2014 International Power Electronics Conference

(IPEC-Hiroshima 2014 ECCE-ASIA)

Hiroshima, Japan
18-21 May 2014

Pages 1634-2412

| IEEE Catalog Number: | CFP14CPB-POD |
| ISBN: | 978-1-4799-2706-7 |

**Copyright © 2014 by the Institute of Electrical and Electronic Engineers, Inc
All Rights Reserved**

Copyright and Reprint Permissions: Abstracting is permitted with credit to the source. Libraries are permitted to photocopy beyond the limit of U.S. copyright law for private use of patrons those articles in this volume that carry a code at the bottom of the first page, provided the per-copy fee indicated in the code is paid through Copyright Clearance Center, 222 Rosewood Drive, Danvers, MA 01923.

For other copying, reprint or republication permission, write to IEEE Copyrights Manager, IEEE Service Center, 445 Hoes Lane, Piscataway, NJ 08854. All rights reserved.

***This publication is a representation of what appears in the IEEE Digital Libraries. Some format issues inherent in the e-media version may also appear in this print version.

IEEE Catalog Number: CFP14CPB-POD
ISBN 13:   978-1-4799-2706-7

Additional Copies of This Publication Are Available From:

Curran Associates, Inc
57 Morehouse Lane
Red Hook, NY 12571 USA
Phone: (845) 758-0400
Fax: (845) 758-2633
E-mail: curran@proceedings.com
Web: www.proceedings.com

TABLE OF CONTENTS

A NOVEL CONTROL SCHEME FOR THREE-LEVEL FULL-BRIDGE CONVERTER ACHIEVING LOW THD OUTPUT VOLTAGE............66
Liu, Jilong ; Xiao, Fei ; Chen, Wei ; Yang, Guorun

PARALLEL CONNECTED THREE PHASE INVERTERS BASED ON MODULAR DESIGN AND DISTRIBUTED CONTROL............72
Xiao, Fei ; Chen, Wei ; Liu, Jilong ; Wang, Hengli

EFFICIENCY INVESTIGATIONS OF A 3KW T-TYPE INVERTER FOR SWITCHING FREQUENCIES UP TO 100 KHZ............78
Anthon, Alexander ; Zhang, Zhe ; Andersen, Michael A.E. ; Franke, Toke

MINIATURIZATION OF THE BOOST-UP TYPE ACTIVE BUFFER CIRCUIT IN A SINGLE-PHASE INVERTER............84
Watanabe, Hiroki ; Koiwa, Kazuhiro ; Itoh, Jun-ichi ; Ohnuma, Yoshiya ; Miyawaki, Satoshi

TESTING FACILITY USING LARGE CAPACITY INVERTER............92
Ishimaru, Yusuke ; Adachi, Mitsuo ; Tsukakoshi, Masahiko ; Nakamura, Ritaka ; Masuda, Hiroyuki ; Ogashi, Yoshihiro ; Tsuboi, Yuichi

PERFORMANCE EVALUATION UNDER THE ACTUAL OPERATING CONDITION OF A LARGE CAPACITY VSI INVERTER FOR STEEL MILL APPLICATIONS............97
Mamun, Mostafa ; Yoshizawa, Daisuke ; Mukunoki, Makoto

A SOFT-SWITCHING SINGLE-PHASE UNIFIED POWER QUALITY CONDITIONER............105
Jiang, Maoh-Chin ; Chang, Kai-Chi ; Lu, Kao-Yi ; Shih, Bing-Jyun ; Liu, Tai-Chun

NOVEL THREE-PHASE PWM AC-AC CONVERTERS SOLVING COMMUTATION PROBLEM............110
Khan, Ashraf Ali ; Shin, Hyunhak ; Cha, Honnyong ; Kim, Heung-Geun

EXPERIMENTAL INVESTIGATION OF NORMALLY-ON TYPE BIDIRECTIONAL SWITCH FOR INDIRECT MATRIX CONVERTERS............117
Sung, Kyungmin ; Iijima, Ryuji ; Nishizawa, Shinichi ; Norigoe, Isami ; Ohashi, Hiromichi

VISUALIZATION OF PWM WAVEFORMS OF OUTPUT VOLTAGE AND INPUT CURRENT FOR A DIRECT MATRIX CONVERTER............123
Asai, Inami ; Takeshita, Takaharu

SPACE VECTOR MODULATION BASED ON VIRTUAL INDIRECT CONTROL FOR HIGH FREQUENCY AC-LINKED MATRIX CONVERTER............130
Inoue, Keita ; Shioda, Masashi ; Katade, Motohumi ; Goto, Akira ; Morishita, Shin ; Itoh, Junichi ; Koiwa, Kazuhiro

A FUNDAMENTAL VERIFICATION OF A SINGLE-PHASE TO THREE-PHASE MATRIX CONVERTER WITH A PDM CONTROL BASED ON SPACE VECTOR MODULATION............138
Nakata, Yuki ; Itoh, Jun-ichi

STEADY STATE CHARACTERISTICS OF THE BOOST-TYPE MATRIX CONVERTER FOR STAND-ALONE POWER SOURCE............146
Nagano, Y. ; Yamamura, N. ; Ishida, M. ; Hirokado, K.

DESIGN PROCEDURE FOR OUTPUT CURRENT CONTROL AND DAMPING CONTROL OF MATRIX CONVERTER............152
Takahashi, Hiroki ; Itoh, Jun-ichi

A NOVEL LCL FILTER PARAMETER DESIGN METHOD BASING ON RESONANT FREQUENCY OPTIMIZATION OF THREE-LEVEL NPC GRID CONNECTED INVERTER............160
Li, Ning ; Wang, Yue ; Niu, Ruigen ; Guo, Wei ; Lei, Wanjun ; Wang, Zhao'An

DESIGN AND ANALYSIS OF ISOLATED BI-DIRECTIONAL DC/DC CONVERTER USING QUASI-RESONANT ZVS............166
Noh, Yong-Su ; Won, Chung-Yuen ; Oh, Min-Seok ; Jeon, Jin-Yong ; Jung, Yong-Chae

AN ACTIVE-CLAMPING ZVS FLYBACK CONVERTER WITH INTEGRATED TRANSFORMER............172
Lin, Jing-Yuan ; Lo, Yu-Kang ; Chiu, Huang-Jen ; Wang, Chao-Fu ; Lin, Chien-Yu

PFM AND PWM HYBRID CONTROLLED LLC CONVERTER............177
Yamamoto, Junichi ; Zaitsu, Toshiyuki ; Abe, Seiya ; Ninomiya, Tamotsu

DISCUSSIONS ON VARIOUS VOLTAGE EQUALIZERS FOR EDLCS USING CW CIRCUIT............183
Khant, Hlaing Kyi Pyar ; Matsui, Keiju ; Hasegawa, Masaru ; Yasubayashi, Mikio ; Umeno, Masayoshi ; Ooishi, Eiji

ISOLATION SYSTEM WITH WIRELESS POWER TRANSFER FOR MULTIPLE GATE DRIVER SUPPLIES OF A MEDIUM VOLTAGE INVERTER............191
Kusaka, Keisuke ; Orikawa, Koji ; Itoh, Jun-ichi ; Morita, Kazunori ; Hirao, Kuniaki

STUDY AND IMPLEMENTATION OF A 15-W POWER AMPLIFIER FOR PIEZOELECTRIC ACTUATOR............199
Lo, Yu-Kang ; Chiu, Huang-Jen ; Liu, Yu-Chen ; Lin, Chung-Yi ; Cheng, Shih-Jen ; Yang, CS

ISOLATED VOLTAGE-BOOSTING CONVERTER............204
Hwu, K.I. ; Jiang, W.Z. ; Shieh, Jenn-Jong

HIGH VOLTAGE CONVERSION RATIO CASCADE BOOST CONVERTER WITH DC SNUBBER............208
Lee, Yuang-Shung ; Yu, Ling-Chia ; Chou, Tzu-Han

DESIGN-ORIENTED ANALYSIS OF RESONANCE DAMPING AND HARMONIC COMPENSATION FOR LCL-FILTERED VOLTAGE SOURCE CONVERTERS............216
Wang, Xiongfei ; Blaabjerg, Frede ; Loh, Poh Chiang

STATE-SPACE AVERAGE MODELING OF BIDIRECTIONAL DC-DC CONVERTER FOR BATTERY CHARGER USING LCLC FILTER..224
Moon, Sang-Ho ; Jou, Sung-Tak ; Lee, Kyo-Beum

A NEW SVPWM STRATEGY FOR INPUT SWITCHED MULTILEVEL CONVERTER..........................230
Xiong, Li ; Prasanna, U.R. ; Bilal, Akin ; Rajashekara, Kaushik

ESD RELIABILITY INFLUENCE OF A 60 V POWER LDMOS BY THE FOD-BASED (& DOTTED-OD) DRAIN...236
Chen, Shen-Li ; Lee, Min-Hua

ENHANCED TRANSVERSE-FLUX MOTOR WITH TORUS COILS...240
Tanaka, Junya ; Sakai, Kazuto

THE INFLUENCE OF MAGNETIC PROPERTIES OF PERMANENT MAGNET ON THE PERFORMANCE OF IPMSM FOR AUTOMOTIVE APPLICATION..246
Yoshioka, S. ; Morimoto, S. ; Sanada, M. ; Inoue, Y.

CHARACTERISTICS OF INTERIOR PERMANENT MAGNET SYNCHRONOUS MOTOR WITH IMPERFECT MAGNETS..252
Shinagawa, Syuhei ; Ishikawa, Takeo ; Kurita, Nobuyuki

STUDY OF STATOR STRUCTURE TO IMPROVE RELUCTANCE TORQUE FOR IPMSM WITH CONCENTRATED WINDING...258
Morikawa, R. ; Sanada, M. ; Morimoto, S. ; Inoue, Y.

DEVELOPMENT AND VERIFICATION OF ENERGY-ACCURATE SIMULATION MODELS FOR PERMANENT MAGNET SYNCHRONOUS MOTORS IN AUTOMATION SYSTEMS....................264
Blank, Frederic ; Roth-Stielow, Jorg

COMPARISON OF THE RESISTANCE- AND INDUCTANCE-BASED SALIENCY OF A PMSM DUE TO A SHORT-CIRCUITED ROTOR WINDING...270
Graus, Johannes ; Rambetius, Alexander ; Hahn, Ingo

DESIGN AND OPTIMIZATION OF HIGH-SPEED SWITCHED RELUCTANCE MOTOR USING SOFT MAGNETIC COMPOSITE MATERIAL...278
Gaing, Zwe-Lee ; Kuo, Kuan-Yi ; Hu, Jia-Sheng ; Hsieh, Min-Fu ; Tsai, Ming-Hsiao

INFLUENCE OF PULSE WIDTH MODULATION (PWM) ON THE IRON LOSSES OF ELECTRICAL STEEL.......283
Boehm, Andreas ; Hahn, Ingo

INVESTIGATION ON IRON LOSS CHARACTERISTICS IN STAR-CONNECTION AND DELTA-CONNECTION UNDER THREE PHASE PWM INVERTER EXCITATION....................................289
Odawara, Shunya ; Fujisaki, Keisuke ; Fukuhara, Shuhei

OPTIMIZATION ON ARRANGEMENT OF PERMANENT MAGNETS FOR MAGNETIC LEVITATION SYSTEM FOR THIN STEEL PLATE (FUNDAMENTAL CONSIDERATION ON LEVITATION PROBABILITY)...294
Ishii, Hirotaka ; Hasegawa, Shinya ; Narita, Takayoshi ; Oshinoya, Yasuo

EFFECT OF A MAGNETIC FIELD FROM THE HORIZONTAL DIRECTION ON A MAGNETICALLY LEVITATED STEEL PLATE (FUNDAMENTAL CONSIDERATIONS ON THE SHAPE ANALYSIS OF ULTRATHIN STEEL PLATE)..299
Kurihara, Takeshi ; Hasegawa, Shinya ; Narita, Takayoshi ; Oshinoya, Yasuo

NOVEL MAGNETIC STRUCTURE OF INTEGRATED DIFFERENTIAL-MODE AND COMMON-MODE INDUCTORS TO SUPPRESS DC SATURATION..304
Umetani, Kazuhiro ; Tera, Takahiro ; Shirakawa, Kazuhiro

A NOVEL CONTROL METHOD IN FLUX-WEAKENING REGION FOR EFFICIENT OPERATION OF INTERIOR PERMANENT MAGNET SYNCHRONOUS MOTOR.................................312
Ueda, K. ; Morimoto, S. ; Inoue, Y. ; Sanada, M.

IMPLEMENTATION OF THE MTPA AND MTPV CONTROL WITH ONLINE PARAMETER IDENTIFICATION FOR A HIGH SPEED IPMSM USED AS TRACTION DRIVE.........................318
Nguyen, Quoc Khanh ; Petrich, Matthias ; Roth-Stielow, Jorg

CORRECTION OF REFERENCE FLUX FOR MTPA CONTROL IN DIRECT TORQUE CONTROLLED INTERIOR PERMANENT MAGNET SYNCHRONOUS MOTOR DRIVES.................................324
Shinohara, Atsushi ; Inoue, Yukinori ; Morimoto, Shigeo ; Sanada, Masayuki

VOLTAGE REGULATION AND MAXIMUM OUTPUT POWER TRACKING OF A 4.5KW PERMANENT-MAGNET SYNCHRONOUS GENERATOR...330
Chang, Yuan-Chih ; Chang, Hsiu-Feng ; Dai, Wei-Fu ; Wu, Chun-Wei

A NOVEL FLUX-WEAKENING CONTROL METHOD BASED ON SINGLE CURRENT REGULATOR FOR PERMANENT MAGNET SYNCHRONOUS MOTOR..335
Fang, Xiaocun ; Hu, Taiyuan ; Lin, Fei ; Yang, Zhongping

PREDICTIVE CURRENT CONTROL METHOD IN INDUCTION MOTOR SPEED SENSORLESS DRIVE..........341
Wei, Sun ; Yong, Yu ; Dianguo, Xu ; Jin, Xu ; Li, Ding

REAL-TIME IMPLEMENTATION OF AN ONLINE MODEL PREDICTIVE CONTROL FOR IPMSM USING PARALLEL COMPUTING ON FPGA..346
Leuer, Michael ; Bocker, Joachim

AN INTEGRAL SLIDING-MODE CONTROLLER FOR ENERGY EFFICIENCY IMPROVEMENT IN AC POWER SOURCE SUPPLIED AC MACHINE DRIVES..351
Shieh, Hsin-Jang ; Chen, Ying-Zuo

PERFORMANCE IMPROVEMENT OF ULTRA-HIGH-SPEED PMSM DRIVE SYSTEM BASED ON DTC BY USING SIC INVERTER..356
Togashi, Ryo ; Inoue, Yukinori ; Morimoto, Shigeo ; Sanada, Masayuki

MATHEMATICAL MODEL FOR HIGH-EFFICIENCY CONTROL OF PERMANENT-MAGNET SYNCHRONOUS MOTOR IN STATOR FLUX LINKAGE SYNCHRONOUS FRAME 363

Inoue, Tatsuki ; Inoue, Yukinori ; Morimoto, Shigeo ; Sanada, Masayuki

WIDE-SPEED-RANGE OPERATION OF DTC-BASED PMSM DRIVE SYSTEM USING MTPF CONTROL 370

Inoue, Yukinori ; Ichiya, Takahiro ; Morimoto, Shigeo ; Sanada, Masayuki

AN INDUSTRIAL LOW-VOLTAGE INVERTER FOR PRM CONTROL 376

Nakamura, M. ; Oka, T. ; Oishi, K.

OPTIMAL PULSE PATTERN DETERMINATION BASED ON PULSE HARMONIC MODULATION 383

Furukawa, Kimihisa ; Ajima, Toshiyuki ; Miyazaki, Hideki

METHOD FOR AUTO-TUNING OF CURRENT AND SPEED CONTROLLER IN IPMSM DRIVE SYSTEM BASED ON PARAMETER IDENTIFICATION 390

Tadokoro, D. ; Morimoto, S. ; Inoue, Y. ; Sanada, M.

COMPARATIVE STUDY OF PWM STRATEGIES FOR THREE-PHASE OPEN-END WINDING INDUCTION MOTOR DRIVES 395

Zhu, B. ; Prasanna, U.R. ; Rajashekara, K. ; Kubo, H.

10MW,3.3MWH ENERGY STORAGE SYSTEM CONSISTING OF 4000 FLYWHEELS CONTROLLED BY ICT NETWORK FOR SHORT CYCLE POWER FLUCTUATION COMPENSATION 403

Kato, Koji ; Ishigma, Satoru ; Nakajima, Yoichiro ; Arai, Haruki ; Ueda, Tetsuya ; Iwata, Tetsuki ; Ito, Yoichi ; Sugao, Kazumi

VERSATILE POWER TRANSFER STRATEGIES OF PV-BATTERY HYBRID SYSTEM FOR RESIDENTIAL USE WITH ENERGY MANAGEMENT SYSTEM 409

Choi, Seong-Chon ; Sin, Min-ho ; Kim, Dong-Rak ; Won, Chung-Yuen ; Jung, Yong-Chae

HIGH-EFFICIENCY AND COST-MINIMIZATION METHOD OF ENERGY STORAGE SYSTEM WITH MULTI STORAGE DEVICES FOR GRID CONNECTION 415

Haga, Hitoshi ; Shimao, Toshihiro ; Kondo, Seiji ; Kato, Koji ; Itoh, Youichi ; Arimatsu, Kenji ; Matsuda, Katsuhiro

BIDIRECTIONAL DC-DC CONVERTER WITH MULTIPLE SWITCHED-CAPACITOR CELLS 421

Lee, Yuang-Shung ; Huang, Hsin-Wei ; Chou, Tzu-Han

SWITCHED-CAPACITOR CHARGE EQUALIZATION CIRCUIT FOR SERIES-CONNECTED BATTERIES 429

Hsieh, Yao-Ching ; Cai, Zheng-Xiu ; Wu, Wen-Zhe

PERFORMANCE ANALYSIS OF UNITL-H6 INVERTER WITH SIC MOSFETS 433

Barater, Davide ; Buticchi, Giampaolo ; Concari, Carlo ; Franceschini, Giovanni ; Gurpinar, Emre ; De, Dipankar ; Castellazzi, Alberto

MAXIMUM POWER POINT TRACKING OF GRID-TIED PHOTOVOLTAIC POWER SYSTEMS 440

Lee, Ya-Ting ; Chiu, Chian-Song ; Chiu, Tse-Wei

A NEW VOLTAGE TYPE MAGNETICALLY COUPLED T-SOURCE INVERTER 446

Tran, Q.V. ; Low, K.S.

A HIGH EFFICIENCY HYBRID 7-LEVEL INVERTER WITH SINGLE DC SOURCE 452

Yanhong, Zhang ; Kazuya, Ogura ; Oi, Kazunobu

OPTIMAL IDLING CONTROL STRATEGY FOR THREE-PORT FULL-BRIDGE CONVERTER 458

Jiang, Yongjie ; Liu, Fuxin ; Ruan, Xinbo ; Wang, Lipeng

FILTER DESIGN FOR THREE-LEVEL GRID-CONNECTED INVERTER WITH LOW SWITCHING FREQUENCY 465

Ren, Kangle ; Zhang, Xing ; Wang, Fusheng ; Tu, Yunwu ; Wang, Lingxiang ; Deng, Lirong

A NOVEL EFFICIENT T TYPE THREE LEVEL NEUTRAL-POINT-CLAMPED INVERTER FOR RENEWABLE ENERGY SYSTEM 470

Wu, Wenlong ; Wang, Fei ; Wang, Yong

A NOVEL NEUTRAL POINT VOLTAGE AUTOMATIC BALANCING CARRIER-BASED MODULATION STRATEGY OF THREE-LEVEL NPC CONVERTER 475

Li, Ning ; Wang, Yue ; Niu, Ruigen ; Guo, Wei ; Lei, Wanjun ; Wang, Zhao'An

A HIGH VOLTAGE GAIN SWITCHED-COUPLED-INDUCTOR QUASI-Z-SOURCE INVERTER 480

Ahmed, Furqan ; Cha, Honnyong ; Kim, Su-Han ; Kim, Heung-Geun

A NOVEL CONTROL STRATEGY TO SUPPRESS DC CURRENT INJECTION TO THE GRID FOR THREE-PHASE PV INVERTER 485

Zhang, Tao ; He, Guofeng ; Chen, Min ; Xu, Dehong

CLC FILTER DESIGN OF A FLYBACK-INVERTER FOR PHOTOVOLTAIC SYSTEMS 493

Shin, Yesl ; Lee, June-Hee ; Lee, June-Seok ; Lee, Kyo-Beum

THREE-PHASE INVERTER TOPOLOGIES FOR GRID-CONNECTED PHOTOVOLTAIC SYSTEMS 498

Ozkan, Ziya ; Hava, Ahmet M.

A THREE-PORT TOPOLOGY COMPARISON FOR A LOW POWER STAND-ALONE PHOTOVOLTAIC SYSTEM 506

Mira, Maria C. ; Knott, Arnold ; Andersen, Michael A.E.

EFFECT OF CONVENTIONAL GRID-VOLTAGE FEEDFORWARD ON THE OUTPUT IMPEDANCE OF A THREE-PHASE PHOTOVOLTAIC INVERTER 514

Messo, T. ; Jokipii, J. ; Suntio, T.

POWER AMPLIFIER SUITABLE FOR PHOTOVOLTAIC CELL BOOSTER 522

Kohama, Teruhiko ; Sogawa, Yuki ; Tsuji, Satoshi

REALIZATION STUDY OF INTERLEAVED PV MICROINVERTER BY QUADRATURE-PHASE-SHIFT SPWM CONTROL 526

Hsieh, Hung-I ; Hsieh, Guan-Cyun ; Hou, Jiaxin

CURRENT SENSORLESS MPPT METHOD FOR A PV FLYBACK MICROINVERTERS USING A DUAL-MODE 532
Lee, June-Hee ; Lee, June-Seok ; Lee, Kyo-Beum

A NOVEL METHOD OF SUPPRESSING INRUSH CURRENTS OF SQUIRREL-CAGE INDUCTION MACHINE USING MATRIX CONVERTER IN WIND POWER GENERATION SYSTEMS 538
Yamada, Hiroaki ; Hanamoto, Tsuyoshi

NONLINEAR PITCH CONTROL DESIGN FOR LOAD REDUCTION ON WIND TURBINES 543
Xiao, Shuai ; Yang, Geng ; Geng, Hua

DEVICE LOADING OF MODULAR MULTILEVEL CONVERTER MMC IN WIND POWER APPLICATION 548
Popova, L. ; Pyrhonen, J. ; Ma, K. ; Blaabjerg, F.

A NOVEL OPTIMAL DESIGN OF DFIG CROWBAR RESISTOR DURING GRID FAULTS 555
Hu, Sheng ; Zou, XuDong ; Kang, Yong

DC-VOLTAGE REGULATION OF A FIVE LEVELS NEUTRAL POINT CLAMPED CASCADED CONVERTER FOR WIND ENERGY CONVERSION SYSTEM 560
Merahi, Farid ; Mekhilef, Saad ; Berkouk, El Madjid

A REACTIVE POWER SHARING METHOD BASED ON VIRTUAL CAPACITOR IN ISLANDING MICROGRID 567
Xu, Haizhen ; Zhang, Xing ; Liu, Fang ; Shi, Rongliang ; Yu, Changzhou ; Zhao, Wei ; Yu, Yong ; Cao, Wei

STORAGE CAPACITY PERFORMANCE FOR HYBRID PV/DIESEL SYSTEM IN SABAH MALAYSIA 573
Hidayat, Nabil M ; Kari, Mat Nasir ; Mohd Arif, Mohd Johari

NEW TECHNIQUES FOR MEASURING ISLANDED MICROGRID IMPEDANCE CHARACTERISTICS BASED ON CURRENT INJECTION 577
Hou, Lixiang ; Liu, Baoquan ; Shi, Hongtao ; Yi, Hao ; Zhuo, Fang

A GENERAL FRAMEWORK TO DESIGN OPERATION MODES OF DC MICROGRIDS WITHOUT COMMUNICATION LINKS 582
Pan, Miao ; Shen, Na ; Yang, Geng ; Morita, Kazunori ; Ogura, Kazuya ; Wu, Weiyang

IMPLEMENTATION DESIGN OF THE CONVERTER-BASED GALVANIC ISOLATION FOR LOW VOLTAGE DC DISTRIBUTION 587
Mattsson, A. ; Vaisanen, V. ; Nuutinen, P. ; Kaipia, T. ; Lana, A. ; Peltoniemi, P. ; Silventoinen, P. ; Partanen, J.

PEAK DETECTION METHOD USING TWO-DELTA OPERATION FOR SINGLE VOLTAGE SAG 595
Lee, Woo-Cheol ; Lee, Taeck-Kie

LINE LOSS MINIMIZATION IN RADIAL DISTRIBUTION SYSTEM USING MULTIPLE STATCOMS AND STATIC CAPACITORS 601
Miyazaki, Kensuke ; Takeshita, Takaharu

A NOVEL CONTROL METHOD FOR INDIVIDUAL DC VOLTAGE BALANCING IN H-BRIDGE CASCADED STATCOM 609
Xu, Rong ; Yu, Yong ; Yang, Rongfeng ; Qu, Lizhi ; Sun, Wei ; Xu, Dianguo

RESEARCH ON THE CONTROL STRATEGY OF STATCOM BASED ON MODULAR MULTILEVEL CONVERTER 614
Zhang, Wei ; Gao, Qiang ; Su, Bonan ; Jin, Miaoxin ; Xu, Dianguo ; Liu, Jianyu

FAULT DIAGNOSIS IN LARGE FORMAT LIFEPO4 ESS APPLICATION THROUGH DWT-BASED MRA 619
Kim, Jonghoon

COMPARISON OF DIFFERENT IGBT BASED DESIGNS OF POWER ELECTRONIC TRANSFORMER 624
Wang, Xinyu ; Ouyang, Shaodi ; Liu, Jinjun ; Meng, Fei ; Javed, Riffat

SEMI-ADAPTIVE HARMONIC CONTROL FOR POWER BALANCING DEVICE FOR AC TRACTION 629
Akagi, Masataka ; Tsuruta, Hironori ; Oso, Hiroshi

RESEARCH OF EFFICIENT MAIN POWER EQUIPMENT USING SIC POWER DEVICE 634
Shinbo, Mitsuo ; Sonoda, Hideki ; Ishida, Takahito ; Abiko, Hiroshi ; Shibanuma, Kenichi ; Chiba, Yoshinori

A HIGH PERFORMANCE CONTROL STRATEGY FOR THREE-LEVEL NPC EMU CONVERTERS 640
Song Kejian ; Wu Mingli ; Wang Hui ; Agelidis, Vassilios Georgios

A DESIGN OF INRUSH CURRENT IDENTIFICATION SYSTEM FOR HIGH-SPEED TRAIN'S TRACTION TRANSFORMER 647
Yu, Weikai ; Liu, Xiankai ; Zhang, Yuzhuo ; Cao, Yuan ; Ma, Weigang ; Hei, Xinhong ; Huang, Zhenhui ; Jiang, Dawang

CURRENT SOURCE INVERTER BASED CASCADED SOLID STATE TRANSFORMER FOR AC TO DC POWER CONVERSION 651
Roy, Sudhin ; De, Ankan ; Bhattacharya, Subhashish

EVALUATION OF HIGH VOLTAGE 15 KV SIC IGBT AND 10 KV SIC MOSFET FOR ZVS AND ZCS HIGH POWER DC -DC CONVERTERS 656
Moballegh, Shiva ; Madhusoodhanan, Sachin ; Bhattacharya, Subhashish

THE DIRECT YAW-MOMENT CONTROL TO FOLLOW THE NEUTRAL STEERING PATH REGARDLESS OF VELOCITY 664
Jang, Young-Jin ; Nam, Kwang-Hee

NEXT-GENERATION IGBT MODULE STRUCTURE FOR HYBRID VEHICLE WITH HIGH COOLING PERFORMANCE AND HIGH TEMPERATURE OPERATION 671
Morozumi, Akira ; Gohara, Hiromichi ; Momose, Fumihiko ; Saito, Takashi ; Nishimura, Yoshitaka ; Mochizuki, Eiji ; Takahashi, Yoshikazu

INTEGRATION OF PLUG-IN ELECTRIC VEHICLES IN POWER SYSTEMS USING CHARGING MODE SWITCHING 677
Wen-Tai Li ; Wen, Chao-Kai ; Chen, Jung-Chieh ; Teng, Jen-Hao ; Ting, Pangan

A NOVEL COMPENSATION METHOD FOR A MOTOR PHASE CURRENT SENSOR OFFSET ERROR VARIED DURING A VSI-MOTOR DRIVE 682

Tamura, Hiroshi ; Noto, Yasuo ; Ajima, Toshiyuki ; Itoh, Jun-ichi

INVESTIGATION OF CALCULATION METHOD OF LOSSES IN PWM INVERTER WITH VOLTAGE BOOSTER USING BOTH DC LINK VOLTAGE CONTROL AND FLUX WEAKENING CONTROL 689

Imakiire, Akihiro ; Hikita, Masayuki ; Yamamoto, Kichiro ; Yonemori, Ryo

DYNAMIC AND STEADY-STATE BEHAVIOR OF A PARALLELING THREE-PHASE AC-TO-DC CONVERTER WITH REDUCED DC BUS CAPACITOR 694

Kamnarn, Uthen ; Kanthaphayao, Yutthana ; Chunkag, Viboon

REACTIVE POWER LOSS OPTIMIZATION METHOD FOR BI-DIRECTIONAL ISOLATED DC-DC CONVERTERS 702

Wen, Huiqing

POWER SUPPLY FOR A WIRELESS SENSOR NETWORK: AIRLINER FLIGHT TEST CASE STUDY 707

Durand Estebe, P. ; Boitier, V. ; Bafleur, M. ; Dilhac, J.-M. ; Berhouet, S.

A CONFIGURABLE THREE-PHASED INVERTER FOR TEACHING POWER ELECTRONICS 712

Kern, Ansgar

A BACHELOR-STUDENT PROJECT: BUCK-BOOST OPERATION OF AN INTEGRATED H-BRIDGE FOR VARIABLE-SPEED ENERGY STORAGE SYSTEMS USING MEASUREMENT COILS IN THE STATOR OF A DC-MACHINE 718

De Belie, Frederik ; Darba, Araz ; Melkebeek, Jan

DEVELOPMENT OF A WEB-BASED REMOTE EXPERIMENT SYSTEM FOR ELECTRICAL MACHINERY LEARNERS 724

Ishibashi, Makoto ; Fukumoto, Hisao ; Furukawa, Tatsuya ; Itoh, Hideaki ; Ohchi, Masashi

DEVELOPMENT OF POWER MEASUREMENT SYSTEM IN SIMULATED MICRO GRID SYSTEM FOR EDUCATION 730

Hira, Yuki ; Furukawa, Tatsuya ; Yakabe, Seichiro ; Fukumoto, Hisao ; Itoh, Hideaki ; Ohchi, Masashi

POWER ELECTRONIC TECHNOLOGIES FOR FLEXIBLE DC DISTRIBUTION GRIDS 736

De Doncker, Rik W.

2.5KV, 200KW BI-DIRECTIONAL ISOLATED DC/DC CONVERTER FOR MEDIUM-VOLTAGE APPLICATIONS 744

Matsuoka, Yuji ; Wada, Keiji ; Nakahara, Mizuki ; Takao, Kazuto ; Kyungmin Sung ; Ohashi, Hiromichi ; Nishizawa, Shinichi

POWER-LOSS BREAKDOWN OF A 750-V, 100-KW, 20-KHZ BIDIRECTIONAL ISOLATED DC-DC CONVERTER USING SIC-MOSFET/SBD DUAL MODULES 750

Akagi, Hirofumi ; Yamagishi, Tatsuya ; Tan, Nadia M.L. ; Kinouchi, Shin-ichi ; Miyazaki, Yuji ; Koyama, Masato

DESIGN CONSIDERATIONS OF A 15KV SIC IGBT ENABLED HIGH-FREQUENCY ISOLATED DC-DC CONVERTER 758

Tripathi, Awneesh ; Mainali, Krishna ; Patel, Dhaval ; Kadavelugu, Arun ; Hazra, Samir ; Bhattacharya, Subhashish ; Hatua, Kamalesh

COMMON-MODE CURRENTS IN MULTI-CELL SOLID-STATE TRANSFORMERS 766

Huber, Jonas E. ; Kolar, Johann W.

SINGLE-STAGE RECONFIGURABLE DC/DC CONVERTER FOR WIDE INPUT VOLTAGE RANGE OPERATION IN HEVS 774

Zeljkovic, Sandra ; Reiter, Tomas ; Gerling, Dieter

A TWO STAGE DC/DC CONVERTER WITH WIDE INPUT RANGE FOR EV 782

Peng Wen ; Changsheng Hu ; Haitao Yang ; Longlong Zhang ; Cheng Deng ; Yashun Li ; Dehong Xu

INTERMEDIATE AND LIGHT LOAD EFFICIENCY IMPROVEMENT OF A HIGH-POWER DENSITY BIDIRECTIONAL DC-DC CONVERTER IN HYBRID ELECTRIC VEHICLES WITH MR FLUID GAP INDUCTOR 790

Ahmed, Furqan ; Su-Han Kim ; Cha, Honnyong ; Kim, Dong-Hun ; Heung-Geun Kim

REGENERATIVE CONTROL OF BI-DIRECTIONAL DC-DC CONVERTER CONTROLLING VARIABLE DC-LINK FOR FCEV 796

Il-Kuen Won ; An-Yeol Ko ; Do-Yun Kim ; Chung-Yuen Won ; Young-Ryul Kim

LARGE DRIVING RANGE INCREASE OF SERIES CHOPPER BASED POWER TRAIN USING MOTOR TEST BENCH 801

Hosoyamada, Yu ; Takeda, Masashi ; Motoi, Naoki ; Kawamura, Atsuo

THE POWER ELECTRONICS PROGRAM AT BEIJING JIAOTONG UNIVERSITY 807

Fei Lin ; Zhongping Yang ; Zheng, T.Q.

EFFORTS FOR POWER ELECTRONICS EDUCATION IN A START-UP COMPANY 811

Hattori, Fumiya ; Imaoka, Jun ; Ishitobi, Manabu ; Nagai, Shinichiroh ; Yamamoto, Masayoshi

EDUCATION FOR THE ENGINEERS OF TRACTION POWER SUPPLY DIVISION IN EAST JAPAN RAILWAY COMPANY 817

Takino, Toshiaki ; Iwakami, Tetsuro

SUCCESSFUL ONLINE EDUCATION - GECKOCIRCUITS AS OPEN-SOURCE SIMULATION PLATFORM 821

Musing, Andreas ; Kolar, Johann W.

AN ELECTRIC VEHICLE PROJECT FOR ECO-RUN RACE 829

Yamagata, Shinichi ; Oda, Yoshinori ; Tanai, Masanobu ; Sung, Kyungmin

MULTI-LOOP CONTROLLER DESIGN FOR DIODE-ASSISTED BUCK-BOOST VOLTAGE SOURCE INVERTER 835

Yan Zhang ; Jinjun Liu ; Xiaolong Ma ; Junjie Feng

VOLUME 2

REAL-TIME SIMULATION OF WIND TURBINE CONVERTER-GRID SYSTEMS 843
Shah, Shahil ; Vieto, Ignacio ; Nian Heng ; Sun, Jian

TECHNOLOGIES FOR MITIGATING FLUCTUATION CAUSED BY RENEWABLE ENERGY SOURCES 850
Katoh, Shuji ; Ohara, Shinya ; Itoh, Tomomichi

RELIABILITY-ORIENTED ENERGY STORAGE SIZING IN WIND POWER SYSTEMS 857
Zian Qin ; Liserre, Marco ; Blaabjerg, Frede ; Poh Chiang Loh

A MULTI-LEVEL VIRTUAL CONDUCTOR AS A BACKBONE OF A DC POWER ROUTING SYSTEM 863
Ramadan, Husam A. ; Imamura, Yasutaka ; Kawachi, Konosuke ; Yang, Sihun ; Shoyama, Masahito

SEMI-NUMERICAL METHOD FOR LOSS-CALCULATION IN FOIL-WINDINGS EXPOSED TO AN AIR-
GAP FIELD 868
Leuenberger, D. ; Biela, J.

LOSS REDUCTION OF LAMINATED CORE INDUCTOR USED IN ON-BOARD CHARGER FOR EVS 876
Tera, Takahiro ; Taki, Hiroshi ; Shimizu, Toshihisa

FEASIBLE EVALUATIONS OF COUPLED MULTILAYER CHIP INDUCTOR FOR POL CONVERTER 883
Imaoka, Jun ; Kimura, Shota ; Itoh, Yuki ; Yamamoto, Masayoshi ; Suzuki, Michiaki ; Kawano, Kenji

OPTIMAL INDUCTOR DESIGN FOR 3-PHASE VOLTAGE-SOURCE PWM CONVERTERS CONSIDERING
DIFFERENT MAGNETIC MATERIALS AND A WIDE SWITCHING FREQUENCY RANGE 891
Burkart, Ralph M. ; Uemura, Hirofumi ; Kolar, Johann W.

COMPARATIVE ANALYSIS OF INDUCTOR CONCEPTS FOR HIGH PEAK LOAD LOW DUTY CYCLE
OPERATION 899
Leibl, Michael ; Kolar, Johann W.

INITIAL POSITION ESTIMATION FOR IPMSMS USING COMB FILTERS AND EFFECTS ON VARIOUS
INJECTED SIGNAL FREQUENCIES 907
Suzuki, Toshiki ; Tomita, Mutuwo ; Hasegawa, Masaru ; Doki, Shinji

ADAPTIVE SIGNAL INJECTION METHOD COMBINED WITH EEMF BASED POSITION SENSORLESS
CONTROL OF IPMSM DRIVES 914
Ohnuma, Takumi ; Makaino, Yuki ; Saitoh, Ryoh

STUDY OF LOW SPEED SENSORLESS DRIVES FOR SPMSM BY CONTROLLING ELLIPTICAL
INDUCTANCE 919
Maekawa, Sari ; Hinata, Toshifumi ; Suzuki, Nobuyuki ; Kubota, Hisao

SUPPRESSION OF INJECTION VOLTAGE DISTURBANCE FOR HIGH FREQUENCY SQUARE-WAVE
INJECTION SENSORLESS DRIVE WITH REGULATION OF INDUCED HIGH FREQUENCY CURRENT
RIPPLE 925
Dongouk Kim ; Yong-Cheol Kwon ; Seung-Ki Sul ; Jang-Hwan Kim ; Rae-Sung Yu

APPLICATION TREND OF SALIENCY-BASED SENSORLESS DRIVES 933
Yamazaki, Akira ; Ide, Kozo

SWITCHING-LEVEL SIMULATION MODEL OF MMC-BASED BACK-TO-BACK CONVERTER FOR
HVDC APPLICATION 937
Byung Moon Han ; Jong kyou Jeong

POWER-CELL SWITCHING-CYCLE CAPACITOR VOLTAGE CONTROL FOR THE MODULAR
MULTILEVEL CONVERTERS 944
Wang, Jun ; Burgos, Rolando ; Boroyevich, Dushan ; Bo Wen

A COMPARISON OF MODULAR MULTILEVEL ENERGY CONVERSION PROCESSES: DC/AC VERSUS
DC/DC 951
Kish, Gregory J. ; Lehn, Peter W.

A NOVEL TOPOLOGY OF WIND POWER PLANT SUITABLE FOR DC POWER TRANSMISSION
SYSTEMS 959
Nishikata, Shoji ; Tatsuta, Fujio ; Suzuki, Katsumi

AN IMPEDANCE-BASED APPROACH TO HVDC SYSTEM STABILITY ANALYSIS AND CONTROL
DEVELOPMENT 967
Liu, Hanchao ; Shah, Shahil ; Sun, Jian

TOPOLOGY EVALUATION OF SLOTLESS BEARINGLESS MOTORS WITH TOROIDAL WINDINGS 975
Steinert, Daniel ; Nussbaumer, Thomas ; Kolar, Johann W.

WINDING ARRANGEMENT IN SINGLE-DRIVE BEARINGLESS MOTOR WITH RADIAL GAP 982
Sugimoto, Hiroya ; Tanaka, Seiyu ; Chiba, Akira ; Rahman, M.A.

DEVELOPMENT OF A ONE-AXIS ACTIVELY REGULATED BEARINGLESS MOTOR WITH A
REPULSIVE TYPE PASSIVE MAGNETIC BEARING 988
Asama, Junichi ; Watanabe, Daisuke ; Oiwa, Takaaki ; Chiba, Akira

CONTROL CHARACTERISTICS OF 8/10 AND 12/14 BEARINGLESS SWITCHED RELUCTANCE MOTOR 994
Zhenyao Xu ; Dong-Hee Lee ; Jin-Woo Ahn

BASIC CHARACTERISTIC OF A TWO-UNIT OUTER ROTOR TYPE BEARINGLESS MOTOR WITH
CONSEQUENT POLE PERMANENT MAGNET STRUCTURE 1000
Takemoto, Masatsugu

VOLTAGE RIPPLE ELIMINATION IN INDUCTOR-LESS AC-TO-AC CONVERTERS FOR MULTI-POLE PERMANENT MAGNET SYNCHRONOUS GENERATORS 1006
Tanaka, Koutaro ; Fujita, Hideaki

A NEW SVM METHOD TO REDUCE COMMON-MODE VOLTAGE IN DIRECT MATRIX CONVERTER 1013
Huu-Nhan Nguyen ; Hong-Hee Lee

EXPERIMENTAL VERIFICATION OF HIGH FREQUENCY LINK DC-AC CONVERTER USING PULSE DENSITY MODULATION AT SECONDARY MATRIX CONVERTER 1021
Itoh, Jun-ichi ; Oshima, Ryo ; Takahashi, Hiroki

LOSS ANALYSIS AND DESIGN METHOD FOR HIGH EFFICIENCY MATRIX CONVERTER 1028
Koiwa, Kazuhiro ; Goh Teck Chiang ; Itoh, Jun-ichi

CAPACITOR CLAMPED MULTI-LEVEL MATRIX CONVERTER 1036
Raju, Siddharth ; Mohan, Ned

EUROPEAN TRENDS AND TECHNOLOGIES IN TRACTION 1043
Drofenik, Uwe ; Canales, Francisco

CO-PHASE POWER SUPPLY SYSTEM FOR HSR 1050
Qunzhan Li ; Wei Liu ; Zeliang Shu ; Shaofeng Xie ; Fulin Zhou

THE APPLICATION OF ELECTRONIC FREQUENCY CONVERTER TO THE SHINKANSEN RAILYARD POWER SUPPLY 1054
Shimizu, Toshimasa ; Kunomura, Ken ; Kai, Masahiko ; Onishi, Mitsuru ; Masuzawa, Hiroshi ; Miyajima, Hiroki ; Otsuki, Midori ; Tsuruma, Yoshinori

APPLICATION EXAMPLES OF ENERGY SAVING MEASURES IN JAPANESE DC FEEDING SYSTEM 1062
Suzuki, Takashi ; Hayashiya, Hitoshi ; Yamanoi, Takashi ; Kawahara, Keiji

LITHIUM ION BATTERY APPLICATION IN TRACTION POWER SUPPLY SYSTEM 1068
Teshima, Masato ; Takahashi, Hirotaka

INTEGRATED ISOLATION AND VOLTAGE BALANCING LINK OF 3-PHASE 3-LEVEL PWM RECTIFIER AND INVERTER SYSTEMS 1073
Boillat, David O. ; Kolar, Johann W.

VOLTAGE STEP-UP CONVERTER BASED ON MULTISTAGE STACKED BOOST ARCHITECTURE (MSBA) 1081
Rufer, Alfred ; Barrade, Philippe ; Steinke, Gina

COMPARISON OF CASCADED MULTILEVEL CONVERTER TOPOLOGIES FOR AC/AC CONVERSION 1087
Ilves, Kalle ; Bessegato, Luca ; Norrga, Staffan

EVALUATION OF ISOLATED THREE-PHASE AC-DC CONVERTER USING MODULAR MULTILEVEL CONVERTER TOPOLOGY 1095
Nakanishi, Toshiki ; Itoh, Jun-ichi

SELF-DECOUPLED DUAL PICK-UP COILS WITH LARGE LATERAL TOLERANCE FOR ROADWAY POWERED ELECTRIC VEHICLES 1103
Choi, Su Y. ; Lee, Sung W. ; Lee, Eun S. ; Jeong, Seog Y. ; Gu, Beom W. ; Rim, Chun T.

CONTACTLESS POWER TRANSFER SYSTEM SUITABLE FOR LOW VOLTAGE AND LARGE CURRENT CHARGING FOR EDLCS 1109
Kudo, Takahiro ; Toi, Takahiro ; Kaneko, Yasuyoshi ; Abe, Shigeru

EXCITATION SYSTEM BY CONTACTLESS POWER TRANSFER SYSTEM WITH THE PRIMARY SERIES CAPACITOR METHOD 1115
Nozawa, Ryosuke ; Kobayashi, Ryota ; Tanifuji, Hikaru ; Kaneko, Yasuyoshi ; Abe, Shigeru

DESIGN OF FERRITE CORES OF INDUCTIVE POWER COLLECTION COILS FOR MOVING VEHICLES 1122
Shimode, Daisuke ; Murai, Toshiaki ; Sawada, Tadashi

TORQUE/CURRENT RATIO IMPROVEMENT AND VIBRATION REDUCTION OF SWITCHED RELUCTANCE MOTORS USING MULTI-STAGE STRUCTURE 1128
Matsui, Ryota ; Nakao, Noriya ; Akatsu, Kan

IMPROVEMENT OF EFFICIENCY BY STEPPED-SKEWING ROTOR FOR SWITCHED RELUCTANCE MOTORS 1135
Sugiura, Makoto ; Ishihara, Yuji ; Ishikawa, Hiroki ; Naitoh, Haruo

A SINGLE PHASE SRM DRIVEN BY COMMERCIAL AC POWER SUPPLY 1141
Aiso, Kohei ; Nakao, Noriya ; Akatsu, Kan

FAST ANALYTICAL MODEL OF SWITCHED RELUCTANCE MACHINE 1148
Smaka, Senad ; Masic, Semsudin ; Cosovic, Mirsad

DETAILED ANALYSIS AND A GENERAL DESIGN PROCEDURE OF DAMPED LCL FILTERS IN THREE PHASE VOLTAGE SOURCE CONVERTERS 1155
Baoquan Liu ; Shaohui Zhong ; Yixin Zhu ; Hao Yi ; Fang Zhuo

70 KHZ, 15 KW SILICON-CARBIDE MOSFET INVERTER FOR INDUSTRIAL INDUCTION HEATING SYSTEMS 1160
Komeda, Shohei ; Tsuboi, Yoshiki ; Fujita, Hideaki

A STUDY ON EFFICIENCY IMPROVEMENT OF HIGH-FREQUENCY CURRENT OUTPUT INVERTER BASED ON IMMITTANCE CONVERSION ELEMENT 1166
Suzuki, Shun ; Shimizu, Toshihisa

HIGH-SPEED SWITCHING METHOD OF MOSFET USING VOLTAGE BOOST AUXILIARY CIRCUIT FED BY GATE DRIVE POWER SUPPLY 1173
Noguchi, Toshihiko ; Murata, Munehiro

OPERATING STRATEGY FOR BI-DIRECTIONAL LLC RESONANT CONVERTER WITH SEAMLESS OPERATION 1179
Abe, Seiya ; Yamamoto, Junichi ; Zaitsu, Toshiyuki ; Ninomiya, Tamotsu

NEGATIVE SEQUENCE CURRENT INJECTION CONTROL ALGORITHM COMPENSATING FOR UNBALANCED PCC VOLTAGE IN MEDIUM VOLTAGE PMSG WIND TURBINES 1185
Jayoon Kang ; Daesu Han ; Suh, Yongsug ; Byoungchang Jung ; Jeongjoong Kim ; Jonghyung Park ; Youngjoon Choi

OPTIMIZATION OF AN OFF-GRID HYBRID SYSTEM FOR SUPPLYING OFFSHORE PLATFORMS IN ARCTIC CLIMATES 1193
Kalogera, Maria ; Bauer, Pavol

ACTIVE DAMPING CONTROL OF LLCL FILTERS FOR THREE-LEVEL T-TYPE GRID CONVERTERS 1201
Alemi, Payam ; Lee, Dong-Choon

DEVELOPING A NEW TOPOLOGY FOR THE DC-DC CONVERTER USED IN FUEL CELL-ELECTRIC DOUBLE LAYER CAPACITOR HYBRID POWER SOURCE SYSTEM FOR MOBILE DEVICES 1207
Tosaka, Shuhei ; Yamanaka, Tatsuya ; Katayama, Noboru ; Hayase, Masanori ; Dowaki, Kiyoshi ; Kogoshi, Sumio

MULTIPLE OUTPUT CHARGER BASED ON PHASE SHIFT FULL BRIDGE CONVERTER WITH NOVEL TIME DIVISION MULTIPLE CONTROL TECHNIQUE 1214
Van-Long Tran ; Woojin Choi

DC-BREAKER FOR A MULTI-MEGAWATT BATTERY ENERGY STORAGE SYSTEM 1220
Demetriades, Georgios D. ; Hermansson, Willy ; Svensson, Jan R ; Papastergiou, Konstantinos ; Larsson, Tomas

ENERGY MANAGEMENT METHOD USING THE IIR FILTER FOR PEMFC-SUPERCAPACITOR HYBRID POWER SOURCE 1227
Yamanaka, Tatsuya ; Katayama, Noboru ; Tosaka, Shuhei ; Kogoshi, Sumio

ADVANCED TORQUE AND CURRENT CONTROL TECHNIQUES FOR PMSMS WITH A REAL-TIME SIMULATOR INSTALLED BEHAVIOR MOTOR MODEL 1234
Tanabe, Ryo ; Akatsu, Kan

COMPENSATION OF THE CURRENT MEASUREMENT ERROR WITH PERIODIC DISTURBANCE OBSERVER FOR MOTOR DRIVE 1242
Yamaguchi, Takashi ; Tadano, Yugo ; Hoshi, Nobukazu

RAPID AND STABLE SPEED CONTROL OF SPMSM BASED ON CURRENT DIFFERENTIAL SIGNAL 1247
Kitajima, Jun ; Ohishi, Kiyoshi

PARALLEL CONNECTED MULTIPLE DRIVE SYSTEM USING SMALL AUXILIARY INVERTER FOR NUMBERS OF PMSM 1253
Nagano, Tsuyoshi ; Itoh, Jun-chi

A TRANSFORMER INRUSH REDUCTION TECHNIQUE FOR LOW-VOLTAGE RIDE-THROUGH OPERATION OF RENEWABLE CONVERTERS 1261
Hsin-Chih Chen ; Ping-Heng Wu ; Cheng, Po-Tai

A CELL CAPACITOR ENERGY BALANCING CONTROL OF MODULAR MULTILEVEL CONVERTER CONSIDERING THE UNBALANCED AC GRID CONDITIONS 1268
Jung, Jae-Jung ; Shenghui Cui ; Kim, Sungmin ; Sul, Seung-Ki

FAULT CURRENT LIMITATION USING THYRISTOR BASED DEVICES 1276
*Komatsu, Wilson ; Giaretta, Antonio Ricardo ; de Miranda, Rubens Domingos ; Jardini, Jose Antonio ; Casolari, Ronaldo Pedro ;
Vasquez-Arnez, Ricardo Leon ; Hojo, Toshiaki ; Carvalho, Eden Luiz ; Maezono, Paulo Koiti*

DC-DC BOOST CONVERTER BASED MSHE-PWM CASCADED MULTILEVEL INVERTER CONTROL FOR STATCOM SYSTEMS 1283
Law, Kah Haw ; Dahidah, Mohamed S.A.

NOVEL PRINCIPLE FOR FLUX SENSING IN THE APPLICATION OF A DC + AC CURRENT SENSOR 1291
Schrittwieser, L. ; Mauerer, M. ; Bortis, D. ; Ortiz, G. ; Kolar, J.W.

UTILIZING VOLTAGE MEASUREMENT OF FET SWITCH FOR MPPT OF DC ENERGY SOURCE 1299
Kimura, Noriyuki ; Niijima, Koji ; Morizane, Toshimitsu ; Omori, Hideki

HIGH FREQUENCY TRANSFORMER BASED ON A COUPLED INDUCTOR TOPOLOGY WITH DIELECTRIC ISOLATION 1303
Amanci, Adrian Z. ; Dawson, Francis P. ; Ruda, Harry E.

CONCEPT AND EXPERIMENTAL EVALUATION OF A NOVEL DC- 100MHZ WIRELESS OSCILLOSCOPE 1309
Lobsiger, Yanick ; Ortiz, Gabriel ; Bortis, Dominik ; Kolar, Johann W.

INTRODUCTION AND EFFECTIVENESS OF STATCOM TO THE INDEPENDENT POWER SYSTEM OF JR EAST 1317
Omi, Masataro ; Kotegawa, Ryo ; Ando, Masato ; Masui, Takeshi ; Horita, Yasuhisa

THE ANALYSIS OF TIME-VARYING RESONANCES IN THE POWER SUPPLY LINE OF HIGH SPEED TRAINS 1322
Chu, Xi ; Lin, Fei ; Yang, Zhongping

FUZZY FEED-FORWARD CHARGE/DISCHARGE CONTROL OF STATIONARY ENERGY STORAGE SYSTEMS FOR DC ELECTRIC RAILWAYS 1328
Kikuchi, Takuya ; Taga, Hironori ; Takagi, Ryo

TRAIN GROUP CONTROL FOR ENERGY-SAVING DC-ELECTRIC RAILWAY OPERATION 1334
Watanabe, Shoichiro ; Koseki, Takafumi

TRANSFORMER-LESS UNIFIED POWER FLOW CONTROLLER USING THE CASCADE MULTILEVEL INVERTER 1342
Fang Zheng ; Shao Zhang ; Shuitao Yang ; Gunasekaran, Deepak ; Karki, Ujjwal

A NEW POWER FLOW CONTROLLER USING SIX MULTILEVEL CASCADED CONVERTERS FOR DISTRIBUTION SYSTEMS......1350

Tsuruta, Ryoji ; Hosaka, Tatsuya ; Fujita, Hideaki

A PROPOSAL OF MODULAR MULTILEVEL CONVERTER APPLYING THREE WINDING TRANSFORMER......1357

Tamada, Shunsuke ; Nakazawa, Yosuke ; Irokawa, Shoichi

BACK-TO-BACK SYSTEM FOR FIVE-LEVEL CONVERTER WITH COMMON FLYING CAPACITORS......1365

Hasegawa, Isamu ; Urushibata, Shota ; Kondo, Takeshi ; Hirao, Kuniaki ; Kodama, Takashi ; Hui Zhang

HARMONIC MODELING OF A VEHICLE TRACTION CIRCUIT TOWARDS THE DC BUS......1373

Haghbin, Saeid ; Karvonen, Andreas ; Thiringer, Torbjorn

AC/DC CONVERTER BASED ON INSTANTANEOUS POWER BALANCE CONTROL FOR REDUCING DC-LINK CAPACITANCE......1379

Tokumasu, Akira ; Taki, Hiroshi ; Shirakawa, Kazuhiro ; Wada, Keiji

MODULAR CONVERTER ARCHITECTURE FOR MEDIUM VOLTAGE ULTRA FAST EV CHARGING STATIONS: DUAL HALF-BRIDGE-BASED ISOLATION STAGE......1386

Vasiladiotis, Michail ; Bahrani, Behrooz ; Burger, Niklaus ; Rufer, Alfred

NEW INTERLEAVED CURRENT-FED RESONANT CONVERTER WITH SIGNIFICANTLY REDUCED HIGH CURRENT OUTPUT FILTER FOR EV AND HEV APPLICATION......1394

Moon, Dongok ; Park, Junsung ; Choi, Sewan

15 PHASE INDUCTION MOTOR DRIVE WITH 1:3:5 SPEED RATIOS USING POLE PHASE MODULATION......1400

Umesh B S ; Sivakumar K

MATHEMATICAL MODEL OF NOVEL WOUND-FIELD SYNCHRONOUS MOTOR SELF-EXCITED BY SPACE HARMONICS......1405

Aoyama, Masahiro ; Noguchi, Toshihiko

DUAL PURPOSE NO VOLTAGE WINDING DESIGN FOR THE BEARINGLESS AC HOMOPOLAR AND CONSEQUENT POLE MOTORS......1412

Severson, Eric ; Nilssen, Robert ; Undeland, Tore ; Mohan, Ned

HARVESTING ENERGY FROM SHIP ROLLING USING AN ECCENTRIC DISK REVOLVING IN A HULA-HOOP MOTION......1420

Yu-Jen Wang

LOAD-INDEPENDENT CURRENT OUTPUT OF INDUCTIVE POWER TRANSFER CONVERTERS WITH OPTIMIZED EFFICIENCY......1425

Zhang, Wei ; Wong, Siu-Chung ; Tse, Chi K. ; Chen, Qianhong

VOLTAGE CONTROL OF INDUCTIVE CONTACTLESS POWER TRANSFER SYSTEM WITH COAXIAL CORELESS TRANSFORMER FOR DC POWER DISTRIBUTION......1430

Miiura, Yushi ; Ojika, Satoshi ; Ise, Tomofumi

CONTACTLESS HIGH POWER TRANSFORMER TECHNOLOGIES FOR RAILWAY VEHICLES......1438

Kondo, Keiichiro ; Yamamoto, Kohei ; Kitazawa, Satochi

TWO-SWITCH VOLTAGE EQUALIZER BASED ON HALF-BRIDGE CONVERTER WITH MULTI-STACKED CURRENT DOUBLERS FOR SERIES-CONNECTED BATTERIES......1444

Uno, Masatoshi ; Kukita, Akio

OPTIMAL ENERGY STORAGE SYSTEM PLANNING FOR MICROGRIDS WITH CONTRACT CAPACITY CONSTRAINT......1452

Shu-Hung Liao ; Jen-Hao Teng ; Yung-Ching Huang ; Dong-Jing Lee

OPTIMAL ZERO SEQUENCE INJECTION IN MULTILEVEL CASCADED H-BRIDGE CONVERTER UNDER UNBALANCED PHOTOVOLTAIC POWER GENERATION......1458

Yu, Yifan ; Konstantinou, Georgios ; Hredzak, Branislav ; Agelidis, Vassilios G.

SIMPLE METHOD FOR MEASURING OUTPUT IMPEDANCE OF A THREE-PHASE INVERTER IN DQ-DOMAIN......1466

Jokipii, Juha ; Messo, Tuomas ; Suntio, Teuvo

ANALYSIS AND DESIGN OF POWER MANAGEMENT SCHEME FOR AN ON-BOARD SOLAR ENERGY STORAGE SYSTEM......1471

Jiang, W. ; Yu, F.Y. ; Lin, Z.Y. ; Wu, G.F. ; Chen, H. ; Hashimoto, S

LVRT CONTROL STRATEGY OF CSC-DPMSG-WGS UNDER UNBALANCED GRID FAULTS......1476

Meiqin Mao ; Yong Ding ; Shiting Weng ; Liuchen Chang

A NEW CURRENT CONTROL DROOP STRATEGY FOR VSI-BASED ISLANDED MICROGRIDS......1482

Shoeiby, B. ; Davoodnezhad, R. ; Holmes, D.G. ; McGrath, B.P.

POWER EXCHANGE USING PFC FOR MICRO GRID......1490

Sakai, Tomoyasu ; Takeda, Takashi ; Yukita, Kazuto ; Goto, Yasuyuki ; Ichiyanagi, Katsuhiro ; Morita, Hiroshi

DETERMINATION OF ROTOR TEMPERATURE FOR AN INTERIOR PERMANENT MAGNET SYNCHRONOUS MACHINE USING A PRECISE FLUX OBSERVER......1501

Specht, Andreas ; Wallscheid, Oliver ; Bocker, Joachim

MONITORING CRITICAL TEMPERATURES IN PERMANENT MAGNET SYNCHRONOUS MOTORS USING LOW-ORDER THERMAL MODELS......1508

Huber, Tobias ; Peters, Wilhelm ; Bocker, Joachim

ROBUST CURRENT CONTROL INSENSITIVE TO GAIN DEVIATION AND OFFSET OF INVERTER DC-LINK CURRENT SENSOR FOR SPMSM......1516

Matsuura, Kei ; Ando, Itaru ; Ohishi, Kiyoshi ; Matsuhashi, Masataka

AUTO-TUNING METHOD OF INDUCTANCES FOR PERMANENT MAGNET SYNCHRONOUS MOTORS......1522

Nomura, Naofumi ; Higuchi, Shinichi

AN IMPEDANCE-BASED STABILITY ANALYSIS METHOD FOR PARALLELED VOLTAGE SOURCE CONVERTERS 1529
Wang, Xiongfei ; Blaabjerg, Frede ; Loh, Poh Chiang

DYNAMIC CHARACTERISTICS AND STABILITY COMPARISONS BETWEEN VIRTUAL SYNCHRONOUS GENERATOR AND DROOP CONTROL IN INVERTER-BASED DISTRIBUTED GENERATORS 1536
Jia Liu ; Miura, Yushi ; Ise, Toshifumi

EMBEDDED LIMITATIONS AND PROTECTIONS FOR DROOP-BASED CONTROL SCHEMES WITH CASCADED LOOPS IN THE SYNCHRONOUS REFERENCE FRAME 1544
D'Arco, Salvatore ; Guidi, Giuseppe ; Suul, Jon Are

VIRTUAL SYNCHRONOUS GENERATOR CONTROL WITH DOUBLE DECOUPLED SYNCHRONOUS REFERENCE FRAME FOR SINGLE-PHASE INVERTER 1552
Hirase, Yuko ; Noro, Osamu ; Yoshimura, Eiji ; Nakagawa, Hidehiko ; Sakimoto, Kenichi ; Shindo, Yuji

CONTACTLESS DC CONNECTOR BASED ON GAN LLC CONVERTER FOR NEXT GENERATION DATA CENTERS 1560
Hayashi, Yusuke ; Toyoda, Hajime ; Ise, Toshifumi ; Matsumoto, Akira

ANALYSIS OF MIS-INTERRUPTION OF SEMICONDUCTOR BREAKER IN DC POWER FEEDING SYSTEM 1567
Murai, Kensuke ; Kanai, Yasuyuki ; Asakimori, Koki ; Babasaki, Tadatoshi

A RELIABLE ELECTRONIC CHOKE WITH NO NEED OF GAIN ADJUSTMENT FOR WIRE COMMUNICATION SYSTEM 1575
Katsuki, Akihiko ; Nakamura, Tatsuya ; Mizuki, Tatsuya ; Shibahara, Kohei ; Abe, Tomohiko ; Ikeda, Tomohiko ; Maeyama, Shigetaka

DESIGN OF NEW CONTROL STRATEGIES FOR A FOUR-LEG THREE-PHASE INVERTER TO ELIMINATE THE NEUTRAL CURRENT UNDER UNBALANCED LOADS 1580
Zhao-Qin Guo ; Panda, Sanjib Kumar ; Prasanna, I.V.

RESEARCH TRENDS OF MODULAR MULTILEVEL CASCADE INVERTER (MMCI-DSCC)-BASED MEDIUM-VOLTAGE MOTOR DRIVES IN A LOW-SPEED RANGE 1586
Okazaki, Yuhei ; Matsui, Hitoshi ; Hagiwara, Makoto ; Akagi, Hirofumi

AN INPUT SWITCHED MULTILEVEL INVERTER FOR OPEN-END WINDING INDUCTION MOTOR DRIVE 1594
Zhu, B. ; Jia, Y. ; Prasanna, U.R. ; Rajashekara, K. ; Kubo, H.

VARIABLE CARRIER FREQUENCY MIXED PWM TECHNIQUE BASED ON CURRENT RIPPLE PREDICTION FOR REDUCED SWITCHING LOSS 1601
Kubo, Hajime ; Yamamoto, Yasuhiro

SLIDING MODE PWM FOR EFFECTIVE CURRENT CONTROL IN SWITCHED RELUCTANCE MACHINE DRIVES 1606
Manolas, Iakovos ; Papafotiou, Georgios ; Manias, Stefanos N.

EXPERIMENTAL VERIFICATION OF AN EMC FILTER USED FOR PWM INVERTER WITH WIDE BAND-GAP DEVICES 1613
Itoh, Jun-ichi ; Araki, Takahiro ; Orikawa, Koji

PACKAGING FOR SIC POWER DEVICE 1621
Funaki, Tsuyoshi

SOLID STATE TRANSFORMER AND MV GRID TIE APPLICATIONS ENABLED BY 15 KV SIC IGBTS AND 10 KV SIC MOSFETS BASED MULTILEVEL CONVERTERS 1626
Madhusoodhanan, Sachin ; Tripathi, Awneesh ; Patel, Dhaval ; Mainali, Krishna ; Kadavelugu, Arun ; Hazra, Samir ; Bhattacharya, Subhashish ; Hatua, Kamalesh

VOLUME 3

GENERALIZED MODULAR MULTILEVEL CONVERTER AND MODULATION 1634
Hui Liu ; Loh, Poh Chiang ; Blaabjerg, Frede

AVERAGE POWER CONTROL OF DC BUS VOLTAGES OF CASCADED H-BRIDGE MULTILEVEL CONVERTERS 1639
Lee, Chia-Tse ; Chen, Hsin-Chih ; Ching-Wei Wang ; Ching-Hsiang Yang ; Cheng, Po-Tai

ANALYSIS AND COMPARISON OF HIGH POWER SEMICONDUCTOR DEVICE LOSSES IN 5MW PMSG MV WIND TURBINES 1646
Kihyun Lee ; Kyungsub Jung ; Seunghoo Song ; Suh, Yongsug ; Changwoo Kim ; Hyoyol Yoo ; Sunsoon Park

APPLICATION OF MODULAR MATRIX CONVERTER TO WIND TURBINE GENERATOR 1654
Inomata, Kentaro ; Hara, Hidenori ; Morimoto, Shinya ; Fujii, Junji ; Takeda, Kotaro ; Yamamoto, Eiji

FREE MOTION MECHANICAL POWER FACTOR; COMPARISON BETWEEN ROBOTS IN DIFFERENT STRUCTURE AND COORDINATE 1660
Mizoguchi, Takahiro ; Nozaki, Takahiro ; Ohnishi, Kouhei

ANALYSIS OF SETTLING BEHAVIOR AND DESIGN OF CASCADED PRECISE POSITIONING CONTROL IN PRESENCE OF NONLINEAR FRICTION 1665
Ruderman, Michael ; Iwasaki, Makoto

FIELD AND BENCH TEST EVALUATION OF RANGE EXTENSION CONTROL SYSTEM FOR ELECTRIC VEHICLES BASED ON FRONT AND REAR DRIVING-BRAKING FORCE DISTRIBUTIONS 1671
Fujimoto, Hiroshi ; Harada, Shingo ; Goto, Yuichi ; Kawano, Daisuke ; Sato, Koji ; Matsuo, Yusuke

VIBRATION SUPPRESSION OF INTEGRATED RESONANT AND TIME DELAY SYSTEM BY REFLECTED WAVE REJECTION 1679
Saito, Eiichi ; Oboe, Roberto ; Katsura, Seiichiro

THRUST CHARACTERISTICS IMPROVEMENT OF A CIRCULAR SHAFT MOTOR FOR DIRECT-DRIVE APPLICATIONS 1685
Omura, Mototsugu ; Shimono, Tomoyuki ; Fujimoto, Yasutaka

DESIGN OF A BEARINGLESS FLUX-SWITCHING SLICE MOTOR 1691
Gruber, Wolfgang ; Radman, Karlo ; Schob, Reto.T.

PROPOSAL OF A PERMANENT MAGNET HYBRID TYPE AXIAL MAGNETICALLY LEVITATED MOTOR 1697
Kurita, Nobuyuki ; Ishikawa, Takeo ; Takada, Hiromu ; Suzuki, Genri

COMPARISON OF HIGH SPEED BEARINGLESS DRIVE TOPOLOGIES WITH COMBINED WINDINGS 1701
Mitterhofer, Hubert ; Mrak, Branimir ; Gruber, Wolfgang

HIGH-SPEED MAGNETICALLY LEVITATED REACTION WHEEL DEMONSTRATOR 1707
Zwyssig, Christof ; Baumgartner, Thomas ; Kolar, Johann W.

STABILIZED SUSPENSION CONTROL CONSIDERING ARMATURE REACTION IN A D-Q AXIS CURRENT CONTROL BEARINGLESS MOTOR 1715
Ooshima, Masahide ; Kumakura, Yoshito

ANALYSIS AND DESIGN OF A HIGH-FREQUENCY ISOLATED DUAL-TANK LCL RESONANT AC-DC CONVERTER 1721
Du, Yimian ; Bhat, Ashoka K.S.

VERIFICATION OF LLC RESONANT CONVERTER APPLIED A CURRENT-BALANCING HIGH-FREQUENCY TRANSFORMER WITH MULTI-OUTPUT WINDINGS 1728
Araki, Jun ; Shinozaki, Ikki ; Funato, Hirohito ; Ogasawara, Satoshi ; Murakami, Daichi ; Hirota, Yukitsugu ; Mihara, Teruyoshi ; Mouri, Masayuki ; Okazaki, Fumihiro

LIGHT-LOAD EFFICIENCY IMPROVEMENT STRATEGY FOR LLC RESONANT CONVERTER UTILIZING A STEP-GAP TRANSFORMER 1734
Huang, Wen-Nan ; Lee, Shiu-Hui ; Chen, Ching-Guo

A NOVEL ACCURATE PRIMARY SIDE CONTROL (PSC) METHOD FOR HALF-BRIDGE (HB) LLC CONVERTER 1738
Jae-Bum Lee ; Kim, Chong-Eun ; Jae-Hyun Kim ; Cheol-O Yeon ; Young-Do Kim ; Moon, Gun-Woo

A SIMPLE CONTROL SCHEME FOR IMPROVING LIGHT-LOAD EFFICIENCY IN A FULL-BRIDGE LLC RESONANT CONVERTER 1743
Kim, Jae-Hyun ; Kim, Chong-Eun ; Lee, Jae-Bum ; Young-Do Kim ; Han-Shin Youn ; Moon, Gun-Woo

POWER CONDITIONER FOR STABILIZING POWER DISTURBANCE CAUSED OF WIND TURBINE GENERATOR SYSTEM 1748
Saga, Yasunao ; Fujii, Kansuke ; Yoda, Kazuyuki

A FRONT-TO-FRONT (FTF) SYSTEM CONSISTING OF MULTIPLE MODULAR MULTILEVEL CASCADE CONVERTERS FOR OFFSHORE WIND FARMS 1761
Sasongko, Firman ; Hagiwara, Makoto ; Akagi, Hirofumi

MODELLING, DESIGN AND CONTROL OF GRID CONNECTED CONVERTER FOR HIGH ALTITUDE WIND POWER APPLICATION 1775
Adhikari, Jeevan ; Rathore, Akshay K. ; Panda, S K

PRACTICAL STUDY OF A HIGH STEP-DOWN CONVERTER 1781
Jinno, Masahito ; Su, Hong-Wei ; Tsai, Jiung-Lin ; Matsuo, Hirofumi

GENERALIZED MODELING AND OPTIMIZATION OF A BIDIRECTIONAL DUAL ACTIVE BRIDGE DC-DC CONVERTER INCLUDING FREQUENCY VARIATION 1788
Jauch, Felix ; Biela, Jurgen

BALANCED DISCHARGING OF POWER BANK WITH BUCK-BOOST BATTERY POWER MODULES 1796
Moo, Chin-Sien ; Wu, Tsung-Hsi ; Hou, Chih-Hao ; Hsieh, Yao-Ching

Y-SOURCE IMPEDANCE-NETWORK-BASED ISOLATED BOOST DC/DC CONVERTER 1801
Siwakoti, Yam P. ; Town, Graham E. ; Loh, Poh Chiang ; Blaabjerg, Frede

MULTI-PHASE DC-DC CONVERTER WITH RIPPLE-LESS OPERATION FOR THERMO-ELECTRIC GENERATOR 1806
Kimura, Noriyuki ; Niijima, Koji ; Morizane, Toshimitsu ; Omori, Hideki

POSITION SENSORLESS START-UP METHOD OF SURFACE PERMANENT MAGNET SYNCHRONOUS MOTOR USING NONLINEAR ROTOR POSITION OBSERVER 1811
Hanamoto, Tsuyoshi ; Yamada, Hiroaki ; Okuyama, Yoshihiro

SENSORLESS CONTROL OF PMSM FOR THE WHOLE SPEED RANGE USING TWO-DEGREE-OF-FREEDOM CURRENT CONTROL AND HF TEST CURRENT INJECTION FOR LOW SPEED RANGE 1816
Seilmeier, Markus ; Piepenbreier, Bernhard

ELLIPSE-TRAJECTORY-ORIENTED VECTOR CONTROL FOR ENERGY EFFICIENT/WIDE-SPEED-RANGE DRIVES OF SENSORLESS PMSM 1824
Shinnaka, Shinji ; Amano, Yuki

DEVELOPMENT OF POSITION SENSORLESS CONTROL FOR PERMANENT-MAGNET SYNCHRONOUS GENERATOR DRIVE 1832
Chang, Yuan-Chih ; Lin, Chia-Yu ; Dai, Wei-Fu ; Wu, Chun-Wei

CONTROL OF A 750KW PERMANENT MAGNET SYNCHRONOUS MOTOR 1837
Liping Zheng ; Dong Le

REGIONAL SMART GRID OF ISLAND IN CHINA WITH MULTIFOLD RENEWABLE ENERGY 1842
Xu Cai ; Zheng Li

STABILIZING SMALL ISLAND POWER SYSTEM WITH RENEWABLES BY USE OF POWER CONDITIONING SYSTEMS - JAPANESE ISLAND SYSTEM CASE - 1849
Baba, Jumpei

POWER ELECTRONICS SOLUTIONS APPLIED TO A VARIETY OF DEMONSTRATIVE MICROGRID PROJECTS 1855
Ueda, Yoshinobu

MOVING TOWARDS THE SMART GRID: THE NORWEGIAN CASE 1861
Fosso, Olav B. ; Molinas, Marta ; Sand, Kjell ; Coldevin, Grete H.

POWER ELECTRONICS TECHNOLOGY IN SMART GRID PROJECTS -APPLICATIONS AND EXPERIENCES- 1868
Kobayashi, Takenori

EV AND HEV MOTOR DEVELOPMENT IN TOSHIBA 1874
Arata, Masanori ; Kurihara, Yoshihiro ; Misu, Daisuke ; Matsubara, Masakatsu

MOTOR STATOR WITH THICK RECTANGULAR WIRE LAP WINDING FOR HEVS 1880
Ishigami, Takashi ; Tanaka, Yuichiro ; Homma, Hiroshi

COMPARISON STUDY OF VARIOUS MOTORS FOR EVS AND THE POTENTIALITY OF A FERRITE MAGNET MOTOR 1886
Matsuhashi, Daiki ; Matsuo, Keisuke ; Okitsu, Takashi ; Ashikaga, Tadashi ; Mizuno, Takayuki

OPTIMAL FIELD EXCITATION CONTROL OF A CLAW POLE MOTOR FOR HYBRID ELECTRIC VEHICLE 1892
Azuma, M. ; Hazeyama, M. ; Morita, M. ; Kuroda, Y. ; Daikoku, A. ; Inoue, M.

A WIDE SPEED RANGE HIGH EFFICIENCY EV DRIVE SYSTEM USING WINDING CHANGEOVER TECHNIQUE AND SIC DEVICES 1898
Takatsuka, Yushi ; Hara, Hidenori ; Yamada, Kenji ; Maemura, Akihiko ; Kume, Tsuneo

PERFORMANCE COMPARISON OF A GAN GIT AND A SI IGBT FOR HIGH-SPEED DRIVE APPLICATIONS 1904
Tuysuz, Arda ; Bosshard, Roman ; Kolar, Johann W.

WIDE-BAND GAP DEVICES IN PV SYSTEMS - OPPORTUNITIES AND CHALLENGES 1912
Sintamarean, C. ; Eni, E. ; Blaabjerg, F. ; Teodorescu, R. ; Wang, H.

POWER ELECTRONICS EQUIPMENTS APPLYING NOVEL SIC POWER SEMICONDUCTOR MODULES 1920
Mino, Kazuaki ; Yamada, Ryuji ; Kimura, Hiroshi ; Matsumoto, Yasushi

EMI PREDICTION METHOD FOR SIC INVERTER BY THE MODELING OF STRUCTURE AND THE ACCURATE MODEL OF POWER DEVICE 1929
Maekawa, Sari ; Tsuda, Junichi ; Kuzumaki, Atsuhiko ; Matsumoto, Shuhei ; Mochikawa, Hiroshi ; Kubota, Hisao

SYSTEM INTEGRATION OF GAN TECHNOLOGY 1935
Ferreira, J.A. ; Popovic, J. ; van Wyk, J.D. ; Pansier, F.

POWER LOSSES OF MULTILEVEL CONVERTERS IN TERMS OF THE NUMBER OF THE OUTPUT VOLTAGE LEVELS 1943
Kashihara, Yugo ; Itoh, Jun-ichi

A LARGE CAPACITY 3-LEVEL IEGT INVERTER 1950
Yoshizawa, Daisuke ; Mukunoki, Makoto ; Omote, Kenichiro ; Hayashi, Makoto ; Isida, Takashi

VIBRATION SUPPRESSING CONTROL METHOD OF ANGULAR TRANSMISSION ERROR OF CYCLOID GEAR FOR INDUSTRIAL ROBOTS 1956
Yoshioka, Takashi ; Hirano, Yosei ; Ohishi, Kiyoshi ; Miyazaki, Toshimasa ; Yokokura, Yuki

AN ADVANCED POSITION CONTROL OF OVERHEAD CRANE BY SWAY SUPPRESSION METHOD EMULATING NATURAL DAMPING 1962
Kurabayashi, Toshiyuki ; Yang Chuan ; Murakami, Toshiyuki

A ROBOTIC CANE FOR WALKING ASSISTANCE 1968
Shimizu, Kyohei ; Smadi, Issam ; Fujimoto, Yasutaka

HAND POSITION ESTIMATION IN BINOCULAR VISUAL SPACE USING LINEAR APPROXIMATION OF KINEMATICS 1974
Komada, Satoshi ; Turpin, Santiago ; Hashimoto, Kento ; Yashiro, Daisuke ; Hirai, Junji

CONTACT STATE RECOGNITION BASED ON HAPTIC SIGNAL PROCESSING FOR ROBOTIC TOOL USE 1978
Matsuzaki, Ryohei ; Okuma, Jun ; Sakaino, Sho ; Tsuji, Toshiaki

RECENT TECHNICAL TRENDS IN MAGNETIC MATERIALS 1984
Wajima, Kiyoshi ; Toda, Hiroaki ; Kosaka, Takashi ; Marukawa, Yasuhiro ; Ishihara, Chio

MULTI-DOMAIN CO-SIMULATION WITH NUMERICALLY IDENTIFIED PMSM INTERWORKING AT HILS FOR ELECTRIC PROPULSION 1990
Park, Gyeong-Jae ; Jung, Hochang ; Kim, Yong-Jae ; Jung, Sang-Yong

RECENT TECHNICAL TRENDS IN PMSM 1997
Morimoto, Shigeo ; Asano, Yoshinari ; Kosaka, Takashi ; Enomoto, Yuji

RECENT TECHNICAL TRENDS IN SRM AND FSM 2004
Kano, Yoshiaki

RECENT TECHNICAL TRENDS IN VARIABLE FLUX MOTORS 2011
Toba, Akio ; Daikoku, Akihiro ; Nishiyama, Noriyoshi ; Yoshikawa, Yuichi ; Kawazoe, Yosuke

A GENERAL DISCRETE TIME MODEL TO EVALUATE ACTIVE DAMPING OF GRID CONVERTERS WITH LCL FILTERS 2019
Parker, S.G. ; McGrath, B.P. ; Holmes, D.G.

ANALYSIS AND REDUCTION OF POWER LOSSES IN PV CONVERTERS FOR GRID CONNECTION TO LOW-VOLTAGE THREE-PHASE THREE-WIRE SYSTEMS 2027
Amma, Ryosuke ; Fujita, Hideaki

DESIGN OF GRID CONNECTED PWM CONVERTERS CONSIDERING TOPOLOGY AND PWM METHODS FOR LOW-VOLTAGE RENEWABLE ENERGY APPLICATIONS 2034
Kantar, Emre ; Hava, Ahmet M.

PERFORMANCE OF DEAD TIME COMPENSATION METHODS IN THREE-PHASE GRID-CONNECTION CONVERTERS 2042
Mannen, Tomoyuki ; Fujita, Hideaki

D-S DIGITAL CONTROL FOR THREE-PHASE BI-DIRECTIONAL INVERTERS 2050
Wu, T.-F. ; Chang, C.-H. ; Lin, L.-C.

EXPECTATIONS OF NEXT-GENERATION POWER DEVICES FOR HOME AND CONSUMER APPLIANCES 2058
Kanouda, Akihiko ; Shoji, Hiroyuki ; Shimada, Takae ; Okubo, Toshikazu

APPLICATION TREND AND FORESIGHT OF SIC POWER DEVICES TO AIR CONDITIONERS 2064
Kamikura, Mamoru ; Murata, Yuichiro ; Kutsuki, Tomohiro ; Saito, Katsuhiko

RECENT TECHNICAL TRENDS AND FUTURE PROSPECTS OF IGBTS AND POWER MOSFETS 2068
Ogura, Tsuneo

RECENT DEVELOPMENT AND FUTURE PROSPECTS OF POWER SIC DEVICES 2074
Nakamura, T. ; Nakano, Y. ; Aketa, M. ; Hanada, T.

RECENT ADVANCES AND FUTURE PROSPECTS ON GAN-BASED POWER DEVICES 2075
Ueda, Tetsuzo

SCALING AND BALANCING OF MULTI-CELL CONVERTERS 2079
Kasper, Matthias ; Bortis, Dominik ; Kolar, Johann W.

HYBRID MODULATED UNIVERSAL SOFT-SWITCHING CURRENT-FED DC/DC CONVERTER FOR WIDE VOLTAGE REGULATION FOR PV/FUEL CELLS/BATTERY APPLICATIONS 2087
Moorthy, Radha Sree Krishna ; Rathore, Akshay Kumar

HIGH EFFICIENCY POWER CONVERTERS FOR BATTERY ENERGY STORAGE SYSTEMS 2095
Kawakami, Noriko ; Iijima, Yukihia ; Li, Haiqing ; Ota, Satoru

IMPLEMENTATION OF BRIDGELESS CUK POWER FACTOR CORRECTOR WITH POSITIVE OUTPUT VOLTAGE 2100
Yang, Hong-Tzer ; Chiang, Hsin-Wei

A NOVEL SYNCHRONOUS RECTIFIER METHOD FOR A LLC RESONANT CONVERTER WITH VOLTAGE-DOUBLER RECTIFIER 2108
Murata, Koji ; Kurokawa, Fujio

LATEST DEVELOPMENTS IN INCREASING THE POWER DENSITY OF TRACTION DRIVES 2113
Bakran, Mark-M. ; Marz, Andreas ; Laska, Bernd ; Krafft, Eberhard ; Korner, Olaf ; Nagel, Andreas

CATENARY AND STORAGE BATTERY HYBRID SYSTEM FOR ELECTRIC RAILCAR SERIES EV-E301 2120
Kono, Y. ; Shiraki, N. ; Yokoyama, H. ; Furuta, R.

TECHNOLOGY FOR ENERGY-SAVING RAILWAY OPERATION THROUGH POWER-LIMITING BRAKES—A CASE STUDY AT AN URBAN RAILWAY 2126
Koseki, Takafumi ; Watanabe, Shoichiro ; Hamazaki, Yasuhiro ; Kondo, Keiichiro ; Hasegawa, Tomonori ; Mizuma, Takeshi

AN OVERVIEW ON BRAKING ENERGY REGENERATION TECHNOLOGIES IN CHINESE URBAN RAILWAY TRANSPORTATION 2133
Yang, Zhongping ; Xia, Huan ; Wang, Bin ; Lin, Fei

TRACTION INVERTER THAT APPLIES COMPACT 3.3 KV / 1200 A SIC HYBRID MODULE 2140
Ishikawa, Katsumi ; Yukutake, Seigo ; Kono, Yasuhiko ; Ogawa, Kazutoshi ; Kameshiro, Norifumi

POWER ELECTRONIC-BASED PROTECTION FOR DIRECT-CURRENT POWER DISTRIBUTION IN MICRO-GRIDS 2145
Tseng, K.J. ; Luo, Guomin

A CONCEPT OF HIGH POWER DC/DC CONVERTER WITH DOUBLE LOW POWER OUTPUTS 2152
Hojo, Masahide ; Nishioka, Tomoya ; Yamanaka, Kenji

PERFORMANCE EVALUATION FOR GRID IMPEDANCE BASED ISLANDING DETECTION METHOD 2156
Liu, Ning ; Aljankawey, A.S. ; Diduch, C.P. ; Chang, L. ; Mao, Meiqin ; Yazdkhasti, Pegah ; Su, Jianhui

IDENTIFYING NATURAL DEGRADATION/AGING IN POWER MOSFETS IN A LIVE GRID-TIED PV INVERTER USING SPREAD SPECTRUM TIME DOMAIN REFLECTOMETRY 2161
Li, Qian ; Khan, Faisal H.

CONTROL METHOD FOR INDUCTIVE POWER TRANSFER WITH HIGH PARTIAL-LOAD EFFICIENCY AND RESONANCE TRACKING 2167
Bosshard, R. ; Kolar, J.W. ; Wunsch, B.

STANDARD MODELS FOR SMART GRID SIMULATIONS 2175
Noda, Taku ; Nagashima, Tomohiro ; Sekisue, Takayuki ; Kabasawa, Yuichiro ; Kato, Shinji ; Sekiba, Yoichi ; Tokuda, Hirokazu ; Kounoto, Masaaki

MODEL DEVELOPMENT FOR MOTOR DRIVE SYSTEM SIMULATIONS 2183
Ishikawa, Hiroki ; Abe, Takashi ; Kato, Toshiji ; Kubota, Yutaka ; Shimomura, Junichi ; Kohno, Yusuke ; Ikeda, Masahiro ; Umeda, Nobuhiro ; Kimura, Noriyuki ; Shigematsu, Koichi ; Inoue, Yukinori

PRACTICAL SIMULATION EXAMPLES OF AUTOMOTIVE AND POWER SUPPLY SYSTEMS 2189
Abe, Takashi ; Fukushima, Kentaro ; Sekisue, Takayuki ; Shigematsu, Koichi ; Ichihara, Junichi ; Kato, Toshiji ; Ishikawa, Hiroki ; Kouno, Yusuke ; Konoto, Masaaki ; Saito, Ryoji ; Nishida, Yasuyuki

ADMITTANCE MATRICES OF VOLTAGE SOURCE CONVERTERS FOR DISTRIBUTED GENERATORS 2195
Lian, K.L. ; Huang, T.D.

FPGA-BASED SIMULATION OF POWER ELECTRONICS USING ITERATIVE METHODS 2202
Zhang, Huiguo ; Sun, Jian

GALLIUM ARSENIDE IC TECHNOLOGY FOR POWER SUPPLIES ON CHIP 2208
Pala, Vipindas ; Peng, Han ; Hella, Mona ; Chow, T.Paul

SILICON ON NANOCRYSTALLINE AND MICROCRYSTALLINE DIAMOND STACKING STRUCTURE FOR POWER SUPPLY ON CHIP 2212
Yamada, Takatoshi ; Hasegawa, Masataka

A NOVEL LOAD REGULATION TECHNIQUE FOR POWER-SOC WITH PARALLEL CONNECTED POLS 2216
Abe, Seiya ; Matsumoto, Satoshi ; Hidaka, Akira ; Rikitake, Jungo ; Ninomiya, Tamotsu

MATRIX-POL ARCHITECTURE FOR INTEGRATED POWER SUPPLY 2222
Ishizuka, Yoichi ; Shibahara, Ryota ; Ninomiya, Tamotsu ; Tanaka, Kiminori ; Abe, Seiya

ON-CHIP BUCK CONVERTER WITH SPIRAL FERRITE INDUCTOR AND REDUCING IR DROP IN 3D STACKED INTEGRATION 2228
Fuketa, Hiroshi ; Shinozuka, Yasuhiro ; Ishida, Koichi ; Takamiya, Makoto ; Sakurai, Takayasu

DCM ANALYSIS OF A SINGLE SIC SWITCH BASED ZVZCS TAPPED BOOST CONVERTER 2232
Choi, Bo H. ; Lee, Eun S. ; Kim, Ji H. ; Rim, Chun T.

EFFECT OF INPUT AND OUTPUT TERMINAL SOURCES ON DYNAMIC BEHAVIOR OF SWITCHED-MODE CONVERTERS 2240
Suntio, T. ; Viinamaki, J. ; Jokipii, J. ; Messo, T. ; Sitbon, M. ; Kuperman, A.

A FULLY SOFT-SWITCHED MULTIPHASE DC-DC CONVERTER WITH REDUCED SWITCH COUNT FOR HIGH POWER APPLICATION 2247
Kim, Minjae ; Yang, Daeki ; Choi, Sewan

A STATIC CHARACTERISTIC ANALYSIS OF PROPOSED BI-DIRECTIONAL DUAL ACTIVE BRIDGE DC-DC CONVERTER 2252
Nagata, Shun ; Takasaki, Mika ; Furukawa, Yutaka ; Hirose, Toshiro ; Ishizuka, Yoichi

HYBRID BATTERY CHARGING SYSTEM COMBINING OBC WITH LDC FOR ELECTRIC VEHICLES 2260
Kim, Seonghye ; Kang, Feel-soon

TRANSIENT BEHAVIOR OF THE DUAL ACTIVE BRIDGE CONVERTER IN HIGH EFFICIENT ENERGY CONVERSION SYSTEM 2266
Aoyama, Kohei ; Motoi, Naoki ; Tsuruta, Yukinori ; Kawamura, Atsuo

STATE-OF-CHARGE ESTIMATION FOR LITHIUM-ION BATTERY PACK USING RECONSTRUCTED OPEN-CIRCUIT-VOLTAGE CURVE 2272
Chun, Chang Yoon ; Seo, Gab-Su ; Yoon, Sung Hyun ; Cho, Bo-Hyung

SYSTEM DESIGN OF ELECTRIC ASSISTED BICYCLE USING EDLCS AND WIRELESS CHARGER 2277
Itoh, Jun-ichi ; Noguchi, Kenji ; Orikawa, Koji

STUDY ON LOW-LOSS GATE DRIVE CIRCUIT FOR HIGH EFFICIENCY SERVER POWER SUPPLY USING NORMALLY-OFF SIC-JFET 2285
Katoh, Kaoru ; Ishikawa, Katsumi ; Hatanaka, Ayumu ; Ogawa, Kazutoshi ; Akiyama, Satoru ; Ogawa, Takashi ; Yokoyama, Natsuki ; Maru, Naoki ; Takahashi, Osamu ; Nishisu, Koji

A SHORT CIRCUIT PROTECTION METHOD BASED ON A GATE CHARGE CHARACTERISTIC 2290
Horiguchi, Takeshi ; Kinouchi, Shin-ichi ; Nakayama, Yasushi ; Oi, Takeshi ; Urushibata, Hiroaki ; Okamoto, Shoji ; Tominaga, Shinji ; Akagi, Hirofumi

HIGHLY RELIABLE 1200-V P-TYPE MOSFET FOR LEVEL-SHIFT CIRCUIT USED IN DRIVER IC 2297
Sakurai, Naoki ; Hakutou, Takuma ; Yura, Masashi

A NEW LEVEL UP SHIFTER FOR HVICS WITH HIGH NOISE TOLERANCE 2302
Akahane, Masashi ; Jonishi, Akihiro ; Yamaji, Masaharu ; Kanno, Hiroshi ; Tanaka, Takahide ; Nishio, Haruhiko ; Sumida, Hitoshi

OUTPUT RIPPLE MINIMIZATION OF SINGLE-STAGE POWER FACTOR CORRECTED BI-DIRECTIONAL BUCK AC/DC CONVERTER 2310
Veerasamy, Balaji ; Kitagawa, Wataru ; Takeshita, Takaharu

THREE-PHASE ISOLATED FULL-BRIDGE BOOST PFC WITH FLYBACK PASSIVE AUXILIARY CONVERTER 2318
Meng, Tao ; Yu, Shuai ; Ben, Hongqi ; Wei, Guo ; Sun, Shaohua

CONTROL AND EXPERIMENT OF A MODULAR PUSH-PULL PWM CONVERTER FOR A BATTERY ENERGY STORAGE SYSTEM 2323
Hagiwara, Makoto ; Akagi, Hirofumi

ACTIVE FRONT-END TOPOLOGY FOR 5 LEVEL MEDIUM VOLTAGE DRIVE SYSTEM WITH ISOLATED DC BUS 2330
Oka, Toshiaki ; Kusunoki, Hironobu ; Tsukakoshi, Masahiko ; Kleinecke, John ; Daskalos, Mike

A DUAL ACTIVE BRIDGE DC-DC CONVERTER WITH OPTIMAL DC-LINK VOLTAGE SCALING AND FLYBACK MODE FOR ENHANCED LOW-POWER OPERATION IN HYBRID PV/STORAGE SYSTEMS 2336
Poshtkouhi, Shahab ; Trescases, Olivier

NOVEL MODULAR MULTIPLE-INPUT BIDIRECTIONAL DC-DC POWER CONVERTER (MIPC) 2343
Hintz, Andrew ; Prasanna, Udupi.R. ; Rajashekara, Kaushik

SINGLE-SWITCH PWM CONVERTER INTEGRATING VOLTAGE EQUALIZER FOR PHOTOVOLTAIC MODULES UNDER PARTIAL SHADING..2351
Uno, Masatoshi ; Kukita, Akio

NEW DC RAIL SIDE SOFT-SWITCHING PWM DC-DC CONVERTER WITH VOLTAGE DOUBLER RECTIFIER FOR PV GENERATION INTERFACE...2359
Sayed, Khairy ; Kwon, Soon-Kurl ; Nishida, Katsumi ; Nakaoka, Mutsuo

MODELING METHOD OF STRAY MAGNETIC COUPLINGS IN AN EMC FILTER FOR A SIC SOLAR INVERTER..2366
Masuzawa, Takashi ; Hoene, Eckart ; Hoffmann, Stefan ; Lang, Klaus-Dieter

DC BUS VOLTAGE EMI MITIGATION IN THREE-PHASE ACTIVE RECTIFIERS USING A VIRTUAL NEUTRAL FILTER...2372
Parker, S.G. ; Segaran, D.S. ; Holmes, D.G. ; McGrath, B.P.

EFFECTS OF TRANSFORMER STRUCTURES ON THE NOISE BALANCING AND CANCELLATION MECHANISMS OF SWITCHING POWER CONVERTERS...2380
Hsieh, Hung-I ; Shih, Sheng-Fang

A NOVEL TECHNIQUE FOR REDUCING LEAKAGE CURRENT BY APPLICATION OF ZERO-SEQUENCE VOLTAGE...2385
Ayano, Hideki ; Murakami, Kouhei ; Matsui, Yoshihiro

AC-CHOPPERS USING INSTANTANEOUS VOLTAGE CONTROL TECHNIQUE TO SOLVE VOLTAGE SAG PROBLEMS..2392
Khomfoi, Surin

VOLTAGE REGULATION IN DISTRIBUTION SYSTEM USING THE COMBINED DVR.....................2400
Nakamura, Sota ; Aoki, Mutsumi ; Ukai, Hiroyuki

NONLINEAR CONTROL OF THREE-PHASE FOUR-WIRE DYNAMIC VOLTAGE RESTORERS FOR DISTRIBUTION SYSTEM..2406
Jeong, Seon-Yeong ; Nguyen, Thanh Hai ; Lee, Dong-Choon ; Kim, Jang-Mok

VOLUME 4

DISTURBANCE CALCULATION BASED ON SPACE VECTOR DOT PRODUCT: APPLICATIONS TO COMPENSATORS...2413
de Carvalho, Kelly Caroline Mingorancia ; Ama, Naji Rajai Nasri ; Komatsu, Wilson ; Martinz, Fernando Ortiz ; Figueredo, Ricardo Souza ; Matakas, Lourenco

PROPOSAL OF 6TH RADIAL FORCE CONTROL BASED ON FLUX LINKAGE..2421
Kanematsu, Masato ; Miyajima, Takayuki ; Fujimoto, Hiroshi ; Hori, Yoichi ; Enomoto, Toshio ; Kondou, Masahiko ; Komiya, Hiroshi ; Yoshimoto, Kantaro ; Miyakawa, Takayuki

AIR GAP CONTROL OF MULTI-PHASE TRANSVERSE FLUX PERMANENT MAGNET LINEAR SYNCHRONOUS MOTOR BY USING INDEPENDENT VECTOR CONTROL...2427
Hwang, Seon-Hwan ; Bang, Deok-Je ; Kim, Ji-Won

MODIFIED DIRECT INSTANTANEOUS TORQUE CONTROL OF SWITCHED RELUCTANCE MOTOR WITH HIGH TORQUE PER AMPERE AND REDUCED SOURCE CURRENT RIPPLE................................2433
Suryadevara, Rohit ; Fernandes, B.G.

CONTROL OF WOUND FIELD SYNCHRONOUS MOTOR INTEGRATED WITH ZSI............................2438
Tajima, G. ; Kosaka, T. ; Matsui, N. ; Tonogi, K. ; Minoshima, N. ; Yoshida, T.

A NOVEL IPMSM MODEL FOR ROBUST POSITION SENSORLESS CONTROL TO MAGNETIC SATURATION..2445
Matsumoto, Atsushi ; Hasegawa, Masaru ; Doki, Shinji

MOTOR DRIVE SYSTEM USING NONLINEAR MATHEMATICAL MODEL FOR PERMANENT MAGNET SYNCHRONOUS MOTORS...2451
Iwaji, Yoshitaka ; Nakatsugawa, Junnosuke ; Sakai, Toshifumi ; Aoyagi, Shigehisa ; Nagura, Hirokazu

SENSORLESS-ORIENTED DESIGN OF IPMSM...2457
Kano, Yoshiaki

NOISE REDUCTION METHOD BY INJECTED FREQUENCY CONTROL FOR POSITION SENSORLESS CONTROL OF PERMANENT MAGNET SYNCHRONOUS MOTOR...2465
Taniguchi, Shun ; Yasui, Kazuya ; Yuki, Kazuaki

FORCE SENSORLESS BILATERAL CONTROL USING A DYNAMICAL ASYMMETRIC COMPENSATOR.....................2470
Hama, Ryota ; Imai, Jun ; Takahashi, Akiko ; Funabiki, Shigeyuki

DESIGN OF M-IPD CONTROLLER OF MULTI-INERTIA SYSTEM USING DIFFERENTIAL EVOLUTION.....................2476
Ikeda, Hidehiro ; Tsuyoshi, Hanamoto

A GUIDE TO DESIGN DISTURBANCE OBSERVER BASED MOTION CONTROL SYSTEMS...................2483
Sariyildiz, Emre ; Ohnishi, Kouhei

IDENTIFICATION OF TWO-MASS MECHANICAL SYSTEMS USING TORQUE EXCITATION: DESIGN AND EXPERIMENTAL EVALUATION..2489
Saarakkala, Seppo E. ; Hinkkanen, Marko

INDUCTOR LOSS CALCULATION OF COUPLED INDUCTORS FOR HIGH POWER DENSITY BOOST CONVERTER...2497
Itoh, Yuki ; Kimura, Shota ; Imaoka, Jun ; Yamamoto, Masayoshi

1.2KW DUAL-ACTIVE BRIDGE CONVERTER USING SIC POWER MOSFETS AND PLANAR MAGNETICS.................2503
De, D. ; Castellazzi, A. ; Lamantia, A.

ANALYSIS OF HYSTERESIS AND EDDY-CURRENT LOSSES FOR A MEDIUM-FREQUENCY TRANSFORMER IN AN ISOLATED DC-DC CONVERTER..2511
Nakahara, Mizuki ; Wada, Keiji

EXPERIMENTAL VERIFICATION OF CAPACITIVE POWER TRANSFER USING ONE PULSE SWITCHING ACTIVE CAPACITOR FOR PRACTICAL USE..2517
Kitabayashi, Tatsuaki ; Funato, Hirohito ; Kobayashi, Hiroya ; Yamaichi, Katsuya

A SINGLE-STAGE HIGH-PF DRIVER FOR SUPPLYING A T8-TYPE LED LAMP......................2523
Cheng, Chun-An ; Chang, Chien-Hsuan ; Cheng, Hung-Liang ; Chung, Tsung-Yuan

ELIMINATION OF ELECTROLYTIC CAPACITOR IN AC-DC SYSTEM OF LED DRIVER..........2529
Mustapa, Rijalul Fahmi ; Hidayat, Nabil M ; Tukiman, Rahayu

A NOVEL BRIDGELESS BOOST HALF-BRIDGE ZVS-PWM SINGLE-STAGE UTILITY FREQUENCY AC-HIGH FREQUENCY AC RESONANT CONVERTER FOR DOMESTIC INDUCTION HEATERS...................2533
Mishima, Tomoakzu ; Nakagawa, Yuki ; Nakaoka, Mutsuo

APPLICATION OF VIRTUAL VALIDATION SYSTEM FOR INVERTER HEAT PUMP SYSTEM.........2541
Kanamori, Masaki ; Noda, Koji ; Endo, Takahisa ; Suzuki, Nobuyuki

TEST SETUP FOR ACCELERATED TEST OF HIGH POWER IGBT MODULES WITH ONLINE MONITORING OF VCE AND VF VOLTAGE DURING CONVERTER OPERATION...............................2547
de Vega, Angel Ruiz ; Ghimire, Pramod ; Pedersen, Kristian Bonderup ; Trintis, Ionut ; Beczckowski, Szymon ; Munk-Nielsen, Stig ; Rannestad, Bjorn ; Thogersen, Paul

DESIGN OF HIGH-SPEED IGBT-BASED SWITCHING MODULES FOR PULSED POWER APPLICATIONS.........2554
Kluge, Andreas ; Goehler, Lutz ; Gueldner, Henry ; Trompa, Thomas ; Mory, David ; Segsa, Karl-Heinz

COMPARATIVE SUITABILITY EVALUATION OF REVERSE-BLOCKING IGBTS FOR CURRENT-SOURCE BASED CONVERTER..2562
De, Ankan ; Roy, Sudhin ; Bhattacharya, Subhashish

NEW REVERSE-CONDUCTING IGBT (1200V) WITH REVOLUTIONARY COMPACT PACKAGE.........2569
Takahashi, K. ; Yoshida, S. ; Noguchi, S. ; Kuribayashi, H. ; Nashida, N. ; Kobayashi, Y. ; Kobayashi, H. ; Mochizuki, K. ; Ikeda, Y. ; Ikawa, O.

AN IMPROVED MODULATED CARRIER CONTROL OF SINGLE-PHASE CCM BOOST PFC CONVERTER.........2575
Kim, Hyejin ; Cho, Bo-Hyung ; Choi, Hangseok

MODIFIED INTERLEAVED CURRENT SENSORLESS CONTROL FOR THREE-LEVEL BOOST PFC CONVERTER WITH ASYMMETRIC LOADS...2580
Chen, Hung-Chi ; Liao, Jhen-Yu

A NOVEL CRITICAL-CONDUCTION-MODE BRIDGELESS INTERLEAVED BOOST PFC RECTIFIER.........2587
Cao, Guoen ; Kim, Hee-Jun

ANALYSIS AND DESIGN OF A PUSH-PULL SINGLE-STAGE FLYBACK POWER FACTOR CORRECTOR.........2593
Lo, Yu-Kang ; Chiu, Huang-Jen ; Liu, Yu-Chen ; Lin, Chung-Yi ; Cheng, Shih-Jen ; Yang, CS

LINEAR OVER-MODULATION STRATEGY FOR CURRENT CONTROL IN PHOTOVOLTAIC INVERTER.........2598
Park, Yongsoon ; Sul, Seung-Ki ; Hong, Ki-Nam

DESIGN OF DECENTRALIZED VOLTAGE CONTROL FOR PV INVERTERS TO MITIGATE VOLTAGE RISE IN DISTRIBUTION POWER SYSTEM WITHOUT COMMUNICATION.....................................2606
Lee, Tzung-Lin ; Yang, Shih-Sian ; Hu, Shang-Hung

STABILITY ANALYSIS AND ACTIVE DAMPING FOR LLCL-FILTER BASED GRID-CONNECTED INVERTERS..2610
Huang, Min ; Blaabjerg, Frede ; Loh, Poh Chiang ; Wu, Weimin

INTEGRATED COMMON AND DIFFERENTIAL MODE FILTER APPLIED TO A SINGLE-PHASE TRANSFORMERLESS PV MICROINVERTER WITH LOW LEAKAGE CURRENT...........................2618
Figueredo, Ricardo Souza ; de Carvalho, Kelly Caroline Mingorancia ; Matakas, Lourenco

DESIGN AND INTEGRATION OF INTERPHASE INDUCTORS FOR INTERLEAVED THREE PHASE VOLTAGE-SOURCE-INVERTERS IN DC-FED MOTOR DRIVE SYSTEMS................................2626
Zhang, Xuning ; Boroyevich, Dushan ; Burgos, Rolando

A NOVEL TRANSFORMER MODEL USING MAGNETIC CIRCUIT...2632
Nakamurame, Fuminori ; Ise, Toshifumi

HARDWARE-IN-THE-LOOP SIMULATION OF A MACHINE MODEL WITH REAL-TIME ANIMATION.........2638
Xiaojie Zhuang ; Hibino, Shinya ; Harakawa, Masaya ; Terabe, Ryosuke ; Ozaki, Takayuki ; Nagano, Tetsuaki

DEVELOPMENT OF REAL TIME DIGITAL SIMULATOR FOR SELF-COMMUTATED SVC TO SUPPRESS VOLTAGE FLICKER...2644
Terao, Yutaka ; Shishida, Yasuhiro ; Tsuruma, Yoshinori ; Ishizuka, Tomotsugu ; Aoyama, Fumio ; Yoshino, Teruo ; Kato, Yutaka ; Belanger, Jean

OPERATIONAL ASPECTS AND POWER ARCHITECTURE DESIGN FOR A MICROGRID TO INCREASE THE USE OF RENEWABLE ENERGY IN WIRELESS COMMUNICATION NETWORKS...........................2649
Kwasinski, Alexis ; Kwasinski, Andres

P+ MULTIPLE RESONANT CONTROL FOR OUTPUT VOLTAGE REGULATION OF MICROGRID WITH UNBALANCED AND NONLINEAR LOADS..2656
Kyungbae Lim ; Jaeho Choi ; Juyoung Jang ; Junghum Lee ; Jaesig Kim

130MVA-STATCOM FOR TRANSIENT STABILITY IMPROVEMENT...2663
Imanishi, Takao ; Nagatomo, Yoshinobu ; Iwasaki, Shinya ; Masaki, Kenji ; Fujii, Toshiyuki ; Ieda, Jun

IMPROVED DROOP CONTROLLER FOR MICROGRID INVERTER CONSIDERING THE LINE IMPEDANCE MISMATCHING...2668
Du Yan ; Liuchen Chang ; Meiqin Mao ; Jianhui Su ; Ning Liu

SUPPRESSION CONTROL METHOD FOR IRON LOSS OF MATRIX MOTOR UNDER FLUX WEAKENING UTILIZING INDIVIDUAL WINDING CURRENT CONTROL 2673
Hijikata, Hiroki ; Akatsu, Kan ; Miyama, Yoshihiro ; Arita, Hideaki ; Daikoku, Akihiro

PERFORMANCE ANALYSIS OF A NEW CONCENTRATEDWINDING INTERIOR PERMANENT MAGNET SYNCHRONOUS MACHINE UNDER FIELD ORIENTED CONTROL 2679
Nguyen, D. ; Dutta, R. ; Fletcher, J. ; Rahman, F. ; Lovatt, Howard

ONLINE PARTICLE SWARM OPTIMIZATION FOR SENSORLESS IPMSM DRIVES CONSIDERING PARAMETER VARIATION 2686
Song, Z.Q. ; Xiao, D. ; Rahman, M.F.

A DTC-PWM CONTROL SCHEME OF PMSM BASED ON 12-SECTORS DIVISION AND SPEED INFORMATION 2693
Yunchang Kwak ; Jin-Woo Ahn ; Dong-Hee Lee

CONTROL OF POWER FLOW BETWEEN THE WIND GENERATOR AND NETWORK 2700
Stumpf, Peter ; Nagy, Istvan ; Vajk, Istvan

ADVANCES IN NANOGRID TECHNOLOGY AND ITS INTEGRATION INTO RURAL ELECTRIFICATION IN INDIA 2707
Mishra, Santanu ; Ray, Olive

STUDY AND IMPLEMENTATION OF SEVEN-LEVEL INVERTER USING COUPLED INDUCTOR AND SWITCHED-CAPACITOR 2714
Yi-Chun Lin ; Jiann-Fuh Chen ; Wen-Chien Hsu ; Sheng-Kai Kao

CASCADED MULTILEVEL CONVERTER BASED BIDIRECTIONAL INDUCTIVE POWER TRANSFER (BIPT) SYSTEM 2722
Bac Xuan Nguyen ; Vilathgamuwa, D.M. ; Foo, Gilbert ; Ong, Andrew ; Sampath, Prasad K. ; Madawala, Udaya K.

UNDERSAMPLING CONTROL OF A BIDIRECTIONAL CASCADED BUCK+BOOST DC-DC CONVERTER 2729
Rosekeit, Martin ; Joebges, Philipp ; Lelie, Markus ; Sauer, Dirk Uwe ; De Doncker, Rik W.

SUB-MICROSECOND RESPONSE DIGITAL CONTROLLER FOR POL 2737
Nonaka, Hirotaka ; Ishizuka, Yoichi ; Mii, Kenji ; Takenami, Fumiaki ; Kanemoto, Daisuke

GAIN CONTROLLED HIGH EFFICIENCY POWER FACTOR CORRECTION CIRCUIT 2745
Yonezawa, Yu ; Nakao, Hiroshi ; Sasaki, Tomotake ; Matsui, Yoshinobu ; Nakashima, Yoshiyasu ; Kaneko, Junji ; Shimamori, Hiroshi ; Yoshino, Yukio ; Hisato, Hosoyama ; Atsushi, Manabe ; Motizuki, Shun ; Yamashita, Shigeharu

DESIGN OF QUASI-RESONANT FLYBACK CONVERTER CONTROL IC WITH DCM AND CCM OPERATION 2750
Kai-Hui Chen ; Tsorng-Juu Liang

LOAD TRANSIENT RESPONSE IMPROVEMENT BASED ON PID CONTROL 2754
Yau, Y.T. ; Hwu, K.I.

AN ACTIVE-CLAMPING FORWARD CONVERTER WITH NON-LINEAR STEP-DOWN CONVERSION 2758
Jing-Yuan Lin ; Yu-Kang Lo ; Huang-Jen Chiu ; Chao-Fu Wang ; Chien-Yu Lin

SWITCHING LOSS MINIMIZATION OF 3-PHASE INTERLEAVED BIDIRECTIONAL DC-DC CONVERTER 2763
Eui-Cheol Nho ; Jae-Hun Jung ; Hak-Soo Kim ; In-Dong Kim ; Heung-Geun Kim ; Tae-Won Chun

MODIFIED THREE-PHASE THREE-LEVEL DC-DC CONVERTER -ADOPTING ASYMMETRICAL DUTY CYCLE CONTROL 2768
Yue Chen ; Xuling Chen ; Liu, Fuxin ; Ruan, Xinbo

DEADBEAT CONTROL OF POWER LEVELING UNIT WITH BIDIRECTIONAL BUCK/BOOST DC/DC CONVERTER 2775
Hamasaki, Shin-ichi ; Mukai, Ryosuke ; Yano, Yoshihiro ; Tsuji, Mineo

DESIGN OF OPTIMIZED ON-OFF CONTROL TO IMPROVE EFFICIENCY OF PARALLELED CONVERTER SYSTEM 2781
Kohama, Teruhiko ; Sogawa, Yuki ; Tsuji, Satoshi

EFFICIENCY IMPROVEMENTS IN A SINGLE ACTIVE BRIDGE MODULAR DC-DC CONVERTER WITH SNUBBER CAPACITANCE OPTIMISATION 2787
Ting, Yeh ; de Haan, Sjoerd ; Ferreira, Jan A.

A WIRELESS POWER TRANSFER SYSTEM OPTIMIZED FOR HIGH EFFICIENCY AND HIGH POWER APPLICATIONS 2794
Bani Shamseh, Mohammad ; Kawamura, Atsuo ; Yuzurihara, Itsuo ; Takayanagi, Atsushi

NON-ITERATIVE LCL FILTER DESIGN FOR THREE-PHASE TWO-LEVEL VOLTAGE-SOURCE PWM CONVERTERS 2802
Byung-Geuk Cho ; Seung-Ki Sul

DSP-BASED INTERLEAVED BUCK POWER FACTOR CORRECTOR 2810
Yu-Chen Liu ; Tsan Chen ; Po-Jung Tseng ; Yu-Kang Lo ; Huang-Jen Chiu

THE AVERAGE MODEL OF A THREE-PHASE THREE-STAGE POWER ELECTRONIC TRANSFORMER 2815
Shaodi Ouyang ; Liu, Jinjun ; Wang, Xinyu ; Wang, Xiaojian ; Fei Meng ; Riffat, Javid

A MULTI-CARRIER PWM FOR AC-DC-AC CONVERTER WITHOUT DC LINK ELECTROLYTIC CAPACITOR 2821
Chung-Chuan Hou ; Hsin-Ping Su

A DECOUPLING OFFSET-BASED PWM CONTROL FOR A MULTILEVEL INVERTER UNDER DC VOLTAGE UNBALANCE 2826
Nho Van Nguyen ; Tam Khanh Tu Nguyen ; Lee, Hong-Hee

?-? PARETO OPTIMIZATION OF 3-PHASE 3-LEVEL T-TYPE AC-DC-AC CONVERTER COMPRISING SI AND SIC HYBRID POWER STAGE...2834
Uemura, Hirofumi ; Krismer, Florian ; Okuma, Yasuhiro ; Kolar, Johann W.

PRACTICAL INVESTIGATION OF THE GATE BIAS EFFECT ON THE REVERSE RECOVERY BEHAVIOR OF THE BODY DIODE IN POWER MOSFETS 2842
Lindberg-Poulsen, Kristian ; Petersen, Lars Press ; Ouyang, Ziwei ; Andersen, Michael A.E.

AN ONLINE VCE MEASUREMENT AND TEMPERATURE ESTIMATION METHOD FOR HIGH POWER IGBT MODULE IN NORMAL PWM OPERATION 2850
Ghimire, Pramod ; de Vega, Angel Ruiz ; Beczkowski, Szymon ; Munk-Nielsen, Stig ; Rannested, Bjorn ; Thogersen, Paul Bach

EVALUATION ON IRON LOSS CHARACTERISTICS IN SERIES CONNECTION AND PARALLEL CONNECTION OF LOADS WITH INVERTER EXCITATION 2856
Odawara, Shunya ; Fujisaki, Keisuke

LOSS AND THERMAL MODEL FOR POWER SEMICONDUCTORS INCLUDING DEVICE RATING INFORMATION 2862
Ma, K. ; Bahman, A.S. ; Beczkowski, S.M. ; Blaabjerg, F.

IMPROVING RELIABILITY OF IGBT SURFACE ELECTRODE FOR 200 C OPERATION......................2870
Nishimura, Tomohiro ; Ikeda, Yoshinari ; Hokazono, Hiroaki ; Mochizuki, Eiji ; Takahashi, Yoshikazu

INFLUENCE OF CARRIER FREQUENCY ON IRON LOSS TAKING ACCOUNT OF DEAD TIME EFFECT 2874
Kogi, Ryosuke ; Odawara, Shunya ; Fujisaki, Keisuke

DECREASE OF SIC-BJT DRIVER LOSSES BY ONE-STEP COMMUTATION..2881
Barth, Henry ; Hofmann, Wilfried

POWER PROFILE BASED SELECTION AND OPERATION OPTIMIZATION OF PARALLEL-CONNECTED POWER CONVERTER COMBINATIONS...2887
Vogt, T. ; Peters, A. ; Frohleke, N. ; Bocker, J. ; Kempen, S.

A NOVEL POWER LOSS CALCULATION METHOD FOR IGBTS IN POWER CONVERTERS VIA CHAOTIC SPWM CONTROL 2893
Boyu Wang ; Li, Hong ; Xiaojie You ; Trillion Zheng

LOSS ANALYSIS AND SOFT-SWITCHING CHARACTERISTICS OF FLYBACK-FORWARD HIGH GAIN DC/DC CONVERTER WITH GAN FET 2899
Zhang Yajing ; Zheng, Trillion Q. ; Li Yan

INSULATED METAL SUBSTRATE FOR POWER MODULES USING ANODIC OXIDE FILM OF ALUMINUM 2904
Tokuyama, Takeshi ; Kusukawa, Jumpei ; Nakatsu, Kinya

A FAST-TRANSIENT-RESPONSE BUCK CONVERTER WITH SPLIT-TYPE III COMPENSATION AND CHARGE-PUMP CIRCUIT TECHNIQUE 2910
Chen, Jiann-Jong ; Wei-Ting Hsu ; Jih-Hua Yu ; Hwang, Yuh-Shyan ; Cheng-Chieh Yu

ADVANTAGES OF LOW PARASITIC INDUCTANCE PACKAGES OF POWER MOSFET FOR SERVER POWER APPLICATIONS..2914
Wonsuk Choi ; Dongkook Son ; Dongwook Kim

MODULAR INTEGRATION OF A MATRIX CONVERTER...2920
Solomon, Adane Kassa ; Skuriat, Robert ; Castellazzi, Alberto ; Wheeler, Pat

A MODULAR NANOSECOND PULSE GENERATION SYSTEM FOR PLASMA-ASSISTED IGNITION 2926
Peng Gao ; Fletcher, John ; O'Byrne, Sean

DEVELOPMENT OF A SINGLE SWITCH CELL FOR MODULAR NANOSECOND PULSE GENERATION SYSTEMS...2932
Peng Gao ; Fletcher, John ; O'Byrne, Sean

ADVANTAGE OF SUPER JUNCTION MOSFET FOR POWER SUPPLY APPLICATION..............................2939
Tabira, K. ; Watanabe, S. ; Shimatou, T. ; Watashima, T. ; Takenoiri, S.

STUDY ON AN ACCURATE CALCULATION OF THE CONDUCTED EMI NOISE OF THE POWER CONVERTERS 2944
Omata, Shinpei ; Shimizu, Toshihisa

AN EXACT DISCRETE-TIME MODEL CONSIDERING DEAD-TIME NONLINEARITY FOR AN H-BRIDGE GRID-CONNECTED INVERTER 2950
Xie, Ruiliang ; Hao, Xiang ; Yang, Xu ; Chen, Wenjie ; Huang, Lang ; Chao Wang

THEORETICAL ANALYSIS OF THE DUALITY PRINCIPLE APPLIED TO INTERLEAVED TOPOLOGIES....................2954
Caris, M.L.A. ; Huisman, H. ; Duarte, J.L.

A NEW IMPEDANCE MEASUREMENT METHOD BASED ON HIGH FREQUENCY COMPENSATION 2960
Yue, Xiaolong ; Zhuo, Fang ; Hao Yi

NUMERICAL AND EXPERIMENTAL INVESTIGATION OF PARASITIC EDGE CAPACITANCE FOR PHOTOVOLTAIC PANEL 2967
Wenjie Chen ; Xiaomei Song ; Hao Huang ; Xu Yang

VEHICLE INTERIOR NOISE CONTROL OF ULTRA-COMPACT ELECTRIC VEHICLE (FUNDAMENTAL CONSIDERATION USING RECTANGULAR ENCLOSURE)..2972
Kato, Taro ; Kato, Hideaki ; Oshinoya, Yasuo ; Suzuki, Ryosuke ; Hasegawa, Shinya

CONSIDERATION FOR THE PROPAGATION PATH OF CONDUCTIVE NOISE IN AIR CONDITIONERS..............2977
Tokiwa, Tsuyoshi ; Kanamori, Masaki ; Endo, Takahisa ; Iida, Mikiya ; Ogasawara, Satoshi ; Yizhanyi Tang

IRON LOSS EVALUATION OF IRON POWDER CORE SUITABLE FOR INDUCTOR USED IN POWER CONVERTERS 2983
Mori, Tomohiro ; Igarashi, Kazunori ; Kanagawa, Kinji ; Yamashita, Nobuyuki ; Shimizu, Toshihisa ; Bizen, Yosio

OPTIMIZED TUNING METHOD OF STATIONARY FRAME PROPORTIONAL RESONANT CURRENT CONTROLLERS 2988

Martinz, Fernando Ortiz ; de Carvalho, Kelly Caroline Mingorancia ; Ama, Naji Rajai Nasri ; Komatsu, Wilson ; Matakas, Lourenco

INSTANTANEOUS POWER THEORY APPLIED TO POWER CONDITIONING UNDER DISTORTED MAINS VOLTAGES: A MATLAB/SIMULINK APPROACH 2996

Nicolae, Petre-Marian ; Popa, Lucian-Dinut ; Nicolae, Marian-Stefan ; Nicolae, Ileana-Diana

THE RESEARCH ON RELIABILITY AND REAL-TIME OF THE SCHEME OF PROCESS LAYER GOOSE NETWORK IN SMART SUBSTATION BASED ON ARTIFICIAL COBWEB TOPOLOGY STRUCTURE 3002

Liu, Xiaosheng ; Zhu, Honglin ; Xu, Dianguo ; Li, Yanxiang

EFFICIENCY IMPROVEMENT OF A SELF-START TYPE PERMANENT MAGNET SYNCHRONOUS MOTOR 3007

Saikusa, H. ; Arikawa, S. ; Higuchi, T. ; Yokoi, Y. ; Abe, T.

CONSIDERATION OF OPTIMAL NUMBER OF POLES AND FREQUENCY FOR HIGH-EFFICIENCY PERMANENT MAGNET MOTOR 3012

Misu, Daisuke ; Matsushita, Makoto ; Takeuchi, Katsutoku ; Oishi, Koji ; Kawamura, Mitsuhiro

BASIC STUDY ON THE SUITABLE STRUCTURE OF A PERMANENT MAGNET SYNCHRONOUS MOTOR WITH A POWDER MAGNETIC CORE 3018

Hashimoto, Shizuka ; Sanada, Masayuki ; Morimoto, Shigeo ; Inoue, Yukinori

CHARACTERISTICS OF A HALF-WAVE RECTIFIED BRUSHLESS SYNCHRONOUS GENERATOR 3024

Hirakawa, Yuki ; Higuchi, Tsuyoshi ; Yokoi, Yuichi ; Abe, Takashi

MODELING OF WOUND ROTOR SYNCHRONOUS MACHINES CONSIDERING HARMONICS, GEOMETRIC SALIENCIES AND SATURATION INDUCED SALIENCIES 3029

Rambetius, Alexander ; Luthardt, Sven ; Piepenbreier, Bernhard

DESIGN AND COMPARISON OF HIGH FREQUENCY TRANSFORMERS USING FOIL AND ROUND WINDINGS 3037

Iyer, Kartik V ; Robbins, William P ; Mohan, Ned

A METHOD TO CALCULATE THE PERFORMANCE OF LINEAR INDUCTION MOTORS USING SIMPLE TWO-PHASE MODEL 3044

Hirahara, Hideaki ; Yamamoto, Shu ; Ara, Takahiro ; Shimizu, Toshihisa

AN ESP DOWNHOLE PARAMETERS MONITORING SYSTEM BASED ON CURRENT LOOP TRANSMISSION METHOD 3050

Jin Miaoxin ; Zhang Wei ; Gao Qiang ; Xu Dianguo

BENDING MAGNETIC LEVITATION CONTROL FOR THIN STEEL PLATE (EXPERIMENTAL CONSIDERATION USING SLIDING MODE CONTROL) 3055

Yonezawa, Hikaru ; Narita, Takayoshi ; Oshinoya, Yasuo ; Marumori, Hiroki ; Hasegawa, Shinya

TRANSFORMER WINDING LOSSES WITH ROUND CONDUCTORS FOR DUTY-CYCLE REGULATED SQUARE WAVES 3061

Iyer, Kartik V ; Robbins, William P ; Basu, Kaushik ; Mohan, Ned

SIMULATION OF RESIN MOLDED TYPE SENSOR IN POLE SWITCH FOR POWER DELIBERY SYSTEMS 3067

Furukawa, Tatsuya ; Muta, Shoichiro ; Fukumoto, Hisao ; Itoh, Hideaki ; Ohchi, Masashi

ROBUST STARTUP CONTROL OF SENSORLESS PMSM DRIVES WITH SELF-COMMISSIONING 3072

Lin, Chiao-Chien ; Tzou, Ying-Yu

POSITION SENSORLESS CONTROL OF PMSM WITH A LOW-FREQUENCY SIGNAL INJECTION 3079

Nimura, Tomohiro ; Doki, Shinji ; Fujitsuna, Masami

A COMPARISON OF DIFFERENT SENSORLESS POSITION ACQUISITION METHODS AT LOW SPEEDS FOR A PERMANENT MAGNET SYNCHRONOUS MACHINE IN VEHICLE APPLICATIONS 3085

Lehmann, Oliver ; Zehelein, Matthias ; Schuster, Johannes ; Roth-Stielow, Jorg

STABILITY COMPARISON OF IPMSM SENSORLESS VECTOR CONTROL SYSTEMS USING EXTENDED EMF 3093

Tsuji, Mineo ; Mizusaki, Hiroshi ; Hamasaki, Sin-ichi

INDUCTION MACHINE BASED FLYWHEEL SPEED ESTIMATION AT STAND-BY MODE 3099

Liu, Rongqiang ; Xu, David

SYMMETRICAL SIGNALING SYSTEM FOR SENSOR-LESS SRM DRIVE 3106

Yamamoto, Kenji ; Takahashi, Hisashi ; Ushiro, Nobumasa ; Shirasawa, Koki

DIGITAL INTEGRATORS FOR CONDITION MONITORING: A DC AND MULTITONE SIGNAL ANALYSIS 3111

Peretti, L.

AUDIBLE NOISE REDUCTION METHOD IN IPMSM POSITION SENSORLESS CONTROL BASED ON HIGH-FREQUENCY CURRENT INJECTION 3119

Tauchi, Yuki ; Kubota, Hisao

A NOVEL DESIGN FOR INDUCTION MOTOR FLUX ESTIMATION USING IMPULSIVE OBSERVER 3124

Peng Wang ; Yan Li ; Jianwen Zhang ; Xu Cai ; Zhengzhi Han

LOAD TORQUE AND INERTIA SIMULATION BASED ON DOUBLE-STATOR PERMANENT-MAGNET SYNCHRONOUS MOTOR 3129

Zhe Wang ; Mingyan Wang ; Ben Guo ; Chai Feng

INDEPENDENT SPEED AND POSITION CONTROL OF TWO PERMANENT MAGNET SYNCHRONOUS MOTORS FED BY A FOUR-LEG INVERTER 3134

Kubo, Yuji ; Moroi, Takayuki ; Kouki, Matsuse ; Kubota, Hisao ; Rajashekara, Kaushik

MINIMIZATION OF STATOR CURRENTS FOR MONO INVERTER DUAL PARALLEL PMSM DRIVE SYSTEM 3140
Yongjae Lee ; Ha, Jung-Ik

PERFORMANCE COMPARISON OF INVERTER AND DRIVE CONFIGURATIONS WITH OPEN-END AND STAR-CONNECTED WINDINGS 3145
Neubert, Markus ; Koschik, Stefan ; De Doncker, Rik W.

INPUT CURRENT HARMONICS REDUCTION CONTROL FOR ELECTROLYTIC CAPACITOR LESS INVERTER BASED IPMSM DRIVE SYSTEM 3153
Abe, Kodai ; Ohishi, Kiyoshi ; Haga, Hitoshi

NONCONTACT GUIDE SYSTEM FOR TRAVELING ELASTIC STEEL PLATES (THEORETICAL STUDY ON THE SHAPE OF TRAVELING STEEL PLATE) 3159
Sakaba, Kouichi ; Hasegawa, Shinya ; Narita, Takayoshi ; Oshinoya, Yasuo

ACTIVE SEAT SUSPENSION FOR ULTRA-COMPACT VEHICLE (FUNDAMENTAL CONSIDERATION ON ELECTROMYOGRAM WHEN FALL FROM THE BUMP) 3162
Mashino, Masahiro ; Sunaga, Keita ; Hasegawa, Shinya ; Ishida, Masaki ; Kato, Hideaki ; Oshinoya, Yasuo

ADAPTIVE CURRENT TRACKING OF THREE-PHASE ACTIVE POWER FILTER USING BACKSTEPPING CONTROL 3168
Yunmei Fang ; Juntao Fei ; Shixi Hou ; Weili Dai

FAST IDENTIFICATION OF RESONANCE CHARACTERISTIC FOR 2-MASS SYSTEM WITH ELASTIC LOAD 3174
Ming Yang ; Liang Hao ; Dianguo Xu

AUTONOMOUS NAVIGATION SYSTEM BASED ON COLLISION DANGER-DEGREE FOR UNMANNED GROUND VEHICLE 3179
Yasuno, Takashi ; Tanaka, Daiki ; Kuwahara, Akinobu

A HIGH-PERFORMANCE BIDIRECTIONAL DC-DC CONVERTER FOR DC MICRO-GRID SYSTEM APPLICATION 3185
Shu-Wei Kuo ; Yu-Kang Lo ; Huang-Jen Chiu ; Shih-Jen Cheng ; Chung-Yi Lin ; Yang, CS

VOLUME 5

IMPROVEMENT IN EFFICIENCY OF LED LIGHTING SYSTEM 3190
Hwu, K.I. ; Jiang, W.Z. ; Jenn-Jong Shieh

COMPARISON AND EVALUATION OF VIBRATION-BASED PIEZOELECTRIC POWER GENERATORS 3194
Basari, Amat A. ; Awaji, Sosuke ; Hashimoto, Seiji ; Kasai, Makoto ; Suto, Kenji ; Kumagai, Shunji ; Kasai, Makoto ; Suto, Kenji ; Wei Jiang ; Shuren Wang

BATTERY SELECTION FOR HYBRID ENERGY SYSTEMS AND THERMAL MANAGEMENT IN ARCTIC CLIMATES 3200
Kalogera, Maria ; Bauer, Pavol

100KW PV PCS WITH NATURAL CONVECTION COOLING FOR OUTDOOR INSTALLATION 3207
Jin, Yasuhiro ; Matsuoka, Kazumasa ; Takahashi, Takehiro ; Takahashi, Nobuhiro

A NEW PLL BASED ON FAST POSITIVE AND NEGATIVE SEQUENCE DECOMPOSITION ALGORITHM WITH MATRIX OPERATION UNDER DISTORTED GRID CONDITIONS 3213
Shaohua Sun ; Hongqi Ben ; Tao Meng ; Jinyong Zhang

PERFORMANCE IMPROVEMENT OF PHOTOVOLTAIC POWER GENERATION SYSTEMS USING ON-OFF CONTROL METHODS 3218
Kenji, Matsumoto ; Nomura, Shinichi

LOW VOLTAGE PV POWER INTEGRATION INTO MEDIUM VOLTAGE GRID USING HIGH VOLTAGE SIC DEVICES 3225
Chattopadhyay, Ritwik ; Bhattacharya, Subhashish ; Foureaux, Nicole C. ; Silva, Sidelmo M. ; Braz Cardoso, F. ; de Paula, Helder ; Pires, Igor A. ; Cortizio, Porfirio C. ; Moraes, Lenin ; de S.Brito, Jose A.

A NOVEL GLOBAL MAXIMUM POWER POINT TRACKING METHOD FOR PHOTOVOLTAIC GENERATION SYSTEM OPERATING UNDER PARTIALLY SHADED CONDITION 3233
Jing-Hsiao Chen ; Yu-Shan Cheng ; Shun-Chung Wang ; Huang, Jia-Wei ; Liu, Yi-Hua

AN APPLICATION OF Z-SOURCE CONVERTER TO BATTERIES CHARGE WITH A PHOTOVOLTAIC SYSTEM 3239
Razik, H. ; Zitouni, Y. ; Maret, C.

PCS WITH SCANNING-TYPE MPPT CONTROL FOR INDUSTRIAL GRID-CONNECTED PV POWER GENERATION SYSTEM 3244
Itako, Kazutaka

FEASIBLE METHOD OF CALCULATING LEAKAGE REACTANCE OF 9-WINDING TRANSFORMER FOR HIGH-VOLTAGE INVERTER SYSTEM 3249
Fukumoto, Hisao ; Furukawa, Tatsuya ; Itoh, Hideaki ; Ohchi, Masashi

HIGH POWER HVDC-DC CONVERTERS FOR THE INTERCONNECTION OF HVDC LINES WITH DIFFERENT LINE TOPOLOGIES 3255
Schon, Andre ; Bakran, Mark-M.

CHARACTERIZATION OF A CURRENT SHUNT AND AN INDUCTIVE VOLTAGE DIVIDER FOR PMU CALIBRATION 3263
Kon, Saytaro ; Yamada, Tatsuji

DISTRIBUTED SERIES/HYBRID-SHUNT COMPENSATION FOR HARMONIC MITIGATION IN COMMERCIAL FACILITIES......3270

Diniz, Rogerio Azevedo ; Pires, Igor A. ; Franca, Gleisson J. ; Cardoso, Braz J.

ROBUST CONTROL DESIGN FOR THE VOLTAGE TRACKING LOOP OF A DVR......3278

Ferrari, Bruno Augusto ; Ama, Naji Rajai Nasri ; de Carvalho, Kelly Caroline Mingorancia ; Martinz, Fernando Ortiz ; Matakas, Lourenco

MULTI-PORT SOLID STATE TRANSFORMER FOR INTER-GRID POWER FLOW CONTROL......3286

Roy, Sudhin ; De, Ankan ; Bhattacharya, Subhashish

REACTIVE POWER CONTROL STRATEGY BASED ON DC CAPACITOR VOLTAGE CONTROL FOR ACTIVE LOAD BALANCER IN THREE-PHASE FOUR-WIRE DISTRIBUTION SYSTEMS......3292

Tint Soe Win ; Hisada, Yoshihiro ; Tanaka, Toshihiko ; Hiraki, Eiji ; Okamoto, Masayuki ; Lee, Seong Ryong

VOLTAGE SAG RIDE-THROUGH PERFORMANCE OF VIRTUAL SYNCHRONOUS GENERATOR......3298

Alipoor, Jaber ; Miura, Yushi ; Ise, Toshifumi

CONTROL OF DISTRIBUTED GENERATION SYSTEMS UNDER UNBALANCED VOLTAGE CONDITIONS......3306

Kabiri, R. ; Holmes, D.G. ; McGrath, B.P.

STABILITY ANALYSIS OF GRID-CONNECTED INVERTERS WITH LCL-FILTER BASED ON HARMONIC BALANCE AND FLOQUET THEORY......3314

Jing Bian ; Hong Li ; Zheng, Trillion Q.

COMPARATIVE EVALUATION OF PASSIVE DAMPING TOPOLOGIES FOR PARALLEL GRID-CONNECTED CONVERTERS WITH LCL FILTERS......3320

Beres, Remus ; Wang, Xiongfei ; Blaabjerg, Frede ; Bak, Claus Leth ; Liserre, Marco

STUDY AND IMPLEMENTATION OF A SEPIC LED DRIVER WITH ADJUSTABLE OUTPUT VOLTAGE......3328

Po-Jung Tseng ; Yu-Chen Liu ; Yu-Kang Lo ; Chiu, Huang-Jen ; Yun-Chu Chiu

AN INTERLEAVED SINGLE-STAGE LLC RESONANT CONVERTER USED FOR MULTI-CHANNEL LED DRIVING......3333

Chang, Chien-Hsuan ; Cheng, Chun-An ; Jinno, Masahito ; Cheng, Hung-Liang

A NOVEL TYPE OF WIRELESS V2H SYSTEM WITH BIDIRECTIONAL RESONANT SINGLE-ENDED INVERTER......3341

Fukuoka, Hiroki ; Iga, Yuichi ; Omori, Hideki ; Morizane, Tosimitsu ; Kimura, Noriyuki ; Nakaoka, Mutuo

DESIGN AND IMPLEMENTATION OF AN INTERLEAVED BCM BOOST PFC CONTROL IC......3346

Kuan-Hsien Chou ; Tsorng-Juu Liang ; Kai-Hui Chen ; Ji-Shiang Lee

LOW CAPACITIVE INDUCTORS FOR FAST SWITCHING DEVICES IN ACTIVE POWER FACTOR CORRECTION APPLICATIONS......3352

Hernandez, Juan C. ; Petersen, Lars P. ; Andersen, Michael A.E.

TEMPERATURE-ROBUST LC3 LED DRIVER WITH LOW THD, HIGH EFFICIENCY, AND LONG LIFE......3358

Lee, Eun S. ; Choi, Bo H. ; Cheon, Jun P. ; Kim, Bong C. ; Rim, Chun T.

OPTIMIZING REPULSIVE LORENTZ FORCES FOR A LEVITATING INDUCTION COOKER......3365

Zingerli, Claudius M. ; Nussbaumer, Thomas ; Kolar, Johann W.

DESIGN OF A MODULAR RESONANT CONVERTER FOR 25KV-8A DC POWER SUPPLY OF RF CAVITIES......3371

Siemaszko, Daniel ; Pittet, Serge ; Aguglia, Davide ; de Mallac, Louis

A NOVEL TRANSFORMER-LESS INTERLEAVED FOUR-PHASE HIGH STEP-DOWN DC CONVERTER WITH LOW SWITCH VOLTAGE STRESS......3379

Ching-Tasi Pan ; Chen-Feng Chuang ; Chia-Chi Chu ; Hao-Chien Cheng

EFFICIENCY IMPROVEMENT OF POWER SUPPLY WITH TRANSIENT CURRENT CIRCUIT USING DIGITAL CONTROL......3386

Takashita, Haruomi ; Shoyama, Masahito ; Yonezawa, Yu ; Nakashima, Yoshiyasu

ULTRA HIGH STEP-DOWN CONVERTER......3392

Yau, Y.T. ; Hwu, K.I.

DIGITAL CONTROL OF PWM INVERTER USING ULTRA HIGH SPEED NETWORK FOR FEEDBACK SIGNALS WITH COMMUNICATION DISTURBANCE OBSERVER BASED ON ROCKET I/O PROTOCOL......3397

Saito, Ryo ; Tsuchida, Kazuo ; Yokoyama, Tomoki

100 KHZ DC CHOPPER DIGITALLY GATE CONTROLLED WITH PARTIAL TURN- OFF SWITCHING USING SIC-MOSFET AND FPGA......3403

Tsuruta, Yukinori ; Kawamura, Atsuo

VARIABLE CARRIER DEADBEAT CONTROL WITH DIGITAL HYSTERESIS METHOD USING SOC-FPGA FOR UTILITY INTERACTIVE INVERTER......3410

Ohashi, Shunsuke ; Yoshida, Morito ; Yokoyama, Tomoki

A SPACE VECTOR MODULATION STRATEGY FOR THREE-LEVEL OPERATION BASED ON DUAL TWO-LEVEL VOLTAGE SOURCE INVERTERS......3417

Kumsuwan, Yuttana ; Srirattanawichaikul, Watcharin

INVESTIGATION ON THE PARALLEL OPERATION OF ALL-GAN POWER MODULE AND THERMAL PERFORMANCE EVALUATION......3425

Cheng, Stone ; Po-Chien Chou

FULL SILICON CARBIDE BOOST CHOPPER MODULE FOR HIGH FREQUENCY AND HIGH TEMPERATURE OPERATION......3432

Pettersson, Sami ; Kicin, Slavo ; Holm, Toni ; Bianda, Enea ; Canales, Francisco

DEVELOPMENT OF ULTRAHIGH VOLTAGE SIC POWER DEVICES..........3440

Fukuda, Kenji ; Okamoto, Dai ; Harada, Shinsuke ; Tanaka, Yasunori ; Yonezawa, Yoshiyuki ; Deguchi, Tadayoshi ; Katakami, Shuji ; Ishimori, Hitoshi ; Takasu, Shinji ; Arai, Manabu ; Takenaka, Kensuke ; Fujisawa, Hiroyuki ; Takei, Manabu ; Matsumoto, Kazushi ; Ohse, Naoyuki ; Ryo, Mina ; Ota, Chiharu ; Takao, Kazuto ; Mizukami, Makoto ; Kato, Tomohisa ; Izumi, Toru ; Hayashi, Toshihiko ; Nakayama, Koji ; Asano, Katsunori ; Okumura, Hajime ; Kimoto, Tsunenobu

HIGH SWITCHING PERFORMANCE OF 1.7KV, 50A SIC POWER MOSFET OVER SI IGBT FOR ADVANCED POWER CONVERSION APPLICATIONS..........3447

Hazra, Samir ; De, Ankan ; Bhattacharya, Subhashish ; Lin Cheng ; Palmour, John ; Schupbach, Marcelo ; Hull, Brett ; Allen, Scott

CONTROL METHOD FOR FIVE LEVEL CONVERTER WITH COMMON FLYING CAPACITORS TO AVOID VOLTAGE LEVEL SKIP..........3455

Wei Yan ; Hui Zhang ; Ogura, Kazuya ; Urushibata, Shota

LOW-COMPLEXITY ANALYTICAL APPROXIMATIONS OF SWITCHING FREQUENCY HARMONICS OF 3-PHASE N-LEVEL VOLTAGE-SOURCE PWM CONVERTERS..........3460

Burkart, Ralph M. ; Kolar, Johann W.

DYNAMIC VOLTAGE BALANCING ALGORITHM FOR MODULAR MULTILEVEL CONVERTER WITH THREE-LEVEL FLYING CAPACITOR SUBMODULES..........3468

Dekka, Apparao ; Wu, Bin ; Zargari, Navid R.

MODULAR MEDIUM VOLTAGE DRIVE FOR DEMANDING APPLICATIONS..........3476

Dujic, Drazen ; Wahlstroem, Jonas ; Marrero Sosa, Juan Alberto ; Fritz, Dominik

ASYMMETRICAL FAULT RIDE-THROUGH OF THREE-PHASE PV SYSTEMS USING FOUR-WIRE DC-AC CONVERTERS..........3482

Iyer, Shivkumar ; Bin Wu ; Yunwei Li ; Singh, B.N.

OPERATION MODE ANALYSIS FOR SOLVING THE PARTIAL SHADOW IN A NOVEL PV POWER GENERATION SYSTEM..........3489

Qi Zhang ; Xiangdong Sun ; Yanru Zhong ; Lie Guo ; Matsui, Mikihiko

ANALYSIS OF PARTIAL POWER PROCESSING DISTRIBUTED MPPT FOR A PV POWERED ELECTRIC AIRCRAFT..........3496

Marzouk, Ahmad Diab ; Fournier-Bidoz, Sebastien ; Yablecki, Jessica ; McLean, Kenneth ; Trescases, Olivier

IMPACTS OF RECTIFIER CIRCUIT LOADS ON ISLANDING DETECTION OF PHOTOVOLTAIC SYSTEMS..........3503

Yoshida, Yoshiaki ; Suzuki, Hirokazu

INDUCTION MOTOR MADE OF SMC..........3509

Morimoto, Masayuki ; Inamori, Mamiko

ESTIMATION AND COMPARISON OF THE WINDAGE LOSS OF A 60 KW SWITCHED RELUCTANCE MOTOR FOR HYBRID ELECTRIC VEHICLES..........3513

Kiyota, Kyohei ; Kakishima, Takeo ; Chiba, Akira

DEVELOPMENT OF HIGH-POWER PMASYNRM USING FERRITE MAGNETS FOR REDUCING RARE-EARTH MATERIAL USE..........3519

Sanada, Masayuki ; Morimoto, Shigeo ; Inoue, Yukinori

CONSIDERATION OF 10KW IN-WHEEL TYPE AXIAL-GAP MOTOR USING FERRITE PERMANENT MAGNETS..........3525

Sone, Kodai ; Takemoto, Masatsugu ; Ogasawara, Satoshi ; Takezaki, Kenichi ; Hino, Wataru

POWER CONTROL METHOD FOR MULTI-PARALLEL DC DISTRIBUTION SYSTEM THROUGH THE EQUIVALENT CIRCUIT MODEL..........3532

Seok-Jin Hong ; Soo-Cheol Shin ; Hee-Jun Lee ; Chung-Yuen Won ; Taeck-Kie Lee

A COMMUNICATION-LESS DISTRIBUTED VOLTAGE CONTROL STRATEGY FOR A MULTI-BUS AC ISLANDED MICROGRID..........3538

Wang, Yanbo ; Yongdong Tan ; Chen, Zhe ; Wang, Xiongfei ; Tian, Yanjun

AN ENHANCED LOAD POWER SHARING STRATEGY FOR LOW-VOLTAGE MICROGRIDS BASED ON INVERSE-DROOP CONTROL METHOD..........3546

Yixin Zhu ; Fang Zhuo ; Baoquan Liu ; Hao Yi

ADDING VIRTUAL RESISTANCE IN SOURCE SIDE CONVERTERS FOR STABILIZATION OF CASCADED CONNECTED TWO STAGE CONVERTER SYSTEMS WITH CONSTANT POWER LOADS IN DC MICROGRIDS..........3553

Mingfei Wu ; Lu, Dylan D.C.

EXPANSION OF OPERATING RANGE AND IMPROVEMENT OF TORQUE RESPONSE OF PMSM DRIVE BY USING MODEL PREDICTIVE CONTROL..........3557

N/A

NONLINEAR MODEL PREDICTIVE TORQUE CONTROL OF A LOAD COMMUTATED INVERTER AND SYNCHRONOUS MACHINE..........3563

Almer, Stefan ; Besselmann, Thomas ; Ferreau, Joachim

MODEL PREDICTIVE CURRENT CONTROL FOR PMSM CONSIDERING NUMBER OF SWITCHING OPERATIONS..........3568

Zanma, Tadanao ; Yasumura, Yuji ; Liu, KangZhi

PREDICTIVE INDIRECT MATRIX CONVERTER FED TORQUE RIPPLE MINIMIZATION WITH WEIGHTING FACTOR OPTIMIZATION..........3574

Uddin, Muslem ; Mekhilef, Saad ; Rivera, Marco ; Rodriguez, Jose

HIGH-POWER DENSITY HYBRID CONVERTER TOPOLOGIES FOR LOW-POWER DC-DC SMPS..........3582

Radic, Aleksandar ; Ahssanuzzaman, S.M. ; Mahdavikhah, Behzad ; Prodic, Aleksandar

COUPLED INDUCTOR BASED CURRENT-FED SWITCHED INVERTER FOR LOW VOLTAGE RENEWABLE INTERFACE 3587

Nag, Soumya Shubhra ; Mishra, Santanu Kumar

A SEMI-ISOLATED MULTI-INPUT CONVERTER FOR HYBRID PV/WIND POWER CHARGER SYSTEM 3592

Cheng-Wei Chen ; Kun-Hung Chen ; Chen, Yaow-Ming

HFL PV MICRO-INVERTER WITH FRONT-END CURRENT-FED CONVERTER AND HALF-WAVE CYCLOCONVERTER 3598

Nayanasiri, D.R. ; Vilathgamuwa, D.M. ; Maskell, D.L.

COMPREHENSIVE STUDY ABOUT STABILITY ISSUES OF MULTI-MODULE DISTRIBUTED SYSTEM 3604

Liu, Fangcheng ; Liu, Jinjun ; Zhang, Haodong ; Xue, Danhong ; Dou, Qinyun

CHARACTERISTICS STUDY OF NEURAL NETWORK AIDED DIGITAL CONTROL FOR DC-DC CONVERTER 3611

Maruta, Hidenori ; Motomura, Masashi ; Kurokawa, Fujio

ZERO CURRENT SWITCHING CURRENT-FED PARALLEL RESONANT PUSH-PULL (CFPRPP) CONVERTER 3616

Moorthy, Radha Sree Krishna ; Rathore, Akshay Kumar

CHARACTERISTICS OF TRANSMISSION CARRIER IN A NEW WIRE COMMUNICATION SYSTEM BY THE USE OF HIGH-RIPPLE DC-DC CONVERTER 3624

Katsuki, Akihiko ; Mizuki, Tatsuya ; Shibahara, Kohei ; Morita, Kosuke ; Masutomo, Kazufumi ; Maeyama, Shigetaka

5MHZ PWM-CONTROLLED CURRENT-MODE RESONANT DC-DC CONVERTER USING GAN-FETS 3630

Hariya, Akinori ; Yanagi, Hiroshige ; Ishizuka, Yoichi ; Matsuura, Ken ; Tomioka, Satoshi ; Ninomiya, Tamotsu

DESIGN AND PERFORMANCE EVALUATION OF DIGITAL CONTROL FOR LLC SERIES RESONANT DC-TO-DC CONVERTERS 3638

Pidaparthy, Syam Kumar ; Choi, Byungcho ; Jang, Jinhaeng

EXPERIMENTAL VERIFICATION OF NOISELESS SAMPLING FOR BUCK CHOPPER CIRCUIT WITH CURRENT CONTROL 3646

Takeuchi, Shun ; Wada, Keiji

CONTROL CHARACTERISTICS IMPROVEMENT OF FULL-BRIDGE DC-DC CONVERTER WITH SNUBBER CAPACITOR 3652

Domoto, Kazuhide ; Ishizuka, Yoichi ; Abe, Seiya ; Ninomiya, Tamotsu

DCM CONTROL METHOD OF BOOST CONVERTER BASED ON CONVENTIONAL CCM CONTROL 3659

Le Hoai Nam ; Orikawa, Koji ; Itoh, Jun-ichi

TECHNICAL ASSESSMENT OF LOAD COMMUTATION SWITCH IN HYBRID HVDC BREAKER 3667

Hassanpoor, Arman ; Hafner, Jurgen ; Jacobson, Bjorn

CONTROL OF HEXAGONAL MODULAR MULTILEVEL CONVERTER FOR 3-PHASE BTB SYSTEM 3674

Hamasaki, Shin-ichi ; Okamura, Kazuki ; Tsubakidani, Takashi ; Tsuji, Mineo

A SYNTHESIZED CAPACITORS VOLTAGE CONTROL FOR MODULAR MULTILEVEL CONVERTER IN HVDC APPLICATION 3680

Rongfeng Yang ; Shunke Sui ; Binbin Li ; Wei Wang ; Dianguo Xu

OPERATING PHASE AND FREQUENCY SELECTION OF LOW FREQUENCY AC TRANSMISSION SYSTEM USING CYCLOCONVERTERS 3687

Achara, Pichetjamroen ; Ise, Toshifumi

FAST ACTING DC CIRCUIT BREAKER FOR HVDC TRANSMISSION LINE BASED ON DC/DC CHOPPER 3695

Liangyi Tang ; Bin Wu ; Yaramasu, Venkata ; Weirong Chen ; Athab, Hussain S.

1700V SI-IGBT AND SIC-SBD HYBRID MODULE FOR AC690V INVERTER SYSTEM 3702

Haining Wang ; Ikawa, O. ; Miyashita, S. ; Nishimura, T. ; Igarashi, S.

SWITCHING SIMULATION OF SIC HIGH-POWER MODULE WITH LOW PARASITIC INDUCTANCE 3707

Yamamoto, Takashi ; Hasegawa, Kohei ; Ishida, Masaaki ; Takao, Kazuto

SWITCHING PERFORMANCE OF PARALLEL-CONNECTED POWER MODULES WITH SIC MOSFETS 3712

Colmenares, Juan ; Peftitsis, Dimosthenis ; Nee, Hans-Peter ; Rabkowski, Jacek

BUILT-IN RELIABILITY DESIGN OF A HIGH-FREQUENCY SIC MOSFET POWER MODULE 3718

Jianfeng Li ; Gurpinar, Emre ; Lopez-Arevalo, Saul ; Castellazzi, Alberto ; Mills, Liam

EXPERIMENTAL SWITCHING FREQUENCY LIMITS OF 15 KV SIC N-IGBT MODULE 3726

Kadavelugu, Arun ; Bhattacharya, Subhashish ; Ryu, Sei-Hyung ; Van Brunt, Edward ; Grider, Dave ; Leslie, Scott

SELECTION OF SUITABLE CARRIER-BASED PWM METHOD FOR MODULAR MULTILEVEL CONVERTER 3734

Ciftci, Baris ; Erturk, Feyzullah ; Hava, Ahmet M.

CONTROL AND EXPERIMENT OF A 380-V, 15-KW MOTOR DRIVE USING MODULAR MULTILEVEL CASCADE CONVERTER BASED ON TRIPLE-STAR BRIDGE CELLS (MMCC-TSBC) 3742

Kawamura, Wataru ; Hagiwara, Makoto ; Akagi, Hirofumi

A POWER ELECTRONIC TRANSFORMER WITH SINUSOIDAL VOLTAGES AND CURRENTS USING MODULAR MULTILEVEL CONVERTER 3750

Sahoo, Ashish Kumar ; Mohan, Ned

VARYING AND UNEQUAL CARRIER FREQUENCY PWM TECHNIQUES FOR MODULAR MULTILEVEL CONVERTERS 3758

Konstantinou, Georgios ; Darus, Rosheila ; Pou, Josep ; Ceballos, Salvador ; Agelidis, Vassilios G.

COMPARISON OF PHASE-SHIFTED AND LEVEL-SHIFTED PWM IN THE MODULAR MULTILEVEL CONVERTER 3764

Darus, Rosheila ; Konstantinou, Georgios ; Pou, Josep ; Ceballos, Salvador ; Agelidis, Vassilios G.

A SINGLE-PHASE POWER CONDITIONER WITH A BUCK-BOOST-TYPE POWER DECOUPLING CIRCUIT 3771
Yamaguchi, Shota ; Shimizu, Toshihisa

A NOVEL ASYMMETRICAL FLC-BASED MPPT TECHNIQUE FOR PHOTOVOLTAIC GENERATION SYSTEM 3778
Yi-Hsun Chiu ; Yu-Shan Cheng ; Yi-Hua Liu ; Shun-Chung Wang ; Zong-Zhen Yang

A NOVEL CURRENT LINK DISTRIBUTED MPPT PV SYSTEM - OVERALL SYSTEM PROTOTYPING AND EVALUATION 3784
Mikihiko ; Toru ; Akira ; Xiang-Dong Sun ; Byung-Gyu Yu

POWER FLOW CONTROL AND MPPT PARAMETER SELECTION FOR RESIDENTIAL GRID-CONNECTED PV SYSTEMS WITH BATTERY STORAGE 3789
Chokchai, Chuenwattanapraniti

A MAXIMUM POWER POINT TRACKING METHOD WITH RIPPLE CURRENT ORIENTATION 3796
Moo, Chin-Sien ; Wu, Gwo-Bin

OUTPUT CHARACTERISTICS OF A SURFACE PERMANENT MAGNET-TYPE VERNIER MOTOR - COMPARISON OF TEST RESULTS AND CALCULATION 3801
Kataoka, Yasuhiro ; Takayama, Masakazu ; Anazawa, Yoshihisa ; Matsushima, Yoshitarou

TOPOLOGY OPTIMIZATION FOR SKEW OF SPMSM BY USING MULTI-STEP PARALLEL GA 3809
Kitagawa, Wataru ; Takeshita, Takaharu

LOSS MINIMIZATION DESIGN USING MAGNETIC EQUIVALENT CIRCUIT FOR A PERMANENT MAGNET SYNCHRONOUS MOTOR 3815
Sato, Daisuke ; Itoh, Jun-ichi

THE PROPOSAL OF A NEW MOTOR WHICH HAS A HIGH WINDING FACTOR AND A HIGH SLOT FILL FACTOR 3823
Makita, Shinji ; Ito, Yasuhide ; Aoyama, Tomohiro ; Doki, Shinji

VARIABLE LEAKAGE FLUX INTERIOR PERMANENT MAGNET SYNCHRONOUS MACHINE FOR IMPROVING EFFICIENCY ON DUTY CYCLE 3828
Minowa, Masanao ; Hijikata, Hiroki ; Akatsu, Kan ; Kato, Takashi

HISTORY AND TRENDS OF CONVERTER TECHNOLOGY FOR DC AND AC TRANSMISSION IN JAPAN 3834
Yoshino, Teruo

ACCURATE OUTPUT POWER CONTROL OF CONVERTERS FOR MICROGRIDS BASED ON LOCAL MEASUREMENT AND UNIFIED CONTROL 3842
Meiqin Mao ; Zheng Dong ; Yong Ding ; Liuchen Chang

IMPEDANCE-BASED ANALYSIS OF ACTIVE FREQUENCY DRIFT ISLANDING DETECTION METHOD FOR GRID-TIED INVERTER SYSTEM 3850
Wen, Bo ; Boroyevich, Dushan ; Burgos, Rolando ; Shen, Zhiyu ; Mattavelli, Paolo

DEVELOPMENT OF 200-MVAR CLASS THYRISTOR SWITCHED CAPACITOR SUPPORTING FAULT RIDE-THROUGH 3857
Ohtake, Asuka ; Fei Zhang ; Fujimoto, Takafumi ; Nakayama, Naoyuki

DETAILED ANALYSIS AND DESIGN OF A THREE-PHASE PHASE-MODULAR ISOLATED MATRIX-TYPE PFC RECTIFIER 3864
Cortes, Patricio ; Fassler, Lukas ; Bortis, Dominik ; Kolar, Johann W. ; Silva, Marcelo

AN ENERGY SAVING DRIVE METHOD OF AN INDUCTION MOTOR WITH THE SUPPRESSION OF SUDDEN ACCELERATION AND DECELERATION 3872
Asano, Yuji ; Inoue, Kaoru ; Kotera, Keito ; Kato, Toshiji

FIELD ORIENTED CONTROL OF SENSORLESS LINEAR INDUCTION MOTOR USING MATRIX CONVERTER 3877
Sayed, Mahmoud A. ; Mohamed, Essam Ebaid ; Mohamed, Tarek Hassan ; Takeshita, Takaharu

A STATOR-EQUATION-BASED REDUCED-ORDER OBSERVER FOR POSITION-SENSORLESS VECTOR CONTROL SYSTEM OF DOUBLY-FED INDUCTION MACHINES 3885
Smiththisomboon, Somrat ; Suwankawin, Surapong

INPUT CURRENT RIPPLE ANALYSIS OF INVERTER FED DUAL THREE-PHASE AC MOTORS 3893
Dahono, Pekik Argo ; Satria, Andri

OFFLINE EXTRACTION OF INDUCTION MACHINE PARAMETERS FOR CONTROL STRATEGY SYNTHESIS 3898
Koschik, Stefan ; Bauer, Florian ; De Doncker, R.W.

HIGH CURRENT PLANAR TRANSFORMER FOR VERY HIGH EFFICIENCY ISOLATED BOOST DC-DC CONVERTERS 3905
Pittini, Riccardo ; Zhe Zhang ; Andersen, Michael A.E.

HIGH VOLTAGE-GAIN INTERLEAVED BOOST DC-DC CONVERTER DISCARDED ELECTROLYTIC CAPACITOR 3913
Nha, Quang Trong ; Huang-Jen Chiu ; Yu-Kang Lo ; Pham Phu Hieu

PARALLEL BI-DIRECTIONAL DC-DC CONVERTER FOR ENERGY STORAGE SYSTEM 3920
Ouchi, Takayuki ; Kanoda, Akihiko ; Takahashi, Naoya

CHARGING SCENARIO OF SERIAL BATTERY POWER MODULES WITH BUCK-BOOST CONVERTERS 3928
Jhen-Yu Jian ; Chu-Shen Chang ; Moo, Chin-Sien ; Hau-Chen Yen

COMPARATIVE THERMAL PERFORMANCE EVALUATION OF SIC MOSFETS AND SI MOSFET FOR 1.2 KW 300 KHZ DC-DC BOOST CONVERTER AS A SOLAR PV PRE-REGULATOR 3933
Taekyun Kim ; Minsoo Jang ; Agelidis, Vassilios G.

TOLERANCE ANALYSIS OF A CONSTANT-ON TIME CURRENT-MODE VOLTAGE REGULATOR WITH ADAPTIVE VOLTAGE POSITION FEATURE 3938

Chih Wei Chen ; Dan Chen ; Shin Shiung Wang

FPGA-BASED DIGITAL-CONTROLLED POWER CONVERTER DESIGNED WITH UNIVERSAL INPUT MEETING 80 PLUS PLATINUM EFFICIENCY CODE AND STANDBY POWER CODE FOR SEVER POWER APPLICATIONS..3942

Lai, Yen-Shin ; Ho, Kung-Min

STATIC AND DYNAMIC ANALYSES OF DIGITAL PEAK CURRENT MODE DC-DC CONVERTER 3950

Kajiwara, Kazuhiro ; Kurokawa, Fujio ; Shibata, Yuichiro

EXTENDED DISCRETE CONTROL OF CLASS E AMPLIFIER IN ORDER TO ACHIEVE NOMINAL OPERATION..3955

Suetsugu, Tadashi ; Xiuqin Wei ; Kuga, Shotaro

ADAPTIVE POWER EFFICIENCY CONTROL BY COMPUTER POWER CONSUMPTION PREDICTION USING PERFORMANCE COUNTERS 3959

Kawaguchi, Shinichi ; Yachi, Toshiaki

Author Index

Generalized Modular Multilevel Converter and Modulation

Hui Liu, Poh Chiang Loh, Frede Blaabjerg
Department of Energy Technology
Aalborg University
Aalborg, Denmark
hui@et.aau.dk, pcl@et.aau.dk, fbl@et.aau.dk

Abstract— Modular multilevel converter (MMC) has gained popularity recently with its modulation, capacitor voltage balancing and circulating current issues widely discussed. Contributing to this effort, a study is presented here to show how the MMC topology can be derived from the viewpoint of two series converters regulating a power grid. The generalized topology derived and notated as GMMC can then be altered to create various types of MMC, including the traditional topology that is presently well-known. This effort has not been previously discussed, and may smoothen the understanding of MMC operation. For controlling the GMMC, a simple modulation scheme is also presented, where the goal is to achieve the desired performance at a minimized complexity. Simulation results are presented to illustrate the operation of the proposed topology.

Keywords— Modular multilevel converter, modulation, triplen-powerd modular multilevel converter, hybrid-source-powered modular multilevel converter.

I. INTRODUCTION

A Modular Multilevel Converter (MMC) shown in Fig. 1, was invented by Siemens in 2001 [1], but did not gain much attention until the past few years. Its usage has since been directed at the medium to high voltage level with its first commercial application identified as the Trans Bay Cable (TBC) project in the US. That project involves High Voltage DC (HVDC) transmission at 400-MW rated power, 200-kV dc-link voltage and 216 sub-modules in each of the MMC arms [2], [3]. With its smooth completion, other HVDC projects subsequently follow like the 2×1000-MW link with 320-kV dc-link voltage from France to Spain built by Siemens for the INELFE project [3], [4]. These developments, no doubt, generate a greater interest in MMC topology especially with its transformer-less modular design that allows easy transportation and flexible scaling up of the power level [5]. Its large number of voltage levels "protected" by sizable redundancy is also an attractive feature that will generate low harmonics, and hence support a filter-less design [6].

The mentioned advantages of MMC have, so far, outweighed its disadvantages, mainly related to its large number of switches [7]. That, in turn, leads to intensive discussions about its modulation schemes, capacitor balancing techniques [8], circulating current related

issues [9] and others. Contributing to this ongoing effort and a better understanding of MMC in general, a study is proposed here to demonstrate how a Generalized MMC (GMMC) topology can be developed from two series converters regulating voltage for a power grid. Modifications of the GMMC can then be pursued to develop various variations of the MMC topology, which can collectively be controlled by the generalized modulation scheme which is proposed in this paper. Finally, simulation results are demonstrated the topologies.

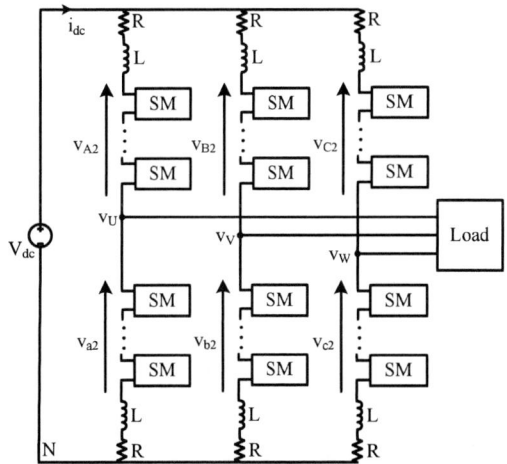

Fig. 1. Traditional Modular Multilevel Converter (MMC).

II. TWO-SERIES-CONVERTER REPRESENTATION

Fig. 2 (a) shows the single-line representation of two series three-phase converters regulating a power grid with two power sources notated as v_{X1} and v_{Y1}, respectively ($X = A, B$ or C and $Y = a, b$ or c). The converters are usually accompanied by their respective transformers, which are theoretically not necessary if each phase of the two converters has its own dc storage, rather sharing a single storage per converter. This has, in fact, been done in [10], where a cascaded multilevel converter has been used for implementing a series dynamic compensator.

978-1-4799-2706-7/14 $31.00 © 2014 IEEE

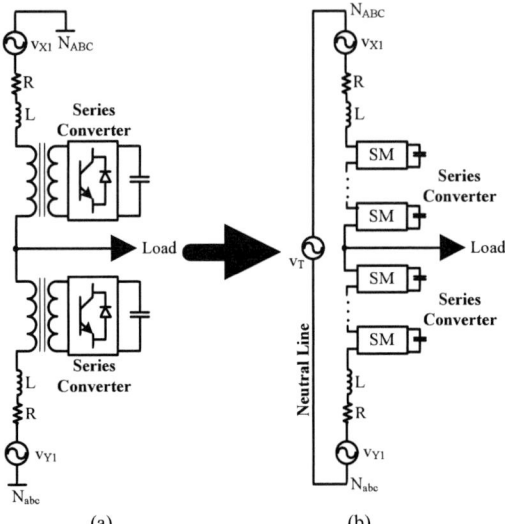

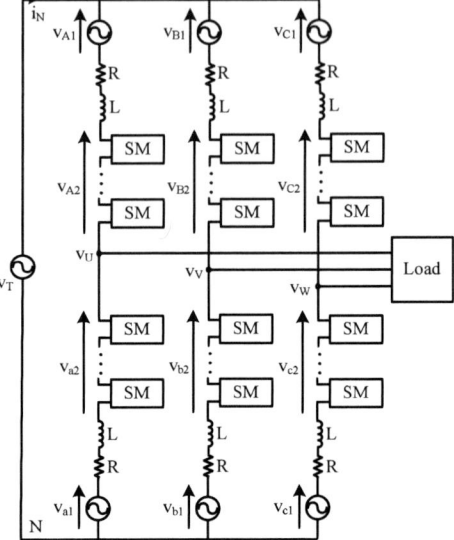

Fig. 2. Two series converters regulating a power grid: (a) two series converter; (b) multilevel converter.

Fig. 3. Three-phase Generalized MMC.

With the removal of the transformers, the single-line diagram shown in Fig. 2 (a) changes to that shown in Fig. 2 (b), where two other features are also observed. The first feature is the realization of each converter with a series string of sub-modules, which has traditionally been practiced for raising the converter voltage level and waveform quality. The second feature introduced is the arbitrary source v_T tying the neutral points of the two three-phase sources. Strictly, these points need not be at the same potential in practice, and can indeed be linked in a three-phase four-wire system. The arbitrary source added is thus not contradicting any circuit laws if its frequency is defined to be triplen multiple of the three-phase output frequency f_{out}.

III. GMMC TOPOLOGY

Expanding the single-line diagram shown in Fig. 2 (b) to its three-phase representation, the GMMC topology is derived and shown in Fig. 3. Sources of this GMMC, when set to $v_{X1} = 0$, $v_{Y1} = 0$ and $v_T = V_{dc}$, give rise to the existing MMC topology shown in Fig. 1, which is strictly not the only possibility. A second triplen-powered topology can, for example, be developed by setting $v_{X1} = 0$, $v_{Y1} = 0$ and v_T to an ac source with a frequency of $3k \times f_{out}$, where k is an integer. The resulting converter is shown in Fig. 4 (a). Instead of only a single source, topologies with more than one source can similarly be developed by, for example, setting $v_{X1} = 0$, $v_{Y1} \neq 0$ and $v_T = V_{dc}$ in order to obtain the converter shown in Fig. 4 (b). These converters in Fig. 1 and Fig. 4 will subsequently be used to illustrate the modulation theories and topological choices made for the sub-modules in each MMC arm.

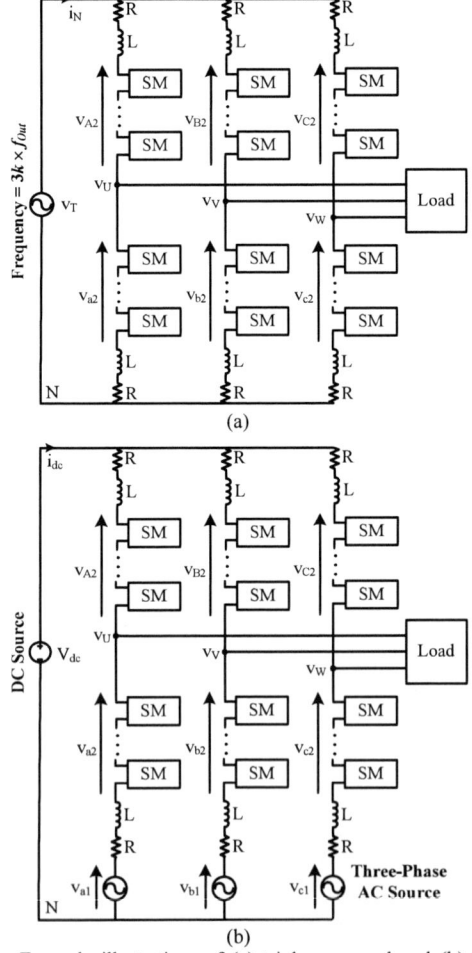

Fig. 4. Example illustrations of (a) triplen-powered and (b) hybrid-source-powered MMCs.

IV. GENERALIZED MODULATION

Using the notations indicated in Fig. 3, (1) to (3) can be written, where subscripts "0" and "*ph*" have been added to indicate zero-sequence and non-zero-sequence components, respectively. In the expressions, voltage drop across filter L of each arm (accompanied R is for representing losses) has been ignored, which in general, is fine for a multilevel system, where the harmonics are low, and hence does not require a large L [11]. The eventual expressions derived in (3) are, in fact, the generalized criteria governing the modulating reference selection for all topologies simplified from the GMMC. To illustrate this, (3) is applied to the example converters shown in Fig. 1 and Fig. 4, whose details are presented as follows.

$$v_T = \{v_{X1} + v_{X2}\} + \{v_{Y1} + v_{Y2}\} \quad (1)$$

in which, $\{X, Y\} = \{A, a\}, \{B, b\}$ or $\{C, c\}$ are from the same phase and (1)

$$v_{T,0} = \{(v_{X1,ph} + v_{X1,0}) + (v_{X2,ph} + v_{X2,0})\} \quad (2)$$
$$+\{(v_{Y1,ph} + v_{Y1,0}) + (v_{Y2,ph} + v_{Y2,0})\}$$

$$\Rightarrow \begin{cases} v_{X1,ph} + v_{X2,ph} = -(v_{Y1,ph} + v_{Y2,ph}) \\ v_{T,0} = (v_{X1,0} + v_{X2,0}) + (v_{Y1,0} + v_{Y2,0}) \end{cases} \quad (3)$$

A. Traditional MMC

Related conditions are summarized as $v_{T,0} = V_{dc}$ and $v_{X1} = v_{Y1} = 0$. Expressions from (3) then become $v_{X2,ph} = -v_{Y2,ph}$ and $V_{dc} = v_{X2,0} + v_{Y2,0}$. That means the two arms per phase must be modulated by the same sinusoidal references, but with opposite polarities (triplen offset V_{off} can, no doubt, be added to increase the linear modulation range). The references must also include dc components, which can theoretically be any combination of $v_{X2,0}$ and $v_{Y2,0}$, so long as they add up to V_{dc}. However, if the same sub-module like in Fig. 5 (a) is used for both arms, it is wiser to have equal dc components, leading to the reference expressions in (4), where $0 \leq M \leq 1.15$, $M|V_M|$ and φ are the modulation index, amplitude and phase of the sine components.

$$\begin{cases} v_{X2,Ref} = M|V_M|cos(2\pi f_{out}t + \varphi) + \frac{V_{dc}}{2} + V_{off} \\ v_{Y2,Ref} = -M|V_M|cos(2\pi f_{out}t + \varphi) + \frac{V_{dc}}{2} - V_{off} \end{cases} \quad (4)$$

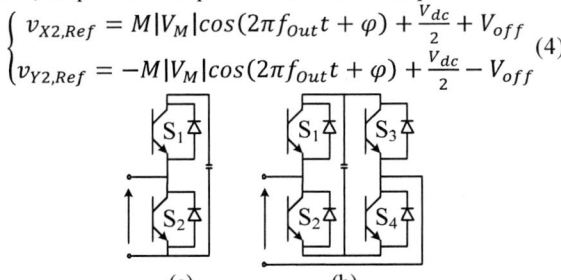

(a) (b)

Fig. 5. Possible sub-module topologies for MMC: (a) half bridge; (b) full bridge.

B. Triplen-Powered MMC

The new circuit conditions are summarized as $v_{T,0} \neq V_{dc}$ and $v_{X1} = v_{Y1} = 0$, which lead to $v_{X2,ph} = -v_{Y2,ph}$ and $v_{T,0} = v_{X2,0} + v_{Y2,0}$ according to (3). The only difference here is the triplen voltage $v_{T,0}$, which should again be shared between the two arms per phase if the same type of sub-module is used. The modulating references demanded are thus given by (5), where no dc components are included. Equation (5) should therefore

be realized by the sub-module in Fig. 5 (b), instead of that in Fig. 5 (a), where a dc offset is always present.

$$\begin{cases} v_{X2,Ref} = M|V_M|cos(2\pi f_{out}t + \varphi) + \frac{v_{T,0}}{2} + V_{off} \\ v_{Y2,Ref} = -M|V_M|cos(2\pi f_{out}t + \varphi) + \frac{v_{T,0}}{2} - V_{off} \end{cases} \quad (5)$$

C. Hybrid-Source-Powered MMC

The mixed source conditions are summarized as $v_{T,0} = V_{dc}$, $v_{X1} = 0$, $v_{Y1,ph} = |V_s|cos(2\pi f_s t + \varphi_s)$ and $v_{Y1,0} = 0$, which when is substituted to (3), gives rise to $v_{X2,ph} = -(|V_s|cos(2\pi f_s t + \varphi_s) + v_{Y2,ph})$ and $V_{dc} = v_{X2,0} + v_{Y2,0}$. The modulating references demanded are hence written as (6), where $0 \leq k \leq 1$ is a fractional number for dividing V_{dc} between the two arms per phase. Here, k should not be equal to 0.5 since the sinusoidal components of the two references are not the same. Instead, it should be set according to (6), where $max( )$ is a function that returns the peak value of the variable enclosed by the parentheses.

$$v_{X2,Ref} = M|V_M|cos(2\pi f_{out}t + \varphi) + kV_{dc} + V_{off}$$
$$v_{Y2,Ref} = -M|V_M|cos(2\pi f_{out}t + \varphi)$$
$$-|V_s|cos(2\pi f_s t + \varphi_s)$$
$$+(1 - k)V_{dc} - V_{off} \quad (6)$$

where

$$k = max(1.15|V_M|cos(2\pi f_{out}t + \varphi))$$
$$\div [max(1.15|V_M|cos(2\pi f_{out}t + \varphi)$$
$$+|V_s|cos(2\pi f_s t + \varphi_s))$$
$$+max(1.15|V_M|cos(2\pi f_{out}t + \varphi))]$$

D. References for Sub-Modules

Regardless of the MMC topology considered, references for sub-modules are computed by dividing (4), (5) or (6) by the number of sub-modules per arm (n_X for upper arm and n_Y for lower arm). The dc-link voltage of each sub-module can then be determined as $max\left(\frac{v_{X2,Ref}}{n_X}\right)\Big|_{M=1.15}$ for the upper arm and $max\left(\frac{v_{Y2,Ref}}{n_Y}\right)\Big|_{M=1.15}$ for the lower arm.

E. Carriers for Sub-Modules

Each sub-module has its own carrier [12], which must be phase-shifted from others by $2\pi/n_X$ for the upper arm and $2\pi/n_Y$ for the lower arm.

V. SIMULATION RESULTS

For evaluation, the traditional MMC and the hybrid-source-powered MMC shown in Fig. 1 and Fig. 4 (b) were simulated using the parameters tabulated in Table I, Table II respectively. The parameters chosen only represent an example case that can better be explained. Other combinations can certainly be used, where necessary.

Setting the expected output and the DC source values as in the first two rows in Table I, the other parameters can be obtained as shown in the Table and the results are displayed in Fig. 6. As it can be seen, the output line voltages, currents, are balanced and sinusoidal. Both the voltages and currents are the same for the upper and lower arms.

TABLE I
SIMULATION SPECIFICATIONS FOR THE TRADITIONAL MMC

AC Output v_U	$\|V_M\| = 4800$ V(rms), $f_{Out} = 50$ Hz $\varphi = 180°$ (3-phase)
DC Source V_{dc}	$7200\sqrt{2}$ V
Sub-Module DC-Link	1140 V, 3000 µF (same for all)
No. of Sub-Modules	$n_X = 12$ (upper), $n_Y = 12$ (lower)
Arm Filter L, R	1 mH, 0.5 Ω

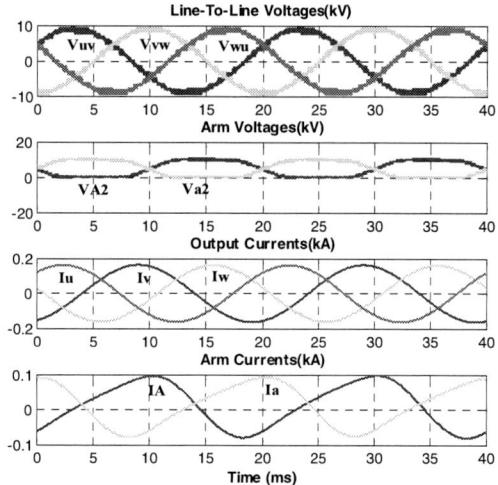

Fig. 6. Simulated results obtained from traditional MMC shown in Fig. 1.

TABLE II
SIMULATION SPECIFICATIONS FOR THE HYBRID-SOURCE-POWERED MMC

AC Source v_{a1}	$\|V_s\| = 2760$ V(rms), $f_s = 50$ Hz $\varphi_s = 0$ (3-phase)
AC Output v_U	$\|V_M\| = 4800$ V(rms), $f_{Out} = 50$ Hz $\varphi = 180°$ (3-phase)
k from (6)	2/3
DC Source V_{dc}	$7200\sqrt{2}$ V
Sub-Module DC-Link	1140 V, 3000 µF (same for all)
No. of Sub-Modules	$n_X = 12$ (upper), $n_Y = 6$ (lower)
Arm Filter L, R	1 mH, 0.5 Ω

With the ac values given in the first two rows in Table II, k from (6) is determined as 2/3 for the upper arm, which can then be used to compute the required dc source voltage as $V_{dc} = 4800\sqrt{2} \times 3/2 = 7200\sqrt{2}$ V (see fourth row). With $k = 2/3$, the number of sub-modules needed for the upper arm is also twice that of the lower arm if the same sub-module components and voltages are used. That leads to the fifth and sixth rows of Table I. Using these parameters, Fig. 7 shows the obtained results, which clearly shows that the output line voltages and currents are balanced and sinusoidal. The line voltage amplitude is also noted to be 1.33 times higher than that achievable by the traditional MMC in Fig. 1, whose k is 0.5. This is, no doubt, contributed by source v_{a1}, whose effect can be seen from the second plot in Fig. 7. To be more specific, v_{a1} causes the lower arm voltage to be

more sinusoidal with negative values. It thus looks different from the upper arm voltage.

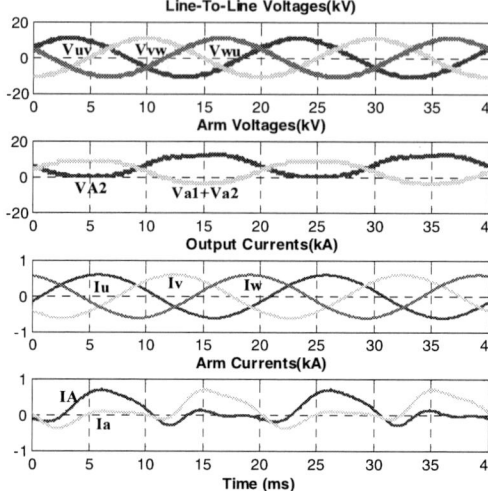

Fig. 7. Simulated results obtained from hybrid-source-powered MMC shown in Fig. 4 (b).

VI. CONCLUSIONS

An equivalent representation has been established in the paper, which leads to the development of the GMMC and its generalized modulation. Various MMC topologies can subsequently be derived with the tranditional and hybrid-source powered variations which are already proven in simulation. The concepts discussed here are expected to be interested in medium and high voltage applications like HVDC, where multilevel converters seem to be alternative for convert source converters.

REFERENCES

[1] R. Marquardt, A. Lesnicar, J. Hildinger, "Modulares Stromrichterkonzept für Netzkupplungsanwendung bei hohen Spannungen," ETG-Fachtagung, Bad Nauheim, Germany, 2002.

[2] CIGRE working group B4.101, "Trans bay cable-world's first HVDC system using multilevel voltage sourced converter," 2010.

[3] J. Z. Xu, C. Y. Zhao, W. J. Liu, and C. Y. Guo, "Accelerated Model of Modular Multilevel Converters in PSCAD/EMTDC," IEEE Trans. on Power Delivery, vol. 28, no. 1, pp. 129-136, Jan. 2013

[4] Siemens HVDC project: http://www.energy.siemens.com/us/en/power-transmission/hvdc/hvdc-plus/references.htm#content=2014%20INELFE%2C%20France-Spain

[5] M. Glinka, R. Marquardt, "A New AC/AC-Multilevel Converter Family Applied to a Single-phase Converter," IEEE Trans. on Industrial Electronics, vol. 52, no. 3, pp. 662-669, June, 2005.

[6] A. Lesnicar, and R. Marquardt, "An Innovative Modular Multilevel Converter Topology Suitable for a Wide Power Range," Proc. of IEEE Power Tech Conference, vol. 3, Bologna, Italy, June 2003, [CD-ROM].

[7] U. N. Gnanarathna, A. M. Gole, and R. P. Jayasinghe, "Efficient Modeling of Modular Multilevel HVDC Converters (MMC) on Electromagnetic Transient Simulation Programs," IEEE Trans. on Power Delivery, vol. 26, no. 1, pp. 316-324, Jan. 2011.

[8] E. Solas, G. Abad, J. A. Barrena, S. Aurtenetxea, A. Carcar, L. Zajac, "Modular Multilevel Converter With Different Submodule Concepts-Part I: Capacitor Voltage Balancing Method," IEEE Trans. on Industrial Electronics, vol. 60, no. 10, pp. 4525-4535, Jul. 2010.

The 2014 International Power Electronics Conference

[9] Q. R. Tu, Z. Xu, L. Xu, "Reduced switching-frequency modulation and circulating current suppression for modular multilevel converters," *IEEE Trans. on Power Electronics*, vol. 26, no. 3, pp. 2009-2017, 2011.

[10] A. M. Massoud, S. Ahmed, P. N. Enjeti, and B. W. Williams, "Evaluation of a multilevel cascaded-type dynamic voltage restorer employing discontinuous space vector modulation," *IEEE Trans. on Industrial Electronics*, vol. 57, no. 7, pp. 2398-2410, Jul. 2010.

[11] S. Allebrod, R. Hamerski, R. Marquardt, "New transformerless, scalable modular multilevel converters for HVDC transmission," *Proc. of IEEE Power Electronics Specialists Conference (PESC)*, pp. 174-179, Rhodes, Greece, June 2008.

[12] B. P. Mcgrath and D. G. Holmes, "Natural current balancing of multicell current source converters," *IEEE Trans. on Power Electronics*, vol. 23, no. 3, pp. 1239-1246, May 2008.

The 2014 International Power Electronics Conference

Average Power Control of DC Bus Voltages of Cascaded H-Bridge Multilevel Converters

Chia-Tse Lee, Hsin-Chih Chen, Ching-Wei Wang, Ching-Hsiang Yang, Po-Tai Cheng
Center for Advanced Power Technologies
Department of Electrical Engineering
National Tsing Hua University
Hsinchu 30013, TAIWAN.

Abstract—This paper presents an average power balancing control technique for the modular multilevel cascaded converter (MMCC) based on single-star bridge cells (SSBC) in the static synchronous compensator (STATCOM) applications. Detailed power flow analyses of the MMCC-SSBC converter, from the bottem level of individual H-bridge modules, then the middle level of per-phase circuit, to the top level of the three-phase circuit, are performed. By utilizing the outcome of such power flow analyses, the proposed method can precisely control all the DC bus voltages of the MMCC-SSBC while performing the reactive power compensation, even under extreme conditions like voltage sags, thus enhance the low voltage ride through functionalities of the STATCOM, which is very critical for renewable energy deployment. The proposed control algorithm is verified by a MMCC-SSBC test bench in the laboratory under normal and fault grid conditions.

Index Terms—STATCOM, distributed energy resources, Modular multilevel cascaded converter, average power balancing.

I. INTRODUCTION

Distributed energy resources (DERs) have attracted great attention in recent years [1] as a potent solution for the $CO2$ reduction. However, their distributed nature poses a significant challenge for stablizing the power system. The static synchronous compensator (STATCOM) has been proven as an effective solution to for the voltage control by injecting or drawing VAR [2], [3], [4], [5], [6], [7]. The modular multi-level cascaded converter (MMCC) based on single-star bridge cells (SSBC) is among the most utilized circuit topology for STATCOM applications.

The balancing of DC bus voltages of MMCC-SSBC is the most fundamental issue for operating this circuit. This paper conducts a very detailed analysis into the power flow of the MMCC-SSBC, from the bottom level of each H-bridge modules, then the middle level of per-phase circuit, to the top level of three-phase circuit. Based on the outcome of the analysis, the proposed average power balancing (APB) technique directly manage the power flow in all three levels with both feed-forward and feedback designs. The advantages of the power flow based approach become critical in the case of low voltage ride-through (LVRT) operation because the unbalanced grid voltages are factored into the power flow calculation. Details of the power flow analysis are presented, and test restuls based on a scaled-down laboratory prototype are given to validate the performance of the proposed method.

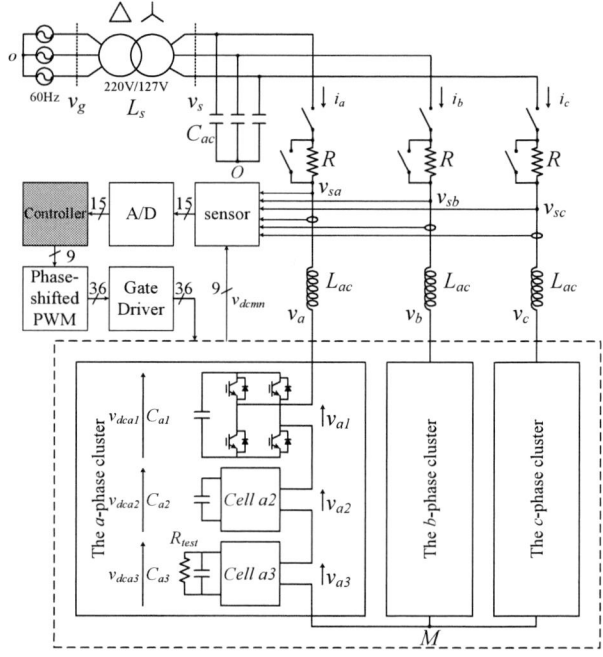

Fig. 1. The system configuration of the STATCOM based on MMCC-SSBC.

II. SYSTEM CONFIGURATION

Fig. 1 shows the circuit diagram of the MMCC-SSBC and the implementation of the proposed voltage balancing control, Fig. 2 shows the overall control block diagram of this paper. The converter's output voltages (v_{sm}, where $m = a, b, c$), phase currents (i_m, where $m = a, b, c$), and DC bus voltages (V_{dcmn}, where $m = a, b, c$ and $n = 1, 2, 3$) are taken by sensors and then converted into digital format by analog-to-digital (A/D) converters.

In the circuit topology, each cluster connects with the grid source of each phase. Thus every bridge cell in the same cluster performs the single phase AC modulation with the separated DC capacitors and the DC voltages will contain the double line frequency ripples. In order to eliminate the double line frequency ripples, the signal processing uses moving averaging filters (MAF) [8]. Fig. 3 shows DC capacitor voltages(V_{dcmn}) are processed by the the MAF and used to calculate the average values for the voltage balancing control. Note that the proposed voltage balancing control is operated at

978-1-4799-2706-7/14 $31.00 © 2014 IEEE   1639

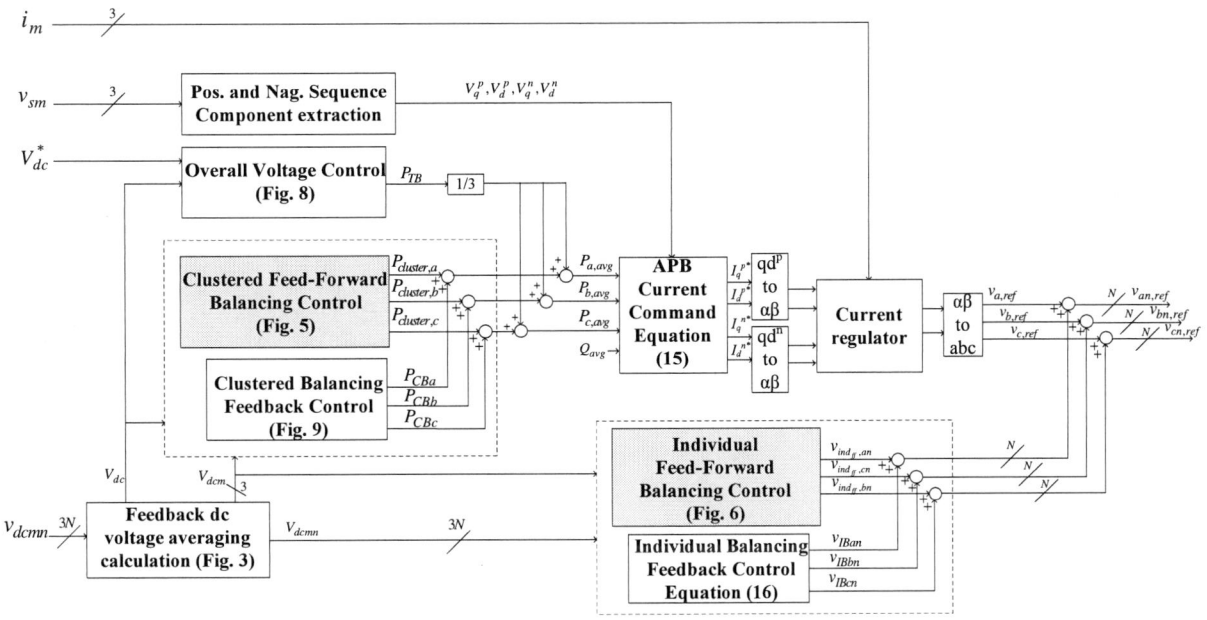

Fig. 2. The system overall control block of the STATCOM based on APB.

$$\begin{bmatrix} v_{sa} \\ v_{sb} \\ v_{sc} \end{bmatrix} = F \cdot \begin{bmatrix} v_\alpha \\ v_\beta \end{bmatrix} = F \cdot \left(T \begin{bmatrix} V_q^p \\ V_d^p \end{bmatrix} + T^{-1} \begin{bmatrix} V_q^n \\ V_d^n \end{bmatrix} \right)$$

$$\begin{bmatrix} i_{sa} \\ i_{sb} \\ i_{sc} \end{bmatrix} = F \cdot \begin{bmatrix} i_\alpha \\ i_\beta \end{bmatrix} = F \cdot \left(T \begin{bmatrix} I_q^p \\ I_d^p \end{bmatrix} + T^{-1} \begin{bmatrix} I_q^n \\ I_d^n \end{bmatrix} \right) \tag{1}$$

$$Where\ T = \begin{bmatrix} \cos(\omega t) & \sin(\omega t) \\ -\sin(\omega t) & \cos(\omega t) \end{bmatrix},\ T^{-1} = \begin{bmatrix} \cos(\omega t) & -\sin(\omega t) \\ \sin(\omega t) & \cos(\omega t) \end{bmatrix},\ F = \begin{bmatrix} 1 & 0 \\ -\frac{1}{2} & -\frac{\sqrt{3}}{2} \\ -\frac{1}{2} & \frac{\sqrt{3}}{2} \end{bmatrix}$$

TABLE I
SYSTEM PARAMETERS OF THE MMCC-SSBC.

parameters	Symbol	Value
Line-to-line rms voltage	v_g	220(V)
Rated reactive power	Q_R	1.0(kVAR)
Cascaded cell number	N	3
Loss test resistor	R_{test}	1000(Ω)
Background inductor	L_s	1.0(mH)
		0.78(%)
AC filter inductor	L_{ac}	6.8(mH)
		5.30(%)
AC filter capacitor	C_{ac}	3.3(μF)
Nominal DC voltage	V_{dc}^*	80.0(V)
DC bus capacitor	C_{mn}	840(μF)
Unit capacitance constant	H	24.2(msec)
Switching frequency	f_{sw}	2.0(kHz)
Sampling frequency	f_{sp}	12.0(kHz)
Switching dead time	T_{dt}	1.0(μsec)
Energy feedforward time value	T_{ff}	500(μsec)

synchronous frame. Equation (1) shows the transformation of voltages and currents. TABLE I lists the corresponding system specification of the circuit topology.

III. OPERATION PRINCIPLE OF THE PROPOSED VOLTAGE BALANCING CONTROL

The voltage balancing control is one of the most important issue of MMCC converter. The voltage balancing control of MMCC can be separated into three layers, which are overall voltage balancing control, clustered voltage balancing control, and individual voltage balancing control.

- Overall voltage control: Control the average voltage (V_{dc}) to voltage command.
- Clustered voltage balancing control: Control the voltages between each cluster (V_{dcm}) to V_{dc}.
- Individual voltage balancing control: Control the voltages between the cascaded cell of each phase (V_{dcmn}) to V_{dcm}.

Chou et al. [9] presented a voltage balancing control method by regulating the average power of each phase. Based on Chou's method, this paper presents a feed-forward clustered and individual voltage balancing controls to accelerate the transient response.

Fig. 4 shows the generic model of H-bridge converter, R_{mn} (where $m = a, b, c$, $n = 1, 2, 3$) expresses the power loss (i.e switching loss) of each individual cell which is considered in this paper. As Fig. 4, the power loss and the energy stored in

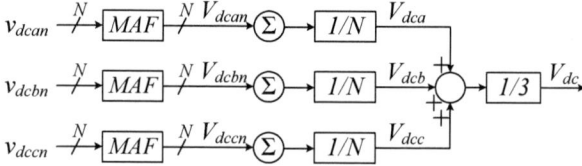

Fig. 3. The calculations for feedback DC averaging values.

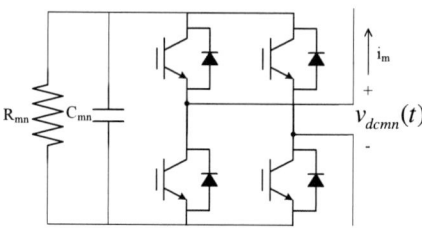

Fig. 4. The generic model of each H-bridge.

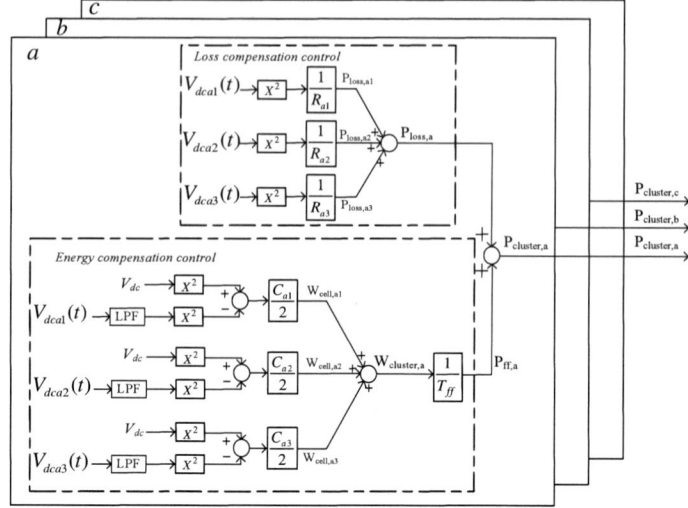

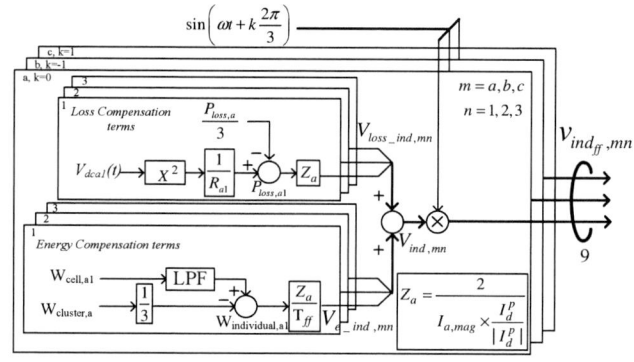

Fig. 5. The control block of the proposed clustered feed-forward balancing control (cluster $m = a, b, c$).

capacitor are two factors which will influence the DC voltage. In this paper, the feed-forward of clustered voltage balancing control and individual voltage balancing control are presented based on Fig. 4.

A. The proposed feed-forward terms of clustered voltage balancing control

Fig. 5 shows the control block diagram of proposed clustered feed-forward voltage balancing control. As above, two compensations are included in Fig. 5, which are loss compensation term and energy balance term.

1) The loss compensation term of clustered voltage control: As Fig. 4, the power loss of each individual bridge cell can be expressed as resistors ($R_{m,n}$) in each H-bridge circuit. Thus, the overall power loss of each phase can be calculated as

$$P_{loss,m} = \frac{V_{dcm1}^2(t)}{R_{m1}} + \frac{V_{dcm2}^2(t)}{R_{m2}} + \frac{V_{dcm3}^2(t)}{R_{m3}}, \text{ where } m=a, b,$$
(2)

2) The energy balance term of clustered voltage control: As above, the clustered voltage balancing control regulates each clustered voltage (V_{dcm}) to V_{dc}. Thus, the proposed method calculates the required energy ($W_{cell,mn}$, $n = 1, 2, 3$) of each bridge from clustered voltage to average voltage as the feed-froward term to better the transient response.

Equation (3) shows the controller takes the V_{dc}, $V_{dc,m1}$, $V_{dc,m2}$, and $V_{dc,m3}$ to calculate the compensating energy of each cell for the same phase, then using Equation (4) to get the overall compensating energy of each phase.

$$\begin{cases} W_{cell,m1} = \frac{1}{2}C_{m1}V_{dc}^2 - \frac{1}{2}C_{m1}V_{dcm1}^2(t_0) \\ W_{cell,m2} = \frac{1}{2}C_{m2}V_{dc}^2 - \frac{1}{2}C_{m2}V_{dcm2}^2(t_0) \\ W_{cell,m3} = \frac{1}{2}C_{m3}V_{dc}^2 - \frac{1}{2}C_{m3}V_{dcm3}^2(t_0) \end{cases}$$
(3)
, where m=a, b, c

$$W_{cluster,m} = W_{cell,m1} + W_{cell,m2} + W_{cell,m3}$$
(4)

Moreover, the controller employs a low-pass filter (LPF) to take off the switching noise and user should pre-define the

Fig. 6. The control block of the proposed individual feed-forward balancing control (Cluster $m = a, b, c$; cell $n = 1, 2, 3$).

compensating time (T_{ff}) to transfer the energy command ($W_{cluster}$) into power command (P_{ff}).

$$P_{ff,m} = \frac{W_{cluster,m}}{T_{ff}}, \text{ where } m=a, b, c$$
(5)

After the loss compensation term and energy balance term have been calculated, the total clustered compensating power can be calculated as Equation (6) then the APB controller [9] will regulate the converter's output currents to balance the clustered voltage.

$$P_{cluster,m} = P_{loss,m} + P_{ff,m}$$
$$where \quad m = a,b,c$$
(6)

B. The proposed feed-forward terms of individual voltage balancing control

Fig. 6 shows the control block diagram of proposed feed-forward calculation for individual voltage balancing control. The individual voltage balancing control is for regulating the DC bus voltage of each bridge cell (V_{dcmn}) to their clustered average voltage (V_{dcm}), where the converter's output phase

978-1-4799-2706-7/14 $31.00 © 2014 IEEE   1641

current can be defined as Equation (7) and the output feed-forward voltage terms of individual voltage balancing control are shown as Equation (8).

$$\begin{cases} i_a = I_{a,mag}\sin(\omega t) \\ i_b = I_{b,mag}\sin(\omega t - \frac{2\pi}{3}) \\ i_c = I_{c,mag}\sin(\omega t + \frac{2\pi}{3}) \end{cases} \quad (7)$$

$$\begin{cases} v_{ind_{ff},an} = V_{ind,an}\sin(\omega t) \\ v_{ind_{ff},bn} = V_{ind,bn}\sin(\omega t - \frac{2\pi}{3}) \\ v_{ind_{ff},cn} = V_{ind,cn}\sin(\omega t + \frac{2\pi}{3}) \end{cases} \quad (8)$$

As Fig. 6, two compensations are considered in this feed-forward term, which are power loss compensation term and energy balance term.

1) The loss compensation term of individual voltage control: Fig. 4 shows the equivalent circuit of each H-bridge cell, and R_{mn} expresses the power loss of the converter. The clustered voltage balancing controller draws in the compensating power ($P_{loss,m}$) and separates into each cell averagely. Since the power loss of each bridge are difference, the required compensating power of each bridge can be calculated as

$$\begin{cases} P_{ind,loss,m1} = -\frac{P_{loss,m1}}{3} + \frac{V_{m1}^2(t)}{R_{m1}} \\ P_{ind,loss,m2} = -\frac{P_{loss,m1}}{3} + \frac{V_{m2}^2(t)}{R_{m2}} \\ P_{ind,loss,m3} = -\frac{P_{loss,m1}}{3} + \frac{V_{m3}^2(t)}{R_{m3}} \end{cases} \quad (9)$$
, where m=a, b, c

Note that, the individual voltage balancing control regulates the average power by controlling the converter's output voltage of each H-bridge. Thus, the Z_m is defined as Equation (10), which is for transformation from the compensating power command to the output voltage command. Since the system is STATCOM compensation, the positive-sequence reactive current (I_d^p) is taken to generate the output voltage command of individual voltage control.

$$Z_a = \frac{2}{I_{a,mag} \cdot \frac{I_d^p}{|I_d^p|}}; \; Z_b = \frac{2}{I_{b,mag} \cdot \frac{I_d^p}{|I_d^p|}}; \; Z_c = \frac{2}{I_{c,mag} \cdot \frac{I_d^p}{|I_d^p|}} \quad (10)$$

Hence, the voltage regulator of loss compensation can be expressed as

$$V_{loss\_ind,mn} = P_{ind,loss,m1} \cdot Z_m \text{ , where m=a, b, c} \quad (11)$$

2) The energy balance term of individual voltage control: The energy balance compensation is also one of issue for individual voltage control. After the clustered voltage balancing control draws in the required energy ($W_{cluster,m}$, where $m = a, b, c$) to compensate the energy stored and separates into each bridge cell, then the individual voltage balancing controller calculates the regulation of energy to balance the voltage, which can be expressed as Fig. 7 and Equation (12).

$$\begin{cases} W_{individual,m1} = W_{cell,m1} - \frac{W_{cluster,m}}{3} \\ W_{individual,m2} = W_{cell,m2} - \frac{W_{cluster,m}}{3} \\ W_{individual,m3} = W_{cell,m3} - \frac{W_{cluster,m}}{3} \end{cases} \quad (12)$$

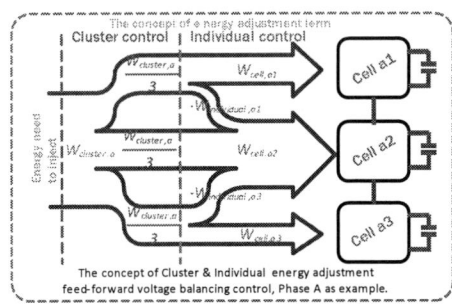

Fig. 7. The relationships of capacitor energy at three different layers.

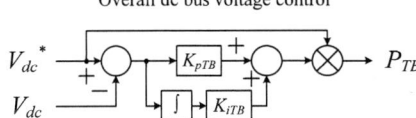

Fig. 8. Overall DC bus voltage feedback control.

Note that, the required compensating energy command should be divided by compensating time (T_{ff}) to compensating power command then Z_m is used to transfer the output power command ($W_{individual,mn}$) to voltage regulation command ($V_{e\_ind,mn}$).

After all the feed-forward terms have been calculated, the amplitude of total voltage command are generated as

$$V_{ind,mn} = V_{loss\_ind,mn} + V_{e\_ind,mn} \quad (13)$$
, where m=a, b, c and n=1,2,3

Then, the output feed-forward voltage terms of individual voltage balancing control are shown as Equation (8) and given to their H-bridge voltage command.

IV. THE AVERAGE POWER BALANCING CONTROL

The average power balancing (APB) controller has been presented in [9], which takes all positive- and negative-sequence voltages and currents to regulate the average power of each phase for DC bus voltage balancing control. The definitions of positive- and negative-sequence voltages and currents are shown in Equation (1). Since the APB method considers all positive- and negative-sequence components, the DC bus voltage will be balanced even voltage sag occurs.

A. Overall voltage controller

Fig. 8 shows the control block diagram of overall average voltage control, the controller generates the overall required active power command (P_{TB}) by a PI control then using to adjust the average voltage (V_{dc}) to voltage command (V_{dc}^*).

The 2014 International Power Electronics Conference

Clustered Balancing Control

V_{dca}, V_{dcb}, V_{dcc}, V_{dc}

Fig. 9. Clustered DC bus voltage feedback control.

TABLE II
CONTROLLER PARAMETERS

Variables	Symbol	Value
Moving averaging filter	MAF	33points at $4kHz$
Overall DC bus voltage control	K_{pTB}	0.4(A/V)
	K_{iTB}	4.0(A/V·sec)
Clustered balancing control	K_{pCB}	0.4(A/V)
	K_{iCB}	4.0(A/V·sec)
Individual balancing control	K_{IB}	2.0(V/V)

B. Clustered voltage balancing control

Fig. 9 shows the control block diagram of clustered voltage balancing control, the controller calculates the difference between clustered average voltages (V_{dca}, V_{dcb}, V_{dcc}) and average voltage (V_{dc}) then using a PI control to get the required power of each phase (P_{CBa}, P_{CBb}, P_{CBc}).

After all the power command of clustered voltage balancing control has been decided, the total required compensating power of each phase can be expressed as

$$P_{m,avg} = \frac{1}{3}P_{TB} + P_{cluster,m} + P_{CB,m} \tag{14}$$
, where m=a, b, c and n=1,2,3

Moreover, the overall reactive power command (Q_{avg}) should be considered since the system is STATCOM operation. Then, [9] presented the Equation (15) to adjust the average power of each clustered and control overall output reactive power by positive- and negative-sequence current injection. After the current command has been decided, a current regulator is for the converter's output current control.

C. Individual voltage balancing control

The Equation (16) shows a useful individual voltage balancing control by taking the voltage error then using the proportional gain (K_{IB}) to generate the compensating voltage command, which has been presented in [7], and the feedforward terms of individual voltage balancing control have been explained at section III-B. Hence, the final voltage command of each H-bridge cell can be calculated as

$$\begin{cases} v_{an,ref} = v_{a,ref} + v_{ind_{ff}an} + v_{IBan} \\ v_{bn,ref} = v_{b,ref} + v_{ind_{ff}bn} + v_{IBbn} \quad , where\ n=1,2,3 \\ v_{cn,ref} = v_{c,ref} + v_{ind_{ff}cn} + v_{IBcn} \end{cases} \tag{17}$$

V. TEST RESULTS

The proposed APB balancing control method is tested in the laboratory prototype of MMCC-SSBC. The balanced and unbalanced grid voltages are applied to verify the DC bus voltage balancing and the reactive power output capability of the MMCC-SSBC under the proposed control technique. In addition, the feed-forward control terms, including the energy balance and the loss compensation are also tested. The

laboratory prototype of the MMCC-SSBC is as given in Fig. 1, where the number of cascaded bridge cells per phase is N = 3. The system parameters of the prototype testbench are given in TABLE I, and the control parameters are given in TABLE II.

A. Operations under Balanced Grid Voltages

The proposed method is first tested under balanced grid voltages. Fig. 10 show that the converter is operated in inductive VAR operation(1kVAR) and test the effect of loss compensation term. Otherwise, Fig. 11 show that the converter is operated in capacitive VAR operation(1kVAR). According to the hardware test result of Fig. 10 and Fig. 11, all the DC bus voltage is precisely converge to the target value when the loss compensation term is activated. From the test result, it also could affirm that the feedback control with loss compensation term is superior to only feedback control. Ultimately, the benefit of the loss compensation term is proved in inductive VAR operation and capacitive VAR operation respectively.

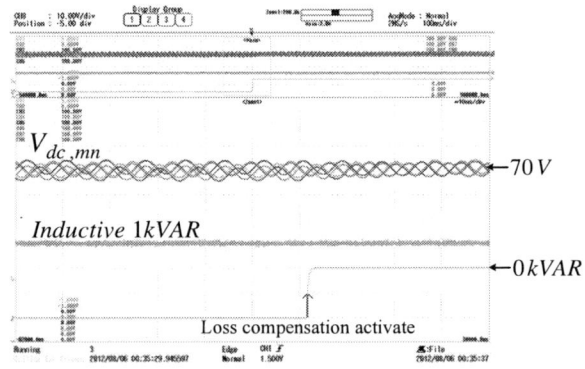

Fig. 10. Experimental test waveform of rated inductive VAR operation before and after the loss compensation term is activated (X axis: 10 ms/div; Y axis: 10 V/div and 1 kVAR/div).

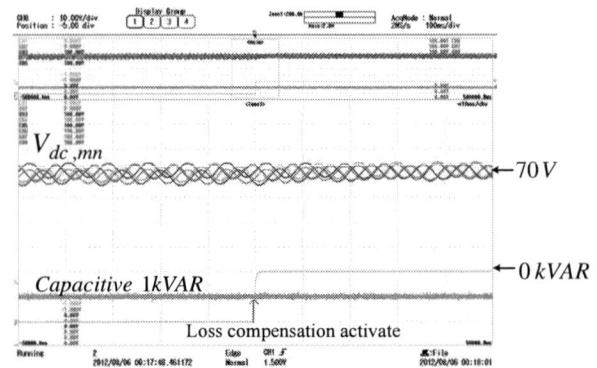

Fig. 11. Experimental test waveform of rated capacitive VAR operation as the loss compensation term is activated (X axis: 10 ms/div; Y axis: 10 V/div and 1 kVAR/div).

The 2014 International Power Electronics Conference

$$
\begin{bmatrix} I_q^{p*} \\ I_d^{p*} \\ I_q^{n*} \\ I_d^{n*} \end{bmatrix} = \begin{bmatrix} \frac{V_q^p}{2}+\frac{V_q^n}{2} & \frac{V_d^p}{2}-\frac{V_d^n}{2} & \frac{V_q^n}{2}+\frac{V_q^p}{2} & \frac{V_d^n}{2}-\frac{V_d^p}{2} \\ -\frac{V_q^p}{2}-\frac{V_q^n}{4}+\frac{\sqrt{3}V_d^n}{4} & \frac{V_d^p}{2}+\frac{\sqrt{3}V_q^n}{4}+\frac{V_d^n}{4} & \frac{V_q^n}{2}-\frac{V_q^p}{4}+\frac{\sqrt{3}V_d^p}{4} & \frac{V_d^n}{2}+\frac{\sqrt{3}V_q^p}{4}+\frac{V_d^p}{4} \\ -\frac{V_q^p}{2}-\frac{V_q^n}{4}-\frac{\sqrt{3}V_d^n}{4} & \frac{V_d^p}{2}-\frac{\sqrt{3}V_q^n}{4}+\frac{V_d^n}{4} & \frac{V_q^n}{2}-\frac{V_q^p}{4}-\frac{\sqrt{3}V_d^p}{4} & \frac{V_d^n}{2}-\frac{\sqrt{3}V_q^p}{4}+\frac{V_d^p}{4} \\ -\frac{3V_d^p}{2} & \frac{3V_q^p}{2} & -\frac{3V_d^n}{2} & \frac{3V_q^n}{2} \end{bmatrix}^{-1} \begin{bmatrix} P_{a,avg} \\ P_{b,avg} \\ P_{c,avg} \\ Q_{avg} \end{bmatrix} \quad (15)
$$

$$
v_{IBan} = \begin{cases} K_{IB} \cdot (V_{dca} - V_{dcan}) \cdot \sin \omega t & \text{if } Q_{avg} > 0, \text{ inductive VAR,} \\ (-1) \cdot K_{IB} \cdot (V_{dca} - V_{dcan}) \cdot \sin \omega t & \text{if } Q_{avg} < 0, \text{ capacitive VAR.} \end{cases}
$$

$$
v_{IBbn} = \begin{cases} K_{IB} \cdot (V_{dcb} - V_{dcbn}) \cdot \sin(\omega t - 120^o) & \text{if } Q_{avg} > 0, \text{ inductive VAR,} \\ (-1) \cdot K_{IB} \cdot (V_{dcb} - V_{dcbn}) \cdot \sin(\omega t - 120^o) & \text{if } Q_{avg} < 0, \text{ capacitive VAR.} \end{cases} \quad (16)
$$

$$
v_{IBcn} = \begin{cases} K_{IB} \cdot (V_{dcc} - V_{dccn}) \cdot \sin(\omega t + 120^o) & \text{if } Q_{avg} > 0, \text{ inductive VAR,} \\ (-1) \cdot K_{IB} \cdot (V_{dcc} - V_{dccn}) \cdot \sin(\omega t + 120^o) & \text{if } Q_{avg} < 0, \text{ capacitive VAR.} \end{cases}
$$

Fig. 12 and Fig. 13 are used to verify the benefit of the energy balance term both in inductive and capacitive operation. For Fig. 11 and Fig. 12, the clustered and individual feedback control are turned off at the happening of first trigger signal so the DC bus voltage start to diverge gradually. At the happening of second trigger signal, the energy balance term is activated so all the DC bus voltage start to converge to its target value at the pre-determinate time($0.05s$). As a result, After the end of second trigger signal, the energy balance term is terminated. Relatively, the clustered and feedback control are activated at the end of second trigger signal. Ultimately, all the DC bus voltage are steadily regulated at the target value.

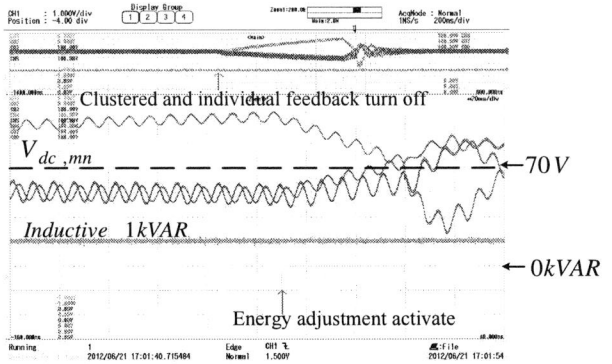

Fig. 12. Experimental test waveform of rated inductive VAR operation as the energy balance term is activated (X axis: 20 ms/div; Y axis: 20 V/div and 1 kVAR/div).

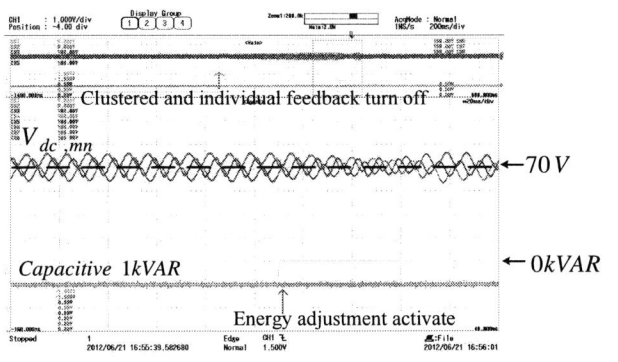

Fig. 13. Experimental test waveform of rated capacitive VAR operation as the energy balance term is activated (X axis: 20 ms/div; Y axis: 20 V/div and 1 kVAR/div).

B. Operation under Unbalanced Grid Voltages

Fig. 14 show that the feedback control with loss compensation term can regulate all the DC voltages at the commanded value under inductive VAR operation, even the voltage source sag. Otherwise, Fig. 15 confirm that all the DC voltages also reach target value in capacitive VAR operation(1kVAR).

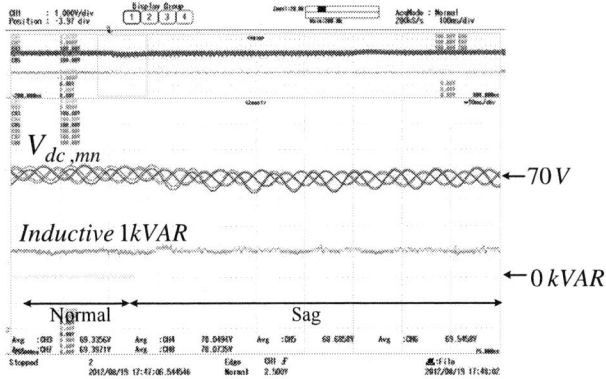

Fig. 14. Experimental test waveform of rated inductive VAR operation as the feed-forward voltage balancing control is activated under unbalanced grid (X axis: 10 ms/div; Y axis: 10 V/div and 1 kVAR/div).

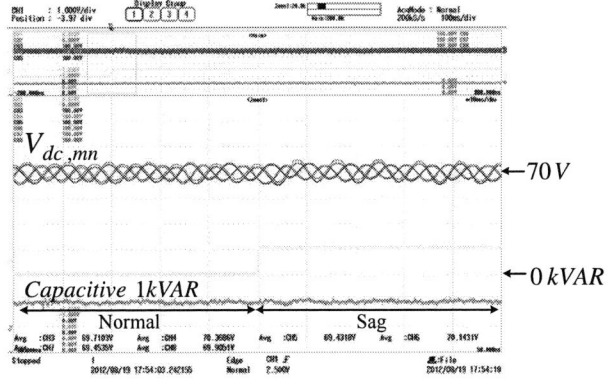

Fig. 15. Experimental test waveform of rated capacitive VAR operation as the feed-forward voltage balancing control is activated under unbalanced grid (X axis: 10 ms/div; Y axis: 10 V/div and 1 kVAR/div).

VI. CONCLUSION

This paper proposes a DC bus voltage balancing and VAR output control technique for the MMCC-SSBC based on the power flow analysis of the converter system. By including the unbalanced grid conditions into the power flow analysis, the proposed method can manage the DC bus voltages of the MMCC-SSBC even under grid faults, and thus enables the low voltage ride-through operation by providing the crucial reactive power support. The proposed method combines the feed-forward and the feedback approaches in order to accomplish fast and accurate voltage balancing and VAR control. Such effectiveness is validated by the laboratory test results under normal grid conditions and fault conditions.

REFERENCES

[1] S. R. Bull, "Renewable energy today and tomorrow," *Proceedings of the IEEE*, vol. 89, no. 8, pp. 1216–1226, 2001.

[2] H. Akagi, S. Inoue, and T. Yoshii, "Control and performance of a transformerless cascade pwm statcom with star configuration," *Industry Applications, IEEE Transactions on*, vol. 43, no. 4, pp. 1041–1049, 2007.

[3] N. S. Choi, G. C. Cho, and G. H. Cho, "Modeling and analysis of a static var compensator using multilevel voltage source inverter," in *Industry Applications Society Annual Meeting, 1993., Conference Record of the 1993 IEEE*. IEEE, 1993, pp. 901–908.

[4] F. Peng and J. Lai, "A static var generator using a staircase waveform multilevel voltage-source converter," in *Proc. Rec. Power Quality Conf*, 1994, pp. 58–66.

[5] F. Z. Peng, J.-S. Lai, J. W. McKeever, and J. VanCoevering, "A multilevel voltage-source inverter with separate dc sources for static var generation," *Industry Applications, IEEE Transactions on*, vol. 32, no. 5, pp. 1130–1138, 1996.

[6] J.-S. Lai and F. Z. Peng, "Multilevel converters-a new breed of power converters," *Industry Applications, IEEE Transactions on*, vol. 32, no. 3, pp. 509–517, 1996.

[7] H. Akagi, "Classification, terminology, and application of the modular multilevel cascade converter (mmcc)," *Power Electronics, IEEE Transactions on*, vol. 26, no. 11, pp. 3119–3130, 2011.

[8] M. Hagiwara, R. Maeda, and H. Akagi, "Negative-sequence reactive-power control by a pwm statcom based on a modular multilevel cascade converter (mmcc-sdbc)," *Industry Applications, IEEE Transactions on*, vol. 48, no. 2, pp. 720–729, 2012.

[9] S.-F. Chou, B.-S. Wang, S.-W. Chen, C.-T. Lee, P.-T. Cheng, H. Akagi, and P. Barbosa, "Average power balancing control of a statcom based on the cascaded h-bridge pwm converter with star configuration," in *Energy Conversion Congress and Exposition (ECCE), 2013 IEEE*. IEEE, 2013, pp. 970–977.

Analysis and Comparison of High Power Semiconductor Device Losses in 5MW PMSG MV Wind Turbines

Kihyun Lee, Kyungsub Jung,
Seunghoo Song, and Yongsug Suh
Department of Electrical Engineering
Chonbuk National University
Jeonju, Korea
lkh0120@jbnu.ac.kr, mineg200@jbnu.ac.kr,
thdsh@jbnu.ac.kr, and ysuh@jbnu.ac.kr

Changwoo Kim, Hyoyol Yoo,
and Sunsoon Park
Dawonsys Co.
Siheong, Korea
ckd9537@dawonsysi.com, hyyoo@dawonsys.com,
and sspark@dawonsys.com

Abstract— **This paper provides a comparison of high power semiconductor devices in 5MW-class Permanent Magnet Synchronous Generator (PMSG) Medium Voltage (MV) wind turbines. High power semiconductor devices of press-pack type IGCT, module type IGBT, press-pack type IEGT, and press-pack type IGBT of both 4.5kV and 6.5kV are considered in this paper. Benchmarking is performed based on neutral point clamped 3-level back-to-back type voltage source converter supplied from grid voltage of 4160V. The feasible number of semiconductor devices in parallel is designed through the loss analysis considering both conduction and switching losses under the given operating conditions of 5MW-class PMSG wind turbines, particularly for the application in offshore wind farms. The loss analysis is confirmed through PLECS simulations. The comparison result shows that press-pack IGCT type semiconductor device has the highest efficiency.**

Keywords—Losses, medium voltage, multi-level system, three-level neutral point clamped inverter, voltage source converter (VSC), power semiconductor, wind turbine.

I. INTRODUCTION

In the multi-MW wind turbine market, the maximum power rating of a commercial wind turbine has been increased more than 5MW with a view to generate more power from wind power sites [1]. Power electronic converters in medium-voltage level are generally realized as multi-level (ML) voltage source converters (VSC) instead of 2L-VSCs in order to improve the performance factors regarding switch power losses, harmonic distortion, and common mode voltage/current [2]. In the family of multilevel inverters, the three-level topology,

called Neutral Point Clamped (NPC) inverter, is one of the few topologies that have received a reasonable consensus in the high power community [3]. These NPC inverters have also been implemented successfully in the industrial applications for high power drives and wind turbines [4].

In the multi-MW wind turbine systems, there are many different types of power converter topologies and high-power switching devices in use. In particular, the recent development of high power semiconductor technology has resulted in a wide variety of practical power devices. The benchmarking of these topologies and its optimal power switches is important for industry to select the most feasible solution in product development of wind turbines. In general, comparison of various power switches involves a great deal of engineering work considering many different aspects of device characteristics. Therefore, it is necessary to pick critical performance factors on which the comparison for device is made so that the selection of the most feasible power device can be made with a meaningful engineering work and insight.

This paper investigates the utilization of four most feasible high power switching devices for 5MW PMSG wind turbine systems; press-pack type IGCT, module type IGBT, press-pack type IGBT, and press-pack type IEGT. The power converter topology of 3L-NPC VSC is selected as a main platform for the device comparison. Power loss dissipated in the semiconductor device is one of the most important performance factors in high power drives considering the total system efficiency and the requirement on cooling system. In addition, power loss

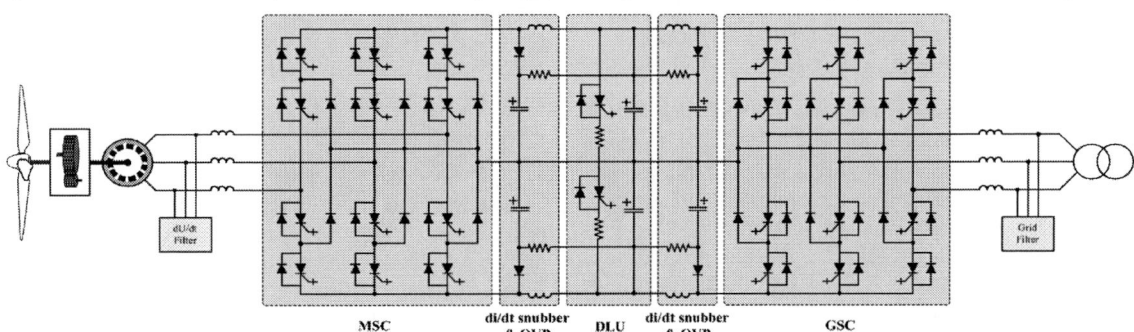

Fig. 1. Three-level neutral-point-clamped (3L-NPC) back-to-back configuration for 5MW PMSG MV wind turbine.

978-1-4799-2706-7/14 $31.00 © 2014 IEEE

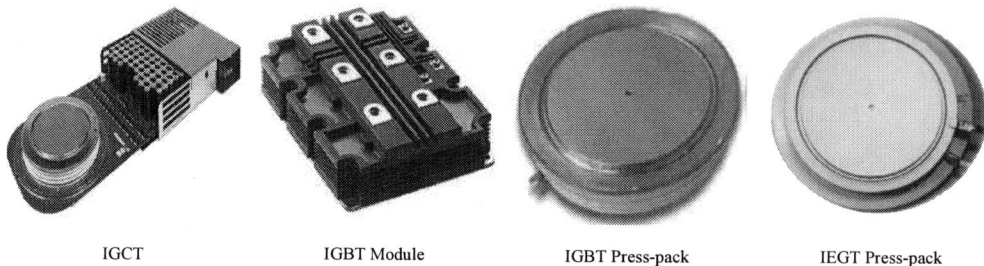

IGCT   IGBT Module   IGBT Press-pack   IEGT Press-pack

Fig. 2. Target power semiconductor devices for MV wind turbines [8]-[11].

TABLE I
CHARACTERISTIC VALUES OF TARGET POWER SEMICONDUCTORS [8]-[12]

Device	IGCT	Module type IGBT		Press-pack IGBT	Press-pack IEGT	Press-pack Diode
Manufacturer	ABB	ABB		Westcode, IXYS	Toshiba	ABB
Code	5SHY 42L6500	5SNA 0750G650300		T2400GB45E	ST2100GXH22A	5SDF 10H6004
Blocking Voltage	6.5 kV	6.5 kV		4.5 kV	4.5 kV	6.5 kV
I_{TGQM}	3800A	-		-	-	-
$I_{C,nom}$ / I_{CM}	-	750A / 1500A		2400A / 4800A	2100A / 5500A	-
$I_{F(AV)M}$	-	-		-	-	1100A
Part	GCT-part	IGBT-part	Diode-part	IGBT-part	-	Diode-part
V_{TO} (Max.)	1.88 V	2.0 V	2.5 V	1.49 V	3.0 V	1.5 V
R_T (Max.)	0.56 mΩ	2.5 mΩ	1.3 mΩ	1.05 mΩ	1.0 mΩ	0.6 mΩ
E_{on} (Max.)	3.1 J	6.4 J	-	15 J	18.4 J	-
E_{off} (Max.)	44 J	5.3 J	2.7 J	14 J	17 J	5 J
Meas. condition	4kV / 3800A	3.6kV / 750A		2.8kV / 2400A	3.0kV / 2100A	2.9kV / 1000A
$T_{vj\_max.}$	125℃	125℃		125℃	125℃	125℃
$R_{th(j-c)}$	8.5 K/kW	11 K/kW	21 K/kW	5.2 K/kW	5.25 K/kW	12 K/kW
$R_{th(c-h)}$	3 K/kW	9 K/kW	18 K/kW	3 K/kW	3 K/kW	3 K/kW
$R_{th(h-a)}$	6 K/kW	10 K/kW	10 K/kW	6 K/kW	6 K/kW	6 K/kW

analysis gives an engineering insight into the cost-effective system design [5]. This paper compares four different high power switching devices with respect to the power loss dissipation. 6.5kV and 4.5kV class devices are considered in the loss calculation of 3L-NPC VSCs. The loss distribution among several switching devices in the converter including the snubber is also explained in this paper.

This paper is structured in four main sections. Section II describes the target power semiconductor devices under comparison. Section III discusses the MV 3L-NPC VSCs and model of semiconductors to calculate the losses. Section IV presents the simulation results of 3L-NPC VSCs in 5MW PMSG MV wind turbines. Finally, Section V provides an analysis and comparison of target power semiconductors.

II. POWER SEMICONDUCTOR DEVICES UNDER COMPARISON

A simplified schematic of the 3L-NPC VSCs for 5MW PMSG MV wind turbines is presented in Fig. 1 [6]-[7]. Semiconductor devices commonly used in high power converters are IGBTs (in a module or press-pack package), press-pack IEGTs, and IGCTs for the 3L-NPC converters as shown in Fig. 2. Recent technology development of 4.5 kV and 6.5 kV IGCTs, IGBTs, and IEGTs has been enabling a substantial improvement of MV converters in many aspects [13]-[15]. These four

major types of semiconductor devices are considered in this paper. Major operating characteristics of target semiconductor devices are summarized in Table I and used in the loss analysis throughout this paper. Important characteristics for the loss calculation are threshold voltage (V_{TO}), slope resistance (R_T) as a function of the collector / anode current, turn-on energy (E_{on}), turn-off energy (E_{off}), maximum operating junction temperature (T_{vj_max}), and thermal resistance (R_{th}).

A. Medium-voltage press-pack IGCT

MV IGCT press-pack devices are mainly used in high power industrial applications owing to advantageous features such as press-pack housing cases, a higher thermal/power cycling capability, and an explosion-free failure mode [5]. Recently, 10kV IGCT device has been introduced and its switching capability has been confirmed [16]. In this paper, 6.5kV/3800A press-pack IGCT (ABB 5SHY42L6500) is considered for the loss analysis. As for the anti-parallel and neutral-point diode, 6.5kV/1100A FRD (ABB 5SDF10H6004) is employed.

B. Medium-voltage module type IGBT

Module type IGBT is widely accepted in the market of power range below 3-4 MW approximately. MV IGBT having the blocking voltage of 6.5kV has been developed by several manufacturers and employed in many industrial applications. The continuous switching current capability of 6.5kV IGBT modules has reached around 750A. In this

978-1-4799-2706-7/14 $31.00 © 2014 IEEE   1647

paper, 6.5kV/750A module type IGBT (ABB 5SNA0750G650300) is considered for the loss analysis.

C. Medium-voltage press-pack type IGBT

Recently developed press-pack type IGBT devices combine the advantages of IGBTs with those of press-pack cases. Thus, press-pack IGBTs have become a competition for IGCTs in medium and high power industrial applications such as MV drives for wind turbines [17]. In this paper, 4.5kV/2400A press-pack IGBT (Westcode, IXYS T2400GB45E) is considered for the loss analysis.

D. Press-pack type IEGT

As a high-voltage and large-capacity power semi-conductor device replacing a conventional GTO thyristor, Injection Enhanced Gate Transistor (IEGT) has been developed in recent years and applied in a practical use, which was led by Toshiba and GE Company. High-voltage, large-capacity, and full-controlled power device of IEGT is intended to combine the advantages of IGBT devices and GTO devices. Press-pack type 4.5kV IEGTs have been placed on the market, mainly for the use in medium voltage converters [18]-[19]. In this paper, 4.5kV/2100A press-pack type IEGT (Toshiba ST2100GXH22A) is considered for the loss analysis.

III. MEDIUM-VOLTAGE VSCS FOR 5MW PMSG WIND TURBINES

A. Medium-voltage 3L-NPC VSCs

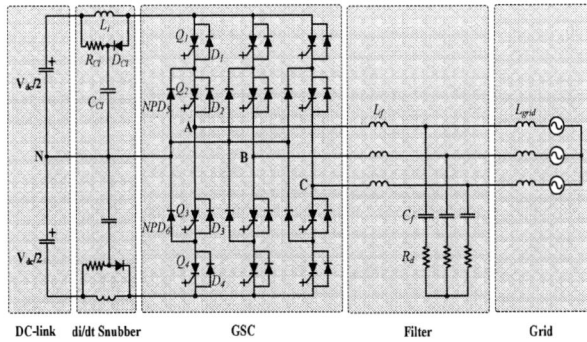

Fig. 3. Three-level neutral-point-clamped VSC under loss analysis.

Figure 3 shows the 3L-NPC VSCs for loss analysis of selected high power semiconductors. Each leg of the VSC consists of two neutral-point clamped diodes, four switches, and four anti-parallel diodes. The DC-bus voltage is split into three-levels by two series connected capacitors. The middle point of two capacitors N can be defined as a neutral point. The output voltage v_{AN} has three states; $V_{dc}/2$, 0, and $-V_{dc}/2$ in each leg, which are produced by specific conduction paths depending on output current direction and output voltage polarity. Due to the relatively low switching frequency, 2nd-order LC-filter system has been employed at the grid side of converter to meet the harmonic constraint of grid code [20]. In addition, di/dt snubber is essential for all IGCT converters to achieve the required di/dt characteristics during switching on.

B. Model of semicondcutors for calculating the losses

The losses of power semiconductor device are approximated by analytical expressions in terms of voltage and currents. Figure 4 shows the simplified loss estimation model for power semiconductor devices.

1) Conduction losses

The total semiconductor device loss P_t consists of the conduction loss P_{cond} and switching loss $P_{switching}$;

$$
\begin{aligned}
P_t &= P_{cond} + P_{switching} \\
&= P_{cond} + P_{on} + P_{off}
\end{aligned}
\tag{1}
$$

Conduction loss of each power semiconductor depends on the instantaneous on-state voltage $v_{sw}(t)$ and the instantaneous switching current $i(t)$ passing through it. A forward on-state voltage of power semiconductor device, $v_{sw}(t)$ can be modeled using a first-order linear approximation comprised of a threshold voltage v_{on} and a series resistance R_{on} as follows;

$$
v_{sw}(t) = v_{on} + R_{on} \cdot i(t)
\tag{2}
$$

The total conduction loss in power semiconductors can be expressed as;

$$
\begin{aligned}
P_{cond} &= \frac{1}{T} \int v_{sw}(t) \cdot i(t) dt \\
&= \frac{1}{T} \int \{v_{on} + R_{on} \cdot i(t)\} i(t) dt \\
&= v_{on} I_{avg} + R_{on} (I_{rms})^2
\end{aligned}
\tag{3}
$$

where T is the fundamental period of the converter [21].

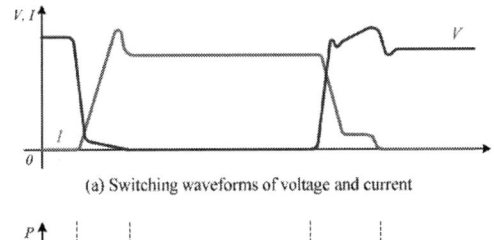

(a) Switching waveforms of voltage and current

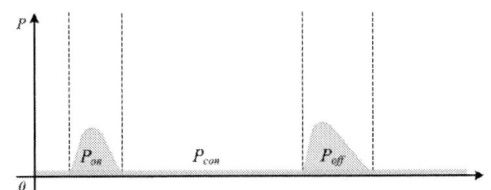

(b) Simplified power losses estimation on switching status

Fig. 4. Simplified device switching waveforms and its power losses.

2) Switching losses

Switching loss of power semiconductor device is determined by the total commutation time in which the device is turned on/off, and by the voltage $v(t)$ and current $i(t)$ across the device. The energy dissipated during commutation is E_{on} and E_{off} for turn on and off, respectively, and is provided by the device manufacturers on their datasheet. The average switching power loss $P_{switching}$ over a complete fundamental period T may be determined by summing all the commutations of the device during the respective interval of time. Each

978-1-4799-2706-7/14 $31.00 © 2014 IEEE

switching loss for turn-on and turn-off can be expressed as;

$$P_{on} = E_{on} \times f_{sw}$$
$$= \frac{V_{on(measure)}}{V_{test}} \times \frac{I_{on(measure)}}{I_{test}} \times E_{on(spec)} \times f_{sw} \quad (4)$$

$$P_{off} = E_{off} \times f_{sw}$$
$$= \frac{V_{off(measure)}}{V_{test}} \times \frac{I_{off(measure)}}{I_{test}} \times E_{off(spec)} \times f_{sw} \quad (5)$$

Equation (4) and (5) represent the linear approximation of actual switching loss for turn-on and turn-off based on the specific values ($E_{on(spec)}$ and $E_{off(spec)}$) provided by manufacturers. This linear approximation gives a fairly good accuracy particularly at the vicinity of a manufacturer's test point (V_{test} and I_{test}) [5].

C. Snubber circuit losses

Converters employing IGCTs need a *di/dt* limiting inductance to meet the required *di/dt* characteristics during switching on transients. This *di/dt* limiting inductor (L_i) usually necessitates an additional over voltage protection snubber or clamping circuitry as shown in Fig. 3. This snubber circuitry dissipates additional power loss and gives a rise to an important loss factor.

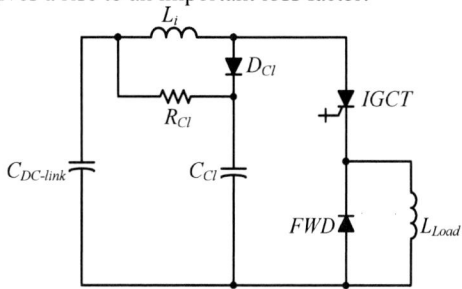

Fig. 5. OVP clamp and *di/dt* snubber circuit of IGCT in the upper-half part of 3L-NPC VSC.

Figure 5 presents an equivalent circuit of *di/dt* limiting inductor, over voltage protection, and one particular IGCT being subject to switching transients in Fig. 3. At the instant of switching off, the stored magnetic energy from the on-state current of the IGCT is given by;

$$E_{R_{Cl}} = \frac{1}{2} L_i \cdot i(t)^2 \quad (6)$$

This stored energy is mainly dissipated in the snubber resistor (R_{Cl}) or fed back to charge the dc link capacitor ($C_{DC-link}$) [22]. In this paper, the total stored energy in the *di/dt* limiting inductor is considered to be snubber circuit power loss since loss analysis is performed to compare the full functionality of different power semiconductor switching devices on an equal basis. Therefore, clamp circuit loss P_{cl} can be expressed as;

$$P_{cl} = E_{R_{Cl}} \times f_{sw}$$
$$= \frac{1}{2} L_i \cdot i(t)^2 \times f_{sw} \quad (7)$$

where f_{sw} is an effective switching frequency which one particular IGCT is subject to during one fundamental period.

IV. SIMILATION RESULTS

The simulation is performed based on the parameters of 5MW MV VSCs as specified in Table II. Since the converter is a 3L-NPC type connected to the ac line of 4160V, the nominal dc-link voltage is chosen to be 7kV. The switching frequency adopted for the grid-side converter is set to 1020 Hz. This switching frequency is selected to be 17 times the fundamental frequency. The selection of switching frequency is done compromising the switching loss and the harmonic content of ac input current.

TABLE II
SIMULATION PARAMETERS OF 5MW MV 3L-NPC VSC

Parameter	Symbol	Value	Per unit
Output Power	$P_{rated-out}$	5 MW	1.0
Frequency	f_{sw}	60 Hz	1.0
Grid side inductance	L_{grid}	1.56 mH	0.17
Grid side input voltage	V_{LL}	4.16 kV	1.0
Grid side input current	I_{AC_input}	708 A	1.0
Switching frequency	f_{GSC_PWM}	1020 Hz	-
DC-link voltage	$V_{DC-link}$	7 kV	-
DC-link capacitance	$C_{DC-link}$	2.6 mF	-
AC filter inductance	L_f	1.5 mH	0.16
AC filter capacitance	C_f	0.35 mF	0.45

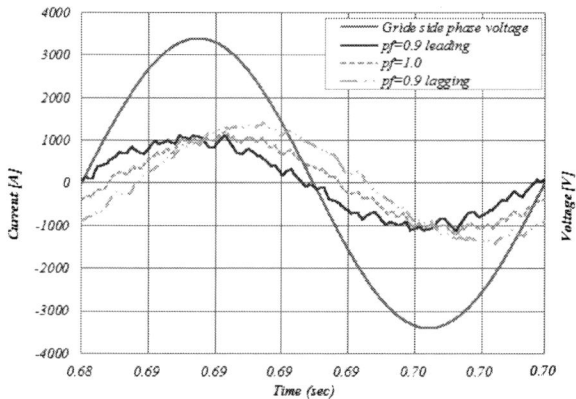

Fig. 6. Waveforms of ac input current at the converter pole under rectifier operating mode (Positive current flowing into converter from grid).

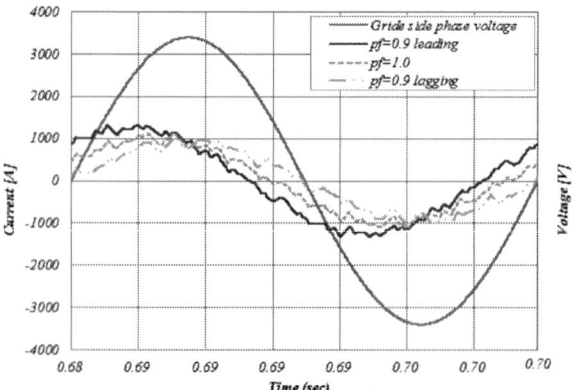

Fig. 7. Waveforms of ac input current at the converter pole under inverter operating mode (Positive current flowing into grid from converter).

The 2014 International Power Electronics Conference

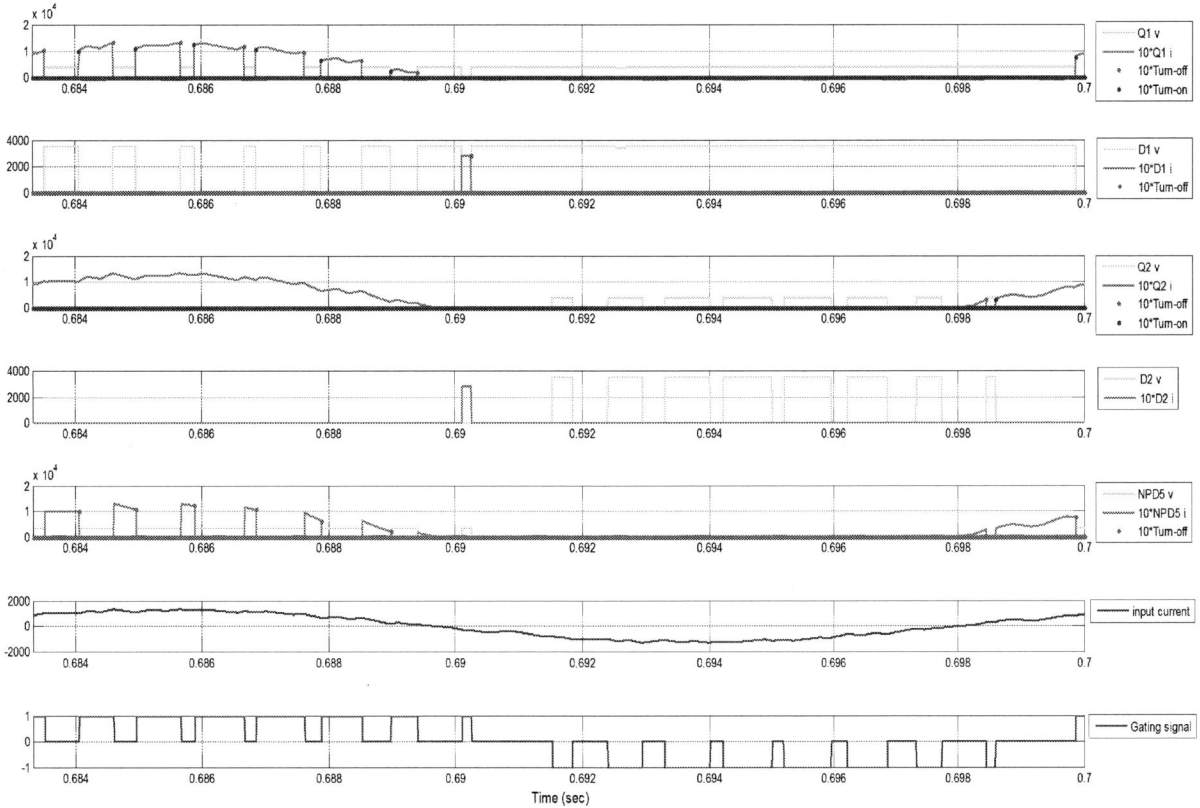

Fig. 8. Waveforms of switching voltage and current in the upper side of each phase-leg under inverter operating mode (pf=0.9 leading condition).

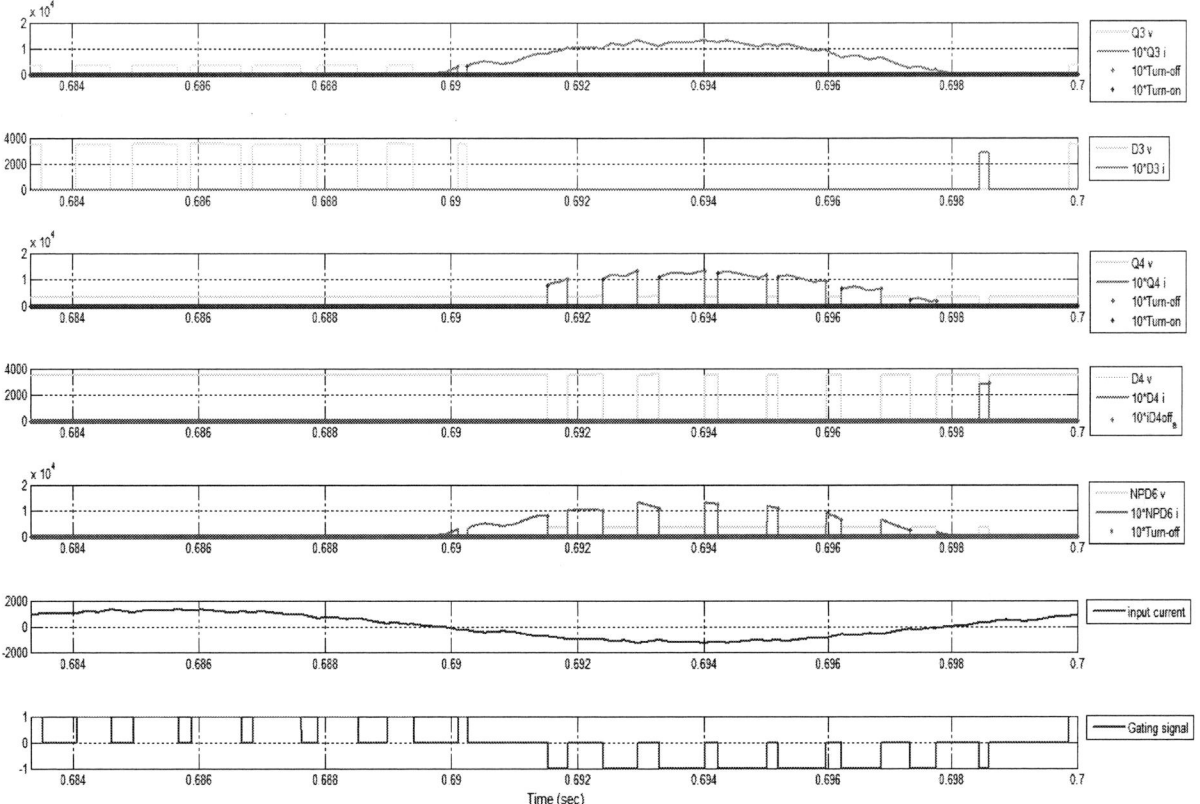

Fig. 9. Waveforms of switching voltage and current in the lower side of each phase-leg under inverter operating mode (pf=0.9 leading condition).

978-1-4799-2706-7/14 $31.00 © 2014 IEEE   1650

In Fig. 6 - 7, ac input currents at the converter pole under the three different power factor conditions (0.9 leading condition, 1.0, and 0.9 lagging condition) are given with a respect to grid phase voltage. It is noted that the amplitude of ac input current for the case of 0.9 leading condition under inverter operating mode is at largest among three conditions. This is due to the fact that the grid side *LC*-filter adds further leading power factor so that the input current at the converter pole requires more leading angle to generate 0.9 leading power factor at the ac input, i.e. up stream of grid side *LC*-filter.

Figure 8 and 9 show waveforms of switching current and voltage for 6.5kV IGCTs in each phase-leg during one ac line period under the power factor of 0.9 leading condition. The converter operates under the inverter operating mode, i.e. power flows from the converter into the grid. In order to obtain the maximum semiconductor losses for the worst cases, the simulation has been performed under the condition of maximum ac input current; line under-voltage of 90% and power factor of 0.9 leading condition, 1.0, and 0.9 lagging condition. The switching voltage and current are sampled at a switching instant from the simulation waveform for each semiconductor device. These sampled voltage and current values are then used to calculate the switching losses based on the specified loss values ($E_{on(spec)}$ and $E_{off(spec)}$) given in the datasheet of power semiconductors. The calculation of switching loss is done in a linear manner as described in (4) and (5).

V. COMPARISON OF LOSSES

The power losses of semiconductor devices in 5MW PMSG MV wind turbines have been summarized in Fig. 10 - 13 under 0.9 leading power factor condition of inverter operating mode. The power losses of four different target semiconductor devices are compared including the snubber losses for the case of IGCT. Power loss numbers in Fig. 10 - 13 represent values for the complete 3-phase legs in a total system of 5MW grid side converter as depicted in Fig. 3. In general, turn-on losses of anti-parallel diode and neutral-point clamp diode are very small, so they are ignored in this paper. For the case

of module type IGBT of 6.5kV/750A, it is required to employ two devices in parallel (n_p=2) to meet the converter operating specification for ac input current of 708A as shown in Table II. This parallel operation can be implemented either by paralleling devices or paralleling two converter systems. In this paper, it is assumed that the current is shared equally between these two parallel devices of IGBT modules.

In Fig. 10 - 11, the switching devices show symmetrical power losses in one leg of the converter, i.e. the outer devices and inner devices have almost same losses, respectively. The junction temperature of each switching device has been computed based on the thermal resistances given in Table I. The ambient temperature, i.e. the temperature of in-let cooling water, has been assumed to be 40 degree. It is interesting to note that the inner anti-parallel diodes (D_2 and D_3) do not have switching off losses. This is due to the modulation characteristic of 3L-NPC VSCs, that is Q_2 and Q_3 being left turn on at the commutation instant of D_2 and D_3. In Fig. 11, press-pack diode of 5SDF10H6004 is commonly employed as antiparallel and neutral-point clamp diode in the cases of IGCT, press-pack IGBT, and press-pack IEGT, for the sake of fair comparison. In the case of module type IGBT, anti-parallel diode part integrated in the package of 5SNA0750G650300 is also utilized as a neutral-point clamp diode. As for neutral-point clamp diode of NPD$_5$ and NPD$_6$, two anti-parallel diodes of 5SNA0750G650300 are paralleled (n_p=2).

Figure 12 provides total loss distribution regarding conduction, switching, and snubber losses in four devices (Q_1, Q_2, Q_3, and Q_4) of four different types of power semiconductors. It is noted from the graph that IGCT is subject to largest switch turn-off loss among four kinds of power devices. On the contrary, IGCT has the smallest switch turn-on loss in the group. The total power loss dissipated in the snubber circuitry for IGCT converter as shown in Fig. 5 is obtained to be around 4.92kW based on Eq. (6) and (7). Even with this additional power loss factor due to snubber circuitry, the switch turn-on loss of IGCT is relatively smaller than those of other power devices as shown in Fig. 12.

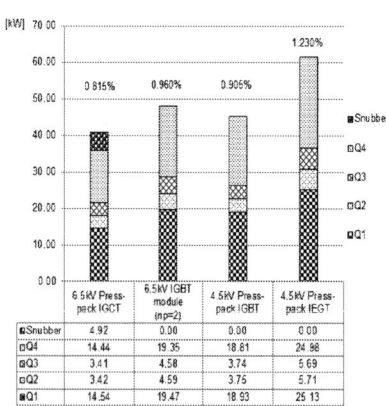

Fig. 10. Total loss distribution in four devices (Q1 - Q4) of four different types of power semiconductors (Total power loss value for three-phase).

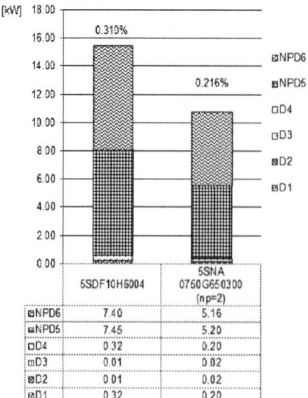

Fig. 11. Total loss distribution in six devices (D1 - D4, NPD5 - NPD6) of two different types of diodes; 5SDF10H6004 and 5SNA0750G650300 (Total power loss value for three-phase).

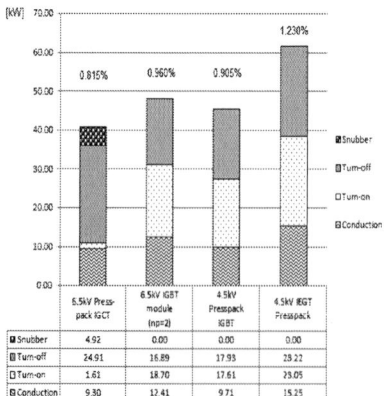

Fig. 12. Total loss distribution regarding conduction, switching, and snubber losses in four devices (Q1 - Q4) of four different types of power semiconductors (Total power loss value for three-phase).

The 2014 International Power Electronics Conference

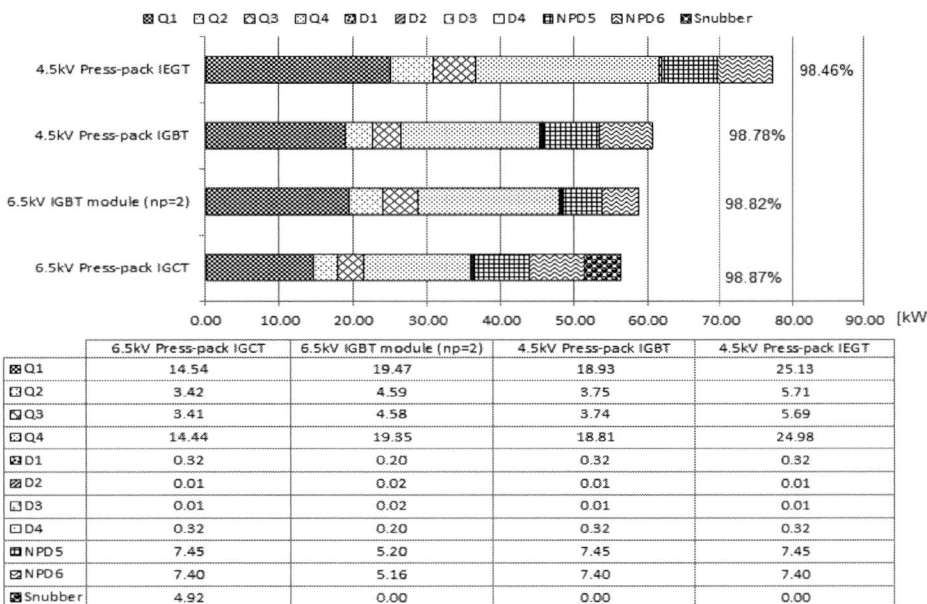

	6.5kV Press-pack IGCT	6.5kV IGBT module (np=2)	4.5kV Press-pack IGBT	4.5kV Press-pack IEGT
Q1	14.54	19.47	18.93	25.13
Q2	3.42	4.59	3.75	5.71
Q3	3.41	4.58	3.74	5.69
Q4	14.44	19.35	18.81	24.98
D1	0.32	0.20	0.32	0.32
D2	0.01	0.02	0.01	0.01
D3	0.01	0.02	0.01	0.01
D4	0.32	0.20	0.32	0.32
NPD5	7.45	5.20	7.45	7.45
NPD6	7.40	5.16	7.40	7.40
Snubber	4.92	0.00	0.00	0.00

Fig. 13. Total loss distribution regarding conduction, switching, and snubber losses of four different types of power semiconductors (Total power loss value for three-phase).

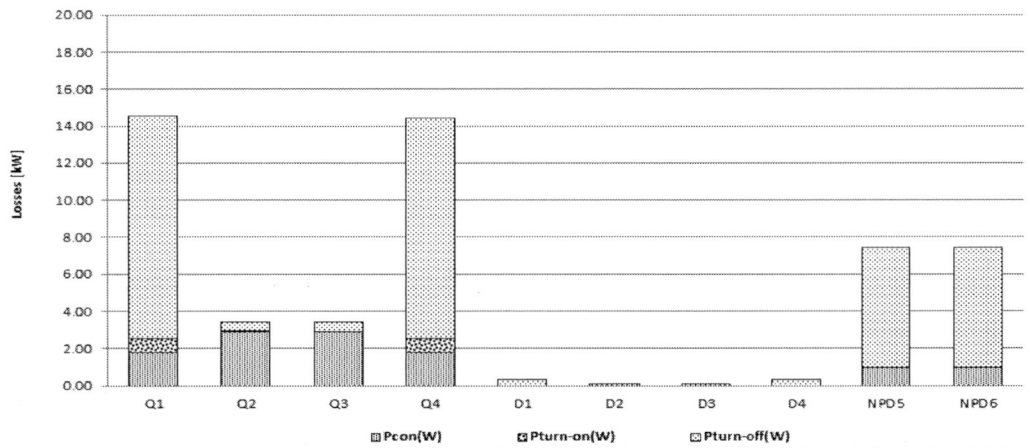

Fig. 14. Total loss distribution in semiconductor devices of three phase-leg under inverter operating mode employing 6.5kV IGCT (*pf*=0.9 leading condition).

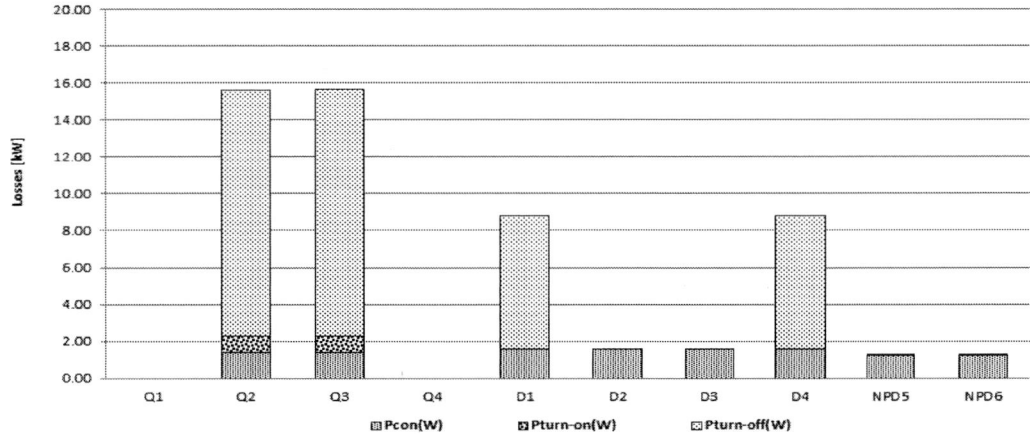

Fig. 15. Total loss distribution in semiconductor devices of three phase-leg under rectifier operating mode employing 6.5kV IGCT (*pf*=0.9 lagging condition).

978-1-4799-2706-7/14 $31.00 © 2014 IEEE

The total losses of MV 3L-NPC VSC for 5MW PMSG wind turbines have been described in Figure 13. It shows that press-pack IGCT has the lowest loss value of 56kW (1.13%) among all four kinds of power semiconductor platforms including the snubber losses. In contrary, press-pack type IEGT has the highest loss value of 77kW (1.54%). Module type IGBT exhibits the loss value of 58kW (1.18%) which is close to those of press-pack IGCT. Press-pack IGBT has the loss value of 61kW (1.22%). Power loss of module type IGBT corresponds to the case of two devices in parallel (n_p=2). In this paralleling of devices or converters, the equal sharing of current becomes a quite important design task. It may further complicate a control algorithm or mechanical concept as compared to the single device approach such as IGCT, press-pack IGBT, and press-pack IEGT.

Figure 14 - 15 describe total loss distribution at worst case under inverter operating mode and rectifier operating mode for the target power semiconductor switches. Under the inverter operating mode, Q_1, Q_4, NPD_5, and NPD_6 are subject to most of power loss. Under the rectifier operating mode, Q_2, Q_3, D_1, and D_4 are subject to most of power loss. As noted in Fig. 6 - 7, the amplitude of ac input current at the converter pole changes as the power factor is varied from 0.9 leading to 0.9 lagging. Higher amplitude of ac input current at the converter pole naturally results in higher power losses in power semiconductor switches. In addition, different power factor angle also changes the loss distribution pattern among power semiconductor devices in each phase leg due to commutation property in 3L-NPC VSC.

VI. CONCLUSION

In this paper, loss analysis of 6.5kV/3800A IGCT, 4.5kV/2400A press-pack type IGBT, 4.5kV/2100A press-pack type IEGT, and 6.5kV/750A module type IGBT for 5MW PMSG MV wind turbine employing a back-to-back 3L-NPC voltage source converter is presented. The switching frequency is set to 1020Hz, under the grid side input voltage of 4.16kV. The press-pack type IGCT has been found to have the lowest device losses, i.e. highest efficiency, among four candidates with the additional snubber loss of IGCT being taken into consideration. The module type IGBT requires to put two devices in parallel in order to meet the given operating specification of ac input current of 708A. This paralleled devices structure may complicate the mechanical and cooling system which are very critical functional elements in multi-MW wind turbine systems

ACKNOWLEDGMENT

This work was supported by the National Research Foundation of Korea (NRF) grant funded by the Korea government (MSIP) (No. 2010-0028509).

REFERENCES

[1] BTM Consult ApS: World Market Update 2008, Forecast 2009-2013

[2] B. Wu, High-Power Converters and AC Drives, Piscataway, NJ, *IEEE Press*, 2006, ISBN 978-0-471-731-9

[3] A. Nabae, I. Takahashi, and H. Akagi, "A new neutral point clamped PWM inverter," *IEEE Trans. Ind. Applicat.*, vol. IA–17, pp. 518–523, Sept./Oct. 1981.

[4] Information about ACS1000 medium voltage drive product at http://www.abb.com/motor&drives.

[5] Y. S. Suh, J. Steinke, and P. Steimer, "Efficiency comparison of voltage source and current source drive system for medium voltage applications," *IEEE Transactions on Industrial Electronics*, vol. 54, no. 5, pp. 2521–2531, October 2007.

[6] S.Kouro, M. Malinowski, K. Gopakumar, L. G. Franquelo, J. Pou, J. Rodriguez, B. Wu, M. A. Perez and J. I. Leon, "Recent Advances and Industrial Applications of Multilevel Converters," *IEEE Trans. Ind. Electron.*, vol. 57, no. 8, Aug. 2010.

[7] A. Zuckerberger, E. Suter, C. Schaub, A. Klett, and P. Steimer, "Design, simulation and realization of high power NPC converters equipped with IGCTs", *Proc. of the Thirty-Third IAS Annual Meeting*, vol. 2, pp.865–872, 1998.

[8] "Asymmetric Intergrated Gate-Commutated Thyristor 5SHY 42L6500", Datasheet, Doc. No. 5SYA1245-03 Dec. 12, ABB Switzerland Ltd., Online: www.abb.com.

[9] "IGBT Moduled 5SNA 0750G650300", Datasheet, Doc. No. 5SYA 1600-02 04-2012, ABB Switzerland Ltd., Online: www.abb.com.

[10] "Insulated Gate Bi-Polar Transistor, Type T2400GB45E", Datasheet, T2400GB45E November. 2011, IXYS, Online: www.westcode.com.

[11] "Toshiba Silicon N-Chnnel IEGT ST1200GXH24A", Datasheet, 2009-09-07, Toshiba, Online: http://www.semicon.toshiba.co.jp.

[12] "Fast Recovery Diode 5SDF 10H6004", Datasheet, Doc. No. 5SYA1109-03 Jan. 10, ABB Switzerland Ltd., Online: www.abb.com.

[13] S. Bernet, "State of the art and developments of medium voltage converters an overview," *in Proc. PELINCEC 2005, Warsaw, Poland*, 2005.

[14] J. Rodriguez, S. Bernet, B. Wu, J. Pontt, and S. Kouro, "Multilevel voltage-source-converter topologies for industrial medium-voltage drives," *IEEE Trans. on Industrial Electron.*, vol. 54, no. 6, pp. 2930–2945, Dec. 2007.

[15] S. Bernet, "Recent developments of high power converters for industry and traction applications," *IEEE Trans. on Power Electron.*, vol. 15, no. 6, pp. 1102–1117, Nov 2000.

[16] S. Bernet, E. Carroll, P. Streit, O. Apeldoorn, P. Steimer, and S. Tschirley, "10 kV IGCTs," *Industry Application Magazine*, vol. 11, no. 2, pp. 53–61, March/April 2005.

[17] Alvarez, R. Filsecker, and S. Bernet, "Characterization of a new 4.5 kV press pack SPT+ IGBT for medium voltage converters," *Proc. of the 1st IEEE Energy Conversion Congress and Exposition, 2009. ECCE 2009, San Jose (CA)*, 20-24 Sept. 2009, pp. 3954 - 3962.

[18] Kon, K. Nakayama, S. Yanagisawa, J. Miwa, and Y. Uetake, "The 4500V-750A planar gate press pack IEGT," *in Proc. ISPSD'98*, pp. 81-84, 1988.

[19] K. Ichikawa, M. Tsukakoshi, and R. Nakajima, "Higher efficiency three-level inverter employing IEGTs," *in Proc. 19th Annu. IEEE APEC*, 2004, vol. 3, pp.1663-1668.

[20] A. Rockhill, M. Liserre, R. Teodorescu, and P. Rodriguez, "Grid-filter design for a multi megawatt medium-voltage voltage-source inverter," *IEEE Trans. on Industrial Electronics*, vol. 58, no. 4, pp. 1205–1217, 2011.

[21] L. Clotea and A. Forcos, "Power Losses Evaluation of Two and Three-Level NPC Inverters considering Drive Applications," *in Proc. OPTIM, 2012*, pp.929-934.

[22] M. Buschendorf, J. Weber, and S. Bernet, "Comparison of IGCT and IGBT for the use in the modular multilevel converter for HVDC applications," *in Proc. 9th Int. Multi-Conf. SSD*, 2012, pp. 1–6.

The 2014 International Power Electronics Conference

Application of Modular Matrix Converter to Wind Turbine Generator

Kentaro Inomata*, Hidenori Hara*, Shinya Morimoto*, Junji Fujii**, Kotaro Takeda**, Eiji Yamamoto**

*Yaskawa Electric Corporation, Corporate R&D Center, 12-1 Otemachi, Kokurakita-ku, Kitakyushu 803-8530 Japan
** Yaskawa Electric Corporation, System Engineering Center, 13-1 Nishimiyaichi 2 Chome, Yukuhashi 824-8511 Japan
E-mail: inomata@yaskawa.co.jp

Abstract— **This paper introduces a new method of generator brake torque control during Fault Ride Through (FRT) drive for a multi level matrix converter of wind power system. It may be exist that a generator speed become over speed during FRT drive. In this case, a generator has to be fixed brake torque. In the past, generator brake torque control during FRT drive for a matrix converter is impossible. However, in this paper, concept of brake torque control method and compatible method of the torque control and FRT control are introduced. Then, fundamental simulation indicates effecting of the propose method. It is expected that safety FRT drive is realization by the proposed method.**

Keywords— AC-AC conversion, Fault Ride Through, Matrix converter, Wind power generation system.

I. INTRODUCTION

Recent years, many countries have installed renewable energy plant such as wind turbines and further more rapidly. The penetration level of wind turbine is getting larger which lead to emerging requirements to connect to grid.

One of requirement is Fault Ride Through (FRT) capability which enables the power generation system to be connected to the grid during a sudden power failure. This is because grid network can be unstable if distributed power plants are all together disconnected from the grid due to voltage and frequency fluctuation.

During under voltage condition, the wind turbine must remain connected to the grid for a given time duration and contribute to the power system stability by supplying reactive power. Fig.1.1 shows the fault clearing times as well as voltage drop requirements in some country [1]. Under voltage drop, the reactive current has to be injected/ absorbed as shown in Fig.1.2 [2].

Variable speed wind turbine generator system by a partial-rating power converter and Doubly Fed Induction Generator (DFIG) is the most popular wind energy conversion system. For FRT capability, it is necessary to limit large transient current in DFIG based system. There are many papers that report technical solutions [3] [4].

On the other hand, adopting full-scaled power converter system has advantages of enhancing wind turbine system performance as well as FRT capability compared to partial-scaled converter based on DFIG. This system totally decouples the generator from the grid and can control generator individually over the entire speed range [5].

Usually, full-scaled back-to-back converter with DC link is the most common power converter topology. As an alternative for wind power application, matrix converter (MC) was introduced [6].

Authors presented the power conversion system employing 3.3kV medium voltage full scaled matrix conversion system for large wind turbine [7].

AC-AC direct conversion system can achieve high efficiency and low maintenance. On the other hand, direct PWM switching between the grid and AC generator can be sensitive to grid disturbance compared to the conventional AC-DC-AC system.

In addition, AC-AC conversion system has drawback that not to be able to conserve the wind turbine speed during FRT. In other words, the conventional system has a way to consume the accelerating energy with controlling brake resistor equipped with DC line.

The paper describes new techniques to improve the AC-AC direct conversion system for controlling the wind generator during FRT.

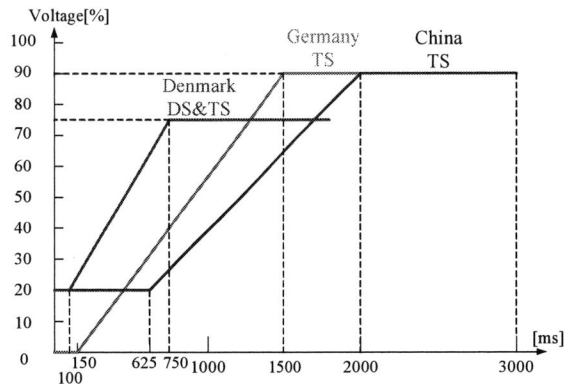

Fig. 1.1. Limit curve for voltage pattern at the grid connection.

978-1-4799-2706-7/14 $31.00 © 2014 IEEE

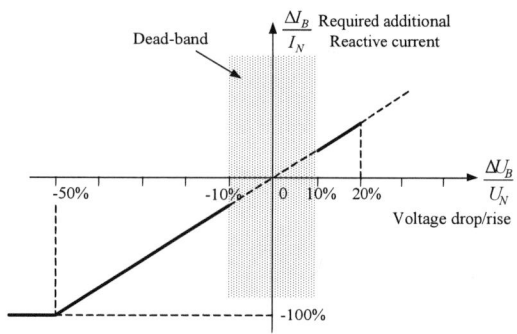

Fig. 1.2. Eon Code : The principle of voltage support in the event of grid faults (reactive power demand)

II. TECHNICAL WORK PREPARATION

Wind power system voltage is expected to increase gradually from standard 690 V to Medium Voltage (MV) as the power ratings of the system increase. The advantages of MV are lower conduction loss between the generator and the power converter and reduction of volume and weight of nacelle, and cost [8].

MV matrix converters have been developed based on the basic MC topology [9]. It provides an attractive solution to the market which needs high power drives with regenerative power capability such as paper winder, large power pump, steel mill, and wind power system. It is compact in size, has low harmonics in input and output waveforms.

A basic module for the modular drive is Single-Phase output Matrix Converter (SPMC) as shown in Fig. 2.1. Basically this module is a three-phase input single-phase output MC. It generates single-phase output voltage according to the reference of three-phase output voltage. Power factor of input current is controlled by adjusting conduction ratio of the each switch according to the input voltage waveforms [10]-[11].

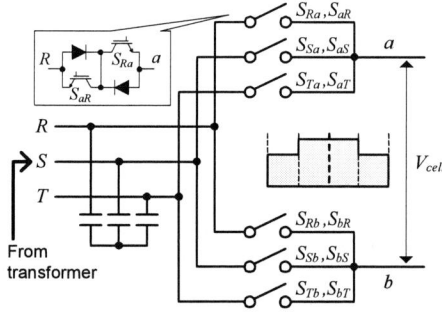

Fig. 2.1. Simplified single-phase output matrix converter.

Fig. 2.2 shows overall configuration of the High Power Matrix Converter (HPMC) at the 3.3 kV including a multi-winding transformer which steps down secondary voltage from 3300 V to 630 V. The HPMC at 3.3kV is composed by three One-Cell-Block Matrix Converters (OCBMCs). And, a OCBMC is composed by three SPMC. Phase shifts of +20, 0, and -20 degree in the secondary windings improve input current harmonics. The standard low voltage MC requires LC

filters to reduce PWM harmonics in the grid but in the modular MC topology, reactors are eliminated by using leakage inductance portion of the transformer. PWM timing of the power cells are shifted by 120 degree with each other to achieve smooth output voltage step change. For 6.6 kV line-line voltage drive, six power cells are connected in series.

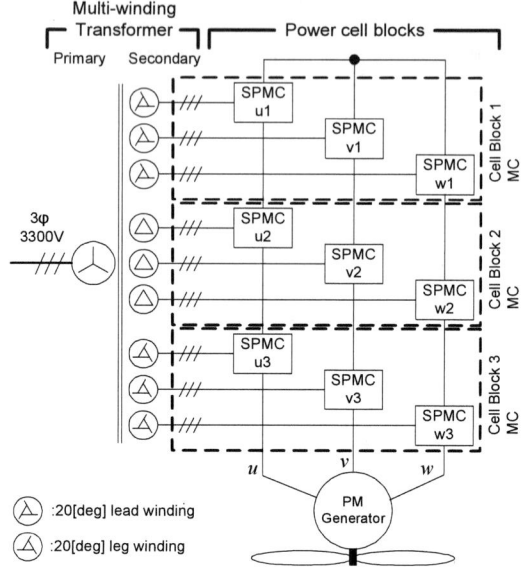

Fig. 2.2. The overall configuration of the High Power Matrix Converter for wind power system.

III. TECHNICAL WORK PREPARATION

Reference [12] shows a method of FRT drive for a MC. It is possible that a generator is considered as a current source. Therefore, grid current can be controlled by MC operating as current circuit.

It assumes that the virtual current circuit shown by Fig. 3.1 is connected to between a generator and grid, instead of a MC. Amount of grid current can be controlled by virtual Current Source Converter (virtual CSC). And, phase angle of grid current can be controlled by virtual Current Source Inverter (virtual CSI).

In basic MC, if flow of current is from w-phase to S-phase and from R-phase to u-phase, switching of basic MC is shown in Fig. 3.2. Therefore, generalized equations of transforming from current circuit switching to basic MC switching can be shown in (3.1) and (3.2).

In OCBMC, if flow of current is from w-phase to S-phase and from R-phase to u-phase, switching of OCBMC is shown in Fig. 3.3. Therefore, generalized equations of transforming from current circuit switching to OCBMC switching can be shown in (3.3) and (3.4).

The concept of a method of FRT drive for a MC was explained. However, if power by wind is bigger than power for FRT drive, the generator accelerates. If the accelerated generator's speed is dangerous, generating brake torque during FRT drive is necessary.

978-1-4799-2706-7/14 $31.00 © 2014 IEEE

The 2014 International Power Electronics Conference

$$
\begin{bmatrix}
S_{Rjua} & S_{Sjua} & S_{Tjua} \\
S_{Rjub} & S_{Sjub} & S_{Tjub} \\
S_{Rjva} & S_{Sjva} & S_{Tjva} \\
S_{Rjvb} & S_{Sjvb} & S_{Tjvb} \\
S_{Rjwa} & S_{Sjwa} & S_{Tjwa} \\
S_{Rjwb} & S_{Sjwb} & S_{Tjwb}
\end{bmatrix}
=
\begin{bmatrix}
S_{Nu} & 0 \\
0 & S_{uP} \\
S_{Nv} & 0 \\
0 & S_{vP} \\
S_{Nw} & 0 \\
0 & S_{wP}
\end{bmatrix}
\begin{bmatrix}
S_{RjP} & S_{SjP} & S_{TjP} \\
S_{RjP} & S_{SjP} & S_{TjP}
\end{bmatrix}
\;(3.4)
$$

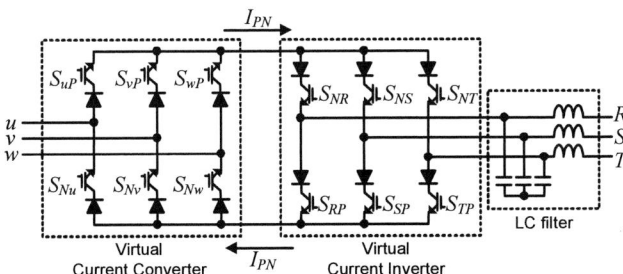

Fig. 3.1. A virtual current circuit.

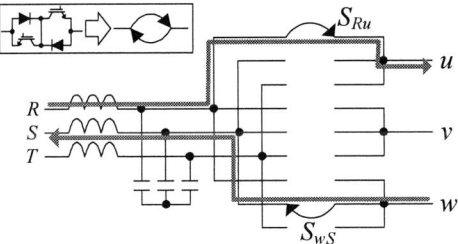

Fig. 3.2. Example of basic matrix converter switching.

IV. PROPOSED METHOD

In this chapter, a method of brake torque control during FRT drive for a matrix converter is proposed. First of all, a new virtual current source inverter switching method for brake torque generating is shown. Next, brake torque control and generator speed limitation control methods are shown.

A. New virtual current source inverter switching

In method of FRT drive for a matrix converter, a generator is considered current source. However, a generator has resistor in series to inductor. Therefore, conduction loss is raised by flowing generator current, and then brake torque is raised by conduction loss. As a result, a generator torque control is realized by the operation of amount of generator current.

Next, the operation method of amount of generator current is explained. In Fig. 4.1, the virtual CSI has three switching patterns. Fig. 4.1(b) and (c) show switching patterns for grid current controlling. And, Fig. 4.1(a) shows the switching pattern for extra brake torque generating. Current flowing at switching pattern of Fig. 4.1(a) generates extra brake torque.

Figure 4.2 shows a current space vector of virtual CSI. In Fig. 4.2, the switching pattern of Fig. 4.1(a) creates zero-vector. Therefore, amount of extra generator current change in proportion with a time of zero-vector. Duty of zero-vector (D_o) is defined by (4.1). I_{bra} shows amount of current for brake torque, and $\|I_{RST}^*\|$ shows norm of grid current reference.

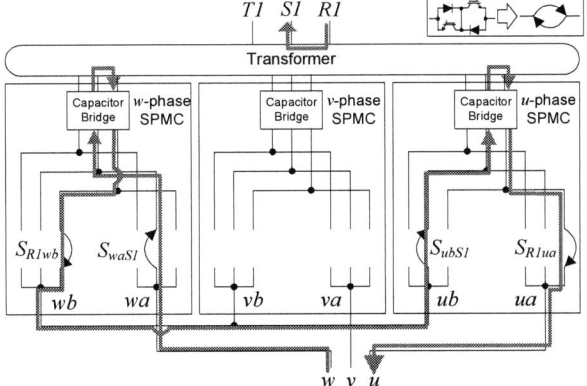

Fig. 3.3. Example of one cell block matrix converter switching.

$$
\begin{bmatrix}
S_{uR} & S_{uS} & S_{uT} \\
S_{vR} & S_{vS} & S_{vT} \\
S_{wR} & S_{wS} & S_{wT}
\end{bmatrix}
=
\begin{bmatrix}
S_{uP} \\
S_{vP} \\
S_{wP}
\end{bmatrix}
\begin{bmatrix}
S_{NR} & S_{NS} & S_{NT}
\end{bmatrix}
\qquad (3.1)
$$

$$
\begin{bmatrix}
S_{Ru} & S_{Su} & S_{Tu} \\
S_{Rv} & S_{Sv} & S_{Tv} \\
S_{Rw} & S_{Sw} & S_{Tw}
\end{bmatrix}
=
\begin{bmatrix}
S_{Nu} \\
S_{Nv} \\
S_{Nw}
\end{bmatrix}
\begin{bmatrix}
S_{RP} & S_{SP} & S_{TP}
\end{bmatrix}
\qquad (3.2)
$$

$$
\begin{bmatrix}
S_{uaRj} & S_{uaSj} & S_{uaTj} \\
S_{ubRj} & S_{ubSj} & S_{ubTj} \\
S_{vaRj} & S_{vaSj} & S_{vaTj} \\
S_{vbRj} & S_{vbSj} & S_{vbTj} \\
S_{waRj} & S_{waSj} & S_{waTj} \\
S_{wbRj} & S_{wbSj} & S_{wbTj}
\end{bmatrix}
=
\begin{bmatrix}
S_{uP} & 0 \\
0 & S_{Nu} \\
S_{vP} & 0 \\
0 & S_{Nv} \\
S_{wP} & 0 \\
0 & S_{Nw}
\end{bmatrix}
\begin{bmatrix}
S_{NRj} & S_{NSj} & S_{NTj} \\
S_{NRj} & S_{NSj} & S_{NTj}
\end{bmatrix}
\;(3.3)
$$

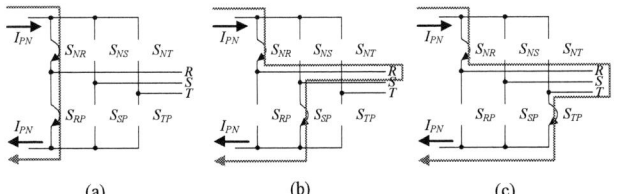

Fig. 4.1. Example of virtual current source inverter switching.

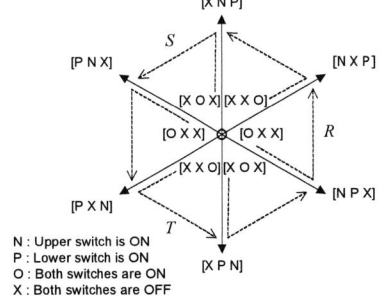

N : Upper switch is ON
P : Lower switch is ON
O : Both switches are ON
X : Both switches are OFF

Fig. 4.2. The switching pattern of virtual current source inverter.

978-1-4799-2706-7/14 $31.00 © 2014 IEEE

Figure 4.3 shows a part of the current space vector. *a*-vector is a vector of forward phase angle in the triangle. The θ_a is defined angle between grid current vector and *a*-vector in the triangle. Therefore, duties of *a*-vector and *b*-vector (D_a and D_b) are shown in (4.2) and (4.3) by θ_a and D_o.

Then, duties of current vector are compared with saw-tooth wave, as shown in Fig. 4.4. S_o, S_a and S_b are made by the comparing. S_o, S_a and S_b show the selection signal of three current vectors. For example, if S_a is true, the *a*-vector is used. The rotation of selecting current vectors is the rotation of "o", "a" and "b". The three signals and virtual CSI switches relationship is shown in Table. 4.1. In Table. 4.1, "T" shows true (switch is ON), and "F" shows false (switch is OFF).

A making flow of virtual CSI switching is shown in Fig. 4.5. A HPMC switching of FRT with brake torque is constructed by using the proposed virtual CSI switching and the virtual CSC switching. And, making method of CSC switching is introduced by [12].

$$D_o = \frac{I_{bra}^*}{I_{bra}^* + \left\| I_{RST}^* \right\|} \tag{4.1}$$

$$D_a = (1 - D_o)\left(\frac{1}{2} + \frac{\sqrt{3}}{2}\tan\left(\frac{\pi}{6} - \theta_a \right) \right) \tag{4.2}$$

$$D_b = (1 - D_o)\left(\frac{1}{2} - \frac{\sqrt{3}}{2}\tan\left(\frac{\pi}{6} - \theta_a \right) \right) \tag{4.3}$$

B. Brake Torque Control and Generator Speed Limitation Control

In before section, it is shown that the duty and switching pattern for generating I_{bra} are decided. Next, estimating method of I_{bra} is indicated. Loss by FRT drive is shown by (4.5). In (4.5), W_{FRT_loss} is made by copper loss of generator side and active power of grid side. W_{bra_loss} shows extra loss by I_{bra}. If extra current I_{bra} is generated, like Fig. 4.1(a), extra loss W_{bra_loss} is generated, too. W_{bra_loss} is made by copper loss of generator side ((4.6)). So, total loss of proposed FRT drive is shown by (4.4). And, extra brake torque τ_{bra} is generated, like equation (4.7). R is value of generator's resistor, ω_m is generator rotation speed, $\|E_{RST}\|$ is norm of input voltage, $\Delta\theta$ is power factor angle.

$$W_{loss} = W_{FRT\_loss} + W_{bra\_loss} \tag{4.4}$$

$$W_{FRT\_loss} = \left\| I_{RST} \right\|^2 R + 3\left\| I_{RST} \right\| \cdot \left\| E_{RST} \right\| \cos \Delta\theta \tag{4.5}$$

$$W_{bra\_loss} = I_{bra}^{*\,2} R \tag{4.6}$$

$$W_{bra\_loss} = \tau_{bra}\omega_m \tag{4.7}$$

Therefore, by (4.6) and (4.7).

$$I_{bra} = \sqrt{\frac{\tau_{bra}\omega_m}{R}} \tag{4.8}$$

TABLE I
RELATIONSHIP OF SWITCHES

	S_{NR}	S_{NS}	S_{NT}	S_{RP}	S_{SP}	S_{TP}
$-\pi/6 < \theta < \pi/6$	T	F	F	S_o	S_a	S_b
$\pi/6 < \theta < \pi/2$	S_a	S_b	S_o	F	F	T
$\pi/2 < \theta < 5\pi/6$	F	T	F	S_b	S_o	S_a
$5\pi/6 < \theta < 7\pi/6$	S_o	S_a	S_b	T	F	F
$7\pi/6 < \theta < 3\pi/2$	F	F	T	S_a	S_b	S_o
$3\pi/2 < \theta < 11\pi/6$	S_b	S_o	S_a	F	T	F

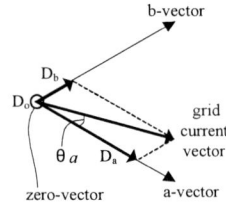

Fig. 4.3. A part of the current space vector.

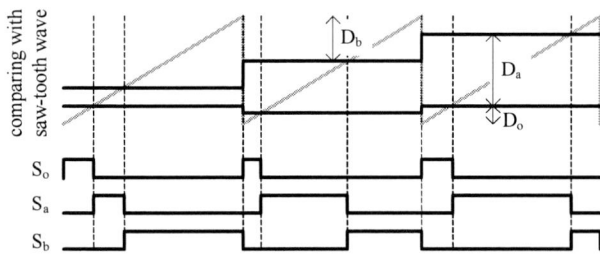

Fig. 4.4. The comparing with saw-tooth wave.

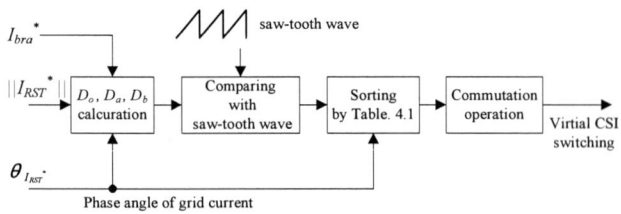

Fig. 4.5. A making flow of virtual CSI switching.

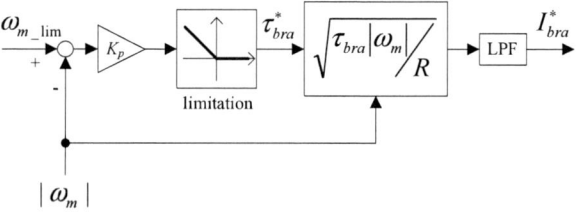

Fig. 4.6 A block diagram of brake torque control and generator speed limitation control

In (4.8), I_{bra}^* can is estimated by τ_{bra}^* and ω_m. So, brake torque control is realized by using (4.8).

And then, reference of brake torque τ_{bra}^* is decided as no exceeding limit of generator speed. So, a block diagram of brake torque control and generator speed limitation control is shown in Fig. 4.6. ω_{m_lim} is limit value of generator speed. K_P is proportional gain. Next, resolving method of value of K_P is explained. Transfer function of Fig. 4.6 is shown in (4.9). In (4.9), J is inertia of mechanical system, and D is viscosity coefficient. Because D is much smaller than J, D can be

978-1-4799-2706-7/14 $31.00 © 2014 IEEE

neglected. Therefore, K_P can be decided by using control response frequency ω_n, like (4.10). A block of "LPF" shows low pass filter processing. The low pass filter has from 5 to 10 times bandwidth at ω_n.

$$G = \frac{K_P}{sJ + D + K_P} \approx \frac{K_P/J}{s + K_P/J} \qquad (4.9)$$

$$K_P = \omega_n J \qquad (4.10)$$

A block diagram of FRT control which can generate brake torque is shown in Fig. 4.7. Processing of the block of "Making CSC switching" is shown in reference [12]. Processing of the block of "Concentration of switchings" is equations (3.1)-(3.4). If MC is basic MC, equations (3.1), (3.2) are used. If MC is HPMC, equations (3.3), (3.4) are used.

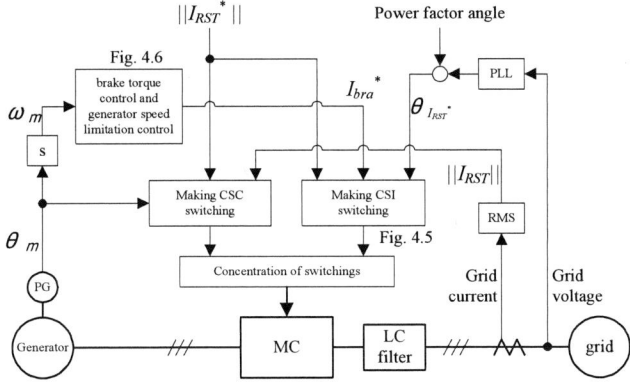

Fig. 4.7 A block diagram of FRT control which can generate brake torque

V. SIMULATIONS

FRT on MC model is simulated by MATLAB Simulink and SimPowerSystems. Simulation parameters are shown in table II. In their simulations, interior permanent magnet synchronous generator model was used. The simulation results of brake torque control for basic MC and HPMC are introduced.

TABLE II
SIMULATION PARAMETER

parameter	value
Generator d-axis inductance	4.5 [mH]
Generator q-axis inductance	3.25 [mH]
Generator resistance	0.1 [Ohm]
Generator back EMF constant	6.25 [V/(rad/s)]
Generator pole number	84
Generator inertia	1000 [kgm³]
Generator coefficient of viscosity	0.001 [Nm/(rad/s)]
Generator mechanical rating speed	2 [Hz]
ω_{m_lim}	2.05 [Hz]
Input rating voltage	3300 [V]
Frequency of input voltage	60 [Hz]
Saw-tooth wave frequency	4000 [Hz]
Reference of grid current's amount	230 [A]
K_P	12000
Frequency of LPF of Fig. 4.6	10 [Hz]
Power factor angle	+90 [deg]

A. Simulation Result on Basic MC

First of all, basic MC's simulation result is explained. The voltage during FRT is 25% (825V). Generator mechanical initial speed is 2 Hz. External load torque of 5000 Nm is applied as load of wind.

Fig. 5.1 shows comparison of generator speed (ω_m) with and without proposed method on basic MC. Fig.5.2 shows comparison of current's amount with and without proposed method on basic MC. Fig. 5.2(a) shows generator current. Fig. 5.2(b) shows grid current. Fig. 5.3 shows Result of power factor angle on grid on basic MC.

In Fig. 5.1, generator speed without proposed method increases little by little. As a result generator limit speed is exceeded. However, generator speed with proposed method can be controlled at around ω_{m_lim}. And, in Fig.5.2(a), amount of generator current with proposed method increases from point of ω_m exceeding ω_{m_lim}. But, grid side current with proposed method is constant. And, power factor angle remains 90 degree. Therefore, brake torque control and generator speed limitation control are realized in basic MC.

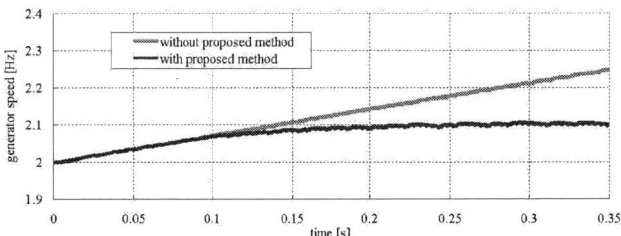

Fig. 5.1. The comparison of generator speed with and without proposed method on basic MC.

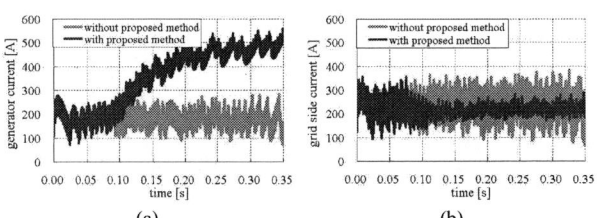

Fig. 5.2. The comparison of current's amount with and without proposed method on basic MC.

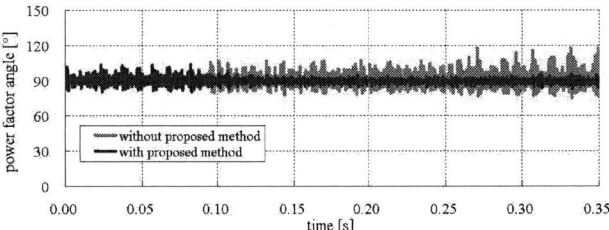

Fig. 5.3. Result of power factor angle on grid on basic MC.

B. Simulation Result on High Power MC

Next, simulation result of the proposed method on HPMC is shown. The voltage during FRT is zero. Initial generator's mechanical speed is 2 Hz, and external load torque is 7500 Nm. Other parameters are based on table II.

Fig. 5.4 shows comparison of generator speed with and without proposed method on HPMC. Fig. 5.5 shows comparison of current's amount with and without proposed method on HPMC ((a) is generator side, (b) is grid side). Fig. 5.6 shows result of grid side power factor angle on HPMC.

In Fig. 5.4, generator speed without the proposed method increased gradually. Finally, the speed arrived at dangerous speed. However, generator speed with the proposed method was limited at around ω_{m_lim}. And, in Fig. 5.5(b), amount of grid side current always was controlled at 230 A. But, with the proposed method, generator current increased from about 0.1 second in Fig. 5.5(b). The increasing was effect by zero-vector of the proposed method. And, because power factor angle of grid current always was +90 degree, it fitted of reference of grid current phase angle. It is clear that the proposed method can use on HPMC.

So, advantageous effect of the proposed method on both of basic MC and HPMC is clear by the simulation results.

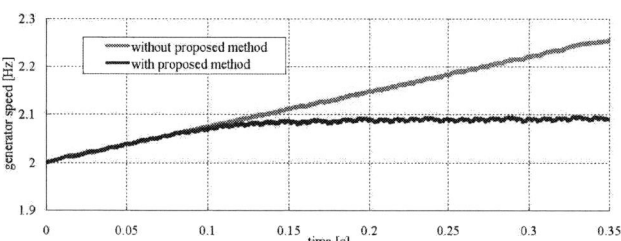

Fig. 5.4. The comparison of generator speed with and without proposed method on HPMC.

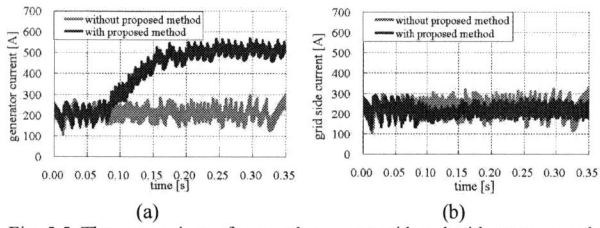

Fig. 5.5. The comparison of current's amount with and without proposed method on HPMC.

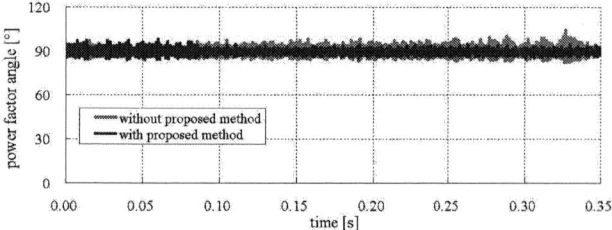

Fig. 5.6. Result of power factor angle on grid on HPMC.

VI. CONCLUSION

In this paper, brake torque control method during FRT on a matrix converter without external machine for brake torque was proposed. If wind power is bigger than power for FRT drive, as a result, generator speed exceeds limit speed. However, by the brake torque control, generator speed can remain in limit speed or acceleration of generator can decrease. Detail of the matrix converter's switching for generating brake torque and the generator speed limitation control method were explained in this paper. The effect of the proposed method on basic MC and HPMC was confirmed by simulations.

If this proposed method was applied on matrix converters, external special machine for generating brake torque isn't demanded or the downsizing of special machine is possible.

REFERENCES

[1] Florin lov, Anca Daniela Hansen et al. "Mapping of grid faults and grid Codes," Riso-National Laboratory, R-1617(EN), July 2007.

[2] E.ON Netz GmbH, "Netzanschlussregeln für Hoch- und Höchstspannung," 1 April 2006.

[3] Luna.Á, Rolán.A, Medeiros.G, Rodrígue.P, Teodorescu.R, "Control strategies for DFIG wind turbines under grid fault conditions," in Proc. the 35th Annual Conference of the IEEE Industrial Electronics Society (IECON 2009), Nov. 2009, pp.3886-3891.

[4] Lasantha Meegahapola, Tim Littler, Damian Flynn, "Decouple-DFIG Fault Ride-Through Strategy for Enhanced Stability Performance During Grid Faults," IEEE Transactions on Sustainable Energy, pp.152-162, Oct. 2010.

[5] Marwan Rosyadi1, S.M.Muyeen, Rion Takahashi, Junji Tamura, "Low Voltage Ride-Through Capability Improvement of Wind Farms using Variable Speed Permanent Magnet Wind Generator," in Proc. 2011 International Conference on Electrical Machines and Systems (ICEMS 2011), 20-23 August 2011, pp.1-6.

[6] L.Zhang, C.Watthanasarn, "A Matrix Converter Excited Double-Fed Induction Machine as a Wind Power Generator," in Proc. 7th International conference on power electronics and variable speed drives, 21-23 September 1998, pp.532-537.

[7] E.Yamamoto, H.Hara, T.Uchino, M.Kawaji, T.Kume, J.Kang, H.Krug, "Development of MCs and its Applications in Industry," IEEE Industrial Electronics Magazine, vol 5, No1, pp.4-12, 2011.

[8] W. Erdman, M. Behnke, "Low Wind Speed Turbine Project Phase II: The Application of Medium-Voltage Electrical Apparatus to the Class of Variable Speed Multi-Megawatt Low Wind Speed Turbines," National Renewable Energy Lab Report, NREL/SR-500-38686, November 2005.

[9] E.Yamamoto, H.Hara, T.Uchino, M.Kawaji, T.Kume, J.Kang, H.Krug, "Development of MCs and its Applications in Industry," IEEE Industrial Electronics Magazine, vol 5, No1, 2011, pp.4-12.

[10] J.Oyama, X.Xia, T.Higuchi, and E.Yamada, "Displacement Angle Control for Matrix Converter," in Proc. 28th Power Electronics Specialists Conference (PESC 1997), June 1997, pp.1033-1039.

[11] S.Ishii, E.Yamamoto, H.Hara, E.Watanabe, A.M.Hava, and X.Xia, "A vector controlled high performance matrix converter – induction motor drive," in Proc. the 4th International Power Electronics Conference (IPEC Tokyo 2000), Apr 2000, pp.235-240.

[12] K.Inomata, H.Hara, S.Morimoto, J.Fujii, T.Takeda, E.Yamamoto, E.Watanabe, "Enhanced Fault Ride Through Capability of Matrix Converter for Wind Power System," in Proc. the 39th Annual Conference of the IEEE Industrial Electronics Society (IECON 2013), Nov 2013.

The 2014 International Power Electronics Conference

Free Motion Mechanical Power Factor; Comparison Between Robots in Different Structure and Coordinate

Takahiro Mizoguchi and Takahiro Nozaki
Keio University
Graduate School of Science and Technology
Yokohama, Kanagawa, Japan
mizobuu@sum.sd.keio.ac.jp, takahiro@sum.sd.keio.ac.jp

Kouhei Ohnishi
Keio University
Department of System Design Engineering
Yokohama, Kanagawa, Japan
ohnishi@sd.keio.ac.jp

Abstract—This paper shows behavior of power factor in mechanical system in the case of work space motion. XY table and two link manipulator are employed to demonstrate the difference between work space power factor and joint space power factor. It turns out that power factor in work space and joint space becomes different even though the same motion is performed. In addition, power factor during free motion is investigated. Experimental results supports that structure of the system is relevant to the efficiency of the motion.

Keywords—Power Factor, Mechanial System, Analogy, Motion Control

I. INTRODUCTION

In recent years, development of technology enables humans to consider quality of support. As well as quality of the support, efficiency of the robot motion needs to be considered, since amount of the fossil fuel is limited[1]. To reduce the consumption of fuels; car energy is supported by electric batteries, urban development are based on precise control of distribution of electric energy [2]. System today are required to be energy efficient [3], [4].

However, robotics today heads opposites of consideration of energy consumption. The focus of research usually claims precision of task execution. Robots are position controlled regardless of disturbance reduction force generated during operation, although reduction force is generated by increasing current consumption[5]. In fact, automation robots operated in factories around the world are controlled this way. Examination of efficiency of the motion takes a big role not only as a research, but also for society. However, only few researches had addressed this problem [6], [7], [8].

In prior researches, authors have pointed out that energy behavior can be observed in a form of power in mechanical system. When a robot is in contact with environment, there always be energy transmission as shown in Fig. 1.

In addition, behavior of power factor in mechanical system resembles to that of electrical system and analyses can be conducted in the same manner. On research [9] power factor of the mechanical system is compared with that of electrical system in very limited situation. In that particular situation,

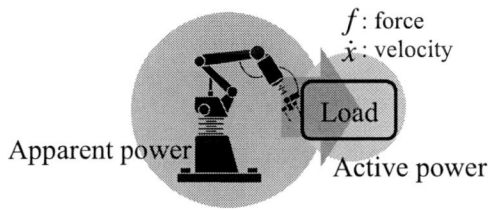

Fig. 1.   Image of power factor in mechanical system.

behavior of two systems resembles to each other. More additional situations are considered in research [10].

In this study, we will discuss about free motion case and robot with multi degree of freedom case. Non contact motion is unique case that exists only in mechanical system. Equivalent electrical system in non contact case is short circuit. It is not natural to consider about power factor of short circuit in electrical system. On the other hand, free motion in mechanical system is likely to occur. This disconnection between electrical system and mechanical event causes difficulty in analyses. In addition, power factor in multi degree of freedom robot case is considered as well. Power factor in work space eventually includes Jacobian matrix in it, which causes power factor to differ between joint space and work space. Power factor behavior in both joint space and work space is obtained in this paper for future consideration on power factor research considering relationship between mechanical structure of robot and trajectory of it.

II. POWER FACTOR IN MECHANICAL SYSTEM

This section explain a concept of power factor in mechanical system. Power factor is one index to represent energy transmission efficiency in electrical system. In alternating current case, two kinds of working electrical power are defined; "active power" and "apparent power". Active power is the product of the current and voltage of the circuit. Apparent power, on the other hand, is the capacity of the circuit for performing work in a particular time. Power factor expresses the ratio of active power over apparent power of the system. Knowing power factor helps us to know sufficient amount of voltage or current to perform particular work.

Based on analogy between quantities in electrical system

978-1-4799-2706-7/14 $31.00 © 2014 IEEE   1660

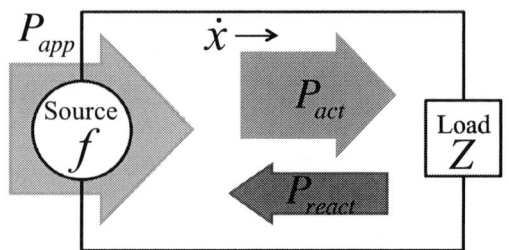

Fig. 2. Analogic representation of power in mechanical system.

and mechanical system, the concept of power factor should also be able to be applied for mechanical system. Fig. 1 represents apparent power and active power in mechanical system. Robot should have had apparent power to do the work, while only some amount of apparent power is used in a motion as a active power. The part of power, which is not used for the motion, is called reactive power. Relationship between active power, apparent power and reactive power is shown on Fig. 2 as analogic representation with electrical circuit. Force provides effort to perform motion, velocity is flow in the system, and energy is transferred to the load as shown in figure. In contact motion case, load is present as impedance Z.

A. Active power

As in electrical system, active power in mechanical system should explain the amount of power used in the work. From the analogy, active power in mechanical system can be defined as

$$p = f\dot{x}, \tag{1}$$

$$P_{act} = \frac{1}{T} \int_0^T p\,\mathrm{dt}. \tag{2}$$

In real system, active mechanical power P_{act} should be available from the integral of instantaneous mechanical power p. f and $\dot{x}$ are instant force and instant velocity respectively.

B. Apparent power

In contrast to active power P_{act}, a system should have some amount of power they apply for work. This power does not necessary be the same amount as active power used for the work. By considering electrical system, apparent power of mechanical system P_{app} is defined as

$$f_{rms} = \sqrt{\frac{1}{T} \int_0^T f^2 \mathrm{dt}}, \tag{3}$$

$$\dot{x}_{rms} = \sqrt{\frac{1}{T} \int_0^T \dot{x}^2 \mathrm{dt}}, \tag{4}$$

$$P_{app} = f_{rms}\dot{x}_{rms}. \tag{5}$$

Subscript rms defines root mean square (RMS) value.

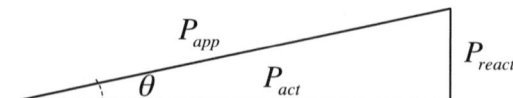

Fig. 3. Triangle of power in mechanical system.

C. Power factor

By applying the concept of active power and apparent power in mechanical system, we should also be able to define power factor of the mechanical system. That explains how much of the applied power is really used in the work. As an electrical system, power factor should be defined by the fraction of active power and apparent power.

$$PF = \cos\theta = \frac{P_{act}}{P_{app}} \tag{6}$$

PF is power factor. In electrical system, θ in power factor defines phase difference of current and voltage, thus, θ should be defined by a phase difference of velocity and force in mechanical system.

Fig. 3 represents the relationship of power in power factor analysis. Power in the system is identified as right triangle. θ is an angle between apparent power and active power. From the relation, reactive power is derived by,

$$P_{react} = \sqrt{P_{app}^2 - P_{act}^2}. \tag{7}$$

D. Distorted wave cases

In general, force and velocity in the motion are not sinusoidal, namely distorted waves. Therefore Fourier transformation is employed to break down distorted wave into combination of sinusoidal waves. In addition, in practical cases, information we can obtain from robot are always discrete. Thus, discrete Fourier transformation (DFT) is conducted. By DFT, distorted force and velocity can be expressed as,

$$f_d(t) \simeq F_0 + \sum_{n=1}^{N} F_{mn} \sin(n\omega t + \theta_{fn}), \tag{8}$$

$$\dot{x}_d(t) \simeq \dot{X}_0 + \sum_{n=1}^{N} \dot{X}_{mn} \sin(n\omega t + \theta_{\dot{x}n}). \tag{9}$$

Subscript d, m, n, f, and $\dot{x}$ signify distorted wave, magnitude, nth data, force, and velocity. N is maximum number of data. θ, ω are phase angle and angular velocity. RMS value for distorted wave is derived from RSS (root sum of squares) of RMS in each harmonic component,

$$F_n = \frac{F_{mn}}{\sqrt{2}}, \tag{10}$$

$$\dot{X}_n = \frac{\dot{X}_{mn}}{\sqrt{2}}, \tag{11}$$

$$F_{rms,d} = \sqrt{F_0^2 + F_1^2 + \cdots + F_N^2}, \tag{12}$$

$$\dot{X}_{rms,d} = \sqrt{\dot{X}_0^2 + \dot{X}_1^2 + \cdots + \dot{X}_N^2}. \tag{13}$$

978-1-4799-2706-7/14 $31.00 © 2014 IEEE

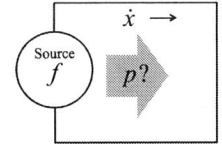
(a) Free motion analogy.

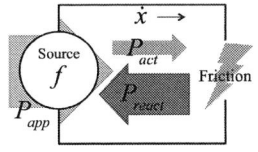
(b) Considering friction.

Fig. 4.   Circuit representation of free motion.

III.   POWER FACTOR IN WORK SPACE

This section explains concept of work space power factor and joint space power factor. When considering energy transfer to the environment from output of the total system, work space power factor is important. On the other hand, when considering efficiency of actuator to the output of the system or environment, joint space power factor should be considered.

A.   Coordinate transformation

Equation (14) shows transformation of joint angle to work space.

$$\mathbf{x} = \mathbf{T}(\boldsymbol{\theta}), \tag{14}$$
$$\mathbf{x} = [x_1, x_2, \cdots, x_b]^T, \tag{15}$$
$$\boldsymbol{\theta} = [\theta_1, \theta_2, \cdots, \theta_a]^T. \tag{16}$$

$\mathbf{T}$ is transformation function. $\mathbf{x}$ is position in work space. $\boldsymbol{\theta}$ signify joint angles. Subscripts a represent number of joints in manipulator, and b represent number of degree in work space. When a exceeds b, a manipulator is called redundant. Velocity in work space is expressed by,

$$\dot{\mathbf{x}} = \mathbf{J}\dot{\boldsymbol{\theta}}. \tag{17}$$

$\mathbf{J}$ is Jacobian matrix. Similarly, force in work space can be expressed by using transpose inverse of Jacobian matrix as,

$$\mathbf{f} = \mathbf{J}^{-T}\boldsymbol{\tau}, \tag{18}$$
$$\mathbf{f} = [f_1, f_2, \cdots, f_b]^T, \tag{19}$$
$$\boldsymbol{\tau} = [\tau_1, \tau_2, \cdots, \tau_b]^T. \tag{20}$$

Assuming no interference between axis in work space, power factor in work space can be expressed separately.

IV.   FREE MOTION CONSIDERATION

This section gives comments on how to consider free motion in power factor analysis. In power factor analysis of the mechanical system, consideration of force in the system has very important role. When the system is in free motion, unlike contact motion, load of the system is not present as shown in Fig. 4(a). In electrical system, such a circuit is called short circuit and has infinite admittance. It is not natural to consider about power factor in short circuit in electrical system. On the other hand, short circuit is likely to occur in mechanical system. This is the difficulty of power factor in mechanical system.

To define free motion power factor, this paper focuses on physical meaning of power and energy flow in the system.

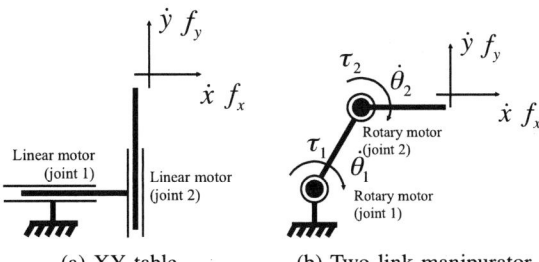

(a) XY table.   (b) Two link manipurator.

Fig. 5.   Experimental system.

Energy flow during the motion is shown in Fig. 4(b). As shown in figure, energy during free motion is not likely to transferred to anywhere, unless there exists friction. Therefore free motion should be constructed by reactive power at the most. If active power presents in the free motion, it is most likely to represent energy transferred to friction.

Reactive power factor is derived from,

$$RPF = \frac{P_{reac}}{P_{app}}. \tag{21}$$

By examining reactive power factor of the system, efficiency of power during free motion can be determined.

V.   EXPERIMENT

Experiments are performed in order to examine two things:

1) Power factor difference due to structural difference of the system
2) Power factor difference between work space and joint space

XY table (XY) and two link manipulator (2 link) are compared to clarify power factor difference due to structure of the system. In addition, work space power factor and joint space power factor are compared in two link manipulator to show that power factor differs between work space and joint space. Figure 5 shows experimental systems. Fig. 5(a) is XY table composed of two sets of linear motors. Joint space and work space of the system has common axis in XY table. Fig. 5(b) is two link manipulator composed of two sets of rotary motors. Joint space and work space in two link manipulator requires coordinate transformation. Transformation function from joint space to work space is,

$$\mathbf{T}(\theta_1, \theta_2) = \left[\begin{array}{c} l_1 \sin(\theta_1) + l_2 \sin(\theta_1 + \theta_2) \\ l_1 \cos(\theta_1) + l_2 \cos(\theta_1 + \theta_2) \end{array} \right]. \tag{22}$$

Table I shows parameters used in the experiment. Velocity is calculated from psudo-derivation from position encoder output.

A.   Power factor in two different structured systems

To examine the power factor difference between different structured systems, the same motion was applied to XY table and two link manipulator. Circular motion given to the system

978-1-4799-2706-7/14 $31.00 © 2014 IEEE

TABLE I.   EXPERIMENTAL PARAMETERS.

XY	Mass of linear motor 1	3.0	[kg]
	Mass of linear motor 2	0.6	[kg]
	Position gain	900	[1/s^2]
	Velocity gain	60	[1/s]
2 link	Inertia of link 1	0.0005	[kgm^2]
	Inertia of link 2	0.0003	[kgm^2]
	Length of link 1 (l_1)	0.09	[m]
	Length of link 2 (l_2)	0.09	[m]
	Position gain	300	[1/s^2]
	Velocity gain	35	[1/s]

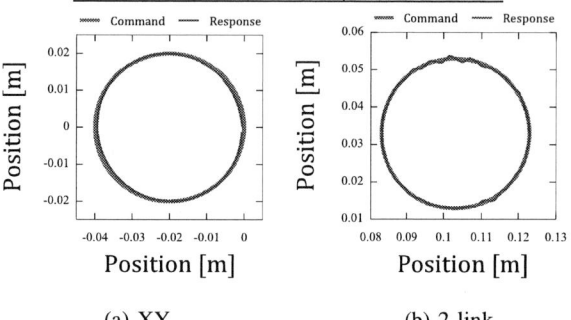

(a) XY.   (b) 2 link.

Fig. 6.   Circular motion experimental results.

and position response of each robots are shown in Fig. 6. Both systems are controlled by PD control with DOB. Responses are both acceptable.

Table II shows power factor and reactive power factor result of circular motion in both robots. As explained in section IV, reactive power factor is also derived from the responses. Though the same commands are given to both systems, power factor results have considerable difference. From the view point of power factor, XY table has low value meaning less energy is transferred to load during free motion, which is most likely to be friction in free motion case. On the other hand, two link manipulator has higher very high power factors (in meaning of absolute value), high power factor tends to signify damping and/or viscosity in the load, thus we can conclude that two link manipulator had higher friction force and energy was transferred to friction during free motion.

Such a tendency is easier to observe by examining reactive power factor. XY table has high reaction power factor and two link manipulator has smaller reactive power factor.

B. Comparison between work and joint space

Work space power factor and joint space power factor are compared in second experiment. The circular motion is the same as shown in Fig. 6(b). Total of four power factors are calculated. They are power factor for x axis, y axis in work space and for joint 1 and joint 2 in joint space.

Table III shows power factor results for both work space case and joint space case. From the viewpoint of reactive power factor, joint space response has much better value, meaning energy is not transferred to the load during free motion.

From two experiments, it is concluded that there exists a difference between work space and joint space power factor.

TABLE II.   POWER FACTOR COMPARISON BETWEEN DIFFERENT STRUCTURES.

		Power factor	Reactive power factor
XY	X axis	-0.103	0.994
	Y axis	-0.212	0.977
2 link	X axis	0.972	0.234
	Y axis	-0.862	0.506

TABLE III.   POWER FACTOR COMPARISON BETWEEN WORK AND JOINT SPACES.

		Power factor	Reactive power factor
Work space	X axis	0.972	0.234
	Y axis	-0.862	0.506
Joint space	Joint 1	0.0542	0.998
	Joint 2	0.423	0.905

Robot structure needs to be considered carefully in order to perform efficient motion from the view point of energy transmission. It is better that axis of joint space and work space to be identical.

VI.   CONCLUSION

This paper took care of power factor in mechanical system especially focusing on free motion and work space cases. We have conducted acceleration reference approach to well explain the free motion power factor, however based on expriment, it turned out approach explained in this study did not have significant effect expressing power factor in free motion. Additionally, this study explained that power factor differs between work space and joint space. Experiment result supported the explanation. Further research about both topics is necessary.

ACKNOWLEDGEMENT

This research was supported in part by the Ministry of Education, Culture, Sports, Science and Technology of Japan under Grant-in-Aid for Scientific Research (S), 25220903, 2013.

REFERENCES

[1] B. Ganji, A. Z. Kouzani, and H. Khayyam : "Look-ahead Intelligent Energy Management of A Parallel Hybrid Electric Vehicle," *IEEE/ASME Trans. Mechatron.*, Vol. 16, No. 6, pp. 1002–1010, 2011.

[2] H. Fujimoto and H. Sumiya : "Range extension control system of electric vehicle based on optimal torque distribution and cornering resistance minimization," *Proceedings of the 37th Annual Conference on IEEE Industrial Electronics Society*, pp.3858–3863, 2011.

[3] M. Hardt, K. Kreutz-Delgado, and J. W. Helton : "Minimal energy control of a biped robot with numerical methods and a recursive symbolic dynamic model," *Proceedings of the 37th IEEE Conference on Decision and Control 1998.*, Vol. 1, pp.413–416, 1998.

[4] C. Santacruz and Y. Nakamura : "Walking motion generation of humanoid robots: Connection of orbital energy trajectories via minimal energy control," *IEEE International Conference on Fuzzy Systems (FUZZ) 2011*, pp.2335–2341, 2011.

[5] D. Verscheure, B. Demeulenaere, J. Swevers, J. De Schutter, and M. Diehl : "Time-Optimal Path Tracking for Robots: A Convex Optimization Approach," *IEEE Transactions on Automatic Control*, Vol. 54, No. 10, pp. 2318–2327, 2009.

[6] C. Changrak, D. Chatzigeorgiou, R. Ben-Mansour, and K. Youcef-Toumi : "Design and analysis of novel friction controlling mechanism with minimal energy for in-pipe robot applications," *IEEE International Conference on Robotics and Automation (ICRA) 2012*, pp.4118–4123, 2012.

[7] W. R. Provancher, S. I. Jensen-Segal, and M. A. Fehlberg : "ROCR: An Energy-Efficient Dynamic Wall-Climbing Robot," *IEEE/ASME Trans. Mechatron.*, Vol. 16, No. 5, pp. 897–906, 2011.

[8] A. Vergnano, C. Thorstensson, B. Lennartson, P. Falkman, M. Pellicciari, F. Leali, and S. Biller : "Modeling and Optimization of Energy Consumption in Cooperative Multi-Robot Systems," *IEEE Transactions on Automation Science and Engineering*, Vol. 9, No. 2, pp. 423–428, 2012.

[9] T. Mizoguchi, T. Nozaki, and K. Ohnishi : "Power Factor in Mechanical System," *IEEE International Conference on Mechatronics 2013*, pp. 576–581, 2013.

[10] T. Mizoguchi, T. Nozaki, and K. Ohnishi : "Power Factor Analyses in Mechanical System Focusing on Trajectory and Environment," *IEEE Internation Simposium on Industrial Electronics 2013*, pp. 1–6, 2013.

Analysis of Settling Behavior and Design of Cascaded Precise Positioning Control in Presence of Nonlinear Friction

Michael Ruderman and Makoto Iwasaki
Department of Computer Science and Engineering
Nagoya Institute of Technology
Nagoya, Japan
ruderman.michael@nitech.ac.jp

Abstract—In several robotics and machinery applications the design of positioning control constitutes a trade-off between the requirements posed on a fast transient response and accuracy in settling. Mostly, neither 'universal' control gains can be found equally suitable for both objectives, so that often gain-scheduling strategies are used, in particularly for the inner velocity loop. However, this can be cumbersome from a systematic and robust control design point of view, and often no analytical solutions are available to ensure the fast and accurate settling. In this paper, the design of precise positioning control is presented which uses an additional feed-forward friction observer (FFFO) in the inner velocity loop. The FFFO approach allows efficiently compensating for nonlinear friction and can be appointed as a plug-in, after designing the surrounding feedback control. Here, the standard cascaded P-PI control is taken as a reference control system. The design of cascaded positioning control is discussed and analyzed in view of the settling behavior. Exposing in details the closed-loop dynamics we discuss the shortcomings of cascaded P-PI and PI-PI feedback regulators in presence of nonlinear presliding friction. The proposed control strategy is evaluated experimentally on a linear stage with drive velocity of 500 mm/s and micrometer positioning accuracy.

Keywords—*Motion control, dynamic friction, precise positioning system, feedforward friction observer*

I. INTRODUCTION

A fast transient response and short settling time are nearly always the opposite objectives of precise positioning control used in robotics and machinery. In several applications such as data storage devices, machine tools, manufacturing tools for electronics components, and industrial robots the required specifications in motion performance, e.g. response and settling time and trajectory and settling accuracy, should be sufficiently achieved [1]. Mostly, neither 'universal' control gains can be found equally suitable for both, fast transient and settling behavior within the specified error band. Usually, the gain-scheduling strategies (see [2], [3] for survey) are used in various industrial applications, in particular when designing the inner velocity control loop. However, these can be cumbersome in design and analysis and often involve several ad hoc steps. Among other related challenges, the linearization gain scheduling depends on

intuitive rules of thumb and extensive simulations for evaluation of stability and performance [3].

For the most part, the settling response of positioning control is affected by nonlinear presliding friction whose appearance is characterized by hysteresis in displacement after each motion reversal (see [4] for details on friction dynamics). Several control strategies have been proposed to overcome the motion nonlinearities, in particular friction. So a gain-scheduling control of systems with dynamic friction has been proposed in [5], where a gain-scheduled linearized friction estimator has been combined with the PI velocity control, and the reference velocity appears as the scheduling variable. However, only the simulation results have been shown, and the overall control design bases on the assumption that the dynamic friction complies the LuGre-modeled behavior. Adjacently, the gain-scheduling control for mechatronic systems with position-dependent dynamics has been proposed in [6]. Here, the presented global gain-scheduling controller is based on transforming the poles, zeros, and gains of local controllers, designed in fixed operating points, into the varying state-space matrices with position as scheduling parameter. This technique proved to be efficient for position-dependent modal system characteristics, but is doubtful in case of nonlinear friction, due to the lack of an appropriate control design for the fixed operating points. An efficient adaptive friction compensation [7] can be used also for positioning, however, assuming the friction uncertainties basically in the Coulomb friction term and not position-dependent. Another possible way to compensate for presliding friction is to observe explicitly the corresponding presliding state and to inject it to the control signal as recently proposed in [8]. The problem of positioning settling performance in presence of nonlinear friction has been also explicitly addressed in [9]. A split initial value compensation has been proposed using the mode-switching control in vicinity to the target position.

Even if it is not immediately evident, an habitual use of integral control term does not really contribute to improvement of the settling behavior in presence of nonlinear friction. On contrary, the so-called hunting limit cycles can be induced through the integral control action, and the linear feedback control tends to micro-oscillations about the target position. The phenomenon of friction-induced hunting limit cycles has been explicitly addressed in [10], while taking the LuGre and static switch friction models for analysis. However, to the best of our knowledge, the problem of positioning settling has not been fully explored up to the present, above all due to the complex and not completely understood presliding friction mechanisms.

The aim of this contribution is to analyze the settling behavior of common positioning controls with cascaded P-PI or PI-PI structures in presence of nonlinear friction. Furthermore, we show a possible way for compensating the nonlinear friction by adding the feed-forward friction observer (FFFO) recently proposed in [11]. It is worth noting that the recent work addresses the feedback control loop only, that is without considering the feed-forward control dynamics (see [1] for 2-DoF control structure). As implication, a transient overshoot in the recent positioning is deliberately allowed. In the following, the closed-loop dynamics in presence of nonlinear presliding friction is analyzed in details in Section II. After giving the main details on the experimental linear stage system in Section III, the positioning control is addressed in Section IV. The experimental control evaluation is described in Section V. Some concluding remarks are drawn in Section VI.

II. ANALYSIS OF SETTLING BEHAVIOR

Consider a simple one degree-of-freedom motion system described by the following equation

$$m\,\ddot{x}(t) = u(t) - F\big(\dot{x}(t), z(t)\big)\,, \qquad (1)$$

with moving mass m and friction F. Assume that the latter is dynamic nonlinear map of the relative velocity $\dot{x}$ and internal state z which represents a relative displacement after each motion reversal. It is worth noting, that the assumed friction function complies with generalized phenomenological friction model structure established in [4]. Further, assume that the system is forced by a scalar control value u which is the output of inner (velocity) feedback control loop.

A. P-PI control loop dynamics

Substituting the conventional PI velocity control

$$u(t) = K_p\left(\dot{x}_r(t) - \dot{x}(t)\right) + K_i \int \left(\dot{x}_r(t) - \dot{x}(t)\right)dt \quad (2)$$

and the surrounding P positioning control

$$\dot{x}_r(t) = P\big(x_r(t) - x(t)\big) \qquad (3)$$

into (1), one obtains the closed-loop dynamics of the total cascaded control system as

$$m\,\ddot{x} + K_p\,\dot{x} + (K_i + PK_p)\,x + PK_i \int x\,dt +$$
$$+ F(\dot{x}, z) = P\left(K_p\,x_r + K_i \int x_r\,dt\right)\,. \quad (4)$$

Note that in (4), and further on, we omit the time argument for the sake of simplicity. Taking the time derivative of (4) and assuming $\dot{x}_r = 0$ for the reference,

the latter due to positioning task we are interested in, one obtains the control system dynamics as

$$m\,\dddot{x} + K_p\,\ddot{x} + (K_i + PK_p)\,\dot{x} + PK_i\,x + \dot{F}(\dot{x}, z) =$$
$$= PK_i\,x_r\,. \quad (5)$$

Using convenient tools of linear control theory [14], e.g. Routh's stability criterion, one can easily determine the stable control gains on the left-hand side of (5), when zero or linear friction dynamics is assumed. Note that the first case, i.e. $\dot{F} = 0$, corresponds to the constant Coulomb friction at unidirectional motion, and the second case, i.e. $\dot{F} = D\ddot{x}$, constitutes the linear viscose friction with the corresponding coefficient D. For both, the steady-state convergence, i.e. $x = x_r$, can be guarantied in time predefined by an appropriate control gain selection.

B. Presliding friction dynamics

Now let us examine the friction dynamics in more details, in particular in the range close to the controlled position settling, that is in presliding. Taking the total derivative with respect to the time one can easily obtain

$$\dot{F}(\dot{x}, z) = \frac{\partial F}{\partial \dot{x}}\ddot{x} + \frac{\partial F}{\partial z}\dot{x}\,. \qquad (6)$$

It can be seen, that even if the first right-hand side term in (6) can be neglected at settling, the second term remains significant at non-zero velocities. Note that the above neglecting of the first term is due to low accelerations and presliding friction as a predominantly function of the relative displacement i.e. $\partial F/\partial \dot{x} \ll \partial F/\partial z$. It means

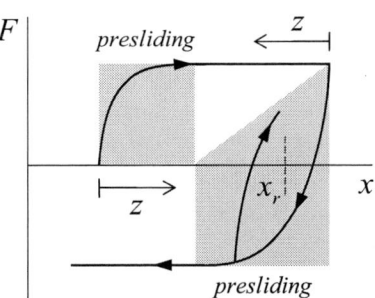

Fig. 1. Presliding friction curves at motion onset and motion reversals

that whenever at settling motion the impact of friction derivative contributes substantially to the control loop dynamics. Note that the impact becomes even larger as closer the system to the motion reversal is. The latter can be explained by means of presliding friction curves schematically shown in Fig. 1. Recall that the bound between presliding and sliding phases is not explicitly defined and is rather characterized by zero friction dynamics (see [12] for details).

Hardly to expect that a fast feedback positioning control provides zero transient overshoot on a large moving distance. Otherwise, an underdetermined control will lead to a untimely motion stop which has a similar impact of presliding friction since a novel motion onset should be enforced for reaching the reference position. Generally, we assume that a transient overshoot is present even if it is infinitesimal small, however beyond the sensing resolution of the feedback control. That is an inevitable motion reversal occurs after which the friction curve undergoes a nonlinear hysteresis pathway as exemplarily shown in Fig. 1. It is evident that within presliding the $\partial F/\partial z$ term is considerable and above all non-constant, that reveals the control performance of system (5) as less efficient. Doubtless, one can try to reduce the impact of friction dynamics by increasing the K_i and P control gains. However, the latter can be bounded by actuator limits and also worsen the overall control performance.

C. PI-PI control loop dynamics

When applying the outer PI positioning control

$$\dot{x}_r = P(x_r - x) + I \int (x_r - x)\, dt \qquad (7)$$

to the same (inner) velocity control (2) one obtains the overall closed-loop dynamics as

$$m\,\dddot{x} + K_p\,\dddot{x} + (K_pP + K_i)\,\ddot{x} + (K_pI + K_iP)\,\dot{x} + \\ + K_iI\,x + \ddot{F}(\dot{x}, z) = r(\dddot{x}_r, \dot{x}_r, x_r)\,. \qquad (8)$$

Note that the right-hand side of (8), which is

$$r(\dddot{x}_r, \dot{x}_r, x_r) = K_pP\,\dddot{x}_r + (K_pI + K_iP)\,\dot{x}_r + K_iI\,x_r, \qquad (9)$$

constitutes the reference-based excitation of the closed-loop dynamics. Here again, one can recognize that the nonlinear friction term in (8) disturbs the (exponential) convergence of the output position towards the reference value. Important to note is that, however, the second-order friction term in (8) has to be considered. It means that if the presliding friction dynamics complies $\dot{F} = const$, no frictional impact on the closed-loop behavior has to be expected. However, this cannot be generally assumed in view of the presliding hysteresis curvature as in Fig. 1.

In order to analyze the impact of presliding friction dynamics in more details consider

$$\ddot{F}(\dot{x}, z) = \frac{d}{dt}\dot{F}(\dot{x}, z) = \frac{\partial F}{\partial \dot{x}}\ddot{x} + \frac{\partial F}{\partial z}\ddot{x}\,. \qquad (10)$$

It is evident that even if the first (high-order) right-hand side term in (10) can be neglected, this for the same reasons as in Section II-B, the second term yields substantial at motion onsets and motion reversals. Neglecting the hight-order terms in (8) the eigendynamics of the closed control loop can be approximated by

$$\left(K_pP + K_i + \frac{\partial F}{\partial z}\right)\ddot{x} + (K_pI + K_iP)\,\dot{x} + K_iI\,x = 0. \qquad (11)$$

It can be easily recognized that (11) constitutes the second-order dynamic system of the form

$$c(z)\,\ddot{x} + d\,\dot{x} + k\,x = 0\,, \qquad (12)$$

where the stiffness k and damping d are determined by the corresponding control gains. At the same time, the inertia c is not longer constant and depends on the relative displacement z within presliding operation range. Recall that ones the nonlinear friction saturates the $\partial F/\partial z$ term becomes zero and $c = K_pP + K_i$. However, the inertial

term during presliding cannot be fully determined by the control gain selection, that has a substantial impact on the positioning response. This will be exposed in the following by analyzing $(x, \dot{x})$ trajectories of the control system (12). Recall that $(x, \dot{x})$ represents the simplified dynamics of the closed-loop control system that, however, allows well to demonstrate qualitatively the impact of presliding friction on the positioning settling.

D. Motion trajectories at controlled settling

Now, let us analyze and compare the trajectories of control system (12), once with and once without impact of presliding friction. Recall that the impact of presliding friction reflects in the $\partial F/\partial z$ term which influences significantly the dynamics (12) at motion onsets and motion reversals. The following control parameters are assumed for the numerical simulation: $K_p = 0.01$, $K_i = 0, 1$, $P = 100$, and $I = 1000$. Note that the selected control parameters, excepting the integral positioning term I, are the same as used by the experimental evaluation described in Section V. The simulated presliding term is assumed to be invers to the relative displacement so that $\partial F/\partial z \sim z^{-1}$. Here it is worth noting that another functional maps, e.g. exponential one, are equally thinkable to represent $\partial F/\partial z$ which would comply with presliding hysteresis curves. Most important is that $\partial F/\partial z$ possesses a relative high value, or even $\rightarrow \infty$, at $z \rightarrow 0$, and is then rapidly decreasing as z exceeds a relative small presliding distance. Since (12) constitutes a free system, its dynamic response is simulated by setting an initial value $x(0) \neq 0$. Recall that this is equivalent to applying a constant position reference. Two cases are considered $x(0) = 0.1$ and $x(0) = 0.01$ that means a micro-positioning for relative displacements differing in one order of magnitude. The $(x, \dot{x})$ trajectories are shown in Fig. 2 for both cases. It can be seen that for $\partial F/\partial z = 0$

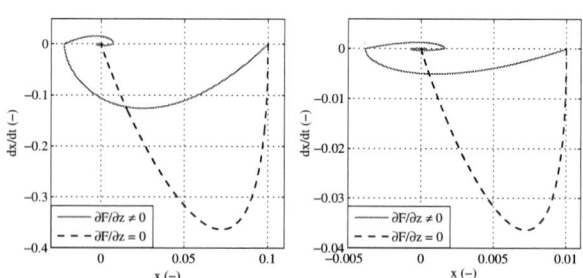

Fig. 2. Simulated trajectories of system (12) with $(\partial F/\partial z \neq 0)$ and without $(\partial F/\partial z = 0)$ impact of presliding friction

a pure linear behavior results in a direct acceleration and deceleration trajectory without positioning overshoot. On contrary when $\partial F/\partial z \neq 0$, several oscillating cycles occur before the trajectory approaches the equilibrium state. Further, it can be seen that the relative (percental) positioning overshoot becomes even larger when the relative displacement decreases, i.e. in case $x(0) = 0.01$. At the same time, the reached maximal velocity magnitude decreases, which results in a clearly slower positioning. The latter becomes particularly evident when inspecting the position time series in Fig. 3. For both relative displacements, $x(0) = 0.1$ and $x(0) = 0.01$, the linear case ($\partial F/\partial z = 0$) provides a fast exponential convergence to the idle state. In case $\partial F/\partial z \neq 0$ the control system exhibits a long-term nonlinear oscillation pattern, obviously due to the impact of presliding friction. It should be stressed that the convergence time increases dramatically by the decreased positioning distance. The flat extrema of oscillating pattern indicate the appearance of stick-slip behavior in view of the integral control action.

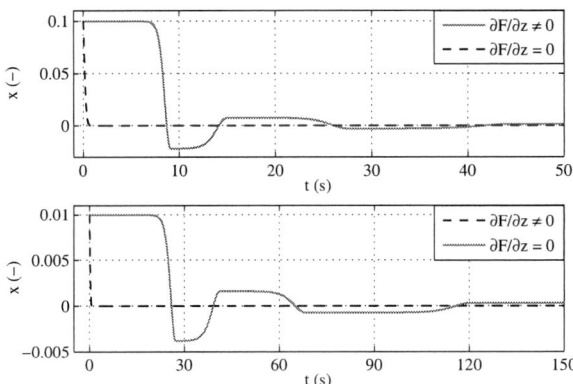

Fig. 3. Simulated position settling of system (12) with ($\partial F/\partial z \neq 0$) and without ($\partial F/\partial z = 0$) impact of presliding friction

From the performed analysis it can be concluded that smaller the considered micro-positioning range is, less efficient is the linear feedback regulation even in presence of integral control terms. Further it seems obvious that the incorporation of integral control term into the outer position loop does not allow efficiently to compensate for nonlinear presliding friction during the settling.

III. LINEAR STAGE SYSTEM

The linear stage system used is this work is schematically represented in Fig. 4. This constitutes a standard industrial linear guide actuator based on the ball-screw transmission, often also denoted as table drives or linear axes. The moving carrier with a payload is actuated by the BLDC (brushless direct current) motor directly coupled to the ball-screw shaft. The key data of the system are taken from the technical specification and listed in Table I. The angular motor displacement and corresponding angular velocity are captured by the motor-embedded 16-bit serial encoder connected to the motor drive amplifier. Accordingly, the theoretical sensor-based resolution of the linear motion is about 0.3 μm. However, the real positioning repeatability is limited by the mechanical backlash in ball-screw, which nominal value accounts for 20 μm. The control implementation and evaluation are performed using the DS1104CLP dSpace realtime board with the sampling time set to 500 μs.

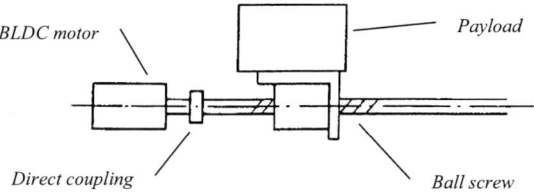

Fig. 4. Schematic representation of linear stage system

TABLE I. TECHNICAL SPECIFICATIONS OF LINEAR STAGE SYSTEM

maximal motor torque	1.9	Nm
rated motor torque	0.64	Nm
rated motor current	1.5	A
ball-screw pitch lead	20	mm
maximal speed of table drive	1480	mm/s
maximal stroke of table drive	400	mm

IV. POSITIONING CONTROL

A. Control structure

The applied positioning control is shown in Fig. 5. This constitutes a common cascaded P-PI structure extended by the FFFO to compensate for nonlinear dynamic friction. Note that the FFFO is driven (see [11] for details) by the output of position control, which in turn constitutes the velocity reference value. The observed friction disturbance $\tilde{f}$, used in FFFO, is provided by the observer whose inputs are the control value u and measured relative velocity $\dot{x}$. Examining the control structure it becomes evident that the applied FFFO scheme can be easily plugged in a common positioning control, which can be designed separately. This appears as particularly suitable for several industrial applications, where the standard PI-type cascaded controls represent a common solution implemented on the dedicated hardware.

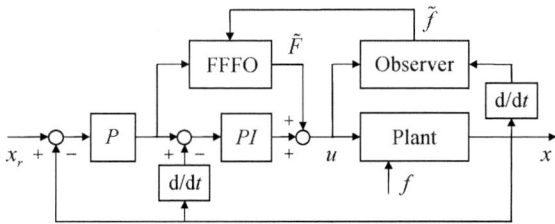

Fig. 5. Block diagram of cascaded positioning control with FFFO

B. Feed-forward friction observer (FFFO)

The feed-forward friction observer (FFFO), first proposed in [13] and elaborated in [11], allows to combine both, the model-based friction feed-forwarding and observer-based estimation of friction disturbances. In the following, the applied FFFO is briefly summarized for convenience of the reader. For more details, however, we refer to [13], [11].

The unknown time-variant friction part $f(t)$ can be assumed as an input disturbance so that the plant dynamics will be captured by

$$m\,\ddot{x}(t) + F\big(\dot{x}(t)\big) + f(t) = u(t)\,. \qquad (13)$$

After determining the plant description, i.e. $\tilde{m}$ and $\tilde{F}(\cdot)$ terms[1], the friction disturbance can be observed by

$$\tilde{f}(t) = u(t) - \tilde{m}\,\ddot{x}(t) - \tilde{F}\big(\dot{x}(t)\big)\,. \qquad (14)$$

It is evident, that the computed disturbance observation will equally contain the (spurious) high-frequent components, which are mainly due to the time derivative of the measured output value and high-order harmonics which are not captured by the simplified model (14). However, these are not interfering the output of FFFO because of the inherent low-pass characteristics of the modeled dynamic friction, as explained in details in [11].

The feed-forward friction observer is given by

$$\dot{\tilde{F}}(t) + \frac{B\,|\dot{x}_{in}|}{|S(\dot{x}_{in})|}\,\tilde{F}(t) = \dot{\tilde{z}}_1(t) + B\,\dot{x}_{in} + L\tilde{f}(t)\,. \qquad (15)$$

Here, the attributes of 2SEP dynamic friction model (see [12], [11] for details) are the weighting parameter B, dynamic presliding friction state $\tilde{z}_1$, and static Stribeck characteristic curve $S(\cdot)$. Note that the input velocity value $\dot{x}_{in}$ depends on the placement of FFFO within the control loop. Here it constitutes the output of P-control, according to Fig. 5.

As described in [11] the observation gain

$$L = \frac{B\,|\dot{x}_{in}|}{|S(\dot{x}_{in})|} \qquad (16)$$

brings the observer excitation on the same scale as the output friction force (see left- and right-hand sides in (15)), thus ensuring the computed friction at steady-state to contain both parts

$$\tilde{F} = S(\dot{x}_{in}) + \tilde{f}\,, \qquad (17)$$

i.e. after the transient response. Note, that the observation gain L is computed analytically and does not require any design parameters to be selected. For further details on FFFO properties the reader is refereed to [11].

C. Linear feedback control

The linear feedback control, i.e. the P-PI one, is designed in two consecutive stages, starting by the inner velocity loop and following by the outer positioning loop. Having a linear plant approximation $G(j\omega) = X(j\omega)/U(j\omega)$, which is in simple case a first-oder time delay transfer element, the PI velocity control gains can be easily assigned using e.g. pole-placement, bode diagrams, and other convenient techniques of the classical control theory [14]. Note, that besides the actuator limits the admissible control gains can be bounded by the impact of dead-time (see e.g. [15] for details), which is frequently encountered in digital control systems. Note, that a possibly large selection of integral gain argues for a better cancelation of the steady-state friction but not necessarily of the presliding one. After designing and evaluating the inner velocity loop an appropriate position control gain P can be easily selected taking into account (i) the maximal relative velocity of the plant, and (ii) acceptable control overshoot before settling. It is worth to recall, that an additional integral term in the outer positioning loop does not improve the cancelation of nonlinear friction as has been analyzed in Section II.

V. Experimental evaluation

The designed P-PI cascaded control and the same control extended by FFFO (see block diagram in Fig. 5) have been evaluated on the linear stage system described in Section III. Before the control design and evaluation, the system has been identified once to determine the friction model parameters used in FFFO and once to obtain a linear plant approximation required for tuning the feedback control gains. Two different motion experiments have been accomplished to identify the system parameters. First, a low-gain closed loop position tracking motion has been performed so as to determine the Coulomb friction level and initial hysteresis stiffness of 2SEP friction model (see [12] for details). Second, the frequency-response measurements at different excitation levels have been done in the range 1–100 Hz to determine the residual nonlinear and linear system parameters. For more details on the frequency-domain system identification in presence of nonlinear friction we refer to [16].

The measured frequency response function at higher excitation amplitude[2] is exemplarily shown in Fig. 6 versus the linear plant approximation $G(j\omega)$. The determined, based thereon, feedback control parameters are as following: $K_p = 0.01$, $K_i = 0.1$, and $P = 100$.

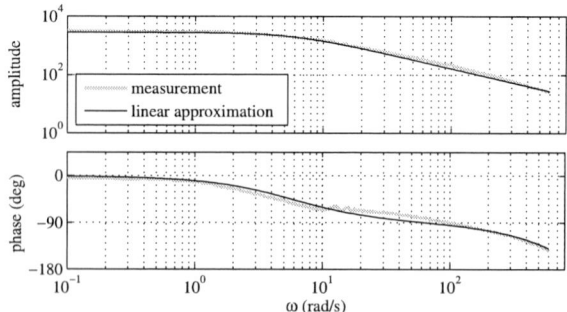

Fig. 6. Measured frequency response function and linear plant approximation G at higher excitation amplitude

The motion profile shown in Fig. 7 has been taken for the control evaluation. This constitutes the bidirectional positioning of different distances with a constant velocity

---

[1]Here the dynamic friction, in the input-output notation, is explicitly denoted as a function of relative velocity only. The z argument, as in (1), is deliberately skipped since this constitutes an internal friction state.

[2]Note that higher the excitation level is, lower is the impact of nonlinear friction (see e.g. [11]). Therefore, the sufficiently excited frequency response of the system with friction nonlinearities approaches the corresponding linear transfer function characteristics

set to ± 500 mm/s. Note, that the latter has been selected so as to protect the mechanical structure of linear stage system (compare with the data in Table I). Two independent experiments, i.e. with different initial positions, have been repeated, each one when using the P-PI and then P-PI-FFFO controller. The close-ups of settling behavior at 250 mm and 75 mm reference (see Fig. 7 in the middle and below) disclose the inferior performance of the P-PI positioning control. It is evident that a slow settling after the transient overshoot relates to the nonlinear friction in presliding and cannot be efficiently compensated by a linear feedback control, as discussed before in Section II. On the contrary, the P-PI-FFFO control is free of a slow settling and, at the same time, has only a slightly higher overshoot than the P-PI control.

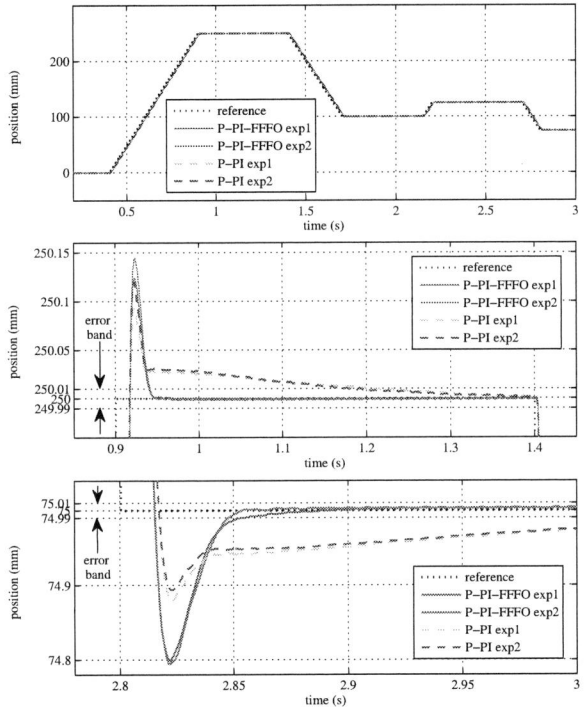

Fig. 7. Experimental positioning response using P-PI and P-PI-FFFO controllers. The overall positioning trajectory above and close-ups of settling response in the middle and below

VI. CONCLUSIONS

In this paper, we have analyzed the settling behavior of precise positioning control in presence of nonlinear friction. The design of cascaded P-PI feedback control has been addressed, also using the plugged-in feed-forward friction observer (FFFO), which allows compensating for dynamic nonlinear friction on the level of inner velocity loop. The proposed method has been described in context of a linear stage system with one degree-of-freedom which offers a micrometer positioning accuracy and, at the same time, provides the set drive velocity of 500 mm/s. We have shown by theoretical analysis and experimental evaluation that a pure integral control action

in feedback cannot be sufficient for the fast compensation of nonlinear friction at the controlled positioning settling.

The experimental control evaluation confirmed the analyzed shortcomings of cascaded feedback control without explicit friction compensation. We suppose that this is not matter of an appropriate control gain selection, since the presliding friction mechanisms generally antagonize the principles of linear feedback regulation. However, a significant improvement in settling behavior of the positioning control could be demonstrated when applying the FFFO compensation method.

REFERENCES

[1] M. Iwasaki, K. Seki, and Y. Maeda, "High-precision motion control techniques: A promising approach to improving motion performance," *IEEE Industrial Electronics Magazine*, vol. 6, no. 1, pp. 32–40, 2012.

[2] K. Astrom, T. Hagglund, C. Hang, and W. Ho, "Automatic tuning and adaptation for PID controllers - a survey," *Control Engineering Practice*, vol. 1, no. 4, pp. 699–714, 1993.

[3] W. J. Rugh and J. S. Shamma, "Research on gain scheduling," *Automatica*, vol. 36, no. 10, pp. 1401–1425, 2000.

[4] F. Al-Bender and J. Swevers, "Characterization of friction force dynamics," *IEEE Control Systems Magazine*, vol. 28, no. 6, pp. 64–81, Dec. 2008.

[5] C. Vivas, F. Rubio, and C. Canudas-De-Wit, "Gain-scheduling control of systems with dynamic friction," in *IEEE 41st Conference on Decision and Control*, vol. 1, 2002, pp. 89–94.

[6] B. Paijmans, W. Symens, H. Van Brussel, and J. Swevers, "A gain-scheduling-control technique for mechatronic systems with position-dependent dynamics," in *American Control Conference*, 2006, pp. 2933–2938.

[7] B. Friedland and Y.-J. Park, "On adaptive friction compensation," *IEEE Transactions on Automatic Control*, vol. 37, no. 10, pp. 1609–1612, 1992.

[8] M. Ruderman and M. Iwasaki, "Control of pre-sliding friction using nonlinear state observer," in *IEEE 13th International Workshop on Advanced Motion Control (AMC)*, 2014, p. n.n.

[9] Y. Maeda, M. Wada, M. Iwasaki, and H. Hirai, "Improvement of settling performance by mode-switching control with split initial-value compensation based on input shaper," *IEEE Transactions on Industrial Electronics*, vol. 60, no. 3, pp. 979–987, 2013.

[10] R. H. A. Hensen, M. J. G. Van de Molengraft, and M. Steinbuch, "Friction induced hunting limit cycles: A comparison between the LuGre and switch friction model," *Automatica*, vol. 39, no. 12, pp. 2131–2137, 2003.

[11] M. Ruderman, "Tracking control of motor drives using feed-forward friction observer (FFFO)," *IEEE Transactions on Industrial Electronics*, vol. 61, no. 7, pp. 3727–3735, 2014.

[12] M. Ruderman and T. Bertram, "Two-state dynamic friction model with elasto-plasticity," *Mechanical Systems and Signal Processing*, vol. 39, no. 1–2, pp. 316–332, 2013.

[13] M. Ruderman and T. Bertram, "Feed-forward friction observer (FFFO) for high-dynamic motion control," in *IEEE 20th Mediterranean Conference on Control and Automation*, 2012, pp. 1013–1018.

[14] G. F. Franklin, J. D. Powell, and E. A. Naeini, *Feedback Control of Dynamic Systems*, 6th ed. Prentice Hall, 2009.

[15] M. Ruderman and T. Bertram, "Variable proportional-integral-resonant (PIR) control of actuators with harmonic disturbances," in *IEEE International Conference on Mechatronics (ICM)*, 2013, pp. 846–851.

[16] M. Ruderman and T. Bertram, "FRF based identification of dynamic friction using two-state friction model with elasto-plasticity," in *IEEE International Conference on Mechatronics (ICM)*, 2011, pp. 230–235.

Field and Bench Test Evaluation of Range Extension Control System for Electric Vehicles Based on Front and Rear Driving-Braking Force Distributions

Hiroshi Fujimoto and Shingo Harada
The University of Tokyo
5-1-5, Kashiwanoha, Kashiwa, Chiba, Japan
fujimoto@k.u-tokyo.ac.jp
Koji Sato and Yusuke Matsuo
Ono Sokki Co.,Ltd.
1-16-1, Hakusan-midoriku, Yokohama, Kanagawa, Japan

Yuichi Goto and Daisuke Kawano
National Traffic Safety and Environment Laboratory
7-42-27, Jindaiji-higashimachi, Chofu, Tokyo, Japan

Abstract—Electric vehicles (EVs) have a disadvantage in that the cruising distance per charge is short. This paper proposes a model-based range extension control system (RECS) for EVs. The proposed system optimizes the front and rear driving-braking force distributions by considering the slip ratio of the wheels and the motor loss. The optimal distribution depends solely on the vehicle acceleration and velocity. Therefore, this system is effective not only at constant speeds but also in acceleration and deceleration modes. Bench tests were conducted for more precise evaluation and to realize experimental results with high reproducibility. The effectiveness of the proposed system was verified through field and bench tests.

Keywords—*bench test, driving and braking force distribution, electric vehicle, range extension control system*

I. INTRODUCTION

Nowadays, EVs are receiving attention because of environmental concerns such as global warming, exhaustion of fossil fuels, and air pollution. In addition, EVs have remarkable advantages in motion control compared with internal combustion engine vehicles (ICEVs) [1]:

1) The response to the driving-braking force by the motor is much faster than that of engines (about 100 times).
2) Development of in-wheel motors enables the individual control of each wheel.
3) The generated torque can be measured precisely from the motor current.
4) Smooth braking torque can be generated by regeneration.

Research is actively ongoing on traction control [2], [3] and stability control [4], [5] to utilize the above advantages.

One reason that is preventing EVs from spreading is that its mileage per charge is shorter than that of conventional ICEVs. In order to solve this problem,

wireless power transfer for moving vehicles [6], [7], [8] is being researched. As another approach, ultracapacitors are being utilized for energy storage systems to improve the energy regeneration [9], [10], [11]. Research is also being carried out to improve the efficiency of motors [12]. In order to realize high-efficiency motor control, Inoue et al. examined torque and angular velocity patterns that maximize efficiency during acceleration and deceleration [13]. Yuan and Wang utilized the independent characteristics of traction motors to develop a torque distribution method for decreasing EV energy consumption where two motors with the same efficiency characteristics are used [14].

The authors' research group previously proposed the range extension control system (RECSs) [15], [16], which does not involve changes to the vehicle structure such as an additional clutch [14] or the motor type. Instead, the RECS extends the cruising range of a vehicle by motion control. Previously, the RECS was evaluated in terms of the acceleration and deceleration on a straight road [17]. The effectiveness of the proposed system was only verified for operation at low speeds. It has not been verified for operation at high speeds, where the ratios of the driving resistance and motor iron loss to the total loss are relatively large. Therefore, experiments on operation at high speeds are necessary for more appropriate evaluation of the RECS. In this study, a bench test was performed to realize high reproducibility of the results along with a field test to evaluate the proposed system [16]. The effectiveness of the proposed system was verified through the field and bench tests.

II. EXPERIMENTAL VEHICLE AND VEHICLE MODEL

A. Experimental Vehicle

This study used the original electric vehicle "FPEV–2 Kanon," which was developed in-house. This vehicle has four outer rotor–type in-wheel motors. Since these are direct drive–type motors, the reaction force from the

(a) FPEV–2 Kanon.

(b) dSPACE AutoBox.

(c) Front motor.

(d) Rear motor.

Fig. 1. Experimental vehicle.

TABLE I. VEHICLE SPECIFICATIONS.

Vehicle mass M	854 kg
Wheelbase l	1.715 m
Distance from CG to front/rear axles l_f, l_r	l_f:1.013 m l_r:0.702 m
Gravity height h_g	0.51 m
Front wheel inertia J_{ω_f}	1.24 Nms^2
Rear wheel inertia J_{ω_r}	1.26 Nms^2
Wheel radius r	0.302 m

road is directly transferred to the motor without backlash from the reduction gear.

Fig. 1 shows the experimental vehicle. The dSPACE AutoBox (DS1103) was used for real-time data acquisition and control. Table I and Table II show the specifications of the vehicle and in-wheel motors. Fig. 2 presents the efficiency map of the front and rear in-wheel motors. In this study, higher torque operation points where the motor torque was greater than one and half times the rated torque were not used for the evaluation. Since the front and rear motors installed in the vehicle were different, their efficiency maps were also different. Therefore, the cruising range can be extended by employing the difference in efficiency.

Fig. 3 illustrates the power system of the vehicle. A lithium-ion battery was used as the power source. The voltage of the main battery was 160 V (ten battery modules were connected in series). The voltage was

TABLE II. SPECIFICATIONS OF IN-WHEEL MOTORS.

	Front	Rear
Manufacturer	TOYO DENKI SEIZO K.K.	
Type	Direct drive system Outer rotor type	
Rated torque	110 Nm	137 Nm
Maximum torque	500 Nm	340 Nm
Rated power	6.0 kW	4.3 kW
Maximum power	20.0 kW	10.7 kW
Rated speed	382 rpm	300 rpm
Maximum speed	1113 rpm	1500 rpm

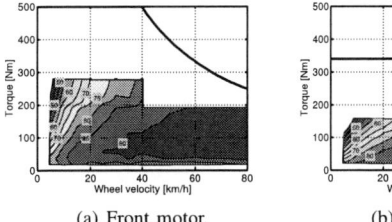
(a) Front motor.   (b) Rear motor.

Fig. 2. Efficiency maps of front and rear motors.

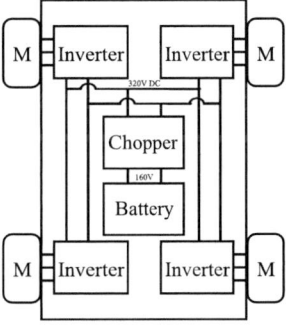

Fig. 3. Electric power system of vehicle.

boosted to 320 V by a chopper. In this study, the chopper loss was not evaluated because it was independent of the torque distribution.

B. Vehicle Model

The four wheel–drive vehicle model is described here. The wheel rotation is expressed by (1). For straight driving, the driving-braking forces of the right and left wheels are equal. Therefore, the vehicle dynamics is

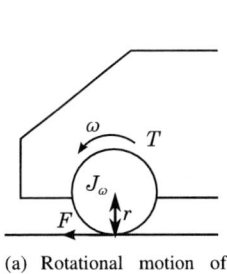
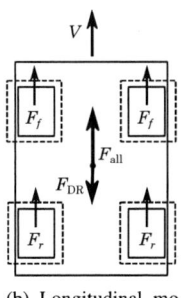

(a) Rotational motion of wheel   (b) Longitudinal motion of vehicle

Fig. 4. Vehicle model.

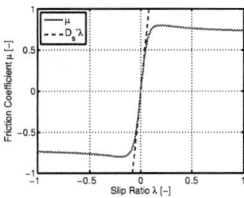

Fig. 5. Example of μ–λ curve.

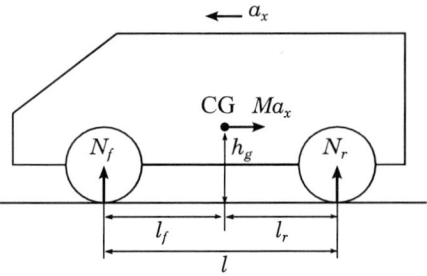

Fig. 6. Load transfer model.

expressed by (2) and (3).

$$J_{\omega_j}\dot{\omega}_j = T_j - rF_j, \quad (1)$$

$$M\dot{V} = F_{\text{all}} - F_{\text{DR}}, \quad (2)$$

$$F_{\text{all}} = 2\sum_{j=f,r} F_j, \quad (3)$$

where ω_j is the wheel angular velocity, V is the vehicle speed, T_j is the motor torque, F_{all} is the total driving-braking force, F_j is the driving-braking force of each wheel, M is the vehicle mass, r is the wheel radius, J_{ω_j} is the wheel inertia, and F_{DR} is the driving resistance. The subscript j represents f or r (f stands for front, and r represents rear).

Next, the slip ratio λ_j is defined as

$$\lambda_j = \frac{V_{\omega_j} - V}{\max(V_{\omega_j}, V, \epsilon)}, \quad (4)$$

where $V_{\omega_j} = r\omega_j$ is the wheel speed and ϵ is a small constant to avoid zero division. $\lambda_j > 0$ means driving, and $\lambda_j < 0$ means braking. The slip ratio λ is known to be related with the coefficient of friction μ, as shown in Fig. 5 [18]. In region $|\lambda| \ll 1$, μ is nearly proportional to λ. By using the normal forces of each wheel N_j during longitudinal acceleration with a_x and the slope of the curve, the driving force of each tire is expressed as

$$F_j = \mu_j(\lambda_j)N_j(a_x) \approx D_s'\lambda_j N_j(a_x), \quad (5)$$

where D_s' is the normalized driving stiffness.

The normal forces of each wheel during the longitudinal acceleration process are calculated as follows:

$$N_f(a_x) = \frac{1}{2}\left(\frac{l_r}{l}Mg - \frac{h_g}{l}Ma_x\right), \quad (6)$$

$$N_r(a_x) = \frac{1}{2}\left(\frac{l_f}{l}Mg + \frac{h_g}{l}Ma_x\right), \quad (7)$$

where N_f and N_r are the front and rear normal forces, respectively, l_f and l_r are the distances from the center of gravity to the front and rear axles, respectively, and h_g is the height of the center of gravity. The acceleration direction is defined as positive when the vehicle is accelerating.

C. Driving-Braking Force Distribution Model

During straight driving, the required total driving-braking force can be distributed to each wheel. Since the

EV motors were assumed to be independently controlled in this study, the driving-braking force distribution has an extra degree of freedom. By introducing the front and rear driving-braking force distribution ratio k, the driving-braking forces can be formulated based on the total driving-braking force F_{all} and the distribution ratio k as follows [16]:

$$F_j(k) = \frac{1}{2}\gamma_j(k)F_{\text{all}}, \quad (8)$$

$$\gamma_j(k) = \begin{cases} 1-k & (j=f) \\ k & (j=r) \end{cases}. \quad (9)$$

The distribution ratio k varies from 0 to 1. $k=0$ means that the vehicle is a front-driven system, and $k=1$ means that it is rear-driven only. Note that, even if the driving force F_j is zero, the torque T_j is not always zero according to (1).

D. Modeling of Inverter Input Power

The slip ratio and motor loss can be considered to derive the distribution ratio that minimizes the inverter input power. Neglecting the inverter loss and mechanical loss of the motor, the inverter input power P_{in} is expressed as

$$P_{\text{in}} = P_{\text{out}} + P_c + P_i, \quad (10)$$

where P_{out} is the sum of the mechanical output of each motor, P_c is the sum of the copper loss of each motor, and P_i is the sum of the iron loss of each motor. P_{out} is given by

$$P_{\text{out}} = 2\sum_{j=f,r} \omega_j T_j. \quad (11)$$

In the modeling of the copper loss P_c, iron loss was neglected for simplicity. Suppose that the magnet torque is much greater than the reluctance torque and that the q-axis current is much greater than the d-axis current; then, the sum of the copper loss of the permanent magnetic motors P_c is expressed as

$$P_c = 2\sum_{j=f,r} R_j i_{qj}^2, \quad (12)$$

where R_j is the armature winding resistance of the motor and i_{qj} is the q-axis current of the motor. Then, the following relationship between the q-axis current and torque is obtained:

$$i_{qj} = \frac{T_j}{K_{tj}} = \frac{T_j}{p_{nj}\Psi_j}, \quad (13)$$

where K_{tj} is the torque coefficient of the motor, p_{nj} is the number of pole pairs, and Ψ_j is the interlinkage magnetic flux. Therefore, the copper loss P_c is given by

$$P_c = 2\sum_{j=f,r} \frac{R_j T_j^2}{K_{tj}^2}. \quad (14)$$

In this study, the equivalent circuit model [19] was used to examine the iron loss. Fig. 7 shows the d- and q-axis equivalent circuits of the permanent magnetic motor.

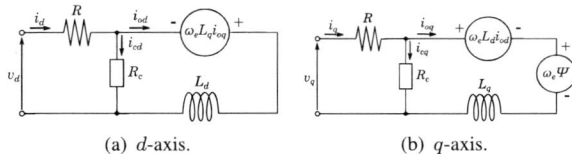

(a) d-axis.   (b) q-axis.

Fig. 7.   Equivalent circuit of PMSM.

From the circuits, the sum of iron loss P_i is expressed as

$$P_i = 2 \sum_{j=f,r} \frac{\omega_{ej}^2}{R_{cj}} \left\{ (L_{dj} i_{odj} + \Psi_j)^2 + (L_{qj} i_{oqj})^2 \right\},$$

(15)

where ω_{ej} is the electrical angular velocity of the motor, R_{cj} is the equivalent iron loss resistance, L_{dj} is the d-axis inductance, L_{qj} is the q-axis inductance, i_{odj} is the difference between the d- and q-axis current i_{dj}, i_{qj} and the d- and q-axis components of the iron loss current i_{cdj}, i_{cqj}, respectively [19]. In (15), the armature reaction of the d-axis $\omega_e L_d i_{od}$ is neglected since it is much smaller than the electromotive force of the magnet $\omega_e \Psi$. In the modeling of the iron loss, ω_{ej} was approximated as $p_{nj} V / r$ for simplicity since the slip ratio of each wheel was small. Under this condition, P_i is approximated by

$$P_i \approx 2 \frac{V^2}{r^2} \sum_{j=f,r} \frac{p_{nj}^2}{R_{cj}} \left\{ \left(\frac{L_{qj}}{K_{tj}} \right)^2 T_j^2 + \Psi_j^2 \right\}.$$

(16)

The equivalent iron loss resistance R_{cj} is expressed as

$$\frac{1}{R_{cj}(\omega_{ej})} = \frac{1}{R_{c0j}} + \frac{1}{R'_{c1j} |\omega_{ej}|}.$$

(17)

In (17), the first and second terms on the right-hand side represent the eddy current loss and hysteresis loss, respectively [20]. By applying $\omega_{ej} = p_{nj} V / r$, R_{cj} is expressed as $R_{cj}(V)$.

From the above equations, P_{in} is expressed as

$$P_{in} = P_{out} + P_c + P_i$$

$$= 2 \sum_{j=f,r} \omega_j T_j + 2 \sum_{j=f,r} \frac{R_j T_j^2}{K_{tj}^2}$$

$$+ 2 \frac{V^2}{r^2} \sum_{j=f,r} \frac{p_{nj}^2}{R_{cj}} \left\{ \left(\frac{L_{qj}}{K_{tj}} \right)^2 T_j^2 + \Psi_j^2 \right\}.$$

(18)

III. OPTIMIZATION OF FRONT AND REAR DRIVING-BRAKING FORCE DISTRIBUTIONS

A. Derivation of Optimal Distribution Ratio

The optimal driving-braking force distribution ratio that minimizes the input power of the inverter is derived here. To derive the optimal distribution ratio, the inertia force of each wheel is neglected in (1) because $J_{\omega_j} \dot{\omega}_j \ll r F_j$ under a high μ load. As noted in [21], the denominator of (4) can be approximated to V when $|\lambda_j| \ll 1$. Therefore, T_j and ω_j can be approximated as

$$T_j = r F_j,$$

(19)

$$\omega_j = \frac{V}{r} (1 + \lambda_j).$$

(20)

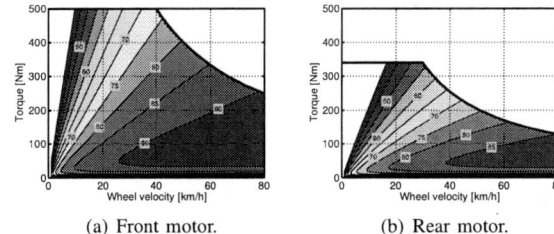

(a) Front motor.   (b) Rear motor.

Fig. 8.   Motor efficiency (calculated).

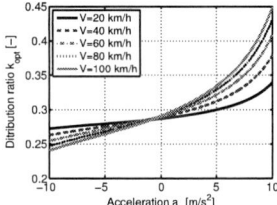

Fig. 9.   Optimal distribution ratio k_{opt}.

By applying the above approximation, P_{in} is obtained as $P_{in}(k)$ [17]. Since $P_{in}(k)$ is a quadratic function of k, the optimal distribution ratio k_{opt} satisfies $\partial P_{in} / \partial k |_{k=k_{opt}} = 0$. Therefore, k_{opt} is derived as a function of V and a_x:

$$k_{opt}(V, a_x) =$$

$$\frac{\dfrac{V}{D'_s N_f(a_x)} + \dfrac{r^2 R_f}{K_{tf}^2} + \dfrac{V^2}{R_{cf}(V)} \left(\dfrac{L_{qf}}{\Psi_f} \right)^2}{\dfrac{V}{D'_s} \sum_{j=f,r} \dfrac{1}{N_j(a_x)} + r^2 \sum_{j=f,r} \dfrac{R_j}{K_{tj}^2} + V^2 \sum_{j=f,r} \dfrac{1}{R_{cj}(V)} \left(\dfrac{L_{qj}}{\Psi_j} \right)^2}.$$

(21)

B. Numerical Calculation

Fig. 8 shows the calculation results of the experimental vehicle's motor efficiencies when R_{cj} was 300 Ω and R'_{c1f} and R'_{c1r} were 0.13 and 0.053 Ωs/rad, respectively. From Fig. 2 and Fig. 8, the modeling error of the motor efficiency was within $\pm 5\%$ for most evaluated operation areas. Moreover, the front motor had a higher global efficiency than the rear motor because the former can have a much smaller internal diameter than the latter. Therefore, the number of turns of the motor windings and the teeth shape can be optimized for the front motor design.

Fig. 9 shows the calculated k_{opt}. Under a high μ load, the normalized driving stiffness D'_s was set to 12. k_{opt} increased with the acceleration and decreased with increased deceleration. This is mainly because of the influence of the variation in the slip ratio due to load transfer and copper loss. On the other hand, k_{opt} increased with the vehicle velocity. The range of k_{opt} was 0.2–0.45. This is because the front motor had higher efficiency than the rear motor in a wide area of the efficiency map, as shown in Fig. 2.

(a) Wheel with bearing.

(b) Test vehicle and RC-S.

Fig. 10. Bench test environment.

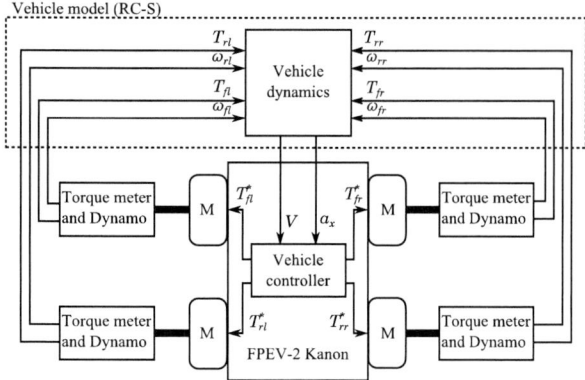

Fig. 11. Block diagram of experiment environment.

IV. EXPERIMENT

A. Test Field and Test Bench

A test field for vehicles owned by the National Traffic Safety and Environment Laboratory in Japan was used for the field test. This test field has a 1350 m long straight road, a low μ load, and a slope. This field allows experiments to be performed under various driving conditions. In this study, no-slope and high μ load conditions were employed for the evaluation.

In the bench test, the Real Car Simulation Bench (RC-S) owned by Ono Sokki Co.,Ltd. was used. Fig. 10 shows the bench test environment. In the experiments using RC-S, driving shafts were directly connected to dynamometers through a bearing wheel, which is different from the case of a chassis dynamometer. Fig. 10(a) shows the bearing wheel. By changing the vehicle model of RC-S, experiments can be conducted under various load conditions. In addition, RC-S can control dynamos with a faster response than a chassis dynamometer using rollers, which have greater inertia. Therefore, RC-S is suitable for bench test of electric vehicles driven by motors. In this research, the test bench was very useful because the experiments were not influenced by changes in the wind and load conditions.

Fig. 11 shows the block diagram of the experimental environment using RC-S. The motor torque of each wheel was measured by a torque meter and input to the vehicle model of RC-S. The velocity and acceleration of the vehicle were calculated by vehicle dynamics model in RC-S. In order to control the motors, these values were input to the vehicle controller.

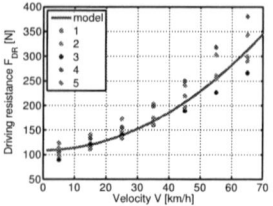

Fig. 12. Comparison of model and measurement results of driving resistance.

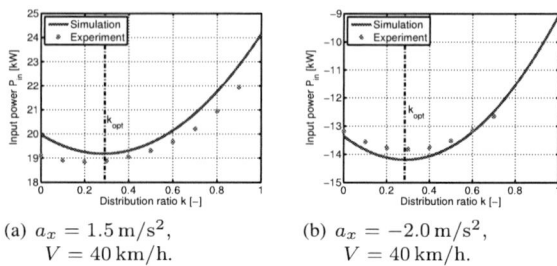

(a) $a_x = 1.5\,\mathrm{m/s^2}$, $V = 40\,\mathrm{km/h}$.

(b) $a_x = -2.0\,\mathrm{m/s^2}$, $V = 40\,\mathrm{km/h}$.

Fig. 13. Experimental result of P_{in}.

B. Driving Resistance

For the simulation and bench test, the driving resistance of the test vehicle was measured in the test field. The driving resistance F_{DR} can be determined by

$$F_{\mathrm{DR}}(V) = \mu_0 M g + \frac{1}{2}\rho C_d A V^2, \qquad (22)$$

where μ_0 is the rolling friction coefficient, ρ is the air density, C_d is the drag coefficient, and A is the frontal projected area. ρ and A were determined to be 1.205 kg/m^3 and 1.2 m^2, respectively. μ_0 and C_d were 1.28×10^{-2} and 0.863, respectively. These values were obtained empirically. Fig. 12 shows the measured and calculated driving resistance. The measurements were taken five times. As shown in the figure, the model described by (22) matched the measured values.

C. Input Power for Change in Distribution Ratio

Fig. 13 shows the experimental results of P_{in} when the distribution ratio was changed. This experiment was conducted using RC-S. The inverter input power P_{in} can be calculated as

$$P_{\mathrm{in}} = V_{\mathrm{dc}} \sum_{j=f,r} I_{\mathrm{dc}j}, \qquad (23)$$

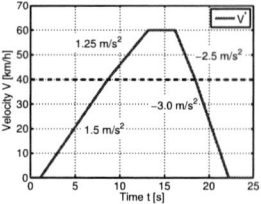

Fig. 14. Reference vehicle speed.

978-1-4799-2706-7/14 $31.00 © 2014 IEEE

The 2014 International Power Electronics Conference

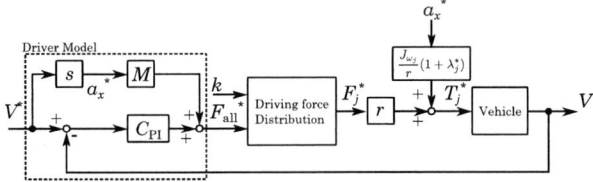

Fig. 15. Vehicle speed control system.

where V_{dc} is the inverter input voltage and $I_{\mathrm{dc}j}$ is the front and rear inverter input currents. Fig. 13 shows the results when a_x and V were 1.5 m/s^2 and 40 km/h, respectively, and -2.0 m/s^2 and 40 km/h, respectively. These conditions were simulated by RC-S. In Fig. 13, the rigid lines represent the calculation results of the computer simulation; here, the approximations presented above were not applied. The value at k_{opt} calculated by (21) is shown as a dashed line. Fig. 13 indicates that P_{in} is a convex function of k. Therefore, a k that minimizes P_{in} exists. In the simulation, although there were errors caused by the approximations of the torque, wheel angular velocity, copper loss, and iron loss, the case of $k = k_{\mathrm{opt}}$ showed almost equal minimum values, as shown in Fig. 13. The experimental results indicate that k_{opt} can mostly minimize the input power, although a little error remains present. Therefore, the approximations assumed in this study were appropriate.

D. Pattern Driving

To demonstrate the effectiveness of the proposed system, the driving cycle was evaluated with both the test field and test bench. Fig. 14 shows the driving cycle, which comprised two-step acceleration, cruising, and two-step deceleration. The accelerations were 1.5 and 1.25 m/s^2, the maximum vehicle speed was 60 km/h, and the decelerations were -2.5 and -3.0 m/s^2. The cases of $k = 0, 0.1, 0.2, 0.3, 0.4, 0.5,$ and k_{opt} were evaluated. In the bench test, the driving resistance was set to the value measured in the field test.

Fig. 15 shows the vehicle velocity control system for determining the vehicle velocity pattern in Fig. 14 during the field test. This system comprised a feedforward controller and a feedback controller. These controllers corresponded to the driver model. The input was the vehicle velocity reference V^*, and the average of all f the wheel velocities was used as the vehicle velocity V in the field test. The value calculated with the vehicle model was used in the bench test. These controllers generated the total reference driving-braking force F_{all}^*. Then, F_{all}^* was distributed to the reference front and rear driving-braking forces F_j^* based on (8) and (9). Represented by the slip ratio, the reference front and rear torques T_j^* are given by

$$T_j^* = rF_j^* + \frac{J_{\omega j} a_x^*}{r}(1 + \lambda_j^*), \qquad (24)$$

where the second term of the right-hand side represents the compensation for the inertia torque of the wheels [21]. In order to consider the stability of the vehicle velocity

control system, the reference acceleration a_x^* and slip ratio λ_j^* were substituted for their measured values. Because $J_{\omega j} a_x^*/r$ was much smaller than rF_j^*, the second term did not have a large effect. Therefore, λ_j^* was simply set to 0.05, 0, and -0.05 during acceleration, cruising, and deceleration, respectively.

The vehicle velocity controller $C_{\mathrm{PI}}(s)$ was a proportional-integral (PI) controller that was designed by the pole placement method. The plant of the vehicle velocity controller is given by

$$\frac{V}{F_{\mathrm{all}}} = \frac{1}{Ms}. \qquad (25)$$

The pole of vehicle velocity controller was set to -5 rad/s.

Fig. 16 shows the vehicle speed control system for the experiments using RC-S. The inverters and motors of the real vehicle and vehicle model in RC-S represent the actual vehicle plant in Fig. 15. The vehicle model comprised the equations given in section 2.

Fig. 17 shows the experimental results for the vehicle motion in the field and bench tests; the results of each test when $k = k_{\mathrm{opt}}$ are shown. Fig. 17(a) shows the vehicle velocity. In each test, the vehicle velocity followed the reference, similar to the simulation results. This figure also shows the distribution ratio. The optimal distribution ratio k_{opt} increased during acceleration and decreased during deceleration. This result matched the previous calculations. Fig. 17(c) and Fig. 17(d) show the front and rear driving-braking forces, respectively. The total driving-braking force F_{all} was distributed based on k. In addition, the absolute values of the driving force in the simulation and bench test were equal to that of the field test. Therefore, the driving resistance model was appropriate, and the test bench realized the same load as the field test.

Fig. 18 shows the energy consumption in each experimental test. The energy consumptions during acceleration, cruising, and deceleration are shown separately. In order to confirm the reproducibility of the experimental results, the average values and standard deviations, shown as error bars, were calculated for the field and bench tests, which were carried out 12 and 8 times, respectively. In the computer simulation and RC-S, the energy consumption and regenerated energy during each driving section were minimized and maximized by the proposed system. In the field test, the effectiveness of the proposed system with regard to the total energy consumption was not clear because of the large dispersion for the data. However, Fig. 18(a) and Fig. 18(e) clearly show the effectiveness in the case of large acceleration because the variations in the wind and load conditions on the energy consumption was relatively small with a large torque. In the computer simulation results, the high-speed operation showed worse efficiency than the experimental results. This is because the simulation model had greater iron loss than the actual values. A comparison of the simulation and two tests showed that their energy consumptions roughly agreed. Thus, the proposed system achieved 9 % and 8 %

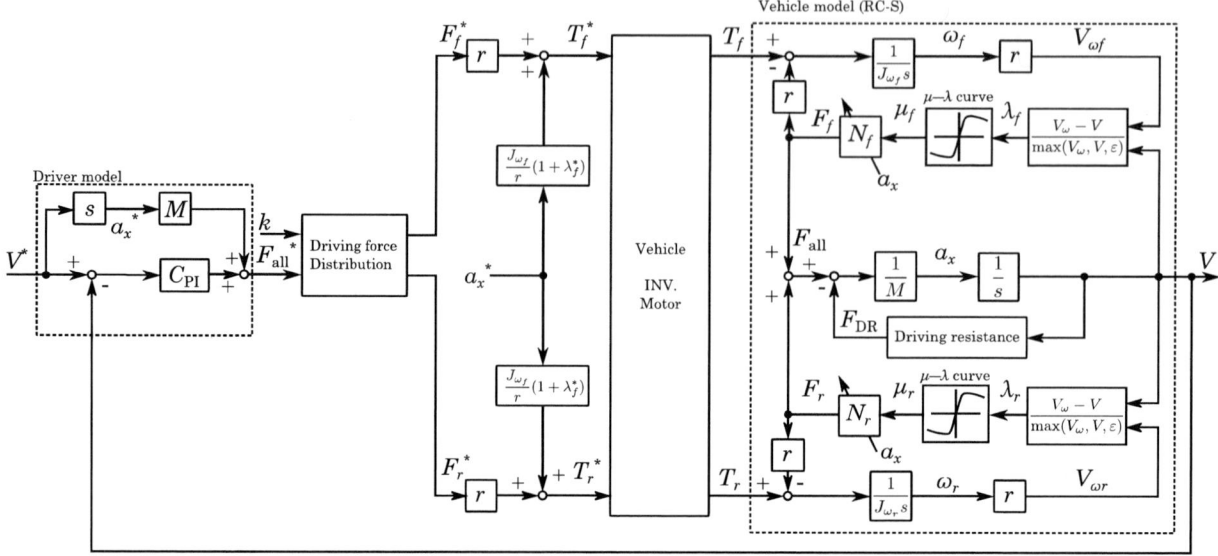

Fig. 16. Vehicle speed control system and vehicle model of RC-S.

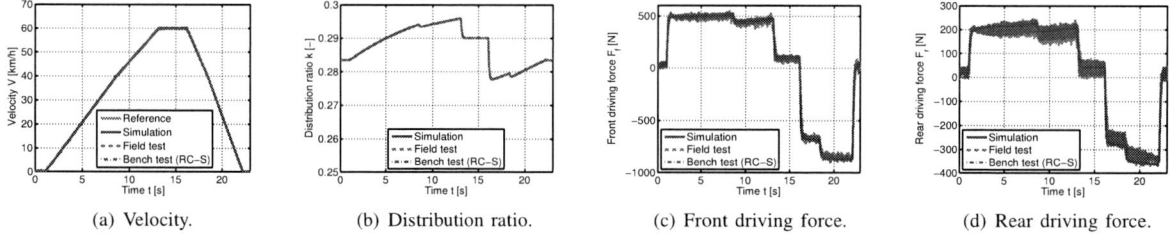

(a) Velocity. (b) Distribution ratio. (c) Front driving force. (d) Rear driving force.

Fig. 17. Experimental results related to vehicle motion of driving pattern ($k = k_\mathrm{opt}$).

decreases in the energy consumption during the bench and field tests, respectively, compared with $k = 0.5$.

V. CONCLUSION

This paper proposes a model-based range extension control system for electric vehicles that optimizes the front and rear driving-braking force distributions. The slip ratio of the wheels and the copper and iron losses of the motors are considered to minimize the energy consumption. Because the proposed distribution method depends on only the vehicle acceleration and velocity, the distribution ratios during the acceleration or deceleration processes can be optimized.

A bench test was carried out to realize results with high reproducibility. The results of a computer simulation, actual field test, and bench test were compared. The simulation and experimental results confirmed the effectiveness of the proposed system. The simulation and bench test results on the energy consumption matched the field test results.

Therefore, this study verified that the proposed system can extend the cruising range of electric vehicles and accurately measure the energy consumption.

ACKNOWLEDGMENT

This research was partly supported by the Industrial Technology Research Grant Program from the New Energy and Industrial Technology Development Organization(NEDO) of Japan (No. 05A48701d) and by a grant from the Ministry of Education, Culture, Sports, Science and Technology (No. 22246057).

REFERENCES

[1] Y. Hori: "Future Vehicle Driven by Electricity and Control—Research on Four-Wheel-Motored: "UOT Electric March II"", IEEE Trans. IE, Vol. 51, No. 5, pp. 954–962 (2004)

[2] K. Maeda, H. Fujimoto, and Y. Hori: "Four-Wheel Driving-Force Distribution Method Based on Driving Stiffness and Slip Ratio Estimation for Electric Vehicle with In-Wheel Motors", 8th IEEE Vehicle Power and Propulsion Conference, pp. 1286–1291 (2012)

[3] T. Hsiao: "Robust Estimation and Control of Tire Traction Forces", IEEE Trans. Veh. Technol., Vol. 62, No. 3, pp. 1378–1383 (2013)

[4] J. Kang, J. Yoo, and K. Yi: "Driving Control Algorithm for Maneuverability, Lateral Stability, and Rollover Prevention of 4WD Electric Vehicles with Independently Driven Front and Rear Wheels", IEEE Trans. Veh. Technol., Vol. 60, No.7, pp. 2987–3001 (2011)

[5] K. Nam, H. Fujimoto, and Y. Hori: "Advanced Motion Control of Electric Vehicles Based on Robust Lateral Tire Force Control via Active Front Steering", IEEE/ASME Trans. Mechatron., Vol. 19, No. 1, pp. 289–299 (2014)

978-1-4799-2706-7/14 $31.00 © 2014 IEEE

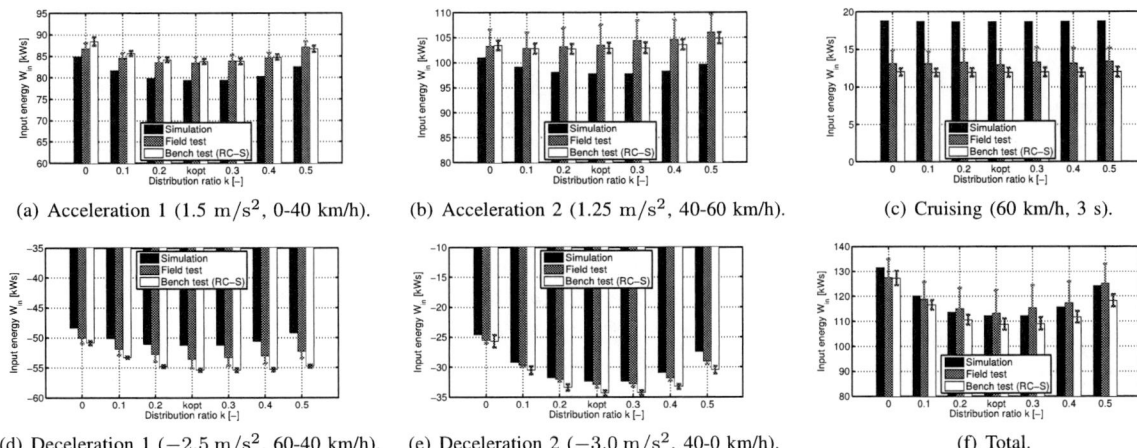

(a) Acceleration 1 (1.5 m/s², 0-40 km/h). (b) Acceleration 2 (1.25 m/s², 40-60 km/h). (c) Cruising (60 km/h, 3 s).

(d) Deceleration 1 (−2.5 m/s², 60-40 km/h). (e) Deceleration 2 (−3.0 m/s², 40-0 km/h). (f) Total.

Fig. 18.   Experimental results of pattern driving (comparison of energy consumption).

[6] T. Imura and Y. Hori: "Maximizing Air Gap and Efficiency of Magnetic Resonant Coupling for Wireless Power Transfer Using Equivalent Circuit and Neumann Formula", IEEE Trans. IE, Vol. 58, No. 10, pp. 4746–4752 (2011)

[7] S. Chopa and P. Bauer: "Driving Range Extension of EV With On-Road Contactless Power Transfer—A Case Study", IEEE Trans. IE, Vol. 60, No. 1, pp. 329–338 (2013)

[8] J. Shin, S. Shin, Y. Kim, S. Ahn, S. Lee, G. Jung, S. J. Jeon, and D. H. Cho: "Design and Implementation of Shaped Magnetic-Resonance-Based Wireless Power Transfer System for Roadway-Powered Moving Electric Vehicles", IEEE Trans. IE, Vol. 61, No. 3, pp. 1179–1192 (2014)

[9] P. J. Grbovic, P. Delarue, P. L. Moigne, and P. Bartholomeus: "The Ultracapacitor-Based Regenerative Controlled Electric Drives with Power-Smoothing Capability", IEEE Trans. IE, Vol. 59, No. 12, pp. 4511–4522 (2012)

[10] J. Cao and A. Emadi: "A New Battery/UltraCapacitor Hybrid Energy Storage System for Electric, Hybrid, and Plug-In Hybrid Electric Vehicles", IEEE Trans. Power Electron., Vol. 27, No. 1 pp. 122–132 (2014)

[11] M. Montazeri, M. Soleymani, and S. Hashemi: "Impact of Traffic Conditions on the Active Suspension Energy Regeneration in Hybrid Electric Vehicles", IEEE Trans. IE, Vol. 60, No. 10, pp. 4546–4553 (2013)

[12] H. Toda, Y. Oda, M. Kohno, M. Ishida, and Y. Zaizen: "A New High Flux Density Non-Oriented Electrical Steel Sheet and Its Motor Performance", IEEE Trans. Magn., Vol. 48, No. 11, pp. 3060–3063 (2012)

[13] K. Inoue, K. Kotera, Y. Asano, and T. Kato: "Optimal Torque and Rotating Speed Trajectories Minimizing Energy Loss of Induction Motor under Both Torque and Speed Limits", in Proc. Power Electronics and Drive Systems, 2013 IEEE 10th International Conference, pp. 1127–1132 (2013)

[14] X. Yuan and J. Wang: "Torque Distribution Strategy for a Front- and Rear-Wheel-Driven Electric Vehicle", IEEE Trans. Veh. Technol., Vol. 61, No. 8, pp. 3365–3374 (2012)

[15] H. Fujimoto and H. Sumiya: "Range Extension Control System of Electric Vehicle Based on Optimal Torque Distribution and Cornering Resistance Minimization", in Proc. 37th Annual Conference of the IEEE Industrial Electronics Society, pp. 3727–3732 (2011)

[16] H. Fujimoto, S. Egami, J. Saito, and K. Handa: "Range Extension Control System for Electric Vehicle Based on Searching Algorithm of Optimal Front and Rear Driving Force Distribution", in Proc. 38th Annual Conference of the IEEE Industrial Electronics Society, pp. 4244–4249 (2012)

[17] S. Harada and H. Fujimoto: "Range Extension Control System for Electric Vehicle on Acceleration and Deceleration Based on Front and Rear Driving/Braking Force Distribution Considering Slip Ratio and Motor Loss", in Proc. 39th Annual Conference of the IEEE Industrial Electronics Society, Vienna, Austria, pp. 6624–6629 (2013)

[18] H. B. Pacejka and E. Bakker: "The Magic Formula Tyre Model", Vehicle System Dynamics: International Journal of Vehicle Mechanics and Mobility , Vol. 21, No. 1, pp. 1–18 (1992)

[19] S. Morimoto, Y. Tong, Y. Takeda, and T. Hirasa: "Loss Minimization Control of Permanent Magnet Synchronous Motor Drives", IEEE Trans. IE, Vol. 41, No. 5, pp. 511–517 (1994)

[20] C. Kaido: "Effects of Cores on the Characteristics of an Induction VCM", IEEE Trans. Magnetics in Japan, Vol. 9, No. 6, pp. 110–116 (1994)

[21] H. Fujimoto, J. Amada, and K. Maeda: "Review of Traction and Braking Control for Electric Vehicle", in Proc. The 8th IEEE Vehicle Power and Propulsion Conference, Seoul, Korea, pp. 1292–1299 (2012)

Vibration Suppression of Integrated Resonant and Time Delay System by Reflected Wave Rejection

Eiichi Saito
System Design Engineering
Keio University
3-14-1 Hiyoshi, Kohoku,
Yokohama 223-8522, Japan
Email: saito@katsura.sd.keio.ac.jp

Roberto Oboe
Management and Engineering
University of Padova
Stradella S. Nicola 3,
Vicenza 36100, Italy
Email: roberto.oboe@unipd.it

Seiichiro Katsura
System Design Engineering
Keio University
3-14-1 Hiyoshi, Kohoku,
Yokohama 223-8522, Japan
Email: katsura@sd.keio.ac.jp

Abstract—The vibration suppression on a resonant system, based on the reflected wave rejection, has been recently proposed. In the method, the resonant system is modeled through a wave equation, in order to consider high order vibrations and vibration suppression is achieved by eliminating a reflected wave from the resonant system. However, it is impossible to implement such method when there are input and output time delays, because it would require the implementation of the inverse of a time delay. This paper proposes a new vibration suppression scheme for a resonant systems, affected by time delay, by using a modified reflected wave rejection. The proposed reflected wave rejection can be applied even in presence of time delays, as long as some conditions on the value of such delays are satisfied. The validity of the proposed method is confirmed by simulations and experimental results.

I. INTRODUCTION

Recently, motion control technologies have been brilliantly developed, and fast and accurate force/position control has been achieved. However, a fast response often excites mechanical resonances, leading the system to vibrate. In particular, there are applications in which the vibration problem can not be avoided, due to the mechanical construction of the controlled system (e.g. using reduction gears in robots). Moreover, it is also known that a time delay in the control system often leads the system to vibrate and to be unstable [1]–[5]. The time delay often appears in practical applications, due to the use of communication system or to the presence of a sensor delay. Therefore, it is important to consider vibration suppression and time delay compensation for realizing advanced motion control systems. This paper focuses on vibration suppression on a motion control system, in presence of both resonant behavior and time delay.

Several vibration control and time delay compensation methods have been proposed so far. As for the vibration suppression methods, there are state feedback control [6], resonant ratio control [7], H_∞ control [8], and so on. Concerning the time delay compensation methods, there are Smith predictors [9], communication disturbance observers [10], and others. Regarding the vibration suppression, the authors proposed the use of a novel wave rejection scheme [11]. In such solution, the resonant system is modeled as a wave equation, in order

to consider high order vibrations, and vibration is suppressed by eliminating the reflected wave (which is inducing the vibration) from the resonant system. In addition, it has been shown how a transfer function of the resonant system without the reflected wave can be represented as a time delay. In other words, the waves only travels from the actuator to the load side of the system. Hence, the resonant system can be regarded as an equivalent time delay system, which can be treaded with standard control techniques. In this regard, the authors already proposed the vibration suppression scheme of integrated resonant and time delay system by using reflected wave rejection and wave compensator, which is based on the communication disturbance observer [12]. In a typical scenario of application of the proposed method, the reflected wave rejection must be implemented at remote side (i.e. after the equivalent delay, at the plant side). On the other hand, it is difficult to obtain a good performance in case the reflected wave rejection is to be implemented at a controller side, because it is difficult to implement inverse system of time delay. This is a typical scenario of an actual delay between controller and resonant plant to be controlled. To solve the problem, this paper proposes a new vibration suppression method for systems containing both resonances and time delay, by using reflected wave rejection. Concretely speaking, the reflected wave is modified to contain the time delay, and the input and output time delay can be buffered in time delays in a novel reflected wave rejection. Accordingly, the proposed method can be used even if there are those time delays as long as specific conditions on the value of the time delays (i.e. input and output delays and the propagation time of the wave) are satisfied.

II. VIBRATION SUPPRESSION ON RESONANT SYSTEM BASED ON REFLECTED WAVE REJECTION [11]

In this section, the suppression of vibrations on a resonant system, based on the reflected wave rejection, is explained. At first, the modeling of resonant system by using wave equation is described. Next, the vibration suppression scheme, based on the reflected wave rejection, is introduced.

The 2014 International Power Electronics Conference

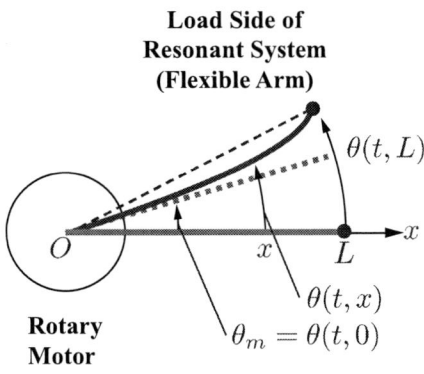

Load Side of Resonant System (Flexible Arm)

$\theta(t, L)$

O

x L

$\theta(t, x)$

$\theta_m = \theta(t, 0)$

Rotary Motor

Fig. 1. Modeling of the resonant system based on wave.

A. Modeling of Resonant System

A system dealt with in this paper is shown in Fig. 1, where $\theta_m(t)$, $\theta(t, x)$ and L denote the motor position, the displacement at x in t, the length of the system, respectively. The control goal for the system is that the load position $\theta(t, L)$ corresponds to the position command by controlling the actuator. In our method, the resonant system is modeled as a wave equation, which is one of a distributed parameter model, in order to consider high order resonances. The wave equation is represented as

$$\frac{\partial^2 \theta(t, x)}{\partial t^2} = c^2 \frac{\partial^2 \theta(t, x)}{\partial x^2} \qquad (1)$$

where c stands for the propagation velocity of the wave. The boundary conditions for the resonant system shown in Fig. 1 are represented as

$$\theta(t, 0) = \theta_m(t) \qquad (2)$$
$$\frac{\partial \theta(t, L)}{\partial x} = 0. \qquad (3)$$

(2) means that the motor position, and it is regarded as an input for a load side of the resonant system which is modeled as the wave equation. It is noted that the motor position is not affected by the reaction force from the load side of the resonant system because the motor implements the robust acceleration control based on disturbance observer (DOB) [13]. (3) means the we have a free end at $x = L$. By using these boundary conditions, a transfer function from the motor position to the load position can be derived [14], and the transfer function is represented as

$$\theta(s, L) = \frac{2e^{-T_w s}}{1 + e^{-2T_w s}} \theta_m \qquad (4)$$

where s and T_w denote the Laplace operator and the propagation time of the wave, which is expressed as

$$T_w = \frac{L}{c}. \qquad (5)$$

The block diagram of the transfer function is shown in Fig. 2. In Fig. 2, the part enclosed by dashed-line denotes the reflected wave which includes the wave traveling to $x = 0$ and the wave reflected at $x = L$. The negative feedback is a cause of vibration on the resonant system.

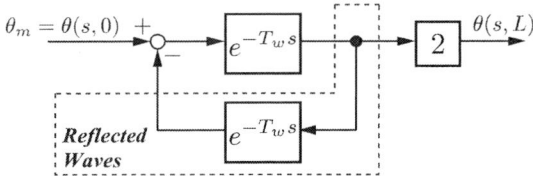

$\theta_m = \theta(s, 0)$

$e^{-T_w s}$

$e^{-T_w s}$

2

$\theta(s, L)$

Reflected Waves

Fig. 2. Block diagram of the resonant system modeled by wave equation.

B. Reflected Wave Rejection

In this part, reflected wave rejection for vibration suppression is explained. The transfer function can be transformed as a following equation:

$$\theta(s, L) = e^{-T_w s}(\theta_m + \theta_{rfl}) \qquad (6)$$

where θ_{rfl} denotes the reflected wave, and it is represented as

$$\theta_{rfl} = \theta_m - e^{-T_w s}\theta(s, L). \qquad (7)$$

The reflected wave is a cause which induces the vibration on the system. To eliminating the reflected wave from resonant system, the reflected wave is estimated and compensated by reflected wave rejection. the reflected wave is estimated as

$$\hat{\theta}_{rfl} = \frac{g_r}{s + g_r}\left[\theta_m - e^{-T_{wn} s}\theta(s, L)\right] \qquad (8)$$

where v, g_r, and T_{wn} stand for the estimated reflected wave, the cut-off frequency of the reflected wave rejection, and the nominal value of the propagation time, respectively. The block diagram of the reflected wave rejection is shown in Fig. 3. In Fig. 3, g_{pd} and $\ddot{\theta}^{cmp}$ denote , the cut-off frequency of the pseudo derivative and the compensation value in acceleration dimension, respectively. In the reflected wave rejection, the reflected wave is estimated through the low-pass filter. Then, the compensation value is calculated by using the inverse system of motor with DOB. Finally, the reflected wave is canceled out by the estimated reflected wave. If the cut-off frequency of the reflected wave rejection is high enough, the transfer function from acceleration reference to the load position reduces to:

$$\frac{\theta(s, L)}{\ddot{\theta}^{ref}} = \frac{1}{s^2}e^{-T_w s}. \qquad (9)$$

From (9), it turns out that the vibration on resonant system is suppressed, because there is no time delay in the denominator of the transfer function. In addition, the resonant system without reflected wave can be regarded as an equivalent time delayed system. According to the above, in our method, the position controller with a communication disturbance observer [10], which is a time delay compensation method, are implemented in the outer loop of Fig. 3.

However, there is a problem in the conventional method, when there are input and output delays (e.g. communication delay and sensing delay). In this case, in fact, the reflected wave rejection can not be implemented because the inverse systems of the time delays, which means the use of future value, are needed and it are difficult to implement them. If

978-1-4799-2706-7/14 $31.00 © 2014 IEEE

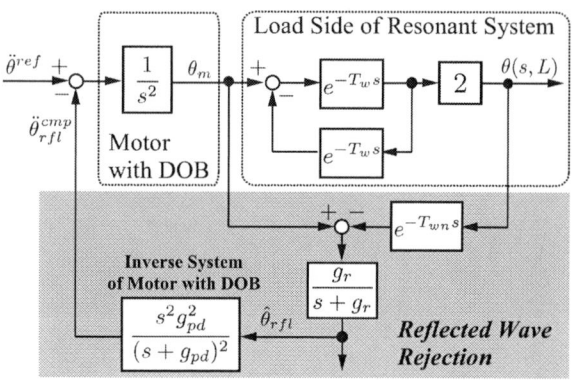

Fig. 3. Block diagram of the conventional reflected wave rejection.

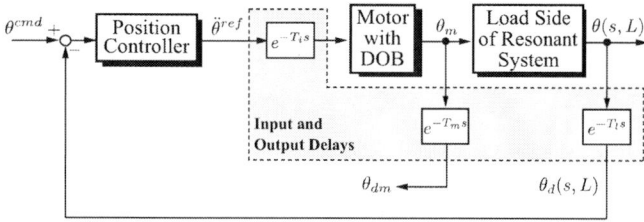

Fig. 4. Integrated resonant and time-delayed system dealt with in this paper.

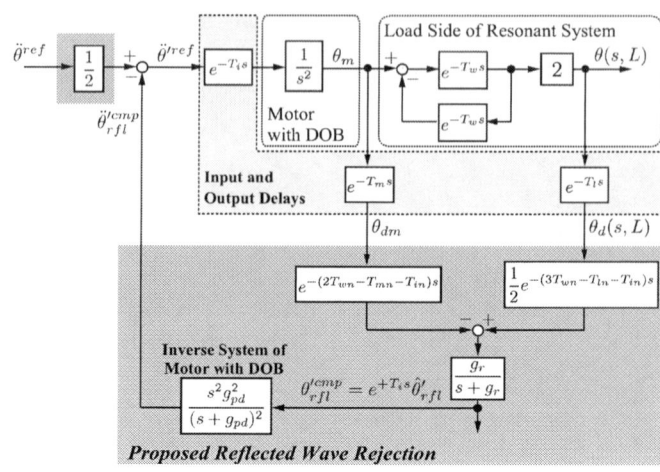

Fig. 5. Block diagram of proposed reflected wave rejection.

the inverse systems are approximated to 1, the approximation causes degradation of the reflected wave rejection.

III. REFLECTED WAVE REJECTION IN INTEGRATED RESONANT AND TIME-DELAYED SYSTEM

As mentioned before, it is difficult to implement the reflected wave rejection if there are input and output delays. Therefore, this paper proposes a novel reflected wave rejection scheme under existence of these delays. The block diagram of integrated resonant and time delayed system is shown in Fig. 4, where T_i, T_m, T_l, θ_{dm} and $\theta_d(s, L)$ are the input delay, the output delay at motor side, the output delay at load side, and the delayed motor position, and the delayed load position, respectively.

First of all, the transfer function is transformed into

$$
\begin{aligned}
\theta(s, L) &= 2e^{-T_w s} \frac{1}{1 + e^{-2T_w s}} \theta_m \\
&= 2e^{-T_w s} \left(1 - e^{-2T_w s} + e^{-4T_w s} \right. \\
&\quad \left. - e^{-6T_w s} + \cdots \right) \theta_m. \\
&= 2e^{-T_w s} \left[\theta_m - e^{-2T_w s} \theta_m + e^{-4T_w s} \times \right. \quad (10) \\
&\quad \left. \left(1 - e^{-2T_w s} + e^{-4T_w s} \cdots \right) \theta_m \right]. \quad (11)
\end{aligned}
$$

In the aforementioned transformation, the relation of geometric series $1/1 + r = 1 - r + r^2 - r^3 \cdots$ is used. Furthermore, by using the relation of geometric series and (6), (11) is transformed into

$$
\theta(s, L) = 2e^{-T_w s} \left[\theta_m - e^{-2T_w s} \theta_m + \frac{1}{2} e^{-3T_w s} \theta(s, L) \right] \quad (12)
$$

Here, let's re-define a reflected wave as the following equation:

$$
\theta'_{rfl} = -e^{-2T_w s} \theta_m + \frac{1}{2} e^{-3T_w s} \theta(s, L). \quad (13)
$$

The re-defined reflected wave corresponds to a wave at the end of negative feedback, i.e. $e^{-T_w s} \theta(s, L)$. It can be noticed that the reflected wave defined in previous section denotes the summation of the wave at the end of negative feedback and the wave reflected at $x = L$. By using (13), the relation between θ_m, $\theta(s, L)$ and θ'_{rfl} is represented as

$$
\theta(s, L) = 2e^{-T_w s} \left[\theta_m + \theta'_{rfl} \right]. \quad (14)
$$

As well as the conventional reflected wave rejection, the reflected wave θ'_{rfl} will be canceled out by the feedforward of the estimated reflected wave. However, because there are input and output delays, the compensation value for eliminating the reflected wave from the resonant system must include the inverse system of the time delays. In the proposed method, the compensation value including the inverse systems are calculated as

$$
\begin{aligned}
\theta'^{comp}_{rfl} &= e^{+T_i s} \hat{\theta}'_{rfl} \\
&= \frac{g_r}{s + g_r} \left(-e^{-(2T_{wn} - T_{mn} - T_{in})s} \theta_{dm} \right. \\
&\quad \left. + \frac{1}{2} e^{-(3T_{wn} - T_{ln} - T_{in})s} \theta_d(s, L) \right) \quad (15)
\end{aligned}
$$

where $\hat{\theta}'_{rfl}$, T_{in}, T_{mn} and T_{ln} denote the estimated reflected wave, and the nominal time delays, respectively. It is noted that, in order to calculate the above compensation value, following conditions must be satisfied,

$$
\begin{aligned}
2T_w &> T_m + T_i \quad (16) \\
3T_w &> T_l + T_i. \quad (17)
\end{aligned}
$$

If the time delays satisfy the above conditions, the proposed reflected wave rejection can be implemented. Compared with the application range of the conventional method against value of time delay, it is found that the application range is extended.

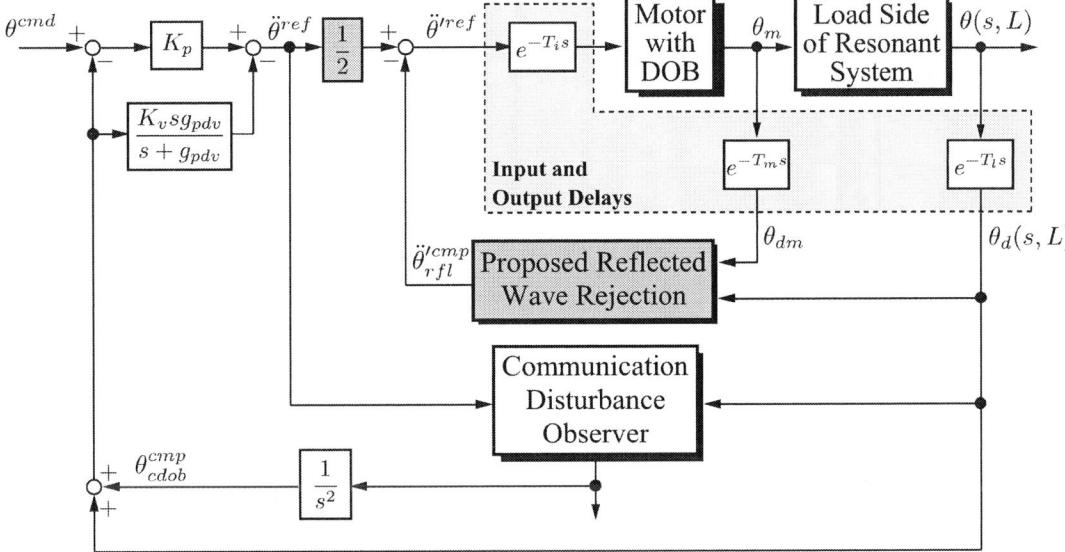

Fig. 6. Whole block diagram of the proposed position control system.

The block digram of the proposed reflected wave rejection is shown in Fig. 5. In Fig. 5, $\ddot{\theta}_{rfl}^{\prime cmp}$ denotes the compensation value in acceleration dimension, it is calculated as

$$\ddot{\theta}_{rfl}^{\prime cmp} = s^2 \frac{g_{pd}^2}{(s + g_{pd})^2} e^{+T_i s} \theta_{rfl}^{cmp}. \tag{18}$$

In (18), the inverse system of the motor with DOB is approximately implemented by using the pseudo derivation. Finally, the acceleration reference injected to the motor is calculated as

$$\ddot{\theta}^{\prime ref} = \frac{1}{2} \ddot{\theta}^{ref} - \ddot{\theta}_{rfl}^{cmp}. \tag{19}$$

It is noted that 1/2 in the right hand of (19) means that cancellation of the superposition of the wave at $x = L$, because, at $x = L$, the wave which travels from $x = 0$ and the wave which is reflected at $x = L$ are immediately superposed (This is meaning of the block "2" in Fig. 2). If the cut-off frequencies in the proposed reflected wave rejection are ideally high value, the transfer function from the acceleration reference to the load position is represented as

$$\frac{\theta(s, L)}{\ddot{\theta}^{ref}} = \frac{1}{s^2} e^{-(T_w + T_i)s}. \tag{20}$$

From (20), it is found that the vibration is suppressed because the time delay has been omitted from the denominator of the transfer function.

By using the proposed reflected wave rejection, the resonant system with input and output time delay can be regarded as an equivalent time delay system described as (20). Then, the position control system of (20) is constructed by using the position controller with a communication disturbance observer (CDOB) [10] which is one of a time delay compensation

method. The acceleration reference is calculated as

$$\begin{aligned}\ddot{\theta}^{ref} &= K_p \left(\theta^{cmd} - \theta_d(s, L) - \theta_{cdob}^{cmp} \right) \\ &\quad - K_v \frac{s g_{pdv}}{s + g_{pdv}} (\theta_d(s, L) + \theta_{cdob}^{cmp}) \end{aligned} \tag{21}$$

where θ^{cmd}, K_p, K_v, g_{pdv} and θ_{cdob}^{cmp} denote the position command, the position gain, the velocity gain, the cut-off frequency for calculation of velocity, the compensation value of CDOB, respectively. In the proposed system, the P control with a velocity minor loop is used as a position controller. The whole block diagram of position control of the integrated resonant and time delay system is shown in Fig. 6. Finally, if the cut-off frequency of the communication disturbance observer is enough large value, the transfer function from position command θ^{cmd} to the load position $\theta(s, L)$ is represented as

$$\frac{\theta(s, L)}{\theta^{cmd}} = \frac{K_p}{s^2 + K_v s + K_p} e^{-(T_w + T_i)s}. \tag{22}$$

It is found that because there is no time delay in the denominator of the full-closed transfer function, vibration on the integrated resonant and time delay system is suppressed.

IV. NUMERICAL RESULTS

A. Simulation Setup

In order to verify the validity of the proposed method, simulation of position control on a three-mass resonant system is conducted. Plant and control parameters used in this simulation are shown in Table I. The control goal of the simulation is that the load position corresponds to the position command (a step of 0.1 rad) without vibration. Performance of the proposed method is compared with that of the method based on the conventional reflected wave rejection. The difference between the proposed and conventional methods is only in the structures of the reflected wave rejections, but the other control parameters are same.

TABLE I.   SIMULATION PARAMETERS.

Parameter	Description	Value
T_s	Sampling time	0.1 ms
T_i	Input delay	7.5 ms
T_m	Output delay at motor	7.5 ms
T_l	Output delay at load	7.5 ms
T_w	Propagation time of wave	40 ms
K_{tn}	Nominal thrust force coefficient	3.0 N/A
M_n	Nominal mass of motor	0.245 kg
w_1	First order resonance freq.	39.5 rad/s
w_2	Second order resonance freq.	103.0 rad/s
K_p	Position gain	900
K_v	Velocity gain	80
g_{dis}	Cut-off freq. of DOB	500 rad/s
g_r	Cut-off freq. of reflected wave rejection	500 rad/s
g_{cdob}	Cut-off freq. of CDOB	1000 rad/s
g_{pd}	Cut-off freq. of pseudo derivation for reflected wave rejection	500 rad/s
g_{pdv}	Cut-off freq. of pseudo derivation for position controller	500 rad/s
$d\theta, d\theta(s,L)$	Position sensor resolutions	65536 pulse/rev

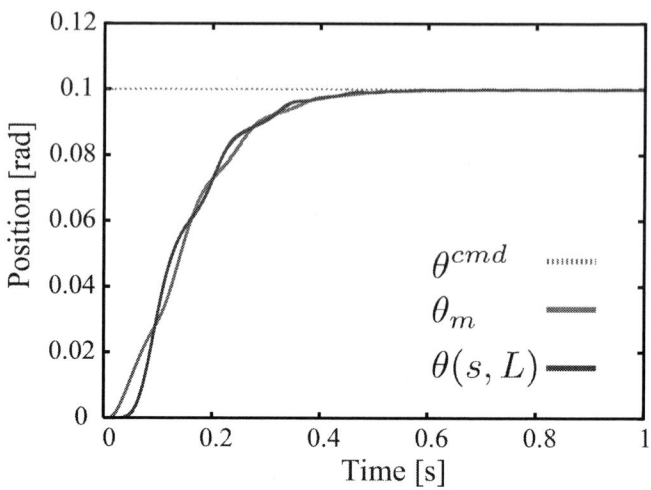

Fig. 8.   Simulation results of the proposed reflected wave rejection.

TABLE II.   EXPERIMENTAL PARAMETERS.

Parameter	Description	Value
T_s	Sampling time	0.1 ms
T_i	Input delay	5.0 ms
T_m	Output delay at motor side	5.0 ms
T_l	Output delay at load side	20 ms
T_w	Propagation time of wave	9.6 ms
K_{tn}	Nominal thrust force coefficient	3.0 N/A
M_n	Nominal mass of motor	0.245 kg
w_1	First order resonance freq.	104.0 rad/s
L	Length of flexible arm	0.3 m
K_p	Position gain	10000.0
K_v	Velocity gain	400
g_{dis}	Cut-off freq. of DOB	1000 rad/s
g_r	Cut-off freq. of reflected wave rejection	200 rad/s
g_{cdob}	Cut-off freq. of CDOB	1000 rad/s
g_{pd}	Cut-off freqs. of pseudo derivation for reflected wave rejection	200 rad/s
g_{pdv}	Cut-off freqs. of pseudo derivation for calculation of velocity	1000 rad/s

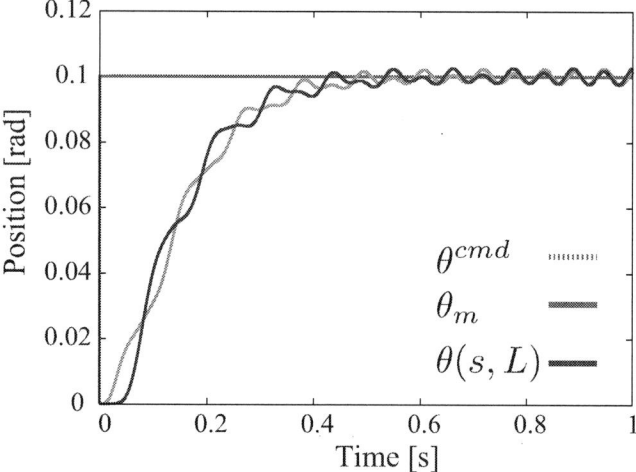

Fig. 7.   Simulation results of the conventional reflected wave rejection.

B. Simulation Results

The simulation results of the conventional and proposed methods are shown in Figs. 7 and 8. From the results of the conventional method, the vibration can be observed because the performance of the reflected wave rejection degrades due to the existence of the delays. On the other hand, it is found that the residual vibration is well suppressed with the proposed approach.

V.   EXPERIMENTS

A. Experimental Setup

To verify the effectiveness of the proposed method, a position control of flexible arm is performed. Experimental setup is shown in Fig. 9. The control goal is that the tip position of a flexible arm $\theta(s,L)$ corresponds to the position command θ^{cmd} without vibrations. The flexible arm is mounted on the direct drive rotary motor (encoder resolution: 20 bit/rev). The load position $\theta(s,L)$ is obtained by the vision sensor (frame rate: 100 Hz). Equivalent resolution of vision sensor is 1.09×10^{-4} rad. Control program is written by C language under Real time application interface 3.8 (RTAI 3.8) on Linux.

Experimental parameters are shown in Table II. Output delay at load side includes processing time of vision sensor. The input and output delays without the image processing delay is artificially generated in the computer. The jitters of input and output delays are not considered in this experiment. In this experiment, a step position command ($\theta^{cmd} = 0.05$ rad) is applied to the control system. It is noted that, because output delay T_l is larger than propagation time of wave T_w, the conventional reflected wave rejection shown in Fig. 3 can not be implemented in this experiment.

B. Experimental Results

The experimental results of the proposed method is shown in Fig. 10. From Fig. 10, the residual vibration on load side is well suppressed by the proposed method although there are input and output time delays. However, it is observed that there is steady state error because of the friction acting on motor

The 2014 International Power Electronics Conference

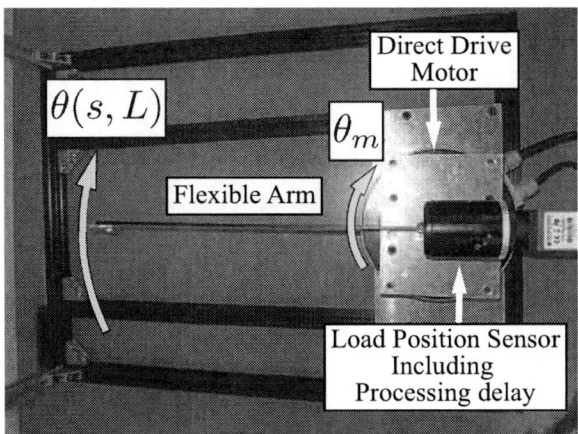

Fig. 9.   Experimental Setup.

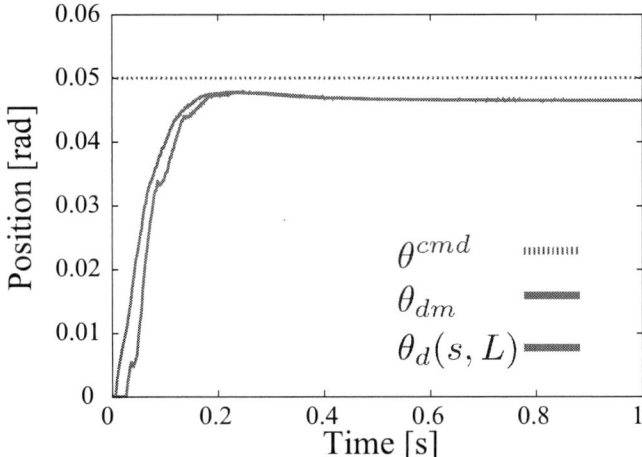

Fig. 10.   Position response of proposed method.

and degradation of performance of disturbance rejection when the CDOB is used. Although the disturbance compensator is not implemented in this system in order to be simplified, the elimination of steady state error is possible by using the method about disturbance rejection in CDOB [15], [16].

Then, validity of the proposed reflected wave rejection is confirmed.

VI.   CONCLUSIONS

This paper proposed a vibration suppression for time-delayed resonant systems, by using reflected wave rejection. Conventional method, based on the reflected wave rejection, could not handle input and output delays such as communication delay and sensing delay, because the conventional reflected wave rejection needed the non-delayed motor position information. On the other hand, by modifying the calculation method of the reflected wave, the proposed method was able to properly control the system, provided that the delays satisfy the conditions (16) and (17). The validity of the proposed method was verified by both simulations and experimental results on

position control of the flexible arm with input and output time delays.

ACKNOWLEDGMENT

This research was supported by the Ministry of Education, Science, Sports and Culture, Grant-in-Aid for JSPS Fellows, 24-5462, 2013.

REFERENCES

[1]   J. P. Richard : "Time-delay systems : an overview of some recent advances and open problems," *Automatica*, Vol. 39, No. 10, pp. 1667–1694, 2003.

[2]   H. Morioka, A. Sabanovic, A. Uchibori, K. Wada and M. Oka : "Application of time-delay-control in variable structure motion control systems," *The IEEE International Symposium on Industrial Electronics Society, ISIE '05–PUSAN*, pp. 1313–1318, June, 2001.

[3]   K. Gu and S-I. Niculescu: "Survey on Recent Results in the Stability and Control of Time-Delay Systems," *ASME Journal of Dynamic Systems, Measurement, and Control*, Vol. 125, No. 2, pp. 158–165, 2003.

[4]   M. Bowthorpe: "Smith Predictor-Based Robot Control for Ultrasound-Guided Teleoperated Beating-Heart Surgery," *IEEE Journal of biomedical and Health Informatics*, Vol. 18, No. 1, pp. 157–166, January, 2014.

[5]   R. Yang, G. Liu, P. Shi, C. Thomas, and M. V. Basin: "Predictive Output Feedback Control for Networked Control Systems," *IEEE transactions on Industrial Electronics*, Vol. 61, No. 1, pp. 512–520, January, 2014.

[6]   S. H. Song, J. K. Ji, S. K. Sul, and M. H. Park : "Torsional Vibration Supprresion Control in 2-mass System by State Feedback Speed Controller," *Proceedings of 2nd IEEE Conference on Applications, '93–VANCOUVER*, Vol. 1, pp. 129–134, September 13–16, 1993.

[7]   Y. Hori, H. Sawada, and C. Yeonghan : "Slow Resonance Ratio Control for Vibration Suppression and Disturbance Rejection in Torsional System," *IEEE Transactions on Industrial Electronics*, Vol. 46, No. 1, pp. 162–168, February, 1999.

[8]   S. Morimoto and Y. Takeda : "Two-Degrees-of-Freedom Speed Control of Resonant Mechanical System Based on H_∞ Control Theory," *Transaction of IEE Japan*, Vol. 116–D, No. 1, pp. 65–70, January, 1996.

[9]   O. J. M. Smith : "A Controller to Overcome Dead Time," *The International Society of Automation Journal*, Vol. 6, No. 2, pp. 28–33, February, 1959.

[10]   K. Natori and K. Ohnishi : "A Design Method of Communication Disturbance Observer for Time-Delay Compensation, Taking the Dynamic Property of Network Disturbance Into Account," *IEEE Transactions on Industrial Electronics*, Vol. 55, No. 5, pp. 2152–2168, May, 2008.

[11]   E. Saito, S. Katsura : "Vibration Suppression of Resonant System by Using Wave Compensator," *The 37th Annual Conference of the IEEE Industrial Electronics Society, IECON '11-MELBOURNE*, pp. 4105–4110, November 7–10, 2011.

[12]   E. Saito, S. Katsura : "Compensation of Integrated Resonant and Time Delay System by Using Wave Compensator," *Journal for Control, Measurement, Electronics, Computing and Communications, Automatika*, Vol. 54, No. 1, pp. 28–38, April, 2013.

[13]   K. Ohnishi, M. Shibata, and T. Murakami : "Motion Control for Advanced Mechatronics," *IEEE/ASME Transactions on Mechatronics*, Vol. 1, No. 1, pp. 56–67, March, 1996.

[14]   E. Kreyszig: "Advanced Engineering Mathematics," Wiley, 2005

[15]   K. Natori, R. Oboe and K. Ohnishi : "Robust Time Delayed Control Systems with Communication Disturbance Observer," *The 33th Annual Conference of the IEEE Industrial Electronics Society, IECON '07-TAIPEI*, pp. 316–321, November, 2007.

[16]   A. A. Rahman and K. Ohnishi : "Robust time delayed control system based on communication disturbance observer with inner loop input," *The 36th Annual Conference of the IEEE Industrial Electronics Society, IECON '10-TAIPEI*, pp. 1621–1626, November, 2010.

Thrust Characteristics Improvement of a Circular Shaft Motor for Direct-Drive Applications

Mototsugu Omura, Tomoyuki Shimono and Yasutaka Fujimoto

Division of Electrical and Computer Engineering

Yokohama National University

Yokohama, Japan

Email: omura-mototsugu-kc@ynu.jp, shimono@ynu.ac.jp, fujimoto@ynu.ac.jp

Abstract—**This paper presents a new circular shaft motor (CSM) model for development of haptic forceps system which has multi–degrees–of–freedom (MDOF). The prototype of CSM has issues; it is difficult to construct the stator as designed and the thrust characteristics is not enough high. The proposed CSM model includes trapezoidal magnets which are magnetized in axial direction. This structure can facilitate construction of the stator and decrease thrust ripple. In addition, by increasing the number of coil turns and shortening the magnetic gap between magnets and coils, the thrust characteristic can be improved. In this paper, the structure of proposed CSM is described. Then, the improvement of thrust characteristic is verified by magnetic field analysis. Finally, a prototype of the proposed model is developed, then thrust characteristic of the prototype of the proposed model is verified by primary experiment.**

I. INTRODUCTION

Linear actuators are utilized for the various industrial applications. A linear actuator has many advantages since it is a direct–drive system. Because a linear actuator has almost no friction and no backlash, it has a high controllability. This aspect is suitable for the industrial system which requires the high accuracy position control. However, it is predicted that the industrial systems will become more huge and complex in order to realize various motions. Therefore, a linear actuator is required to attain the high thrust capacity and various motions. In papers [1] ~ [5], various tubular linear actuators are developed and the design of parameters is optimized to improve the thrust characteristic. In order to obtain the high thrust density, a helical–shaped linear actuator is introduced [6], [7]. Moreover, the two–degrees–of–freedom actuator is reported [8]. Industrial applications will be sustained by the advancement of the linear motion technology.

In addition, the linear actuator has another aspect of high back–drivability based on direct–drive. High back–drivability system is suitable for the realization of human-system interaction. From this advantage, the researches of linear systems are reported for medical and welfare system. In article [9], the haptic forceps system installing linear actuators is presented. This system has two pairs of linear actuators as a master system and a slave system. This system can transmit vivid touching sensation between a master system and a slave system. This technique is very useful for the surgical robot application. In order to construct the surgical robot application installing this

technique, the realization of a multi–degrees–of–freedom system is desired.

Various researches about the linear system for realizing a multi–degrees–of–freedom (MDOF) system have been reported. Some studies on the combination use of the motor and the push–pull cable have been reported [10], [11]. Besides, a MDOF haptic endoscopic surgery robot was presented [12]. Although this surgery robot can realize the flexible allocation of the linear actuators, the haptic transmission performance is low. For the surgical robot, the system which has both MDOF and good haptic transmission performance is required. The development of such system remains as a key issue for the surgical robot application.

In order to develop the surgical system which has both MDOF and good haptic transmission performance, the development of a new direct–drive motion system is expected. For the surgical robotic application, not only thrust motion but also the rotational motions around the fixed fulcrum point based on direct–drive are required. For realization of these motions, a circular shaft motor (CSM) is proposed [13]. A CSM is a kind of tubular linear actuator. A CSM can realize the circumference motion in the direct–drive mechanism. The combination use of CSMs can realize a rotational motion around a fulcrum point in the MDOF system. However, the prototype of CSM has a problem that it is difficult to construct the stator. Because of repulsive force between magnets, iron pieces between magnets are rotated in the stator. Therefore, the construction of the stator as designed is very difficult. In addition, the value of thrust constant is insufficient for applying to the MDOF system.

In this paper, the new CSM model is presented. The proposed CSM model includes the trapezoidal magnets which are magnetized in axial direction. By using these magnets, magnetic flux can be generated in the normal direction without using iron pieces as same as that of the prototype of CSM. This stator structure can facilitate the construction of the stator and can decrease thrust ripple. In addition, by increasing the number of coil turns, the thrust constant is improved to satisfy the requirement in development of the MDOF direct–drive motion system.

II. STRUCTURE OF PROPOSED CSM

Fig. 1 and Fig. 2 show structures of the conventional model of a CSM and the proposed model of a

978-1-4799-2706-7/14 $31.00 © 2014 IEEE

The 2014 International Power Electronics Conference

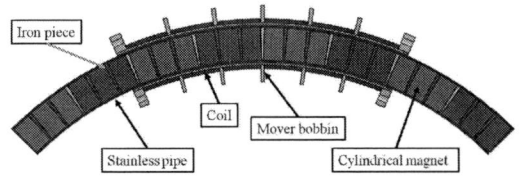

Fig. 1: A conventional model of CSM

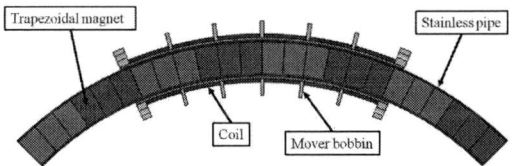

Fig. 2: A proposed model of CSM

CSM. In the conventional model of CSM, Nd–Fe–B magnets in cylindrical shape are inserted. These magnets are magnetized in the axial direction. Iron pieces are sandwiched between each magnet so as to keep the stator parts in the circular shape. However, because the repulsive force between magnet poles are generated, the conventional model of a CSM has a problem that iron pieces are rotated in the stator pipe. Hence, constructing the stator is very difficult as designed. In the proposed model of a CSM, instead of iron pieces and cylindrical shape magnets, trapezoidal magnets are inserted in the stator. The trapezoidal magnets are magnetized in axial direction. This structure of the proposed model makes it easier to construct the stator than the conventional model of CSM. As a result, the decrease of the thrust ripple is expected in real machine.

In addition, the thrust capacity in the conventional model of a CSM is not sufficient for a MDOF direct–drive motion system. In order to improve the thrust capacity, the new CSM model has more coil turns than the conventional model of CSM and shortens the length of magnetic gap. However, because the sufficient pipe thickness is required for bending a pipe, the magnetic gap has been increased. Therefore, the relation between the thrust capability and the pipe thickness will be investigated to determine the design parameter.

III. EVALUATION OF THRUST CHARACTERISTICS

The thrust constant is evaluated in the tangential direction of circular arc in a CSM. Fig. 3 shows simple model of a CSM. The initial angle is set on y axis in Fig. 3. The positive direction of the angle θ is determined in the counterclockwise direction. Thrust force in tangential direction $F_{t\theta}$ can be obtained from the thrust force based on the world coordinate as equation (1).

$$F_{t\theta} = -F_x \cos\theta - F_y \sin\theta \qquad (1)$$

TABLE I: Specifications of stator part and mover part

Stainless pipe	
Material	SUS304
Outside diameter [mm]	19.00
Pipe wall thickness [mm]	0.2500~2.000
Bend radius [mm]	175
Permanent magnet	
Material	Nd–Fe–B
Angle [deg]	11.25
Mover bobbin	
Material	ABS resin
Diameter of central hole [mm]	20.00
Mover thickness [mm]	1.000
Number of mover slots	6

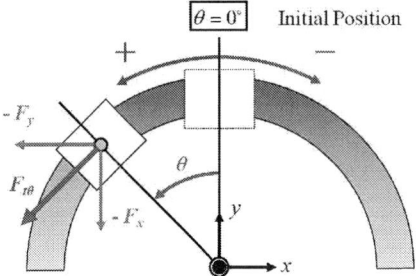

Fig. 3: The model of CSM

In paper [13], the thrust constant is calculated from the average value of thrust force at each current value. In addition, it is confirmed that the relation between thrust force and current is almost linear. Thrust constant is defined as a value of the thrust force generated when the current 1A is fed. Therefore, the thrust constant is calculated by the average value of the thrust force generated when the current 1A is fed at each angle. Thrust constant $K_{t\theta}$ is expressed in terms of step number N in equation (2).

$$K_{t\theta} = \frac{1}{N} \sum_{j=1}^{N} F_{t\theta j} \qquad (2)$$

In addition, the thrust ripple is also evaluated in the magnetic field analysis. From the thrust data, thrust ripple γ is derived by equation (3).

$$\gamma = \frac{(F_{t\theta max} - F_{t\theta min})/2}{K_{t\theta}} \times 100 \qquad (3)$$

IV. MAGNETIC FIELD ANALYSIS IN THRUST CHARACTERISTIC

In this section, the thrust characteristic of the proposed model is compared with that of the conventional model by magnetic field analysis.

A. Model condition

The thrust constant is influenced greatly by the length of the magnetic gap. Thus, the new CSM model is designed to decrease the length of magnetic gap so as to obtain the objective thrust constant. As follows, the

978-1-4799-2706-7/14 $31.00 © 2014 IEEE

TABLE II: Stator design for verifying thrust constant

Case	1	2	3	4	5
Pipe outside diameter [mm]	19	19	19	19	19
Pipe thickness [mm]	2.0	1.5	1.0	0.5	0.25
Magnet diameter [mm]	14	15	16	17	18
Magnetic gap [mm]	4.0	3.5	3.0	2.5	2.0

TABLE III: Simulation condition

Motion	
Step Number	41
Time step [s]	5.000×10^{-4}
Movement angle [deg/step]	0.5625
Pole pitch angle [deg]	11.25
Current	
Amplitude [A]	$\sqrt{2}$
Current frequency [Hz]	50.00
Phase of the U–phase [deg]	0
Coil resistance [ω]	4.007
Coil turns at each slot	400

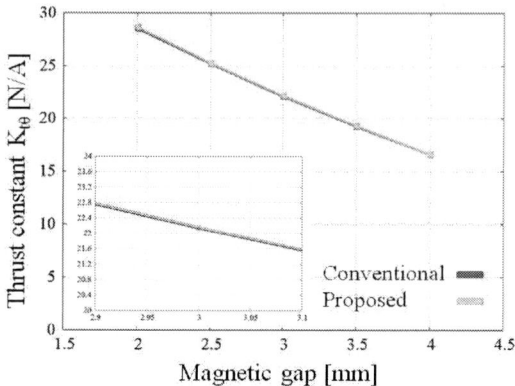

Fig. 4: Relation between thrust constant and magnetic gap

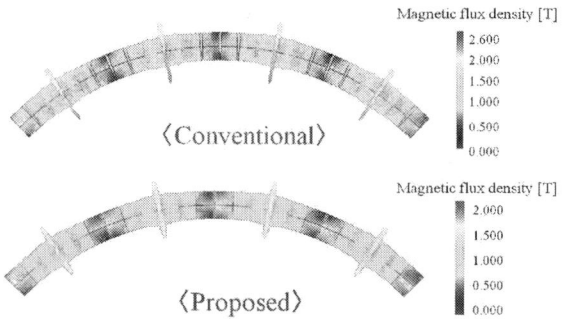

Fig. 5: Magnetic field distribution

influence on the thrust force by the length of magnetic gap and the effect of decrease of thrust ripple in both models is investigated by the magnetic field analysis. Both the conventional model and the proposed model are described in Fig. 1 and Fig. 2. The structure's difference between the conventional model and the proposed model are two points in the stator inside parts. The conventional model has iron pieces and uses cylindrical shape magnets that are magnetized in the axial direction. On the other hand, the proposed model has no iron pieces and includes the trapezoidal magnets that are magnetized in the axial direction. The mover part of the proposed model has the same one as those of the conventional model. The model parameter is shown in Table II.

B. Simulation condition

The simulation is performed as follows. Firstly, initial angle of the mover is set at $0°$. The mover moves in constant angular at each step. Three-phase AC is used as the drive current. Thrust force in the world coordinate is observed at each step. The thrust force in tangential direction is derived by equation (1). In addition, thrust ripple is evaluated by standard deviation in equation (3). The detailed simulation condition is shown in Table III.

C. Result on thrust constant

In order to realize the MDOF direct–drive motion system with CSMs for surgical application, higher thrust capacity is required for the CSM itself. Therefore, the relationship between thrust constant and magnetic gap is investigated. Fig. 4 shows the relation between the thrust constant and the length of magnetic gap. The thrust constant is derived by equation (2). From Fig. 4, both models are little different in the value of thrust constant. Fig. 4 confirms that the thrust force monotonically decreases with increasing the length of magnetic gap. From this result, the CSM has almost same characteristic in conventional linear motor about thrust constant value.

D. Result on thrust ripple

Fig. 5 shows magnetic flux density distribution for both models. The magnetic flux of the conventional model is interlinked at narrow area. The amount of interlinked magnetic flux is different between inside of coils and outside of coils. On the other hand, in the proposed model, the almost same amount of the magnetic flux is interlinked at each angle of the coils. Fig. 6 shows the comparison result of the thrust force generated at each step between the conventional model and the proposed model in the case 3. From Fig. 6, it is shown that thrust force waveform of the proposed model is very similar to that of the conventional model. Thrust constant at each model is shown in equation (4).

$$K_{t\theta\mathrm{con}}^{sim} = 22.13 [\mathrm{N/A}]$$
$$K_{t\theta\mathrm{pro}}^{sim} = 22.18 [\mathrm{N/A}] \qquad (4)$$

In this simulation, $K_{t\theta\mathrm{con}}^{sim}$ expresses the thrust constant of the conventional model and $K_{t\theta\mathrm{pro}}^{sim}$ expresses the thrust constant of the proposed model. In the thrust constant, it is shown that the proposed model has almost same value as the conventional model. In addition, the thrust ripple is calculated by simulation result based on equation (3).

The 2014 International Power Electronics Conference

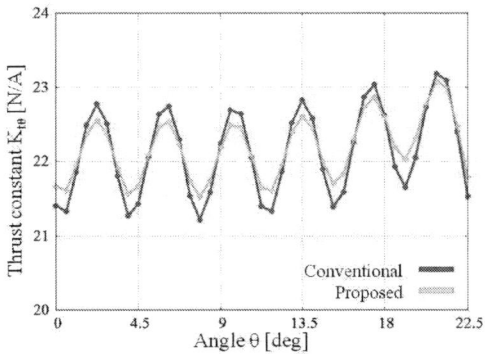

Fig. 6: Thrust force at each angle

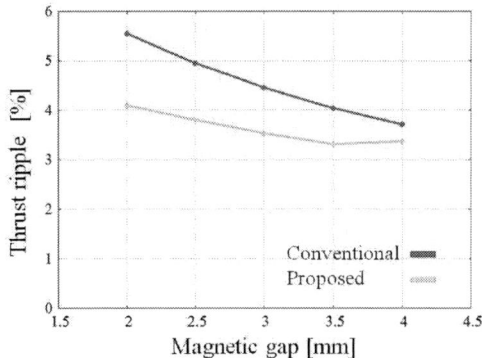

Fig. 7: Thrust ripple

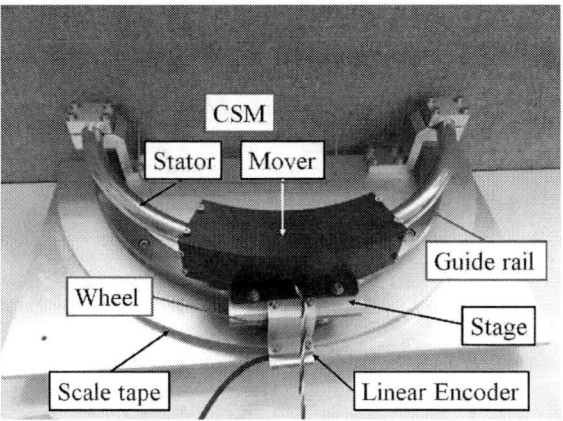

Fig. 8: Proposed CSM system

TABLE IV: Design parameter of prototype based on proposed model

Stainless pipe	
Material	SUS304
Outside diameter [mm]	19.0
Pipe wall thickness [mm]	1.20
Bend radius [mm]	175
Permanent magnet	
Material	Nd-Fe-B N-45SH
Diameter [mm]	16.0
longer Length [mm]	11.9
Cutting angle [deg]	1.88
Mover	
Number of mover slots	6
Coil turns at each slot	400
Gap length from magnet to coil [mm]	3.00
Linear stage size	
Radius of the magnetic tape [mm]	225
Angle range of movement [deg]	-52.0~52.0
Area [mm^2]	46.0 × 23.0
Optical linear encoder	
Manufacturer	RENISHAW
Model number	RGH24Y30A30A
Limit of resolution [μm]	0.100

The thrust ripple of each model is shown in equation (5).

$$\gamma_{con}^{sim} = 4.46\%$$
$$\gamma_{pro}^{sim} = 3.53\% \tag{5}$$

In this simulation, γ_{con}^{sim} expresses the thrust ripple of the conventional model and γ_{pro}^{sim} expresses the thrust ripple of the proposed model. As can be seen, the proposed model is about 1% smaller than the conventional model in the thrust ripple. From this result, the effect of decreasing the thrust ripple is confirmed in the proposed model.

V. PRIMARY EXPERIMENT

In this section, a developed CSM system based on proposed model is described. In order to calculate thrust constant of the proposed CSM, thrust test is performed as primary experiment of the proposed CSM system. The thrust ripple is also verified from these experimental results.

A. Proposed CSM system

Fig. 8 shows a proposed CSM system. This CSM system is developed based on the case 3 of the simulation

model. A same linear platform as a conventional CSM system is used for this system [13]. Thrust force is affected by the static friction force generated between linear guide and wheels. Therefore, thrust constant is calculated in consideration of the static friction force. Table IV shows the specifications of the prototype of CSM system based on proposed model.

B. Equipments and Conditions

Fig. 9 shows experimental setup for measurement of thrust force. In the experiment, an aluminum rod is attached to side of the mover. Firstly, a motor driver feeds alternate current to the coils in mover. Then, the aluminum rod contacts a load cell in order to measure generated thrust force. The thrust force is observed by an output meter. Table V shows the details of experimental equipment for the thrust force measurement.

Experimental condition is shown as follows. Firstly, the initial angle of the mover is set every 2.25° from 0°

978-1-4799-2706-7/14 $31.00 © 2014 IEEE

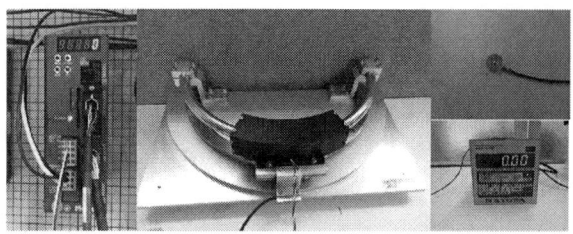

Motor driver	Proposed new CSM system	Load cell
		Output meter

Fig. 9: Experiment setup for thrust force measurement

TABLE V: Specifications of the experimental equipments

Motor driver	
Manufacturer	Servoland corporation
Model number	SVFM2-DSP
Rated current [A]	0.800
Maximum rated current [A]	2.00
Load cell	
Manufacturer	Kyowa Electronic Instruments
Model number	LMA-50A
Rating capacity [N]	50.0

to 22.5°. This range is equivalent to twice of pole pitch angle. Initial angle is determined by the reading value of the linear optical encoder. Setting various initial angles can investigate the thrust characteristic at each angle. The verification of thrust characteristic at each angle leads to verify the thrust ripple. Secondly, the alternate current is given to the mover. The current is fed in every 0.20A from 0.20A to 2.0A In order to investigate effect to the thrust characteristic by static friction force, small amplitude current in every 0.02A from 0.02A to 0.1A are also given to the mover. Under these conditions, the thrust force is measured three times at each current value. In order to consider the static friction force, the valid data is defined more than the minimum current that the mover can move to 2.0A for calculation of the thrust constant. The thrust constant can be calculated by using least square method based on valid data.

C. Result

Fig. 10 shows the thrust characteristic between the feeding current I and thrust force $F_{t\theta}$ at initial angle $\theta = 0°$. The red line is measured data of the thrust force by the experiment. The green line is approximated linear line of these measured data. As indicated in Fig. 10, the thrust characteristic between the current I and thrust force $F_{t\theta}$ is almost linear as well as a conventional CSM system. From this result, the thrust constant can be mathematically calculated by the least square method. The thrust constant of the proposed CSM system at $\theta = 0°$ is shown as follows.

$$K_{t\theta\text{pro}}^{exp}|_{\theta=0°} = 20.89[\text{N/A}] \qquad (6)$$

Then, thrust constant set at other initial angle is calculated as well as when initial angle is set at $\theta = 0°$. Fig. 11 shows the thrust constant at each angle. The red line

expresses the thrust constant at each angle. The green line expresses the average value of the thrust constant. From Fig. 11, the thrust constant of proposed CSM system is described as follows.

$$K_{t\theta\text{pro}}^{exp} = 20.54[\text{N/A}] \qquad (7)$$
$$(K_{t\theta\text{con}}^{exp} = 7.50[\text{N/A}])$$

$K_{t\theta\text{pro}}^{exp}$ expresses thrust constant of the proposed C-SM system and $K_{t\theta\text{con}}^{exp}$ expresses thrust constant of the conventional CSM system. In general, thrust constant is proportional to the feeding current. The proposed CSM system has twice coil turns of the conventional CSM system. In contrast, the proposed CSM system can generate about three times higher thrust force than the conventional system. Therefore, compared with the conventional CSM system, it is found that this model can realize improvement of the thrust constant even if increase of coil turns is taken into account. In addition, the thrust constant expresses the thrust force generated by the feeding current 1.0A. Therefore, based on equation (3), thrust ripple can be verified from Fig. 11. The thrust ripple is verified by the measured thrust constant as follows.

$$\gamma_{\text{pro}}^{exp} = 3.94\% \qquad (8)$$
$$(\gamma_{\text{con}}^{exp} = 4.85\%)$$

$\gamma_{\text{pro}}^{exp}$ expresses of the proposed CSM system and $\gamma_{\text{con}}^{exp}$ expresses of the conventional CSM system. This experimental data is equivalent to the result when step number is set to 11 in the simulation. Compared with the conventional CSM system, thrust ripple of the proposed CSM system is about 1% lower than that of the conventional CSM system. This result proves the improvement of the thrust ripple by the proposed CSM system.

VI. CONCLUSION

In this paper, the new CSM model is presented. The proposed model is designed in order to realize a multi–degrees–of–freedom direct–drive motion system. Firstly, the new structure of CSM using the trapezoidal magnet is indicated. Secondly, the evaluation index of thrust characteristics is expressed. Thirdly, the relation between magnetic gap and thrust constant is investigated by magnetic field analysis. In addition, the effect of decreasing thrust ripple in the proposed model is verified. Finally, the prototype of CSM based on the proposed model is developed, then the thrust characteristic of proposed CSM system is investigated by the primary experiment. From the experimental results, it can be said that thrust ripple can be decreased by the proposed CSM system.

ACKNOWLEDGMENT

This research was supported in part by Japan Society for the Promotion of Science under Grant-in-Aid for Young Scientists (A), 23676046. The authors would like to thank Mr. Noritaka Ishiyama, the Representative Director, President of GMC Hillstone Co. Ltd. for his valuable advice on the development of a circular shaft motor.

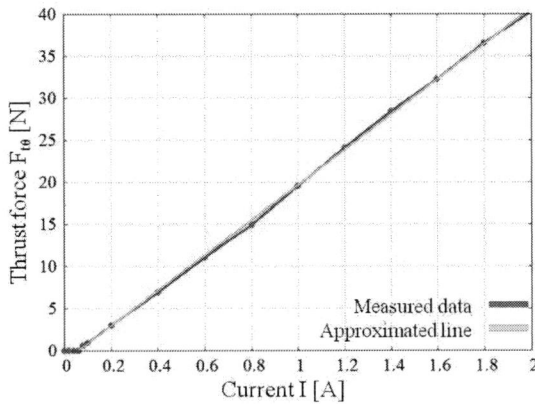

Fig. 10: Thrust characteristic of CSM at $\theta = 0°$

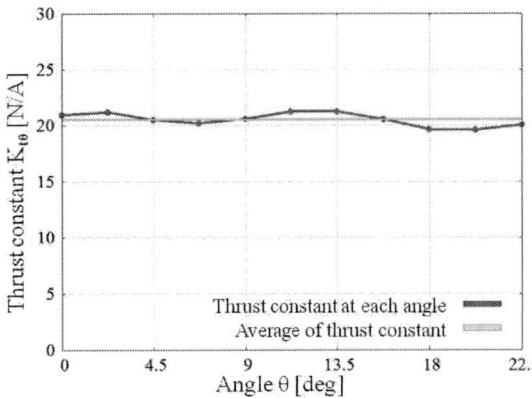

Fig. 11: Thrust characteristic of CSM at each angle

REFERENCES

[1] S.T. Boroujene, J. Milimonfared, and M. Ashabani, "Design, Prototyping, and Analysis of a Novel Tubular Permanent–Magnet Linear Machine," *IEEE Transactions on Magnetics*, Vol. 45, No. 12, pp. 5405–5413, 2009.

[2] R.D. Stefano and F. Marignetti, "Tubular Linear Permanent Magnet Actuator with Fractional Slots", *IEEJ Journal of Industry Applications*, Vol. 1, No. 3, pp. 172–177, 2012.

[3] J. Wang, G W. Jewell, and D. Howe, "A General Framework for the Analysis and Design of Tubular Linear Permanent Magnet Machines", *IEEE Transactions on Magnetics*, Vol. 35, No. 3, pp. 1986–2000, 1999

[4] J. Wang, D. Howe, and G W. Jewell, "Analysis and Design Optimization of an Improved Axially Magnetized Tubular Permanent–Magnet Machine," *IEEE Transactions on Energy Conversion,* Vol. 19, No. 2, pp 289–295, JUN, 2004.

[5] S. T Boroujeni, J. Milimonfared, and M. Ashabani, "Design, Prototyping, and Analysis of a Novel Tubular Permanent–Magnet Linear Machine" *IEEE Transactions on Magnetics,* Vol. 45, No. 12, pp. 5405–5413, DEC, 2009.

[6] I.A. Smadi, H. Omori, and Y. Fujimoto, "Development, Analysis, and Experimental Realization of a Direct–Drive Helical Motor," *IEEE Transactions on Industrial Electronics*, Vol. 59, No. 5, pp. 2208–2216, 2012.

[7] Y. Fujimoto, T. Kominami, and H. Hamada, "Development and Analysis of a High Thrust Force Direct–Drive Linear Actuator," *IEEE Transactions on Industrial Electronics*, Vol. 56, No. 5, pp. 1383–1392, 2009.

[8] T. T. Overboom, J. W. Jansen, E. A. Lomonova, and F. J. F. Tacken, "Design and Optimization of a Rotary Actuator for a Two–Degree–of Freedom $z\phi$–Module," *IEEE Transactions on Industry Applications*, Vol. 46, No. 6, pp. 2401–2409, 2010.

[9] K. Ohnishi, S. Katsura, and T. Shimono, "Motion Control for Real–World Haptics," *IEEE Industrial Electronics Magazine*, pp. 16–19, 2010.

[10] V. Agrawal, W.J. Peine, and B. Yao, "Model of Transmission Characteristics Across a Cable–Conduit System," *IEEE Transactions on Robotics*, Vol. 26, No. 5, pp. 914–924, 2010.

[11] S. Hyodo, Y. Soeda, and K. Ohnishi, "Vertification of Flexible Actuator From Position and Force Transfer Characteristic and Its Application to Bilateral Teleoperation System," *IEEE Transactions on Industrial Electronics*, Vol. 56, No. 1, pp. 36–42, 2009.

[12] H. Tanaka, K. Ohnishi, H. Nishi, T. Kawai, Y. Morikawa, S. Ozawa, and T. Furukawa, "Implementation of Bilateral Control System Based on Acceleration Control Using FPGA for Multi–DOF Haptic Endoscopic Surgery Robot," *IEEE Transactions on Industrial Electronics*, Vol. 56, No. 3, pp. 618–627, 2009.

[13] M. Omura, T. Shimono, and Y. Fujimoto, "Development of a Half–Circle–Shaped Tubular Permanent Magnet Machine", *Proceedings of the 39th Annual Conference of the IEEE Industrial Electronics Society, (IECON2013)*, 2013.

Design of a Bearingless Flux-Switching Slice Motor

Wolfgang Gruber, Karlo Radman
Institute of Electrical Drives and Power Electronics
Johannes Kepler University
Linz, Austria
wolfgang.gruber@jku.at

Reto. T. Schöb
Levitronix GmbH
Zurich, Switzerland

Abstract— **This work introduces a novel bearingless slice motor design, the bearingless flux-switching slice motor. In contrast to state-of-the-art bearingless slice motors the rotor in this new design does not possess any permanent magnets. This offers advantages for disposable rotor devices for instance in the medical industry and extends the range of bearingless slice motors towards high temperature applications. The force and torque generation of the novel drive design is outlined, the used control scheme is presented and the prototype is described. Conducted measurements conclude the paper.**

Keywords— *bearingless motor, magnetic levitation, self-bearing drive, flux-switching*

I. INTRODUCTION

Using an air gap field with permanent magnetic bias flux and a disk-shaped rotor, it becomes possible to stabilize three degrees of freedom (the two tilting and the axial displacement) passively by reluctance forces. The active control of the remaining three degrees of freedom (the radial movement and the rotation) by the stator currents completes a fully magnetically levitated drive. Such systems are called bearingless slice motors and were first introduced in 1996 [1]. Steady development led to industrial applications of these drives, mainly as pumps in the medical and semiconductor industry [2], [3].

However, in state-of-the-art bearingless slice motors, the permanent magnets exciting the air gap bias flux are located in the rotor [4]. Thus, the rotor cross section of bearingless motors shows a strong affinity to rotors of brushless permanent magnet synchronous machines.

Especially in high temperature or high speed applications and in disposable devices (where the rotor has to be replaced frequently), permanent magnet free rotors would be favourable. Using a magnet-free rotor setup leads to reduced manufacturing costs (the state-of-the-art rare earth magnet materials have increased in price significantly in the last years) and enhanced thermal and mechanical robustness. First studies concerning bearingless slice motors without magnetic material in the rotor, also called bearingless reluctance slice motors, were published recently in [5] and [6]. The motor variant considered in this work, the bearingless flux-switching slice drive, constitutes an additional promising possibility to realize such a system.

II. BEARINGLESS FLUX-SWITCHING MOTOR

The main constructional characteristic of the flux-switching motor, depicted in Fig. 1, is the placement of the permanent magnets. They are located in the midst of the stator teeth and separate them electromagnetically. The flux lines of each permanent magnet close mainly over the neighbouring stator teeth and, thus, do not penetrate the whole air gap, but only influence the air gap region close to the permanent magnet. Unfortunately, there is also a considerable amount of fringing flux closing over the outer side of the stator. To keep this kind of stray flux low, a small air gap length (compared to the magnet width) is favourable and necessary.

However, the air gap flux density in flux-switching machines is often higher than in comparable other electric machines, leading to an increased torque capacity [7]. This is due to the flux concentration capability of the construction. The surfaces of the stator teeth normal to the magnetization direction collect permanent magnetic flux and concentrate it towards the much smaller area of the stator teeth adjoining the air gap.

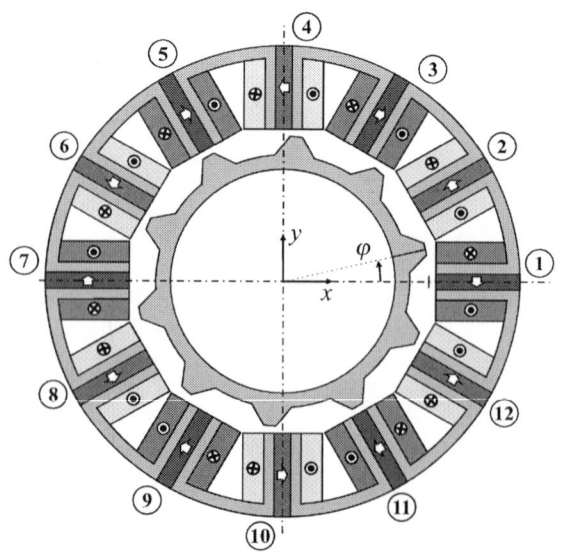

Fig. 1. Cross section of a flux-switching motor with twelve stator and ten rotor teeth. The arrows indicate the direction of the magnetization.

III. Force and Torque Generation

Due to the saliency of the rotor, the permanent magnetic flux Ψ_{PM} linked with the stator coils changes its direction, resulting in an induced stator back-electromotive force. This is a necessity for linear permanent magnetic torque generation, because

$$T_z(\varphi) = \frac{\partial \Psi_{PM}}{\partial \varphi} i_s \tag{1}$$

holds true with the rotor angle φ and stator coil current i_s.

Additionally, the superposition of the stator field with the permanent magnetic field also leads to a creation of bearing forces by field strengthening and weakening, respectively. Both principles are depicted in Fig. 2 and lead to the possibility of bearingless motor operation for this kind of machine. Unfortunately the force generation of one coil features a nonlinear correlation with the referring coil current. This fact complicates the necessary decoupling of x- and y-force, necessary for stable levitation. However, a series connection of two opposing coils in a bearingless flux-switching motor linearizes the force to phase current relationship. A closer look into the force and torque generation of the flux switching motor is given in [8]. In this work also theoretical design consideration led to the result, that a bearingless flux-switching motor with ten rotor and twelve stator teeth offers favourable operational behaviour. Thus, it was decided to build such a system as depicted in Fig. 1. Due to the announced series connection of opposing coils the overall bearingless slice motor consists of six phases, three for force generation (indicated with yellow coils in Fig. 1) and another three for torque generation (represented by orange coils in Fig. 1).

Because of the linear behaviour the following description of the radial suspension forces (F_x, F_y) and drive torque (T_z)

$$\begin{pmatrix} F_x(\varphi) \\ F_y(\varphi) \\ T_z(\varphi) \end{pmatrix} = T_m(\varphi) \begin{pmatrix} i_1 & i_2 & i_3 & i_4 & i_5 & i_6 \end{pmatrix}^T \tag{2}$$

holds true. i_1 to i_6 represent the six stator phase currents and $T_m(\varphi)$ stands for the so called current-force-matrix. The characteristic normalized curves of a force and a torque phases are depicted in Fig. 3. The direction of the tangential force F_t and the normal force F_n refer to the axis of the considered phase (defined by the two opposing coils connected in series).

For a proper decoupling of suspension forces and drive torque the inverse relation $K_m(\varphi)$, also referred to as force-current matrix, has to be found. It yields

$$\begin{pmatrix} i_1 & i_2 & i_3 & i_4 & i_5 & i_6 \end{pmatrix}^T = K_m(\varphi) \begin{pmatrix} F_x \\ F_y \\ T_z \end{pmatrix}. \tag{3}$$

A reasonable computation for this matrix is given by

$$K_m(\varphi) = T_m(\varphi)^T \left(T_m(\varphi) T_m(\varphi)^T \right)^{-1}, \tag{4}$$

additionally minimizing the copper losses [9].

IV. Winding System

It was already mentioned that the considered bearingless flux-switching slice motor features six phases. Energizing all six phases independently leads to

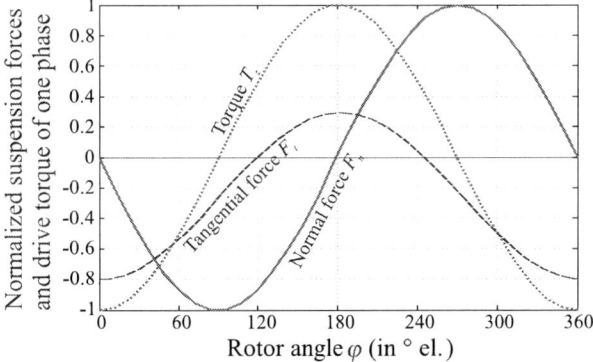

Fig. 3. Bearing force and motor torque (acting on the rotor) with proper connection of opposing coils at constant flux linkage over the rotor angle (only the fundamental wave is considered).

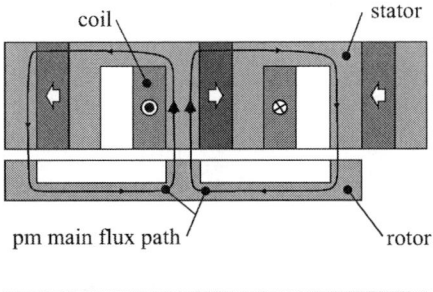

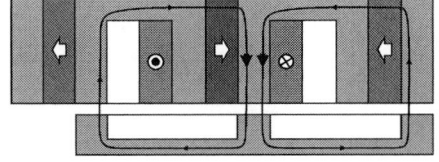

Fig. 2. Basic principles regarding the generation of drive torque (left) and suspension force (right). For sake of simplicity the flux-swiching motor is illustrated in a planar and segmented way.

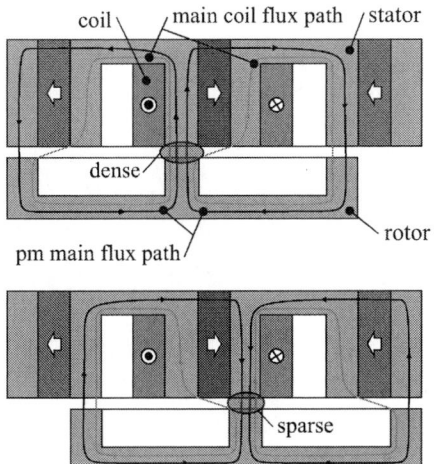

978-1-4799-2706-7/14 $31.00 © 2014 IEEE

the need for a power electronics with six full bridges (and therefore 24 power switches) and six current sensors to control the phase currents properly. To save power electronics effort, often a connection of the phases is implemented. Especially a connection in star is common and allows the usage of half bridges instead of full bridges, reducing the necessary power switches by half to twelve and the necessary current sensors to five. In previous works with six phase bearingless slice motors a further simplification in the winding system has been proven to be feasible. A double star connection, featuring three phases each [10], [11], was implemented.

By the help of connection matrices, that have to be multiplied with $T_m(\varphi)$ and represent the according phase connections, three different $K_m(\varphi)$-matrices (for the three different winding systems) can be deduced. The entries of these matrices help to evaluate the operational behaviour of the system. Figure 4 shows the capability of force generation in x-direction and drive torque generation in dependence of the rotor angle for all three described winding systems. These curves are normalized. Thus, an amplitude of 2 means, that the overall system is capable

to create twice the force of a singe phase.

Figure 4 incorporates all three considered winding systems. In the first orbital section with a mechanical rotor angle from 0 to 120° no winding connection is used. In the following area the behaviour of a single star connection of the six phases is illustrated. Finally, the last sector gives the results of a double star connection. It can be seen that a single star connection does not deteriorate the performance of the drive, whereas a double star connection worsens especially the bearing force generation. Hence, this composition is not feasible. The reason for this is the unique non centred force orbit, resulting in the need of a common mode current component in the three force creating phases. Thus, the single star winding connection was chosen for actual implementation.

V. CONTROL SCHEME

With the help of (4), Nonlinear feedback transformation is used to decouple the bearing forces and the drive torque in the control system accordingly. This method is proven for bearingless slice motors [5], [6], [10]. Figure 5

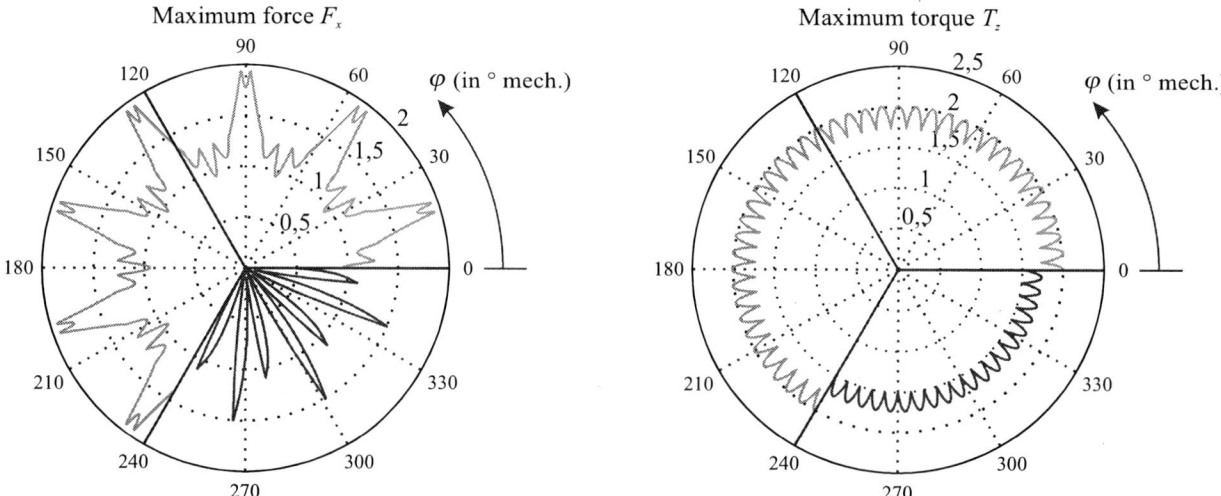

Fig. 4. Orbital plots of the producible maximum suspension force (left) and the generateable drive torque (right) over the rotor angle for three winding systems. The orange curve represents the unconnected six phases system, cyan the single star point connection and blue the double star point connected configuration.

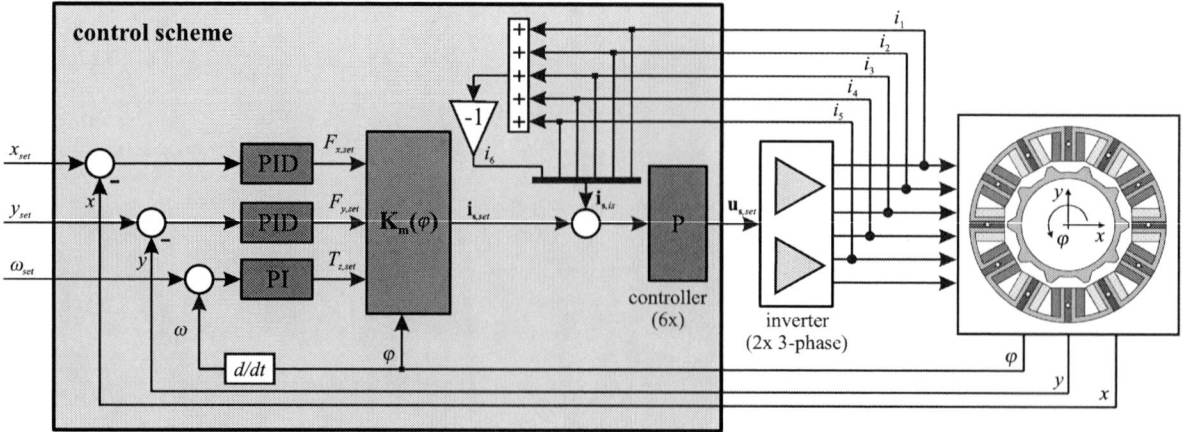

Fig. 5. Utilized nonlinear feedback control scheme block diagram.

978-1-4799-2706-7/14 $31.00 © 2014 IEEE   1693

illustrates this control scheme with outer PID position and PI speed controller. Faster nested P current controllers impress the necessary currents into the phases with the help of a standard half bridge voltage inverter.

VI. FINITE ELEMENT OPTIMIZATION

A 3D finite element optimization of the geometry illustrated in Fig. 1 was conducted with the software package Maxwell 3D from Ansys. Previous works [12], [13] optimized the geometric parameters of standard (non bearingless) flux-switching motors with respect to torque generation. From these papers, favorable geometric parameters are known. For instance, a recommendation to set the rotor to stator diameter ratio between 0.55 and 0.6 is given. Additionally, it is favorable to have similar values for the stator tooth width, the magnet width and the stator slot width; whereas the rotor tooth width should be about 1.4 times bigger than the stator tooth width.

Beside the maximization of the motor torque, also the increase of the passive stabilizing stiffnesses and the achievable bearing forces are optimization criteria. During the finite element simulations it turned out that saturation effects severely restrict the parameter choice, especially of the stator geometry. Due to the limited design space, saturation effects occur in the iron of the stator teeth (at higher current densities or torque creation, respectively) as depicted in Fig. 6. Hence, the geometry was adopted accordingly. The main geometric and electromagnetic characteristic parameters are visible in

TABLE I
MAIN GEOMETRIC PARAMETERS OF THE PROTOTYPE

Variable	Description	Value	Unit
d_{so}	stator iron outer diameter	266	mm
d_{ro}	rotor outer diameter	150	mm
h_z	axial rotor and stator length	10	mm
δ	magnetic air gap	3	mm
w_{pm}	magnet width	8,5	mm
l_{pm}	magnet length	54	mm
laminated iron material		M330-35A	
magnetic material		NdFeB N38	

TABLE II
ELECTROMAGNETIC CHARACTERISTIC OF THE PROTOTYPE

Variable	Description	Value	Unit
k_r	radial stiffness	-41.6	N/mm
k_z	axial stiffness	7.3	N/mm
k_φ	tilt stiffness	21.4	Nm/rad
J_{max}	max. current density	6	A/mm²
F_{max}	max. bearing force	30	N
T_{max}	max. drive torque	1.5	Nm

Fig. 7. The referring geometric and electromagnetic values after the optimization process are summarized in Table I and II.

VII. PROTOTYPE

Figure 8 pictures the manufactured prototype motor. The laminated stator iron stack is held by an aluminum

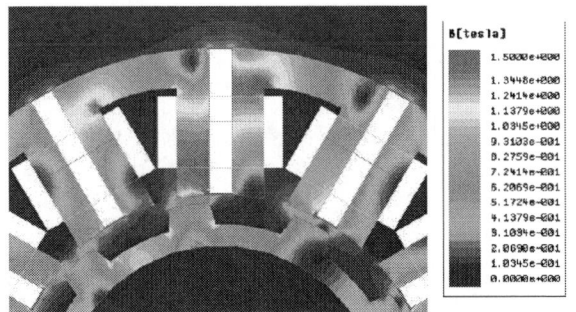

Fig. 6. Flux density plot at maximum current density, obtained from a 3D finite element simulation: the stator teeth are close to saturation.

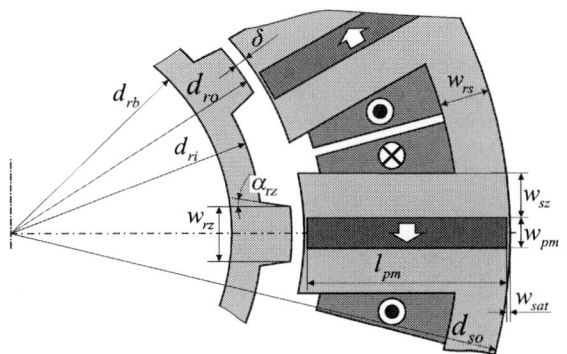

Fig. 7. Geometry parameters of the flux-switching bearingless slice motor varied in the optimization run.

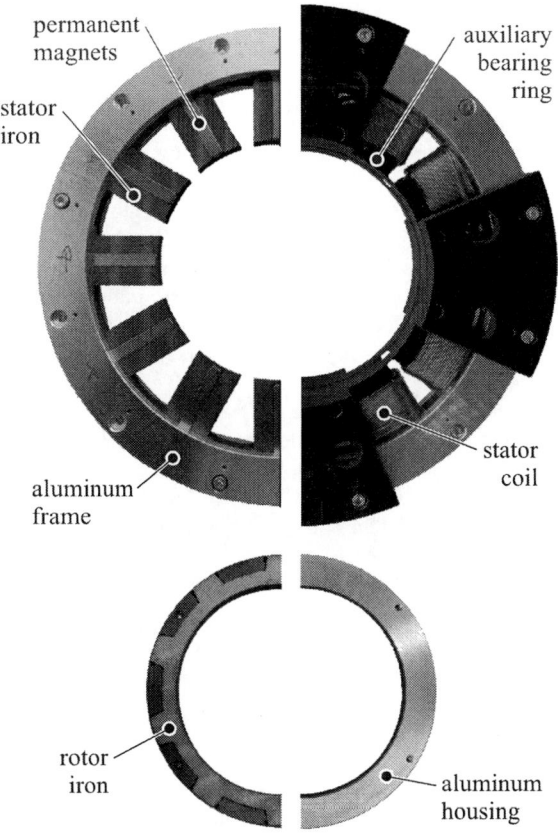

Fig. 8. Picture of fully assembled and unassambled stator and rotor.

978-1-4799-2706-7/14 $31.00 © 2014 IEEE

frame to strengthen the structure. The permanent magnets are glued into pockets that separate the stator elements. The externally wound stator coils are easily slipped on the stator teeth and connected accordingly. Hall and eddy current sensors are placed on the inside of the stator and in-between the stator teeth to measure the radial deflection and the rotor angle. An aluminum housing covers the rotor and acts as target for the eddy current sensors.

The used power electronics inverter is composed of two three-phase power modules. The rated phase voltage and current of the inverter are 300V and 15A, respectively. The nonlinear control scheme for proper decoupling of force and torque is implemented on a 32-bit digital signal processor, which is placed in the integrated control electronics of the inverter.

VIII. MEASUREMENTS

A. Static Measurements

The prototype system was mounted on a test bench with x-y-z-table, load cell, torque sensor and load machine. This setup is quite universal and allows the measurement of a lot of different values. Generally, the conducted measurements show good correlation with the finite element simulation results.

Exemplarily, the force orbit and the torque characteristic at constant magneto-motive force are depicted in Fig. 9 and 10. The force orbit is quite unique due to its offset in tangential direction. Both the force and the torque curves differ only by about 10% from the expected values.

Dynamic Measurements

Additionally, measurements were conducted under bearingless operation to check the functionality of the control scheme and its performance. The reaction of the bearingless motor for a step change of the reference position in x-direction is illustrated in Fig. 11. It can be seen that after about 80ms the new position is stably reached. Due to the used controller parameters an overshot and oscillation is observed. A small cross coupling between x- and y-axis is present, because there is also a reaction in y-position signal. This measurement was conducted at zero speed.

Furthermore, also a disturbance feedforward x-force step was implemented, again at zero speed. The reaction of the bearingless motor is visible in Fig. 12. The error is compensated and the original position signal reached after 100ms.

Finally, rotor orbit measurements are presented in Fig. 13 at two different rotor speeds. For an operation below critical speed it is typical that the rotor orbit rises with higher speed due to unbalance effects.

IX. CONCLUSION

This work sheds light on the design of a novel bearingless slice motor type, the bearingless flux-switching slice motor. Starting with theoretical consideration about the basic design parameters and the winding system, this research work concludes in a

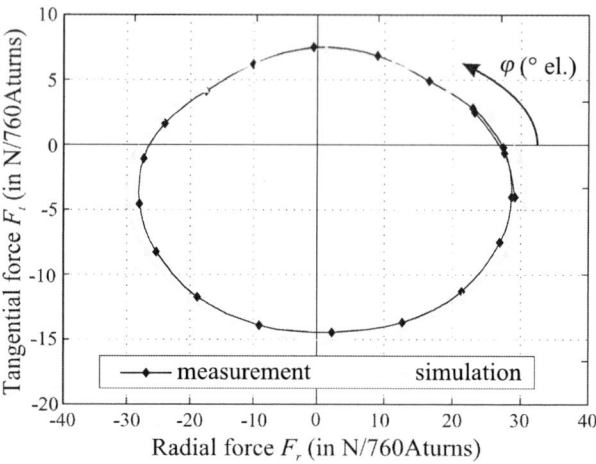

Fig. 9. Measurement of the suspension force orbit of one phase at constant magneto-motive force.

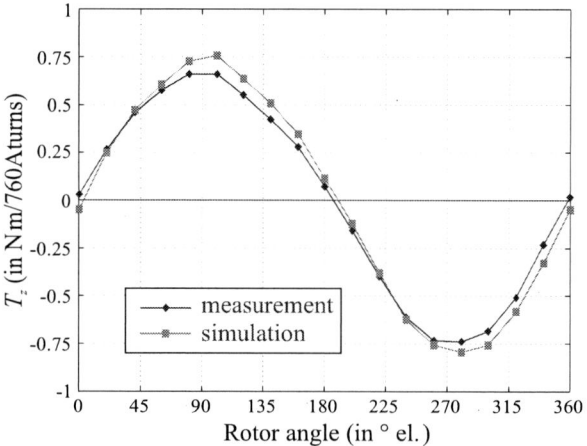

Fig. 10. Measured torque characteristic of one phase.

prototype drive. Several static and dynamic measurements show the functionality and potential of the bearingless flux-switching slice motor for industrial applications.

Ongoing work the efficiency of the system will be tested. Concerning this issue the utilization of the

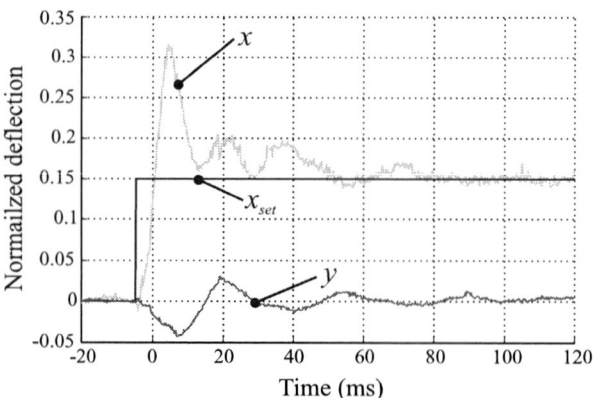

Fig. 11. Position reference step response in x-direction.

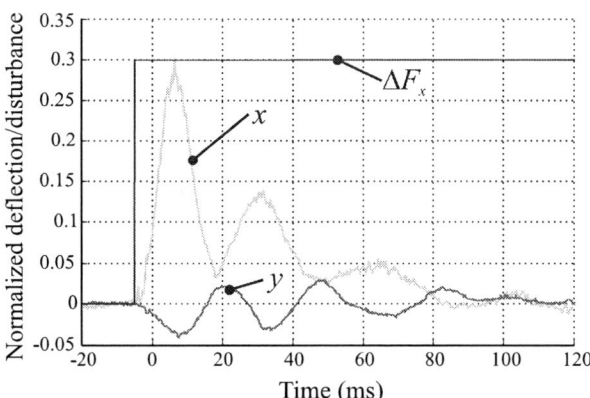

Fig. 12. Feedforward disturbance of a step force in x-direction.

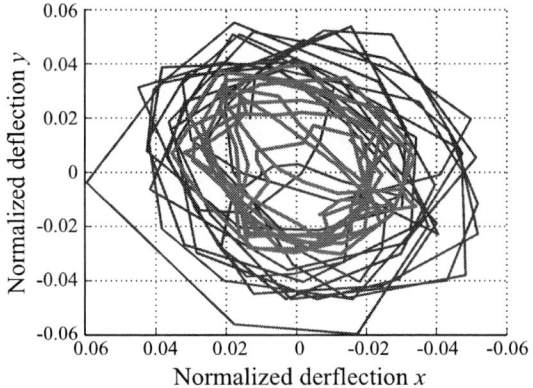

Fig. 13. Rotor position orbit at 300rpm (red) and 500rpm (blue).

aluminum rotor housing is very disadvantageous, because of the heteropolar magnetic air gap field. Thus, a lot of eddy current losses are created in the aluminum. In an updated version the measurement system (which is in need of the aluminum target) will be updated and placed on the inside of the rotor. This measure will avoid these losses and boost efficiency.

ACKNOWLEDGMENT

Parts of this work were supported by the Linz Center of Mechatronics (LCM) GmbH, a K2-centre of the COMET program of the Austrian Government. The authors thank the Austrian and Upper Austrian Government for their support.

REFERENCES

[1] R. Schöb, N. Barletta, "Principle and application of a bearingless slice motor", Proc. 5th Int. Symp. on Magnetic Bearings (ISMB), pp. 333-338, 1996

[2] J. Asama, T. Fukao, A. Chiba, M. A. Rahman, T. Oiwa, "A design consideration of a novel bearingless disk motor for artificial hearts", Proc. 1st IEEE Energy Conversion Congress and Exposition (ECCE), pp. 1693-1699, 2009

[3] T. Nussbaumer, K. Raggl, P. Boesch, J. W. Kolar, "Trends in integration for magnetically levitated pump systems", Proc. Power Conversion Conf. (PCC), pp. 1551-1558, 2007

[4] T. Nussbaumer, P. Karutz, F. Zürcher, J. W. Kolar, "Magnetically levitated slice motors—an overview", IEEE Trans. on Industry Applications, vol. 47, no. 2, pp. 754-766, 2011

[5] W. Gruber, W. Briewasser, M. Rothböck, R. Schöb, "Bearingless slice motor concepts without permanent magnets in the rotor", Proc. IEEE Int. Conf. on Industrial Technology (ICIT), pp. 259-265, 2013

[6] W. Gruber, M. Rothböck, R. Schöb, "Design of a novel homopolar bearingless slice motor with reluctance rotor", Proc. 5th IEEE Energy Conversion Congress and Exposition (ECCE), 2013

[7] E. Hoang, A. H. Ben-Ahmed, J. Lucidarme, "Switching flux permanent magnet poly-phase synchronous machines", Proc. 7th European Conf. on Power Electronics and Applications, pp. 903-908, 1997

[8] W. Gruber, W. Bauer, K. Radman, W. Amrhein, R. T. Schöb, "Considerations regarding bearingless flux-switching slice motors", Proc. 1st Brazilian Workshop on Magnetic Bearings, 2013

[9] S. Silber, "Power optimal current control scheme for bearingless pm motors", Proc. 7th Int. Symp. on Magnetic Bearings (ISMB), 2000.

[10] W. Gruber, T. Nussbaumer, H. Grabner, W. Amrhein, "Wide Air Gap and Large-Scale Bearingless Segment Motor With Six Stator Elements", IEEE Trans. on Magnetics, vol. 46, no. 6, pp. 2438-2441, June 2010

[11] H. Grabner, S. Silber, W. Amrhein, "Bearingless torque motor - modeling and control", Proc. 13th Int. Symp. on Magnetic Bearings (ISMB), 2012

[12] Z. Q. Zhu, Y. Pang, D. Howe; S. Iwasaki, R. Deodhar, A. Pride, "Analysis of electromagnetic performance of flux-switching permanent-magnet machines by nonlinear adaptive lumped parameter magnetic circuit model", IEEE Trans. on Magnetics, vol. 41, no. 11, pp. 4277-4287, 2005

[13] Z. Q Zhu, Y. Pang, J. Chen, Z. P. Xia, D. Howe, "Influence of design parameters on output torque of flux-switching permanent magnet machines", Proc. 2008 IEEE Vehicle Power and Propulsion Conference (VPPC), pp. 1-6, 2008

Proposal of a Permanent Magnet Hybrid Type Axial Magnetically Levitated Motor

Nobuyuki Kurita, Takeo Ishikawa, Hiromu Takada, Genri Suzuki

Division of Electronics and Informatics
Gunma University
Kiryu Gunma Japan
nkurita@gunma-u.ac.jp

Abstract— A permanent magnet hybrid type axial magnetically levitated motor is proposed. The motor consists of a spherical permanent magnet and an axial type bearing-less motor that has tilt control function. The rotor has two permanent magnets on one side, and the motor stator has eight poles, which include eight concentrated windings. The operating principle was investigated by numerical analysis. It was verified that the proposed motor can control the translational motion, inclinational motion and rotational motion independently.

Keywords— *Axial flux motor, Bearingless motor, Magnetically levitation technology, Passive magnetic bearing.*

I. INTRODUCTION

A magnetically levitated motor (maglev motor) can support and rotate a rotor by using magnetic force with no mechanical contact and, therefore, has many advantages over conventional mechanical bearings [1], [2]. In order to accomplish complete magnetic levitation of the rotor, however, it is necessary to control five axes actively. Two radial magnetic bearings and one axial magnetic bearing are usually required, which gives rise to major issues, such as a complicated levitation control system and associated electronics that result in enlarged equipment size.

In order to down size a maglev motor and to simplify the control system, several types of bearing-less motors (BℓM) have been proposed and researched, such as a radial-BℓM [3]-[4] and an axial-BℓM [5]-[6]. To increase passive stability, the rotor shape should be a disk shape for radial-BℓM or a long cylindrical shape for axial-BℓM. Consequently, permanent magnet (PM) size and stator surface area will be decreased due to these rotor shapes. Additionally, these geometries are associated with low rotational torque and weak suspension force.

To solve this problem and increase PM size and stator surface area, several types of axial maglev motors that have inclination motion control function were researched. Shimbo et al. [7] and Chongk-wanyuen el al. [8] used an axial-BℓM that has a PM synchronous motor on the upper side and a reluctance motor on the lower side. Osa et al. [9] and Takada et al. [10] used an identical stator on both sides of the rotor. Since these axial-BℓMs required two stators to control the tilt motion, the number of amplifiers required for levitation control increases, and

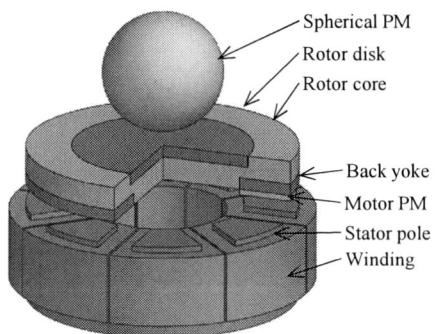

Fig. 1. Schematic of proposed maglev motor.

the levitation control becomes more complex.

As an alternative, the research proposed in this paper details a PM hybrid type axial maglev motor, shown schematically in Fig. 1. The maglev motor consists of a spherical PM and an axial-BℓM which has a tilt control function. A spherical permanent magnet was selected in order to minimize the negative torque produced by the upper PM. The number of amplifiers required is only eight this is half as many as compared with the double stator type magnetically levitated axial gap motor [10]. Moreover, the structure of the proposed motor is very simple.

The operating principle is described below, as is an FEM magnetic analysis that indicates the proposed maglev motor can control each degree of freedom independently. Based on this theoretical framework, an experimental setup was fabricated. Magnetically levitation was achieved and the control performance of the maglev motor was also investigated.

II. OPERATING PRINCIPLE

The proposed maglev motor consists of a spherical PM and a conventional axial-BℓM. The rotor of the axial-BℓM has two PMs, and the stator has eight salient poles with a concentrated winding on each pole. The flux density distribution produced in the PM lower-side airgap is described by (1).

$$B_{pm}(\theta, t) = B_{PM} \cos(\omega t - \theta), \qquad (1)$$

where B_{PM} (T) is the maximum value of the rotor PM flux

The 2014 International Power Electronics Conference

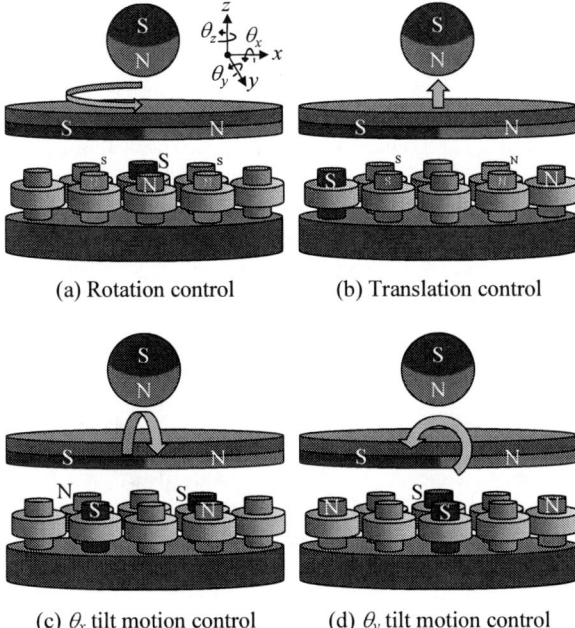

(a) Rotation control   (b) Translation control

(c) θ_x tilt motion control   (d) θ_y tilt motion control

Fig. 2. Magnetic pole arrangements.

density, ω (rad/s) is rotation speed, t (sec) is time, and θ (rad) is angle.

To control the rotor rotation θ_z, a two-pole rotating magnetic field of phase difference ψ (rad) is generated by the stator windings. The flux density $B_{\theta z}$ of the rotational control current can be written as (2)

$$B_{\theta z}(\theta, t) = B_{\Theta Z} \cos(\omega t - \theta + \psi),\qquad(2)$$

where $B_{\Theta Z}$ is the maximum value of the flux density produced by the motor current.

A schematic of magnetic pole arrangements is shown in Fig. 2 (a). The spherical PM attracts the rotor upward. The near-side poles (magnetized to N poles) attract the S pole of the rotor PM, while the far-side salient poles (magnetized to S poles) attract the N pole of the rotor PM. These magnetic attractive forces result in a rotating torque in a counter-clockwise direction.

To control the axial translation motion z, the stator windings should produce a two-pole magnetic field which has the same phase as the PM magnetic field. The flux density B_z created by the axial direction control current thus described as (3)

$$B_z(\theta, t) = B_Z \cos(\omega t - \theta),\qquad(3)$$

where, B_Z is the maximum value of the flux density produced by the axial direction control current. Fig. 2 (b) shows the magnetic pole arrangements of the rotor PM and the axial direction control current. Here, the left side poles are magnetized to S poles at the same time the right side poles are magnetized to N poles, so a repulsive force acts between the rotor and the stator, resulting in an upward suspension force acting on the rotor.

Meanwhile, to control the tilt motion θ_x and θ_y, a four-pole magnetic field should be generated by the control current. The flux density $B_{\theta x}$ and $B_{\theta y}$ for θ_x and θ_y tilt motion control are given by (4)

$$\begin{aligned} B_{\theta x}(\theta, t) &= B_{\Theta} \sin(\omega t - 2\theta) \\ B_{\theta y}(\theta, t) &= B_{\Theta} \cos(\omega t - 2\theta) \end{aligned},\qquad(4)$$

where B_{Θ} is the maximum value of the flux density produced by the tilt control current. In Fig. 2 (c), the left near side pole and the right far side pole are magnetized to N pole, and the left far side pole and the right near side pole are magnetized to S pole. In this configuration, an attractive force acts on the rotor at the near side and a repulsive force acts in far side. As a consequence of these forces, a restoring torque acts in θ_x direction. A restoring torque for θ_y direction is produced in the same way. Additionally, radial directional translation motions are controlled passively.

III.   MAGNETIC FIELD ANALYSIS

In order to verify that rotation control, translation control, and inclination control are not cross-coupled, an FEM magnetic field analysis was carried out; principle sizes of the model are listed in Table I. The rotor was located at the geometric center of the spherical PM and motor stator. The airgap between the rotor and the stator was 3 mm. By applying each control current (rotation control current $I_{\theta z}$, axial directional translation motion control current I_z, and tilt control current $I_{\theta x}$, $I_{\theta y}$) to each coil according to (2), (3) and (4), the rotation torque $\tau_{\theta z}$, Translation suspension force F_Z and restoring torque $\tau_{\theta x}$, $\tau_{\theta y}$ were calculated.

When the phase difference of the stator pole and rotor pole is $\psi = \pi/2$, the relationships between rotational torque $\tau_{\theta z}$ and each control current are as shown in Fig. 3 (a). Specifically, a linear relationship exists between the rotational torque $\tau_{\theta z}$ and rotation control current $I_{\theta z}$, but the rotational torque created by translation control current and tilt control current are negligibly small.

TABLE I
PRINCIPLE SIZE OF AN FEM ANALYSIS MODES

Rotor		
Disk diameter	ϕ30	mm
Disk thickness	1	mm
Back yoke OD	ϕ45	mm
Back yoke ID	ϕ27	mm
Back yoke thickness	2	mm
PM thickness	1	mm
Motor stator		
OD	ϕ45	mm
ID	ϕ27	mm
Pole height	13	mm
Slot width	4	mm
Winding number	129	turns/pole
Spherical PM size	ϕ30	mm

OD: Outer diameter, ID: Inner diameter

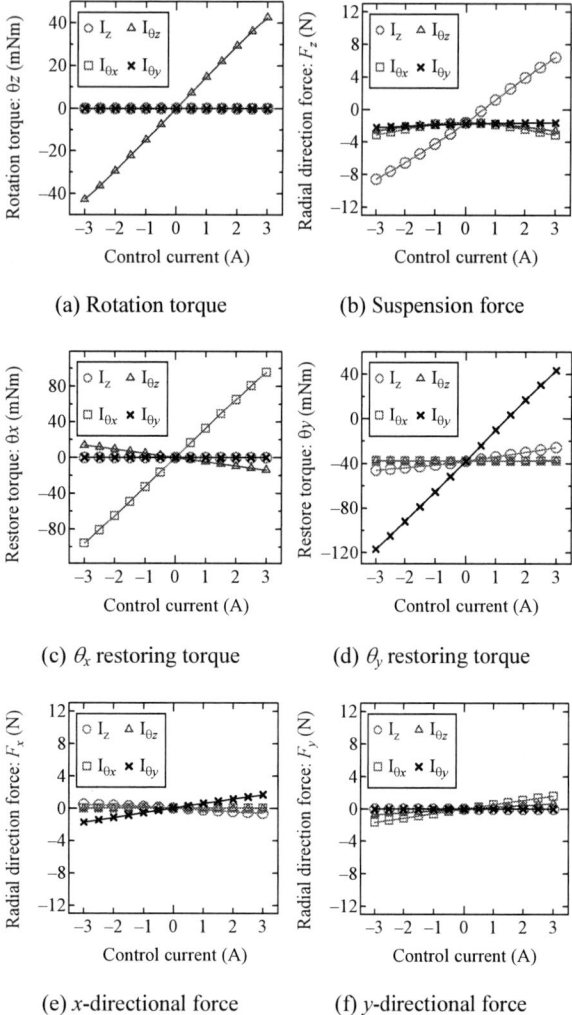

(a) Rotation torque

(b) Suspension force

(c) θ_x restoring torque

(d) θ_y restoring torque

(e) x-directional force

(f) y-directional force

Fig. 3. Analytical result of interference of control current.

Fig. 3(b) shows the relationships between the axial direction suspension force F_Z and each control current. When the upper and lower airgaps are both 3 mm, the resulting attractive force by the motor PMs is stronger than the attractive force of the spherical PM by 1.8 N. The suspension force F_Z is approximately proportional to the translation motion control current I_z. With a low control current of ±1.5 A, the suspension force produced by the rotation and both tilt control currents are negligible. However, when current of more than ±1.5 A is applied, the rotation and θ_x tilt control current produce a suspension force. This force is due to the interaction of the flux produced by the spherical PM on the flux produced by the rotor PM, which results in the flux density distribution in the lower side airgap varying from the ideal sinusoidal wave.

The relationship between restoring torque $\tau_{\theta x}$ and each control current are shown in Fig. 3(c). Restoring torque is mainly produced by θ_x tilt control current $I_{\theta x}$, however, the translation motion control current $I_{\theta z}$ does contribute some torque. Similarly, Fig. 3(d) shows a similar

relationship between restoring torque $\tau_{\theta y}$ and each control current. The variation of the flux density distribution of the motor PM caused by the flux of the spherical PM produced a constant torque of 37 mNm in the θ_y direction. As seen for the other tilt direction, the restoring torque is mainly produced by θ_y tilt control current $I_{\theta y}$, but some contribution also comes from the translation motion control current I_z. Those interferences of torque and force are again due to the variation of the flux density distribution of the motor PM. We postulate that a closed loop magnetic circuit for the upper side PM would decrease the mutual interference of torque and force.

Finally, Fig. 3(e) and (f) shows the relationship between radial directional force and the each control current. Interferences of torque and force are observed. In this case x-directional force produced by translation motion control current I_z and y-directional force produced by the rotation control current $I_{\theta z}$ are thought to be due to the variation of the flux density distribution of the motor PM. Meanwhile, the x-directional force produced by tilt control current $I_{\theta y}$ and and y-directional force produced by tilt control current $I_{\theta z}$ are thought to be due to the structure of the maglev motor. Forces in the radial direction were not observed in the double stator type maglev motor [11] because they were canceled due to the symmetric stator design. It is difficult to reduce these forces in the proposed motor without using an additional electromagnet; however, the magnitude of this force is small enough such that it does not pose a significant problem for magnetic levitation control.

IV. EXPERIMENTAL RESULTS

In order to confirm levitation control performance, a simple experimental setup was designed and fabricated. To control the axial translation motion, the tilt, and the rotation independently, eight sets of linear amplifiers are required. Rotor levitation control with 0 min^{-1} rotation was achieved. Once the rotor was levitated stably, an impulse disturbance was applied, and the rotor displacement and control current were recorded.

As shown in Fig. 4(a), the response of the axial controller is fast: the rotor vibration dropped to less than less than ±5% of the maximum displacement within about 0.02 sec in response to an impulse input which moved the rotor about 0.1 mm in the z-direction. Moreover, since there was almost no rotor vibration in the inclination direction (θ_x, θ_y) by the z-direction perturbation, the axial direction translation control does not appear to affect the control performance of the tilt control.

The response of the tilt controller was measured to be similarly quick for an impulse tilting input of about 0.4 deg in the θ_x-direction, the rotor's rotational vibration became less than ±5% of the maximum tilt within about 0.03 sec. It was also verified that the tilt control also does not affect the control performance of the axial translation control.

Fig. 4(c) shows the rotor behavior in response to an impulse input which moved the rotor about 1.0 mm in the

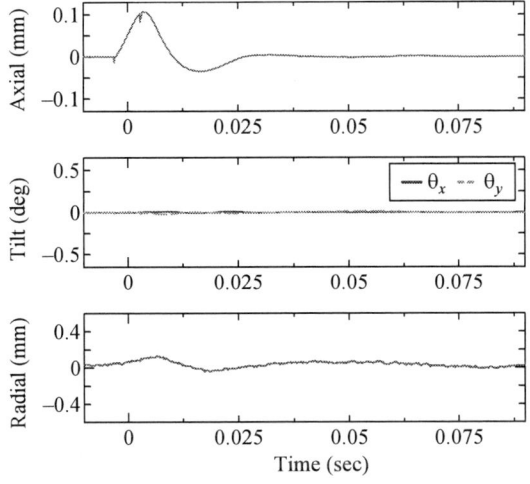

(a) Disturbance input: Axial direction z

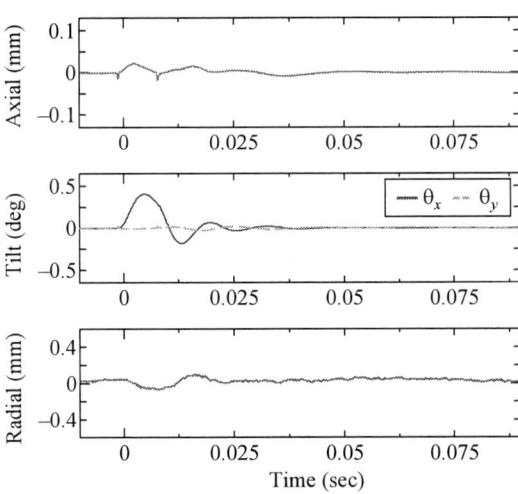

(b) Disturbance input: Tilt θ_x

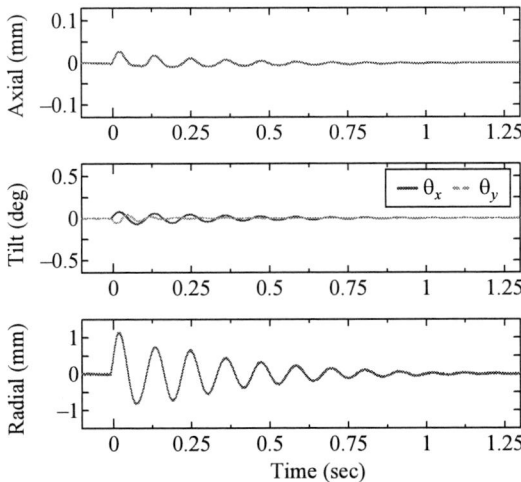

(c) Disturbance input: Radial direction x

Fig. 4. Impulse response results.

radial direction. In this case, the radial vibration took nearly 0.9 sec to fall below ±5% of the maximum displacement. The slow settling time of the radial direction control is a consequence of the radial direction being stabilized passively by the restoring force produced by the axial attractive force (rather than employing an active stabilization control).

V. CONCLUSION

A permanent magnet hybrid type axial magnetically levitated motor was proposed. FEM magnetic analysis showed that there was no significant interference between rotational torque, suspension force, and restoring torque, indicating that these four axes are therefore able to be controlled independently. The impulse response measured in the experimental test rig showed quick response for the actively controlled degrees of freedom (z and θ_x, θ_y) and showed that the translation motion and the inclination motion can be controlled independently.

As future work, the levitated rotation performance will be investigated, and the test rig will be modified to explore motor performance for specific applications such as use as an artificial heart and/or ultrapure water pump.

REFERENCES

[1] Yohji Okada, Kenzo Nonami, "Research Trends on Magnetic Bearings", JSME International Journal, Series C, Vol. 46, No. 2 (2003), pp. 341-342

[2] Akira Chiba, Tadashi Fukao, et al., "Magnetic Bearings and Bearingless Drives", Newnes (2005)

[3] Yohji OKADA, Tetsuo OHISHI, Kazutada DEJIMA "General Solution of Levitation Control of a Permanent Magnet(PM)-Type Rotating Motor", JSME International Journal Series C, Vol.38, No.3 (1995)

[4] Reto SCHOEB, Natale BARLETTA "Principle and Application of a Bearingless Slice Motor", JSME International Journal Series C, Vol.40, No.4 (1997)

[5] Yohji Okada, Satoshi Ueno, Tetsuo Ohishi, Takashi Yamane, Chit Chiow Tan, "Magnetically Levitated Motor for Rotary Blood Pumps", Artificial Organs, Vol. 21, Issue 7 (1997), pp. 739–745.

[6] Satoshi Ueno, Yohji Okada, "Characteristics and control of a bidirectional axial gap combined motor-bearing", IEEE/ASME Transactions on Mechatronics, Vol. 5, Issue 3 (2000), pp. 310-318.

[7] Keisuke Shimbo, Ichiro Tomita, Osamu chikawa, Chikara Michioka, Akira Chiba, Tadashi Fukao: "Axial Gap Length and the Maximum Torque of Shaftless Axial Gap Bearingless Motors", IEEJ Annual meeting 1997, Paper No. 1218 (1997) (in Japanes)

[8] Kijja Chongkwanyuen, Osamu chikawa, Chikara Michioka, Akira Chiba, Tadashi Fukao: "Inclination Control of Axial Gap Bearingless Motors", IEEJ Annual meeting 1997, Paper No. 1219 (1997) (in Japanese)

[9] Masahiro Osa, Toru Masuzawa, Eisuke Tatsumi, "Miniaturized Axial Gap Maglev Motor with Vector Control for Pediatric Artificial Heart", Journal of JSAEM, Vol. 20, No. 2, pp. 397-403, (2012)

[10] Hiromu Takada, Nobuyuki Kurita, Takeo Ishikawa, "Proposal of a Double Stator Type Magnetically Levitated Axial Gap Motor", IEEJ Industry Applications Society Conference 2012, Paper No. Y-114 (2012) (in Japanese)

Comparison of High Speed Bearingless Drive Topologies with Combined Windings

Hubert Mitterhofer, Branimir Mrak
Linz Center of Mechatronics
Altenbergerstrasse 69, 4040 Linz
hubert.mitterhofer@lcm.at

Wolfgang Gruber
Institute for Electrical Drives and
Power Electronics
Johannes Kepler University Linz
Altenbergerstrasse 69, 4040 Linz

Abstract—For high speed bearingless disk drives, certain topologies are advantageous. The authors have published works on a bearingless disk drive for high speeds which is characterized by a slotless stator and a toroid winding set. Several different variations of this setup are imaginable. The ones which this work focuses on are the variation of the number of phases, of the number of coils, and of the applied coil connection. A comparison of the constructed and tested setup with the possible variations is presented in the course of this work.

I. INTRODUCTION

Bearingless drives are a special form of magnetically levitated device where the active magnetic bearing function is provided by the motor itself. The winding system can be carried out as separated or combined windings. The former features actual suspension windings for conducting the bearing currents which are physically separated from the motor windings. The latter instead superposes the two current components in the control scheme and uses only one common winding set for conducting the resulting currents. This increases the compactness of the system while reducing the mechanical complexity [1].

As a bearingless unit typically only stabilizes either radial or axial deflections of the rotor, additional bearings are usually necessary in order to fully stabilize a rotor. A disk drive represents the most compact overall design since it allows using the passive stability in axial and tilt direction which is inherently provided by a flat magnetic rotor and its stator [2].

In earlier works, a prototype high speed drive was built as a bearingless disk drive with 5 phases supplying a combined winding system (cf. Fig. 1). The rotor position is stabilized passively in axial and tilt direction whereas the radial position signal from eddy current sensors is used to actively control the radial degrees of freedom. A detailed system description can be found in [3]. Since the drive was designed to run at up to 100 000 rpm, a slotless stator was chosen which reduces the occurring iron losses [4]. Additionally, a special toroid winding form was introduced to bearingless drives in [5] and [6] which reduces the necessary copper length and facilitates the manufacturing process. In this scheme which is displayed in Fig. 2, each phase is connected to two coils which are wound in opposing winding sense. Recently, a project was published [7] using a similar general drive idea and

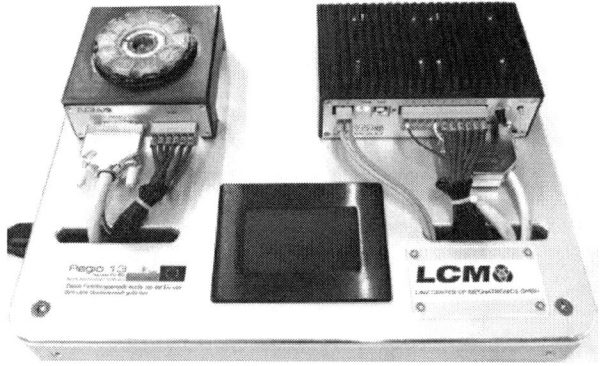

Figure 1. High speed disk drive prototype used for testing

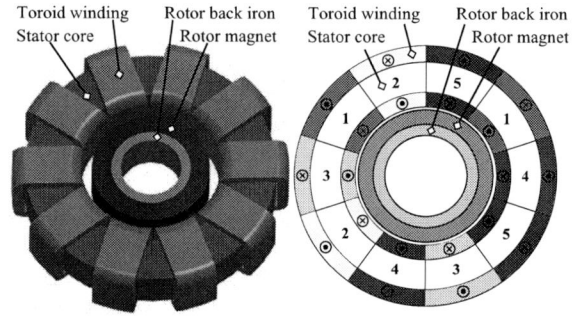

Figure 2. Winding scheme of the realized 5 phase double coil prototype.

a winding structure which applies 6 phases with one single coil for each phase. This calls for a comparison and a more detailed investigation of the advantages of each scheme. Therefore, several such variations are investigated in this paper with an explanation of the evaluation criteria, simulation and comparison of six designs and presentation of the measurement results of the constructed prototype.

II. COMPARISON SETUP

For all investigated types, the stator and rotor dimensions are identical with the values of the prototype drive given in Table I. As the physical air gap width δ is also not modified, the available winding space varies only slightly due to the different coil separators isolation space requirements for designs with different coil numbers. For the 3D

Table I
KEY DATA AND STIFFNESS VALUES OF PROTOTYPE DRIVE

	Parameter	Value
d_{ro}	Outer rotor diameter	32 mm
d_{so}	Outer stator diameter	60 mm
δ	Physical air gap width	0.5 mm
δ_m	Magnetic air gap width	4.5 mm
l_{fe}	Axial motor length	10 mm

FE simulations, no certain wire gauge was specified, it was rather assumed that the entire available winding space is filled with a copper fill factor of $k_{cu} = 0.48$ as this was the fill factor obtained in the constructed prototype. The current density peak value for the simulations was set to $6\,^{\text{A}}\!/\text{mm}^2$.

A. Evaluation criteria

As shown in Fig. 3, the presented drive with passive stabilization in axial and tilt direction needs to produce active bearing forces in the two radial degrees of freedom and, of course, the motor torque. Therefore, the radial

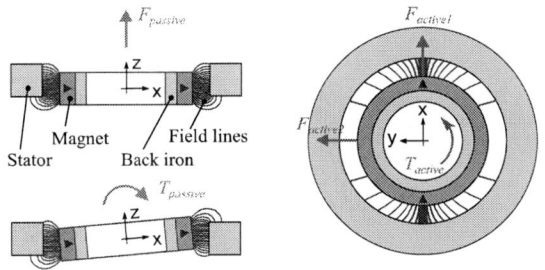

Figure 3. Passive stabilization of axial and tilt displacement, active creation of radial forces and torque.

force and torque capacity of the different designs constitute the main comparison values. However, there is no standard definition of these values for bearingless drives as e.g. the torque coefficients for conventional electrical drives. This is due to the fact that, here, forces and torque have to be created simultaneously and that certain winding properties have quite different effects on the torque and force capacity. One way of obtaining a suitable evaluation criterion can be to look at the achievable force values for *one* distinct direction (either F_x or F_y) while producing *no* force in the other direction and also *no* torque T_z. Therefore, the rotor angle φ dependent matrix $\mathbf{T}(\varphi)$ has to be found, defining the relationship between the vector of phase currents for the m phases

$$\mathbf{i} = \begin{bmatrix} i_1 \\ \vdots \\ i_m \end{bmatrix} \tag{1}$$

and the radial force / torque vector

$$\mathbf{Q}(\varphi) = \begin{bmatrix} F_x(\varphi) \\ F_y(\varphi) \\ T_z(\varphi) \end{bmatrix} \tag{2}$$

as

$$\mathbf{Q}(\varphi) = \mathbf{T}(\varphi)\mathbf{i}. \tag{3}$$

The j-th column of $\mathbf{T}(\varphi)$ holds the angle dependent force and torque value which is created when only the j-th phase is powered with a constant current value while the rotor rotates about φ.

1) General criteria: It may seem tempting to use this relationship for creating evaluation criteria for bearing force

$$t_F = \frac{\sum\limits_{i=1}^{2} \sum\limits_{j=1}^{m} rms(\mathbf{T}(\varphi)_{i,j})}{2m} \tag{4}$$

and torque capacity

$$t_T = \frac{\sum\limits_{j=1}^{m} rms(\mathbf{T}(\varphi)_{3,j})}{m} \tag{5}$$

respectively. Unfortunately, the two terms mentioned above would not yield a good basis for comparison since it is not guaranteed in that case, that only one of the components of $\mathbf{Q}(\varphi)$ is created exclusively. This however, is one of the key features of a bearingless drive as we want to be able to produce torque and forces independent from each other and we also want the force and torque indices to reflect this property. We, therefore, have to turn to the inverse relationship of (3), expressed as

$$\mathbf{i} = \mathbf{K}(\varphi)\mathbf{Q}(\varphi) \tag{6}$$

which links the demanded output $\mathbf{Q}(\varphi)$ to the necessary phase currents $\mathbf{i}$. The matrix $\mathbf{K}(\varphi)$ is calculated by inversion of $\mathbf{T}(\varphi)$. For this inversion, additional criteria such as the star-connection of the motor phases or a minimum winding loss criterion are necessary since usually, $\mathbf{T}(\varphi)$ is not a square matrix. The detailed process of obtaining and inverting the matrix $\mathbf{T}(\varphi)$ is described in [8]. Applying the currents

$$\mathbf{i}_{Fx}(\varphi) = \mathbf{K}(\varphi) \begin{bmatrix} 1 \\ 0 \\ 0 \end{bmatrix} \tag{7}$$

$$\mathbf{i}_{Fy}(\varphi) = \mathbf{K}(\varphi) \begin{bmatrix} 0 \\ 1 \\ 0 \end{bmatrix} \tag{8}$$

$$\mathbf{i}_{Tz}(\varphi) = \mathbf{K}(\varphi) \begin{bmatrix} 0 \\ 0 \\ 1 \end{bmatrix} \tag{9}$$

ensures that, in the respective situation, only the demanded force or torque component is produced without exciting any other components. Applying the currents in the appropriate ratios according to (7)-(9) with the highest current not exceeding the current specification, gives the mentioned maximum exclusive torque or maximum exclusive force in one distinct direction. Calculating the rms-values of $\mathbf{i}_{Fx}(\varphi)$, $\mathbf{i}_{Fy}(\varphi)$ and $\mathbf{i}_{Tz}(\varphi)$ and then taking the mean value over all phases leads to more suitable comparison factors for bearingless drives which were introduced in [9], calculating force and torque constants as

$$k_F = \frac{2m}{\displaystyle\sum_{i=1}^{m}\sum_{j=1}^{2} rms(\mathbf{K}(\varphi)_{i,j})} \qquad (10)$$

and

$$k_T = \frac{m}{\displaystyle\sum_{i=1}^{m} rms(\mathbf{K}(\varphi)_{i,3})}, \qquad (11)$$

respectively. These constants can be interpreted as the reciprocal mean value of the minimum rms current necessary to create 1 N of F_x or F_y force or 1 N m of T_z torque. The constants, therefore, have the units $\mathrm{N/A}$ and $\mathrm{Nm/A}$, respectively. They provide a very adequate comparison value for the different designs, yet for k_F, there are two potential flaws. The first one comes from the fact that only distinct force directions (F_x and F_y) are regarded and evaluated. The additional use of the rms value of $\mathbf{K}(\varphi)$ means that there may be certain rotor angles φ where only little forces and torques can be created but which stay unnoticed.

The second weakness of this criterion is that the forces and torques used for the calculation of $\mathbf{T}$ and $\mathbf{K}$ are normalized for 1 A of magneto motive force, thus eliminating the link to the available copper cross section of the respective motor design.

2) Start-up criterion: From this conclusion, we can take one step further and rethink the meaning of the radial force components that the force factor k_F is based on. These components, $F_x(\varphi)$ and $F_y(\varphi)$, stand for the force in *x*- and *y*-direction for all rotor angles, respectively. This means, however, that all other positions apart from the *x*- and *y*-direction are *not* regarded even though they may show different maximum force values. This is of critical importance for the starting moment, when the rotor sticks to the stator due to the reluctance forces and needs to be lifted out of this resting position into the center position. Therefore, this criterion will be named *start-up* criterion. Forces of *any* direction may be required, since the resting position may be in *any* point around the stator surface, with, theoretically, *any* rotor orientation. We must, therefore, look at every force direction and, for each one, evaluate the force capacity over the rotor angle φ. For doing this, we can no longer use the rms-value of $\mathbf{K}(\varphi)$ and we also cannot take the mean value of F_x and F_y capacity. As the resulting term is not simple to interpret, a more intuitive graphic approach will be chosen for evaluating the different designs.

B. Evaluated designs

As mentioned in the introduction, several design variations are imaginable for the toroid disk drive setup. The chosen variations feature a phase number of either 5, 6 or 8 phases with either one *single coil* or a set of a *double coil* connected in anti-series (the two connected coils are would in opposite sense). These resulting six variations are hence called 5ps (5 phases, single coil), 5pd (5 phases, double coil) and further 6ps, 6pd, 8ps and 8pd. Fig. 4 and

Fig. 5 exemplarily show the winding schemes of the 5ps and 8pd topology.

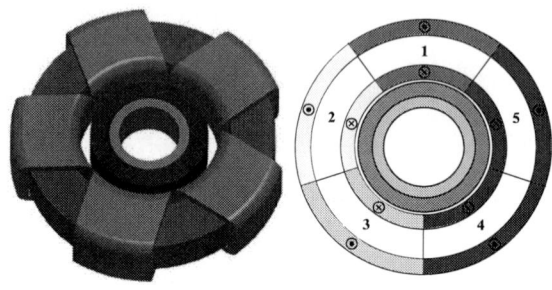

Figure 4. Exemplary setup and winding scheme of the 5 phase single coil (5ps) setup. The 6ps, and 8ps setups are structured accordingly.

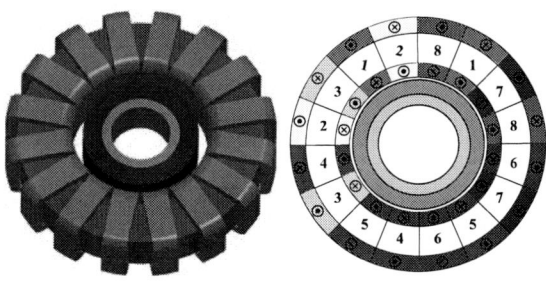

Figure 5. Exemplary setup and winding scheme of the 8 phase double coil (8pd) setup. The 5pd and 6pd setups are structured accordingly.

III. COMPARISON RESULTS

A. Validation of Simulation Approach through Measurement

For obtaining the comparison data, the targeted setups were magneto-statically simulated using a 3D-FE solver. Of course, all of the simulations use the same stator and rotor geometries, rely on the same nonlinear stator iron characteristics and also allow the same current density in the winding. Only the available copper cross section A_{Cu} per phase varies with the chosen phase number due to the different number of isolation walls between the coils.

During the simulations, 1 phase was energized with a constant current density of $6\,\mathrm{A/mm^2}$ while the permanent magnet rotor was rotated in order to obtain the respective values of achievable forces and torque dependent on φ. The resulting vectors are normalized to 1 A of magento motive force and are then used to calculate $\mathbf{T}(\varphi)$ and follow the calculation of evaluation criteria outlined in section (II-A).

Since two of the designs, the 5pd and the 6ps version, were already available as prototypes, the same procedure as in the simulation was also carried out at a test bench. Fig. 6 shows the comparison of the overall achievable F_x, F_y and T_z values when all phases can be optimally energized while only respecting the limitation due to

978-1-4799-2706-7/14 $31.00 © 2014 IEEE

exclusiveness (no other force or torque created) and the star-connection of the system. The quantitative mismatch of up to 19 % can be explained with the ideality of the simulation including the ideal filling of the available winding space. In the real prototype, the winding space can only hold a whole number of winding turns which, at a fixed current density for simulation and real prototype, can lead to a certain discrepancy in available copper cross section and thus, of the current linkage. Additionally, a certain imprecision of the angular winding distribution and of the measurement process adds to the mentioned error. However, except for a certain phase shift of about 10° notable in the figures of the 6ps topology, the characteristics of the simulated data fits the measurement quite well. This decent qualitative agreement justifies the further use of simulation results for the comparison of the remaining designs.

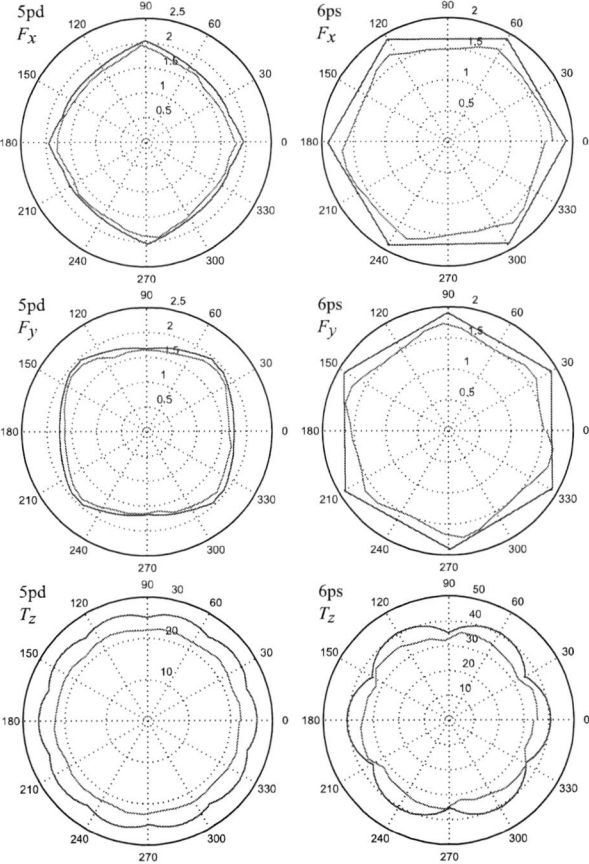

Figure 6. Simulation (blue) based and measurement (red) based exclusive force (F_x and F_y in N in top and middle row) and exclusive torque creation (T_z in mN m in bottom row) for the 5pd and 6ps topology.

B. Torque Comparison

The maximum torque calculation as it was used for the comparison of simulated and measured data in the bottom row of Fig. 6 is now used for comparing the 6 designs under consideration. Fig. 7 gives the six torque orbits

where the single coil designs are displayed in dashed lines and the double coil designs are shown in solid lines.

The single coil versions using 6 or 8 phases are clearly the best choices concerning torque creation. Both have similar peak values whereas the torque ripple is more prominent in the 6 phase design. Next in line is the 5ps version and only then, the double coil variants follow. It is interesting to note that an increase in the number of phases is beneficial for the single coil but detrimental for the double coil designs.

Both effects, the superiority of the single coil arrangements and the torque capacity reduction with rising phase number in the double coil arrangements can be explained with a look at Fig. 8 where the winding schemes of both 8 phase designs are given exemplarily. The single coil design can emulate a quasi-full pitch coil, producing only torque and yielding a winding factor of 1, by having e.g. positive current in phase 1 and negative current in 5. Contrary to that, the two connected coils of each phase in the double coil design are short-pitched and thus have a very small winding factor.

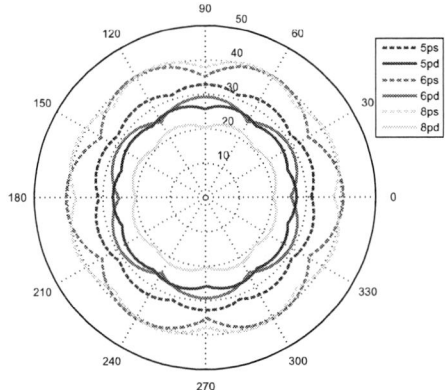

Figure 7. Maximum torque in mN m over rotor angle in degree for all 6 topologies when no radial forces are created.

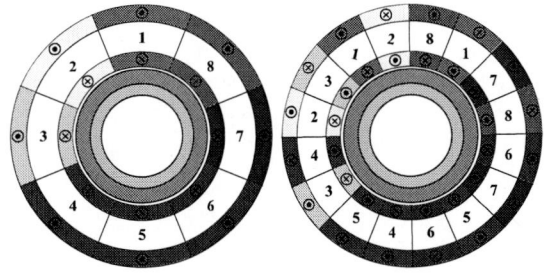

Figure 8. Winding scheme of 8ps (left) and 8pd (right)

Table II holds the torque capacity constant k_T, given in (11), the mean torque value for a current density of 6 A/mm² over one rotor period and the torque ripple peak to peak value relative to the mean torque.

C. Force Comparison

Let us finally take a look at the radial bearing forces. As proposed in (10) and given in Table II, a force

Table II
Overall comparison

Topology	A_{Cu} per phase	Torque constant k_T	Mean torque	Torque ripple pp in % of mean value	Force constant k_F	Minimum producible radial force
5ps	21.67 mm²	0.339 mN m/A	32.32 mN m	4.9 %	18.9 mN/A	1.73 N
5pd	21.18 mm²	0.286 mN m/A	26.6 mN m	4.9 %	18.2 mN/A	1.63 N
6ps	17.98 mm²	0.399 mN m/A	38.8 mN m	14.0 %	21.6 mN/A	1.65 N
6pd	17.48 mm²	0.295 mN m/A	27.9 mN m	14.0 %	23.4 mN/A	1.74 N
8ps	13.36 mm²	0.550 mN m/A	39.7 mN m	7.8 %	31.0 mN/A	1.76 N
8pd	12.87 mm²	0.307 mN m/A	21.4 mN m	7.8 %	28.9 mN/A	1.57 N

constant k_F can be consulted for evaluating the different designs. Due to the mentioned potential flaw of these factors, Fig. 9 considers the *start-up* criterion mentioned in section II-A2. There are two main striking points about this comparison. The first one is that the fear of a direction where significantly less radial force could be produced than in the principle *x*- and *y*-axis, thus creating a "dead spot" from which the rotor could not be levitated if the active bearing characteristic was designed exclusively according to k_F, was needless. The second one is that even though the range between minimum and maximum force value is quite different for the 6 designs, the minimum producible force is similar.

D. General Remark

It is important to notice in the comparison in Table II that the factors k_T and k_F are normalized values. They do not take into account that there is less winding space available in the designs with higher phase numbers or two coils instead of one coil per phase. In contrast to that, the *mean torque* and *minimum producible radial force* do reflect this difference in copper volume as they are based on results for identical current density. So while the k_T and k_F are suitable for making a principle topological comparison, the mean torque and minimum force values should be used for judging the actual potential of a design. The 8pd topology, for instance, shows significantly higher k_T and k_F factors than the 5pd design while it performs worse than the 5pd for equal current density.

IV. CONCLUSION

The results of the topology comparison are summarized in Table II, allowing several interesting conclusions.

The choice of appropriate comparison values is essential. For the case of the motor torque, the proposed torque constant k_T should not be taken as the only evaluation criterion. Since the available copper cross section is not taken into account but instead, only the values normalized to 1 A of magento motive force are used, the resulting values are of theoretical nature. The actually achievable mean torque is a better choice when looking at the real performance of a certain design. The same is true for the bearing forces where the constant k_F does not take

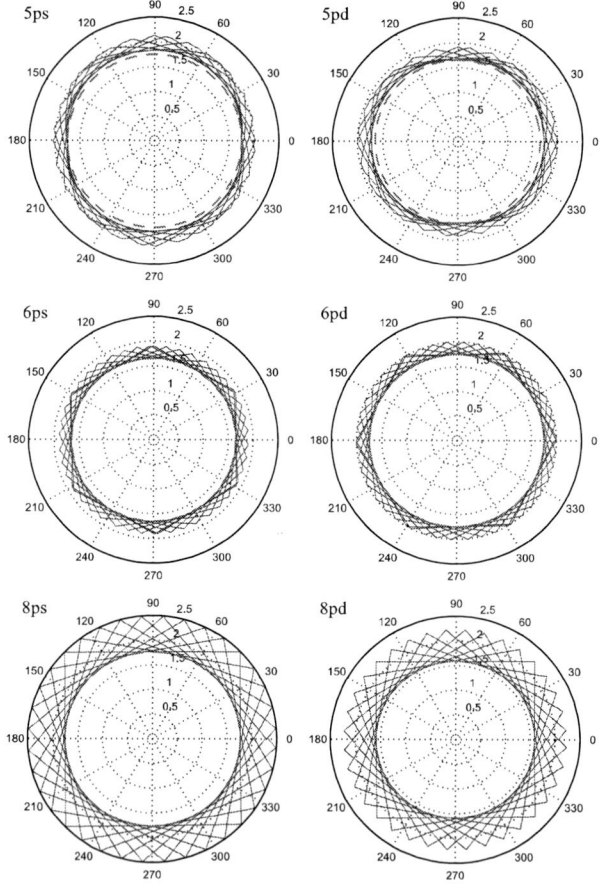

Figure 9. Overlay of the force orbits in N for evaluating of the *start-up* criterion for all designs. Every line represents the maximum achievable force in one direction over all rotor angles φ. The force directions are evaluated in steps of $10 - 15°$. The resulting minimum producible force range is displayed with a red dashed line.

the winding space into account and is thus, misleading. Generally speaking, the topologies with higher number of phases and or coils show a higher ratio of k_T to mean torque and of k_F to minimum radial force.

Once the criteria for the comparison are chosen and the simulation and post-processing is done, it becomes quite evident, that the 5pd topology chosen for the prototype drive is not the ideal choice. Even without increasing the phase count, the 5ps topology would constitute a better choice.

On the overall comparison, the 8-phase single coil arrangement would constitute the best choice of the compared topologies for both, torque and force creation. This conclusion seems plausible as this winding scheme is the only one where both, the 2-pole field for torque creation and the 4-pole field for force creation can be reproduced perfectly due to the appropriate coil placement.

However, an 8-phase machine also demands at least 8 half bridges in the used power electronics circuit. This constitutes a major disadvantage since power electronic circuits with 8 half bridges are very uncommon outside of laboratories. Despite the fact that the 6-phase designs need an additional half bridge compared to the 5-phase arrangements, they may both be fed by two very common and widely available 3-phase inverters [10]. With this advantage taken from the 5-phase arrangements, the 6-phase single coil design could be selected as a good compromise between simplicity and performance.

ACKNOWLEDGMENT

This work is part of the research project "Sustainable and resource saving electrical drives through high energy and material efficiency" and is sponsored within the program of the European Union "Regionale Wettbewerbsfaehigkeit OOe 2007-2013 (Regio 13)" by the European Regional Development Fund and the Province of Upper Austria. Initial parts of the project were funded by the LCM K2-centre of the COMET program of the Austrian and Upper Austrian Government. The authors thank all involved partners for their support.

REFERENCES

[1] K. Raggl, T. Nussbaumer, and J. Kolar, "Comparison of winding concepts for bearingless pumps," *Proc. ICPE 2007*, 2007.

[2] R. Schob and N. Barletta, "Principle and application of a bearingless slice motor," *Proc. 5th International Symposium on Magnetic Bearings*, 1996.

[3] H. Mitterhofer, W. Gruber, and W. Amrhein, "On the high speed capacity of bearingless drives," *IEEE Transactions on Industrial Electronics*, vol. 61, no. 6, pp. 3119–3126, 2014.

[4] N. Bianchi, S. Bolognani, and F. Luise, "High speed drive using a slotless pm motor," *IEEE Transactions on Power Electronics*, vol. 21, no. 4, pp. 1083–1090, 2006.

[5] H. Mitterhofer, W. Gruber, and W. Amrhein, "Towards high speed bearingless drives," *Proc. 12th International Symposium on Magnetic Bearings*, 2010.

[6] H. Mitterhofer and W. Amrhein, "Design aspects and test results of a high speed bearingless drive," *Proc. 9th Power Electronics and Drive Systems Conference*, 2011.

[7] D. Steinert, T. Nussbaumer, and J. Kolar, "Concept of a 150krpm bearingless slotless disc drive with combined windings," *Proc. International Electric Machines and Drives Conference*, 2013.

[8] S. Silber, "Power optimal current control scheme for bearingless PM motors," *Proc. 7th International Symposium on Magnetic Bearings*, 2000.

[9] S. Silber, W. Amrhein, H. Grabner, and R. Lohninger, "Design aspects of bearingless torque motors," *Proc. 13th International Symposium on Magnetic Bearings*, 2012.

[10] H. Grabner, S. Silber, and W. Amrhein, "Bearingless torque motor - modeling and control," *Proc. 13th International Symposium on Magnetic Bearings*, 2012.

High-Speed Magnetically Levitated Reaction Wheel Demonstrator

Christof Zwyssig
Celeroton Ltd.
Zurich, Switzerland
christof.zwyssig@celeroton.com

Thomas Baumgartner
Celeroton Ltd.
Zurich, Switzerland
thomas.baumgartner@celeroton.com

Johann W. Kolar
Power Electronic Systems Laboratory
ETH Zurich
Zurich, Switzerland
kolar@lem.ee.ethz.ch

Abstract— **Reaction wheels (RWs) for small satellites with active magnetic bearings allowing for ultra-high-speed operation show advantages in angular momentum density over ball bearing RWs with limited speed according to scaling laws developed in this paper. A reaction wheel demonstrator design based on a novel dual hetero-/homoploar, slotless, self-bearing, permanent-magnet synchronous motor concept with a rotational speed of 250 000 rpm is investigated. The design includes the rotor dynamics, mechanical stress analysis, electromagnetics, power electronics and control, and the sensor concept. The experimental setup ready for experimental verification is presented.**

Keywords— *reaction wheel, self-bearing, active magnetic bearing, high-speed.*

I. INTRODUCTION

Single reaction wheels (RWs), 3 or more RWs combined into reaction wheel assemblies (RWAs) or integrated power and attitude control systems (IPACS) using an RWA also for energy storage, are used for attitude control of satellites in orbit flight. Changes in rotational speed vary the angular momentum of an RW, which causes the spacecraft to counter-rotate through conservation of total angular momentum. Usually, ball bearings are employed in RWs [1], but also magnetic bearing reaction wheels have been developed, usually for larger satellites, in a few cases also for smaller satellites [2]. If an RW for small spacecrafts is designed, typically, the maximum rotor diameter is limited. Thus, a very high rotational speed has to be selected to maximize the angular momentum (and energy storage capacity in case of IPACS). Today, the maximum rotational speed is typically limited by the employed bearings and the required lifetime, to typically 5 000 to 10 000 rpm, also in the case of magnetic bearings.

Therefore, in this paper a RW demonstrator for small satellites based on a high-speed, magnetically levitated electrical drive system is presented, which allows to increase the rotational speed from currently achieved 10 000 rpm to over 200 000 rpm.

First, RW requirements and scaling laws are identified, showing that the only way to higher performance density RWs is an increase in rotational speed. Then, the electromagnetic concept and design based on two different slotless self-bearing motors, a heteropolar and a homopolar motor, is described in detail. The main aspects of the mechanical design, the stress analysis and rotordynamic design, are also presented. Finally, the experimental setup including reaction wheel demonstrator and power and control electronics is shown and first measurement results are presented.

II. REACTION WHEEL REQUIREMENTS AND SCALING

Most space vehicles designed today require some form of attitude control. Some typical applications are:

- Orientation of solar arrays
- Orientation of optical payloads such as telescopes
- Orientation of scientific instruments
- Orientation of high gain antennas to a ground based station or another spacecraft
- Orientation of orbit thrusters

A typical attitude system consists of an attitude measurement system, a control algorithm and a set of RWs. The measurement system is used to determine the actual attitude. The control algorithm compares the actual attitude with the desired attitude and determines the control input to obtain the desired attitude. Finally, the RWs apply the control torque to the spacecraft.

The design of an attitude control system is mainly determined by the space mission requirements such as the accuracy and stability of the attitude control, the spacecraft's mass and the nature of disturbance torques. Since the functionality of the attitude control system is essential for the overall space mission, the lifetime of the control actuators is crucial. Typical design lifetimes of satellites are 5 to 15 years [3]. Therefore, if using ball bearings, the actuators speeds are limited to around 10 000 rpm.

978-1-4799-2706-7/14 $31.00 © 2014 IEEE

A. Reaction disk scaling laws

When the reaction disk is operated at the stress limitation (optimal operation conditions), the angular momentum L (which is the main RW performance criterion) can be expressed by

$$L = I\omega = \frac{1}{2}m_{rotor}r^2\omega = \frac{1}{2}\pi\rho\frac{r^5}{x}\omega$$
$$= \pi\sqrt{\frac{2}{3+\nu}}\sqrt{\sigma_{max}\rho}\cdot\frac{r^4}{x} \tag{1}$$

where I is the inertia, ω the angular rotational speed, m_{rotor} the weight, r the radius, x the factor of radius to length, ρ the density, ν the Poisson's ratio, and σ_{max} the tensile strength of the inertia disk material.

From this it can be seen that if L is to be maximized, the radius r has to be selected as large as possible. For a given L, this results in a thin disk (small value of x). Typically, the maximum diameter of the rotor is limited by the geometrical constraints of the spacecraft and the rotor dynamical behavior. For fixed rotor proportions (constant x), all rotor dimensions scale according to above equation with

$$r \propto \sqrt[4]{L} \tag{2}$$

as a function of angular momentum storage capacity L. With this relationship, the rotor mass scaling can be derived to

$$m_{rotor} \propto \sqrt[4]{L^3} . \tag{3}$$

This result indicates that the rotor mass does not scale linearly to the momentum capacity L, i.e. smaller reaction wheels feature less angular momentum per mass than bigger ones. This is one of the main limitations and challenges when designing reaction wheels for small spacecrafts as small reaction wheels intrinsically are less mass efficient. Therefore, an optimization of the momentum capacity per mass is essential, i.e. an operation at the stress limitation becomes increasingly important when decreasing spacecraft size to achieve an acceptable performance.

The fact that smaller reaction wheels feature lower momentum capacity per mass is also validated by the market study presented in [4].

In order to utilize the rotor material, the rotational speed has to be adapted to the size of the rotor, i.e. the rotational speed scales with

$$\omega \propto \frac{1}{\sqrt[4]{L}} . \tag{4}$$

This means that reaction wheels which are designed for smaller angular momentum storage capacities have to be designed for a higher rotational speed.

B. Electrical machine scaling laws

Beside the high-inertia rotor also an electric machine is necessary to accelerate or decelerate the rotor. The size required for the machine is mainly given by the torque requirement of the attitude control system. The torque density (torque T devided by machine volume $m_{machine}$) obtained from an electric machine is widely independent of its size and rotational speed [5]. Thus, a constant torque density and torque per machine mass can be assumed for the scaling

$$\frac{T}{m_{machine}} \propto const. \tag{5}$$

The machine power P can be derived by multiplying the torque with the rotational speed. Thus, the power density increases linearly with the rotational speed

$$\frac{P}{m_{machine}} \propto \omega \tag{6}$$

In most applications of electrical machines, the power P has to meet the requirements of the application. Thus, if the rotational speed and with it also the power density is increased, the system mass can be reduced. Therefore, a high rotational speed is desirable. If this machine scaling is applied to a RW and if the acceleration time

$$t_{acc} = \frac{L}{T} \tag{7}$$

is kept constant, a machine mass scaling linear to the angular momentum results

$$m_{machine} \propto L . \tag{8}$$

Consequently, the mass ratio between the machine and the rotor scales to

$$\frac{m_{machine}}{m_{rotor}} \propto \sqrt[4]{L} \tag{9}$$

This means that for smaller reaction wheels the size of the machine decreases faster than the size of the rotor. The overall power of the system scales to

$$P \propto \sqrt[4]{L^3} . \tag{10}$$

Consequently, small reaction wheels feature more power per angular momentum capacity than bigger ones. Based on this scaling law, it can be concluded that a high rotational speed is especially desirable for RW used in IPACS since high power densities can be achieved.

C. Bearing requirements

All RWs RWAs and IPACS require bearings to support the rotor. The bearings are a critical part as they influence the RW design and limit performance of the RW. Critical bearing specifications are:

- Rotational speed of the rotor
- Lifetime
- Weight and inertias of the rotor
- Load transmission
- Size and weight of the bearing
- Atmosphere requirements

Most RWs employ ball bearings because of simplicity, compactness, low weight and high bearing load capability. However, ball bearings show a limited lifetime depending on rotational speed and lubricant life

978-1-4799-2706-7/14 $31.00 © 2014 IEEE

[6]. In section II.A it is derived that smaller angular momentum (L) RWs show in lower angular momentum density (L/m), and the only way to counteract and optimize this (to increase L/m) is to increase the rotational speed. As a consequence, the ball bearing becomes an increasing challenge, and is usually defining the performance limit, in small RWs. A further disadvantage of ball bearings is they usually require a pressurized atmosphere to prevent outgassing of the lubricant, and therefore containment is required.

To overcome the limitations and drawbacks of ball bearings at high rotational speeds and vacuum requirement, a novel magnetically levitated high-speed machine has been developed in [7]. This is the basis of RW demonstrator presented in this paper.

D. Summary of scaling laws

RWs employing ball bearings are compared to magnetically levitated RWs based on the high-speed, slotless, self-bearing, permanent-magnet synchronous machine presented in [7]. The total system mass (including machine, electronics, housing, etc.) is calculated based on realistic reference designs for ball bearing RWs and for the high-speed self-bearing machine. Both systems employ the same reaction disk shape, machine model (mass proportional to torque), power electronics model (mass proportional to power), and a constant acceleration time of 20 s is assumed. The control electronics are constant in both systems, however the mass of the self-bearing control is much higher than the ball bearing control (only motor control needed here). The ball bearing RW rotational speed is limited to a (high) value of 8000 rpm.

The results of the system scaling are shown in Figure 1. In order to give a comparison with state of the art RW systems employing ball bearings, typical reference designs are also shown in the total mass plot. For this, the systems RW1, RW35, RW90 and RW150 from Astro- und Feinwerktechnik Adlershof GmbH [8], which feature a similar acceleration time, are added to the plot. The total mass plot show in in Figure 1 shows that the mass of the commercially available systems is underestimated by approximately 20%. However, the general trend of the total mass versus the angular momentum is correctly rendered in the scaling. The red diamond shown in the plot shows the prototype system employing the machine presented in [7] when extended with a high inertia titanium reaction disk.

The comparison shows that above an angular momentum capacity of roughly 0.1 Nms (typical RW performance required in a satellite with total mass of 10-30 kg) the magnetically levitated system can be realized with a lower system mass. The mass saving is about 50% in the range of 1 Nms. Given the higher rotational speed and the identical torque requirement, the resulting power is much higher in the magnetically levitated system. This is a major disadvantage when used in an RWA, but a major advantage in IPACS because the RWs can be used as high power sources.

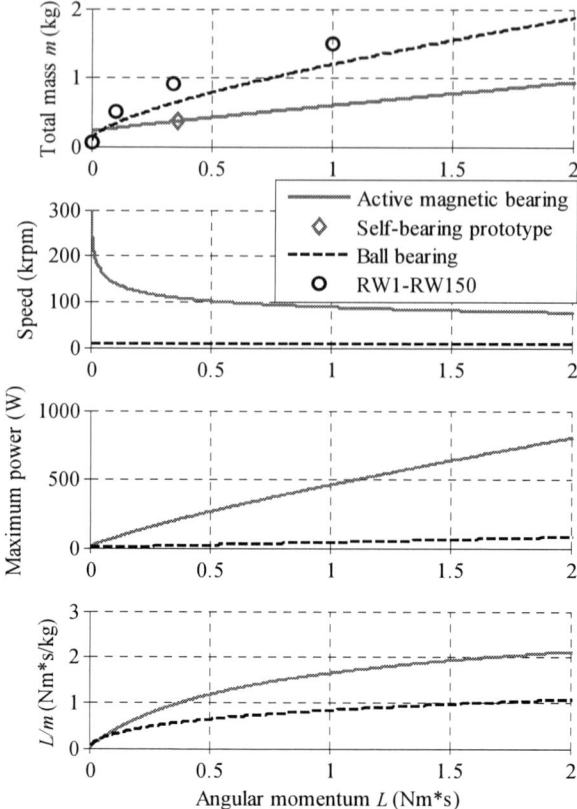

Figure 1: Reaction wheel system scaling. Typical RWs employing ball bearings are compared to RWs employing high-speed self-bearing motors. Additionally, commercially available systems employing ball bearings (RW1-RW150) are also shown for comparison.

III. REACTION WHEEL DEMONSTRATOR DESIGN

A. Dual hetero-/homopolar self-bearing motor concept

In the proposed novel dual hetero-/homopolar slotless self-bearing motor concept depicted in Figure 1, the bearing forces and drive torque are generated by two slotless, self-bearing, permanent-magnet synchronous motors employing an ironless rotor. Such slotless, self-bearing concepts have also been presented in [9] and [10]. There a disk motor (small length compared to diameter) with one active radial magnetic bearing and a passive stabilization of the tilting are utilized. Using disk motors for RWs would require integrating the permanent magnet(s) in the inertia disk. Because of the limited mechanical strength of magnets, the stress limited operation would be at much lower speeds than if a separate, high strength inertia disk can be used. In contrary to the disk motor concepts, the proposed demonstrator employs two individual self-bearing motors with two small rotor diameters in two active magnetic bearing parts instead of one large diameter disk motor. This separation of inertia disk and magnetic bearing allows integrating a specially designed reaction disk in between the two self-bearing motors and therefore going to stress limited operation at much higher rotational speeds –essential to realize RWs for small spacecraft.

A similar motor concept has also been presented in

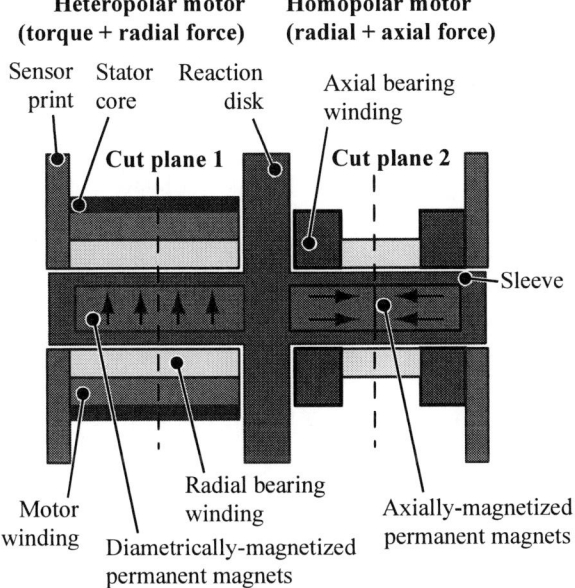

Figure 2: Novel dual hetero-/homopolar slotless self-bearing motor concept.

[7], where two heteropolar self-bearing motors, and an additional axially magnetized magnet plus an axial bearing coil control all six degrees of freedom. In contrary to this concept, the proposed demonstrator employs one heteropolar self-bearing motor (the same as in [7]) but replaces the second heteropolar self-bearing motor with a homopolar self-bearing motor. In the homopolar motor the radial bearing forces and the axial bearing force can be controlled with the same magnetic field generated by two axially magnetized magnets. This has the advantage that no extra magnet is required for the axial bearing force only, and the rotor can be made short for high bending modes, allowing high rotational speeds. Furthermore, in contrary to the concept in [7], the two motors are not placed right beside each other, but they are holding the application, in this case the inertia disk, from two ends. This is a further advantage concerning rotordynamics and further increases the first bending mode.

The application of forces and torque is divided up into the two motors: Motor torque and radial bearing forces are applied in the heteropolar motor, axial and radial bearing forces are applied in the homopolar motor. All six-degrees of freedom are controlled actively: The displacement in x-and y-direction, and the torque/tilting around the x- and y- axis is controlled by two radial bearings separated in z-direction (separated into the heteropolar motor and the homopolar motor). The torque around the z-axis is controlled by the heteropolar motor, the displacement in z-axis is controlled by the homopolar motor. The torque and force generation by Lorentz forces (by interaction of magnetic field and current) is explained in more detail in the following subsections.

The novel slotless self-bearing motor concept is completed with two PCB based eddy current displacement and hall effect rotation angle sensors. This sensor concept has been presented in [11] and [7].

Compared to state-of-the-art reluctance type active magnetic bearings, this concept has various advantages allowing for high rotational speed such as

- Very short axial length and therefore shifting of bending modes to high frequencies
- Simple mechanical rotor construction resulting in low mechanical stresses at high speed
- Low high frequency losses due to slotless motor topology
- High force control bandwith
- Feasible for miniaturization due to low mechanical complexity (especially on rotor side)

B. Electromagnetic force generation

1) Heteropolar motor

The force and torque generation is depicted in Figure 3, which is a cross section view of the cut plane 1 indicated in Figure 2. In the heteropolar motor the torque is generated with the motor winding sitting in between the bearing winding and the stator core. The motor winding has one pole-pair as the magnet. A first current in axial direction, together with a magnetic field in radial direction, leads to a force vector in azimuthal direction. Combining two currents in opposing axial directions displaced by 180° leads to a force pair that generates a torque vector in z-axis.

The bearing forces in x- and y- direction are generated by a bearing winding sitting between the air gap and the motor winding. The bearing winding has two pole-pairs, as for force generation the winding poles pair number has to be the rotor pole-pair number +/-1. The four current vectors of a two pole-pair winding are displaced by 90°, and together with the magnetic field that changes angular direction every 90° as well this leads to forces in the same direction, leading to a bearing force e.g. in –y direction.

The vast amount of generated torque and forces is based on Lorentz force generation, only a small, in the design negligible amount of reluctance force is present. The heteropolar slotless motor design and calculation of Lorentz based torque and forces with analytical models (and some FE calculations for reluctance forces) has been presented in [12] and [7] in detail. The same models and principles are used for the heteropolar motor in this paper. Therefore, linear relationships between the winding currents and the forces and torques are assumed.

2) Homopolar motor

The homopolar motor has two axially magnetized magnets pointing towards each other, and contains no back iron. This results in a field distribution as shown in Figure 4, which is homopolar, meaning there is no change in field if the rotor is rotating (assuming ideal isotropic conditions in the magnet and a fully concentric rotor design).

The axial force generation is depicted in Figure 4. The axial force is generated by two separate coils which are wound in azimuthal turns around the rotor. The axial bearing coils are placed in a radius and z-location where they see mainly a radial magnetic field. This, together with the azimuthal currents, results in an axial force.

The radial force generation is depicted in Figure 5, which is a cross section view of the cut plane 2 indicated in Figure 2 and Figure 4. The radial forces are generated by a winding which essentially is the same as the motor winding of the heteropolar motor: a winding with one pole-pair sitting in between the axial bearing coils and also seeing mainly a radial field. The axial currents in opposing directions displaced by 180°, together with the homopolar field that changes angular direction also every 180°, leads to force vectors pointing in the same direction and therefore a bearing force e.g. in –y direction.

For the homopolar motor, the generated torque and forces consists uniquely of Lorentz forces, therefore the equations and models in [12] and [7] can be applied for the homopolar motor in this paper, and all forces are proportional to current.

C. Specifications

The specifications for the RW demonstrator are not derived from a defined mission or satellite, but are chosen to allow for the verification of the feasibility of a RW based on the slotless, self-bearing permanent-magnet synchronous machine similar to the one firstly presented in [7]. Therefore, in this design, dimensions, rotational speed and motor torque are inputs parameters into the design, the angular momentum L, the acceleration time and the stored energy are results from the design. In a design for a defined satellite or mission, design input parameters would be resulting specifications, and vice versa results would be input parameters.

The machine described therein allows for rotational speeds up to 500 000 rpm. However, the rotational speed in a high-speed RW is limited by the stresses in the reaction disk. In the RW demonstrator presented in this paper, the reaction disk diameter is chosen to 48 mm, the

disk material is titanium grade 5. According to the mechanical stress calculation (Figure 6) this results in a maximum rotational speed of 250 000 rpm. The rotor length is chosen such that the first bending mode is approximately at double this maximal speed (Figure 7). This results in a total rotor length of 49 mm, of which 17 mm can be used for the heteropolar motor and 17 mm for the homopolar motor. The rotor diameter within the motors results from stress analysis of the permanent magnets and results in 8 mm.

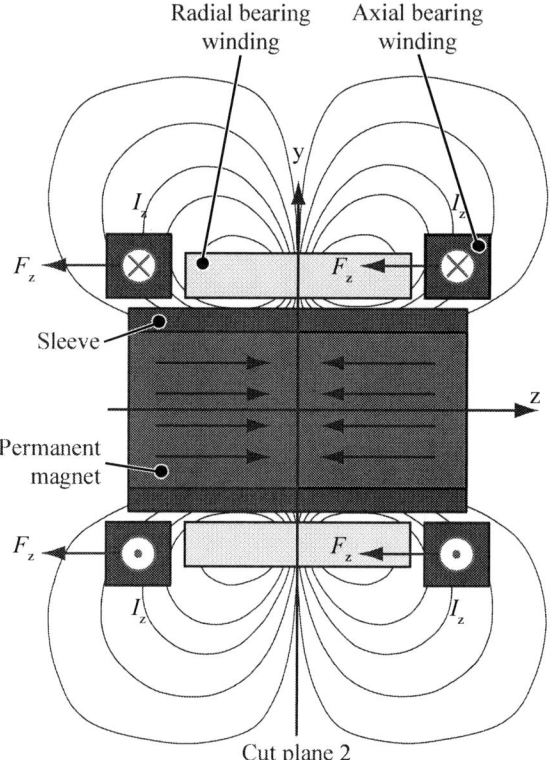

Figure 4: Axial force generation in the homopolar motor.

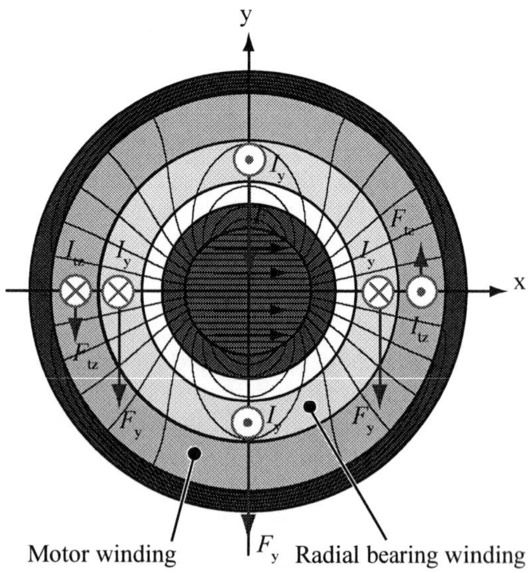

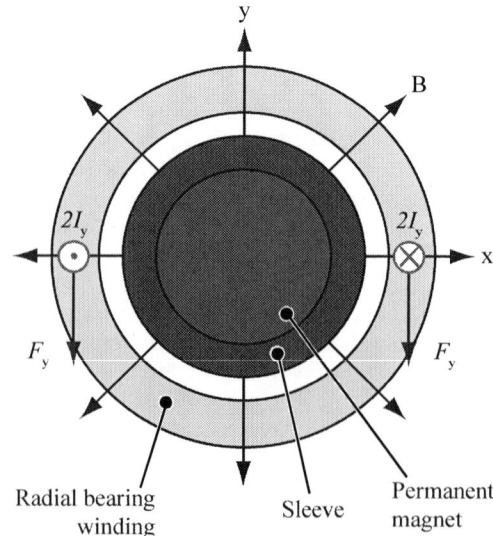

Figure 3: Torque and radial force generation in the heteropolar motor (cut plane 1 of Figure 2).

Figure 5: Radial force generation in the homopolar motor (cut plane 2 of Figure 2 and Figure 4).

This results in a rated motor torque of 7.8 Nmm, a maximal motor power of 345 W, an acceleration time of 45 s, a maximal angular momentum of 0.355 Nms, an and a maximal stored kinetic energy of 4.65 kJ. The motor back EMF, phase current and inverter signals for one phase of the motor at rated speed and power are depicted in Figure 8.

IV. EXPERIMENTAL VERIFICATION

A. Experimental Setup

Based on the concept and specifications presented in the previous sections, a demonstrator RW has been realized. In addition to the already presented parts in Figure 2 it consists of two motor casings where the heteropolar and the homopolar motor are built into, a central casing where these two motors are attached to including a holder, with which the entire setup can be mounted onto a baseplate. A photo of the demonstrator is shown in Figure 9.

In Figure 10, the power electronics to control the demonstrator are shown. The power electronics consist of two three-phase channels for the radial bearing windings, a single-phase channel for the axial bearing winding and a three-phase channel for the motor winding. It is controlled by a DSP/FPGA digital control platform, which also demodulates the rotor position sensor signals. The power electronics and control are similar to the system presented in [7] and [13].

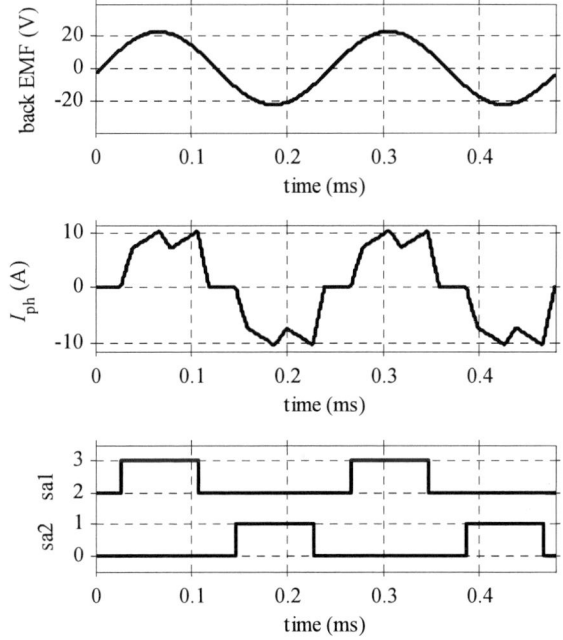

Figure 8: Simulated motor back EMF, phase current and PAM inverter switching signals for one phase at rated speed (250 000 rpm) and rated power (345 W).

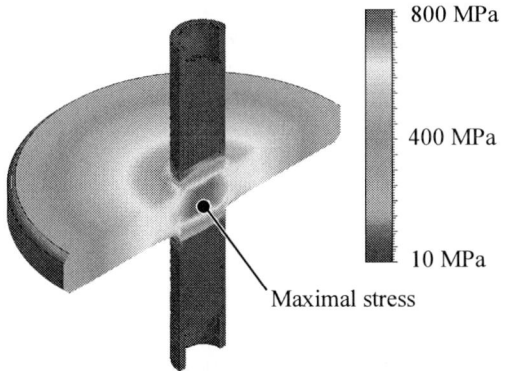

Figure 6: Mechanical stresses (von Mises) in the rotor at an overspeed of 250 000 rpm with the maximal stress of of 808 MPa in the center of the rotor.

Figure 7: First bending mode of the rotor at 544 000 rpm.

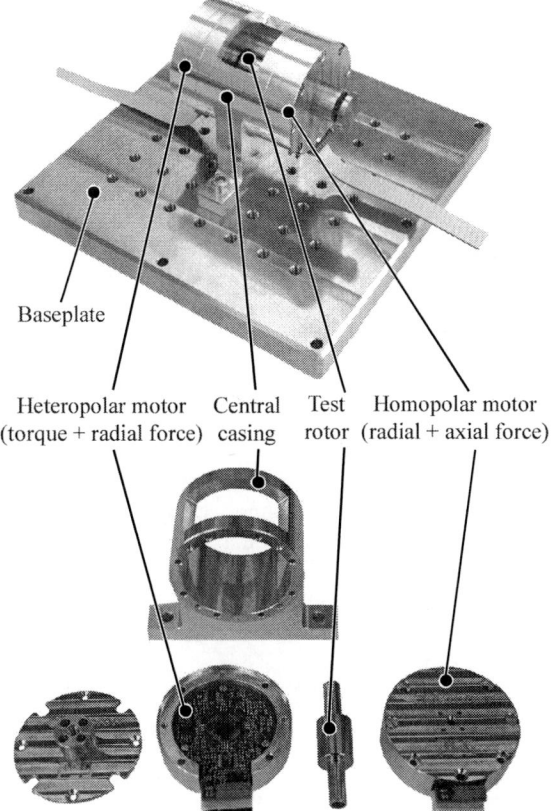

Figure 9: Novel dual hetero-/homoploar, slotless, self-bearing, permanent-magnet synchronous machine reaction wheel demonstrator.

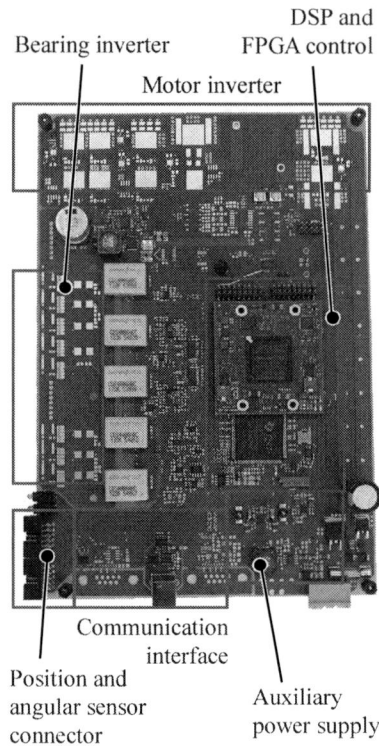

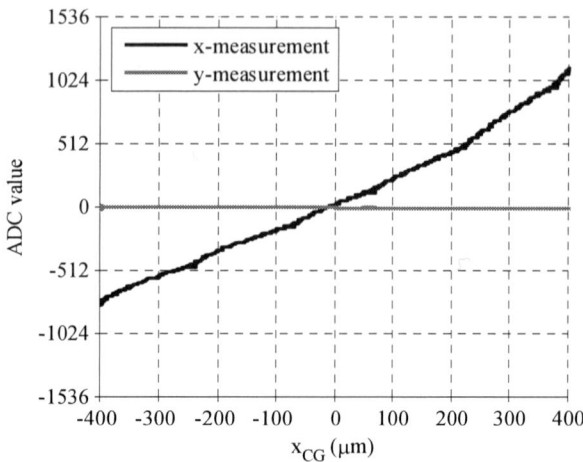

Figure 11: Radial rotor position measurement obtained by the PCB eddy current displacement sensor: Radial rotor position measurement when varying the rotor position xCG (yCG =0). The reference measurement is obtained by an external optical displacement sensor (Keyence LK-H022).

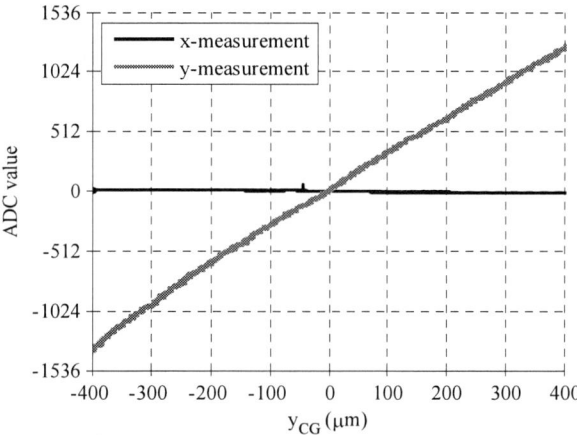

Figure 12: Radial rotor position measurement obtained by the PCB eddy current displacement sensor: Radial rotor position measurement when varying the rotor position yCG (xCG =0). The reference measurement is obtained by an external optical displacement sensor (Keyence LK-H022).

Figure 10: Power and control electronics to drive and control novel dual hetero-/homoploar, slotless, self-bearing, permanent-magnet synchronous motor reaction wheel demonstrator.

B. Measurements

The PCB based eddy current displacement sensors have been tested together with the demodulation electronics and the RW demonstrator rotor. The signals for a displacement in x-direction (Figure 11) and y-direction (Figure 12) show a good linearity no cross coupling between x-displacement and y-measurement and vice versa, and have a resolution of about 1 to 2 μm. The next steps in the experimental verification will be the closure of the current control loop, and the position control loop subsequently.

V. CONCLUSIONS

Smaller satellites require smaller angular momentum (L) reaction wheels (RWs), which results in lower angular momentum density (L/m), especially if the speed is limited as with ball bearings. The only way to increase L/m is to increase the rotational speed, use high strength materials for the inertia disks, and operate closer to or at the stress limit of these materials. Active magnetic bearings are the only alternative for higher rotational speeds and vacuum requirement.

Scaling law comparison of active magnetic bearing RWs and ball bearing RWs shows that above an angular momentum of 0.05 Nms magnetic bearing RWs perform better (higher L/m) but at a much higher maximum power demand.

A novel dual hetero-/homoploar, slotless, self-bearing, permanent-magnet synchronous machine concept has been presented. The application of forces and torque is therefore divided up into two motors: Motor torque and radial bearing forces are applied in the heteropolar motor, axial and radial bearing forces are applied in the homopolar motor. All six-degrees of freedom are controlled actively.

The concept is ideally suited for minimizing the size of RWs or integrated power and attitude control systems (IPACS) for small satellites, as it allows for an increase in rotational speed and can run in vacuum, both not possible with today's state-of-the-art ball bearing RWs and IPACS. A RW demonstrator design has been presented for a rotational speed of 250 000 rpm. The demonstrator is realized in hardware and ready for experimental verification.

In the current implementation of the self-bearing RW demonstrator, the mass of the total system is at least ten times the mass of the rotor. This means the mass overhead is considerable. This is a consequence of the additional components such as the bearing windings, the bearing power electronics, the rotor position sensor and the control electronics. If this mass overhead can be reduced, the magnetically levitated RW becomes even more competitive compared to RWs employing ball bearings. Therefore, future research has to be focused on a higher integration of system components such as the integration of bearing windings and rotor position sensor (self-sensing magnetic bearing).

REFERENCES

[1] R. Varatharajoo and R. Kahle, "A review of conventional and synergistic systems for small satellites," *Aircraft Engineering and Aerospace Technology*, vol. 77, no. 2, pp. 131–141, 2005.

[2] M. Scharfe, T. Roschke, E. Bindl, D. Blonski, "Design And Development Of A Compact Magnetic Bearing Momentum Wheel For Micro And Small Satellites", *Conference on small satellites*, 2001.

[3] W. Ley, K. Wittmann and W. Hallmann, *Handbook of Space Technology*. John Wiley & Sons, 2009.

[4] R. Votel and D. Sinclair, "Comparison of control moment gyros and reaction wheels for small earth-observing satellites," *Proceedings of the AIAA/USU Conference on Small Satellites*, SSC12-X-1, 2012.

[5] C. Zwyssig, J. W. Kolar, and S. Round, "Megaspeed drive systems: Pushing beyond 1 million r/min," *IEEE/ASME Transactions on Mechatronics*, vol. 14, no. 5, pp. 564–574, Oct. 2009.

[6] J. Fausz, B. Wilson, C. Hall, D. Richie, and V. Lappas, "Survey of technology developments in flywheel attitude control and energy storage systems," *Journal of Guidance, Control, and Dynamics*, vol. 32, no. 2, pp. 354–365, 2009.

[7] T. Baumgartner, R. Burkart, and J. W. Kolar, "Analysis and Design of a 300-W 500 000-r/min Slotless Self-Bearing Permanent-Magnet Motor," *IEEE Transactions on Industrial Electronics*, Vol. 61, No. 8, pp. 4326-4336, August 2014.

[8] Online: http://www.astrofein.com/astro-und-feinwerktechnik-adlershof/produkte/raumfahrt-eng/8/reaktionsraeder/

[9] D. Steinert, T. Nussbaumer, J. W. Kolar, "Concept of a 150 krpm Bearingless Slotless Disc Drive with Combined Windings," *Proceedings of the IEEE International Electric Machines and Drives Conference* (IEMDC 2013), Chicago, USA, May 12-15, 2013.

[10] H. Mitterhofer and W. Amrhein, "Design aspects and test results of a high speed bearingless drive," in Proceedings of the 9th IEEE International Conference on Power Electronics and Drive Systems (PEDS 2011), Dec. 2011, pp. 705–710.

[11] A. Muesing, C. Zingerli, P. Imoberdorf, and J. W. Kolar, "PEEC based numerical optimization of compact radial position sensors for active magnetic bearings," in Proceedings of the 5th IEEE International Conference on Integrated Power Systems (CIPS 2008), March 2008, pp. 1–5.

[12] A. Looser, T. Baumgartner, J. W. Kolar, and C. Zwyssig, "Analysis and measurement of three-dimensional torque and forces for slotless permanent-magnet motors," IEEE Transactions on Industry Applications, vol. 48, no. 4, pp. 1258–1266, July–Aug. 2012.

[13] T. Baumgartner and J. W. Kolar, "Multivariable state feedback control of a 500 000 rpm self-bearing motor," in Proc. IEEE IEMDC, May 2013, pp. 347–353.

Stabilized Suspension Control Considering Armature Reaction in a d-q Axis Current Control Bearingless Motor

Masahide Ooshima and Yoshito Kumakura
Department of Electronic Systems Engineering
Tokyo University of Science, Suwa
Chino, JAPAN
moshima@rs.suwa.tus.ac.jp

Abstract-This paper presents a novel control strategy to stably support the rotor shaft when a bearingless motor (BELM) based on d-q axis current control is driven under loaded condition. In this type BELM, the magnitude and direction of the suspension force are unfortunately varied depending on the rotor rotation, which causes the unstable rotor levitation during the motor operation. Thus, the compensation method of the suspension force has been proposed by the authors earlier and its effectiveness has been verified by a prototype machine. However, it is possible that the magnitude and the direction of suspension force are varied with armature reaction under loaded condition. It causes the serious problem in the rotor levitation. In this paper, thus, it is obviously found how the compensation coefficients are conducted to realize the stable rotor levitation even under loaded condition. The effectiveness of the updated compensation coefficients is confirmed in simulation based on a machine model designed by Finite Element Method (FEM) software and the experimental test results.

Keywords— Armature reaction, bearingless motor, permanent magnet synchronous motor, stable rotor levitation .

I. INTRODUCTION

Recently, the bearingless motors (BELMs) are receiving attention in some special cases, particularly, for high vacuum and the other environment where the lubrication oil cannot be used due to some of its advantageous features as compared to the conventional motor drives [1]-[8]. In the BELMs, the rotor shaft is supported without mechanical contact in the bearing journal. It is maintenance free and hence the longevity of the motor is increased. The functions of motor and magnetic rotor levitation are successfully integrated so that the rotor shaft can be made shorter and at the same time it requires less number of inverters, controller and electric wires as compared to a motor with magnetic bearings. Thus, the overall size and cost of the bearingless machine is considerably reduced as compared to the machine with magnetic bearing. Furthermore, in the BELM as the rotor shaft is shorter, there is no fear to decrease the critical speed of the BELM by the rotational axis bend.

The BELM based on d-q axis current control, which has been proposed by the authors, is one of the BELMs with the integrated winding [5]-[7]. The advantageous feature of the proposed BELM is as follows. 1) The motor structure is just the same as the conventional brushless dc motors. The stator winding is short-pitched and hence, it is quite simple. 2) The control method is similar to that of the interior permanent magnet synchronous motor (IPMSM), i.e., in the proposed BELM, the rotational torque is controlled by the q-axis current and the suspension force is controlled by the d-axis current (the field-weakening or field-strengthening controls). Thus, the control method is also quite simple and the general-use 3-phase inverter can be employed to control the torque and suspension force.

The motor structure, principle of the torque and suspension force generations, control strategy have been proposed by the authors [6]. A prototype machine was built and it was confirmed by the experimental test that the rotor shaft was stably suspended without mechanical contact on the bearing journals. However, it was done only under no load condition. In this paper, hence, a novel control strategy to stably support the rotor shaft even under loaded condition is proposed. The bearingless motor performance with the stabilized control technique under loaded condition is tested by a simulation software using Finite Element Method (FEM). It is found that the suspension force ripple and the error of the force direction are remarkably decreased with the proposed identification method of the compensation coefficients. Furthermore, the experimental test of rotor levitation is carried out using a prototype machine. The compensation coefficients are updated by the proposed identification method based on the simulated results by FE analysis and they are practically input to the control system. The suspension performance is compared with that when the conventional compensation coefficients are applied in the control system. From the result, the validation of the updated compensation coefficients derived by the proposed identification method is confirmed.

978-1-4799-2706-7/14 $31.00 © 2014 IEEE

The 2014 International Power Electronics Conference

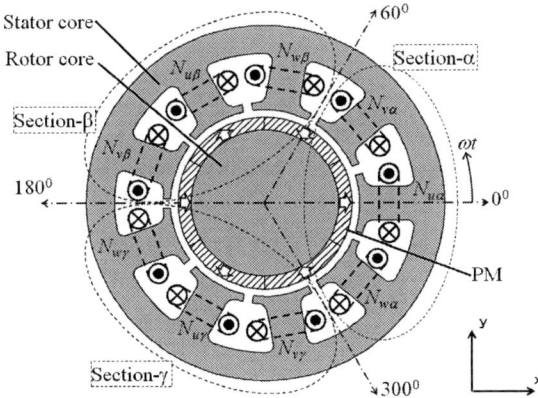

Fig. 1. Cross-section of the proposed BELM.

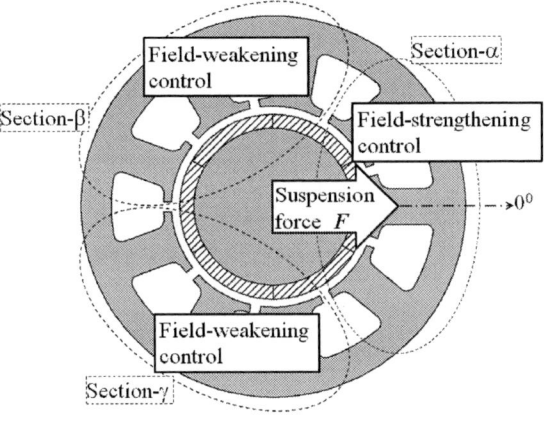

Fig. 2. Principle of the suspension force generation in the proposed BELM.

II. STRUCTURE AND PRINCIPLE OF THE PROPOSED BEARINGLESS MOTOR

Figure 1 shows the cross section of the proposed BELM. As there is only one stator winding wound on one stator core, the structure of the BELM is simple and also the winding arrangement looks like conventional blushless dc motor. The stator core is classified into three sections as the section-α, the section-β and the section-γ. In the section-α, the three-phase three-wire windings $N_{u\alpha}$, $N_{v\alpha}$ and $N_{w\alpha}$ are wound; in the section-β, $N_{u\beta}$, $N_{v\beta}$, $N_{w\beta}$ are wound; in the section-γ, $N_{u\gamma}$, $N_{v\gamma}$ and $N_{w\gamma}$ are wound. All these windings are short-pitch and simple. Each section is controlled by separate general-use 3-phase inverters. Totally, three 3-phase inverters are needed to drive the proposed BELM. The number of inverters is much than that of the conventional BELM. However, the capacity per an inverter may be decreased to 1/3 times that of the conventional motor. Hence the total volume and cost of the inverters are almost same as the conventional motors.

The suspension force is regulated to suspend the rotor of BELM. Fig. 2 shows the principle of the suspension force generation in the proposed BELM. The suspension

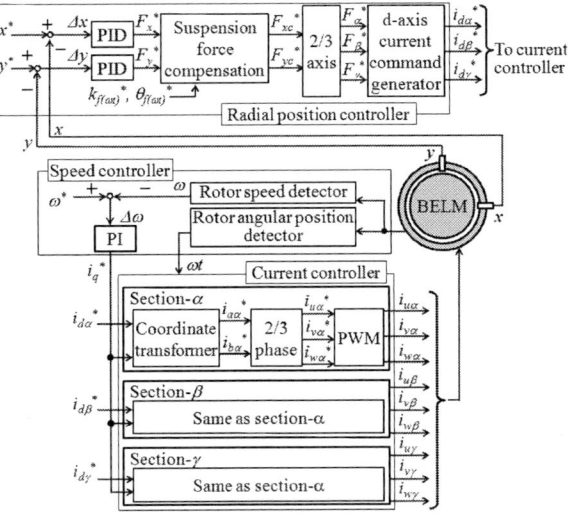

Fig. 3. Control system of BELM based on d-q axis current control.

force is generated by unbalanced flux density in air-gap with controlled d-axis currents $i_{d\alpha}$, $i_{d\beta}$ and $i_{d\gamma}$ in each section. For example, in section-α the field-strengthening control is done and then the air-gap flux density is increased; in the section-β and section-γ the field-weakening control is done and then the air-gap flux density is decreased. By the net vector sum in three sections, thus the suspension force is generated in the x-positive direction. By these controlled d-axis currents, the suspension force can be successfully generated in the arbitrary radial direction.

III. CONTROL SYSTEM AND SUSPENSION FORCE COMPENSATION

Figure 3 shows the control system configuration of the proposed BELM. In the motor controller, the rotor speed control method is same as the field oriented control of the conventional ac motors. The difference $\Delta\omega$ between the detected rotor speed ω and command speed ω^* is input to the proportional-integral (PI) controller, and q-axis current command i_q^* is output of the PI speed controller which is the input to the current controllers.

In the radial position controller, the detected rotor positions x and y on the x- and y- axes with eddy-current type gap sensors are input. The differences Δx and Δy between the detected rotor positions x, y and the commands x^*, y^* are input to the proportional-integral-derivative (PID) controller, and the suspension force commands F_x^* and F_y^* in the x- and y- axes coordinate are determined. In the block of suspension force compensation, F_x^* and F_y^* are significantly compensated [5]-[7]. Because the magnitude and direction of suspension force are undesirably varied depending on the rotor angular position due to the rotation of d-axis direction. The magnitude and direction compensation coefficients $k_{f(\omega t)}$ and $\theta_{f(\omega t)}$ are estimated based on the FE

978-1-4799-2706-7/14 $31.00 © 2014 IEEE   1716

The 2014 International Power Electronics Conference

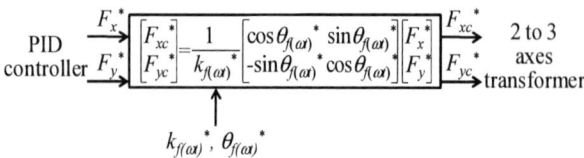

Fig. 4. Block diagram of suspension force compensation.

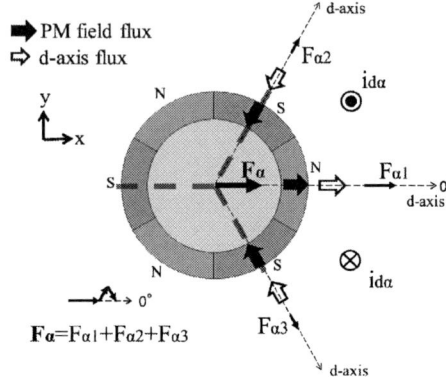

Fα=Fα1+Fα2+Fα3

Fig. 5. Suspension force at no load. (Section-α)
(At 0 degree)

analyzed results of the suspension force and then, the compensated suspension force commands F_{xc}^* and F_{yc}^* are obtained by the equations shown in Fig. 4. These commands F_{xc}^* and F_{yc}^* are transformed into F_α^*, F_β^* and F_γ^* in the α-, β- and γ- axes coordinate. Then, the d-axis current commands $i_{d\alpha}^*$, $i_{d\beta}^*$ and $i_{d\gamma}^*$ are determined to be proportional to the suspension force commands, respectively.

In the current controller, the current is independently controlled in each section. In the section-α, for example, the current commands i_q^* and $i_{d\alpha}^*$ are transformed into 2-phase the current commands $i_{a\alpha}^*$ and $i_{b\alpha}^*$ in the stationary coordinate. Then $i_{a\alpha}^*$ and $i_{b\alpha}^*$ are transformed into $i_{u\alpha}^*$, $i_{v\alpha}^*$ and $i_{w\alpha}^*$. The winding currents $i_{u\alpha}$, $i_{v\alpha}$ and $i_{w\alpha}$ are regulated to follow the current commands in the Pulse Width Modulation (PWM) block. The current controllers of the sections -β and -γ are same as that of the section-α.

IV. UPDATE OF COMPENSATION COEFFICIENTS UNDER LOADED CONDITION

A. Control Method at No Load

When the proposed BELM is operated at no load and particularly positioned at 0^0, the suspension force in each section is generated along the α-, β- and γ-axis, respectively. Fig. 5 shows the suspension force of the α-section at the rotor position of 0^0 when it is commanded in the x-positive direction.

However, when the rotor is at an angular position except for 0^0, the suspension force does not agree its command. Because the d-axis is also rotated in

accordance with the rotor rotation. Hence, the authors have proposed the ripple reduction control method of the suspension force. In the proposed method, the magnitude and direction of the suspension force command is successfully compensated with the rotor angular position. The effectiveness of the proposed method has been verified by the FE analysis and the experimental results [5][6].

B. Influence of Armature Reaction and Update Method of Compensation Parameters

When the proposed BELM is driven under loaded condition, the q-axis flux is added on the BELM magnetic field and it causes the variation of the magnitude and direction of the suspension force. As a result, the rotor fluctuation may be seriously increased and the rotor levitation may be unstable. Fig. 6 shows the suspension force under loaded condition at the rotor angular position of 0^0, in which the suspension force is commanded in the x-positive direction. It is described only in the section-α. The resultant flux of the permanent magnet (PM) field flux and the q-axis flux leads the PM field flux to the rotor rotational direction due to armature reaction. If the d-axis flux is superimposed on the resultant flux under this condition, the suspension force F_α is consequently generated by the resultant vector sum of $F_{\alpha1}$, $F_{\alpha2}$ and $F_{\alpha3}$ as described in Fig. 6. The direction of suspension force F_α leads to the rotor rotational direction, i.e., it is not oriented to the x-positive direction although the suspension force is commanded to its direction.

In order to overcome this problem, a new identification method of the compensation coefficient is proposed as follows; Fig. 7 shows the flux distribution when the rotor angular position is at –φ [deg] (φ >0). The resultant flux is oriented by the vector sum of the PM field flux and the q-axis flux as described in Fig. 7. In this situation, the d-axis flux is superimposed on this magnetic field. As a result, the suspension force F_α is generated by the resultant vector sum of $F_{\alpha1}$, $F_{\alpha2}$ and $F_{\alpha3}$ just in the x-positive direction, i.e., it agrees to the suspension force command. From this consideration, the compensation coefficients are justly updated as the reference of the magnitude and direction at –φ [deg] as shown in the equations (1) and (2). It introduces the stable magnetic levitation of the rotor shaft.

$$k_{f(\alpha)} = \frac{F_{(\alpha)}}{F_{(-\varphi)}}, \tag{1}$$

$$\theta_{f(\alpha)} = \theta_{(\alpha)} - \theta_{(-\varphi)}, \tag{2}$$

where the angular position –φ is obtained from the direction $\theta_{(\alpha)}$ of the suspension force under loaded condition. It is the angular position when the $\theta_{(\alpha)}$ is equal to zero.

978-1-4799-2706-7/14 $31.00 © 2014 IEEE   1717

The 2014 International Power Electronics Conference

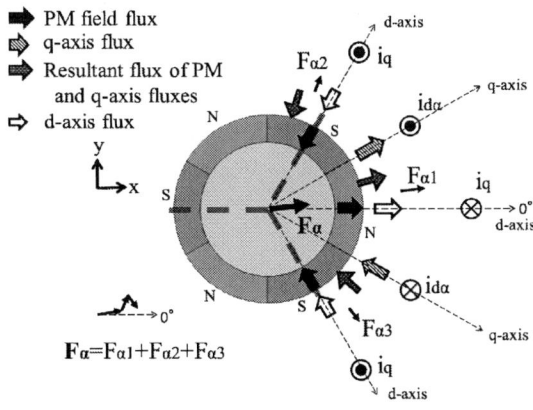

Fα=Fα1+Fα2+Fα3

Fig. 6. Suspension force under loaded condition. (Section-α)

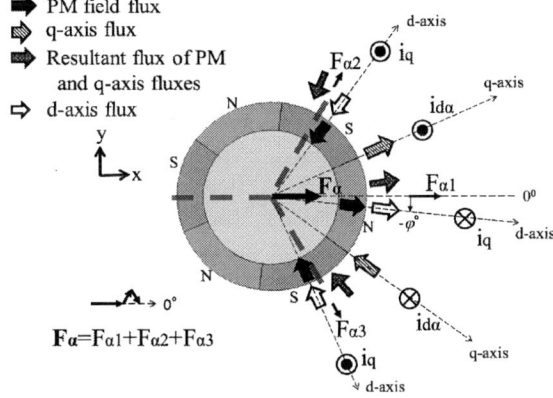

Fα=Fα1+Fα2+Fα3

Fig. 7. Reference of compensation coefficients under loaded condition. (at -φ deg)

In addition, the magnitude of the suspension force is increased due to the influence of armature reaction when the BELM based on d-q axis current control is driven under loaded condition. Hence, the magnitude of suspension force should be compensated. The authors have proposed the compensation method in [8]. The magnitude is successfully compensated in accordance with the q-axis current and then, its validation has been verified by the experimental tests using a prototype machine. In this paper, its compensation method is properly applied also in the proposed BELM based on d-q axis current control.

The examples of the updated $k_{f(\omega t)}$ and $\theta_{f(\omega t)}$ in a prototype machine are conducted below. Fig. 8 and Table 1 show the dimension and specification of a prototype machine, respectively. It is an interior permanent magnet (IPM) synchronous motor. The thin several PMs are buried just below the rotor iron surface. Hence, the saliency ratio is relatively low. The motor performance is almost same as that of the surface-mounted permanent magnet (SPM) type motor.

Fig. 9 shows the FE analyzed results of the magnitude $F_{(\omega t)}$ and direction $\theta_{(\omega t)}$ of suspension force at the

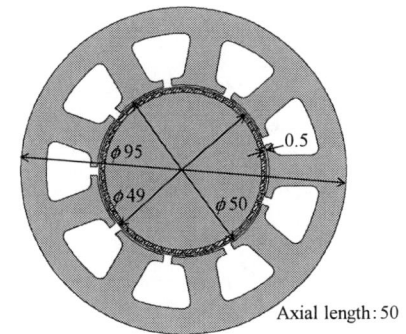

Axial length: 50

Fig. 8. Dimension of FE model.

TABLE I
SPECIFICATION OF FE MODEL

Rotor core, Stator core	Silicon steel
PM	Nd-Fe-B
Stator winding	128 turns

rotational speed of 4,500 r/min, under 77 % loaded condition in the FE machine model before the compensation is implemented. They are computed by FEM using a simulation software (JMAG Designer, Ver.12.0, JSOL Corporation). The suspension force of 30 N is commanded. The compensation coefficients $k_{f(\omega t)}$ and $\theta_{f(\omega t)}$ are updated based on Fig. 9 as follows; in Fig. 9, the $\theta_{(\omega t)}$ is zero at -6⁰, therefore, the magnitude and direction at -6⁰ (−φ = -6⁰) are set as the references $F_{(-\varphi)}$ and $\theta_{(-\varphi)}$, respectively. Fig. 10 shows the updated compensation coefficients $k_{f(\omega t)}$ and $\theta_{f(\omega t)}$, which are calculated by (1) and (2) based on the $F_{(\omega t)}$ and $\theta_{(\omega t)}$ in Fig. 9.

C. Verification by FEM

The authors verify the validation of updated compensation coefficients by the simulated results of magnitude and direction of the suspension force by FE analysis. Figs. 11 (a) and (b) show the computed results of the magnitude and direction of the suspension force under loaded condition by FEM using a simulation software. They are examples of FE model shown in Fig. 7 and it is under 77% loaded condition at the rotational speed of 4,500 r/min. The suspension force of 30 N is commanded. Fig. 11 (a) shows the computed results when the compensation coefficients at no load are applied even under loaded condition. Namely the compensation coefficients are determined as the reference of the suspension force at the rotor angular position of 0⁰ as shown in Fig. 5. On the other hand, Fig. 11 (b) shows those when the compensation coefficients under loaded condition are justly applied. From Fig. 11 (a), the suspension force ripple is not changed in comparison with that before the compensation is implemented. The error of suspension force direction is decreased a little bit. Thus, the effectiveness of the compensation cannot be detected. On the other hand, it is seen in Fig. 11 (b) that

978-1-4799-2706-7/14 $31.00 © 2014 IEEE   1718

The 2014 International Power Electronics Conference

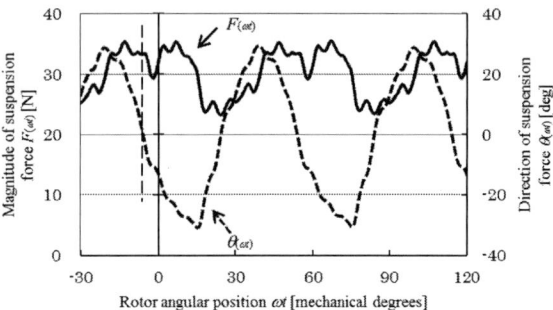

Fig. 9. Magnitude and direction of suspension force under loaded condition without compensation.

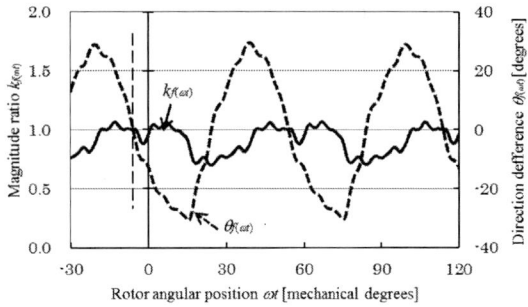

Fig. 10. Updated compensation coefficients under loaded condition.

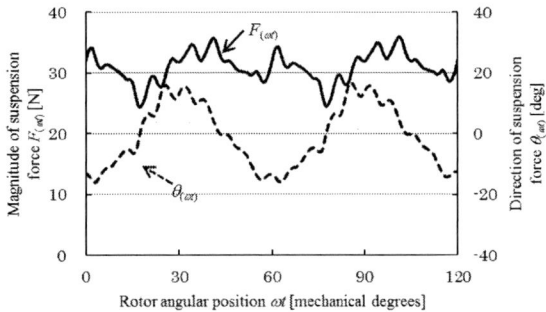

(a) With the compensation coefficients at no load even under loaded condition.

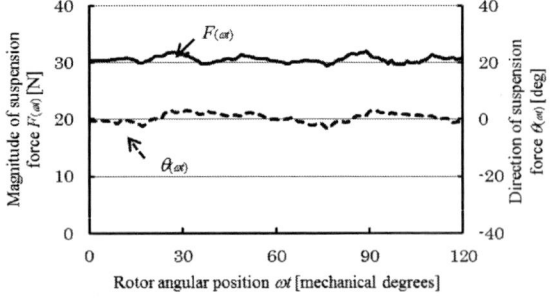

(b) With the compensation coefficients under loaded condition.

Fig. 11. Magnitude and direction of suspension force with compensation coefficients.

the ripple and direction variation of the suspension force are considerably reduced. Both of the magnitude and direction are correctly compensated and the suspension force is completely generated according to its command. Hence, the validation of the proposed compensation method is verified by the simulated results in Fig. 11 (a) (b).

V. EXPERIMENTAL TEST RESULT

In order to confirm the validation of the compensation method and the updated compensation coefficients under loaded condition, a prototype machine is built and the rotor levitation tests are carried out. Fig. 12 shows the structure of the prototype BELM based on d-q axis current control. It is a lateral type. The right side of the shaft is mechanically supported by the ball bearing and the position of its left side is actively controlled by the electromagnetic force. The touchdown bearing is equipped around the left side of the shaft. The shaft is touched down at standstill and emergency when the trouble unfortunately occurs in the machine or the controller during the operation. The clearance between the shaft and the inside surface of the touchdown bearing is 100 μm. The gap sensors are equipped in two perpendicular axes x and y and the rotor radial position is detected. The rotary encoder is in the right side of the shaft. Fig. 8 shows the dimension of the prototype bearingless motor. The small permanent magnets are buried just below the rotor core surface. Fig. 13 shows the photo of the experimental setup to measure the motor and magnetic suspension performances under loaded condition. The PM synchronous generator is connected to the prototype BELM by the mechanical coupling. The output power of the prototype BELM is 600 W at the rotational speed of 4,500 r/min.

The rotor levitation tests under loaded condition are carried out using a prototype machine. Fig. 14 (a) shows the rotor radial displacements x and y when the compensation coefficients $k_{f(\omega t)}$ and $\theta_{f(\omega t)}$ are not updated. Fig. 14 (b) shows those when the coefficients are correctly updated as shown in Fig. 10 and they are practically input to the control system. In these figures, the machine is driven at the rotational speed of 1,500 r/min, under 50% loaded condition. The suspension force is constantly generated in the y-positive direction against the gravity of the rotor shaft as the experimental setup is a lateral type. It is seen in Figs. 14 (a) and (b) that the rotor radial displacements are reduced by employing the updated compensation coefficients. In particular, the decrease in the y-direction displacement is remarkable because the displacement in the y-direction was relatively large before the compensation coefficients are updated as shown in Fig. 14 (a) since the rotor shaft gravity is constantly applied in the y-negative direction. The validation of the update of compensation coefficients is obviously verified by the test results.

978-1-4799-2706-7/14 $31.00 © 2014 IEEE

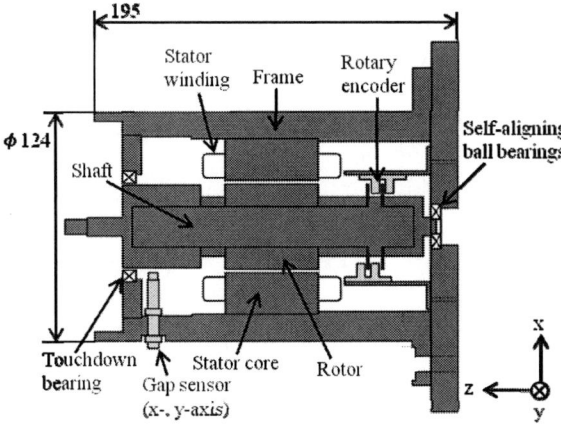

Fig. 12. Longitudinal-section of the prototype machine.

Fig. 13. Photo of the prototype machine.

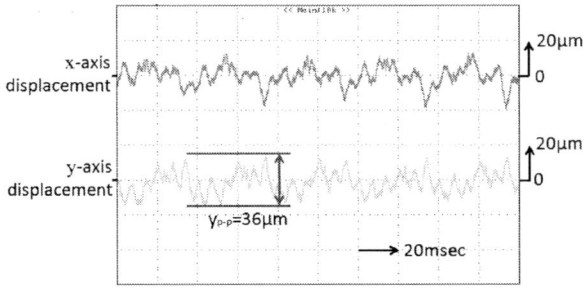

(a) Compensation coefficients are not updated.

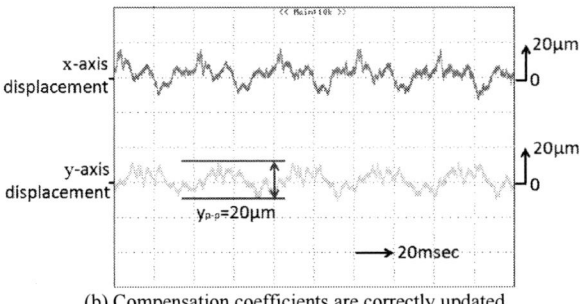

(b) Compensation coefficients are correctly updated.

Fig. 14. Rotor radial displacements in x- and y-axes.

VI. CONCLUSION

The stabilized control method of the rotor position under loaded condition in the BELM based on d-q axis current control has been presented. In this paper, the validation of the proposed method has been verified by a simulation software of FE analysis using a machine model. Furthermore, the rotor levitation tests under loaded condition are carried out using a prototype machine. The compensation coefficients $k_{f(\omega t)}$ and $\theta_{f(\omega t)}$ are successfully updated by the proposed identification method. The rotor radial displacements are reduced, particularly the displacement in the y-axis is remarkably reduced to 20 μm from 36 μm by the updated compensation coefficients. The validation of the updated compensation coefficients are verified by the test results. In the next stage, it will be found how the compensation coefficients are correctly updated when the motor load is varied during the operation.

ACKNOWLEDGMENT

The authors would like to thank Mr. Karasawa who was a former graduate student at the Tokyo University of Science, Suwa and Mr. Yajima who is a student at the Tokyo University of Science, Suwa.

REFERENCES

[1] Akira Chiba, Tadashi Fukao, Osamu Ichikawa, Masahide Oshima, Masatsugu Takemoto, and David G Dorrell: "Magnetic Bearings and Bearingless Drives", *Newnes*, ISBN 0 7506 5727 8, 2005.

[2] W. Gruber, W. Amrhein, M. Haslmayr, "Bearingless Segment Motor With Five Stator Elements—Design and Optimization", *IEEE Transactions on Industry Applications*, Vol. 45, No. 4, 2009.

[3] Andreas Binder and Gabriel Munreanu, "Bearingless PM Levitation Systems" *Proceedings of the 20 12 International Conference on Electrical Machines and Systems*, Special Lecture SL-4, @Sapporo, CDROM, 2012.

[4] Zhenyao Xu, Dong-Hee Lee, and Jin-Woo Ahn, "Suspending Force Control of a Novel 12/14 Hybrid Stator Pole Type Bearingless SRM", *Proceedings of the 2012 International Conference on Electrical Machines and Systems*, LS2B-3, @Sapporo, CDROM, 2012.

[5] Masahide Ooshima, Syunsuke Kobayashi and M. Nasir Uddin, "Magnetic Levitation Tests of a Bearingless Motor Based on d-q Axis Current Control", *IEEE Industry Application Society 2012 Annual Meeting*, 2012-IACC-212, CDROM, 2012.

[6] Syunsuke Kobayashi, Masahide Ooshima and M. Nasir Uddin, "A Radial Position Control Method of Bearingless Motor Based on d-q Axis Current Control", *IEEE Transactions on Industry Applications*, vol.49, No.4, pp.1827-1835, 2013.

[7] Masahide Ooshima, Toshiki Karasawa and M. Nasir Uddin, "Stabilized Control Strategy Under Loaded Conditions in a d-q Axis Current Control", *IEEE Industry Application Society 2012 Annual Meeting*, 2013-IACC-315, CDROM, @Orlando, 2013.

[8] Masahide Oshima, Satoru Miyazawa, Akira Chiba, Fukuzo Nakamura, Tadashi Fukao "Parameter Measurements and Radial Position Control Characteristics of a Permanent Magnet Type Bearingless Motor Under Loaded Condition", *The Transaction of The Institute of Electrical Engineers of Japan, A Publication of Industry Applications Society*, vol.120-D, No.8/9, pp.1015-1023, 2000 (in Japanese).

Analysis and Design of a High-Frequency Isolated Dual-Tank LCL Resonant AC-DC Converter

Yimian Du
Department of Electrical and Computer Engineering
University of Victoria
Victoria, B.C., Canada
duyimian@ece.uvic.ca

Ashoka K.S. Bhat
Department of Electrical and Computer Engineering
University of Victoria
Victoria, B.C., Canada
bhat@ece.uvic.ca

Abstract—An integrated single-phase single-stage high-frequency (HF) isolated dual-tank LCL-type series resonant ac-dc converter with fixed frequency, phase-shift control strategy is presented. The circuit configuration combines a diode rectifier, boost converter and two identical half-bridge dc-dc LCL resonant converters. A high power factor (PF) and low total harmonic distortion (THD) is achieved by discontinuous current mode (DCM) operation of the front-end power factor correction (PFC) circuit. The output voltage is regulated by fixed-frequency, phase-shift between two identical half-bridge LCL resonant converters. Soft-switching operation is achieved for all the switches. The operation of intervals and steady-state analysis are presented briefly. Design example of a 100 W proposed converter is given together with its simulated and experimental results for different input voltages.

I. INTRODUCTION

Single-phase single-stage HF isolated ac-dc converter asks for following features: functions of HF isolation, PFC and output voltage regulation in one single-stage; a high PF and low THD at the ac input side; simple control strategy and high efficiency. Efforts to achieve all the expected functions in a single-stage are reported in [1-3]. A high PF is achieved in [1,2] by operating the ac input side in DCM and the output voltage regulation is achieved by a wide variation in switching frequency. These configurations require a wide variation (almost 2:1) in switching frequency for power control. Wide variation in switching frequency for power control makes filter design difficult while introducing higher losses at increased frequencies. In order to avoid variable frequency control, [3] employs a fixed-frequency, secondary-side phase-shift control for output voltage regulation, but a low PF and high THD occurs when a large phase-shift is applied.

To improve PF and THD at the ac input side, an integrated dual-tank LCL-type series resonant ac-dc converter is proposed in this paper (Fig. 1) that uses a fixed frequency phase-shift control for output voltage regulation. The proposed circuit configuration combines a diode rectifier, boost converter and two identical half-bridge LCL resonant dc-dc converters. The concept of dual-tank used in dc-dc converters [4] is utilized in realizing the proposed integrated ac-dc converter for the first time. A high PF and low THD is obtained by DCM

in the front-end PFC circuit. The proposed converter output voltage can be controlled by fixed-frequency phase-shift between the two identical half-bridge converters. Also, a LCL-type series resonant converter guarantees the expected soft-switching operation. The targeted applications are in small scale wind energy conversion systems (WECS) employing synchronous generators [5] and ac-dc power supplies.

II. PROPOSED CONVERTER AND ITS OPERATION

In the proposed configuration Fig. 1), a dual-switch boost converter (D_{r1}, D_{r2}, S_1, S_2, L_1) is integrated with a half-bridge resonant converter (S_1, S_2, C_1, C_2, L_{r1}, C_{r1}, T_1). The other half-bridge resonant converter (S_3, S_4, C_1, C_2, L_{r2}, C_{r2}, T_2) shares the dc bus capacitor (C_{bus}) with the first half-bridge converter. Two identical HF transformers (T_1, T_2) are connected in series on secondary-side. So it forms the desired dual-tank configuration. An external inductor (L_t) is connected in parallel with the terminal of secondary-side of T_1/T_2 to achieve LCL-type series resonant circuit. The magnetizing inductances of each transformer can be considered as parts of the paralleled inductor. A high PF and THD are achieved by DCM in L_1 over the entire line frequency cycle at the ac input side. Since T_1 and T_2 are connected in series on secondary-side, i_{rT1} and i_{rT2} are always identical ($i_{rT1} = i_{rT2}$).

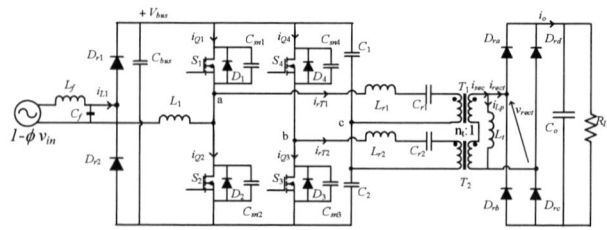

Fig. 1. Proposed HF isolated dual-tank LCL-type series resonant ac-dc converter.

Fixed-frequency complementary gating signals (v_{gs1}, v_{gs2}) and (v_{gs3}, v_{gs4}) of 50% duty cycle and with a small dead-time between them are applied to the switches (S_1,S_2) and (S_3,S_4), respectively. The output voltage can be controlled by the phase-shift θ between the two half-bridge converters. When the minimum input voltage is applied, θ is set as zero, the outputs of two bridges are added together on the secondary-side to obtain twice of

978-1-4799-2706-7/14 $31.00 © 2014 IEEE

output from single tank circuit. If the ac input voltage rises, θ needs to be increased to regulate the output voltage.

Due to the symmetry of the proposed circuit, only the operation of the positive half cycle of ac input is described. There are 11 operation intervals in one HF switching period. The key HF waveforms and interval equivalent circuits are shown in Fig. 2 and Fig. 3, respectively. To simplify the operation and analysis, following assumptions are made: (a) All semiconductors and passive components are ideal. (b) Leakage inductance of HF transformer is part of resonant inductor. (c) Effects of all snubber capacitors (C_{sn1} to C_{sn4}) are ignored. Capacitor C_{bus}, C_1, C_2 and C_o are assumed to be large enough so that the dc bus voltage V_{bus} and output voltage V_o can be regarded as constant values.(c) Effects of small dead-gaps between two groups of complementary signals are neglected.(d) The parallel inductance (L_p) is assumed to include magnetizing inductances(L_m) of T_1/T_2, and the external inductance (L_t) [6].

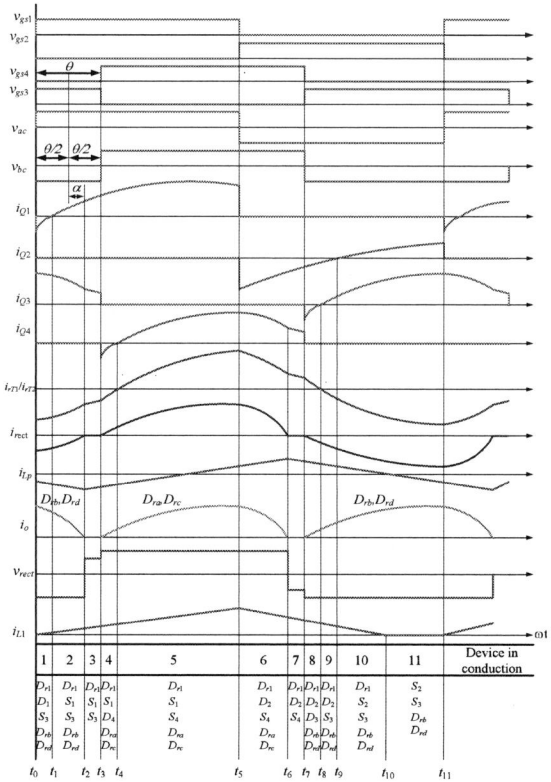

Fig. 2. Steady-state waveforms of the proposed converter in one HF cycle.

Interval 1 (t_0-t_1) (Fig. 3a): This interval starts when v_{gs1} is applied at $t = t_0$, v_{gs3} remains ON (v_{gs2} and v_{gs4} are OFF). During the interval, the power is injected into the circuit through D_{r1} from the ac source. The boost converter current i_{L1} starts to increase linearly from zero. The voltage between a and c is, $v_{ac} = V_{bus}/2$, and between b and c is $v_{bc} = -V_{bus}/2$. D_1 and S_3 conduct while the two identical HF resonant currents i_{rT1} and i_{rT2} are negative during the interval. This will allow ZVS turn-on for S_1 in

the next interval. HF diodes rectifier input current i_{rect} is negative and rectifier diodes D_{rb} and D_{rd} are conducting delivering power to the load. Also, the current through parallel inductor i_{Lp} increases linearly. This interval ends when D_1 current goes to zero.

Interval 2 (t_1-t_2) (Fig. 3b): At $t = t_1$, S_1 is turned-ON with ZVS and starts to conduct. The other conducting devices are the same as those in Interval 1. At $t = t_2$, i_{Lp} reaches its peak value, i_{rect} reaches zero and current flowing through D_{rb} and D_{rd} goes to zero turning-OFF with zero-current switching (ZCS). This interval ends at $t = t_2$.

Interval 3 (t_2-t_3) (Fig. 3c): On the primary-side of HF transformers, all conducting situations are the same as the previous interval. On the secondary-side i_{Lp} starts decreasing and since $i_{rect} = 0$, all the rectifier diodes are in OFF state and load power is supplied by the filter. This interval ends at $t = t_3$, when v_{gs3} is removed and v_{gs4} is applied.

Interval 4 (t_3-t_4) (Fig. 3d): At $t = t_3$, since v_{gs3} is removed and v_{gs4} is applied, i_{Q3} becomes zero. On the primary-side, i_{Q1} flows through S_1 and D_4 also conducts. v_{bc} changes polarity so that $v_{bc} = v_{ac} = V_{bus}/2$. At the same time, i_{rect} changes its polarity so that D_{ra} and D_{rc} start to conduct, and delivering power to the load. i_{Lp} continues to decrease linearly from the peak value and D_{r1} continues to conduct allowing i_{L1} to increase linearly. At the end of this interval, i.e., at $t = t_4$, i_{rT1}/ i_{rT2} reach zero and $i_{Q4} = 0$.

Interval 5 (t_4-t_5) (Fig. 3e): At $t = t_4$, resonant currents i_r become positive; S_4 is turned-ON with ZVS and starts to conduct. The conducting devices are the same as those in Interval 4. i_{Lp} naturally changes its polarity during this interval. D_{ra} and D_{rc} are conducting delivering power to the load. This interval ends at $t = t_5$, when v_{gs1} is removed and v_{gs2} is applied. Boost inductor current i_{L1} reaches its peak value.

Interval 6 (t_5-t_6) (Fig. 3f): At $t = t_5$, since v_{gs1} is removed and v_{gs2} is applied i_{Q1} becomes zero. On the primary-side, Q_4 continues to conduct, current i_{Q4} flows through S_4 and D_2 also conducts. Voltage v_{ac} changes polarity so that $v_{ac} = -V_{bus}/2$. Current i_{L1} starts decreasing linearly and i_p increases in negative direction. Output rectifier diodes conducting are the same as interval 5. This interval ends at $t = t_6$, when i_{rect} reaches zero and i_{Lp} reaches its positive peak value.

Interval 7 (t_6-t_7) (Fig. 3g): On the primary-side of HF transformers, all conducting situations are the same as the previous interval. On the secondary-side, i_{Lp} starts decreasing towards zero and since $i_{rect} = 0$, all the rectifier diodes are in OFF state and load power is supplied by the filter capacitor C_o. This interval ends at $t = t_7$, when v_{gs4} is removed and v_{gs3} is applied.

Interval 8 (t_7-t_8) (Fig. 3h): At $t = t_7$, since v_{gs4} is removed and v_{gs3} is applied i_{Q4} becomes zero and D_3 starts conducting. D_2 continues to conduct. Voltage v_{bc} changes polarity so that $v_{bc} = v_{ac} = -V_{bus}/2$. At the same time, i_{rect} changes its polarity (becomes negative) so that D_{rb} and D_{rd} start to conduct delivering power to the load. Currents i_p and i_{L1} (D_{r1} continues to conduct) continue to decrease linearly towards zero. This interval ends at $t = t_8$ when i_r reaches zero and current through D_3 also goes to zero.

Interval 9 (t_8-t_9) (Fig. 3i): At $t = t_8$ resonant currents i_r become negative; S_3 is turned-ON with ZVS and starts to conduct. The other conducting devices are the same as those in Interval 8. Rectifier diodes D_{ra} and D_{rd} continue to conduct delivering power to the load. This interval ends at $t=t_9$, when current through D_2 (i_{Q2}) reaches zero.

Interval 10 (t_9-t_{10}) (Fig. 3j): At $t = t_9$, S_2 is turned-ON with ZVS and starts to conduct. The other conducting devices are the same as those in Interval 9. This interval ends at $t = t_{10}$, i_{L1} reaches zero turning-off D_{r1}.

Interval 11 (t_{10}-t_{11})(Fig. 3k): At $t = t_{10}$, since i_{L1} reaches zero turning-off D_{r1}, line current is in DCM since D_{r2} is also not conducting. S_2 and S_3 are conducting and energy stored in the circuit is still delivered to the load through D_{rb} and D_{rd}. Current i_{Lp} reaches zero and changes direction during this interval. This interval ends at $t = t_{11}$, when v_{gs2} is removed and v_{gs1} is applied. This completes the operation for one HF switching cycle and next cycle begins.

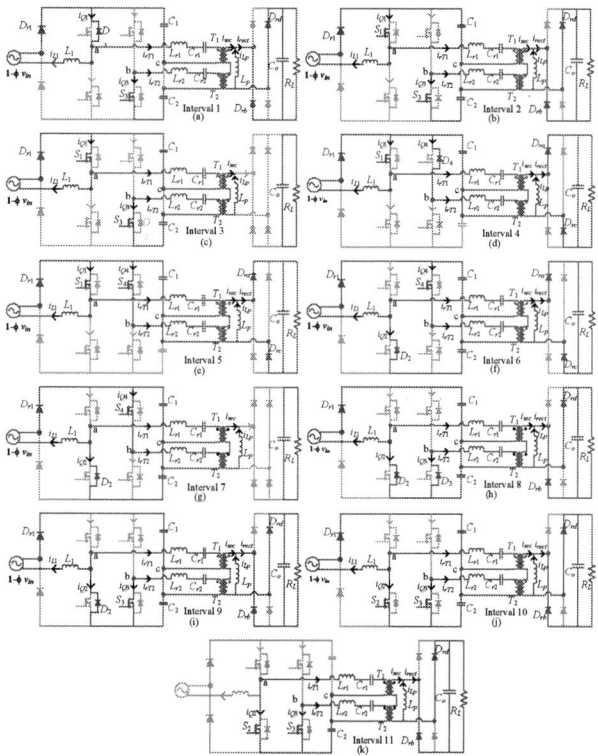

Fig. 3. Steady-state equivalent circuits of the proposed converter in one HF cycle.

III. STEADY-STATE ANALYSIS

To simplify the analysis, following assumptions are made: (a) All semiconductors and passive components are ideal. (b) Leakage inductances of HF transformers are parts of resonant inductors. (c) Effects of all snubber capacitors (C_{sn1} to C_{sn4}) are ignored. (d) Effects of small dead-gaps between two groups of complementary signals are neglected. (e) Filter capacitors C_{bus}, C_1, C_2 and C_o are assumed to be large enough so that the dc bus voltage V_{bus} and output voltage V_o can be regarded as constant values.

For analysis purpose, the single-stage converter can be viewed as two separate parts: (i) front-end PFC circuit and (ii) dual-tank LCL-type series resonant dc-dc converter. The PFC circuit operates in DCM and its analysis is well known [6,7]. The ac input power P_{in} is given by

$$P_{in} = \frac{\left(\sqrt{2}V_{in}\right)^2 D^2 y_1(\rho)}{2\pi f_s L_1} \tag{1}$$

where

$$y_1(\rho) = -\frac{2}{\rho} - \frac{\pi}{\rho^2} + \frac{2}{\rho^2\sqrt{1-\rho^2}}\left[\frac{\pi}{2} - tan^{-1}\left(\frac{-\rho}{\sqrt{1-\rho^2}}\right)\right] \tag{2}$$

and $\rho = \sqrt{2}V_{in}/V_{bus}$, V_{bus} is the voltage across the dc bus capacitor C_{bus}. V_{in} is the rms value of ac input voltage. $D = 0.5$ is the duty cycle of the shared switches (S_1/S_2). The value of boost inductor (L_1) can be determined by (1) and (2).

The steady-state analysis of the dual-tank LCL-type series resonant dc-dc converter is done using Fourier series analysis [8]. In order to combine the two identical magnetizing inductances L_{m1} and L_{m2} of two identical HF transformers with the external paralleled inductor L'_t (Fig. 4(a)), it is necessary to simplify this equivalent circuit before starting to analyze the dual-tank dc-dc converter. The steps of transformation used to simplify the equivalent circuit in time domain are shown as Fig. 4(a) to (c).

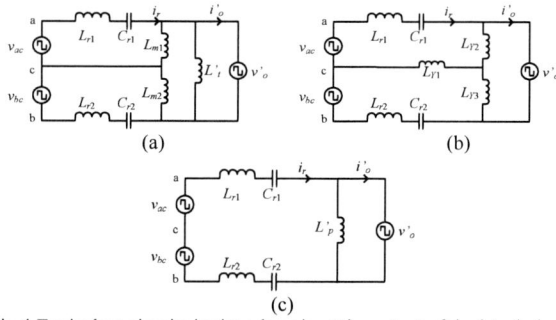

Fig.4 Equivalent circuits in time domain at the output of dual-tank dc-dc resonant converter. (a) Delta connection for L_{m1}, L_{m2} and L'_t before transformation; (b) Y-connection after the Δ-Y transformation; (c) Simplified equivalent circuit after transformation and neglecting L_{Y1} (large value).

The original connection of the equivalent circuit in time domain is shown in Fig. 4(a). As can be seen, L_{m1}, L_{m2} and L'_t are connected in delta. In order to simplify the analysis, the delta-connection is transformed to Y-connection as shown in Fig. 4(b). The transformation equations are given by

$$L_{Y1} = \frac{L_{m1}L_{m2}}{L_{m1}+L_{m2}+L'_t} \tag{3}$$

$$L_{Y2} = \frac{L_{m1}L'_t}{L_{m1}+L_{m2}+L'_t} \tag{4}$$

$$L_{Y3} = \frac{L_{m2}L'_t}{L_{m1}+L_{m2}+L'_t} \tag{5}$$

978-1-4799-2706-7/14 $31.00 © 2014 IEEE

Since $L_{m1} = L_{m2} = L_m$ are large compared to L'_t, L_{Y1} is also large and therefore, we can neglect that current flowing through L_{Y1}. Hence, it can be considered as open circuited. On the other hand, $L'_p = L_{Y2} + L_{Y3} = 2L_mL'_t/(2L_m+L'_t)$. Simplified equivalent circuit obtained is shown in Fig. 4(c). The following analysis is based on the equivalent circuit shown in Fig. 4(c). Since two half-bridge resonant converters are identical, $L_{r1} = L_{r2} = L_r$, and $C_{r1} = C_{r1} = C_r$. In the following presentation, L_r and C_r will be used to simplify the circuit analysis.

All parameters on the secondary-side of HF transformers are reflected to primary-side, which are denoted by the superscript "'". To get generalized design curves, all the parameters are normalized using the base values: $V_B = V_{bus,min}$, $Z_B = (L_r/C_r)^{1/2}$, $I_B = V_B/Z_B$, where $L_r = L_{r1} = L_{r2}$, and $C_r = C_{r1} = C_{r2}$. The dual-tank LCL resonant dc-dc converter gain is defined as $M_f = V'_o/V_{bus}$. $V'_o = n_tV_o$, where n_t is the two identical transformers primary-to-secondary turns ratio. The normalized switching frequency is given by $F = \omega_s/\omega_r = f_s/f_r$ where resonant frequency $\omega_r = 2\pi f_r = 1/(\sqrt{L_rC_r})$ and switching frequency, $f_s = \omega_s/(2\pi)$. The normalized values and the nth harmonic components are denoted by subscript "0" and "n" respectively. Hence all normalized n^{th} harmonic reactance components are given by $X_{Lr,n0} = nF$, $X_{Cr,n0} = -1/(nF)$, $X_{s,n0} = X_{Lr,n0} + X_{Cr,n0} = nF - 1/(nF)$, $X'_{Lp,n0} = knF$ where $k = L_p/L_r$. $X_{eq,n0} = 2X_{s,n0}X'_{Lp,n0}/(2X_{s,n0} + X'_{Lp,n0})$.

The nth harmonic phasor equivalent circuit used for analysis is shown in Fig 5(a). Two HF voltage sources (v_{ac}, v_{bc}) which generates square wave with amplitude of $V_{bus}/2$ are placed on the left, with phase angle of $\theta/2$ and -$\theta/2$, respectively (center of angle θ is taken as the origin). The third HF voltage source (v'_o) placed on the right is the diode rectifier input voltage reflected to primary-side of HF transformers. It can be assumed as an approximate square wave with amplitude of V'_o and phase angle at -α. L'_p includes the external parallel inductor and the magnetizing inductors of T_1 and T_2.

The normalized n^{th} harmonic square wave voltage sources in phasor domain are given by

$$\bar{V}_{ac,n0} = \frac{2}{n\pi}\angle(\frac{n\theta}{2} - \frac{\pi}{2}) \tag{6}$$

$$\bar{V}_{bc,n0} = \frac{2}{n\pi}\angle(-\frac{n\theta}{2} - \frac{\pi}{2}) \tag{7}$$

$$\bar{V}'_{o,n0} = \frac{4M_f}{n\pi}\angle(-n\alpha - \frac{\pi}{2}) \tag{8}$$

Note that two voltage sources on the left can be considered as a single equivalent voltage source (Fig. 5(b)), given by

$$\bar{V}_{eq,n0} = \frac{4}{n\pi}\sin\left(\frac{n\pi}{2}\right)\sin\left(\frac{n\delta}{2}\right)\angle - \pi/2 \tag{9}$$

where $\delta = \pi - \theta$ is the pulse width of equivalent voltage source.

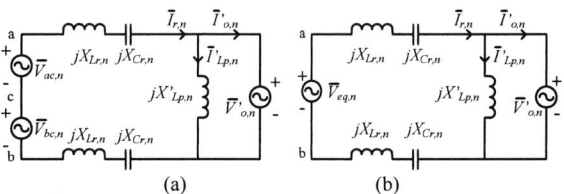

Fig.5. n^{th} harmonic phasor equivalent circuit, all parameters are referred to primary-side.

The Superposition theorem is used to analyze the circuit. The corresponding equivalent circuits are shown in Fig. 6.

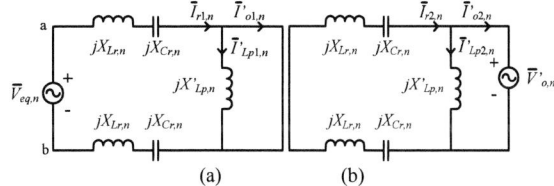

Fig.6. Equivalent circuit by using Superposition: (a) output voltage source short circuited; (b) input voltage sources short circuited.

According to Fig. 6(a) (output voltage source is short circuited), phasor currents for n^{th} harmonic are given by

$$\bar{I}'_{o1,n0} = \bar{I}_{r1,n0} = \frac{4\sin(n\pi/2)\sin(n\delta/2)\angle - \pi}{2n\pi X_{s,n0}} \tag{10}$$

$$\bar{I}'_{Lp1,n0} = 0 \tag{11}$$

According to Fig. 6(b) (input source is short circuited retaining the output source), n^{th} harmonic currents are

$$\bar{I}_{r2,n0} = \frac{\bar{V}'_{o,n0}}{2jX_{s,n}} = -\frac{4M_f}{2n\pi X_{s,n0}}\angle(-n\alpha - \pi) \tag{12}$$

$$\bar{I}'_{Lp2,n0} = \frac{\bar{V}'_{o,n0}}{jX_{Lp,n0}} = \frac{4M_f}{n\pi X_{Lp,n0}}\angle(-n\alpha - \pi) \tag{13}$$

$$\bar{I}'_{o2,n0} = \frac{\bar{V}'_{o,n0}}{jX_{eq,n0}} = -\frac{4M_f}{n\pi X_{eq,n0}}\angle(-n\alpha - \pi) \tag{14}$$

Hence the normalized tank resonant current and reflected HF rectifier input current in phasor $(\bar{I}_{r,n0} = \bar{I}_{r1,n0} + \bar{I}_{r2,n0})$ and time domain are given by

$$\bar{I}_{r,n0} = \frac{4}{2n\pi X_{s,n0}}\sin\left(\frac{n\pi}{2}\right)\sin\left(\frac{n\delta}{2}\right)\angle(-\pi)$$
$$-\frac{4M_f}{2n\pi X_{s,n0}}\angle(-n\alpha - \pi) \tag{15}$$

$$i_{r,0}(t) = -\frac{4}{\pi}\sum_{n=1,3}^{\infty}\frac{\sin\left(\frac{n\pi}{2}\right)\sin\left(\frac{n\delta}{2}\right)\cos(n\omega_s t)}{2nX_{s,n0}}$$
$$+\frac{4M_f}{\pi}\sum_{n=1,3}^{\infty}\frac{\cos(n\omega_s t - n\alpha)}{2nX_{s,n0}} \tag{16}$$

The normalized n^{th} harmonic reflected HF rectifier input phasor current is given by $(\bar{I}'_{o,n0} = \bar{I}'_{o1,n0} + \bar{I}'_{o2,n0})$,

$$\overline{I}'_{o,n0} = \frac{4}{n\pi}\left(\frac{\sin(n\pi/2)\sin(n\delta/2)\angle-\pi}{2X_{s,n0}} - \frac{M_f\angle(-n\alpha-\pi)}{X_{eq,n0}}\right) \quad (17)$$

The corresponding equation in time domain is given by

$$i'_{o,0}(t) = -\frac{4}{\pi}\sum_{n=1,3}^{\infty}\frac{\sin\left(\frac{n\pi}{2}\right)\sin\left(\frac{n\delta}{2}\right)\cos(n\omega_s t)}{2nX_{s,n0}}$$
$$+\frac{4M_f}{\pi}\sum_{n=1,3}^{\infty}\frac{\cos(n\omega_s t-n\alpha)}{nX_{eq,n0}} \quad (18)$$

The normalized n^{th} harmonic resonant capacitor voltage equation in phasor domain is given by

$$\overline{V}_{cr,n0} = \frac{4X_{Cr,n0}}{2n\pi X_{s,n0}}\Big(\sin(n\pi/2)\sin(n\delta/2)\angle(-3\pi/2)$$
$$- M_f\angle(-n\alpha-3\pi/2)\Big) \quad (19)$$

The corresponding time domain equation is

$$v_{Cr,0}(t) = \frac{4}{\pi}\sum_{n=1,3}^{\infty}\frac{\sin(n\pi/2)\sin(n\delta/2)\sin(n\omega_s t)X_{Cr,n0}}{2nX_{s,n0}}$$
$$-\frac{4M_f}{\pi}\sum_{n=1,3}^{\infty}\frac{\sin(n\omega_s t-n\alpha)X_{Cr,n0}}{2nX_{s,n0}} \quad (20)$$

To find α, the normalized output current (16) is equated to zero at $\omega_s t=\alpha$. The Newton-Raphson method is used to compute α. The initial guess value for α is found using the fundamental component ($n = 1$, $\alpha = \alpha_1$), given by:

$$\cos(\alpha_1) = \frac{M_f}{\sin(\delta/2)}\left(1+\frac{2X_{s,10}}{X'_{Lp,10}}\right) \quad (21)$$

In order to find the average value of output current, the normalized load current J is obtained first by averaging the nth harmonic component

$$J_n = \frac{1}{\pi}\left[-\int_0^\alpha i'_{o,n0}d(\omega_s t)+\int_\alpha^\pi i'_{o,n0}d(\omega_s t)\right] \quad (22)$$

Therefore, the normalized load current J is obtained as

$$J = \frac{8}{\pi^2}\sum_{n=1,3,...}^{\infty}\frac{\sin(n\alpha)}{2n^2X_{sn,0}}\sin\left(\frac{n\pi}{2}\right)\sin\left(\frac{n\delta}{2}\right) \quad (23)$$

In this case, V_{bus} rises with θ increased due to the fixed duty cycle of gating signals applied to S_1/S_2. In order to find the relationship between V_{bus} and θ, power balance concept is employed for the analysis. The proposed ac-dc converter output power is given by

$$P_o = (I'_o)^2 R'_L = (JI_B)^2 n_t^2 R_L \quad (24)$$

Assuming an ideal case (100% of efficiency), $P_{in}= P_o$, we obtain

$$\frac{(\sqrt{2}V_{in})^2 D^2 y_1(\rho)}{2\pi L_1 f_s} = (JI_B)^2 n_t^2 R_L \quad (25)$$

Hence V_{bus} can be calculated based on (2), (23) and (25).

IV. DESIGN EXAMPLE

To illustrate design procedure, a design example is provided as follows: peak ac input voltage varies between 60V, 40Hz to 80V, 60Hz (e.g., in a WECS); converter output power is 100W, and dc output voltage is 100V. Switching frequency is 100 kHz.

Based on the analysis, in the frond-end dual-switch boost converter part, the boost inductor L_1 is determined by the minimum ac input voltage. The value can be found based on (1) and (2), so $L_1=40\,\mu H$.

Using the Fourier series analysis, several design curves under different normalized switching frequency F are plotted for the example with $k=20$ and $\theta=0$, and are shown in Fig. 7. As can be seen, a smaller F brings higher M_f, J, resonant capacitor voltage, and kVA/kW, but lower rms tank current. Also F needs to be greater than 1 due to operation in lagging power factor mode to achieve ZVS. Hence, the proposed converter should be designed at minimum ac input voltage, and the optimal design parameters are chosen as $F=1.1$, $J=0.5$, $M_f=0.955$, $V_B=120V$.

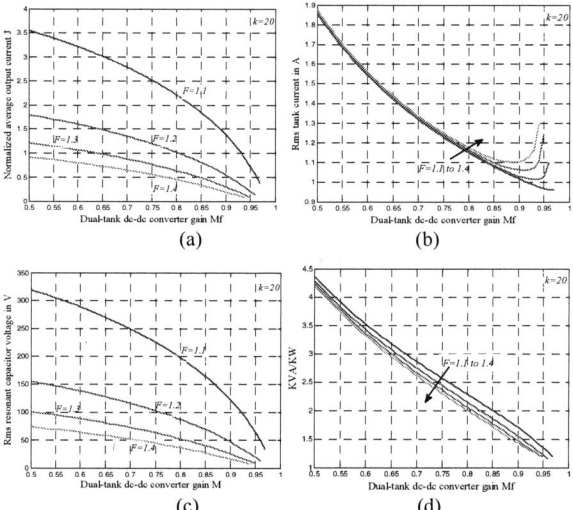

(a)　　　　(b)

(c)　　　　(d)

Fig.7. Design curves for different normalized switching frequency F, for $k=20$ and $\theta =0$ (i.e. $\delta=\pi$): (a) normalized average output current J; (b) rms tank current I_r;(c) rms tank capacitor voltage V_{Cr}; and (d) kVA/kW; vs M_f.

In the design example, $R_L = V_o^2/P_o = 100\Omega$, $I_o = V_o/R_L = 1A$, $V'_o = M_fV_B = 114.6V$, $n_t = V'_o/V_o = 1.146$. The resonant inductance and capacitance can be calculated as [5,8]

$$L_{r1} = L_{r2} = \frac{1}{2}\left(\frac{M_f V_B^2 JF}{2\pi f_s P}\right) = 60.2\,\mu H$$

$$C_{r1} = C_{r2} = 2\left(\frac{PF}{2\pi f_s M_f V_B^2 J}\right) = 50.9\,nF$$

So $L'_p = k*L_{r1/2} = 1.2mH$ and $L_p = L'_p/n_t^2 = 923\mu H$. When the ac input voltage increases, θ needs to be increased to keep the output voltage constant.

978-1-4799-2706-7/14 $31.00 © 2014 IEEE

V. SIMULATION AND EXPERIMENTAL RESULTS

To verify the analysis, PSIM 6.0 is used to simulate the proposed converter. Simulated waveforms are shown in Fig. 8. Note that all HF waveforms are observed from peak areas of LF waveforms. When $\sqrt{2}V_{in}$= 60V, 40 Hz, zero phase-shift is applied (Fig. 8(a)), the dc bus voltage with a small ripple factor is twice the peak ac input voltage (120 V) as expected. Harmonic spectra of line current are also given in Fig. 8(a). It is observed that the input ac side power factor is 0.99 with 12% of line current THD. High power factor with low THD is obtained due to the DCM operation of front-end dual-switch boost converter. All switches work in ZVS mode, since the integrated dual-tank dc-dc converter stage operates at above resonance or lagging power factor mode. Other key waveforms agree with the theory as well.

When $\sqrt{2}V_{in}$= 80V, 60 Hz, 108° of phase-shift is applied to regulate the output voltage at 100 V, corresponding to the value obtained with minimum input voltage. In Fig. 8(b), the dc bus voltage reaches about

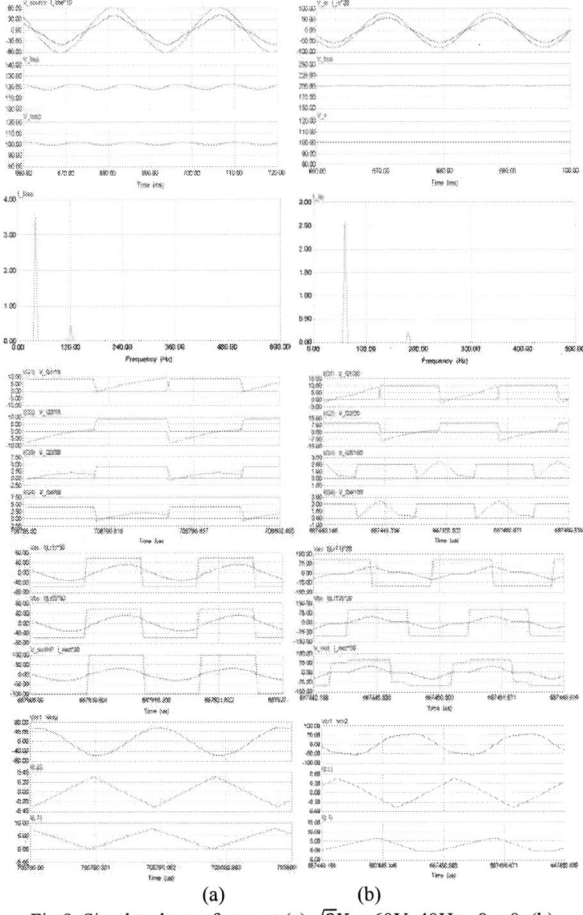

(a)   (b)

Fig.8. Simulated waveforms at (a) $\sqrt{2}V_{in}$=60V, 40Hz , θ =0; (b) $\sqrt{2}V_{in}$=80V, 60Hz, θ =108°: Top to bottom: line voltage (v_{in}) and current (i_{in}), bus voltage (V_{bus}) and output voltage (V_o); harmonic spectra of line current; Voltage across and current through switches; inverter output voltages (v_{ac}, v_{bc}) and tank currents(i_{rT1}, i_{rT2}), diode rectifier input voltage and current(v_{rect}, i_{rect}); resonant capacitor voltages (v_{cr}),current through L_p(i_{Lp}),and boost current(i_{L1}) through L_1.

200V, which brings a high power factor (unity) and low line current THD (4%) at the ac input side. All switches work in ZVS mode because the dual-tank LCL dc-dc converter still operates in above resonance mode. Other key waveforms agree with the theory as well.

A 100W prototype is built and the corresponding experimental waveforms are captured, shown in Fig. 9. The experimental components details used are shown in Table I. The gating signals are generated by TMS320F2812 DSP board and the open loop control is used in the experiment for demonstration purpose. The proposed converter is tested under minimum ($\sqrt{2}V_{in}$= 60V, 60 Hz) and maximum input ($\sqrt{2}V_{in}$= 80 V, 60 Hz) voltage conditions. Note that 60Hz ac input source is used due to non-availability of 40 Hz supply.

TABLE I
COMPONENTS USED IN EXPERIMENTAL CONVERTER

Components	Modules/Values
Dr1/Dr2	RHRP 1560
L_1	L_1 = 40 μH
S_1/S_2	G22N60S
S_3/S_4	G20N50S
HF transformer cores	TOKIN ETD44
D_{ra} to D_{rd}	MUR 460
L_{r1}/L_{r2}, C_{r1}/C_{r2}, L_t	65μH, 50nF, 1 mH

When $\sqrt{2}V_{in}$= 60 V is applied, the phase-shift angle θbetween two tanks is zero. Because of DCM operation of front-end dual-switches boost converter, 0.99 of power factor and 9.3% of THD at input line current are obtained as shown in Fig. 9(a). The dc bus voltage is 120V due to 0.5 duty cycle of the boost converter so the amplitudes of v_{ac} and v_{bc} are both 60V. The resonant currents are in lagging power factor operation, achieving ZVS mode for all switches. The amplitude of v_{rect} = 100 V same as the output voltage. The measured converter efficiency is 94.4% at full-load. The waveforms (Fig. 9(a)) of voltage across C_{r2}, current through L_t, and current through boost inductor $L_1$are captured. All of them match the theoretical values and simulated results.

When $\sqrt{2}V_{in}$ =80V is applied, the phase-shift angle θbetween two tanks is adjusted to 108°, shown in Fig 9(b).At the input side, we still obtain a high power factor (0.99) and low THD (5%) for ac input current. The amplitudes of v_{ac} and v_{bc} (107 V) are half of the dc bus voltage. The tank currents still lag v_{ac} and v_{bc} and all switches operating in ZVS mode. The amplitude of v_{rect} = 100 V equals the output voltage. The converter efficiency is 89.3% at full-load with an input of 80 V peak. The waveforms (Fig. 9(b)) of voltage across C_{r2}, current through L_t, and current through boost inductor $L_1$are captured. All of them match the theoretical values and simulated results.

Table II shows the comparison of theoretical values obtained from the Fourier series analysis with simulated and experimental values. It is observed that most of the theoretical values are close to those simulated and experimental ones.

978-1-4799-2706-7/14 $31.00 © 2014 IEEE

TABLE II
COMPARISON OF THEORETICAL, SIMULATION AND EXPERIMENTAL RESULTS

	$\sqrt{2}V_{in}$=60V, 40Hz			$\sqrt{2}V_{in}$=80V, 60Hz		
	Pred.	Simn.	Expt.	Pred.	Simn.	Expt.
V_o (V)	100	100	96	100	98.7	100
$I_{in,rms}$(A)	2.35	2.5	2.3	1.76	1.9	1.98
V_{dc} (V)	120	120	117	243	230	215
$I_{r,rms}$	0.97	0.98	0.95	1.39	1.35	1.29
$I_{Lp,rms}$	-	0.18	0.15	-	0.3	0.35
$I_{L1,pk}$	-	7.76	7	-	8.07	8.4
$V_{cr,rms}$	34	32	31.1	47.4	40.8	40.1
θ	0°	0°	0°	110°	108°	108°
PF	0.99	0.99	0.99	unity	unity	0.99
THD		12.3	9.3%		4%	5%
efficiency			94.4%			89.3%

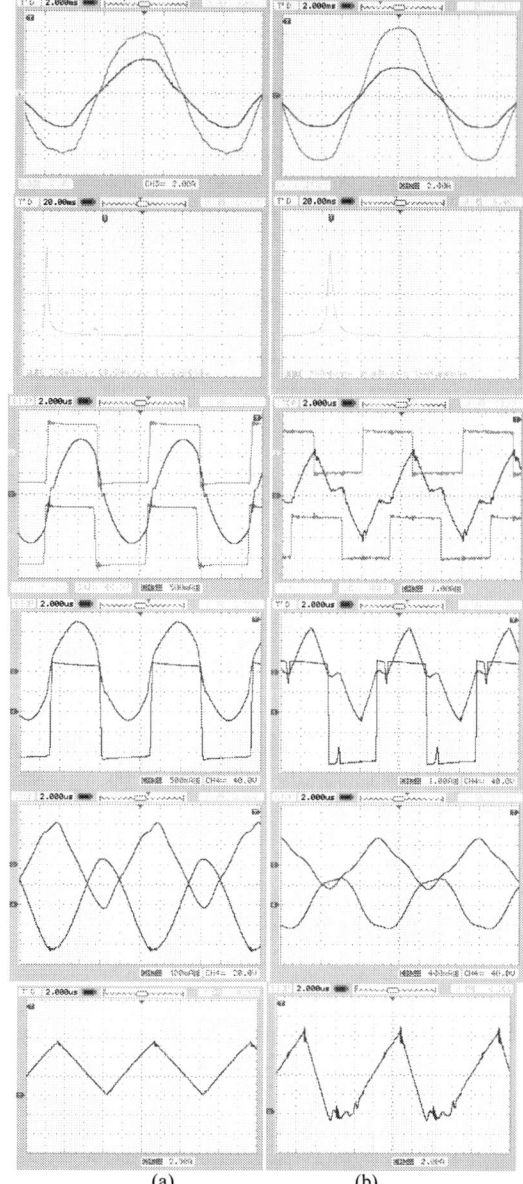

(a) (b)

Fig.9. Experimental results at(a)$\sqrt{2}V_{in}$=60V, 60Hz , θ = 0; (b) $\sqrt{2}V_{in}$=80V, 60Hz, θ = 108°:Top to bottom: line voltage (ch1, 40V/div) and current (ch3, 2A/div), 2ms/div; FFT spectrum of line current, 0.5A/div, 25Hz/div for (a) and 50Hz/div for (b); v_{ac} (ch1, 100V/div), v_{bc} (ch2, 100V/div,), i_{rT1} (ch3, 0.5A/div for (a) and 1 A/div for (b)), 2µs/div; v_{rect}(ch4, 40V/div) and i_{rect} (ch3, 0.5 A/div for (a) and 1 A/div for (b)); v_{cr1}(ch4, 20 V/div for (a) and 40 V/div for (b)) and i_{Lp} (ch3, 0.1A/div for (a) and 0.4 A/div for (b)); boost current i_{L1} (2.5A/div for (a) and 2 A/div for (b)) through L_1, 2µs/div.

VI. CONCLUSIONS

A new single-phase single-stage HF isolated dual-tank LCL-type series resonant ac-dc converter is proposed in this paper. Functions of HF isolation, PFC and output voltage regulation are in one single-stage. Two identical half-bridge HF isolated LCL-type resonant converters are employed to form the dual-tank configuration. The fixed-frequency, phase-shift control is employed to regulate output voltage. The proposed converter obtains high PF and low THD at ac input side. Also ZVS mode for all switches is guaranteed in entire operating range. Three identical converters will be used in interleaved configuration for small scale WECS applications [5].

REFERENCES

[1] H.L. Cheng, K.H. Lee, Y.C. Li, and C.S. Moo, "A novel single-stage high-power-factor high-efficiency ac-to-dc resonant converter", *IEEE PECon 2008*, pp.1135–1140, 2008.

[2] H.L. Cheng, Y.C. Hsieh, and C.S. Lin. "A novel single-stage high-power-factor ac/dc converter featuring high circuit efficiency". *IEEE Trans. on, Industrial Electronics*, vol. 58, no. 2, pp. 524–532, 2011.

[3] Y. Du and A.K.S Bhat, "Analysis and design of a single-stage high-frequency isolated ac-dc converter", *3rd international conf. onAdvances in Electrical & Elctronics*,pp. 52-58, 2012.

[4] Li, X., and A.K.S. Bhat, "Analysis and design of high-frequency isolated dual-bridge series resonant DC/DC converter", *IEEE Trans. on Power Electronics*, vol. 25, no. 4, , pp. 850-862, 2010.

[5] Y. Du, "A high-frequency isolated integrated resoant ac-dc conveters for PMSG based wind energy conversion system," Ph.D. Dissertation,University of Victoria, 2013.

[6] K.H. Liu and Y.L. Lin. "Current waveform distortion in power factor correction circuits employing discontinuous-mode boost converters", *IEEE PESC*, pp. 825–829, 1989.

[7] Bhat, A.K.S., and R. Venkatraman, "A soft-switched full-bridge single-stage ac-to-dc converter with low line-current harmonic distortion,"*IEEE Trans. on Ind. Electronics*, vol. 52, no. 4, pp. 1109-1116, Aug. 2005.

[8] A.K.S Bhat, "Analysis and design of a fixied-frequency LCL-type series-resonant converterwith capacitive output filter,"*Proc. Inst. Elec. Eng. Circuit, Dev.Syst.*, vol144, no. 2, pp 97-103, Apr. 1997.

Verification of LLC Resonant Converter Applied a Current-Balancing High-Frequency Transformer with Multi-Output Windings

Jun Araki, Ikki Shinozaki, Hirohito Funato
Department of Electrical and Electronic Engineering
Utsunomiya University
Utsunomiya, Tochigi, Japan
funato@cc.utsunomiya-u.ac.jp

Satoshi Ogasawara, Daichi Murakami
Department of System Science and Informations
Hokkaido University
Sapporo, Hokkaido, Japan
oga@ist.hokudai.ac.jp

Yukitsugu Hirota, Teruyoshi Mihara, Masayuki Mouri, Fumihiro Okazaki
Calsonic Kansei Corporation

Abstract—DC-DC converters of hybrid vehicles and electric vehicles are required to become smaller and lighter for building roomy and light weight EV. High-frequency switching can be used to miniaturize DC-DC converter. However, the influence of the skin effect and the eddy current become large. Parallel operation of converter is often for current applications using multiple winding transformers in order to reduce of skin effect. However, it is difficult to equalize the each current of converter due to unbalance of winding resistance, drop voltage of diodes and so on. In this paper, the parallel LLC converter applied a current-balancing high-frequency transformer is proposed. The proposed converter can realize nearly equal current even if the stray impedance of secondary circuit of LLC are unbalanced.

Keywords—*DC/DC converter, LLC, current balancer, parallel*

I. INTRODUCTION

Recently, electric vehicles(EVs) has been popularized. An EV is equipped with an auxiliary power supply of 12V composed of a battery and a DC-DC converter, which is a DC-DC converter for supplying 12V power to electrical devices from a high-voltage battery. To build a roomy and lightweight EV, miniaturization of the DC-DC converter is required[1]. Since an EV requires high output power at low output voltage, the DC-DC converter must supply high output current.

High-frequency switching is very effective to miniaturize the DC-DC converter because of the reduction in magnetic flux in a high-frequency transformer. This reduction depends on the time integral of voltage. Next-generation power semiconductor devices, such as those based on SiC and GaN, have been developed in recent years[2]. The switching frequency of such devices may increase in the future; however, iron loss in the magnetic core and the skin effect in the winding conductor would also increase. A copper bar is used as the winding

conductor in high current transformers. Because of this, the influence of the skin effect and the eddy current are large. In order to reduce their influence, the secondary conductor is sometime divided into multi-output windings.

On the other hands, diodes connected in parallel are often used in DC-DC converters to produce large output current. Although each diode is connected to a divided output winding, differences in the forward voltage of the diodes produce a current imbalance in the output windings. To equalize the diode currents, a current balancer with multiple magnetic cores is proposed in [3]. However, the current balancer is bulky and has a complex magnetic structure.

Authers have proposed a current-balancing high-frequency transformer(CB-HFT) with multi-output windings[4]. This transformer has a simple structure with a pot core that has multiple magnetic paths. Next, parallel LLC resonant converter is proposed. The effectiveness of the proposed converter is verified through simulations and experiments.

II. CONVENTIONAL TRANSFORMERS WITH MULTI-OUTPUT WINDINGS

A. Conventional Transformer with Multi Output Windings

Fig. 1 shows a conventional high-frequency transformer with multi-output windings. From the law of equal ampere-turns , the sum of secondary winding currents is equal to N times the primary current. Each secondary winding can step down the primary voltage V_1 to V_1/N. However, there is no current-balancing function in the secondary windings.

B. Principle of Current Balance

Fig. 2 shows the circuit diagram of CB-HFT. This transformer has *n* cores. Primary windings, which are connected in

978-1-4799-2706-7/14 $31.00 © 2014 IEEE

series, are wound N turns around each core. The transformer has n secondary windings. Each secondary wire passes through n-1 cores. The total secondary current I_2 is defined as the sum of the current I_{2i} of each secondary winding:

$$I_2 = \sum_{i=1}^{n} I_{2i} \tag{1}$$

The law of equal ampere-turns with regard to the each core gives the following system of equations:

$$
\begin{aligned}
I_1 \times N &= I_{22} \times 1 + I_{23} \times 1 + \cdots + I_{2n} \times 1 \\
I_1 \times N &= I_{21} \times 1 + I_{23} \times 1 + \cdots + I_{2n} \times 1 \\
&\quad\vdots \\
I_1 \times N &= I_{21} \times 1 + I_{22} \times 1 + \cdots + I_{2(n-1)} \times 1
\end{aligned}
\tag{2}
$$

This simultaneous equation problem can be expressed in matrix form:

$$
\begin{bmatrix} I_1 \\ I_1 \\ I_1 \\ \vdots \\ I_1 \\ I_1 \end{bmatrix}
=
\begin{bmatrix}
0 & 1 & 1 & \cdots & 1 & 1 \\
1 & 0 & 1 & \cdots & 1 & 1 \\
1 & 1 & 0 & \cdots & 1 & 1 \\
\vdots & \vdots & \vdots & \ddots & \vdots & \vdots \\
1 & 1 & 1 & \cdots & 0 & 1 \\
1 & 1 & 1 & \cdots & 1 & 0
\end{bmatrix}
\begin{bmatrix} I_{21} \\ I_{22} \\ I_{23} \\ \vdots \\ I_{2(n-1)} \\ I_{2n} \end{bmatrix}
\tag{3}
$$

Solving these simultaneous equation, the current in each secondary winding is

$$I_{2i} = I_{21} = I_{22} = \cdots = I_{2n} = \frac{I_2}{n} \tag{4}$$

Therefore, each secondary current can be balanced.

C. Transformarion ratio

Summing all equations in Eq(2) gives.

$$I_1 \times N = \frac{n-1}{n} \sum_{i=1}^{n} I_{2i} = \frac{n-1}{n} I_2 \tag{5}$$

In a general transformer, the relation between the primary current I_1 and the secondary current I_2 is expressed by

$$I_1 N_1 = I_2 N_2 \tag{6}$$

where N_1 and N_2 are number of turns at the primary and secondary windings, respectively. Comparison of Eq(5) to (6) yields the transformation ratio of the proposed CB-HFT (Fig.2)

$$N_1 : N_2 = N : \frac{n-1}{n} \tag{7}$$

Note that number of turns on the secondary side is not each 1, even if windings in secondary side is wound one turn.

D. Structure of the CB-HFT

Fig.3 shows a structure of the CB-HFT with a single pot core. The pot core has three symmetrical magnetic paths ($n=3$) due to its three notches. Each magnetic path corresponds to a magnetic core in Fig.2. Since the same primary winding in Fig.2 is wound on to the left side of each core, the windings can share this part. Therefore, the primary winding can be wound around the center leg of the pot core.

Each secondary winding in Fig.2 passes through $n-1n$-1 cores and is wound 2/3 of a revolution around pot core as shown in Fig.3.

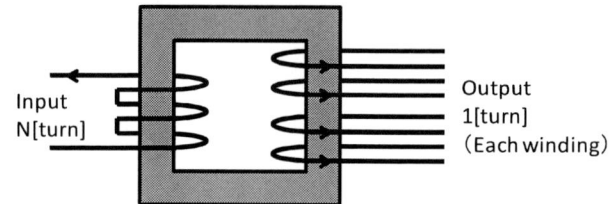

Fig. 1.   Traditional multi output transformer

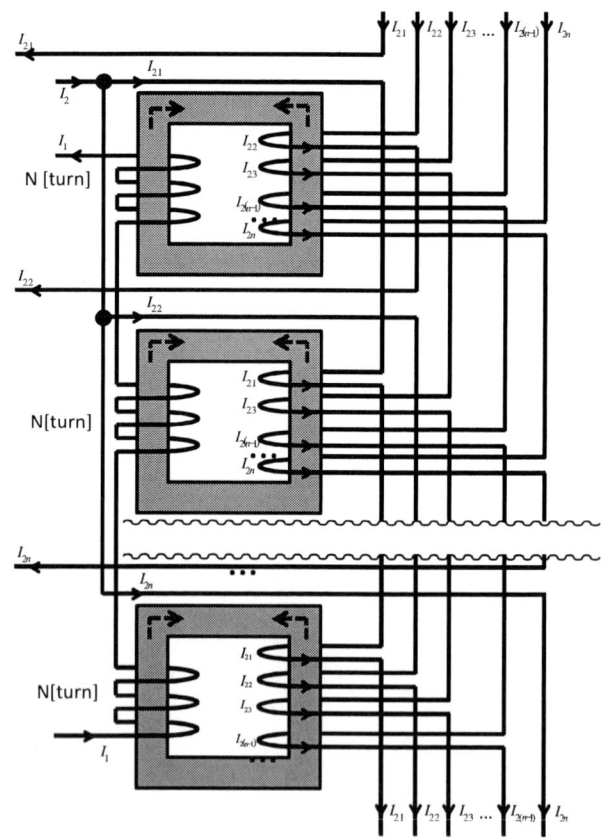

Fig. 2.   Circuit diagram of CB-HFT

The 2014 International Power Electronics Conference

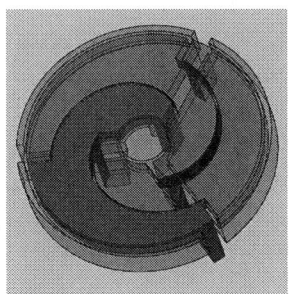

Fig. 3.   Three-dimensional diagram of the pot-core CB-HFT(n=3)

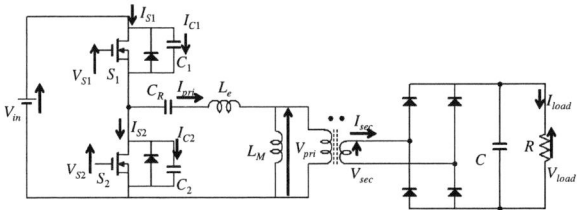

Fig. 4.   Basic LLC converter

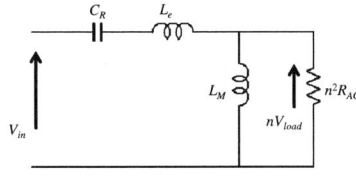

Fig. 5.   AC Equivalent Circuit

III.   PARALLEL CONNECTING OF LLC RESONANT CONVERTER WITH THE CB-HFT

A. Basic design of LLC Converter

Fig.4 shows a basic LLC converter, and Fig.5 shows a AC equivalent circuit of the basic LLC converter. From Fig.5, the series impedance Z_s and the parallel impedance Z_p can be derived as following equations;

$$Z_s = \frac{1}{j\omega C_R} + j\omega L_e \tag{8}$$

$$Z_p = \frac{j\omega n^2 L_M R_{AC}}{j\omega L_M + n^2 R_{AC}} \tag{9}$$

therefore, the voltage-gain $T = V_{load}/V_{in}$ can be calculated as;

$$T = \frac{1}{(1 + \frac{L_e}{L_M} - \frac{1}{\omega^2 L_M C_R}) + j(\frac{\omega L_e}{n^2 R_{AC}} - \frac{1}{\omega n^2 C_R R_{AC}})} \tag{10}$$

From Eq(11)-Eq(14), the voltage-gain $|T|$ can be changed to the Eq(15).

$$\text{Resonance frequency}: f_R = \frac{1}{2\pi\sqrt{L_e C_R}} \tag{11}$$

$$\text{Normalized frequency}: f_n = \frac{f_{SW}}{f_R} \tag{12}$$

$$\text{Quality factor}: Q = \frac{\omega_R L_e}{n^2 R_{AC}} = \frac{1}{\omega_R n^2 C_R R_{AC}} \tag{13}$$

$$\text{Inductance ratio}: m = \frac{L_M}{L_e} \tag{14}$$

$$|T| = \frac{1}{\sqrt{\left\{1 + \frac{1}{m}(1 - \frac{1}{f^2{}_n})\right\}^2 + \left\{Q(f_n - \frac{1}{f_n})\right\}^2}} \tag{15}$$

Fig.6 shows the voltage-gain curve. LLC converter can be designed to select the appropriate gain curve by defining quality factor Q in order to realize soft-switching[5]. In this paper, the quality factor Q is selected 0.26, and the switching frequency f_{SW} is selected 200kHz in this simulation. The simulation parameters of LLC converter are defined by Eq(11)-Eq(14).

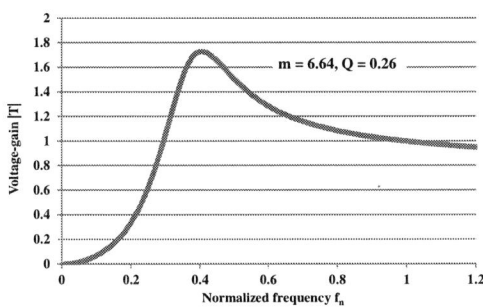

Fig. 6.   voltage-gain curve

B. Parallel Connecting of LLC Converter

In this section, effectiveness of the LLC resonant converter with the CB-HFT is verified by simulations. The LLC resonant converter with normal transformer is also simulated for comparison with the LLC resonant converter with the CB-HFT. Fig.7 and Fig.8 shows parallel connecting of LLC converter with normal transformer and CB-HFT respectively. Transformer in Fig.8 has 3 cores. Therefore, the winding ratio of CB-HFT is $N : 2/3$ according to Eq(7), and the CB-HFT has 3 secondary windings. Each output winding passed through two cores.

Table.I and Table.II show circuit parameters of Fig.7 and Fig.8 respectively. Unbalance resistance is connected to only one output secondary winding to realize imbalanced condition. Fig.9 shows simulation results of Fig.7 and Fig.8 respectively. Fig.9(a) shows the simulation result of the LLC resonant converter with normal transformer. Fig.9(b) shows the LLC resonant converter with the CB-HFT. In Fig.9(a), it is clear that ZVS and voltage step-down function from 400V to 12V are realized. However, secondary winding current I_{sec1}, I_{sec2}, I_{sec3} produce imbalance. On the other hand, in Fig.9(b), it is clear that ZVS and voltage step-down function from 400V to

978-1-4799-2706-7/14 $31.00 © 2014 IEEE   1730

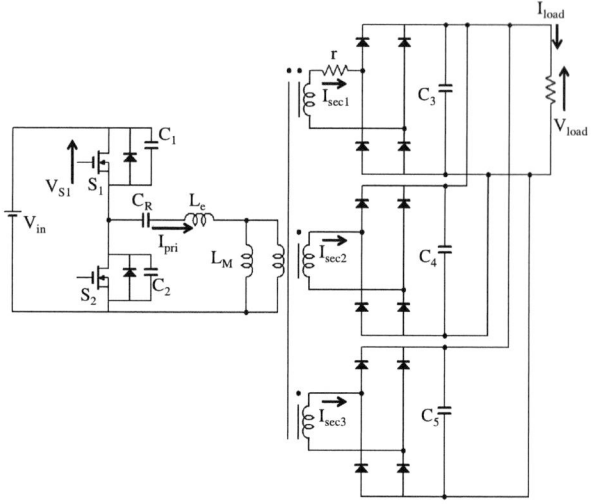

Fig. 7. Parallel LLC converter with normal transformer

TABLE I. PARAMETERS OF LLC CONVERTER WITH NORMAL TRANSFORMER

Parameters	Value
Input Voltage [V_{in}]	400 [V]
Output Voltage [V_{load}]	12 [V]
Output Current [I_{load}]	150 [A]
Switching Frequency [f_{SW}]	200 [kHz]
Transformer Winding Ratio [n]	17 : 1
Resonant Capacitance [C_R]	160 [nF]
Resonat Inductance (Leakage Inductance) [L_e]	4 [uH]
Magnetizing Inductance [L_M]	26 [uH]
Parallel Capacitance [C_1, C_2]	3 [nF]
Output Capacitance [$C_3 \sim C_5$]	100 [uF]
Unbalance Resistance [r]	1[mΩ]
Dead Time	0.5[usec]

12V are realized with balanced secondary current I_{sec1}, I_{sec2}, I_{sec3}. In this simulation, it is confirmed that the CB-HFT can perform both transformer function and current balancing function, even when the CB-HFT is applied to an LLC resonant converter.

IV. EXPERIMENTS OF THE LLC RESONANT CONVERTER WITH THE CB-HFT

In this section, effectiveness of the LLC resonant converter with the CB-HFT is verified through experiment.

A. Basic Characteristic

In the experiments, the same circuit shown in Fig.8 is used. Table.III shows circuit parameters. Fig.10 shows experimental result of V_{S1}, V_{S2}, V_{load} and I_{pri}. In Fig.10(a), it is clear that V_{load} is stepped down from 282V to 14V , and I_{pri} becomes sine wave by resonance.

Fig.11 shows experimental waveforms of voltage and current of S_1. From this figure, it is clear that ZVS switching is successfully realized.

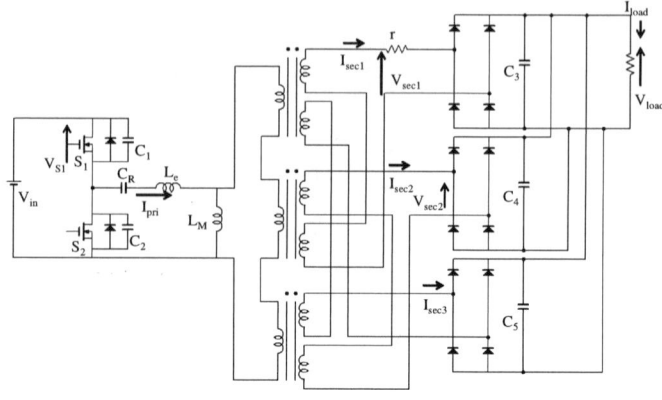

Fig. 8. circuit of the LLC converter with CB-HFT

TABLE II. PARAMETERS OF LLC CONVERTER WITH CB-HFT

Parameters	Value
Input Voltage [V_{in}]	400 [V]
Output Voltage [V_{load}]	12 [V]
Output Current [I_{load}]	150 [A]
Switching Frequency [f_{SW}]	200 [kHz]
Transformer Winding Ratio [n]	11 : 2/3
Resonant Capacitance [C_R]	160 [nF]
Resonat Inductance (Leakage Inductance) [L_e]	4 [uH]
Magnetizing Inductance [$L_{M1} \sim L_{M3}$]	26 [uH]
Parallel Capacitance [C_1, C_2]	3 [nF]
Output Capacitance [$C_3 \sim C_5$]	100 [uF]
Unbalance Resistance [r]	1[mΩ]
Dead Time	0.5[usec]

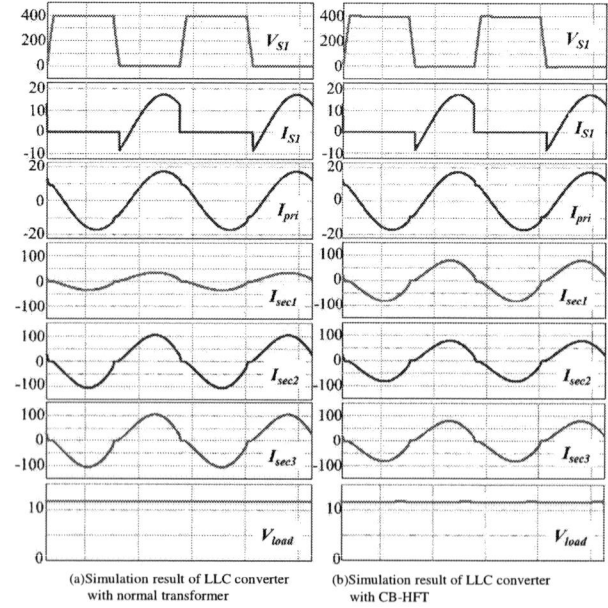

(a)Simulation result of LLC converter with normal transformer

(b)Simulation result of LLC converter with CB-HFT

Fig. 9. Simulation results

978-1-4799-2706-7/14 $31.00 © 2014 IEEE

B. Current-balancing Characteristic

Fig.12 (a) and (b) shows experimental result of transformer currents I_{sec1} and I_{sec2} using normal transformer and CB-HFT respectively. The unbalance resistance r is connected to only secondary winding to make an imbalanced condition. I_{sec1} is current of secondary winding that unbalance resistance r is connected. I_{sec2} is current of secondary winding that unbalance resistance r is not connected. In Fig.12(a), secondary winding current is unbalanced. On the other hand, in Fig.12(b), secondary winding current is balanced. From these experiments, the CB-HFT is confirmed that it can perform both transformer function and current balancing function, even when the CB-HFT is applied to the LLC resonant converter.

Fig.13 shows the experimental result for the current-balancing characteristic. The horizontal axis represents the unbalance resistance and the vertical axis shows the current-balancing ratio defined as;

$$Br = \frac{I_{sec1}}{I_{sec2}} \times 100\% \qquad (16)$$

Br represents the ratio of the current in the imbalanced output winding to the maximum value of the current of the balanced output windings. If the output current is fully balanced, the ratio becomes 100%.

When unbalance resistance is 0.1Ω, the current-balancing ratio is 100%. On the other hand, Br is 76% for LLC converter with normal transformers using the same unbalance resistance.i Therefore, it is possible to improve current-balancing ratio because of using the CB-HFT. However, the current balance is not perfect for highly unbalanced situation.

V. CONCLUSIONS

This paper presents a novel parallel LLC converter using CB-HFT. The proposed converter realizes current balancing function for unbalance of paralleled components. The effectiveness of the proposed converter are verified through simulations and experiments.

REFERENCES

[1] Wolfgang Schmit "DCDC Converter for Hybrid Vehicle Applications", *CIPS Integrated Power Systems* , pp. 1-3, 2008.

[2] Arun Kadavelugu, Seunghun Baek, Sumit Dutta, Subhashish Bhattacharya, Mrinal Das, Anant Agarwal, James Scofield "High-frequency Design Considerations of Dual Active Bridge 1200V SiC MOSFET DC-DC Converter", *APEC 2011 Applied Power Electronics Conference and Exposition* , pp. 314-320, 2011

[3] Jianfeng Wang, Junming Zhang, Xinke Wu, Yangyu Shi, Zhaoming Qian, "A Novel High Efficiency and Low-Cost Current Balancing Method for Multi-LED Drier", *ECCE 2011 Energy Conversion Congress and Exposition* ,pp. 2296-2301, 2011.

[4] Daichi Murakami, Satoshi Ogasawara, Masatsugu Takemoto, Yukitsugu Hirota, Takahiro Wakabayashi, Hirohumi Okazaki, Yoshiyuki Kikuchi "Development of Multi-Output-Winding High-Frequency Transformer Having Current Balancing Function", *The Papers of Technical Meeting on SPC, VT, HCA, IEEE Japan*, SPC-12-170, VT-12-021, HCA-12-055, pp. 7-12, 2012(in japanese).

[5] Han-sol Chang, Byung-Hun Lee, Hyung-Nam Park, Jae-Du La, Young-Seok Kim,"Design of Dual Output LLC Resonant Converter for a High-Power LED Lamp", *ICEMS 2012 The 15th International Conference on Electrical Machines and Systems* , 2012.

TABLE III. EXPERIMENT PARAMETERS OF LLC CONVERTER WITH CB-HFT

Parameters	Value
Input Voltage [V_{in}]	282 [V]
Output Voltage [V_{load}]	14 [V]
Output Current [I_{load}]	7 [A]
Switching Frequency [f_{SW}]	50 [kHz]
Transformer Winding Ratio [n]	7 : 2/3
Resonant Capacitance [C_R]	27 [nF]
Resonat Inductance (Leakage Inductance) [L_e]	310 [uH]
Magnetizing Inductance [$L_{M1} \sim L_{M3}$]	1545 [uH]
Parallel Capacitance [C_1, C_2]	0.5 [nF]
Output Capacitance [$C_3 \sim C_5$]	66 [uF]
Unbalance Resistance [r]	$0 \sim 1.1$ [Ω]
Dead Time	2[usec]

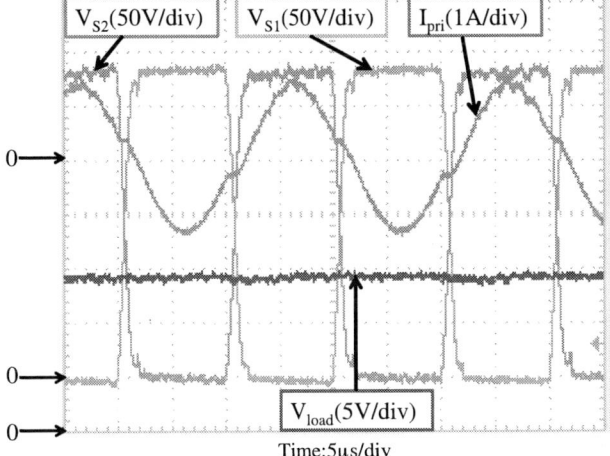

Fig. 10. Experimental result of V_{S1}, V_{S2}, V_{load} and I_{pri}

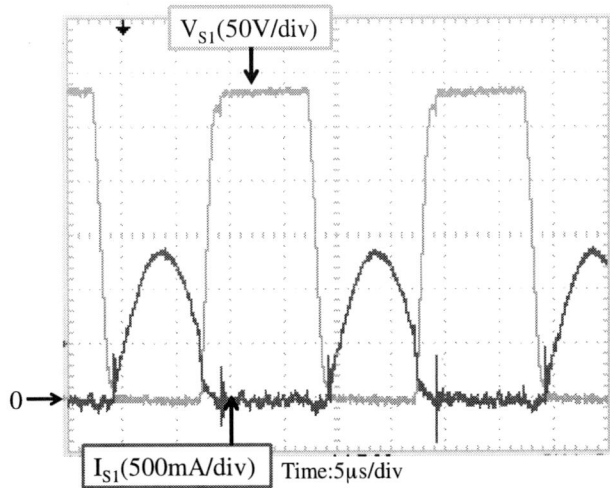

Fig. 11. experimental result of V_{S1} and I_{S1}

The 2014 International Power Electronics Conference

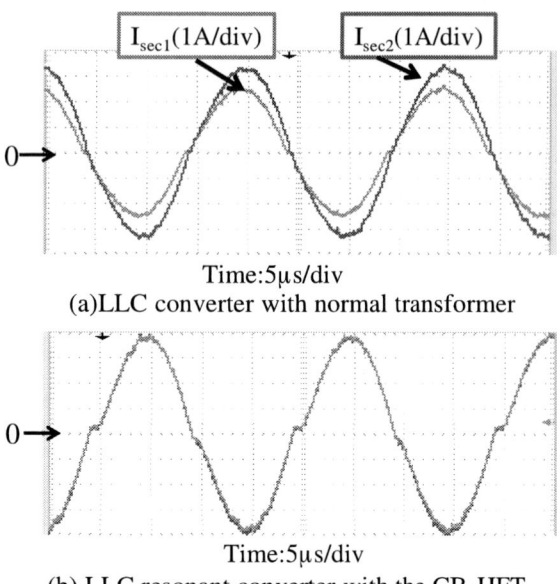

Time:5μs/div
(a)LLC converter with normal transformer

Time:5μs/div
(b) LLC resonant converter with the CB-HFT.

Fig. 12. Experimental result of I_{sec1} and I_{sec2}

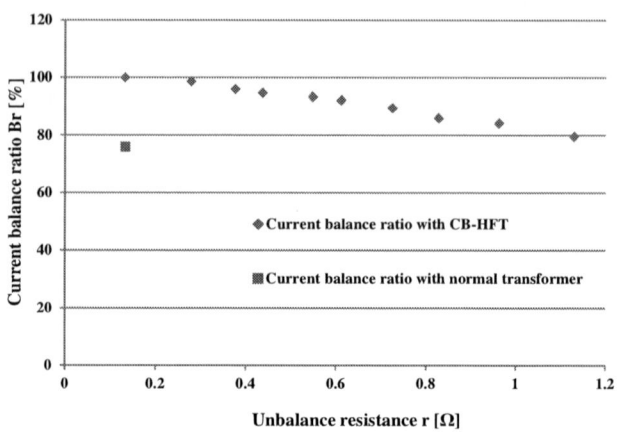

Fig. 13.   Current-balancing characteristic

Light-Load Efficiency Improvement Strategy for LLC Resonant Converter Utilizing a Step-Gap Transformer

Wen-Nan Huang*, Shiu-Hui Lee** and Ching-Guo Chen*

* Chicony Power Technology Co., Ltd
**National Taipei University of Technology

Abstract— **An inductance-based control strategy, for further optimize parameter design which is conducted through inductance adaptive variation on LLC resonant converter, is proposed. A simple switch concept is applied to set the desired magnetizing inductance, the main control factor, to achieve light load efficiency improving and overall required switching frequency range narrowing. Evaluation for showing the enhancement effect of this control method is verified theoretically in a deriving mathematic model based on the relation between magnetizing inductance and air gap of LLC transformer. Two platforms using LLC resonant converter demonstrates the effectiveness of this control strategy experimentally.**

I. INTRODUCTION

With further demand of energy saving for external power conversion products, such as the requirement from affiliation 80 PLUS [1], and directives of market standard depicted from Eup [2], efficiency index for power supply product at standby mode and extremely light load is still a challenge for circuitry designers and researchers. Topology of LLC resonant converter is featured with high power density, high efficiency with ZVS (zero voltage switching), etc. and it is drawing more and more attention and put into consideration on application and implementation. Several strategies have been adopted trying to upgrade low load efficiency by less cost adding, e.g. skip/burst mode gate driving pattern with control bandwidth decreasing or asymmetrical switch for on/off duration [3-5]. However, in some applications, acoustic noise and stability at light load would be another trade-off case needs to be addressed and compromised while switches' gate-drive trigger had been set ahead. Thus, a simple control method with easily fabricated magnetic core structure is proposed to make use the magnetizing inductance variation for further improve the efficiency adaptively. Comparing to conventional transformer design, inductance has been usually limited at rated operation load condition for preventing core saturation in original set dimension, and meanwhile, narrowing on the overall operation frequency range for resonant response is the other consideration to constrain the desired inductance while the proposed strategy can extend the usage setting of inductance at specific light load range in resonant mode.

II. MAIN CONCEPT OF CONTROL STRATEGY

For LLC resonant operation, magnetizing inductance is proposed herein as the control factor of this adaptive control strategy. A simple relation is applied and depicted as Fig.1. Magnetizing inductance Lm is taken to move resonant operation into a further improving state which can let the power loss of the converter go downward at set loading. As Fig.1(b), a close two-state inductor is designed to provide the advantage for efficiency through the enlargement of the magnetizing inductance which is meant that less magnetizing current to be drawn, hence, less power consumption is generated, especially during light load.

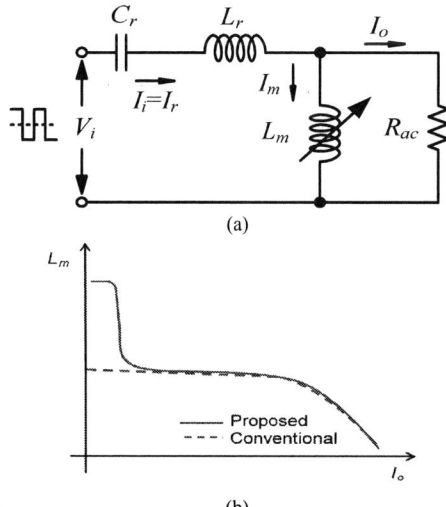

Fig. 1 Equivalent circuit and inductance set. (a) LLC resonant converter circuit model. (b) Magnetizing inductance under different load current.

According to relation of LLC resonant conversion, when the magnetizing inductance can be set as Fig.1(b), the effect of narrowing on converter switching frequency band can be evaluated as Fig.(2) which is plotted by converter gain w.r.t. frequency ratio, defined by operation switching frequency f_o divided by resonant frequency f_r.

978-1-4799-2706-7/14 $31.00 © 2014 IEEE

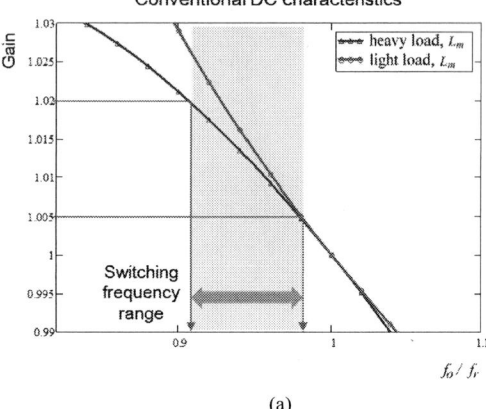

Conventional DC characteristics

(a)

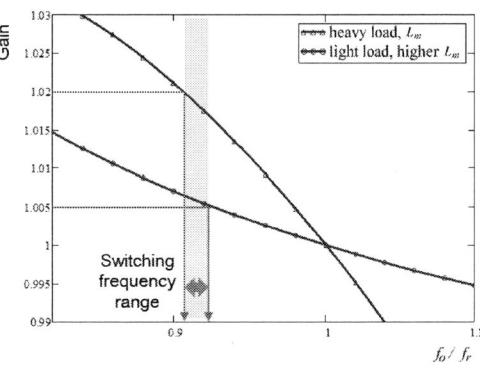

Proposed (variable magnetizing inductance)

(b)

Fig. 2 The switching frequency band. (a) Conventional gain response w.r.t. frequency ratio. (b) Proposed gain response w.r.t. frequency ratio.

The design of inductance features are given as follows. The adaptive variation of inductance is realized by setting the air gap of the transformer and its depiction is shown as Fig. 3. It is so called a step-gap transformer. The a, b, and c are scaling factors. l_e is effective core path length. A_e is the effective core area. The reluctance of original core without air gap is expressed as

$$R_m = \frac{l_e}{\mu_0 \, \mu_c \, A_e} \tag{1}$$

where μ_0 is the permeability of air, and μ_c is the relative permeability of the core material. The reluctance of original core with air gap is expressed as

$$R_1 = (1-b) \, R_m \tag{2}$$

The reluctance while saturation point is not reached at convex region is

$$R_{2a} = \frac{b-a}{c} \, R_m \tag{3}$$

The reluctance while saturation point is reached at convex region is

$$R_{2b} = \frac{b-a}{c} \, \mu_c \, R_m \tag{4}$$

The air gap reluctance at convex region is

$$R_3 = \frac{a}{c} \, \mu_c \, R_m \tag{5}$$

The air gap reluctance at concave region is

$$R_4 = \frac{b}{1-c} \, \mu_c \, R_m \tag{6}$$

The equivalent reluctance in R_{2a} status is

$$R_{eqa} = R_1 + \frac{(R_{2a} + R_3) \, R_4}{(R_{2a} + R_3) + R_4} \tag{7}$$

The equivalent reluctance in R_{2b} status is

$$R_{eqb} = R_1 + \frac{(R_{2b} + R_3) \, R_4}{(R_{2b} + R_3) + R_4} \tag{8}$$

According to equation (7) .The magnetizing inductance at first state is

$$L_{m1} = \frac{N^2}{R_{eqa}} \tag{9}$$

where N is number of turns. According to equation (8) .The magnetizing inductance at second state is

$$L_{m2} = \frac{N^2}{R_{eqb}} \tag{10}$$

We can easy obtain first corner current while L_{m1} starts turning from first state to second state.

$$I_{corn1} = \frac{B_{sat} \, R_{eqa} \, c \, A_e}{N} \tag{11}$$

The second corner current while L_{m2} reaches at saturation is

$$I_{corn2} = \frac{B_{sat} \, R_{eqb} \, A_e}{N} \tag{12}$$

Derivation result to obtain the relation among air gap and inductance is given in Fig. 4. A simple control concept is presented to set the desired magnetizing inductance which is the main control factor to achieve light load efficiency upgrade and let the overall required switching frequency range to be reduced.

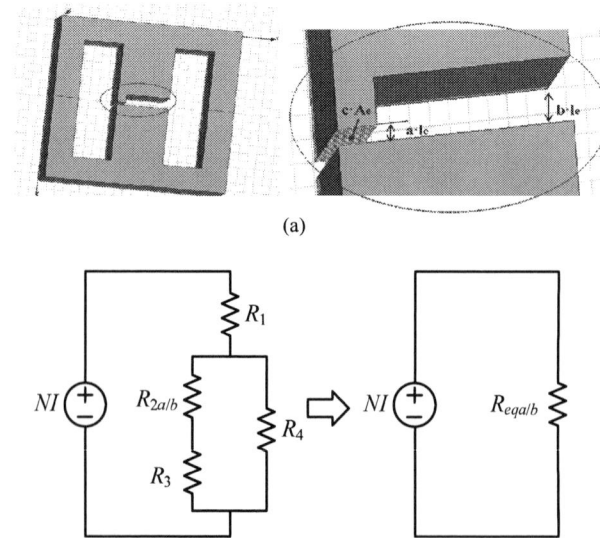

(a)

(b)

Fig. 3 Illustration of variable inductance. (a) Core structure. (b) Magnetic circuit.

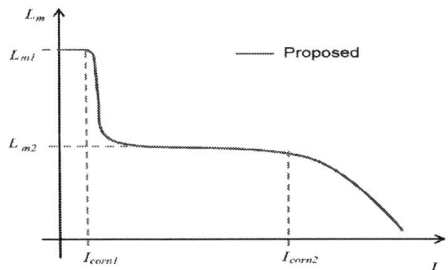

Fig. 4. Derivation of variable inductance.

III. EXPERIMENTAL RESULTS

Platform of 550W power supply without skip mode control is chosen for demonstration on improving effect while taking the inductance-based adaptive control strategy for implementation. The parameters of EE42/20 core and step-gap are as below Table I and Table II. As shown of Fig. 5, the design information of the applied core for the LLC resonant transformer is given.

TABLE I
PARAMETERS OF EE42/20 CORE

Core material	JPP95
Initial permeability [μ_i]	3300
Effective area [A_e]	235mm^2
Effective length [l_e]	97.8mm
Saturation flux density [B_{sat}]	450mT

TABLE II
PARAMETERS OF STEP-GAP

Parameters	JPP95
N	30 turns
$a\ l_e$	0.02mm
$b\ l_e$	0.38mm
$c\ A_e$	39.167mm^2

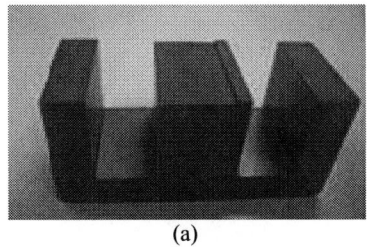

(a)

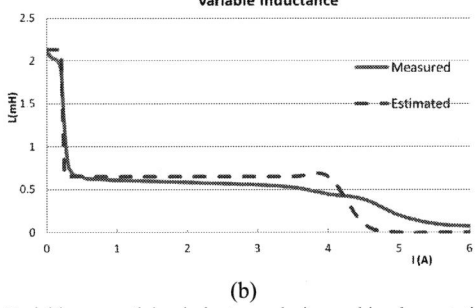

(b)

Fig. 5 Variable magnetizing inductance design and implementation. (a) Core for implement. (b) Verification on inductance estimation.

The experiment result is highlighted as below Table III. Efficiency at light load and the required switching frequency band are verified to be improved and decreased, respectively. Efficiencies at 10% and 20% rated load conditions are increasing observably while the switching frequency range is narrowed by 4.8 kHz.

TABLE III
THE COMPARISON OF OPERATION EFFECT

Load	Conventional		Proposed	
	Efficiency [%]	fo [kHz]	Efficiency [%]	fo [kHz]
10%	87.04%	53.8	88.13%	48.7
20%	91.18%	50.1	91.35%	49.5
50%	92.69%	46.5	92.65%	47.3
100%	91.09%	43.7	91.07%	44.2

Another platform for demonstration is 400W power supply with skip mode control. The design of variable magnetizing inductance for the LLC resonant transformer is shown as Fig. 6 shows the gating signal and output voltage of LLC resonant converter while using the ETD49 core without step-gap at no load condition. Fig. 7 shows the gating signal and output voltage of LLC resonant converter while using the ETD49 core with step-gap at no load condition. Comparing the gating signal frequency of LLC resonant converter with and without step-gap, we can conclude that the ETD49 core without step-gap requires more input current/power at same load condition. Fig. 8 shows the efficiency of LLC resonant converter with and without step-gap. LLC resonant converter with step-gap has a much better performance at no and light load.

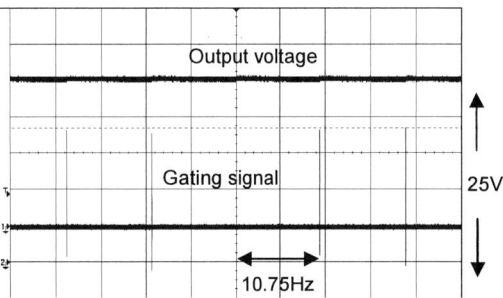

Fig. 6 The gating signal and output voltage of LLC resonant converter while using the ETD49 core without step-gap at no load condition.

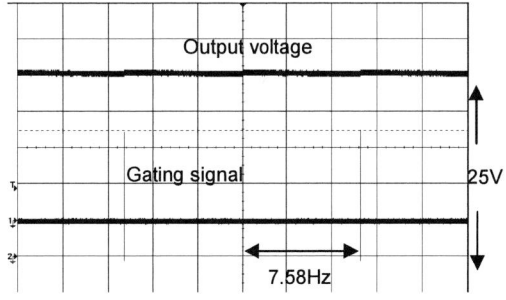

Fig. 7 The gating signal and output voltage of LLC resonant converter while using the ETD49 core with step-gap at no load condition.

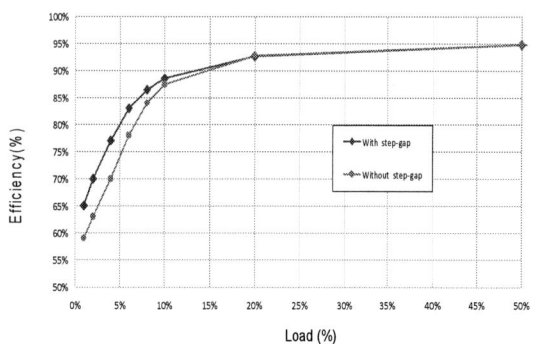

Fig. 8 The efficiency of LLC resonant converter with and without step-gap.

IV. CONCLUSION

In this design, an inductance-based control strategy is proposed to enhance the LLC resonant operation with respect to output loading current adaptively. The operation principle of the presented approach is implemented and verified through a step-gap transformer core design performing a close two-state variable inductance in an LLC resonant converter platform. Whatever with or without skip mode control, the switching frequency range is reduced. The efficiency is increased at first-state inductance region to improve the light load performance. The proposed control strategy helps on optimizing for the operation of LLC resonant converter at light load condition and makes the adaptive design of this energy conversion possible and based on this the further high average efficiency and power density of LLC resonant converter can be achieved.

REFERENCES

[1] Ecova Plug Load Solutions, "*80 PLUS® Certified Power Supplies and Manufacturers,*" from website: http://www.plugloadsolutions.com/80PlusPowerSupplies.aspx.

[2] The European Council for an Energy Efficient Economy, "Lot 6: Standby and off-mode," from website: http://www.eceee.org/Eco_design/products/standby.

[3] Y. K. Lo, S. C. Yen and C. Y. Lin, "A High-Efficiency AC-to-DC Adaptor with a Low Standby Power Consumption," *IEEE Transactions on Industrial Electronics*, vol. 55, no. 2, pp. 963-965, Feb. 2008.

[4] X. Hong, S. Wang, Z. Lu and S. Ye, "High efficiency soft-switched step-up dc-dc converter with hybrid mode LLC+C resonant tank," *IEEE 2010 Applied Power Electronics Conference*, 2010, pp. 1358-1364.

[5] Y. Fang, D. Xu, Y. Zhang, F. Gao, L. Zhu and Y. Chen, "Standby Mode Control Circuit Design of LLC Resonant Converter," *IEEE Power Electronics Specialists Conference*, 2007, pp.726-730.

[6] W.N. Huang, Y.W Tsai and S.H. Lee, *"Power Factor correction Apparatus,"* Taiwan Patent M434364, Jul. 2012

A Novel Accurate Primary Side Control (PSC) Method for Half-Bridge (HB) LLC Converter

Jae-Bum Lee, Chong-Eun Kim, Jae-Hyun Kim, Cheol-O Yeon, Young-Do Kim, and Gun-Woo Moon

E-mail: leejb83@angel.kaist.ac.kr

Abstract—Until now, several researches have been progressed on the primary side control (PSC) methods which decrease the size and cost of the overall system. However, all of them have been applied to the flyback converter, and it is difficult to apply them to the half-bridge (HB) LLC converter due to the large voltage across the secondary leakage inductor of the transformer. In this letter, a new PSC method for the HB LLC converter is proposed to obtain accurate output voltage. In the proposed method, the output voltage is regulated by obtaining the voltage across the primary side of the transformer when the external resonant inductor voltage becomes 0V. At this time, since the voltage across the transformer secondary leakage inductor is small, the proposed method can accurately regulate the output voltage. A 400V input and 20V/85W output laboratory prototype is built and tested to verify the effectiveness of the proposed PSC method.

Keywords—*Half-bridge (HB) LLC converter, high-power adaptor applications, primary side control (PSC).*

I. INTRODUCTION

Recently, as the portable data-processing equipment such as the mobile phones, tablet PCs, and laptop computers has grown explosively and their energy consumption has also been increased, the energy saving has been an essential issue. Meanwhile, since the consumers tend to prefer a compact device, the manufacturers have an effort to minimize its size. For these reasons, the power systems for those devices have been required to be small and achieve a high efficiency. Among the power systems, the adaptors for the laptop computers are strongly required to have a high power density in excess of 5W/in^3 and average efficiency above 85% because of the energy efficiency standards like the energy star program.

At a high power level above 70W for the laptop computers, the conventional adaptors have been developed based on two-stage structure to comply with the IEC 1000-3-2 harmonic standards and obtain a high efficiency [1], [2]. This two-stage configuration consists of the power factor correction (PFC) stage and dc/dc stage. Since the boost converter has many advantages such as a direct control of the line current and low input current ripple, it has been widely utilized in the PFC stage [3], [4]. For the dc/dc stage, the conventional half-bridge (HB) LLC converter is the most attractive candidate because of low voltage stresses on the primary switches and no transformer dc-offset current [5]-[7]. In addition, since it is generally designed in resonant or below region to achieve a full zero-voltage-switching (ZVS) of the primary switches and minimize the switch

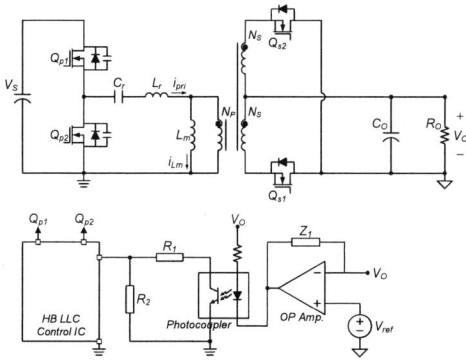

Fig. 1. Schematic of HB LLC converter with conventional SSC method.

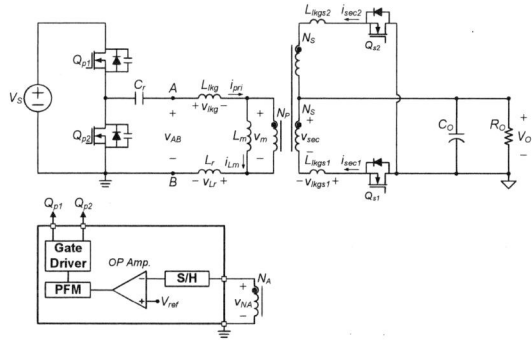

(a) Simplified block diagram.

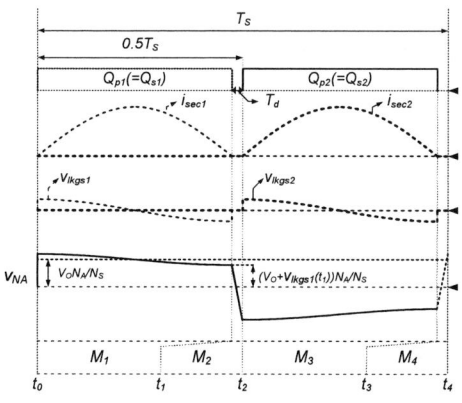

(b) Operational key waveforms.

Fig. 2. Conventional PSC method for HB LLC Converter.

turn-off losses, it can operate at a high switching frequency, which enables the size of reactive components to be decreased [8]. Due to these many advantages such as a high power density and efficiency, the HB LLC converter has been widely used in the dc/dc stage for high-power adaptor applications.

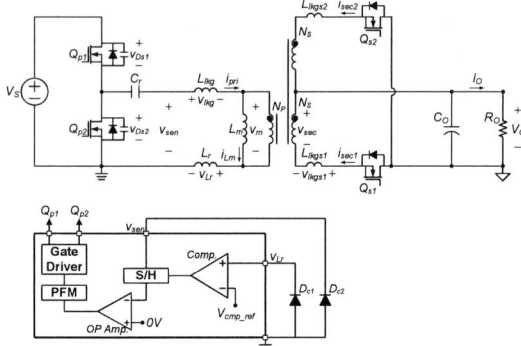

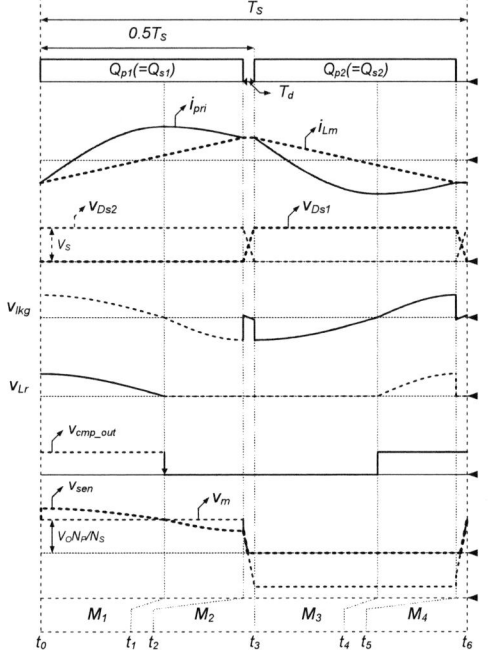

Fig. 3. Block diagram of HB LLC converter with PSC method.

Fig. 4. Key waveforms of HB LLC converter with PSC method.

The HB LLC converter is generally implemented by the secondary side control (SSC), using a HB LLC controller in the primary side, operational amplifier in the secondary side, and photocoupler in between the primary and secondary side to obtain a constant output voltage as shown in Fig. 1. However, since the SSC method requires three ICs with different grounds, it is difficult to integrate them into a single-chip, which increases the size and cost of the overall system. Moreover, the current transfer ratio (CTR) of the photocoupler often changes with the temperature, which degrades the reliability on the control [9], [10].

To overcome the problems of the SSC method, several researches on the primary side control (PSC) method have been proceeded [9]-[14]. By eliminating the photocoupler and secondary feedback circuitry, PSC methods can decrease the size and cost of the overall system compared with the SSC method. However, all of them have been applied only to the flyback converter, and it is difficult to apply them to the HB LLC converter. The reason is stated below. Fig. 2(a) and 2(b) show the

simplified block diagram and operational key waveforms of the HB LLC converter with the conventional PSC methods, respectively. In the conventional PSC methods, when the secondary current $i_{sec1}(t)$ reaches 0A at time t_1, the auxiliary winding voltage $v_{NA}(t_1)$ is obtained by the sample-and-hold (S/H) circuit, and the voltage control loop enables $v_{NA}(t_1)$ to follow a reference value V_{ref}. At time t_1 when the primary and secondary switches Q_{p1} and Q_{S1} are turned off, the voltage $v_{sec}(t_1)$ across the secondary side of the transformer and $v_{NA}(t_1)$ become $V_O+v_{lkgs1}(t_1)$ and $N_A(V_O+v_{lkgs1}(t_1))/N_S$, respectively. However, in the HB LLC converter, the voltage $v_{lkgs1}(t_1)$ across the transformer secondary leakage inductor L_{lkgs1} becomes large due to steep slope of $i_{sec1}(t)$ at time t_1 caused by sinusoidal secondary current as shown in Fig. 2(b). Since $v_{NA}(t_1)$ follows V_{ref}, i.e., $V_{ref}=N_AV_O/N_S$, the output voltage cannot be accurately regulated. Therefore, it is difficult to apply the conventional PSC methods to the HB LLC converter due to large $v_{lkgs1}(t_1)$.

In this letter, a new PSC method for the HB LLC converter is proposed as shown in Fig. 3. In the proposed PSC method, the output voltage is regulated by obtaining $v_{sen}(t)$ when the external resonant inductor voltage $v_{Lr}(t)$ becomes 0V. At this time, since the voltage across the transformer secondary leakage inductor is small, the proposed method can accurately regulate the output voltage.

II. PROPOSED METHOD

A. Principle of Proposed PSC Method

Fig. 3 and 4 respectively show the simplified block diagram and operational key waveforms of the HB LLC converter with the proposed PSC method under the following assumptions.

1) All parasitic components except for those specified in Fig. 3 are ignored.
2) The transformer secondary leakage inductors L_{lkgs1} and L_{lkgs2} are ignored. The analysis including L_{lkgs1} and L_{lkgs2} is discussed in Section II-C.
3) The drain-source on-state resistance $R_{ds,on}$ of the secondary switches Q_{s1} and Q_{s2} is small enough to be ignored.
4) The output voltage V_O is constant.

As shown in Fig. 3, the external resonant inductor L_r is located beside the primary ground to easily sense $v_{Lr}(t)$, and two schottky diodes D_{c1} and D_{c2} are used to enable two sensing voltages $v_{Lr}(t)$ and $v_{sen}(t)$ to have positive value that is appropriate for the controller IC. As shown in Fig. 4, the HB LLC converter is regulated by the pulse-frequency modulation (PFM) with a 50% fixed duty cycle, and its operation can be divided into two half cycles t_0–t_3 and t_3–t_6. Since two half cycles have symmetric operation, the first half cycle is only explained.

Mode 1 [t_0–t_2]: The primary switch Q_{p1} is turned on at time t_0. Since the secondary switch Q_{s1} is turned on, the magnetizing inductor current $i_{Lm}(t)$ is linearly increased with a slope of $N_PV_O/N_S/L_m$, and the transformer leakage inductor L_{lkg}, external resonant inductor L_r, and resonant capacitor C_r begin to resonate. Thus, the primary current

978-1-4799-2706-7/14 $31.00 © 2014 IEEE

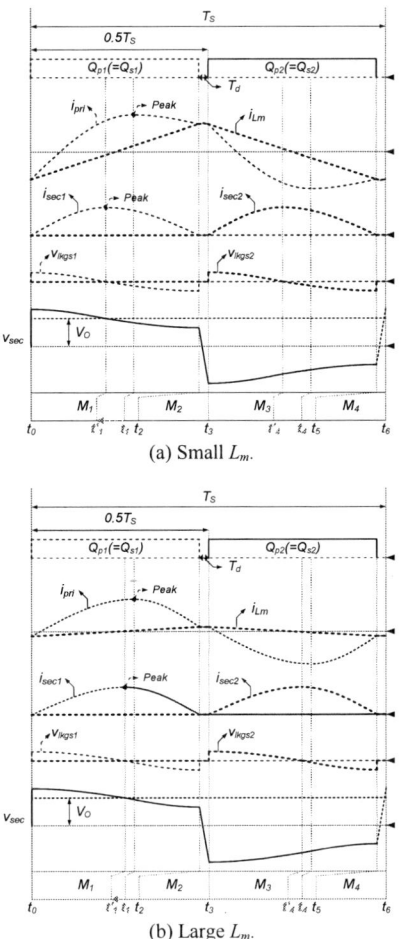

(a) Small L_m.

(b) Large L_m.

Fig. 5. The current and voltage waveforms according to L_m.

$i_{pri}(t)$ has sinusoidal shape. The voltages $v_{lkg}(t)$ and $v_{Lr}(t)$ across L_{lkg} and L_r can be expressed as follows:

$$v_{lkg}(t) = L_{lkg}\frac{di_{pri}(t)}{dt} \qquad (1)$$

$$v_{Lr}(t) = L_r\frac{di_{pri}(t)}{dt}. \qquad (2)$$

Based on (1) and (2), it can be seen from Fig. 4 that $v_{lkg}(t_1)$ and $v_{Lr}(t_1)$ become 0V when $i_{pri}(t)$ reaches its peak value at time t_1. Since $v_{lkg}(t_1)$ and $v_{Lr}(t_1)$ become 0V, $v_{sen}(t_1)$ becomes N_PV_O/N_S. By using the S/H circuit, $v_{sen}(t_1)$ can be obtained, and the voltage control loop enables $v_{sen}(t_1)$ to follow V_{ref}, i.e., $V_{ref}=N_PV_O/N_S$. Therefore, the output voltage can be accurately regulated by obtaining $v_{sen}(t_1)$ since $v_{lkg}(t_1)$ and $v_{Lr}(t_1)$ become 0V. Due to D_{c1} and D_{c2}, $v_{Lr}(t)$ and $v_{sen}(t)$ below 0V are clamped to 0V.

Mode 2 $[t_2–t_3]$: At time t_2, $i_{pri}(t)$ becomes equal to $i_{Lm}(t)$, and Q_{p1} is turned off. The peak magnetizing inductor current $i_{Lm}(t_2)$, considered as a current source, discharges the output capacitor of Q_{p2}. Thus, $v_{Ds2}(t)$ is linearly decreased and reaches zero. At this time, Q_{p2} is turned on, and the ZVS of Q_{p2} can be achieved.

B. Implementation of Proposed PSC Method

As mentioned in Section II-A, since $v_{lkg}(t_1)$ and $v_{Lr}(t_1)$

become 0V, $v_{sen}(t_1)$ only includes the information about the output voltage. Therefore, $v_{sen}(t_1)$ should be obtained by detecting the point that $v_{Lr}(t)$ becomes 0V. As shown in Fig. 4, $v_{Lr}(t)$ is compared with 0V. When $v_{Lr}(t)$ decreases and reaches 0V at time t_1, the output $v_{cmp\_out}(t_1)$ of the comparator is changed from high value to low value. By using the S/H circuit, $v_{sen}(t_1)$ can be obtained at only falling edge of $v_{cmp\_out}(t)$ [12]. Based on the simplified block diagram in Fig. 3, the proposed method can be a single-chip solution.

C. Effect of Transformer Secondary Leakage Inductor

In this section, since the transformer secondary leakage inductors L_{lkgs1} and L_{lkgs2} have a strong influence on the regulation accuracy, the analysis considering L_{lkgs1} and L_{lkgs2} is discussed. Fig. 5(a) and 5(b) show the current and voltage waveforms according to L_m, considering L_{lkgs1} and L_{lkgs2}. During mode 1, $i_{Lm}(t)$ is linearly increased with a slope of $N_PV_O/N_S/L_m$, and L_{lkg}, L_r and C_r resonate. Therefore, $i_{Lm}(t)$ and $i_{pri}(t)$ can be expressed as follows:

$$i_{Lm}(t) = \frac{N_PV_O}{N_SL_m}(t-t_0) - \frac{N_PV_O}{4N_SL_mF_S} \qquad (3)$$

$$i_{pri}(t) = i_{pri}(t_0)\cos(\omega_O(t-t_0)) \\ + \left(\frac{V_S - v_{Cr}(t_0) - N_PV_O/N_S}{Z}\right)\sin(\omega_O(t-t_0)), \qquad (4)$$

where $v_{Cr}(t_0)$ and $i_{pri}(t_0)$ are the initial values of the resonant capacitor voltage $v_{Cr}(t)$ and primary current $i_{pri}(t)$, F_S is the switching frequency, and the characteristic impedance Z and resonant angular frequency ω_O are defined as $[(L_{lkg}+L_r)/C_r]^{0.5}$ and $1/[(L_{lkg}+L_r)C_r]^{0.5}$.

Based on (1) and (2), when $i_{pri}(t)$ reaches its peak value at time t_1, both $v_{lkg}(t)$ and $v_{Lr}(t)$ become 0V, and t_1 can be expressed as follows:

$$t_1 = t_0 + \frac{1}{\omega_O}\left[\pi + \tan^{-1}\left(\frac{V_S - v_{Cr}(t_0) - N_PV_O/N_S}{i_{pri}(t_0)Z}\right)\right]. \qquad (5)$$

Meanwhile, the secondary current $i_{sec1}(t)$ during mode 1 is the difference between $i_{pri}(t)$ and $i_{Lm}(t)$, and $i_{sec1}(t)$ can be expressed as follows:

$$i_{sec1}(t) = \frac{N_P}{N_S}\left(i_{pri}(t) - i_{Lm}(t)\right). \qquad (6)$$

As shown in Fig. 5(a) and 5(b), since $i_{sec1}(t)$ reaches its peak value at time not t_1 but t'_1, the voltage $v_{lkgs1}(t_1)$ across L_{lkgs1} becomes negative, and t'_1 can be expressed as follows:

$$t'_1 = t_1 - \frac{1}{\omega_O}\sin^{-1}\left[\frac{N_PV_O}{N_SL_m\omega_O\sqrt{i_{pri}^2(t_0) + \left(\frac{V_S - v_{Cr}(t_0) - N_PV_O/N_S}{Z}\right)^2}}\right]. \qquad (7)$$

From (7), as the transformer magnetizing inductor L_m is decreased, it can be seen from Fig. 5(a) that t'_1 deviates farther from t_1 and $v_{lkgs1}(t_1)$ becomes more negative due to steep slope of $i_{sec1}(t)$ at time t_1. In the proposed method, since $v_{sec}(t_1)$ and $v_{sen}(t_1)$ respectively become $V_O+v_{lkgs1}(t_1)$ and $N_P(V_O+v_{lkgs1}(t_1))/N_S$, designing large L_m indicates that $v_{lkgs1}(t_1)$ is small and accurate output voltage is obtained.

Meanwhile, in the HB LLC converter with the conventional PSC methods, when $i_{sec1}(t)$ reaches 0A at

TABLE I
CIRCUIT PARAMETERS OF HB LLC CONVERTER WITH PSC METHODS

Rated power	85W	Input voltage, V_S	400V
		Output voltage, V_O	20V
Main switches (Q_1, Q_2)		IPP60R600CP	
External resonant inductor (L_r)		31.05μH	
Resonant capacitor (C_r)		47nF	
Main transformer	Core	PQ2620	
	$N_P : N_S : N_S = 20 : 2 : 2$		
	L_m: 2.37mH, L_{lkg}: 5.5μH		
Secondary switches (SR_1, SR_2)		BSC077N12NS3G (120V, 98A, $R_{ds,on}$: 7.7mΩ)	

TABLE II
MEASURED OUTPUT VOLTAGE OF PSC METHODS ACCORDING TO OUTPUT LOAD

Output load	Output Voltage		Output load	Output Voltage	
	Conventional PSC Method	Proposed PSC Method		Conventional PSC Method	Proposed PSC Method
10%	20.135V	20.051V	60%	20.401V	20.110V
20%	20.191V	20.061V	70%	20.454V	20.123V
30%	20.242V	20.072V	80%	20.507V	20.136V
40%	20.294V	20.084V	90%	20.560V	20.148V
50%	20.348V	20.095V	100%	20.614V	20.162V

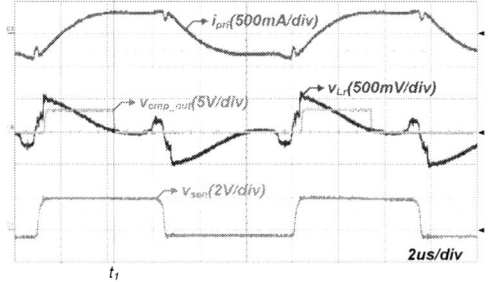 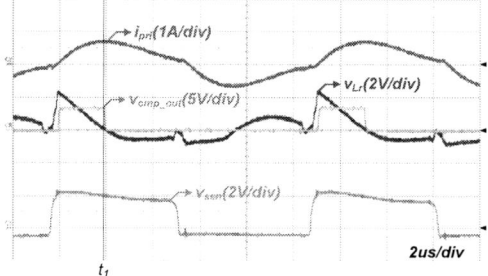

(a) At 10% load conditions.   (b) At full load conditions.

Fig. 6. Experimental key waveforms of i_{pri}, v_{Lr}, v_{cmp_out}, and v_{sen} at 10% and full load conditions.

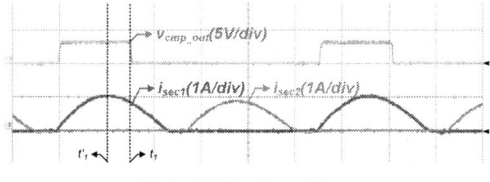 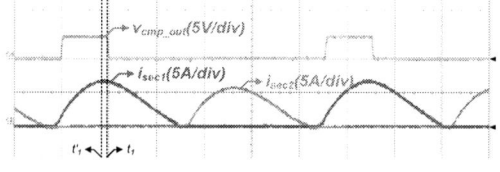

(a) At 10% load conditions.   (b) At full load conditions.

Fig. 7. Experimental key waveforms of v_{cmp_out}, i_{sec1}, and i_{sec2} at 10% and full load conditions.

time t_2 as shown in Fig. 5(a) and 5(b), $v_{NA}(t_2)$ is obtained, i.e. $v_{NA}(t_2)=N_A(V_O+v_{lkgs1}(t_2))/N_S$. Since $v_{lkgs1}(t_2)$ is larger than $v_{lkgs1}(t_1)$, the output voltage in the conventional PSC methods can be less accurately regulated than that in the proposed PSC method.

III. EXPERIMENTAL RESULTS

To verify the validity of the proposed PSC method for the HB LLC converter, a 85W prototype converter with the specification of V_S=400V and V_O=20V has been built, and a PIC33FJ16GS502 micro control unit (MCU) is utilized. The design parameters utilized in this experiment are presented in Table I.

Fig. 6(a) and 6(b) show the experimental key waveforms of the HB LLC converter with the proposed

PSC method at 10% and full load conditions. As can be seen from $i_{pri}(t)$ and $v_{Lr}(t)$ in Fig. 6(a) and 6(b), $v_{Lr}(t_1)$ becomes 0V when $i_{pri}(t)$ reaches its peak value at time t_1. In addition, from $v_{cmp\_out}(t)$ in Fig. 6(a) and 6(b), it can be seen that $v_{cmp\_out}(t_1)$ is changed from high value to low value. Moreover, from $i_{sec1}(t)$ in Fig. 7(a) and 7(b), it can be seen that $i_{sec1}(t)$ reaches its peak value at time not t_1 but t'_1. However, since L_m is large enough, the slope of $i_{sec1}(t)$ is gradual at time t_1, and $v_{lkgs1}(t_1)$ is small.

Table II shows the measured output voltage of the conventional and proposed PSC methods according to the load conditions. As shown in this table, it can be seen that the proposed method has the improved output voltage regulation capability within 0.81% by minimizing $v_{lkgs1}(t_1)$ compared the output voltage regulation capability of the conventional method within 3.1%.

IV. CONCLUSIONS

This paper introduces a new PSC method for the HB LLC converter. The proposed method has the output voltage regulation capability within 0.81% by minimizing the voltage across the transformer secondary leakage inductor. The validity of this study is confirmed by the experimental results. It is suitable for high-power adaptor applications employing the HB LLC converter for the tablet PCs and laptop computers.

ACKNOWLEDGMENT

This work was supported by the National Research Foundation of Korea(NRF) grant funded by the Korea government(MSIP) (No.2010-0028680).

REFERENCES

[1] Y. Panov and M. M. Jovanovic, "Performance Evaluation of 70-W Two-Stage Adaptors for Notebook Computers," in *Proc. IEEE APEC*, 1999, pp. 1059-1065.

[2] S. W. Choi, B. W. Ryu, and G. W. Moon, "Two-Stage AC/DC Converter Employing Load-Adaptive Link-Voltage-Adjusting Technique with Load Power Estimator for Notebook Computer Adaptor," in *Proc. IEEE ECCE*, 2009, pp. 3761-3767.

[3] L. H. S. C. Barreto, M. G. Sebastiao, L. C. de Freitas, E. A. Alves Coelho, V. J. Farias, and J. B. Vieira, "Analysis of a Soft-Switched PFC Boost Converter Using Analog and Digital Control Circuits," *IEEE Trans. Ind. Electron.*, vol. 52, no. 1, pp. 221–227, Feb. 2005.

[4] J. P. R. Balestero, F. L. Tofoli, R. C. Fernandes, G. V. Torrico-Bascope, and F. J. M. de Seixas, "Power Factor Correction Boost Converter Based on the Three-State Switching Cell," *IEEE Trans. Ind. Electron.*, vol. 59, no. 3, pp. 1565–1577, Mar. 2012.

[5] B. Yang, F. C. Lee, A. J. Zhang, and G. Huang, "LLC Resonant Converter for Front End DC/DC Conversion," in *Proc. IEEE APEC*, 2002, pp. 1108-1112.

[6] D. Y. Kim, C. E. Kim, and G. W. Moon, "High-Efficiency Slim Adapter With Low-Profile Transformer Structure," *IEEE Trans. Ind. Electron.*, vol. 59, no. 9, pp. 3445–3449, Sep. 2012.

[7] I. O. Lee and G. W. Moon, "The k-Q Analysis for an *LLC* Series Resonant Converter," *IEEE Trans. Power Electron.*, vol. 29, no. 1, pp. 13–16, Jan. 2014.

[8] K. Jin, X. Ruan, M. Yang, and M. Xu, "A Hybrid Fuel Cell Power System," *IEEE Trans. Ind. Electron.*, vol. 56, no. 4, pp. 1212–1222, Apr. 2009.

[9] J. Fang, Z. Lu, Z. Li, and Z. Li, "A New Flyback Converter with Primary Side Detection and Peak Current Mode Control," in *Proc. IEEE ICCCAS*, 2002, pp. 1707-1710.

[10] J. Shen and T. Liu, "Constant Current LED Driver Based on Flyback Structure with Primary Side Control," in *Proc. IEEE PEAM*, 2011, pp. 260-263.

[11] X. Xie, J. Wang, C. Zhao, Q. Lu, and S. Liu, "A Novel Output Current Estimation and Regulation Circuit for Primary Side Controlled High Power Factor Single-Stage Flyback LED Driver," *IEEE Trans. Power Electron.*, vol. 27, no. 11, pp. 4602–4612, Nov. 2012.

[12] Jianwen Shao, "A Highly Accurate Constant Voltage (CV) and Constant Current (CC) Primary Side Controller for Offline Applications," in *Proc. IEEE APEC*, 2013, pp. 3311-3316.

[13] H. H. Chou, Y. S. Hwang, and J. J. Chen, "An Adaptive Output Current Estimation Circuit for a Primary-Side Controlled LED Driver," *IEEE Trans. Power Electron.*, vol. 28, no. 10, pp. 4811–4819, Oct. 2013.

[14] Y. C. Li and C. L. Chen, "A Novel Primary-Side Regulation Scheme for Single-Stage High-Power-Factor AC-DC LED Driving Circuit," *IEEE Trans. Ind. Electron.*, vol. 60, no. 11, pp. 4978–4986, Nov. 2013.

The 2014 International Power Electronics Conference

A Simple Control Scheme for Improving Light-Load Efficiency in a Full-Bridge LLC Resonant Converter

Jae-Hyun Kim[1], Chong-Eun Kim[2], Jae-Bum Lee[1], Young-Do Kim[2], Han-Shin Youn[1], Gun-Woo Moon[1]

[1]Department of Electrical Engineering, KAIST, 335, Gwahangno, Yuseong-Gu, Daejeon, Republic of Korea
[2]Samsung Electro-Mechanics, 314, Maetan3-Dong, Yeongtong-Gu, Suwon, Gyunggi-Do, Republic of Korea

Abstract— In this paper, a load adaptive phase-shift control is proposed for the full-bridge LLC resonant converter to improve light-load efficiency. The proposed method reduces the effective duty ratio and the magnetizing current. Therefore, the core loss on transformer and the turn-off switching loss on switches are significantly reduced. To confirm the validity of this study, the prototype with 320~400V DC input, 12V/40A DC output is experimented.

Keywords— *LLC resonant converter, light load efficiency, phase shift control*

I. INTRODUCTION

Today, as consumer electronic devices and IT equipment have grown explosively, the high efficiency in power supplies is becoming more and more important in order to suppress global warming. Consequently, the efficiency requirement, defined by Energy Star [1], 80 Plus incentive program [2], Climate Saver Computing Initiative [3], and European Code of Conduct [4], is focused on the light load as well as the heavy load. Therefore, manufacturers for power supply have a significant challenge to improve conversion efficiencies of their products.

The LLC resonant converter has drawn much attention for high efficiency due to excellent soft switching characteristics such as zero-voltage switching of primary-side switches and zero-current switching of secondary-side rectifiers [5]-[7]. However, the light-load efficiency is significantly degraded by load-independent losses such as the switching loss and the gate driving loss in semiconductor devices, the power loss in controller ICs, and the core loss in magnetic components [8]-[10].

Meanwhile, many literatures have been presented in order to reduce load-independent losses. The resonant gate-drive circuit [11]-[13] can reduce the gate driving loss by recycling the gate charge of a switch. The adaptive gate driving voltage methods [14]-[15] can reduce the gate charge by decreasing the gate driving voltage V_G of the gate driver, which results in low power consumption in the gate driver. The work in [16] can reduce the loss on controller IC. Meanwhile, burst mode operation with hysteresis band has been widely used for reducing core loss and switching loss [17]. However, it

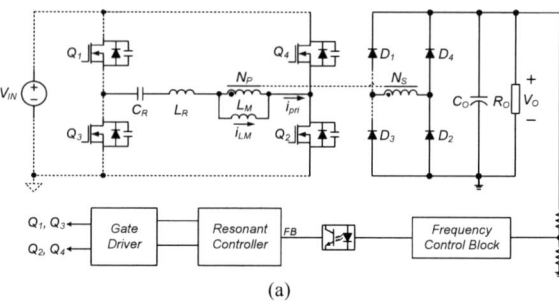

Fig. 1. The circuit diagram of the full-bridge LLC resonant converter

leads to poor dynamic performance because the operating frequency changes over a wide range according to the load variation, and poor electromagnetic interference (EMI) and audible noise are other issues [9]. In addition, large hysteresis band induces a large output voltage ripple, and small band leads to low burst efficiency. In other words, the burst mode operation degrades performance in the main power stage.

This paper presents a simple control method to reduce power loss under light load conditions for the full-bridge LLC resonant converter. The proposed method can reduce core loss of transformer greatly by load adaptive phase-shift control without additional power devices. Moreover, since the operation under heavy load conditions is the same as conventional one, there are no adverse features. In addition, the since burst operation is not used, performance in the main power stage is not degraded.

II. PROPOSED METHOD

A. Motivation and Concept of the Proposed Method

Fig. 1 shows the circuit diagram of the full-bridge LLC resonant converter. It is composed of the primary switches $Q_1 \sim Q_4$, resonant capacitor C_R, resonant inductor L_R, magnetizing inductance L_M, transformer T, rectifier diodes D_1 and D_2, output capacitor C_O, and output resistance R_O. N_P and N_S are the primary and secondary turns of the transformer, respectively. Generally, it uses the variable frequency control where the duty ratio is fixed as 0.5 and the switching frequency f_S is changed to

978-1-4799-2706-7/14 $31.00 © 2014 IEEE   1743

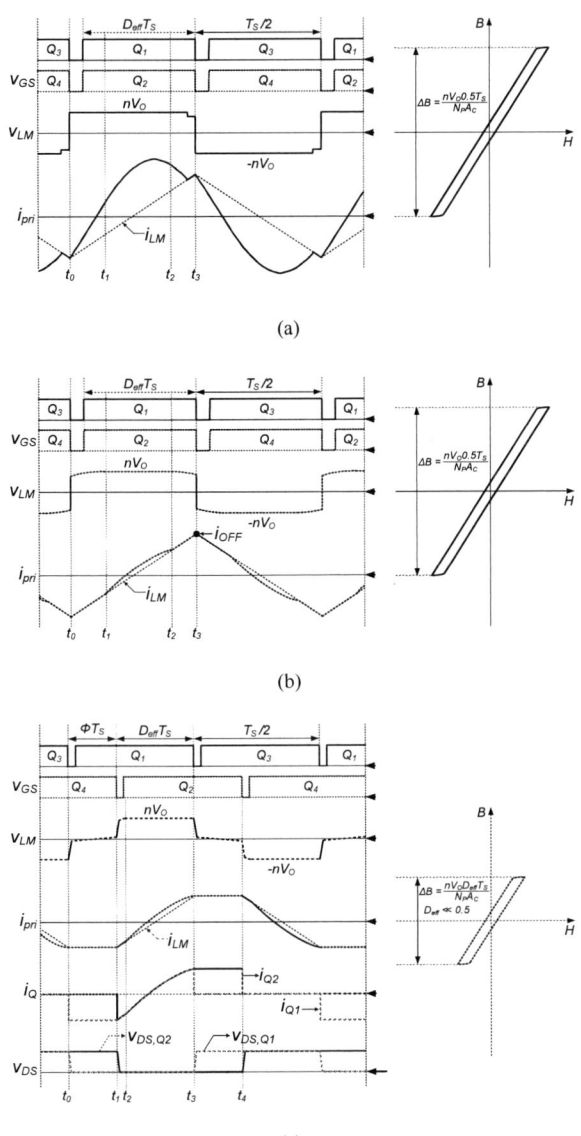

(a)

(b)

(c)

Fig. 2. Key waveforms of the full-bridge LLC resonant converter using conventional frequency control under light load conditions

regulate the output voltage V_O.

In order to describe the core loss of the transformer in the full-bridge LLC resonant converter, the key waveforms under heavy load conditions and light load conditions are presented in Fig. 2(a) and Fig. 2(b), respectively. The operational principle in detail is presented in [18]. Typically, the LLC resonant converter is designed to operate at the resonant frequency f_R under heavy load conditions for high efficiency, as shown in Fig. 2(a). In other words, the powering period T_{PWR}, when the input power transfers to the output, is almost the same as a half of the switching period $T_S/2$. Meanwhile, the core loss P_{core} is given by

$$P_{core} = c\left(\Delta B\right)^a f_s^b V_C, \qquad (1)$$

where a, b, c are the coefficients of the utilized core

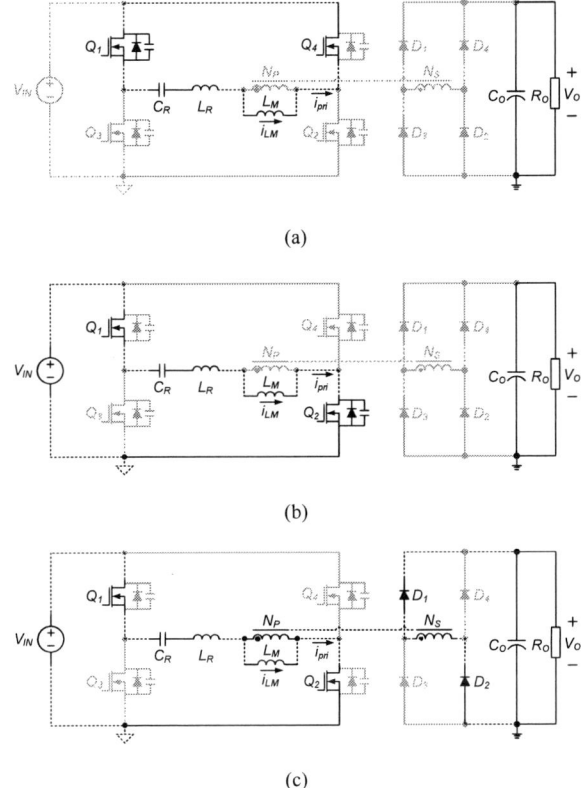

(a)

(b)

(c)

Fig. 3. Key waveforms of the full-bridge LLC resonant converter using conventional frequency control under light load conditions

material, ΔB is the ac flux density, f_s is the switching frequency, and V_C is the volume of the transformer core. In addition, ΔB can be expressed as

$$\Delta B = \frac{V_{LM} D_{eff} T_S}{N_P A_C}, \qquad (2)$$

where V_{LM} is the voltage of L_M, D_{eff} is the effective duty ratio, T_S is the switching period, and A_C is the cross sectional area. Since V_{LM} is induced by nV_O during T_{PWR} or $T_S/2$ under heavy load conditions, D_{eff} is equal to 0.5. Thus, (2) can be expressed as

$$\Delta B = \frac{nV_O 0.5 T_S}{N_P A_C}. \qquad (3)$$

On the other hand, under light load conditions, T_{PWR} is much shorter than that of the heavy load conditions. Instead, the freewheeling period T_{FR}, when the input current freewheels in the primary side and no input power transfers to the output, has large portion of the switching period, as shown in Fig. 2(b). Meanwhile, V_{LM} is induced by nV_O unnecessarily during T_{FR} as well as during T_{PWR}. In other words, V_{LM} is induced by nV_O during $T_S/2$ and then D_{eff} is equal to 0.5 like the heavy load conditions. Therefore, ΔB is the same as (3) and it cannot be reduced under light load conditions. Consequently, P_{core} is almost constant regardless of the load conditions, which leads to large portion of the power loss under light load conditions. On the other hand, if V_{LM} becomes zero during T_{FR}, ΔB and P_{core} can be reduced significantly.

978-1-4799-2706-7/14 $31.00 © 2014 IEEE

The 2014 International Power Electronics Conference

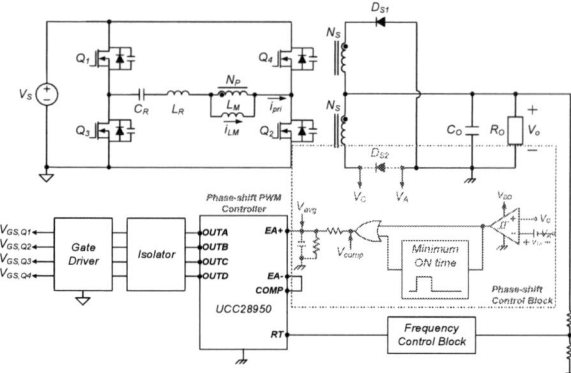

Fig. 4. Block diagram of the proposed control circuit

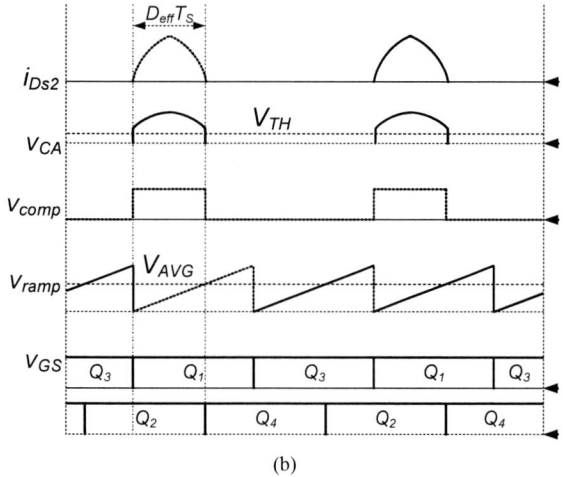

(b)

Fig. 5. Key waveforms of the proposed control circuit

In the proposed method, the phase-shift control is used under light load conditions whereas the frequency control is used under heavy load conditions. As shown in Fig. 3, the proposed control makes it possible that V_{LM} becomes low voltage during T_{FR} under light load conditions. Therefore, since D_{eff} becomes much lower than 0.5, ΔB and P_{core} can be reduced greatly.

B. Operational Principle of the Proposed Method

Fig. 2(c) shows the key waveforms of the full-bridge LLC resonant converter using the proposed method. Fig. 3 shows the equivalent circuit of each operational mode. For convenience, it is assumed that Moreover, the parasitic components are neglected except for the output capacitance of the primary switches.

Mode 1(t_0~t_1) : When Q_3 is turned off at t_0, Mode 1 begins. After, the output capacitance of Q_1 is discharged and $V_{DS,Q1}$ decreases to zero voltage. When Q_1 is turned on at this time, ZVS of Q_1 can be achieved. As shown in Fig. 3(a), V_{LM} is equal to $-V_{CR}$ which is the voltage of C_R. Unlike the conventional frequency control shown in Fig. 2(b), small voltage is induced to V_{LM} during this freewheeling period. In this mode, the input current freewheels in the primary side of the transformer.

TABLE I
SPECIFICATIONS AND PARAMETERS OF PROTOTYPE

Item	Parameter
Input voltage	V_S=320~400V
Nominal input voltage	$V_{S,nom}$=400V
Output voltage	V_O=48V
Turns of transformer	N_P:N_S:N_S =25:3:3
Magnetizing inductance	L_M=800µH
Resonant inductance	L_R=30µH
Resonant capacitance	C_R=47nF
Primary switches	IPP60R190C6
Secondary diodes	V60100C
Controller	UCC28950

Mode 2(t_1~t_2) : At t_1, Q_4 is turned off. The output capacitance of Q_4 is discharged and $V_{DS,Q2}$ decreases to zero voltage. When Q_2 is turned on at this time, ZVS of Q_2 can be achieved. Since V_{LM}=V_{IN}-V_{CR} and it is smaller than nV_O, V_{LM} is insufficient to turn on D_1 and D_2. Thus, D_1 and D_2 are turned off, as shown in Fig. 3(b).

Mode 3(t_2~t_3) : When V_{LM} become nV_O and it is sufficient to turn on D_1, Mode 3 begins. In this mode, L_R and C_R resonate and the input power transfers to the output, as shown in Fig. 3(c).

C. Implementation of the proposed method

The implementation of the proposed method is described in Fig. 4 and its key waveforms are shown in Fig. 5. The cathode-to-anode voltage is directly sensed without sensing resistor. When the cathode voltage V_C is larger than the threshold voltage V_{TH}, the output voltage of the comparator, V_{comp} become high level. On the contrary, when V_C is smaller than V_{TH}, V_{comp} become low level. The period of high level of V_{comp} is equal to T_{CON}. In addition, V_{comp} is scaled by voltage divider and averaged by RC filter. This averaged value V_{avg} is connected to the EA+ pin of the phase-shift PWM controller UCC28950. The scale factor can be set to be equal to the peak value of the internal ramp voltage v_{ramp} in the UCC28950. Phi can be determined by comparing v_{avg} with v_{ramp}. Therefore, $D_{eff}T_S$ is equal to T_{CON}. In order to ensure ZVS of the primary switches, the minimum on-time is set so that minimum D_{eff} can be limited. Meanwhile, to regulate the output voltage tightly, feedback control is used through RT pin of UCC28950.

III. EXPERIMENTAL RESULTS

To confirm the validity of the proposed method, 480W prototype of the server power supply is implemented. The parameters of LLC resonant converter are listed in Table I. The conventional method uses just variable frequency control, but the proposed method uses phase-shift control and variable frequency control. Fig. 6 shows the experimental waveforms using conventional control. It can be seen that T_{PWR} is small. Fig. 7 shows the experimental waveforms using the proposed control. It is confirmed that the proposed method is operated as

978-1-4799-2706-7/14 $31.00 © 2014 IEEE

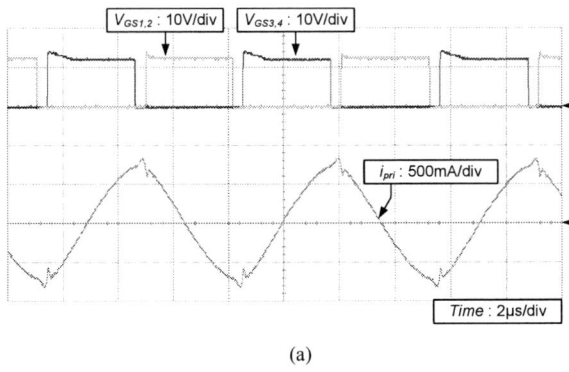

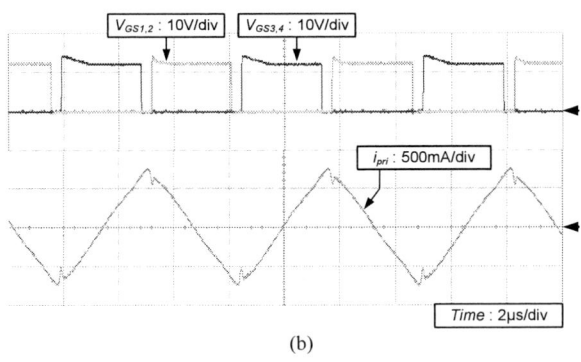

Fig. 6. Experimental waveforms of the conventional full-bridge LLC resonant converter under (a) 10% load condition (b) 2% load condition

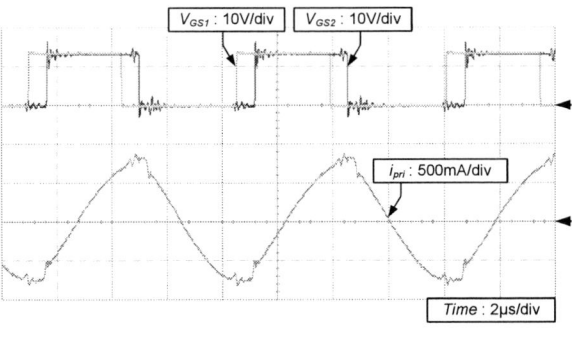

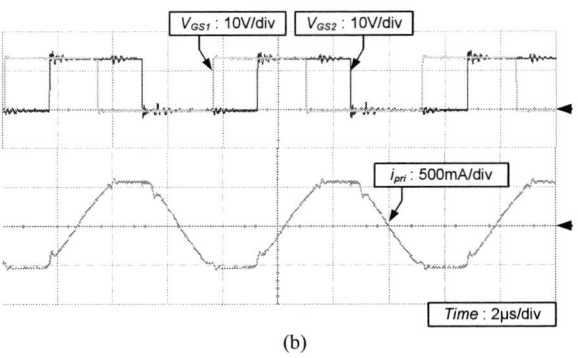

Fig. 7. Experimental waveforms of the proposed phase-shift full-bridge LLC resonant converter under (a) 10% load condition (b) 2% load condition

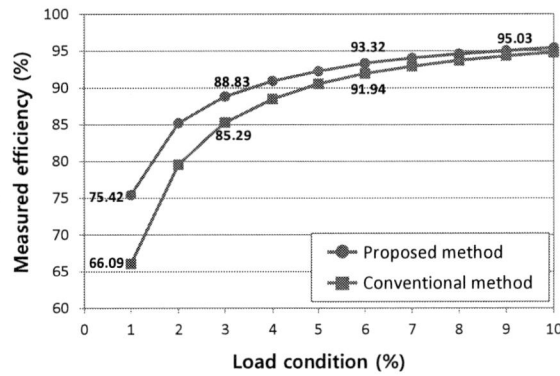

Fig. 8. Measured Efficiency

expected. Fig. 8 shows the measured efficiency. It is demonstrated that the efficiency under light load conditions can be improved by using the proposed method. Especially, the efficiency at 1% load condition is greatly improved by 9%.

IV. CONCLUSIONS

This paper introduces a load adaptive phase-shift control for full-bridge LLC resonant converter that achieves low power consumption under light load conditions. Compared to the conventional variable frequency control, the proposed method reduces core loss on transformer and turn-off switching loss on primary switches. The validity of this study is confirmed by the experimental results. It is suitable for the power system applications requiring high efficiency under light load conditions such as server computer, PC, LED TV and so on.

ACKNOWLEDGMENT

This work was supported by the National Research Foundation of Korea(NRF) grant funded by the Korea government(MSIP) (No.2010-0028680).

REFERENCES

[1] Energy Star Program. [Online]. Available: http://www.energystar.gov

[2] 80 Plus Incentive Program. [Online]. Available: http://www.80plus.org

[3] Climate Savers Computing Initiative. [Online]. Available: http://www.climatesaverscomputing.org

[4] Power Integrations, Inc., San Jose, CA, Green Room page. [Online]. Available: http://www.powerint.com/greenroom/index.html.

[5] B. Lu, W. Liu, Y. Liang, F. C. Lee, and J. D. van Wyk, "Optimal design methodology for LLC resonant converter," in Proc. 21st Annu. IEEE Appl. Power Electron. Conf. Expo., Mar. 2006, pp. 553-558.

[6] F. C. Lee, S. Wang, P. Kong, C. Wang, and D. Fu, "Power architecture design with improved system efficiency, EMI and power density," in Proc. IEEE Power Electron. Spec. Conf., Jun. 2008, pp. 4131-4137.

[7] W. Feng, D. Huang, P. Mattavelli, and F. C. Lee, "Digital implementation of driving scheme for synchronous rectification in LLC resonant converters," in Proc. IEEE Eur. Convers. Congr. Expo., Sep. 2010, pp. 256-263.

[8] Y. Jang, and M. M. Jovanovic, "Light-Load Efficiency Optimization Method," in *IEEE Trans. on Power Electronics*, vol. 25, no. 1, pp. 67-74, Jan. 2010.

[9] L. Huber, and M. M. Jovanovic, "Methods of Reducing Audible Noise Caused by Magnetic Components in Variable-Frequency-Controlled Switch-Mode Converters," in *IEEE Trans. on Power Electronics*, vol. 26, no. 6, pp. 1673-1681, Jun. 2011.

[10] B. Wang, X. Xin, S. Wu, H. Wu, and J. Ying, "Analysis and Implementation of LLC Burst Mode for Light Load Efficiency Improvement," in Applied Power Electronics Conference(APEC'09), pp. 58-64, 2009.

[11] W. Eberle, Y-F. Liu, and P. Sen, "A New Resonant Gate-Drive Circuit With Efficient Energy Recovery and Low Conduction Loss," in *IEEE Trans. on Power Electronics*, vol. 55, no. 5, pp. 2213-2221, May. 2008.

[12] Z. Yang, S. Ye, and Y-F. Liu, Z. Du, "A New Dual-Channel Resonant Gate Drive Circuit for Low Gate Drive Loss and Low Switching Loss," in *IEEE Trans. on Power Electronics*, vol. 23, no. 3, pp. 1574-1583, May. 2008.

[13] Z. Yang, S. Ye, and Y-F. Liu, "A New Resonant Gate Drive Circuit for Synchronous Buck Converter," in *IEEE Trans. on Power Electronics*, vol. 22, no. 4, pp. 1311-1320, Jul. 2007.

[14] M. D. Mulligan, B. Broach, and T. H. Lee, "A Constant-Frequency Method for Improving Light-Load Efficiency in Synchronous Buck Converters," in *IEEE Power Electronics Letters*, vol. 3, no. 1, pp. 24-29, Mar. 2005.

[15] O. Abdel-Rahman, J. A. Abu-Qahouq, L. Huang, and I. Batarseh, "Analysis and Design of voltage Regulator With Adaptive Fet Modulation Scheme and Improved Efficiency," in *IEEE Trans. on Power Electronics*, vol. 23, no. 2, pp. 896-906, Mar. 2008.

[16] J. H. Kim, J. K. Kim, J. B. Lee, J. W. Kim, and G. W. Moon, "Light-load efficiency improvement using load adaptive gate driving method," in Proc. IEEE ECCE Asia Downunder, pp. 240-244, Jun. 2013.

[17] Texas Instruments, Dallas. UCC25600: 8-pin high-performance resonant mode controller. [Online]. Available: Http://www.ti.com/lit/ds/symlink/ucc25600.pdf

[18] ST Microelectronics, Application note AN2450: An introduction t o LLC resonant half-bridge converter. [Online]. Available: http://www.st.com/web/en/resource/technical/document/application_note/CD00174208.pdf

Power Conditioner for Stabilizing Power Disturbance caused of Wind Turbine Generator System

Yasunao Saga, Kansuke Fujii and Kazuyuki Yoda

Development Division, Power Supply Group
Fuji Electric Co., Ltd.
Hyogo, Japan
saga-yasunao@fujielectric.co.jp

Abstract— **In Japan, Feed in Tariff has started since July in 2012. Moreover only solar power is not attractive energy, a wind turbine generator is attractive energy from the point of view to reduce double carbon oxide emission. However, to connect the wind turbine generator system to the grid, the independent power producer (IPP) must compensate the power disturbance caused of the wind turbine. Thus, a power conditioner for stabilizing power disturbance is necessary. The authors have developed a new power conditioner using batteries in this paper. The control technique, specifications, circuit configuration and prototype test results are described.**

Keywords— *Power conditioner, Battery, control and three-level topology.*

I. INTRODUCTION

A wind turbine system is one of the attractive renewable energies, because the cost of power generation is quite lower than that of the solar power. In addition, it is possible to build a few 100 MW wind farm, because the capacity of the wind turbine can be increased up to a few MW [6]. The problem of the wind turbine system is that the power disturbance caused of the wind turbine influences the quality of the grid; i.e. frequency and voltage [5], because the capacity of the wind turbine generator system is quite larger than that of the normal mega solar system. Therefore, electric power companies, who are responsible for stabilizing the grid, order the IPP to install the power conditioners for stabilizing the power disturbance caused of the wind turbine.

Fig. 1 shows a typical power conditioning system for wind turbine generator system. The power disturbance caused of the wind turbine is absorbed to the battery by using the power conditioner. The battery can be lead-acid battery, Li-ion battery, NAS battery [1] and so on. These batteries can be charged and discharged more times than the battery for stand-by use.

The authors have developed a new power conditioner for stabilizing power disturbance. Its circuit topology is Advanced T-type NPC three-level circuit [2], which uses a Reversed blocking IGBT (RB-IGBT) [3]. In this paper, as a control technique of the power conditioner, how to

enable the fault ride through capability and how to control output power is described. After that, the specifications and characteristics of the power conditioner are described. Finally, as the experimental results of the prototype (500-kVA and 600-kVA converter), the efficiencies and transient response are discussed.

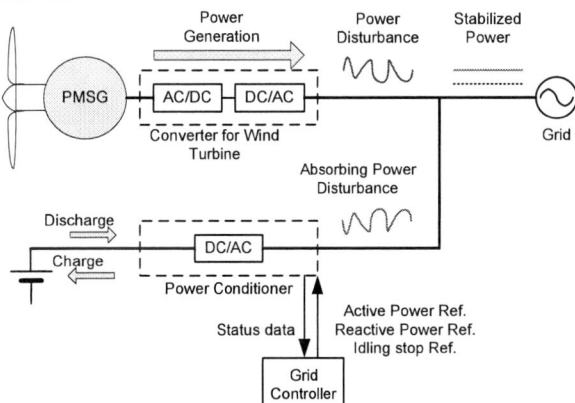

Fig. 1. Power Conditioning System for Wind turbine generator system

II. CONTROL TECHNIQUES

A. Inverter control for enabling Fault Ride Through

Inverter control consists of d-q current reference calculation and current control. Fig. 2 shows an inverter-control diagram except for the calculation of the d-q current reference ($i_p{}^*$ and $i_q{}^*$). In this inverter, a phase reference ($\cos\theta$), which is necessary for changing the current reference d-q axis to abc axis, is calculated from the grid voltage (v_{suvw}). The v_{suvw} is filtered with a Band-pass filter to eliminate a high harmonic component, and the phase of the v_{suvw} is set forward to compensate the phase delay caused of the output filter. In the phase reference calculation, $\cos\theta$ is calculated with the following equation [4].

$$\cos\theta = \frac{1}{3A}\left\{ v_{fu} + \left(-\frac{v_{fv}}{2} - \frac{\sqrt{3}v_{fv90}}{2} \right) + \left(-\frac{v_{fw}}{2} + \frac{\sqrt{3}}{2}v_{fw90} \right) \right\} \quad (1)$$

, where A is amplitude of the $\cos\theta$ without

standardization, v_{fv90} and v_{fw90} are 90 degree delayed with v-phase and w-phase of v_{fuvw} respectively. The amplitude is the same with the amplitude of the v_{fu}, when the grid voltage is balanced. Though the detailed calculation about cosθ is not described (refer [4]), cosθ is in phase of v_{fu}, which is u-phase voltage of the filtered grid voltage, even if two-phase short circuit occurs. This means that the inverter can flow a balanced current without changing phase, even if the phase of the grid voltage is suddenly changed because of two-phase short circuit.

In addition, to flow a balanced current, the trapezoidal base voltage (v_{tuvw}) is added to the output of the current control (Δv_{iuvw}). The v_{tuvw} is calculated by using cos3θ, which is calculated with cosθ, and each amplitude of three trapezoidal base voltages is changed according to the grid voltage (v_{suvw}).

Finally, the sum of the v_{tuvw} and Δv_{iuvw} ($v_{iuvw}*$) is used in a pulse width modulation unit, which generates gate signals of the inverter by comparing a triangle-wave carrier.

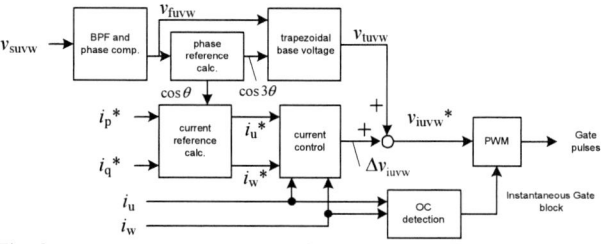

Fig. 2. Inverter current control block diagram for enabling fault ride through capability

Fig. 3 shows experimental results of the small scaled model (5 kW) of the power conditioner. During the three-phase short-circuit, 15 % of the grid voltage remains. When the grid voltage increases or decreases, the inverter current increases. However, the instantaneous gate block keeps the inverter current below an over current level. Thus, the inverter can achieve the FRT requirement.

In addition, during the grid fault (two-phase short-circuit), the inverter current is balanced as shown in Fig. 4, even if the grid voltage is unbalanced. This verifies the inverter current control proposed in this paper works well.

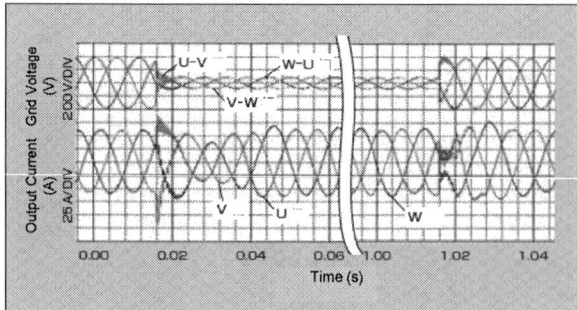

Fig. 3. Experimental result during three-phase short-circuit

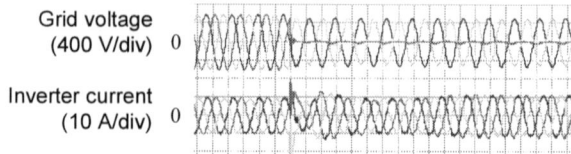

Fig. 4. Experimental result during two-phase short-circuit

B. Active power control

Usually, the power conditioner receives the active power reference from a grid controller. The grid controller is responsible for a battery management, grid frequency compensation and grid voltage compensation. The power conditioner generates the active power according to the reference, when the battery voltage is inside the specified range, which depends on the characteristics of the battery. When the battery is outside the range, the power conditioner rejects the active power reference as shown in Fig. 5. If the DC voltage is near the lower output limit or upper output limit, active power is limited to nearly zero. Furthermore, if the DC voltage reaches under voltage level or upper voltage level, the power conditioner stops the operation.

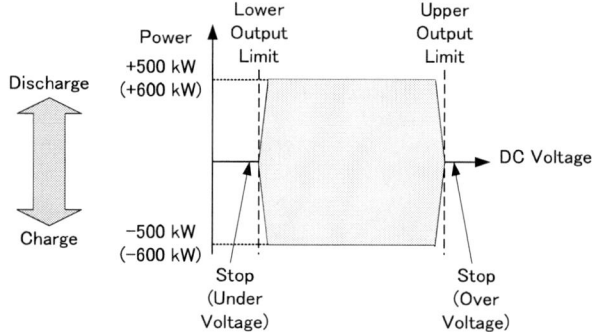

Fig. 5. Active power profile vs. DC input voltage

C. Reactive power control

To stabilize the grid voltage, the power conditioner possesses two reactive power control modes.

1) Constant power factor mode

The power conditioner changes the reactive power according to the active power and the power factor. The power factor can be set from Lead 0.8 to Lag 0.8.

2) Variable reactive power mode

In this mode, the power conditioner generates the reactive power according to the reactive power reference from the grid controller. The grid controller gives the reference as an analog current. If the apparent power according to the reference is beyond the capacity of the power conditioner, the reactive power is limited

D. Idling stop mode

A wind turbine sometimes stops due to week wind. During this time, a power conditioner does not need to output. However, the conventional power conditioner continued switching due to the lifetime of the magnet contactor shown in Fig. 7. This switching deteriorated the stand-by losses.

This control mode is developed to reduce the stand-by losses. During this mode, magnet contactors do not open, and the switching of the power conditioner stops. Thus, the stand-by loss occurs only in the AC filter. The effect of this control will be discussed in the prototype test section.

III. POWER CONDITIONER

A. Specifications

Table I shows the specifications of the power conditioner. The differences between the two models are capacity, AC voltage and DC voltage range. The two models are selected according to the battery voltage range.

The outlook of the power conditioner is shown in Fig. 6. From the bottom of the right-side cabinet, the DC cable is connected to the DC input. And, the AC cable is connected through the bottom of the left-side cabinet. On the center cabinet, the operator can monitor the power conditioner by using touch panel and operate by using the ON/OFF switch.

TABLE I
SPECIFICATIONS

Model	PVI650-3/500	PVI800-3/600
Capacity	500 kVA	600 kVA
AC voltage	210 V ± 10 %	270 V ± 10 %
DC voltage during operation	345 V – 650 V	480 V – 800 V
Frequency	50 / 60 Hz (± 6 %)	
Power factor	Lead 0.8 to Lag 0.8	
THD	Less than 5 %	
Size	W 2400 mm, D 900 mm, H 1950 mm	
Weight	2000 kg	
Ope. Temp.	-5 °C to 40 °C	

Fig. 6. Outlook of the power conditioner

B. Circuit configuration

The circuit configuration is shown in Fig. 7. The power conditioner has a DC input, a DC breaker, two power units, two magnet conductors and an AC breaker. Each power unit is controlled by using each output current and the all PWM pulses of the units are generated by using one carrier. When the power conditioner starts operation, the all units generate the AC voltage in phase with the grid voltage. After that, the magnet conductors close. By this start up method, the surge current during start up can be reduced.

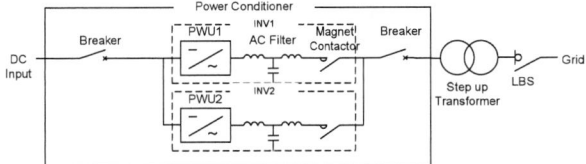

Fig. 7. Circuit configuration of the power conditioner

The power unit and LCL filter of the inverter is shown in Fig. 8. The capacity of this unit is 250 kVA in case of PVI650-3/500. In the power unit, the AT-NPC inverter is configured by using AT-NPC IGBT modules [7]. To enlarge the capacity of the power unit, five AT-NPC IGBT modules are connected in parallel for each phase. The differences of the current among four modules are limited less than 10 %.

As shown in Fig. 8, the neutral point of the LCL filter is connected to the DC-link voltage in this inverter. Actually, this inverter is always connected to AC mains through step up transformer. Therefore, the electromagnetic disturbance voltage of the inverter including the transformer is not limited according to the standard (CISPR11). However, EMC problem can be happened in the actual field caused of the coupling capacitance of the transformer. Thus, to maintain the electromagnetic disturbance voltage of the inverter below the limit of Group 2 (> 75kVA), the neutral point of the LCL filter is connected to the DC-link voltage.

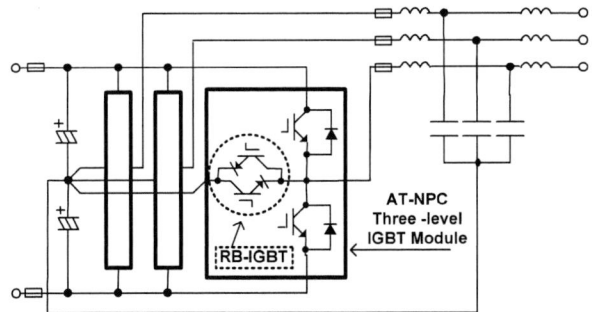

Fig. 8. Converter circuit configuration

IV. THE PROTOTYPE TEST

In this prototype test section, the 500-kVA and 600-kVA prototypes were evaluated. To avoid from describing the similar results, in the 500-kVA prototype test, only the efficiency results are discussed. In the 600-kVA prototype test, the efficiency and transient response are described.

A. 500 kVA Prototype test

In this section, the test results obtained by using a 500 kVA prototype converter are described.

The efficiency curves are shown in Fig. 9 and Fig. 10. The efficiency according to the DC input voltages during charge mode is shown in Fig. 9. In addition, the efficiency during discharge mode is shown in Fig. 10. The efficiency during discharge mode including internal control power source reaches about 97.6 %, in case of minimum DC input voltage. The efficiency during charge mode is about 0.2 % lower than that of the discharge mode. The reason of achieving the highest efficiency in case of DC 345V is that the switching loss becomes minimum and the conduction loss is not changed according to the DC input voltage.

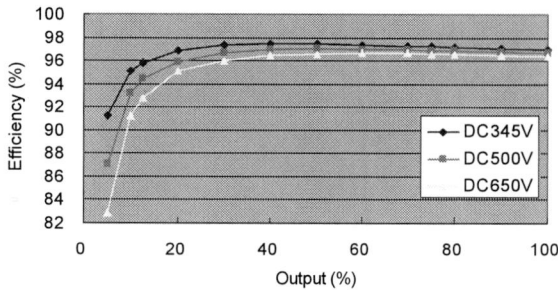

Fig. 9. Efficiency curves in charge mode

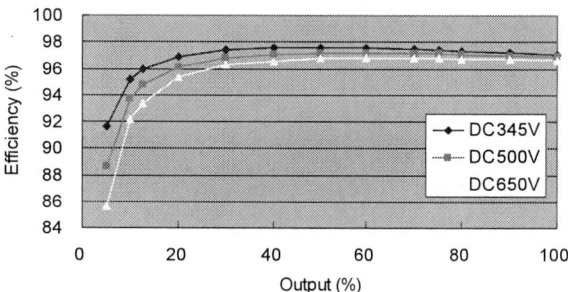

Fig. 10. Efficiency curves in discharge mode

B. 600 kVA Prototype test

In this section, the test results obtained by using a 600 kVA prototype converter are described.

The efficiency curves are shown in Fig. 11 (Charge mode) and Fig. 12 (Discharge mode). The efficiency during discharge mode including internal control power source reaches about 97.8 %, in case of minimum DC input voltage. The efficiency during charge mode is almost equal to the efficiency during discharge mode.

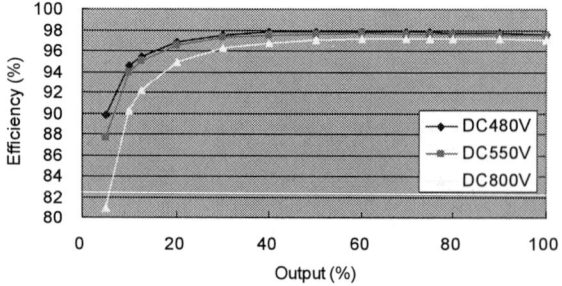

Fig. 11 Efficiency curves in charge mode

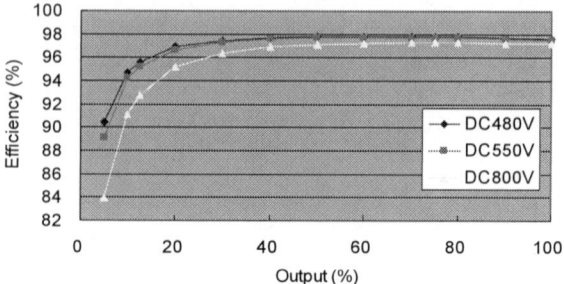

Fig. 12 Efficiency curves in discharge mode

Fig. 13 shows the measurement results of stand-by losses. Fig. 13(a) shows the losses when active and reactive power references are zero. Fig. 13(b) shows the stand-by loss during the idling stop mode. By changing idling stop mode, the stand-by loss is reduced to 5.7%.

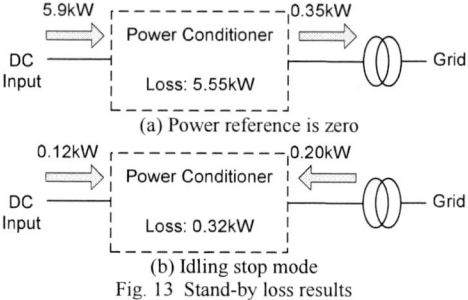

Fig. 13 Stand-by loss results

Fig. 14 shows the transient characteristic according to changing the active power reference. The active power reference is given through the analog current input. In the figure, the output voltages, the output currents and the active power reference are shown respectively. The response time is about 5 ms. This response time satisfies the transient characteristic (less than 30 ms) without any problems.

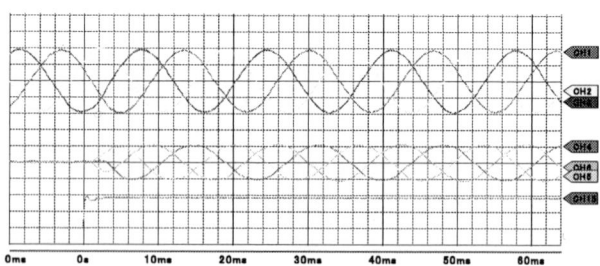

Fig. 14. Output waveforms according to changing active power reference (Ch1-3: Output voltage; 200 V/div, Ch4-6: Output current; 1500A/div, Ch15: Active power reference; 50 %/div)

Fig. 15 and Fig. 16 show the transient waveforms after recovering from idling stop mode to charge mode and discharge mode respectively. In this test, the active power reference is changed to 100 % before canceling the idling stop mode. In case of charge mode, the drop of the grid voltage occurs, but the power conditioner can flow the current properly.

978-1-4799-2706-7/14 $31.00 © 2014 IEEE   1751

The 2014 International Power Electronics Conference

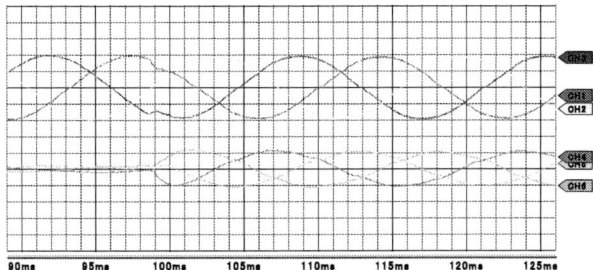

Fig. 15 Output waveforms during changing idling stop mode to charge mode (Ch1-3: Output voltage; 200V/div, Ch4-6: Output current; 1500A/div)

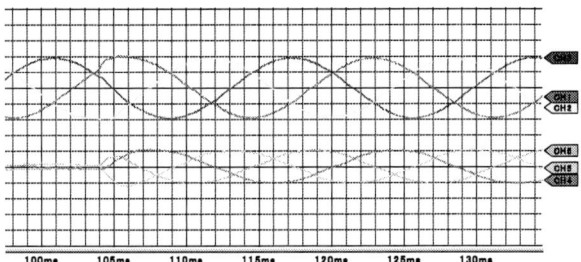

Fig. 16 Output waveforms during changing idling stop mode to discharge mode (Ch1-3: Output voltage; 200V/div, Ch4-6: Output current; 1500A/div)

V. CONCLUSIONS

In this paper, the power conditioner for stabilizing the power disturbance caused of the wind turbine generator is developed. This power conditioner achieves the fault ride through capability by adopting phase references which are calculated from the grid voltage. In addition, the PQ control method and specifications of the developed 500-kVA and 600-kVA power conditioner are described. As the prototype test results of the 500-kVA converter, the efficiency curves during charge mode and discharge mode are shown, and the influence to the efficiency from the DC input voltage is discussed. Furthermore, the efficiency curves of the 600-kVA prototype are described. Finally, the transient response according to the active power reference and transient waveforms after recovering idling stop mode are shown respectively. These waveforms show that the power conditioner possesses the enough performance from the point of view of stabilizing the power disturbance caused of the wind turbine generator.

REFERENCES

[1] M. Kamibayashi, K. Nichols, T. Oshima, "Development update of the NAS battery," Conference Proceedings on Transmission and Distribution Conference and Exhibition 2002, Vol. 3, pp. 1664-1668, 2002

[2] K. Fujii, T. Kikuchi, H. Koubayashi, K. Yoda, "1-MW advanced T-type NPC converters for solar power generation system," Conference Proceedings on Power Electronics and Applications (EPE), pp. 1-10, 2013.

[3] H. Nakazawa, M. Ogino, H. Wakimoto, T. Nakajima, Y. Takahashi, D. H. Lu, "Hybrid isolation process with deep diffusion and V-groove for reverse blocking IGBTs," Conference Proceedings on ISPSD, 2011.

[4] K. Fujii, N. Kanao, T. Yamada, Y. Okuma, "Fault Ride Through Capability for Solar Inverters," Conference Proceedings on Power Electronics and Applications (EPE), pp. 1-9, 2011.

[5] N. Kawakami, Y. Iijima, "Overview of battery energy storage systems for stabilization of renewable energy in Japan," Conference Proceedings on Renewable Energy Research and Applications (ICRERA), pp. 1 - 5, 2012.

[6] Global Wind Report Annual Market Update 2012, http://www.gwec.net/wp-content/uploads/2012/06/Annual_report_2012_LowRes.pdf

[7] K. Komatsu, M. Yatsu, S. Miyashita, S. Okita, H. Nakazawa, S. Igarashi, Y. Takahashi, Y. Okuma,Y. Seki, T. Fujihira, "New IGBT modules for advanced neutral-point-clamped 3-level power converters," Conference Proceedings on Power Electronics Conference (IPEC), pp. 523 - 527, 2010.

978-1-4799-2706-7/14 $31.00 © 2014 IEEE

Gap in pagination due to withheld paper.

Pages 1753-1760

The 2014 International Power Electronics Conference

A Front-to-Front (FTF) System Consisting of Multiple Modular Multilevel Cascade Converters for Offshore Wind Farms

Firman Sasongko, Makoto Hagiwara and Hirofumi Akagi
Department of Electrical and Electronic Engineering
Tokyo Institute of Technology
NE-11, 2-12-1, O-okayama, Meguro, Tokyo, JAPAN
E-mail: akagi@ee.titech.ac.jp

Abstract—**This paper presents a front-to-front (FTF) system based on modular multilevel cascade converters (MMCC). It is called an FTF system because the ac sides of the MMCCs are connected together making a front-to-front configuration via a medium-frequency transformer for voltage matching and galvanic isolation. The system configuration is applicable to dc power collections. Moreover, it is suitable as a power converter for multi-terminal dc power networks since it can handle dc faults inherently without using costly dc circuit breakers. Simulated results using a "PSCAD/EMTDC" software package verify the operating principles and control method of the FTF system.**

Keywords—*Medium-voltage high-power dc-dc converter, modular multilevel cascade converter, multi-terminal dc power network.*

I. INTRODUCTION

Offshore wind energy reserves an enormous potential for future large-scale sustainable energy resources. Offshore winds have several favorable characteristics compared to onshore winds, e.g. higher wind speeds, less turbulence, and large areas availability, thus leading to a higher energy yield and higher capacity factor. Moreover, many problems associated with installation of onshore wind farms, such as acoustic noises, visual impacts, and land conflicts, are less relevant to offshore wind farm projects. Recently, a substantial shift towards more large offshore wind farms has been made [1], [2]. Most offshore wind farms today are less than 30 km from shore and using ac interconnections for power collection as well as transmission to inland grids. Since submarine ac cables can produce large charging current, the upper limits of transmission voltage and distance are restricted. Future offshore wind farms are likely to be further away from shore to increase the size and to reduce visual impacts. However, as the distance from shore increases, the requirements for the wind turbines and their foundations, the transmission distance and its capacity, and also the collection layout and grid interconnection will need to be designed to suit the far offshore conditions [1]–[9].

For longer distance and higher power transmission, high-voltage direct current (HVDC) transmission is a preferred option for future large-scale offshore wind farms. Since the space requirement is critical in offshore installation, an HVDC based on line-commutated converters (LCC-HVDC) is unfavorable because it requires heavy and bulky auxiliary filters. On the other hand, an HVDC based on voltage-sourced converters (VSC-HVDC) offers attractive advantages such as fast control of active and reactive power, fewer auxiliary filters, and reactive power support during grid faults. However, the VSC-HVDC system suffers from higher converter cost and losses. Nevertheless, to comply with grid requirements imposed to the integration of wind farms to grid networks, the VSC-HVDC may be the preferred solution as the converters for offshore power transmissions [2]–[6], [9]–[13]. For high-power and high-voltage applications, compared to two-level or three-level converters, the modular multilevel converter (MMC) is more attractive since it has a modular structure and redundant operation, control and power/voltage rating flexibility, and also low voltage- and current-harmonic contents which comply with the power quality standard [9], [14]–[16].

The concept of using dc power systems for power collection and transmission could be an alternative solution for future offshore wind farms [5], [6], [8], [10], [11], [17]–[19]. The dc power systems show higher reliability and flexibility than the conventional ac power systems. Moreover, the transmissible power is not limited by the transmission length as in the ac power systems. In order to realize dc power transmissions, medium- and high-voltage dc-dc converters are required. Fig. 1 shows an example of a large wind farm configuration based on dc-dc layout. Each power generation unit can be a single connection of a wind turbine (WT) or multiple series-parallel connections of wind turbines (WTs). Furthermore, interconnection of distant renewable offshore power generators to the inland grid via a multi-terminal dc network will improve the system reliability and stability [20]–[22]. Since the multi-terminal dc transmission requires a robust protection system to deal with dc faults, instead of using costly dc breakers, other alternatives have to be found [14], [23], [24]. The conventional MMC is incapable to handle dc faults in such that when a dc-side short circuit occurs, the fault current continues to flow through the anti-parallel diodes across the switching devices. However, the anti-parallel diodes can not withstand the large surge current and should be equipped by using press-pack thyristors in parallel with the diodes. Fast interruption of the

978-1-4799-2706-7/14 $31.00 © 2014 IEEE

The 2014 International Power Electronics Conference

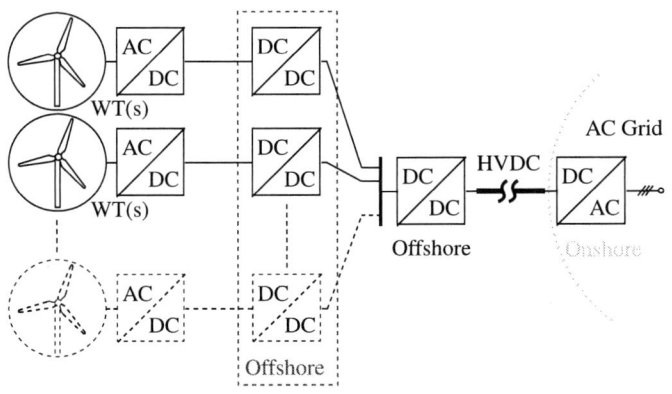

Fig. 1. Large offshore wind farm configuration based on dc-dc layout.

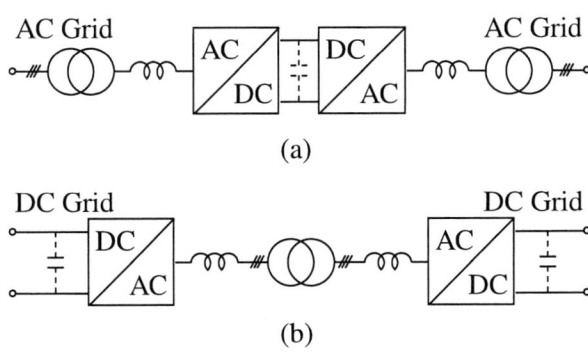

Fig. 2. Basic configurations of back-to-back (BTB) and front-to-front (FTF) systems. (a) BTB system between two ac grids. (b) FTF system between two dc grids.

fault currents is essential for the grid reliability.

This paper presents a front-to-front (FTF) system as a building block for the interconnection of several dc power generation based on multi-terminal HVDC. The dc power collection in one cluster of power generation is achieved by interconnecting several converters at their ac sides, forming front-to-front systems consisting of modular multilevel cascade converters based on double-star chopper-cells (MMCC-DSCC) and double-star bridge-cells (MMCC-DSBC). In this paper the MMCC-DSCC will be called as DSCC while the MMCC-DSBC will be called as DSBC for simplicity. The circuit configuration for an FTF system based on MMCC will be explained in the next section. The FTF system features the flexibility of dc power collection and inherent dc faults protection. The circuit operation of the DSCC is also explained in particular. Later in this paper, the control strategy of the MMCC-based FTF system is presented. The control method for the MMCC will be separated into two categories, the power-flow control and the capacitor-voltage balancing-control. The power-flow control enables a bi-directional power flow for the FTF system while the capacitor-voltage balancing-control ensure the capacitor voltages in all cells maintained at the desired value. Finally, simulation results are presented to verify the proposed circuit configuration and control method for the front-to-front system under steady-state, transient, and fault conditions.

II. MMCC-BASED FTF POWER COLLECTION

The interconnection of two ac grids having different phases or frequencies to improve power system reliability can be fulfilled by using a back-to-back (BTB) system. Fig. 2(a) shows the basic configuration of a BTB system as an interface between two ac networks. The BTB systems can be found in several applications such as HVDC transmissions, frequency changers and asynchronous power-flow controllers.

As opposed to the BTB system, a front-to-front (FTF) system ties two dc grids by using two bi-directional power converters via a transformer. Fig. 2(b) shows the basic configuration for an FTF system. One of the applications of the FTF system is the dc power collection. The dc end of the

converters are connected either to the power-collecting side or transmission side. The transmission-side converter will have a higher dc voltage rating than the collecting-side converter. In this case, the voltage elevation can be done within each converter and/or by adjusting the transformer ratio. For dc transformer applications, only a few literatures address the use of FTF system [24], [25]. However, no literature has discussed the three-phase FTF system configuration using several MMC converters at the collecting side yet.

Several generator types have been used for variable-speed wind turbine applications. The market share of doubly-fed induction generator (DFIG) for wind turbine application is approximately 50%. However it may not be suitable for future high-power offshore application since it needs gearbox, slip rings and brushes which need regular maintenances. It is reported that the low avaibility of wind turbines in offshore wind farms is mainly due to slip-rings and gearbox failures combined with low access to the offshore sites [1], [9]. A direct-drive multi-pole permanent magnet synchronous generator (PMSG) for variable-speed wind turbine offers better reliability and noise reduction since no slip rings and gearbox are used, thus is believed to be suitable for future multi-megawatt offshore generator [9], [26]–[30]. Although the multi-pole PMSG produces less losses and has ride-through capability, it has larger size and weight compared to DFIG. A good compromise between size and reliability may be achieved by using a single-stage gearbox with PMSG which enable the generator to operate at medium-speed operation [9], [27].

The PMSGs are usually equipped with multipulse diode rectifiers with dc choppers, or back-to-back (BTB) converters either using two-level or neutral point clamped (NPC) topology for ac grid interconnection or used in standalone power generation [9], [27], [29]–[33]. For multi-megawatt applications, the NPC converters are the most adopted topology because of their good performances to extract the maximum power from wind as well as to comply with stringent grid requirements. For offshore applications where the power converters should be highly reliable and less maintenance, the diode rectifier may become more preferred option than the VSC-based rectifier. However, the diode rectifier produces high harmonic distortion

978-1-4799-2706-7/14 $31.00 © 2014 IEEE

The 2014 International Power Electronics Conference

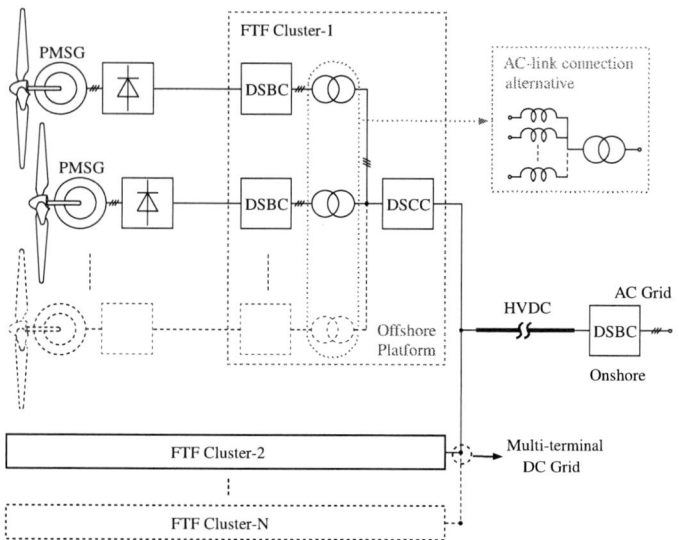

Fig. 3. System configuration overview for offshore wind power collection based on multiple MMCCs.

on the generator side leading to increased generator losses and torque pulsations. Nevertheless, multipulse diode rectifiers have higher efficiency and reliability, and also have lower cost compared to VSC-based rectifiers.

Fig. 3 shows the proposed dc power collection based on a front-to-front system applicable for the interconnection of offshore wind-farms to the inland grids. The power from the wind turbines are collected first and then a large MMCC system steps up the voltage to the transmission level. Offshore medium-voltage dc-dc power converters are not needed since the dc-output voltage from each turbine is directly connected to a large power collection and transmission system. This power collection and transmission design could be advantageous since a two-voltage-level system is employed [6]. The idea of using series connection of wind turbines may improve efficiency and remove completely the offshore platforms [18], [19]. However, the insulation level of the equipments represents a major practical issue for the proposed configuration.

A combination of a multi-pole permanent magnet synchronous generator (PMSG) with a three-phase six-pulse diode rectifier for dc power generation shows a promising candidate for offshore wind farm applications. The multi-pole permanent magnet generator enables the variable-speed power generation to operate at its maximum power coefficient over a wide range of wind speed, while at the same time achieving reliability improvement and reducing maintenance expenses since the gearbox and slip-ring parts can be eliminated. Moreover, although a variable dc voltage will be produced by the three-phase six-pulse diode rectifier, the FTF system can still extract maximum power from wind by varying the input power based on the speed signal from the generator [10], [30].

The proposed system consists of a transmission-side converter and power-collecting-side converters. For offshore wind farm applications, the system could be located in the offshore platform several kilometers away from the wind farm location.

The proposed system collects the distributed dc power from wind turbines and transmits the power to the inland grid directly through the HVDC transmission without intermediate converters for medium-voltage elevation. Furthermore, the system offers bi-directional power flow which can provide the wind towers with emergency power from inland grid instead of using diesel generators when on-site maintenance works are needed. The double-star bridge-cells (DSBC) and double-star chopper-cells (DSCC) are applicable to the front-to-front application since they have five-terminal circuits to interconnect the dc terminals to the three-phase ac circuit [34]. However, since the dc-link voltages from wind turbines vary widely depending on wind conditions, the DSBC configuration is more suitable for the power-collecting converter as it can produce a constant ac voltage [35]. Either the DSCC or DSBC configuration can be used for the transmission-side converter. Note that when DSCC converter configuration is utilized, ac circuit breakers should be put at the ac-link of the converter for protective measure. On the contrary, the DSBC can be utilized without the need of ac breakers. Nevertheless, the DSCC configuration offers less switching devices, compared to the DSBC configuration.

The system can be expanded into several clusters which may be located in different areas or platforms. Each cluster can have any number of branches of the power-collecting side (lower-voltage side) which connected at the ac link of the transmission-side converter. Each collecting-side converter may have a different power rating depending on the generated power at that point. With this configuration, the transmission-side converter will have a power rating equal to the total power rating in the collecting side. The frequency of the transformer used here can be 50/60 Hz or higher to reduce the size of the transformer and passive components. Moreover, the galvanic isolation can be achieved by using a single transformer with line inductances or a multi-winding transformer. The flexibility of the proposed configuration is very favorable for interconnecting large offshore wind farms or other distributed power generations using renewable energy sources such as photovoltaic.

For handling dc faults, the proposed system can use the safety-procedure operation of the FTF system with or without the common ac breakers, depending on the DSBC-DSCC or DSBC-DSBC combination. When a dc fault on the transmission side occurs, all converters in each end will be turned off, avoiding short-circuit current to flow into the converters. Since the fault is handled directly by the converter operation, the rating of the ac circuit breaker may be lower than the nominal load current. Moreover, when a dc fault occurs in one of the power-collecting feeders, other healthy feeders may still operate normally if the DSBC configuration is used at the feeders. The DSBC configuration can block fault current without turning off the other converters. This feature improves the reliability of the dc power-collecting system.

Fig. 4 shows the basic circuit diagram of an FTF system using two modular multilevel cascade converters (MMCCs). The MMCC can be seen as the building block converter for each

978-1-4799-2706-7/14 $31.00 © 2014 IEEE   1763

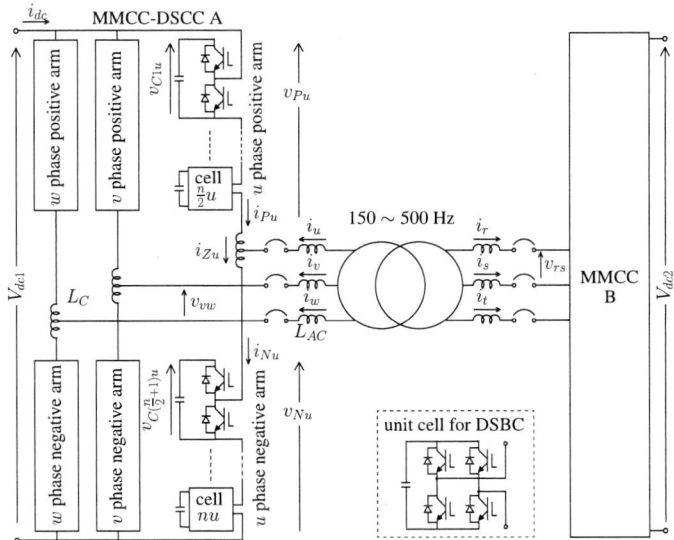

Fig. 4. Circuit diagram of front-to-front system based on two modular multilevel cascade converters (MMCCs).

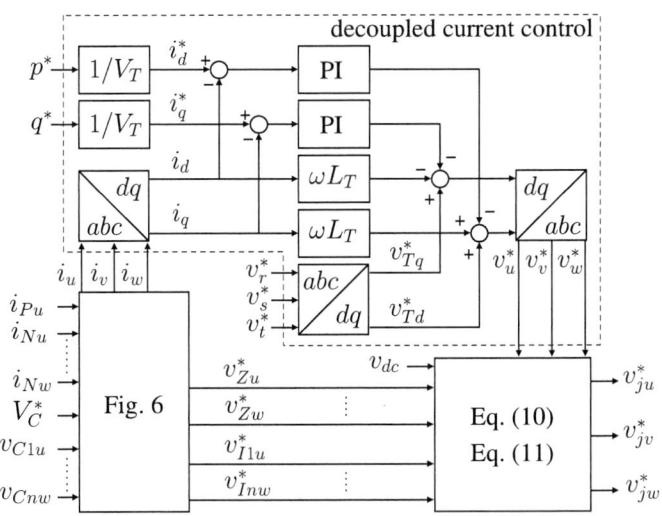

Fig. 5. Overall control block diagram for the collecting-side converters.

FTF cluster in the multi-terminal dc power network. Two three-phase MMCCs are connected together at their ac sides via a transformer for galvanic isolation. Each phase leg composed of series-connected chopper-cell or bridge-cell circuits and a center-tapped inductor. The number of cells in each MMCC can be adjusted to withstand the respective system voltage at their dc sides. In this case, the transmission-side MMCC will have more cells than the collecting-side MMCCs to withstand high-power and high-voltage transmission operation. Moreover, for multiple collecting-side MMCCs operation, the number of cells in each converter may differ depending on the power rating of each converter. Since the ac-side voltages nearly resemble sinusoidal waveforms for large number of the cells, the interface inductances can be made smaller, thus reducing the size and cost. The coupled inductances and the interface inductances can be reduced further by applying a higher frequency operation.

In this paper, only the DSCC circuit configuration will be further discussed, since the DSBC circuit configuration and control system are almost the same [35]. The DSCC was firstly introduced as modular multilevel converter (MMC) [36] and then named as modular multilevel cascade converter double-star chopper-cells (MMCC-DSCC) [34] as one of the family members of the MMCC which applicable for grid interconnections and motor drives. Each DSCC has three legs and two arms per leg for three-phase implementation. The output voltages for the upper and lower arms in the u-phase leg can be expressed as

$$v_{Pu} = \frac{V_{dc}}{2} - v_u, \tag{1}$$

$$v_{Nu} = \frac{V_{dc}}{2} + v_u, \tag{2}$$

where V_{dc} and v_u are the dc and u-phase output voltage respectively. The upper- and lower-arm voltages for v-phase and w-phase can be expressed in a similar way. Within one

leg of the DSCC, dc and ac currents will flow in each arm and can be written as

$$i_{Pu} = i_{Zu} - \frac{i_u}{2}, \tag{3}$$

$$i_{Nu} = i_{Zu} + \frac{i_u}{2}, \tag{4}$$

where i_{Zu} and i_u are the circulating and u-phase output current respectively. The arm currents for v-phase and w-phase can be derived in a similar manner. Equations (3) and (4) show that the arm currents consist of two independent variables, i.e., the circulating current and the output phase current. Since a center-tapped inductor is used instead of two single inductors in each leg, the voltage drop caused by the ac components of the arm currents will cancel each other out [34].

III. CONTROL STRATEGY

A. Power Flow Control

Similar to other VSC topologies, the DSCC can act as an inverter or a rectifier, thus capable of injecting or absorbing power to or from the grid respectively. The power flow will be controlled indirectly by controlling the currents in rotating dq frame, i.e., controlling the d-axis current for active power and the q-axis current for reactive power respectively. The decoupled current control as depicted in Fig. 5 is implemented and the control response depends on the gain parameters used. Since this control method needs a voltage reference for synchronization, the DSCC converter at the transmission side should produce a voltage reference for all converters using power control at the collecting side.

The ac voltage command in the transmission-side DSCC can be written as

$$v_r^* = \sqrt{2}V_T \sin \omega t, \tag{5}$$

$$v_s^* = \sqrt{2}V_T \sin(\omega t - \frac{2\pi}{3}), \tag{6}$$

$$v_t^* = \sqrt{2}V_T \sin(\omega t + \frac{2\pi}{3}), \tag{7}$$

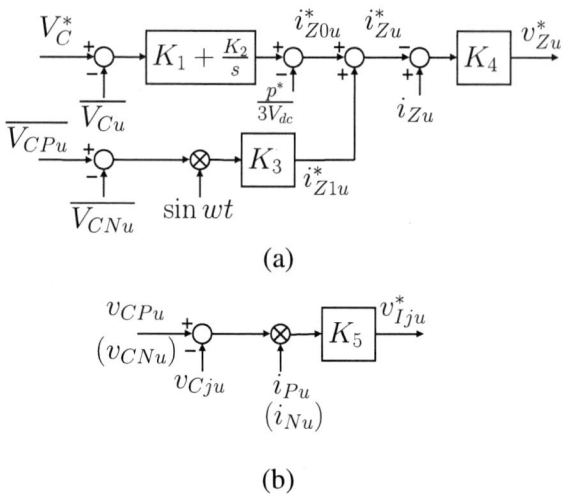

(a)

(b)

Fig. 6. Capacitor voltage balancing control block diagram.

where V_T is the line-to-line rms voltage at the ac-link of transmission-side DSCC and ω is the desired ac frequency of the ac-link. By using these voltage commands as a feedforward control for the decoupled current control, only the line currents need to be acquired for feedback purpose without the need of a phase-locked loop (PLL) and ac voltage sensors.

B. Capacitor Voltage Balancing Control

Controlling the capacitor voltages for the modular multi-level converter comprising many floating dc capacitors is one of the common issues to be solved. The capacitor voltage in each individual cell may deviate from the desired value because of the disturbances such as parasitic resistances, transients, and harmonics. An appropriate control method should be applied to regulate the capacitor voltages.

Since the upper- and lower-arm voltages and currents have dc and ac components as expressed in (1)-(4), the upper- and lower-arm average power will also have dc and ac part as follows:

$$P_{Pu} = \int_{t_0}^{t_0+T} v_{Pu} i_{Pu}\, dt$$
$$= \frac{VI\cos\theta}{2\sqrt{3}} + \frac{V_{dc}I_{Z0}}{2} - \frac{VI_{Z1}}{\sqrt{3}}, \quad (8)$$

$$P_{Nu} = \int_{t_0}^{t_0+T} v_{Nu} i_{Nu}\, dt$$
$$= \frac{VI\cos\theta}{2\sqrt{3}} + \frac{V_{dc}I_{Z0}}{2} + \frac{VI_{Z1}}{\sqrt{3}}, \quad (9)$$

where V, I, and θ are the line-to-line rms voltage, the line rms current and the power factor angle respectively, while I_{Z0} and I_{Z1} are dc and fundamental ac component of the circulating current respectively. The first term of each equation above is related to ac power, the second term is related to dc power and the third term is related to the circulating power in each arm. Note that the dc power within one leg is directly related to the dc component of the circulating current. The total dc

power is equal to the summation of dc powers in all legs. By controlling the dc term of the circulating current, we can regulate the average capacitor voltage in each leg [37].

Because of the circuit topology nature, the ac power will fluctuate between the upper and lower arms and can cause imbalance in each arm capacitor voltage. This problem can be solved by controlling the third term of (8) and (9), i.e., by controlling the fundamental component of the circulating current which has the same phase as the output voltage. Fig. 6(a) shows the control block diagram for balancing arm average capacitor voltages.

To ensure each capacitor voltage within one arm is equal to the average value, the individual capacitor control has to be applied. When the arm current flows into the cell, the capacitor will be charged if the cell is activated. Likewise, when the arm current flows out from the cell, the capacitor will be discharged if the cell is activated. The arm current will not flow into or out from the capacitor if the cell is turned off, thus no charging/discharging condition occurs. By using these condition states as a rule, the capacitor voltage in each cell can be maintained at the desired value. Fig. 6(b) shows the individual-balancing-control block diagram.

The overall control for all cell output voltages in the upper arm of u-phase (for $j : 1 \sim \frac{n}{2}$) is given by

$$v_{ju}^* = v_{FF} - \frac{2v_u^*}{n} + v_{Zu}^* + v_{Iju}^*. \quad (10)$$

With similar derivation, all cell output voltages in the lower arm of u-phase (for $j : \frac{n}{2}+1 \sim n$) is given by

$$v_{ju}^* = v_{FF} + \frac{2v_u^*}{n} + v_{Zu}^* + v_{Iju}^*, \quad (11)$$

where v_{FF} is the feedforward control to maintain the voltage at the dc side which is equal to V_{dc}/n; v_u^* is the power control output; v_{Zu}^* is the circulating-current-control output; and v_{Iju}^* is the individual-control output. Each cell output voltage command should then be divided by the corresponding cell voltage to produce the modulation for each cell. Fig. 5 shows the overall control block diagram for each DSCC. Note that for transmission-side converter, the voltage-reference command should be used instead of the decoupled current control.

IV. FTF SYSTEM PERFORMANCES

This paper uses a front-to-front system using three DSCC-based converters shown in Fig. 4 for simulation circuit configuration. The power-collection side uses two DSCCs while the transmission-side has a single DSCC. The DSBC-DSCCs combination can also be applied with minor changes. Table I summarizes the circuit parameters with all the dc-side voltages to be 13.2 kV and 6.6 kV/150 Hz on the ac link. In the actual system, the rated power of the transmission-side converter should be equal to the total rated power of all collecting-side converters. The ac-link frequency could be decided based on a good compromise between switching losses and ac-link harmonic contents. The higher ac-link frequency will produce

978-1-4799-2706-7/14 $31.00 © 2014 IEEE

TABLE I.   CIRCUIT PARAMETERS FOR SIMULATION

Parameter	Symbol	Value
Rated power	P	10 MW
Nominal dc voltage	V_{dc}	13.2 kV
Ac-link voltage reference	V_S^*	6.6 kV
Ac-link frequency reference	f_S^*	150 Hz
Transformer voltage ratio		1 : 1
Cell count per leg	n	16
Dc capacitor	C	3 mF
Dc capacitor voltage reference	V_C^*	1.65 kV
Unit capacitance constant	H	20 ms at 1.65 kV
Ac-link inductor	L_{AC}	0.37 mH (8%)
Center-tapped inductor	L_C	1.1 mH (23.8%)
Switching method		Phase-shifted PWM
PWM carrier frequency	f_C	1350 Hz
Equivalent switching frequency	nf_C	21.6 kHz
Dead time		4 μs

on a three-phase 6.6-kV, 10-MW, 150-Hz base

the higher switching losses but also less harmonic contents. In the following discussions, only 150 Hz of frequency is used for the system-operation verification under several conditions. The phase-shifted sinusoidal pulse-width modulation (PWM) technique is applied and 16 triangular-carrier signals with the frequency f_C of 1350 Hz are phase-shifted each other by 22.5^o. With this technique, all chopper-cells will have equal switching and conduction power losses. Note that the equivalent switching frequency for each leg is 21.6 kHz for 1350-Hz carrier frequency and 16 carrier-signals operation. The simulation is conducted by using "PSCAD/EMTDC" software package and a fully digital control method is implemented with a dead time of 4 μs.

A. System Performance Under Steady-State

In the first case, the system works under steady-state condition with both collecting-side converters deliver power to the transmission-side converter with unequal power distribution with the total of 9 MW at the transmission-side converter. Fig. 7 shows simulated waveforms for FTF system with three DSCCs under steady-state condition. Both the collecting-side DSCCs use decoupled current control with the transmission-side DSCC as the reference. The results show that the line currents distribution in all three converters has the same ratio as the power command for both power directions. Note that the currents i_{u1} and i_{u2} from the collecting-side DSCCs have the same phase while the current i_r from the transmission-side DSCC is 180° out of phase from the other two. The individual capacitor voltage in each cell within one arm is well balanced. The capacitor voltage ripple in each converter has different magnitude due to the power distribution difference.

B. System Performance Under Transient

Power-command alterations for an offshore wind farm are needed to extract the maximum power from the varying wind speed. The power command should follow the maximum-power curve based on the corresponding wind speed. In

transient case, the system runs under power transient from one of the collecting-side converters. The power flow from one of the collecting-side converters increases from 2 MW to 4 MW in 10 ms and finally to 3 MW resulting in power transient on the transmission-side converter from 7 MW to 9 MW and finally to 8 MW. Figs. 8 shows simulated waveforms for a transient operation in the FTF system. As we can see from the results, the line currents follow the power transient smoothly. The average capacitor voltage in each arm within one leg is balanced although the power-flow transients happen. Moreover, the individual capacitor voltage in each cell within one arm has only a slight variation from the arm average capacitor voltage. From these results, it is obvious that the capacitor balancing control is very effective to maintain the capacitor voltages within acceptable values under steady-state and transient conditions.

C. System Performance Under DC Fault

DC short circuits may occur both on the transmission side or the power-collection feeders. In the case of a dc fault occurrence on the transmission side, all the converters in the FTF system should be turned off to stop the fault currents to flow from the wind farm. When dc faults occur on one or several of the power-collection feeders, the corresponding collecting-side converters, in this case only the DSBC can be used, should be turned off to allow the other healthy feeders to continue the operation. Note that the transmission-side converter (either DSCC or DSBC) can still be operating unaltered. However, when the DSCC configuration is used for collecting-side converters, all the converters including the transmission-side converter need to be turned off for safety measure.

In this section, only the FTF system using DSCCs is simulated under a dc fault on one of the collecting-side feeders. Figs. 9 shows simulated waveforms for the FTF system performance under a dc fault occurrence. The simulated waveforms show that when a dc fault occurs on the feeder 2 at $t = 0$, the FTF system can act fast to block the fault current. Since all of the ac-link voltages are zero when the fault occurs, there is no power flow in the FTF system. The overshoot on a dc current from the faulty feeder is occurred because of a discharging process of the reserve energy on the coupled inductors within the converter. This current amplitude depends on the impedances within the converter arms and the dc cable impedance from the converter to the fault location. The discharging time and current-overshoot amplitude will need to be compromised by adjusting the values of the coupled inductors, thereby the ac-link frequency, to get the permissible amplitude of fault current. Moreover, the dc currents on the other converters are blocked almost instantaneously without overshoot as in the faulty feeder. Note that the dc capacitor voltages are slightly different from the reference value after the fault.

The 2014 International Power Electronics Conference

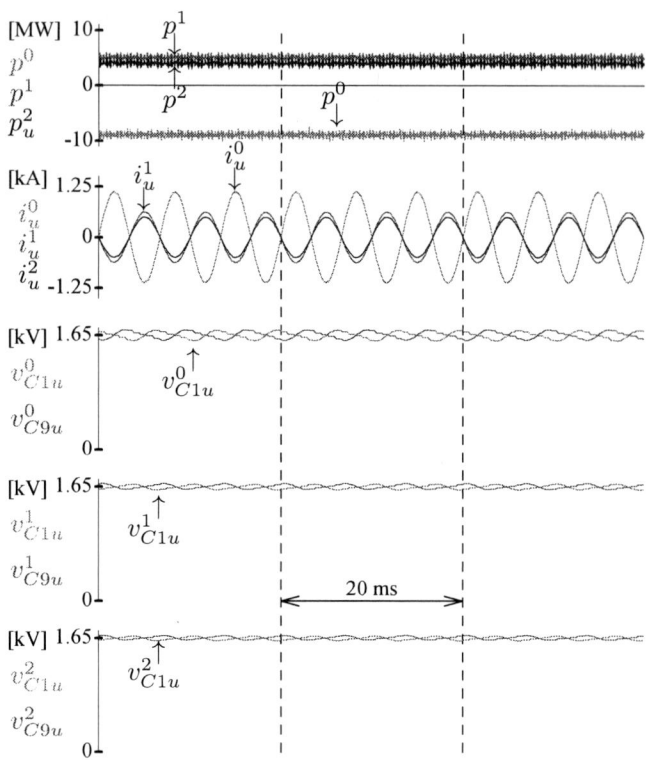

Fig. 7. Simulation results for the transmission-side (superscript 0) and the collecting-side DSCCs (superscript 1 and 2) under steady-state condition.

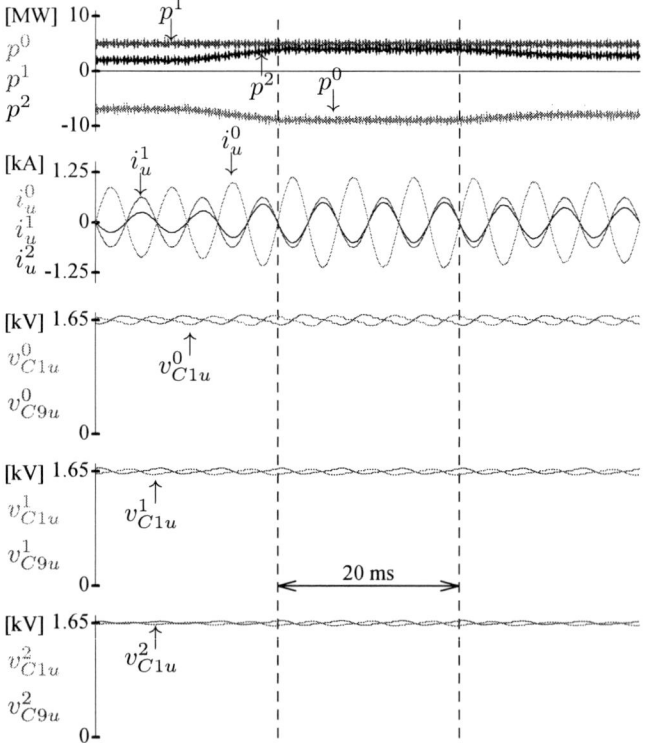

Fig. 8. Simulation results for the transmission-side (superscript 0) and the collecting-side DSCCs (superscript 1 and 2) under transient condition.

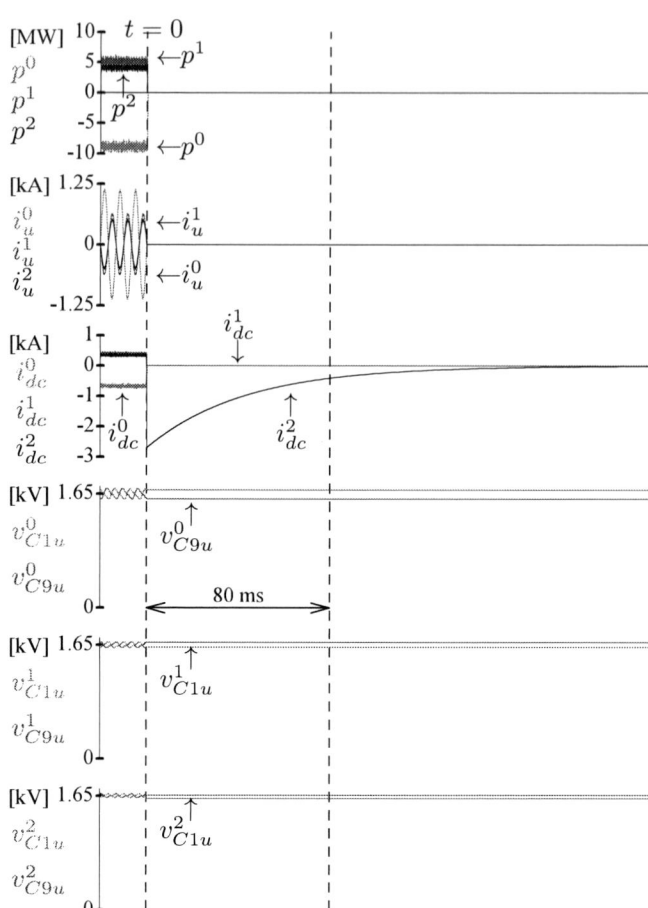

Fig. 9. Simulation results for the transmission-side (superscript 0) and the collecting-side DSCCs (superscript 1 and 2) under fault condition.

V. CONCLUSIONS

A front-to-front (FTF) system based on modular multilevel cascade converters (MMCC) topology has been presented in this paper. The system configuration can be implemented for a medium-/high-voltage high-power dc power collection. A dc-dc power converter based on FTF system has been proposed for multi-terminal dc power network applications. Several DSCCs or DSBCs can be connected together at their ac sides to produce a higher total output power to be transmitted. The simulated results show the effectiveness of the system configuration and control system under steady and transient conditions for dc power collection. The power sharing in the collecting-side MMCCs can easily be controlled by implementing the decoupled current control in the collecting-side MMCCs while the transmission-side MMCC acts as a voltage reference. Furthermore, the system can handle dc faults inherently by turning off the operation of the converters, thus leading to a fast fault protection without the need of dc circuit breakers. To maintain the capacitor voltages in all cells within each converter, the capacitor balancing control is indispensable requirement for stable operation of the converters.

978-1-4799-2706-7/14 $31.00 © 2014 IEEE

REFERENCES

[1] J. Kaldellis and M. Kapsali, "Shifting towards offshore wind energyrecent activity and future development," *Energy Policy*, vol. 53, pp. 136–148, Feb. 2013.

[2] I. Erlich, F. Shewarega, C. Feltes, F. W. Koch, and J. Fortmann, "Offshore wind power generation technologies," *Proceedings of the IEEE*, vol. 101, no. 4, pp. 891–905, Apr. 2013.

[3] T. Ackermann, "Transmission systems for offshore wind farms," *IEEE Power Engineering Review*, pp. 23–27, 2002.

[4] P. Bresesti, W. L. Kling, R. L. Hendriks, and R. Vailati, "HVDC connection of offshore wind farms to the transmission system," *IEEE Transactions on Energy Conversion*, vol. 22, no. 1, pp. 37–43, Mar. 2007.

[5] P. McKeever, "Next generation HVDC network for offshore renewable energy industry," *10th IET International Conference on AC and DC Power Transmission (ACDC 2012)*, pp. 11–11, 2012.

[6] C. Meyer, M. Höing, A. Peterson, and R. W. D. Doncker, "Control and design of dc grids for offshore wind farms," *IEEE Transactions on Industry Applications*, vol. 43, no. 6, pp. 1475–1482, 2007.

[7] G. Quinonez-Varela, G. Ault, O. Anaya-Lara, and J. Mcdonald, "Electrical collector system options for large offshore wind farms," *IET Renewable Power Generation*, vol. 1, no. 2, pp. 107–114, 2007.

[8] H. J. Bahirat, B. A. Mork, and H. K. Hø idalen, "Comparison of wind farm topologies for offshore applications," *Power and Energy Society General Meeting, 2012 IEEE*, pp. 1–8, 2012.

[9] F. Blaabjerg and K. Ma, "Future on power electronics for wind," *IEEE Journal of Emerging and Selected Topics in Power Electronics*, vol. 1, no. 3, pp. 139–152, 2013.

[10] J. Robinson, D. Jovcic, and G. Joós, "Analysis and design of an offshore wind farm using a MV dc Grid," *IEEE Transactions on Power Delivery*, vol. 25, no. 4, pp. 2164–2173, 2010.

[11] F. Deng and Z. Chen, "Operation and control of a dc-grid offshore wind farm under dc transmission system faults," *IEEE Transactions on Power Delivery*, vol. 28, no. 3, pp. 1356–1363, 2013.

[12] N. M. Kirby, L. Xu, M. Luckett, and W. Siepmann, "HVDC transmission for large offshore wind farms," *Power Engineering Journal*, vol. 16, no. 3, pp. 135–141, Jun. 2002.

[13] V. G. Agelidis, G. D. Demetriades, and N. Flourentzou, "Recent advances in high-voltage direct-current power transmission systems," *IEEE International Conference on Industrial Technology, 2006. ICIT 2006.*, pp. 206–213, Dec. 2006.

[14] A. M. Abbas and P. W. Lehn, "PWM based VSC-HVDC systems - a review," *PES '09. IEEE Power & Energy Society General Meeting, 2009.*, pp. 1–9, Jul. 2009.

[15] J. Glasdam, J. Hjerrild, L. H. Kocewiak, and C. L. Bak, "Review on multi-level voltage source converter based HVDC technologies for grid connection of large offshore wind farms," *IEEE International Conference on Power System Technology (POWERCON) 2012*, pp. 1–6, Oct. 2012.

[16] S. Kouro, M. Malinowski, K. Gopakumar, J. Pou, L. G. Franquelo, B. Wu, J. Rodriguez, M. A. Pérez, and J. I. Leon, "Recent advances and industrial applications of multilevel converters," *IEEE Transaction on Industrial Electronics*, vol. 57, no. 8, pp. 2553–2580, 2010.

[17] S. Lundberg, "Evaluation of wind farm layouts," *EPE Journal*, vol. 16, pp. 14–21, 2006.

[18] N. Holtsmark, H. J. Bahirat, M. Molinas, B. A. Mork, and H. K. Hø idalen, "An all-dc offshore wind farm with series-connected turbines : an alternative to the classical parallel ac model ?" *IEEE Transaction on Industrial Electronics*, vol. 60, no. 6, pp. 2420–2428, 2013.

[19] E. Veilleux and P. W. Lehn, "Interconnection of direct-drive wind turbines using a series-connected dc grid," *IEEE Transactions on Sustainable Energy*, vol. 5, no. 1, pp. 139–147, 2014.

[20] W. Lu and B.-T. Ooi, "Premium quality power park based on multi-terminal HVDC," *IEEE Transactions on Power Delivery*, vol. 20, no. 2, pp. 978–983, Apr. 2005.

[21] L. Xu, B. W. Williams, and L. Yao, "Multi-terminal dc transmission systems for connecting large offshore wind farms," *2008 IEEE Power and Energy Society General Meeting - Conversion and Delivery of Electrical Energy in the 21st Century*, pp. 1–7, Jul. 2008.

[22] J. Zhu and C. Booth, "Future multi-terminal HVDC transmission systems using voltage source converters," *45th International Universities Power Engineering Conference (UPEC) 2010*, pp. 1–6, Aug. 2010.

[23] R. Marquardt, "Modular multilevel converter topologies with dc-short circuit current limitation," *8th International Conference on Power Electronics ECCE Asia*, pp. 1425–1431, Jun. 2011.

[24] S. Kenzelmann, A. Rufer, M. Vasiladiotis, D. Dujic, F. Canales, and Y. de Novaes, "A versatile dc-dc converter for energy collection and distribution using the modular multilevel converter," *Proceedings of the 2011-14th European Conference on Power Electronics and Applications (EPE 2011)*, pp. 1–10, Aug. 2011.

[25] S. Kenzelmann, D. Dujic, F. Canales, Y. de Novaes, and A. Rufer, "Modular dc/dc converter: comparison of modulation methods," *15th International Power Electronics and Motion Control Conference (EPE/PEMC) 2012*, pp. LS2a.1–1–LS2a.1–7, Sep. 2012.

[26] E. Spooner and a.C. Williamson, "Direct coupled, permanent magnet generators for wind turbine applications," *IEE Proceedings - Electric Power Applications*, vol. 143, no. 1, p. 1, 1996.

[27] M. Liserre, R. Cárdenas, M. Molinas, and J. Rodríguez, "Overview of multi-MW wind turbines and wind parks," *IEEE Transaction on Industrial Electronics*, vol. 58, no. 4, pp. 1081–1095, 2011.

[28] H. Polinder, J. A. Ferreira, B. B. Jensen, A. B. Abrahamsen, K. Atallah, and R. A. Mcmahon, "Trends in wind turbine generator systems," *IEEE Journal of Emerging and Selected Topics in Power Electronics*, vol. 1, no. 3, pp. 174–185, 2013.

[29] M. Chinchilla, S. Arnaltes, and J. C. Burgos, "Control of permanent-magnet generators applied to variable-speed wind-energy systems connected to the grid," *IEEE Transactions on Energy Conversion*, vol. 21, no. 1, pp. 130–135, Mar. 2006.

[30] E. Haque, M. Negnevitsky, and K. M. Muttaqi, "A novel control strategy for a variable-speed wind turbine with a permanent-magnet synchronous generator," *IEEE Transactions on Industry Applications*, vol. 46, no. 1, pp. 331–339, 2010.

[31] A. Faulstich, J. K. Steinke, and F. Wittwer, "Medium voltage converter for permanent magnet wind power generators up to 5 MW," *2005 European Conference on Power Electronics and Applications*, pp. 1–9, 2005.

[32] J. Wang, D. D. Xu, B. Wu, and Z. Luo, "A low-cost rectifier topology for variable-speed high-power PMSG wind turbines," *IEEE Transactions on Power Electronics*, vol. 26, no. 8, pp. 2192–2200, 2011.

[33] S.-h. Song, S.-i. Kang, and N.-K. Hahm, "Implementation and control of grid connected ac-dc-ac power converter for variable speed wind energy conversion System," *APEC '03. Eighteenth Annual IEEE Applied Power Electronics Conference and Exposition, 2003.*, vol. 00, no. C, 2003.

[34] H. Akagi, "Classification , terminology , and application of the modular multilevel cascade converter (MMCC)," *IEEE Transactions on Power Electronics*, vol. 26, no. 11, pp. 3119–3130, Apr. 2011.

[35] N. Thitichaiworakorn, H. Akagi, and M. Hagiwara, "Experimental verification of a modular multilevel cascade inverter based on double-star bridge-cells," *IEEE Transactions on Industry Applications*, vol. 50, no. 1, pp. 509–519, Jan. 2014.

[36] A. Lesnicar and R. Marquardt, "An innovative modular multilevel converter topology suitable for a wide power range," *IEEE Bologna PowerTech Conference*, 2003.

[37] H. Fujita, M. Hagiwara, and H. Akagi, "Power flow analysis and dc-capacitor voltage regulation for the MMCC-DSCC," *IEEJ Transactions on Industry Applications*, vol. 132, no. 6, pp. 659–665, Dec. 2012.

Gap in pagination due to withheld paper.

Pages 1769-1774

The 2014 International Power Electronics Conference

Modelling, Design and Control of Grid Connected Converter for High Altitude Wind Power Application

Jeevan Adhikari, *Student Member, IEEE*, Akshay K. Rathore, *Senior Member, IEEE* S K Panda, *Senior Member, IEEE*
Department of Electrical and Computer Engineering, National University of Singapore, Singapore

Abstract—**High altitude wind based renewable energy generating system can be connected to a distribution level grid. The generated power at high altitude above the ground is transmitted at medium voltage DC to the ground based station. Thus, transmitted power is interfaced with the distribution grid at the ground station. This paper presents the power electronic converter (PEC) rated at 100 kW HAWP application that converts medium voltage DC to three phase distribution level grid voltage. The proposed converter topology consists of a neutral point clamped (NPC) three level DC-DC converter followed by three phase grid connected two level inverter. The designed power electronic converter uses four high voltage (HV) rating power semiconductor switches for buck converter before inversion to three phase AC distribution voltages. The active and passive components selection for two stage conversion is presented in the paper. The grid side current is controlled using quadrature axes current control method and inverter switches are switched using space vector modulation (SVM) technique. Simulations of the proposed PEC and control of the inverter are carried out using software programs PSIM-9 and MATLAB. The designed converter converts the 8 kV DC transmission voltage to 415 V grid side voltage with current total harmonic distortion (THD) of about 1.2%.**

Key Words: High altitude wind power (HAWP), three level neutral point clamped (NPC) DC-DC converter, grid-connected inverter, total harmonic distortion (THD)

I. INTRODUCTION

Wind and solar energy are two major renewable energy sources those have potential to reduce the number of fossil-fuel based power generating system. Solar power has low energy density and conventional wind power (CWP) harvesting system has low capacity factor and needs huge infrastructure constructions [1], [2]. High altitude wind power (HAWP) harvesting system generates wind power at low cost and high capacity factor [3], [4]. HAWP harvesting system using light gas filled blimp/aerostat is enlightened in [3], [5]. Fig 1 shows a prototype of blimp supported HAWP generation system developed by Altaeros Energy [6]. A light weight airborne wind turbine drives a permanent magnet synchronous generator (PMSG) at high altitude above the ground. Thus, the generated power is transformed into medium voltage DC for efficient transmission and to minimize the weight of the power transmission cable. An electro-mechanical tether is used to transmit power to the ground based station where it is interfaced with a distribution grid. Power electronic converter (PEC) topology for harvesting HAWP using blimp/aerostat is

illustrated in Fig. 2. The PEC topology consists of a rectifier and a DC-DC converter in an air-borne unit and a grid connected converter at the ground based station. Detail study of DC-DC converters for blimp supported HAWP application is explained in [7], [8].

Fig. 1: A prototype of HAWP generating system developed by Altaeros Energy [6]

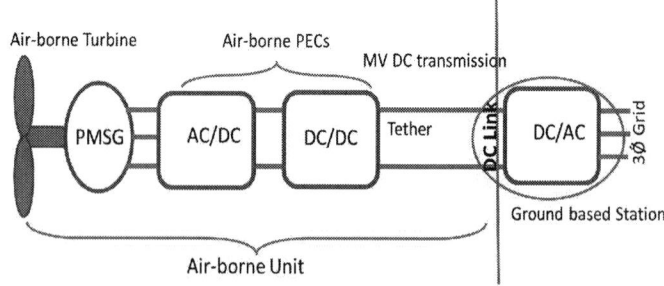

Fig. 2: Electrical architecture for harvesting high altitude wind power generating system

Various inverter topologies and their design, modelling and limitations are explained in [9]- [10] for industrial use. The multi-level inverters explained in [9]- [10] are used to drive high power motor drive system, compressor pumps and other electrical loads. Grid connected inverters used for wind farms are described in [11]- [14]. In [11]- [14], and the PECs designed for different power level for on-shore/off-shore wind farms. The close loop current control using different modulation techniques for the designed converter are explained as

978-1-4799-2706-7/14 $31.00 © 2014 IEEE

well. However, grid connected converter for HAWP has not been explored yet. In contrast to conventional inverters, the converters for HAWP should be capable of converting medium voltage DC to three phase distribution level grid voltages. For 100 kW HAWP application, optimal transmission voltage is 8 kV [4] which gives maximum power-to-weight (P/W) ratio for an air-borne unit. Transmission voltage at the ground station acts as DC link voltage for the converter which is interfaced to distribution grid voltage at 415 V (RMS line-line). A HAWP harvesting system considered in the paper is a distributed renewable generation system connected to the distribution grid.

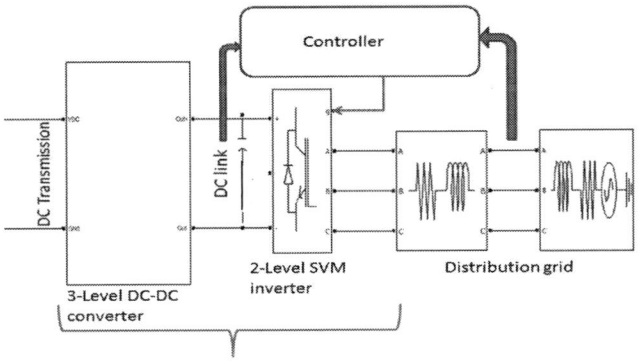

Fig. 3: The converter topology for interfacing HAWP to distribution grid

The paper introduces the converter that transforms 8 kV DC link voltage to 415 V grid voltage connected to three phase power distribution system. The proposed converter consists of three level zero voltage switching (ZVS) isolated DC-DC converter and two level SVM inverter. The isolation provided by DC-DC converter protects the grid side power system from unwanted power signals from lightening. The converter consists of four high voltage switches due to the use of three level isolated buck converter instead of using isolated full bridge DC-DC buck converter. Two level inverter is used to transform the output of 3-level buck converter into 3-phase AC with controlled grid side current. The inverter is switched using SVM technique where total harmonic distortion (THD) of the grid side current lies within industrial standard. The paper explains the design of the converter, selection of devices depending on the rating of the switches and control of grid connected inverter to get current distortion with in the permissible limits. Fig. 3 shows the proposed converter for HAWP application for grid interface.

The description and design of the converter are given in Sections II and III. Sections IV and V summarize the modeling and the control of the converter. Switching strategy for the inverters with space vector modulation is explained in Section VI. Simulation results to verify the design of the converter is shown in Section VII.

II. DESCRIPTION OF THE CONVERTER

The specifications of the ground based converter are:

Power Rating: 100 kW
Input Voltage: 8000 V (MV DC transmission voltage)
Intermediate DC link voltage to inverter: 700 V
Output Voltage: 415 V (L-L RMS)

The proposed converter consists of multilevel isolated buck converter followed by two level inverter. Transforming 8000 V to 700 V using an isolated full bridge DC-DC converter requires series connection of the switches (with high voltage rating) in the primary side in order to withstand high input voltage. The series connections of switches (high voltage rating) reduce the reliability of the converter and increase switching and conduction losses. In addition, higher rating switches limit the switching frequency of the converter. So, three level ZVS based isolated buck converter is better choice which reduces the switch voltage stress to half of the input voltage and increases the switching frequency due to soft switching characteristics. A three level isolated buck converter as explained in [17] is used for step down operation as shown in the Fig. 4.

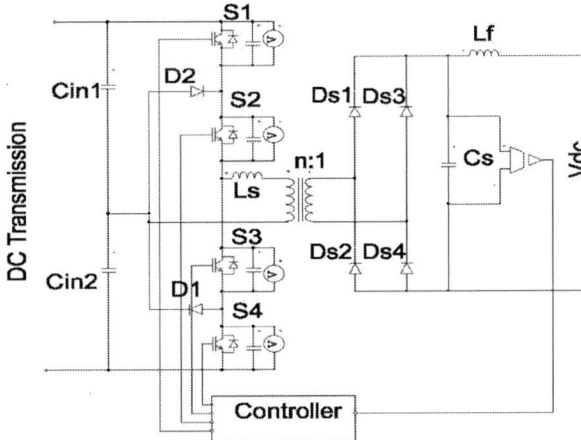

Fig. 4: NPC three level DC-DC converter for HAWP

The input voltage to the inverter is reduced to 700 V using multilevel buck converter. This allows the use of two level SVM inverter which gives reduced THD, increased switching frequency and reduced filtering requirements. Different multilevel inverters are explained in [9]- [10] which suffer hardware design difficulty, voltage balancing issues etc. Two level inverter has reduced numbers of switches, capacitors and diodes, so use of multilevel inverter is skipped by reducing DC link voltage to a lower value. Schematic circuit diagram for two level inverter for HAWP application is shown in Fig. 5.

III. DESIGN OF THE CONVERTER

This section gives ratings of switches, diodes and passive components for three level DC-DC converter and two level inverter. Table I gives the components' ratings for DC-DC converters for buck operation. The devices selected for step down operation are shown in Table II. The use of three level topology reduces the switch stress and capacitor voltage stress to half. The complete buck operations using multilevel

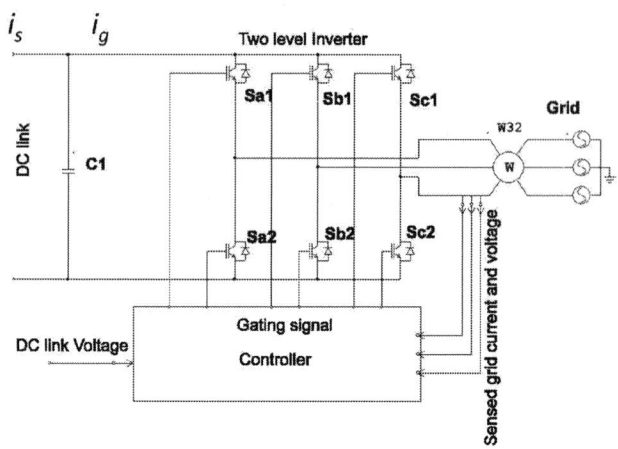

Fig. 5: Two level inverter for interfacing HAWP to distribution grid

topology is explained in [17]. The conduction of antiparallel diode just before turning on the switch facilitates ZVS of all HV switches.

The DC-DC converter uses four Infenion IGBT (FZ200R65KF2) and two split input capacitors. The three level neutral point clamped topology eliminates series connection of HV switches. Similar step down operation can be done using full bridge isolated DC-DC converter which requires eight (4*2) FZ200R65KF2 switches with HF transformer.

TABLE I: Device rating for multilevel isolated DC-DC buck operation

Parameters	Rating	Parameters	Rating
Switch voltage (V)	4000	Rectifier diode voltage (V)	700
Peak switch current (A)	85	Rectifier diode peak current (A)	280
Av. switch current (A)	35	Rectifier diode av. current (A)	160
Clamp diode voltage (V)	4000	Filter capacitor voltage (V)	700
Av. clamp diode current (A)	25	Filter capacitance (mF)	13.3
Peak clamp diode current (A)		Inductor peak current (V)	85
Maximum voltage across inductor (V)	8000	HF transformer KVA	105
		Input capacitor voltage (V)	4000
Leakage inductance (uH) Filter	332	Input capacitance (uF)	625
Turns ratio	3.42:1	inductance (mH)	3

Two level inverter interfaces 700 V dc link voltage to 415 V AC grid. The inverter consists of 3 legs with 2 switches in each leg. Ratings of the switches are given in Table III along with the selected modules' details. Infenion fast IGBT modules are selected for the inverter application.

IV. MODELING OF GRID SIDE INVERTER

The schematic diagram of grid side converter is shown in Fig 3. Modelling of the grid side converter is carried out by modelling the DC link side and grid connected inverter.

TABLE II: Name of switch and diode selected multilevel isolated DC-DC buck operation

	Primary Side	Secondary Side
IGBT Module	FZ200R65KF2	
Voltage	6600 V	
Clamp Diode	DD250S65K3	
Voltage	6500 V	
Rectifier diode		BSM300GA120DN2
Voltage		1200 V

TABLE III: Two level inverter devices ratings

	Inverter switch
Voltage (V)	700
RMS current (A)	135
Peak current (A)	184
Selected module	FS200R12KT3

DC link capacitor stabilizes the DC link voltage. Current flowing through the DC link capacitor is given by:

$$C\frac{dv_{dc}}{dt} = i_s - i_g \qquad (1)$$

where v_{dc} is DC link voltage, i_s is current of DC link after buck converter, i_g is current of DC link at the grid side and C is the DC link capacitance.

Considering that the grid side inverter does not consume power, eqn 1 can be expressed as:

$$C\frac{dv_{dc}}{dt} = \frac{P_s}{v_{dc}} - i_g \qquad (2)$$

where P_s is generated output power by an air-borne electric generator which is equal to electromagnetic power ($P_e = T_e w_s$), T_e is electrical torque of the generator, w_s is angular frequency of rotation of the air-borne turbine.

The voltage equation per phase between the inverter leg and the grid is given by:

$$e_a = v_a + iR_a + L_a\frac{di_a}{dt} \qquad (3)$$

where e_a is grid side phase voltage, v_a is output voltage of one of the leg of inverter, L_a and R_a are equivalent series inductance and resistance between grid and inverter leg per phase.

Similar voltage equations can be written for phase-b and phase-c as expressed in eqn. 3. Three rotating vectors can be transformed into two stationary vectors by $\alpha - \beta$ transformation. The stationary vectors are transformed into rotating $d - q$ frame by park transformation. Three phase grid voltages (e_a, e_b and e_a) are represented by single rotating voltage e_s, which rotates at angle θ along the rotating frame. e_s is aligned along d-axis of rotating frame which results the q-axis component of rotating voltage zero. The $d - q$ transformation of three voltage equations results the following equations:

$$v_d = L_d \frac{di_d}{dt} + R_s i_d - w_r L_q i_q + e_s \qquad (4)$$

$$v_q = L_q \frac{di_q}{dt} + R_s i_q + w_r L_d i_d \qquad (5)$$

where v_d and v_q are quadrature axes voltages of the inverter, L_d and L_q are the inductances in $d-q$ axes, i_d and i_q are the currents in $d-q$ frame, w_r is angular frequency of grid.

Active power flow through the inverter is controlled by d-axis current control and reactive power is controlled by q-axis current. The equations for active (P) and reactive (Q) power are given by:

$$P = 1.5(V_d I_d + V_q I_q) = 1.5 V_d I_d \qquad (6)$$

$$Q = 1.5(V_d I_q - V_q I_d) = 1.5 V_d I_q \qquad (7)$$

Thus, P can be controlled by I_d and Q can be controlled by I_q as expressed in eqns. 6 and 7.

V. CONTROL OF GRID SIDE INVERTER

The converter consists of two stages: buck stage and inversion stage. The buck operation is controlled to maintain DC link voltage to reference DC link voltage determined by the generated power by HAWP system. The grid side inversion is controlled using quadrature current control method and it is carried out for the following purposes:

1) Regulation of DC link voltage which value should be greater than L-L grid voltage. DC link voltage is controlled by controlling d-axis current.
2) Control of reactive power flowing into the grid by controlling the q-axis current.

Grid side phase voltages are sensed and fed into phase lock loop (PLL) to get the orientation of the rotating voltage vector. Three grid side currents are converted into corresponding $d-q$ axis currents using the angle obtained from PLL. Using quadrature axes currents; quadrature axes voltages V_d and V_q are calculated as shown in Fig 6 and 7.

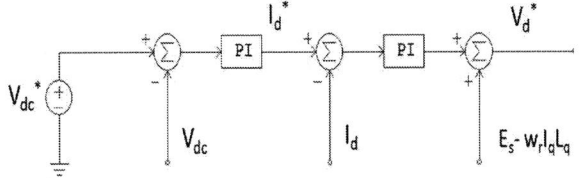

Fig. 6: Control block diagram of d-axis current generating V_d

The quadrature axis converter voltages are converted into $a-b-c$ frame by inverse park transformation. Three reference voltages V_a, V_b and V_c are used to generate six switching signals using space vector techniques.

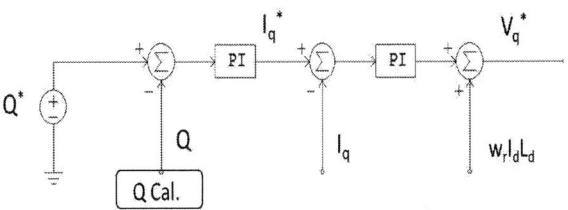

Fig. 7: Control block diagram of q-axis current generating V_q

VI. SWITCHING OF GRID SIDE TWO LEVEL INVERTER

Space vector modulation has been extensively used for high power and high voltage application. SVM utilizes the DC link voltage better than sine PWM method of switching and generates low current ripple. Moreover, implementation of SVM in digital signal processor (DSP) is easier than other modulation techniques [15]- [16]. In two level inverter, eight different switching vectors are possible (six active vectors and two zero vectors).

VII. SIMULATION RESULTS

The 3-level NPC DC-DC converter of the proposed grid connected converter allows zero voltage switching of the IGBT switches ($S_1 - S_4$) due to the conduction of anti-parallel diode ($D_1 - D_4$) just before turning on IGBTs. The ZVS operations of switches are illustrated in Fig. 8. Soft-switching characteristic enables the converter to operate at higher switching frequency that reduces the size of magnetic used. The DC-DC converter transforms 8 kV transmission voltage into 700 V DC link voltage as demonstrated in Fig. 9.

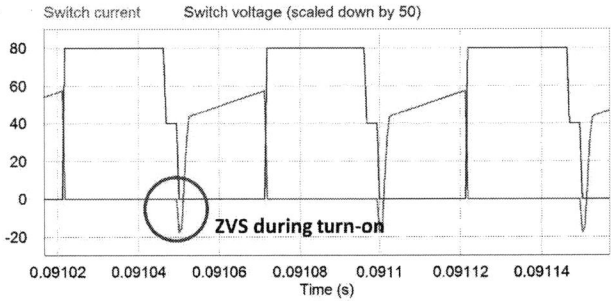

Fig. 8: Switch transient of three level DC-DC converter

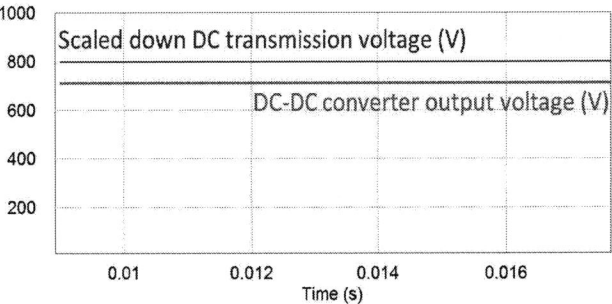

Fig. 9: Input/output voltages of three level DC-DC converter

The 2014 International Power Electronics Conference

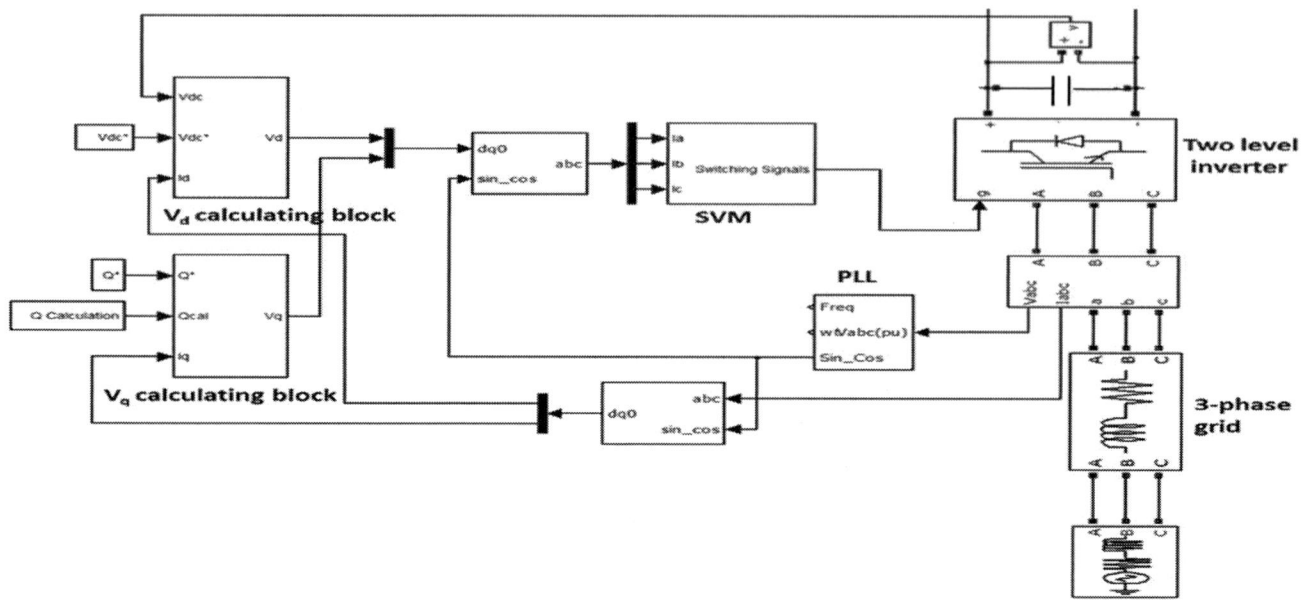

Fig. 10: Complete control diagram of grid side inverter in HAWP application

The output of the DC-DC converter is fed to an inverter for interfacing HAWP with distribution grid. The complete block diagram for interfacing the grid and inverter is shown in Fig 10. Two level inverter is controlled using $d-q$ current control strategy and switched using SVM technique. Fig. 11a shows the three phase sinusoidal voltage at grid side while slightly distorted grid side current is pointed out in Fig 11b. However, injected grid side current is roughly sinusoidal.

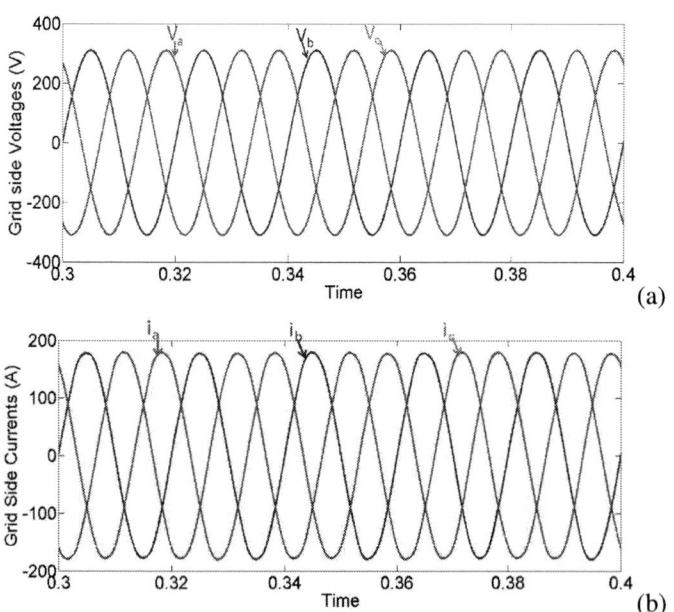

Fig. 11: Grid side phase voltage and phase current of grid connected inverter for HAWP

The converter switching signals are controlled to maintain unity power factor of the drawn current as shown in the Fig 12. However, reactive power can be injected into the grid

changing the reference current I_q in the controller loop when grid demands reactive power from the converter.

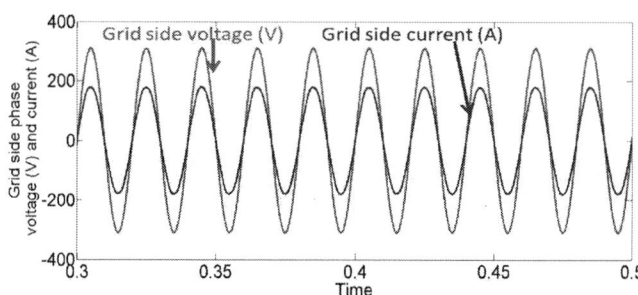

Fig. 12: Converter injecting the power into the grid at unity power factor

Fig 13 and 14 show active and reactive power drawn by the distribution grid. The grid side rotating voltage phasor V_s is aligned along with V_d. So, the reactive power extracted by the grid is zero.

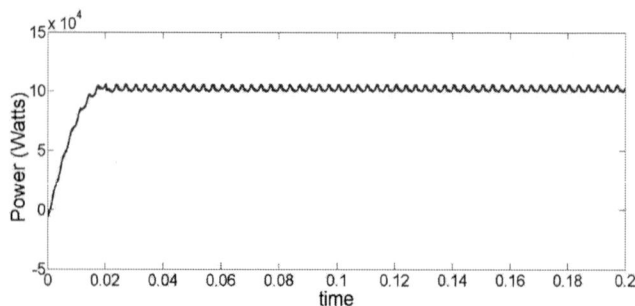

Fig. 13: Active power flowing to the grid from HAWP system

Since the HAWP generating system is interfaced into the distribution grid, the converter should not inject non-linear

978-1-4799-2706-7/14 $31.00 © 2014 IEEE   1779

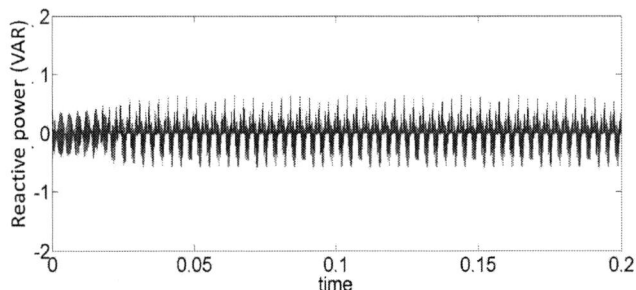

Fig. 14: Reactive power flowing to the grid from HAWP system

current into the grid. The proposed converter controls the injected current harmonics with in the industrial standard limit. Fig 15 shows the harmonic spectrum of grid side current. Fifth, seventh and eleventh harmonics are injected into the grid current, but the value of total harmonic distortion is within the industrial limit. The value of RMS phase current drawn by the grid is approximately 130 A with total harmonic distortion of 1.2%.

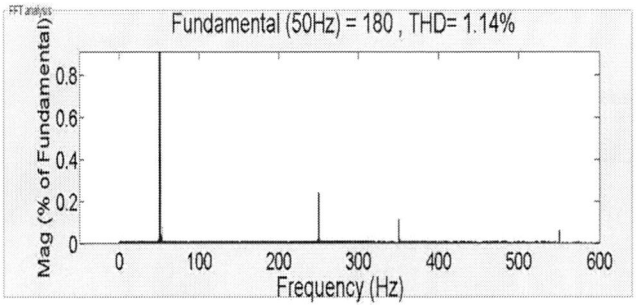

Fig. 15: THD of the grid side current

VIII. CONCLUSION

HAWP generating systems can be interfaced into the distribution grid as a source of renewable generation. The converter proposed in the paper is used to connect 8 kV DC transmission line of HAWP system to a 415 V distribution grid. The paper has presented the design of the converter with cascaded isolated three level DC-DC converter followed by two level space vector modulated inverter. Design and description of 3-level DC-DC converter and 2-level inverter has been carried out in Sections II and III. Current and voltage ratings of the active and passive devices for the complete converter are listed. In addition, for the calculated ratings suitable power semiconductor device modules are itemized in the paper. The converter uses four high voltage rating switches during buck operating. The use of buck converter facilitates the use of low voltage rating IGBT switches for two level inversion operation. The inverter is controlled using quadrature current control method and switched using SVM techniques. The proposed converter delivers 100 kW active power to the grid at unity power factor with grid side current THD of 1.2%.

REFERENCES

[1] Renewables 2011 and 2012 Global status Report, REN21 Renewable Energy Policy Network for the 21st Century

[2] G. M. Masters, Renewable and Efficient Electric Power Systems (Wiley, 2004).

[3] Adhikari, J.; Panda, S.K.; Rathore, A.K., "Harnessing high altitude wind power using light gas filled blimp," Industrial Electronics Society, IECON 2013 - 39th Annual Conference of the IEEE , vol., no., pp.7163,7168, 10-13 Nov. 2013

[4] Kolar, J.W.; Friedli, T.; Krismer, F.; Looser, A.; Schweizer, M.; Steimer, P.; Bevirt, J., "Conceptualization and multi-objective optimization of the electric system of an Airborne Wind Turbine," Industrial Electronics (ISIE), 2011 IEEE International Symposium on , vol., no., pp.32,55, 27-30 June 2011

[5] White, N.; Tierno, N.; Garcia-Sanz, M., "A novel approach to airborne wind energy: Design and modeling," Energytech, 2011 IEEE , vol., no., pp.1,6, 25-26 May 2011

[6] http://www.altaerosenergies.com/

[7] Adhikari, J.; Rathore, A.K.; Panda, S.K., "Comparison of ZVS based isolated DC-DC converters for high altitude wind power application," Innovative Smart Grid Technologies - Asia (ISGT Asia), 2013 IEEE , vol., no., pp.1,6, 10-13 Nov. 2013

[8] Adhikari, J.; Rathore, A.K.; Panda, S.K., "Modular interleaved ZVS current fed isolated DC-DC converter for harvesting high altitude wind power," Industrial Electronics Society, IECON 2013 - 39th Annual Conference of the IEEE , vol., no., pp.7187,7192, 10-13 Nov. 2013

[9] Abu-Rub, H.; Holtz, J.; Rodriguez, J.; Ge Baoming, "Medium-Voltage Multilevel ConvertersState of the Art, Challenges, and Requirements in Industrial Applications," Industrial Electronics, IEEE Transactions on , vol.57, no.8, pp.2581,2596, Aug. 2010

[10] Malinowski, M.; Gopakumar, K.; Rodriguez, J.; Perez, M.A., "A Survey on Cascaded Multilevel Inverters," Industrial Electronics, IEEE Transactions on , vol.57, no.7, pp.2197,2206, July 2010

[11] Xie Lei; Xie Da; Zhang Yanchi, "Three-level inverter based on direct power control connecting offshore wind farm," Sustainable Power Generation and Supply, 2009. SUPERGEN '09. International Conference on , vol., no., pp.1,6, 6-7 April 2009

[12] Qiang Zhang; Lewei Qian; Chongwei Zhang; Cartes, D., "Study On Grid Connected Inverter Used in High Power Wind Generation System," Industry Applications Conference, 2006. 41st IAS Annual Meeting. Conference Record of the 2006 IEEE , vol.2, no., pp.1053,1058, 8-12 Oct. 2006

[13] Vilathgamuwa, D.M.; Jayasinghe, S. D G; Madawala, U.K., "Space vector modulated cascade multi-level inverter for PMSG wind generation systems," Industrial Electronics, 2009. IECON '09. 35th Annual Conference of IEEE , vol., no., pp.4600,4605, 3-5 Nov. 2009

[14] Chongming Qiao; Smedley, K.M., "Three-phase grid-connected inverters interface for alternative energy sources with unified constant-frequency integration control," Industry Applications Conference, 2001. Thirty-Sixth IAS Annual Meeting. Conference Record of the 2001 IEEE , vol.4, no., pp.2675,2682 vol.4, Sept. 30 2001-Oct. 4 2001

[15] Keliang Zhou; Danwei Wang, "Relationship between space-vector modulation and three-phase carrier-based PWM: a comprehensive analysis [three-phase inverters]," Industrial Electronics, IEEE Transactions on , vol.49, no.1, pp.186,196, Feb 2002

[16] Fei Wang, "Sine-triangle versus space-vector modulation for three-level PWM voltage-source inverters," Industry Applications, IEEE Transactions on , vol.38, no.2, pp.500,506, Mar/Apr 2002

[17] Pinheiro, J.R.; Barbi, I., "The three-level ZVS PWM converter-a new concept in high voltage DC-to-DC conversion," Industrial Electronics, Control, Instrumentation, and Automation, 1992. Power Electronics and Motion Control., Proceedings of the 1992 International Conference on , vol., no., pp.173,178 vol.1, 9-13 Nov 1992

Practical Study of a High Step-down Converter

Masahito Jinno[†]   Hong-Wei Su[†]   Jiung-Lin Tsai[†]   Hirofumi Matsuo[††]

[†]Department of Electrical Engineering
I-Shou University
Kaohsiung, Taiwan

[††]Nagasaki University
Nagasaki, Japan

Abstract— In order to face the coming era of DC power supplying, it is needed to establish the technology of high step-down converter for the electrical appliances' control circuits, especially for dynamic appliances like air conditioner, refrigerator and freezer.

For a forward converter suitable for high step-down[1]~[2], mathematical derivations, including steady-state analysis in time domain and static characteristic analysis in frequency domain, are performed for establishing precise design criteria. Also for the converter, the theoretical analysis is verified by the experiment. Furthermore, voltage stress of component is improved for practical use of this converter.

Keywords— *DC power supplying, high step-down, forward converter, voltage stress*

I. INTRODUCTION

In recent years, owing to the speedy rise of energy conservation awareness and the encouragement policies of government to development of renewable energy, new energy, such as PV, wind, fuel cell, etc., is rapidly developed. On the other hand, thinking of traditional AC power supplying has gradually changed to that of DC power supplying for reducing transmission loss. Accordingly in the developed countries, DC micro-grid linking new energy is experimentally introduced to data centers, offices, factories and small areas of residence. These DC micro-grids will become popular in the near future, and will probably construct the complete DC grid in the future.

To cope with DC power supplying, in power electronics field, appropriate DC to DC or DC to AC SMPS(switched mode power supply) for each electronic appliance or electric machine is absolutely needed, as shown in Fig. 1.

In SMPSs, single switch forward converter[3]~[6] is one of the basic circuits, where the switch is generally a power MOSFET. The forward converter, as shown in Fig. 2, is basically buck type with isolation transformer. The circuit configuration of this converter is very simple; besides, the turn ratio of isolation transformer can be chosen to enlarge the effect of step-down. Since the DC voltage for data center is 400V_{DC}, and that for residence is 300V_{DC}, single switch which should sustain such high voltage will cause large conduction loss in conduction state and high voltage stress at switching transient. For high input voltage, forward converter with two MOSFET switches [7]~[8], as shown in Fig. 3, is usually adapted to reduce voltage stress of each MOSFET switch. In this way, MOSFET of lower blocking voltage will be used to reduce conduction loss.

The relation concerning output voltage E_o, input voltage E_i, duty ratio D and turn ratio of transformer N in traditional two-switch forward converter can be expressed as $E_o=(E_i/N)\cdot D$, where $N=N_1/N_2$. In this case, $D\leq0.5$. If high step-down is required to perform with this general forward converter, it is necessary to lower duty ratio or to increase turn ratio of transformer. Lowering duty ratio of MOSFET switch will decrease the utilization of switch and increase the higher harmonics of input current. On the other hand, high turn ratio of transformer is disadvantageous to magnetic coupling, thus larger leakage inductance will cause higher voltage surge.

In this paper, a novel two-switch forward converter is proposed for high step-down solution. In the proposed converter, conversion ratio can be greatly expanded without increasing turn ratio of transformer. For practical use of this converter, expansion of conversion ratio, circuit operation and design criteria are analyzed.

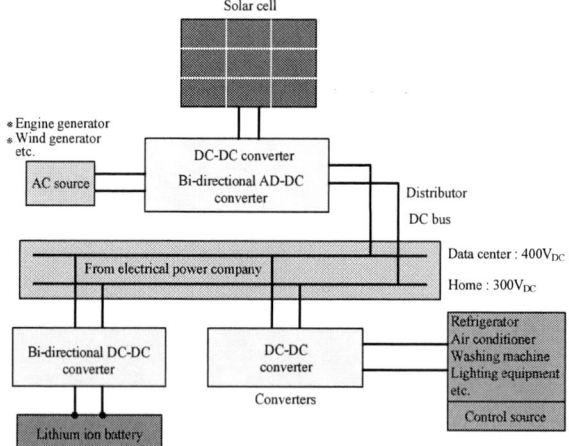

Fig. 1. High voltage DC power distribution system.

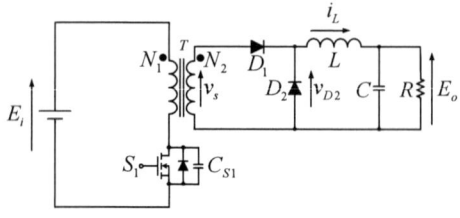

Fig. 2. Single-switch forward converter.

978-1-4799-2706-7/14 $31.00 © 2014 IEEE

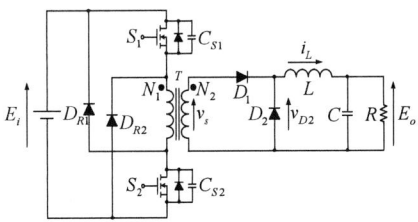

Fig. 3. Traditional two-switch forward converter.

II. OPERATION PRINCIPLE

A. Circuit Configuration and Definitions of Element Symbols

The circuit configuration of the high step-down converter is shown in Fig. 4. Definitions of symbols used in this figure are as follows:

E_i : input voltage
i_i : input current
S_1 : high-side MOSFET switch
S_2 : low-side MOSFET switch
C_{S1}: parasitic capacitance of S_1
C_{S2}: parasitic capacitance of S_2
D_{R1} : diode connected with drain of S_1 for energy recovery
D_{R2} : diode connected with source of S_2 for energy recovery
N_1 : primary winding turns of transformer
N_2 : secondary winding turns of transformer
N : turn ratio of transformer , where $N=N_1/N_2$
v_s : secondary voltage of transformer
D_1 : rectification diode
D_2 : flywheeling diode
v_{D2} : reverse voltage across D_2
L_1 : separate inductance for high step-down
i_{L1} : current flowing through L_1
L_2 : output inductance
i_{L2} : current flowing through L_2
n : inductance ratio , where $n=L_1/L_2$
C : output capacitance
R : load resistance
E_o : output voltage

In this circuit, L_1 and L_2 should be magnetically non-coupled for high step-down. Hence the circuit configuration is different from that of conventional tapped inductor[9]~[10], as shown in Fig. 5.

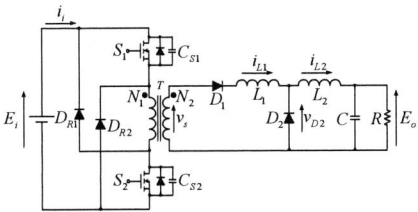

Fig. 4. Proposed two-switch forward converter.

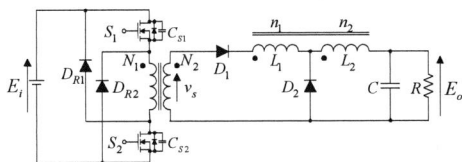

Fig. 5. Two-switch forward converter with tapped-inductor.

B. Circuit Operation

When both S_1 and S_2 are in on-state and i_{L1} is lower than i_{L2}, load is powered by i_{L2} only. As i_{L1} reaches and keeps equal to i_{L2} in on-state of S_1 and S_2, load is powered by v_s through L_1 and L_2. In off-state of S_1 and S_2, energy stored in L_1 and the magnetizing inductance L_{m1} of transformer will recover to E_i through D_{R1} and D_{R2}. And load is powered again by i_{L2}. In off-state of S_1 and S_2 , as i_{L1} decreases to and keeps equal to zero before i_{Lm1} reaches zero, load is still powered by i_{L2}. After magnetizing current i_{Lm1} reaches zero in off-state of S_1 and S_2, D_{R1} and D_{R2} stop conducting, but load is still powered by i_{L2}.

When D_2 conducts, i.e., $i_{L1} \neq i_{L2}$, load is powered by i_{L2}. While in off-state of D_2, i.e., $i_{L1}=i_{L2}$, i_{L2} rises linearly and energy is stored in L_2. The ratio of the duration of $i_{L1}=i_{L2}$ to one switching period will be the effective duty ratio of the converter, denoted as D_{D1} in this paper. It should be noted that D represents duty ratio of S_1 and S_2, while D_{D1} represents effective duty ratio of circuit performance.

III. STEADY STATE AND STATIC ANALYSIS

A. Steady State Analysis

For simplicity of steady state analysis, every element shown in Fig. 4 is thought to be ideal. Equivalent circuits corresponding to their operation states are shown in Fig. 6 to Fig. 10, respectively. As previously mentioned, large N is not preferred to achieve high step-down of converter. Accordingly, for state 3 as shown in Fig. 8, it can be easily derived that L_2V_s will not be smaller than L_1E_o, where $V_s(=E_i/N)$ represents peak value voltage of secondary winding. Key waveforms of the circuit are shown in Fig. 11. In this figure, v_{GS1} and v_{GS2} represent each drive voltage of S_1 and S_2, respectively, while v_{DS1} and v_{DS2} represent each drain-source voltage of S_1 and S_2, respectively.

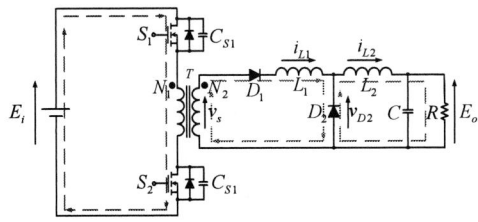

Fig. 6. Equivalent circuit of state 1.

978-1-4799-2706-7/14 $31.00 © 2014 IEEE

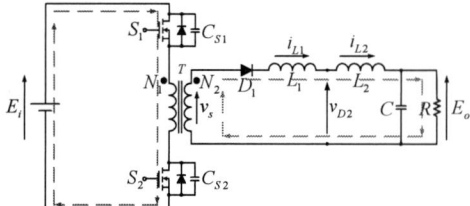

Fig. 7. Equivalent circuit of state 2.

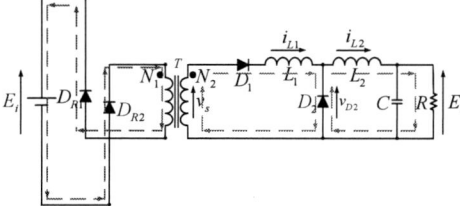

Fig. 8. Equivalent circuit of state 3.

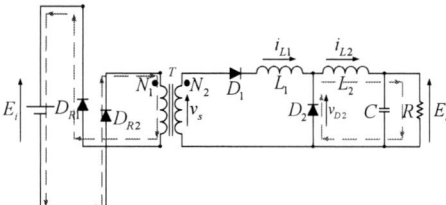

Fig. 9. Equivalent circuit of state 4.

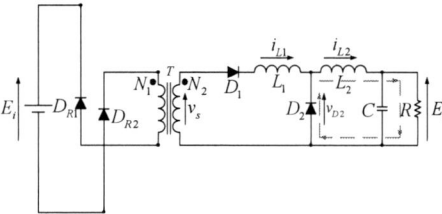

Fig. 10. Equivalent circuit of state 5.

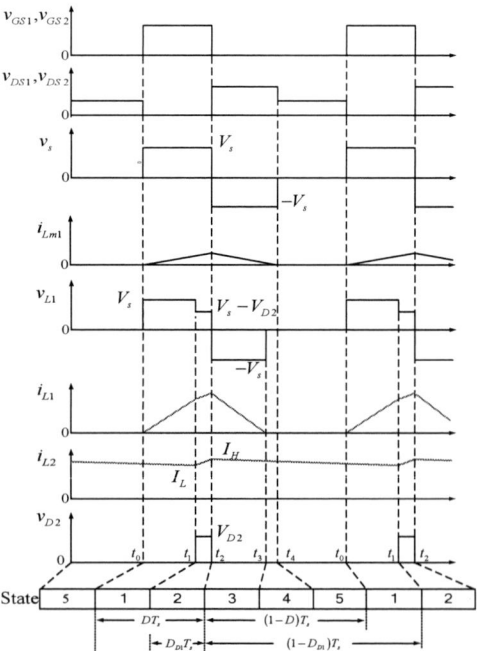

Fig. 11. Schematic diagram of waveform in the proposed converter.

From steady state analysis, the actual duty ratio D_{D1} can be obtained as follows.

$$D_{D1} = \frac{E_o(1+n)}{V_s + nE_o} \qquad (1)$$

In state 2, $v_{D2} = V_{D2}$, V_{D2} can be obtained as follows.

$$V_{D2} = \frac{V_s + nE_o}{1+n} \qquad (2)$$

From Eqs. (1) and (2), E_o can be obtained as follows.

$$E_o = V_{D2} \cdot D_{D1} \qquad (3)$$

The maximum value I_H and the minimum value I_L of i_{L2} can be expressed as follows:

$$I_H = \frac{E_o}{R} + \frac{\Delta I_{L2}}{2} \qquad (4)$$

$$I_L = \frac{E_o}{R} - \frac{\Delta I_{L2}}{2} \qquad (5)$$

where ΔI_{L2} is expressed as follows.

$$\Delta I_{L2} = \frac{V_{D2} \times D_{D1} \times (1 - D_{D1})}{L_2 f_s} \qquad (6)$$

The correlation between D_{D1} and D can be derived as follows.

$$D = \frac{L_1 f_s I_o N}{E_i} - \frac{1}{2n} + \frac{D_{D1}}{2} + \frac{1+n}{2n[1 + n(1 - D_{D1})]} \qquad (7)$$

And the conversion ratio M can be obtained as follows.

$$M = \frac{D_{D1}}{N(1 + n - nD_{D1})} \qquad (8)$$

On the other hand, M_r, conversion ratio of traditional two-switch forward converter, is expressed as follows.

$$M_r \triangleq \frac{E_o}{E_i} = \frac{D}{N} \qquad (9)$$

The difference value between M and M_r is given as follows.

$$\Delta = M - M_r$$

$$= \frac{1}{N} \frac{DV_s(-n + nD) - I_o L_1 f_s (1 - \frac{1}{2}\sigma_{L2})(1 + nD)}{(1 + n - nD)V_s + nI_o(1 - \frac{1}{2}\sigma_{L2})L_1 f_s} \qquad (10)$$

where σ_{L2} denotes the ripple current ratio of i_{L2}.
In Eq. (10), certainly $N > 0$, $n > 0$, $0 \le D \le 1$; besides, σ_{L2} is generally required to be lower than 0.05. Since $(1 - 0.5\sigma_{L2}) > 0$, it is seen that Δ is absolutely negative. It means that under the same turn ratio of transformer, the conversion ratio of the proposed converter is definitely lower than that of the traditional two-switch forward converter.

B. Static Analysis

The circuit shown in Fig. 4 can be analyzed with the state-space averaged method for efficiency investigation. There are two operation states for the CCM (continuous current mode) of i_{L2}. One $(0 \sim D_{D1}T_s)$ is derived from Fig. 7, the other$(D_{D1}T_s \sim T_s)$ is derived from Fig. 8, Fig. 9, Fig. 10 and Fig. 6. The equivalent circuits of these two operation states are shown in Fig. 12 and Fig. 13, respectively. In these figures, r_1 and r_2 represent the equivalent resistance corresponding to the power loss of each equivalent circuit, respectively.

$[0 \sim D_{D1}T_s]$:

This state starts when i_{L1} equals i_{L2}, and ends when S_1 and S_2 are simultaneously turned off. The equivalent circuit is expressed as Fig. 12.

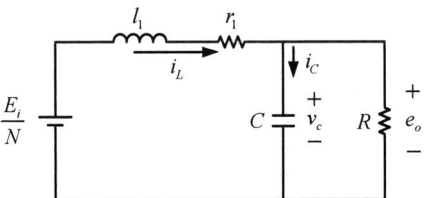

Fig. 12. Equivalent circuit of state 2 in Fig. 11 for static analysis.

If the mutual inductances of L_1 and L_2 can be neglected, in this duration, l_1 and r_1 can be expressed as follows.

$$l_1 = L_1 + L_2 = (n+1)L_2 \qquad (11)$$

$$r_1 = r'_{PT} + r_{D1} + r_{L1} + r_{L2} \qquad (12)$$

$[D_{D1}T_s \sim T_s]$:

This state starts at $t=t_2$ when both S_1 and S_2 are simultaneously turned off, and ends at $t=t_1$ when i_{L1} becomes equal to i_{L2} in on-state of both S_1 and S_2. The equivalent circuit is expressed as Fig. 13.

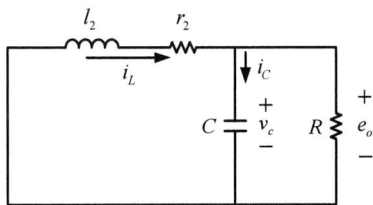

Fig. 13. Equivalent circuit of the duration except state 2 in Fig. 11 for static analysis.

In this duration, l_2, r_2 and i_L can be expressed as follows.

$$l_2 = L_2 \qquad (13)$$

$$r_2 = r_{D2} + r_{L2} \qquad (14)$$

$$i_L = i_{L2} \qquad (15)$$

In this way, equivalent circuit for static characteristic can be obtained as Fig. 14.

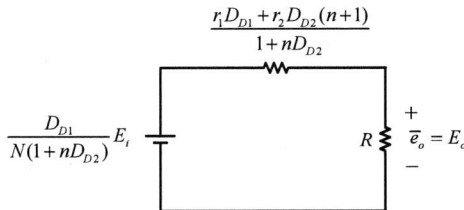

Fig. 14. Equivalent circuit for static characteristic.

The symbol definitions for the aforesaid analysis are expressed in TABLE I.

TABLE I
SYMBOL DEFINITIONS

Symbol	Definition
r_{PT}	Equivalent resistance of total losses in primary side
$R_{S1(on)}$、 $R_{S2(on)}$	Conduction resistance of S_1 or S_2
r_{DR1}、 r_{DR2}	Equivalent forward resistance of D_{R1} or D_{R2}
r_{D1}、 r_{D2}	Equivalent forward resistance of D_1 or D_2
r_{L1}、 r_{L2}	Equivalent resistance for power dissipation of L_1 or L_2
r_{N1}	Equivalent resistance for transformer primary winding
r_{Ei}	Equivalent internal resistance of source E_i

where,

$$r_{PT} = R_{S1(on)} + R_{S2(on)} + r_{N1} + r_{Ei} \qquad (16)$$

$$r'_{PT} = \frac{1}{N^2} r_{PT} \left(\because N = \frac{N_1}{N_2} \right) \qquad (17)$$

Output impedance is derived as follows.

$$Z_o = \frac{r_1 D_{D1} + r_2 D_{D2}(n+1)}{1 + n D_{D2}} \qquad (18)$$

where,

$$D_{D2} = 1 - D_{D1} \qquad (19)$$

Applying the state-space averaged method, the relation of conversion ratio vs. output impedance can be obtained as follows.

$$M = \frac{D_{D1}}{\left(Z_o / R + 1 \right) N \left(1 + n - n D_{D1} \right)} \qquad (20)$$

Besides, theoretical efficiency η can be expressed as follows.

$$\eta = \frac{1}{1 + Z_o / R} \qquad (21)$$

IV. DESIGN FLOW AND EXPERIMENTAL RESULTS

The structure of the proposed high step-down forward converter is intuitively similar to that of the traditional two-switch forward converter. However, their circuit operations are completely different. To verify the correctness of the theoretical analysis, the experiments which especially focus on the circuit operations of L_1 and L_2 were carried out.

A. Design Flow

As seen from Eq. (8), increasing N will be helpful for step-down. However, since the parasitic elements of the transformer will affect the circuit performance of this converter, N should not be too large even for higher step-down. For this research, N is set to be 3, and the design flow is shown in Fig.15.

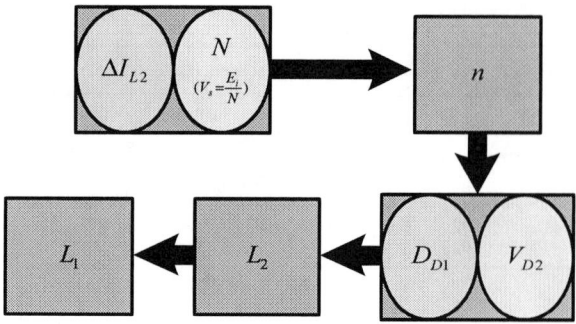

Fig. 15. Design flow chart.

B. Specifications and Component Values

In the near future, the power supply of residence will be switched from AC voltage to $300V_{DC}$. In other words, through DC to DC converters, the $300V_{DC}$ bus bar will provide $300V_{DC}$ for air conditioner, $140V_{DC}$ for refrigerator, freezer and washing machine, $48V_{DC}$ for lighting equipment, computer, television, etc. On the other hand, the control voltage of these appliances, for example, $5V_{DC}$, will also be supplied by this $300V_{DC}$ busbar through a high step-down converter. For the experimental implement, the electrical specifications of this high step-down converter are shown in TABLE II, and the component values are shown in TABLE III.

TABLE II
SPECIFICATIONS OF THE PROTOTYPE

Symbol	Definition	Value
E_i	Input voltage	300V
f_s	Switching frequency	100kHz
E_o	Output voltage	5V
I_o(max)	Maximum output current	1A

TABLE III
COMPONENT VALUES OF THE PROTOTYPE

Symbol	Definition	Model/Value
S_1, S_2	High side and low side switches	IRF730
D_{R1}, D_{R2}	Energy recovery diodes	ER304
D_1, D_2	Rectification diode, flywheeling diode	HER303
N	Turn ratio of transformer	3
L_1	Middle inductor	0.387mH
L_2	Output inductor	1.005 mH
C	Output capacitor	100μF

C. Experimental Results

The experimental results which are needed to verify the circuit operation agree well with the theoretical analysis. Under the same electric specifications, compared with the traditional two-switch forward converter, the MOSFETs of the proposed converter are really driven by pulse with larger duty ratio. Fig. 16 shows the experimental waveforms of the proposed converter.

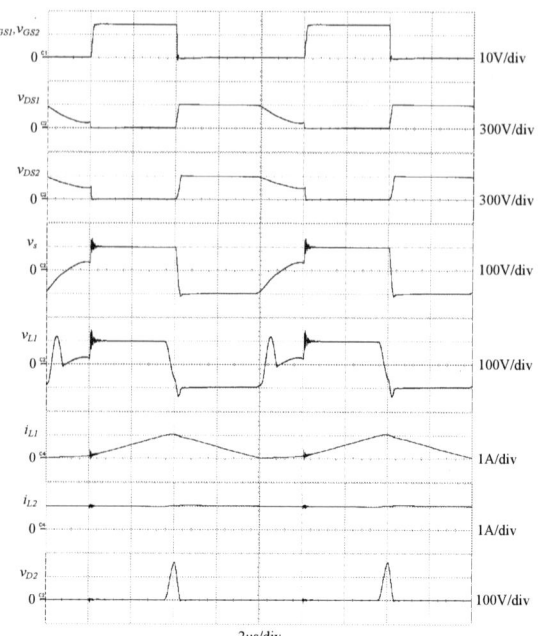

Fig. 16. Experimental waveforms in the proposed high step-down converter. (I_o=1A)

D. Voltage Surge across Flywheeling Diode

In theory, the reverse voltage across the flywheeling diode of the proposed converter is smaller than that of the traditional two-switch forward converter. However, the voltage surge across the flywheeling diode incurred by parasitic elements will cause the deviation of duty ratio between simulation and experiment.

An example of the comparison between the simulated and the experimental result is shown in TABLE IV, where $V_{D2(peak)}$ represents the peak value of the reverse voltage across D_2. The experimental waveforms of i_{D2} and v_{D2} are shown in Fig. 17.

TABLE IV
AN EXAMPLE OF THE COMPARISON BETWEEN THE
SIMULATED AND THE EXPERIMENTAL RESULT
(E_o=5V, I_o=1A)

Simulated results		Experimental results	
$V_{D2(peak)}$(V)	D(%)	$V_{D2(peak)}$(V)	D(%)
73.59	44.60	165.8	39.5

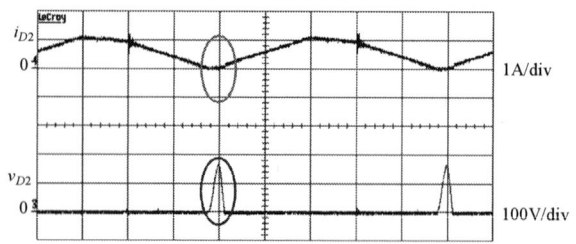

Fig. 17. Experimental waveforms under full load condition.(E_o=5V, I_o=1A)

Compared with the illustration waveform v_{D2} shown in Fig. 11, actually, t_1 will delay due to the storage time of the flywheeling diode, and t_2 will also delay due to the parasitic elements of the lead connected with this diode. The delay waveform of v_{D2} and the waveform of i_{D2} are shown in Fig.18, while the delay waveform of v_{D2} and the waveform of v_{GS2} are shown in Fig.19. Therefore, under the condition of the same output voltage, actual average voltage of v_{D2} in the OFF state of D_2 will be different from the theoretical average voltage. In the high step-down converter, since D_{D1} is quite smaller than 1, the forward conduction voltage of D_2, V_F, cannot be neglected. Accordingly Eq. (3) should be modified as follows.

$$E_o = V_{D2} \cdot D_{D1} - V_F (1 - D_{D1}) \qquad (22)$$

Thus, Eq. (22) can be derived as follows.

$$E_o = V_{D2,eq.} \cdot D_{D1} \qquad (23)$$

where,

$$V_{D2,eq.} = V_{D2} - \frac{V_F (1 - D_{D1})}{D_{D1}}. \qquad (24)$$

Accordingly, the comparison between the simulation and the experiment results are shown in TABLE V. In TABLE V, the simulated results obtained from Eq. (3) can also be confirmed with Eq. (23).

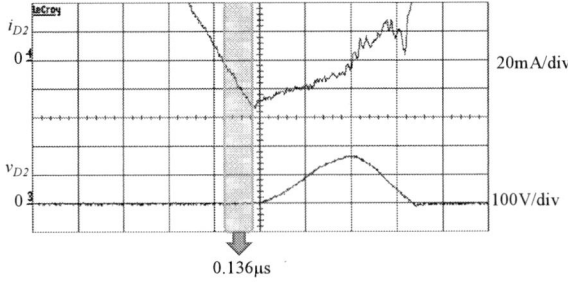

Fig. 18. Experimental waveforms of i_{D2} and v_{D2} in the proposed high step-down converter.

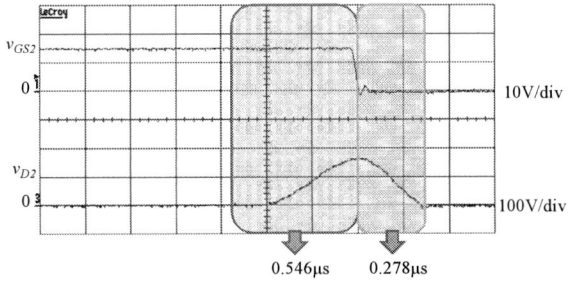

Fig. 19. Experimental waveforms of v_{GS2} and v_{D2} in the high step-down converter.

TABLE V
THE RELEVANT DATA DURING OFF-STATE OF D_2 (I_o=1A)

E_o (V)	Simulated results		Experimental results			
	D_{D1} (%)	V_{D2} (V)	D_{D1} (%)	$V_{D2(peak)}$ (V)	V_{D2} (V)	$V_{D2,eq.}$ (V)
2	2.75	72.75	5.64	86.2	51.84	34.74
3	4.11	73.03	5.96	117.7	66.14	49.04
4	5.46	73.31	6.40	146.9	77.92	63.90
5	6.79	73.59	6.74	165.8	89.77	73.41
6	8.12	73.87	7.20	168.9	98.28	82.68
7	9.44	74.14	8.02	169.7	101.3	87.87

E. Utilization of RCD Snubber

Use of RCD snubber is made to suppress the voltage surge across D_2, as shown in Fig. 20. Under the condition of fixed snubber resistance R_s, the suppression effect of the voltage surge vs. snubber capacitance C_s can be shown in TABLE VI. The experimental waveforms of i_{D2} and v_{D2} are shown in Fig. 21. From TABLE VI and Fig. 21, it is clear that $V_{D2(peak)}$ is significantly reduced. For example, $V_{D2(peak)}$ is greatly reduced from 165.8V which is shown in TABLE V to 68V when C_s of 12nF is used in this snubber.

Owing to the utilization of RCD snubber, $D_{D1}T_s$ will slightly increase. However the product of $V_{D2,eq.}$ and D_{D1} is correctly equal to E_o. By use of RCD snubber, experimental waveforms shown in Fig. 22 become more similar to those illustration waveforms shown in Fig. 11, especially for the experimental waveform of v_{D2}.

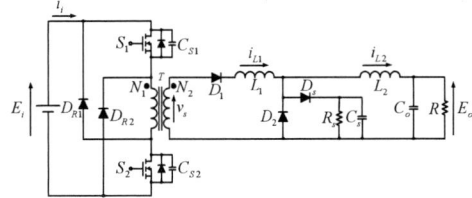

Fig. 20. A high step-down converter with RCD snubber.

TABLE VI
EFFECT OF THE CAPACITANCE IN RCD SNUBBER
(E_o=5V, I_o=1A, R_s=22kΩ)

	D_{D1}(%)	$V_{D2(peak)}$ (V)	$V_{D2,eq.}$(V)
C_s=2.2nF	10.58	72.7	47.40
C_s=3.3nF	10.66	71.0	47.01
C_s=6.8nF	10.76	68.6	46.82
C_s=12nF	10.86	68.0	46.04

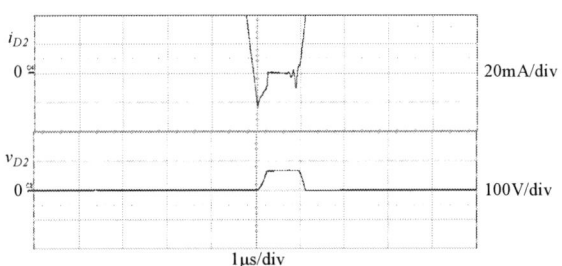

Fig. 21. Experimental waveforms of i_{D2} and v_{D2} in the high step-down converter with RCD snubber.
(E_o=5V, I_o=1A, R_s=22kΩ, C_s=12nF)

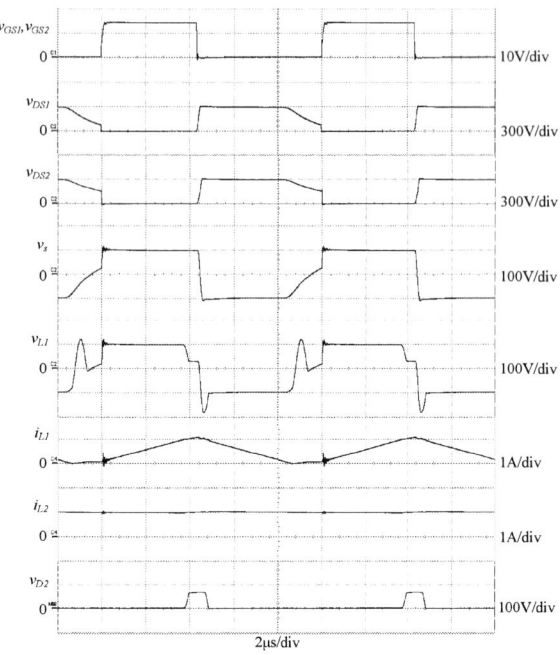

Fig. 22. Experimental waveforms in the high step-down converter with RCD snubber.
(E_o=5V, I_o=1A, R_s=22kΩ, C_s=12nF)

V. CONCLUSIONS

For the proposed high step-down forward converter, the feasibility and appropriateness have been confirmed with theoretical analysis and experimental verification. Furthermore, through detail investigation on voltage surge across the flywheeling diode, a fitful solution has been found, and the applicability is also confirmed with the comparison of the simulation and the experiment results. The research results are concluded as follows.

1. Under the same step-down condition, duty ratio of MOSFETs in the proposed converter will be larger than that in the traditional two-switch forward converter. The utilization ratio of these MOSFETs is accordingly improved.

2. The voltage across the flywheeling diode in the proposed converter can be basically reduced. However, voltage surge will occur due to the reverse recovery time of this diode. RCD snubber is proved to be effective to suppress the voltage surge.

REFERENCES

[1] Masahito Jinno, Hong-Wei Su, Hirofumi Matsuo, Daisuke Ueno "Characteristic Investigation of a High Step-down Converter," *The 10th Taiwan Power Electronics Conference & Exhibition*, 2011, pp.142-146.

[2] J. Zhao and M. Sekine, "A Novel Two-Switch Forward Converter for High Step-Down Conversion," *IEEE Telecommunications Energy Conference*, 2009, pp. 1-5.

[3] Masahito Jinno, "Efficiency Improvement for SR Forward Converters With LC Snubber," *IEEE Transactions on Power Electronics*, vol. 16, no. 6, pp. 812-820, 2001.

[4] Masahito Jinno and Wen-Lune Wu, "LC Snubber Combination Forward DC-DC Converter Employing Synchronous Rectifier," *International Power Electronics Conference*, 2000, pp. 889-894.

[5] Masahito Jinno and Wen-Lune Wu, "Efficiency Improvement for Forward DC-DC Converter Employing Synchronous Rectifier," *IEEE Power Electronics Specialists Conference*, 2000, pp. 1516-1521.

[6] Masahito Jinno, Po-Yuan Chen, Kun-Chih Lin, "An Efficient Active LC Snubber for Forward Converters," *IEEE Transactions on Power Electronics*, vol. 24, no.6, pp. 1522-1531, 2009.

[7] Jianping Xu, Xiaohong Cao, Qianchao Luo, "An improved two-transistor forward converter," *PEDS*, 1999, pp. 225-228.

[8] M. chen, D. Xu and M. Matsui, "Study on Magnetizing Inductance of High Frequency Transformers in the Two-Transistor Forward Converter," *IEEE PCC-Osaka* 2002, pp. 597-602.

[9] C. F. JIN, and T. Ninomiya, "Single-Stage Power-Factor-Correction Converter Using Tapped Inductor," *IEEE Power Electronics Specialists Conference*, 2004, pp. 1520-1524.

[10] K.Yao, M.Ye, M. Xu, and F.C. Lee, "Tapped-Inductor Buck Converter for High-Step-Down DC-DC Conversion," *IEEE Transactions on Power Electronics*. vol. 20. no. 4, pp. 775-780, July 2005.

Generalized Modeling and Optimization of a Bidirectional Dual Active Bridge DC-DC Converter including Frequency Variation

Felix Jauch, Jürgen Biela
Laboratory for High Power Electronic Systems, ETH Zurich
Email: jauchf@ethz.ch
URL: http://www.hpe.ee.ethz.ch

Abstract—The paper presents a novel modeling approach of the power flow in a bidirectional dual active bridge DC-DC converter. By using basic superposition principles, the mathematical distinction of cases is avoided in the modeling process of high-frequency transformer currents for different types of modulation. The generalized model is used in an optimization of converter losses of a 3.3 kW electric vehicle battery charger with an input voltage of 400 V and a battery voltage range of 280 V to 420 V. Besides the commonly used control variables such as phase-shift and clamping intervals, also the variation of switching frequency is considered in the optimization process. The optimal modulation including frequency variation leads to an increase of converter efficiency up to 8.6 % using IGBTs and 17.8 % using MOSFETs in the most critical point compared to phase-shift modulation at fixed switching frequency.

Index Terms—DC-DC Converter, Dual Active Bridge, Frequency Variation, Modeling, Optimization

I. Introduction

During the last few decades the environmental impact of petroleum-based transportation infrastructure gained more and more significance. Fossil fuel-powered vehicles lead to large emissions of CO_2 and other pollutions. In order to reduce those impacts, electric vehicles will play an important role in our future transportation infrastructure as the use of renewable energy sources is constantly increasing.

For charging the batteries of electric vehicles or storage systems in general, suitable power electronic systems are necessary. Usually, a basic two-stage approach comprising a boost Power Factor Correction (PFC) rectifier and a subsequent high-frequency isolated DC-DC converter is used for a charging system connected to the low-voltage AC grid. Additionally, for implementing Vehicle-2-Grid (V2G) concepts, the converter systems feature bidirectional power flow capability. Suitable DC-DC converters comprise the Dual Half-Bridge (DHB) [1], Dual Active (Full-)Bridge (DAB) [2] or resonant DC-DC converters [3].

For the DAB, several modulation methods like phase-shift modulation [2], triangular and trapezoidal current mode modulation [4] have been investigated. Further adjustments of these methods have been presented in [5]. Usually, for each of these modulation methods, piecewise linear equations are used to describe the currents where several mathematical cases depending on the control variables have to be distinguished. A general optimization of converter losses becomes relatively complex and demands high computational power.

In this paper, a novel generalized modeling approach of the power flow is presented to include all possible modulation

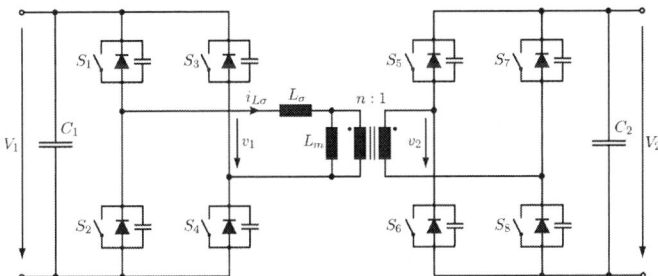

Fig. 1. Topology of the Dual Active Bridge (DAB) DC-DC converter with primary full-bridge with switches S_1, S_2, S_3, S_4 and secondary full-bridge with switches S_5, S_6, S_7, S_8 connected to a two-winding transformer with leakage inductance L_σ and magnetizing inductance L_m.

methods at once. The general power flow equation is then used in an optimization procedure to find the optimal modulation, which leads to highest converter efficiency over the whole operating range. Besides the commonly used control variables like phase-shifts and clamping intervals, also the switching frequency is considered to control a DAB [6].

First, in section II the DAB converter topology with its modulation methods is introduced. The mathematical derivation of a novel generalized power flow equation depending on the control variables is presented in section III. Then, section IV shows the optimization of converter losses of a DAB prototype system. Finally, the optimal modulation including frequency variation is compared to conventional modulation methods with respect to converter efficiency.

II. Topology and Modulation

In the following, first the topology of the DAB DC-DC converter with its operating principle is shortly explained. Afterwards, commonly used modulation methods with their operating limits are summarized. These are the phase-shift modulation as well as the trapezoidal and the triangular current mode modulation.

A. Dual Active Bridge DC-DC Converter

Fig. 1 shows the converter topology of a DAB DC-DC converter. The converter consists of a primary and a secondary full-bridge with unidirectional switches. The two full-bridges are connected to the windings of a two-winding transformer and generate high-frequency (HF) square-wave voltages with amplitudes of the DC port voltages V_1, V_2. The converter is operated by phase-shift control where the control variables are

978-1-4799-2706-7/14 $31.00 © 2014 IEEE

The 2014 International Power Electronics Conference

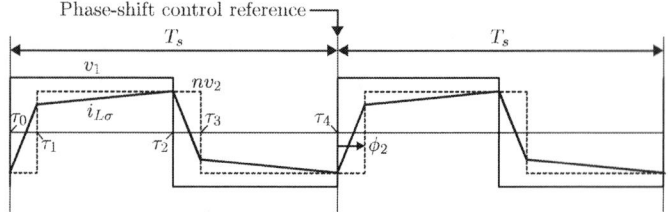

Fig. 2. High-frequency voltages v_1, nv_2 applied to the transformer windings in *phase-shift modulation* and resulting leakage inductance current $i_{L\sigma}$.

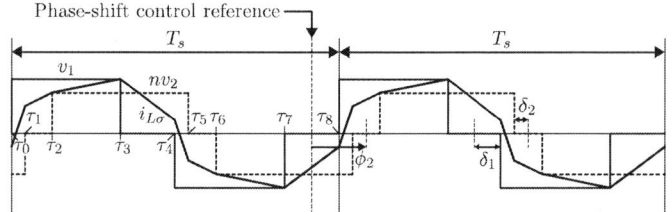

Fig. 3. High-frequency voltages v_1, nv_2 applied to the transformer windings in *general trapezoidal current mode modulation* and resulting leakage inductance current $i_{L\sigma}$.

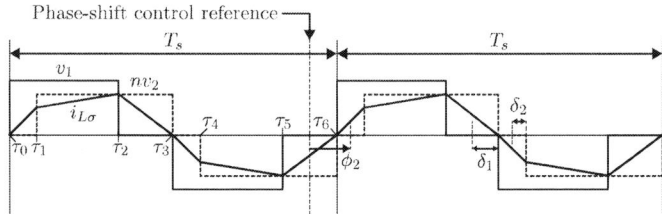

Fig. 4. High-frequency voltages v_1, nv_2 applied to the transformer windings in *trapezoidal current mode modulation* and resulting leakage inductance current $i_{L\sigma}$.

the clamping intervals ($v_1 = 0$ and/or $v_2 = 0$) and the phase-shifts of the HF voltages. In Fig. 7a the basic equivalent circuit of the converter is given where the full-bridges are modeled by HF voltage sources. The primary referred leakage inductance L_σ of the transformer acts as decoupling and energy transfer element between the square-wave voltages. The magnetizing inductance L_m of the transformer is neglected.

By applying a positive or negative voltage $v_{L\sigma}$ across the leakage inductance L_σ (see Fig. 7a), the current waveform $i_{L\sigma}$ can be controlled during the switching cycle T_s as shown for different modulation methods in Fig. 2, Fig. 3, Fig. 4, Fig. 5, Fig. 6.

B. Phase-Shift Modulation

In phase-shift modulation, square-wave voltages without clamping intervals are applied to the transformer windings as shown in Fig. 2. The power transferred from primary to secondary side is controlled by the phase-shift ϕ_2 between the two voltages v_1, nv_2 and given by

$$P_{12} = -\frac{V_1 n V_2 \phi_2 (\pi - |\phi_2|)}{\pi \omega_s L_\sigma} \qquad (1)$$

with $\phi_2 \in [-\pi, \pi]$. V_1, V_2 are the input and output voltage, $\omega_s = 2\pi/T_s$ the angular switching frequency, $n = N_1/N_2$ the turns ratio and L_σ the leakage inductance of the transformer.

The maximum transferable power is

$$P_{12,max} = \pm\frac{\pi V_1 n V_2}{4\omega_s L_\sigma}. \qquad (2)$$

The soft-switching range of the modulation where Zero-Voltage-Switching (ZVS) can be achieved is strongly dependent on the voltage ratio V_1/nV_2 as well as the power level P_{12} [2]. Especially for voltage ratios $V_1/nV_2 \ll 1$ and $V_1/nV_2 \gg 1$ at low loads, ZVS cannot be maintained. Disadvantages like limited soft-switching range and high RMS transformer currents can be overcome by using the trapezoidal current mode modulation explained in the next section.

C. Trapezoidal Current Mode Modulation

In trapezoidal current mode modulation, square-wave voltages with clamping intervals are applied to the transformer windings as shown in Fig. 3 for the general modulation mode.

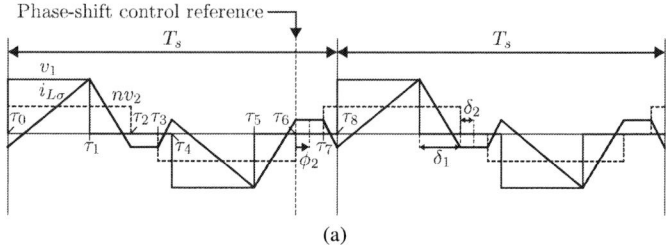

(a)

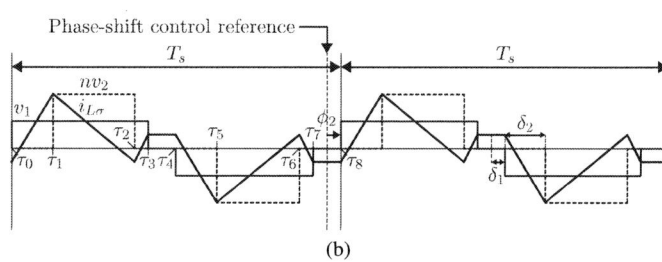

(b)

Fig. 5. High-frequency voltages v_1, nv_2 applied to the transformer windings in *general triangular current mode modulation* and resulting leakage inductance current $i_{L\sigma}$ for $V_1 > nV_2$ (a) and $V_1 < nV_2$ (b).

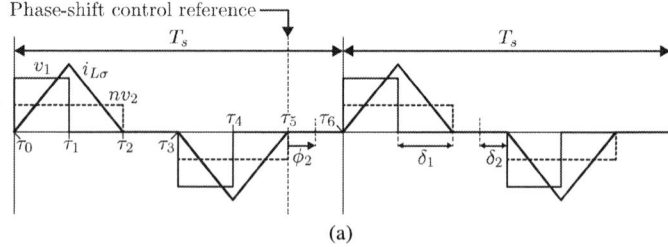

(a)

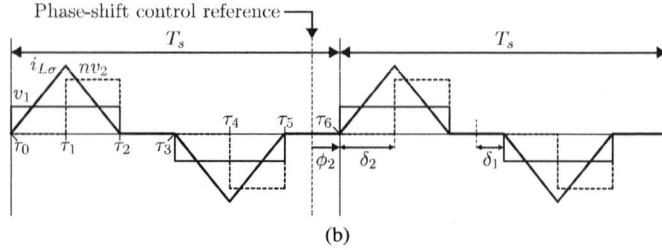

(b)

Fig. 6. High-frequency voltages v_1, nv_2 applied to the transformer windings in *triangular current mode modulation* and resulting leakage inductance current $i_{L\sigma}$ for $V_1 > nV_2$ (a) and $V_1 < nV_2$ (b).

978-1-4799-2706-7/14 $31.00 © 2014 IEEE

By setting $\tau_0 = \tau_1$ and $\tau_4 = \tau_5$ in Fig. 3 the trapezoidal current mode modulation given in Fig. 4 is obtained. The transferred power is then described by

$$P_{12} = -\text{sign}(\phi_2)\frac{V_1 n V_2(\pi\,|\phi_2| - 2\phi_2^2 + 2\delta_1\delta_2)}{\pi\omega_s L_\sigma} \qquad (3)$$

with $\delta_1 = f(\phi_2) \in [0, \pi/2]$, $\delta_2 = f(\phi_2) \in [0, \pi/2]$ and $\phi_2 \in [-\pi, \pi]$. The maximum transferable power is

$$P_{12,max} = \pm\frac{\pi V_1^2 n^2 V_2^2}{2\omega_s L_\sigma (V_1^2 + n^2 V_2^2 + V_1 n V_2)}. \qquad (4)$$

The current $i_{L\sigma}$ reaches zero at switching instants τ_0, τ_3, τ_6 where Zero-Current-Switching (ZCS) is possible [4]. At $\tau_1, \tau_2, \tau_4, \tau_5$ ZVS is possible as far as the minimum commutation current needed for the resonant transition is reached. Nevertheless, the modulation method cannot be applied for low output power. This leads to the triangular current mode modulation with a seamless transition between the modulation methods.

D. Triangular Current Mode Modulation

In triangular current mode modulation, also square-wave voltages with clamping intervals are applied to the transformer windings as can be seen from Fig. 5 for the general modulation mode. Considering $\tau_3 = \tau_4$ and $\tau_7 = \tau_8$ in Fig. 5a as well as $\tau_2 = \tau_3$ and $\tau_6 = \tau_7$ in Fig. 5b the triangular current mode modulation shown in Fig. 6 is obtained. The transferred power can then be written as

$$P_{12} = -\frac{V_1 n V_2 \phi_2(\pi - 2\delta_1)}{\pi\omega_s L_\sigma}, \qquad V_1 > nV_2 \qquad (5)$$

$$P_{12} = -\frac{V_1 n V_2 \phi_2(\pi - 2\delta_2)}{\pi\omega_s L_\sigma}, \qquad V_1 < nV_2 \qquad (6)$$

with $\delta_1 = f(\phi_2) \in [0, \pi/2]$, $\delta_2 = f(\phi_2) \in [0, \pi/2]$ and $\phi_2 \in [-\pi, \pi]$. The maximum transferable power is

$$P_{12,max} = \pm\frac{\pi n^2 V_2^2 (V_1 - nV_2)}{2\omega_s L_\sigma V_1}, \qquad V_1 > nV_2 \qquad (7)$$

$$P_{12,max} = \pm\frac{\pi V_1^2 (V_1 - nV_2)}{2\omega_s L_\sigma nV_2}, \qquad V_1 < nV_2. \qquad (8)$$

The current $i_{L\sigma}$ reaches zero at switching instants $\tau_0, \tau_2, \tau_3, \tau_5, \tau_6$ where ZCS is possible [4]. At instants τ_1, τ_4 ZVS is possible. Depending on the voltage ratio, the modulation shown in Fig. 6a with $V_1 > nV_2$ or the modulation shown in Fig. 6b with $V_1 < nV_2$ is applied. Power transfer in case of $V_1 = nV_2$ is not possible in triangular current mode.

III. MODELING OF POWER FLOW

For designing and controlling the DAB converter system, the mathematical description of the power flow depending on the control variables is essential. The optimization procedure shown in Fig. 11 requires the calculation of the power flow P_{12} for evaluating the power equality constraint.

The well-known approach uses piecewise linear equations for the transformer leakage inductance current where several mathematical cases depending on the phase-shifts and the clamping intervals have to be distinguished. Due to the mathematical complexity, especially for high port numbers in multi-port converters [7], the following analysis uses basic

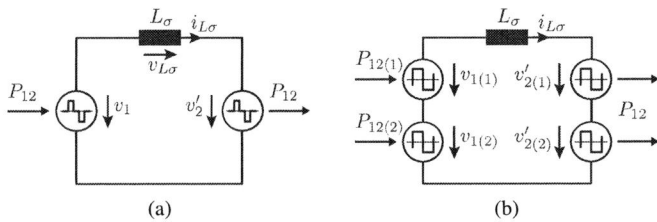

Fig. 7. Simplified circuit of a two-port converter applying square-wave voltages with clamping intervals (a) and the equivalent four-port circuit applying square-wave voltages without clamping intervals at the ports (b).

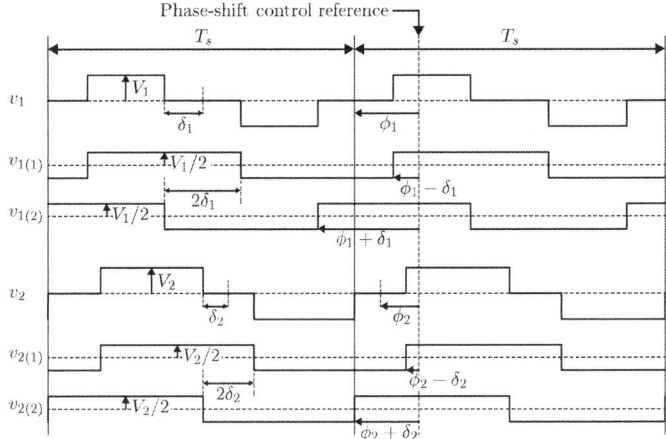

Fig. 8. Square-wave voltages v_1, v_2 with clamping intervals $2\delta_1, 2\delta_2$ and the underlying square-wave voltages $v_{1(1)}, v_{1(2)}, v_{2(1)}, v_{2(2)}$ without clamping intervals and duty cycles 50 % which add up to v_1 and v_2 respectively.

superposition principles to find the general analytical formula for the power flow. With this approach, there is no need for mathematical distinction of cases.

The mathematical analysis of the power flow is based on the primary side referred equivalent circuit of the converter topology shown in Fig. 7a. The full-bridges are modeled by HF square-wave voltage sources v_1, v_2' with clamping intervals as shown in Fig. 7a.

The power flow over one switching cycle $T_s = 2\pi/\omega_s$ between two ports (from a first port 1 to a second port 2) applying square-wave voltages with clamping intervals as shown in Fig. 7a is based on the well-known power flow equation [8] (power from primary port p to secondary port s)

$$P_{ps} = \frac{V_p n V_s}{\omega_s L_\sigma}(\phi_p - \phi_s)\left(1 - \frac{|\phi_p - \phi_s|}{\pi}\right). \qquad (9)$$

There, two square-wave voltages with 50 % duty cycles, amplitudes V_p, V_s and phases $\phi_p, \phi_s \in [-\pi, \pi]$ are applied across the windings of a two-winding transformer with primary referred leakage inductance L_σ, negligible large magnetizing inductance and turns ratio $n = N_p/N_s$. The phase angles are measured against a given reference, a positive angle defines a leading signal and a negative angle a lagging signal with respect to the reference.

The two-port circuit with clamping intervals given in Fig. 7a can be modeled by the equivalent four-port circuit shown in Fig. 7b where only square-wave voltages without clamping intervals and duty cycles of 50 % occur. This is done by splitting up voltage v_1 with clamping interval into a sum $v_{1(1)}+v_{1(2)}$ of two voltages with 50 % duty cycle, no clamping interval and a phase-shift of $2\delta_1$ against each other as depicted in Fig. 8. Analogously, this is done for the voltage v_2. The power transferred from port 1 to port 2 is then given by

$$
P_{12} = \underbrace{\frac{1}{T_s}\int_0^{T_s} v_{1(1)}i_{L\sigma}\, d\tau}_{P_{12(1)}} + \underbrace{\frac{1}{T_s}\int_0^{T_s} v_{1(2)}i_{L\sigma}\, d\tau}_{P_{12(2)}} \quad (10)
$$

with the two power shares of voltage sources $v_{1(1)}, v_{1(2)}$ (see Fig. 7b). The leakage inductance current $i_{L\sigma}$ is split up into three parts $i_{L\sigma(I)}, i_{L\sigma(II)}, i_{L\sigma(III)}$ which are obtained by applying the superposition principle as shown in Fig. 9 by selectively short-circuiting voltage sources. In this way, the power exchange of source $v_{1(1)}$ with sources $v_{1(2)}, v'_{2(1)}, v'_{2(2)}$ is described. The power share $P_{12(1)}$ in (10) can then be written as

$$
P_{12(1)} = \underbrace{\frac{1}{T_s}\int_0^{T_s} v_{1(1)}i_{L\sigma(I)}\, d\tau}_{P_{12(1)(I)}} + \underbrace{\frac{1}{T_s}\int_0^{T_s} v_{1(1)}i_{L\sigma(II)}\, d\tau}_{P_{12(1)(II)}}
$$

$$
+ \underbrace{\frac{1}{T_s}\int_0^{T_s} v_{1(1)}i_{L\sigma(III)}\, d\tau}_{P_{12(1)(III)}}. \quad (11)
$$

Analogously, the second power share $P_{12(2)}$ is described. From Fig. 9 and (11) it is concluded, that the power shares $P_{12(1)(I)}, P_{12(1)(II)}, P_{12(1)(III)}$ are given by (9). This is also the case for the power shares $P_{12(2)(I)}, P_{12(2)(II)}, P_{12(2)(III)}$. By summing up all the power shares, the resulting power transferred per switching cycle from port 1 to port 2 applying square-wave voltages v_1, v_2 with clamping intervals $2\delta_1, 2\delta_2$ and phases ϕ_1, ϕ_2 as shown in Fig. 8 is thus given as

$$
P_{12} = \frac{V_1 n V_2}{4\omega_s L_\sigma}\Bigg[((\phi_1-\delta_1)-(\phi_2-\delta_2))\left(1-\frac{|(\phi_1-\delta_1)-(\phi_2-\delta_2)|}{\pi}\right)
$$

$$
+((\phi_1-\delta_1)-(\phi_2+\delta_2))\left(1-\frac{|(\phi_1-\delta_1)-(\phi_2+\delta_2)|}{\pi}\right)
$$

$$
+((\phi_1+\delta_1)-(\phi_2-\delta_2))\left(1-\frac{|(\phi_1+\delta_1)-(\phi_2-\delta_2)|}{\pi}\right)
$$

$$
+((\phi_1+\delta_1)-(\phi_2+\delta_2))\left(1-\frac{|(\phi_1+\delta_1)-(\phi_2+\delta_2)|}{\pi}\right)\Bigg]. \quad (12)
$$

In general, the proposed superposition method for deriving analytical power flow equations can be applied to any number of ports which are connected in series in multi-port converters as for instance shown in [9] for an isolated three-phase bidirectional AC-DC converter.

IV. OPTIMIZATION OF CONVERTER LOSSES

The general power flow equation (12) is used in an optimization of converter losses of a DAB prototype system designed

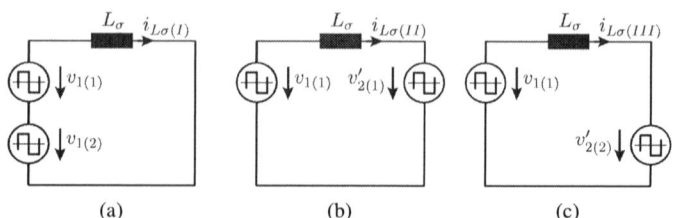

Fig. 9. Leakage inductance current $i_{L\sigma} = i_{L\sigma(I)} + i_{L\sigma(II)} + i_{L\sigma(III)}$ obtained by applying the superposition principle with three parts (a), (b), (c) by selectively short-circuiting voltage sources.

TABLE I
PARAMETERS OF THE PROTOTYPE SYSTEM.

Input voltage	V_1	400 V
Battery voltage	V_2	280 V … 420 V
Output power	P_2	3.3 kW
Switching frequency	f_s	20 kHz … 50 kHz
Transformer turns ratio	n	8/7
Transformer leakage inductance	L_σ	
Phase-shift modulation	$L_{\sigma(1)}$	181 µH
Triangular/Trapezoidal current modulation	$L_{\sigma(2)}$	158 µH
Optimized control variables modulation	$L_{\sigma(3)}$	158 µH
Transformer magnetizing inductance	L_m	neglected

for a nominal output power of 3.3 kW. The loss model includes the main loss shares as conduction and switching losses of the semiconductor devices, skin and proximity effect losses of the transformer windings as well as the core losses of the transformer.

A. Prototype System

As a prototype system to find optimal control variables, a 3.3 kW electric vehicle battery charger to connect to a fixed 400 V DC-link with an output voltage range of 280 V to 420 V of a lithium-ion battery is considered. The primary and secondary switching devices are chosen to be 650 V IGBTs of type IKW50N65F5 for a first converter solution and 650 V MOSFETs of type IPW65R019C7 for a second converter solution, both from Infineon [10]. The system parameters are listed in detail in Table I.

B. Loss Models

For optimizing the DAB converter system, its main losses occurring in the semiconductor devices as well as in the transformer have to be modeled. For the semiconductor devices, both IGBT and MOSFET loss models based on datasheet parameters are used and given in the following.

1) IGBT Losses: For calculating the conduction losses, the typical output characteristic $i_C = f(v_{CE})$ of the IGBT and the typical diode forward current as a function of forward voltage $i_D = f(v_D)$ of the anti parallel diode at a junction temperature of $T_{j,max}-25\,°C = 150\,°C$ from the datasheet are considered. Given the current i_C flowing through the IGBT and the current i_D through the diode over the interval $[0, T_s]$ of a switching period, the conduction losses of an IGBT co-pack device can be calculated by

$$
P_{c,S} = \frac{1}{T_s}\int_0^{T_s} i_C(\tau)\cdot v_{CE}(i_C(\tau))\, d\tau, \quad (13)
$$

$$P_{c,D} = \frac{1}{T_s} \int_0^{T_s} i_D(\tau) \cdot v_D(i_D(\tau)) \, d\tau, \qquad (14)$$

$$P_c = P_{c,S} + P_{c,D}. \qquad (15)$$

When optimizing and designing a converter system, the details of the gate drives and the parasitics of the commutation path are usually unknown, so that only approximations of switching losses can be performed. For estimating the switching losses of an IGBT co-pack device, the following assumptions are made: For a device which is turned on

- diode losses at zero-voltage turn-on are neglected,
- IGBT losses $E_{on} = f(i_C)$ at turn-on including diode reverse recovery losses are taken from datasheet and are linearly scaled with voltage according to datasheet.

For a device which is turned off

- diode losses at turn-off are neglected,
- IGBT losses $E_{off} = f(i_C)$ at turn-off are taken from datasheet and are linearly scaled with voltage according to datasheet.

For soft-switching in terms of ZVS at the switching instant τ_s, the switching losses of an IGBT co-pack device are approximated by

$$P_s = f_s \cdot E_{off}(i_C(\tau_s)) \frac{v_{CE}(\tau_s)}{V_{CE}}, \qquad (16)$$

whereas for hard-switching in terms of forced diode commutation, the switching losses are estimated according to

$$P_s = f_s \cdot E_{on}(i_C(\tau_s)) \frac{v_{CE}(\tau_s)}{V_{CE}} \qquad (17)$$

with V_{CE} being the collector-emitter voltage where switching losses were measured according to the datasheet. For soft-switching in terms of ZCS for small $i_C(\tau_s)$, losses according to (16) and (17) become negligible small.

2) MOSFET Losses: Also for MOSFETs, the typical output characteristic $i_D = f(v_{DS})$ at a junction temperature of $T_{j,max} - 25\,°C = 125\,°C$ from the datasheet can be used to calculate the conduction losses. Given the current i_D flowing through the MOSFET over the interval $[0, T_s]$ of a switching period, the conduction losses are given by

$$P_c = \frac{1}{T_s} \int_0^{T_s} i_D(\tau) \cdot v_{DS}(i_D(\tau)) \, d\tau. \qquad (18)$$

The approximation of switching losses of a MOSFET device is based on the following assumptions: For ZVS conditions when stored energy in the output capacitance is transferred from one MOSFET to another, difference between released and absorbed energies are negligible small. In other words, losses caused during the commutation of the inductive current are neglected, $P_s = 0$. For hard-switching, two loss effects are modeled, these are

- dissipation of energy $E_{oss} = f(v_{DS})$ stored in the output capacitance at turn-on,
- body diode reverse recovery losses occurring in turn-on device squarely scaled with voltage and linearly scaled with current [11].

Switching losses at time instant τ_s are then estimated using

$$P_{oss} = f_s \cdot E_{oss}(v_{DS}(\tau_s)), \qquad (19)$$

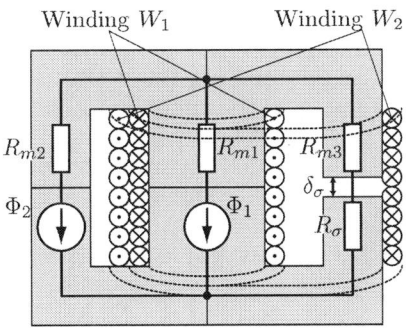

Fig. 10. 2D drawing of the transformer including reluctance model consisting of four U-cores with primary winding W_1 wound around the inner leg and secondary winding W_2 around the inner and the right-hand sided leakage leg. By inserting an air gap of length δ_σ in the leakage leg, the leakage inductance L_σ can be set.

TABLE II
TRANSFORMER PARAMETERS OF THE PROTOTYPE SYSTEM.

Magnetic core	2x AMCC-32 VITROPERM 500F
Primary winding W_1	24 turns, litz wire 945 strands, 0.071 mm
Secondary winding W_2	21 turns, litz wire 945 strands, 0.071 mm

Air gap length δ_σ
  Phase-shift modulation
  Triangular/Trapezoidal current modulation
  Optimized control variables modulation
	
	1.6 mm
	1.8 mm
	1.8 mm

$$P_{rr} = f_s \cdot Q_{rr} \left(\frac{v_{DS}(\tau_s)}{V_R} \right)^2 \frac{i_D(\tau_s)}{I_F} \cdot v_{DS}(\tau_s), \qquad (20)$$

$$P_s = P_{oss} + P_{rr}, \qquad (21)$$

with the curve $E_{oss} = f(v_{DS})$ and the reverse recovery charge Q_{rr} measured for reverse voltage V_R and forward current I_F from the datasheet. For soft-switching in terms of ZCS for small $i_D(\tau_s)$, losses according to (20) become negligible small, hence losses are mainly described by (19).

3) Transformer Losses: Besides the losses in the semiconductor devices, the transformer losses in terms of core and winding losses are modeled. The transformer is built with four U-cores of size AMCC-32 [12] with material VITROPERM 500F [13], which exhibits a relatively high saturation flux density of 1.2 T and is therefore ideally suited for switching frequencies in the range of a few 10 kHz. For the windings, litz wire with 945 strands of diameter 0.071 mm is used. On the primary side, there are 24 turns, whereas the secondary winding consists of 21 turns. The primary winding W_1 is directly wound on the inner leg with the secondary winding W_2 around the inner leg and the outer leakage leg as shown in Fig. 10. By inserting an air gap of length δ_σ in the leakage leg, the leakage inductance L_σ can be set. Practically, distributed air gaps are used in order to reduce losses induced by the fringing field. The leakage inductance L_σ is determined in such a way, that the maximum input power of 3.3 kW can be transferred at the lowest switching frequency of 20 kHz and the lowest battery voltage of 280 V for the considered modulation method. The transformer parameters are summarized in Table II.

In the loss model, the core losses per volume are calculated by applying the improved Generalized Steinmetz Equation (iGSE) [14]. The skin and proximity effect losses per unit length in litz wires for each current harmonic are determined according to [15]. The external magnetic field strength for evaluating proximity effect losses is derived by a 1D approximation using the Dowell method [16].

4) Auxiliary Losses: Besides the load dependent loss shares shown in the previous sections, a constant loss share for gate drives, control, sensing and fans of 8 W is considered.

C. Optimization Procedure

The optimal control variables in terms of clamping intervals δ_1, δ_2, phase-shift ϕ_2 and switching frequency f_s are numerically determined by minimizing the total converter losses (semiconductor losses P_{sw}, transformer losses P_{tr} and auxiliary losses P_{aux}) subject to power flow constraint. The optimization procedure is shown in Fig. 11. For given output voltage $V_2 \in [280\,\text{V}, 420\,\text{V}]$ and reference output power $P_{12}^* \in [0.33\,\text{kW}, 3.3\,\text{kW}]$, the optimization routine calculates the optimal control variables. The optimization problem is stated as

$$\min_{x}\left[P_{sw} + P_{tr} + P_{aux}\right] \text{ with respect to } x = \begin{bmatrix} \delta_1 \\ \delta_2 \\ \phi_2 \\ f_s \end{bmatrix} \quad (22)$$

with

$$x_{lb} = \begin{bmatrix} 0 \\ 0 \\ -\pi \\ 20\,\text{kHz} \end{bmatrix}, \quad x_{ub} = \begin{bmatrix} \pi/2 \\ \pi/2 \\ \pi \\ 50\,\text{kHz} \end{bmatrix} \quad (23)$$

where x denotes the vector of control variables which is restricted to lower and upper bounds x_{lb}, x_{ub} respectively. The equality constraint is given by setting the power transfer $P_{12} = P_{12}^*$ using (12).

D. Optimization Results

Calculated relative converter efficiencies applying phase-shift modulation, combined triangular/trapezoidal current mode modulation and modulation with optimized control variables including frequency variation are shown in Fig. 13 and Table III for an IGBT solution and in Fig. 14 and Table IV for a MOSFET solution.

For phase-shift modulation, it can be seen that for the MOSFET solution in the soft-switching area higher efficiencies are achieved than for the IGBT solution (compare Fig. 14a to Fig. 13a). This is mainly due to the fact, that IGBT turn-off losses cannot be substantially reduced by using ZVS. In the hard-switching region, ZVS is lost and forced diode commutations occur. The switching losses in this region are strongly dependent on the characteristics of the anti parallel diode of the IGBT and the body diode of the MOSFET respectively.

To improve efficiencies, especially in the hard-switching region, the combined triangular/trapezoidal current mode modulation can be used. There, also for low output power in the areas of low and high output voltages, soft-switching (ZCS combined with ZVS) can be achieved.

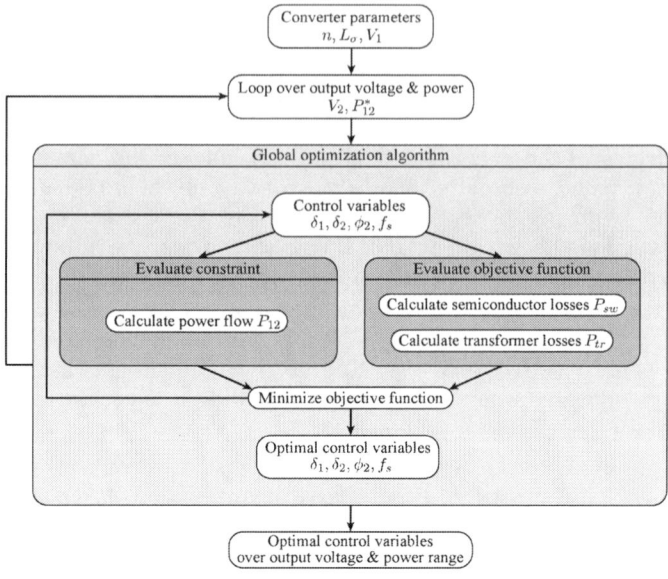

Fig. 11. Visualization of the optimization procedure to find optimal control variables $\delta_1, \delta_2, \phi_2, f_s$ by minimizing the DAB converter losses subject to power flow constraint. P_{12}^* represents the reference value for the transferred power.

TABLE III
EFFICIENCIES FOR DIFFERENT MODULATION METHODS APPLYING IGBTS. AVERAGE EFFICIENCY IS CALCULATED OVER THE WHOLE OUTPUT VOLTAGE/POWER RANGE.

Modulation method	Calculated efficiency	
	Peak	Average
Phase-shift modulation	98.0 %	96.9 %
Triangular/Trapezoidal current modulation	98.2 %	97.5 %
Optimized control variables modulation	98.2 %	97.6 %

TABLE IV
EFFICIENCIES FOR DIFFERENT MODULATION METHODS APPLYING MOSFETs. AVERAGE EFFICIENCY IS CALCULATED OVER THE WHOLE OUTPUT VOLTAGE/POWER RANGE.

Modulation method	Calculated efficiency	
	Peak	Average
Phase-shift modulation	99.0 %	97.6 %
Triangular/Trapezoidal current modulation	99.0 %	98.3 %
Optimized control variables modulation	99.1 %	98.5 %

Considering also frequency variation, efficiencies can be slightly increased compared to triangular/trapezoidal current mode modulation. Soft-switching is achieved in the whole operating range: ZCS and ZVS for the IGBT solution and only ZVS for the MOSFET solution. The modulation modes, which are found by the optimization procedure, are given in Fig. 13c and Fig. 14c. Fig. 12 shows the resulting switching frequencies found by the optimization, both for the IGBT and the MOSFET solution. Especially for the MOSFET solution, the frequency is varied over a wide area of the operating range. With decreasing power, frequency can be increased to lower the power transfer and achieve high efficiencies at the same time. Nevertheless, at low input powers, it is more attractive to decrease the switching frequency and change the modulation mode when necessary (see Fig. 14c).

978-1-4799-2706-7/14 $31.00 © 2014 IEEE   1793

TABLE V
CHARGING EFFICIENCIES OF A 14.2 kWh LITHIUM-ION BATTERY PACK FROM 10 % TO 90 % STATE-OF-CHARGE FOR DIFFERENT MODULATION METHODS APPLYING IGBTs AND MOSFETs.

Modulation method	Charging efficiency	
	IGBTs	MOSFETs
Phase-shift modulation	98.0 %	99.0 %
Triangular/Trapezoidal current modulation	98.1 %	99.0 %
Optimized control variables modulation	98.1 %	99.0 %

For a 14.2 kWh battery pack with 11.5 Ah lithium iron phosphate (LiFePO$_4$) cells [17], efficiencies of a typical charging process from 10 % to 90 % state-of-charge (battery voltage from 385 V to 418 V) with constant input power $P_1 = 3.3$ kW are calculated and given in Table V for different modulation methods applying IGBTs and MOSFETs. It can be seen, that the charging efficiencies at maximum input power do not differ much for the same semiconductor technology applying different modulation methods.

V. CONCLUSION

A novel modeling approach of the power flow in a bidirectional DAB DC-DC converter is presented. By using basic superposition principles, the mathematical distinction of cases is avoided in the modeling process of HF transformer currents for different types of modulation. The generalized model is used in an optimization of converter losses for an IGBT and a MOSFET solution considering all types of modulations, also including variation of frequency for converter control. For a 3.3 kW electric vehicle battery charger, the efficiency increases up to 8.6 % using IGBTs and 17.8 % using MOSFETs in the most critical point compared to phase-shift modulation at fixed switching frequency.

ACKNOWLEDGMENT

The authors would like to thank Swisselectric Research and the Competence Center Energy and Mobility (CCEM) very much for their strong financial support of the research work.

REFERENCES

[1] H. Fan and H. Li, "High Frequency High Efficiency Bidirectional DC-DC Converter Module Design for 10 kVA Solid State Transformer," in *Proc. 25th Applied Power Electronics Conference and Exposition (APEC)*, 2010, pp. 210–215.

[2] M. N. Kheraluwala, R. W. Gascoigne, D. M. Divan, and E. D. Baumann, "Performance Characterization of a High-Power Dual Active Bridge DC-to-DC Converter," *IEEE Transactions on Industry Applications*, vol. 28, no. 6, pp. 1294–1301, 1992.

[3] R. L. Steigerwald, "A comparison of half-bridge resonant converter topologies," *IEEE Transactions on Power Electronics*, vol. 3, no. 2, pp. 174–182, 1988.

[4] N. Schibli, "Symmetrical multilevel converters with two quadrant DC-DC feeding," Ph.D. dissertation, Swiss Federal Institute of Technology Lausanne (EPFL), 2000.

[5] F. Krismer, "Modeling and Optimization of Bidirectional Dual Active Bridge DC-DC Converter Topologies," Ph.D. dissertation, ETH Zurich, 2010.

[6] G. Guidi, M. Pavlovsky, A. Kawamura, T. Imakubo, and Y. Sasaki, "Improvement of light load efficiency of Dual Active Bridge DC-DC converter by using dual leakage transformer and variable frequency," in *Energy Conversion Congress and Exposition (ECCE)*, 2010, pp. 830–837.

[7] F. Jauch and J. Biela, "An Innovative Bidirectional Isolated Multi-Port Converter with Multi-Phase AC Ports and DC Ports," in *Proc. 5th EPE Joint Wind Energy and T&D Chapters Seminar*, 2012.

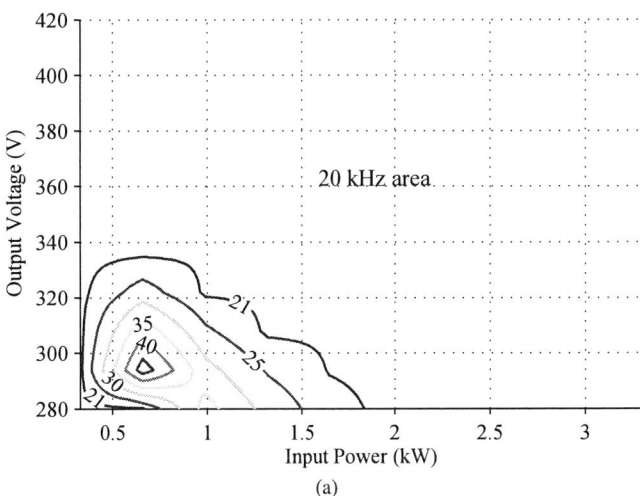

(a)

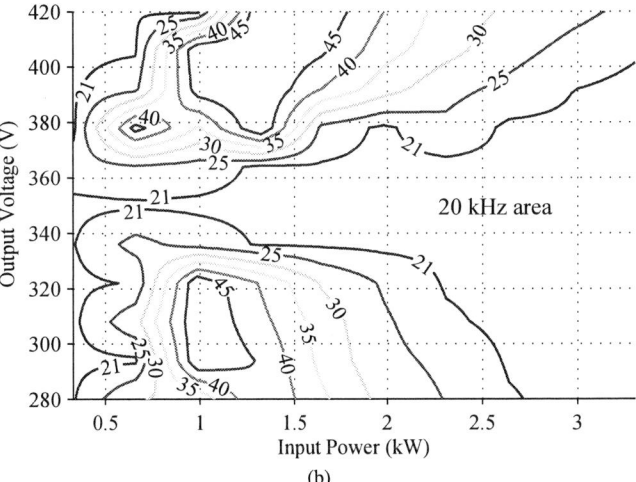

(b)

Fig. 12. Switching frequencies in kHz for optimized control variables modulation for an input power range of 10 % to 100 % of maximum input power and an output voltage range of 280 V to 420 V using IGBTs (a) and using MOSFETs (b).

[8] A. M. Ari, L. Li, and O. Wasynczuk, "Modeling and Analysis of N-Port DC-DC Converters using the Cyclic Average Current," in *Proc. 27th Applied Power Electronics Conference and Exposition (APEC)*, 2012, pp. 863–869.

[9] F. Jauch and J. Biela, "Modelling and ZVS control of an isolated three-phase bidirectional AC-DC converter," in *15th European Conference on Power Electronics and Applications (EPE)*, 2013, pp. 1–11.

[10] [Online]. Available: http://www.infineon.com

[11] N. Mohan, T. M. Undeland, and W. P. Robbins, *Power Electronics: Converters, Applications, and Design*. John Wiley & Sons, 2002.

[12] [Online]. Available: http://www.hitachi-metals.co.jp

[13] [Online]. Available: http://www.vacuumschmelze.de

[14] K. Venkatachalam, C. Sullivan, T. Abdallah, and H. Tacca, "Accurate Prediction of Ferrite Core Loss with Nonsinusoidal Waveforms using only Steinmetz Parameters," in *Proc. 8th IEEE Workshop on Computers in Power Electronics*, June 2002, pp. 36–41.

[15] J. Mühlethaler, "Modeling and Multi-Objective Optimization of Inductive Power Components," Ph.D. dissertation, ETH Zurich, 2012.

[16] P. Dowell, "Effects of Eddy Currents in Transformer Windings," *Proceedings of the Institution of Electrical Engineers*, vol. 113, no. 8, pp. 1387–1394, 1966.

[17] A. Vezzini, "Lithiumionen-Batterien als Speicher für Elektrofahrzeuge, Teil 1: Technische Möglichkeiten heutiger Batterien," *Bulletin SEV/AES*, vol. 3, pp. 19–23, 2009.

978-1-4799-2706-7/14 $31.00 © 2014 IEEE

The 2014 International Power Electronics Conference

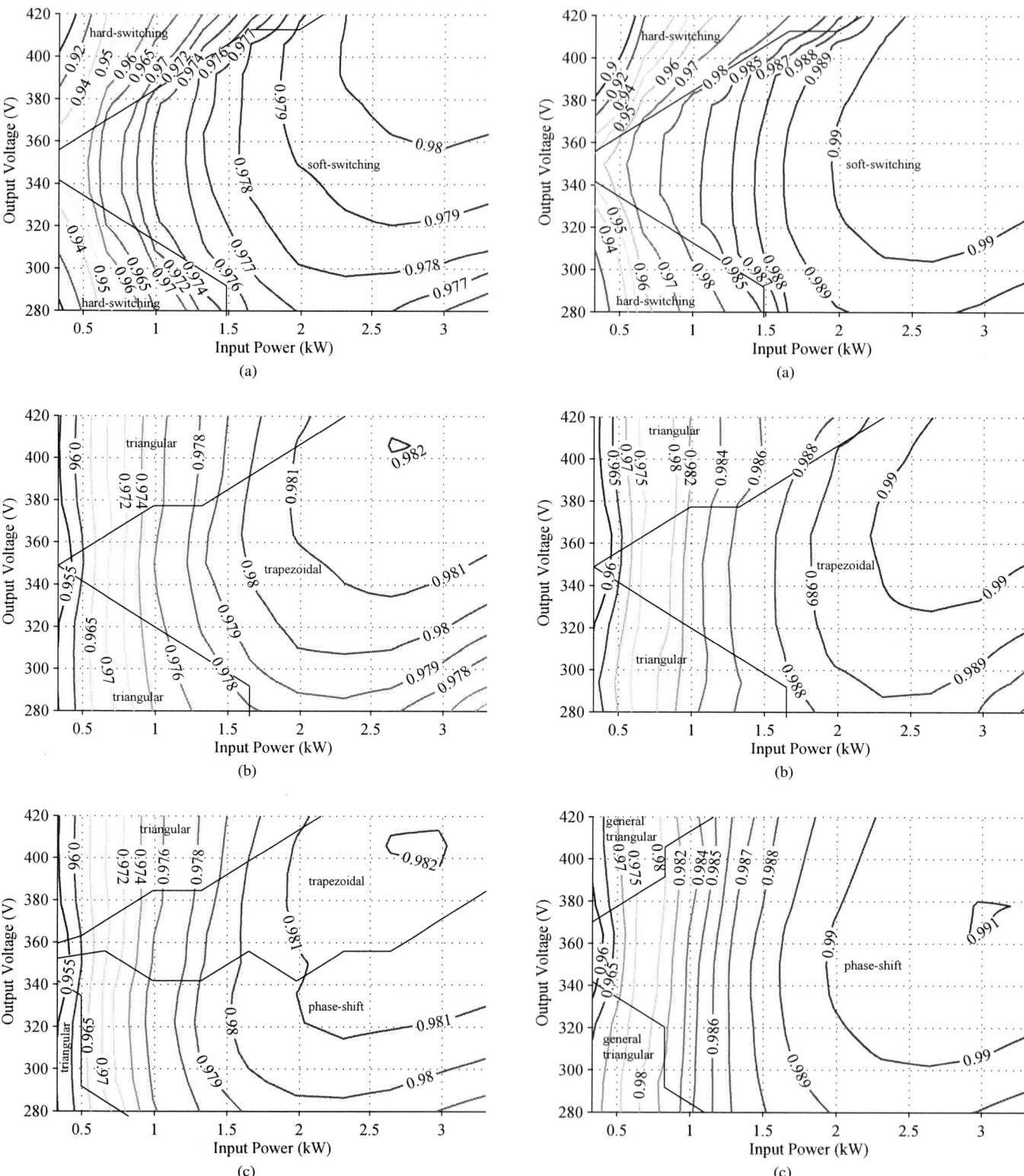

Fig. 13. Relative converter efficiencies applying IGBTs for an input power range of 10 % to 100 % of maximum input power and an output voltage range of 280 V to 420 V for phase-shift modulation (a) and triangular/trapezoidal current mode modulation (b) both with fixed switching frequency at 20 kHz and for modulation with optimized control variables with variable switching frequency from 20 kHz to 50 kHz (c). For phase-shift modulation in (a), soft- and hard-switching areas are given, whereas with modulations in (b) and (c) soft-switching is always achieved. In (c), also the modulation mode found by the optimization is given.

Fig. 14. Relative converter efficiencies applying MOSFETs for an input power range of 10 % to 100 % of maximum input power and an output voltage range of 280 V to 420 V for phase-shift modulation (a) and triangular/trapezoidal current mode modulation (b) both with fixed switching frequency at 20 kHz and for modulation with optimized control variables with variable switching frequency from 20 kHz to 50 kHz (c). For phase-shift modulation in (a), soft- and hard-switching areas are given, whereas with modulations in (b) and (c) soft-switching is always achieved. In (c), also the modulation mode found by the optimization is given.

Balanced Discharging of Power Bank with Buck-Boost Battery Power Modules

Chin-Sien Moo, Tsung-Hsi Wu, Chih-Hao Hou
Department of Electrical Engineering
National Sun Yat-sen University
Kaohsiung, Taiwan
mooxx@mail.ee.nsysu.edu.tw

Yao-Ching Hsieh
Department of Electrical Engineering
National Dong Hwa University
Hualien, Taiwan
ychsieh@mail.ndhu.edu.tw

Abstract— The operation of a battery power bank with the buck-boost type battery power modules (BPMs) is studied. All BPMs in the power bank are collaboratively to cope with the load requirements but substantially are operated individually. They can be scheduled to discharge currents from batteries in accordance with their state-of-charges (SOCs) and the operating modes of the converters. The operations of BPMs are analyzed to derive the current distribution equations and then to figure out the discharging strategy accordingly. Experimental results demonstrate that excellent performance on charge equalization can be achieved during the discharging processes. In addition, a fault-tolerance function can be included to isolate those with completely exhausted or damaged batteries. These features are helpful to maintenance and management of a battery power system.

Keywords— Battery power modules (BPMs), Charge equalization, Discharging strategy, State-of-charge (SOC).

I. INTRODUCTION

Battery power has been now widely used in high-voltage and high-power applications, such as electric vehicles (EVs) and distributed energy resources. To fulfill the system requirements, a number of battery cells are connected in series to reach the demanded high-voltage and in parallel to supply a higher power or a sustainable working cycle. With such a power pack or bank, however, serious problematic imbalance in state-of-charges (SOCs) may occur to the batteries, especially to those in series due to intrinsic discrepancies or different initial states. The charge imbalance among batteries will be magnified along with the times of charging or discharging cycles, and eventually causes over-charge or over-discharge to the batteries. To solve the problems, a conventional method is to introduce a battery management system with charge equalization [1-3]. The charge equalization technique and over charge/discharge protection circuits along with the battery management system, can relieve the problematic charge imbalance, but may bring about other problems including additional losses, production cost, and inefficient utilization of the battery power [4-6].

Alternatively, the concept of battery power module (BPM) was proposed, in which battery cells or packs are associated with power electronic converters [7-9].

A power bank can be assembled by a number of BPMs connected in series to achieve the required high load voltage [10], and in parallel for high power and energy applications. The BPMs connected in series are with an identical output current but share a portion of the aggregated load voltage [11-13]. When attached in parallel, the BPMs are with the same output voltage but share the load current cooperatively. On the other hand, the battery cells or packs in all BPMs can be controlled individually with such a configuration. As a result, charge equalization is not critical issue since charging or discharging currents of the BPMs with unequal SOCs can be scheduled in accordance with the corresponding SOCs [14]. In other words, the BPMs are associated with full power balance capability. Nevertheless, the utilization of battery power can be further improved if charge equalization is included in the discharging process. In this research, the effort is focused on balanced discharging of a power bank formed by BPMs with buck-boost converters.

II. CONFIGURATION OF POWER BANK WITH BUCK-BOOST BPMs

Fig. 1 illustrates the configuration of a battery power bank constructed by a number of $n \times m$ BPMs, which are connected first in series and then in parallel. Each BPM consists of a battery set with an associated bidirectional buck-boost converter. The battery power bank is formed by m BPM queues in parallel. Each queue is composed of n BPMs in series to meet the load voltage requirement, V_o. The BPMs with bidirectional buck-boost converters can be functioned for either charging or discharging. For discharging operation, the active power switch denoted by M_i in the i-th BPM plays the role of the main power switch for buck-boost conversion and the auxiliary active power switch S_i performs synchronous rectification conducting the free-wheeling current of the inductor.

All BPMs are controlled by microprocessors to regulate the discharging currents. With this configuration, isolation of the exhausted or damaged batteries can be easily implemented simply by removing the gate signals of the active power switches in the corresponding BPMs. The remaining BPMs will still supply a lighter load collaboratively without shutting the system down.

978-1-4799-2706-7/14 $31.00 © 2014 IEEE

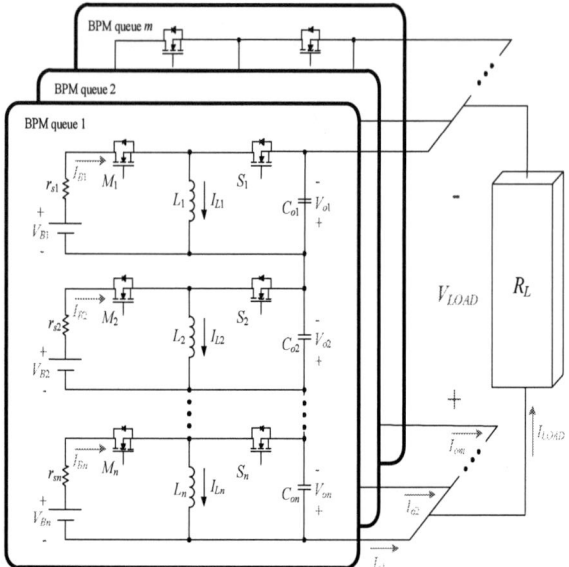

Fig. 1. Configuration of power bank with buck-boost BPMs.

Fig. 2. Buck-boost BPMs with parallel-series configuration.

III. CIRCUIT OPERATION

To simplify the analysis, all BPMs are with identical circuit parameters. The filter capacitors at output terminals are large enough so that the output voltages are assumed to be constant at the steady state. The buck-boost type BPMs may be operated at the continuous conduction mode (CCM) or the discontinuous conduction mode (DCM), depending on the continuity of the inductor current which is determined by the designed circuit parameters as well as the load condition. Different operation modes will make diverse effects on the batteries.

In the power bank, all BPM queues are with the same output voltage but contribute different currents. The queue which has the largest sum of battery voltages is defined as the master queue to govern the output voltage and the remnant queues are the slave queues.

Fig. 2 illustrates the configuration of a battery power bank constructed by m BPM queues in parallel. Theoretically, the equivalent load resistance, R_L, can be equivalent to m identical resistance, R_{Li}, of m BPM queues connected in parallel.

$$R_L = \frac{1}{\displaystyle\sum_{i=1}^{m} \frac{1}{R_{Li}}} \tag{1}$$

A. CCM operation of parallel-series configuration

The average output currents of all BPMs connected in series are the same. For the CCM operation, the average current discharged from the battery is proportional to the duty-ratio of the corresponding buck-boost converter in a BPM.

$$I_{B1}:I_{B2}\cdots:I_{Bi}\cdots:I_{Bn} = \frac{d_1}{1-d_1}:\frac{d_2}{1-d_2}:\cdots:\frac{d_i}{1-d_i}:\cdots:\frac{d_n}{1-d_n} \tag{2}$$

where I_{Bi} is the average battery current and d_i is the duty-ratio of the buck-boost converter in the i-th BPM.

Equation (2) indicates that the battery with a higher battery voltage can contribute a larger current by designating a larger duty-ratio to the associated converter, meaning that charge equalization can be achieved by adjusting the duty-ratios of BPMs.

For BPMs with parallel-series configuration, the contribution of a BPM queue to output current is related to the battery internal resistances, r_{si}.

$$I_{oi} = \sum_{i=1}^{n} \frac{V_{Bi} d_i (1-d_i)}{R_{Li}(1-d_i)^2 + d_i^2 r_{si}} \tag{3}$$

where I_{oi} is the output current of the i-th queue. The portion of the equivalent load resistance for each queue can be expressed as

$$R_{Li} = R_L \left(1 + \frac{1}{I_{oi}} \sum_{k=1,k\neq i}^{m} I_{ok} \right) \tag{4}$$

Then, the output current of a BPM queue in Eq. (3) can be rewritten as

$$I_{oi} = \sum_{i=1}^{n} \frac{V_{Bi} d_i (1-d_i) - R_L(1-d_i)^2 \displaystyle\sum_{k=1,k\neq i}^{m} I_{ok}}{R_L(1-d_i)^2 + r_{si} d_i^2} \tag{5}$$

With the same duty-ratio, the queue which has the higher sum of battery voltages will output more currents to the

The 2014 International Power Electronics Conference

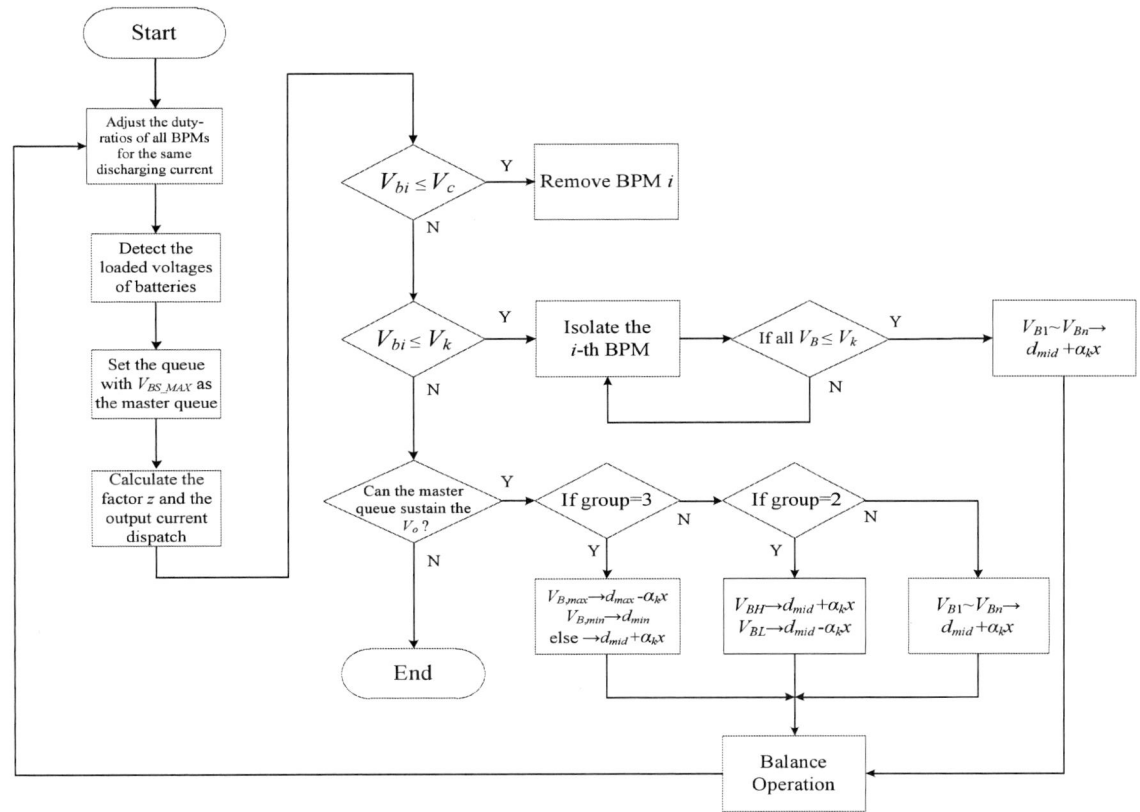

Fig. 3. Control flowchart of balanced discharging.

load. Therefore, the balanced discharging can be achieved by adjusting the duty-ratios of all BPMs.

B. DCM operation of parallel-series configuration

The relationship between battery currents and duty-ratios under DCM operation are as follows

$$I_{B1}:I_{B2}\cdots:I_{Bi}\cdots:I_{Bn}=\frac{d_1}{d_1'}:\frac{d_2}{d_2'}:\cdots:\frac{d_i}{d_i'}:\cdots:\frac{d_n}{d_n'} \qquad (6)$$

where d_i' is the time ratio for the inductor current decreasing from its peak to zero in a cycle. Then, the relations among battery voltages can be expressed as

$$V_{B1}:V_{B2}\cdots:V_{Bi}\cdots:V_{Bn}=\frac{1}{d_1 d_1'}:\frac{1}{d_2 d_2'}:\cdots:\frac{1}{d_i d_i'}:\cdots:\frac{1}{d_n d_n'} \qquad (7)$$

Equations (6) and (7) indicate that the BPM with a higher battery voltage has the smaller d_i', revealing that charge equalization may automatically be performed when all BPMs are operated with a same duty-ratio at the DCM.

IV. DISCHARGING STRATEGY

Fig. 3 illustrates the discharging strategy for the parallel-serial BPMs. The goals of the discharging strategy are not only to perform charge equalization but to provide load voltage regulation. The adjustable range of the duty-ratio of the buck-boost converter is designated between d_{max} and d_{min}. For balanced discharging, the BPM with the largest battery voltage is operated at the

maximum duty-ratio, d_{max}, and the BPM with the lowest battery voltage is set at the minimum duty-ratio, d_{min}.

The discharging process is divided into two cyclical stages, the detection stage and the balance stage. During the detection stage, all batteries are operated with the same discharging current, which can cope with the load requirements. With an identical current, the measured terminal voltages reveal the battery capacities more convincingly. At the balance stage, BPMs are operated with the designated duty-ratios to relieve the charge imbalance among batteries. The detection stage and the balance stage are interchanged alternately during the whole discharging process.

To schedule the discharging currents, all BPMs are categorized into most three groups in accordance with the measured battery voltages. The BPMs in a group are operated with an identical duty-ratio. As time goes by, the voltage differences between the batteries will be decreased. The voltage differences among batteries can be smaller and smaller. Eventually, charge equalization can be achieved.

For a BPM queue with m BPMs which are functioning, the ratio between the k-th battery's voltage and the sum of battery voltages in the queue can be express as

$$a_k = \frac{V_{Bk}}{\sum_{i=1}^{m} V_{Bi}} \qquad (8)$$

where V_{Bi} is the battery voltage of the i-th BPM.

978-1-4799-2706-7/14 $31.00 © 2014 IEEE   1798

The charge equalization in a queue can be achieved by considering the factor α_k for adjusting the duty-ratios of BPMs. The master queue can be selected by sorting the sum of battery voltages of all queues. The master queue not only governs the output voltage but loads larger output current than the slave queues do. The output current of each slave queue is distributed by the sum of its battery voltages. A factor z_k is introduced for reasonably dispatching the portion of load current.

$$z_k = \frac{V_{Sk}}{\sum\limits_{i=1}^{m} V_{Si} - V_{SM}} \qquad (9)$$

where V_{Sk} is the summation of battery voltages in k-th queue, and V_{SM} represents that voltage of the master queue.

The BPM which has the lowermost SOC is operated at d_{min}, and the associated battery supplies the smallest discharging current. As the battery has been completely exhausted, the associated converter is stopped to isolate the battery from the battery power bank.

In general, a battery's SOC can be estimated by the measured terminal voltage. The higher the voltage of the battery, the higher the SOC is. Fig. 4 illustrates the battery voltage curves for various discharging currents. The battery voltages decline more quickly under the larger discharging currents, and decrease drastically since the knee points, V_k, and thereafter to the cut-off voltage, V_c. A battery is considered to be exhausted and has to be isolated when has reached the cut-off voltage. With the diverse discharging currents, the knee point voltages can be slightly different to each other. In this research, the knee point is chosen at 12.5 V according to the circuit parameters.

V. EXPERIMENTAL TESTS

An implementation example of a battery power bank is built by two BPM queues connected in parallel. Each BPM queue consists of 4 buck-boost BPMs connected in series. The BPMs are numbered by 1 to 4 in the first queue and 5 to 8 in the second queue. Table 1 lists the circuit parameters of the battery power bank.

Fig. 5 shows the current sharing between the master and slave queues during the detection and balance stages. In the case of Fig. 5(a), all BPMs are operated at an identical duty-ratio at the detection stage. The output current is allocated equally by the two queues to ensure both queues are detected under the same discharging current. The Fig. 5(b) describes the current sharing during the balance stage with an output current of 1 A in the experiment. In this case, the sum of battery voltages of Queue 2 is higher than that of Queue 1, and hence is selected as the master queue to govern the load voltage. For balancing the battery capacities, the master queue delivers an average current of 0.73 A to the load, which is much higher than the slave queue does.

Fig. 6 shows an experimental result of the balanced discharging process. In the experiment, 8 lithium-ion battery packs are deliberately charged to be with different SOCs before discharging. At the beginning, all BPMs are operated at the detection stage with a same duty-ratio, drawing an identical current from batteries for measuring the corresponding loaded voltages. During the following balance stage, the battery currents are scheduled in accordance with their measured loaded voltages by controlling the duty-ratios of the corresponding buck-boost converters. At around the 490th second after the experiment has been started, the battery voltages of BPMs 2 and 4, which are initially with relatively low SOCs, reach the knee point. At this moment, these two BPMs are isolated temporarily.

Thereafter, only BPMs 1 and 3 are activated in the first queue. As the time goes by, the battery voltages of BPMs 5, 6, and 8 are getting closer to each other. About 1800 seconds later, the battery voltages of BPMs 3, 5, 6, and 8 reach the knee point almost at the same time. Then, the load is supplied only by BPM 7. After 2400 seconds, all battery voltages are at the same level and all BPMs are operated with a same duty-ratio. When all the batteries reach the cut-off voltage, the battery power bank completes the discharging process.

Fig. 7 shows the load voltage variation of the battery power bank with open-loop control. The drastic changes are caused by the switching of the discharging stage and the detection stage. The variation in the load voltage is less than 3.5 V over the discharging process.

TABLE I
CIRCUIT PARAMETERS

Rated output current, I_{LOAD}	1 A (0.217 C)
Rated output voltage, V_o	48 V
Rated output power	48 W
Cut-off voltage of battery pack	10 V
Knee point voltage of battery pack	12.5 V
Maximum battery discharge current	7 A (1.5 C)
Switching frequency, f_S	20 kHz
Inductor, L_i	200 μH
Output capacitor, C_o	470 μF

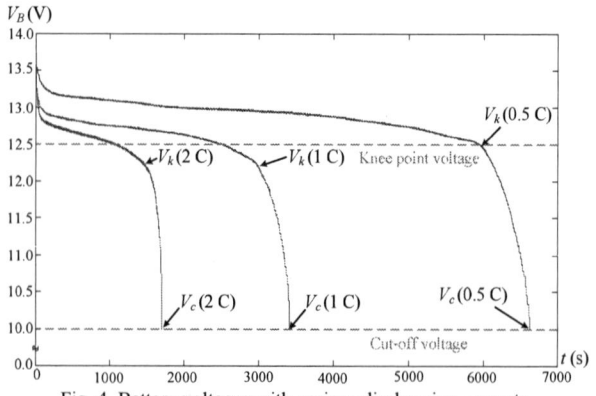

Fig. 4. Battery voltages with various discharging currents.

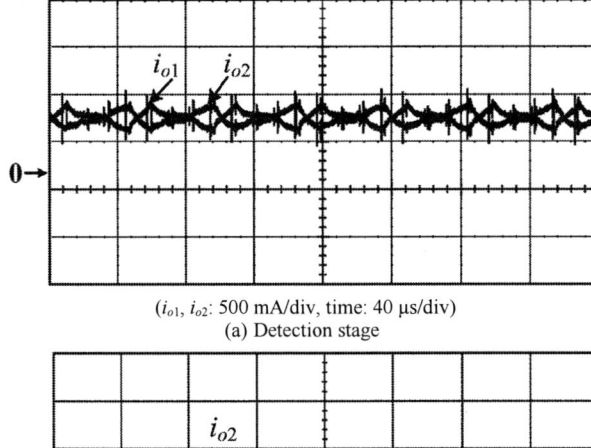

(i_{o1}, i_{o2}: 500 mA/div, time: 40 μs/div)
(a) Detection stage

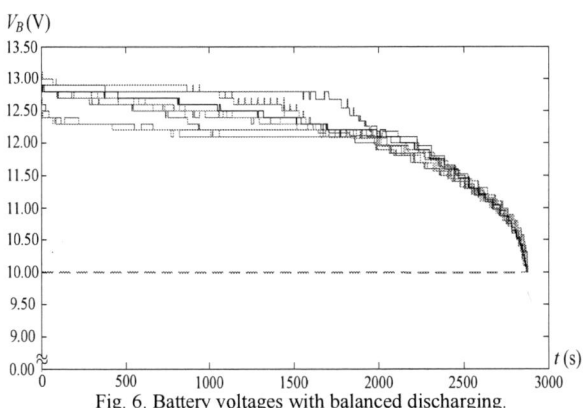

(i_{o1}, i_{o2}: 500 mA/div, time: 20 μs/div)
(b) Balance stage

Fig. 5. Current sharing between the master and slave queues during the detection and balance stages.

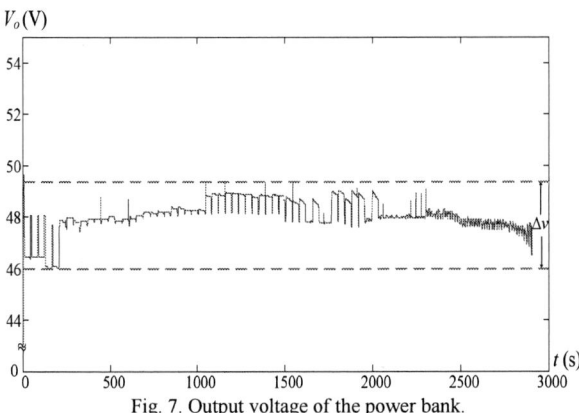

Fig. 6. Battery voltages with balanced discharging.

Fig. 7. Output voltage of the power bank.

VI. CONCLUSIONS

The configuration of a battery power bank with buck-boost type battery power modules connected in parallel-series has been illustrated. With such a connection, all BPMs, which are mutually affected to each other, have to be operated collaboratively to meet the demanding load voltage and current. In practice, charge equalization is not essential for the battery power bank with bidirectional BPMs. Nevertheless, to further improve the battery power utilization, a balanced discharging strategy is proposed for discharging the batteries in the BPMs in accordance with the corresponding SOCs.

Experimental results on a laboratory battery power bank have demonstrated that the BPMs can virtually be operated individually, so that balanced discharging can be achieved with a satisfied output voltage regulation.

REFERENCES

[1] X. F. Wang, S. Y. Yang, N. J. Park, K. J. Lee, and D. S. Hyun, "A three-port bidirectional modular circuit for li-ion battery strings charge/discharge equalization applications," in *Proc. IEEE PESC*, pp. 4695-4698, June 2008.

[2] C. H. Kim, M. Y. Kim, and G. W. Moon, "A modularized charge equalizer using a battery monitoring IC for series-connected li-ion battery strings in electric vehicles," *IEEE Trans. on Power Electronics*, vol. 28, no. 8, pp. 3779-3787, Aug. 2013.

[3] Y. S. Lee and G. T. Cheng, "Quasi-resonant zero-current-switching bidirectional converter for battery equalization applications," *IEEE Trans. on Power Electronics*, vol. 21, no. 5, pp. 1213-1224, Sep. 2006.

[4] W. C. Lee, D. Drury, and P. Mellor, "Comparison of passive cell balancing and active cell balancing for automotive batteries," in *Proc. IEEE VPPC*, pp. 1-7, Sep. 2011.

[5] Z. G. Kong, C. B. Zhu, R. G. Lu, and S. K. Cheng, "Comparison and evaluation of charge equalization technique for series connected batteries," in *Proc. IEEE PESC*, pp. 1-6, June 2006.

[6] S. M. Lukic, J. Cao, R. C. Bansal, F. Rodriguez, and A. Emadi, "Energy storage systems for automotive applications," *IEEE Trans. on Industrial Electronics*, vol. 55, pp. 2258-2267, June 2008.

[7] C. S. Moo, K. S. Ng, and Y. C. Hsieh, "Parallel operation of battery power modules," *IEEE Trans. on Energy Conversion*, vol. 23, no. 2, pp. 701-707, June 2008.

[8] K. H. Lin, L. R. Yu, C. S. Moo, and C. Y. Juan, "Analysis on parallel operation of boost-type battery power modules," in *Proc. IEEE PEDS*, pp. 809-813, Apr. 2013.

[9] C. S. Moo, T. H. Wu, K. S. Ng, Y. C. Hsieh, and C. Y. Juan, "Battery power module with tri-port DC-to-DC converter," in *Proc. IEEE AE*, pp. 1-4, Sep. 2011.

[10] C. S. Moo, J. Y. Jian, T. H. Wu, L. R. Yu, and C. C. Hua, "Battery power system with arrayed battery power modules," in *Proc. IEEE ICSSE*, pp. 437-441, July 2013.

[11] C. S. Moo, K. S. Ng, and J. S. Hu, "Operation of battery power modules with series output," in *Proc. IEEE ICIT*, pp. 1-6, Feb. 2009.

[12] C. H. Hou, C. T. Yen, T. H. Wu, and C. S. Moo, "A battery power bank of serial battery power modules with buck-boost converters," in *Proc. IEEE PEDS*, pp. 211-216, Apr. 2013.

[13] L. R. Yu, Y. C. Hsieh, W. C. Liu, and C. S. Moo, "Balanced discharging for serial battery power modules with boost converters," in *Proc. IEEE ICSSE*, pp. 449-453, July 2013.

[14] W. Hong, K. S. Ng, J. H. Hu, and C. S. Moo, "Charge equalization of battery power modules in series," in *Proc. IEEE IPEC*, pp. 1568-1572, June 2010.

Y-Source Impedance-Network-Based Isolated Boost DC/DC Converter

Yam P. Siwakoti, Graham E. Town
Department of Engineering
Macquarie University
NSW 2109, Australia
yam.siwakoti@mq.edu.au, graham.town@mq.edu.au

Poh Chiang Loh, Frede Blaabjerg
Department of Energy Technology
Aalborg University
Pontoppidanstræde 101, 9220 Aalborg, Denmark
pcl@et.aau.dk, fbl@et.aau.dk

Abstract—A dc-dc converter with very high voltage gain is proposed in this paper for any medium-power application requiring a high voltage boost with galvanic isolation. The proposed converter topology can be realized using only two switches. With this topology a very high voltage boost can be achieved even with a relatively low duty cycle of the switches, and the gain obtainable is presently not matched by any existing impedance network based converter operated at the same duty ratio. The proposed converter has a Y-source impedance network to boost the voltage at the intermediate dc-link side and a push-pull transformer for square-wave AC inversion and isolation. The voltage-doubler rectifier provides a constant dc voltage at the output stage. A theoretical analysis of the converter is presented, supported by simulation and experimental results. A 250 W down-scaled prototype was implemented in the laboratory to demonstrate the feasibility and performance of the proposed converter topology.

I. INTRODUCTION

In recent years, the development of high-voltage-gain isolated dc-dc converters has become an important topic of research due to the boom in distributed power generation. The low and wide varying output voltage from various distributed generators, *e.g.* PV, fuel cells and small scale wind turbines (20 - 150 V) requires a high-step-up dc-dc converter to boost the voltage to much higher link dc-voltage (200 - 600 V) needed for interfacing to the utility grid (110 V_{ac}, 230 V_{ac} or 400 V_{ac}) [1], [2].

In addition, the trends towards dc electrification in residential and industrial applications are also increasing due to the emergence of the technologies of distributed dc power systems and dc micro-grids. Direct current (dc) power systems have many advantages compared to their alternating current (ac) counterparts, which is motivating many large IT companies to retrofit existing ac supply systems with dc systems. The trend towards dc power distribution highlights the need for highly efficient dc-dc boost converters for power conversion and conditioning [3]- [5].

Further, an isolated dc-dc converter with a high boost capability is also a prime requirement of many industrial applications, *e.g.* for Uninterruptible Power-supply Systems (UPS), telecommunications, the automobile industry, etc. [6]

In all such applications the boost stage is usually a critical point in high-efficiency converter design due to the high current at the input side and the high voltage at the output side. The high current drawn from the source leads to higher switching and conduction losses in the semiconductor switches and therefore reduces the efficiency of the converter. In addition, the high voltage at the output side often poses a threat to the safety requirements, so galvanic isolation is often required for flexibility of system reconfiguration and for meeting safety requirements. This requirement challenges power electronic researchers and engineers to implement a very-high-boost isolated dc-dc converter.

Various isolated as well as non-isolated boost converter topologies are presented in the literature to achieve a high step-up voltage gain [7], [8]. Generally, this is achieved by implementing one or more of the following techniques:

1) operating the converter at higher duty cycle;
2) multilevel topology;
3) cascading of converters;
4) voltage multiplier cells;
5) high-turns-ratio transformer;
6) high-turns-ratio coupled inductor;
7) impedance source network.

The selection and implementation of one or more of these techniques depends on the power density, efficiency, cost, reliability and isolation requirements of the system. Operating the converter at an extreme duty cycle reduces the efficiency and increases the stress on the switching device. On the other hand, voltage boost using a voltage multiplier and a multilevel/cascading structure requires many active and passive devices, which not only reduces the efficiency and power density but also increases the cost of the system. Voltage boost by implementing a high-turns-ratio transformer is not a good practice as it increases the size and cost. Higher boost can be possible with a coupled inductor, however higher boost requires a large turns ratio, which increases a leakage inductance and causes a problem with large voltage spikes in the switching devices [7]. Design of an isolated boost converter using a Z-source/quasi Z-source impedance

978-1-4799-2706-7/14 $31.00 © 2014 IEEE

network is implemented in [9], however the boost is restricted by the losses and the EMI associated with the higher shoot-through duty cycle.

In this paper, we propose a very high gain Y-source impedance network based isolated dc-dc converter using a Y-source impedance network [10], [11] with higher gain, and more tuning parameters. The proposed topology is versatile and gives the flexibility to choose the range of shoot-through duty cycle and the turns ratio of the coupled inductor. It also eliminates the requirement for a large turns ratio of the isolation transformer or coupled inductor. The topology of the proposed converter is presented in Section II with its operating principle. Simulation and experimental results are presented in Section III and a conclusion in Section IV.

II. PROPOSED CONVERTER TOPOLOGY AND PRINCIPLE OF OPERATION

The proposed converter consists of a Y-source impedance network whose primary function is to boost the voltage at the dc-link; a push-pull isolation transformer which galvanically isolates the input from the output; and a Voltage-Doubler Rectifier (VDR). The circuit diagram of the proposed converter topology is depicted in Fig. 1.

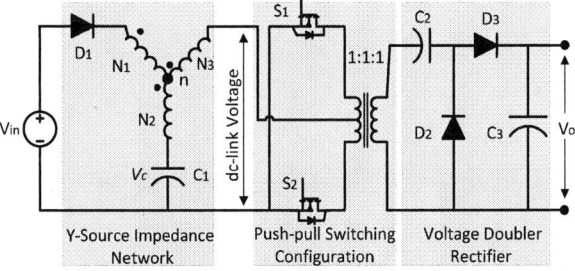

Fig. 1. Proposed Y-source impedance-network-based isolated dc-dc converter.

The voltage boost in the converter is mainly achieved by using a Y-source network which primarily consists of three windings (N_1, N_2, N_3) on the same core, and a capacitor and diode at the input current path. It has two states of operation: an active state and a shoot-through state. The impedance network stores energy during the shoot-through state which is transferred to the load during the active state. Here 'shoot-through state' refers to the condition where both switches S_1 and S_2 operate at the same time. This condition is normally forbidden in a push-pull converter. However, this state is utilized here to boost the input voltage at the dc-link voltage, which is then fed to the push-pull transformer via switch S_1 or S_2. Only one switch operates at a time during normal operating conditions, *i.e.* when the input voltage is equal to the desired dc-link voltage. The input diode conducts all the time and no switching takes place in the impedance network. The whole input voltage appears across the dc-link and the output voltage of the converter becomes

$$V_O = 2V_{in} \qquad (1)$$

However, when the dc-link voltage drops below the desired value, the converter operates in the shoot-through mode. The shoot-through is created by turning on both switches at the same time. Since the fluxes generated in the core due to the two primary windings are equal in magnitude and opposite in direction, the net flux in the core is zero. This zero-flux condition creates magnetic shoot-through and consequently electrical shoot-through at the dc-link side. During this shoot-through mode the diode D_1 is reverse biased which makes it open circuited and current $I_{in}=0$. Active state follows the shoot-through state; the capacitor C_1 charges and the dc-link voltage appear across the primary winding of the transformer. The voltage at the dc-link during this mode is given as

$$\hat{V}_{DC-link} = \frac{V_{in}}{[1-Kd_{st}]} \qquad (2)$$

where $K = \frac{N_1+N_3}{N_3-N_2}$ is the winding factor of the integrated magnetics and d_{st} is the overlap time period of switches S_1 and S_2 as shown in Fig. 2, and expressed in terms of the total duty cycle (D) as

$$d_{st} = 2D - 1 \qquad (3)$$

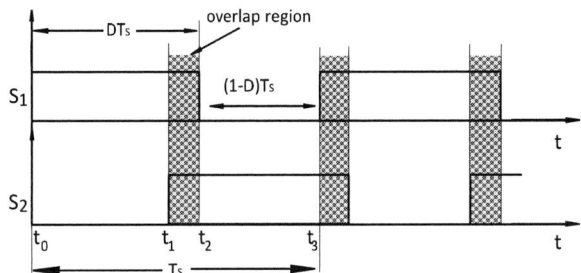

Fig. 2. Switching sequence of S_1 and S_2 in boost operation (D ≥ 50%).

Now (2) can be written as a function of the total duty cycle and K as

$$\hat{V}_{DC-link} = \frac{V_{in}}{[1-K(2D-1)]} \qquad (4)$$

Theoretically, any gain can be possible at the dc-link of the Y-source impedance network by varying the shoot-through duty cycle and K ($K \geq 2$). This large voltage-gain capability of the converter helps to reduce the large turns-ratio requirement of the isolation transformer, which improves the power density of the converter as well as reducing cost. For this work, a 1:1:1 transformer has been implemented simply for providing isolation in the push-pull circuit. Regardless of the final design chosen for the coupled inductor and transformer, their leakage inductances must be kept small by following common practices which is well documented in [12].

With the gains introduced by the Y-source network, transformer and VDR combined, the overall converter output voltage, V_o, and gain, $G_V(K, D)$, can be written as (5), where N_{PP} represents the transformer turns ratio, which for the 1:1:1 transformer is $N_{PP} = 1$,

$$V_o = 2\hat{v}_{dc-link} = \frac{2N_{PP}V_{in}}{[1-K(2D-1)]}$$

$$G_V(K,D) = \frac{2N_{PP}}{[1-K(2D-1)]} \quad (6)$$

The overall gain can thus be varied by changing K, D and N_{PP}, which can be deduced from (5) and is shown in Fig. 3 for different K and D with N_{PP} set to unity. Another tuning freedom shown more clearly in Table I is the possibility of generating a particular winding factor K, and hence gain G_V, by various winding combinations (N_1, N_2, N_3) of the coupled inductor. The eventual combination adopted should ideally help the converter to avoid large turns ratio for the coupled inductor and transformer, which would otherwise be difficult to realize. Such flexibilities are not available in other isolated impedance-source converters.

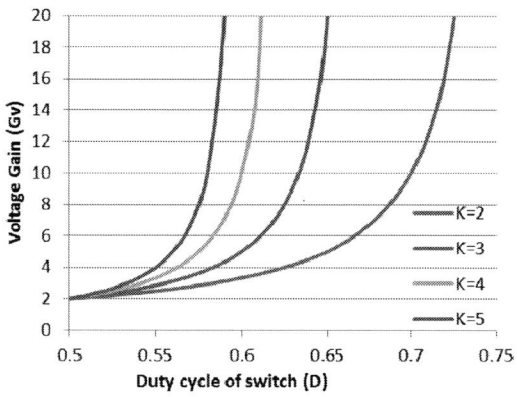

Fig. 3. Theoretical voltage gain of the proposed converter for different winding factors, K.

TABLE I
GAIN OF PROPOSED CONVERTER VERSUS WINDING FACTOR AND TURNS RATIO

K	Range of D	Voltage Gain (G_V)	Possible Turns Ratios $N_1 : N_2 : N_3$
2	$0.5 < D < 3/4$	$2[1-2(2D-1)]^{-1}$	1:1:3, 2:1:4, 1:2:5, 3:1:5, 4:1:6
3	$0.5 < D < 4/6$	$2[1-3(2D-1)]^{-1}$	1:1:2, 3:1:3, 2:2:4, 1:3:5, 4:2:5
4	$0.5 < D < 5/8$	$2[1-4(2D-1)]^{-1}$	2:1:2, 1:2:3, 5:1:3, 4:2:4, 8:1:4
5	$0.5 < D < 6/10$	$2[1-5(2D-1)]^{-1}$	3:1:2, 2:2:3, 1:3:4, 7:1:3, 6:2:4

Since, the integrated magnetics connect directly to the switches S_1, S_2 and D_1, its coupling must be tight to minimize leakage inductances seen at its windings. The latter can cause large switching transients, and reduces the voltage gain and efficiency of the system.

III. SIMULATION AND EXPERIMENTAL RESULTS

Simulation was performed using PLECS, using the parameters listed in Table II, which are also the experimental parameters.

TABLE II
PARAMETERS AND COMPONENT VALUES OF THE CONVERTER

Parameter/Description	Value/Part Number
Power rating	230 W
Input voltage range, V_{in}	48 V - 115 V
Output voltage, V_o	230 V
Capacitance, C_1, C_2 and C_3	470 µF @ 400V Kemet
Turns ratio of coupled inductor	5:1:3 (80:16:48) on *C055863A2* MPP core from Magnetics Inc.
Winding factor, K	4
Permitted ranges of D and d_{st}	$0.5 < D < 0.625,\ 0 < d_{st} < 0.25$
Isolation transformer	1:1:1 (30:30:30) on *0W48613TC* ferrite core from Magnetics Inc.
Switching frequency, f_s	6.1 kHz
Switching devices, S_1 and S_2	V_{DS} = 650 V, I_{DS} = 47 A CoolMOS *SPW47N60C3*
Diode, D_1	*C3D25170H*–Silicon Carbide Schottky Diode Z-Rec™ Rectifier
Diodes, D_2 and D_3	*C3D20060D*–Silicon Carbide Schottky Diode Z-Rec™ Rectifier

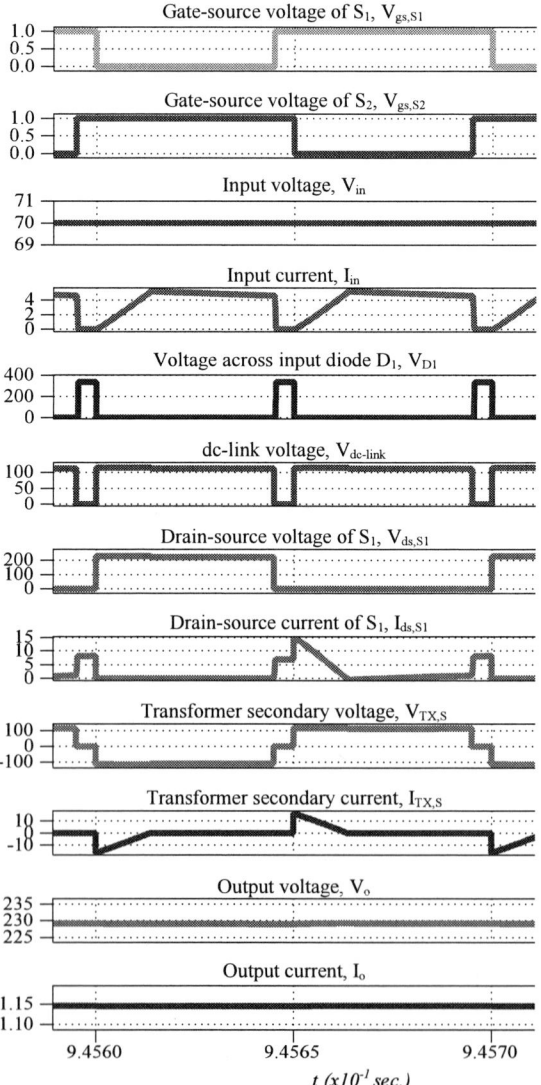

Fig. 4. Simulation results of Y-source impedance-network-based isolated dc-dc converter at d_{st} = 0.1 and K = 4.

Fig. 4 shows the simulated result of the proposed converter with an input voltage of 70 V and D = 0.55 (or d_{st} = 0.1). The peak dc-link voltage and output voltage are indeed close to the values of 116 V and 230 V computed from (4) and (5), respectively. The simulation results have an overall voltage gain of 3.33, which is in agreement with the gain computed using (5). Also anticipated is the doubled dc-link voltage stress experienced by the switches, when they are in the push-pull active state.

A 230 W prototype was designed in the laboratory (see Fig. 5) to measure the performance of the converter. Different measurements were taken at different values of d_{st} for different input voltages (see Table II) to regulate the output voltage to 230 V.

Fig. 6 shows the experimental waveform of the input/output voltage and current of the converter at d_{st} = 0.15, which gives a computed gain of 5. The respective waveforms are obtained after reaching thermal stability. The output voltage is close to the calculated value and the waveforms are similar to the simulated ones. All other intermediate waveforms and values of the converter are in agreement with the theoretical values and simulated results as shown in Fig. 7.

Fig. 5. The experimental setup of proposed converter.

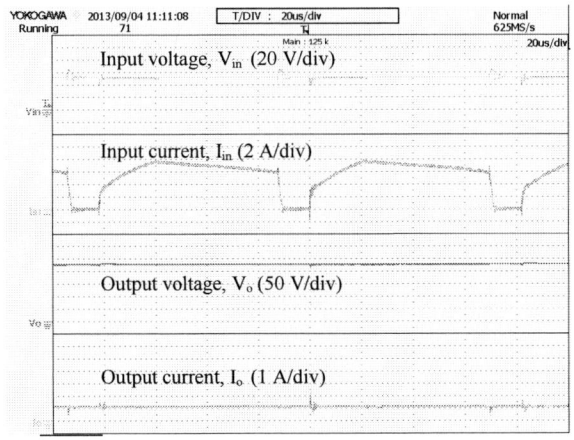

Fig. 6. Experimental waveforms of the input/output voltage and current of the converter obtained at d_{st} = 0.15.

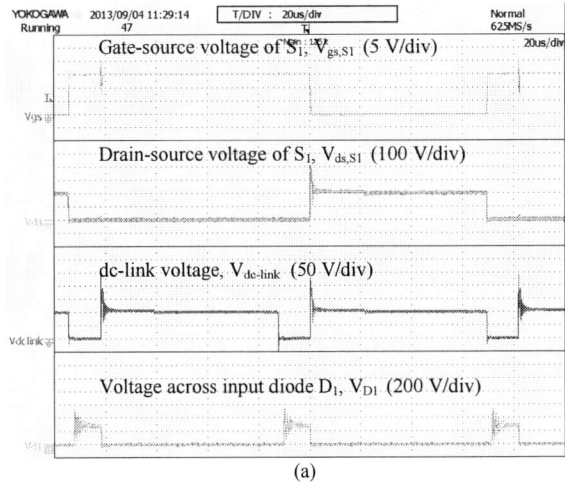

(a)

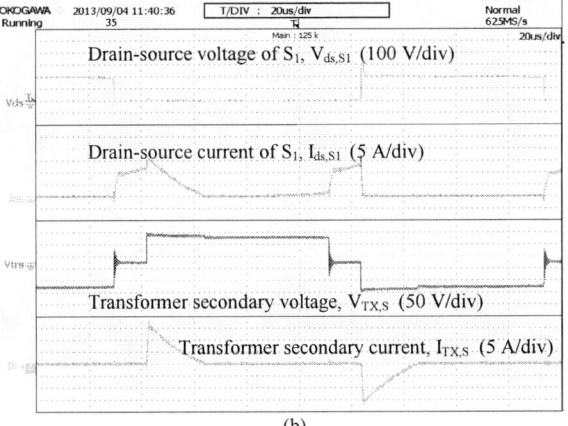

(b)

Fig. 7. Experimental waveforms of the proposed converter when D = 0.575 (d_{st} = 0.15) and K = 4.

Fig. 8 shows the measured efficiency of the converter over a range of input voltages with D varied to keep the output voltage constant at 230 V. The maximum efficiency recorded is 96.1%, which happens at D = 0.5, or zero shoot-through time. It falls to 91.1% at D = 0.575, or 15% shoot-through time in a switching period. The drop in efficiency is likely related to the large shoot-through current, whose accompanied losses can be reduced by using better graded wires and busbars to implement the converter.

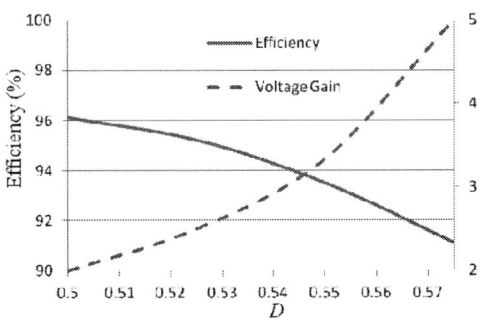

Fig. 8. Measured efficiency and voltage gain of the converter when K = 4 and 0.5 ≤ D ≤ 0.575.

IV. CONCLUSION

We have proposed a high voltage gain isolated dc-dc converter based on a Y-source impedance network. The shoot-through state of the converter can be varied in a narrow range to obtain a very wide range of voltage gain. This reduces the otherwise large turns-ratio required of the isolation transformer. Furthermore, the integrated magnetics improve the power density and efficiency of the converter as well as reducing the cost and size. The operation and performance of the proposed converter have been verified by both simulation and experimental testing under a variety of operating conditions.

REFERENCES

[1] F. Blaabjerg, Z. Chen and S. B. Kjaer, "Power Electronics as Efficient Interface in Dispersed Power Generation Systems," *IEEE Trans. Power Electron.*, vol. 19, no. 5, pp. 1184-1194, Sept. 2004.

[2] E. H. Ismail, M. A. Al-Saffar, A. J. Sabzali and A. A. Fardoun, "A Family of Single-Switch PWM Converters With High Step-Up Conversion Ratio," *IEEE Trans. on Circuits and Systems*, vol. 55, no. 4, pp. 1159-1171, 2008.

[3] Y. P. Hsieh, J. F. Chen, T. J. Liang and L. S. Yang "A Novel High Step-Up DC–DC Converter for a Microgrid System," *IEEE Trans. Power Electron.*, vol. 26, no. 4, pp. 1127-1136, April 2011.

[4] Y. P. Siwakoti and G. Town, "Performance of Distributed DC Power System using Quasi Z-Source Inverter Based DC/DC Converters," in *Proc. Applied Power Electronics Conference APEC2013*, pp. 1946-1953.

[5] An Electric Power Research Institute (EPRI) white paper on DC Power Production, Delivery and Utilization, June 2006.

[6] http://www.emersonnetworkpower.com/en-US/About/newsroom/NewsReleases/Pages/Key-Applications-400VDC-Power-Technology.aspx

[7] W. Li and X. He, "Review of Nonisolated High-Step-Up DC/DC Converters in Photovoltaic Grid-Connected Applications," *IEEE Trans. Ind. Electron.*, vol. 58, no. 4, pp. 1239-1250, April 2011.

[8] Y. Du, S. Lukic, B. Jacobson and A. Huang, "Review of high power isolated bi-directional DC-DC converters for PHEV/EV DC charging infrastructure," in *Proc. Energy Conversion Congress and Exposition (ECCE)*, pp. 553,560, 17-22 Sept. 2011.

[9] D. Vinnikov and I. Roasto, "Quasi-Z-Source-based Isolated DC/DC Converters for Distributed Power Generation," *IEEE Trans. Ind. Electron.*, vol. 58, no. 1, pp. 192-201, Jan. 2011.

[10] Y. P. Siwakoti, P. C. Loh, F. Blaabjerg and G. E. Town, "Y-Source Impedance Network," in *Proc. Twenty-Ninth Annual IEEE Applied Power Electronics Conference and Exposition (APEC)*, Fort Worth, TX, March 16-20, 2014.

[11] Y. P. Siwakoti, P. C. Loh, F. Blaabjerg and G. E. Town, "Y-Source Impedance Network," *IEEE Trans. Power Electron. (Letter)*, vol. 29, no. 7, pp. 3250-3254, Jul. 2014.

[12] Y. P. Siwakoti, P. C. Loh, F. Blaabjerg and G. E. Town, "Effects of Leakage Inductances on Magnetically-Coupled Impedance-Source Networks," *IEEE Trans. Power Electron. (Letter)* (Accepted).

Multi-phase DC-DC Converter with Ripple-less Operation for Thermo-Electric Generator

Noriyuki Kimura*, Koji Niijima*, Toshimitsu Morizane* and Hideki Omori*

*Electrical and Electronic Systems Engineering Dept.,
Osaka Institute of Technology, Osaka, Japan
Email: n.kimura@ieee.org

Abstract-**This paper presents the possibility to make the dc current ripple to zero. The simulation results show that the ripple-less operation is possible theoretically. The fast response in the current control is also shown by using the simple digital control algorithm. The control for the load change is investigated in simulation. The simulation results show that the output current loop is more suitable than the output voltage loop. Experimental results show that the ripple cannot be suppressed substantially because of the delay in control system. Therefore, the suppression of the effect of the delay in control is next step.**

I. INTRODUCTION

Recently many dc output distributed energy source have been developed. Some of these have low dc voltage output and need to step up the voltage at the dc side.

Thermo-Electric Generator (TEG) is used to generate electricity from the temperature difference [1,2]. Well known materials are kinds of semiconductor. The materials of p-type and n-type are connected as shown in Fig. 1. It can regenerate the electricity from the wasted heat, such as exhausted water of the boiler or exhausted gas from the combustion engine. The output voltage is low and usually a boost type chopper is connected. The voltage-current (V-I) characteristics of TEG is as shown in Fig. 1. And the equivalent circuit can be expressed with series connected voltage source and the constant inner resistance as shown in Fig. 2.

The dc-dc converter provides the ability of maximum power point tracking (MPPT) for some dc energy source. The current ripple of the dc-dc converter may cause the decrease in output power.

To reduce the current ripple, using larger inductance or higher switching frequency are the simple solutions. However the increase of the cost or the switching loss limits these solutions. Another solution is using multi-phase converter circuit [3].

In this paper, we propose the ripple-less dc-dc converter using the 2-phase dc-dc boost converter with the constant duty ratio and output voltage ratio in the critical current mode.

II. PROPOSED CIRCUIT AND STEADY STATE SIMULATION

The proposed 2-phase dc-dc boost converter circuit is shown in Fig. 3. We have compared the ripple ratio with the conventional single chopper circuit in simulation.

The conventional single chopper circuit is shown in Fig. 4. The inductance Ld is main parameter to suppress the current ripple. Simulation results are shown in Fig. 5 when the inductance Ld is 0.05mH. Triangular carrier frequency is set to be 5kHz. The generator open voltage is 10V and the terminal voltage at maximum power point is 5V, that is half of open voltage.

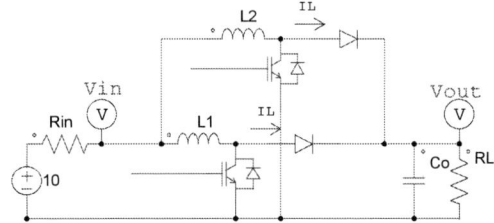

Fig.3. Proposed 2-phase boost converter

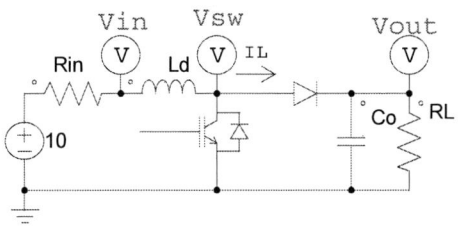

Fig. 4 Conventional single chopper circuit

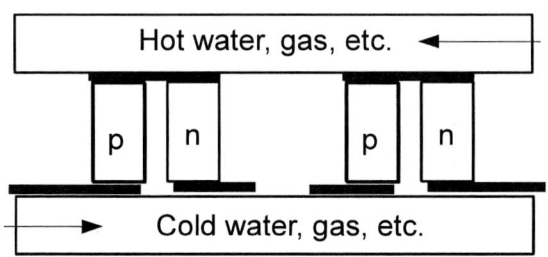

Fig. 1 Structure of thermo-electric generator

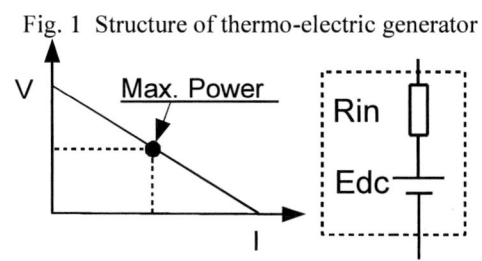

(a)V-I characteristics (b) Equivalent circuit
Fig. 2 Characteristics of thermo-electric generator

978-1-4799-2706-7/14 $31.00 © 2014 IEEE

In Fig. 5, there is a large current ripple in the input current I_{in}. It affects the input voltage to the chopper and the input power from the thermo-electric generator largely. Then the energy extracted from the TEG is reduced largely.

Fig. 6 shows the steady state simulation of the proposed dc-dc converter using 2-phase boost chopper. It is operated in the CRM and the input current has almost no ripple. The output voltage is fixed at $V_o=2V_i$. That means the duty factor is fixed to 0.5 in steady state. The current can be changed by changing the switching frequency. Merits are that the current ripple is theoretically zero and the soft switching is realized at switch-on. Demerits are that the output voltage is fixed to 2Vi and the switching frequency becomes high at the small current.

Simulation results of the proposed circuit with TEG are shown in Fig. 7. The inductances of the proposed circuit, L_1 and L_2, are 0.05mH. This parameter does not affect the ripple ratio theoretically. It affects the switching frequency. The switching frequency changes depending on the current amplitude. It is around 10kHz in this simulation. In this case, the output power is almost at the maximum power point and kept constant.

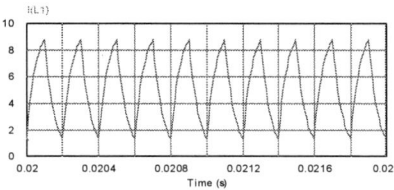

(a) Input current

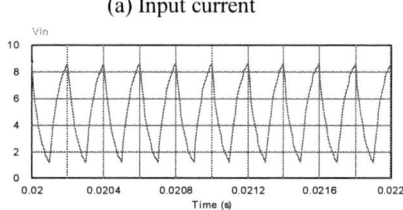

(b) Input voltage

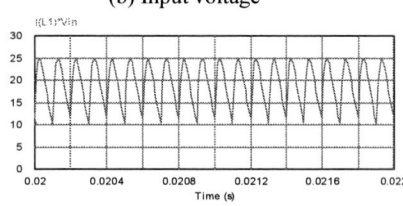

(c) Input power
(output power from thermo-electric generator)

Fig. 5 Simulation results of single chopper

(Vs=10V, Rin= 1 ohm, Ld=0.05mH, Rd=0.1 ohm)

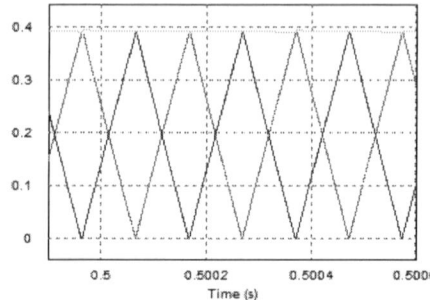

Fig. 6 Simulation results of steady state

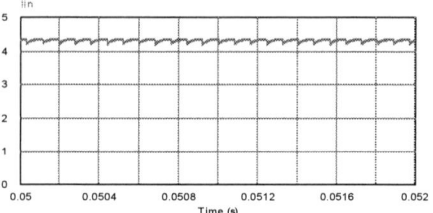

(a) Input current

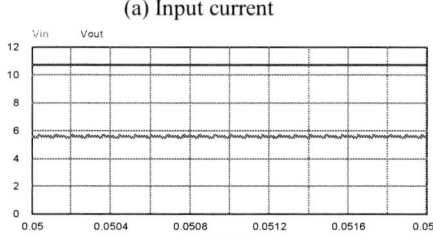

(b) Input and output voltage

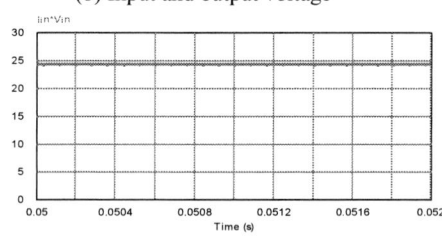

(c) Input power
(output power from thermo-electric generator)

Fig. 7 Simulation results of double chopper

(Vs=10V, Rin= 1 ohm, Ld=0.05mH, Rd=0.1 ohm)

III. CONTROLLER

To correspond to the load change or the parameter change of the energy source, the feedback control is added. Schematic diagrams of the controllers and the simulation results are shown in Fig. 8.

One method uses the output voltage as the input signal as shown in Fig. 8(a). Another method uses the output current as the input signal as shown in Fig. 8(b).

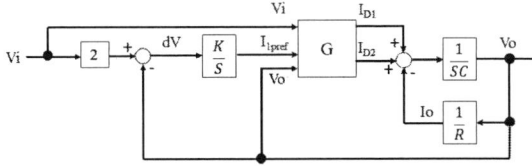

(a) Controller 1

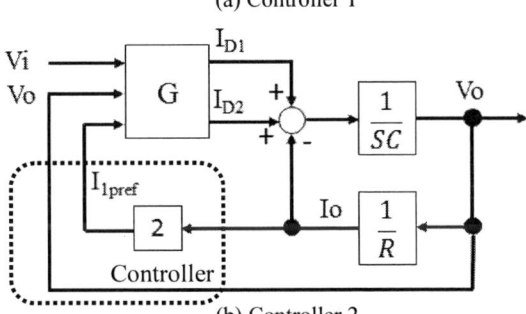

(b) Controller 2

Fig. 8 Schematic diagram of controller

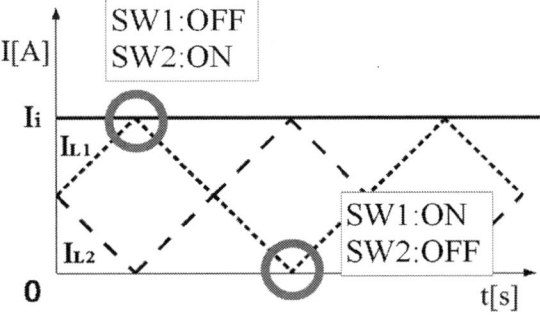

Fig. 9 Basic control of proposed boost converter

The on switching signal of two chopper circuit is shifted 180 degree. The control method of this circuit has the following characteristics.

1. In steady state, the duty factor is controlled to be 0.5. In other words, the output voltage is controlled to be the twice of the input voltage.
2. The switching frequency is changed to control the current, which is inversely proportional to the switching frequency. The smallest output current may be limited by the highest switching frequency.
3. Zero current soft switching (ZCS) can be achieved at the switch-on timing.
4. The inductance of the inductor can be much smaller than that of the single converter at the same ripple ratio limit.

We have designed the controller to have larger merits in the proposed circuit. Here, we have compared two controllers. One measures the output current and controls the input current half of it. Another measures the output voltage and controls it twice of the input voltage. Both controllers output the current reference I_{ref}.

We make one of the inductor currents as the master, e.g. the chopper 1 current I_{L1}, and another slave, e.g. the chopper 2 current I_{L2}. The sequence of the control is as follows.

1. The switch 1 is switched off and the switch 2 is switched on, when the current I_{L1} reaches the current reference.
2. The switch 1 is switched on and the switch 2 is switched off, when the current I_{L1} reaches zero.

IV. EXPERIMENTAL RESULTS OF STEADY STATE

Experimental results of steady state are shown in Fig. 10. There is certain amount of ripple in total input current. It is also shown that the decreasing rate of the inductor current is larger. It means that the output voltage is larger than twice of the input voltage. Another difference from the simulation results is longer zero current period.

We investigated the possibility of the effect of the delay in control. Simulation results with the delay in the control are shown in Fig. 11. The large ripple is seen in the total input current as same as the experimental results.

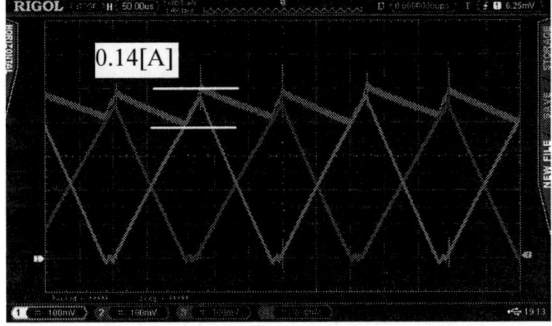

Fig.10 Experimental results of steady state

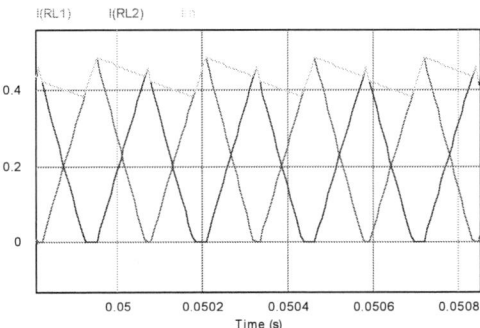

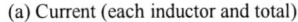

(a) Current (each inductor and total)

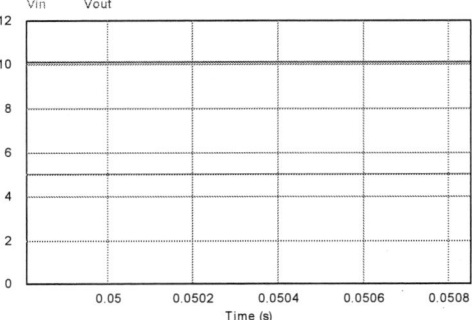

(b) Voltage (input and output)

Fig. 11 Simulation results of controller 2 with delay

The delay cause the excess of the peak current value and it results in the larger output voltage than twice of the input voltage as shown in Fig. 11 (b).

V. SIMULATION RESULTS OF CURRENT CONTROL

Fig. 12 shows the simulation results of the proposed circuit when the current reference is changed in step manner. Each inductor current of the parallel boost circuits are I_{L1}, I_{L2}. Each current is controlled to limit the peak value at the current reference. Input current is the total of each current and is the ripple-less dc current at the steady state. The value of the dc current is equal to the current reference I_{ref}.

When the current reference is changed, the response is within one switching period and very fast. In case of current increase, it is seen that the change from t_1 to t_2 is smaller than the half cycle. On the other hand, in case of the current decrease, more than one cycle is thought to be necessary to change from t_3 to t_4, since the current has to be decreased much more than the reference value.

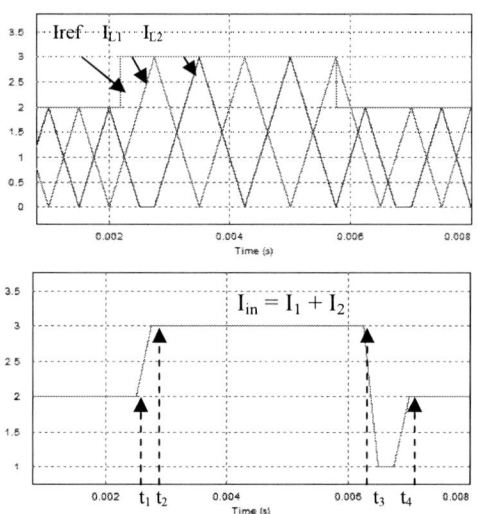

Fig. 12 Simulation results of proposed boost converter

To correspond to the load change or the parameter change of the energy source, the feedback control is added. Simulation results are shown in Fig. 13 and 14.

In simulation, the load changes from 5 ohm to 10 ohm at the time 0.5 sec. Simulation results in Fig.13(a) and Fig.14(a) show that the inductor total current decreases.

It is clearly shown that the controller 2, which uses the output current as the input signal, is faster than the controller 1, which uses the output voltage as the input signal. The reason is that the change of the current is faster than the voltage change.

VI. CONCLUSION

We have investigated the ripple-less dc-dc converter with two-phase boost chopper. The simulation results show that the ripple-less operation is possible

theoretically. The fast response in the current control is also shown by using the simple digital control algorithm.

Experimental results show that the ripple cannot be suppressed substantially because of the delay in control system. Therefore, the suppression of the effect of the delay in control is next step.

The control for the load change is investigated in simulation. The simulation results show that the output current loop is faster than the output voltage loop.

REFERENCES

[1] Singh, B.; Tan, L.; Date, A.; Akbarzadeh, A. Power Generation from Salinity Gradient Solar Pond Using Thermoelectric Generators for Renewable Energy Application. In Proceeding of the 2012 IEEE International Conference on Power and Energy (PECon), Kota Kinabalu, Malaysia, 2-5 December 2012; pp. 89-92, 2012.

[2] Hiroaki Yamada, Koji Kimura, Tsuyoshi Hanamoto, Toshihiko Ishiyama, Tadashi Sakaguchi and Tsuyoshi Takahashi, "A Novel MPPT Control Method of Thermoelectric Power Generation with Single Sensor", Appl. Sci. 2013, vol.3(No.2), pp. 545-558; doi:10.3390/app3020545, 2013.

[3] Akihiro Toru, Hitoshi Haga, Seiji Kondo:" Comparison between Phase Number and Operation Mode of Multiphase Boost Chopper for Reducing Input Current Ripple", IEE-Japan Trans. IA, Vol.132, No.2, p.250-257 (2012)

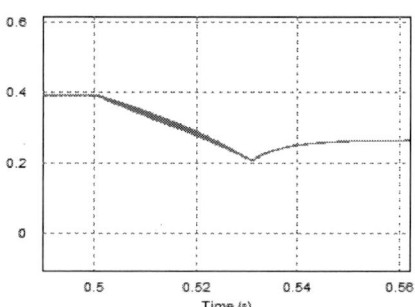

(a) Simulation results of total current

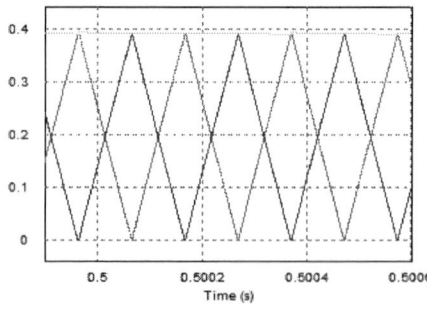

(b) Simulation results of currents just after load change

Fig. 13 Simulation results of controller 1

The 2014 International Power Electronics Conference

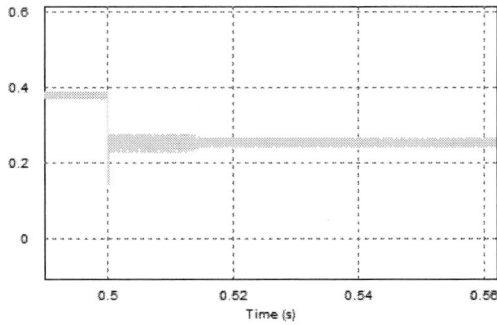

(a) Simulation results of total current

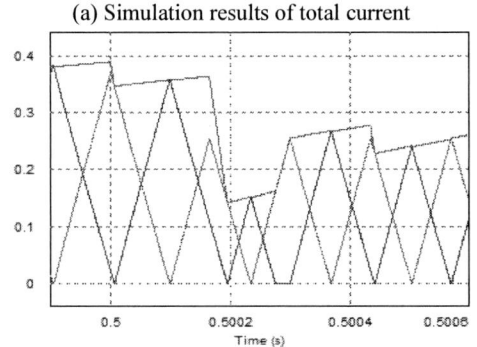

(b) Simulation results of currents just after load change

Fig. 14 Simulation results of controller 2

Position Sensorless Start-up Method of Surface Permanent Magnet Synchronous Motor Using Nonlinear Rotor Position Observer

Tsuyoshi Hanamoto[*], Hiroaki Yamada[*] and Yoshihiro Okuyama[**]
[*]Graduate School of Life Science and Systems Engineering,
Kyushu Institute of Technology, Wakamatsu-ku, Kitakyushu, JAPAN 808-0196
Email: hanamoto@life.kyutech.ac.jp
[**] Technology Research Laboratory, Shimadzu Corporation,
Hikaridai Seika-cho, Kyoto, JAPAN

Abstract— **This paper presents a new start-up method from standstill to a sensorless control for surface permanent magnet synchronous motor (SPMSM). To estimate the rotor position of SPMSM from the measured induced voltage, a DC current adds to the motor and nonlinear observer is introduced to the motor torque equation while the motor is vibrating. To achieve the start-up smoothly, first order induced voltage observer also calculates the rotor position during the vibration and the estimation method is replaced after a certain period of time. The validity of the proposed method is verified by experiments using DSP based voltage source inverter.**

Keywords— *permanent magnet motors, estimation of motor speed and rotor position, nonlinear observer, particle swarm optimization*

I. INTRODUCTION

Recently, motor controls are being widely used in various industrial applications, for example, as an actuator in robots and numerical control machines, and as drive units in electric vehicles and plug-in hybrid vehicles. Specifically, AC motors such as induction motors and permanent magnet synchronous motors (PMSM) have certain advantages over DC motors. They are maintenance free, have high energy density, and attain high performances needed to employ vector control.

In addition, sensorless speed controls in AC motors have been proposed in response to reductions in weight, size, and total cost requirements [1–3]. Two types of the sensorless control methods are popular, one is the based on the estimating induced voltage and the other is based on high-frequency signal injection method [4-6]. The former one can't use when the motor is standstill because the induced voltage doesn't induce theoretically. The latter one estimates even if the zero-speed condition, but it is not used for the surface permanent magnet synchronous motors (SPMSMs) and in some cases the imposing high frequency signal causes of the torque ripple to deteriorate the system performance.

In this paper, we consider the speed control system that has large moment of inertia and small Coulomb's friction, for example the system of turbo molecular pumps (TMPs). In this case, it is difficult to impose high-frequency signal to keep standing still because the Coulomb's torque is too small, so that the estimating method based on the induced voltage type is employed.

Here, a position estimator for SPMSMs at start-up is proposed. To estimate the initial position of a SPMSM, DC excitation is used at start-up. While the rotor of SPMSM is vibrating during DC excitation, the rotor position is immediately estimated from the induced voltage. We proposed an estimation method using the nonlinear observer because the motor model consists of nonlinear terms involving the Coulomb's friction torque and trigonometric function [7]. The modified Euler method is employed to solve the observer. In addition, the motor parameters and observer gains require an estimate of the rotor position so that a particle swarm optimization (PSO) can be employed [8, 9]. To achieve the speed control system, the first order induced observer replaces the nonlinear observer because the induced voltage can't measure after the motor rotates. Combining the two observer output property the smooth start-up from standstill is succeeded. The validity of the proposed method is verified by experiments using DSP based voltage source inverter.

II. PROPOSED ESTIMATOR OF SPMSM

A. Position estimating method of SPMSM

The torque and angular speed equations of a SPMSM in the three-phase model are described as follows [4].

$$
\left.
\begin{aligned}
& J\frac{d\omega_m}{dt} + D\omega_m + T_L = T_e \\
& \frac{d\theta_e}{dt} = p\omega_m \\
& T_e = -p\Phi_r\left\{ i_a \sin\theta_r + i_b \sin\left(\theta_r - \frac{2}{3}\pi\right) + \right. \\
& \qquad\qquad\qquad \left. i_c \sin\left(\theta_r + \frac{2}{3}\pi\right)\right\}
\end{aligned}
\right\}, \quad (1)
$$

where J: inertia, D: viscous friction coefficient, T_L: Coulomb friction torque expressed as $T_L = T_{La}sign(\omega_m)$, T_{La}: amplitude of T_L, i_a, i_b, i_c: armature current of the three phase, ω_m : motor angular speed, θ_r : electrical rotor position from the u-axis, p : number of pole pairs, and Φ_r : flux linkage. The angle θ_r is the complement of θ_e as $\theta_r = \pi/2 + \theta_e$ when the DC excitation is added to the b-c phase; it is easy to understand that the stop position is converged to zero.

First, we consider the estimation method of the rotor position from the standstill. A new nonlinear observer is introduced as follows. When the DC current I_b flows from the b-phase to the c-phase, T_e is described as follows:

$$T_e = -p\Phi_r \left\{ I_b \sin\left(\theta_r - \frac{2}{3}\pi \right) - I_b \sin\left(\theta_r + \frac{2}{3}\pi \right) \right\}$$
$$= -\sqrt{3}\Phi I_b \sin\theta_e. \tag{2}$$

The motor torque equation is rewritten as follows;

$$\frac{d}{dt}\begin{bmatrix} \omega_m \\ \theta_e \end{bmatrix} = \begin{bmatrix} -\dfrac{D}{J}\omega_m + \dfrac{-\sqrt{3}\Phi I_p \sin\theta_e}{J} - \dfrac{T_L sign(\omega_m)}{J} \\ p\omega_m \end{bmatrix}, \tag{3}$$

and the induced voltage of the a-phase e_a is given as follows:

$$e_a = -\frac{d\phi_a}{dt} = -\Phi\frac{d\theta_e}{dt} = -\Phi p \omega_m \cos\theta_e. \tag{4}$$

An estimated value of e_a can be calculated if the angular speed and the rotor position of SPMSM are estimated as follows.

$$\hat{e}_a = -\Phi p \hat{\omega}_m \cos\hat{\theta}_e. \tag{5}$$

As $\hat{e}_a$ includes $\hat{\omega}_m$ and $\hat{\theta}_e$ shown in (5), the estimated rotor position is convergence to the real value when the error of measured and calculated induced voltage goes to zero. Then the observer theory is applied to the motor torque equation.

$$\frac{d}{dt}\begin{bmatrix} \hat{\omega}_m \\ \hat{\theta}_e \end{bmatrix} =$$
$$\left. \begin{bmatrix} -\dfrac{D}{J}\hat{\omega}_m + \dfrac{-\sqrt{3}\Phi I_p \sin\hat{\theta}_e}{J} - \dfrac{T_L sign(\hat{\omega}_m)}{J} \\ p\hat{\omega}_m \end{bmatrix} + \boldsymbol{g}\varepsilon \right\},$$
$$\varepsilon = e_a - \hat{e}_a \tag{6}$$

where, observer gains are defined as $\boldsymbol{g} = \begin{bmatrix} g_1 & g_2 \end{bmatrix}^T$, and the initial values are assumed as $\begin{bmatrix} \hat{\omega}_{m0} & \hat{\theta}_{e0} \end{bmatrix}^T = \begin{bmatrix} 0 & 0 \end{bmatrix}^T$. $\hat{\theta}_r$ is calculated simply the following relationship $\hat{\theta}_r = \pi/2 + \hat{\theta}_e$. As (6) are nonlinear equations, the modified Euler method is used for the calculation and the observer gains are searched using particle swarm optimization (PSO). In this paper, we apply PSO to search the appropriate gains of the nonlinear observer and to estimate the motor parameters.

This observer is effective the start-up but when the motor starts to rotate $\hat{\theta}_r$ can't estimate because induced voltage is not able to be measured for ordinary voltage source inverter (VSI). Next, we consider the observer for rotating. Here, the first order induced observer is employed. The voltage equations of $\alpha\beta$ stationary coordinate and the model of induced voltage are described as follows,

$$\left. \begin{array}{l} L_a \dfrac{di_\alpha}{dt} + R_a i_\alpha = v_\alpha - e_\alpha \\ \dfrac{de_\alpha}{dt} = e_{\alpha 0} \\ \dfrac{de_\alpha}{dt} = 0 \end{array} \right\} \tag{7}$$

where L_a : armature inductance, R_a : armature resistance,

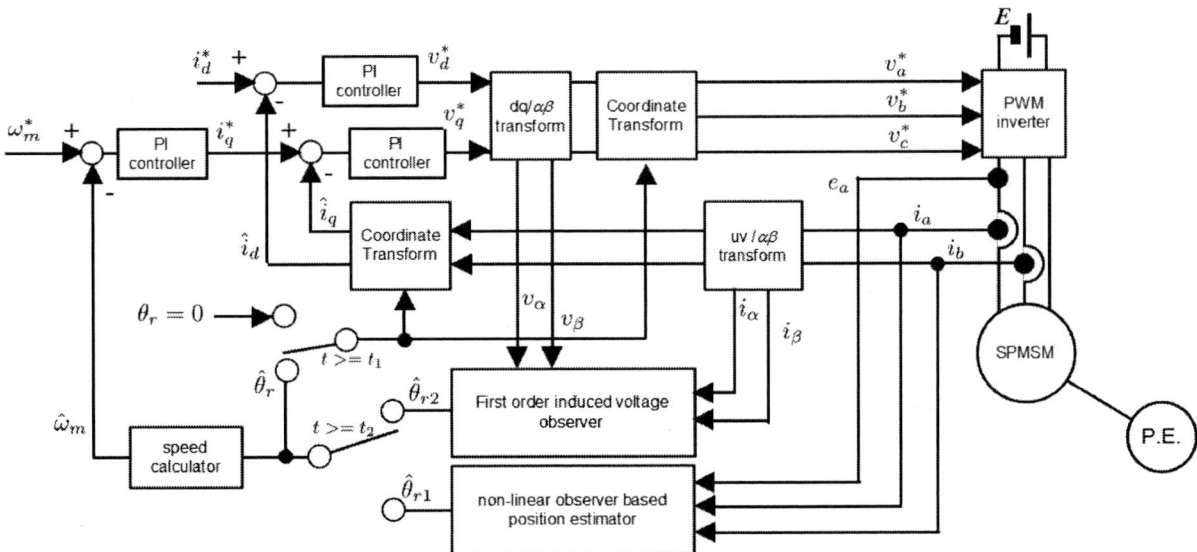

Fig. 1 Configuration of the proposed system

TABLE I.	OPERATOIN OF THE PROPOSED SENSORLESS CONTROL METHOD	
MODE	Control method	estimated variables
1	DC excited current add to the motor, not control	using nonliner observer
2	closed loop speed control	using nonliner observer ,but observer gains set to zero
		using the 1st order observer but not user for the control
3	closed loop speed control	using the 1st order observer

i_α: α axis current, v_α: α axis terminal voltage, e_α: α axis induced voltage, $e_{\alpha 0}$: differential component of $e_{\alpha 0}$. (7) shows α axis and β axis equation is obtained as the same manner. Full order observer is applied to estimate induced voltage for (7), where $x(k) = \begin{bmatrix} i_\alpha & e_\alpha & e_{\alpha 0} \end{bmatrix}^T$, and observer gain $g_e = \begin{bmatrix} g_{e1} & g_{e2} & g_{e3} \end{bmatrix}^T$

The observer for (7) is given by the usual observer theory,

$$\hat{x}(k+1) = (A - gc)\hat{x}(k) + bv_\alpha(k) + g_e i_\alpha(k) . \quad (8)$$

Using the result of (8) $\hat{e}_\alpha$ and the β axis result $\hat{e}_\beta$ are able to be estimated. Then the rotor position and the motor angular speed are estimated as follows,

$$\hat{\theta}_r = -\frac{\hat{e}_\alpha}{\hat{e}_\beta} . \quad (9)$$

$$\hat{\omega}_m = \frac{1}{p} \frac{\left(\hat{\theta}_r(k) - \hat{\theta}_r(k-1) \right)}{\Delta T} . \quad (10)$$

Where, ΔT : sampling period, $\hat{\theta}_r(k)$, $\hat{\theta}_r(k-1)$: k-th and (k-1)th value of $\hat{\theta}_r$. In the experimental system, LPF (low pass filter) is applied to (10) for decreasing the noise component.

Finally, we consider how to combine the two observers and achieve the sensorless control from stand-still. Here we define the transition time t_1 and t_2 ($t_1 < t_2$) indicated in Fig.1 and Fig. 2.

When t=0s the DC excitation is started. We call during $0 < t < t_1$ as MODE 1. At t= t_1 the mode change to the speed control mode (MODE 2) from the DC excitation mode. During t_1 to t_2 the system is controlled using the output of the nonlinear observer ($\hat{\theta}_r$, $\hat{\omega}_m$), but the observer gains set to zero in (6). This means the estimated results use no correction because this duration is not DC excitation mode so the output of (6) just calculates the rotor position from the calculated input torque. At the same time first order induced voltage

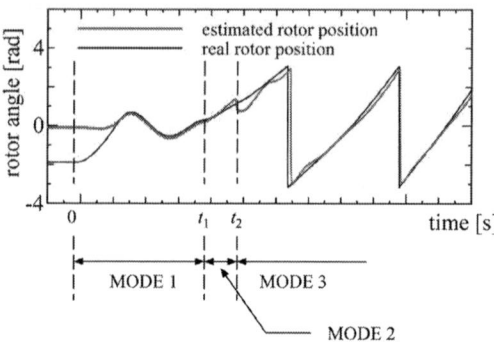

Fig. 2 Operation of the proposed sensorless control method from standstill

observer starts to calculate the rotor position $\hat{\theta}_{r2}$ and the estimated values converge at the end of the transition time t_2.

At t= t_2 the estimated rotor position is replaced to the result of the first order observer completely (MODE 3). And the sensorless speed control based on the induced voltage observer is achieved. When the property observer gains are selected the position error is almost disappeared. Table I and Fig. 2 show the explanation of each Mode.

B. Particle swarm optimization

In this paper, we apply PSO to search the appropriate gains of the nonlinear observer and to estimate the motor parameters. The PSO is one of the optimization technique and kind of evolutionary computation technique. All solutions in the PSO can be represented as particles in a swarm. Each particle (agent) has a position and velocity vector and each position coordinate represents a parameter value. The PSO which use the concept of velocity described in (11) and (12) can easily get over the range of initial parameters,

$$v_i^{k+1} = \omega \cdot v_i^k + c_1 \cdot rand() \cdot \left(pbest_i - x_i^k \right) + c_2 \cdot rand() \cdot \left(gbest - x_i^k \right), \quad (11)$$

$$x_i^{k+1} = x_i^k + v_i^{k+1}, \quad (12)$$

where, x_i is the i-th particle, v_i is the velocity for particle i, $pbest_i$ means the previous best position of the i-th particle, $gbest$ is the best particle in the population, c_1 and c_2 are the acceleration constant, $rand()$ is the random function and ω is the inertia weight factor described in (13) where k is generation. By selecting appropriate ω, c_1, c_2, it enables the particle to move toward the optimum position.

$$\omega = \omega_{max} - \frac{\omega_{max} - \omega_{min}}{k_{max}} k . \quad (13)$$

We define an evaluation function F_{ev} as (14) to minimize the estimate error of the induced voltage, where $\hat{e}_a$ is estimated induced voltage from (5),(6) and e_a is the measured one.

$$F_{ev} = 1 / \sum_{k=1}^{n} \sqrt{\left(\hat{e}_a(k) - e_a(k) \right)^2} . \quad (14)$$

TABLE II.   SYSTEM PARAMETERS

Symbol	Meaning	value
I_b	DC excited current	2.0 A
p	number of pole pairs	4
T_s	control period	1.0×10^{-4} s
J	inertia	4.26×10^{-5} kgm^2
D	viscous friction coefficient	2.05×10^{-4} Nm/ (rad/s)
Φ	flux linkage	7.39×10^{-2} Wb
T_{La}	Amplitude of Coulomb friction torque	6.57×10^{-3} Nm

TABLE III.   PARAMETERS OF PSO FOR OBSERVER GAIN ESTIMATION

Parameters	Specification
Number of iterations	500
Swarm size	30
Dimension	2
C_1, C_2	2.0
$\omega_{max}, \omega_{min}$	0.9, 0.4

Fig. 3 Experimental system

III. EXPERIMENTAL RESULTS

The experimental system is shown in Fig. 3, which consists of a 400W SPMSM, a voltage source PWM inverter and DSP based control unit. A Pulse encoder, mounted on the shaft of the SPMSM, is used just for measurement and not used for the rotor position estimation and the speed control. TI microcontroller TMS320F28035 is employed for the estimation and the vector control of the proposed system. This device is 32 bit microcontrollers with high performance integrated peripherals designed for real-time control applications. In the system, the carrier frequency and the control frequency is set to 10kHz and nonlinear observer is performed every 2ms.

TABLE II lists system parameters and estimated parameters of the SPMSM searched using PSO. Table III shows the parameter of PSO for searching nonlinear

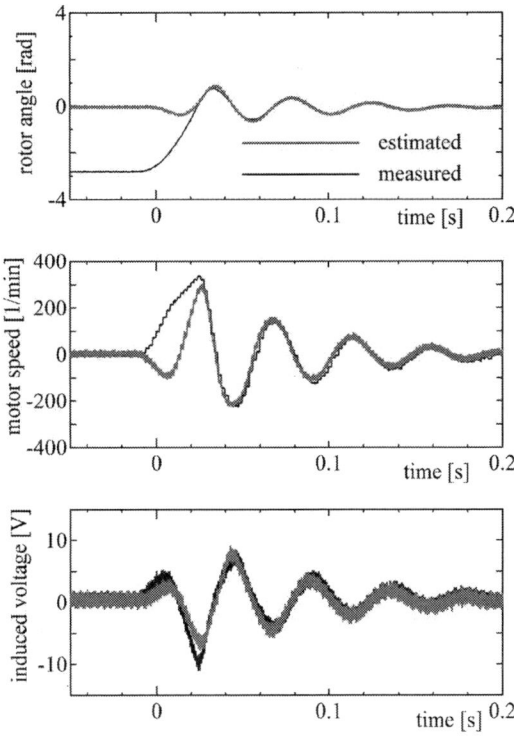

Fig. 4 Experimental results at DC current add to b-c phase when the initial position θ_{eini}=-2.8 rad.

observer gains and the DC current flows between the b-phase and the c-phase for the estimation. Different gains are searched because the gains depend on the initial position. After several experiments, we decided the gain for this system as follows; $\boldsymbol{g} = \begin{bmatrix} 1.76 & -1.27 \times 10^{-2} \end{bmatrix}^{\mathrm{T}}$.

Fig. 4 shows the estimation results when the initial position is $\theta_{eini} = -2.8$ rad. The top graph shows the measured and estimated rotor position; the red (black) curves are the estimated (measured) values. The next two graphs show the motor speed and induced voltage estimated results, respectively. From the figure, though the value of the estimated starts from zero, the estimated value immediately converges to the measured (real) position.

The waveform of rotor position, motor speed and phase currents at start-up from standstill are shown in Fig. 5 and Fig. 6. The speed command set as ω_m^*=300min^{-1} (31.4 rad/s). Fig. 5 shows the results when θ_{eini}=-1.9 rad and Fig. 6 shows θ_{eini}=2.7 rad. In both experiments transition time is set as t_1=80ms and t_2=100ms. The initial vibration direction depends on the initial position, but the vector speed control is achieved immediately within t_2 smoothly.

IV. CONCLUSIONS

We proposed a new rotor position estimator for a SPMSM at start-up from standstill. Combining a nonlinear observer and a first order induced observer the sensorless speed control is achieved in short time

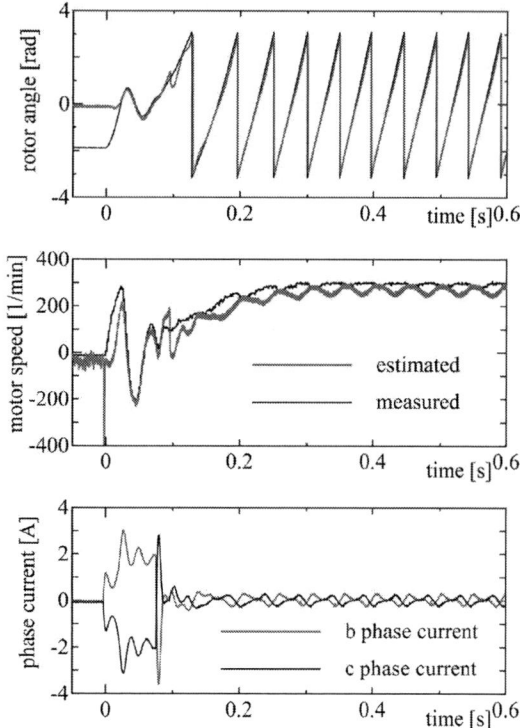

Fig. 5 Experimental results at start-up from standstill when the initial position θ_{eini}=-1.9 rad.

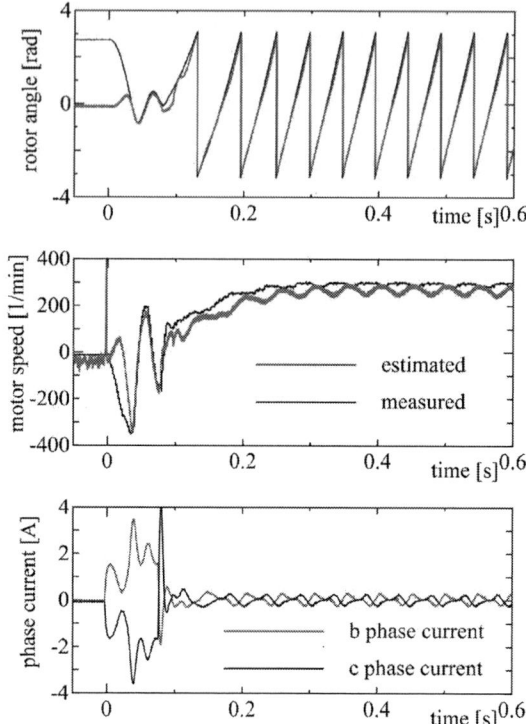

Fig. 6 Experimental results at start-up from standstill when the initial position θ_{eini}= 2.8 rad

smoothly. Experimental results demonstrated the validity of the proposed method. Proposed method needs motor mechanical parameters like the moment of inertia and load torque so that this method is useful in case of the constant load torque at the start-up duration. Now our experimental system is the ordinary servo motor system, to apply the proposed method to the large moment inertia and small Coulomb's friction system is the next step.

REFERENCES

[1] P. Acarnley, J. Watson , "Review of Position-Sensorless Operation of Brushless Permanent-Magnet Machines", IEEE Trans. Ind. Election, Vol.53, No. 2, pp.352-362 (2006)

[2] S. Ichikawa, M. Tomita, S. Doki, S. Okuma, "Sensorless Control of Permanent-Magnet Synchronous Motors Using Online Parameter Identification Based on System Identification Theory", IEEE Trans. Ind. Election, Vol.53, No. 2, pp.363-372 (2006)M. Young, "The PWM strategy on DC-DC converter," *IEEJ Journal of Industry Applications*, vol. 28, no. 15, pp. 123-129, 1989.

[3] T. Hanamoto, A.Ghaderi and T.Tsuji, "RTLinux Based Speed Control System of SPMSM with an Online Real Time Simulator", IEEJ Tans. Ind. App, Vol.126, No.4, pp 453-458 (2006)G. Eason, B. Noble, and I. N. Sneddon, "On certain integrals of Lipschitz-Hankel type involving products of Bessel functions," *IEEE Trans. on Power Electronics*, vol. 247, no. 8, pp. 529-551, 1995.

[4] Tsuyoshi Hanamoto, Ahmad Ghaderi, Teppei Fukuzawa and Teruo Tsuji, "Sensorless Control of Synchronous Reluctance Motor Using Modified Flux Linkage Observer with an Estimation Error Correct Function", Proc of ICEM2004, 249 (CD-ROM)

[5] Myoungho Kim, Seung-Ki Sul,"An Enhanced Sensorless Control Method for PMSM in Rapid Accelerating Operation",Proc. of IECON-2010,pp.2761-2767

[6] K. Ide, H. Iura, M. Inazumi: " Hybrid Sensorless Control of IPMSM Combining HFIM and Back EMF Method", proc. of IECON-2010,pp.2230-2235

[7] T.Hanamoto, H.Yamada, R.Kawano and Y.Okuyama, "Rotor position estimator using non-linear observer of surface permanent magnet synchronous motor", Proc. of IEEE PEDS 2013, pp.306-310,(2013)

[8] J.Kenney and R.Eberhart: "Particle Swarm Optimization" Proc. of 1995 IEEE International Conference on Neural Networks, pp.1942-1948(1995)

[9] T.Hanamoto, M.Takenouchi and H.Ikeda, "Vibration Suppression Control of 3-mass Resonance System Using Particle Swarm Optimization for Design of Coefficient Diagram Method", Journal of the Japan society of applied electromagnetics and mechanics, vol. 19, supplement, 2011.09 (S16-S20) (2011).

The 2014 International Power Electronics Conference

Sensorless Control of PMSM for the whole speed range using Two-Degree-of-Freedom current control and HF test current injection for low speed range

Markus Seilmeier and Bernhard Piepenbreier, *Senior Member, IEEE*

Chair of Electrical Drives and Machines, University of Erlangen-Nuremberg, Germany

Email: markus.seilmeier@fau.de

Abstract—**In this paper an innovative sensorless Two-Degree-of-Freedom current control scheme for the whole speed range is proposed. It consists of a model based dynamic feed forward control to set the reference response and model reference tracking controllers providing disturbance rejection and the position error signals needed for sensorless control. For low and zero speed operation test current injection is used to gain a position error signal. A flatness based test signal pre-control provides compensation for secondary saliencies like cross-saturation and higher harmonics. For mid and high speed operation it is shown how the model based dynamic feed forward control can be modified to obtain a high quality position error signal from the tracking controller. Like this no additional model based estimator evaluating back-EMF information is needed. The effectiveness of the proposed method is proven by experimental results.**

Index Terms—**permanent magnet motors, sensorless control, saliency, test current injection**

I. INTRODUCTION

Field oriented control of PMSM without use of a position sensor is still a field of intensive research. For zero and low speed operation the PMSM has to be excited by a test signal, otherwise the rotor position is not observable [1]. Basically test voltage or test current injection can be used [2]. Methods using HF voltage injection and evaluation of the related HF current response are state-of-the-art. The voltage injection methods can be categorized as to the shape of the test signal: sinusoidal or pulse injection (PWM injection) methods [3]. Moreover one can distinguish between alternating [4], [5] and rotating test signal injection [6], [7]. Low and zero speed sensorless control is challenging, because the machine has to provide a resistive or inductance saliency in the desired current operating range [8]. The control performance suffers from secondary saliencies like cross-saturation [9] and higher harmonic saliencies [10], [11]. For mid and high speed operation the fundamental frequency model is the basis for estimating the electric rotor position of the PMSM. These methods are often referred to as back-EMF based methods, because the back-EMF is the dominant quantity which carries position information. A lot of different methods have already been proposed. One can basically distinguish between methods which use a full or a reduced order model of the PMSM. Full order model means that the complete state space model of the machine consisting of currents, speed and position as state variables is used to create an estimator. In case of reduced-order models one omits

the speed and position differential equations of the full order model and just uses the differential equations for the currents in order to find an estimator. Designing a state observer based on the full order model of the PMSM is fairly complex, because the state space model is nonlinear [12], [13]. A lot of methods exist which use the reduced order model like flux estimators which evaluate the flux position in the machine [3], observers using Active Flux Concept [14] or Extended-EMF [15]. A survey of those methods is given in [16].

In this paper test current instead of voltage injection is used for low speed operation. An extended control scheme is needed in order to provide zero steady-state control error in case of superimposed sinusoidal HF reference signals. Using test current injection proper separation of the high frequency and fundamental components is provided by the controller itself in case of zero control error. In transient operation, however, an interference of fundamental and HF components might occur which deteriorates sensorless position estimation. To overcome this limitation a Two-Degree-of-Freedom current control scheme is used in this paper. The (fundamental frequency) reference response is set by a model based dynamic feed forward control which has already been successfully used for compensation of higher harmonic currents in PMSM [17]. Model reference tracking controllers are used for disturbance rejection and in case of sensorless control to provide the position error signal which drives the position and speed observer. Like this the fundamental and HF components are also decoupled in case of transient operation. Thus a high performance position error signal is provided without need for additional filtering. The innovation in this paper is that the PI+resonant-controller used in [18] is modified that it directly provides a DC error signal without need for further demodulation and delay compensation. In order to further enhance control performance a flatness based test-signal pre-control which inherently includes compensation for secondary saliencies is used as proposed in [18]. In order to further enhance the dynamics a speed estimate by evaluating the back-EMF of the machine is used [18].

The innovation for mid and high speed operation is that in contrast to conventional methods no additional estimator is needed. By suitable modification of the feed forward control a high quality position error signal can be gained from the fundamental tracking controllers which is used to drive the

978-1-4799-2706-7/14 $31.00 © 2014 IEEE

position and speed observer. The effectiveness of the proposed novel sensorless control method is validated by test bench measurement results for an IPMSM which shows strong saturation effects.

II. PMSM MODEL

Superimposing a test signal whose frequency is significantly higher than the fundamental frequency of the machine requires separate modeling of the fundamental and test signal frequency behavior of the machine. To do so the actuating signals and currents are split up into fundamental (^{f}) and high frequency (^{HF}) components, whereby $\underline{X}_{dq} = [X_d, X_q]^T$ represents voltages or currents in field oriented coordinates:

$$\underline{X}_{dq} = \underline{X}_{dq}^f + \underline{X}_{dq}^{HF} \tag{1}$$

A. Non-ideal fundamental frequency model considering secondary inductance saliencies

In practice machines typically do not only feature one single sinusoidal saliency. Secondary saliencies may occur due to the machine geometry and due to saturation effects [19]:

$$\begin{bmatrix} u_d^f \\ u_q^f \end{bmatrix} = \underline{R}_{dq} \begin{bmatrix} i_d^f \\ i_q^f \end{bmatrix} + \underbrace{\begin{bmatrix} l_{dd} & m_{dq} \\ m_{qd} & l_{qq} \end{bmatrix}}_{\underline{L}_{dq}} \frac{d}{dt} \begin{bmatrix} i_d^f \\ i_q^f \end{bmatrix} + \omega_{el} \underbrace{\begin{bmatrix} \Psi_d^\gamma \\ \Psi_q^\gamma \end{bmatrix}}_{\underline{\Psi}_{dq}^\gamma} \tag{2}$$

$\underline{R}_{dq} = diag(R)$ is the diagonal resistance matrix. The matrix of the differential inductances $\underline{L}_{dq}$ consists of the differential self-inductances l_{dd} and l_{qq} and the differential mutual inductances $m_{dq} = m_{qd}$ [20] which can occur due to cross saturation and higher harmonic components in the inductances. In case of nonlinear material characteristics one can not distinguish whether a contribution to the flux is caused by the d- or q-axis current. To account for this the new parameters Ψ_x^γ were defined in [19]. Ψ_x^γ denotes a change of flux with the rotor position γ which causes an induced voltage in the machine axis $x \in \{d, q\}$ and depends on the PM flux, differential and absolute inductances [19]. All inductance and flux parameters are defined to be current and position dependent.

B. Extension of the Non-ideal Model for HF related resistive components (HF model)

Recent papers show that also HF resistive saliencies can be used for self-sensing [21]. The underlying physical effect is that additional AC losses occur in case of HF injection, which depend on the currents and the rotor position. To deal with this phenomenon additional AC resistances are introduced as to [22]. Test signal injection based sensorless control is just used for low and zero speed operation, thus the motional EMF is neglected by setting $\omega_{el} = 0$.

$$\begin{bmatrix} u_d^{HF} \\ u_q^{HF} \end{bmatrix} = \begin{bmatrix} r_{dd}^{HF} & r_{dq}^{HF} \\ r_{qd}^{HF} & r_{qq}^{HF} \end{bmatrix} \begin{bmatrix} i_d^{HF} \\ i_q^{HF} \end{bmatrix} + \begin{bmatrix} l_{dd}^{HF} & m_{dq}^{HF} \\ m_{qd}^{HF} & l_{qq}^{HF} \end{bmatrix} \frac{d}{dt} \begin{bmatrix} i_d^{HF} \\ i_q^{HF} \end{bmatrix} \tag{3}$$

All parameters are defined to be position and current dependent. Therefore the model includes all kind of saliencies

that may occur in a PMSM. In general the inductance and resistance parameters depend on the injection frequency and amplitude [22]. To indicate the frequency for which the parameters are valid the superscript (^{HF}) is introduced. The experiments carried out in [22] indicated the existence of secondary diagonal elements r_{dq}^{HF} and r_{qd}^{HF} in the resistance matrix $\underline{R}_{dq}^{HF}$. As to [23] the secondary diagonal elements are equal.

III. TWO-DEGREE-OF-FREEDOM CURRENT CONTROL SCHEME

The sensorless current control scheme has to fulfill various requirements:

- The HF signal should not affect the fundamental current reference response and vice versa (proper separation)
- Provide zero steady state control error for fundamental and high frequency current reference values/ trajectories
- Enable demodulation of the HF sine or cosine part of the $\hat{q}$-axis control signal (error signal for HF injection based sensorless control)
- Compensate for non-ideal characteristics of the machine like secondary saliencies

To fulfill these requirements the Two-Degree-of-Freedom control scheme in Fig. 1 is used which features the following properties:

- Two-Degree-of-Freedom Control is a model based control scheme. That is a model of the control plant is included in the controller.
- The reference response and disturbance rejection can be set independently from each other.
- The reference response is determined by the model based dynamic feed forward control, e. g. by means of suitable eigenvalue assignment.
- Modeling errors and disturbances are rejected by the (model reference) tracking controllers. Suitable internal models are needed to ensure zero steady-state control error.

For proper separation of fundamental and high frequency current components the fundamental frequency reference response is set by a model based dynamic feed forward control. The feed forward control provides the actuating signals which are directly fed to the plant input and the model currents which are used as reference for the tracking controllers. The tracking controllers ensure zero steady state control error in case of model parameter errors and provide the position error signals for sensorless control.

A. Feed Forward Control of (Fundamental) Reference Response

The state space description of the PMSM model (2) is used for the derivation of the model based dynamic feed forward control for the fundamental currents:

$$\frac{d}{dt} \underline{i}_{dq}^f = -\underline{L}_{dq}^{-1} \underline{R}_{dq} \underline{i}_{dq}^f + \underline{L}_{dq}^{-1} \underline{u}_{dq}^f - \omega_{el} \underline{L}_{dq}^{-1} \underline{\Psi}_{dq}^\gamma \tag{4}$$

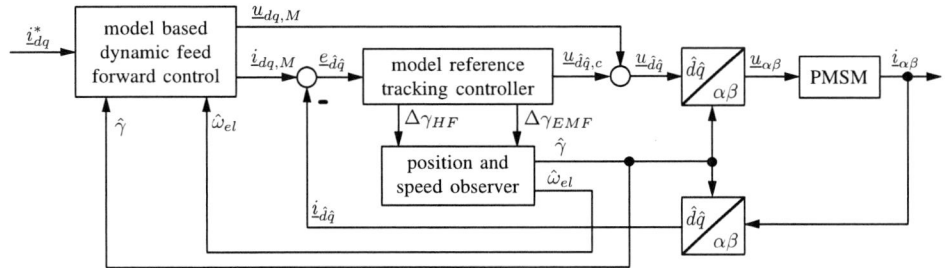

Fig. 2.   Two-Degree-of-Freedom current control scheme including sensorless position and speed estimation.

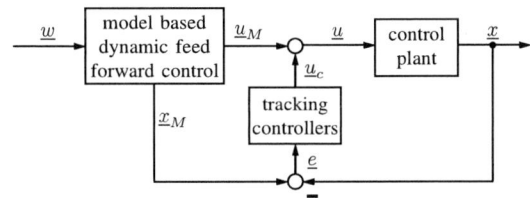

Fig. 1.   Two-Degree-of-Freedom control scheme consisting of a model based dynamic feed forward control which generates control signals $\underline{u}_M$ from the reference values $\underline{w}$. The model state vector $\underline{x}_M$ is used as reference value for the tracking controllers which generate the control signals $\underline{u}_c$ based on the control error $\underline{e} = \underline{x}_M - \underline{x}$.

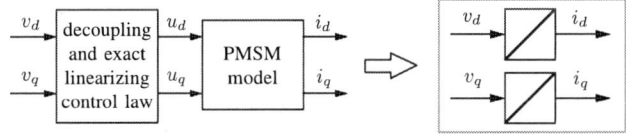

Fig. 3.   Principle of exact state linearization technique, also called feedback linearization. The decoupling and exact linearizing control law is a nonlinear coordinate transformation providing a linear state space description which just consists of integrators (grey box).

Using exact linearization [24], [25], also called feedback linearization, of the nonlinear system one yields linear system dynamics in new coordinates without any approximation. The decoupling and exact state linearizing control law is:

$$\begin{bmatrix} u_d^f \\ u_q^f \end{bmatrix} = \underbrace{\begin{bmatrix} R & 0 \\ 0 & R \end{bmatrix}}_{\underline{R}_{dq}} \begin{bmatrix} i_d^f \\ i_q^f \end{bmatrix} + \underbrace{\begin{bmatrix} l_{dd} & m_{dq} \\ m_{qd} & l_{qq} \end{bmatrix}}_{\underline{L}_{dq}} \begin{bmatrix} v_d \\ v_q \end{bmatrix} + \omega_{el} \underbrace{\begin{bmatrix} \Psi_d^\gamma \\ \Psi_q^\gamma \end{bmatrix}}_{\underline{\Psi}_{dq}^\gamma}$$
(5)

The decoupled system description using the decoupling control law (5) just consists of two integrators, as shown in Fig. 3, whereby the new system input $\underline{v}_{dq} = [v_d, v_q]^T$ is defined:

$$\underline{v}_{dq} = \frac{d}{dt} \underline{i}_{dq}^f$$
(6)

The system description in new coordinates is unstable and thus needs to be stabilized using constant state feedback, as depicted in Fig. 4. Constant state feedback is sufficient, because the integrators force constant errors (reference values are DC signals in steady state) $\underline{e}_{dq,M} = i_{dq}^* - i_{dq,M}$ to zero. The desired reference response can be set by suitable choice of k_x e. g. using eigenvalue assignment. It is worth pointing out that the feed forward control does not affect stability of the control system.

B. Flatness Based Test Signal Pre-Control

In order to obtain a position error signal for low and zero speed operation a high frequency cosine shaped test current is superimposed to the fundamental current:

$$i_{\hat{d}}^{HF} = \hat{I}_{HF} \cos(\omega_{HF} t)$$
(7)

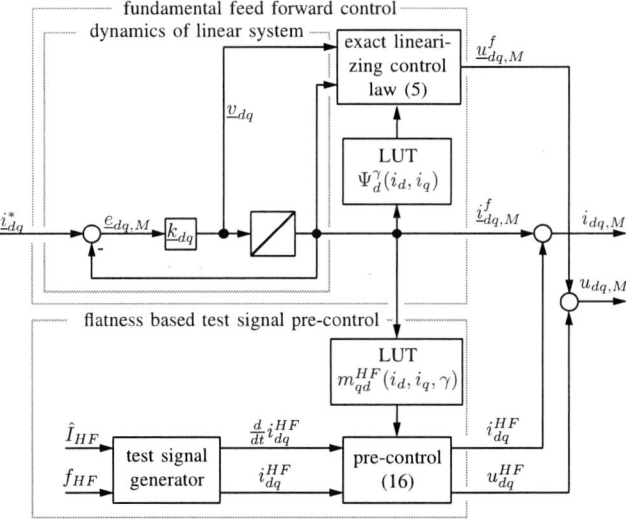

Fig. 4.   Model based dynamic feed forward control consisting of a model based dynamic feed forward control for fundamental current and flatness-based test signal pre-control. Dynamics of the fundamental feed forward control are set by suitable choice of the state feedback $\underline{k}_{dq} = [k_d \ k_q]^T$. The reference value from the speed or torque controller is $\underline{i}_{dq}^*$. The model state $\underline{i}_{dq,M}$ is the reference value for the tracking controller.

$$\frac{d}{dt} i_{\hat{d}}^{HF} = -\omega_{HF} \hat{I}_{HF} \sin(\omega_{HF} t)$$
(8)

The feed forward control presented in the previous subsection is not able to provide zero steady-state control error in case of superimposing an alternating test current reference. In contrast to the fundamental reference values, the test current reference trajectory is known beforehand in an analytical way as to (7). Using the concept of differential flatness [26] the test current pre-control can be found in a very elegant way. The system S

978-1-4799-2706-7/14 $31.00 © 2014 IEEE

1818

is given by the state space description:

$$S: \quad \underline{\dot{x}} = \underline{f}(\underline{x}, \underline{u}) \tag{9}$$

The system S is a *flat system*, if a (differential) parametrization of the input vector $\underline{u}$ and the state vector $\underline{x}$ exists which just depends on the flat output $\underline{y}_f$ and a finite number of its time derivatives:

$$\underline{x} = \underline{\psi}_x(\underline{y}_f, \underline{\dot{y}}_f, \underline{\ddot{y}}_f, \ldots) \tag{10}$$

$$\underline{u} = \underline{\psi}_u(\underline{y}_f, \underline{\dot{y}}_f, \underline{\ddot{y}}_f, \ldots) \tag{11}$$

The flat output can in general depend on the state vector $\underline{x}$, the input vector $\underline{u}$ and a finite number of its time derivatives:

$$\underline{y}_f = \underline{\Phi}(\underline{x}, \underline{u}, \underline{\dot{u}}, \underline{\ddot{u}}, \ldots) \tag{12}$$

Applying theory of flatness to the HF current control problem using the model (2) one yields for the flat output:

$$\underline{y}_f = \begin{bmatrix} y_{f,1} \\ y_{f,2} \end{bmatrix} = \begin{bmatrix} i_d^{HF} \\ i_q^{HF} \end{bmatrix} \tag{13}$$

The differential parametrization of the state vector and the input is:

$$\underline{x} = \underline{\psi}_x(\underline{y}_f) = \underline{y}_f \tag{14}$$

$$\underline{u} = \underline{\psi}_u(\underline{y}_f, \underline{\dot{y}}_f) = \begin{bmatrix} u_d^{HF} \\ u_q^{HF} \end{bmatrix} =$$

$$= \begin{bmatrix} r_{dd}^{HF} y_{f,1} + r_{dq}^{HF} y_{f,2} + l_{dd}^{HF} \dot{y}_{f,1} + m_{dq}^{HF} \dot{y}_{f,2} \\ r_{qd}^{HF} y_{f,1} + r_{qq}^{HF} y_{f,2} + m_{qd}^{HF} \dot{y}_{f,1} + l_{qq}^{HF} \dot{y}_{f,2} \end{bmatrix} \tag{15}$$

(15) is the inverse system S^{-1} of (9) and provides an algebraic calculation formula for the HF actuating signal trajectory based on the knowledge of the test signal trajectory and its first time derivative. In this paper the test current was chosen to be injected into the $\hat{d}$-axis as to (7) whilst the $\hat{q}$-axis current is kept zero ($y_{f,2} = \dot{y}_{f,2} = 0$). This results in:

$$\begin{bmatrix} u_d^{HF} \\ u_q^{HF} \end{bmatrix} = \underline{\psi}_u(\underline{y}_f, \underline{\dot{y}}_f) = \begin{bmatrix} r_{dd}^{HF} y_{f,1} + l_{dd}^{HF} \dot{y}_{f,1} \\ r_{qd}^{HF} y_{f,1} + m_{qd}^{HF} \dot{y}_{f,1} \end{bmatrix} \tag{16}$$

The actuating signal (16) is added to the fundamental actuating signal generated by the fundamental feed forward control as to Fig. 4. In case there are no initial, model, parameter or position estimation errors, the real system tracks the desired reference trajectory perfectly.

C. Model Reference Tracking Controllers

In practice model errors and a misalignment of the estimated rotor angle will cause control errors which have to be compensated for by use of suitable tracking controllers. The resonant controller used in [18] provides zero steady state control error in case of superimposing an alternating test current:

$$F_{res}(s) = k_{res} \cdot \frac{s}{s^2 + \omega_{HF}^2} \tag{17}$$

The resonant controller (17) provides AC actuating signals with frequency ω_{HF}. However, for sensorless control a DC error signal is desired. Quadrature amplitude demodulation and compensation for delay times caused by the discrete time

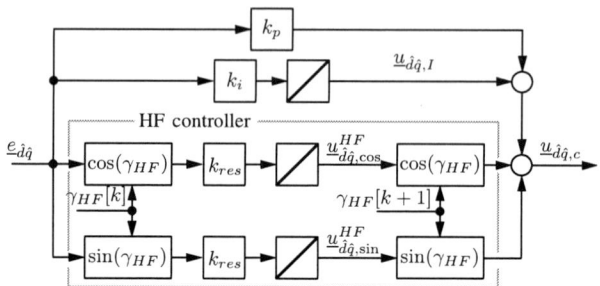

Fig. 5. Model reference tracking controller consisting of the fundamental PI-type controller and the HF controller in parallel.

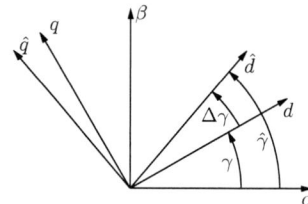

Fig. 6. Coordinate Systems: The orthogonal axes of the stator fixed two phase equivalent coordinates are denoted by α and β. The field oriented coordinate system consists of the orthogonal d- and q-axis whereby the angle of the d-axis with respect to the α-axis is given by γ. The estimated field oriented coordinate system which is misaligned by the angle $\Delta\gamma = \hat{\gamma} - \gamma$ are given by $\hat{d}$ and $\hat{q}$.

implementation of the control is needed [18]. In this paper the controller is modified in order to directly provide DC signals for sensorless control. The HF controller depicted in Fig. 5 has identical input/ output behaviour as the controller (17), however featuring direct demodulation of sine and co-sine components of the HF actuating signals and delay time compensation by using $\gamma_{HF}[k + 1]$ instead of $\gamma_{HF}[k]$ for the back transformation [27]. Comparing equivalence with Fourier series expansion one finds that the structure contains a calculation of the sine and cosine Fourier coefficients for the HF frequency.

IV. PRINCIPLE OF TEST CURRENT INJECTION BASED SENSORLESS CONTROL FOR LOW SPEED RANGE

In this paper inductance saliency based sensorless control using an alternating test current is adressed. To derive a position error signal the system description (3) is transformed to the estimated $\hat{d}\hat{q}$ coordinate system which is misaligned by the angle $\Delta\gamma = \hat{\gamma} - \gamma$ as to Fig. 6. The coordinate transformation from the field oriented into the estimated coordinates and its inverse are given by:

$$\underline{X}_{dq} = \underline{T}_{\Delta\gamma} \underline{X}_{\hat{d}\hat{q}} \tag{18}$$

$$\underline{X}_{\hat{d}\hat{q}} = \underline{T}_{\Delta\gamma}^{-1} \underline{X}_{dq} \tag{19}$$

$$\underline{T}_{\Delta\gamma} = \begin{bmatrix} \cos(\Delta\gamma) & -\sin(\Delta\gamma) \\ \sin(\Delta\gamma) & \cos(\Delta\gamma) \end{bmatrix} \tag{20}$$

$$\underline{T}_{\Delta\gamma}^{-1} = \underline{T}_{\Delta\gamma}^T \tag{21}$$

A. Error signals for alternating injection in the $\hat{d}$-axis

Without loss of generality the cosine shaped alternating test current is chosen to be injected into the estimated $\hat{d}$-axis as to (8). The HF component of the estimated $\hat{q}$-axis current is controlled to zero by a suitable controller. It was shown in [18] that the sine component of $u_{\hat{q}}^{HF}$ carries position error information. Therefore $u_{\hat{q}}^{HF}$ is split up as follows:

$$u_{\hat{q}}^{HF} = u_{\hat{q},cos}^{HF} \cos(\omega_{HF}t) + u_{\hat{q},sin}^{HF} \sin(\omega_{HF}t) \quad (22)$$

As to [28] one yields for the sine coefficient:

$$u_{\hat{q},sin}^{HF} = \left(\Delta l \sin(2\Delta\gamma) - m_{qd}^{HF} \cos(2\Delta\gamma)\right)\omega_{HF}\hat{I} \quad (23)$$

If only the main saliency given by $\Delta l = \frac{l_{dd}^{HF} - l_{qq}^{HF}}{2}$ occurs ($m_{qd}^{HF} = 0$) driving $u_{\hat{q},sin}^{HF}$ to zero provides zero steady state position error, as can be seen from:

$$\Delta\gamma = \frac{1}{2}\arcsin\left(\frac{u_{\hat{q},sin}^{HF}}{\Delta l \omega_{HF}\hat{I}}\right) \quad (24)$$

The position error information gets lost, if the main saliency vanishes ($\Delta l = 0$). In case secondary saliencies $m_{qd}^{HF} \neq 0$ occur a steady-state position estimation error results:

$$\Delta\gamma_{err} = \frac{1}{2}\arctan\left(\frac{2m_{qd}^{HF}}{l_{dd}^{HF} - l_{qq}^{HF}}\right) \quad (25)$$

The steady-state error (25) just depends on inductance parameters. This is a crucial advantage of test current injection based sensorless control, because no secondary resistive saliencies have to be decoupled in case of inductance saliency based sensorless control. To compensate for the error (24) a compensation voltage has to be provided. This is already inherently included in the flatness based test signal pre-control as to (16). It is obvious that the position error signal (24) does not include polarity information, thus an initial start-up method is needed.

V. POSITION AND SPEED ESTIMATION BASED ON BACK-EMF EVALUATION

For zero speed the back-EMF is zero and thus can not be used to gain a position error signal. Therefore back-EMF based position estimation can only be used for mid and high speed operation. A speed estimate, however, can always be obtained from evaluating the fundamental frequency actuating signals of the tracking controller.

A. Position error signal from back-EMF evaluation

In order to obtain a high quality position error signal and a speed estimate the parameter Ψ_q^γ is set to zero in the fundamental frequency feed forward control. Like this, in case there are no model and parameter errors the fundamental tracking controllers just have to compensate for the induced voltage $\omega_{el}\Psi_q^\gamma$. If there is no position error ($\Delta\gamma = 0$) the actuating signals of the controllers are:

$$\begin{bmatrix} u_{d,c} \\ u_{q,c} \end{bmatrix} = \omega_{el} \begin{bmatrix} 0 \\ \Psi_q^\gamma \end{bmatrix} \quad (26)$$

Considering position estimation errors $\Delta\gamma \neq 0$ the actuating signals of the $\hat{d}\hat{q}$-controllers can be found using (19) and (26):

$$\begin{bmatrix} u_{\hat{d},c} \\ u_{\hat{q},c} \end{bmatrix} \approx \underline{T}_{\Delta\gamma}^{-1} \begin{bmatrix} u_{d,c} \\ u_{q,c} \end{bmatrix} = \omega_{el} \begin{bmatrix} \Psi_q^\gamma \sin(\Delta\gamma) \\ \Psi_q^\gamma \cos(\Delta\gamma) \end{bmatrix} \quad (27)$$

Equation (27) is just approximately fulfilled for $\Delta\gamma \neq 0$, because the model implemented in the feed forward control is a model for correct alignment ($\Delta\gamma = 0$). In case of misalignment ($\Delta\gamma \neq 0$) the controllers will also have to compensate for those errors. However, this is not a severe problem, because with increasing speed the back-EMF is dominant. There are two options to gain a position error signal from (27):

1) Using just the actuating signal $u_{\hat{d},c}$ in the estimated $\hat{d}$-axis provides:

$$\Delta\gamma = \arcsin\left(\frac{u_{\hat{d},c}}{\omega_{el}\Psi_q^\gamma}\right) \quad (28)$$

The obtained error signal is within the interval $\Delta\gamma \in [-\pi/2, \pi/2]$. The current dependent parameter Ψ_q^γ has to be provided e. g. by a look-up table and ω_{el} can either be used from the speed estimate $\hat{\omega}_{el,EMF}$ or $\hat{\omega}_{el}$ from the position and speed observer. How to obtain those quantities is shown later.

2) Using both actuating signals in the estimated $\hat{d}$- and $\hat{q}$-axis yields:

$$\Delta\gamma = \arctan\left(\frac{u_{\hat{d},c}}{u_{\hat{q},c}}\right) \quad (29)$$

The error signal (29) is also within the interval $\Delta\gamma \in [-\pi/2, \pi/2]$. Using atan2 instead of arctan provides $\Delta\gamma \in [-\pi, \pi]$, because the sign of the nominator and denominator are also evaluated. This provides the following error signal:

$$\Delta\gamma = \text{atan2}\left(u_{\hat{d},c}\, sign(\omega_{el}\Psi_q^\gamma),\ u_{\hat{q},c}\, sign(\omega_{el}\Psi_q^\gamma)\right) \quad (30)$$

In contrast to the error signal obtained from HF injection the back-EMF based error signal (30) includes polarity information.

The aim of sensorless control is to keep the position estimation error very small even in transient operation. If this is the case, the trigonometric error functions can be linearized for small errors $\Delta\gamma \approx 0$ resulting in less computational effort. Using $\sin(x) \approx x$ and $\tan(x) \approx x$ for small x provides:

$$\Delta\gamma = \frac{u_{\hat{d},c}}{\omega_{el}\Psi_q^\gamma} \quad (31)$$

$$\Delta\gamma = \frac{u_{\hat{d},c}}{u_{\hat{q},c}} \quad (32)$$

The accuracy of the back-EMF based position estimation mainly depends on the knowledge of the parameter Ψ_d^γ which is implemented in the feed forward control. The induced voltages $\omega_{el}\Psi_d^\gamma$ and $\omega_{el}\Psi_q^\gamma$ are the dominant quantities in mid and high speed operation. Therefore errors in the other parameters do not affect the position estimation performance in a severe manner. The precise knowledge of Ψ_d^γ is of similar importance as the parameter m_{qd}^{HF} in case of HF test signal injection based sensorless operation in the low speed range.

978-1-4799-2706-7/14 $31.00 © 2014 IEEE

B. Speed estimate based on back-EMF evaluation

The speed estimation is based on (27). There are two options to gain a speed estimate:

1) Using the actuating signal in the estimated $\hat{q}$-axis:

$$\hat{\omega}_{el,EMF} = \frac{u_{\hat{q},c}}{\Psi_q^\gamma \cos(\Delta\gamma)} \qquad (33)$$

2) Using both actuating signals in the estimated $\hat{d}$- and $\hat{q}$-axis: Like this the position estimation error $\Delta\gamma$ can be eliminated by summing up the squared actuating signals of the $\hat{d}$- and $\hat{q}$-axis:

$$u_{\hat{d}}^2 + u_{\hat{q}}^2 = \omega_{el}^2 (\Psi_q^\gamma)^2 \left(\sin^2(\Delta\gamma) + \cos^2(\Delta\gamma) \right) \qquad (34)$$

Resulting in:

$$\hat{\omega}_{el,EMF} = \frac{\pm\sqrt{u_{\hat{d}}^2 + u_{\hat{q}}^2}}{\Psi_q^\gamma} \qquad (35)$$

The speed estimate based on (35) has always two solutions. The sign and thus the rotational direction can be found using (33), $\Psi_q^\gamma > 0$ (valid for PMSM) and the assumption $\Delta\gamma \in\ ] -\pi/2, \pi/2[$ resulting in:

$$\hat{\omega}_{el,EMF} = sign(u_{\hat{q},c}) \frac{\sqrt{u_{\hat{d}}^2 + u_{\hat{q}}^2}}{\Psi_q^\gamma} \qquad (36)$$

Due to the sign function in the speed estimate (36) undesired ripple in the speed estimate occurs for near zero speed operation, because the actuating signal $u_{\hat{q},c}$ might quite often change its sign. Therefore this speed signal is not well suited for low speed operation. The error signal (33), however, requires the precise knowledge of $\Delta\gamma$. Assuming a small position error $\Delta\gamma \approx 0$ both speed estimates (33) and (36) lead to:

$$\hat{\omega}_{el,EMF} \approx \frac{u_{\hat{q},c}}{\Psi_q^\gamma} \qquad (37)$$

The precision of the estimation depends on the knowledge of the parameter Ψ_q^γ. However, this is not a severe problem, because the position and speed observer introduced in the next section ensures zero steady state speed error.

VI. Position and Speed Estimation

For low and zero speed operation the error signal (24) and for mid and high speed operation the error signal (31) or (32) can be used. A transition function is needed to perform a smooth change between HF injection based $\Delta\gamma_{HF}$ and the back-EMF based position error signal $\Delta\gamma_{EMF}$. The transition method proposed in [29] is used as a basis for this paper. In order to prevent the need for additional anti-windup methods the transition function is placed at a different position in the block diagram resulting in less computational effort, Fig. 7. The transition function is shown in Fig. 8.

A mechanical model in state space description is the basis for observer design:

$$\dot{x} = Ax + Bu \qquad (38)$$

$$\dot{\gamma} = \omega_{el} \qquad (39)$$

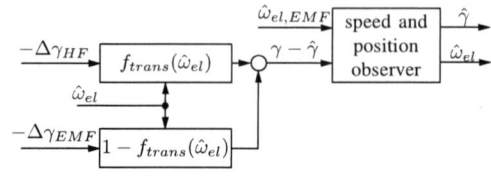

Fig. 7. Luenberger observer providing position and speed estimate based on back-EMF based speed estimate $\hat{\omega}_{el,EMF}$ and position error signal from evaluation of HF injection for low and zero speed and back-EMF for mid and high speed. In the transition region both error signals are evaluated based on the weighting factor from the respective transition function.

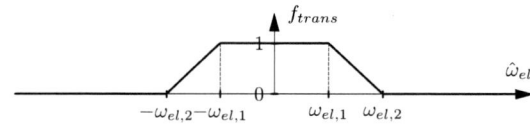

Fig. 8. Transition functions provide speed dependent weighting factors for HF injection and back-EMF based position error.

The state variable $x = \gamma$ is the electric rotor position and the input $u = \omega_{el}$ is the (electric) angular speed. Moreover $A = 0$ and $B = 1$.

Due to parameter errors the speed estimate $\hat{\omega}_{el,EMF}$ provided by evaluation of the back-EMF won't match the real speed ω_{el} resulting in steady state speed estimation errors. To guarantee zero steady state speed estimation error the speed estimation error $\Delta\omega_{el} = \omega_{el} - \hat{\omega}_{el,EMF}$ is introduced and used as an additional observer state variable, resulting in the extended model for the observer:

$$\dot{x}_e = A_e x_e + B_e u_e \qquad (40)$$

$$\frac{d}{dt}\begin{bmatrix} \gamma \\ \Delta\omega_{el} \end{bmatrix} = \underbrace{\begin{bmatrix} 0 & 1 \\ 0 & 0 \end{bmatrix}}_{A_e}\begin{bmatrix} \gamma \\ \Delta\omega_{el} \end{bmatrix} + \underbrace{\begin{bmatrix} 1 \\ 0 \end{bmatrix}}_{B_e}\hat{\omega}_{el,EMF} \qquad (41)$$

The extended state vector is $x_e = [\gamma, \Delta\omega_{el}]^T$. The differential equation $\Delta\dot{\omega}_{el} = 0$ is a disturbance model for constant disturbances. The measurement equation for the observer is:

$$y = C_e x_e \qquad (42)$$

$$y = \underbrace{\begin{bmatrix} 1 & 0 \end{bmatrix}}_{C}\begin{bmatrix} \gamma \\ \Delta\omega_{el} \end{bmatrix} \qquad (43)$$

The Luenberger type observer is given by the following state space equations:

$$\dot{\hat{x}}_e = A_e \hat{x}_e + B_e u_e + L(y - \hat{y}) \qquad (44)$$

$$\hat{y} = C\hat{x}_e \qquad (45)$$

Introducing the observation error $\varepsilon = x - \hat{x}$ provides the observation error dynamic, which needs to be stabilized by suitable choice of the gain vector L, e. g. using eigenvalue assignment:

$$\dot{\varepsilon} = (A - LC)\varepsilon \qquad (46)$$

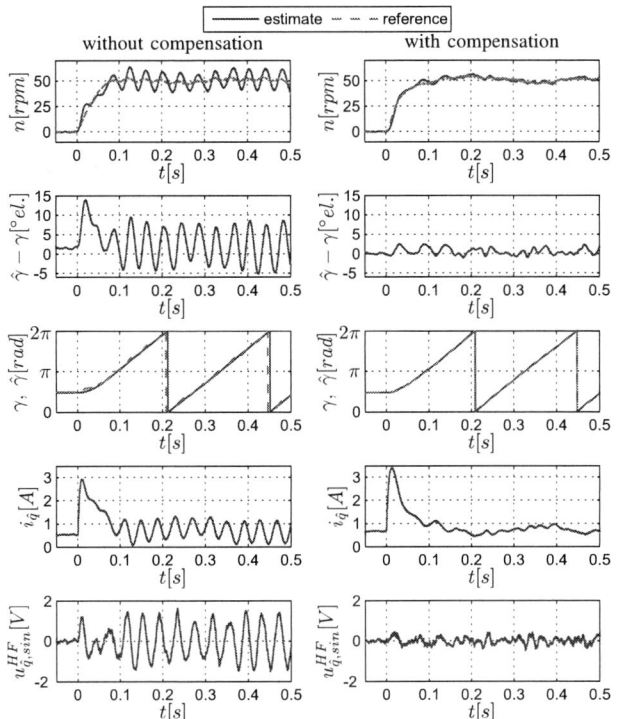

Fig. 9. Speed step response from standstill to $50rpm$ using HF test current injection without (left) and with (right) compensation for m_{qd}^{HF}.

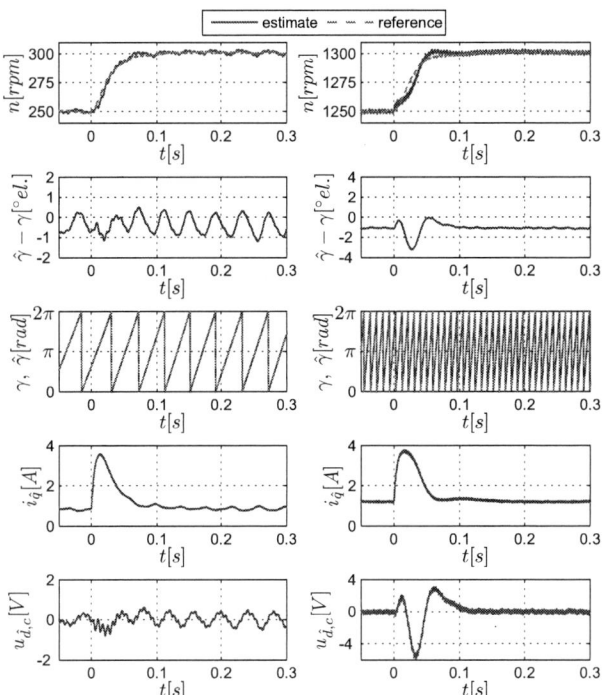

Fig. 10. Speed step response from $250rpm$ to $300rpm$ and $1250rpm$ to $1300rpm$ using back-EMF information based on the error signal from the fundamental frequency actuating signal $u_{\hat{d},c}$ from the tracking controller.

VII. EXPERIMENTAL RESULTS

For the experiments an IPMSM which shows significant 6th harmonic and saturation effects was used. The PMSM model considering higher harmonics [30] was parametrized using the methods proposed in [22] and [31]. The test signal frequency was $f_{HF} = 1kHz$ and the transition frequencies were chosen to be $\omega_{el,1} = 2\pi 8.3Hz$ and $\omega_{el,2} = 2\pi 16.6Hz$.

The speed step response for HF injection based low speed operation with and without compensation for $m_{qd}^{HF}(i_d, i_q, \gamma)$ is shown in Fig. 9. One can see that without compensation significant oscillations occur in the position estimation error $\Delta\gamma = \hat{\gamma} - \gamma$, the current $i_{\hat{q}}$ and in the speed estimate. If the compensation is used a very smooth position and speed estimate can be observed. The voltage $u_{\hat{q},sin}^{HF}$ related to the position error is nearly perfectly driven to zero and the position estimation error is less than three degrees electric position.

The speed step response from $250rpm$ to $300rpm$ and from $1250rpm$ to $1300rpm$ using back-EMF based sensorless control are shown in Fig. 10. The speed estimate as well as the position estimate are of good quality. However, a small steady state position estimation error is observed which might be caused by imperfect knowledge of the parameter Ψ_d^γ. The position estimation error is less than $3°$ el. rotor position.

The speed reversal tests from from $-250rpm$ to $250rpm$ and from $-1250rpm$ to $1250rpm$ are shown in Fig. 11. The current limit was chosen to be $i_{\hat{q},max} = 10A$ which is 125% rated current of the IPMSM. In this case the saliency Δl is very

small. For the speed reversal from $-250rpm$ to $250rpm$ the maximum position estimation error is a bit more than $10°$ el. rotor position. In case of the speed reversal from $-1250rpm$ to $1250rpm$ the maximum position error is approximately $6°$ el. rotor position. In both cases the speed estimate is of very good quality and no oscillations occur in the currents.

VIII. CONCLUSION

In this paper an innovative Two-Degree-of-Freedom current control scheme for sensorless control is proposed. Using Two-Degree-of-Freedom current control proper separation of fundamental reference response and HF test signal is achieved, thus providing very dynamic sensorless operation. The accuracy of inductance saliency based sensorless control using test current injection just depends on the knowledge of the position and current dependent coupling inductance m_{qd}^{HF}. The compensation for secondary saliencies is already inherently included in the flatness based test signal pre-control. The tracking controller used in this paper features direct demodulation of the HF sine and cosine components needed for sensorless control and phase shift compensation due to the discrete-time implementation of the control algorithm. A high quality position error signal for mid and high speed range is obtained from the fundamental tracking controllers. Like this no additional observer is needed for back-EMF based sensorless control. The accuracy of the proposed back-EMF based method mostly depends on the knowledge of the parameter Ψ_d^γ. The performance of the speed and position observer

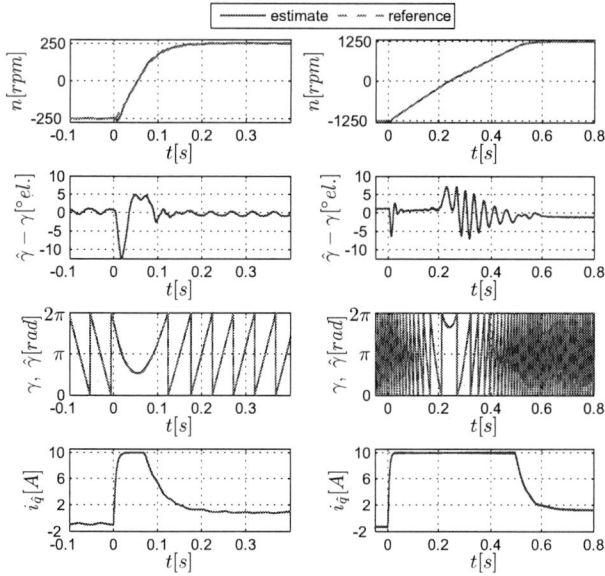

Fig. 11. Speed reversal test from $-250rpm$ to $250rpm$ and $-1250rpm$ to $1250rpm$ using back-EMF evaluation for mid and high speed operation and HF test current evaluation for low speed range. The current limit is 125% rated current of the IPMSM. In this case the saliency Δl nearly vanishes.

is enhanced by a speed estimate gained from back-EMF evaluation. Measurement results prove proper functioning of the proposed method.

REFERENCES

[1] D. Zaltni, M. N. Abdelkrim, M. Ghanes, and J. P. Barbot, "Observability analysis of pmsm," in *Internation Conference on Signals, Circuits and Systems*, 2009.

[2] L. A. S. Ribeiro, M. W. Degner, F. Briz, and R. D. Lorenz, "Comparison of carrier signal voltage and current injection for the estimation of flux angle and rotor position," in *IEEE Industry Applications Conference*, 1998.

[3] R. Bojoi, M. Pastorelli, J. Bottomley, and P. Giangrande, "Sensorless control of pm motor drives - a technology status review," in *IEEE Workshop on Electrical Machines Design Control and Diagnosis (WEMDCD)*, 2013.

[4] M. Linke, R. Kennel, and J. Holtz, "Sensorless speed and position control of synchronous machines using alternating carrier injection," in *Electric Machines and Drives Conference, 2003. IEMDC'03.*, 2003.

[5] J. Holtz, "Acquisition of position error and magnet polarity for sensorless control of pm synchronous machines," in *IEEE Transactions on Industry Appliactions*, 2008.

[6] P. L. Jansen and R. D. Lorenz, "Transducerless position and velocity estimation in induction and salient ac machines," in *IEEE Transactions on Industry Applications*, 1995.

[7] S. Kim and S.-K. Sul, "High performance position sensorless control using rotating voltage signal injection in IPMSM," in *Proceedings of the 14th European Conference on : Power Electronics and Applications (EPE 2011)*, 2011, pp. 1–10.

[8] S. C. Yang and R. D. Lorenz, "Comparison of resistance-based and inductance-based self-sensing control for surface permanent magnet machine using high frequency signal injection," in *2011 IEEE Energy Conversion Congress and Exposition.* IEEE, 2011, pp. 2701–2708.

[9] T. Frenzke, "Impacts of cross-saturation on sensorless control of surface permanent magnet synchronous motors," in *2005 European Conference on Power Electronics and Applications.* IEEE, 2005.

[10] D. Reigosa, P. Garcia, D. Raca, F. Briz, and R. D. Lorenz, "Measurement and Adaptive Decoupling of Cross-Saturation Effects and Secondary Saliencies in Sensorless-Controlled IPM Synchronous Machines," in *IEEE Industry Applications Annual Meeting*, 2007.

[11] A. Piippo and J. Luomi, "Inductance harmonics in permanent magnet synchronous motors and reduction of their effects in sensorless control," in *Proceedings of the XVII International Conference on Electric Machines (ICEM 2006)*, 2006.

[12] J. Solsona, M. I. Valla, and C. Muravchik, "Nonlinear control of a permanent magnet synchronous motor with disturbance torque estimation," in *IEEE Transactions on Energy Conversion, Vol. 15*, 2000.

[13] S. Beineke, J. Schirmer, J. Lutz, H. Wertz, A. Bähr, and J. Kiel, "Implementation and applications of sensorless control for synchronous machines in industrial inverters," in *Symposium on Sensorless control for Electrical Drives (SLED)*, 2010.

[14] I. Boldea and S. C. Agarlita, "The active flux concept for motion-sensorless unified ac drives: A review," in *International Conference on Electrical Machines and Power Electronics (ACEMP)*, 2011.

[15] S. Morimoto, K. Kawamoto, M. Sanada, and Y. Takeda, "Sensorless control strategy for salient-pole pmsm based on extended emf in rotating reference frame," in *IEEE Transactions on Industry Applications*, 2002.

[16] O. Benjak and D. Gerling, "Review of position estimation methods for ipmsm drives without a position sensor part i: Nonadaptive methods and part ii adaptive methods," in *XIX International Conference on Electrical Machines (ICEM)*, 2010.

[17] M. Seilmeier, S. Arenz, B. Piepenbreier, and I. Hahn, "Model based closed loop control scheme for compensation of harmonic currents in PM-synchronous machines," in *SPEEDAM 2010.* IEEE, 2010, pp. 1–6.

[18] S. Ebersberger, M. Seilmeier, and B. Piepenbreier, "Flatness Based Sensorless Control of PMSM using Test Current Signal Injection and Compensation for Differential Cross-Coupling Inductances at Standstill and Low Speed Range," in *4th Symposium on Sensorless Control for Electrical Drives*, 2013.

[19] M. Seilmeier, S. Ebersberger, and B. Piepenbreier, "PMSM Model for Sensorless Control Considering Saturation Induced Secondary Saliencies," in *4th Symposium on Sensorless Control for Electrical Drives*, 2013.

[20] J. Melkebeek and J. Willems, "Reciprocity relations for the mutual inductances between orthogonal axis windings in saturated salient-pole machines," *IEEE Transactions on Industry Applications*, vol. 26, no. 1, pp. 107–114, 1990.

[21] S.-C. Yang and R. D. Lorenz, "Surface permanent magnet synchronous machine self-sensing position estimation at low speed using eddy current reflected asymmetric resistance," in *Proceedings of the 14th European Conference on : Power Electronics and Applications (EPE 2011)*, 2011.

[22] M. Seilmeier, S. Ebersberger, and B. Piepenbreier, "Identification of high frequency resistances and inductances for sensorless control of PMSM," in *4th Symposium on Sensorless Control for Electrical Drives*, 2013.

[23] L. Alberti, N. Bianchi, M. Morandin, and J. Gyselinck, "Finite-element analysis of electrical machines for sensorless drives with signal injection," in *Energy Conversion Congress and Exposition (ECCE), 2012 IEEE*, 2012, pp. 861–868.

[24] J. Slotine, *Applied nonlinear Control.* New Jersey: Prentice Hall, 1991.

[25] A. Isidori, *Nonlinear control systems.* London: Springer Verlag, 1995.

[26] M. Fliess, J. Levine, P. Martin, and P. Rouchon, "Flatness and defect of non-linear systems: introductory theory and examples," *Internation Journal of Control*, vol. 61, Issue 6, pp. 1327–1361, 1995.

[27] M. Seilmeier, A. Boehm, I. Hahn, and B. Piepenbreier, "Identification of time-variant high frequency parameters for sensorless control of PMSM using an internal model principle based high frequency current control," in *XXth International Conference on Electrical Machines*, 2012.

[28] M. Seilmeier, S. Ebersberger, and B. Piepenbreier, "HF test current injection based self-sensing control of PMSM for low and zero speed range using Two-Degree-of-Freedom current control," in *5th Symposium on Sensorless Control for Electrical Drives*, 2014.

[29] J. Hong, S. Jung, and K. Nam, "An incorporation method of sensorless algorithms: Signal injection and back emf based methods," in *International Power Electronics Conference (IPEC)*, 2010.

[30] M. Seilmeier, S. Ebersberger, and B. Piepenbreier, "PMSM Model for Sensorless Control Considering Saturation Induced Secondary Saliencies," in *4th Symposium on Sensorless Control for Electrical Drives*, 2013.

[31] M. Seilmeier and B. Piepenbreier, "Identification of steady-state inductances of PMSM using polynomial representations of the flux surfaces," in *Industrial Electronics Conference (IECON), Vienna*, 2013.

Ellipse-Trajectory-Oriented Vector Control for Energy Efficient/Wide-Speed-Range Drives of Sensorless PMSM

Shinji Shinnaka

Dept. of Electrical Engineering
Kanagawa University
3-27-1 Rokkakubashi, Kanagawaku, Yokohama, Japan
shinnaka@kanagawa-u.ac.jp

Yuki Amano

Kokusan Denki Co., Ltd.
3744 Ooka, Numazu, Japan

Abstract—**This paper proposes a new sensorless vector control method for energy-efficient and wide-speed-range drives of permanent-magnet synchronous motors. The proposed sensorless method is endowed with all of the following functions: 1) wide-speed-range drive under voltage limit, 2) current limitation, 3) high-efficiency. These functions are realized in a very simple manner and can operate adaptively in all sensorless driving modes including motoring/regenerating, steady/transient states. The usefulness of the proposed method is verified by extensive experiments.**

Keywords—*current limitation, efficiency, sensorless drive, voltage limitaion, wide-speed-range drive.*

I. INTRODUCTION

Applications of sensorless drives of permanent-magnet synchronous motors (PMSMs) are expanding in areas such as industry, home-appliance, automobile etc. Sensorless drives of PMSMs basically require the phase and speed estimates of the rotor N-pole. In addition to the estimation function, sensorless-drive apparatuses that are applied in wide-speed range over the rated speed should have simultaneously the following functions/properties,

  (a) wide-speed range drive in the circumstance of the voltage limit
  (b) current limitation/over-current protection
  (c) efficient drive minimizing losses
  (d) simple and consistent implementation of all functions including estimating one.

An approach to the above request will be the direct one that has been taken in the sensor-used vector control [1]. However, this approach cannot improve the implementation simplicity, conversely will worsen it in phase/speed estimation circumstance.

A potential approach will be improvements of the conventional sensorless methods that attain simultaneously items (b) to (d) in speed range below the rated speed. These methods accomplish the simplicity by determining directly the phase of the stator current via no estimation of the rotor phase. They can be classified into non-parametric [2]-[5] and parametric [2], [6]-[11] methods. A typical parametric method estimates the rotor flux, the back EMF, or the extended back EMF by using

a constant observer inductance different from the actual value, and is called TOVC (trajectory-oriented vector control) method [2], [6], [9].

This paper newly proposes E-TOVC (ellipse-trajectory-oriented vector control) method that can attain simultaneously all items (a) to (d). The E-TOVC method changes dynamically the observer inductance that can take large negative values according to the stator voltage under the voltage limit.

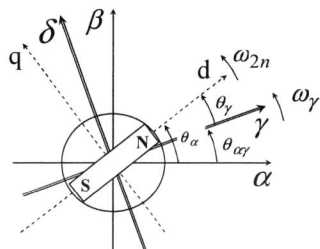

Fig. 1. Phase of rotor N-pole in $\gamma - \delta$ rotating reference frame rotating at instant speed ω_γ with instant phase $\theta_{\alpha\gamma}$.

II. MATHEMATICAL MODEL

Consider a $\gamma - \delta$ rotating reference frame where the orthogonal $\gamma - \delta$ coordinates rotate at an instant speed ω_γ, with an instant γ-axis phase $\theta_{\alpha\gamma}$ as shown in Fig. 1. Rotating polarity is defined such that the direction from the principal axis (γ-axis) to the secondary axis (δ-axis) is positive. Note that all of the following 2x1 vector signals related to PMSMs are defined in the $\gamma - \delta$ rotating reference frame.

Using the D-matrix defined in (4), the electromagnetic characteristics of PMSMs can be described as [1]

$$v_1 = R_1 i_1 + D(s, \omega_\gamma)\phi_1 \tag{1}$$

$$\phi_1 = \phi_i + \phi_m \tag{2}$$

$$\phi_i = [L_i I + L_m Q(\theta_\gamma)]i_1 \tag{3}$$

with

$$D(s, \omega_\gamma) = sI + \omega_\gamma J \tag{4}$$

978-1-4799-2706-7/14 $31.00 © 2014 IEEE

$$Q(\theta_\gamma) = \begin{bmatrix} \cos 2\theta_\gamma & \sin 2\theta_\gamma \\ \sin 2\theta_\gamma & -\cos 2\theta_\gamma \end{bmatrix} \tag{5}$$

$$\phi_m = \Phi\, u(\theta_\gamma) = \Phi \begin{bmatrix} \cos\theta_\gamma \\ \sin\theta_\gamma \end{bmatrix} \; ; \; \Phi = \text{const} \tag{6}$$

$$s\theta_\gamma = \omega_{2n} - \omega_\gamma \tag{7}$$

where 2x1 vectors v_1, i_1, ϕ_1 are the voltage, the current and the flux of the stator, respectively; 2x1 vectors ϕ_i, ϕ_m are the components of the stator flux ϕ_1 – more precisely, ϕ_i indicates the flux evolved directly by the stator current i_1 and ϕ_m is the flux due to the rotor magnet; I is a 2x2 identity matrix; J is a 2x2 skew symmetric matrix such as

$$J = \begin{bmatrix} 0 & -1 \\ 1 & 0 \end{bmatrix}; \tag{8}$$

ω_{2n} is the rotor electrical speed; R_1 is the stator resistance; L_i, L_m are the in- and mirror-phase inductances having a relation with d- and q-inductances such as

$$\begin{bmatrix} L_i \\ L_m \end{bmatrix} = \frac{1}{2}\begin{bmatrix} 1 & 1 \\ 1 & -1 \end{bmatrix}\begin{bmatrix} L_d \\ L_q \end{bmatrix}; \tag{9}$$

symbol "s" indicates a differential operator d/dt.

III. TOVC METHOD

A. Basic System Structure of TOVC System

One of the basic structures of the sensorless vector control system based on the TOVC method (simply refer to as TOVC system in the following) can be illustrated as in Fig. 2(a) [2], [6], [9]. The phase converter and the vector rotator in the system are defined such as

$$S = \sqrt{\frac{2}{3}} \begin{bmatrix} 1 & 0 \\ \dfrac{-1}{2} & \dfrac{\sqrt{3}}{2} \\ \dfrac{-1}{2} & \dfrac{-\sqrt{3}}{2} \end{bmatrix} \tag{10}$$

$$R(\hat{\theta}_{\alpha\gamma}) = \begin{bmatrix} \cos\hat{\theta}_{\alpha\gamma} & -\sin\hat{\theta}_{\alpha\gamma} \\ \sin\hat{\theta}_{\alpha\gamma} & \cos\hat{\theta}_{\alpha\gamma} \end{bmatrix}. \tag{11}$$

Note that γ-current command is always set at zero in the TOVC system.

The phase-speed estimator of the system, which plays a role of determining phase and speed of the $\gamma - \delta$ rotating reference frame, is illustrated as in Fig. 2 (b). It consists of two basic blocks of phase-error estimator and phase synchronizer.

A kind of rotor flux, back EMF or extended back EMF observers using motor parameters is implemented in the phase-error estimator, and a generalized integral-type PLL method in the phase synchronizer [2], [6], [9]. The observers have the ability that estimates correctly the

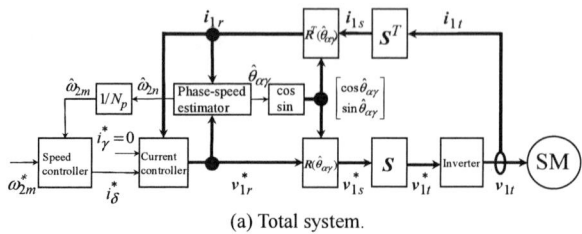

(a) Total system.

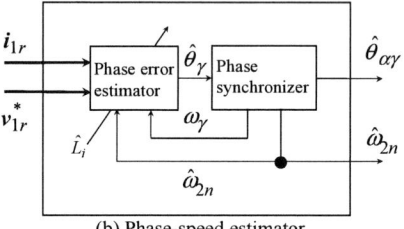

(b) Phase-speed estimator.

Fig. 2. A basic structure of sensorless vector control systems based on TOVC method.

physical quantities and rotor phase, when the correct motor parameters are employed for the observers.

B. Principle and Basic Characteristics of TOVC Method

Consider the situation that the stator current of the system in Fig. 2 is controlled to follow a current command. In the situation, the following theorems 1-3 hold [2], [6], [9].

[Theorem 1 (parameter error theorem)]

If an inductance $\hat{L}_i$ is used in the observer (i.e. phase-error estimator) instead of the actual q-inductance L_q, and the difference between observer-used and actual stator resistances can be negligible, for example, due to high speed, then the steady-state stator current of the sensorless control system is putted on the trajectory governed by

$$\Phi i_d - (\hat{L}_i - L_d)i_d^2 - (\hat{L}_i - L_q)i_q^2 = 0, \tag{12}$$

where i_d, i_q are d-current and q-current, respectively.

[Theorem 2 (trajectory theorem I)]

If an inductance $\hat{L}_i = L_d$ is used in the observer (i.e. phase-error estimator) instead of actual q-inductance L_q, and the difference between observer-used and actual stator resistances can be negligible, for example, due to high speed, then the steady-state stator current of the sensorless control system is putted on the parabolic trajectory governed by

$$\Phi i_d - 2L_m i_q^2 = 0. \tag{13}$$

[Theorem 3 (trajectory theorem II)]

If the zero inductance $\hat{L}_i = 0$ is used in the observer (i.e. phase-error estimator) instead of actual q-inductance L_q, and the difference between observer-used and actual stator resistances can be negligible, for example, due to high speed, then the steady-state stator current of the

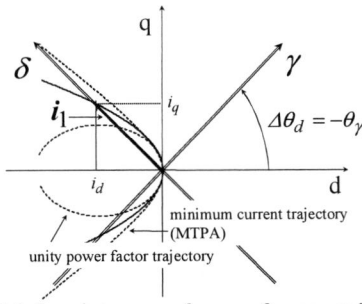

Fig.3. Relation between reference frames and current trajectory.

sensorless control system is putted on the ellipse trajectory of the unity power-factor governed by

$$\Phi i_d - 2L_m i_q^2 = 0. \tag{14}$$

Exploiting theorems 1 to 3, the TOVC method that allows efficient and sensorless drive with current limitation function is established, and can be summarized as follows [2], [6], [9].

[TOVC method]
First, treat a salient-pole PMSM with two inductances of L_d, L_q as a non-salient-pole PMSM with a single constant inductance $\hat{L}_i$ such as

$$0 \le \hat{L}_i \le L_i. \tag{15}$$

Secondly, control γ-current to be zero, and realize the observer (i.e. phase-error estimator) of the phase-speed estimator with a single constant inductance $\hat{L}_i$ as in (15). Then the steady-state stator current is putted on the trajectory governed by (12) that can achieve a kind of efficient drives.

Fig. 3 shows the relations between reference frames and current trajectories. Note that 1) the stator current is always on δ-axis; 2) the phase of γ, δ-axes is lead/lag to that of d,q-axes according to positive/negative polarity of δ-current.

In the case that a single observer inductance of $\hat{L}_i = L_d$ is used, the stator current is putted on the parabolic trajectory in (13) [2], [6], [9], which is close to the minimum current (MTPA) trajectory that attains the minimum copper loss, i.e.

$$0 = \Phi i_d - 2L_m i_q^2 \\ \approx \Phi i_d + 2L_m(i_d^2 - i_q^2) \ ; \ 2L_m i_d << \Phi \tag{16}$$

Two trajectories using motor parameters in TABLE I where N_p means the number of pole pairs are depicted in Fig. 4. It is confirmed that the parabolic trajectory according to theorem 2 is very close to the minimum current (MTPA) one in rage of the rated current.

IV. E-TOVC METHOD

The conventional TOVC method mentioned in the above does not take the voltage limit into account. In

TABLE 1. CHARACTERISTICS OF TEST MOTOR (SST4-20P4AEA-L).

R_1	2.201 [Ω]	rated torque	2.2[Nm]
L_i	0.026845 [H]	rated speed	183 [rad/s]
L_m	-0.005655[H]	rated current	1.7 [A, rms]
Φ	0.24 [Vs/rad]	rated voltage	163 [V, rms]
N_p	3	moment of inertia	0.0016 [kgm²]
rated power	0.4 [kW]	effective resolution of encorder	4×1024 [p/r]

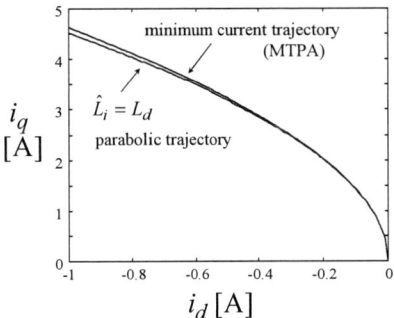

Fig.4. Parabolic and minimum current trajectories.

other words, it cannot be used over the rated speed where the voltage limit practically exists. This chapter improves the conventional TOVC method so that the improved one can extend speed-region, and attain the most efficient drive even in the extended speed-region under the voltage limit. The improved TOVC method is referred to as E-TOVC (ellipse-trajectory-oriented vector control) method in the following.

A. Additional Characteristics
The characteristics summarized in the following theorems play a key role to get the aforementioned performances.

[Theorem 4]
Consider the situation where theorem 1 holds. If observer inductance $\hat{L}_i$ satisfies the condition of $\hat{L}_i \le L_q$, then there exists the following relation between observer inductance $\hat{L}_i$ and the phase difference $\Delta\theta_d$ (refer to Fig. 3),

$$\hat{L}_i = L_q + |\sin \Delta\theta_d| \left(2L_m |\sin \Delta\theta_d| - \frac{\Phi}{|i_\delta|} \right) \ ; |\Delta\theta_d| \le \frac{\pi}{2} \tag{17}$$

< proof >
Under the condition of $\hat{L}_i \le L_q$, the following relation holds from Fig. 3

$$i_d = -i_\delta \sin \Delta\theta_d \le 0 \ ; \ |\Delta\theta_d| \le \frac{\pi}{2}. \tag{18}$$

If theorem 1 holds, then the following relation also holds [2], [6], [9]

$$\Delta\theta_d = \sin^{-1} \frac{-i_d}{i_\delta}$$

978-1-4799-2706-7/14 $31.00 © 2014 IEEE

$$= \sin^{-1}\left(\frac{\Phi - \sqrt{\Phi^2 + 8L_m(\hat{L}_i - L_q)i_\delta^2}}{4L_m i_\delta}\right). \qquad (19)$$

Arranging (19) in terms of $\hat{L}_i$ using (18) yields

$$\hat{L}_i = L_q + \frac{i_d(2L_m i_d + \Phi)}{i_\delta^2}$$

$$= L_q + \sin\Delta\theta_d\left(2L_m \sin\Delta\theta_d - \frac{\Phi}{i_\delta}\right) \qquad (20)$$

$$= L_q + \left|\sin\Delta\theta_d\right|\left(2L_m\left|\sin\Delta\theta_d\right| - \frac{\Phi}{\left|i_\delta\right|}\right)$$

(20) implies the theorem.

[Theorem 5]

Consider the situation where theorem 2 holds. In the current range of

$$\left(\frac{4L_m\left|i_\delta\right|}{\Phi}\right)^2 \ll 1, \qquad (21)$$

the sinusoidal value of the phase difference is nearly proportional to δ-current such as

$$\sin\Delta\theta_d \approx \frac{L_q - L_d}{\Phi}i_\delta = \frac{-2L_m}{\Phi}i_\delta \qquad . \qquad (22)$$

< proof >

Applying the condition of $\hat{L}_i = L_d$ employed in theorem 2 to theorem 4 yields

$$\hat{L}_i = L_q + x\left(2L_m x - \frac{\Phi}{\left|i_\delta\right|}\right) = L_d \qquad (23)$$

with

$$\left.\begin{array}{c} x \equiv \left|\sin\Delta\theta_d\right| \\ 0 < x < 1 \end{array}\right\}. \qquad (24)$$

Arrange (23) in terms of positive sinusoidal value x and taking positive one yields

$$x = \frac{\dfrac{\Phi}{\left|i_\delta\right|}\cdot\dfrac{1}{2L_m} + \sqrt{\left(\dfrac{\Phi}{\left|i_\delta\right|}\cdot\dfrac{1}{2L_m}\right)^2 + 4}}{2}. \qquad (25)$$

Using the condition in (21) and negative polarity of L_m, (25) can be rearranged and approximated as follows,

$$x = \frac{\dfrac{\Phi}{\left|i_\delta\right|}\cdot\dfrac{1}{2L_m} + \dfrac{\Phi}{\left|i_\delta\right|}\cdot\dfrac{1}{\left|2L_m\right|}\sqrt{1 + 4\left(\dfrac{2L_m\left|i_\delta\right|}{\Phi}\right)^2}}{2}. \qquad (26)$$

$$\approx \frac{-2L_m\left|i_\delta\right|}{\Phi} = \frac{(L_q - L_d)\left|i_\delta\right|}{\Phi} \geq 0$$

(26) implies the theorem.

QED

If the strategy that the observer inductance is set at $\hat{L}_i = L_d$ in lower speed-region without practical voltage limit is taken, then theorems 4, 5 and (17) derive the

minimum/maximum values of both absolute sinusoidal value $\left|\sin\Delta\theta_d\right|$ and observer inductance $\hat{L}_i$ such as

$$\frac{-2L_m\left|i_\delta\right|}{\Phi} \leq \left|\sin\Delta\theta_d\right| < 1 \qquad (27)$$

$$\left(1 - \frac{\Phi}{L_d\left|i_\delta\right|}\right)L_d < \hat{L}_i \leq L_d \quad . \qquad (28)$$

The current trajectories governed by (12) with (28) result in "ellipses" except for $\hat{L}_i = L_d$. The name of "E-TOVC" is derived from this principal fact.

B. Self-tuning of Observer Inductance

The proposed E-TOVC method is established based on theorems 4, 5, (27), and (28), and can attain the expected performance. The E-TOVC method is characterized by control of the phase difference $\Delta\theta_d$ (i.e. stator current phase, refer to Fig. 3) in wide region over the rated speed through self-tuning observer inductance $\hat{L}_i$.

Observer inductance self-tuning algorithms directly based on theorems 4, 5 are summarized as follows.

[Self-tuning algorithm I (output-gain type)]

$$u(k) = \begin{cases} 0 & ; \quad \left\|v_1^*(k)\right\| < c_v(k) \\ 1 & ; \quad \left\|v_1^*(k)\right\| \geq c_v(k) \end{cases} \qquad (29a)$$

$$x'(k) = \alpha_1 x'(k-1) + (1-\alpha_1)u(k) \; ; \; 0 < \alpha_1 < 1 \quad (29b)$$

$$x(k) = \frac{-2L_m\left|i_\delta(k)\right|}{\Phi} + \left(1 + \frac{2L_m\left|i_\delta(k)\right|}{\Phi}\right)x'(k) \qquad (29c)$$

$$\hat{L}_i(k) = L_q + x(k)\left(2L_m x(k) - \frac{\Phi}{\left|i_\delta(k)\right| + \Delta_0}\right) \; ; \; \Delta_0 > 0 \qquad (29d)$$

[Self-tuning algorithm I (input-gain type)]

$$u(k) = \begin{cases} 0 & ; \quad \left\|v_1^*(k)\right\| < c_v(k) \\ 1 & ; \quad \left\|v_1^*(k)\right\| \geq c_v(k) \end{cases} \qquad (30a)$$

$$x'(k) = \alpha_1 x'(k-1) + (1-\alpha_1)\left(1 + \frac{2L_m\left|i_\delta(k)\right|}{\Phi}\right)u(k)$$

$$; \; 0 < \alpha_1 < 1 \qquad (30b)$$

$$x(k) = \frac{-2L_m\left|i_\delta(k)\right|}{\Phi} + x'(k) \qquad (30c)$$

$$\hat{L}_i(k) = L_q + x(k)\left(2L_m x(k) - \frac{\Phi}{\left|i_\delta(k)\right| + \Delta_0}\right) \; ; \; \Delta_0 > 0 \qquad (30d)$$

In (29) and (30), signals at a time instant $t = kT_s$ is simply expressed by integer k instead of kT_s. The role of (29a) and (30a) is to detect the voltage limitation, in other words, to check whether the voltage command reaches

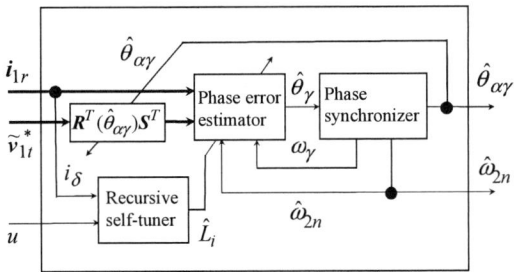

Fig.5. Configuration of new phase-speed estimator based on E-TOVC method.

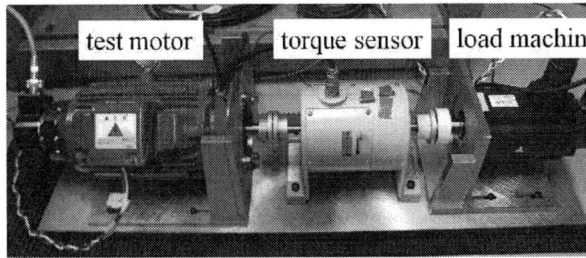

Fig. 6. Test system setup.

the allowed maximum voltage or not. If the norm of the voltage command is smaller than the voltage limit, then $u(k) = 0$ is set. Otherwise, $u(k) = 1$.

In the self-tuning algorithms in (29) and (30), firstly the absolute sinusoidal value $x \equiv |\sin \Delta\theta_d|$ associated with phase difference $\Delta\theta_d$ is self-tuned; secondly observer inductance $\hat{L}_i$ is adjusted according to the self-tuned value. Design parameter α_1 determines the tuning speed. A small positive constant Δ_0 is simply introduced in order to avoid the "zero-division".

Considering the minimum and maximum of the observer inductance shown in (28), it is possible to tune directly the observer inductance. Direct self-tuning algorithms for the observer inductance are summarized as follows.

[Self-tuning algorithm II (output-gain type)]

$$u(k) = \begin{cases} 0 & ; \quad \left\| v_1^*(k) \right\| < c_v(k) \\ 1 & ; \quad \left\| v_1^*(k) \right\| \ge c_v(k) \end{cases} \quad (31a)$$

$$x'(k) = \alpha_1 x'(k-1) + (1-\alpha_1)u(k) ; \quad 0 < \alpha_1 < 1 \quad (31b)$$

$$\hat{L}_i(k) = L_d - \frac{\Phi}{|i_\delta(k)| + \Delta_0} x'(k) ; \quad \Delta_0 > 0 \quad (31c)$$

[Self-tuning algorithm II (input-gain type)]

$$u(k) = \begin{cases} 0 & ; \quad \left\| v_1^*(k) \right\| < c_v(k) \\ 1 & ; \quad \left\| v_1^*(k) \right\| \ge c_v(k) \end{cases} \quad (32a)$$

$$x'(k) = \alpha_1 x'(k-1) - (1-\alpha_1)\frac{\Phi}{|i_\delta(k)| + \Delta_0} u(k) \quad (32b)$$

$$; \quad 0 < \alpha_1 < 1, \quad \Delta_0 > 0$$

$$\hat{L}_i(k) = L_d + x'(k) \quad (32c)$$

C. Phase-speed Estimator Based on E-TOVC Method
  C-1 Recursive Self-tuner
A new phase-speed estimator based on the proposed E-TOVC method is shown in Fig. 5. One of the self-tuning algorithms in (29)-(32) is installed in the block of "recursive self-tuner".
  C-2 Phase-error Estimator

An observer using a single varying inductance $\hat{L}_i$ is implemented in the phase-error estimator similarly to the conventional TOVC method. For example, the following minimum order D-state flux observer is applied [2], [12].

[D-state flux observer]

$$D(s, \hat{\omega}_{2n})\widetilde{\phi}_1 = G[v_1 - R_1 i_1] - |\hat{\omega}_{2n}|\widetilde{\phi}_m \quad (33a)$$

$$\hat{\phi}_m = \widetilde{\phi}_1 - G\hat{\phi}_i \quad (33b)$$

$$\hat{\phi}_i = \hat{L}_i i_1 \quad (33c)$$

$$G = I - \mathrm{sgn}(\hat{\omega}_{2n}) g J \quad ; \quad g > 0 \quad (33d)$$

$$\hat{\theta}_\gamma = \tan^{-1}(\hat{\phi}_{m\delta}/\hat{\phi}_{m\gamma}) \quad (33e)$$

As a voltage signal, voltage-limited command $\widetilde{v}_{1t}^*$ in the three-phase reference frame is used. The signal is inputted into the phase-error estimator through the reference-frame conversion using $R^T(\cdot)S^T$ (refer to (10), (11)).
  C-3 Phase Synchronizer
A generalized integral-type PLL method is implemented in the phase synchronizer similarly to the conventional TOVC method. The PLL method is summarized as follows [2], [12].

[Generalized Integral-type PLL Method]

$$\omega_\gamma = \hat{\omega}_{2n} = C(s)\hat{\theta}_\gamma \quad (34a)$$

$$\hat{\theta}_{\alpha\gamma} = \frac{1}{s}\omega_\gamma \quad (34b)$$

$$C(s) = \frac{C_N(s)}{C_D(s)} = \frac{c_{nm}s^m + c_{nm-1}s^{m-1} + \cdots c_{n0}}{s^m + c_{dm-1}s^{m-1} + \cdots c_{d0}} \quad (34c)$$

The sensorless drive (E-TOVC) system in Fig. 2(a) that employs the new phase-speed estimator in Fig. 5 instead of Fig. 2(b) can overcome the problem of voltage limit and allows wide speed range drives with efficiency.

V. EXPERIMENTS

A. Experiment System and Design Parameter
In order to examine properties and performance of the proposed E-TOVC system, extensive experiments were carried out using the equipment illustrated in Fig. 6. Test motor is a 400-W SP-PMSM (IPMSM, SST4-20P4AEA-

L) made by Yaskawa Electric Corporation (refer to TABLE I for characteristics). A rotor-mounted encoder is just for monitoring of the actual rotor phase and speed, and is not used for control.

The E-TOVC system with the load machine was realized as in Fig. 2(a), but with the phase-speed estimator in Fig. 5. The stator current feedback loop is designed so that its bandwidth can be 2,000-rad/s in consideration of control period of 100-μs (equivalent to 10-kHz). The voltage limit was set intentionally at a small value of $c_v = 170$ -V in consideration of the performance of the load machine.

The design parameters of the recursive self-tuner are selected as

$$\alpha = 0.999, \quad \Delta_0 = 0.01. \tag{35}$$

The phase-error estimator is realized based on the D-state observer with observer gain of $g = 1$ in (33). The phase controller $C(s)$ in (34c) constructing the phase synchronizer was designed as

$$C(s) = 150 + \frac{1500}{s}, \tag{36}$$

so that PLL bandwidth of 150-rad/s can be attained.

B. Current Control

The speed controller was taken off from the E-TOVC system with the load machine so that the stator current can be directly controlled. The speed of the test motor is controlled by the load machine. This system allows the examination and evaluation of the basic functions at varying speed including the detection of voltage limit, the self-tuning of the observer inductance, d- and q-currents control corresponding to the flux weakening.

In consideration of the low voltage limit, the motor speed is controlled between 10 and 120-rad/s with acceleration of ± 20 rad/s^2. γ- and δ-currents of the test motor are controlled constant of 0, 3-A (rated current), respectively.

Fig.7 shows a result. Fig. 7(a) indicates, from the top, the actual and estimated rotor speeds ω_{2m}, $\hat{\omega}_{2m}$ (two waveforms are almost the same, and seen as a single trapezoidal curve), the voltage limit c_v =170-V, the voltage command norm $\|v_1^*\|$, the self-tuned observer inductance normalized by actual d-inductance $\hat{L}_i / L_d$, and its theoretical minimum value $\hat{L}_{i\,\min} / L_d$. It is observed that even after the stator voltage reaches the voltage limit at speed of about 90-rad/s, the speed keeps increasing up to about 120-rad/s and the observer inductance quickly goes down to the theoretical minimum below zero. Concurrently, d-current negatively increases, and q-current decreases as shown in Fig. 6(b). The change of d- and q-currents, which are obtained through conversion of γ- and δ-currents, implies the stator current entered into the mode of so-called "flux weakening."

When the speed goes down from the top speed, the observer inductance quickly goes up. When the stator voltage becomes less than the voltage limit, the observer

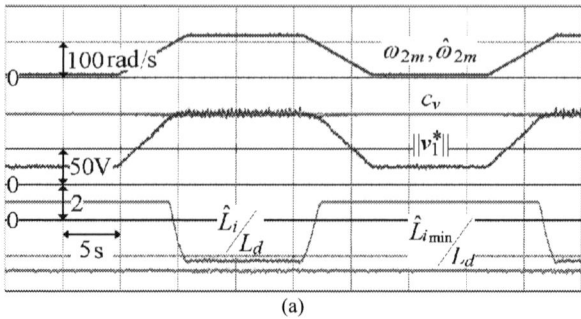

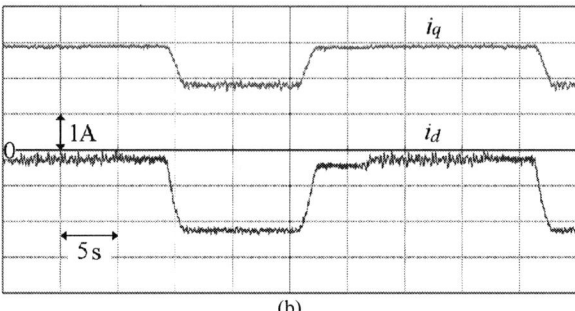

Fig.7. Experimental result of current control.

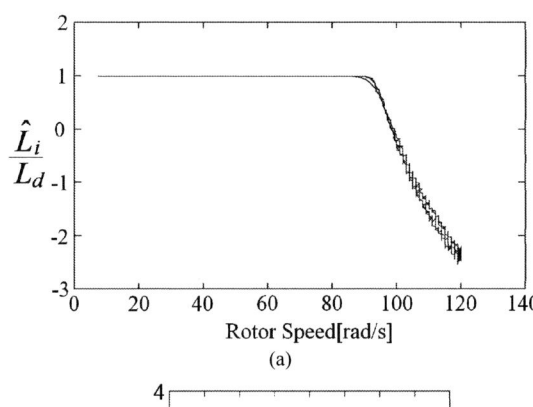

(a)

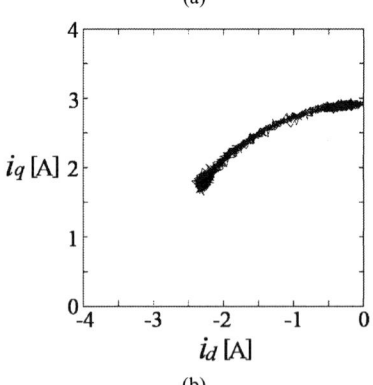

(b)

Fig.8. Experimental result of current control.

inductance converges to d-inductance that guarantees the minimum copper loss drive. Concurrently, d- and q-currents go out of the flux-weakening mode and converge to the standard values on the minimum current trajectory (MTPA trajectory).

Fig. 8(a) shows the normalized observer inductance $\hat{L}_i / L_d$ vs the rotor mechanical speed, which is converted from the time-waveform in Fig. 7(a). The waveform difference of $\hat{L}_i / L_d$ between increasing and decreasing speeds is associated with the modes of motoring and regenerating. Fig. 8(b) shows the space trajectory of the stator current in the dq-synchronous reference frame, which is converted from the time-waveform in Fig. 7(b). The space trajectory clearly shows the stator current enters and goes out of the flux weakening mode. The responses in Figs. 7 and 8 are almost ideal.

C. Speed Control under 50% rated torque

In order to examine the applicability of the E-TOVC method to speed controls and variable current commands, the system with a speed controller in Fig. 2(a) was prepared. The E-TOVC is a kind of the "norm-based current controls", which controls the stator current norm with polarity instead of the direct control of d- and q-currents [13]. It is pointed out by [13] that a speed control system constructed above the norm-based current control system turns out to be nonlinear, but its stability can be guaranteed if the employed linear speed controller is designed based on the Popov's stability theorem. The linear speed controller of the system in Fig. 2(a) was designed so that its time-constant can be about 0.1-s based on the design method by [13]. In consideration of the low voltage limit, the motor speed command between 10 and 120-rad/s with acceleration of ± 20 rad/s² was assigned under the 50% rated torque by the load machine.

Fig.9 shows a result. The meaning of the waveforms is the same as that of Fig. 7 except for the upper ones indicating the speed command and its actual response. As the speed command increases, the speed response and the voltage command norm also increase as in Fig. 9(a). Even after the voltage command norm reaches the voltage limit, the speed response continues to increase according to its command. The normalized observer inductance $\hat{L}_i / L_d$ corresponding to these varying signals shows almost ideal response similarly to Fig. 7(a). It approaches to its theoretical minimum value as observed in Fig. 9(a).

Note that the theoretical minimum value of normalized observer inductance varies according to magnitude of the stator current as analyzed in (28). The stator current in Fig. 9(b) shows a good agreement with the waveforms in Fig. 9(a).

Fig. 10 shows the space trajectory of the stator current and the current limit circle in the dq-synchronous reference frame. It is well known that the optimum current minimizing the copper loss under the constant load torque and the voltage limit exists on the intersecting point of the constant torque and the voltage limit ellipse trajectories. Since the voltage limit ellipse converges to the point of ($i_d = -\Phi / L_d$, $i_q = 0$) with speed increase, the optimum current slides to the left on the constant torque trajectory with speed increase. Fig. 10 shows that the optimal current control under the voltage limit is attained by the E-TOVC method.

Figs. 9 and 10 verify that the proposed E-TOVC method operates appropriately in the speed control mode,

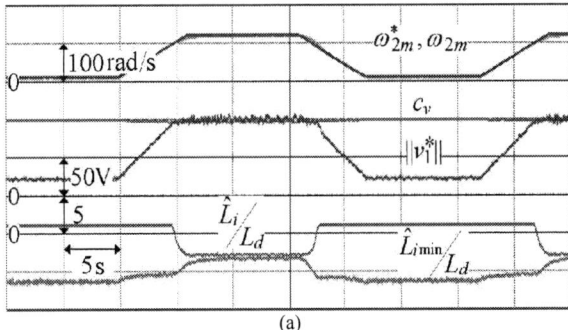

(a)

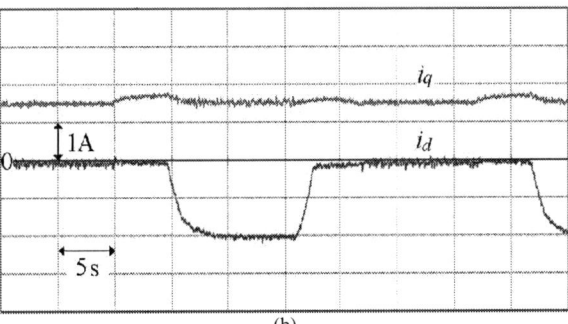

(b)

Fig.9.　Experimental result of speed control (50% load).

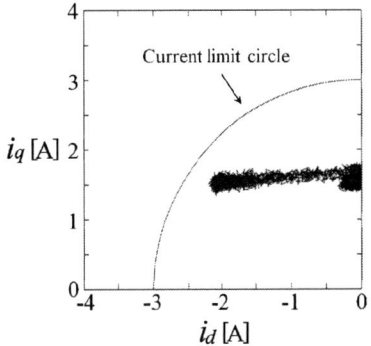

Fig.10.　Experimental result of speed control (50% load).

and in high change of the stator currents under the voltage limit.

D. Speed Control with 100% rated torque

In order to examine the characteristics and performance where both functions of voltage and current limitations operate simultaneously, experiments of speed control under about rated load torque were carried out.

Figs. 11 and 12 show a result. The meaning of the waveforms is the same as that of Figs. 9 and 10. The speed response followed the command up to about 70-rad/s and attained the maximum speed of about 90-rad/s. In other words, it cannot follow up the assigned speed command of 120-rad/s. At the maximum speed, both the stator voltage and current reach their limits. The reached maximum current of 3-A is specified by limiting simply the γ - current command. The speed increasing over about 90-rad/s is suspended by the limited stator current rather than the limited voltage. As the speed decreases,

978-1-4799-2706-7/14 $31.00 © 2014 IEEE　　　1830

The 2014 International Power Electronics Conference

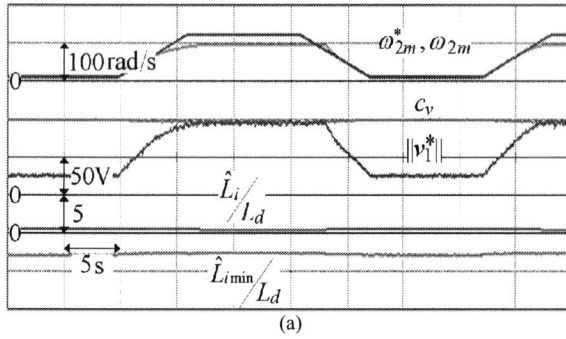

(a)

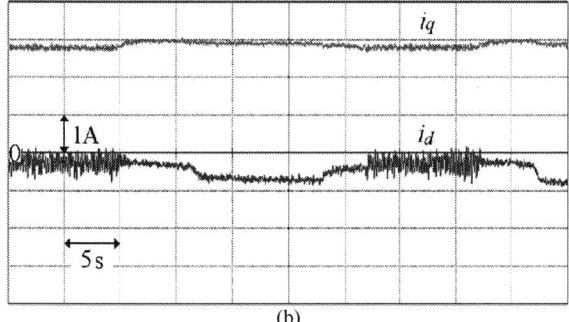

(b)

Fig.11. An experimental result (100% load).

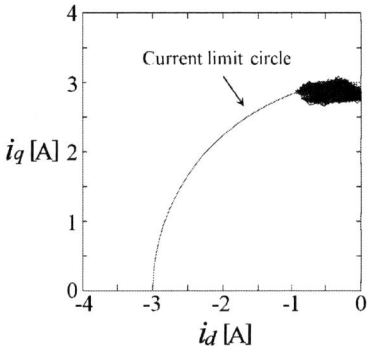

Fig.12. Current trajectory (100% load).

the stator current converges again to the optimum value on the minimum current trajectory (MTPA trajectory).

Figs. 11 and 12 verify that the proposed E-TOVC method operates appropriately in the circumstance where both functions of voltage and current limitations operate simultaneously.

VI. CONCLUSITONS

This paper proposed newly the E-TOVC method for sensorless PMSM drives, which is endowed with all of the following functions: 1) wide-speed-range drive under voltage limit, 2) current limitation. 3) high-efficiency. These functions are realized in a very simple manner and can operate adaptively in all sensorless driving modes including motoring/regenerating, steady/transient states. The usefulness of the proposed method was verified by extensive experiments.

REFERENCES

[1] S.Shinnaka, "Vector Control of Permanent-Magnet Synchronous Motor (From Principle to High-end Technologies)," ISBN978-4-88554-972-4, Denpashinbun-sha, 2008.

[2] S.Shinnaka, "Vector Control of Permanent-Magnet Synchronous Motor (Significance of Sensorless Drive Technologies)," ISBN978-4-88554-973-1, Denpashinbun-sha, 2008.

[3] S.Shinnaka, "A New Simple Power-Factor-Based Vector Control Method for Sensorless Drive of Permanent-Magnet Synchronous Motors, Quasi-Optimal Current Control Insensitive to Motor Parameter Variation, " *IEEJ Journal of Industry Applications*, vol. 127, no. 10, pp. 1070-1080, 2007.

[4] K.Yamanaka and T.Ohnishi, "Sensorless Phase-Tracking Control System for Permanent-Magnet Synchronous Motors," *IEEJ Journal of Industry Applications*, vol. 129, no. 4, pp .432-437, 2009.

[5] S.Shinnaka, "A New Power-Factor-Based Vector Control Method for Sensorless Drive of Permanent-Magnet Synchronous Motors, Simplification and Conversion to Voltage Reference Frame," *IEEJ Journal of Industry Applications*, vol. 130, no. 2, pp. 215-227, 2010.

[6] S.Shinnaka, "A New Unified Analysis of Estimate Errors by Model-Matching Phase-Estimation Methods for Sensorless Drive of Permanent-Magnet Synchronous Motors and New Trajectory-Oriented Vector Control, Part I," *IEEJ Journal of Industry Applications*, vol. 127, no. 9, pp. 950-961, 2007

[7] A.Matsumoto, M.Hasegawa, and K.Matsui, "Position Sensorless Control of IPMSMs Based on a Novel Flux Model Suitable for Maximum Torque Control," *IEEJ Journal of Industry Applications*, vol.132, no.1, pp.67-77, 2012.

[8] K.Tobari, K.Sakamoto, D.Maeda, and T.Endo, "Maximum Torque Control Technique Suitable for Sensorless Permanent Magnet Synchronous Motor Drives," *Proc. of the 2006 JIAS Conf.*, vol. 1, pp.389-392, 2006.

[9] S.Shinnaka, "A New Unified Analysis of Estimate Errors by Model-Matching Phase-Estimation Methods for Sensorless Drive of Permanent-Magnet Synchronous Motors and New Trajectory-Oriented Vector Control, Part II," *IEEJ Journal of Industry Applications*, vol. 127, no. 9, pp. 962-972, 2007.

[10] H.Hida, Y.Tomigashi, and K.Kishimoto, "Position Sensorless Vector Control for Permanent Magnet Synchronous Motors Based on Maximum Torque Control Frame," *IEEJ Journal of Industry Applications*, vol. 127, no. 12, pp. 1190-1196, 2007.

[11] T.Ohnuma, S.H.Jung, S.Doki and S.Okuma, "Maximum Torque Control with Inductance Setting of Extended EMF Observer", *IEEJ Journal of Industry Applications*, vol.130, no.2, pp. 158-165, 2010.

[12] S.Shinnaka, "New D-State-Observer-Based Vector Control for Sensorless Drive of Permanent-Magnet Synchronous Motors", *IEEE Trans. on Industry Applications*, vol. 41, no. 3, pp. 825-833, 2005.

[13] S.Shinnaka, "A New Current Control Method for Energy-Efficient/Wide-Speed-Range Drive of Permanent Magnet Synchronous Motor", *IEEJ Journal of Industry Application*, vol.125, no. 3, pp. 212-220, 2005.

Development of Position Sensorless Control for Permanent-Magnet Synchronous Generator Drive

Yuan-Chih Chang, Chia-Yu Lin, Wei-Fu Dai and Chun-Wei Wu

Elegant Power Application Research Center (EPARC), Department of Electrical Engineering
National Chung Cheng University
Min-Hsiung, Chia-Yi 62102, Taiwan, ROC
Email:ycchang@ccu.edu.tw

Abstract-This paper develops the position sensorless control of a 5kW permanent-magnet synchronous generator (PMSG) drive based on the extended EMF method. With the progress of motor manufacturing, motor design, digital control units and power electronics converters, the permanent-magnet synchronous generator is widely used in electric vehicles, hybrid electric vehicles, flywheel energy storage system (FESS) and wind power generators. The main object of this paper is to establish a PMSG drive suitable for discharging control of the FESS. Position sensorless methods are required in the operating environment that is not convenient for installation of the position sensor. Since the PMSG in FESS is operated at high and medium speeds, the extended electromotive force (EMF) method is adopted. In deriving the extended-EMF of the PMSG, the polarity of winding currents is opposite to the motor mode. The developed sensorless control in this paper is applied to wide speed range. The estimated rotor position is used in the space-vector based current control scheme. In general, the position sensorless control is implemented in the rotor reference frame. In this paper, it is applied to the space-vector based current control, which requires less mathematical calculation. The position estimation and current control are digitally realized by the microprocessor. The generator voltages and currents are measured to implement the digital control of PMSG drive. Some measured waveforms verify the generating performance of the PMSG drive.

Keywords— Position sensorless control, permanent magnet synchronous generator, space-vector based current control, flywheel energy storage system.

I. INTRODUCTION

Permanent-magnet synchronous machines (PMSM) [1] can be found in many industrial applications since it inherently has higher power density and better torque generating performance. According to the embedding method of the permanent magnet, PMSM can be classified into surface-mounted type (SPMSM) [2] and interior type (IPMSM) [3]. PMSM are extensively implemented in hybrid vehicles [4], elevator [5], wind generation [6] and flywheel energy storage system [7]. The generating performance of the PMSG is affected by the current control scheme. Traditional current control approaches of PMSM include fixed frequency [8], hysteresis [9] and predictive [10].

Position sensorless control [11-16] is employed in the circumstance that assembling the position sensor is not convenient. The decision of the sensorless method is based on the specific application of the PMSM. The operating speed range, permanent magnet type and square-wave or sine-wave driving should be taken into account. The sensorless control algorithms in the literatures are summarized as: (i) parameter identification [11]; (ii) back-EMF method [3,12]: this method can be easily implemented in square-wave PMSM and is not applicable at low speed; (iii) observer based techniques: include sliding mode [13], Kalman filter [14] and adaptive [15]; (iv) saliency-based algorithm [16]: this method is first implemented in the induction motor, with the progress of high-frequency injection, this algorithm is both applicable to IPMSM and SPMSM. In the application of generator, the voltage equations should be modified according the polarity of generator current. In the literature, the position sensorless control is implemented in the rotor reference frame. This paper achieve the position sensorless control with space-vector based current control, which requires less mathematical calculation and reduces the operating time of the microcontroller.

The PMSG implemented in this paper is surface-mounted type. The main object of this PMSG drive is being applied to the flywheel energy storage system (FESS). Since the PMSG is operated at high and medium speed range in the FESS, the extended EMF method is proposed to fulfill the sensorless control. In this study, the system configuration including the current control and sensorless schemes is introduced. The modeling and parameter estimation are performed for the design of current control and sensorless control schemes. The derivation of the space-vector based current control and extended-EMF sensorless control is accomplished. The position estimation and current control are implemented by the microcontroller. The estimation accuracy of the rotor speed and rotor position is first verified. Then the performance of the proposed current control is validated by the measured current waveforms. Finally, the DC-link voltage is well regulated under output power and speed variations.

978-1-4799-2706-7/14 $31.00 © 2014 IEEE

The 2014 International Power Electronics Conference

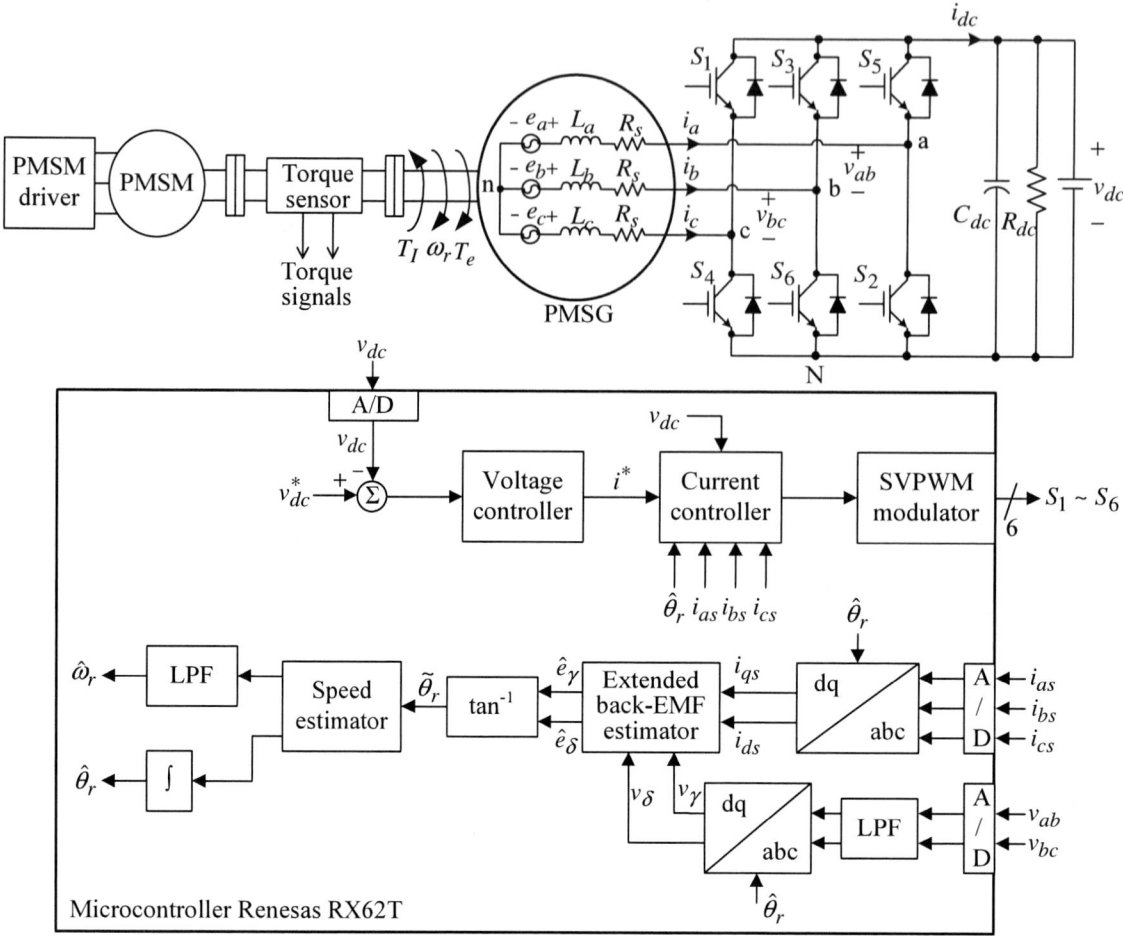

Fig. 1. System configuration of the developed sensorless PMSG drive.

II. System Configuration of the PMSG Drive

A. System Configuration and Dynamic Models

The system configuration of the developed sensorless PMSG drive is shown in Fig. 1. A permanent-magnet synchronous motor (PMSM) with its drive is set up as the prime mover of the PMSG. The torque sensor is equipped to measure the input torque of the PMSG. Three-phase winding currents and two line voltages (v_{ab} , v_{bc}) are sensed and properly filtered to estimate the extended EMF of the PMSG. The estimated rotor speed and position are used to implement the proposed current control scheme. All the control algorithms are digitally realized by the microcontroller Renesas RX62T.

The voltage equations of the PMSG can be expressed in the rotor reference frame (dq-frame) as [1]:

$$\begin{bmatrix} v_d \\ v_q \end{bmatrix} = \begin{bmatrix} R_s + pL_d & -\omega_r L_q \\ \omega_r L_d & R_s + pL_q \end{bmatrix} \begin{bmatrix} -i_d \\ -i_q \end{bmatrix} + \begin{bmatrix} 0 \\ \omega_r \lambda_m' \end{bmatrix} \tag{1}$$

where v_d, v_q are the d- and q-axis voltages, i_d, i_q are the d- and q-axis currents, L_d, L_q are the d- and q-axis inductances, λ_m' is the flux linkages established by the permanent magnet, ω_r is the rotor electrical speed and

p represents the differential operator (d/dt).The mechanical equation of the PMSG is:

$$T_I = T_e + J\left(\frac{2}{P}\right)\frac{d\omega_r}{dt} + B\left(\frac{2}{P}\right)\omega_r \tag{2}$$

where T_I is the input torque, T_e is the electromagnetic torque of PMSG, J is the inertia, B is the damping coefficient and P is the pole numbers of PMSG.

B. Parameter Estimation and Current Control Scheme

To develop the current control scheme and position sensorless algorithm, the parameter estimation of the PMSG is performed in advance. Because the windings of the PMSG are Y-connected with isolated neutral, the winding resistance and inductance are measured between line to line with the LCR-meter. The d- and q-axis inductances can be obtained from the inductance profile as $L_d = 1.896 \, \text{mH}$ and $L_q = 2.131 \, \text{mH}$. The PMSG is found to be the surface-mounted type from the slight difference of d- and q-axis inductances. The winding resistance is found as $R_s = 0.48\Omega$.

The back-EMF of the PMSG can be represented as the function of rotor speed and rotor position:

978-1-4799-2706-7/14 $31.00 © 2014 IEEE　　　1833

$$e_a(\theta_r) = k_e \omega_r \cos\theta_r \tag{3}$$

where k_e is back-EMF constant, ω_r is rotor speed and θ_r is rotor position. The back-EMF constant can be measured by driving the PMSG at no-load condition under different speeds. From the measured results, $k_e = 99.79$ V/krpm is obtained.

The proposed current control scheme and its switching mechanism shown in Fig. 2 are based on the space-vector PWM. The three-phase current waveforms are divided into six sections. There are only two power switches under PWM operation in each section, the others are kept turn-off. The duty ratios in section I can be derived [17] as:

$$\begin{bmatrix} D_{S4} \\ D_{S2} \end{bmatrix} = \frac{1}{v_{dc}T}\begin{bmatrix} (L_a + L_b)\Delta i_a + L_b\Delta i_c \\ L_b\Delta i_a + (L_b + L_c)\Delta i_c \end{bmatrix} + \begin{bmatrix} 1 - e_{ab}/v_{dc} \\ 1 + e_{bc}/v_{dc} \end{bmatrix} \tag{4}$$

where L_a, L_b, L_c are the three-phase winding self-inductances, Δi_x is the current variation of phase-x, T is the switching period and v_{dc} is the DC-link voltage. The duty ratios in sections II to VI can be found via the same procedure.

C. Voltage Control Scheme

The traditional PI-controller is chosen for the voltage regulation. The controller parameters are chosen as $K_{PV} = 0.2$, $K_{IV} = 1.2$.

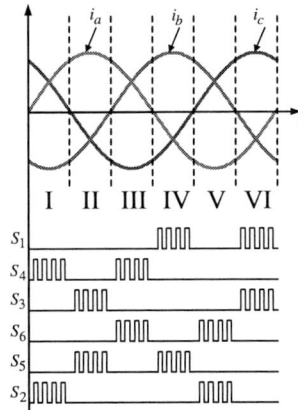

Fig. 2. Time sections and the switching mechanisms of the space-vector based current control scheme.

III. POSITION SENSORLESS CONTROL SCHEME

The relationships of the stator frame, the actual rotor frame (dq-frame) and the estimated rotor frame ($\gamma\delta$-frame) are demonstrated in Fig. 3(a). In the extended-EMF method, the voltage equations in (1) are rewritten as:

$$\begin{bmatrix} v_d \\ v_q \end{bmatrix} = \begin{bmatrix} R_s + pL_d & -\omega_r L_q \\ \omega_r L_q & R_s + pL_d \end{bmatrix}\begin{bmatrix} -i_d \\ -i_q \end{bmatrix} + \begin{bmatrix} 0 \\ E_{ex} \end{bmatrix}$$
$$E_{ex} = \omega_r[(L_q - L_d)i_d + \lambda'_m] + (L_d - L_q)pi_q \tag{7}$$

where E_{ex} is the extended-EMF of PMSG. Since the actual rotor position is not available, the voltage equations in (7) should be transformed to the estimated

frame as:

$$\begin{bmatrix} v_\gamma \\ v_\delta \end{bmatrix} = \begin{bmatrix} R_s + pL_d & -\omega_r L_q \\ \omega_r L_q & R_s + pL_d \end{bmatrix}\begin{bmatrix} -i_\gamma \\ -i_\delta \end{bmatrix} + \begin{bmatrix} e_\gamma \\ e_\delta \end{bmatrix} \tag{8}$$

$$\begin{bmatrix} e_\gamma \\ e_\delta \end{bmatrix} = E_{ex}\begin{bmatrix} -\sin\tilde{\theta}_r \\ \cos\tilde{\theta}_r \end{bmatrix} + (\hat{\omega}_r - \omega_r)L_d\begin{bmatrix} i_\delta \\ -i_\gamma \end{bmatrix} \tag{9}$$

where $\tilde{\theta}_r \equiv \theta_r - \hat{\theta}_r$ is the rotor position error. In a SPMSM with $L_d \cong L_q$, the E_{ex} in (7) can be regarded as constant at steady state. Assume that $\hat{\omega}_r \cong \omega_r$ at steady state, the estimated extended-EMF in (8) can be obtained:

$$\begin{bmatrix} \hat{e}_\gamma \\ \hat{e}_\delta \end{bmatrix} = \begin{bmatrix} v_\gamma \\ v_\delta \end{bmatrix} + \begin{bmatrix} R_s + pL_d & -\hat{\omega}_r L_q \\ \hat{\omega}_r L_q & R_s + pL_d \end{bmatrix}\begin{bmatrix} i_\gamma \\ i_\delta \end{bmatrix} \tag{10}$$

Therefore, $\hat{e}_\gamma$ and $\hat{e}_\delta$ can be obtained via the extended back-EMF estimator shown in Fig. 3(b).

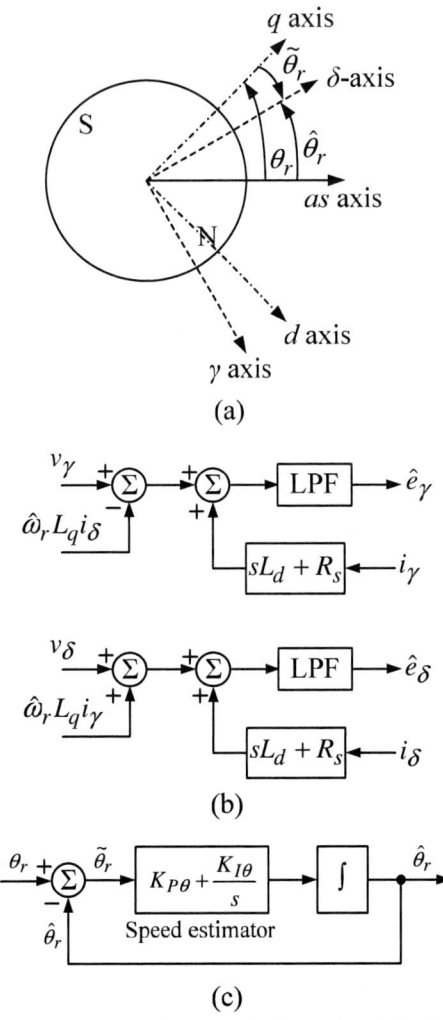

(a)

(b)

(c)

Fig. 3. The position sensorless control scheme: (a) relationship of the stator frame, actual rotor frame and the estimated rotor frame; (a) block diagram of the extended back-EMF estimator; (b) equivalent block diagram of the rotor position estimation scheme.

978-1-4799-2706-7/14 $31.00 © 2014 IEEE

After the estimation of $\hat{e}_\gamma, \hat{e}_\delta$ and assuming $\hat{\omega}_r \cong \omega_r$, the rotor position error can be found from (9):

$$\tilde{\theta}_r = \tan^{-1}\left(\frac{-\hat{e}_\gamma}{\hat{e}_\delta}\right) \tag{11}$$

In the speed estimator diagram shown in Fig. 1, the rotor position error is regulated by the PI controller:

$$G_\theta(s) = K_{P\theta} + \frac{K_{I\theta}}{s} \tag{12}$$

This algorithm is represented by the equivalent block diagram shown in Fig. 3(c). The tracking transfer function of this closed-loop system can be derived as:

$$H_\theta(s) \equiv \frac{\hat{\theta}_r}{\theta_r} = \frac{K_{P\theta}s + K_{I\theta}}{s^2 + K_{P\theta}s + K_{I\theta}} \tag{13}$$

It is obvious that the tracking error will be regulated to zero at steady state via the PI controller.

IV. EXPERIMENTAL RESULTS

The ratings of the established 3φ 8-poles PMSG drive are summarized in Table I. The controllers and low-pass filters implemented in the rotor position estimation algorithms are listed in Table II. The estimation errors at different speeds are compared in Table III. From the results one can find that the speed estimation error is less than 0.2% for all speed range. To verify the estimation accuracy of the rotor position, the estimated rotor position and the Hall signal is compared in Fig. 4. The zero degree of the estimated rotor position is always center at the high-side edge of the Hall signal, which indicates that the estimated rotor position is correct.

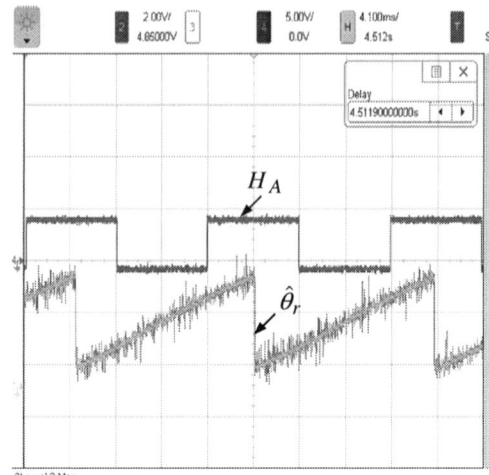
Fig. 4. The estimated rotor position and the Hall signal.

The winding currents of the PMSG are measured to verify the control performance of the proposed current control algorithm. Fig. 5(a) shows the waveforms at $\omega_r = 900$rpm, $P_{dc} = 1.1$kW. Fig. 5(b) shows the waveforms at $\omega_r = 1500$rpm, $P_{dc} = 2.2$kW. The constant output power is controlled by the current controller. The waveforms with different output power at $\omega_r = 1800$rpm are shown in Fig. 6.

For the PMSG drive in FESS, the output voltage should be regulated under output power variation and speed variation. The voltage regulation is verified by the measured DC-link voltage and winding currents in Fig. 7. It can be found that the output voltage is well regulated by the voltage controller under power variation and speed variation.

TABLE I
SPECIFICATIONS OF THE PMSG DRIVE

rated speed	1800 rpm	DC-link voltage	380Vdc
rated power	5 kW	rated voltage (driver)	220Vrms
rated voltage	3φ 220V	rated current (driver)	13.1Arms

TABLE II
PARAMETERS OF THE POSITION ESTIMATION ALGORITHMS

rotor position estimator	$G_\theta(s) = 46.5 + \dfrac{180}{s}$
LPF of the extended-EMF estimator	$\dfrac{3141.6}{s+3141.6}$, $f_c = 500\,\text{Hz}$
LPF of $\hat{\omega}_r$	$\dfrac{31.42}{s+31.42}$, $f_c = 5\,\text{Hz}$

TABLE III
ESTIMATION ERROR OF THE ROTOR SPEED

estimation error	rotor speed	estimation error	rotor speed
0.7 rpm	1803 rpm	1.7 rpm	907 rpm
0.1 rpm	1504 rpm	0.3 rpm	605 rpm
1.8 rpm	1205 rpm		

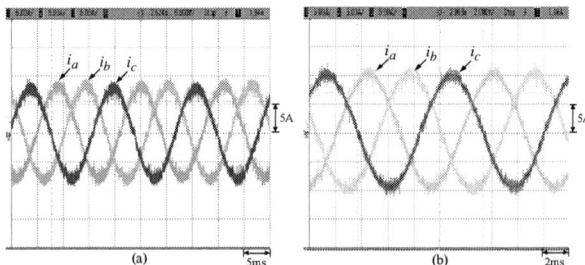
Fig. 5. Measured winding currents at different speeds and output powers: (a) $\omega_r = 900$rpm, $P_{dc} = 1.1$kW; $\omega_r = 1500$rpm, $P_{dc} = 2.2$kW.

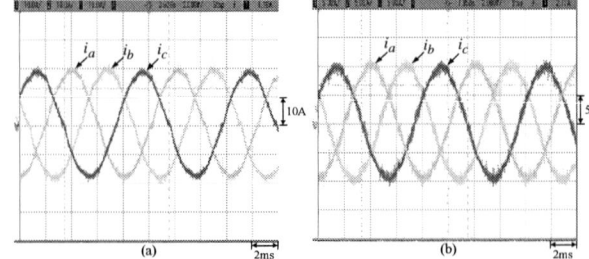
Fig. 6. Measured winding currents at $\omega_r = 1800$rpm: (a) $P_{dc} = 5$kW; (b) $P_{dc} = 2.7$kW.

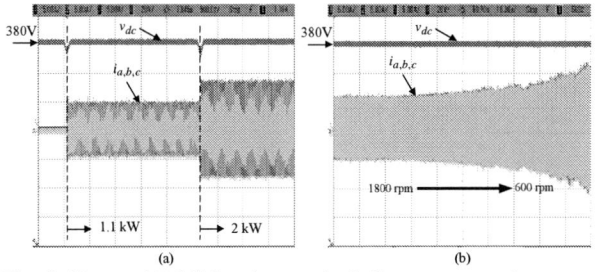

Fig. 7. Measured DC-link voltage and winding currents under output power and speed variations: (a) output power variation at ω_r = 1800rpm; (b) speed variation at P_{dc} = 1.1kW.

V. CONCLUSIONS

A 5kW PMSG drive with position sensorless control is developed. This PMSG drive combines the position sensorless control with the space-vector based current control scheme. The estimation accuracy of the rotor speed and rotor position is verified. The measured winding currents validate the control effectiveness of the proposed current control algorithm. The developed sensorless control method can be successfully achieved at wide speed and output power range. The output voltage is well regulated by the voltage controller.

ACKNOWLEDGMENT

This research was supported by the National Science Council, Taiwan, ROC, under the Grant of NSC 102-2221-E-194-029.

REFERENCES

[1] P. C. Krause, O. Wasynczuk and S. D. Sudhoff, *Analysis of Electric Machine and Drive System.* New York: Wiley, John & Sons, Inc., 2002.

[2] L. Romeral, J.C. Urresty, J.-R. Riba Ruiz and A. G. Espinosa, "Modeling of surface-mounted permanent magnet synchronous motors with stator winding interturn faults," *IEEE Trans. Ind. Electron.*, vol. 58, no. 5, pp. 1576-1585, 2011.

[3] Z. Chen, M. Tomita, S. Doki and S. Okuma, "An extended electromotive force model for sensorless control of interior permanent-magnet synchronous motors," *IEEE Trans. Ind. Electron.*, vol. 50, no. 2, pp. 288-295, 2003.

[4] H.C.M. Mai, F. Dubas, D. Chamagne and C. Espanet, "Optimal design of a surface mounted permanent magnet in-wheel motor for an urban hybrid vehicle, in *Proc. IEEE VPPC*, 2009, pp. 481-485.

[5] S. Cicale, L. Albini, F. Parasiliti and M. Villani, "Design of a permanent magnet synchronous motor with grain oriented electrical steel for direct-drive elevators," in *Proc. IEEE ICEM*, 2012, pp. 1256-1263.

[6] S.M. Dehghan, M. Mohamadian and A.Y. Varjani, "A new variable-speed wind energy conversion system using permanent-magnet synchronous generator and Z-source inverter," *IEEE Trans. Energy Convers.*, vol. 24, no. 3, pp. 714-724, 2009.

[7] C. Huynh, L. Zheng and P. McMullen, "Thermal performance evaluation of a high-speed flywheel energy storage system," in *Proc. IEEE IECON*, 2007, pp. 163-168.

[8] M. N. Uddin, T. S. Radwan, G. H. George and M. A. Rahman, "Performance of current controllers for VSI-fed IPMSM drive," *IEEE Trans. Ind. Appl.*, vol. 36, no. 6, pp. 1531-1538, 2000.

[9] A. Lekshmi, R. Sankaran and S. Ushakumari, "Comparison of performance of a closed loop PMSM drive system with modified predictive current and hysteresis controllers," in *Proc. IEEE ICEMS*, 2008, vol. 1, no. 1, pp. 2876-2881.

[10] J. Weigold and M. Braun, "Predictive current control using identification of current ripple," *IEEE Trans. Ind. Electron.*, vol. 55, no. 12, pp. 4346-4353, 2008.

[11] S. Ichikawa, M. Tomita, S. Doki, and S. Okuma, "Sensorless control of permanent-magnet synchronous motors using online parameter identification based on system identification theory," *IEEE Trans. Ind. Electron.*, vol. 53, no. 2, pp. 363-372, 2006.

[12] F. Genduso, R. Miceli, C. Rando and G. R. Galluzzo, "Back EMF sensorless-control algorithm for high-dynamic performance PMSM," *IEEE Trans. Ind. Electron.*, vol. 57, no. 6, pp. 2092-2100, 2010.

[13] Z. Chen, M. Tomita, S. Doki and S. Okuma, "New adaptive sliding observers for position- and velocity-sensorless controls of brushless DC motors," *IEEE Trans. Ind. Electron.*, vol. 47, no. 3, pp. 582-591, 2000.

[14] M. C. Huang, A. J. Moses and F. Anayi, "The comparison of sensorless estimation techniques for PMSM between extended Kalman filter and flux-linkage observer," in *Proc. IEEE APEC*, 2006, vol. 2, pp. 654-659.

[15] J. Lee, J. Hong, K. Nam, R. Ortega and L. Praly, "Sensorless control of surface-mount permanent-magnet synchronous motors based on a nonlinear observer," *IEEE Trans. Power Electron.*, vol. 25, no. 2, pp. 290-297, 2010.

[16] E. de M Fernandes, A. C. Oliveira, C. B. Jacobina and A. M. N. Lima, "Comparison of HF signal injection methods for sensorless control of PM synchronous motors," in *Proc. IEEE APEC*, 2010, pp. 1984-1989.

[17] Y. C. Chang, J. T. Chan, J. C. Chen and J.G. Yang, "Development of permanent magnet synchronous generator drive in electrical vehicle power system," in *Proc. IEEE VPPC*, 2012, pp. 115-118.

The 2014 International Power Electronics Conference

Control of a 750kW Permanent Magnet Synchronous Motor

Liping Zheng* and Dong Le
Calnetix Technologies, LLC
Cerritos, CA, USA
*lzheng@calnetix.com

Abstract— **Permanent magnet synchronous motors have been widely used due to their high performance and high efficiency. In this paper, we talk about the control of a 750kW permanent magnet synchronous generator which is used for a hybrid turbocharger for a marine application. The controller outputs regulated 700V dc bus voltage with a voltage variation of less than 5% under 100% load transient condition to ensure that the inverter which relies on this 700V input will provide stable three-phase ac power output. The system overview, control methodology and control simulation using Matlab/Simulink is provided in detail. The tests and simulation results are also provided and compared to show the validation of the simulation model and the performance of the generator control and dc bus regulation.**

Keywords— *DC Bus Voltage Regulation, Motor Control, Permanent Magnet Synchronous Motor, Sensorless*

I. INTRODUCTION

Permanent magnet synchronous motors (PMSM) are getting widely used in many industrial applications. This has been made possible with the advent of high performance permanent magnets with high energy density and high operating temperature, providing the PMSM with industry leading power density and efficiency. The sensorless control is also very popular for high performance PMSM control. The position sensors or rotational transducers not only increase cost, maintenance, and complexity but also impair robustness and reliability of the drive system. Long cable length between the variable speed drive (VSD) and motor also makes it less attractive to use position sensors. Various sensorless control methods have been developed to provide high performance control [1-4].

This paper describes a new sensorless control method for high performance and very stable regulated voltage output. A method for initial position and speed estimation is also provided. In addition, the new control is not sensitive to the parameters variation such as induced back electromagnetic force (EMF), winding resistance and inductance.

II. SYSTEM OVERVIEW

The overall system diagram is shown in Fig. 1. The PMSM which is shown in Fig. 2 is attached to the turbocharger of a diesel engine. The specification of PMSM is shown in Table I. The inverter is used to convert the power from 700Vdc voltage to three-phase 440V /60 Hz ac grid. The VSD/converter, which is the main focus of this paper, is used to convert generator high frequency ac power to 700V dc power.

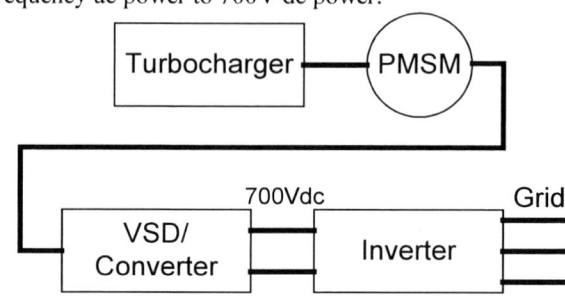

Fig. 1. Overall system diagram.

Fig. 2. Picture of the prototype PMSM.

TABLE I
SPECIFICATION OF THE PMSM

Name	Value
Nominal Speed	9,500 rpm
Nominal Power	750 kW
D-axis Inductance	18 µH
Q-axis Inductance	18 µH
Line-line resistance	0.95 mΩ

The prototype of the designed converter, shown in Figure 3, includes three sections. The left section is the input section, which has a programmable logic controller (PLC) and display, a main contactor, and the pre-charge circuits. The middle section has line reactors which are used to reduce switching harmonics and also to provide load sharing between three parallel switching devices. The right section is the main power converting section which includes main control circuits, pulse-width

978-1-4799-2706-7/14 $31.00 © 2014 IEEE   1837

modulation (PWM) switching IGBT bridges and dc link capacitors.

Fig. 3. Prototype of the designed controller.

III. SYSTEM CONTROL SCHEME

The simplified control scheme is shown in Fig. 4. The block bc/qd is used to convert three-phase currents to qd-axis currents. The current regulation block is used to regulate d-axis and q-axis currents independently. The PWM block is a space vector pulse width modulation block, which converts q-axis and d-axis voltage signals to the switching on-off time of each IGBT. The catch spin & initial angle detection block is used to detect generator speed and initial angle during start-up.

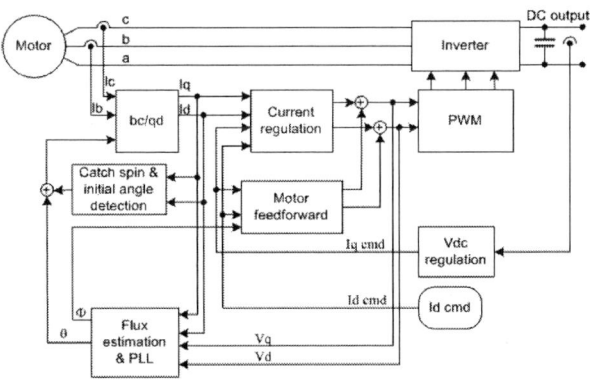

Fig. 4. Simplified control scheme.

During normal operation, d-axis current command (Id cmd) is set to zero, and q-axis current command (Iq cmd) is controlled by the output of the Vdc regulation block. If the DC bus voltage is lower than the voltage setting of 700V, the vdc regulation block will output negative Iq command. If the DC bus voltage is higher than the voltage setting of 700V, the Vdc regulation block will automatically output positive Iq command. A negative Iq command will generate power from the motor while a positive Iq command will automatically do motoring to convert input electric power to kinetic energy.

A. Dq0 transformation

Park's transformation is used to convert stationary reference frame signals to orthogonal rotational reference frame signals. The three-phase abc signals and qd-axis used in the motor/generator controller are shown in Fig. 5.

The d-axis is 90 electrical degrees behind of q-axis. The angle, θ, is the angle between the q-axis of the rotating axis and a-axis of the stationary abc-axis. The phase current is positive when motoring, and negative when generating.

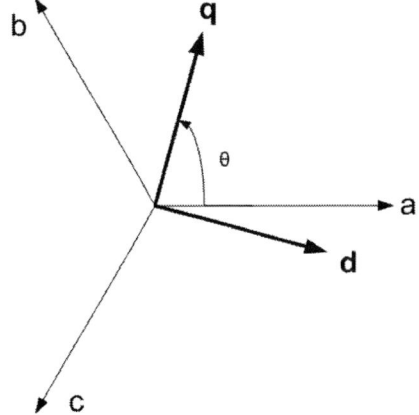

Fig. 5. Three-phase and qd axis.

The relationship between the abc-axis and the dq-axis is shown in the following equations, where S represents any of the variables (current, voltage, flux linkage,...).

$$S_a = S_q \cos(\theta) + S_d \sin(\theta)$$
$$S_b = S_q \cos(\theta - \frac{2\pi}{3}) + S_d \sin(\theta - \frac{2\pi}{3})$$
$$S_c = S_q \cos(\theta + \frac{2\pi}{3}) + S_d \sin(\theta + \frac{2\pi}{3})$$

(1)

$$S_q = \frac{2}{3}\left[S_a \cos(\theta) + S_b \cos(\theta - \frac{2\pi}{3}) + S_c \cos(\theta + \frac{2\pi}{3}) \right]$$
$$S_d = \frac{2}{3}\left[S_a \sin(\theta) + S_b \sin(\theta - \frac{2\pi}{3}) + S_c \sin(\theta + \frac{2\pi}{3}) \right]$$

(2)

B. Motor feed forward

The voltage equations of PMSM in the rotational reference frame can be expressed as [5]:

$$V_q = R_s I_q + \frac{d\lambda_q}{dt} + \lambda_d \frac{d\theta_r}{dt}$$
$$V_d = R_s I_d + \frac{d\lambda_d}{dt} - \lambda_q \frac{d\theta_r}{dt}$$

(3)

where V_q, V_d, I_q, I_d, λ_q, and λ_d are q-axis and d-axis components of voltage, current and flux linkage respectively. θ_r is the rotor angle.

At steady state, equ. (3) will yield to (4), which can be used as feed forward equations.

$$V_q = R_s I_q + \omega_e L_d I_d + E_f$$
$$V_d = R_s I_d - \omega_e L_q I_q$$

(4)

C. Catch-spin operation

For sensorless control, it is still challenging to accurately detect the initial frequency and angle of the spinning machine for flying catch. There is much literature talking about initial speed detection [6-8]. The method developed here is based on the theory that the change of current through inductance is proportional to

978-1-4799-2706-7/14 $31.00 © 2014 IEEE

the applied voltage and time, and inversely proportional to the inductance.

The typical schematic of the 2-level PWM output and the motor/grid is shown in Fig. 6, where switches S1-S6 are power switching devices. The line inductances L1-L3 are used to reduce current harmonics and are optional.

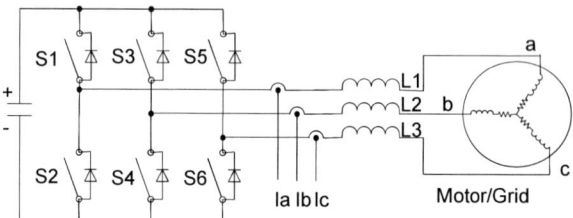

Fig. 6. Typical schematic of the 2-level PWM output and the motor/grid.

Assuming the motor has the three-phase open circuit voltage as shown below:

$$V_a = V_m \cos(\omega t + \theta)$$
$$V_b = V_m \cos\left(\omega t + \theta - \frac{2\pi}{3}\right) \tag{5}$$
$$V_c = V_m \cos\left(\omega t + \theta + \frac{2\pi}{3}\right)$$

If the bottom three switches (S2, S4 and S6) close for a period of time Δt, the final current flow through phase a, b, and c will be

$$I_a = \frac{\Delta t}{L_s} V_m \cos(\omega t + \theta)$$
$$I_b = \frac{\Delta t}{L_s} V_m \cos\left(\omega t + \theta - \frac{2\pi}{3}\right) \tag{6}$$
$$I_c = \frac{\Delta t}{L_s} V_m \cos\left(\omega t + \theta + \frac{2\pi}{3}\right)$$

From (6), at time t=0, we have,

$$V_m = \sqrt{I_a^2 + \frac{(I_b - I_c)^2}{3}}$$
$$\theta = \tan^{-1}\left(\frac{I_b - I_c}{\sqrt{3} I_a}\right) \tag{7}$$

The frequency (ω) can be easily calculated from V_m based on the known back EMF constant.

After initial estimation, further refining of the speed and angle is required to accurately estimate the speed and angle.

D. Flux Estimation and PLL

Flux estimation is the key part of sensorless motor control. The performance of flux estimation directly affects the system performance of the motor control. Virtual flux estimation together with phase lock loop (PLL) is used to provide reliable position and speed estimation. The virtual flux estimator uses q and d axes components of voltage command and current feedback to estimate the position and speed. By using PLL, the flux can be tracked smoothly, thus the position noise due to arc-tangent function is greatly reduced.

IV. SIMULATION

The control scheme has been verified using Matlab/Simulink simulations. Fig. 7 shows the simulation model.

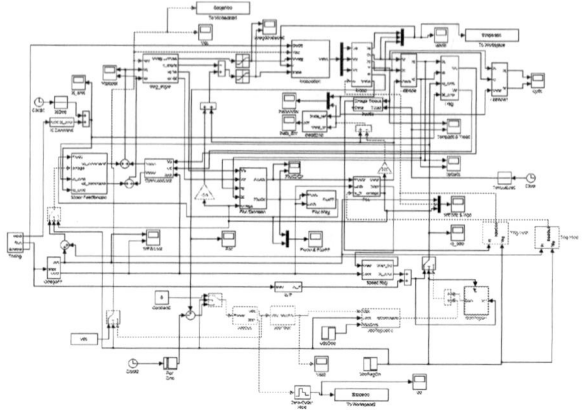

Fig. 7. Simulation model.

Fig. 8 and Fig. 9 show the simulated dc bus voltage response and phase current waveforms when a step load of 0% to 100% applied at the time of 0 seconds. The results show that the dc bus voltage dip is below 5%.

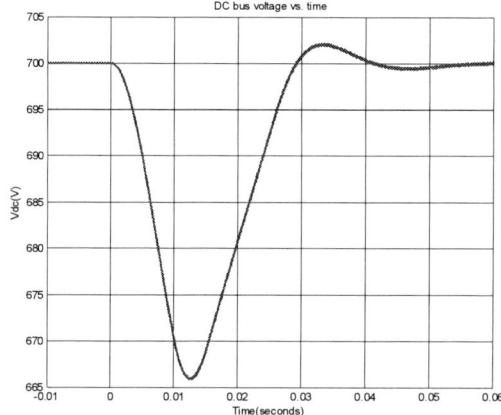

Fig. 8. DC bus voltage overshoot when step load from 0 kW to 750 kW at 10,000 rpm.

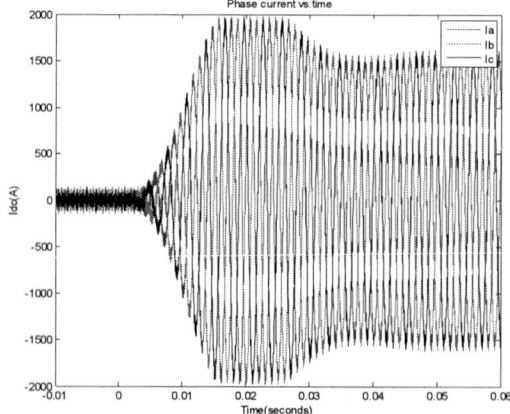

Fig. 9. Phase current waveforms when step load from 0 kW to 750 kW at 10,000 rpm.

978-1-4799-2706-7/14 $31.00 © 2014 IEEE

Fig. 10 and Fig. 11 show the dc bus voltage response and phase current waveforms when a step load of 100 % to 0% is applied at the time of 0 seconds. The result also shows that the dc bus voltage overshoot is below 5%.

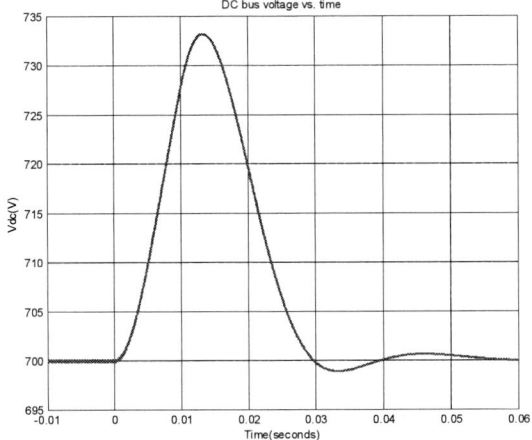

Fig. 10. DC bus voltage overshoot when step load from 750 kW to 0 kW at 10,000 rpm.

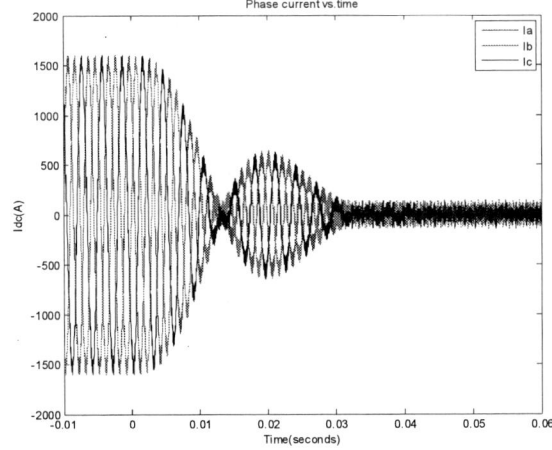

Fig. 11. Phase current waveforms when step load from 750 kW to 0 kW at 10,000 rpm.

V. TEST RESULTS

Catch-spin performance was tested at different initial speed conditions (shown in Table II). From these results we can see that the speed error is less than 5%. The detected speed error is lower at higher speeds because of the higher current signal to noise ratio at higher speeds.

TABLE II
CATCH SPIN RESULTS

Actual Speed (rpm)	Speed Initial Est (rpm)	Speed Refined (rpm)	Refined Error (%)
10000	8962	10117	1.2
10000	8475	10146	1.5
10000	8795	10032	0.3
10000	8847	10160	1.6
10000	8757	10141	1.4
5400	5300	5601	3.7
5113	5295	4889	-4.4
4900	5061	5138	4.8
4745	5166	4870	2.6

Fig. 12 shows the phase current waveform when catch-spinning and then boosting the dc bus voltage to the rated voltage of 700V.

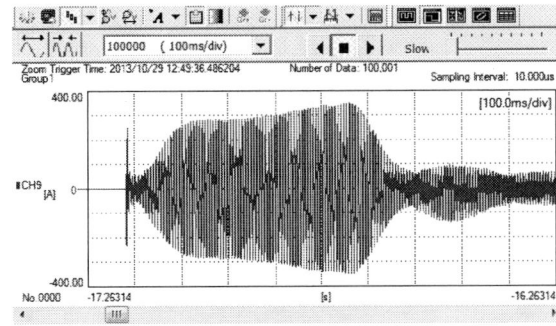

Fig. 12. Current waveform when catch-spinning and then boosting dc bus voltage to the rated 700V.

The DC bus voltage variations under transient load conditions were tested at 580 kW load condition and compared with simulated results. Fig. 13 to Fig. 15 show simulation results and actual test results of the step load response of the dc bus voltage and phase current when the external load changes from 0 kW to 580 kW. The results show that the simulation and actual test results match very well.

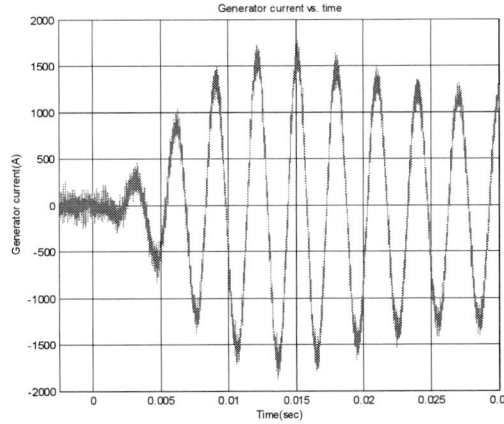

Fig. 13. Measured phase current when the load changed from 0 to 580 kW at 10,000 rpm.

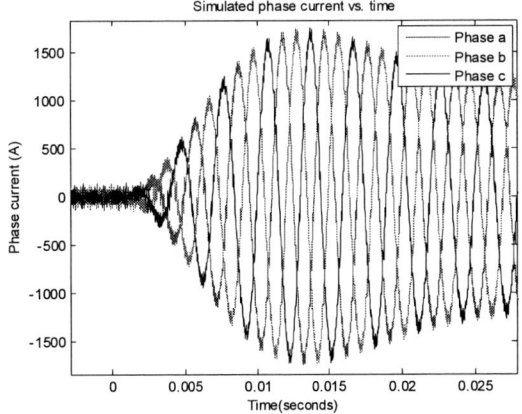

Fig. 14. Simulated phase current when load changes from 0 to 580 kW at 10,000 rpm.

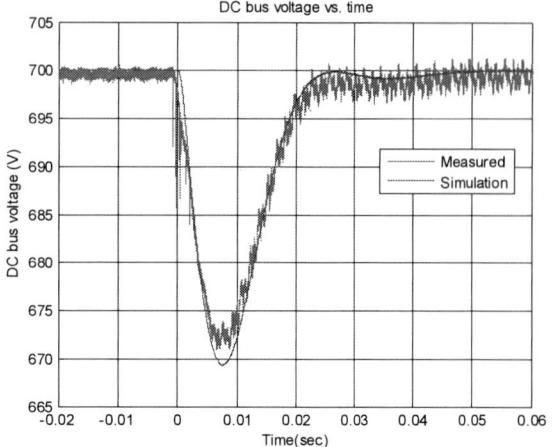

Fig. 15. DC bus voltage dip when step load from 0 to 580kW at 10,000 rpm.

Fig. 16 shows the measured and simulated dc voltage waveforms when the 580 kW load is removed at time 0. Fig. 15 and Fig. 16 show that transient response of the dc bus voltage is less than 5%.

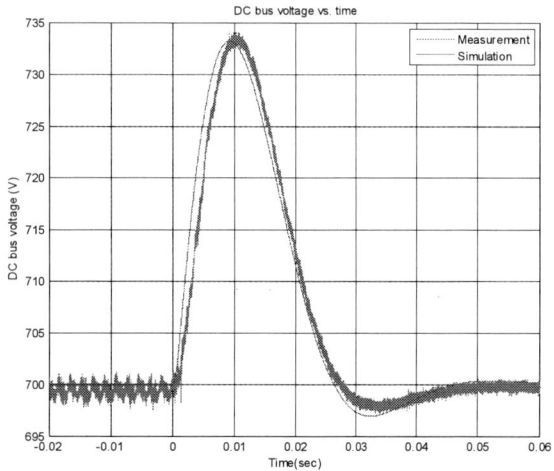

Fig. 16. DC bus voltage overshoot when step load from 580kW to 0 kW at 8,000 rpm.

VI. CONCLUSION

The control of a 750kW permanent magnet synchronous generator which is used for marine hybrid turbocharger applications has been proposed to meet the tough requirement of less than 5% dc bus voltage variation under transient load condition. The system overview, control methodology, and control simulation using Matlab/Simulink has been conducted to provide simulation results that meet system performance requirements. Comparison of the tests and simulation results show the validation of the simulation model and the promising performance of the generator control and dc bus voltage regulation, meeting the performance requirements of the system.

REFERENCES

[1] B. Bae, S. Sul, J. Kwon, and J. Byeon, "Implementation of Sensorless Vector Control for Super-High-Speed PMSM of Turbo-Compressor," *IEEE Trans. on Industry Applications*, vol. 39, no. 3, pp. 811-818, 2003.

[2] J. X. Shen, Z. Q. Zhu, and D. Howe, "Improved Speed Estimation in Sensorless PM Brushless AC Drives," *IEEE Trans. on Industry Applications*, vol. 38, no. 4, pp. 1072-1080, 2002.

[3] I., S. Tomita, M. Doki, S. Okuma, S. "Sensorless Control of Permanent-Magnet Synchronous Motors Using Online Parameter Identification Based on System Identification Theory" *IEEE Trans. Ind. Applications*, 2006, vol.53, no.2, pp.363-372, 2006.

[4] J. H. Kim, S. Lee, R. Y. Kim, and D. S. Hyun, "A Sensorless Control using Extended Kalman Filter For an Interior Permanent Magnet Synchronous Motor Based on an Extended Rotor Flux," *IEEE 38th Annual Conf. on Industrial Electronics Society*, Oct. 2012.

[5] Chee-mun Ong, *Dynamic Simulation of Electric Machinery Using Matlab/Simulink*, Prince Hall PTR, 1997.

[6] M. Tursini, R. Petrella, and F. Parasiliti, "Initial Rotor Position Estimation Method for PM Motors," *IEEE Trans. Ind. Applications*, vol.39, no.6, pp.1630-1640, 2003.

[7] P. B. Schmidt, M. L. Gasperi, G. Ray, and A. H. Wijenayake, "Initial rotor angle detection of nonsalient pole permanent magnet synchronous machine," *IEEE-IAS Annual Meeting*, pp. 459–463, New Orleans, 1997.

[8] T. Noguchi, K. Yamada, S. Kondo, and I. Takahashi, "Initial Rotor Position Estimation Method of Sensorless PM Synchronous Motor with No Sensitivity to Armature Resistance," *IEEE Trans. on Industry Electronics*, vol. 45, no.1, pp. 118-125, 1998.

Regional Smart Grid of Island in China with Multifold Renewable Energy

Xu Cai
Wind Power Research Center & SEIEE
Shanghai Jiao Tong University
Shanghai, China
xucai@sjtu.edu.cn

Zheng Li
College of Information Science & Technology
Donghua University
Shanghai, China
lizheng@dhu.edu.cn

Abstract— A series of challenges have been put forward for safe operation and stable control of the power grid with integration of renewable energy, such as wind, photovoltaic, etc., and also novel loads such as electric vehicles. It stimulates the technology development of smart grid. This paper introduces a demonstration project of an island smart grid in China. The renewable energy resources on this island are mainly wind, photovoltaic and biomass energy. The penetration level of the renewable energy will up to 40% in 2015 and the full load will be powered by the renewable energy in the future. And the whole island will build the power supply and service system for electric vehicles, the green farm energy supply system characterized by biomass make full use of, the generation system combined wind power and energy storage system, the micro grid characterized by building photovoltaic, and industrial micro grid with induction motor load and power energy storage.

Keywords—smart grid, renewable energy, island grid, demonstration project

I. Introduction

After the concept of smart grid was presented, research and applications of smart grid technology are developed in many countries. Italian power companies installed and renovate 3,000 smart meters, and established the smart metering grid in 2001. And the first commercial network using smart grid technology adopted the intelligent measurement system was built in Enel S.p.A in 2005. In the same year, EU started to construct Smart Grid Technology Platform[1] and then presented a series of special project in FP5 projects[2] . While, the first smart grid of America was built in Colorado Boulder City of Colorado and the exemplary application of smart grid for the client was built in Southern California in 2008[3,4]. In Asian, China had built some small exemplary projects in succession in Zhang North County , SSTEC of Tianjin ,and Expo Park of Shanghai . Furthermore, the Chinese Ministry of Science and Technology has launched a technological support program named "The integrated comprehensive exemplary project of area - based smart grid with a high proportion of the intermittent energy sources" in 2013, planning to build four exemplary projects in China up to 2015. They have the characteristics of large-scale use of renewable energy, large regional feature, low-carbon of the grid and comprehensive interactive multi-link. In Japan, Tokyo

Electric Power Company realized gradually real-time measurement and automatic control for the 6kV medium voltage feeders within the reticular structure in system through a fiber-optic communication network, becoming the basis of future smart distribution grid. In South Korean, government established the exemplary project of comprehensive smart grid in Jeju. It integrates the new information technology into the satellite positioning technology, and realizes online real-time monitoring for the electrical demand and supply. It can be a testing base for the future construction of the smart grid. Singapore also built an experimental smart micro-grid in Jurong Island, to research, test and verify the new smart grid technology.

This paper will introduce the meta synthesis of smart grid technology in an island in China, including wind power integration and consumption in the whole island, the electric vehicle energy supply and service system, the industrial park micro grid with induction motor load and power energy storage, the large building energy system with high density photovoltaic integration, the green farm with biomass energy application, and the household energy system featured by small wind, photovoltaic and energy storage integration. All above compose the island general demonstration project.

II. Construction Scheme of the Smart Grid

A. Description of the Island

The demonstration island has geographically relatively closed, modest area, relative independent power grid, and five kinds of renewable energy resources. The strategic orientation of Shanghai ecological island makes it an ideal place for smart grid demonstration

The island, the national third biggest island, covers the area of 1200 km^2, which accounts for about 1/5 of the total area of Shanghai. Located in China's coastal zone and the middle of mouth of the Yangtse River Junction, it is strategic reserves of Shanghai. Thus the ecological environment is relatively good, and the ecological resources are well protected.

Project Supported by the National Science and Technology Support Program (2013BAA01B00)

978-1-4799-2706-7/14 $31.00 © 2014 IEEE

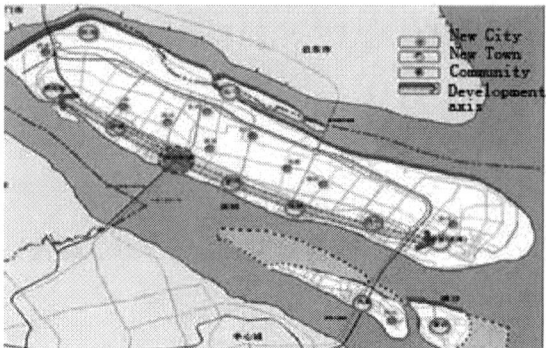
Fig. 1 The demonstration island description

The whole island is divided into five functional regions, which are science and technology innovation gateway landscape region, central forest region, agricultural region, exhibition region and new town region. There are 5 types of renewable energy which accounts for 40% of comprehensive energy consumption of the whole society in 2015. Among these, wind energy is 207.5MW, photovoltaic is 51MW, biomass energy is 9MW. Further, up to 2020, onshore wind energy will be 460MW, offshore wind energy 2450MW, tide energy 240MW, photovoltaic 105MW, and biomass energy 12MW. Renewable energy installed capacity will be up to approximately 3200 ~ 4200MW, and island saturated load will be approximately 1200 ~ 1500MW. From the perspective of energy balance, the island can power the load by itself. Regarded as an isolated grid, it is a typical island consuming large scale renewable energy.

B. Structure of the smart grids

The power grid of the whole island is divided into three layers.

The first layer is high voltage transmission network with voltage level 220 kV and 110 kV. An 800 MW gas-fired power plant, the large scale offshore and onshore wind farms are directly integrated to the 110 kV transmission network. The dispatch able and controllable of power generation on this layer are realized by forecasting the wind power of large scale wind farms and meeting the forecast error through adjusting the output of gas-fired power plant.

The second layer is 35 kV distribution network. The small-scale wind farms and photovoltaic stations connect to it. The large-capacity energy storage system is used to balance the generation power and load demand, and smooth the fast wind power fluctuation to some extent.

And then, the third layer is micro grid and user layer. For building area, large-scale photovoltaic power system accesses to the building power grid with multi-points. The APF and SVG are adopted for dynamic reactive power compensation and harmonic control. The small-scale energy storage system is used to balance the power supply and demand. For industrial area, the power energy storage devices connect to the industrial park power grid, and then constitute the micro grid with the photovoltaic generation system located on the plant roof. Through micro grid, the impact by induction motor on the local

power grid can be eliminated and the photovoltaic power is consumed locally. For the green farm, pumps are driven by wind power directly for irrigation. The CCHP units use the biogas turned from local biomass to provide electric power and heating. Meanwhile, an energy storage system is used to ensure the reliability of power supply for core loads. Thus a high degree autonomy micro grid is composed and then accesses to the distribution grid. For community household, photovoltaic power and the energy supply system with an energy storage system which can operate under on-grid or off-grid conditions are used and are set to achieve the maximum energy self-balance. In addition, the energy supply system of electric vehicles and renewable energy vehicles around the island is composed of the charging station, charging and discharging integration station, hydrogen refueling station and its information service system.

The smart grids information platform is adopted for coordinated operation of the above-mentioned system.

C. The Recent Construction Plan

The smart grid construction is a long process, in accordance with three steps of recent phase (before 2015), mid phase (before 2020) and long-term phase (before 2050). The recent phase will access 3 large-scale onshore wind farms, one hybrid wind-energy storage power plant, and one small-scale photovoltaic station. Three micro grids will also be constructed in recent phase, which are the industrial park micro grid with induction motor load and NaS battery energy storage system, the building micro grid with photovoltaic, ground source heat pump and lithium battery energy storage system, the green farm micro grid with biomass energy respectively.

III. KEY SUBSYSTEMS

Six subsystems are designed in the regional smart grid.

A. Connecting large wind farms to main grid

Wind power resources offshore and onshore all over the island is 2910MW. 7 large-scale off-shore wind farms and 6 large-scale on-shore wind farms will be built around island. Wind power forecasting system is installed for the power generation control of wind farms cluster. The power forecasting information is interactive with EMS system of the grid at real time. The power prediction error is balanced by the gas power plant on the island.

1) MTDC network for integrating offshore wind farms into transmission grid

Seven offshore wind farms with the capacity of 2450 MW will be commissioned in the north and south of the island. They are connected to island transmission grid via MTDC system, as shown in Fig. 2.

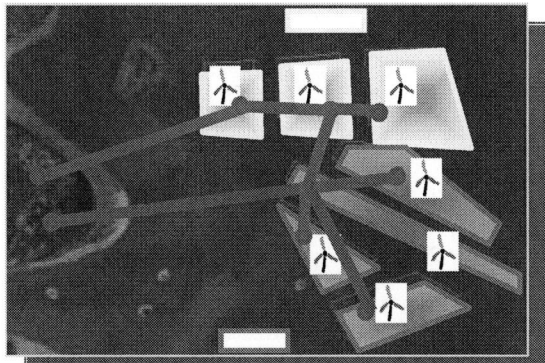

Fig. 2 DC grid for offshore wind farm integration

2) Dispersed access of onshore wind farms to main grid
6 onshore wind farms with the capacity of 460 MW will be built in the northern land of the island. The output voltage of the wind farm is lifted to 110 kV for the transmission level. At present, there are 3 large wind farms with total installed capacity of 156MW.

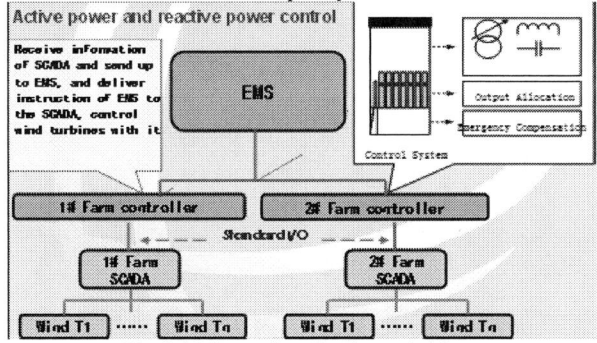

Fig. 3 AC integration of onshore wind farm

B. Distributed hybrid wind-energy storage generation system

A number of small-scale wind farms will be constructed on the island. These wind farms and energy storage systems will constitute hybrid generation systems and then integrated to the 35 kV distribution grid, as shown in Fig. 4. Now there is a 19.5 MW hybrid wind-energy storage power plant, which is composed of 13 wind turbines rated at 1.5 MW and is connect to two buses. In the first phase, 7 wind turbines on bus 1# coordinated by 2MW*2h energy storage system supply about 3 MW local load. In this case, two tasks will be taken by the energy storage device: one is inhibiting the fast fluctuation of wind power and two is the balance of power generation and demand.

C. Industrial park micro grid

The power supply system of industrial park in island contains a large number of induction motors. When the motor starts, it will have great impact on the grid. Added the volatility of renewable energy, the power quality of the grid must be focused on. It is a solution that integrating a power energy storage device coordinates with roof photovoltaic, APF and SVG devices to compose a micro grid. It is a high autonomy system.

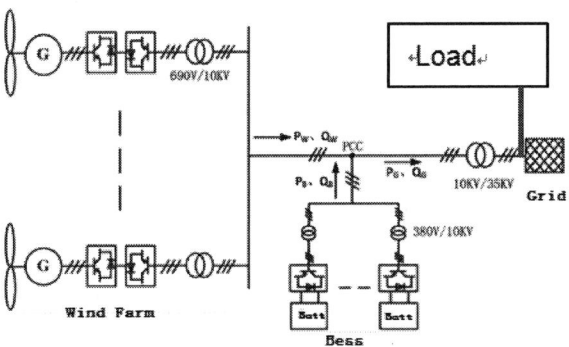

Fig. 4 Hybrid wind-energy storage generation system

D. Green farm micro grid

Many modern farms are constructed on the island. The farm makes full use of its output of straw, cow dung to produce biogas. After purification, the biogas is used by micro CCHP gas turbine for power generation and heating. The CCHP, the park photovoltaic, small wind turbine and energy storage system constitute micro energy network and connect to the distribution network.

E. Household energy system

The residential community on the island will use household photovoltaic energy system. Its topology is shown in Fig. 5. The whole system is composed of three-stage converters. The photovoltaic battery is connected to the common DC bus via boost converter and lithium battery is connected to the common DC bus via buck-boost converter. Therefore, the photovoltaic battery and lithium battery can access to the public grid or isolated to supply the AC load of household via full-bridge inverter on the common DC bus.

According to different operating conditions, the PV battery may work under MPPT mode (maximum power point track) or constant voltage mode (control common DC bus voltage). [5-6] When the boost converter works in the MPPT mode, it aims to collect the solar energy to the full extent. And when the boost converter works in the CV mode, it aims to stabilize the voltage of dc bus. The lithium battery works under buck mode (lithium battery charging control) or boost mode (control common DC bus voltage). So when the converter works in boost mode, it aims to stabilize the voltage of dc bus. And when the Buck-Boost converter works in buck mode, it aims to control the battery charging to prevent over-charging. Load-side full bridge converter works under grid connected mode or isolated inverter mode. When the full-bridge inverter works in grid-connected mode, it aims to control the current feeding back to the grid. Otherwise, when it works in island mode, it aims to control the output voltage. Depending on every converter's working mode, the whole system can work in the following four different modes: photovoltaic power generation mode, feeding energy back to grid mode, power supplying by battery mode, power supplying by grid mode, which can be shown in Fig.6. The Control diagram of each converter is shown in Fig.7. [7-8] The key problem in this system is to judge the system operation situations according to practical states of PV battery, lithium battery as well as household load and then to control each converter coordinately. The target of control is to ensure the system high efficiency and stability, meanwhile household UPS load.

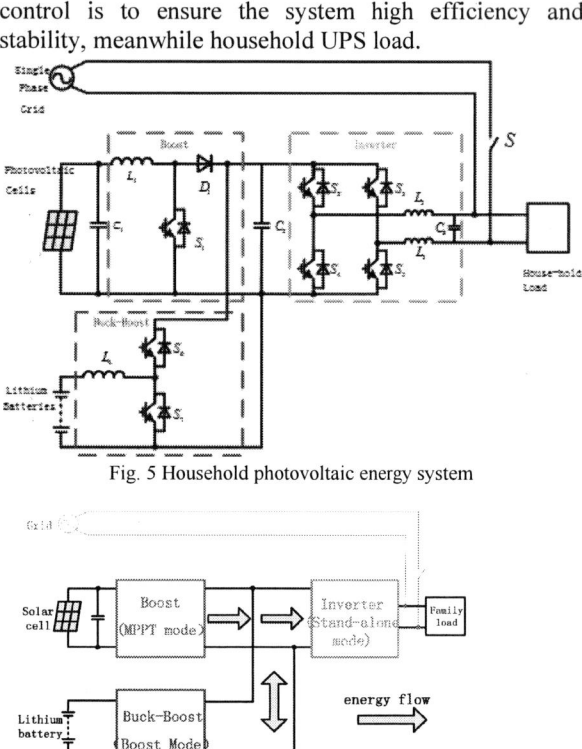

Fig. 5 Household photovoltaic energy system

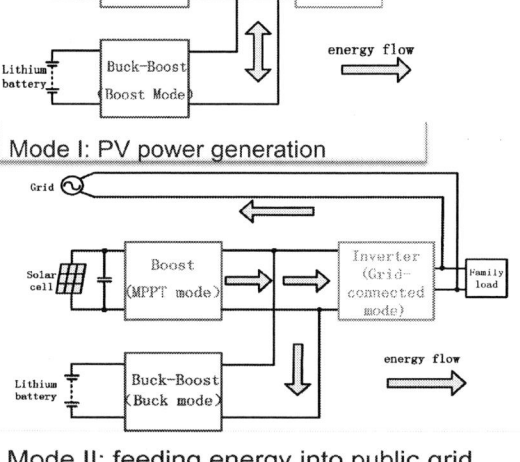

Mode I: PV power generation

Mode II: feeding energy into public grid

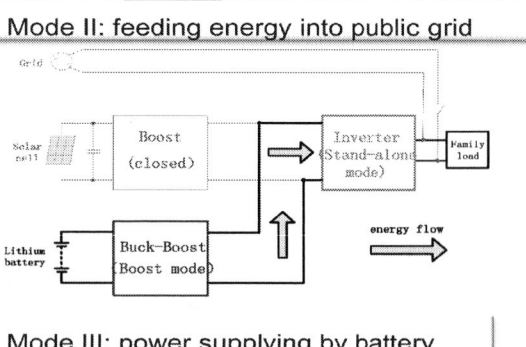

Mode III: power supplying by battery

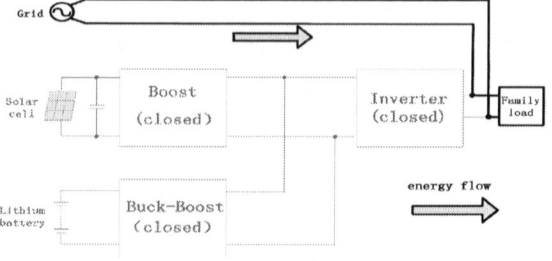

Mode IV: power supplying by grid

Fig. 6. Mode of on/off-grid integrated photovoltaic power generation

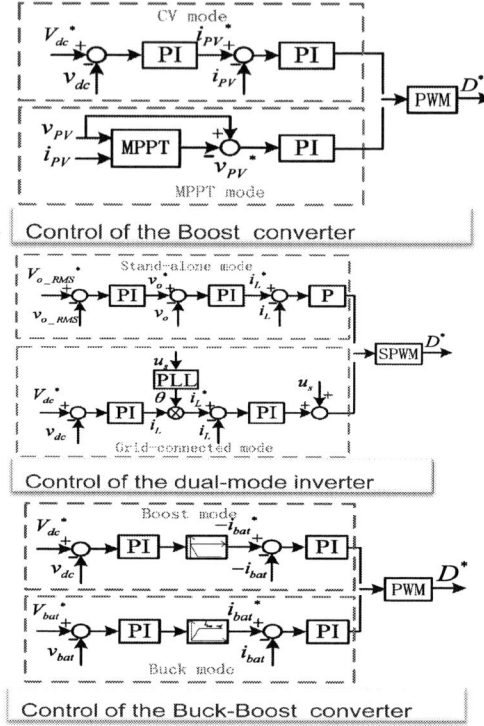

Fig. 7. Control diagram of each converter

To validate the control strategy presented above. A 5KW prototype is built. The maximum output power of solar cells in this system is 5KW. The range of MPPT voltage is 175V~450V. The maximum charging and discharging power of lithium battery is 5KW. The maximum power of family loads is 5KW too. The power electronic switch tube used in the prototype is Mitsubishi IPM: PM100RL1A120.

Fig. 8(a) shows the waveforms when the system works in mode I. The voltage of DC bus is 480V. Output voltage of the inverter is 220V/50Hz. The charging current of lithium battery is 10A. The output current of PV is 20A. So, the whole output power of PV is 4KW. Among the 4KW output power, 2KW is used to charge the lithium battery. The rest power is used to power the family loads. Fig. 8(b) is a partial enlarged view of the Fig. 8(a). Fig. 8(c) shows the waveforms when the system works in mode III. The power of family loads and lithium battery discharging is 2KW. Fig. 8(d) and Fig. 8(e) shows the switching process between mode I and mode III. Fig. 8(d) shows the switching process from mode I to mode III. Fig. 8(e) shows the switching process from mode III to mode I.

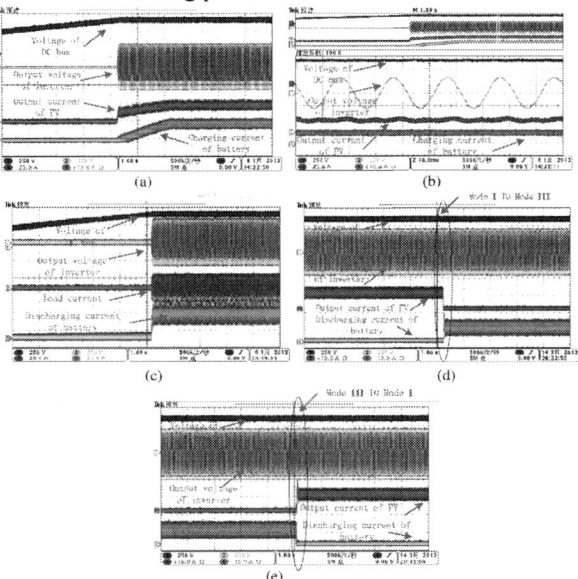

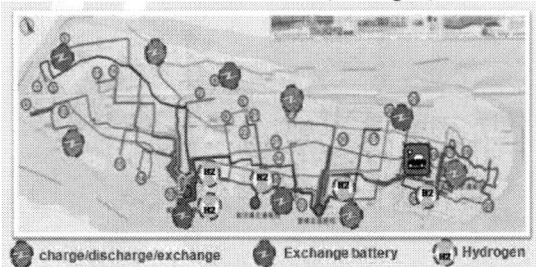

Fig. 8. Experimental results

F. Electric vehicle energy and service system

Only renewable vehicles will be used in island such as electric cars, hydrogen cars. The buses are running mainly along the south island. At the eastern entrance of the island, there is a large car parking area. The private cars will be exchanged by electric vehicles here. Visitors can drive electric vehicle traveling around the island. Multiple battery replacement stations and the power stations will be built on the island (see Fig. 9).

Fig. 9 Power supply system of island renewable vehicles

IV. CONTROL STRATEGY AND SIMULATION

A power flow monitoring center is needed to provide centralized control of on- and off-shore wind farms together with gas power plant which are directly connected to the power transmission network, in this way the island smart grid can be regarded as a virtual power plant. Its net power output becomes controllable, as shown in Fig. 10. Four control objectives are included: (1) Wind turbine is controlled to smooth wind power output of seconds level. (2) The reactive power compensator of the step-up substation connected to wind farm is controlled along with the wind turbine to achieve voltage control. (3) Gas power plant is controlled to balance wind power fluctuations of minutes level, the set point is the difference between scheduling command and predicted

wind power output. (4) Virtual power plant cooperates with the superior scheduling center to achieve integrated energy management and scheduling optimization.

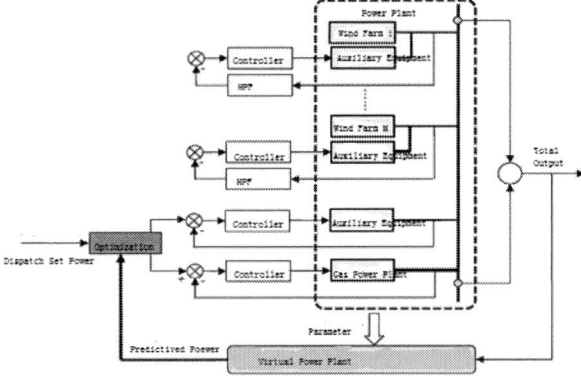

Fig. 10 Control architecture of island smart grid

The island smart grid can operate under on-grid and off-grid conditions. According to the island smart grid architecture (shown in Fig. 11) the simulation model is built in RSCAD. Under on-grid condition, the power flow at the PCC can be controlled while maintaining the renewable energy power plants stable. Fig. 12 shows the power flow at the PCC with random speed and Fig. 13 is the voltage magnitude and frequency curve at the PCC. Under off-grid condition, the frequency and voltage of island smart grid can be stabilized with a wind power penetration level of 40%, as shown in Fig. 14 and 15.

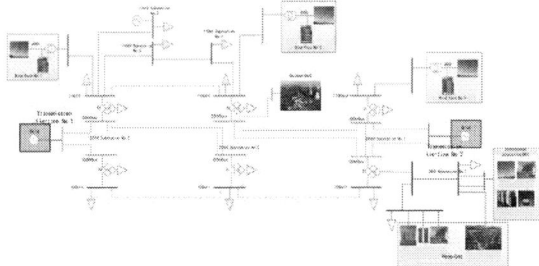

Fig. 11 Island smart grid architecture

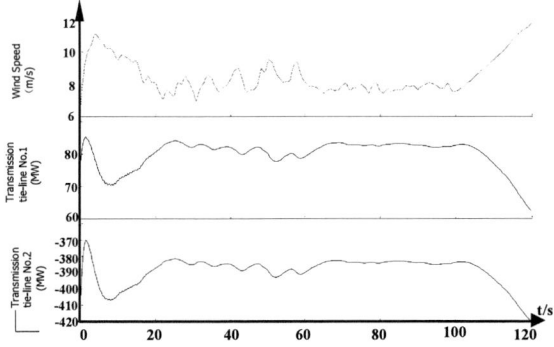

Fig. 12 Power flow at the PCCs
+: flow into island smart grid

978-1-4799-2706-7/14 $31.00 © 2014 IEEE

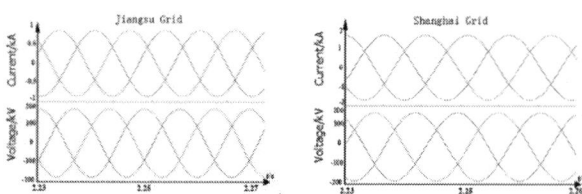

Fig. 13 Current and voltage waveforms at the PCCs

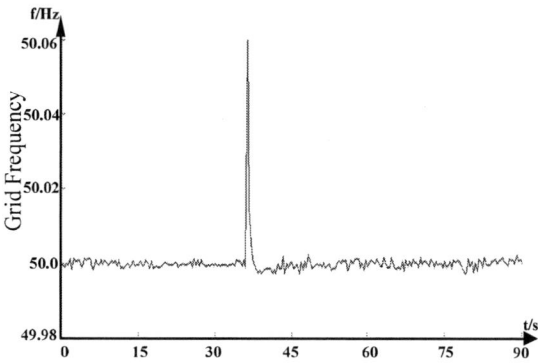

Fig. 14 Frequency waveform from on-grid to off-grid operation

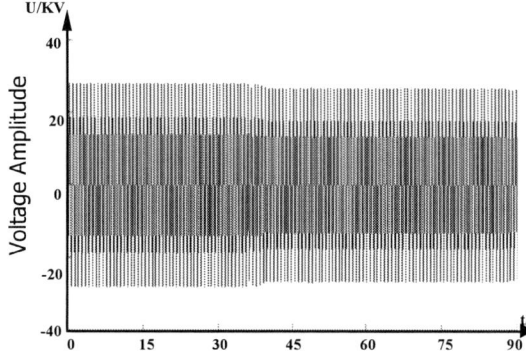

Fig. 15 Voltage amplitude from on-grid to off-grid operation

The fluctuation rate of wind power output is reduced due to coordinated operation of wind farms and energy storage system, which conduces to frequency and voltage stability in the micro-grid, as shown in fig. 16 and 17. The energy storage system can also participate in the scheduling command response and inertia response control to improve the command response speed and the transient support ability of the wind farms.

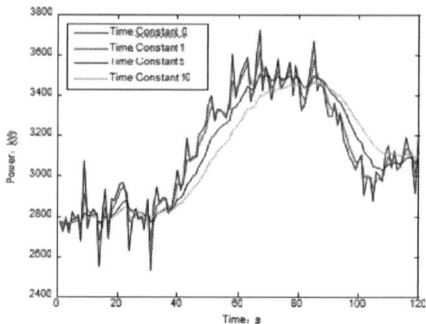

Fig. 16. Bess output simulation in LPF

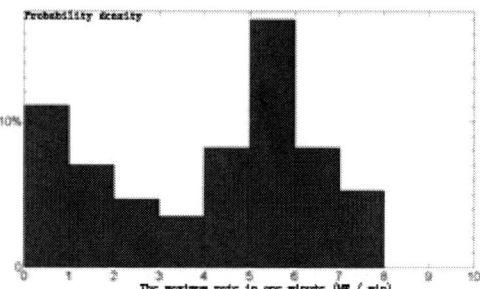

(a)　Battery energy storage system not connected to the grid

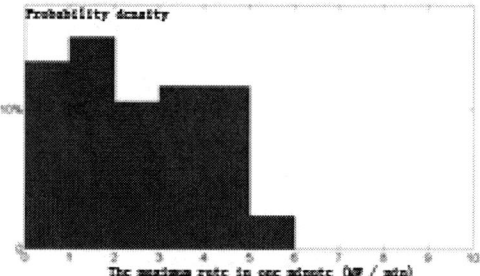

(b)　Battery energy storage system connected to the grid

Fig. 17. Comparison of the fluctuation rate of active wind power in one minute before and after filtering

V. CONCLUSION

The construction scheme of island smart grid with abundant renewable resources is introduced. The renewable energy penetration level in the island will reach 40% in 2015. The full renewable energy use all over the island will be achieved in the future.

REFERENCES

[1] European Commission. European technology platform smart grids: vision and strategy for Europe's electricity networks of the future [EB/OL].2008-10-10.
http://ec.europa.eu/research/energy/pdf/smartgrids_en.pdf
.

[2] Lessons learned from European research FP5 projects. EUR21970, Towards Smart Power Networks.
[3] U. S. Department of Energy. National Energy Technology Laboratory. Modern grid initiative: a vision for modern grid [EB/OL]. 2008-10-10.
http:// www.netl.doe.gov/moderngrid/docs/.

[4] Amin M, Schewe P F. Preventing blackouts: building a smarter power grid[J]. Scientific American, 2008(8): 60-67. or http://www.scientificamerican.com/article.cfm.

[5] Liu B, Duan S, Cai T. Photovoltaic DC-building-module-based BIPV system—Concept and design considerations[J]. Power Electronics, IEEE Transactions on, 2011, 26(5): 1418-1429.

[6] Zhang L, Sun K, Xing Y, et al. A modular grid-connected photovoltaic generation system based on DC bus[J]. Power Electronics, IEEE Transactions on, 2011, 26(2): 523-531.

[7] Kerekes T, Liserre M, Teodorescu R, et al. Evaluation of three-phase transformerless photovoltaic inverter topologies[J]. Power Electronics, IEEE Transactions on, 2009, 24(9): 2202-2211.

[8] Sun K, Zhang L, Xing Y, et al. A distributed control strategy based on DC bus signaling for modular photovoltaic generation systems with battery energy storage[J]. Power Electronics, IEEE Transactions on, 2011, 26(10): 3032-3045.

Stabilizing Small Island Power System with Renewables by use of Power Conditioning Systems
- Japanese Island System Case -

Jumpei BABA
The University of Tokyo, Graduate School of Frontier Sciences
Department of Advanced Energy
3-1, Hongo 7, Bunkyo-ku, Tokyo, Japan

Abstract - **Recently, the capacity of renewable power sources installed in small remote islands, whose power grid is isolated from large scale grid, increases because of the environmental affairs and the generation cost and so on. But in the small remote island grid, the power fluctuation caused by renewable power sources affects power grid stability and operation. In this paper, several stabilizing methods based on Japanese case by use of power conditioning system (PCS) and related technology are reviewed.**

Keywords— Remote Island Grid, Energy Storage Systems, Power System Operation, Renewable Energy Sources

I. INTRODUCTION

In Japan, there are about 400 islands (the number depends on the definition) around 4 main islands and about 50 isolated remote island grids. In the isolated remote island system, diesel engine generators are used as main power supply, however, the fuel oil becomes very expensive. As a result, the generation cost is much higher than that of large scale utility. For example, it is reported that the generation rate of the remote island grid company of Okinawa Electric Power Company, that is responsible to supply electricity in Okinawa prefecture without main Okinawa island, is about 35Yen/kWh (2012)[1], and this is much higher than the residential electricity rate (26Yen/kWh). This rate is calculated from whole cost including small and medium scale island grids, the generation rate of the small islands may be much higher value than that of the medium/large scale island grids because the fuel logistics cost is high and low-grade fuel oil is difficult to use in small scale diesel engine. Compared with a Japanese Feed in Tariff (FIT) of 36-38Yen/kWh for PV system[2], and of 23Yen/kWh for WT (20kW<) (2013), the generation rate of the small island system may be expensive. This cost is not singular because average retail price of electricity in Hawaii is reported as 36.58 Cent/kWh (About 37Yen/kWh) in December 2013[3]. Capacity of renewable energy sources increases in small remote island grids to reduce the generation cost.

Fig. 1 shows the relation between the number of island grids in Japan and its total generation capacity [4]. As shown in Fig.1, the capacity varies from 100kW to over 100MW. It is difficult to determine the border that divides the capacity into "small" and "large", but, the system whose capacity is less than 10MW may be "small" grid. In the following, 10MW is as the border.

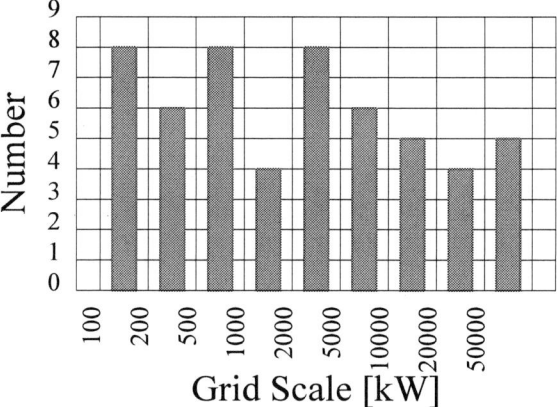

Fig. 1. Relation between generation capacity and number of Japanese Islands [4]

There are a number of diesel generation unit ratings in remote island systems. In Japan, at least two generation units should be in operation in case of accidents. Most diesel generation system cannot operate at below half load for long duration because of incomplete combustion. As a result, there is an acceptable minimum demand to realize stable system operation. Fig.2 shows the relation between the acceptable minimum and maximum power of each remote island grids belongs to a Japanese utility company when the half load is assumed to be the lower limit [4]. Green bar is the minimum generation power in the case smallest two generation units are in operation, and blue bar is that in the case smallest generation unit is in operation.

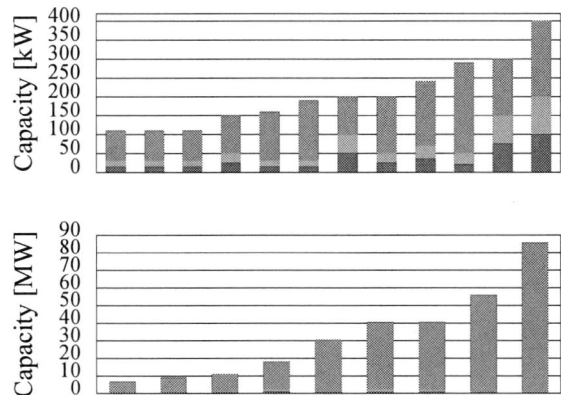

Fig. 2. Relation between generation capacity and acceptable operation power range [4]

Fig.3 shows the relation between the number of unit of each capacity and total grid generation capacity in the remote island system with the capacity under 400kW [4]. The color of dot shown in Fig.3 indicates the number of the unit of a certain capacity.

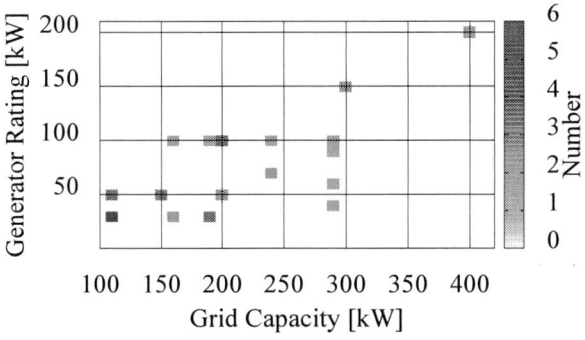

Fig. 3. Relation between generation capacity and generator ratings, number of units [4]

As shown in Fig.2 and Fig.3, rather large remote island systems have the variety of unit capacity and the number of units so that the wide generation range is acceptable. On the other hand, small remote island system is limited to narrow operation range. In case of large penetration of renewable power sources to small remote island grids, fluctuation of the demand including the loads and the renewables can become so large that operating point of diesel generator exceeds the maximum or minimum limit.

Not only static operating point but also lack of dynamic characteristics, power change rate, is the important issue. The power change rate of diesel engine generator is usually higher than that of thermal power plant, however, in small remote island grids, the inertia of the grid is so small that the grid frequency fluctuates improperly. The generator paralleling time from cold start is several to ten minutes, and the time is much shorter than that of thermal power plant. The diesel generator is easy to use on-off operation, and this is useful for stable

small grid operation. But, too much on-off operation brings the lack of compressed air to start up diesel engines and expense of start-up cost.

To assist power fluctuation compensation and other problems that affect operation of small remote island grid, to use several kinds of equipment with PCS is proposed and some of them are in operation. In the following several examples of Japanese island grids that are stabilized by the system with PCS are introduced.

II. DEMONSTRATION PROJECT OF LARGE PENETRATION OF RENEWABLES IN ISOLATED ISLAND GRID

From 2009, Agency Natural Resources and Energy, Japan start the funding program "demonstration project of large penetration of renewables in isolated remote island grid". The aim of this project is to establish the stabilizing method for small and medium remote island power system with large penetration of renewable power sources, mainly PV. 10 projects are accepted for this funding program. Table I shows the demonstration site and installed renewable power sources and energy storage systems [5][6]. Fig.4 shows the location of the test sites listed in Table I. All test sites locate south west part of Japan.

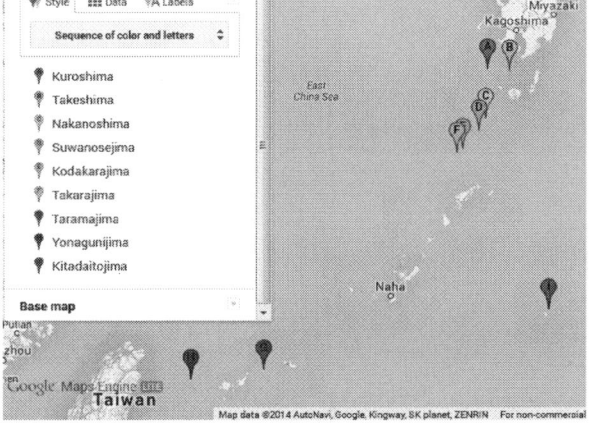

Fig. 4. Location of test sites

These projects are divided into two groups; small scale island grid (Kuroshima – Takarajima) and middle scale island grid (Miyakojima – Kitadaitojima).

There are six targets in the demonstration project;

1. PV short-term power fluctuation compensation (Takeshima, Nakanoshima, Takarajima, Miyako-jima)
2. PV power leveling : long-term power fluctuation compensation (Suwanosejima, Kodakarajima)
3. PV power shift : daytime to night (Kuroshima)
4. Frequency stabilization (Miyakojima, Yonagunijima, Taramajima, Kitadaitojima)
5. Scheduled output of large scale PV (Miyakojima)
6. Optimal multi-layer BESS control (Miyakojima)

TABLE I
SITE AND INSTALLED RENEWABLE POWER SOURCE AND ENERGY STORAGE SYSTEMS IN DEMONSTRATION PROJECT

Site	Maximum Demand (kW)	Newly Installed		Existing Renewable	Lead Acid BESS (kWh)	LiB (kWh)	Li ION Capacitor (kW-kWh)	NAS BESS (kW)
		PV (kW)	WT (kW)					
Kuroshima	193	60	10	-	256	66	-	-
Takeshima	83	7.5	-	-	-	33	-	-
Nakanoshima	193	15	-	-	80	-	-	-
Suwanosejima	78	10	-	-	80	-	-	-
Kodakarajima	71	7.5	-	-	80	-	-	-
Takarajima	125	10	-	-	80	-	-	-
Miyakojima	50000	4000	-	4200(WT)	-	200	-	4000
Yonagunijima	2160	150	-	1200(WT)	-	-	200 - 4.7	-
Taramajima	1160	250	-	280(WT)	-	-	300 - 7.2	-
Kitadaitojima	860	100	-	40(PV)	-	-	100 - 2.9	-

Fuji Electric made the systems without the Miyako-jima one. Detail of Miyakojima project will be introduced in the other paper. The system made by Fuji Electric consists of existing diesel engine generators, renewable energy sources and energy storage systems (ESS). In the small remote island systems, battery ESS (BESS), based on Un-interruptible Power Supplies (UPS) for the system with gas engine generation system, is used in order to compensate the steep power fluctuation caused by load and renewables. Fig.4 shows the schematic diagram of the BESS.

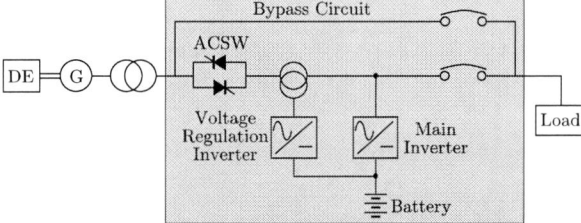

Fig. 4. Schematic diagram of BESS used in small island system

As shown in Fig.4, BESS consists of the main parallel inverter, the series inverter for voltage regulation, high speed AC switch and the battery. The BESS has an "absorber function" that detects instantaneous power fluctuation and compensates them. Fig. 5 shows the schematic diagram of the absorber function.

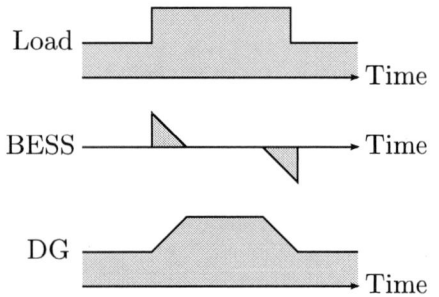

Fig. 5. Schematic diagram of absorber function

Central grid controller measures the power flow at the diesel engine generators and renewable generation systems. When the grid controller detects steep power change, it makes main inverter of BESS to compensate the fluctuation by charging/discharging the battery. Power change rate of diesel generators is limited, and as a result, stable power grid operation is realized. The BESS system compensates power fluctuation in several to 10 msec. Fig. 6 shows the simulation results of frequency fluctuation with and without instantaneous power compensation function. As shown in Fig. 6, deviation of frequency is about 0.1Hz with compensation and 1.5Hz without compensation. The instantaneous power compensation function is useful to suppress frequency fluctuation.

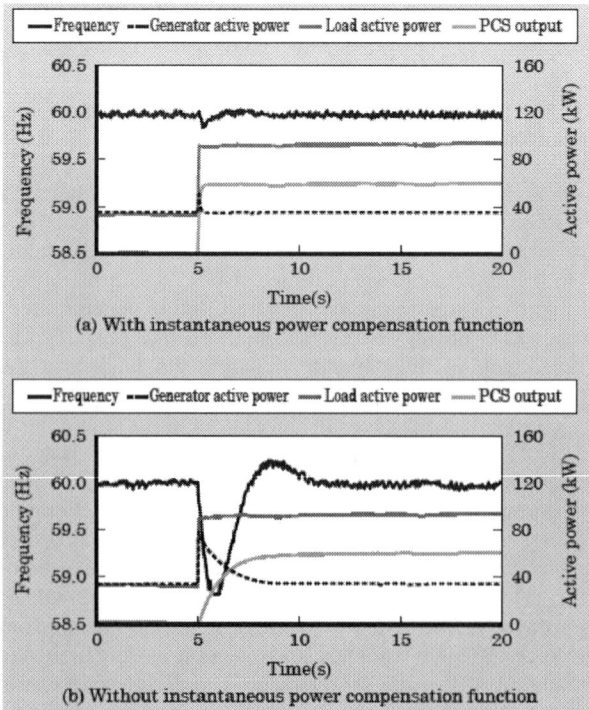

Fig. 6. Effect of instantaneous power compensation function [5]

In the middle scale island grid, Li-ION capacitors provided by FDK are installed for grid frequency stabilization. Fig.7 shows the photo of the Li-ION capacitor module. Table II shows the specifications of the Li-ION capacitor module.

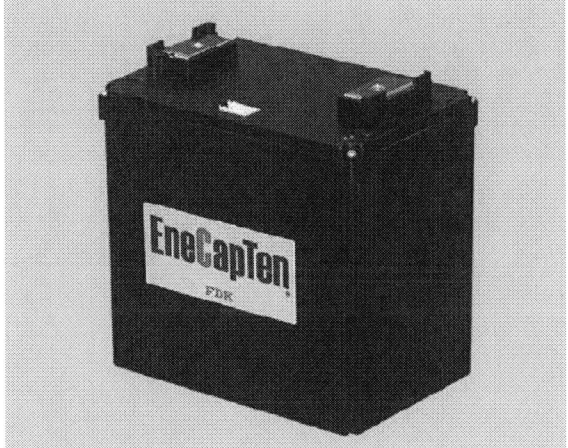

Fig. 7. Photo of Li-ION capacitor module [5]

TABLE II SPECIFICATION OF LI-ION CAPACITOR MODULE

Rated Voltage	DC45V (27 to 45 V)
Rated Current	20A
Initial static capacitance	200F or higher (for 1A discharge)
Initial internal DC resistance	19 m or less (for 100A discharge)

Volumetric energy density of Li-ION capacitor for a certain rated power is lower than that of battery. For efficient use of the energy capacity of Li-ION capacitor, the fast frequency detector (Δ f sample comparison frequency detection [7]) is used to make the power reference for Li-ION capacitors. The frequency detection method can follow the grid frequency in approximately 30ms. Fig. 8 shows the example measurement results of the fast frequency detector.

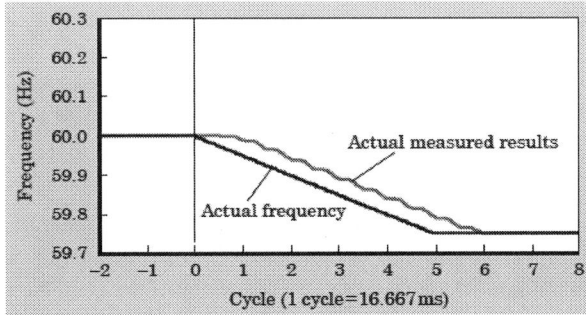

Fig. 8. Example measurement results of fast frequency detector [5]

Li-ION capacitor system responds to frequency fluctuation before response of governor-free function of existing diesel generators.

III. STABILIZING BY BESS – YONAGUNI ISLAND GRID-

In 2002, the "Hybrid System" (the system with WTs and BESS) was started in operation in Yonaguni-Jima (Island), because large WT (600kW x 2) are installed to the grid. Compared with maximum demand of the grid (around 2MW), the capacity of WTs is too large to operate only with diesel generators. Same systems have been installed in Hateruma-Jima, Tarama-Jima, Aguni-Jima, Tonashiki-Jima. The system installed in Yonaguni has the largest BESS capacity of all, 800kWh (400kWh x 2, PCS 300kW x 2) [7].

Fig.9 shows the system configuration of the hybrid system in Yonaguni. The hybrid system consists of a lead acid BESS, a monitoring system and a generator control system. Diesel generators are operated in AFC and Auto mode. The generator operated in AFC mode regulates the grid frequency by its governor and reference from system controller. When the power margin of AFC mode generator will not be enough for frequency stabilization, system controller gives start/stop signal to Auto mode generator. The system also compensates fast power fluctuation by use of BESS.

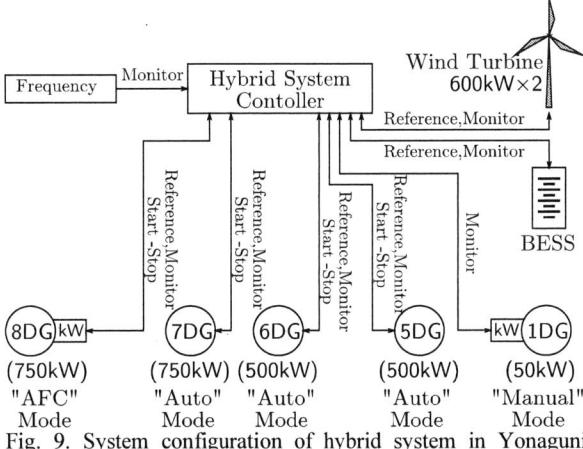

Fig. 9. System configuration of hybrid system in Yonaguni Island

One of the largest disturbances for the remote grid is cut out of WTs. When the wind brows above the cut out velocity, WTs stop their power generation immediately to protect mechanical components. In such conditions, diesel generators increase their output immediately to compensate large supply loss. This operation needs steep generator power increase, so in most system, diesel generators are operated at low partial load when WTs generate at near ratings. Low partial load operation of diesel generators makes their efficiency low. To manage this problem, one generator (AFC generator) operation is allowed in Yonaguni system. When WT cut out occurs, BESS compensates the loss of supply power immediately, and at the same time, system controller orders Auto mode generator to start. BESS have enough capacity to supply power until Auto mode generator is in parallel. As mentioned previously, startup time of diesel generator is short, and this enables cut out compensation with rather small energy capacity BESS. Fig.10 shows the schematic of hybrid system operation during WT cut out.

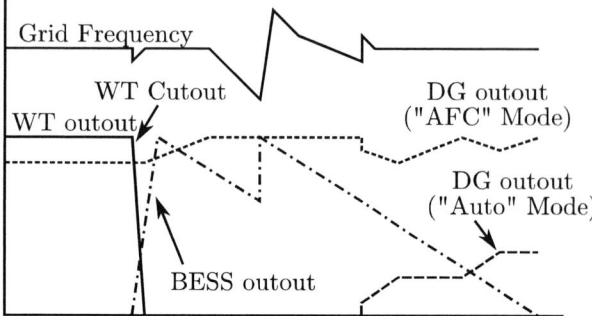

Fig. 10. WT cut out operation of hybrid system

Thanks to the hybrid system, WT can generate more than the half of demand. Fig.11 shows the daily load curve and WT generation curve on 2004/03/05. As shown in Fig.11, about 70% of power is supplied by WT for whole day.

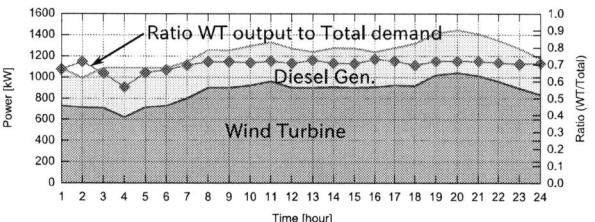

Fig. 11. Daily load and WT generation curve on 2004/03/05

Capacity factor of the WT is reported 26.5% (2002). The generated energy is reported 2.75GWh on 2004, and this energy is about 30% to the total electricity consumption of the island.

IV. STABILIZING BY FLY WHEEL– HATERUMA ISLAND GRID-

The capacity of BESS decreases gradually, especially, repetitive wide SOC change accelerates degradation of BESS. Redundant design and cell replacement should be planed for long term operation of BESS. The environment of small remote islands is severe for BESS; salt damage, long black out and so on. For long time, many projects to find the way to use other ESSs have been done. One of the projects is the fly wheel (FW) ESS project in Hateruma Island. Fig. 12 shows the configuration of Hateruma island grid [8].

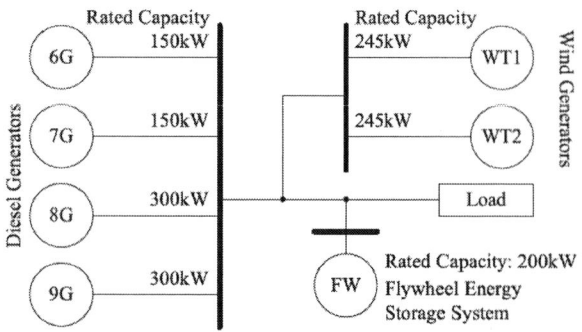

Fig. 12. Power system configuration of Hateruma grid

The 490kW WT system is connected to the grid whose maximum demand is about 600kW. To stabilize the system, FW- ESS (200kW-7.2MJ) is also installed. FW system consists of 8 fly wheels whose capacity is 30kW-900kJ (1500-6000 rpm), and is divided into two banks. Each FW is connected to a PCS (INV, 37kW), and DC side of 4 PCSs are connected to DC side of a PCS (100kVA, 95kW) which is connected to the grid. Fig. 13 shows the configuration of Hateruma FW system.

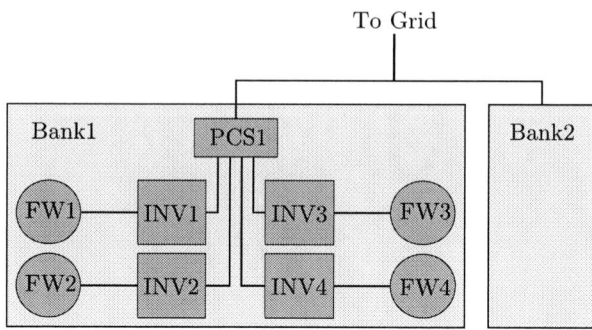

Fig. 13. Configuration of Hateruma FW system

Compared with BESS, the capacity of this FW-ESS is quite small. To make good use of the FW-ESS, control scheme has been studied by use of simulations. In the study, the control strategy of the FW-ESS is discussed based on the measured response of the diesel generators and the measured power fluctuation of wind turbines. The FW-ESS compensates only short periodic power fluctuations caused by WTs and loads, because of small energy capacity of the ESS. As a result, $\Delta P + \Delta f$ with SOC regulation control method has been proposed and implemented to the system. Fig.14 shows the block diagrams of $\Delta P + \Delta f$ control method [9].

In the $\Delta P + \Delta f$ control method, the reference for FW-ESS is made by detected short period power fluctuation of WT output (ΔP) and medium period grid frequency fluctuation (Δf).

FW-ESS compensates the fast power fluctuations (-several seconds) of WTs by ΔP compensation, and rather slow power fluctuations (-10 seconds) that cannot followed by DGs, by Δf compensation. To avoid the shortage of FW-ESS's stored energy, SOC control loop (Center Frequency Control) based on detected rotating speeds of FW is also added in the controller. To determine the parameters of the controller, the response of diesel generators to frequency deviation should be taken into account not to operate at upper/lower limits of the generators. The parameters are determined by simulations so that the 9G that has a very fast response to frequency deviation does not operate beyond the lower/higher limits.

Fig. 15 shows the test result when one WT (150kW) tripped due to its output limitation control (at dotted ellipse). FW-ESS compensates the fast power fluctuation caused by WTs output power, and stabilizes the system in the case of one WT trip by fast response to the steep power change.

978-1-4799-2706-7/14 $31.00 © 2014 IEEE

The 2014 International Power Electronics Conference

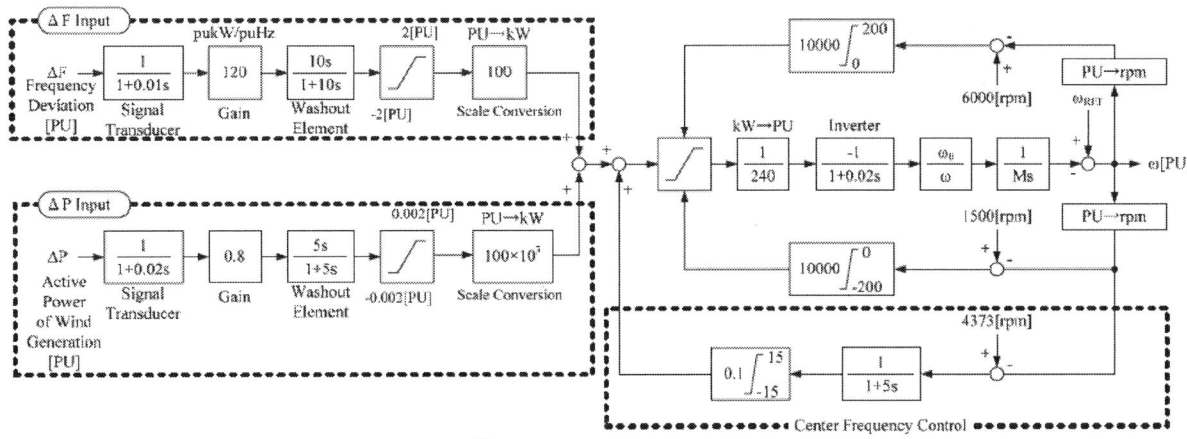

Fig. 14. Block diagram of $\Delta P + \Delta f$ control method [9]

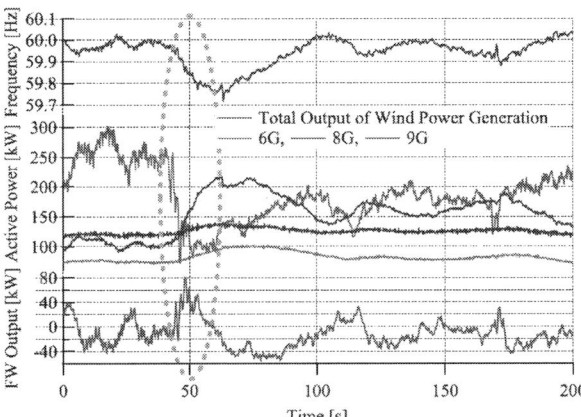

Fig. 15. Result of FW-ESS power compensation against WT trip [9]

After short duration of high FW-ESS power output, FW-ESS decreases its output power as 9G increases its output power. And then, FW-ESS starts to charge energy. After energy charging state, FW-ESS operates normally.

The generated electricity is reported 375MWh during 2010.04-2010.12, and this is about 12.5% of total electricity consumption of the island (3009MWh). It is also reported that the capacity factor of WT is 11.6% and availability factor is 95.4%.

V. CONCLUSION

In this paper, brief overview of Japanese isolated remote island grid is introduced. Some Japanese projects that use PCS and ESS are also introduced. As shown in those examples, PCS and ESS are key technologies to introduce a large amount of renewable to the small remote isolated island system. ESS with small energy capacity such as FW or capacitor is shown to be useful for this kind of system. The capacity of renewable energy sources will increase in isolated remote islands grid because of generation cost reduction and so on. Isolated remote island grids will need more power/energy capacity for grid stabilization, so more PCS and ESS will be introduced in remote island grids.

REFERENCES

[1] Okinawa Electric Power Company, "IR information 2014", pp. 10-11 (in Japanese)

[2] Agency Natural Resources and Energy, Japan, http://www.meti.go.jp/english/press/2013/0329_01.html

[3] U.S. Energy Information Administration, "Electric Power Monthly with Data for December 2013", p.123, February 2014

[4] J. Baba, T. Yoshihara, M. Imanaka, Y. Onda, A.Yokoyama, Y. Kuniba, N. Higa and S. Asato, "Feasibility study on Power System Stabilization using Seawater Desalination Plant in Small Island Power System", IEEJ Technical Meeting, PE-11-005,PSE-11-022, SPC-11-059, 2011 (in Japanese)

[5] T. Kojima and Y.Fukuya, "Microgrid System for Isolated Islands", Fuji Electric Review, vol. 57, No. 4, pp.125-130, 2011

[6] Y.Fukuya, M.Toi and S.Sato, "Development of high-speed frequency detection device for power system", IEEJ Technical Meeting, PPR-10-37, 2010 (in Japanese)

[7] Ministry of Economy, Trade and Industry Japan, Handout of Next Generation Transmission and Distribution System Committee WG 1, No. 7-2, pp. 13-22, 2010 (in Japanese)

[8] A. Sakuma, "New technology development in Okinawa Electric Power Company in 2005", *Electrical Review*, vol.91, no. 1, pp266-268, 2006 (in Japanese)

[9] K. Yamashita, O. Sakamoto, Y. Kitauchi, T. Nanahara, T. Inoue, T. Shiohama, H.Fukuda, " Development of Frequency stabilizing Scheme for Integrating Wind Power Generation into an Isolated Grid", Electrical Engineering in Japan, Vol. 180, Issue 1, pp. 24–35, 15 July 2012

Power Electronics Solutions Applied to A Variety of Demonstrative Microgrid Projects

Yoshinobu Ueda

PUS Products Engineering Division
Meidensha Corporation
Tokyo, Japan
ueda-yo@mb.meidensha.co.jp

Abstract— **Three different energy storage systems by use of power electronics technologies are presented. These systems were applied to demonstrative microgrid projects which have concepts different one another. Concepts of the projects, requirements on energy storage systems and demonstration results are introduced.**

Keywords— *Energy storage system, microgrid, renewable energy.*

I. INTRODUCTION

As an application of power electronics technology, it is believed that an energy storage system (ESS) which is composed of converters and batteries or other devices plays a very important role in microgrids together with output control of d.c. or a.c. generators and load control like variable speed drives.

Especially , because an inertia of a microgrid is relatively small compared with fluctuations of individual load and generator in islanding operation when the microgrid is isolated from the utility grid, frequency control is severely affected by those fluctuations. Therefore, ESS which can control demand and supply in a microgrid flexibly and quickly, is a strong candidate for a countermeasure of frequency stabilization.

In this paper, concepts, requirements on ESS and demonstration results of three demonstrative microgrid projects are described.

II. CONCEPTS OF MICROGRIDS AND REQUIREMENTS ON ESSS

We categorized microgrids into three types. The first one is a small-scale independent electric power supply system which can be use for electrifying areas which are difficult to connect to existing bulk utility grids. For this kind of microgrid, simple configuration and operation are expected because operators cannot be experts of the system and supply of materials for maintenance is not sufficient. ESS is requested to work autonomously without any control server.

The second one is a commonly-used microgrid. Multiple generators, loads and ESSs are joined in the microgrid and they are connected to a control server (often called "Energy Management System (EMS)") via communication network.

The server supervises and controls demand and supply in a microgrid in order to maximize electric or economic efficiency, or to minimize environmental load. An ESS in this kind of microgrid has two roles. One is a high-speed demand and supply adjustor for contributing to the optimal operation of the microgrid by following output commands from the control server. The other is a power quality conditioner as autonomously adjusting its output (local-following control) especially in islanding operation.

The last one is a large-scale renewable energy power plant such as Mega-solar and wind farm. A renewable energy (RE) power plant connected to the utility grid may not be a literal microgrid, however, if the capacity of the plant is large, its output fluctuation may affect frequency control of the grid. An ESS with a large-scale RE power plant can decrease the impacts on the grid. For the system, the compatibility between performance and economic efficiency is very important, that is to say, power and energy capacity of an ESS should be minimized while requirements from the utility grid are conformed. In order to realize the system, a large amount of information, such as weather forecast, state of generators and ESSs, are utilized to calculate output commands for each storage device of ESS. Therefore, output control server which decide the total output of the system and controller such as programmable logic controller which calculate real-time output commands of storage devices are required.

Characteristics of the three types of microgrids are compared in Table 1.

TABLE I
COMPARISON OF THE THREE TYPES OF MICROGRIDS

Item	Independent	Islanding with High Power Quality	Large-scale RE Power Plant
System Requirement	Simple Configuration and Operation	Compatibility of efficient operation and power quality	Minimize ESS Capacity
Control Strategy	Autonomous	Control Server and Local Control by ESS	Output Command Distribution among ESSs
ESS Control	Autonomous	Both External Command and Local Control	Target from Server and Local Control by Controller.

978-1-4799-2706-7/14 $31.00 © 2014 IEEE

III. STAND-ALONE POWER SUPPLY SYSTEM PURELY WITH RENEWABLE ENERGIES

A. Outline of The Project

A small-scale independent electric power supply system composed of micro hydro power (MH) and photovoltaic (PV) which does not consume any fossil fuel in operation is one of the promising solutions to electrify areas which are difficult to be connected to existing bulk utility grids. It is expected that the both resources complement each other especially in a region which has rainy and dry seasons. However, MH has a poor response against output fluctuations while generation from PV is significantly fluctuated by weather conditions. Therefore, it is considered to be difficult to keep frequency without stabilization by ESS.

In order to demonstrate the microgrid, we constructed a verification system for output stabilization technology of PV by use of an electric double layer capacitor (EDLC) and executed experiments from FY2007 to FY2010 in Lao P.D.R. as an international joint proof project of New Energy and Industrial Technology Development Organization (NEDO), Japan [1-4].

Figure 1 shows the system configuration of the Lao project. MH (110kW) is the main power source. PV (40kW) connects to the a.c. main bus with a grid-connection inverter. EDLC (+/-40kW, 30s) connects to the d.c. side of the PV inverter with a DC/DC converter (chopper) and stabilizes frequency fluctuations from short-term variations of PV output and loads which cannot be followed by MH. Generated electricity is boosted to 22kV by the outdoor transformer and is distributed to villages by way of transmission lines which have a length of 20km in total. A photo of the electrical room is shown in Fig. 2. EDLC is in the rightmost panel and chopper for EDLC is in the next panel.

B. Control Strategy

An electric double layer capacitor (EDLC) was selected as the storage device because repetitive operation was required and large electricity for efficient operation of microgrid was not so important because all resources are renewable. Two stabilization controls are implemented to EDLC. The first one is ΔP control. In ΔP control, EDLC detects output fluctuations

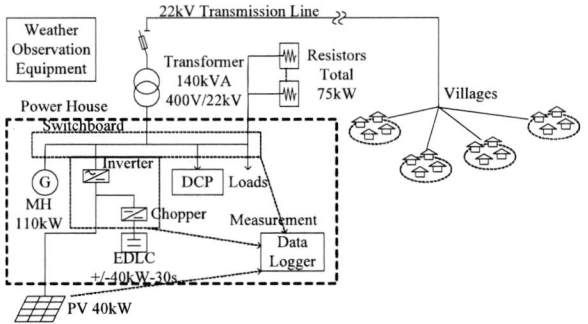

Fig. 1. System configuration of the Lao project.

Fig. 2. Photograph of the Electrical Room. EDLC is in the rightmost panel and Chopper is in the next panel.

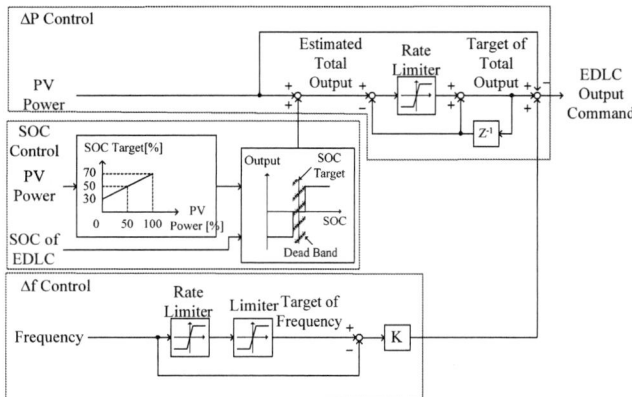

Fig. 3. Control Diagram of EDLC.

of PV directly and cancel the part which exceeds the allowable change-rate of MH. The other one is Δf control. EDLC detects a frequency deviation between the target of frequency and a present value and output in proportional to the deviation. The two controls can be activated simultaneously. The control diagram is shown in Fig. 3.

C. Results

Fig. 4 shows test results of ΔP control and ΔP+Δf control. PV outputs were suddenly stopped by opening the MCCB of PV manually. In the case of ΔP control, EDLC output decreased linearly just after the initial reaction, frequency decreased to 49.1Hz and recovery of the frequency didn't start until 70seconds after the stopping of PV. The response time of EDLC was observed to be less than 10 milliseconds. In the case of ΔP+Δf control, decreasing rate of EDLC output gradually got smaller than that of ΔP control, dropping of frequency is slower and an initiation of frequency recovery is faster compared with ΔP control. The amounts of discharged electricity from EDLC were 1,360 kJ in ΔP control and 1,031 kJ in ΔP+Δf control, respectively. Increment ratio of the electricity was only 2.2% after corrected the difference of PV outputs before stopping (38.1kW and 32.8kW, respectively).

The 2014 International Power Electronics Conference

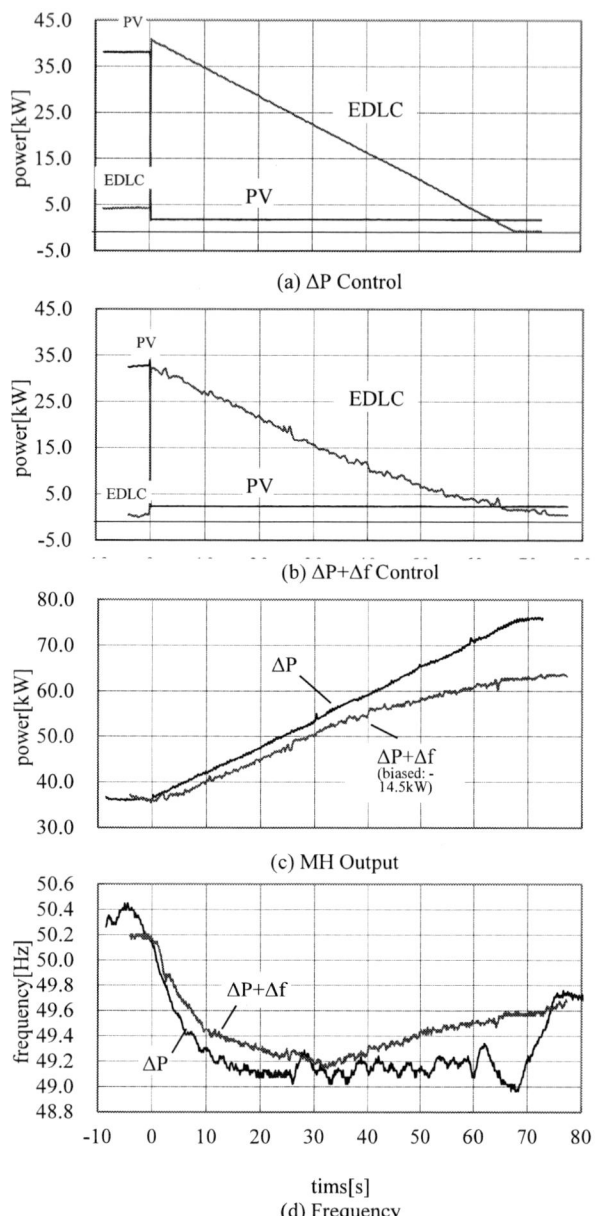

(a) ΔP Control

(b) ΔP+Δf Control

(c) MH Output

(d) Frequency

Fig. 4. Test Results of ΔP Control and ΔP+Δf Control at Stopping of PV.
(a) Power Output in ΔP Control, (b) Power Output in ΔP+Δf Control, (c)
MH Output of Both Controls and (d) Frequency Change in Both Controls.

Long term operations in both control modes were executed. In ΔP control, frequency was often observed to be less than 49Hz because of sudden load changes while no frequency drop was observed inΔP+Δf control.

Therefore, it was concluded that the performance of ΔP+Δf control is better than ΔP control.

IV. MICROGRID CAPABLE OF ISLANDING OPERATION FUNCTION WITH HIGH POWER QUALITY BY USE OF MULTIPLE GENERATORS AND ESSs

A. Outline of The Project

The concept of microgrid described in this section can be considered as of a commonly-used microgrid.

In order to demonstrate the microgrid especially focusing on power quality, verification facilities were installed in Hangzhou, China and experiments were worked out from FY2007 to FY2010 [5-7].

Fig. 5 shows the system configuration of the Hangzhou project. Generators in the system are PV and Diesel engine generator (DG) with capacities of 120kW respectively.

ESSs are EDLC (+/-100kW, 2s) and lead-acid battery (BAT, 100kW, 50kWh) which can operate in a coordinate way autonomously. A power quality conditioner and a voltage dip compensator are installed as countermeasures against higher harmonics, voltage dip, and so on. A photo of the EDLC system is shown in Fig. 6.

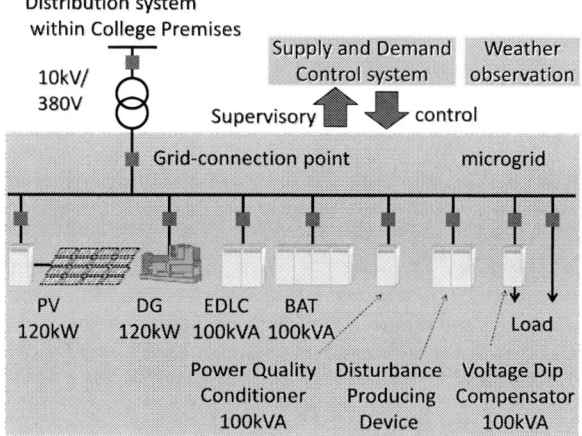

Fig. 5. System configuration of the Hangzhou project.

Fig. 6. Photograph of EDLC System. Incoming Panel, Inverter Panel and EDLC Panel (a Door is Open) from Left.

978-1-4799-2706-7/14 $31.00 © 2014 IEEE

B. Control Strategy

Supply and demand control system forecasts PV generation and load demand. Then calculate optimal operation pattern of DG and BAT. They are operated in accordance with the pattern while actual PV generation and load demand follow the forecasted pattern. If there is a difference between forecasts and results, the system gradually modifies the pattern and sends commands to DG and BAT.

BAT and EDLC compensate the difference between planned and actual power flow at grid-connection point directly measuring the power flow. BAT and EDLC share outputs by use of coordination control. No communication is required for the coordination control. The conceptual image of the control is shown in Fig. 7. EDLC covers shorter time period and BAT covers longer one. Compensated power flow which slowly changes is followed by DG from command from the system. Therefore, BAT outputs the sum of command from the system and coordination control in grid-connected operation of the microgrid. Voltage and frequency are referred in islanding operation of the microgrid.

C. Results

Fig. 8 shows an experimental result of the microgrid. When frequency changes during islanding operation caused by a change of load or PV output, EDLC reacts at first and then gradually reduces its output and BAT increases its output as to fill the gap at the same time. DG covers only changes longer than 30 seconds or 1 minute. Frequency fluctuation was suppressed within 50Hz +/-0.3Hz against PV fluctuation of 20-30kW.

V. OUTPUT FLUCTUATION SUPPRESSION SYSTEM FOR LARGE-SCALE RENEWABLE ENERGY POWER PLANT

A. Ouline of the Project

There is a possibility that large-scale PV systems have some adverse impact on a power grid because the PV system output may fluctuate greatly. Therefore, there is concern that the progress of large-scale PV systems will be limited. To reduce the impact on the power grid, technologies to control power plant output using ESS is required. To demonstrate a large-scale PV power plant with ESS, "Verification of Grid Stabilization with Large-scale PV Power Generation" project was conducted from FY2006 to FY2010 in Wakkanai City, Hokkaido, Japan [8, 9]. Fig.9 shows the system configuration of the Wakkanai project. 5.02MW of PV and 1.5MW of Sodium Sulfur (NAS[*1]) batteries were installed in the system. Inverter for 1MW NAS battery is shown in Fig. 10.

B. Control Strategy

Output management system decides the target of the total output of the system from the weather forecast and the state of charge (SOC) of each battery and sends it to the system stabilizing controller. The controller monitors outputs of PV and SOCs of NAS batteries and calculates output commands of NAS batteries and sends it to the inverters of batteries at

---

*1 NAS® is trademark of NGK INSULATORS, LTD.

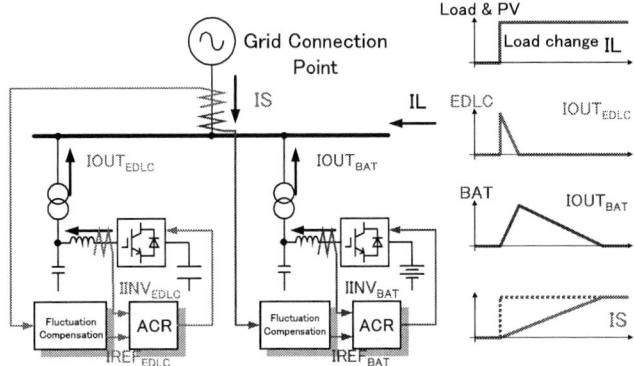

Fig. 7. Conceptual Diagram of the Coordination Control between BAT and EDLC.

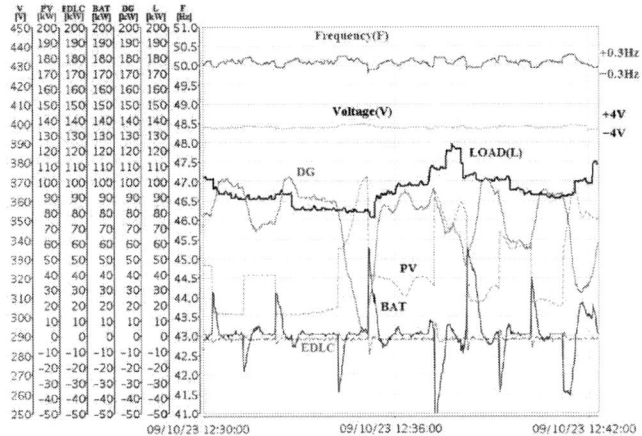

Fig. 8. Waveforms of Frequency, Voltage and Powers at PV Fluctuation in Islanding Operation of the microgrid.

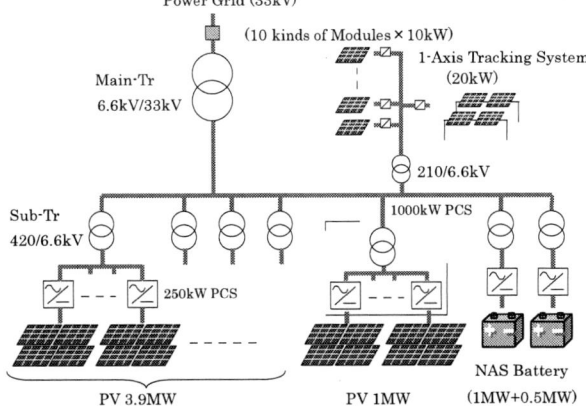

Fig. 9. System configuration of the Wakkanai project.

100ms interval. It was measured that the inverters of batteries react about 300ms after the output change of PV.

The total output command of NAS batteries are calculated from the difference between the total output of the system and actual output by use of a conventional proportional-integral control.

The 2014 International Power Electronics Conference

Fig. 10. Photograph of 1MW Inverter for NAS Battery. NAS Battery Stands behind the Inverter.

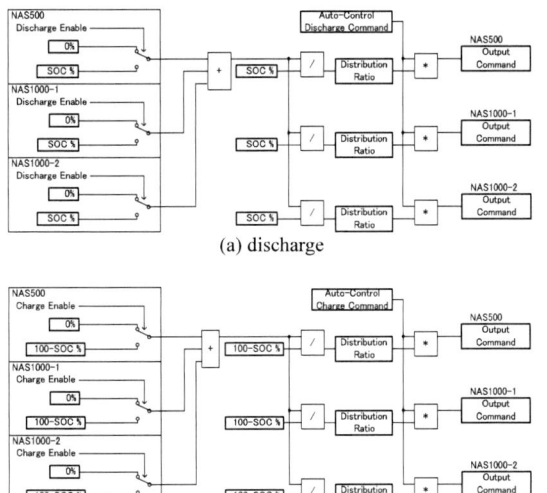

(a) discharge

(b) charge

Fig. 11. Output Distribution Calculation among NAS batteries.

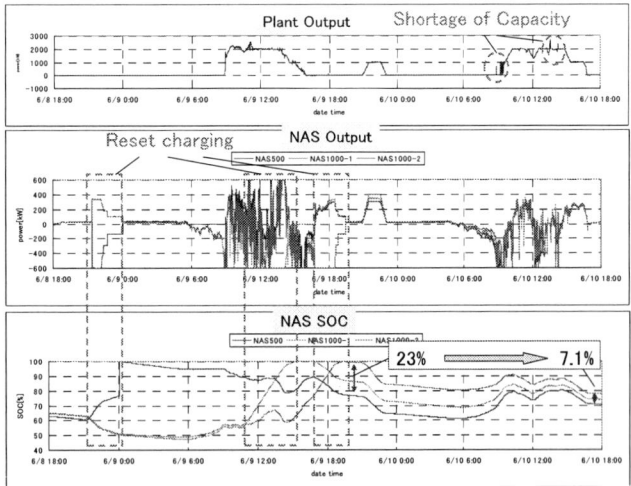

Fig. 12. Outputs of NAS Batteries and Changes of SOC.

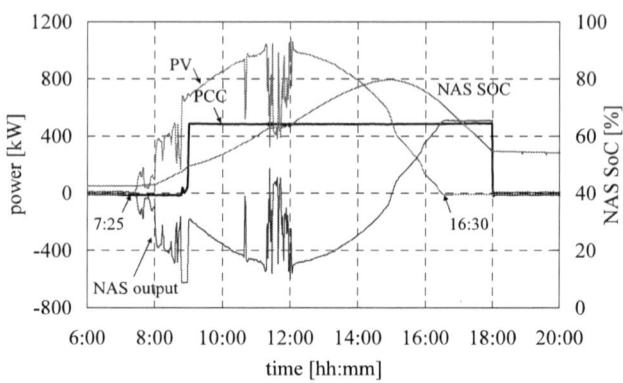

Fig. 13. An Example of the Scheduled Operation.

NAS batteries can be controlled independently as three 500kW-units. Therefore, the system stabilizing controller determine the number of controllable units and distribute output commands considering SOC of each unit in order to equalize SOC. That is, if total output command is discharge, a unit with higher SOC preferentially discharges, and if total output command is charge, a unit with less SOC preferentially charges. The distribution method is shown in Fig. 11. The output distribution can be extended for larger system easily.

C. Results

SOC equalizing test result is shown in Fig. 12. Difference among SOCs of three units was 23% just after the finish of reset charging for one unit at the center of the horizontal axis. After one-day operation, the difference was reduced to 7.1%.

Scheduled operation test result is shown in Fig. 13. Power flow at PCC (total output of the system) was kept constant for about nine hours. 500kW NAS battery was used for this experiment.

VI. CONCLUSIONS

Three kinds of microgrids and ESSs were proposed and demonstrated. An ESS in a stand-alone microgrid is requested to work autonomously in order to keep frequency within allowable range without any control server. An ESS in "the microgrid" which aims at compatibility between high power quality and efficient operation is requested for both autonomous operation to keep power quality and operation in accordance with a command from a control server for efficient operation of the microgrid. An ESS in a large-scale RE power plant should be a hierarchic structure that enables detailed control to make the best use of capacity of the ESS.

ACKNOWLEDGMENT

The demonstrative microgrid projects were supported by NEDO.

REFERENCES

[1] Y. Ueda, S. Oyabu, N. Wakasugi, H. Fujimori, M. Toguchi, I. Takara, H. Matsuda, and M. Shimabuku, "Photovoltaic, Micro-hydraulic and Electric Double Layer Capacitor, Hybrid Electric Power Supply Project in LAO P.D.R. – Test Operation Results",

Proc. Annual Conference of IEEJ-PES, No. 231, pp. 26-11 - 26-12 (2010) (in Japanese).

[2] Y. Ueda, S. Oyabu, K. Higa, T. Takara, L. Kakefuku and M. Shimabuk, "Hybrid Electric Power Supply Project in LAO P.D.R. - Verification Experiment", Joint Workshop of PE, PSE, SPC of IEEJ, PE-11-002 / PSE-11-019 / SPC-11-056 pp. --- (2011) (in Japanese).

[3] Y. Ueda, S. Oyabu, K. Higa, T. Takara, L. Kakefuku and M. Shimabuku, "Hybrid Electric Power Supply Project in LAO P.D.R. - Comparison of Frequency Stabilizing Effects from Various Control Methods of EDLC -", Proc. of Annual Conference of IEEJ- IAS, No.1-11, pp.I-133 – I-134 (2011) (in Japanese).

[4] Y. Ueda, S. Oyabu, K. Higa, H. Gushiken, L. Kakefuku and M. Shimabuku, "Frequency Stabilizing Experiments by Use of Electric Double Layer Capacitors for a Small-Scale Independent Grid with Renewable Energies", Proc. CIGRE SC C4 2012 Hakodate Colloquium (2012).

[5] H. Esaki, H. Sugihara, S. Suzuki, S. Maehira, E. Shimoda, K. Morino, S. Liu and Q. Han, "Simulation of supply-demand control in Micro-Grid with fluctuationg natural power supply", ICEE2008, No. O-175 (2008).

[6] S. Suzuki, S. Maehira, K. Morino, E. Shimode, H. Sugihara, H. Esaki, S. Liu and L. Zheng, "Site test of Power System Stabilizer in Micro-Grid into which a large amount of PV Power generation system are introduced", Proc. CIGRE SC C4 2009 Kushiro Colloquium (2009).

[7] Y. Ueda, "Development of a Power System Stabilizer Suitable for Promotion of Renewable Energies into Small-scale Networks", Electrical Review, 2011-3 (2011) (in Japanese).

[8] Y. Ueda, S. Suzuki, T. Ito, S. Miwa, R. Hara and H. Kita, "Development of an Output Management System for the "Wakkanai Mega-Solar" Project", Proc. CIGRE SC C4 2009 Kushiro Colloquium (2009).

[9] M. Niiyama, A. Kuwayama, Y. Saito, H. Kita, Y. Ueda and K. Yamaguchi, "Development of Output Control Technologies for a Large-Scale PV System with Battery Storage-Wakkanai Mega-Solar Project", Proc. 2012 CIGRE Session, C2-112 (2012).

Moving towards the Smart Grid:
The Norwegian Case

Olav B. Fosso, Marta Molinas, Kjell Sand
Department of Electric Power Engineering
Norwegian University of Science and Technology (NTNU)
Trondheim, Norway
Corr. Author email: olav.fosso@ntnu.no

Grete H. Coldevin
The Norwegian Smartgrid Centre
Trondheim, Norway
Email: grete.coldevin@smartgrids.no

Abstract- **Encompassing the global developments towards more sustainable and environment-friendly energy solutions for the future, Norway has been developing its own Smart Grid strategy. This strategy follows a path defined by the specific characteristics of the Norwegian energy system and the societal context. This article presents the Norwegian Smart Grid case by collecting the experiences and actions taken by industry, academic and research sectors. The role of power electronics technologies in smart grid research and in industry-driven innovation is also addressed in the paper.**

I. INTRODUCTION

The Smart Grid is a term coined to give a name to a wide range of solutions for the electricity grids of the future. Experience is already showing that each country and even regions are responding to the driving forces by developing their own versions of Smart Grids, depending on the national/regional characteristics of energy systems, communication systems, geography, topography, industrial and societal contexts. Consequently, some issues might well be common on a global scale and will find solutions through international development and cooperation while others instead will need tailored solutions to fit the national/regional context.

In Norway, the Smart grid priorities are partly driven by the regulator (e.g. the requirement in Norway to implement smart meters by 2019-01-01 [1] and partly by new technologies and challenges such as distributed generation (e.g. small hydro, PV), adoption of electrical vehicles, new challenging electric devices such as induction stoves, fiber-to-home communication, smart phones etc. Given these peculiarities, Norway can in some areas import Smart grid solutions and knowledge from international cooperation while in others it will need to develop tailored solutions to fulfill its own needs. At the same time, some nationally developed solutions, technologies and software might have potential for the international market.

In support to this emerging vast potential for development in the Smart Grid sector, Norway has given priority to this topic through dedicated R&D programs by the Research Council of Norway (RCN), academia and industry [2]. This has resulted in a set of joint multi-sector initiatives to face the challenges brought along with the Smart Grid.

This article will present Smart Grid initiatives at different levels in an attempt to portrait the Norwegian approach to Smart Grid and expose it to a discussion. The Norwegian Smart Grid Center [3], the National Smart Grid Laboratory, the various actions taken by the Power Industry in Norway ("living labs", demonstration projects like the "Demo Norway" [4], and other smart grid prototype systems) and the strategic positioning of R&D institutions and the industry partners in the Norwegian Smart Grid Centre to respond to the upcoming call of the Research Council of Norway aiming at establishing a select numbers of Centres for Environment-friendly Energy Research (FME). A future FME within the smart grid area will conduct concentrated, focused and long-term research of high international calibre in joint cooperation between innovative enterprises and prominent research groups. The proposal for a Smart Grid FME will be presented as a major step towards the active and intelligent grid in Norway.

Smart grids development will be different in different parts of the world depending on national or regional energy system characteristics and challenges. In Norway historically inexpensive hydro power has been the main energy source for electricity generation. Thus, today 98-99% of the total electricity generation in Norway is hydro power based. Due to this fact, it is not on the Norwegian smart grid agenda to convert fossils based electricity generation to renewable generation. But as Norway is committed to fulfill the so-called European Renewables Directive [15], 67% of the total energy use in Norway should be based on renewable energy by 2020 which requires an increase in renewable electricity generation which should be used to substitute the use of fossils based energy in industry (on-shore and off-shore) and in transport. Incentive schemes in terms of a common green certificate market with Sweden thus motivates for more renewable electricity generation which in part will be based on intermittent energy sources like wind, PV and small hydro power plants without any reservoir capacity.

The Norwegian power system and electricity use has several characteristics different from most countries giving specific challenges and opportunities within the Smart Grid context:

- Large part of electricity in the domestic sector used for space and water heating offers much flexibility for demand response and demand side management schemes.

- Large availability of hydropower plants with reservoirs which are fast and easy to control. These offer low-cost balancing services. (Most new production is small scale distributed generation without storage.)
- Quickly growing use of purely battery based electric vehicles due to very good incentives (tax exempt, free parking, free use of toll roads and bus lanes etc.) [5].
- Significant part of the LV distribution system is of type 230 Volt IT system (230 V line voltage) different from the 400 Volt line voltage systems in most of Europe.
- Weak grids with approx. 40% of the supply terminals weaker than the standardized EMC reference impedance give more severe voltage quality problems when connecting EVs, PVs etc. than many countries [6].
- Well-developed broadband communication to homes and increased use of fiber-to-home communication provided by power utilities.
- Well-developed electricity markets. There are multi-national markets [7] with significant volumes for day-ahead, intra-day and balancing with participation of producers and consumers.

In total, Norway's power system and markets are well positioned for a future smarter and more renewable power and energy system, but some barriers such as weak grids in parts of the LV system needs to find their cost efficient and smart solutions.

II. THE NORWEGIAN SMART GRID CENTER

The Norwegian Smart Grid Centre was established in 2010 on the basis of a recommendation of the Ministry of Petroleum and Energy in its national strategy process for defining future Energy R&D in Norway (Energi21 process [17]). NTNU and SINTEF answered to this challenge and became the locus of coordinating national research, demonstration, laboratory, education, standardisation and information activities to optimise the use of resources and avoid uncoordinated parallel activities [3]. The Norwegian Smart Grid Centre has currently 47 members from universities, research bodies, supply industry, transmission and distribution companies as well as infra-structure providers within telecommunication. Some of these members are large companies operating worldwide while others are smaller niche companies dealing with more specialized issues in the Norwegian system [18]. As ICT security, reliability and privacy of data are essential also service providers within these disciplines are well represented. The Smart Grid Centre is a strategic partnership, organized as a membership organization, where the purpose is to coordinate initiatives, exchange information and be a promoter of initiatives of national interests in the field on Smart Grid. The overall goal is to develop a platform for demonstration and R&D activities of international quality in Norway. Sub-goals are:

1. Establish a national roadmap of Smart Grid in Norway,
2. Develop national demo sites run by network operators and a national Laboratory facility at NTNU/Sintef,
3. Promote activities within ICT security and reliability in Smart Grids,
4. Synthesize knowledge and promote robust solutions for Advanced Metering Systems (AMS) and,
5. Standardization and interoperability as critical issues for a successful implementation of smart grid solutions.
6. Contribute to the competiveness of the emerging Norwegian Smart Grid industry

III. SMART GRID NATIONAL LABORATORY

In the area of Smart Grids, several research laboratories have been operating in Norway in a rather uncoordinated manner for several years. Each of them could offer a limited range of services and testing capabilities with limited access. The concept of a National Smart Grid Laboratory did not exist until the Norwegian Research Council decided to open a dedicated call for applying for funding to build National Infrastructures in the year 2012. As a result of that, a brand new Smart Grid National Laboratory was granted by the Norwegian Research Council to a consortium composed by NTNU, SINTEF, Narvik University College (NUC), and NCE SMART under the leadership of NTNU [8]-[11]. This consortium has the mandate to establish in the period 2014-2018 a one-site Smart Grid laboratory facility equipped with advanced infra- and control structures with high modularity to simulate in as realistic setting as possible occurring scenarios in real life. One of the objectives of this laboratory is the testing of new equipment, functions and control strategies to gain insight on their operation before they are implemented in a real application. Although the central laboratory facility will be located in Trondheim, there exist several other connected facilities and demonstration sites that will be linked to the central laboratory by a high speed communication system by which remote access to the facilities and databases will be given.

The Smart Grid National Laboratory project has just started in March 2014 and will run for the next 5 years. This laboratory is built around an existing facility comprised of the renewable energy laboratory, the energy

Fig.1. Distributed generation system emulator set with controllers from the laboratory

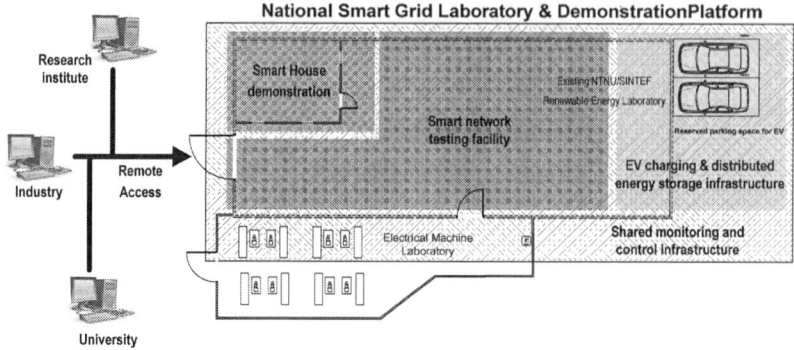

Fig.2. Lay-out of the laboratory facility with planned extension and upgrade

storage laboratory, the converter and control activities connected to the power electronics laboratory, the electrical installation laboratory and the PV solar panels at the campus [12]. The existing laboratories include: The present NTNU/SINTEF Renewable Energy Laboratory with its high power rating of 150 kVA and its wide range of network components is well suited for modelling both transmission and distribution voltage networks with a variety of electrical generators. The lab includes equipment to emulate distribution and transmission networks as well as energy storage technologies. An example of distributed generation emulation at the NTNU/SINTEF renewable lab is shown in Fig. 1. The other two partners of the project, NUC and NCE SMART, currently do not have local physical facilities to conduct medium/large scale experiments and developing large infrastructure locally would be unfeasible in a short term perspective. A schematic lay-out of the new smart grid laboratory including the extension and upgrading is shown in Fig. 2.

The Smart Grid National Laboratory project is subdivided into 6 sub-projects covering a wide range of potential smart grid developments in Norway and worldwide. The 6 sub-projects to be carried out are described in the following:

Subproject 1: Smart House demonstration

A Smart House demonstration covering a surface area of 25-30 m^2 will represent a controlled testing environment for prosumer products (e.g. smart appliances, smart meters, microgeneration etc.) and home automation (e.g. shut down loads during peak hours, programmed appliance operations). The space will be furnished as a standard apartment equipped with modern kitchen appliances (induction stove, oven, and refrigerator), laundry appliances (washing machines, dryers), units for thermal conditioning (water boiler, climate control). The resemblance to a living space will create realistic testing conditions and is expected to be very attractive for students especially at BS and MS level. The installation will be provided with a centralized control unit based on Labview connected to each plug and capable of

monitoring electrical parameters and operate switching actions.

Subproject 2: EV charging and distributed energy storage infrastructure

The charging station for EV functional unit will extend the infrastructure possibilities to testing of EV, electrochemical batteries and impact of EV on the grid. A rectifier unit will provide a controllable DC voltage for direct DC charging and battery testing. A connection of the rectifier to the Energy Storage Laboratory will be arranged in order take advantage of an area already equipped for safe storage and operation of Electro-chemical batteries. An external space adjacent to the laboratory will be reserved for parking two EVs and will be equipped with plugs and accessories for standard AC charging (230 V) and for fast DC charging. In addition, lithium battery packs will be added to the present Energy Storage Laboratory.

Subproject 3 & 4: Smart network testing facility and Real-time digital simulator

This functional unit will extend the capabilities of the current laboratory and will integrate the current equipment. This part of the infrastructure aims to establish a low voltage (400V) network system which can be operated as a distribution or transmission grid with high penetration of renewable units, storage units and power electronics converters. Advanced features of this functional unit will be the rapid prototyping and the Power Hardware in the loop (P-HIL) based on an OPAL RT platform. A third OPAL RT unit will be added to the two already available at NTNU/SINTEF. Moreover, a high bandwidth power converter (e.g. 50-100 kHz, 40-50 kW) will act as a power interface for the OPAL-RT platform. The new OPAL-RT platform and this converter together will allow to fully exploit the potentialities of P-HIL with one main use as a high fidelity grid emulator. The resulting network facility will be well suited for a range of purposes like: testing advanced Smart Grid technology for short term management and islanded (Microgrid) operation in future smart distributions systems - it can also be used for testing equipment and concepts for continuous stand-alone operation as a microgrid in remote/isolated areas (i.e. new grids to be started in developing countries, islands, remote settlements etc.), and it can be used for testing connection of independent (house-hold) microgrids as a bottom-up approach for establishing a local, expandable power system for developing regions without an existing transmission network.

Subproject 5: Physical Extension and Shared monitoring and control infrastructure

The planned research infrastructure will be located in the shared NTNU/SINTEF facilities in Trondheim at the Gløshaugen campus and will represent an extension and a restructuring of the existing Renewable Energy Laboratory. The research infrastructure will be accessible electronically with protected access through internet connection.

A shared control and monitoring infrastructure will ensure a pertinent integration of the previously described functional units to set up and test use cases. In addition, this level will ensure the possibility to test the impact of communication delay and of communication failures on the electrical system. The backbone of this structure will be based on NI equipment and LabView technologies interconnected via Ethernet cables. NTNU and SINTEF have already positively tested the potentialities of LabView for these applications with the installation of 4 cRIO units to create a system for distributed measurement.

Subproject 6: Remote access and database design

An additional feature of the Smart Grid Laboratory infrastructure as a national resource will be the remote accessibility through Internet in order to conduct remote use cases and experiments. In this context, research partners or industrial customers will be allowed to setup experiments in the physical facility but execute tests and monitor the results remotely. To make this possible, all the partners in this project and the ones to be incorporated in future will be provided with remote access to make possible to virtually run experiments and obtain data and information from its databases.

This will benefit from the already high-bandwidth internet connection existing between research institutes in Norway (UNINETT). The physical infrastructure will be located in Trondheim but part of the budget is allocated to NUC Narvik and NCE SMART (as pilot remotely located partners) to equip their facilities with terminal points for remote access (e.g. PC, Labview licenses etc). Conversely, NCEs simulation and scenario center infrastructure can be utilized from NTNU/SINTEF and NUC. All metering, sensor and experiment data will be stored in a common high performance database through a uniform data storage interface. The data will be accessible for aggregation, simulation and control purposes through a uniform XML based interface.

A. The relevance of Power Electronics in the Smart Grid:

The Smart Grid National Laboratory infrastructure is built around the need for a more flexible and active distribution grid. Most of this flexibility and active properties of the network will be enabled by *power electronics technologies* as the core high speed actuator for applications integrating several required functionalities such as monitoring/metering, protection, communication and control. It will be the high frequency operability of power electronics units that will enable the integration of the above required functionalities into one single device for the Smart Grid.

The six sub-projects of the previously described Smart Grid National Laboratory are covering the most critical areas in which innovation enabled by power electronics technologies are expected to happen. The smart house sub-project will give room to explore active demand response and aggregation behaviour of loads. Metering, protection and communication features needed at the load ends will most likely be integrated in single devices in which power electronics will provide the fast switching, compactness and intelligent/flexible properties required to fulfil the demands of real-time management of the smart grid. Power electronics, by their embedded fast digital signal processing features, low losses, high power density/compactness of the new generation semiconductor devices (SiC, GaN, etc), and higher frequency switching possibilities, provides the Smart grid with enormous flexibility to explore innovative integrated solutions with high potential for cost reduction compared to the state of the art solutions. Power electronics decidedly emerges as the basic building block for the vast range of needs and application that the Smart Grid demands.

Sensors and actuators are examples of areas where most innovations in smart grid will emerge. Real-time monitoring capabilities, fast communication and actuation could be integrated in single and compact devices where power electronics will play a fundamental role. Vast research opportunities and synergies with industry needs will be opened by the new laboratory in which power electronics will be central in all sub-projects and where synergies with already existing devices and functionalities can be exploited (renewable energy interfaces, load interfaces, circuit breakers, etc). Some of the local power electronics industries in Norway are already involved in some of these research tasks through the Smart Grid Center or through ongoing projects with NTNU/SINTEF (Eltek, Siemens, ABB, Wärtsilä) [18].

IV. THE SMART GRID LIGHTHOUSE

The Smart Grid Lighthouse is an initiative by the Norwegian University of Science and Technology (NTNU) aimed at highlighting and coordinating the research and education activities related to Smart Grid within the University. The Lighthouse integrates multidisciplinary activities covering the areas of electrical engineering, control systems, computer science, telecommunication as well as societal aspects, economics and new business models. The university allocated significant strategic funding to the lighthouse projects by assigning PhD/Post.Doc positions equivalent of 3 mill EURO in addition to the normal funding obtained through the large research projects funded by the Norwegian Research Council and the European Union [13]. The research projects involve a number of PhD-students and Master students performing their research in close collaboration with researchers from NTNU's close collaborative partner SINTEF. The curriculums of the involved disciplines have also been through restructuring to streamline both the ICT- and the energy related education. A number of new courses within smart energy systems are developed additionally to the upgrading of existing courses. A new course addressing the multidisciplinary aspects of smart energy systems is developed and offered as a continued education course for the power industry [13].

V. DEMO NORWAY

Full scale demonstration projects connected to real power systems are necessary to properly develop, test and verify Smart Grids solutions. Immature and high-risk solutions are best studied and tested in laboratories while the more mature cases and cases which include the behaviour or human response of customers need to be tested in demonstration projects that are linked to real power systems with real customers.

One of the goals of the Norwegian Smart Grid Center is to establish national demos and laboratories for the purpose of developing, testing and verifying Smart Grid technology and services. Demo Norway is the result of coordinated development of complementary demos at individual sites ("Living Labs") with modern off-grid laboratory facilities at the research institutions NTNU/SINTEF.

The living labs of Demo Norway are planned to be connected to the National Smart Grid Laboratory via remote access. Connection between the demos and the off-grid laboratories will be established for exchange of data and to be able to run research projects in a relevant setting. The demos are designed in a way to be useful also for the supply industry both for component and services and in this way they are also incubators for research and business development.

In the Norwegian Smartgrid Centre there are 6 such living labs which are geographically distributed across

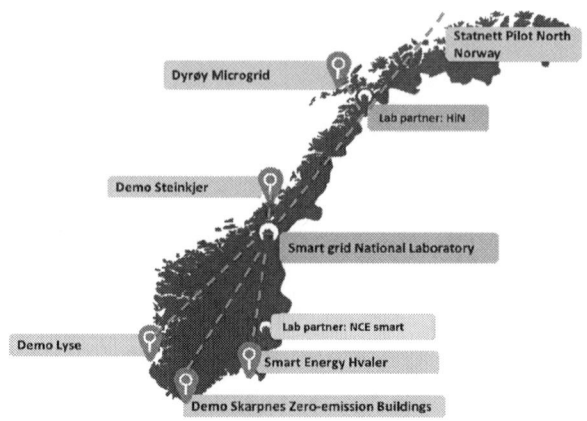

Fig. 3. Demonstrations facilities of the Demo Norge distributed across Norway

Norway according to what is shown in Fig. 3. A brief description of the focused areas of each of these Demos is given in the following:

1. In **Demo Steinkjer** energy companies, vendors, researchers, customers and governmental bodies can test smart meters, system services and other products on 800 end users consisting of ordinary households, commercial companies and industry. The main focus of the demo project is to develop commercial products and services for the next generation of smart networks as well as to prepare for the smart meter roll-out required by the regulator. Flexibility of end users and value added services for the distribution companies will especially be developed and tested in this project. Dynamic tariffs are one of the incentive schemes being tested. Both the technologies involved and the customer response are addressed. A high-level identification of information security threats of the AMI pilot in Demo Steinkjer have been investigated primarily concerning the smart meter and its communication with the main system of the Distribution System Operator (DSO). A number of information security threats have been identified, and will be further analyzed in a complete risk analysis to prioritize risks and possible measures – for more information see [16].

2. In **Smart Energy Hvaler** the living lab consists of all the 6800 customers in the local community. With the implemented infrastructure, the demo will focus on development and testing of enhanced network utilization and end user flexibility. Both Steinkjer and Hvaler are currently the common platform for testing in the Industry driven project called DeVid supported by the Research Council of Norway. This project is considered to be the power companies' main research project within Smartgrid. The main element in DeVID project is testing and verification of Smart grid use

cases for smart operation and planning of the distribution system. The availability of new smart meter data has already given improved information on end user load profiles and thus changed the decision base for network planning. The old load profiles have often given a conservative peak load estimate. Bringing the new data into the planning process have thus reduced the needs for grid reinforcements and saved money.

3. In **Demo Lyse** the aim is to demonstrate ICT-infra structures and architectures. This demo will provide the foundation for services to the customers as for example welfare technology. The grid company Lyse will install AMS-meters at all their end users together with a unit facilitating additional services. The information of this unit may be displayed on for example an iPad for the customers and the information will be embedded in the control systems of Lyse. In Demo Lyse one of the subprograms is called "Smart power" where 40 network customers have been recruited to test solutions for energy efficiency and energy control utilizing a smart house control concept. In this concept a fiber communication based AMI infrastructure is used for controllable household devices interaction.

4. **Demo Dyrøy** is a small micro grid test bed which will be a laboratory for system operators. The micro grid of this Demo will supply a remote district heating system and the demo will be a laboratory for testing technical aspects and solutions for future private and public customers. The energy sources to be used for distributed generation in the micro grid are: Wind, solar and bio fuels (combined heat and power - CHP solution).

5. **Demo Statnett R&D Pilot project North Norway** will address development and testing of principles and procedures which contribute to the planning and

operation of the power system from a system operator's (TSO) point of view. Statnett is the Norwegian Transmission System Operator (TSO). In this demo close to real time reliability and risk assessment solutions are tested and verified. Wide area monitoring systems including phasor measurement units (PMUs) both for monitoring and power system damping control are important elements in the demo. The demo also addresses TSO load management for system balancing purposes by playing on the DSO loads in the northern region of the Norwegian power system.

6. In **Demo Skarpnes** 40 houses with so-called passive or zero-emission building standard of which 37 will produce electricity with solar cells forms the demo site. Norwegian authorities plan to introduce the passive house standard as a requirement for new buildings from 2015. To be able to dimension the electricity grid in areas dominated by passive houses it is necessary to learn how load and generation profiles in passive houses differ from those of traditional houses. In 2020 the building regulations might be even stricter, as there are plans to introduce zero energy buildings as a requirement for new residential buildings from this year. This will enhance the need for new modeling tools for grid planning, tools that must take into account the electricity production from distributed electricity generation as well as new load profiles. In the research project "Electricity usage in Smart Village Skarpnes" data from smart electricity meters will be used to monitor load profiles and electricity production in the new passive and zero energy houses at Skarpnes as well as the production from the PV-systems.

The site offers a unique opportunity to study a larger number of buildings to prepare for the future, and the project aims to harvest experience and data that can be used to plan the future electricity grid. Expansions and up-grades of the electricity grid are expensive and careful and accurate planning will be of great value for the grid

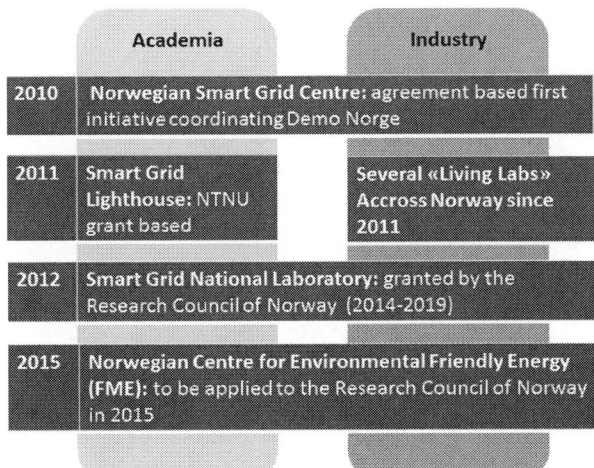

Fig. 4. Timeline overview of major Smart Grid Initiatives in Norway

Fig. 5. All houses equipped with solar cells, solar collectors and energy storage at Demo Skarpnes

operators. Fig. 5 shows an artist vision of a passive house under development in this Demo.

VI. Norwegian Smart Grid Centre for Environmental Friendly Energy (FME)

This in an initiative built on the foundation of the several joint Smart Grid actions taken since 2010 and described in the previous sections of this paper. A chronological overview describing the nature of the smart grid initiatives preceding this Innovation Centre is given in Fig. 4. The Research Council of Norway is responsible for funding a select number of Centres for Environment-friendly Energy Research (FME), which seeks to develop expertise and promote innovation through focus on long-term research in the area of environment-friendly energy. The FME call is scheduled for 2015. The total budget allocation from the Research Council over the life span of eight years for each FME will be 10 and 20 MNOK per year (1,2 - 10,7 MEUR per year in eight years). The host institution and partners responsible for an FME proposal must contribute with at least the same amount as RCN. The ambition of the Norwegian Smart Grid Centre is to establish an advanced research centre for intelligent distribution systems of world class that builds on the experiences gained from the academic, industry and research institutions with the aim to open the roads to research based innovation. A Smart Grid FME will, by its scope, perspective and nature, embrace all the preceding initiatives in Smart Grid in Norway. A Smart Grid FME will encourage enterprises and network operators to innovate by placing stronger emphasis on long-term research and by making it attractive for enterprises that work on the international arena to establish R&D activities in Norway. The FME will promote the development of research groups that are on the cutting edge of international research and are part of strong international networks. Further, the FME will stimulate researcher training in fields of importance to the user partners and encourage the transfer of research-based knowledge and technology. In Norway, the next 10-20 years massive investments will be done within the distribution system on automation, monitoring and control. Integration of dispersed generation, introduction of advanced meeting systems and utilization of local storage will give an active distribution system and pose a lot of challenges and opportunities. A major motivation of the initiative is: by systematic and long-term involvement of the suppliers of components, decision support systems and services and match these with research / education and distribution grid operators in order to develop a new range of products and services and by this enhance competitiveness of the industry to meet these future challenges and benefit from the opportunities.

References

[1] http://www.nve.no/no/Kraftmarked/Sluttbrukermarkedet/AMS/

[2] www.energi21.no

[3] http://smartgrids.no/

[4] Demo Norway is a collection of smart grid pilot projects coordinated by the Norwegian Smartgrid Centre with more than 10.000 pilot customers involved.

[5] European Association for Battery, Hybrid and Fuel Cell Electric Vehicles (AVERE) (2012-09-03). "Norwegian Parliament extends electric car initiatives until 2018". AVERE.

[6] IEC/TR 60725 Consideration of reference impedances and public supply network impedances for use in determining the disturbance characteristics of electrical equipment having a rated current ≤ 75 A per phase

[7] http://www.nordpool.com/

[8] http://www.ntnu.edu/

[9] http://www.sintef.no/home/SINTEF-Energy-Research/

[10] http://www.ncesmart.com/Pages/default.aspx

[11] http://www.hin.no/nor/hovedside/forskning/fou-grupper/elektromekaniske-systemer

[12] www.sintef.no/energylab

[13] http://www.item.ntnu.no/projects/smartgrid/overview

[14] http://www.ntnu.no/elkraft/videreutd_kurs (ET6010 Energy Systems of the future - In Norwegian)

[15] Directive 2009/28/EC of the European Parliament and of the Council of 23 April 2009 on the promotion of the use of energy from renewable sources and amending and subsequently repealing Directives 2001/77/EC and 2003/30/EC

[16] I. A. Tøndel, M. G. Jaatun, M. B. Line , Security Threats in Demo Steinkjer, SINTEF ICT and Telenor Report, 2012-09-12

[17] http://www.energi21.no/prognett-energi21/Home_page/1253955410599

[18] http://smartgrids.no/senteret/medlemmer/

The 2014 International Power Electronics Conference

Power Electronics Technology in Smart Grid Projects -Applications and Experiences-

Takenori Kobayashi
Transmission & Distribution Systems Division
Toshiba Corporation
Kawasaki, Japan

Abstract— For the sustainable development of our highly electrified societies, power electronics technology is playing an important role. Especially in the field of electric power systems, it is considered as one of the essential technologies which make power grids more efficient, more reliable and more environment-friendly. Those are the goals of so-called smart grids. In this paper, expected roles and applications of power electronics in smart grids are first introduced. Then recent technological achievements in actual smart grid projects are presented.

Keywords—Battery energy storage system (BESS), Field demonstration, Lithium-ion battery SCiB^{TM}, Tyrister voltage regulator (TVR)

I. INTRODUCTION

Power electronics is one of the core technologies that constitute a smart grid. Toward realization of sustainable and highly electrified societies, power electronics technology will play an important role to expand renewable energy introduction and to promote efficient

energy use by controlling electric power in the grids.

A smart grid concept is illustrated in Fig. 1, where the following elements are considered as the key actors of smart grids:

- Intermittent renewable energy sources such as photovoltaic (PV) and wind power (WP),
- Controllable energy storage,
- Regional energy management such as micro energy management system (μEMS),
- Demand-side energy management such as building energy management system (BEMS) and home energy management system (HEMS),
- Advanced metering infrastructure including smart meters.

In order for them to work cooperatively as a smart grid, power electronics is used everywhere in the grid, i.e. in generation, consumption, transmission and distribution. Using power electronics, large-scale intermittent renewable generation can be integrated into the grids.

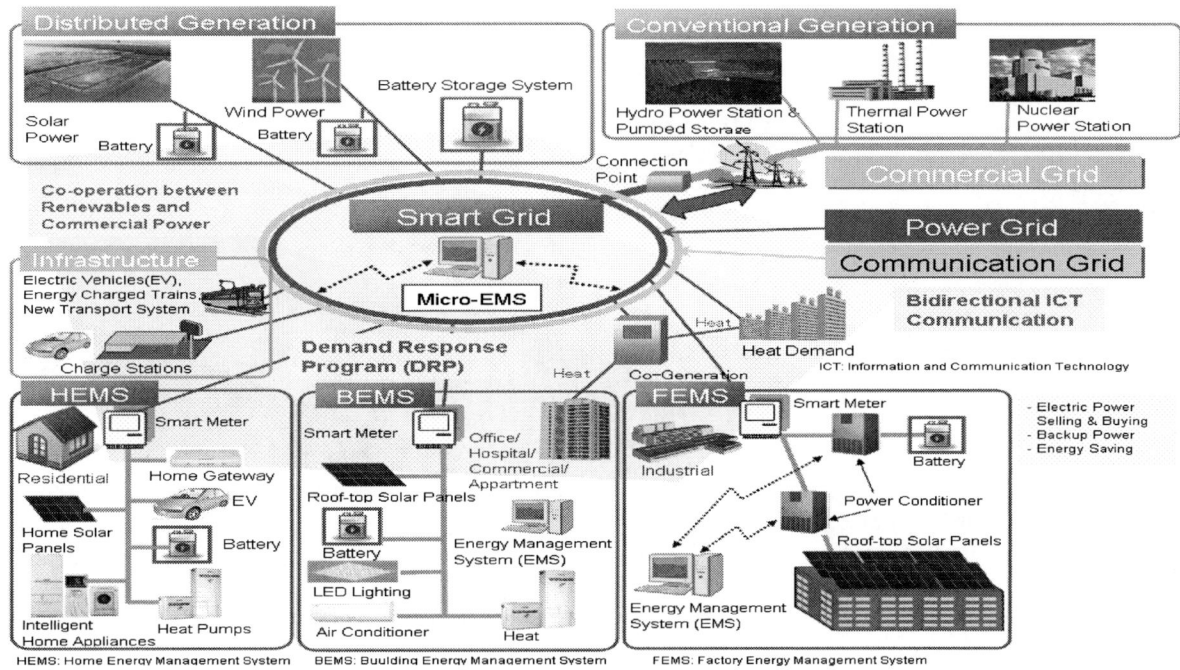

Fig. 1. Overview of smart grid concept

978-1-4799-2706-7/14 $31.00 © 2014 IEEE   1868

Efficient use of electric energy can be achieved by installing highly efficient facilities using the latest power electronics technology.

In this paper, battery energy storage system and thyristor voltage regulator are introduced as power-electronics-based apparatus promised in smart grids. Their applications and experiences in actual smart grid projects are then presented.

II. ROLES OF POWER ELECTRONICS IN SMART GRIDS

A. DC/AC and AC/DC Power Conversion

Many kinds of DC devices are used in smart grids such as PV cells, battery cells, and fuel cells. Since existing power system are AC grids, if DC devices supply their power to the AC grids, power electronics technology is used to convert DC power to AC power.

Power electronics equipment for this purpose is called power conditioning system (PCS), which consists of inverter and grid connection equipment. For an energy storage system, the inverter of the PCS should be bi-directional so that the energy storage system can both charge from and discharge to the grid.

B. Measures to Expand Introduction of Renewable Generation

When a large amount of intermittent renewable generation such as PV and WP are installed in a distribution network, power flows on the distribution lines behave in different manners. Not only power flow directions but also unexpected and sudden changes in power flow occur due to output fluctuations from the renewable generation. Accordingly, voltage deviations from the regulated range may also occur. Hence, fast and precise control of both active and reactive power is expected for power electronics technology to manage power flows and voltages.

Battery energy storage systems (BESS) are the most promising apparatus for this purpose since they can provide active and reactive power in four (4) quadrants. If only voltage regulation is focused, thyristor voltage regulator (TVR) or static var compensator (STATCOM or SVC) is applicable and effective because they can only provide reactive power.

C. Cooperative Operations with Supervisory Energy Management Systems

Many power-electronics-based apparatus work locally and independently. However, in the smart grids, coordinated operation of individual apparatus is expected to achieve the total grid optimization. For this purpose, supervisory energy management systems such as μEMS have been introduced in several smart grid projects.

Using information and communication technology (ICT), μEMS monitors local information of the grid in real time, and control individual apparatus if needed. Being a key control device, power-electronic-based apparatus is requested to work cooperatively with a supervisory control system.

D. Other Practical Requiements

The roles of power electronics discussed above are mainly focused on its functionality, i.e. responsiveness, controllability and operability. These are the technological requirements from the smart grid viewpoint.

On the other hand, from a practical and commercial viewpoint, the following requirements for the power electronics equipment are also important:

- Lower initial cost,
- Downsizing and weight saving,
- Tolerance to overvoltage and overcurrent,
- Compliance with standards and regulations,
- Higher efficiency with lower losses,
- Reliability for long-time use,
- Durability ,
- Maintainability.

The first four (4) are related to the viability of facility introduction. The latter four (4) are related to the viability of operation continuity.

III. BATTERY ENERGY STORAGE SYSTEM (BESS)

A. Necessity of BESS

An electric power system cannot maintain stable operation if imbalance between demand and supply become larger. In existing power systems, the change in demand is forecasted from the past experience, and the actual gap from the forecasted demand is compensated by controlling output from thermal and hydroelectric power plants in real time.

However, if installed capacity of intermittent renewable generation increases beyond the limit, the power system may not keep its stable operation. In such a case, BESS would be a promising solution because of its fast and precise control response, short construction time and decentralized nature with a compact footprint compared to other electric energy storage.

B. Basics of Battery Chemistry

Figure 2 shows available battery types for stationary use classified by their chemistry. BESS is attracting a lot of attentions from people who are involved in smart grid businesses and technologies. One of the key drivers of this trend is the recent development of lithium-ion batteries.

In case of using lithium-ion batteries for smart grid applications, safety, life span, and cost are the issues to be solved. Toshiba has developed a lithium-ion battery named SCiB™[1], using lithium titanate oxide in its anode. SCiB™ has excellent characteristics suitable for BESS as shown in Fig. 3.

C. Configuration of BESS

Figure 4 illustrates a typical configuration of BESS using SCiB™. The specification of the PCS used in Fig. 4 is shown in Table I. The PCS rated 500kW is equipped with best-in-class power conversion efficiency, which is important performance metric for BESS because major portion of charge and discharge losses of BESS comes from the AC/DC and DC/AC conversion.

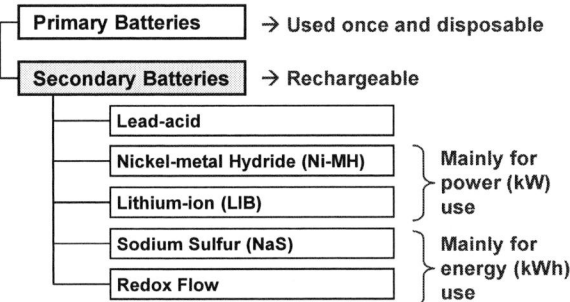

Fig. 2. Battery Types by Chemistry

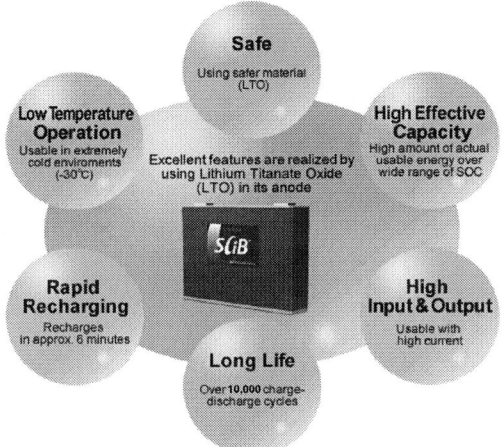

Fig. 3. Features of SCiB™

IV. THYRISTOR VOLTAGE REGULATOR (TVR)

A. Necessity of TVR

Voltages in distribution systems are normally regulated between their upper and lower limits. Recent years, many PV generators have been introduced and connected to the distribution system. When the outputs of PV increase, reverse power flow would occur, and the voltage at the end of the distribution line would exceed the upper limit.

Without taking any measures, PV generation have to be reduced, so that power which could be generated by PV is not fully utilized. A measure to prevent such unfavorable situations is to install TVR, which can improve the voltage profile by controlling reactive power.

B. Advantages of TVR

At present, step voltage regulator (SVR) is widely applied in order to keep the voltages on the distribution lines within specified range. While SVR uses mechanical contactors for voltage control, TVR uses thyristor switches. Hence, TVR has the following advantages compared with SVR.

- Response time is much faster than SVR. Response time of SVR is 1 minute while that of TVR is 200ms at maximum.

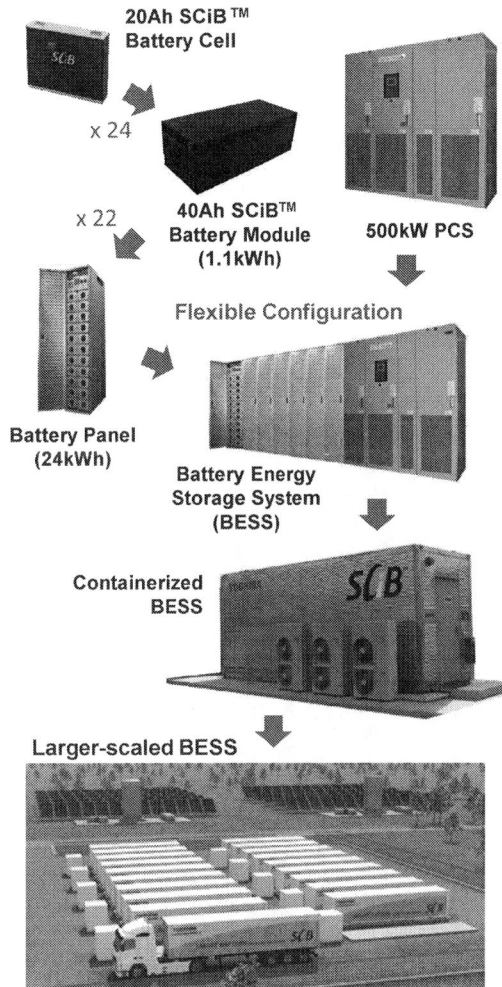

Fig. 4. Basic configuration of SCiB™ based BESS

TABLE I
SPECIFICATION OF 500kW BIDIRECTIONAL PCS FOR BESS

Rated Power	+/- 500 kW
	+/- 360 kvar
DC Voltage	450 ~ 800 VDC
Rated Voltage	310 VAC
Frequency	50/60 Hz ± 3%
Maximum Efficiency	98.5 %
Dimension	1900(W) x 700(D) x 1950(H) [mm]

- Maintenance is much easier due to no mechanical contactors for taps. SVR has limitation of tap change 10,000 times for example, while TVR has no limitation.

C. Configuration of TVR

An example of TVR installation is shown in Fig. 5, where the TVR is installed between two poles. A control panel is beside the transformer and radiators. Section

978-1-4799-2706-7/14 $31.00 © 2014 IEEE   1870

switches are used for connecting or bypass the TVR. Table II shows specification of the TVR.

Figure 6 shows the main circuit of TVR where output of a smaller tap is 100 V and a larger tap is 200 V. By selecting the combination of thyristors to be switched on, the voltage magnitude to be adjusted is decided. For example, if S3R and S4R are switched on, -300 V is selected. As shown in Fig. 6, seven (7) tap positions can be selected and here the voltage step is 100 V. MCCBs in the circuit are used for thyristor protection.

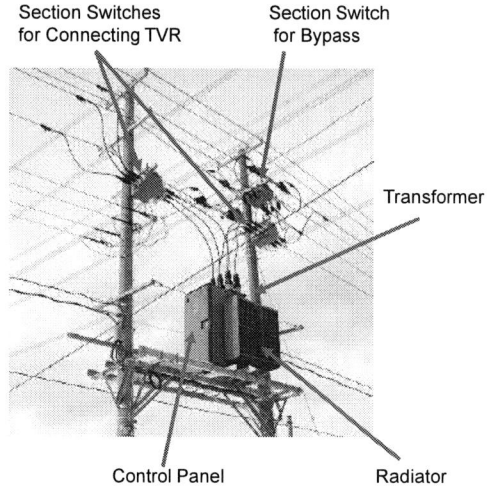

Fig. 5. TVR installed between two poles.

TABLE II
SPECIFICATION OF TVR

Item		Specification
Rated Line Capacity		3000 kVA
Range of Voltage Setting		6300 V ~ 6900 V
Range of Voltage Adjustment		+ 300 V, - 300 V
One Tap Voltage		100 V (7 steps)
Mass		2300 kg
Dimension	Depth	1555 mm
	Width	1410 mm
	Height	2335 mm

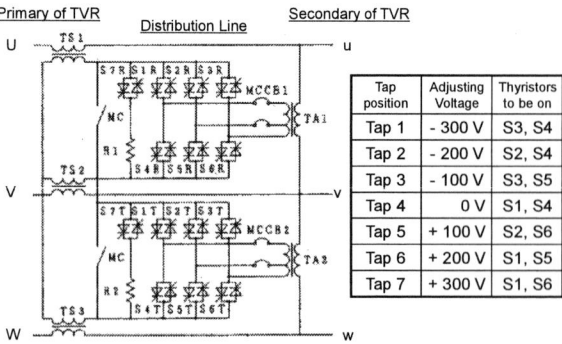

Tap position	Adjusting Voltage	Thyristors to be on
Tap 1	- 300 V	S3, S4
Tap 2	- 200 V	S2, S4
Tap 3	- 100 V	S3, S5
Tap 4	0 V	S1, S4
Tap 5	+ 100 V	S2, S6
Tap 6	+ 200 V	S1, S5
Tap 7	+ 300 V	S1, S6

Fig. 6. Main circuit of TVR

V. FIELD EXPERIENCES FROM ACTUAL SMART GRID PROJECTS

A. High-output BESS for PV smoothing

High-output BESS for PV smoothing is being operated since September 2010 at Toshiba Fuchu Complex, Tokyo, Japan. As shown in Fig.7, 400kW multiple PV generation, 300kW-60kWh BESS using 4.2Ah SCiB™ cells are installed. Figure 8 shows the system configuration where µEMS is implemented for the battery control.

These are in-house demonstration facilities for research and development, where battery system control and long-term evaluation of PV and batter characteristics have been investigated.

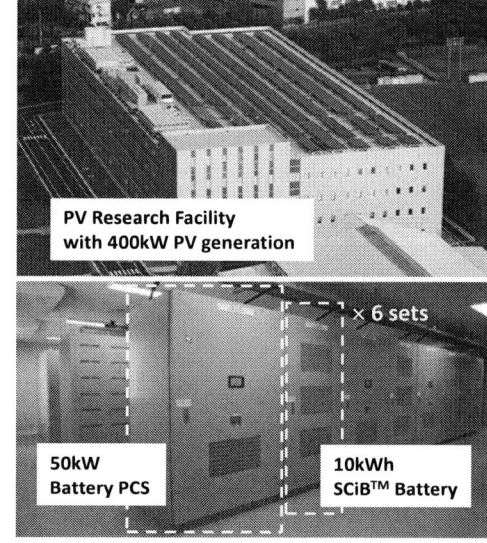

Fig. 7. Facility overview.

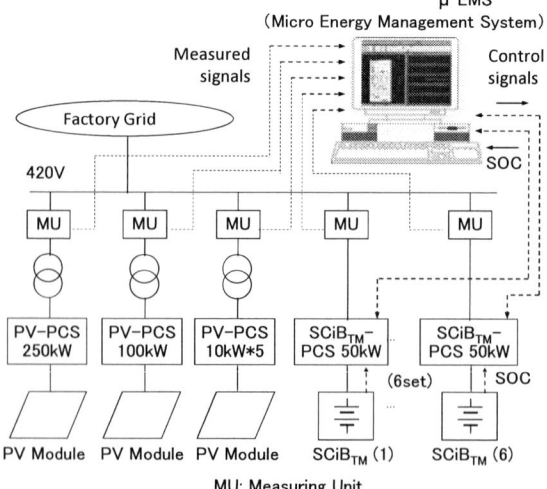

Fig. 8. System configuration.

Test results of PV smoothing are shown in Fig.9. In response to the wide fluctuations of PV outputs due to weather change, BESS outputs are appropriately controlled by μEMS to suppress the fluctuation. In Fig.9, the maximum magnitude of BESS output reaches 300kW which is 5C rate charging for the 60kWh SCiB™.

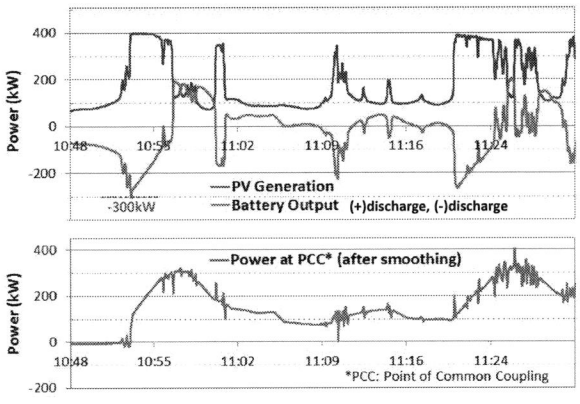

Fig. 9. Test results of PV smoothing (Sept.17, 2012).

B. Coordinated control of BESS and TVR with PV Generation and EV Charger

A smart grid system has been delivered for ACEA SpA. at Raffinerie substation, Rome, Italy. ACEA is Italy's power and water supply public utility.

The system started operation on December 31, 2011, aiming efficient use of PV generation for EV (electric vehicle) and stable operation of distribution network by the advanced voltage management. The overview of the system is shown in Fig.10 where a coordinated control of BESS and TVR are instructed by μEMS. The BESS shown in Fig. 11 is Toshiba's first BESS using 20Ah SCiB™ cells, which are of the current product lineup.

Two control modes are implemented in voltage control. One is "auto tap control" for MV voltage fluctuation suppression, and the other is "manual tap control" which allows operator to adjust the output voltage freely.

Figure 12 shows field test results. Power flow at the interconnection point can be managed with the fluctuation suppression control by BESS, while EV charging is conducted by using PV-generated power.

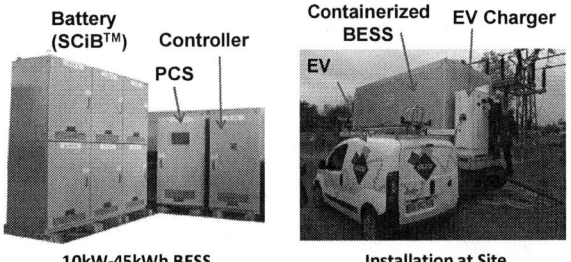

Fig. 11. BESS and its installation at site.

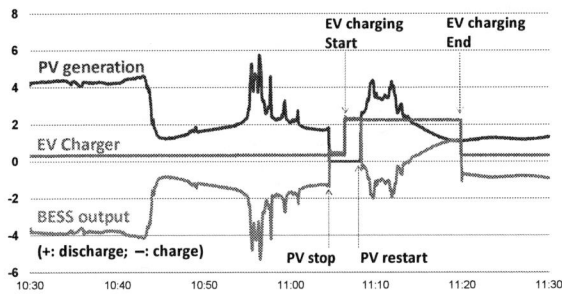

Fig. 12. Test results of fluctuation suppression control by BESS.

C. EV Charging Station Integrated with BESS and PV Generation

Plug-in Ecosystem that integrates EV chargers with BESS and PV generation, a similar concept of the project for ACEA, started operation in February 2013 at Clay Terrace, a major shopping mall in Indiana, US. The project is initiated by Energy System Network (ESN), with the participation of Duke Energy, Simon Property Group, Itochu Corporation, and Toshiba.

The integrated system offer customers access to EV chargers in the mall parking lot, one quick charger and two standard chargers, and makes effective use of renewable energy to provide shoppers with a fast, convenient and environmental service while they visit the mall. As shown in the project overview in Fig. 13, PV energy stored in BESS are used for EV charging as well as providing power to the mall during peak-demand time, without drawing on the distribution grid.

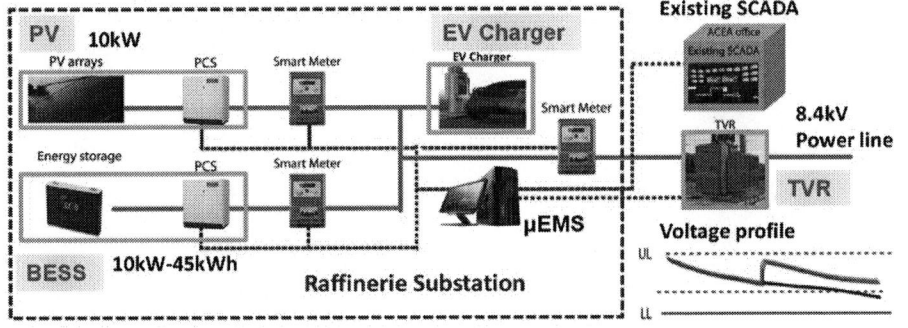

Fig. 10. Overview of ACEA smart grid project.

978-1-4799-2706-7/14 $31.00 © 2014 IEEE

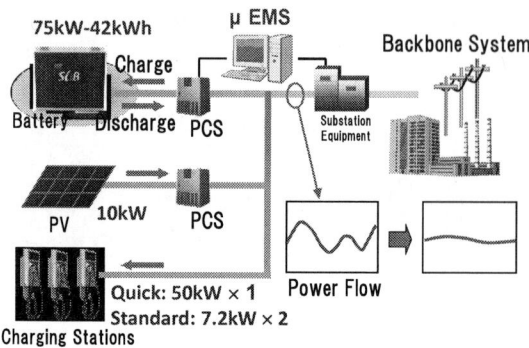

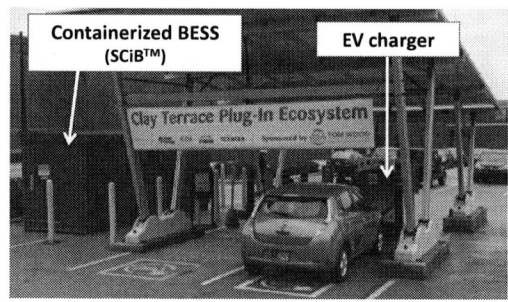

Fig. 13. Overview of Clay Terrace Plug-In Ecosystem Project.

Fig. 14. 300kW-100kWh BESS for the YSCP.

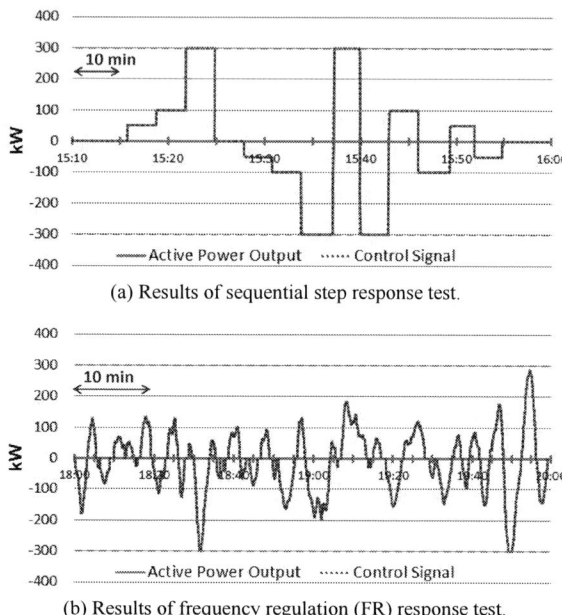

(a) Results of sequential step response test.

(b) Results of frequency regulation (FR) response test.

Fig. 15. Test results of experimental operation (+: discharge; –: charge)

D. Large-scale BESS for Frequency Regulation and Renewable Integration

Toshiba developed and installed a large-scale BESS rated 300kW-100kWh at Battery SCADA Demonstration Center[2][3], Yokohama, Japan, as shown in Fig. 14. The demonstration projects is a part of the Yokohama Smart City Project (YSCP), one of the largest smart city demonstration projects in Japan, and is partly funded by Ministry of Economy, Trade, and Industry (METI) of Japan.

The BESS is high-output type with high C rate, designed in accordance with the basic configuration illustrated in Fig. 4. Therefore it can be considered as a small prototype model of over-megawatt BESS, which is used for the grid support applications such as frequency regulation (FR) and renewable integration. Since it was commissioned in October 2012, various cases of experimental operations have been carried out in order to evaluate performance such as controllability, operability and reliability of the BESS.

Figure 15 demonstrates some of those experimental operation results. From Fig. 15(a), fast and precise responses for both charge and discharge operation are verified within ±300kW of the PCS rated power.

In Fig. 15(b), continuous operation of the BESS in response to FR signals is conducted. Historical regulation signals in the PJM Regulation Market, available on PJM website[4], are used as *dummy* control order for the BESS. Through the FR response test, long-term operability and following accuracy of the BESS have been verified.

VI. CONCLUDING REMARKS

In this paper, actual smart grid projects Toshiba has experienced are introduced from the viewpoint of power electronics applications. Among various power-electronics-based apparatus, battery energy storage systems are chosen and featured to demonstrate how energy storage will play important role in the power grids where large-scale renewable generation is integrated.

Without power electronics technology, battery energy storage systems cannot work; and without battery energy storage systems, smart grids cannot be established. The author is looking forward to further progress in power electronics technology, making the power grids more efficient, more reliable and more ecnviroment-frriendly.

REFERENCES

[1] http://www.scib.jp/en/index.htm
[2] E.Isono, Y.Ebata, T.Isogai, and H.Hayashi, "Development of battery aggregation technology for smart grid," PowerTech (POWERTECH), 2013 IEEE Grenoble.
[3] E.Isono, Y.Ebata, T.Isogai, and H.Hayashi, "Battery SCADA demonstration system in YSCP," Electricity Distribution (CIRED 2013), 22nd International Conference and Exhibition on.
[4] http://www.pjm.com/

The 2014 International Power Electronics Conference

EV and HEV motor development in TOSHIBA

Masanori Arata[1], Yoshihiro Kurihara[2], Daisuke Misu[1]

TOSHIBA Corporation: Power and Industrial System R&D Center[1]/Automotive Systems Division[2]
Power Systems Company[1]/Social Infrastructure Systems Company[2]
Yokohama[1]/Kawasaki[2], JAPAN

Masakatsu Matsubara

Toshiba Industrial Products and Systems Corporation
EV Motor Development Group, EV Div.
Asahi-cho, Mie, JAPAN

Abstract- **Hybrid Vehicles (HEVs) and Electric Vehicles (EVs) improve the efficiency of a drive and reduce CO2 emissions by regenerative braking and operating the engine under optimum conditions. It is desirable to decrease the motor size and increase the maximum torque to minimize the driving units. For these purposes, motors employing permanent magnets have been commonly used recently. For HEVs and EVs motors, Toshiba developed the reluctance torque largely employed new motor named PRM, and has been applying to various vehicles. This paper describes the current motor development status in TOSHIBA.**

Keywords— EV Motor,HEV Motor, Electromagnetic Noise, Memory Motor.

I. INTRODUCTION

Hybrid Vehicles (HEVs) and Electric Vehicles (EVs) improve the efficiency of a drive and reduce CO^2 emissions by regenerative braking and operating the engine under optimum conditions. Toshiba has started making motors, inverters and their control system at early stage in the history of HEVs development.

We started HEVs and EVs motor drive business from an induction motor and its control. In 1980, Toshiba developed one motor HEVs system for a commuter Bus. And Toshiba started to develop a permanent magnet motor for HEVs and EVs in middle of 1990s. Developments in improvement of efficiency and reduction of size have been continuously carried out.

This paper describes outline of Toshiba's HEVs and EVs motor and highlights the unique feature of the permanent magnet reluctance motor(PRM), silent motor technology, current and future high efficient motor developments.

II. EV/HEV MOTOR DEVELOPMENT IN TOSHIBA

The hybrid systems for HEVs and EVs have three main system structures according to its combination the engine with the electric motor.

The series hybrid system as shown in Fig.1 is only motor driven system. Structure of series hybrid system is simple and low cost because of the engine is used only for battery charge.

The parallel hybrid system as shown in Fig.2 is motor and engine driven system. The parallel hybrid system is expected to improve fuel efficiency by utilization of the motor in low speed range and high torque. It is forecasting that HEVs market of the parallel hybrid system will be increased in the future. Because of the system cost and fuel cost will be economical.

The series parallel hybrid system as shown in Fig.3 is combined system the series hybrid system and the parallel hybrid system.

Toshiba had started series production of parallel HEVs drive system for a bus in 1991. Thin motor had requested to install between the engine and the transmission without changing the power-train layout.

On commercial vehicle experiences, we had developed a permanent magnet motor for series parallel hybrid system in 1996. We started its series production since 2004 for passenger vehicle.

Toshiba has over 20 years of experience in series productions several HEVs as shown in Fig.4. And also we released a permanent magnet motor for a HEVs truck in 2006.[2] Our technology covers all types of HEVs and EVs drive systems.

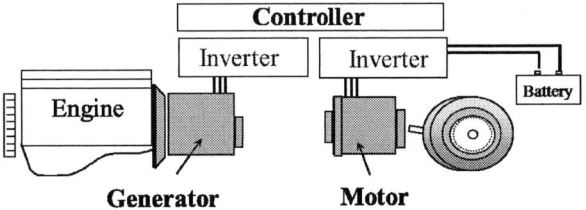

Fig. 1. Series hybrid system

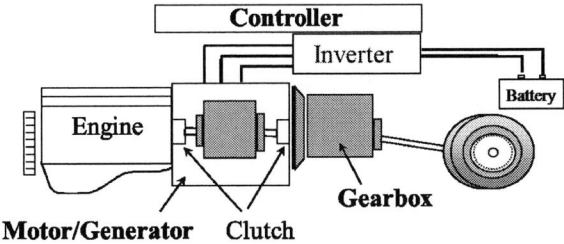

Fig. 2. Parallel hybrid system

978-1-4799-2706-7/14 $31.00 © 2014 IEEE

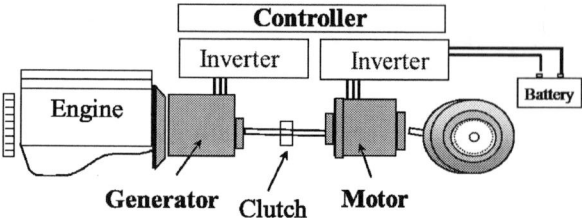

Fig. 3. Series parallel hybrid system

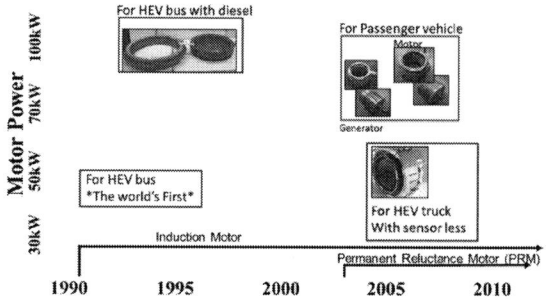

Fig. 4. Toshiba Production of HEV systems

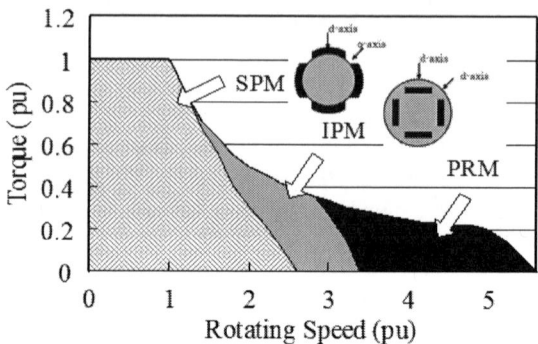

Fig. 5. Performance of Motors at Variable speed.

III. PERMANENT MAGNET RELUCTANCE MOTOR

The motor driving systems of hybrid electric vehicle (HEV) or electric vehicle (EV) must operate at variable speed ranges of up to 1:5. So, it would be desirable to save energy if the total efficiency for large speed range is high. Permanent Magnet (PM) motors using NdFeB magnets having high power and high efficiency are applied to many drive systems these days. But PM motors must control the air-gap flux to keep the voltage constant by the flux weakening for constant-power operation, because the voltage induced by PM increases in proportion to speed.

However, it is difficult for an Interior Permanent magnet Motor (IPM) to operate at a constant-power speed range exceeding 1:3, as shown in Fig.5. And it results in poor efficiency by increase of the flux weakening current in high-speed region. In addition, there is still a possibility for breakdowns of capacitors and/or power devices of an inverter due to excess voltage in case the loss of flux- weakening control. The authors have been developing the PRM to resolve those defects of the IPM by largely employing the reluctance torque by changing the magnet position and magnetic circuit design.[1] Increase of reluctance torque leads to decrease of permanent magnet amounts and smaller back EMF. They allow a large variable speed range over 1:5, smaller flux weakening current and higher efficiency at high speed operating region.

Fig.6 shows a typical cross section of the PRM. Air holes inside a rotor create magnetic flux flow controller increasing the difference d-axis and q-axis reactance.

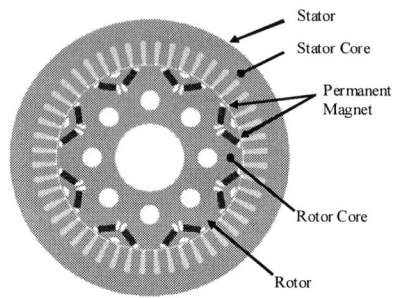

Fig. 6. An example of cross section of the PRM.

IV. SILENT MOTOR TECHNOLOGY

During the motor development for hybrid passenger's car, three type noises took place due to the very high power density of the motor, fairly narrow air gap between the rotor and the stator and different stator support from well-developed industrial motors. Those noise problems are well dissolved at present. The motors designed on the same philosophy are very silent and customs welcomed them. [3]

The first noise phenomenon is the resonance between electromagnetic force and structural natural frequency. The large noise took place when the electromagnetic force space mode and frequency coincide with the structural natural frequency. Through electromagnetic force analysis and noise evaluations, the rotor pole number and stator slot number combinations are changed to avoid the resonance because structural conditions were not changeable due to mechanical support strain. Trough those frequency and mode shifting, noise was lowered significantly. [4][5]

The second noise phenomenon is the sub-harmonics frequency noise due to the electromagnetic force excited by rotor dynamic eccentric motion and circulation currents in parallel circuit. Fig.7 shows the noise measured in the mock up test of rotor eccentric motion. Besides the fundamental harmonics numbers of multiples of four, integral-order noise of rotating speed are distinct as shown like dashed oval A, B and C. And the integral harmonics are outstanding around harmonics numbers of multiples of eight such as 8th, 16th, 24th and 32nd harmonics. Those tendencies are same as the improved motor for the first noise phenomenon.

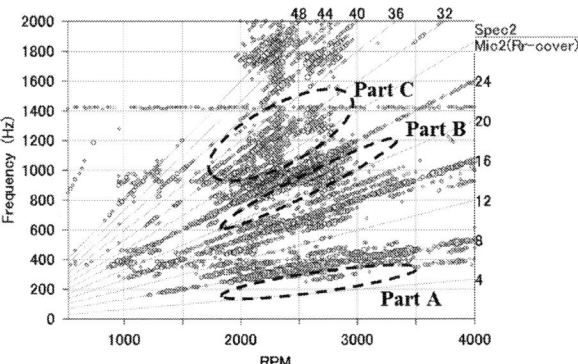

Fig. 7. Noise at the mock up test of 2Y adjacent pole connection with eccentricity

A countermeasure of winding scheme change of the skip-pole connection is presented to avoid the circulation current even where rotor eccentric motion takes place as shown in Fig.8.

Fig.9 shows the Campbell plot of noise with skip pole connection with eccentricity. The sub-harmonics frequency noise almost disappears It is very effective for this phenomenon and the motor noise improves largely.

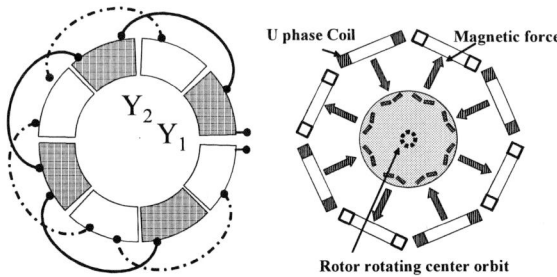

Fig. 8. 2Y-skip pole connection schematic diagram and Electromagnetic force to eccentric rotor motion

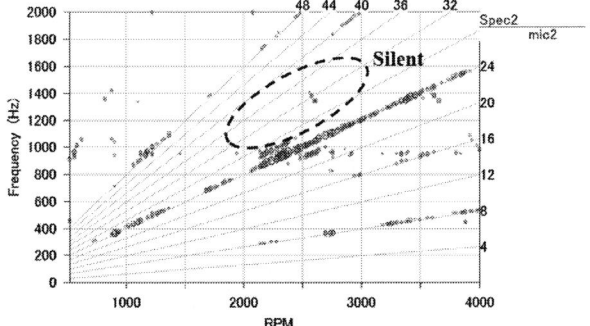

Fig. 9. Noise of 2Y-skip pole connection with eccentricity

The third noise phenomenon is the motor torsional resonance noise due to the resonance of slot ripple torque and a motor torsional natural frequency. The noise took place when the slot ripple torque axial distributions by specified skew scheme coincide with the motor axial natural frequency. This phenomenon can be avoided if the rotor skew phase is changed and optimized. The 4-step V skew as shown schematically in Fig.10 is presented as a countermeasure keeping the torque ripple compensation and the required maximum torque.

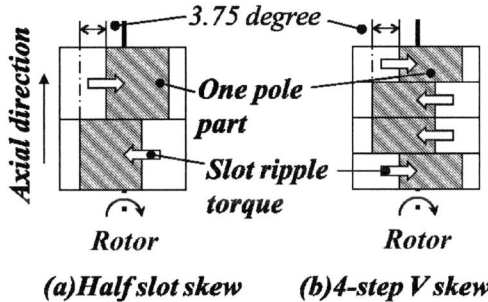

Fig. 10. Schematic draw of a half slot skew and 4-step V skew

Fig.11 shows the FFT analysis result of the 4-step V skew motor's noise in arbitrary unit at 48th order of rotating speed. The noisy part around 3,000rpm to 4,000rpm is suppressed largely as expected. It is very effective to change electromagnetic force axial distribution from axial torsional natural frequency mode. After adoption of a new skew scheme, the motor noises were also largely improved. [6][7]

After applying those technologies and accomplishment of large improvement in noise, the PRM are applied for a hybrid SUV. [8]

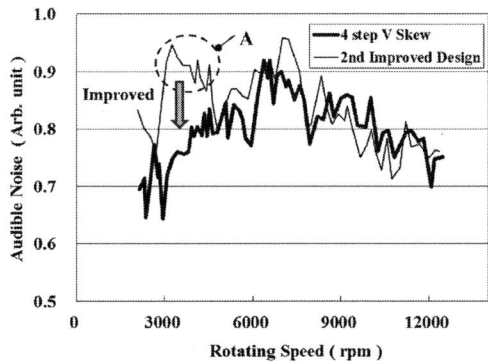

Fig. 11. 48th harmonics Noise of rotating speed of the 4-step V skew motor

V. DESIGN IMPROVEMENT FOR PASSENGER VEHICLE IN SERIES PRODUCTION MOTOR

The series parallel hybrid system is appropriate for a passenger vehicle because it requires high torque and high efficiency in the wide speed range. In the system, motor and engine are combined to operate at each high efficiency speed range. It is preferable for the system to reduce motor loss without major change in dimension.

Design improvement studies were conducted by changing motor design parameters. Table. 1 shows the main specification of the Toshiba's PRM motors for passenger vehicle in series production. The first model is adopted silent motor technologies explained in previous chapter and it had been in mass production since 2004.

Design parameters of first model were reviewed to achieve the motor losses reduction. They were optimized to be better in iron loss, copper loss, motor torque and power. Winding turn number and parallel wire number are selected as design parameters to keep the motor dimension same.

Combinations of three numbers in winding turn and five numbers in parallel wire number were considered and examined. Three winding model shown in TABLE 2 are selected to be studied.

These three Model (1) and (2), (3) are studied first from view point of thermal performance and efficiency without changing the stator slot space. Fig.12 shows design results for maximum torque delivering currents and their current density at maximum torque operation. The currents delivering maximum torque of these three models are smaller than the first model. The current densities of the Model (2) and (3) increase from the first model because diameter of copper wires became smaller with increasing number of turn. The Model (3) is highest in current density because of thin copper wire. It could not satisfy thermal performance specification requires.

The Model (1) and the Model (2) were compared and evaluated from view point of motor losses at operating rotating speed. Fig.13 shows the result of loss calculation for these two models at low speed. The copper loss and the core (iron) loss increase in the Model (2). But the inverter loss decreases in them. The total loss of the Model (2) is a little better than the Model (1). Fig.14 shows the result of loss calculation at high speed. In this high speed, the core loss becomes dominant and the total loss of the Model (1) is better than the Model (2).

It was important and high priority for this passenger vehicle's motor to decrease the power loss at high speed compared to low speed because the specification requires high efficiency in the high speed range. We have finally decided the Model (1) as improved model.

TABLE 1.
Main specification of the PRM for passenger vehicle

		First model	Improvement model
Performance	Max.Torque	210Nm	225Nm
	Max.Power	61kW	92kW
	Voltage	DC220V	DC400V
Size	Pole number	8 pole	8 pole
	Outer diameter	φ236mm	φ236mm
	Total length	160mm	160mm

TABLE2
Winding Model

	Turns	Parallel winding
Model (1)	A	c (a + 3para)
Model (2)	B (A + 1turn)	b (a + 1para)
Model (3)	C (A + 2turns)	a

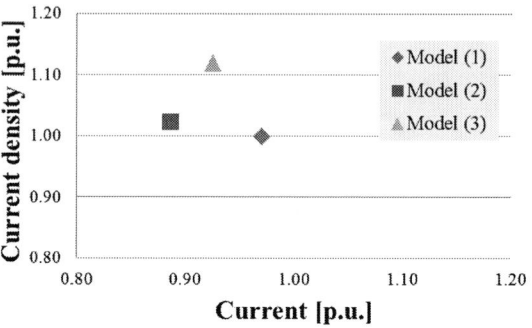

Fig. 12. Motor current density study

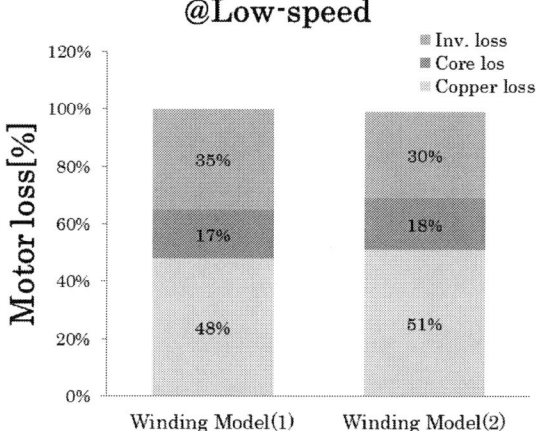

Fig. 13. Loss calculation result at Low speed range

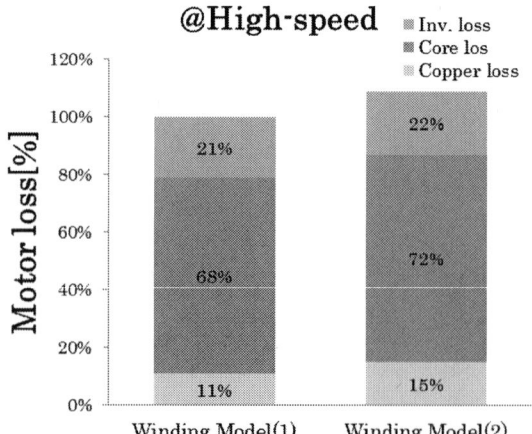

Fig. 14. Loss calculation result at High speed range

VI. NEXT GENERATION MOTOR DEVELOPMENT

The PRM achieves large torque and high efficiency at low speed using reluctance torque. And it also realizes high efficiency at high speed by reducing field weakening current. From the view point of reducing power consumption, it is desirable to enlarge the maximum torque and high efficiency in high speed operation. But it is difficult to achieve both requirements for the motor with constant magneto motive force.

Toshiba has been developing a new technology next to the PRM, which controls the magneto motive force of a permanent magnet depending on the load and speed of the motor. This next generation motor is the variable magneto motive force memory motor. It is a kind of so called "memory motor". Fig.15 shows the rotor of the conventional motor and the memory motor. The conventional motor adopts constant magnetized magnets only. On the other hand, the memory motor adopts two types of magnets, constant magnetized magnet, and variable magnetizing magnet; variable magnet. The variable magnetizing magnet is magnetized by increasing d-axis currents as shown in Fig.16. The d-axis current flow in an armature coil arise a magnetic field in direction to change magnetization of a variable magnets like light black arrows in fig.15. If negative d-axis current is increased, the polarity of the variable magnet will be reversed like dark black arrows in fig.15. In this state, the total combined magnetic flux of the constant magnetized magnet and the variable magnet can be decreased remarkably. Next, if positive d-axis current increases, the polarity of the variable magnetized magnet will be reversed. Then it will be magnetized in the initial magnetizing direction.[9] It can achieve the higher efficiency over a wide speed range than the conventional motor as shown in Fig.17. Because of the memory motor can decrease the magneto motive force of a permanent magnet together with the rotation speed increase. It decreases flux-weakening current resulting in the reduction of copper losses and iron core losses in wide speed range.

Toshiba have already manufactured and tested the principle model. In order to verify that it could change the magnetic flux as a function of rotation speed by using the d-axis current. After confirming the principle of memory motor, the HEVs and EVs motor real size model was constructed to evaluate the effect on iron loss decrease in high speed range by using variable magnetization. It has 8-poles and delivers about 60kW and its maximum rotation speed is 12000rpm.

The result of evaluation is shown in Fig.18 and 19. Fig.18 is relation of d-axis magnetization current in armature winding and induced terminal voltage. The 100% induced voltage means magnet motive force is maximum. Fig.18 shows the magnet motive force range which is obtained by using magnetization current. According to the result of performance test, the magnet

motive force changing range is approximately 12%

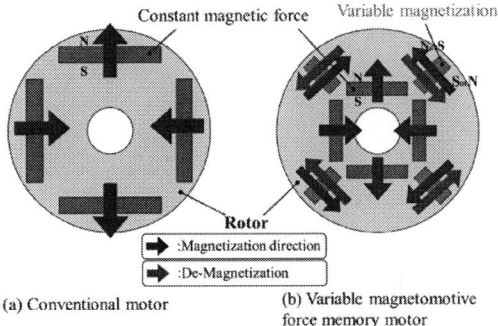

(a) Conventional motor   (b) Variable magnetomotive force memory motor

Fig. 15. Variable magneto motive force memory motor

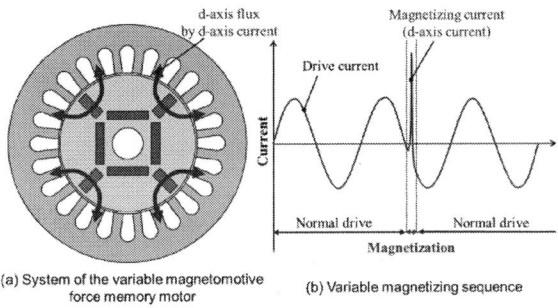

(a) System of the variable magnetomotive force memory motor

(b) Variable magnetizing sequence

Fig. 16. Variable magnetizing by d-axis current

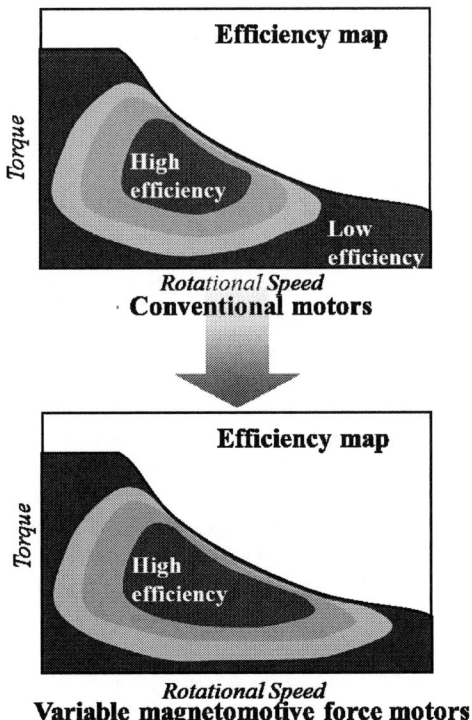

Fig .17. Comparison image with the conventional motor

Fig.19 shows no-load iron loss performance depending on rotation speed. As an effect of magneto motive force change, the memory motor could achieve to decrease no-load iron loss in entire speed range.

Toshiba will continue to confirm the possibility of applying the variable magneto motive force memory motor to apply EVs and HEVs. [10]

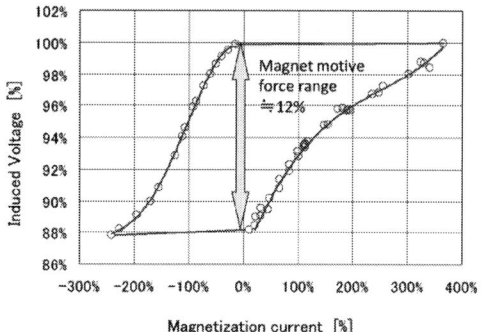

Fig .18. Magnet motive force range

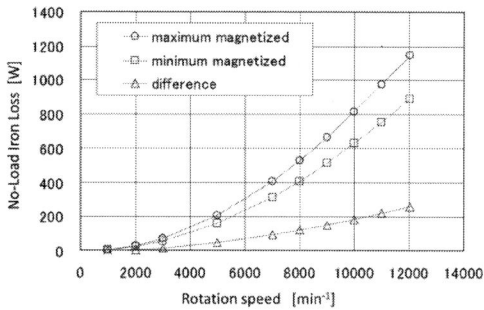

Fig .19. Magnet motive force range

VII. CONCLUSION

The EV/HEV motor development history in Toshiba is described in this paper. Since 1991 for commercial vehicle, and since 2004 for passenger vehicle, Toshiba started the series production for HEVs and EVs motors which are applied parallel and series parallel hybrid systems respectively. To improve efficiency and reduce the size of motors, the PRM was developed and applied to series parallel system through silent motor technology.

For the series parallel system, improvements have been conducted in torque and efficiency in the wide speed range by design parameter study without its major change in dimensions. The result of parameter study for winding design, the improved model had achieved approximately 12 % reduction in motor loss at high speed range compared to the first model.

To lower the power consumption and achieve higher efficiency, Toshiba have developed the variable magneto motive force memory motor and built a prototype for EV.

According to the result of performance test, the magnet motive force range is approximately 12%. And Toshiba have confirmed the memory motor could achieve to decrease no-load iron loss in entire speed range.

REFERENCES

[1] M. Arata and K. Sakai, "Development of a permanent magnet reluctance motor for automobile use," Journal of IEEJ, vol. 128, No.4, pp.231-234, 2008.

[2] M. Arata, N. Takahashi, K. Hagiwara, etal., "Large torque and high efficiency permanent magnet reluctance motor for a hybrid truck," The 22nd International Battery, Hybrid and Fuel Cell Electric Vehicle Symposium and Expo (EVS22), D5, Yokohama (Japan), 2006.

[3] M. Arata, N. Takahashi, M. Fujita, M. Mochizuki, T. Araki, T.Hanai, "Noise Lowering for a Large Variable Speed Range Use Permanent Magnet Motor by Frequence Shift of Electromagnetic Forces Causing Structural Resonance", JPE, vol. 12, No. 1, pp.67-74(2012)

[4] M. Arata, M. Mochizuki, T. Araki, T. Hanai, M. Matsubara, "Decrease of asynchronous rotation-frequence noise and vibration caused by electromagnetic force inside the motor for a hybrid-vehicle", Print ISBN: 978-1-61284-167-0, 14th European Conference on Power Electronics and Applications - EPE 2011, 160, Birmingham(UK)(2011)

[5] M. Arata, M. Mochizuki, T. Araki, T. Hanai, M. Matsubara, "Decrease in Sub-harmonics Frequency Noise Induced by Rotor Eccentric Motion and its Induction Currents between Parallel Stator Circuits", IEEJ Trans. IA, Vol.133, No.10, 2013, PP995

[6] M. Arata, N. Takahashi, M. Mochizuki, T. Araki, T.Hanai, "Torsional Resonance Noise Reduction by Motor Torque Phase Adjustment", The 26th International Electric Vehicle Symposium (EVS26), E4, LA(USA)(2012).

[7] M .Arata, N. Takahashi, M. Mochizuki, T. Araki, T. Hanai, "Permanent Magnet Motor Torsional Resonance Noise Reduction by Rotor Skew Phase Adjustment", IEEJ Trans. IA, Vol.133, No.10, 2013, PP1003

[8] H. Hisada, T. Taniguchi, K. Tsukamoto, et al., "AISIN AW New Full Hybrid Transmission for FWD vehicles", SAE 2005 World Congress, 2005-01-0277, 2005

[9] D. Misu, M. Arata, N. Takahashi, Y. Hashiba, T. Tokumasu, M. Mochizuki "Variable Magneto motive Force Memory Motor for Electric Vehicles" , Society of Automotive Engineers of Japan(2011),2011-39-7257

[10] Y. Hashiba, N. Takahashi, M. Matsushita, D. Misu, K. Yuuki, M. Takabatake "Examination of the no-load iron loss reduction effect by variable-magnet motive-force memory motor", The institute of electrical engineers of Japan(2013), pp.20, No.5, 2013

Motor Stator with Thick Rectangular Wire Lap Winding for HEVs

Takashi Ishigami, Yuichiro Tanaka, Hiroshi Homma
Manufacturing Technology Research Center
Hitachi, Ltd., Yokohama Research Laboratory
Yokohama-shi, Japan
takashi.ishigami.mk@hitachi.com

Abstract— A stator structure with rectangular wire distributed winding exhibiting high productivity and high design flexibility was designed for HEV motors. We devised a method of forming thick rectangular wire diamond coils and determined the best conditions for high density coil forming. A terminal connection structure was also devised to achieve a short coil end. The slot-fill rate of a prototype stator was 80.5% and its total coil end was 45 mm, which is roughly the same motor performance of a stator with segmented coil wave winding. The key advantage here is that the proposed stator has fewer coils and welding points than the stator with segmented coil wave winding and that the possible number of conductors per slot is higher than four, which is the limit of the stator with segmented coil wave winding.

Keywords— *motor stator, rectangular wire, lap winding, hybrid electric vehicle.*

I. INTRODUCTION

Motors for hybrid electric vehicles (HEVs) must be small enough to be mounted on vehicle bodies and must generate a high enough torque at low rotating speed. Rectangular wire coils are currently being used for reducing copper loss and improving the heat dissipation of automotive rotating machinery.

For motors utilizing rectangular winding wire as a stator coil, a stator structure with divided cores and concentrated wound coils is widely used for HEVs. Such a stator is superior in terms of downsizing and productivity, but its performance, especially with cogging torque, noise, and output torque, is inferior to that of a stator with distributed wound coils. A stator structure with distributed wound rectangular wire coils that is sufficiently small and has a high enough productivity must therefore be developed.

The most popular motor stator with rectangular wire distributed windings that has already been mass produced is shown in Fig. 1. First, rectangular wire bars are bent into a hairpin shape and inserted into the slots of the stator in the axial direction. The ends of the hairpin coils are then bent and welded so that the connected coils form a wave shape. These types of stators have been used as alternators and EPS motors [1] [2] [3]. Motors with stators featuring these "segmented coil wave windings" are very small and deliver an excellent performance, but the stator has too many segment conductors and welding points. For example, in the case of 72 slots (with four conductors per slot), there are 147 segment conductors and 288 terminals. Moreover, for the conductors in one slot, six conductors is essentially the limit in mass production. In other words, this stator structure does not have high electrical design flexibility.

Given the background described above, in the present research, we developed a motor stator for HEVs that has few conductor parts and few terminals and that has high electrical design flexibility. We also came up with a basic process for its production.

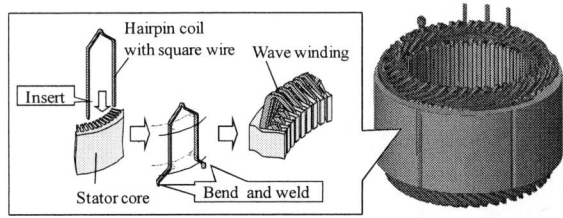

Fig. 1. Motor stator with segmented coil wave winding.

II. MOTOR STATOR WITH RECTANGULAR WIRE LAP WINDING

We devised a stator structure featuring rectangular wire lap winding (Fig. 2) to solve the problems outlined in the previous section. A diamond coil is wound by a continuous single wire, so a stator with this type of coil has fewer conductor parts and coil terminals than a stator with segmented coil wave winding. Furthermore, when the number of slots per pole per phase (NSPP) is 2, the number of connection points can be further reduced to half by continuously winding the neighboring two coils. Also, the numbers of turns of a coil can be changed fairly easily. All of this adds up to a more flexible electrical design.

978-1-4799-2706-7/14 $31.00 © 2014 IEEE

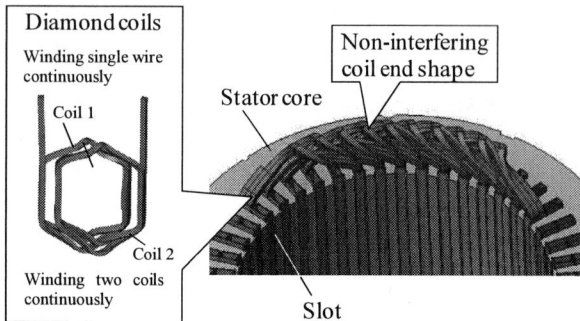

Fig. 2. Motor stator with rectangular wire lap winding.

III. PRODUCTION PROCESS OF STATOR WITH RECTANGULAR WIRE LAP WINDING

Figure 3 shows the basic production process of a stator with rectangular wire lap winding that we devised. First, truck-shaped base coils are wound (Fig. 3 (1)). At this point, a wire with a self-adhesive layer is used as the coil material. Next, an electric current flows through the base coils and the two coils, and four pieces of insulation paper are adhered to each other with the occurring Joule heat (Fig. 3 (2)). After that, the straight parts of the coils that are inserted into the stator slots are gripped and then spread, resulting in diamond-shaped base coils (Fig. 3 (3)). This shape is advantageous because the coil ends of the neighboring coils do not interfere with each other, so they can be small. Finally, the diamond coils are inserted into the slots from the inner part of the stator core (Fig. 3 (4)).

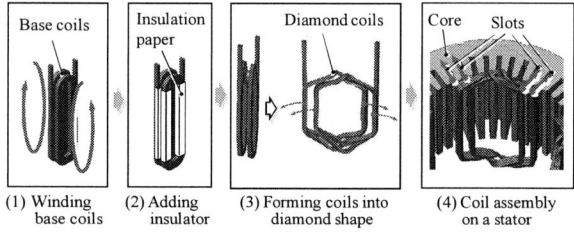

Fig. 3. Motor stator with rectangular wire lap winding.

In this research, we grappled with the following two issues in the manufacturing process of the stator.
(1) High precision forming of diamond coils

In the target stator of this research, we used an electric wire that was quite thick compared with the base coil (coil size: 20 mm × 102 mm; cross-sectional area of conductor: 2.47 mm × 3.32 mm). High-precision forming technology was needed to avoid spring-back and to obtain sufficiently small non-interfering diamond coils.
(2) Small terminal connection structure

A diamond coil has two terminal lines outside of its wound part. If the terminal lines are connected above the coil end, the coil end that includes the terminal line

connections will be higher than that of a stator with segmented coil wave winding (Fig. 4). We therefore designed a small terminal connection structure for the stator with a segmented lap winding. Our target was a coil end height of 45 mm, which is shorter than that of a stator with segmented coil wave winding.

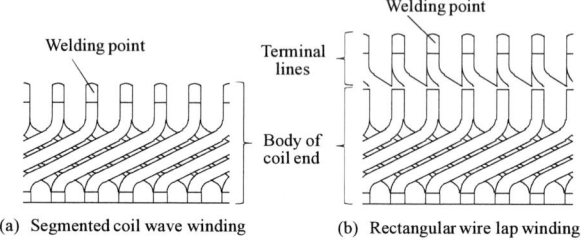

Fig. 4. Problem facing the terminal connection.

IV. PRODUCTION METHOD OF HIGH-PRECISION DIAMOND COIL

To create sufficiently small coil ends that have no interference, we investigated high-precision forming technologies for the thick rectangular wire diamond coils.
(1) Winding of base coils without swellings

The cross section area of the target base coil is large (2.47 mm × 3.32 mm) compared to the coil size (20 mm × 102 mm), so there are some swellings left on the coils if the coils are wound by a conventional method using a nozzle and spindle. Base coils before forming need to be wound with no swellings to ensure high precision, so we adopted a system that pushes an electric wire around a winding form with rollers (Fig. 5 (a)). We were able to obtain base coils without swellings by setting the pressure of the rollers to 1010N and setting their rotation speed to 3 rpm (Fig. 5 (b)).

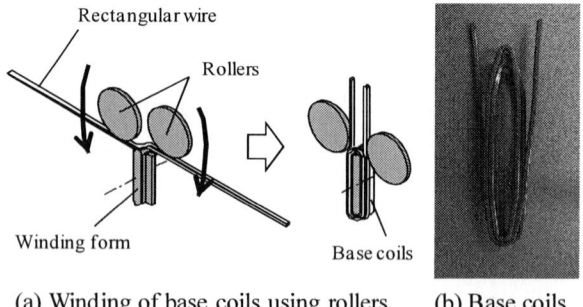

Fig. 5. Winding of base coils without swellings.

(2) Improvement in the accuracy of the coil shape

As shown in Fig. 6, the forming of base coils into a diamond shape is done by one forming in the circumferential direction and another in the radial direction. The resultant shape is a combination of these two forming directions. In order to raise the precision of the diamond coils, the following three conditions were set.

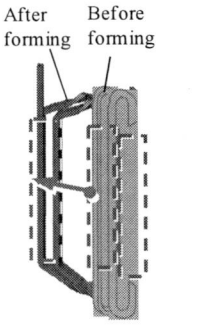 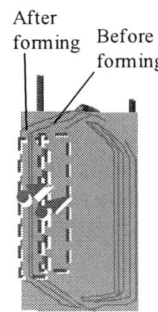

(a) Forming in circumferential direction (b) Forming in radial direction

Fig. 6. Two forming directions for diamond coils.

(2.1) Selection of forming process

There are two forming processes stemming from combinations of forming in the circumference direction and in the radial direction (Fig. 7). In the case of forming process 1 (Fig. 7 (1)), the deviation from the designed shape was 5 mm because two straight parts of the coil were twisted and were not formed enough during the forming in the radial direction. In contrast, in the case of forming process 2 (Fig. 7 (2)), the deviation from the designed shape was less than 1 mm because the coils were sufficiently formed in the two different directions. We therefore selected forming process 2 for the high-precision forming of diamond coils.

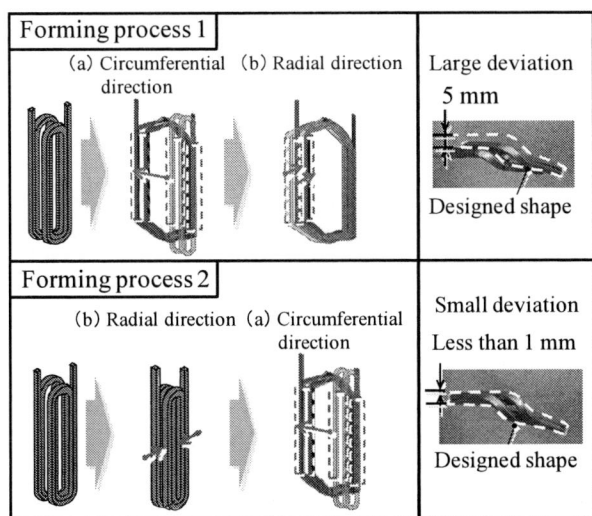

Fig. 7. Two forming processes.

(2.2) Adjustment of Ri size of the base coil

If the diamond coil's radius of the bend section "Rt" is large, the coil ends swell in the radial direction and adjacent coils interfere with each other. We investigated the relation between Ri and Rt and clarified that Rt can be reduced by reducing Ri. When Ri is 2 mm, we can avoid the interference of adjacent coils (Fig. 8).

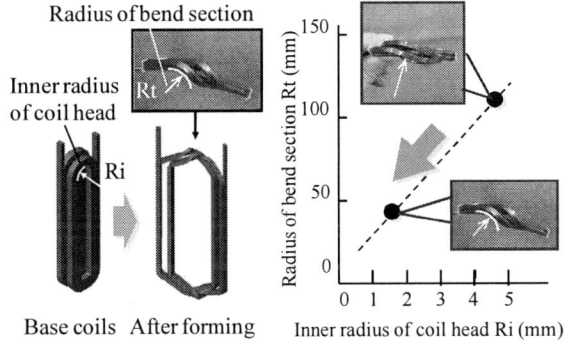

Fig. 8. Relation between Ri and Rt.

(2.3) Forming coil ends in the radial direction

When the base coils are spread, the coil ends are formed in a straight shape on the shortest course. We therefore formed the coil ends in the radial direction to avoid interference with adjacent coils by forming dies (Fig. 9).

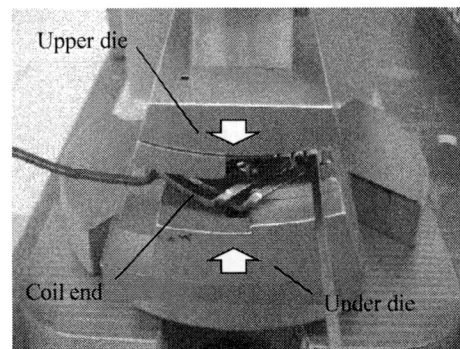

Fig. 9. Forming of coil ends in radial direction.

V.   STRUCTURE OF TERMINAL CONNECTION

We devised the terminal connection structure shown in Fig. 10 to create a coil end height smaller than that of segmented coil wave winding. The inner terminal lines are bent at a right angle and drawn radially over the body of the coil end. The outer terminal lines are bent along the outside of the body of each coil end in a spiral arrangement. An electrode is then introduced to the tops of the two terminal lines from outside the stator and welded by TIG welding. We can achieve a small coil end by connecting the terminal lines on a right-angled surface of the motor axis.

978-1-4799-2706-7/14 $31.00 © 2014 IEEE   1882

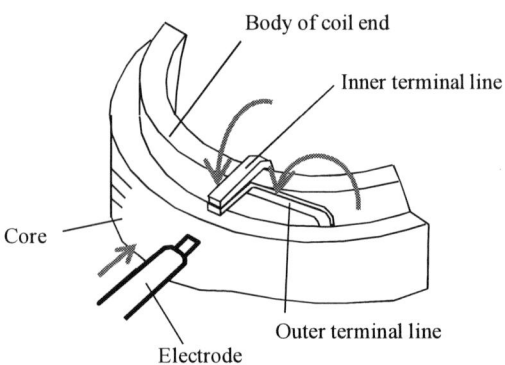

Fig. 10. Proposed small terminal connection structure.

VI. PROTOTYPE MOTOR STATOR

We fabricated a prototype stator using the precision coil forming technologies described in section 3 and installed the small terminal connection structure discussed in section 5. Figure 11 shows the diamond coils after forming and Fig. 12 shows the prototype stator assembled using diamond coils. Using the developed coil forming technology enabled us to obtain diamond coils that had no interference with adjacent coils. At this point, the length of the body of the coil ends was 42 mm. Moreover, as a result of adopting the devised terminal connection structure, the total height of the coil ends, including terminal connections, became 45 mm (terminal connection side: 24 mm, opposite side: 21 mm). This is actually under our target coil end height.

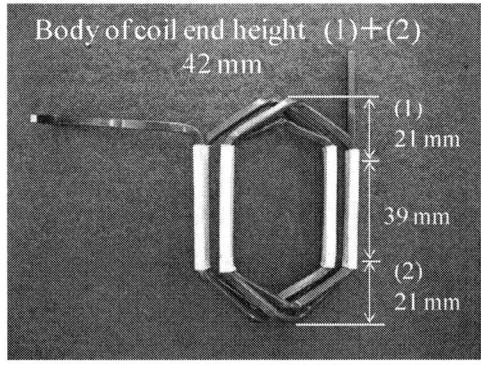

Fig. 11. Prototype diamond coils.

(1) Evaluation of stator structure

Table I shows a comparison of a stator with segmented coil wave winding and the prototype stator with rectangular lap winding. In this comparison, each core was the same size, had a conductor cross section that was the same size, and had the same number of conductors per slot. The stator with segmented coil wave windings had 9 kinds of conductor bars for a total of 147 conductor bars and 288 terminal lines. The prototype stator with

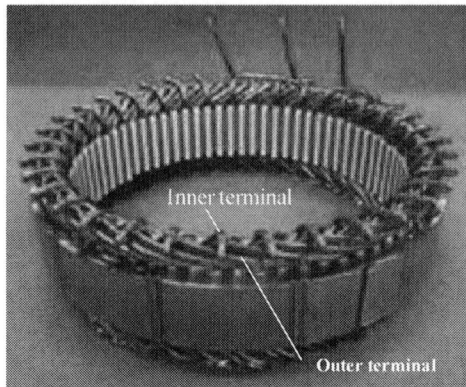

Terminal connection side

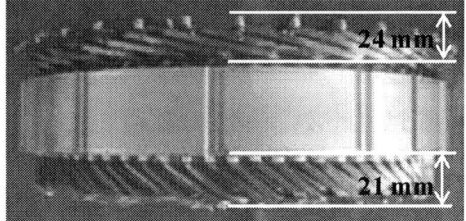

Opposite side of terminal connection

Fig. 12. The prototype stator.

rectangular wire lap winding had only one kind of coil for a total of just 36 coils and only 72 terminal lines.

TABLE I
COMPARISON OF RECTANGULAR WIRE LAP WINDING AND
SEGMENTED COIL WAVE WINDING

Winding type	Rectangular wire lap winding	Segmented coil wave winding
Type of coils	1	9
Total number of coils	36	147
Number of terminals	72	288
Possible number of conductors per slot	2n ($\geqq$4)	2,4
Slot fill rate	80.5%	80.5%

The stator with segmented coil wave winding needs two kinds of basic hair-pin conductors, four kinds of half-conductors that are used for terminal lines or neutral lines, and three kinds of jumper conductors that connect between the inner and outer layers of wave coils (Fig. 13).

In contrast, the stator with rectangular wire lap winding does not need so many kinds of coils because its electrical circuitry is made by forming and connecting the terminal lines. We therefore predicted that the stator with the rectangular wire lap winding would have superior productivity, and indeed, results showed that it had better design flexibility than the stator with segmented coil wave windings (whose possible number of conductors per stator slot was two or four).

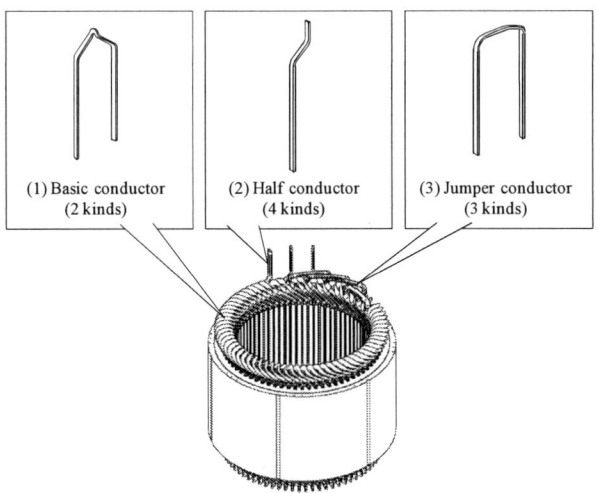

Fig. 13. Segmented conductors for wave winding.

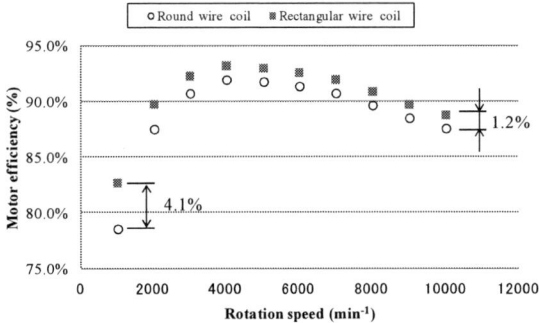

Fig. 14. Torque-characteristic curve.

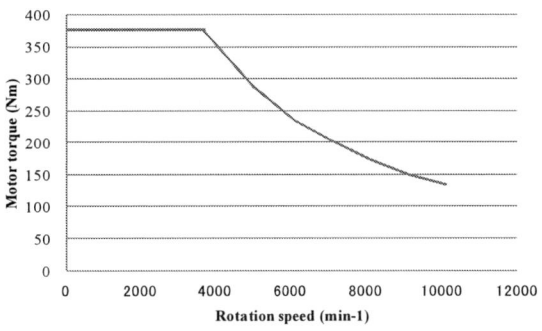

(2) Evaluation of motor performance

The coil end height of the prototype rectangular wire lap winding stator was 45 mm, which is shorter than that of the segmented coil wave winding stator (46 mm). Both stators had the same slot-fill rate and so the motor of each had nearly the same performance. The efficiency of the prototype motor (with rectangular wire lap windings) was calculated based on targeted torque-characteristic curve shown in Fig. 14. Efficiency El of the motor at each rated rotation speed is indicated by the "□" marks in Fig. 15.

The slot-fill rates for each type of coil lead us to assume that the cross-sectional area of the round-wire distributed winding coils is 0.79 times that of the rectangular-wire lap winding coils. The coil lengths for each type of coil are also assumed to be nearly the same. The copper loss of the motor with round-wire distributed windings (Wc) was thus calculated by using the measured copper loss of the prototype motor with rectangular-wire lap windings (Wl) by

$$Wc = Wl \diagup 0.79, \qquad (1)$$

and efficiency E_C of the motor with round-wire distributed windings at various rotating speeds ("×" marks in Fig. 15) was calculated by

$$E_c = (Po - (Wc - Wl)) \diagup Pi, \qquad (2)$$

where Po is the input energy and Pi is the output energy.

The copper loss of the stator with rectangular-wire lap windings was estimated to be around 20% smaller than that of the stator with round-wire concentric windings. It follows that the efficiency of the rectangular-wire lap-winding motor was 4.1% higher at low rotation speed (1000 min⁻¹) and 1.2% higher at high rotation speed (10000 min⁻¹) than that of the round-wire concentric-winding motor.

VII. FUTURE WORK

One problem remains for the mass production of motor stators with rectangular wire lap winding: realizing the automatic assembly of diamond coils. The process of assembling a stator with rectangular wire lap winding by hand is shown in Fig. 16. In the stator with lap winding, all the diamond coils are arranged annularly and adjacent coils overlap each other. Therefore, in the early stages of assembly, diamond coils can be un-interferingly inserted into the stator slots while the insertion points are shifted by one slot (as shown in Fig. 16 ((state 1) to (state 3)). However, after the state shown in Fig. 16 (state 3), no more coils can be inserted because some parts of the coils are already inserted inside the stator slots. These already inserted coil parts have to be pulled out (Fig. 16 (state 4)), and the straight parts of un-inserted coils that should be located outside the stator slots must be passed below the pulled-out coil parts and inserted into the slots.

It is extremely difficult to perform these actions with an automatic machine, and so we devised an automatable production process, as shown in Fig. 17. In this process, first, all coils are arranged annularly in the stator core (Figs. 17 (1)), and next, all the outer parts of these coils are inserted radially into the stator slots at the same time (Fig. 17 (2)). In the final step, all the inner parts of these coils left inside the stator are inserted radially into the stator slots at the same time (Fig. 17 (3)). The assembly of diamond coils can be achieved automatically in this

The 2014 International Power Electronics Conference

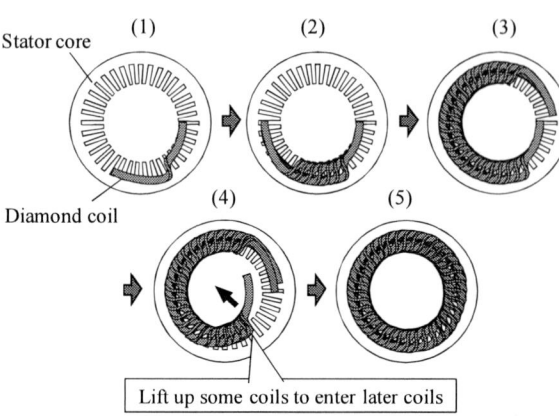

Fig. 16. Assembly of diamond coils in stator core by hand.

way because there is no complicated insertion of coils. We are planning to develop an experimental coil assembling machine to verify the effectiveness of this production process. Also, for thin wire diamond coils, we have developed a machine that forms all the coils into a diamond shape inside the stator core and inserts them into the slots at the same time [4].

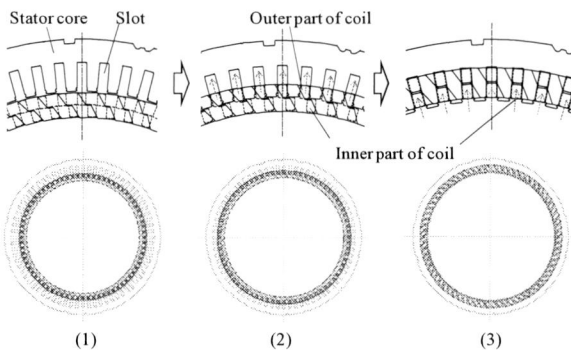

Fig. 17. Concept of automatic assembly of diamond coils.

VIII. CONCLUSION

We designed a stator structure with rectangular wire distributed winding featuring high productivity and high design flexibility for HEV motors and clarified its manufacturing technologies.

(1) We designed a motor stator structure with rectangular wire lap winding for HEV motors.

(2) We obtained base coils with no swellings by using rollers that pressurize coils during winding. We were also able to obtain non-interfering diamond coils by specifying a forming process, adjusting the Ri size of the base coil, and forming the coil end in the radial direction.

(3) We designed a small coil end structure for rectangular wire lap winding. The coil end height of the trial stator, 45 mm, was smaller than that of a stator with segmented

coil wave winding.

(4) The trial stator with rectangular wire lap winding had 36 coils and 72 terminal lines—in other words, fewer conductor parts and terminal lines than that of the segmented coil wave winding stator. Moreover, the possible number of conductors per slot was even higher than 4.

(5) Our future work for the mass production of stators with rectangular wire lap winding will focus on automatic coil assembly to the stator core.

IX. REFERENCES

[1] A. Umeda, "A Technology to Improve the Space Factor of Alternator Stator and its Applications," *2006 Motor Technology Symposium C-1*, 2006 (in Japanese).

[2] M. Meaki, "Development of Rectangular Wires for Automotive Electric Equipment," *2006 Motor Technology Symposium C-1*, 2006 (in Japanese).

[3] M. Ohashi and T. Takahashi, "2-Drive Motor Control Unit for Electric Power Steeling," *2013 JSAE Annual Congress (Autumn) on 10.25.2013*.

[4] T. Ishigami, Y. Tanaka, and H. Homma, "Development of Motor Stator with Rectangular Wire Lap Winding," *T.IEE Japan*, vol. 132, no. 10, pp. 976–982, 2012 (in Japanese).

The 2014 International Power Electronics Conference

Comparison Study of Various Motors for EVs and the Potentiality of a Ferrite Magnet Motor

Daiki Matsuhashi, Keisuke Matsuo, Takashi Okitsu, Tadashi Ashikaga, Takayuki Mizuno

System Technology Research Department, Research & Development Group

MEIDENSHA CORPORATION

2-1-1 ohsaki shinagawa-ku Tokyo 141-6029 Japan

matsuhashi-d@mb.meidensha.co.jp

Abstract—**This paper presents a comparison study of various motors for application to an electric vehicle (EV) and the potentiality of a ferrite magnet motor as a candidate for rare-earth-less motor for an EV. The rotor shape of the proposed ferrite magnet motor is based on a spoke-type shape. Further, we developed two types of prototypes: one has a large magnetic torque equivalent to a conventional rare-earth permanent magnet motor for an equal volume, and another has a wide variable-speed range suitable for an EV.**

Keywords— *electric vehicle, ferrite magnet, rare-earth-less motor, spoke-type rotor*

I. INTRODUCTION

Recently, the development and popularization of electric vehicles (EVs) and hybrid EVs (HEVs) have accelerated owing to energy and environmental problems. We are furthering the research and development of several motors for EVs/HEVs from the 1990s [1]. In July 2009, Mitsubishi Motors Corporation started regular distribution of their electric vehicles named "i-MiEV" for the first time worldwide. In this vehicle, interior permanent magnet synchronous motors (IPMSMs) supplied by Meidensha Corporation are adopted as traction motors in which rare-earth (Nd–Fe–B) magnets are applied (Fig. 1).

Currently, an IPMSM has mainly been adopted as a traction motor for EVs/HEVs, and it is obvious that rare-earth magnets have greatly contributed to the development of EVs/HEVs. However, there are some disadvantages for rare-earth magnets, such as scarcity, concern about a stable supply, and price fluctuation of the raw materials, namely Nd and Dy. Fig. 2 shows the price transitions of Nd and Dy over the past several years. As shown in Fig. 2, these prices have fluctuated sharply in recent years, and the cost of permanent magnets accounted for almost half of an IPMSM when these prices were the highest. Therefore, it is important to develop alternative motors that are independent of rare-earth magnets.

II. COMPARISON OF VARIOUS MOTORS

As alternative technologies for rare-earth permanent magnet motors, rare-earth-less or -free permanent magnet motors have developed rapidly [2]–[4]. Table I summarizes a comparison of various electric motors, and the typical features of each motor are as follows:

(i) IPMSM using a rare-earth permanent magnet (R-IPMSM)

An R-IPMSM has outstanding performance such as high efficiency, high power density, and drivability over a wide variable-speed range. Therefore, it is the most suitable for the specifications of an EV.

(ii) Induction motor (IM)

An IM is a motor that has been developed and utilized for many years. An IM has outstanding performance such as a wide variable-speed range and low drag loss. Further, an IM also has some advantages such as productivity and the toughness of the structure. On the other hand, the torque density and efficiency over the entire operating region are inferior to a PMSM.

Fig. 1. IPMSM for i-MiEV.

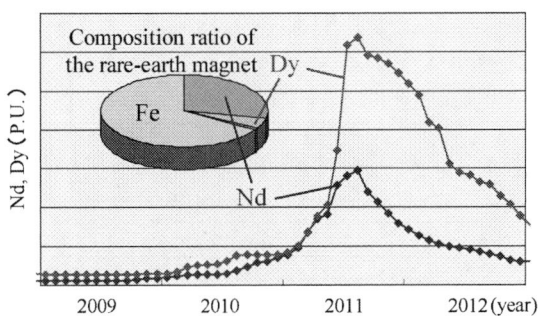

Fig. 2. Price transitions of Nd and Dy.

978-1-4799-2706-7/14 $31.00 © 2014 IEEE

1886

TABLE I COMPARISON OF VARIOUS ELECTRIC MOTORS

motor type	maximum efficiency	variable speed range	power/weight ratio	cost
(ⅰ) R-IPMSM	◎	○	◎	△
(ⅱ) IM	△	◎	△	○
(ⅲ) HSY	○	○	○	△
(ⅳ) SynRM	△	△	×	○
(ⅴ) F-IPMSM	○	○	○	◎

(ⅲ)   Hybrid-excitation-type   permanent   magnet synchronous motor (HSY) [5]

An HSY equips an excitation coil separately from the armature coil in order to control a magnetic flux. An HSY could gain high efficiency over a wide variable-speed range by properly controlling a magnetic flux in each operating region. However, the volume and cost of the entire motor are disadvantages.

(ⅳ) Synchronous reluctance motor (SynRM)

An SynRM has the advantage of low cost because it does not require a permanent magnet or an extra magnetization power supply. However, the decreasing output power in the high-speed region due to the low power factor and low torque density are major disadvantages. As a result, the cost would increase because an increase in the physical size of the motor or power supply capacity is necessary for obtaining the same output power as a PMSM.

(ⅴ) IPMSM using a ferrite magnet (F-IPMSM)

The maximum torque performance is inferior to an R-IPMSM according to the difference in the residual flux density of the magnet. However, an F-IPMSM has advantages in the cost of the materials and its efficiency in the high-speed and low-load region because it requires less field-weakening current.

In this way, there are advantages and disadvantages for each motor. At present, the mainstream traction motor for an EV/HEV is an IPMSM using a rare-earth magnet, but other types of motors may be applicable depending on the requirement specifications.

III.   STUDY OF A FERRITE MAGNET MOTOR

The ability to drive over a wide variable-speed range is a required characteristic for the traction motor for an EV. Regarding the rotational speed-torque characteristics, the output characteristics under the limitation of a terminal voltage and an armature current are largely dependent on the motor parameter expressed as follows [6]:

$$\Psi_{d\min} = \Psi_a - L_d I_{d\max} \qquad (1)$$

where Ψ_a is the flux linkage due to the permanent magnet, and $L_d I_{dmax}$ is the maximum d-axis armature reaction flux linkage.

Fig. 3 shows the relationship between the motor characteristics and Ψ_{dmin}. In the case where Ψ_a is large enough, i.e., the magnetic torque is sufficiently secured, the motor generates a large maximum torque. However, the induced voltage increases owing to the increasing rotational speed. Further, field-weakening control becomes difficult, and high-speed drive becomes impossible. On the other hand, the maximum torque becomes smaller, and the output power in the high-speed region decreases in the case where there is poor Ψ_a, i.e., the reluctance torque is mainly used. When the ratio of Ψ_a and $L_d I_{dmax}$ is the same, the widest constant output power range could be achieved.

We manufactured two types of prototype F-IPMSMs; one is designed to create a large Ψ_a, and the other is designed to let Ψ_{dmin} be approximately zero.

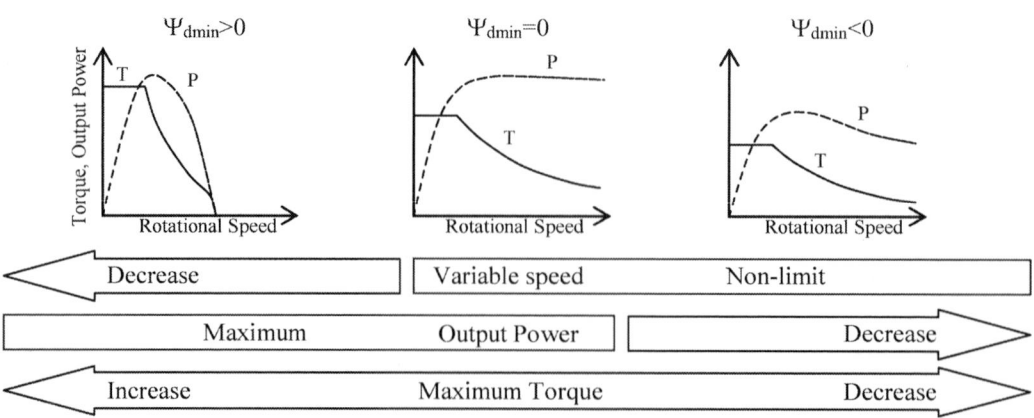

Fig. 3. Relationship between the motor characteristics and Ψ_{dmin}.

The 2014 International Power Electronics Conference

TABLE II SPECIFICATIONS OF THE 1ST F-IPMSM

Number of poles	8
Number of slots	36
Outer diameter of stator [mm]	ϕ 175
Stack length [mm]	50
Rated speed [min⁻¹]	1,500
Residual flux density [T]	1.30 (Nd-Fe-B) 0.45 (Ferrite)

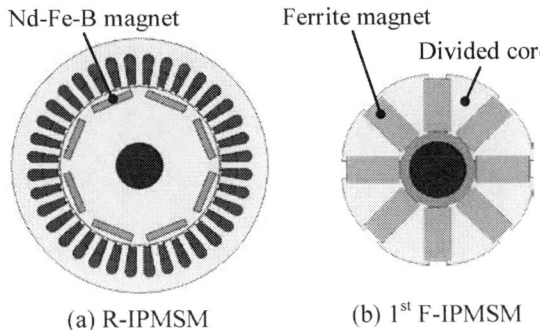

(a) R-IPMSM   (b) 1ST F-IPMSM

Fig. 4. Stator and rotor structure of the first prototype.

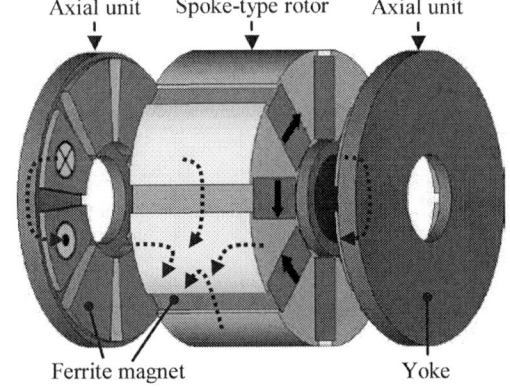

Fig. 5. General view of the rotor of the 1ST F-IPMSM.

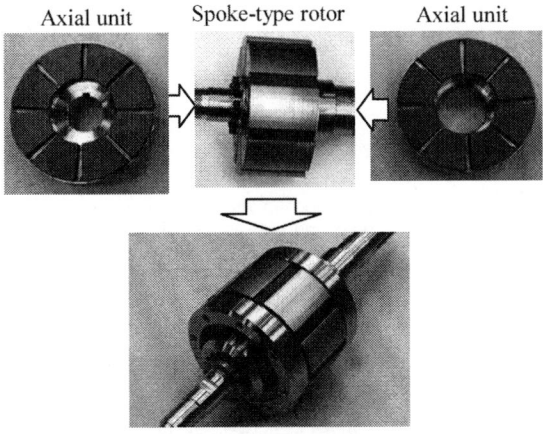

Fig. 6. Photographs of the 1ST prototype F-IPMSM rotor.

A. First Prototype [7]

First, we focused on generating an equivalent torque with a conventional R-IPMSM under equal design restrictions and specifications when designing the first prototype (1ST F-IPMSM). A common stator was applied to each IPMSM. Table II lists the specifications of the 1ST F-IPMSM. The residual flux density of the ferrite magnet is approximately one-third of the rare-earth magnet. Briefly, it is necessary to increase its surface area by about three times that of a rare-earth magnet to generate the same magnetic flux density in the air gap, as in an R-IPMSM. However, it is difficult to achieve this for an equal volume. Therefore, the spoke-type rotor in Fig. 4 was selected to obtain a surface area for the ferrite magnet that is as wide as possible, and a divided rotor core was adopted to decrease the leakage of magnetic flux. In addition, axial units that consist of magnets magnetized in the axial direction and yokes are located at each end of the divided rotor core. Fig. 5 shows a general view of the rotor of the 1ST F-IPMSM. Here, the axial length of the axial units is less than that of the armature coil end. The flux path indicated by the dotted lines in Fig. 5 flows in the axial units, and it works to increase the

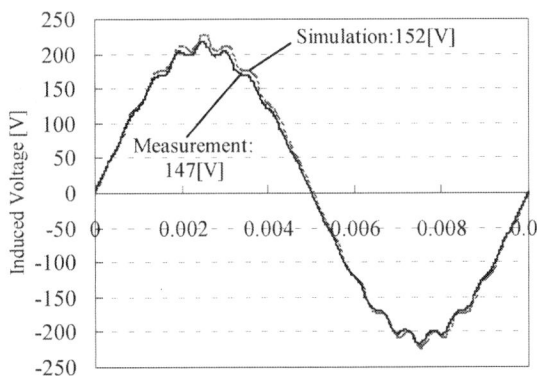

Fig. 7. Comparison of the induced voltage waveforms (simulated and measured results).

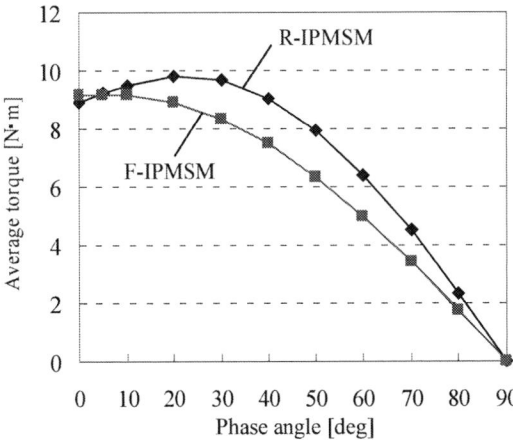

Fig. 8. Relationship between the average torque and the phase angle at the rated current (measured results).

magnetic flux density of the magnetic poles. Fig. 6 shows photographs of the 1st F-IPMSM prototype. The axial units serve as the end plates that press down the spoke magnets and also function as the components that provide balance correction; they were designed so that the rotor might not become large.

Fig. 7 shows a comparison of the induced voltage waveforms, and Fig. 8 shows the measured results of the relationship between the average torque and the phase angle at the rated current. As shown in Fig. 7, the expected value is mostly acquired, even though the measured induced voltage is approximately 3% lower than simulated. As shown in Fig. 8, the F-IPMSM generates almost the same torque as the R-IPMSM. However, this prototype has some weaknesses, i.e., the productivity, the weakness to centrifugal force, and poor reluctance torque generation.

B. Second Prototype [8]

In order to improve the weaknesses of the 1st F-IPMSM, we endeavored to use the reluctance torque effectively and a non-divided rotor core, considering the productivity and driving in the high-speed region when designing the second prototype motor (2nd F-IPMSM). Table III lists the specifications of the 2nd F-IPMSM, and Fig. 9 shows the stator and rotor structure of the 2nd F-IPMSM and conventional R-IPMSM. Here, we applied the same stator to each IPMSM as in designing 1st F-IPMSM. The rotor core of the 2nd F-IPMSM between the

magnetic poles is divided to reduce the leakage of magnetic flux, but the core by the side of the shaft is not divided. Ferrite magnet B is placed to reduce the leakage of magnetic flux, and ferrite magnet C is located at the center of the magnetic pole to lower the d-axis inductance and cause Ψ_{dmin} to be approximately zero.

Fig. 10 shows the analysis results of the relationship between the average torque and the phase angle at the same current. By properly placing ferrite magnet C, the torque of the F-IPMSM increases by approximately 14%

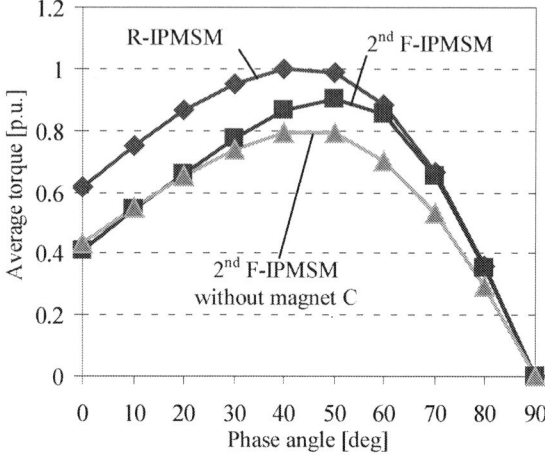

Fig. 10. Relationship between the average torque and the phase angle at the rated current (analysis results).

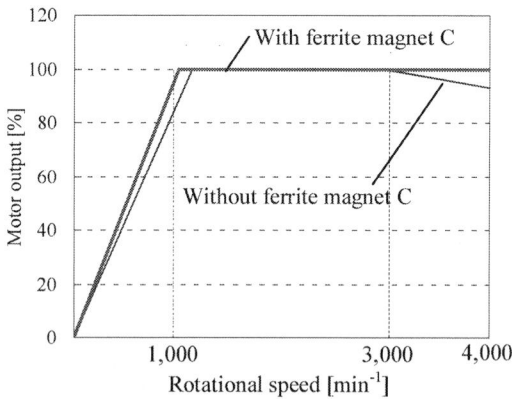

Fig. 11. Comparison of the constant output power range (analysis results).

TABLE III SPECIFICATIONS OF THE 2ND F-IPMSM

Number of poles	6
Number of slots	36
Outer diameter of stator [mm]	ϕ220
Stack length [mm]	108
Base speed [min^{-1}]	1,000
Maximum speed (electrical) [min^{-1}]	4,000
Maximum speed (mechanical) [min^{-1}]	7,200
Residual flux density [T]	1.30 (Nd-Fe-B) 0.45 (Ferrite)

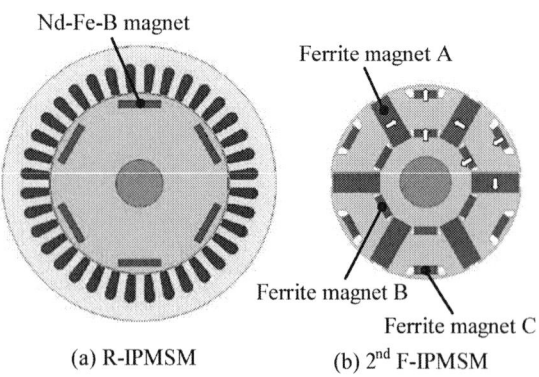

(a) R-IPMSM   (b) 2nd F-IPMSM

Fig. 9. Stator and rotor structure of the second prototype.

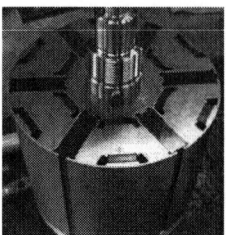

Fig. 12. Photograph of the 2nd prototype F-IPMSM rotor.

978-1-4799-2706-7/14 $31.00 © 2014 IEEE

The 2014 International Power Electronics Conference

compared with the F-IPMSM without ferrite magnet C, and this torque is equal to 90% of that of an R-IPMSM. Fig. 11 shows a comparison of the constant output power range. As shown in Fig. 11, the 2nd F-IPMSM has a wider constant output power range with the addition of ferrite magnet C.

Fig. 12 shows a photograph of the 2nd prototype F-IPMSM rotor. The bridges of the rotor core located among ferrite magnets A and B were designed to be unbreakable at a centrifugal force that is 180% of the electrical maximum speed. When the thickness of this bridge is increased, the intensity will increase; however, the leakage flux also simultaneously increases, causing a reduction in torque.

Fig. 13 shows a comparison of the measured current values. The current value of the F-IPMSM at 1,000 min^{-1} is approximately 14% larger than that of the R-IPMSM to obtain the same torque, but it is approximately 42% lower at 4,000 min^{-1}. This is because of the decrease in current during field-weakening operation.

Fig. 14 shows the measured results for the efficiency. The maximum efficiency of the F-IPMSM is lower than that of the R-IPMSM, especially in the lower-speed and large-torque region. However, the efficiency of the F-IPMSM is higher than that of the R-IPMSM in high-speed region because the field-weakening current is small.

We considered the irreversible demagnetization of the ferrite magnets because the coercivity of the ferrite magnet is low. Fig. 15 shows the simulation results of the demagnetizing field distribution. Under the conditions of maximum current and maximum field-weakening current phase angle, a large demagnetizing field is observed on the stator side of ferrite magnet A. We evaluated the influence of this demagnetizing field by the decreasing

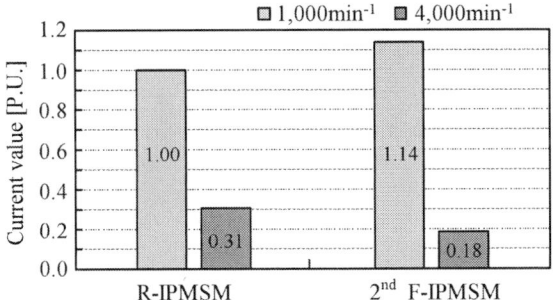

Fig. 13. Comparison of the measured current value.

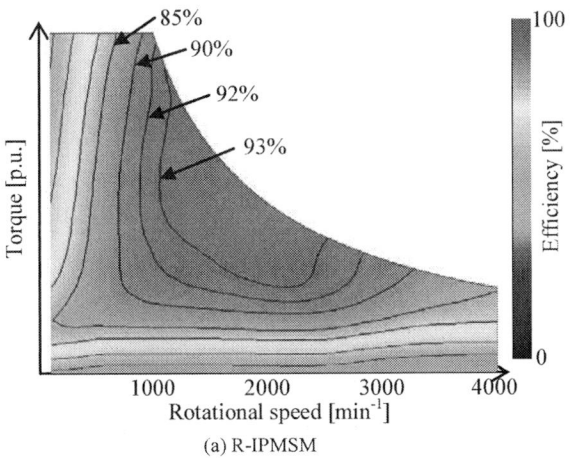

(a) R-IPMSM

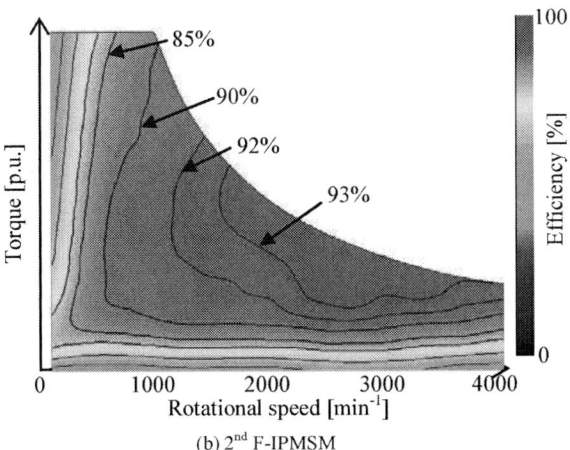

(b) 2nd F-IPMSM

Fig. 14. Efficiency map of the proposed motor (measured results).

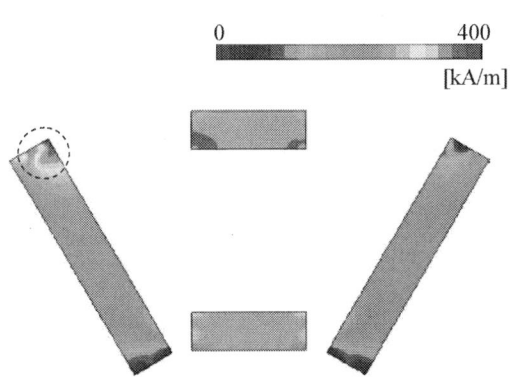

Fig. 15. Demagnetizing field distribution (analysis results).

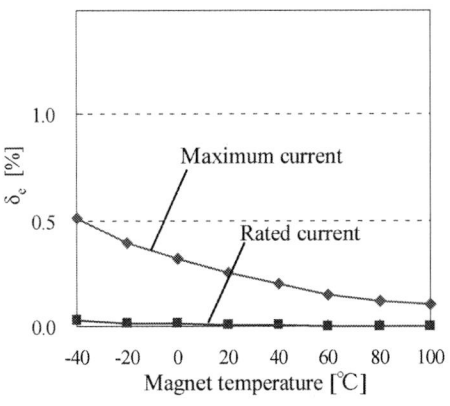

Fig. 16. Decreasing rate of induced voltage (analysis results).

978-1-4799-2706-7/14 $31.00 © 2014 IEEE

rate of the induced voltage given by

$$\delta_e = \frac{E_a - E_b}{E_a} \times 100 \, [\%] \qquad (2)$$

where E_a is the fundamental effective value of a no-load induced voltage before demagnetizing, and E_b is the fundamental effective value of a no-load induced voltage after demagnetizing. Fig. 16 shows that the decreasing rate of induced voltage depends on the magnet temperature. When the 2nd F-IPMSM is driven by the rated current, a decrease in the induced voltage is not observed. However, a decrease in the induced voltage of approximately 0.5% would be caused if the maximum current is added under the condition of the largest phase angle. This is not a large demagnetization that causes a problem, and the influence on the characteristics is small; however, this is one issue that should be solved.

IV. CONCLUSION

We have developed an IPMSM using ferrite magnets as a candidate for a rare-earth-less motor. In the first prototype, ferrite magnets magnetized in the axial direction were located on each side of the spoke-type rotor so that it could generate a torque that is approximately equivalent to a conventional R-IPMSM. Moreover, we manufactured a second prototype that aimed to increase the torque by increasing the reluctance torque and drive with a constant output power over a wide speed range. Even though the residual flux density of the ferrite magnet is approximately one-third of that of a rare-earth magnet, the 2nd F-IPMSM could generate a torque that is equivalent to 90% of that of an R-IPMSM and gain a wider constant output power range.

As the result of these evaluations, the F-IPMSM could achieve an output power density equivalent to an R-IPMSM, even though the maximum torque with the same current is slightly inferior, and gain a wide variable-speed range suitable for the specifications of an EV. Therefore, we suggest that the ferrite magnet motor may be applicable to EVs.

ACKNOWLEDGMENT

The authors would like to deeply thank Prof. Ichiro Miki, Meiji University, for his contributions to the discussion during the research.

REFERENCES

[1] T. Abe, T. Ashikaga, H. Oguri, Y. Nakano, K. Nagata, A. Morikawa, and M. Oyadomari, "Development and commercialization of motor and inverter for electric vehicle, "i-MiEV"," EVS-25 Shenzhen, China, Nov. 5–9, 2010.

[2] S. Chino, T. Miura, M. Takemoto, and S. Ogasawara, A. Chiba, and N. Hoshi, "Fundamental characteristics of a ferrite permanent magnet axial gap motor with segmented rotor structure for the hybrid electric vehicle," in *Proc. 2011 IEEE Energy Conversion Congress and Exposition (ECCE 2011)*, Phoenix, AZ, 2011, pp. 2805–2811.

[3] S. Ooi, S. Morimoto, M. Sanada, and Y. Inoue, "Performance evaluation of a high-power-density PMASynRM with ferrite magnets," *IEEE Trans. Ind. Appl.*, vol. 49, no. 3, pp. 1308–1315, May–June 2013.

[4] T. Kosaka, T. Hirose, and N. Matsui, "Brushless synchronous machines with wound-field excitation using SMC core designed for HEV drives," in *Proc. of 2010 Int. Power Electron. Conf.*, Sapporo, 2010, pp. 1794–1800.

[5] T. Mizuno, K. Nagayama, T. Ashikaga, and T. Kobayashi, "Basic principles and characteristics of hybrid excitation type synchronous machine," *Electr. Eng. Jpn.*, vol. 117, no. 5, pp. 110–123, 1996.

[6] S. Morimoto and Y. Takeda, "Generalized analysis of operating limits on PM motors and suitable machine parameters for constant-power operation," *Electr. Eng. Jpn.*, vol. 123, no. 3, pp. 55–63, May 1998.

[7] K. Matsuo, T. Okitsu, D. Matsuhashi, and I. Miki, "The development of high torque IPMSM using ferrite magnet," in *Proc. JIASC 2012*, chiba, 2012, vol. 3, pp. 175–178.

[8] T. Okitsu, S. Ota, K. Matsuo, and D. Matsuhashi, "The development of spoke type IPMSM using ferrite magnet," in *Proc. JIASC 2013*, yamaguchi, 2013, vol. 3, pp. 99–100.

Optimal Field Excitation Control of a Claw Pole Motor for Hybrid Electric Vehicle

M. Azuma, M. Hazeyama, M.Morita, Y. Kuroda, A. Daikoku, M. Inoue

Advanced Technology R & D Center, Mitsubishi Electric Corporation

8-1-1 Tsukaguchi-Honmachi, Amagasaki, Hyogo 661-8661 JAPAN

Abstract- **Permanent magnet synchronous motors (PMSMs) with rare-earth permanent magnets (PMs) are most popular for automotive applications in which high torque density and high efficiency are required. Because the rare-earth materials have risks of lacking stable supply, rare-earth-free motors such as claw pole motors are widely studied. This paper presents the driving characteristics of a claw pole motor with no use of heavy rare-earth materials. In this motor the magnetic flux density in the air gap can be controlled arbitrarily by controlling field current. This paper shows that the variable field excitation can contribute to enlarge high efficiency range compared with fixed field excitation. Moreover, the electricity consumption driven in the Japan's fuel economy criteria (JC08 cycle) as EV and HEV applications are discussed.**

Keywords— Automotive electric motor, claw pole motor, field excitation control, rare-earth-free motor

I. INTRODUCTION

Recently the electrically-driven vehicles such as hybrid electric vehicles (HEVs) and electric vehicles (EVs) have become the key technologies of reducing greenhouse gas emissions[1]. In these vehicles, the rare-earth permanent magnet motors, using Nd-Fe-B magnets including rare-earth material such as neodymium (Nd), dysprosium (Dy) or terbium (Tb), are commonly used because of high magnetic density and high temperature capability. However, increasing demands of rare-earth material such as Dy and Tb will lead undersupply of these resources in the future. To solve this problem, rare-earth-free motors have been studied in automotive technologies.

There have been several rare-earth free motors which were investigated as conventional power trains [2]. As one example, induction motors (IMs) are commonly used for many application such as electric train. IMs are generally adopted longer axial length and distributed winding in order to increase torque capability. Thus, there have been studied for EV application by using IMs. Although, IMs seem not to be suitable for HEV application. Assuming mild HEV, motors are required to be sandwiched between the transmission and the engine. In this system, the requirements of the motors are shorter axial length and high torque density. By paying attention to this point, we have focused on claw pole motors (CPMs). CPMs can increase number of magnetic poles in the rotor easily because it is not necessary to substantially change the manufacturing. In addition, CPMs are able to field magnetomotive force arbitrarily by using control field current. As another type of the changing field magnetomotive force, there are also wound-field synchronous motors. Although, it has difficultness to increase rotor poles. Therefore, CPMs is expected to adopt HEV application.

CPMs are able to control field magnetic force by means of controlling filed current arbitrary. When the field current is energized, the field flux is emerged. As the field magnetomotive force is increased, the more field current increases. In automotive application, the motor requires to keep high efficiency and high torque density. Although, the efficiency map of conventional PMs are fixed field magnetomotive force. In such as PMs, it is necessary to control magnetic flux weakening in high rotation to reduce high line voltage. Thus, copper loss is increased and iron loss is kept highly. On the other hand, claw pole motors are easily to reduce magnetic flux not to use flux weakening control. It is possible to achieve the effect of magnetic flux weakening by decreasing field current. Therefore, claw pole motors are able to keep high efficiency in any operational points.

Our final target was achieved that the torque density was more than 5Nm/kg, motor efficiency was more than 90% in order to be close to PMSM performance. We achieved at the first stage that torque density was 3Nm/kg, motor efficiency was 80% which was based on alternators in the 1st prototype (CPM1). However, CPM1 was required to reduce weight and size to be close to PMSM performance at the same torque. Based on the results of CPM1, the 2nd prototype (CPM2) was made in order to improve efficiency and torque density. Firstly, with the goal of high efficiency, the surfaces of the claws were changed from solid core to laminated core for the purpose of decreasing eddy current loss in the rotor. Secondary, targeting of high output density, each claw was inserted ferrite magnets to decrease magnetic saturation. As a result, we achieved that torque density was 4Nm/kg, motor efficiency was 89% [3].

As a way of more improvements, we made the 3rd prototype (CPM3) so that the rotor diameter of CPM3 was larger than that of CPM2 to decrease magnetic saturation between these claws. As a result, we achieved the final targets, 5Nm/kg of torque density and 90% of motor efficiency which are nearly equal to the performance of PM motors[4].

978-1-4799-2706-7/14 $31.00 © 2014 IEEE

The 2014 International Power Electronics Conference

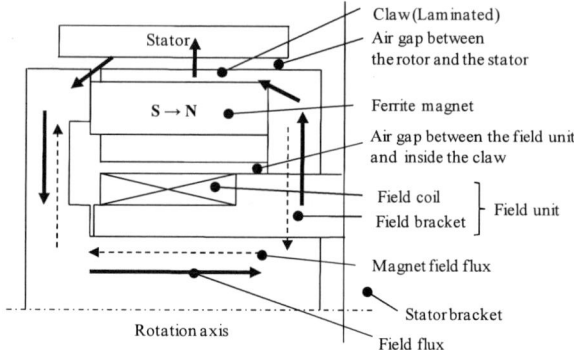

Fig. 1. Magnetic circuit of CPM2 and CPM3
(Cross section of A-A' plane)

TABLE I
SPECIFICATIONS OF PROPOSED CPM2 AND CPM3

Item	Unit	CPM3	CPM2
Nd-Fe-B	-	None	
Dy, Tb	-	None	
Number of poles	-	32	24
Ferrite magnet weight ratio	-	1.3	1
Rotor outer diameter	mm	Φ190	Φ185
Rotor stack length	mm	70	
Core type	-	Laminated	
Stator outer diameter	mm	Φ260	
Stator stack length	mm	40	
Winding type	-	Concentrated	

II. SPECIFICATIONS OF PROPOSED LAMINATED CPM WITH FERRITE MAGNTETS

In this paper, it presents that the validity of the variable field excitation control of CPM3 which can be controlled field magnetomotive force arbitrary. The difference of the motor efficiency between variable field excitation and fixed field excitation is shown using FEA analysis. Moreover, it is shown that the simulation of the EV and HEV application in the case of JC08 cycle.

Figure 1 shows magnetic circuit of CPM2 and CPM3. As shown in Fig.1, there are two air gaps in this motor. One of them is between stator and rotor. Another is between inside the claw and field unit. When the field coil is energized, it supplies the field coil without slip rings throughout the stator bracket. The field flux passes between field coil and inside the rotor. Moreover, the ferrite magnets are located between each claw. Here, it is noted that the magnetic polarity of the ferrite magnets was opposite to the field flux. The opposite polarity can suppress the flux leakage in the rotor so that magnet field flux cancels flux leakage in the rotor. Therefore, it is expected to increase maximum torque while it suppress magnetic saturation.

Table I shows specifications of proposed CPM2 and CPM3. As a way of increasing maximum torque, the number of rotor poles was increased from 24 to 32 so that CPM3 could decrease magnetic saturation in the rotor. Thus, the magnet weight of CPM3 was increased 1.3 times than that of CPM2. The reason for this is that the increased number of ferrite magnets could suppress flux leakage effectively by increased pole number.

Moreover, the diameter of the rotor was increased so that CPM3 could increased air gap flux density. The stack length of the rotor was longer than that of the stator so that the axial flux could pass through the rotor. In addition, the concentrated winding was aimed to decrease copper loss compared to the distributed winding. Moreover, these claws were adopted laminated core so that the core could decrease eddy current loss. Most eddy current loss was generated on the surface of the claws. The other parts that faced stator core was composed of solid core.

Figure 2 shows cross-section of CPM3. As shown in Fig.2, ferrite magnets could be located between each claw so as to decrease magnetic saturation. Once again, the polarity of the claws are generated N and S alternately when the field coil is energized. CPM3 can be controlled optimally by means of vector control as well as field excitation control. Figure 3 and 4 show overview of the field unit and rotor. The field unit consisted of the field coil and the field bracket.

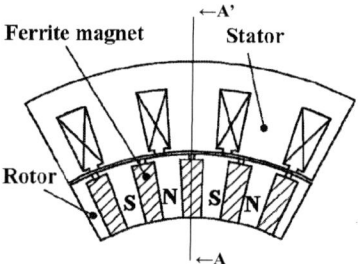

Fig. 2. Cross-section of CPM3

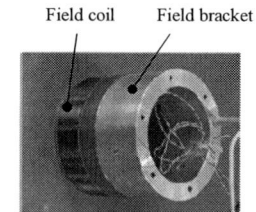

Fig. 3. Overview of the field unit

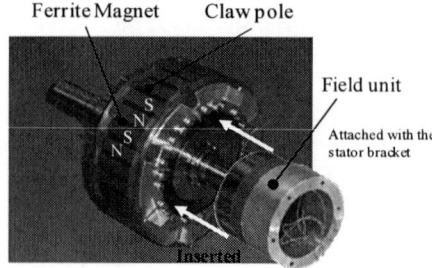

Fig. 4. Overview of the rotor

The field unit was inserted inside the rotor. It is noted that the field unit was attached with the stator bracket. Thus, field unit was supplied field magnetomotive force while it was fixed with stator bracket.

The specification shown above, CPM3 has an advantage that field loss can be decreased arbitrary by controlling field current optimally.

III. ANALYSIS CONDITION OF THE PROPOSED CPM

A. Efficiency map of the proposed CPM3 in FEA

We evaluated two conditions in order to compare these loss in FEA. Condition I and II are assumed for assist of acceleration (1000min⁻¹, 65Nm) and steady state (3000 min⁻¹, 25Nm). In these conditions, the effect of the efficiency were compared variable field excitation with fixed field excitation. The assist of acceleration required high magnetomotive force so that the motor can output required high torque. The steady state requires low field excitation so that the field excitation decrease iron loss in the high rotational speed.

As previous described, field excitation is necessary to control arbitrary in each operational points. Here, it is shown that the motor efficiency map are compared variable field excitation with fixed field exaction.

B. Electricity consumption for EV application

Using motor efficiency map in FEA results, electricity consumption was simulated in the case of JC08 cycle. TABLE II shows the simulated condition for EV application. In this application, the electricity consumption was calculated by all operational points in JC08 cycle. In this TABLE, car weight was considered 1300kg included buttery weight. The gear ratio was set at 6. Besides, the gear efficiency was considered 1.0. The conditions of two field excitation were considered variable and fixed conditions. In the simulation, the maximum value of the fixed field excitation was set 4500AT so that the motor could output maximum torque in JC08 cycle. For example, when the gear ratio is 6, maximum torque is 90Nm.

C. Electricity consumption for HEV application

Using motor efficiency map in FEA results, electricity consumption was simulated when the maximum output was 10kW assistance. TABLE III shows the simulated condition of 10kW assistance for HEV. The simulation was taken four conditions into consideration. These conditions compared the loss dependence of the gear control and field excitation.

Two conditions in terms of the mechanical gear were compared. The variable gear was considered from 2 to 10. The gear could change so as to minimize the motor loss. The fixed gear ratio was set at 6. In this simulation, the gear efficiency was 1.0. In the field condition, the conditions of field excitation were variable and fixed. The variable excitation was set optimally within 4500AT. It is for the reason that the maximum value could output maximum torque.

TABLE II
Simulated condition for EV

	Case1	Case2
Car weight	1300 [kg]	
Gear condition	Fixed	
Gear ratio	6	
Gear efficiency	1.0	
Field excitation	Variable (0 - 4500[AT])	Fixed (4500[AT])

TABLE III
Simulated condition of 10kW assistance for HEV

	Case3	Case4	Case5	Case6
Car weight	1300 [kg]			
Gear condition	Variable	Fixed	Variable	Fixed
Gear ratio	2 - 10	6	2 - 10	6
Gear efficiency	1.0		1.0	
Field excitation	Variable (0 - 4500[AT])		Fixed (4500[AT])	

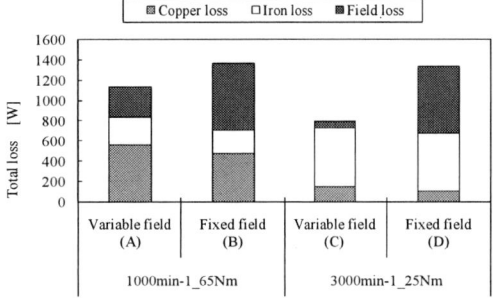

Fig. 5. Comparison of total loss with variable field excitation control and fixed field excitation

IV. ANALYTICAL RESULT

A. Analysis Results of the Proposed CPM

Figure 5 shows comparison of total loss with field excitation control and fixed field excitation. As for condition I, the total loss was decreased 230W with variable field excitation control. The field loss was decreased 360W, nevertheless, the copper loss and iron loss were increased 80W and 50W. It is considered that the armature current was increased so as to output the same torque. Thus, the armature current was increased. Field flux was also decreased. In the iron loss, it could be seen that the iron loss was increased though the fixed field excitation control was decreased. It suggests that alternating magnetic field was increased because the current amplitude was increased.

In condition II, condition II could be found that the total loss was decreased 540W by variable field excitation control. Compared to fixed field excitation control, copper loss and iron loss were increased as well as condition I.

Here, we assume that field loss of B and D are zero, these conditions are close to PMs performance (B', D'). In the condition I, total loss of the fixed field without field loss (B') is less than that of the variable field (A). Thus, PMs is superior to CPM3 in this point. In the condition II, total loss of the fixed field without field loss (D') is close to that of the variable field. Thus, CPM3 is close to PMs performance.

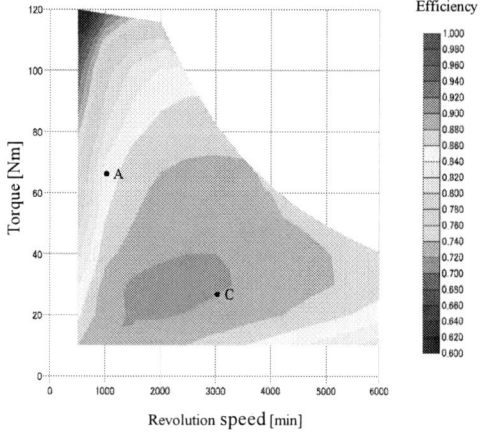

Fig. 6. Efficiency map with variable field excitation

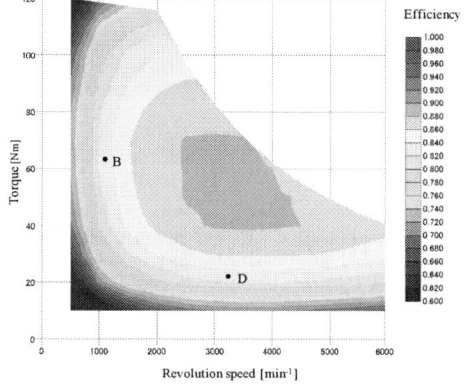

Fig. 7. Efficiency map with fixed field excitation

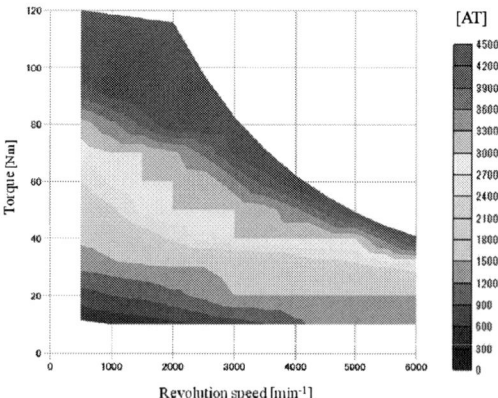

Fig. 8. Distribution of the field magnetomotive force

Figure 6 and 7 show efficiency map of CPM3 with variable and fixed field excitation control. In the case of variable field excitation control, maximum efficiency was distributed in low speed (1,000 - 3,000 min^{-1}) and low torque (20 - 40Nm). It is considered that not only decreased field loss but also the iron loss were decreased by reduced magnetic force in the rotor. On the other hand, in the fixed field excitation control, maximum efficiency was distributed in the domain of the middle torque condition. Then, the efficiency of the low torque operational range was distributed about 60% without field excitation control. i.e. 10Nm at 1000min^{-1}.

By contrast, the efficiency could be increased by field excitation control in the domain of the lower torque (around 10Nm). In this range, not only copper loss but also iron loss could be decreased. Thus, the efficiency with field excitation was higher than fixed excitation in the middle speed and low torque region. In the HEV application, it has been required the high efficiency at middle speed and low torque. Therefore, it is suitable for CPM3 to adopt HEV due to controlling field flux.

Figure 8 shows the distribution of the field magnetomotive force. It found that there were different value in each operational point. In the lower torque range, lower field magnetomotive force was selected, because the reduction of field loss was effective to keep high efficiency nevertheless copper loss and iron loss were increased. On the other hand, it can be seen that high torque and output line need maximum field magnetomotive force.

B. *Electricity consumption for EV application*

When the required torque or required revolution speed was zero, the value of the field excitation kept to be energized at fixed field excitation. For this reason is that required operational points need high responsiveness. In the case of low responsiveness, it was not enough to drive in JC08 cycle. Thus, it is necessary to keep high magnetomotive force.

Figure 9 shows comparison of electric consumption in JC08 cycle assumed for EV. Here, electricity consumption was normalized by maximum value shown in HEV results later. As shown in this figure, it was found that the dominant loss was field loss.

On the other hand, the field loss could be reduced by using variable field excitation control. It is considered that both iron loss and field loss could be increased by using low field electric magnetomotive force. Therefore, high efficiency caused by variable field excitation control could keep in the low torque range.

Figure 10 shows frequency distribution of the field magnetomotive force in JC08 cycle. As shown in this figure, the distribution of the field magnetomotive force was biased low field magnetomotive force range, which could be seen less than 2250AT. It is the reason that the frequency of the JC08 cycle was occupied stop state. Furthermore, it found that the copper loss was increased while iron loss was decreased when the variable field excitation was controlled. As a result, it is shown that CPM3 had effective field excitation by variable excitation.

The 2014 International Power Electronics Conference

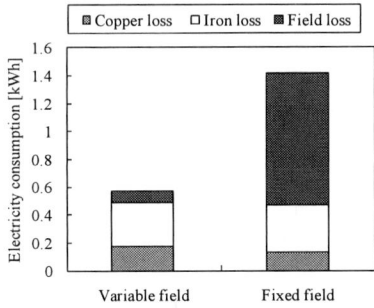

Fig. 9. Comparison of electricity consumption in JC08 cycle assumed for EV

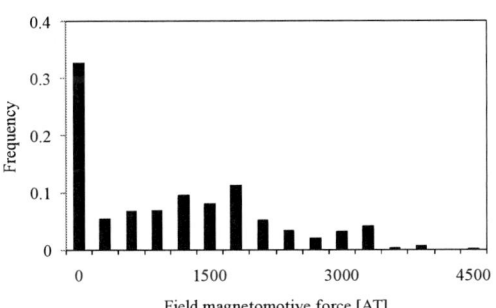

Fig. 10. Frequency distribution of field magnetomotive force in JC08 cycle

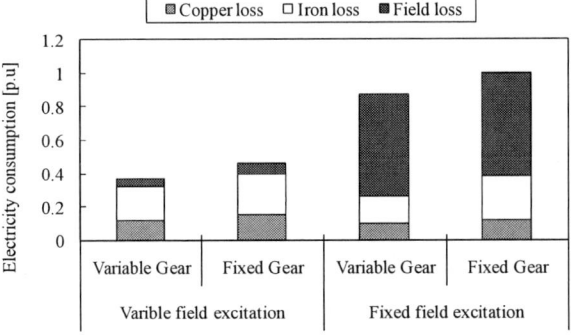

Fig. 11. Comparison electricity consumption in JC08 cycle assumed for HEV

C. Electricity consumption for HEV application

Figure 11 compares the electricity consumptions of the motor assuming the usage for HEV driven in JC08 cycle. The electricity consumption was calculated in each operational point based on JC08 cycle. The efficiency of the operational point was calculated by FEA results. Here, electricity consumption was normalized so as to compare the fixed condition in gear ratio and field excitation. Electricity consumption was consisted of iron loss, field loss and copper loss shown in Fig.11. It was not effective for CPM3 to apply high output. The result implies that CPM3 is suitable for lower assist. As shown in Fig.11, the electricity consumption of field excitation control could be found a half value of the fixed field excitation regardless of any gear ratio.

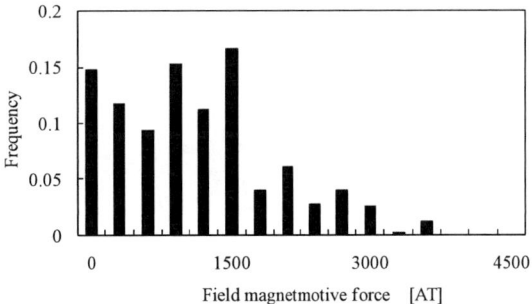

(a) Variable gear with variable field excitation

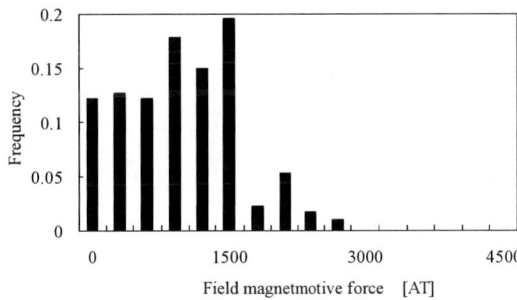

(b) Fixed gear with variable field excitation

Fig.12. Frequency distribution of the field magnetomotive force in JC08 cycle

On the other hand, copper loss and iron loss were increased slightly nevertheless field loss was decreased by controlling variable field excitation. It is because that the armature current increased to keep same torque due to decreasing field magnetomotive force. In comparison fixed gear with variable gear such as CVT, iron loss was decreased by at least 10% less than fixed gear. The variable gear ratio contributed to keep high efficient range in any output by using variable field excitation.

Figure 12 shows the frequency distribution of the field magnetomotive force in JC08 cycle. As shown in Fig.12, both variable and fixed gear were distributed low magnetomotive force. Especially, field magnetomotive force of the variable gear ratio was distributed lower range intensively than that of fixed gear. The difference between variable and fixed gear means that the variable gear ratio was selected optimally so that CPM3 can output to keep high efficiency.

Figure 13 shows characteristics of the electricity consumption as a function of maximum output of CPM3. As shown in Fig.13, the consumption of the variable gear was less than that of the fixed gear. It is considered that variable gear could minimize motor loss in any operational points compared with fixed gear. Moreover, it is found that there were different slopes in low output and high output. This is because the operational points in JC08 cycle were distributed low output. As can be seen in Fig.6, high efficiency was distributed low output range. Thus, CPM3 could keep high efficient drive in JC08 cycle.

978-1-4799-2706-7/14 $31.00 © 2014 IEEE   1896

The 2014 International Power Electronics Conference

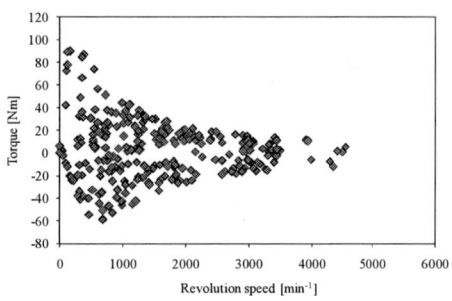

(a) Fixed gear with variable field excitation

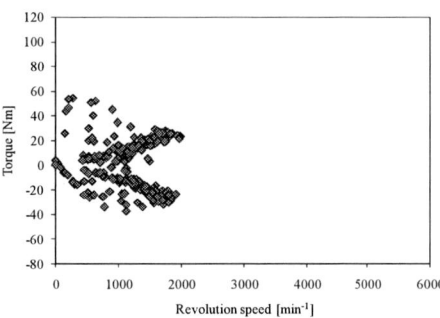

(b) Variable gear with variable field excitation

Fig. 13. Distribution of the operational points with fixed gear and variable gear in JC08 cycle.

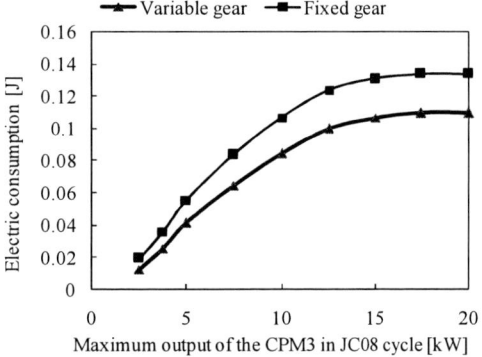

Fig. 14. Characteristics of the electricity consumption as a function of maximum output of CPM3

IV. CONCLUSION

In this paper the driving characteristic of a claw pole motor was studied, in which the magnetic flux density in the air gap can be controlled easily by controlling the field current. To increase the efficiency of the claw pole motor, using the variable field excitation is effective compared with using the fixed field excitation, especially in the lower torque region. The electricity consumptions assuming EV and HEV applications driven in JC08 cycle were simulated and found that the claw pole motor could keep maximum efficiency optimally by using both variable field excitation and variable gear. Therefore, it is suggested that the field excitation control is effective for automotive applications.

ACKNOWLEDGMENT

A part of this work was conducted in the Li-EAD project supported by the New Energy and Industrial Technology Development Organization (NEDO) in Japan.

REFERENCES

[1] C. C. Chan, R. Zhang, K. T. Chau, and J. Z. Jiang, "A novel brushless PM hybrid motor with a claw-type rotor topology for electric vehicles", Proc. 13th Int. Electric Vehicle Symp., Osaka, Japan, vol. II, pp. 579–584, Oct. 1996.

[2] S. Matsumoto, "Advancement of hybrid vehicle technology", Power Electronics and Applications, 2005 European conference,7 pp.-P.7.

[3] Y. Kuroda, M. Morita, M. Hazeyama, M. Azuma and M. Inoue, " Improvement of a claw pole motor using additional ferrite magnets for hybrid electric vehicles", International Conference on Electrical Machines 2010.

[4] M. Azuma, M. Morita, M. Hazeyama, Y. Kuroda, A. Daikoku and M. Inoue, "Fundamental Characteristics of a Claw Pole Motor Using Additional Ferrite Magnets for HEV", Electric Vehicle Conference (IEVC), 2012

[5] Lorilla, L. M.; Keim, T. A.; Lang, J. H.; Perreault, D. J.: "Topologies for Future Automotive Generators – Part II: Optimization", Vehicle Power and Propulsion, 2005 IEEE Conference, pp.831-837 (2005)

[6] S. Kupper and G. Henneberger, "Numerical procedures for the calculation and Design of Automotive Alternators", MAGNETICS IEEE TRANSACTION , vol.33, No.2, pp2022-2025, March 1997.

[7] K. Aoki et al., "Development of integrated motor assist hybrid system: Development of the 'Insight,' a personal hybrid coupe," presented at the SAE Technical Paper Series, Paper,#2000-01-2216, Jun 2000.

978-1-4799-2706-7/14 $31.00 © 2014 IEEE

A Wide Speed Range High Efficiency EV Drive System Using Winding Changeover Technique and SiC Devices

Yushi Takatsuka[*], Hidenori Hara[*], Kenji Yamada[†], Akihiko Maemura[†], Tsuneo Kume[*]

[*] Corporate R&D Center
YASKAWA ELECTRIC CORPORATION
12-1 Otemachi, Kokura-kita, Kitakyushu 803-8530 JAPAN
[†]EV Powertrain Business
YASKAWA ELECTRIC CORPORATION
13-1 Nishimiyaichi 2, Yukuhashi, Fukuoka 824-8511 JAPAN

Abstract— The Silicon Carbide (SiC) based devices will become the mainstream in the motor drive technology in the near future. These next generation power devices give benefits of lower losses, higher voltage, and high temperature operation in the power conversion system, realizing its downsizing and high power density. The winding change over technique, on the other hand, is useful to extend the speed control range with high efficiency of ac motors. This paper describes the effects of using high power SiC MOSFET and SiC SBD, together with the electronic winding changeover technique in the EV motor drives.

Keywords— *EV drive system, SiC device, Winding changeover technique*

I.   INTRODUCTION

In many applications, it is often needed to extend the speed control range of an electric motor. In permanent magnet AC motors, the maximum speed is limited by the available voltage for a given counter EMF (electromotive force) value. Extension of the speed range can be achieved to some extent by using the field weakening principle, changing machine parameters and inverter rating.

In some electric vehicles, boost converter technique is used to have higher dc bus voltage to keep output voltage higher than counter EMF value. However, the total efficiency from high voltage battery to electrical motor via boost converter and inverter cannot be improved because of two steps of semiconductor switching circuits.

The electrical winding changeover technique is an effective way to extend the speed control range [1]-[3], and it has been successfully applied to commercially leased EV models.

Another important point to be considered for motor drive system is to make the power losses in the converter as small as possible. With a wide range of applications including power supplies, electric vehicles, industrial equipment, and electric appliances, if these devices are operated at very high efficiencies, they will contribute significantly to the reduction of energy consumption. However, the characteristic of Si (silicon) power devices,

currently widely used in power electrics, have almost reached their theoretical limitations, and further improvements are becoming difficult to achieve. Fortunately, long awaited new power switching devices using Silicon Carbide (SiC) and Gallium Nitride (GaN) have become available.

This paper describes an electronic winding changeover technique to achieve high efficiency motor drive system for electric vehicles. The active driving mode is selected from two winding connection modes, so that the drive parameters in each mode are optimally controlled to obtain high efficiency. By combining individually optimized two winding modes, even wider high-efficiency areas on the torque-speed plane are achieved. In addition to the basic outstanding performance of the winding changeover system, SiC power switching devices are employed both for the main inverter and the winding changeover units, leading to the further improvement in total efficiency.

II.   APPROACH TO HIGH EFFICIENCY

In this section, the approach to high efficiency is examined.

A.   Theory Behind Electronic Winding Changeover Technique

Typical torque–Speed characteristics of IPM motor for electric vehicle is shown in Fig. 1. As the motor speed increases, the counter EMF of the motor approaches and tends to exceed the maximum inverter output voltage. Then, it will not be able to provide enough current for the required torque if no countermeasure is taken.

In order to extend the operating speed range, the electronic winding changeover method as shown in Fig. 2 has been proposed. The motor is a three phase open-wye connected type with center tap in each phase. The center taps U1, V1, W1 and end terminals U2, V2, W2 are connected to ac input terminals of diode bridges DB1 and DB2, respectively. The dc output terminals of DB1 and

The 2014 International Power Electronics Conference

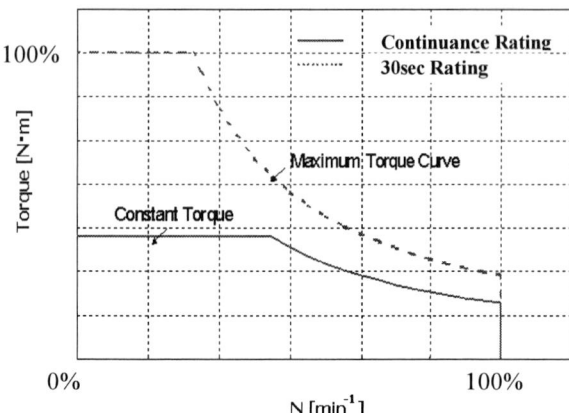

Fig. 1. Torque vs. speed characteristics of required motor

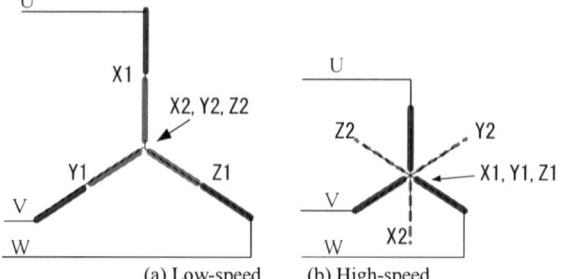

(a) Low-speed   (b) High-speed

Fig. 3. Voltage representation across windings for the proposed technique

Fig. 2. Electronic winding changeover scheme

TABLE I
COMPARISON OF SEMICONDUCTOR PHYSICAL PROPERTIES

Characteristics	Si	SiC
Band Gap Energy[eV]	1.12	2.20 ~ 3.02
Relative Permittivity[p.u.]	11.8	9.6 ~ 10.0
Electron Mobility[cm^2/Vs]	1350	370 ~ 900
Break Down Voltage[MV/cm]	0.3	1.2 ~ 2.0
Drift Saturation Velocity[cm/s]	1×10^7	2×10^7
Thermal Conductivity[W/(cm·K)]	1.5	4.9

B. Advantage of SiC power device

Physical Properties of Si and SiC devices are compared in Table I. The SiC device can be operated at higher temperature environment because of its wide band gap and high thermal conductivity. The practically maximum operating junction temperature (Tj max) for SiC device can be 200 degree C or higher, whereas that for Si device is 150 degrees C or less. Moreover, the high breakdown-voltage property can realize thinner construction leading to the lower conduction resistance for the same voltage rating.

III. PROPOSED TECHNIQUE AND DEDICATED SiC POWER DEVICE

In this section, more details of physical phenomena related to the winding changeover technique, which plays important roles both in extended speed control range and higher motor efficiency, as well as the specially developed SiC power device for this particular application.

A. Proposed circuit of winding changeover system

The circuit of Fig. 2 is applied to the IPM motor drive. In order to extend the high speed operating range, the number of stator turns is changed from higher number NSL to a lower number NSH at the speed of □BL in Fig. 4. The reduction in NS reduces the counter EMF by the ratio NSH / NSL, thereby allowing the speed to be increased for the same counter EMF value. Thus, the dual torque characteristics for the motor as shown in Fig.4 are realized.

DB2 are connected to the power switching devices SW1 and SW2. When SW1 is turned on in this configuration, terminals U1, V1, and W1 of the motor are shorted forming a wye connection using only one half of the motor windings. Since the counter EMF is relatively low, this selection is suitable for high speed operation. When SW2 is turned on, on the other hand, terminals U2, V2, and W2 are shorted forming a wye connection using all the windings, resulting in high counter EMF, which is suitable for low speed operation. The snubber capacitor voltage helps eliminate transient voltage across the outgoing switch. Under steady state operating condition, one switch is turned on while the other switch is turned off. The voltage across the open windings during high-speed and low-speed operation can be visualized using the vector representation shown in Fig. 3. Each half of the winding is designed to be symmetric and is of equal number of turns. Fig. 3(b) shows that the induced voltage across the open terminals X2, Y2, and Z2 do not exceed the applied voltage across X1, Y1, and Z1.

978-1-4799-2706-7/14 $31.00 © 2014 IEEE

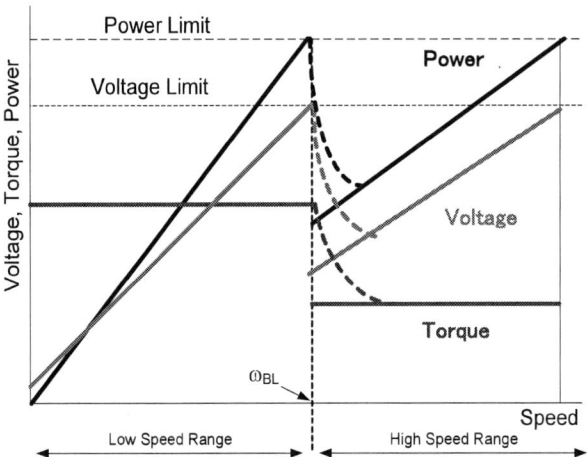

Fig. 4. Dual torque characteristics achievable using the proposed technique in IPM motor.

The motor parameters used in motor control algorithm in inverter controller are changed according to the set of winding being used at the time of transition from either high-speed to low-speed or vice versa. There is no dead time between turn-off of one switch and turn-on of the other switch to prevent "torque shock" and assure comfortable ride.

B. Shock-less switching control

Waveforms at the instant of winding changeover action in the test running of actual electric vehicle are shown in Fig. 5. Since dead-time-less switch control and fast change of the motor control algorithm are applied to this system, the motor current becomes continuous. Driver cannot feel any shock at the instant of winding changeover action because of the smooth transition.

C. Dedicated SiC power device

The newly developed SiC switching power device used in the drive system employs the trench-gate MOSFET (TMOS) structure [4]. It can reduce the On-resistance (Ron) to 1/3 of that for the conventional double-diffused

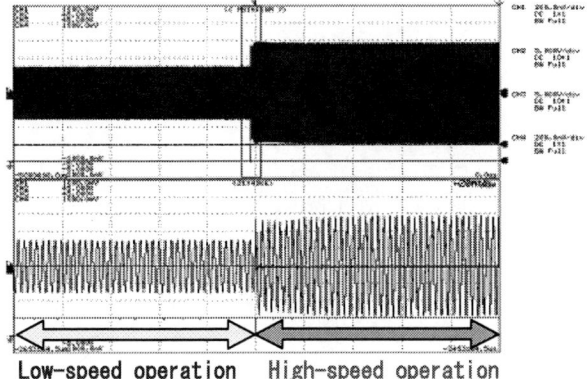

Upper : 200A/div. , 500msec./div.
Lower : 200A/div. , 20msec./div.
Fig. 5. Actual motor current waveform at the winding changeover instant.

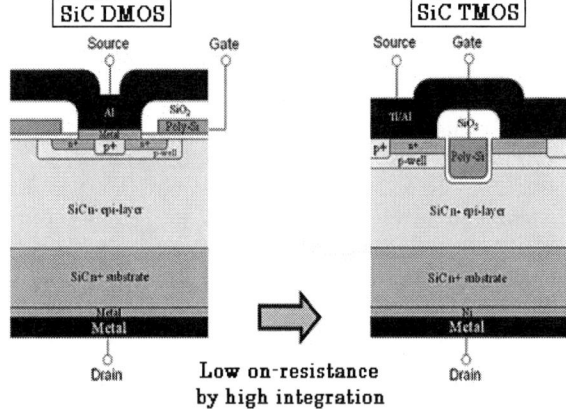

Fig. 6. Structure of SiC-DMOS and SiC TMOS

Fig. 7. Winding Changeover SiC power module for all-in-one integrated motor drive
(600V / 400A / Containing two switching circuits)

MOSFET (DMOS). Fig. 6 compares the cross sections of the DMOS and TMOS structures. The channel area of TMOS is formed vertically along a groove (trench), which is made vertically against the surface of a semiconductor substrate. In this configuration it is possible to reduce Ron by improving the channel density. Use of the SiC SBD for rectifiers realized high temperature tolerance and lower losses.

The dedicated power modules of inverter unit and winding changeover unit were developed with these SiC devices. Fig. 7 shows the external view of the winding changeover module. It realized downsizing and ultra high-efficiency of the drive system because the SiC devices are possible to operate at a high temperature over 200 degrees C, and they carry large current. Compared with the case of with conventional Si devices, the volume of less than 1/3 and the conduction-loss of less than 1/2 were attained.

IV. DEVELOPMENT OF ALL-IN-ONE INTEGRATED MOTOR DRIVE SYSTEM

A prototype of IPM motor drive system employing the winding changeover circuit and newly developed SiC power devices was produced. Specification and detail of the prototype are described in this section.

A. Specifications of prototype

Table II shows the specifications of the prototype motor drive system. The maximum motor speed is 12,000 min^{-1} and the maximum output is 65kW. The rated input voltage value depends on the specifications of the battery but it is set to 350Vdc in this examination.

Fig. 8 shows the block diagram of the drive system. The inverter unit controls the motor current, while the winding changeover unit controls the switches (SW1 and SW2 in Fig. 2) to selectively short circuit the open-wye terminals or center taps of the motor. The motor is integrated with the drive section, which further consists of the inverter unit and winding changeover unit. The terminal temperatures become equal to the motor stator winding temperature because each terminal is directly connected to the windings which protrude from the motor in the unit inside. In case of conventional silicon power device used, it is necessary to separate motor parts and semiconductor parts. However, SiC power device can be connected directly thanks to hight temperature operation capability.

B. Efficiency map

Efficiencies of the power converter section were repeatedly calculated for various speed and torque conditions, and plotted on the torque-speed coordinate plane as an efficiency map shown in Fig. 9 (a). Both of the inverter and winding changeover units were included in this evaluation. The maximum efficiency as high as 98% was obtained owing to the low loss nature of the MOSFET SiC devices.

The efficiency map for the whole integrated drive system including the motor was obtained as well by the similar process as shown in Fig. 9 (b). Here, the maximum efficiency as high as 95 % including the motor was obtained.

It should be noted that there are two areas of high efficiencies in the drive system efficiency map. It proves the fact that the winding changeover method not only extends the speed control range of ac motors, but also significantly improves the efficiencies around the most frequently used speed ranges for town driving as well as the highway cruising.

TABLE II
SPECIFICATION OF PROTOTYPE MODEL

Characteristics	Specifications
Maximum speed [min^{-1}]	12,000
Maximum output torque [Nm]	120
Continuous output power [kW]	30
Maximum output power [kW]	65
Rated input voltage [Vdc]	350

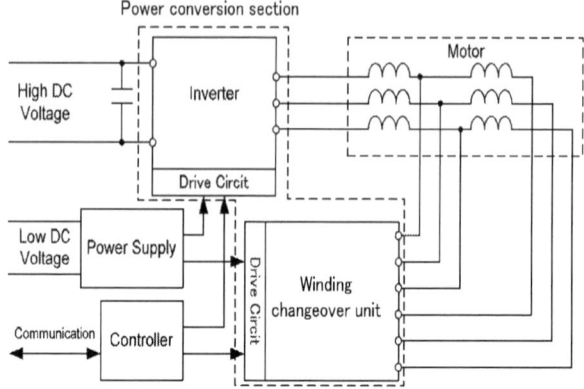

Fig. 8. Block diagram of that drive system

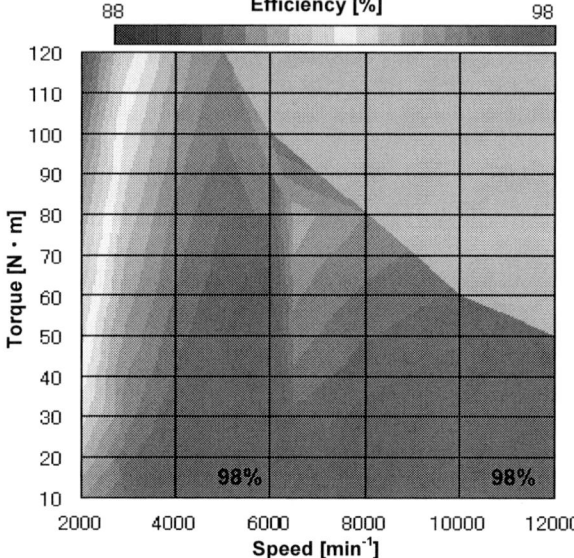

(a) Inverter and winding changeover components

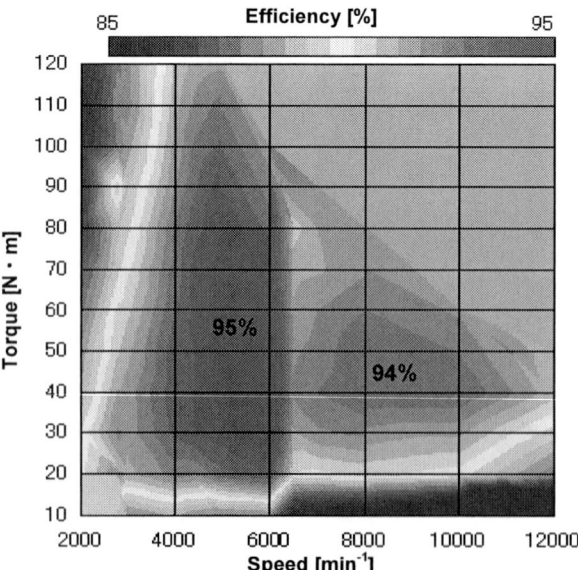

(b) Motor drive system
Fig. 9. Efficiency map

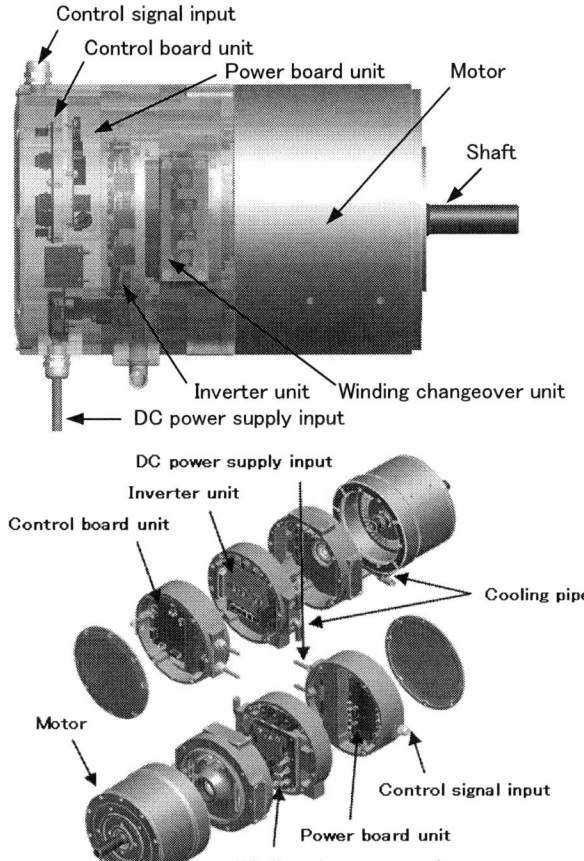

Fig. 10. Exploded view of the drive system (CAD drawing)

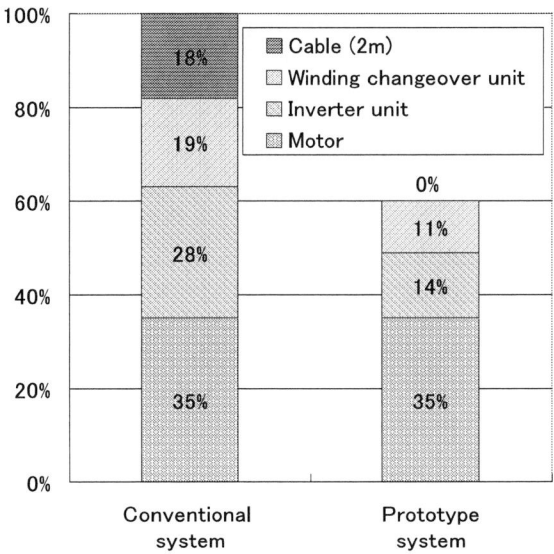

Fig. 11. Volume comparison with the conventional system

C. Outline of the prototype

Fig. 10 shows external and exploded CAD drawn view of the drive system. It realized all-in-one integrated motor drive system by using SiC power devices.

The inverter unit and winding changeover unit are densely packaged under condition of high temperature tolerant devices with water cooling. Although the winding changeover method requires a fairly large number of wire or bus-bar connections, the compact and dense packaging helps reduce the total volume of conductors leading to even less weight and lower heat dissipation.

V. EVALUATION OF PROTOTYPE

The prototype model introduced in the previous section was thoroughly tested, and its results and observation are described in this section.

A. Effect of downsizing

The volume of each functional components of the prototype are evaluated and compared with those in the conventional model as shown in Fig. 11. The total volume of the prototype is 25.8 liters (φ280mm×L419mm), and its power density including the motor is 2.5kW/liter as for max power of 65kW. The

model used here as the conventional system is an EV Drive system with the same specifications including the power ratings. However, Si devices are used for all switching function, and the inverter unit and the winding changeover unit are separately packaged. It is observed from Fig. 11 that both inverter unit and winding changeover unit are downsized by 40%-50%. As for the power cable, there is virtually no need for the cable since the prototype is a single integrated unit. Fig. 12 shows external of the prototype all-in-one integrated motor drive system.

B. Motor current waveform

The motor current waveform when winding changeover switch was shifted from low-speed mode to high-speed mode is shown in Fig.13. The motor counter EMF becomes half when the winding changeover unit switches from low speed winding to high speed winding. Therefore, it is necessary to increase the motor current in order to keep the same output power.

In the lower trace of waveform, the moment of switching is enlarged. It shows that the motor current of high speed winding is stabilized in about 3ms. In this way, it runs without torque fluctuation by completing the winding changeover in a short time.

C. Measured efficiency

Actual efficiency with prototype is shown in Fig.14. It was measured at 6,000min⁻¹ which is the rated speed. The power conversion efficiency for inverter unit and winding changeover unit is as high as 98%, whereas the total system efficiency including the motor is as high as 95%. These measured data meet well with the calculated values shown in Fig. 9.

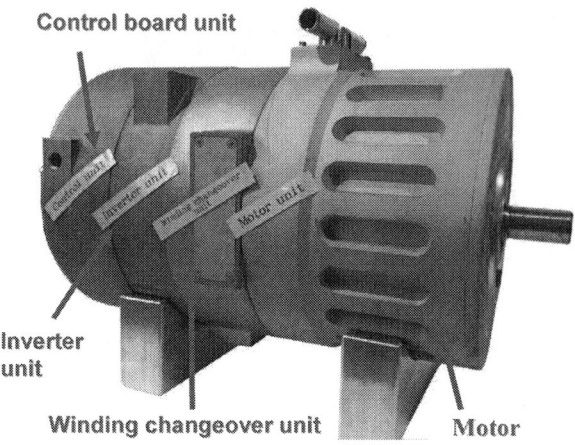

Fig. 12. External of the prototype all-in-one integrated motor drive system

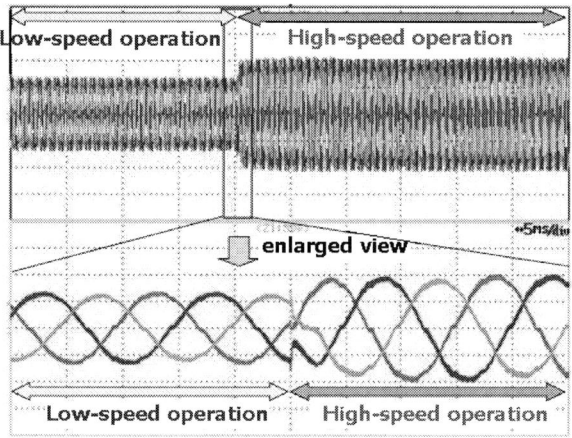

Fig. 13. Motor current waveform at winding changeover operation

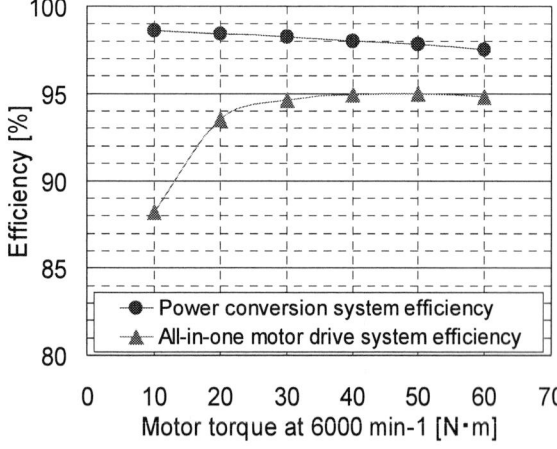

Fig. 14. Measured efficiency of all-in-one integrated motor drive system

VI. CONCLUSIONS

The motor drive system employing two novel techniques, SiC power devices and electronic winding changeover was introduced. It was verified through the preliminary calculation and actual measurement using a prototype that the specially designed SiC power module for this purpose played the key role in improving the overall efficiency and down sizing of the drive system. It was also found that the electronic winding changeover method not only extend the operating speed range, but also improves the motor efficiencies in the most frequently used seed ranges, one for the city driving and the other for the highway cruising.

REFERENCES

[1] Tsuneo J. Kume, Mahesh M. Swamy, Mitsujiro Sawamura, Kenji Yamada, Ikuma Murokita, " A Quick Transition Electronic Winding Changeover Technique for Extended Speed Ranges," CD-ROM IEEE Power Electronics Specialists Conference 2004, Aachen, Germany.

[2] Mahesh M. Swamy, Tsuneo J. Kume, Akihiko Maemura, Shinya Morimoto, "Extended High Speed Operation via Electronic Winding Change Method for AC Motors",CD-ROM IEEE Industry Applications Society Annual Meeting 2004, Seattle, USA.

[3] Akihiko Maemura, Shinya Morimoto, Kenji Yamada, Toshihiro Sawa, Tsuneo J. Kume, Mahesh M. Swamy. "A Novel Method for Extending Stroke Length in Moving Magnet Type Linear Motor Drive System by Employing Winding Changeover Technique", CD-ROM IEEJ International Power Electronics Conference 2005, Niigata, Japan.

[4] ROHM Co., Ltd. "The Industry's First SiC Trench MOSFET and Schottky Barrier Modules for Vehicle Motors", ROHM news 2010-10-04.

Performance Comparison of a GaN GIT and a Si IGBT for High-Speed Drive Applications

Arda Tüysüz, Roman Bosshard, and Johann W. Kolar
Power Electronic Systems Laboratory
ETH Zurich
Zurich, Switzerland
tuysuz@lem.ee.ethz.ch

Abstract— GaN power switches enable better switching characteristics compared to state-of-the-art power transistors that are widely used today. Due to their lower switching losses, GaN switches may lead to new horizons in key application areas of power electronics such as photovoltaic converters, high-speed electrical drives and contactless power transfer. However, this technology has not yet diffused fully into the industry. Therefore, today only limited experimental data is available on those switches.

In this paper, a synchronous buck converter is designed using two 600 V, 15 A GaN GIT switches developed by Panasonic. Guidelines for an optimum PCB layout are given. Both electrical and calorimetric power loss measurements are shown. Finally, a comparison is made between the GIT and a similarly rated Si IGBT for high-speed electrical drive applications where the higher switching frequencies enabled by the use of GaN is shown to reduce the rotor losses in two typical types (slotted and slotless stator) of high-speed permanent-magnet electric machines.

Keywords— *GaN, switching loss, loss measurements, motor drives*

I. INTRODUCTION

Increasing the switching frequency of power electronic converters has been a long-term trend, driven mainly by smaller volume and/or higher power density requirements. Particularly converters employing a transformer or inductors benefit from higher switching frequencies. Typical examples are photovoltaic inverters, DC-DC converters used in distributed photovoltaic maximum power point trackers [1], telecommunication power supplies [2], automotive industry [3] and inductive power transfer applications [4].

Additionally, in electrical drives increasing the switching frequency of the inverter decreases the ripple of the machine current, leading to lower losses in the electrical machine. In [5] it is shown that by increasing the pulse-width-modulation (PWM) frequency of the inverter, the rotor losses of a permanent-magnet (PM) motor can be decreased. However, the increasing switching losses of the inverter may lead to a lower overall efficiency. This method may be used in applications such as high-speed spindles or drills below a few kW, where the machine compactness is of highest priority and additional losses in the converter can be tolerated.

However, the switching frequency of a converter cannot be increased beyond a certain limit, above which the switching losses decrease the converter efficiency to unacceptable values and the increased heatsink sizes decrease the power density. Moreover, the safe operation of the converter is endangered due to increased junction temperatures of the switches.

So far Gallium Nitride (GaN) devices have been employed mainly for radio-frequency or microwave applications [6]. However, their low on-state resistances and recently higher breakdown voltages make them an interesting option for power electronic switches [7]. In [8], a commercially available 200 V GaN transistor is used to build a switched capacitor voltage doubler that achieved 94.4% efficiency at 480 W output power while switching at 893 kHz. The same GaN transistor is tested in [9] and its switching characteristics are reported under hard and soft switching conditions. The performance of GaN and Si switches are compared in a 1 MHz 150 W dual active bridge DC-DC converter [10] where Si switches are only considered for the secondary side bridge.

Until recently, there have been no commercially available GaN power switches in the market with blocking voltages higher than 200 V [6]. However, today the performance of high voltage GaN switch prototypes is reported in an increasing number of publications. In [11], advances in GaN power switches are reviewed, and a novel cost-effective, normally-off Gate-Injection-Transistor (GIT) is presented. The specific on-state resistance and the off-state breakdown voltage of this device are reported as 2 mΩcm^2 and 700 V. Furthermore, six GITs forming a three-phase inverter are integrated in one IC which is used for driving a motor with up to 20 W of output power. Discrete GaN GITs with a blocking voltage of 650V are used to build a three-phase motor drive in [12]. The efficiency of this inverter is reported to be over 98% between 100 W and 900 W output power at 200 V DC link voltage and 6 kHz switching frequency. A similar inverter built using discrete GaN GIT switches is reported to have an efficiency of 99.3% at 1500 W output power when the switching frequency is 4 kHz in [13]. A boost converter for photovoltaic maximum power point tracking is built in [6]. A 600 V 15 A GaN GIT and a Si IGBT with similar ratings are compared for the realization of the transistor and a 600 V 15 A GaN and 600 V 16 A SiC Schottky diode are comparatively evaluated for implementing the freewheeling diode of the boost converter. The GaN GIT is shown to lead to higher efficiencies compared to the Si IGBT whereas the GaN and SiC Schottky diodes show similar performance. However, the lower production costs of the GaN diode is argued to result in the future in very competitive prices compared to the SiC devices. A 1 kW, 1 MHz resonant DC-DC converter built using GaN GITs is shown in [14] where the factors limiting the performance (overheating components) are reported to be the inductive components and not the transistors. A motor drive inverter using different 600 V GaN switches is shown in [15], and the reduction of the output filter size with increasing PWM frequency is highlighted. The possibility of using GaN power switches in automotive applications is reviewed in [3].

In this work, the performance of 600 V 15 A GaN GITs that are developed by Panasonic is evaluated experimentally in a DC-DC converter. The same GITs are also used in [6]. Due to the high switching speed of the GaN devices, switching losses are very difficult to determine accurately by measuring the device voltages and currents simultaneously, due to highly demanding bandwidth and synchronization requirements. Therefore, in this work a calorimeter is used for measuring the total heat generated by the converter system, resulting in a direct power loss measurement. Furthermore, using a DC-DC converter topology enables highly accurate electrical DC input and output power measurements for evaluating the switch performance, which are also carried out in addition to the calorimetric measurements.

Moreover, in this work the effect of the PCB layout on the converter performance is stressed and guidelines for an optimum PCB layout are given. Finally, the measured GaN GIT performance is compared to the performance of a Si IGBT in a motor drive application to show the benefits of the GaN switches.

II. THE PROTOTYPE CONVERTER

A. Topology selection

The main goal of this work is to evaluate the performance of the GaN switches experimentally and to assess the possible performance increase GaN switches may offer in a practical application. For doing so, a DC-DC buck converter is built using two GaN switches in a half bridge configuration.

There are several reasons for choosing this topology. Firstly, measuring the switch current and the voltage across the switch to determine the switching losses requires high measurement bandwidth and a very good synchronization between the current and voltage measurements. However, in a DC-DC converter, input and output power of the converter can be easily measured with high accuracy and the switching losses can be deduced using a model for the other lossy elements (in this case an inductor), eliminating the need for challenging measurement bandwidth and synchronization.

Secondly, the half bridge configuration is a building block that is used in converters for applications such as

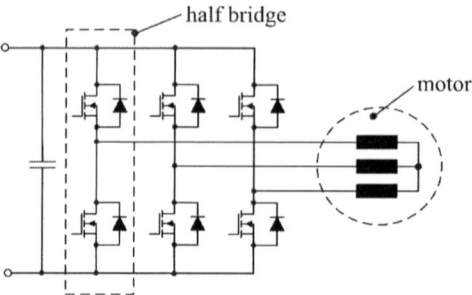

Fig. 1. A three-phase drive inverter, comprising three half bridge circuits (bridge legs), each of which can be seen as a bidirectional DC-DC converter.

bi-directional DC-DC converters and motor drives. As an example, Fig. 1 shows a half bridge topology in a 3-phase 2-level inverter, which is widely used for AC motor drives. The abovementioned applications may benefit from the better switching performance GaN power switches are offering. However, the switching losses are also affected to a large extent by the actual circuit board layout and the layout parasitics. As no current and voltage measurements are needed directly on the switches, the circuit board layout can be optimized as in a final converter design. Therefore, the results taken on a half bridge configuration with an optimized circuit board layout are directly applicable to converters for a wide range of applications. The space requirement for two gate driver circuits instead of one makes the board layout more challenging, but also more realistic with respect to these applications.

Finally, in a half bridge topology, the parasitic couplings limit the switching speeds due to the false-firing (or spurious triggering) problem [16], [17]. For example when the upper switch turns on very fast, the parasitic capacitive coupling between the drain and the gate of the lower switch may cause the lower switch to turn on inadvertently, leading to a temporary cross conduction (short circuit of the DC link) which decreases the efficiency and may even lead to the destruction of the converter. In this work, this problem is taken into account as a half bridge topology is used.

Due to the reasons listed above, a half bridge topology was chosen to evaluate the switch performance while reflecting the situation in converters used for a wide range of applications.

Fig. 2 shows the half bridge converter built using 600 V, 15 A GaN GIT switches.

B. Gate drive circuit

Two identical gate drive circuits fed with isolated DC-DC converters are used for driving the power switches. A LM5114 gate driver IC from Texas Instruments is selected because its high current source/sink capability allows fast switching of the GaN switches. Another useful feature of this IC is the independent source and sink outputs, enabling different impedances for the turn on and turn off paths and making it possible to control rise and fall times independently. The gate drive circuit is shown in Fig. 3. For turning the switch on, 3.5 V is applied between its gate and source

978-1-4799-2706-7/14 $31.00 © 2014 IEEE

The 2014 International Power Electronics Conference

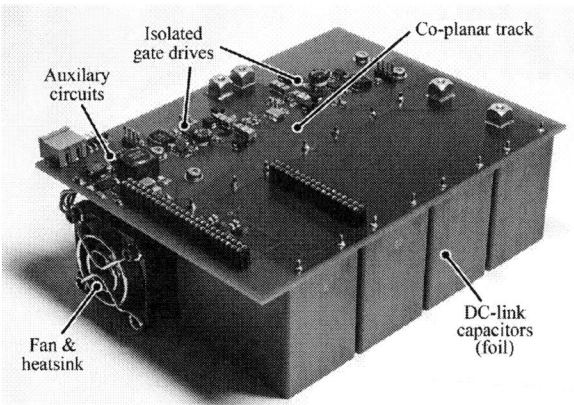

Fig. 2. Photograph of the half bridge circuit using GaN GIT switches.

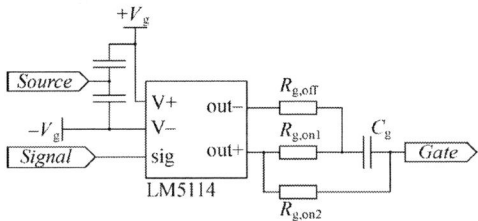

Fig. 3. Schematic of the gate drive circuit. $+V_g$ and $-V_g$ voltages are generated by an isolated DC-DC converter, and may have different amplitudes (e.g., +3 V, -5 V) with respect to the source potential of the power switch.

via an R-C network. $R_{g,on1}$ along with C_g determines the maximum gate current and its duration whereas $R_{g,on2}$ is used to adjust the continuous gate current to keep the device in conduction. Increasing the maximum gate current leads to higher switching speeds and hence to lower switching losses. However, due to the parasitic inductance of the commutation loop, very high switching speeds may lead to voltage spikes on the switches. As the GITs used in this work are not avalanche rated, the voltage across the device should never exceed the rated voltage, not even during the switching transients. Furthermore, the low threshold voltage (around 1.2 V) of the GITs and a very fast switching may lead to unintentional turn on as described in [16] and [17]. A very low or zero $R_{g,off}$ ideally reduces the impedance between the gate and source when the device is off. However, it has been found experimentally that a resistance lower than 15 Ω leads to unwanted oscillations in the gate circuit due to the parasitic inductances and capacitances.

In the off-state a negative voltage is applied between the gate and source of the switch to avoid unintentional turn on. However, a negative gate voltage increases the voltage drop on the switch during its reverse current conduction (during the dead time, where both switches of the half bridge configuration are turned off), leading to higher losses [18]. Therefore the dead time has to be as short as possible. On the other hand, the minimum

necessary dead time in a half bridge configuration depends on the output current. As the investigation of a load dependent dead time is beyond the scope of this work, here the dead time is set to 150 ns, which is found to be adequate for output currents above 1 A.

The values of $R_{g,off}$, $R_{g,on1}$, $R_{g,on2}$, and C_g leading to the shortest switching times without causing oscillations exceeding the safe operating range according to the datasheet were experimentally determined to be 15 Ω, 5 Ω, 40 Ω, and 47 nF respectively.

C. Board layout considerations

For a fast switching converter, special attention must be paid to the printed circuit board (PCB) layout. Along with the package of the power semiconductors, the board design defines the parasitics that affect the performance of the converter, mainly due to the strong dependency of the switching losses on the parasitic elements. The GITs used in this work have a standard TO-220 package, which defines an important portion of the actual parasitics on the board that cannot be further reduced (the lead inductance of a TO-220 package can be up to 12 nH according to [19]). On the other hand, the board layout is optimized for minimum additional parasitics while using discrete elements for power switches, gate drivers and gate resistors.

Two important aspects to consider when designing the board are the gate loop layout, and the commutation loop layout [20], and they are discussed in the following sections.

1) Gate loop

The gate drive loop, i.e., the path of the gate current while turning the power switch on or off is shown in Fig. 4(a). Any parasitic inductance of this loop limits the switching speed of the power switch and leads to higher switching losses. A photo of the gate drive circuit is shown and the turn on path is highlighted in Fig. 4(b). When the switch is turned on, the current flows from the capacitor bank through the gate driver IC and the R-C network to the gate pin on the top layer of the PCB; and from the source pin back to the capacitor bank through the middle layer (the one that is closest to the top layer) of the PCB. The return path on the middle layer is designed as a plane instead of a thin track and connected to the top layer using several parallel connected vias to reduce the inductance as much as possible. Similarly, the gate current path while turning the switch off is shown in Fig. 4(c).

It can be seen in Fig. 4 that the gate loop design provides a return path for the current immediately beneath its flow path, minimizing the parasitic inductance. The arrangement of components given in Fig. 4 shows that a much smaller loop is not possible using discrete elements for the gate driver, the gate resistors and the gate capacitor. In such a configuration, the pins of the power switch's package are contributing significantly to the overall parasitic inductances.

978-1-4799-2706-7/14 $31.00 © 2014 IEEE

The 2014 International Power Electronics Conference

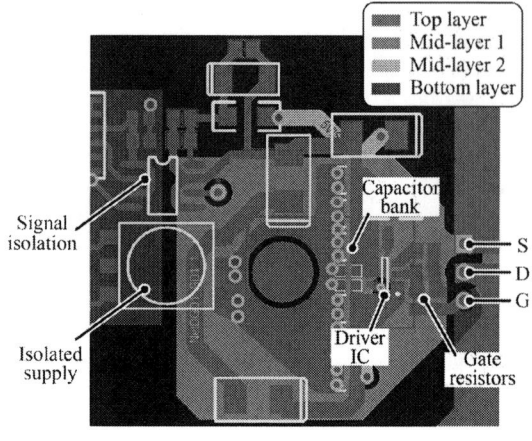

(a)

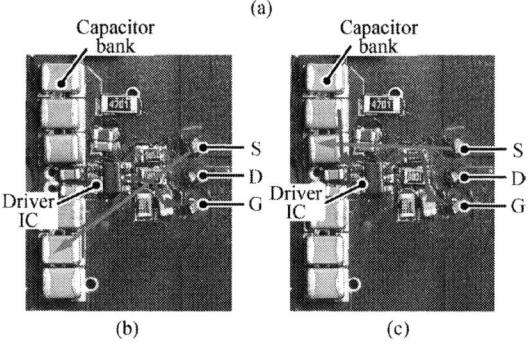

(b)   (c)

Fig. 4. (a) The gate drive circuit PCB layout. (b) Gate current path during turn on. (c) Gate current path during turn off. The colors of the arrows in (b) and (c) denote the board layer where the current flows.

2) Commutation loop

The commutation loop is the area covered by the path from the DC link capacitor bank through the upper and lower switch back to the capacitor bank, as shown in Fig. 5(a). A large commutation loop leads to higher voltage overshoot, decreased input voltage capability, and higher EMI [20].

The capacitor bank is made of foil capacitors that can be seen in Fig. 2, due to their low parasitic inductances. Four capacitors are connected in parallel for decreasing the parasitic inductance L_C of the capacitor bank further. The large package size and the pin positions of the foil capacitors lead to a large commutation loop area. Therefore, a small ceramic capacitor with a damping resistor is placed on the board close to the pins of the switches, as shown in Fig. 5(c). The commutation loop current path is arranged in co-planar tracks as shown in Fig. 5(b), in order to minimize the parasitic inductance.

III. THE MEASUREMENT SETUP

The total losses of a converter can be measured directly using thermal measurements with a calorimeter. As this is a direct loss measurement, the accuracy is independent of the actual power being processed by the converter. This enables more accurate measurements when compared to electrical input and output power

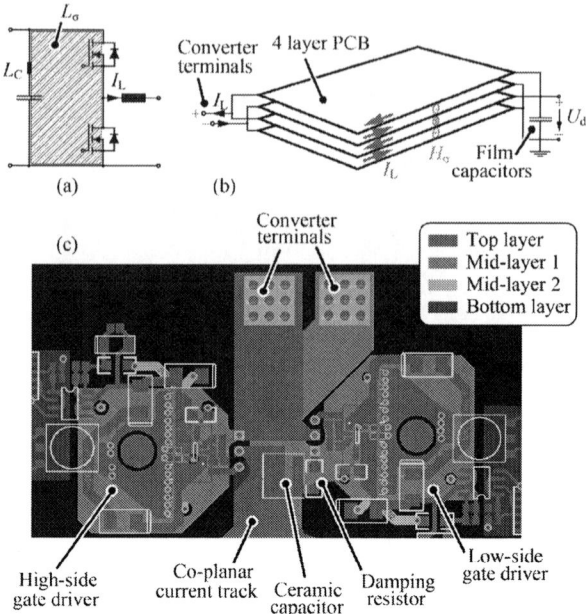

Fig. 5. (a) Commutation loop where L_C denotes the stray inductance of the capacitors and L_σ is the parasitic inductance introduced by the circuit board and the packages of the switches. (b) A co-planar track topology used to minimize the track inductance. (c) Circuit board layout showing the positioning of the two switches and their gate drive circuits.

measurements. On the other hand, calorimetric measurements are time consuming, as the thermal time constant of a large calorimeter like the one used in this work can be in the order of hours, meaning that each measurement may take considerable time. Electrical measurements, on the other hand, can be taken as soon as the converter reaches a given operating point.

The measurement setup is shown in Fig. 6. An 11 mH air cored inductor and a 220 µF electrolytic capacitor is used to build a synchronous buck converter with the half bridge board. Yokogawa WT3000 power analyzer is used to measure the DC input and output powers as well as the separate power inputs for the gate drive circuits and the auxiliary power that supplies the signal processor board and the fan on the heat sink.

The difference of the electrical input – output power measurements gives the total losses of the converter, including the switching and conduction losses of the power switches as well as the losses of the inductor. As an air core inductor is used with a sufficiently large inductance, the output current can be assumed to be DC. Hence, the inductor losses can be calculated using its DC resistance, which is measured to be 507 mΩ (including the cabling of the test setup) using a voltage source with a current limit, two Agilent 34410A multimeters and a Burster 1282 10 mΩ shunt resistor. The accuracy of the multimeters are ± 0.6% for the current measurement and ± 0.04% for the voltage measurement. The shunt resistor has 20 ppm accuracy.

978-1-4799-2706-7/14 $31.00 © 2014 IEEE

IV. MEASUREMENT RESULTS

Table I shows the measurement results taken by both the power analyzer and the calorimeter where f_{sw} is the switching frequency, P_{out} is the output power, I_{out} is the inductor current, P_{aux} is the supply of the signal processor and the cooling fan, T_j is the estimated junction temperature based on a thermal model of the switches and the heat sink, P_{cal} is the power measurement read by the calorimeter and P_{el} is the difference of the total electrical input and output power measurements after subtracting the estimated power losses of the inductor and the cables of the set setup. The calorimeter and the electrical power analyzer are connected together, i.e., electrical and thermal measurements are taken simultaneously. The input voltage is 400 V and the duty cycle is 0.5 for all the measurements described in this work.

Results of the calorimetric loss measurements are also plotted in Fig. 7 and Fig. 8. In Fig. 7, the estimated junction temperature difference between the maximum and the minimum switching frequencies is minimum (15 °C) for 3.5 A and maximum (25 °C) for 6.5 A.

A junction temperature rise from 75 °C to 100 °C increases the on-state resistance of the GaN GIT by only 10%; therefore this effect is neglected when estimating the switching energies. Assuming that the only frequency dependent loss component in the half bridge converter is the switching loss of the GITs, the total switching energies can be extracted as 140, 128, 118 and 93 μJ based on the calorimeter and 162, 141, 115 and 100 μJ based on the electrical power measurements at 6.5, 5.5, 4.5 and 3.5 A, respectively (all at 400 V). The estimated junction temperatures are not constant for these operating points, however, the results of double pulse measurements given in [6] shown that the switching losses of the GaN GITs do not change significantly with temperature, especially at drain current levels considered in this work.

V. APPLICATION EXAMPLE: HIGH-SPEED MOTOR DRIVES

High-speed drives is an emerging topic gaining increasing popularity both in academia and industry due to the higher power densities enables by high rotational speeds. In [21], two typical high-speed machines are analyzed and the asynchronous harmonics in the air gap flux are shown to induce eddy current losses in the rotor. This decreases the machine efficiency. Furthermore, the rotors of these machines contain permanent magnets which demagnetize at elevated temperatures. In standard low-speed surface-mount permanent-magnet machines, there is no retaining sleeve on the rotor and the magnets can be segmented to limit the eddy current losses. However, for the high-speed machines such as the ones shown in [21], segmenting the magnets would only lead to shifting the losses from the magnets to the sleeve and segmenting the sleeve is not possible as a one piece metallic sleeve is needed to provide sufficient mechanical strength.

The rotor losses can be decreased by increasing the inverter switching frequency and having a more sinusoidal machine current [21]. However, the inverter switching frequency cannot be increased above a certain level due to the increasing switching losses. In certain applications where the machine compactness is of highest importance, lower inverter efficiency can be tolerated to save on the machine losses. However, the cooling capability and the maximum allowed junction temperature of the semiconductor switches of the inverter set an upper limit on the maximum inverter losses. For this reason, in this section the possibility of exploiting the performance of GaN GITs in high-speed drives is analyzed and the performance of the GIT is compared to a state-of-the-art Si IGBT.

The conceptual drawings of the two types of typical high-speed motors analyzed in [21] are shown in Fig. 9. The actual off-the-shelf machines analyzed in that work are rated at 600 W and 3.5 A, therefore not suitable to be

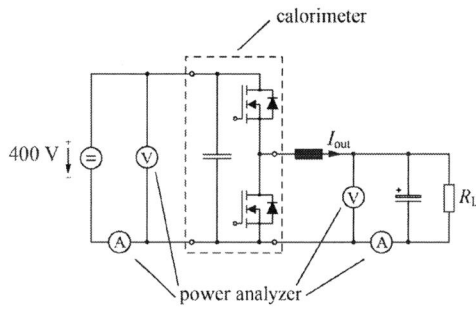

Fig. 6. The measurement setup. The power analyzer measures the input and output powers as shown. Two additional channels of the power analyzer are used to measure the gate and auxiliary (signal processor and fan) powers separately (not shown). The half bridge losses are also measured in a calorimeter. The output inductor is not placed in the calorimeter, therefore its losses need to be estimated and subtracted from the electric power measurements for a comparison with the calorimeter.

Table I. Measurement results.

f_{sw} (kHz)	I_{out} (A)	P_{out} (W)	P_{aux} (W)	T_j (°C)	P_{el} (W)	P_{cal} (W)
25	3.50	689.3	7.17	48	12.36	11.02
25	4.50	882.9	7.26	57	13.48	12.14
25	5.50	1076.2	7.26	64	13.16	13.58
25	6.50	1268.2	7.25	76	14.11	15.81
50	3.50	684.0	7.26	44	14.89	13.47
50	4.51	876.9	7.27	60	15.47	15.36
50	5.51	1068.9	7.25	70	16.58	16.42
50	6.52	1259.0	7.26	83	18.22	19.15
80	3.50	678.6	7.27	58	18.50	17.09
80	4.51	872.0	7.27	67	20.12	18.48
80	5.51	1057.8	7.26	78	21.26	21.26
80	6.50	1246.9	7.24	94	24.85	23.88
100	3.50	678.1	7.24	60	19.52	17.92
100	4.50	867.6	7.29	69	21.37	21.04
100	5.51	1055.8	7.27	81	23.46	23.32
100	6.51	1241.6	7.27	95	25.85	26.16
120	3.50	674.1	7.27	64	22.01	19.87
120	4.52	865.0	7.25	74	24.32	23.49
120	5.50	1048.8	7.20	87	26.67	25.49
120	6.50	1233.0	7.22	101	29.63	29.11

The 2014 International Power Electronics Conference

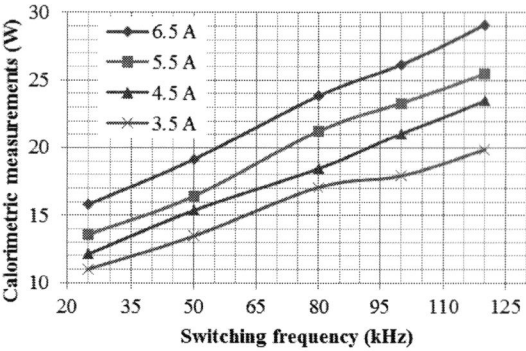

Fig. 7. Results of calorimetric loss measurements vs. switching frequencies. The measured values include the half bridge converter losses (switching and conduction losses, gate supply and auxiliary power) but exclude the inductor losses. If the temperature variations within measurements with the same output currents are neglected, the slope of the curve gives the switching energy of the GIT. The estimated junction temperatures for each operating point can be found in Table I.

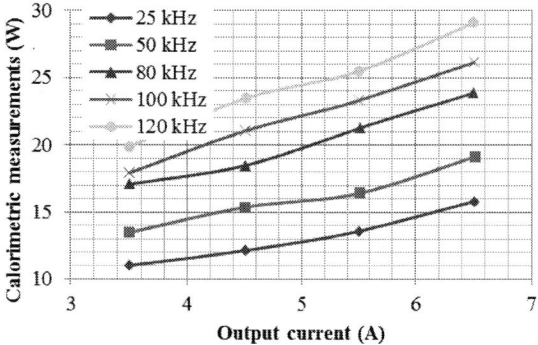

Fig. 8. Results of calorimetric loss measurements vs. different output currents. The measured values include the half bridge converter losses (switching and conduction losses, gate supply and auxiliary power) but exclude the inductor losses. The estimated junction temperatures for each operating point can be found in Table I.

considered as a direct application to the GITs analyzed in this work. Consequently, based on the actual machines of [21], two fictitious machines are designed by scaling the diameter of the machines to increase the rated power while keeping the nominal current density in the windings constant. The number of winding turns is adjusted to match the machine voltage with the 400 V DC link voltage at which the GITs have been tested. As the focus is on the relative machine loss change with respect to inverter switching frequency, a detailed thermal analysis is omitted in a first step. Mechanical constraints such as the stresses in the rotor or rotordynamic constraints are also neglected for the same reason. The parameters of the resulting slotted and slotless machines are given in Table II.

After the machines are designed, the phase current waveform is calculated analytically assuming a 2-level, 3-phase inverter operating with space-vector-modulation, taking the DC link voltage, the machine back EMF,

Table II. Parameters of the fictitious machines.

	Slotted	Slotless
Rated speed (rpm)	100,000	
Rated power (W)	3140	
Rated current (A)	7	
Flux linkage (mWb)	20.6	20.3
Phase inductance (μH)	320	155
Phase resistance (Ω)	0.54	0.21

machine inductance and the inverter switching frequency into account. The phase current waveforms for the slotless machine for 25 kHz and 200 kHz inverter switching frequency can be seen in Fig. 10.

Once the current waveform is generated, the core (stator iron) and rotor (eddy current) losses in the machine are calculated using a 2-D time-transient finite element (FE) model in which the calculated phase currents are impressed in the machine windings. The copper losses are calculated by the machine phase resistance and the RMS value of the calculated phase current, neglecting any skin and proximity effects. This is a reasonable assumption as generally litz wires are used in high-speed electric machines to limit skin and proximity effect losses.

A 600 V, 15 A Si IGBT with antiparallel diode in a TO-220 package (IKP15N60T) is selected to be compared to the GaN GITs analyzed in this work. This IGBT is selected because the current and voltage ratings as well as the package and the internal antiparallel diode make it a direct replacement for the analyzed GaN GITs.

Two inverters, one built with Si IGBT/antiparallel diodes and the other with GaN GITs are assumed. From the machine current calculated above and the switching signals, the currents through the IGBT/diodes (for Si inverter) and through the GaN GITs (for GaN inverter) are calculated separately. Currents at switching instants are calculated similarly. Finally, voltage drops over the Si IGBTs and the diodes as well as the switching energies for the IGBTs are read from the IGBT datasheet to estimate the total losses of the Si inverter for a given machine and operating point. For the GaN inverter, the on-state resistance is read from the datasheet (which is verified in [6]) and the switching energies are extracted from the measurements presented in this work.

The datasheet for the IGBT contains information on the switching and conduction losses at different junction temperatures. In [6], on-state resistance of the GIT is measured also at different temperatures. However, as the measurements in this work are carried out in a fixed ambient temperature the junction temperature of the GITs could not be controlled and varied between 45 °C and 100 °C. The average of estimated junction temperatures for all the measurement points is around 70 °C. Therefore, in order to make a fair comparison, also a junction temperature of 70 °C is assumed when reading the on-state resistance of the GIT and voltage drops over the Si IGBT and the antiparallel diode as well as the switching energies for the IGBT from datasheets.

978-1-4799-2706-7/14 $31.00 © 2014 IEEE

The 2014 International Power Electronics Conference

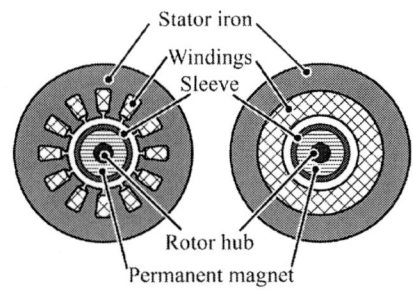

Fig. 9. Conceptual drawing of the slotted (left) and the slotless (right) types of high-speed permanent-magnet synchronous motors as shown in [21].

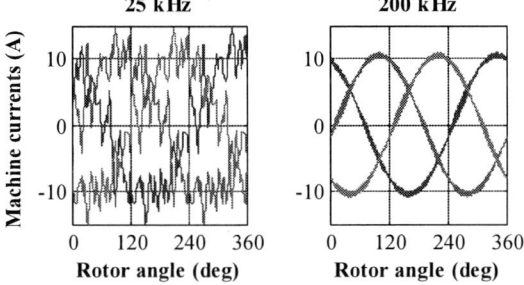

Fig. 10. Phase currents of the slotless machine at 25 kHz (left) and 200 kHz (right) inverter switching frequency.

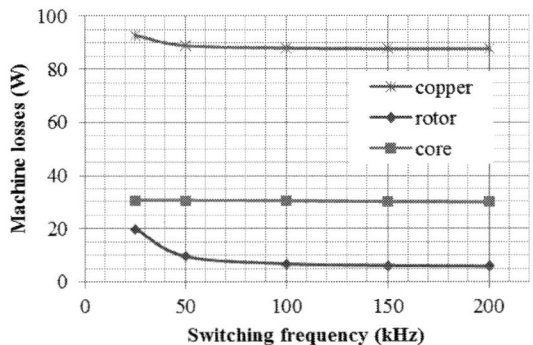

Fig. 11. Losses of the slotted machine vs. inverter switching frequency. If ideal (pure sinusoidal) phase currents are assumed, the copper losses are 80 W and the rotor losses are 5.5 W.

Different components of the machine losses are shown with respect to the inverter switching frequency in Fig. 11 for the slotted machine and in Fig. 12 for the slotless machine. It is clearly visible that for both machines, the rotor losses can be decreased significantly by increasing the inverter switching frequency. The slotted machine has more core losses due to higher flux density in the stator core and also due to its lossier stator core material compared to that of the slotless machine. Nevertheless, in both of the machines the iron losses do not change strongly with the inverter switching frequency because of the weak armature reaction. As the RMS value of the phase current also does not change much, the copper losses also stay almost constant beyond 50 kHz. The rotor losses approach zero in the slotless machine whereas they cannot be decreased below 5.5 W in the

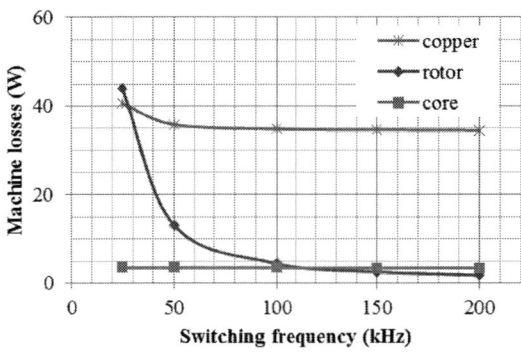

Fig. 12. Losses of the slotless machine vs. inverter switching frequency. If ideal (pure sinusoidal) phase currents are assumed, the copper losses are 31 W and the rotor losses are 0.3 W.

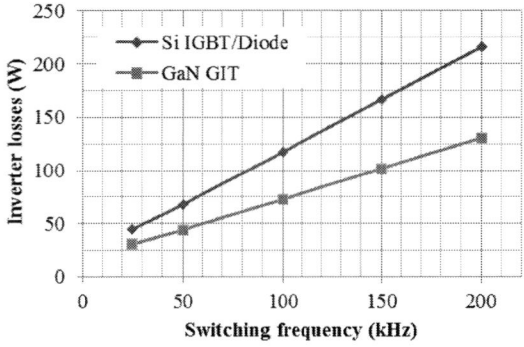

Fig. 13. Losses of the GaN and Si inverters vs. switching frequency. The machine is running at 100 krpm and 300 mNm.

slotted machine. This amount of rotor losses is caused by the slot harmonics of the slotted machine, and would be present even if no harmonics are present in the phase currents of the machine.

Fig. 13 shows the estimated inverter losses when the slotless machine is driven by the Si inverter or the GaN inverter. The GaN inverter clearly outperforms its Si counterpart for this application. In other words, the GaN GITs analyzed in this work can be used advantageously in high-speed motor drives to push the switching frequency above limits that are today given by the state-of-the-art Si devices and lead to decreased rotor losses in the machines.

VI. CONCLUSIONS AND OUTLOOK

In this work, the performance of 600 V, 15 A GaN GIT power switches is investigated experimentally. Guidelines are summarized for an optimum design of the PCB layout. The test converter is a synchronous buck converter which comprises a half bridge configuration with an external inductor and a capacitor. The losses are determined under different operating conditions by electrical input and output power measurements as well as calorimetric measurements. The switching energies of the GITs are calculated from these experiments. When compared to measuring the switch voltage and currents

978-1-4799-2706-7/14 $31.00 © 2014 IEEE

simultaneously to determine the switching energies, the method presented in this paper avoids the high bandwidth and synchronization requirements. Furthermore, as there is no current and voltage measurement directly on the transistors, the PCB layout of the half bridge circuit is optimized for minimum parasitics. As the actual switching losses depend on the PCB parasitics, the results of the measurements can be directly applied to converter systems that employ half bridge structures, e.g., bi-directional DC-DC converters or AC motor drives. On the other hand, as the measurements are done in a fixed ambient temperature, the junction temperature is not the same for the measurements presented in this work. In this work, this effect is neglected as the dependency of the switching energies and the on-state resistances of the GaN GITs are reported to be relatively insensitive to temperature variations.

High-speed electric drives are shown as a possible application for the GaN power transistors. Higher switching frequencies enabled by the use of GaN are shown to reduce the rotor losses in high-speed permanent-magnet electric machines and lead to higher efficiencies and/or higher power densities in machines.

In this work, the gate resistor and capacitor values are determined experimentally to keep the oscillations in the gate loop within safe operating limits and to avoid spurious triggering. Using packages that are better suited for high switching speeds (i.e., packages with smaller parasitic inductances) will enable better utilization of the GaN GITs.

ACKNOWLEDGEMENTS

The authors would like to thank David Eriksson and Mehmet Onur Gulbahce for their valuable contributions in building the hardware and carrying out the measurements. Furthermore, they would like to thank Panasonic for providing the sample GITs used in this work.

REFERENCES

[1] M. Kasper, D. Bortis, J. W. Kolar, "Classification and Comparative Evaluation of PV Panel Integrated DC-DC Converter Concepts," *IEEE Trans. on Power Electronics*, to be published (Early access, retrieved online Oct. 2013).

[2] U. Badstübner, A. Stupar and J. W. Kolar, "Sensitivity of Telecom DC-DC Converter Optimization to the Level of Detail of the System Model," *IEEE 26th Ann. Appl. Pow. Electr. Conf. and Expo (APEC)*, pp. 585 – 592, Mar. 2011

[3] T. Kachi, M. Kanechika and T. Uesugi, "Automotive Applications of GaN Power Devices," *IEEE Comp. Semico. Int. Circ. Symp.* (CSICS), pp. 1-3, Oct. 2011.

[4] R. Bosshard, J.W. Kolar, J. Mühlethaler, I. Stevanović, B. Wunsch, and F. Canales, "Modeling and η-α-Pareto Optimization of Inductive Power Transfer Coils for Electric Vehicles," *IEEE J. Emerg. Sel. Topics Pow. Electron,* (accepted for publication), 2014.

[5] L. Schwager, A. Tüysüz, C. Zwyssig and J. W. Kolar, "Modeling and Comparison of Machine and Converter Losses for PWM and PAM in High Speed Drives," *IEEE Trans. on Industry Applications*, to be published (Early access, retrieved online Oct. 2013)

[6] A. Hensel, C. Wilhelm and D. Kranzer, "Application of a new 600 V GaN Transistor in Power Electronics for PV

systems," *15th Int. Pow. Electr. and Mot. Cont. Conf., (EPE-PEMC ECCE Europe)*, Sep. 2012.

[7] T. Tanaka, T. Ueda and D. Ueda, "Highly Efficient GaN Power Transistors and Integrated Circuits with High Breakdown Voltages," *10th Int. Conf. Solid State Integ. Circ. Techn. (ICSICT)*, pp. 1315-1318, Nov. 2010.

[8] M. J. Scott, K. Zou, J. Wang, C. Chen, M. Su and L. Chen, "A Gallium Nitride Switched-Capacitor Circuit Using Synchronous Rectification," *IEEE Trans. on Ind. Appl.*, vol. 49, no. 3, pp. 1383-1392, May/Jun. 2013.

[9] M. Danilovic, Z. Chen, R. Wang, F. Luo, D. Boroyevich, and P. Mattavelli, "Evaluation of the Switching Characteristics of a Gallium-Nitride Transistor," *IEEE Ener. Conv. Cong. and Expo. (ECCE USA)*, pp. 2681-2688, Sep. 2011.

[10] D. Costinett, H. Nguyen; R. Zane and D. Maksimovic, "GaN-FET Based Dual Active Bridge DC-DC Converter," *26th IEEE Appl. Pow. Elect. Conf. and Expo. (APEC)*, pp. 1425-1432, Mar. 2011

[11] S. Tamura, Y. Anda, M. Ishida, Y. Uemoto, T. Ueda, T. Tanakaa and D. Ueda, "Recent Advances in GaN Power Switching Devices," *IEEE Comp. Semi. Integ. Circ. Symp. (CSICS)*, pp.1-4, Oct. 2010.

[12] T. Morita, S. Tamura, Y. Anda, M. Ishida, Y. Uemoto, T. Ueda, D. Tanaka and D. Ueda, "99.3% Efficiency of Three-Phase Inverter for Motor Drive Using GaN-based Gate Injection Transistors," *26th IEEE Appl. Pow. Electr. Conf. and Exp. (APEC)*, pp. 481-484, Mar. 2011.

[13] M. Ishida, T. Ueda, T. Tanaka and D. Ueda, "GaN on Si Technologies for Power Switching Devices," *IEEE Trans. on. Elect. Dev.*, vol. 60, no. 10, Oct. 2013.

[14] A. H. Wienhausen and D. Kranzer, "1 MHz Resonant DC/DC-Converter Using 600 V Gallium Nitride (GaN) Power Transistors ," *Materials Science Forum*, vol. 740-742, pp. 1123 – 1127, Jan. 2013.

[15] K. Shirabe, M. Swamy, J.-K. Kang, M. Hisatsune, Y. Wu, D. Kebort and J. Honea, "Advantages of High Frequency PWM in AC Motor Drive Applications," *IEEE Ener. Conv. Cong. and Expo. (ECCE USA)*, pp. 2977-2984, Sep. 2012.

[16] J. Wang, and H. S-h Chung, "Impact of Parasitic Elements on the Spurious Triggering Pulse in Synchronous Buck Converter," *IEEE Ener. Conv. Cong. and Expo. (ECCE USA)*, pp. 480-487, Sep. 2013.

[17] Z. Zhang, W. Zhang, F. Wang, L. M. Tolbert, and B. J. Blalock, "Analysis of the Switching Speed Limitation of Wide Band-gap Devices in a Phase-leg Configuration," *IEEE Ener. Conv. Cong. and Expo. (ECCE USA)*, pp. 3950-3955, Sep. 2012.

[18] H. Umegami, F. Hattori, Y. Nozaki, M. Yamamoto, and O. Machida, "A Novel High-Efficiency Gate Drive Circuit for Normally Off-Type GaN FET,", *IEEE Trans. on Ind. Appl.* pp. 593-599, vol. 50, no. 1, Jan/Feb 2014.

[19] Linear Technology, "High Power No R_{SENSE}™ Current Mode Synchronous Step-Down Switching Regulator," LTC1775 datasheet, retrieved online Mar. 2014.

[20] D. Reusch, and J. Strydom, "Understanding the Effect of PCB Layout on Circuit Performance in a High-Frequency Gallium-Nitride-Based Point of Load Converter," *IEEE Trans. on Pow. Elect.*, pp. 2008-2015, vol. 29, no. 4, Apr. 2014.

[21] L. Schwager, A. Tüysüz, C. Zwyssig, and J. W. Kolar, "Modeling and Comparison of Machine and Converter Losses for PWM and PAM in High Speed Drives," *IEEE 20th Int. Conf. Electr. Mach. (ICEM)*, pp. 2441-2447, Sep. 2012.

Wide-band gap devices in PV systems - opportunities and challenges

C. Sintamarean, E. Eni, F. Blaabjerg, R. Teodorescu, and H. Wang

Department of Energy Technology, Center of Reliable Power Electronics
Aalborg University
Aalborg, Denmark
ncs@et.aau.dk, epe@et.aau.dk, fbl@et.aau.dk, ret@et.aau.dk, hwa@et.aau.dk

Abstract— The recent developments in wide band-gap devices based GaN and SiC is showing a high impact on the PV-inverter technology, which is strongly influenced by efficiency, power density and cost. Besides the high efficiency of PV inverters, also the mechanical size, the compactness and simple structure have an important role in the cost reduction. To increase the efficiency of PV systems, most of solutions for PV inverters have moved to three-level (3L) structures reaching typical efficiencies of 98% due to low switching losses of 600V Si IGBT or MOSFET and reduced core losses in the filter. With the appearance of SiC 1200V MOSFETs, it becomes possible to return to more simple two-level (2L) structure with comparable efficiency but high potential to reduce the overall cost. This paper deals with a comparison study between a Si-based 3L-Diode Neutral Point Clamped (DNPC) and a SiC-based 2L-Full Bridge (FB) three-phase PV-inverter topologies in terms of efficiency, thermal loading distribution and costs. Moreover the above mentioned PV-inverters are built and tested in laboratory in order to validate the obtained results.

Keywords— cost analyse, efficiency measurement, high switching frequency operation, thermal loading.

I. INTRODUCTION

The improvement of PV systems is directly related with the evolution of power semiconductor devices. Silicon-based power devices are challenged to handle the market demands in terms of high power device requirements for many applications [1].

Nowadays, the power device requirements include higher blocking voltages, higher switching frequency, higher efficiency, higher power density, higher reliability and lower cost. In order to achieve this goal, the development of power electronic devices based on Wide Band-Gap (WBG) semiconductors, like Silicon Carbide (SiC) and gallium nitride (GaN), are interesting. Their recent commercialization is expected to revolutionize a part of the power electronics industry in the future. The integration on the actual commercial market [1] of the WBG-devices is emphasized with red dotted line in Fig.1. It is worth to mention that, according to their rated power (available on the market today), the main applications that can be covered by this field are PV inverters, adjustable speed drives, small pumps and also automotive. Modern power electronics devices become more demanding by going beyond the capabilities offered by silicon devices. Silicon devices are limited by the maximum junction temperature (T_J=135°C), breakdown voltage, switching frequency and power density [1]. These limitations can be exceeded by WBG - devices

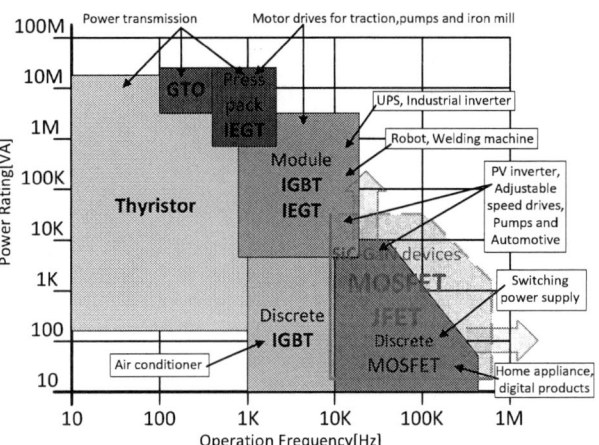

Fig. 1. The integration on the market of SiC/GaN-based applications

based on silicon-carbide. The capabilities of these devices can be translated into higher efficiency, lower volume or both [2]. The higher breakdown field of silicon-carbide allows the development of switches with high voltage blocking capability, when compared to the classic silicon switches [1] [3]. In order to take advantage of the high breakdown voltage and low switching losses of these devices, the 3L-BS NPC (bipolar switch) has been proposed as a dedicated topology for SiC-devices. This topology came as an alternative to the 3L-DNPC. Moreover, PV-inverter companies already started to produce converters based on these 3L-BS NPC topology by using SiC-devices (ex: SMA with STP 20000TLHE-10) [4]. In order to reduce the costs, the clamping to the neutral point is achieved by using Si IGBTs. This converter has achieved the highest efficiency on the market of 98.5% [5]. Even the converter has a higher efficiency compared with Si-based ones, the main disadvantage is the relative high cost. Therefore, in order to reduce the overall cost and to have a comparable efficiency, the return to 2L-Full Bridge with split capacitor has been proposed as an alternative. This paper deals with a comparison study of the above mentioned PV-Inverter topologies in terms of efficiency, thermal loading distribution and cost.

II. WIDE BAND-GAP DEVICES

Nowadays, the development of silicon (Si) semiconductor technology has almost achieved its theoretical maturity, exploiting through the limits the silicon material. In 1950's it was announced that the

WBG-semiconductors will be used in the next generation of power electronics devices once Si will be exploited through its physical limits.[6] These WBG-devices have superior electro-thermal properties and are likely candidates to replace Si in some applications. The main advantages of WBG-devices in comparison with the Si based ones are [1][7][8]:

- Perform in a higher temperature operation levels without losing their electrical characteristics (slightly variation of forward and reverse characteristics) due to the wider band-gap. Therefore they can perform well in extreme conditions, where Si devices cannot be used.
- Higher breakdown voltage due to the higher electric breakdown field of the material. Moreover, due to this characteristic much higher doping levels can be achieved. Thus a thinner semiconductor chip is used to achieve the same breakdown voltage like in Si. This results in a lower on-resistance which implies lower conduction losses.
- Higher power density of the device may be achieved due to higher thermal conductivity of WBG-semiconductor (exception GaN).
- These devices can operate at higher switching frequency (lower switching losses) due to the higher saturation electron drift velocity, thinner semiconductor chip and higher thermal conductivity.

Even the WBG-devices have the above mentioned advantages, when compared with Si-based ones, there are some challenges, which have to be overpassed in order to take fully advantage of this semiconductor material. The WBG-semiconductor inside the device cannot be exploited through its rated limits due to the present unavailability of high temperature packaging techniques which now are in the process of being developed. To understand the fully potential of this material it is worth to mention that the device may require a packaging with performances of power density of 1000 W/cm^2 and a temperature of 300°C or more. Meanwhile, the currently available packaging technique has a power density of 280

W/cm^2 and a maximum allowed temperature of less than 125°C [9]. In terms of short-circuit capability, the performance of commercialized SiC-devices is lower compared with Si-devices. Due to the wider band-gap and higher thermal conductivity of SiC-material, there is a huge potential to improve the short-circuit capability of this devices, thus a lot of research is focused on this direction [10].

When discussing about the technology maturity it can be stated that the first SiC-diodes were commercially available with more than 10 years ago and various SiC switches in the last 3-4 years. The first low voltage GaN devices have just recently started to be available on the market.

A comparison study between SiC and GaN devices in terms of on-resistance versus breakdown voltage is shown in Fig. 2. It can be noticed that below 1 kV GaN is preferred due to the reduced channel resistance. Above this value SiC is leading because of the necessary derating of GaN devices (safety margin for stable operation) [7]. Considering their main advantages, currently it seems that GaN transistors are suitable-dedicated for (voltage and current levels up to 600V and 30A) low power high frequency integrated circuits, while SiC is rather suited for discrete devices or modules with breakdown voltage up to 1.7 kV and current up to 100A.

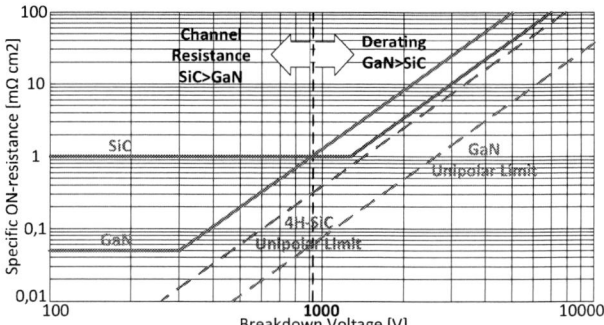

Fig. 2. WBG-devices SiC vs. GaN in terms of rating [7]

TABLE I.   ASSOCIATED Si-SiC DEVICES FOR 1200 V

Device type/ ratings for Tj=100°C	Discrete Devices 1200V, 20A			One-Leg Module 1200V, 100A			Three-Phase Module 1200V, 50A		
Si-IGBT	IHW20T120-Infineon			FF100R12YT3-Infineon			FS50R12KT4B-Infineon		
SiC-MOSFET	C2M0080120D-CREE			CAS100H12AM-CREE			CCS050M12CM-CREE		
Rth(jc) [K/W]	Si		SiC	Si		SiC	Si		SiC
Transistor.-F.Diode	0.85	1.3	0.6	0.6	0.28	0.48	0.19	0.37	0.54	0.81	0.37	0.42
Threshold Voltage	5.8V		3V	5.8V		2.4V	5.8V		2.3V
Device Cost Euro	5.7		13.5	110		374	138		372
Euro/A	0.29		0.67	0.55		1.87	0.46		1.24
SiC/Si	2.3X			3.4X			2.7X		
Sw. test conditions	800V, 20A, Tj=150°C			600V, 100A, Tj=125°C			600V, 50A, Tj=150°C		
Gate resistance [Ω]	35	6.8		6.8	5		15	20	
Switching loss	Si	SiC	Si/SiC	Si	SiC	Si/SiC	Si	SiC	Si/SiC
Turn-on Eon [mJ]	2	0.28	6.9X	9.5	2	6.5X	5.8	1.1	8.2X
Turn-off Eoff [mJ]	1.8	0.27		9	1.4		4.5	0.6	
Revers rec. Err [mJ]	included			6	0.4		3.6	-	

Nowadays, there is a wide range of SiC-devices available on the market, from discrete to module devices.

Table I presents a comparison study between Si and SiC devices which have the same power ratings in terms of: junction to case thermal impedance, threshold voltage, switching losses and costs. It can be seen that the costs of

SiC device is around 2-3 times higher and the switching losses are around 6-8 times lower when comparing with Si ones. Moreover, the junction to case thermal impedance and threshold voltage are lower in the SiC case. All these advantages have a positive impact on the switching frequency operation of these devices.

III. WBG-DEVICES INPACT IN PV-INVERTER SYSTEMS

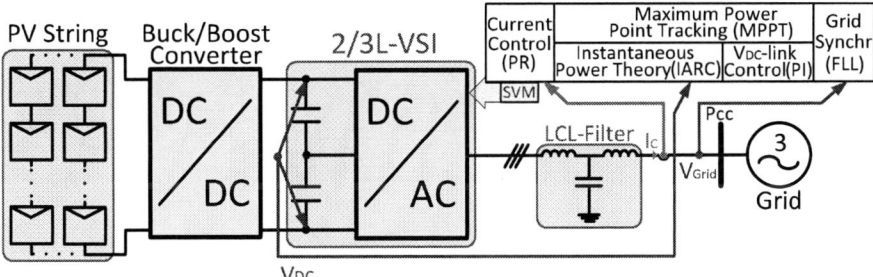

Fig. 3. Grid connected PV-system

Power electronics is the key technology which enables the photovoltaic (PV) system to be connected to the grid. Fig. 3 presents a typical grid connected PV-system which consists of: the PV-string, the buck-boost DC-DC converter, the voltage source inverter (VSI) and the output LCL-filter. Depending on the PV-system power, the use of the DC-DC converter stage is optional. Moreover the main control blocks required to achieve the required system functionality are also shown in Fig. 3.

Due to the high price of the solar electricity-technology, the use of high efficiency inverters is an important requirement.

For inverters with silicon (Si) based devices, the switching frequency is limited (typical to 16 kHz) by the switching losses leading to increased magnetics and cooling requirement [11], [12]. To increase the efficiency of PV-systems, most of the solutions for Si based PV-inverters have moved to 3L structures reaching typical efficiencies up to 98% due to low switching losses of 600V Si IGBT or MOSFET and reduced core losses in the filter. The 3L-DNPC inverter is a good option as it enables the use of lower voltage rated switches (600 V) and also reduces the filtering requirements [12]. This PV-inverter topology (Fig. 4 a) is ea. used by Danfoss Solar in their products. The higher breakdown field of silicon-carbide allows the development of switches with a high blocking voltage capability when compared to silicon switches [3]. With the appearance of SiC 1200V MOSFETs, it becomes possible to return to a more simple 2L structure (Fig. 4 c) with a comparable efficiency but with a high potential to reduce the overall cost. It is worth to mention that according to the rated power available on the market today, WBG devices can be used in PV-inverter applications up to 3 kW for GaN and 50 kW for SiC respectively [14], [15]. In order to take advantage of the high breakdown voltage and the

low switching losses of these devices, the 3L-BS NPC has been proposed as a dedicated topology (Fig. 4 b) for SiC-devices. This topology came as an alternative to the 3L-DNPC. Moreover, the PV-inverter companies already started to produce converters based on these topologies (3L-BS NPC) by using SiC-devices (ex: SMA with STP 20000TLHE-10)[4], [5]. In order to reduce the costs, the clamping to the neutral point is achieved by using Si IGBTs. This converter has obtained the highest efficiency on the market of 98.5% [5]. Even the converter has a higher efficiency when compared with the ones based on Si-devices, the main disadvantage is the high cost. Therefore, in order to reduce the overall cost and still to have a comparable efficiency, the return to 2L-Full Bridge (Fig. 4 c) with split capacitor has been proposed as an alternative.

Table II presents a comparison of the following PV-inverter topologies: 3L-DNPC based on Si-IGBTs, 3L-BS NPC based on SiC-MOSFETs (main switches) and Si-IGBTs (clamping switches) and 2L-FB with split capacitor based on SiC-MOSFETs. When SiC switches were not available, the 3L-NPC topology won a fair place on the market due to the advantages offered over 2L-inverters. Compared with 2L-inverter, the main disadvantages of the 3L ones are related to the double count of switches associated with the drive and protection circuitry and the double voltage requirement in the DC-link along with the need for voltage balancing loop. [16] Three-level inverters exhibit problems such as unequal loss distribution within the switching devices, which leads to unequal switch temperature distribution. The uneven loss distribution among the switches increases once with an increase of the switching frequency and it has a negative impact seen from a reliability point of view.

978-1-4799-2706-7/14 $31.00 © 2014 IEEE

The 2014 International Power Electronics Conference

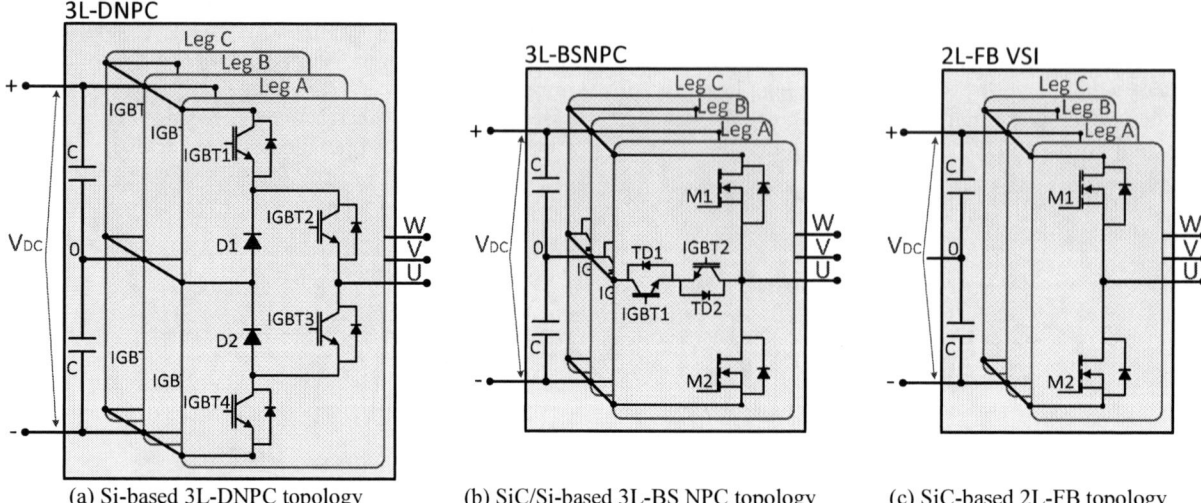

(a) Si-based 3L-DNPC topology   (b) SiC/Si-based 3L-BS NPC topology   (c) SiC-based 2L-FB topology

Fig. 4. Three-phase PV-inverter topologies: 3L-DNPC (a), 3L-BS NPC (b) and 2L-FB (c)

Fig. 5 presents the thermal loading distribution across the converter devices (for a 25 kW grid connected PV-inverter) when the rated active power is injected into the grid. Three different cases are studied considering the converter topologies presented in Fig. 4.

Fig. 5 (a) shows the 3L-DNPC converter devices thermal loading at nominal power. As it was expected due to the current stress repartition and the higher switching frequency, the outer IGBTs chip temperature is higher than the inner IGBTs.

The thermal loading of 3L-BS NPC topology devices is presented in Fig. 5 (b). For the clamping IGBT it is plotted also its freewheeling diode temperature, because in this topology the current stress of the neutral point clamping IGBT is the same with the freewheeling diode. Even they have the same current stress, the temperature of the IGBT and its freewheeling diode differs considerably due to the thermal impedance, which in the case of the diode is doubled. Finally, the thermal loading difference within the devices will have a negative impact

on the converter reliability [17]. This problem can be avoided by using a 2L-topology. According to Fig. 5 (c) the 2L-FB topology has a better thermal loading distribution among the devices.

With the introduction of SiC devices, the 2L inverter is again a good competitor for the 3L ones, due to its simplicity [16]. The disadvantages like the output filter size can be overcome by selecting a proper design. The size of the output filter components is strongly dependent on the current ripple and decreases, when the switching frequency increases [11]. As the SiC devices operate with much lower switching losses, the switching frequency can be increased. Having lower losses, the overall size and weight can be reduced due to the smaller heat-sink. A major factor contributing to heat-sink reduction is also the high temperature operation. By using a high switching frequency, a smaller DC-link capacitance value is required. Moreover, film capacitors can be used thus overcoming the limited life-time problem of the electrolytic ones. [18]

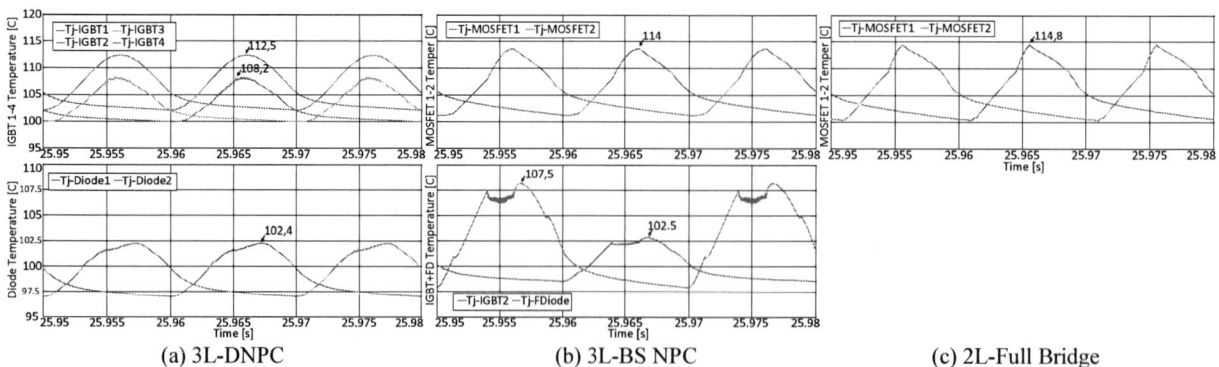

(a) 3L-DNPC   (b) 3L-BS NPC   (c) 2L-Full Bridge

Fig. 5. Thermal loading distribution across PV-Inverter devices at nominal load operation for: 3L-DNPC (a), 3L-BS NPC (b) and 2L-FB (c)

TABLE II. COMPARISON BETWEEN 3-PHASE PV-INVERTER TOPOLOGIES USING Si AND SiC TECHNOLOGIES

	Three-Phase PV-Inverter Topologies		
	3-Level NPC Si	3L-BSNPC SiC-Si	2L-Full Bridge SiC
Switches	12 + 6 diodes ☹	12 ☺	6 ☺
Gate Drivers	12 ☹	12 ☹	6 ☺
PWM Algorithm	Complex ☹	Complex ☹	Simple ☺
PCB Size (L x l)	Higher ☹	Higher ☹	Lower ☺
Output Filter Size	Decreased ☺	Decreased ☺	Decreased ☺
THD	Decreased ☺	Decreased ☺	Increased ☹
Efficiency	Moderate ☺	Higher ☺	Higher ☺
Size and Weight	Higher ☹	Moderate ☺	Lower ☺
Switch Stress	Uneven ☹	Uneven ☹	Even ☺
DC Link Balancing	Needed ☹	Needed ☹	Not needed ☺
Operating Temperature	Low ☹	High ☺	High ☺
DC Link Capacitor Bank	Bigger ☹	Moderate ☺	Smaller ☺
Protection	Complex ☹	Complex ☹	Simple ☺

IV. GRID CONNECTED PV-INVERTER HARDWARE DESIGN

This paper deals also with the comparison of two associated Si and SiC based PV-inverters in terms of efficiency and cost. The Si based PV-inverter relies on 3L-DNPC topology and the SiC based one on 2L-Full Bridge topology. The main target is to achieve a competitive cost with 3L-DNPC converter and a higher efficiency.

The proposed PV-system consists of an inverter connected to the three phase grid through a passive LCL-filter (Fig. 7). A more detailed description of the grid connected PV-inverter control design has been presented in [19]. The design of 25 kVA grid connected PV-inverter will be performed in agreement with the power rating parameters given in Table III.

According to Table III, the maximum DC-link voltage is 1 kV and the nominal output current is 37 A (RMS). The chosen device modules are dedicated for the implemented inverter topologies. When dealing with 3L-DNPC topology at this power levels, the 3-phase power modules are not available on the market. Therefore, for implementing the 3L-DNPC inverter, single phase Si-IGBT modules are used. Thus three modules have to be used, one for each phase leg. For implementing the 2L-FB PV-inverter, a 3-phase SiC-MOSFET module (available on the market from CREE) is chosen.

Fig. 6 shows the 3L-DNPC (a) and 2L-FB (b) PV-inverters hardware implementation. The PV-inverter hardware implementation can be divided in the following main sections as follows:

- 1. – 3 phase - output power
- 2. – DC - input power
- 3. – Heatsink
- 4. – DC - link capacitors: electrolitic Fig. 6 (a) and film Fig. 6 (b)
- 5. – Device modules: IGBT Fig. 6 (a) and MOSFET Fig. 6 (b)
- 6. – Gate Driver circuit
- 7. – DSP control

TABLE III. CONVERTER DESIGN RAITINGS

2L/3L PV-inverter Specifications		
Rated power	S=25 kVA	
Conv. Output phase voltage	V_N = 230 V (RMS) (325 V peak)	
Max. Output current	I_{max} = 37 A (RMS) (52 A peak)	
Max. DC-link Voltage	V_{DC-max} = 1000 V	
Switching Frequency	3L-DNPC	2L-FB
	f_{sw} = 16 kHz	f_{sw} = 50 kHz
Heatsink thermal resistance		
Thermal Resistance	Si-based 3L-DNPC	SiC-based 2L-FB
	R_{th}=0.11 K/W	R_{th}=0.12 K/W
Device Power Ratings		
Device Type	Single-Phase IGBT module Infineon-F3L50R06W1E3_B11	3-Phase MOSFET module CREE-CCS050M12CM2
Power Ratings Voltage/Current	$V_{(BR)CE}$=600 V / I_C=50 A	$V_{(BR)DS}$=1200 V / I_D=52 A
Max allowed. J/C Temperature	T_J= 135 °C T_C= 80 °C	T_J= 135 °C T_C= 100 °C
DC-Link and output filter specifications		
	3L-D NPC	2L-FB
DC-Link Capacitance	32uF	10uF
Total Filter Inductance	0.52 mH	0,32 mH

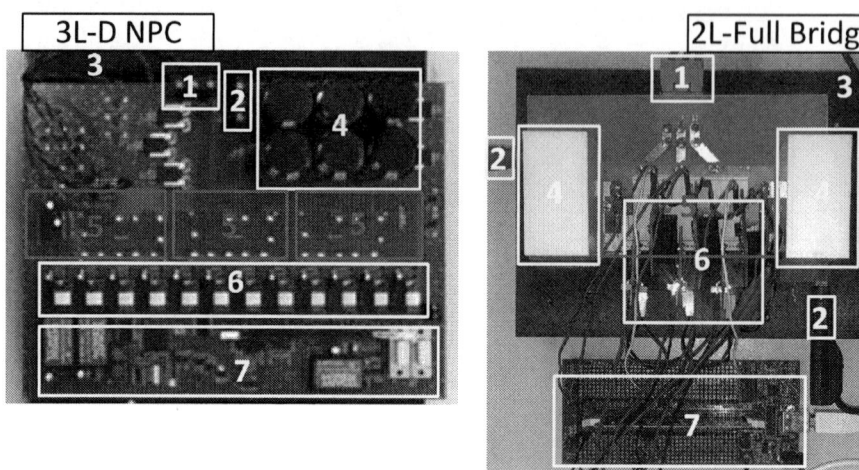

(a) Si-based 3L-DNPC   (b) SiC-based 2L-FB

Fig. 6. Hardware design of the 3L-DNPC (a) and 2L-FB (b) PV-inverters

V. SI VS. SIC PV-INVERTER COST ANALYZES

A hardware costs analyze of the Si-based 3L-DNPC and the SiC based 2L-HB PV-inverters has been performed in parallel in order to emphasize the cost-rentability. The PV-Inverters cost analyzes considers the following components used in hardware implementation:

- Device Module – as it was mentioned above the main target is to compare a Si-based 3L-DNPC versus a SiC-based 2L-HB PV-inverters. Thus Si-IGBT and SiC-MOSFET modules with similar power ratings have to be used. The selected device modules are dedicated for the implemented inverter topologies. For implementing the 3L-DNPC inverter, single phase Si-IGBT modules from Infineon are used. Thus three modules have to be used, one for each phase leg. For implementing the 2L-FB PV-inverter, a 3 phase SiC-MOSFET module from CREE has been used.

- Gate-Driver (GD) Circuit – the GD circuit has been custom made and consists of the following components: GD (Avago Technologies), Insulated dc/dc converter (Recom Power) and linear regulators (Texas Instruments). The driver price for Si and SiC is shown in Table IV.

- Printed Circuit Board – The PCB has been designed in Altium Designer software. The PCB has been produced by PCBCART company. The price mainly varies according with the number of layers and surface (L x l [cm^2]).

- Heat-Sink – The heatsink has been ordered from HS-Marston company. Mainly the price varies according with the cooling method, heatsink thermal impedance and power dissipation. The heatsink Zth has been calculated according with the maximum allowed junction and case temperature of the device for the rated power operation.

- DC-Link Capacitor – The DC-Link Capacitance has been calculated for a maximum voltage ripple of 5%. It is worth to mention that the capacitance of the SiC inverter is three times lower than the one of the Si inverter. The capacitors have been ordered from Vishay Roederstein manufacturer.

- LCL- filter – the output filter has been custom made by Trafox. The filter inductance of the SiC inverter is with almost 40 % lower when compared with Si one.

TABLE IV. CONVERTER COST ANALYZE HARDWARE COMPARISON

		Three-Phase PV-Inverter			
		3L-DNPC			2L-HB
No.	Main Components	Price [€]	Manufacturer		Price [€]
1	Device Module	(3 x) 60	Infineon IGBT	CREE MOSFET	360
2	Gate Driver Circuit	(12 x) 24.2	Custom made		(6 x) 23.5
3	PCB	23 (30 x 20)	PCBCART Dimension L x l [cm^2]		12.5 (20 x 15)
4	Heat-sink	60	HS-Marston		55
5	DC-Link Capacitors	52	Vishay Roederstein		24
6	LCL-Filter	154	Polylux-Trafox		102
	Total Price:	755.6 €		694.5 €	

The price of the main components used in hardware development of the mentioned PV-Inverters, is shown in Table IV. The price of 3L-DNPC is with 8.8 % higher than 2L-FB SiC. The overall cost reduction is mainly due to: a reduced output filter size (40 % lower inductance), a smaller DC-link capacitance value (70 % lower capacitance), the half number of gate drivers and a smaller PCB dimension. The price difference it will have a positive impact in the price reduction of the PV-inverter technology.

VI. EXPERIMENTAL RESULTS

The proposed laboratory setup consists of a PV-inverter connected to the grid through a LCL-filter and a galvanic isolator transformer.

The schematic of the laboratory setup is introduced in Fig. 7. The input power to the inverter is provided by the DC-source Magna-Power electronics DP series II.

The desired system controllability is achieved by implementing the designed control algorithm using a DSP board from Texas Instruments. The current controllability has been achieved by using proportional resonant (PR) controllers, for the grid synchronization a frequency locked loop (FLL) was implemented and for the active-reactive power injection has been used the instantaneous active and reactive control (IARC).

The PV-inverters are tested at the same conditions. The main purpose is to measure the PV-inverter efficiency for different active power levels injected into the grid. The efficiency is measured for different power ratings from 0 to 25 kW with steps of 2.5 kW by using Yokogawa WT3000 power analyser.

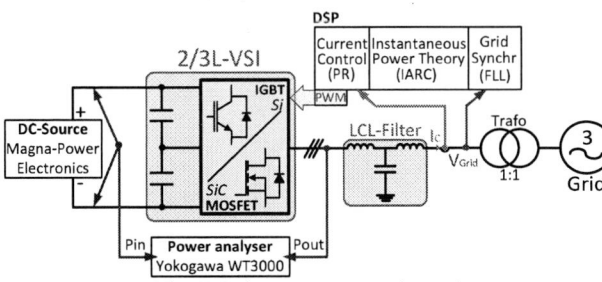

Fig. 7. Laboratory setup schematic

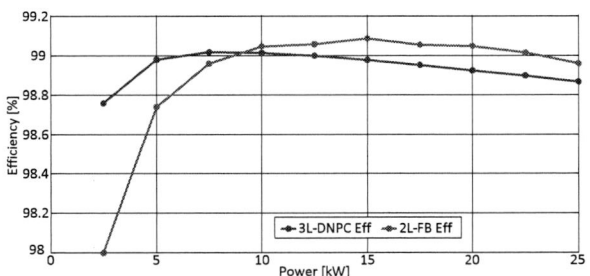

Fig. 8. Efficiency measurements curve of 3L- DNPC and 2L-FB PV-inverters

By analyzing the obtained results presented in Fig. 8, the efficiency of 2L-FB (50 kHz operation) is higher than 3L-DNPC (16 kHz operation) for the injected power higher than 9 kW. It is worth to mention that the efficiency it was measured only for the inverter devices. The LCL-filter was not considered.

VII. CONCLUSIONS

The design and hardware implementation of two associated Si and SiC-based PV-inverters have been performed. A hardware cost analyzes comparison of the Si-based 3L-DNPC and SiC based 2L-HB PV-inverters have been performed in order to investigate the cost-rentability.

Even the price of SiC-MOSFET module is higher than Si-IGBT module, the reduced complexity of SiC-based 2L-FB topology and its higher switching frequency operation (when compared to the Si-based 3L-DNPC topology) have a positive impact in cost reduction of the PV-inverter. The overall cost reduction is mainly due to: a reduced output filter size (40 % lower inductance), a smaller DC-link capacitance value (70 % lower capacitance), half number of gate drivers and a smaller PCB dimension.

Moreover, the efficiency has been measured in the laboratory for both inverters. The obtained results show that the efficiency of 2L-FB (50 kHz operation) is higher than 3L-DNPC (16 kHz operation) for injected power higher than 9 kW. It offers higher efficiency while exhibiting increased reliability (due to the decreased number of switches and a better thermal loading distribution across the converter devices), reduced complexity, reduced weight/size and a competitive cost.

As a final conclusion it can be stated that SiC-based 2L-FB inverter is a good candidate to replace the Si-based 3L-DNPC inverter in the application of three-phase PV-inverters for +10kW range in a cost-effective way.

REFERENCES

[1] B.J. Baliga, Silicon Carbide Power Devices, World Scientific Publishing Co. Pte. Ltd. 2006.

[2] E. Cilio, B. McPherson, R. Schupbach, A. Lostetter, J. Garrett, "A Novel High Density 100kW Three-Phase Silicon Carbide (SiC) Multichip Power Module (MCPM) Inverter", Proc. of APEC, 2007 IEEE, pp. 666-672.

[3] M. Shen, S. Krishnamurthy, M, Mudholkar, "Design and Performance of a High Frequency Silicon Carbide Inverter", Proc. of ECCE, 2011 IEEE, pp. 2044-2049.

[4] G. Deboy, R Rupp, R. Mallwitz, H. Ludwing,"New SiC JFET Boost Performance of Solar Inverters", Issue 4 Power Electronics Europe 2011.

[5] The Solar Power Magazine-Photon International, July 2012.

[6] W. Shockley, "Introductory Remarks in Silicon Carbide, A High Temperature Semiconductor", Pergamon Press, 1960

[7] N. Kaminski,and O. Hilt, "SiC and GaN Devices-Competition or Coexistance?", Proc. of CIPS 2012, March 6, 2012, Nuremberg-Germany,pp 1-11.

[8] B. Ozpineci and L.M. Tolbert, "Comparison of Wide-Bandgap semiconductors for Power Electronics Applications", US Department of Energy-Oak Ridge National Laboratory, December 12,2003.

[9] Y. Liu, Power Electronics Packaging, ISBN 978-1-4614-1052-2, Springer, 2012.

[10] K. Yano, Y. Tanaka, T. Yatsuo, A. Takatsuka, and K. Arai, "Short-Circuit Capability of SiC Buried-Gate Static Induction Transistors: Basic Mechanism and Impacts of Channel Width on Short-Circuit Performance", IEEE Transactions on Electronic Devices, Vol. 57, No. 4, April 2010.

[11] M. Hudson, R. Behnike, R. West, S. Gonzalez, J. Ginn, "Design Considerations for Three-Phase Grid Connected Photovoltaic Inverters", Photovoltaic Specialists Conference, 2002 IEEE, pp.1396-1401.

[12] Y. Kim, H. Cha, B. Song, Y. Lee," Design and Control of a Grid-Connected Three-Phase 3-Level NPC Inverter for Building Integrated Photovoltaic Systems", Proc. of IEEE Innovative Smart Grid Technologies (ISGT), pp. 1-7.

[13] R. Inzunza, H. Yamaguchi, E. Ikawa, T. Sunmiya, Y Fujiii, A. Satoh, "Design and Development of a 500kW Utility-Interactive Switch-Clamped Three-Level Photovoltaic Inverter", Proc. of Power Electronics and ECCE Asia (ICPE & ECCE), 2011,pp.1627-1631.

[14] G. Wang, F. Wang, G. Magai, Y. Lei, A. Huang, M. Das, "Performance Comparison of 1200V 100A SiC MOSFET and 1200V 100A Silicon IGBT", Proc. of IEEE ECCE, pp. 3230-3234, 2013.

[15] www.transphormusa.com

[16] H. Preckwinkel, D. Krishna, N. Fröhleke, J. Böcker, "Photovoltaic Inverter with High Efficiency over a Wide Operation Area – A Practical Approach", Proc. of IECON 2011, pp.912-917.

[17] F. Blaabjerg, K. Ma and D. Zhou, "Power electronics and reliability in renewable energy systems", Proc. of IEEE International Symposium on Industrial Electronics (ISIE), pp. 19-30, July 2012.

[18] H. Wang, H. Chung, W. Liu, F. Blaabjerg," Long lifetime DC-link voltage stabilization module for smart grid application", Advances in Power System Control, Operation and Management (APSCOM 2012), 9th IET International Conference, pp. 1-6, 2012

[19] C. Sintamarean, F. Blaabjerg and H. Wang, "Comprehensive Evaluation on Efficiency and Thermal Loading of Associated Si and SiC based PV Inverter Applications", Proc. of IECON ,November 2013.

Power Electronics Equipments Applying Novel SiC Power Semiconductor Modules

Kazuaki Mino   Ryuji Yamada   Hiroshi Kimura   Yasushi Matsumoto

Corporate R&D Headquarters
Fuji Electric Co., Ltd.
1, Fuji-machi, Hino-city, Tokyo, Japan

Abstract—**New SiC (Silicon Carbide) power semiconductor modules were developed and applied to the power electronics equipments. 25% of loss in the motor drive inverter is reduced by applying hybrid modules which are composed of conventional Silicon IGBTs (Si-IGBT) and SiC Schottky Barrier Diodes (SiC-SBD). In case of a 20kW inverter for solar photovoltaic (PV) generation, 99% of main circuit efficiency and reduction of size to 25% are realized by applying all-SiC modules using SiC-MOSFETs and SiC-SBDs. Furthermore, we developed a boost chopper using the all-SiC modules for high power PV inverter (Mega solar) application. By applying the boost choppers, the input voltage fluctuation is reduced and 25% of power density is increased.**

Keywords— SiC, MOSFET, PV inverter, motor drive inverter

I.   INTRODUCTION

To realize dramatically reduced size, weight, and loss of power electronics, technical innovation of the power semiconductors including package technologies is indispensable.   The characteristics of Silicon power semiconductors have been improved. However they are now close to the limit. Therefore, further research of the power semiconductors using the wide band gap materials, such as SiC and GaN, is required [1][2][3].

In this paper, characteristics of a hybrid module and an all-SiC module are introduced. The hybrid module consists of Si-IGBTs and SiC-SBDs and the all-SiC module consists of SiC-MOSFETs and SiC-SBDs, respectively. The hybrid modules are applied to a general purpose inverter for motor drive and the all-SiC modules are applied to the PV inverters. This paper also describes their performance.

II.   CHARACTERISTICS OF SiC POWER MODULES

A. Hybrid Module (Si-IGBT/SiC-SBD)

Fig. 1 illustrates the internal circuit of the hybrid power integrated module (PIM). As shown in Fig. 1, six SiC-SBDs are used as free wheeling diodes (FWD). The SiC-SBD chips have been developed in collaboration with National Institute of Advanced Industrial Science and Technology (AIST) [1]. The 6th-generation IGBT chips made by Fuji Electric are applied for the switching devices. The package of the hybrid

PIMs is the same to the conventional one in order to maintain compatibility (Fig. 2). In the actually used operating range, the forward voltage drop of the SiC-SBDs is almost same to that of the conventional Si-PNDs (PN junction diodes) [4]. Moreover, since the forward voltage drop characteristic of the SiC-SBDs has strong positive temperature dependency, a current can be easily balanced in case of a parallel connection even using many devices.

Fig. 3 shows the reverse recovery waveforms of the FWD and Fig. 4 shows the turn-on waveforms of the Si-IGBT, respectively. Since pouring behavior of a minority carrier does not occur, the peak reverse current flowing to the SBD, which is a uni-polar device, can be suppressed at the reverse recovery timing. Furthermore, this causes also reduced the peak turn-on current of the Si-IGBTs. 30% of the reverse recovery loss and 46% of the turn-on loss are reduced compared to the conventional PIM, respectively.

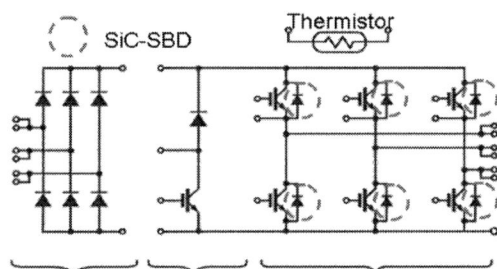

Rectifier   Brake Chopper   Inverter
Fig. 1. Internal circuit diagram of the hybrid PIM.

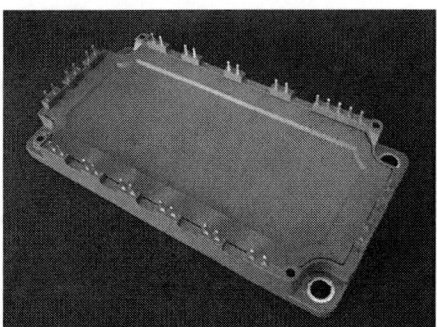

L:122 * W:62 * H:17 (mm^3)
Fig. 2. Appearance of the hybrid PIM (1200V/50A).

The 2014 International Power Electronics Conference

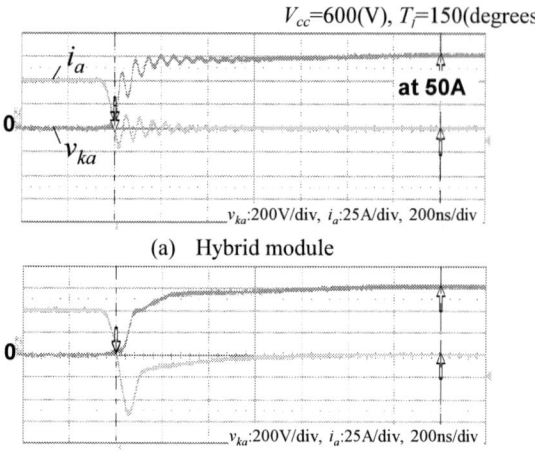

V_{cc}=600(V), T_j=150(degrees)

(a) Hybrid module

v_{ka}:200V/div, i_a:25A/div, 200ns/div

(b) Conventional Si module

Fig. 3. Reverse recovery waveforms of the FWD.

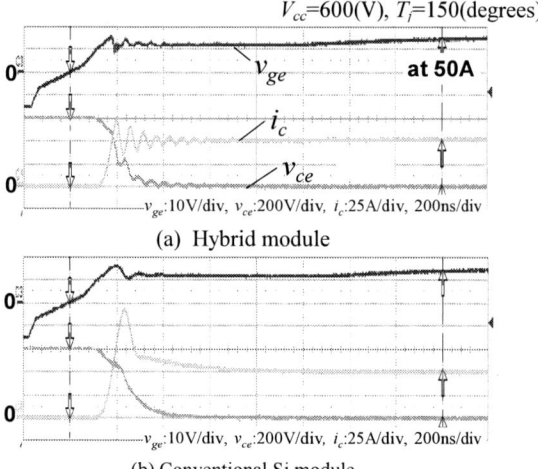

V_{cc}=600(V), T_j=150(degrees)

(a) Hybrid module

v_{ge}:10V/div, v_{ce}:200V/div, i_c:25A/div, 200ns/div

(b) Conventional Si module

Fig. 4. Turn-on waveforms of the Si-IGBT.

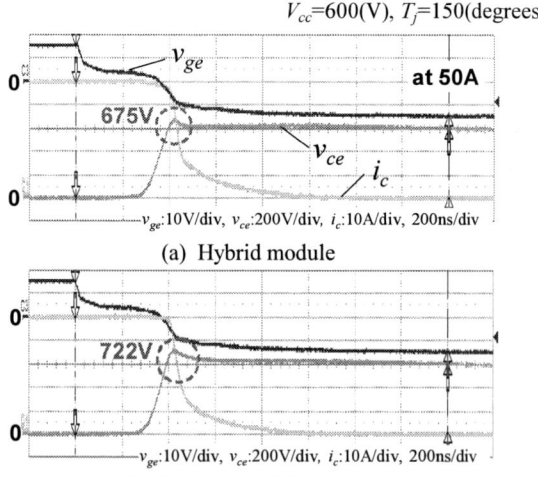

V_{cc}=600(V), T_j=150(degrees)

(a) Hybrid module

v_{ge}:10V/div, v_{ce}:200V/div, i_c:10A/div, 200ns/div

(b) Conventional Si module

Fig. 5. Turn-off waveforms of the Si-IGBT.

Fig. 5 shows the turn-off waveforms of the Si-IGBT. The peak voltage of the hybrid PIM at the turn-off is slightly lower than that of the conventional PIM.

B. All-SiC Module (SiC-MOSFET/SiC-SBD)

SiC-IEMOS (Implantation and Epitaxial Metal Oxide Semiconductor) chips are used for the MOSFETs. They are developed in collaboration with AIST [2].

As well as the SiC-SBDs, the charcteristics of the SiC-MOSFETs has positive temperature dependency. This makes parallel operation easy. From Fig. 6, it is obvious that very low on resistance (RonA) and enough threshold voltage (Vth) to realize normally-off are obtained. The normally-off state can be maintained even at high temperature such as 200 degrees centigrade. Fig. 7 and Fig. 8 show the turn-off and turn-on waveforms of the SiC-MOSFET, respectively. Both turn-off and turn-on switching periods are remarkably short compared to those of the Si-IGBTs. It is noted that the switching behaviors little change depending on temperature. Therefore, both turn-off and the turn-on losses are almost constant. The turn-off loss and turn-on loss at the DC-link voltage V_{cc} of 600V and the junction temperature T_j of 150 degrees centigrade are reduced to 40% and 31% compared to those of the Si-IGBTs, respectively.

Fig. 9 illustrates internal structures of conventional and new packages. In order to realize high power density, the newly developed structure has copper pins attached to each chip electrode instead of using conventional wire-bond [5]. To realize high temperature operation and high power density, low thermal-resistance is obtained by using the direct copper bonding (DCB) board which has thick Cu and Silicon-Nitride (Si$_3$N$_4$) boards [6]. Moreover, high heat resistant epoxy molding structure is applied instead of the conventional silicone-gel filled structure, which results in reduced distortion in the junction of the chip circumference. Multiple parallel connections of many small chips are carried out by the new structure. The structure enables large capacity and small size. Fig. 10 shows the photo of a newly developed 2 in 1 type

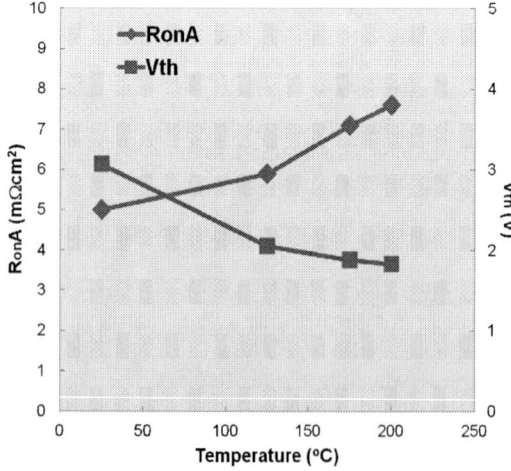

Fig. 6. Temperature dependencies of RonA and Vth.

978-1-4799-2706-7/14 $31.00 © 2014 IEEE

The 2014 International Power Electronics Conference

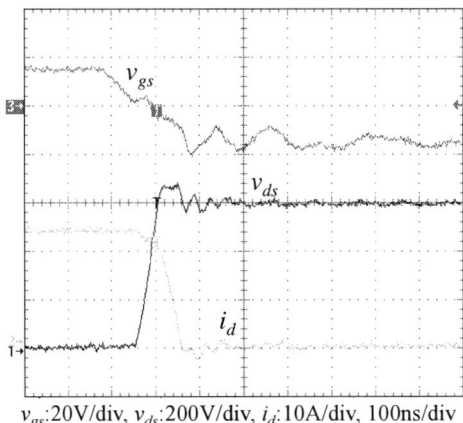

v_{gs}:20V/div, v_{ds}:200V/div, i_d:10A/div, 100ns/div

Fig. 7. Turn-off waveforms of the SiC-MOSFET at V_{cc}=600V and T_j=200°C.

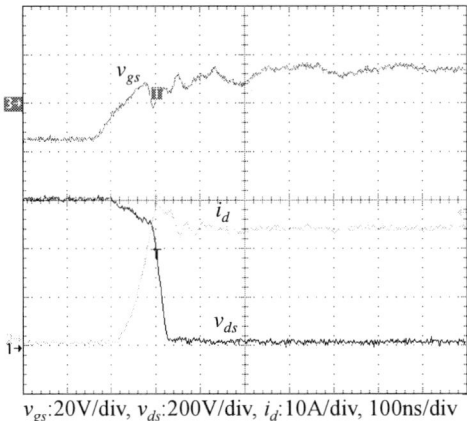

v_{gs}:20V/div, v_{ds}:200V/div, i_d:10A/div, 100ns/div

Fig. 8. Turn-on waveforms of the SiC-MOSFET at V_{cc}=600V and T_j=200°C.

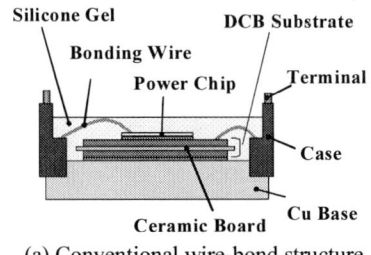

(a) Conventional wire-bond structure.

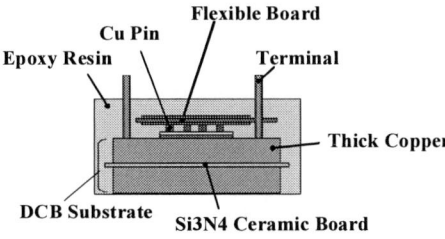

(b) Newly developed structure.

Fig. 9. Comparison of structure schematics of the packages.

module (1200V/100A). As shown in Fig. 11, the foot-print size of the newly developed module is half compared with that of the conventional one.

III. POWER ELECRONICS EQUIPMENTS APPLYING THE SiC POWER DEVICES

A. Moter Drive Inverter Applying SiC-SBDs

In order to increase efficiency of motor drive inverters for application of production facilities and air conditioners in factories, we developed the inverters applying the SiC-SBDs. Fig. 12 shows the photos of the inverter (400V/11kW) and its main power unit. By applying the hybrid modules explained in the previous chapter, the reverse recovery loss and the turn-on loss can be reduced. As shown in Fig. 13, 25% of the total loss is reduced compared to the conventional one. We recommend using the inverter with a permanent magnet synchronous motor to obtain significant reduction of energy consumption.

Fig. 10. Appearance of the newly developed 2 in 1 type all-SiC package (1200V/100A).

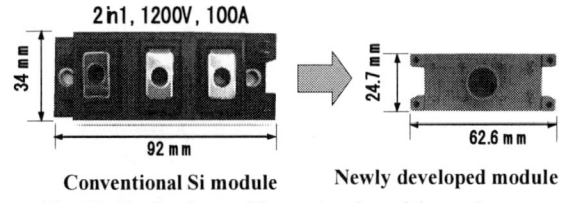

Fig. 11. Comparison of foot-print size of the packages.

Fig. 12. Appearance of the newly developed inverter and its main power unit.

B. 20kW PV Inverter Applying SiC-MOSFETs and SiC-SBDs

In order to confirm the validity of the all-SiC modules, we developed a 20kW PV inverter applying the SiC-MOSFETs and SiC-SBDs. We use the advanced neutral-point-clamped 3-level inverter as shown in Fig. 14 [7][8]. A bidirectional switch used in the inverter consists of two 2 in 1 type modules. Therefore, all the modules in the main circuit are the same. The switching frequency is 20 kHz which is decided by considering a trade-off between the size of output filter and the switching losses. Fig. 15 shows the photos of the developed PV inverter (300V/20kW output) and its main power unit including the output filter. By applying the all-SiC modules explained in the previous chapter, the remarkable size of the output filter and the heat sink is significantly reduced. The overall size of the PV inverter is 25% of the conventional one. Fig. 16 shows the three-phase output voltage waveforms in case of the individual operation. Although we use the small size output filter, the output voltages are sinusoidal and almost no ripple. Moreover, the main circuit efficiency is 99%, which is significantly improved compared to the conventional products.

C. Boost Chopper Appling SiC-MOSFETs and SiC-SBDs for high power PV (Mega Solar) inverters

PV inverters have to be designed with consideration of the input voltage fluctuation. In order to reduce the input voltage fluctuation and to increase power density, we developed a boost chopper. For instance, in case of a high power PV inverter system without the boost chopper as shown in Fig. 17 (a), the rated output voltage is AC270V which is close to the

limit to output at the minimum input voltage of DC450V. To obtain the output power of 250kW, the output current of AC534A is then required. On the other hand, the boost chopper is inserted between solar cells and the inverter in Fig. 17 (b). A conventional 250kW inverter using Si-IGBTs, which is shown in Fig.17 (a), is applied with the newly developed boost chopper. The boost chopper increases the minimum inverter input voltage from 450V to DC800V. Therefore, the output voltage of the PV inverter system can also be increased from AC270V to AC480V. It should be noted that the output

(a) 20kW PV inverter

(b) Main power unit.

Fig. 15. Appearance of the newly developed 20kW PV inverter.

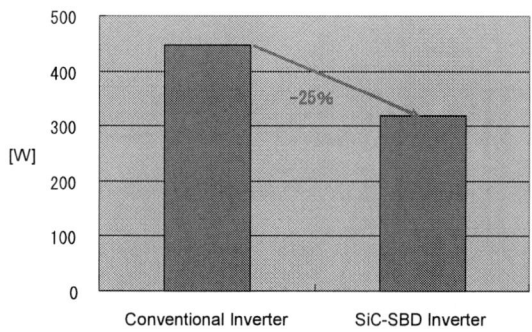

Fig. 13. Total loss comparison of the inverters

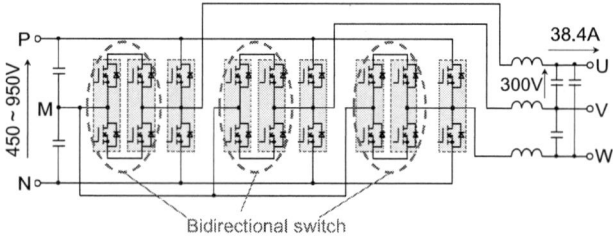

Fig. 14. Circuit configuration of the 20kW PV inverter using the all-SiC modules.

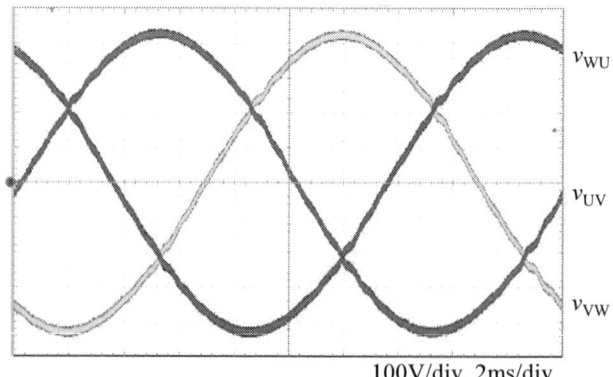

100V/div, 2ms/div

Fig. 16. Output voltage waveforms of the 20kW PV inverter in case of the individual operation.

current is almost same compared to the system without the boost chopper. Therefore, the output power can be increased from 250kW to 500kW by applying the boost chopper.

The circuit configuration and the photo of the boost chopper are shown in Fig. 18 and Fig. 19, respectively. The main circuit consists of four-phase chopper and it is controlled by interleaved manner. The 2 in 1 type all-SiC modules are employed and the switching frequency is 20kHz. The frequency of an output voltage ripple is high, which is 80kHz. Therefore, the film capacitors can be used instead of electrolytic capacitors. When the output voltage from solar cells is higher than DC800V, the upper side SiC-MOSFETs are always on in order to reduce the switching loss. The rated

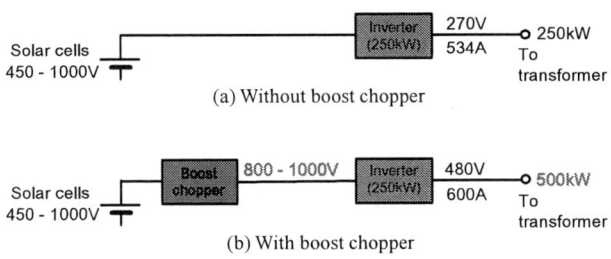

(a) Without boost chopper

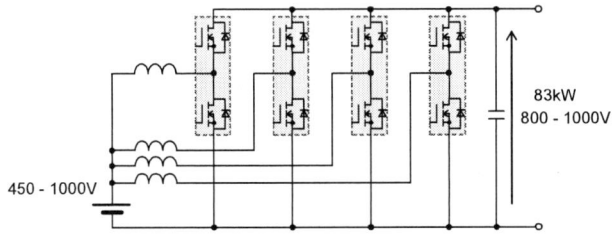

(b) With boost chopper

Fig. 17. Hihg power PV inverter system with and without the boost chopper.

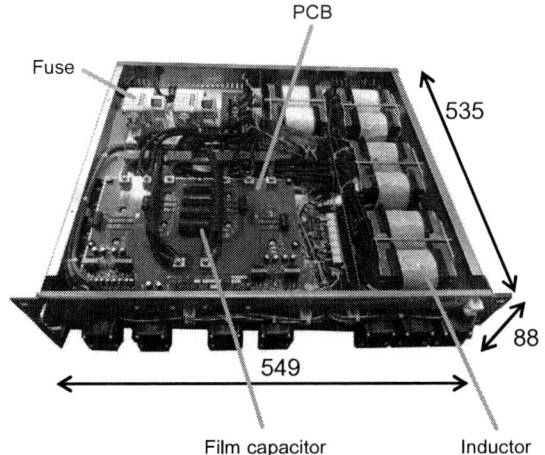

Fig. 18. The circuit configuration of the boost chopper.

Fig. 19. The Photo of the 83kW boost chopper.

output power is 83kW and therefore twelve boost choppers will be used in a 1 MW system.

Although the total volume is increased by 1.6 times by adding the boost chopper, the output power can be increased by factor 2. Therefore, 25% of the total power density is increased compared to the system without the boost chopper. Furthermore, the maximum efficiency of the system is increased from 98.4% to 98.8%.

IV. CONCLUSIONS

The motor drive inverter applying the hybrid modules and the PV inverters applying the all-SiC modules were developed. By using the SiC power semiconductors, it is clear that the power electronics equipments become smaller and more efficient. Therefore, package and circuit technologies to derive superior SiC characteristics contribute to significant improvement in power electronics equipments.

ACKNOWLEDGMENT

This work was supported by an AIST R&D Initiative Program on the Innovative Industrial Technology (FY 2009-2011), under a title of "Demonstration on Mass Production Technology for SiC Power Devices". The authors would like to thank members of the Advanced Power Electronics Research Center of AIST for their technical support.

REFERENCES

[1] T. Tsuji, A. Kinoshita, N. Iwamuro, K. Fukuda, T. Tsuyuki, and H. Kimura, "Experimental demonstration of 1200V SiC-SBDs with lower forward voltage drop at high temperature," *Proceedings of ICSCRM* 2011, Tu-P-27, September 2011, pp. 205-208.

[2] S. Harada, Y. Hoshi, Y. Harada, T. Tsuji, A. Kinoshita, M. Okamoto, Y. Makifuchi, Y. Kawada, K. Imamura, M. Gotoh, T. Tawara, S. Nakamata, T. Sakai, F. Imai, N. Ohse, M. Ryo, A. Tanaka, K. Tezuka, T. Tsuyuki, S. Shimizu, N. Iwamuro, Y. Sakai, H. Kimura, K. Fukuda, and H. Okumura, "High performance SiC-IEMOSFET/SBD module," *Proceedings of ICSCRM* 2011, MO-3A-1, September, 2011, pp. 52-57.

[3] N. Ikeda, Y. Niiyama, H. Kambayashi, Y. Sato, T. Nomura, S. Kato, and S. Yoshida, " GaN Power Transistors on Si Substrates for Switching Applications," *Proceedings of the IEEE*, vol.98, July 2010, pp. 1151-1161.

[4] H. Mine, Y. Matsumoto, R. Yamada, K. Mino, H. Kimura, Y. Kondo, Y. Ikeda, "Characteristics of the power electronics equipments applying the SiC power devices" *International Conference on Power Engineering and Renewable Energy (ICPERE)*, Bali, 3-5 July, 2012, pp. 1-6.

[5] Y. Ikeda, Y. Iizuka, Y. Hinata, M. Horio, M. Hori, and Y. Takahashi, "Investigation on Wirebond-less Power Module Structure with High-Density Packaging and High reliability," *IEEE 23rd International Symposium on Power Semiconductor Devices and ICs (ISPSD)*, San Diego, 23-26 May, pp. 272-275, 2011.

[6] M. Horio, N. Nashida, Y. Iizuka, Y. Ikeda, and Y.Takahashi, "New Power Module Structure with Lower Thermal Impedance and High Reliability for SiC Devices," *Proceedings of PCIM Europe*, May, pp. 229-234, 2011.

[7] A. Nabae, I, Takahashi, H. Akagi, "A New Neutral-Point-Clamped PWM Inverter," *IEEE Transactions on Industry Applications*, Vol. IA-17, Issue 5, pp. 518-523, 1981.

[8] K. Komatsu, M. Yatsu, S. Miyashita, S. Okita, H. Nakazawa, S. Igarashi, Y. Takahashi, Y. Okuma, Y. Seki, and T. Fujihira, "New IGBT modules for advanced neutral-point-clamped 3-level power converters," *International Power Electronics Conference (IPEC)*, Sapporo, June 21-24, pp. 523-527, 2010.

Gap in pagination due to withheld paper.

Pages 1925-1928

The 2014 International Power Electronics Conference

EMI prediction method for SiC inverter by the modeling of structure and the accurate model of power device

Sari Maekawa, Junichi Tsuda, Atsuhiko Kuzumaki,
Shuhei Matsumoto, Hiroshi Mochikawa
TOSHIBA CORPORATION FUCHU OPERATIONS
1, Toshiba-Cho, Fuchu-Shi, Tokyo, Japan 183-8511
sari1.maekawa@toshiba.co.jp

Hisao Kubota
Graduate School of Science and Technology,
Meiji University
1-1-1 Higashimita, Tamaku, Kawasaki, 214-8571,JAPAN
kubota@isc.meiji.ac.jp

Abstract— **In recent years, the switching speed is increased accelerately. And, the increase of EMI by high dv/dt is a problem. In this paper, the Tri-phase 400 V$_{rms}$ inverter for system interconnections which used SiC-JFET is analyzed. And, it is shown that noise terminal voltage is analyzable with an error of ±15 dB by highly precise modeling.**

Keywords— *EMI, analysis, power electronics, SiC.*

I.   INTRODUCTION

In recent years, to reduce a switching loss, a switching speed is dramatically increased. The increase of Electro Magnetic Interference (EMI) by high gradient of time may causes improper operation of neighbor devices.

For the above reason, it is necessary to reduce EMI by a noise filter. In order to estimate EMI in a design phase before a trial production, highly precise EMI analysis technology is required. Estimating EMI for power electronics are studied by high precise analytical modeling [1-6], and the modeling including parasitic components with electromagnetic-field analysis in recent years [1-18].

In this paper, Tri-phase 400 V$_{rms}$ inverter for system interconnections with SiC-JFET is analyzed. The frequency range is from 150 kHz to 30 MHz specified by CISPR11 Class A [19]. The study achieved that the noise terminal voltage is analyzable with an error of ±15 dB by high precise modeling.

II.   COMPOSITION OF THE INVERTER OF ANALYSIS OBJECT

A.   Composition of an EMI evaluation system

Fig.1 is the evaluation system of noise terminal voltage, and Fig.2 is experimental setup for measuring conducted EMI. The evaluation system consists of a DC power supply, an inverter, a normal mode filter, and load resistance. And, in order to measure noise terminal voltage, LISN is added between an inverter and load resistance. The inverter of this paper is targeting system interconnections, such as photovoltaic generation. Output

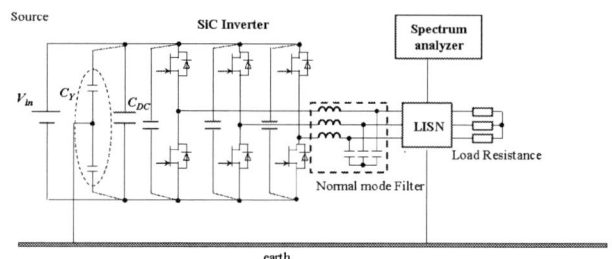

Fig. 1. Evaluation system configuration of conducted.

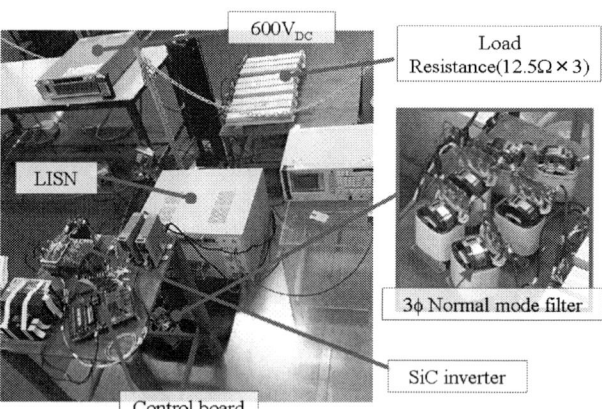

Fig. 2. Experimental setup for measuring conducted EMI.

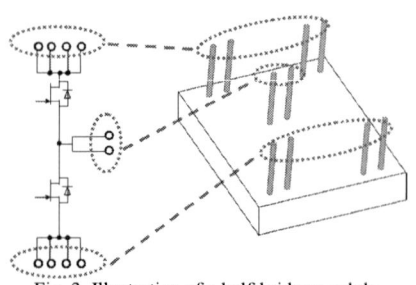

Fig. 3. Illustration of a half-bridge module.

978-1-4799-2706-7/14 $31.00 © 2014 IEEE

The 2014 International Power Electronics Conference

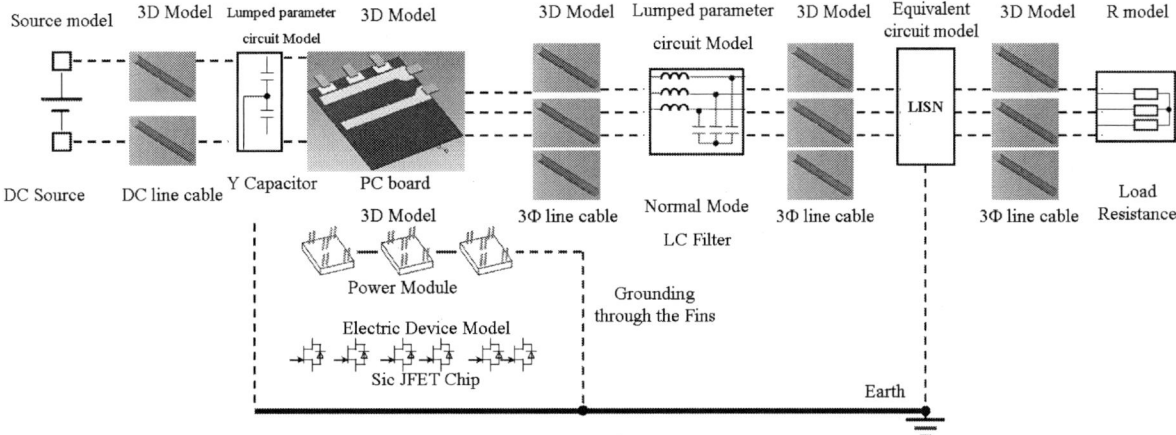

Fig. 4. Modeling structure of the evaluation system.

rating is tri-phase 400V$_{rms}$ and 10 kW.

B. Composition of a main circuit

Main circuit composition is 3-phase inverter of two levels. The inverter uses three half bridge modules shown by Fig.3.

III. MODELING OF THE INVERTER IN EMI ANALYSIS

A. The outline of modeling

Fig.4 shows the modeling composition of the evaluation system.

The parasitic components generated according to the structure of the circuit (ex. printed circuit board, power module). For this reason, a parasitism components is extracted from these by conducting electromagnetic-field analysis of the 3-D model.

The device model which reproduced the switching characteristic is used for the SiC chip mounted in a power module.

The frequency characteristic of the impedance of a reactor and a capacitor is reproduced in the equivalent circuit by a lumped constant.

B. Modeling of a printed circuit board, a power module, and a cable

The pattern of the DC section and the UVW section is modeled in the printed circuit board. Furthermore, the composition between the layers of a pattern is also modeled. As for a power module, an internal circuit pattern is modeled. And, the layer composition to the

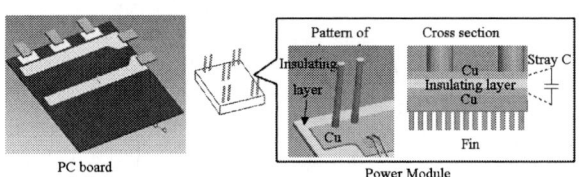

(a) Model of the circuit board   (b) Model of the power module
Fig. 5. Illustration of the (a) PC board and (b) power module of the inverter.

radiating fin is modeled. Since a radiating fin is grounded to Earth in Fig.4, the modeling of the stray capacitance between a SiC device and a fin is important. The 3-D model of a printed circuit board is shown in Fig.5 (a), and the 3-D model of a power module is shown in (b). The parasitic components are extracted from each model and it outputs as an equivalent circuit of *LCR* as shown in Fig.7. Furthermore, they are combined on a circuit simulator (Fig.6). In the equivalent circuit, the resistance *R*, stray capacitance *C*, and inductance *L,M* between each node of a model are taken into consideration.

In addition, the parasitic components is extracted with BEM(Boundary Element Method) by Q3DExtractor of ANSYS. Circuit analysis is using Simplorer of ANSYS.

TABLE.I shows the parasitism components of Sic module. The parasitism components shown in TABLE.I is the stray capacitance between a installation part of a fin

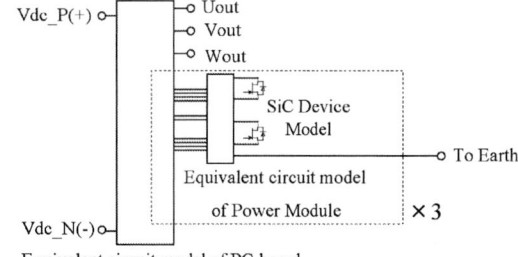

Fig. 6. Reflect on the circuit simulator.

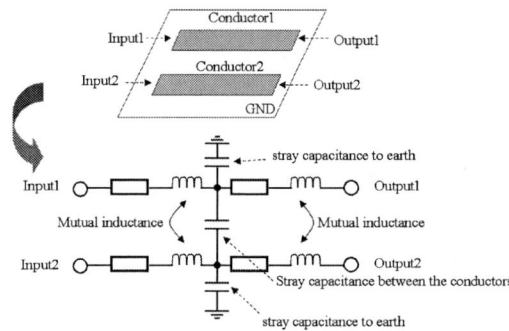

Fig. 7. Equivalent circuit of the parasitic components.

978-1-4799-2706-7/14 $31.00 © 2014 IEEE   1930

TABLE. I. Comparison of the stray capacitances of analysis and measurement

Part	Analysis Capacitance [pF]	Measurement Capacitance [pF]
$V_{DC}P$ to Fin	79	-
$V_{DC}N$ to Fin	66	-
Output to Fin	74	-
Sum	219	202

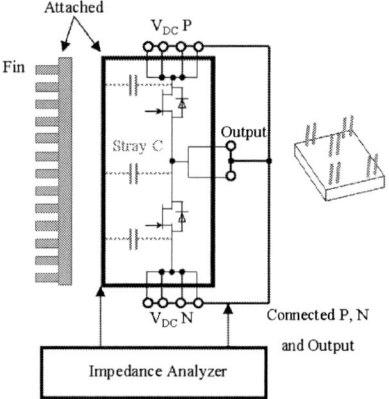

Fig. 8, Experimental setup for measurement stray capacitances of SiC module

and each part. The measuring method of stray capacitance is shown in Fig. 8. The terminal of $V_{DC}P$, $V_{DC}N$ and Output are connected, and the stray capacitance is measured with the impedance analyzer. The output capacitance C_{ds} of SiC JFET is very large compared with stray capacitances, when V_{ds} is small. Therefore, it is difficult to measure the stray capacitances of each part correctly.

C.   Modeling of a capacitor and reactor

The normal mode filter which consists of a reactor and a capacitor is connected to the output of an inverter. The frequency characteristic of L, C, and R which constitute a filter is important in the frequency range (150kHz-30MHz) of the CISPR11 class A. For this reason, the model is using the lumped constant circuit.

Fig.9 shows the lumped constant circuit and value of the normal mode filter (reactor, capacitor). Fig.10 (a) shows the frequency characteristic of the reactor. In contrast, Fig.10 (b) shows that for capacitor. The frequency characteristic is reproducible. Other capacitors are modeled similarly. The resistor for loads shown by Fig.1 is expressed only by R.

D.   Modeling of LISN

LISN is a circuit which measures noise terminal voltage. The equivalent circuit shown by the maker is used. Fig.11 shows the equivalent circuit of LISN in an evaluation system. The terminal point to Earth are LISN, Y capacitor of DC part, and the radiating fin of a power module, as shown in Fig.4.

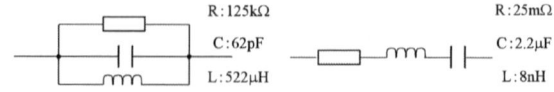

(a) Reactor   (b) Capacitor

Fig. 9. Equivalent circuit of the (a) reactor and (b) capacitor.

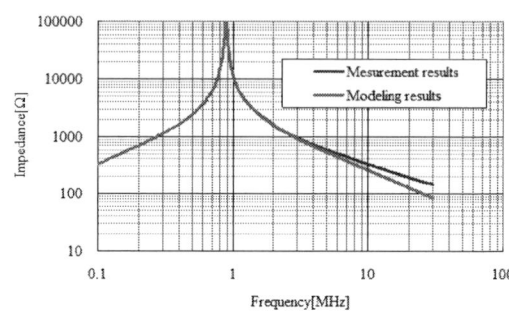

(a) Reactor

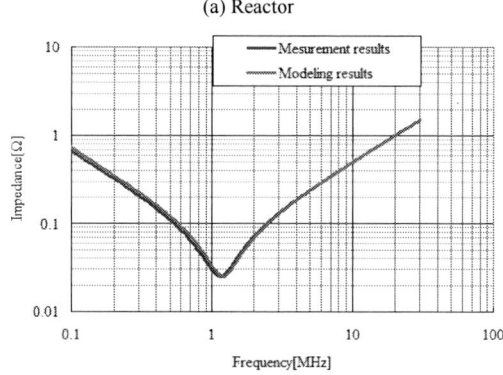

(b) Capacitor

Fig.10. Impedance vs frequency plot for the (a) reactor and (b) capacitor.

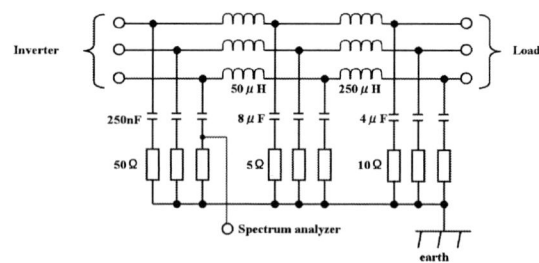

Fig. 11. Equivalent circuit of the LISN.

E.   Modeling of SiC power device

In this paper, a SiC power device is switched on condition of high dv/dt. For this reason, the switching characteristic of a SiC is important, and the characteristic is expressed by the equivalent circuit model shown in Fig.12. Although a power device is JFET, the characteristic of normaly on of JFET is expressed in the MOSFET model with negative gate threshold voltage. The resistance and inductance which are connected to gate, drain, and source are parasitic components in a chip. The parasitic components in a power module are

The 2014 International Power Electronics Conference

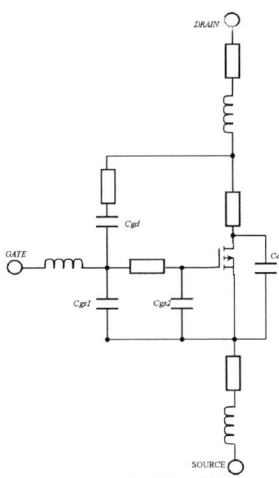

Fig. 12. Equivalent circuit of the SiC power devices.

expressed by (II. *B*).

The switching characteristic is expressed by modeling V_{gs}-I_d and V_{ds}-I_d as a static characteristic, and modeling the characteristic of C_{gs}, C_{gd}, and C_{ds} as dynamic characteristics. The accuracy of the waveform of V_{ds} is improved by changing each capacitance by ON/OFF of SiCJFET [20,21]. Each capacitance has determined from the experimental result of the switching waveform.

The reverse-recovery characteristics of a diode are as important as the switching characteristic, and it is expressing them using the device level diode dynamic-characteristics model of ANSYS Simplorer [22]. As a result, the characteristic of recovery-current is reproducible.

Fig.13 shows the switching waveform of SiCJFET which modeled the switching characteristic (static and

dynamic) and reverse-recovery characteristics.

The experiment uses the double pulse circuit shown by Fig.14(a)(b). Parasitism components, such as stray inductance and stray capacitance are added to the simulation circuit to compare in Fig.14(a). SiCJFET uses the module shown by Fig.3.

In Fig.13, the dv/dt of V_{ds} is reproduced by modeling of a static characteristic and dynamic characteristics.

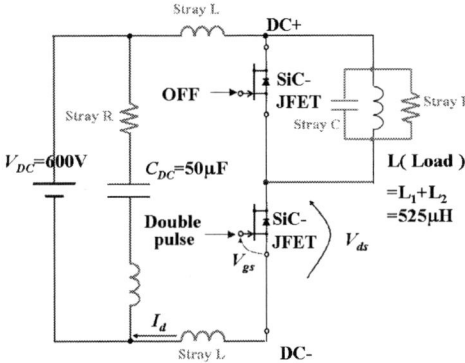

(a) Equivalent circuit for simulation

(b)Experimental setup

Fig. 14. Configuration of the double-pulse test circuit.

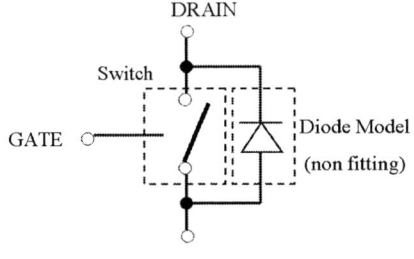

Fig. 15. Simulation and experimental results of the noise voltage in frequency

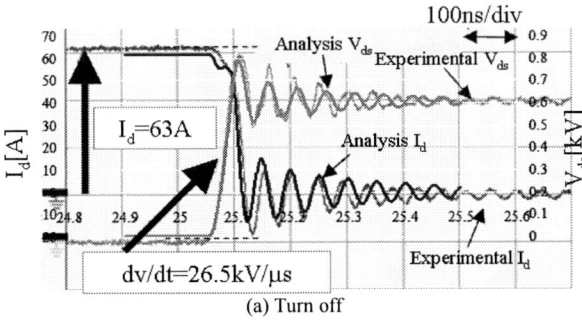

(a) Turn off

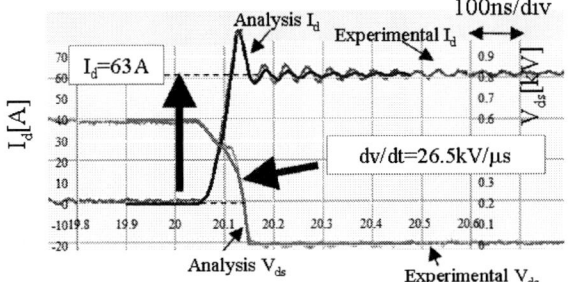

(b) Turn on

Fig. 13. Reproducibility of switching characteristics at (a) turn off and (b) turn on.

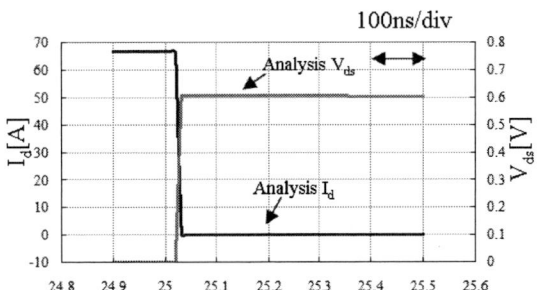

Fig. 16. Characteristics of the ideal switch model (OFF).

978-1-4799-2706-7/14 $31.00 © 2014 IEEE   1932

Although I_d and V_{ds} waveform at the time of turn-on are greatly influenced by the recovery characteristic in a double pulse examination, the accuracy of estimation by a simulation is good.

In contrast, when it does not model the switching characteristic and reverse recovery characteristics shown in Fig.15, a turn-off becomes a waveform as shown in Fig. 16. This uses the simple switch model. In a simple switch model, since the characteristics of capacitance differ, dv/dt of V_{ds} is higher than the experimental result.

IV.   SIMULATION RESULTS

A.   Estimation of noise terminal voltage

As a result of our analysis, noise terminal voltage is obtained by a model with following items. The model consists of equivalent circuits of parasitism components in each part of our test inverter, reactors and capacitors modeled as lumped constant circuit, and characteristics of SiC power devices.

Fig.17 shows Fast Fourier Transform (FFT) of simulation and experimental results of noise terminal voltage in the case of modeling above all components. Compared to an experimental result, our proposed method achieves to estimate EMI with an error of ±15 dB in frequency range from 150 kHz to 30 MHz specified in the CISPR11 class A.

B.   The influence of accurate model of power device

The power device of an inverter is SiC and its dv/dt is high. Fig.18 shows the difference arising from a power device model. The simply power device model is a model shown in Fig. 15 and Fig.16. The accurate model is reproducing static characteristics, dynamic characteristics and reverse recovery characteristics. The difference has appeared from the frequency of over 5 MHz. Since dynamic characteristics is not reproduced enough, the simply power device model has quick switching.

As a result, the frequency of dv/dt became high and the difference appeared in EMI.

V.   CONCLUSION

(1)This paper investigates EMI analysis for SiC inverter.
(2)This paper modeled following items: parasitism

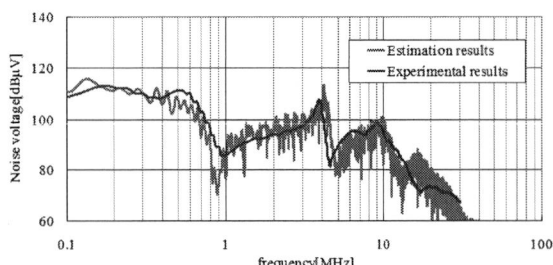

Fig. 17. Simulation and experimental results of the noise
voltage in frequency

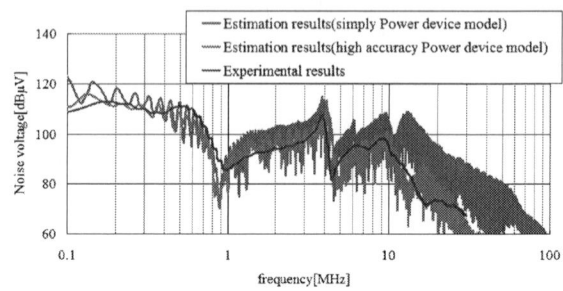

Fig. 18. Simulation and experimental results of the noise
voltage in frequency

components of our test inverter circuit, frequency characteristics of passive components, and switching characteristics of SiC power device.
(3)As a result, our proposed method achieves to estimate EMI with an error of ±15 dB in frequency range from 150 kHz to 30MHz specified in CISPR11 class A.
(4)This paper considers a modeling method for three circuit components in EMI analysis for SiC inverter. The study verifies an accuracy of our model by experiments of switching test, EMI measurement, and impedance measurement.

REFERENCES

[1] R. Kraus, P. TÜrkes, J. Sigg : "Physics-based models of power semiconductor devices for the circuit simulator SPICE", *Power Electronics Specialists Conference*, 1998. PESC 98 Record. 29th Annual IEEE, Vol. 2, pp. 1726-1731(1998)

[2] Allen R. Hefner, Daniel M. Diebolt : "An experimentally verified IGBT model implemented in the Saber circuit simulator", *IEEE Transactions on Power Electronics*, Vol. 9, No. 5, pp. 532-542 (1994)

[3] N. Okada, T. Kikuma, M. Takasaki, K. Kodani, A. Kuzumaki : " Development of Device Model for Inverter Simulation Program" , *Industrial Application Conference of IEEJ*, No.1-131, pp. 515-520 (2008) (in Japanese)

[4] A. Hatanaka, T. Kawashima, A. Mishima : "Analysis technique of EMC for on-board power supply with recovery diode model" , *Industrial Application Conference of IEEJ*, No. 1-132, pp. 525-528 (2008) (in Japanese)

[5] Y. Iwata, S. Tominaga, H. Fujita, F. Akagi, T. Horiguchi, S. Kinouchi, T. Oi, H. Urushibata : "Investigation of noise and switching-energy loss by using a precise MOSFET model" , *Industrial Application Conference of IEEJ*, No. 1-135, pp. 615-618 (2011) (in Japanese)

[6] A. Mishima, T. Kawashima : "Switching analysis methods using Power Device Model and Magnetic Field Coupling System", *Industrial Application Conference of IEEJ*, No. 1-S7-5, pp. 95-100 (2003) (in Japanese)

[7] Y. Koyama, M. Tanaka, H. Akagi : "Modeling and Analysis for Simulation of Common-Mode Noises Produced by an Inverter-Driven Air Conditioner", *Industrial Application Conference of IEEJ*, No.1-O2-5, pp.175-180 (2009) (in Japanese)

[8] T. Shimizu, G. Kimura, J. Hirose : "High Frequency Leakage Current Caused by the Transistor Module and Its Suppression Technique", *T. IEE Japan*, Vol. 116-D, No. 7, pp. 758-766 (1966) (in Japanese)

[9] J.-S.Lai, X. Huang, E. Pepa, S.Chen, and T.W.Nehl : "Inverter EMI Modeling and Simulation Methlogies", *Proceedings of 29th Annual Conference of the IEEE Industrial Electronics Society*, Vol. 2, pp. 1 533-1539 (2003)

[10] B. Revol, J. Roudet, J.L. Schanen, and P. Loizelet : "Fast EMI Prediction method for three phase inverter based on Laplace

Transforms", *Proceedings of 34th IEEE Annual Power Specialists Conference*, Vol. 3, pp. 1133-1138 (2003)

[11] M. Tamate, T. Sasaki, A. Toba : "Quantitative Estimation of Conducted Emission from an Inverter System", *T. IEE Japan*, Vol. 128-D, No. 3, pp. 193-200 (2008) (in Japanese)

[12] S. Ogasawara : "Modeling and Simulation of EMI in Power Electronics systems", *Industrial Application Conference of IEEJ*, 1-S11-2, pp.69-74 (2004) (in Japanese)

[13] A. Okuno, S. Ogasawara : "Simulation in Power Electronic Systems-Characteristics of General-Purpose Simulators and System Modeling Methods-", *T. IEE Japan*, Vol.122-D, No. 9, pp. 893-898 (2002) (in Japanese)

[14] Y. Kondo, M. Izumichi : "VHF conduted emission simulation of power electronic devices", *Industrial Application Conference of IEEJ*, No. 1-40, pp. 205-208 (2012) (in Japanese)

[15] [15] T. Koga, K. Shigematsu, S. Hasumura: " A Study of common mode current reduction in PWM inverter with core modeling and circuit simulation" , Annual Conference of IEEJ, No. 4-019, p. 33 (2009) (in Japanese)

[16] T. Chida, A. Mishima, T. Kamezawa, K. Mou, S. Ibori : "Analysis of conducted emission in general purpose inverter", Industrial Application Conference of IEEJ, No. 1-38, pp. 299-300 (2007) (in Japanese)

[17] G. Xun, J. A. Ferreira : "Investigation of Conducted EMI in SiC JFET Inverters Using Separated Heat Sinks", *IEEE Transactions on Industrial Electronics*, Vol. 61, No. 1, pp. 115-125 (2014)

[18] W. Junsheng, D. Gerling, S. P. Schmid : "Prediction of conducted EMI in power converters using numerical methods", Proceedings of Power Electronics and Motion Control Conference (EPE/PEMC) 2012 , DS1a.3-1 - DS1a.3-6 (2012)

[19] IEC CISPR 11 Edition.5.0: Industrial, scientific and medical equipment – Radio-frequency disturbance characteristics – Limits and methods of measurement, IEC Standard, May, 2009.

[20] T. Sekisue(ANSYS JAPAN): "Parameter for SIMPLORER's IGBT device model", ANSYS JAPAN user support documents.(2010) (in Japanese)

[21] J. Aurich, T. Barucki, : "Fast dynamic model family of semiconductor switches", Power Electronics Specialists Conference, 2001. PESC 2001 Record. 32th Annual IEEE, Vol. 1, pp. 67-74 (2001)

[22] T. Sekisue(ANSYS JAPAN): "Parameter fitting for SIMPLORER's device model starting with diode characteristics", ANSYS JAPAN user support documents. (2009)(in Japanese)

System Integration of GaN Technology

J.A.Ferreira, J. Popovic
Electrical Power Processing
Delft University of Technology
Delft, The Netherlands
j.a.ferreira@tudelft.nl

J.D.van Wyk
University of Johannesburg
Johannesburg, South Africa
daanvw@uj.ac.za

F. Pansier
NXP Semiconductors
Nijmegen, The Netherlands
frans.pansier@nxp.com

Abstract— Gallium nitride (GaN) power semiconductor technology offers a potential for significant performance increase in power electronic converters. This potential cannot be fully exploited if GaN devices are used as drop-in replacement for silicon devices in existing systems.

This paper investigates the switching limits influenced by the device output parasitic capacitance and parasitic inductance of the commutation cell capacitor. The trade-offs between thermal management and high frequency switching of GaN devices in power converters are explored. Finally, an outlook on technology needs for 3D integration of GaN converters for achieving high switching frequencies and power densities is given.

Keywords— GaN power semiconductors, system integration, switching loss, high frequency switching, thermal management.

I. Introduction

GaN power semiconductor technology requires a paradigm shift in power electronics. By increasing the switching speed and voltage rating by an order of magnitude compared to silicon, GaN devices could significantly improve performance in existing systems and enable new applications. However, the full potential of GaN devices cannot be reached by using GaN devices as a drop-in replacement for silicon. We need to look further than devices into the converter and system level and revisit the way we design and build power electronic converters to better suit and fully exploit the new GaN power semiconductors. The electromagnetic volume of the GaN switch has become negligible in comparison to the electromagnetic volume of the interconnects, the passives, the cooling structure – ie. the rest of the power converter circuit. This has a profound influence on measured device characteristics, topology and packaging technologies for optimal use of GaN devices. We need to understand the influence of the parasitic inductances and capacitances on the behaviour of the devices and the implications thereof on the circuit design and operation. Furthermore, the influence of thermal management techniques on high frequency switching needs to be understood. Finally, technology platforms and system integration concepts able to properly accommodate these high frequency, high power density converters are needed.

II. Influence of parasitics on switching loss

Due to the faster switching and higher di/dt and dv/dt of GaN devices circuit "parasitics" move into the pH and fF range, especially when utilizing the higher voltage ratings of the GaN devices at extended frequencies. This puts limits on high frequency operation for hard switching topologies. Device packaging as well as PCB layout become critical and dominate the circuit behaviour. Not all parasitic inductances are detrimental to circuit operation and in some cases parasitic capacitances can assist reducing switching losses. To exploit the full potential of the new devices it is necessary to gain an understanding of the behaviour in the circuit topology environment.

The most common circuit topology is the buck or boost dc/dc converter and the phase arm of a dc/ac converter that operates as a buck/boost converter depending on power flow direction. They operate in hard switching, continuous current mode whereby the diode current is commuted by turning on the switch while the diode is conducting. This implies that the stored charge in both the top and bottom switches is discharged through the switch channel while the supply voltage appears across the switch. As a result, substantial current overshoot is observed and large turn-on losses occur. The stored charge can divided into current amplitude related charge and voltage related charge (capacitive). A typical voltage and current switching waveforms are shown in Figure 1. Notice that the drain source voltage has overshoot and displays oscillatory behaviour due to the parasitic inductance in the commutation path.

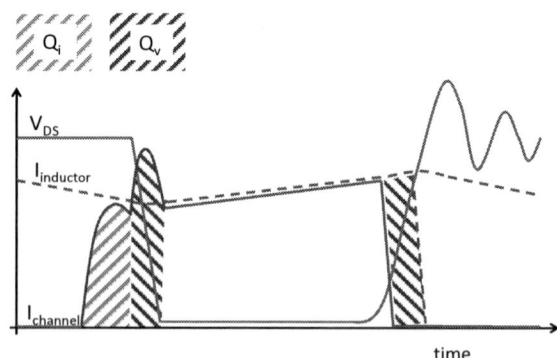

Figure 1 Influence of storage charge on channel current

Zero-voltage switching (ZVS) topologies are most promising for pushing the switching frequencies with GaN. This avoids the capacitive discharge condition at device turn-on. A requirement is that current reversal should take place prior to turn-on so that the switch capacitance voltage is reset by resonance. This can achieved by discontinuous mode operation, using the resonant the pole topology or one of the full resonant converter topologies as shown in Figure 2.

By careful design and accurate timing control of the switches the turn-on losses can be minimized. However, the turn-off losses are still present and the question is how the circuit parasitics, especially the parasitic commutation inductance and the parallel switch capacitance influence the switching waveforms and limit the maximum possible switching frequency. Not only the inductance of the layout but also the parasitic inductance of the capacitor need to be taken into account.

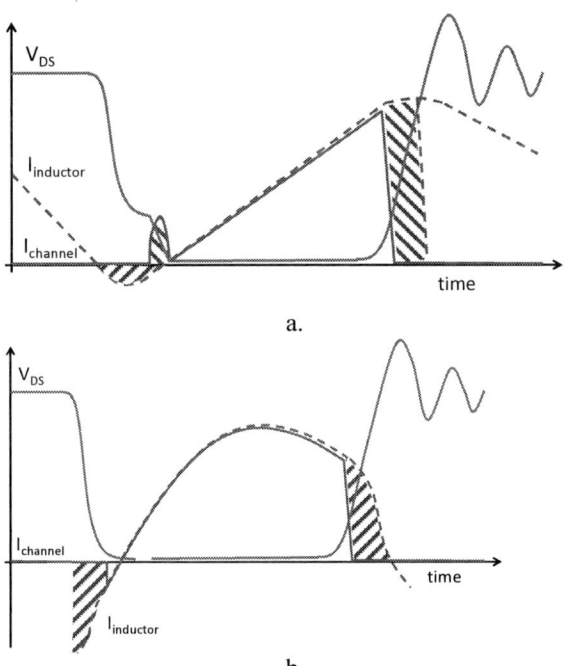

a.

b.

Figure 2 Zero-voltage switching a. discontinuous mode operation b. resonant topologies

A very different environment can be created in a switched circuit for which the load impedance is matched to the voltage and current that is switched (Figure 3). If the true switching losses of GaN switches have to be determined, it is essential to switch in an electromagnetically controlled environment, and not in the constructions on PCBs, such as is currently being used. These set-ups couple to the surroundings and are different every time, especially at the nanosecond and sub-nanosecond times possible. Furthermore, the only way to determine the actual switching characteristics of the device, and not the switching as influenced by the circuit, is to switch a perfect resistive load. Again in the nanosecond region, the only way to achieve an actual resistive load, is to switch a matched transmission line. The switching time determined in this way can then be used with confidence for switching loss calculation. In such a case the circuit parasitics are not playing any role, but the internal capacitance becomes the limiting factor.

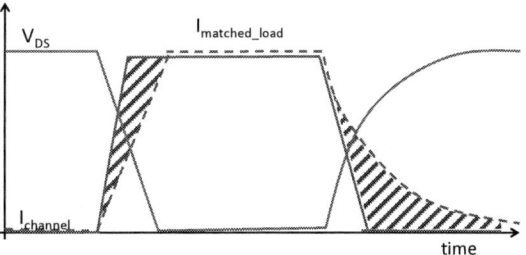

Figure 3 Inductance free switching into a matched impedance load

III. ANALYSIS OF TURN-OFF LOSSES AND VOLTAGE OVERSHOOT

To investigate the switching limits, an analytical model of the switching losses was developed and the boost converter circuit with specifications in Table 1 used as a case study. The model includes the turn-off loss component due to the current fall time (Figure 2) and the loss component (not dissipated in the channel) caused by the voltage ringing between the device capacitance and parasitic inductance. The turn-on loss is assumed to consist of the capacitive discharge component only.

The switching losses and switch drain-source voltage overshoot as function of (a) turn-off time, (b) switch capacitance and (c) commutation loop inductance through the capacitor were modelled and the results are shown in Figure 4 to Figure 6. Figure 4 shows the component of turn-off losses in the channel due to the current fall while Figure 6 shows the total turn-off losses, including the component originating from the parasitic ringing between the switch output capacitance and parasitic inductance of the capacitor. It can be seen that the current fall time has a large influence on this component of turn-off losses, especially for lower capacitance ranges, as can be expected.

TABLE 1 PARAMETERS OF THEORETICAL CASE STUDY OF BOOST CONVERTER.

Output voltage (capacitor voltage)	$V_s = 400V$
Output voltage ripple	$\Delta V_s / V_s = 5\%$
Transistor current at turn-off	$I_t = 15A$
Switching frequency	$f_s = 1MHz$
Current fall time	$t_f = 2ns, 10ns, 20ns$
Output capacitance of transistor	$C_t = 10pF \dots 1nF$
Inductance of output capacitor	$L_C = 100pH \dots 10nH$

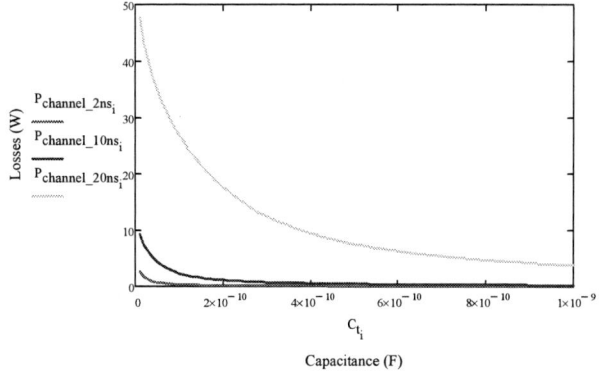

Figure 4 Turn-off losses through channel due to current fall time, t_f [2]

978-1-4799-2706-7/14 $31.00 © 2014 IEEE

It can be seen from Figure 6 that the switch capacitance has a large influence on the losses and peak device voltage. Below certain value of the capacitance the losses increase since the voltage increase becomes faster and soft switching is lost. Furthermore, as the capacitance increases the peak voltage becomes higher increasing the turn-off loss component caused by ringing.

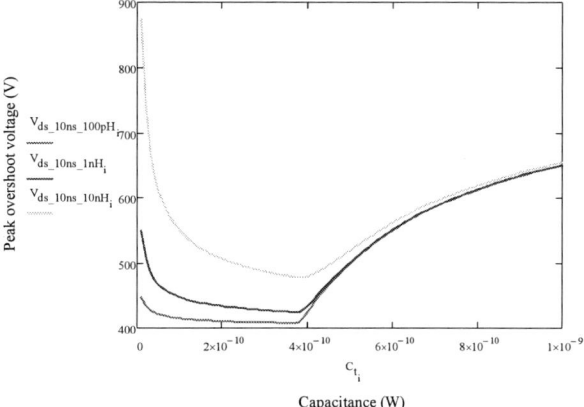

Figure 5 Simulated voltage peak overshoot as function of switch capacitance and output capacitor inductance

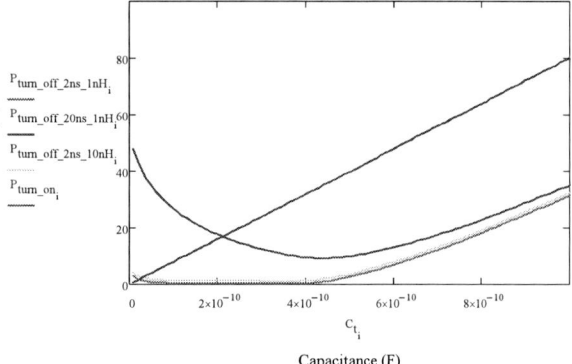

Figure 6 Calculated losses in GaN transistor as function of capacitance and current fall times

It can be seen that if ZVS is not achieved, the turn-on losses dominate the total loss. In so-called BCM valley switching (Figure 2a), where the voltage is resonantly brought to a value much lower than the output voltage, the turn-off losses can become higher than the turn-on losses, especially for larger current fall times. In a topology commonly used in consumer power supply, flyback operating in BCM-VS, increased switch voltage rating of GaN would be beneficial in reducing the turn-on voltage by increasing the reflected voltage but would come at the cost of the increased switch turn-off voltage.

The influence of the variation of the parasitic inductance on the turn-off loss is very small at the upper range of the current fall times (i.e. 20ns), as can be seen in Figure 6, and becomes larger as the current fall times become smaller, the output current increases and switching frequencies increase. Nevertheless, in this case, the effect is quite small.

It should be noted that the switching loss picture would look quite different for low voltage, high current switching waveforms. The influence of the parasitic

inductance would be dominant by influencing the current fall time and voltage overshoot.

Table 2 shows the calculated values of the inductance value of the output capacitor for the given voltage ripple, as a function of geometry (Figure 7). Parasitic inductance was calculated by considering the current path in a capacitor as a current sheet [3][4]. The current sheet has a width of w and encloses a rectangular area h by l. In the calculation, the effects of high frequency operation on current distribution were not taken into account. The table shows the calculation for the required capacitance value from the case study. As frequency increases, the required capacitance value decreases and as such the parasitic inductance.

TABLE 2 CAPACITANCE INDUCTANCE VALUES AS FUNCTION OF GEOMETRY

Capacitors (52nH, 630V)	Capacitance density (F/m³)	Ratio l/w	ESL (nH)
Ceramics 1 (X7R)	1.1	0.1	0.59
		1	6.42
	5.3	0.1	0.60
		1	6.46
Film (Polypropylene)	0.03	0.1	0.61
		1	5.86
	0.11	0.1	0.60
		1	5.97

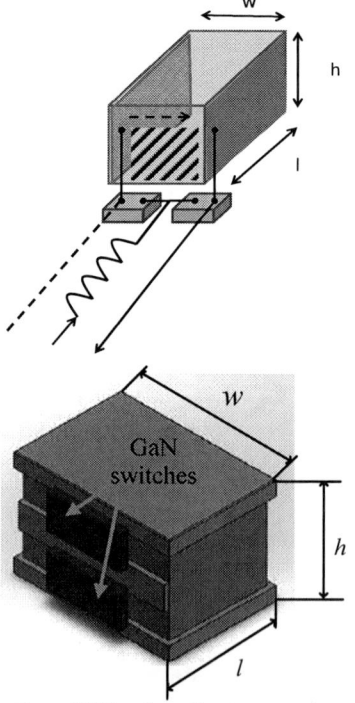

Figure 7 ESL values of output capacitor

A. Parasitics in GaN devices and circuits

In the previous section, the influence of the parasitic inductance and device capacitance was shown at the hand of a simple analytical model. In this section we will take a look at the available GaN devices and the effect of their parasitics in published literature. It should be noted that the model in the previous section does not explicitly take into account the common source inductance, but since this inductance considerably influences the current fall time, the effect is still included.

Low voltage GaN HEMT devices (< 200V) have been commercially available for a couple of years [5]. The isolated lateral device structure devices allows the device to be used as flip chip bare die resulting in a very low parasitic inductance of the package compared to discrete SMT or through-hole power packages (e.g. 0.7nH for a LFPAK package vs 0.15nH for a LGA package) [5][6]. Since the common source inductance of the package is negligible, the dominating parasitics become the commutation inductance of the PCB layout. While in discrete SMT packages (e.g LFPAK) the majority of the switching loss comes from the package common source inductance (> 60%), the commutation inductance in case of chip-size LGA package is responsible for almost 70% of the dominant turn-off switching losses in a hard switched synchronous buck POL converter [7]. The SIP approach of monolithic multi-switch GaN device and driver enables the switching frequency to go up to 5MHz due to a very low commutation inductance design of 0.2nH [7][8].

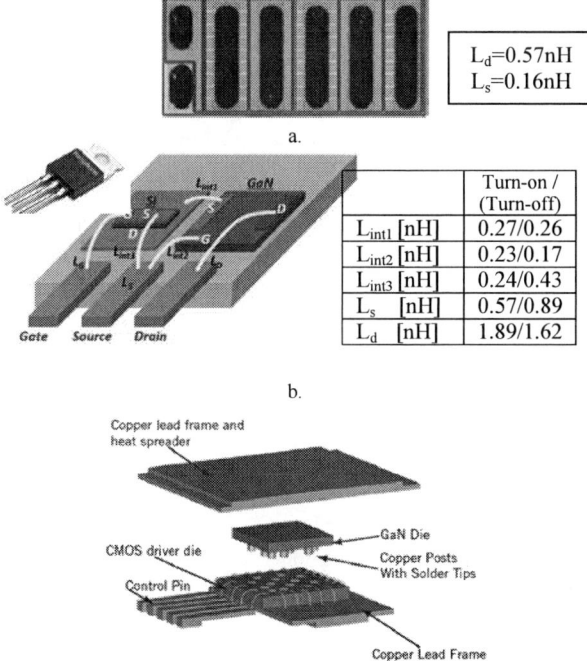

	Turn-on / (Turn-off)
L_{int1} [nH]	0.27/0.26
L_{int2} [nH]	0.23/0.17
L_{int3} [nH]	0.24/0.43
L_s [nH]	0.57/0.89
L_d [nH]	1.89/1.62

Figure 8 GaN packages and parasitic inductances a. LGA package of LV GaN b. TO-220 c. QFPAK [7][10][11]

High voltage (600V) normally-off GaN HEMT devices are currently available as samples in a cascode structure (HV normally-on GaN HEMT and a low voltage silicon MOSFET) [9]. They are currently mostly supplied in a TO-220 package [9] which has large parasitic inductances. It is shown that the critical inductances of the package almost double the dominant turn-on switching energy in a 400V/10A synchronous buck converter [10]. This is due to the common source inductance of the package only, on top of that comes the contribution of the commutation inductance. Furthermore, the reverse recovery of the Si diode in the bottom switch of the

cascode increases the current overshoot and transition time thus increasing turn-on losses of the top switch significantly. An alternative package for the cascode GaN HEMT is demonstrated by [12] where the low voltage MOSFET is placed directly onto the HV GaN HEMT in a flip-chip configuration. This minimises the two most critical inductances of the cascode configuration identified. The intrinsic current source driving mechanism of the cascode configuration reduces the GaN HEMT channel current fall time, the effect of which were shown in the previous section.

The effect of the output capacitance of the device on the switching loss was analysed in the previous section. Parasitic capacitances of the circuit can become comparable to the parasitics of devices (C_{ds}) increasing the switching loss in hard switching topologies.

Reference [13] shows more than 25% total loss increase in a three-phase SiC inverter due to parasitic capacitive coupling between the device pads and the heat sink (Figure 9a). In a phase-leg these capacitances become effectively parallel to the device drain-source capacitance. This becomes even more critical as the switching frequency increases. The drain-source voltage overshoot and oscillatory behaviour due to the parasitic inductance in the commutation path and total parasitic capacitance can be noticed in Figure 9b. The large current peak in the turn-on current is due to the capacitive discharge of the complementary device being turned off. If there is no parasitic coupling capacitance to the heat sink, this current peak is approximately halved.

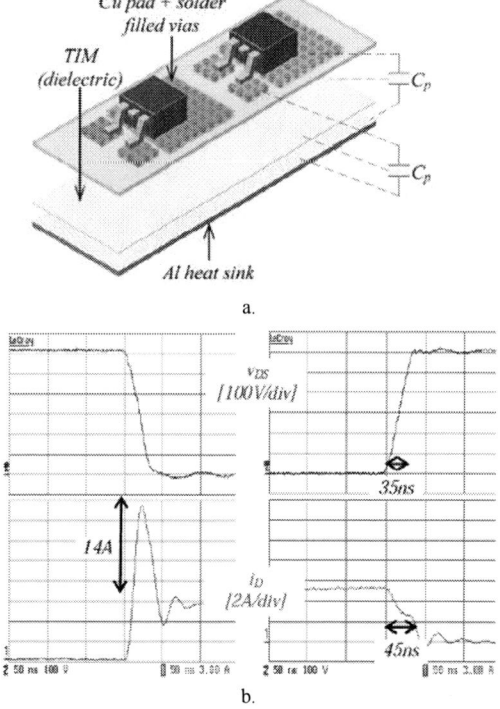

Figure 9 Parasitic capacitive coupling to PCB embedded heatsinks a. coupling b. switching waveforms

In case of the GaN cascode configuration packaged in a TO-220 package, the parasitic coupling capacitance to the heat sink from the drain terminal with a 3 cm^2 copper pad around the drain pad in a phase arm configuration would give an equivalent parasitic drain-source capacitance of 65pF for the top device (in case of through hole mounted components). In case of surface-mount components and the same size of the drain pad for all three terminals of the phase arm the equivalent capacitances would be in the range of 50pF. Looking at the reported device capacitances of HV cascode GaN-HEMT of 25pF-56pF $C_{o(er)}$ [9][14] it is obvious that the parasitic capacitances would contribute significantly to the performance.

As shown in Section II the influence of this parasitic coupling depends on the way that the transistor is switched, so it depends on the topology and switching mode. In case of synchronous buck topology, the parasitic capacitance would increase the switching losses of the control switch. Reference [15] shows that the turn-on losses of the control switch in a buck converter with a GaN schottky diode is dominated by the junction capacitance of the bottom switch (Schottky diode or GaN HEMT) and the parasitic coupling capacitance is in the same range as this junction capacitance. In the case of phase leg configuration (i.e. inverter) the turn-on loss of the bottom device will also be increased due to increasing current peak because of the charging of the equivalent parasitic capacitance parallel to the top switch. For LLC topology, the use of GaN is advantageous due to its low output capacitance which enables both the magnetizing current and dead time to be lowered consequently decreasing both conduction and switching losses (as high as 35-40%) in the primary side devices for GaN devices with half the output capacitance of the Si counterpart [15][16]. Adding a parasitic capacitance in the same range as that of the device can impair this advantage. In a VHF synchronous buck converter implemented with GaN-on-SiC HEMT even 3pF of additional parasitic capacitance at the switching node at 40MHz results in a 2% efficiency drop or 70% increase in the total loss under hard switching conditions [17] due to the capacitive discharge through the channel. Again, the parasitic capacitance of the circuit is comparable to the output capacitance of the devices. On the other hand, ZVS enables operation of up to 40MHz with efficiencies above 90%.

IV. THERMAL MANAGEMENT VS HIGH FREQUENCY OPERATION TRADE-OFFS

The physical construction of the die and package, electrical and thermal properties of GaN devices compared to Si result pose challenges to the way heat is dissipated, from the chip to the board level. For the same power rating the surface area of GaN devices is several times smaller than that of their Si counterparts while the thermal conductivity of GaN is in the same range. Furthermore, there tends to be a trade-off between good thermal management and high frequency performance. On the package level, the chip size packages offer low parasitic inductances and thus potential for good high frequency performance while the thermal management becomes challenging due to the high junction-to-board thermal resistance. On the other hand, through hole and

SMD packages with copper drain pads have low package thermal resistance but suffer from large parasitic inductances. Such trade-offs will be explored in this section.

1) Board level thermal management

a) Heat spreading and thermal vias

Given the above discussion, the thermal management on the board level must be designed carefully with GaN devices. In case of packaged devices (such as TO-220) the thermal resistance junction-to-case (R_{th_JA}) for the GaN device is twice that of R_{th_JA} of a super-junction MOSFET with the same R_{dson} (1.55K/W vs. 0.8K/W) [9]. In assemblies where the case-to-ambient thermal resistance (R_{th_CA}) is dominant in the heat path (PCB with natural convection cooling, no heat sinks) this does not have a large influence, but in designs where the board to ambient thermal resistance is lower this difference plays a role.

In case of chip-size packages, such as LGA, due to the large junction-to-board thermal resistance (15-35K/W) caused mostly by the solder bumps the importance of the board level thermal management is even more critical. Due to the very small die surface area, the effectiveness of heat spreading on a standard PCB stack-up is limited. For a small die, even the thermal resistance of an "infinite" heat spreader (inversely proportional to the die length) is prohibitively high. The junction-to-ambient thermal resistance (R_{th_JA}) of the device given by the manufacturer [5] in the data sheet is for a copper heat spreader area around the device 100X larger than the surface area of the package which is already in the flattening area of the thermal resistance. R_{th_JA} values for the 4X2mm device (with infinite heat spreading) are 67K/W for 35µm thickness of the copper layer and 50K/W for 70µm thickness. Larger copper thicknesses are not recommended since it is very difficult to make reliable pads for the components. In case of chip-size devices, the thermal spreading should be done on the drain contact of the top device and source contact of the bottom device while the surface area of the switching node pad should be as low as possible to ensure that the equivalent parasitic coupling capacitances are low.

Other than decreasing the heat spreading area, which as we seen has its limitations, one can use thermal vias underneath the device connecting the device pad to the bottom side of the PCB. For a stack-up with two 70 µm Cu layers the thermal resistance can be lowered to ~42K/W. Here the limiting factor is in-plane resistance of the stack-up. The effectiveness of vias will also be dependent on their distance from the device, given the limitations of heat spreading. Furthermore, Lastly, this type of PCB embedded heat sink causes parasitic capacitive coupling, the effects of which have been described in the previous sections.

The thermal resistance can be further reduced if the board is mounted on a metal heat spreader. R_{th_JA} of the PCB with thermal vias (100vias/cm^2, 0.5mm diameter, 70µm plating) and 1mm Al heat spreader with a 0.1mm layer of adhesive tape for isolation is 27K/W compared to R_{th_JA} of convectively cooled PCB with 100mm^2 copper pad around the device which is 77K/W. Furthermore, it can be seen that even with the thermal vias and external

978-1-4799-2706-7/14 $31.00 © 2014 IEEE

heat spreader the thermal performance of this arrangement is limited by R_{th_JB}.

High R_{th_JB} limits the power handling of flip chip devices if bottom side cooling is employed. For a flip chip die, R_{th_JC} is several times (~5X) smaller then R_{th_JB}, making top-side cooling a preferred choice for high power application. However, very small and brittle dies make clamping the heatsink directly and applying uniform force over the top die surface difficult without suitable thermal interface material. This is especially the case when a single heatsink is used for several devices. On the other hand, the thermal interface material will introduce additional thermal resistance which can be as large as R_{th_JB}. With PCB heat spreading and additional thermal conducting material around the device the total thermal resistance ($R_{th\_JC} + R_{th\_CH}$) can be 50% lower, when compared to bottom side cooling ($R_{th\_JB} + R_{th\_BH}$), but at the expense of more difficult mounting. Such a construction can make placing decoupling capacitor close to the devices more difficult. On the other hand, top side cooling will free the bottom side of the PCB for components, including decoupling capacitors.

b) Higher thermal conductivity substrates

As shown above, conventional FR4 PCBs even when enhanced with heat spreading and thermal vias run into their limits and higher thermal conductivity circuit carriers may need to be used. However, not only thermal conductivity of the circuit carrier is important but also the suitability for high frequency layout, in terms of number of layers, carrier thickness etc. Circuit carriers superior from the thermal conductivity viewpoint can work adversely with respect to high frequency performance.

Reference [7] compares the use of PCB and alumina DBC in high power density GaN based POL modules. While the use of a DBC decreases the thermal resistance and thus increases the power handling capability by 50% compared to the PCB substrate, the losses increase 25% (efficiency decreases 2.5%). This is due to the distance of the Cu layers in a DBC vs PCB (0.38mm vs 0.12mm) which results in a considerably larger loop inductance of the DBC design since the second layer acts as a shield resulting in a field cancellation and thus reduction in parasitic inductance. The effect would be even larger for thicker DBCs (0.635mm is a standard).

Another alternative used in power electronic circuits is an insulated metal substrate (IMS) which is a circuit carrier consisting of a top copper layer bonded to a very thin dielectric layer and a thick (1mm) bottom Al heat spreader. The thermal behaviour is superior compared to a PCB (due to thin dielectric and thick heat spreader) but since it is a single layer board it limits the flexibility in component placement and only accommodates SMD devices. Furthermore, due to the very thin dielectric the capacitive coupling to the bottom heat spreader is pronounced resulting in the already discussed increase in switching loss.

2) System thermal management considerations

Pushing the switching frequency with GaN devices can lead to a shift in the volume distribution between electrically and thermally active construction parts in the converter. High frequency operation will lead to

miniaturisation of electrical parts, especially passives, and thermal management parts could take up a significant portion of the total volume [18]. Approaches for a more effective multifiunctional use of converter parts (heat conduction, electrical interconnection, mechanical support) are needed [19].

Due to their fast switching properties and low on-resistance the losses in semiconductors can be reduced significantly which makes the shift in loss distribution between semiconductors and passive components and makes the losses in passives even more decisive for high efficiencies. The DC link capacitor losses are the same as the switch losses in a GaN inverter due to high frequency currents [20]. Reference [21] reports an LLC 1kW, 1MHz demonstrator where the temperature increase on the transformer is much higher than that of the transistors. In VHF GaN converters semiconductors could contribute less than 20% of the total loss [22]. The bottleneck to pushing the switching frequency further shifts passive components, especially magnetics. This can result in the need to change the way that converters are cooled. In the power range of interest for GaN, most converters, with the exception of very high power density converters where the increase in passive losses is acceptable for size reduction and more advanced cooling is needed [23], passives have little dedicated cooling. Cooling of passives can become critical at high frequencies and new thermal management approaches for passives (and/or converter as a whole) are needed. One can think of 3D heat spreading, integrating thermally active parts into components, use of thermally conductive materials for overmoulding and housing of passive assemblies etc.

Due to the much smaller surface area of GaN devices and thus limitations in heat spreading, spreading the heat over multiple sources can be desirable. Here one can think about multi-phase or multi-level topologies and architectures. This approach can also enable further increase in switching frequency without compromising component stresses and switching losses [24]. In applications with a wide range operating conditions, modular based converter architectures aside from heat distribution benefit can also bring a higher efficiency operation through each cell operating under a narrow range of operating conditions [25].

V. OUTLOOK ON TECHNOLOGIES FOR 3D SYSTEM INTEGRATION

GaN devices are still in their early phase of development and the focus is on their switching performance but in order to take a full advantage of their potential in terms of increasing the level of integration and power density, we need to look at technology platforms and system integration concepts. An optimal concept for circuit integration would place the switching and electromagnetic functions in the 3D space and at the same time distribute the heat in an optimal manner. A technology platform that gives more freedom to place the power devices, voltage and current sensors and create and shape the fields associated with inductors, capacitors and transformers is desirable. The technology should be able to accommodate trade-offs between high frequency switching and thermal management. Furthermore, it should be able to enable integration of passives and in order to ensure market adoption, manufacturability has to be taken into account.

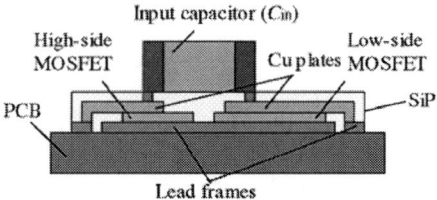

a.

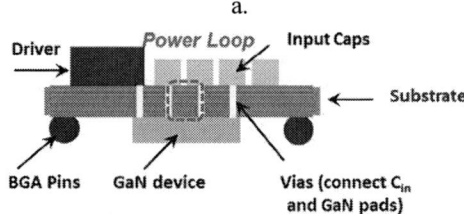

b.

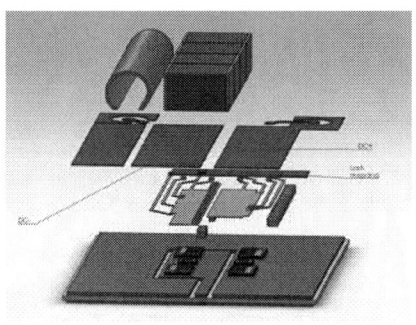

c.

Figure 10 3D commutation cell SiP structures for low parasitic inductance a. SiP with decoupling capacitor on top [24] b. GaN SiP with decoupling capacitor on the opposite side of the substrate [8] c. DBC/PCB integration concept with laminated PCB bus bar structure on top of the chips [27]

For hard switching topologies the commutation cell (switch, diode and capacitors) are the basic building block. The majority of the research effort in enabling pushing the switching frequency (further) into the MHz range has been directed in developing technology concepts aimed at reducing the parasitic inductance of the commutation cell, for POL converters and power modules. Figure 10a shows a SiP implementation with >40% decrease in the loop parasitic inductance due to the position of the decoupling capacitor on top of the SiP package [26]. However, this prohibits the use of the top side for heat removal. Similar reasoning holds for placing the capacitors directly below the devices on the opposite side of the circuit carrier, as IR has demonstrated in their SiP approach where the decoupling capacitors are mounted on top of the GaN SiP reducing the parasitic inductance to 0.2nH (Figure 10a). If the capacitors are placed next to the device instead, the inductor could be integrated on the bottom of the carrier, reducing the size and increasing power density with the trade-off of single-sided thermal management and an increase in the parasitic loop inductance. Reference [8] shows an implementation of a 5MHz POL converter with the inductor integrated in LTCC technology on the bottom of the carrier, increasing

the power density to 800W/in^3 but also increasing the parasitic inductance by 40%. Reference [27] presents a low inductance hybrid DBC/PCB integration concept where the power switches are placed on the DBC and the low inductance is achieved by a bus bar structure using laminated PCB process, achieving the commutation loop inductance of 0.57nH.

MHz conversion at low voltages is state-of-the-art with Si (POL) converters. By utilizing faster GaN devices in SiP like structures with low parasitic inductances, the frequency is being pushed further into the MHz range. For very high frequency conversion (>10MHz) the passives could be integrated in PCB resulting in very high power densities, provided that suitable magnetic and dielectric materials are available for very high frequency operation. An alternative is utilising air-core magnetics for light weight because of small heatsinks and elimination of magnetic material. This will naturally result in a penalty in power density.

High voltage GaN devices can enable MHz operation of line voltage converters. Using Si devices the same topology and switching frequency, and applying the system integration concept based on multilayer PCB technology with SMT passives and automated assembly the power density of the commercial HID ballast operating at 80kHz was doubled (Figure 11) [28]. GaN could allow for the increase in switching frequency while using similar system integration concept provided that topology and thermal management considerations from previous sections are taken into account.

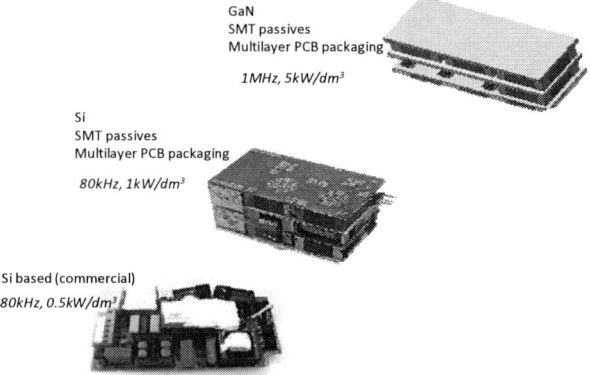

Figure 11GaN enabled MHz operation of line voltage converters

VI. CONCLUSIONS

In order to fully exploit the potential of GaN a detailed understanding of the device behaviour and its limits in the environment is mandatory. Hard switching topologies reach their limits as switching frequency increases as circuit "parasitics" become dominant and circuit design becomes a battle against nH and pF that can be pushed with clever designs but reaches the limit at some point. Determining exact switching characteristics in nanosecond times should be done in defined electromagnetic surroundings. Zero-voltage switching is desirable for avoiding capacitive discharge at turn-on, but at high switching frequencies turn-off losses and the parasitic inductance of the commutation cell capacitor pose the limit.

GaN devices require careful thermal management design due to their small size and high heat flux densities. There are trade-offs between thermal management techniques and high frequency switching and the circuit designer should to be aware of these issues. System integration technologies that can optimally accommodate electromagnetic and thermal functions are needed to fully exploit the potential of GaN in enabling the increase of switching frequency and power densities.

REFERENCES

List only one reference per reference number according to the following samples:

[1] J.D. Van Wyk, "An Experimental Study of switching GaN FETs in a coaxial Tranmission Line", Conference Proceeding of 2014 IEEE EPE ECCE-Europe.

[2] W. McMurray, "Optimum Snubbers for Power Semiconductors", IEEE Transactions on Industry Applications, Volume:IA-8 , Issue: 5, 1972 , pp593- 600

[3] C. R. Sullivan and A. M. Kern, "Capacitors with fast current switching require distributed models", in Proc. IEEE Power Electron. Spec. Conf. ,Vancouver, BC, Jun. 2001, vol. 3, pp. 1497–1503

[4] Frederick W. Grover, "Inductance Calculations: Working Formulas and Tables", Dover Publications, Inc., New York, 1946, pp.70-71

[5] EPC Power Conversion, http://epc-co.com

[6] J. Wurfl, GaN Power Devices (HEMT):Basics, Advantages and Perspectives, ECPE Seminar on WBG devices, 2013.

[7] D. Reusch, "High Frequency, High Power Density Integrated Point of Load and Bus Converters" PhD thesis, Virginia Polytechnic Institute and State University, http://scholar.lib.vt.edu/theses/available/etd-04162012-151740/unrestricted/Reusch_4_15_12_Final.pdf

[8] Shu Ji et al."High-Frequency High Power Density 3-D Integrated Gallium-Nitride-Based Point of Load Module Design," IEEE Transactions on Power Electronics,vol.28,no.9,pp.4216,4226,Sept. 2013

[9] http://www.transphormusa.com/

[10] Liu, Z.; Huang, X.; Lee, F.; Li, Q., "Package Parasitic Inductance Extraction and Simulation Model Development for High Voltage Cascode GaN HEMT," Power Electronics, IEEE Transactions on , vol.PP, no.99, pp.1,1, 0

[11] http://www.gansystems.com/

[12] Xiucheng Huang; Qiang Li; Zhengyang Liu; Lee, F.C., "Analytical Loss Model of High Voltage GaN HEMT in Cascode Configuration," Power Electronics, IEEE Transactions on , vol.29, no.5, pp.2208,2219, May 2014

[13] Josifovic, I.;Popovic-Gerber, J.;Ferreira, J.A.,"Improving SiC JFET Switching Behavior Under Influence of Circuit Parasitics" IEEE Transactions on Power Electronics,vol.27,no.8,pp.3843,3854, Aug. 2012

[14] GaN Transistors for Efficient Power Conversion, Power Conversion Publications, January 2012.

[15] Huang, X.; Liu, Z.; Li, Q.; Lee, F.C., "Evaluation and application of 600V GaN HEMT in cascode structure," Power Electronics, IEEE Transactions on , vol.PP, no.99, pp.1,1, 0

[16] W. Zhang, Z. Xu, Z. Zhang, F. Wang, L.M. Tolbert, B.J. Blalock "Evaluation of 600 V Cascode GaN HEMT in Device Characterization and All-GaN-Based LLC Resonant Converter", ECCE 2013.

[17] Rodriguez, M.; Zhang, Y.; Maksimovic, D., "High frequency PWM Buck converter using GaN-on-SiC HEMTs," Power Electronics, IEEE Transactions on , vol.PP, no.99, pp.1,1, 0

[18] Yi Wang; De Haan, S. W H; Ferreira, J.A., "Potential of improving PWM converter power density with advanced components," Power Electronics and Applications, 2009. EPE '09. 13th European Conference on , vol., no., pp.1,10, 8-10 Sept. 2009

[19] Popovic, J.; Ferreira, J. A., "An approach to deal with packaging in power electronics," Power Electronics, IEEE Transactions on , vol.20, no.3, pp.550,557, May 2005

[20] R. Mitova "Evaluation of 600V GaN devices for industrial applications", ECPE Workshop on WBG devices, 2-3 May 2013.

[21] D. Kranzer, "Application of GaN Transistors in Hard Switched and Resonant Power Electronics" ECPE Workshop on WBG devices, 2-3 May 2013.

[22] Saito, W. et. Al. "Demonstration of 13.56-MHz class-E amplifier using a high-Voltage GaN power-HEMT," Electron Device Letters, IEEE , vol.27, no.5, pp.326,328, May 2006

[23] Gerber, M.; Ferreira, J.A.; Hofsajer, I.W.; Seliger, N., "High density packaging of the passive components in an automotive DC/DC converter," Power Electronics, IEEE Transactions on , vol.20, no.2, pp.268,275, March 2005

[24] M. Maerz, "Parasitics in Power Electronics" ECPE Workshop „Future Trends for Power Semiconductors", January 2012.

[25] Rivas, J.M.; Wahby, R.S.; Shafran, J.S.; Perreault, D.J.; "New Architectures for Radio-Frequency DC–DC Power Conversion," IEEE Transactions on Power Electronics,vol.21,no.2,pp.380- 393, March 2006

[26] Hashimoto, T. et al.,"A System-in-Package (SiP) With Mounted Input Capacitors for Reduced Parasitic Inductances in a Voltage Regulator," IEEE Trans on Power Electronics,vol.25,no.3,pp.731-740, March 2010

[27] E. Hoene et. al. "Packaging Very Fast Switching Semiconductors", CIPS 2014 - 8th International Conference on Integrated Power Electronics Systems

[28] Josifovic, I.; Popović-Gerber, J.; Ferreira, J.A.; Van Casteren, D. H J, "Multilayer SMT high power density packaging of electronic ballasts for HID lamps," Energy Conversion Congress and Exposition (ECCE), 2010 IEEE , vol., no., pp.1275,1282, 12-16 Sept. 2010

978-1-4799-2706-7/14 $31.00 © 2014 IEEE

Power Losses of Multilevel Converters in Terms of the Number of the Output Voltage Levels

Yugo Kashihara
Energy and Environmental Science
Nagaoka University of Technology
Nagaoka, Niigata, Japan
kasihara@stn.nagaokaut.ac.jp

Jum-ichi Itoh
Electrical Engineering
Nagaoka University of Technology
Nagaoka, Niigata, Japan
itoh@vos.nagaokaut.ac.jp

Abstract— This paper presents loss calculation methods of which the multilevel converters have several number of the output voltage level. The multilevel converters of the flying capacitor topology and the active neutral point clamp topology are evaluated in terms of the high efficiency in this paper. In addition, the power losses of the multilevel converters are discussed using two power devices. As a result, in case of the MOSFET device, the active neural point clamp topology is better than the flying capacitor topology, regardless of the number of level. The power loss of the two-level inverter is lower than other topologies.

Keywords— *Inverter, Multilevel converter, Power loss, Photo voltaic.*

I. Introduction

In terms of system integration on power electronics system, it is important to choose suitable circuit topology according to the purpose of the system. Multilevel converters are one of the good options in order to obtain high efficiency. In general, multilevel converters are applied in medium-voltage applications, such as large power motor drives and 6.6-kVA power transmission lines because by comparing between the multilevel converters and the conventional two-level converters, the multilevel converter can reduce the voltage stress of a switching device to 1/(n-1) of the DC input voltage and also reduce the harmonic component of the output voltage. Hence, applications of multilevel converters have been actively investigated [1-5].

However, in order to achieve high efficiency using low conduction loss devices such as MOSFETs and size reduction of the output filter, the multilevel converters have been applied in low-voltage applications, such as uninterrupted power supplies (UPSs) and power converters for photo voltaic cells (PVs) [6].

In order to achieve high efficiency when multilevel converter is concerned, this factor need to be considered seriously, i.e., the suitable circuit topologies selection and the number of output voltage level. Although there are many topologies for multilevel converters, however in general, this converter can be categorized into separated two topologies, which are the diode clamp (DCLMP) topology and the flying capacitor (FC) topology [1], [2]. Besides that, these two topologies have shared the same waveforms at the same levels. Thus, it is difficult to select the circuit topology in the multilevel topologies. In addition, the number of the voltage levels of multilevel converter is decided depending on the application.

Numbers of studies have demonstrated the power losses of a multilevel converter in terms of number of the output voltage level [7], [8]. Those studies analyzed the power losses of the multilevel converters using mathematical expression. This method calculates the power loss depending on the device parameters and the circuit structure only. Thus, it is possible to design the multilevel converters depending on the application in terms of some parameters such as efficiency, volume, cooling performance, cost, reliability and so on. However, previous studies did not evaluate and compare power losses of the multilevel converter topologies in terms of number of the output voltage levels.

This paper presents several numbers of loss calculation methods with regard to the type of the multilevel converters, and regardless of the number of level. In addition, the power losses of the multilevel converters are discussed using two kinds of the power devices which are MOSFET and IGBT. Thus, the best topology and output voltage level of the multilevel converter can be selected based on its application. First, the loss calculation methods of the FC topology and the active neural point clamp (ANPC) topology are discussed [1], [3] because these topologies do not require the voltage balance circuit at DC link capacitor. In addition, the power losses of the two multilevel topologies using two power devices are discussed. The power losses of the two multilevel topologies using MOSFETs from three-level to eleven-level of output voltage levels are calculated based on the mathematical expression. On the other hand, the power losses of the two multilevel topologies using IGBTs from two-level to five-level of output voltage levels are calculated. In addition, power losses of the conventional two-level inverter and the diode clamp (DCLMP) topology are also discussed with simulation results in the case of IGBT. Finally, power loss characteristics are compared in terms of the numbers of the output voltage level. From the point of efficiency, in the case of MOSFET, the ANPC topology shows better results than other types of multilevel converters, regardless of the number of level. On the other hand, in

978-1-4799-2706-7/14 $31.00 © 2014 IEEE

the case of IGBT, the two-level converter shows the best result compared to other types of multilevel converters, regardless of the number of level.

II. MULTILEVEL CONVERTER TOPOLOGY

A. Flying capacitor topology

Figure 1 shows the single phase generalized FC topology [1]. The number of the flying capacitors and switches in the generalized FC topology increases in proportion to the number of levels. Outputs step waveform of FC topology is sum of the voltages of the flying capacitor and DC smoothing capacitor.

B. Active neutral point clamp topology

Figure 2 shows the single phase generalized ANPC topology. The ANPC topology combines the DC and FC topologies into one converter. Due to switching devices of the ANPC topology has two switching frequencies, the ANPC topology can be separated into two cells as shown in the Figure 2. Switching frequency of the Cell 1 switches is the carrier frequency. On the other hand, switching frequency of the Cell 2 switches is same to the output frequency. Thus, switching loss of the Cell 2 switches is low. In addition, circuit structure of the Cell 1 is similar to FC topology. Thus, number of the flying capacitors in the Cell 1 increases in proportion to the number of levels. On the other hand, number of the semiconductors in the Cell 2 increases in proportion to the number of levels. However, in the Cell 2, a high voltage rating device can be used instead of many low voltage rating devices.

III. CALCULATION METHODS OF THE TWO GENERALIZED MULTILEVEL CONVERTER TOPOLOGIES

These multilevel converter topologies are assumed to be operated under ideal condition. The power losses are calculated under ideal condition, i.e., no current and voltage ripples in the capacitors. The voltage fluctuation in the flying capacitor occurs only during the switching cycle. In addition, the applied voltage of the switches fluctuates during the switching cycle. However, there are two switches that apply low voltage and high voltage for the same switching pattern. Thus, the power loss by voltage ripple is counterbalanced.

Semiconductor loss is separated into the switch-side loss and the FWD-side loss [6]. The power losses of the switch-side P_{sw} and the FWD-side P_D are given by

$$P_{sw} = P_{con\_sw} + P_{switch} + P_{nl\_sw}, \quad\quad\quad\quad\quad (1)$$

$$P_D = P_{con\_D} + P_{rec} + P_{nl\_D}, \quad\quad\quad\quad\quad\quad (2)$$

where P_{con_sw} is the conduction loss of the switch-side, P_{switch} is the switching loss, P_{nl_sw} is the no-load loss, P_{con_D} is the conduction loss of the FWD-side, P_{rec} is the recovery loss, and P_{nl_D} is the no-load loss of the FWD-side.

The conduction loss is separated into the switch-side loss and the FWD-side loss. In addition, if the switching

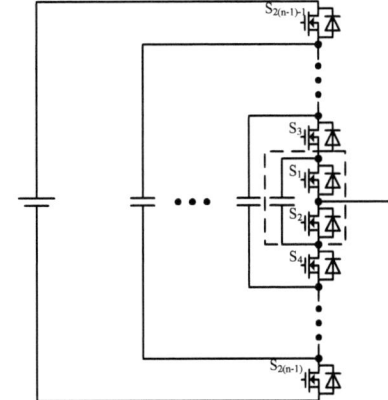

Fig.1. Single phase generalized flying capacitor converter.

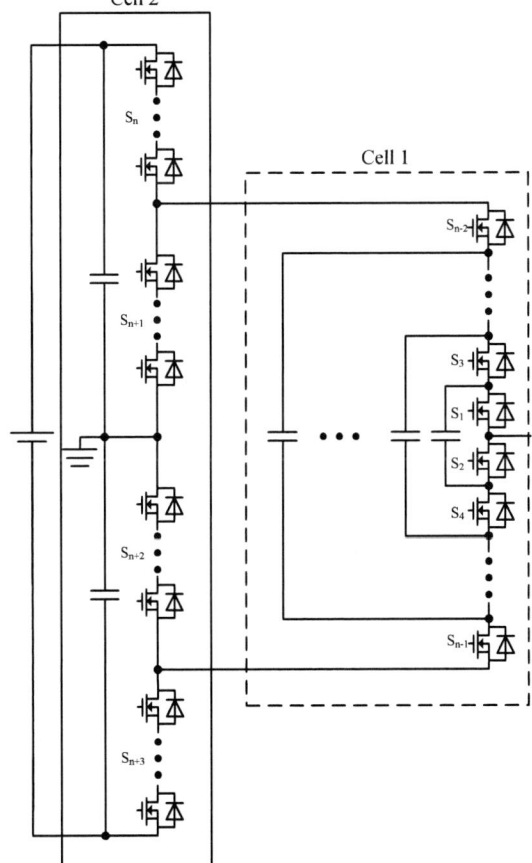

Fig.2. Single phase generalized active neutral point clamp converter.

device of the two-level converter is a MOSFET, both the positive and negative currents will flow into the switch side due to low on-resistance. We assume that positive current flows into the switch-side and negative current flows into the FWD-side. The conduction losses of the switch-side and the FWD-side are calculated from on-voltage of the semiconductor and switch current. Thus, conduction loss of the semiconductor P_{con} is given by

$$P_{Con} = \frac{1}{2\pi}\int_{\alpha}^{\beta} v_{on} i_{sw} dx, \quad\quad\quad\quad\quad (3)$$

$$v_{on} = r_{on} i_{sw} + v_0, \quad\quad\quad\quad\quad\quad\quad (4)$$

$$i_{sw} = \lambda I_m \sin(\theta + \phi), \quad\quad\quad\quad\quad\quad (5)$$

where v_{on} is the on-voltage, i_{sw} is the switch current, α and β are phase angle during current flowing, r_{on} is the on-resistance, v_0 is the on-state voltage when $I_m\sin\theta$ equals to approximately 0 A, λ is the duty ratio command, θ is the power factor, and ϕ is the phase angle.

Using switching characteristics from a data sheet, the switching loss and recovery loss are given by

$$P_{switch} = \frac{E_{dc}}{n-1}(e_{on}+e_{off})f_c\frac{1}{2\pi}\int_x^y i_{out}d\theta, \dots\dots\dots\dots(6)$$

$$P_{rec} = \frac{E_{dc}}{n-1}(e_{rec})f_c\frac{1}{2\pi}\int_x^y i_{out}d\theta, \dots\dots\dots\dots(7)$$

where E_{dc} is the input voltage, e_{on} is the turn-on energy per switching from datasheet, e_{off} is the turn-off energy from switching at datasheet, I_{out} is the output current, f_c is the carrier frequency, n is output voltage level, x and y are the phase angles while the current is passed, e_{rec} is the recovery energy per switching from datasheet.

The no-load loss occurs as a result of the parasitic capacitance of switching devices. When an input voltage is applied to the switching devices, the parasitic capacitance of the drain-source of the switching device charges the voltage in the MOSFET. The parasitic capacitance for the IGBT is on the collector-emitter of the switching device. When the voltage of the floating capacitor is discharged, the no-load loss occurs at the on resistance of the switching device. The no-load loss is given by

$$P_{nloss} = \frac{1}{2}C_p\Delta V_{sw}^2 f_c, \dots\dots\dots\dots\dots(8)$$

where C_p is the parasitic capacitance of the switching device, and ΔV_{sw} is the applied voltage of the switching device.

A. Power loss of the generalized FC topology

This section explains the power loss expression of the generalized FC topology. The switching pulse pattern of all switches in the FC topology is same. Thus, semiconductor loss per one switch is same.

The conduction loss $P_{FC\_con\_sw}$ on the switch-side and conduction loss $P_{FC\_con\_FWD}$ on the FWD-side can be given by

$$P_{FC\_con\_sw} = \left(\frac{1}{8}+\frac{1}{3\pi}a\cos\phi\right)r_{on}I_m^2 + \left(\frac{1}{2\pi}+\frac{1}{8}a\cos\phi\right)v_0I_m, \dots\dots(9)$$

$$P_{FC\_con\_FWD} = \left(\frac{1}{8}-\frac{1}{3\pi}a\cos\phi\right)r_{on}I_m^2 + \left(\frac{1}{2\pi}-\frac{1}{8}a\cos\phi\right)v_0I_m. \dots\dots(10)$$

Therefore, the switching loss depends on the current flows through the switches and the number of switches. The switching loss P_{FC_sw} is given by

$$P_{FC\_sw} = \frac{1}{(n-1)\pi}\frac{E_{dc}I_m}{E_{dcd}I_{md}}(e_{on}+e_{off})f_c \dots\dots\dots\dots(11)$$

In addition, the recovery loss P_{FC_rec} is given by

$$P_{FC\_rec} = \frac{1}{(n-1)\pi}\frac{E_{dc}I_m}{E_{dcd}I_{md}}e_{rr}f_c \dots\dots\dots\dots(12)$$

Finally, the no-load loss P_{FC_nl} is given by

$$P_{FC\_nl} = \frac{1}{2}C_p\left[\frac{E_{dc}}{(n-1)}\right]^2\frac{f_c}{2} \dots\dots\dots\dots(13)$$

Thus, total semiconductor loss per one switch in the n-level FC topology is given by

$$P_{FC\_Loss\_semi} = 2(n-1)\left(P_{FC\_con\_Sw}+P_{FC\_con\_FWD}+P_{FC\_switch}+P_{FC\_rec}+P_{FC\_nl}\right). \dots\dots\dots\dots(14)$$

B. Power loss of the generalized ANPC topology

This section explains the power loss expression of the generalized ANPC topology (Figure 2). The conduction loss $P_{ANPC\_con\_Cell1\_sw}$ on the switch side can be given by

$$P_{ANPC\_con\_Cell1\_sw} = \frac{1}{2\pi}\left(\left[\frac{1}{4}\sin2\phi-\frac{1}{2}\phi+\frac{4}{3}a\cos\phi\right]r_{on}I_m^2\right. \\ \left.+\left[1+\left(\frac{\pi}{2}a-1\right)\cos\phi\right]v_0I_m\right) \dots\dots\dots(15)$$

On the other hand, the conduction loss $P_{ANPC\_con\_Cell1\_FWD}$ on the switch side is given by

$$P_{ANPC\_con\_Cell1\_FWD} = \frac{1}{2\pi}\left(\left[-\frac{1}{4}\sin2\phi+\frac{1}{2}\phi-\frac{4}{3}a\cos\phi+\frac{\pi}{2}\right]r_{on}I_m^2\right. \\ \left.+\left[1+\left(1-\frac{\pi}{2}a\right)\cos\phi\right]v_0I_m\right) \dots(16)$$

The conduction loss in Cell2 is obtained by the same formula that is used to calculate the conduction loss in Cell 1. However, the current that flows into the Cell 2 switches is different from the current that flows into the Cell 1 switches because S_n and S_{n+2} are turned on when the output voltage command is positive and S_{n+1} and S_{n+3} are turned on when the output voltage command is negative.

Therefore, the conduction loss $P_{ANPC\_con\_Cell2\_swA}$ for the switch side of S_n and S_{n+3} is given by

$$P_{ANPC\_con\_Cell2\_swA} = \frac{a}{2\pi}\left[\left(\frac{1}{6}\cos2\phi+\frac{2}{3}\cos\phi+\frac{1}{2}\right)r_{on}I_m^2\right. \\ \left.+(-\sin\phi+(\pi+\phi)\cos\phi)\frac{1}{2}v_0I_m\right] \dots\dots\dots(17)$$

The conduction loss $P_{ANPC\_con\_Cell2\_FWDA}$ for the FWD side of S_n and S_{n+3} is given by

$$P_{ANPC\_con\_Cell2\_FWDA} = \frac{a}{12\pi}\left[8\sin\left(\frac{\phi}{2}\right)^4 r_{on}I_m^2 + 3(-\sin\phi+\phi\cos\phi)v_0I_m\right]. \dots\dots\dots\dots(18)$$

Likewise, the conduction loss for the switch side of S_{n+1} and S_{n+2} is given by

$$P_{ANPC\_con\_Cell2\_swB} = \frac{1}{2\pi}\left[\left[\left(\frac{\pi}{2}+\frac{\phi}{2}-\frac{1}{4}\sin2\phi\right)+a\left(\frac{1}{6}\cos2\phi+\frac{2}{3}\cos\phi+\frac{1}{2}\right)\right]r_{on}I_m^2\right. \\ \left.+\left[(\cos\phi+1)-a\left(\frac{\pi}{2}\cos\phi-\frac{1}{2}\sin\phi+\frac{1}{2}\phi\cos\phi\right)\right]v_0I_m\right] \\ , \dots\dots\dots\dots(19)$$

and the conduction loss for the FWD side of S_{n+1} and S_{n+2} is given by

$$P_{ANPC\_con\_Cell2\_FWDB} = \frac{1}{2\pi}\left[\left[\left(\frac{\phi}{2}-\frac{1}{4}\sin2\phi\right)+a\left(\frac{1}{6}\cos2\phi-\frac{2}{3}\cos\phi+\frac{1}{2}\right)\right]r_{on}I_m^2\right. \\ \left.+\left[-1+\cos\phi-\frac{1}{2}a(\sin\phi-\phi\cos\phi)\right]v_0I_m\right]. \\ \dots\dots\dots\dots(20)$$

Thus, the switching loss of the switches in Cell 1 is proportional to the applied voltage and current. Therefore, the switching loss of Cell 1 depends on the current flows through the switches and the numbers of switch. The Cell 1 switching loss $P_{ANPC\_switching\_Cell1}$ is given by

$$P_{ANPC\_switching\_Cell1} = \frac{1}{(n-1)\pi}\frac{E_{dc}I_m}{E_{dcd}I_{md}}(e_{on}+e_{off})f_c. \dots\dots\dots\dots(21)$$

The recovery loss $P_{ANPC\_rec\_Cell1}$ is given by

$$P_{SA\_rec\_Cell1} = \frac{1}{(n-1)\pi}\frac{E_{dc}I_m}{E_{dcd}I_{md}}e_{rr}f_c. \dots\dots\dots\dots(22)$$

The switching loss in Cell 2 depends on the output frequency (50 Hz).As a result, the switching loss in Cell

2 is lower than that in Cell 1, which is approximately zero, and therefore can be ignored.

No-load loss in the Cell1 $P_{ANPC\_nl\_Cell1}$ is calculated by

$$P_{ANPC\_nl\_Cell1} = \frac{1}{2} C_P \left[\frac{E_{dc}}{(n-1)} \right]^2 f_c \quad \text{.............................(23)}$$

On the other hand, the no-load loss in Cell 2 is also approximately zero, and therefore can be ignored based on the switching loss in the Cell 2.

Thus, total semiconductor loss per one switch in the n-level FC topology is given by

$$
\begin{aligned}
P_{ANPC\_Loss\_semi} = 2&\left[\frac{(n-3)}{2} + 1 \right] \Big(P_{ANPC\_con\_sw\_Cell1} + P_{ANPC\_con\_FWD\_Cell1} \\
&+ P_{ANPC\_switch\_Cell1} + P_{ANPC\_rec\_Cell1} + P_{ANPC\_nl\_Cell1} \Big) \\
+ 2&\left[\frac{(n-1)}{2} \right] \Big(P_{ANPC\_con\_sw\_Cell2A} + P_{ANPC\_con\_FWD\_Cell2A} \\
&+ P_{ANPC\_con\_sw\_Cell2B} + P_{ANPC\_con\_FWD\_Cell2B} \Big)
\end{aligned}
$$

$$\text{.......................(24)}$$

C. Experimental verification

Table 1 shows the converter specifications and device parameters. This section discusses the validity of mathematically calculated losses based on the experimental results. Thus, the three-level FC inverter and the five-level ANPC inverter are designed based on converter specifications and device parameters.

Figure 3 shows the experimental waveforms of the prototypes of the single-phase three-level FC inverter and the single-phase five-level ANPC inverter for a 3.3 kW load. Both inverters show a perfect sinusoidal waveform without distortion of the output current, respectively. In addition, a three-step waveform of the output voltages of the three-level FC inverter is shown, Figure 3 (a). On the other hand, a five-step waveform of the output voltages of the five-level ANPC inverter is shown as well, Figure 3 (b).

Figure 4 shows the no-load loss comparison between the calculation and simulation results of the both multilevel inverters. Note that the parametric capacitance of the MOSFET is measured by LCR meter (5 V, 10 kHz). Both, the calculation results of no-load loss and the experimental results show a good agreement. In addition, the error ratio is under 2.2 %.

Figure 5 shows the power loss comparison between the calculation results and experimental results of the both inverters. Both, the calculation results and the experimental results show a good agreement. In addition, the error ratio is under 6 %. The validity of the loss

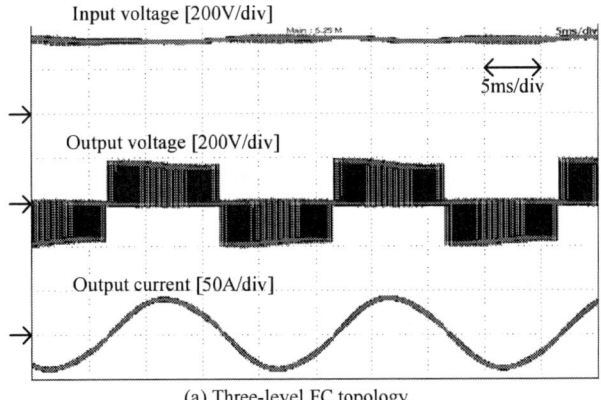

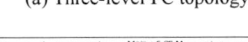

(a) Three-level FC topology.

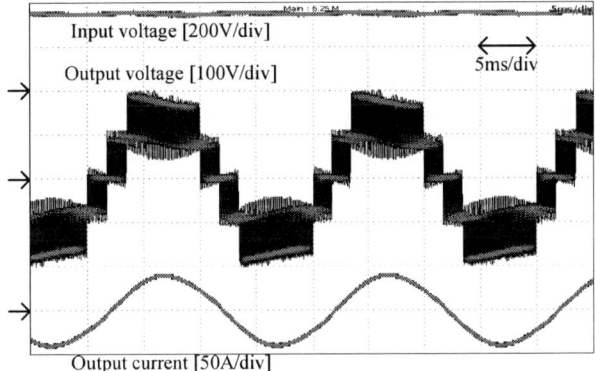

(b) Five-level ANPC topology.
Fig. 3. Experimental waveforms.

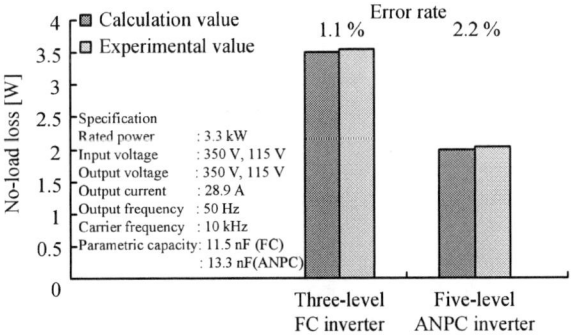

Fig. 4. No-load loss comparison.

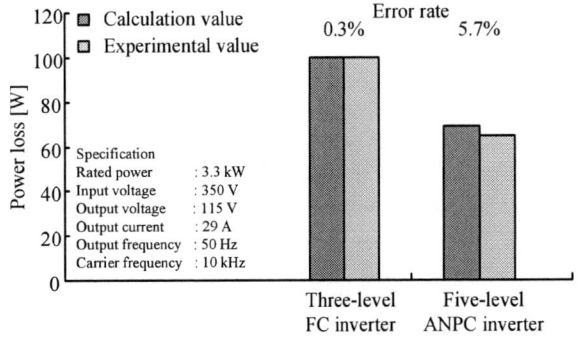

Fig. 5. Power loss comparison.

TABLE 1
SPECIFICATION

Rated power		3300 W	Output frequency	50 Hz
Input voltage		350 V	Output voltage	115 V
Modulation index		0.93	Output current	29 A
Switching device	FC	MOSFET:IXFB170N30P(IXYS)		
	ANPC	Cell1	MOSFET:IRFP4668pBF(IR)	
		Cell2	MOSFET:IXFB170N30P(IXYS)	
Flying capacitor		LGU2W101MELZ (Panasonic)		
		100 µF 450 V		
DC smoothing capacitor		FXA2G472YE (Hitachi)		
		4700 µF 400 V		

calculation method for the both generalized multilevel inverters is confirmed by the experimental results. The power loss comparison results as shown in Figure 5 include the wire resistance, equivalent series resistance (ESR) of the flying capacitor, and ESR of the DC smoothing capacitor. The measurement results of those parameters are as follow, 13.5 mΩ (wire resistance of the three-level FC), 14.9 mΩ (wire resistance of the five-level ANPC), 21.0 mΩ (ESR of the flying capacitor), and 19.8 mΩ (ESR of the DC smoothing capacitor).

IV. POWER LOSS EVALUATION IN TERMS OF THE NUMBER OF LEVELS

This section discusses the power loss in terms of the number of levels using two kinds of the power devices, which are MOSFET and IGBT. First, in the case of MOSFET, the power losses from three-level to eleven-level are calculated based on the mathematical expression. On the other hand, in the case of IGBT, the power losses from two-level to five-level are calculated similar to the case of MOSFET.

A. MOSFET

Figure 6 shows the scatter plot of the on-resistance and the breakdown voltage of the MOSFET. The MOSFETs are selected from five different semiconductor manufactures which are Infineon, IR, IXYS, Renesas, and TOSHIBA. The criterions of selecting devices are the range of breakdown voltage from 60 V to 300 V and the range of continuous drain current from 50 A to 100 A. In Figure 6, we assume that the on-resistance of the MOSFET increases in proportional to the breakdown voltage of the MOSFET. Thus, a line in the Figure 6 is calculated by approximate expression based on the hypothesis situation. In addition, power losses of the both multilevel converters the range of the number of levels from three-level to eleven-level are calculated based on the line. Note that the 10-kW application of three-phase inverter for PV is considered. On the other hand, ANPC topology is considered for two conditions. First condition is Cell 2 devices of the ANPC topology 1 use high-voltage rating devices. Second condition is Cell 2 devices of the ANPC topology 2 use the same devices rating of the Cell 1 devices.

Figure 7 shows the power loss characteristics of the multilevel converters that are structured from three-level to eleven-level. From Figure 7 (a), the conduction loss of ANPC topology 2 is the same to the conduction loss of FC topology. On the other hand, the switching loss of the ANPC topology is half of the switching loss of the FC topology in the Figure 7 (b). Thus, the power loss of the ANPC topology is lower than the power loss of the FC topology in the Figure 7 (c).

Table 2 shows the relationship between number of switch per one switching state and total on-voltage. Power loss of the multilevel inverter decreases in inverse proportion to the number of level. The condition that

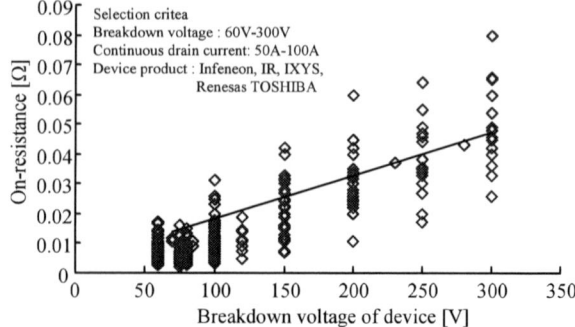

Fig.6. Scatter plot of on-resistance and breakdown voltage of MOSFET.

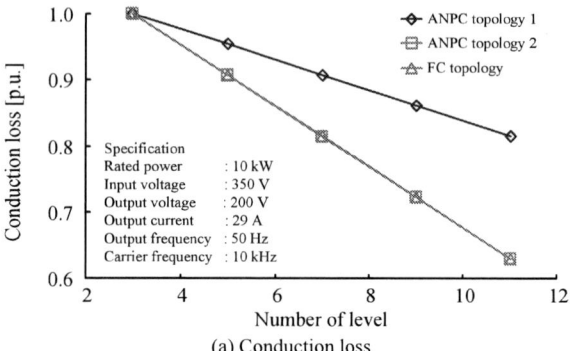

(a) Conduction loss

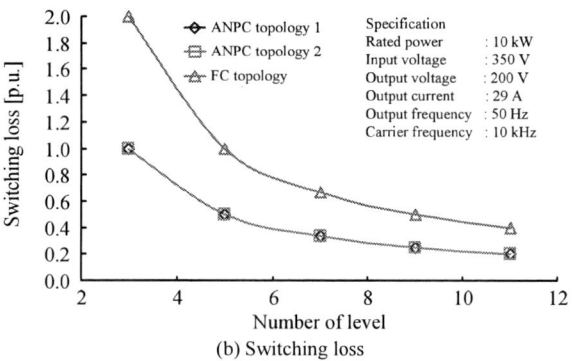

(b) Switching loss

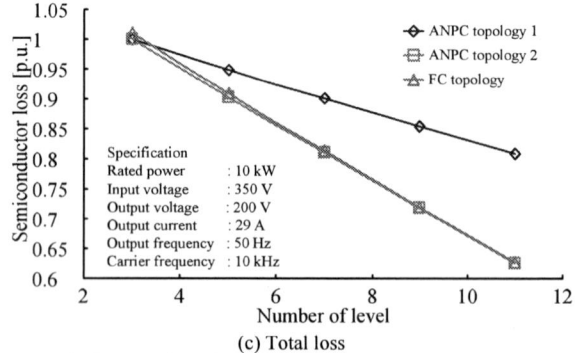

(c) Total loss

Fig.7. Power loss characteristics of multilevel converters.
1 [p.u.] of the figure 7 (a) is normalized value by total conduction loss of the three-level ANPC topology 1. On the other hand, 1 [p.u.] of the figure 7 (b) are normalized by total switching loss of the three-level ANPC topology 1. Finally, 1 [p.u.] of the semiconductor losses of figure 7 (c) are normalized value by total loss of the three-level ANPC topology 1. Cell 2 devices of the ANPC topology 1 use high-voltage rating devices. Cell 2 devices of the ANPC topology 2 use same rating devices of the Cell 1 devices.

TABLE 2
RELATIONSHIP BETWEEN LEVEL AND NUMBER OF SWITCH AND TOTAL ON-RESISTANCE

Level	Topology	Number of switch per one switching state	Total on-voltage per one switching state (on-resistance)
2	Two-level	1	v_{on} (r_{on})
3	FC, DCLMP, ANPC2 (ANPC1)	2 (2)	$2v_{on}$ ($2r_{on}$)
5		4 (3)	$4v_{on}$ ($4r_{on}$)
7		6 (4)	$6v_{on}$ ($6r_{on}$)
9		8 (5)	$8v_{on}$ ($8r_{on}$)
11		10 (6)	$10v_{on}$ ($10r_{on}$)

power loss decreases in inverse proportion to the number of level is discussed based on Figures 6 and 7. Multilevel converter increases the number of semiconductor device per on switching state in proportion to the number of level in the table 2. Thus, if the total resistance of semiconductor per on switching state of n-level inverter is lower than the total resistance of the (n-1)-level inverter, the n-level inverter can achieve low semiconductor loss. The condition that power loss decreases in inverse proportion to the number of level is given by

$$v_{on\_nl} < \frac{1}{(n-1)} v_{on\_2l}, \quad\quad\quad\quad\quad\quad\quad (22)$$

Where v_{on_nl} is on-voltage of semiconductor of n-level inverter, v_{on_2l} is on-voltage of semiconductor of 2-level inverter. For example, the level of inverter is assumed to change from three-level to five-level. If the total on-resistance of the five-level inverter is achieved by referring equation (22), the power loss of the five-level inverter is reduced. On the other hand, if the five-level inverter is not achieved by referring equation (22), the power loss of the five-level inverter is larger than the three-level inverter. Thus, it is not effective to change the level from three-level to five-level. However, the power loss of three-level inverter can be reduced if low on-resistance MOSFET which is lower than existing MOSFET. Note that the level of inverter is held the same level. Thus, the power loss of three-level inverter can be reduced.

B. IGBT

Figure 8 shows the scatter plot of the on-voltage and the breakdown voltage of the IGBT device. The IGBTs are selected from five different semiconductor manufactures which are ABB, Fuji electric, Infineon, IR, MITUBISHI, and Renesas. The criterions of selecting devices are the range of breakdown voltage from 600 V to 1800 V and the range of continuous drain current from 550 A to 1800 A. In Figure 8, we assume that the on-voltage of the IGBT increases in proportion to the breakdown voltage of the IGBT. Thus, a line in the Figure 8 is calculated by approximate expression based on the hypothesis situation. In addition, this section discusses the power loss among four topologies which are ANPC topology, FC topology, DCLMP topology, and conventional two-level topology. It is because the

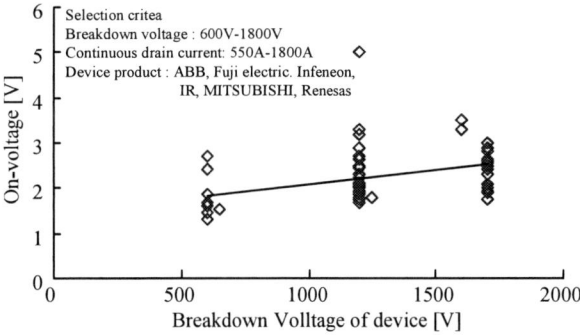

Fig.8. Scatter plot of on-voltage and breakdown voltage of MOSFET.

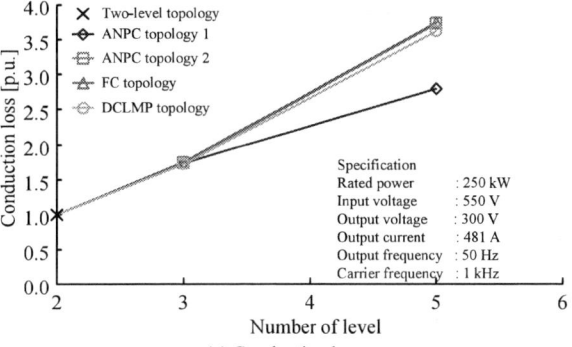

(a) Conduction loss

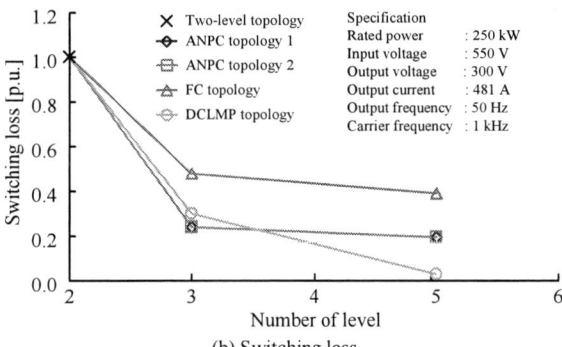

(b) Switching loss

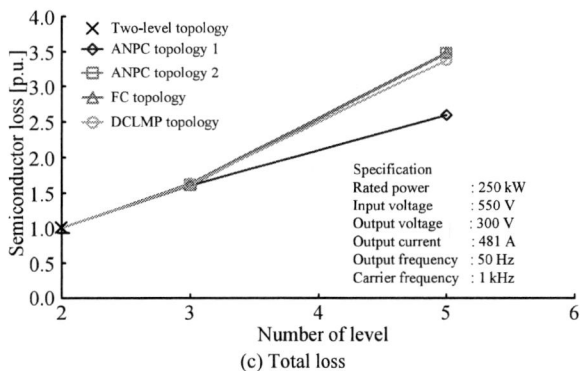

(c) Total loss

Fig.9. Power loss characteristics of multilevel converters.
1 [p.u.] of the figure 9 (a) is normalized value by total conduction loss of the two-level topology. On the other hand, 1 [p.u.] of the figure 9 (b) are normalized by total switching loss of the two-level topology. Finally, 1 [p.u.] of the semiconductor losses of figure 9 (c) are normalized value by total loss of the two-level ANPC topology 1. Cell 2 devices of the ANPC topology 1 use high-voltage rating devices. Cell 2 devices of the ANPC topology 2 use same rating devices of the Cell 1 devices.

conventional two-level topology and the DCLMP topology are applied in medium-voltage application. The power losses of four multilevel converters, the range of the number of levels from two-level to five-level are calculated based on the line. Note that the application of four topologies is 250-kW three-phase inverter for PV. In addition, the two-level topology is calculated by mathematical expression [8]. On the other hand, DCLMP topology is calculated by simulation software (PSIM).

Figure 9 shows the power loss characteristics of multilevel converters that are structured from two-level to five-level. From Figure 9 (a), the conduction loss of the two-level inverter is the lowest than other topologies. On the other hand, the conduction loss of ANPC topology 2 is the same to the conduction loss of FC topology. From Figure 9 (b), the switching loss of the five- level DLMPC inverter is the lowest than other topologies. From Figure 9 (d), the power loss of the two-level inverter is lower than other topologies.

Power loss of the multilevel inverter increases in proportion to the number of level. This result is different in the case of MOSFET. It is because there is little change in on-voltage of the IGBT which is the range of breakdown voltage from 600 V to 1800 V. It is difficult to achieve the criteria of selecting devices by equation (22). Thus, in case of the IGBT, it is not effective for power converter in medium-voltage application to increase the number of levels in terms of the power loss.

V. CONCLUSION

This paper discussed power losses of two multilevel topologies in terms of numbers of the output voltage level using two power devices. In case of the MOSFET, power loss of the multilevel inverter decreases in inverse proportion to the number of level. In addition, the ANPC topology is better results comparing to the FC topology, regardless of the number of level. On the other hand in case of the IGBT, power loss of the multilevel inverter increases in proportion to the number of level. The two-level converter shows the best result compared to other types of multilevel converters, regardless of the number of level.

REFERENCES

[1] F. Z. Peng : "A Generalized Multilevel Inverter Topology with Self Voltage Balancing", IEEE Transactions on industry applications, Vol.37, No.2, pp. 2024-2031, 2001

[2] A. Nabae, I. Takahashi, H. Akagi, "A new neutral-point-clamped PWM inverter", IEEE Trans.Industry Applications, Vol.IA-17, pp.518-523, 1981.

[3] Barbosa, P.; Steimer, P.; Steinke, J.; Meysenc, L.; Winkelnkemper, M.; Celanovic, N: "Active Neutral-point-Clamped Multilevel Converter", Power Electronics Specialists Conference, 2005. PESC '05. IEEE 36th 16-16 June 2005 Page(s):2296 – 2301, 2005.

[4] Gateau, G., Meynard, T.A., Foch, H.: "Stacked multilcell converter (SMC) : properties and design", Power Electronics Specialists Conference, IEEE 32nd Annual, 2001.

[5] ABB RESEARCH LTD. : P2009-525717A

[6] Lin Ma, Tamas Kerekes, Remus Teodorescu, Xinmin Jin, Dan Floricau, Marco Liserre : "The High Efficiency Transformer-less PV Inverter Topologies Derived From NPC Topology", EPE 2009-Barcelona , pp.1-10, 2009

[7] M. Kamaga, Y. Sato, K. Sung, H. ohashi: "An investigation of power device loss in multilevel converters", EDD-08-73, SPC-08-160, 2008.

[8] Yugo kashihara, Jun-ichi Itoh, "The performance of the multilevel converter topologies for PV inverter", International Conference on Integrated Power Electronics Systems 2012, Nuremberg, Germany, 2012.

[9] J. W. Kolar, J Biela and J, Minibock : "Exploring the Pareto Front of Multi –Objectice Single-Phase PFC Rectifier Design Optimization -99.2% Efficiency vs. 7kW/d m3 Power Density", IPEMC 2009-China, 2009.

A Large Capacity 3-level IEGT Inverter

Daisuke Yoshizawa, Makoto Mukunoki, Kenichiro Omote, Makoto Hayashi, Takashi Isida

Toshiba Mitsubishi-Electric Industrial Systems Corporation

* 1 Toshiba-cho, Fuchu-shi, Tokyo, 183-8511, JAPAN

YOSHIZAWA.daisuke@tmeic.co.jp

Abstract – **Many large capacity motor drive equipments have been used in the steel plants. High-efficiency motor drive equipments are required in the industrial fields. On the other hand, the replacement demand of cycloconverter to VSI (Voltage Source Inverter) is also increasing. Normally, the VSI such as IGBT (Insulated Gate Bipolar Transistor) / IEGT (Injection Enhanced Gate Transistor) inverter has low-harmonics and high-efficiency. So, it has spread widely now. In this paper, it is introduced high-efficiency and large capacity 3-level IEGT inverter for steel plant application. It is realized 6MVA rated output capacity. With the series line-up of 3-level IEGT inverter, it is possible to combine with the early development of 9MVA IEGT inverter. Moreover, it has the flexibility to connect with a common converter and several numbers of inverters.**

Keywords— 3-level Inverter, IEGT, steel plant, motor drive

I. INTRODUCTION

High-efficiency and low-harmonics motor drive equipments are required in the industrial fields. Normally, the VSI has high-efficiency and low-harmonics. So, it has spread widely now. However, the cycloconverter which has been applied in 1980's are still operating now in the steel plants. The cycloconverter has relatively low power factor and high-harmonics compared with VSI. So, an equipment size becomes larger for introducing a harmonics filter.

IGBT/IEGT VSI has better speed control accuracy with speed response and load response, and high power factor. Recently, the replacement demand of cycloconverter to IGBT/IEGT VSI is increasing. Therefore, it has been developed high -efficiency and large capacity IEGT inverters for steel plant applications. It is realized 6MVA rated output capacity. 1 bank of this inverter has a capacity of 6MVA. So, by connecting 2 banks in parallel it is possible to obtain 12MVA.

Moreover, front side maintenance is possible for 6MVA, and back side space is unnecessary. It is performed space-saving of the installation area containing maintenance space area. The installation area was miniaturized to about 57% compared with the conventional drive equipment. If the maintenance space area is included, it miniaturized a maximum of 36%.

By including 9MVA with this line-up, it is possible to realize a large capacity inverter rating from 6MVA to 36MVA.

II. CHARACTERISTICS OF IEGT

It is applied 4500V/3000A rating IEGT in this development. This semiconductor device is configured with 21 pieces of IEGT chips installed inside the press pack. Fig.1 shows the outline of the 4500V/3000A IEGT.

Fig.1 Outline of the 4500V/3000A IEGT

TABLE I
CHARACTERISTICS OF IEGT

Items	Characteristics	Conditions
DC voltage	3000V	Failure rate 100FIT
Collector-emitter voltage	4500V	VGE=-15V Tj=125℃
Collector maximum turn-off current	3000A	Vcp=4500V,Vcc=3000V, VGE=±15V,Tj≦125℃,
Collector-emitter Saturation voltage	3.4V (typ)	Ic=1500A,VGE=15V, Tj=125℃
Turn-on loss	15J	Vcc=3000V,Ic=1500A, Tj=125℃,VGE=±15V
Turn-off loss	10.5J (Vce:3.4V)	

Table I shows the characteristic of IEGT. Although this is high voltage device, the Collector-emitter saturation voltage of the device is about 3.4V (typ). Moreover, a switching loss is low even if DC voltage is high. It is suitable as a device applied to a large capacity inverter.

978-1-4799-2706-7/14 $31.00 © 2014 IEEE

Fig.2 shows the example of the turn-off waveform. It is confirmed that the turn-off is carried out more than 3000VDC. Surge voltage is suppressed about 4100V by low inductance structure.

In order to reduce the surge voltage at turn-off, reduction of the inductance is important. In this development, it is confirmed low inductance structure by 3D-CAD simulation. And finally it is realized high current turn-off.

Fig.3 shows the inductance structure by 3D-CAD simulation. A current loop changes by the turn-off device (mode 1 is turn-off of IEGT1, mode 2 is turn-off of IEGT2). Bus-bar and heat sink in each mode has modeled, and the target inductance has realized shown below.

Moreover, the gate pulse timing is adjusted so that a turn-on loss can be reduced. Usually, if a turn-on is possible by low gate resistance, a turn-on loss is reduced. However, the recovery loss of the diode is increased. The capacity of the gate resistance value also has a limit.

Therefore, after turn-on by high gate resistance value, it turns on once again by low gate resistance value. Thus, the turn-on loss is reduced.

Fig.4 shows the gate voltage (VGE), collector current (Ic) and collector-emitter voltage (Vce). At a certain mirror period of IEGT, IEGT is turned on by low gate resistance value. The turn-on loss was reduced. Fig.5 shows the turn-on waveform of the IEGT in which gate timing is adjusted.

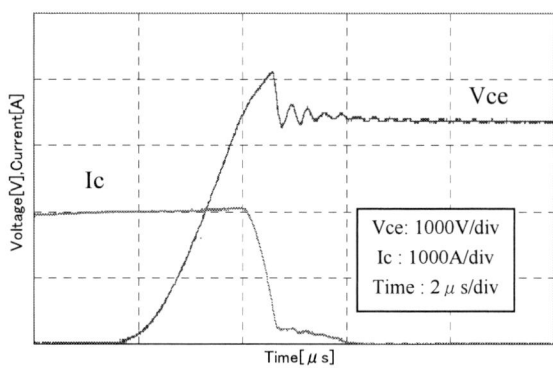

Fig.2 IEGT turn-off experimental waveform

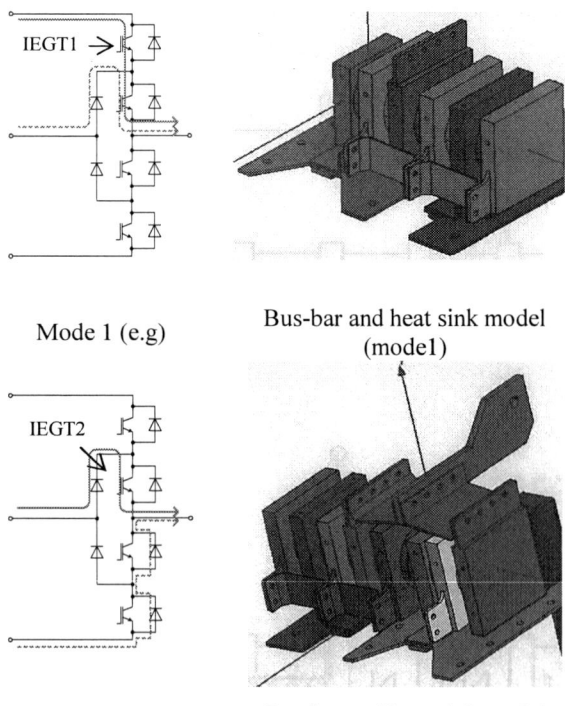

Mode 1 (e.g)

Bus-bar and heat sink model (mode1)

Mode 2 (e.g)

Bus-bar and heat sink model (mode2)

Fig 3. Inductance simulation by 3D-CAD

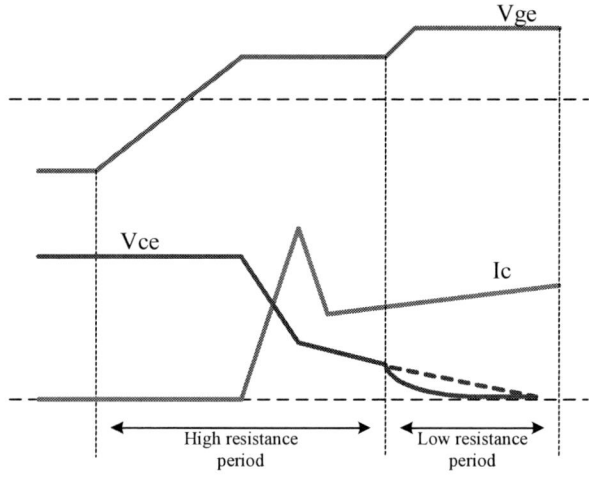

Fig.4 Gate voltage and IEGT turn-on waveform

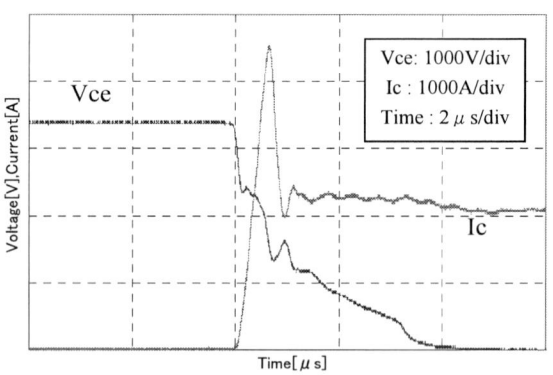

Fig.5 IEGT turn-on experimental waveform

III. SYSTEM CONFIGURATION AND SPECIFICATION

Fig.6 shows the configuration of a 3-level IEGT inverter circuit. Both inverter and converter are configured with a power unit. The new type IEGT and the pair of freewheel diodes, coupling diodes are included to a power unit.

978-1-4799-2706-7/14 $31.00 © 2014 IEEE

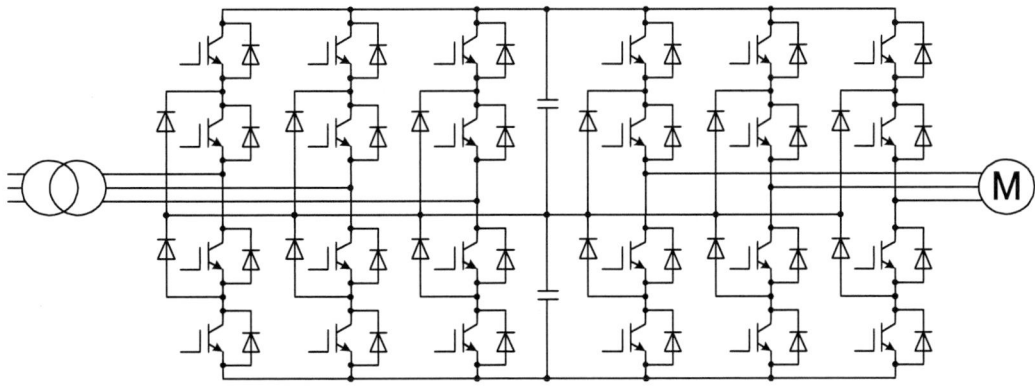

Fig.6 Configuration of a 3-level IEGT inverter circuit

Table II shows the specification of the 3-level IEGT inverter equipment. The developed equipment has two kinds of panel sizes. The outline of the equipment is W: 3000 x H: 2300 x D: 750 or W: 1500 x H: 2300 x D: 1500.

The installation area was miniaturized to about 57% compared with the conventional drive equipment. If the maintenance space area is included, it miniaturized a maximum of 36%.

TABLE II
SPECIFICATION OF THE EQUIPMENT

Items	Specification
Main circuit	3-level circuit
Rated power	6MVA
AC output voltage	3650Vrms
AC output current	950Arms(over load 150%-1min)
Output frequency	0〜75Hz
Cooler type	water cooling
Efficiency	99%
Outline (mm)	W3000×H2300×D750 or W1500×H2300×D1500
Weight	2800kg

Fig.8 shows the common converter type. Four set of inverter and two set of converter are combined. For example, in the case of induction motor, inverter capacity becomes smaller than the converter capacity controlled with the power factor 1. For the steel plant, the motor rating can be various. If it is possible to perform a converter in common to two or more inverters, the drive equipment can be widely applied.

Fig.9 shows the configuration example of a common converter.

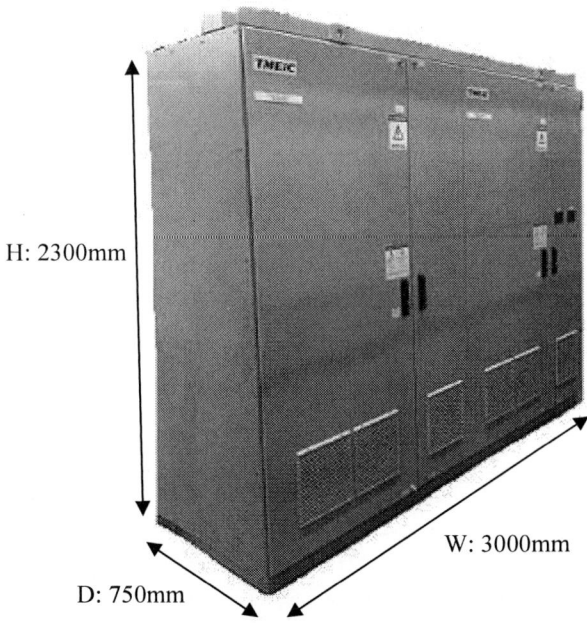

Fig.7 drive equipment appearance

Fig.7 shows the appearance of the drive equipment. Fig.7 shows only an inverter and a converter. However, a cooling panel and a control panel are installed separately. It is possible to combine with 9MVA which is developed previously [1].

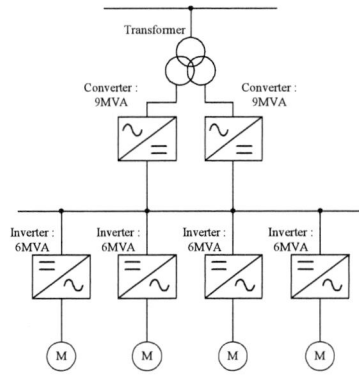

Fig.8 Common converter type

978-1-4799-2706-7/14 $31.00 © 2014 IEEE

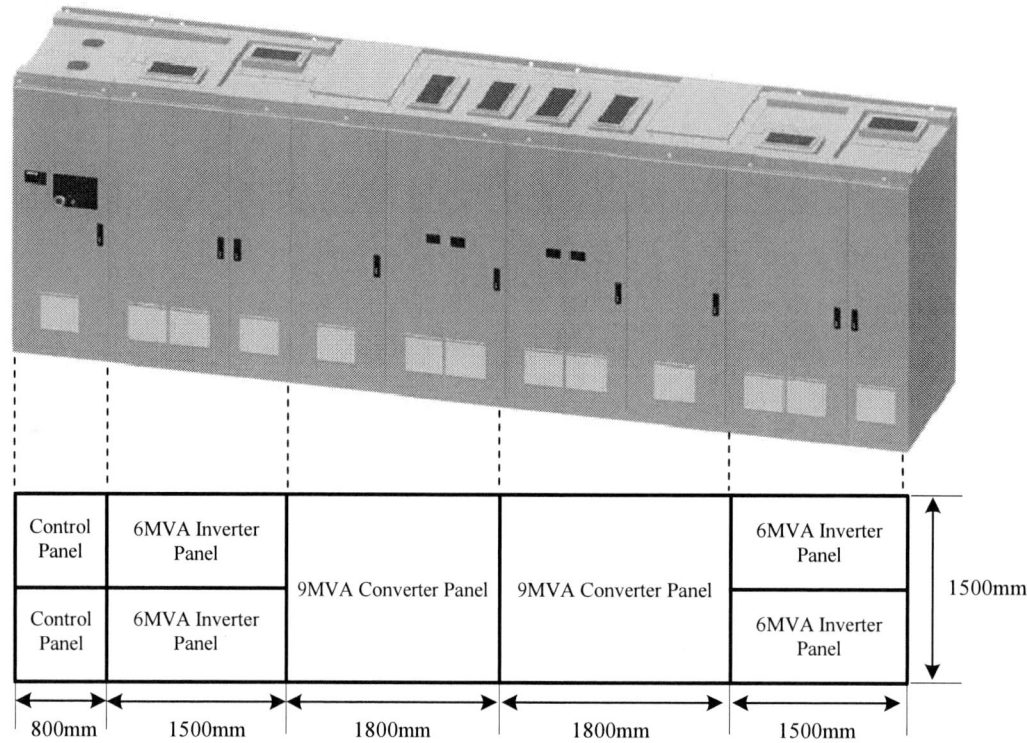

Control Panel	6MVA Inverter Panel	9MVA Converter Panel	9MVA Converter Panel	6MVA Inverter Panel	
Control Panel	6MVA Inverter Panel			6MVA Inverter Panel	1500mm
800mm	1500mm	1800mm	1800mm	1500mm	

Fig.9 Configuration example of a common converter

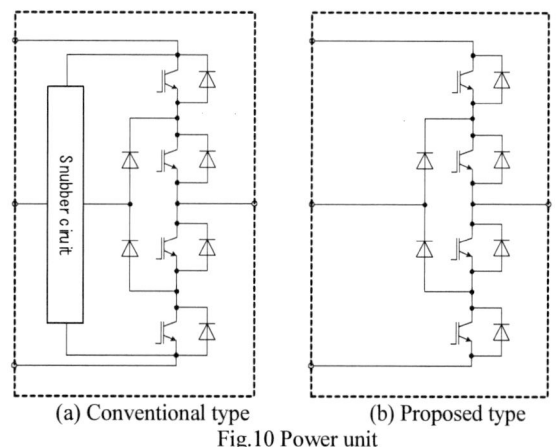

(a) Conventional type   (b) Proposed type
Fig.10 Power unit

Fig.11 shows the appearance of the developed power unit. IEGT/DIODE/Gate drive board/Gate power supply transformer are mainly built as one stack. The newly developed aluminum heat sink provides for a IEGT power unit weight reduction of approximately 33%.

And all of the DC capacitors are realized oil less. It has been applied the resin mold type capacitor without an oil leakage. Moreover, the resin mold capacitor is an insulator. So, the high density package of the equipment is possible to realize.

It is realized snubberless by low inductance structure and low heat-thermal-resistance type heat sink application. Fig.10 shows the configuration of the conventional power unit and the proposed power unit.

Although the snubber circuit is applied in the conventional power unit, it is realized snubberless in the power unit proposed. So, the power unit is simplified and maintenance has been improved. At the time of a device failure, the whole unit is exchanged so that it is possible to shorten the MTTR (Mean Time To Repair) of the equipment.

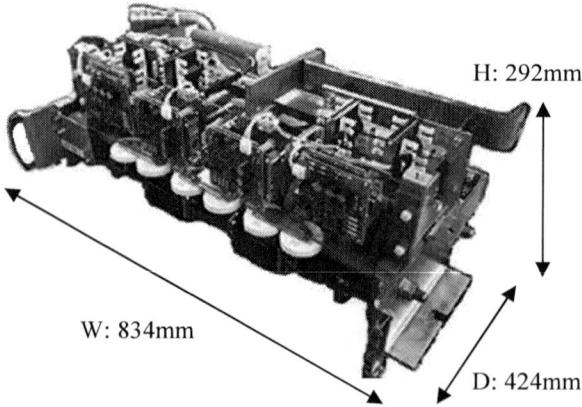

H: 292mm

W: 834mm

D: 424mm

Fig.11 power unit appearance

IV. COOLING CAPABILITY

Heat thermal resistance can be reduced by using aluminum material about 40% rather than the conventional aluminum heat sink by improving the shape and the internal flow channel to the heat sink of aluminum. Thus, it is possible to realize downsizing of the equipment.

The heat thermal resistance ratio of the conventional heat sink and the proposed heat sink is shown in Fig.12. The proposed aluminum heat sink is applied to this equipment [2], [3]. Therefore, 6MVA is realized.

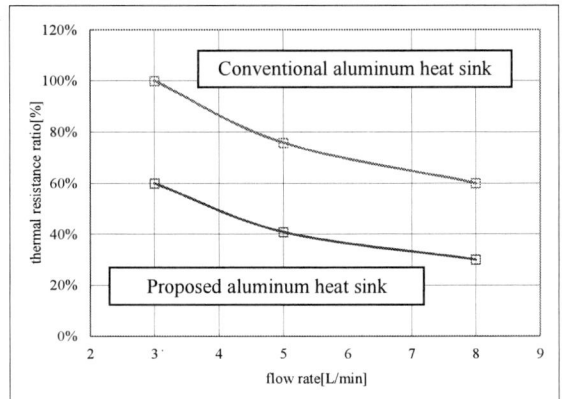

Fig.12 Comparison of the heat thermal resistance

V. OUTLINE OF THE EQUIPMENT

The developed equipment has two kinds of panel sizes. One composition is front maintenance type (Fig.13). Front maintenance is possible for this composition. The maintenance space of the back side is unnecessary.

Another one composition is double side maintenance type (Fig.14). Although the maintenance space of the double side is needed with this composition, it is possible to make the panel width smaller. Rated output capacity of 6MVA is possible for both of the composition. It enables these compositions to satisfy various demands.

VI. TEST RESULT

In the type test circuit, most of the electric power is circulated through load inductance and only the switching loss is supplied by the power supply. By this test, it is possible to perform verification tests for semiconductor devices and other parts as an actual test condition. Thus, the large capacity equipment can be tested at the factory. Fig.15 shows the configuration of the type test circuit.

The rated output voltage and current test was carried out as the type test. The type test condition is shown below.

*Output voltage: 3650Vrms (100%)

*Output current: 950Arms (100%), overload 150%-1min

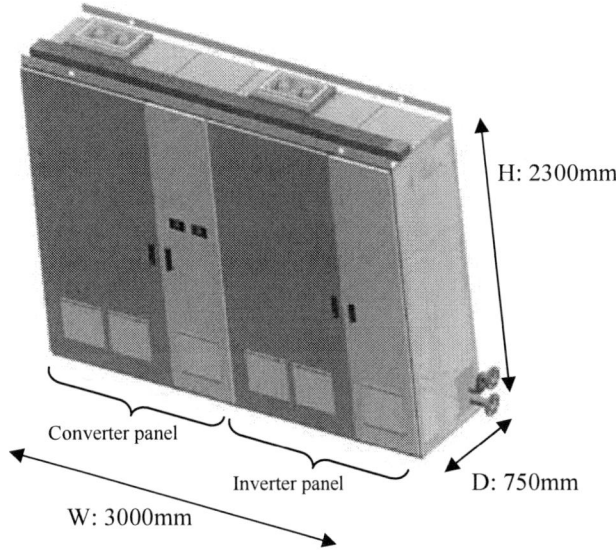

Fig.13 Front maintenance type

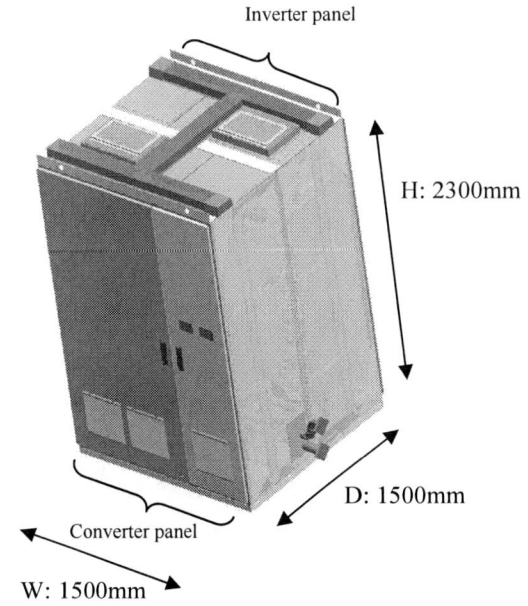

Fig.14 Double side maintenance type

Fig.16 and Fig.17 show the line to line output voltage waveform and the output phase current waveform at the rated power respectively. From the test results, it is confirmed five levels of line to line output voltage. Moreover, the equipment efficiency is confirmed 99% at the rated power (Fig.18).

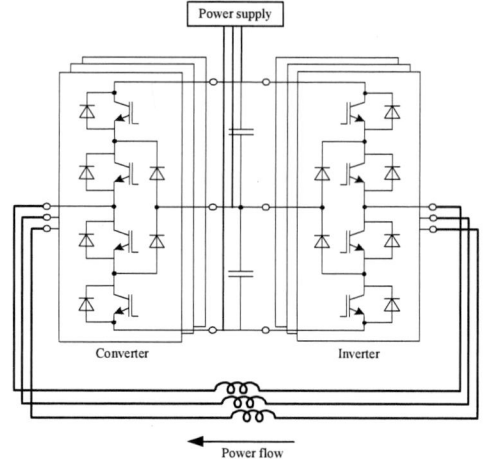

Fig.15 Configuration of the type test circuit

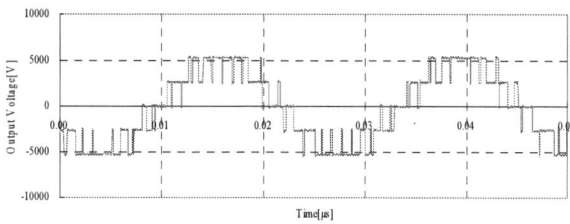

Fig.16 Line to line output voltage waveform

Fig.17 output phase current waveform

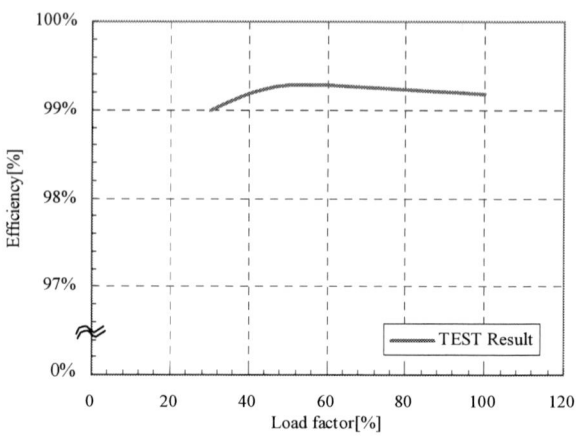

Fig.18 Equipment efficiency

REFERENCES

[1] M.A. Mamun, D. Yoshizawa, M. Mukunoki, "Performance Evaluation of a Large Capacity 3-Level IEGT Inverter", *IEEE ECCE Asia*, Melbourne, Australia, pp.201-207, Jun. 3-6, 2013.

[2] R. Nakajima, M. Mukunoki and K. Omote, "Application of an aluminum material to the large IEGT stack", *The Institute of Electrical Engineers of Japan (IEEJ)*, Hiroshima, Japan, session 4-154, pp. 265, March 2012.

[3] Z. Jun, R. Nakajima, M. Mukunoki and K. Omote, "Development of the new power conversion module for press-packed power device", *IEEJ Industry Applications Society Conference (JIASC)*, Yamaguchi, Japan, session 1-104, pp.419-420, August 2013.

VII. CONCLUSION

A 3-level IEGT inverter of 6MVA rated power is developed to apply in the steel plants. The new type IEGT is also able to successfully use for this equipment. And it is realized snubberless by low inductance structure and low heat-thermal-resistance type heat sink application. Moreover, this equipment efficiency is confirmed 99% at the rated power.

It can respond to various motors by combining with 9MVA drive equipments. This drive equipment is expected to apply in the steel plants in near future.

Vibration Suppressing Control Method of Angular Transmission Error of Cycloid Gear for Industrial Robots

Takashi Yoshioka, Yosei Hirano, Kiyoshi Ohishi,
Toshimasa Miyazaki and Yuki Yokokura
The Department of Electrical Engineering,
Nagaoka University of Technology,
Nagaoka, 940-2188, JAPAN
marui29@stn.nagaokaut.ac.jp, s123164@stn.nagaokaut.ac.jp, ohishi@vos.nagaokaut.ac.jp
miyazaki@vos.nagaokaut.ac.jp, yokokura@vos.nagaokaut.ac.jp

Abstract—**This paper proposes a method for suppressing the speed vibration caused by the angular transmission error in the cycloid gear used in industrial robots. It is important for industrial robots to have high accuracy. The arms of robot, however, are caused to vibrate by cycloid gears that have an angular transmission error. The proposed method for compensating for the speed vibration uses a new extended state observer that is based on the model of a cycloid gear with angular transmission error. A compensation current is produced from the torque vibration, which is converted by the estimated speed vibration. The speed vibration is suppressed using this compensation current. The experimental results show that the proposed system suppresses the vibration of load speed.**

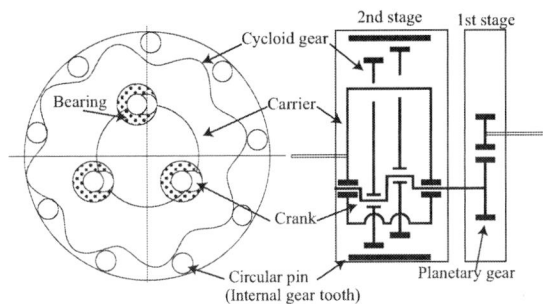

Fig. 1. Mechanical structure of cycloid gear.

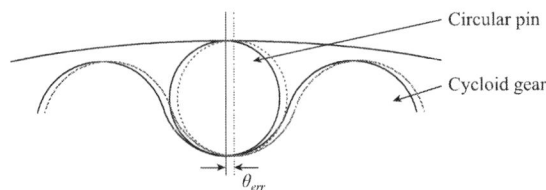

Fig. 2. Mechanism of angular transmission error of cyclid gear.

I. INTRODUCTION

In many cases, an actuator for a joint drive applies a slowdown machine, which is connected to a servo motor for the purpose of downsizing and producing a high torque output. Therefore, the cycloid gear is used for the basic axles of many industrial robots because the cycloid gear has the high torsional rigidity, a high slowdown ratio, small size, low backlash and a high torque density. However, because of an assembling error, a production error and a mechanism of the cycloid gear, the cycloid gear produces an angular transmission error. The angular transmission error has a periodicity depending on the rotary speed of the input motor. The robot arm vibrates because of the angular transmission error, which may reduce its trace precision.

To overcome this problem, this paper proposes a new vibration suppression method caused by the angular transmission error of the cycloid gear. A number of vibration-restraint laws have been proposed for harmonic drive gearings [1]–[4]. On the other hand, this paper focuses on the cycloid gear. First, considering the mechanism of the cycloid gear, this paper proposes the angular transmission error model for the cycloid gear. This paper shows that the speed vibration frequency is proportional to the number of the circular pin. Second, this paper proposes a vibration suppression control system using the extended state observer.

II. MODELING OF CYCLOID GEAR

A. Mechanical Structure

Fig. 1 shows the mechanical structure of a cycloid gear. The cycloid gear has the planetary gears and cycloid gearing structure in the first and second stages, respectively. At first, the input torque is conducted to the planetary gears which have the shaft in the first stage. In the next step, the shaft drives the two cycloid gears in the second stage. The rotational axis of the cycloid gears is decentering as with a crank shaft. Moreover, the cogs of the cycloid gears mesh with the circular pins, which are installed in around the gears. The number of circular pins is larger than the number of the cogs. When the cycloid gears make a round, the output gear rotates so that the toque is amplified.

The angular transmission error of the cycloid gear in the case of the cycloid gear has a pitch error. The black line

978-1-4799-2706-7/14 $31.00 © 2014 IEEE　　1956

The 2014 International Power Electronics Conference

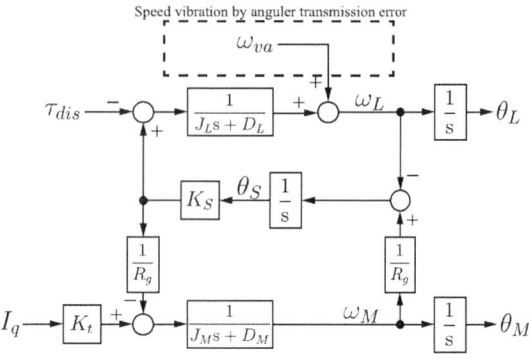

Fig. 3. Model of cycloid gear with angular transmission error.

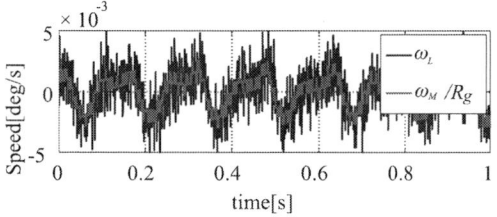

(a) Experimental results using cycloid gear.

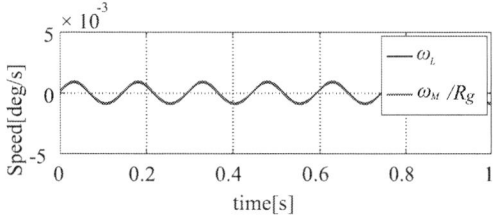

(b) Numerical simulation results of proposed model.

Fig. 4. Speed ripple waveform on condition of constant speed ω_L=7.3min^{-1}.

shown in Fig. 2 is the ideal mesh of the gear in the case of the circular pin does not have the pitch error to the cycloid gear. However, in the case of the circular pin has the pitch error to the cycloid gear shown as the red line of Fig. 2, the circular pin meshes with the cycloid gear having a meshing error θ_{err}. The meshing error occurs at each meshing with a circular pin, therefore, the vibration frequency caused the angular transmission error is proportional to the number of circular pins N_p.

B. Model of Cycloid Gear

Fig. 3 represents the block diagram of the cycloid gear. Fundamentally, the cycloid gear is modeled as the two inertia resonance system [5][6]. In general, the cycloid gear is approximated as elastic bodies. The vibration occurs due to the elasticity of the gears. In this system, I_q, τ_{dis}, ω, and θ denote the q-axis current, load torque, velocity, and position response, respectively. The subscript M and L represents "motor side" and "load side." J, D, K_s, and K_t denote the inertia, viscosity, stiffness, and torque constant of the motor, respectively.

As with another type of gears, the angular transmission error between the input and output of the cycloid gear occurs. The vibration of the load side velocity ω_L appears due to the

angular transmission error. The velocity components of the vibration ω_{va} are expressed as

$$\omega_{va} = \sum_{k=1}^{n} A_{\omega k} \sin\left(\frac{N_p k}{R_g}\omega_M t + \phi_k\right) \quad (1)$$

where $A_{\omega k}$, N_p, R_g, and ϕ_k denote the vibration amplitude, number of the circular pins, gear ratio, and vibration phase, respectively. As shown in Fig. 3, the vibration components stimulate the load side mechanism. This paper aims to reduce the velocity vibration, which occurs in the plant system.

Fig. 4 shows the motor-side speed vibration and the load-side speed vibration obtained by the numerical simulation and the experiments using the actual cycloid gear. The fundamental harmonic $k = 1$ is considered in the numerical simulation. Fig. 4 (a) show that the load-side speed vibration amplitude and the phase are coincident with motor-side speed waveform normalized by the gear ratio R_g. Fig. 4 (b) also show that the load-side speed vibration amplitude and the phase are coincident with motor-side speed waveform. From the results, this paper confirms that the proposed model of the angular transmission error shown in Fig. 3 is coincident with the actual plant system.

III. Vibration Suppression Control Based on Extended State Observer

A. Design of Vibration Suppression Control System

This paper focuses on the fundamental speed vibration $k = 1$ and the second-order speed vibration $k = 2$. Thus, the proposed extended state observer estimates the speed vibration of N_p and $2N_p$ times frequency of load-side speed. Focusing on the fundamental speed vibration $k = 1$, (1) is expressed as (2). Here, ω_d expressed as (3), and the speed vibration amplitude and the vibration phase are defined as A_ω and ϕ', respectively. The proposed extended state observer estimates the speed vibration ω_{va} having the frequency variation due to the load-side speed variation.

To design the extended state observer, this paper derives the second-order differential of the speed vibration $\ddot{\omega}_{va}$. $\ddot{\omega}_{va}$ is expressed as (4) because ω_{va} is the periodic function.

$$\omega_{va} = A_\omega \sin(\omega_d t + \phi') \quad (2)$$

$$\omega_d = \frac{N_p}{R_g}\omega_M \quad (3)$$

$$\ddot{\omega}_{va} = -\omega_d^2 A_\omega \sin(\omega_d t + \phi') = -\omega_d^2 \omega_{va} \quad (4)$$

The second-order speed vibration $k = 2$ is also derived same way. The fundamental speed vibration and the second-order speed vibration are defined as $\hat{\omega}_{va1}$ and $\hat{\omega}_{va2}$, respectively. The fundamental vibration frequency and the second-order vibration frequency are defined as ω_{d1} and ω_{d2}, respectively. $\hat{\omega}_{va}$ is expressed as the sum of the fundamental speed vibration $\hat{\omega}_{va1}$ and the second-order speed vibration $\hat{\omega}_{va2}$.

$$\hat{\omega}_{va} = \hat{\omega}_{va1} + \hat{\omega}_{va2} \quad (5)$$

Fig. 5 shows a block diagram of the method for suppressing the speed vibration caused by angular transmission error. The

978-1-4799-2706-7/14 $31.00 © 2014 IEEE

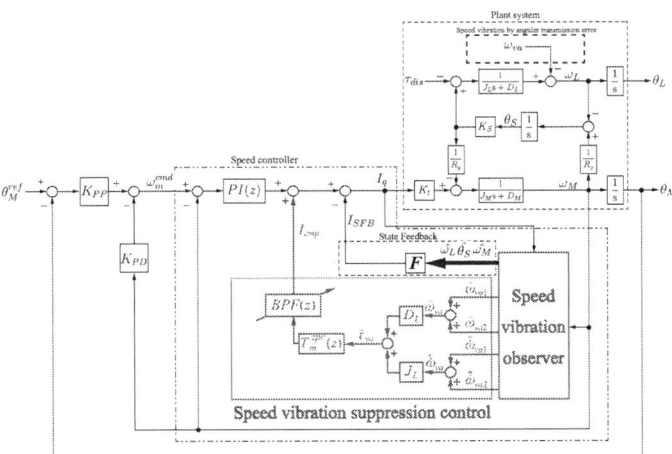

Fig. 5. Proposed speed vibration suppression control system.

speed vibration is estimated by the extended state observer expressed as

$$\dot{x}_0 = \hat{A}x_0 + \hat{B}u_0 \tag{6}$$

$$y_0 = \hat{C}x_0 \tag{7}$$

$$\hat{A} = \begin{bmatrix} -\frac{D_M+J_M g_1}{J_M} & -\frac{K_S}{J_M R_g} & 0 & 0 & 0 & 0 & 0 & 0 \\ \frac{1-R_g g_2}{R_g} & 0 & -1 & 0 & 0 & 0 & 0 & 0 \\ -g_3 & \frac{K_S}{J_L} & -\frac{D_L}{J_L} & -\frac{D_L}{J_L} & -1 & \frac{D_L}{J_L} & -1 & -\frac{1}{J_M} \\ -g_4 & 0 & 0 & 1 & 0 & 0 & 0 & 0 \\ -g_5 & 0 & 0 & -\omega_{d1}^2 & 0 & 0 & 0 & 0 \\ -g_6 & 0 & 0 & 0 & 0 & 1 & 0 & 0 \\ -g_7 & 0 & 0 & 0 & 0 & -\omega_{d2}^2 & 0 & 0 \\ -g_8 & 0 & 0 & 0 & 0 & 0 & 0 & 0 \end{bmatrix},$$

$$\hat{B} = \begin{bmatrix} \frac{K_t}{J_M} & g_1 \\ 0 & g_2 \\ 0 & g_3 \\ 0 & g_4 \\ 0 & g_5 \\ 0 & g_6 \\ 0 & g_7 \\ 0 & g_8 \end{bmatrix}, \hat{C} = \begin{bmatrix} 1 & 0 & 0 & 0 & 0 & 0 & 0 & 0 \\ 0 & 1 & 0 & 0 & 0 & 0 & 0 & 0 \\ 0 & 0 & 1 & 0 & 0 & 0 & 0 & 0 \\ 0 & 0 & 0 & 1 & 0 & 0 & 0 & 0 \\ 0 & 0 & 0 & 0 & 1 & 0 & 0 & 0 \\ 0 & 0 & 0 & 0 & 0 & 1 & 0 & 0 \\ 0 & 0 & 0 & 0 & 0 & 0 & 1 & 0 \\ 0 & 0 & 0 & 0 & 0 & 0 & 0 & 1 \end{bmatrix},$$

$$x_0 = \begin{bmatrix} \hat{\omega}_M \\ \hat{\theta}_S \\ \hat{\omega}_L \\ \hat{\omega}_{va1} \\ \hat{\dot{\omega}}_{va1} \\ \hat{\omega}_{va2} \\ \hat{\dot{\omega}}_{va2} \\ \hat{\tau}_{dis} \end{bmatrix}, u_0 = \begin{bmatrix} I_q \\ \omega_M \end{bmatrix}, y_0 = \begin{bmatrix} \hat{\omega}_M \\ \hat{\theta}_S \\ \hat{\omega}_L \\ \hat{\omega}_{va1} \\ \hat{\omega}_{va1} \\ \hat{\omega}_{va2} \\ \hat{\omega}_{va2} \\ \hat{\tau}_{dis} \end{bmatrix}$$

where g_1 to g_8 denote the pole of the extended state observer. The subscript $va1$ and $va2$ denote the primary and secondary components of the vibration. The extended state observer estimates the speed vibration, the speed vibration time derivative value, and the two-inertia state variable.

Fig. 6 shows the frequency response of the proposed extended state observer. This figure shows the frequecny characteristics from the speed vibration ω_{va} caused by the angular transmission error to the estimated speed vibration $\hat{\omega}_{va}$. Here, the fundamental vibration frequency ω_{d1} and the second-order vibration frequency ω_{d2} are set as 100rad/s and 200rad/s, respectively. Fig. 6 shows that the bode plot of the estimated speed vibration becomes 0dB and 0deg at the designed vibration frequency.

To compensate the velocity vibration, the vibration components of the velocity are converted to the torque $\hat{\tau}_{va}$ calculated

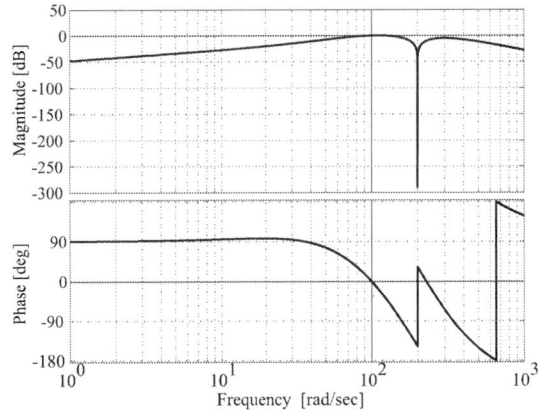

(a) Bode plot of transfer function from ω_{va1} to $\hat{\omega}_{va1}$.

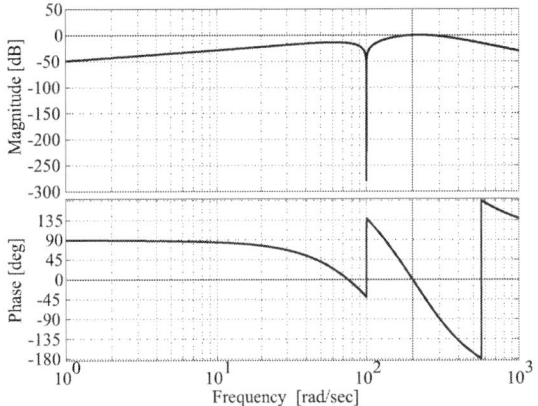

(b) Bode plot of transfer function from ω_{va2} to $\hat{\omega}_{va2}$.

Fig. 6. Frequency characteristics of proposed extended state observer.

as

$$\hat{\tau}_{va} = (J_L s + D_L)\hat{\omega}_{va}$$
$$= J_L(\hat{\dot{\omega}}_{va1} + \hat{\dot{\omega}}_{va2}) + D_L(\hat{\omega}_{va1} + \hat{\omega}_{va2}). \tag{8}$$

In the next step, the compensation current I_{cmp} for suppressing the vibration is calculated from the torque $\hat{\tau}_{va}$ as follows:

$$I_{cmp}(s) = T_m(s)\hat{\tau}_{va}(s) \tag{9}$$

where $T_m(s)$ denotes the transfer function to $I_{cmp}(s)$ from $\tau_{va}(s)$. $T_m(s)$ is derived as

$$T_m(s) = \frac{I_{cmp}(s)}{\tau_{va}(s)} = \frac{G_t(s)}{G_i(s)}. \tag{10}$$

$G_t(s)$ and $G_i(s)$ denote the transfer functions given by [7].

$$G_t(s) = \frac{\omega_L(s)}{\tau_{va}(s)} \tag{11}$$

$$G_i(s) = \frac{\omega_L(s)}{I_{cmp}(s)}. \tag{12}$$

In order to implement the suppression method to digital systems, (10) is discretized as follows:

$$T_m(z) = \frac{-7.1394(z^2 - 1.993z + 0.9959)}{z + 0.9986} \cdot \frac{1}{z}. \tag{13}$$

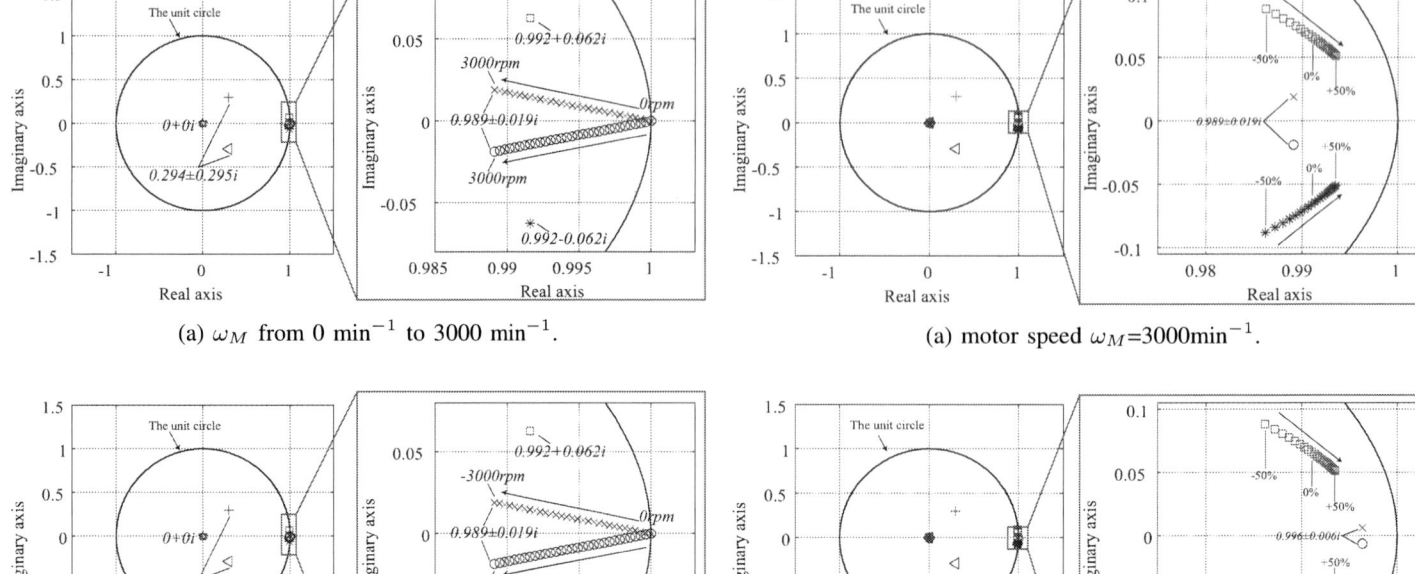

(a) ω_M from 0 min^{-1} to 3000 min^{-1}.

(a) motor speed ω_M=3000min^{-1}.

(b) ω_M from 0 min^{-1} to -3000 min^{-1}.

(b) motor speed ω_M=1000min^{-1}.

Fig. 7. Root locus of transfer function from ω_M^{ref} to ω_M on condition of motor speed reference change.

Fig. 8. Root locus of transfer function from ω_M^{ref} to ω_M on condition with variation of load-side inertia moment J_L from -50% to +50%.

In (13), the unit delay has to be implemented because the transfer function T_m is not proper. However, (13) cannot be used due to internal stability of the feed forward compensator. Accordingly, $T_m(z)$ is redesigned by ZPC [8]. The redesigned transfer function $T_m^{ZPC}(z)$ is derived as

$$T_m^{ZPC}(z) = \frac{-7.1394(z^2 - 1.993z + 0.9959)}{1 + 0.9986} \cdot \frac{1}{z^2}. \quad (14)$$

Since the compensation current I_{cmp} generated by (10) and (14) includes observation noise, the bandpass filter ($BPF(z)$) is used to remove the noise of the compensation current. In this paper, $BPF(z)$ is given by

$$BPF(s) = \frac{\frac{\omega_{d1}}{Q_1}s}{s^2 + \frac{\omega_{d1}}{Q_1}s + \omega_{d1}^2} + \frac{\frac{\omega_{d2}}{Q_2}s}{s^2 + \frac{\omega_{d2}}{Q_2}s + \omega_{d2}^2} \quad (15)$$

Finally, the compensation current is added to the q-axis current reference as shown in Fig. 5.

B. Stability analysis of Vibration Suppression Control System

To extract the compensation signal of the angular transmission error, the proposed motion control system uses the bandpass filter. The center frequency of the bandpass filter is changed with the motor-side velocity. In other words, the transfer function of the proposed control system is changed with the operating point. Therefore, this paper carries out the stability analysis. This paper analyzes the stability of the discretized motion control system using the zero-order hold on the Z-plane.

Fig. 7 shows the root locus of the transfer function from ω_M^{ref} to ω_M on the condition of the motor speed reference change. Fig. 7 (a) shows the analysis results of the forward rotation, and Fig. 7 (b) shows the analysis results of the reverse rotation. In the case which the motor-side velocity is not zero, all roots are located in the unit circle. The conjugate complex roots move to the center of the unit circle with increase in the motor-side velocity. However, in the case of the zero speed, the conjugate complex roots are located on the unit circle. Hence, to prevent that the motion control system becomes unstable, this paper disables the proposed vibration suppression method in the low speed condition of less than 20min^{-1}.

Considering the application for the industrial robot, this paper confirms the stability of the proposed method against the load-side inertia variation. Here, the stability of the motion control system is confirmed on the two conditions, ω_M =3000min^{-1} and ω_M=1000min^{-1}. In this analysis, the load-side inertia moment J_L is changed from -50% to +50%. Fig. 8 shows the root locus of the transfer function from ω_M^{ref} to ω_M on the condition with the variation of the load-side inertia moment J_L from -50% to +50%. The analysis results show that all roots are located in the unit circle. Therefore, the motion control system is stable against ± 50% load-side inertia variation.

978-1-4799-2706-7/14 $31.00 © 2014 IEEE

The 2014 International Power Electronics Conference

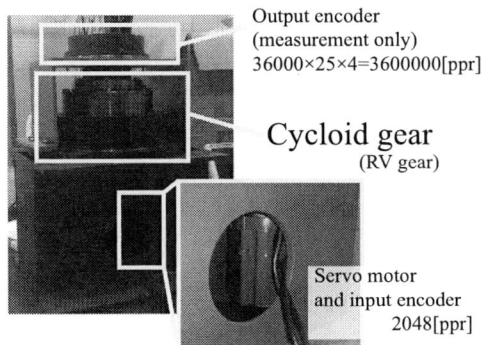

Output encoder
(measurement only)
36000×25×4=3600000[ppr]

Cycloid gear
(RV gear)

Servo motor
and input encoder
2048[ppr]

Fig. 9. Experimental setup.

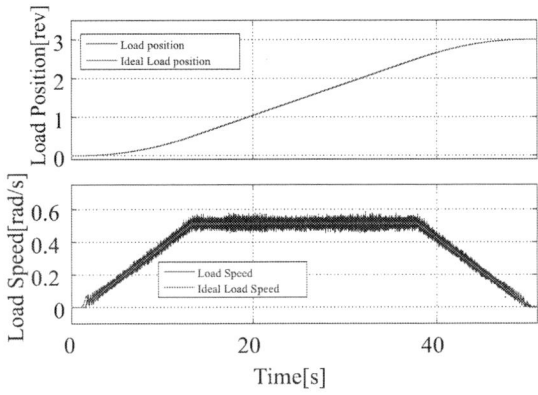

Fig. 10. Experimental results of load position θ_L and load speed ω_L with proposed speed vibration suppression control.

IV. EXPERIMENT

A. Experimental Setup

Fig. 9 shows the experimental setup. The servo motor drives the input side of the cycloid gear. Furthermore, the high accuracy rotary encoder is mounted on the output side, in order to observe the angular transmission error.

B. Experimental Results

Fig. 10 shows the overview of the experimental results. The tested motion is consisted from three intervals, the acceleration time, constant velocity time and the deceleration time.

First, this paper shows the experimental results of the constant velocity time. Fig. 11 (a) shows the experimental results of the load speed and the FFT analysis without vibration suppression method. The experimental results show that the vibration occurs due to the angular transmission error of the cycloid gear. On the other hand, Fig. 11 (b) represents the experimental results of the proposed vibration suppression method. Compared with the conventional system, the experimental results show that the proposed system suppresses the vibration of the load speed. The vibration of both frequencies of 52 times and 104 times that of the ideal load speed decreases.

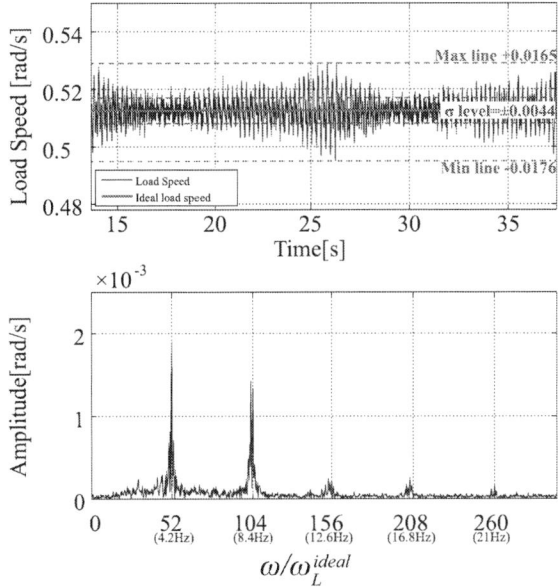

(a) Without suppression of speed vibration.

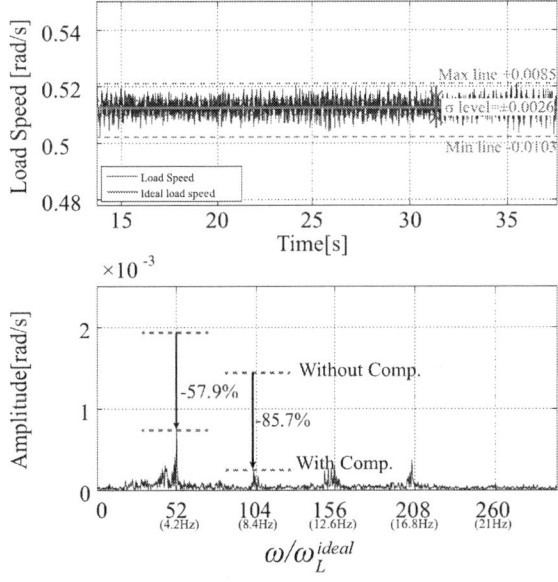

(b) With suppression of speed vibration.

Fig. 11. Experimental results of constant speed response.

Fig. 12 (a) shows the experimental results of the load position error and the FFT analysis without vibration suppression method. The experimental results show that the vibration occurs due to the angular transmission error of the cycloid gear. On the other hand, Fig. 12 (b) represents the experimental results of the proposed vibration suppression method. The vibration of both frequencies of 52 times and 104 times that of the ideal load speed decreases. Compared with the conventional system, the experimental results show that the proposed system suppresses the vibration of the load position error.

Second, this paper shows the experimental results of the

978-1-4799-2706-7/14 $31.00 © 2014 IEEE

The 2014 International Power Electronics Conference

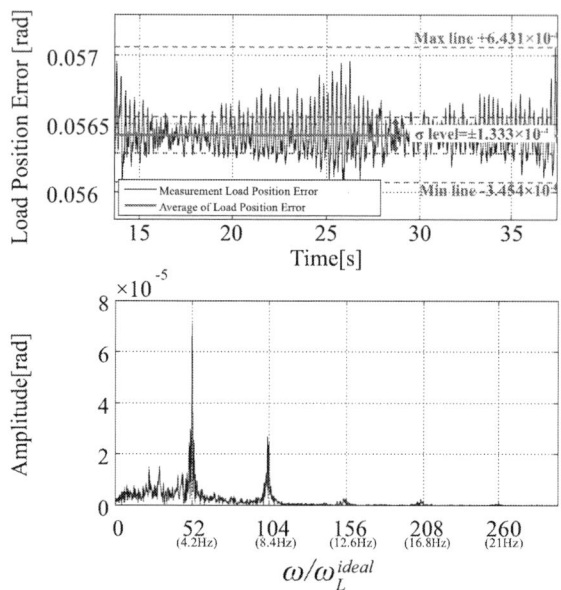

(a) Without suppression of speed vibration.

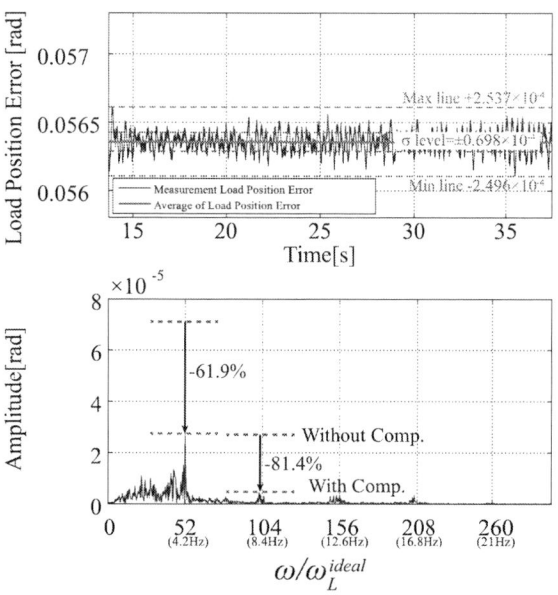

(b) With suppression of speed vibration.

Fig. 12. Experimental results of load position error.

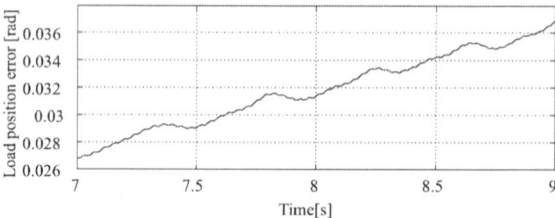

(a) Without suppression of speed vibration.

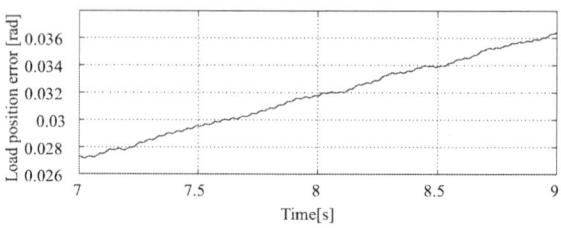

(b) With suppression of speed vibration.

Fig. 13. Experimental results of load position error response on condition of constant acceleration.

error due to the angular transmission error of the cycloid gear is reduced. In this paper, the validity of the proposed method was verified by the comparative experiment. The proposed vibration suppression method is useful for industrial robots.

REFERENCES

[1] T. Miyazaki and K. Ohishi, "Robust Speed Control System Considering Vibration Suppression Caused by Angular Transmission Error of Planetary Gear", IEEE/ASME Trans. on Mechatronics, vol. 7, Jun. 2002, pp. 235–244.

[2] P. S. Gandhi and F. H. Ghorbel, "Closed-Loop Compensation of Kinematic Error in Harmonic Drives for Precision Control Applications", IEEE Trans. Control Systems Technology, vol. 10, Nov. 2002, pp. 759–768.

[3] Iwasaki. M, Yamamoto. M., Hirai. H, Okitsu. Y, Sasaki. K, Yajima. T, "Modeling and compensation for angular transmission error of harmonic drive gearings in high precision positioning", IEEE/ASME International Conference on Advanced Intelligent Mechatronics, July. 2009, pp. 662–667.

[4] Iwasaki. M, Yamamoto. M., Hirai. H, Okitsu. Y, Sasaki. K and Yajima. T, "Compensation for synchronous component of angular transmission errors in harmonic drive gearings", IEEE International Workshop on Advanced Motion Control, July. 2010, pp. 361–365.

[5] Jing Zhang, Bingkui Chen and Sung-Ki Lyu, "Mathematical model and analysis on cycloid planetary gear", Second International Conference on Mechanic Automation and Control Engineering (MACE), vol. 1, July. 2011, pp. 400–403.

[6] Zhang YingHui, He WeiDong and Xiao JunJun, "Dynamical Model of RV Reducer and Key Influence of Stiffness to the Nature Character," Third International Conference on Information and Computing (ICIC), June. 2010, pp. 192–195.

[7] T. Miyazaki, S. Otaki, S. Tungpataratanawong, and K. Ohishi, "High Speed Motion Control Method of Industrial Robot Based on Dynamic Torque Compensation and Two-Degrees-of-Freedom Control System," T.IEEJ, vol.123-D, No.5, May, 2003, pp.525–532 (in Japanese).

[8] M. Tomizuka, "Zero Phase Rrror Tracking Algorithm for Digital Control," ASME Journal of Dynamic Systems, Meas and Control, vol.113, 1989, pp. 6–10.

acceleration time. Fig. 13 shows the experimental results of the load position error. In the case of the acceleration, Fig. 13 shows that the proposed system suppresses the vibration of the load position error.

V. CONCLUSIONS

This paper proposes the new vibration suppression method using the extended state observer, which is capable of estimating the velocity vibration. The vibration components of the velocity are converted to the compensation current by the transfer function designed by ZPC. In order to suppress the vibration, the compensation current is added to the current reference of the servo motor. As a result, the load position

An Advanced Position Control of Overhead Crane by Sway Suppression Method Emulating Natural Damping

Toshiyuki Kurabayashi,Yang Chuan
Department of System Design Engineering
Keio university
Yokohama city Japan
Email: kurabayashi@sum.sd.keio.ac.jp

Toshiyuki Murakami
Department of System Design Engineering
Keio university
Yokohama city Japan
Email: mura@sum.sd.keio.ac.jp

Abstract—**Overhead crane is widely used to convey the payload due to its convenience. However it is hard to control the crane because of its underactuated structure. The most important control requirement is to make the trolley position converge to target position accurately with suppressing payload sway. This makes it possible to achieve the purpose of overhead crane which is to convey payload to target position stably. In the practical implementation, however, the achievement of both conveying trolley to target position and suppressing payload sway is not always easy. In this paper, emulating natural damping method is proposed. The natural damping is defined as energy damping of crane system caused by trolley friction. Not only trolley but also payload energy is damped by natural damping effect. The proposed approach brings to achieve both trolley position convergence and payload sway suppression.**

I. INTRODUCTION

Overhead crane has been widely used for transportation due to its convenience. For example, it is mainly used in the factory, port and under the construction. Generally, overhead crane is operated by humans but there are automatic controls which are controlled by machines. Though convey performance always changes by operator's skill and condition in human control, automatic control can make stable result and also decrease employment cost. In this paper, automatic control is focused on. As a problem of overhead crane, it is difficult to control because it is underactuated system. Underantuated system is defined the system which input is less than output. Crane system has 1 trolley input but there are 2 outputs which are trolley position and payload sway angle. It is required that to converge trolley position to target position accurately and to suppress payload sway are achieved at the same time because the purpose of overhead crane is to convey payload to target position.

Many methods are proposed to solve this problem. As a feedforward control, Smith [1] proposed input shaping method. This control method has widely used in cranes. They make command signal to reduce vibration by convolving a sequence of impulses. But this method is week to the disturbance because this method is feedforward control. To compensate sway angle disturbance, H.Ishino [2] proposed sway angle disturbance observer(SADOB). SADOB can compensate disturbance and gives system robustness. But trolley position convergence term and sway suppress term are often interfered and there are some case that stable result is not obtained.

Ning Sun [3] proposed model-free output feedback control method. This method designs virtual model and drop whole crane energy to virtual energy. Payload sway can be suppressed by this method, but it is difficult to design parameter values.

In this paper, emulating natural damping method is proposed. Natural damping is defined as energy damping of crane system caused by trolley friction. Not only trolley but also payload energy is damped by natural damping. In actual situation, if there is no control force, energy of whole system is damped by natural damping. But convergence capability of energy cannot be best by actual natural damping. Natural damping capability depends on friction coefficient of trolley and mass proportion between trolley and payload. In this paper, optimal natural damping moving is implemented with control. There are 2 steps in proposed method. First, proper parameters which can converge whole energy quickly are analyzed by using stability index[4]. Second, natural damping is reproduced by control with using the proper parameters. This paper focuses on the experimental evaluations.[5]

II. MODELING

In this section, the modeling of overhead crane is introduced. In chapter II-A, model equation of overhead crane is introduced. In chapter II-B, 2 equation of the model equation, which are used in this paper is introduced.

A. Derivation of model equation of overhead crane

Crane system is shown in Fig.1. The dynamics of crane system is described as follows.

$$\ddot{x} = N_1 + N_3 \mu_1 \cos \theta_x \tag{1}$$

$$\ddot{y} = N_2 + N_3 \mu_2 \sin \theta_x \sin \theta_y \tag{2}$$

$$\ddot{\theta}_x = \frac{\sin \theta_x N_1 - \cos \theta_x \sin \theta_y N_2}{l}$$
$$+ \frac{(\mu_1 - \mu_2 \sin^2 \theta_2) \cos \theta_x \sin \theta_x N_3 + V_5}{l} \tag{3}$$

$$\ddot{\theta}_y = \frac{-(\cos \theta_y N_2 + \mu_2 \sin \theta_x \sin \theta_y \cos \theta_y N_3 + V_6)}{\sin \theta_x l} \tag{4}$$

$$\ddot{l} = -\cos \theta_x N_1 - \sin \theta_x \sin \theta_y N_2$$
$$- (1 + \mu_1 \cos^2 \theta_x + \mu_2 \sin^2 \theta_x \sin^2 \theta_y) N_3 + V_7 \tag{5}$$

978-1-4799-2706-7/14 $31.00 © 2014 IEEE

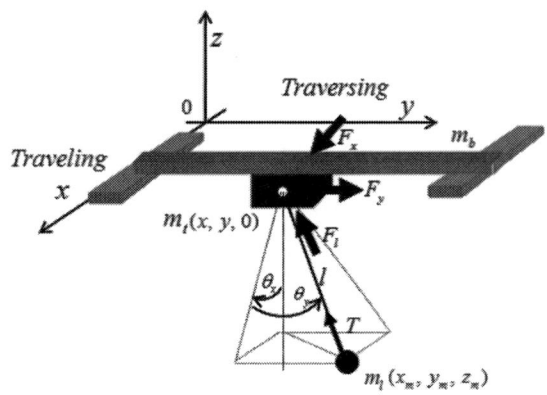

Fig. 1.   Crane model

TABLE I.   PARAMETERS ABOUT CRANE

Variables	Definition
u_1	$\frac{F_x}{m_t+m_b}$
u_2	$\frac{F_y}{m_t}$
u_3	$\frac{F_R}{m_p}$
T_1	$\frac{T_x}{m_t}$
T_2	$\frac{T_y}{m_t+m_b}$
T_3	$\frac{T_R}{m_p}$
μ_1	$\frac{m_p}{m_t}$
μ_2	$\frac{m_p}{m_t+m_b}$
m_p	Load mass [kg]
m_t	Trolley mass [kg]
m_b	Boom mass [kg]
F_x	Force driving the cart with boom [N]
F_y	Force driving the cart [N]
F_R	Force controlling the length of lope [N]
T_x, T_y, T_R	Friction coefficients
g	Gravity acceleration

where N and V are defined as follows.

$$N_1 = u_1 - T_1\dot{x} \tag{6}$$
$$N_2 = u_2 - T_2\dot{y} \tag{7}$$
$$N_3 = u_3 - T_3\dot{l} \tag{8}$$
$$V_5 = \cos\theta_x \sin\theta_x\dot{\theta}_2^2 l - 2\dot{l}\dot{\theta}_x + g\cos\theta_x\cos\theta_y \tag{9}$$
$$V_6 = 2\dot{\theta}_y(\cos\theta_x\dot{\theta}_x l + \sin\theta_x\dot{l}) + g\sin\theta_y \tag{10}$$
$$V_7 = \sin^2\theta_x\dot{\theta}_2^2 l + g\sin\theta_x\cos\theta_y + \dot{\theta}_1^2 l \tag{11}$$

Using state vector $\boldsymbol{q} = [x \ \ y \ \ \theta_x \ \ \theta_y \ \ l]^T$ and input vector $\boldsymbol{u} = [F_x \ \ F_y \ \ F_z]^T$, Eq.(5) can be rewritten as Eq.(12)

$$\boldsymbol{M}(\boldsymbol{q})\ddot{\boldsymbol{q}} + \boldsymbol{d}(\boldsymbol{q}, \dot{\boldsymbol{q}}) = \boldsymbol{f}(\boldsymbol{q}) + \boldsymbol{u} \tag{12}$$

$\boldsymbol{M}$ is inertia matrix, $\boldsymbol{d}$ is Coriolis vector and $\boldsymbol{f}$ is gravity vector.

B. Remarkable model equation

Model equation of overhead crane is expressed as Eq.(12). If Eq.(12) is calculated, 5 equations are obtained. 3 of them describe the relationship between input and trolley moving of

x, y, z direction respectively. 2 of them describe the relationship between trolley moving and payload sway of x, y direction respectively. In this paper, y direction is focused. y direction equation of Eq.(12) are expressed as Eq.(13)-Eq.(14).

$$(m_t + m_p)\ddot{y} = F_y - k_y^{fric}(m_t + m_p)\dot{y}$$
$$+ m_p l(\dot{\theta}_y^2\sin\theta_y - \ddot{\theta}_y\cos\theta_y) \tag{13}$$
$$\cos\theta_y\ddot{y} + l\ddot{\theta}_y = -g\sin\theta_y + T_{dis} \tag{14}$$

k_y^{fric} is friction coefficient of trolley and T_{dis} is sway angle disturbance. Eq.(13) describe the relationship between input F_y and trolley moving of y direction. Eq.(14) describe the relationship between trolley moving and payload sway of y direction.

III. ANALYSIS OF NATURAL DAMPING

In this section, natural damping is analyzed. Natural damping is defined as energy damping of whole crane system caused by trolley friction. Natural damping is occurred by model equations Eq.(13),(14). Model equations Eq.(13),(14) are re-expressed as Eq.(15),(16) without any control force($F_y = 0$).

$$(m_t + m_p)\ddot{y} = -k_y^{fric}(m_t + m_p)\dot{y}$$
$$+ m_p l(\dot{\theta}_y^2\sin\theta_y - \ddot{\theta}_y\cos\theta_y) \tag{15}$$
$$\cos\theta_y\ddot{y} + l\ddot{\theta}_y = -g\sin\theta_y + T_{dis} \tag{16}$$

Theorem 1: Natural damping caused by Eq.(15),(16) suppress the payload sway and trolley position y converges to 0. This property is also indicated when there are any sway angle disturbances.

Proof: Whole system energy $E(t)$ is express as Eq.(17).

$$E(t) = \frac{1}{2}(m_t + m_p)\dot{y}^2 + \frac{1}{2}m_p(l\dot{\theta}_y)^2$$
$$+ m_p l\dot{\theta}_y\dot{y}\cos\theta_y + m_p gl(1 - \cos\theta) \tag{17}$$

Derivative of $E(t)$ is expressed as Eq.(18).

$$\dot{E}(t) = (m_t + m_p)\dot{y}\ddot{y} + m_p l(\ddot{\theta}_y\cos\theta_y - \dot{\theta}_y^2\sin\theta_y)\dot{y}$$
$$+ m_p l\dot{\theta}_y\left(\cos\theta_y\ddot{y} + l\ddot{\theta}_y + g\sin\theta_y\right) \tag{18}$$

Eq.(18) is also expressed as Eq.(19) by Eq.(16).

$$\dot{E}(t) = (m_t + m_p)\dot{y}\ddot{y} + m_p l(\ddot{\theta}_y\cos\theta_y - \dot{\theta}_y^2\sin\theta_y)\dot{y}$$
$$+ m_p l\dot{\theta}_y T_{dis} \tag{19}$$

Eq.(15) is substituted to Eq.(19) and Eq.(20) is obtained.

$$\dot{E}(t) = -(m_t + m_p)k_y^{fric}\dot{y}^2 + m_p l\dot{\theta}_y T_{dis} \tag{20}$$

Limit of $\dot{E}(t)$ is expressed as Eq.(21) with a condition of $\lim_{t\to\infty} T_{dis} = 0$.

$$\dot{E}(\infty) = -(m_t + m_p)k_y^{fric}\dot{y}^2 \leq 0 \tag{21}$$

From Eq.(15), unless payload sway exist, $\dot{y}$ does not always 0. Therefore, Eq.(22) can be said from Eq.(21).

$$E(\infty) = 0 \tag{22}$$

Eq.(23),(24) is also obtained from Eq.(22).

$$\lim_{t \to \infty} \theta_y = 0, \ \lim_{t \to \infty} \dot{\theta}_y = 0, \ \lim_{t \to \infty} \ddot{\theta}_y = 0 \tag{23}$$

$$\lim_{t \to \infty} \dot{y} = 0, \ \lim_{t \to \infty} \ddot{y} = 0 \tag{24}$$

Here, integrate of Eq.(15) is expressed as Eq.(25).

$$\dot{y} = -k_y^{fric} y - \frac{m_p}{m_t + m_p} l\dot{\theta}_y \cos \theta_y \tag{25}$$

Getting the limit of both side, Eq.(26) is obtained.

$$k_y^{fric} \lim_{t \to \infty} y = -\lim_{t \to \infty} \left(\dot{y} - \frac{m_p}{m_t + m_p} l\dot{\theta}_y \cos \theta_y \right) \tag{26}$$

Eq.(26) is also expressed as Eq.(27) by Eq.(23),(24).

$$k_y^{fric} \lim_{t \to \infty} y = 0 \tag{27}$$

Eq.(28) can be said from a condition of $k_y^{fric} > 0$.

$$\lim_{t \to \infty} y = 0 \tag{28}$$

∎

IV. PROPOSAL

In this section, proposed method is introduced. Natural damping suppresses the whole crane energy but there is a demerit in actual situation. The demerit is that convergence speed of whole crane energy depends on some parameters which are trolley and payload mass and friction coefficient of trolley. Therefore the convergence capability is changed by actual crane situations.

To solve this problem, Emulating best performance natural damping method is proposed. With proposed method, payload sway is suppressed without moving over the target position. In chapter IV-A, proposed input is introduced. In chapter IV-B, convergent method to target position is introduced. In chapter IV-C, determine method of parameters is introduced.

A. Proposed input

From *Theorem 1*, proposed natural damping control input $\ddot{y}_n$ is determined as Eq.(29).

$$\ddot{y}_n = -k_{yn}^{fric} \dot{y} + m_n l(\dot{\theta}_y^2 \sin \theta_y - \ddot{\theta}_y \cos \theta_y) \tag{29}$$

m_n is nominal value defined as $m_n = \frac{m_{pn}}{m_{tn} + m_{pn}}$. m_{tn} and m_{pn} are nominal trolley and payload mass respectively. m_n get the range of $0 < m_n < 1$. k_{yn}^{fric} is a nominal value of trolley friction coefficient.

Here, low-pass filter(LPF) is introduced to reduce the effect of noise and Eq.(29) is expressed as Eq.(30).

$$\ddot{y}_n = \frac{g_n}{s + g_n} \left\{ -k_{yn}^{fric} \dot{y} + m_n l(\dot{\theta}_y^2 \sin \theta_y - \ddot{\theta}_y \cos \theta_y) \right\} \tag{30}$$

g_n is cutoff frequency of LPF. Eq.(30) is transformed as Eq.(31).

$$\ddot{y}_n = \frac{g_n}{s + g_n} g_n \left(k_{yn}^{fric} y + m_n l\dot{\theta}_y \cos \theta_y \right) \\ - g_n \left(k_{yn}^{fric} y + m_n l\dot{\theta}_y \cos \theta_y \right) \tag{31}$$

Finally, proposed natural damping control input $\ddot{y}_n$ is determined as Eq.(31). Block diagram of proposal is shown in Fig.2.

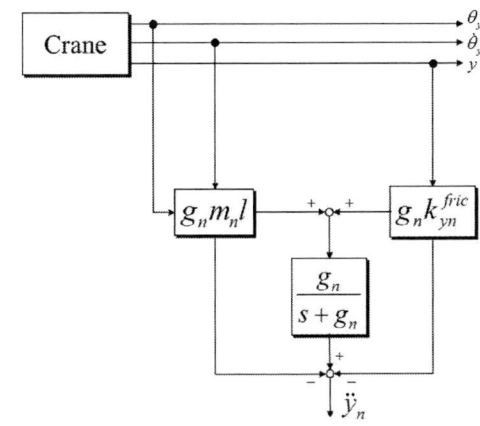

Fig. 2. Block diagram of proposed natural damping control input

B. Convergent method to target position

In actual situation, not only suppressing the payload sway but also conveying to the target position is required. In this chapter, convergent method to target position is considered. Here, Hamiltonian H is introduced. If Hamiltonian H is satisfied with the conditions of both positive definite function and zero condition-detectability, it is known that there is an input to stabilize the system. Zero condition-detectability is defined as Eq.(32).

$$(\mathbf{u}, \dot{\mathbf{q}}) \equiv 0 \Rightarrow (\mathbf{q}, \mathbf{p}) \to 0 \tag{32}$$

$\mathbf{u}$ is an input vector, $\mathbf{q}$ is an vector (y, θ) and $\mathbf{p}$ is a momentum vector. Hamiltonian H of crane is defined as Eq.(33).

$$\begin{aligned} H &= E(t) \\ &= \frac{1}{2}(m_t + m_p)\dot{y}^2 + \frac{1}{2}m_p(l\dot{\theta}_y)^2 \\ &\quad + m_p l\dot{\theta}_y \dot{y} \cos \theta_y + m_p gl(1 - \cos \theta) \end{aligned} \tag{33}$$

Hamiltonian H expressed as Eq.(33) does not be satisfied with the conditions of both positive definite function and zero condition-detectability. Input to stabilize the system cannot be given. However it is known input which stabilizes the system is derived with choosing positive definite function $U(t)$. In this case, Input u to stabilize the system is given as Eq.(34).

$$\mathbf{u} = -\frac{\partial U}{\partial \mathbf{q}} + \overline{u} \tag{34}$$

Here, positive definite function $U(t)$ is chosen as Eq.(35).

$$U = \frac{1}{2}k_p(y^{cmd} - y)^2 \tag{35}$$

k_p is a proportion gain and y^{cmd} is chosen that is satisfied with the condition of Eq.(36)

$$\lim_{t \to \infty} y^{cmd} = P_d \tag{36}$$

P_d is a target position. A new Hamiltonian $\overline{H}$ is defined as Eq.(37).

$$\overline{H} = H + U \qquad (37)$$

Hamiltonian $\overline{H}$ expressed as Eq.(37) is satisfied with the conditions of both positive definite function and zero condition-detectability. Therefore control input F_y is determined as Eq.(38) by Eq.(34).

$$F_y = (m_t + m_p)\left\{ k_p(y^{cmd} - y) + \ddot{y}_n \right\} \qquad (38)$$

C. Determine method of parameters

In this chapter, parameters are analyzed. Parameters are determined by using stability index[4]. Relationship between trolley command y^{cmd}, trolley position y and payload sway angle θ_y is expressed as Eq.(39),(40) because control force is given as Eq.(38).

$$\ddot{y} = k_p\left(y^{cmd} - y\right) - k_{yn}^{fric}\dot{y}$$
$$+ m_n l(\dot{\theta}_y^2 \sin\theta_y - \ddot{\theta}_y \cos\theta_y) \qquad (39)$$
$$\cos\theta_y \ddot{y} = -l\ddot{\theta}_y - g\sin\theta_y \qquad (40)$$

Eq.(39) and (40) can be approximated as Eq.(41) and (42).

$$\ddot{y} = k_p\left(y^{cmd} - y\right) - k_{yn}^{fric}\dot{y} - m_n l\ddot{\theta}_y \qquad (41)$$
$$\ddot{y} = -l\ddot{\theta}_y - g\theta_y \qquad (42)$$

Eq.(43) can be obtained to substitute Eq.(41) to Eq.(42).

$$k_p l \ddot{y}^{cmd} + g k_p y^{cmd} = (1 - m_n) l \ddddot{y} + k_{yn}^{fric} l \dddot{y}$$
$$+ (k_p l + g) \ddot{y} + (k_p l + g) \ddot{y}$$
$$+ g k_{yn}^{fric} \dot{y} + g k_p y \qquad (43)$$

Transfer function $G(s)$ from trolley command y^{cmd} to response y is expressed as Eq.(44) from Eq.(43).

$$G(s) = \frac{k_p l s^2 + g k_p}{(1 - m_n) l s^4 + k_{yn}^{fric} l s^3 + (k_p l + g) s^2 + g k_{yn}^{fric} s + g k_p} \qquad (44)$$

Characteristic polynomial $P(s)$ is expressed as Eq.(45) by Eq.(44).

$$P(s) = (1 - m_n) l s^4 + k_{yn}^{fric} l s^3 + (k_p l + g) s^2$$
$$+ g k_{yn}^{fric} s + g k_p \qquad (45)$$

Stability indexes $\gamma_1, \gamma_2, \gamma_3$ of Eq.(45) is expressed as Eq.(46)-(48)[4].

$$\gamma_1 = \frac{g k_{yn}^{fric\,2}}{k_p (k_p l + g)} \qquad (46)$$

$$\gamma_2 = \frac{(k_p l + g)^2}{g l k_{yn}^{fric\,2}} \qquad (47)$$

$$\gamma_3 = \frac{l k_{yn}^{fric\,2}}{(1 - m_n)(k_p l + g)} \qquad (48)$$

Stability indexes are designed as Eq.(49) to improve tracking capability.

$$\gamma_1 = 2.0, \gamma_2 = 2.0, \gamma_3 = 2.0 \qquad (49)$$

Therefore, k_p, k_{yn}^{fric} and m_n are designed as Eq.(50)-(52).

$$k_p = \frac{g}{3l} \qquad (50)$$

$$k_{yn}^{fric} = \sqrt{\frac{8g}{9l}} \qquad (51)$$

$$m_n = \frac{2}{3} \qquad (52)$$

At last, block diagram of whole system is shown in Fig.3.

V. SIMULATION

In this section, simulation result is shown. In chapter V-A, conventional method is introduced. In chapter V-B, simulation result of comparison between conventional method and proposed method is shown.

A. Conventional method

Lyapnov controller[2] is introduced as a conventional method. It is expressed as Eq.(53).

$$\ddot{y}^{lya} = \frac{l\left(k_1 \theta_y + k_2 \dot{\theta}_y\right) - g\sin\theta_y - \hat{T}^{dis}}{\cos\theta_y} \qquad (53)$$

Controller is used expressed as Eq.(54).

$$\ddot{y}^{ref} = k_p'\left(y^{cmd} - y\right) - k_d'\dot{y} + \ddot{y}^{lya} \qquad (54)$$

k_p' is proportional gain and k_d' is derivative gain.

B. Simulation result

In this chapter, 2 patterns simulation shown as follow are implemented.

- Comparison of conventional method and proposed method by trajectory and oscillation without sway angle disturbance

- Comparison of conventional method and proposed method by trajectory and oscillation with sway angle disturbance

1) Comparison of conventional method and proposed method without sway angle disturbance: Trolley position, velocity and payload sway angle are shown in Fig.5(a),(b) and (c) respectively.

2) Comparison of conventional method and proposed method with sway angle disturbance: Robustness to sway angle disturbance is simulated. Disturbance is defined as table.III. Trolley position, velocity and and payload sway angle are shown in Fig.6(a),(b) and (c) respectively. With conventional method, trolley moves over to compensate disturbance when disturbance is added to payload. But trolley does not move over and compensate disturbance in proposed method with same situation. From this result, it can be said that proposed method is robustness to sway angle disturbance.

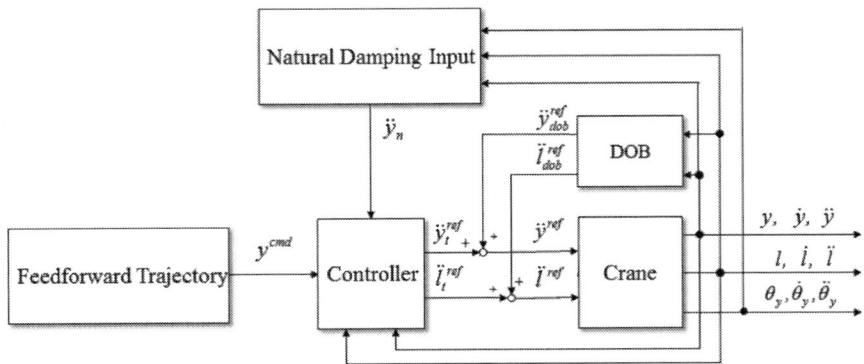

Fig. 3. Block diagram

TABLE II. PARAMETERS VALUE IN SIMULATION

Parameter	Value	Definition
m_t	2.0[kg]	Payload mass
m_p	1.0[kg]	Trolley mass
l	1.0[m]	Rope length
k_1	1.0	Lyapunov gain1
k_2	0.1	Lyapunov gain2

TABLE III. DISTURBANCE IN SIMULATION

Disturbance	Value	Term
Wind	1.0[Nm]	$1.0 - 3.0$[sec]

VI. EXPERIMENT

In this section, experimental result is shown. Experiment is implemented by machine shown in Fig.4. Trolley is conveyed by 2 links system and sway angle is measured by PSD camera. Experimental condition is expressed as table.IV. Trolley position, velocity and payload sway angle are shown in Fig.7(a),(b) and (c) respectively. From Fig.7(c), it can be said that proposed method can suppress payload sway than conventional method. Proposed method can suppress payload sway without trolley moving over largely. There is little interference between feedforward trajectory y^{cmd} and sway suppress command $\ddot{y}_n$. However conventional method moves trolley over largely to suppress payload sway. Therefore there is large interference between feedforward trajectory y^{cmd} and sway suppress command $\ddot{y}^{lya}$. From these reasons, it can be considered that proposed method has better result than conventional method.

VII. CONCLUSION

In this paper, emulating natural damping method was proposed. Proposed method has 3 merit. First, payload sway is suppressed rapidly. Second, trolley converges to initial position as a feature of proposed method. Therefore, proposed method has less interference to position convergence control. Third, proposed method is robustness to sway angle disturbance. As a future work, adaptation of proposed method to the case of rope length varying in conveying will be implemented. With this

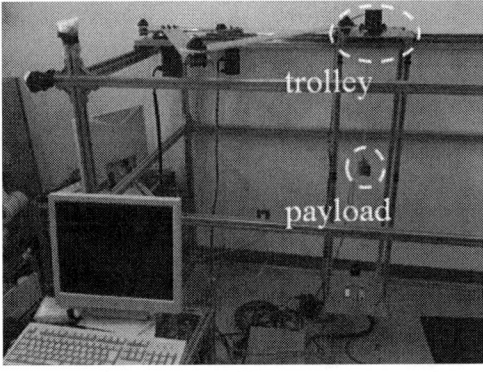

Fig. 4. Experimental machine

TABLE IV. PARAMETERS VALUE IN EXPERIMENT

Parameter	Value	Definition
m_t	2.0[kg]	Payload mass
m_p	1.0[kg]	Trolley mass
l	0.6[m]	Rope length
k_1	1.0	Lyapunov gain1
k_2	0.1	Lyapunov gain2
k'_p	4.0	Proportional gain of Lyapnov method
k'_d	4.0	Derivative gain of Lyapnov method

work, proposed method will be able to be used more situation.

REFERENCES

[1] Smith, O. J. M.: "Feedback control systems," New York, McGraw-Hill series in control systems engineering, pp. 1-694, 1958

[2] H.Ishino,T. Murakami: "Anti Swinging Control of Automatic Crane System with Variable Rope", 8th France-Japan and 6th Europe-Asia Congress on Mechatronics, USB proceedings, Nov, 2010

[3] Ning Sun, Yongchun Fang, Xiuyun Sun, and Zhekui Xin: "An energy exchanging and dropping-based model-free output feedback crane control method", Mechatronics, Vol.23, pp. 549-558, September 2013

[4] Franklin, C. F., J. D. Powell, and Abbas Emami-Naeini: "Feedback Control of Dynamic Systems", Addison-Wesley,1994

[5] T.Kurabayashi: "Position Convergence and Sway Suppress Method of Overhead Crane by Emulating Natural Damping" The 31th Annual Conference of the Robotics Society of Japan, 1K2-01, July 2013

The 2014 International Power Electronics Conference

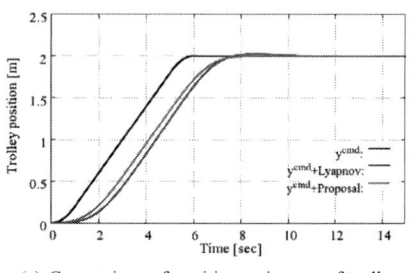 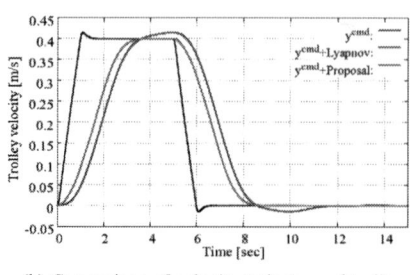 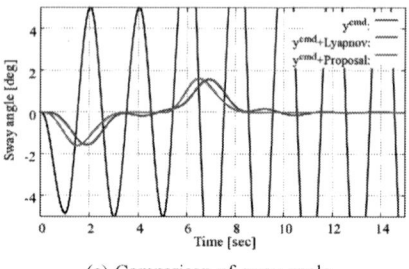

(a) Comparison of position trajectory of trolley   (b) Comparison of velocity trajectory of trolley   (c) Comparison of sway angle

Fig. 5.   Comparison of trolley trajectory and sway angle of payload(Simulation)

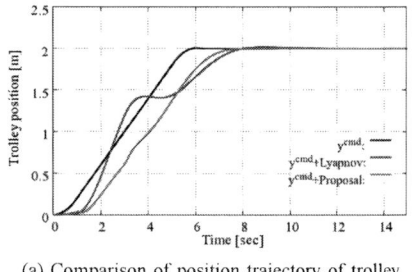 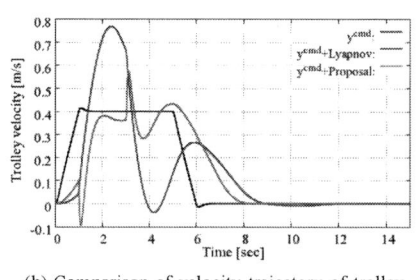 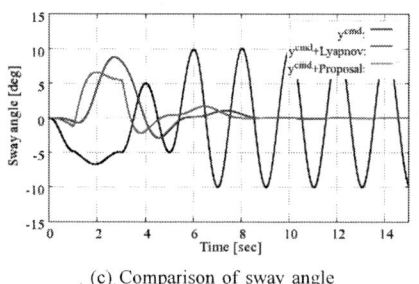

(a) Comparison of position trajectory of trolley   (b) Comparison of velocity trajectory of trolley   (c) Comparison of sway angle

Fig. 6.   Comparison of trolley trajectory and sway angle of payload with disturbance(Simulation)

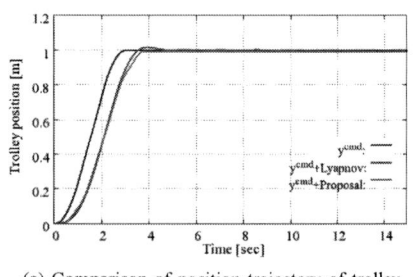 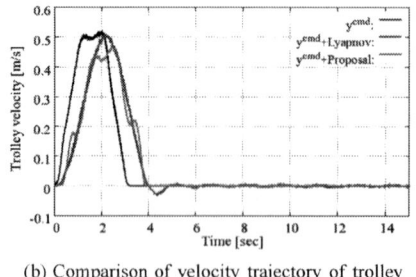 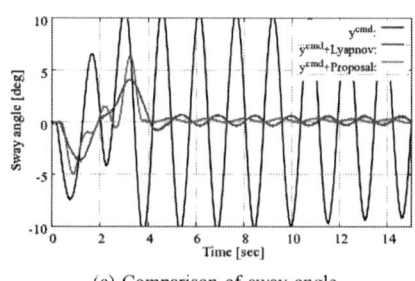

(a) Comparison of position trajectory of trolley   (b) Comparison of velocity trajectory of trolley   (c) Comparison of sway angle

Fig. 7.   Comparison of trolley trajectory and sway angle of payload(Experiment)

978-1-4799-2706-7/14 $31.00 © 2014 IEEE

A Robotic Cane for Walking Assistance

Kyohei Shimizu
Yokohama National University
Japan
shimizu-kyohei-jb@ynu.jp

Issam Smadi
Jordan University of Science and
Technology
Jordan
iasmadi@just.edu.jo

Yasutaka Fujimoto
Yokohama National University
Japan
fujimoto@ynu.ac.jp

Abstract - **New design and control of a robotic cane have been proposed in this paper. Our robotic cane has been designed so that it moves according to user's small force applied to the cane and it stably supports the user when large force is applied by the user. We confirmed our control and estimation of human force method in 2D and 3D simulation. In this paper, an overview of the design and a robust control method are presented.**

Keywords—robotic cane, inverted pendulum, walking assist, omni-directional wheel, nonlinear control.

I. INTRODUCTION

Many studies regarding mobility-assisting devices for the disabled/limited people have been conducted recently. Honda Motor Co., Ltd., has been developing a device that assists motion of human hip joints by two geared motors controlled based on information obtained by pressure sensors equipped on thighs [1]. Cyberdyne Inc. released robotic suit HAL, which assists motion of lower limbs using geared motors and electromyography sensors [2]. There are several walking assist robots with omni-directional wheels; walker-type [3]-[5] and cane-type [6][7]. Some of them have static stability due to multiple wheels [3]-[6]. In general, cane-type assist robots are simpler and lighter than walker-type.

In this paper, new design and control of a robotic cane will be proposed. The concept is similar to a device presented in [7]. However, no actual development has been reported yet regarding a model in [7]. Target functions of our robotic cane are as follows:

1. Ease of handling by a user. The device moves according to user's small force applied to the device.
2. Ability of supporting a user. The device stays stably and supports a user when he/she applies large force to the device.
3. Ability of guiding user's motion. The device guides standing motion of a user.

In order to realize these functions, we have designed a light-weight, high-performance robotic cane. In this paper, an overview of the design and a robust control method are presented.

II. DESIGN OF ROBOTIC CANE

Our robotic cane consists of an active omni-directional wheel and a rod as shown in Fig. 1. The detailed design of the omni-directional wheel is shown in Fig. 2 and Fig.

3. The wheel contains six minor sub-wheels. It employs a differential at the center of the wheel. Two inputs of the differential are driven by two geared motors. If the motors rotate in the same direction at the same speed, the major wheel rotates with respect to the rod but the sub-wheels keep stopping. The cane travels along the *x*-direction of Fig. 1. On the other hand, if the motors rotate in the opposite direction at the same speed, the major wheel keeps stopping but sub-wheels rotate. In this case, the cane travels along the *y*-direction.

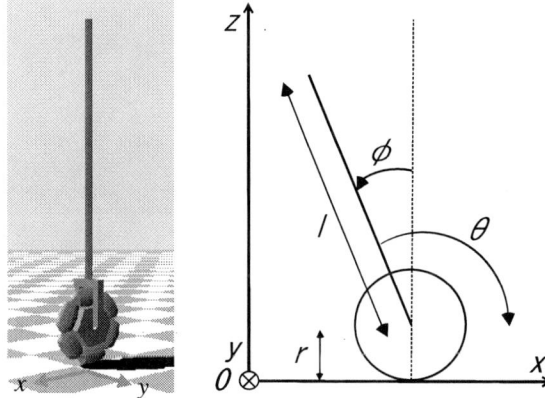

Fig. 1. A robotic cane with an active omni-directional wheel.

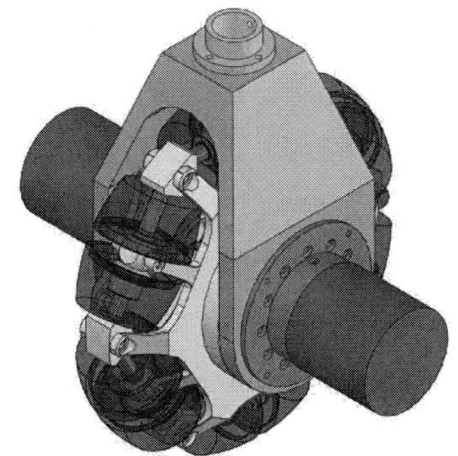

Fig. 2. A detailed design of an active omni-directional wheel.

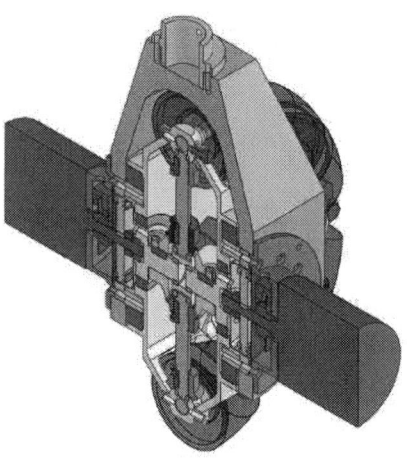

(a) A front cross-sectional view.

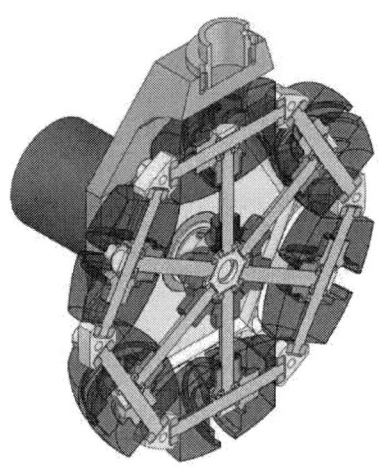

(a) A side cross-sectional view.

Fig. 3. A cross-sectional view of the omni-directional wheel.

III. MODEL AND CONTROL OF ROBOTIC CANE

The major wheel angle θ_{Pitch} and minor wheel angle θ_{Roll} are obtained as an average and a differential of angles of the two geared motors placed right and left side, θ_R and θ_L.

$$\theta_{Pitch} = \frac{\theta_R + \theta_L}{2}$$

$$\theta_{Roll} = G\frac{\theta_R - \theta_L}{2}$$

(1)

where G is the ratio of bevel gears at the differential. In our design $G = 2$.

Let the vertical axis be the z-axis of the world coordinates system. Then xz-plane and yz-plane are defined as the sagittal and lateral plane, respectively. The cane can be approximately modeled as a wheeled inverted pendulum in each plane. We employ two-dimensional inverted pendulum models.

By selecting ϕ, the angle of rod with respect to vertical axis, and the wheel angle θ as generalized coordinates of the inverted pendulum. The equation of motion is derived from Lagrange's equation as follows.

$$\frac{d}{dt}\left(\frac{\partial L}{\partial \dot{q}}\right) - \frac{\partial L}{\partial q} = \tau_{all} - d$$

(2)

$$\begin{bmatrix} H_{11} & H_{12} \\ H_{21} & H_{22} \end{bmatrix}\begin{bmatrix} \ddot{\phi} \\ \ddot{\theta} \end{bmatrix} + \begin{bmatrix} b_1 \\ b_2 \end{bmatrix} = \begin{bmatrix} 0 \\ \tau \end{bmatrix} - \begin{bmatrix} d_1 \\ d_2 \end{bmatrix}$$

(3)

where q is the generalized coordinates vector, H_{ij} is an element of the inertia matrix, b_i is a nonlinear terms, τ is the actuation torque, and d_i is the generalized load torque applied by a user.

Note that this is a trivially underactuated system. There are several approaches to control such nonlinear underactuated systems. Authors have been proposing a nonlinear control method based on differential geometry for underactuated systems with two-degree-of-motion-freedom [8]. The method is as follows.

Let u be a virtual input and determine τ by the following equation.

$$\tau = \left(H_{22} - \frac{H_{12}H_{21}}{H_{11}}\right)u - \frac{H_{21}b_1}{H_{11}} + b_2$$

(4)

Using this equation, the original system (4) can be rewritten as follows.

$$\dot{z} = \begin{bmatrix} \dot{\phi} \\ \ddot{\phi} \\ \dot{\theta} \\ \ddot{\theta} \end{bmatrix} = \begin{bmatrix} \dot{\phi} \\ -\dfrac{b_1}{H_{11}} \\ \dot{\theta} \\ 0 \end{bmatrix} + \begin{bmatrix} 0 \\ -\dfrac{H_{12}}{H_{11}} \\ 0 \\ 1 \end{bmatrix}u = f(z) + g(z)u$$

(5)

Then, we define an output y by

$$y = h(z) = \int_0^\phi \frac{H_{11}}{H_{12}}d\phi + \theta$$

(6)

The high-order differentiations of the output y are calculated as follows.

$$\dot{y} = L_f h(z) + L_g h(z)u$$
$$\ddot{y} = L_f^2 h(z) + L_g L_f h(z)u + \delta_2$$
$$y^{(3)} = L_f^3 h(z) + L_g L_f^2 h(z)u + \delta_3$$
$$y^{(4)} = L_f^4 h(z) + L_g L_f^3 h(z)u + \delta_4$$

(7)

where L_f and L_g are Lie derivatives. Thanks to the selection of the output function y as in (6), the condition $L_g h(z) = 0$, $L_g L_f h(z) = 0$, $\delta_2 = 0$, and $\delta_3 = 0$ holds. In addition, we can assume that $L_g L_f^2 h(z) \cong 0$ and $\delta_4 \cong 0$.

Hence, the following controller which is derived from the last equation in (7) can stabilize the system.

$$v = \lambda_0\left(y_r - y\right) + \lambda_1\left(\dot{y}_r - \dot{y}\right) + \lambda_2\left(\ddot{y}_r - \ddot{y}\right) + \lambda_3\left(y_r^{(3)} - y^{(3)}\right)$$
$$= L_f^4 h + L_g L_f^3 h(z)u$$

(8)

where λ_0, λ_1, λ_2, and λ_3 are coefficients of a Hurwitz polynomial. Now y, $\dot{y}$, $\ddot{y}$ and $y^{(3)}$ are new coordinates and y_r is its reference. The system (5) is approximately transformed into the following system.

$$\frac{d}{dt}\begin{bmatrix} y \\ \dot{y} \\ \ddot{y} \\ y^{(3)} \end{bmatrix} = \begin{bmatrix} \dot{y} \\ \ddot{y} \\ y^{(3)} \\ \lambda_0(y_r - y) + \lambda_1(\dot{y}_r - \dot{y}) + \lambda_2(\ddot{y}_r - \ddot{y}) + \lambda_3(y_r^{(3)} - y^{(3)}) \end{bmatrix} \tag{9}$$

Although this controller does not guarantee global stability, it realizes a wider stability region than linear controllers.

IV. ESTIMATION OF HUMAN FORCE

A. Sensorless Human Force Estimation

In order to control the cane according to the force imposed by a user, we have to measure or estimate it. A nonlinear disturbance observer [9] is applicable for this purpose. A state space representation of the nonlinear disturbance observer is described as follows.

$$\dot{\xi} = -K\xi + K^2 \frac{\partial L}{\partial \dot{q}} + K\left(\frac{\partial L}{\partial q} + \tau_{all}\right) \tag{10}$$

$$\hat{d} = \xi - K\frac{\partial L}{\partial q}$$

where $\hat{d}$ is estimated disturbance, ξ is state variable of the observer, L is the Lagrangian, and K is gain of the observer.

The human force (f_x, f_z) in Fig. 4 appears in the generalized coordinates through a transformation as follows.

$$\begin{bmatrix} d_1 \\ d_2 \end{bmatrix} = \begin{bmatrix} -l\cos\phi - r & -l\sin\phi \\ r & 0 \end{bmatrix}\begin{bmatrix} f_x \\ f_z \end{bmatrix} \tag{11}$$

where l is length of the rod and r is radius of the wheel.

If the inversion of the above transformation exists, we can compute estimation of human force by solving (11) and utilizing the nonlinear disturbance observer (10) as follows.

$$\hat{f}_x = \frac{1}{r}\hat{d}_2 \tag{12}$$

$$\hat{f}_z = -\frac{1}{rl\sin\phi}\left(r\hat{d}_1 + (l\cos\phi + r)\hat{d}_2\right) \tag{13}$$

However, the estimator (13) is not applicable when angle of the rod is around zero. Thus we adopt a conditional recursive least squares algorithm instead.

$$z_n = rl\sin\phi$$

$$x_n = -r\hat{d}_1 - (l\cos\phi + r)\hat{d}_2$$

$$k_n = \frac{P_n z_n}{\rho + P_n z_n^2} \tag{14}$$

$$\begin{cases} \hat{f}_{zn+1} = \hat{f}_{zn} - k_n(z_n f_z - x_n) \\ P_{n+1} = \frac{1}{\rho}(P_n - k_n z_n P_n) \end{cases} \quad \text{if } |x_n| > \varepsilon$$

where ρ is the forgetting factor and ε is a threshold for updating estimation $\hat{f}_z$ and P_n is the covariance matrix.

Once the human force is estimated, reference for angle of the rod, ϕ_r, can be obtained from the equation of force equilibrium:

$$\begin{bmatrix} -lmg\sin\phi_r \\ 0 \end{bmatrix} = \begin{bmatrix} 0 \\ \tau \end{bmatrix} - \begin{bmatrix} -l\cos\phi_r - r & -l\sin\phi_r \\ r & 0 \end{bmatrix}\begin{bmatrix} \hat{f}_x \\ \hat{f}_z \end{bmatrix} \tag{15}$$

Solving this equation for ϕ_r, we have

$$\phi_r = \sin^{-1}\left(\frac{\tan\psi}{1 + \tan^2\psi}\left(\frac{r}{l} + \sqrt{\left(1 - \frac{r^2}{l^2}\right)\tan\psi + 1}\right)\right) \tag{16}$$

where

$$\tan\psi = \frac{\hat{f}_x}{\hat{f}_z - mg}. \tag{17}$$

The reference of the rotation angle of the wheel θ_r holds the following relation with the rod reference angle ϕ_r.

$$\theta_r = \frac{l}{r}\sin\phi_r \tag{18}$$

The references ϕ_r and θ_r are transformed to y_r, $\dot{y}_r$, $\ddot{y}_r$ and $y_r^{(3)}$ by (6)(7) and given in (8).

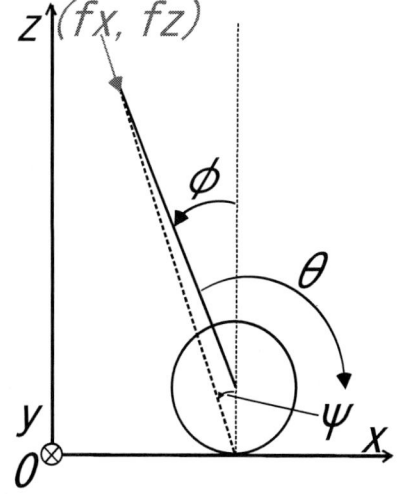

Fig. 4. Inverted pendulum model with human force.

B. Human Force Measurement

Using disturbance observer only small imposed force is estimated. However, several problems regarding stability and difficulty of estimating a large imposed force occurred. Therefore, a load cell is added to measure the imposed force that is parallel to the rod, as follows.

$$\begin{bmatrix} f_x \\ f_z \end{bmatrix} = \begin{bmatrix} \cos\phi & -\sin\phi \\ \sin\phi & \cos\phi \end{bmatrix} \begin{bmatrix} f_x{}' \\ f_z{}' \end{bmatrix} \tag{19}$$

where f_x' and f_z' are human force orthogonal and parallel to the rod, respectively, as shown in Fig. 5.

We can measure the force f_z' directly by the load cell. On the other hand, f_x' can be estimated as a similar manner described in the previous section. By solving (11) and (19) for f_x' we have

$$\hat{f}_x{}' = -\frac{1}{l}(\hat{d}_1 + \hat{d}_2). \tag{20}$$

We can obtain the estimation of f_x and f_z by substituting (20) into (19) as follows.

$$\hat{f}_x = -\frac{1}{l}(\hat{d}_1 + \hat{d}_2)\cos\phi - f_z{}'\sin\phi$$

$$\hat{f}_z = -\frac{1}{l}(\hat{d}_1 + \hat{d}_2)\sin\phi + f_z{}'\cos\phi \tag{21}$$

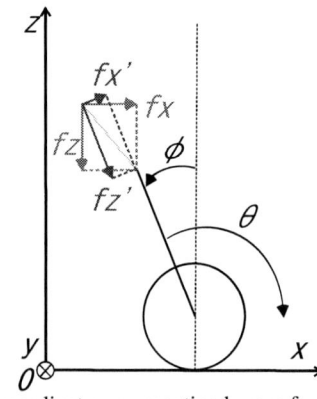

Fig. 5. Two coordinates representing human force.

V. NUMERICAL SIMULATION

A. 2D simulation

The proposed control and estimation methods of human force described in the sections III and IV are applied to a two-dimensional simulation model. Fig. 6(a) and (b) show angle and contact position response of the cane using the sensorless method described in the section IV. A. Human force $f_x = 1$N and $f_z = -2$ N were applied after 3s. Fig. 6(c) shows results of the estimation for the human force. It can be seen that these estimations converge to the actual values.

Fig. 7(a), (b), and (c) show angle and contact position response of the cane, and estimated imposed force in the case of using load cell. Fig. 8(a), (b) and (c) are angle and position response of the cane, and estimated human

force in the case of using load cell. Human force $f_x = 10$N and $f_z = -20$ N were applied after 3s. As a result, we confirmed that its stability is higher than the case of estimation and it can estimate more large force.

B. 3D simulation

The proposed control method described in the section III and IV are applied to a three-dimensional simulation model [10]. The same control strategy is adopted in both sagittal and lateral plane. Fig. 9 shows an initial condition response of the cane with initial pitch angle $\phi(0)=10$deg. From Fig. 9(a), (b), angles and position of the rod stably converge to the upright position.

Fig. 10 shows angle and position response of the cane under the existence of human force of $f_x = 0.3$N and $f_z = -0.5$ N. This simulation we set suitable reference directly without disturbance observer. It can be seen there are little deviation, and it has some swing in lateral plane, but rod angle and position are converged to the reference.

VI. CONCLUSION

In this paper, a robotic assistive cane that extends a function of conventional cane was proposed. A design and a control of the robotic cane were presented. It consists of an omni-directional wheel driven by two geared motors and a rod. The wheel has two-degree-of-motion-freedom. Nonlinear control and human force estimation were proposed and validated via simulations. An experimental setup is under the development. We will finish the equipment and conduct preliminary experiment.

REFERENCES

[1] T. Hirata, et al., "Motion assisting device, control method therefor, and rehabilitation method," *US Patent Application*, pub. no. US2012/0215140A1, 2012.

[2] H. Kawamoto and Y. Sankai, "Power assist method based on phase sequence and muscle force condition for HAL," *Advanced Robotics*, vol. 19, no. 7, 2005.

[3] C. Zhu, et al., "A new type of omnidirectional wheelchair robot for walking support and power assistance," *proc. IEEE/RSJ IROS*, pp. 6028-6033, 2010.

[4] R. Tan, et al., "Adaptive controller for omni-directional walker: Improvement of dynamic model," *proc. IEEE Int. Conf. on Mechatronics and Automation*, pp. 325-330, 2011.

[5] G. Lee, T. Ohnuma, and N. Y. Chong, "Design and control of JAIST active robotic walker," *J. Intelligent Service Robotics*, vol. 3, no. 3, pp. 125-135, 2010.

[6] K. Wakita, et al., "Human-walking-intention-based motion control of an omnidirectional-type cane robot," *IEEE/ASME Trans. Mechatronics*, vol. 18, no. 1, pp. 285-296, 2013.

[7] Y. Ota, et al., "Robotic cane devices," *US Patent Application*, pub. no. US2013/0041507A1, 2013.

[8] I. A. Smadi, "Nonlinear control of mechanical systems with application to monowheel robot," Ph.D. thesis, Yokohama National University, 2009.

[9] I. A. Smadi and Y. Fujimoto, "On nonlinear disturbance observer based tracking control for Euler-Lagrange systems," *JSME J. System Design and Dynamics*, vol. 3, no. 3, pp. 330-343, 2009.

[10] Y. Fujimoto and A. Kawamura, "Simulation of an autonomous biped walking robot including environmental force interaction," *IEEE Robotics and Automation Magazine*, vol. 5, no. 2, pp. 33-42, 1998.

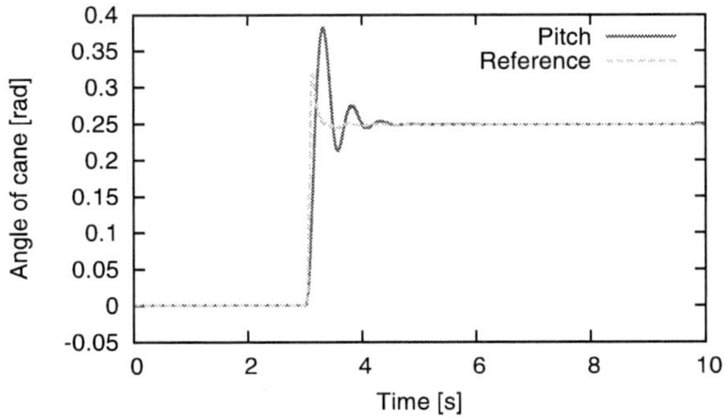

(a) Response of angle of the rod.

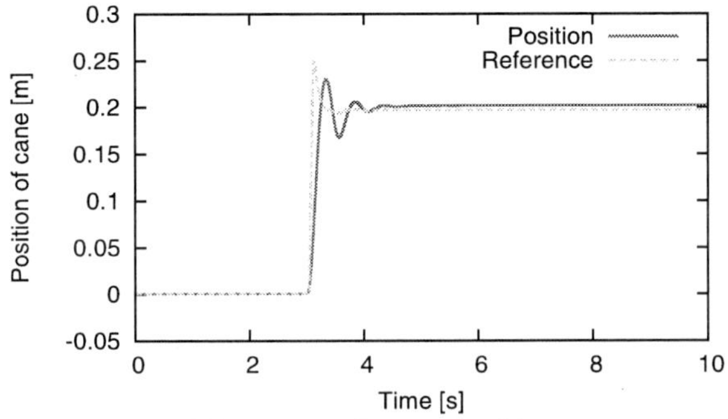

(a) Response of angle of the rod.

(b) Response of contact position.

(b) Response of contact position.

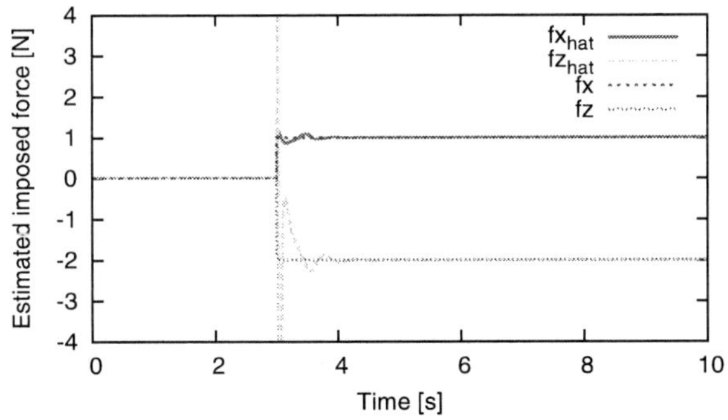

(b) Estimated imposed force.

(c) Estimated imposed force.

Fig. 6. 2D simulation of the robotic cane with sensorless human force estimator (f_x =1N and f_z = -2N).

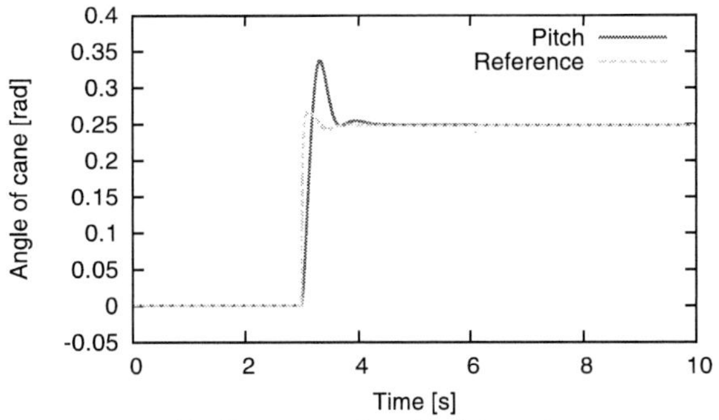

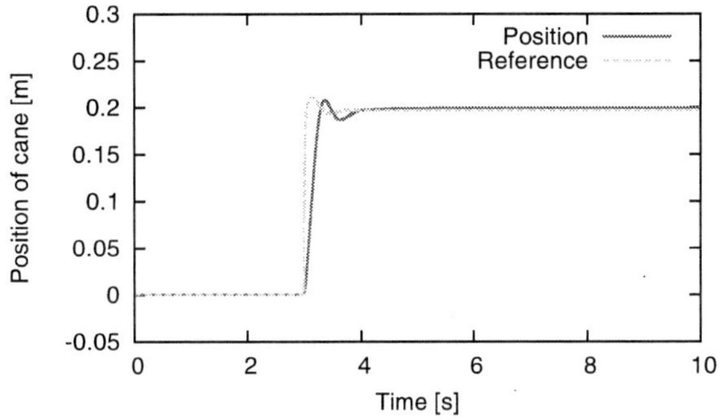

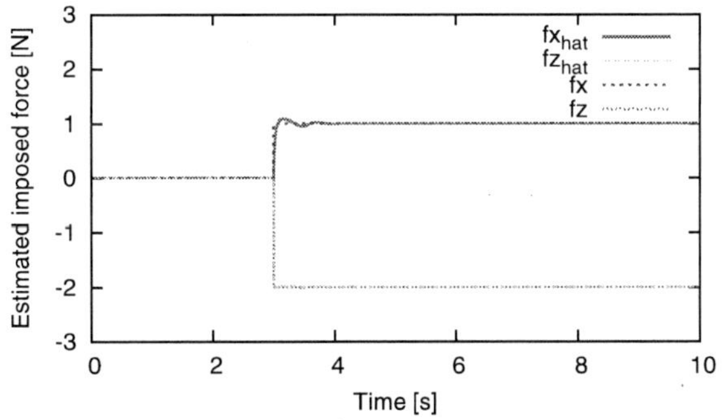

Fig. 7. 2D simulation of the robotic cane with human force measurement by load cell (f_x =1N and f_z =-2N).

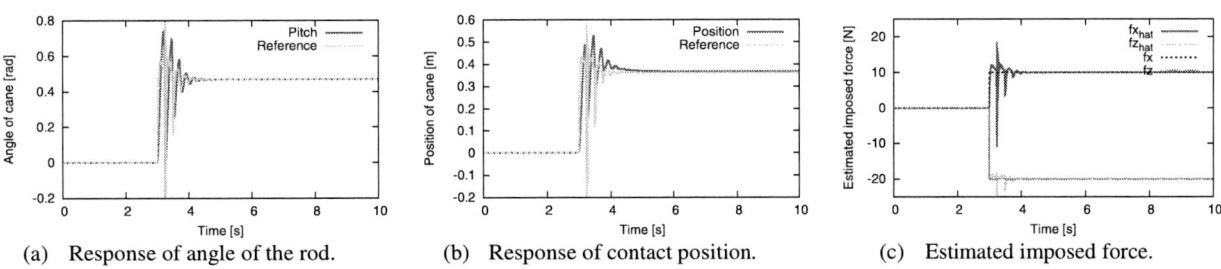

(a) Response of angle of the rod.

(b) Response of contact position.

(c) Estimated imposed force.

Fig. 8. 2D simulation of the robotic cane with human force measurement by load cell (f_x =10N and f_z =-20N).

The 2014 International Power Electronics Conference

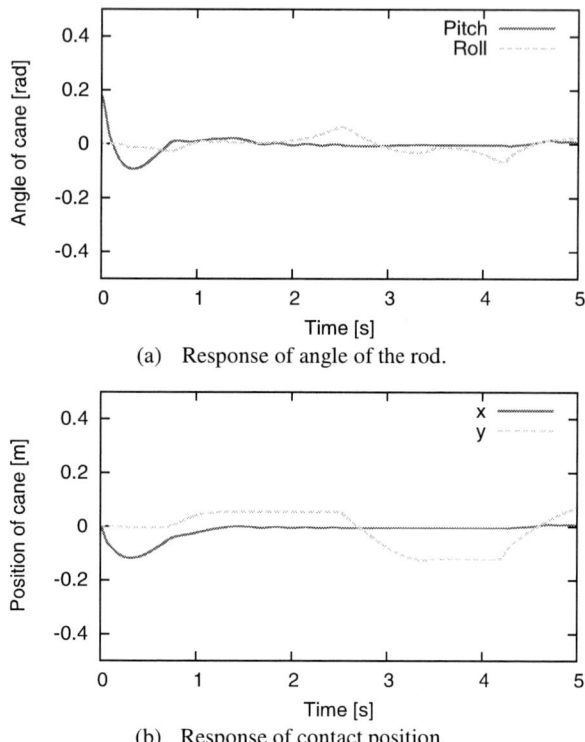

(a)   Response of angle of the rod.

(b)   Response of contact position.

Fig. 9.   3D simulation of initial condition response of the robotic cane without human force.

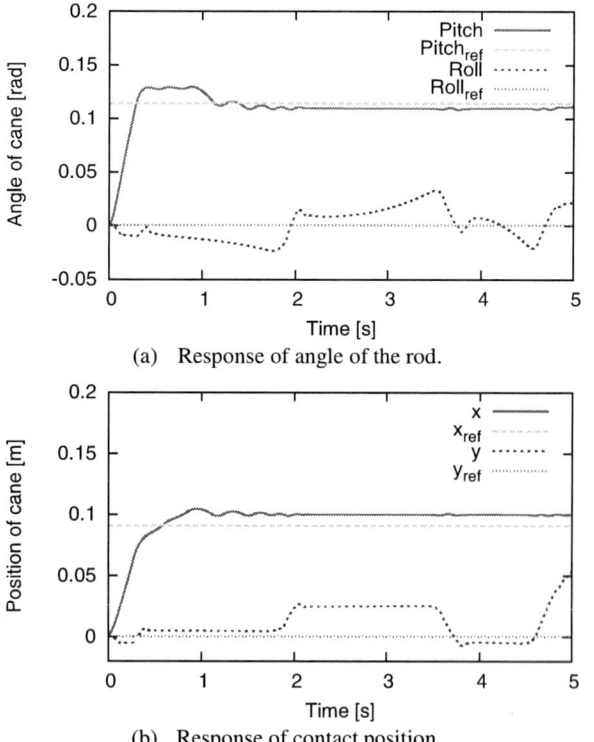

(a)   Response of angle of the rod.

(b)   Response of contact position.

Fig. 10.   3D simulation of the robotic cane with human force (f_x =0.3N and f_z =-0.5N).

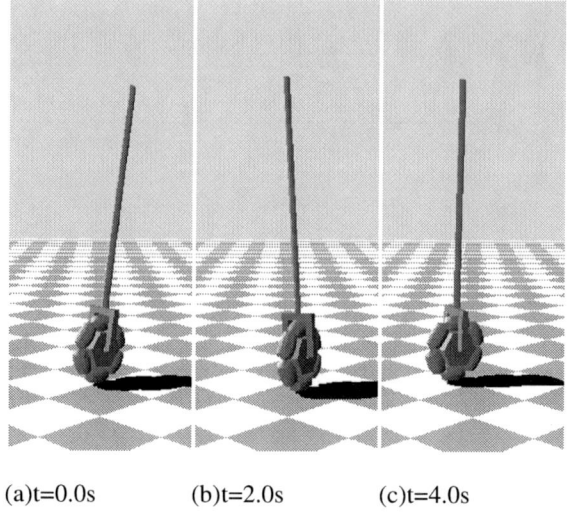

(a)t=0.0s   (b)t=2.0s   (c)t=4.0s

Fig. 11 Snapshots of 3D model of initial condition response.

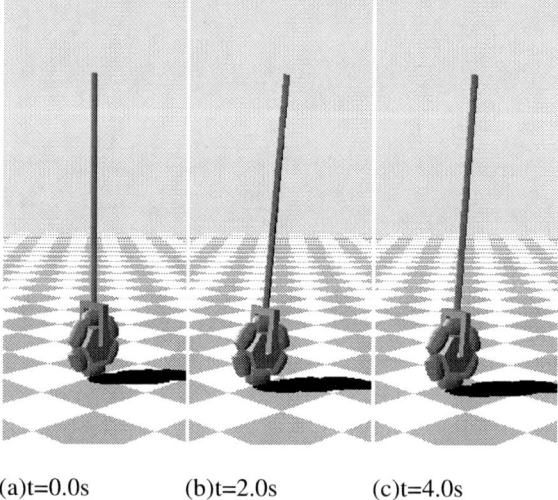

(a)t=0.0s   (b)t=2.0s   (c)t=4.0s

Fig. 12. Snapshots of 3D model of the robotic cane with human force.

978-1-4799-2706-7/14 $31.00 © 2014 IEEE   1973

Hand Position Estimation in Binocular Visual Space using Linear Approximation of Kinematics

Satoshi Komada
Graduate School of
Engineering
Mie University
Tsu, Mie, Japan
komada@elec.mie-u.ac.jp

Santiago Turpin
Launch Department
Ford
Spain

Kento Hashimoto
Graduate School of
Engineering
Mie University
Tsu, Mie, Japan

Daisuke Yashiro
Graduate School of
Engineering
Mie University
Tsu, Mie, Japan
yashiro@elec.mie-u.ac.jp

Junji Hirai
Graduate School of
Engineering
Mie University
Tsu, Mie, Japan
hirai@elec.mie-u.ac.jp

Abstract— A linear visual servoing is based on linear approximation of kinematics made by a constant matrix. The combination of binocular visual space and configuration of a human upper limb and human eyes realizes the linear approximation. In this research, a method of compensating the image-processing delay for the linear visual servoing is proposed. A delay compensation method of image features from the image Jacobian matrix and the joint angle displacements has been proposed. In order to apply this method to the linear visual servoing, a method of compensating delay based on the linear approximate expression is proposed. This method compensates delay by estimating the hand position in the binocular visual space from the linear approximation of kinematics and joint angle displacements. Moreover, a parameter derivation method of the constant matrix is also proposed in order to reduce the estimation error. Accuracy of the proposed method is evaluated through simulations.

Keywords— *binocular visual space, image-processing delay, kinematics, linear visual servoing*

I. INTRODUCTION

Vision sensors can acquire much information visually without contacting environments. One of control methods using vision sensors is a visual servo. The visual servo has a dynamic visual feedback including a vision sensor into a feedback loop directly, which reacts reflectively to surrounding environments.

Linea visual servo (LVS)[1][2][3] uses a linear approximation of kinematics which consists of a constant matrix instead of image Jacobian matrix. In the system of human structure arm and eyes, the inverse kinematics from hand position in binocular visual space to joint angles is approximated linearly. Since there are no parameters in the transformation, LVS can acquire a robust system against parameter error.

Image processing period of general image processor is longer than control period. In visual servo system, there are problems on time delay by image processing and difference between image processing period and control period. Therefore, realization of precise and high-speed visual servo is difficult because of time delay of image

processing.

Methods to overcome this problem are hardware approaches to develop high-speed image processors and software approaches to estimate the delay due to image processing. 1ms high-speed visual processor[4] has been realized high performance visual servo. High-speed image processing system using high-speed camera and FPGA[5] has also been developed. However, these methods are expensive and are not general.

Estimated image feature has been applied to visual servoing as one of the latter approach[6][7]. Delay of image features of a manipulator is compensated by an image Jacobian matrix from joint angle displacement to image feature displacement of a hand.

This paper proposes a delay compensation method in the binocular visual space using the linear approximation of manipulator kinematics with the same size of human arm and eyes. Image-processing dclay is compensated by estimating displacement of the hand position from the linear approximation and the joint angle displacement of the manipulator. This method uses the constant matrix instead of the image Jacobian matrix which is a function of joint angles. Image-processing delay compensation is performed by a simulation.

Moreover, a parameter derivation method of the constant matrix is also proposed in order to reduce the estimation error. Conventionally, the parameter is decided so that the least-square error is minimized for a certain range of motion because the parameter is used for inner loop of visual servoing. Moreover, the method is affected by modeling error of manipulator kinematics because the method is based on the model. The proposed method updates the matrix according to the estimation error which makes the parameter so that the estimation error becomes small. Simulation is performed in order to show usefulness of the proposed method.

II. BINOCULAR VISUAL SPACE AND JOINT SPACE[1]

A. *Model of Cameras and Arm*

In this research, the structure of an arm and cameras

similar to those of humans as shown in Fig. 1 is used. The stereo camera can rotate perpendicularly (Tilt) and horizontally (Pan). The manipulator has 2 and 1 degree of freedom in the shoulder joint and the elbow joint, respectively.

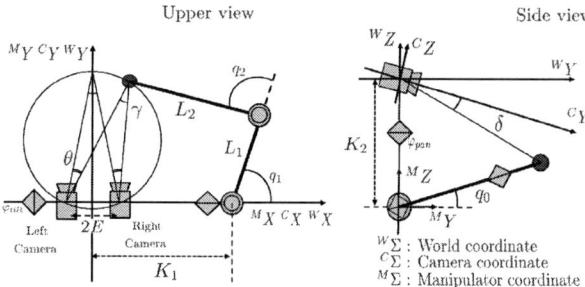

Fig. 1 Model of hand-eye system.

B. Binocular Visual Space

The binocular visual space is the coordinate system defined by the angle of vergence γ and the horizontal and vertical viewing directions θ and δ as shown in Fig. 1.

In the stereo camera model shown in Fig. 2, when the horizontal angles of left and right cameras are α_L and α_R, and the perpendicular angle is φ_{tilt}, the image features of the hand of the manipulator $\left({}^I p_L^x, {}^I p_L^z\right)$ and $\left({}^I p_R^x, {}^I p_R^z\right)$ on the image planes are changed into the hand position of the manipulator $v_p = (\gamma, \theta, \delta)^T$ in the binocular visual space as follows:

$$v_p = \begin{bmatrix} \gamma \\ \theta \\ \delta \end{bmatrix} = \begin{bmatrix} \alpha_L - \alpha_R \\ (\alpha_L + \alpha_R)/2 \\ \varphi_{tilt} \end{bmatrix} + \begin{bmatrix} \left({}^I p_L^x - {}^I p_R^x\right)/f \\ \left({}^I p_L^x + {}^I p_R^x\right)/2f \\ \left({}^I p_L^z + {}^I p_R^z\right)/2f \end{bmatrix} \quad\text{.............. (1)}$$

where f is focal length of cameras. Here, it is assumed that ${}^I p_L^x, {}^I p_R^x, {}^I p_L^z, {}^I p_R^z \ll f$. Moreover, the approximation such as $\tan^{-1}\left({}^I p_L^x/f\right) \cong {}^I p_L^x/f$ are used.

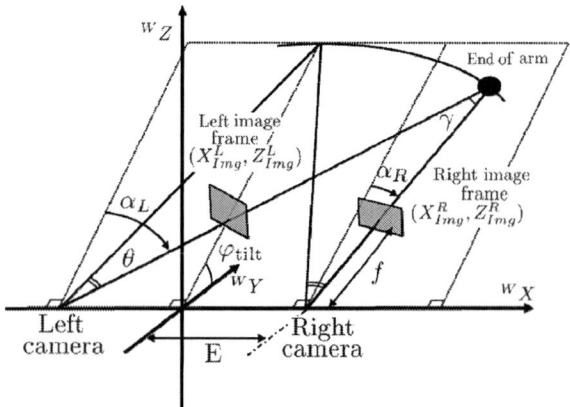

Fig. 2. Model of the active stereo camera.

C. Linear Approximation of Kinematics

In Fig. 2, when the head angle is fixed with $\varphi_{pan} = 0$, the hand position of the manipulator v_p in the binocular visual space is changed into the world-coordinates $\left({}^W x_p, {}^W y_p, {}^W z_p\right)$ as follows.

$$ {}^W x_p = E\sin(2\theta)/\sin(\gamma) $$
$$ {}^W y_p = {}^W \bar{y}_p \cos(\delta) \qquad\qquad\text{.................................... (2)} $$
$$ {}^W z_p = {}^W \bar{y}_p \sin(\delta) $$

where ${}^W \bar{y}_p = E\{\cos(\gamma) + \sin(2\theta)\}/\sin(\gamma)$

From (2) and the relation between the world coordinate and joint coordinate, the inverse dynamics from the hand position in binocular visual space to the joint angle is approximated as the linear equation as follows[3]:

$$ q \approx Rv_p + C \qquad\qquad\text{..................................... (3)} $$

Here, $q = (q_0, q_1, q_2)^T$ is the joint angle vector. R and C are constant matrices.

If we would like to find relation between displacement of joint angle Δq and displacement of hand position in binocular visual space Δv_p, C in (3) disappears as follows:

$$ \Delta q \approx R\Delta v_p \qquad\qquad\text{....................................... (4)} $$

III. ESTIMATION OF HAND POSITION IN BINOCULAR VISUAL SPACE

In this chapter, the hand position of the manipulator in binocular visual space is estimated using the linear approximation of manipulator kinematics to compensate the time delay according to image processing.

A. Time series of image feature

In Fig. 3, the subscript i expresses the i-th image-processing cycle. The image data obtained at the present time t is behind from the actual value by one image-processing cycle. Moreover, the same value is maintained until the next processing is finished. Therefore, the image feature is delayed from 1 to 2 image sampling period. Here, the delay time ΔT is defined as shown in Fig. 3.

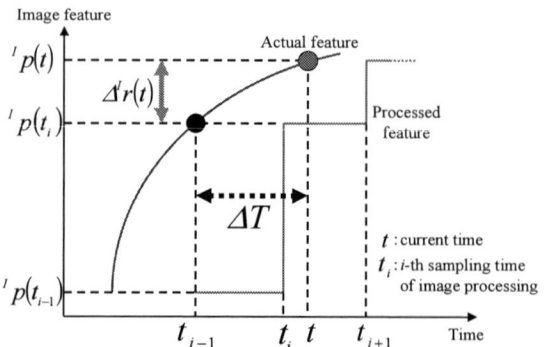

Fig. 3. Time series of image feature.

978-1-4799-2706-7/14 $31.00 © 2014 IEEE

$$\Delta T = t - t_{i-1} \quad\text{.. (5)}$$

The hand position of the manipulator in the binocular visual space obtained from (1) using the delayed image data also includes image-processing delay. If the delayed hand position is used in the linear visual servoing, it causes deterioration of control performance.

B. Estimation of Manipulator Hand in Binocular Visual Space

Based on the time series of image features shown in Fig.3, the hand position of the manipulator in the binocular visual space is estimated.

The displacement of the hand position in binocular visual space between t_{i-1} and t is obtained from (4).

$$\Delta \hat{v}_p(t) = v_p(t) - v_p(t_{i-1})$$
$$= R^{-1}\Delta q(t) \quad\text{...................................... (6)}$$

Here, $\wedge$ expresses an estimated value, $\Delta q(t)$ is the displacement of the joint angle between t_{i-1} and t.

$$\Delta q(t) = q(t) - q(t_{i-1}) \quad\text{.. (7)}$$

Finally, current hand position can be estimated by the joint displacement $\Delta q(t)$ using (6) and the delayed hand position in binocular visual space $v_p(t_i)$ as follows

$$\hat{v}_p(t) = v_p(t_i) + \Delta \hat{v}_p(t)$$
$$= v_p(t_i) + R^{-1}\Delta q(t) \quad\text{..................................... (8)}$$

When there is offset error in the detected joint angle, accuracy of the conventional method is deteriorated because the joint angle is included in the image Jacobian matrix. Accuracy of the proposed method is not affected by it because R is expressed by a constant matrix.

IV. PARAMETER ADJUSTMENT OF R

A. Background

The parameters R and C in (3) are decided by the least-square method for a certain range of motion. Since the parameter is used for inner loop of visual servo, the accuracy is not so important. In this paper, the parameter is used for hand position estimation in binocular visual space as shown in (8). It is better to obtain an accurate parameter so that the estimation error makes as small as possible. Therefore, this paper proposes a parameter estimation method for each operating point.

In order to reduce estimation error of (8), the parameter R is updated during motion of manipulator. If we formulate the square matrix from $\Delta q(t)$ and $\Delta v_p(t)$, R can be derived as follows:

$$R = \Delta Q \Delta V_p^{-1} \quad\text{(3)}$$

However, it is affected by sensor noise easily if the motion of manipulator in binocular visual space is small. If model of manipulator is utilized to avoid sensor noise, modelling error causes parameter error of R. Therefore, real time parameter update based on estimation error is

utilized.

B. Parameter Estimation

In order to reduce estimation error of (8), the parameter R^{-1} is updated during motion of manipulator. Initially, the estimation error vector is defined.

$$e_p(t_i) = \Delta \hat{v}_p(t_{i-1}) - \Delta v_p(t_i)$$
$$= \psi \Delta q(t_{i-1}) - \Delta v_p(t_i) \quad\text{................................. (9)}$$

where $\psi = R^{-1}$ is introduced. The displacements are defined as follows:

$$\Delta q(t_{i-1}) = q(t_{i-1}) - q(t_{i-2}) \quad\text{................................... (10)}$$

$$\Delta v_p(t_i) = v_p(t_i) - v_p(t_{i-1}) \quad\text{.....................................(11)}$$

The estimation error is defined by (9).

$$E(t_i) = \frac{1}{2} e_p^T(t_i) e_p(t_i) \quad\text{................................... (12)}$$

Parameter ψ is updated so that the estimation error shown in (12) makes small.

$$\psi(t_{i+1}) = \psi(t_i) - G \frac{\partial E(t_i)}{\partial \psi(t_i)}$$
$$= \psi(t_i) - G\Delta q^T(t_{i-1})\left(\psi^T(t_i)\Delta q(t_{i-1}) - \Delta v_p(t_i)\right) \quad\text{........ (13)}$$

where G is a gain matrix. Obtained $\psi(t_{i+1})$ is given to R^{-1} in (8).

V. SIMULATION

A. Simulation Setup

Simulation conditions are shown in Fig. 4 and Table 1. In this simulation, since the manipulator only operates on horizontal plane ${}^W X - {}^W Y$, $q_0 = 0$. The angles of the cameras are $\alpha_L = \alpha_R = 0$ to be parallel stereo camera and $\varphi_{pan} = 0$ and $\varphi_{tilt} = 45$ degrees in order to observe the hand of the manipulator inside the image plane.

The following constant matrix is used for estimation of hand position in binocular visual space.

$$R = \begin{bmatrix} -1.60 & -1.45 \\ 6.03 & 0.71 \end{bmatrix} \quad\text{................................ (9)}$$

It is derived from the least-square method so that the joint angle error is minimized based on the linearized equation shown in (3) for combinations of the joint angle $q_0 = 0$, $q_1 = 40, 45, \ldots, 60$, $q_2 = 110, 115, \ldots, 130$[deg] and corresponding hand position in binocular visual space v_p.

The joint angle of the manipulator is moved as follows:

$$q_1(t) = 45 - 5 \cdot \sin\left(\frac{\pi}{6}t\right)$$
$$\quad\text{.. (10)}$$
$$q_2(t) = 80 - 5 \cdot \sin\left(\frac{\pi}{6}t\right)$$

In the parameter update of R, the following gain is utilized.

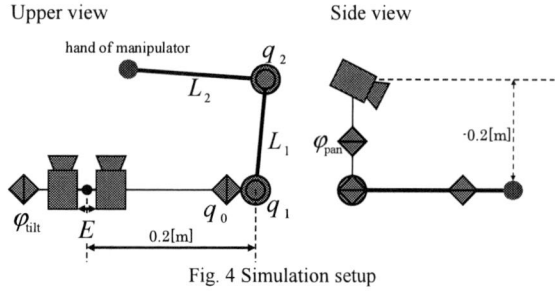

Fig. 4 Simulation setup

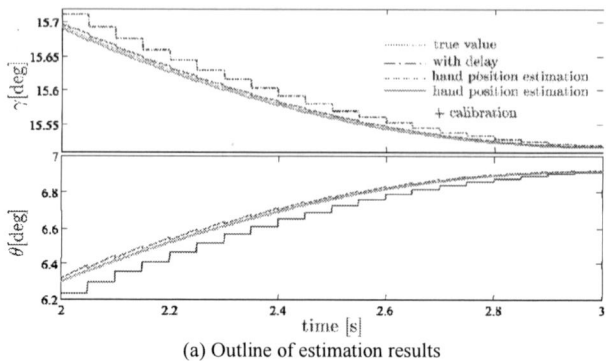

(a) Outline of estimation results

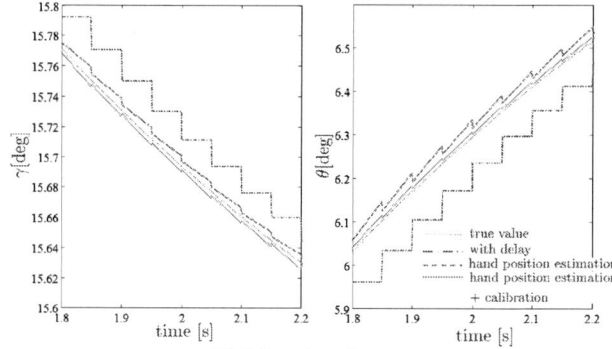

(b) Enlarged results

Fig. 5. Hand position estimation

TABLE I
SIMULATION PARAMETERS

Link length [m]	L_1=0.25 , L_2=0.38
Base-line length [m]	2E=0.07
Position of camera [m]	W=200, K=200
Angle of camera [deg]	φ_{pan} =0, φ_{tilt} =45
Image processing Period [ms]	50
Focal length [mm]	8
Size of a image plane [pixel^2]	640 × 480
Pitch of pixel [mm]	0.01375

$$ G = \begin{bmatrix} 1.7 & 0 \\ 0 & 0.002 \end{bmatrix} \quad\dots\dots\dots\dots\dots\dots\dots (11) $$

B. Simulation Results

In order to show influence of parameter error for the estimation, the length of W is increased by 10 mm. Fig. 5 shows the simulation result, where the hand position similar to the true value is obtained by the hand position estimation. However, there is estimation error due to the parameter error of W. By applying the parameter calibration for R^{-1}, the error is decreased.

VI.   CONCLUSION

In this paper, as a compensation method of image-processing delay suitable for the linear visual servoing, the estimation method of hand position in binocular visual space is proposed using the linear approximation from hand position in binocular visual space to joint angles. Although image Jacobian matrices are functions of joint angles used for conventional image feature estimation, the linear approximation is the constant matrix. Therefore, the proposed method is robust against the offset error of the joint angles. The simulation result by the proposed method shows that estimated hand position is similar to the actual one.

The parameter of the matrix used for the delay compensation is updated so that the estimation error becomes small. Simulation result shows improvement of estimation error due to a calibration error.

REFERENCES

[1] T. Mitsuda, N. Maru, K. Fujikawa, and F. Miyazaki, "Binocular visual servoing based on linear time-invariant mapping", *Advanced Robotics*, vol.11, no.5, pp.429-443, 1997

[2] K. Namba, N. Maru, "Positioning Control of the Arm of the Humanoid Robot by Linear Visual Servoing", *Proceedings of the 2003 IEEE, International Conference on Robotics and Automation*, vol. 3, pp. 3036-3041 , 2003

[3] S. Mukai, N. Maru, "Redundant Arm Control by Linear Visual Servoing Using Pseudo Inverse Matrix", *Proceedings of the 2006 IEEE/RSJ International Conference on Intelligent Robots and Systems*, pp. 1237-1242, 2006

[4] Y. Nakabo and M. Ishikawa, ``Visual Impedance Using 1ms Visual Feedback System," *Proc. of IEEE Int. Conf. Robotics and Automation (ICRA '98)*, pp. 2333--2338, 1998.

[5] J. Oaki, ``Application of 1,000 fps High Speed Image Processing to Active Camera System - Tracking of Unspecified Moving Object under Ordinary Illumination and Cluttered Background -," *Journal of the Robotics Society of Japan*, vol. 23, no. 3 , pp. 282--285, 2005. (in Japanese)

[6] S. Komada, M. Yoshida, and T. Hori, "Visual Servoing of Robots using Estimated Image Features," *IEEJ Journal of Industry Applications*, vol. 123, no. 10, pp. 1200-1205, 2003 (in Japanese)

[7] S. Higashi, S. Komada, M. Ishida, and T. Hori: ``Obstacle Avoidance of Redundant Manipulators on Visual Servo System Using Estimated Image Features," *Proc. International Workshop on Advanced Motion Control*, pp. 165-170, 1998.

[8] I. Kinbara, S. Komada, and J. Hirai, "Visual Servo of Active Camera and Manipulator with Simple On-line Calibration for Estimated Image Feature," *in Proc. 9th International Workshop on Advanced Motion Control*, pp. 636-640, 2006

Contact State Recognition Based on Haptic Signal Processing for Robotic Tool Use

Ryohei Matsuzaki, Jun Okuma, and Sho Sakaino
Department of Electrical and Electronic Systems
School of Science and Engineering, Saitama University
Saitama, Japan

Toshiaki Tsuji
Department of Electrical and Electronic Systems
School of Science and Engineering, Saitama University
Saitama, Japan / JST PRESTO

Abstract—**Robotic tool use is a challenging issue, which can be a key for extending the ability of robots in human environment. This study deals with a technique for contact state recognition of robotic tool use. A contact model and a method to calculate the parameters of the model from the haptic information are proposed. Some kinetic information is estimated from the model parameters and utilized for the tool use.**

Keywords—force sensing, haptics, robotic tool use

I. Introduction

Many robots are working in factories today since robotics technologies have been developed well. However, still the use of robots are limited because most of existing robots in practice are specialized for a single task. A technique to control robots for a universal appliance is desirable for applying robots in human environment. Studies of tool use deal with essential issues for dexterous motion in complicated tasks.

Robotic tool use is a quite popular study in cognitive robotics [1], [2]. There are some studies on development of control architectures, too. Kemp and Edsinger examined detection of task relevant feature of human tools by vision [3]. Nabeshima and Kuniyoshi proposed a method to achieve sustainable sensory-motor coordination for robotic tool-use [4].

Development of controllers for robotic tool-use systems requires much human labor. Control architectures for various tools are all different, while the tools with similar contact state during the tool use have similar control architecture. It infers that the controller design can be simplified if the controllers are schematized. The contact states between multiple general objects has been defined by Mason *et al.* [5] depending on the kinetic degree of freedom based on the relative attitude of each object. Extending the definition of the contact states, the authors have categorized tools into four groups and showed the control architecture of them [6].

One of the largest issues for robotic tool use is the position error at the tip of the tool. When a robot tries to accomplish a dexterous contact motion with objects, position error at the tip of a robot arm often becomes an issue. Keeping smooth touch control between two objects is a challenging theme especially when both objects have high stiffness [7], and the problem becomes more serious when position errors exist. Especially, tool use robots often confront with this problem because the

kinematics of the robot tool varies with the tool change. Using vision sensors is a common way to update the kinematics of the robot holding tool [8], while vision sensors still have issues: slight error of the vision sensor often cause large error in force dimension when the tool and the object both have high stiffness; and the vision sensors have some blind area in most cases especially around the point of contact. To solve this issue, this study proposes a method to update the kinematic information of tool held by a robot and compensate the error.

This paper first proposes the model of contact. Then, a method to calculate the parameters of the model from the haptic information is proposed. Results of experimental verifications are shown after the control architecture.

II. Identification of Properties of Tools Based on Tactile Sensations

A. Basic concept

When using tools to perform sophisticated work, it is necessary to gain understanding regarding the position of the leading edge of such tools, as well as its width and center of gravity. Recognizing properties of such tools is therefore very important. However, it is not always possible to correctly identify properties of tools. The position of the leading edge of a tool with respect to the grip section, for instance, varies depending on how such tool is held. This is because holding onto a tool always at the same position and attitude is a difficult thing to achieve, since there is some backlash in the grip section of tools used by humans. This is particularly the case when multiple tools are used, as tools are switched from one hand to another and each time hands are switched to hold these tools, the position of the leading edge and attitude vary, causing problems such as the center of gravity shifting and offset acting on force sensors, which make it difficult to manipulate such tools well. Humans are capable of compensating for such changes by using tactile information. Humans are believed to be conducting active sensing [9], [10] of properties of an object held in their hand, by getting such objects to come into contact with environment. The fact that people with impaired vision are able to handle kitchen knives and spatulas without any visual information suggests that it is possible to estimate properties based on tactile information and to compensate positional errors.

978-1-4799-2706-7/14 $31.00 © 2014 IEEE

The method for estimating properties of an object held in hand based on the position and force information gained from active sensing is described below. First, a contact model of robotic tool use is proposed based on the equilibrium of force moment. Properties of tools that can be identified by using the model, as well as the compensation method will then be described. A spatula is used to provide descriptions, since it is a tool with high stiffness. Furthermore, the robot used for in this research had six degrees of freedom, capable of detecting force and moment at the tip of the arm. An ordinary robot arm also could have been used for this purpose, since a robot arm that is capable of holding and securing a tool is assumed for our purpose.

B. Force equilibrium

First, the force equilibrium on the tool is considered by an example in Fig. 1. In this example, the tool is fixed on a gripper on the tip of the robot arm. The force on the gripper F^o is measured by a 6-axis force sensor as well as M^o, the force moment on the gripper. Suppose external force F^e acts on the tool, the force equilibrium on the tool is presented as follows:

$$F^e + F^o = 0 \tag{1}$$

On the other hand, the force moment equilibrium on the tool is represented as follows:

$$(P^e - P^o)F^e + M^o = 0 \tag{2}$$

Here, P^e denotes the contact point, the point external force acts, and P^o denotes the position of the force sensor. In case the external force acts as distributed pressure force, P^e represents the center of pressure.

As shown in the previous studies[11]-[13], the contact point P^e can be calculated from force and moment information. By developing (1) and (2), the following equations are acquired.

$$F^e = -F^o \tag{3}$$

$$P_y^e = \frac{F_y^e}{F_z^e}(P_z^e - P_y^e) + P_y^o + \frac{M_x^o}{F_z^e} \tag{4}$$

where the subscripts x, y and z denote axes in Cartesian coordinates. Equation (4) represents that P^e exists on a straight line along the external force vector F^e, as shown in Fig. 2. This study assumes that the kinematic information of the tool is known, while it may contain some errors as described above. Then, the contact point P^e is determined as the intersection of the line represented by (4) and the surface of the tool. Additionally, M^e, the moment from the environment at the center of the contact surface P^{eo} is derived as follows:

$$M^e = (P^e - P^{eo}) \times F^e. \tag{5}$$

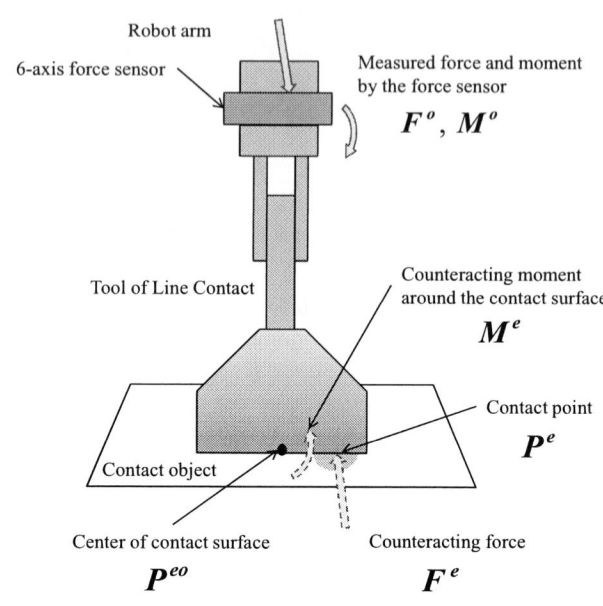

Fig. 1.   Force and moment on the tool

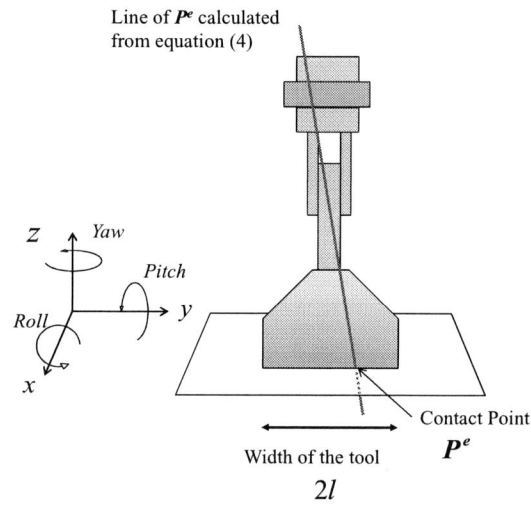

Fig. 2.   Center of Pressure

C. Proposal of contact model

A model of the attitude and moment for active sensing is shown in Fig. 3. For the purpose of simplification, the tool was considered two dimensionally, with the tool connected to environment with a line. This was expanded to three dimensions, however, for the experiment in which the tool was actually used. In case of a completely rigid body, when the tool is to be tilted the moment acting from environment initially increases in the Area ① of Fig. 3 but with no changes in attitude and once the moment that exceeds lF_z^e is applied the tilting starts to occur as shown in Area②. l in this instance represents the length from the center of the leading edge of the tool to the very edge of it. This can be estimated using (2) and (4) as well. Equation (4) shows how the center of pressure P^e changes with the increase in the moment. Since the range of actual $\boldsymbol{P}^e$ is dependent on the width of the tool we have in this case:

$$|\boldsymbol{P}^e - \boldsymbol{P}^{eo}| \le l \tag{6}$$

while M_{max}, the upper limit of the reaction force moment from environment would be:

$$M_{max} = lF_z^e \tag{7}$$

Here, M is the force moment about the point $\boldsymbol{P}^{eo}$. Note that a force moment in a two-dimensional space is represented by a scalar value. Since the balance with the reaction moment from environment cannot be maintained when $M > M_{max}$ in (4), the spatula starts to tilt. The reaction force moment the tool is subjected by environment at this moment does not exceed beyond the upper limit. Descriptions were provided thus far with the assumption that the tool was a complete rigid body for the purpose of simplicity, but tools of such completely rigid bodies do not exist in reality. Descriptions are provided for tools of non-rigid bodies below. A model of the attitude and moment for active sensing on a tool of non-rigid body is shown in Fig. 4. The major difference from a tool of complete rigid body is in Area ① of the diagram. Since the tool is not a completely rigid body, distortions occur with the tool even when $M \le M_{max}$ or less and the attitude begins to change. Assuming the rotational stiffness of the spatula in such an instance to be K [Nm/rad], then the reaction force moment the tool is subjected by environment would be expressed as:

$$M = K\theta \tag{8}$$

When the force moment M increases and exceeds M_{max}, the attitude of the tool starts to change significantly like in Fig. 4. As shown in the figure, the tilting of the tool results in the decrease of M from M_{max}. The attitude of the tool as it is in contact with the environment with its contact surface positioned in parallel to the environment was set to 0 rad thus far and the moment the tool was subjected by the environment was considered to be 0 Nm. In reality, however, physical errors of the inclination in the attitude of the tool, as well as impact of the offset on the force sensor occur, when a tool is held in a manner as described in the introduction. A model for the attitude moment with this taken into consideration is shown in Fig. 5. This graph represents a parallel moving of the graph in Fig. 4, by the error in the inclination of the attitude, as well as the offset for the sensor.

D. Estimation method for properties of tools

The properties of the tool that can be understood using the proposed model, as well as the estimation method thereof, are described below. When active sensing is performed using actual equipment, $\boldsymbol{F}^o(t)$, $M^o(t)$ and $\theta(t)$ can be given by sensors as positional and force information. Furthermore, (3) to (5) can be used to derive $\boldsymbol{F}^e(t)$, as well as the z direction component of this force $F_z^e(t)$, and $M^e(t)$. Since the theoretical moment M^o becomes largest at this point a, M^a can be derived using the following equation:

$$M_a = \max(M^o(t)) \tag{9}$$

Letting the time t at which (9) holds up to be t_a, θ_a, the attitude θ at point a, can be derived using the following equation:

$$\theta_a = \theta(t_a) \tag{10}$$

Point a is where the tool starts to tilt and the moment thereof can be derived using (7), but it is actually expressed by the following equation, since in reality the offset of the force sensors must be taken into consideration.

$$M_a = lF_z^e + M_{offset} \tag{11}$$

Similarly, since moment M^o is the smallest at point b we get:

$$M_b = \min(M^o(t)) \tag{12}$$

Letting time t for this instance to be t_b, θ_b, the attitude at point b can be derived using the following equation:

$$\theta_b = \theta(t_b) \tag{13}$$

By taking into consideration the offset with (7) we get:

$$M_b = -lF_z^e + M_{offset} \tag{14}$$

where we get the following based on (11) and (14):

$$M_a - M_b = lF_z^e + lF_z^e \tag{15}$$

The width of the tool $2l$ can therefore be calculated in the following manner:

$$2l = \frac{M_a - M_b}{F_z^e} \tag{16}$$

Next we draw our attention on the line derived by connecting points a and b. This line is expressed by (8). The environment and the tool are in contact with each other in practically parallel condition in Area ①, which is $\theta_b < \theta \le \theta_a$.

In reality the graph becomes rounded at the borderline between areas ① and ②, which potentially leads to error in the inclination of the graph with respect to the line that connects points a and b, but these are ignored as they are quite minimal.

The offset that applies to the force sensors in the initial attitude M_{offset} and the inclination of the initial attitude with

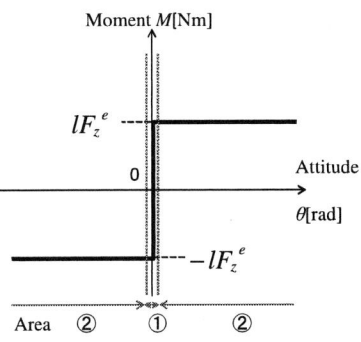

Fig. 3.   Model for attitude and moment (rigid body)

respect to the environment θ_{offset} were derived next. From (11) and (14) we get:

$$M_{offset} = \frac{M_a + M_b}{2} \tag{17}$$

Similarly, the inclination of the initial attitude θ_{offset} can be calculated by using the following equation:

$$\theta_{offset} = \frac{\theta_a + \theta_b}{2} \tag{18}$$

θ_{offset} and M_{offset} therefore are derived as the center point c of a and b. At the time point c is reached, the leading edge of the tool is actually in contact with the environment in parallel, in other words with the inclination θ of 0 rad and the moment due to the environment M becomes 0 Nm . The width of the tool $2l$, rotational stiffness K, error in attitude of tool θ_{offset} and offset of force sensors M_{offset} can be derived as properties of the tool, based on the above.

III. CONTROL SYSTEM

A. General Structure

The robot is controlled by a hybrid control system composed of the force control or position control assigned to each axis. The block diagram is shown in Fig. 6. Selection matrix S switches the control target force and position. Although there is a need to switch the control systems corresponding to each category of tools, the classification of each control is switched by only changing numeric values of the selection matrix S. Therefore, most part of the control system of the robot has been unified.

By giving the trajectory to this control system, the robot will be able to use a tool. The robot arm used in this study has six degrees of freedom. Reference value of acceleration $\ddot{x}^{ref}$ of the tip of the arm for control robot is given by

$$\ddot{x}^{ref} = S(K_p(x^{cmd} - x) + K_v(\dot{x}^{cmd} - \dot{x})) \\ + (I - S)(K_f(f^{cmd} - f)) \tag{19}$$

where x^{cmd}, $\dot{x}^{cmd}$, and f^{cmd} are command value of position, velocity, and force of the tip of the arm, x and f are value

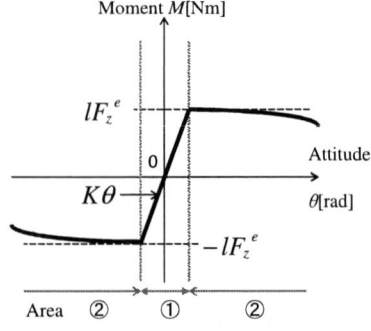

Fig. 4. Model for attitude and moment (non-rigid body)

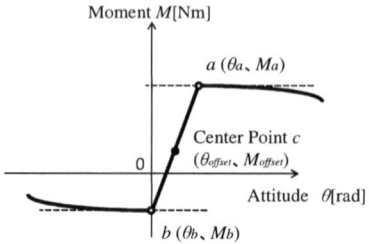

Fig. 5. Model for attitude and moment (with offset)

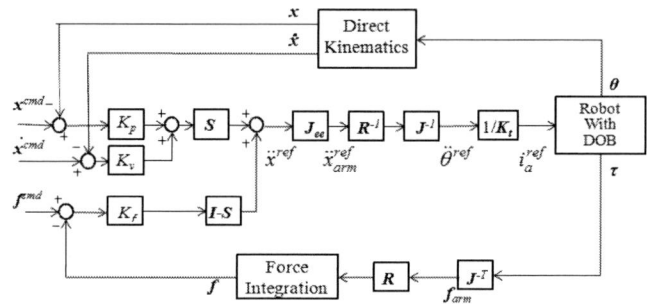

Fig. 6. Block diagram of robot control system

of position and force of the tip of the arm. Position gain K_p, velocity gain K_v, and force gain K_f, are matrix, given by

$$K_p = diag[K_{pp}, K_{pp}, K_{pp}, K_{pr}, K_{pr}, K_{pr}] \tag{20}$$
$$K_v = diag[K_{vp}, K_{vp}, K_{vp}, K_{vr}, K_{vr}, K_{vr}] \tag{21}$$
$$K_f = diag[K_{fp}, K_{fp}, K_{fp}, K_{fr}, K_{fr}, K_{fr}]. \tag{22}$$

K_{pp}, K_{vp}, K_{fp} are position, velocity, and force gain of the direction of the x, y, and z axes. K_{pr}, K_{vr}, K_{fr} are position, velocity, and force gain for rotation of the direction of the R, P, and Y. I is identity matrix. Selection matrix S is a diagonal matrix, shown in (23). x comprises six components of the position x, y, z and attitude of the R, P, Y. f comprises six components of the force and moment.

$$S = diag[S_x, S_y, S_z, S_R, S_P, S_Y] \tag{23}$$

Diagonal components switch the control target force and position. When diagonal component is the maximum, 1.0, the control target is position. When diagonal component is the minimum, 0.0, the control target is force.

In the next place, we give a description of the variables of the block diagram (Fig. 6), except hybrid control. Reference value of acceleration $\ddot{x}^{ref}_{arm}$ of the tip of the each arm is calculated from Jacobian matrix J_{ee} and reference value of acceleration $\ddot{x}^{ref}$ of the end effector, comprises the tip of three arms. Rotation matrix R coordinates global coordinate into each arm coordinates. Jacobian matrix J associates reference joint angular acceleration $\ddot{\theta}^{ref}$ with acceleration of tip of arm. The robot is under acceleration control based on disturbance observer (DOB) [14]. "Force Integration" calculates external force and moment f of end effector from force and moment f_{arm} of tip of each arm.

IV. Experiment

A. Experiment equipment

The parallel link manipulator [15], [16], which adopts a twin drive system that can reduce the impact of friction from the motor on respective joints, was used for the experiment. The system is shown in Fig. 7. This parallel manipulator was comprised of three combinations of robot arms having three degrees of freedom. The leading edges of the three arms were arranged together with an end effector, as shown in the figure. A spatula was secured there for the purpose of our experiment. An ordinary robot arm could have been used for our purpose, since the control system of this research assumed a robot arm with six degrees of freedom and capable of detecting force and moment at the leading edge of the arm. Control parameters are shown in Table I. The grasping mechanism was installed on the end effector via force sensors as shown in Fig. 8 a) in our experiment and the system was made to grip on a metallic spatula shown in Fig. 8 b) to perform our experiment.

A scene from the experiment is shown in Fig. 9.

Fig. 7. Parallel link manipulator

a) Grasping mechanism

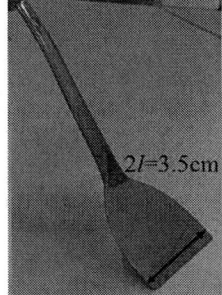

b) Spatula

Fig. 8. Experiment equipment

Fig. 9. A scene from active sensing

TABLE I. CONTROL PARAMETERS

Control gain parameter	Variable	Value
Position gain	K_{pp}	100.0
Velocity gain	K_{vp}	12.0
Position gain for rotation	K_{pr}	80.0
Velocity gain for rotation	K_{vr}	10.0
Force gain	K_{fp}	0.6
Force gain for rotation	K_{fr}	10.0
Cutoff frequency of DOB	G_{dob}	60.0

B. Parameter identification experiment using active sensing

Active sensing was performed to conduct an experiment for identifying the properties of the tool held by the robot. The operation involved the issuing of a force instruction to the robot to apply constant pressure at 3.0 N to the environment, while manipulating the robot to tilt the spatula in the direction of R by ± 0.15 rad. x, y, P and Y axes are position controlled and target values were all set to remain at respective initial positions with the value of 0. The instruction and response values for attitude in the R direction, as well as response values for the moment are shown as experiment results in Fig. 10, while the relationship between the attitude and the moment is shown in Fig. 11. The experiment resulted in a significant manifestation of hysteresis characteristics for the proposed model. Data for tilting the attitude of the tool is comprised of ① and ③ with larger changes of moments, while the data for reverting the tilted attitude is comprised of ② and ④ with smaller changes of moments. With this model, the properties of the tool are derived by using data for tilting of the attitude in ① and ③, since significant tilting is presumed from the condition where the entire leading edge of the tool is in contact with the environment in our model. First of all, the starting points for tilting (attitude [rad], moment [N·m]) were (-0.040, -0.023) and (0.026, 0.105). Since the center of these two points was (-0.007, 0.041), the initial attitude was inclined to -0.007 rad and the offset of 0.041 Nm was on the force sensors. Since the amplitude of the measured moment was 0.064 Nm, furthermore, the width of the tool was identified as 0.042 m by using (16).

The 2014 International Power Electronics Conference

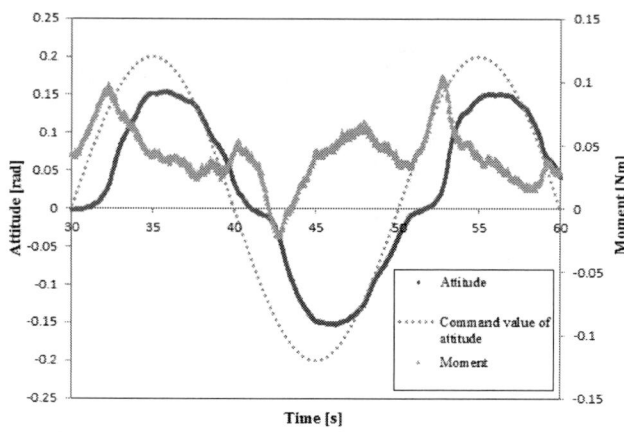

Fig. 10. Attitude and moment in R direction

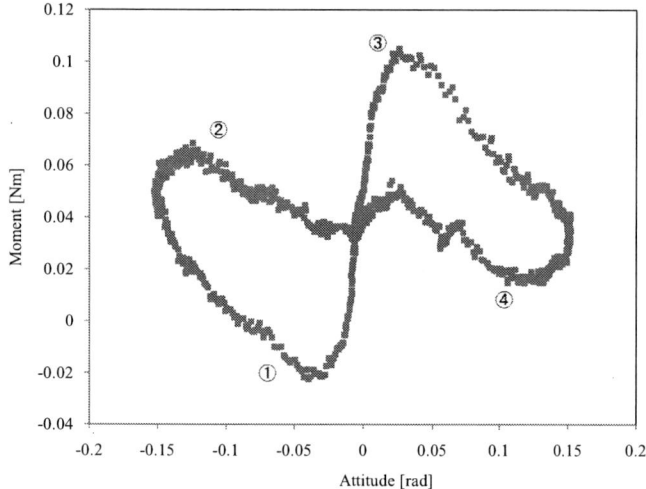

Fig. 11. Attitude and moment

V. Conclusion

This research was aimed to achieve robots that can use multiple tools, in order to enhance the versatility of robots. When using tools to perform tasks, it is necessary to gain understanding regarding the properties of such tool. Recognizing parameters of the tool is therefore very important. Position and force information obtained by performing active sensing of the tool in relation to a flat environment potentially includes information on the properties of the tool. A method for a robot to automatically recognize characteristics of tools based solely on the position and force information was therefore proposed. Active sensing was performed to conduct an experiment for identifying the properties of the tool held by the robot. The experimental result led to the identification of the width of the tool, initial attitude error, as well as offset of force sensors. One of the future works is to confirm the effectiveness of the method by conducting an experiment that involves the use of the proposed method in skilled tool use control.

Acknowledgments

This work was supported in part by PRESTO grant from Japan Science and Technology Agency(JST).

References

[1] A. Stoytchev: "Behavior-grounded representation of tool affordances," *Proc. IEEE Int. Conf. on Robotics and Automation*, pp. 3060–3065, (2005).

[2] C. Nabeshima, Y. Kuniyoshi and M. Lungarella: "Adaptive Body Schema for Robotic Tool-use," *Advanced Robotics*, Vol. 20, No. 10, pp. 1105–1126, (2006).

[3] Charles C. Kemp and Aaron Edsinger: "Robot Manipulation of Human Tools: Autonomous Detection and Control of Task Relevant Features," *Proc. 5th IEEE International Conference on Development and Learning (ICDL-06)*, (2006).

[4] C. Nabeshima and Y. Kuniyoshi: "A Method for Sustaining Consistent Sensory-Motor Coordination under Body Property Changes Including Tool Grasp/Release," *Advanced Robotics*, Vol. 24, No. 5–6, pp. 687–717, (2010).

[5] M. T. Mason, and J. K. Salisbury :Robot Hands and the Mechanics of Manipulation, Cambridge, MA, MIT Press (1985).

[6] R. Matsuzaki, M. Kamibayashi, S. Sakaino, T. Tsuji: "Classification of a Hybrid Control System for Robotic Tool Use," *Proc. IEEE Int. Conf. on Mechatronics(ICM2013)*, pp. 712–717, (2013).

[7] N. Shimada, T. Yoshiaoka, K. Ohishi, M. Miyazaki: "Smooth touch control between position control and force control for industrial robots." *Proc. Annual Conf. on IEEE Industrial Electronics Society(IECON2010)*, pp. 1949–1954, (2010).

[8] V. Tinkhanoff, U. Pattacini, L. Natale, G. Metta: "Exploring affordances and tool use on the iCub," *Proc. IEEE Int. Conf. on Humanoid Robotics*, (2013).

[9] N. Mimura and Y. Funahashi: "Parameter Identification of Contact Conditions by Active Force Sensing," *Proc. IEEE Int. Conf. on Robotics and Automation*, Vol. 3, pp. 2645–2650, (1994).

[10] T. Yamada, A. Tanaka, M. Yamada, Y. Hunahashi, and H. Yamamoto: "Identification of Contact Conditions by Active Force Sensing - Estimated Parameter Uncertainty and Experimental Verification -," *Journal of Robotics and Mechatronics*, Vol. 23, No. 1, pp. 44–52, (2011).

[11] T. Tsuji, Y. Kaneko, and S. Abe: "Whole-body Force Sensation by Force Sensor with Shell-shaped End-effector," *IEEE Trans. on Industrial Electronics*, Vol. 56, No. 5, pp. 1375–1382, (2009).

[12] T. Tsuji, J. Arakawa: "Real-time personal identification based on haptic information," *Proc. IEEE Int. Conf. on Robotics and Automation*, pp. 1354–1359, (2011).

[13] N. Kurita, S. Sakaino, T. Tsuji: "Whole-Body Force Sensation by Force Sensor with End-Effector of Arbitrary Shape," *Proc. IEEE Int. Conf. on Intelligent Robots and Systems*, pp. 5428–5433, (2012).

[14] T. Murakami, F. Yu, K. Ohnishi :"Torque Sensorless Control in Multidegree-of-Freedom Manipulator," *IEEE Trans. on Ind. Electron.*, Vol. 40, No. 2, pp. 259–265, (1993).

[15] N. Hayashida, T. Yakoh, T. Murakami, and K. Ohnishi : "A Sensorless Bilateral Robot Manipulator Based on Twin Drive System," The Japan Society for Precision Engineering, Vol. 67, No. 11, pp. 1843–1838, (2001). (in Japanese)

[16] T. Tsuji, K. Ohnishi, A. Sabanovic: "A controller design method based on functionality," *IEEE Trans. Ind. Electron.*. Vol. 54, No. 6, pp. 3335–3343, (2007).

Recent Technical Trends in Magnetic Materials

Kiyoshi Wajima
Instrument System R&D Div., Process Research Labs.
Nippon steel & Sumitomo metal Corporation
Futtsu, Chiba, Japan
wajima.p7z.kiyoshi@jp.nssmsc.com

Yasuhiro Marukawa
Magnetic Materials Company Planning Dept.
Hitachi Metals, Ltd.
Shibaura Minato-ku, Tokyo, Japan
yasuhiro.marukawa.zv@hitachi-metals.com

Hiroaki Toda
Electrical Steel Research Dept., Steel Research Lab.
JFE Steel Corporation
Kurashiki, Okayama, Japan
h-toda@jfe-steel.co.jp

Chio Ishihara
Powder Metal Development Dept.
Hitachi Chemical Co., Ltd.
Matsudo, Chiba, Japan
c-ishihara@hitachi-chem.co.jp

Takashi Kosaka
Dept. of Computer Science and Engineering
Nagoya Institute of Technology
Nagoya, Japan
kosaka@nitech.ac.jp

Abstract— **Permanent magnets and magnetically soft materials are indispensable constituent materials for motors and expected to play an important role in realizing compact and high efficiency motors. This paper overviews recent technological trends in developments of N_dF_eB sintered magnets, ferrite magnets, electrical steel sheets and soft magnetic composites.**

Keywords— *permanet magnet motors, permanent magnets, electrical steel sheets, soft magnetic composites.*

I. INTRODUCTION

The requirement of global environmental protection and economical use of energy promotes developments of smaller and more efficient motors. For this reason, permanent magnet (PM) motor is widely used for compact and high efficiency motor such as traction motor of hybrid/electric vehicle since it realizes the best efficiency and torque density economically at this moment. Permanent magnets and magnetically soft materials are fundamental and important materials for improving drive performances in most of motors. Therefore, a survey of their recent technological developments and trends is reported in this paper.

Rare earth metals are essential elements for high performance permanent magnets. At first, their current situation from a viewpoint of material resource is overviewed and subsequently, the recent developments in NdFeB sintered magnets and Ferrite magnets are reviewed. In response to the requirements for reducing size and improving efficiency of motors, new Non-oriented (NO) electrical steel sheet products with improved properties are presented. High silicon steels for high frequency applications have been manufactured with practical process and their evaluation result as reactors

are demonstrated. The magnetic properties of improved soft magnetic composites are introduced. As an application example, 3-dimensional magnetic circuit motors are shown.

II. PERMANENT MAGNETS

A. Rare-earth Material Source

Triggered by the Senkaku Islands incident in September 2010, prices of rare-earth (RE) metals have soared and it has affected permanent magnet industry. This section reports the current situation regarding RE resources as well as the situation and issues of N_dF_eB sintered magnets and ferrite magnets.

United States was a major supplier of RE resources when N_dF_eB sintered magnets were invented in 1982 [1], however, after 2000, the permanent magnet industry was highly dependent on China for RE resources. Although RE prices have been essentially determined based on supply and demand relationship, they have soared due to change of Chinese government policy and the Senkaku Islands incident. As an example, a heavy RE, Dysprosium (D_y) price was more than \$3,000/kg at its peak time. Recently, the prices have been stabilized because of restart of production of RE resources outside of China as well as establishment of alternative technologies. However, the permanent magnet industry has been still dependent on China for D_y and therefore, it has been important to reduce D_y contents for N_dF_eB sintered magnets.

B. Recent Development of N_dFB Sintered Magnets

In order to elevate coercivity of N_dF_eB sintered magnets, generally, the heavy RE such as D_y and Terbium

(T_b) have been added to the magnets for their high temperature use. Recent research found that the coercivity of N_dF_eB sintered magnets was determined at grain boundaries so that the boundary diffusion process of D_y and T_b on surface of permanent magnet has been proposed and employed as one of methods for reduction of D_y while maintaining higher magnetic properties [2][3]. Fig. 1 has illustrated the magnetic properties when D_y boundary diffusion technology has been applied [4]. Moreover, current research has tended to be targeting to increase the coercivity without D_y. As an example, it has found that the coercivity could be elevated by segregating main phase that was strong magnetic phase by non magnetic grain boundaries. Coupled with effectiveness of small additives and improvement of manufacturing processes, less Dy materials reduced by 2% are proposed. Fig. 2 has shown the magnetic properties of "Low D_y Series" [4].

C. Recent Development of Ferrite Magnets

Magnetic properties of S_r-ferrite magnet were significantly improved by replacing a part of S_r and F_e with L_a and C_o. In addition, by adding C_a on L_a and C_o, the ferrite magnet with further improved properties ($C_aL_aC_o$-M type ferrite magnets) has been successfully commercialized. Fig. 3 has illustrated magnetic properties of various kinds of ferrite magnets. Furthermore, these high performance materials also have shown improvement of temperature coefficient compared with conventional materials. Fig. 4 has summarized temperature coefficient of conventional materials (NMF-6 Series) and high performance Ferrite

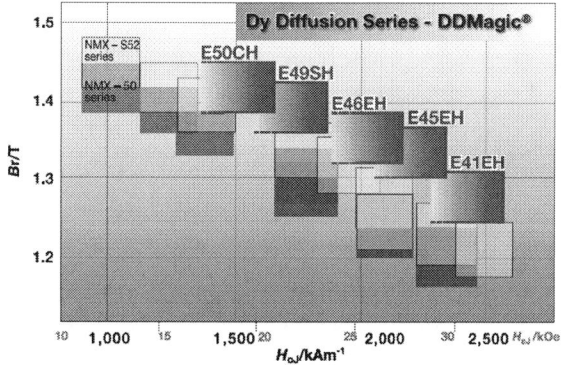

Fig. 1. Magnetic properties of D_y diffusion DDMagic®.

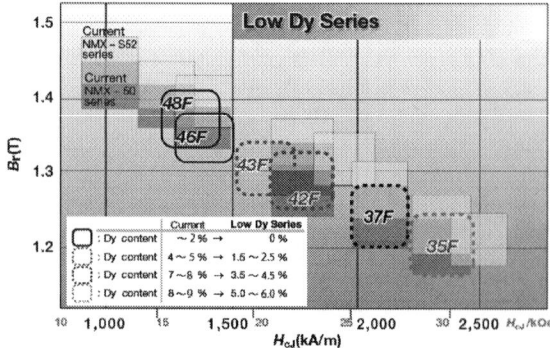

Fig. 2. Magnetic properties of Low D_y Series.

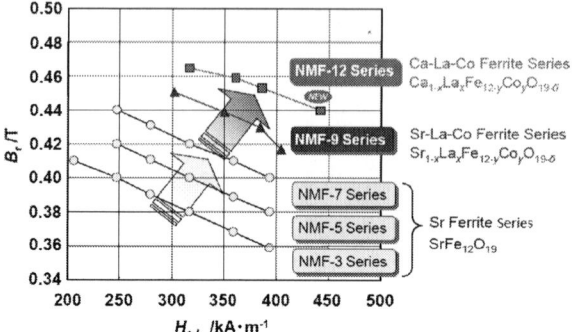

Fig. 3. Magnetic characteristics of various ferrite magnets.

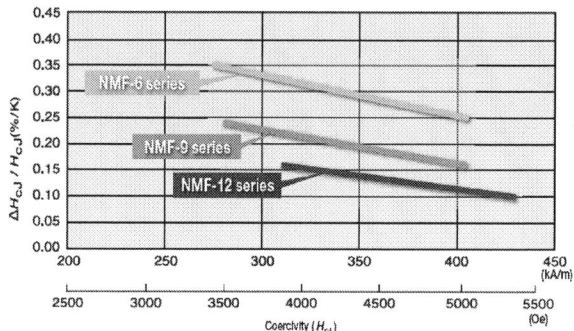

Fig. 4. Temperature coefficient of coercivity.

Magnet (NMF-9 Series, NMF-12 Series) [4]. Some applications where N_dF_eB sintered magnets has been used, have returned to reuse high performance ferrite magnets due to price elevation of RE resources.

III. ELECTRICAL STEEL SHEETS

A. Electrical Steel Sheets Suitable for Compact and High Efficiency Motors

Non-oriented electrical steel (NO) is widely used as motor cores since it economically meets requirements for size reduction and efficiency improvement of motors. Fig. 5 depicts the required properties of magnetically soft materials for the cores employed in hybrid/electric vehicle traction motors [5][6]. Low loss and high magnetizing properties in the most frequent operating range of the traction motors are essential for economical use of energy. As the traction motors need to have a high torque for starting, higher magnetic flux density is required. High induction properties of core material contribute not only high torque but also size reduction. In addition to that, high-processability at core stamping process is important because it makes the airgap between stator and rotor cores narrow, resulting in the increases in the magnetic flux density and the motor torque. As one of trends in motor design, it is well-known to expand the maximum operating speed because it enables to reduce the motor size. In this case, the motor cores are exposed to high frequency magnetic excitations and hence, their materials must have low iron loss properties at high frequency. The rotors of high speed motors also receive a large centrifugal force and therefore, the core materials must have also sufficient mechanical strength to endure it.

Fig. 6 illustrates magnetizing characteristics of electrical steel sheet and major influential factors for them. Since these factors also affect iron loss of electrical steel sheet, it is very crucial to optimize the balance of them to realize target properties. For example, silicon is a typical alloying element. When the percentage of constituent of silicon is increased, the eddy current loss is suppressed because electrical resistivity of steel sheet is increased. On the other hand, it deteriorates saturation flux density at the same time. Thus, electrical steel sheets for high efficiency motors are realized by controlling the content of additive elements and size/orientation of grains in a well-balanced manner.

The magnetic properties of the developed high-efficiency electrical steel sheets are shown in Fig. 7. These product series have the improved magnetic flux density B50 (the flux density under a magnetizing force of 5000A/m), in comparison with those of conventional grade NO series. These new series have higher B50 values at the same value of iron loss W10/400 (the iron loss under excitation of 1.0T at 400Hz). It implies that higher motor torque can be expected by applying these products.

Fig. 8 depicts the magnetic properties of a thin-gauge

electrical steel sheet series for high-frequency use, which is designed to have low iron loss under high-frequency excitation. The thin-gauge electrical steel sheet with 0.2mm thickness presents lower W10/400 values than those of conventional 0.35mm thickness products, while inhibiting the drop of its magnetic flux density B50 caused by its thin thickness.

Fig. 9 shows the magnetic and mechanical properties of a high tensile strength electrical steel sheet suitable for a rotor core used in high-speed machine. Its yield strength is twice as high as those of conventional JIS grade products or more, while suppressing the increase in the W10/400 value.

In response to the requirements of improved properties for reducing size and improving efficiency of motors, electrical steel sheet products have been being developed. To design motors of target performance economically, it is important to select the most suitable type of electrical steel sheet and utilize them with appropriate utilization technologies.

B. High Silicon Steel Sheet (6.5% Si Steel Sheet)

The demand for high frequency reactors has been expanding with the spread of power electronics technologies for electric power conversion. Essential properties of core material for high frequency reactors have been low iron loss at high frequency, high saturation flux density for downsizing and low magnetostriction for reducing

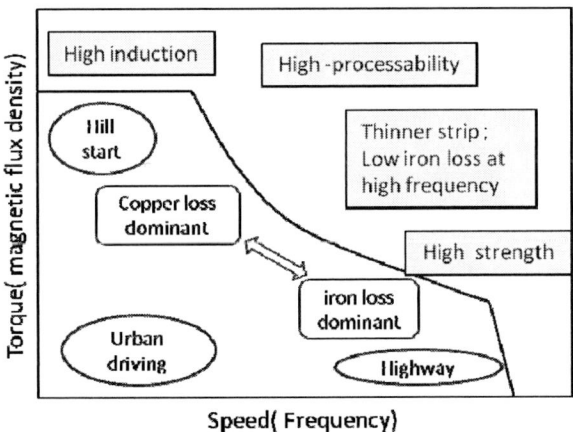

Fig. 5. Required properties of electrical steel in torque-speed characteristic of traction motor.

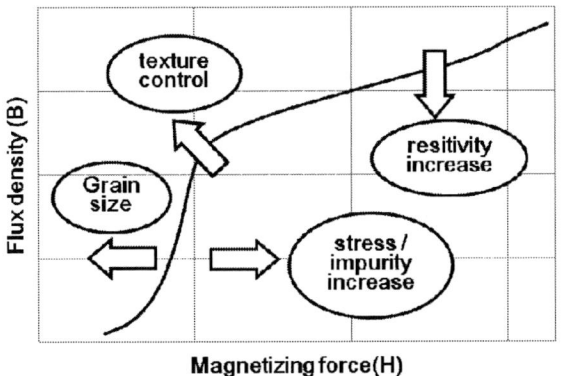

Fig. 6. Magnetizing characteristics of electrical steel sheet and principal influential factors.

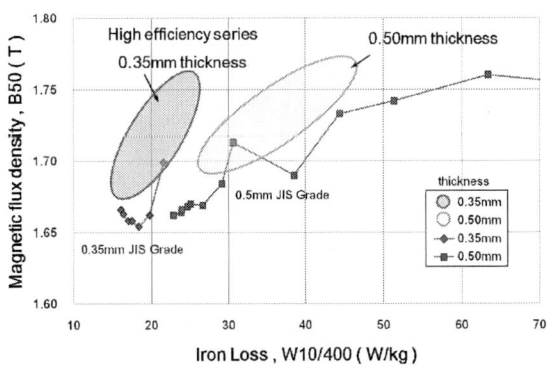

Fig. 7. Magnetic properties of high-efficiency electrical steel sheets.

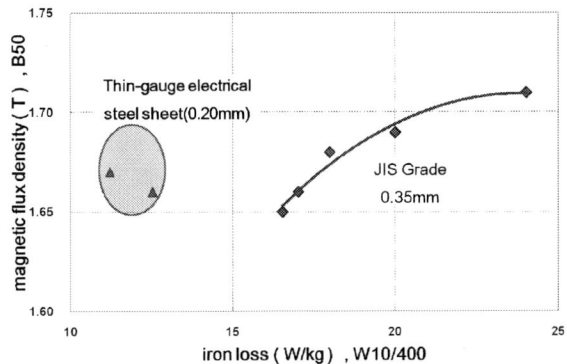

Fig. 8. Magnetic properties of thin gauge electrical steel sheets.

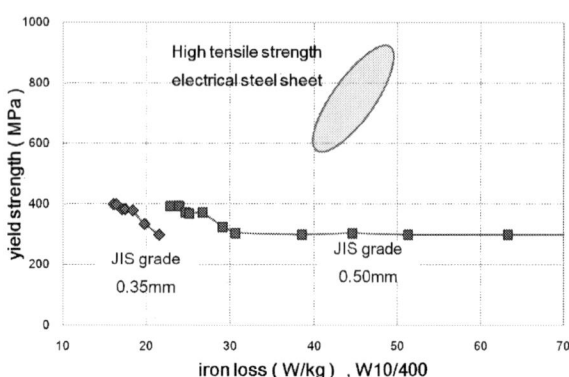

Fig. 9. Magnetic properties of high tension electrical steel sheets.

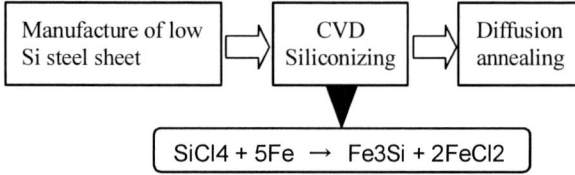

Fig. 10. Manufacturing process of 6.5% Si steel.

audible noise. 6.5% S_i steel has been considered to be the optimal material for satisfying the above-mentioned requirements. Adding S_i to electrical steel sheet has increased its resistivity and decreased its iron loss. It has been well known that its magnetic permeability has taken the maximum value and its magnetostriction reached nearly zero under S_i content of 6.5% since 1950s [7]. Although Sendust, Permalloy and C_o-based amorphous have been well known as low magnetostriction and high permeability material, their saturation magnetization has been low and therefore, the size of reactors has tended to be larger. On the other hand, 6.5% S_i steel has exhibited nearly zero magnetostriction, relatively high saturation magnetization and resistivity. Hence, it has been being considered to be an excellent soft magnetic material for high frequency applications.

However, the ductility of steel sheets decreases as S_i content increases. The material with S_i content more than 4% shows remarkable embrittlement making cold rolling difficult. For this reason, the production of high S_i steel sheet with more than 4% S_i content had been considered to be unsuitable for industrial use.

To solve the problem, the production technology for 6.5% S_i steel sheets using chemical vapor deposition (CVD) method has been developed [8]. Fig. 10 shows the principle of the manufacturing process. The cold-rolled low S_i steel sheet is prepared as the base material. S_i is deposited on the surface of the base material by the

chemical reaction between F_e and S_iC_{l4} gas in an inert gas atmosphere. Then, S_i diffuses into the interior of the steel sheet by high temperature annealing. 6.5% S_i steel sheets with 0.1mm thickness are commercially produced by this method. Furthermore, gradient high S_i steel having the S_i content distribution (gradient) in the sheet thickness direction has been developed [9]. As shown in Fig. 11, the gradient high S_i steel sheet has a concentration distribution pattern in which S_i concentration increases continuously from the sheet center to the surface layer. It has a 6.5% Si composition in the sheet surface layer.

Table I appears examples of the typical magnetic properties of 6.5% S_i steel sheet and gradient high S_i steel sheet [9]. The table also shows the typical magnetic properties of thin-gauge grain-oriented silicon steel sheet (thin-gauge GO) and F_e-based amorphous which is representative material for high frequency applications. The F_e-based amorphous shows the lowest iron loss over the entire frequency because the sheet thickness is extremely thin and its resistivity is high. However, since its saturation magnetization is low and the magnetostriction is large, this material has problems in terms of downsizing and audible noise. On the other hand, thin-gauge GO has high saturation magnetization, but its iron loss is higher than that of the 6.5% S_i steel sheet. In comparison with these materials, 6.5% Si steel and gradient high S_i steel present the excellent balance of iron loss and saturation magnetization being suitable for the high frequency reactor application. Fig. 12 demonstrates the comparison of iron losses between 6.5% Si steel sheet and gradient high S_i steel sheet with 0.1mm thickness. 6.5% Si steel displays lower iron losses at frequencies below 5 kHz, whereas gradient high Si steel shows excellent iron losses at higher frequencies higher than 5 kHz. The reason why is because the eddy current loss of gradient high Si steel is lower by optimizing Si distribution through the thickness. Therefore, 6.5% Si steel is mainly suitable for the applications where the switching frequency is several kHz or

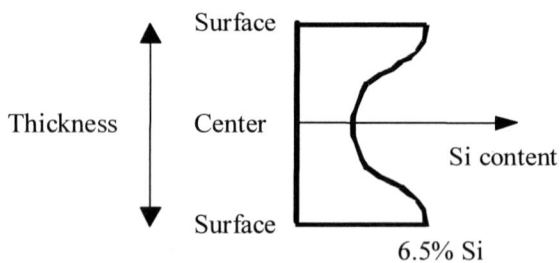

Fig. 11. S_i profile of gradient high S_i steel sheet.

TABLE I TYPICAL MAGNETIC PROPERTIES OF REPRESENTATIVE MATERIALS FOR HIGH FREQUENCY APPLICATIONS

Materials	Thickness (mm)	Saturation magnetization (T)	Iron loss W/kg					Magnetostriction at 400Hz, 1.0T ($\times 10^{-6}$)
			$W_{10/50}$	$W_{10/400}$	$W_{5/2k}$	$W_{1/10k}$	$W_{0.5/20k}$	
6.5%Si steel (JNEX Core)	0.10	1.8	0.5	5.7	11.3	8.3	6.9	0.1
Gradient high Si steel (JNHF Core)	0.10	1.9	1.1	10.1	11.2	7.1	5.0	—
Thin-gauge grain oriented 3%Si steel	0.10	2.0	0.7	6.4	20.0	18.0	14.0	−0.8
Fe-based amorphous	0.025	1.5	0.1	1.5	8.1	3.0	3.3	27.0

less. In contrast, gradient high *Si* steel is suitable for the applications with frequencies range of 10 kHz or high.

Fig. 13 shows the comparison of flux density dependency of audible noise between 6.5% *Si* steel, thin-gauge GO and F_e-based amorphous for high frequency reactor application. The thickness of 6.5% S_i steel and thin-gauge GO is 0.1mm and that of amorphous is 0.025mm. The reactors are energized using PWM voltage waveform with a fundamental frequency of 50 Hz and a carrier frequency of 16 kHz. The audible noise (A scale) of the reactors is measured at a position 10 cm far from the core. 6.5% *Si* steel exhibits an extremely low audible noise, reflecting better magnetostriction characteristic of the material. The results also show that the differences in noise between 6.5% *Si* steel and other materials increase as the flux density becomes higher. Therefore, it is evident that 6.5% *Si* steel sheet is the optimal material to achieve low audible noise and low loss for high frequency reactor application.

IV. SOFT MAGNETIC COMPOSITES (SMC)

Recently, reviews and applications of SMC, which is made of powdered metals (PM), have been being promoted for such as various types of sensors, actuators, motor cores and so forth as control devices and drive units. There seems to be two reasons as this background. Firstly, magnetic properties of soft magnetic materials have been improved drastically. Thanks to the technical improvement of various compacting techniques [10], the magnetic properties as same as ingot steel have been available. Secondly, automobiles have dramatically promoted the applications of high performance control units employing power electronics and motor drives. Fig. 14 depicts the conceptual figure of composite elements of SMC. It is produced by following procedures. At first, the surface of iron powder (nearly 100μm) is insulated with inorganic oxidative product. Secondly, it mixes a little organic binder. After that, it is compressed with high pressure and is processed by heat treatment under the temperature that does not destroy the inorganic insulator and the organic binder. Consequently, it has a good property in terms of low eddy current loss in high frequency applications. In general, iron loss is expressed as a sum of two factors; one is eddy current loss W_e and the other is hysteresis loss W_h as given in (1)[11].

$$W = W_e + W_h = \frac{k_e B_m^{\,2} t^2 f^2}{\rho} + k_h B_m^{\,1.6} f \qquad (1)$$

where, k_e, k_h are eddy current loss and hysteresis loss coefficients, f is frequency, B_m is magnetic flux density, ρ is specific resistivity, t is thickness of material, respectively. According to (1), as SMC can minimize powder size being equivalent of t, relatively low eddy current loss is realized.

Fig. 15 illustrates applicable range of SMC under AC magnetic field. SMC have some aspects which compensate no good region of the both magnetic materials such as the laminated steel and ferrite cores. Fig. 16 shows relations of the core loss and magnetic flux density between various SMC and laminated sheets. The developed SMC material shown in Fig. 16 has high magnetic flux

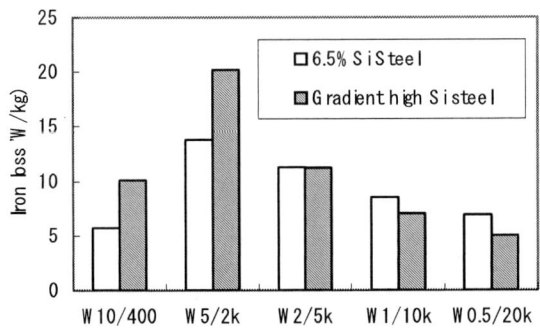

Fig. 12. Iron loss of 6.5% *Si* steel and gradient high *Si* steel.

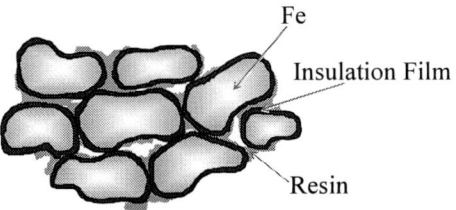

Fig. 14. Conceptual figure of composite elements of SMC.

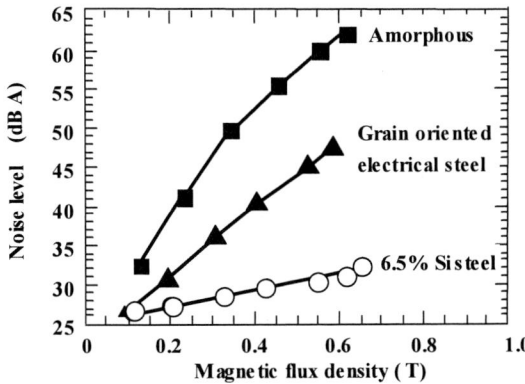

Fig. 13. The effect of core materials on audible noise for high frequency reactor application.

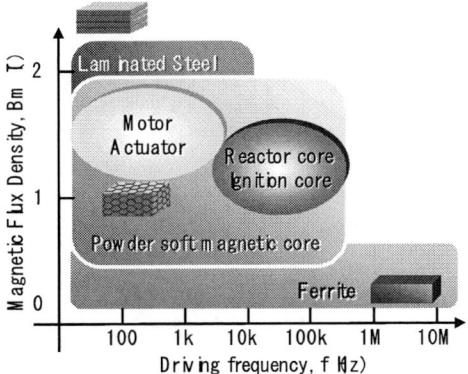

Fig. 15. Applicable range of soft magnetic material under AC magnetic field.

density and low core loss compared with other conventional SMC. Furthermore, it has almost the same magnetic properties in comparison with laminated sheet (0.35mm).

As one of its application examples, Fig. 17 demonstrates Claw-teeth motor® whose feature is a use of 3 dimensional magnetic circuits [12]. Fig. 18 shows another application example, axial-type motor employing SMC, which also has 3-dimensional magnetic circuit [13], is actively developed. It is expected that the applicable range of SMC will expand much more in near future.

V. CONCLUSIONS

This paper has presented some survey of recent technical trends in magnetic materials for power electronics and motor drive applications. The properties of permanent magnets and soft magnetic materials have been improved continuously, in response to pursuit of compact and high-efficient motors such as hybrid/electric vehicle traction motors. It is expected that PM motors accomplishes further development by these magnetic materials in cooperation with the utilization technologies that maximize the excellent performance of materials.

ACKNOWLEDGMENT

The development of Claw-teeth® motor is the collaborative work of Hitachi, Ltd. and Hitachi Industrial Equipment Systems Co., Ltd. We also express thanks to corporation of Höeganäs AB regarding on improvement of insulated iron powders for SMC.

REFERENCES

[1] Japan Patent, No. 1431617
[2] H. Nakamura, K. Hirota, M. Shimao, T. Minowa, and M. Honshima, "Magnetic properties of extremely small NdFeB sintered magnets", *IEEE Trans. on Mag.* Vol. 41, No.10, pp. 3844-3846 (2005)
[3] Japan Patent, No. 4241900
[4] Hitachi-Metals, Ltd Catalog, No. HG-A27-A (Dec. 2013)
[5] C. Kaido, "Latest technologies of electrical steel sheet-Development of electric automobile and materials," *CMC*, pp. 120-129 (1999)
[6] M. Yabumoto, C. Kaido, T. Wakisaka, T. Kubota and N. Suzuki, "Electrical steel sheet for traction motors of hybrid/electric vehicles," *Nippon STEEL TECHNICAL REPORT*, No. 87, pp. 57-61 (2003)
[7] R. M. Bozorth, "Ferromagnetism", D. Van Nostrand Co. Inc. (1951)
[8] T. Yamaji, M. Abe, Y. Takada, K. Okada, and T. Hiratani, "Magnetic properties and workability of 6.5% silicon steel sheet manufactured in continuous CVD siliconizing line," *J. Magn. Magn. Mater*, Vol. 133, pp.187-189 (1994)
[9] M. Namikawa, H. Ninomiya, and T. Yamaji, "High Silicon Steel Sheets Realizing Excellent High Frequency Reactor Performance," *JFE TECHNICAL REPORT*, No.6, pp.12-17 (Oct. 2005)
[10] Saito, Iwakiri, Kagaya, "The properties of sintered material by high density method combined with compaction by using die wall lubricant and warm compacting," *Hitachi powdered metal technical report*, Vol.2, pp. 28-33 (2003)
[11] Kato, "magnetism and magnetic materials for engineer," *NIKKAN KOUGYOU SHINBUN*, p.67 (1991)
[12] Inagaki, Ishihara, Enomoto, Ito, "Evaluation result of motor properties with SMC core (No.4)," Abstract of autumn presen-tation (2009) of Japan Society of Powder and Powder Metallurgy, p.105 (2009)
[13] S. Ogasawara, M. Takemoto, "Development of ferrite magnet motor for hybrid automobile without rare-earth material", Innovation Japan (2010)

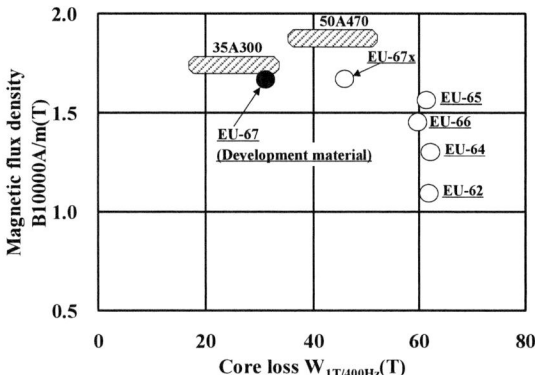

Fig. 16. Core loss and magnetic flux density between various SMC and laminated sheets.

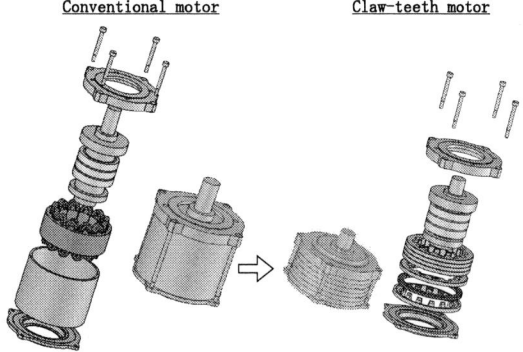

Fig. 17. Comparison of motor structure: conventional type and Claw-teeth type employing SMC core.
(Claw-teeth® is registered brand of Hitachi, Ltd.)

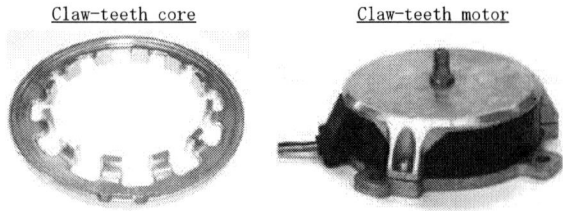

Fig. 18. Photographs of Claw-teeth® core and Claw-teeth® motor.

The 2014 International Power Electronics Conference

Multi-Domain Co-simulation with Numerically Identified PMSM interworking at HILS for Electric Propulsion

Gyeong-Jae Park*, Hochang Jung[†], Yong-Jae Kim[‡], Sang-Yong Jung*

*School of Electronic and Electrical Engineering
Sungkyunkwan University, Suwon 440-746, Korea
syjung@ece.skku.ac.kr
[†]Diesel/Hybrid Research and Development Center
Korea Automotive Technology Institute, Cheonan 330-912, Korea
[‡]Department of Electrical Engineering
Chosun University, Gwangju 501-759, Korea

Abstract— **In this paper, design and analysis of interior permanent magnet synchronous motor for electric vehicle bus propulsion is performed based on Finite Element Method. As the increase of the advanced application which requires precise control, the inverter is regarded as requisite for controlling electric machines. In addition multi-domain co-simulation is important to consider the inter-linking effect of the system as well as the conventional sinusoidal input analysis to investigate the characteristics of electric machines, specifically. Therefore, Multi-domain co-simulation is performed to consider the effect of inverter current, containing time harmonics, to the characteristics of interior permanent magnet synchronous motor in system circumstance. Simulated result is validated by comparing it with experimental result. Lastly, virtual driving test using Hardware-in-the-loop simulation is performed with co-simulated data to verify the reliability of multi-domain co-simulation model in virtual driving system.**

Keywords—Multi-domain, Co-simulation, Hardware-in-the-Loop Simulation, Electric propulsion.

I. INTRODUCTION

Axle-mounted motor of EV bus requires properties such as high torque, wide operating speed range, high torque density, high efficiency and durability. Interior Permanent Magnet Synchronous Motor (IPMSM) is the proper type to fulfill requirements. IPMSM is mostly utilized as the motor for eco-friendly vehicle on the virtue of its high power density, efficiency and high torque characteristic. Furthermore, concentrated winding application guarantees higher torque and wider speed range operation in various industry systems compared to another winding types[4].

As mentioned, the motor for propulsion system is required to be controlled in variable speed and wide operating speed range. Therefore using inverter, which can control motor in certain intended condition by vector control method, is regarded as requisite. As a side effect, time harmonics occurred by switching is contained on the current flows into the motor. Thus, time harmonics should be considered for the accurate motor analysis in the system. Multi-domain co-simulation has advantage on

considering interworking effect, such as time harmonics of current, between parts of system.

For co-simulation, two models can be considered, the Finite element method (FEM) model and the map-based equivalent model. The first method using FEM motor model connects the motor model to inverter circuit directly. In this case, instantaneous current is produced considering the load condition (motor) under certain simulation time and the effect of time harmonics and nonlinearity of motor model is considered. This method is the most precise method to consider motor-inverter linking effect, but takes an excessive analysis time. The second one using map-based equivalent model is the look-up table model that contains the performance data of motor according to the rotation angle, the magnitude and the phase angle of input current analyzed by FEM. Since the result according to input signal and rotating angle comes out from the database, this model guarantees rapid response. Accuracy of this model is highly affected by the intervals of input signal to compose look-up table and the omitted values between the analyzed one is approximated by interpolation method. Hence, for the analysis which requires accuracy such as characteristic analysis, FEM model is used. For the analysis that connects the motor to the inverter and the thorough system, map-based equivalent model is used owing to its rapid response according to input signal.

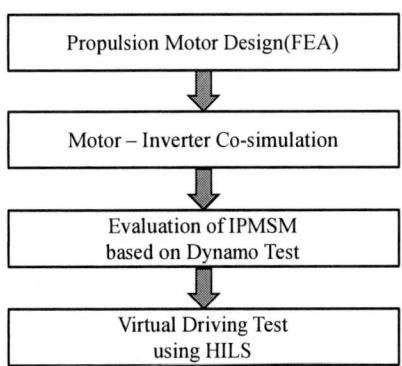

Fig. 1. Multi-domain co-simulation procedure

978-1-4799-2706-7/14 $31.00 © 2014 IEEE

In this paper, design and characteristic analysis of IPMSM for electric propulsion system is conducted with conventional method using sinusoidal input signal. Then, characteristic analysis of IPMSM is performed through multi-domain co-simulation with FEM motor model. Its result is compared to the one by conventional analysis method and the prototype model. Lastly using map-based equivalent motor model for co-simulation, Hardware-in-the-Loop Simulation (HILS) is performed to validate the proper operation of motor in system circumstance.

II. ELECTROMAGNETIC CHARACTERISTIC AND DESIGN OF IPMSM FOR ELECTRIC PROPULSION SYSTEM

A. Electromagnetic characteristic of IPMSM

Since permanent magnet is adapted instead of field winding, Permanent Magnet Synchronous Motor (PMSM) produces a field flux in constant magnitude and there is no copper loss. Therefore, PMSM guarantees high efficiency, high power density and rapid response. In case of IPMSM, permanent magnet is located at the inside of rotor core. Relative permeability of permanent magnet is close to that of air. Hence, reluctance of the flux path through the permanent magnet gets greater, which increases the difference of reluctance. This gives rise to the saliency of the motor to have advantage on the torque density, producing reluctance torque by reluctance difference as well as alignment torque by the field and armature flux.

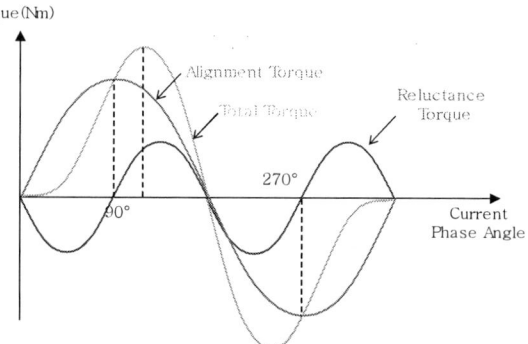

Torque (Nm)

Fig. 2. Torque Characteristic of IPMSM

For the designed model, difference between inductances of d-axis and q-axis get greater as the layers of internal magnet is increasing which leads to enhancing saliency (reluctance torque). Flux linkage in d-q axis is described as (1) and the conventional voltage and torque equations of IPMSM in d-q reference are shown in (2) and (3). The torque equation can be separated into alignment torque by field and armature flux and reluctance torque by reluctance difference between d- and q-axis.

$$\lambda_{ds} = L_{ds} i_{ds} + \lambda_f$$
$$\lambda_{qs} = L_{qs} i_{qs}$$
(1)

$$V_{ds} = R_s i_{ds} + L_{ds} \frac{di_{ds}}{dt} - \omega_r L_{qs} i_{qs}$$

$$V_{qs} = R_s i_{qs} + L_{qs} \frac{di_{qs}}{dt} + \omega_r (L_{ds} i_{ds} + \lambda_f)$$
(2)

$$T_e = \frac{3}{2} \frac{P}{2} (\lambda_{ds} i_{qs} - \lambda_{qs} i_{ds})$$

$$= \frac{3}{2} \frac{P}{2} [\lambda_f i_{qs} + (L_{ds} - L_{ds}) i_{ds} i_{qs}]$$
(3)

Where, P is the number of poles, i_{ds} and i_{qs} are current in d-q reference, R_s is phase resistance and ω_r is angular velocity. λ_f is linkage flux by permanent magnet and λ_{ds}, λ_{qs} are linkage flux obtained by non-linear FEM. The first term of the torque at (3) is the alignment torque by the effect of magnet and armature flux and the second term is about the reluctance torque caused by the difference of the reluctance.

The characteristics region of IPMSM can be separated into three regions according to the control strategies under voltage and current constraints. In the first region, constant torque region, maximum torque is retained as a constant value until the base speed. To produce constant maximum torque, constant current is applied and voltage is increased proportional to the rotating speed of IPMSM. For effective operation, maximum torque per ampere (MTPA) control method is used to produce more torque with the limited current.

In the second region, constant power region, output power is remain constant using flux weakening control. As the rotating speed increases, Back Electro-Motive Force (BEMF) is increased. Since voltage has its constraint, BEMF is matched with voltage constraint on the base point and output torque is decreased dramatically after base point. Using flux weakening control, flux contributes to decrease BEMF by applying negative flux to the magnet by adjusting the phase angle of current. This leads to wider speed range of motor while satisfying the voltage constraint.

In the third region, characteristic region, maximum torque per voltage (MTPV) control produces higher torque than flux weakening control with equal voltage. Current is reduced while the voltage is maintaining, thus, power is reduced as well.

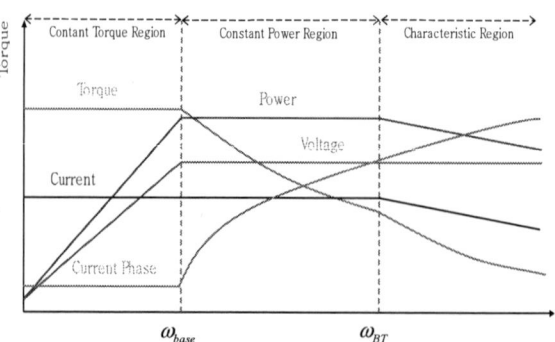

Fig. 3. Torque-speed characteristic of IPMSM.

B. Design of IPMSM using FEM

Specification of the motor mounted on the electric vehicle should be considered for performance requirement and spatial constraint condition. Particularly for the case of axle-mounted system, spatial constraint condition is harshly strict. Therefore tooth concentrated winding, which guarantees identical magneto motive force with less winding, is applied. 8 poles and 36 slots are selected for stator and rotor is designed to produce maximum torque per 2-Dimensional area with V-shape double layered magnet. Specification and shape of designed motor are shown in TABLE I and Fig. 4.

TABLE I
SPECIFICATION OF INTERIOR PERMANENT MAGNET SYNCHRONOUS
MOTOR

Section	Design Parameters	Specification	Unit
Performance	Torque	144	[Nm]
	Speed	4000	[r/min]
General	No. of Pole and Slot	8 / 36	
	No. of Phase	3	
	Air-Gap	1	[mm]
Stator	Outer Diameter	320	[mm]
	Core Material	Silicon Steel	
Rotor	Permanent Magnet Property	NdFeB (Br=1.16~1.19)	[T]

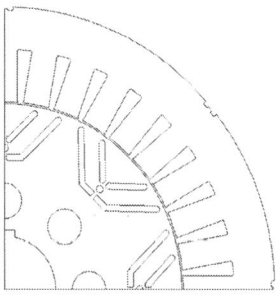 

Fig 4. Configuration of IPMSM

When the current to satisfy the performance is applied at the rated point, the result is shown as Fig. 5.

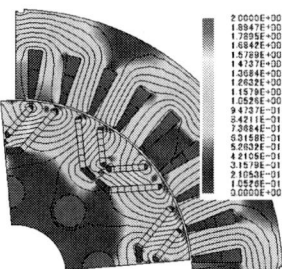 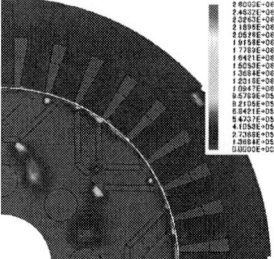

(a) Magnetic Flux Density (b) Torque

Fig 5. Characteristics of designed IPMSM

C. Iron Loss calculation of IPMSM

For accurate characteristic analysis of IPMSM, reliable loss should be calculated by numerical analysis. To consider non-linearity of magnetic steel sheet, FEM is performed for loss calculation. Particularly flux wave locally distorted in rotor is separated as a n-th order harmonics in current and total iron loss is calculated by summation of losses by n-th order harmonics of current, as shown in (3). Also, eddy current loss of permanent magnet, W_{PM}, is calculated by the n-th order component of eddy current density as shown in (4).

$$W_i = \int_{iron} \sum_n K_e D \cdot (nf)^2 \cdot \{B_{r,n}^2 + B_{\theta,n}^2\} dv$$
$$+ \int_{iron} \sum_n K_h D \cdot (nf) \cdot \{B_{r,n}^2 + B_{\theta,n}^2\} dv \quad (3)$$

$$W_{PM} = \sum_n \left(\int_{mag} \frac{|J_n|^2}{2\sigma} dv \right) \quad (4)$$

where, f is frequency, K_e, K_h are coefficients of eddy current loss and hysteresis loss. D is density of magnetic steel sheet, σ is conductivity of permanent magnet and $B_{r,n}$, $B_{\theta,n}$ are radial and circumferential magnetic flux density by n-th order harmonics.

The model considered in this paper operates at high speed. Thus iron loss, highly affected by machine's rotating speed, is relatively higher than copper loss. Furthermore, when the current containing time harmonics is applied to motor, iron loss increases even more owing to the n- factor of the time harmonics in (3).

III. INVERTER-IPMSM CO-SIMULATION

A. Inverter- IPMSM co-simulation model

There are two motor models for multi-domain co-simulation, FEM model and map-based equivalent model. For the case of FEM model, motor model is directly connected to the inverter circuit. This gives rise to the accurate analysis on the instantaneous effect of input current, owing to time harmonics, to the motor through FEM analysis while consuming excessive simulation time. Co-simulation model using FEM motor model is shown in Fig. 6. DC voltage from the battery is supplied to the motor by inverter switching with space vector pulse width modulation. Corresponding current flows into the motor and the performance and characteristic analysis of IPMSM by the input current is performed with FEM.

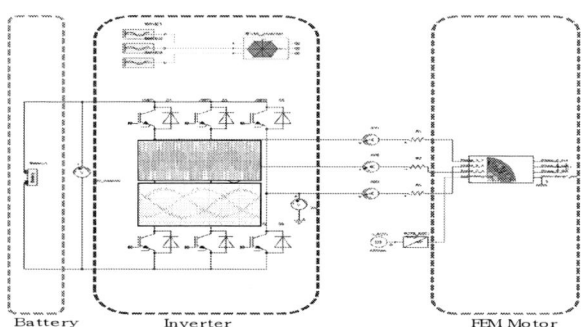

Fig. 6. Co-simulation model using FEM motor model

Map-based equivalent model is a data map of input signal and output performance. Data is obtained as the form of the control parameter by FEM analysis according to the magnitude and the phase angle of input current and the rotation angle. Map-based equivalent model consists of these data as a form of map and put out the control parameter according to the input signal. Accuracy of this model is highly depends on the data interval when initially mapping the data and interpolation is applied to the point between the data. Thus, this analysis is generally less accurate but faster calculate time than FEM model analysis as long as it has omitted value between mapping information. Owing to its rapid response to the input signal, the performance variation due to the input signal can be observed and not only the characteristic of inverter-motor simulation model but also the characteristic of the system that applies inverter-motor application can be investigated. The Map-based equivalent model which consists of the parameters according to the magnitude and phase angle of input current is shown in Fig. 7.

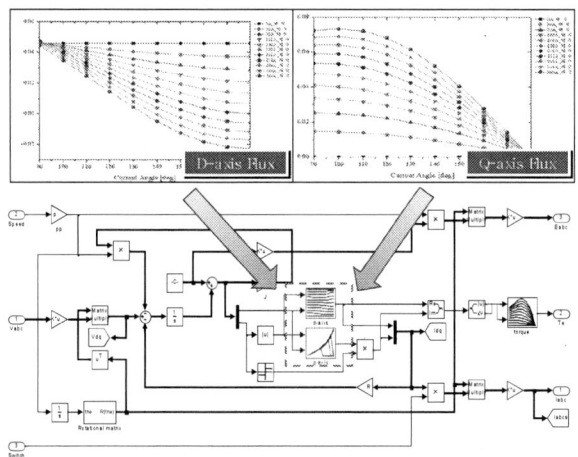

Fig. 7. Map-based equivalent model

B. Vector-Control of IPMSM

For IPMSM for EV propulsion system, vector control by inverter is regarded as requisite to enable variable speed control in wide speed range. For accurate vector control, inverter sensors the position of the rotor magnet and reflect it to the operation algorithm.

Vector control is the control method driving motor considering current and voltage constraints as shown in Fig 8. Voltage and current constraints are calculated as (5)-(7).

$$V_{ds}^2 + V_{qs}^2 \leq V_{smax}^2 \tag{5}$$

$$(\omega_e L_d)^2 (i_{ds} + \frac{\lambda_f}{L_d})^2 + (\omega_e L_q)^2 i_{qs}^2 \leq V_{smax}^2 \tag{6}$$

$$I_{ds}^2 + I_{qs}^2 \leq I_{smax}^2 \tag{7}$$

where, $V_{ds}, V_{qs}, I_{ds}, I_{qs}$ are voltage and current.

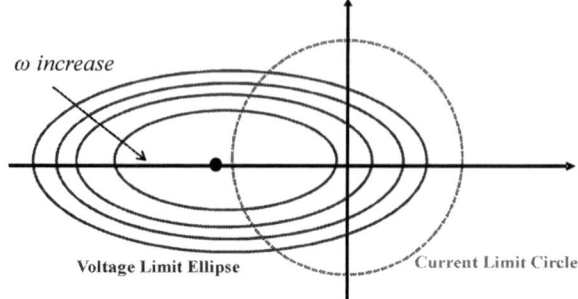

Fig. 8. Current and voltage constraints of IPMSM

In low speed region, torque is only controlled by current since voltage constraint ellipse is large enough. Motor is controlled at the point of contact between torque curve and current constraint circle which can produce maximum torque per ampere (MTPA). As the speed increases, voltage constraint ellipse decreases according to (5)-(6). If controlling motor at MTPA, torque decreases rapidly. Therefore applying flux weakening control which is changing phase of current in d-q reference decreases reduction of torque. The operating condition of IPMSM is shown in Fig. 9.

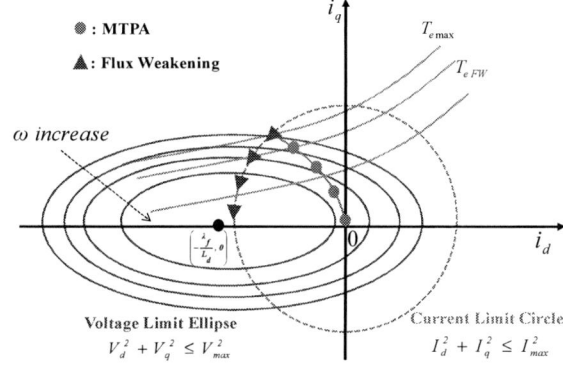

Fig. 9. Operating condition of IPMSM

C. Space Vector Pulse Width Modulation(SVPWM)

Inverter used for EV propulsion motor adopts SVPWM to control torque. SVPWM is the methodology converting 3-phase reference voltage into space vector on the complex domain. The concept of SVPWM is shown in Fig. 10.

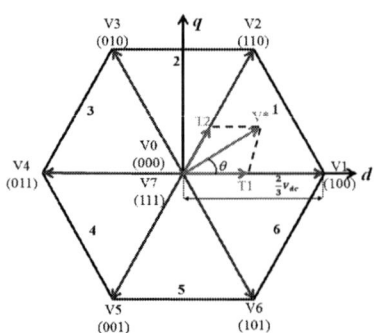

Fig 10. Conceptual diagram of SVPWM

The 2014 International Power Electronics Conference

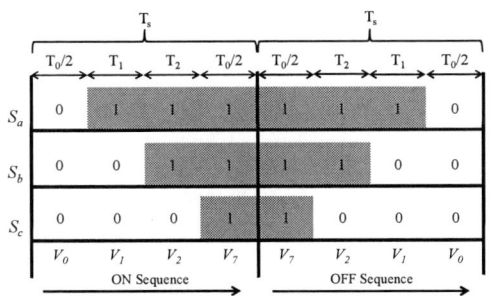

Fig. 11. Applying time control of SVPWM

SVPWM controls the applied time of reference voltage to make average value of reference voltage in time period to be identical with intended voltage as shown in (5) and Fig. 11.

$$\int_0^{T_s} V^* dt = \int_0^{T_1} V_1 dt + \int_{T_1}^{T_1+T_2} V_2 dt + \int_{T_1+T_2}^{T_s} V_{0,7} dt \quad (5)$$

D. Result analysis of IPMSM-inverter co-simulation

Numerical investigation on the adaptively built-up input topology has been performed, and its results are emphasized in terms of running performance compatibility between the steady-state current source and the time-harmonic voltage source. As shown in Fig. 13, electric machine fed by inverter proving the voltage excitation with the time-harmonics has the exclusive characteristics in terms of torque and current harmonics, which are quite different to the results from the steady-state ideal current excitation.

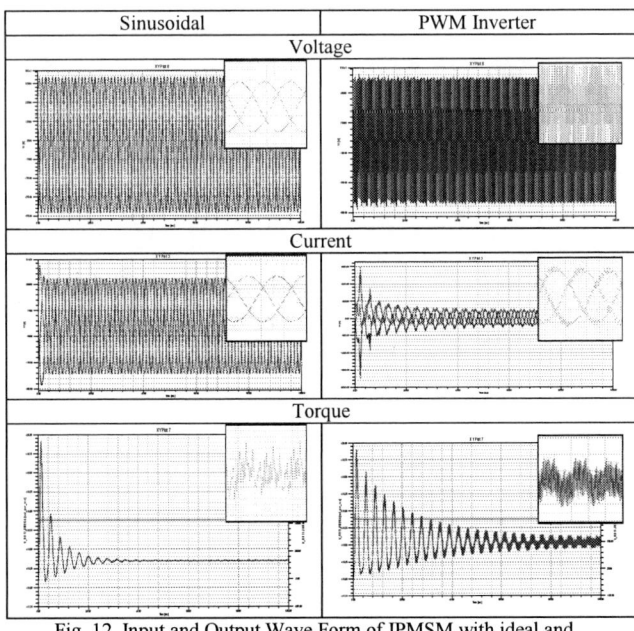

Fig. 12. Input and Output Wave Form of IPMSM with ideal and inverter-fed current

The average torque of motor when using inverter has smaller than the one using ideal sinusoidal input current owing to the voltage drop on the switch in the inverter and the harmonics of the current.

Flux density and iron loss distributions are shown in

Fig. 13. It is observed that the flux density of IPMSM is affected by harmonics when co-simulation considering time harmonics is conducted. When co-simulation is conducted, not the all point shows the higher flux density. However the variation of flux density is more frequent and severe. This gives rises to the fluctuation of air gap flux and the more iron loss is produced as a result. It is observed that iron loss of co-simulated model has higher distribution than that of conventional sinusoidal analysis in Fig. 13.

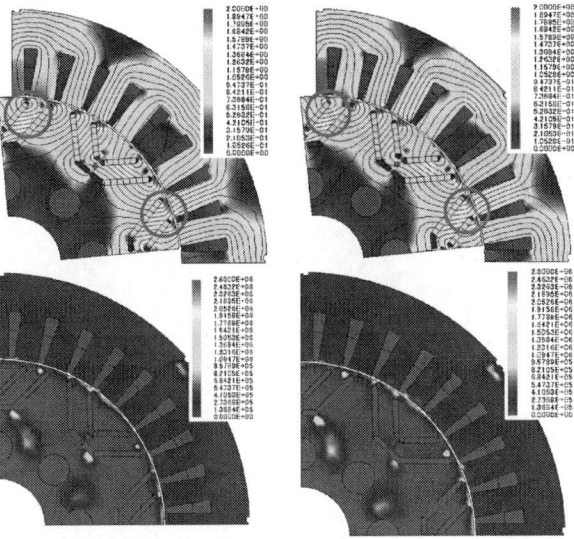

(a) Conventional   (b) Co-simulation

Fig. 13. Flux density distribution and Iron loss density of IPMSM with conventional and co-simulation analysis

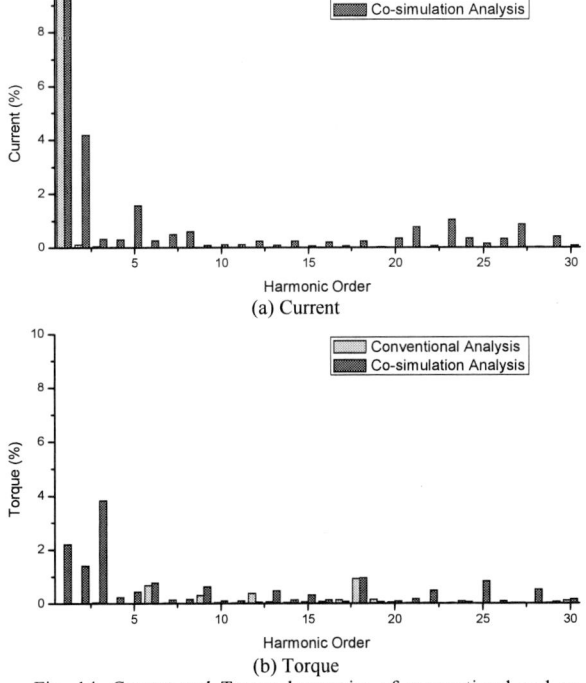

(a) Current

(b) Torque

Fig. 14. Current and Torque harmonics of conventional and co-simulated model

978-1-4799-2706-7/14 $31.00 © 2014 IEEE

As a result of the harmonics analysis in Fig. 14, current from the inverter contains time harmonics. All harmonic components are manifested in percent of the fundamental component which is the 1st harmonics for the current and the DC component for the torque. The multiples of 6th-order harmonic component are produced by the configuration and input condition. As shown in Fig. 14, other than the multiples of 6th harmonic component have been increased when co-simulation conducted. It can be interpreted that those harmonics of torque are caused by the time harmonics from the current.

IV. COMPARATIVE ANALYSIS OF SIMULATED AND EXPERIMENTED RESULTS

As shown in Fig. 15, experiment is performed with dynamometer and designed motor.

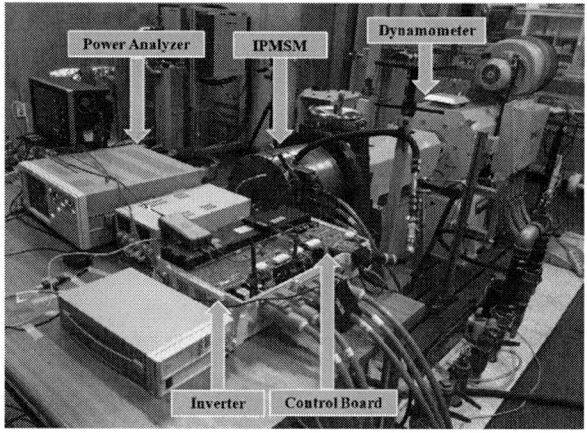

Fig. 15. Layout of dynamometer experiment

Co-simulated and experimentally obtained data is shown in Table II. Since data obtained from the experiment has wide measuring time interval and intervening element such as filters owing to its experimental circumstance, obtained experimental torque ripple is difficult to compare with the one from the simulation. The average torque and experimental efficiency of the motor are compared with the one from the simulation instead to validate the result of the co-simulation, because the effect of the harmonics to the efficiency has been investigated through the co-simulation result.

TABLE II
SIMULATED AND EXPERIMENTAL RESULTS OF IPMSM

Section	Conventional Simulation	Co-simulation	Experiment
Speed	4000 rpm	4000 rpm	4000 rpm
Avg. Torque	148.3 Nm	147.3 Nm	146.7 Nm
Total Loss	1533 W	1767 W	4982 W
Motor Efficiency	97.6 %	97.2 %	95.3 %

The comparison between the simulations and experimental results is shown in Table II. As a result, measured torque and efficiency of the experimental result has close value to the simulation result and the co-simulation result has closer value than the conventional simulation result. It emphasizes that the co-simulation result is accurate. The efficiency from the conventional simulation and the co-simulation has difference owing to the harmonics as explained above.

V. VIRTUAL DRIVING TEST USING HILS

HILS is simulation method that conducted to validate the performance of single part or the whole system before assembling. Therefore it has advantage on lower cost and shorter development time than actual assembly test relatively. Especially for the industrial applications, such as automotive, using various ECUs and modules, it is generalized to perform HILS instead of assembling parts and test, which leads to enhancing reliability, moving up the release point and reducing budget of development.

The virtual driving test using HILS is performed replacing the performance of motor and inverter by data from multi-domain co-simulation. This test can consider various driving circumstances, system efficiency and fuel efficiency as well.

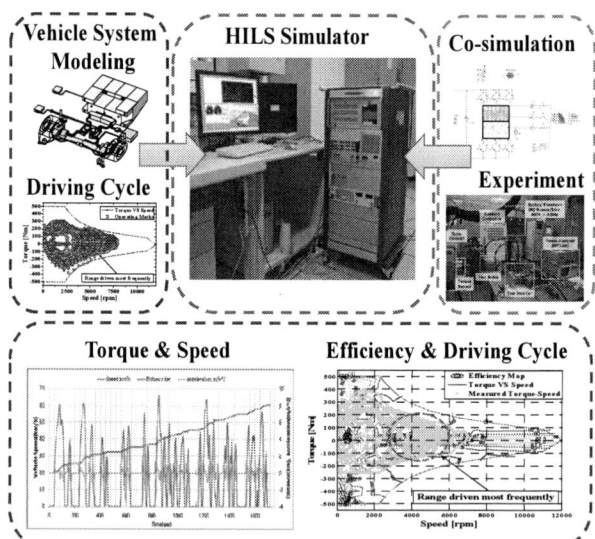

Fig. 16. Virtual driving test using HILS considering driving circumstances

As shown in Fig. 16, operation of co-simulation model is validated in propulsion system through HILS. The system connects to the motor demands the torque based on the accumulated torque reference by the driving test, and the motor is required to produce demanded torque. The performance and efficiency of the motor has been compared to the result of the driving test. As a result, it is observed that the result from the HILS test is analogous to the one from the driving test. It emphasizes that the designed co-simulation model of motor can operate properly in system circumstance.

The 2014 International Power Electronics Conference

VI. CONCLUSION

In this paper, axle-mounted IPMSM for EV bus is designed and analyzed with FEM. From the designed model, control variables are obtained and the map of inductance in various magnitudes and phases of current is composed using those variables. Then, inverter is employed to the map-based IPMSM and co-simulation is performed. The co-simulation result contains the effect of time harmonics compared to ideal current source. However, it is highly similar to the experimented result with slight difference by mechanical loss, which is not taken account in co-simulation circumstance, and it validates the reliability of multi-domain co-simulation. Lastly, virtual driving test using HILS is performed with co-simulation model, which proves the reliability of co-simulation model in virtually assembled vehicle system.

ACKNOWLEDGMENT

This work was supported by the Basic Science Research Program through the National Research Foundation of Korea (NRF) funded by the Ministry of Education, Science and Technology (NRF-2013R1A1A1A05011966) and the Human Resources Development (No. 20114030200030) of the Korea Institute of Energy Technology Evaluation and Planning (KETEP) grant funded by the Korea government Ministry of Trade, industry & Energy.

REFERENCES

[1] G. Ugalde, G. Almandoz, J. Poza, A. Gonzalez, "Computation of iron losses in permanent magnet machines by multi-domain simulations," *Power Electronics and Applications, 2009. EPE '09. 13th European Conference on*, pp. 1-10, 1989.

[2] Chang-Sun Yoo, Young-shin Kang, Bum-Jin Park, "Hardware-In-the-Loop simulation test for actuator control system of Smart UAV," *Control Automation and Systems (ICCAS), 2010 International Conference on*, pp. 1729-1732, 2010.

[3] B. Tabbache, Y. Aboub, K. Marouani, A. Kheloui, M.E.H. Benbouzid, "A simple and effective hardware-in-the-loop simulation platform for urban electric vehicles," *Renewable Energies and Vehicular Technology (REVET), 2012 First International Conference on*, pp. 251-255, 2012.

[4] Y. Honda, T. Nakamura, T. Higaki, Y. Takeda, "Motor design considerations and test results of an interior permanent magnet synchronous motor for electric vehicles," *Industry Applications Conference, 1997. Thirty-Second IAS Annual Meeting, IAS '97., Conference Record of the 1997 IEEE*, vol 1. pp. 75-82, 1997

[5] T. Murata, U. Kawatsu, J. Tamura, T. Tsuchiya, "Modeling and simulation technique of two quadrant chopper and PWM inverter-fed IPMSM drive system and its application to Hybrid Vehicle," *Electrical Machines and Systems (ICEMS), 2010 International Conference on*, pp. 706-711, 2010

[6] H. Jung, D. Kim, C.-B. Lee, J. Ahn, S.-Y. Jung, "Numerical and Experimental Design Validation for Adaptive Efficiency Distribution Compatible to Frequent Operating Range of IPMSM," *Magnetics, IEEE Transactions on*. Vol. 50 Issue 2, 2014

978-1-4799-2706-7/14 $31.00 © 2014 IEEE

Recent Technical Trends in PMSM

Shigeo Morimoto*, Yoshinari Asano**, Takashi Kosaka*** and Yuji Enomoto****

*Osaka Prefecture University, Sakai, Japan
**Daikin Industries, Ltd. Environmental Technology Laboratory, Kusatsu, Japan
***Nagoya Institute of Technology, Nagoya, Japan
****Hitachi, Ltd. Hitachi Research Laboratory, Hitachi, Japan

Abstract— This paper overviews the recent technical trends in permanent magnet synchronous motors (PMSMs), especially the rare-earth less PMSMs in Japan. The rare-earth less PMSMs including the PMSMs with reduced rare-earth PMs, permanent magnet assisted synchronous motors with ferrite magnets and axial gap PMSMs with ferrite magnets are reviewed.

Keywords— *permanent magnet syanchronous motor, rare-earth less motor, axial gap PMSM, PMASynRM*

I. INTRODUCTION

Permanent magnet synchronous motor (PMSM) is widely used because it has many advantages, such as maintenance-free operation, high controllability, robustness against the environment, high efficiency and high power factor operation. Recently, there is a great demand for energy saving because of environmental problem, and thus the highly efficient motors are demanded in various fields. The developments of high performance rare-earth permanent magnets (usually NdFeB) in addition to such environmental problem have expanded the application fields of PMSMs such as the FA, the home appliances, and the automobiles [1]. Especially, the requirements of electric motors for traction applications such as hybrid electric vehicles (HEVs) and electric vehicles (EVs) include high torque density, high power density, a wide constant power speed range, and high efficiency. In such applications, PMSMs with rare-earth PMs are usually employed.

However, rare-earth PMs are high in cost and there is

concern about the stable supply of rare-earth materials. Therefore, their use should be reduced, and thus the most recent research is development of PMSMs with less or no rare-earth permanent magnet material. This paper reviews the recent technical trends in PMSM, especially the rare-earth less PMSM in Japan.

II. REDUCTION OF RARE-EARTH MATERIAL IN RARE-EARTH PMSMs

A. Reduction of Dysprosium

The heavy rare-earth element such as Dysprosium (Dy) has to be added to NdFeB sintered magnets in order to increase the coercivity of NdFeB sintered magnets. However, Dy is very rare and very expensive. On the other hand, as shown in Table I, the usage amount of rare-earth elements in the magnet continues to increase for improving the motor efficiency till several years before. In particular, it is important to reduce the usage amount of heavy rare-earth elements from the viewpoint of resource problem. Therefore, it is highly desirable to reduce the amount of Dy. Recently, the grain boundary diffusion-type NdFeB sintered magnets were developed, where the amount of heavy rare-earth element was reduced by about 60% and remanence was improved by 11%. Such rare-earth permanent magnets were applied to the IPMSM for compressor drive of air conditioner. Fig. 1 shows the compressor motor with the grain boundary diffusion-type NdFeB sintered magnets [3]. The magnet arrangement was changed to the V-shape because the

TABLE I
TREND OF USAGE AMOUNT OF RARE EARTH ELEMENTS IN COMPRESSOR MOTORS[2].

Year	1996	2001	2007	2009
Rotor structure				
Weight of magnet	1.0 (reference)	0.9	1.1	1.3
Weight ratio of heavy rare earth [wt%]	6%	Over 6%	Over 6%	3%
Weight of hevy rare earth	1.0 (reference)	1.2	1.6	0.7

978-1-4799-2706-7/14 $31.00 © 2014 IEEE

The 2014 International Power Electronics Conference

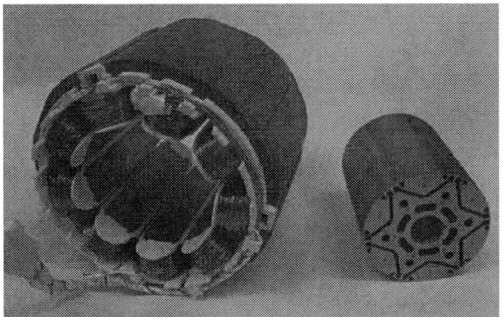

Fig. 1. Newly developed IPMSM using the grain boundary diffusion-type Nd-Fe-B sintered magnets for compressor motor.

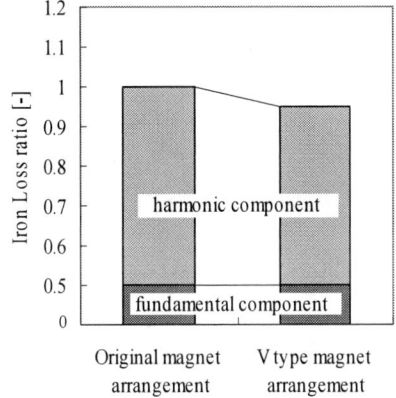

Fig. 2. A comparison of iron loss of magnet arrangement.

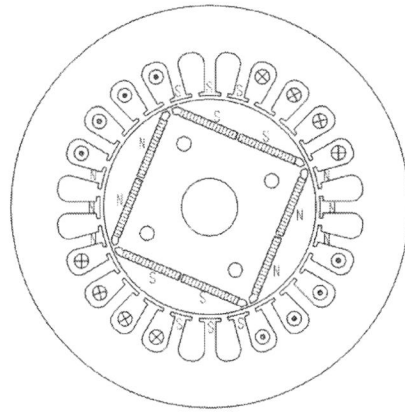

(a) conventional IPMSM at rotor position of 45 degrees

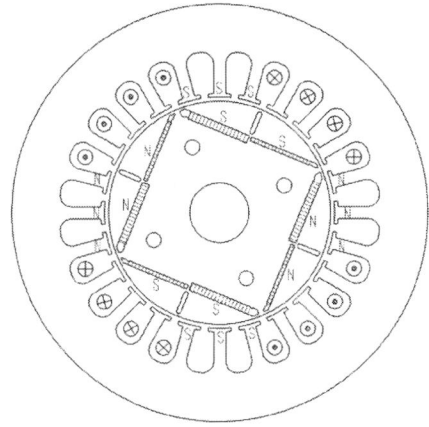

(b) newly designed IPMSM at rotor position of 45 degrees

Fig. 3. Sectional views of the conventional and the newly designed IPMSMs under 120 degrees current conduction with 30 degrees of the current-phase-lead angle.

edge of the grain boundary diffusion-type magnet is difficult to be demagnetized and the magnet surface area can be increased. Furthermore, the V-shape magnet arrangement can reduce higher flux harmonic components by enlarging the area of rotor core and creates a gradual distribution of magnetic flux components which prevents the flux from saturation [4].

Fig. 2 shows a comparison of iron loss between a conventional motor and a motor with a V-shaped magnet arrangement. Fig. 2 also shows that the V-shaped magnet arrangement decreases the higher harmonic flux, which achieves a reduction in iron loss of about 5%.

In the newly developed IPMSM using the grain boundary diffusion-type NdFeB sintered magnets, the amount of the heavy rare earths elements were reduced by about 60% and the efficiency was improved by 1.4-1.7%.

B.   Design Study on Volume Reduction of PM

A volume reduction of magnet itself undoubtedly contributes to reducing rare-earth materials being included in NdFeB sintered magnets. As an example, Ref. [5] has reported design study on permanent magnet volume reduction of IPMSM for compressor drive in air-conditioner applications.

Fig. 3 shows the conventional IPMSM with 4-pole and the newly designed IPMSM with less volume of permanent magnet. In both cases, the motors are driven as a brushless DC motor under 120 degrees current

conduction with 30 degrees of the current-phase-lead angle and the rotor position is at electrical 45 degrees. In the target compressor drive, the IPMSM is exposed to relatively high temperature environment around 60-120°C. Since a coercivity of NdFeB sintered magnets decreases with the temperature rise, the thickness of permanent magnet has to be designed so as to prevent its irreversible demagnetization taking into account of current-phase-lead angle control range. In the conventional IPMSM, the permanent magnet per pole is divided into two pieces by a center bridge. Each piece of permanent magnets designed has uniform thickness with 2mm, by which no irreversible demagnetization occurs even if the current-phase-lead angle reaches 90 degrees under the maximum current amplitude condition. On the other hand, each piece of permanent magnets in the newly designed IPMSM is not uniform as shown in Fig. 3(b). One has thickness with 2mm, the other has 1mm that is a half thickness of other piece. As a result, the newly designed IPMSM enables to reduce 25% volume of permanent magnet compared to the conventional

978-1-4799-2706-7/14 $31.00 © 2014 IEEE   1998

TABLE II
COMPARISONS OF DEMAGNETIZING CURRENT FOR TWO MOTORS

Current-phase-lead Angle [elec. deg]	Demagnetizing current level [A]	
	Conventional IPMSM	Newly designed IPMSM
90	40	17
45	50	50

IPMSM. An air slit located at the center between two pieces of permanent magnet plays a role of flux barrier. The flux barrier prevents from the irreversible demagnetization of this permanent magnet on the assumption that the current controller limits the current-phase-lead angle less than 45 degrees. Table II summarizes comparisons of test results of demagnetizing current levels for different current-phase-lead angles in two IPMSMs. As it can be seen from the table, the newly designed IPMSM ensures enough level of demagnetizing current same with that in the conventional IPMSM under 45 degrees of current-phase-lead angle. Under such constraint of current-phase-leading angle, the proposed motor has the same performance as the conventional IPMSM while the amount of rear-earth magnets is reduced by 25%.

III. PERMANENT MAGNET ASSISTED SYNCHRONOUS RELUCTANCE MOTOR WITH FERRITE MAGNETS

The rare-earth less PMSM for traction applications are examined in recent years. Such motors must have a competitive size, power density and efficiency with respect to those in the rare-earth PMSMs. Here permanent magnet assisted synchronous motors (PMASynRMs) using ferrite magnets instead of rare-earth magnets are introduced.

A. PMASynRM with Multiple Flux Barriers

A PMASynRM with ferrite magnets that has the same power density and the same maximum efficiency as a rare-earth PMSM for HEV application has been proposed [6]. Fig. 4 shows the PMASynRM with ferrite magnets, which are inserted into the multiple flux barriers. The rotor has a center rib to provide mechanical strength at high speed of 10,000 r/min. The flux barriers of the first and second layers are tapered to reduce the effect of armature reaction on the PMs and to prevent irreversible demagnetization. The rotor structure including the shape and thickness of ferrite magnets, the shape of flux barriers, and the width of center ribs and outer bridges were designed taking into account the irreversible demagnetization of ferrite magnets and the mechanical strength.

The calculated efficiency map of the proposed PMASynRM is shown in Fig. 5. The efficiency tends to increase as the speed increases. The proposed PMASynRM is over 90% efficient across a wide operating range, with a maximum efficiency of 97.2%.

A prototype PMASynRM was manufactured. The

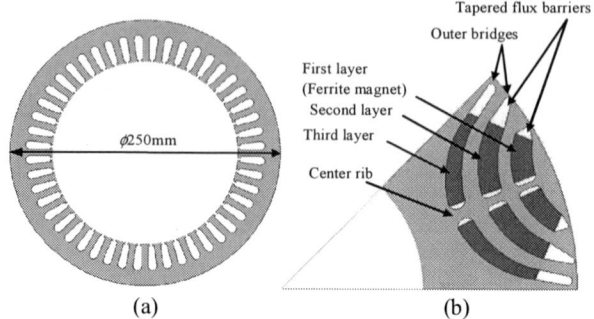

Fig. 4. PMASynRM with ferrite magnets for automotive applications. (a) Stator. (b) Rotor.

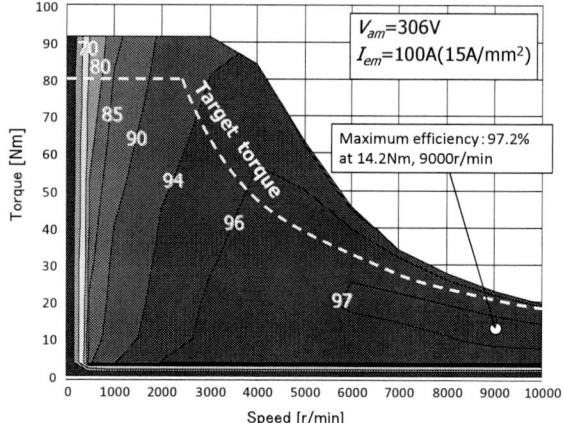

Fig. 5. Efficiency map of PMASynRM with ferrite magnets (analysys results).

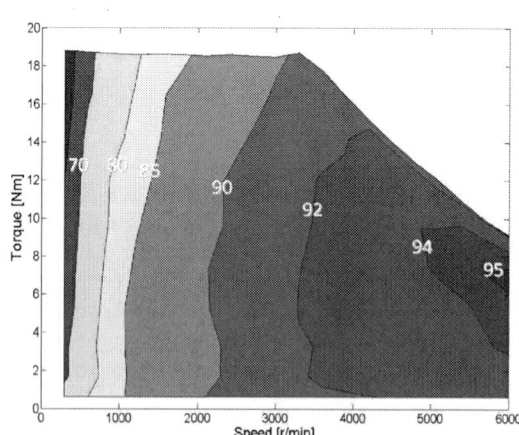

Fig. 6. Efficiency map of prototype PMASynRM with ferrite magnets (experimental resuls).

stack length of the prototype PMASynRM is 20 mm, which is a half of the analysis model, and the current limit was set to the rated current ($7.5A/mm^2$). The experimental efficiency map is shown in Fig. 6. Because the ratio of iron loss is comparatively small in the PMASynRM with ferrite magnets, the efficiency in high-speed region becomes high. In general, the efficiency of the rare-earth PMSM with large magnet flux-linkage tends to decrease by d-axis current for the flux-weakening control at high-speed and light-load condition.

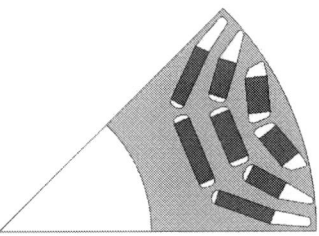

Fig. 7. Rotor structure of PMASynRM with rectangular ferrite magnets.

On the other hand, the proposed PMASynRM with small magnet flux-linkage has the feature that the decrease of efficiency at high-speed and light-load condition is small compared to the rare-earth IPMSM. As the electromotive force at no load is less than the limited voltage, the uncontrolled generator mode operation following unexpected inverter shutdowns never occurs in the PMASynRM with ferrite magnets.

In consideration of the productivity and practical use, a less-expensive rotor structure with excellent productivity has been examined [7]. The rotor structure of the PMASynRM with ferrite magnets is shown in Fig. 7. This rotor uses rectangular ferrite magnets in place of arc-shaped magnets (see Fig. 4 (b)) in consideration of the cost and manufacturing concerns. The performance of PMASynRM with rectangular ferrite magnets was evaluated by the FEA and experiments, and it was confirmed that its performance is compatible with the PMASynRM with the arc-shaped ferrite magnets.

By incorporating ribs and appropriately designing the shape of the ferrite magnets, flux barriers, center ribs, and outer bridges, a PMASynRM with ferrite magnets was designed that satisfies the mechanical strength requirements in the high-speed region and resists demagnetization.

B. PMASynRM with Double Gap Structure

As a one of candidates of ferrite permanent magnet motor with high torque density, PMASynRM employing double gap structure is proposed [8]. This machine makes its airgap surface area enlarge so that high torque is essentially expected. Fig. 8 demonstrates several examined machine structure and a standard IPMSM with single airgap using NdFeB sintered magnets for drive performance comparisons. Based on the general torque expression of IPMSM in (1), magnetic pole configuration suitable for the double gap structure is examined.

$$T = P\{\phi_a + (L_d - L_q)i_d\}i_q \qquad (1)$$

According to (1), to make inductance difference $L_q - L_d$ large is considered firstly in order to gain larger torque. SynRM with a simple structure and single airgap shown in Fig. 8(a) is treated as a beginning structure for design investigations and subsequently, rotor salient pole geometry, gap configuration and ferrite magnet arrangement are refined step by step as shown in Figs. 8(b), (c), (d) and (e).

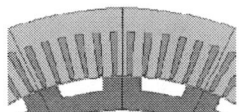

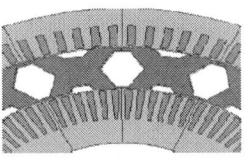 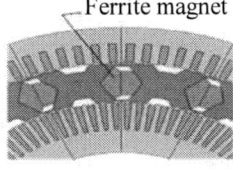

(a) salient type SynRM with single airgap

(b) segment type SynRM with single airgap

(c) Honeycomb type SynRM with double airgap

(b) ferrite magnet embedded honeycomb type PMASynRM with double airgap

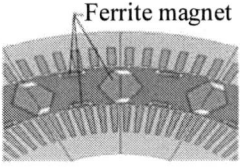 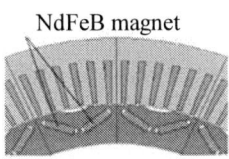

(e)PMASynRM shown in Model (d) + ferrite magnets adding between poles

(f) stantadrd IPMSM using NdFeB

Fig. 8. Sectional views of the examined motor structures.

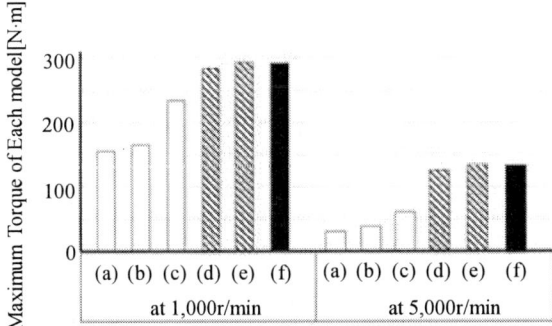

Fig. 9. Maximum torque comparisons between examined 6 motors shown in Fig. 8.

Fig. 9 illustrates comparison of the maximum torque of the treated six structures shown in Fig. 8. The maximum torque on each motor is computed by 2 dimensional FEA (2D-FEA) and the comparison is done under the uniform conditions such as the outer diameter of stator core of 265mm, the stack length of motor of 50mm, the pole numbers of 16, the airgap length of 0.6mm and the maximum current density of 25A/mm^2. As can be seen from the figure, the segment type SynRM (Fig. 8(b)) has superior maximum torque capability compared to the simple salient pole type one (Fig. 8(a)). This is because the segment type SynRM makes q-axis flux path short so that the increase in q-axis inductance makes the inductance difference $L_q - L_d$ large. The torque of SynRM with double gap structure (Fig. 8 (c)) is larger

978-1-4799-2706-7/14 $31.00 © 2014 IEEE   2000

than that with conventional single gap structure (Fig. 8 (a) and (b)). Moreover, it is shown that adding ferrite magnets improves the maximum torque performance in spite of motor speed and consequently, the maximum torque performance of PMASynRM shown in Fig. 8 (e) is the best. Its torque performance is competitive to the target rare-earth IPMSM for HEV.

IV. AXIAL-GAP PMSMs WITH FERRITE MAGNETS

The axial gap configuration is expected to have higher torque than the radial gap configuration in case of motors with short axis. Applying this effect, the research on substituting rare earth magnets with ferrite magnets to obtain the same motor's performance has been carrying out. The axial gap PMSMs are developed as a solution for rare-earth less PMSM. In this section, the research on high-efficiency technology of vehicle traction motors, home appliances motors and industry motors are introduced.

A. Axial-gap PMSM with Segmented Rotor

First, the development of a 50kW motor used for HEV traction system is introduced. The ferrite magnets are applied to HEV traction motor system to replace rare earth magnets. Due to the energy density of ferrite magnets is one tenth of the rare earth magnets, the outer diameter of the motor is increased to obtain larger areas of magnets [9].

Fig. 10 shows the structure of the axial gap motor. The motor contains one flat disk-shaped rotor with two stators arranged at the both sides. Fig. 11 shows the details of the rotor. On the rotor surface, ten ferrite permanent magnets are placed to have ten-pole excitation. Soft magnetic composite (SMC) iron cores are installed between magnets. These iron cores generate reluctance torque as an inset permanent magnet motor. The outside of the rotor is fixed by a stainless steel ring. The novel segmented rotor structure provides the following advantages: the reluctance torque can be effectively used and it is possible to prevent the irreversible demagnetization in the ferrite PMs.

Fig. 12 shows the efficiency map of the prototype when the rotational speed changes from 300 to 5,100 r/min. The load test was carried out with various current densities and current phase angles, and the efficiency was measured. As shown in Fig. 12, the efficiency is high in the light load area where the HEV is most frequently driven, and the maximum measured efficiency is 92.5% at 1,500 r/min. The maximum average torque is 289.3 Nm at the fundamental speed of 1,700 r/min. Thus, the maximum output power of 51.5 kW can be generated. The prototype demonstrates that the axial gap motor can offer the same output power with the high-performance radial gap type PMSM of the second-generation Toyota Prius.

Fig. 10. Outline of a ferrite permanent magnet axial gap motor.

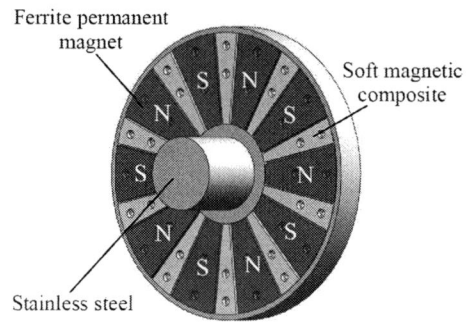

Fig. 11. Rotor configuration of segmented rotor structure.

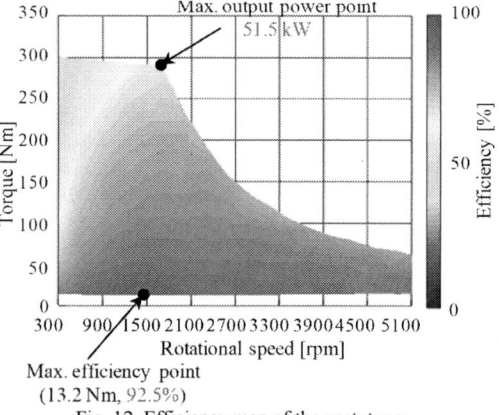

Max. efficiency point
(13.2 Nm, 92.5%)

Fig. 12. Efficiency map of the prototype.

B. Axial-Gap PMSM with Amorphous Core

Second, the experiences on improving motor efficiency for home appliances and industry applications are introduced. The same concept with the HEV traction motor by increasing the magnets area in the axial-gap PMSMs is employed. The difference is that the two-rotor and one-stator structure is used here. Furthermore, low iron loss magnetic materials are applied to stator cores to increase motor efficiency.

Amorphous alloy exhibits high permeability and extremely low iron loss compared to magnetic steel sheets. Therefore, the amorphous alloy is expected to make a contribution on motor affiance improvement. On the other hand, the thickness of amorphous alloy sheet is one tenth of magnetic steel sheets. It is difficult

The 2014 International Power Electronics Conference

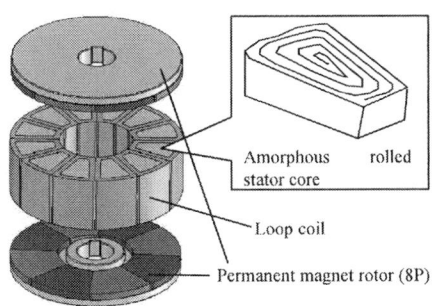

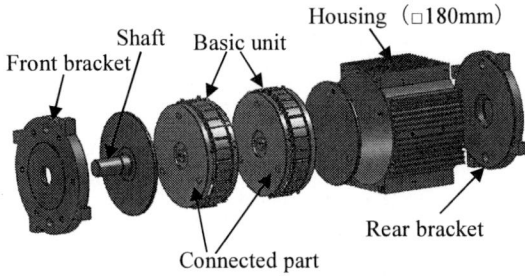

Fig. 15. Axial-type PMSM of multistage structure.

Fig. 13. Structure of the proposed axial-type PMSM.

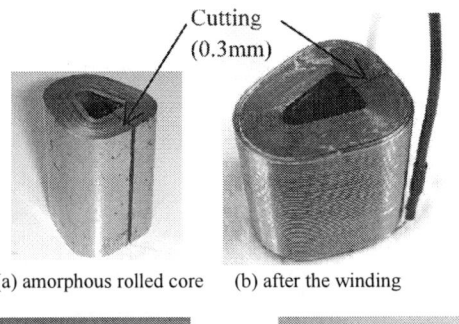

(a) amorphous rolled core   (b) after the winding

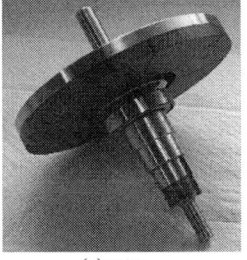 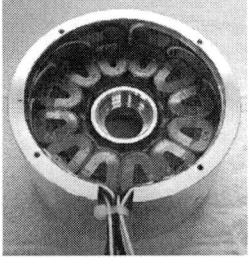

(c) rotor   (d) stator

Fig. 14. Structure of the prototype axial-gap PMSM.

Left : conventional industrial 11kW induction motor
Right : New motor developed
(a)   comparison of motor size

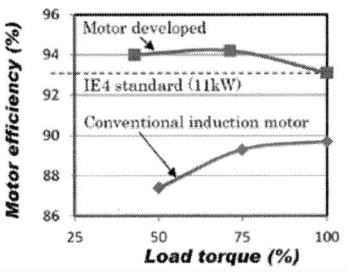

(b)   comparison of motor efficiency (11kW)

Fig. 16. Structure of the proposed axial-type PMSM.

to process it to complicated core shapes for motor applications because of the hardness of amorphous alloy. An axial flux PMSM, which applied rolled amorphous cores to the stator, was proposed as a solution for amorphous alloy's motor application [10]. Fig. 13 shows the structure of the proposed axial-type PMSM with the amorphous rolled stator care, where ferrite magnets are mounted on the rotor surface. Fig. 14 shows the structure of the prototype axial-gap PMSM. The flow path of eddy current in the rolled core can be cut off by removing one section of the amorphous rolled core as shown in Fig. 14(a), which made it possible to prevent the generation of eddy current loss in the rolled core. The evaluation results of the prototype axial-gap PMSM with ferrite magnets verified that the proposed structure can meet the target efficiency of 85 % with a size smaller than φ100mm x 60mm. Furthermore, applying amorphous cut cores or increasing motor speed has being under discussion to improve motor efficiency [11].

The developed high efficiency technology has been applied to large output industry motors. Fig 15 shows one

example to construct axial-gap PMSM with ferrite magnets with a limit dimension [12]. Several axial gap motors are superimposed at axial direction to increase output power.

Fig. 16 shows the test results on an 11kW industrial motor that used axial-gap structure [13]. This motor's efficiency meets the requirements of IEC-IE4 regulations.

V. CONCLUSIONS

This paper overviewed the recent technical trends in permanent magnet synchronous motors (PMSMs), especially the rare-earth less PMSMs in Japan. The performance of PMSMs is greatly improved by not only the further progress of electromagnetic material technology such as the permanent magnets and soft magnetic materials, but also the innovative machine structure. The application of PMSMs will be expanded to various fields by such advances.

978-1-4799-2706-7/14 $31.00 © 2014 IEEE   2002

REFERENCES

[1] S. Morimoto, "Trend of permanent magnet synchronous machines," *IEEJ Trans.*, Vol. 2, pp. 101-108, 2007.

[2] K. Ohyama, "The resources problem and its correspondence in the viewpoint of magnet users," The 33rd motor technology symposium of Japan Management Association, C-2-3, 2013 (in Japanese)

[3] H. Domeki, N. Matsui, K. Ohyama, Y. Shimogaki, "Performance improvement of reluctance torque assisted motor by advances in magnetic materials -general remarks-," *Proc. of JIASC 2012*, pp. III-27-III-30, 2012.

[4] H. Kamiishida, N. Tomioka, K. Iida, K. Yuasa, Y. Kataoka, A. Yamagiwa, K. Aota, "Development of high efficiency swing type compressor using new interior permanent magnet synchronous motor," *International Compressor Engineering Conference at Purdue*, pp. 12-15, July 2010.

[5] M. Sato, S. Kaneko, M. Tomita, S. Doki and S. Okuma, "Discussion and proposal for reduction of rare-earth magnet of IPM motor by attention to the relation between magnet amount and its erasing," *The 2012 Annual Meeting Record IEE Japan*, vol. 5, no. 5-017, pp. 28-29 (2012) (in Japanese)

[6] S. Ooi, S. Morimoto, M. Sanada, Y. Inoue, "Performance evaluation of a high power density PMASynRM with ferrite magnets," *IEEE Transactions on Industry Applications*, vol. 49, no. 3, pp. 1308-1315, 2013.

[7] M. Obata, S. Morimoto, M. Sanada, Y. Inoue, "A study of high efficiency PMASynRM with ferrite magnets for vehicle drive," *The 2013 IEEJ Industry Applications Society Conference*, nol. 3, pp. 135-138 (2013) (in Japanese)

[8] S. Kusase, T. Maekawa and K. Kondoh, "Study of double gap motor for rare-earth free", *The 2012 Annual Meeting Record IEE Japan*, vol. 5, no. 5-031, pp. 53-54 (2012) (in Japanese)

[9] S. Chino, T. Miura, M. Takemoto, S. Ogasawara, A. Chiba, N. Hoshi, "Fundamental characteristics of a ferrite permanent magnet axial gap motor with segmented rotor structure for the hybrid electric vehicle," *Proc. of ECCE2011*, pp. 2805-2811, 2012.

[10] H. Amano, Y. Enomoto, M. Ito, H. Itabashi, S. Tanigawa, R. Masaki, "Examination of applying amorphous rolled core to permanent magnet synchronous motors," *IEEJ Trans. on IA*, vol. 130, no. 5, pp. 632-638, 2010.

[11] Z. Wang, Y. Enomoto, M.Ito, R. Masaki, S. Morinaga, H. Itabashi, and S. Tanigawa, "Development of a permanent magnet motor utilizing amorphous wound cores," *IEEE Trans. on Magn.*, vol. 46, no. 2, pp. 570-573, 2010

[12] Y. Enomoto, Z. Wang, R. Masaki, K. Soma, "Development of a large capacity axial gap type motor with iron-based amorphous core" Journal of the Japan Society of Applied Electromagnetics and Mechanics, vol. 21, no. 2, pp. 314-319, 2013.

[13] Hitachi, Ltd. News Release
http://www.hitachi.com/New/cnews/120411.html
"Highly efficient industrial 11kW permanent magnet synchronous motor without rare-earth metals –Realizing IE4 class efficiency standard with a smaller motor–"

Recent Technical Trends in SRM and FSM

Yoshiaki Kano

Dept. of Information and Computer Engineering
Toyota National College of Technology
Toyota, Japan
kano@toyota-ct.ac.jp

Abstract— **This paper overviews recent technical trends of research and development of non rare-earth permanent magnet motors. In particular, switched reluctance motors (SRMs) made of recent developed core materials and flux switching machines (FSMs) are reviewed.**

Keywords—Non rare-earth, reluctance motor, wound-field synchronous motor, flux-switching motor

I. INTRODUCTION

Interior permanent magnet synchronous motors (IPMSMs) are widely used in many applications such as electric vehicles and compressor drives of air conditioner because of its superior power density and high efficiency [1]. However, the IPMSM contains rare-earth magnets with rare-earth materials such as Neodymium but also Dysprosium. The price and the supply of rare earth materials have been recognized as one the major problem for mass production. Therefore, the continuous research and development of permanent magnet machines with less or no rare-earth materials would be very important [2], [3].

This paper overviews recent technical trends of research and development of non rare-earth permanent magnet motors. In particular, switched reluctance motors (SRMs) made of recent developed core materials and switching flux machines (FSMs) are reviewed.

II. COMPARISON OF IM, IPMSM, AND SRM-DRIVES FOR EV APPLICATIONS

Permanent magnet motors, induction motors and switched reluctance motors are the main candidates for HEV or EV traction applications which extensively have been subject of research in both academia and industry [4]. Table I shows the comparison of IM, IPMSM and SRM-drives for EV applications. The specific target application is the 2003 Prius IPMSM as discussed in [5]. In order to make direct comparison, the three different machines share the same axial core length (84mm), and stator outer diameter (269mm). The comparison of motor performance at 1500 and 6000 r/min at maximum power are also shown in table I. It is interesting to note that the change in dominant losses at base speed (copper loss) to full speed (iron losses). The SRM has higher iron losses due to the increased frequency of the flux, and the copper losses are high at 1500 r/min. Therefore, in order to produce a traction drive system based on SRM, the major

TABLE I
COMPARISON OF IPMSM, IM, AND SRM

Parameter		IPMSM	SRM	IM
No. of poles/slots		8/48	12/18	8/48
Outer stator diam. [mm]		269.0		
Inner rotor diam. [mm]		111.0		
Outer rotor diam. [mm]		160.5	170.0	180.0
Axial core length [mm]		84.0		
Air-gap length [mm]		0.73	0.3	1.5
Total weight [kg]		31.16	26.71	36.25
Total material cost [US$]		242.2	74.2	143.8
Base speed 1500 r/min	Torque [Nm]	303	297	294
	Efficiency[%]	91.3	85.2	83.1
	Copper loss [W]	4328	7653	8591
	Iron loss [W]	198	404	148
Full speed 6000 r/min	Torque [Nm]	45.6	52.1	50.8
	Efficiency[%]	96.1	88.2	95.2
	Copper loss [W]	219	306	730
	Iron loss [W]	953	4074	439

Laminated steel[a] :1.3 US$/kg, Copper[b]:6.6 US$/kg, Nd-Fe-B[c]:132 US$/kg
[a]Lamination estimated to be approximately double bulk steel cost
[b]London Metal Exchange, June 2010
[c]W.T.Benecki, The Permanent Magnet Industry Outlook, Great Western Minerals Group, June 2008

points of consideration are:
1) Increase of torque/copper loss at low speed
2) Iron loss reduction at high speed

From above viewpoints, magnetic characteristics of iron cores are particularly important for the design of SRM. The following section reviews several remarkable research and development of SRM made of recent developed core materials [6]-[19].

III. SWITCHED RELUCTANCE MOTORS

A. SRM designed for HEVs using super core

The interior permanent magnet synchronous motor (IPMSM) is the most popular electric motor for HEV applications. It is very desirable to replace the IPMSM with a rare-earth magnet free material SRM as an alternative. So, many researchers have attempted to reach an equivalent motor in this respect. In this subsection, some feasibility studies on SRM applying to hybrid electric vehicles drive applications are reported [6-13].

The design target is the motor with a maximum torque of 400Nm, a maximum power of 50kW, a maximum torque density more than 45Nm/L with similar

The 2014 International Power Electronics Conference

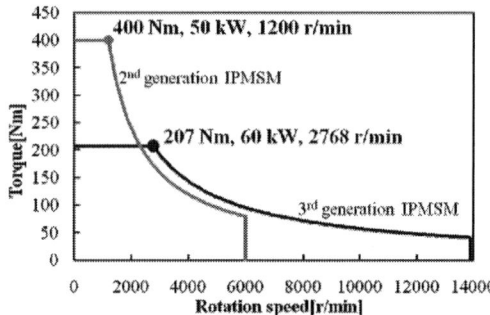

Fig.1. Torque vs. speed characteristic of 2nd and 3rd generation IPMSMs for HEV[11].

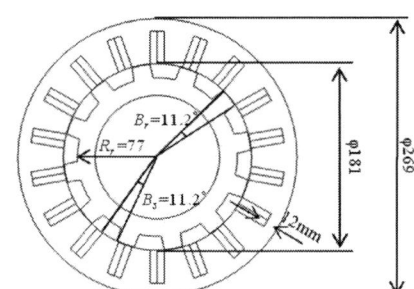

(a) Construction of designed SRM

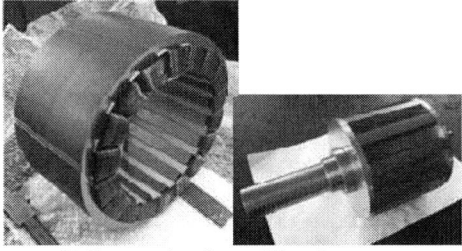

(b) Photographs of assemblies of test motor
Fig.2. Structure of designed SRM[6],[9].

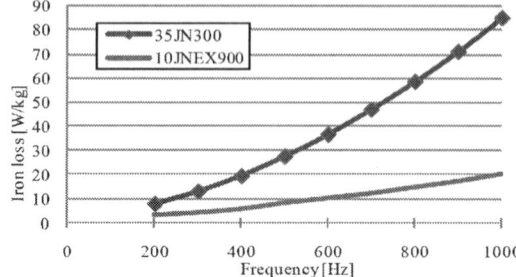

Fig. 3. Comparison of iron loss at $1.0T^{[11]}$.

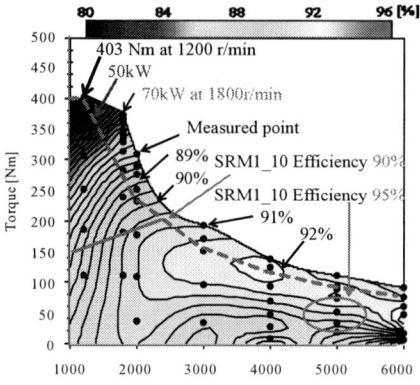

(a) SRM with 35JN300

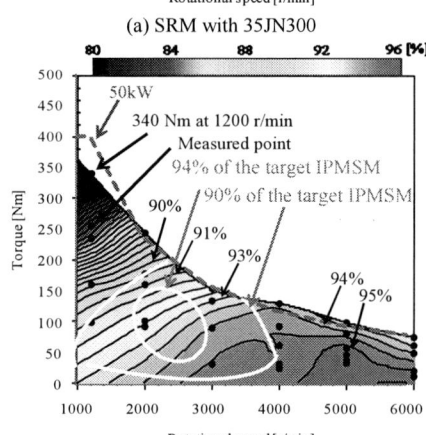

(b) SRM with 10JNEX900
Fig. 4. Comparison of measured efficiency maps of two prototype motors[6].

restrictions and specifications in IPMSM used for the second generation Toyota Prius commercialized in 2003. The target torque-speed envelop characteristics are shown in Fig.1.

Fig.2 shows the structure of the designed 50kW SRM (SRM1) [6-9]. The SRM1 has 18 stator poles, 12 rotor poles, and three phase windings. The outer diameter 269mm and the total axial length 156mm of the design machine are identical with those of IPMSM. To satisfy the target efficiency (95% of maximum efficiency), the designed SRM has high silicon steel, which is so called Super Core, 10JNEX900 having a thickness of 0.1mm. Fig.3 shows a comparison of iron loss at 1.0T of super core (10JNEX900) and conventional non-oriented Si steel with a thickness 0.35mm (35JN300). The figure indicates that the iron loss of super core 10JNEX900 is less than one third of that of 35JN300 at high frequency. In [6], two types of SR motors, one is made of 35JN300, the other is made of super core 10JNEX900, are compared by experiment.

Fig.4 shows the measured efficiency maps of the two types of test motors under the condition that the DC-bus voltage and maximum current limitation are respectively, set to 500V and 380A.

From the figure, it can be seen that high efficiency area appears in low-torque high speed region compared to the SRM with 35JN300. In addition, compared to the IPMSM, the efficiency of the SRM with 10JNEX900 is better in low torque area. Generally, since the motor for vehicles is frequently operated in low-torque region, it is reported that the SRM is suitable for HEV in terms of motor efficiency under the frequent operating regions. Moreover, the maximum motor efficiency of 95.4% and the maximum power of 50kW can be achieved.

However, the maximum torque at 1,200r/min is 340Nm, i.e., 85% of the target torque of 400Nm. The discrepancy may cause by a three dimensional leakage flux, a lamination stacking factor setting, BH characteristics of the iron after process and etc. In [8,12],

978-1-4799-2706-7/14 $31.00 © 2014 IEEE

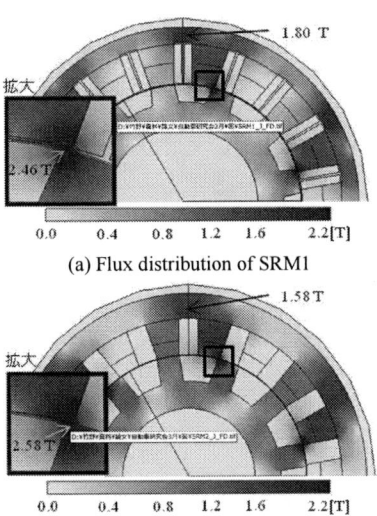

(a) Flux distribution of SRM1

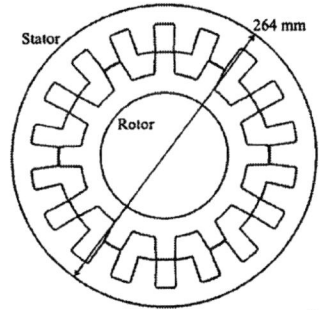

(b) Flux distribution of SRM2

Fig. 5. Flux distribution under the maximum torque condition[12].

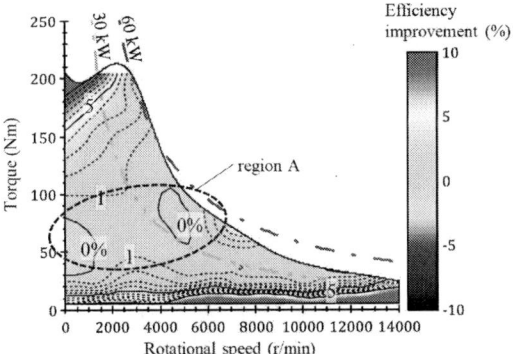

Fig. 6. Construction of designed SRM[11].

TABLE II
DESIGN SPECIFICATIONS OF SRM[11]

	target IPMSM	Designed SRM
Number of poles	8	18/12
Outer diameter of stator	264 mm	264 mm
Motor axial length	108 mm	108 mm
Coil-end length	29 mm	10.5 mm
Iron stack length	50 mm	87 mm
Weight	22.2 kg	25.2 kg
Slot fill factor	56%	56%
DC side voltage	650 V	650 V
Current density	18.8 A/mm^2	23.9 A/mm^2
RMS Current	144 A	137 A
Max torque	207 Nm	211 Nm
Max power	60 kW	61 kW
Max power density	10.2 kW/l	10.4 kW/l
Max torque density	35 Nm/l	36 Nm/l
Power-weight ratio	2.7 kW/kg	2.4 kW/kg
Torque-weight ratio	9.3 Nm/kg	8.4 Nm/kg
Copper loss	6.0 kW	6.3 kW
Iron loss	2.7 kW	1.3 kW
Efficiency	87.4 %	89.0 %

Fig. 7. Efficiency difference of the IPMSM and the SRM[13].

these differences are considered to design the second stage prototype machine SRM2. Fig.5 shows the flux density distributions of the SRM1 and the SRM2 under the maximum torque condition. It is noted that the magnetic saturation of the stator yoke is eased by a change in the stator yoke width. As a result, the maximum toque of the SRM2 is reached 447Nm by the 3D-FEM simulation.

Fig.6 and Table II shows the construction of SRM of 60kW and its design specifications. The design target is the motor with a maximum torque of 207Nm, a maximum power of 60kW with similar restrictions and specifications in IPMSM used for the third generation Toyota Prius commercialized in 2009. The target torque-speed envelop characteristics are also shown in Fig.1.

From the table, the maximum torque is 211Nm and the maximum power reaches 61kW, which met the target requirement for HEV drive. However, the current density and the weight are slightly increased. Thus, the maximum torque density and power density are approximately 10% smaller than that of the IPMSM.

Fig.7 shows the measured efficiency difference of the target IPMSM and the designed SRM [13]. From the figure, it is found that the increases of motor efficiency of 5% can be achieved in low torque area less than 20Nm.

B. SRM made of permendur

The performance of a SRM made of permendur which has extremely high saturation flux density and very low

core loss have been investigated in [14, 15]. Two types of SR motors, one is made of conventional non-oriented Si steel with a thickness of 0.35mm, the other is made of permendur with a thickness of 0.2mm, are compared by simulation and experiment. Fig.8 shows the iron loss and B-H curves of core material. The figure indicates that the saturation flux density of permendur is larger than that of conventional non-oriented Si steel by 40% or more. Fig.9 shows the constructions of the trial SR motor. The axial core length is 40mm, and stator outer diameter is 136mm. The SRM has a 12/8 configuration.

Fig.10 shows the comparison of torque vs. speed characteristics of the trial SR motors. The maximum torque of the SRM made of permendur is greater than that of the conventional Si steel by over 40%.

Fig.11 shows the efficiency characteristics of the both motors. From the figure, it found that the efficiency of the SRM made of permendur is higher than that of non-oriented Si steel over all the operating range.

C. SRM with segmental rotors using grain oriented magnetic steel

The SRMs with segmental rotors have been introduced in recent years [16], indicating that the machine is able to make much better utilization of the magnetic geometry, and thereby gives much greater torque density than a conventional SRM. Shortening flux paths and columned composite rotors result that windage losses and iron

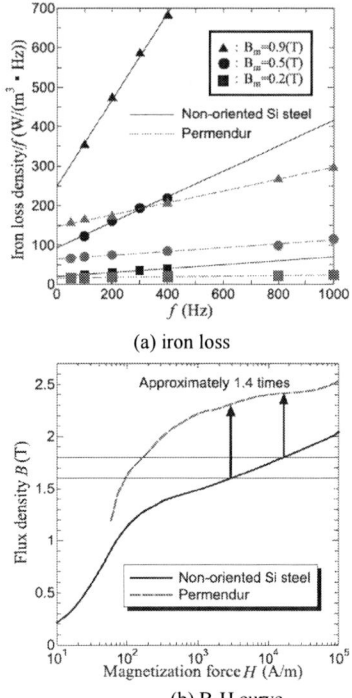

(a) iron loss

(b) B-H curve

Fig. 8. Comparison of B-H curves and iron losses of core materials[15].

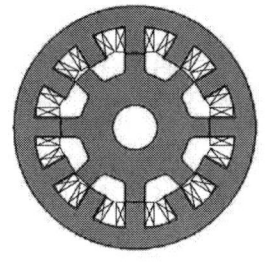

Stack length:	40 mm
Stator diameters :	136 mm
Rotor diameters :	83.2 mm
Gap :	0.2 mm
Number of windings/phase:	140 turns
Winding resistance/phase :	0.37 Ω

Fig. 9. Construction of the trial 12/8-pole SRM [15].

losses are greatly decreased at high speed. However, it had some problems of complexity for manufacturing and weakness of mechanical strength.

In [17], a novel segment type SRM in which the segment cores are embedded in aluminum rotor block in order to increase the mechanical strength and easy manufacturing as well as to improve the performance and reduce the vibration and acoustic noise is proposed. Fig.12 shows the construction of the proposed segment type SRM with 6 stator poles and 4 rotor segment cores. The stator has full pitch windings. The novel segment type SRM increases in the average torque by 40% in comparison with the VR type SRM of same size. The vertical force for one pole reduces by 76%. On the other hand, the novel segment type SRM increases in the average torque by 2.7% and reduces in the vertical force for one pole by 4.8% comparing with the conventional

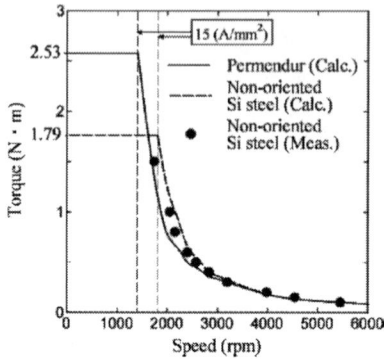

Fig.10. Torque vs. speed characteristics[15].

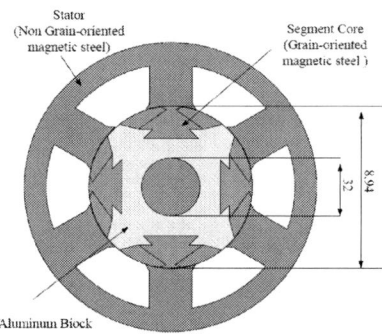

Fig.11. Efficiency characteristics[15].

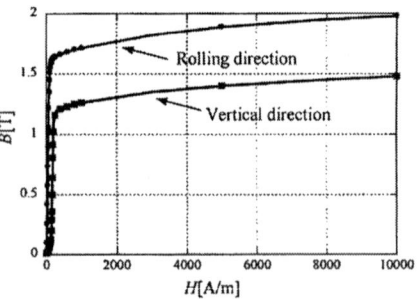

Fig.12. Construction of segment type SRM in [17].

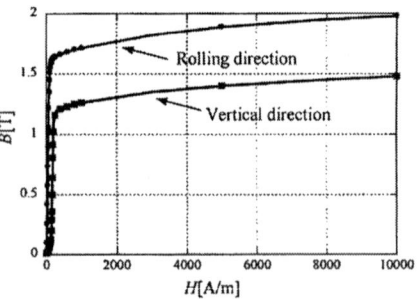

Fig.13. B-H curves of grain-oriented magnetic steel[18].

segment type SRM. In order to increase maximum torque and efficiency, T.Abe proposed a segment type SRM using grain-oriented magnetic steel. The grain-oriented steel is applied to the segmental rotor cores as shown in Fig.12 [18]. The rotor core is divided into three peaces to fit the rolling direction of the core to flux path in the rotor

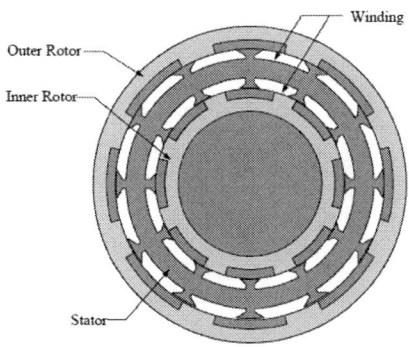

Fig.14. Construction of dual rotor segment type SRM in [19].

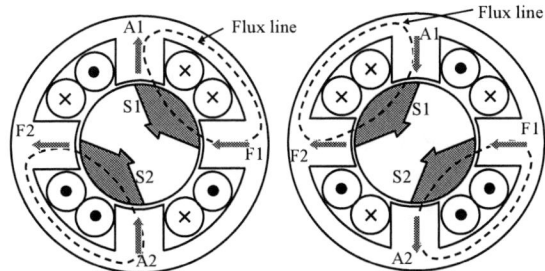

(a) First alignment position (b) Second alignment position
Fig.15. Flux distribution in stator teeth with field excitation only.

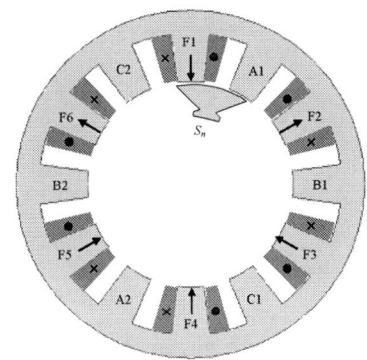

Fig.16. Wound-Field three-phase FSM with segmental rotors.

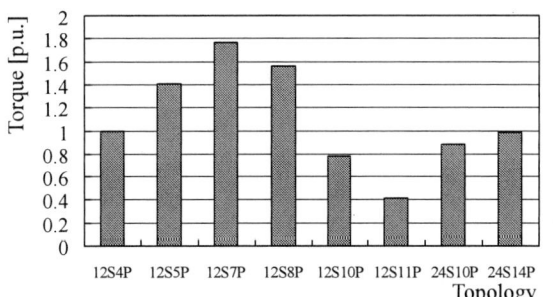

Fig.17. Torque capability of various topologies.

core. By using grain-oriented steel for segmental cores, the average torque is increased by 4% and the efficiency is also increased by 5%. Fig.13 shows the B-H curves of grain-oriented magnetic steel for rolling and vertical directions.

The conventional segment type SRMs with full pitched windings have overlapping and substantially longer end-windings, which reduce the electric loading and fault-tolerant ability, and also make them impractical for applications which combine a short lamination stack length with a large pole pitch. In order to shorten length of end-windings, T.Abe also proposed a dual rotor segment type novel SRM [19]. The configuration of the proposed SRM is shown in Fig.14.

IV. FLUX SWITCHING MACHINES

A. Wound-field three-phase FSM with segmental rotors

Recent research has shown that, through use of segmental rotor configurations, the performance of flux switching machines (FSM) is enhanced [20]. Use of rotor segments to achieve flux switching is illustrated on a basic 4/2 arrangement shown in Fig.15 with four stator teeth (F1, F2, A1, A2) and two rotor segments (S1, S2) with constant DC field excitation in coils around F1 and F2. For the two aligned positions shown, coupling of coils on the stator teeth is through segments. The motion of the rotor from position (a) to (b) not only varies flux in the armature teeth A1 and A2, but also changes its polarity. This results in the armature coils experiencing a bipolar AC magnetic field, with induction of EMF.

In practice, the concept is implemented on a stator with an even number of teeth, with half the number of teeth designated as field and the other half as armature. The field teeth alternate in position with the armature teeth, with the polarity of each field tooth arranged to be opposite that of the next field tooth. The wound-field segmented-rotor FSM have following advantages compared to conventional wound filed excited synchronous motors.

1. Ease of cooling of all active parts such as armature coil and field coil.
2. Dispensation of the brush gear.

To provide further attractive characteristics, the three phase topologies has been proposed by Barrie.C. Mecrow et al. [20]. Fig.16 shows the topology of three phase machine. Six of the twelve stator teeth, F1 to F6, are

wound as field coils and excited with DC current. These field coils produce three N poles interspersed between three S poles. The remaining six stator teeth, A1, A2, B1, B2, C1, and C2, contain six armature coils.

Fig.17 shows the torque capability at optimum segment spans for various topologies. If the topologies with an odd number of segments are discounted on account of the potential for unequalised radial forces, the 12/8 topology appears to give the highest torque capability.

B. Field-Excitation FSM

Fig.18 shows a new design of 24slot-10pole field excitation FSM (FEFSM) [21]. Similar to the wound-field segmented-rotor FSM, both field excitation coils and armature coils are allocated at stator side, but coil span is different. On the other hand, the rotor consists of only single piece iron, becoming more robust and more suitable for high speed operation coupled with reduction gear compared to conventional IPMSM. The field excitation can be used to control flux with variable flux

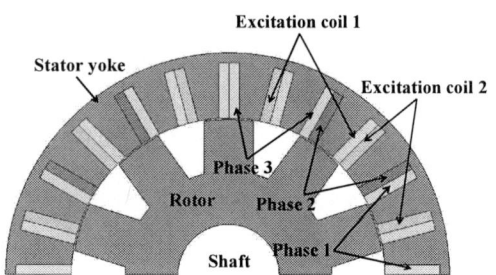

Fig.18. 3-phase 24-slot-10pole field-excitation FSM.

TABLE III
FEFSM DESIGN RESTRICTIONS AND SPECIFICATIONS FOR HEV APPLICATIONS[22].

Item	IPMSM	WFFSM
Max. DC-bus voltage inverter [V]	650	
Max. inverter current [A$_{rms}$]	Conf.	260
Max. current density in armature wdg. [A/mm^2]	Conf.	21
Max. current density in excitation coil [A/mm^2]	NA	21
Stator outer diameter [mm]	264	
Motor core stack length [mm]	70	
Shaft diameter [mm]	30	
Air gap length [mm]	0.8	
Maximum speed [r/min]	12,400	20,000
Maximum torque [N·m]	333	>210
Reduction gear ratio	2.478	4
Max. axle torque via reduction gear [N·m]	825	>840
Max. power [kW]	123	>123
Power density [kW/kg] (estimated)	3.5	>3.5

capabilities. From these viewpoints, FEFSM is recently received much attention as a candidate of non rare-earth PM vehicle motors. In this section, a study example trying to apply FEFSM to hybrid electric vehicles drive application is introduced [22-23].

The design target is the motor with a maximum torque of 210Nm with reduction gear ratio of 4:1, a maximum power of 123kW, a maximum power density more than 3.5kW/kg, and a maximum speed of 20,000r/min with similar restrictions and specifications in IPMSM used for LEXUS RX400h. Table III shows the target specifications for the IPMSM. The deterministic design optimization approach is used to treat design parameters defined in rotor, armature and FEC repeatedly until the target performances are achieved, under maximum current density 21A/mm^2 for both armature and FEC.

Fig.19 shows the construction of the designed 123kW FEFSM and photographs of rotor and stator of the fabricated test machine.

Fig.20 illustrates the measured torque and power versus speed curves under the condition that the DC-bus voltage, armature current and the field winding's ampere-turns limitation are respectively, set to 650V, 260A$_{rms}$ and 2200AT. At base speed 5,000r/min, the torque obtained is 182.1Nm as the maximum and the maximum power of 98kW is achieved at 5800r/min. Thus, the obtained maximum power is approximately 20% smaller than the target value of 123kW. The discrepancy may cause by a three dimensional leakage flux, BH characteristics of the iron after process and etc. However, the total weight of finally designed motor is 23.2kg and thus, the maximum power density is 4.2kW/kg, which meet the target requirement for the HEV drive.

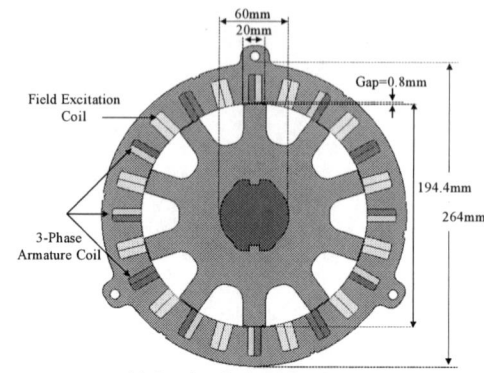

(a) Sectional view of prototype

(b) Photographs of assemblies of test motor
Fig.19. Construction of designed FEFSM[23].

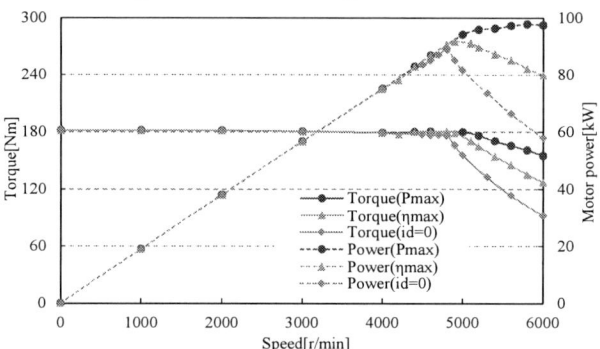

Fig.20. Speed vs. maximum torque and power characteristics. [23]

V. CONCLUSIONS

A review of non rare-earth PM motors has been presented including switched reluctance motors made of recent developed core materials and flux switching machines.

ACKNOWLEDGMENT

The author wishes to thank Prof. Akira Chiba of Tokyo Inst. of Tech., Prof. Masayuki Morimoto of Tokai Univ., Prof. Shigeo Morimoto of Osaka Prefecture Univ., Prof. Tsuyoshi Higuchi of Nagasaki Univ., Prof. Shoji Shimomura of Shibaura Inst. of Tech., Prof. Yasukazu Sato of Yokohama National Univ., Mr. Motoyasu Mochizuki of Toshiba Industrial Products and Systems Co., Mr. Yuichi Yoshikawa of Panasonic Corporation, and Mr. Masayuki Nashiki of Nagoya Univ. for their valuable advise and assistance during the course of this investigation.

REFERENCES

[1] R. Mizutani: "The present state and issues of the motor employed in Toyota HEVs", *Proc. of the 29thSymposium on Motor Technology in Techno-Frontier*, 2009, pp.E3-2-1-E3-2-20.

[2] Mineral Resource Information Center affiliated to Japan Oil, Gas and Metals National Corporation: "Metal resources report", vol. 36, no.1, 2006, pp.11-16.

[3] T. Kosaka, T. Hirose and N. Matsui: "Brushless synchronous machine with wound-field excitation using SMC core designed for HEV drives", *Proc. The 2010 International Power Electronics Conference*, (IPEC 2010), Sapporo (Japan), June 2010.

[4] T.Jahns and V.Blasko, "Recent advances in power electronics technology for industrial and traction machine drives", *Proc. Of the IEEE, vol.89, no.6, pp.963-975, jun. 2001.*

[5] Mitch Olszewski, "Evaluation of the 2007 Toyota Camry Hybrid Synergy Drive System", Oak Ridge National Laboratory, U.S. Department of Energy, USA, 2009

[6] M. Takeno,N.Hoshi, A. Chiba, M. Takemoto, and S. Ogasawara, "A comparison of high power and high efficiency machines of 50kW SRM designed for HEVs", Proceedings of the 2011 IEE-Japan Industry Applications Society Conference, Vol. III, No.3-88, pp.III-407-III-412 (2011) (in Japanese)

[7] M. Takeno, Y. Takano, A. Chiba, N. Hoshi, M. Takemoto, and S. Ogasawara, "A test result of a 50kw switched reluctance motor designed for a hybrid electric vehicle," in Power and Energy Society General Meeting, 2011 IEEE , July 2011, pp. 1–2.

[8] M. Takeno, A. Chiba, N. Hoshi, S. Ogasawara, M. Takemoto, and M. Rahman, "Test results and torque improvement of the 50-kw switched reluctance motor designed for hybrid electric vehicles," IEEE Transactions on Industry Applications, vol. 48, no. 4, pp. 1327 –1334, July-Aug. 2012.

[9] M.Takeno, N.Hoshi, A.Chiba, M.Takemoto, and S.Ogasawara, "Test results at 1200-6000r/min of the Switched Reluctance Motor deigned for hybrid vehicles", 2011 National Convention Record, IEE Japan, Vol. 5, No.5-002, pp.2-3 (2011)(in Japanese)

[10] Y. Takano, M. Takeno, N. Hoshi, A. Chiba, M. Takemoto, S. Ogasawara, and M. Rahman, "Design and analysis of a switched reluctance motor for next generation hybrid vehicle without pm materials," in Power Electronics Conference (IPEC), 2010 International, June 2010, pp. 1801–1806.

[11] K. Kiyota, and A. Chiba, "Design and analysis of a 60kW Switched Reluctance Motor for Hybrid Electric Vehicles", Proceedings of the 2011 IEE-Japan Industry Applications Society Conference, Vol. III, No.3-87, pp.III-401-III-406 (2011)

[12] K.Kiyota, T.Kakishima, H.Sugimoto, and A. Chiba, "A comparison of analysis and test results of a 50kW switched reluctance motor and a design of the second prototype of SRM", The papers of Technical Meeting on Vehicle Technology, IEE Japan, VT-11-014, pp.53-58 (2011)(in Japanese)

[13] K.Kiyota, T.Kakishima, H.Sugimoto, and A. Chiba, "Test results of a 60kW Switched Reluctance Motor for Hybrid Electric Vehicle," The papers of Technical Meeting on Vehicle Technology, IEE Japan, VT-13-024, pp.55-60 (2013)(in Japanese)

[14] Y.Hasegawa, K.Nakamura, and O.Ichinokura, "Optimization of a Switched Reluctance Motor Made of Permendur", IEEE Trans. Magn., vol.46, No.6, pp.1311-1314 (2010)

[15] Y.Hasegawa, K.Nakamura, and O.Ichinokura, "Experimental Evaluation of Characteristics of SR Motor Made of Permendur", IEEJ Trans. on Industry Applications, vol.132, No.4, pp.458-463 (2012)

[16] B.C. Mecrow, J.W. Finch, E.A EI-Kharashi and A.G Jack, "Switched reluctance motors with segmental rotors", IEE *Proc. Of Elect. Power Appl., Vol.140, pp.245-254, 2002*

[17] J. Oyama, T.Higuchi, T.Abe, N.Kifuji, "Novel Switched Reluctance Motor with Segment Core Embedded in Aluminum Rotor Block", IEEJ Trans., IA, Vol.126, No.4, pp. 385-389, 2006

[18] M.Matsumoto, Y.Matsuo, T.Higuchi, nad T.Abe, "Design Analysis of Dual Rotor Segment Type Novel Switched Reluctance Motor", Proc. of the 2011 Japan Industry Applications Society Conference, 3-89, pp.413-416, 2011

[19] O.Kaneki, Y.Matsuo, T.Higuchi, T.Abe, Y.Miyamoto, and M.Ohto, "Characteristics of A Novel Segment Type Switched Reluctance Motor using Grain-Oriented Magnetic Steel", Proceedings of the 2011 IEE-Japan Industry Applications Society Conference, No.Y-124, pp.Y-123 (2011) (in Japanese)

[20] A. Zulu, B.C. Mecrow, M. Armstrong, "TOPOLOGIES FOR WOUND-FIELD THREE-PHASE SEGMENTED-ROTOR FLUX-SWITCHING MACHINES", Proc. of Energy Conversion Congress and Exposition, ECCE2010, pp. 1617-1622

[21] E. Sulaiman, T.Kosaka, N.Matsui, "A New Structure of 12Slot-10Pole Field-Excitation Flux Switching Synchronous Machine for Hybrid Electric Vehicles", Proc. of EPE2011, CD-ROM, 2011

[22] Y.Kuwahara, T.Kosaka, N.Matsui, Y.Kamada, and H.Kajiura, "Experimental Drive Characteristics of Field-Excitation Flux Switching Motor for HEV Drives", Proceedings of the 2012 IEE-Japan Industry Applications Society Conference, Vol. III, No.3-22, pp.III-133-III-136 (2012) (in Japanese)

[23] Y.Kuwahara, T.Kosaka, N.Matsui, Y.Kamada, and H.Kajiura, "Drive Performance Evaluation of Wound Field Flux Switching Motor for HV Drives", The paper of Technical Meeting on "Vehicle Technology", IEE Japan, VT-13-023, pp.49-54(2013) (in Japanese)

Recent Technical Trends in Variable Flux Motors

Akio Toba
Fuji Electric Co., Ltd.
Fuji-machi 1, Hino, Tokyo 191-8502,
JAPAN

Akihiro Daikoku
Mitsubishi Electric Corporation
Tsukaguchi-Honmachi 8-1-1,
Amagasaki, Hyogo 661-8661,
JAPAN

Noriyoshi Nishiyama
Panasonic Corporation
Yagumo-naka-machi 3-1-1, Moriguchi,
Oasaka 570-8501, Japan

Yuichi Yoshikawa
Panasonic Corporation
Morofuku 7-1-1, Daito,
Osaka 574-0044, JAPAN

Yosuke Kawazoe
Yaskawa Electric Corpolation
Nishimiyaichi 2-13-1, Yukuhashi,
Fukuoka 824-8511, JAPAN

Abstract— This paper reviews recent technical trends in variable flux motors. Permanent magnet motor is a major component to realize high torque density with keeping high efficiency. However, when high speed revolution is required for further miniaturization, as well as the large torque in the low speed region, the flux yielded by the permanent magnet encounters a serious trade-off, i.e. the former needs less flux and the latter requires much. The solution for this trade-off is varying the flux. In general, the flux of the permanent magnet is constant, and many efforts are under development to overcome this as to be shown in this paper.

Keywords— *review, permanent-magnet, trends, variable-flux*

I. INTRODUCTION

Increasing the maximum revolution is a basic strategy to minimize the motor size with keeping its power output and reducing its torque. This strategy is the most effective way to reduce the size and weight of the power-train in the automotive utilizing the electric drive system. This can be confirmed by the trend of the maximum revolution of the motors implemented in the electric vehicles (EV) or the hybrid electric vehicles (HEV) as shown in Fig. 1.

When the increase of the maximum motor-revolution comes with a wide range of the constant power, i.e. a large torque is needed in the low speed region as shown in Fig. 2, a tradeoff occurs with the air-gap flux of the motor. That is, large air-gap flux contributes to the increase of the torque, while it disturbs the high-speed rotation by making the terminal voltage high. This problem is critical in the permanent magnet motors because the magnetization of its permanent magnet is constant in general. In this case, field weakening control is usually employed. However, it accompanies the increase of the current, which results in the increase of the copper and iron losses. Therefore, it is ideal if the air-gap flux can be adjusted corresponding to the rotational-speed. Various efforts are done toward the realization of this ideal characteristic.

This paper reviews several remarkable researches and developments of variable flux motors.

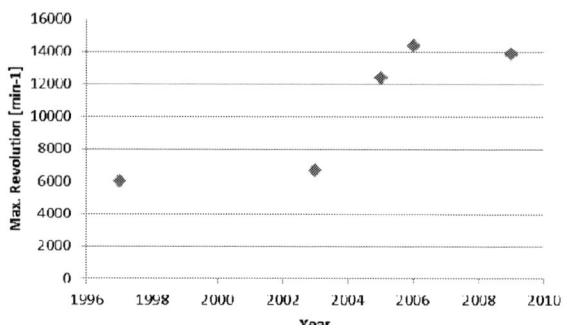

Fig. 1. Trend of the maximum revolution of the traction motors in commercialized HEVs.

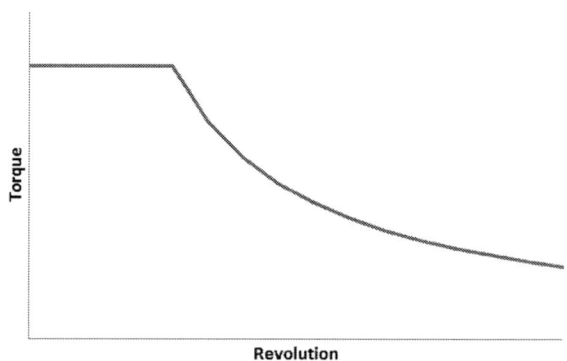

Fig. 2. Typical characteristics of torque vs. revolution of the motor for EV and HEV.

II. VARIABLE FLUX MOTORS

A. Hybrid-excitation motor [1]

Figure 3 shows the hybrid-excitation motor, in which the flux can be controlled by the hybrid usage of a permanent magnet and electric magnets. As can be seen in Fig. 3, the rotor has two salient pole parts stacked axially and a ring-shaped permanent magnet between them. The stator has armature coils outside the rotor in the radial direction, and two end-plates with field coils in both sides of the axial direction. The motor has three-dimensional magnetic paths including the end-plates, and these plates are made by soft magnetic composite (SMC).

The three-dimensional flux paths and the principle of the field strength control are shown in Fig. 4. Figure 4 (a) shows the flux paths without field coil excitation. The black arrows indicate the flux path which links to the armature coils via the air gap between rotor and stator, and the pink arrows indicate the leakage flux path which goes through the back yoke of the stator and does not link to the armature coils. Figure 4 (b) shows the field weakening excitation condition. The flux path excited by the field coil, drawn as the pink arrows in Fig. 4 (b), is the same as the leakage flux path in Fig. 4 (a). When the magnetic flux density of the back yoke becomes higher and saturated, the amount of magnetic flux linked to the armature coils is decreased and the magnetic field is weakened. Figure 4 (c) shows the field strengthening excitation condition. The flux path excited by the field coil, drawn as the green arrows in Fig. 4 (c), goes through the salient poles of the rotor and links to the armature coils via air gap and come back to the SMC field-pole shaped in the end-plate. Since the direction of the flux excited by the field coil is the same as the flux excited by the permanent magnet, the magnetic field strengthened.

A prototype machine of the hybrid-excitation motor has been designed and the characteristics of the motor have been simulated. Specifications of the prototype are as follows: the maximum output power is 123 [kW], and the maximum rotational speed is 20,000 [r/min].

Figure 5 shows typical characteristics of the prototype. The

variation of the flux-linkage of the armature coil is shown in Fig. 5 (a). The origin of the vertical axis points the flux linkage without field coil excitation and the scale of the vertical axis means the flux variation rates [%]. The data is calculated by monitoring the open-circuit induced voltage of

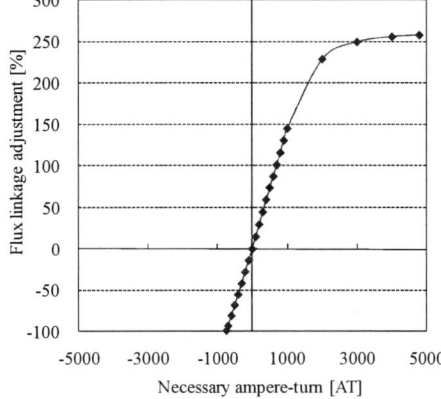

(a) Variation of flux-linkage by the current of the field toroidal coil.

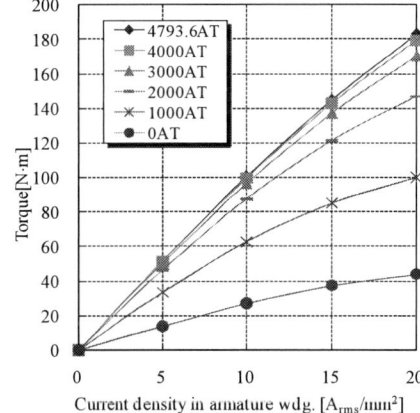

(b) Torque vs. armature current with various field current.

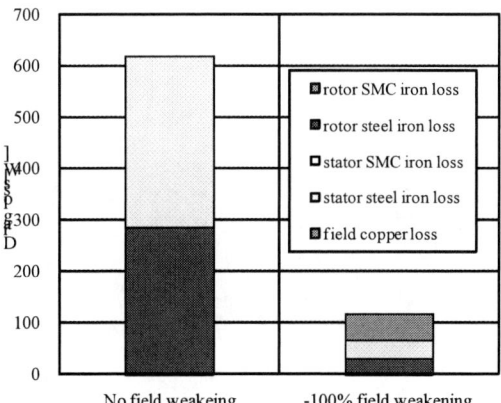

(c) Comparison of drag loss with / without field weakening.

Fig. 5. Results of the prototype machine of hybrid-excitation motor.

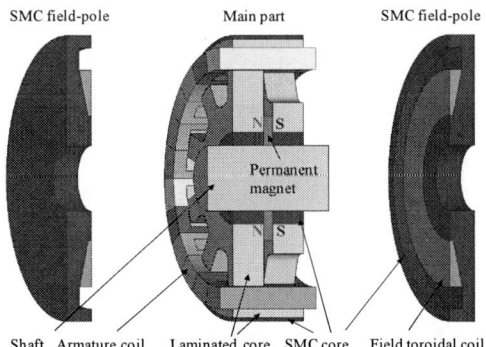

Fig. 3. An overview of hybrid-excitation motor.

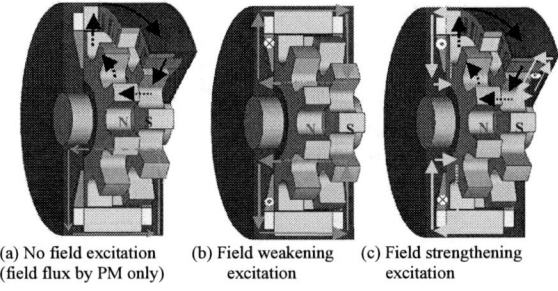

(a) No field excitation (field flux by PM only)
(b) Field weakening excitation
(c) Field strengthening excitation

Fig. 4. The field control principle of hybrid-excitation motor.

the armature coil by rotating the rotor. It can be realized that the flux-linkage can be adjusted with the range of -100% to 250%. Here, -100% means that the flux-linkage of the armature coils can be canceled out by exciting the field coil with -750 [AT].

Figure 5 (b) shows the torque vs. armature current with various filed current. With the same current density in the armature coil, it is confirmed that the torque can be increased by increasing the field current. The amount of the torque is almost proportional to the field current up to 2000 [AT], while the increasing rate of the torque falls down with further field current because of the saturation of the magnetic flux path.

Figure 5 (c) shows the effect of the magnetic field control. The vertical axis indicates the drag loss, which is the loss of the motor with no load, in the maximum rotational speed. The drag loss reaches 619 [W] without field control, while it is reduced to 64 [W] with the field weakening.

In this study, a down-scaled prototype is manufactured and the accuracy of the simulated result has been confirmed [1].

Through these studies, it is proved that the proposed machine has the ability of the flux variation and thus the suitable performance can be obtained to fit the required characteristics shown in Fig. 2.

B. Claw-pole motor for EV/HEV traction [2]

Claw-pole motors are widely used as the alternator in automotive that usually have small capacity as a few kilo-watts. Claw-pole motor is a kind of synchronous motor with field excitation coil. The proposed motor is the one aiming the application to the traction for EV or HEV, which has typically several-ten kilo-watts.

Figure 6 and figure 7 show the overview of the rotor and the cross section of the prototype of the proposed claw-pole motor, respectively. The number of poles is 32, and a ferrite magnet is inserted in every gap between two poles. There is also an excitation coil, usually set inside of the rotor but placed at the stator in this prototype, which eliminates the slip-rings necessary for power-feeding to the rotating field coil. The major flux contributing to the motor torque, drawn by the red arrows in Fig. 7(b), is made by exciting the field coil, while the flux excited by the ferrite magnets acts to relax the magnetic saturation in the rotor core, as shown by the purple dashed arrows in Fig. 7(b).

One unique point of this type of the motor is that the flux

can be optimized by controlling the current of the excitation coil and the d- and q- axis currents of the stator coil. Figure 8 shows the efficiency maps of the prototype with and without the optimization of the excitation flux. In the case without controlling the field current, the amount of the field current is fixed to the minimum value required for driving with JC08 mode. As to be seen, the overall efficiency is high with the optimization, and it is remarkable that high efficiency is accomplished in the low-speed and low-torque region with the optimization, that is especially important in the EV or HEV.

Figure 9 shows the comparison of the total losses at the two typical drive-conditions with the variable field excitation control and the fixed field excitation. As for the condition I, the torque is 65 [Nm] and the rotational speed is 1000 [r/min]. In this case, the total loss has decreased by 230[W] with variable field excitation control. The breakdown of the loss difference is that the field loss has decreased by 360[W], on the other hand the copper and the iron losses have increased by 80[W] and 50[W], respectively. The reason is that the armature current with the variable field excitation control has increased compared to that with the fixed field control so as to keep the same output torque. As for condition II, the torque is 25 [Nm] and the rotational speed is 3000 [r/min]. In this case, it can be found that the total loss with the variable field excitation control has decreased by 540[W] compared to that

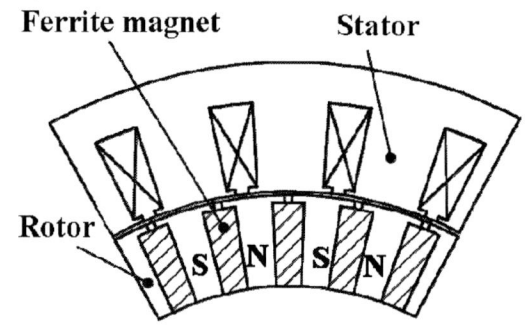

(a) Cross-section of the claw-pole motor with the plane normal to the rotational axis.

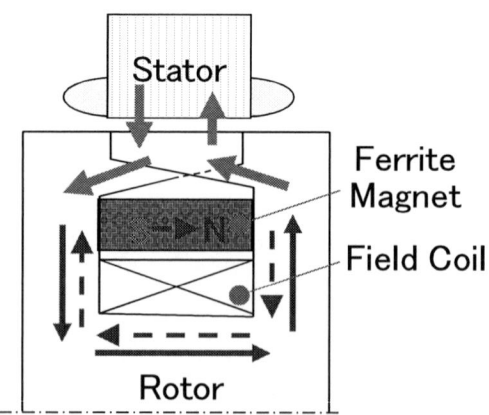

(b) Cross-section of the claw-pole motor with the plane including rotational axis.

Fig. 7. Cross section and flux path of the claw-pole motor.

Fig. 6. An overview of the rotor of claw-pole motor for HEV usage.

with the fixed field excitation control. The breakdown of the loss difference is that the field loss has decreased by 580[W], while the copper and the iron loss have both increased similar to the condition I.

The above-mentioned results suggest that the strength of the magnetic field, decided in the minimum value required for driving through JC08 mode, is too strong in terms of minimizing losses. Therefore, the variability of the magnetic field strength is very effective for raising the efficiency of EV or HEV.

C. Variable-Magnetization Permanent Magnet Motor [3],[4]

The magnetization of the permanent magnets in a motor is constant in general, and this is regarded to be desirable in terms of the operation stability. In the presenting motor, however, the magnetization of permanent magnets is designed to be variable to enhance the overall performance.

Figure 10 depicts the basic ideas of the rotor structure having permanent magnets with variable-magnetization. Two types of layouts are depicted, i.e. the series type and the parallel type. In both types, the flux linkage can be variable. In a rotor, two kinds of permanent magnets are placed, which have high and low coercive forces, respectively. The NdFeB

magnet is a typical example for the former, while the alnico magnet is the one for the latter.

Figure 11 indicates the magnetic characteristics of permanent magnet for variable-magnetic force. The coercive forces of NdFeB and alnico magnets are about 1,000kA/m and 100kA/m, respectively. Typical maximum magnetic field that can be provided by the stator coil is about 500kA/m. Therefore, by choosing an adequate NdFeB magnet for the one having the high coercive force, irreversible demagnetization can be avoided for it. For the other type of the permanent magnet having the lower coercive force, the magnetic force of the coil current is enough to change its magnetization. Therefore, the total flux of the two types of the permanent magnet can be controlled by adding pulsation

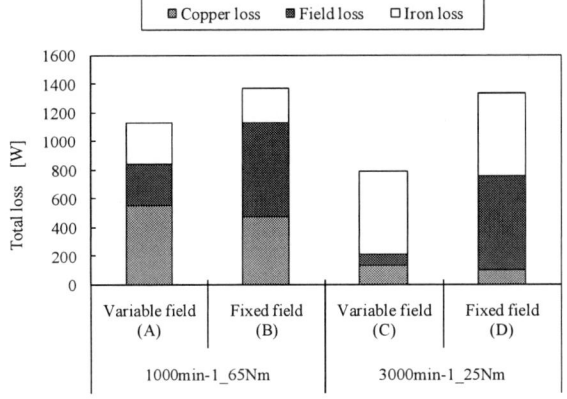

Fig. 9. Comparison of total loss with variable field excitation control and fixed field excitation

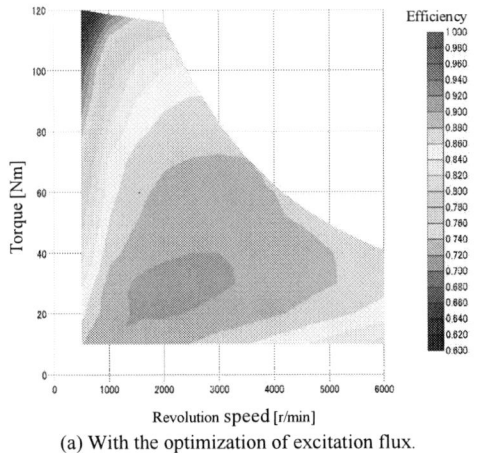

(a) With the optimization of excitation flux.

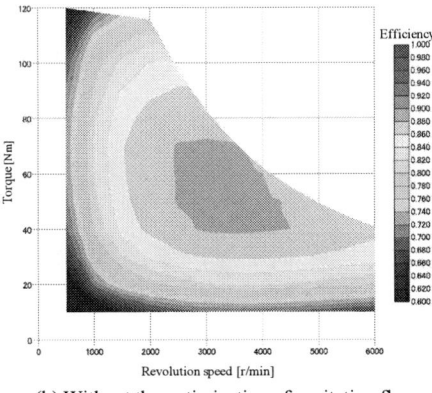

(b) Without the optimization of excitation flux.

Fig. 8. Efficiency maps of the prototype machine of the claw-pole motor.

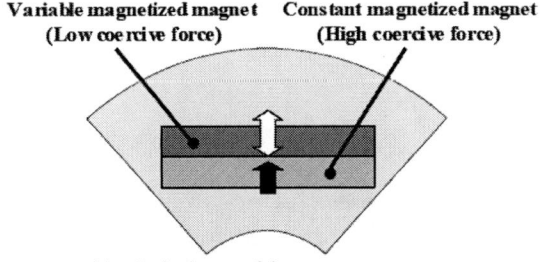

(a)   Series layout of the permanent magnets.

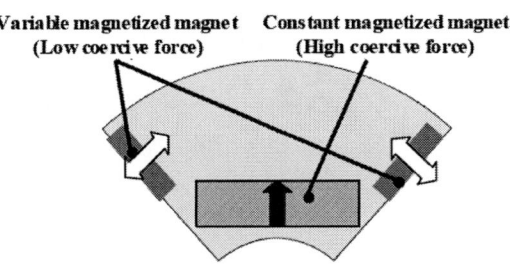

(b)   Parallel layout of the permanent magnets.

Fig. 10. Structures of the rotor having permanent magnets with variable-magnetization.

current of the stator coil amid motor drive operations.

Figure 12 depicts the change of operating point for the variable-magnetized magnet. At first, the operating point of the permanent magnet is at the point A. When the permanent magnet is demagnetized by negative d-axis current of the stator coil, the operating point is moved to the point B. If the negative d-axis current to the stator coil is further increased, the operating point goes through the point C and reaches to the point D. Then, by removing the negative d-axis current, the operating point is moved from the point D to the point E along the recoil permeability. That is, the magnet has negative magnet flux.

In this situation, for the case of the parallel magnet layout as shown in Fig. 10(b), the total flux linkage can be decreased because of the magnetic flux difference between the permanent magnets of the constant and variable magnetization. Therefore, the flux-weakening current can be much smaller compared to the conventional constant-magnetization PM motor, since the back EMF becomes low. This brings a merit that the amount of the expensive Dy in the NdFeB magnet can be decreased due to the lower demagnetization force by the coil current.

This type of the motor has already been in the commercial use for home laundry machines that have two-mode operations of the motor, i.e. washing and dewatering. The washing operation requires low-speed and high-torque, while the dewatering needs high-speed and low-torque. Figure 13 shows the overview of the motor. The motor is an outer-rotor type and the magnetic poles are configured by placing the permanent magnets of high and low coercive force alternately. This can be regarded as another type of the series layout of the permanent magnets described in Fig. 10(a).

D. Permanent magnet motor with variable-magnetization and pole-change [5]

With the above described motors, they are all aiming to overcome the trade-off between low-speed-high-torque and high-speed-low-torque characteristics. To pursue this more, there is another item to remark; the pole-change. Figure 14 illustrates typical driving characteristics of electric vehicles.

Sakai et al. proposed a novel PM motor called Three-torque Mode Pole-Changing PM motor (3M-PC PM motor) that can change the number of poles and generate three different types of torque. As shown in Fig. 14, when the motor operates with 8 poles in the low-speed area, it produces permanent magnet (PM) torque. When the motor operates with 4 poles in the medium-speed area, it produces PM torque and reluctance torque. Then, at the high-speed area, the motor produces only reluctance torque.

Figure 15 shows the configuration of 3M-PC PM motor. The rotor of the 3M-PC PM motor has a salient core and PMs with low coercive force embedded in the iron core. When all the variably magnetized magnets have the same direction of

(a) An overview.

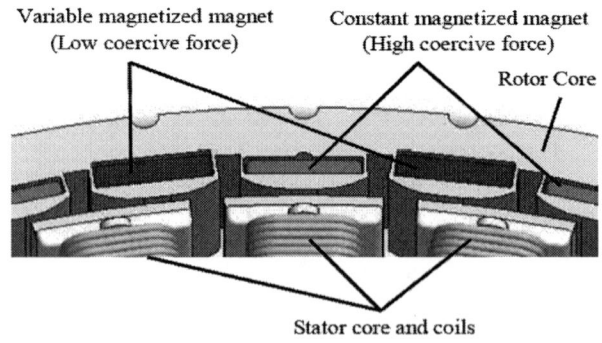

(b) Detailed structure of the rotor and the stator.

Fig. 13. Permanent magnet motor with variable-magnetization for home laundry.

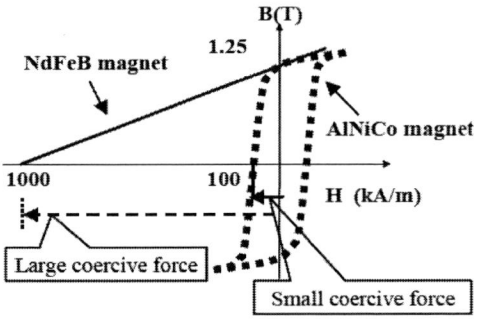

Fig. 11. Magnetic characteristics of permanent magnet for variable-magnetic force.

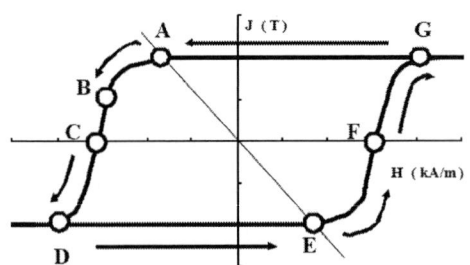

Fig. 12. Change of operating point for the variable-magnetized magnet.

magnetization, magnetic poles with the reverse polarity to the magnets are formed in the rotor core between the adjacent permanent magnet poles. Thus, the rotor forms 8 poles (Fig. 16(a)). When the adjacent permanent magnets of the rotor has opposite polarity, the rotor has 4-poles (Fig. 16(b)). Finally, if all the permanent magnets of the rotor has no magnetization, the rotor works as a 4-pole reluctance one (Fig. 16(c)). Figure 17 shows the winding connections for pole changing of the stator. As can be seen, the connection of the armature winding can be switched between 8-pole and 4-pole.

The change of the motor operation among the areas shown in Fig. 14 can be accomplished as follows. At the low-speed area, the rotor and the stator both has 8-pole configuration. Then at first, the armature winding is changed from 8- to 4-poles, followed by the d-axis pulse current to change the number of the rotor poles from 8 to 4. With this, the motor operation can be shifted to the medium-speed area. Next, by applying an adequate negative d-axis current, the permanent magnets of the rotor can be demagnetized and the motor operation enters to the high-speed area.

Figure 18 shows the distribution of the magnetic flux density in the motor for the three pole-configurations. When all the PMs are magnetized with the same polarity, the magnetic flux forms an 8-pole distribution, as shown in Fig. 18(a). When the PMs are magnetized in the polarity opposite to their adjacent poles, the magnetic flux forms a 4-pole distribution, as shown in Fig. 18(b). When all PMs are demagnetized, the magnetic flux resulting from the excitation current forms a 4-pole distribution, as shown in Fig. 18(c).

Figure 19 shows the variable characteristics of the induced voltage. It can be confirmed that the frequency for the 4-pole-IPM mode is just one half of the one for the 8-pole-PM mode. Figure 20 shows the torque characteristics during rotation with the same armature current. As to be seen, the torque decreases corresponding to the modes with the speed areas. Figure 21 shows the core loss of the stator in each pole mode. The results indicate that the core loss in the 4-pole mode decreases by 21% of that in the 8-pole mode. At the moment, all those results are obtained from the FEM analysis. Experimental evaluations of this type of the motor will show more details.

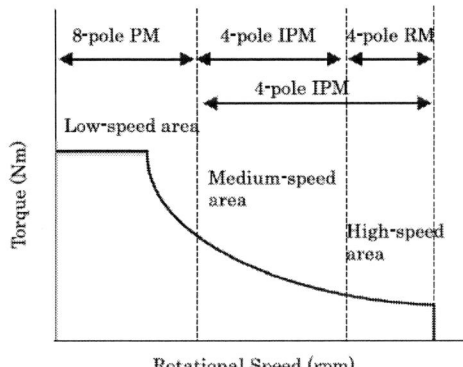

Fig.14. Driving characteristics typical of electric vehicles

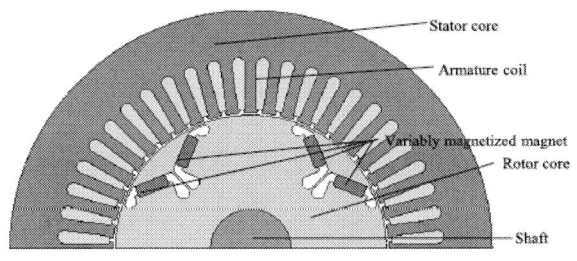

Fig.15. Configuration of a 3M-PC PM motor

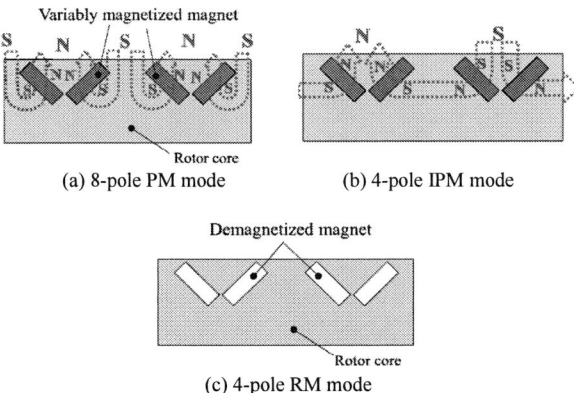

Fig.16. Change in poles and torque modes in a 3M-PC-PM motor

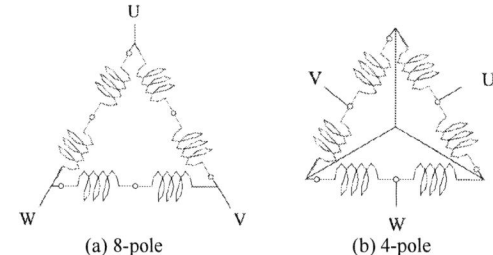

(a) 8-pole   (b) 4-pole

Fig.17. Winding connections of a 3M-PC PM motor

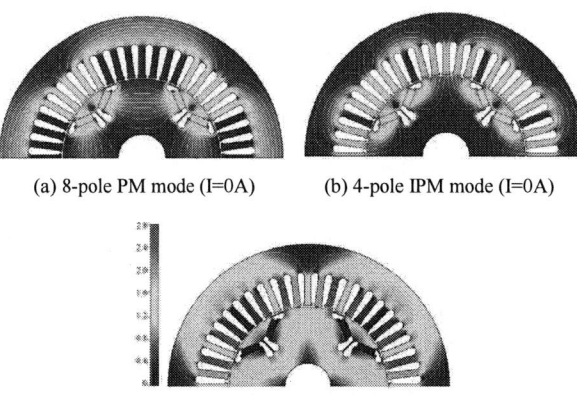

Fig.18. Flux distribution in each mode of 3M-PC-PM motor

E. PERMANENT MAGNET MOTOR WITH WINDING CHANGEOVER METHOD

The winding change motor which has dual characteristics in one motor is developed. It is possible to change machine parameters such as winding resistance, inductance, and magnetic flux, by changing the winding connections.

1) Constitution of Winding Changeover Method

Figure 22 shows the circuit that was used for electronic

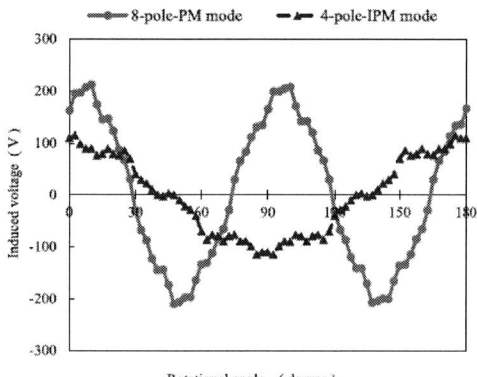

Fig.19. Variation of induced voltage by pole changing

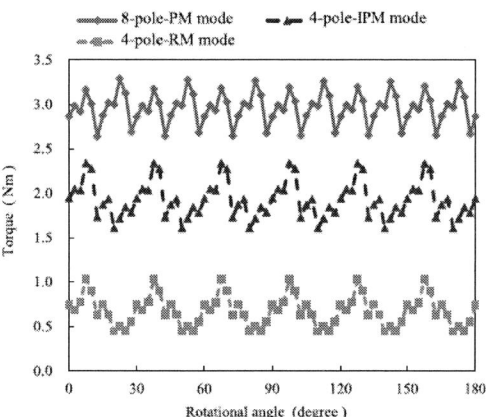

Fig.20. Torque characteristics of a 3M-PC PM motor

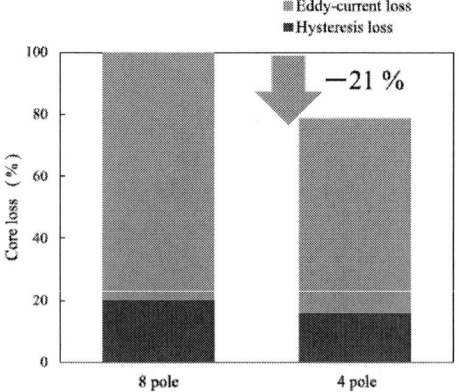

Fig.21. Reduction of core loss in 3M-PC PM motor by pole changing

winding change method[6]. The motor is an IPM motor and has specially wound coils with center taps. By short-circuiting the center tap of the three-phase winding to form the neutral point, the number of turns of the winding is decreased. This leads to the step change of the emf, resulting in a wider constant power operating range in high-speed region.

The switches SW1 and SW2 shown in Fig. 22 are complementary, i.e. their ON/OFF states are always opposite to each other. The combination of a diode bridge and a switching device (S1 and S2) works as a switch to short-circuiting or opening of the three-phase circuit in the motor. By utilizing the semiconductor switch, the change-over can be performed quickly.

The voltage across the open windings during high-speed and low-speed operation can be visualized using the vector representation shown in Fig.23. Figure 23 (b) shows that the induced voltage across the open terminals X2, Y2, and Z2 do not exceed the applied voltage across X1, Y1, and Z1.

Figure 24 summarizes the torque, voltage and power characteristics to the speed with the winding change over. As to be seen, the voltage can be suppressed below the limitation in both of the low and high speed operations. Torque becomes one half at the high speed operation.

The motor parameters used in the motor control algorithm of the inverter controller are changed according to the set of the winding at the time of transition from either high-speed to low-speed or vice versa. There is no dead time between turn-off of one switch and turn-on of the other switch to prevent "torque shock" for pursuing comfortable vehicle ride.

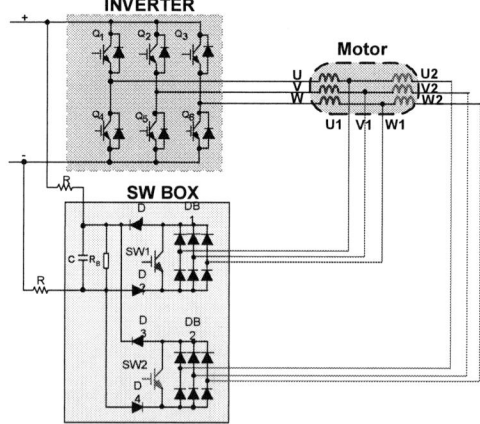

Fig. 22. Schematic of winding change circuit

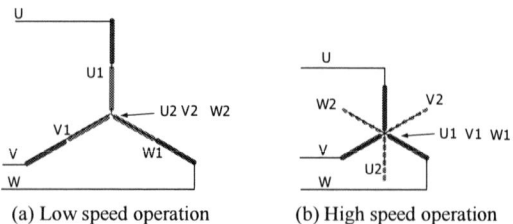

(a) Low speed operation   (b) High speed operation

Fig. 23. Voltage representation across open windings

Waveforms at the instant of winding changeover action in the test running of actual electric vehicle are shown in Fig.25. Since no-dead-time switch control and fast change of the motor control algorithm are applied to this system, the motor current becomes continuous despite the winding changeover action. The driver of electric vehicle cannot feel any shock at the instant of winding changeover action because of the stable torque condition.

Figure 26 shows the efficiency map measured from the inverter to the motor with a winding change unit. This motor drive system has wide high efficiency area. Two high efficiency areas can be recognized in the map, which correspond to city area and highway for a vehicle.

2) Development Example

Figure 27 shows the drive system employing the winding change motor / inverter, which is designed for an electric vehicle. The vehicle equipped with the system has been released. The specifications of the system are as follows: maximum power 75kW(102ps) / 5,200-12,000min-1, maximum torque 150N·m (15.3kgfm) / 0~2,800 min-1.[7]

III. CONCLUSIONS

This paper reviews several remarkable proposals of variable flux motors. Although it was regarded to be only an idea a few years ago, there are already examples of having been commercialized as described in this paper. Its practical applications have just started and more improvements and refinements of this concept can be expected in coming years.

REFERENCES

[1] T. Kosaka, T. Hirose, and N. Matsui, "Some Considerations on Experimental Drive Characteristics of Less Rare-Earth HEM", Proc. of the 2011 JIASC, Vol.I, No.I-O6-2, pp.I-85 - I-90, 2011.

[2] M. Azuma, M. Hazeyama, M. Morita, Y. Kuroda, A. Daikoku, and M. Inoue, "Driving Characteristics of a Claw Pole Motor Using Field Excitation for Hybrid Electric Vehicles", The Papers of Technical Meeting on Vehicle Technology IEEJ, No. VT-12-013, pp. 49-54, 2012.M. Young, "The PWM strategy on DC-DC converter," *IEEJ Journal of Industry Applications*, vol. 28, no. 15, pp. 123-129, 1989.

[3] I. Nitta, S. Maekawa, and T. Shiga, "Serial type Variable-Magnetic-Force Motor", The 2010 Annual Meeting Record IEEJ, No. 5-013, pp.20-21, 2010.

[4] K. Sakai, K. Yuki, Y. Hashiba N. Takahashi K. Yasui and L. Kovundhikulrungsri, "Principle and basic characteristics of variable-magnetic-force memory motors", IEEJ Trans. IA, Vol.131, No.1, pp.8,53-60, 2011.

[5] K. Sakai, N. Yuzawa, H. Hashimoto, "Permanent Magnet Motors Capable of Pole Changing and Three-Torque-Production Mode using Magnetization", IEEJ Journal of Industry Applications, Vol.2, No.6, pp.269–275, 2011.

[6] Akihiko Maemura, "Extended High Speed Operation via Electronic Winding Change Method for Interior Permanent Magnet Motors", 2010 JIASC, 2-S8-6.

[7] Mazda News Releases : "Mazda to Lease 'Demio EV' Electric Vehicle in Japan from October," http://www.mazda.com/publicity/release/2012/201207/120706a. html, 2012/07/06

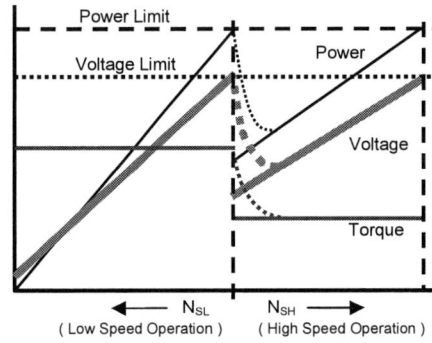

Fig. 24. Dual torque characteristics when proposed method is applied to IPM motor.

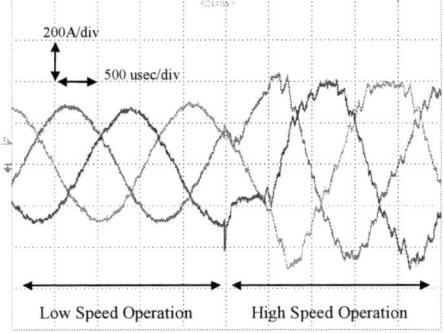

Fig. 25. Actual motor current waveform at the winding changeover instant.

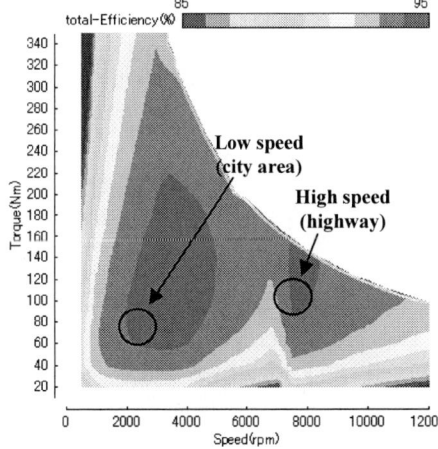

Fig.26. The total Efficiency map which is measured from inverter to motor with winding change unit.

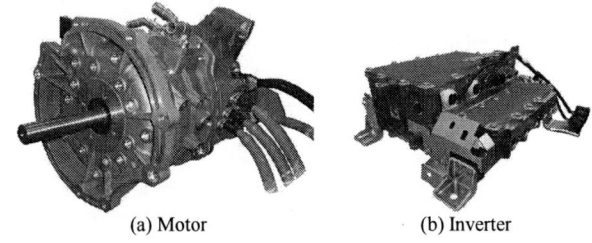

(a) Motor   (b) Inverter

Fig. 27. The winding change motor / inverter for EV.

A General Discrete Time Model to Evaluate Active Damping of Grid Converters with LCL Filters

S. G. Parker   B. P. McGrath   D. G. Holmes

School of Electrical and Computer Engineering
RMIT University, Melbourne, Australia

stewart.parker@student.rmit.edu.au   brendan.mcgrath@rmit.edu.au   grahame.holmes@rmit.edu.au

Abstract — **While LCL filters offer significant benefits for grid connected converters, they always require management of their inevitable filter resonance. Many active damping strategies have been proposed to resolve this issue, using various feedback combinations of the converter side, filter capacitor and grid side currents. However, these strategies are often presented as continuous time formulations that ignore the impact of converter PWM delays, and are also not usually evaluated on an exactly comparable basis. This makes it hard to select between them to find the most suitable alternative for a specific context.**

This paper presents a generalised discrete time model for a grid connected VSI with an LCL filter, fully accounting for all practical converter delays and second order effects. The model is arranged to allow any active damping strategy to be readily integrated and evaluated under identical operating conditions, so that a comprehensive comparative evaluation of its relative damping performance can be easily performed. Experimental and simulation results are presented to compare the relative performance of several well known active damping strategies, to illustrate the capability and flexibility of the generalised model.

Keywords— LCL Filter, Active Damping, Discrete Time, Grid Connected.

I. INTRODUCTION

Grid connected converters need series line filters to attenuate the high frequency harmonic currents that are produced by the switching action of their converter. Increasingly, inductive-capacitive-inductive (LCL) filters are being used for this purpose since they provide superior harmonic attenuation for the same volume, weight and cost as a simple first order inductive (L) filter [1][2]. However LCL filters introduce additional control complexity since they are resonant, with a pair of open loop system poles on the closed loop stability boundary that must be damped to ensure system stability [1]-[3].

The preferred approach to achieve this damping is an active feedback system, where the control system uses feedback of specific dynamic state variables to emulate a virtual damping resistor [3][4]. Feedback of the filter capacitor current is the most conventional approach, and provides a simple and robust active damping strategy [1][2][5]. Alternative active damping strategies achieve damping action by feeding back different state variables. For example converter side current feedback provides inherent damping action, irrespective of whether it is the primary control target [6] or used only for damping [7]. Other approaches have proposed a weighted combination of converter and grid side currents as the feedback variable [8], or hybrid combinations of weighted average current and capacitor current feedback [9]. Of course, each alternative claims some degree of additional benefit.

Choosing between these damping alternatives is quite challenging, since their performance is rarely presented on an exactly comparable basis, and their gains are often set without taking account of the critical constraint of sampling and PWM transport delays [1][4]. This paper now addresses this issue, by presenting a generalised model structure that allows any form of capacitor current type active damping to be modelled and evaluated on an exactly comparable basis. The model uses a discrete time formulation that fully accounts for sampling and PWM transport delays. Comparable gains are set using matched closed loop pole locations, to achieve equivalent transient or resonant damping performance. Evaluations of several active damping strategies with matching simulation and experimental results are presented to confirm the validity and usefulness of the generalised model.

II. DISCRETE TIME MODEL OF AN LCL SYSTEM

Fig. 1 shows the topology of a grid connected converter incorporating an LCL filter with converter and grid side inductances L_1 and L_2, and a filter capacitance C_f. All active damping control strategies appropriate for this paper are anticipated to be implemented digitally, and measure one or more of the LCL filter currents (i.e. i_1, i_2 or i_c) and set the reference command signals for a

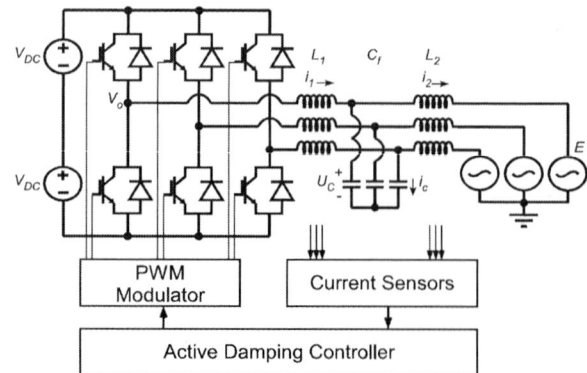

Fig. 1: Three phase grid connected voltage source inverter topology, with an LCL filter and active damping control.

$$\mathbf{A_d} = \begin{bmatrix} \dfrac{L_1 + L_2 \cos(\omega_{res}T)}{L_1 + L_2} & \dfrac{-\sin(\omega_{res}T)}{\omega_{res}L_1} & \dfrac{L_2 - L_2 \cos(\omega_{res}T)}{L_1 + L_2} \\[2ex] \dfrac{\sin(\omega_{res}T)}{\omega_{res}C_f} & \cos(\omega_{res}T) & \dfrac{-\sin(\omega_{res}T)}{\omega_{res}C_f} \\[2ex] \dfrac{L_1 - L_1 \cos(\omega_{res}T)}{L_1 + L_2} & \dfrac{\sin(\omega_{res}T)}{\omega_{res}L_2} & \dfrac{L_2 + L_1 \cos(\omega_{res}T)}{L_1 + L_2} \end{bmatrix}, \ \mathbf{B_d} = \begin{bmatrix} \dfrac{V_{DC}L_1\omega_{res}T + V_{DC}L_2 \sin(\omega_{res}T)}{L_1\omega_{res}(L_1 + L_2)} \\[2ex] \dfrac{V_{DC}L_2 - V_{DC}L_2 \cos(\omega_{res}T)}{L_1 + L_2} \\[2ex] \dfrac{V_{DC}\omega_{res}T - V_{DC}\sin(\omega_{res}T)}{\omega_{res}(L_1 + L_2)} \end{bmatrix}, \ \mathbf{C_d} = \begin{bmatrix} 0 \\ 0 \\ 1 \end{bmatrix}^T \quad (1b)$$

regularly sampled pulse width modulator. Hence a discrete time dynamic model of the physical converter and LCL filter is required to properly assess the control system performance. This model can be readily derived by applying a Zero-Order-Hold transformation [10] to the well-known continuous time state space equation of the converter with an LCL filter shown in Fig. 1, to yield [5]:

$$\begin{aligned} \mathbf{x_d}(k+1) &= \mathbf{A_d}\mathbf{x_d}(k) + \mathbf{B_d}m(k) \\ i_2(k) &= \mathbf{C_d}\mathbf{x_d}(k) \end{aligned} \quad (1)$$

where the state vector is $\mathbf{x_d}(k) = \begin{bmatrix} i_1(k) & U_c(k) & i_2(k) \end{bmatrix}^T$, the system output is the grid current $i_2(k)$ and the input is the PWM command signal $m(k)$. The $\mathbf{A_d}$, $\mathbf{B_d}$ and $\mathbf{C_d}$

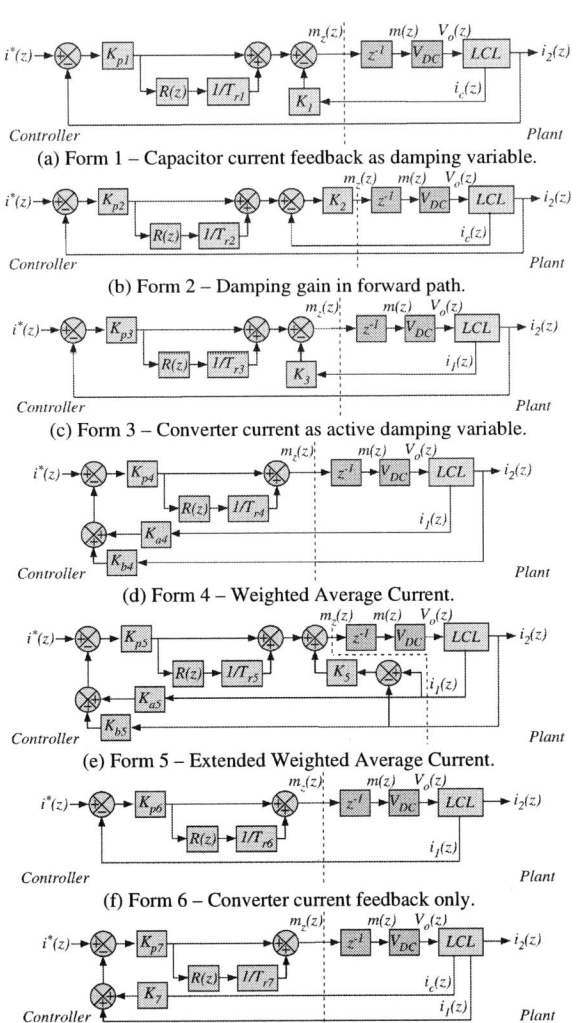

(a) Form 1 – Capacitor current feedback as damping variable.

(b) Form 2 – Damping gain in forward path.

(c) Form 3 – Converter current as active damping variable.

(d) Form 4 – Weighted Average Current.

(e) Form 5 – Extended Weighted Average Current.

(f) Form 6 – Converter current feedback only.

(g) Form 7 – Converter current feedback with additional capacitor current active damping.

Fig. 2: Control system block diagrams of each active damping form.

state space matrices in (1a) are given in (1b) above, with V_{DC} defined as half the DC link voltage (see Fig. 1), T is the sampling period (equal to a half-carrier interval) and the resonant frequency ω_{res} is defined as:

$$\omega_{res} = \sqrt{(L_1 + L_2)/L_1 L_2 C_f} \quad (2)$$

Note that all parasitic resistances in the LCL filter have been neglected to show the worst case damping scenario.

The asymmetric regularly sampled PWM process introduces a one sample period transport delay [1]. This can be accounted for by defining a new input modulation signal $m_z(k)$, and delaying the actual modulation command $m(k)$ by one sample period, viz.:

$$m(k) = m_z(k-1) \quad (3)$$

Hence, $m(k)$ becomes an additional state variable that can be incorporated into (1). System analysis then proceeds by defining a general closed loop state space equation for various active damping control strategies.

III. GENERALISED DISCRETE TIME CLOSED LOOP MODEL

Fig. 2 shows the controller architectures of several capacitor current feedback active damping strategies that have been reported in the literature [6]-[18]. The dynamic converter model defined in (1) is represented by the inverter gain V_{DC} and the *LCL* filter blocks, while the z^{-1} block accounts for the PWM transport delay defined in (3). In each case a proportional plus resonant (PR) controller is used which has a proportional gain K_{px} and a resonant reset time T_{rx}, where x denotes the active damping architecture being considered (e.g. 1 to 7). The term $R(z)$ is a discrete representation of the resonant filter required for the PR controller and is defined as [1]:

$$R(z) = \frac{\sin(\omega_0 T)}{2\omega_0} \frac{z^2 - 1}{z^2 - 2\cos(\omega_0 T)z + 1} \quad (4)$$

To combine this controller with the inverter model in (1), the resonant dynamics defined in (4) must be expressed in state space form [10], given by:

$$\begin{aligned} \mathbf{x_r}(k+1) &= \mathbf{A_r}\mathbf{x_r}(k) + \mathbf{B_r}u(k) \\ m_r(k) &= \mathbf{C_r}\mathbf{x_r}(k) + D_r u(k) \end{aligned} \quad (5a)$$

where $\mathbf{x_r}(k) = \begin{bmatrix} w_1(k) & w_2(k) \end{bmatrix}^T$ are the resonant filter state variables, $u(k)$ is input signal to the resonant filter $R(z)$ (see Fig. 2), ω_0 is the grid frequency, and:

$$\mathbf{A_r} = \begin{bmatrix} 2\cos(\omega_0 T) & -1 \\ 1 & 0 \end{bmatrix}, \quad \mathbf{B_r} = \begin{bmatrix} 1 \\ 0 \end{bmatrix}, \quad D_r = \frac{\sin(\omega_0 T)}{2\omega_0},$$

$$\mathbf{C_r} = \begin{bmatrix} \dfrac{\cos(\omega_0 T)\sin(\omega_0 T)}{\omega_0} & -\dfrac{\sin(\omega_0 T)}{\omega_0} \end{bmatrix}. \quad (5b)$$

Each system in Fig. 2 is now analysed by combining equations (1), (3) and (5) in accordance with each specific system architecture. To illustrate, consider Form 1 in Fig. 2(a), which shows capacitor current feedback through the damping gain K_1, and has the control law:

$$u(k) = K_{p1}\left(i^*(k) - i_2(k)\right) \quad (6a)$$

$$\begin{aligned} m_z(k) &= u(k) + m_r(k)/T_{r1} - K_1 i_c(k) \\ &= u(k) + m_r(k)/T_{r1} - K_1[i_1(k) - i_2(k)] \end{aligned} \quad (6b)$$

Observe that in the second line of (6b) the capacitor current has been expressed in terms of the LCL filter state variables $i_1(k)$ and $i_2(k)$ to facilitate the formation of a closed loop state space equation. Combining (1), (3), (5) and (6) yields after some algebra a 6th order closed loop state space equation for the converter with the reference input $i^*(k)$, viz.:

$$\begin{aligned} \mathbf{x_{cl}}(k+1) &= \mathbf{A_{cl}}\mathbf{x_{cl}}(k) + \mathbf{B_{cl}}i^*(k) \\ i_2(k) &= \begin{bmatrix} 0 & \mathbf{C_d} & 0 \end{bmatrix}\mathbf{x_{cl}}(k) \end{aligned} \quad (7a)$$

where $\mathbf{x_{cl}}(k) = \begin{bmatrix} m(k) & \mathbf{x_d}(k) & \mathbf{x_r}(k) \end{bmatrix}^T$. The $\mathbf{A_{cl}}$ and $\mathbf{B_{cl}}$ state space matrices are defined as follows:

$$\mathbf{A_{cl}} = \begin{bmatrix} 0 & -\mathbf{I_{cl}} - (1 + D_r/T_{rx})\mathbf{J}_{cl} & \mathbf{C_r}/T_{rx} \\ \mathbf{B_d} & \mathbf{A_d} & 0 \\ 0 & -\mathbf{J_{cl}}\mathbf{B_r} & \mathbf{A_r} \end{bmatrix} \quad (7b)$$

$$\mathbf{B_{cl}} = \begin{bmatrix} (1 + D_r/T_{rx})K_{px} \\ 0 \\ \mathbf{B_r}K_{px} \end{bmatrix} \quad (7c)$$

where the vectors $\mathbf{I_{cl}}$ and $\mathbf{J_{cl}}$ vectors contain the controller feedback gains. For Form 1 these vectors are:

$$\mathbf{I_{cl}} = \begin{bmatrix} K_1 & 0 & -K_1 \end{bmatrix} \quad \mathbf{J_{cl}} = \begin{bmatrix} 0 & 0 & K_{p1} \end{bmatrix} \quad (8)$$

Applying this principle to all of the control system architectures of Fig. 2 reveals that equation (7) is in fact a generalised formulation, and it is only the vectors $\mathbf{I_{cl}}$ and $\mathbf{J_{cl}}$ that change. This allows for the rapid exploration of closed loop system behaviour, irrespective of the controller type. This paper will now apply the developed model to compare the behavior of the systems in Fig. 2.

IV. GAIN SELECTION PROCEDURE

The closed loop system responses for the architectures of Fig. 2 will be explored considering the LCL converter with parameters given in Table I. Controller gains are selected using the principles in [1], in which the proportional gain K_{px} and reset time T_{rx} are set so as to maximise the closed loop system bandwidth (from [1]

this occurs at $\omega_c = 0.36\omega_{res}$). Then for architectures with an independent damping gain K_x, discrete root locus methods are utilized to maximize system damping. E.g. for Form 1 the gains are [1]:

$$K_{p1} = \omega_c(L_1 + L_2)/V_{DC} \quad T_{r1} = 10/\omega_c \quad (9)$$

Identification of these gains now allows for the root locus of closed loop poles to be determined using the eigenvalues of the $\mathbf{A_{cl}}$ matrix in (7b) as a function of K_1. This is shown in Fig. 3 with the optimal K_1 value selected based on positioning the LCL resonant poles at the maximum damping locations. Fig. 4 shows a PSIM simulation of the LCL system with the Form 1 control strategy and gains calculated using the procedure outlined above ($K_{p1} = 0.06$, $K_1 = 0.083$, $T_{r1} = 0.0021$), which clearly shows a rapid transient response and well damped LCL resonance, consistent with established literature [1].

V. GENERALISED EVALUATION OF ACTIVE DAMPING

A. Form 2 - Damping Gain in Forward Path

The Form 2 controller [11]-[13] is a subtle variation on Form 1, where the damping gain K_2 is moved from the feedback path, into the forward path, this is shown in Fig. 2(b). In terms of the control law this means that the resultant output of the PR controller is also multiplied by the damping gain, giving the control law:

$$u(k) = K_{p2}\left(i^*(k) - i_2(k)\right) \quad (9a)$$

$$m_z(k) = K_2 u(k) + K_2 m_r(k)/T_r - K_2[i_1(k) - i_2(k)] \quad (9b)$$

And thus feedback gain vectors:

$$\mathbf{I_{cl}} = \begin{bmatrix} K_2 & 0 & -K_2 \end{bmatrix} \quad \mathbf{J_{cl}} = \begin{bmatrix} 0 & 0 & K_{p2}K_2 \end{bmatrix} \quad (10)$$

Equivalent gains (to achieve matching pole locations) can be identified by comparing the coefficients of $i_2(k)$ and $i_c(k)$ or $i_1(k)$ from (9) with (7). Note that the coefficients of the reference current do not impact the closed loop poles. The equivalent gains are given by:

$$K_{p2} = K_{p1}/K_1 \quad K_2 = K_1 \quad T_{r2} = T_{r1} \quad (11)$$

With these gains Form 2 is, in fact, entirely equivalent to Form 1. The pole zero map and transient response are identical (see Fig. 3 and Fig. 4). It offers no advantage or disadvantage over Form 1.

B. Form 3 - Converter current active damping

Form 3 uses converter side current as the inner active damping loop [7], instead of capacitor current (Fig. 2(c)). This is due to the desire to have the current sensor placed in the converter current path for protection. The control law is derived by replacing the capacitor current with

TABLE I - LCL CONVERTER SYSTEM PARAMETERS

$L_1 = 3mH$	$L_2 = 1mH$	$C_f = 7.5\mu F$
$V_{DC} = 325V$	$E = 415V_{llrms}$	$\omega_{res} = 13.3 krads^{-1}$
$f_{sw} = 10kHz$	$f_{samp} = 20kHz$	$T = 50\mu s$

The 2014 International Power Electronics Conference

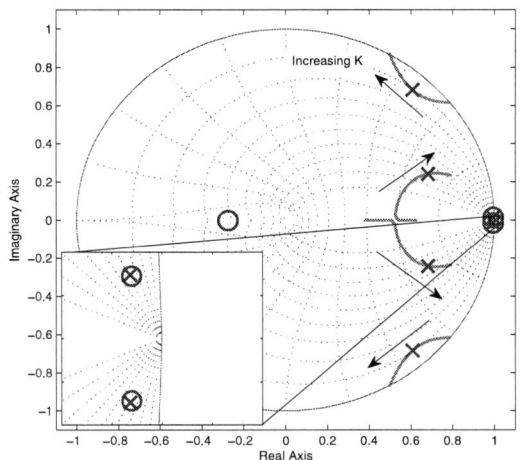

Fig. 3: Root locus and closed loop poles and zeros of Form 1.

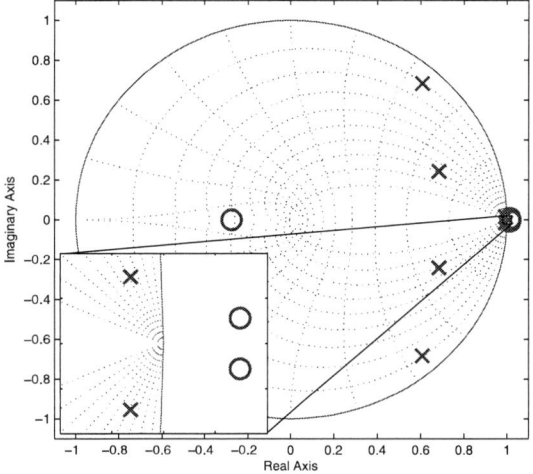

Fig. 5: Closed loop poles and zeros of Form 3.

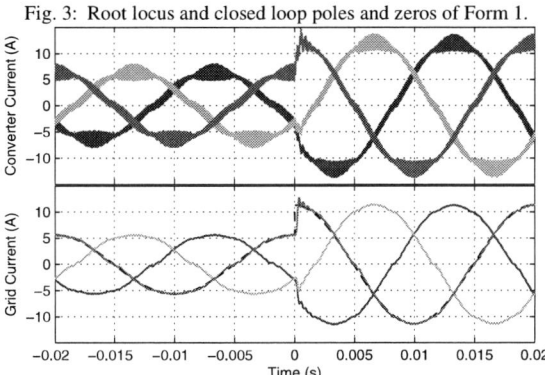

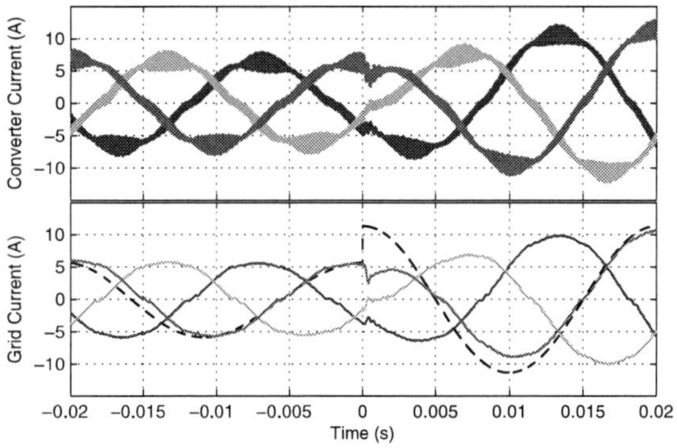

Fig. 4: Converter and grid current step response, Form 1 and Form 2.

Fig. 6: Converter and grid current step response of Form 3.

converter current in (6), giving:

$$u(k) = K_{p3}\left(i^*(k) - i_2(k)\right) \qquad (12a)$$

$$m_z(k) = u(k) + m_r(k)/T_{r3} - K_3 i_1(k)$$
$$= u(k) + m_r(k)/T_{r3} - K_3\left[i_2(k) - i_c(k)\right] \qquad (12b)$$

And the feedback gain vectors are:

$$\mathbf{I_{cl}} = \begin{bmatrix} K_3 & 0 & 0 \end{bmatrix} \qquad \mathbf{J_{cl}} = \begin{bmatrix} 0 & 0 & K_{p3} \end{bmatrix} \qquad (13)$$

The Form 3 controller can be designed to achieve identical closed loop pole locations as for the Form 1 control system, as shown in Fig. 5, using:

$$K_{p3} = K_{p1} - K_1 \qquad K_3 = K_1 \qquad (14a)$$

$$T_{r3} = \left(K_{p1} - K_1\right)T_{r1}/K_{p1} \qquad (14c)$$

Comparison of Figs. 3 and 5 shows that a key difference between the Form 1 and 3 controllers is the formation of a pair of non-minimum phase zeros. These zeros are responsible for a degradation of the transient response due to undershoot as shown in Fig. 6.

C. Form 4 – Weighted Average Current

Also known as WAC [14][15], Form 4 creates a feedback variable of the weighted summation of i_1 and i_2 for the PR controller (refer to Fig. 2(d)). The literature states this form controls a novel state where the LCL resonant poles are cancelled by a zero pair. To reflect form 4 into the model, the control law is given as:

$$u(k) = K_{p4}\left(i^*(k) - K_{a4}i_1(k) - K_{b4}i_2(k)\right) \qquad (15a)$$

$$m_z(k) = u(k) + m_r(k)/T_{r4} \qquad (15b)$$

Giving the feedback vectors:

$$\mathbf{I_{cl}} = \begin{bmatrix} 0 & 0 & 0 \end{bmatrix} \qquad \mathbf{J_{cl}} = \begin{bmatrix} K_{a4}K_{p4} & 0 & K_{b4}K_{p4} \end{bmatrix} \qquad (16)$$

The weighting factors are provided in the literature as:

$$K_{b4} = 1 - L_1/(L_1 + L_2) \qquad K_{a4} = 1 - K_{b4} \qquad (17a)$$

The proportional gain and resonant time constant are matched to Form 1 to achieve the same system bandwidth, viz:

$$K_{p4} = K_{p1} \qquad T_{r4} = T_{r1} \qquad (17b)$$

As identified in [16], the resonant poles still exist in the pole zero map (Fig. 7) when the grid current is the observed variable, and that they are placed on the stability boundary. This observation is reflected in the simulated waveform (see Fig. 8), where the highly underdamped resonance is evident at the step change.

D. Form 5 – Extended Weighted Average Current

In order to overcome the marginal stability of Form 4, additional capacitor current feedback may be applied [16]. This structure is shown in Fig. 2(e). The control law is developed from (15) with additional capacitor current feedback, viz:

$$u(k) = K_{p5}\left(i^*(k) - K_{a5}i_1(k) - K_{b5}i_2(k)\right) \qquad (18a)$$

978-1-4799-2706-7/14 $31.00 © 2014 IEEE

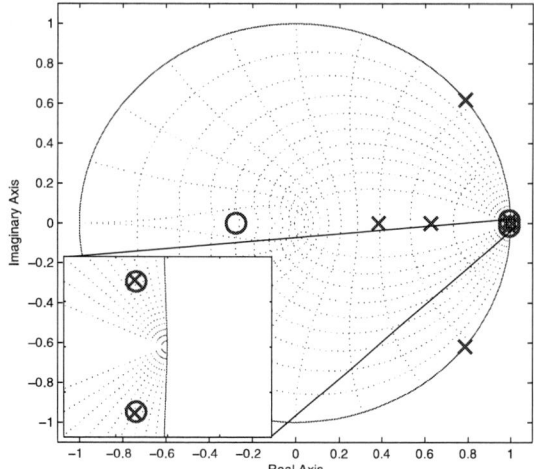

Fig. 7: Closed loop poles and zeros of Form 4.

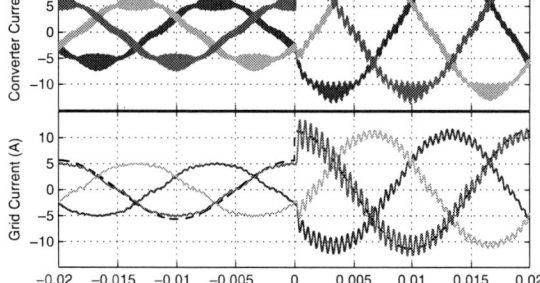

Fig. 8: Converter and grid current step response of Form 4.

$$m_z(k) = u(k) + m_r(k)/T_{r5} + K_5 i_c(k)$$
$$= u(k) + m_r(k)/T_{r5} + K_5[i_1(k) - i_2(k)] \quad (18b)$$

Thus the feedback gain vectors are an extension of (16):

$$\mathbf{I_{cl}} = [K_5 \quad 0 \quad -K_5] \quad (19a)$$
$$\mathbf{J_{cl}} = [K_{a5}K_{p5} \quad 0 \quad K_{b5}K_{p5}] \quad (19b)$$

To achieve the equivalent resonance damping to Form 1, the total feedback of i_c should be equal, i.e.

$$K_5 = K_1 - K_{a5}K_{p1} \quad (20)$$

The other gains remain the same as Form 4, i.e. (17).

With this variation the poles are well damped (Fig. 9) and the transient response is fast (Fig. 10). What is not clear from the pole zero map, but can be seen in the transient response is that the grid current does not track the reference due to the weighted feedback into the PR controller, thus there is steady state error in i_2.

E. Form 6 – Converter Current Feedback Only

The converter side current can be used as the main feedback variable, as it will provide a degree of active damping [6][17][18]. This form is shown in Fig. 2(f). The control law for Form 6 is:

$$u(k) = K_{p6}(i^*(k) - i_1(k))$$
$$= K_{p6}(i^*(k) - i_2(k) - i_c(k)) \quad (21a)$$
$$m_z(k) = u(k) + m_r(k)/T_{r6} \quad (21b)$$

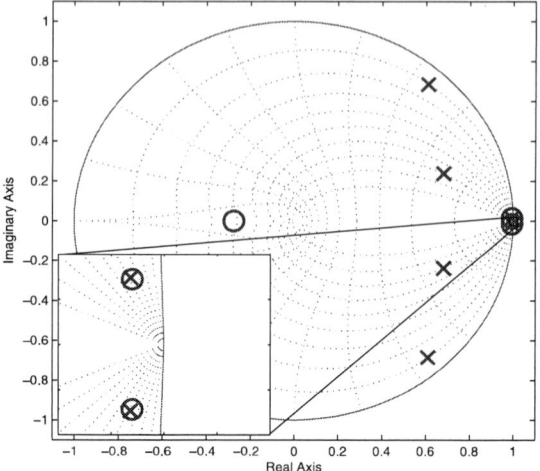

Fig. 9: Closed loop poles and zeros of Form 5.

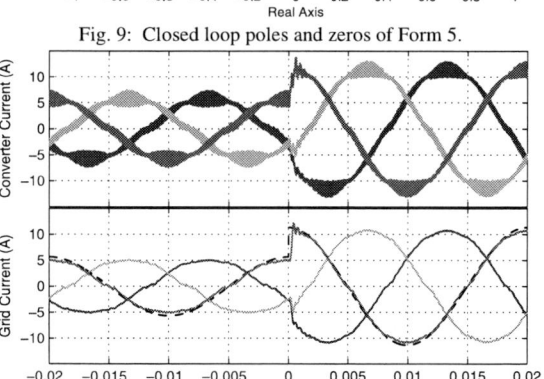

Fig. 10: Converter and grid current step response of Form 5.

and the gain vectors are:

$$\mathbf{I_{cl}} = [0 \quad 0 \quad 0] \qquad \mathbf{J_{cl}} = [K_{p6} \quad 0 \quad 0] \quad (22)$$

This form has only a single degree of freedom; this can be viewed as Form 1, with the limitation $K_6 = K_{p6}$.

Thus, the system can not have the same closed loop poles. There are, however, two options for a fair comparison; either the same bandwidth or the same LCL resonant pole damping.

1) Gain Option 1 – Same System Bandwidth

The same bandwidth is obtained with the proportional gain and resonance time constant from Form 1, viz:

$$K_{p6} = K_{p1} \qquad T_{r6} = T_{r1} \quad (23)$$

The reduced resonance damping of this system can be seen in Fig. 11, where the closed loop poles are closer to the unit circle stability boundary. The transient results show a greater amount of resonance, with longer a damping time (see Fig. 12(a)).

2) Gain Option 2 – Same Resonance Damping

The same resonant pole damping can be obtained by using a root locus to place the resonant poles (see Fig. 11). This gives a gain of:

$$K_{p6} = 0.092 A^{-1} \quad (24)$$

The extra gain required to match the pole damping characteristic results in a larger current overshoot seen in the transient results of Fig. 12(b).

978-1-4799-2706-7/14 $31.00 © 2014 IEEE

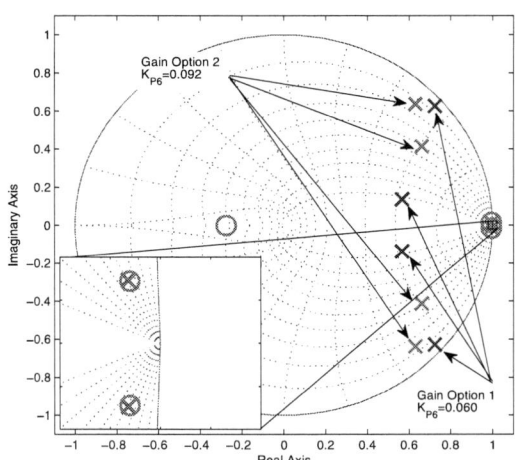

Fig. 11: Closed loop poles and zeros of Form 6 for matched bandwidth ($K_{mA}=0.060$) and matched damping ($K_{mA}=0.092$).

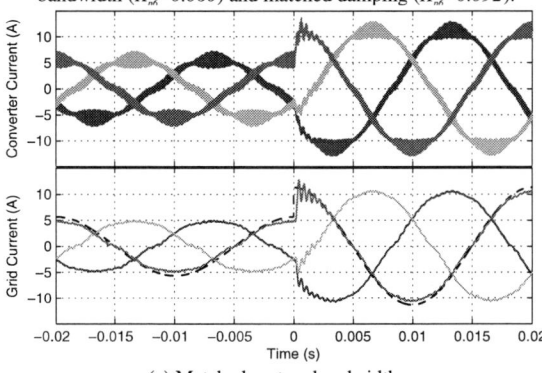

(a) Matched system bandwidth.

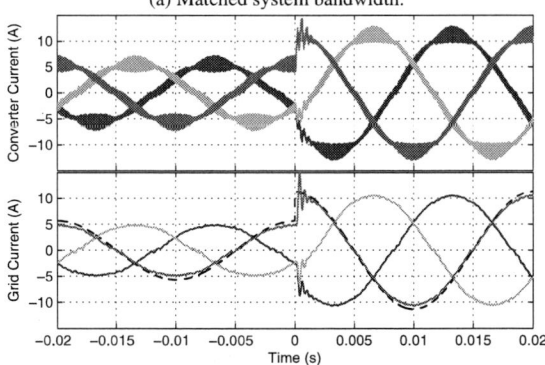

(b) Matched resonance damping.

Fig. 12: Converter and grid current step response of Form 6.

As evident from the transient results, for Form 6, the grid current does not track the reference as it is not part of the feedback path. For Form 6 there is also evidently a trade off between transient performance (rise time, settling time) and resonance damping.

F. Form 7 – Converter Current Feedback with Additional Active Damping

Form 7 extends Form 6 to provide better resonant pole damping by including additional capacitor current feedback [6]. The control diagram for this form is shown in Fig. 2(g). The control law is developed by taking (21) from Form 6 and adding capacitor current feedback into the PR controller, viz:

$$
\begin{aligned}
u(k) &= K_{p7}\left(i^*(k) - i_1(k) - K_7 i_c(k)\right) \\
&= K_{p7}\left(i^*(k) - i_2(k) - (1 + K_7)i_c(k)\right)
\end{aligned} \tag{25a}
$$

$$
m_z(k) = u(k) + m_r(k)/T_{r7} \tag{25b}
$$

while the gain vectors are:

$$
\mathbf{I_{cl}} = \begin{bmatrix} 0 & 0 & 0 \end{bmatrix} \tag{26a}
$$

$$
\mathbf{J_{cl}} = \begin{bmatrix} K_{p7}(1 + K_7) & 0 & -K_{p7}K_7 \end{bmatrix} \tag{26b}
$$

1) Gain Option 1 – From Literature

Reference [6] provides a formula to calculate a value for K_7, to give critically damped resonant poles, i.e.

$$
K_7 \approx \frac{2\zeta L_1 \omega_{res}}{K_{p7} V_{DC}} - 1 \tag{27}
$$

with $\zeta = 0.707$. However this equation was derived based on a continuous time model and results in stable, well damped, closed loop poles for the continuous time pole zero map, Fig. 13(a). When applied to the discrete time model used in this work, the result is a prediction of an unstable system, Fig. 13(b).

The transient response of a system using these gains is clearly unstable (see Fig. 14(a)). However, an unstable LCL system may be stabilised in practice by parasitic resistances in the system, not modelled here.

2) Gain Option 2 – Equivalent Poles

As the system contains the same number of degrees of freedom in the controller as Form 1, the same poles can be achieved, using:

$$
K_{p7} = K_{p1} \qquad K_7 = K_1 / K_{p1} - 1 \tag{28}
$$

The transient result of gain option 2, Fig. 14(b), shows a stable, well damped, response. However, the grid current does not track the reference, and has a steady state error.

G. Evaluation Summary

Three main issues have been identified from this comparative evaluation. Firstly, many forms do not directly regulate the grid current. In order to control real and reactive power flow (and power factor) the current flowing into the grid must be controlled. The forms which do not regulate the grid current directly (Forms 3, 4, 5 and 6) will require current reference compensation, introducing more complexity.

Secondly, if any two currents are measured, the third current can be calculated and used as the active damping feedback variable. For example Form 3 shows that using converter current as the active damping variable can result in poor transient performance, despite its ability to damp the LCL resonance. In this case the capacitor current can be simply calculated internally and used as the active damping variable, ensuring identical operation to Form 1.

Thirdly, some gain selection strategies presented in the literature result in unacceptable responses. Primarily, this is because they have focused on the poles and zeros produced in the converter current transfer function, and have not appreciated that it is the poles and zeros of the grid current transfer function that actually define the

The 2014 International Power Electronics Conference

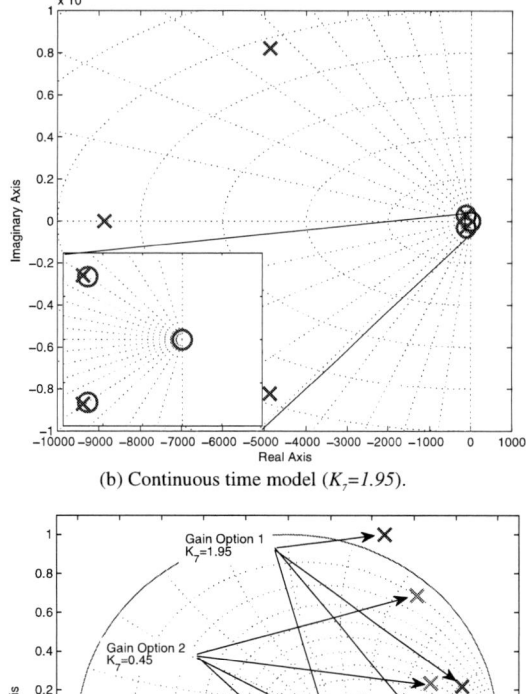

(b) Continuous time model ($K_7=1.95$).

(b) Discrete time model.

Fig. 13: Closed loop poles and zeros of form 7 for gains from literature ($K_7=1.95$) and matched damping ($K_7=0.41$).

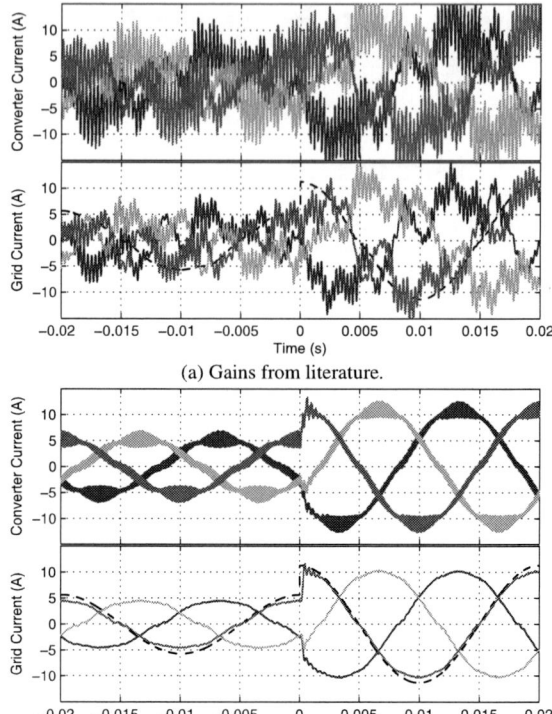

(a) Gains from literature.

(b) Matched resonance damping.

Fig. 14: Converter and grid current step response of form 7.

active damping requirements. Secondly, many studies ignore PWM transport delay. In this paper it is shown how this delay severely limits the controller gains, dramatically alters the root locus paths [1][3], and must always be taken into account.

The best controller forms from this analysis are Form 1 (and equivalent Form 2). It provides the fastest transient performance, best resonance damping and directly regulates the grid current without steady state error. However the best placement of current measurement transducers is in both inductor current paths, to provide overcurrent protection.

If the system design is limited to a single current transducer then feedback of the converter side current will provide a degree of active damping, with a trade off of transient performance. Further damping may be obtained by approximating the capacitor current, from the converter current, using a lower pass or notch filter to remove the fundamental frequency component [6]. This may reduce dynamic performance; the performance of these filters is beyond the scope of this evaluation.

A summary of this evaluation is provided Table II.

VI. EXPERIMENTAL VERIFICATION

The system architectures described in this paper have been tested on an experimental platform to validate the generalised model and simulation investigations. The experimental system includes a 5kVA active rectifier coupled to a California Instruments CX30 grid emulator with parameters detailed in Table 1. Fig. 15 shows the transient response of the experimental system for the form 1 and form 6 control strategies. The dashed line in the experimental result waveforms is the current reference, stepping from $4A_{rms}$ to $8A_{rms}$ capacitive.

The experimental step response of form 1 is shown in Fig. 15(a). Very little LCL resonance oscillation is visible during the transient step change and the grid current is tracking the reference signal. This is in good agreement with the simulation results shown in Fig. 4.

Both the equivalent transient response gain ($K_{p6} = 0.060$) and equivalent damping gain ($K_{p6} = 0.092$) have been tested experimentally for form 6 (Fig. 15(b) and 15(c) respectively). Neither system show any great

TABLE II
SUMMARY OF COMPARATIVE EVALUATION

Form	Stable	Track i2	Damping	Transient	Sensors
1	Y	Y	Good	Good	2
2	Y	Y	Good	Good	2
3	Y	N	Good	Bad	2
4	Y	N	Bad	Good	2
5	Y	N	Good	Good	2
6 ($K_{p6}=K_{p1}$)	Y	N	Poor	Good	1
6 ($K_{p6}=0.092$)	Y	N	Good	Poor	1
7 ($K_7=1.95$)	N	N/A	N/A	N/A	2
7 ($K_7=0.41$)	Y	N	Good	Good	2

difference in LCL resonance damping as the parasitic resistances in the system provide a degree of damping above that from the controller. However the higher proportional gain still leads to larger overshoot and oscillations. It is also evident the grid current does not track the reference signal. These results are in agreement with the simulations presented in Fig. 12.

VII. CONCLUSION

This paper has presented a generalised discrete time model of an LCL filtered converter for active damping control strategies that fully accounts for PWM transport and sampling delays. The model enables the analysis of any active damping controller architecture, and has been applied to strategies that employ either the capacitor or converter current for damping or primary control functions. Despite the equivalent gains being used between forms, they showed variation in transient performance, LCL resonance damping and grid current tracking. The best forms, using one or two current sensors, were identified and experimentally verified.

REFERENCES

[1] S. G. Parker, B. P. McGrath and D. G. Holmes, "Regions of active damping control for LCL filters," in Proc. IEEE Energy Conversion Congress and Exposition (ECCE), Raleigh, US, 2012, pp. 53-60.

[2] D. Ricciuto, M. Liserre, T. Kerekes, et al., "Robustness analysis of active damping methods for an inverter connected to the grid with an LCL-filter," in Proc. IEEE Energy Conversion Congress and Exposition (ECCE), Phoenix, US, 2011, pp. 2028-2035.

[3] J. Dannehl, F. W. Fuchs, S. Hansen, et al., "Investigation of active damping approaches for PI-based current control of grid-connected pulse width modulation converters with LCL filters," IEEE Trans. Ind. Appl., vol. 46, no. 4, pp. 1509-1517, Jul. /Aug. 2010.

[4] J. He and Y. W. Li, "Generalized closed-loop control schemes with embedded virtual impedances for voltage source converters with LC or LCL filters," IEEE Trans. Power Electron., vol. 27, no. 4, pp. 1850-1861, Apr. 2013.

[5] X. Wang, X. Ruan, C. Bao, et al., "Design of the PI regulator and feedback coefficient of capacitor current for grid-connected inverter with an LCL filter in discrete-time domain," in Proc. IEEE Energy Conversion Congress and Exposition (ECCE), Raleigh, US, 2012, pp. 1657-1662.

[6] Y. Tang, P. C. Loh, P. Wang, et al., "Exploring inherent damping characteristics of LCL-filters for three-phase grid-connected voltage source inverters," IEEE Trans. Power Electron., vol. 27, no. 3, pp. 1433-1443, Mar. 2012.

[7] K. H. Ahmed, A. M. Massoud, S. J. Finney, et al., "Optimum selection of state feedback variables PWM inverters control," in Proc. IET Conference on Power Electronics, Machines and Drives (PEMD), York, UK, 2008, pp. 125-129.

[8] N. He, D. Xu, Y. Zhu, et al., "Weighted average current control in a three-phase grid inverter with an LCL filter," IEEE Trans. Power Electron., vol. 28, no. 6, pp. 1785-2797, Jun. 2013.

[9] J. Xu and S. Xie, "Optimization of weighted current control for grid-connected LCL-filtered inverters," in Proc. IEEE Energy Conversion Congress and Exposition Asia (ECCEAsia), Melbourne, Australia, 2013, pp. 1170-1175.

[10] G. F. Franklin, J. D. Powell and M. Workman, Digital Control of Dynamic Systems, 3rd ed.: Addison Wesley Longman, Inc., 1998.

[11] E. Twining and D. G. Holmes, "Grid current regulation of a three-phase voltage source inverter with an LCL input filter," IEEE Trans. Power Electron., vol. 18, no. 3, pp. 888-895, May 2003.

[12] Y. W. Li, "Control and resonance damping of voltage-source and current-source converters with LC filters," IEEE Trans. Ind. Electron., vol. 56, no. 5, pp. 1511-1521, May 2009.

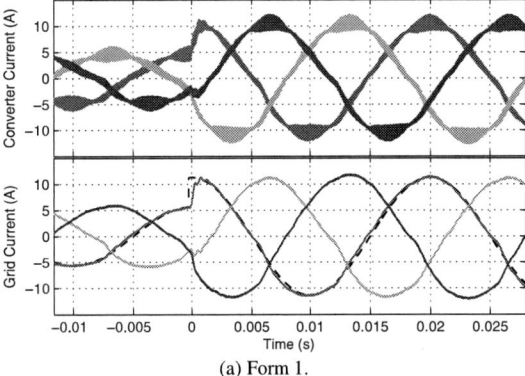

(a) Form 1.

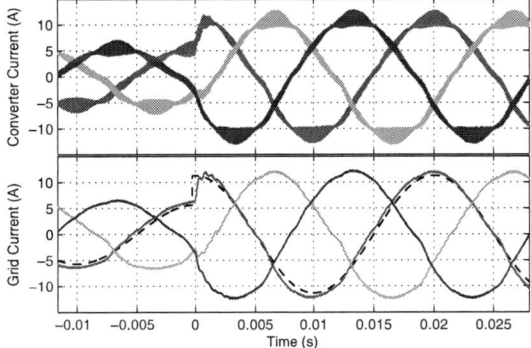

(b) Form 6 – Matched system bandwidth ($K_{p6} = 0.060$).

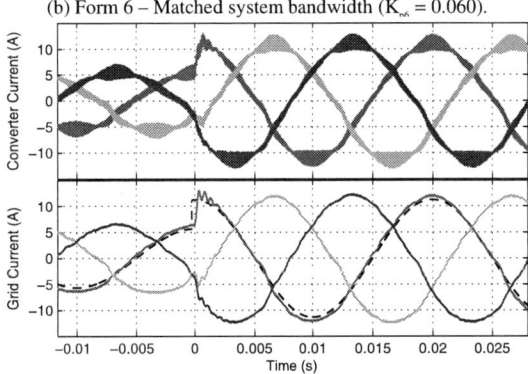

(c) Form 6 – Matched resonance damping ($K_{p6} = 0.092$).

Fig. 15: Experimental results. Reference shown as dashed line.

[13] P. C. Loh and D. G. Holmes, "Analysis of multiloop control strategies for LC/CL/LCL-filtered voltage-source and current-source inverters," IEEE Trans. Ind. Appl., vol. 41, no. 2, pp. 644-654, Mar./Apr. 2005.

[14] G. Shen, Z. Xuancai, Z. Jun, et al., "A new feedback method for PR current control of LCL-filter-based grid-connected Inverter," IEEE Trans. Ind. Electron., vol. 57, no. 6, pp. 2033-2041, 2010.

[15] N. He, D. Xu, Y. Zhu, et al., "Weighted average current control in a three-phase grid inverter with an LCL filter," IEEE Trans. Power Electron., vol. 28, no. 6, pp. 1785-2797, Jun. 2013.

[16] J. Xu and S. Xie, "Optimization of weighted current control for grid-connected LCL-filtered inverters," in Proc. IEEE Energy Conversion Congress and Exposition Asia (ECCEAsia), Melbourne, Australia, 2013, pp. 1170-1175.

[17] Y. Tang, P. C. Loh, P. Wang, et al., "Generalized design of high performance shunt active power filter with output LCL filter," IEEE Trans. Ind. Electron., vol. 59, no. 3, pp. 1443-1452, Mar. 2012.

[18] L. Huang, B. Li, Z. Lu, et al., "PR controller for grid-connected inverter control using direct pole placement strategy," in Proc. IEEE International Symposium on Industrial Electronics (ISIE), Hangzhou, China, 2012, pp. 469-474.

Analysis and Reduction of Power Losses in PV Converters for Grid Connection to Low-Voltage Three-Phase Three-Wire Systems

Ryosuke Amma, and Hideaki Fujita, *Senior Member, IEEE*
Department of Electrical and Electronic Engineering
Tokyo Institute of Technology, Tokyo, Japan

Abstract—This paper discusses analysis and reduction of power losses in PV converter systems using conventional Si-IGBTs as well as SiC-MOSFETs. The PV converter system discussed in this paper consists of a boost converter and two half-bridge PWM inverters, considering grid connection to three-phase three-wire low-voltage power systems in Japan. Power losses in Si-IGBTs and SiC-MOSFETs are measured and theoretically discussed from experimental result in a buck converter benchmark. The power losses in PV converters using Si-IGBTs and SiC-MOSFETs are also measured and separated into the components in each part. As a result, it is clarified that the dominant power loss occurs in the two half-bridge inverters, and their switching power losses are larger than the on-state losses. Moreover, application of SiC-MOSFETs to both boost converter and half-bridge inverters makes it possible to improve the overall power conversion efficiency from 93.5% to 95.5% in the Euro-eta definition.

Keywords—*Grid-connected converters, low-voltage three-phase three-wire systems, PV converters, switching power losses*

I. INTRODUCTION

Recently, solar power generation systems have widely been installed. PV converters are used to adapt the dc electric power generated by the PV modules to the ac utility power grid. Due to the demand for miniaturization of the systems, transformer-less PV converters are widely adopted especially in PV systems whose rated power is around 50-kW or less. On the other hand, PV converters are required to reduce the leakage currents caused by the fluctuation of the common mode voltage or the potential of the PV modules. Researches on PWM methods and topologies effective in suppression of the common mode voltage have been conducted [1][2].

In the case of PV converters for grid connection to single-phase systems, Heric converters [3] introduce an additional bidirectional switch on the ac side, H5 converters [4] are equipped with a decoupling switch on the dc side, and ZCC converters [5] are composed of two bidirectional buck converters and an H-bridge PWM converter connected in series. These converters make it possible not only to suppress the common mode voltage but also to reduce the switching power losses.

Different topologies and control methods are applied to PV converters for three-phase low-voltage distribution systems depending on their grounding systems in terms of

suppression of the common mode voltage. A PV converter for three-phase four-wire systems achieves a maximal conversion efficiency as high as 98.0% [6]. On the other hand, PV converters for three-phase three-wire 200-V systems have a low power efficiency around 93%, because a relatively high dc-link voltage is required.

Nowadays, it is widely recognized that silicon carbide (SiC) has outstanding electric properties, and the researches and developments of SiC power devices have been well advanced [7]. Today, Silicon Carbide MOS-FETs (SiC-MOSFETs) withstanding 1200 V are available, and they are applicable to PV converters for low-voltage three-phase three-wire systems. Applying them to PV converters instead of convetional Silicon IGBTs (Si-IGBTs) is expected to enable higher performance.

This paper discusses the power losses in PV converters for grid connection to low-voltage three-phase three-wire systems in Japan and the effects on reducing power losses obtained by replacing Si-IGBTs with SiC-MOSFETs. At first, the power losses in a buck converter equipped with Si-IGBTs and SiC-MOSFETs are measured to analyze the differences in power losses between Si-IGBTs and SiC-MOSFETs. Moreover, for evaluating the power losses in PV converters, two prototypes of PV converters are designed and assembled by using conventional Si-IGBTs and SiC-MOSFETs. The power losses are measured and separated into the components in each part of the PV converter prototypes to analyze the tendency of the power losses and their factors. As a result, it has been clarified that the PV conversion systems using SiC-MOSFETs can improve the overall Euro-eta efficiency from 93.5% to 95.5%.

II. PV CONVERTERS FOR LOW-VOLTAGE THREE-PHASE THREE-WIRE SYSTEMS IN JAPAN

Fig. 1 shows the circuit configuration of a PV converter for grid connection to the three-phase three-wire systems in Japan. The PV converter consists of a boost converter and two half-bridge inverters. The boost converter boosts the input PV voltage up and provides a required dc-link voltage to the capacitors. Then, two half-bridge inverters perform dc-to-ac power conversion [8].

The midpoint of the dc link is grounded to the S-phase terminal which is connected to the earth to suppress the common mode voltage or to sustain the potential of the

978-1-4799-2706-7/14 $31.00 © 2014 IEEE

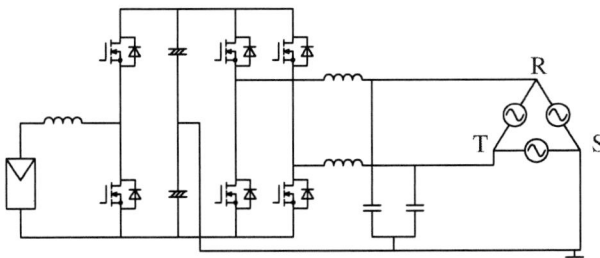

Fig. 1. Circuit configuration of a PV converter for low-voltage three-phase three-wire systems in Japan.

PV module at a stable level. In this case, the line-to-line grid voltage is applied between the output terminal of the inverters and the midpoint of the dc-link. As a result, the dc-link voltage V_{dc} need to be higher than the peak-to-peak value of the line-to-line grid voltage, as follows:

$$V_{dc} > 2\sqrt{2}V_s, \tag{1}$$

where V_s is the rms line-to-line grid voltage.

PV converters in Europe are generally connected to three-phase four-wire low-voltage systems whose neutral point is grounded to the earth, and usually use PV modules having an open voltage around 800 V. Then the dc-link voltage should be higher than the peak-to-peak value of the phase voltage, given by,

$$V_{dc} > \frac{2\sqrt{2}}{\sqrt{3}}V_s. \tag{2}$$

Thus, the dc-link voltage is about 653V when the rms line-to-line voltage of the grid is 400 V. Furthermore, the PV module voltage is controlled to be about 80% of the open voltage when maximum power point tracking (MPPT) operation is adapted. Therefore, the MPPT voltage is estimated as 640 V. In this case, the boost converter can be operated with a relatively low boost ratio with a low duty factor, and then, it is possible to reduce the power losses of switching devices in the boost converter and the core loss in the boost dc inductor. For these reasons, the latest three-phase four-wire PV converters realize a power conversion efficiency as high as 98%.

By contrast, the grid voltage is only 200 V in three-phase three-wire systems in Japan. Thus, the grid current is almost double of that in 400-V PV systems in Europe. However, the dc-link voltage has to be boosted up to 565 V even when the grid voltage is only 200 V. Therefore, the switching power loss is significant as well as the on-state loss. Moreover, the open-circuit voltage of PV modules is generally designed to be 600 V or below in Japan, due to the safety regulation. Accordingly, the output voltage of PV modules is in a range from 400 V to 500 V under MPPT operating conditions. For these reasons, the low-voltage PV system in Japan causes a relatively large power losses and has a conversion efficiency as low as 93%.

Thus, the power losses in the PV converters for the low-voltage three-phase three-wire systems in Japan are larger than those for other systems in principle, due to lager switching power losses caused by the high dc-link voltage and the continuous boost operation.

III. POWER LOSS CHARACTERISTICS OF SiC-MOSFETs AND Si-IGBTs

A. Loss measurement of switching devices

Fig. 2 shows the circuit configuration for buck converter benchmarks to evaluate the power loss characteristics of Si-IGBTs and SiC-MOSFETs, and Table I is its circuit parameters. Two buck converter circuits are assembled by using conventional Si-IGBT modules (CM100DY-24NF: Mitsubishi, 1200 V, 100 A) and SiC-MOSFET modules (BSM120D12P2C005: ROHM, 1200 V, 120 A) as switching devices. A dc input capacitor C is installed on the dc side of the buck converter to keep the dc input voltage at a constant level, whose capacitance is 3400 μF. On the ac side, L is an ac output inductor of 3 mH for smoothing the output current. Input power P_{in} and output power P_{out} are measured by using a digital power analyzer (WT1800: Yokogawa, 6ch, 5 MHz, 0.15%), and the power losses in switching devices P_{loss} are calculated by subtracting P_{out} from P_{in}. The power losses are measured at 100 points of combination of V_{in} and I_{out}, where V_{in} is varied in a range from 100 V to 500 V and I_{out} is adjusted from 0 to 30 A.

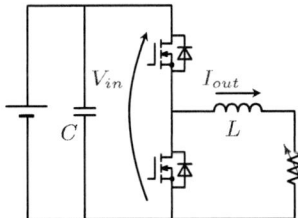

Fig. 2. Circuit configuration for evaluating power loss characteristics of switching devices

TABLE I. CIRCUIT PARAMETERS OF THE BUCK CHOPPER.

inductor L	3 mH
capacitor C	3400 μF
switching frequency f_{sw}	10 kHz
on-delay time T_d	3.2 μs

B. Analysis of the power losses in the switching devices by means of curve fitting technique

Assuming ripples in V_{in} and I_{out} are negligible, the power losses in switching devices P_{loss} can be expressed as a function of V_{in} and I_{out}. Based on this assumption, P_{loss} is assumed as a quadratic function of V_{in} and I_{out} in this paper, as follows:

$$P_{loss} = \begin{pmatrix} 1 \\ V_{in} \\ V_{in}^2 \end{pmatrix}^{\mathrm{T}} \begin{pmatrix} a_{00} & a_{01} & a_{02} \\ a_{10} & a_{11} & a_{12} \\ a_{20} & a_{21} & a_{22} \end{pmatrix} \begin{pmatrix} 1 \\ I_{out} \\ I_{out}^2 \end{pmatrix}. \tag{3}$$

The power loss characteristics of switching devices can be evaluated based on the elements of the above matrix. Each

TABLE II.   ELEMENTS OF THE MATRIX

(a) Si-IGBT (CM100DY-24NF: Mitsubishi, 1200 V, 100 A)

	1	I_{out}	I_{out}^2
1	8.0×10^{-15}	1.0	1.4×10^{-2}
V_{in}	1.3×10^{-2}	5.2×10^{-3}	1.1×10^{-17}
V_{in}^2	1.1×10^{-5}	1.7×10^{-19}	1.2×10^{-20}

(b) SiC-MOSFET
(BSM120D12P2C005: ROHM, 1200 V, 120 A)

	1	I_{out}	I_{out}^2
1	1.4×10^{-1}	3.6×10^{-13}	1.8×10^{-2}
V_{in}	1.4×10^{-16}	3.1×10^{-3}	2.6×10^{-5}
V_{in}^2	2.5×10^{-5}	2.2×10^{-19}	1.8×10^{-8}

TABLE III.   POWER LOSSES DERIVING FROM THE ELEMENTS($V_{in} = 500$ V, $I_{out} = 30$ A)

(a) Si-IGBT (CM100DY-24NF: Mitsubishi, 1200 V, 100 A)

	1	I_{out}	I_{out}^2
1	8.0×10^{-15}	$\mathbf{3.0 \times 10^{1}}$	$\mathbf{1.3 \times 10^{1}}$
V_{in}	$\mathbf{6.5}$	$\mathbf{7.8 \times 10^{1}}$	5.0×10^{-12}
V_{in}^2	2.8	1.3×10^{-12}	2.7×10^{-12}

(b) SiC-MOSFET
(BSM120D12P2C005: ROHM, 1200 V, 120 A)

	1	I_{out}	I_{out}^2
1	1.4×10^{-1}	1.1×10^{-11}	$\mathbf{1.6 \times 10^{1}}$
V_{in}	7.0×10^{-14}	$\mathbf{4.7 \times 10^{1}}$	$\mathbf{1.2 \times 10^{1}}$
V_{in}^2	$\mathbf{6.3}$	1.7×10^{-12}	4.1

element can be determined by two-dimensional curve fitting technique, in which the squared error between measured and calculated losses is minimized. Table II shows matrixes of the elements of Si-IGBTs and SiC-MOSFETs obtained by the curve fitting technique.

C. Physical meaning of the elements

In general, the loss in switching devices is separated into three components: on-state losses, off-state losses, and switching power losses. Off-state loss is caused by leakage currents flowing through an off-state switching device and negligibly small. On-state loss is brought about by currents flowing through an on-state switching device. It is independent from input voltage V_{in} but depends only on output current I_{out}. Consequently, the elements from a_{00} to a_{02} can be considered as terms representing on-state losses. On the other hand, switching power losses occur in turn-on and turn-off processes, and it is well known that they are proportional to the switching frequency f_{sw}. In addition, switching power losses depend on both input voltage V_{in} and output current I_{out}, so that switching power losses are considered to be caused by the elements of a_{10} through a_{22}. In those elements, a_{10} and a_{20} are independent from output current I_{out}. The element a_{10} is considered to indicate the losses deriving from the reverse recovery charge and a tail current. Note that the tail current appears only in IGBTs, but not in MOSFETs. Furthermore, a_{20} represents the losses deriving from the output capacitance [9]. The other elements, that is, a_{11}, a_{12}, a_{21}, and a_{22} can be considered as the power losses caused by a rise time t_r, a fall time t_f, a reverse recovery time t_{rr} and a Miller effect duration [10]. Therefore, the power losses in switching devices expressed in (4) are rewritten as follows:

$$P_{loss} = \boldsymbol{V}_{in}^{\mathrm{T}} \begin{pmatrix} 0 & V_f & R_{on} \\ Q_{rr}f_{sw} & K_1 f_{sw} & K_2 f_{sw} \\ C_{oss}f_{sw} & K_3 f_{sw} & K_4 f_{sw} \end{pmatrix} \boldsymbol{I}_{out}, \quad (4)$$

where Q_{rr}, C_{oss}, V_f and R_{on} are the reverse recovery charge, the output capacitance, the forward voltage drop and resistance. Furthermore, K_1 [s], K_2 [s/A], K_3 [s/V] and K_4 [s/(A·V)] are the elements representing the power losses affected by a rise time t_r, a fall time t_f, a reverse recovery time t_{rr} or a Miller effect duration and so on.

D. Comparison of the power loss characteristics

Table III shows the power losses deriving from each elements indicated in Table II under the condition where the input voltage V_{in} and the output current I_{out} are 500 V and 30 A, respectively. The elements larger than 5 W are emphasized by bold type face in Table III.

In the elements of SiC-MOSFETs, a_{01} is negligibly small and a_{02} is large compared with those of Si-IGBTs. These tendencies agree with the characteristics of MOS-FETs, in which the forward voltage drop V_f appearing in IGBTs and diodes is not observed, because synchronous rectification is applied to the buck converter. Resistance R_{on} is larger than in IGBTs. Each element representing on-state losses is consistent with the device characteristics presented in the data-sheet.

In the elements indicating switching power losses, a_{10} is not observed in SiC-MOSFETs. This corresponds to the fact that the power losses caused by reverse recovery charge Q_{rr} and the tail current do not exist in SiC-MOSFET modules due to the effect of the anti-parallel connected SiC Schottky-Barrier-Diode (SBD). Conversely, a_{20} is larger in SiC-MOSFETs because the output capacitance C_{oss} of SiC-MOSFETs is larger than that of Si-IGBTs. Regarding a_{11}, a_{12}, a_{21} and a_{22}, the sum of them in SiC-MOSFETs is smaller than that in Si-IGBTs. This means that SiC-MOSFETs are faster than Si-IGBTs in switching speed.

E. Characteristics of on-state losses and switching power losses

Fig. 3 and Fig. 4 show on-state and switching power losses of SiC-MOSFET modules and Si-IGBT modules under the condition that the output power is varied and input voltage V_{in} is fixed at 500 V.

The on-state loss in Si-IGBTs is almost proportional to the output power. Considering from above, the forward voltage drop V_f has a stronger influence rather than resistance R_{on}. On the other hand, only the resistance R_{on} causes the on-state loss in SiC-MOSFETs, so that the on-state loss is proportional to square of the output power. Because the loss caused by the forward drop V_f

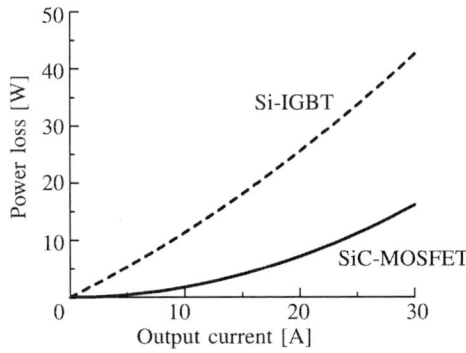

Fig. 3. Characteristics of on-state losses

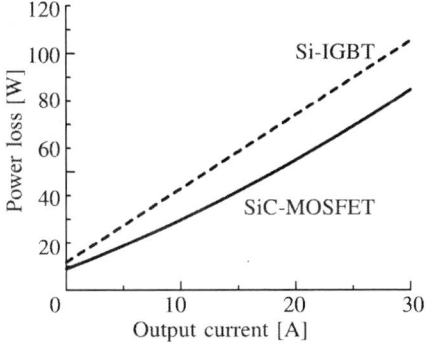

Fig. 4. Characteristics of switching power losses

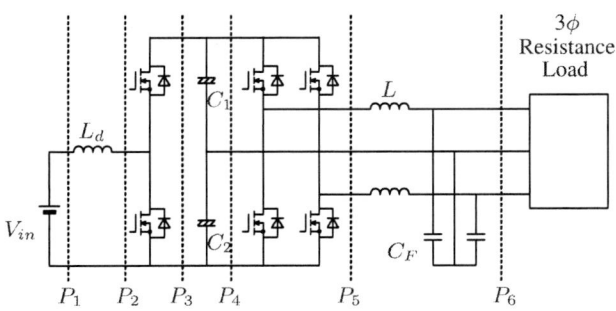

Fig. 5. Experimental system.

TABLE IV. SPECIFICATIONS AND CIRCUIT PARAMETERS OF THE EXPERIMENTAL SETUP

rated power	10 kW
ac main	200 V, 50 Hz
ac inductor L	1.0 mH (4.5%)
filter capacitor C_F	3 μF
dc capacitor C_1, C_2	6800 μF
boost inductor L_d	1.0 mH
dc input voltage V_{in}	490 V
dc-link voltage V_{dc}	590 V
switching frequency f_{sw}	10 kHz
on-delay time	3.2 μs

does not exist in the on-state loss of SiC-MOSFETs, the on-state loss in SiC-MOSFET is small especially in low-output range, compared with that in Si-IGBTs. The switching power loss of SiC-MOSFETs is smaller than that in Si-IGBTs at a light load condition due to its small reverse recovery charge Q_{rr}. Si-IGBTs have a switching power loss almost proportional to output power, while the switching power loss of SiC-MOSFETs increases by the square of the output power.

The switching power losses are tend to be higher in both switching devices rather than the on-state losses in the buck converter operation. Both switching power loss and on-state loss are reduced by replacing the Si-IGBT modules with SiC-MOSFETs, and the effect on power loss reduction is significant in the on-state loss rather than the switching power loss. Furthermore, the effect is remarkable especially in low-output range operation.

IV. EXPERIMENTAL SETUP

Fig. 5 shows the experimental system configuration of a 200-V, 10-kW PV converter, and its specifications and circuit parameters are shown in Table IV. The experimental system consists of a boost converter and two half-bridge inverters just like the most popular circuit configuration in Japan called as transformer-less PV converters. Two main circuits are constructed for the following comparison: one uses Si-IGBTs (CM100DY-24NF: Mitsubishi, 1200 V, 100 A), and the other employs SiC-MOSFETs (BSM120D12P2C005: ROHM, 1200 V,

120 A). The switching frequencies for both circuits are set to the same frequency of 10 kHz, and the dead time is 3.2 μs. The inverters are connected to a three-phase resistive load instead of the power grid, and the voltage reference is set to be 200-V, 50 Hz three-phase ac voltage. A low-loss core (Finemet: Hitachi metal) is used for the ac and dc inductors. Two electrolytic capacitors are employed to compose the dc link. The dc-link voltage is set to 590 V which is slightly higher than the required dc-link voltage of 565 V, taking the voltage drops in switching devices and filter circuits into account.

V. MEASUREMENT AND ANALYSIS OF LOSSES

A. Measurement of losses and evaluation of efficiency

A power analyzer (WT1800: Yokogawa) is also used to measure the power through the broken lines in Fig. 5. The losses in each part can be obtained from the difference of the corresponding measured power. The curve fitting technique is applied to the measured power losses to analyze the relationship against the current and/or output power. The total circuit losses are measured by subtracting the output power P_6 from the input power P_1, and the conversion efficiency of the system is also evaluated by P_1 and P_6. Moreover, the European efficiency η_{EU} is defined by

$$\eta_{EU} = 0.03\eta_5 + 0.06\eta_{10} + 0.13\eta_{20} + 0.1\eta_{30} + 0.48\eta_{50} + 0.2\eta_{100}, \quad (5)$$

where η_a indicates the conversion efficiency at an $a\%$ load operation [11].

B. Separation of the switching power losses and the on-state losses

The power losses in switching devices can be divided into switching power losses and on-state losses. In this paper, the switching power losses are calculated by subtracting the calculated on-state losses from the total device losses. The on-state power losses are obtained from the V-I characteristics at a junction temperature T_j of 25°C on the datasheet. The forward voltage drop V_f and resistance R_{on} are considered for Si-IGBT, while the on-state resistance is only taken into account for SiC-MOSFETs. The average on-state power losses P_{cond} are simply given by

$$P_{cond} = V_f I_{ave} + R_{on} I_{rms}^2, \qquad (6)$$

where I_{ave} indicates the average of the absolute values of the current, and I_{rms} means the rms values of the current. The current flows through only one of the two switching devices in each leg depending on the switching of them, and the average and rms current values are assumed to be constant in one switching cycle, as those flowing through the inductor connected to the output terminals of the legs. Accordingly, substituting the average and rms values of the current flowing through the inductor into (6) leads to the sum of the on-state losses of two devices composing a half-bridge leg.

VI. EXPERIMENTAL RESULT

A. Experiment waveforms

Fig. 6 is the circuit schematic showing the measurement points of the following experiment waveforms, and Fig. 7 is the experimental waveforms of the system using SiC-MOSFETs operated around the rated power condition. The half-bridge inverters provided almost sinusoidal line-to-line voltages v_{rs} and v_{ts} to the resistive load, resulting in three-phase sinusoidal currents. The dc-capacitor voltages v_{C1} and v_{C2} were fluctuated at the line frequency, because the S-phase current flows thorough the dc capacitors. The current flowing through the dc inductor of the boost converter is a almost constant value.

B. Power losses and conversion efficiencies of experimental systems

Fig. 8 and Fig. 9 show the breakdowns of the power losses in the experimental systems using Si-IGBTs and

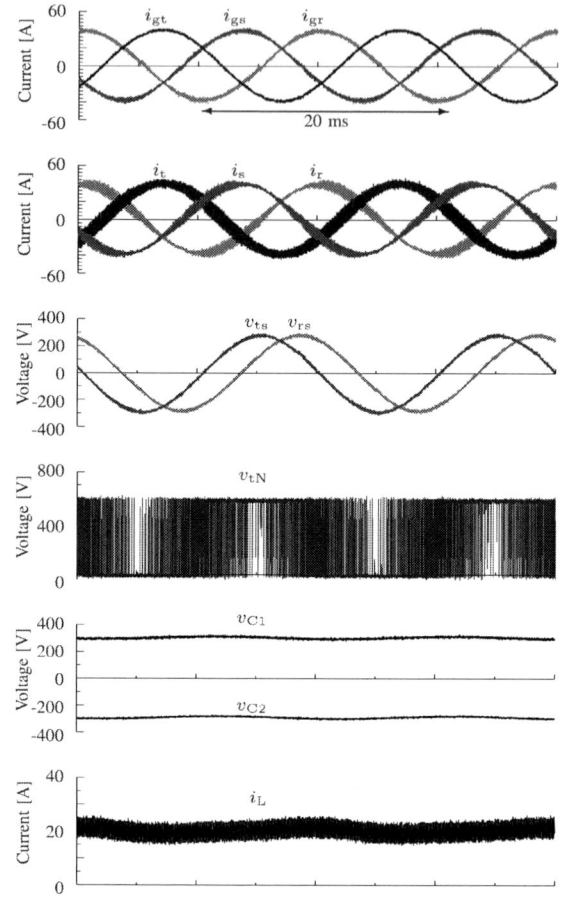

Fig. 7. Experiment waveforms of the system using SiC-MOSFETs outputting approximately rated power

using SiC-MOSFETs, respectively. The losses of inverters are much larger than the other losses because the current flowing in the inverters is larger than that in the boost converter due to the low output ac voltage of 200 V compared with the input dc voltage of 490 V. Both systems exhibit the same power losses in the passive elements, such as the dc inductors, the dc capacitors and the ac filters. On the other hand, the system using SiC-MOSFETs has reduced power losses in the inverters and the boost converter, compared with the Si-IGBT based system. Consequently, the losses in the system using SiC-MOSFETs are lower than those in the system of Si-IGBTs, especially in a low-power range.

Fig. 10 shows the switching power losses and the on-state losses in the switching devices composing the inverters of the system using Si-IGBTs, and those of the system using SiC-MOSFETs are shown in Fig. 11. In both systems, the switching power losses are dominant because the dc-link voltage is relatively high in these experiments. This tendency observed in the experiment agrees with the characteristics of the power losses obtained from the buck converter benchmark. The on-state loss in the systems using Si-IGBTs is almost proportional to the output power, because the built-in potentials in IGBTs and

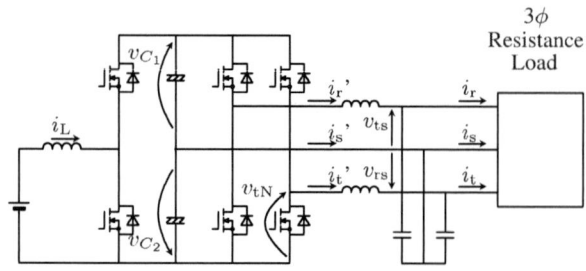

Fig. 6. Measurement points of experiment waveforms

978-1-4799-2706-7/14 $31.00 © 2014 IEEE

The 2014 International Power Electronics Conference

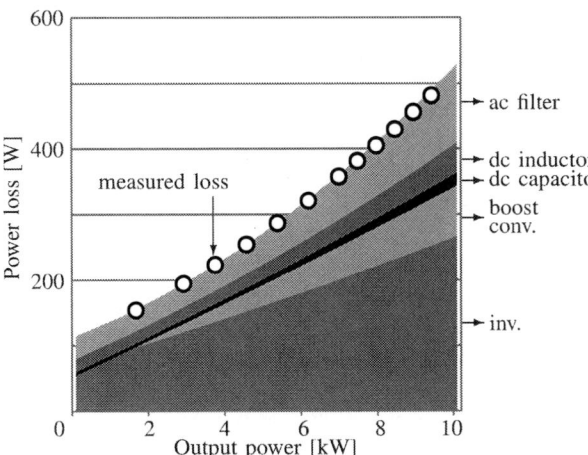

Fig. 8. Breakdown of the losses in the experimental system using Si-IGBTs.

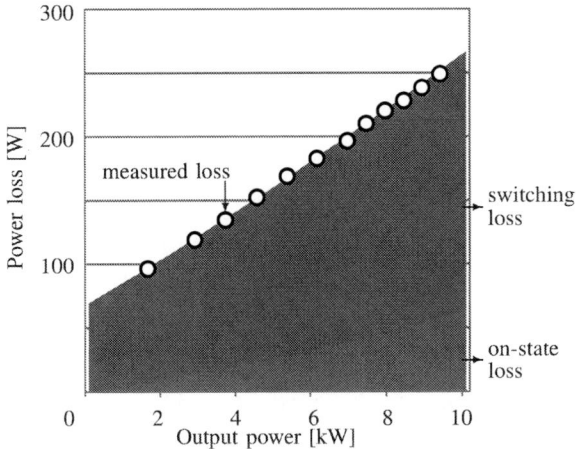

Fig. 10. The switching power losses and on-state losses of the Si-IGBT inverter.

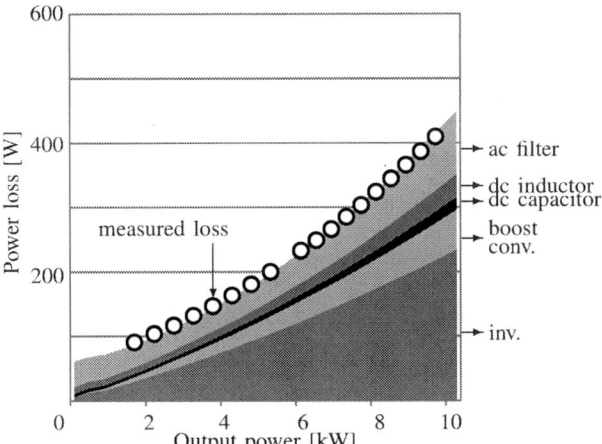

Fig. 9. Breakdown of the losses in the experimental system using SiC-MOSFETs.

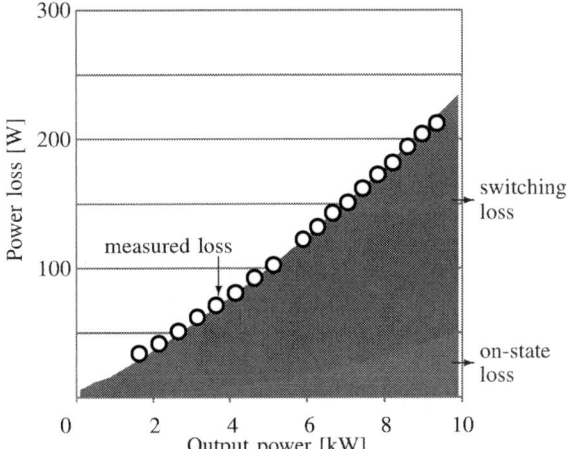

Fig. 11. The switching power losses and on-state loss of the SiC-MOSFET inverter.

diodes are much greater than their ohmic voltage drop. The system using Si-IGBTs has a switching power loss as large as 80 W at an output power of zero. On the other hand, the on-state power loss in the systems with SiC-MOSFETs is proportional to the square of the output power, and exhibits almost no switching power loss under the zero output power condition because there are neither reverse recovery charge Q_{rr} in the SiC-SBDs nor tail current in MOSFETs.

By replacing Si-IGBTs with SiC-MOSFETs, the switching power loss and the on-state loss become less than 90% and 80%, respectively, under a rated power operating condition. When the load is 50% of the rated power, the switching power loss and the on-state loss in the system using SiC-MOSFETs reduce to less than 65% and 45%, compared with the system using Si-IGBTs. Accordingly, compared with the system using Si-IGBTs, the losses in the system using SiC-MOSFETs are reduced in the whole range, especially in low-output operating range. In particular, the reduction of the on-state losses in the SiC-MOSFET system is remarkable. These tendencies coincide with the characteristics of power losses obtained

by the buck converter based benchmark.

Fig. 12 shows power conversion efficiencies of the experimental systems. In the case of the system using Si-IGBTs, the conversion efficiency reaches the maximal value of 95.1% at the 8 kW, and is higher than 94% in a range from 3 kW to the rated power. On the other hand, the system using SiC-MOSFETs attains the maximal efficiency as high as 96.2% at the 5 kW, and its conversion efficiency exceeds 95% in a range from 2 kW to the rated power of 10 kW. Table V shows the European efficiencies of the systems. Replacing Si-IGBTs with SiC-MOSFETs makes it possible to improve the efficiency by 2%. In the system using SiC-MOSFETs, the losses in low output operation are reduced, leading to notable improvement in the European efficiency.

TABLE V.   EUROPEAN EFFICIENCY OF THE EXPERIMENTAL SYSTEMS

Si-IGBT	93.5%
SiC-MOSFET	95.5%

978-1-4799-2706-7/14 $31.00 © 2014 IEEE   2032

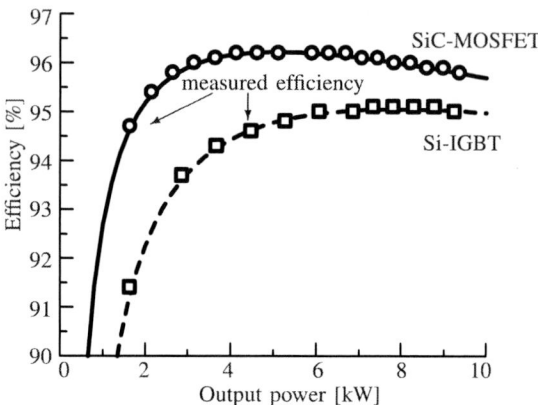

Fig. 12. Conversion efficiency of the experimental systems

VII. CONCLUSION

This paper has discussed the possibility of SiC-MOSFETs to improve the power conversion efficiency in PV converters for grid connection to the low-voltage three-phase three-wire systems in Japan. Experimental prototypes of PV converters using Si-IGBTs and SiC-MOSFETs are assembled and evaluated in experiments. As a results, it is revealed that the losses in the inverters are dominant and that the switching power losses are large, rather than the on-state losses. And it is also clarified that replacing Si-IGBTs with SiC-MOSFETs reduces both the switching power losses and the on-state losses, especially under low output operating conditions. As a result, it is also confirmed that SiC-MOSFETs make it possible to improve their European efficiency from 93.5% to 95.5%.

REFERENCES

[1] Y.-S. Lai and F.-S. Shyu: "Optimal common-mode voltage reduction pwm technique for inverter control with consideration of the deadtime effects-part i: Basic development," *IEEE Tansactions on Industry Applications*, vol. 40, no. 6, pp. 1605-1612, Nov./Dec. 2004, (2004)

[2] Z. Özkan, A. M. Hava: "A survey and Extension of High Efficiency Grid Connected Transformerless Solar Inverters with Focus on Leakage Current Characteristics," in *Energy Conversion Congress and Exposition (ECCE)*, 10.1109/ECCE.2012.6342322 ,pp.3453-3460, (2012)

[3] H. Schmidt, C. Siedle, and J. Ketterer, "Inverter for converting an electric direct current into an alternating current or an alternating voltage,", *European Patent*, EP1369985, DE1221592, (2003)

[4] M. Victor, F. Greizer, S. Bremicker, and U. Hübler, "Circuit arrangement having a dual coil for producing alternating voltage or an alternating current,", *International Patent*, WO/2007/073946, (2003)

[5] H. Fujita, M. Mabuchi, Y. Tsubota, and T. Mizogami, "Solar Power Conditioners Using Bidirectional Chopper Circuits Connected in Series," IEEJ Transactions on Industry Application, Vol.132, No.1, pp.50-57, (2012), in Japanese

[6] Johnson, B. ; Krein, P. ; Zheming Zheng ; Lentine, A, "A single-stage three-phase AC module for high-voltage photovoltaics," Applied Power Electronics Conference and Exposition (APEC), (2012)

[7] J.A. Cooper, JR. and A. Agarwal, "SiC power-switching devices—The second electronics revolution?," Proc. of IEEE, Vol.90, No.6, pp.956-968, (2002)

[8] H. W. Van Der Broeck and J. D. Van Wyk, "A comparative investigation of a three-phase induction machine with a component minimized voltage-fed inverter under different control options," IEEE Transactions on Industry Application, vol. IA-20, no. 2, pp309-320, (1984)

[9] Loceday H. Mweene, Chris A. Wright, and Martin F. Schlecht, "A 1 kW 500 kHz Front-End Converter for a Distributed Power Supply System," IEEE Transactions on Power Electronics, vol. 6, no. 3, pp398-407, (1991)

[10] H. Fujita, "Switching Loss Analysis of a Three-Phase Solar Power Conditioner Using a Single-Phase PWM Control Method," IEEE Energy Conversion Congress and Exposition (ECCE), (2009)

[11] H. Haeberlin, L. Borga, M.Kaempfer, U. Zwahlen, "New Tests at Grid-Connected PV Inverters : Overview over Test Result and Measured Value of Total Efficiency η_{tot}," 21st European Photovoltaic Solar Energy Conference, (2006)

Design of Grid Connected PWM Converters Considering Topology and PWM Methods for Low-Voltage Renewable Energy Applications

Emre Kantar
Electrical and Electronics Engineering Department
Middle East Technical University
Ankara, Turkey
emre.kantar@metu.edu.tr

Ahmet M. Hava
Electrical and Electronics Engineering Department
Middle East Technical University
Ankara, Turkey
hava@metu.edu.tr

Abstract- **This paper assesses the feasibility three-phase PWM three-level voltage source converters (VSCs), namely NPC-type VSC (3L-NPC) and T-type VSC (3L-T) as alternatives to standard three-phase PWM two-level VSC (2L-VSC) for low-voltage grid interface applications. The topological performances are compared through a design algorithm regarding varying switching frequency, load condition, modulation index, and converter side inductance under distinct pulse-width modulation methods. Design of the LCL filter is also addressed. With the proposed design algorithm, a wide range of power levels (100kW-7MW) can be spanned and in the end the optimum topology and its design details are determined by comparing the return on investment (the pay-off time of total cost of ownership) of each design in the major energy markets.**

Keywords— **Grid-connected VSC, LCL filter, two-level VSC, three-level NPC, three-level T.**

I. INTRODUCTION

As renewable energy projects are spreading all over the world rapidly, three-phase grid connected pulse-width modulation (PWM) voltage source converters (VSCs) are getting more popular due to enhanced controllability, high efficiency, and reliable performance with long life. The connection to the electric utility is provided through LCL line filters in modern PWM VSCs ensuring a more compact system and better harmonic attenuation compared to the conventional L filters. Power range of grid connected three-phase PWM VSC based renewable energy systems covers 100kW-7MW and such systems are classified according to the power levels of an application. For instance, solar energy (PV) applications (of solar farms, involving central inverters) are effective for the power ranges starting from typically 100 kW up to 1MW whereas; wind power converters are generally designed for 1-7MW power scale nowadays. Hence, particularly for wind power applications the two-level (2L) VSC topology reaches the design limits due to the restricted technology of power semiconductors that can be switched up to only a few kHz. Thus, utilization of three-level (3L) VSC topologies in place of 2L-VSC in low-voltage high power applications is favored.

In the literature, there exist several performance comparison studies concerning 2L-VSC and 3L-VSC topologies. In [1]-[6], extensive information about 2L-VSC design is provided. In [7]-[9], 3L-T topology has been examined in comparison with 2L-VSC and 3L-NPC-VSC (neutral point clamped) in terms of efficiency and cost.

This paper provides a head to toe methodology by presenting a complete design algorithm that the literature still lacks. The design algorithm spans a wide power scale of grid connected PWM converters to unveil the optimum topology delivering the lowest cost solution to achieve the same objective. The design algorithm starts with LCL-filter design and its stability analysis. Then, a topological (2L, 3L-NPC, and 3L-T) comparison under distinct PWM methods, varying load, modulation index (m_i), switching frequency (f_{sw}) and converter side inductance (L_c) is provided by regarding the efficiency and total harmonic distortion (THD) specifications of the application. With the proposed design method, the selection criterion for the optimum topology is based essentially on the pay-off time of the total cost of ownership (TCO) (i.e. return on investment, ROI) of each design. To do so, a figure of merit defined as 'operational efficiency' is used to estimate the ROI of each design in the major energy markets. Thus, an easy forward design method becomes possible.

II. SYSTEM MODELING AND GRID INTERFACE

Fig. 1 demonstrates a three-phase 2L-VSC connected to the grid via an LCL-filter. Fig. 2 introduces one leg diagrams of 2L-VSC, 3L-NPC and 3L-T VSC topologies in (a), (b) and (c) respectively. All three converter legs of Fig. 1 shall be replaced with the 3L leg modules of Fig. 2, to obtain the full schemes of 3L-NPC and 3L-T type VSCs.

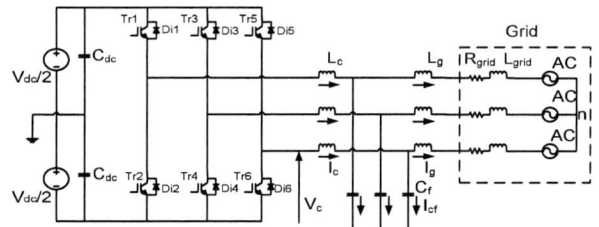

Fig. 1. Grid connected three-phase 2L-VSC system.

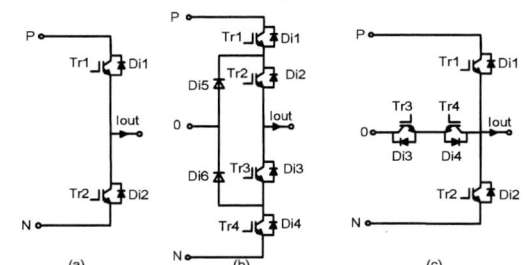

Fig. 2. One converter leg (a) 2L (b) 3L-NPC (c) 3L-T.

978-1-4799-2706-7/14 $31.00 © 2014 IEEE

LCL-filters ensure a more compact system and a better harmonic attenuation compared to the conventional L filters. However, amplification of undesired harmonics at the resonant frequency (f_{res}) of the LCL-filters may occur. Fig. 3 depicts the resonance effect of the designed LCL-filter with the 'undamped' label. Resonance shall be avoided with the employment of inherent or active or passive damping methods [1]-[5] since the rapid phase transition occurring at f_{res} as in Fig. 3 causes instability of current controllers. With the utilization of damping methods [1]-[5], the resonance impact can be suppressed as seen in Fig. 3 at the magnitude response labeled with 'damped' waveform and the rapid phase transition at f_{res} is softened. The LCL-filter design and stability analysis are covered in detail in Section IV.

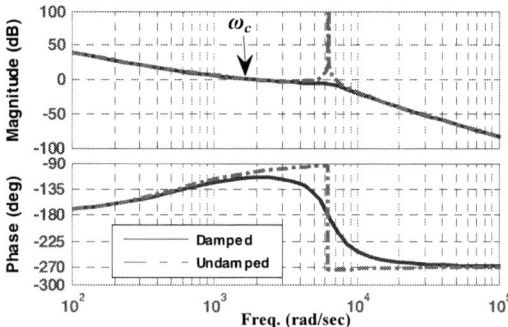

Fig. 3. Bode plot of the open-loop VSC system with LCL-filter.

III. CONVERTER DESIGN ALGORITHM

In the literature, converter design algorithms consider only the preferred methodology (topology, PWM method, set of filter parameters, current control technique, etc.) under the specified working conditions (rated power, f_{sw}, full load, etc.). However, the design algorithm in this paper (Fig. 4) can span a wide range of power levels under different PWM methods, VSC topologies, load conditions, modulation indices, switching frequencies, and set of filter parameters. Thus design results for different systems with different performances can be evaluated and compared, allowing a better determination of a suitable system.

The converter design process is divided into four main parts, namely semiconductor selection part, LCL-filter design part, switching frequency adjustment part, and topological comparison part. In the first part, suitable semiconductors are selected for 2L, 3L-NPC and 3L-T VSC topologies regarding the input parameters such as power rating of the VSC (P_{rated}), DC bus voltage (V_{dc}), the grid frequency (f_g), grid voltage (V_g), power factor (PF), and modulation index (m_i). In the second part, LCL-filter design is provided based on 2L topology including filter stability analysis. Third, f_{sw} for 3L-NPC and 3L-T is optimized so that all topologies could afford a unique solution providing the same quality on the grid-side current (in THD_i and ripple aspect). However, each topology delivers the same quality output under different energy efficiency values. TCO considers initial cost, operating cost, and economical value of a product over a determined life time. Thus, in the final part, TCO of each topology is compared with the accumulated profit over the specified life time. Then, the topology having the minimum ROI is selected to be the optimum solution. Owing to this design algorithm, ROI becomes the only parameter to decide on the optimum topology providing the same unique solution. The following sections comprehensively examine the design algorithm and clarify all of the design steps with elaborated illustrations throughout a thorough design example and cost analysis.

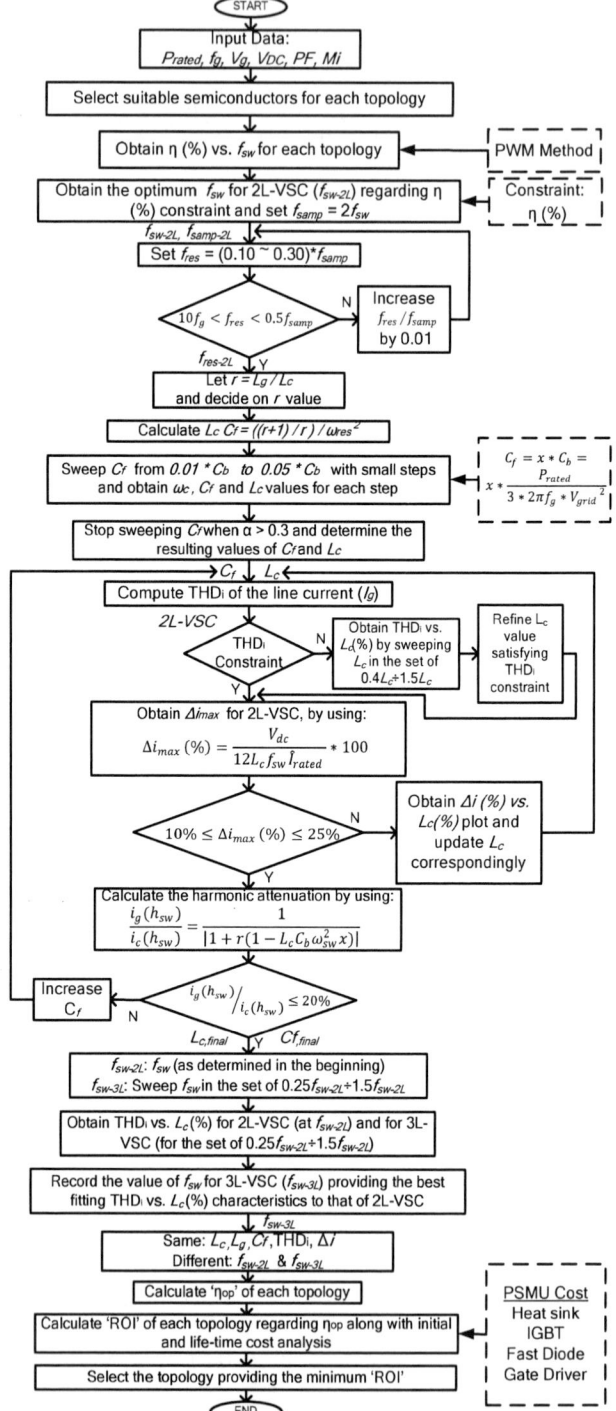

Fig. 4. Complete design flow diagram (dashed boxes provide supplementary information for the related step).

IV. DESIGN EXAMPLE

In this section, the performance analysis of the three VSC topologies will be elaborated and compared through a design example. To exemplify the process, a 1 MW grid-connected PWM converter employing space-vector PWM (SVPWM) will be considered in separate subsections. The input data of the design algorithm in Fig. 4 is provided in Table I.

TABLE I: SYSTEM PARAMETERS

V_g	Grid Line Voltage	690 $V_{rms,1\text{-}l}$	V_{dc}	DC Bus Voltage	1070 V
f_g	Grid Frequency	50 Hz	P_{rated}	Rated Power	1 MW
M_i	Modulation Index	1.05	PF	Power Factor	0.95~1
Z_b	Base Impedance	0.4761 Ω	$\eta(\%)$	Efficiency Constraint	98~99%

A. Designation of f_{sw} and Semiconductor Loss Comparison

For the first step of the design algorithm, suitable semiconductors are set to be determined for each topology. 1700V/1800A IGBTs were selected for 2L and 3L-T topologies. Besides, 1200V/1800A switches were picked for the active bidirectional switch part in 3L-T. 1200V/1800A IGBTs and fast diodes are opted for 3L-NPC.

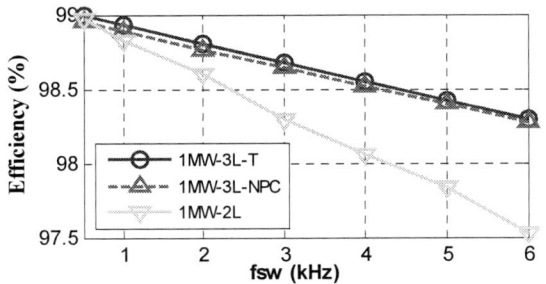

Fig. 5. η (%) vs. f_{sw} plot of three VSC topologies under SVPWM.

For the second step of design algorithm in Fig. 4, efficiency η (%) vs. switching frequency f_{sw} plot is obtained for each topology by using the parameters provided in Table I. In efficiency calculations, semiconductor losses along with LCL-filter losses are regarded. LCL-filter loss is reflected to the η curve which accounts for an increase of 0.5% of P_{rated} in total losses at full load. Details of the filter loss will be investigated in subsection D-2. Optimum f_{sw} for 2L-VSC ($f_{sw\text{-}2L}$) is selected as 2.5 kHz and it yields η (%) ≥98.5% which is compatible with the efficiency constraint of 1MW system in Table I. Afterwards, $f_{sw\text{-}2L}$ will be adjusted for 3L-NPC and 3L-T topologies so that the output current injected to the grid (I_g) could have the same THD$_i$ and ripple characteristic for all topologies. Yet, prior to these steps, a brief topological comparison will be held at f_{sw}=2.5 kHz.

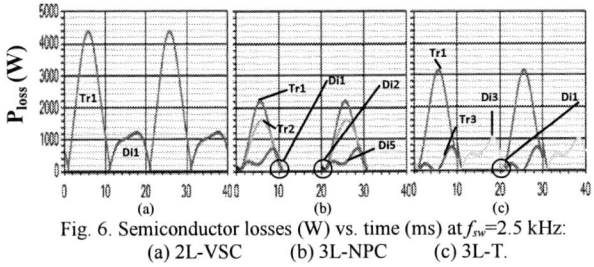

Fig. 6. Semiconductor losses (W) vs. time (ms) at f_{sw}=2.5 kHz:
(a) 2L-VSC   (b) 3L-NPC   (c) 3L-T.

The time behavior of total semiconductor losses of each semiconductor (symmetric components in each leg are omitted) for each topology are shown in Fig. 6 (at f_{sw}=2.5 kHz). The characteristics of only one leg of the VSC are shown, as the other two are identical. In Fig. 6, the peak stress distributed on each semiconductor in 2L topology is higher than 3L topologies as the heat is concentrated in a few numbers of switches. Yet, in 3L topologies the peak stresses

are distributed among more transistors and diodes resulting in reduction of the total stress on each component. Thus, better thermal performance is realized (Fig. 7). Besides, the peak stress on each semiconductor is the lowest in 3L-NPC topology providing the most stable thermal performance as shown in Fig. 7. Consequently, volume and cost of the heat sink to be used in 3L-NPC and 3L-T is less than that of 2L.

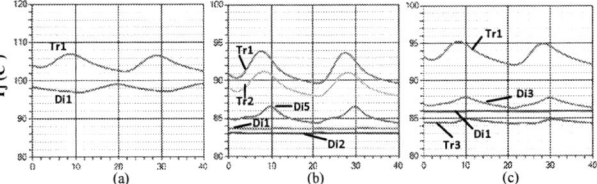

Fig. 7. Junction temperature (T_j(C°)) vs. time (ms) at f_{sw}=2.5 kHz:
(a) 2L-VSC   (b) 3L-NPC   (c) 3L-T.

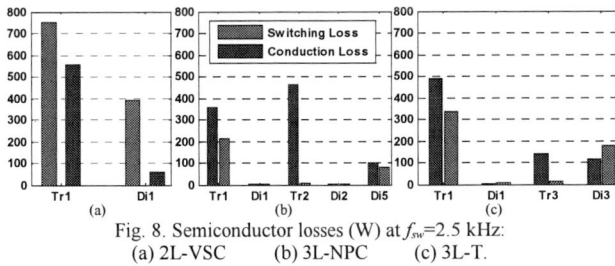

Fig. 8. Semiconductor losses (W) at f_{sw}=2.5 kHz:
(a) 2L-VSC   (b) 3L-NPC   (c) 3L-T.

3L-VSCs (T-type and NPC-type) are more favorable than 2L-VSCs in terms of reduced switching losses. The commutation voltage of all semiconductors in 3L-NPC is $V_{dc}/2$ rather than V_{dc} as in 2L-VSC. However, 3L-T is composed of both $V_{dc}/2$ (Tr3-4) and V_{dc} (Tr1-2) rated semiconductors. Hence, the total switching loss extent of 3L-T is lower than that of 2L-VSC; conversely, higher than that of 3L-NPC as can be seen in Fig. 8. Although the efficiency curves of NPC and T-type are very flat, nearly parallel and they intersect at 9 kHz in Fig. 5, this intersection point can move to the left by significant amount for higher V_{dc} values making the NPC-type favorable over the T-type beyond the intersection point. In terms of conduction losses, with fewer elements along the current path yielding less losses, the 2L-VSC, T-type and NPC-type converters are ordered from the best to the worst in a sequence. Evaluating the total losses and determining the efficiency from here, it becomes obvious at frequencies less than 9 kHz the T-type is more favorable. However, considering the efficiency constraint, f_{sw} cannot exceed 5-6 kHz which becomes a switching frequency constraint. Thus, in a more confined f_{sw} range, still the ordering from the best to worst in terms of efficiency is T-type, NPC-type and 2L-VSC.

After completing the topological comparison part, the design algorithm proceeds by determining the sampling frequency (f_{samp}). f_{samp} can be selected as equal to f_{sw} (single-update) or twice of f_{sw} (double-update) [6]. Preferred sampling is double-update ($f_{samp} = 2*f_{sw}$) throughout this paper that yields f_{samp} for 2L-VSC ($f_{samp\text{-}2L}$) to be 5 kHz.

B. LCL-Filter Design, Control and Stability Analysis

The LCL-filter design part aims to predict the dynamic behavior of the system and turn the stabilization into an easy task. Therefore, f_{res} should be set to a certain frequency to rule the control actions properly. In the algorithm, f_{res} is chosen as 10-30% of f_{samp} as suggested in [3], [5]. However, designation of this proportion depends on the power level

978-1-4799-2706-7/14 $31.00 © 2014 IEEE

and the preferred LCL resonance damping technique. For instance, at low power designs (100-500kW) where active damping is generally preferred, 10% should be the starting value in the algorithm. On the other hand, 20% should be the onset value in higher power applications (>500 kW) where either inherent damping characteristics of the LCL-filter or passive damping methods are generally benefited [5], [6]. Thus, f_{res} should be assigned as $0.2*f_{samp}$ since converter side current feedback is preferred (inherent damping) as the current control technique in the 1MW system in this paper. Hence, f_{res} of 2L-VSC design (f_{res-2L}) was found as 1 kHz satisfying $10f_g<f_{res}<f_{sw}=0.5f_{samp}$ constraint.

$$\omega_{res} = \sqrt{\frac{L_c+L_g}{L_cL_gC_f}} \qquad (1)$$

Next, the proportion (r) between L_c and grid-side inductance (L_g) is selected as $r=1$ to maximize the filter attenuation extent and minimize the filter size [4]. Then, ω_{res} formula in (1) is used to calculate L_cC_f multiplication because f_{res} and r are predetermined parameters at this phase in the algorithm. Hence, if the filter capacitance (C_f) is determined, then L_c could be found. Yet, rather than picking a specific C_f value, a set of C_f values will be spanned and the optimum L_c, C_f pair is chosen by regarding the stability analysis presented in the following paragraph. To do so, C_f value is swept from 1% to 5% of base capacitance (C_b) so that the reactive power absorption (x) by C_f could be limited in 1-5% to deliver a power factor (PF) within 0.95-0.99.

The distance between ω_{res} (ω_{res} and f_{res} are used alternatingly) and zero dB gain crossover frequency (ω_c) in the magnitude response of the open loop system (Fig. 3) plays the key role on ensuring LCL-filter stabilization by achieving an adequate phase margin. For this reason, ω_c must be fixed sufficiently below ω_{res} regardless of damping extent and technique. The correlation in between can be represented as $\omega_c=\alpha\omega_{res}$. Setting $\alpha\leq0.3$ (i.e. $\omega_c\leq0.3\omega_{res}$) not only succeeds an adequate phase margin (50°) and a sufficient damping ratio ($\zeta=0.707$) but also avoids the rapid phase transition securely [6].

In the design procedure, C_f is varied from 1% to 5% of C_b until $\alpha>0.3$ condition arises. ω_c differs for each combination of C_f and L_c values, and the values fulfilling $\alpha\leq0.3$ requirement are opted as the raw filter parameters. The raw LCL-filter parameters were acquired as $L_c=153\mu H$ and $C_f=332\mu F$ in the light of design steps implied in Fig. 4. The bode diagram in Fig. 3 was obtained for these raw filter components.

C. THD and Ripple Performance of the LCL-Filter

The LCL-filter should comply with the stringent grid codes specified by IEEE 519-1992. Thus, THD_i of I_g should be monitored carefully and a reasonable maximum THD_i limit should be set in the design phase according to the type and susceptibility of the application.

However, the raw LCL-filter components may fail to meet the THD_i aim set by the design rules. In this case, L_c should be refined to attain the desired THD_i performance. For this purpose, THD_i vs. L_c (%) characteristics should be obtained by spanning L_c value up and down sufficiently (from $0.4*L_c$ to $1.5*L_c$ is recommended) as in Fig. 9 and L_c value should be tuned with regard to the intersection point of L_c (%) and desired THD_i value. After the proper refinement of L_c, design algorithm proceeds to current ripple analysis as

depicted in Fig. 4. In the design example, $L_c=153\mu H$ (10%) results in 1.04% THD_i that satisfies the specified constraint ($THD_i\leq2\%$) for 1MW-VSC. However, any L_c that is bigger than 8-10% of the base inductance (L_b) does not enhance THD_i performance considerably as can be seen in Fig. 9.

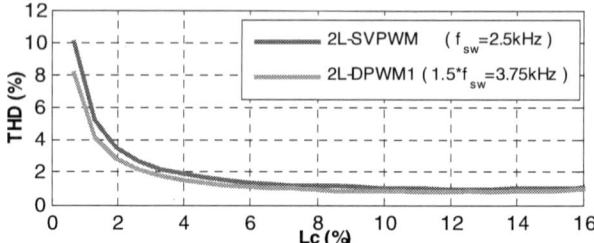

Fig. 9. THD_i vs. L_c (%) for 2L-VSC.

In addition to THD_i restriction, IEEE 519-1992 compels the worst case peak-to-peak converter side current ripple (Δi_{max}) to be confined into 10-25% of peak rated load current ($\hat{I}_{rated}$). Hence, the ripple suppressing capability of L_c must always be adequate to confine $\Delta i_{max(\%)}$ within 10-25% under varying grid conditions. However, Δi_{max} is independent of the load; whereas, highly dependent on the grid voltage (V_g). Thus, varying V_g impact on Δi_{max} can be investigated by considering varying modulation index (m_i). m_i is defined as the peak line-to-neutral V_g over half of V_{dc} with a definition range of 0-1.15 for SVPWM [8]. V_{dc} value is kept constant and V_g is altered correspondingly to practice varying m_i case. Then, Δi_{max} expression is derived as a function of m_i in (2) under SVPWM [10].

$$\Delta i_{max} \cong \frac{V_{dc}}{3L_cf_{sw}}(1-m_i)m_i \qquad (2)$$

Theoretical studies in the literature state that Δi_{max} becomes the largest at $m_i=1/\sqrt{3}$ under SVPWM. In the design example, the simulated system generated the maximum ripple as 21.3% at $m_i=1/\sqrt{3}$ rather than at the rated $m_i=1.05$ as can be seen in Fig. 10(a). If L_c design were made by regarding only rated m_i, L_c would be insufficient to suppress the ripple in case of V_g change. Besides, the boundary region between linear-modulation region and over-modulation region in SVPWM is somewhat uncertain and probably unstable. As a result, Δi_{max} rises substantially near 1.15 as demonstrated in Fig. 10(a). Hence, maximum m_i should be limited to 1.05-1.1 to have enough margin for the control in SVPWM method.

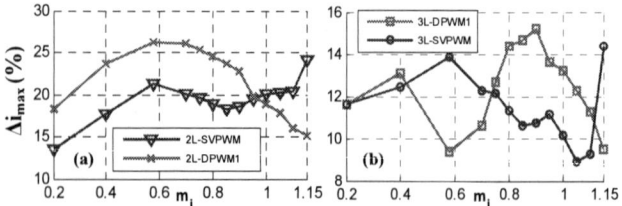

Fig. 10. Δi_{max} (%) vs. m_i plot under SV and DP (a) 2L-VSC (b) 3L-VSC.

Fig. 9 and 10 also contains analyses for discontinuous PWM1 (DPWM1). However, these waveforms will be used to enlighten the influence of the employed PWM pattern on ripple characteristics in subsection D-1.

Substituting $m_i=1/\sqrt{3}$ on (2) yields $\Delta i_{max\%}$ formula as shown in (3). The corresponding converter parameters in

Table I have been inserted in (3) and $\Delta i_{max\%}$ was found as 20.7% (in high correlation with the simulations 21.3%).

$$\Delta i_{max}(\%) = \frac{V_{dc}}{12 L_c f_{sw} I_{rated}} * 100 \qquad (3)$$

Designed filter components satisfied the ripple constraint (21.3% < 25%) defined in Fig. 4. If the given constraint could not be met, Δi (%) vs. L_c (%) plot for 2L-VSC should be obtained as shown in Fig. 11(a) by using (3) as described in Fig. 4. Then, L_c should be refined properly. However, if L_c is updated, algorithm goes back to the THD$_i$ calculation stage and iterations last until fulfilling the constraint (Fig. 4).

Design algorithm states the next step as the analysis of the current ripple attenuation (from converter side to grid side) of the LCL-filter. The relationship between the harmonics generated by the converter and injected into the grid can be found by using (4).

$$\frac{i_g(h_{sw})}{i_c(h_{sw})} = \frac{1}{|1 + r(1 - L_c C_b \omega_{sw}^2 x)|} \qquad (4)$$

The primary objective of the LCL-filter is to confine the current ripple on L_c within 10-25%. Then, with the addition of $L_g C_f$ branch, minimum 80% additional current ripple attenuation with respect to the ripple on the converter side (10-25%) is aimed. To deduce, the final target of the LCL-filter is to reduce the current ripple on the grid-side to 2-5% of I_{rated} in total. Thereby, maximum 0.2 (20%) harmonic injection is intended to satisfy $i_g(h_{sw})/i_c(h_{sw})$ constraint in Fig. 4. With the corresponding parameters substituted on (4), the total harmonic injection to the grid was found to be 0.1 (Fig. 11(b)) that complies with the defined constraint.

If the aimed constraint could not be fulfilled, $i_g(h_{sw})/i_c(h_{sw})$ vs. r plot should be acquired as shown in Fig. 11(b) by employing (4) for various r. Then, minimum r providing sufficient harmonic attenuation should be opted by means of Fig. 11(b). Nevertheless, increase in r makes the filter become too bulky and costly. Therefore, increasing C_f to enhance the attenuation capability of the LCL-filter is recommended rather than increasing r. Yet, there is a trade-off between reactive power absorption and size & cost of the filter. In this paper, augmenting C_f is preferred in the design algorithm regarding power levels and design specifications.

After satisfying harmonic attenuation constraint, the final values of the filter components are unveiled as shown in Fig. 4. In the design example, final L_c and C_f values have fulfilled THD$_i$, ripple and harmonic attenuation constraints so any refinement in filter components was not needed.

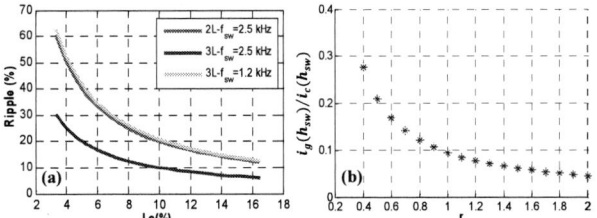

Fig. 11. (a) Ripple (%) vs. L_c (%) (b) Harmonic attenuation vs. r.

D. f_{sw} Synchronization of 3L Topologies to 2L for SVPWM

After completing the filter design phase regarding 2L-VSC, the analysis is extended by involving 3L-NPC and 3L-T type topologies. As a remark, the effective f_{sw} in 3L topologies is almost twice of f_{sw-2L}. Therefore, THD$_i$ performance of 3L topologies is expected to prevail over 2L topology for the same f_{sw}. As mentioned before, the target of the design algorithm is to provide the same output waveform quality (in THD$_i$ and ripple aspect) under each topology by using the same LCL-filter parameters. For this reason, f_{sw} for 3L topologies (f_{sw-3L}) is varied with reasonable frequency steps to cover the set of f_{sw} values ($0.25 f_{sw-2L} \div 1.5 f_{sw-2L}$). Then, the optimum f_{sw-3L} is chosen providing the closest THD$_i$ performance to that of 2L-VSC under varying L_c (%).

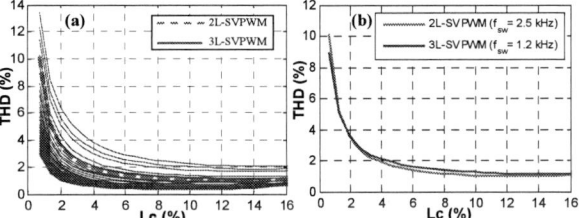

Fig. 12. THD$_i$(%) vs. L_c under SVPWM
(a) f_{sw-2L}=2.5kHz, f_{sw-3L}=$0.25 f_{sw-2l} \div 1.5 f_{sw-2L}$ (b) f_{sw-2L}=2.5kHz, f_{sw-3L}=1.2 kHz.

Fig. 12(a) depicts THD$_i$ vs. L_c (%) plots for 2L-VSC at f_{sw-2L} and 3L-VSC under the set of $0.25 f_{sw-2L} \div 1.5 f_{sw-2L}$. Then, f_{sw} iteration steps were adjusted sensitive enough to match the 2L and 3L plots. Fig. 12(b) shows the best fitting THD$_i$ curve of 3L-VSC to the THD$_i$ curve of 2L-VSC (at f_{sw-3L}=1.2 kHz). Consequently, f_{sw-3L} is found to be approximately half of the f_{sw-2L} as expected. In [8], (3) is derived for the 3L-NPC topology, and the constant term in the denominator is '24'. Thus, the determined f_{sw-3L} yields almost the same value with 2L-VSC ($24 * f_{sw-3L} \equiv 12 * f_{sw-2L}$) in the denominator of (3) and the ripple characteristics of 3L-VSC were found to be very close to 2L-VSC as depicted in Fig. 11(a). Besides, the determined f_{sw-3L} (i.e. $f_{samp-3L}$) has to satisfy the stability constraints regarded for 2L-VSC in subsection B. Otherwise, filter design procedure should be repeated by refining 10-30% proportion between f_{res} and f_{samp} so that f_{res} of 3L-VSC (f_{res-3L}) could fulfill the constraints eventually.

1) Impact of the PWM Pattern on THD and Ripple

The flow of the design algorithm is interrupted in this subsection to indicate that PWM modulation methods have an influence not only on THD$_i$ and Δi_{max} performance of I_g, but also on total semiconductor loss extent significantly. Thereby, the impact of the preferred PWM pattern is reflected to design algorithm on f_{sw-2L} designation part in Fig. 4. In this paper, two of the most popular PWM methods containing zero sequence signal injection, namely SVPWM and DPWM1 are examined for 2L-VSC and 3L-VSC under "*equal switching loss*" principle. To do so, f_{sw} is increased by 50% in DPWM1 method so that the switching count and therefore the switching losses could remain the same while ripple of DPWM1 becomes less compared to SVPWM [10].

The preferred PWM method influences η(%) vs. f_{sw} of 2L-VSC (i.e. optimum f_{sw-2L}) as depicted in Fig. 13(a). Picking f_{sw-2L} as 2.5 kHz for SVPWM and 3.75 kHz for DPWM1 verifies the theoretical approach and ensures "*equal switching loss*". Moreover, in 2L-VSC the selected f_{sw-2L} for DPWM1 method delivers slightly better THD$_i$ performance against L_c variation than that of SVPWM as shown in Fig. 9.

SVPWM at low m_i and DPWM1 at high m_i have been widely used in industry. SVPWM loses its advantage around $m_i \approx 0.8$ since the ripple increases as m_i increases. Therefore; use of DPWM1 methods beyond $m_i > 0.8$ is suggested in

[10]. The simulation analysis of 2L-VSC in Fig. 10(a) depicts that DPWM1 exhibited worse ripple performance than that of SVPWM up to $m_i \approx 0.9$. However, a substantial improvement in DPWM1 has been recognized beyond that point and DPWM1 was found to be the optimum PWM pattern in ripple aspect for 2L-VSC at the rated m_i=1.05.

However, 2L and 3L topologies shows opposite reactions to m_i variation at DPWM1 case as can be seen in Fig. 10(b). In DPWM1 case, 3L topologies performed the lowest Δi_{max} at m_i=1/√3; whereas, 2L-VSC had its peak as shown in Fig. 10. Yet, the ripple variation in 3L-VSC was confined to 9-15% yielding much lower ripple compared to 2L-VSC varying in a narrower envelope.

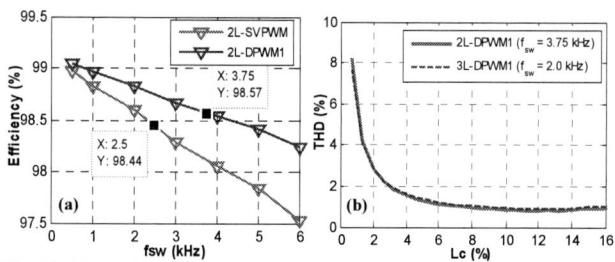

Fig. 13. (a) η (%) vs. f_{sw} (SV and DP) (b) $f_{sw\text{-}3L}$=2 kHz $f_{sw\text{-}2L}$=3.75 kHz (DP).

Finally, to equalize the output current waveform quality (THD$_i$) of all topologies under DPWM1 operation and same efficiency, the switching frequency values are adjusted. As a result, the same THD$_i$ for 2L SVPWM (Fig 9, 2.5kHz), DPWM1 (Fig 9 and 13(b), 3.75kHz) and for 3L SVPWM (Fig. 12(b), 1.2kHz) and DPWM1 (Fig. 13(b), 2kHz) is obtained for all cases.

2) Efficiency and Filter Loss Variation versus Load

LCL-filter loss is another significant loss mechanism in grid-connected converters apart from semiconductor loss. Therefore, a reasonable filter loss should be assumed according to power level of the design. For instance, applications higher than 300-400 kW presume less than 0.5% total filter loss at full load. As the power rating decreases, this percentage can be increased up to 1-2%. In the LCL-filter, the losses are predominantly inductor losses. The core losses mainly occur in L_c and the ohmic losses occur both in L_c and L_g. Thus, in the design, only inductor losses are considered for the filter loss calculations as in (5). LCL-filter loss (P_{filter}) contributes as an offset in efficiency characteristics of 1MW-VSC which accounts for a 0.5% decrease at full load.

$$P_{filter} = P_{fe} + P_{cu} = k_1 + k_2 * I_{load}^2 \quad (5)$$

where k_1 stands for core loss, P_{fe} (W), P_{cu} (W) is copper loss, and k_2 (W/A^2) is a constant term.

k_2 is calculated by subtracting predetermined P_{fe} (assumed at full load) from P_{filter} and the result is divided by the square of full load current. Afterwards, (5) can be employed under various load current (I_{load}) accordingly.

In Fig. 14(a), with the LCL-filter loss included, η (%) of 2L-VSC is plotted by means of (5) for two distinct filter designs against varying I_{load} including light load and overload conditions. In the first design, efficiency degradation owing to the filter loss is evaluated by assuming P_{fe} constituting 25% of the total filter loss (0.5% of P_{rated}) at full load. Yet, the second design assesses the case that is less insensitive to I_{load} variations as opposed to first design. P_{fe} is

constant regardless of I_{load}; whereas, P_{cu} is highly dependent on I_{load} due to its second order characteristics. Therefore, in both cases a substantial degradation in efficiency compared to the case without filter losses is observed in Fig. 14(a) as I_{load} increases. However, the deflection is much sensitive to increase in I_{load} in P_{cu}=75% case. Conversely, I_{load} increase does not affect the shape of the efficiency curve in P_{cu}=25% case significantly and it contributes more like offset. Typically, the inductors are designed such that the core losses are about 25-40% of the total inductor losses, such that the light and no-load losses are maintained low.

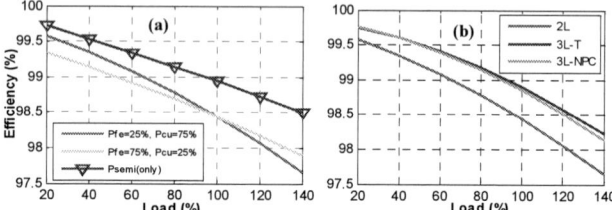

Fig. 14. With the inclusion of LCL-filter loss, the η (%) vs. load (%) characteristics (a)2L-VSC (b) All topologies under P_{fe}=25%.

After encompassing filter losses, η vs. load characteristics for all topologies were gathered into a single plot in Fig. 14(b) for topological comparison. Conduction losses increase faster (quadratic dependency) than switching losses (linear dependency) with increasing load current [9]. This impact can be realized in Fig. 14(b), particularly as the load current passes beyond full load range. Up to 60-70% load, total losses of 3L-T and 3L-NPC are very close; however, beyond that point, the conduction losses of 3L-NPC rises considerably since four components are present in one leg in the main current path. Therefore, T-type is more favorable near rated loads and above. Additionally, 2L-VSC has shown the worst η (%) characteristic amongst all topologies. The flow of the design algorithm proceeds with operational efficiency analysis utilizing $f_{sw\text{-}2L}$ and $f_{sw\text{-}3L}$ found in D.

E. Figure of Merit: Operational Efficiency

Converters operating in renewable energy systems do not always perform at rated load due to intermittent characteristic of nature. Energy derived from photovoltaic (PV) and wind sources varies with time, day of the week, season, weather and similar natural factors. Particularly, wind turbines usually operate 40% of the time annually. They generate small amount or no power about 60% of the time [11]. Thereby, topological comparison regarding the rated load is not a fair assessment. In the literature, there are several operational efficiency definitions such as 'Euro Efficiency 'and 'CEC Efficiency'. However, these criteria are more suitable to assess PV applications. Yet, a novel operational efficiency definition including wind turbine applications is set to be developed in this section. Moreover, Euro and CEC Efficiency definitions will be provided as the alternative definitions in the Appendix section.

To develop an operational efficiency definition to analyze the wind turbine applications, power output vs. wind speed characteristics is required. MWT62/1.0 (Mitsubishi Wind Turbine Generator) is selected as a reference wind turbine generator. Its rated power is 1MW and the DC/AC converter is connected to the low-voltage grid at 690 V line-to-line.

In wind turbines the term 'capacity factor' (C.F) is widely used. The capacity factor is the actual output over a period of time as a proportion of a wind turbine or plant's maximum capacity. The C.F versus wind speed characteristics provided in MWT62/1.0 datasheet is shown in Fig. 15(a) and it is elaborated in Fig. 15(b). Then, these curves are converted into rectangular boxes without changing the area under the curve to ease the numerical calculations. The rectangular form assumes that the capacity factor is steady for the corresponding wind speed intervals.

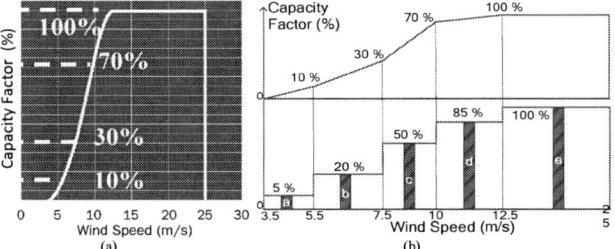

Fig. 15. (a) Power curve of MWT62/1.0 (b) C.F vs. wind speed (m/s).

The red-hatched slices named as a-e in Fig. 15(b) depict the portions of which wind class is benefited throughout the operation time (i.e. when wind speed is between 3.5-25m/s). Depending on the width of these slices, the equivalent C.F is determined by weighing 5%, 20%, 50%, 85%, and 100% capacity factors correspondingly as given in (6). The maximum operation time of a wind turbine is assumed as 60% of the time annually in this paper as depicted in (7).

$$C.F(\%) = a * (5\%) + b * (20\%) + c * (50\%) + d * (85\%) + e * (100\%) \tag{6}$$

$$a + b + c + d + e = 0.6 \ (\equiv 60\%) \tag{7}$$

Wind speed is classified according to possible wind strengths in Table II. Then, the incidences of these wind classes are weighed such that resulting outputs could provide 20%, 30%, and 40% C.Fs. $\mathcal{C}_1$-$\mathcal{C}_5$ coefficients represent the weight of each wind class benefited during 60% operation time. The sum of the wind speed coefficients is unity (i.e. $\sum \mathcal{C}_n$ =1). Thus, a-e constants should be normalized by multiplying $\mathcal{C}_1$-$\mathcal{C}_5$ with 0.6 respectively. Hence, equivalent capacity factors representing the 60% operation time per annum are embodied.

TABLE II: WIND SPEED CLASSIFICATION

C.F (%)	Wind Speed Coefficients ($\mathcal{C}_n$)				
	$\mathcal{C}_1$ 3.5-5.5 m/s	$\mathcal{C}_2$ 5.5-7.5 m/s	$\mathcal{C}_3$ 7.5-10 m/s	$\mathcal{C}_4$ 10-12.5 m/s	$\mathcal{C}_5$ 12.5-25 m/s
20 %	0.30	0.40	0.09	0.11	0.10
30 %	0.15	0.25	0.30	0.05	0.25
40 %	0.05	0.30	0.05	0.05	0.55

$$\eta_{op} = \mathcal{C}_1 \cdot \eta_{5\%} + \mathcal{C}_2 \cdot \eta_{20\%} + \mathcal{C}_3 \cdot \eta_{50\%} + \mathcal{C}_4 \cdot \eta_{85\%} + \mathcal{C}_5 \cdot \eta_{100\%} \tag{8}$$

Wind speed classification as given in Table II, and an efficiency curve as in Fig. 14(b) are necessary to compute the operational efficiency (η_{op}) of the wind turbine. Operational efficiencies can be calculated by using (8) where $\eta_{5\%}$, $\eta_{20\%}$, $\eta_{50\%}$, $\eta_{85\%}$, and $\eta_{100\%}$ are the efficiencies at 5%, 20%, 50%, 85%, and 100% load, respectively. As a remark, operation at 20% C.F means that the converter employed in the wind turbine operates at 20% load. The wind data in [11] were benefited to determine $\mathcal{C}_1$-$\mathcal{C}_5$ in Table II. The calculated operational efficiencies of each topology for 20%, 30%, and 40% C.F are tabulated in Table III. At all capacity factors,

3L-T topology prevailed over the other topologies; whereas, 2L-VSC delivered the worst efficiency performance.

TABLE III: COMPARISON OF η_{op} (%) UNDER C.F

C.F (%)	η_{2L}	η_{3L-NPC}	η_{3L-T}
20 %	99.38	99.59	99.60
30 %	99.17	99.43	99.45
40 %	98.90	99.22	99.24

In the upcoming section, ROI (i.e. pay-off time) of each topology regarding C.F and the energy price in major energy markets will be investigated.

V. OPERATIONAL COST EVALUATION

Operation schedule of a plant is the key point in ROI analysis. TCO contains initial cost, operating cost, and economical value of a plant over a determined life time. Therefore, the overall profit gained along the life time of the plant is compared with TCO to present a cost evaluation criterion. Then, topological comparison is made in terms of ROI of each design regarding the energy price in a corresponding energy market and the topology providing the minimum ROI is selected to be the optimum solution.

A. Initial Cost Analysis

In this section, the current market quotations are considered for the components and energy prices (all in $) to perform cost calculations. IGBTs with 600V, 1200V, and 1700V were priced as 9ct/A, 17 ct/A, and 23ct/A per IGBT, respectively. Furthermore, 1200V fast diodes employed in 3L-NPC were determined as 5 ct/A per diode, respectively. Gate drive units were accounted for $10 per channel.

The heat sink design has been made regarding the required minimum sink to ambient resistance value ($R_{th(s-a)}$) found by the detailed temperature analysis provided in this paper. The LCL-filter is designed for once and it is the same for each topology; therefore cost analysis of it will be redundant. Additionally, cost of any system elements common in each topology is also omitted for simplicity since they would just contribute as an offset to calculations. Thus, the components comprising Table IV can be named as power semiconductor main unit (PSMU) and TCO comparison based on ROI of each topology is reduced to replacing PSMUs in the system where the remaining VSC components are the same.

TABLE IV: INITIAL COST ANALYSIS OF PSMU

1MW System	2L-VSC	3L-NPC	3L-T
IGBTs	$2484	$3672	$4320
Fast Diodes	$0	$540	$0
Gate Driver	$60	$120	$120
Heat Sink	$2551	$1275	$1275
TOTAL	$5095	$5607	$5715

The initial PSMU costs of 3L topologies are slightly higher than that of 2L due to the higher semiconductor and gate driver cost. The savings in the heat sink cost has reduced the gap between the PSMU costs of 2L and 3L topologies. Besides, the initial PSMU cost of NPC-type is marginally cheaper than that of T-type.

B. Accumulated Cost Savings Using Capacity Factor

In this section, instead of calculating the overall TCO of each topology, the cost savings (difference in TCOs) owing to the improved efficiency in 3L-NPC and 3L-T topologies in comparison to 2L topology will be determined. To do so, the operational efficiency of each topology under 20%, 30% and 40% C.F is made use of (Table III) considering 1 year, 5 years, and 10 years of operation time. Then, the resulting kWh difference is utilized to calculate the cost savings in one of the high price energy markets, such as Japan. The net cost paid per kWh in industry was 17.9ct/kWh in Japan in 2011 [11]. The overall cost savings including a possible annual interest rate in electricity can be calculated by employing (9).

$$s_0 = \sum_{i=1}^{12n} \frac{s_i}{12(1+\frac{k}{12})^i} \qquad (9)$$

where s_0 is the cost savings in present, s_i is cost savings in month i, k is annual interest rate, and n is number of years.

A 2L grid connected VSC operating at 20% C.F (1752h annually) has 2169 W more losses than 3L-T as depicted in Table III. Thus, 3800 kWh less energy losses of 3L-T achieves a cost saving of s_i=$681 in each single year in Japan. Then, by means of calculated s_i, the overall energy cost savings at present value, s_0 was calculated regarding the operation time of 1 year, 5 years, and 10 years. The annual interest rate is taken as 3% (k=0.03). Eventually, the resulting s_0 was tabulated in Table V. Likewise, the same approach was adopted for the comparison between 2L-VSC and 3L-NPC and resulting outcomes are presented in the Table V as well. Similarly, the cost savings of the systems operating at 30% and 40% C.F were also considered.

TABLE V: S_0 VS. OPERATION TIME IN JAPAN

Japan	2L vs. T			2L vs. NPC		
	1 y	5 y	10 y	1 y	5 y	10y
20%	$670	$3155	$5871	$643	$3029	$5636
30%	$1306	$6153	$11448	$1235	$5819	$10827
40%	$2122	$9999	$18606	$1975	$9307	$17317

Although 3L topologies cost $500-600 more than 2L topology, 3L topologies pay-off the initial PSMU cost difference just in one year in high energy price markets like Japan. However, C.F and total operation time have a strong impact ROI. For instance, if the system were operating at 40% C.F for ten years, 3L-T would save $18606 over 2L; whereas, 3L-NPC would save $17317 over 2L in Japan. Thus, T-type would become the most optimum solution for the implied conditions. Conversely, NPC and T-type barely pays-off in one year at low C.F (20%) in low energy price markets as the USA (7ct/kWh [11]).

The initial PSMU cost of 3L-T is expected to be lower than 3L-NPC because 3L-NPC has a higher total volt-ampere (VA) per volume index due to additional diodes. However, T-type has become popular recently and its production volume is much lower than the NPC-type settled for years in industry. As the area of utilization of 3L-T increases, the initial PSMU cost of 3L-T will decrease in the long term and the advantages brought by 3L-T over NPC will become more evident.

VI. CONCLUSIONS

This study provides a complete design analysis for grid-connected PWM converters by combining the LCL-filter design procedure with topological comparison. Hence, a step by step method ending with optimized filter elements along with the best topological choice based on the minimum ROI

was presented. THD$_i$ and efficiency performances of 3L topologies were found to be superior to 2L topology under the same f_{sw}. Thus, 3L-VSCs are economically feasible in low-voltage applications especially in high energy-cost markets. In addition, the reduction in the commutation voltage has provided lower temperature rises in 3L-VSCs in return reducing the cost and size of the heat sinks. Besides, the 3L-T was found to be the better choice for low-voltage applications if efficiency and costs are the main concern.

APPENDIX

In addition to 1MW-VSC design, converters at different power ratings are evaluated regarding Euro and CEC Efficiency definitions given in (10) and (11), respectively.

$$\eta_{EURO} = 0.03 \times \eta_{5\%} + 0.06 \times \eta_{10\%} + 0.13 \times \eta_{20\%} + 0.1 \times \eta_{30\%} + 0.48 \times \eta_{50\%} + 0.2 \times \eta_{100\%} \qquad (10)$$

$$\eta_{CEC} = 0.04 \times \eta_{10\%} + 0.05 \times \eta_{20\%} + 0.12 \times \eta_{30\%} + 0.21 \times \eta_{50\%} + 0.53 \times \eta_{75\%} + 0.05 \times \eta_{100\%} \qquad (11)$$

TABLE VI: COMPARISON OF η_{op} (%) UNDER EURO AND CEC

Power Level	EURO			CEC		
	η_{2L}	$\eta_{3L\text{-}NPC}$	$\eta_{3L\text{-}T}$	η_{2L}	$\eta_{3L\text{-}NPC}$	$\eta_{3L\text{-}T}$
250kW	98.88	99.09	99.19	98.82	99.05	99.18
1 MW	99.18	99.45	99.47	99.13	99.44	99.46
3 MW	99.33	99.44	99.48	99.31	99.43	99.48

The ordering from the best to worst in terms of efficiency is still T-type, NPC-type and 2L-VSC under both definitions for all power levels as shown in Table VI. η_{EURO} and η_{CEC} performances were found to be very close to η_{op} performance of 1MW-VSC under 30% C.F (Table III). Besides, the design procedure stated in Fig. 4 can also be adopted for 250kW and 3MW-VSC designs and the resulting ROI of each topology in each design can be compared by using either of η_{op}, η_{EURO}, η_{CEC} definitions.

REFERENCES

[1] M. Liserre, F. Blaabjerg, and S. Hansen, "Design and control of an LCL-filter-based three-phase active rectifier,"IEEE. Trans. Industrial Applications, vol. 41, pp. 1281-1291, 2005.

[2] J. Dannehl, F.W. Fuchs, S. Hansen, and P.B. Thogersen, "Investigation of Active Damping Approaches for PI-Based Current Control of Grid-Connected Pulse Width Modulation Converters with LCL Filters," IEEE. Trans. Industrial Applications, vol. 46, pp. 1509-1517, 2010.

[3] S. G. Parker, B. P. McGrath, and D. G. Holmes, "Regions of Active Damping Control for LCL Filters" IEEE Trans. Ind. App., 2012-IPCC-463.

[4] Y. Tang, P. C. Loh, P. Wang, F. H. Choo and F. Gao, "Exploring inherent damping characteristics of LCL-filters for three-phase grid-connected VSI," IEEE Trans. Power Elec., vol. 27, no. 3, pp. 1433-1443, Mar. 2012.

[5] E. Kantar, S. N. Usluer, and A. M. Hava, "Design and Performance Analysis of a Grid Connected PWM-VSI System," Electrical and Electronics Engineering (ELECO), 2013 8th International Conf., pp. 220-224, Nov. 2013.

[6] E. Kantar, S. N. Usluer, and A. M. Hava, "Control Strategies for Grid Connected PWM-VSI Systems," Electrical and Electronics Engineering (ELECO), 2013 8th International Conf., pp. 157-161, Nov. 2013.

[7] M. Schweizer and J. W. Kolar, "Design and implementation of a hybrid efficient three-level T-type converter for low-voltage applications," IEEE Trans. Power Electron., vol. 28, no. 2, pp. 899–907, Feb. 2013.

[8] A. A. Rockhill, M. Liserre, R. Teodorescu, and P. Rodriguez, "Grid filter design for a multi-megawatt medium-voltage voltage source inverter," IEEE Trans. Ind. Electron, vol. 58, no. 4, pp. 1205–1217, Apr. 2011.

[9] R. Teichmann and S. Bernet, "A comparison of three-level converters versus two-level converters for low-voltage drives, traction, and utility applications," IEEE Trans. Ind. Applicat., pp. 855-865, June 2005.

[10] A. M. Hava and N. O. Cetin, "A Generalized Scalar PWM Approach With Easy Implementation Features for Three-Phase, Three-Wire VSI in IEEE Trans. Power Electron., vol. 24, no. 5, pp. 1385-1395, May 2011.

[11] National Wind Watch (2001). Retrieved December 15, 2013, from https://www.wind-watch.org/

Performance of
Dead Time Compensation Methods
in Three-Phase Grid-Connection Converters

Tomoyuki Mannen and Hideaki Fujita
Department of Electrical and Electronic Engineering
Tokyo Institute of Technology
Tokyo, JAPAN

Abstract—This paper discusses effect of the power factor on voltage error caused by the dead time in voltage-source grid-connection PWM converters. The theoretical analysis derives the voltage error considering output capacitances of the switching devices and brings out that the voltage error depends on the current ripple amplitude through the converter which is changed by the power factor. The analytical result reveals the compensation characteristics of three approximation compensation methods and turn-off transition-based compensation method from the viewpoint of the power factor. The compensation characteristics of four different compensation methods are compared with those with each other in experiments using a 200-V, 5-kW three-phase grid-connection converter. As a result, the approximation compensation methods exhibit a good compensation performances at a specific power factor. In contrast, the turn-off transition-based compensation method has a better compensation performance allover the range of power factor.

Keywords—dead time, grid connection converters, power factor, current ripples, phase angle.

I. INTRODUCTION

Recently the switching frequency of voltage-source PWM converters tends to become higher to reduce the volume of passive components and to improve the controllability, according to the improvement in switching performance, especially in wide-band gap semiconductor devices [1]. Voltage-source converters need a dead time, which is the blanking period where both upper and lower switching devices are turned off to avoid the short circuit of its dc link. As the switching frequency increases, the dead time might cause a remarkable voltage error according to the increasing rate of the dead time to the switching cycle [2]. The voltage error of the dead time may cause a current distortion in grid-connection converters and a torque ripple in motor drive systems.

Recent grid-connection converters are equipped with a high-order switching-ripple filter which effectively suppress grid current ripples in spite of using relatively small ac inductors [3]. However, large current distortion is caused by the voltage error and has become remarkable in the converters equipped with the small inductors because of their low impedance. In voltage-source PWM converters, current feedback control usually has the capability of mitigating the current distortion caused by the dead time. When the small inductors are applied to the switching-ripple filter, current feedback control is not expected to

mitigate the current distortion, because of the stability problem.

To reduce the voltage error caused by the dead time in voltage-source PWM converters, various compensation methods have been proposed [4]–[7]. The most popular compensation methods are two-level approximation compensation method [4], [5], linear approximation compensation [6] and three-level approximation compensation [7], [8]. These approximation-based compensation methods (ACMs) usually calculate the compensation voltage from the detected converter current, assuming that the amplitude of the current ripple is constant or negligible.

Besides, compensation methods detecting zero current clamping [8] and compensation methods paying attention to the output capacitance of switching devices [9], [10] are proposed Reference [9] proposes a feed forward compensation method using a look-up table containing the compensating voltage derived from simulation results of the dynamic response of charging and discharging the output capacitance. Reference [10] has reported detailed waveforms of the converter voltage during the dead time and implies that the voltage error is strongly related with the current ripple because of the existence of the output capacitance in switching devices. The result in [10] also suggests that it is difficult for conventional ACMs to calculate and compensate the voltage error accurately when the power factor or the dc-link voltage in the converter changes. For example, detecting the turn-off currents or the output voltage edges would make it possible to solve such the problem [11]. However, it is difficult to be implemented in a popular microcontroller because it needs a specially designed detection circuit using a high-speed A/D converter and a FPGA.

Authors have proposed turn-off transition-based compensation method (TTCM), whose compensating voltage is derived from the analytic result considering output capacitances of the switching devices [12]. Theoretical analysis in [12] reveals functions of the voltage error caused by the dead time and the turn-off current of three-phase PWM converters. The feature of TTCM is that the compensating voltage is derived from the analytic result of the voltage error and the turn-off current. TTCM can accurately compensate the voltage error and effectively suppress the current distortion regardless of the power factor. Since turn-off current estimation is employed in the TTCM, it is easy to implement this method into a popular microcontroller, instead of high-speed A/D

978-1-4799-2706-7/14 $31.00 © 2014 IEEE

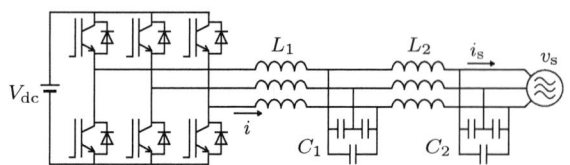

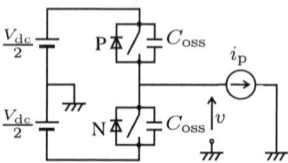

Fig. 1. Circuit configuration of a three-phase grid-connection converter equipped with a cascaded switching-ripple filter.

Fig. 2. Single-phase equivalent circuit for analysis of voltage error caused by the dead time.

TABLE I. CIRCUIT PARAMETERS OF THE EXPERIMENTAL CIRCUIT

L_1	0.3 mH			C_1	3 μF
L_2	0.1 mH			C_2	0.22 μF
f_{sw}	20 kHz	T_{sw}	50 μs	T_{DT}	3 μs
V_{dc}	330 V	f_s	50 Hz		

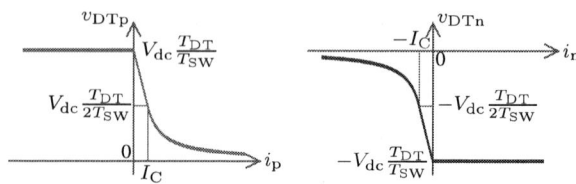

(a) Commutation from P to N (b) Commutation from N to P

Fig. 3. Average voltage error against the turn off currents.

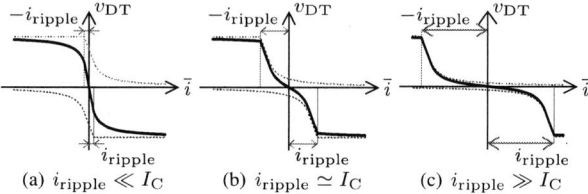

(a) $i_{ripple} \ll I_C$ (b) $i_{ripple} \simeq I_C$ (c) $i_{ripple} \gg I_C$

Fig. 4. Average voltage error against the converter current $\bar{i}$ under three different current ripple conditions.

converters and FPGAs.

The dead time compensation for grid connection converters has been discussed in [8], [12], on the assumption of a unity power factor, considering applications to PV converters. Grid-connection converters are operated under various conditions of power factor and modulation index, according to applications such as PV converters, STATCOMs, active power filters and so on. For example, the power factor is almost zero in case of STATCOMs. It is unable to define the power factor of active power filters because the converter current contains only harmonic components. Moreover, the dc voltage of the active power filter is higher than other converters even when the grid voltage is the same. Thus, the current ripple through the converter in active power filter is large, compared with other applications.

This paper discusses the compensation characteristics of voltage error caused by dead time in case of application of conventional ACMs and TTCM to grid-connection converters. Theoretical analysis derives the turn-off current of the voltage-source PWM converter operated as a grid-connection converter. The analytical results reveals the voltage error caused by the dead time depends on the power factor of the grid-connection converter. It is difficult for conventional ACMs to calculate an accurate compensating voltage because they do not consider the effect of the power factor. On the other hand, TTCM can effectively suppress the current distortion at any power factor because it calculates the compensating voltage using the turn-off current estimation.

Experimental results are shown to compare and to evaluate the performance of voltage error compensation and source-current harmonic distortion in conventional two-level, linear and three-level ACMs and TTCM. The linear and three-level ACMs exhibit a lower harmonic distortion or a better voltage error compensation than the two-level ACM around a unity power factor. On the other hand, two-level ACM makes it possible to reduce harmonic distortion compared with the other ACMs under low power factor conditions. Moreover, it is also confirmed that TTCM can effectively compensate the voltage error in all operation range, and thus, suppress the harmonic components contained in the source current compared with the conventional ACMs

II. EXPERIMENTAL SETUP

Fig.1 shows the circuit configuration of the experimental three-phase grid-connection converter rated at 5 kVA, and its circuit parameters are listed in Table.I. The main circuit of the converter is a conventional three-phase bridge PWM converter using a dual in-line package intelligent power module (DIP-IPM: PS21869, 600 V, 50 A, Mitsubishi). The converter is equipped with a high order ripple filter on its ac side, which consists of ac inductors L_1 and L_2 and filter capacitors C_1 and C_2. Although a relatively large current ripple of 4 A in peak value flows through L_1, the filter circuit makes it possible to eliminate current ripples flowing out to the grid effectively. The switching frequency was set to be $f_{sw} = 20$ kHz ($T_{sw} = 50$ μs), and the dead time was $T_{DT} = 3$ μs in the following experiments.

III. ANALYSIS OF VOLTAGE ERROR

A. Voltage error against the current ripples

Fig.2 shows a single-phase equivalent circuit considering parasitic capacitances C_{oss}. In Fig.2, IGBTs are represented by parallel connection of an ideal switch, an ideal diode and parasitic capacitance C_{oss}. Turn-off currents of the upper and lower switching devices, i_p and i_n are defined as follows:

$$i_p = \bar{i} + i_{ripple} \tag{1}$$
$$i_n = \bar{i} - i_{ripple}, \tag{2}$$

where i_{ripple} is the peak value of the current ripples.

Fig.3 shows three different cases of the average voltage error as a function of turn-off currents. The average voltage error in the commutation from switch P to switch

N is summarized as follows:

$$v_{\mathrm{DTp}}(i_{\mathrm{p}}) = \begin{cases} V_{\mathrm{dc}}\dfrac{T_{\mathrm{DT}}}{T_{\mathrm{SW}}} & (i_{\mathrm{p}} < 0) \\ V_{\mathrm{dc}}\dfrac{T_{\mathrm{DT}}}{T_{\mathrm{SW}}}\left(1 - \dfrac{i_{\mathrm{p}}}{2I_{\mathrm{C}}}\right) & (0 \le i_{\mathrm{p}} \le I_{\mathrm{C}}) \\ V_{\mathrm{dc}}\dfrac{T_{\mathrm{DT}}}{2T_{\mathrm{SW}}}\dfrac{I_{\mathrm{C}}}{i_{\mathrm{p}}} & (I_{\mathrm{C}} < i_{\mathrm{p}}) \end{cases} , \quad (3)$$

where $I_{\mathrm{C}} = 2C_{\mathrm{oss}}V_{\mathrm{dc}}/T_{\mathrm{DT}}$. As the same manner, the switching transition from switch N to switch P also causes a similar voltage error given by

$$v_{\mathrm{DTn}}(i_{\mathrm{n}}) = \begin{cases} V_{\mathrm{dc}}\dfrac{T_{\mathrm{DT}}}{2T_{\mathrm{SW}}}\dfrac{I_{\mathrm{C}}}{i_{\mathrm{n}}} & (i_{\mathrm{n}} < -I_{\mathrm{C}}) \\ -V_{\mathrm{dc}}\dfrac{T_{\mathrm{DT}}}{T_{\mathrm{SW}}}\left(1 + \dfrac{i_{\mathrm{n}}}{2I_{\mathrm{C}}}\right) & (-I_{\mathrm{C}} \le i_{\mathrm{n}} \le 0) \\ -V_{\mathrm{dc}}\dfrac{T_{\mathrm{DT}}}{T_{\mathrm{SW}}} & (0 < i_{\mathrm{n}}) \end{cases} .$$
$$(4)$$

Fig.4 shows the relation between the average converter current $\bar{i}$ in a switching cycle and the voltage error v_{DT} derived from

$$v_{\mathrm{DT}}(\bar{i}) = v_{\mathrm{DTp}}(\bar{i} + i_{\mathrm{ripple}}) + v_{\mathrm{DTn}}(\bar{i} - i_{\mathrm{ripple}}). \quad (5)$$

The voltage error strongly depends on the amplitude of the current ripples. In other words, the voltage error caused by the dead time is affected by the inductance of the ripple filter, dc-link voltage, switching frequency, and power factor.

B. The converter current ripple amplitude

The converter current i can be calculated by using only a relatively simple multiply-accumulate operation, when a three-phase PWM control method based on a triangular carrier signal is applied. Here, the grid voltage is assumed to be three-phase balanced voltage source, and the switching ripple filter is considered as a simple inductor L. When the three-phase voltage references are $v_{\mathrm{a}}^* > v_{\mathrm{b}}^* > v_{\mathrm{c}}^*$, the turn-off currents of three-phase PWM converters are obtained by

$$i_{\mathrm{ap}} = \left(\frac{v_{\mathrm{a}}^*}{3} - \frac{v_{\mathrm{b}}^*}{6} - \frac{v_{\mathrm{c}}^*}{6} - \frac{v_{\mathrm{sa}}}{4} - \frac{v_{\mathrm{sa}}v_{\mathrm{a}}^*}{2V_{\mathrm{dc}}}\right)\frac{T_{\mathrm{sw}}}{L} + i_{\mathrm{a0}} \quad (6)$$

$$i_{\mathrm{an}} = \left(\frac{v_{\mathrm{a}}^*}{3} - \frac{v_{\mathrm{b}}^*}{6} - \frac{v_{\mathrm{c}}^*}{6} - \frac{3v_{\mathrm{sa}}}{4} + \frac{v_{\mathrm{sa}}v_{\mathrm{a}}^*}{2V_{\mathrm{dc}}}\right)\frac{T_{\mathrm{sw}}}{L} + i_{\mathrm{a0}} \quad (7)$$

$$i_{\mathrm{bp}} = \left(\frac{v_{\mathrm{b}}^*}{6} - \frac{v_{\mathrm{c}}^*}{6} - \frac{v_{\mathrm{sb}}}{4} - \frac{v_{\mathrm{sb}}v_{\mathrm{b}}^*}{2V_{\mathrm{dc}}}\right)\frac{T_{\mathrm{sw}}}{L} + i_{\mathrm{b0}} \quad (8)$$

$$i_{\mathrm{bn}} = \left(\frac{v_{\mathrm{b}}^*}{2} - \frac{v_{\mathrm{a}}^*}{3} - \frac{v_{\mathrm{c}}^*}{6} - \frac{3v_{\mathrm{sb}}}{4} + \frac{v_{\mathrm{sb}}v_{\mathrm{b}}^*}{2V_{\mathrm{dc}}}\right)\frac{T_{\mathrm{sw}}}{L} + i_{\mathrm{b0}} \quad (9)$$

$$i_{\mathrm{cp}} = \left(-\frac{v_{\mathrm{sc}}}{4} - \frac{v_{\mathrm{sc}}v_{\mathrm{c}}^*}{2V_{\mathrm{dc}}}\right)\frac{T_{\mathrm{sw}}}{L} + i_{\mathrm{c0}} \quad (10)$$

$$i_{\mathrm{cn}} = \left(\frac{2v_{\mathrm{c}}^*}{3} - \frac{v_{\mathrm{a}}^*}{3} - \frac{v_{\mathrm{b}}^*}{3} - \frac{3v_{\mathrm{sc}}}{4} + \frac{v_{\mathrm{sc}}v_{\mathrm{c}}^*}{2V_{\mathrm{dc}}}\right)\frac{T_{\mathrm{sw}}}{L} + i_{\mathrm{c0}}, (11)$$

where i_{a0}, i_{b0}, i_{c0} are initial values of the converter current at the beginning of the corresponding switching cycle or at the apex of the triangular carrier signal. The voltage references v_{a}^*, v_{b}^* and v_{c}^* for the grid-connection converters can approximately be assumed to equal the source voltages v_{sa}, v_{sb} and v_{sc}. The source voltage v_{sa} and the voltage reference v_{a}^* are given by

$$v_{\mathrm{sa}} = \sqrt{2}V_{\mathrm{s}}\sin\theta \quad (12)$$
$$v_{\mathrm{a}}^* = \sqrt{2}V_{\mathrm{s}}\sin\theta + v_0, \quad (13)$$

where θ is the phase angle of the source voltage and v_0 is the zero sequence voltage of the converter. Substituting

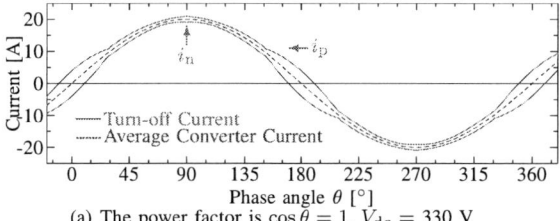

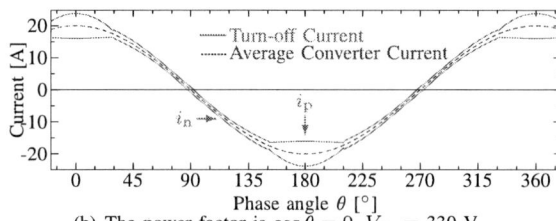

(b) The power factor is $\cos\theta = 0$, $V_{\mathrm{dc}} = 330$ V

Fig. 5. The converter current ripple amplitude of grid-connection converters against the phase angle.

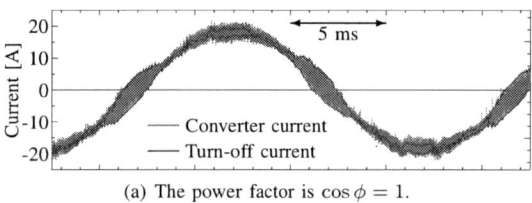

(a) The power factor is $\cos\phi = 1$.

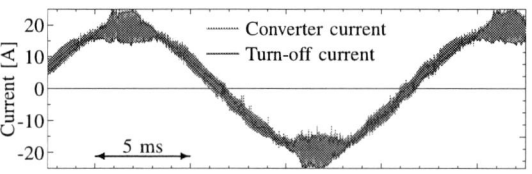

(b) The power factor is $\cos\phi = 0$.

Fig. 6. Measured converter current and turn-off current waveforms of grid-connection converters.

(12) and (13) into (6) through (11), the amplitude of the current ripple in a-phase is represented by

$$i_{\mathrm{ripple\_a}} = \begin{cases} \left\{(v_0 + V_{\mathrm{dc}})\sin\theta - 2\sqrt{2}V_{\mathrm{s}}\sin^2\theta\right\}\dfrac{\sqrt{2}V_{\mathrm{s}}}{4V_{\mathrm{dc}}}\dfrac{T_{\mathrm{sw}}}{L} \\ \qquad\qquad (30° \le \theta < 150°) \\ \left\{\begin{array}{l} 6v_0\sin\theta + \sqrt{3}V_{\mathrm{dc}}|\cos\theta| \\ \quad -6\sqrt{2}V_{\mathrm{s}}\sin^2\theta \end{array}\right\}\dfrac{\sqrt{2}V_{\mathrm{s}}}{4V_{\mathrm{dc}}}\dfrac{T_{\mathrm{SW}}}{L} \\ \qquad (-30° \le \theta < 30°, 150° \le \theta < 210°) \\ \left\{(v_0 - V_{\mathrm{dc}})\sin\theta - 2\sqrt{2}V_{\mathrm{s}}\sin^2\theta\right\}\dfrac{\sqrt{2}V_{\mathrm{s}}}{4V_{\mathrm{dc}}}\dfrac{T_{\mathrm{SW}}}{L} \\ \qquad\qquad (-150° \le \theta < -30°) \end{cases} .$$
$$(14)$$

The above equation implies that the amplitude of the current ripple should be considered as the function of the source-voltage phase angle θ or a function of time. Thus, a conventional ACMs causes a compensation error because they assume a constant ripple amplitude in the converter current.

Fig.5 shows envelopes of the theoretical turn-off currents i_{p} and i_{n} in a grid-connection converter. Here, the average converter current $\bar{i}$ is assumed to be a sinusoidal current waveform of 14.4 A in rms value and the dc voltage is to be $V_{\mathrm{dc}} = 330$ V. The turn-off currents i_{p} and i_{n} are calculated considering a simple ac filter composed

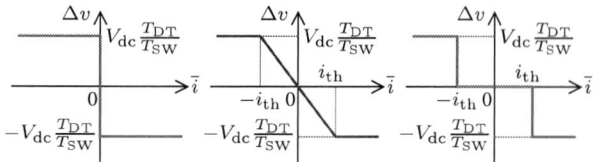

(a) Two-level ACM   (b) Linear ACM   (c) Three-level ACM

Fig. 7. Compensating voltages Δv in conventional compensation methods.

only an inductor $L = 0.4$ mH.

Fig.5(a) shows the turn-off currents in a case of a unity power factor, and Fig.5(b) presents a case that the power factor is zero. In both cases, a relatively large current ripple appears in the converter current contained around $\theta = 0°$ and $\theta = 180°$. On the other hand, the current ripple is small in a phase angle range $30° \leq \theta < 150°$ and $210° \leq \theta < 330°$. Thus, the amplitude of the current ripple depends on the phase angle of the source voltage, but is independent from the power factor or the phase angle of the source current.

Paying attention to the zero-crossing of the source current, the amplitude of the current ripple is large around the zero-crossing in Fig.5(a), while it is small in Fig.5(b). For this reason, the voltage error caused by dead time in Fig.5(a) is also different from that in Fig.5(b) because the voltage error is the function of the current ripple amplitude as shown in (4) and Fig.5.

Fig.6 shows the current envelopes of the measured converter and turn-off currents. Here, the source current and the dc voltage are set to the same as those in Fig.5. Fig.6(a) is the waveforms in case of the power factor $\cos\phi = 1$, and Fig.6(b) is those in case of the power factor $\cos\phi = 0$. The turn-off currents i_p and i_n are sampled at the instants of the commutation of the switching devices in the corresponding phase leg. Note that the converter current ripple is larger than the turn-off current in ranges $30° \leq \theta < 150°$ and $210° \leq \theta < 330°$. The converter current ripple is caused by the commutation not only in the corresponding phase leg but also the other two phase leg. Therefore, the other two phase legs induce larger current ripples than the corresponding phase leg in these phase angle ranges. The sampled turn-off currents are well agree with (14) and the theoretical plots in Fig.5.

IV. DEAD TIME COMPENSATION METHODS

A. Conventional approximation compensation methods

Fig.7 shows the compensating voltage Δv in conventional approximation compensation methods (ACMs). The conventional ACMs calculates the compensating voltage Δv from a simple approximation of the voltage error caused by the dead time as a function of only the average converter current $\bar{i}$ on the assumption of a constant amplitude of current ripples. The voltage reference should be modified from v^* to $v^* - \Delta v$ as a feed forward compensation, and then the PWM controller generates gating pulses for the corresponding switching devices using a triangular carrier signal, a space vector modulation method, or so on. Thus, it is easy to implement the conventional ACM into popular microcontrollers because the sampled converter current or the current reference can be used as the average converter current $\bar{i}$.

Fig.7(a) shows the compensating voltage of a two-level ACM [4]. The compensating voltage is directly calculated from the direction of the average converter current $\bar{i}$ as given by

$$\Delta v = \begin{cases} V_{\text{dc}} \frac{T_{\text{DT}}}{T_{\text{SW}}} & (\bar{i} < 0) \\ -V_{\text{dc}} \frac{T_{\text{DT}}}{T_{\text{SW}}} & (0 < \bar{i}) \end{cases}. \tag{15}$$

Fig.7(b) illustrates the compensating voltage of a linear ACM [6], which is given by

$$\Delta v = \begin{cases} V_{\text{dc}} \frac{T_{\text{DT}}}{T_{\text{SW}}} & (\bar{i} < -i_{\text{th}}) \\ -V_{\text{dc}} \frac{T_{\text{DT}}}{T_{\text{SW}}} \frac{\bar{i}}{i_{\text{th}}} & (-i_{\text{th}} < \bar{i} < i_{\text{th}}) \\ -V_{\text{dc}} \frac{T_{\text{DT}}}{T_{\text{SW}}} & (i_{\text{th}} < \bar{i}) \end{cases}, \tag{16}$$

where i_{th} is a threshold current for calculating the compensating voltage. Fig.7(c) shows the compensating voltage of a three-level ACM [7], [8], given by

$$\Delta v = \begin{cases} V_{\text{dc}} \frac{T_{\text{DT}}}{T_{\text{SW}}} & (\bar{i} < -i_{\text{th}}) \\ 0 & (-i_{\text{th}} < \bar{i} < i_{\text{th}}) \\ -V_{\text{dc}} \frac{T_{\text{DT}}}{T_{\text{SW}}} & (i_{\text{th}} < \bar{i}) \end{cases}. \tag{17}$$

In linear and three-level ACMs, the threshold current might be set to around the current ripple amplitude near the zero-crossing of the average converter current $\bar{i}$ to obtain an accurate compensating voltage. In this case, the threshold current is usually set to a fixed value because the ripple current amplitude is assumed to be constant. The compensation characteristic becomes worse when the threshold current deviates from the amplitude of the current ripple.

B. Turn-off transition-based compensation method

Turn-off transition-based compensation method (TTCM) calculates the compensating voltage from the voltage error based on the theoretically derived equation considering the effect of output capacitances in switching devices and the estimated turn-off currents. The turn-off current can be estimated by substituting the voltage references v_a^*, v_b^*, v_c^* into (6)–(11). This estimation requires only a relatively simple multiply-accumulate operation which is also easily calculated in microcontrollers. TTCM can calculate an appropriate compensating voltage Δv even when the amplitude of the converter current ripple changes because it applies the estimated turn-off current. Thus, TTCM is expected to achieve better compensation characteristics regardless of the phase angle of the converter current or the power factor unlike conventional ACMs.

Aforementioned, turn-off currents i_p and i_n are the functions of voltage phase angle θ in case of grid-connection converters, as shown in Fig.5. Thus, the turn-off currents can easily be calculated from the current ripple amplitude i_{ripple} in (14) instead of the turn-off current estimation using (6)–(11). In case of grid-connection converters, the dc voltage V_{dc} and the source voltage V_s are assumed to be constant, and the zero-sequence voltage v_0 is usually zero. Then, the function of $\sin\theta$ and $\cos\theta$ are only required for calculation of the current ripple amplitude i_{ripple} in (14), which can be obtained from a look up table against the phase angle of the source voltage.

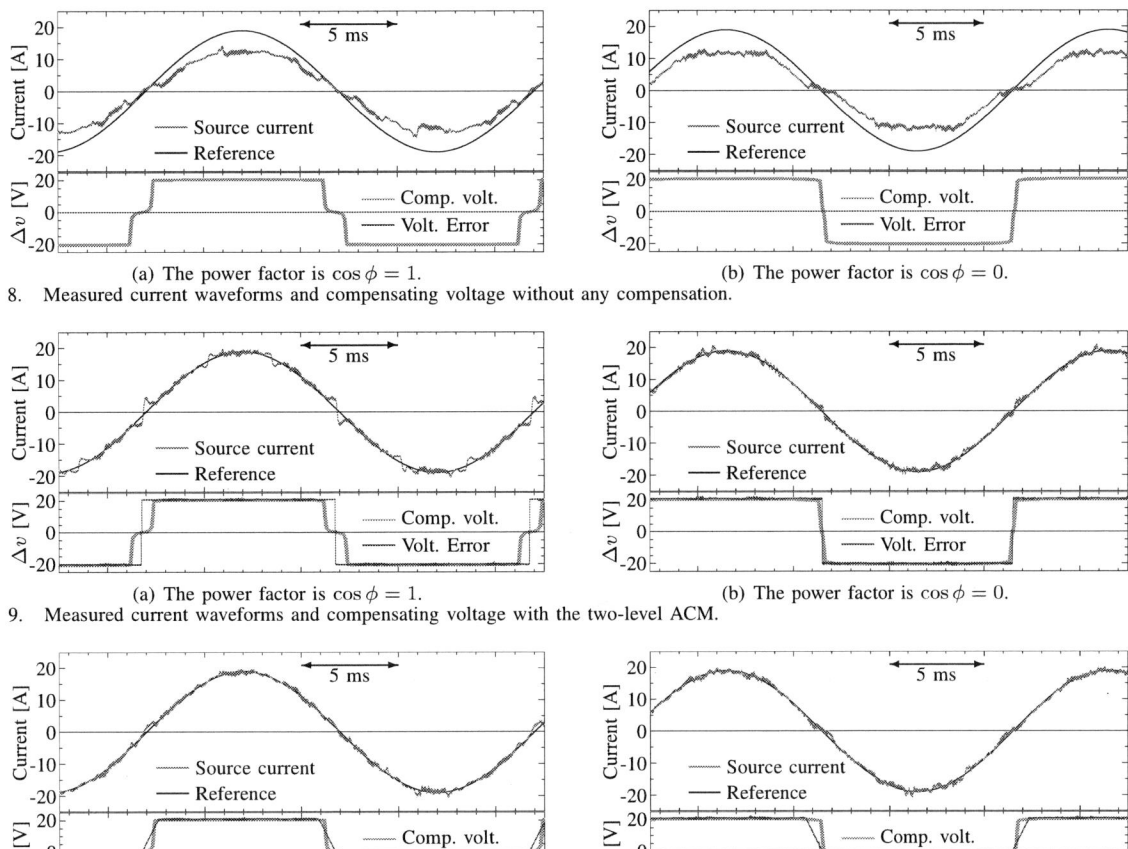

(a) The power factor is $\cos\phi = 1$. (b) The power factor is $\cos\phi = 0$.

Fig. 8.   Measured current waveforms and compensating voltage without any compensation.

(a) The power factor is $\cos\phi = 1$. (b) The power factor is $\cos\phi = 0$.

Fig. 9.   Measured current waveforms and compensating voltage with the two-level ACM.

(a) The power factor is $\cos\phi = 1$. (b) The power factor is $\cos\phi = 0$.

Fig. 10.   Measured current waveforms and compensating voltage with the linear ACM.

V.   EXPERIMENTAL RESULTS

Fig.8–12 show experimental waveforms of the grid-connection converter shown in Fig.1 and its reference in the grid-connection converter with and without voltage error compensation. In the following experiments, a sinusoidal three-phase balanced current reference was provided to the converter, the amplitude of which was set to 14.4 A. A dc power supply is connected to the dc side of the converter to maintain the dc voltage at a constant level of 330 V. The converter was connected to a three-phase power grid with a line-to-line rms voltage of 200 V and a frequency of 50 Hz. The converter is controlled using a simple current feedback with a proportional gain of 4.0 V/A.

Fig.8 shows the measured current waveforms without any voltage error compensation. Fig.8(a) is the waveform in case of a unity power factor. The source current i_s was smaller than its reference i_s^* in amplitude due to the voltage error caused by the dead time. The source current well followed its reference near the zero-crossing of i_s^* because the current ripple was relatively large. As shown in Fig.4(c), the voltage error v_{DT} is small in case of large current ripples. Therefore, the source current can follow its reference only near zero-crossings. Fig.8(b) is the waveform when the source current reference i_s^* was set to lead by 90° from the grid voltage. The source current i_s had a trapezoidal wave shape. Around the zero-crossing,

the source current was distorted because the current ripple is small, as shown in Fig.5(b). Then, the voltage error can be considered as shown in Fig.4(a), and a large voltage error would appear around the zero-crossing. For this reason, Fig.8(b) had a serious current distortion compared with Fig.8(a). These waveforms imply that the voltage error strongly depends on the power factor of the converter current.

Fig.9 shows experimental source current waveforms when a conventional two-level ACM is applied. Fig.9(a) is the waveform in case of a unity power factor. The amplitude of the source current followed its reference. However, a large current distortion occurred near the converter current zero-crossings of each phase. There was a large current ripple near the zero-crossing of the converter current and, the voltage error like Fig.4(c) occurred. In this case, the compensating voltage calculated by two-level ACM is larger than the real voltage error, and thus, a relatively large current distortion appeared near the zero-crossing. Fig.9(b) is the waveform when the power factor was set to $\cos\phi = 0$. The source current distortion in Fig.9(b) is very small, compared with that in Fig.9(a). There is a small current ripple near the zero-crossing of the converter current and the voltage error like Fig.4(a) occurs. In this case, the two-level ACM can compensate effectively the voltage error caused by the dead time. These results imply that the current phase angle affects

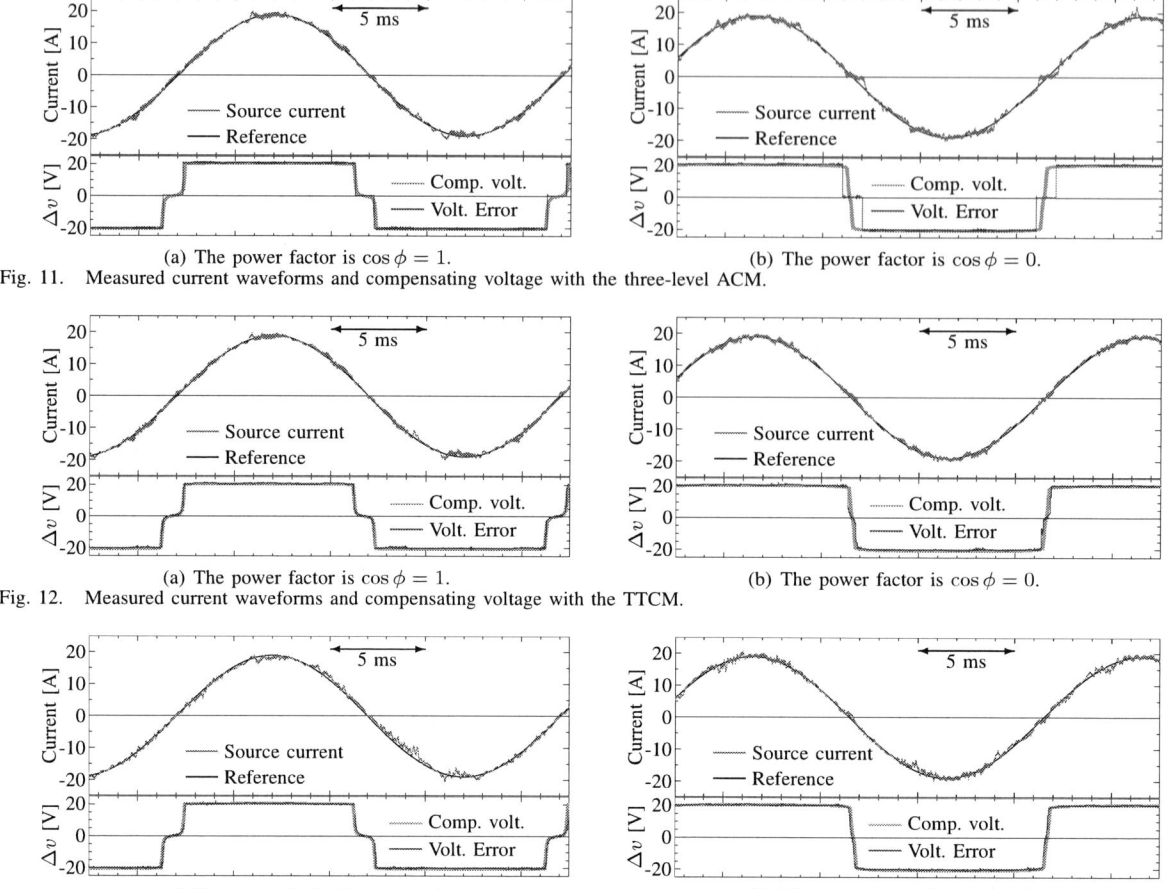

(a) The power factor is $\cos\phi = 1$.   (b) The power factor is $\cos\phi = 0$.

Fig. 11. Measured current waveforms and compensating voltage with the three-level ACM.

(a) The power factor is $\cos\phi = 1$.   (b) The power factor is $\cos\phi = 0$.

Fig. 12. Measured current waveforms and compensating voltage with the TTCM.

(a) The power factor is $\cos\phi = 1$.   (b) The power factor is $\cos\phi = 0$.

Fig. 13. Measured current waveforms and compensating voltage with the TTCM with (14).

the compensation characteristics of the two-level ACM.

Fig.10 shows experimental source current waveforms when a conventional linear ACM is applied. Here, the threshold current of the linear ACM was set to be $i_{\text{th}} = 5.3$ A in order to minimize the harmonic current at a power factor of $\cos\phi = 1$. The source current contains not so large current distortion as that of two-level ACM regardless of the power factor. The current distortion occurs near the zero-crossing of the source current, which the compensating voltage changes linearly, because of overcompensation in case of a power factor of $\cos\phi = 1$ shown in Fig.10(a) or undercompensation in case of a power factor of $\cos\phi = 0$ shown in Fig.10(b).

Fig.11 shows experimental source current waveforms when a conventional three-level ACM is applied. Here, the threshold current of the three-level ACM was set to be $i_{\text{th}} = 3.5$ A in order to minimize the harmonic current at the unity power factor. Fig.11(a) is the waveform in case of the unity power factor, which is less distorted sinusoidal waveform near the zero-crossing. Fig.11(b) is the waveform in case of a power factor of $\cos\phi = 0$. The source current waveform has a large distortion near the zero-crossing because the compensating voltage is not adequate. In this case, there is a small current ripple near the zero-crossing and a large voltage error caused by dead time like Fig.4(a). The inadequate compensating voltage of the three-level ACM is caused by the unsuitable thresh-

old current because the threshold current was adjusted under the unity power factor condition where the current ripple is large.

Fig.12 shows experimental source current waveforms when the TTCM is applied. The source current in Fig.12 contains less current distortion in both cases of $\cos\theta = 1$ as shown in Fig.12(a) and $\cos\theta = 0$ in Fig.12(b) because the TTCM can adjust the compensating voltage according to the phase angle. The TTCM exhibits almost the same compensation characteristics in Fig.12(a) and Fig.12(b), and the power factor does not affect to the compensation performance of the voltage error caused by dead time.

Fig.13 shows experimental source current waveforms of the TTCM where the current ripple estimation method is replaced with the simplified equation in (14). The waveform in Fig.13 also contains less current distortion regardless of the power factor. The compensation characteristics of the TTCM with (14) are almost the same as those of the TTCM using (6)–(11). This result clarifies that the on-line turn-off current calculation is not required in case of application of TTCM to grid-connection converters.

Fig.14 shows the relationship between the current phase angle ϕ and harmonic distortion included in the source current. In this paper the phase angle ϕ is defined as the phase angle of the source current against the source voltage. Therefore, a positive phase angle means that the

The 2014 International Power Electronics Conference

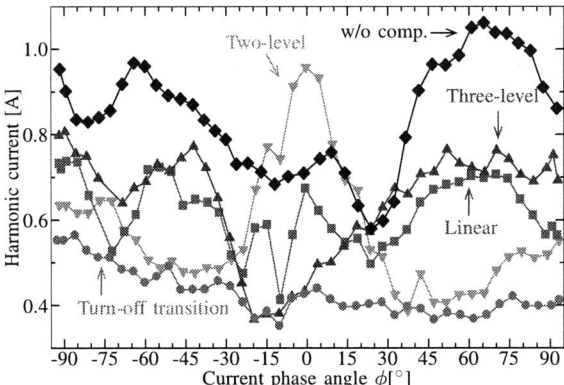

Fig. 14. Harmonic current amplitude included in the source current against the phase angle.

source current leads the source voltage, and a negative phase angle represents a lagging source current. In other words, the converter acts as a inductor in case of $\phi > 0$, and its behaves as a capacitor in case of $\phi < 0$, paying attention to the current direction in Fig.1. A three-phase balanced sinusoidal is provided as the current reference, the rms value of which is set to $I_S^* = 14.4$ A. The threshold current was set to be $i_{th} = 5.3$ A for the linear ACM, and $i_{th} = 3.5$ A for the three-level ACM in order to minimize the harmonic current at $\phi = 0°$. These threshold current are fixed in these experiments.

The amplitude of the harmonic current is varied according to the phase angle. In case of no compensation, the harmonic current reaches 1 A at $\theta = \pm 60°$, while it is about 0.7 A at $\theta = 0°$. The two-level ACM effectively suppresses the harmonic current less than 0.6 A where the phase angle is $\phi < -30°$ and $\phi > 30°$. However, it causes a large harmonic current of 1.0 A at $\theta = 0°$, which is larger than in case of no compensation.

The linear and three-level ACM can reduce the harmonic current in a wide phase angle range $-90° < \theta < 90°$. However, the two-level ACM has a better compensation performance than the linear or three-level ACM in ranges $\theta < -30°$ and $\theta > 30°$. The three-level ACM can suppress the harmonic current effectively compared with the linear ACM. In the experiments, the converter current ripple was as large as 3 A near the zero-crossing of the converter current, and thus, the three-level ACM is the almost suitable among these three ACMs. The linear and three-level ACMs require an appropriate threshold current setting to obtain a good compensation performance. The compensating performance would be improved in the linear and three-level ACMs, if their threshold currents could be adjusted to minimize the harmonic current and/or the current ripple amplitude at each phase angle ϕ. The phase angle strongly affects the compensating performance of the three-level ACM compared with that of the linear ACM because the compensating voltage of the three-level ACM steeply changes at the threshold current i_{th}.

The harmonic current of TTCM is the smallest among the four compensation methods regardless of the phase angle. The compensating performance of TTCM around $\phi = 0°$ is as well as that of the three-level ACM which the threshold current of is adjusted to minimize the harmonic current at $\phi = 0°$. TTCM can suppress the harmonic

current more effectively than two-level ACM, even if the phase angle is in ranges $\phi < -30°$ and $\phi > 30°$. The phase angle has a small effect to the compensations characteristic of TTCM because TTCM estimates the turn-off current or the current ripple amplitude and calculates the compensating voltage online.

Fig.15 and Fig.16 shows experimental demonstration of harmonic current injection. In this experiment, harmonic current was used as the current reference for the converter, which was a sinusoidal current with a frequency of 350 Hz and an rms value of 14.4 A. A dc power supply was also used to sustain the dc voltage of the converter, and the current feedback control gain was set to 4.0 V/A.

Fig.15(a) is the measured waveform without any compensation. The source current amplitude does not follow the reference. It seems a trapezoidal wave shape which is distorted around tops and zero-crossings. Fig.16(a) is the measured frequency spectrum in case of no compensation. The source current contains 1750 Hz harmonic component of 0.6 A and 2150 Hz harmonic component of 0.3 A. These harmonic components were induced by the voltage error caused by dead time.

Fig.15(b) is the measured waveform with two-level ACM and Fig.16(b) is its frequency spectrum. The source current amplitude can follow the reference. However, it seems the source current distortion occurs in all area because the frequency of the source current is high. The 1750 Hz harmonic component in the source current is reduced to 0.4 A and the 2150 Hz component is suppressed effectively.

Fig.15(c) is the measured waveform with TTCM and Fig.16(c) is its frequency spectrum. The source current contains small distortion compared with the other compensation method because TTCM can compensate the voltage error caused by dead time regardless of the phase angle of the source voltage. The all harmonic components are reduced to less than 0.2 A. The TTCM can effectively suppress the current distortion even if the current reference consists of the harmonic components.

VI. CONCLUSION

This paper discusses the compensating performance of various compensation methods for the voltage error caused by the dead time in grid-connection converters. The compensating performances of four different methods are compared with each other from the view point of theory and experiments. The current ripple analysis in three-phase grid-connection converters reveals the phase angle of the source voltage affects the current ripple. The current ripple amplitude around the zero-crossing of the average converter current decides the voltage error which occurs in a cycle of the source voltage because the voltage error caused by the dead time changes steeply around the zero-crossing of the current when the switching devices turn off. The operating power factor should be considered to select the most effective compensation method because the power factor affects the current ripple amplitude around the zero-crossing and the voltage error. On the other hand, the compensating performance of TTCM remains at any power factor.

978-1-4799-2706-7/14 $31.00 © 2014 IEEE

The 2014 International Power Electronics Conference

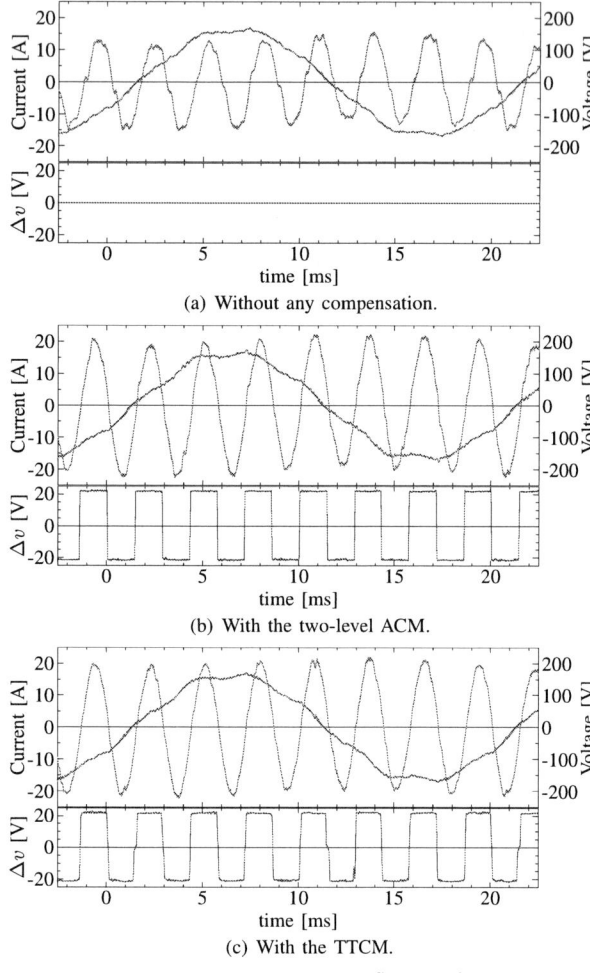

(a) Without any compensation.

(b) With the two-level ACM.

(c) With the TTCM.

------ Source current ------ Source voltage

Fig. 15. Measured current and voltage waveforms when the current reference consists of the 350 Hz harmonic component.

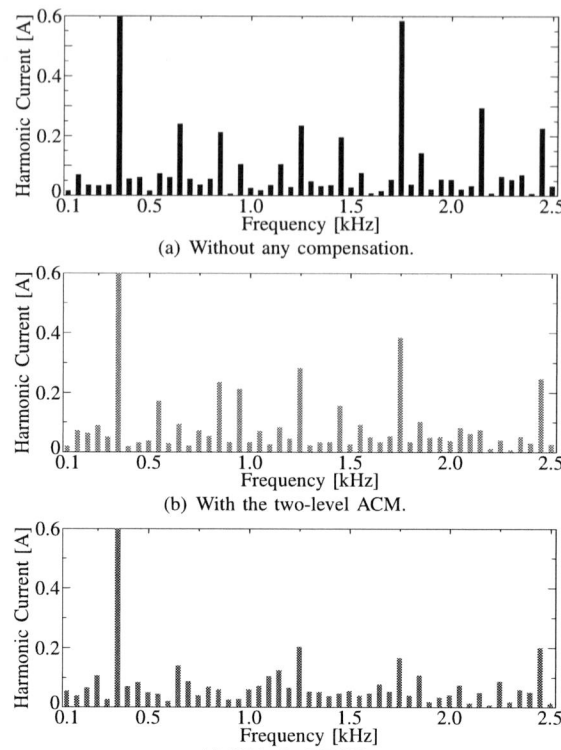

(a) Without any compensation.

(b) With the two-level ACM.

(c) With the TTCM.

Fig. 16. Measured frequency spectra of the source current when the current reference consists of the 350 Hz harmonic component.

The compensation characteristics of conventional approximation-based compensation methods and turn-off transition-based compensation method are compared with each other in the experiments under grid-connection converters. As a result, it is confirmed the two-level ACM has good compensation characteristics in case of the small current ripple, in ranges $\phi < -30°$ and $\phi > 30°$, and the linear and three-level ACM exhibit good compensation performance around the specific power factor where the threshold current is adjusted at an appropriate level. TTCM can effectively suppress the harmonic current at any power factor, compared with the other conventional ACMs. When the current reference including harmonic components, the current distortion of TTCM is the smallest among the four different compensation methods. The TTCM might be suitable for the operation which power factor changes or the harmonic compensation of active power filters because a better performance could be achieved under any operating condition.

REFERENCES

[1] T. Friedli, S. D. Round, D. Hassler, J. W. Kolar: "Design and Performance of a 200-kHz All-SiC JFET Current DC-Link Back-to-Back Converter," IEEE Trans.IA, Vol. 45, No. 5, pp. 1868–1878, (2009)

[2] A.C. Oliveira, C.B. Jacobina, A.M.N. Lima: "Improved Dead-Time Compensation for Sinusoidal PWM Inverters Operating at High Switching Frequencies," IEEE Trans.IE, Vol. 54, No. 4, pp. 2295-2304, (2007)

[3] J. Muhlethaler, M. Schweizer, R. Blattmann, J. W. Kolar, A. Ecklebe: "Optimal Design of LCL Harmonic Filters for Three-Phase PFC Rectifiers," IEEE Trans.PE, Vol. 28, No. 7, pp. 3114–3125, (2013)

[4] L. Bang-sup, K. Kyung-seo, P. Min-ho: "The Analysis and Compensation of Dead Time Effects in Pwm Inverters," IEEE IECON '88, vol. 3, pp. 667–671, (1988)

[5] A. Imura, T. Takahashi, M. Fujitsuna, T. Zanma, S. Doki: "Dead-Time Compensation in Model Predictive Instantaneous-current Control," IEEE IECON 2012, pp. 5037–5042, (2012)

[6] A.C. Oliveira, C.B. Jacobina, A.M.N. Lima, E.R.C. da Silva: "Dead-time compensation in the zero-crossing current region," IEEE PESC '03, pp. 1937–1942 vol. 4, (2003)

[7] J. M. Schellekens, R. a. M. Bierbooms, J. L. Duarte: "Dead-time compensation for PWM amplifiers using simple feed-forward techniques," IEEE ICEM 2010, pp. 1–6, (2010)

[8] M.A. Herrán, J.R. Fischer, S.A. González, M.G. Judewicz, D.O. Carrica: "Adaptive Dead-Time Compensation for Grid-Connected PWM Inverters of Single-Stage PV Systems," IEEE Trans.on Power Electron., vol. 28, no. 6, pp. 2816–2825, (2013)

[9] J.M. Schellekens, R.A.M. Bierbooms, J.L. Duarte: "Dead-Time Compensation for PWM Amplifiers using Simple Feed-forward Techniques," IEEE ICEM 2010, pp. 1–6, (2010)

[10] M. Miyazaki, Senior Member, Y. Hayashi, T. Fukumoto: "An Estimation of Voltage Control Error in PWM Inverter Taking Instantaneous Phase Current Value into Account," IEEJ IASC 2011, 1–26, (2011) (in Japanese)

[11] M. Ogawa, S. Ogasawara, M. Takemoto: "A feedback-type dead-time compensation method for high-frequency PWM inverter –Delay and pulse width characteristics–,' IEEE APEC 2012, pp. 100–105, (2012)

[12] T. Mannen, H. Fujuta: "Dead Time Compensation Method Based on Current Ripple Estimation,' IEEE ECCE 2013, pp. 775–782, (2013)

978-1-4799-2706-7/14 $31.00 © 2014 IEEE

D-Σ Digital Control for
Three-Phase Bi-Directional Inverters

T.-F. Wu

Elegant Power Electronics Applied Research Lab.
(EPEARL)
Department of Electrical Engineering
National Tsing Hua University
Hsin-Chu 30013, Taiwan, ROC
Email: tfwu@ee.nthu.edu.tw

C.-H. Chang and L.-C. Lin

Elegant Power Application Research Center (EPARC)
Department of Electrical Engineering
National Chung Cheng University
Min-Hsiung, Chia-Yi 62102, Taiwan, ROC
Email: tfwu@ee.ccu.edu.tw

Abstract—This paper presents a division-summation (D-Σ) digital control for a three-phase bi-directional photovoltaic inverter with wide inductance variation. The bi-directional inverter fulfilling grid connection and rectification with power factor correction has been implemented in the laboratory. The proposed D-Σ approach summarizes the inductor-current variations over one switching cycle to derive control laws directly, which can overcome limitations of d-q transformation. With the digital control, the inverter can deal with wide inductance variation, reducing core size significantly. In the design and implementation, the inductances corresponding to various inductor currents are measured and tabulated into a single-chip microcontroller for tuning loop gain cycle by cycle, ensuring stable operation. Measured results from a 10 kVA 3ϕ bi-directional inverter have been presented to confirm the feasibility of the discussed control approaches.

Keywords—D-Σ digital control, leakage ground current, inductance variation and inverter.

I. INTRODUCTION

Renewable power generation systems grow rapidly. By nature, renewable power is not continuous and reliable. It will be converted into dc form and buffered with energy storage elements. This brings dc-driving opportunities for electric appliance and equipment which are mostly supplied with dc voltage sources. However, the distributed generation (DG) systems require bi-directional inverters to control the power flow between dc bus and ac grid, and to regulate the dc bus to a certain range of voltages.

Three-phase inverters with grid-connection mode or rectification mode (with power factor correction (PFC)) can adapt discontinuous renewable power to usable dc distribution systems, which have been widely studied. The major concerns of an inverter design include component selection, control scheme and soft-switching topology. Component selection is always a primary task in the growing demand for higher efficiency and smaller size, especially in high power applications. The others are

also important issues for achieving fast dynamics, low current distortion and low EMI.

For grid-connected inverters, there are several applications, such as regular grid-tied inverter (for injecting power into grid) [1]-[3], rectifier (drawing power from grid) [4], [5], reactive power compensator (i.e., Static Synchronous Compensator, STATCOM) [6]-[8], and harmonic current compensator (*i.e.*, Active Power Filter, APF) [9], [10], where their current controls are essentially based on space-vector pulse width modulation (SVPWM). In the design procedure of a conventional controller, a dynamic model of inverter and its state equations in the a-b-c frame are derived first, and the state equations will be transformed to the d-q frame to find a reference voltage vector for SVPWM [11]-[13]. However, the three-phase inductances are assumed constant for simplifying state equations and d-q transformation, while the inductance varies with current level, resulting in inductor current fluctuation at high power applications. Moreover, grid voltage harmonics and three-phase voltage imbalance will complicate the d-q transformation, limiting its wide applications [14]-[16]. Even though some approaches can compensate harmonics [15], [16], the orders of the compensated harmonics are still finite [16], requiring extra effort to design a proper controller for specific orders. Additionally, many attempts have been devoted to reducing current harmonics with grid-voltage compensation and current prediction [1], [3], [11]-[13]. However, their compensators and predictors cannot be designed readily, and their inductance varying with current level has not been considered yet. Another approach, namely one-cycle control (OCC) [17], uses the dual-buck concept to derive the control law without the conventional SVPWM and d-q transformation. Its controller is mainly realized by an integrator with reset, linear and logic components, but no DSP, microprocessor and software. However, inductance variation with current level has not been considered in the controller design yet.

In our previous research, two-phase modulation (TPM) based D-Σ digital control for a 10 kVA 3φ bi-directional inverter with wide inductance variation overcoming some of the aforementioned limitations has been designed and implemented [18]. The proposed D-Σ approach, which is based on the method developed in [18], summarizes the inductor-current variations over one switching period to derive the control laws, reducing switching loss. With the enhancement work of previous research, SVPWM-based D-Σ control laws are derived and expressed in a general form, covering universal four-quadrant operations: grid connection, rectification, STATCOM, and APF. The TPM-based control laws are also derived again they are expressed in a general form, while the parameters for grid-connection and rectification modes are separated into two tables. Additionally, a D-Σ transformation is presented to simplify the derivation procedure. Experimental results from a 10 kVA 3φ inverter are used to confirm the analysis and discussion of the proposed control approaches.

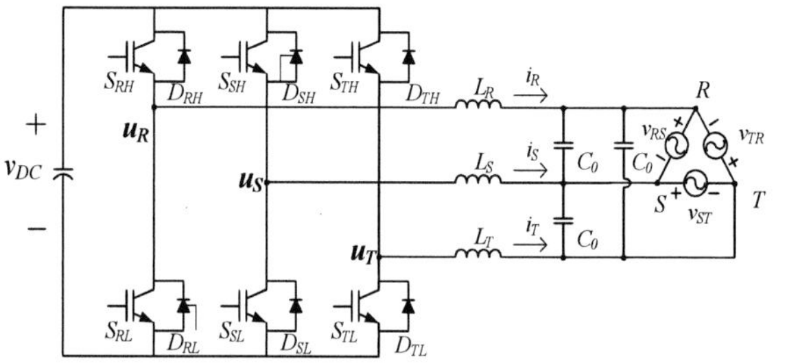

Fig. 1. Circuit diagram of a three-phase six-switch bi-directional inverter.

II. D-Σ DIGITAL CONTROL

For a digital control system, only can the total inductor-current variation be obtained every switching cycle. However, with the D-Σ control, one switching cycle is divided into several time intervals and each time interval is corresponding to a portion of total current variation. Thus, it is difficult to find the duty-ratio control laws directly due to the unknown current variation. The individual current variations are summarized to obtain the total variation over one switching cycle to derive the control laws.

Power circuit diagram of a three-phase six-switch bi-directional inverter is shown in Fig. 1. From Kirchhoff's Current Law (KCL), the three-phase currents satisfy the following equation:

$$i_R + i_S + i_T = 0 \tag{1}$$

where current i_R, i_S and i_T are the inductor currents shown in Fig. 1. The differential form of (1) can be written as:

$$\frac{di_R}{dt} + \frac{di_S}{dt} + \frac{di_T}{dt} = 0 \tag{2}$$

Without considering inductance variation, the three phase inductances are treated as constant, and the following equation derived from (2) will hold:

$$L_R \frac{di_R}{dt} + L_S \frac{di_S}{dt} + L_T \frac{di_T}{dt} = 0 \tag{3}$$

However, equation (3) is no longer valid for a 3φ inverter of which its inductance varies with current widely. That is,

$$L_R \frac{di_R}{dt} + L_S \frac{di_S}{dt} + L_T \frac{di_T}{dt} \neq 0 \tag{4}$$

With the D-Σ digital control, the control laws are derived based on (1), (2) and the state equations of the inverter, and they can adapt to wide inductance variation. In the following, the control laws based on the TPM and the SVPWM for the inverter operated in grid-connection mode are derived and presented.

A. TPM-based D-Σ Control

The proposed D-Σ control based on the TPM includes two control laws for grid-connection and rectification modes, since the two switching ways are different. In the following derivation, the two control laws will narrow down to a general form, while the parameters for the two modes are separated into two tables.

(A) Grid-Connection Mode

The state equations of the 3φ inverter shown in Fig. 1 are described as

$$u_R - L_R \frac{di_R}{dt} - v_{RS} + L_S \frac{di_S}{dt} - u_S = 0 \tag{5}$$

and

$$u_T - L_T \frac{di_T}{dt} + v_{ST} + L_S \frac{di_S}{dt} - u_S = 0 \tag{6}$$

where u_R, u_S, and u_T stand for the switching state voltages which change with the states of the switches. The above switching-state voltages, u_{RS} and u_{ST}, can be also expressed in a matrix form

$$\begin{bmatrix} u_{RS} \\ u_{ST} \end{bmatrix} = \begin{bmatrix} L_R & -L_S \\ L_T & L_S + L_T \end{bmatrix} \begin{bmatrix} \dfrac{di_R}{dt} \\ \dfrac{di_S}{dt} \end{bmatrix} + \begin{bmatrix} v_{RS} \\ v_{ST} \end{bmatrix} \tag{7}$$

where

$$u_{RS} = u_R - u_S$$

and

$$u_{ST} = u_S - u_T$$

Division(D):

In the two-phase modulation, two of the three legs are switching for the two lower line currents and one is fixed to unity duty ratio for the highest line current, which can be equivalent to a dual-buck converter [17]. Therefore, one line period is divided into six regions according to the zero-crossing points of the line currents, as shown in Fig. 2. In region 0°~60°, the gate driving signals can be determined from switching state voltages u_{RS}, u_{ST}, and u_{TR}, and their corresponding switching sequences are shown in Fig. 3. In Fig. 3, only the upper arms of phases R and T are switched for the positive line currents, and the lower arm of phase S is fixed on all the time, treating as a common path for the line currents of phases R and T. The switching sequence consists of three time intervals: T_0, T_x, and T_y, and each of which has its corresponding state of switching voltage: $u_{RS,0}$, $u_{ST,0}$, $u_{RS,x}$, $u_{ST,x}$, $u_{RS,y}$, and $u_{ST,y}$. Thus, equation (7) should be divided into three portions to cover the three switching state voltages, which can be expressed in three different matrixes corresponding to the three time intervals, T_0, T_x, and T_y.

Interval T_0:

$$\begin{bmatrix} u_{RS,0} \\ u_{ST,0} \end{bmatrix} = \begin{bmatrix} L_R & -L_S \\ L_T & L_S+L_T \end{bmatrix} \begin{bmatrix} \Delta i_{R,0} \\ \Delta i_{S,0} \end{bmatrix} \frac{1}{T_0} + \begin{bmatrix} v_{RS} \\ v_{ST} \end{bmatrix} \quad (8)$$

Interval Tx:

$$\begin{bmatrix} u_{RS,x} \\ u_{ST,x} \end{bmatrix} = \begin{bmatrix} L_R & -L_S \\ L_T & L_S+L_T \end{bmatrix} \begin{bmatrix} \Delta i_{R,x} \\ \Delta i_{S,x} \end{bmatrix} \frac{1}{T_x} + \begin{bmatrix} v_{RS} \\ v_{ST} \end{bmatrix} \quad (9)$$

Interval Ty:

$$\begin{bmatrix} u_{RS,y} \\ u_{ST,y} \end{bmatrix} = \begin{bmatrix} L_R & -L_S \\ L_T & L_S+L_T \end{bmatrix} \begin{bmatrix} \Delta i_{R,y} \\ \Delta i_{S,y} \end{bmatrix} \frac{1}{T_y} + \begin{bmatrix} v_{RS} \\ v_{ST} \end{bmatrix} \quad (10)$$

where

$$T_0 = T - T_x - T_y,$$

T is the switching period, v_{RS} and v_{ST} are the ac grid line-to-line voltages, and v_{DC} is the dc-bus voltage. Note that Δi_R and Δi_S include two current variations, current reference variation $i_{v(\cdot)}$ ($= I_{ref}(n+1) - I_{ref}(n)$) and current error $i_{e(\cdot)}$ ($= I_{ref}(n) - i_{fb}(n)$), as shown in Fig. 4. Current variation $i_{v(\cdot)}$ is determined from current reference model to insure inductor current tracking the sinusoidal current reference precisely.

Summation(Σ):

In a digital controller, the values of $\Delta i_{R,n}$ and $\Delta i_{S,n}$ are unknown or difficult to be obtained so as the duty-ratio control laws for determining T_0, T_x, and T_y cannot be derived directly from (8) to (10). Only can the overall inductor current variation over one switching period T be sensed or calculated by the current reference model. The overall current variation can be calculated by the summation of all of the individual current variations in each time interval. These are expressed as follows:

Interval T_0:

$$\begin{bmatrix} \Delta i_{R,0} \\ \Delta i_{S,0} \end{bmatrix} = -\begin{bmatrix} \frac{L_S+L_T}{L_{total}^2} & \frac{L_S}{L_{total}^2} \\ \frac{-L_T}{L_{total}^2} & \frac{L_R}{L_{total}^2} \end{bmatrix} \begin{bmatrix} v_{RS} \\ v_{ST} \end{bmatrix} T - \begin{bmatrix} \frac{L_S+L_T}{L_{total}^2} & \frac{L_S}{L_{total}^2} \\ \frac{-L_T}{L_{total}^2} & \frac{L_R}{L_{total}^2} \end{bmatrix} \begin{bmatrix} u_{RS,0} \\ u_{ST,0} \end{bmatrix} T \quad (11)$$

Interval T_x:

$$\begin{bmatrix} \Delta i_{R,x} \\ \Delta i_{S,x} \end{bmatrix} = -\begin{bmatrix} \frac{L_S+L_T}{L_{total}^2} & \frac{L_S}{L_{total}^2} \\ \frac{-L_T}{L_{total}^2} & \frac{L_R}{L_{total}^2} \end{bmatrix} \begin{bmatrix} v_{RS} \\ v_{ST} \end{bmatrix} T - \begin{bmatrix} \frac{L_S+L_T}{L_{total}^2} & \frac{L_S}{L_{total}^2} \\ \frac{-L_T}{L_{total}^2} & \frac{L_R}{L_{total}^2} \end{bmatrix} \begin{bmatrix} u_{RS,x} \\ u_{ST,x} \end{bmatrix} T_x \quad (12)$$

Interval T_y:

$$\begin{bmatrix} \Delta i_{R,y} \\ \Delta i_{S,y} \end{bmatrix} = -\begin{bmatrix} \frac{L_S+L_T}{L_{total}^2} & \frac{L_S}{L_{total}^2} \\ \frac{-L_T}{L_{total}^2} & \frac{L_R}{L_{total}^2} \end{bmatrix} \begin{bmatrix} v_{RS} \\ v_{ST} \end{bmatrix} T - \begin{bmatrix} \frac{L_S+L_T}{L_{total}^2} & \frac{L_S}{L_{total}^2} \\ \frac{-L_T}{L_{total}^2} & \frac{L_R}{L_{total}^2} \end{bmatrix} \begin{bmatrix} u_{RS,y} \\ u_{ST,y} \end{bmatrix} T_y \quad (13)$$

where

$$L_{total}^2 = L_R L_S + L_S L_T + L_T L_R .$$

By summarizing the above three state equations, we can have

$$\begin{bmatrix} \Delta i_R \\ \Delta i_S \end{bmatrix} = -\begin{bmatrix} \frac{L_S+L_T}{L_{total}^2} & \frac{L_S}{L_{total}^2} \\ \frac{-L_T}{L_{total}^2} & \frac{L_R}{L_{total}^2} \end{bmatrix} \begin{bmatrix} v_{RS} \\ v_{ST} \end{bmatrix} T - \begin{bmatrix} \frac{L_S+L_T}{L_{total}^2} & \frac{L_S}{L_{total}^2} \\ \frac{-L_T}{L_{total}^2} & \frac{L_R}{L_{total}^2} \end{bmatrix} \begin{bmatrix} u_{RS,0} & u_{RS,x} & u_{RS,y} \\ u_{ST,0} & u_{ST,x} & u_{ST,y} \end{bmatrix} \begin{bmatrix} T_0 \\ T_x \\ T_y \end{bmatrix} \quad (14)$$

where

$$\Delta i_R = \Delta i_{R,0} + \Delta i_{R,x} + \Delta i_{R,y}$$

and

$$\Delta i_S = \Delta i_{S,0} + \Delta i_{S,x} + \Delta i_{S,y} .$$

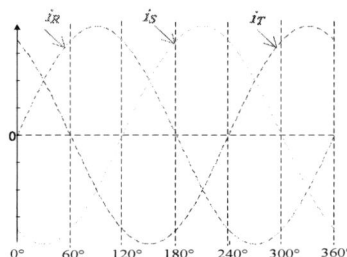

Fig. 2. Six regions in one line period divided according to the zero-crossing points of line currents i_R, i_S and i_T.

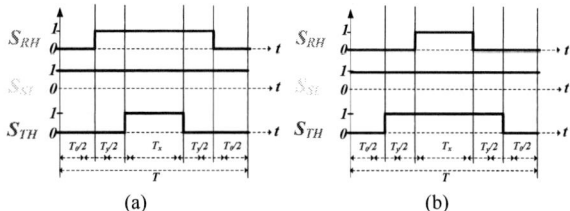

Fig. 3. Symmetrical switching sequence based on the TPM in region $0°\sim60°$ with (a) $D_{RH} > D_{TH}$ (type A) and (b) $D_{RH} < D_{TH}$ (type B) in grid-connection mode.

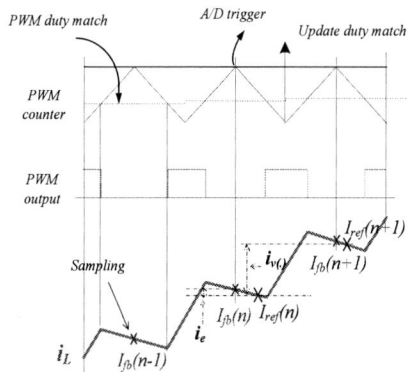

Fig. 4. Illustration of inductor current i_L varying with time and its corresponding PWM signal.

In (14), switching voltages $u_{RS,0}$ and $u_{ST,0}$ are zero so as equation (14) can be simplified as:

$$\begin{bmatrix} \Delta i_R \\ \Delta i_S \end{bmatrix} = -\begin{bmatrix} \dfrac{L_S+L_T}{L_{total}^2} & \dfrac{L_S}{L_{total}^2} \\ \dfrac{-L_T}{L_{total}^2} & \dfrac{L_R}{L_{total}^2} \end{bmatrix}\begin{bmatrix} v_{RS} \\ v_{ST} \end{bmatrix} T - \begin{bmatrix} \dfrac{L_S+L_T}{L_{total}^2} & \dfrac{L_S}{L_{total}^2} \\ \dfrac{-L_T}{L_{total}^2} & \dfrac{L_R}{L_{total}^2} \end{bmatrix}\begin{bmatrix} u_{RS,x} & u_{RS,y} \\ u_{ST,x} & u_{ST,y} \end{bmatrix}\begin{bmatrix} T_x \\ T_y \end{bmatrix} \tag{15}$$

From (15), the time intervals, T_x and T_y, can be determined as:

$$\begin{bmatrix} T_x \\ T_y \end{bmatrix} = \begin{bmatrix} u_{RS,x} & u_{RS,y} \\ u_{ST,x} & u_{ST,y} \end{bmatrix}^{-1}\left\{\begin{bmatrix} L_R & -L_S \\ L_T & L_S+L_T \end{bmatrix}\begin{bmatrix} \Delta i_R \\ \Delta i_S \end{bmatrix} + \begin{bmatrix} v_{RS} \\ v_{ST} \end{bmatrix}T\right\}. \tag{16}$$

The control law (duty ratio) for the TPM can be then determined as:

$$\begin{bmatrix} D_x \\ D_y \end{bmatrix} = \begin{bmatrix} u_{RS,x} & u_{RS,y} \\ u_{ST,x} & u_{ST,y} \end{bmatrix}^{-1}\left\{\frac{1}{T}\begin{bmatrix} L_R & -L_S \\ L_T & L_S+L_T \end{bmatrix}\begin{bmatrix} \Delta i_R \\ \Delta i_S \end{bmatrix} + \begin{bmatrix} v_{RS} \\ v_{ST} \end{bmatrix}\right\}. \tag{17}$$

In region 0°~60°, it consists of two switching sequences, and they will lead to different switching voltages and duty ratio expressions for phases R and T. In Fig. 3(a), D_{RH} is equal to the sum of D_x and D_y, and D_{TH} is equal to D_x. By taking into account the values of four switching voltages: $u_{RS,x}=v_{DC}$, $u_{ST,x}=-v_{DC}$, $u_{RS,y}=v_{DC}$, and $u_{ST,y}=0$, the duty-ratio control laws can be expressed as:

$$\begin{bmatrix} D_{RH} \\ D_{SL} \\ D_{TH} \end{bmatrix} = \begin{bmatrix} \dfrac{(L_R+L_S)\Delta i_R + L_S\Delta i_T}{v_{DC}T} \\ 0 \\ \dfrac{(L_T+L_S)\Delta i_T + L_S\Delta i_R}{v_{DC}T} \end{bmatrix} + \begin{bmatrix} \dfrac{v_{RS}}{v_{DC}} \\ 1 \\ -\dfrac{v_{ST}}{v_{DC}} \end{bmatrix}, \tag{18}$$

and D_{RL}, R_{SH}, and D_{TL} are set to zero. On the other hand, by considering the switching voltages and duty-ratio expressions in Fig. 3(b), the duty-ratio control laws can be obtained to be identical to (18). For general form expressions, the corresponding parameters to each region based on type A are listed in Table I, in which the definition of type A is that the duty ratio of one arm corresponding to the line current rising from zero at the start of the region is $D_x + D_y$. Additionally, it can be

observed that each control law or duty ratio has two inductance variables. In other words, inductance variation has been taken into account by the controller to tune the duty ratios cycle by cycle corresponding to different current levels. Thus, the control laws shown in (18) can handle wide inductance variation. Note that the inductance varying with its inductor current can be measured offline and tabulated into the memory of a microcontroller, or it can be measured with a self-learning program online.

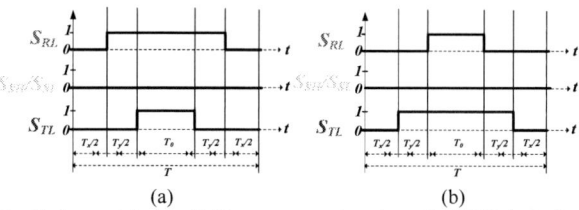

(a)   (b)

Fig. 5. Symmetrical switching sequence based on the TPM in region 0°~60° with (a) $D_{RH} > D_{TH}$ (type A) and (b) $D_{RH} < D_{TH}$ (type B) in rectification mode.

(B) Recfication Mode

When the bi-directional inverter is operated in rectification mode with power factor correction, the inverter acts like a boost converter which is just the complementary operation of a buck converter in grid-connection mode. Fig. 5 shows two symmetrical switching sequences, types A and B, in region 0°~60°. It can be seen that only are the switches S_{RL} and S_{TL} PWM controlled, while the others are all turned off over this region. Since the line current of phase S will go through the diode of the lower arm, the switches of phase S can be operated without PWM control. Note that time interval T_0 is changed with T_x, maintaining switching-state voltages $u_{RS,0}$ and $u_{ST,0}$ zero. Thus, again based on the TPM, the general form of the control laws for the rectification mode can be obtained as the same as (17), in which the corresponding parameters to each region can be referred to Table II.

TABLE I
THE PARAMETERS IN THE GENERAL EXPRESSION (17) BASED ON THE TPM FOR GRID-CONNECTION MODE.

Parameter Region	Switching-State Voltage				Duty Ratio					
	$u_{RS,x}$	$u_{ST,x}$	$u_{RS,y}$	$u_{ST,y}$	D_{RH}	D_{RL}	D_{SH}	D_{SL}	D_{TH}	D_{TL}
I: 0°~60°	v_{DC}	$-v_{DC}$	v_{DC}	0	$D_x + D_y$	0	0	1	D_x	0
II: 60°~120°	v_{DC}	0	0	v_{DC}	1	0	0	D_x	0	$D_x + D_y$
III: 120°~180°	0	v_{DC}	$-v_{DC}$	v_{DC}	D_x	0	$D_x + D_y$	0	0	1
IV: 180°~240°	$-v_{DC}$	v_{DC}	$-v_{DC}$	0	0	$D_x + D_y$	1	0	0	D_x
V: 240°~300°	$-v_{DC}$	0	0	$-v_{DC}$	0	1	D_x	0	$D_x + D_y$	0
VI: 300°~360°	0	$-v_{DC}$	v_{DC}	$-v_{DC}$	0	D_x	0	$D_x + D_y$	1	0

TABLE II
THE PARAMETERS IN THE GENERAL EXPRESSION (17) BASED ON THE TPM FOR RECTIFICATION MODE.

Parameter Region	Switching-State Voltage				Duty Ratio					
	$u_{RS,x}$	$u_{ST,x}$	$u_{RS,y}$	$u_{ST,y}$	D_{RH}	D_{RL}	D_{SH}	D_{SL}	D_{TH}	D_{TL}
I: 0°~60°	v_{DC}	$-v_{DC}$	v_{DC}	0	0	$1-D_x$	0	0	0	$1-D_x-D_y$
II: 60°~120°	v_{DC}	0	0	v_{DC}	0	0	$1-D_x-D_y$	0	$1-D_x$	0
III: 120°~180°	0	v_{DC}	$-v_{DC}$	v_{DC}	0	$1-D_x-D_y$	0	$1-D_x$	0	0
IV: 180°~240°	$-v_{DC}$	v_{DC}	$-v_{DC}$	0	$1-D_x$	0	0	0	$1-D_x-D_y$	0
V: 240°~300°	$-v_{DC}$	0	0	$-v_{DC}$	0	0	0	$1-D_x-D_y$	0	$1-D_x$
VI: 300°~360°	0	$-v_{DC}$	v_{DC}	$-v_{DC}$	$1-D_x-D_y$	0	$1-D_x$	0	0	0

B. SVPWM-based D-Σ Control

In the SVPWM, all of the three legs are switched with complementary operations, and the region transition is based on the zero-crossing point of line voltages, instead of line currents, as shown in Fig. 6. Since the current directions are not limited to regions, e.g. the line currents of each region are always positive or negative in the TPM, the inverter based on the SVPWM can employ a general form of duty-ratio control law to cover grid-connection and rectification modes. Fig. 7 shows a general switching sequence in region 0°~60°, which is composed of two non-zero vectors, $V_1(100)$ and $V_2(110)$, and two zero vectors, $V_0(000)$ and $V_7(111)$. As identical to D-Σ derivation in the TPM, the general form of T_x and T_y can be also solved directly as follows:

$$\begin{bmatrix} T_x \\ T_y \end{bmatrix} = \begin{bmatrix} u_{RS,x} & u_{RS,y} \\ u_{ST,x} & u_{ST,y} \end{bmatrix}^{-1} \left\{ \begin{bmatrix} L_R & -L_S \\ L_T & L_S + L_T \end{bmatrix} \begin{bmatrix} \Delta i_R \\ \Delta i_S \end{bmatrix} + \begin{bmatrix} v_{RS} \\ v_{ST} \end{bmatrix} T \right\}. \tag{19}$$

The corresponding parameters for each region are listed in Table III.

Compared with TPM, the SVPWM-based D-Σ control can use a parameter table to cover all current applications, since the switching way of the SVPWM can change the direction of inductor current any time. It is easy to change phase shift angle for current reference.

TABLE III
THE PARAMETERS IN THE GENERAL EXPRESSION (19) BASED ON THE SVPWM.

Parameter Region	$u_{RS,x}$	$u_{ST,x}$	$u_{RS,y}$	$u_{ST,y}$	x	y	V_x	V_y
I: 0°~60°	v_{DC}	0	0	v_{DC}	1	2	$V_1(100)$	$V_2(110)$
II: 60°~120°	0	v_{DC}	$-v_{DC}$	v_{DC}	2	3	$V_2(110)$	$V_3(010)$
III: 120°~180°	$-v_{DC}$	v_{DC}	$-v_{DC}$	0	3	4	$V_3(010)$	$V_4(011)$
IV: 180°~240°	$-v_{DC}$	0	0	$-v_{DC}$	4	5	$V_4(011)$	$V_5(001)$
V: 240°~300°	0	$-v_{DC}$	v_{DC}	$-v_{DC}$	5	6	$V_5(001)$	$V_6(101)$
VI: 300°~360°	v_{DC}	$-v_{DC}$	v_{DC}	0	6	1	$V_6(101)$	$V_1(100)$

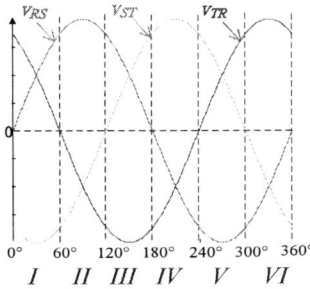

Fig. 6. Six regions in one line period divided according to the zero-crossing points of line voltage v_{RS}, v_{ST} and v_{TR}.

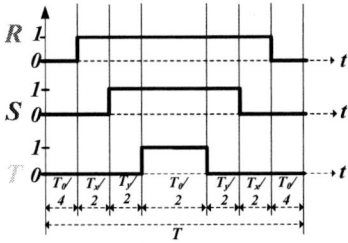

Fig. 7. Symmetrical switching sequence based on the SVPWM in region 0°~60° with vector $V_1(100)$ and $V_2(110)$.

III. D-Σ TRANSFORMATION

The above derivation procedure of the proposed D-Σ approach can be simplified through a transformation equation, which can be expressed as follows:

$$\begin{bmatrix} u_{RS} \\ u_{ST} \end{bmatrix} T = \begin{bmatrix} u_{RS,0} & u_{RS,x} & u_{RS,y} \\ u_{ST,0} & u_{ST,x} & u_{ST,y} \end{bmatrix} \begin{bmatrix} T_0 \\ T_x \\ T_y \end{bmatrix}. \tag{20}$$

For time interval T_0, its related switching-state voltages $u_{RS,0}$ and $u_{ST,0}$ are equal to zero in all regions, so that they can be ignored in this matrix. The D-Σ transformation equation can be then expressed as

$$\begin{bmatrix} u_{RS} \\ u_{ST} \end{bmatrix} T = \begin{bmatrix} u_{RS,x} & u_{RS,y} \\ u_{ST,x} & u_{ST,y} \end{bmatrix} \begin{bmatrix} T_x \\ T_y \end{bmatrix}. \tag{21}$$

As mentioned above, equation (7) is the original mesh equation, from which the state equation can be written as follows:

$$\begin{bmatrix} \Delta i_R \\ \Delta i_S \end{bmatrix} = -\begin{bmatrix} \dfrac{L_S + L_T}{L_{total}^2} & \dfrac{L_S}{L_{total}^2} \\ \dfrac{-L_T}{L_{total}^2} & \dfrac{L_R}{L_{total}^2} \end{bmatrix} \begin{bmatrix} v_{RS} \\ v_{ST} \end{bmatrix} T + \begin{bmatrix} \dfrac{L_S + L_T}{L_{total}^2} & \dfrac{L_S}{L_{total}^2} \\ \dfrac{-L_T}{L_{total}^2} & \dfrac{L_R}{L_{total}^2} \end{bmatrix} \begin{bmatrix} u_{RS} \\ u_{ST} \end{bmatrix} T, \tag{22}$$

By substituting (21) into (22) and inverting the matrix of the switching-state voltages, the control laws for T_x and T_y can be derived directly as

$$\begin{bmatrix} T_x \\ T_y \end{bmatrix} = \begin{bmatrix} u_{RS,x} & u_{RS,y} \\ u_{ST,x} & u_{ST,y} \end{bmatrix}^{-1} \left\{ \begin{bmatrix} L_R & -L_S \\ L_T & L_S + L_T \end{bmatrix} \begin{bmatrix} \Delta i_R \\ \Delta i_S \end{bmatrix} + \begin{bmatrix} v_{RS} \\ v_{ST} \end{bmatrix} T \right\}. \tag{23}$$

It can be observed that the control law is the same as those shown in (16) and (19), the general forms for the TPM and SVPWM, while its derivation procedure is much more simplified.

TABLE IV
SYSTEM PARAMETERS OF THE DESIGNED INVERTER.

Parameters	Symbols	Values
DC-bus voltage	v_{DC}	360 ~ 400 V
AC grid voltage	v_{RS}, v_{ST} & v_{TR}	220 V$_{rms}$(nominal)
Maximum rated power	P_{max}	10 kVA
Line frequency	f_l	60 Hz
Inductors	L_R, L_S & L_T	300μH ~ 2 mH
Output filter capacitor	C_o	5 uF
Power switch	IGBT HGTG40N60A4	$V_{CE(on) typ.}$ = 1.6 V, V_{CES} = 600V, and $I_{C(TC=25℃)}$ = 75 A
Power diode (silicon carbide)	CREE C3D20060D	$V_{F(TJ=25℃) typ.}$ = 1.5 V Zero-Recovery Time
Switching frequency	f_s	20 kHz
Power factor	PF	0 ~ 1 leading or lagging

IV. EXPERIMENTAL RESULTS

The proposed D-Σ control was confirmed by a 10 kVA three-phase inverter. Based on the analysis, parameter determination of the power stage is summarized in Table IV. The range of dc-bus voltage is specified from 360 V to 400 V and the switching frequency is 20 kHz. The nominal phase voltage is 220 V_{rms} and the line frequency is 60 Hz. The inverter inductance varies from 2 mH to 300 µH per phase. With the variation, the core size can be reduced by 5 times. The power diodes are realized with silicon carbide, which have no reverse-recovery time. The harmonic components of the output currents were measured up to 17th harmonic with a power analyzer WT1600.

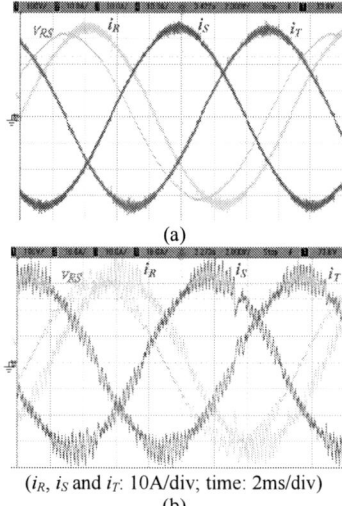

(i_R, i_S and i_T: 10A/div; time: 2ms/div)

(b)

Fig. 8. Measured waveforms of the three-phase inductor currents using the proposed TPM-based D-Σ control (a) with and (b) without considering wide inductance variation in grid-connection mode at 10 kW.

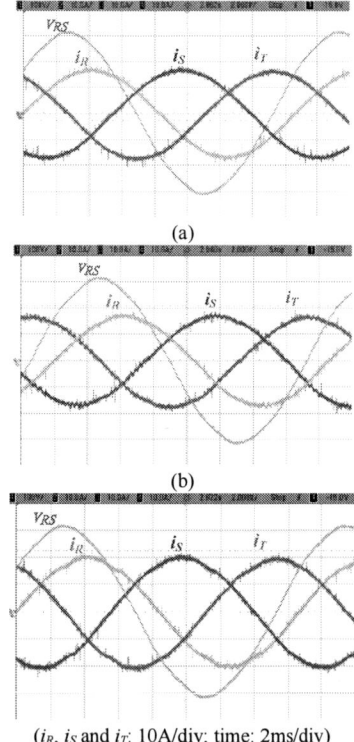

(i_R, i_S and i_T: 10A/div; time: 2ms/div)

(c)

Fig. 9. Measured waveforms of the three-phase inductor currents using the proposed SVPWM-based D-Σ control (a) with and (b) without considering wide inductance variation in grid-connection mode at 5 kW and (c) at 6 kW.

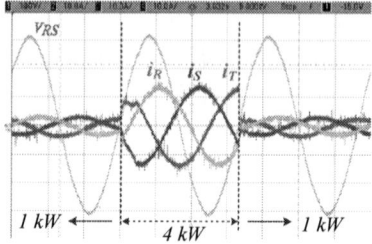

(i_R, i_S and i_T: 10A/div; time: 5ms/div)

(a)

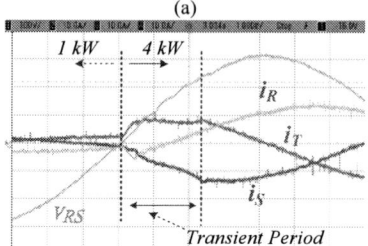

(i_R, i_S and i_T: 10A/div; time: 1ms/div)

(b)

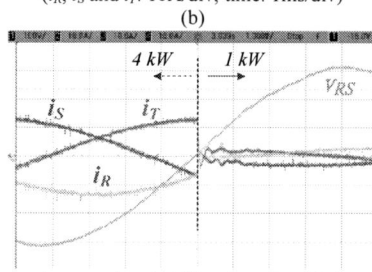

(i_R, i_S and i_T: 10A/div; time: 1ms/div)

(c)

Fig. 10. Measured waveforms of the three-phase inductor currents using the proposed SVPWM-based D-Σ control under step load change and its expansions: (a) from 1 kW ←→ 4 kW, (b) from 1 kW → 4 kW, and (c) from 4 kW → 1 kW.

A. Test with Wide Inductance Variation and Dynamic Response

Fig. 8 shows the current waveforms with and without considering wide inductance variation at 10 kW. The tests were based on the proposed control laws (TPM) and the inverter was operated in grid-connection mode. With the consideration of wide inductance variation, the inverter can track sinusoidal current reference precisely, which employs only a unity-gain controller ($K_p = 1$) to compensate the current errors. While, those without the consideration yield sub-harmonic oscillation. The inverter requires other well-designed controllers to compensate the lack of inductance-variation consideration.

Figs. 9(a) and 9(c) show current waveforms by using the SVPWM-based D-Σ control at 5 kW and 6kW, respectively. The current waveform at 6 kW has a little distortion, and the three-phase currents do not reach current references near the peak. Compared with TPM, the maximum open-loop duty ratio of SVPWM switching sequence is $T_0/2$ more than that of TPM, so as the inverter with the SVPWM-based control requires higher dc bus voltage. In the test, the dc-bus voltage is fixed at 380 V. The maximum open-loop duty ratios of TPM and SVPWM are 311/380 (0.818) and 1/2 + 311/380/2 (0.909), respectively, where 311 is the peak value of 220 V_{RMS} AC grid. Thus, the duty ratio of SVPWM might exceed unity when considering closed-loop term and dead time under heavy load. The solution is to increase the dc-bus voltage to reduce the open-loop duty ratio.

Fig. 9(b) shows current waveform without considering wide inductance variation at 5 kW. It can be seen that the three-phase currents have a little sub-harmonic oscillation, and it might become larger at full load.

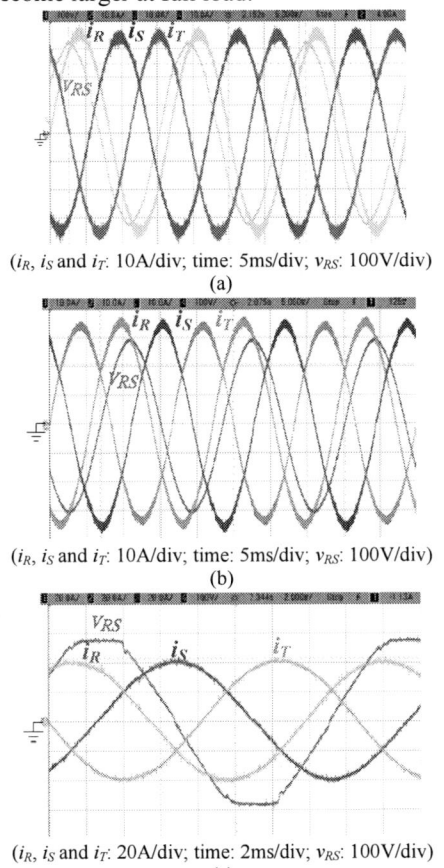

(i_R, i_S and i_T: 10A/div; time: 5ms/div; v_{RS}: 100V/div)

(a)

(i_R, i_S and i_T: 10A/div; time: 5ms/div; v_{RS}: 100V/div)

(b)

(i_R, i_S and i_T: 20A/div; time: 2ms/div; v_{RS}: 100V/div)

(c)

Fig. 11. Measured waveforms of the three-phase inductor currents and line voltage v_{RS} in (a) grid-connection, (b) rectification, and (c) grid-connection modes with distorted voltage source at 10 kW.

For dynamic test, Fig. 10 shows line current transient response to step power change from 1 kW to 4 kW and from 4 kW back to 1 kW. It can be seen that the transient response time from 1 kW to 4 kW is about 4.6 ms, and the inductor current of phase R reaches the reference rapidly, as shown in Fig. 10(b). In any regions, the range of the maximum open-loop duty ratio is from 0.85 to 0.91 and from 0.91 to 0.85, according to the proposed SVPWM-based control law. Thus, it will decrease the tracking speed. For the transient from 4 kW to 1 kW, three-phase currents reach their steady state at the same time, and the response time is about 0.3 ms.

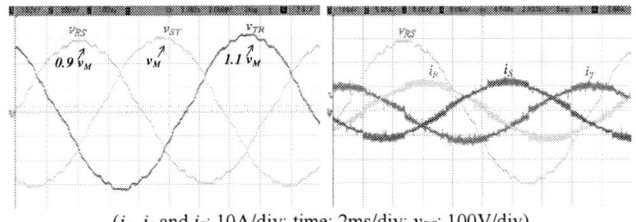

(i_R, i_S and i_T: 10A/div; time: 2ms/div; v_{RS}: 100V/div)

(a)   (b)

Fig. 12. Measured waveforms of (a) unbalanced three-phase line voltages, and (b) inductor currents in grid-connection mode at 1 kW.

B. Test with Grid-Voltage Harmonics

Fig. 11 shows waveforms of three-phase inductor currents and line voltage v_{RS} in grid-connection and rectification modes at 10 kW. By considering wide inductance variation, the inverter can track sinusoidal reference currents precisely and stably. The test is based on a Δ-connected source system so that its line voltages lead inductor currents (line currents) by 30°. Fig. 11(c) shows that the inverter can maintain sinusoidal current waveform under distorted grid voltage.

For three-phase imbalance tests, a programmable AC source, Chroma 61512, is used to provide an unbalanced AC grid. The test conditions regard the imbalance range, $v_{RS} = 0.9*220 \ V_{AC}$, $v_{ST} = 220 \ V_{AC}$, and $v_{TR} = 1.1*220 \ V_{AC}$, which is shown in Fig. 12(a). Fig 12(b) shows the measured three-phase inductor currents, in which the inductor currents are still maintained sinusoidal waveforms, verifying the feasibility of the proposed control approach.

C. Comparison between TPM and SVPWM

Fig. 13 shows measured waveforms of the three-phase inductor currents from the inverter with the TPM-based and SVPWM-based D-Σ control by considering wide inductance variation in grid-connection mode and at 800 W. It can be seen that the region transition under the TPM-based D-Σ control is obvious and results in current distortion, while it becomes smooth with the SVPWM-based D-Σ control.

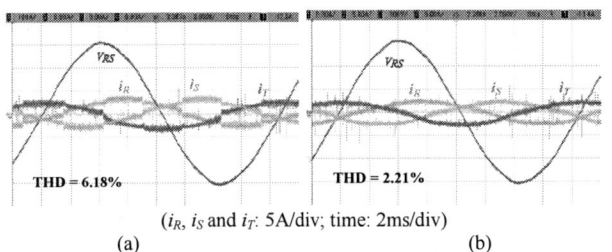

(i_R, i_S and i_T: 5A/div; time: 2ms/div)

(a)   (b)

Fig. 13. Measured waveforms of the three-phase inductor currents from the inverter with (a) the TPM-based D-Σ control and (b) the SVPWM-based D-Σ control by considering wide inductance variation in grid-connection mode and at 800 W.

Table V lists a performance comparison between TPM and SVPWM. With SVPWM, current ripple can be reduced to around half of the TPM one, and THD is much lower due to smooth region transition, especially under low power conditions. The SVPWM-based D-Σ control can use a general form and a parameter table to cover all current applications, such as grid connection, rectification, STATCOM, and APF. While, the TPM-based D-Σ control requires different parameter tables for the four operations. For the merit of TPM, the two-phase modulation has only four switchings in one switching cycle, which is lower than the SVPWM (with six switchings), resulting in lower switching loss. Finally, the inverter with SVPWM requires higher dc-bus voltage, since its maximum open-loop duty ratio is higher than that of TPM. The TPM and SVPWM have their merits and limitations. Designers can select properly according to the system requirements, *e.g.* low power with SVPWM and high power with TPM.

TABLE V.
THE PERFORMANCE COMPARISON BETWEEN TPM AND SVPWM.

Modulation / Performance	TPM	SVPWM
Current ripple	Larger	Half of TPM
THD	< 3 % (3 kW up) > 5% (below 1kW)	< 3%
Switching loss	Low (4 times a switching cycle)	High (6 times a switching cycle)
A general form covers all current applications	No (every mode requires a parameter table)	yes
Maximum open-loop duty-ratio	$v_{AC,peak}/v_{DC}$	$0.5+v_{AC,peak}/2v_{DC}$

V. CONCLUSIONS

A D-Σ digital controlled three-phase inverter considering wide filter-inductance variation and taking into account line voltage distortion has been presented in the paper. The control laws for both TPM and SVPWM have been derived in detail. With the proposed D-Σ approach, the inverter can track sinusoidal current reference precisely and stably, and can accommodate grid voltage harmonics and wide filter-inductance variation. This paper has also presented a D-Σ transformation, which can save the traditional d-q transformation and obtain the control laws directly. The tradeoff between TPM-based and SVPWM-based D-Σ digital controls has been discussed and presented. Experimental results have verified the discussed performance and feasibility of the proposed control approaches.

REFERENCES

[1] K. H. Ahmed, A. M. Massoud, S. J. Finney, and B. W. Williams, "A Modified Stationary Reference Frame-Based Predictive Current Control With Zero Steady-State Error for LCL Coupled Inverter-Based Distributed Generation System," IEEE Trans. on Industrial Electronics, vol. 58, no. 4, pp. 1359-1370, April. 2012.

[2] R. Li, Z. Ma, and D. Xu, "A ZVS Grid-Connected Three-Phase Inverter," IEEE Trans. on Power Electronics, vol. 27, no. 8, pp. 3595-3604, Aug. 2012.

[3] J. R. Massing, M. Stefanello, A. G. Hilton, and H. Pinheiro, "Adaptive Current Control for Grid-Connected Converters with LCL Filter," IEEE Trans. on Industrial Electronics, vol. 59, no. 12, pp. 4681-4693, April. 2012.

[4] J. W. Kolar, and T. Friend, "The Essence of Three-Phase PFC Rectifier Systems Part I," IEEE Trans. on Industrial Electronics, 2012.

[5] M. A. Chaudhari, H. M. Suryawanshi, and M. M. Renge, "A Three-Phase Unity Power Factor Front-End Rectifier for AC Motor Drive," IET Trans. on Power Electronics, vol. 5, no. 1, pp. 1-10, 2012.

[6] A. Yazdani, M. L. Crow, and J. Guo, "An Improved Nonlinear STATCOM Control for Electric Arc Furnace Voltage Flicker Mitigation," IEEE Trans. on Power Delivery, vol. 24, no. 4, pp. 2284-2290, 2009.

[7] J. A. Barrena, L. Marroyo, M. A. R. Vidal, and J. R. T. Apraiz, "Individual Voltage Balance Strategy for PWM Cascaded H-Bridge Converter-Based STATCOM," IEEE Trans. on Industrial Electronics, vol. 55, no. 1, pp. 21-29, 2008.

[8] A. Luo, C. Tang, Z. Shuai, X.-Y. Xu, and D. Chen, "Fuzzy-PI-Based Direct-Output-Voltage Control Strategy for the STATCOM Used in Utility Distributed Systems," IEEE Trans. on Industrial Electronics, vol. 56, no. 7, pp. 2401-2411, 2009.

[9] R. L. D. A. Riberio, C. C. D. Azevedo, and R. M. D. Sousa, "A Robust Adaptive Control Strategy of Active Power Filter for Power-Factor Correction, Harmonic Compensation, and Balancing of Nonlinear Loads," IEEE Trans. on Power Electronics, vol. 27, no. 2, pp. 718-730, Feb. 2012.

[10] W. H. Choi, M. C. Wong, and Y. D Han, "Adaptive DC-Link Voltage-Controlled Hybrid Active Power Filters for Reactive Power Compensation," IEEE Trans. on Power Electronics, vol. 27, no. 4, pp. 1758-1772, April. 2012.

[11] M. Prodanovic and T. C. Green, "Control and Filter Design of Three-Phase Inverter for High Power Quality Grid Connection," IEEE Trans. on Power Electronics, vol. 18, no. 1, pp. 373-380, Jan. 2003.

[12] Q. Zeng and L. Chang, "An Advanced SVPWM-Based Predictive Current Control for Three-Phase Inverters in Distribution Generation Systems," IEEE Trans. on Industrial Electronics, vol. 55, no. 3, pp. 1235-1246, March. 2008.

[13] J. Castello, R. Garcia-Gil, G. Garcera, and E. Figueres, "An Adaptive Robust Predictive Current Control for Three-Phase Grid-Connected Inverters," IEEE Trans. on Industrial Electronics, vol. 58, no. 5, pp. 3537-3546, Aug. 2011.

[14] D.N. Zmood and D.G. Holmes, "Stationary Frame Harmonic Reference Generation for Active Filter Systems," IEEE Trans. on Industrial Applications, vol. 38, no. 6, pp. 1591-1599, Nov./Dec. 2002.

[15] P. Mattavelli, "A Close-Loop Selective Harmonic Compensation for Active Filters," IEEE Trans. on Industrial Applications, vol. 37, no. 1, pp. 81-89, Jan./Feb. 2001.

[16] M. Liserre, R. Teodorescu, and F. Blaabjerg, "Multiple Harmonics Control for Three-Phase Grid Converter Systems with the Use of PI-RES Current Controller in A Rotating Frame," IEEE Trans. on Power Electronics, vol. 21, no. 3, pp. 836-841, May. 2006.

[17] T.-T. Jin, L.-H. Li, and K. M. Smedley, "A Universal Vector Controller for Four-Quadrant Three-Phase Power Converters," IEEE Trans. on Circuits and System I: Regular Papers, vol. 54, no. 2, pp. 377-390, 2007.

[18] T.-F. Wu, C.-H. Chang, L.-C. Lin and Y.-R. Chang, "Two-Phase Modulated Digital Control for Three-Phase Bidirectional Inverter With Wide Inductance Variation," IEEE Trans. on Power Electronics, vol. 28, no 4, pp. 1598-1607, April. 2013.

Expectations of Next-Generation Power Devices for Home and Consumer Appliances

Akihiko Kanouda, Hiroyuki Shoji, Takae Shimada, Toshikazu Okubo

Hitachi Research Laboratory
Hitachi, Ltd.
7-1-1, Omika-cho Hitachi city, Japan
akihiko.kanoda.zz@hitachi.com

Abstract— This paper describes expected effects and characteristics required for next-generation power devices to be used in home appliances and consumer electronics. This paper focuses on the challenges presented by the IH cooking heater, energy storage systems, and solar power generation system and presents the circuits that will solve these challenges. Also, requirements for power devices used for the circuits are specified. The authors also show the effect of reducing the recovery loss when a SiC diode is applied to the boost converter. Moreover, the efficiency characteristics are compared for using the SJ-MOSFET and IGBT in the boost converter. The experimental results revealed that SJ-MOSFET improves the efficiency by 0.3 percentage point more than the IGBT. By using the SiC diode and SJ-MOSFET, efficiency of the boost converter is 98.7 % at rating 5.7 kW and output DC 380 V.

Keywords— *IH cooking heater, Energy storage, Photovoltaic power generation, Next-generation power devices*

I. INTRODUCTION

To help prevent climate change, home appliances and consumer electronics have been made to use less energy, use fewer resources, and be recycled more [1]. On the other hand, to survive in the global markets for these products, Japanese companies have to develop cost-competitive high-performance products with limited development personnel and resources while retaining technological superiority. In not only home appliances and consumer electronics but also power electronics,

equipment performance is significantly affected by power devices. In recent years, equipment performance is expected to be improved by applying the next-generation high-speed low-loss power devices such as a wide band gap semiconductor SiC (Silicon Carbide) or GaN (Gallium Nitride), which have started to be commercialized [2]. This paper describes the requirements and expectation for the new power devices for home appliances and consumer electronics.

Fig. 1 shows the main devices of interest. Of the numerous home appliances and consumer electronics, this paper focuses on the induction heating (IH) cooking heater, energy storage system, and residential photovoltaic (PV) power conditioning system.

II. IH COOKING HEATER

Among home appliances, IH cooking heater requires especially large electric power, 2.5-3 kW per one heater. Therefore, its performance is significantly affected by the conversion circuit using power devices and its control method. Some IH cooking heaters can heat all-metal pans and others can heat only iron and stainless steel pans [3].

Fig. 2 shows a general IH circuit to heat iron or stainless steel pans. Since Cf has a relatively small capacity, Cf voltage (DC link voltage) has a waveform that pulsates at twice the line frequency. Using and insulated gate bipolar transistor (IGBT) for Q1 and Q2, the single ended push-pull (SEPP) series resonant inverter is constituted. Using resonance of the heating

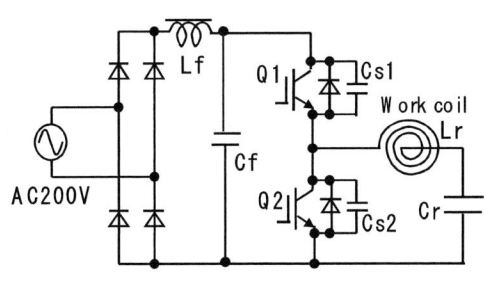

Fig. 1 Home appliances and consumer electronics to which next-generation power devices are to be applied

Fig. 2 SEPP inverter for IH cookers

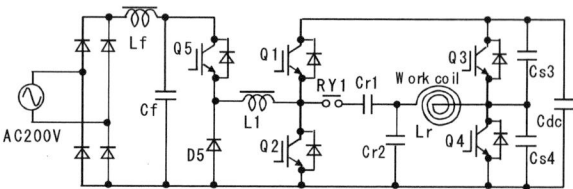

Fig. 3 Buck-Boost -Full-Bridge inverter for all-metal IH cookers [4]

coil (work coil) Lr and resonant capacitor Cr, high frequency AC magnetic field is enerated. In order to keep resonance, Q1 and Q2 are switched alternately at a frequency from 20 kHz to 30 kHz. The pans are heated with eddy current loss caused by the AC magnetic field. Output power can be adjusted with pulse frequency modulation (PFM) of Q1 and Q2, which are always switched under soft-switching conditions.

On the other hand, the above SEPP inverter circuit has a significantly different power circuit from the all-metal corresponding IH cooking heater to heat both copper or aluminum pans and iron or stainless steel pans using the same heater. Fig. 3 schematically shows the IH cooking heater that can heat the all-metal pans [4]. This circuit can heat aluminum or copper pans made of a non-magnetic low resistance material and iron or stainless steel pans made of a high resistance magnetic material. For aluminum or copper pans, the frequency of the eddy current generated in the pan and the skin resistance are increased, and the magnetic field is increased by increasing the current and number of turns of the heating coil. Furthermore, for the lightweight aluminum pan, buzz noise is caused by the heating coil current being changed due to the pulsation of the DC link voltage. The DC link voltage needs to be smoothed in order to prevent this noise. However, if the DC link voltage is smoothed by enlarging the capacity of the capacitor of the rectifier circuit, the input current harmonics are increased. Therefore, a power factor correction (PFC) function is required. Moreover, since the frequency characteristics of the series resonant circuit rapidly change near the resonant frequency, it is difficult to use PFM similar to those of iron or stainless steel pans. Therefore, pulse amplitude modulation (PAM) control is used for aluminum and copper pans. On the other hand, the optimum value of the resonant frequency and the number of turns of the heating coil are different for aluminum and iron pans. Therefore, the number of turns of the heating coil is fixed, and the amplitude of the voltage applied to the resonant circuit is changed by changing the topology of the inverter in accordance with the material of the pans. The resonant and the switching frequencies are changed by changing the capacitance of the resonant capacitor. IGBTs that have breakdown voltage of 600 V are used for Q1 - Q5 in Fig. 3. Q1 and relay RY1 are turned off when an aluminum or copper pan is heated. The H-bridge buck-boost converter circuit is configured with Q1, Q5, D5, L1, and Q2. The circuit performs PAM control for output adjustment and PFC control. Q3, Q4, Lr, and Cr2 operate as the SEPP inverter.

On the other hand, RY1 turns on when an iron or stainless steel pan is heated and a full-bridge inverter circuit that has a boost function is configured. Q2 is included in both the inverter and boost converter.

Table I lists the specifications of power devices required for the next generation IH cooking heater. Currently, 600V IGBT is widely used.

Because hard switching operation area is included in switching devices of the all-metal compatible IH cooking heater shown in Fig. 3 (Q1, Q2, Q5), SiC-MOSFET with excellent fast turn-off characteristics and SiC-SBD with excellent reverse recovery characteristics are desired for switching devices and anti-parallel diodes, respectively. On the other hand, switching devices Q3 and Q4 of the inverter are controlled in soft switching operation. For these switching devices, lower on-voltage (lower on-resistance) characteristics are more important than the fast reverse recovery characteristics.

TABLE I   Requirements for power devices to be used in next-generation IH cooking heaters

Power stage	Switching mode	Power devices	Requirements for power devices
Buck-Boost converter	Both soft and hard switching	Switching device	・Breakdown Voltage : > 600V ・Lower cost ・Higher speed turn off characteristics ex: SiC-MOSFET
		Anti-parallel diode	・Breakdown Voltage : > 600V ・Lower cost ・Excellent reverse recovery characteristics ex: SiC-SBD
SEPP inverter	Soft switching (Current resonant)	Switching device	・Breakdown Voltage : > 600V ・Lower cost ・Lower ON voltage (ON resistance)
		Anti-parallel diode	・Breakdown Voltage : > 600V ・Lower cost ・Lower ON voltage (ON resistance)

Since inverters of the IH cooking heater for iron and stainless steel pans are controlled in soft switching operation, the demand for power devices is the same. However costs still need to be reduced, the IGBT will be replaced by the next-generation power devices after its price has reduced to close to that of the current IGBT.

III. ENERGY STORAGE SYSTEM

The power storage system is used not only to supply backup power during a power failure but also to absorb the fluctuation of the electric power generated by wind or PV power, to stabilize the grid, or to shift the peak power consumption. Lead batteries are mainly used for the large capacity storage system because they are relatively inexpensive. On the other hand, Li-ion battery has higher energy density and smaller size than the lead battery.

In the energy storage system, fine control is needed for voltage and/or current between output (DC bus side) and the battery side in order to stabilize output line voltage and to keep longer life of the battery. Therefore, a bi-directional DC-DC converter is used. There is a bi-directional converter of two types of isolated and non-isolated.

As a circuit example of the power storage system, Fig. 4 shows a circuit of a bi-directional isolated DC-DC converter used to charge and discharge the storage battery [5]. V1 is the DC bus side DC link voltage, and its rated voltage is 300 V to 400 V. By connecting V1 to the power conditioning system or to an inverter that is provided separately, it is possible to charge and discharge the battery with the commercial AC line or to supply power from the battery to the home appliances.

Both the IGBT (H1, H2) and the power MOSFET (H3, H4) are used in the high voltage (HV) side legs. Zero voltage switching (ZVS) is promoted by using Lr. Cr is used for cutting direct current. N1 is the HV side winding of the transformer. N21 and N22 are center-tapped windings of the low voltage (LV) side. Power MOSFET S1, S2, transformer windings N21, N22, and smoothing inductor L are connected to the LV side storage battery. The main feature of this circuit is it has an active clamping circuit composed of power MOSFET S3, S4, and Cc. The active clamp circuit can suppress the surge voltage generated across S1 and S2 when operating boost mode and buck mode and can use a power MOSFET that has a lower breakdown voltage than a converter without an active clamp circuit. For the HV-side bridge, in the boost mode operation, reverse recovery characteristics of the body diode of the power MOSFET is a problem.

In the buck mode operation, the switching loss of the IGBT at turn-off is another problem. To solve these problems, a hybrid bridge configuration of the power MOSFET and IGBT is chosen. The circuit shown in Fig. 4 is the preferred circuit topology, especially when V2 has a relatively lower voltage battery.

Fig. 5 shows another bi-directional isolated DC-DC converter called a dual active bridge (DAB) [6]. A DAB converter is a circuit that has a full-bridge to both the DC bus side and the battery side, connected by a transformer. The power flow through the transformer can be controlled by changing the phase of the switching between those full-bridges. DAB is chosen when battery voltage V2 is higher than in the circuit in Fig. 4. Since DAB has the same problem as the HV side in the circuit in Fig. 4, hybrid full-bridges of MOSFET and IGBT are also effective for the DAB converter. IGBTs are used for H1 - H2 and S1 - S2 legs, and MOSFETs are used for H3 - H4 and S3 - S4 legs in Fig. 5.

Meanwhile, a non-isolated DC-DC converter is also used for storage applications. Fig. 6 shows a bi-

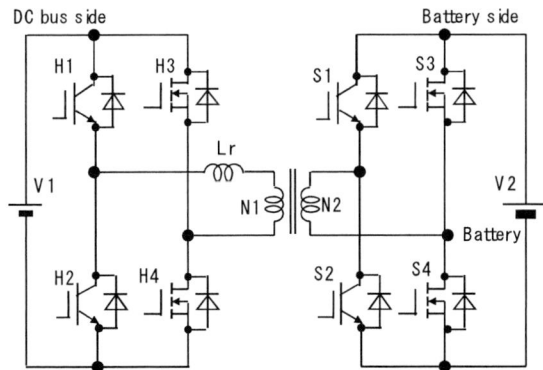

Fig. 5   DAB bi-directional isolated DC-DC converter

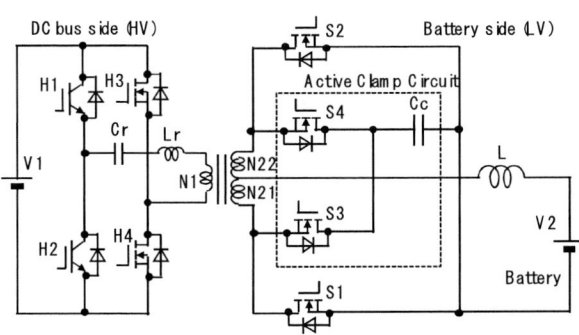

Fig. 4   Bi-directional isolated DC-DC converter
with active clamp circuit [5]

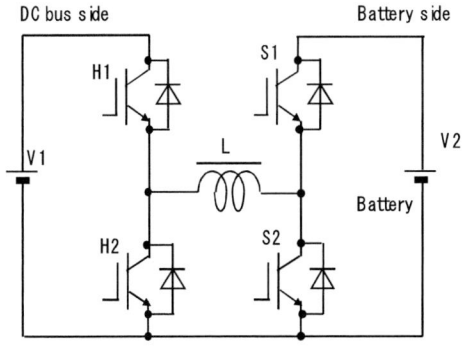

Fig. 6   Bi-directional non-isolated H-bridge DC-DC converter

TABLE II Requirements for power devices applied for bi-directional DC-DC converter in energy storage systems

Topology	Power stage	Switching mode	Requirements for power devices
Isolated DAB	DC bus side full bridge, Battery side full bridge	Phase shift ZVS	• Breakdown voltage: 600V (DC bus voltage ≒ 380V) • Lower switching loss when turned-off • Lower on voltage (Lower on resistance) • Excellent reverse recovery characteristics of anti-parallel diode (Rectfier mode)
Isolated full-bridge/current doubler with active clamp	DC bus side full bridge		
	Battery side	Active clamp ZVS	• Breakdown voltage depends on battery voltage • Lower on resistance
Non-isolated H-bridge	DC bus side leg	Hard switching (PWM)	• Breakdown voltage: 600V (DC bus voltage ≒ 380V) • Lower switching loss • Lower on voltage (Lower on resistance) • Excellent reverse recovery characteristics of body diode
Reversible chopper	Battery side leg	Hard switching (PWM)	• Breakdown voltage depends on battery voltage • Lower switching loss • Lower on voltage (Lower on resistance) • Excellent reverse recovery characteristics of body diode

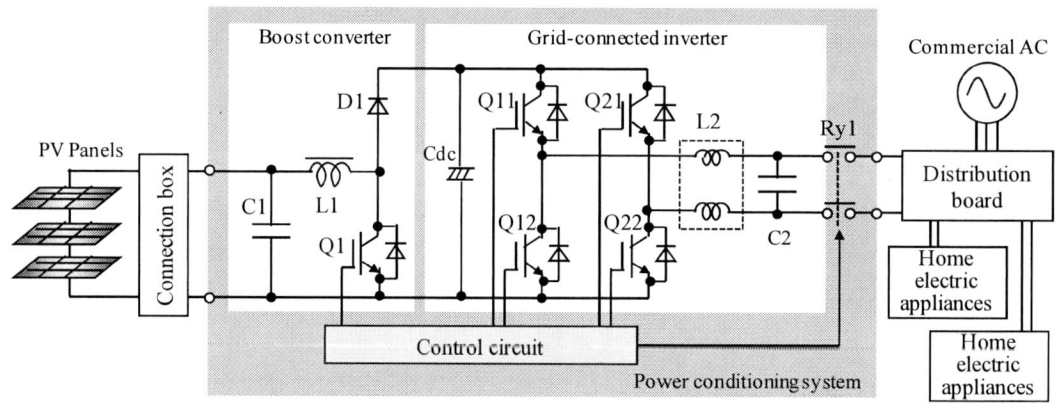

Fig. 7 Overall configuration of a residential PV power conditioning system

directional non-isolated H-bridge DC-DC converter. Four IGBTs or MOSFETs are used in this circuit. The H-bridge converter is able to charge and discharge continuously by changing smoothly between buck and boost modes depend on the voltage relationship of the DC bus side and battery side. The efficiency of the converter maximizes when the DC bus voltage is equal to the battery voltage. Since the circuit is simpler and has higher efficiency than that of the isolated, H-bridge is suitable for the power storage applications.

Also the reversible chopper circuit that removed the leg of S1 and S2 from the circuit in Fig. 6 is used when battery voltage is always lower than the DC bus voltage.

In the energy storage use, power passes twice through the DC-DC converter: once for the charge time, and again for the discharge time. Therefore, power loss of the DC-DC converter must be reduced.

Table II lists the requirements for power devices used in the storage systems. Of these, for the HV side in Fig. 4 and the DAB method, a next-generation power device must have lower switching loss at turn-off, excellent reverse recovery characteristics of the body diode, lower on-resistance, and 600 V class breakdown voltage since

both power MOSFET and IGBT contain problems. Also, specifications of power devices of the LV side in Fig. 4 vary in accordance with the voltage of V2. However, higher current, lower voltage switching, and lower on-resistance are required.

IV. PV POWER CONDITIONING SYSTEM

Fig. 7 shows an example of a residential PV power system configuration. Terminal voltage of the series-connected PV panels is approximately 250 V rated and is connected to the boost converter of the power conditioner through the connection box. As the power generated by the PV panels is always maximized, maximum power point tracking (MPPT) control varies the operating point by calculating the input power of boost converter.

Cdc voltage, which is the output of the boost converter (DC link voltage), is applied to the full-bridge grid-connected inverter. The inverter forms an alternating current waveform of a sine wave synchronized to the grid by the pulse width modulation (PWM) control. Converted commercial AC power can not only be used in

TABLE III   Requirements for power devices which applied for next generation power conditioning system.

Power stage	Switching mode	Power devices	Requirements for power devices
Boost converter Grid-connected inverter	Hard switching (PWM)	Switching devices	·Breakdown Voltage : > 600V ·Lower cost ·Lower on resistance (Lower on voltage) ·Higher speed /Lower switching loss ·IGBT → SJ-MOSFET → SiC-MOSFET
		Boost diode, Anti-parallel diodes	·Breakdown Voltage : > 600V ·Lower cost ·Excellent reverse recovery characteristics ex: Si-SBD → SiC-SBD

residential properties but also to sell reverse power flow back to the grid.

Future power conditioning systems will need to have lower cost, lighter weight for easy to installation, and higher efficiency. In response, requirements collectively required for the power devices are listed in Table III. First, for the boost converter, the reactor (L1) must be lightened. To reduce the volume and weight of the reactor, higher switching frequency is needed, so the loss of the switching devices must be reduced. Because of smaller switching loss at higher speed and lower on-resistance than the conventional IGBT, SiC MOSFET or SJ (Super junction) MOSFET is expected to apply Q1 to the boost converter. At the same time, excellent reverse recovery characteristics of SiC schottky barrier diode (SBD) can significantly reduce the recovery loss. Thus, SiC SBD should be adopted for D1.

Also, for Q11-Q22, which are used for the grid-connected inverter, the adoption of the SiC-MOSFET or SJ-MOSFET is expected to replace the conventional IGBT. Using a higher frequency PWM carrier reduces the copper loss of the reactor L2, making L2 smaller and lighter. This can be achieved at the same time as reducing recovery loss and conduction loss of the power devices.

Table IV lists waveforms of diode D1 of the boost converter used in PV power conditioning system when the switching device Q1 is turned on. In this situation, 8.0 A flows D1 in the forward direction before Q1 is turned on, but 12.8 A peak recovery current flows on the conventional boost converter that used IGBT and a silicon diode. In contrast, by using the SJ-MOSFET and SiC diode, the recovery current can be reduced to 4.5 A peak, and recovery loss can be also reduced to 81 % that of the conventional technology.

Fig. 8 compares the efficiency characteristics of experimental results using the SJ-MOSFET and IGBT used in the boost converter in a PV power conditioning system. RJH60F5DPK was used as the IGBT, and TK65L60V was used as the SJ-MOSFET.

In this comparison, the same SiC diode is used for the boost diode D1, and only Q1 is different. Maximum efficiency for IGBT is 98.5 % at 3.4 kW output and 98.4 % at rated 5.7 kW output condition. On the other hand, maximum efficiency for SJ-MOSFET is 98.75 % at

TABLE IV   Comparison of power devices for Boost converter

	Conventional	Developed
Q1	IGBT	SJ -MOSFET
D1	Silicon diode	SiC diode
Current wave of diode	8.0A ··· 49ns 0 ID -12.8A	8.0A ··· 25ns 0 ID -4.5A
Recovery loss	98.2μJ	18.6μJ (-81%)

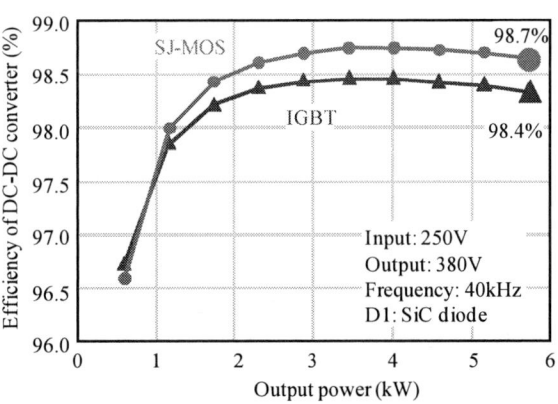

Fig. 8 Efficiency comparison of boost converter for PV power conditioning system.

3.4 kW and 98.7 % at the rated output. From these results, SJ-MOSFET can increase efficiency by 0.3 percentage point more than IGBT at rated power. This is equivalent to a loss-reduction of 18 W. The efficiency can increase more than 0.1 percent at rated power if the output voltage can lower to 320 V.

V.   CONCLUSIONS

This paper described challenges for home appliances and consumer electronics and summarized the requirements for next generation power devices needed to

978-1-4799-2706-7/14 $31.00 © 2014 IEEE   2062

meet these challenges. An IH cooking heater, power storage system, and solar power system were focused on because they require especially large power. In particular, for the photovoltaic power conditioning system, the effect of reducing the recovery loss when applying the SiC diode for boost converter was explained. The authors also compared experimentally efficiency characteristics in the case of using the SJ-MOSFET or IGBT in the boost converter. The results revealed that the SJ-MOSFET improves efficiency by 0.3 percentage point more than the IGBT. The SiC diode and SJ-MOSFET can obtain 98.7 % efficiency for the boost converter in the condition of rated 5.7 kW and output DC 380 V. The required characteristics of power device usually change in accordance with the application, but they also change depending on the circuit system.

The circuit developers need not only to tell the power device developers about problems but also to learn about which commercially available power devices have the best cost performance.

The circuit system will need to be reviewed constantly while the evolution of the power devices is monitored.

REFERENCES

[1] Investigating R&D Committee on Information Apparatus Energy System Trends in Homes and Offices, "Technology Trends in Energy Systems of Information Apparatuses in Home and Offices," *IEEJ Technical Report*, No. 1251, 2012.

[2] A. Kanouda, H. Shoji, and T. Shimada "Challenges and Expectations for the Next-Generation Power Devices in Home Appliances and Consumer Electronics," *IEEJ Proceedings of the 2013 Industry Applications Society Conference*, 4-S10-2, pp. IV-65-68, 2013.

[3] H. Shoji, "The Latest Technology in the Induction Heating Appliance," *The Journal of The Institute of Electrical Engineers of Japan*, vol. 132, No. 8, pp. 545-547, 2012.

[4] H. Shoji, J. Uruno, and M. Isogai, "Buck-Boost-Full-Bridge Inverter for All-Metal Induction-Heating Cookers," *IEEJ Proceedings of the 2011 Industry Applications Society Conference*, 1-110, pp. I-513-516, 2011.

[5] T. Shimada, H. Shoji, and K. Taniguchi, "Turbo Acceleration Method Expanding Boost Operating Range for Bi-directional Isolated DC-DC Converter," *IEEJ Proceedings of the 2011 Industry Applications Society Conference*, 1-51, pp. I-293-296, 2011.

[6] T. Matsuda, G. Guidi, A. Kawamura, T. Imakubo, and T. Jikumaru, "Manufacture of High Efficiency Isolated DC-DC Converter and Its Performance Evaluation ," *IEEJ Proceedings of the 2012 Industry Applications Society Conference*, 1-92, pp. I-391-394, 2012.

Application trend and foresight of SiC power devices to air conditioners

Mamoru Kamikura, Yuichiro Murata
Advanced Technology Center
Mitsubishi Electric Corporation
Hyogo, Japan
Kamikura.Mamoru@ay.MitsubishiElectric.co.jp

Tomohiro Kutsuki, Katsuhiko Saito
Shizuoka Works
Mitsubishi Electric Corporation
Shizuoka, Japan

Abstract— **Large energy-saving effect can be expected for air conditioners that occupy the highest power consumption rate of home appliance, by introducing the SiC. In this paper, the transition of power devices, the characteristics, and the issues in widespread use, and foresight of the next generation power devices, SiC, are reported.**

Keywords— *Air conditioners, inverters, AC/DC converters, SiC.*

I. Introduction

Recently there have been considerable interests in energy-saving appliance. Fig. 1 shows an example of power consumption rate of home appliance. From the figure, air conditioners occupy more than 50% of the power consumption rate of home appliance. Therefore application of the energy-saving technology to air conditioners would be very effective. In order to lower the power consumption, application of wide band gap materials, such as Silicon Carbide (SiC), to inverters and AC/DC converters in air conditioner is expected.

In this paper technical transition of power devices, which are key components in the power-saving technology, characteristics of SiC devices, an example of the application of SiC devices to air conditioners, and foresight of SiC devices are reported.

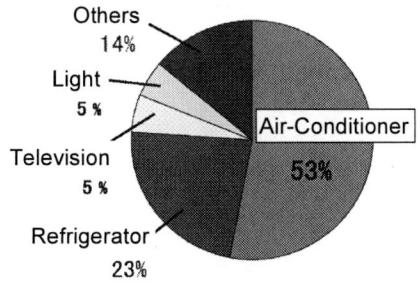

Fig. 1. Power consumption rate of home appliance.

II. Technical transition of power device

Fig. 2 shows the transition of power loss of an insulated gate bipolar transistor (IGBT). It is indicated

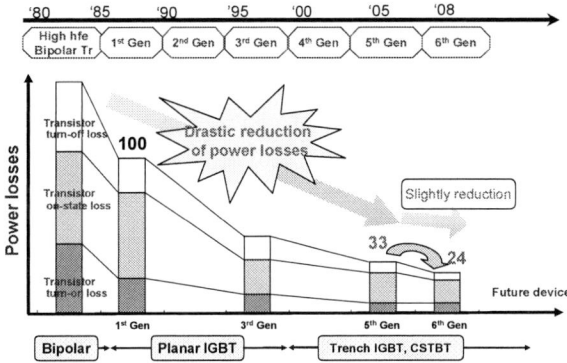

Fig. 2. Transition of power loss of IGBTs

that the power loss of 5th generation IGBT in 2005 is less than 33% compared to that of 1st generation IGBT in 1985. However the improvement of the power loss of the 6th generation IGBT compared to that of 5th generation IGBT is supposed to be little. Therefore the improvement of the performance by introducing SiC devices to air conditioners is strongly expected.

There are some merits in SiC devices compared to Si devices. Band gap of SiC is 3 times larger than that of Si, so SiC devices can function at much higher temperature than Si devices. In addition, breakdown voltage of SiC is 10 times higher than that of Si, so the insulating films of SiC devices could be 10 times thinner than those of Si devices. Furthermore, the density of impurity of SiC devices could be 100 times larger than that of Si devices. It indicates that resistance of SiC devices could be much lower than that of Si devices. Thus SiC devices have superior characteristics to Si devices.

Fig. 3 shows the comparison of the power loss of an SiC device with that of an Si device. The Si device is defined as combination of an Si IGBT and an Si recovery diode, on the other hand the hybrid SiC device is defined as combination of an Si IGBT and an SiC recovery diode, and the full SiC device is defined as combination of an SiC metal-oxide-semiconductor field-effect-transistor (MOSFET) and an SiC recovery diode. In comparison of the Si device and the hybrid SiC device, the switching loss in the hybrid SiC device is lower because of the

978-1-4799-2706-7/14 $31.00 © 2014 IEEE

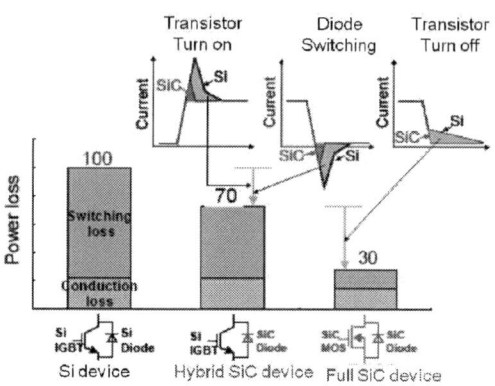

Fig. 3. Comparison of power loss of Si with that of SiC.

lower recovery current of the SiC diode. In comparison of the hybrid SiC device and the full SiC device, the full SiC device is much lower in switching loss because of the lower tail current in the turn off process. Low power loss in switching devices is suitable for high frequency drive, so it is concluded that the full SiC device is the most suitable for high frequency drive.

SiC devices are superior to Si devices from the viewpoint of power loss. However SiC has three challenges. First, the cost of SiC is much higher than Si. Second, threshhold voltage is lower, so SiC device is subject to wrongly be turned on. Third SiC is hard but brittle, so it is difficult to process. Therefore it is important to overcome these challenges to benefit the maximum advantage of SiC devices.

III.   Application of SiC devices to air conditioners

In 2010, Mitsubishi Electric Corporation introduced SiC Schottky barrier diode (SBD) to dual-in-line (DIP) intelligent power module (IPM) in inverters of air conditioners for the first time to the best of authors' knowledge. Fig. 4 shows the outline view of the DIP IPM. Fig. 5 shows the circuit block diagram of the DIP IPM. The recovery diodes are SiC SBDs. Fig. 6 shows the circuit block diagram of air conditioners. The DIP IPM is installed in the inverter circuit. Fig. 7 shows the comparison of the wave forms in turn on process of Si DIP IPM and that of hybrid SiC DIP IPM. From the figure, the recovery current of SiC SBD is much smaller than that of Si diode. By introducing the hybrid SiC DIP IPM, switching loss of the air conditioner was lowered by 60% than that launched in 2009 as shown in Table I.

Fig. 4. Outline view of DIP IPM.

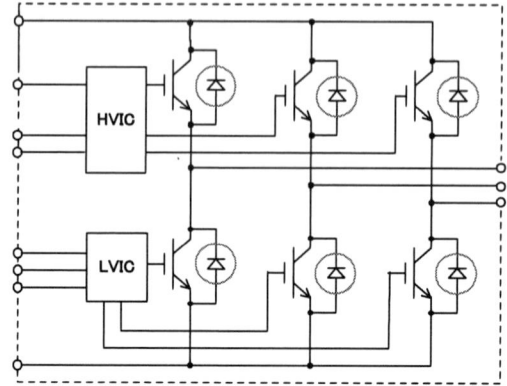

Fig. 5. Circuit block diagram of DIP IPM.

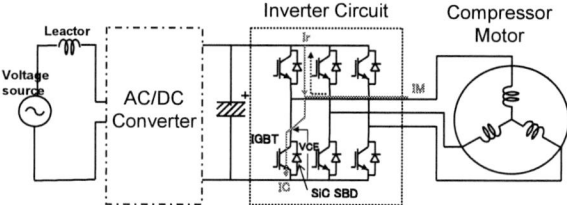

Fig. 6. Circuit block diagram of air conditioners.

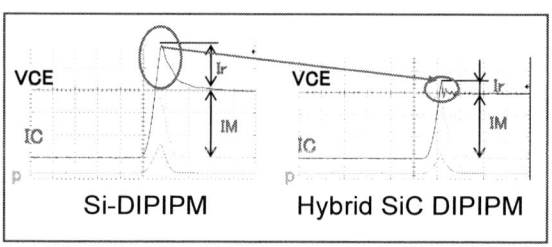

Fig. 7. Comparison of the wave forms in turning on of Si-DIP IPM and that of hybrid SiC DIP IPM.

TABLE I
COMPARISON OF SWITCHING LOSS COMPARED WITH SI-DIP IPM

	Hybrid SiC DIP IPM (%)	Rate of power loss Improvement (%)
POWER LOSS OF IGBT TURNING ON	43	57
POWER LOSS OF FRD SWITCHING	25	75
POWER LOSS OF SWITCHING	40	60

Application of SiC devices to air conditioners is not expected to inverters only, but also AC/DC converters. If SiC devices are introduced to AC/DC converters and AC/DC converters are driven in high frequency, the reactor and the noise filter would be reduced in size and simplified. Therefore AC/DC converters of SiC are attractive. It is supposed that the strong candidate for SiC application is to bridgeless AC/DC converters and interleaved AC/DC converters.

Fig. 8 shows the circuit block diagram of an inverter including a bridge-less AC/DC converter. Mitsubishi Electric Corporation have introduced to air conditioners

The 2014 International Power Electronics Conference

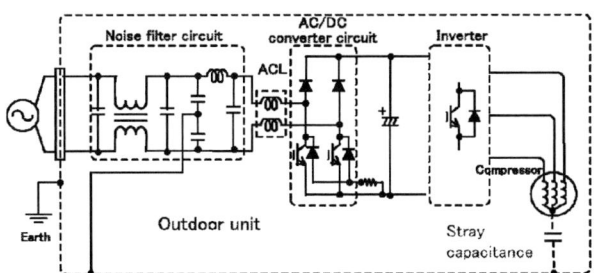

Fig. 8. Circuit block diagram of inverters including bridgeless AC/DC converter.

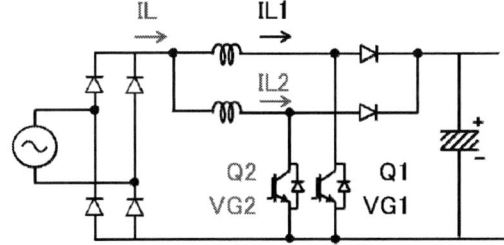

Fig. 9. Circuit block diagram of interleaved AC/DC converter.

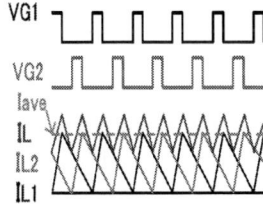

Fig. 10. Timing chart of interleaved AC/DC converter control system.

from the viewpoint of the low power loss and the harmonic current suppression. The bridgeless AC/DC converter fluctuates the AC line voltage directly during the switching process, so the electromagnetic noise would increases severely when it functions in high frequency drive. The main reasons of the electromagnetic noise are switching noise of the converters and the inverters, and the stray capacitance of the compressor motor.

Fig. 9 shows the circuit block diagram of an interleaved AC/DC converter. Fig. 10 shows the timing chart of its control system. In this converter the current is commuted by diode bridge first, and then chopped by two switching devices (Q1 and Q2) which turn on and off alternatively. The interleaved converter generates less electromagnetic noise than that of the bridgeless AC/DC converter's. Fig. 11 shows comparison of the conducted noise of the bridgeless AC/DC converter and the interleaved AC/DC converter, when the noise filters were removed. The interleaved AC/DC converter generated less electromagnetic noise than that of bridgeless AC/DC converter in the frequency range from 0.15 to 30 MHz. It is supposed that the interleaved AC/DC converter generated less electromagnetic noise even if in high frequency drive, so interleaved AC/DC converters would be suitable to introduce SiC devices.

Fig. 12 shows the comparison of the power loss of the interleaved AC/DC converter consisting of Si IGBTs and Si diodes with that consisting of SiC MOSFETs and SiC SBDs. It is indicated that 50% lower power loss were achieved by introducing SiC devices to the interleaved AC/DC converter.

Fig. 13 shows the wave forms in turn on process of the interleaved AC/DC converter. On the other hand, Fig. 14 shows the wave form in turn off process of the interleaved AC/DC converter. In Fig. 13 the peak current of the Si device was larger because of higher recovery current of Si diode compared to that of SiC SBD. It indicates that the SiC device had much less turn on loss than that of the Si device. In Fig. 14 turn off time of the converter of SiC is less than that of the converter of Si, so turn off loss of the converter of SiC is less than that of the converter of Si.

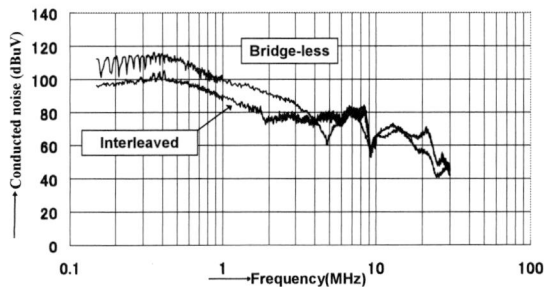

Fig. 11. Comparison of conducted noise of bridgeless AC/DC converter and interleaved AC/DC converter.

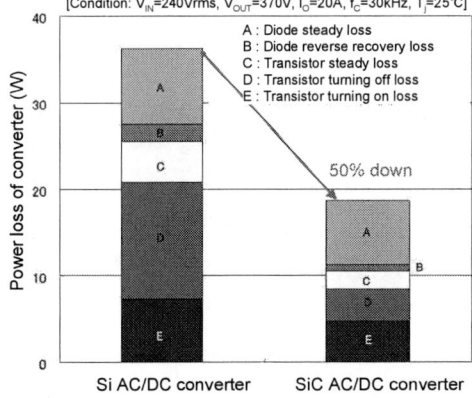

Fig. 12. Comparison of the power loss of interleaved AC/DC converter.

978-1-4799-2706-7/14 $31.00 © 2014 IEEE

The 2014 International Power Electronics Conference

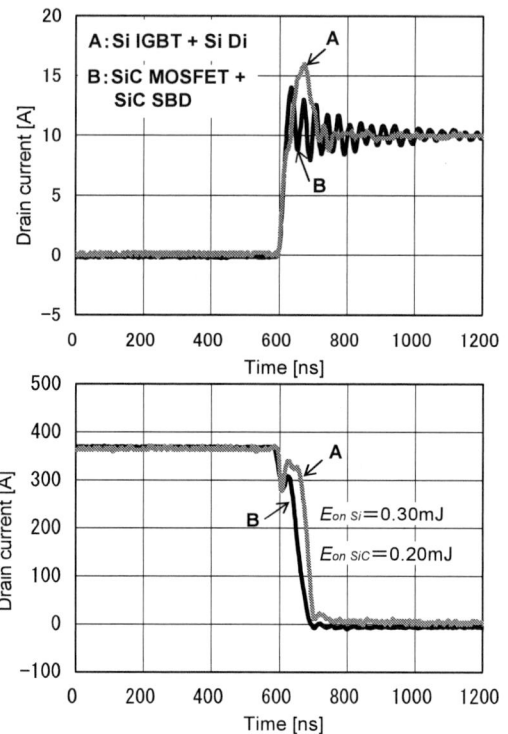

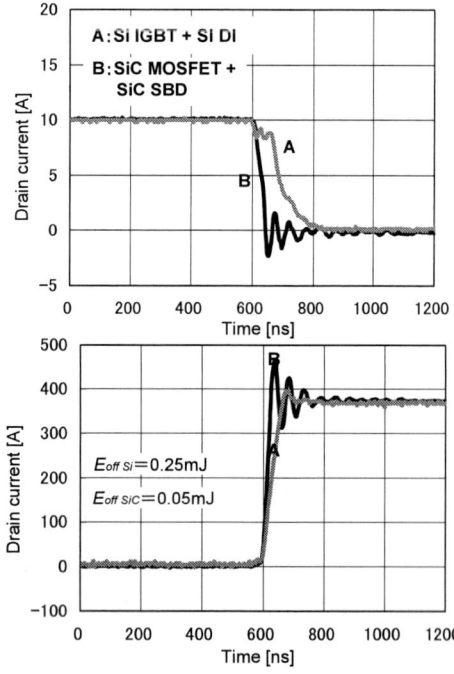

Fig. 13. Wave forms of turning on process of Si and SiC.

Fig. 14. Wave forms in turning off process of Si and SiC.

highest rate in home appliances, by introducing the SiC. The transition of power devices, the features and the characteristics of the next generation power devices (SiC), the issue in widespread use and expectations of SiC are reported. We confirmed that a converter with SiC has much less turn on loss than that with Si using an interleaved AC/DC converter.

References

[1] G. Majumdar , M. Fukunaga, T. Ise, "Trends of Intelligent Power Module," *IEEJ Trans*, vol.2, issue 2, pp. 143-153, 2007.

[2] G. Majumdar, "Power Module Technology for Home Power Electronics," *Proc. on IEEE International Power Electronics Conference*, pp.773-777, 2010.

[3] T. Nakamura, M. Sasagawa, Y. Nakano, T. Otsuka, and M. Miura, "Large current SiC power devices for automobile applications," *Proc. on IEEE International Power Electronics Conference*, pp.1023-1026, 2010.

[4] J. Biela, M. Schweizer, S. Waffler, B. Wrzecionko, J.W. Kolar, "SiC vs. Si – Evaluation of potentials for performance improvement of power electronics converter systems by SiC power semiconductors," *IEEE Trans. on Industrial Electronics*, vol.58, no. 7, pp.2872-2882, 2011.

[5] B. Ozpineci, L.M. Tolbert, "Characterization of SiC Schottky diodes at different temperatures," *IEEE Power Electronics Letters*, vol.1, no.2, pp.54-57, 2003.

IV.   Conclusion

Large energy-saving effect can be expected for air conditioners, those power consumption occupy the

978-1-4799-2706-7/14 $31.00 © 2014 IEEE

Recent Technical Trends and Future Prospects of IGBTs and Power MOSFETs

Tsuneo Ogura

Discrete Semiconductor Division, Semiconductor & Storage Products Company,
Toshiba Corporation
Kawasaki city, Japan

Abstract— **In this paper, recent technical trends and future prospects of IGBTs and power MOSFETs is presented. Device technologies mainly for reducing power loss are discussed. This is because the reduction in power loss of these power devices is important for home and consumer appliances. Firstly, historical main road maps of these device technologies are introduced. Next, proposed future road maps and distinguishing results are also introduced. And, a comparison with IGBTs, power MOSFETs and SiC-MOSFETs is discussed. Finally, I will conclude that Si-power devices and SiC-power devices will coexist in home and consumer appliances by taking advantage of each characteristic in the near future.**

Keywords— *power semiconductor device, IGBT, power MOSFET*

I. INTRODUCTION

In recent years, wide-band-gap (WBG) power semiconductor devices, such as SiC and GaN power devices, are expected as next-generation power devices from the requirement of energy-saving in home and consumer appliances. This is because these WBG power devices have a great potential as compared to Si power devices that are widely used today. This potential includes the characteristics of low on-state loss, high switching speed and high temperature operation, resulting from the advantage in the physical properties over Si materials. There is also the background that the improvements of Si power semiconductor devices, promoted over more than 30 years, have close to their limit that is determined by the physical properties of Si materials. Therefore, Si power devices are believed to have a few rooms of performance improvement.

However, Si power devices, such as IGBTs and power MOSFETs are now progressing steadily with these characteristics. And it is considered that these devices play an important role in home and consumer appliances in the near future. So, in this paper I will discuss outlook and recent technical trends and future prospects of IGBTs and power MOSFETs, and will compare with WBG power devices.

A general correlation of main characteristics of power devices and main characteristics required from home and consumer appliances are the same both in Si power devices and WBG power devices (Fig. 1). Low power-loss, high temperature operation and high ruggedness in power devices are required to achieve compact & light weight, low noise, low cost, high reliability and high efficiency for home and consumer appliances. Among them, it is particularly important to realize low power-loss in power devices without sacrificing the other properties.

In addition, discussions in this paper are not limited to home and consumer appliances, because the discussions may be considered as common technologies in other appliances, such as general purpose, automotive, electric train, information and communication.

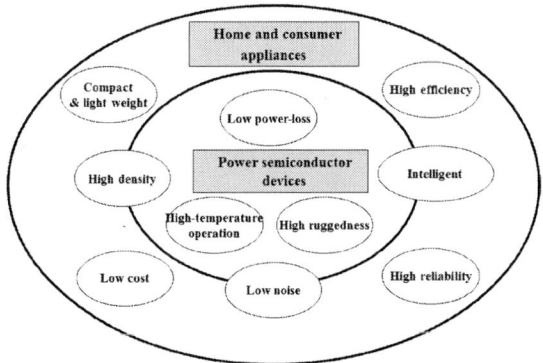

Fig.1 General correlation of characteristics of power devices and characteristics required from home and consumer appliances.

II. TECHNICAL TREND OF IGBTS

First, I will introduce main application products of home and consumer appliances of IGBTs (Fig. 2). IGBTs are widely used in these products, such as air conditioners, refrigerators, washing machines, photovoltaic power conditioners, digital cameras, microwave ovens, induction heaters and rice cookers. Many types of circuit configurations, such as inverter, voltage resonant, current resonance and pulse-current generator circuits, are used for these appliances. In this regard, the circuits for home and consumer appliances are vivid contrast to those for industrial and automotive appliances which have mainly inverter circuits for driving motors. Therefore, a performance required to IGBTs for home and consumer appliances, is different at each appliances in detail design level. However, low power-loss as a basic characteristic is always required for these circuits. Therefore, though it is not limited to home and consumer appliances, I will describe trend toward to lower loss of IGBTs in this section.

978-1-4799-2706-7/14 $31.00 © 2014 IEEE

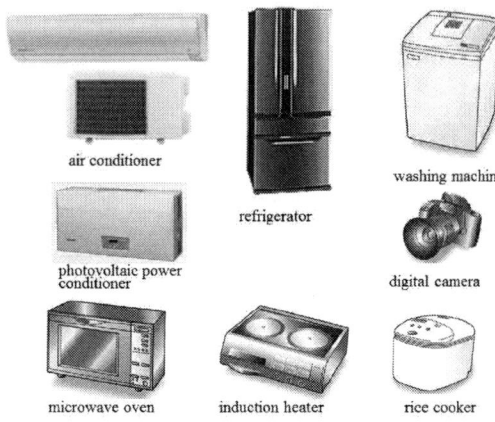

Fig.2 Application examples of IGBTs for home and consumer appliances.

To realize low power-loss in bipolar devices such as IGBTs, it is basically important to improve the trade-off relation between the reduction of on-state voltage by increasing accumulated carriers and the reduction of turn-off switching loss by high-speed discharge of accumulated carriers. Therefore, in the beginning of development stage, the reduction of turn-off switching loss by high-speed discharge of accumulated carriers was an important development challenge. In this view point, the injection efficiency control of a p-collector layer and the reduction of carrier lifetime in an n-base layer were most important issues. In the next step of development stage, the reduction of on-state voltage was a major problem, because the basic structure of IGBTs hardly accumulates carriers in the n-base layer near a p-base layer. Therefore, a structure with electron injection enhancement effect (IE-effect) has been developed [1], and a significant reduction in on-state voltage was realized (Fig. 3).

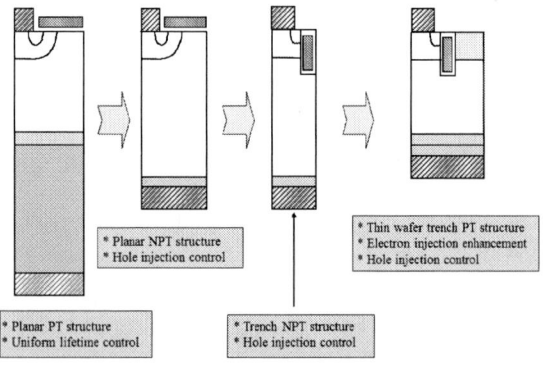

Fig. 3 Historical progress in device structures of IGBTs.

Latest trend to realize low on-state voltage, a narrow mesa structure for further increase in electrons at an emitter side was proposed as shown in Fig. 4(a). It was theoretically shown that a minimum value of on-state voltage of IGBTs is realized at a case of integration of the both sides of channel of the trench surface [2]. To realize this condition, the mesa width of two trench gates located next to each other was calculated to about 20 ~ 40 nm.

And, on-state voltage at 600 V IGBTs was expected to reduce to 2 mΩcm^2 which is about 30 % reduction compared to the latest IGBTs. Also, at this condition, it was expected that turn-off switching was also faster than that of the latest IGBTs, because only electron current contributes to conduction by suppressing injection amount of holes from the p-collector.

A reduction in on-state voltage by the narrow mesa width has been experimentally verified [3]. A 40 % reduction of the mesa width has realized 80 % of on-state voltage compared to that of the latest IGBTs. However, it is considered that the effect of narrow mesa structure is not proportional to a reduction rate in on-state voltage, but is gradually saturated. And ultimately, a minimum value of on-state voltage will lead to the assumption of the theoretical analysis described above.

Furthermore, since a layout of gate electrode and emitter electrode at the surface of emitter is difficult at the condition of fine mesa width, a fine pitch structure only at the trench bottom (partially narrow mesa structure) is also reported [4]. Figure 4(b) shows a schematic diagram of an ultra-fine pitch IGBT having a minimum width of 30 nm mesa. By fining the mesa width, because carrier density at the emitter side is increased, on-state voltage is reduced theoretically and experimentally. However, it is noted that due to carrier density increase at the emitter side, delay time at turn-off switching is increased and resulting switching loss increase.

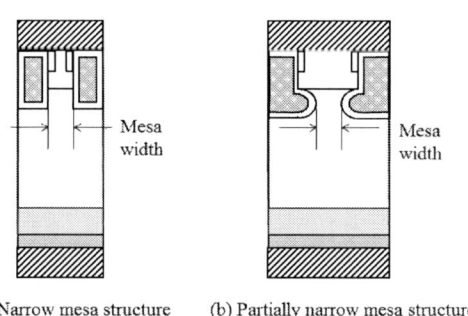

(a) Narrow mesa structure   (b) Partially narrow mesa structure

Fig.4 Recent progress in device structures of IGBTs.

In order to miniaturization of the mesa width, it is considered that an adoption of the process developed for the LSI is a powerful tool. It has been reported that on-state voltage can be reduced by reducing a structure design pitch in accordance with a certain rule likely to LSI process miniaturization rule [5]. However, a special structure design is required for IGBTs, such as a trench gate structure, an oxide film thickness and a collector layer profile. Therefore, it should be pointed out that a proper modification of the LSI process is an important issue for next generation IGBTs.

As described above, an introduction of fine pitch process technology is important for next generation of IGBTs. Possibility of reduction in on-state voltage is shown by the fine pitch IGBT, and development has been in progress. However, it is considered that there are many problems to be solved, such as processing technology, reliability of oxide film, increase in switching loss due to carrier accumulation and ruggedness. Therefore, the reduction in on-state voltage should be developed in parallel without sacrifice of these characteristics.

Next, I will discuss a latest trend for reverse conductive type IGBTs (RC-IGBTs), which have already been developed for home and consumer appliances. Because the reduction in on-state voltage is progress as described above, a downsizing of chip area has been achieved by increasing current density to realize both size and cost reductions. However, a significant reduction in chip size becomes difficult in recent years. Therefore, to realize the reduction in chip area, a new approach of functional integration has been studied. Power ICs with integrated control circuit are those examples, but I will introduce RC-IGBTs which aim to integrate power devices. Power electronics circuits for home and consumer appliances, an inverse parallel connection of IGBT and wheeling diode (FWD) is commonly used. IGBTs do not integrate FWDs in their original device structure, unlike MOSFETs. Therefore, a special structure of adding an n-cathode layer for FWDs has been developed as shown in Fig.5. This special integrated structure can reduce chip area compared with the conventional two-chip configuration of IGBTs and FWDs.

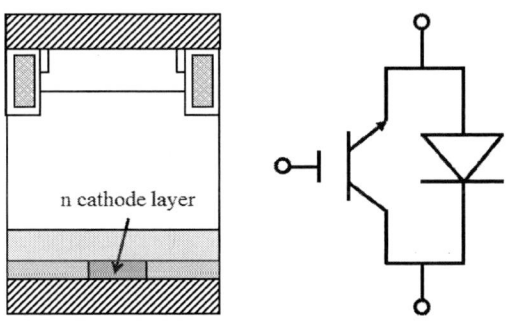

Fig.5 Structure cross section of reverse conducting IGBT and its equivalent circuit.

RC-IGBTs are already applied to soft switching circuits. Figure 6 shows switching waveforms of IGBTs in a current resonance circuit and a voltage resonance circuit to achieve soft switching. The former is used for microwave ovens; the latter is used for IH cooking heaters, and so on. Because high-frequency switching is essentially required for these applications, these soft switching circuits to reduce in switching loss are used. In these resonant circuits, high-frequency operation of more than 20 kHz has been realized. In addition to this, it is a merit to form a resonant circuit without using an additional circuit with a large inductance. For these circuits, a device design of RC-IGBTs is easier than that for the hard switching circuit, because over load condition at switching transient is light and the current flowing through FWDs is small. Therefore, RC-IGBTs for the soft switching circuit have developed ahead of that for the hard switching circuit.

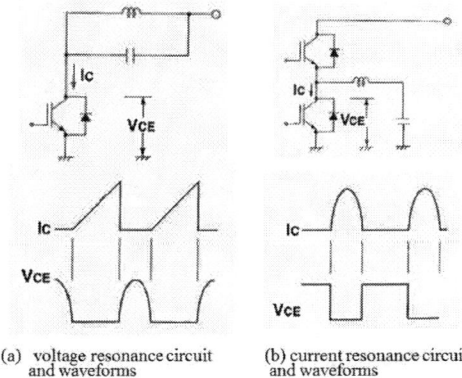

(a) voltage resonance circuit and waveforms   (b) current resonance circuit and waveforms

Fig.6 Circuits and switching waveforms of reverse conducting IGBTs in current resonance circuit and voltage resonance circuit.

Thereafter, RC-IGBTs for the conventional hard switching are also developed. It was reported that, as a loss reduction in a 600 V RC-IGBT, increase in rated current of about 33 % to 150 % has been realized at three types of packages. Miniaturization of these packages has achieved with downsizing by integrated chips [6].

For an RC-IGBT, because an IGBT and a FWD are integrated in one chip, a device and package design that takes into account both influence of heat and electrical effects is important. It is believed that there still remain many problems to be solved for wide range of applications; RC-IGBTs shall continue to contribute significantly to realize small size and high density of power electronics systems.

TECHNICAL TREND OF POWER MOSFETS

First, I will introduce main application products of home and consumer appliances of power MOSFETs (Fig.7). Power MOSFETs are widely used in these products, such as LCD-TVs, air conditioners, photovoltaic power conditioners, LED lightings, printers and power tools. Many types of circuit configurations, such as AC/DC converters, PFC (Power Factor Correction) circuits, DC/DC converters and DC/AC converters, are used for these applications. Like IGBT case, low power-loss as basic characteristics is always required for these circuits. Therefore, though it is not limited to home and consumer appliances, I will mainly describe trend toward to lower loss of power MOSFETs in this section.

Power MOSFETs are switching elements which are most widely used in breakdown voltage range from 10 V to around 600 V. Here, devices having breakdown voltage from approximately 20 V to 30 V are called low-voltage MOSFETs (LV-MOSFETs). And LV-MOSFETs are used widely in DC/DC converter circuits to obtain a DC voltage required in power circuits of home and consumer appliances. The device of this voltage region, on-resistance is determined mainly by channel resistance rather than n-type drain layer resistance, because of narrow and high concentration of n-type drain layer. For this reason, reduction of channel resistance is an important issue. Also, reduction of gate capacitance for

978-1-4799-2706-7/14 $31.00 © 2014 IEEE   2070

high-speed switching is an important issue. Therefore, fine patterning and etching process developed for the LSI fabrication process have been applied to realize low channel resistance and low gate capacitance.

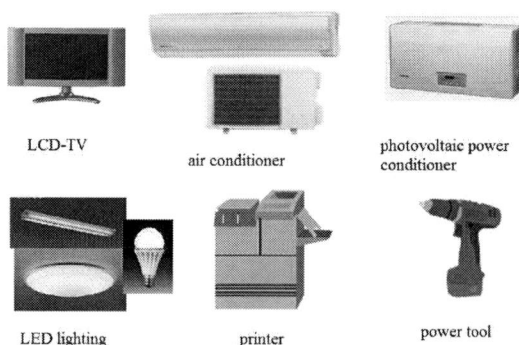

Fig.7 Application examples of power MOSFETs for home and consumer appliances.

At initial stage, planar gate structures were developed. But now trench gate structures have become a mainstream, because channel density is easily increased by fining of trench pitch (Fig. 8). Features, compared to IGBTs with trench structure, are advanced fine trench pitch because of shallow trench depth and thin oxide. Further, in power MOSFETs, because feedback capacitance is dominant over switching unlike IGBTs, special device structure with trench structure is important.

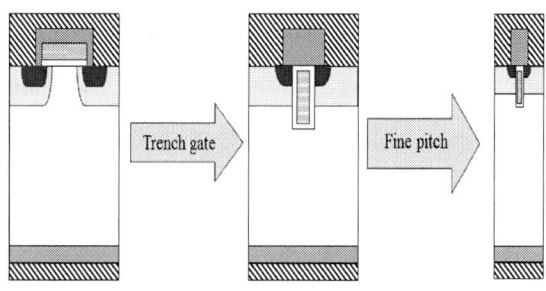

Fig. 8 Historical progress in device structures of low-voltage MOSFETs.

Latest two types of trench gate structure for LV-MOSFETs are showed in Fig. 9. Figure 9(a) shows a LV-MOSFET with the breakdown voltage of 35 V which has reduced feedback capacitance by increasing the thickness of the oxide film in the trench bottom [7]. A 27% reduction from 762 pF to 207 pF at drain voltage of 0 V and a 55 % reduction from 163 pF to 89 pF at drain voltage of 30 V have been realized when compared with the conventional structure. As the result, conversion efficiency of 1.5 % higher than that of the conventional structure in switching operation of 300 kHz in a DC/DC converter circuit from 20 V to 1.5 V was realized. Figure 9(b) shows a LV-MOSFET with the breakdown voltage of 25 V which has a buried source electrode in the bottom of trench [8]. A feedback capacitance can be reduced to 85 % as compared with the conventional structure

without buried source electrode, thereby achieving a higher switching characteristic of two times or more.

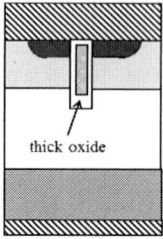 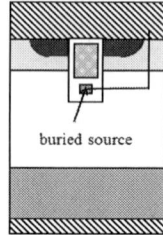

(a) thick bottom oxide structure   (b) buried source electrode

Fig.9 Recent progress in device structure of LV-MOSFETs.

In addition, to improve the trade-off relation between switching speed and on-resistance, a combination structure of trench and planar structure has also been developed in recent years (Fig. 10). This planar channel structure can reduce feedback capacitance drastically. Also this planar structure can be fabricated using fully LSI fine patterning process, so it has an advantage on the performance improvement by miniaturization. And the trench structure is used to form drain electrode at the bottom of the chip which can realize high heat dissipation by double-sided cooling [9].

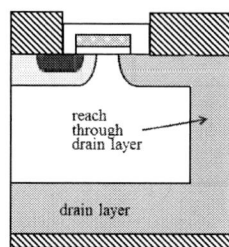

Fig.10 Structural cross section of new planar MOSFET.

Next, power MOSFETs having breakdown voltage from 500 V to 600 V are called high-voltage MOSFETs (HV-MOSFETs). And HV-MOSFETs are used widely in the primary side of AC/DC converters to obtain from AC power supply to DC voltage which is usually required for home and consumer appliances. Because this type of converters always operates, improvement of conversion efficiency by reducing on-resistance of HV-MOSFETs is very important for energy conservation.

In the device of this voltage range, because of low-concentration and wide n-type drain layer, on-state resistance is mainly determined by n-type drain layer resistance rather than channel resistance. Therefore, instead of the fine pitch of gate structure, a design of the n-type drain layer is important. However, on-resistance at conventional planar MOSFETs has approached a theoretical limit (about 70 mΩcm^2 in 600 V class MOSFETs) by the latter half in the 1990's.

At that stage, Super-Junction MOSFETs (SJ-MOSFETs) have been developed as shown in Fig. 11. To

978-1-4799-2706-7/14 $31.00 © 2014 IEEE

realize the SJ-MOSFET structure, it is necessary to form a p/n-pillar layer in the horizontal direction. By adopting this pillar structure, it is possible to increase the concentration of the n-drain layer. Moreover, it is possible to reduce on-state resistance by decreasing the pitch of the p/n-pillar layer (Fig. 12). Therefore, a technique for a narrow p/n-pitch width is a major issue, and the development of process technology is a key point. The basic and original method for forming p/n-pillar layer is repeating epitaxial growth and implantation method (so-called multi-epitaxial method). Using this method, it is believed that control of the epitaxial layer thickness and control of the concentration of the p/n-pillar layer is easy, but it is necessary to increase the number of epitaxial growth steps to obtain a narrow pitch of the p/n-pillar layer. Therefore, there is a limit from the viewpoint of production cost. And, various process techniques have been studied to solve this problem.

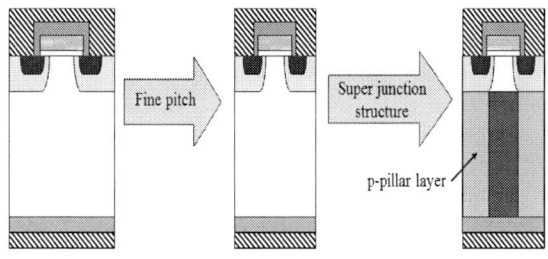

Fig. 11 Progress in device structures of high-voltage MOSFETs.

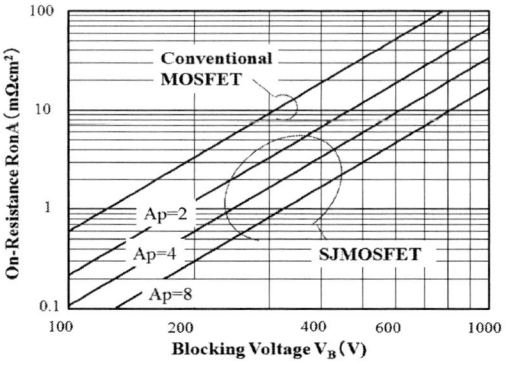

Fig.12 Ideal On-Resistance of DMOSFETs and SJMOSFETs as a parameter of pillar aspect ratio (Ap).

Recently, a method of forming the p/n-pillar layer using deep trench filled by epitaxial growth for the p-pillar layer has been developed. Because of simplification of process, this method (so-called single-epitaxial method) is expected to be a next generation process for SJ-MOSFETs. To obtain good performance by using this method, it is important to form a uniform epitaxial p-pillar layer in the trench without crystal defects. Recently, epitaxial p-pillar layer of 37 μm depth in the trench has been formed, and on-resistance has been reduced to about 1/10 of a conventional MOSFET. This SJ-MOSFET has the on-resistance of 7.8 mΩcm² at the breakdown voltage of 600 V class [10].

Figure 13 shows a road map of on-resistance reduction in SJ-MOSFETs. The on-resistance of current production level is about 20 mΩcm², but that of research level is 7.8 mΩcm². However, the progress of on-resistance reduction has not progressed from 2008. It is considered that difficulty in process has remarkably increased, but development of new technology is desired in the future.

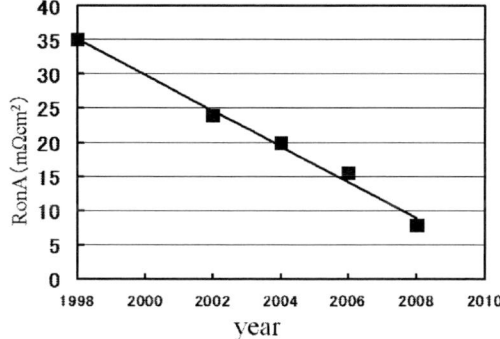

Fig. 13 Roadmap for deduction in on-resistance of SJ-MOSFETs at research level.

In addition to this, reduction of the switching noise has become important for high-switching speeding condition. Increase in dv/dt due to fast switching causes noise. A new structure which can reduce noise even at high dv/dt condition is shown in Fig.14 [11]. A dummy p-base layer can control internal parasitic capacitances of the MOSFET. Therefore, it is possible to use in high dv/dt condition, realizing high switching speed without sacrificing noise problem.

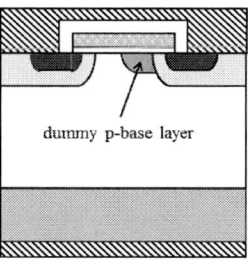

Fig.14 Structural cross section of new HV-MOSFET.

III. CONCLUSIONS

Finally, I will compare on-state voltage between IGBTs, power MOSFETs and SiC-MOSFETs at the breakdown voltage of 600 V which is commonly used in home and consumer appliances. The relationship between on-state voltage and chip size is shown in Fig. 15. In IGBTs, the internal 0.7 V potential due to the pn-junction were added to the on-resistance. In addition, the relation for MOSFETs, SJ-MOSFETs and SiC-MOSFETs can be expressed only by changing parameter of on-resistance. The on-resistance of latest SJ-MOSFET was considered as 20 mΩcm², and the limit value of the on-resistance was estimated as 5 mΩcm². In SiC-MOSFETs, on-resistance of 2 mΩcm² of current production situation is plotted and almost theoretical limit value of on-resistance of 0.2 mΩcm^2 is also plotted in the figure.

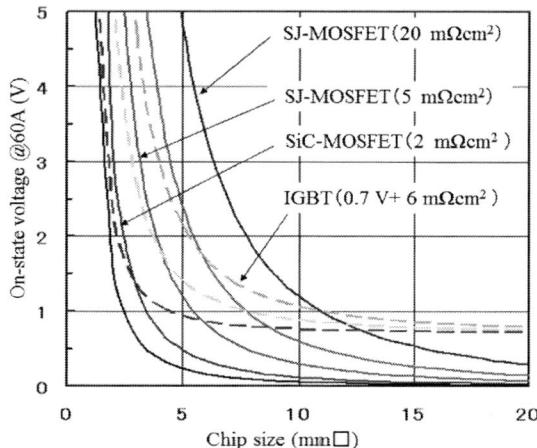

Fig.15 Relationship between on-state voltage and chip size with IGBTs and power MOSFETs.

When comparing the on-state voltage of 60 A at the chip size of 5 mm□, the on-state voltage of a latest IGBT is about 2 V, that of a latest SJ-MOSFET is about 5 V and that of a latest SiC-MOSFET is about 0.5 V. Ignoring switching losses assumed low switching frequency driving, the turn of low loss is SiC-MOSFET, IGBT and SJ-MOSFET. However, because of high cost of SiC-MOSFETs, the chip size of SiC-MOSFETs should be reduced. Assuming that chip size of SiC-MOSFETs is 1/4 compared with that of IGBTs, it is considered that SiC-MOSFETs do not have many advantages compared with IGBTs.

However, SiC-MOSFETs have a merit of lower switching loss than IGBTs. Therefore, it is necessary to apply SiC-MOSFETs to suitable systems by comprehensively considering their characteristics. The problem is that the on-resistance of current situation has stopped at 10 times larger than that of theoretical limit of SiC-MOSFETs. If this theoretical limit is achieved, the on-resistance of the 60 A will be significantly decreased to about 0.05 V at the chip size of 5 mm□ and about 0.2 V at the chip size of 1/4.

Difficulty degree of process technology to develop next generation IGBTs and power MOSFETs for home and consumer appliance is increasing; however, technologies for total loss reduction in these devices will progress in the future. In the near future, it is believed that Si-power devices and SiC-power devices will coexist in home and consumer appliances by taking advantage of each characteristic.

In recent years, the growth of power consumption by industrial usage has stopped in Japan, but that for home and consumer appliances has extended year by year. It considered that this is caused by the growth of numbers of air-conditioners and lightings and high functionality of refrigerators and TV sets. Also, I believe that the use of energy storage at home, including solar power, will rapidly increase in the future. For these reasons, energy efficiency at home and consumer appliances has become an increasingly important issue. Therefore further development towards power electronics with higher energy efficiency and power devises with lower total loss are strongly expected.

REFERENCES

[1] M.Kitagawa, I.Omura, S.Hasegawa, T.Inoue and A.Nakagawa, "A 4500 V Injection Enhanced Insulated Gate Bipolar Transistor (IEGT) Operating in a Mode Similar to a Thyristor", IEDM'93, pp.679-682.

[2] A.Nakagawa, "Theoretical Investigation of Silicon Limit Characteristics of IGBT", ISPSD'06, pp.5-8, 2006.G. Eason, B. Noble, and I. N. Sneddon, "On certain integrals of Lipschitz-Hankel type involving products of Bessel functions," *IEEE Trans. on Power Electronics*, vol. 247, no. 8, pp. 529-551, 1995.

[3] S.Honda, Y.Haraguchi, A.Narazaki, T.Terashima and Y.Terasaki, "Next Generation 600V CSTBTTM with an Advanced Fine Pattern and a Thin Wafer Process Technologies", ISPSD'12, pp.149-152, 2012.

[4] M.Sumitomo, J.Asai, H.Sakane, K.Arakawa. Y.Higuchi and M.Matsui, "Low loss IGBT with Partially Narrow Mesa Structure (PNM-IGBT)", ISPSD'12, pp.17-20, 2012.

[5] M.Tanaka and I. Omura, "Scaling Rule for Very Shallow Trench IGBT toward CMOS Process Compatibility", ISPSD'12, pp.177-180, 2012.

[6] H.Ruthing, F.Hille, F.-J.Niedernostheide, H.J.Schulze and B.Brynner, "600 V Reverse Conducting (RC-) IGBT for Drives Applications in Ultra-Thin Wafer Technology", ISPSD'07, pp.27-30, 2007.

[7] M.Darwish, C.Yue, K.H.Lui, F.Giles, B.Chan, K.Chen, D.Pattanayak, Q.Chen, K.Terrill and K.Owyang, "A New Power W-Gated Trench MOSFET (WMOSFET) with High Switching Performanc", ISPSD'03, pp.24-27, 2003.

[8] P.Goarin, G.E.J.Koops, R.v. Dalen, C.L.Cam and J.Saby, "Split-gate Resurf Stepped Oxide (RSO) MOSFETs for 25V applications with record low gate-to-drain charge", ISPSD'07, pp.61-64, 2007.

[9] G.Loechelt, G.Grivna, L.Golonka, C.Hoggatt, H.Massie, F.D.Pestel, N.Martens, S.Mouhoubi, J.Roig, T.Colpaert, P.Coppens, F.Bauwens and E.D.Backer, "A High-Speed Silicon FET for Efficient DC-DC Power Conversion", ISPSD'12, pp.85-88, 2012.

[10] J.Sakakibara, Y.Noda, T.Shibata, S.Nogami, T.Yamaoka and H.Yamaguchi, "600V-class Super Junction MOSFET with High Aspect Ratio P/N Columns Structure", ISPSD'08, pp.299-302, 2008.

[11] W.Saito, S.Aida, S.Koduki amd M.Izumisawa, "Improvement of Switching Trade-off Characteristics between Noise and Loss in High Voltage MOSFETs", ISPSD'11, pp.316-319, 2011.

Recent development and future prospects of Power SiC devices.

T. Nakamura, Y. Nakano, M. Aketa, T. Hanada
Power Electronics R&D Unit, ROHM Co., Ltd.
21 Saiin Mizosaki-cho, Ukyo-ku, 615-8585 Kyoto, Japan

Abstract- **Silicon Carbide (SiC) devices have the potential to reduce energy losses in high power applications. However SiC devices have yet to achieve ideal performance levels. The SiC diodes and MOSFETs with advanced trench structures succeeded in improving performance by reduction of the internal electric field. In addition, transfer mold type power modules using SiC devices demonstrated high temperature operation and high power density.**

I. ADVANCED SiC DEVICES

In an effort to reduce the electric field in critical areas, a double-trench structure was developed. Simulation results reveal that the electric field at the gate trench bottoms is reduced from 2.66MV/cm to 1.66MV/cm by use of this structure when there is 600V applied from drain to source. A very low specific on-resistance of 0.79 and 1.41mΩcm^2 is achieved for each device, respectively

It is one of the most important subjects to lower V_F in SiC SBD. There are two kinds of step which lower V_F. One is reducing Vth. We succeeded in reducing Vth drastically by the proposal of the trench SBD [1]. It is that a step reduces resistance to one more which reduces V_F. This time, we succeeded in reducing resistance drastically by making a SiC substrate thin to 50um. The resistance comparison of conventional SiC SBD and SBD fabricated in this work is shown in Fig.1.

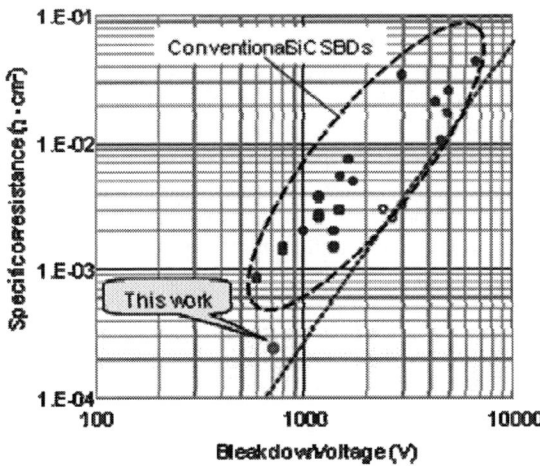

Fig. 1. Comparison between the conventional SBDs and the SBD on 50um SiC.substrate.

II. ULTRA SMALL SiC MODULES

To make the most of SiC potentials, we have developed a transfer-molded design with a new encapsulation resin which is able to be used at 200°C and above. We succeed to develop the SiC module which is dramatically smaller than case-type modules by using the new technologies.

Work has been done on the development of a single switch position module with the aforementioned molding techniques using trench MOSFETs in parallel with dimensions 40.0mm x 16.5mm x 7.0mm. Switching operation at 600V, 575A has been confirmed.

A three phase inverter setup is realized using six of the single switch position modules and an integrated gate driver board in a configuration shown in Fig. 2. The single modules are designed in such a way that the positive and negative terminals of each are at different heights. This facilitates tight bus bar interconnection with the positive and negative busses overlapping. Partial cancellation of mutual inductances is achieved through this setup. Furthermore the compact bus bar arrangement makes for a small system foot print and reduction in parasitic inductance, as well.

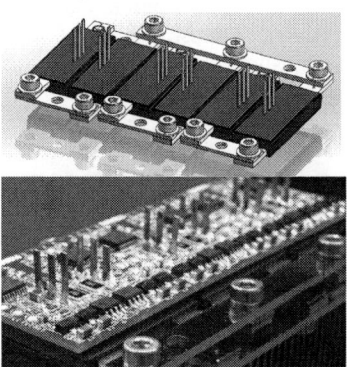

Fig. 2. 3 phase module configuration illustration and final assembly with gate driver.

REFERENCES

[1] M. Aketa, Y. Yokotsuji, M. Miura and T. Nakamura, Materials Science Forum Vols. 717-720 (2012) pp. 933

Recent Advances and Future Prospects on GaN-based Power Devices

Tetsuzo Ueda

Power Electronics Development Center, Automotive & Industrial Systems Company, Panasonic Corporation
3-1-1 Yagumo-nakamachi, Moriguchi-shi, Osaka 570-8501, JAPAN
E-mail: ueda.tetsuzo@jp.panasonic.com

Abstract— **Recent advances of GaN Gate Injection Transistors (GITs) and their applications to power switching systems are reviewed. The GITs are fabricated on cost effective Si substrates in a large diameter up tp 6inch, which exhibit normally-off operations by the use of p-AlGaN over an AlGaN/GaN hetero-junction. The GIT is free from the current collapse up to 600V enabling reliable operation of the power switching systems. The inverter for motor drive using GITs enables high efficiency of 99.3% at 1.5kW which is higher than that by conventional Si-IGBT (Insulated Gate Bipolar Transistor). The GITs are also applied various circuits in a power supply taking advantages of the high frequency operations. A GaN-based LLC resonant converter used in isolated DC-DC converter enables 1MHz operation with high efficiency of 96.4%, while conventional Si devices cannot be operated at such high frequencies. The above results demonstrate very promising potentials of high-voltage normally-off GaN GITs.**

Keywords—**GaN, Gate Injection Transistor, Inverter, DC-DC converter.**

I. INTRODUCTION

Gallium Nitride (GaN)-based power transistors are very promising for switching applications taking advantages of the material's high electron velocity and high breakdown strength. High carrier density can be achieved at the AlGaN/GaN hetero-interface by its unique polarization-induced electric field, which leads to low on-state resistances with high breakdown voltages. Recent progress of epitaxial growth of GaN on large diameter Si substrates by novel metal-organic chemical vapor deposition (MOCVD) technologies enables high performance GaN devices at low fabrication cost [1-4]. Power switching systems with far better performances than the conventional ones would be possible with the GaN power devices. Figure 1 shows various power switching applications and those suitable for GaN. Since the breakdown voltage is limited for GaN on Si up to 1000V with the present epitaxial growth technologies, GaN is the best suitable for the applications at higher frequencies and with the operating voltage less than 1000V.

This work is partly supported by the New Energy and Industrial Technology Development Organization (NEDO), Japan, under the Strategic Development of Energy Conservation Technology Project.

In this paper, GaN-based power devices on Si developed at Panasonic are reviewed. Highly efficient power switching systems using the GaN devices are also presented, which demonstrate superior potential of those new devices.

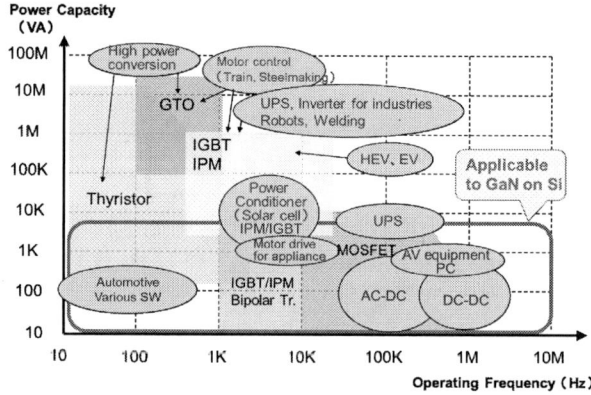

Fig.1 Potential applications of GaN power switching devices.

II. NORMALLY-OFF GaN GATE INJECTION TRANSISTORS ON SI SUBSTRATES

A. Epitaxial Growth of GaN on Large Diameter Si Substrates

Epitaxial growth of GaN on Si had suffered from large mismatch of the lattice constant and the thermal expansion coefficient, which had resulted in rough surface and/or cracks in the film, so far. The epitaxial structure as shown in Fig.2 grown by MOCVD relaxes the mismatch and enables mirror surface free from cracks over a large diameter Si substrate. Here, an AlN initial layer prevents the inter-diffusion of Ga and Si which would cause the rough surface. An AlN/GaN superlattice interlayers enable the growth of GaN by relaxing the strain between Si and GaN. Figure 3 shows the photograph of the AlGaN/GaN heteroepitaxial films over a 6-inch Si substrate. Note that the motilities are over 2000 cm^2/Vs indicating the good crystalline quality. Although off-state breakdown voltages of lateral GaN transistors are linearly increased by the extension of the gate-drain spacing L_{gd}, they are saturated up to a certain value on the conductive Si substrate. The saturated value is increased by the thickness of the GaN since the vertical electric fields determines the breakdown voltage for longer L_{gd}. Figure 4 shows the measured off-state breakdown voltages of GaN transistors as a function of the Lgd for the epitaxial GaN with the thickness of 5.4μm. The breakdown voltage can be increased further by the extension of the Lgd in case the substrate is floating. The maximum breakdown voltage as the substrate is grounded is 1150V which satisfies the requirement for the operation at the drain bias less than1000V.

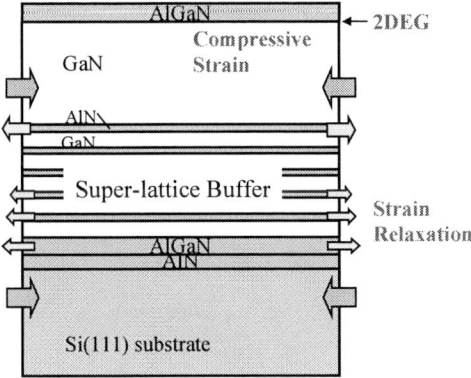

Fig.2 Schematic cross section of an epitaxial structure of GaN on a Si.

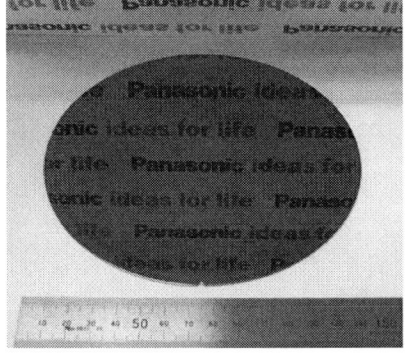

Fig.3 Photograph of the fabricated GaN-GITs over a 6 inch Si substrate

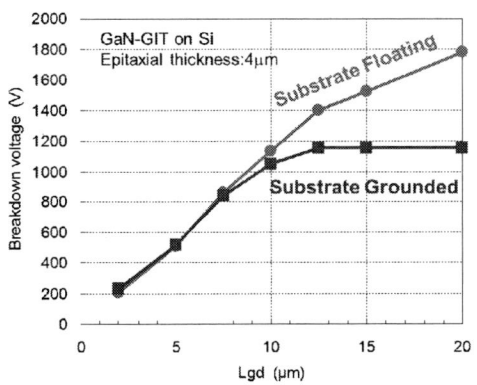

Fig.4 Off-state breakdown voltages of AlGaN/GaN hetero-junction field effect transistors on Si for various gate-drain distances.

B. Novel Normally-off Gate Injection Transistors (GITs)

Normally-off operation is strongly desired for power switching transistors for the safe operation of the systems. Since the polarization in GaN-based materials produces high carrier density at AlGaN/GaN under the gate, achieving the normally-off operation of the GaN transistors has been very difficult. A new normally-off device structure called Gate Injection Transistor (GIT) is proposed, where p-AlGaN gate is placed over AlGaN/GaN [2]. The p-AlGaN lifts up the potential enabling the normally-off operation. Injection of holes from the p-gate forms the equal numbers of election by so-called conductivity modulation resulting in high drain current with low on-state resistance. Figure 5 shows a schematic cross section of the GIT describing its operating principle. At present, 65mΩ GIT with 600V of the blocking voltage, of which the DC performances are summaried in Table 1, is available. Here, $R_{on}Q_g$ (R_{on}: on-state resistance, Q_g: gate charge) is a figure- of-merit for high speed switching. The fabricated GIT exhibits the low $R_{on}Q_g$ of 715mΩnC which is one thirteenth lower than that by the state-of-the-art Si devices. Faster switching than conventional Si devices can be achied by the presented GITs. So far, the phenomena called current collapse, in which on-state resistance is increased after the application of high drain voltages, has been the most serious technical issue for GaN transistors because of the capture of electrons in the device structure. The collapse is no longer observed up to 600V of the off-state drain voltage in the improved GITs as summarized in Fig.6. The GITs free from the current collapse enable reliable operation of power switching systems.

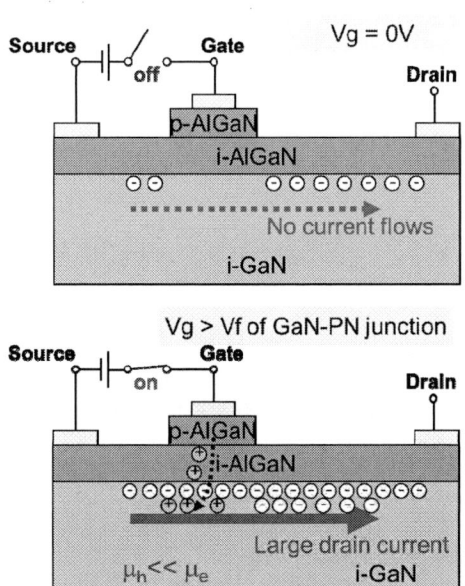

Fig.5 A schematic illustration of a GIT structure and its basic operation.

978-1-4799-2706-7/14 $31.00 © 2014 IEEE

Table 1 Summary of DC performances of fabricated GITs on Si.

Threshold voltage Vth	1.2V (Normally-off)
Blocking voltage BVds	600V
Raiting current (continuous) Id	15A
On-state resistance Ron	65mΩ
Gate charge Qg	11nC

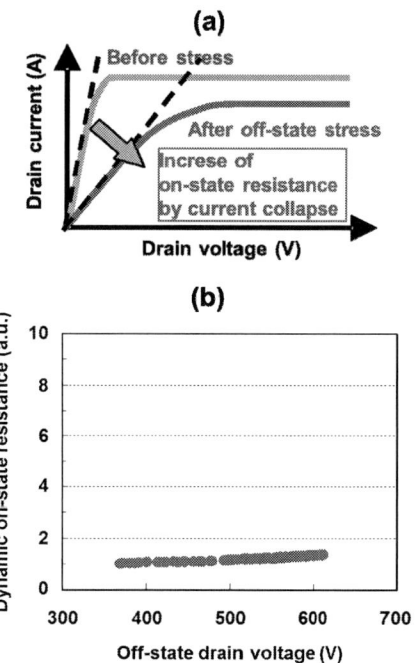

Fig.6 (a) Current collapse in GaN power transistors and (b) measured increase of on-state resistances after off-state stress in GITs on Si for various off-state drain voltages.

III. HIGHLY EFFICIENT POWER SWITCHING SYSTEM USING GATE INJECTION TRANSISTORS

A. GaN-based Inverter for Motor Drive

Inverter systems are widely used for motor drive in various applications, by which the total operating loss can be greatly reduced. Since the transistor requires flowing bidirectional currents, conventional Si-based insulated gate bipolar transistor (IGBT) requires so called fast recovery diode (FRD) in parallel with it. The lateral GIT can flow the bidirectional current by a single device so that the total number of the devices in the GaN-based system can be half from that in the IGBT-based one. The technical advantages of the use of GaN transistor instead of IGBT in the inverter system are summarized in Fig.7. The current-voltage characteristics of the GaN device are free from the forward voltage off-set which is seen for the IGBT and FRD. Then, the operating loss is fully reduced in the GaN-based inverter as long as the on-state resistance of the GaN transistor is

reduced. Figure 8 shows the conversion efficiency of the GaN-based inverter using GITs on Si substrates for various output power [5]. The values for conventional IGBT-based inverter are also plotted in the same figure for comparison. The efficiencies for the GaN-based inverter are higher than those by IGBTs, of which the highest value is 99.3% at the output of 1.5kW. The low on-state resistance and the diode-free operation by the GITs remarkably reduce the operating loss in the wide range of the output power.

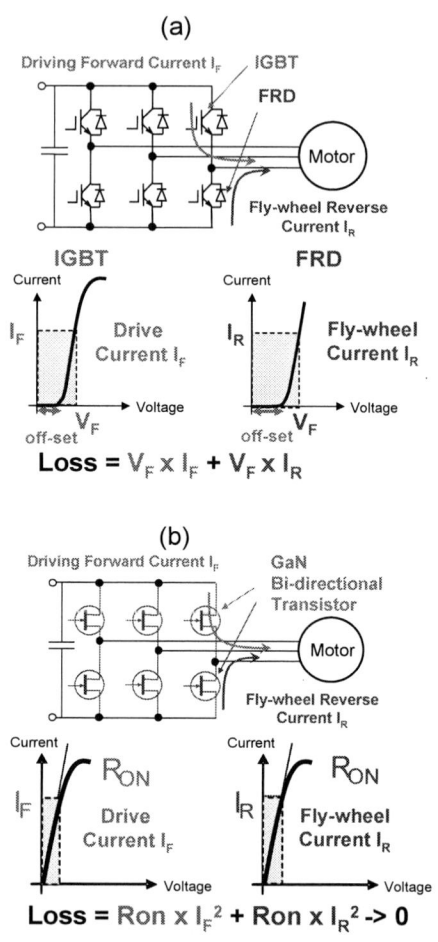

Fig.7 Operations and forward and reverse conduction losses in inverter systems by (a) IGBTs with FRDs, (b) GaN transistors.

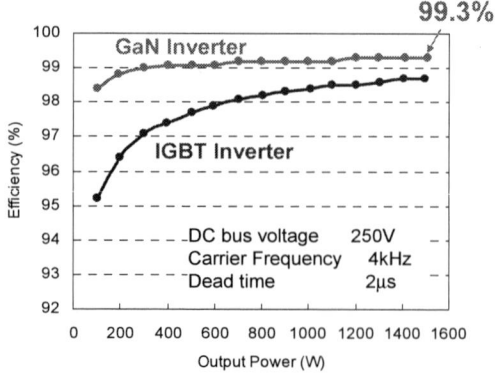

Fig.8 Power conversion efficiency of GaN inverter for various output power as compared with those by conventional IGBT-based one.

B. 1MHz Operation of GaN-based Resonant LLC Converter

Power supplies are one of the promising applications of GaN transistors, since the high frequency operation can reduce the size of the system with smaller passive components. An isolated DC-DC converter with a transformer in it is commonly used in power supplies. The GaN GITs are applied to a resonant LLC converter which is a typical circuit topology of the isolated DC-DC converter as shown in Fig.9. The measured performance as shown in Fig. 10 confirms the successful operation at 1MHz with high efficiency of 96.4% at 1kW output. Such high frequency operation enables the reduction of the system size by half from the conventional one. Since the performance could not be achieved by conventional Si power devices so far, the results demonstrate the great potential of the GaN transistors for power supply applications.

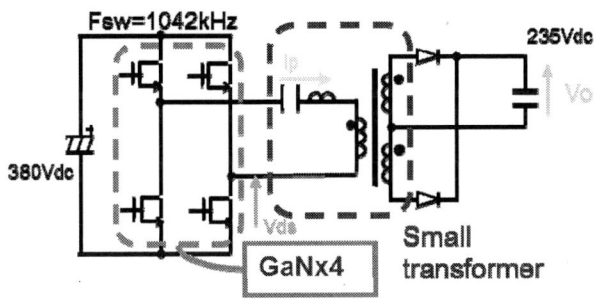

Fig.9 Circuit diagram of a resonant LLC converter used as an isolated DC-DC converter

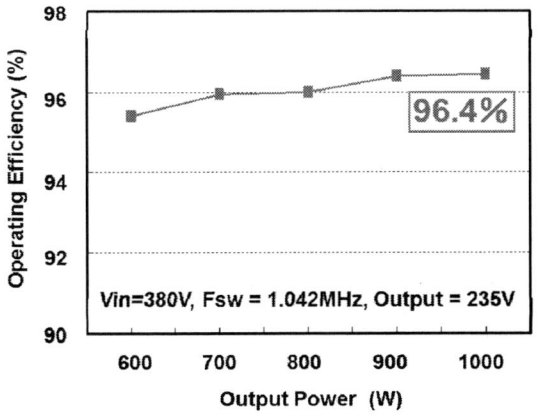

Fig.10 Operating efficiencies of a fabricated resonant LLC converter using GITs on Si at 1MHz for various output powers.

In order to extract the full potential of GaN transistors, peripheral technologies such as high speed gate drivers and design platforms need to be established in addition to the improvement of the device performances. Panasonic's "Drive-by-microwave" technology using electro-magnetic resonant coupling to transfer the power and signal for the isolated gate driving with faster switching speed [6]. Integrated design platform is also developed for the GaN transistors including the simulation of the thermal distribution, noise and device parameters [7]. These technologies would further improve the performances of the switching systems.

IV. CONCLUSIONS

Satte-of-the-art performances of high-voltage GaN GITs and their applications to power switching systems are reviewed. The normally-off GITs are fabricated on cost-effective 6-inch Si substrates. The developed GIT is free from the current collapse up to 600V enabling reliable operation of the power switching systems. Highly efficient operations of GaN-based inverter for motor drive and a resonant LLC converter at 1MHz, a typical circuit topology of isolated DC-DC converter, using the GaN GITs are presented. These performances are far better than those by conventional Si power devices. The GaN GITs are thus very promising for future power switching systems with high efficiencies.

ACKNOWLEDGMENTS

The author would like to acknowledge Dr. D. Ueda, Dr. M.Kubo and Dr. T. Tanaka and other members of Panasonic Corporation for their help and advices throughout the research works on GaN devices.

REFERENCES

[1] M. Hikita, M. Yanagihara, K. Nakazawa, H. Ueno, Y. Hirose, T. Ueda, Y. Uemoto, T. Tanaka, D. Ueda, and T. Egawa, "350V/150A AlGaN/GaN power HFET on Silicon substrate with source-via grounding (SVG) structure," *IEEE Trans. Electron Device*, vol.52, no.9, pp1963-1968, 2005.

[2] Y. Uemoto, M. Hikita, H. Ueno, H. Matsuo, H. Ishida, M. Yanagihara, T. Ueda, T. Tanaka, D. Ueda, "Gate Injection Transistor (GIT)—A Normally-Off AlGaN/GaN Power Transistor Using Conductivity Modulation,"*IEEE Trans. Electron Device*, vol.54, no.12, pp3393-3399, 2007.

[3] T. Ueda, T. Tanaka, and D. Ueda, "GaN Transistors for Power Switching and Millimeter-wave Applications," *International Journal of High Speed Electronics and Systems*, vol.19, pp. 145-152, 2009,

[4] M.Yanagihara, Y.Uemoto, T. Ueda, T.Tanaka, and D.Ueda, "Recent advances in GaN transistors for future emerging applications" *Physica Status Solidi(a)*, Vol. 206, pp.1221-1227, 2009.

[5] T. Morita, S. Tamura, Y. Anda, M. Ishida, Y. Uemoto, T. Ueda, T. Tanaka, and D. Ueda, "99.3% Efficiency of Three-Phase Inverter Using GaN-based Gate Injection Transistors," *Proc. 26th IEEE Applied Power Electronics Conf. and Expo.(APEC2011)*, Fort Worth, USA, March 2011, pp.481-483.

[6] S.Nagai, N.Negoro, T.Fukuda, N.Otsuka, H.Sakai, T.Ueda, T.Tanaka, and D. Ueda, "A DC-Isolated Gate Drive IC with Drive-by-microwave Technology for Power Switching Devices," *Digest of Technical Papers 2012 ISSCC*, San Francisco, USA, February 2012, pp. 404 – 406.

[7] K.Mizutani, H. Ueno, Y.Kudoh, S.Nagai, K.Inoue, N.Otsuka, T.Ueda, T. Tanaka, and D.Ueda, "Integrated Power Design Platform Based on Modeling Dynamic Behavior of GaN Devices," *IEEE IEDM Tech. Dig.*, San Francisco, USA, December.

Scaling and Balancing of Multi-Cell Converters

Matthias Kasper, Dominik Bortis and Johann W. Kolar
Power Electronic Systems Laboratory
ETH Zurich, Physikstrasse 3
Zurich, 8092, Switzerland
kasper@lem.ee.ethz.ch

Abstract—In this paper, the potential of the multi-cell approach for power electronic converters with efficiencies and power densities beyond the barriers of state-of-the-art systems is discussed. Based on fundamental scaling laws the benefits of splitting a system into multiple converter cells are derived in terms of lower volume and/or higher power density for a given cooling capacity. In addition, the conditions for equal current and/or voltage balancing of multi-cell systems is reviewed. The advantages of the mulit-cell systems are examined in more detail based on the example of a DC-DC boost converter realized with either parallel- or series-interleaved boost cells. It is shown, that the multi-cell systems can offer lower switching and conduction losses and/or an improved voltage spectrum depending on the choice of the switching frequency relative to a single system. Furthermore, the effects of parasitic capacitances on unwanted ground currents are investigated for both configurations.

Index Terms—Multi-cell converters, scaling laws, series-interleaving, parallel-interleaving

I. INTRODUCTION

The development of power electronic converter systems towards more efficient, compact and cost effective systems is nowadays to a large extent driven either by the performance improvement of power electronic components or by a higher level of integration. These improvement processes, however, evolve only over longer periods of time that often span decades; the development and market introduction of wide-bandgap semiconductors which started about two decades ago could serve as an example here. In contrast, the development of new topologies is able to shift the system performance (e.g. efficiency, power density and system cost) to new levels in a much shorter time. However, many newly developed topologies are based on adding components to standard topologies with added components [1] which improves individual performance aspects of the basic system but also leads to a higher system complexity and often reduced reliability since the failure rate increases with increasing component count [2].

In this paper, a multi-cell (MC) topology approach is presented which allows to break the barriers of traditional single-stage converter systems by employing basic converters as individual converter cells. The paper is structured as follows: First, general scaling laws of power electronic converters are derived in Section II in order to provide the basis for general statements about the advantages of MC converters. The operation and balancing of multi-cell systems is discussed and reviewed in general in Section III. In a third step, the exemplary multi-cell realization of a DC-DC boost converter in either parallel or series configuration is presented in Section IV and its performance is comparatively evaluated against a single stage converter system. In Section V

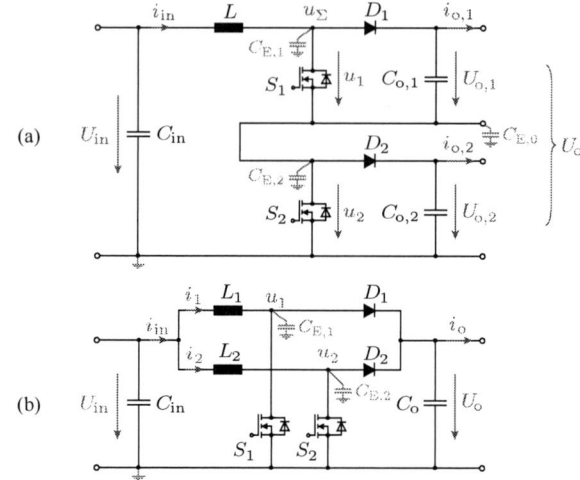

Fig. 1: Multi-cell converter realizations of a DC-DC boost converter: (a) series interleaving of two boost stages and (b) parallel interleaving of two boost stages.

the leakage current caused by parasitic ground capacitances is assessed for both multi-cell converter realizations in comparison to a single boost converter. Finally, the findings are summarized in Section VI.

II. SCALING LAWS OF MULTI-CELL CONVERTERS

Multi-cell topologies can in general be classified as converter systems consisting of two or more subsystems that are connected in one of the following configurations: input-series output-parallel (ISOP), input-series output-series (ISOS), input-parallel output-series (IPOS) or input-parallel output-parallel (IPOP) [3]. As an example of multi-cell converters a standard DC-DC boost converter is shown in **Fig. 1(a)** realized as series-interleaved MC converter (ISOS) and in **Fig. 1(b)** as a parallel-interleaved MC converter (IPOP). These configurations allow to either share the input current between the converter cells (i.e. IPOP) or distribute the output voltage between the converter cells (i.e. ISOS). As a result of splitting either the current or the voltage among the converter cells, the system power is also split in such way that each cell transfers only a fraction of the total power.

The concept of splitting the overall system into smaller subsystems with relatively low power rating leads to benefits that can be leveraged to improve one or more of the performance criteria mentioned in Section I, as will be shown in the following paragraph.

978-1-4799-2706-7/14 $31.00 © 2014 IEEE

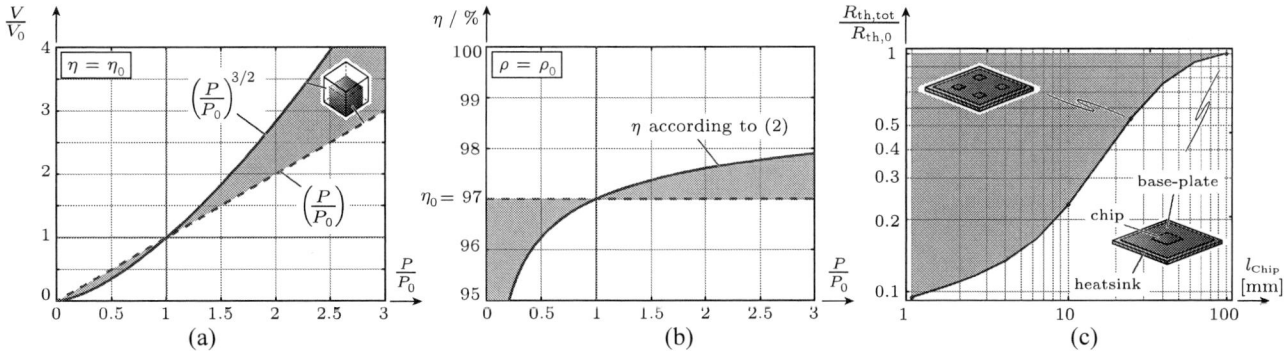

Fig. 2: General scaling laws of power electronic converters: (a) scaling of converter volume V with the converter power P at a constant conversion efficiency η and (b) required scaling of converter efficiency with the converter power at a constant power density ρ, both under the assumption of a constant heat dissipation per converter surface area; (c) reduction of the total thermal resistance of semiconductors by distributing the total chip area to multiple chips with shorter edge lengths.

As derived in Appendix A the volume V of a power electronic system with output power P scales compared to a reference system (V_0, P_0) with

$$\frac{V}{V_0} = \left(\frac{P}{P_0}\right)^{3/2} \tag{1}$$

under the assumption of a constant efficiency (i.e. $\eta = \eta_0$). This is visualized in **Fig. 2(a)**. As a result, the total boxed volume of MC converters with multiple converter cells scales advantageously compared to a single converter with the same total power rating as the MC converter.

With a similar approach (cf. Appendix A), assuming a constant power density (i.e. $\rho = P/V = \rho_0 = P_0/V_0$), it can be found that the efficiency of a converter system has to be scaled with the rated system output power by

$$\eta = \frac{\eta_0 \cdot \left(\frac{P}{P_0}\right)^{1/3}}{1 + \eta_0 \cdot \left(\left(\frac{P}{P_0}\right)^{1/3} - 1\right)} \tag{2}$$

since only a fixed amount of losses (dissipated as heat) can be extracted per surface area. This relationship is depicted for $\eta_0 = 97\%$ in **Fig. 2(b)** and shows that, for example, doubling the system power while keeping the same power density requires to increase the efficiency to $\eta = 97.6\%$.

By splitting the system into lower rated subsystems due to parallel or series interleaving, the semiconductor ratings can be reduced, resulting in a smaller silicon area of the employed chips. Due to the better heat-spreading of smaller chips on a (comparably large) base-plate, the total thermal resistance $R_{\text{th,tot}}$ of the lower-rated chips reduces compared to the thermal resistance $R_{\text{th,0}}$ of the full-rated semiconductor, as shown in **Fig. 2(c)**. The values were determined with FEM simulations for a chip structure based on a TO-247 package with $350\,\mu\text{m}$ thick silicon, $2\,\text{mm}$ thick copper base-plate, $40\,\mu\text{m}$ thick phase-change material ($\lambda = 0.3474\,\text{W}/(\text{m} \cdot \text{K})$) and a $5\,\text{mm}$ thick heat-sink connected to a reference temperature of $T_{\text{amb}} = 40\,^{\circ}\text{C}$. The copper area was chosen to be ten times larger than the silicon area. As the thermal resistance decreases with decreasing chip length, the overall heat-sink volume can be decreased by applying the MC approach.

As a result of the above mentioned fundamental scaling laws, which are independent of the employed converter topology, the MC approach offers advantages in terms of converter efficiency and/or power density and potentially also costs since lower rated semiconductors and/or smaller heat-sinks can be employed.

III. BALANCING AND CONTROL OF MC CONVERTERS

The balancing of currents and/or voltages is an important issue in multi-cell converter systems as the design of the converter cells relies on an equal current and/or voltage sharing among the cells such that the overall system power is equally distributed. Thus, any conditions that lead to a violation of the power sharing might cause an overloading and ultimately a destruction of individual converter cells possibly resulting in a failure of the system. Therefore, current and/or voltage sharing among the converter cells has to be guaranteed for steady state condition and transients. Also the influence of component mismatches, such as slightly different inductance values of parallel connnected converter cells, on the system balancing needs to be addressed.

The operation of multi-cell systems with common-duty-ratio control, where all converter cells are operated with the same duty cycle, relies upon the natural balancing capabilities of a multi-cell topology and is thus only feasible for IPOS and ISOP systems [3]–[5]. For those topologies any component mismatch leads to slightly unequal sharing conditions but not to a runaway situation. In general, ISOS and IPOP are considered to have no natural balancing mechanism and thus require additional control means to guarantee a balanced operation (even though also for the ISOS converter some rebalancing mechanisms could be found [6]). For the IPOP converter it is sufficient to control an equal sharing of the output current as the input current will then also be equally shared [7]. Different control schemes such as droop methods and active control schemes have been published and reviewed in literature [8]–[11]. For ISOS converter structures it was found that controlling the output voltage sharing does not ensure input voltage sharing due to the presence of a right-half-plane pole and thus control efforts should focus on input voltage sharing [12].

IV. MULTI-CELL BOOST CONVERTERS

In this chapter, the operation and the scaling benefits of series-interleaved and parallel-interleaved DC-DC boost converters are investigated.

978-1-4799-2706-7/14 $31.00 © 2014 IEEE

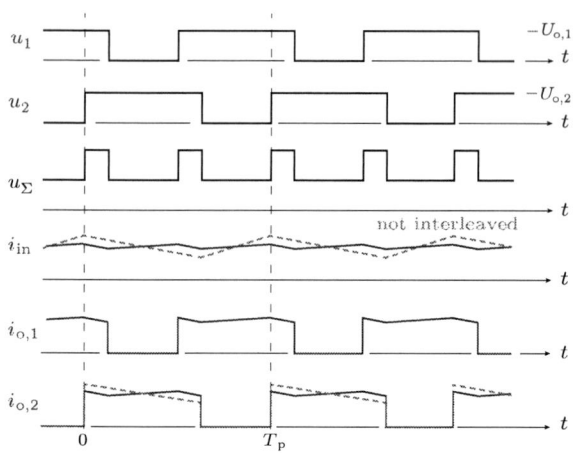

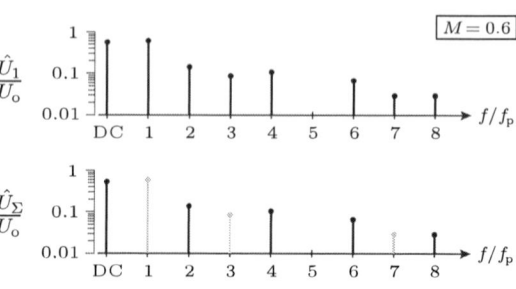

Fig. 4: Comparison of the harmonic spectrum of a single boost converter (top) with the harmonic spectrum of two series-interleaved boost converters (bottom) at a modulation index of $M = 0.6$. Harmonics shown in grey are canceled in the spectrum of the interleaved converters.

Fig. 3: Schematic waveforms of the operation of two series-interleaved DC-DC boost converters. Dashed lines denote the waveforms for operation without phase shift, i.e. non-interleaved operation.

A. Series-interleaved boost converters

In the series-interleaved boost converter (cf. **Fig. 1(a)**), with N_s series connected converter cells, the total output voltage U_o is shared among the output capacitors $C_{o,1}, C_{o,2}...C_{o,N_s}$ of the individual converter cells, such that $U_{o,i} = U_o/N_s$. The system is advantageously operated with a phase-shifted (interleaved) modulation scheme with a phase shift of $\delta = 2\pi/N_s$ as shown in **Fig. 3**. This results in an effective switching frequency of the total switch node voltage $u_\Sigma = \sum_{i=1}^{N} u_i$ of $f_{sw,eff} = N_s \cdot f_{sw}$ with f_{sw} being the switching frequency of one converter cell. Hence, the peak-to-peak current ripple Δi_{in} of the input current i_{in} can be calculated in dependency of the modulation index $M = U_{in}/U_o$ by introducing an effective modulation index $M_{eff} = (M \text{ modulo } 1/N_s)$ as

$$\Delta i_{in} = \frac{U_o}{L \cdot f_{sw}} \cdot M_{eff}\left(M_{eff} - \frac{1}{N_s}\right) . \tag{3}$$

The maximum value of Δi_{in} can be found for $M_{eff} = 0.5/N_s$ as

$$\Delta i_{in,max} = \frac{U_o}{4N_s^2 f_{sw}L} \tag{4}$$

yielding a $\propto 1/N_s^2$ decrease of the ripple amplitude in dependency of the number of converter cells.

Furthermore, the harmonic spectrum of the multi-cell converter voltage u_Σ can be derived from a single converter system by considering only the harmonics with orders that are multiples of the number of cells N_s, as visualized in **Fig. 4**.

Semiconductors: The losses caused by semiconductors can be divided into conduction and switching losses and their dependency on the number of converter cells is described in the following.

Conduction losses: For standard MOSFETs the fundamental relation between the blocking voltage U_{DS} and the lowest achievable on-state resistance $R_{DS,on}$ of a device is determined by the so-called silicon limit [14], which can be expressed for a given semiconductor area A_{Si} as

$$R_{DS,on,(1)} \cdot A_{Si} = 8.3 \cdot 10^{-9} \cdot U_{DS}^{2.5}[\Omega\text{cm}^2] = k_{Si} \cdot U_{DS}^{2.5}[\Omega\text{cm}^2] . \tag{5}$$

This relation holds true for MOSFETs where the $R_{DS,on}$ is mainly influenced by the resistance of the drift region of the device, i.e. only for devices with blocking voltages larger than around 50 V [15]. In the multi-cell system with series-interleaved boost converters, the required blocking voltage of $U_{DS} = U_o$ is divided to N_s series connected switches, i.e. each switch has to be capable of blocking a voltage U_{DS}/N_s, as shown in **Fig. 5(a)**. It can either be assumed that the chip area of each semiconductor is equal to the chip area of the full-rated switch (i.e. N_s devices with A_{Si}, (2) in **Fig. 5(a)**) or that the total chip area A_{Si} is equally distributed among the semiconductors (i.e. N_s devices with A_{Si}/N_s, (3) in **Fig. 5(a)**). For option (2) the total resistance can be calculated as

$$R_{DS,on,N,(2)} = N_s \cdot \frac{1}{A_{Si}} \cdot k_{Si} \cdot \left(\frac{U_{DS}}{N_s}\right)^{2.5}[\Omega\text{cm}^2]$$
$$= \frac{1}{\sqrt{N_s} \cdot N_s} \cdot \frac{1}{A_{Si}} \cdot k_{Si} \cdot U_{DS}^{2.5}[\Omega\text{cm}^2] . \tag{6}$$

whereas the total resistance of option (3) equals

$$R_{DS,on,N,(3)} = N_s \cdot \frac{N_s}{A_{Si}} \cdot k_{Si} \cdot \left(\frac{U_{DS}}{N_s}\right)^{2.5}[\Omega\text{cm}^2]$$
$$= \frac{1}{\sqrt{N_s}} \cdot \frac{1}{A_{Si}} \cdot k_{Si} \cdot U_{DS}^{2.5}[\Omega\text{cm}^2] . \tag{7}$$

Both equations can be interpreted as a shift of the silicon limit towards lower specific on-state resistances [16], which can be expressed as

$$R_{DS,on,N,(2)} = \frac{R_{DS,on,(1)}}{\sqrt{N_s} \cdot N_s} \text{ and } R_{DS,on,N,(3)} = \frac{R_{DS,on,(1)}}{\sqrt{N_s}} . \tag{8}$$

This relationship is visualized in **Fig. 5(b)**. The fundamental limits of wide bandgap materials such as GaN and SiC can be shifted in the same manner, since their on-state resistance for a given semiconductor area also increases more than quadratically with the break-down voltage (i.e. $R_{DS,on,GaN} \propto U_{DS}^{2.5}$ and $R_{DS,on,6H\text{-}SiC} \propto U_{DS}^{2.6}$) [17].

Switching losses (Option 1): The first option of calculating switching losses considers the overlapping of voltage and current across the transistor (NB: mainly applicable to circuits with IGBTs). The switching losses of N_s series connected switches can be compared to a single switch with full blocking voltage by assuming equal rates of du/dt and di/dt for all switches [13]. The switching losses of the single full-rated switch are

$$P_{Sw,loss,1} = E_{sw} \cdot f_{sw,1} = \frac{1}{2} \cdot (T_{r,i} + T_{f,u}) \cdot I_i \cdot V_{DC} \cdot f_{sw,1} . \tag{9}$$

978-1-4799-2706-7/14 $31.00 © 2014 IEEE

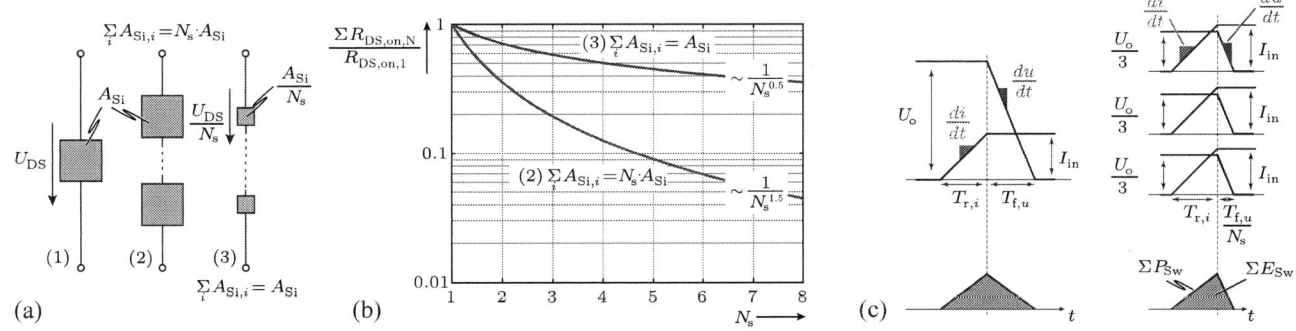

Fig. 5: Scaling laws of conduction and switching losses for series-interleaved boost converters: (a) replacing a single semiconductor with a blocking voltage of U_{DS} (1) with N_s semiconductors with blocking voltages of U_{DS}/N_s where each has either the same silicon area A_{Si} as the full-rated semiconductor (2) or where the total chip area is equal to A_{Si}; (b) improvement of the total $R_{DS,on}$ for scenarios (2) and (3) with the number of series converter cells; (c) reduction of switching losses in a series-interleaved multi-cell system with the same du/dt and di/dt as the single converter system [13].

As can be seen from (4), the switching frequency of a system with N_s cells can be scaled by $1/N_s^2$ while keeping the same current ripple amplitude for a certain inductance. Since the voltage across one switch is equal to U_o/N_s and the same du/dt is assumed, time $T_{f,u}$ decreases by a factor of $1/N_s$. Thus, the overall switching losses of N_s series connected switches become

$$
\begin{aligned}
P_{Sw,loss,Ns} &= N_s \cdot E_{sw,cell} \cdot \frac{f_{sw,1}}{N_s^2} \\
&= N_s \cdot \frac{1}{2} \cdot \left(T_{r,i} + \frac{T_{f,u}}{N_s} \right) \cdot I_{in} \cdot \frac{U_o}{N_s} \cdot \frac{f_{sw,1}}{N_s^2} \\
&= \frac{1}{2N_s^2} \cdot \left(T_{r,i} + \frac{T_{f,u}}{N_s} \right) \cdot I_{in} \cdot U_o \cdot f_{sw,1} \, . \quad (10)
\end{aligned}
$$

Neglecting $T_{f,u}/N$ (compared to $T_{r,i}$) in first step or assuming low values of $T_{r,i}$ (i.e. low values of the input current I_{in}), an improvement of the switching losses of

$$
P_{Sw,loss,Ns} \approx \frac{P_{Sw,loss,1}}{N_s^2} \cdots \frac{P_{Sw,loss,1}}{N_s^3} \quad (11)
$$

can be found.

Switching losses (Option 2): The second option of calculating switching losses considers the energy stored in the parasitic capacitances of the transistor and the diode of a half-bridge. The energy stored in a parasitic non-linear capacitance ($C_{T,OSS}$ or $C_{D,OSS}$) can be calculated by introducing an energy-equivalent capacitance

$$
C_{OSS,E,eq}(U_{DS}) = \frac{2 \cdot E_{OSS}(U_{DS})}{U_{DS}^2} = \frac{2 \int_0^{U_{DS}} v \cdot C_{OSS}(v) \mathrm{d}v}{U_{DS}^2} \, ; \quad (12)
$$

furthermore, a charge-equivalent capacitance

$$
C_{OSS,Q,eq}(U_{DS}) = \frac{Q_{OSS}(U_{DS})}{U_{DS}} = \frac{\int_0^{U_{DS}} C_{OSS}(v) \mathrm{d}v}{U_{DS}} \, . \quad (13)
$$

can be defined for the switch and the diode. This allows to calculate the energy $E_{on,1}$ lost per switching cycle in a system

with only one converter cell (i.e. $N_s = 1$) to be

$$
\begin{aligned}
E_{on,1} = &\frac{1}{2} \cdot C_{T,OSS,E,eq,1}(U_o) \cdot U_o^2 \\
&- \frac{1}{2} \cdot C_{D,OSS,E,eq,1}(U_o) \cdot U_o^2 \\
&+ C_{D,OSS,Q,eq,1}(U_o) \cdot U_o^2
\end{aligned} \quad (14)
$$

(cf. [18]) since the turn-off transition of the MOSFET can be regarded as loss-less (ZVS) and thus $E_{off,1} = 0$. The above equation can be simplified since the contribution of $C_{D,OSS,E,eq,1}$ is typically small compared to the other terms and thus negligible, such that

$$
E_{on,1} = \frac{1}{2} \cdot C_{eff,1}(U_o) \cdot U_o^2 \quad (15)
$$

by introducing an effective capacitance

$$
C_{eff,1}(U_o) = C_{T,OSS,E,eq,1}(U_o) + 2 \cdot C_{D,OSS,Q,eq,1}(U_o) \, . \quad (16)
$$

Hence, the switching losses for a single cell system are

$$
P_{Sw,loss,1} = E_{on,1} \cdot f_{sw,1} = \frac{1}{2} \cdot C_{eff,1}(U_o) \cdot U_o^2 \cdot f_{sw,1} \, . \quad (17)
$$

In a system with N_s converter cells the voltage across each switch is only U_o/N_s. Thus, in the same manner as before, the power dissipated in the switches of a multi-cell system can be calculated as

$$
\begin{aligned}
P_{Sw,loss,Ns} &= N_s \cdot E_{on,N} \cdot \frac{f_{sw,1}}{N_s^2} \\
&= \frac{1}{N_s} \cdot \frac{1}{2} \cdot C_{eff,Ns}(U_o/N_s) \left(\frac{U_o}{N_s} \right)^2 f_{sw,1} \\
&= \frac{1}{2N_s^3} \cdot C_{eff,Ns}(U_o/N_s) \cdot U_o^2 \cdot f_{sw,1} \quad (18)
\end{aligned}
$$

and/or

$$
P_{Sw,loss,Ns} = \frac{1}{N_s^3} \cdot \frac{C_{eff,Ns}(U_o/N_s)}{C_{eff,1}(U_o)} \cdot P_{Sw,loss,1} \, . \quad (19)
$$

Depending on how the values of the effective capacitances of the employed low voltage switches (i.e. at $U_{DS} = U_o/N_s$) compare to those of the higher voltage switch (i.e. at $U_{DS} = U_o$) a significant improvement of the switching losses can be achieved in the multi-cell system.

The 2014 International Power Electronics Conference

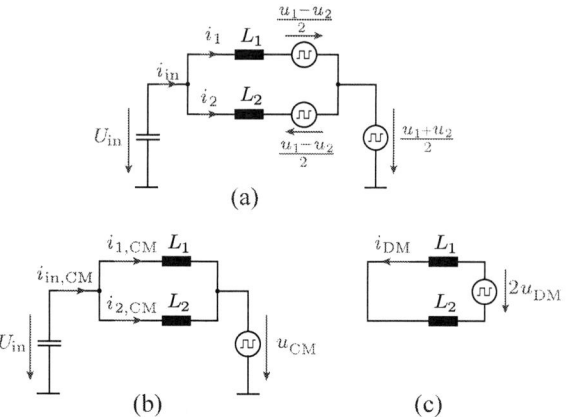

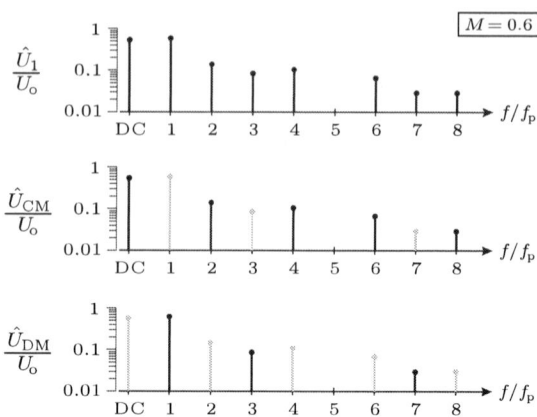

Fig. 6: Equivalent circuit diagram of two parallel-interleaved boost converters: (a) replacement of switch node voltages with rectangular voltage sources and splitting into (b) common mode and (d) differential mode equivalent circuits.

Fig. 8: Comparison of the harmonic spectrum of a single boost converter (top) to the common mode voltage spectrum (middle) and the differential mode voltage spectrum (bottom) of two parallel-interleaved boost converters.

B. Parallel-interleaved boost converter

In the parallel-interleaved boost converter (cf. **Fig. 1(b)**) with N_p parallel connected boost converters, the DC value of the input current I_{in} of the system is equally shared among the converter cells, i.e. average input current values of $I_1 = I_2 = ... = I_{Np} = I_{in}/N_p$ are occuring. The operation of the individual converter cells is phase shifted with the same phase shift $\delta = 2\pi/N_p$ as for the series interleaved system. This mode of operation allows to derive the equivalent circuit of **Fig. 6(a)** for a system with two parallel-interleaved boost converters. In this circuit the switches are replaced with rectangular voltage sources in order to accurately model the influence of the switch node voltages u_1 and u_2 on the input currents i_{in}, i_1 and i_2. The rectangular voltage sources can be divided into a common mode voltage component (cf. **Fig. 6(b)**)

$$u_{CM} = \frac{u_1 + u_2}{2} \tag{20}$$

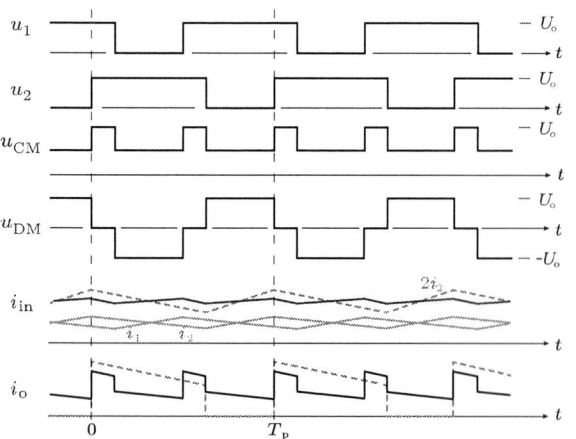

Fig. 7: Schematic waveforms of the operation of two parallel-interleaved phase shifted boost converters. Dashed lines denote the waveforms for operation without phase shift.

and a differential mode voltage component [19] (cf. **Fig. 6(c)**)

$$u_{DM} = \frac{u_1 - u_2}{2} . \tag{21}$$

Based on these equivalent circuits it can be concluded, that the input current of the system i_{in} is only influenced by the common mode voltage component u_{CM} (hence $i_{in} = i_{in,CM}$), whereas the current i_{DM}, driven by the differential mode voltage component u_{DM}, circulates only between the boost stages and does not contribute to the power transfer from the source to the load. The schematic waveform of those quantities (cf. **Fig. 7**) illustrates the similarity between the common mode voltage u_{CM} of the parallel-interleaved boost converters and the voltage u_Σ of the series-interleaved boost converters of **Fig. 3**. Both exhibit an effective switching frequency of $f_{sw,eff} = N_p \cdot f_{sw}$ or $f_{sw,eff} = N_s \cdot f_{sw}$, respectively. The harmonic spectrum of multiple parallel-interleaved boost converters can also be derived from the harmonic spectrum of a single boost converter. The spectrum of the common mode voltage u_{CM} is basically identical to the spectrum of the single converter but contains only harmonics with orders that are multiples of the cell number N_p, as shown in **Fig. 8**. The spectrum of the differential mode voltage u_{DM} contains the remaining harmonics, i.e. those harmonics of the original spectrum that are not present in the spectrum of u_{CM}. It is important to note, that the common mode voltage exhibits the same spectrum as the voltage u_Σ of the series-interleaved boost converters for the same number of converter cells.

Stored energy and converter volume: The peak-to-peak current ripple of the inductor current in any of the parallel interleaved boost stages can be calculated for a given modulation index M as

$$\Delta i_i = \frac{U_o \cdot M \cdot (1 - M)}{f_{sw} \cdot L} \tag{22}$$

whereas the DC value of the inductor current equals

$$I_i = \frac{I_{in}}{N_p} \tag{23}$$

under the assumption of equal power sharing between the parallel boost stages. The total peak energy stored in the inductors of a

The 2014 International Power Electronics Conference

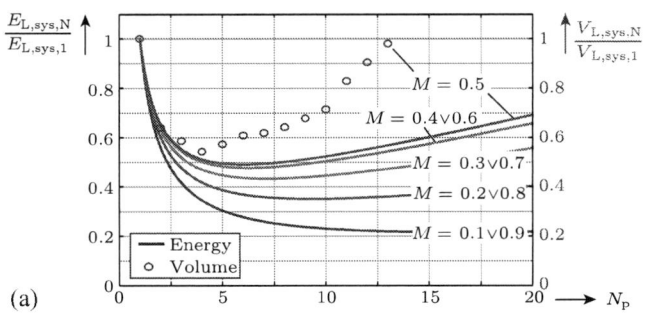

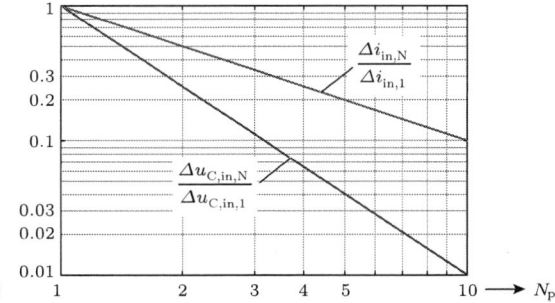

(a) (b)

Fig. 9: Scaling laws of parallel-interleaved boost converters. (a) Dependency of the normalized total peak energy stored in the inductors ($E_{\mathrm{L,sys,N}}/E_{\mathrm{L,sys,1}}$) with the number of converter cells and the modulation index. The total normalized inductor volume ($V_{\mathrm{L,sys,N}}/V_{\mathrm{L,sys,1}}$) of parallel-interleaved boost converters is shown for $M = 0.5$ for efficiency-constrained designs with N87 core material and litz wires (cf. [20]). (b) Reduction of the normalized system input current ripple and normalized input voltage ripple with the number of converter cells.

system with N_{p} parallel-interleaved boost converters is

$$E_{\mathrm{L,sys,N}} = N_{\mathrm{p}} \cdot \frac{1}{2} \cdot L \cdot \left(I_i + \frac{\Delta i_i}{2} \right)^2 . \quad (24)$$

The energy stored in a system with interleaved boost converters can now be compared to the energy in single boost converter system as depicted in **Fig. 9(a)** for $U_{\mathrm{o}} = 400\,\mathrm{V}$, $L = 200\,\mu\mathrm{H}$, $f_{\mathrm{sw}} = 100\,\mathrm{kHz}$ and $I_{\mathrm{in}} = 15\,\mathrm{A}$. It has to be pointed out that for this and the following considerations the inductance L of each converter cell is equal to the inductance of a single boost converter (i.e. resulting in an increased relative ripple of the inductor currents). The result shows that for each modulation index an optimum number of parallel-interleaved boost converters can be found where the total peak energy is minimized. The relation between stored energy in inductive components and their corresponding volume has been studied in [20] where it was shown that the inductor volume is largely proportional to the stored energy as long as low frequency losses dominate compared to high-frequency losses. The total inductor volume of parallel-interleaved boost converters with efficiency-constrained inductor designs is also shown in **Fig. 9(a)** for a modulation index of $M = 0.5$. Thus, for a fixed switching frequency, minimizing the total energy also minimizes the overall inductor volume until high frequency losses predominate.

The peak-to-peak current ripple of the input current can be derived for a given number of parallel boost converters N_{p} and an effective modulation index $M_{\mathrm{eff}} = (M \bmod 1/N_{\mathrm{p}})$ as

$$\Delta i_{\mathrm{in}} = \frac{U_{\mathrm{o}} \cdot N_{\mathrm{p}}}{f_{\mathrm{sw}} \cdot L} \cdot M_{\mathrm{eff}} \left(\frac{1}{N_{\mathrm{p}}} - M_{\mathrm{eff}} \right) . \quad (25)$$

with a maximum value at $M_{\mathrm{eff}} = 0.5/N_{\mathrm{p}}$ of

$$\Delta i_{\mathrm{in,max}} = \frac{U_{\mathrm{o}}}{4 f_{\mathrm{sw}} N_{\mathrm{p}} L} . \quad (26)$$

The ripple of the input current introduces a voltage ripple on the input capacitor C_{in} which can be calculated with the relation of $u = \int i \mathrm{d}t / C$ as

$$\Delta u_{\mathrm{C,in,max}} = \frac{\Delta i_{\mathrm{in,max}}}{4 N_{\mathrm{p}} f_{\mathrm{sw}} C_{\mathrm{in}}} = \frac{U_{\mathrm{o}}}{16 N_{\mathrm{p}}^2 f_{\mathrm{sw}}^2 C_{\mathrm{in}} L} \quad (27)$$

(assuming a constant current current drawn from the voltage source powering the converter system). These scaling laws are

shown in **Fig. 9(b)** normalized to the values of a single boost converter system.

Switching losses (Option 1): The switching losses can be calculated by considering the overlap of voltages and currents of the switches of the multi-cell converter, as shown in **Fig. 10**. Based on the result of (26) the switching frequency of each stage can be reduced by a factor of $1/N_{\mathrm{p}}$ compared to the switching frequency of a single boost converter to obtain the same peak-to-peak amplitude of the input current. By assuming the same rates of $\mathrm{d}u/\mathrm{d}t$ and $\mathrm{d}i/\mathrm{d}t$ as in the single converter system the switching losses can be found to be

$$\begin{aligned} P_{\mathrm{Sw,loss,Np}} &= N_{\mathrm{p}} \cdot E_{\mathrm{sw,cell}} \cdot \frac{f_{\mathrm{sw,1}}}{N_{\mathrm{p}}} \\ &= N_{\mathrm{p}} \cdot \frac{1}{2} \cdot \left(\frac{T_{\mathrm{r},i}}{N_{\mathrm{p}}} + T_{\mathrm{f},u} \right) \cdot \frac{I_{\mathrm{in}}}{N_{\mathrm{p}}} \cdot U_{\mathrm{o}} \cdot \frac{f_{\mathrm{sw,1}}}{N_{\mathrm{p}}} \\ &= \frac{1}{2 N_{\mathrm{p}}} \cdot \left(\frac{T_{\mathrm{r},i}}{N_{\mathrm{p}}} + T_{\mathrm{f},u} \right) \cdot I_{\mathrm{in}} \cdot U_{\mathrm{o}} \cdot f_{\mathrm{sw,1}} . \quad (28) \end{aligned}$$

An upper boundary of the switching loss reduction can be found for low output voltage values, i.e. negligible times $T_{\mathrm{f},u}$, thus the switching loss reduction lies in the range of

$$P_{\mathrm{Sw,loss,Np}} = \frac{P_{\mathrm{Sw,loss,1}}}{N_{\mathrm{p}}} \ldots \frac{P_{\mathrm{Sw,loss,1}}}{N_{\mathrm{p}}^2} \quad (29)$$

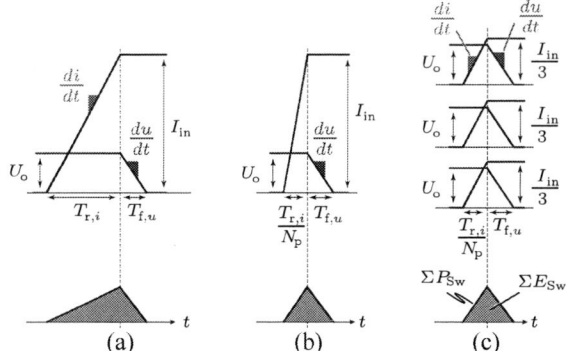

(a) (b) (c)

Fig. 10: Comparison of switching losses between a single boost converter (a) and parallel-interleaved boost convertes (c) for the same rates of $\mathrm{d}u/\mathrm{d}t$ and $\mathrm{d}i/\mathrm{d}t$. The single boost converter can only reach the same level of switching losses as the parallel-interleaved converter, if the rate of $\mathrm{d}i/\mathrm{d}t$ is increased by a factor of N_{p} (b).

978-1-4799-2706-7/14 $31.00 © 2014 IEEE

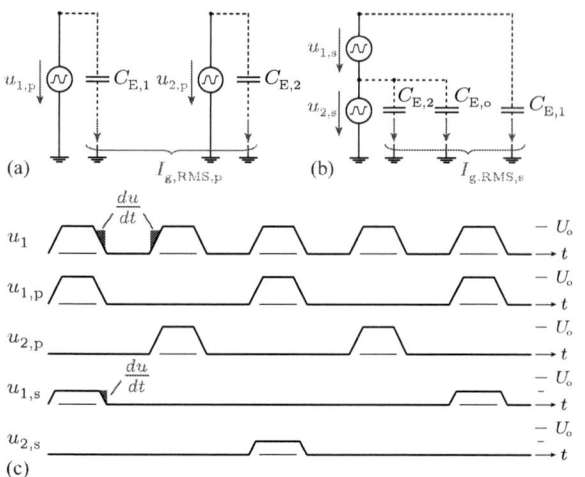

Fig. 11: Equivalent circuit model for ground currents of (a) the parallel-interleaved boost converter and (b) the series interleaved boost converter. The waveforms of the switch-node voltages of the parallel-interleaved system ($u_{1,p}$, $u_{2,p}$) and series-interleaved system ($u_{1,s}$, $u_{2,s}$) are compared to those of a single boost converter (u_1) in (c) under the assumption of same du/dt in all systems.

Switching losses (Option 2): The calculation of the switching losses based on the energy stored in the parasitic transistor and diode capacitances yields for the system with parallel-interleaved boost converters

$$
\begin{aligned}
P_{\text{Sw,loss,Np}} &= N_p \cdot E_{\text{on,Np}} \cdot \frac{f_{\text{sw,1}}}{N_p} \\
&= \frac{1}{2} \cdot C_{\text{eff,Np}}(U_o) \cdot U_o^2 \cdot f_{\text{sw,1}}
\end{aligned}
\tag{30}
$$

and/or

$$
P_{\text{Sw,loss,Np}} = \frac{C_{\text{eff,Np}}(U_o)}{C_{\text{eff,1}}(U_o)} \cdot P_{\text{Sw,loss,1}} .
\tag{31}
$$

Since the switches and diodes in the parallel-interleaved boost converters have to be rated for the same voltage but for a lower current than the switch in the single boost converter, the employed Silicon area can be smaller and thus also the parasitic capacitances will be smaller and the capacitive switching losses will decrease.

V. GROUND CAPACITANCES

The parasitic ground capacitances at the switch-nodes of the converters are a source for unwanted common mode currents which necessitate the application of common-mode filters, e.g. in PFC boost rectifiers, that substantially increase the volume of the converter. The voltage of the switch-node can be modeled with a trapezoidal wave voltage source that exhibits a certain du/dt, as shown in **Fig. 11(a)** for the parallel-interleaved boost converter and in **Fig. 11(b)** for the series-interleaved boost converter with parasitic capacitances as defined in **Fig. 1**. A comparison of the voltage waveforms between the single boost converter (u_1), the parallel-interleaved ($u_{1,p}$, $u_{2,p}$) and the series-interleaved boost converter ($u_{1,s}$, $u_{2,s}$) is depicted in **Fig. 11(c)**. Based on the fundamental relationship of $I_C = C \cdot du/dt$ the RMS value of the ground current of a single boost converter system through capacitor C_E can be calculated as

$$
I_{\text{g,RMS,1}} = \sqrt{2k} \cdot C_E \cdot \frac{du}{dt} \text{ with } k = \frac{U_o \cdot f_{\text{sw}}}{\frac{du}{dt}} .
\tag{32}
$$

The switching frequency of the parallel-interleaved boost converters can be reduced by $1/2$ (for $N_p = 2$) for an equal current ripple amplitude (cf. (26)) leading to

$$
k_p = \frac{U_o \cdot f_{\text{sw}}}{\frac{du}{dt} \cdot 2} = \frac{k}{2} .
\tag{33}
$$

Under the assumption of $C_{E,1} = C_{E,2} = C_E$ the RMS value of the ground current can be found as

$$
I_{\text{g,RMS,p}} = \sqrt{2} \cdot \sqrt{2k_p} \cdot C_E \cdot \frac{du}{dt} = \sqrt{2k} \cdot C_E \cdot \frac{du}{dt} .
\tag{34}
$$

For the series-interleaved system the switching frequency can be reduced by $1/4$ (for $N_s = 2$) for an equal current ripple amplitude (cf. (4)) and in combination with an voltage amplitude of only $U_o/2$ for each equivalent source results

$$
k_s = \frac{\frac{U}{2} \cdot f_{\text{sw}}}{\frac{du}{dt} \cdot 4} = \frac{k}{8} .
\tag{35}
$$

By super-position and again assuming $C_{E,1} = C_{E,2} = C_E$ the equation for the ground current yields

$$
\begin{aligned}
I_{\text{g,RMS,s}} &= \sqrt{2k_s \cdot \left((C_{E,o} + 2C_E)\frac{du}{dt} \right)^2 + 2k_s \cdot \left(C_E \frac{du}{dt} \right)^2} \\
&= \frac{1}{2} \cdot \sqrt{k} \cdot \sqrt{(C_{E,o} + 2 \cdot C_E)^2 + C_E^2} \cdot \frac{du}{dt} .
\end{aligned}
\tag{36}
$$

By relating the ground currents of multi-cell systems to the single boost converter it can be found that

$$
\frac{I_{\text{g,RMS,p}}}{I_{\text{g,RMS,1}}} = 1
\tag{37}
$$

$$
\frac{I_{\text{g,RMS,s}}}{I_{\text{g,RMS,1}}} = \frac{1}{2\sqrt{2}} \sqrt{ \left(2 + \frac{C_{E,o}}{C_E} \right)^2 + 1 } .
\tag{38}
$$

This means that the RMS ground current of the series-interleaved system can only reach the same value as the parallel-interleaved system (and thus as the single boost converter) if $C_{E,o} = (\sqrt{7} - 2) \cdot C_E \approx 0.64 \cdot C_E$. This is rather unlikely since capacitance $C_{E,o}$ denotes the capacitance of the entire "upper" converter cell to ground, which is amongst others defined by the physical size of the entire converter cell. Thus, the series-interleaved boost converter system will potentially exhibit larger RMS ground currents.

The RMS value of the ground currents does not allow to extract any information about the frequency and the amplitude of the ground currents. Thus, in order to draw any conclusions for the design and volume of a common mode filter a more in depth analysis of the ground currents becomes indispensable, including a Fourier analysis to obtain the current spectrum for the different converter concepts. This will be investigated in a future publication in a more detailed and comprehensive way.

VI. CONCLUSION

Based on the investigation of the scaling laws it can be summarized that both the series-interleaved as well as the parallel-interleaved approach offer considerable advantages in terms of conduction losses, switching losses and/or harmonic distortion compared to a single converter system. The parallel-interleaved concept also shows the same RMS values of the parasitic ground currents as the single converter. The series-interleaved converter,

978-1-4799-2706-7/14 $31.00 © 2014 IEEE

however, exhibits most likely larger ground currents (depending on $C_{E,o}$) since the converter cells are stacked and thus their potential referenced to ground also depends on the switching states of the lower converter cells in the converter stack.

The analysis of voltage and/or current balancing in multi-cell systems reveals that a common-duty cycle operation of all converter cells is applicable to ISOP and IPOS converters due to their self-balancing capabilities. For ISOS and IPOP converters the input voltage or input current sharing, respectively, needs to be ensured by an additional controller.

In conclusion, this paper has demonstrated that splitting a power electronic converter into multiple converter cells which share the system power equally can increase the overall system efficiency and power density. The benefits of multi-cell systems can be leveraged in many dimension e.g. to improve the harmonic spectrum of the converters and/or to obtain a reduction of the switching and conduction losses, depending on the choice of switching frequency of the individual cells compared to the switching frequency of the single converter. Since standard components with lower voltage and/or current ratings can be employed, the costs for a multi-cell system also decreases.

APPENDIX A
DERIVATION OF GENERAL SCALING LAWS

The efficiency of a power electronic system with input power P_1, output power P and losses P_L is defined as

$$\eta = \frac{P}{P_1} = \frac{P}{P + P_L} \Rightarrow P_L = \frac{1 - \eta}{\eta} P . \tag{39}$$

In the most simplified approach the power electronic system is considered to be a cube with surface A which scales with the volume V of the cube by $A = 6V^{2/3}$. The heat dissipation capability p_L of a system can be defined by relating the losses of a system to its surface area,

$$p_L = \frac{P_L}{A} = \frac{1 - \eta}{\eta} \cdot \frac{P}{6V^{2/3}} . \tag{40}$$

This allows to derive fundamental scaling laws since any scaled system needs to have the same value of p_L as the reference system (denoted by subscript "$_0$"),

$$\frac{1 - \eta}{\eta} \cdot \frac{P}{6V^{2/3}} = \frac{1 - \eta_0}{\eta_0} \cdot \frac{P_0}{6V_0^{2/3}} . \tag{41}$$

Rearranging this equation yields

$$\frac{P}{P_0} = \frac{(1 - \eta_0)}{(1 - \eta)} \cdot \frac{\eta}{\eta_0} \cdot \left(\frac{V}{V_0}\right)^{2/3} . \tag{42}$$

By assuming a constant efficiency, i.e. $\eta = \eta_0$, the relation

$$\frac{V}{V_0} = \left(\frac{P}{P_0}\right)^{3/2} \tag{43}$$

can be found, whereas the condition of a constant power density, i.e. $\rho = P/V = \rho_0 = P_0/V_0$, leads to

$$\eta = \frac{\eta_0 \left(\frac{P}{P_0}\right)^{1/3}}{1 + \eta_0 \left(\left(\frac{P}{P_0}\right)^{1/3} - 1\right)} . \tag{44}$$

REFERENCES

[1] S. Waffler and J. Kolar, "Comparative Evaluation of Soft-Switching Concepts for Bi-Directional Buck+Boost DC-DC Converters," in *Proc. of the International Power Electronics Conference (IPEC)*, pp. 1856–1865, 2010.

[2] IEC TR 62380, "Reliability data handbook - Universal model for reliability prediction of electronics components, PCBs and equipment," IEC, Tech. Rep., 2004.

[3] R. Giri, V. Choudhary, R. Ayyanar, and N. Mohan, "Common-Duty-Ratio Control of Input-Series Connected Modular DC-DC Converters with Active Input Voltage and Load-Current Sharing," *IEEE Trans. Ind. Appl.*, vol. 42, no. 4, pp. 1101–1111, 2006.

[4] W. van der Merwe and T. Mouton, "Natural Balancing of the Two-Cell Back-to-Back Multilevel Converter with specific Application to the Solid-State Transformer Concept," in *Proc. of the 4th IEEE Conf. on Industrial Electronics and Applications (ICIEA)*, pp. 2955–2960, 2009.

[5] H. Mouton, J. Enslin, and H. Akagi, "Natural Balancing of Series-Stacked Power Quality Conditioners," *IEEE Trans. Power Electron.*, vol. 18, no. 1, pp. 198–207, 2003.

[6] J. van der Merwe and H. du T Mouton, "An Investigation of the Natural Balancing Mechanisms of Modular Input-Series-Output-Series DC-DC Converters," in *Proc. of the IEEE Energy Conversion Congress and Exposition (ECCE)*, pp. 817–822, 2010.

[7] W. Chen, X. Ruan, H. Yan, and C. Tse, "DC/DC Conversion Systems Consisting of Multiple Converter Modules: Stability, Control, and Experimental Verifications," *IEEE Trans. Power Electron.*, vol. 24, no. 6, pp. 1463–1474, 2009.

[8] R. Ayyanar, R. Giri, and N. Mohan, "Active Input-Voltage and Load-Current Sharing in Input-Series and Output-Parallel Connected Modular DC-DC Converters Using Dynamic Input-Voltage Reference Scheme," *IEEE Trans. Power Electron.*, vol. 19, no. 6, pp. 1462–1473, 2004.

[9] I. Batarseh, K. Siri, and H. Lee, "Investigation of the Output Droop Characteristics of Parallel-Connnected DC-DC Converters," in *Power Electronics Specialists Conference, PESC '94 Record., 25th Annual IEEE*, pp. 1342–1351, 1994.

[10] S. Luo, Z. Ye, R.-L. Lin, and F. Lee, "A Classification and Evaluation of Paralleling Methods for Power Supply Modules," in *Proc. of the 30th IEEE Power Electronics Specialists Conference (PESC)*, vol. 2, pp. 901–908, 1999.

[11] M. Li, C. Tse, H. Iu, and X. Ma, "Unified Equivalent Modeling for Stability Analysis of Parallel-Connected DC/DC Converters," *IEEE Trans. Circuits Syst. II*, vol. 57, no. 11, pp. 898–902, 2010.

[12] Y. Huang, C. Tse, and X. Ruan, "General Control Considerations for Input-Series Connected DC/DC Converters," *IEEE Trans. Circuits Syst. I*, vol. 56, no. 6, pp. 1286–1296, 2009.

[13] J. Huber and J. Kolar, "Optimum Number of Cascaded Cells for High-Power Medium-Voltage Multilevel Converters," in *Energy Conversion Congress and Exposition (ECCE), 2013 IEEE*, pp. 359–366, 2013.

[14] F. Udrea, "Future and Expected Applications of Si Power Devices," in *ECPE Workshop Future Trends for Power Semiconductors, Zurich*, 2012.

[15] J. Lutz, H. Schlangenotto, U. Scheuermann, and R. W. De Doncker, *Semiconductor Power Devices: Physics, Characteristics, Reliability*. Springer, 2011.

[16] J. Kolar, "What are the Big Challenges in Power Electronics," in *Proc. of 8th International Conference on Integrated Power Electronic Systems (CIPS) Keynote presentation*, 2014.

[17] M. Henini and M. Razeghi, *Optoelectronic Devices: III Nitrides*. Elsevier Science & Technology, 2004.

[18] F. Krismer, "Modeling and Optimization of Bidirectional Dual Active Bridge DC-DC Converter Topologies," Ph.D. dissertation, Power Electronic Systems Laboratory, ETH Zurich, 2008.

[19] S. Utz and J. Pforr, "Current-Balancing Controller Requirements of Automotive Multi-Phase Converters with Coupled Inductors," in *Proc. of the IEEE Energy Conversion Congress and Exposition (ECCE)*, pp. 372–379, 2012.

[20] R. Burkart, H. Uemura, and J. W. Kolar, "Optimal Inductor Design for 3-Phase PWM Converters Considering Different Magnetic Materials and a Wide Switching Frequency Range," in *Proc. of the International Power Electronics Conference (IPEC)*, 2014.

Hybrid Modulated Universal Soft-switching Current-fed DC/DC Converter for Wide Voltage Regulation for PV/Fuel cells/Battery Applications

Radha Sree Krishna Moorthy
Electrical and Computer Engineering
National University of Singapore
Singapore 117583
a0107273@nus.edu.sg

Akshay Kumar Rathore, *Senior Member, IEEE*
Electrical and Computer Engineering
National University of Singapore
Singapore 117583
eleakr@nus.edu.sg

Abstract—A soft-switching extended secondary universal current-fed converter has been proposed, analyzed, and designed. The proposed converter accommodates wide source voltage range to cover several sources to interface through extended secondary circuit and hybrid modulation. Proposed fixed frequency hybrid modulation and design achieves soft-switching of semiconductor devices under all operating conditions. It is therefore suitable to be universally adopted for solar photovoltaic (PV), fuel cells, and battery applications. Current-fed technology is suitable for such sources (low voltage high current). Experimental results are illustrated to verify the claims and to show performance of the converter over wide input voltage range of 20 to 60 V with load variation.

Keywords— *Hybrid modulation, Current-fed converter, Soft-switching, Voltage regulation.*

I. INTRODUCTION

The concept of localized distributed generation utilizing non-conventional energy sources such as solar panels, fuel-cell, etc. is gaining significant attention owing to environmental concerns and growing energy demand. Storage is still desired for grid stability and to overcome intermittency of the renewable sources having discontinuous and unsecured output. Front end dc/dc converters are necessary to boost the source voltage with required isolation [1-2]. It has been a challenge to design a universal converter that is promising for interfacing different sources with different operating voltage range. Also, maintaining soft-switching of semiconductor devices at high-frequency (HF) while achieving high efficiency is another challenge. HF operation offers obvious merits of compact and light weight system. Voltage-fed converters have several issues like rectifier diode ringing, duty cycle loss, etc. [3] and are not suited for high gain and high current applications. Soft-switching can be achieved using additional components [4] but at the cost of increasing the topology's complexity with limited soft-switching range [5].

Alternatively, current-fed converters are preferred for such high gain applications [6-9] because current-fed topologies overcome above mentioned issues and allows low input current ripple. However, current-fed topologies suffer from high voltage spike across the semiconductor devices at turn-off and needs snubber circuit [10]. Active-clamp snubs the spike while assisting in zero voltage switching (ZVS) [11-14]. Magnetizing inductance assisted design has been proposed in [15-16] to maintain ZVS with wide variations in load and input voltage.

In this paper, a modified secondary universal current-fed converter is proposed that operates with ZVS and zero current switching (ZCS) and achieves output voltage regulation with wide input voltage and load variation. The converter is claimed to be universal as it can accommodate input voltage variations of 1:3 compared to conventional converters. Also, the range can be extended by slight modifications in secondary circuit. Thus it is seen as a potential candidate for interfacing solar panels (20–42V), fuel-cells (22–41V) and batteries (24/36/48/60 V) with load. To achieve load voltage regulation, hybrid fixed frequency modulation is proposed.

Hybrid fixed frequency modulation consists of (a) primary circuit duty cycle modulation (fixed secondary circuit duty cycle) that is active for PV, fuel cells, and batteries voltage below 42 V and (b) secondary circuit duty cycle modulation that is active for input voltage above 42V, i.e., 48/60 V batteries or two series connected solar panels. For input voltage of 40 to 60 V, duty cycle of primary circuit devices is fixed at 55% while the duty cycle of the secondary devices is varied to regulate the load voltage, i.e., full or partial utilization of secondary.

The objectives and layout of the paper are as follows. Steady-state operation of the proposed converter has been explained in Section II. Converter design has been illustrated in Section III. Simulation and experimental results are demonstrated in Section IV to support the claims and verify the proposed operation and design.

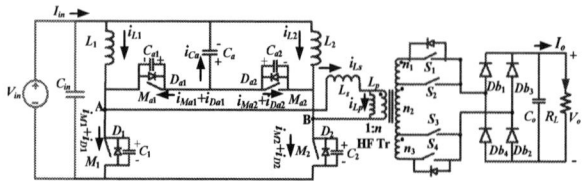

Fig. 1. Proposed current-fed universal dc/dc converter.

II. STEADY STATE OPERATION AND ANALYSIS

The steady-state waveforms of the proposed converter are shown in Fig. 2 for V_{in} = 50 V. The following assumptions are made to study the operation and analysis: a) Boost inductors L_1 and L_2 are large to maintain constant current through them, b) Clamp capacitor C_a is large to maintain constant voltage across it, c) All the components are ideal, d) Series inductance L_s and parallel inductance L_p includes the leakage and magnetizing inductance of the transformer, respectively.

Gating signals of the main switches M_1 and M_2 are phase shifted by 180° with an overlap that varies with duty cycle. Auxiliary switches M_{a1} and M_{a2} are operated complementary to respective main switches. The duty cycle of the main devices is always kept above 50% to prevent under-utilization of the main devices. The duty cycle of the secondary devices $S_1 \sim S_4$ is varied with input voltage. The secondary switches are operated with a duty cycle d_1 for S_1, S_4 and d_2 for S_2, S_3. Gating signals of devices S_1, S_4 are phase-shifted by 180°. The same is true for pair S_2, S_3. The equivalent circuits for different intervals of operation in a half HF cycle are shown in Fig. 3. For the other half HF cycle the intervals repeat in the same sequence.

A. Interval 1 (Fig. 3(a): $t_0 < t < t_1$): In this interval, the switch M_1 and body diode of switch M_2 are conducting. Switch M_2 can be gated for ZVS turn-on. Energy is stored in the boost inductors L_1 and L_2. Output capacitor C_o transfers power to the load. Rectifier diodes are reverse-biased and blocking the voltage $V_o/2$. Constant current $-I_{Lp,peak}$ flows through the series and the parallel inductors and is given by

$$i_{Lp,peak} = \frac{V_{in}}{2f_s(L_s + L_p)} \tag{1}$$

Voltage across the clamp capacitor V_{Ca} is given by

$$V_{Ca} = \frac{DV_{in}}{1-D} \tag{2}$$

Voltage across the auxiliary switches is given by

$$V_{Ma1} = V_{Ma2} = \frac{V_{in}}{1-D} \tag{3}$$

Switch currents are $i_{M1} = I_{in}/2 + I_{Lp,peak}$ and $i_{D2} = I_{in}/2 - I_{Lp,peak}$.

B. Interval 2 (Fig. 3(b): $t_1 < t < t_2$): At $t = t_1$, M_1 is turned-off and secondary switch S_1 is turned-on. Boost inductor current gets divided in proportion to the capacitances and starts charging and discharging C_1 and C_{a1}, respectively. Rectifier diodes remain reverse biased. Final values are: $v_{M1}(t_2) = V_o/(n_1+n_2+n_3)$ and $v_{Ma1} = [V_{in} +V_{Ca} - V_{M1}]$ respectively.

C. Interval 3 (Fig. 3(c): $t_2 < t < t_3$): The charging and discharging of snubber capacitors still continues till the main switch voltage increases to $V_{in} + V_{Ca}$. Positive voltage $[V_{M1}-V_o/(n_1+n_2+n_3)]$ appears across L_s and the series inductor current starts increasing in the positive direction. The current through the series and the parallel inductor is given by

$$i_{Ls} = -I_{Lp,peak} + \frac{(V_{M1}-V_o/(n_1+n_2+n_3))}{L_s}(t-t_2) \tag{4}$$

$$i_{Lp} = -I_{Lp,peak} + \frac{V_o/(n_1+n_2+n_3)}{L_p}(t-t_2) \tag{5}$$

The current through M_2 given by

$$i_{M2} = \frac{I_{in}}{2} + i_{Ls}(t-t_2) \tag{6}$$

At the end of the interval, C_{a1} discharges completely i.e., $v_{Ca1}(t_3)=0$ and C_1 charges to $v_{C1}(t_3) = V_{in} + V_{Ca} = V_{in}/(1-D)$.

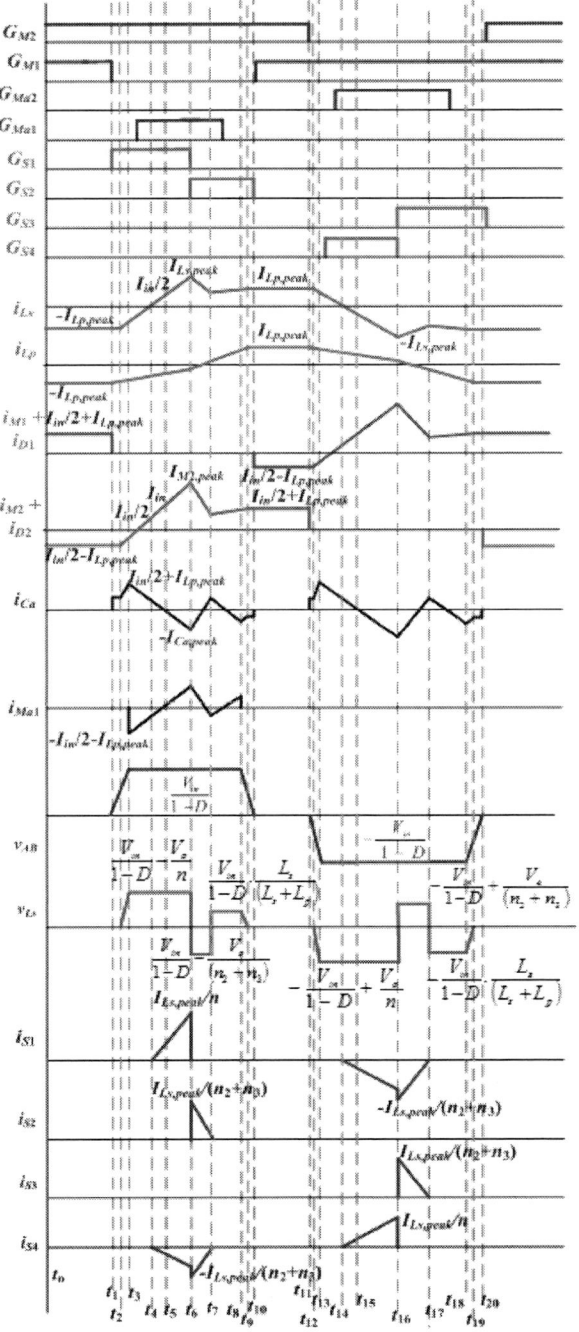

Fig. 2. Steady-state operating waveforms of the proposed converter.

D. Interval 4 (Fig. 3(d): $t_3 < t < t_4$): In this interval, the body diode D_{a1} of auxiliary switch M_{a1} starts conducting and M_{a1} can now be gated for ZVS turn-on. Switch M_2 begins to conduct with ZVS when the current $I_{in}/2+i_{Ls}$ becomes positive. The series inductor current i_{Ls} increases with a slope of $[V_{in}+V_{Ca}-V_o/(n_1+n_2+n_3)]$. The currents through the various components are given by

$$i_{Ls} = i_{Ls}(t_3) + \frac{(V_{in}+V_{Ca}-V_o/(n_1+n_2+n_3))}{L_s}(t-t_3) \qquad (7)$$

$$i_{Lp} = i_{Lp}(t_3) + \frac{V_o/(n_1+n_2+n_3)}{L_p}(t-t_3) \qquad (8)$$

$$i_{M2} = i_{M2}(t_3) + \frac{(V_{in}+V_{Ca}-V_o/(n_1+n_2+n_3))}{L_s}(t-t_3) \qquad (9)$$

$$i_{Ca} = I_{Ca,peak} - \frac{(V_{in}+V_{Ca}-V_o/(n_1+n_2+n_3))}{L_s}(t-t_3) \qquad (10)$$

Final values: $i_{Ca}(t_4) = I_{in}/2$ and $i_{Ls}(t_4) = 0$.

E. Interval 5 (Fig. 3(e): $t_4 < t < t_5$): In this interval, current i_{Ls} increases above zero with the same slope and switch M_2 is turned-on with ZVS. Current i_{Lp} also increases with the same slope while the clamp capacitor current continues to decrease.

$$i_{Ca} = \frac{I_{in}}{2} - i_{Ls} \qquad (11)$$

$$i_{M2} = i_{M2}(t_4) + i_{Ls} \qquad (12)$$

Final values: $i_{Ca}(t_5)=0$ and $i_{Ls}(t_5)= I_{in}/2$ and $i_{M2}(t_5)= I_{in}$.

F. Interval 6 (Fig. 3(f): $t_5 < t < t_6$): Auxiliary switch M_{a1} starts conducting with ZVS. The clamp capacitor current decreases linearly. Current i_{Ls} increases beyond $I_{in}/2$ with the same slope and i_{Lp} continues to increase with the same slope. The current through M_2 and clamp capacitor C_a are given by

$$i_{M2} = i_{M2}(t_4) + i_{Lp}(t-t_5) \qquad (13)$$

$$i_{Ca} = \frac{I_{in}}{2} - i_{Lp}(t-t_5) \qquad (14)$$

At the end of interval, i_{Ls} and i_{M2} reaches peak values $I_{Ls,peak}$ and $I_{M2,peak}$ respectively while i_{Ca} reaches its negative peak.

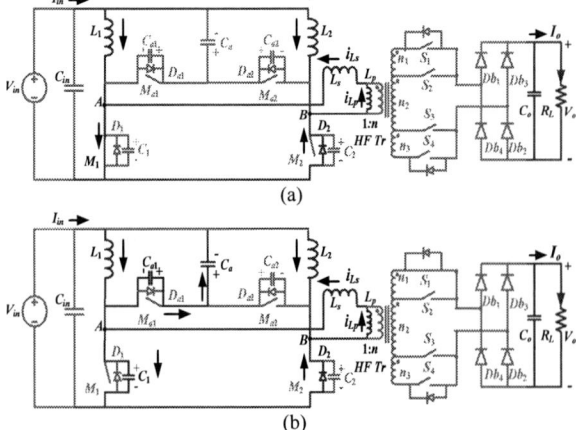

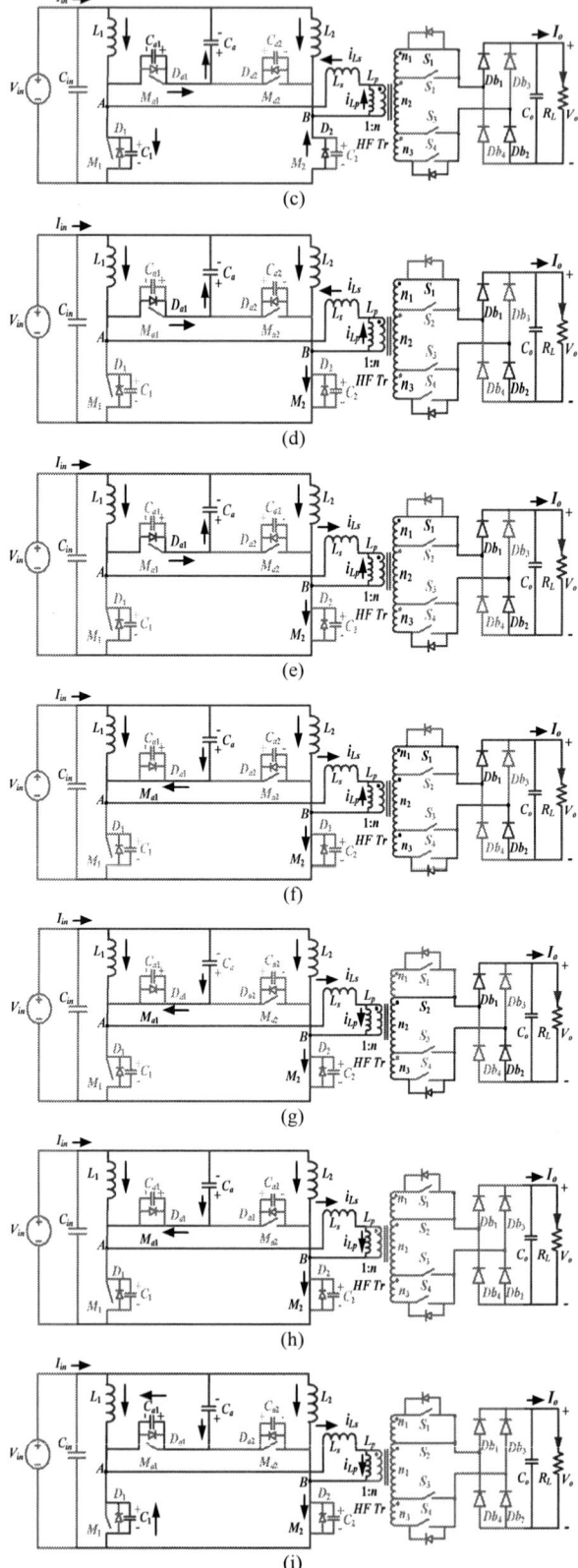

Fig. 3. Equivalent circuits representing the different intervals of operation of the proposed converter.

G. Interval 7 (Fig. 3(g): $t_6 < t < t_7$): At $t = t_6$, the switch S_1 is turned-off and S_2 is turned-on. Negative voltage $[(V_{in}+V_{Ca}-V_o/(n_2+n_3))/L_s]$ appears across L_s and i_{Ls} starts decreasing. The slope of current i_{Lp} changes to $(V_o/(n_2+n_3))/L_p$. Power is transferred to load through secondary switch S_2, body diode of switch S_4 and rectifier diodes D_{b1} and D_{b2}. The current through L_s, L_p, and M_2 is given by

$$i_{Ls} = i_{Ls}(t_6) + \left(\frac{V_{in}+V_{Ca}-V_o/(n_2+n_3)}{L_s}\right)(t-t_6) \tag{15}$$

$$i_{Lp} = i_{Lp}(t_6) + \left(\frac{V_o/(n_2+n_3)}{L_p}\right)(t-t_6) \tag{16}$$

$$i_{M2} = i_{M2}(t_6) + i_{Ls} \tag{17}$$

At the end of this interval, $i_{Ls} = i_{Lp}$ and no energy is transferred to the load.

H. Interval 8 (Fig. 3(h): $t_7 < t < t_8$): During this interval, the series and the parallel inductor currents are given by

$$i_{Ls} = i_{Lp} = i_{Lp}(t_7) + \frac{V_{Lp}}{L_p}(t-t_7) \tag{18}$$

The main switch current i_{M2} is given as

$$i_{M2} = \frac{I_{in}}{2} + i_{Ls}(t-t_7) \tag{19}$$

At the end of the interval, switch M_{a1} is turned-off.

I. Interval 9 (Fig. 3(i): $t_8 < t < t_9$): At $t=t_7$, the auxiliary switch M_{a1} is turned-off. The series inductor L_s resonates with the snubber capacitors C_1 and C_{a1} during this short interval and discharges and charges C_1 and C_{a1} respectively. The resonant frequency is given by

$$f_r = \frac{1}{2\pi\sqrt{L_s(C_1+C_{a1})}} \tag{20}$$

The leakage inductor current is given by

$$i_{Ls} = i_{Ls}(t_8)\cdot\cos(\omega_r(t-t_7)) \tag{21}$$

The main switch current i_{M2} is given by

$$i_{M2} = i_{M2}(t_8)\cdot\cos(\omega_r(t-t_7)) \tag{22}$$

The voltage across switches M_{a1} and M_1 are given by

$$v_{Ma1} = i_{Ma1}(t_8)\cdot\sqrt{\frac{L_s}{(C_1+C_{a1})}}\cdot\sin(\omega_r(t-t_7)) \tag{23}$$

$$v_{M1} = \left(\frac{V_{in}}{1-D}\right) - v_{Ma1} \tag{24}$$

J. Interval 10 (Fig. 3(i): $t_9 < t < t_{10}$): The series inductor current charges and discharges C_{a1} and C_1 resonantly. This period is very small and the resonant frequency is given by (20). At the end of the interval current flowing through the series and the parallel inductor becomes $I_{Lp,peak}$ and the secondary switch S_2 is turned-off with ZCS. Final values: $v_{M1}(t_9)=0$; $v_{Ma1}(t_9) = V_{in}/(1-D)$; $i_{Ls}(t_{10}) = i_{Lp}(t_{10}) = I_{Lp,peak}$.

III. CONVERTER DESIGN

The converter design procedure is illustrated and explained with a design example in this Section. The specifications are provided in Table I.

A. Maximum duty ratio of main switches: Based on switch voltage rating, D_{max} is computed for minimum input voltage $V_{in}=20$ V and full load using

$$D_{max} = 1-\left(\frac{V_{in}}{V_{SW}}\right) \tag{25}$$

For $V_{SW,max} = 100$ V, $D_{max} = 0.8$.

TABLE I SPECIFICATIONS OF THE CONVERTER.

Input voltage V_{in}	20 to 60 V
Output voltage V_o	180 V
Output power P_o	200 W
Switching frequency f_s	100 kHz

B. Static voltage gain: The output voltage for the two different operating interval i.e., $V_{in} = 20 - 40$ V and $41 - 60$ V can be given by (26) and (27), respectively.

$$V_o = \frac{V_{in}}{(1-D)}\cdot(n_1+n_2+n_3) \tag{26}$$

$$V_o = \frac{V_{in}}{(1-D)^2}\cdot[d_1(2n_1+n_2)+d_2(n_1+n_2)] \tag{27}$$

Where, n_1, n_2, and n_3 are the turns ratio of the multi-winding transformer and d_1 and d_2 are the duty cycles of the secondary switches S_1 and S_2 respectively.

C. Inductor values L_s and L_p and turns ratio: Minimum input voltage and full load condition is used to calculate the inductance values using

$$L_s = \frac{R_L}{f_s}\left[\frac{\left(V_{in}/V_o\right)^2}{1+(L_s/L_p)} - \frac{V_{in}/V_o\cdot(1-D_{max})}{(n_1+n_2+n_3)}\right] \tag{28}$$

For a rated power P_o and switching frequency f_s, the inductance values depend on the transformer turns ratio $n_1+n_2+n_3$, inductor ratio L_p/L_s and the maximum duty cycle D_{max} given in (25). The ratio L_p/L_s determines ZVS range for a particular turns ratio n. ZVS can be maintained over wide range with lower L_p/L_s ratio. For a given value of n, the rms current through the switch decreases as the L_p/L_s ratio increases. Hence, the ratio L_p/L_s should be a trade-off between the ZVS range and rms current through the switch. It has been observed that decrease in switch rms current becomes negligible for $L_p/L_s > 20$. Therefore, an optimal inductance ratio is selected i.e., $L_p/L_s = 20$ that aids in maintaining ZVS for $V_{in} =20$ to 60 V and low conduction losses in the converter. The external inductor value to be connected in parallel on the primary side of the transformer is given by

$$L_{p,ex} = \frac{L_m}{\left(\frac{L_m}{L_p}-1\right)} \tag{29}$$

Where, L_m is the magnetizing inductance of the transformer referred to the primary.

Transformer turns ratio is selected so that D is maintained above 0.5 to regulate load voltage with input voltage variation while ensuring $L_s > 0$ using (28). Turns ratio condition for primary modulation is

$$n > (1-D_{max})\cdot\frac{V_o}{V_{in}}\cdot(1+L_s/L_p) \tag{30}$$

Turns ratio limitation for secondary modulation is

$$n_2 + n_3 > (1-D)\frac{V_o}{V_{in}}(1+L_s/L_p) \tag{31}$$

Where $n = n_1+n_2+n_3$. The value of n is chosen to be 2. From the above equations, we get $n_1 = n_3 = 0.6$ and $n_2 = 0.8$. For $V_{in} = 60$ V, effective turns ratio is 1.4 as only 2/3 of the secondary winding is utilized. For the selected turns ratio, the variation in duty cycle D of main switches and d_1 and d_2 of the secondary switches is shown in Table II.

TABLE II   VARIATION IN DUTY CYCLE FOR DIFFERENT OPERATING CONDITIONS

Output power P_o	Input voltage V_{in}	Duty cycle of the primary switches D	Duty cycle of the secondary switches	
			d_1	d_2
200 W (full load)	20 V	0.8	0.255	-
	40 V	0.55	0.505	-
	50 V	0.55	0.172	0.306
	60 V	0.55	-	0.505
40 W (20% load)	20 V	0.769	0.286	-
	40 V	0.54	0.517	-
	50 V	0.55	0.075	0.402
	60 V	0.55	-	0.505

D. Inductors' ratings: The rms current through the series inductor for $V_{in} = 20$ V is given by

$$I_{Ls,rms} = \sqrt{I_{in}^2 \cdot \left(\frac{2T_{DR}}{3T_s}\right) + I_{Lp,peak}^2} \tag{32}$$

The rms current through the parallel inductor is given by

$$I_{Lp,rms} = I_{Lp,peak}\left(1 - \frac{4T_{DR}}{3T_s}\right)^{1/2} \tag{33}$$

Where T_{DR} is rectifier diode conduction time and is given by

$$T_{DR} = \frac{nV_{in}}{V_o f_s\left(1+\dfrac{L_s}{L_p}\right)} \tag{34}$$

Using (1), the peak parallel inductor current $I_{Lp,peak} = 2.88$ A for $V_{in} = 60$ V. $I_{Lp,rms}$ is computed to be 4.1 A for $V_{in} = 20$ V. Peak series inductor current is 10.6 A for $V_{in} = 20$ V.

E. Current ratings of primary switches: RMS current through main and auxiliary switches is computed for $V_{in} = 20$ V using (35) and (36).

$$I_{SW,rms} = \sqrt{\left(\frac{I_{in}}{2}\right)^2 \cdot D + I_{Lp,peak}^2 \cdot D + I_{in}^2 \cdot \frac{T_{DR}}{T_s} + I_{in} \cdot I_{Lp,peak} \cdot \left(D - 1 + \frac{2T_{DR}}{T_s}\right)} \tag{35}$$

$$I_{auxsw,rms} = (I_{in} + 2I_{Lp,peak})[(1-D)/24]^{1/2} \tag{36}$$

The values of $I_{SW,rms}$ and $I_{auxsw,rms}$ computed are 6.33 A and 1.48 A for $V_{in} = 20$ V respectively. Peak current through the main switch for $V_{in} = 20$ V is $I_{SW,peak} = 3I_{in}/2 + I_{Lp,peak} = 15.5$ A.

F. RMS current through the secondary switches: For $V_{in} = 20$ to 40 V, the rms current through the secondary switches S_1 and S_4 operating with duty ratio d_1 is given by

$$I_{S1,rms} = I_{S4,rmw} = \sqrt{\frac{2(I_{in} + I_{Lp,peak}) \cdot d_1}{3(n_1+n_2+n_3)^2}} \tag{37}$$

Where, $I_{Lp,peak}$ is given by (1). The rms current through the secondary switches was computed to be 2.2

A. For $V_{in} = 50$ V, rms current through the secondary switches is given by

$$I_{S1,rms} = I_{S4,rms} = \sqrt{\frac{d_1 \cdot I_{Ls,peak}^2}{3n^2} + \frac{I_{Ls,peak}^2}{3(n_2+n_3)^2} \cdot (d_1 + 0.45d_2)} \tag{38}$$

$$I_{S2,rms} = I_{S3,rms} = \frac{I_{Ls,peak}}{(n_2+n_3)} \cdot \sqrt{0.45d_2/3} \tag{39}$$

Where $I_{Ls,peak}$ is given by

$$I_{Ls,peak} = \left[\frac{V_{in}}{1-D} - \frac{V_o}{n}\right] \cdot \frac{d_1 T_s}{L_s} - I_{Lp,peak} \tag{40}$$

The rms current through the secondary switches S_1 and S_2 for $V_{in} = 50$ V are 1.49 A and 1.03 A respectively.

G. Auxiliary clamp capacitor: The rms current through the auxiliary clamp capacitor for $V_{in} = 20$ V is given by

$$I_{Ca,rms} = I_{Ca,peak}\sqrt{\frac{2(1-D)}{3}} \tag{41}$$

Rms current through the auxiliary clamp capacitor is 2.08 A for $V_{in} = 20$ V. The value of the clamp capacitor was computed for $V_{in} = 20$ V using

$$C_a = \frac{I_{Ca,peak}\sqrt{2(1-D)/3}}{4\pi f_s \Delta V_{Ca}} \tag{42}$$

The frequency of clamp capacitor current is 2x the switching frequency i.e., 200 kHz.

H. Condition for ZVS: To achieve ZVS of the main switches, energy stored in the leakage inductance should be sufficient to charge and discharge C_{a1} and C_1, respectively and is given by

$$L_s I_{Ls,peak}^2 \geq (C_{a1} + C_1) \cdot \left(\frac{V_{in}}{1-D}\right)^2 \tag{43}$$

IV.   SIMULATION AND EXPERIMENTAL RESULTS

Simulation results using PSIM 9.1 are shown in Fig. 4 for $V_{in} = 50$V, full load (200 W). It should be observed that as a result of hybrid modulation, secondary devices are modulated to modify the effective transformer turns ratio. Simulation waveforms match closely with theoretical operating waveforms as in Fig. 2. The negative current or anti-parallel body diode conduction before the switch conduction demonstrates the ZVS of the primary side main and auxiliary devices in Fig. 4(a). Fig. 4(b) shows the device voltage is zero before the device starts conducting showing ZVS turn-on of secondary side switch S_1. Fig. 4(c) shows the current through S_2 goes zero and device voltage is still zero resulting in ZCS turn-off of other secondary switch. It is clear that the converter devices undergo soft-switching as explained and claimed.

Experimental prototype of the proposed converter rated at 200 W is designed and developed in lab. Experimental results for $V_{in} = 20$ V, 50 V, and 60 V at rated load are demonstrated in Figs. 5-7, respectively. Fig. 5(a) clearly illustrates ZVS of main switches. Voltage across the devices becomes zero, i.e., $v_{ds} = 0$ before the gating signals v_{gs} is applied. Also, the body diode conduction prior to the switch conduction verifies ZVS operation. It should be observed from Fig. 5(b) that body diode conduction precedes auxiliary switch ensuring its ZVS turn-on. Fig. 5(c) shows that bipolar transformer voltage v_{AB} appearing for duration (1-D) when a device is off and is equal to device voltage. It should be noticed that leakage inductance current is continuous unlike conventional converters. Fig. 5(d) shows parallel inductor current. It adds to the value of peak leakage inductance current increasing the stored energy to improve ZVS range. It helps discharging the device capacitance before the device is gated for its turn-on. It helps achieve ZVS over wide input voltage and load variations.

Fig. 6(a) and (b) illustrate the ZVS of the main and auxiliary switches for V_{in} = 50 V illustrating the proposed hybrid modulation. For higher voltages, D reduces thereby prolonging the duration for which v_{AB} appears and this can be observed from Fig. 6(c). Additionally, an increase in parallel inductor current with input voltage can also be noticed from Fig. 6(c). Fig. 6(d) and 6(e) show the current through and voltage across secondary switches. For the period d_1T_s, switch S_4 is gated followed by switch S_3 for d_2T_s. Thus the turns ratio is varied from ($n_1+n_2+n_3$) to (n_2+n_3) to adjust the effective gain to regulate load voltage. It should be observed that while switch voltage is zero, i_{S3} turns-off naturally followed by the rise in corresponding switch voltage confirming its ZCS turn-off.

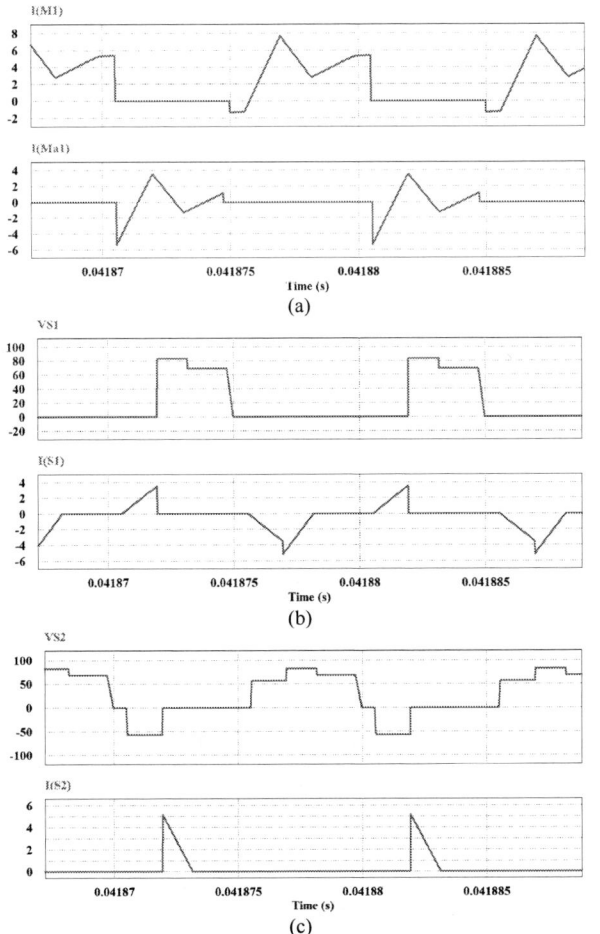

Fig. 4. Simulation results for V_{in} = 50 V, full load (200 W).

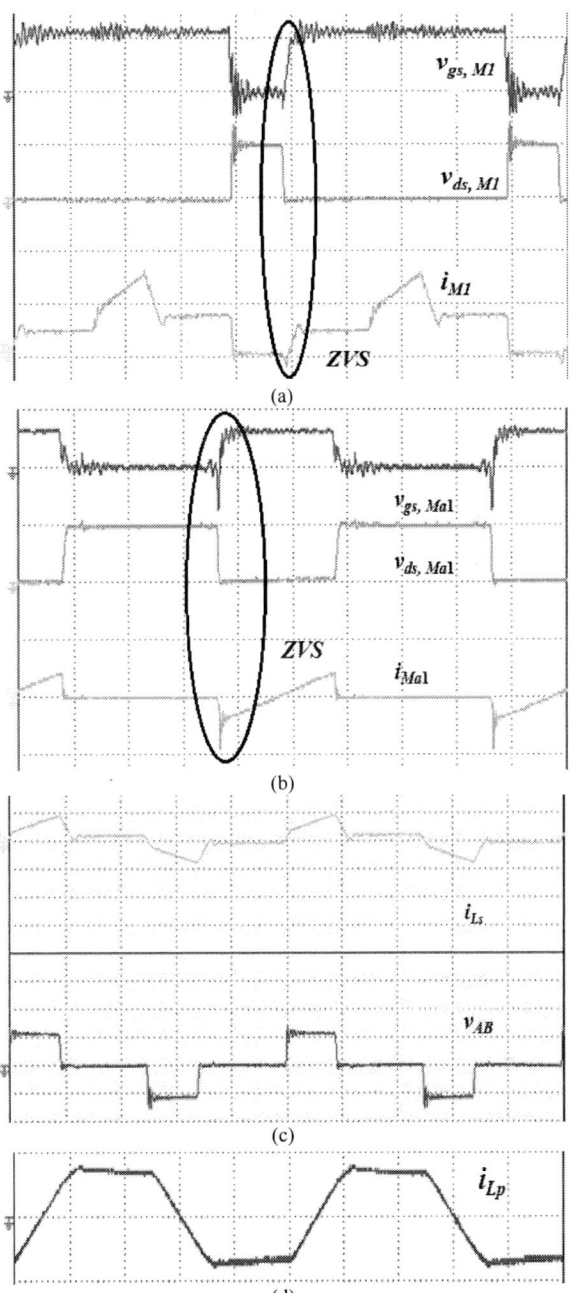

Fig. 5. Experimental waveforms at V_{in} = 20 V and full load (200 W): (a) $v_{gs,M1}$ (10 V/div, 2 µs/div), $v_{ds,M1}$ (100 V/div, 2 µs/div) and i_{M1} (20 A/div, 2 µs/div), (b) $v_{gs,Ma1}$ (20 V/div, 1 µs/div), $v_{ds,Ma1}$ (200 V/div, 1 µs/div) and i_{Ma1} (10 A/div, 1 µs/div), (c) i_{Ls} (10 A/div, 2 µs/div) and v_{AB} (100 V/div, 2 µs/div) and (d) i_{Lp} (2 A/div, 2 µs/div).

Fig. 7 demonstrates the same for V_{in} = 60 V at rated load. Owing to the fixed duty cycle of primary side devices, duration for which v_{AB} appears is the same as that of V_{in} = 50 V as seen from Fig. 7(c). Secondary modulation regulates the load voltage with fixed duty cycle of primary switches at 0.55 for V_{in} = 41 to 60 V. Generally, with change in load at fixed input voltage, duty cycle variation to regulate output voltage is not significant.

The 2014 International Power Electronics Conference

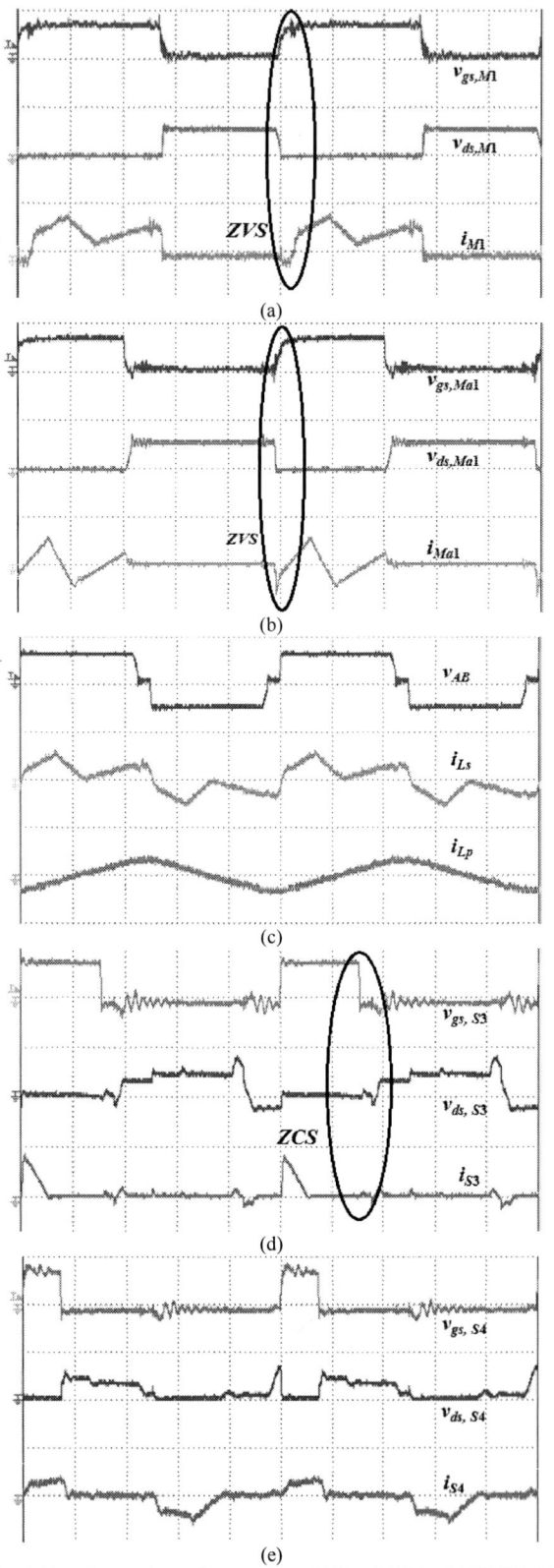

(a)

(b)

(c)

(d)

(e)

Fig. 6. Experimental waveforms at V_{in} = 50 V and full load (200 W): (a) $v_{gs,M1}$ (20 V/div, 2 μs/div), $v_{ds,M1}$ (200 V/div, 2 μs/div) and i_{M1} (10 A/div, 2 μs/div) (b) $v_{gs,Ma1}$ (20 V/div, 1 μs/div), $v_{ds,Ma1}$ (200 V/div, 1 μs/div) and i_{Ma1} (5 A/div, 1 μs/div) (c) v_{AB} (200 V/div, 2 μs/div), i_{Ls} (10 A/div, 2 μs/div) and i_{Lp} (10 A/div, 2 μs/div), (d) $v_{gs,S3}$ (20 V/div, 2 μs/div), $v_{ds,S3}$ (200 V/div, 2 μs/div) and i_{S3} (5 A/div, 2 μs/div) and (e) $v_{gs,S4}$ (20 V/div, 2 μs/div), $v_{ds,S4}$ (200 V/div, 2 μs/div) and i_{S4} (10 A/div, 2 μs/div).

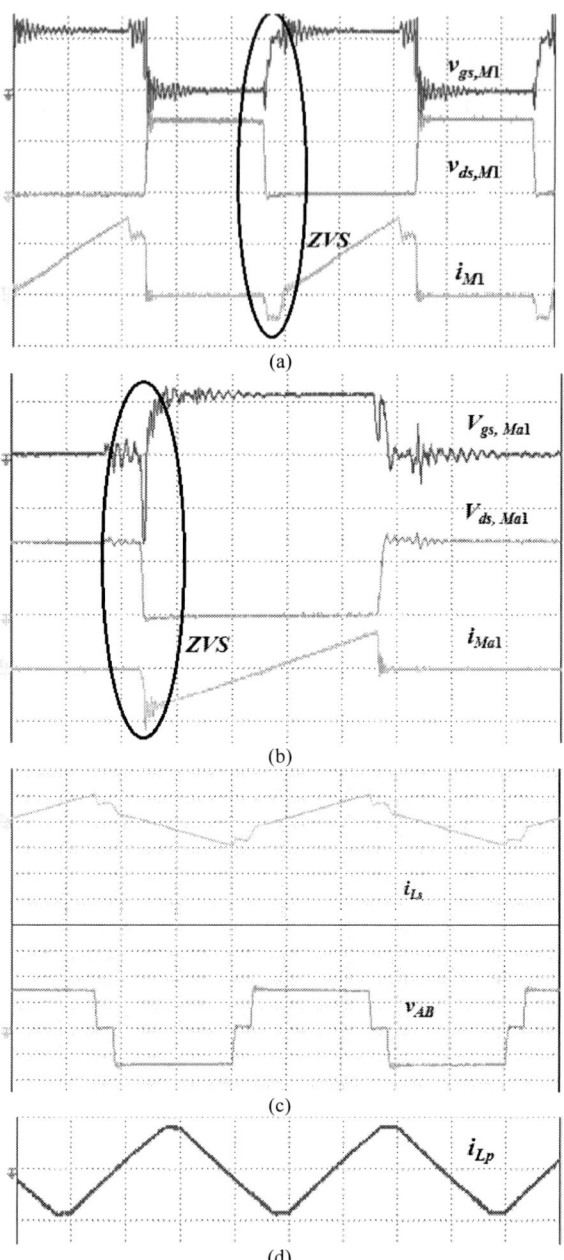

(a)

(b)

(c)

(d)

Fig. 7. Experimental waveforms at V_{in} = 60 V and full load (200 W): (a) $v_{gs,M1}$ (10 V/div, 2 μs/div), $v_{ds,M1}$ (100 V/div, 2 μs/div) and i_{M1} (5 A/div, 2 μs/div), (b) $v_{gs,Ma1}$ (10 V/div, 1 μs/div), $v_{ds,Ma1}$ (100 V/div, 1 μs/div) and i_{Ma1} (5 A/div, 1 μs/div), (c) i_{Ls} (5 A/div, 2 μs/div) and v_{AB} (100 V/div, 2 μs/div) and (d) i_{Lp} (5 A/div, 5 μs/div).

Experimental results for V_{in} = 60 V, 20% load (40 W) are shown in Fig. 8. The experimental results coincide with the steady-state operating waveforms shown in Fig. 2 validating the claims. It is clear that the main, auxiliary and the secondary switches S_2 and S_3 operate with ZVS and ZCS respectively for wide variations in input voltage and load. Also, load voltage is regulated effectively. Peak leakage current $I_{Ls,peak}$ and so its stored energy is increased by external parallel inductor. It aids in holding ZVS for wide source voltage and load power range. Additional energy helps discharge the device capacitance.

978-1-4799-2706-7/14 $31.00 © 2014 IEEE

The 2014 International Power Electronics Conference

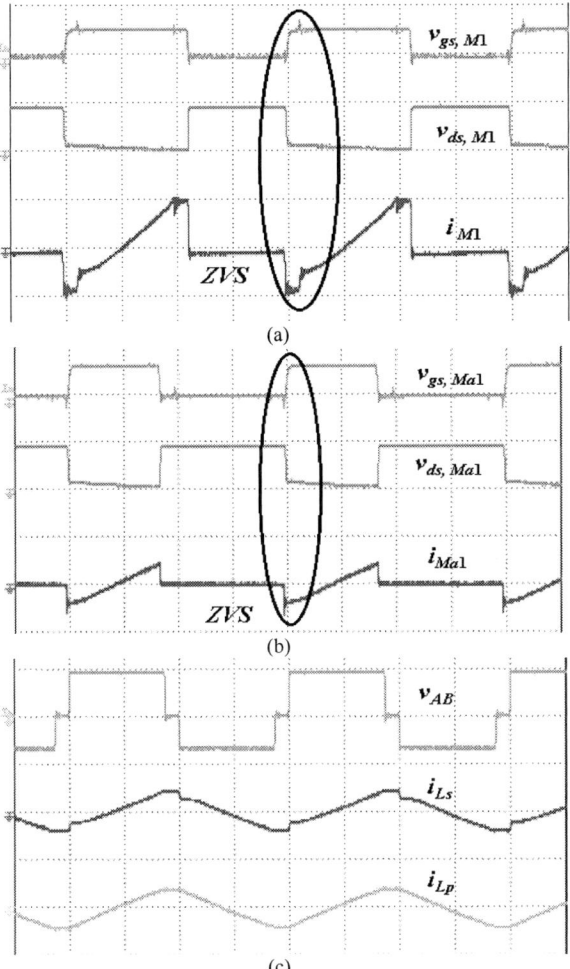

(a)

(b)

(c)

Fig. 8. Experimental waveforms at V_{in} = 60 V and 20% load (40 W): (a) $v_{gs,M1}$ (20 V/div, 5 μs/div), $v_{ds,M1}$ (100 V/div, 5 μs/div) and i_{M1} (10 A/div, 5 μs/div) (b) $v_{gs,Ma1}$ (20 V/div, 5 μs/div), $v_{ds,Ma1}$ (100 V/div, 5 μs/div) and i_{Ma1} (20 A/div, 5 μs/div) and (c) v_{AB} (100 V/div, 5 μs/div), i_{Ls} (20 A/div, 5 μs/div) and i_{Lp} (20 A/div, 5 μs/div).

Hybrid modulation ensures load voltage regulation over entire range by utilizing the transformer effectively. It has been demonstrated that the proposed converter maintains soft-switching and load voltage regulation under all operating conditions simultaneously.

V. Summary and Conclusion

An extended secondary universal current-fed converter that accommodates wide input voltage variation (1:3) is proposed. The range can be extended by modifying secondary circuit design. Magnetizing inductance assists in maintaining ZVS of the primary switches over this range. The secondary switches operate with ZCS thus minimizing the overall switching losses. Inherent current sharing property of two inductor topology and high L_p/L_s ratio reduces the current stress on the devices compared to soft-switching voltage-fed PWM and resonant converters. Proposed hybrid modulation enhances transformer utilization and ensures load voltage regulation over wide input voltage and load variation. Above all, a single converter can be used for different

sources such as batteries, fuel-cells, solar PV etc. and therefore, can be used in several applications like micro-grid and distributed generation. Steady-state analysis and design of the converter have been reported. Performance has been demonstrated through experimental results to verify the claims.

REFERENCES

[1] J. Wang, F. Z. Peng, J. Anderson, A. Joseph and R. Buffenbarger, "Low cost fuel cell converter system for residential power generation," *IEEE Trans. Power Electron.*, vol. 19, no. 5, pp. 1315-1322, 2004.

[2] R.-J. Wai and R.-Y. Duan, "High-efficiency power conversion for low power fuel cell generation system, " *IEEE Trans. Power Electron.*, vol. 20, no. 4, pp. 847-856, 2005.

[3] H. Tao, A. Kotsopoulos, J. L. Duarte and M. A. M. Hendrix, "Transformer-coupled multiport ZVS bidirectional dc-dc converter with wide input range," *IEEE Trans. Power Electron.*, vol. 23, no. 2, pp. 771-781, 2008.

[4] A. J. Mason, D. J. Tschirhart and P. K. Jain, "New ZVS phase shift modulated full-bridge converter topologies with adaptive energy storage for SOFC applications," *IEEE Trans. Power Electron.*, vol. 23, no. 1, pp. 332-342, 2008.

[5] S. Jung, Y. Bae, S. Choi and H. Kim, "A low cost utility interactive inverter for residential fuel cell generation," *IEEE Trans. Power Electron.*, vol. 22, no. 6, pp. 2293-2298, 2007.

[6] U. R. Prasanna, A. K. Rathore and S. K. Mazumder, "Novel zero-current-switching current-fed half-bridge isolated DC/DC converter for fuel-cell-based applications," *IEEE Trans. Industry applications*, vol. 49, no. 4, pp. 1658-1668, 2013.

[7] S. Lee, J. Park and S. Choi, "A three-phase current-fed push-pull dc-dc converter with active clamp for fuel cell applications," *IEEE Trans. Power Electron.*, vol. 26, no. 8, pp. 2266-2277, 2011.

[8] D. S. Oliveira and I. Barbi, "A three-phase ZVS PWM dc-dc converter with asymmetrical duty cycle associated with a three-phase version of the hybrid rectifier," *IEEE Trans. Power Electron.*, vol. 20, no. 2, pp. 354-360, 2005.

[9] U. R. Prasanna and A. K. Rathore, "Analysis, design and experimental results of novel snubberless bidirectional naturally clamped ZCS/ZVS current-fed half-bridge dc/dc converter for fuel cell vehicles," *IEEE Trans. Industrial Electronics*, vol. 60, no. 10, pp. 4482-4491, 2013.

[10] S.-K. Han, H.-K. Yoon, G.-W. Moon, M.-J. Youn, Y.-H. Kim and K.-H. Lee, "A new active clamping zero-voltage switching PWM current-fed half-bridge converter," *IEEE Trans. Power Electron.*, vol. 20, no. 6, pp. 1271-1279, 2005.

[11] R. L. Andersen and I. Barbi, "A ZVS-PWM three-phase current-fed push-pull dc-dc converter," *IEEE Trans. Industrial Electronics*, vol. 60, no. 3, pp. 838-847, 2013.

[12] S. -J. Jang, C. -Y. Won, B. -K. Lee and J. Hur, "Fuel cell generation system with a new active clamping current-fed half-bridge converter," *IEEE Trans. Energy Conversion*, vol. 22, no. 2, pp. 332-340, 2007.

[13] E.-S. Park, S. J. Choi, J. M. Lee, and B. H. Cho, "A soft-switching active-clamp scheme for isolated full-bridge boost converter," in *Proc. 19th IEEE-APEC*, 2004, pp. 1067-1070.

[14] T. -F. Wu, J. -C. Hung, J. -T. Tsai and C. -T. Tsai, "An active-clamp push-pull converter for battery sourcing applications," *IEEE Trans. Industry Applications*, vol. 44, pp. 196-204, 2008.

[15] A. K. Rathore, A. K. S. Bhat and R. Oruganti, "Analysis, design and experimental results of wide range ZVS active-clamped L-L type current-fed dc/dc converter for fuel cells to utility interface," *IEEE Trans. Industrial Electronics*, vol. 59, pp. 473-485, 2012.

[16] P. Xuewei, U. R. Prasanna, and A. K. Rathore, "Magnetizing-inductance-assisted extended range soft-switching three-phase AC-link current-fed dc/dc converter for low dc voltage applications," *IEEE Trans. Power Electron.*, vol. 28, no. 7, pp. 3317-3328, 2013.

978-1-4799-2706-7/14 $31.00 © 2014 IEEE

High Efficiency Power Converters for Battery Energy Storage Systems

Noriko Kawakami , Yukihia Iijima, Haiqing Li, Satoru Ota
Power Electronics Department, Power Electronics Systems Division
Toshiba Mitsubishi-electric Industrial Systems Corporation (TMEIC)
Tokyo, Japan
KAWAKAMI.noriko@tmeic.co.jp

Abstract— **Renewable energy sources such as wind-turbine and photovoltaic power generators may make the power grid unstable due to their output fluctuations. Battery energy storage systems (BESSs) are being considered as a countermeasure for this issue. "BESSs" require bidirectional ac-to-dc and dc-to-ac converters, so improvement of the converter efficiency is desired to improve the efficiency of BESSs. This paper presents the performance of 100-kVA and 500-kVA converters for BESSs in practical use, and introduces the development and the verification test results of a real-scale (500 kW) single-star bridge cell (SSBC) based modular multilevel cascade converter (MMCC) which is expected to be a new converter topology for BESSs.**

Keywords— *Battery energy storage systems, battery storage plants, frequency stability, multilevel converters, power systems, pulse-width modulation and two-level converters*

I. INTRODUCTION

In Japan, photovoltaic (PV) power generation systems have been strongly promoted. By 2012, the accumulated installations amounted to approximately 6.6 GW, and national targets for PV installation capacity have been set at 28 GW by 2020 and 53 GW by 2030. However, the output power of PV systems depends on the weather conditions and the natural environment, which cause large fluctuations in the output power, leading to insufficient frequency adjustment capacity of the power grid. Battery energy storage systems (BESSs) are being considered as a countermeasure for this issue, and the installation capacity of BESSs is expected to increase in the future. BESSs require bidirectional ac-to-dc and dc-to-ac converters. In current actual applications, BESS converters typically have two-level or three-level topologies [1] [2] [3].

On the other hand, "a modular multilevel cascade converter based on single-star bridge cell (MMCC-SSBC) [4]" has been applied to a 200-V, 10-kW, 3.6-kWh laboratory prototype BESS, and its control strategy and tactics have been discussed for the laboratory prototype [5] [6], and a fault-tolerant control scheme has been proposed and demonstrated [7]. In the SSBC (single-star bridge cell)-based MMCC topology, each bridge cell can control the state-of-charge (SOC) of the

battery units independently, and ac filters can be eliminated from the MMCC because the ac terminal waveform of the MMCC has very low harmonics. Thus, MMCC is one suitable topology for BESS converters. A 500-kW, 238-kWh BESS employing an SSBC-based MMCC with six bridge cells per phase was developed and verification test was conducted as the first step toward actual applications [8].

In this paper, the performances of 100-kVA and 500-kVA converters in practical use for BESSs are described, and the development and the verification test results of full-scale demonstration equipment employing SSBC-based MMCC are introduced.

II. CONVERTERS USED IN PRACTICAL APPLICATIONS

In this chapter, the performance, circuit configurations, and test results of 100-kVA and 500-kVA converters which have been put to practical use are described [3].

A. Specifications and Circuit Configuration

Table I summarizes the technical specifications of the 500-kVA converter and the 100-kVA converter, whose external views are shown in Figure 1. Figure 2 shows the

TABLE I. SPECIFICATIONS OF 100-KVA AND 500-KVA CONVERTERS

Item	Specifications	
	500-kVA PCS	*100-kVA PCS*
DC Voltage Range	450 V~800 V	320 V~550 V
Rated ac Voltage	300 V	210 V
Rated Output Capacity	500 kVA	100 kVA
Frequency	50/60 Hz	
Main Circuit configuration	3-Level	2-Level
Isolation Transformer	External	Internal
Maximum efficiency	98.5%	96.4%
Cooling Method	Forced-air-cooling	Self-cooling
Installation Location	Container or indoor	
Dimensions	1900W×700D ×1950H	1400W×800D ×1950H

The 2014 International Power Electronics Conference

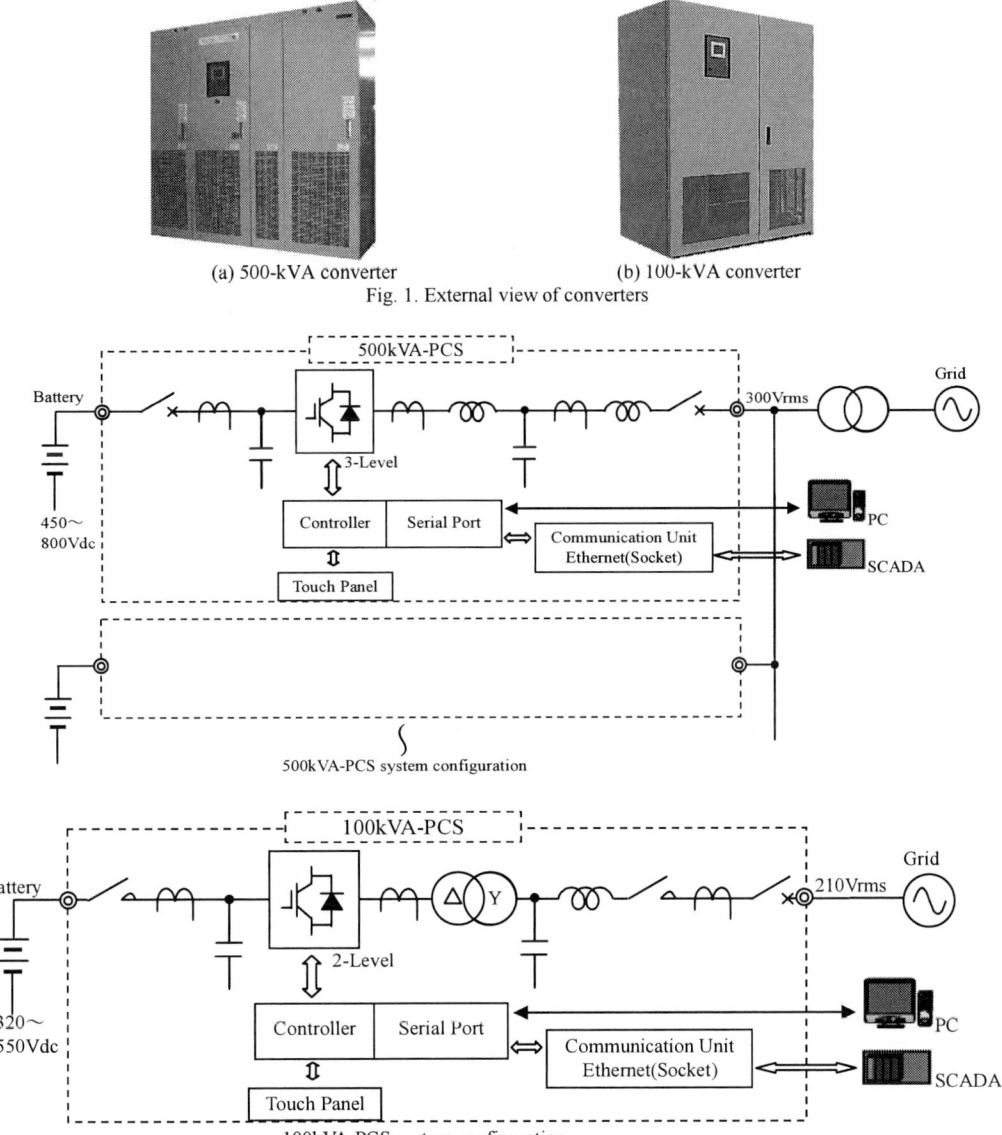

(a) 500-kVA converter   (b) 100-kVA converter

Fig. 1. External view of converters

Fig. 2. Main circuit configurations of 500-kVA converter and 100-kVA converter

circuit configurations. The 500k-VA converter employs a three-level conversion circuit, for the purpose of reducing the total loss of semiconductors and improving the output waveform. As a result, the size of harmonic filter is reduced. An ac filter is used to improve the output waveform. Beside the LC filter, a reactor for parallel connection is also mounted. In addition, both dc and ac circuit breakers are equipped, so that the PCS can be separated from the grid rapidly by opening the circuit breaker automatically.

In the 100-kVA converter, a two-level conversion circuit is used. In order to achieve high efficiency, the latest 6th generation IGBTs [9] are used. In addition, a three-phase Δ/Y inverter-transformer is used, thus eliminating the necessity of the outside isolation transformer. Furthermore, an efficient cooling system without cooling fans [10] is equipped, which is the first

fan-less converter at this capacity class. Removing the fan is beneficial in noise reduction, reliability improvement and decrease of the maintenance cost. Moreover, it also plays a partial role in improving efficiency of the converter by saving its auxiliary power consumption.

B. Test Results

Through factory tests, it has been verified that the total harmonic distortion of the output current of converter was less than 5%. The 2~40th order harmonics also met the requirements of IEEE standard 519.

The maximum efficiency of 500-kVA converter not including transformer is 98.5%, and that of 100-kVA converter including transformer is 96.4%. These efficiencies are the highest level in their respective capacity classes.

978-1-4799-2706-7/14 $31.00 © 2014 IEEE

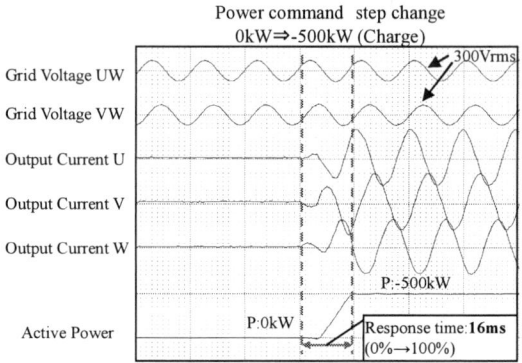

Fig. 3. Step response of active power command
(0 kW to charge 500 kW)

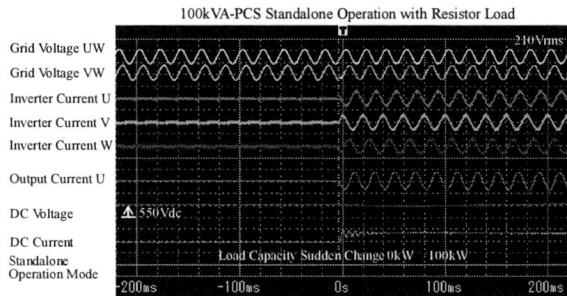

Fig. 4. Experiment results of stand-alone operation of 100-kVA PCS
(Load capacity changes from 0 kW to 100 kW)

Figure 3 shows the test results when the active power command step changes from 0-kW to 500-kW (charge operation) for the 500-kVA converter. In shown in the figure, the output of PCS changes smoothly within 16ms (500kVA-PCS) without an overshoot. In factory test, the soft-start time is set to 20ms.

Both 500-kVA and 100-kVA converters have standalone operation mode. Figure 4 shows the test results of standalone operation of the 100-kVA converter with changing resistive loads from 0 kW to 100 kW.

TABLE II. SPECIFICATIONS OF BATTERY ENERGY STORAGE SYSTEM

Items	Actual equipment	Verification test equipment
Power capacity	1000 kW	500 kW
Storage energy capacity	1000 kWh	238kWh
Circuit topology	MMCC-SSBC Modular Multilevel Cascade Converter based on Single-Star Bridge Cells	
Cascade number	6	6
Capacity of unit cell	56 kVA	28 kVA
DC voltage range	250 V - 380 V	280 V - 380 V Nominal voltage 331.2V
Output AC voltage of converter	1500 V	1500 V
Grid side AC voltage	6600 V	6600 V
Power device	600-V, 600-A IGBT	
Configuration of each bridge cell	1-series, 2-parallel, and 4-arm structure	1-series, 1-parallel, and 4-arm structure

III. REAL SCALE VERIFICATION EQUIOMENT EMPLOYING MMCC TOPLOGY

In this chapter, the outline of the development and the verification test results of a real scale MMCC for BESSs are described.

A. Specifications and Circuit Configuration

Table 2 summarizes the main technical specifications of actual equipment that is expected to be introduced in the secondary side of distribution substations in the future, as well as verification test equipment. The capacity of verification test equipment was chosen with sufficient capacity to evaluate technical issues in actual applications.

Figure 5 shows the circuit configuration of the verification test equipment. Six cells, each composed of an H-bridge IGBT unit and a battery unit, were cascade-connected for each of the U, V, and W phases to form a SSBC-based MMCC.

To ensure sufficient fault-ride-through (FRT) capability, the impedance of the converter transformer was determined to be 8% by numerical simulations. Considering 10% grid voltage fluctuations and a 5% negative phase sequence voltage, we chose 1500 V as the secondary side voltage of the transformer so that a rated power of 500 kW can be output even with the minimum battery voltage. Figure 6 shows the entire outside view of the verification converter including converter panels, a transformer, a circuit breaker for 6600 V side, an auxiliary power supply, and a control panel. Figure 7 shows the inside view of converter panels.

B. Verification test results

The verification test was carried out by taking four steps, an H bridge IGBT unit test, Inductor load current test, downscaled grid model test, and real scale distribution line test. In this paper, the representative test results using the real scale distribution line test are shown. This verification test equipment was installed at the Akagi testing center of the Central Research Institute of the Electric Power Industry (CEPRI) in Japan. The performance test of the entire BESS including full scale battery units was conducted using the 6.6-kV real-scale

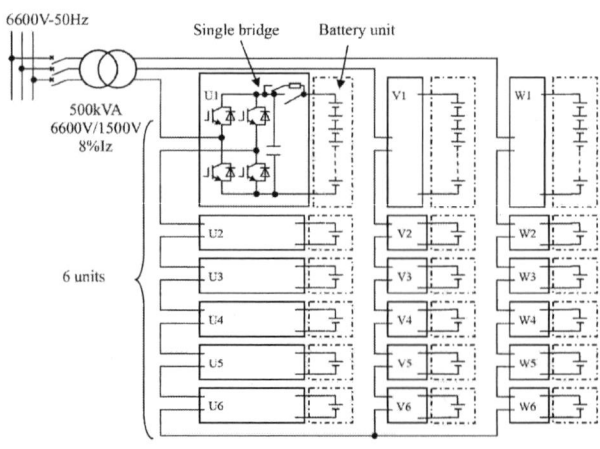

Fig. 5. Main Circuit Configuration

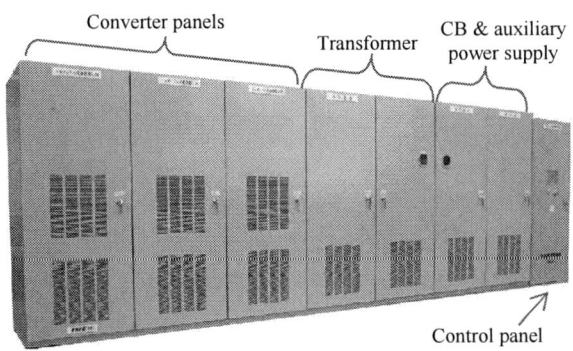

Fig. 6. Outside view of the developed equipment

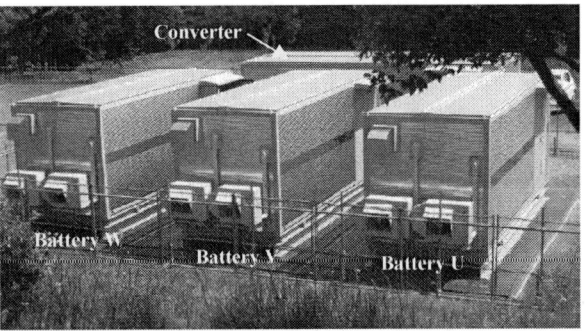

Fig. 8. Overall external view of the BESS at Aakagi testing center

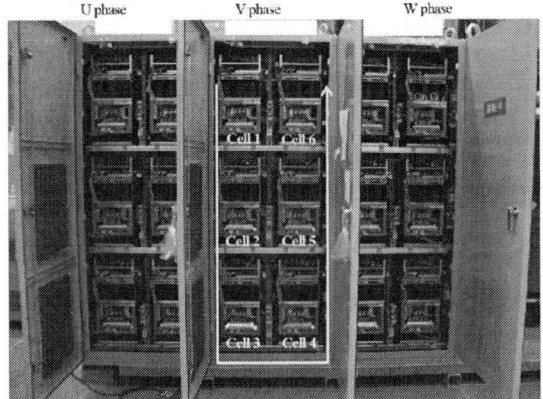

Fig. 7. Inside view of the converter panels

distribution line. Figure 8 shows an overall external view of the equipment installed at the Akagi testing center. The batteries were housed in an outdoor enclosure provided individually for each phase. A total of three outdoor enclosures were used for batteries. The converter was housed in one other enclosure.

Figure 9 shows the system circuit configuration of Akagi testing center. At the Akagi testing center, tests were conducted with two test circuit configurations. In test system 1, the distribution line of the Akagi testing center was connected to the 66-kV power grid of Tokyo Electric Power Company. With this test circuit configuration, basic performance tests, such as a continuous charge/discharge test and harmonic measurement, were confirmed. In test system 2, a 6.6-kV 1600-kVA programmable ac voltage source that can vary the phase and voltage was used. With this test circuit configuration, continuous operation performance is confirmed in the presence of disturbances on the grid side; the items that were confirmed included the fault-ride-through capability, phase and frequency jump and distorted voltage.

Figure 10 shows step response waveforms when the power reference of BESS is changed from charge 500 kW to discharge 500 kW. A response time is within 20-ms, a sufficiently quick response for BESS. Figure 9 shows a current waveform of 500 kW charging operation and its harmonic analysis.

Figure 11 shows a current waveform of 500 kW charging operation and its harmonic analysis. Total harmonic distortion below 20 kHz achieved 1.37% without an ac filter.

Figure 12 shows waveforms of an anti-islanding performance test. The output of BESS was adjusted so

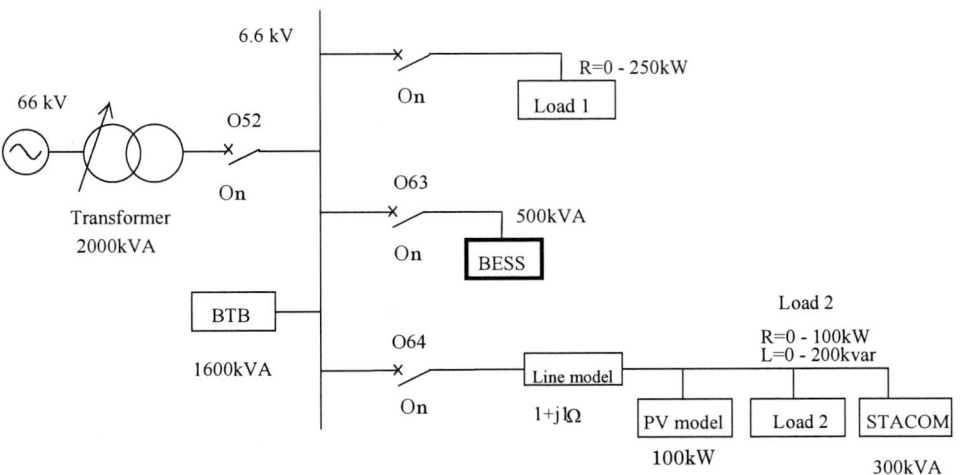

Fig. 9. Circuit configuration of a real scale distribution line test

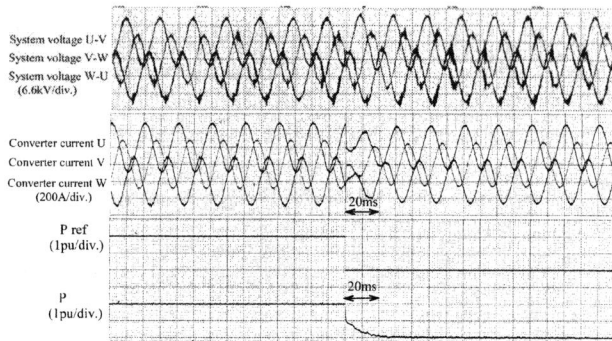

Fig. 10. Power reference step response test
(charge 500 kW to discharge 500 kW)

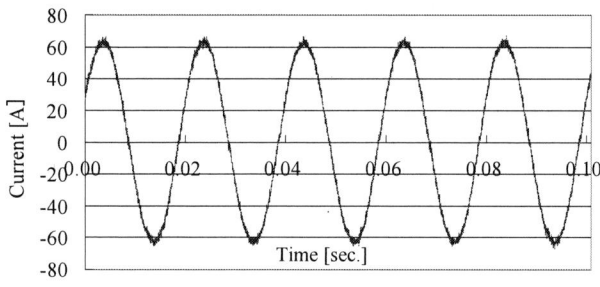

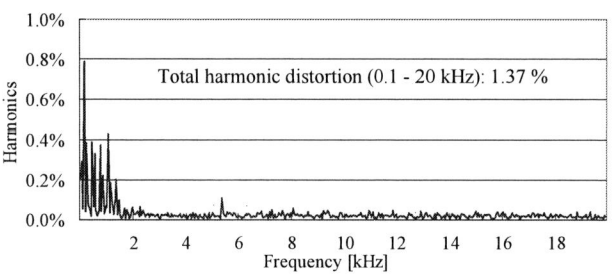

Fig. 11. Current waveform at 500 kW discharge and its harmonics

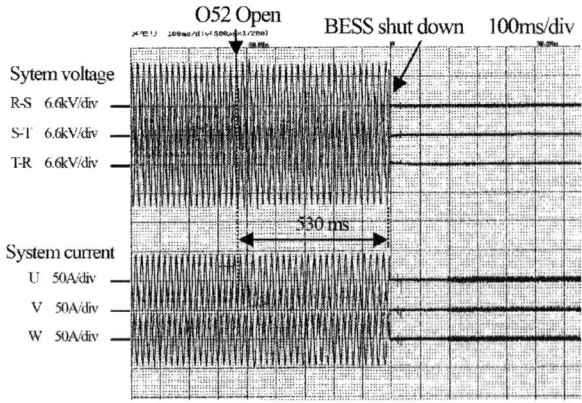

Fig. 12. Anti-islanding performance test

IV. CONCLUSIONS

The performances of 100-kVA and 500-kVA converters in practical use for BESSs were described, and the development and the verification test results of real-scale demonstration equipment employing a SSBC-based MMCC topology were introduced. These high efficiency converters for BESSs will contribute to stable operation of distribution lines with high penetration of photovoltaic generation systems in the future.

REFERENCES

[1] N. Kawakami, Y. Iijima, Y. Sakanaka, M. Fukuhara, K. Ogawa, M. Bando, T. Matsuda, "Development and field experiences of NAS battery Inverter for Power Stabilization of a 51 MW Wind Farm," The 2010 International Power Electronics Conference, pp.1837–1841, June 2010.

[2] T. Tanaka, I. Tominaga, N. Kawakami, "Development of high efficiency PCS for storage batteries," 2011 Japan Industry Application Society Conference Record, pp.1-421 - 1-422, September 2011. (in Japanese)

[3] H. Li, Y. Iijima, N. Kawakami, "Development of Power Conditioning System (PCS) for Battery Energy Storage Systems," the 5th Annual International Energy Conversion Congress and Exhibition for the Asia/Pacific region (ECCE Asia 2013), pp. 1295 -1299, 2013.

[4] H. Akagi, "Classification, Terminology, and Application of the Modular Multilevel Cascade Converter (MMCC)," IEEE Transactions on Power Electronics, vol.26, No.11, pp.3119-3130 , 2011.

[5] L. Maharjan, T. Yamagishi, H. Akagi, "Active-Power Control of Individual Converter Cells for a Battery Energy Storage System Based on a Multilevel Cascade PWM Converter," IEEE Transactions on Power Electronics, vol.27, No.3, pp.1099-1107, 2012

[6] L. Maharjan, S. Inoue, H. Aakagi, J. Asakura, "State-of-Charge (SOC)-Balancing Control of a Battery Energy Storage System Based on a Cascade PWM Converter," IEEE Transactions on Power Electronics, Vol.24, No.6, pp.1628-1636, June 2009.

[7] L. Maharjan, T. Yamagishi, H. Aakagi, J. Asakura, "Fault-Tolerant Operation of a Battery-Energy-Storage System Based on a Multilevel Cascade PWM Converter," IEEE Transactions on Power Electronics, vol. 27, No.3, pp. 1099-1107, March 2012

[8] N. Kawakami, S. Ota, H. Kon, S. Konno, H. Akagi, H. Kobayashi, N. Okada, "Development of a 500-kW Modular Multievel Cascade Converter for Battery Energy Storage Ssystems," IEEE Energy Conversion Congress & Expo (ECCE 2013), pp.3375 -3381, 2013

[9] Catalog, "Power Modules", Mitsubishi Electric Corporation, Tokyo, Japan, 2011. [Online]. Available: http://www.mitsubishielectric.com/

[10] T. Takahashi, E. Ikawa, R. Inzunza, T. Aambo, "100-kW High-Power PV PCS with No Cooling Fans," the 5th Annual International Energy Conversion Congress and Exhibition for the Asia/Pacific region (ECCE Asia 2013), pp. 1300 -1305, 2013.

that the power flow of circuit breaker O52 was zero. Then, O52 was opened to make islanding mode. BESS detected the islanding situation and shut down in 0.53 sec.

The 2014 International Power Electronics Conference

Implementation of Bridgeless Cuk Power Factor Corrector with Positive Output Voltage

Hong-Tzer Yang and Hsin-Wei Chiang
Research Center for Energy Technology and Strategy
Department of Electrical Engineering
National Cheng Kung University, Tainan 70101, Taiwan
htyang@mail.ncku.edu.tw; eszs125689@gmail.com;

Abstract- **A single-phase, bridgeless Cuk AC/DC power factor correction (PFC) rectifier with positive output voltage is proposed in this paper. For low output voltage product applications, the rectifier is designed to convert high input voltage to low output voltage. Due to no bridge-diodes required and thus decreased input conduction losses, the proposed rectifier efficiency can be improved. The proposed rectifier operates in discontinuous conduction mode (DCM) and the current-loop circuit is hence not needed. Also, only a single switch is used in the rectifier to simplify the control circuit design. A simple translation method to have the positive output voltage in the Cuk converter is presented in the rectifier to reduce the component counts and cost as well. The operational principles, steady-state analysis, and design procedure of the proposed rectifier are addressed in detail in this paper. Simulation and experimental results obtained from a 150 W-rated prototype circuit with input 90 V_{rms} -130 V_{rms} , 60Hz, and output 48 V_{dc} have verified the validity of the proposed rectifier.**

I. INTRODUCTION

In recent years, switched-mode power supply technologies have developed rapidly. Most switched-mode power supplies for electronic products are used to transfer AC to DC source in different applications. The use of a transformer, a bridge rectifier, and capacitors can achieve a DC output voltage easily, but the input current is seriously distorted.

Therefore, the PFC converters are critically required for AC/DC conversion [1]. A variety of circuit topologies have been developed for the PFC applications. The conventional PFC converter is a full-bridge rectifier followed by a boost converter, as shown in Fig. 1. The converter is widely used, because of its simplicity. However, due to boosting behavior of the converter, the output voltage is always greater than the input voltage. In many applications such as low-voltage and low-power supplies, it is desired to have the output voltage lower than the peak of input voltage. A buck-type converter is thus required.

The buck converter is seldom used in the PFC application. Since as the input current of the buck converter is discontinuous, it would lose control when the line input voltage is lower than the output voltage [2]. Also, to filter the input current, additional passive filter must be used at the buck converter input [2]. A buck PFC rectifier is recently proposed in [3] and [4] for voltage

step-down applications. However, the input line current cannot follow the input voltage around the zero crossing of the input line voltage.

Also, the buck PFC converter leads to increased total harmonic distortion (THD) and reduced power factor [4]. Therefore, in such applications, converters like buck-boost, single-ended primary-inductor converter (SEPIC) or Cuk converter are often used next to a full-bridge rectifier, as shown in Figs. 2-4 [5]-[13], to have a PFC converter with low output voltage.

All the converters mentioned above can be used in DCM or continuous conduction mode (CCM). There is no need of any control circuit, while operating in DCM to shape the input current sinusoidally, since these converters have intrinsic PFC characteristics at fixed duty ratio [14]. However, the drawbacks of buck-boost converter operating in DCM are high current stress on semiconductor devices and discontinuous input current, which increases the THD.

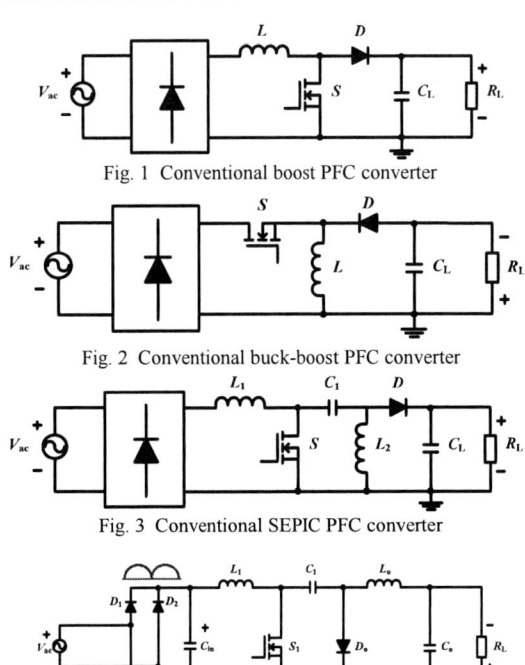

Fig. 1 Conventional boost PFC converter

Fig. 2 Conventional buck-boost PFC converter

Fig. 3 Conventional SEPIC PFC converter

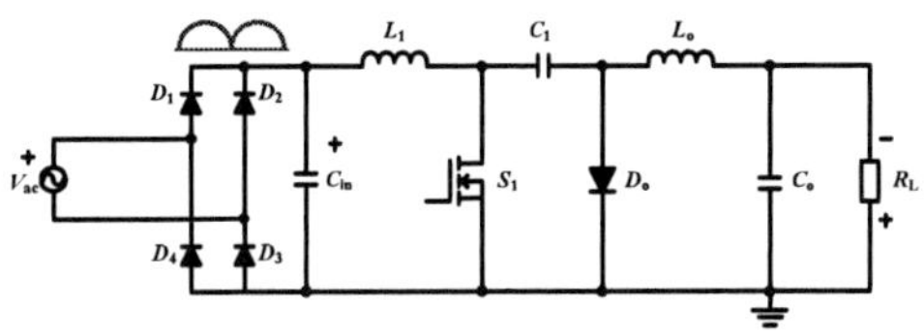

Fig. 4 Conventional Cuk PFC converter

SEPIC and Cuk converters are, therefore, alternative choices, because their input currents are continuous, while operating in DCM with the output voltage lower than the input voltage. Similar to the boost converter, the SEPIC converter has the disadvantage of discontinuous output current, resulting in relatively high output ripple [15].

A Cuk converter offers several advantages in PFC applications, such as easy implementation of transformer isolation, natural protection against inrush current occurring at start-up or overload current, lower input current ripple, and less electromagnetic interference (EMI) associated with the DCM topologies [14] and [16].

Unlike the SEPIC converter, the Cuk converter has both continuous input and output currents with a low current ripple. Thus, for applications requiring low current ripples at the both input and output ports of the converter, the Cuk converter seems to be a better candidate in the basic converter topologies.

In practical applications the DCM operation of the Cuk converter significantly increases the conduction losses, due to the increased current stress on the circuit components. However, using CCM for low-power applications, it requires extra components to achieve PFC performance [1]. As a result, additional circuit cost is increased. This leads to DCM operation of the Cuk converter limiting its use only in low-power applications (< 300 W) [17].

A conventional PFC Cuk rectifier is shown in Fig. 4. The current flows through the two rectifier bridge diodes and the power switch (S_1) during the switch ON-time, and through two rectifier bridge diodes and the output diode (D_o) during the switch OFF-time. Thus, during each switching cycle, the current flows through three power semiconductor devices.

As a result, a significant conduction loss, caused by the forward voltage drop across the bridge diodes, degrades the converter's efficiency, especially at low line input voltage. To reduce the conduction losses, the number of semiconductor devices must be reduced in the current path. Some methods to reduce conduction losses in Cuk and SEPIC converters are proposed in [15], [18] and [19].

In [15] and [18], the control circuits are complex using two main switches in the Cuk PFC topology. The bridgeless SEPIC converter introduced in [19] consists of two SEPIC converters, each of which is used for a half-line cycle. Thus, the number of devices and the cost are increased. Similar to the boost converter, the SEPIC converter has the disadvantage of discontinuous output current, resulting in a relatively high output ripple.

Besides, in [15], [18] and [19], the Cuk and SEPIC converters used have negative output voltages. Therefore, additionally required is an inverse amplifier circuit to translate the negative into the positive voltage [20]. The additional inverse amplifier circuit thus increases the cost required for the Cuk or SEPIC converters.

II. PROPOSED BRIGELESS CUK PFC RECTOFIER WITH POSITIVE OUTPUT VOLTAGE

Fig. 5 shows the proposed initial bridgeless Cuk PFC rectifier, which has a negative output voltage, like the existing Cuk PFC rectifier. As noted, for this circuit an inverting circuit to transfer the negative to the positive output voltage is still required for analog feedback control, as shown in Fig. 6.

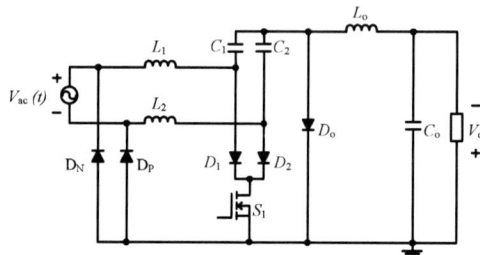

Fig. 5 Proposed Bridgeless Cuk power factor correction rectifier with negative output voltage

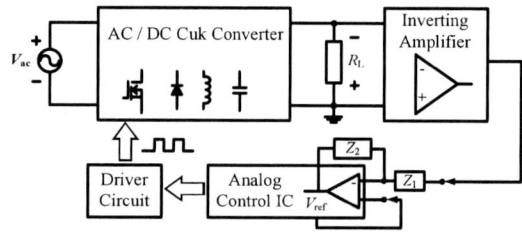

Fig. 6 Blocking Diagram of the conventional Cuk PFC circuit (with negative output voltage)

To obtain the positive output voltage without the inverting amplifier circuit, we transfer the polarity of all the components in Fig. 5 into those as shown in Fig. 7, and obtain the proposed bridgeless Cuk PFC rectifier in Fig. 8. Thus, the feedback control circuit is simpler and the cost can also be reduced, as compared with the conventional feedback control circuit, as shown in Fig. 6.

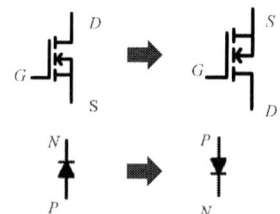

Fig. 7. Transferring the polarity of all components in the proposed initial topology in Fig. 5

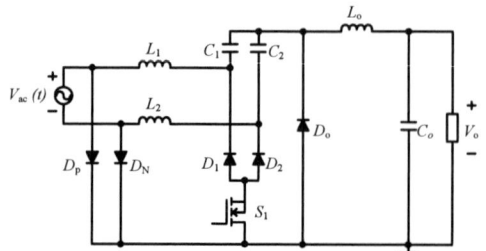

Fig. 8. Proposed Bridgeless Cuk PFC rectifier with positive output voltage

Before analyzing the proposed rectifier, the analysis supposes that the converter is operating at a steady state in addition to the following assumptions:

1. Both of the ON-state resistance R_{DS_ON}, parasitic capacitances of the main switch S_1 and the forward voltage drops (V_d) of the diodes are neglected.

2. The input capacitances are large enough such that during a switching period (T_s) their voltages are considered to be constant.

3. The output capacitor C_o is sufficiently large that the capacitor voltage is considered to be constant.

4. The proposed converter is operated in the DCM.

5. Due to symmetry of the circuit, it is sufficient to analyze the circuit during the positive half cycle of the input voltage.

A. Principles of Operation

Mode I [t_0 - t_1]: This mode starts when switch S_1 is turned ON, as shown in Figs. 9 and 10. Input inductors L_2 starts to charge linearly in slope of $V_{ac}(t)/L_2$ and diode D_p is forward biased by the indictor current i_{L2}. The voltage across L_o is equal to $V_{ac}(t)$, thus i_{Lo} increases linearly in slope of $V_{ac}(t)/L_o$. The inductor currents of L_2 and L_o during this mode are given by

$$\frac{di_{L_n}}{dt} = \frac{V_{ac}(t)}{L_n}, \text{n} = 2, \text{o}. \tag{1}$$

Accordingly, the peak current through the active switch S_1 is given by

$$I_{s1,pk} = \frac{V_m}{L_e} D_1 T_s \tag{2}$$

where V_m is the amplitude of the input voltage $v_{ac}(t)$, D_1 is the switch duty cycle, and L_e is the paralleled inductance of inductors L_1, L_2 and L_o.

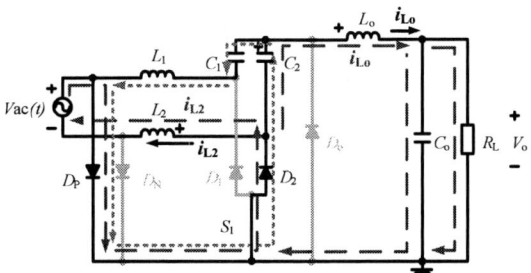

Fig. 9 The equivalent circuit in mode I (Switch S_1 is turned ON)

Mode II [t_1 - t_2]: This mode starts when switch S_1 is turned OFF and diode D_o is turned ON, simultaneously, as shown in Fig. 11 and Fig. 12. Input inductor L_2 starts to discharge linearly in slope of V_o/L_2 and diode D_p is

forward biased by the indictor current i_{L2}. The voltage across L_o is equal to V_o, thus i_{Lo} decreases linearly in slope of V_o/L_o. Note that diode D_o is turned OFF at zero current. The inductor currents of L_2 and L_o during this mode are given by

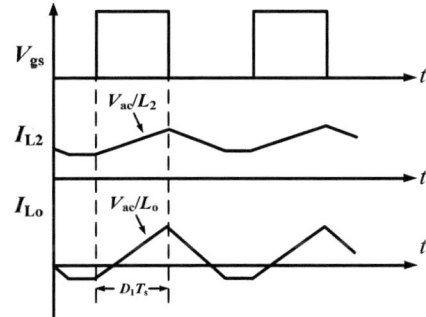

Fig. 10 Theoretical DCM waveforms during one switching period T_s in mode I

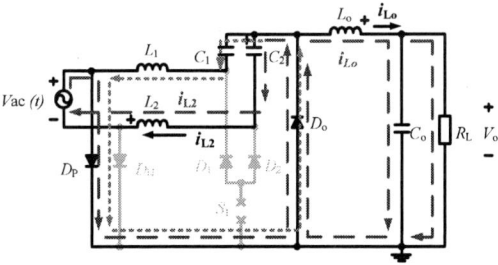

Fig. 11 The equivalent circuit in mode II (Switch S_1 is turned OFF)

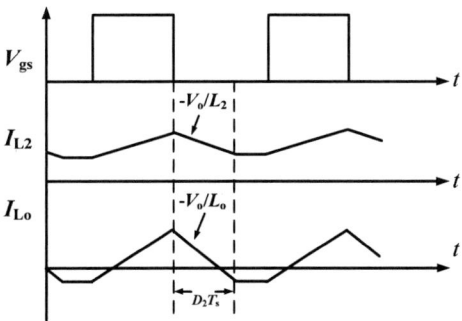

Fig. 12 Theoretical DCM waveforms during one switching period T_s in mode II (Switch S_1 is turned OFF)

$$\frac{di_{L_n}}{dt} = \frac{-V_o}{L_n}, \text{n} = 2, \text{o}. \tag{3}$$

Mode III [t_2 - t_3]: During this interval, only diode D_p conducts to provide a path for i_{L2}. Accordingly, the inductors L_2 and L_o in this interval behave as constant current source. Thus, the voltage of inductors (L_2 and L_o) is zero. Capacitor C_2 is being charged by the inductor current i_{L2} and the energy of capacitor C_o is released to load. This is a freewheeling mode. The theoretical waveforms in this mode are shown in Figs. 13 and 14.

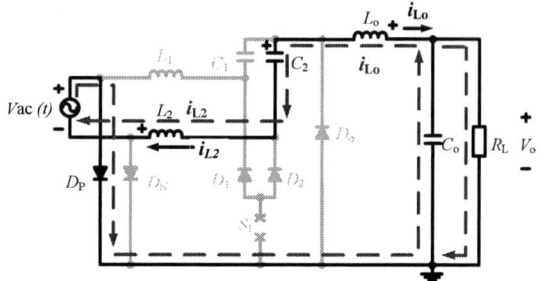

Fig. 13 The equivalent circuit in mode III (Switch S_1 is turned OFF)

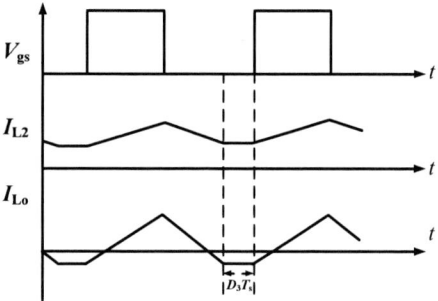

Fig. 14 Theoretical DCM waveforms during one switching period T_s in mode III (Switch S_1 is turned OFF)

This mode lasts until the start of a new switching period. The turn-OFF time of the switch and the output diode is given by

$$t_{\text{off}} = T_s - t_{\text{on}} - t_{\text{don}} \tag{4}$$

where t_{on} is the conducting interval of switch S_1 and that of the output diode D_o is t_{don}.

According to Equations (2) and (3), the normalized length of Mode II period can be obtained as follows:

$$D_2 = \frac{D_1}{M}\sin \omega t \tag{5}$$

where ω is the line angular frequency, and M is the voltage conversion ratio ($M = V_o/V_m$).

B. Analysis and Design of the Proposed Cuk Power Factor Correction Rectifier

1) Design of Input Inductors

Maximum input current is calculated by the output power and efficiency of the converter as follows [18]:

$$I_{\text{in}} = I_m \sin \omega t = \frac{2P_o}{\eta\, I_m}\sin \omega t \tag{6}$$

By the relationship of input ripple current (ΔI_{L2}), as shown in Fig. 10, and input voltage (V_{ac}) in positive half

cycle, the values of input indictor (L_1 and L_2) can be obtained below:

$$L_2 = \frac{V_{\text{ac}}(t)\, D_1}{2\,\Delta I_{L2}\, f_s} \tag{7}$$

2) Voltage Conversion Ratio M

The voltage conversion ratio M in terms of the rectifier parameters can be obtained by applying the power-balance principle below [20]:

$$P_{in}(t) = \frac{2}{T}\int_0^{\frac{T}{2}} v_{\text{ac}}(t)\, i_{\text{ac}}(t)\, dt \tag{8}$$

According to the DCM switch network large-signal model [21], as shown in Fig. 15. It is be noted that the average input current can be given as follows.

$$< i_{\text{ac}}(t) >_{T_s} = < i_{L2}(t) >_{T_s} = \frac{V_{\text{ac}}(t)}{R_e} \tag{9}$$

where the R_e is defined as the input resistance and given by

$$R_e = \frac{2L_e}{D_1^2\, T_s} \tag{10}$$

Evaluating (6) by using (7) and applying the power balance between the input and output ports, the voltage conversion ratio can be given by

$$M = \frac{V_o(t)}{V_m} = \sqrt{\frac{R_L}{2R_e}} \tag{11}$$

3) Boundaries Between CCM and DCM

To operate in DCM, the following inequality must be satisfied as follows:

$$D_2 \leqslant 1 - D_1 \tag{12}$$

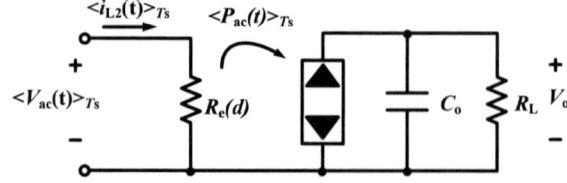

Fig. 15 Large signal model [21] of the proposed Cuk PFC rectifier

Substituting Equation (5) into (12) and applying (10) and (11), the following condition for DCM is obtained:

$$K_e \leqslant K_{e\_crit} = \frac{1}{2(M + \sin(\omega t))^2} \quad (13)$$

where the parameter K_e is expressed as follows:

$$K_e = \frac{2L_e}{R_L T_s} \quad (14)$$

It is obvious that from Equation (13) the value of $K_{e\text{-crit}}$ depends on the line angle ωt. Thus, the minimum and maximum parameters of K_{e_crit} can be obtained, respectively:

$$K_{e\_crit(min)} = \frac{1}{2(M+1)^2} \quad \text{and}$$

$$K_{e\_crit(max)} = \frac{1}{2M^2} \quad (15)$$

Therefore, for $K_e < K_{e\_crit(min)}$, the proposed bridgeless Cuk PFC rectifier with positive output voltage always operates in DCM.

4) Selection of Input Capacitors

In the Cuk converter as PFC, voltages of the input capacitors C_1 and C_2 should be nearly constant value within the switching period T_s and follow the input voltage profile within a line period T_L. Also, input capacitors C_1 and C_2 should not cause low-frequency oscillations with the converter inductors. Thus, the energy transfer capacitors C_1 and C_2 are determined based on inductor L_1, L_2 and L_o values such that the line frequency (f_L) should be well below the switching frequency (f_s). And a better initial approximation for choosing the resonant frequency (f_r) is given by [14] and [15].

$$f_L < f_r < f_s \quad (16)$$

where $f_r = \dfrac{1}{2\pi\sqrt{C_1(L_1 + L_o)}} \quad (17)$

5) Design of Output Capacitor C_o

Output ripple frequency of the converter is two times the input frequency. In the worst case, the output current during the half period of ripple frequency is provided by the output capacitor. Therefore, C_o can be obtained as follows:

$$C_o = \frac{P_o}{4f_L V_o \Delta V_o} \quad (18)$$

III. EXPERIMENTAL RESULTS

By following the specification given in Table I and the design procedure described above, components and values in the proposed power-stage circuit are used in Table II for verification of the proposed rectifier.

TABLE I
Specification of the proposed rectifier

Specifications	
INPUT VOLTAGE V_{IN}	90-130 V_{RMS}
OUTPUT VOLTAGE V_{OUT}	48 V_{DC}
RATED POWER P_{OUT}	150 W
SWITCHING FREQUENCY F_s	100 kHz

TABLE II
Component parameters used in the proposed circuit

Device	Component	Part/Value
POWER SWITCH	MAIN SWITCHES S_M	IXFH3650P
DIODES	INPUT DIODE $D_P D_N$	MBR20200CT
	SERIES DIODE $D_1 D_2$	STTH6003CW
	OUTPUT DIODES D_o	
INDUCTOR	INPUT INDUCTORS L_1 AND L_2	1×10^{-3} H
	OUTPUT INDUCTOR L_o	22×10^{-6} H
CAPACITOR	INPUT CAPACITORS C_1 C_2	1×10^{-6} F/400V
	OUTPUT CAPACITOR C_o	2×10^{-3} F/100V
CONTROL IC	UC 3525 (VOLTAGE MODE)	

The measurement results show that the main switch turns on under ZCS condition and D_o turns off under ZCS condition, as shown in Fig. 16(a), Fig. 16(b), Fig. 17(a) and Fig. 17(b), respectively. Therefore, reverse recovery problem of the main diodes is resolved by employing the proposed Cuk rectifier in DCM.

The measurement results of the input voltage and current waveforms are shown in Fig. 18 to Fig. 21 for 20%, 50%, 80%, and full loads, respectively.

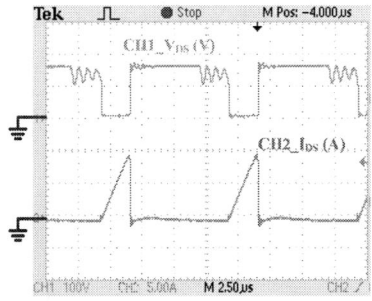

Fig. 16(a) Measurement waveforms of the MOSFET input voltage and current at full load (150 W)

The 2014 International Power Electronics Conference

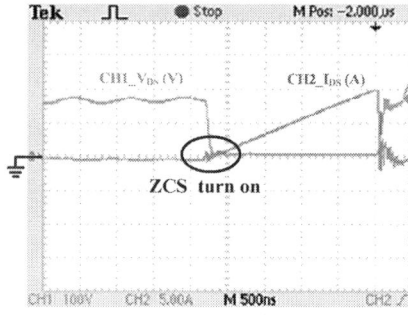

Fig. 16(b) Measurement waveforms of the MOSFET input voltage and current at full load (150 W)

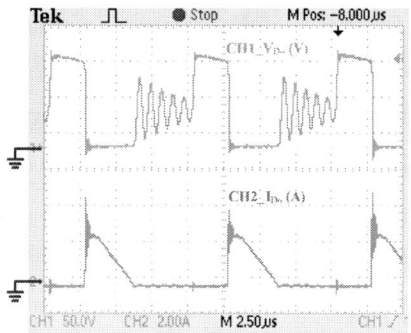

Fig. 17(a) Measurement of the output diode voltage and current waveforms at full load (150 W)

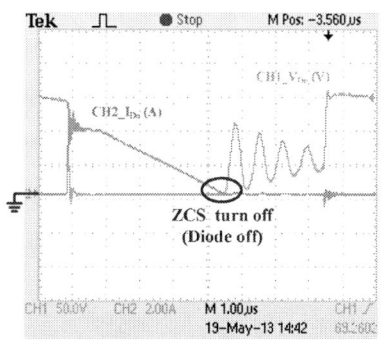

Fig. 17(b) Measurement of the output diode voltage and current waveforms at full load (150 W)

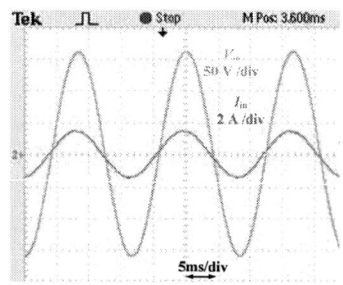

Fig. 18 Measurement waveforms of the input voltage and current for 20% load (30 W)

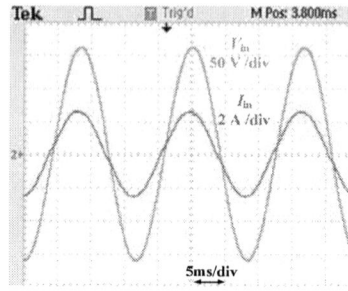

Fig. 19 Measurement waveforms of the input voltage and current for 50% load (75 W)

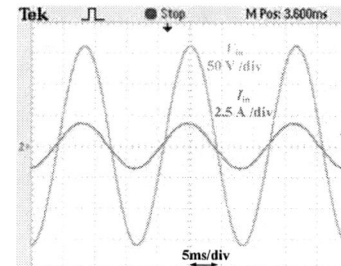

Fig. 20 Measurement waveforms of the input voltage and current for 80% load (120 W)

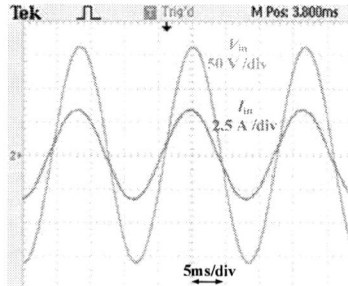

Fig. 21 Measurement waveforms of the input voltage and current for full load (150 W)

According to Fig. 22, it can be observed that the IEC 61000-3-2 Class D limits are well met by the proposed Cuk rectifier. The measured results of the power factor and the THD values of the input current for different loads of 20% to full loads (V_{in}=110 V_{rms}) are shown in Table III. The PF is about 0.9961 and the THD in percentage of the input line current is 2.66% for the full power output of 150 W.

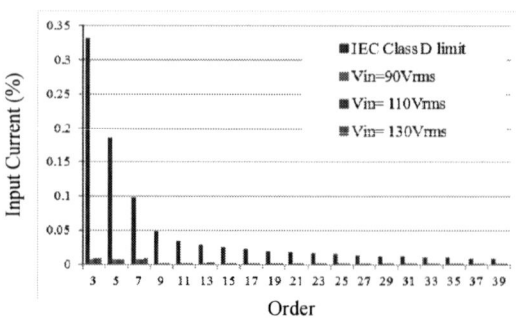

Fig. 22 Measured current harmonics complying with IEC 61000-3-2 Class D limits at full load

TABLE III
Measured THD and PF values for different loads
(V_{in}=110 V$_{rms}$)

Load (%)	THD (%)	Power Factor
20%	7.66	0.9908
30%	6.27	0.991
40%	4.50	0.9912
50%	3.04	0.9918
60%	2.97	0.9923
70%	2.85	0.9932
80%	2.78	0.994
90%	2.75	0.9958
Full Load	2.66	0.9961

Since the proposed rectifier has the functions of ZCS turn-on and ZCS turn-off (for the output diode), the switch and output diode losses can be decreased at full load, significantly. The measured efficiencies of the prototype rectifier at line voltages of 90 V$_{rms}$, 110 V$_{rms}$, and 130 V$_{rms}$, respectively, and 20% to full loads are shown in Fig. 23. It reveals that all the efficiencies of different loads for the line voltage of 110Vrms are above 91%. The best efficiency 96.2% is obtained at high line voltage of 130Vrms for full load, and the worst efficiency of 89.7% is observed at low line voltage (90 V$_{rms}$) for 20% load.

Efficiency comparisons with the Cuk circuit proposed in [4] and the conventional one are displayed in Fig. 24. It shows that the proposed circuit has higher efficiency than the ones for comparisons in addition to the advantage of positive output voltage.

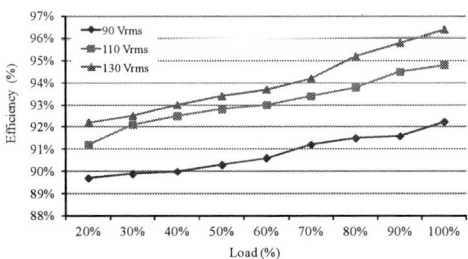

Fig. 23 Efficiencies of the proposed bridgeless Cuk PFC rectifier for different input voltages and loads

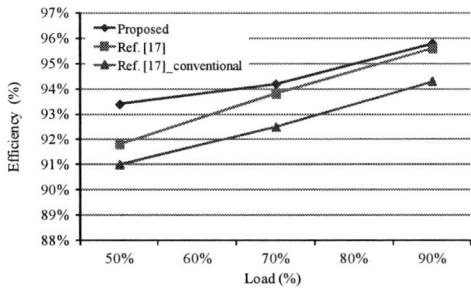

Fig. 24 Efficiency comparisons of the proposed bridgeless Cuk PFC rectifier with that proposed in **Error! Reference source not found.** and the conventional one

IV. CONCLUSIONS

In this paper, the Cuk PFC rectifier with positive output voltage has been proposed and experimentally verified. The experimental results have shown good agreements with the predicted waveforms analyzed in the paper. The power factor of the circuit is able to keep above 0.99 at all the specified input and output conditions. Moreover, with higher efficiency the proposed topology is able to obtain high power factor that can apply to most of the consumer electronic products for 150 W in the market. Above the power factor of 0.98 and satisfaction of the IEC 61000-3-2 requirements can be easily achieved by the proposed circuit. Also, by only using single switch, the implemented system control circuit is simple with the high power factor achieved by applying any PWM control IC. Moreover, in the proposed Cuk topology, no extra circuit is not required to transfer the original negative output voltage to the positive one, as needed in the traditional Cuk circuit. Convenience of using the Cuk rectifier can thus be obtained.

V. ACKNOWLEDGMENTS

The financial supports by National Science Council of R.O.C. under Grants NSC 102-3113-P-006-015 and NSC 102-3113-P-194-002 are greatly acknowledged.

REFERENCES

[1] "Power Factor Correction (PFC) Handbook", Rev. 4, *on semiconductor®*, Feb. 2011.

[2] M. Mahdavi and H. Farzanehfard, "Bridgeless SEPIC PFC Rectifier With Reduced Components and Conduction Losses", *IEEE Trans. Industrial Electronics*, Vol. 58, No.9, Sep. 2011.

[3] D.S.L. Simonetti, J. Sebastian, and J.Uceda, "The discontinuous conduction mode Sepic and Cuk power factor preregulators: analysis and design", *IEEE Trans. Ind. Electron.*, 44, (5), pp. 630-637, 1997.

[4] M. Mahdavi and H. Faarzaneh-fard, "Bridgeless Cuk power factor correction rectifier with reduced conduction losses", *IET Power Electron.*, Vol. 5, lss. 9, pp. 1733-1740, 2012.

[5] R. Itoh, K. Ishizaka, and H. Okada, "Single-phase buck rectifier employing voltage reversal circuit for sinusoidal input current waveshaping", *IEEE Proc. Electronics, Power,* Appl., 146, (6), pp. 707-712, 1999.

[6] R. Oruganti and M. Palaniapan, "Inductor Voltage Control of Buck-Type Single-Phase", *IEEE Trans. Power Electronics*, Vol. 15, No. 2., Mar. 2012.

[7] M. A. Al-Saffar, E. H. Ismail, and A. J. Sabzali, "Integrated Buck–Boost–Quadratic Buck PFC", *IEEE Trans. Power Electronics*, Vol. 24, No. 12, Dec. 2009.

[8] L. Petersen, "Input-Current-Shaper Based on a Modified SEPIC Converter with Low Voltage Stress", *IEEE PESC Conf. Thirty-second*, Vol. 2, pp. 666-671, Jun. 2001.

[9] P. Scalia, "A Double-Switch Single-Stage PFC Offline Switcher Operating in CCM with High Efficiency and Low Cost", *Twenty-ninth IEEE PESC Conf.*, Vol. 2, pp. 1040-1047, May 1998.

[10] C. Jingquan, D. Maksimovic, and R. Erickson, "A New Low-

Stress Buck-Boost Converter for Universal-Input PFC Applications", *Sixteenth IEEE APEC Conf.*, Vol. 1, pp. 343-349, Mar. 2001.

[11] W. Huai and I. Batarseh, " Comparison of Basic Converter Topologies for Power Factor Correction", *Southeastcon '98. IEEE Proceedings*, 24-26, pp. 348-353, Apri. 1998.

[12] P.F. de Melo, R. Gules, E.F.R. Romaneli, and R.C. Annunziato, "A modified SEPIC converter for high-power-factor rectifier and universal input voltage applications", *IEEE Trans. Power Electron.*, 25, (2), pp. 310-321, 2010.

[13] T. Ching-Jung and C. Chern-Lin, "A novel ZVT PWM Cuk power-factor Corrector", *IEEE Trans. Ind. Electron.*, 46, (4), pp. 780-787, 1999.

[14] C. Jingquan, D. Maksimovic, and R.W. Erickson, "Analysis and design of a low-stress buck-boost converter in universal-input PFC applications", *IEEE Trans. Power Electron.*, 21, (2), 2006.

[15] R. Martinez and P.N. Enjeti, "A high performance single-phase AC to DC rectifier with input power factor correction", *IEEE Trans. Power Electron.*, 11, (2), pp. 311–317, 1996.

[16] M. Brkovic and S. Cuk, "Input current shaper using Cuk converter", *Proc. Int. Telecommun. Energy Conf.*, pp. 532-539, 1992.

[17] Y. S. Roh, Y. J. Moon, J. G. Gong, and C. Yoo, "Active power factor correction (PFC) circuit with resistor-free zero-current detection", *IEEE Trans. Power Electron.*, Vol. 26, No. 2, pp. 630-637, Feb. 2011.

[18] A. A. Fardoun, E. H. Ismail, and A. J. Sabzali, "New Efficient Bridgeless Cuk Rectifiers for PFC Applications", *IEEE Trans. Power Electron.*, Vol. 27, No. 7, July 2012.

[19] A. J. Sabzali, E. H. Ismail, M. A. Al-Saffar, and A. A. Fardoun, "New Bridgeless DCM Sepic and Cuk PFC Rectifiers With Low Conduction and Switching Losses", *IEEE Trans. Ind. Appl.*, 47, (2), pp. 873-881, 2011

[20] *Application Hint 85 MIC2295 and MIC6211 Cuk Converter, Buck/Boost Inverter, MICREL*®, Oct. 2010.

[21] *Fundamental of Power Electronics*, Second Edition, *Springer*, R. W. Erickson and D. Maksimovic, 2011.

A Novel Synchronous Rectifier Method for a LLC Resonant Converter with Voltage-doubler Rectifier

Koji Murata and Fujio Kurokawa
Nagasaki University
Nagasaki, Japan
bb52212202@cc.nagasaki-u.ac.jp

Abstract— **This paper presents a new synchronous rectifier method for a LLC resonant converter with voltage-doubler. The pulse width of secondary rectifier cannot be obtained in LLC resonant converter based on the primary side current because primary current include not only the current that is transferred to the secondary side but also the magnetizing current. In the proposed method, the pulse width of synchronous rectifier is obtained from the voltage across the secondary capacitor in the voltage-doubler rectifier.**

Keywords— Resonant, synchronous rectifier, voltage doubler.

I. INTRODUCTION

In recent years, the resonant converters have been drawing attention to minimize the power conversion loss. Among the resonant converter, LLC resonant converter has been popular because the LLC resonant converter can achieve the zero voltage switching in the whole load range. The secondary rectifier achieves zero current switching in wide operating region. A lot of papers about LLC resonant converter have been reported [1]-[9].

One of the central research targets in LLC resonant converter is the synchronous rectification. The phase of voltage across secondary transformer doesn't coincide with the phase of secondary current [1], [2].

In LLC resonant converters, the primary current includes the current that is transferred to the secondary side and the magnetizing current. It is difficult to determine the pulse width of synchronous rectifier in secondary side by detecting the primary current of LLC resonant converter, while the secondary current detection is not desirable in the view point of power loss. So the synchronous rectifier methods by detecting the voltage across the switches have been proposed [1]-[4].

In ref. [5], [6], the phase of secondary current is obtained by cancelling the magnetizing current through primary side. Although it is the most direct way to detect secondary current to achieve synchronous rectifier, the power loss by the detecting secondary current is not negligible.

In ref. [7], the open loop synchronous rectification is proposed in order to obtain both its simplicity and large improvement on efficiency over the diode rectifier. The synchronous rectifier method is proposed for LLC resonant converter with voltage-doubler [11]. In this method, the detection circuit is only one while it is required two detection circuits in LLC resonant converter with the center-tap structure. The number of turn of detection circuit is only one. Furthermore, the current used for detection is transferred to the load, which helps decrease the power loss. However, the complex circuits are needed for detection circuits.

The center-tap type rectifier shown in Fig. 1 is widely used in the secondary side of LLC resonant converters. The LLC resonant converter with voltage-doubler also has the advantage of ZVS capability for primary switches and ZCS for secondary switches. The LLC resonant circuit allows boost and step down operations by changing switching frequency.

In this paper, a new synchronous rectifier method for a LLC resonant converter with the voltage-doubler rectifier is proposed. It is well known that LLC resonant converter with the center-tap structure can achieve Zero Current Switching (ZCS) of secondary side switches for the sake of capacitive filter in output side. On the other hands, the LLC resonant converter with the voltage-doubler rectifier can also obtain advantages of primary Zero Voltage Switching (ZVS) capability for primary switches for whole load range, Zero Current Switching for secondary switches, and it can boost and step down the voltage. The purpose of the proposed method is to realize the synchronous rectification using simple control circuit without the current sensing resistor. The LLC resonant converter with voltage doubler circuit has the capacitor in the voltage doubler rectifier circuit. The output current can be calculated by using the voltage across the capacitor in the voltage-doubler circuit. Only the differentiator is used. The operating characteristics of synchronous rectification for LLC resonant converter with voltage-doubler in the proposed synchronous rectifier are shown.

978-1-4799-2706-7/14 $31.00 © 2014 IEEE

II. OPERATION PRINCIPLE

Figure 2 shows the configuration of LLC resonant converter with voltage-doubler rectifier. The half bridge configuration is used in the primary side that consists of switches S_1 and S_2. The resonant circuit consists of the resonant capacitor C_r, resonant inductor L_r and magnetizing inductor L_m. In the secondary side, the switches S_3, S_4 and capacitor C_f constitute voltage-doubler rectifier. C_o is the output smoothing capacitor and R is the load. i_{Lr}, i_{s3} and i_{s4} indicate the primary side current, the secondary current that goes through switches S_3 and S_4, respectively. v_{cr} and v_{cf} indicate the voltage across the primary resonant capacitor and secondary capacitor, respectively. The number of turn ratio is n:1.

Figure 3 shows the simplified bock diagram of proposed control circuit for the secondary synchronous rectifier. The voltage across the secondary capacitor C_f is detected. The phase of differentiator of voltage across the capacitor C_f coincides with the phase of secondary current so that the phase of secondary current is directly detected with simple differential circuit and comparator. Thus, the optimum turn-off timing for synchronous rectifier S_3 and S_4 are obtained. Figure 4 describes the theoretical waveforms of LLC resonant converter with the voltage-doubler rectifier. v_{gs1} and v_{gs2} are driving signals for primary switches. v_{gs3} and v_{gs4} are driving signals for the synchronous rectification in the secondary side.

State 1(t_0-t_1)

At t_0, the secondary current i_3 starts to flow, while the primary resonant current i_{Lr} continues to flow in a negative direction. Then the switch S_1 is turned on with ZVS condition. This interval is the resonant interval between the resonant inductance and resonant capacitor. The equivalent circuit in this interval is shown in Fig. 5(a).

State 2(t_1-t_2)

At t_1, the primary resonant current i_{Lr} becomes positive. For the sake of magnetizing current flowing though primary side, the start of primary and secondary current is different. Thus, the primary current and resonant capacitance voltage cannot be directly utilized for turn-on of secondary synchronous rectification.

State 3(t_2-t_3)

The operation during interval t_2-t_3 is shown in Fig. 5(c). The energy is transferred from primary to secondary side in this period. At the end of this interval, the primary resonant current decreases to the magnetizing current. Before t_3, the differential of the voltage across the capacitor decrease. So the driving signal for switch S_3 become low. At t_1, the current through switch S_3 becomes zero.

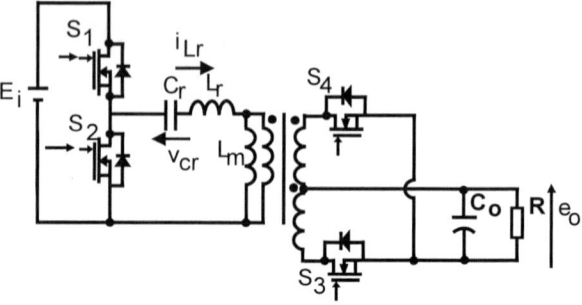

Fig. 1. Configuration of LLC resonant converter with center-tap structure

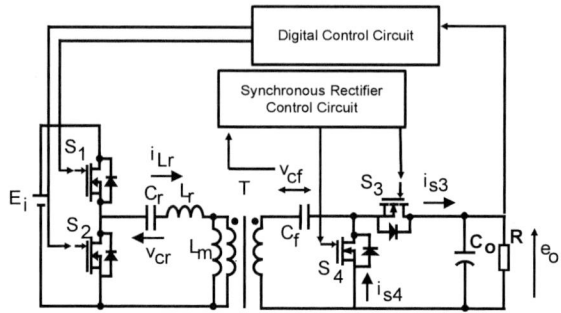

Fig. 2. Configuration of LLC resonant converter with voltage-doubler rectifier.

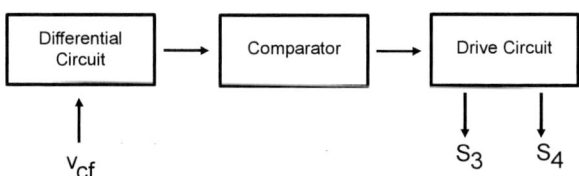

Fig. 3. Block diagram of proposed control circuit for synchronous rectifier.

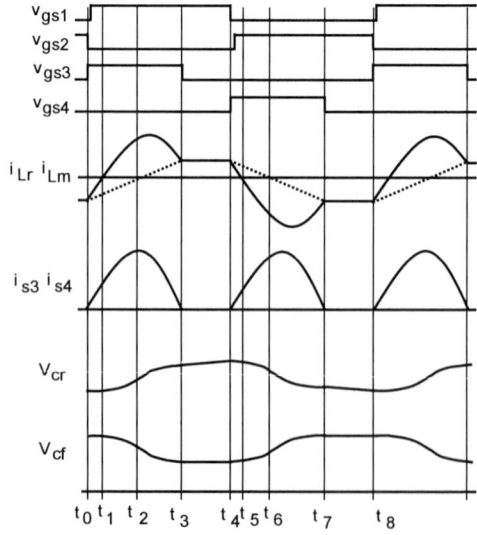

Fig. 4. Operating principle of with synchronous rectifier for LLC resonant converter with voltage-doubler rectifier.

978-1-4799-2706-7/14 $31.00 © 2014 IEEE

State 4(t3-t4)

At t_3, the secondary current becomes zero, while the primary resonant current i_{Lr} still flows in a positive direction. Therefore, the turn-on and turn-off instants of driving signals for secondary synchronous rectifiers cannot be directly obtained from the primary current. Since the center-tapped LLC resonant converter has no resonant capacitor in the secondary side, the current transformer is needed to directly detect the secondary current. In this paper, LLC resonant converter with voltage doubler rectifier is employed. As shown in Fig. 4, the voltage across the resonant capacitor, v_{cr} increases gradually by the magnetizing current i_{Lm}.

On the other hand, the voltage across secondary capacitor, v_{cf} is almost constant since there is no current flowing through in the secondary side during this interval.

Thus, the switch S_3 is automatically turned off based on differential of v_{cf}. During t_3-t_4, both secondary switches S_3 and S_4 are turned off. During t_4-t_8, the energy is not transferred to the load. The energy is transferred. At t_4, the switch S_4 is turned on and the resonant interval between C_r, C_f and L_r starts. During t_4-t_7, the energy is transferred to the capacitor C_f while the energy is not transferred to the load. At t_7, the primacy current decrease to the magnetizing current. During t_7-t_8, C_r, L_r and L_m resonance. The energy is not transferred from the primary side to the secondary side.

The operation during t_4-t_8 is shown in Figs. 5(e) through (h). The driving scheme is same as the driving for the switch S_3, which is based on the voltage across capacitor v_{cf}.

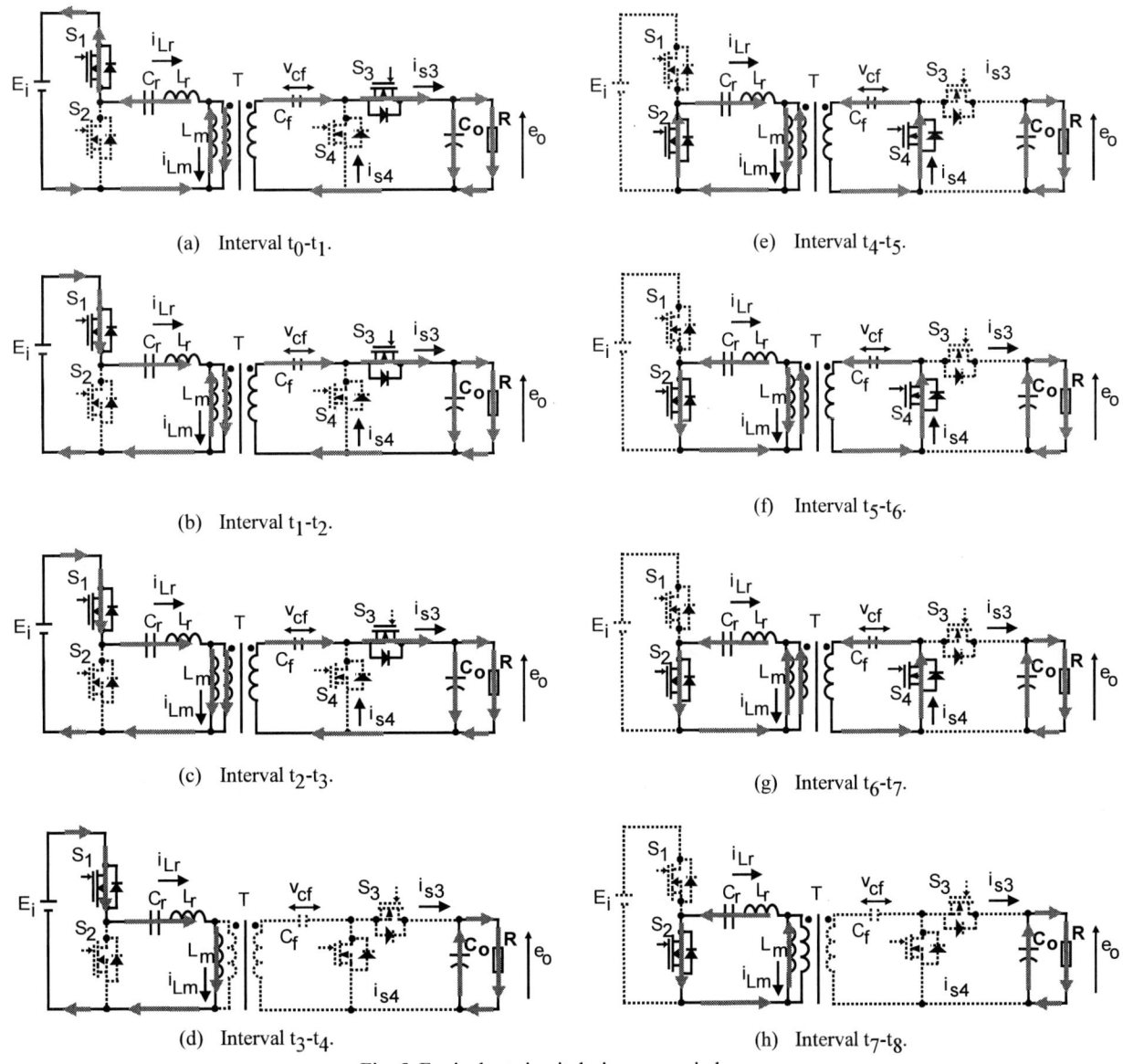

(a) Interval t_0-t_1.

(b) Interval t_1-t_2.

(c) Interval t_2-t_3.

(d) Interval t_3-t_4.

(e) Interval t_4-t_5.

(f) Interval t_5-t_6.

(g) Interval t_6-t_7.

(h) Interval t_7-t_8.

Fig. 5. Equivalent circuit during one period.

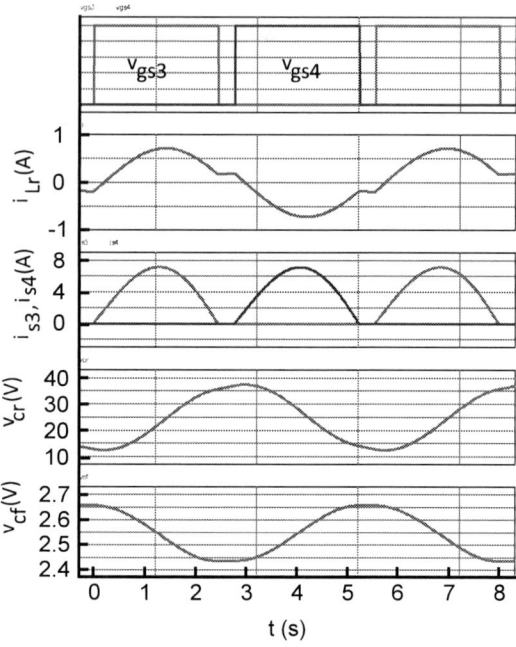

Fig. 6. Waveforms of current in primary, secondary capacitor and switches.

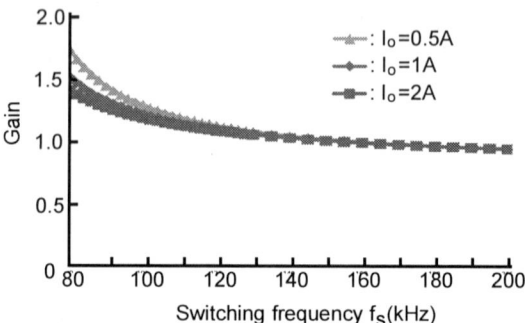

Fig. 7. Output voltage-frequency characteristic of LLC resonant converter with voltage-doubler rectifier.

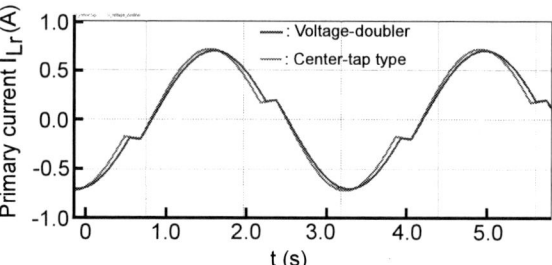

Fig. 8. Waveforms of primary current.

As mentioned in the above discussion, the primary side current includes the current that is transferred to the secondary side and magnetizing current as the center-tap type LLC resonant converters have. The current through the capacitor C_f is obtained based on the voltage across it. Thus, the synchronous rectification is achieved in switches S_3 and S_4. Figure 6 shows the waveform of current through primary side and secondary side, and voltage across the primary resonant capacitor and secondary capacitor in voltage-douber rectifier obtained from simulator PSIM.

III. STATIC CHARACTERISTICS

Figure 7 shows the frequency-gain characteristics of LLC resonant converter with the voltage-doubler rectifier. The resonant capacitor C_r is 47nF, resonant inductor L_r is 14μF, magnetizing inductor L_m is 170μF, and the capacitor C_f is 50μF. The resonant frequency f_{r1} and f_{r2} is 196kHz and 54kHz, respectively. The comparison of the primary side resonant current obtained from the center-tap rectifier and voltage-doubler rectifier is shown in Fig. 8. The blue line and red line show the primary current of LLC resonant converter with the center-tap structure and that with the voltage-doubler rectifier, respectively. Only the number of ratio between two structures is different so that the output voltage is same in both converters. The number of ratio of transformer in the LLC resonant converter with the voltage doubler is one half of the center-tap type. As it can be seen, the capacitor C_f in the secondary side has a little impact on resonant interval. The LLC resonant converter with the voltage-doubler rectifier is regulated by the Pulse FrequencyModulation (PFM) control. The horizontal axis

is normalized with the series resonant frequency. The series resonant frequency can be expressed as

$$f_{r1} = \frac{1}{2\pi\sqrt{C_{r2} * L_r}} \tag{1}$$

$$f_{r2} = \frac{1}{2\pi\sqrt{C_r(L_r + L_m)}} \tag{2}$$

where

$$C_{r2} = \frac{n^2 C_r C_f}{C_r + n^2 C_f} \tag{3}$$

The capacitor C_f has less impact on the series resonant frequency as expressed in eq. (1). Equation (2) shows the resonant frequency during interval t_3-t_4, and t_7-t_8. Therefore, the frequency-gain characteristics are almost same as that from LLC resonant converters with a center-tap structure. The output voltage is regulated by PFM (Pulse Frequency Modulation) control.

IV. CONCLUSIONS

A new synchronous rectifier method for a LLC resonant converter with voltage-doubler is proposed. In the LLC resonant converter, the phase of the primary side current doesn't match the phase of secondary current. In the proposed method, the turn-off timing of secondary rectifier of LLC resonant converter is obtained from voltage across the secondary capacitor. Its detection circuit has only differentiator and no complex circuit.

ACKNOWLEDGMENT

This work is supported in part by the Grant-in-Aid for Scientific Research (No.24360112) of JSPS (Japan Society for the Promotion of Science) and the Ministry of Education, Science, Sports and Culture.

REFERENCES

[1] Dianbo Fu, Ya Liu, Fred C. Lee, and Ming Xu, "A novel driving scheme for synchronous rectifiers in LLC resonant converters", IEEE Transactions on Power Electronics, vol. 24, no. 5, pp. 1321-1329, May 2009.

[2] Dianbo Fu, Ya Liu, Fred C. Lee and Ming Xu, "An improved novel driving scheme of synchronous rectifiers for LLC resonant converters," in Proc. Applied Power Electronics Conference and Exposition (APEC), pp. 510 - 516, Feb. 2008.

[3] Weiyi Feng, Fred C. Lee, Paolo Mattavelli, and Daocheng Huang, "A universal adaptive driving scheme for synchronous rectification in LLC resonant Converters," IEEE Trans. on Power Electronics, vol. 27, no. 8, pp. 3775-3781, Aug. 2012.

[4] Weiyi Feng, Daocheng Huang, Paolo Mattavelli, Dianbo Fu, Fred C. Lee, "Digital implementation of driving scheme for synchronous rectification in LLC resonant converter," Energy Conversion Congress and Exposition (ECCE), pp. 256-263, Sep. 2010.

[5] Chen Zhao, LI Bao-hong, Jing Cao, Yue Chen, Xinke Wu, and Zhaoming Qian, "A novel primary current detecting concept for synchronous rectified LLC resonant converter," in Proc. IEEE Energy Conversion Congress and Exposition (ECCE), pp. 766-770, Sep. 2009.

[6] Xinke Wu, Baohong Li, Zhaoming Qian, Rongxiang Zhao, "Current driven synchronous rectifier with primary current sensing for LLC converter," in Proc. IEEE Energy Conversion Congress and Exposition (ECCE), pp. 738-743, Sep. 2009.

[7] Jing Wang and Bing Lu, "Open loop synchronous rectifier driver for LLC resonant converter," in Proc. APEC, pp. 2048-2051, Mar. 2013.

[8] Seiya Abe, Toshiyuki Zaitsu, Junichi Yamamoto, and Shinji Ueda, "Adaptive driving of synchronous rectifier for LLC converter without signal sensing," in Proc. IEEE Applied Power Electronics Conference and Exposition (APEC), pp. 1370-1375, Mar. 2013.

[9] Weiyi Feng, Paolo Mattavelli, Fred C. Lee, and Dianbo Fu, "LLC converters with automatic resonant frequency tracking based on synchronous rectifier (SR) gate driving signals," in Proc. IEEE Applied Power Electronics Conference and Exposition (APEC), pp. 1-5, Mar. 2011.

[10] Dong Wang, Liang Jia, Jizhen Fu, Yan-Fei Liu, Paresh C Sen, "A new driving method for synchronous rectifiers of LLC resonant converter with zero-crossing noise filter," in Proc. IEEE Energy Conversion Congress and Exposition (ECCE), pp. 249-255, Sep. 2010.

[11] Junming Zhang, Jiawen Liao, Jianfeng Wang, and Zhaoming Qian, "A current-driving synchronous rectifier for an LLC resonant converter with voltage-doubler rectifier structure," IEEE Trans. on Power Electronics, vol. 27, no. 4, pp. 1894-1904, Apr. 2012.

Latest Developments in Increasing the Power Density of Traction Drives

Mark-M. Bakran, Andreas März
University of Bayreuth
Department of Mechatronics
Bayreuth, Germany
bakran@uni-bayreuth.de

Bernd Laska, Eberhard Krafft,
Olaf Körner, Andreas Nagel
Siemens AG, Traction Drives
Nürnberg, Germany
bernd.laska@siemens.com

Abstract-**Volume and weight of the propulsion equipment of a traction drive are most important features to provide more efficient transportation in the future. It will be shown how the converter technology for traction drives has improved its performance in the past decades and new developments are presented promising a further increase in power density. The machine and corresponding gear and suspension system is even more weight sensitive, and latest measures to increase the performance will be presented.**

Keywords— power density, traction, efficiency

I. INTRODUCTION

There are some basic features distinguishing a railway traction drive from a standard industrial drive system:

- Environmental conditions
  temperature, humidity, shock, vibration
- Supply system conditions
  voltage level, "weak" line, line voltage tolerances
- Restrictions on space e.g. low floor installations for EMUs,
- Restrictions on axle weight load

The latter two aspects are the driving force regarding innovation in the drive system. A better power-per-volume ratio will allow more efficient vehicle design with more space for passengers but also more flexibility regarding, for example space requirements for cross-border operation due to safety and control systems. A better power-per-weight ratio is the path to go in order to provide less axle load and the basis for a more efficient vehicle design. The overall benefits of a higher power density can only be seen on a vehicle level. As an example the next generation of Intercity trains for Germany will show how innovation in the drive system resulted in a completely new train design, where the distributed system with transformer and converter on separate cars could be condensed into specific power cars.

II. POWER DENSITY AND TRAIN DESIGN

Though this paper focuses on the drive system, this paragraph should first show the effect on system level. How huge the benefits on the rail -car design is, depends on many issues, e.g. axle load limit, power requirement, train concept – length of cars. One impressive example is

shown in fig. 1. Here, the traction -drive could be designed as to concentrate the drive components on one car. The benefits are found in a less complex interface between neighboring cars and a higher integration level, for example, the combination of transformer cooler and converter cooler as well as the integration of the auxiliary and traction converter in one unit.

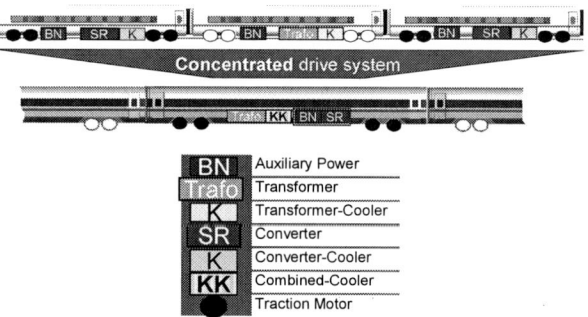

Fig. 1. Concentrated drive system for an EMU.

This example is to demonstrate that the main aim of increased power density is always to gain higher flexibility and to achieve better overall optimization of a train.

III. TRACTION CONVERTER

After a short summary of past developments this paragraph will analyze current developments, which will improve the converter power density and also try to show the major obstacles the power electronics development faces.

A. Development in the last two decades

The driving forces towards higher power density in the past two decades have been new and improved semiconductors, but also the corresponding simplification of the circuit concepts. The first big step was the replacement of the GTO by first generation IGBTs

Specific for Europe is the distinction between AC – single -system and multi -system trains and locomotives including 3 kV DC. The introduction of the IGBT achieved about 40% increase in power density for AC-

978-1-4799-2706-7/14 $31.00 © 2014 IEEE

traction, while for multi-system operation a nearly 80% increase could be achieved. This significant progress is thanks to the combination of the IGBT itself and the new voltage class 6.5 kV, which for the first time allowed designing a two level converter to be operated directly on the 3 kV DC line. Thus, the previously necessary chopper with high volume chokes or 3-level structures could be omitted.

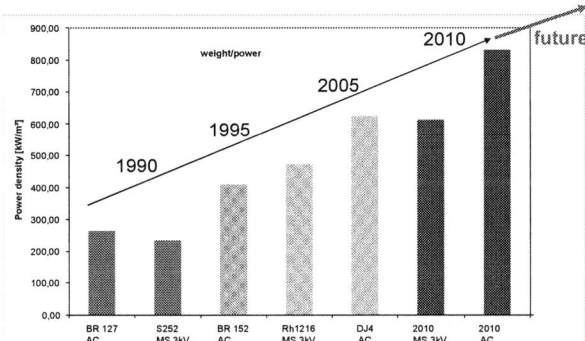

Fig. 2. Past development of power density for high power locomotive converters

Another significant step could be observed with the second generation of IGBTs. A major improvement in the losses was achieved by thinning the IGBT with the introduction of a field-stop design.

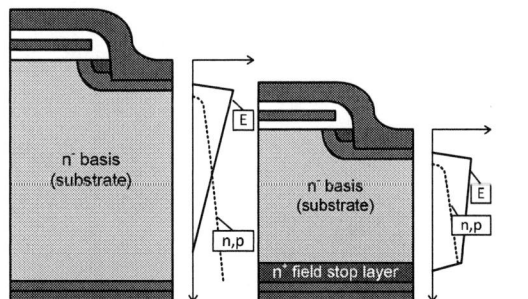

Fig. 3. From NPT-IGBT (left side) to FS-IGBT (right side) and reduction in thickness

B. Recent Developments

The introduction of the latest generation of trench IGBTs together with an optimization of the charge carrier profile presents the state of the art in reducing losses and thus improving the power per silicon area delivered.

This development has also shown the limits, which especially traction application faces.

- The diode performance does not match the IGBT performance. This applies first to switching losses, where turn-on IGBT losses dominate because of insufficient dynamic diode characteristics and this applies also to a reduced surge-current capability of the diode.
- The load-cycling limits of the converter output power. With the available higher allowable junction temperature, the requirement on load cycling rises

faster than joining technology has improved in this respect.

- The module packaging, especially the ampacity of the load terminal limits the performance, because the increased output current cannot be handled thermally.

Consequently future developments have to overcome these limitations to achieve a further increase in power density. Especially on the power cycling and high temperature capability, there has been a lot of development lately. The use of sinter layers between chip and ceramic substrate and also substrate to base-plate promise an increase of cycling capability of a factor of more than 10, see [24, 25]. This LTJT (low temperature joining technology) on the bottom side of the chip has to be combined with either copper bonding on top or also a sintered contact on top of the chips. This combination promises an increased power cycling capability of more than 20, while at the same time working at a 25K higher $T_{j,max}$.

C. Future Developments

Before considering wide-band gap devices, one should analyze the still existing potential of Si-based semiconductors.

For high-power high-voltage applications the RC-IGBT or BIGT is a very promising concept [14]. The combination of diode and IGBT area will improve the surge current performance significantly. Furthermore the low fundamental frequency thermal cycling between IGBT and diode can be omitted with the RC-IGBT.

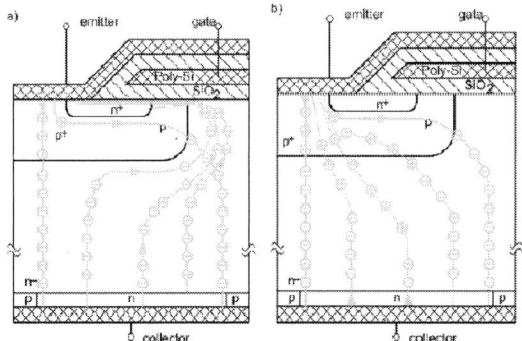

Fig. 4. Structure of the RC-IGBT and charge carrier flow in diode mode:
a) open gate-channel, b) closed gate-channel

As shown in fig. 4 the reverse side of the RC-IGBT is structured with an n-doped area, which produces a pin-diode structure. This has the additional feature that depending on the switching state of the MOS-channel a diode with high emitter efficiency (off) or low efficiency (on) can be. This produces a diode with either low V_F or with low Q_{rr}. In this diode controlled operation mode the gate-drive is capable of combining the low on-state losses with low switching losses by turning the diode for a short period into low Q_{rr} mode before turn-on of the opposite IGBT.

This evolutionary semiconductor concept promises an improved performance for HV-IGBTs. It is especially suited for applications, where the operation at low fundamental frequencies determines the performance. Regarding traction this applies dominantly to locomotives with a high starting tractive effort, which has to be provided continuously.

As a consequence a power per silicon area increase of about 20% can be expected [18].

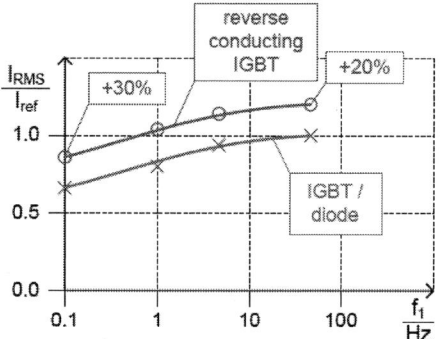

Fig. 5. Output power increase with respect to fundamental frequency with RC-IGBT compared to conventional IGBT and diode

This shows, that innovation based on silicon will continue to deliver improvements in power density.

For applications, which profit from higher switching frequencies, the material of choice is SiC. From fig. 6 one can observe that SiC will outperform Si with regard to chip area and also efficiency. This gap widens for higher switching frequencies and is substantial in the 1700 V class.

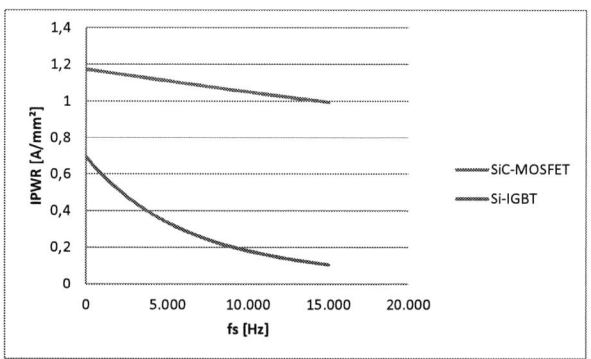

Fig. 6. Comparison of output current per chip area for 1700V Si-IGBT compared to 1700 V SiC MOSFET

IV. DRIVE SYSTEM DESIGN

The advantage of an increased switching frequency has to be evaluated with respect to the complete drive train. Basically three effects on system level have to be considered:

1. An increased switching frequency will lead to a lower harmonic content in the current, which corresponds to decreased copper and iron losses.
2. Reduced losses in the machine can be used to shrink the size and thus increase the power density.

3. An increased switching frequency is able to maintain the same torque pulsation also for a machine design with lower winding number. The lower winding number can be used to provide full braking up to the maximum motor speed. This way the recuperation energy during braking can be increased. However, this depends on the mission profile and whether by this approach the pneumatic brake can be omitted or reduced.

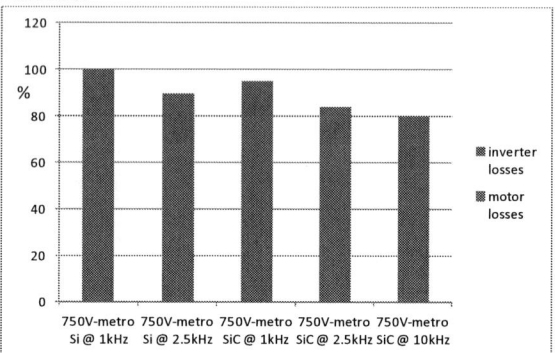

Fig. 7. Reduction of motor and inverter losses achieved with higher switching frequency.

From fig. 7 it can be seen, that the effect of using an elevated switching frequency in the motor side inverter is relatively limited. The motor losses for a low number of turns (winding) can be reduced by about 15%, with 10 kHz switching frequency by approx. 25%. By using SiC this can be achieved with inverter losses, that are slightly decreased despite the high switching frequency.

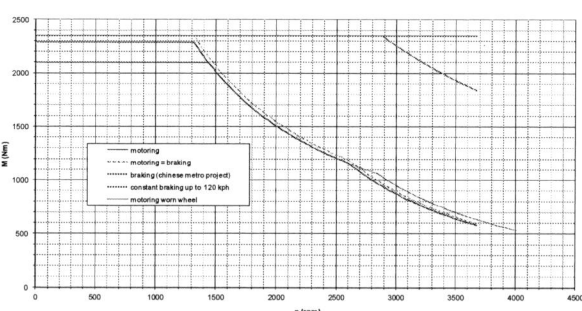

Fig. 8. Torque-speed characteristic of different motor designs for a metro. Blue: typical torque for motoring mode, dashed-red: Braking torque identical to motoring torque, green: Braking torque with increased performance, red: braking torque until maximum speed.

In fig. 8 typical torque-speed characteristics of a traction motor are shown. The standard motor design is characterized by the constant power region, which is limited by the maximum infeed power. The braking torque can basically be chosen higher, however this only lead to a better recuperation if the power can be fed back into the line. The ideal braking torque would stay constant up to maximum speed.

By using a mission profile of a metro with peak power the additionally fed back energy can be calculated. This is demonstrated in fig. 9. However, this requires a feed-back power that is factor 2 higher than the infeed-power.

The benefit of the high switching frequency of a SiC-based inverter can be found for this design, because the motor with the high braking torque has a low turn-number and only the high switching frequency can still provide a low distortion current with a low harmonics on the torque and low motor losses. However, for standard metros this goal can usually not be achieved because of the limited line current, which prohibits the feed-back of the regenerative power.

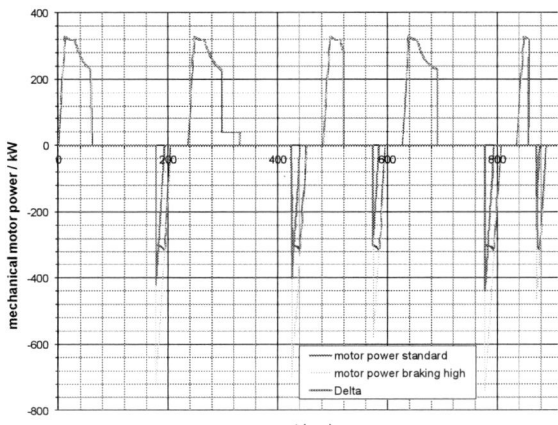

Fig. 9. Power at motor terminals for a metro mission profile. Blue: Standard profile with regenerating power equal to motoring power, green: braking torque = 2 motoring torque, red: additionally recuperated power by omitting the pneumatic brake.

In fig. 10 it is shown, that a metro, that uses the pneumatic brake at normal operation can profit from an increased recuperation performance. However, this will usually happen at the cost of additional losses due to the higher power operation at recuperation. Consequently inverters with a higher switching frequency add an additional degree of freedom in the drive system design and can thus improve overall losses in a typical metro application.

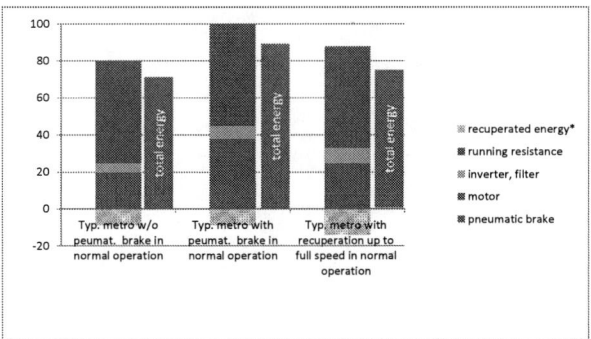

Fig. 10. Potential for loss reduction in a metro. The standard metro without pneumatic brake shows the lowest losses (left). The pneumatic brake used in normal operation increases the losses (middle), the full recuperation up to maximum speed can omit these losses (right). *) Recuperated energy: Only if power supply system allows full recuperation, otherwise at least partially dissipated in braking resistors.

V. AUXILIARY CONVERTER

Auxiliary converters are a vital subsystem of rolling stock and are key subsystems for a high availability of traction power. The main differences to the traction converter can be found in two aspects:

1. The insulation requirements between the high voltage supply system and the auxiliary supply is often realized by using a medium frequency transformer within a dc-dc converter
2. To fulfill the necessary voltage quality and to reduce harmonics, an output filter has to be used.

Both aspects lead to the effect that an increased switching frequency can significantly shrink the size and weight of the passive components. Therefore, SiC based semiconductors on the one hand or multi-level structures on the other hand promise a further increase in power density.

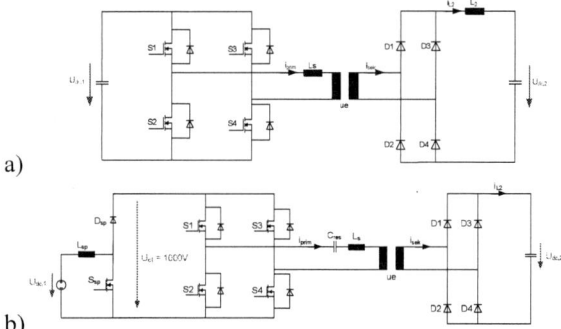

Fig. 11 Basic dc-dc converters with 1700 V SiC MOSFET
a) hard switched converter with full voltage range
b) resonant switched converter and boost converter for constant voltage transfer ratio

In [20] an auxiliary converter with SiC is demonstrated and shows that the inverter will exhibit smaller losses, a smaller filter and need less cooling. The essential ingredient to achieve this is a low inductive dc-link circuit.

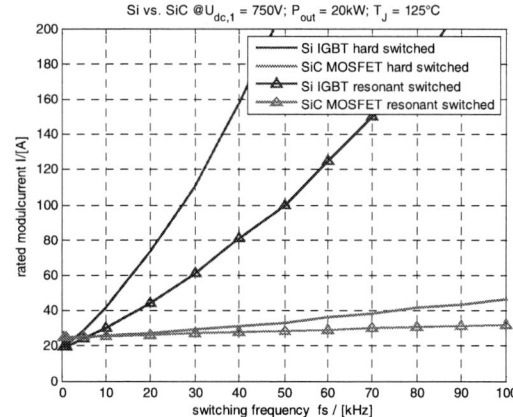

Fig. 12 Calculation of necessary chip area (equals nominal module current) as function of the switching frequency 1700V Si-IGBT [21] vs. 1700V SiC MOSFET [22]

The dc-dc converter as shown in fig. 11 promises the highest potential in weight saving. The primary side dc-voltage has a nominal value of 750 V (for LRV), for higher values like 1500 V or 3000 V a series stacking of the dc-dc converters is used. Therefore, the availability of the 1700 V class in SiC is essential for a modular converter concept based on SiC.

The competing topologies are the hard switched dc-dc converter, which copes with the large input voltage variation by adjusting the duty ratio and the resonant topology, which uses a boost converter to achieve a constant dc-link voltage. This topology as shown in fig. 11 b) has the advantage that the resonant converter does not have to operate at a variable frequency, which becomes quite a design constraint, when the voltage range is high and the load also varies from nearly no – load to maximum load. These are typical conditions for an auxiliary converter in railways.

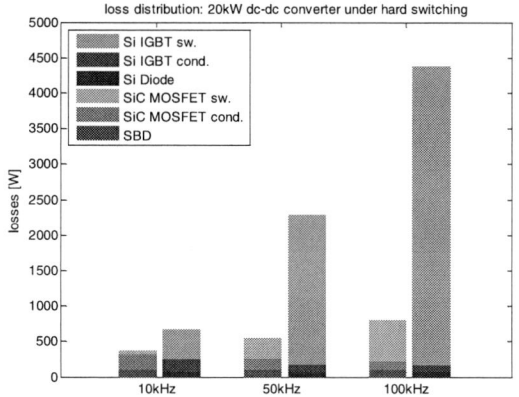

Fig. 13 Switching and conduction losses as function of the switching frequency 1700V Si-IGBT [21] vs. 1700V SiC MOSFET [22]

From fig. 12 one can derive that a SiC based converter can operate at 50 kHz without a too high penalty on needed chip area, while a Si based converter is limited to a few kHz.

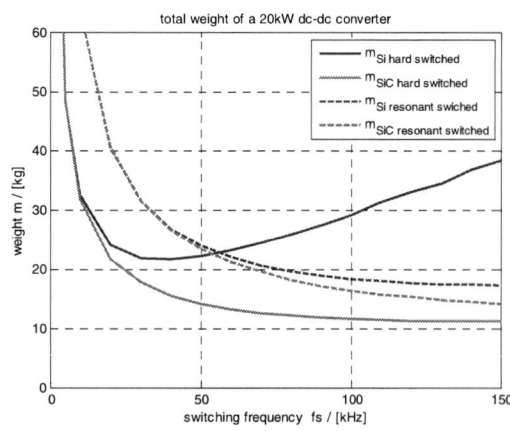

Fig. 14 Comparison of weight for a 20kW dc-dc converter for auxiliary supply. Hard switching operation vs. soft switching and 1700V Si-IGBT vs. 1700V SiC MOSFET

The increase in needed chip area with frequency is due to the switching losses. On a system level, this results in a

bigger heat sink, which starts to dominate the total converter weight at some point, see fig 14.

As consequence it is demonstrated in fig. 14 that SiC can lead to a weight reduction of about 30%, when operating a hard switched dc-dc converter. With soft switching a change to SiC doesn´t have such a big impact on weight savings. The weight minimum of the Si based dc-dc converter at about 30 kHz is due to the calculation of the heat-sink weight, which increases significantly for the high switching losses. For a complete comparison on system-level it will be decisive how the cost situation is affected by a penalty cost for weight. Today's situation in traction suggests, that a value of 10€ per kg is a good estimate to evaluate the benefit weight saving solutions vs. investment cost.

VI.   TRACTION MOTOR AND DRIVE TRAIN

Traction motor and the mechanical drive train are an integral part of a powered bogie. The powered bogie contributes considerably to the total vehicle weight especially to the weight of a locomotive (about 40%).

Lighter powered bogies for EMU or locomotives help to achieve development goals for lighter and more energy efficient vehicles or allow installation of additional equipment on a multi system locomotive e.g. several train control systems. Two examples of recent vehicle developments will be given below: the new EMU Desiro City and the new Vectron locomotive family.

A.   Recent Developments

Half suspended drive of EMU Desiro City:

An important new component of the EMU Desiro City is the compact and light weight bogie SF 7000 [15] with inner bearings which replaces the SF 5000 bogie (outside wheelset bearings) of the EMU Desiro UK.

Fig. 15 Desiro City traction motor. Self ventilated induction motor.

Along with the bogie a new self ventilated traction motor (fig. 15) and drive train was developed to offer the same performance within the smaller envelope of the compact SF 7000. The small high power traction motor is

978-1-4799-2706-7/14 $31.00 © 2014 IEEE

fully suspended by the bogie frame. A short coupling transfers the torque to the wheelset shaft mounted gear unit with two stages. The gear unit allows space for the coupling and is built very compact especially around the first gear stage. Moreover, the traction motor end shield at the drive end is designed to give room for the coupling as well.

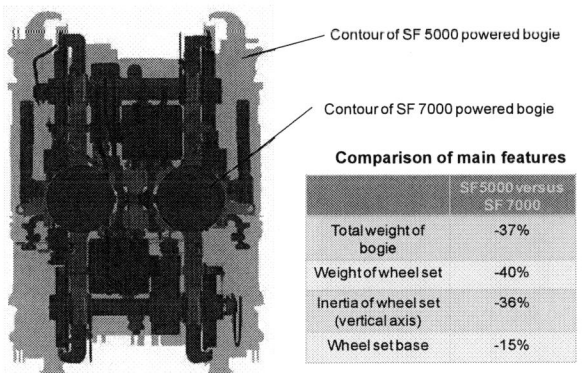

Fig. 16   Comparison of the powered bogies SF 5000 (predecessor) and SF 7000 for EMUs.

Because of the inner bearings, the traction motor, the coupling and the gear unit has to fit into a lateral space of less than 1000 mm whereas conventional bogies (outside bearings) offer lateral space up to 1300 mm. Sometimes this space is decreased by brake calipers and disks. In order to allow all the available lateral space within the SF 7000 for the drive components, tread brakes were chosen for this powered bogie. For service brake applications, motor braking is used as much as possible; therefore the use of the tread brakes is limited.

The SF 7000 powered bogie weighs approx. one third less compared to the SF 5000 (fig. 16). The wheelset base was shortened by 15 %. In terms of whole life cycle costs, the SF 7000 lowers the total vehicle weight, the energy consumption and the impact on the track (vertical forces and lateral forces in curves) which leads to lower track access charges.

Half suspended drive of Vectron locomotive:
The Vectron locomotive family is the successor of the Siemens Eurosprinter locomotives (fig. 17):
- Eurosprinter ES64U4 – high power universal locomotive, 6400 kW, max. speed 230 kph, SF 1 bogie with fully suspended drives.
- Eurosprinter ES64F4 - high power freight locomotive, 6400 kW, max. speed 140 kph, SF 2 bogie with nose suspended drives.

For the Vectron locomotive family, offering a maximum power of 6400 kW, the SF4 bogie with half suspended drives was developed which allows a maximum speed of 200 kph [16]. The bogie can also be equipped with fully suspended drives for maximum speed above 200 kph.
The pinion hollow shaft drive shown in fig. 17 consists

of a fully suspended traction motor and a wheelset shaft mounted gear unit. This configuration leads to lower unsprung masses compared to the nose suspended drive of the ES64F4. The unsprung mass is slightly higher compared to the fully suspended drive of the ES64U4.

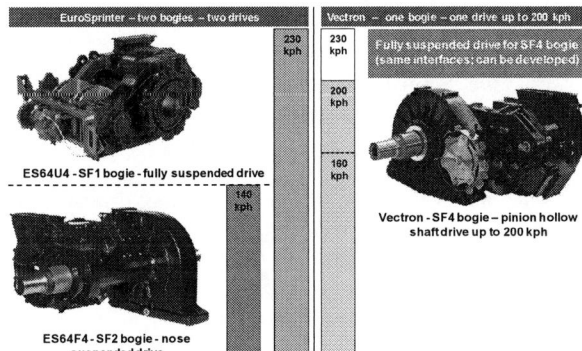

Fig. 17   Development of the drives for the Siemens locomotives from EuroSprinter (predecessor) to the new Vectron locomotive family.

The pinion hollow shaft drive shown in fig. 17 consists of a fully suspended traction motor and a wheelset shaft mounted gear unit. This configuration leads to lower unsprung masses compared to the nose suspended drive of the ES64F4. The unsprung mass is slightly higher compared to the fully suspended drive of the ES64U4.

The drive (motor and gear unit) is supported at three points within the bogie. Two supports are located at the head stock of the bogie which allow a horizontal pendular movement of the drive, the lateral movement is damped by a lateral motor damper. The third support is at the center bogie cross member. The gear unit is supported by a rod like suspension extending vertically from the pinion area of the gear unit to the traction motor end shield. There is no link of the gear unit directly to the bogie frame; therefore the pinion hollow shaft drive has the same interfaces to the bogie frame like a future fully suspended drive.

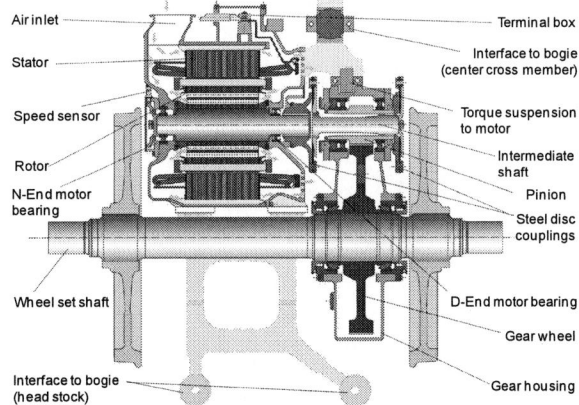

Fig. 18 Pinion hollow shaft drive for Vectron locomotive.

To install the large traction motor and the coupling between the bogie suspended motor and the wheel set shaft mounted gear unit, the pinion hollow shaft drive was chosen. The coupling unit with its two steel discs and the intermediate shaft takes up little lateral space because the shaft itself is located within the hollow pinion of the gear unit and the steel discs are very compact in terms of lateral space. With a conventional coupling unit located entirely between motor and gear unit, little space would be left for the traction motor. The former nose suspended drive and fully suspended drive for the Eurosprinter locomotive family did not have any coupling located between motor and gear unit, therefore no similar lateral space constraints.

The total mass of the pinion hollow shaft drive (motor, coupling, gear unit) is about 15 % lower compared ES64F4 drive and approx. 30 % lower compared to the fully suspended drive of the ES64U4 (fig. 19). However, the U4 drive (fully suspended cardan hollow drive) incorporates the mechanical brakes as well which increases the weight of the drive.

B. Future Developments

In the last ten to fifteen years, developers took a very close look at the benefits of permanent magnet (PM) traction machines replacing induction motors which are widely used in newer rail vehicles.

PM machine technology omits the inherent rotor losses of an induction machine (IM). The rotor losses account for 25% to 30% of the overall losses of an IM [17].

Moreover, PM machines can be designed with higher poles numbers without lowering the power factor. Smaller stator yokes and winding overhangs lead to higher torque and power density of the PM machine. This can help to reduce the weight of the traction motors used in a vehicle. This results in further reduction of the energy consumed during a duty cycle of a vehicle, if the vehicle speed frequently changes over time, e.g. in metro applications.

The possible application of PM direct drives without gear contributes to this trend since losses in gear units are omitted and the moment of inertia of the rotating parts of the drive is also reduced, resulting in a smaller equivalent mass of a train.

VII. CONCLUSION

A further increase in power density of all traction equipment, main converter, auxiliary converter and drive train is the way forward to deliver a higher value to the rail customer. For auxiliary converters the introduction of SiC based semiconductors will lead to significant weight savings. The traction converters will benefit from future improvements of Si devices. Especially new Si based semiconductor concepts like the RC-IGBT will lead the way for power electronics in the high-voltage and high-power domain of rail vehicles. Traction motor and drive train show significant potential with the mechanical integration into the bogie. Future PM drives with gear but

also with the capability to operate as a direct drive will be the next step towards higher power density.

REFERENCES

[1] A. Kopta, M. Rahimo, S.Eicher, U. Schlapbach; "A landmark in electrical performance of IGBT modules utilizing next generation chip technology", ISPSD 2006, Naples, Italy, Conf. Proceedings CD

[2] S. Iura, A. Narazaki, M. Inoue, S. Fujita, E. Thal; "Development of New Generation 3.3kV IGBT module", PCIM 2006, Nürnberg, Germany, Conf-Proceedings CD

[3] C. Gerster; "Trends in der modernen Antriebstechnik: Integrierte multisystemfähige Antriebssysteme"; SYMPOSIUM Elektrische Fahrzeugantriebe und –ausrüstungen, Dresden, 15. Oct 2006

[4] M. Bakran, M. Helsper, H.-G. Eckel, A. Nagel; "Challenges in using the latest generation of IGBTs in traction converters", EPE 2003, Toulouse, Conf-Proceedings CD

[5] J.-M. Bodson et al.; "ONIX 3000: How to Combine IGBT and Inverter Directly Coupled to a 3kV Catenary"; International Conference Railway Traction System 2001, Capri, pp. 2.39-2.54.

[6] Y. Hagiwara, M. Tanaka, M. Ueno "Technological Development of an IGBT applied Traction System for the Series 700 Shinkansen High-Speed Train"; International Conference Railway Traction System 2001, Capri, pp. 2.1-2.19

[7] M. M. Bakran, H.-G. Eckel; "Traction Converter with 6.5kV IGBT Modules"; EPE 2001, Graz, Conf. Proceedings CD.

[8] B. Kießling, Ch. Thoma; "Europalokomotive BR 189"; ZEVrail Glasers Annalen, Nr. 126 – 9/2002.

[9] Ch. Gerster, M. Meyer; "LCC based Evalutation of Traction Chain Systems for Multi-System Locomotives"; EPE 2003, Toulouse, Conf. Proceedings CD

[10] A. Colasse, A. Dandoy, Ch. Delecluse, R. Maffei, Ph. Thomas; "Development of a Multi-Voltage Locomotive with 6.5kV IGBT"; EPE 2003, Toulouse, Conf. Proceedings CD

[11] M.-M. Bakran; "A Power Electronics View on Rail Transportation"; EPE 2009, Barcelona, Spain, Conf. Proceedings CD

[12] M. Mermet-Guyennet; "Heavy and Light Train Technologies", EPE 2009, Barcelona, Spain, Conf. Proceedings CD

[13] Ch. Gerster; "Technology trends in railway traction"; PCIM 2009, Nürnberg, Germany, Conf. Proceedings CD

[14] L. Storasta et al.; "The next generation 6500V BIGT HiPak Modules"; PCIM 2013, Nürnberg, Germany, Conf. Proceedings CD

[15] J. Hirtenlechner, J. Brandstetter: SF7000 – das innovative Fahrwerkskonzept als Antwort auf Whole Life Cost Modell. ZEV rail 137 (2013), Tagungsband SFT Graz 2013.

[16] R. Paar: The SF4 bogie for Vectron locomotives. Rail Technology Review May 2012; pp. 32-35.

[17] O. Koerner: Permanent magnet machine technology in Traction applications. VDE Conference 2010, Leipzig, Germany, Conf. Proceedings CD.

[18] E. Krafft, R. Hermann, A. Maerz; „Reverse Conducting IGBTs – A new technology setting new benchmarks in traction converters"; EPE 2013, Lille, France, Conf. Proceedings

[19] T. Kobayashi et al; "Energy Saving Operation for Railway Inverter System with SiC Power Module"; PCIM 2012, Nürnberg, Germany, Conf. Proceedings

[20] F. Alkayal, J. Saada; „ Compact three phase inverter in Silicon Carbide technology for auxiliary converter used in railway applications"; EPE2013, Lille, France, Conf. Proceedings

[21] Infineon AG, FF150R17KE4 datasheet, 2012

[22] Rohm, BSM100D17P1C004 preliminary datasheet, 2012

[23] K. Park, S. Peterson, F. Canales: „Auxiliary power supply for LV inverter with 1700V SiC switch"; IECON 2013, Vienna, Austria, Conf. Proceedings

[24] P. Beckedahl: „Breakthrough into the third dimension – Sintered multi layer flex for ultra low inductance power modules"; CIPS 2014, Nurenberg, Germany, Conf. Proceedings

[25] N. Heuck: "Aging of new Interconnect-Technologies of Power-Modules during Power-cycling"; CIPS 2014, Nurenberg, Germany, Conf. Proceedings

Catenary and Storage Battery Hybrid System for Electric Railcar Series EV-E301

Y.Kono, N.Shiraki, H.Yokoyama, R.Furuta

East Japan Railway Company, Transport and Rolling Stock Department, 2-2-2 Yoyogi, Shibuya-ku, TOKYO, JAPAN

Abstract- **East Japan Railway Company (JR East) has developed the catenary and storage battery hybrid train system using a test car for the purpose of through operation service between electrified section and non-electrified section and decreasing environmental impact of diesel trains operating. We will develop commercial train Series EV-E301 applied this hybrid system and start operation in March of 2014. We introduce the catenary and storage battery hybrid system and technical items in this paper.**

Keywords— Hybrid System, Lithium-ion battery, Commercial Train.

I. INTRODUCTION

JR East has decreased negative environmental impact in non-electrified section. As the first step, the operational service of railcar using the hybrid system with diesel engine and the storage battery (Lithium-ion battery) has started since 2007. Furthermore, we have developed a "catenary and storage battery hybrid train system" to realize the through operation service between electrified section and non-electrified section and to further decrease negative environmental impact of diesel railcar. We manufactured the test car called "New Energy Train (NE-Train)" which has the hybrid system is installed and have verified system performance and technical tasks by running test.

This hybrid railcar charges the battery in electrified section. On the other hand, in non-electrified section, it runs by electric power of only storage battery. The storage battery charges at battery charging facilities of station. The running energy can be reduced by charging regenerative brake energy to storage battery on the other hand, diesel car cannot use it.

We introduce the running test result and verification by using new railcar Series EV-E301 applied the "catenary and storage battery hybrid train system".

II. DEVELOPMENT OF HYBRID SYSTEM IN JR EAST

A. Engine and Lithium-ion Hybrid Diesel Railcar

There are a lot of non-electrified lines except for the metropolitan area and the main lines in local areas (27 percent, 2030km of JR East area), where many diesel railcars are operated. Due to the rise of environmental problems in recent years, for the purpose of reducing the negative environmental impact of diesel railcars, JR East developed the hybrid system with a diesel engine and a lithium-ion battery, and began operational service of the world's first hybrid diesel railcar of series Ki-Ha E200 in July, 2007 (as shown in Fig.1 (a)) [1]. In 2010, the railcar of series HB-E300 with the same hybrid system began operational service in Aomori, Akita and Nagano areas (as shown in Fig.1 (b)) [2].

(a) Series Kiha-E200   (b) Series HB-E300

Fig.1 Diesel hybrid vehicle in JR East

B. Catenary and Storage Battery Hybrid Train System

JR East began to develop the car which is able to run by energy of only storage battery for decreasing negative environmental impact of the car in non-electrified section. JR East has gotten the experience of diesel engine hybrid system. In addition, increasing the power output and the capacity of storage battery decreasing has achieved in the field of automobile and industry and the cost of storage battery. Therefore storage battery can be applied to railcar [3].

We started to develop the catenary and storage battery hybrid train system in 2008 as shown in Fig.2.

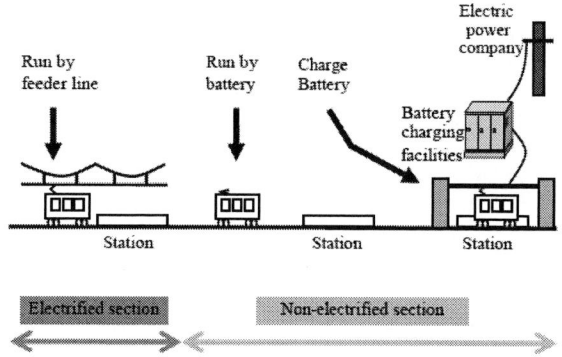

Fig.2 Catenary and storage battery hybrid train system

On electrified sections, the catenary and storage battery hybrid train runs by electric energy received from the catenary through the pantograph, charging the battery at the same time. In non-electrified sections, it runs by electric energy from the storage battery. Battery charging facilities are installed at some stations along the non-electrified section, and the electric energy for running is supplied when the train is stopped.

The purposes of this hybrid system are as follows.
- Running through electrified section and non-electrified section.
- Effective use of regenerative energy.
- Peak cut by leveling off the charging power.

Verification item of the field test by using test car of NE Train are as follows.
- Drive control of an induction motor with power rectifier and storage battery.
- Driving during charging to storage battery by power from catenary line in electrified section.
- Driving by power of storage battery in non-electrified section.
- Fast charging to storage battery at charging facility.

As reference from the result of NE-Train test running, new railcar Series EV-E301 applied the catenary and storage battery hybrid train system will be introduced on the Karasuyama Line (as shown in Fig.3).

Fig.3 Series EV-E301

Series EV-E301 train will run mainly from Utsunomiya to Karasuyama through 2 sections. One is Tohoku Line beween Utsunomiya and Hoshakuji. This section is electrified and this distance is 11.7km. The other is Karasuyama Line between Hoshakuji and Karasuyama. This section is non-electrfied and this distance is 20.4km. There is the facility for charging on board battery at Karasuyama terminal (as shown in Fig.4).

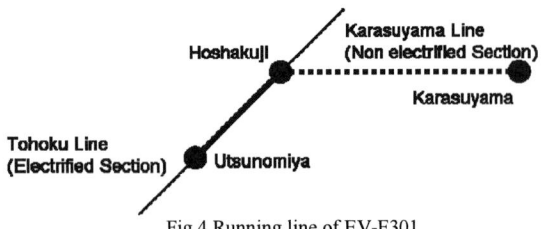

Fig.4 Running line of EV-E301

III. OUTLINE OF THE HYBRID SYSTEM OF SERIES EV-E301

A. Specifications of the Hybrid System
TABLE I shows specifications of the hybrid system of Series EV-E301 and outline of storage battery and

traction rectifier in Fig.5 and Fig.6.

TABLE I
SPECIFICATION OF HYBRID SYSTEM OF SERIES EV-E301

Item	Specification
Voltage of Main Circuit (Voltage of Main Battery)	DC630V
Max Speed	100km/h
Acceleration	0.556m/s^2 (2.0km/h/s)
Propulsion System	3 phase 2 level Converter – Inverter + 3 phase 2 level APS
Main Motor	3 phase induction motor, 95kW
Auxiliary Power Unit	3 phase AC440V 100kVA
Main Storage battery	Lithium-ion battery 190kWh

Fig.5 Storage battery

Fig.6 Traction rectifier

B. Composition of the Hybrid System
This hybrid system is composed of propulsion rectifier, storage battery and traction motors and can drive two traction motors per one system unit. The propulsion rectifier consists of DC to DC (DC-DC) converter that bi-directionally converts between DC 1,500V of catenary line and 630V for storage battery and VVVF (Valuable Voltage Valuable Frequency) inverter that drives traction motors.

Series EV-E301 is comprised of two cars and has four traction motors. Therefore Series EV-E301 consists of two systems which is made up the part under pantograph to traction motor and has high redundancy by dividing each main circuit independently (as shown in Fig.7).

978-1-4799-2706-7/14 $31.00 © 2014 IEEE

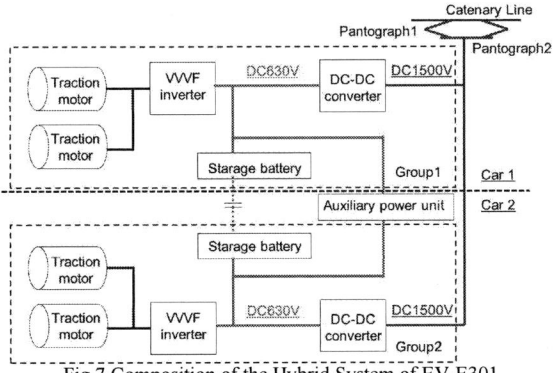

Fig.7 Composition of the Hybrid System of EV-E301

C. Operation of Hybrid System in Electrified Section

The traction circuit of the car connect DC-DC converter, VVVF inverter and storage battery, and receives DC1500V electricity from the catenary line, which is converted into DC 630V by the DC-DC converter and supplied to two traction motors. The storage battery supplies electric power to auxiliary power unit, too. The main circuit also supplies power to storage battery if SOC (state of charge) of storage battery is low.

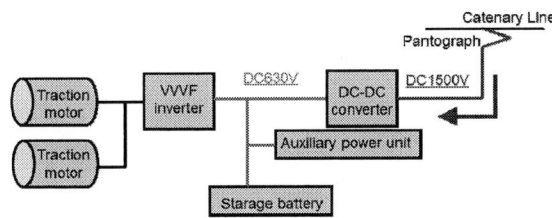

Fig.8 Hybrid system operation in electrified section

D. Operation of Hybrid System in Non-electrified Section

The traction circuit of the car connects VVVF inverter with storage battery directly, drives two traction motors and supplies the energy to auxiliary power unit. The regenerated braking energy is charged to storage battery in braking.

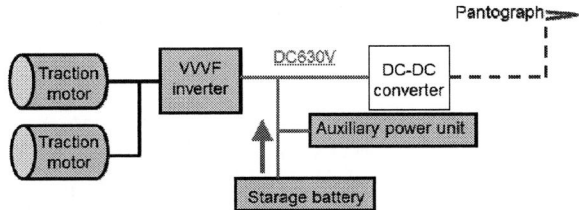

Fig.9 Hybrid system operation in non-electrified section

E. Specification of Storage Battery

Storage energy is considered as one cycle of battery charging form starting station to terminal station including the work of charging. Consequently high energy is required to run over 20 kilometers. However installation space is limited in the case of rolling stock, Lithium-ion battery with high energy density is decided.

Specification of storage battery is decided by the result of running simulation. DC 630V of storage voltage is realized by storage battery unit of 22 direct connection modules with eight cells that consist of 30Ah rated Lithium-ion battery for industry. There are 5 storage battery units in one system unit and storage battery capacity is 190kWh. It is described in chapter V.

IV. TRACTION CIRCUIT CONTROL AND RUNNING RESULT

Control function in electrified section of converter can be divided into four types shown below by difference of running section (electrified or non-electrified) and SOC.

A. Powering Control and Normal Charging

In powering, DC-DC converter controls all electric energy from catenary for traction, auxiliary power unit and storage battery. In the case that storage battery reaches designed upper limit of voltage, energy for traction and auxiliary power unit are supplied from catenary.

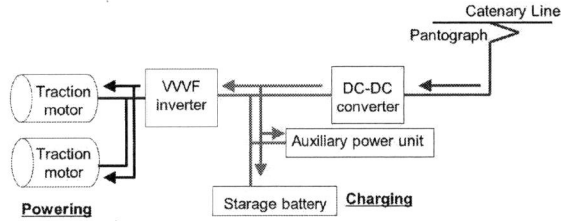

Fig.10 Block diagram of powering control

The running result in powering at 710V of storage battery voltage is shown in Fig.11. The motor current according to torque pattern flowed by powering control and the train accelerated. In addition, 100A of storage battery current flowed by normal charging function and SOC was increase over time. Thus normal charging function was operated accurately as to SOC in powering.

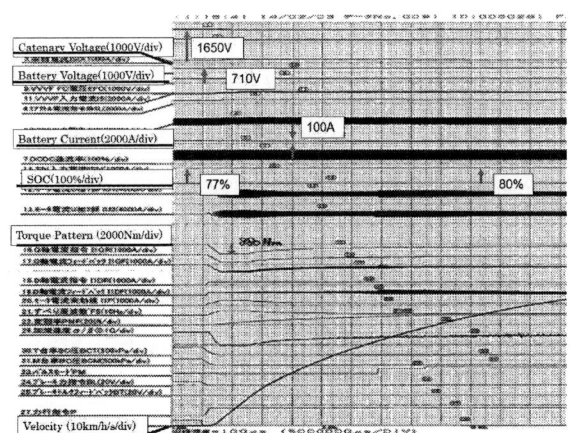

Fig.11 Running result in powering (at 710V of Storage Battery)

The running result in maximum powering is shown in Fig.12. The voltage of storage battery was 730V which reached the designed value to stop charging and charging current is 0A. This case of powering control is the same as normal electric vehicle.

We confirmed good result that starting acceleration at the region of constant torque in low velocity was 2.07 km/h/s,

978-1-4799-2706-7/14 $31.00 © 2014 IEEE

which is within 5% of designed tolerance.

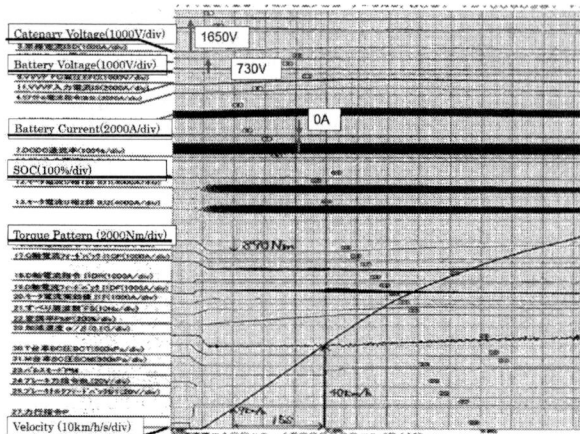

Fig.12 Running result in powering (at 730V of Storage Battery)

B. Powering Control in Non-electrified Section

Energy for traction and auxiliary power unit is supplied from storage battery.

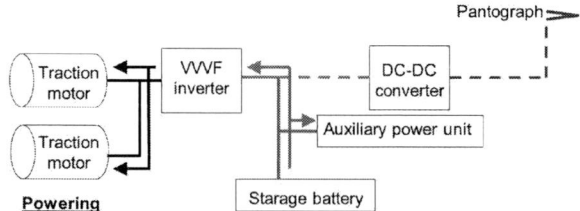

Fig.13 Block diagram of powering control in non-electrify section

The running result in non-electrified section is shown in Fig.14. As a result that output current of DC-DC converter was 0A, about 600A/car of storage battery flowed as to powering control. We confirmed good result that starting acceleration at the region of constant torque in low velocity was 2.01 km/h/s, which is within 5% of designed tolerance same as electrified section.

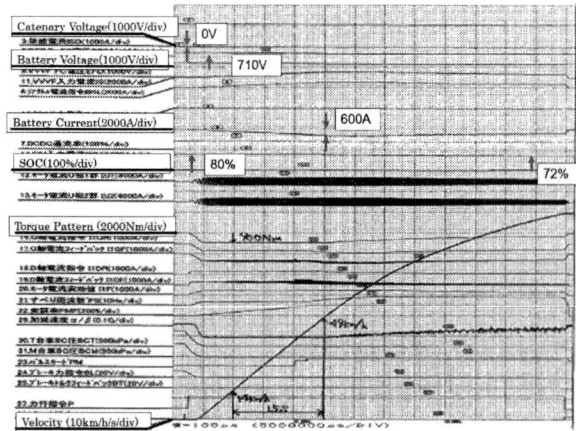

Fig.14 Running result in powering (in non-electrified section)

C. Regenerative Brake Charging Control

In regenerative braking, DC-DC converter carries out the regenerative brake charging control if the sum of regenerative energy (negative value) of VVVF inverter and supply energy (positive value) of auxiliary power unit is negative value. Therefore DC-DC converter return regenerative power to catenary in the case that storage battery reaches designed upper limit of voltage, but it is sent to the storage battery basically.

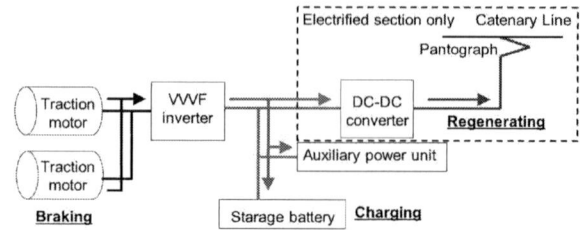

Fig.15 Block diagram of regenerative control

Running result of regenerative brake control in non-electrify section shown in Fig.16. The voltage of storage battery was 680V at the beginning of braking and regenerative energy used for charging of storage battery. On the other hand, running result in electrified section shown in Fig.17 was the case that the voltage of storage battery was 730V which reached the designed value to stop charging and regenerative energy return to catenary line. It was confirmed that the control according to running section and SOC was good.

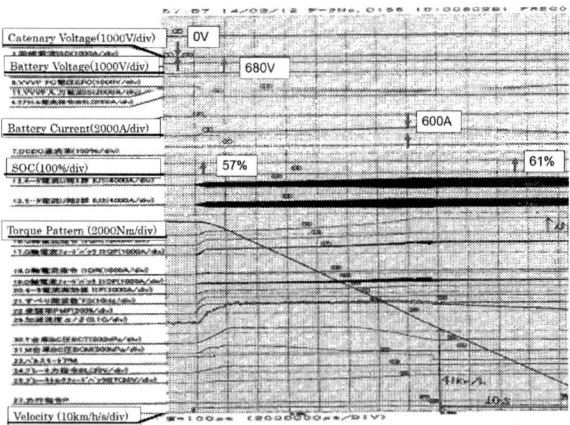

Fig.16 Running result in braking (in non-electrified section)

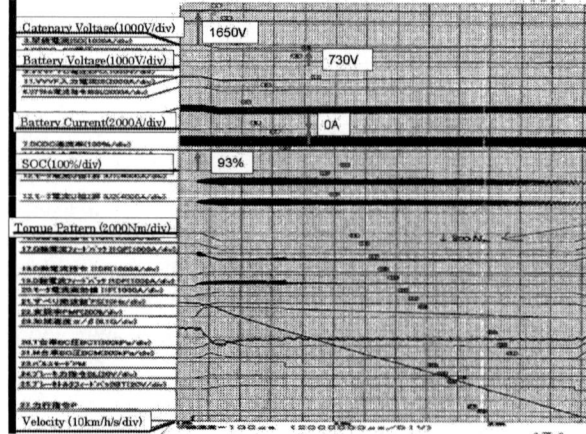

Fig.17 Running result in braking (in electrified section)

D. High Power Charging Control

When battery is charged from charging facility at terminal station, high power charging control is carried out by increasing the charging current in order to decrease the charging time. But catenary may be heat up and cut by flowing large current in stopping at station. Therefore it is necessary that the input current of pantograph is controlled within the bounds of thermal limit of catenary and pantograph.

The detail result of this control is stated in next chapter.

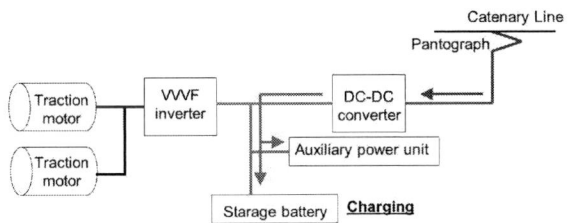

Fig.18 Block diagram of high power charging control

V. DESIGNING STORAGE BATTERY AND ENEGY CONSUMPTION RESULT

A. Designing about Storage Battery Capacity

It is necessary that the storage battery has to supply the energy for traction and auxiliary power unit. Considering commercial operation, designing of storage battery capacity include train delay and deterioration of storage battery itself. This design result leads the mass of storage battery shown in chapter III.

B. SOC Simulation and Running Result

SOC changing include time of high power charging is simulated for confirming that designed storage battery capacity is enough to run. Considering SOC about commercial train, it is important that charging time in high power charging control in addition to accuracy of regenerative brake energy and auxiliary power unit energy consumption. If charging time is long, train operation become to be inefficiency because the shuttle time become to be long.

Considering the worst case, the value of maximum powering energy and minimum regenerative brake energy is selected from the simulation result based on the performance characteristic and line profile and the test result of NE Train. The energy consumption of auxiliary power unit is also based on real value NE Train results in summer.

In simulated commercial running test, simulation result was verified by measuring the SOC changing and energy consumption. Fig.19 and Fig.20 are shown the result per car.

In down line direction, SOC was increased because there was downgrade slope between Kounoyama and Ogane. On the other hand, SOC was highly decreased in this section of up line direction. At the time of stopping the Kobana station, we got the result that decreasing of SOC was low because of braking and downgrade slope. It was confirmed the power consumption of auxiliary

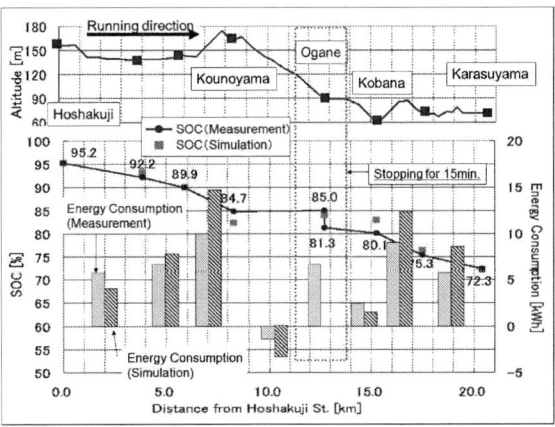

Fig.19 Running result (in down line direction)

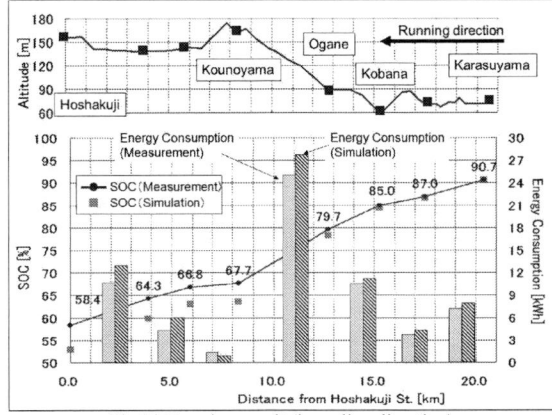

Fig.20 Running result (in up line direction)

power unit was about 28kW/car by using the measured energy consumption in stopping at Ogane station.

Although Karasuyama Line has heavy slope and powering time intends to long cause majority of upgrade slope in up line direction, SOC was 58.4% at the arrival of Hoshakuji station. Analysis of energy consumption of storage battery is shown in TABLE II.

TABLE II
ANALYSIS OF ENERGY CONSUMPTION

Direction	Total Energy Consumption [kWh]	Running Time [min]	Energy Consumption in APU [kWh]	Traction Energy [kWh]
Down	43.5	49	22.7	20.8
UP	61.4	40	18.7	42.7

The capacity of storage battery in this time was suitable for Karasuyama Line. In addition, this result showed accuracy of energy consumption simulation.

C. Consideration and Test Result of Charging Time

Charging points are Karasuyama and Hoshakuji stations, and there is the restrict current depended on catenary and charging facility.

- At Karasuyama Station

Charging power is limited 750kW/train since upper limit of supplied power capacity of high power charge facility. As a result, charging current is limited

978-1-4799-2706-7/14 $31.00 © 2014 IEEE   2124

250A/pantograph.

- At Hosyakuji Station
Charging current is limited 250A/pantograph since upper limit of current capacity of pantograph.

Using internal resistance taken account of storage battery deterioration, the charging time when 30kWh of required energy is shown in Fig.21.

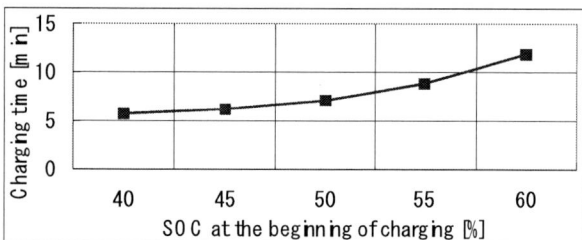

Fig.21 Charging time calculation for energy of 30kWh

In the case that SOC is high at the beginning of charging, charging time is intended to be long since it is influence that narrowing area of charging current occurs for overcharge protection.

High power charging at Karasuyama and Hoshakuji station is shown in Fig.22 and Fig.23.

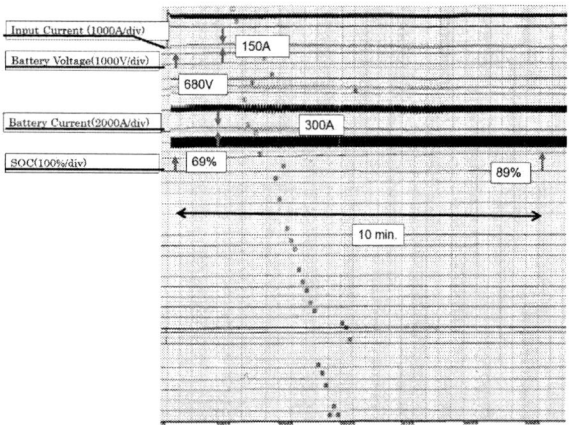

Fig.22 Fast power charging at Karasuyama Station

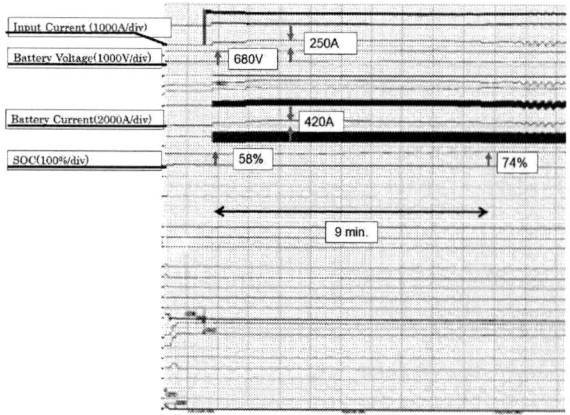

Fig.23 Fast power charging at Hoshakuji Station

At Karasuyama station, the input current from catenary was 150A, which was within 250A of upper current limit. The current of 150A was the explanation of overcharge protection. It was confirmed that SOC changing from 69% to 89% was in 10 minutes.

At Hoshakuji station, input current from catenary was about 250A near the limitation. Reduction of charging current was not occurred for 8 minutes of the start of charging because of the condition that voltage of storage battery was about 680V.

SOC at the beginning of charging was 58% and it was 74% after charging of 30kWh. Charging time in this situation was 9 minutes. This result was shorter than calculation result in Fig.21 because increasing of internal resistance taken account of storage battery deterioration was considered. Therefore it was commendable that study of charging time was validated.

VI. CONCLUSION

A catenary and storage battery hybrid train system to decrease negative environmental impact on non-electrified lines has been developed since 2008. Series EV-E301 train has applied this accomplishment will start commercial use. Development of Series EV-E301 as commercial train is reflected the traction circuit system designing due to change to a train of 2 cars and the result of SOC consideration.

Storage battery capacity designed in this time verified using Series EV-E301. By running test, it was confirmed that the control of hybrid system operated accurately both electrified section and non-electrified section. Moreover, it also was confirmed that consideration of storage battery capacity had high validity by simulated commercial running test.

The optimum volume of storage battery for Karasuyama Line was that additional high power charging facility was not required at the midpoint of non-electrify section. And it can be installed in rolling stock. In the future, for the purpose of developing the vehicle which can run another long distance in non-electrify section, it is necessary that storage battery will be high density. Because charging at the midpoint lead that travel time for destination becomes long and transportation service for passenger become poor. Additionally, it will be desirable that Lithium-ion battery come down in price with application of other industrial category. Hereafter technical knowledge as maintenance is obtained by continuous researching storage battery deterioration.

REFERENCES

[1] N. Shiraki, H. Satoh, S. Arai, "A Hybrid System for Diesel Railcar Series Ki-Ha E200," *INTERNATIONAL POWER ELECTRONICS CONFERENCE -ECCE ASIA- 2010.*

[2] M. NAkagami, "Summary of Series HB-E300 for Resort Hybrid Train in JR-East," ROLLING STOCK & MACHINERY, vol. 18, no. 11, pp. 4-7, 2010.

[3] H. Hirose, K. Yoshida, K. Shibanuma, "Development of Catenary and Storage Battery Hybrid Train System," *ESARS2012.*

Technology for Energy-Saving Railway Operation through Power-Limiting Brakes
---A case study at an urban railway---

Takafumi Koseki, Shoichiro Watanabe
Department of Electrical Engineering and Information
Systems, The School of Engineering
The University of Tokyo
Tokyo, Japan
takafumikoseki@ieee.org

Yasuhiro Hamazaki[1], Keiichiro Kondo[2],
Tomonori Hasegawa[3],Takeshi Mizuma[3]
Shin-Keisei Electric Railway Co. Ltd.[1], Chiba University[2],
National Traffic Safety and Environment Laboratory[3]
Chiba[1,2] and Tokyo[3], Japan

Abstract— An appropriate driver-assistance in order to use regenerative braking effectively is significant as a feasible method of a smart power management for energy-saving and economical solutions in sustainable electric railway operation. It is substantially significant to suppress braking force in high speed so that the braking force is less than corresponding traction force in the field-weakening speed region so that the whole braking force can be supplied by the regenerative electric brakes without active use of ordinary mechanical brakes. This power limitation in high speed is also beneficial to reduce the probability of regenerative braking cancellation caused by occasional high voltage at pantographs. If one limit the braking force, the total traveling time is increased, but this supplemental running time is relatively small, when the limitation of the deceleration is only applied in high speed and this supplemental time can be absorbed in realistic time margin, which is the difference of the initially scheduled conservative running time and actual one. Authors propose this type of the fully used electric braking pattern as "constant-power braking" or "power-limiting braking" and for realistic application to existing rolling stocks, they also propose a minor variation called as discrete approximate constant-power electric braking. The train driver must start braking operation considerably earlier than conventional ordinary train operation, and the braking force must be precisely controlled depending on the actual train speed timely. Since this driving method is, therefore, inherently more complicated than conventional one, a smart onboard driver-assistance or automatic train stopping control is requested for the realization.

Keywords— *electric railway, regenerative brake, energy-saving train operation, assistance*

I. Introduction

Most substations for DC-electrification have no functionality of regenerating power from railway to commercial power network side, in spite that many electric rolling stocks have regenerating brakes, and the technology contributes to ecological mass transportation[1]. Therefore, regenerative electric brakes cannot be used when there are timely no other accelerating trains near the braking train. Consequently, the effective usage of regenerative brake is less than

initial expectation since power squeezing and cancellation of regenerating brakes often occur. There are researches and technical development of regenerative power converters for DC-railway substations, and application of onboard or wayside power storage, *e.g.*, fly wheels, secondary batteries and electric double-layer capacitors[2]. They are, however, relatively expensive solutions. It is also possible to avoid the squeezing and cancellation of regenerative brakes by limiting the peak of braking power with carefully designed train motions. This paper discusses a train operation for better use of regenerative brakes and resultant energy-saving operations from this point of view. We explain our theoretical proposal of a train operation "approximate power constant braking" or "power-limiting braking"[3] to suppress the largest braking power and its practical problems from our experience of vehicle tests at Shinkeisei-Line. An appropriate driver assistance is needed for realization of the proposal in ordinary DC-train services, since the proposed train operation requests substantially difficult train motion control in comparison with conventional ones. The proposed idea will be more useful if it is involved in designs of new automatic train operation systems.

II. Energy-Saving Operation Control of a Train Motion

It is well-known that the combination of maximal powering, the possibly longest coasting, and maximal braking for energy-saving train operations. In fact, many commercial train operations are basically planned in this way[4]. However, the application of constant braking force from high speed operation have the following two problems.

The first problem is that electric braking force is usually smaller at high speed operation, since ordinary traction motor has "field-weakening mode" at high speed for efficient and economically rational designs of power electronic and electric machinery components. If the large braking operation is requested at high speed, mechanical braking system is requested to assist the

978-1-4799-2706-7/14 $31.00 © 2014 IEEE

shortage of the electric braking force. The mechanical braking system causes substantial energy loss.

The second problem is that large regenerated power often causes occasional and rapid rising of the overhead voltage, which results in squeezing or cancellation of regenerating braking operation for protecting power electronic components. Therefore, is is important to intentionally suppress the braking force at high speed for liming braking power by modifying train traveling pattern; the speed-distance profile called a run-curve. Since it can be realized just by the modification of the run-curve without any additional hardware, it may be an economy solution for saving electric energy.

Braking patterns appropriate for energy-saving operation is schematically represented in Fig. 1. The blue solid line in Fig. 1 (b) is conventional braking command with constant deceleration. The dotted green line represents the "approximate power constant" or "power-limiting" brake. The braking force is kept constant and large level at low speed. The regenerated braking power is simply proportional to the speed. The torque is reduced in proportion to in high speed. The green solid line represents the approximate quantized power-limited braking pattern, which is a step-wise internal envelope obtained by dividing the maximal force on traction curve equally to N (=7 steps in this case). This pattern gives a constant braking deceleration in a certain portion range of speed, and it may be appropriate for braking commands to assist a human driver so that one can control braking force under the performance of regenerative brakes.

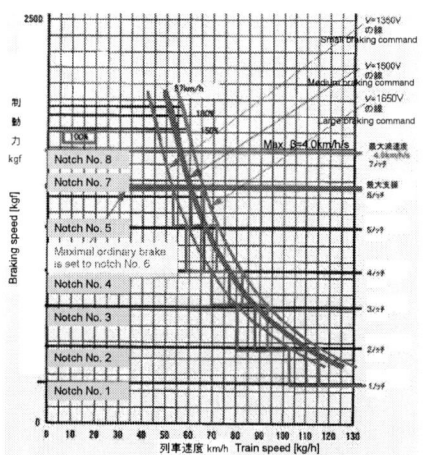

(a) General idea of the power-limited braking patterns.

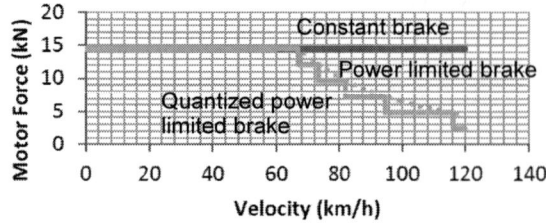

(b) A concrete quantitative implementation to a vehicle test.

Fig. 1 Relationship between train speed and braking-force proposal braking patterns

A train motion and power flow have been calculated based on the block diagram illustrated in Fig. 2 to evaluate the energy-saving effects of the proposed braking pattern in comparison with conventional constant braking deceleration on a simple run-curve of the combination of single powering, coasting, and braking as illustrated in Fig. 3. In this preliminary numerical case study, the following simple conditions have been assumed: The overhead voltage is kept 1500V constant, the maximal running coasting speed is 100km/h, and the inter-station distance is 2.0km/h. The regenerated electric energy in the proposed braking pattern has been 1.25 times as much as one in the conventional constant braking pattern, whereas the proposed pattern needs more running time of 1%. The reason for the small surplus of the running time is that the sensitivity of the suppression of the train performance at high speed to total running time is low in general.

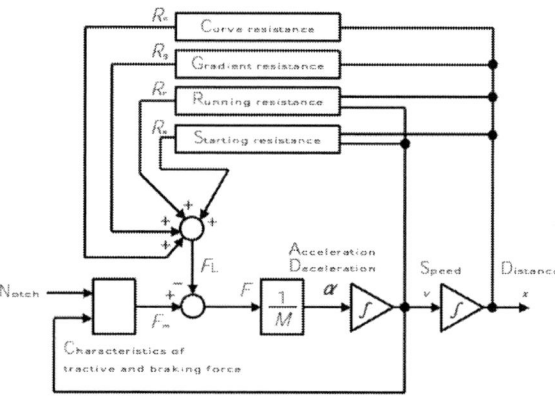

Fig. 2 Numerical calculation of a train motion

III. INTRODUCTION OF QUANTIZED POWER-LIMITING BRAKES AND A FUNDAMENTAL STRATEGY OF DRIVER ASSISTANCE

The inter-station running time must be strictly kept unchanged in real train operation. Therefore, the following factors have to be considered.

(1) The setting of the maximal braking-power limitation shall be decided corresponding to actual margin of the running-time at a certainly early time before the start of braking operation and consequent step-wise braking pattern shall be calculated and presented to a driver at appropriate timing as assistance braking commands for keeping the total inter-station running time.

(2) Since the actual overhead voltage has not been considered in the decision of the maximal braking power, it does not guarantee perfect regenerative braking operation, even though it gives the least possible braking power for keeping requested traveling time.

We propose the following braking strategy by considering the two factors above.

(1) At an appropriate timing during coasting mode, the possibly weakest step-wise braking pattern is calculated so that the train can spend out the actual time-margin in comparison with standard run-curve, which was designed conservatively by assuming the lowest overhead voltage.

(2) After the decision point above, concrete assistance braking commands are timely displayed to a driver by an onboard driving assistance device.

These strategies cannot guarantee successful regenerative braking operations, but it gives the best-effort of brakes for maximizing regenerated electric energy and for minimizing peak of the regenerated power in a deterministic decision process without any inter-train communication on the information of actual powering/regenerating power flows and overhead voltages.

IV. TRAIN OPERATION FOR A BETTER USE OF REGENERATING BRAKES

An experimental trial to apply the proposed braking commands to a real train is explained in this chapter.

A. Field for Vehicle Tests and Fundamental Information for a Driver Assistance

The quantized power-limiting braking pattern described in previous section has been applied to Shinkeisei-Line to confirm its energy-saving effect and its technical problems. Fig. 3 shows a case study at a section, where the inter-station distance is long, and the speed limitations from horizontal and vertical curve profiles are not severe. The test vehicle is a rolling stock of Shinkeisei 8000series, which consists of four motor- and two trailer- cars. The nominal overhead voltage is 1500V, and sixteen motors can produce maximal acceleration of 3.0 km/h/s/ and maximal deceleration of 4.0 km/h/s as illustrated in Fig. 3 (a). An onboard driving assistance device is needed for realizing the proposed braking operation.

B. Proposed Braking Assistance Commands

The fundamental train motion calculation has been executed based on the block diagram illustrated in Fig. 2. Fig. 3 (a) shows a calculated braking notch assistance pattern, whose maximal braking notch number is set to 6, where the largest normal braking notch number is designed to N=7. The horizontal line represents the position of the train where the driver shall change the braking notches.

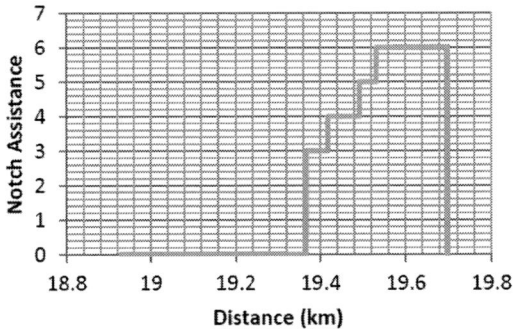

(a) A step-wisely quantized braking assistance command.

(b) Baking curves corresponding to the three braking patterns illustrated in Fig. 1 (a).

Fig. 3 An example of step-wisely quantized braking assistance command with power-limitation and corresponding braking curves o be used for driver-assistance calculated on a standard train-run curve.

C. Experimental Case Study

A vehicle test was carried out on a Sunday in March 2012 on the commercial line at Shinkeisei. An 8000-series train-set, which has a train state recorder at the second car, traveled from eastern to western terminals. The longest inter-station section between Gokoh-station and Tokiwadaira-station will be analyzed here in details.

V. EXPERIMENTAL RESULTS AND DISCUSSIONS

The purposes of the test data analysis are as follows:
(1) numerical verification of the train simulator, and
(2) analysis of practical problems of the proposed approximate power-limited braking pattern and train driver assistance by watching difference of the assistance commend and actual action of a driver.

A. Preliminary test in March 2012

The speed-logs shown in Fig. 4 have been obtained from a train state recorder. The acceleration and distance information have been obtained from low-pass-filtered numerical differentiation and integral of the speed record. Since there were coarse quantization and unexpected jumps, a precise data post-processing was difficult. Fig. 4 (a) and (b) show the comparison of braking commands and actual reaction of a driver. The consequent action of the electric brake is also plotted as dotted line. Unfortunately the driver could not appropriately follow the braking assistance commands, since he was too afraid of overruns at the next station. He started a strong braking action at high speed in spite of the weak braking command at high speed. This action requested large braking force exceeding the ability of the electric braking system and caused the undesired squeezing of regenerative electric brakes immediately. The voltage profile recorded in Fig. 4 (c) shows that the overhead voltage rose immediately after the strong braking action and it caused consequently the mechanical braking operation for compensating the reduction of electric brakes. The recycled energy was limited to just one third times of the initial expectation. In spite of the theoretical advantages in energy-saving effect, the first experimental case study with real human-driver assistance was not successful.

We discussed the following improvements for conducting next experimental trial.

(1) We should carefully explain the driver participating in the experimental trial technical reasons and the possible contribution of the propose braking assistance pattern to energy-saving operation in advance,

(2) we should decide and display the braking assistance commands sufficiently earlier than the braking start, and

(3) we should keep sufficiently long time for one braking step, whereas the shortest braking command interval was just less than two seconds in the previous trial.

B. Improved assistance test in October 2012

In order to realize the sufficiently early assistance as described in (2) above, we have set fixed "assistance-decision point" and prepared two layer of data bases.

The lower layer database consists of one-dimensional series of braking curves depending on different braking force corresponding to braking power limitations, which are calculated through "virtual backward acceleration from terminal point" as shown in Fig. 3 (b).

The upper layer database is a two dimensional database of the index of the braking power limitations, which are function of speed and reserve time when the train is passing the fixed assistance-decision point as shown in Fig. 5.

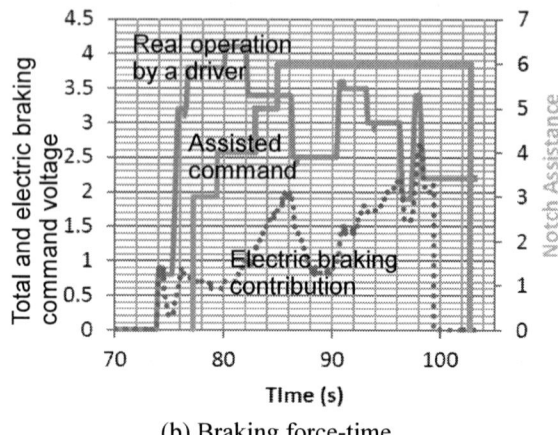

(b) Braking force-time

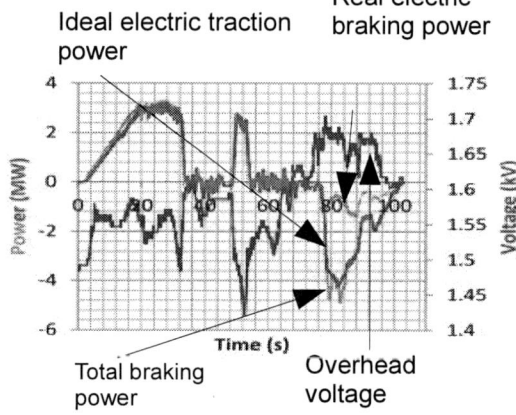

(c) Power flow and overhead voltage

Fig. 4 Braking command, actual braking operation, power flow and overhead voltages in a vehicle test.

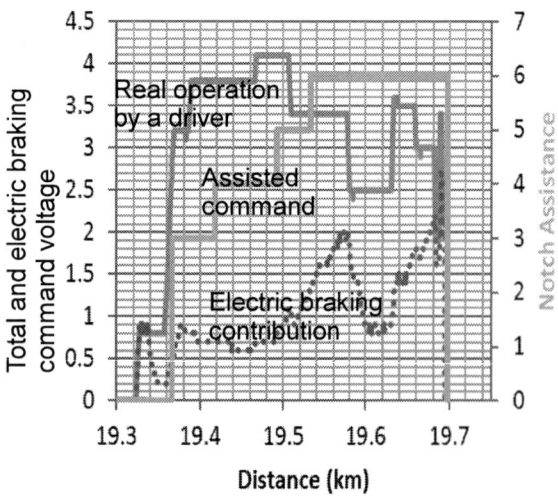

(a) Braking force-Distance

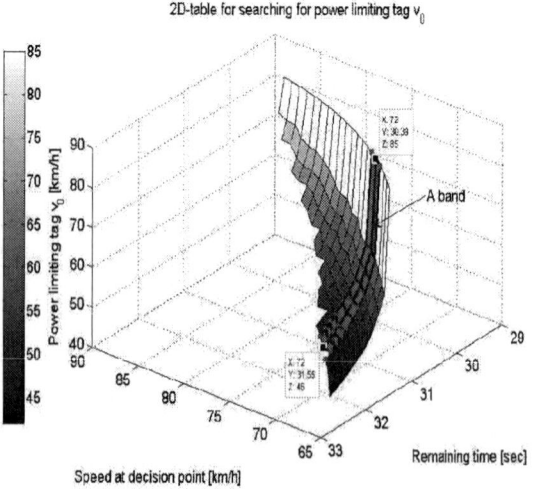

Fig. 5 Two-dimensional database of maximal braking power index dependent on speed and remaining time at an assistance-decision point..

The fundamental idea of the power limiting brake is to avoid squeezing regenerative brake by an explicit limitation of braking command for avoiding spontaneous rising of pantograph voltage caused by large electric braking power. However, it requires braking operation difficult for a human driver: he must start braking action earlier of several hundred meters than conventional ones, and the braking force shall be strengthened depending on the train speed. Therefore, an onboard sophisticated braking assistance as illustrated in Fig. 6 is necessary.

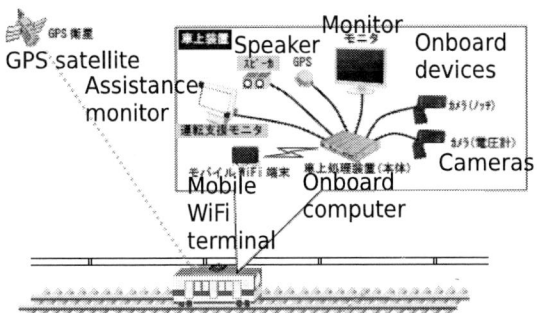

Fig. 6 Configuration of onboard drive assistance.

Based on the real-time data-acquisition of the speed and the reserve time at the assistance-decision point, the onboard driver assistance system gives a vocal braking instruction as well as the visual display as shown in Fig. 7. The vocal instruction is inherently significant since the driver must watch forward direction for safety.

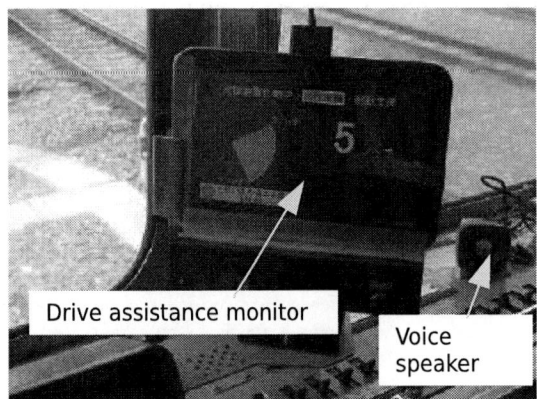

Fig. 7 Driver interface of the onboard drive assistance system.

A train operation status monitor records scheduled, assistance- command, and actual run-curves as shown in Fig. 8. When the train has accelerated faster than the initial schedule, the train has more reserve time than the initial schedule at the assistance-decision point. This allows weaker braking command than the initial schedule. The driver has followed the assistance-command almost well. But alight delay of the braking action can be also observed in Fig. 8.

Figs. 9 show that the proposed quantized power-limiting brake successfully increase the ration of electric brakes, which results in substantially better usage of regenerated electric energy.

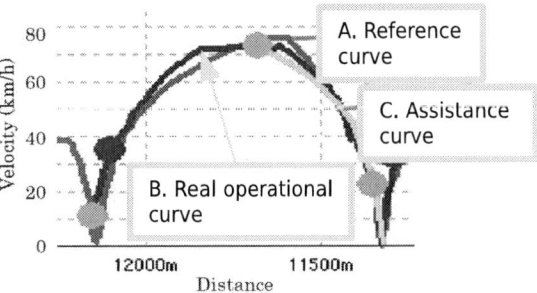

Fig. 8 An example of the driver assistance curve in the vehicle test in October 2012.

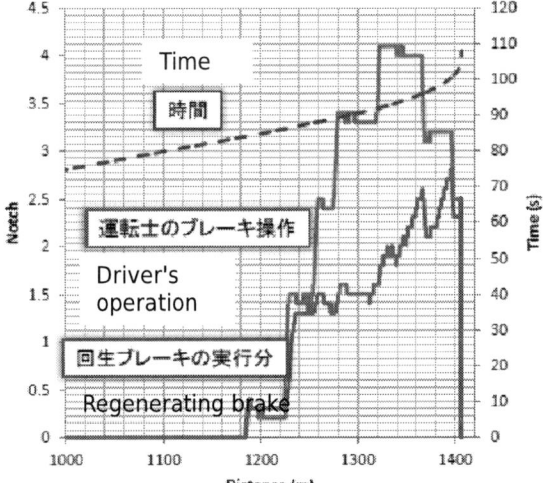

(a) Without assistance.

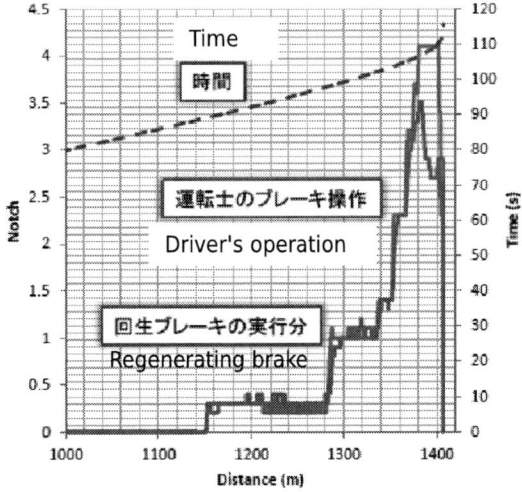

(b) With assistance.

Fig. 9 Regenerating braking action measured during the vehicle test in October 2012.

The train driver told us the following comments after the test.

(1) On the assistance display:

The visual display has become much better than previous preliminary tests. The indication of other lectric loads and energy-saving operation achievement may be useful assistance information.

(2) On the timing of assistance command:

It is very difficult to respond spontaneous vocal instructions timely. Earlier vocal instructions are preferable.

(3) On the assistance contents:

A new assistance command in a section of the arriving station is too late and awful. Also braking command after a neutral command is very unnatural and not preferable.

These comments mean that the intended assistance was almost successful, but the comment (2) means that the response delay to a spontaneous braking command was difficult and earlier instructions are required. Actually, Figs. 10 shows that braking delay requests stronger brake in low speed and such actions is not desirable from a safety point of view. Furthermore, the problem referred in the comment (3) occurs when a strong speed limitation is set before entering an arrival station. requests stronger brake in low speed and such actions is not desirable from a safety point of view. Furthermore, the problem referred in the comment (3) occurs when a strong speed imitation is set before entering an arrival station.

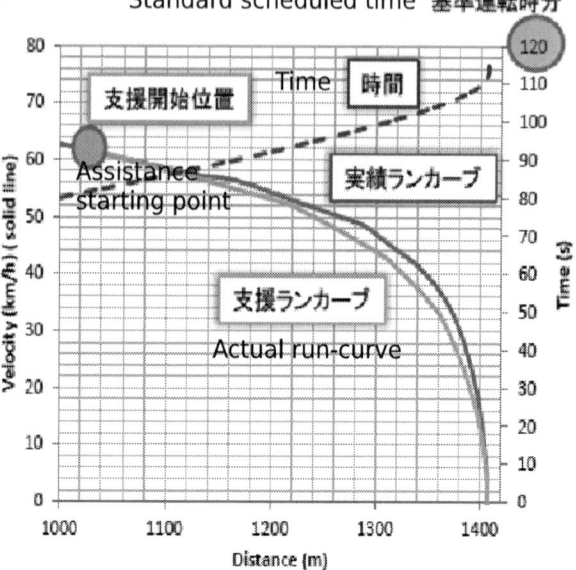

(b) With assistance.

Fig. 10 Run-curves measured during the vehicle test in October 2012.

C. Final assistance test in March 2013: Compensation of braking action delay and limitation of braking action in a section closed to the arriving station

The following two strategies have been added to solve the problems in a vehicle test held in March 2013.

(1) A braking command of a couple of seconds earlier based on prognosis of train running trajectory alculated after deciding assistance command.

(2) Introduction of an intentional braking notch ver-limitation in the braking curve calculation as a irtual backward acceleration from the stopping point.

Fig. 11 shows a method of the compensation of the raking action delay by earlier command based on rain-trajectory prognosis described in (1) above. Figs. 5 show the calculated braking command curve and raking notch actions when a strong speed limitation xists near the approaching station was very difficult or the driver to increase braking force just before train tops, therefore, the driver did not completely follow he braking assistance commands. However, the risk of he over-run has been successfully mitigated by the ntroduction of the intentional braking notch limitation n braking command pre- calculation. The energy-aving effects was, however, lost by this safer braking peration. The evaluation of the energy aspect is iscussed in [5]

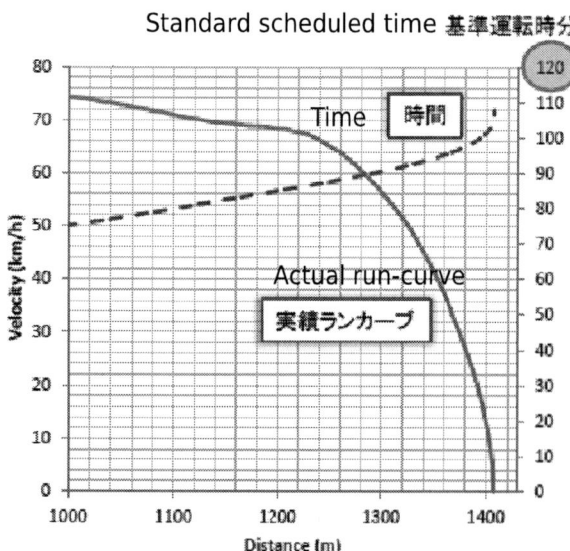

(a) Without assistance.

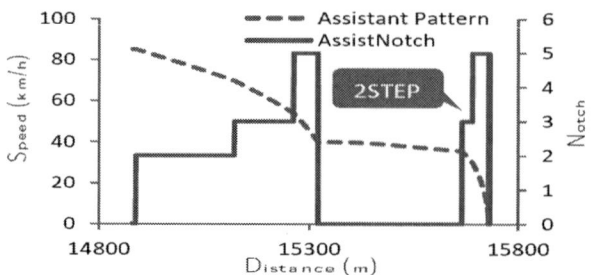

(a) Driver assistance

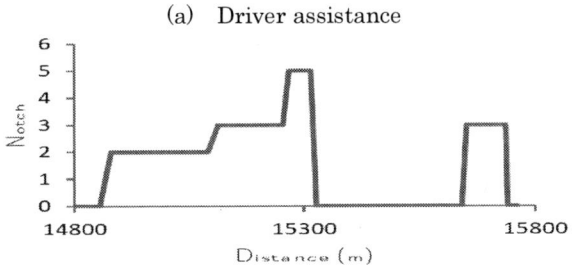

(b) Actual driver operation

Fig. 11 Braking notch reference and action for a section of strong speed limitation before an arriving station.

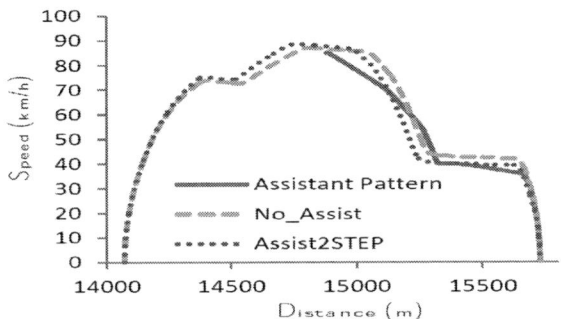

Fig. 12 Effects of the improved driver assistance: measured train run curve.

VI. APPLICATION OF THE ENERGY-SAVING BRAKING PHYLOSOPHY TO AYUTOMATIC TRAIN OPERATION

The necessity of quantized braking commands and difficulty in implementation discussed in chapter V were caused by the assistance of manual train operation. The energy-saving braking operation can be realized substantially simply and straightforwardly if it is applied to fully automatic train operations. In such automatic train operations,

(1) continuous power-limiting braking is possible,

(2) there are no problems of annoying man-machine interface discussed in chapter V,

(3) consistent rational designs and powering and braking train operation are possible, and

(4) vertical track profile significant to energy issue can be explicitly considered in the operational designs.

That is to say, theoretically optimal train operation can be mathematically calculated and implemented to realistic case studies. Also best-effort energy-saving strategy is possible based on the two-layer data base proposed in chapter V, which is flexible to variation of pantograph voltages and train payload in response to the request of keeping traveling time constant.

VII. CONCLUSIONS

We have proposed quantized power-limiting braking pattern and an appropriate driver assistance for more

efficient use of regenerated electric power from an electrically braking train at a DC-electrified section. The motion of an electric rolling stock and its power flow have been numerically calculated for deciding the appropriate driver-assistance commends and the advantages and problems have been investigated though vehicle tests at Shinkneisei-Line as an experimental case study. Fundamental advantage of the proposed method has been confirmed if a driver can react appropriately the numerically derived braking assistance commands. However, the driver could not always follow the assistance-commands successfully, since the proposed operational method is substantially more difficult and is often felt unnatural for a human driver because of considerably earlier braking start and the gradually increased braking forces. Sufficiently earlier decision and presentation of the braking assistance commands to a driver is needed and we have investigated the earlier and carefully designed driver guidance for the following experiments based on two-layer data base of braking index as a function of reserve-time and passing speed of a train at a fixed braking-assistance decision point.

This research is financially supported by Japan Railway Construction, Transportation and Technology Agency, in the framework of its promotion of fundamental technical researches in the field of transportation; the project name "Power management for sustainable economy and energy-saving railway systems (Project No. 2010-04)." We are cordially grateful for this support.

REFERENCES

[1] http://www.mlit.go.jp/k-toukei/search/excelhtml/23/23000000x00012.html, *written in Japanese*

[2] Shimada, M. et al. (2010): "Energy Storage System for Effective Use of Regenerative Energy in Electrified Railways", Hitachi Review, Vol. 59, No.1, pp. 33-38, written in Japanese

[3] Koseki, T., Mizuno, Y., and Mizuma, T. (2012): " The best use of regenerative brakes by adjusting train intervals in DC-electrified railway systems," IEEJ Annual Conference 2011, 5-075, March 2012, Hiroshima, Japan, written in Japanese.

[4] Miyatake, T. (2011) :"A Simple Mathematical Model for Energy-saving Train Scheduling," IEEJ Transactions on IAS, Vol. 131, No. 6, pp. 860-861, written in Japanese

[5] Watanabe, S. and Koseki, T. :"Train Group Control for Energy-Saving DC-Electric Railway Operation," Proceedings of IPEC 2014

An Overview on Braking Energy Regeneration Technologies in Chinese Urban Railway Transportation

Zhongping YANG , Huan XIA, Bin WANG and Fei LIN
School of Electrical Engineering
Beijing Jiaotong University
Beijing, China
zhpyang@bjtu.edu.cn

Abstract— **In order to prevent the failure of regeneration and reduce energy consumption, making full use of regenerative energy has been widely recognized as an important issue by Chinese urban railway transportation community. This paper provides an overview on the technologies of regenerative energy braking in China. It is first explained that why the ground resistor is popular in China in order to prevent the failure of regeneration. Alternative technology using regenerative inverter and ground supercapacitor is then introduced including main circuit, working mechanism and existing problems. Next, the simulation, status quo and experimental results of the ground supercapacitor energy storage system are discussed in detail for a 200kW prototype system developed in China. Finally, future improvements are mentioned for Chinese braking energy regeneration technology.**

Keywords—. **Regenerative Braking, Regenerative Inverter, Super Capacitor, Urban Railway Transportation.**

I. INTRODUCTION

With the continuous and rapid development of Chinese economy, the numbers of passenger cars and urban population significantly increased in recent years. Traffic jam and pollution are now serious social problems in Chinese big cities such as Beijing, Shanghai and Guangzhou. It is a national-wide consensus to mitigate traffic jam and improve air quality through developing energy saving and punctual mass transportation systems such as the urban railway system. Under this basic consideration, China is now leading the world in the development of the urban railway transportation system. By the end of 2012, there are 2,008 km urban railways in operation in 17 Chinese cities. In addition, 80 lines are under construction, whose total length is about 2,000km. And more than 400 lines are being planned with total length of over 14,000km.

The two biggest advantages of the railway transportation over other means of transportations are its small rolling resistance and the capability of braking energy regeneration. However, the regenerated energy is not able to return to the grid through traction substation due to the DC power supply of the urban railway systems. While the train is regenerating the braking energy, if there is no other train absorbing the energy, the pantograph voltage will sharply increase that may cause

the traction motor incapable to work as a generator any more. This situation is the so-called failure of regeneration. Then the train's kinetic energy can only be converted into heat and dissipated in air through mechanical braking or braking resistor. It received a considerable attention in recent years to prevent the regeneration failure, and thus reduce energy consumption and improve train's performance by making full use of the regenerative braking energy [1].

This paper introduces the current situation and problems of the braking energy regeneration technology in Chinese urban railway transportation systems. Then a detailed discussion is provided on the development of regenerative energy storage system using ground supercapacitors. The system will be implemented in Beijing subway Line 10.

II. REGENERATIVE ENERGY RECOVERY TECHNOLOGIES IN CHINA

Urban railway transportation has developed for over 40 years in China since the first line, Line 1 from Beijing railway station to Pingguoyuan opened in 1969. The power supply system of Chinese urban railway transportation system is either 750V DC or 1500V DC. Ground and viaduct lines usually receive electrical power from overhead wires, while underground lines are commonly powered by a third rail. The urban railway systems in China are significantly diversified in recent years including subway, monorail, linear motor metro, LRT (light rail transit) and so on. In terms of size of the car, there are two types in China, type A with 20m long and 3.0m wide, and type B with 18m long and 2.8m wide. Type A railway cars are widely used in Shanghai and Guangzhou. Type B ones mainly operate in Beijing now, while the adoption of type A cars is under discussion and may also be introduced in Beijing in the future. In addition, AC electric driving has totally replaced the DC electric driving in Chinese railway trains since 1990s.

As an important sub-system of urban railway cars, the performance of braking system directly influences driving safety, energy consumption, riding comfort, and temperature rise in tunnels. In Chinese urban railway cars, the electro-pneumatic blending brake is usually adopted, in which the regenerative braking plays a major role and the air braking only serves as an auxiliary brake. As shown in Table 1, there are multiple technologies to absorb the regenerative energy when regeneration failure occurs, such as on-board resistor, ground resistor, the combination of ground resistor and regenerative inverters, regenerative inverters, ground supercapacitors [2]. It is obvious that on-board and ground resistors are only able to dissipate energy instead of the recycle and reuse of the regenerative energy. However, since the way using resistor to consume the regenerative energy is easy to control and maintain with low cost, it is now a major technology in China to prevent the regeneration failure. Therefore, this technology is also briefly reviewed in this paper.

TABLE 1
EXAMPLES OF METRO REGENERATIVE BRAKING ENERGY ABSORBING TECHNOLOGIES IN CHINA

ABSORBING TECHNOLOGIES	APPLIED LINES (YEAR OF OPERATION)		
ON-BOARD RESISTOR	LINE 1 IN BEIJING SUBWAY(1969)	LINE 1 IN SHANGHAI METRO(1993)	LINE 1 IN GUANGZHOU METRO(1997)
GROUND RESISTOR	LINE 4 IN GUANGZHOU METRO(2005)	LINE 6 IN BEIJING SUBWAY(2012)	-
GROUND RESISTOR AND REGENERATIVE INVERTER	LINE 14 IN BEIJING SUBWAY(2013)	LINE 5 IN GUANGZHOU METRO(2009)	LINE 9 IN BEIJING SUBWAY(2011)
REGENERATIVE INVERTER	LINE 10 IN BEIJING SUBWAY(2008)	LINE 14 IN BEIJING SUBWAY (2013)	-
SUPER CAPACITOR	LINE 5 IN BEIJING SUBWAY(2007)	-	-

III. RESISTOR BRAKING

There are two ways of rheostatic braking that are widely used in China, braking using on-board resistor and ground resistor, respectively. It is a long-term controversy on the capacity of the on-board braking resistor in China. The controversy focuses on if the braking resistor is required to absorb all or part of the regenerative energy from the maximum speed to zero speed [3]. As a practical solution, the A type railway cars absorb all the regenerated energy by the on-board resistor such as Line 1 of Guangzhou Metro, while considering the reduction of weight the B type cars only absorb part of the regenerative energy through the on-board resistor such as Line 13 of Beijing Subway. For example, the maximum speed of Line 13 is 80km/h, but the capacity of its braking resistor was determined to absorb the regenerative energy with the initial braking speed of 50km/h. This leads to the reduction of both the space and weight of braking resistor. The difference between the capacities of the type A and type B cars is due to the different understanding on the amount of regenerative

energy that need be absorbed during the regeneration failure.

Line 2 of Chongqing monorail and Line 4 of Guangzhou linear motor metro started in 2005 and 2006, respectively. The ground resistor developed by Hunan Hengxin Electric corporation are being applied in the above two lines for same reasons: (1) from the operation experience of Line 2 of Guangzhou Metro, the on-board brake choppers and resistor weight 500kg in a single car. Thus for the 4M2T formation, the total weight is increased by 2t. But both the monorail and linear motor cars strictly require the lightweight design. The weight of cars will increase if on-board resistor is used. (2) The installation space under floor is limited. (3) Since Chongqing and Guangzhou are hot in summer, it is undesirable to use on-board resistor that will raise the tunnel temperature. In addition to the above two lines, Beijing Fangshan Line, Guangzhou Line 4 and Tianjin Line 1 use ground braking resistor, as shown in Figure 1.

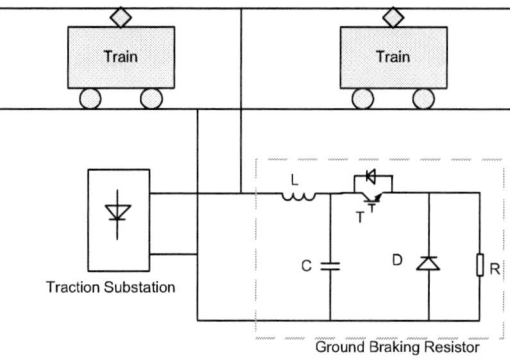

Fig. 1. The schematic diagram of the ground braking resistor

The combination of ground resistor and regenerative inverter has been adopted in Line 1 of Tianjin subway in Oct. 2007. The system absorbs the peak power of regenerative energy by resistor, and converts the left energy from DC to 380V AC in order to supply electric power for lighting, fans and other auxiliary equipment. However, this system is currently not widely applied [4].

IV. INVERTER FEEDBACK APPROACH

Although the rheostatic braking is simple with low-cost, it can only dissipate the regenerated energy into heat, thus fails to reuse the energy and follow the energy saving policy advocated by Chinese government. Therefore, China started the development of the regenerative inverter, which has been in trail operation at the 2nd stage of Beijing subway Line 10 in December 2012 and Line 14 in May 2013 [5].

There are two types of regenerative inverters using SCR and PWM mechanism (IGBT or the other full-controlled power electronic devices is adopted) [6]. Disadvantages of the SCR inverters are the strong high harmonic components and the large distortion of the AC electric power. It is not suitable to directly connect with the gird. The PWM inverter has small harmonic components, small distortion and good dynamic performance. Therefore, Beijing Subway Operation

Corporation decided to develop the PWM regenerative inverter in the future.

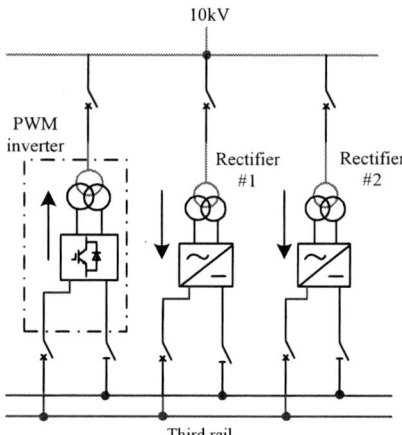

Fig.2 The schematic diagram of PWM regenerative inverter

Figure 2 shows the schematic diagram of the PWM regenerative inverter system that has been applied at the 2nd stage of Beijing subway Line 10. The regenerative inverter system consists of transformer, low-voltage switchgear cabinet and the PWM inverter (see Figures 3 and 4). There are two substations installing the PWM regenerative inverter system. The rated voltage of the AC side is 10kV and the operating range of DC voltage is 750V~1000V. The inverter uses 1700V/2400A IGBT, and the rated power of every inverter is 2MW. Line 10 uses 750V DC power supply. When the DC bus voltage is higher than 880V, the inverter is activated to send the regenerative energy to the grid. The highest allowed voltage is 1000V. And there are two power supply modes, 750V DC and 1500V DC, in Beijing subway system. Different series-parallel connections of the DC-link of the PWM inverter correspond to the two different power supply modes (see Figure 3).

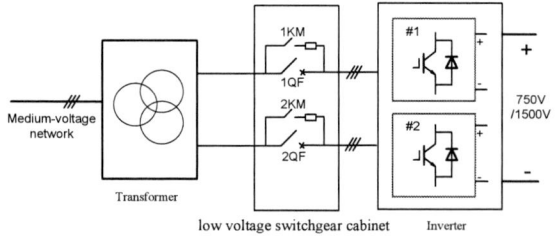

Fig.3 Series-parallel connection of PWM inverter DC-link

Fig.4 Photo of the PWM regenerative inverter system

After the initial operation of the PWM regenerative inverter system, there were several protection malfunctions (caused by improper setting of the overvoltage and overcurrent protection parameters) and communication malfunctions occurring at the beginning. After fixing the malfunctions, the system was able to work properly. Figure 5 shows the statistical results of the feedback power of PWM inverter in a week. It can be seen that the average daily electricity saving is 1772kWh using the PWM regenerative inverter system.

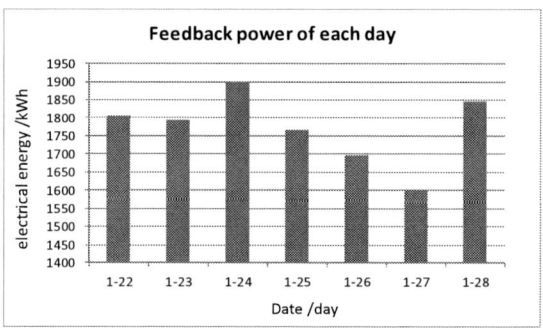

Fig.5 The statistical results of the feedback power in a week

However, the electric power of Chinese urban railway systems is supplied by SGCC (State Grid Corporation of China). SGCC is worrying about the influences of the feedback power to grid stability and other electrical equipment. It is critically important to work with SGCC together to make the PWM regenerative inverter system a general technology that could be widely applied in China [7]. The AC medium voltage network uses decentralized power supply system. Since the peak power of the inverter cannot be completely consumed by the subway system, the left power has to be sent to the grid, which need be proved by SGCC. While for the centralized power supply system, the above problem does not exist because of its large capacity and more loads.

V. STATIONARY SUPER CAPACITOR ENERGY STORAGE

With the rapid development of secondary battery, flywheel, supercapacitor and other energy storage devices in recent years, utilizing energy storage devices to absorb the regenerative energy has been gradually applied in urban railway systems. Total length of 27.6 km of Beijing subway Line 5 was opened in October 2007. 4 SITRAS-

SEC supercapacitor energy storage devices (provided by Siemens) have been installed in 4 substations, which are used to absorb regenerative energy in order to prevent regeneration failure and mitigate fluctuations of power supply voltage. The Line 5 is the first subway line to adopt supercapacitor in China [8].

There are two major reasons for the Line 5 to adopt the stationary ground supercapacitor: 1) supercapacitor works by physical reaction. Compared to Lithium-ion batteries or other secondary batteries, its advantages are long cycle life, high power density, fast charging and discharging, etc. Supercapacitor is suitable for absorbing and releasing the highly dynamic regenerative braking energy; 2) the usage of the stationary ground supercapacitor avoids the space and weight problems compared to the on-board one.

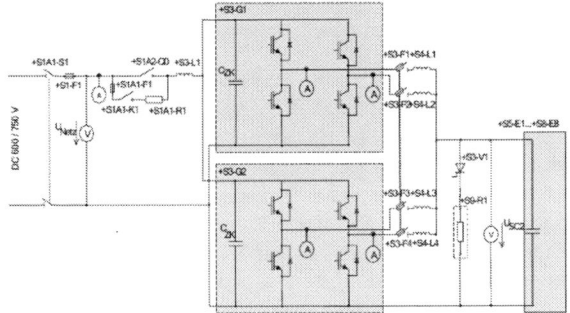

Fig.6 The schematic diagram of the supercapacitor system in Line 5

Figure 6 shows the diagram of the main circuit in the supercapacitor energy storage system of Line 5. The system can be applied to a subway system with a voltage level of 600V or 750V. The bi-directional converter is connected with a 4-bridge arm in parallel. The total capacitance is 94F and the energy storage capacity is 2.5kWh. The system can work by two modes, energy saving mode and constant voltage model. In the energy saving mode, the regenerative braking energy is stored and reused; while in the constant voltage model, the energy storage system is charged by the traction network or release the energy back to the network in order to stabilize the voltage and improve the train performance. Figure 7 shows the SES system from Siemens.

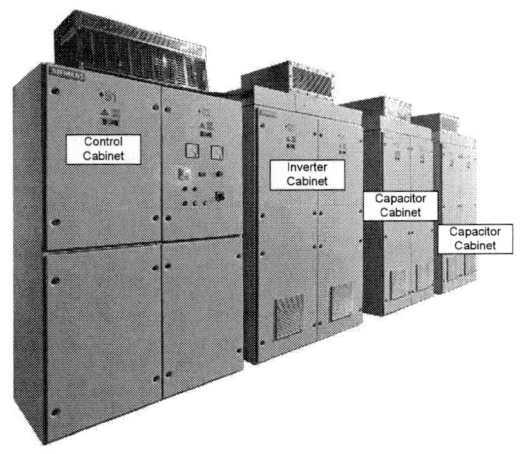

Fig.7 SES system from Siemens

If the similar energy storage system is adopted by other Chinese urban railway lines as Line 5 in Beijing, two issues have to be addressed. One is that how to choose the location for installation and capacity of the supercapacitor; the other is developing home-made energy storage systems [9]. Beijing Subway Operation Corporation is now working with universities, research institutes and other companies to solve the two issues.

The supercapacitor energy storage system innovates the traditional power supply system for urban railway. Developers should consider simultaneously the energy storage capacity, control strategy, charging and discharging power. The subway operation corporations also want to exactly know the energy-saving effect, the maintenance of a constant DC bus voltage and quantitative results on the improvement of railway car performance after the implementation of the system.

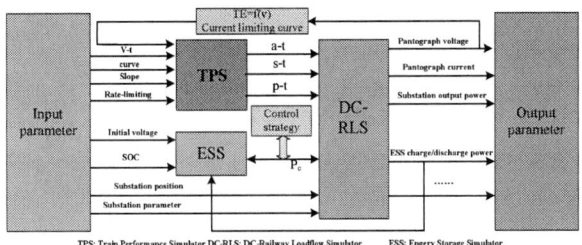

Fig.8 The system simulation flow chart

In Figure 8, the simulation model of the supercapacitor energy storage system is shown consisting of traction calculation module (TPS), DC line power flow simulation module (DC-RLS) and energy storage module (ESS). This simulation model can evaluate energy-saving effect, optimize and configure the capacity and position of the energy storage system. The departure interval, charging and discharging threshold voltages are also represented in the model.

Figure 9 and Figure 10 show the results of the maximum power and energy storage capacity of supercapacitor charging and discharging under 10min/5min departure intervals, respectively [10]. In Figure 9, with a 10min departure interval, the maximum power (800kW) and energy storage capacity (2.5kWh) of the supercapacitor is the largest at station 12. In Figure 10, with a 5min departure interval, energy exchange among the nearby trains becomes more frequent. The maximum power (550kW) and energy storage capacity (2.3kWh) of the supercapacitor reach the largest at station 12 and 13, which are smaller than the results in Figure 9.

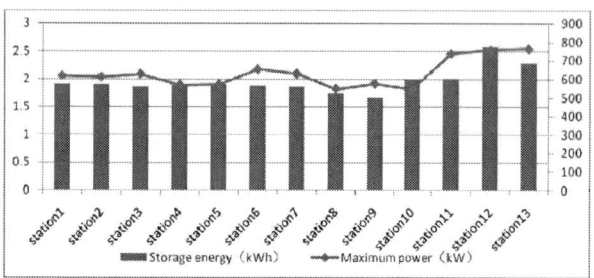

Fig.9 The maximum power and energy storage capacity under 10min departure interval

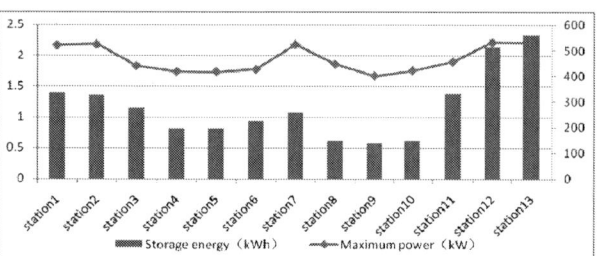

Fig.10 The maximum power and energy storage capacity under 5min departure interval

In order to verify effects of energy saving and constant voltage maintenance of the supercapacitor energy storage system in house, the authors developed a 200kW experimental platform that emulates the subway regenerative energy recovery system. A megawatt-level prototype system will also be developed in the future, which is close to the real practical systems.

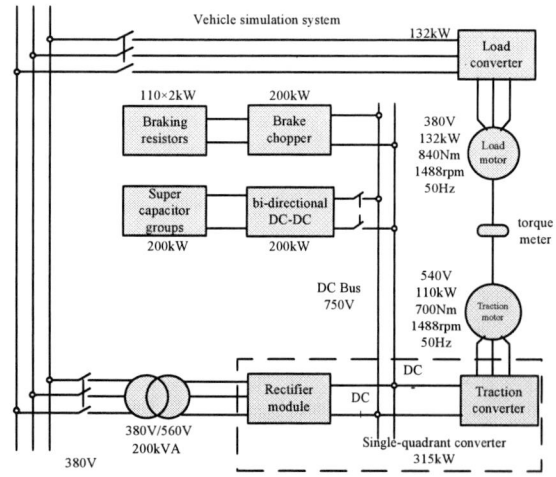

Fig.11 The experimental platform for subway regenerative energy recovery

Figure 11 shows the schematic diagram of 200kW experimental platform. The 380V AC is boosted and rectified to 750V DC through boosting transformer and diode rectifier. The output voltage 750V DC is the DC bus voltage. The traction converter and traction motor emulate the subway traction driving system; while the load motor and load converter emulate the subway train. The supercapacitor energy storage device is connected with the DC bus in parallel.

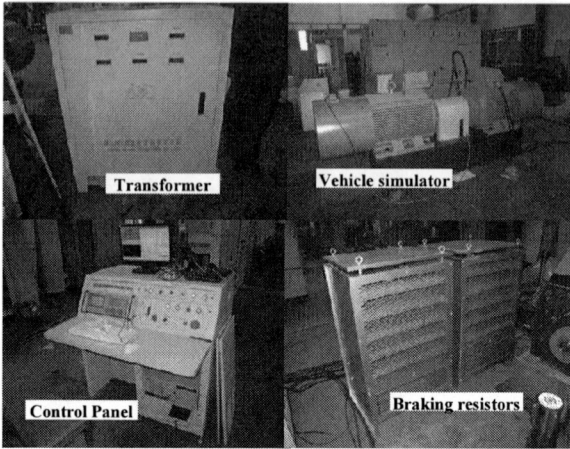

Fig.12 The experimental platform physical appearance

Figure 12 shows the physical appearance of the experimental platform, after adding super capacitor energy storage system, when the simulated train is accelerating, DC line voltage drops, super capacitor group discharges, and supplies the electrical power together with the rectifier; during the braking, energy comes back to DC line, and the voltage start to rise up, super capacitor group start to take in the regenerative energy, and if the braking energy power level is higher than the maximum power super capacitor group can withstand, the remaining energy will be consumed by the braking resistor, the energy flow is shown in figure 13 [11-13].

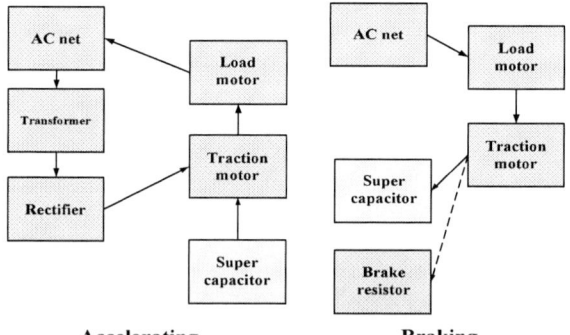

Fig.13 The experimental platform energy flow chart

Figure 14 [14-16] shows the main circuit of the supercapacitor energy storage system. It uses Maxwell 63F/125V module. A bigger module, the 31.5F/500V supercapacitor module consists of 4 series and 2 parallel connection of the 63F/125V modules. The peak power is 200kW and the energy storage capacity is 1kWh. Figure 15 shows the photos of the system.

978-1-4799-2706-7/14 $31.00 © 2014 IEEE   2137

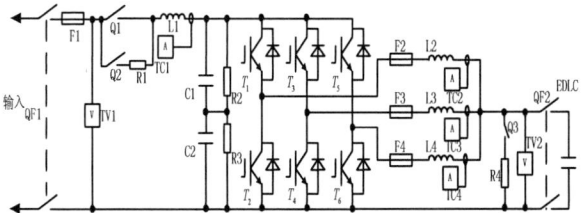

Fig.14 The main circuit of the supercapacitor energy storage system

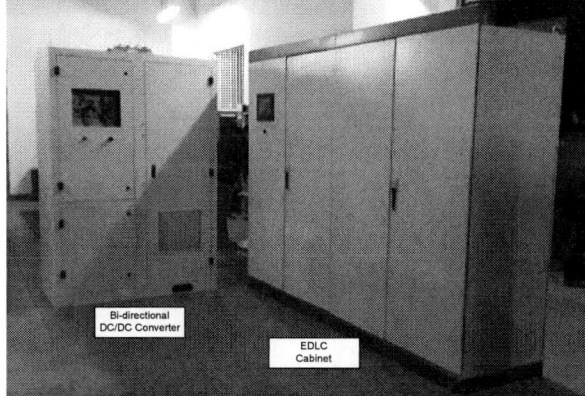

Fig.15 200kW supercapacitor energy storage system

The platform can emulate the traction, coasting and braking of the trains that cause DC bus voltage fluctuations. Figure 16 shows the experimental results of speed and torque. As shown in Figure 17, with the supercapacitor energy storage system, the decline of the DC bus voltage is limited in the first duration because using the power released by supercapacitor can reduce the output power of the diode rectifier. In the second duration, the supercapacitors do not absorb or release energy because DC bus voltage is in its normal operation range. While in the third duration, the supercapacitor absorbs the regenerative braking energy in order to suppress the rise of the DC bus voltage. The authors' group is now developing a megawatt-level prototype system that is expected to perform online test running in October 2014.

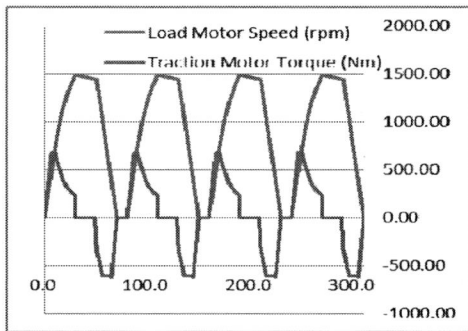

Fig.16 Experimental results of speed and torque

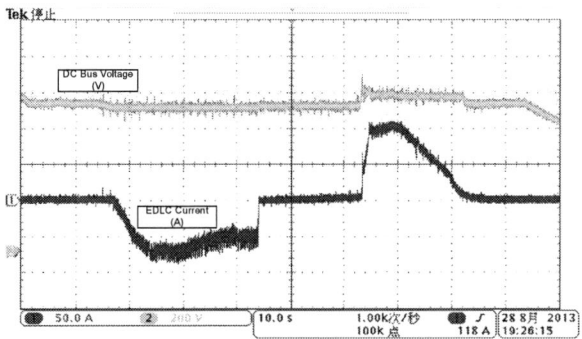

Fig.17 Experimental results of supercapacitor energy storage system

VI. CONCLUSIONS

This paper introduces the regenerative energy recovery technologies in China. For the conventional on-board resistor, it is needed to determine a correct capacity considering the space and weight limitations. Although ground resistor is widely used in China, it will become obsolete due to its incapability of energy saving and reuse. Currently the regenerative inverters and supercapacitors are not widely used. In addition to the cost, the operating companies are asking for the proof of the energy-efficiency and reliability of the technology. Quantitative studies are needed to validate its energy-saving effect, stabilization of the DC bus voltage and improvement of train operation. It is also important for China to develop home-made high performance energy storage devices such as secondary battery, supercapacitor and flywheel.

ACKNOWLEDGMENT

This paper was supported by a grant from the Beijing Laboratory of Urban Rail Transit, People's Republic of China.

REFERENCES

[1] Shi Wei, Fang Yu, "Model Building and Simulation of the Electric Brake Energy of Urban Rail Vehicle", ELECTRIC DRIVE FOR LOCOMOTIVES, Jan. 2011.

[2] Rao Huiming, Yang Jian, Fang Yu, "Test and Research on Resistance Brake of Urban Rail Transit Vehicle", URBAN MASS TRANSIT, Oct. 2010.

[3] Fang Ming, "Analysis on Parameters of Ground Absorption Equipment of Regenerative Braking for Urban Mass Transit Trains", CHINA RAILWAY SCIENCE, Sept. 2008.

[4] Zhao Ronghua, Yang Zhongping, Zheng Qionglin, "Simulation of Ground Rheostatic Braking in URT", URBAN MASS TRANSIT, Sept. 2008.

[5] Xia Jinghui, Zheng Ning, Zuo Guangjie, "Feedback Scheme and Device of Regenerative Braking Power in Invert Subway Vehicle", URBAN MASS TRANSIT, Jun. 2013.

[6] Chen Xiaoli, Yang Jian, "Design and Analysis of Braking Energy Recovery System for Urban Railway Vehicles", ELECTRIC DRIVE, Oct. 2010.

[7] Liu Manzu, Pan Difu, Wang Zhiwei, "Study on energy feedback equipment based on high-frequency PWM rectifier/inverter", ELECTRIC LOCOMOTIVES & MASS TRANSIT VEHICLES, Mar. 2008.

[8] Wang Xuedi, Yang Zhongping, "Study of Electric Double Layer Capacitors to Improve Electric Network Voltage Fluctuation for Urban Railway Transit", ELECTRIC DRIVE, Mar. 2009.

[9] Chen Lang, "Application of Super Capacitors in Urban Rail Transit System", URBAN RAPID RAIL TRANSIT, Mar. 2008.

[10] Zhang Tiejun, Chen Xue, Chen Guangzan, Lin Li, "A Novel Regeneration Braking Power Feedback Set for MRT Power System", HIGH POWER CONVERTER TECHNOLOGY, May 2011.

[11] Shen X, Chen S, Zhang Y, et al. Configuration Method for the Onboard Super-Capacitor Bank of Urban Rail Transit Considering Power and Capacity Constraints[J]. CHINA RAILWAY SCIENCE, 2013, 2: 020.

[12] González-Gil A, Palacin R, Batty P. Sustainable urban rail systems: Strategies and technologies for optimal management of regenerative braking energy[J]. ENERGY CONVERSION AND MANAGEMENT, 2013, 75: 374-388.

[13] Ma Z, Jiang J, Liu S, et al. The Design of Traction Power Battery System for Dual Power Urban Rail Metro[C]. PROCEEDINGS OF THE 2013 INTERNATIONAL CONFERENCE ON ELECTRICAL AND INFORMATION TECHNOLOGIES FOR RAIL TRANSPORTATION (EITRT2013)-VOLUME I. Springer Berlin Heidelberg, 2014: 75-86.

[14] Battistelli L, Fantauzzi M, Iannuzzi D, et al. Generalized approach to design supercapacitor-based storage devices integrated into urban mass transit systems[C]. CLEAN ELECTRICAL POWER (ICCEP), 2011 INTERNATIONAL CONFERENCE ON. IEEE, 2011: 530-534.

[15] Chen Y Q, Zhang J Y, Chen X X, et al. Dynamics Analysis and Power Allocation Strategy Research for a New Hybrid Electric Urban Rail Vehicle[J]. APPLIED MECHANICS AND MATERIALS, 2013, 300: 93-98.

[16] Ciccarelli F, Iannuzzi D, Lauria D. Supercapacitors-based energy storage for urban mass transit systems[C]. Power Electronics and Applications (EPE 2011), PROCEEDINGS OF THE 2011-14TH EUROPEAN CONFERENCE ON. IEEE, 2011: 1-10.

Traction Inverter that Applies Compact 3.3 kV / 1200 A SiC Hybrid Module

Katsumi Ishikawa

Mito Rail System Product Div., Rail System Company
Hitachi, Ltd.
Hitachinaka-shi, Ibaraki-ken, Japan
Katsumi.ishikawa.zw@hitachi.com

Kazutoshi Ogawa

Hitachi Research Labolatory
Hitachi, Ltd.
Hitachi-shi, Ibaraki-ken, Japan
Kazutoshi.Ogawa.ev@hitachi.com

Seigo Yukutake

Hitachi Research Labolatory
Hitachi, Ltd.
Hitachi-shi, Ibaraki-ken, Japan
Seigo.yukutake.et@hitachi.com

Norifumi Kameshiro

Central Research Laboratory
Hitachi, Ltd
Kokubunji-shi, Tokyo-to, Japan
norifumi.kameshiro.nh@hitachi.com

Yasuhiko Kono

Mito Rail System Product Div., Rail System Company
Hitachi, Ltd.
Hitachinaka-shi, Ibaraki-ken, Japan
yasuhiko.kono.fn@hitachi.com

Abstract- A compact 3.3 kV / 1200 A SiC hybrid module which adopts silicon carbide Schottky barrier diodes (SiC-SBDs) and insulated gate bipolar transistors (IGBTs) has been developed. The size of the developed SiC hybrid module is 130 mm × 140 mm, approximately 2/3rd the size of a conventional IGBT module.

Using this SiC hybrid module technology, a new traction inverter for railway applications has been developed. The new inverter has been reduced in weight by up to 40% compared to a conventional IGBT inverter. This has been achieved through the use of SiC modules, active gate control technology, reduced cooling system requirements and lightweight oil-free capacitors. The total energy loss of the new inverter is reduced by approximately 35%, through the use of SiC hybrid modules and active gate control drive circuits.

Keywords- SiC hybrid module, SiC Schottky barrier diodes (SBDs), junction barrier Schottky (JBS), traction inverter, active gate control, total energy loss, turn-on energy loss, recovery energy loss

I. INTRODUCTION

The development of high power semiconductor such as 1.7 kV, 3.3 kV, 4.5 kV and 6.5 kV IGBTs [1-3], and widespread introduction of three-level topologies [4] have led to a drastic increase the market share of PWM controlled IGBT inverters. Moreover, there is a growing demand for more efficient, higher power density and high temperature operation traction inverters. Silicon carbide

(SiC) has been indentified as the material with the potential to replace Si devices (such as IGBTs) in the near term because of its superior material properties such as wider bandgap, higher thermal conductivity and higher critical breakdown field strength. New inverters, in which SiC devices are used, are expected to meet these needs. Junction Barrier Schottky (JBS) structures have especially attracted a great deal of attention in SiC-SBDs, because they feature low-leakage current at high-blocking voltages. Several JBS diodes have been developed [5-7]. We made the 3kV SiC-SBDs with JBS structure, and built a prototype of a 3 kV/200 A SiC hybrid module, combined with Si-IGBTs and SiC-SBDs. By using a SiC hybrid module and a high-speed drive circuit, the turn-on loss could be reduced to about 1/7, and the reverse recovery loss could be reduced to about 1/10, compared with a conventional IGBT module [8].

This paper reports on the reduction of losses within a traction inverter installed with compact 3.3 kV / 1200 A SiC hybrid modules, and active gate control circuit. Test results of the SiC traction inverter are included within this paper.

II. COMPACT 3.3 KV/1200 A SiC HYBRID MODULE

Figure 1 shows the cross-section of an SiC-SBD with JBS structure. This structure contributes to both low on-resistance and low leakage current at a high blocking voltage. The optimized JTE (Junction Termination Extension) structure for the termination structure is adopted to ease the concentration of the electric field.

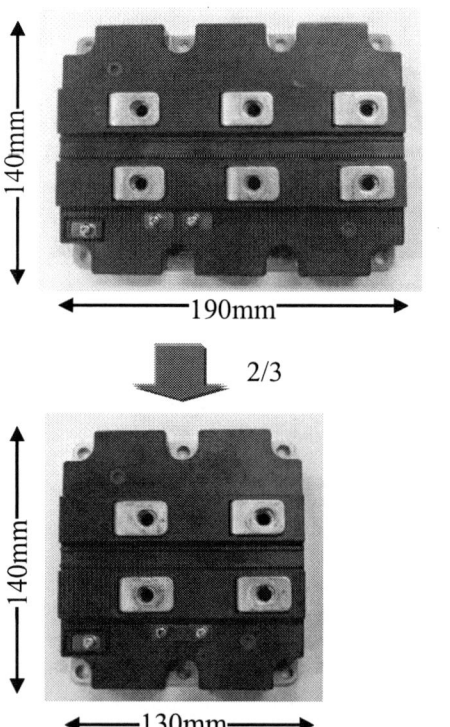

Fig. 1. Cross-sectional structure of SiC-SBD

Figure 2 shows photographs of a conventional IGBT module and SiC hybrid module. The rating voltage and the rating current are 3.3 kV / 1200 A respectively. The size of the conventional IGBT module is 190 mm × 140 mm. The developed SiC hybrid module adopts SiC-SBD and IGBT which has lower on-state voltage (Vce). The size of the developed SiC hybrid module is 130 mm × 140 mm, a reduction of a 1/3rd compared to a conventional module.

Fig. 2. Conventional IGBT module and developed SiC hybrid module

Figure 3 shows the typical reverse I-V characteristics of a SiC-SBD at 25 ℃. Here, we achieved a low leakage current of 0.1 mA or less at 3.3 kV reverse bias.

Figure 4 compares the switching behaviour of the reverse recovery waveforms and turn-on waveforms between a Si-pn diode and a SiC-SBD. Comparison of the two waveforms shows that the reverse recovery energy loss within the SiC-SBD is reduced by approximately 90% in comparison to the Si-pn diode, at 1500 Vdc voltage. Additionally, the turn-on peak current is also reduced because of the reduction in the recovery current. Therefore, the turn-on energy losses are reduced by approximately 50%.

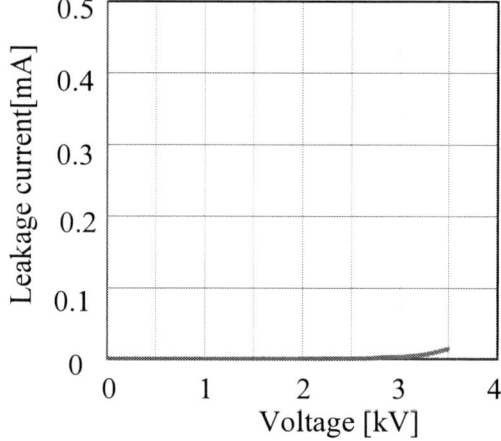

Fig. 3. Leakage current of developed 3.3 kV / 1200 A SiC hybrid module

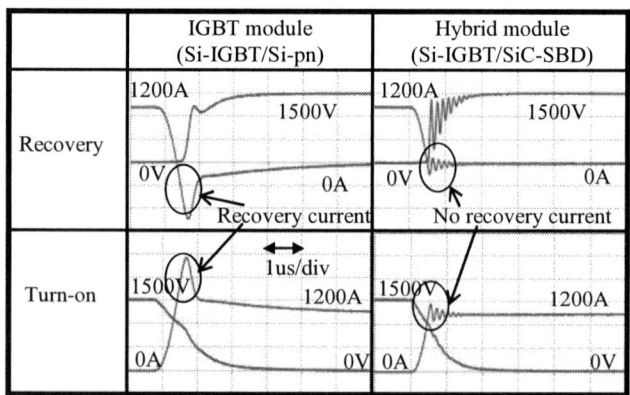

Fig. 4. Switching waveforms of conventional IGBT module and Hybrid module

III. ACTIVE GATE CONTROL OF SiC HYBRID-MODULE

The trade-off of surge voltage and turn-on energy loss has been achieved by using active gate control circuitry. Figure 5 shows the active gate control circuit that has been adopted. Figure 6 shows the gate waveform of the active gate control drive circuit. This turn-on process is divided into the following three stages.

Period ①: Shortening of the turn-on time

Period ②: Surge voltage suppression

Period ③:Turn-on energy loss reduction

For period ①, to shorten the turn-on time, the input capacitance (Cies) of the IGBT is charged at high speed via capacitor Csp. For period ②, the gate current flows through Rg1 and Rg2. Therefore the turn-on di/dt slows, and the surge voltage is suppressed. For period ③, the gate current flows through Rg1 and SW. The gate resistance reduces compared with period ② and therefore the turn-on speed increases, and the turn-on energy loss is reduced.

Figure 7 compares the gate current, the collector voltage, the recovery voltage, and turn-on energy loss waveforms when the fixed gate resistance drive circuit (depicted by the blue line) and the active gate control circuit (depicted by the red line) are used. The switching condition is 1500 V, and 1200 A. As can clearly be seen from the gate current waveform, the active gate control is achieved. It is confirmed that turn-on energy loss is reduced in the period ③, due to the active gate control circuit.

Figure 8 shows the three different trade-off curves between the turn-on energy loss and the surge voltage. The switching condition is 1500 V, and 1200 A. The conventional IGBT module uses the active gate control. When the cases of the same surge voltage are compared, the turn-on energy loss can be reduced by 44% by applying a SiC hybrid module and the active gate control circuit. When the active gate control circuit is compared with the fixed gate resistance, the turn-on energy loss can be reduced by 15%.

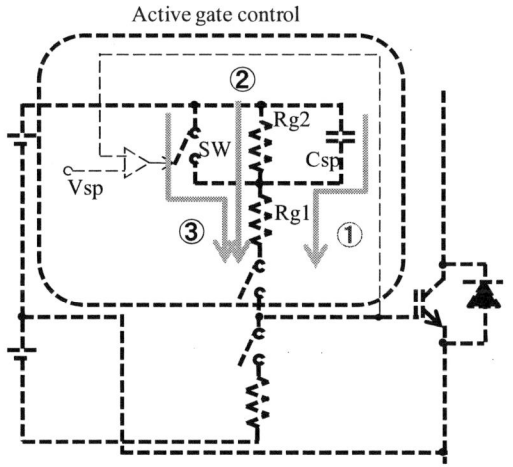

Fig. 5. Active gate control circuit

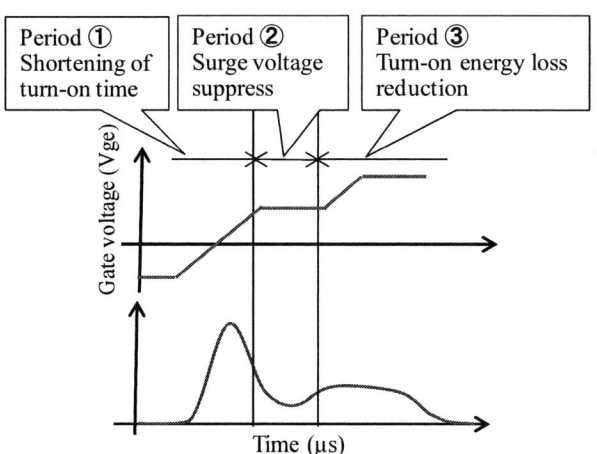

Fig. 6. Gate waveform that applies the active gate drive circuit

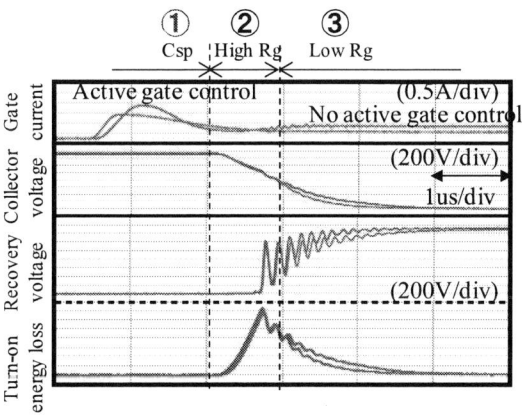

Fig. 7. Switching waveforms of active gate control circuit

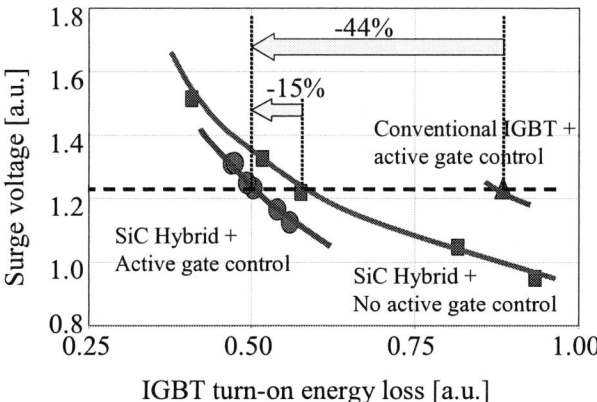

Fig. 8. IGBT turn-on energy loss reduction by using SiC hybrid module and active gate control circuit

IV. COMPACT SiC HYBRID INVERTER

This chapter explains the railway inverter equipped with SiC hybrid module. Table 1 shows the comparison of the circuit topologies of railway inverters. Over 90% railway track is electrified by DC 1500V. There are a lot of ratios of the line voltage of DC 1500V and DC 3000V all over the world. For the line voltage of DC 1500 V, three level inverter configuration adopted with 1.7 kV IGBT or two level inverter configuration adopted with 3.3 kV IGBT are required. Necessary modules of three level inverters are three times compared with two level inverters. By using 3.3 kV SiC hybrid module, two level inverter of high power density can be applied.

Figure 9 shows the conventional IGBT inverter and the developed SiC hybrid inverter. The weight of the developed inverter has been reduced to 60% of a conventional IGBT inverter. Reduction of the weight has been through the use of a compact cooler, lightweight oil-free capacitor, 3.3 kV compact SiC hybrid modules and active gate control technology.

Figure 10 shows the effect of loss reduction in the traction inverter with compact 3.3 kV SiC hybrid modules. The total losses from the devices are estimated at 125 ℃ from the forward I-V characteristics of the IGBT and SiC-SBD, and the dependence of turn-on energy loss, turn-off energy loss and recovery energy loss on current. The total losses are calculated based on the assumption of a commuter unit, operating on a 1500V line voltage, taking in to consideration the powering modes and regenerative brake modes. By applying a SiC hybrid module and active gate control drive circuit, it has been estimated that the total energy loss of the inverter is reduced by about 35%.

Figure 11 shows an example of operational current waveforms driving four when 150 kW motor. It was confirmed that the developed traction inverter is ready for release for the market.

TABLE 1 Comparison of circuit topologies

Line voltage	750V, 600V	1500 V	
Length of Japanese railway track	865.3 km (7.4 %)	10817.1 km (92.6 %)	
Rating voltage	1.7 kV	1.7 kV	3.3 kV
Number of needed modules/leg	2	6	2
Circuit topology	2 level	3 level	2 level

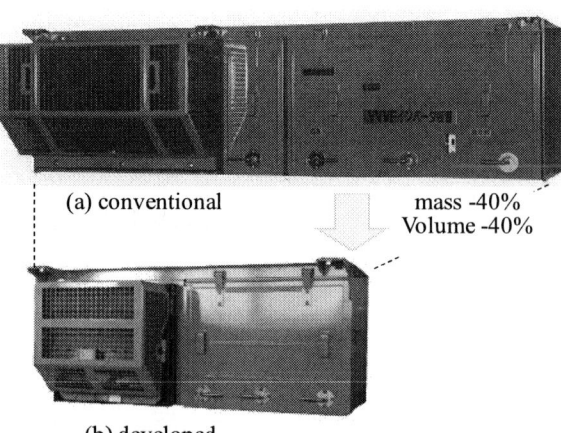

(a) conventional   mass -40% Volume -40%

(b) developed

Fig. 9 Conventional IGBT inverter and developed SiC hybrid inverter

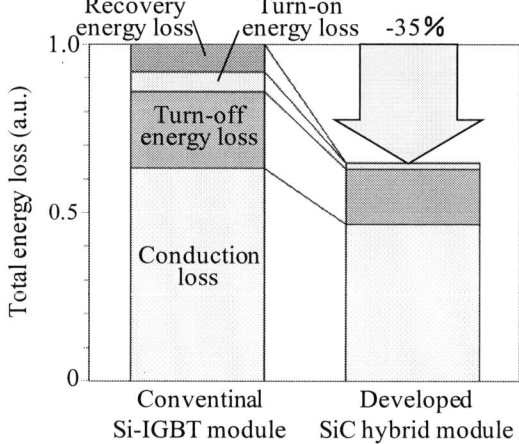

Line voltage	1500V
Type of train	Commuter car
Operating mode	Powering operation Regenerative brake operation

Fig. 10 Simulation results for inverter loss

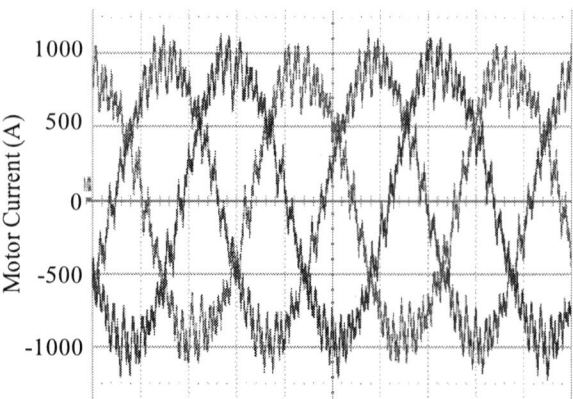

Fig. 11 Current waveforms for motor drive

V. Conclusions

A compact 3.3 kV / 1200 A SiC hybrid module which adopts SiC-SBDs and IGBTs has been developed. By adopting SiC-SBD with the JBS structure, both low on-state voltage drop and low leakage current at a high blocking voltage have been improved. The size of the developed SiC hybrid module is 130 mm × 140 mm, a reduction of a 1/3rd compared to a conventional module.

We examined the active gate control circuit where the trade-off of surge voltage and turn-on energy loss is improved. When the cases of the same surge voltage are compared, the turn-on energy loss can be reduced by 44% by applying a SiC hybrid module and the active gate control circuit.

A traction inverter for railway applications equipped with SiC hybrid module has been developed. The developed inverter has been reduced in weight to approximately 60% of a conventional IGBT inverter, by using 3.3 kV / 1200 A compact SiC hybrid module, active gate control technology, compact cooler and lightweight oil-free capacitors .We estimated that the total energy loss of inverter is reduced by about 35%, by applying an SiC hybrid module and active gate control drive circuit.

References

[1] M. Mori, R, Saitou and T.Yatsuo, " A high power IGBT module for traction moter drive " , in Proceedings. ISPSD1993, pp.287-291(1993)

[2] Y. Hagiwara, M. Tanaka, M. Ueno, " Technological Development of an IGBT applied Traction System for the Series 700 Shinkansen High-Speed Train." , International Conference Railway Traction System 2001, Capri, pp. 2.1-2.19.

[3] M. M. Bakran, H-G. Eckel, "Traction Converter with 6.5kV IGBT Modules", EPE 2001, Graz, Conf. Proceedings CD.

[4] K. Nakata, K. Nakamura, S. Ito and K, Jinbo, "A three level traction inverter with IGBTs for EMU", in Proceedings. IAS, pp667-672 (1994)

[5] P. Brosselard, et al., "High temperature behavior of 3.5 kV 4H-SiC JBS diode", Proc. of ISPSD2007, p. 285

[6] B. A. Hull, et al., "Performance and stability of large-area 4H-SiC 10-kV junction barrier Schottky rectifiers", IEEE Trans. Electron Devices, vol. 55, no. 8, p. 1864

[7] H. Okino, N. Kameshiro, K. Konishi, N. Inada, K. Mochizuki, A. Shima, N. Yokoyama, and R. Yamada, "Electrical Characteristics of Large Chip-size 3.3 kV SiC-JBS Diodes" Materials Science Forum Vols. 740-742, pp 881-886 (2013).

[8] K. Ishikawa, Kazuyoshi Ogawa, H. Onose, N. Kameshiro and M. Nagasu, "Traction Inverter that Applies Hybrif module using 3-kV SiC-SBDs "in Proceedings. IPEC2010, pp3266-3270(2010)

Power Electronic-Based Protection for Direct-Current Power Distribution in Micro-Grids

K.J. Tseng
School of Electrical & Electronic Engineering
Nanyang Technological University
Singapore, Email: ekjtseng@ntu.edu.sg

Guomin Luo
Beijing Jiaotong University
China

Abstract—**DC power distribution can be more effective than AC power distribution in the built environment. However, DC bus protection by using DC circuit breaker (CB) can be costly. This paper proposes a protection method which adopts power electronic-based methods to isolate the faulted segment instead of using CBs. Starting with investigations on the architecture of DC power distribution system for typical data center as an example of DC micro-grid, and the requirements on its protection, the currents of different kinds of faults and the possible ways to eliminate fault currents are discussed. Then, the details of the proposed protection method are introduced. Finally, the effectiveness of the proposed method is demonstrated by simulations and comparisons using the PSCAD platform.**

Index Terms—**Contactor, Data center, DC power distribution, DC bus protection, LVDC, ring-bus microgrid.**

I. Introduction

With the increasing interests in energy efficient data centers with DC distribution, protection of the DC power systems should be further explored. Therefore, DC bus protection issues are studied in this paper, first by reviewing the fundamentals of data center with DC distribution and the available protection techniques. Next, the possible faults of DC bus are introduced and their fault currents are studied. Based on the analysis of possible fault current eliminations, a DC bus protection method is investigated. In this method, contactors that need only switch low current are adopted instead of DC CBs. With the cooperation of fast power device switching operations in converters and special designed capacitor branch, the DC buses can discharge quickly and the fault currents drop sharply. The contactors at the both ends of potential lines switch as soon as the fault currents drop to a low threshold. Once the faulted segment is isolated, the DC system is re-energized and returns back to normal operation. The whole procedure is extremely short in duration such that the power supply to the loads will not be interrupted. Finally, simulations obtained using PSCAD are presented to demonstrate the effectiveness of the proposed method.

II. DC Bus Protection of DC Micro-Grid

A. Data Center as Example of DC Micro-Grid

Data center is a facility to house the computer system and related components. It generally includes DC loads, AC loads, and power supplies which consist of both the utility AC power supply and redundant or back up power supply.

A typical ring-bus DC power distribution system of data center is shown in Fig.1.

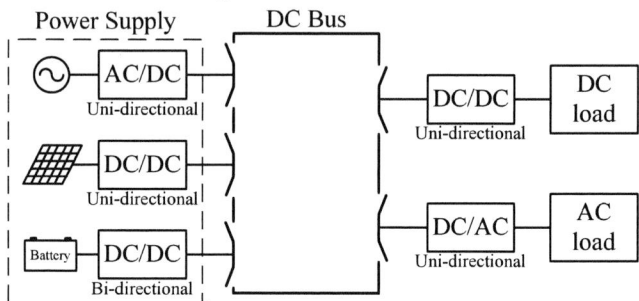

Fig.1 Ring-bus DC distribution system of a data center

1) DC loads: The DC loads in a typical data center are mostly composed of IT loads which include the servers, storage, and telecommunication equipment. These loads are of the highest priority of data center operation. Since power shutdown of computing equipment results in data loss and discontinuity of communication, the power supply for DC loads must be reliable and continuous. Usually, the IT computing loads are supplied by two or more power sources to ensure its reliable operation.

2) AC loads: The AC power is usually consumed by environment control system such as cooling equipment. Heat given out by the electrical devices in data center is a big problem in reliable operation. Higher temperature leads to higher failure rates of equipment. Different kinds of cooling methods are commonly employed, for example, air conditioners, chillers, and fans. Almost all of these methods use AC power supply. However, when compared with IT loads where power loss is strictly not allowable, the cooling system can tolerate a short power discontinuity.

3) Power supply: Data center is usually powered by the utility grid. Although the continuity of power supply from utility grid is highly reliable, it is also possible to be interrupted from time to time, for example, the large scale power outage due to serious faults occurring far away from the data center. To ensure continuous power supply, backup or redundant supply is a must for data center. The backup supply generally includes battery and distributed power generation system such as photovoltaic sources. When the power supply from utility grid is cut off, the battery starts to power the data center immediately. As the same time, the backup power generation starts as soon as possible.

B. Protection Techniques

Traditionally, the commonest practice in DC power system is not to install any protection on the DC side, and the link between AC and DC opens when faults in DC system are detected[5, 6]. However, the separation of DC and AC system interrupts the power supply of DC systems. The short-coming of this DC protection system cannot fulfill the requirement of users in many fields such as data centers.

After further research and in later applications, some protective devices were used on the DC side to perform isolations directly in DC system. Among these devices, fuse and DC CBs are often mentioned in articles and technical reports. A fuse is a type of resistive short which can provide over-current protection. It consists of a fuse link which melts to open the circuit when large current flows through it and a heat-absorbing part which is used to quench the arc. Its operation time depends on many factors such as current magnitude, current rise time, and temperature. However, the fuse can only trip the faulted line. Human intervention is required to clear the fault and re-energize the system after the fault is separated. Unlike fuses that can only operate once, DC circuit breakers can reset with controlled signals to resume normal operation. Since tripping a faulted line with large current is very dangerous, the design of CBs is very important in aspect of arc quenching. Therefore, different current limiting methods were proposed, for example, special design of freewheeling loop[7], improvements on the snubber circuit[8], usage of better arc quenching materials[9], and so on. Such research is still on-going with opportunities for improvement.

A handshake method was subsequently developed [6]. The faulted segment is located and isolated with de-energizing of the DC system. Such method applies fast DC switches to isolate the fault lines in DC side after the power in AC side is cut-off. It does not face the problems of manual recovery in using fuses and dangerous arcs in using CBs. However, the slow discharge of DC system leads to unnecessary outage in LVDC micro-grid. As the loads are connected to the DC bus via power electronics converters, the capacitors in converters can provide a short duration of power supply after fault occurrence. Therefore, such handshake method can be improved for applications in data center by improving the speed of protection. If procedures of system discharging, fault isolation and recovering normal operation of DC system can be completed in extremely short durations, the interruptions of power supply to the loads could be minimized. Since the switches can operate very fast with apposite selection of switches, the high-speed discharge and recovery of DC system is a potential research direction to speed up the protection procedure.

III. ANALYSIS OF FAULT CURRENTS IN DC BUS

To achieve fast discharge and re-energizing of DC bus after fault occurrence, the fault currents of faulted segment

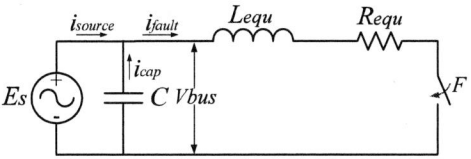

under different conditions should be analyzed in details.

A. Possible Faults

Two types of faults are usually encountered in DC bus

systems: line-to-ground fault and line-to-line fault. The line-to-ground faults occur when either positive or negative line is short-circuited to ground. The line-to-line fault generally short-circuits the positive and negative lines. While line-to-line faults usually are characterized as low fault impedance, the line-to-ground faults often have both high and low fault impedance[10]. The faults occur on DC bus are critical and will affect the stability of all devices connected to the bus. Unlike the faults on DC feeders where most of them are line-to-line faults, the line-to-ground faults are the most common types on DC bus in industrial distribution system such as data center[7].

B. Fault Currents with DC Source

When fault occurs, the current in a DC bus includes two components: the load current and fault current[7]. The load current will not change if the source and load keep the same, while the amplitude of fault current depends on many factors such as locations, impedance of the faulted path, and so on. An equivalent circuit of faulted DC bus is portrayed in Fig.2.

Here, E_s represents the DC source voltage of power supply. In most cases, it is the output voltage of AC/DC converter that is connected to utility grid. In some cases when AC power discontinues, E_s represents the output voltages of backup power sources such as battery. C represents the capacitor in converters on the bus side. It is used to maintain the stability of V_{bus} and is thus also called "smoothing capacitor"[11, 12]. L_{equ} and R_{equ} are the equivalent series inductance and resistance of the fault path. Since the capacitance of DC bus is very small when compared with smoothing capacitor C, they are usually neglected in fault current calculation to avoid complex computation.

Fault occurrence is simulated by the operation of a switch F. Before the fault occurrence, the switch F keeps open. The fault current i_{fault} is zero, and the voltage V_{bus} equals E_s. Once faults happen at t_{fault}, the switch F closes and the fault current i_{fault} is generated. Since the capacitor C is fully charged before the fault, the voltage over smoothing capacitor $V_c(t_{fault+})$ equals E_s. The capacitor C will not discharge and the current i_{cap} is zero at the beginning of the fault. Only the source E_s continues powering the system. Therefore, the transient fault current is

$$i_{fault} = \frac{E_s}{R_{equ}}(1 - e^{-\frac{R_{equ}}{L_{equ}}t}) \qquad (1)$$

According to (1), the magnitude of i_{fault} can be minimized by either increasing the R_{equ} or decreasing the E_s. As the series impedance R_{equ} and L_{equ} of fault path varies with the fault location, fault impedance, and so on, the magnitude of fault current are different and unpredictable under different conditions. Furthermore, connecting a series resistance directly into the faulted path is also dangerous because of the switch operations with large-amplitude DC currents. On the other hand, isolation of power supply was proved to be a good method to eliminate the fault current[6]. With E_s forced to zero, the magnitude of i_{fault} drops greatly according to (1). However, as demonstrated in Fig.2, the smoothing capacitor C is fully charged before fault occurrence. It will discharge and power the bus after the isolation of DC sources. Therefore, strategies to quickly discharge the capacitor and decrease the bus current should be explored.

C. Line-to-Ground Fault Current without DC Sources

In data center, the smoothing capacitors connect the positive and negative lines directly without any connections to earth at the middle point. Therefore, although line-to-ground faults compose the major part of bus faults of LVDC systems, they produce less terrible results when compared with line-to-line faults. For a ring-type DC bus, the equivalent circuit of line-to-ground fault with isolation of DC sources can be revised as portrayed in Fig.3. Since only one line segment is considered, the grounding capacitance C_{cable} of DC cable should also be considered. The voltage V_l is the voltage-to-earth of either positive or negative lines. Due to the direct connection of smoothing capacitors, the voltage over smoothing capacitor equals the bus voltage V_{bus} in normal operation, and the $V_{lpositive}$ and $V_{lnegative}$ have same amplitude $V_{bus}/2$ but opposite directions. When fault is detected, the DC source is isolated immediately, and the smoothing capacitor C starts to discharge with an initial voltage V_{bus}. However, the magnitude of voltage-to-earth V_l of faulted line decreases quickly due to the grounded fault. Accordingly, the magnitude of V_l of the other normal line will increase to balance the voltage V_{bus} in a short duration. The initial voltage $V_l(t_{fault0-}) = V_l(t_{fault0+}) = V_{bus}/2$, but the $V_l(t_\infty) = 0$. Thus, the fault current i_{fault} can be written as

$$i_{fault} = V_l C_{cable} + \frac{V_l}{R_{equ}}(e^{-t} - e^{-\frac{R_{equ}}{L_{equ}}t}) \qquad (2)$$

Since the voltage-to-earth $V_l(t_\infty)$ will decrease to zero with an attenuation coefficient $\exp(-t) - \exp(-R_{equ}/L_{equ} \cdot t)$, the attenuating speed of the fault current should be a constant for the same cable. Generally, the R_{equ} of a DC bus cable is around one thousand time larger than L_{equ}[10]. The magnitude of fault current i_{fault} can decrease quickly after the fault occurrence. When the fault current reduces to a small value, the switches on the faulted bus can open to avoid generation of large-amplitude arcs.

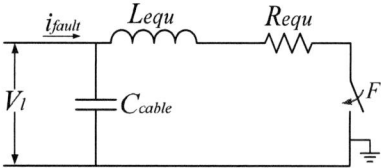

Fig.3 Equivalent circuit of line-to-ground fault without DC source

D. Line-to-Line Fault Current without DC Source

When line-to-line fault happens, the positive and negative lines are connected together. The positive line, negative line and direct connected smoothing capacitor form a loop path as portrayed in Fig.2. The grounding capacitance C_{cable} of DC cable is very small when compared with smoothing capacitor and ignored in modeling line-to-line fault. As the fully charged smoothing capacitor will continue to power the faulted segment, the equivalent circuit of line-to-line fault without DC source can be represented by the model in Fig.4.

The initial voltage of smoothing capacitor is $V_{bus}(t_{fault0+}) = E_s$. Then, without considering the grounding capacitance of cable, the fault current of line-to-line fault is

$$i_{fault} = \frac{E_s}{L_{equ}\xi} e^{-\frac{R_{equ}}{2L_{equ}}t} \sin(\xi t) \qquad (3)$$

where $\xi = \sqrt{\dfrac{1}{L_{equ}C} - \dfrac{R_{equ}^2}{4L_{equ}^2}}$.

From equation (3), the magnitude of fault current depends on many factors such as the value of parameter $L_{equ}\xi$, the attenuation constant R_{equ}/L_{equ}, and the oscillation frequency $\xi/2\pi$. Since the L_{equ} is much smaller than capacitance C and R_{equ}, the attenuation constant is very large. However, such small L_{equ} will lead to a complex coefficient ξ. The magnitude of i_{fault} and oscillation frequency of sinusoidal component may become very large. A possible way to eliminate the magnitude of fault current is to increase the value of L_{equ}. As discussed in Part A, it is very dangerous to connect an additional component in series with the faulted path due to the large DC current flowing in the cable. However, before the isolation of DC sources, the current i_{cap} in the smoothing capacitor branch is very small. Such low current provide a chance to revise the value of L_{equ} by connecting an additional inductance into the fault path to reduce the magnitude of fault current when fault occurs.

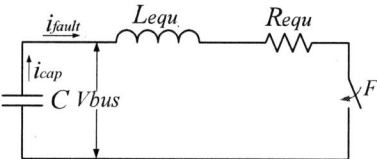

Fig.4 Equivalent circuit of line-to-line fault without DC source

IV. PROPOSED PROTECTION DESIGN

To provide a reliable and continuous power transmission in data center, ring-type DC bus is generally employed in its micro-grid design. The DC bus can be divided into several sections as shown in Fig.5 for bus protection. The positive and negative lines between two converters are defined as one section. The switches at the ends of two nearby sections are controlled by one protection unit. The current signals are measured by current measuring devices in contactors and monitored by the protection units. The contactors are installed as close as possible to the connections of converters to minimize the possibility of faults between two sections. When faults occur in one section, the fault currents are detected by the protection unit located at the both ends of the faulted section and the switchers controlled by the protection units will open to isolate the fault when the system is discharged. For example, if a fault happens in section 2, the switches labeled S_{2*} will open as soon as the current i_{in2*} are low enough.

The aim to discharge the DC system after fault is to avoid dangerous arcs from switch operations. Since only small-magnitude currents will flow through the switches, it is unnecessary to use expensive DC CBs. Contactors which are much cheaper and can trip a circuit at lower power levels are adopted in our research. Contactor usually has a breaking current as low as several amperes. Contactor switching with such a small current is less likely to generate dangerous arcs. However, the small-amplitude bus current can only be reached when the DC bus is greatly discharged. If the discharge duration lasts for a long time, for example, 100

978-1-4799-2706-7/14 $31.00 © 2014 IEEE

milliseconds, the power supply of load will be interrupted. However, power interruption of loads, especially IT equipment, is strictly not allowed. Thus, the discharge procedure must be completed in an extremely short time. According to the discussions in Section II and III, a DC bus protection and fault isolation method is proposed. The details of protection procedure are introduced in following contents.

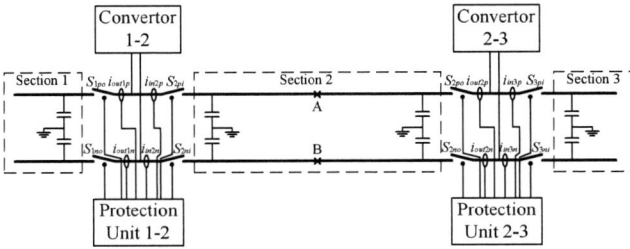

Fig.5 Diagram of proposed protection system

A. Fault Detection

The fault detection of DC bus needs to detect the fault as soon as the fault occurred, quickly identify the faulted line, and find out the fault type with simple judgment.

The currents flowing through all the contactors are monitored continually by protection units. When fault occurs, a fault path is formed and an abnormal current is generated. Since DC currents flow in the buses, the current difference i_d between the in-coming i_{in} and out-going i_{out} currents of a single line is used as the evidence of fault existence. When i_d exceeds a threshold, for example, 5A, the protection units will receive this signal and compare it quickly with i_ds of the other line of the same section and the lines of nearby sections. Take the section 2 as an example, if a line-to-ground fault occurs at point A on positive line, the magnitude of current i_{d2p} will increase. When its magnitude exceeds the threshold, the protection units: unit 1-2 and unit 2-3 will receive this alarm signal and compared i_{d2p} quickly with i_{d2n} in negative line of section 2 and the current differences i_{d1p}, i_{d1n}, i_{d3p} and i_{d3n} of nearby sections 1 and 3. For line-to-ground fault at point A, the magnitude of i_{d2p} of faulted positive line should be larger than the magnitude of

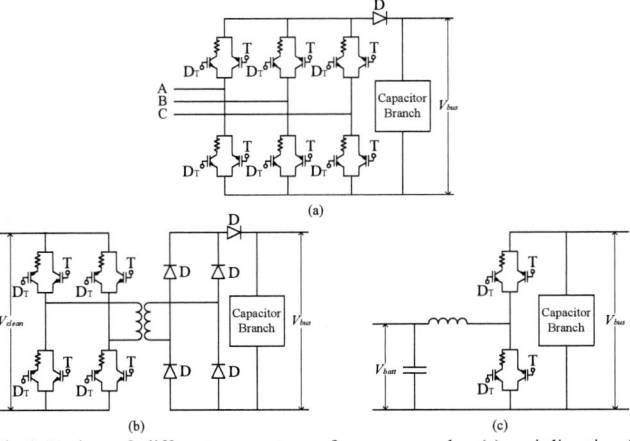

Fig.6 Design of different converters of power supply, (a) uni-directional rectifier, (b) uni-directional DC/DC converter, (c) bi-directional DC/DC converter

i_{d2n} of negative line, and the magnitudes of i_{d2p} and i_{d2n} are much larger than these of i_ds in nearby sections. On the other hand, when a line-to-line fault connects point A and B, the current differences i_{d2p} and i_{d2n} will increase with same

amplitude and they are much greater than those of i_ds in nearby sections. According to the simple comparisons, the fault type and faulted line can be quickly determined.

B. De-energizing the DC Bus

After faults are detected, the bus currents and voltage need to decrease immediately to allow the operation of contactors. As discussed in Section III, in order to reduce the magnitudes of current and voltage as soon as possible, the power supply should be isolated and the DC bus should discharge quickly.

1) Isolation of Power Supply

To isolate the power supply devices such as AC utility grid, backup energy source and battery, the most convenient way is to install CBs at the input or output of the electronics converters. However, another simpler and easier approach is to adopt the switches inside converters for power interruptions[13]. The IGBT switches inside converters can open or close the circuit with extremely high speed and can be used as CBs with suitable controls. Hence, the design of converters needs certain modifications. As portrayed in Fig.1, three types of power supply are usually adopted in data center. The designs of CB functions embedded in these three kinds of converters: uni-directional AC/DC rectifier, uni-directional DC/DC converter, and bi-directional DC/DC converter are discussed here.

Uni-directional AC/DC rectifier is typically three-phase bridge-type rectifier with parallel diodes. If there is a fault on DC bus, the AC/DC conversion will stop by opening all the IGBT switches T. However, the power from AC side will still feed high-current to the faulted DC side because the parallel diodes act as a bridge. Therefore, all the parallel diodes need being replaced with turnoff devices in order to be able to interrupt the current before it gets too high[13]. A typical revised rectifier is shown in Fig.6(a). The parallel diodes which are connected in series with resistances are all replaced by controlled IGBTs D_T. When faults are detected, all the IGBTs, T and D_T will open to isolate the power supply sources. Other re-design concepts of power converters can also be considered.

The same design is adopted in DC/DC converters. When designing uni-directional DC/DC converter, those with an isolation transformer provides much flexibility[13]. As shown in Fig.6(b), the diodes at primary side of the transformer are replaced by the IGBTs with controlled parallel flyback circuit. For the bi-directional DC/DC converter, the same replacement is used to ensure the converter is controllable. The design of bi-directional DC/DC converter is shown in Fig.6(c).

2) Discharge of DC Bus

When faults occur, the power supply sources are isolated immediately by opening all the IGBTs and their flyback circuits in converters. Such action eliminates the increase of fault current to a great extent. However, as discussed in Section III, the smoothing capacitor will continue discharge and power the faulted lines. In order to decrease the magnitude of current as soon as possible, the smoothing capacitor branch is re-designed as in Fig.7.

The 2014 International Power Electronics Conference

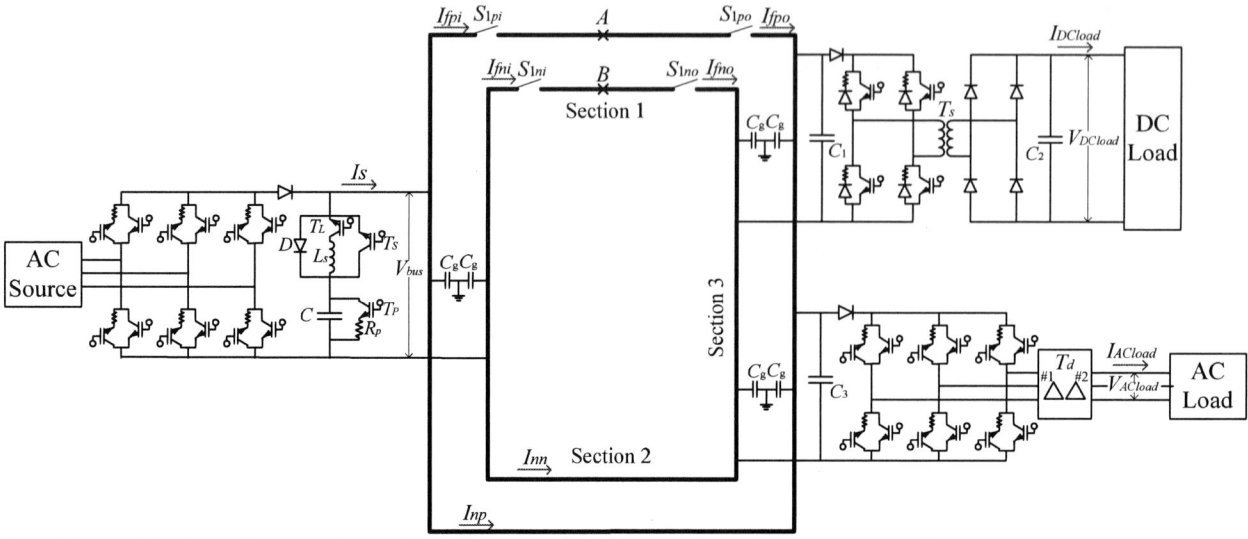

Fig.8 Simulation model of DC distribution system of data center which consist of one source, one DC load and one AC load

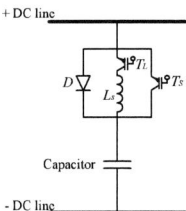

Fig.7 Design of smoothing capacitor branch

In normal operation, the smoothing capacitor is charged and discharged continuously to maintain the stability of DC bus voltage. The currents go through the diode D and IGBT T_S which is closed. When line-to-ground fault is detected, the status of T_S and T_L will keep the same. Because the smoothing capacitor is directly connected between positive and negative lines, it will discharge slowly and the bus voltage V_{bus} will keep unchanged in a short duration. At the same time, the line-to-earth voltage of faulted line decreases quickly with an attenuation coefficient R_{equ}/L_{equ}. On the other hand, if line-to-line fault is detected, according to the equation in (3), adding a large inductance in-series with the capacitor is a simple and effective way to eliminate the fault current. Thus, the IGBT T_S will open and T_L will close at the same time to connect the additional inductance L_s into the smoothing capacitor branch. Usually, the additional inductance is selected to be very large, for example, 10H, to eliminate the sudden change of fault current.

C. Contactor Operation and Re-energize DC Bus

As soon as the magnitude of fault currents reduce to a small value such as 10A, the contactors located at both ends of possible faulted line will open to clear the fault. Take the DC buses in Fig.5 for example, when the fault occurs on the lines of section 2 and the currents flowing in section 2 are small enough, the contactors S_{2pi}, S_{2po}, S_{2ni}, S_{2no} will open at the same time to isolate the faulted area.

After the possible faulted area is isolated, the DC bus is re-energized immediately. The IGBT switches inside the power source converters resume normal operations. The switches in capacitor branch are also reset to their initial status where the T_L is open and the T_S is closed.

All the procedures of DC bus protection perform in an extremely short time. Such a short time will not result in

great discharge of capacitors in electronics converters at the load side. Those capacitors can still provide energy to loads and the power supply of loads will not be interrupted.

V. SIMULATIONS AND RESULTS

To demonstrate the effectiveness of proposed protection method, a small data center with one power supply: AC utility grid, one DC load and one AC load is modeled on the platform of PSCAD. The PSCAD model and its connections are shown in Fig.8. As claimed by some researchers in [14], the LVDC distribution system with a bus voltage around 380V is regarded as most effective and is highly recommended. Thus, the bus voltage V_{bus} between positive and negative lines is set to be 425V. Since three converters are connected to the bus, the ring-type bus is divided into 3 sections. The system parameters of each component of the system are listed in Table I.

The faults are simulated at points A and B that are located at the middle of positive and negative lines of section 1. The performance of proposed method is studied for cases of different faults. In our simulations, the communications between different protection units need extremely short time which is not considered in our analysis.

TABLE I
SYSTEM PARAMETERS OF RING-TYPE DC BUS

Component	Data	Component	Data
DC bus voltage V_{bus}	425V	Unit resistance (cable)	0.6Ω/km
Grounding capacitor C_g	0.05mF	Unit inductance (cable)	0.5mH/km
Rectifier capacitor C	0.1F	Unit capacitance (cable)	0.1μF/km
Bus length (section 1)	60m	Series inductance L_s	10H
Bus length (section 2)	70m	Capacitor C_1	0.1F
Bus length (section 3)	60m	Capacitor C_2	5mF
Transformer T_s turns ratio	20:1	Capacitor C_3	0.1F
Transformer T_d turns ratio	2:1		

A. Line-to-ground Fault

A line-to-ground fault is simulated on the positive bus at the point A. It occurs at 5ms. As soon as the fault current goes through point A to ground, the current difference i_d between I_{fpi} and I_{fpo} increases immediately. When i_d reaches 5A, the fault is detected and all the switches in rectifier open to isolate the power supply. By comparing the i_ds of different lines in same and nearby sections, the potential

978-1-4799-2706-7/14 $31.00 © 2014 IEEE   2149

fault type and line are found out. For line-to-ground fault, switches in smoothing capacitor branch do not operate, and the capacitor C starts to discharge slowly. Due to the ground fault, the voltage-to-earth V_l of faulted line decreases sharply. When the current I_{fpi} decreases to the operation current of contactor which is 10A in our simulations, the contactors in faulted section will open to clear the fault. The simulated DC bus voltage and current in faulted line are shown in Fig.9. As portrayed in Fig.9(b), the magnitude of current I_{fpi} decreases rapidly after the isolation of power supply. The fault current is interrupted in much less than 1milisecond. Because of the slow discharge of smoothing capacitor, the DC bus voltage V_{bus} keeps unchanged during protection.

The bus voltage V_{bus} and fault current I_{fpi} without protection are also portrayed in Fig.9. These signals are marked in grey. Without isolation of the power supply, the fault does not influence the bus voltage V_{bus}, but it induces current oscillations in faulted line which can be harmful to DC buses.

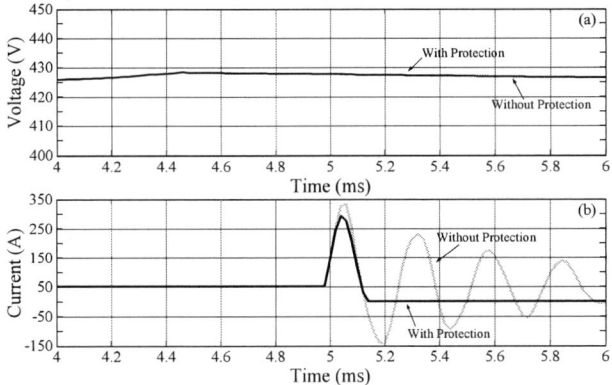

Fig.9 Simulated line-to-ground signals with and without protection, (a) DC

B. Line-to-ground Fault

A line-to-line fault is simulated by a short-circuit fault between point A and B in section 1 in Fig.8. This fault also happens at 5ms. With directly connected smoothing capacitor, the DC bus voltage V_{bus} and current I_{fpi} change greatly. The signals with and without protection are shown in Fig.10. Similar to line-to-ground fault, once the fault is detected, the switches in the rectifier are opened to isolate the power supply. At the same time, the switches T_L closes and T_S opens to connect an additional inductance L_S to eliminate the fault current. As shown in Fig.10(b), the proposed method can clear the fault current I_{fpi} in less than 0.2 milliseconds. Due to the large inductance L_s, the voltage V_{bus} decreases sharply before contactor operation and recovered quickly after the contactors separate the faulted line segment.

One the other hand, if the fault is not cleared immediately, the bus voltage decreases slowly, and the fault current increases to dangerous values such as several thousand amperes.

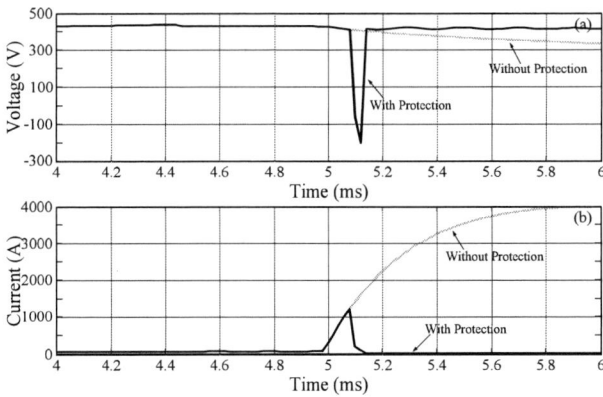

Fig.10 Simulated line-to-line signals with and without protection, (a) DC

VI. COMPARISONS

The proposed protection method is an improvement of the handshake method presented in [6]. Its effectiveness is illustrated by comparing with the handshake method in this section.

The handshake DC bus protection adopts fast DC switches. This method combines the usage of AC CBs and fast DC switchers to protect the multi-terminal DC systems. When faults are detected on the DC buses, all the AC CBs open to discontinue the power supply to DC buses. Once the fault currents decreases and reaches the threshold, 10A, the fast DC switches located at the both ends of potential faulted line will open to isolate the fault. Then, the AC CBs re-close and the smoothing capacitor re-charges. By using our proposed method, no additional DC CBs are required and the cost is greatly reduced by adopting relatively cheap fast DC switches. However, the smoothing capacitor between positive and negative lines is grounded in this method. Therefore, when fault occurs, either line-to-ground fault or line-to-line fault, the smoothing capacitor and the faulted path can form an enclosed loop such that the attenuation of fault current is much slower. Accordingly, it will need longer time before the fault currents reduce to the preset threshold and interruptions of the bus voltage is unavoidable.

Simulations with the same data center model were carried out by using fast DC switches based method. The simulation results of line-to-ground fault and line-to-line fault portrayed in Fig.11 and Fig.12 show that the handshake method requires longer time to isolate the fault segment and the currents in faulted lines reach dangerous levels. Usually, 20 milliseconds are needed to open the switches while our proposed method requires only about 0.2 milliseconds. The largest magnitudes of fault currents in Fig.11 and Fig.12 are around 4000 amperes. They are much larger than the current peaks in Fig.9 and Fig.10.

978-1-4799-2706-7/14 $31.00 © 2014 IEEE   2150

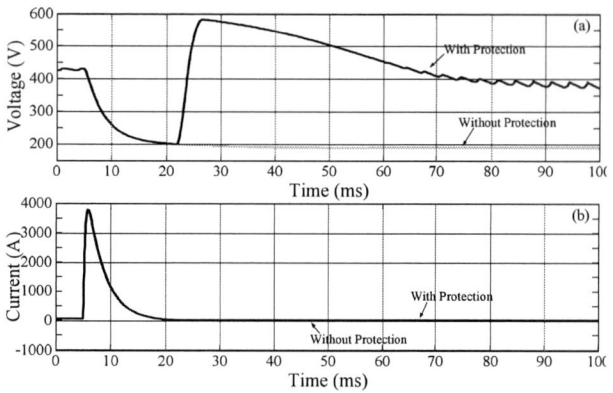

Fig.11 Simulated line-to-ground signals with and without protection by using handshake method, (a) DC bus voltage V_{bus}, (b) current I_{pfi} flowing in

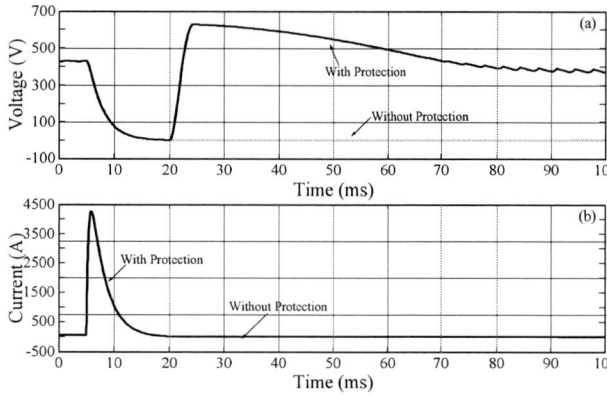

Fig.12 Simulated line-to-line signals with and without protection by using handshake method, (a) DC bus voltage V_{bus}, (b) current I_{pfi} flowing in

VII. CONCLUSIONS

This paper presents a DC bus protection method for DC micro-grid such as a data center without need of CBs. Interruptions of power supply at load side is also avoided. The DC bus faults are cleared by isolating the power supplies to DC bus. IGBT switches in the power source converters act as CBs and provide fast protections. The DC bus voltage and current are reduced in different ways according to the type of faults. Then, contactors are used to trip the faulted segment when fault current is low enough. The simulation results on the simple model of data center illustrate that the proposed method can quickly locate and isolate the faults in less than 1 millisecond. Comparisons between proposed method and previous one provide further demonstrations of the effectiveness of proposed method. Those extensive simulation results show that the contactor and IGBTs based DC protection is reliable and promising in data center DC power distribution applications.

REFERENCES

[1] K. Kumon, "Overview of Next-Generation Green Data Center," *Fujitsu Scientific & Technical Journal*, vol. 48, pp. 177-183, Apr 2012.

[2] J. Koomey, "Worldwide electricity used in data centers " *Environmental Research Letters*, vol. 3, pp. 1-8, 2008.

[3] D. Salomonsson, L. Soder, and A. Sannino, "An Adaptive Control System for a DC Microgrid for Data Centers," *IEEE Transactions on Industry Applications*, vol. 44, pp. 1910-1917, 2008.

[4] J. D. Park, J. Candelaria, L. Ma, and K. Dunn, "DC Ring-Bus Microgrid Fault Protection and Identification of Fault Location," *IEEE Transactions on Power Delivery*, vol. PP, pp. 1-1, 2013.

[5] P. Cairoli, R. A. Dougal, U. Ghisla, and I. Kondratiev, "Power sequencing approach to fault isolation in dc systems: Influence of system parameters," in *IEEE Energy Conversion Congress and Exposition (ECCE)*, 2010, pp. 72-78.

[6] L. Tang and B.-T. Ooi, "Locating and Isolating DC Faults in Multi-Terminal DC Systems," *IEEE Transactions on Power Delivery*, vol. 22, pp. 1877-1884, 2007.

[7] J. D. Park and J. Candelaria, "Fault Detection and Isolation in Low-Voltage DC-Bus Microgrid System," *IEEE Transactions on Power Delivery*, vol. 28, pp. 779-787, 2013.

[8] J. M. Meyer and A. Rufer, "A DC hybrid circuit breaker with ultra-fast contact opening and integrated gate-commutated thyristors (IGCTs)," *IEEE Transactions on Power Delivery*, vol. 21, pp. 646-651, 2006.

[9] F. Luo, J. Chen, X. Lin, Y. Kang, and S. Duan, "A novel solid state fault current limiter for DC power distribution network," in *Twenty-Third Annual IEEE Applied Power Electronics Conference and Exposition, APEC*, 2008, pp. 1284-1289.

[10] D. Salomonsson, L. Soder, and A. Sannino, "Protection of Low-Voltage DC Microgrids," *IEEE Transactions on Power Delivery*, vol. 24, pp. 1045-1053, 2009.

[11] T. Kato, M. Hisada, Y. Suzuoki, and H. Yamawaki, "Feasibility of increase in smoothing-capacitor of battery system for dumping power oscillation in transition to isolated operation of distributed generator," in *IEEE 30th International Telecommunications Energy Conference, INTELEC 2008.* , 2008, pp. 1-7.

[12] A. J. Snyders, P. A. Janse-van Rensburg, H. C. Ferreira, and A. J. Han Vinck, "AC-DC smoothing capacitor current coupling for improved powerline signal reception," in *2011 IEEE International Symposium on Power Line Communications and Its Applications (ISPLC)*, 2011, pp. 341-345.

[13] M. E. Baran and N. R. Mahajan, "Overcurrent Protection on Voltage-Source-Converter-Based Multiterminal DC Distribution Systems," *IEEE Transactions on Power Delivery*, vol. 22, pp. 406-412, 2007.

[14] N. Rasmussen, "AC vs. DC Power Distribution for Data Centers," *Schneider Electric White Paper 63*, pp. 2-4, 2011.

The 2014 International Power Electronics Conference

A Concept of High Power DC/DC Converter with Double Low Power Outputs

Masahide Hojo, Tomoya Nishioka, and Kenji Yamanaka
Institute of Technology and Science
The University of Tokushima
Tokushima, Japan
hojo@ee.tokushima-u.ac.jp

Abstract— This paper proposes a circuit topology to draw additional dc power outputs from a dc transmission line by series-connected circuits. It is based on a conventional concept of HVDC tap but the HVDC tap is duplicated in this paper in order to reduce current fluctuations on the dc transmission line. Inversed switching pattern between the two HVDC tap results in small current fluctuation and realizes additional low dc power outputs. This circuit topology is expected to be applied to a dc transmission system for offshore wind power, etc. In this paper, the operating characteristics are discussed in detail based on simulation study. The condition to achieve the soft-switching is also revealed. In addition to this, operating characteristics of the total system are verified by numerical analysis to investigate the availability of the proposed circuit.

Keywords— *dc/dc converter, dc power supply, high voltage dc transmission line, soft switching*

I. INTRODUCTION

A High-Voltage DC transmission system has been constructed all over the world and now they play an important role to trade high power over a long distance. Most of them employ line-commutated converters. On the other hand, power electronics technologies related to high power applications have been drastically developed with improvement of semiconductor power devices and various advanced converter topologies such as multi-level power converters [1]. In the circumstances, self-commutated power converters are tried to be applied to the dc transmission system. The dc transmission system based on self-commutated converters has a great flexibility of its converter control to trade a reactive power as well as an active power. Moreover, challenges about multi-terminal dc power transmission system [2] can be found in literatures. Today, dc power transmission system is expected to be applied to not only an off-shore wind power system, but also dc microgrid [3]-[4], dc power supply network for telecommunication system [5].

However, each terminal converter consists of remarkable capacity as large as dc transmission line even if the all terminals do not handle high powers.

On the other hand, a concept of HVDC tap has been proposed [6], which consists of relatively low power

converter in series with the dc transmission line as shown in Fig. 1. A series-connected bridge constructed by GTOs and diodes induce an ac component on the main dc line and the ac component is delivered to a diode bridge through a series transformer. However, single-use of this HVDC tap cannot prevent dc current fluctuations on the dc transmission line.

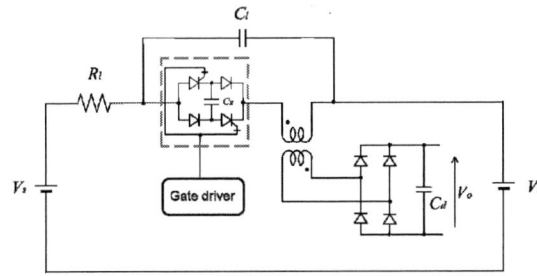

Fig. 1. A simple circuit model with single HVDC tap.

This paper proposes to use double HVDC tap to reduce the dc current fluctuations. When the switching pattern of each HVDC tap is inversed, the dc current fluctuations are expected to be reduced. First, in order to prove the availability of duplicated HVDC tap, this paper investigates its operating characteristics in detail based on simulation study. Secondly, the operating area of the duplicated HVDC tap with achieving the soft-switching is also investigated to reduce the switching losses. And then, the dc current fluctuations are compared between the cases when the switching pattern is inversed or not.

II. OPERATING PRINCIPLES OF THE HVDC TAP

Fig. 1 shows a simple dc power supply system which consists of a HVDC tap introduced in [6]. The dc voltage source V_s indicates a sending-end dc terminal voltage, and V_r is a receiving-end dc voltage. The major dc power is transmitted through the line resistance R_l from V_s to V_r. On the other hand, an additional dc voltage terminal is equipped by the HVDC tap, as shown in Fig. 1. It consists of a dc line capacitor C_l, which does not disconnect the receiving-end but produce a high

978-1-4799-2706-7/14 $31.00 © 2014 IEEE

frequency and small amplitude current fluctuations with an H-bridge. The H-bridge consists of a capacitor C_H, two GTOs and two power diodes. The introduced current fluctuations are delivered to the additional dc voltage terminal V_o through the series-connected transformer.

A. Operating modes of the bridge in the HVDC tap

The operating modes of the HVDC tap are defined by the H-bridge operating modes. Fig. 2 summarizes four switching modes of the H-bridge, where the two GTOs are driven by a same gate signal.

The four switching modes can be explained as follows.

(1) Mode 1: This mode begins at the time when the two GTOs are turned-off, simultaneously. Before this mode, the capacitor C_H has been discharged so that the voltage across it is zero. Then, the capacitor C_H is charged through the two diodes during this mode, as shown in Fig. 2 (a).

(2) Mode 2: After the voltage across the capacitor C_H is developed, the two diodes are turned-off and the H-bridge is opened as shown in Fig. 2 (b). In this mode, the voltage across the H-bridge and transformer is determined by the voltage across the capacitor C_l.

(3) Mode 3: When the two GTOs turn-on at the same time, the capacitor C_H begins to be discharged until the voltage across it becomes zero, as shown in Fig. 2 (c). In this mode, the two diodes keep its off-state.

(4) Mode 4: If the capacitor C_H is discharged, the two diodes are turned-on and the H-bridge is bypassed by the GTOs and diodes.

The dc line current fluctuates by the above behaviors of the H-bridge, therefore the small amount of power is delivered through the series transformer to the rectifier circuit of its secondary side.

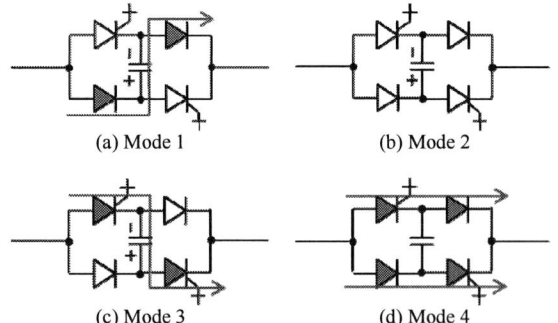

(a) Mode 1   (b) Mode 2

(c) Mode 3   (d) Mode 4

Fig. 2. Mode transitions to charge and discharge the capacitor C_H.

B. Basic performance of the HVDC tap

In order to accomplish the zero voltage turn off of the GTOs, the capacitor C_H should be completely discharged before the beginning of Mode 1. It depends on both the operating frequency of the HVDC tap and the duty ratio of the GTOs. In order to verify the operating area in which such condition is realized, the following tests are

TABLE I
PARAMETERS FOR SIMPLE CIRCUIT MODEL IN FIG. 3

The main dc system	$V_s = 1000$kV
	$V_r = 995$kV
	$R_l = 0.0025\Omega$
	$C_l = 45\mu$F
HVDC tap	$C_H = 10 \mu$F
	$C_d = 50 \mu$F
	ratio of transformer 25 : 1

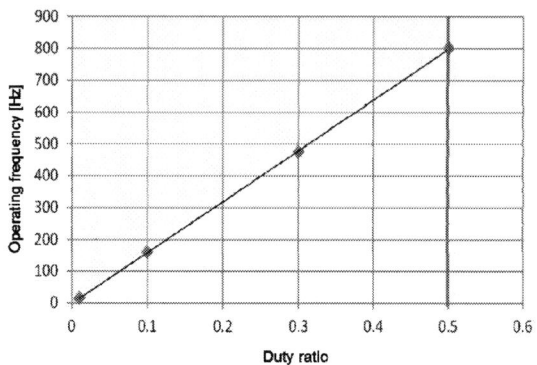

Fig. 3. Operating area to realize the soft-switching of GTOs.

executed by a simple circuit model with single HVDC tap shown in Fig. 1. Circuit parameters are given in Table 1. As indicated in Table 1, a constant voltage drop of 5kV is assumed between the sending and receiving end of the dc transmission line. Although there may be an optimal design of such parameters depending on the operating frequency of the HVDC tap, the parameters are assumed to be constant for the simplicity.

Fig. 3 shows the relationship between the operating frequency of the HVDC tap and duty ratio of the GTOs, which is confirmed by numerical analysis. The soft-switching of the GTOs cannot be accomplished in the colored zone. Under the same circuit parameters, the time to discharge the capacitor C_H is almost constant. Therefore, the higher operating frequency results in requiring higher duty ratio to discharge the capacitor.

Based on the soft-switching conditions, Fig. 4 confirms the operating characteristics of the system from the viewpoints of controllability of the additional terminal voltage and the dc current on the main dc line with comparison of two cases, where the operating frequency is 100Hz or 300Hz, respectively. As the dc currents inherently fluctuate as discussed above, they are evaluated by their averaged values.

Figs. 4 (a) and (b) indicate that, in case of 300Hz, both the output voltage of the sub-circuit and the main dc current become larger than the case of 100Hz. It means that the dc voltage at the additional terminal can be regulated by the operating frequency. On the other hand, they can be varied by regulating the duty ratio of the GTOs. However, in case of 100Hz, the additional dc voltage cannot be decreased under 72V as shown in Fig. 4 (a). Simultaneously, the dc line currents are increased by the duty ratio because the high duty ratio results in bypassing the H-bridge during longer time. Therefore, the output power regulation at the additional dc voltage

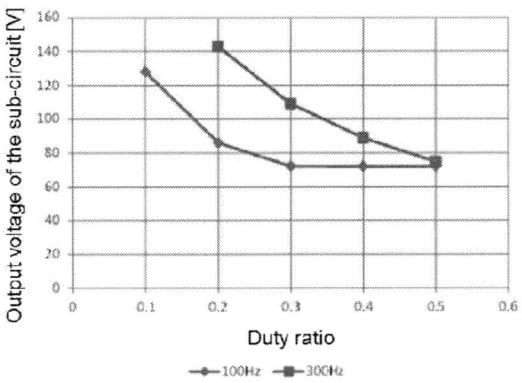

(a) The output voltage of the sub-circuit,

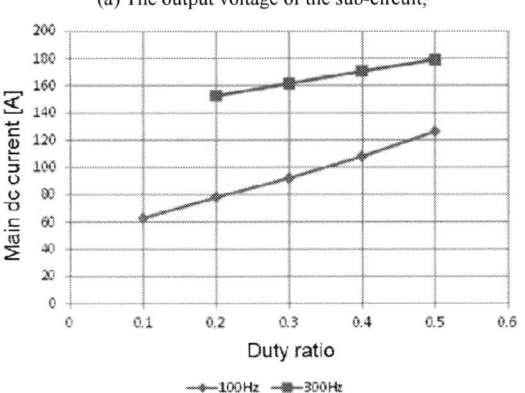

(b) Main dc currents,

Fig. 4. Operating characteristics of the system.

terminal can be realized by modulation of the operating frequency.

Fig. 5 displays the dc line current when the operating frequency is 600Hz and the duty ratio is 0.5. In this case, the soft-switching of the GTOs are accomplished. As shown in Fig. 5, the dc current is highly fluctuated. If it is not acceptable, this can be a drawback of this circuit topology.

If a hard switching of the power device can be considered, the operating area can also be enhanced. Fig. 6 summarizes results in case that the duty ratio is 0.3. The colored zone indicates the soft-switching cannot be accomplished, where the operating frequency is higher than 475Hz. When the operating frequency is higher than 475Hz, the dc line current is almost same but the additional dc terminal voltage can be increased.

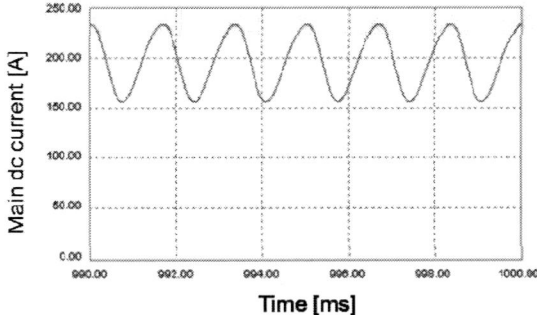

Fig. 5. Fluctuating dc line current when the operating frequency is 600Hz and the duty ratio is 0.5.

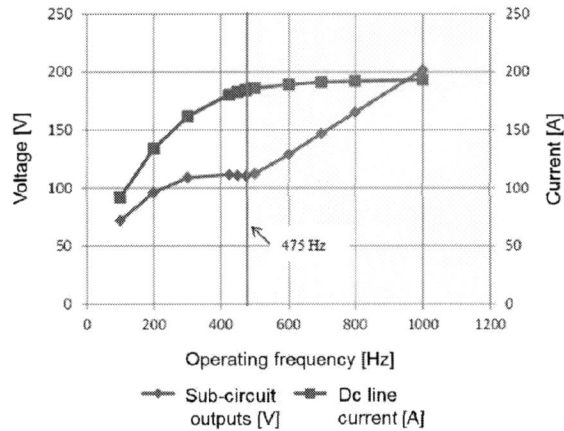

Fig. 6. Relation between the operating frequency and variables.

III. DC POWER SUPPLY WITH DOUBLE HVDC TAPS

As discussed above, single HVDC tap cannot avoid the dc current fluctuation inherently. However, if the HVDC tap is duplicated and their switching pattern is inversed, the dc current fluctuation can be compensated. This paper proposes a circuit topology with double HVDC taps as shown in Fig. 7. The dc voltages V_{oA} and V_{oB} are additional two dc outputs whose voltages are lower than the main dc output voltage V_r.

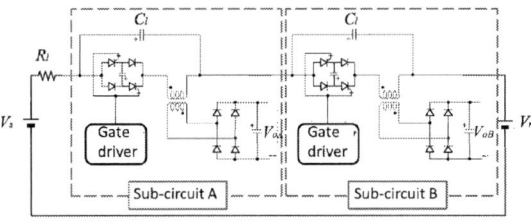

Fig. 7. The proposed circuit topology with double HVDC taps.

To confirm the operating characteristics of the proposed circuit, numerical analysis were executed with same parameters summarized in Table. 1.

Fig. 8 shows the simulation results of the dc line currents with comparison of the inversed or same switching signal. When the same signal is used as shown in Fig. 8 (a), the fluctuation of the dc current is larger than the case of Fig. 5, where single HVDC tap is used. However, the inversed switching pattern can suppress the fluctuations as shown in Fig. 8 (b). Therefore, it can be concluded that the double HVDC tap circuit is useful when their circuit parameters are same.

Fig. 9 summarizes the results of the sub-circuit output voltage when the operating frequency was varied with comparison of the switching signal was inversed or not. The inversed switching signal makes the output voltage higher than the one in case of same switching signal. Fig. 9 also indicates that the output voltage can be regulated by the operating frequency by linear in both cases.

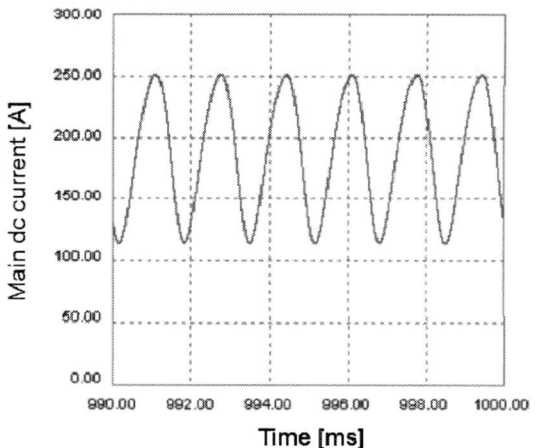

(a) with same signals,

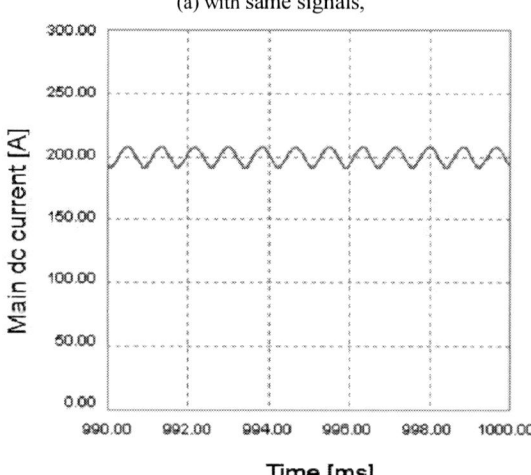

(b) with inversed signals,

Fig. 8. Dc current fluctuations when the switching signal is inversed or not.

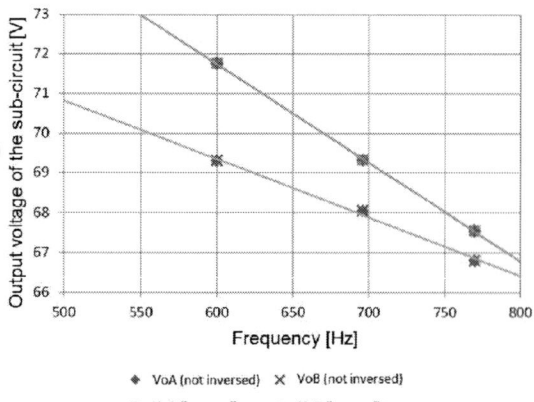

♦ VoA (not inversed)   ✕ VoB (not inversed)
▩ VoA (inversed)   ✚ VoB (inversed)

Fig. 9. Output voltage regulation by frequency control when the switching signal is inversed or not.

IV. CONCLUSIONS

This paper introduces a concept of multi-output dc transmission system for high voltage dc transmission system such as offshore wind power system. The concept of HVDC tap is very useful but, if two HVDC taps are used on the dc line simultaneously and their switching signal are inversed, their inherent fluctuation of the dc current can be reduced drastically. The additional dc outputs are low power rating but the circuit can be compact. It is useful in case that low power is required at the end of dc transmission line.

In future work, the concept should be verified by a small scale experimental setup.

REFERENCES

[1] J. Rodríguez, J.-S. Lai, and F. Z. Peng, "Multilevel inverters: a survey of topologies, controls, and applicatoins," *IEEE Trans. Industrial Electronics*, vol. 49, pp. 724-738, Aug. 2002.

[2] Agustí Egea-Alvarez, Fernando Bianchi, Adria Junyent-Ferré, Gabriel Gross, Oriol Gomis-Bellmunt, "Experi-mental Implementation of a Voltage Control for a Multiterminal VSC-HVDC Offshore Transmission System," *Proceedings of the Innovative Smart Grid Technologies (ISGT) Europe*, 2012.

[3] L. Xu and D. Chen, "Control and operation of a dc microgrid with variable generation and energy storage," *IEEE Trans. Power Delivery*, vol. 26, pp. 2513-2522, Oct. 2011.

[4] D. Salomonsson, L. Soder, A. Sannino, "An adaptive control system for a dc microgrid for data centers," *IEEE Trans. Industry Applications*, vol. 44, pp. 1910-1917, Nov/Dec. 2008.

[5] W.D. Reeve, *dc Power System Design for Telecommunications*, Piscataway: IEEE Press, 2007, p. 5-11.

[6] Maurício Aredes, Robson Dias, Antonio Felipe Da Cunha De Aquino, Carlos Portela and Edson Watanabe, "Going the Distance Power-Electronics-Based Solutions for Long-Range Bulk Power Transmission," *IEEE Industrial Electronics Magazine*, Vol.5, No.1, pp.36-48 (2011).

978-1-4799-2706-7/14 $31.00 © 2014 IEEE

Performance Evaluation for Grid Impedance Based Islanding Detection Method

Ning Liu[*1], A. S. Aljankawey[*2], C. P. Diduch[*2], L. Chang[*2], Meiqin Mao[*1], Pegah Yazdkhasti[*2], Jianhui Su[*1]

[*1]: The Institute of Energy, Hefei University of Technology-Hefei, Anhui, China
[*2]: Department of Electrical and Computer Engineering, University of New Brunswick-Fredericton, NB, Canada
E-mails: n.liuyu@gmail.com, a.aljankawey@unb.ca, diduch@unb.ca, Lchang@unb.ca, mmqmail@163.com,
pegah.yazdkhasti@gmail.com, su_chen@126.com

Abstract— This paper develops an approach for islanding detection based on measurements of the frequency dependent impedance at the point of common coupling (PCC) that exploits the presence of harmonics introduced by the electric power system (EPS) and harmonics introduced by the distributed generator (DG). The approach is new: i) an analytic model for the frequency dependent impedance is derived from the interconnection topology and this may be used as a basis for selecting features of the impedance that change when islanding occurs, and ii) the variation of a frequency dependent feature is used to characterize a non detection zone (NDZ) and iii) the load parameter space is used to compare the NDZ of the grid impedance based scheme with under/over frequency schemes. Although the focus is passive islanding detection, it also encompasses active schemes where certain harmonics are intentionally injected rather than inherent.

Keywords— *Distributed generator, frequency dependent impedance, non detection zone, islanding detection.*

I. INTRODUCTION

Unintentional islanding is an undesirable operating condition for grid connected distributed generation (DG) systems that cause safety hazards and risks to equipment, the electrical power system (EPS), and personnel. To mitigate the risk, IEEE Std. 1547 [1] specifies requirements for islanding detection and prevention to be implemented within DG equipment.

Passive approaches to islanding detection are desirable because i) only measurements at the PCC are required and ii) external disturbances are not injected into the EPS [2-3]. Many of the proposed signals processing schemes are based on heuristics, and the effectiveness has not been demonstrated. Such approaches could be strengthened by incorporating the harmonic grid impedance (HGI) based methodology such as that proposed in [4] where an impedance model is shown to characterize the actual physical interconnection topology. The HGI scheme exploits the presence of harmonic distortion in the current and voltage at the PCC as a basis for computing measures of impedance at frequencies where there is sufficient harmonic content. Unlike signals based schemes that rely on heuristics, the HGI method detects islanding based on an analytic model that reflects the interconnection topology of distributed generators (DGs) with the EPS. The DGs may include power converters or not.

This work is supported by Canadian Natural Sciences and Engineering (NSERC) Research Council, Guangdong Innovative Research Team Program (GIRTP).

This paper develops an approach to characterize the effectiveness of the HGI methodology. The analytic performance evaluation of passive islanding detection methods includes under/over voltage scheme, under/over frequency (UF/OF) method and phase jump method [5-6]. The performance indexes have included parameterization such as the power imbalance space, ΔP versus ΔQ, the load parameter space, and quality factor (Q_f) versus resonant frequency (f_o), as a means for mapping the non detection zones (NDZs) of the various passive methods [7].

The objective of this study is to evaluate the effectiveness of the HGI methodology by analyzing the DG interconnection topology and developing a performance metric for comparison with other passive methods. First, a new approach for performance evaluation of HGI scheme is developed where NDZ is defined. Then the NDZ of the HGI approach and the NDZ of the UF/OF approach are compared using the load parameter space (Q_f versus f_o).

II. THE HARMONIC GRID IMPEDANCE BASED ISLANDING DETECTION METHOD

The essence of HGI based islanding detection method is detecting changes in impedance at the PCC. The DG interconnection topology is modeled by the small signal equivalent circuit shown in Fig. 1. It consists of a local load, Z_L, connected at the PCC to a DG and the EPS through a breaker, B, and grid impedance $Z_{EPS}=R_g+sL_g$. The harmonic distortion of the EPS is represented by an ideal current source, N_{gridi}. At the instant of islanding, the equivalent impedance as seen by the DG at the PCC suddenly changes from $Z_L(s)Z_{EPS}(s)/(Z_L(s)+Z_{EPS}(s))$ to $Z_L(s)$. This results in sudden changes in the harmonic components of V_{pcc} and I_{dg}.

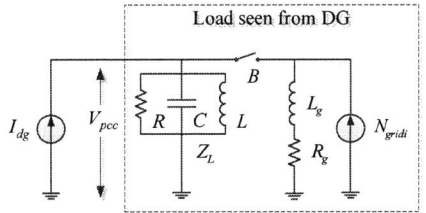

Fig. 1. Harmonic model of grid-connected DG system.

An equivalent transfer function model of the topology in Fig. 1 appears in Fig. 2, where, V_{pcc} denotes the harmonic component of the voltage at the PCC caused by a particular harmonic in I_{dg}, and N_{gridi}. Define the impedance as seen by the DG during islanding operation as $Z_{PCC}^i(s) = G_z^i(s)$ and the impedance as seen by DG during normal operation as $Z_{PCC}^0(s) = G_z^0(s)$.

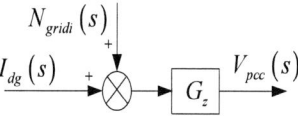

Fig. 2. Transfer function model of grid-connected DG system.

If there exists distinct harmonics in $I_{dg}(s)$ at complex frequencies, $s = j\omega_n$, $n = 1, 2, \dots N$, and if these harmonics are not present in N_{gridi}, then the measurements of the harmonics at V_{pcc} and I_{dg} can be used to compute the magnitude of the impedance at the PCC at these frequencies from,

$$|Z_{PCC}(j\omega_n)| = \frac{|V_{PCC}(j\omega_n)|}{|I_{dg}(j\omega_n)|} = |G_z(j\omega_n)|, \; n = 1, 2\dots$$

If the computed values of $|Z_{PCC}(j\omega_n)|$ align with $|Z_{PCC}^0(j\omega_n)|$ then the DG is operating normally. If the computed values for $|Z_{PCC}(j\omega_n)|$ align with $|Z_{PCC}^i(j\omega_n)|$ then the DG is islanding. The magnitude of the frequency dependent impedance $|Z_{PCC}(j\omega)|$ is shown in Fig. 3. The impedance as seen by the DG changes when islanding occurs, causing a shift in the resonant peak from 280 Hz to 60 Hz. There is also an increase in the impedance of approximately 30dB over the frequency range 0.01 to 280Hz, depending on the value of $\alpha = R_g / L_g$.

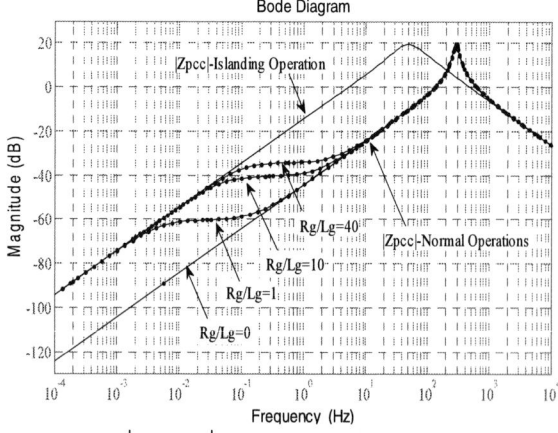

Fig. 3. The $|Z_{PCC}(j\omega)|$ during islanding and normal operations.

The frequency dependent change in impedance serves as the basis for HGI based islanding detection [4]. In the following we extend the analysis by defining a performance metric that allows a NDZ to be computed. The performance metric uses the grid impedance parameter space, based on the grid inductance L_g and the grid resistance-inductance ratio, α. This perspective and the methodology developed in the next section can be extended to other grid impedance based methods.

III. GRID IMPEDANCE PARAMETER SPACE

The change in harmonic impedance after islanding occurs depends on the grid impedance and the local load. Consider the change in the resonant frequency of the impedance at the PCC, define the resonant frequency of the local load under islanding condition as ω_o^j. Consider the worst islanding condition, when the resonant frequency of the load matches the grid frequency, $\omega_o^j = \frac{1}{\sqrt{LC}} = 2\pi \cdot f_o^i$. Define the resonant frequency under normal operation as ω_o^0.

During normal condition, the impedance as seen by the DG is,

$$Z_{PCC}^0(j\omega) = \left[\frac{R_g}{R_g^2 + \omega^2 L_g^2} + \frac{1}{R} + j\left(\omega C - \frac{1}{\omega L} - \frac{\omega L_g}{R_g^2 + \omega^2 L_g^2} \right) \right]^{-1} \quad (1)$$

Consider, $R_g \neq 0, \alpha L_g = R_g, R_g > L_g$, typical of low voltage distribution networks, α is the resistance inductance ration of the grid impedance. At the resonant frequency under normal operation,

$$\text{Im}\left[Z_{PCC}^0\left(j\omega_o^0 \right) \right] = \omega_o^0 C - \frac{1}{\omega_o^0 L} - \frac{\omega_o^0 L_g}{R_g^2 + \omega_o^{02} L_g^2} = 0 \quad (2)$$

The value of ω_o^0 can be calculated from (2) as

$$\left(\omega_o^0 \right)^2 = \frac{L_g^2 + LL_g - R_g^2 LC + \sqrt{\Delta}}{2 L_g^2 LC}$$

$$\Delta = (L_g^2 + LL_g - R_g^2 LC)^2 + 4 R_g^2 L_g^2 LC \quad (3)$$

Equation (3) indicates that $\sqrt{\Delta} > \left| L_g^2 + LL_g - LCR_g^2 \right|$.

Define a function of the grid impedance parameters, R_g and L_g that characterizes the difference in the resonant frequency under normal and islanding operation,

$$f(R_g, L_g) = \left(\omega_o^0 \right)^2 - \left(\omega_o^j \right)^2 =$$

$$\frac{L_g^2 + LL_g - R_g^2 LC + \sqrt{\Delta}}{2 L_g^2 LC} - \left(\omega_o^j \right)^2, \; L_g \neq 0 \quad (4)$$

If $R_g \neq 0, L_g = 0$ then,

$$Z_{PCC}^0(j\omega) = \left[\frac{1}{R_g} + \frac{1}{R} + j\left(\omega C - \frac{1}{\omega L} \right) \right]^{-1} \quad (5)$$

In the case described by (5), $\omega_o^0 = \omega_o^j = \frac{1}{\sqrt{LC}}$, ω_o does not change and islanding cannot be detected based on the size of $f(R_g, L_g)$. Hence, $R_g \neq 0, L_g = 0$ belongs to the NDZ of the HGI approach.

For $L_g > 0$ and $\alpha \in (0,1000)$, Fig. 3 shows a distinct difference in the impedance during normal and islanding operation. Under these conditions, if measurements of voltage and current at the various harmonic frequencies of the EPS fundamental, $\omega = n\omega_o$, are available, then the HGI method will detect islanding under the worst case local load when, $\left(\omega_o^0\right)^2 - \left(n\omega_o\right)^2 > 0$ at frequency, $\omega = n\omega_o$. A measure of the difference between the resonant frequency under normal operation and islanding operation may be given by,

$$f(L_g,\alpha)\big|_{L_g>0} = \left(\omega_o^0\right)^2 - \left(n\omega_o\right)^2 = \frac{1}{2LC} + \frac{1}{2L_gC} - \frac{\alpha^2}{2}$$
$$+ \sqrt{\left(\frac{1}{2LC} + \frac{1}{2L_gC} - \frac{\alpha^2}{2}\right)^2 + \frac{\alpha^2}{LC} - (n \times 2\pi \times 50)^2} \quad (6)$$

The NDZ of HGI method can be defined by the parameter space of $f(L_g,\alpha) < 0$. The parameters, $R_g \neq 0$, $L_g = 0$, fall within the NDZ.

Fig. 4 shows the computed NDZ as a function of L_g and α when $n = 2$ and the rated power of the DG is 5kW and the rate voltage is 220V. The boundary between is defined by $f(L_g,\alpha) = 0$. If L_g and α are located in the shaded region the islanding detection is not possible.

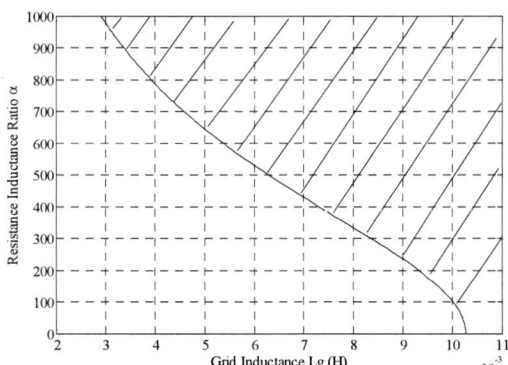

Fig. 4. L_g versus α Grid Impedance Parameter Space, $n = 2$, $R = 9.68\Omega$, $Q_f = 2.5$.

This approach for determining the NDZ for an index is parameterized by the load or grid impedance can be extended to other islanding detection algorithms.

IV. PERFORMANCE AND COMPARISON OF HGI APPROACH

If the grid impedance falls within the detection zone of the HGI approach, then islanding detection is influenced primarily by the local load. The local load parameters, Q_f and f_o^i, have been previously used for evaluating the NDZ in [7] and we will extend this to the HGI approach.

A. UF/OF Approach

The UF/OF passive islanding detection scheme will be used as a baseline for comparison with the HGI approach. Under the worst islanding, the resonant frequency of the load f_o^i matches the grid frequency. Since the frequency of the DG after islanding is f_o^i and if f_o^i is close to the grid frequency then islanding will not trip the UF/OF relay.

If the UF/OF interval is defined by $f_{\min} = 49.3\text{Hz}$, $f_{\max} = 50.5\text{Hz}$, and if $f_{\min} \leq f_o^i \leq f_{\max}$, then UF/OF will not trip and islanding will occur independent of Q_f. The NDZ of the UF/OF approach when parameterized by Q_f and f_o appears as the shaded area in Fig. 5.

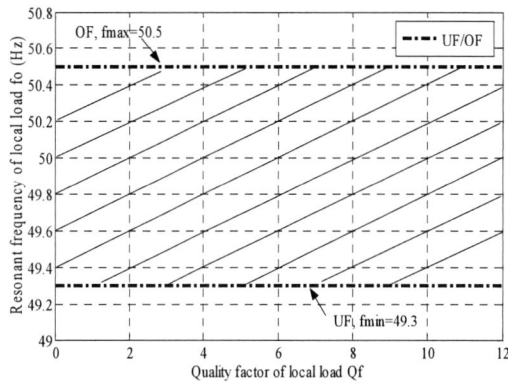

Fig. 5. NDZ of UF/OF approach.

B. HGI Approach

IEEE Std 929–2000 [8] defines the quality factor, Q_f, as 2π times the ratio of the maximum stored energy to the energy dissipated per cycle at a given frequency. For a parallel RLC load, $Q_f = \omega_o^i RC = \frac{R}{\omega_o^i L}$, and $f_o^i = \frac{1}{2\pi\sqrt{LC}}$ is the resonant frequency of the local load, thus, $L = \frac{R}{2\pi f_o^i Q_f}$, $C = \frac{Q_f}{2\pi f_o^i R}$.

The comparison measure stated in (6) may be rewritten in terms of Q_f and f_o^i,

$$f(Q_f,f_o^i) = 2\pi^2\left(f_o^i\right)^2 + \frac{\pi R f_o^i}{L_g Q_f} - \frac{\alpha^2}{2} +$$
$$+ \sqrt{\left(2\pi^2\left(f_o^i\right)^2 + \frac{\pi R f_o^i}{L_g Q_f} - \frac{\alpha^2}{2}\right)^2 + 4\pi^2\alpha^2\left(f_o^i\right)^2 - (n \times 2\pi \times 50)^2} \quad (7)$$

Fig. 6 overlays the NDZ associated with (7) for $n=2$ with the NDZ of the UF/OF in Fig. 5. It shows that if the HGI approach is combined with the UF/OF approach then the NDZ is reduced.

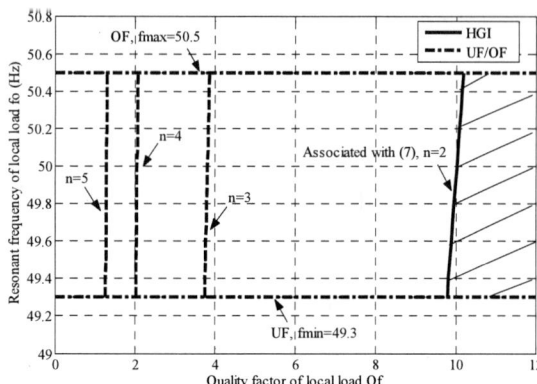

Fig. 6. NDZ of HGI scheme combined with UF/OF scheme, $n = 2, 3, 4, 5$, and $L_g = 1e^{-3}$H, $R = 9.68\Omega$, $\alpha = 100$.

The comparison measure may be augmented to achieve better detection performance. Fig. 6 also shows the load parameter space (Q_f versus f_o^i) that represents the condition, $(2\omega_o)^2 < (\omega_o^0)^2 < (n\omega_o)^2$, $n = 3, 4, 5$, i.e., the space located between the boundary associated with (7) for $n=3,4,5$ and the boundary associated with (7) for $n=2$. For the HGI scheme, if the harmonic orders used for calculating harmonic impedance are smaller than n, then the parameter space will fall within the NDZ. If the measured frequencies available for calculating the harmonic impedances to cover the harmonic order, n, then the parameter space will fall outside of the NDZ. This indicates that the better performance of the HGI method is achieved by selecting a frequency band that includes higher harmonic orders.

As an alternative to the shift in resonant frequency, consider the difference in the magnitude of the impedance at a particular frequency (λ),

$$(\omega_o^0)^2 - (n\omega_o)^2 > 0 \tag{8}$$

$$\left\| Z_{pcc}^0 (j\omega) \right| - \left| Z_{pcc}^i (j\omega) \right\| > \lambda, \quad \omega = m\omega_o \tag{9}$$

Using the load and grid impedance models in Fig. 1 it can be shown that (9) can be written as,

$$4\pi L_g Q_f f_o^i \left| \frac{(m50)^2}{(f_o^i)^2} - 1 \right| > \lambda (R + 2\alpha L_g) \tag{10}$$

The NDZ associated with both (8) and (10) can readily be generated for particular values of n and m. The NDZ may be reduced by choosing an appropriate value for m.

V. VERIFICATION BASED ON SIMULATION

To verify the analytical results derived in this paper, an inverter based grid connected DG system was modeled with MATLAB/Simulink. A dc voltage source is used to represent the photovoltaic array and is connected to a current controlled voltage source inverter. The rated power of this simulated DG is 5kW at the rate voltage of 220V. The EPS is modeled as an ac voltage source connected in series with grid impedance of $R_g = 0.1\Omega$, $L_g = 1$mH. A parallel RLC block is used to represent the

local load with four sets of values, which yield (f_o^i, Q_f) to be A (2.5, 49.3), B (2.5, 50), C (5, 50) and D (8.5, 50.7) respectively, which represent four cases of the local load as depicted in Fig. 7.

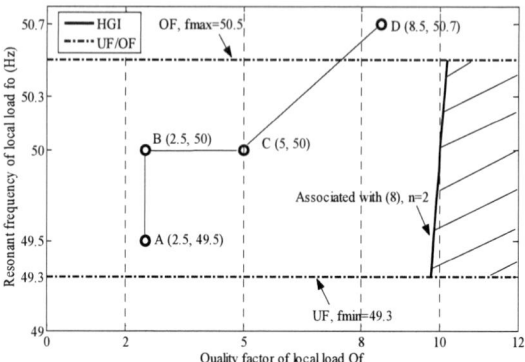

Fig. 7. Cases for verifying the NDZ of HGI scheme combined with UF/OF scheme, $n = 2$, $L_g = 1e^{-3}$H, $R = 9.68\Omega$, $\alpha = 100$.

A. Verification of the NDZ for case A

In case A, system parameters with the quality factor of 2.5 and the resonant frequency of 49.5Hz drive DG system fall into the NDZ of UF/OF but fall out of the NDZ of HGI scheme. The simulation collected transient data for I_{dg} and V_{pcc}, and computed the FFT for each. The magnitude of the impedance is computed from the magnitude of the FFT of V_{pcc} divided by the magnitude of the FFT of I_{dg} at the harmonic frequencies, and appears in Fig. 8 together with the associated bode plots of $Z_{pcc}(s)$ for both normal and islanding condition. It indicates that, in the NDZ of UF/OF scheme, with the chosen frequency band of n=2, HGI scheme can detect islanding by monitoring the frequency dependent change in the impedances at the PCC.

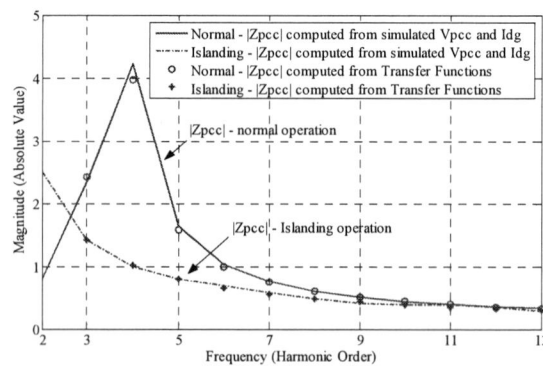

Fig. 8. Simulation results for case A (2.5, 49.5 Hz).

B. Verification of the NDZ for case B

Case B represents one of the most challenging islanding conditions for passive detection algorithms with $Q_f = 2.5$ and $f_o = 50$Hz. Comparing with the case A, the quality factor does not change and the resonant frequency increases to a value equal to the fundamental frequency of EPS. The same simulation process as stated in case A was carried out in this case, the simulation results are depicted in Fig. 9. The islanding condition can be detected by HGI scheme in the NDZ of UF/OF scheme by choosing an appropriate frequency band for calculating the impedances. Case B is in the detection zone of HGI scheme.

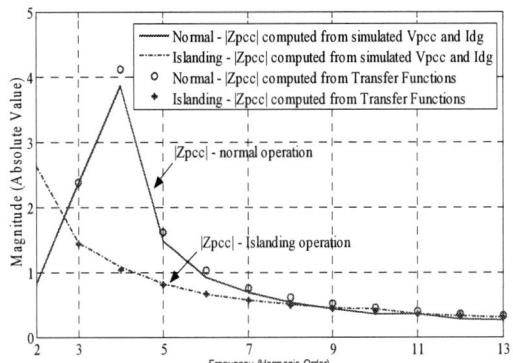

Fig. 9. Simulation results for case B (2.5, 50 Hz).

C. Verification of the NDZ for case C

In Case C, another worst case islanding condition with $Q_f = 5$ and $f_o = 50$Hz is simulated. The quality factor of the local load in this case is higher in comparison with cases A and B. The simulation results, shown in Fig. 10, show that the increase in Q_f produces a reduction in the frequency dependent impedance. The frequency of the peak value is shifted to a lower harmonic order when Q_f increases. Islanding can be detected in the NDZ of UF/OF scheme by choosing an appropriate frequency band.

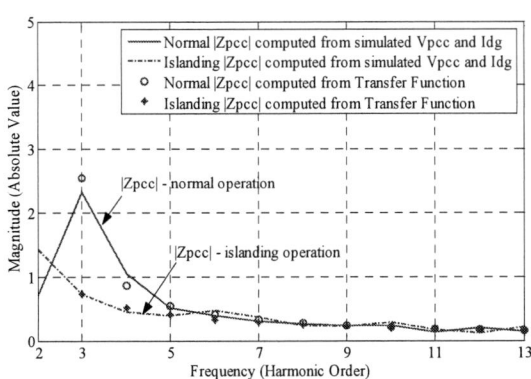

Fig. 10. Simulation results for case C (5, 50 Hz).

D. Verification of the NDZ for case D

Case D represents the islanding condition with $Q_f = 8.5$ and $f_o = 50.7$Hz, which is in the detection zone for both UF/OF and HGI schemes. The simulation results appear in Fig. 11. With a higher quality factor of 8.5, the frequency dependent change in impedances decreases, and the resonant peak shifts to a lower harmonic.

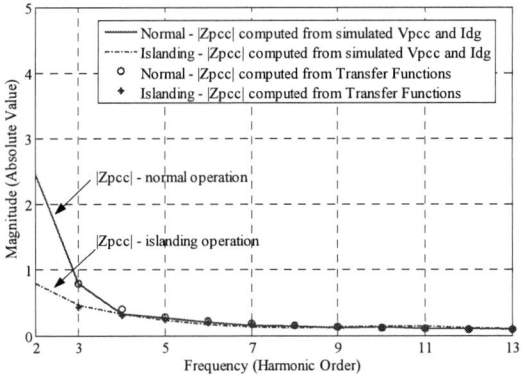

Fig. 11. Simulation results for case D (8.5, 50.7 Hz).

The above simulation results for cases A, B, C and D indicate that, the NDZ of UF/OF scheme can be reduced by in combination with the HGI scheme.

VI. CONCLUSIONS

This paper proposes a new Grid Impedance Parameter Space based on the values of the grid inductance L_g and the resistance inductance ratio α to characterize the NDZ of HGI based islanding detection. The boundaries that separate the NDZ relate to the shift in the resonant frequency after islanding. This methodology may be extended to other comparison measures of grid impedance during islanding and normal operation. The NDZ of the HGI approach has been analyzed using new comparison measures and parameterizations. By combining HGI with UF/OF based islanding detection this work shows that the resulting NDZ can be reduced.

REFERENCES

[1] "IEEE standard for interconnecting distributed resources with electric power systems," *IEEE Std 1547-2003*, pp. 0–1–16, Jun 2003.

[2] J.-H. Kim, J.-G. Kim, Y.-H. Ji, Y.-C. Jung, and C.-Y. Won, "An islanding detection method for a grid-connected system based on the goertzel algorithm," *IEEE Trans. on Power Electronics,* vol. 26, no. 4, pp. 1049–1055, 2011..

[3] J. Yin, C. Diduch, and L. Chang, "Islanding detection using proportional power spectral density," *IEEE Trans. on Power Delivery*, , vol. 23, no. 2, pp. 776–784, 2008.

[4] N. Liu, A. S. Aljankawey, C. Diduch, L. Chang, J. Su, and M. Yu, "A new impedance-based approach for passive islanding detection scheme," *in Power Electronics for Distributed Generation Systems (PEDG), 4th IEEE International Symposium conference (in press)* , 2013.

[5] H. Zeineldin and J. Kirtley, "Performance of the over/under voltage and frequency method with voltage and frequency dependent loads," *IEEE Trans. on Power Delivery*, vol. 24, no. 2, pp. 772–778, 2009.

[6] Ye Zhiyong, A. Kolwalkar, Y. Zhang, Du Pengwei, R. Walling, "Evaluation of Anti-Islanding Schemes Based on Nonedetection Zone Concept," *IEEE Trans. on Power Electronics*, vol. 19, no. 5, pp. 1171-1176, 2004.

[7] L.A.C Lopes, Sun Huili, "Performance Assessment of Active Frequency Drifting Islanding Detection Methods," *IEEE Trans. on Energy Conversion*, vol. 21, No. 1, pp. 171-180, 2006.

[8] IEEE Recommended Practice For Utility Interface Of Photovoltaic (PV) Systems, *IEEE Standard 929–2000*, Apr., 2000.

Identifying Natural Degradation/Aging in Power MOSFETs in a Live Grid-Tied PV Inverter Using Spread Spectrum Time Domain Reflectometry

Qian Li and Faisal H. Khan
Dept. of Electrical and Computer Engineering
University of Utah
Salt Lake City, USA

Abstract—Spread spectrum time domain reflectometry (SSTDR) has been applied to a live PV inverter circuit to measure impedance variations caused by natural degradation in switching devices (MOSFET), and this method was applied without altering the normal operation of the circuit. Therefore, the proposed technique is able to perform condition monitoring – the state of health of the inverter. The experimental results and the corresponding analysis have been included which show that it is possible to determine the various path impedances inside a PV inverter using SSTDR, and thereby, it is possible to detect any natural degradation associated with the power semiconductor devices inside the circuit.

Keywords—condition monitoring, PV inverter, reliability, time domain reflectometry.

I. INTRODUCTION

Grid-tied PV inverter is one of the key components in the PV based renewable energy system. Different kinds of PV inverters exist – ranging from few hundreds of watts to kWs. Regardless of the size, a grid-tied PV system facilitates the maximum power point tracking (MPPT) and necessary dc-ac power conversion with reasonabe high efficiency (95%). Making solar power system more reliable is a way to prevent the loss of potential energy production and reduce the cost of repair [1]. As a result, reliability has become one of the most promising research directions. Unfortunately, PV inverters are responsible for most of the reliability issues in the PV based power systems [1]- [3]; the operational life of PV cells is reported to be more than 20 years while the PV inverter suffers from a much shorter life span.

Existing methods can analyze the reliability of PV inverters in the form of failure rate. In [4], both MIL HDBK 217 and IEC TR procedures have been used to calculate the failure rate. A model framework for decomposing the inverter into subsystems has been suggested in [1]. By acknowledging inverter failures and repairs over the lifetime of the system, the authors of [5] illustrate a systematic method to integrate PV inverter reliability into energy-yield estimation with Markov reliability models. Unfortunately, even though it is possible to estimate the reliability of PV inverters using the methods mentioned above, a failure cannot be predicted (with reasonable accuracy) using these

techniques before it happens. Thus, in this case, we can either change the PV inverter before it breaks down (solution 1) or efforts should be in place to locate the fault after the failure happens (solution 2). However, the first solution is not cost-effective whereas the repair is time intensive thereby causes the loss of potential for energy production [1]. To address this problem, different condition monitoring (CM) techniques have been suggested to give a real-time measurement of component conditions or even to monitor the entire system. These CM techniques can be further divided into two groups: internal sensor based methods and model based methods [8]. For the internal sensor based methods, additional hardware is always needed which makes the solutions more expensive and sometimes even impossible to apply to a commercial PV inverter circuit. Meanwhile, model based methods can put huge calculation stress onto the system as well. For example, some of the CM methods use on-state resistance as the precursor of a degrading MOSFET [9]-[12]. By measuring electrical parameters such as voltages and currents of the circuit, on-state resistance of a switching device can be calculated. However, it is not always possible to detect the small electrical variations caused by the change in the on-state resistance $R_{DS(ON)}$ [8].

A new technique was proposed by one of the co-authors of this paper to monitor the degradation level of power semiconductor devices and electrolytic capacitors using spread spectrum time domain reflectometry (SSTDR). SSTDR was used with individual components while they were disconnected from the circuit. In contrast, this paper applies SSTDR to monitor the natural degradation of a power MOSFET in an energized commercial PV inverter. SSTDR can predict a failure condition and locate the faults without interfering with PV inverters' normal working condition. As can be seen in the following analysis, SSTDR method only needs four accessible points to monitor all the MOSFETs in an H-bridge PV inverter. Its hardware flexibility together with its simplicity in calculation makes SSTDR a promising method to monitor live power circuits.

Among all the components in a PV inverter, semiconductor and soldering failure contribute with 21% of all the failures. Capacitor is also a fragile component; nearly 30% of the failures are the results of capacitor

978-1-4799-2706-7/14 $31.00 © 2014 IEEE

degradation [3], [13]-[17]. The failure mechanisms for a MOSFET can be divided into two groups: intrinsic faults and extrinsic faults. Intrinsic faults include dielectric breakdown, hot carrier injection and electro-migration while extrinsic faults mainly consist of contact migration, wire lift-off, die solder degradation and package de-lamination [18]. When a MOSFET is subjected to thermal cycling, the major failure mechanism is the die-attachment degradation, and it leads to an increase in on-state resistance of a MOSFET.

Our previous SSTDR related studies have been reported in [19]-[21]. In [19], authors verified the feasibility of using SSTDR to identify the impedance variation for IGBTs, MOSFETs and electrolytic capacitors during their aging processes. After measuring different path impedances for an AC-AC converter, a numerical computation process of an impedance matrix was used to locate the aged components and to find out the level of component degradation [20], [21]. In this paper, the on-state resistance of four MOSFETs in a single phase PV inverter has been selected as the precursor of failure, and the equivalent path impedance analysis will be conducted accordingly.

The degradation process in electrolytic capacitors shows a direct relationship with the increase in equivalent series resistance (ESR) of the capacitor. Therefore, an increase in ESR is the indicator of capacitor degradation. As a result, ESR of the input capacitor will be taken into consideration in our equivalent path impedance analysis shown in section III, and a new data processing algorithm will be illustrated in section IV. Meanwhile, experimental results will be presented to verify the correctness of the proposed algorithm.

II. SSTDR TECHNOLOGY AND PRIOR WORK

Reflectometry is a method that has been used to locate faults on electrical wiring, to characterize the impedance of electrical components, and to measure the electrical properties of materials. When using a reflectometry method, a high frequency signal will be sent down the wire. By analyzing the reflected signal, the impedance discontinuity can then be found [19]. As shown in equation (1), reflection coefficient ρ represents the relationship between the impedance of the transmission medium and the terminal impedance,

$$\rho = \frac{Z_t - Z_o}{Z_t + Z_o} \qquad (1)$$

where Z_t is the impedance at the end terminal, and Z_o is the characteristic impedance of the transmission line. From this equation, if the terminal impedance is higher than the impedance of the transmission medium, the reflection coefficient will be positive. In the same way, if the terminal impedance is lower than that of the transmission medium, ρ is negative. Finally, when these two impedances are equal, reflection coefficient ρ will be zero.

When using SSTDR, a sine wave modulated pseudonoise (PN) code is sent down the wire. Though the level of SSTDR incident and reflected signals is buried in noise, they are detectable using cross-correlation which makes it an ideal fault detection method to be applied to live circuits [23], [24]. The ability to pick out the signal when using SSTDR is due to processing gain [24], which is defined as,

$$PG = \frac{T_s}{T_c} = \frac{R_c}{R_s} = \frac{w_{ss}}{2R_s} \qquad (2)$$

where T_s is the duration of one entire SSTDR sequence, T_c is the duration of a PN code chip, R_c is the chip rate in chips/second, R_s is the symbol rate, which is the number of full sequences per second, and w_{ss} is bandwidth of the spread-spectrum signal.

Figure 1 shows the SSTDR test setup for a live system. As presented in this figure, two accessible nodes are needed for each SSTDR test. These two points are connected to the SSTDR hardware through a 30 feet long cat-5 twisted pair cable.

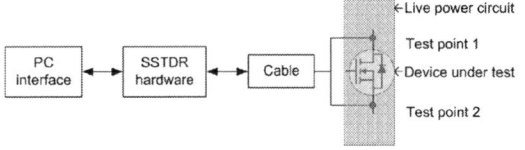

Fig. 1. SSTDR test setup [21].

III. PV INVERTER: CURRENT PATHS AND EQUIVALENT PATH IMPEDANCES

The first step to apply SSTDR to a live circuit is to define the equivalent path impedances for all possible current paths. For the single phase H-bridge PV inverter shown in Figure 2, four accessible nodes are needed in order to characterize all four MOSFETs using SSTDR: two nodes across the DC input and another two nodes on the AC output side. These points are typically available in real applications; therefore, no additional access is necessary.

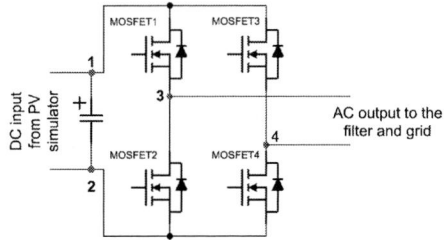

Fig. 2. SSTDR test diagram for PV inverter connected between a PV simulator and grid.

In order to determine the equivalent path impedances under various operating states, the modulation scheme for the PV inverter under test was studied, and two operating states were found.

978-1-4799-2706-7/14 $31.00 © 2014 IEEE

The 2014 International Power Electronics Conference

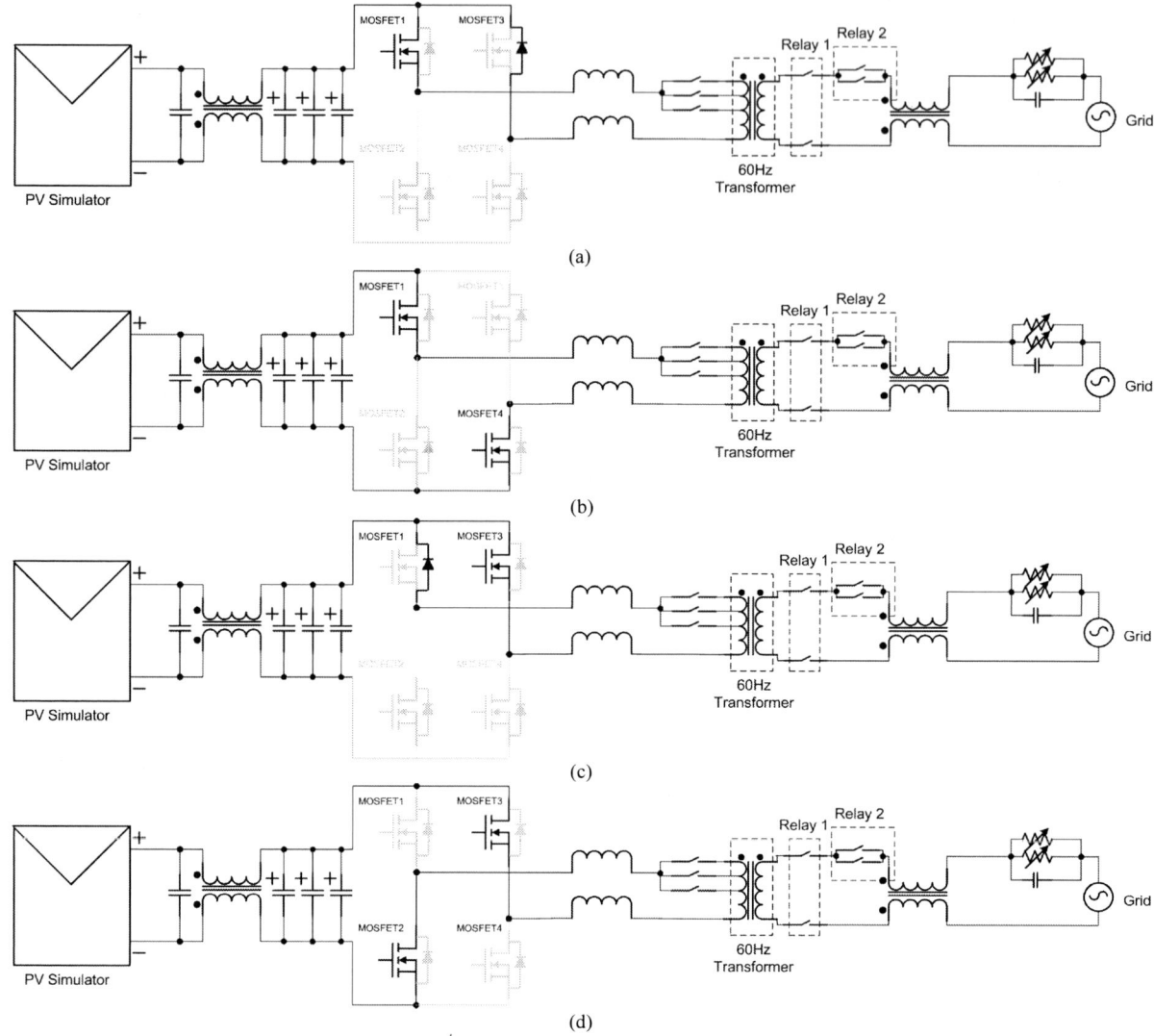

Fig. 3. Inverter equivalent current paths. (a) and (b) – state 1, (c) and (d) – state 2.

State 1: MOSFET1 is ON and MOSFET2 is OFF, MOSFET3 is OFF and MOSFET4 is working in PWM mode;

State 2: MOSFET3 is ON and MOSFET4 is OFF, MOSFET1 is OFF and MOSFET2 is working in PWM mode.

State 1 can be further separated into two different sub-state which are shown in Figures 3(a) and 3(b), respectively. In Figure 3 (a), both MOSFET1 and the free-wheeling diode of MOSFET3 are ON; MOSFET4 will then replace the body diode of MOSFET3 to conduct current as shown in Figure 3(b). In reality, conduction states will change rapidly between these two sub-states when the PV circuit is working in state 1. Because this is a 60 Hz grid-tied inverter, the time duration of state 1 is 1/120 sec.

The analysis of state 2 is similar to the procedure shown for state 1. During state 2, the operating condition will alternate between the two sub-states according to the control signals; the corresponding current paths for these two sub-states are presented in Figures 3(c) and (d).

MOSFET1 will be used as an example for the equivalent path impedance analysis. However, the method described below is applicable to all the switching components as well as the dc bus electrolytic capacitor.

In order to monitor the on-state impedance variation in MOSFET1, the SSTDR hardware was connected across test points 1 and 3 as shown in Figure 2. Because multiple circuit components are connected in parallel with MOSFET 1, it is of paramount importance to isolate the effect of any variation in $R_{DS(ON)}$ of MOSFET1 from other components connected across MOSFET1. Figure 4 shows the equivalent path impedances for states 1 and 2, and these impedance paths lay the foundation of the condition monitoring method proposed in this paper.

978-1-4799-2706-7/14 $31.00 © 2014 IEEE

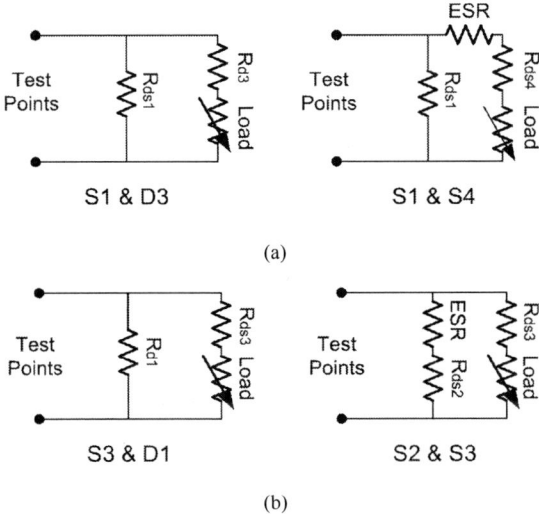

Fig. 4. Equivalent path impedances under (a) state 1; (b) state 2.

IV. EXPERIMENTAL RESULTS AND ANALYSIS

As shown in Figure 4 (a), R_{ds1} is directly connected across the two test nodes during state 1. Because the load impedance is much larger than the on-state resistance of MOSFET1 (which is equal or less than 85mΩ, according to the device datasheet), it can be concluded that the total impedance across these two test nodes during this half cycle is comparatively stable. On the other hand, a big impedance variation is observed between the two working sub-states of state 2. When the free-wheeling diode of MOSFET1 is conducting current along with MOSFET3, the impedance across the test points is the parallel of the resistance of diode 1 and R_{ds3} + the load impedance as shown in Figure 4(b). Meanwhile, when both MOSFET2 and MOSFET3 are ON, the equivalent series resistance (ESR) of the input capacitor will also make a contribution to the total impedance across test points as shown in Figure 4(b).

TABLE I
RELATIONSHIP BETWEEN SSTDR TEST RESULT AND TERMINAL IMPEDANCE

Polarity of SSTDR cross-correlation peak		Amplitude of SSTDR cross-correlation peak
Positive	$Z_t > Z_o$	If $P_a > P_b$, $Z_a > Z_b > Z_o$
Negative	$Z_t < Z_o$	If $|P_a| > |P_b|$, $Z_a < Z_b < Z_o$
Zero	$Z_t = Z_o$	P=0

When SSTDR signal is applied across any impedance, the amplitude and polarity of the SSTDR cross-correlation peak have a certain relationship with the terminal impedance. When the terminal impedance is smaller than the wire impedance, the cross-correlation will show a negative peak. In this case, the smaller the absolute peak value becomes, the bigger the actual impedance is. For the cases when the terminal impedance is bigger than that of the wire, a positive peak can be observed and the amplitude of the positive peak is proportional to the terminal impedance. Table I is a summary of the relationship described above where P_a and P_b are the corresponding SSTDR test results (cross-correlation peaks) of two different terminal impedances Z_a and Z_b, Z_o is the wire impedance, and Z_t is the terminal impedance as defined in section III.

A 700W commercially available PV inverter was modified to verify the proposed technique, and Figure 5 shows this arrangement. Figure 6 shows the plot of SSTDR peak values from continuous scans in our experiment to monitor the impedance of MOSFET1. This result is obtained when the PV inverter is operating in steady state. Because R_{ds1} is always connected directly across the test points as well as in parallel with big impedances in both sub-states during working state 1(as shown in Figure 4(a)), SSTDR cross-correlation peak variation from this region (as shown in Figure 6) can be considered as the indicator of MOSFET1 degradation level. In contrast, SSTDR test data obtained from state 2 (as shown in Figure 6) will not be influenced when only MOSFET1 is experiencing an aging process.

When the SSTDR signal arrives at the test terminals, MOSFET 1 may not be completely ON as the SSTDR system and the inverter under test are not synchronized. This is why we see variations of some level in the SSTDR data during state 1. In addition, any change in load impedance will influence the test result as well. As a consequence, a data analysis algorithm is necessary to determine the level of MOSFET degradation. Using this algorithm, all the cross-correlation results will be rearranged from high to low based on their absolute peak values. Then, the entire data set will be separated into two parts: the first data set with higher absolute peak values (lower corresponding impedances) which represents operating state 1 and the second data set with lower absolute peak values (higher corresponding impedances) which reflects the equivalent impedances during working state 2. In order to further reject the abnormal data which is influenced by disturbances during the experiment, a certain portion of captured data points will be discarded from both high and low sides in each of these data sets.

It is known that the on-resistance of a MOSFET can be greatly influenced by temperature. Therefore, observing the first data set is not sufficient to decide the MOSFET's degradation level. Considering the fact that both of these two data sets will be equally influenced by temperature or other disturbances and only the first data set will be influenced by an increased on-resistance of MOSFET1, the difference between the average values from the first and second data sets will indicate the level of MOSFET1 degradation.

Thus, in the next step of the proposed algorithm, averaged SSTDR cross-correlation peak value of each of the data set will be calculated and the outcome from the first data set will be subtracted from the result of the second data set. The resulting differences will then be documented and compared with the base value which is obtained when the PV inverter circuit is in its initial working condition with no aged components.

an aged MOSFET1 was present in the circuit. This indicates that the technique developed for the grid-tied inverter works when $R_{DS(ON)}$ of a MOSFET increases due to natural degradation.

TABLE II
DATA PROCESSING RESULTS OF TWO DATA SETS

	Test 1(without degradation)	Test 2 (an 25mΩ increase in $R_{ds\,(ON)}$ due to degradation)
Average peak value for the **first** data set	-27897	-27649
Average peak value for the **second** data set	-23467	-23549
Difference	**4430**	**4100**

V. CONCLUSION AND FUTURE WORK

Condition monitoring of a grid-tied PV inverter has been accomplished by applying spread spectrum time domain reflectometry (SSTDR) technique to the live circuit. SSTDR can detect small changes in impedance without interfering with the normal operation of the circuit. The amplitude and polarity of the cross-correlation can be used as the indicator of impedance variations across test points. Thus, after defining the equivalent path impedances of the inverter, the degradation level of the MOSFETs can be determined using the data analysis method proposed in this paper. The authors are presently working on this project to characterize all four MOSFETs by taking a single measurement, and the experimental results will be documented in future publications. We are also investigating the cases when multiple MOSFETs are aged, and the solution will require a new algorithm to analyze the experimental data.

REFERENCES

[1] A.Ristowand and M. Begovic, " Development of a methodology for improving photovoltaic inverter reliability," *IEEE Trans. Ind. Electron.*, vol. 55, no. 7, pp. 2591-2592, Jul. 2008.

[2] F. Chan and H. Calleja, "Design strategy to optimize the reliability of grid-connected PV systems," *IEEE Trans. Ind. Electron.*, vol. 56, no. 11, pp. 4465-4472, Nov. 2009.

[3] T. Messo, J. Jokipii, H. Puukko, and T. Suntio, "Determining the value of DC-link capacitance to ensure stable operation of a three-phase photovoltaic inverter," *Power Electronics, IEEE Transactions on*, vol. 29, no.2, pp. 665-673, Feb. 2014.

[4] S.E. De Leon-Aldaco, H. Calleja, F. Chan, and H. R. Jimenez-Grajales, "Effect of the mission profile on the reliability of a power converter aimed at photovoltaic applications-a case study," *Power Electronics, IEEE Transactions on*, vol. 28, no. 6, pp. 2998-3007, June 2013.

[5] S.V. Khople, A. Davoudi, P. L. Chapman, and A. D. Dominguez-Garcia, "Integrating photovoltaic inverter reliability into energy yield estimation with Markov models," *Proc. 2010 IEEE 12th Workshop on Control and Modeling for Power Electronics (COMPEL)*, Boulder, CO, pp. 1-5, Jun. 2010.

[6] E. Hofreiter and A. M. Bazzi, "Single-stage boost inverter reliability in solar photovoltaic applications," *Power and Energy Conference at Illinois (PECI)*, 2012 IEEE, pp. 1-4, Feb. 2012.

[7] B. Gu, J. Dominic, J. –S. Lai, C.-L. Chen, T. Labella, and B. Chen, "High reliability and efficiency single-phase transformerless inverter for grid-connected photovoltaic systems," *IEEE Trans. Power Electron.*, vol. 28, no. 5, pp. 2235-2245, May 2013.

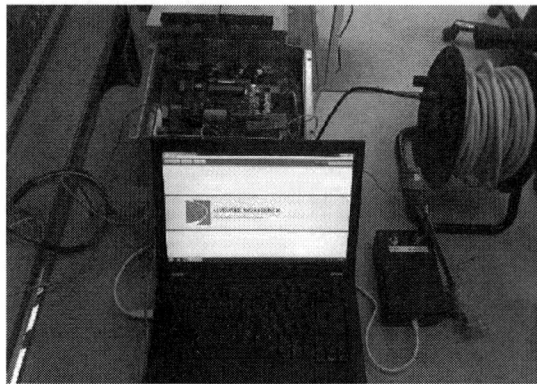

Fig. 5. Test setup using a 700W PV inverter. All four MOSFETs used in the inverter were relocated in order to age them individually.

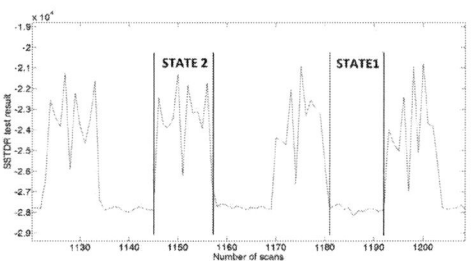

Fig. 6. SSTDR test results across node pairs 1 and 3 during normal operation.

In order to verify the accuracy of the proposed algorithm, two tests were conducted to see if any change in the averaged peak value difference between the two data sets can represent the level of MOSFET aging. In the first test, four new MOSFETs were connected in the inverter under test; SSTDR was applied between test point 1 and 3 to measure the equivalent path impedances. During test 2, a 25mΩ resistor was connected in series with the source terminal of MOSFET1 to mimic the effect of MOSFET aging. SSTDR signal was applied across this combination and the test results were recorded. Table II shows the data processing results for test 1 and 2 using the proposed algorithm. From Table II, it is obvious that when MOSFET1 is aged, the difference between the averaged peak values will decrease. This difference was 4430 for a healthy system, and it dropped to 4100 when

978-1-4799-2706-7/14 $31.00 © 2014 IEEE

[8] S. Yang, D. Xiang, A. Bryant, P. Mawby, L. Ran, and P. Tavner, "Condition monitoring for device reliability in power electronic converters: a review," *IEEE Trans. On Power Electronics*, vol.25, no. 11, pp. 2734-2752, Nov. 2010.

[9] J. Morroni, A. Dolgov, M. Shirazi, T. Zane, and D. Maksimovic, "Online health monitoring in digitally controlled power converters," *in IEEE Power Electron. Spec. Conf.*, pp. 112-118, 2007.

[10] J.R. Celaya, P. Wysocki, V. Vashchenko, S. Saha, and K. Goebel, "Accelerated aging system for prognostics of power semiconductor devices," *in AUTOTESTCON, 2010 IEEE*, pp. 1-6, 2010.

[11] J.R. Celaya, A. Saxena, C. S Kuldarni, S. saha, and K. Goebel, "Prognostics approach for power MOSFET under thermal-stress aging," *Annual Proceedings on Reliability and Maintainability Symposium (RAMS)*, 2012.

[12] J. R. Celaya, A. Saxena, V. Vashchenko, S. Saha, and K. Goebel, "Prognostics of Power MOSFET," *23rd International Symposium on Power Semiconductor Devices and ICs (ISPSD)*, 2011.

[13] K. W. Lee, M. Kim, Y. Yoon, S. B. Lee, and J. Y. Yoo, "Condition monitoring of DC-link electrolytic capacitors in adjustable-speed drives," *in IEEE Tran. on Ind. Applic.*, vol. 44, no. 5, pp. 1606-1613, Sept.-Oct. 2008.

[14] R. Jano and D. Pitica, "Accelerated aging tests for predicting capacitor lifetimes," *Design and Technology in Electronic Packaging (SIITME)*, pp. 63-68, Oct. 2011.

[15] J. R. Celaya, C. Kulkarni, S. Saha, G. Biswas, and K. Goebel, "Accelerated aging in electrolytic capacitors for prognostics," *Reliability and Maintainability Symposium (RAMS)*, pp. 1-5, Jan. 2012.

[16] C. S. Kulkarni, J. R. Celaya, G. Biswas, and K. Goebel, "Accelerated aging experiments for capacitor health monitoring and prognostics," *2012 IEEE AUTOTESTCON*, pp. 356-361, Sept. 2012.

[17] E. Wolfgang, "Example for failures in power electronics systems," presented at ECPE Tutorial 'Rel. Power Electron, Syst.', Nuremberg, Germany, Apr. 2007.

[18] G. Sonnenfeld, K. Goebel and J. R. Celaya, "An agile accelerated aging, characterization and scenario simulation system for gate controlled power transistors," *AUTOTESTCON, 2008 IEEE*, pp. 1088-7725, Sept. 2008.

[19] M. Sultana Nasrin and F. H. Khan, "Characterization of aging process in power converters using spread spectrum time domain reflectometry," *IEEE Energy Conversion Congress and Exposition (ECCE)*, pp. 2142-2148, Sept. 2012.

[20] M. Sultana Nasrin and F. H. Khan, "Use of spread spectrum time domain reflectometry to estimate state of health of power converters," *IEEE 13th Workshop on Control and Modeling for Power Electronics (COMPEL)*, pp. 1-6, Jun. 2012.

[21] M. Sultana Nasrin and F. H. Khan, "Real time monitoring of aging process in power converters using the SSTDR generated impedance matrix," *Applied Power Electronics Conference and Exposition (APEC)*, pp. 1199-1205, March 2013.

[22] J. Celaya, C. Kulkarni, and K. Goebel, "A model-based prognostics methodology for electrolytic capacitors base on electrical overstress accelerated aging," *Proceedings of Annual Conference of the PHM Society*, Sept. 2011.

[23] C. Lo and C. Furse, "Noise-domain reflectometry for locating wiring faults," *IEEE Trans. Electromagn. Compat.*, vol. 47, no. 1, pp. 97-104, Feb. 2005.

[24] P. Smith, C. Furse, and J. Gunther, "Analysis of spread spectrum time domain reflectometry for wire fault location," *IEEE Sensors J.*, vol. 5, no. 6, pp. 1469-1478, Dec. 2005.

Control Method for Inductive Power Transfer with High Partial-Load Efficiency and Resonance Tracking

R. Bosshard*, J. W. Kolar*, and B. Wunsch[†]

*Power Electronic Systems Laboratory, ETH Zürich, Switzerland, Email: bosshard@lem.ee.ethz.ch
[†]ABB Switzerland Ltd., Corporate Research, 5405 Baden-Dättwil, Switzerland

Abstract—**Frequency controlled Inductive Power Transfer (IPT) systems for Electric Vehicle (EV) battery charging applications often suffer from high power losses in partial-load, because the transmitter coil current is not significantly reduced at low output power. Therefore, in this paper a novel control method is presented that exhibits a substantially higher partial-load efficiency, while it also enables full control of the power semiconductor switching conditions. The power flow control is based on the dynamic regulation of the dc-link voltages on both sides of the resonant system with dc-dc-converters. Additionally, a tracking of the resonance with a current zero crossing detection circuit and a PLL makes the switched current an additional degree of freedom, that can be used, e.g., for the minimization of IGBT soft-switching losses due to stored-charge. All calculated results are supported by experimental measurements on an existing 5 kW/52 mm air gap/210 mm coil diameter prototype system with an efficiency of more than 96.5% at maximum power and above 96% down to 20% rated power.**

I. INTRODUCTION

Inductive Power Transfer (IPT) is widely discussed as a battery charging technology for Electric Vehicles (EV), because of the considerable simplification of the charging process due to the contactless transmission of the charging energy. It has been shown in recent publications that with an appropriately designed system a dc-to-dc efficiency comparable to conventional chargers with direct electrical connection can be achieved [1]–[4]. However, additional requirements for the battery charging system arise from the employed battery technology, which have not yet been fully addressed in literature on IPT systems.

For an efficient and fast charging of the lithium-ion batteries, which are commonly used in modern EVs, because of their high energy density, Constant Current/Constant Voltage (CC/CV) charging profiles similar to **Fig. 1** are typically used. Hence, a substantial part of the charging process does not require the full output power from the converter, especially when the State-of-Charge (SoC) is close to the allowed maximum and the current must be reduced to protect the cells from over-charging. Therefore, apart from the efficiency at the nominal point, for a battery charging system also the efficiency in partial-load is of importance. For IPT charging systems with vehicle-to-grid capabilities [4]–[6] the efficiency in partial-load is even more important to make bi-directional energy exchange with the grid financially and ecologically attractive. However, the optimization of a resonant converter system, such as an IPT system, for more than one operating point is highly challenging. Depending on the used control method, reactive currents in the IPT coil windings cause high losses in partial-load which reduce the efficiency dramatically. Therefore, in this paper it is shown that by dynamically adapting the dc-link voltage to the load conditions with two additional dc-dc-converters on both sides of the IPT link, the efficiency can be increased substantially.

In a practical IPT system additional requirements for the control arise from parameter uncertainties. Apart from coil misalignment,

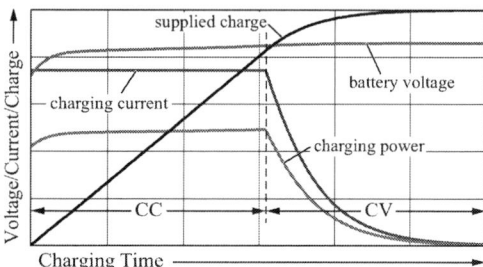

Fig. 1. Typical CC/CV charging profile for EV Li-ion battery cells.

which is widely discussed in literature, also component tolerances and temperature drift are influencing the voltage gain and the resonant frequency of an IPT system. The control method presented in this paper incorporates an automatic tracking of the resonant frequency and dynamic compensation of the varying gain. As an additional feature, the novel control method provides full control over the switching conditions of the transmitter-side power semiconductors. This is particularly important for IPT battery charging systems employing IGBTs as switches to transmit power in the range of several tens of kilowatts. Due to the stored charge of IGBTs [7]–[10] high soft-switching losses occur despite close-to-zero current switching. Because the current at the turn-off instant of the IGBT is a degree of freedom of the presented control method, it enables highest efficiencies for high-power IPT battery charging systems.

The paper is divided into five sections: in Section II insight into the design of a 5 kW IPT system with a focus on controllability is given. Existing control methods are discussed in Section III and a novel control method is presented in Section IV. A comparison of the described existing and the novel control method based on experimental measurements is included. The improved performance of the novel control method is achieved at the cost of additional volume and losses of an additional dc-dc-converter. Therefore, the efficiency requirements and design aspects for the dc-dc-converter are discussed in Section V. Concluding remarks are given in Section VI.

II. IPT SYSTEMS FOR BATTERY CHARGING

In [2], [3], the design of a 5 kW/52 mm air gap/210 mm coil diameter IPT system with a dc-to-dc efficiency of 96.5% was presented. The main specifications of the IPT prototype are given in **Tab. I**. This system is also used for the experiments in this paper. Therefore, in this section a brief overview of the control-related design aspects of the system is given and requirements for an IPT controller are derived.

A. Design Rules for SS-Compensation

As discussed in [3], the design rules for a series-series compensated IPT system as shown in **Fig. 2(a)** for maximum transmission

978-1-4799-2706-7/14 $31.00 © 2014 IEEE

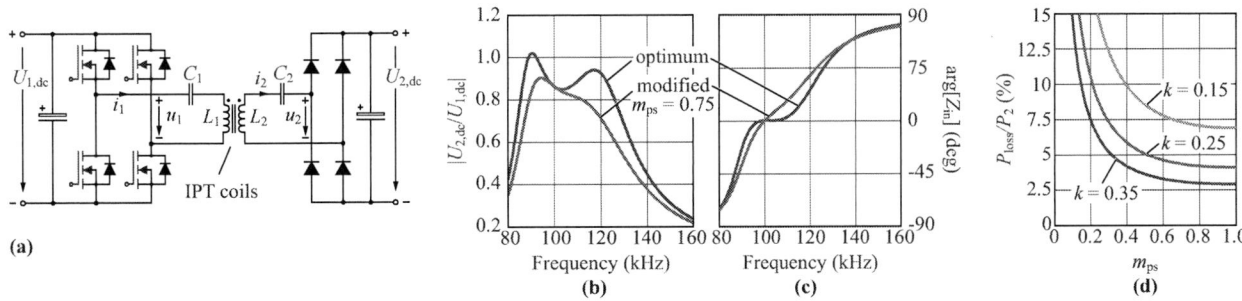

Fig. 2. (a) Equivalent circuit diagram of a series-series compensated IPT system; (b) transfer function from the input to the output dc-link voltage and (c) phase angle of the input impedance of a system designed according to the design rules of [3] and for a system with parameters that were modified with $m_{\mathrm{ps}} = 0.75$ to avoid pole-splitting; (d) total loss factor P_{loss}/P_2 as a function of the factor m_{ps}.

Table I
SPECIFICATIONS OF THE PROTOTYPE IPT SYSTEM IN [3],
WHICH IS USED FOR THE PRESENTED EXPERIMENTS.

Var.	Value	Description
P_2	5000 W	output power
$U_{1,\mathrm{dc}}$	400 V	transmitter-side dc-link voltage
$U_{2,\mathrm{dc}}$	350 V	receiver-side dc-link voltage
U_{batt}	350 V	battery voltage
δ	52 mm	air gap
D_{c}	210 mm	coil diameter
f_0	100 kHz	transmission frequency
α	1.47 kW/dm^2	power density
η	96.5%	max. dc-to-dc efficiency

efficiency at the angular resonant frequency $\omega_0 = 2\pi f_0$ are

$$\omega_0 L_2 \approx \frac{R_{\mathrm{L,eq}}}{k_0} \tag{1}$$

for the reactance of the receiver coil and

$$L_1 = L_2 \left(\frac{U_{1,\mathrm{dc}}}{U_{2,\mathrm{dc}}}\right)^2 \tag{2}$$

for the transmitter coil self-inductance, where k_0 denotes the magnetic coupling at the nominal position of the coils. This design ensures an optimal matching of the receiver coil to the equivalent resistance of the load at the nominal point, which is given by

$$R_{\mathrm{L,eq}} = \frac{8}{\pi^2} \frac{U_{2,\mathrm{dc}}^2}{P_2}. \tag{3}$$

for a diode rectifier with a capacitive output filter [11].

The resonant frequency is set by choosing the capacitances for the transmitter and the receiver side resonant compensation as

$$C_1 = \frac{1}{\omega_0^2 L_1} \quad \text{and} \quad C_2 = \frac{1}{\omega_0^2 L_2}. \tag{4}$$

These design rules are discussed in detail and experimentally verified in [3]. The focus of this paper is the controllability of the system. Therefore, in the next part, the transfer characteristics are discussed and it is shown how they are influenced by capacitor tolerances and temperature drift. Based on the discussion, requirements for a controller are derived.

B. Transfer Characteristics

The transfer function from the input to the output dc-link voltage is shown in **Fig. 2(b)**. For a system design based on the guidelines

above, a pole-splitting occurs and the voltage transfer function exhibits the two peaks that are characteristic for this phenomenon [12], [13]. At the same time, the phase angle of the input impedance seen from the transmitter-side shown in **Fig. 2(c)** has a saddle point at the resonant frequency.

These transfer characteristics are undesired, because they cause

i) hard-switching of the transmitter-side power semiconductors during operation close to the resonant frequency
ii) non-monotonous transfer behavior that may lead to instability of a variable frequency controller

Power losses due to hard-switching (i) of the power semiconductors are of highest concern. Because of the small (or even negative) phase angle between voltage and current at the transmitter-side terminals of the resonant circuit, the current at the switching instants of the power semiconductors can be insufficient to completely discharge the output capacitance of the MOSFET that is turned on. Therefore, Zero Voltage Switching (ZVS) is no longer achieved, because the remaining charge in the output capacitance is dissipated in the MOSFET during the turn-on transition. If IGBTs are used instead, the soft-switching losses due to the stored charge in the junction [8]–[10] similarly impair the performance of the converter, as is discussed in more detail in Section IV.

As shown in later parts of this paper, typically the frequency characteristic of the resonant circuit is exploited to regulate the power flow in the IPT system. In this case, the non-monotonous shape of the voltage transfer function (ii) may cause instability of conventional PID-controllers or could cause convergence to an undesired operating point far above the resonant frequency.

For these reasons a pole-splitting is best avoided by the design process. However, as discussed in [12], for the series-series compensated system a pole-splitting occurs as soon as the magnetic coupling is higher than the limit value

$$k_{\mathrm{lim}} = \frac{R_{\mathrm{L,eq}}}{\omega_0 L_2}. \tag{5}$$

Unfortunately, this coincides with the design rule (1) that is required to reach the physical maximum transmission efficiency. Hence, a pole-splitting can only be avoided at the cost of additional losses in the IPT coils.

A straight-forward solution is to avoid a pole-splitting by reducing the self-inductance of the receiver coil by a factor m_{ps} below the optimum value

$$L_2' = m_{\mathrm{ps}} L_2, \tag{6}$$

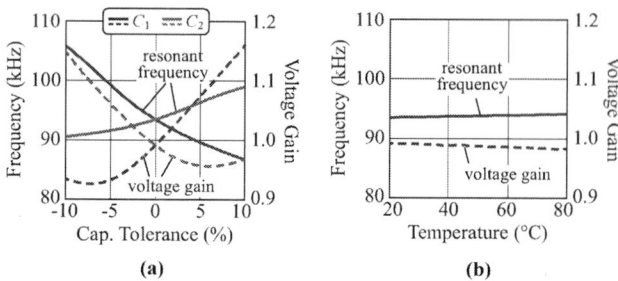

Fig. 3. (a) Influence of capacitor tolerance and **(b)** temperature drift on the IPT system in [3], which must be compensated by the control.

which increases the limit value k_{lim} by m_{ps}^{-1}. In order to provide the same voltage transfer ratio at the resonant frequency,

$$|G_{\text{v}}(\omega_0)| = \left|\frac{U_{2,\text{dc}}}{U_{1,\text{dc}}}\right| = \frac{R_{\text{L,eq}}}{\omega_0 L_{\text{h}}}, \tag{7}$$

the mutual inductance L_{h} must remain constant. This is achieved by adapting the self-inductance of the transmitter coil as $L_1' = L_1/m_{\text{ps}}$. The resonant capacitances are adapted to the corrected self-inductance values according to (4). The resulting voltage transfer function and the phase angle of the input impedance for $m_{\text{ps}} = 0.75$ are shown in **Fig. 2(b)-(c)**.

The price that has to be paid for the improved controllability and reduced switching losses can be calculated as a function of the factor m_{ps} from the total loss factor P_{loss}/P_2 given in [1], [3], which is shown in **Fig. 2(d)**. Owing to the relatively flat minimum of the total loss factor, a correction with $m_{\text{ps}} = 0.75$ causes only few additional losses and is, therefore, acceptable.

Note that the pole-splitting may occur either if the magnetic coupling becomes larger than k_{lim}, e.g., because the IPT coils are placed closer together, or if the equivalent load resistance $R_{\text{L,eq}}$ becomes low enough that k_{lim} increases until it exceeds the magnetic coupling. Because a low equivalent resistance corresponds to a high output power (or a low receiver-side dc-link voltage), the IPT system has to be designed with the discussed design rules for the maximum expected magnetic coupling and the maximum output power. Then, even if during operation the magnetic coupling is reduced below the nominal value or if the output power is decreased, no pole-splitting will occur.

C. Requirements for the Control

The key requirement for an EV battery charging system is a tight control of the battery current according to a CC/CV charging profile similar to **Fig. 1**. Additionally, the battery temperature and the SoC must be constantly monitored and the applied current and voltage must be limited in order to prevent damage of the battery, due to excessive heat generation or over-charging. Preferably, these tasks are mainly executed on the receiver side of the IPT system, because this allows shorter reaction times than what could be achieved via a wireless communication across the air gap.

The task of the transmitter is to operate the IPT system under optimum conditions and the supply of the charging power. For lowest switching losses in the transmitter-side power electronics, a controller for the modified IPT system must be able to guarantee operation in the inductive region of the input impedance (cf. **Fig. 2(c)**). However, the actual resonant frequency is an uncertain parameter which needs to be determined by the control in real time.

For an industrially produced IPT system, the effect of capacitance tolerances shown in **Fig. 3(a)** has to be considered. Given (4), a

mis-match in capacitance has an effect on the resonant frequency. However, also the voltage gain (7) is considerably affected if the transmitter-side and receiver-side resonant circuits are no longer tuned to the same frequency. Additionally, **Fig. 3(a)** shows that depending on which capacitor is inflicted by production tolerances, the voltage gain may be increased or reduced. Thus, a control margin in both directions must be included in the design.

Fig. 3(b) shows an estimate of the influence of the operating temperature on the resonant frequency and the voltage gain. To model the effect of temperature drift, the temperature dependency of typical core material (N87) and of polypropylene film capacitors with metallized plastic film (MKP) have been approximated based on the data given in [14], [15]. A temperature coefficient of $0.0039\,\text{K}^{-1}$ was assumed for the resistivity of the copper litz-wire. A finite element simulation with FEMM[1] was used to determine the self-inductances and the magnetic coupling for the prototype coil used in the experiments. An influence of the operating temperature on the IPT system can be observed, but it is small when compared to the effect of capacitance tolerances. This can be explained by the large air gap which mainly determines the reluctance of the arrangement, while the permeability of the core material has only a small influence. Additionally, the employed MKP film capacitors exhibit a comparably small temperature variance, while other types, e.g., MKN or MKT, would lead to a higher change in capacitance for increased temperatures.

III. EXISTING CONTROL METHODS

A block diagram of a typical IPT battery charging system is shown in **Fig. 4(a)**. A grid rectifier with Power Factor Correction (PFC) is to regulate the dc-link voltage $U_{1,\text{dc}}$, from which the IPT system is supplied. The IPT system is shown schematically in **Fig. 4(a)** as an inverter, a resonant circuit, and a rectifier on the receiver side. At the output of the rectifier, typically a dc-dc-converter is connected, which is used to control the battery current. An output filter is needed between the output of the dc-dc-converter and the battery to reduce the switching frequency ripple. For the regulation of the receiver-side dc-link voltage $U_{2,\text{dc}}$, from which the the dc-dc-converter is supplied, mainly the two different control methods that are discussed in the following are possible.

A. Control of the Switching Frequency

A typical control scheme for IPT systems is the frequency control method, where the transmitter-side inverter switching frequency is used as the actuating variable to regulate the output voltage $U_{2,\text{dc}}$ at the receiver-side. A possible implementation in shown in **Fig. 4(b)**, where the measured difference between $U_{2,\text{dc}}$ and its reference value $U_{2,\text{dc}}^*$ is fed to, e.g., a PI-controller. The controller continuously updates the inverter switching frequency f_{sw} from which the inverter switching signals $s_{1...4}$ are generated by a Pulse Width Modulation (PWM) module. An internal cross-check against a Safe Operating Area (SOA) is implemented in the PI controller to limit the switching frequency to the ZVS region of the resonant circuit that was either calculated or measured prior to operation. However, the switching conditions need to be constantly monitored and adapted, because the resonant frequency may differ from the anticipated value as discussed above.

B. Dual Control / Self-Sustained Oscillating Control

Alternatively, it is possible to determine the resonant frequency f_0 in real time, e.g., sing a current transformer circuit and a comparator

---

[1]Freeware, available at www.femm.info (3.3.2014).

978-1-4799-2706-7/14 $31.00 © 2014 IEEE

The 2014 International Power Electronics Conference

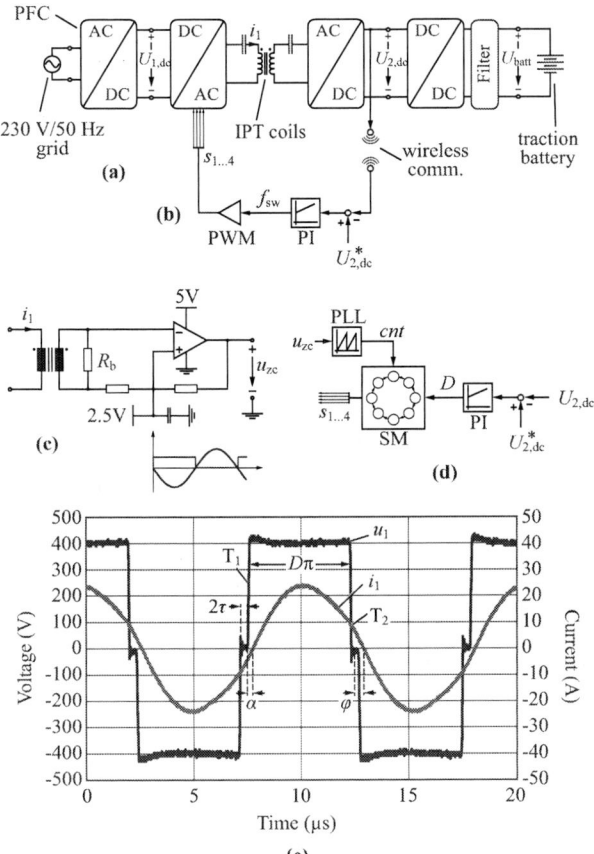

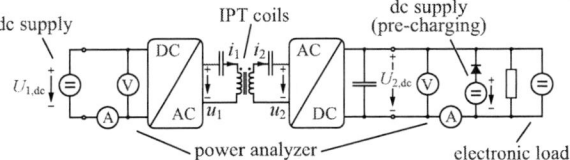

Fig. 5. Measurement setup used in the experiments.

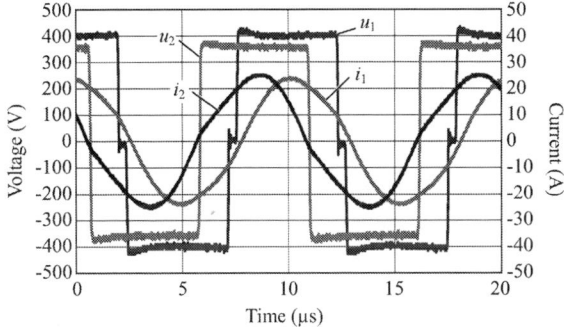

Fig. 6. Waveforms of a measurement at the transmission of 5.7 kW over an air gap of 52 mm. The efficiency at this operating point is 96.5%.

$$\varphi \equiv \underbrace{\frac{\pi}{2}(1 - D)}_{\tau} + \alpha \equiv \arg[\underline{Z}_{\mathrm{in}}]. \qquad (8)$$

This allows to account for variations of the resonant frequency due to component tolerances, temperature drift, or coil misalignment in an automatic fashion with guaranteed soft-switching in the ZVS region of the resonant circuit, which presents a fundamental advantage of the dual control method over the frequency control method [18].

An inherent disadvantage of both of the presented control methods is that the measured value of $U_{2,\mathrm{dc}}$ has be transmitted to the controller on the transmitter side via the wireless communication link. This introduces a time delay in the control loop and results in an increased reaction time of the output voltage regulation, which needs to be compensated by over-dimensioning of the passive components, e.g., the dc-link capacitance of the receiver.

C. Measured Performance of Existing Methods

The measurement setup that was used for these and all the measurements discussed in the following is shown in **Fig. 5**. A dc supply was used to set the transmitter-side dc-link voltage. For the regulation of the dc-link voltage at the receiver side, an electronic load in constant voltage mode was used. To dissipate the transmitted power, load resistors were used additionally to the electronic load. For the pre-charging of the receiver-side dc-link, another dc-supply was used, which is isolated by a series diode as soon as the power transmission is initiated.

Fig. 6 shows measured current and voltage waveforms in the resonant system at the transmission of the maximum output power of 5.7 kW. Waveforms for the dual control method with duty-cycle $D = 0.75$ (4.7 kW) and $D = 0.65$ (3.7 kW) are shown in **Fig. 7(a)-(b)**, respectively. A comparison of **Fig. 6** and **Fig. 7(b)** reveals a fundamental disadvantage of the dual control method (and also of the frequency control method): even though for the reduced output power the amplitude of the receiver coil current i_2 is lower, the amplitude of the current in the transmitter coil i_1 remains almost unchanged.

The calculated rms value of the current in the transmitter coil I_1 and the receiver coil current I_2 are shown in **Fig. 8(a)-(b)**,

Fig. 4. **(a)** Block diagram of a series-series compensated IPT battery charging system that is supplied from the 230 V/50 Hz grid; **(b)** control diagram for the frequency control method; **(c)** zero crossing detection circuit; **(d)** control diagram for the dual control method; **(e)** measured waveform of the transmitter coil current i_1 and the transmitter-side inverter output voltage u_1 for the dual control method.

as shown in **Fig. 4(c)**. The comparator output voltage u_{zc} is used to trigger a Phase-Locked Loop (PLL) that synchronizes an internal counter to the the zero crossings of the current in the transmitter coil. As shown in **Fig. 4(d)**, the PLL counter signal cnt is fed to a State Machine (SM) which generates the inverter switching signals $s_{1\ldots4}$. Based on the zero crossing detection, it is possible to implement the dual control or self-sustained oscillating control method [16]–[18]. In this control scheme, the SM generates the switching signals such that the angular length $D\pi$ of the power interval as defined in the measured waveform in **Fig. 4(e)** can be adjusted by the PI-controller. Additionally, the angle α can be set such that the current during the switching transition T_1 is always sufficient to completely discharge the output capacitance of the MOSFET that is turned-on.

Instead of controlling the switching frequency directly, the angles $D\pi$ and α are controlled and the temporal duration of the individual inverter switching states are derived in real time from the PLL counter. Because the switching frequency is uncontrolled, the resonant system will autonomously converge to the frequency, where the phase of the input impedance $\arg[\underline{Z}_{\mathrm{in}}]$ (cf. **Fig. 2(c)**) is equal to the phase angle φ between the fundamental components of the voltage u_1 and the current i_1. The resulting operating point is given approximately by

978-1-4799-2706-7/14 $31.00 © 2014 IEEE

2170

The 2014 International Power Electronics Conference

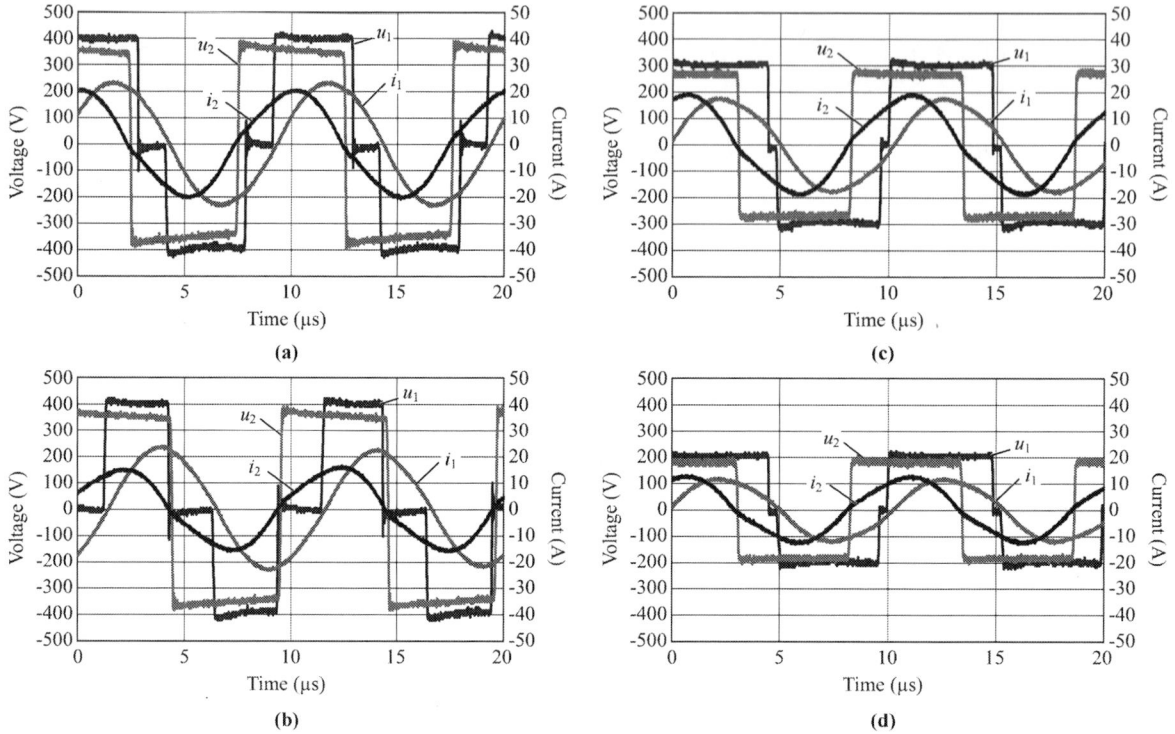

Fig. 7. (a)-(b) waveforms for the dual control method and duty-cycles $D = 0.75$ (4.7 kW) and $D = 0.65$ (3.7 kW), respectively; (c)-(d) waveforms at reduced output power with dc-link voltages controlled at $U_{1,dc} = 300\,V/U_{2,dc} = 265\,V$ (3.2 kW) and $U_{1,dc} = 200\,V/U_{2,dc} = 176\,V$ (1.4 kW).

respectively. For the dual control and the frequency control method, the rms value of the current in the transmitter coil is almost constant, while the current in the receiver coil is reduced. Consequently, the power loss in the transmitter coil also remains almost constant at reduced output power. The calculated power losses for the two control methods are shown in **Fig. 9**(a)-(b). For the calculation of the power losses in the coils $P_{coil,1/2}$ a finite element tool was used as described in [2], [3]. The losses in the resonant capacitors $P_{cap,1/2}$ were estimated based on the manufacturer data given in [15]. For the power semiconductors $P_{semi,1/2}$, conduction losses were calculated based on the device datasheet. Because of the ZVS operation, no switching losses have to be included.

In partial-load, the losses in the transmitter coil $P_{coil,1}$, the transmitter-side resonant capacitor $P_{cap,1}$, and the transmitter-side power semiconductors $P_{semi,1}$ are almost equal for all values of the output power for the dual control and the frequency control method. This characteristic fundamentally impairs the efficiency of the system at reduced load. In order to overcome this limitation, a control method with an improved partial-load efficiency is presented in the next section.

IV. NOVEL CONTROL METHOD FOR HIGH-POWER IPT

As discussed in the previous section, while a deviation from the resonant frequency of the IPT system reduces the transmitted power, the current in the transmitter coil remains almost constant and, therefore, causes high losses in partial-load. A better performance can be achieved if the switching frequency is constantly at the resonant frequency of the IPT link and the two dc-link voltages $U_{1,dc}$ and $U_{2,dc}$ are used to control the output power [4]. For the series-series compensated IPT system, the relation

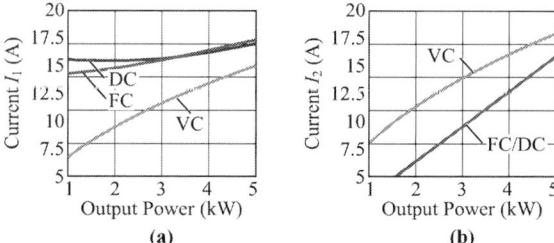

Fig. 8. (a) Calculated transmitter coil current I_1 and (b) receiver coil current I_2 for the frequency control (FC) method, the dual control (DC) method, and voltage control (VC) method (rms value).

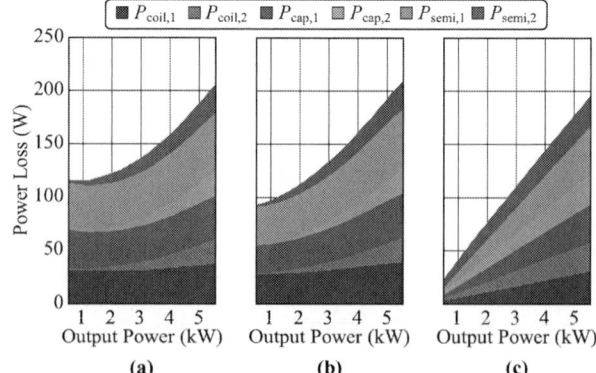

Fig. 9. Calculated power loss as a function of the output power for (a) the frequency control, (b) dual control, and (c) for the proposed voltage control method. Losses of the dc-dc-converter are not included.

978-1-4799-2706-7/14 $31.00 © 2014 IEEE

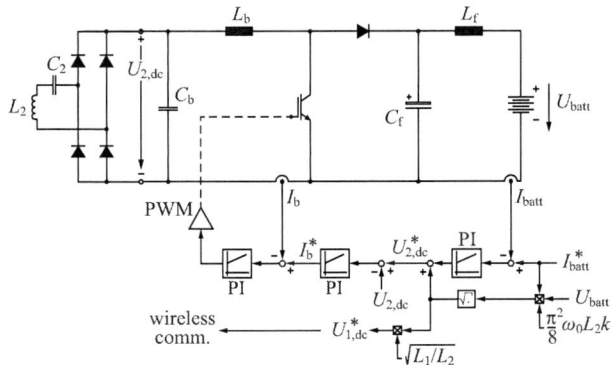

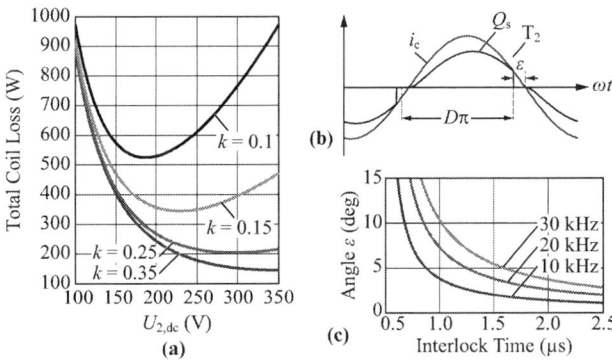

Fig. 10. Control diagram for the voltage control method. All critical parts are implemented within the receiver-side power electronic converter. The reference value $U^*_{1,\mathrm{dc}}$ is the only inherently required communication with the transmitter that has to be transmitted across the air gap.

$$P_2 = \frac{8}{\pi^2}\frac{U_{1,\mathrm{dc}}U_{2,\mathrm{dc}}}{\omega_0 L_{\mathrm{h}}} \qquad (9)$$

holds for the output power P_2 of the IPT system. Therefore, if the dc-link voltages are controlled, the battery current can be regulated according to

$$I_{\mathrm{batt}} = \frac{8}{\pi^2}\frac{1}{\omega_0 L_{\mathrm{h}}}\frac{U_{1,\mathrm{dc}}U_{2,\mathrm{dc}}}{U_{\mathrm{batt}}}, \qquad (10)$$

since the battery is a voltage-impressing element and the battery current results from $I_{\mathrm{batt}} = P_2/U_{\mathrm{batt}}$.

In the following, a possible implementation is discussed in more detail and further improvements of the method for high-power IPT systems that employ IGBTs as switches are shown. Additionally, experimental results are presented that demonstrate the performance, which can be achieved with this control method.

A. Controller Implementation

An implementation of the described method is possible if the grid-side rectifier shown in **Fig. 4(a)** is realized as an active front-end that allows controlling the transmitter-side dc-link voltage or if an additional dc-dc-converter is implemented at the transmitter side. The receiver-side dc-link voltage can be regulated by the dc-dc-converter on the receiver side, which is also shown in **Fig. 4(a)**.

The condition for the maximum efficiency of the IPT system is that the receiver reactance is matched to the equivalent load resistance (3) according to (1). Hence, if the receiver-side dc-link voltage can be controlled according to

$$U^*_{2,\mathrm{dc}} = \sqrt{\frac{\pi^2}{8}\omega_0 L_2 k P_2}. \qquad (11)$$

at reduced output power, the equivalent load resistance is maintained constant and the matching condition is always fulfilled. Therefore, this process could be termed *active impedance matching*. Note that if the equivalent load resistance is maintained constant independently of the output power, the strong load-dependency of the transfer characteristics of the series-series compensated IPT system described, e.g., in [19], is eliminated.

Because of the symmetry of (9), it is sufficient if a feedback controller is used to adapt one of the reference values dynamically to compensate inaccuracies in the analytical calculation shown above. **Fig. 10** shows a possible configuration, where the reference value $U^*_{2,\mathrm{dc}}$ is manipulated by a PI-controller to regulate the battery current.

Fig. 11. (a) Coil losses as a function of the secondary-side dc-link voltage $U_{2,\mathrm{dc}}$ and the magnetic coupling k (coil misalignment); (b) schematic waveforms of IGBT collector current i_{c} and stored charge Q_{s}; (c) turn-off angle ϵ for minimum IGBT soft-switching losses as a function of the interlock time and the transmission frequency.

Also the cascaded control loop for the receiver-side dc-link voltage is shown. Note that **Fig. 10** shows a boost topology for the dc-dc-converter mainly as an example. Depending on the battery voltage and the rating of the employed power semiconductors, also a buck-boost-type converter could be used.

The reference value for the dc-link voltage $U_{1,\mathrm{dc}}$ follows from (2) as

$$U^*_{1,\mathrm{dc}} = \sqrt{\frac{L_1}{L_2}}U^*_{2,\mathrm{dc}}. \qquad (12)$$

The reference value $U^*_{1,\mathrm{dc}}$ is sent through, e.g., a wireless communication channel to the power electronic converter on the transmitter side, where another, local PI-controller regulates the voltage. It is also possible to adapt $U^*_{1,\mathrm{dc}}$ with the PI-controller instead of $U^*_{2,\mathrm{dc}}$. However, then the reaction time of the battery current control loop would be lower, due to the delay that is introduced by the communication across the air gap. In the proposed configuration, all critical parts of the battery charging control are implemented on the receiver, which has a more direct access to the measurements. This is considered the inherently safest and most reliable design.

B. Tracking of the Efficiency Maximum with Misalignment

To calculate the reference values (11) and (12) accurately, an estimate of the magnetic coupling is needed. If a feedback controller is used, also a pre-calculated, approximative value can be used and the control would compensated for the estimation error. However, ideally a method is implemented to estimate the magnetic coupling in real time, e.g., from measurements of the currents in the IPT coils and an equivalent circuit.

If an estimate is available, the reference value $U^*_{2,\mathrm{dc}}$ can be adapted to the magnetic coupling in real time to follow the loss minima shown in **Fig. 11(a)**, which are described by (11). Alternatively, it is also possible to measure the dc input and output power of the system and calculate the power loss online. An optimization algorithm or a look-up table could then be used to determine reference values that lead to minimum total losses in the system.

C. Control of IGBT Soft-Switching Conditions

If the circuit for the detection of the zero crossings and the PLL discussed in Section III for the dual control method are also implemented for the voltage control scheme, another valuable advantage arises: it was shown in [7]–[10] that the stored charge of IGBTs causes high switching losses, even if they are commutated with

978-1-4799-2706-7/14 $31.00 © 2014 IEEE

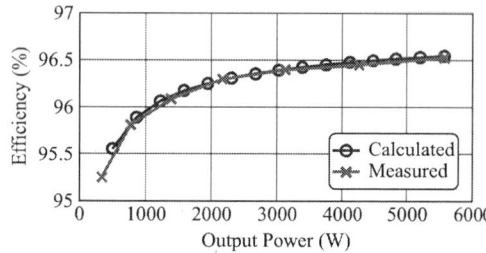

Fig. 12. Calculated and measured dc-to-dc conversion efficiency (incl. losses in the IPT coils, resonant capacitors, and power semiconductors) as a function of the output power at an air gap of 52 mm, using the voltage control method.

almost zero current. For IPT systems with a high power level, e.g., several tens of kWs, for all of the discussed control methods this effect could severely impair the performance, because the IGBTs are always switched close to the zero crossings of the current. Novel SiC modules with better switching performance are often proposed to circumvent this problem, e.g, in [4]. However, due to their high reliability, low cost, and better availability at present time IGBT modules are still clearly preferred in industry. The possibility to minimize the effect of the stored charge, which is discussed in the following is, therefore, a key advantage of the proposed voltage control scheme.

In [9], it was shown that an optimum exists for the current during the turn-off transition of an NPC-bridge leg operated in discontinuous conduction mode in a solid state transformer application. The optimally switched current is given where the charge that is removed from the IGBT by the magnetizing current of the transformer during the zero current interval is equal to the stored charge in the junction of the IGBT, because the turn-on losses of the complementary device are eliminated and the total losses reach a minimum.

Similarly, for the schematically reproduced waveforms of the IPT system in **Fig. 11(b)**, a condition for minimum IGBT switching losses follows for the turn-off angle ϵ. At the switching transition T_2, the charge that is removed from the IGBT by the collector current i_c during the interlock time must be equal to the charge Q_s that is stored in the IGBT junction in order to completely eliminate the turn-on losses of the complementary device. Under the assumption of a sinusoidal current in the transmitter coil and with the stored charge model discussed in [8], [9], the optimum angle ϵ can be calculated numerically. In **Fig. 11(c)** the optimum is shown as a function of the interlock time and for different switching frequencies. It can be seen that for higher switching frequencies a higher angle ϵ is needed to completely remove the charge from the IGBT, which is still higher at the turn-off due to the time constant of the device.

With the zero crossing detection and the PLL, the switched current during the transition T_2 can be controlled by adjusting the angle $D\pi$ in combination with the described control voltage method. If the voltages can be adjusted within a sufficient range, the angle $D\pi$ is an additional degree of freedom of the control method. This allows to minimize the IGBT switching losses significantly by tracking the discussed optimum for the turn-off angle ϵ. This optimization is not possible with the existing control methods that were described in Section III, since there, the phase shift between current and voltage is needed for the power control and is, therefore, not available for another control loop.

D. Measured Performance of the Voltage Control Method

Measured waveforms at two power levels are shown in **Fig. 7(c)-(d)**. The calculated rms currents that results in the transmitter and receiver coil are indicated in **Fig. 8(a)-(b)**, respectively. The current in the transmitter coil can be significantly reduced due to the reduction of the dc-link voltages at lower output power. The current in the receiver coil is slightly increased, but this is needed to drive the system to the global loss optimum, which is also evident from the calculated power losses shown in **Fig. 9(c)**.

Using a Yokogawa WT3000 power analyzer, the dc-to-dc efficiency of the system was measured for different load points. The results of the efficiency measurement at an air gap of 52 mm are shown in **Fig. 12**. As indicated, the measured and calculated performance show an excellent agreement, which validates the used methods.

A more detailed discussion of the used IPT prototype system and the measurement setup can be found in [2], [3], where also thermal and stray field measurements are presented to verify the finite-element based models. In the experiments, the dc-link voltages were adjusted with the dc-supply and the electronic load shown in **Fig. 5**. In a practical system, this would be the task of the dc-dc-converters. However, it is expected that the outcome of the measurements would be the same.

As confirmed by the measurements, the proposed voltage control methods shows an excellent performance, that can not be reached with the existing control methods. However, the price that has to be paid is the additional losses and the volume of the dc-dc-converter that is required on the transmitter side of the system.

V. REQUIREMENTS FOR THE ADDITIONAL DC-DC-CONVERTER

Generally, it can be considered an advantage of the proposed solution that power losses can be shifted from the IPT coils to the power electronic converters, since an efficient cooling of the IPT coils is more complex than the cooling system of a power electronic converter. Due to the stray field of the coils, high eddy current losses would occur in any metal parts. Therefore, the copper windings and the core materials of the IPT coils would need to be mechanically fixed mostly with plastic components, which typically have a comparably low thermal conductivity. The cooling would need to be provided either passively, or by forced air from integrated fans or a compressed air supply. Meanwhile, the components of the power converter could be mounted on a standard heatsink, which could be forced air cooled or even connected to a water-cooling circuit of the EV, which would allow for a highly compact and robust realization [20], [21]. Nevertheless, the over-all system efficiency is still an important performance aspect. Therefore, requirements of the dc-dc-converter are derived in the following.

For the topology shown in **Fig. 13(a)** and equal, bi-directional dc-dc-converters with efficiency $\eta_{dcdc}(P_2)$, for the proposed voltage control method

$$
\begin{aligned}
\eta_{dcdc}(P_2)^2 \cdot \eta_{VC}(P_2) \dots \\
\dots \geq \max\left[\eta_{FC}(P_2), \eta_{DC}(P_2)\right] \cdot \eta_{dcdc}(P_2)
\end{aligned}
\tag{13}
$$

must apply to outperform the frequency control and the dual control methods. In (13), $\eta_{VC}(P_2)$ is the efficiency of the voltage control method as a function of the output power P_2, while $\eta_{FC}(P_2)$ and $\eta_{DC}(P_2)$ are the efficiencies of the frequency control and dual control methods, respectively.

It is assumed that for the frequency control and the dual control methods, a dc-dc converter is implemented on the receiver side to control the battery current locally, which is not strictly needed, but advisable and commonly used for the protection of the battery. Then, the required efficiency of the dc-dc-converter is given as

$$
\eta_{dcdc}(P_2) \geq \frac{\max\left[\eta_{FC}(P_2), \eta_{DC}(P_2)\right]}{\eta_{VC}(P_2)}.
\tag{14}
$$

978-1-4799-2706-7/14 $31.00 © 2014 IEEE

The 2014 International Power Electronics Conference

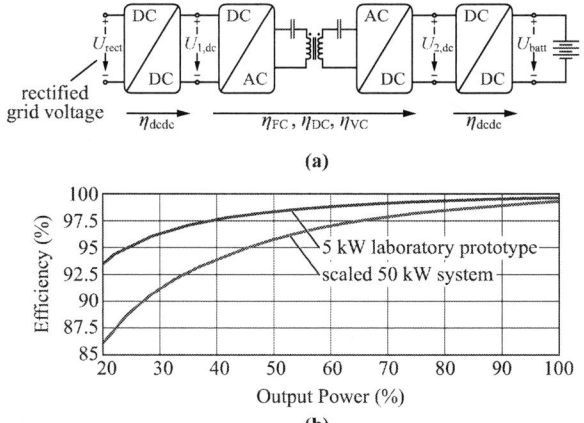

(a)

(b)

Fig. 13. (a) Considered system topology for the derivation of the efficiency requirement for the dc-dc-converters; (b) required minimum efficiency as a function of the output power for the 5 kW laboratory prototype with MOSFETs and a scaled 50 kW high-power IPT system that incorporates 1.2 kV IGBTs.

As shown in **Fig. 13(b)**, for a scaled 50 kW high-power IPT system, which incorporates IGBTs, a large loss reduction is possible with the proposed control method, and the efficiency requirement for the dc-dc-converter is feasible. Particularly, if a modular converter system as shown in [20], [21] is implemented, a high partial-load efficiency is possible if the modules are sequentially activated and deactivated, depending on the output power. For the 5 kW laboratory prototype, where MOSFETs are employed in the IPT system and no switching losses occur due to ZVS, the required efficiency is higher and a realization of the converter is more challenging. However, considering state-of-the-art dc-dc-converters also this task seems achievable [22].

VI. CONCLUSION

In this paper, the two existing control schemes frequency control and dual control are analyzed and compared in detail, using both calculations and measurements. It is shown that a significant disadvantage of both methods is the high reactive current in the transmitter coil during phases of reduced output power and, therefore, high partial-load losses in the transmitter coil.

Based on the results, a novel control method with significantly lower losses in partial-load is derived, where the dc-link voltages on both sides of the IPT system are regulated by dc-dc-converters to control the output power. The novel method includes the zero crossing detection of the dual control method, which allows full control of the power semiconductor switching conditions. It, therefore, enables the minimization of IGBT soft-switching losses, since the output current during the switching transition is an additional degree of freedom of the control. Experimental results demonstrate the excellent performance of the proposed novel method.

The price for the good results is an additional dc-dc-converter, for which the requirements are also derived. Assuming a state-of-the-art converter with 99% efficiency, and given the experimentally verified 96.5% transmission efficiency of the described prototype IPT system, a total conversion efficiency from grid to battery of 95% is expected with the proposed control method, even for high-power systems that incorporate IGBTs.

ACKNOWLEDGMENT

The authors would like to thank ABB Switzerland Ltd. for their funding and for their support regarding many aspects of this research project.

REFERENCES

[1] E. Waffenschmidt and T. Staring, "Limitation of inductive power transfer for consumer applications," in *Proc. 13th European Conf. on Power Electronics and Applications (EPE)*, 2009, pp. 1–10.

[2] R. Bosshard, J. W. Kolar, and B. Wunsch, "Accurate finite-element modeling and experimental verification of inductive power transfer coil design," in *Proc. 29th Appl. Power Electronics Conf. and Expo. (APEC)*, 2014.

[3] R. Bosshard, J. W. Kolar, J. Mühlethaler, I. Stevanovic, B. Wunsch, and F. Canales, "Modeling and η-α-Pareto optimization of inductive power transfer coils for electric vehicles," *IEEE J. Emerg. Sel. Topics Power Electron. (accepted for publication)*, 2014.

[4] B. Goeldi, S. Reichert, and J. Tritschler, "Design and dimensioning of a highly efficient 22 kW bidirectional inductive charger for e-mobility," in *Proc. Int. Exhibition and Conf. for Power Electronics (PCIM Europe)*, 2013, pp. 1496–1503.

[5] U. K. Madawala and D. J. Thrimawithana, "A bidirectional inductive power interface for electric vehicles in V2G systems," *IEEE Trans. Ind. Electron.*, vol. 58, no. 10, pp. 4789–4796, 2011.

[6] R. M. Miskiewicz, A. J. Moradewicz, and M. P. Kazmierkowski, "Contactless battery charger with bi-directional energy transfer for plug-in vehicles with vehicle-to-grid capability," in *Proc. IEEE Int. Ind. Electron. Symp. (ISIE)*, 2011, pp. 1969–1973.

[7] P. Ranstad and H.-P. Nee, "On dynamic effects influencing IGBT losses in soft-switching converters," *IEEE Trans. Power Electron.*, vol. 26, no. 1, pp. 260–271, 2011.

[8] G. Ortiz, H. Uemura, D. Bortis, J. W. Kolar, and O. Apeldoorn, "Modeling of soft-switching losses of IGBTs in high-power high-efficiency dual-active-bridge dc/dc converters," *IEEE Trans. Electron Devices*, vol. 60, no. 2, pp. 587–597, 2013.

[9] J. Huber, G. Ortiz, F. Krismer, N. Widmer, and J. W. Kolar, "η-ρ-Pareto optimization of bidirectional half-cycle DC/DC converter with fixed voltage transfer ratio," in *Proc. 28th Appl. Power Electronics Conf. and Expo. (APEC)*, vol. 1, 2013, pp. 1413–1420.

[10] D. Dujic, G. K. Steinke, M. Bellini, M. Rahimo, L. Storasta, and J. K. Steinke, "Medium-Frequency Soft-Switched Applications," *IEEE Trans. Power Electron.*, vol. 29, no. 2, pp. 906–919, 2014.

[11] R. Steigerwald, "A comparison of half-bridge resonant converter topologies," *IEEE Trans. Power Electron.*, vol. 3, no. 2, pp. 174–182, 1988.

[12] P. E. K. Donaldson, "Frequency hopping in r.f. energy-transfer links," *Electronics & Wireless World*, pp. 24–26, Aug. 1986.

[13] C.-S. Wang, G. A. Covic, and O. H. Stielau, "Power transfer capability and bifurcation phenomena of loosely coupled inductive power transfer systems," *IEEE Trans. Ind. Electron.*, vol. 51, no. 1, pp. 148–157, 2004.

[14] EPCOS, "Ferrites and accessories - data handbook," 2006.

[15] ——, *Film capacitors - data handbook*, 2009.

[16] J. A. Sabate, M. M. Jovanovic, F. C. Lee, and R. T. Gean, "Analysis and design-optimization of LCC resonant inverter for high-frequency AC distributed power system," *IEEE Trans. Ind. Electron.*, vol. 42, no. 1, pp. 63–71, 1995.

[17] H. Pinheiro, P. K. Jain, and G. Joos, "Self-sustained oscillating resonant converters operating above the resonant frequency," *IEEE Trans. Power Electron.*, vol. 14, no. 5, pp. 803–815, 1999.

[18] R. Bosshard, U. Badstübner, J. W. Kolar, and I. Stevanovic, "Comparative evaluation of control methods for inductive power transfer," in *Proc. 1st Energy Research and Applicat. Conf. (ICRERA)*, vol. 1, no. 1, 2012, pp. 1–6.

[19] I. Nam, R. Dougal, and E. Santi, "Optimal design method to achieve both good robustness and efficiency in loosely-coupled wireless charging system employing series-parallel resonant tank with asymmetrical magnetic coupler," *Proc. 5th Energy Conversion Congr. and Expo. (ECCE USA)*, pp. 3266–3276, 2013.

[20] B. Eckardt and M. März, "A 100kW automotive powertrain DC/DC converter with 25kW/dm3 by using SiC," in *Proc. Int. Exhibition and Conf. for Power Electronics (PCIM Europe)*, 2006.

[21] S. Waffler and J. Kolar, "Efficiency optimization of an automotive multiphase bi-directional DC-DC converter," in *Proc. 6th Int. Power Electron. and Motion Control Conf. (ECCE Asia)*, 2009, pp. 566–572.

[22] J. W. Kolar, F. Krismer, Y. Lobsiger, J. Mühlethaler, T. Nussbaumer, and J. Miniböck, "Extreme efficiency power electronics," in *Proc. 7th Int. Conf. Integrated Power Electron. Syst. (CIPS)*, 2012, pp. 1–22.

978-1-4799-2706-7/14 $31.00 © 2014 IEEE

Standard Models for Smart Grid Simulations

Taku Noda
CRIEPI
Yokosuka, Kanagawa, Japan
takunoda@criepi.denken.or.jp

Shinji Kato
Kobe City College of Technology
Kobe, Hyogo, Japan

Tomohiro Nagashima
CRIEPI
Yokosuka, Kanagawa, Japan

Yoichi Sekiba
Denryoku Computing Center
Kawasaki, Kanagawa, Japan

Takayuki Sekisue
ANSYS Japan K. K.
Tokyo, Japan

Hirokazu Tokuda
Fuji Electric Co,. Ltd.
Hino, Tokyo, Japan

Yuichiro Kabasawa
Tohoku Electric Power Co., Inc.
Sendai, Miyagi, Japan

Masaaki Kounoto
Panasonic Corporation
Moriguchi, Osaka, Japan

Abstract— **Cooperative Study Group on Standard Model Development for Power Electronics Simulations, created under the Industry Applications Society of the IEEJ (Institute of Electrical Engineers of Japan), is active to develop standard models for simulations related to power electronics. Under the study group, four working groups (WGs) have been created for different fields of simulation. Among them, Smart Grid WG is developing standard simulation models of components which will form smart grids in the near future. As the government of Japan assumes, a large portion of renewable energy will be provided by photovoltaic (PV) generation systems installed at ordinary houses, buildings and solar farms, all of which are connected to distribution lines. This implies that smart grids in Japan will be established at the distribution and the end-consumer level. Therefore, the components for which the Smart Grid WG is developing standard models include distribution substations, distribution lines, voltage regulation equipment, pole-mounted transformers, PV panels, Li-ion batteries, power conditioning systems, and micro and small wind power generation systems. This paper presents these standard models and also future development plan for those which are still under development at this moment.**

Keywords— *Institute of Electrical Engineers of Japan (IEEJ), Simulations, Smart Grids, and Standard models.*

I. INTRODUCTION

Today, simulations are used for the design of real power-electronics systems and also for solving real-world problems related to power electronics applications. The Industry Applications Society of the Institute of Electrical Engineers of Japan (IEEJ) has launched a series of cooperative study groups on power electronics simulation since 1994. These study groups have been carrying out reviews and investigations of simulation methods,

modeling methods and simulation programs related to power electronics [1], [2]. The current successor of the study groups is Cooperative Study Group on Standard Model Development for Power Electronics Simulations, which was launched in 2012. It is active to develop and to discuss standard models of components often used in power electronics simulations. Under the study group, the following working groups (WGs) have been created for different fields of simulations: Smart Grid WG, Motor Drive WG, Power Supply WG and Automobile WG.

Preliminary results and future plan of the Smart Grid WG, among those four WGs mentioned above, have been presented in [3] and [4]. This paper presents principal results recently obtained by the Smart Grid WG. The Smart Grid WG is developing and discussing standard simulation models of components will form smart grids in the near future. The purpose of developing the standard models is to use in transient simulations related to smart grids. The components for which standard models are developed cover the distribution and the end-consumer level of power systems. This is because the government of Japan assumes that a large portion of renewable energy in the near future will be provided by photovoltaic (PV) generation systems installed at ordinary houses, buildings and solar farms all connected to distribution lines and smart grids in Japan will be established mainly at the distribution and the end-consumer level [5], [6]. Therefore, the components for which the Smart Grid WG is developing standard models include distribution substations, distribution lines, voltage regulation equipment, pole-mounted transformers, PV panels, Li-ion batteries, power conditioning systems (PCSs), and micro and small wind power generation

978-1-4799-2706-7/14 $31.00 © 2014 IEEE

systems. This paper presents these standard models and also future development plan for those which are still under development at this moment.

The authors of this paper are the main contributors, and the following experts also contribute this project:

- T. Abe (Nagasaki University)
- K. Fukushima (CRIEPI)
- J. Ichihara (AZAPA Co. Ltd.)
- M. Inoue (Osaka Prefectural University)
- H. Ishikawa (Gifu University)
- H. Kakigano (Ritsumeikan University)
- T. Kato (Doshisha University)
- N. Kimura (Osaka Institute of Technology)
- Y. Kouno (Toshiba Corp.)
- H. Kubota (Mitsubishi Heavy Industries Ltd.)
- M. Matsui (Tokyo Polytechnic University)
- Y. Nishida (Chiba Institute of Technology)
- K. Shigematsu (Cybernet Systems, Co. Ltd.)
- J. Shimomura (Meidensha Corp.)
- K. Tago (Hitachi, Ltd.)
- N. Umeda (Yaskawa Electric Corp.)

II. Importance of Standard Models

Consider the situation that an engineer is designing a power-electronics product which will be used in distribution systems or end-consumer houses as an element of smart grids. The engineer has to carry out various types of simulations at the design stage, and among them, he/she has to carry out simulations in which the behaviors of the product in a smart grid are analyzed and checked to optimize the main circuit and the controller of the product. To do this kind of simulations, models of the rest of the smart grid to be simulated are needed. But, if the engineer prepared those simulation models by himself/herself, it would take an enormous amount of time for figuring out, for each smart grid component, which kind of equivalent circuit should be used, what parameter values are appropriate, and etc. If the standard models of components of smart grids are fully prepared with appropriate parameter values, the engineer can readily use those models for such simulations and does not have to waste time. Since the standard models should be reliable so as to obtain reasonable simulation results, the models have to be developed and their parameters have to be suggested by experts in this field. This is the reason why the Smart Grid WG has been created to develop the standard models of the components of smart grids.

The standard models can be used not only for the design of a product but also for the coordination of control systems and protective relays of equipment in a smart grid. In this situation, small differences of specific products are not of interest, and thus, the standard models are appropriate for this kind of simulations.

It is often true that power-electronics engineers are not familiar with power distribution systems and distribution engineers are not familiar with power electronics. This also justifies the necessity of the standard models, since they are useful when one has to do simulations including technology areas without enough knowledge.

It should be noted that the Smart Gird WG is not trying to develop "average" models which reproduce average characteristics of existing products. Manufacturers invent and apply new circuit topologies and new control strategies one after another. Thereby, obtaining an average circuit topology, average control strategy and average model parameter values is not an easy task and even useless considering its efforts. The approach taken by the Smart Grid WG is the following. For a smart grid component, the simplest circuit topology and the simplest control strategy, which may be found in text books, are selected for its standard model. And then, its parameter values are adjusted so as to reproduce average characteristics as close as possible. In this regard, the standard models are useful for educational and research purposes, since they use the simplest circuit topologies and control strategies with practical parameter values.

III. Standard Models

Figure 1 shows the components for which standard models are developed by the Smart Grid WG.

A. Distribution Substation

A distribution substation steps down the voltage come from a transmission line into 6.6 kV (in the case of Japan) using transformers and supplies power to distribution lines. Each transformer is equipped with a load ratio control transformer (LRT), and the voltage of the distribution lines connected to the transformer is regulated by tap changes of the LRT.

Fig. 2 shows the standard model developed by the Smart Grid WG. In this model, two voltage sources are used to represent the voltage of each phase. One generates the nominal voltage of the distribution line, and the other represents voltage variations due to the tap changes. The latter is controlled by control blocks which

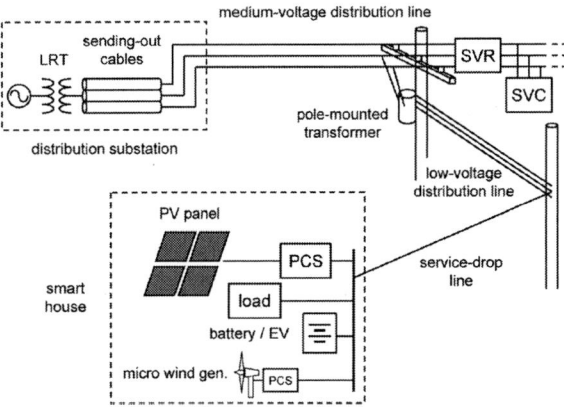

Fig. 1. Components for which standard models are developed.

reproduce a typical control algorithm of the load drop compensator (LDC) used in Japan. The control algorithm which can be considered typical has been investigated by the WG. The impedance seen from the 6.6-kV distribution line, the capacitance of sending-out cables and other substation equipment, and the equivalent grounding resistance of the grounded potential transformer (GPT) are represented by *RLC* elements.

Recommended parameter values and recommended parameter determination procedures will be described in the final report that will be issued by the IEEJ.

B. Distribution Lines and Service-Drop Lines

In Japan, a distribution line consists of 6.6-kV three-phase lines and 200/100-V single-phase three-wire lines. Those two kinds of distribution lines are connected to each other by pole-mounted transformers. A service-drop line is used to connect a low-voltage distribution line to a customer.

The power-frequency characteristics of 6.6-kV lines and 200/100-V lines can be represented by their series impedances. Since their capacitances have little effect, they can be neglected to simulate power-frequency behaviors. The impedance matrix of a 6.6-kV or a 200/100-V line can be calculated by textbook formulas taking the skin effects of wires and ground soil into account [7]. An off-the-shelf program may be used for this calculation. The real part of the impedance matrix is considered the resistance, and the imaginary part divided by $\omega_0 = 2\pi f_0$ (f_0: power frequency) the inductance. A coupled resistance element and a coupled inductance element in a simulation program are respectively used to represent the resistance and the inductance matrix, and their series connection represents the impedance matrix. If a ground wire (shielding wire) exists in a 6.6-kV distribution line, its effects are taken into account in the impedance calculation [8]. The impedance calculation of service-drop lines is not easy, since they are tightly twisted. For service-drop lines, impedance values provided by manufacturers may be used. If the imped-

ance value provided by a manufacturer is a positive-sequence impedance, a series connection of a resistance and a inductance is inserted in each phase.

Fig. 3 shows a candidate of the standard conductor arrangements for 6.6-kV and 200/100-V lines. These arrangements can be used, when specific arrangement data are not available.

When frequency components higher than the power frequency are of interest, the capacitances of lines should be represented. The easiest way is to use a π equivalent in which the capacitance matrix is lumped to both ends of the impedance. The capacitance matrix can be calculated by a textbook formula, or, an off-the-shelf program may be used. If traveling waves on a line are of interest, a constant-parameter line model is used. If a wide range of frequency is of interest as even more advanced simulations, a frequency-dependent line model which reproduces the frequency dependence of the impedance due to the skin effects of conductors and ground soil is used. For these line models, see [8], [9].

C. SVR, SVC and STATCOM

A step voltage regulator (SVR) is a voltage regulation device installed on a distribution line. It is used to compensate a voltage drop when the length of a distribution line is long and the voltage drop is thus large. An SVR consists of an autotransformer and an on-load tap changer (LTC), and the tap position of the LTC is controlled so that the voltage of the load side of the distribution line is kept constant.

Fig. 4 shows the standard model of SVRs developed by the Smart Grid WG. As shown in the figure, an autotransformer equipped with an LTC is modeled by the equivalent circuit consisting of an ideal transformer, winding resistance, and leakage inductance. The change of the tap position is equivalently reproduced by modifying the turn ratio of the ideal transformer, and it is controlled by control blocks which reproduce a control

Fig. 2. Standard model of distribution substations.

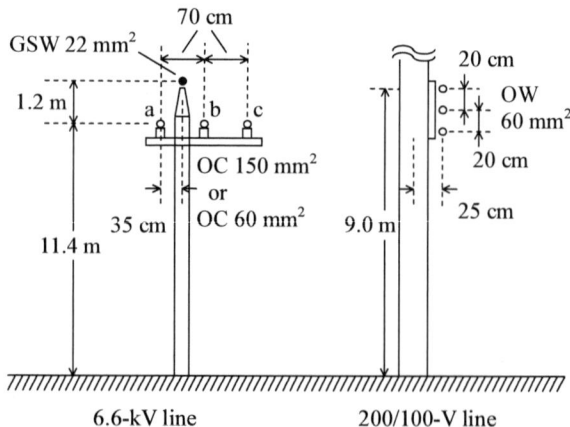

Fig. 3. Candidate of the standard conductor arrangements
for 6.6-kV and 200/100-V lines.

978-1-4799-2706-7/14 $31.00 © 2014 IEEE

algorithm of the LDC. The control algorithm has been investigated by the WG so that it can be considered typical in Japan.

The standard models of SVCs and STATCOMs are still under development and investigation. The Smart Grid WG is active to find out typical circuit topologies, control algorithms and their parameters. The results will be presented in the final report.

D. Pole-Mounted Transformers

A pole-mounted transformer is basically a single-phase transformer which steps down 6.6 kV to 200/100 V. Its primary side is connected to two of the three phases of a 6.6-kV line, and the secondary side is connected to a 200/100-V line or directory to a service-drop line. The secondary side winding has a center tap which is grounded at the distribution pole. When the power-frequency behaviors of a pole-mounted transformer are of interest, the fundamental equivalent circuit, also known as Steinmetz's equivalent circuit, shown in Fig. 5 is adequate as the standard model. It consists of an ideal transformer, winding resistance, leakage inductance and a magnetizing circuit. When high frequency behaviors are of interest, a more detailed model such as the one proposed in [10] is recommended.

E. PV Panels

A PV Panel is a semiconductor device which directly converts sunlight into electricity. Its characteristics are dependent on irradiation and temperature. When PV power generation systems come into wide use, the fluctuations of PV outputs have an essential impact on the power quality of end consumers.

The Smart Grid WG has selected the single-diode model proposed in [11] as the standard model of PV panels. The model, shown in Fig. 6 (a), offers good trade-off between simplicity and accuracy. The value of the photovoltaic current source I_{pv} is given by

$$I_{pv} = \left\{ \frac{R_p + R_s}{R_p} I_{sc,n} + K_I (T - T_n) \right\} \frac{G}{G_n} \qquad (1)$$

where R_p and R_s are the parallel and the series resistance in the model, $I_{sc,n}$ is the nominal short-circuit current, K_I is the short-circuit current/temperature coefficient, T and T_n are the actual and the nominal temperature, and G and G_n are the actual and the nominal irradiation. The V–I curve of the diode in parallel with I_{pv} is given by

$$I_d = \{I_{sc,n} + K_I (T - T_n)\}$$
$$\times \frac{\exp[q(V + R_s I) / a N_s kT] - 1}{\exp[q\{V_{oc,n} + K_V (T - T_n)\} / a N_s kT] - 1} \qquad (2)$$

where $V_{oc,n}$ is the nominal open-circuit voltage, K_V is the open-circuit voltage/temperature coefficient, N_s is the number of PV cells connected in series, q is the electron

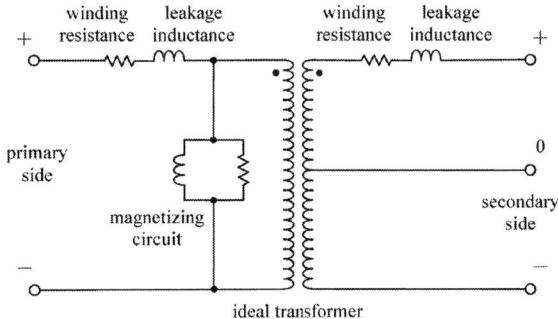

Fig. 5. Fundamental equivalent circuit of pole-mounted transformers.

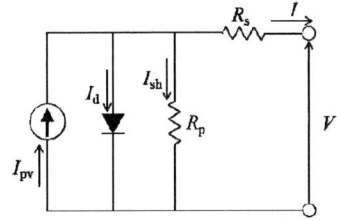

(a) Equivalent circuit.

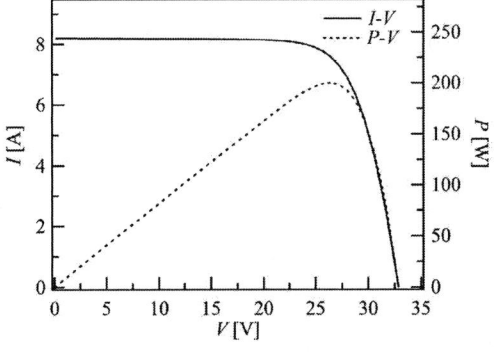

(b) Calculated I–V and P–V curves.

Fig. 6. Standard model of PV Panels and its validation result.

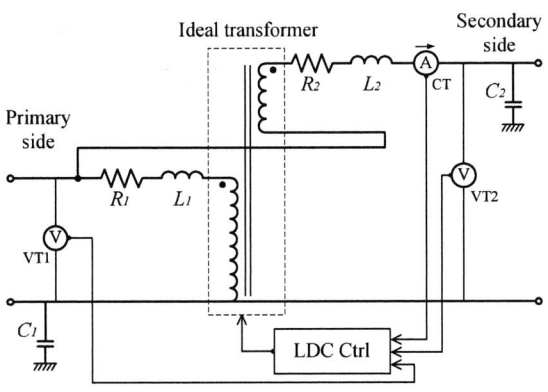

Fig. 4. Standard model of SVRs.

978-1-4799-2706-7/14 $31.00 © 2014 IEEE

charge, a is the diode ideality constant, and k is the Boltzmann constant. Most of these parameters can be obtained from typical PV panel datasheets or by the method proposed in [11]. The availability of parameter values is another point to select this model.

Fig. 6 (b) shows the I–V and the P–V curve calculated by the standard model for an actual PV panel. The calculated maximum-power-point current and voltage values agree well with those shown in the datasheet of the PV panel.

F. PCS for PV Power Generation

A PCS for a PV power generation system converts DC power from PV panels into AC power in order to connect to a residential or a building power wiring. The generated power is primarily supplied to electric appliances in the house or the building. If the generated power is greater than the power consumed by those appliances, then the surplus power is sent back to a distribution line.

The typical configuration of PCSs for PV power generation systems is shown in Fig. 7. It consists of a boost converter and a single-phase full-bridge inverter. Actual PCS products use various worked-out circuit topologies to solve problems like common-mode currents due to stray capacitance. However, the Smart Grid WG has selected the topology shown in Fig. 8 as the standard model, according to the principles mentioned in Section II. Considering that the purpose of the Smart Grid WG is to develop models for distribution- and end-consumer-level simulations, the selected topology should be adequate for these kinds of simulations. If phenomena

inside a PCS are of interest, the topology is of course not adequate. Table I shows specifications of the standard model.

Fig. 9 shows control block diagrams of the standard model. The control blocks shown in Fig. 9 (a) are for the control of the boost converter. Using the AVR (Automatic Voltage Regulator) block, the voltage from the PV panels connected is controlled and adjusted to the command value V_{pv}^* that is sent from the MPPT (Maximum Power Point Tracking) controller. The MPPT controller tracks the maximum power point obtained from the output power P_{pv} from the PV panels. P_{pv} is calculated by the voltage V_{pv} and the current I_{pv} of the PV panels, and V_{pv}^* is obtained by the perturb and observe (P&O) method. The control blocks shown in Fig. 9 (b) are for the control of the inverter. Using the AVR block,

TABLE I

SPECIFICATIONS OF THE STANDARD MODEL OF
PCSs FOR PV POWER GENERATION SYSTEMS.

Category	Item	Specifications and constants
General	Configuration	Boost converter / Inverter. Non-isolated.
	Output ratings	AC 200V 5kW 50Hz(25A) Single phase 2 wire.
	Input ratings	DC 250V (20A)
Inverter	Main circuit	Single phase full bridge. Carrier frequency: 20kHz.
	Control scheme	Current control (Instantaneous ACR)
	Output filter	ACL: 0.4mH × 2 (3%) C: 5μF (Cutoff frequency: 2.5kHz)
DC link	Capacitance	4700μF
	Rated voltage	400V
Boost converter	Main circuit	Boost converter. Carrier frequency: 20kHz.
	Control scheme	Current Control (Instantaneous ACR) Voltage Control (Instantaneous AVR)
	DC reactor	DCL: 1mH

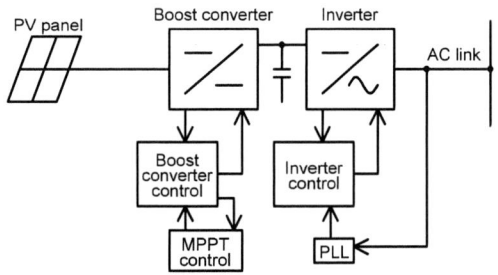

Fig. 7. Typical configuration of PCSs for
PV power generation systems.

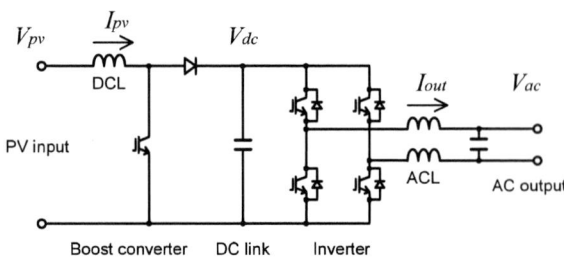

Fig. 8. Standard model of PCSs for PV power generation systems.

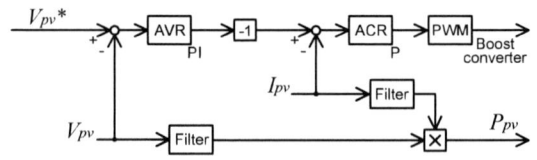

(a) Control blocks for the boost converter.

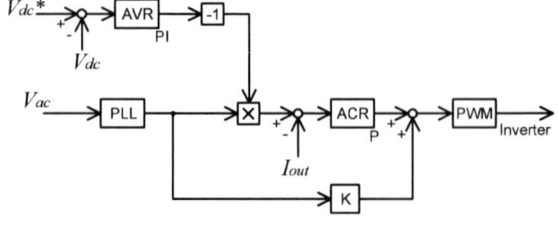

(b) Control blocks for the inverter.

Fig. 9. Control blocks of the standard model of PCSs for
PV power generation systems.

the dc-link voltage V_{dc} is controlled to be a constant value so that all power supplied from the boost converter is converted into AC power.

G. Li-Ion Batteries

Owing to high energy density and ease of maintenance, Li-ion batteries are now used as home-stationing-type energy storages. Normally, the capacity of a home storage battery is about 5 kWh, and it is constituted by a number of cells. The capacity of each cell is 2.5 Ah, and its nominal voltage is 3.7 V. The vehicle-to-home (V2H) concept, which becomes common today, utilizes the batteries of EVs as home storage batteries.

Since the Smart Grid WG focuses on simulations up to tens of seconds, charge-and-discharge performance in a partial charge state and detection of overcharge should be represented by the standard model of Li-ion batteries. As the simplest model that is able to represent these characteristics, the WG currently investigates the equivalent circuit model shown in Fig. 10 (a). To obtain this standard model, information from [12]–[14] has been used. The two parallel R–C branches, whose parameters R_{C1}, C_{C1}, R_{C2} and C_{C2} can be obtained by charge and discharge characteristics, reproduce activation and concentration polarizations. Transient responses of the model are mainly due to these R–C branches.

The relationship between the electromotive force OCV and the state of charge SOC can be described using the nonlinear function $f(\cdot)$ which is expressed by exponential functions in regions close to the fully-charged and fully-discharged states and is linearly interpolated in between.

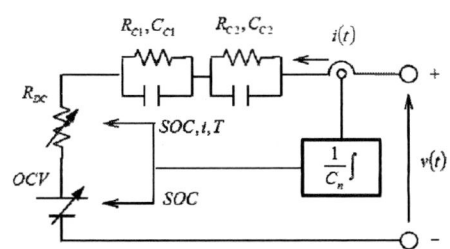

(a) Equivalent circuit.

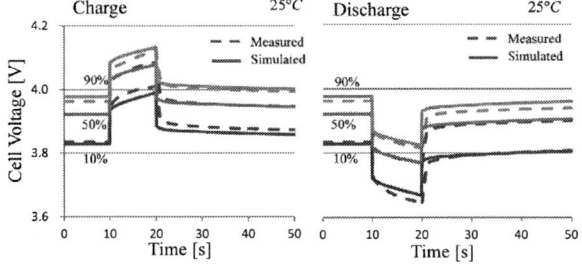

(b) Simulation results.

Fig. 10. Standard model of Li-ion batteries and its preliminary simulation results.

Then, OCV is given by

$$OCV = E_0 - f(SOC) \tag{3}$$

where E_0 is the electromotive force at the fully-charged state. The value of SOC is obtained by the integration of the battery current i:

$$SOC = SOC_0 + \eta \frac{1}{C_{nom}} \int i \cdot dt \tag{4}$$

where η is the charge-and-discharge efficiency, C_{nom} is the rated capacity, and SOC_0 is the initial state of charge. To reproduce the C rate discharging characteristics, the polarization resistance R_{DC} is expressed by the DC resistance R_0 and the current dependent resistance R_1:

$$R_{DC} = \alpha \left\{ R_0 + R_1 \left(1 - \frac{i}{I_{nom}} \right) \right\} \tag{5}$$

where I_{nom} is the nominal current and α is the temperature coefficient whose value is adjusted according to operating regions.

Preliminary simulation results obtained using this standard model is shown in Fig. 10. In the figure, the left-hand-side plots show transient voltage waveforms for pulse current charges with different SOCs, and the right-hand-side plots show those for pulse current discharges. The simulation results closely reproduce corresponding measured results.

H. Micro and Small Wind Power Generation Systems

Wind power generation systems with capacity less than 1 kW and propellers less than 2 m are called "micro" wind power generation systems. Those with capacity less than 20 kW and propellers less than 7 m are called "small" wind power generation systems. A micro or a small wind power generation system consists of a propeller, a generator and a PCS.

Propellers are classified into vertical and horizontal types in terms of rotation axis. For the purpose of electrical transient simulations, however, these classifications are not important, and any type of propeller can be represented by its power coefficient C_p and its projected area A. The power P obtained by a propeller is expressed by

$$P = \frac{1}{2} C_p m v^2 = \frac{1}{2} C_p (\rho A v) v^2 = 2\pi n \tau \tag{6}$$

where m is the mass of air which goes through the propeller per second, v is the velocity of wind, ρ is the density of air, n is the speed of rotation, and τ is torque [15]. In most cases, a permanent magnet synchronous machine (PMSM) is used as the generator owing to its efficiency and ease of maintenance. A propeller is directly mounted to the shaft of the PMSM.

The standard model of micro and small wind power generation systems consist of the following models. First, the input to the generator model mentioned next is calculated by (6). A three-phase dq-frame-based PMSM model is prepared to represent the generator. The electrical output from the generator model is connected to the following PCS model. In the PCS model, the input is connected to a three-phase diode-bridge rectifier, and its output is then connected to a boost chopper. The output of the boost chopper is finally connected to a single-phase or a three-phase pulse width modulation (PWM) inverter so as to connect to a residential or a building power wiring.

As a preliminary simulation, a micro wind power generation system of 1 kW is modeled by the standard model. Fig. 11 shows the simulation case created using XTAP (a transient analysis program developed by CRIEPI). Since this a preliminary simulation, the output from the rectifier model is directly connected to a load resistor of 5 Ω, and the boost chopper model and the PWM inverter model remain unused. Fig. 12 (a) shows the wind speed variation with respect to time, and this is used as the input to the propeller. Fig. 12 (b) shows the characteristics of C_p with respect to the tip speed ratio of the propeller assumed in this simulation. Fig. 13 shows the simulation result. It is observed that the outputs at different stages vary in response to the wind speed variation. The maximum power generated is about 500 W at a wind speed of 11.4 m/s.

IV. SIMULATION SCENARIOS

Since the development of some of the standard models mentioned above has not yet finished, the Smart Grid WG unfortunately cannot start creating simulation cases using the standard models at this moment. However, the Smart Grid WG plans to create simulation cases for the following scenarios:

1. Transients when a PV power generation system or a small/micro wind power generation system rapidly changes its output in a smart house. Also, transients when a Li-ion battery system is applied to mitigate the rapid change.

2. Transients due to rapid changes of loads in a smart house. Also, transients when a Li-ion battery system is applied to mitigate the rapid change.

3. Interactions of different voltage regulation equipment such as LRT in a substation, SVR, SVC, STATCOM and PCSs of PV generation systems in a distribution system. Their coordination by simulations is of great interest.

4. Tests of islanding and FRT operations for the PCSs of PV and other distributed generation systems in a distribution system.

V. CONCLUSION

Cooperative Study Group on Standard Model Development for Power Electronics Simulations, created under the Industry Applications Society of the IEEJ, is

active to develop standard models for simulations related to power electronics. Under the study group, four WGs have been created for different fields of simulation. Among them, Smart Grid WG is developing standard simulation models of components which will form smart grids in the near future. With the assumption that smart grids in Japan will be established at the distribution and the end-consumer level, the Smart Grid WG is developing the standard models of the following components: distribution substations, distribution lines, voltage regulation equipment, pole-mounted transformers, PV panels, Li-ion batteries, power conditioning systems and micro and small wind power generation systems. This paper has presented these standard models and also future plan for those which are still under development.

The final results of the standard model development

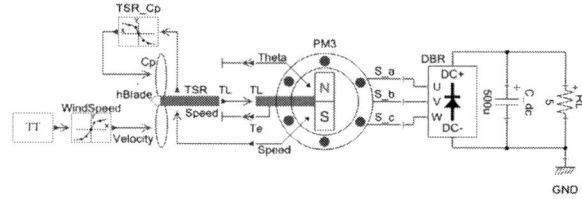

Fig. 11. Preliminary simulation case using the standard model of micro and small wind power generation systems.

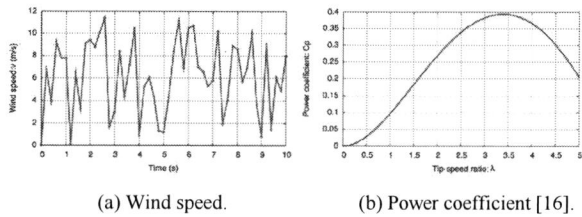

(a) Wind speed.   (b) Power coefficient [16].

Fig. 12. Wind speed and power coefficient assumed in the preliminary simulation case shown in Fig. 11.

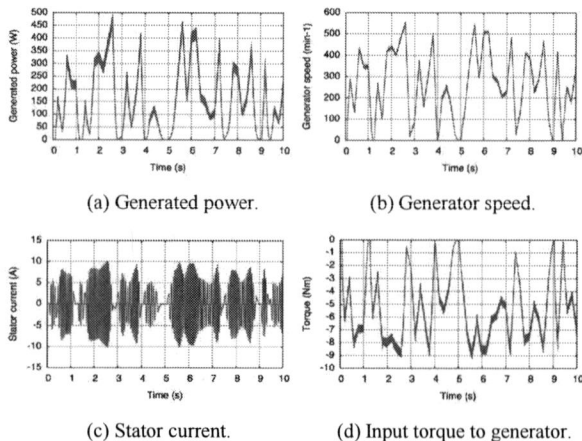

(a) Generated power.   (b) Generator speed.

(c) Stator current.   (d) Input torque to generator.

Fig. 13. Result of the preliminary simulation case shown in Fig. 11.

with recommended model parameters and recommended parameter determination methods will be published in a Technical Report issued by IEEJ in 2015.

REFERENCES

[1] "Technologies for the Simulation of Power Electronic Systems," (title translated into English by the authors) Technical Report, No. 761, IEEJ (Institute of Electrical Engineers of Japan), 2000.

[2] "Modeling and Simulation Techniques of Power Electronic Systems," Technical Report, No. 1114, IEEJ, 2008.

[3] T. Noda, Y. Kabasawa, K. Fukushima, H. Tokuda, H. Ishikawa, J. Ichihara, S. Kato, H. Kakigano, T. Sekisue, T. Kato, N. Kimura, Y. Kuroe, R. Saito, J. Shimomura, and M. Matsui, "Development of Standard Models for Transient Simulations of Smartgrids: Present State and its Plan," 2012 Annual Meeting Record of IEEJ, Vol. 4, Paper No. 4–S13–2, pp. S13 (3)–(6), 2012.

[4] T. Noda, Y. Kabasawa, K. Fukushima, H. Tokuda, H. Ishikawa, J. Ichihara, S. Kato, H. Kakigano, T. Sekisue, T. Kato, N. Kimura, Y. Kuroe, R. Saito, J. Shimomura, and M. Matsui, "Present State and Future Plan of Standard Model Development for Smart Grid Simulations," Proceedings of IEEE COMPEL 2012, Paper # CP-0242, Kyoto, Japan, 2012.

[5] A. Yokoyama, "Toward the Development of a Smarter Grid: Part 1," (title translated into English by the authors), IEEJ Journal, Vol. 130, No. 2, pp. 94–97, 2010.

[6] A. Yokoyama, "Toward the Development of a Smarter Grid: Part 2," (title translated into English by the authors), IEEJ Journal, Vol. 130, No. 3, pp. 163–167, 2010.

[7] T. Noda, "Numerical Techniques for Accurate Evaluation of Overhead Line and Underground Cable Constants," IEEJ Trans. on Electrical and Electronic Engineering, Vol. 3, Issue 5, pp. 549–559, 2008.

[8] H. W. Dommel, "Electro-Magnetic Transients Program (EMTP) Theory Book," Bonneville Power Administration.

[9] A. Martinez-Velasco (Editor), "Power System Transients: Parameter Determination", CRC Press, 2010.

[10] T. Noda, H. Nakamoto, and S. Yokoyama, "Accurate Modeling of Core-Type Distribution Transformers for Electromagnetic Transient Studies," IEEE Trans. on Power Delivery, Vol. 17, No. 4, pp. 969–976, 2002.

[11] M. G. Villalva, J. R. Gazoli, and E. R. Filho, "Comprehensive Approach to Modeling and Simulation of Photovoltaic Arrays," IEEE Trans. on Power Electronics, Vol. 24, No. 5, pp. 1198–1208, 2009.

[12] Y. Inui, Y. Watanabe, and Y. Kobayashi, "Numerical Simulation of Transient Voltage Response of Lithium-Ion Secondary Battery," IEEJ Trans. on Power & Energy, Vol. 126, No. 5, pp. 532–538, 2006.

[13] A. Baba and S. Adachi, "Simultaneous State and Logarithmic Parameter Estimation of Lithium-Ion Batteries using UKF," IEEE Trans. on Industry Applications, Vol. 133, No. 12, pp. 1139–1147, 2013.

[14] E. Kuhn, C. Forgez, P. Lagonotte, and G. Friedrich, "Modeling Ni-mH battery using Cauer and Foster structures," Journal of Power Sources, Vol. 158, No. 2, pp. 1490–1497, 2006.

[15] I. Ushiyama and M. Mino, "Small Wind Turbine," Power-sha, 1980.

[16] T. Matsuzaka and K. Tuchiya, "Study on Stabilization of a Wind Generator Power Fluctuation," IEEJ Trans. on Power and Energy, Vol. 117, No. 5, pp. 625–633, 1997.

Model Development
for Motor Drive System Simulations

Hiroki Ishikawa
Dept. of Electrical, Electronic and Computer Engineering
Gifu University
Gifu, JAPAN
ishikawa@gifu-u.ac.jp

Masahiro Ikeda
Nagasaki Institute of Applied Science

Takashi Abe
Nagasaki University

Nobuhiro Umeda
YASKAWA Electric Corporation

Toshiji Kato
Doshisha University

Noriyuki Kimura
Osaka Institute of Technology

Yutaka Kubota
Mitsubishi Heavy Industries, Ltd.

Koichi Shigematsu
Cybernet Systems Co., Ltd.

Junichi Shimomura
Meidensha Corporation

Yukinori Inoue
Osaka Prefecture University

Yusuke Kohno
TOSHIBA CORPORATION

Abstract—**In the motor drives working group of Cooperative Study Group on Applications of Power Electronics Simulations of IEEJ, standard models for simulation of motor drive systems are discussed. This paper provides a standard motor model of induction motor drive systems with parameters. Parameters of more than 30 induction motors have been gathered, and tendency of the parameters have been also investigated. In this paper, some simulation example are shown and compared with the experimental results.**

Keywords— simulation, standard models, motor drive systems, and induction motors.

I. INTRODUCTION

Motor drive systems (MDSs) are applied to various fields widely, and replacement of other power supply systems with MDSs is advanced because of higher efficiency, faster responses, lower noise and vibration, and so forth. Higher layer systems, such as smart grid and automotive system, include MDSs, and become more under influence of characteristics of MDSs with widespread use of MDSs.

At the design development stage of such higher layer systems especially, computer simulation is one of key technologies for effective design of the systems and confirmation of the system operation. The simulation of an MDS requires a motor model with motor parameters

and a controller configuration with control gains at least. However, engineers except motor and motor drive specialists have less opportunity to get such information. Therefore, not only electric engineers but also engineers in other field often require information about operating characteristics of motors, design techniques of controller, and performances of motor drive systems.

Standard motor parameters and standard models for simulation of MDSs have been discussed in the motor drives working group (MDWG) of Cooperative Study Group on Applications of Power Electronics Simulations in IEE JAPAN (IEEJ) on such demand. Target motors are dc motors, induction motors, permanent magnet synchronous motors, reluctance motors, and so on.

As first report, a standard model and parameters for permanent magnet dc motor drive systems was proposed[1][2]. In the report, parameters of 77 dc motors were gathered and discussed in MDWG. Tendency of motor parameters with respect to rated output power was found and formulated by approximation. The standard controller model and the design for dc motor drive systems were also shown in the report.

The MDWG continues to discuss standard models for other motors drive systems, especially induction motor drive system. This paper reports a standard model and parameters for induction motor drive systems as a progress report.

II. A STANDARD MODEL OF INDUCTION MOTORS

The configuration of an analysis model must to be selected by analyzed phenomena. In case of confirmation of operating principles, equivalent circuit models can be applied. For analysis of details, a model corresponding to a phenomenon is added to the equivalent circuit model.

Figure 1 shows a typical single-phase electric equivalent circuit of induction motors. In the figure, r_1 is primary resistance, r_2 is secondary resistance, X_1 is primary leakage winding reactance, X_2 is secondary leakage winding reactance, R_m' is resistance of excitation winding, X_m' is excitation winding reactance, and s is slip. The equivalent circuit in Fig. 1 has been identified as a basic standard induction motor model in MDWG. The parameters can be obtained by winding resistance measurement, no load test under the rated voltage v_{rated}, and locked rotor test under the rated current i_{rated}.

The relations between winding inductances L_1, L_2, and L_m' and X_1, X_2, and X_m', respectively, are expressed as follows.

$$L_1 = \frac{X_1}{\omega_e}, \tag{1}$$

$$L_2 = \frac{X_2}{\omega_e}, \text{ and} \tag{2}$$

$$L_m' = \frac{X_m'}{\omega_e}. \tag{3}$$

where, ω_e is angular frequency of voltage source v_1. The model equations in Fig. 1 are

$$v_1 = r_1 i_1 + L_1 \frac{di_1}{dt} + \frac{r_2}{s} i_2 + L_2 \frac{di_2}{dt}, \tag{4}$$

$$v_1 = r_1 i_1 + L_1 \frac{di_1}{dt} + R_m' i_m + L_m' \frac{di_m}{dt}, \text{ and} \tag{5}$$

$$i_1 = i_2 + i_m. \tag{6}$$

The slip s, rotational angular speed ω_m, and synchronous angular speed ω_0 have next relation.

$$s = \frac{\omega_0 - \omega_m}{\omega_0} = \frac{\frac{P}{2}\omega_e - \omega_m}{\frac{P}{2}\omega_e} \tag{7}$$

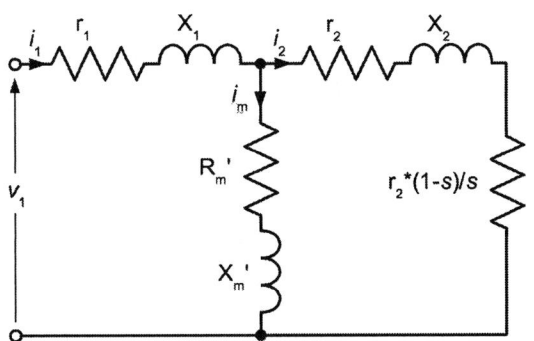

Fig. 1. Single-phase equivalent circuit of induction motors.

where, P is pole number. Torque Tq can be gotten from next equation.

$$Tq = L_m' \cdot \frac{P}{2}\left(i_{q1} i_{d2} - i_{d1} i_{q2}\right) \tag{8}$$

where, i_{d1}, i_{q1}, i_{d2}, and i_{q2} are motor currents on d-q coordinate system.

III. INVESTIGATIONS OF MOTOR PARAMETERS

In MDWG, the opened measurement test results for 19 models[3] have been gathered, and measurement tests in laboratories has been carried out for 13 models. The parameters of 32 induction motors have been calculated from all results and discussed. The rated output range of the extracted models is from 400W to 4.5MW. In order to clarify tendency of the parameters with respect to each rated output power P_O, the calculated parameters have been converted into normalized values by next equations.

$$k = \frac{i_{rated}}{v_{rated}}, \tag{9}$$

$$r_{1PU} = k r_1, \tag{10}$$

$$r_{2PU} = k r_2, \tag{11}$$

$$R_{mPU} = k R_m', \tag{12}$$

$$X_{1PU} = k X_1, \tag{13}$$

$$X_{2PU} = k X_2, \text{ and} \tag{14}$$

$$X_{mPU} = k X_m' \tag{15}$$

where, k is normalized factor of each motor, v_{rated} is the rated voltage of each motor, i_{rated} is the rated current of each motor, and r_{1PU}, r_{2PU}, R_{mPU}, X_{1PU}, X_{2PU}, and X_{mPU} are normalized values of r_1, r_2, R_m', X_1, X_2, and X_m', respectively.

Figure 2 shows the survey results of normalized parameters. The (a) and (b) shows the tendency of r_{1PU} and r_{2PU} with respect to P_O, respectively. It is clear from the tendencies larger P_O decreases r_{1PU} and r_{2PU}. The R_m', X_1, X_2, and X_m' are increased for larger P_O as shown in from (c) to (f). The parameters of the standard motor model in Fig.1 can approximated from the tendencies in Fig.2 by the decision of P_O. The P_O can be designed next equation.

$$P_0 = \omega_{mrated} Tq_{rated}, \tag{16}$$

where, ω_{mrated} is the rated angular velocity of a motor, and Tq_{rated} is the rated motor torque.

In the MDWG, approximate equations for parameters of the standard model will be derived from tendencies of the calculated parameters as shown in Fig. 2, and the validity of the equations will be discussed.

IV. SIMULATION RESULTS WITH THE STANDARD MODEL

The MDWG has started to discuss the difference between experimental results and simulation results by the standard model due to demonstration of the validity and usefulness for the standard model.

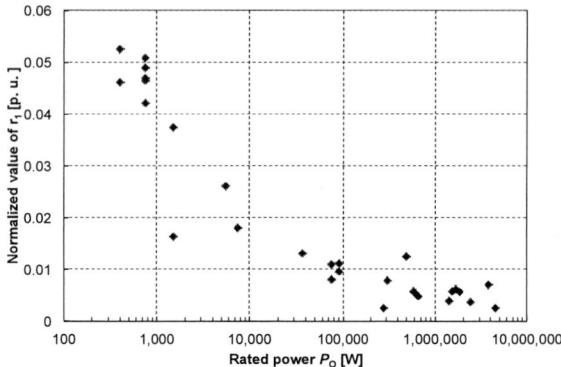

(a) Primary resistance r_1.

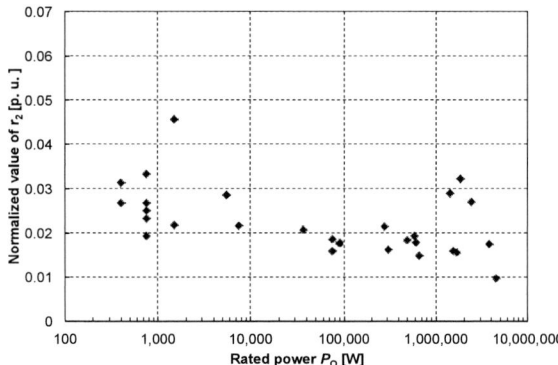

(b) Secondary resistance r_2.

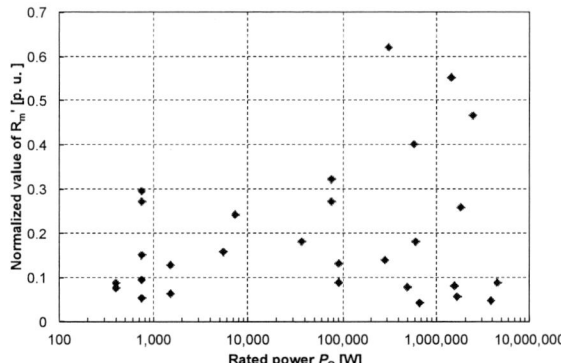

(c) Excitation winding resistance R_m'.

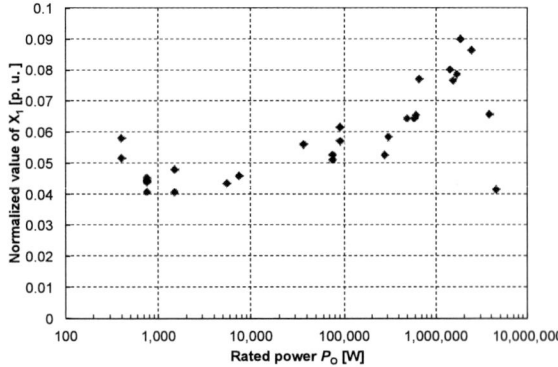

(d) Primary leakage winding reactance X_1.

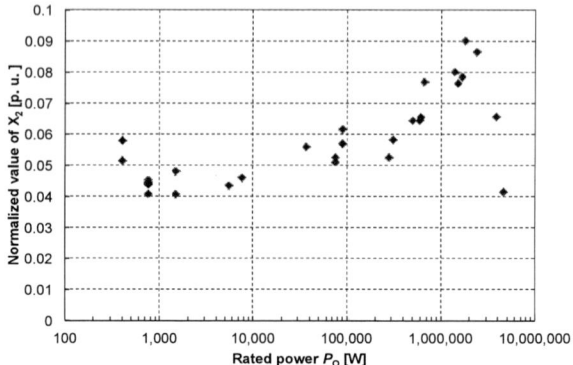

(e) Secondary leakage winding reactance X_2.

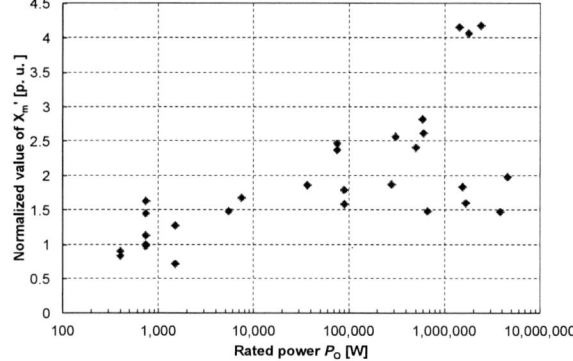

(f) Excitation leakage winding reactance X_m'.

Fig. 2. Parameter calculation results of induction motors.

Figures 3 and 4 show examples of the comparison of simulation results using the standard model with experimental results[4] of a 750W squirrel-cage induction motor. The specifications of the tested motor and parameters of the equivalent circuit for the tested motor are listed in Table I and II, respectively. Inertia J in Table II has been obtained from the results of free-run test for the tested motor.

Figure 3 shows comparison of transient phase current through motor windings at starting under across the line starting and no load condition. The input line-to-line voltage v_{in} has been set to 200V. The blue lines are experimental results, and the red lines are simulation results. The (a) shows comparison of instantaneous phase current and line-to-line voltage waveforms, and (b) shows peak value of phase current and rotational speed at starting. Just after starting, current of simulation is much smaller than that of the experimental results even though phase and magnitude of v_{in} for simulation are almost same condition. As a result, larger torque is generated in the experimental results just after starting, and the difference between the experimental and simulated speed is also observed.

Figure 4 shows the comparison results of the waveforms at $v_{in}=35[V]$ under across the line starting. The blue lines are experimental results, and the red lines are simulation results. In the figure, simulated results of both of current and speed are almost same as the experimental results.

978-1-4799-2706-7/14 $31.00 © 2014 IEEE

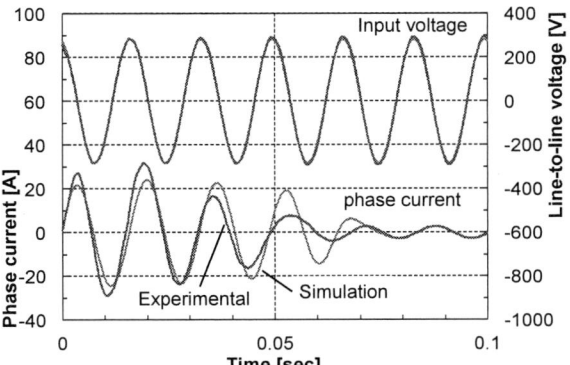

(a) Comparison of current waveform.

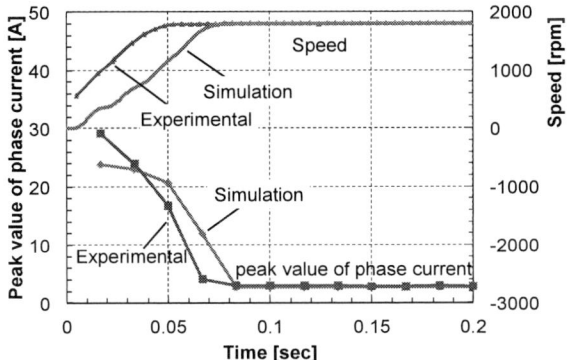

(b) Comparison of peak value of phase current and rotational speed.
Fig. 3. Comparison of transient state between simulation and experimental results at starting under across the line starting at 200V.

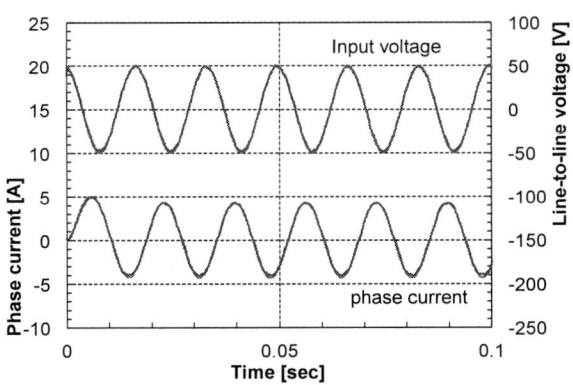

(a) Comparison of current waveform.

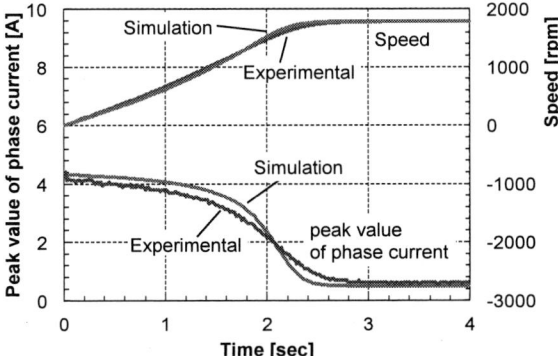

(b) Comparison of peak value of phase current and rotational speed.
Fig. 4. Comparison of transient state between simulation and experimental results at starting under across the line starting at 35V.

The ratio of the experimental current peak to simulated current peak with respect to v_{in} is shown in Fig. 5. Larger v_{in} increases the ratio. One of the reasons is magnetic saturation in motor cores because larger current flows under larger v_{in}. Considerations of magnetic saturation in motor cores and deep bar effect have an effect to correct the transient error[5]. The deep bar effect and magnetic saturation is explained next section. The MDWG also has started discussion to add the considerations to the standard model.

Figure 6 compares characteristics at steady state. The (a) and (b) are slip and phase current characteristics and slip and average torque characteristics, respectively. The simulation results have almost same properties with experimental results under from no-load to full-load condition. The difference between simulation and experimental results, however, is larger under over-load condition. The correction for steady state error will also be discussed in MDWG.

V. CONSIDERATION OF DEEP BAR EFFECT AND MAGNETIC SATURATION

The starting current simulation result of a squirrel-cage induction motor is shown as red lines in Fig. 3 assuming that all parameters are constant. The experimental starting current measured by actually load method is shown as blue line in same Fig. 3. By comparing the starting current by measurement with that by calculation, it can be seen that the maximum value of the measured current peak is about 30% greater than that of simulated current peak just after starting. It is considered that the following two causes make the measured starting current greater than the calculated starting current.

The first cause is a deep bar effect of squirrel-cage induction motor shown by the following formula K_r and K_x. This is because the deep bar type slot is employed for the rotor bars of the squirrel cage induction motor having several tens of kW. The resistance of the rotor bar will increase by the deep bar effect by the following formula K_r, but the leakage reactance of the rotor bar is reduced by the deep bar effect shown as formula K_x. By comparing the magnitude of the leakage resistance and reactance of the rotor bar, the leakage reactance is larger than 10 times and more. Therefore, impedance becomes smaller influenced by reduction of the leakage reactance at starting, and the starting current of squirrel cage induction motor will increase.

$$K_r = \xi \frac{\sinh 2\xi + \sin 2\xi}{\cosh 2\xi - \cos 2\xi}, \qquad (17)$$

$$K_x = \frac{3}{2\xi} \frac{\sinh 2\xi - \sin 2\xi}{\cosh 2\xi - \cos 2\xi}, \qquad (18)$$

where, $\xi = \alpha d$, $\alpha = 2\pi \sqrt{(sf/\rho) \times 10^{-11}}$.

The second cause is due to saturation of the rotor teeth of the squirrel-cage IM. Magnetic flux density of the gap is 0.9T or less. And other magnetic path parts such as

TABLE I
SPECIFICATION OF THE TESTED SQUIRREL-CAGE INDUCTION MOTOR.

Rated power	750 W
Rated line-to-line voltage	200 V
Rated current	3.3 A
Rated speed	1,710 rpm
Pole number	4
Synchronous speed	1,800 rpm
Slip at rated speed	0.05

TABLE II
PARAMETERS OF EQUIVALENT CIRCUIT FOR THE TESTED MOTOR.

r_1	2.55 Ω
r_2	1.47 Ω
L_1 and L_2	7.26 mH
$R_m{}'$	3.14 Ω
$L_m{}'$	155.4 mH
Inertia J	0.003 kg/m^2

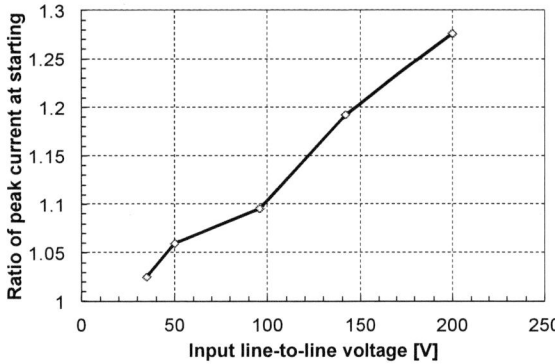

Fig. 5. Ratio of peak current with respect to input voltage under across the line starting.

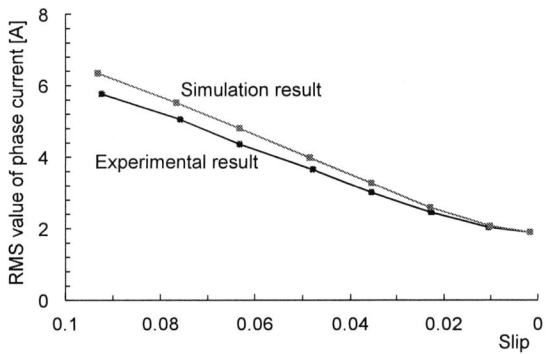

(a) Characteristics of slip and rms value of phase current.

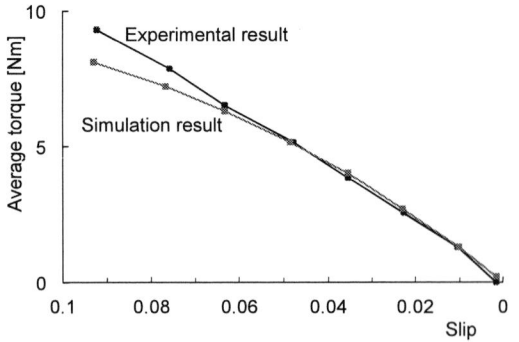

(b) Characteristics of slip and average torque.
Fig. 6. Comparison of steady state characteristics between simulation and experimental results.

stator and rotor core and teeth are limited to less than 1.8T. If the current of about 100% of rated current flows, there is no change in leakage reactance of the rotor teeth in rated speed, but the current of the magnitude of 400-600% of the rated current flows as starting current with slip frequency, leakage reactance of "b1" of Fig. 7 can be regarded as equivalent bridge portion "b" as shown in Fig.8 by saturation of rotor slot tips. As the leakage reactance of the rotor bar at the starting decreases by the above mentioned saturation, naturally the starting impedance becomes small, and the starting current of the squirrel cage IM will increase.

VI. CONCLUSIONS

This paper described a standard model and tendency of motor parameters for induction motor drive systems under discussion in the motor drives working group of Cooperative Study Group on Applications of Power Electronics Simulations in IEE JAPAN. Motor drive systems are applied to various fields widely. Higher layer systems also become more under influence of characteristics of MDSs with wide-spread use of MDSs. Standard motor parameters and standard models for simulation have discussed on such demand. This paper reports a standard model and parameters for induction motor drive systems as a progress report.

In this paper, some parameter tendency of the equivalent circuit for induction motors was clarified. The investigated parameters correlate with the rated power, and parameters of a motor with different output power can be obtained from the tendency approximately. In the MDWG, approximate equations will be derived from the tendency for higher usability.

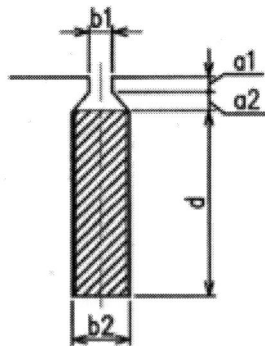

Fig. 7. Configuration of rotor slot.

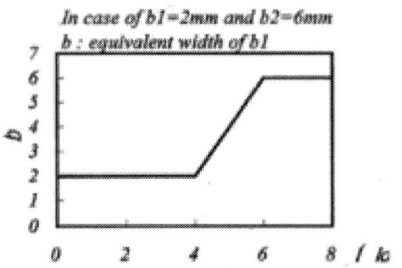

Fig. 8. Relation between b and I/I_0.

978-1-4799-2706-7/14 $31.00 © 2014 IEEE

The validity of the standard model was also verified. Simulated results with the model correspond pretty much to experimental results under not exceeding rated current flowing, such as steady state simulation. The model,however, need to be improved for larger current flowing, such as simulation under over-load condition and just after starting at rated voltage because the model cannot simulate influences on magnetic saturation and deep bar effect. The MDWG will also continue to discuss how to solve the problems.

REFERENCES

[1] Hiroki Ishikawa, Toshiji Kato, Nobuhiro Umeda, Shinji Kato, Kazutami Tago, Takayuki Sekisue, Takashi Abe, Shunsuke Ohashi, Satoshi Ogasawara, Yasuaki Kuroe, Yutaka Kubota, Koichi Shigematsu, Junichi Shimomura, Takeshi Horiguchi, Toshimitsu Morizane, and Kazuya Yasui, "Investigation into Standard Parameters of Motor Drive Systems", *IEEE-COMPEL 2012*, 2012.

[2] H. Ishikawa, T. Kato, N. Umeda, S. Kato, K. Tago, T. Sekisue, T. Abe, S. Ohashi, S. Ogasawara, Y. Kuroe, H. Kubota, K. Shigematsu, J. Shimomura, T. Horiguchi, T. Morizane, and K. Yasui, "Investigation for Standard Parameters of Motor Drive Systems," *The 2012 Annual Meeting Record I. E. E. Japan*, vol. 4, pp. S13(7)-S13(10), 2012 (in Japanese).

[3] "Proposal of Novel Characteristic Estimations for Induction Machines," *Technical report of IEEJ*, No. 435, 1992 (in Japanese).

[4] Hiroki Ishikawa and Yoshihiro Murai, "A New Approach of Characteristics Analysis for 3-Phase Induction Motor," *Conference Record of the 1998 IEEE Industry Applications Conference*, Vol. 1, pp. 73-78, 1998.

[5] Masahiro Ikeda and Takashi Hiyama, "Simulation Studies of the Transients of Squirrel-Cage Induction Motors," *IEEE Trans. on Energy Conversion*, vol. 22, no. 2, pp. 233-239, 2007.

Practical Simulation Examples of Automotive and Power Supply Systems

Takashi Abe
Nagasaki University : Graduate School of Engineering,
Division of Electrical Engineering and Computer Science
1-14 Bunkyo-machi Nagasaki, Japan
abet@nagasaki-u.ac.jp

Kentaro Fukushima
Electric Power Engineering Research Laboratory,
Central Research Institute of Electric Power Industry,
2-6-1 Nagasaka, Yokosuka City, Kanagawa, Japan
k-fuku@criepi.denken.or.jp

Takayuki Sekisue
ANSYS Japan K. K.

Koichi Shigematsu
Cybernet Systems Co., Ltd.

Junichi Ichihara
AZAPA Co., Ltd.

Toshiji Kato
Doshisha University

Hiroki Ishikawa
Gifu University

Yusuke Kouno
Toshiba Corporation

Masaaki Konoto
Panasonic Corporation

Ryoji Saito
Formerly with Origin Electric Co., Ltd.

Yasuyuki Nishida
Chiba Institute of Technology

Abstract— **In the Automotive Working Group and Power Supply Working Group of Cooperative Study Group on Standard Development for power electronics systems in IEEJ, practical simulation examples of Automotive and Power Supply Systems have been discussed. This paper provides the practical simulation examples for the inexperienced engineers of automotive and power electronics, in order to easily understand the practical operation and apply to other simulation areas. Each of the practical simulation examples is shown and the validity is demonstrated.**

Keywords— *Simulation model, Power Supply Systems, Automotive, Practical examples*

I. Introduction

In recent years, renewable energy such as wind turbine generation and photovoltaic generation is installed actively in the world. Especially in Japan, after the huge earthquake of March 11th, this tendency is accelerating, and efficient usage of electrical power is paid attention. Power Electronics technology is one of the key factors of its realization.

Concurrently with this situation, automotive system is also paid attention for efficient usage of the energy and for the emission control. On the other hand, recent innovation in automotive system has brought the extensive application of power electronics technology, which makes the system more complex. In order to reduce time and cost for development and to satisfy environmental correspondence and market demand, computer simulation technique has become an indispensable for this complicate automotive system both in the design and development stage [1][2].

However, automotive system is too complex for beginners and entry learners to understand the meaning of those parameter values. They will be eager to understand the rough operation. Furthermore, their parameter values are different and may be confidential because each automotive system which is on the market is customized and optimized. Therefore, the practical simulation examples of automotive systems are required for education.

The practical simulation examples of power supply systems are also required for the learners to understand the power supply systems. Power supply systems require much knowledge such as, circuit analysis, control system, semiconductor device and so on. Also power supply systems are the combination of each essence. The learners can touch and understand the essence of power supply systems effectively by using the simulation examples [3].

978-1-4799-2706-7/14 $31.00 © 2014 IEEE

This paper shows the practical simulation examples of automotive and power supply systems for the inexperienced engineers, in order to easily understand the practical operation and apply to other simulation areas. A fuel consumption model is reported as the automotive example. This model consists of the driver, engine, transmission, wheels, running resistance and power-network including battery, alternator and electric load. And the behavior of the each model element is discussed. The converter design examples are reported as power supply systems. These models are set to practical parameter values so that one can readily carry out practical simulations.

II. Modeling of Automotive System

A. Multi-Physics Modeling

The automotive system is a multi-domain system that requires many branches of science and engineering, such as mechanical, thermal, magnetic, hydraulic and electrochemical fields, as well as the electrical circuits and control circuits drive them. Therefore, we considered modeling using VHDL-AMS (Very High Speed Integrated Circuit Hardware Description Language-Analog and Mixed Signal) [4]. The VHDL-AMS is a general-purpose circuit description language, which has the possibility of applying equilibrium conditions (Kirchhoff's law) to the multi-domain energy simulation using the same topology of electrical circuits [5][6].

B. Fuel Consumption Model

The whole simulation model for fuel consumption calculation is shown Fig. 1. This model consists models of the driver (drive train), engine, transmission (differential gear, gear box), wheels, shift select, running resistance and power-network including battery, alternator and electric load. The driving pattern is used the JC08 mode as the drive train data. Three physics domains such as the electrical, the translational and the rotational are mixed in this simulation model. The constitution and description using VHDL-AMS for each model element are discussed below.

C. Description of the Each Model Element

(1) Fuel consumption calculation

The ICE (Internal Combustion Engine) model is composed as a map reference model of the engine output torque according to the throttle angle and engine speed which are shown in Fig. 2. The engine output torque is consumed by inertia and internal friction of engine, and operates as a source of vehicle driving.

Also the engine map model is described using fuel consumption per hour contour line as shown Fig. 3. The fuel consumption per hour is derived from the relationship of the engine speed and the engine torque. As shown in Fig. 4, we defined total vehicle distance divided by the total fuel consumption as the final fuel consumption. Where, the total vehicle distance is

obtained by integrating vehicle speed, and the total fuel consumption is obtained by integrating fuel consumption per hour, respectively.

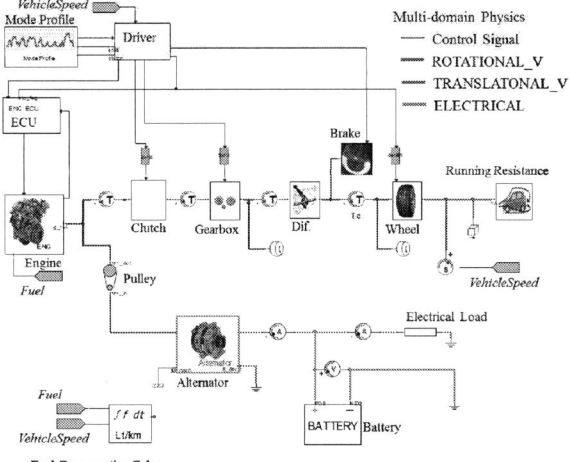

Fig. 1. Simulation model for fuel consumption.

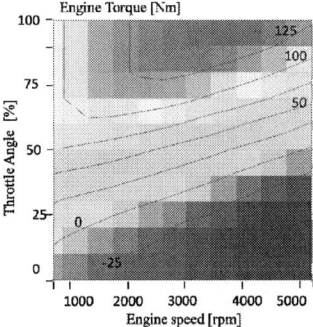

Fig. 2. Engine torque map.

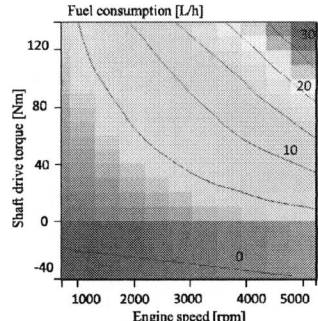

Fig. 3. Fuel consumption map.

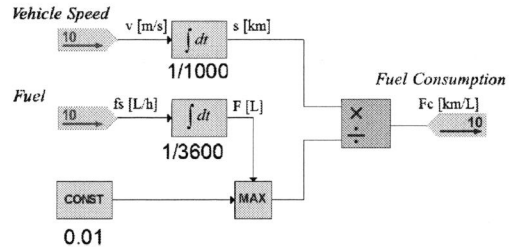

Fig. 4. Fuel consumption calculation.

(2) Wheel model

This wheel model exchanges the Translational _Velocity domain for the Rotational_Velocity domain, as shown List 1 and Fig. 5. Those relationship, velocity and angular velocity, force and torque, are calculated from the radius of the tire.

(3) Alternator model

The modeling of the alternator can be represented using basic equivalent circuit consisted of an averaged winding inductance L_a, a resistence R_a, an electromotive force and a rectifier diode. The electromotive force is controled by regulation voltage, rotational speed and output current map as shown in Fig. 6. Linear interpolation of the output current restrictions is carried out using the map data of the output current characteristic acquired from the measurement results based on the JIS D1615 alternator testing method. Similarly the alternator drew the torque consumed power generation using the efficiency characteristic referred from rotational speed, and it considered as the model by which the energy balance of electricity and a machine is held. The look-up tables of the output current characteristic and efficiency is shown in Fig. 7. This model exchanges the rotational _velocity domain from the engine and gear model for the electrical domain of the battery model, as shown in Fig. 8.

This model operates as an energy converter of the machine and electricity in consideration of efficiency. Hence, the torque consumption does not occur until it reaches the cut-in revolving speed which the output current of an alternator produces. For this reason, when applying to the simulation in consideration of the newest control system that performs electrical power generation control of an alternator actively, the result of mechanical energy loss is evaluated low.

(4) Battery model

The functions of battery which required for automotive fuel consumption estimation are the charge-and-discharge performance in a partial charge state and detection of overcharge, and several dozen of 10ms to seconds cycle behavior should be simulated. Moreover, the simulation should take into consideration the voltage drop according to a charge state until over 10min of time range. The equivalent circuit model which consists of minimum elements which imitate the function needed as a model which imitates these, and a means to derive model parameters from general measurement method and examination were examined.

Figure 9 shows the battery model using the directional equivalent circuit consisted of diodes, 2-level capacitance - resistance parallel circuits and OCV voltage source. The behavior of the polarization resistance R_{DC} and the electromotive pressure OCV according to SOC obtained by current integration are acquired from the quantity of the electrolysis solution in a storage battery, and imitate the charge-and-discharge characteristics. Resistance R_{C1}, R_{C2}, R_{D1}, R_{D2}, and capacitance C_{C1}, C_{C2}, C_{D1}, and C_{D2} are equivalent to activation polarization and concentration polarization imitates the transient response characteristic, and parallel configuration using the diodes considered in order to correspond to the difference in the characteristic of charge and discharge.

Each parameter was extracted to this model from the measurement result using the SBA S0101 lead acid battery life test pattern for idling stop vehicles. On the other hand, this model has adopted the fixed value as R_{DC}, since the SBA S0101 test pattern cannot acquire the nonlinear characteristics of the polarization resistance depending on discharge current quantities. For this reason, the difference from measurement about behavior of the voltage at the high current charge and discharge is expected.

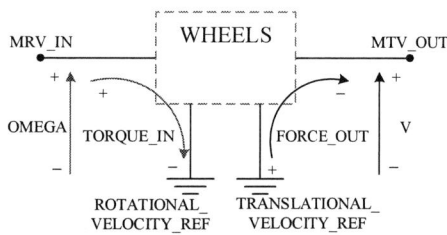

Fig. 5. Domain exchange of wheel model.

```
entity WHEELS is
  generic (
        tire_r : REAL      := 0.6;
      );
  port (
        terminal MRV_IN    : ROTATIONAL_VELOCITY;   -- provides torque
        terminal MTV_OUT   : TRANSLATIONAL_VELOCITY  -- drives vehicle
      );
end entity WHEELS;
architecture SIMPLE of WHEELS is
  quantity OMEGA across TORQUE_IN through MRV_IN;
  quantity V across MTV_OUT;
  quantity FORCE_OUT through TRANSLATIONAL_VELOCITY_REF to
MTV_OUT;
begin
  V       == tire_r * OMEGA;
  FORCE_OUT * tire_r == TORQUE_IN;
end architecture SIMPLE;
```

List. 1. Wheel model.

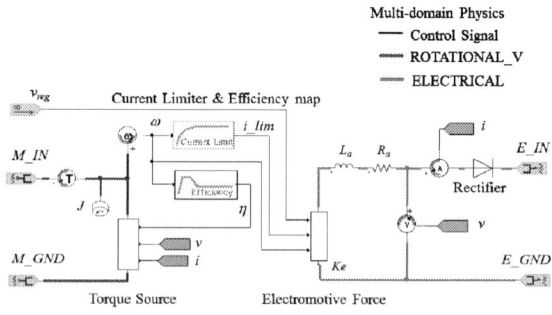

Fig. 6. Alternator model.

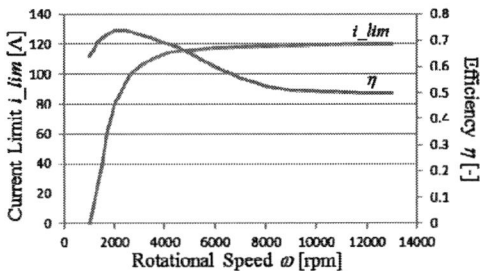

Fig. 7. Current and efficiency characteristics.

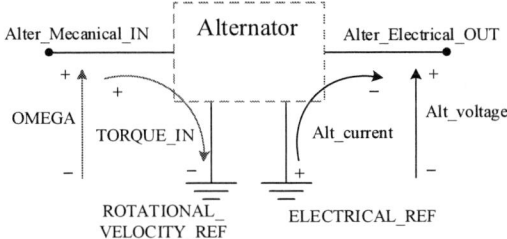

Fig. 8. Domain exchange of alternator model.

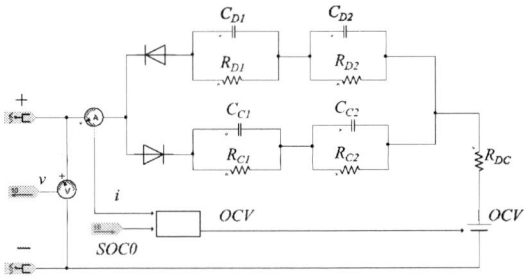

Fig. 9. Battery model.

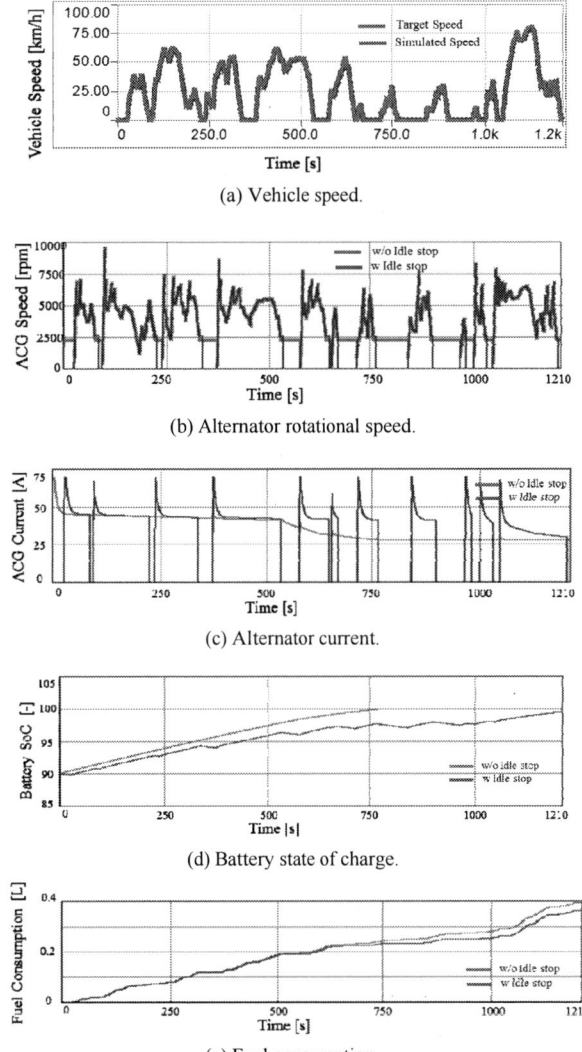

(a) Vehicle speed.

(b) Alternator rotational speed.

(c) Alternator current.

(d) Battery state of charge.

(e) Fuel consumption.

Fig. 10. Result of vehicle fuel consumption simulation.

D. Simulation Results

The simulation results of having presumed the fuel consumption of small/light size passenger car running as mode JC08 using the vehicles simulation model shown in Fig. 1 are shown in Fig. 10. The fuel consumption is 20.3km/L in usual driving mode, 22.1km/L in idling stop driving modes, respectively. From these results, 9% of fuel consumption improvement was obtained.

In the situation which changes the regulation voltage of an alternator and is overcharged during operation, fuel consumption was seen get worse 2~5% by the rise of the acceptance resistance to a battery, and the influence which an electric behavior has on fuel consumption has been shown.

III. Examples of Power Supply Systems

The beginners of power electronics have less image of circuit operation. In order to understand them, it is usable to check the waveforms and operation visually by simulation. Therefore, it is important that simulation examples can readily carry out. For example, the java applets of iPES (Interactive Power Electronics Seminar) are part of the Introductory Course on Power Electronics taught by Prof. Kolar at the ETH Zurich[7]. The interactive and animated applets are used as aid for teaching in the classroom and are displayed using a laptop and a beamer. In addition, the learner might have no idea how parameter values are adequate. If the parameters set to practical values in the examples, the learners can understand the practical condition automatically by using the simulation examples. This thing will be usable for the design engineers in R&D group.

In this paper, the simulation example of a buck DC-DC converter and a boost DC-DC converter are shown.

978-1-4799-2706-7/14 $31.00 © 2014 IEEE

A. Example of Buck DC-DC Converter

Figure 11 shows a circuit diagram of buck converter. Firstly, steady state characteristics are estimated from the specification. The circuit parameter values are shown in Table I. V_o is defined as the voltage of R1. In ideal condition, V_o is written as the bellow:

$$V_o = DV_i \tag{1}$$

Secondly, the inductance is estimated from the inductor current ripple ratio to the rated current. The inductor current ripple is written as below:

$$\Delta I_L = \frac{(1-D)DV_i}{Lf_s} \tag{2}$$

In Fig. 11, ripple current ratio is set as 10 % of rated current.

It is important to realize the dynamic characteristics when a controller of converter is analyzed and designed. Figure 12 shows the frequency response characteristics of the converter. This simulation is used SCAT produced by Keisoku Giken Co., Ltd [8]. Blue line is the simulation result when C2 is not inserted in the circuit of Fig. 11. Red line is the simulation result when C2 is inserted in the circuit of Fig. 11. From Fig. 12, it is observed that the phase margin is improved and the value is 39.3. C2 operates as phase compensation of the converter. Figure 13 shows the transient output voltage waveforms at each case. In this case, output resistance is changed from 2Ω to 5Ω. From this figure, it is observed that voltage response time is improved when C2 is inserted in the Fig. 11 circuit.

B. Example of Boost DC-DC Converter

Figure 14 shows a circuit diagram of boost converter. Steady state characteristics are estimated from the specification. The circuit parameter values are shown in Table II. V_o is defined as the voltage of R1. In ideal condition, V_o is written as the bellow:

$$V_o = \frac{1}{1-D}V_i \tag{3}$$

Secondly, the inductance is estimated from the inductor current ripple ratio to the rated current. The inductor current ripple is written as below:

$$\Delta I_L = \frac{DV_i}{Lf_s} \tag{4}$$

In Fig. 14, ripple current ratio is set as 10 % of rated current.

Here, the parameter values of the controller in Fig.14 are studied two cases written in table III. Fig.15 shows the frequency response characteristics of the converter. This system is narrowly stable at case #1. If the proportional gain is increased, the system will be unstable. On the other hand, DC gain of the system is increased at case #2. Furthermore, the phase margin is 89 degree at 96Hz. The system is operated stable. At both case, it cannot respond to high frequency disturbance. Fig.16 shows the transient output voltage waveforms at each case. In this case, output resistance is changed from 2 ohm to 10 ohm. From this figure, it is observed that the

steady-state deviation is generated slightly at case #1. Furthermore, the stabilization time is improved at case #2.

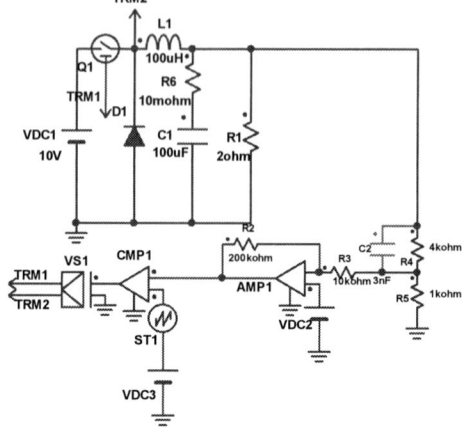

Fig. 11. Circuit diagram of a buck converter.

TABLE I
CIRCUIT PARAMETER VALUES

Parameters	Symbol	Values
Input voltage	V_i	10 V
Output voltage	V_o	5 V
Rated current	I_o	2.5 A
Switching frequency	f_s	100 kHz

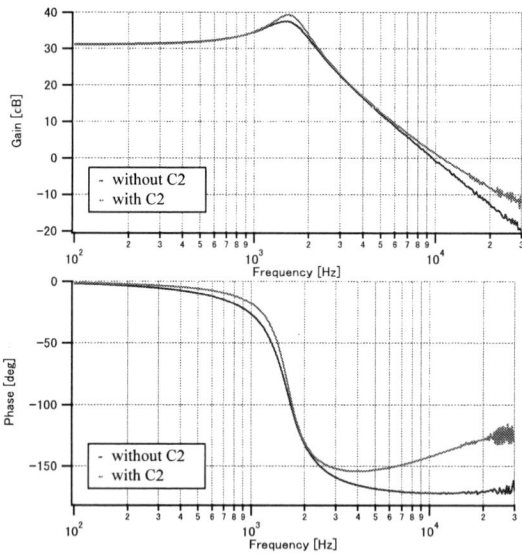

Fig. 12. Frequency response characteristics of the converter.

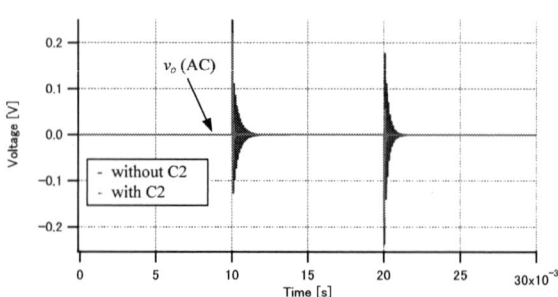

Fig. 13. Simulation result of transient output voltage waveforms.

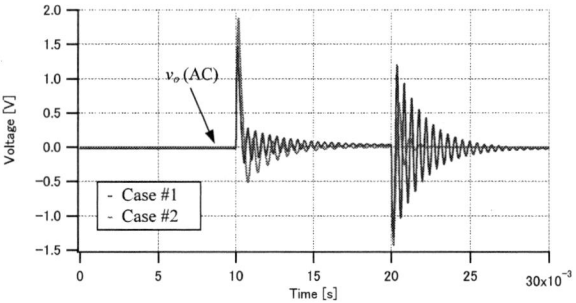

Fig. 16. Simulation result of transient output voltage waveforms.

For the learner for power electronics, they can understand the circuit operation and the meanings of the parameter values by using those examples. For example, the learner will study the characteristics of the circuit when the parameter values are changed. Furthermore, by using these parameters, they can compare the simulation result with experimental result which they make.

IV. Conclusions

This paper shows the practical simulation examples of automotive and power supply systems for beginner to be able to understand those operations, easily. The practical simulation examples are expected to be used for the leaner, and the examples can support to understanding and attract to automotive and power electronic system.

References

[1] D. W. Gao, C. Mi, and A. Emadi, "Modeling and Simulation of Electric and Hybrid Vehicles", Proceedings of the IEEE, Vol. 95, No.4, pp.729-745, 2007
[2] K. Shigematsu, T. Sekisue and K. Tsuji, "Auto-motive System Simulation (Japanese only)", Proc. of the 2006 IEE Japan IAS Conference, Vol.1, No.S9-3, pp.51-56, 2006
[3] T. Tamura, T. Abe, T. Higuchi, and K. Shigematsu, "Development of Power Electronics Education Tool Using a Circuit Simulator," Proceedings of IEEJ Industrial Application Society Conference 2009, pp. I-587-588, Aug. 2009 (in Japanese).
[4] IEEE:"IEEE Standard VHDL Analog and Mixed-Signal Extensions", IEEE Std 1076.1-2007, ISBN: 0-7381-5627-2, 2007
[5] Cooperation study group of modeling and simulation techniques of power electronic systems, "Modeling and simulation techniques of power electronic systems", IEEJ-IA, Vol.1114, 2008
[6] K. Tsuji, Y. Kido, T. Abe, "A Study of Vehicle Energy Management during Warming up Process Using VHDL-AMS Multi-domain Simulation", IEEJ Trans. IA, Vol.131, No.8, pp.985-991, 2011
[7] iPES web site, "http://www.ipes.ethz.ch/ipes/index.html"
[8] Keisoku Giken Co., Ltd. Web site, "http://www.keisoku.co.jp/en/"

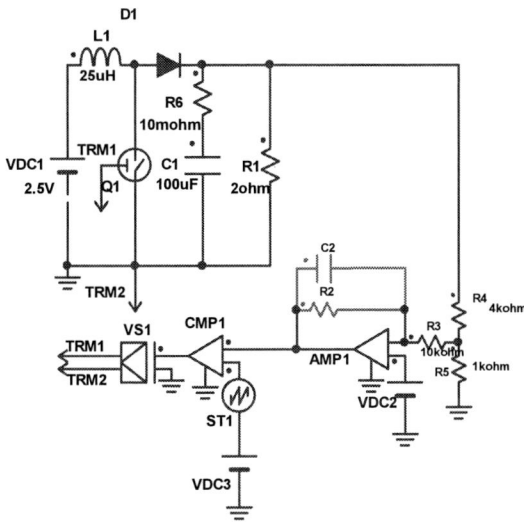

Fig. 14. Circuit diagram of a boost converter.

TABLE II
CIRCUIT PARAMETER VALUES

Parameters	Symbol	Values
Input voltage	V_i	2.5 V
Output voltage	V_o	5 V
Rated current	I_o	2.5 A
Switching frequency	f_s	100 kHz

TABLE III
PARAMETER VALUES AT EACH CASE

Case	R2	C2
#1	5 kΩ	-
#2	100 kΩ	300 nF

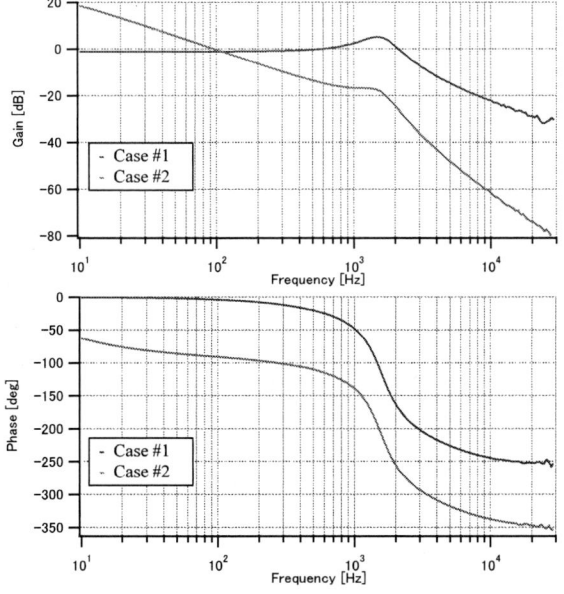

Fig. 15. Frequency response characteristics of the converter.

Admittance Matrices of Voltage Source Converters for Distributed Generators

K. L. Lian
Department of Electrical Engineering
National Taiwan University of Science and Technology
Taipei, Taiwan
ryanlian@mail.ntust.edu.tw

T. D. Huang
Department of Electrical Engineering
National Taiwan University of Science and Technology
Taipei, Taiwan
d10007101@mail.ntust.edu.tw

Abstract—To integrate distributed generators (DGs) with the power grid, interfacing power converters such as voltage source converters (VSCs) are required. However, due to the switching natures, these converters generate harmonics, which may have detrimental effects on the system. To find out how these harmonics may interact with the grid, it is essential to find the admittance matrices of the converters. In this paper, a new proposed method will also be introduced, which can significantly increase the simulation efficiency. A Microgrid example will be presented, and the results are compared with those of brute-force time domain simulation such as PSCAD/EMTDC, demonstrating the validity of the proposed method.

I. INTRODUCTION

Driven by environmental concerns and fluctuations of electric energy cost, the integration of distributed generators (DGs) into distribution networks close to the loads, has gained its importance and emerged as complementary to the conventional central power plants. To integrate these DGs or distributed resources with the grid, interfacing power converters are required. However, due to the switching natures, power converters generate harmonics, which may have detrimental effects on the system. It is therefore essential to predict how harmonics generated by these power converters and to quantify the distortions in voltage and current waveforms at various locations in the power network. There are, in general, three approaches of obtaining an admittance matrix of a power converter. The first approach is done by directly converting the differential equations of a converter into harmonic algebraic equations [1]. However, as shown in [2], the accuracy of this method (called harmonic domain or HD method) depends on the numbers of harmonics being included during the calculation. The second approach is proposed by Lehn and Lian [2]. They showed how to obtain the admittance matrix of a voltage source converter (VSC) [3] directly in the time domain by solving the differential equations piecewise-linearly, and treat the harmonics of interest as state variables [2]. This method, (called the fast time domain or FTD method) is more efficient than the brute-force time domain method and is very accurate. However, the computation time required by the FTD method increases as the switching index (m_f) [4] increases. This is because it requires to evaluate the exponential term, for each switching interval, and the number of switching intervals increases as m_f increases. The third approach is to derive a harmonic domain transfer function (HDTF). The transfer function method leads

to a model useful for both dynamic and steady-state analysis. There are in general two methods of obtaining the HDTF for power converters. One is based on the Floquet theory [5], [6], which essentially converts a time-varying system into a time-invariant system by a coordinate change. However, since the coordinate transformations are usually not easily obtained [5], this method has not been widely used for modeling power converters. The second method is based on the harmonic state space (HSS) theory, proposed by Wereley [7], and has been used for modeling the Thyristor controlled reactor (TCR) [8], DC-DC converters [9], and a diode train [10].

The purpose of this paper is to analytically model a pulse-width modulated (PWM) switch and incorporated it in the HSS method to improve the modeling accuracy of a PWM VSC, and to increase the speed of computation. As the paper shows, the proposed method is suitable for a microgrid system where multiple VSCs are employed.

The rest of the paper is arranged as follows: In Section II, the differential equations of a VSC are described. In Section III, the admittance matrix based on the HSS method is reviewed. Section IV describes the analytical model of a PWM switch and how it can be incorporated into the HSS method to obtain the admittance matrix of a VSC. In Section V, two numerical examples are presented. The first example is a system containing a single VSC. The admittance matrices of the systems are obtained by the HD, FTD and the improved method, respectively, and their computational efficiencies will be compared. In the second example, a microgrid system, which is similar to the Aichi microgrid system [11], and contains eight VSCs, is studied. The admittance matrices of the microgrid system are derived based on the proposed method. An arbitrary sets of input harmonics is applied to the microgrid system, and the corresponding output harmonics are compared with those obtained by PSCAD/EMTDC (a brute-force time domain simulator), HD and FTD methods. All the results are in great consistency, demonstrating the validity of the proposed method. Finally, a conclusion is given in Section VI.

II. ADMITTANCE OF A VOLTAGE SOURCE CONVERTER

For illustration purpose, the admittance matrix of a VSC is derived. Fig. 1 shows the schematic diagram of a VSC. Among these power converters, VSCs are probably the most widely used. This is mainly due to the advances in the voltage and

978-1-4799-2706-7/14 $31.00 © 2014 IEEE

current ratings of self-commutated semi-conductor switches, such as insulated gate bipolar (IGBTs) and gate turn-off thyristors (GTOs). In a smart grid or microgrid system, VSCs with dc-dc converters [12], [13] or back-to-back VSCs [14], [15] are commonly used as interfaces for distributed energy sources such as solar or wind energy conversion systems. Equation (1) is the differential equation [2] describing Fig. 1.

$$\dot{x} = A(t)x + Bu \tag{1}$$

where

$$x = \begin{bmatrix} i_a \\ i_b \\ i_c \\ v_{dc} \end{bmatrix},$$

$$A(t) = \begin{bmatrix} -\dfrac{R_{aa}}{L_t} & -\dfrac{R_{ab}}{L_t} & -\dfrac{R_{ac}}{L_t} & \dfrac{S'_a}{L_t} \\ -\dfrac{R_{ba}}{L_t} & -\dfrac{R_{bb}}{L_t} & -\dfrac{R_{bc}}{L_t} & \dfrac{S'_b}{L_t} \\ -\dfrac{R_{ca}}{L_t} & -\dfrac{R_{cb}}{L_t} & -\dfrac{R_{cc}}{L_t} & \dfrac{S'_c}{L_t} \\ \dfrac{S_a(t)}{C} & \dfrac{S_b(t)}{C} & \dfrac{S_c(t)}{C} & \dfrac{-1}{R_{dc}C} \end{bmatrix},$$

$$B = \begin{bmatrix} \dfrac{L_b+L_c}{L_t} & -\dfrac{L_c}{L_t} & -\dfrac{L_b}{L_t} & 0 \\ -\dfrac{L_c}{L_t} & \dfrac{L_a+L_c}{L_t} & -\dfrac{L_a}{L_t} & 0 \\ -\dfrac{L_b}{L_t} & -\dfrac{L_a}{L_t} & \dfrac{L_a+L_b}{L_t} & 0 \\ 0 & 0 & 0 & \dfrac{1}{C} \end{bmatrix},$$

$$L_t = L_aL_b + L_bL_c + L_aL_c,$$

$$R_{aa} = R_a(L_b + L_c) + R_bL_c,$$

$$R_{ab} = 0, R_{bc} = 0, R_{ca} = 0,$$

$$R_{ac} = (R_bL_c - R_cL_b),$$

$$R_{ba} = (R_cL_a - R_aL_c),$$

$$R_{bb} = R_b(L_a + L_c) + R_cL_a,$$

$$R_{cb} = (R_aL_b - R_bL_a),$$

$$R_{cc} = R_c(L_a + L_b) + R_aL_b,$$

$$S'_a = -(L_b + L_c)S_a(t) + L_cS_b(t) + L_bS_c(t),$$

$$S'_b = -(L_a + L_c)S_b(t) + L_cS_a(t) + L_aS_c(t),$$

$$S'_c = -(L_a + L_b)S_c(t) + L_bS_a(t) + L_aS_b(t).$$

Note that $S_a(t)$, $S_b(t)$, and $S_c(t)$ are the periodic switching functions, which take the values of 1 or 0 to represent switches on or off, respectively. The admittance matrix of a VSC can be obtained from (1). Section III describes how HSS method can apply to (1) to obtain such a matrix.

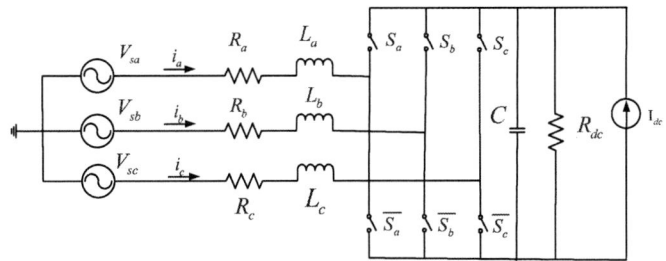

Fig. 1. The schematic diagram of a VSC

III. HSS METHOD

The frequency (or harmonic) response of a linear time invariant (LTI) system can be easily described by its transfer function. A transfer function can be found by inputting the complex exponential signal,

$$u(t) = u_o e^{j\omega t}. \tag{2}$$

The corresponding response is also a complex exponential signal of the same frequency but with possibly different amplitude and phase. However, such a signal does not lead to such a convenient frequency response if it is fed to a linear time periodic (LTP) system. This can be illustrated by a simple LTP system as shown in Fig. 2. As the figure indicates, the output signal is the sum of two complex exponentials with frequencies different from the input.

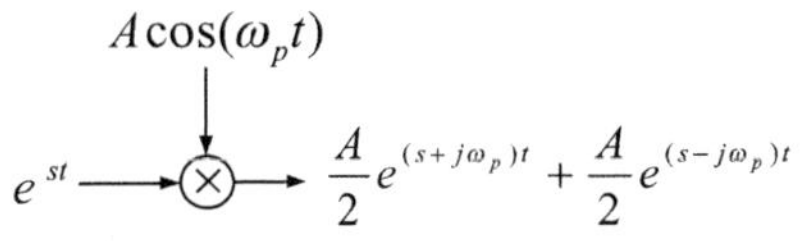

$$e^{st} \longrightarrow \otimes \longrightarrow \frac{A}{2}e^{(s+j\omega_p)t} + \frac{A}{2}e^{(s-j\omega_p)t}$$

$$A\cos(\omega_p t)$$

Fig. 2. A simple LTP system to illustration the input and output relationship

In 1991, N. M. Wereley [7] elegantly circumvented this problem by inputting the complex exponentially modulated periodic (CEMP) signal, instead of a complex exponential signal, to a LTP system, and the output will also be a CEMP signal. A CEMP signal is defined as

$$u(t) = e^{st} \sum_{n=-\infty}^{\infty} u_n e^{jn\omega_p t}. \tag{3}$$

Consequently, a LTP transfer function, also named harmonic transfer function (HTF) can be derived and describes its frequency response. According to [7], the HTF of a periodic system, described by (4) and (5).

$$\dot{x}(t) = A(t)x(t) + B(t)u(t) \tag{4}$$
$$y(t) = C(t)x(t) + D(t)u(t), \tag{5}$$

is found to be

$$H(s) = C_{TP}\left[sI - (A_{TP} - N)\right]^{-1}B_{TP} + D_{TP} \tag{6}$$

where C_{TP}, and D_{TP} are the complex Fourier coefficients of $C(t)$, and $D(t)$ expressed in Toeplitz form. N is a diagonal

matrix having the form of

$$
\begin{bmatrix}
\ddots & & & & & \\
& 2j\omega_p & & & & \\
& & j\omega_p & & & \\
& & & 0 & & \\
& & & & -j\omega_p & \\
& & & & & -2j\omega_p \\
& & & & & & \ddots
\end{bmatrix}
$$

where ω_p is the period of the system.

Since the differential equation of a VSC has the form of (4), its transfer function can be found from (6), and the admittance matrix can be found by setting $s = j\omega$.

The drawback of HD, FTD and HSS methods is that switching times are needed to be solved in order to define the switching functions. For naturally PWM [16], iterative methods are required to solve the switching times. Consequently, the calculation time increases as the m_f increases because the number of switching times need to be solved increases with m_f. The accuracy of both HD and HSS methods depends on the numbers of harmonics being included during the calculation. In general, the accuracy increases as the number of harmonic terms increases. As a rule of thumb, the number of harmonics being included should be more than three times the switching frequency [2]. On the other hand, the accuracy of the FTD method does not depend on the number of harmonics being included. However, it requires to evaluate the exponential term for each switching interval, and the number of switching intervals increases as m_f increases. HD method can be considered as a special case of HSS method because they are equal when s is set to $j\omega$. Moreover, since HSS is defined in terms of s, it can also capture phenomena associated with transients and interhamonics. Table I is the comparison chart among these three methods.

TABLE I. THE SWITCHING TIMES AND INITIAL CONDITIONS AT THE OPERATING POINT

	HD	FTD	HSS
accuracy depends on number of harmonics being included	Yes	No	Yes
the need to evaluate integration or exponential matrices	No	Yes	No
only capable of dealing with integer number of harmonics	Yes	Yes	No

IV. PROPOSED METHOD

The computational time of obtaining the admittance matrices is influenced by two problems — the inherent problems and numerical ones. The inherent problems are essentially the nature of the methods, and yields little room for improvements. On the other hand, problems associated with numerical methods can be eliminated by better or smarter techniques. For instance, the need for iteratively solving for switching times to define the switching functions can be eliminated if one can obtain the analytical expression of the switching functions. Moreover, the model accuracy will be enhanced as it avoids

taking Fast Fourier Transform, which may result in aliasing or leakage problems [17].

The analytical expression of the PWM switching function can be obtained by the method presented in [18] and [19]. In [19] and [20], it has been shown that a pulse train can be obtained either by a 2-D model (Fig. 3(a)) or 3-D model (Fig. 3(b)).

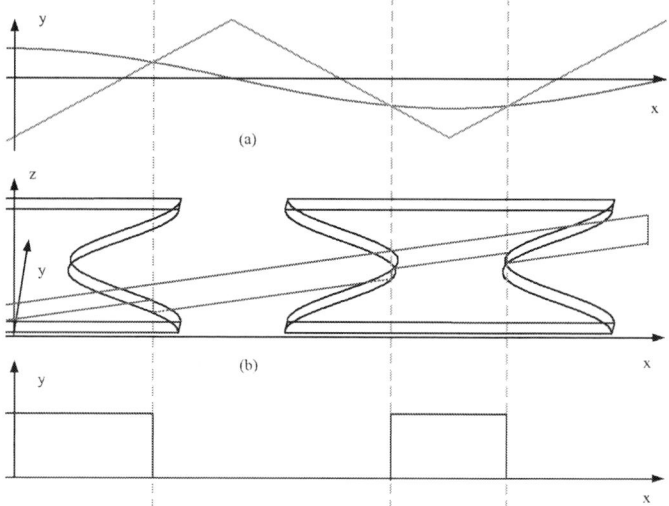

Fig. 3. (a) 2-D model of the PWM modulation; (b) 3-D model of the PWM modulation

For the 2-D model, the modulating signal (v_m) is compared with the carrier waveform (v_c). The intersection points form the switching edge of the pules. In the 3-D model, v_m with a unit height in the z-direction is repeatedly drawn on the x-y plane, running parallel to the y-axis. The repetition period in both x ($x = \omega_c t$, where ω_c is the carrier frequency) and y ($y = \omega_m t$, where ω_m is the modulating frequency) direction is 2π. To generate the PWM signal, a plane with slope of $\frac{\omega_m}{\omega_c}$ is traced perpendicularly to the x-y plane, and is intersected with the modulating-signal wall. The intersection points coincide with those generated in the 2-D model, as shown in Fig. 3. Since the resulting signal has a period of 2π in both x and y directions, it can be represented as a double Fourier series of the form [21]:

$$
\begin{aligned}
F(x, y) = {} & \frac{A_{00}}{2} + \sum_{n=1}^{\infty} A_{0n} \cos(ny) + B_{0n} \sin(ny) + \\
& \sum_{m=1}^{\infty} A_{m0} \cos(mx) + B_{m0} \sin(mx) + \\
& \sum_{m=1}^{\infty} \sum_{n=\pm\infty}^{\pm 1} [A_{mn}(mx + ny) + B_{mn} \cdot \\
& \sin(mx + ny)] \quad (7)
\end{aligned}
$$

The first term of (7) corresponds to the DC offset, the second term to the frequency components of the modulating signal, the third term to the carrier frequency and its harmonics, and the last term is the ensemble of all possible harmonics frequency formed by taking the sum and difference between the modulating and carrier waveform [20], [21].

To find out the Fourier coefficient of (7), one can look at the unit cell, indicated in dashed box in Fig. 3, which is shown in Fig. 4 and conclude that f(x,y) changes from 0 to 1 when

$$x = 2\pi p - \frac{\pi}{2}\left(1 + M\cos\omega_m t\right) \quad p = 0, 1, 2 \cdots \infty, \quad (8)$$

and f(x,y) changes from 1 to 0 when

$$x = 2\pi p + \frac{\pi}{2}\left(1 + M\cos\omega_m t\right) \quad p = 0, 1, 2 \cdots \infty, \quad (9)$$

where M is the modulation index [4].

Under the integration limits defined by (8) and(9),the Fourier coefficients can be obtained [16], [21]

$$A_{mn} + jB_{mn} = \frac{1}{\pi^2}\int_{-\pi}^{\pi}\int_{-\frac{\pi}{2}(1+M\cos y)}^{\frac{\pi}{2}(1+M\cos y)} e^{j(mx+ny)}\,dx\,dy \quad (10)$$

By applying trigonometry identities and Jacob-Anger expansion [16], one can obtain the compact form of analytical expansion of the switching functions of the PWM as shown in (11) .

$$
\begin{aligned}
S(t) \;=\;& 0.5 + 0.5M\cos\left(w_0 t + \theta_0\right) + \frac{2}{\pi}\sum_{m=1}^{\infty}\frac{1}{m}\,\cdot\\
& J_0\left(m\frac{\pi}{2}M\right)\sin\left(m\frac{2}{\pi}\right)\cos\left(m\left(w_c t + \theta_c\right)\right) +\\
& \frac{2}{\pi}\sum_{m=1}^{\infty}\sum_{\substack{n=-\infty\\(n\neq 0)}}^{\infty}\frac{1}{m}J_n\left(m\frac{\pi}{2}M\right)\sin\left[(m+n)\frac{\pi}{2}\right]\cdot\\
& \cos[m\left(\omega_c t + \theta_c\right) + n\left(\omega_0 t + \theta_0\right)] \qquad (11)
\end{aligned}
$$

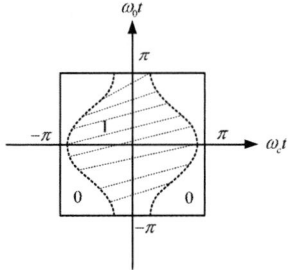

Fig. 4.   The unit cell of the modulating-signal wall

To overcome the above mentioned problems of the existing method, the proposed method is to incorporate the analytical model of the PWM switches based on (7) into the HSS model (i.e. (6)). The advantages of the proposed method are as follows:

1)   As will be shown in Section V-A, the computation time of the proposed method is much smaller than the

existing three methods. In fact, the computation time is almost independent of the switching frequency, as will be demonstrated in Fig. 5.

2)   The proposed method is very suitable for modeling a microgrid system such as Aichi microgrid system, which consists of multiple number of converters since the overall simulation time can be greatly reduced. This will be further illustrated in Section V-B.

3)   The proposed method is HSS method based and models the converters in the "s" domain. Therefore, it has the potential of handling interharmonics emitted by back-to-back converter system connecting two areas operating at different power frequency. Nevertheless, this point is not addressed in this paper. This point will be further addressed in a future paper.

V.   NUMERICAL EXAMPLES

To validate the proposed methods, two cases are studied. In the first case, a single VSC system is studied to compare the simulation time between the proposed method, HD, FTD, and brute-force time-domain methods to calculate particular low order output harmonics because these harmonics can be substantially influenced by interaction between the ac and dc networks [22]. In the second case, a microgrid system, which consists of eight VSCs for interfacing renewable resources are modeled. The output harmonics obtained by the proposed method are compared with the brute-force time domain method to show the validity of the proposed method for simulating a microgrid or smart grid system.

A.   Example 1 – Single VSC system

Fig. 1 shows the schematic diagram of a VSC whose parameters values are extracted from [23]. A set of test input stimuli, as given in Table II, is applied to the admittance matrices of the VSC to investigate how much the low order odd current harmonics will be injected into the system. Note that the superscript of the voltage stimuli in Table II represents the order of harmonics.

TABLE II.   THE PARAMETER VALUES OF FIG. 1

R_a, R_b, R_c (Ω)	0.15, 0.165, 0.18
L_a, L_b, L_c (mH)	0.637, 0.707, 0.7644
V_{sa}^1, V_{sb}^1, V_{sc}^1 (V_{rms})	$110\angle 0°, 110\angle -120°, 110\angle 120°$
V_{sa}^5, V_{sb}^5, V_{sc}^5 (V_{rms})	$8.8\angle 0°, 8.8\angle -120°, 8.8\angle 120°$
V_{sa}^7, V_{sb}^7, V_{sc}^7 (V_{rms})	$6.6\angle 0°, 6.6\angle -120°, 6.6\angle 120°$
C (μF)	4820
θ $(°)$	1.2
M	0.9
I_{dc} (A)	5
$Switching\ freq.\ (Hz)$	180 Hz

Three admittance matrices are obtained based on the HD, FTD and proposed improved methods. Fig. 5 shows the simulation time required by the three methods when m_f varies from 3 to 111. As seen from the figure,the FTD method has the highest efficiency when m_f ratio is below 13. On the other hand, when m_f is greater than 13, the proposed method becomes the most efficient method. In fact, the simulation time remains relatively constant regardless how large m_f is. On the other hand, the simulation times of the FTD and the HD methods increase as m_f increases due to the fact that both of them require finding the switching times. Note that the harmonic being included in the HD and the proposed methods are 50, which is to minimize the truncation error. Fig. 6 shows the admittance matrix, indicated in dashed boxes, obtained by the proposed method.

Figs. 7 to 9 show the corresponding output ac current harmonics, obtained by these models, and PSCAD/EMTDC, respectively. The results are consistent with each other, demonstrating the validity of the proposed method. Consequently, the proposed method is the best choice for simulating system containing multiple converters which operate at high frequencies.

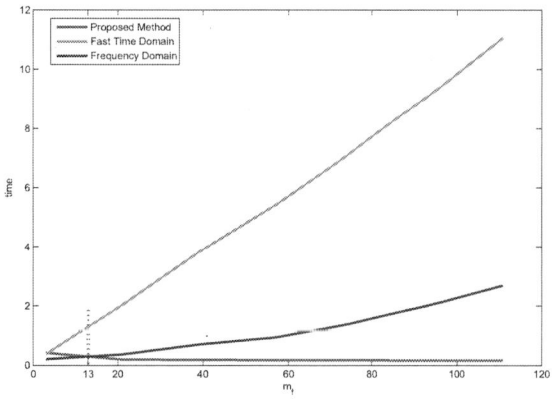

Fig. 5.   Simulation time for HD, FTD and the proposed method

B. Example 2 – A Microgrid System

The system shown in Fig. 10 is a microgrid system, similar to the Aichi microgrid system. The grid connects to eight

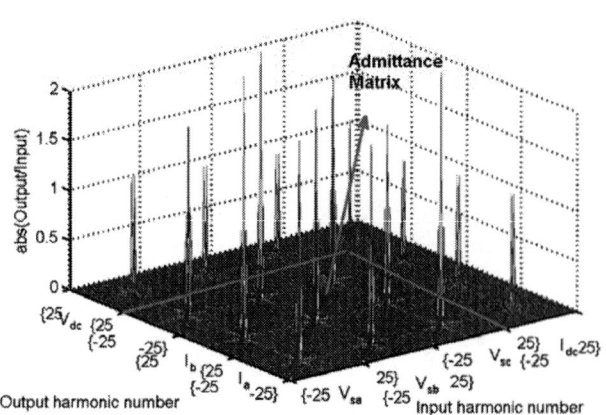

Fig. 6.   Admittance matrix obtained by the proposed method

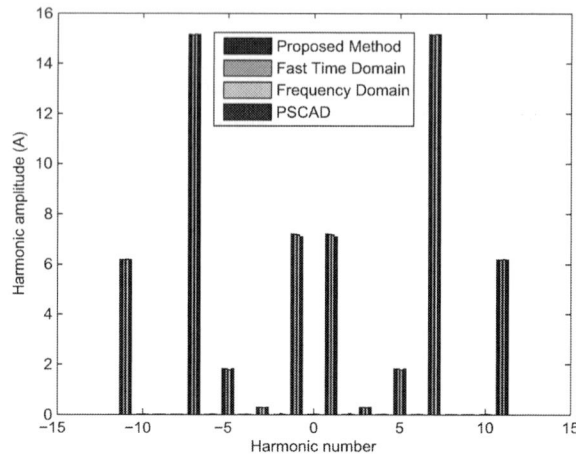

Fig. 7.   AC current harmonics (phase A)

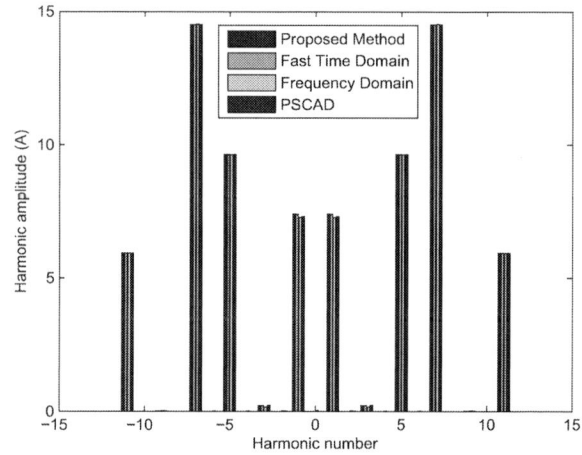

Fig. 8.   AC current harmonics (phase B)

different distributed energy resources (DER) via eight VSCs. It is assumed that there are some even harmonics existing in the dc buses of the VSCs (as indicated in Table III), which are interfaced to the DGs. These even harmonics, modulated by the VSC will in turn inject odd harmonics into the microgrid system. Moreover, it is also assumed that the ac source voltage, V_s also contains some fifth harmonics ($\frac{1}{10}$ of the fundamental component; see Table IV). Tables V and VI summarize the different operating conditions of the individual VSC. They will make PSCAD/EMTDC very hard to simulate as it requires to reduce the simulation time step to be $0.5~\mu s$ in order to obtain accurate results.

Table VII shows the computation times required by the proposed method and PSCAD/EMTDC. The proposed method only requires 6.3 seconds to obtain the results, which is about 61 times faster than the time required by PSCAD/EMTDC.

978-1-4799-2706-7/14 $31.00 © 2014 IEEE   2199

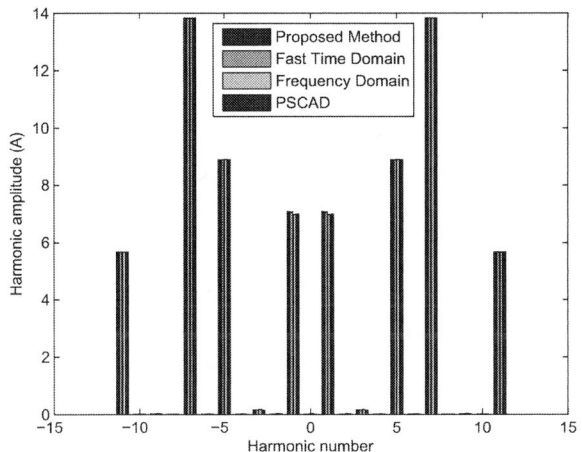

Fig. 9. AC current harmonics (phase C)

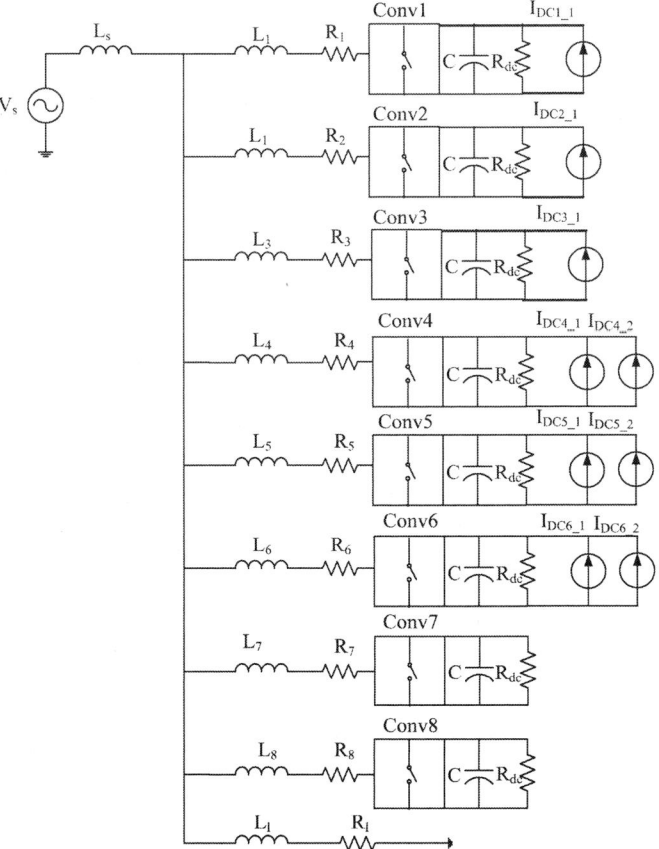

Fig. 10. Aichi microgrid

Figs. 11 and 12 show the total output current harmonics (for phase A) injected into the grid, I_T for the operating conditions specified in TABLEs V and VI, respectively. In order to

TABLE III. THE EVEN HARMONICS INJECTED BY DIFFERENT DGS

	A(rms)	Hz		A(rms)	Hz
I_{DC1_1}	$0.76\angle0°$	120	I_{DC5_1}	$1.14\angle0°$	120
I_{DC2_1}	$0.6\angle0°$	120	I_{DC5_2}	$0.6\angle0°$	240
I_{DC3_1}	$0.38\angle0°$	120	I_{DC6_1}	$1.52\angle0°$	120
I_{DC4_1}	$0.76\angle0°$	120	I_{DC6_2}	$0.76\angle0°$	240
I_{DC4_2}	$0.38\angle0°$	240			

TABLE IV. THE PARAMETER VALUES OF FIG. 10

$V_{sa}^1, V_{sb}^1, V_{sc}^1$	$220\angle0°, 220\angle-120°, 220\angle120°\ Vrms$
$V_{sa}^5, V_{sb}^5, V_{sc}^5$	$11\angle0°, 11\angle-120°, 11\angle120°\ Vrms$

TABLE V. THE FIRST OPERATING CONDITION

	m_f	M	θ (degree)
$Converter_1$	3	0.9	-1.1
$Converter_2$	3	0.875	-1.2
$Converter_3$	9	0.85	-1.3
$Converter_4$	9	0.825	-1.4
$Converter_5$	15	0.8	1.1
$Converter_6$	15	0.775	1.2
$Converter_7$	111	0.75	1.3
$Converter_8$	333	0.725	1.4

TABLE VI. THE SECOND OPERATING CONDITION

	m_f	M	θ (degree)
$Converter_1$	15	0.7	-1.5
$Converter_2$	15	0.8	1.5
$Converter_3$	15	0.9	-1.1
$Converter_4$	33	0.7	1.1
$Converter_5$	33	0.8	-1.4
$Converter_6$	111	0.85	1.4
$Converter_7$	111	0.9	-2.1
$Converter_8$	333	0.75	2.1

show each harmonic more clearly, all the harmonics in the figures are plotted in log scale. As indicated in the figures, the values predicted by the proposed method are consistent with those obtained by PSCAD/EMTDC. The proposed method has proven to be very efficient and accurate, and can be used to evaluate the harmonics injected by a microgrid or to investigate the resonant point in a microgrid. As demonstrated in the case studies, as long as the operating condition of the converter is known, regardless of control methods being used, the output harmonics can be easily obtained.

TABLE VII. THE COMPUTATION TIME COMPARISON

	computation time (sec)
Proposed Method	7.5
PSCAD/EMTDC	454

VI. CONCLUSIONS

This paper has achieved three goals. Firstly, the state-of-the-art methods (HD, FTD, and HSS methods) for modeling power converters are reviewed. Secondly, a method for improving HSS method is proposed. Essentially, an analytical method of modeling a PWM switch is introduced and incorporated in the HSS method to significantly improve the computation time. Finally, the computational speed of all the available methods are compared, and we concluded that for $m_f < 13$,

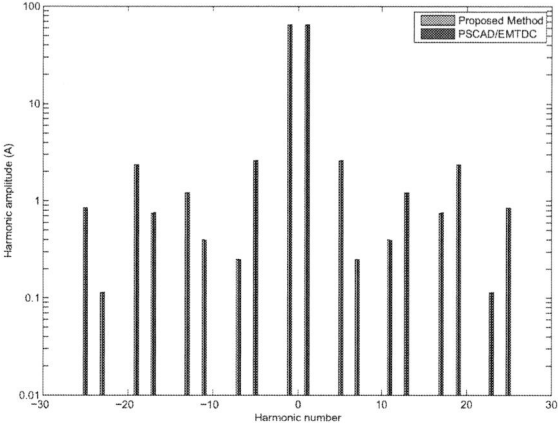

Fig. 11. The total current harmonics injected into the grid (phase A) for operating condition in TABLE V

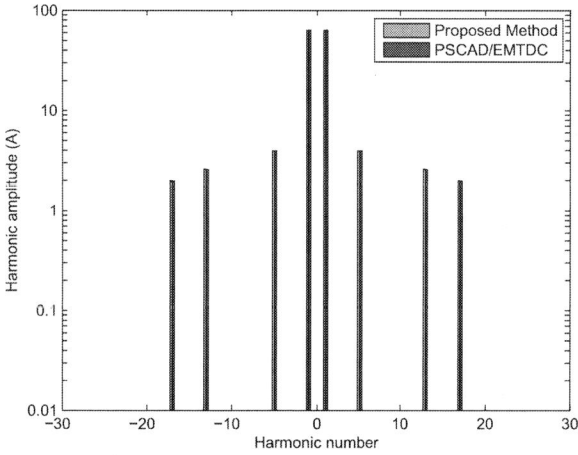

Fig. 12. The total current harmonics injected into the grid (phase A) for operating condition in TABLE VI

FTD method is the most efficient whereas for $m_f > 13$, the proposed improved method is the most efficient. This is very useful for power engineers when undertaking the tasks of simulation studies and system planning of a smart grid or microgrid system.

REFERENCES

[1] E. Acha and M. Medina, *Power Systems Harmonics: Computer Modeling and Analysis*, 1st ed. Wiley & Sons, 2001.

[2] P. Lehn and K. Lian, "Frequency coupling matrix of a voltage-source converter derived from piecewise linear differential equations," *IEEE Transactions on Power Delivery*, vol. 22, no. 3, pp. 1603 –1612, july 2007.

[3] N. G. Hingorani and L. Gyugyi, *Understanding FACTS: concepts and technology of flexible ac transmission systems*. New York: IEEE Press, 2001.

[4] N. Mohan, T. Undeland, and W. Robins, *Power Electronics: Converters, Applications, and Design, 2nd edition*. John Wiley & Sons, 1995.

[5] A. Semlyen and A. Medina, "Computation of the periodic steady state in systems with nonlinear components using a hybrid time and frequency domain methodology," *IEEE Transactions on Power Systems*, vol. 10, no. 3, pp. 1498 –1504, aug 1995.

[6] A. Semlyen, "s-domain methodology for assessing the small signal stability of complex systems in nonsinusoidal steady state," *IEEE Transactions on Power Systems*, vol. 14, no. 1, pp. 132 –137, feb 1999.

[7] N. Wereley, *Analysis and Control of Linear Periodically Time Varying Systems*. MIT Phd. dissertation, 1990.

[8] J. Orillaza and A. Wood, "Reduced harmonic state space model of tcr," in *14th International Conference on Harmonics and Quality of Power*, sept. 2010, pp. 1 –5.

[9] G. Love and A. Wood, "Harmonic state space model of power electronics," in *ICHQP 2008*, 28 2008-oct. 1 2008, pp. 1 –6.

[10] E. Mollerstedt and B. Bernhardsson, "Out of control because of harmonics-an analysis of the harmonic response of an inverter locomotive," *Control Systems, IEEE*, vol. 20, no. 4, pp. 70 –81, aug. 2000.

[11] N. Lidula and A. Rajapakse, "Microgrids research: A review of experimental microgrids and test systems," *Renewable and Sustainable Energy Reviews*, vol. 15, no. 1, pp. 186 – 202, 2011.

[12] R.-J. Wai, W.-H. Wang, and C.-Y. Lin, "High-performance standalone photovoltaic generation system," *IEEE Transactions on Industrial Electronics*, vol. 55, no. 1, pp. 240 –250, jan. 2008.

[13] L. Gertmar, P. Karlsson, and O. Samuelsson, "On dc injection to ac grids from distributed generation," in *European Conference on Power Electronics and Applications*, 0-0 2005, pp. 10 pp. –P.10.

[14] T. Friedli, S. Round, D. Hassler, and J. Kolar, "Design and performance of a 200-khz all-sic jfet current dc-link back-to-back converter," *IEEE Transactions on Industry Applications*, vol. 45, no. 5, pp. 1868 –1878, sept.-oct. 2009.

[15] M. Noroozian, A.-A. Edris, D. Kidd, and A. Keri, "The potential use of voltage-sourced converter-based back-to-back tie in load restorations," *IEEE Transactions on Power Delivery*, vol. 18, no. 4, pp. 1416 – 1421, oct. 2003.

[16] D. G. Holmes and T. A. Lipo, *Pulse Width Modulation for Power Converters - Principels and Practice*, 1st ed. IEEE, 2003.

[17] A. Girgis and F. Ham, "A quantitative study of pitfalls in the fft," *IEEE Transactions on Aerospace and Electronic Systems*, vol. AES-16, no. 4, pp. 434 –439, july 1980.

[18] W. R. Bennett, "New results in the calculation of modulation products," *The Bell System Technical Journal*, vol. 12, pp. 228 –243, april 1993.

[19] S. R. Bowes, "Novel approach to the analysis and synthesis of modulation processes in power converters," *IEE Proceedings (London)*, vol. 122, no. 5, pp. 507 – 513, may 1975.

[20] G. Chang, H.-W. Lin, and S.-K. Chen, "Modeling characteristics of harmonic currents generated by high-speed railway traction drive converters," *IEEE Transactions on Power Delivery*, vol. 19, no. 2, pp. 766 – 773, april 2004.

[21] I. Deslauriers, N. Avdiu, and B. T. Ooi, "Naturally sampled triangle carrier pwm bandwidth limit and output spectrum," *IEEE Transactions on Power Electronics*, vol. 20, no. 1, pp. 100 –106, jan. 2005.

[22] E. Larsen, D. Baker, and J. McIver, "Low-order harmonic interactions on ac/dc systems," *IEEE Transactions on Power Delivery*, vol. 4, no. 1, pp. 493 –501, jan 1989.

[23] C. Sao, P. Lehn, M. Iravani, and J. Martinez, "A benchmark system for digital time-domain simulation of a pulse-width-modulated d-statcom," *IEEE Transactions on Power Delivery*, vol. 17, no. 4, pp. 1113 – 1120, oct 2002.

FPGA-Based Simulation of Power Electronics Using Iterative Methods

Huiguo Zhang
School of Physics and Electronic Engineering
Changshu Institute of Technology
Changshu, 215500, China
E-mail: hweigo@cslg.edu.cn

Jian Sun
Center for Future Energy Systems
Rensselaer Polytechnic Institute
Troy, NY 12180-3590, USA
E-mail: jsun@ecse.rpi.edu

Abstract – **Real-time (RT) simulation is increasingly used in power electronics and power engineering as a tool to replace actual hardware for use in the development and verification of control and complex system designs. New computing architectures and algorithms, such as parallel and distributed computing, hold the promise to meet the speed requirement of such RT simulation. Due to its inherent parallelism at the hardware layer, field-programmable gate array (FPGA) provides an attractive alternative to conventional computing devices such as CPUs and GPUs for RT simulation. On the other hand, much of this advantage is lost when an FPGA is simply used as replacement for a CPU or GPU to execute existing simulation routines. This paper presents our preliminary work on the development of alternative circuit simulation algorithms that best exploit the parallelism and other unique features of FPGA for real-time simulation of power electronic converters and systems. Instead of solving the discretized circuit equations directly by inverting the coefficient matrix, we construct a signal flow diagram corresponding to solving the discretized model by the Jacobi iterative method. By exploiting the time-scale separation properties of typical power electronics circuits and properly selecting the discretizing time step, the response of the signal flow diagram can be made to converge to the original discretized circuit model response within a few iterations. Simulation of buck converter with average current control is used to illustrate and test the proposed method.**

Keywords - Real-Time Simulation, FPGA, Iterative Methods, Simulation

I. INTRODUCTION

Real-time (RT) simulation has been used in the design and testing of electrical power systems such as utility grids and automotive applications. It eliminates the need for actual hardware and helps to reduce the cost and improve design reliability by facilitating extensive testing under different conditions. For simulation of power electronics circuits and systems, the high switching frequency is a major barrier for real-time simulation as very small time step is required to accurately capture the switching transitions [1-3]. Existing RT simulation platforms usually limits the minimum time step to about 10 μs, which is too large for most power electronics circuits.

Field-Programmable Gate Array (FPGA) has the advantages of parallelism at the hardware layer. This coupled with the high clock frequency makes FPGA an attractive alternative for RT simulation applications.

Several papers have dealt with modeling of power electronics circuits for simulation in FPGA and proposed the use of different switch models, such as ideal switch, ADC-based model, device level behavior model, state space switch model [4-7]. Based on the switch model selected, the traditional method is to use nodal analysis to form a set of ordinary differential equations of the circuit under each switching state. Each set of differential equations are then discretized using a selected numerical integration method, such as Euler method and trapezoidal method [4, 6]. The resulting difference equations can be solved by using pre-calculated inverse matrix of the difference equations, as proposed in [4, 7]. However, inverting the coefficient matrix method directly or indirectly using traditional iteration methods requires intensive computation that is not best suited for FPGA implementation. Additionally, the advantage of FPGA due to its inherent ability of distributed and parallel computation is lost when a solution algorithm is implemented based on pure mathematical formulation [8].

This paper presents our preliminary work on the development of alternative circuit simulation algorithms that best exploit the parallelism and other unique features of FPGA for real-time simulation of power electronic converters and systems. Instead of solving the discretized circuit equations directly by inverting the coefficient matrix, we construct a signal flow diagram corresponding to solving the discretized model by the Jacobi iterative method. By exploiting the time-scale separation properties of typical power electronics circuits and properly selecting the discretizing time step, the response of the signal flow diagram can be made to converge to the original discretized circuit model response within a few iterations. The paper is organized as follows. In section II, discretized circuit models are developed for an example buck converter with average current-mode control by using an ideal switch model and trapezoidal discretization method. Section III discuss the solution of the discretized model and investigates the effects of time step on the convergence of an iterative solution method. Section IV presents details of the implemented simulation model, and Section V compares the resulting RT simulation results with offline simulation in SABER. Section VI concludes the work.

II. MODELING OF CIRCUITS

Fig. 1 shows the schematic of a buck converter with an

average current controller used for the study of real-time simulation in this paper.

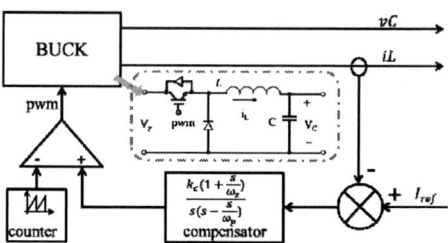

Fig.1. Close loop buck converter

A. Open Loop Buck Converter Model

Eq. (1)-(4) shows the state-space equation for buck converter, where first two-equations are for the ON state of the switch and the remaining two are for the OFF state.

$$\frac{di_L(t)}{dt} = -\frac{v_C}{L} \tag{1}$$

$$\frac{dv_C(t)}{dt} = \frac{i_L(t)}{C} - \frac{v_C(t)}{RC} \tag{2}$$

$$\frac{di_L(t)}{dt} = \frac{v_g - v_C(t)}{L} \tag{3}$$

$$\frac{dv_C(t)}{dt} = \frac{i_L(t)}{C} - \frac{v_C(t)}{RC} \tag{4}$$

Trapezoidal discretization is most commonly used numerical method for solving ODEs and is at the core of most SPICE solvers. Trapezoidal rule (TR) is an implicit method with $O(h^3)$ error and it provides excellent stability. It is suitable for stiff systems, and has better accuracy than both Forward Euler method and Backward Euler method [9].

As for the ODE system: $x'(t) = f(x, t)$, where $x(t_0) = x_0$, the state vector $x(t_{n+1})$ can be obtained based on the previous step value $x(t_n)$, as $x(t_{n+1}) = x_n + \frac{h}{2}[f(x_{n+1}, t_{n+1}) + f(x_n, t_n)]$, where h is the time step interval.

Discretized derivative of state variables for buck converter using trapezoidal method are given by (5)-(6).

$$i_L(t_{n+1}) - i_L(t_n) = \frac{\Delta t}{2}(i_L'(t_{n+1}) + i_L'(t_n)) \tag{5}$$

$$v_c(t_{n+1}) - v_c(t_n) = \frac{\Delta t}{2}(v_c'(t_{n+1}) + v_c'(t_n)) \tag{6}$$

Using above discrete representation of state-variable derivatives, the state-space model of (1)-(4) can be represented in discrete form by (7)-(10).

$$i_L(t_{n+1}) + \frac{\Delta t}{2L}v_C(t_{n+1}) = i_L(t_n) - \frac{\Delta t}{2L}v_C(t_n) \tag{7}$$

$$-\frac{\Delta t}{2C}i_L(t_{n+1}) + \left(1 + \frac{\Delta t}{2RC}\right)v_C(t_{n+1})$$
$$= \frac{\Delta t}{2C}i_L(t_n) + v_C(t_n)\left[1 - \frac{\Delta t}{2RC}\right] \tag{8}$$

$$i_L(t_{n+1}) + \frac{\Delta t}{2L}v_C(t_{n+1})$$
$$= \Delta t * \frac{V_g}{L} + i_L(t_n) - \frac{\Delta t}{2L}v_C(t_n) \tag{9}$$

$$-\frac{\Delta t}{2C}i_L(t_{n+1}) + \left(1 + \frac{\Delta t}{2RC}\right)v_C(t_{n+1})$$
$$= \frac{\Delta t}{2C}i_L(t_n) + v_C(t_n)\left[1 - \frac{\Delta t}{2RC}\right] \tag{10}$$

B. Compensator Model

The buck converter uses the average current control method using a PI compensator with a high-frequency pole to attenuate switching noise from duty-ratio, as shown in (11).

$$H(s) = \frac{1}{s} \cdot \frac{1 + \frac{s}{\omega_z}}{1 + \frac{s}{\omega_p}} \tag{11}$$

For implementation of compensator if (11) on digital platform it needs to be discretized. Impulse invariance and bilinear transformation are standard methods for the discretization of continuous domain transfer function. As the impulse invariance method has frequency mixing problem, it requires high sampling rate. While bilinear method has no such problem and it is equivalent to the Trapezoidal method and is employed here. The discretized version of compensator is given by (12).

$$H(z) = \frac{T}{2} \cdot \frac{1 + z^{-1}}{1 - z^{-1}} \cdot \left(\frac{1 - k_z}{1 - k_p}\right)\left(\frac{\frac{1 + k_z}{1 - k_z} - \frac{1 + k_p}{1 - k_p}}{\frac{1 + k_p}{1 - k_p} + z^{-1}} + 1\right) \tag{12}$$

where $k_z = \frac{2}{\omega_z T}$, $k_p = \frac{2}{\omega_p T}$, T is sample period.

PWM signals for switching are produced by comparing the current feedback signal with the saw tooth waveform produced by an incremental counter.

III. ITERATION METHODS AND CONVERGENCE CONSTRAINTS

The objective of the simulation is to solve (7)-(10), which is corresponds to the ON and OFF states of the switch. The matrix form of the circuit model is given by (5), where S represents the switching state, which is one when the switch is ON and zero otherwise.

$$\begin{bmatrix} 1 & \frac{\Delta t}{2L} \\ -\frac{\Delta t}{2C} & \left(1 + \frac{\Delta t}{2RC}\right) \end{bmatrix}\begin{bmatrix} i_L(t_{n+1}) \\ v_C(t_{n+1}) \end{bmatrix} =$$
$$\begin{bmatrix} 1 & -\frac{\Delta t}{2L} \\ \frac{\Delta t}{2C} & \left(1 - \frac{\Delta t}{2RC}\right) \end{bmatrix}\begin{bmatrix} i_L(t_n) \\ v_C(t_n) \end{bmatrix} + \begin{bmatrix} S \\ 0 \end{bmatrix} * \Delta t * \frac{V_g}{L} \tag{13}$$

At $n+1$ time step, the right side of (13) is the constant values, we got the linear algebraic equations with the form of $Ax = b$, where

$$A = \begin{bmatrix} 1 & \frac{\Delta t}{2L} \\ -\frac{\Delta t}{2C} & \left(1 + \frac{\Delta t}{2RC}\right) \end{bmatrix},$$

$$b = \begin{bmatrix} 1 & -\frac{\Delta t}{2L} \\ \frac{\Delta t}{2C} & \left(1 - \frac{\Delta t}{2RC}\right) \end{bmatrix}\begin{bmatrix} i_L(t_n) \\ v_C(t_n) \end{bmatrix} + \begin{bmatrix} S \\ 0 \end{bmatrix} * \Delta t * \frac{V_g}{L}$$

978-1-4799-2706-7/14 $31.00 © 2014 IEEE

The characteristics of matrix A determine the solution of the equation. It is shown in [10] that if the state-transition matrix A is strictly diagonally dominant matrix, then the Jacobi iteration methods will be convergent. The inductance L and capacitance C of buck converter circuit are the design parameters and beyond control for simulation, however, if time step Δt is chosen properly, matrix A can be made diagonally dominant matrix. Specifically, in our situation, Δt is chosen such that $\frac{\Delta t}{2L} < 1$ and $\frac{\Delta t}{2C} < 1$, the Jacobi iteration method can be used to solve the equation (13).

IV. IMPLEMENTATION

A. Open Loop Buck converter

Eq. (7)-(10) describe the discretized circuit model and it can be represented in the Z-domain by (14)-(17). Notice that (15) and (17) are same.

$$v_C(z) + v_C(z) \cdot z^{-1} = \frac{2L}{\Delta t}[-i_L(z) + i_L(z) \cdot z^{-1}] \quad (14)$$

$$i_L(z) + i_L(z) \cdot z^{-1} = \frac{2C}{\Delta t}\left(1 + \frac{\Delta t}{2RC}\right)$$
$$\cdot v_C(z) - \frac{2C}{\Delta t}\left(1 - \frac{\Delta t}{2RC}\right) \cdot v_C(z) \cdot z^{-1} \quad (15)$$

$$v_C(z) + v_C(z) \cdot z^{-1}$$
$$= \frac{2L}{\Delta t}[-i_L(z) + i_L(z) \cdot z^{-1}] + 2V_g \quad (16)$$

$$i_L(z) + i_L(z) \cdot z^{-1} = \frac{2C}{\Delta t}\left(1 + \frac{\Delta t}{2RC}\right) \cdot v_C(z) -$$
$$\frac{2C}{\Delta t}\left(1 - \frac{\Delta t}{2RC}\right) \cdot v_C(z) \cdot z^{-1} \quad (17)$$

These equations can be simulated by the signal block diagram independently. Fig. 2 shows the signal block diagram for simulation of (14). Similar diagram for (15) and (16) is shown in Fig. 3. Fig. 4 corresponds to the implementation of (17).

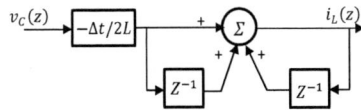

Fig.2. Block diagram of equation (14)

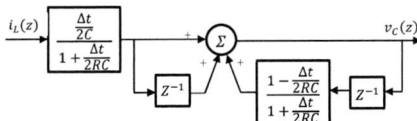

Fig.3. Block diagram of equation (15)(16)

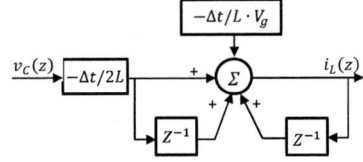

Fig.4. Block diagram of equation (17)

To solve the circuit efficiently, we proposed a circular feedback method. Fig.5 is the signal block diagram for the open loop buck converter based on the proposed method. For the evaluation of state variables, iL and vC,

at any step, their values from previous time step are fed directly to the input, forming the circular signal block diagram. It is verified that such implementation is equivalent to the Jacobi iterative method.

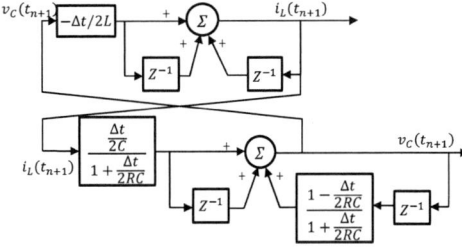

Fig.5. Circular feedback Block diagram

Fig.6 shows the implementation in Matlab. As equations (14) and (15) are describing switch off state, their block diagrams can be lumped to one block as indicated in the block B of Fig.6. Similarly, blocks for switch on state equations can also be lumped into one block. In buck converter case, the coefficient matrix is the same whenever the switch is on or off. This gives additional simplification to the solver. In Fig.6, the block A is the multiplexer served for switch purpose.

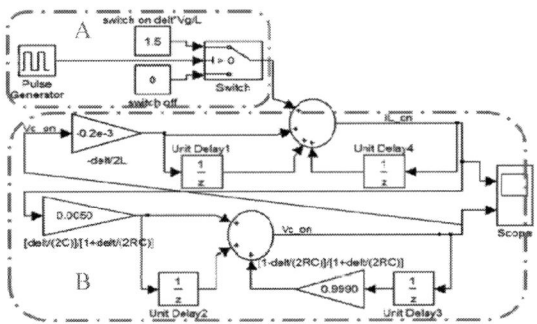

Fig.6. Open loop buck converter

The computing speed can be improved by the pipeline method. The state can be solved in one stage with the time of longest time delay between two neighbor registers.

B. Overall Signal Block Diagram

The compensator described by equation (12) can be solved by the diagram in Fig.7. The PWM can be implemented by counter and comparator.

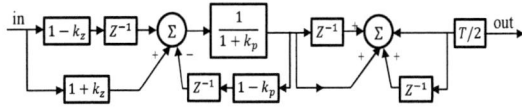

Fig.7. Block diagram of compensator

With the model of open loop converter, the compensator and PWM module are also implemented with signal block diagram. Fig.8 is the overall signal block diagram. We use this diagram to automatically generate the needed Hardware Description Language

(HDL) Code. Then, HDL code is used to generate the corresponding hardware in FPGA.

Fig.9 shows the design procedure using signal flow diagram. The first step is to use the Matlab/Simulink for the verification of model, and to implement it on Matlab/DSP Builder. In the second step, the HDL code is automatically generated and verified in the HDL simulation environment. These steps from the algorithm to HDL generation are falls under the unified model-based procedure. With this approach, the time consuming coding work is greatly reduced and simulation on FPGA can be easily realized.

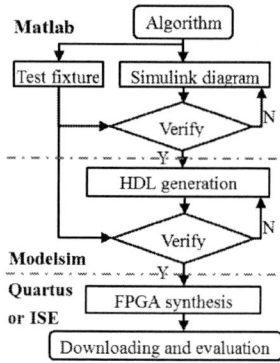

Fig.9. Model based design procedure

Fig.10 shows the overall implementation of the close loop buck converter in Quartus II environment. There are two principal modules in the design. One is frequency divider which provides the 1MHz clock, the other is close loop buck converter. The output signals are clock signal, PWM signal, 10-bit digital signal for inductor current (VGA_R[9..0]), 10-bit digital signal for capacitor voltage (VGA_B[9..0]).

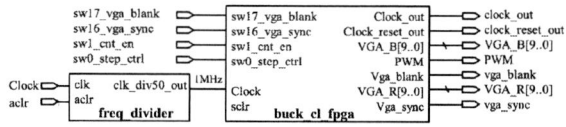

Fig.10. Implementation in FPGA

V. SIMULATION RESULTS

A. Test Scheme

The simulation is carried on the Altera DE2 development and education board based on Cyclone II 2C35 FPGA in a 672-pin package. To evaluate the simulation results, we test the simulation results with an oscilloscope. As the simulation outputs digital signals, we first convert the digital signals through a digital-to-analog converter (DAC) ADV7123, which is located on the DE2 board. Fig.11 shows the signal connection block diagram of the test system.

  a) 50MHz is the clock signal for FPGA
  b) SW0 , SW1 , SW16 , SW17 are toggle switches on DE2 board：SW0 control the reference current switching from 50A to 300A, SW1 used to enable/disable the saw tooth signal.
  c) SW16, SW17 and vga_clk signal control the chip ADV7123 to convert digital signal to analog signal.
  d) R，B represent the inductor current (iL) and the capacitor voltage(vC).

As is shown, we use 10 bits to represent the inductor current and capacitor voltage.

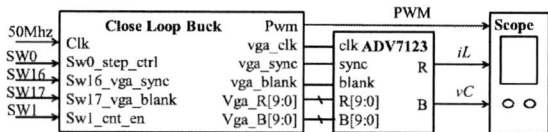

Fig.11. Test diagram

B. Calibration of the signals

As tested, the output voltage range is 0 to 1.27V, this is corresponds to the digital output 0 to 11_1111_1111. The inductor current and capacitor voltage are scaled down to make it possible to represent them by 10 bits. With scaling, the output signal of 1.27V is equal to 4092V real capacitor voltage or 1023A real inductor current.

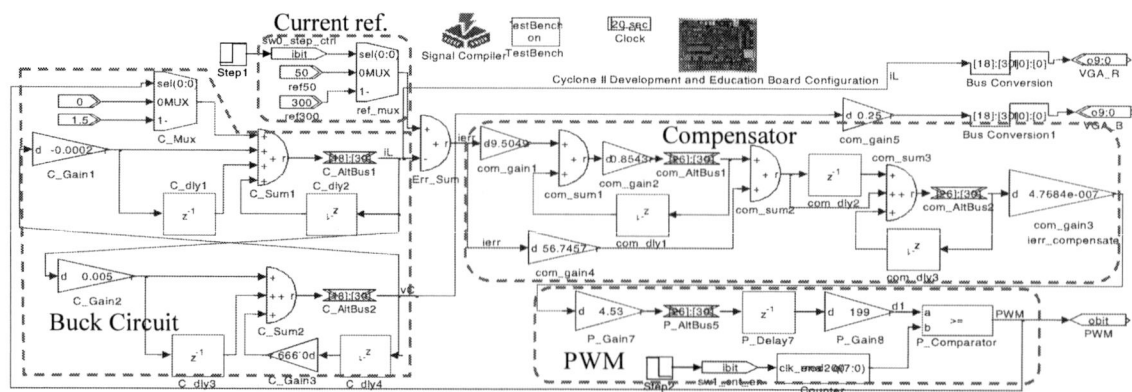

Fig.8. Overall signal block diagram

978-1-4799-2706-7/14 $31.00 © 2014 IEEE   2205

C. Test results

The transient response of buck converter with average current control is simulated offline using SABER as shown in Fig. 12. The current reference is changed from 50 A to 300 A at 30ms. These results are used as reference to compare the real-time simulation results.

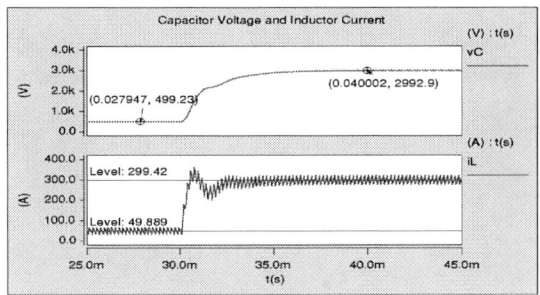

Fig.12 SABER simulation of the buck converter

The measured waveforms of FPGA simulation are shown in Fig.13. Waveform (a) shows the inductor current and capacitor voltage transient when reference current change from 50A to 300A. The steady inductor current signal is measured as 370mV, which represents the real inductor current of about 300A. Similarly, the steady state capacitor signal is measured as 1000mv, which means the real capacitor voltage is about 3kV. The measured waveforms agree with the SABER simulation. However, as the reference current changes from 50A to 300A, some differences can be found in the transient waveform between the FPGA simulation and SABER simulation. One possible reason for the error is that the number of iterations employed are not sufficient for convergence. Waveform (b) is the measured PWM generator output, confirming the switching frequency of 5kHz.

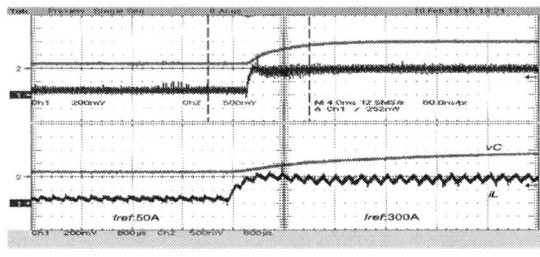

(a) Inductor current and capacitor voltage

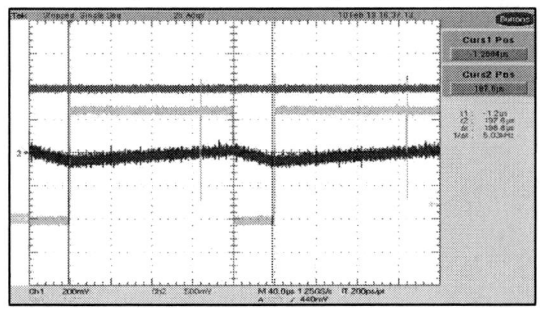

(b) PWM waveform

Fig.13. FPGA measurement results

VI. CONCLUSIONS

This paper realized a discretized circuit solver based on Jacobi iteration method using an example buck converter with the average current control. Signal block diagram is used for the implementation based on the mode-based design. Under the constraints on time step, the convergence of simulation is guaranteed. A detailed buck converter with 5kHz switch frequency was simulated with 1μs time step. The measurements agreed with the offline benchmark simulations on SABER, validating the procedure and signal block diagram solver. This method reduces the coding work of FPGA. However, as the complexity of circuit increases, we need to pay more attention to construct the signal block.

APPENDIX

Buck converter and compensator parameters used in the simulation are shown in Table I.

TABLE I
PARAMETERS OF CONVERTER

Parameter	Symbol	Value
Input voltage	V_g	3750V
Inductance	L	2.5mH
Capacitance	C	100μF
Resistance	R	10 Ω
Sample time	T	1μs
Switching frequency	f_{sw}	5kHz
Compensator coefficient	K_c	4.53
Compensator zero point	ω_z	3kHz
Compensator pole point	ω_p	15.71kHz
Current reference	I_{ref}	50A or 300A

ACKNOWLEDGMENT

This project is supported by Jiangsu Government Scholarship Fund and by the Nature Science Foundation of Jiangsu Province (Grant No. BK2011366). This work was conducted when HZ was a visiting scholar at the Center for Future Energy Systems at RPI.

REFERENCES

[1] Christian Dufour, Sébastien Cense, et al. "Review of state-of-the-art solver solutions for HIL simulation of power systems, power electronic and motor drives," in *Proceedings of* 15th *European Conference on Power Electronics and Applications* (EPE),2013, pp.1-12.

[2] Grégoire, L., J. Bélanger, et al, "FPGA-based real-time simulation of multilevel modular converter HVDC systems," in Proceedings of *ELECTRIMACS Cergy-Pontoise*, France , 6-8th June 2011, pp.1-6.

[3] Dufour, C., T. Ould Bachir, et al, "Real-time simulation of power electronic systems and devices," *Dynamics and Control of Switched Electronic Systems*, pp.475-485, 2012.

[4] Matar, M. and R. Iravani, "FPGA implementation of the power electronic converter model for real-time simulation of electromagnetic transients," *IEEE Trans. on Power Delivery*, vol. 25, no.2, pp. 852-860, 2010.

[5] Parma, G. G. and V. Dinavahi, "Real-time digital hardware simulation of power electronics and drives." *IEEE Trans. on Power Delivery*, vol. 22, no. 2, pp. 1235-1246, 2007.

[6] Myaing, A. and V. Dinavahi, "FPGA-based real-time emulation of power electronic systems with detailed representation of device characteristics," *IEEE Trans. on Industrial Electronics*, vol. 58, no. 1, pp. 358-368, 2011.

[7] Blanchette, H., T. Ould-Bachir, et al, "A State-Space Modeling Approach for the FPGA-based Real-Time Simulation of High Switching Frequency Power Converters," *IEEE Trans. On Industrial Electronics*, vol. 59, no. 12, pp.4555-4567, 2012.

[8] Kiffe, A., S. Geng, et al, "Automated generation of a FPGA-based oversampling model of power electronic circuits," *IEEE 15th International Power Electronics and Motion Control Conference (EPE/PEMC)*, Sept. 2012, pp. 1-8.

[9] Farid N. Najm, *Circuit simulation*, John Wiley & Sons, Inc., Hoboken, New Jersey, 2010, pp. 211-215.

[10] K.R. James, "Convergence of matrix iterations subject to diagonal dominance," *SIAM Journal on Numerical Analysis*, vol. 10, no. 3, pp.478-484,1973.

Gallium Arsenide IC Technology for Power Supplies on Chip

Vipindas Pala, Han Peng, Mona Hella and T. Paul Chow*
Center for Industrial Electronics
Rensselaer Polytechnic Insitute
Troy, NY-USA
chowt@rpi.edu

Abstract— **This research presents a power IC technology platform based on AlGaAs/InGaAs/AlGaAs pseudomorphic field effect transistors (pHEMTs) on a GaAs substrate. A quantitative assessment of a foundry-available 11 V GaAs pHEMT process indicates that due to their superior material properties, the intrinsic figure of merit for pHEMT switching devices show an order of magnitude improvement over the state-of-the-art Silicon NMOS transistors. The characterization results GaAs pHEMTs with a breakdown voltage up to 47V is presented and shown to be comparable to GaN based transistors for power switching applications. An integrated pHEMT DC-DC converter that can switch at frequencies above 100 MHz are demonstrated. A 4.2 V pHEMT buck converter designed for envelope tracking applications achieved 88 % conversion efficiency at 100 MHz.**

Keywords— Power Semiconductors, pHEMT, Power IC

I. INTRODUCTION

AS Power management in consumer electronics and communication systems has evolved, increasingly the power conversion circuits are placed in close proximity to various sub-systems to cater to the fast-changing load requirements (Point of Load or POL). It is advantageous to integrate the POL power converter along with the load in a system-on-chip or a system-in-package implementation. With these converters, if the operating frequency of the switching circuit can be increased to the high MHz range, the size of the passives-especially the inductors, can be reduced to a point so that it is feasible to integrate the entire power converter on to a same IC. An added advantage to this approach is that of improved bandwidth, which enables complex power management strategies to be employed to increase the power efficiency of the system.

The major bottlenecks to extending the switching frequencies of low-voltage power converters are a) switching and drive losses in state-of-the-art power transistors significantly degrade the efficiency when switched above a few MHz and b) the core and hysteresis losses in magnetic inductors at these frequencies make them less attractive. In this paper we propose a Gallium Arsenide power IC platform, which enables transistors that can switch efficiently at 100MHz or above. This approach combines the use of integrated gate drivers on chip along with the power transistor, and the use of air-core inductors that become attractive at these frequencies. The power electronic system is then integrated either on package or on chip – not dissimilar to a microwave monolithic IC (MMIC), commonplace in the RF industry today.

II. GAAS PHEMTS FOR POWER CONVERSION

The advantages of GaAs over silicon are its high electron mobility, high critical field strength, larger bandgap, higher operating temperature and the availability of a semi-insulating material to provide isolation and minimize parasitics,[2]. GaAs based pHEMTs (Pseudomorphic High Electron Mobility Transistors, Fig. 1) are commercially available, and are extensively used for power RF applications [3]. We have designed lateral power devices using a commercially available, 0.5 μm feature size GaAs pHEMT process [4]. The process integrates depletion mode (D-pHEMT) and enhancement mode (E-pHEMT) devices on the same wafer. The availability of low-standby current enhancement mode devices makes the pHEMT process suited to fabrication of integrated power management circuits. An enhancement mode pHEMT device with a gate width of 100 um was characterized to estimate the basic device characteristics. The output curve of the device is shown in Fig. 2 and transfer curve of the device in the linear regime is shown in Fig. 3.

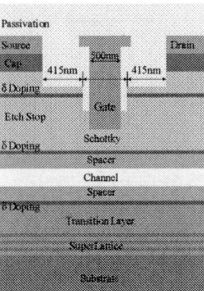

Fig. 1: GaAs pHEMT Device Structure

Based on the transfer curve of the device in the linear regime, a threshold voltage of 0.36 V is observed, which is the gate voltage needed for an output current of 1 mA/mm. The device has a Schottky type gate, as the gate current starts to increase exponentially when the gate voltage is greater than 0. We choose a maximum permissible gate bias depending on this power dissipation limit. If the ratio between the output current and the gate current has to be limited to less than a hundred, the maximum permissible gate bias is 0.88 V from Fig. 3.

978-1-4799-2706-7/14 $31.00 © 2014 IEEE

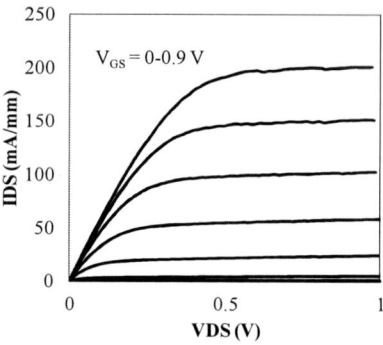

Fig. 2: Output characteristics of the enhancement mode pHEMT device.

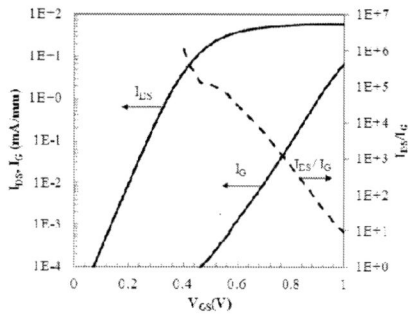

Fig. 3: Gate current, drain current and their ratio as a function of gate voltage.

TABLE I
ON RESISTANCE OF FABRICATED pHEMTs

Gate Width (W)	5 mm	10 mm	20 mm
Current Rating	0.5 A	1 A	2 A
Area	0.12 mm^2	0.24 mm^2	0.48 mm^2
R_{ON}	319 mΩ	168 mΩ	92 mΩ
R_{ON} x W	1.6 Ω.mm	1.7 Ω.mm	1.8 Ω.mm
R_{ON} x Area	38 mΩ.mm^2	40 mΩ.mm^2	44 mΩ.mm^2

The multi-level metallization was used to scale up the device to larger areas. Using an inter-digitated layout using 2 metal layers, and by using a flip-chip packaging technique to minimize the distance from each pad on the PCB, we have fabricated devices with an ON resistance of less than 100 mΩ. The characteristics of the fabricated devices are shown in table 1.

When the gate voltage is set to zero, the maximum operating voltage is determined by the leakage current rather than avalanche breakdown. As shown in Fig. 4, for a current ratio of 10^4 between the OFF state and the ON state, the drain voltage cannot exceed 11 V, and the leakage current is limited to 10 µA/mm.

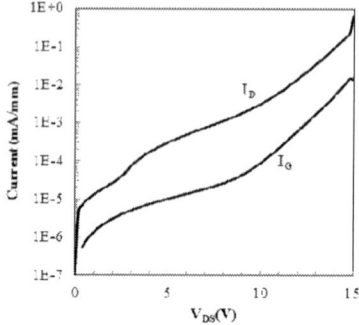

Fig. 4: Leakage current of the TQPED Enhancement mode pHEMT when V_{GS} is 0 V.

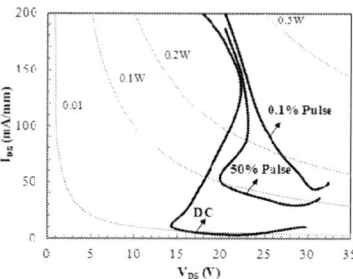

Fig 5: The SOA of the transistor as a function of duty cycle. Also shown are constant power contours.

The SOA characteristics were measured on chip in a poor thermal environment where the devices were allowed to fail due to a thermal runaway [5]. The devices showed good ruggedness for the entire voltage range, as indicated by the safe operating area, shown in Fig 5.

III. EXTENDED DRAIN PHEMTs

Using an optimized version of the process, we have also extended the breakdown voltage of the devices. The optimized process uses a double recessed structure, shown in Fig. 6, to reduce the field crowding near the gate. For this structure, the breakdown voltage increases in proportion to the separation between the gate and drain (L_D), as shown in Fig. 7. For L_D of 2.5 µm, which was the largest drift length in the experiment, a breakdown voltage of 47 V was obtained.

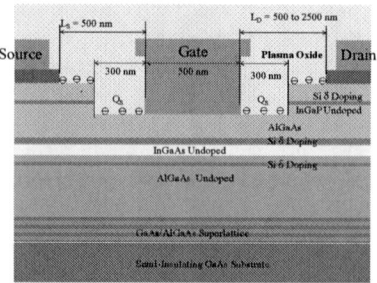

Fig. 6: Structure of the double recessed pHEMT

The ON resistance of the extended drain devices as a function of drift length is shown in Fig. 8. It is seen from Fig. 7 and Fig. 8 that with over a four-fold increase in BV can be obtained with only a 1.8x increase in resistance. This shows that the extrinsic components of the resistance [6], including contact and interconnects are a significant fraction of the total ON resistance of the device.

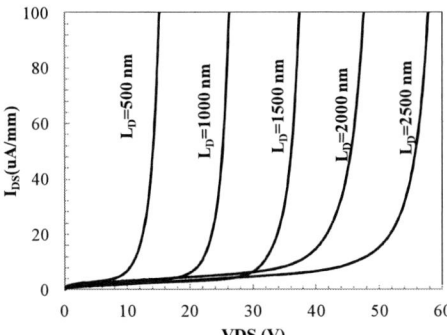

Fig. 7: Breakdown characteristics of the optimized devices with varying L_D at $V_{GS} = 0V$.

A figure of merit (FOM) for high frequency switching devices is the product of the ON resistance and the gate charge (Q_G) of the transistor [7]. A low FOM indicates higher power conversion efficiency for the same frequency or alternatively a higher operating frequency for a given power conversion efficiency for a switching device. As shown in Fig. 9, the fabricated GaAs pHEMTs show close to 5x improvement over commercial Silicon based transistors, and is in the same range as commercialized GaN based transistors. Also shown in Fig. 9 is a theoretical comparison of the three technologies [6] which indicates that for low voltage devices GaAs is the optimal technology platform for low voltage high frequency power electronics.

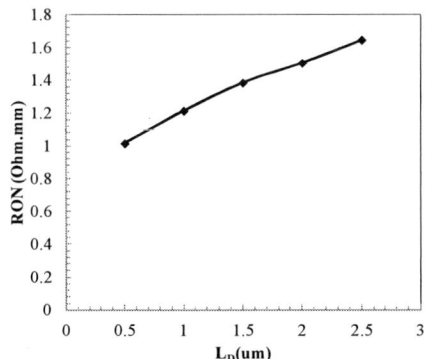

Fig. 8: The ON resistance of the optimized device as a function of L_D.

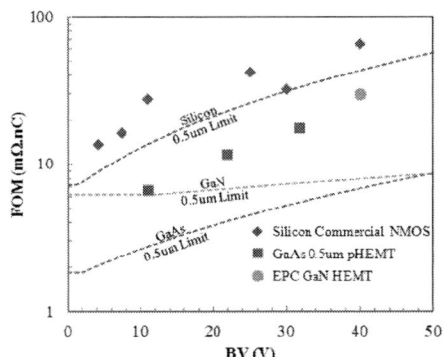

Fig. 9: Comparison of switching FOM between Silicon, GaN and GaAs pHEMT devices.

IV. GaAs Power ICs

As a demonstration vehicle for the GaAs power IC technology, we have built prototype buck type DC-DC converters using this technology. In the design example presented, we have developed a high frequency buck converter with a varying output voltage, designed as a high-bandwidth, envelope-tracking power supply for an RF power amplifier in a cellular handset. The nominal input voltage was 4.2V, which is the Lithium ion battery voltage, and the nominal switching frequency is 100MHz, to enable both a high loop bandwidth, and to enable the use of high-Q air core inductors and capacitors in the nano-Farad range.

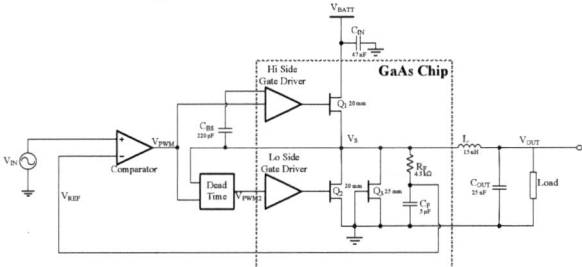

Fig. 10: Schematic of the single phase converter with hysteric control.

The output stage of the buck converter, consisting of switches and drivers integrated on the GaAs pHEMT die, is shown in Fig. 10. The high side switch is an enhancement mode pHEMT that is connected directly to the battery voltage. The low side consists of a synchronous rectifier in parallel with a diode connected pHEMT. Both high side switch and the synchronous rectifier are 11V rated enhancement mode pHEMTs with a gate periphery width of 20 mm. The diode connected pHEMT has a gate periphery width of 25 mm. The diode connected pHEMT is used instead of a Schottky diode because of the small threshold voltage of the enhancement mode pHEMT (0.36 V) which gives a forward voltage drop of 0.5 V for a 25 mm device when the current is 1 A. The diode connected pHEMT is a majority carrier device since the pHEMT does not have a body diode. Therefore this device does not suffer from reverse recovery and the power losses are almost entirely due to the forward voltage drop.

In this implementation we have also integrated the high and low-side gate driver on the same GaAs chip. Enhancement and depletion mode transistors with the same voltage rating as the power device are used to design a three-stage buffer that drive the power transistor. A schematic of the gate driver is shown in Fig. 11. A photograph of the fabricated output stage IC is shown in Fig. 12.

The power IC was flip-chip assembled on a 5mmx5mm laminate on which the GaAs die and the passives were integrated. A 0402 size 15 nH wire-wound surface mount inductor was used as the output inductor. A hysteretic control system was also implemented to generate a variable output. The open loop power efficiency of the output stage for various voltage conversion ratios for an input voltage of 4.5 V and a switching frequency of 100 MHz is shown in Fig. 13. When switched at 100 MHz, the peak measured efficiency of the converter at a steady state output voltage of 3.375 V is 88 % for 34dBm power output. The converter can operate up to 37 dBm output power for at the peak voltage. At lower output voltages, the efficiency degrades. The major reason for this degradation at low voltages in our design is the static gate driver losses in the GaAs driver, which do not scale as a function of output power.

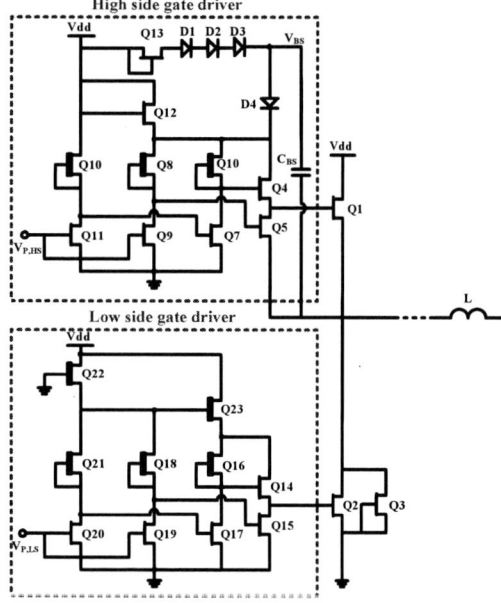

Fig. 11: Single phase GaAs pHEMT DC-DC converter chip with integrated switches and drivers.

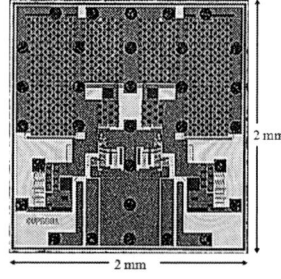

Fig. 12: Die photo of the GaAs pHEMT DC-DC converter chip

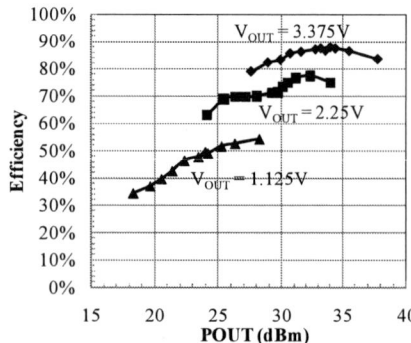

Fig. 13: Measured power efficiency of the converter as a function of output power at 75 %, 50 % and 25 % voltage conversion ratios.

V. Summary

We have demonstrated a power IC technology platform based on enhancement mode GaAs pHEMTs that can enable integrated power ICs that can extend the frequency range of switching circuits to 100MHz or above without compromising the efficiency. GaAs pHEMTs with a breakdown voltage of up to 47V were demonstrated that have a five-fold advantage over silicon NMOS in terms of switching figure of merit and are comparable to GaN based lateral switches. We have also demonstrated a prototype buck converter power IC, with integrated gate drivers that achieved 88% conversion efficiency at 100MHz, which is among the highest report for switching converters in the power and frequency range. On a broader scale GaAs pHEMTs could also be utilized to fabricate DC-DC converters for microprocessors, base band amplifiers, and in other applications where wide bandwidth power supplies could offer huge gains. In a scenario where innovations in silicon based low voltage transistors have saturated, this approach gives a new way of breaking the paradigm and making large leaps in performance.

Acknowledgment

The authors want to thank TriQuint Semiconductor Inc for providing foundry, packagiing and technical support for this work.

References

[1] B.J. Baliga, M.S. Adler, and D.W. Oliver, "Optimum semiconductors for power field effect transistors," *IEEE Electron Dev. Lett.*, vol. 2, pp. 162-164, 1982.

[2] A.J. Atkinson, "Power devices in Gallium Arsenide," *IEE. Proc. Int. Solid-State and Electron. Dev.*, vol. 132, pp. 264-271, 1985.

[3] M. Miller, "Design, performance and application of high voltage GaAs FETs," in *Compound Semiconductor Integrated Circuit Symp.*, 2005, pp. 236-239.

[4] TriQuint Semiconductor Inc. (2010, Mar.) TriQuint Semiconductor Website. [Online].

[5] V. Pala, M. Hella, and T.P. Chow, "Safe operating area of AlGaAs/InGaAs/GaAs HEMT power transistors," in *Int. Symp. Power Semiconductor Devices and ICs*, 2011, pp. 243-246.

[6] V. Pala, H. Peng, P. Wright, M.M. Hella, and T.P. Chow, "Integrated High-Frequency Power Converters Based on GaAs pHEMT: Technology Characterization and Design Examples," *IEEE Trans. Power Electron.*, vol. 27, no. 5, pp. 2644-2656, 2012.

[7] Z.J. Shen, D.N. Okada, F. Lin, S. Anderson, and X. Cheng, "Lateral Power MOSFET for Megahertz-Frequency, High-Density DC/DC Converters," *IEEE Trans. Power Electron.*, vol. 21, no. 1, pp. 11-17, 2006.

Silicon on nanocrystalline and microcrystalline diamond stacking structure for power supply on chip

Takatoshi Yamada and Masataka Hasegawa
Nanotube Research Center, National Institute of Advanced
Industrial Science and Technology
Tsukuba, Japan
takatoshi-yamada@aist.go.jp

Abstract—We proposed a stacking structure of nanocrystalline diamond (NCD) and microcrystalline diamond (MCD) films for an insulating material of silicon-on-insulator (SOI) substrate. NCD was deposited on Si wafer substrates by surface wave plasma CVD and then MCD was deposited on as-grown NCD surface. The break down field of the stacking structure was three times higher than that of MCD layers. The thermal conductivities were in the range from 320 to 500W/mK, which were almost same as MCD layer.

Keywords— *Nanocrystalline diamond, Microcrystalline diamojd, Stacking strcuture, Chemical vapor deposition*

I. INTRODUCTION

Diamond is expected for one of the most appropriate insulating materials for silicon on insulator (SOI) substrates [1-3] since diamond has both high thermal conductivity and high break down field. The thermal conductivity and the break down field of single crystal diamond are 2000 W/mK and 10^7V/cm, respectively. Therefore, chemical vapor deposited (CVD) diamond films were applied for the insulating materials for SOI substrates and improvements of device properties on the SOI were reported using CVD diamond films as the insulating layers [4-5]. Most of reports described polycrystalline CVD diamond films for the insulating materials of SOI substrates, thus the thermal conductivities and the break down fields were far from those of single crystal diamond and theoretical data.

In general, both of thermal conductivity and electrical resistivity of polycrystalline CVD diamond films were changed by grain size as well as quality. We can modulate the grain size and quality by controlling the CVD parameters. Reductions of non-diamond phases, such as graphitic and/or amorphous carbon, increased both the thermal conductivity and the break down fields. However, relationship between the thermal conductivity and the electrical resistivity of polycrystalline diamond films is trade-off. Therefore, controlling of grain sizes of CVD diamond is important technology to use diamond thin films as the insulating materials of SOI substrates.

In order to develop SOI substrates having diamond insulator layers for industrial mass productions, depositions of homogeneous diamond films on large area are requirement. There are two CVD techniques to form homogeneous diamond films on the large area. One is hot filament CVD. Using the hot filament CVD techniques, it is possible to deposit microcrystalline diamond (MCD) films on large area. However, less coalescence in MCD at early growth stage is origin of low break down field. In addition, contaminations in MCD films, which are coming from filaments, play current leak path. Scratching on the substrate surfaces, which are performed for nucleation, makes damages on the interface between MCD films and substrates. These are also current leak paths. The other is surface wave microwave plasma CVD technique. The surface wave microwave plasma CVD enables us to deposit nanocrystalline diamond (NCD) films on large area. NCD deposited by SWPCVD showed electrical resistivity of 10^{10}-$10^{13}\Omega$cm [6] and thermal conductivity of a few tens W/mK [6]. Although the reported thermal conductivities of NCD were one order higher than those of silicon oxide (SiO_2), the reported data were one or two order smaller than that of microcrystalline diamond (MCD) [7].

In this, paper, a stacking structure of the NCD and the MCD for the insulating materials of SOI substrate is proposed in order to improve both of the thermal conductivity and the break down field at the same time. The breakdown fields were evaluated by the measurements of current vs voltage (*I-V*) characteristics. The thermal conductivities were measured by thermo-reflectance method [8-9].

II. EXPERIMENTALS

The stacking structure of NCD and MCD were fabricated on the conductive p-type Si substrate ($10^{-1}\Omega$cm). First, about 200nm of NCD film was deposited by surface wave microwave plasma CVD [10] and then about 6μm of microcrystalline diamond film was deposited on as-grown NCD surface by hot filament CVD. We also prepared MCD films direct deposited on Si substrates.

Before diamond depositions by CVD techniques, seeding processes on Si substrates were carried out in water suspension with dispersed detonation NCD powder of 5 nm in average crystal size. The suspension was spread on the Si substrates using ultra-sonic bathes. Using the seeding processes, nucleation density was achieved from 10^{11} to 10^{12} cm^{-2} [10].

Typical CVD conditions of NCD and MCD were as follows. NCD was deposited by surface-wave microwave plasma CVD. In this study, we used a slot antenna type microwave plasma CVD apparatus as shown in Fig. 1. Microwave is propagated in the microwave guide and emit into CVD apparatus though the slot antenna and the quartz window. Microwave power was 1500W. The substrate temperature was 350ºC. Mixture of (H_2+CH_4+CO_2) was used as a reactant gas and the C/H ratio was 5% [11]. The pressure was 20Pa. Under this CVD condition, electron density near the dielectric plate window was in the order of 10^{11}cm^{-3}, which means generation of the surface-wave plasma [11].

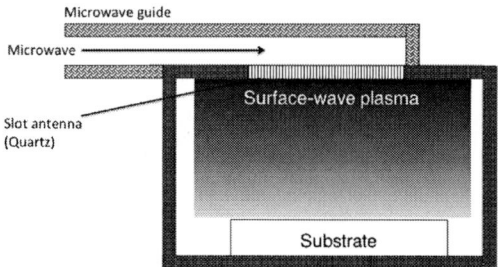

Fig. 1 The surface wave plasma CVD apparatus.

MCD was deposited by hot filament CVD on the NCD films deposited on Si substrates. The tungsten (W) filament temperature was 2200°C. The substrate temperature was about 800°C. The mixture of (CH$_4$+H$_2$) was used as the reactant gas for MCD deposition and C/H ratio was 4.3%. The pressure was 1333Pa.

The breakdown fields were evaluated from the *I-V* characteristics. Ti/Au contacts were formed by an electron beam evaporator after wet chemical oxidization [12] in order to remove surface conductive properties related with hydrogen termination. The wet chemical treatments were carried out by boiling of samples in mixed acid solutions (HNO$_3$+H$_2$SO$_4$) at 200 °C for 60 min [10]. The diameters of contacts were from 100 to 400 μm. The thickness of Ti and Au layers were 30 and 100nm, respectively. The *I-V* characteristics were measured at room temperature in air. The silicon substrates were connected ground potential and the positive bias was applied on metal contacts.

The thermal conductivity was characterized by thermo-reflectance method. For measurements of the thermal conductivity, Si substrates were removed in the mixed solution of HF and HNO$_3$ and then Mo thin films with 100nm in thickness were deposited on both diamond surfaces by radio frequency (RF) sputtering. The detailed measurement method is described in literature [8-9].

The surface morphology of the stacking structure was observed by optical microscopy.

Cross-sectional image of the stacking structure was observed by scanning electron microscopy (SEM). The samples were cut by the laser, in order to observe the interface between diamond layer and Si substrate.

Both the stacking structure and the MCD film were characterized by Raman spectroscopy. A semiconducting laser of 532 nm wavelength was for excitation. The laser spot size was 1 μm in diameter. The depth of characterization was about 1 um from the surface.

Secondary ion mass spectroscopy (SIMS) was used to determine impurities in both CVD diamond films and Si substrates. Oxygen (O$^+$) ions were used to etch diamond films and Si substrates.

III.   RESULTS AND DISCUSSION

Typical surface morphology observed by optical microscopy is shown in Fig. 2. It was confirmed that the polycrystalline film was obtained. The grain sizes on the surface of the stacking structure were about 2 μm.

The cross-sectional image of observed by SEM as shown in Fig. 3 (a) indicates that the MCD was deposited on NCD with thickness of 220 nm. In addition, coalescence of NCD is confirmed. While, less-coalescence at the early growth stage of

MCD layer [Fig. 3 (b)] is observed as was seen the previous reposts. For both the stacking structure and the MCD film, columnar structures are observed. Therefore, growth mechanism of MCD layer for both samples are same. Low nucleation density of diamond using hot filament CVD was reported. It was considered that NCD layer acted as nucleus for MCD layer formation.

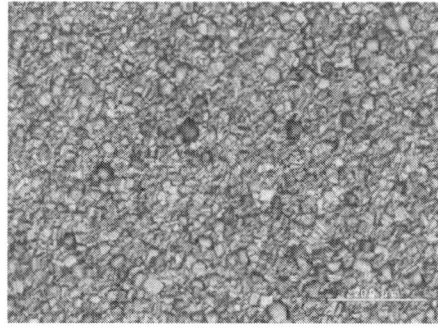

Fig. 2 The surface morphology observed by optical microscopy.

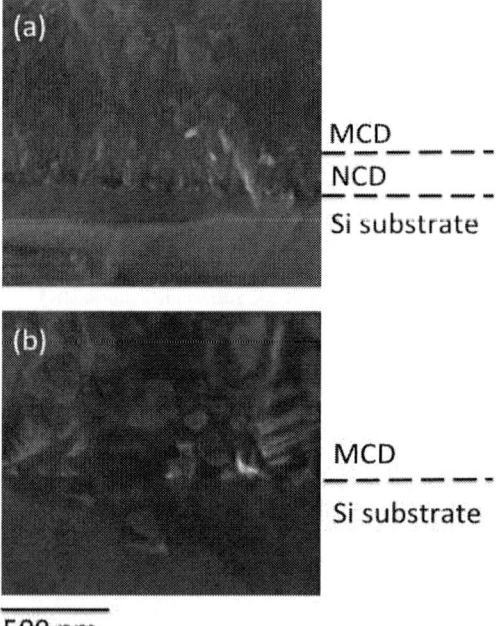

Fig. 3 (a) The cross-sectional images of the stacking structure...Growth of MCD layer on NCD layer is confirmed. (b) The cross-sectional image of the MDC film. Columnar structures are observed.

Typical Raman spectra of the stacking structure and the MCF film are shown in Fig. 4. The strong and sharp peaks due to diamond at 1333 cm^{-1} are observed for the stacking structure. Disordered graphite or amorphous carbon in the stacking structure is not detected. It was considered that the background was attributed to nitrogen impurities. The Raman spectrum of the MCD film is also shown in Fig. 4. These data indicated that the high quality MCD layer was formed on the NCD layer.

The 2014 International Power Electronics Conference

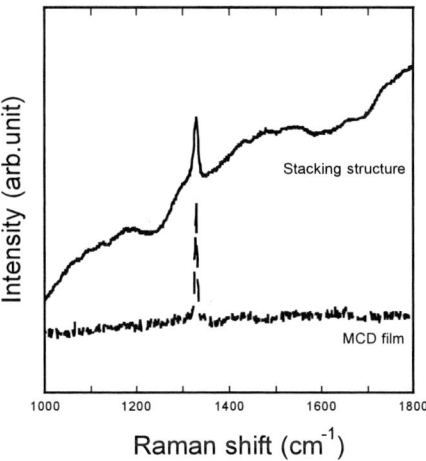

Fig. 4 Raman spectra of the stacking structure and the MCD film. Only peak due to diamond is observed which indicates the high quality CVD diamond film is obtained.

Typical current density (J) vs electric field (E) characteristics obtained from the I-V characteristics are shown in Fig.5. The current densities were calculated by dividing the current by the contact area and the fields were obtained from the voltage and the total thickness of insulating layers. The J-E characteristics (Fig. 5) show linear relationships between the current densities and the electric fields at low electric field regions. While the current densities at high electric field regions are rapidly increased. We defined that the electric field where the current density increased rapidly was the breakdown field. The breakdown fields were calculated to be 1.4×10^5 and 4.5×10^5V/cm for the MCD layer and the stacking structure, respectively. The calculated breakdown fields were one order lower than the reported single crystal diamond, which would be explained by the columnar structures [13]. The grain boundaries exist from the bottom to top surface, which was almost same direction of current flow. Although high quality diamond films were confirmed by Raman spectra as shown in Fig. 5, the grain boundaries are well known defective. Therefore, the grain boundaries played current leakage path.

The breakdown field of stacking structure was three times higher than that of the MCD layer. Although it is not understand the breakdown mechanism up-to-now, discharge and current leakage were considerable explanations. The discharge in voids at interface between MCD and Si substrate is one of the triggers to occur the breakdown. Second was the leakage current at the MCD and Si interface. In SIMS measurements, W contaminations were detected in MCD and Si interfaces as shown in Fig. 6 (a). While no W contamination was observed in NCD and Si interface of the stacking structure [Fig. 6 (b)]. Therefore, it was speculated that the W contaminations acted as leakage current path [14]. During CVD process using hot filaments, evaporated materials from filaments were incorporated at the interface between MCD layers and Si substrates. For the stacking structure case, NCD layer protected the incorporation of filament materials into Si substrates.

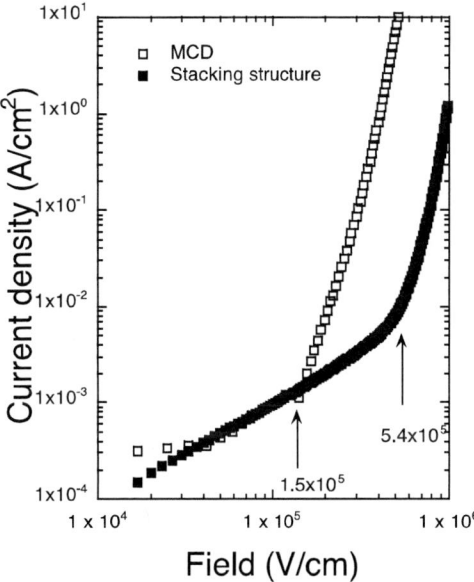

Fig. 4. Current density-electric field (J-E) characteristics of the stacking structure and the MCD layer. The breakdown field is three times higher than the MCD layer.

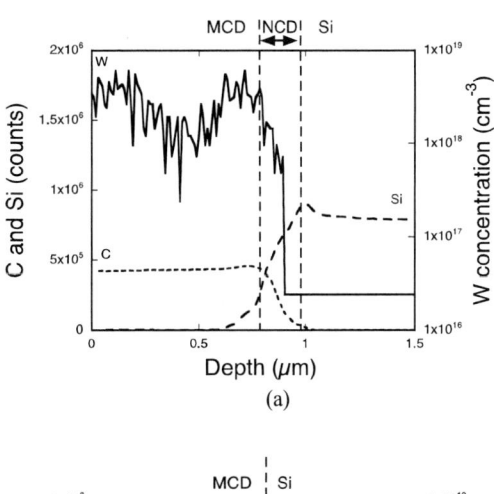

(a)

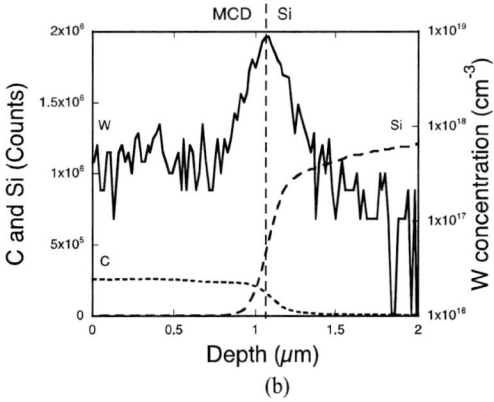

(b)

Fig. 6. SIMS profiles of (a) the stacking structure and (b) MCD film. High W impurity concentration region is observed at the interface between MCD and Si substrate for the MCD film.

978-1-4799-2706-7/14 $31.00 © 2014 IEEE

At low electric fields, almost same slopes are confirmed for both samples. This means that the average resistivities of two samples were almost same and the estimated reisitivities were about $3.7 \times 10^9 \, \Omega$ cm. It was expected that the average resistivity would be increased be formation of the stacking structure compared to the MCD film, since the reported resistivities of NCD films were in the range of $10^{11}\Omega$cm [7, 15-16]. The average resistivity was estimated from the slope in the J-E characteristics. We could estimate the total resistance (R_{total}) of the stacking structure and R_{total} was obtained from equation (1), where R_{MCD} was resistance of MCD layer with 6μm and R_{MCD} was resistance of NCD layer with 220 nm.

$$R_{total} = R_{MCD} + R_{NCD} \qquad (1)$$

Using the resistivity of NCD layer ($10^{11}\Omega$cm), the total resistance of the stacking structure was estimated to be $10^{10} \, \Omega$. The obtained value was almost same as the resistance estimated from the slop of the J-E characteristics ($6 \times 10^9 \Omega$) as shown in Fig. 4. The above discussion could be well the resistivity of the stacking structure. Both higher resistivity and break down filed would be expected by the optimizing the NCD layer [17].

The thermal conductivities of the stacking structure of NCD and MCD are in the range between 320 to 500 W/mK, which are almost the same as the reported thermal conductivities of polycrystalline diamond films having the thickness over 1μm [18-19]. The thermal conductivities of the NCD layers were in the order of 10W/mK [16] and were almost independent of thickness [20]. It is considered that the thermal transport in MCD layer is dominant in the stacking structure of MCD and NCD since the thickness of the MCD layer in the stacking structure is three hundreds times thicker than the NCD layer. It was reported that defective regions in CVD diamond films near the Si substrate (early growth stage of CVD diamond films) affected the thermal conductivity [19]. From the obtained data, it would be expected that further increasing the thermal resistivities by the formation of multilayer structures of MCD and NCD in addition to improving MCD and NCD qualities.

IV. SUMMARY

The stacking structure of MCD and NCD was formed for the insulating layer of SOI substrates in order to improve both of the breakdown fields and the thermal conductivities. The NCD layers were deposited on Si substrates by the surface wave microwave plasma CVD and then the MCD layers were deposited on as-grown NCD layer surfaces by the hot filament CVD. The J-E characteristics showed that the break down field of the stacking structure was three times higher than that of the MCD layer. Thermal conductivities of the stacking structure were from 320 to 500W/mK. This was almost the same as MCD layer. From the obtained results, maintaining of the thermal conductivities of MCD layers was achieved regardless increasing of breakdown fields of MCD layer.

It was expected that combination of surface wave microwave plasma CVD for NCD and hot filament CVD for MCD was the most appropriate technique toward the industrial mass productions of SOI substrates since both CVD techniques enabled us to deposit diamond films on large area.

ACKNOWLEDGMENT

The authors would like to acknowledge Dr. T. Yagi, AIST, Japan for the thermal conductivity measurements.

REFERENCES

[1] A. Aleksov, X. Li, N. Govindaraju, J. M. Gobien, S. D. Wolter, J. T. Prater and Z. Sitar, Diam. Relat. Mater. vol. 14, no. 3-7, pp.308-313, 2005.

[2] M. Lions, S. Saada, B. Bazin, M. –A. Pinault, F. Jomard, F. Andrieu, O. Faynot and P. Bergonzo, Diam. Relat. Mater. vol. 19, no.5-6, pp. 413-417, 2010.

[3] M. Rabarot, J. Widiez, S. Saada, J.-P. Mazellier, C. Lecouvey, J. –V. Roussin, J. Dechamp, P. Bergonzo, F. Andieu, O. Faynot, D. Deleonibus, L. Clavelier, J. P. Roger, Diam. Relat. Mater. vol. 19, no.7-9, pp. 796-805, 2010.

[4] A. Aleksov, J. M. Gobien, X. Li, J. T. Prater and Z. Sitar, Diam. Relat. Mater. vol. 15, no. 2-3, pp. 248-253, 2006.

[5] J.-P. Mazellier, M. Mermoux, F. Andrieu, J. Widiez, J. Dexhamp, S. Saada, M. Lions, M. Hasegawa, K. Tsugawa, P. Bergonzo and O. Faynot, J. Appl. Phys. vol. 110, pp. 08901-XXX, 2011

[6] K. Tsugawa, M. Ishihara, J. Kim, M. Hasegawa and Y. Koga, J. Sur. Sci Soc. Jpn. vol. 30, pp. 267-272, 2009 [in Japanese].

[7] J. E. Greabner, Ch. 7, "Thermal conductivity of diamond", in Diamond: Electronic properties and applications, L. S. Pan, Ed., Kumar Academic Publication, 1995, pp.285-318.

[8] T. Yagi, K. Tamano, Y. Sato, N. Taketoshi, T. Baba and Y. Shigesato, J. Vac. Sci, Technol. A vol. 23, no. 4, pp. 1180-1186, 2005.

[9] N. Taketoshi, T. Baba, E. Schaub and A. Ono, Rev. Sci. Inst. vol. 74, no. 12, pp. 5226-5230, 2003.

[10] K. Tsugawa, S. Kawaki, M. Ishihara, J. Kim, Y. Koga, H. Sakakita, H. Koguchi, and M. Hasegawa, Dia. Rela. Mater. vol. 20, no. 5-6, pp. 833-838, 2011.

[11] K. Tsugawa, M. Ishihara, J. Kim, Y. Koga and M. Hasegawa, Phys. Rev. B, vol. 82, pp.125460-XXX, 2010.

[12] T. Yamada, C. E. Nebel, K. Somu, H. Uetsuka, H. Yamaguchi, Y. Kudo, K. Okano and S. Shikata, Phys. Stat. Sol. (a), vol. 204, no. 9, pp. 2957-2964, 2007.

[13] W. J. P. van Enckevort, Ch. 9 "Physical, chemical, and microstructural characterization and properties of diamond", K. E. Spear and J. P. Dismukes, Eds., Jhon Wiley and Sons, 1994, pp. 307-353.

[14] R. Hessmer, M. Schreck, S. Geier and B. Strizker, Diam. Relat. Mater. vol. 3, no. 5-6, pp. 951-956, 1994.

[15] O. A. Williams, Diam. Relat. Mater. vol. 20, no. 5-6, pp. 621-640, 2011.

[16] O. Auciello and A. V. Sumant, Diam. Rela. Mater. vol. 19, no. 7-9, pp. 699.-718, 2010.

[17] T. Yamada and M. Hasegawa, Phys. Stat. Sold. (a), vol. 210, no.10, pp1998-2001, 2013.

[18] S. Shikata, S. Yun, T. Yamada, H. Umezawa, F. Nakamura, T. Yagi, N. Taketoshi, N. Yamada and T. Baba, Abst. European Conf. SiC. Relat. Mater, TP103, 2010.

[19] E. B. –Grayeli, A. Sood, M. Asheghi, V. Gambin, R. Sandhu, T. I. Feygelson, B. B. Pate, K. Hobart and K. E. Goodson, Appl. Phys. Lett. vol. 102, pp.111907, 2013.

[20] M. Lion, S. Saada, J. –P. Mazellier, F. Andrieu, O. Faynot and P. Bergonzo, Phys. Stat. Sol. RRL, vol. 3, no.6, pp. 205-207, 2009.

A Novel Load Regulation Technique for Power-SoC with Parallel Connected POLs

Seiya Abe
International Centre for the Study of East Asian
Development (ICSEAD)
Kitakyushu, Japan
abe@icsead.or.jp

Akira Hidaka, Jungo Rikitake
Kyushu Institute of Technology
Kitakyushu, Japan

Satoshi Matsumoto
Kyushu Institute of Technology
Kitakyushu, Japan

Tamotsu Ninomiya
International Centre for the Study of East Asian
Development (ICSEAD)
Kitakyushu, Japan

Abstract— **This paper presents the novel load regulation technique for parallel connected POLs which reads for power supply on chip (power-SoC). In power-SoC, many POLs are implemented on the same chip. In this case, the conventional loop control (feedback control) may have some problem such as oscillation. The proposed strategy regulates the output voltage by changing number of the working POLs under fixed duty ratio. The parallel connected POLs system is implemented by means of MATLB/Simulink, and the operating characteristics are confirmed. In addition, the proposed control strategy is also verified experimentally.**

Keywords— POL, Power-Soc, Parallel connection

I. INTRODUCTION

Point of load (POLs) becomes more important through increasing clock speed and the power consumption of MCU. It becomes main stream to put POLs near MCU in order to reduce voltage drop by line resistance and voltage variation with parasitic inductance. Therefore, downsizing of these is more significant [1]. In such a scheme, a power supply on chip (power-SoC) which integrates power devices, control circuits and passive devices on the same chip as shown in Fig. 1. Power-Soc has been attracted attentions of many researchers because it can realize ultimate minimization of the power supply [2-5].

One of the most effective ways to shrink the size of the power supply is to reduce the volume of the passive components such as inductors and capacitors. Increasing the switching frequency of power supply is one of the most promising approaches to do this. Therefore, the switching frequency of more than 10 MHz is required for realizing power-SoC and it will reach several tenth MHz in near future. In such situations, the traditional analog based PWM will face the problems.

In addition, many POLs are connected in parallel on the power-SoC because the downsized POLs can handle smaller power per chip.

Therefore, to develop new control strategies with stable operation at high frequency and parallel connections becomes very important.

Hence, the control technique of the power-SoC based on parallel connections of many POLs is investigated by means of digital control [6].

The number of the parallel connected working POLs according to the required power by detecting the load current has been proposed in previous report [6]. However, the accurate current detection is difficult in downsized POLs and the control technique based on current detecting induces the increasing the number of the parts and this inhibits downsizing. Moreover, the current detection control has some difficulties as follows;

(1) Making multi output voltage
(2) Accurate adjustment of the target output voltage

In this paper, the novel control strategy which detects the output voltage is proposed for the power-SoC based on parallel connected POLs.

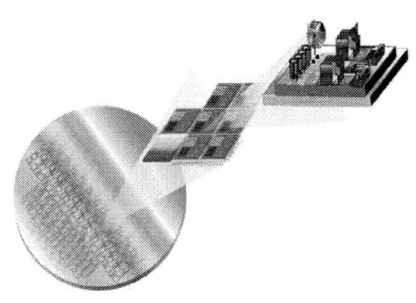

Fig. 1. Power-SoC.

The proposed strategy regulates output voltage by changing number of the working POLs under fixed duty ratio (without loop control). At first, the system operation and the advantage of proposed control strategy are confirmed by means of MATLAB/Simulink. Finally, the proposed technique is verified through prototype POLs.

II. System Configuration and Modeling

The concept of parallel connected POLs based control technique is schematically as shown in Fig. 2. This system does not have feedback loop. Instead of the feedback control, the number of the working POLs which is connected in parallel is changed according to the required power (load current).

For example, smaller number of the parallel connected POLs work at light load as shown in Fig. 2 (a), and larger number of them work at heavy load as shown in Fig. 2 (b). The POL can be drawn like simplified model in steady state operation as shown in Fig. 3. This simplified model consists of the internal resistance which includes the conducting resistance of switches and equivalent series resistance of the smoothing inductor. In single POL case, the output voltage can be derived from Fig. 3 (a).

$$V_o = DV_{in} - r_{DC}I_o \quad (1)$$

As shown in Eq. (1), the output voltage decreases with increasing r_{DC} and Io. In order to reduce the output voltage variation without feedback control, reducing the internal resistance r_{DC} is needed.

In parallel connected case, the output voltage can be derived from Fig. 3 (b).

$$V_o = DV_{in} - \frac{r_{DC}}{n}I_o \quad (2)$$

In this case, the internal resistance is divided number of connected POLs "n". In discrete system, each POL has individual variation of circuit components. However, the individual variation can be minimized in power-SoC system because POLs are fabricated by LSI and MEMS process. This is one of the most promising advantages of power-SoC system.

From Eq (2), the output voltage can be regulated by changing the number of the working POLs "n" without feedback control.

Figure 4 show the analytical results of load characteristics and efficiency characteristics on parallel connected POLs system. The slope of the output voltage variation is gradual with increasing number of POLs as shown in Fig. 4 (a). Moreover, the output voltage can be regulated by properly changing the number of POLs as shown red line in Fig. 4 (a).

Similarly, the slope of the efficiency variation is gradual and the peak is moved to high current side with increasing number of POLs as shown in Fig. 4 (b).

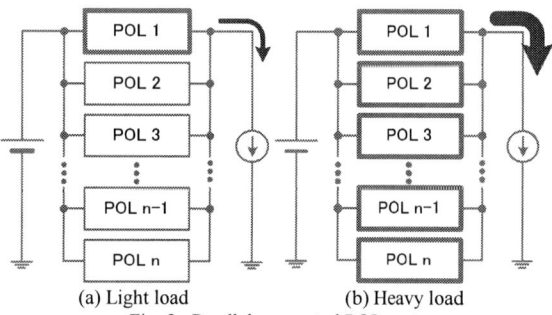

(a) Light load   (b) Heavy load
Fig. 2. Parallel connected POL system.

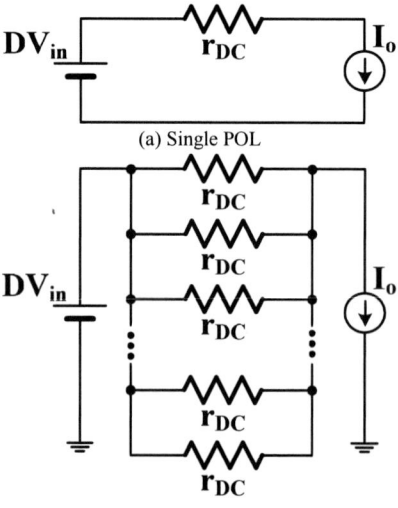

(a) Single POL

(b) Parallel POLs
Fig. 3. Static models for parallel connected POL system.

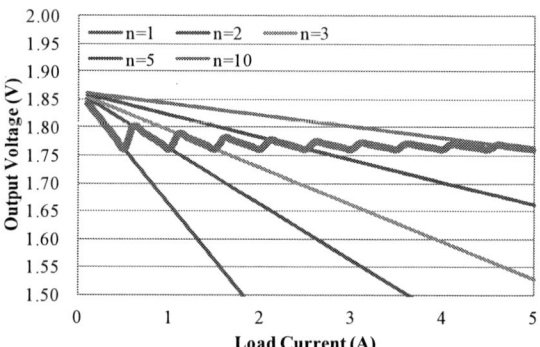

(a) Load characteristics (Analytical results).

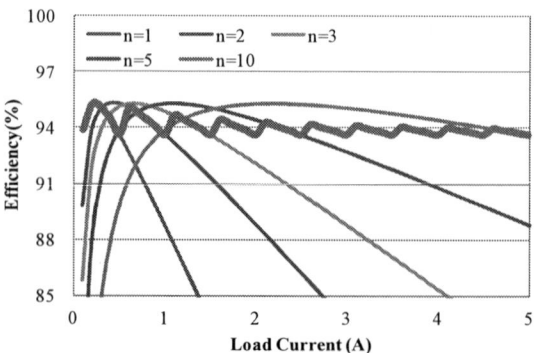

(b) Efficiency characteristics
Fig. 4 Analytical results.

Furthermore, the efficiency can be kept high by properly changing the number of POLs as shown red line in Fig. 4 (b).

The voltage and efficiency variation at switching behavior (changing the number of POLs) can be minimized by increasing total number of POLs

III. SIMULATION BY MATLAB / SIMULINK

The experimental measurements on discrete system instead of power-SoC system are impractical in many POLs case. In order to investigate the power-SoC based on parallel connected POLs before experimental measurements, the implementation of the simulation system is necessary. Here, MATLAB / Simulink is used as a simulation tool [7]. Figure 5 shows the parallel connected POLs system.

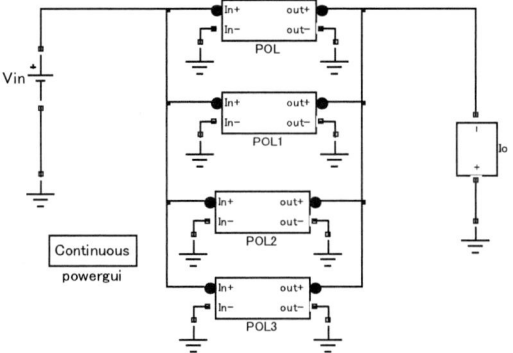

(a) total system

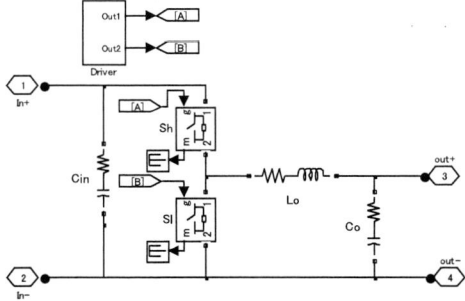

(b) Synchronous buck converter as POL

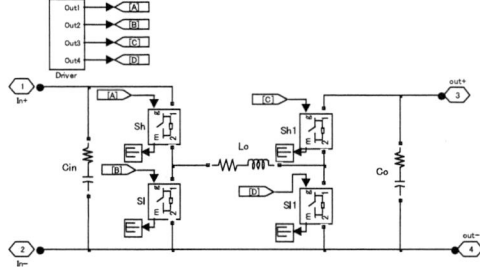

(c) H-bridge converter as POL.
Fig. 5. Parallel connected POL system.

The system consists of voltage source as an input voltage, POLs and current source as a load current as shown in Fig. 5 (a). Figure 5 (b), (c) show the circuit topologies of POLs as examples, respectively. A buck converter with synchronous rectification is usually used for POL application as shown in Fig. 5 (b). The buck converter consists of input capacitor with ESR, two ideal switches, smoothing inductor with ESR and output capacitor with ESR. The gate singles are supplied form the driver.

The boost mode operation is demanded due to the diversification of POL application. The H-bridge converter is one of the most popular topology as shown in Fig. 5 (c). The H-bridge converter is very flexible because there are two operating mode (buck and boost mode) by changing the switching sequence.

In order to evaluate the system operation, the simulated model shown in Fig. 5 is executed. Here, the buck converter with synchronous rectification is used as an example. The circuit parameters and specifications of POL are shown in Table I.

Where, the on resistance of switches and equivalent series resistance of smoothing inductor are included as an internal resistance r_{DC}. In actual simulated works, 10 POLs are connected in parallel, and the number of POLs is switched by load current.

Figure 6 and 7 show the simulation results of load characteristics and efficiency characteristics of buck converter and H-bridge converter, respectively.

In buck converter, the slope of the output voltage variation becomes gradual with increasing number of POLs as shown in Fig. 6 (a). Moreover, the output voltage can be regulated by changing the number of POLs as shown red line in Fig. 6 (a). These results are similar to above mentioned analytical discussions. The slope of the efficiency variation is also gradual and the peak is moved to high current side with increasing number of POLs as shown in Fig. 6 (b). Furthermore, the efficiency can be kept high as shown red line in Fig. 6 (b).

In H-bridge converter case, the slope of the output voltage variation and the efficiency variation are similar to buck converter case as shown in Fig. 7. These results are also similar to above mentioned analytical discussions.

The output voltage variation at switching behavior is larger than buck converter case though the output voltage can be also regulated by changing the number of POLs as shown red line in Fig. 7 (a).

TABLE I
CIRCUIT PARAMETERS AND SPECIFICATIONS

Symbol	Description	Value
Vin	Input Voltage	3.8V
Io	Load Current	0.1-5A (0.5A/POL)
Lo	Smoothing Inductor	1uH
r_{DC}	Internal Resistance	200mOhm
Co	Output Capacitor	10uF
fs	Switching Frequency	10MHz
D	Duty Ratio	0.5

This is because the internal resistance r_{DC} becomes higher than buck converter due to the influence of duty ratio, and the apparent resistance becomes $r_{DC}/(1-D)^2$ in boost operation of H-bridge converter.

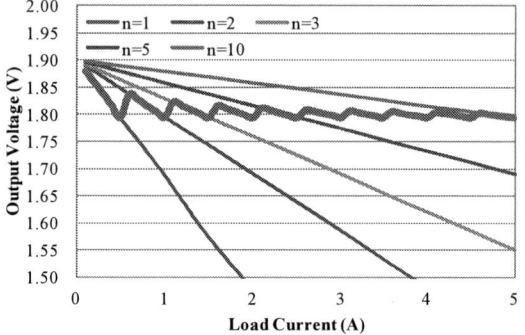

(a) Load characteristics

(b) Efficiency characteristics

Fig. 6. Simulated results (Synchronous buck converter)

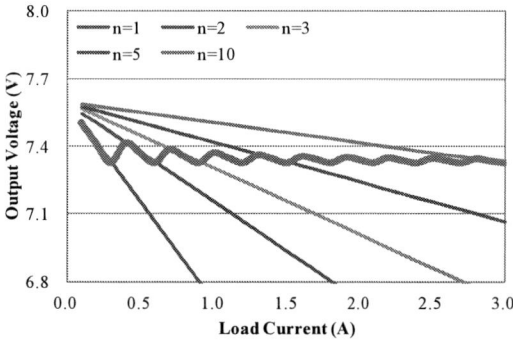

(a) Load characteristics

(b) Efficiency characteristics

Fig. 7. Simulated results (Synchronous buck converter)

IV. PROPOSED CONTROL STRATEGY

A. Principle of proposed techniqe

The proposed control strategy is to use the output voltage detection instead of load current detection to regulate the output voltage. This reduces the number of parts and realizes more accurate output voltage regulation compared with previously proposed current detection technique.

Two features to prevent the unnecessary oscillation of output voltage are introduced

(1) The output voltage sampling are performed several times and compare at AD converter

(2) The switching points, where change the number of the working POLs have hysteresis.

The flowchart of the proposed control strategy is shown in Fig. 8 (a). The 'N' is number of working POLs, Vset is target output voltage and Vout is the output voltage.

Firstly, FPGA reads the output voltage data from AD converter which averages 9 times of digitized data.

Secondly, the averaged data (data_C) is compared with the target voltage and Vset+VN. Where, VN is Vout/N. If data_C is less than target voltage of Vset, the proposed strategy changes the number of the working POLs in order to regulate the output voltage. The number of working POLs is decided by comparing with V12_gap and (Vset–Vout)*N.

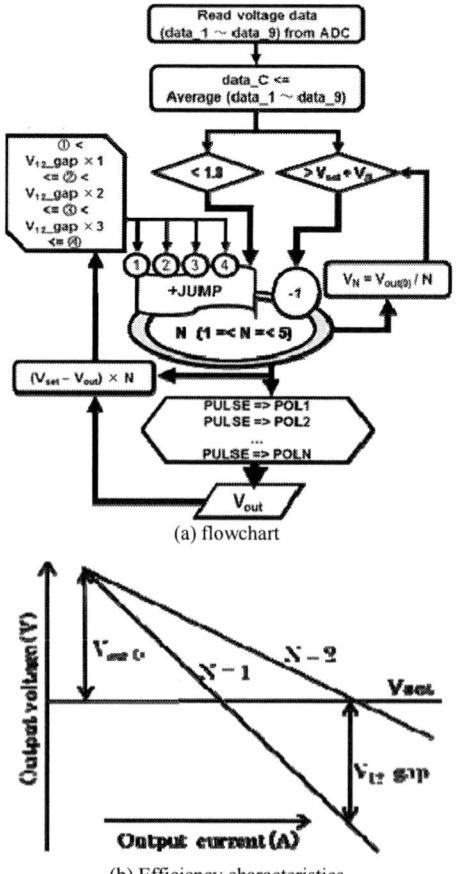

(a) flowchart

(b) Efficiency characteristics

Fig. 8. Proposed control algorism.

978-1-4799-2706-7/14 $31.00 © 2014 IEEE

The definition of V12_gap and Vout(0) are shown in Fig. 8 (b). The horizontal axis is the output current and the vertical axis is the output voltage. As shown in Fig. 7 (b), the load characteristics when the number of the working POL are single and two, respectively. Vout(0) is intersection of two lines. V12_gap is output voltage difference between single POL and two POLs when the output voltage of the two working POLs is equal to Vset.

The control scheme is as follows;

When (Vset–Vout)*N is less than V12_gap, one additional POL starts working, when (Vset–Vout)*N is between V12_gap and twice of V12_gap (2*V12_gap), two additional POLs start working, when (Vset–Vout)*N is between (n-1)*V12_gap and n*V12_gap, n additional POLs start working.

When data_C is over Vset+VN, number of working POLs is reduced one by one.

B. Simulation results

In order to evaluate the proposed control strategy, the control algorism is implemented by means of MATLAB/Simulink. Figure 9 shows the simulated system. In this case, 5 POLs are connected in parallel. An enable port is added to POL in order to switch the number of POLs as shown in Fig. 9 (a). The gate driver is controlled by input signal of the enable port. Moreover, the averaging block is included in order to suppress the influence of the output voltage ripple as shown in Fig. 9 (a). The averaged output voltage is entered into the control block. By using the input signal, the number of working POLs are decided according to proposed control algorithm.

Figure 9 (b) shows the property of control algorithm which is the core of this control strategy. This part consists of function block and switch network block. In the function block "N" and "(Vset–Vout)*N" are calculated by using the averaged output voltage. The switch network is to change the number of working POLs by using calculated signal "N".

The circuit parameters and specifications of simulated system are shown in Table II.

Figure 10 shows the simulation results including the proposed control strategy. The output voltage can be regulated without load current signal as shown red line in Fig. 10 (a). Moreover, the effect of the proposed control strategy is confirmed. The output voltage is slightly higher than target voltage. This is the influence of the hysteresis value to suppress the oscillation problem. Furthermore, the efficiency can be kept high as shown red line in Fig. 10 (b).

C. Experimental results

The picture of the prototype evaluation board is shown in Fig. 11. The five parallel connected buck converters with synchronous rectification are used as an example. The circuit parameters and specifications are shown in Table II.

After the output voltage is detected and digitized by AD converter (ADC0804LCN), digitized data is fed to FPGA (Cyclone III).

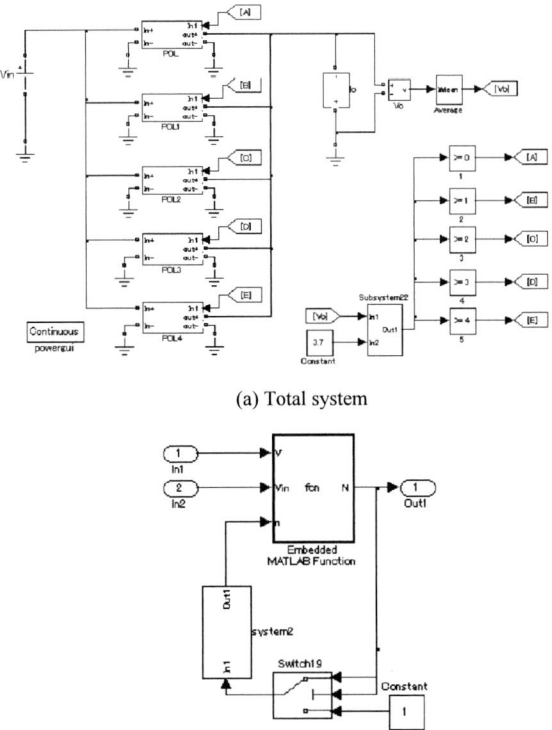

(a) Total system

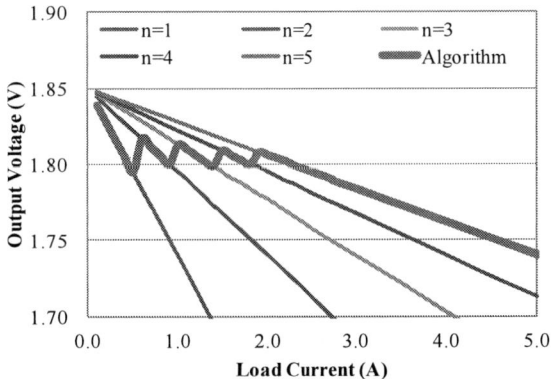

(a) Load characteristics

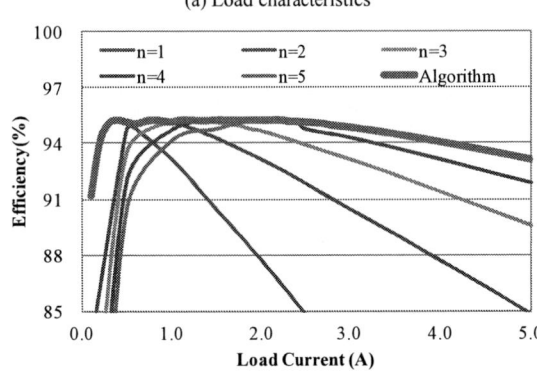

(b) Control algorithm

Fig. 9. Simulated model for proposed control strategy.

(b) Efficiency characteristics

Fig. 10. Simulated results for proposed control strategy

Then, the gate driving signals are generated by FPGA. In this case, the gate driving signals are fed to POLs which will be working.

Dependence of the output voltage on the output current and dependence of the output current on the efficiency are shown in Figs. 12 (a) and (b). The red line indicates the proposed strategy. The other lines indicate these dependences under parallel connection without control.

The proposed strategy can regulate output voltage by only changing the numbers of parallel connected converters at the constant duty ratio without oscillation over wide range. The hysteresis is composed of red line in Fig.12 (a) is effective for suppressing the oscillation of the output voltage. In addition, the proposed strategy enables to keep high efficiency over the wide load current range. The experimental results and simulation results are agreed well. These results show that the proposed control strategy is effective, and the simulated model by means of MATLAB/Simulink is also effective.

V. CONCLUSIONS

In this paper, a new control strategy suitable for power-supply on chip by voltage detecting technique is proposed. The proposed strategy can regulate the output voltage by changing number of the working POLs instead of changing duty ratio. The value of proposed control strategy is confirmed by simulation and experiment.

TABLE II
CIRCUIT PARAMETERS (PROPOSED CONTROL STRATEGY)

Symbol	Description	Value
Vin	Input Voltage	3.7V
Vo	Output Voltage	1.8V
Lo	Smoothing Inductor	10uH
rDC	Internal Resistance	100mOhm
Co	Output Capacitor	470uF
fs	Switching Frequency	300kHz

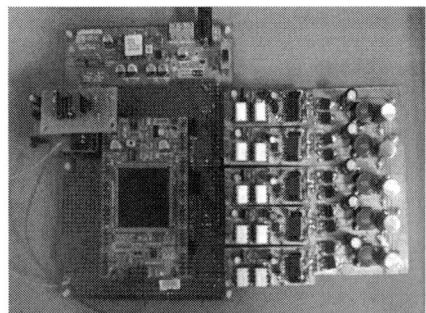

Fig. 11. Prototype evaluation board.

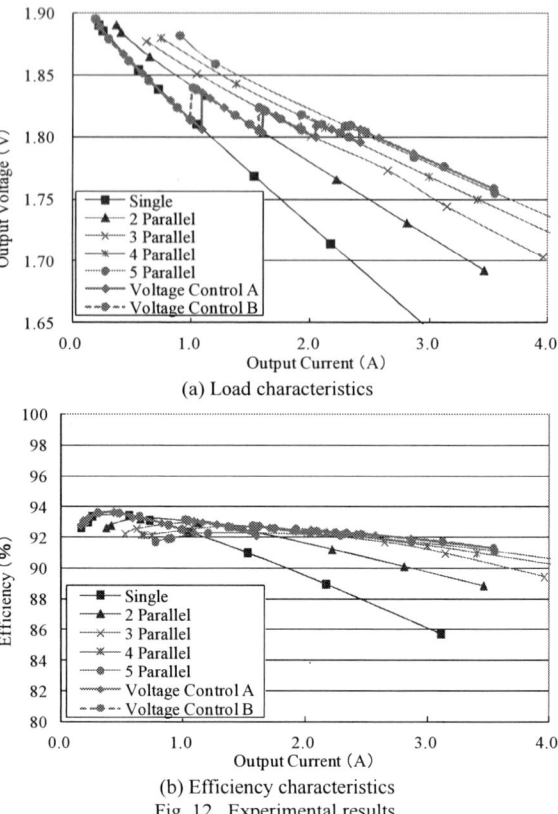

(a) Load characteristics

(b) Efficiency characteristics

Fig. 12. Experimental results.

REFERENCES

[1] http://www.tij.co.jp/analog/jp/docs/analogsplash.tsp?contentId=46073

[2] S. Matsumoto, M. Mino, and T. Yachi: "Integration of a power supply part for a system on silicon", IEICE Trans. Communication and Computer Science, vol.80-A, No.2, pp.276-282, 1997

[3] International Workshop on Power Supply On Chip 2010(PwrSoc' 10) http://www.powersoc.org/index.php

[4] A. Prodic: "High-Pergormance Mixed-Signal Controllers for On-Chip Integrated SMPS" International Workshop on Power Supply On Chip 2010(PwrSoc' 10), Session 1.4,2010

[5] H. Meyvaert and E. Micas: "The importance of fully-integrated CMOS: Cost Effective Integrated DC-DC Converters", International Workshop on Power Supply On Chip 2010(PwrSoc' 10), Session 1.4,2010

[6] T. Yamamoto, J. Rikitake, S. Matsumoto, S. Abe, "A New Control Strategy for Power Supply on Chip Using Parallel Connected DC-DC Converter", IEEE 10th International Conference on Power Electronics and Drive Systems (PEDS) 2013, pp. 109-112.

[7] MATLAB R2011b copyright 19984-2011.The MathWorks, Inc

978-1-4799-2706-7/14 $31.00 © 2014 IEEE

Matrix-POL Architecture for Integrated Power Supply

Yoichi Ishizuka/Nagasaki University
Graduate School of Engineering
Nagasaki Univ.
Nagasaki, Japan
isy2@nagasaki-u.ac.jp

Kiminori Tanaka/Nagasaki University
Graduate School of Engineering
Nagasaki Univ.
Nagasaki, Japan

Ryota Shibahara/Nagasaki University
Graduate School of Engineering
Nagasaki Univ.
Nagasaki, Japan

Seiya Abe
The International Centre for the Study of
East Asian Development
Electronics Research Group for Sustainability
ICSEAD
Kitakyushu, Japan
abe@icsead.or.jp

Tamotsu Ninomiya
The International Centre for the Study of
East Asian Development
Electronics Research Group for Sustainability
ICSEAD
Kitakyushu, Japan
t_ninomiya@icsead.or.jp

Abstract— **In this study, integrated H-bridge converter with the Matrix-POL power supply system is proposed. From the simulation results, the validity of the Matrix-POL is revealed. The results revealed that the fast response to the load current and the voltage change can be done with duty and parallel number control by the proposed system.**

Keywords— Point-Of-Load(POL), Power integrated circuit, Voltage Regulator Modules(VRM)

I. INTRODUCTION

In Fig.1, the estimation of the increased power consumption by information appliance in both of Japan and world is shown. The consumption is expected that it would be increased by 5.2 times over the period of 2006 to 2025 in Japan. Also, it is expected to be 9.4 times over the same period in the world. Japan has the plans to reduce 40% of the power consumption by 2025 [1]. Recently, Micro Processing Unit (MPU) has been required to realize energy conservation without reducing the information processing capacity [2-6]. It can be possible to perform power saving at the system level. Dynamic voltage and frequency scaling (DVFS) and power gating have been studied for MPU power conservation. DVFS is the technique that realizes the low power consumption by the optimal value of the supply voltage and clock frequency of LSI as shown in Fig.2 [7]. The power gating is the technique that suppressing the leak current by interrupting the power supply when the part of the MPU is in de-active mode in Fig.3 [8]. The concepts of these techniques were very effective for power management. But, these techniques require the tracking control for very fast change of the power supply voltage or the load current for point of load (POL) type DC-DC converter. Conventional bulky POL is not enough for achieving fast response because of the effect of the package or line impedance. Therefore, nowadays, the integration of POL with MPU has been paid attention, rapidly. In this paper, as one of the integrated POL study, Matrix-POL power supply system is proposed.

II. MATRIX-POL POWER SUPPLY SYSTEM MODEL

Recently, MPU is incorporating many cores. Each core and, further, SRAM and interface block requires independent supply voltage for energy conservation. Therefore, it is necessary to provide multiple power supply lines in the MPU. Figure 4 shows an example of proposed integrated Matrix-POL power supply system.

Each single POL is constructed with H-bridge converter, which has the functions of bi-directional power transfer and buck-boost voltage conversion as shown in Fig.5.

This example has two-input and m-output terminals. Each output terminal of POL is connected in parallel. The parallel number is dynamically changed by the output voltage regulation feedback control. The connection of the output terminal of each POL is realized with the power gating switches. With this technique, active series regulator can be realized by controlling the parallel number. The rest of the parallel POL is connected to the next output stage. At least, one of the terminals is connected to next output stage.

Figure 6 shows buck-mode operation and Fig.7 shows boost-mode operation of the H-bridge. The both mode is realized with just the gate control of the switches. In this paper, the simulation results are obtained just with buck-mode operation. The duty ratio of each POL in the same stage is synchronized. And the duty ratio of the each stage is adjusted with the voltage reference output (VID) of the each independent power supply blocks of MPU. VID is supplied as feed forward signal. For more detail is mentioned next chapter.

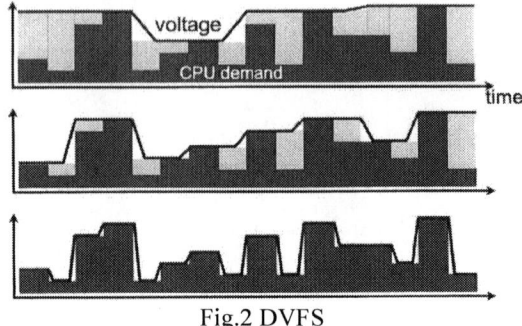

Fig.2 DVFS

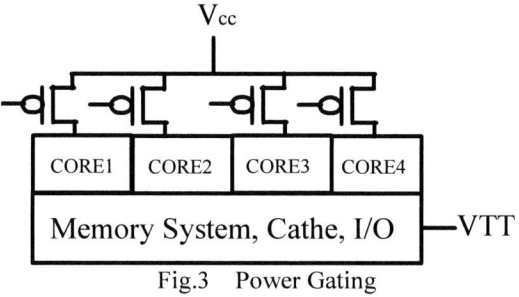

Fig.3   Power Gating

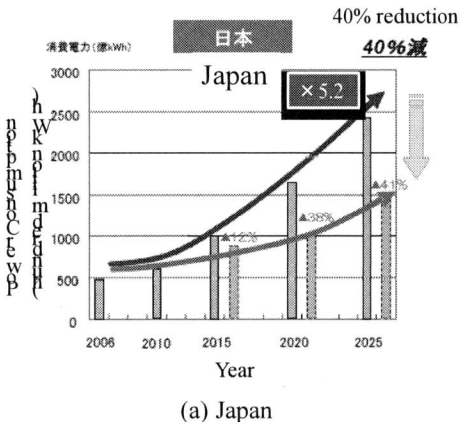

(a) Japan

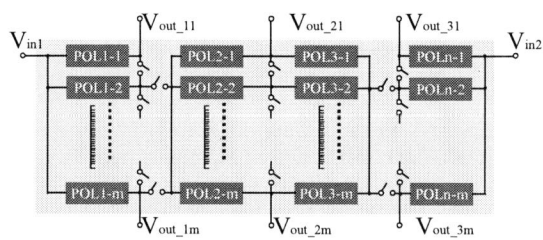

Fig.4 Matrix-POL Power Supply System

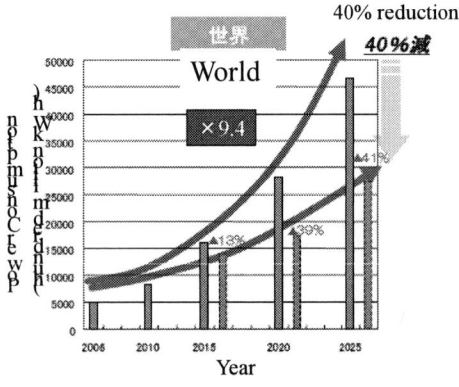

(b) World

Fig.1 Changes in Estimates of Power Consumption by the Information Appliance

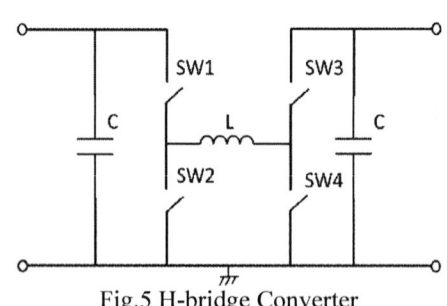

Fig.5 H-bridge Converter

III.   MATRIX-POL POWER SUPPLY SYSTEM MODEL

Figure 8 shows the control algorithm of the Matrix-POL power supply system. In this system, method of controlling the output voltage has two ways. One is the duty ratio and on the other hand is the number of parallel. In steady term, the each output terminal voltage is regulated with the feed-forward controlled duty ratio depends on VID of MPU. And, dynamical output voltage regulation is performed with feed-back controlled parallel number of POL. With this technique, high-speed-response of the output voltage is realized. Figure 9 shows simulation model. The output terminals of each ten paralleled POLs are connected to one output terminal. The gate switch can adeptly change the parallel number for the output terminal. The rest of the POLs which have not be connected to output terminal, are all connected to the input terminal of next output stage.

To confirm the validation of the proposed method, we performed numerical simulation for multi-output condition with Maplesim5. The Matrix-POL is configured one input voltage and three output voltage terminal. To each output voltage terminal, paralleled POL whose duty ratio is same is connected. The parallel number is dynamically changed from one to nine depends on the feed-back signal.

IV.   LOAD CHAGE SIMULATIONS

To confirm the validation of the proposed method, we performed simulation for multi-output condition by Maplesim5. The Matrix-POL is configured one input voltage and three output voltage terminal. To each output voltage terminal, paralleled POL whose duty ratio is same is connected. The parallel number is dynamically changed from one to nine depends on the feed-back signal. The all of the rest parallel POL is connected to next output voltage stage. Figure 9 is simulation circuit. Square block of the figure is the H-bridge. Table I shows simulation parameters. Table II, III and IV shows the look-up-table of switch combination of terminal of output voltage 1, 2 and 3, respectively. Figure 10, 11, and 12 show simulation results of light to heavy load. Figure 13, 14 and 15 show simulation results of heavy to light load. From the results, it confirms that the output voltage is within the operating range of the load in all conditions. But there is also a part of the output voltage meets barely the operating range of the load. It is further study in this regard.

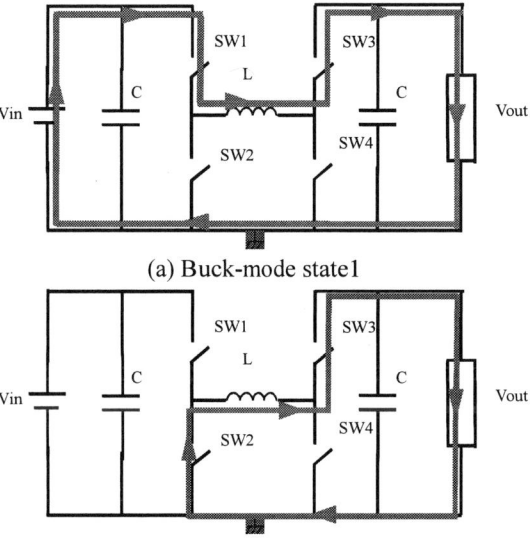

(a) Buck-mode state1

(b)Buck-mode state2

Fig.6 Buck-mode Architecture

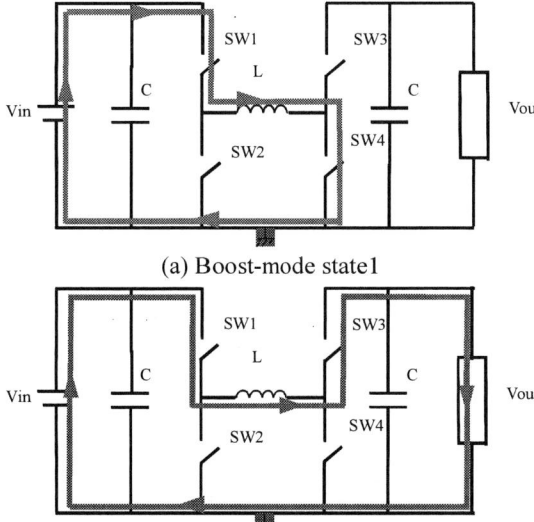

(a) Boost-mode state1

(b) Boost-mode state2

Fig.7 Boost-mode Architecture

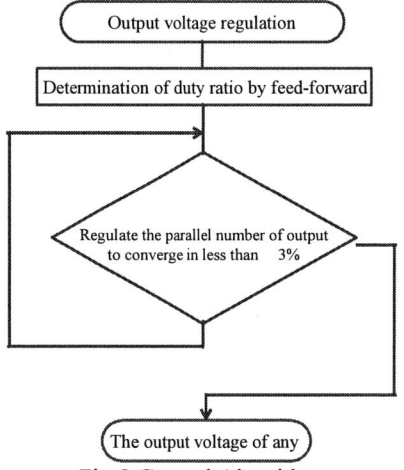

Fig.8 Control Algorithm

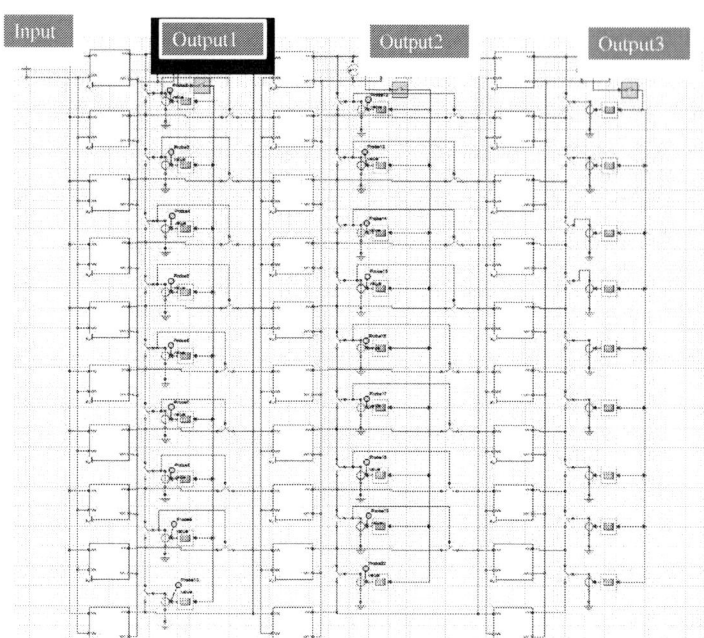

Fig.9 Simulation Circuit

Table I Simulation Parameter

Switching Frequency [MHz]	10
Input Voltage [V]	12
Output Voltage1 [V]	5
Output Voltage2 [V]	3.3
Output Voltage3 [V]	1.5
Load Current(light-load) [A]	0.3
Load Current(heavy-load) [A]	0.6
Input Resistance [Ω]	0.1
Input Capacitor [µF]	0.1
Parasitic Resistance of Input C [Ω]	0.1
Output Capacitor [µF]	0.5
Parasitic Resistance of Output C [Ω]	0.01
Inductor [µH]	0.1
Parasitic Resistance of L [Ω]	0.1
On-Resistance [Ω]	0.01
Number of Plots	5000

Table II Signal of Switches (Output Voltage1)

Output Voltage1	Connection Stage								
	first-second	second-third	third-fourth	fourth-fifth	fifth-sixth	sixth-seventh	seventh-eighth	eighth-ninth	ninth-tenth
0	1	1	1	1	1	1	1	1	0
4.9	1	1	1	1	1	1	1	1	0
4.9249999	1	1	1	1	1	1	1	1	0
4.925	1	1	1	1	1	1	1	1	0
4.949999	1	1	1	1	1	1	1	1	0
4.95	1	1	1	0	0	0	0	0	0
4.9749999	1	1	1	0	0	0	0	0	0
4.975	1	1	1	0	0	0	0	0	0
4.999999	1	1	1	0	0	0	0	0	0
5	1	1	1	0	0	0	0	0	0
5.001	1	1	1	0	0	0	0	0	0
5.0249999	1	1	1	0	0	0	0	0	0
5.025	1	1	0	0	0	0	0	0	0
5.0499999	1	1	0	0	0	0	0	0	0
5.05	1	0	0	0	0	0	0	0	0
5.0749999	1	0	0	0	0	0	0	0	0
5.075	0	0	0	0	0	0	0	0	0
5.1	0	0	0	0	0	0	0	0	0
100	0	0	0	0	0	0	0	0	0

Table III Signal of Switches (Output Voltage2)

Output Voltage	Connection Stage								
	first-second	second-third	third-fourth	fourth-fifth	fifth-sixth	sixth-seventh	seventh-eighth	eighth-ninth	ninth-tenth
0	1	1	1	1	1	1	1	1	0
3.2	1	1	1	1	1	1	1	1	0
3.2249999	1	1	1	1	1	1	1	1	0
3.225	1	1	1	1	1	1	1	1	0
3.249999	1	1	1	1	1	1	1	1	0
3.25	1	1	1	1	1	1	1	1	0
3.2749999	1	1	1	1	1	1	1	1	0
3.275	1	1	1	1	1	1	1	1	0
3.2999999	1	1	1	1	1	1	1	1	0
3.3	1	1	0	0	0	0	0	0	0
3.3000001	1	1	0	0	0	0	0	0	0
3.3249999	1	1	0	0	0	0	0	0	0
3.325	1	1	0	0	0	0	0	0	0
3.3499999	1	1	0	0	0	0	0	0	0
3.35	1	0	0	0	0	0	0	0	0
3.3749999	1	0	0	0	0	0	0	0	0
3.375	0	0	0	0	0	0	0	0	0
3.4	0	0	0	0	0	0	0	0	0
100	0	0	0	0	0	0	0	0	0

978-1-4799-2706-7/14 $31.00 © 2014 IEEE

Table IV Signal of Switches (Output Voltage3)

Output Voltage	Connection Stage								
	first-second	second-third	third-fourth	fourth-fifth	fifth-sixth	sixth-seventh	seventh-eighth	eighth-ninth	ninth-tenth
0	1	1	1	1	1	1	1	1	0
1.4	1	1	1	1	1	1	1	1	0
1.4249999	1	1	1	1	1	1	1	1	0
1.425	1	1	1	1	1	1	1	1	0
1.449999	1	1	1	1	1	1	1	1	0
1.45	1	1	1	1	1	1	1	1	0
1.4749999	1	1	1	1	1	1	1	1	0
1.475	1	1	1	1	1	1	1	1	0
1.4999999	1	1	1	1	1	1	1	1	0
1.5	1	1	0	0	0	0	0	0	0
1.5000001	1	0	0	0	0	0	0	0	0
1.5249999	1	0	0	0	0	0	0	0	0
1.525	1	0	0	0	0	0	0	0	0
1.5499999	1	0	0	0	0	0	0	0	0
1.55	1	0	0	0	0	0	0	0	0
1.5749999	1	0	0	0	0	0	0	0	0
1.575	0	0	0	0	0	0	0	0	0
1.6	0	0	0	0	0	0	0	0	0
100	0	0	0	0	0	0	0	0	0

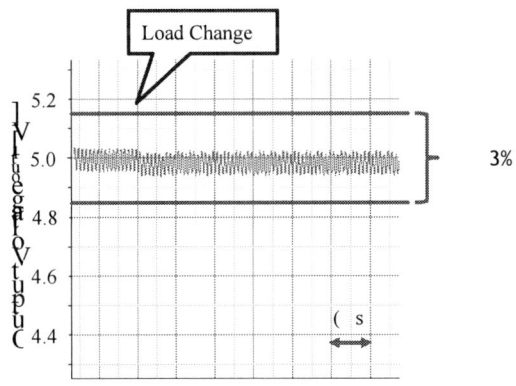

Fig.10 Output Voltage (Light to Heavy)

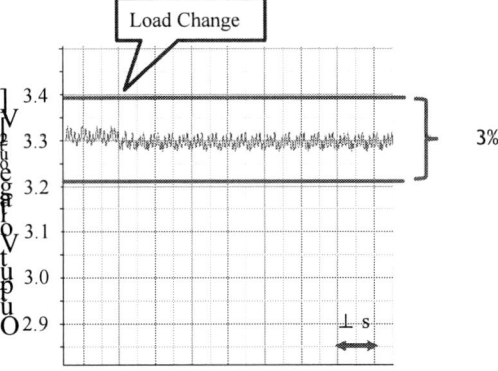

Fig.11 Output Voltage (Light to Heavy)

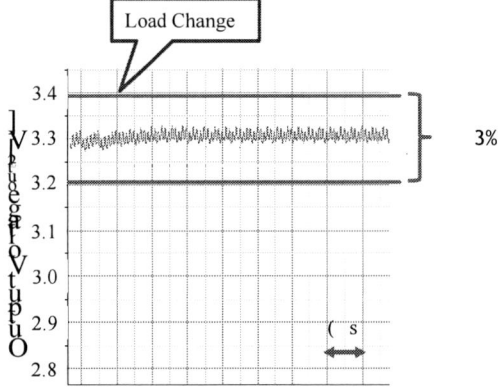

Fig.13 Output Voltage (Heavy to Light)

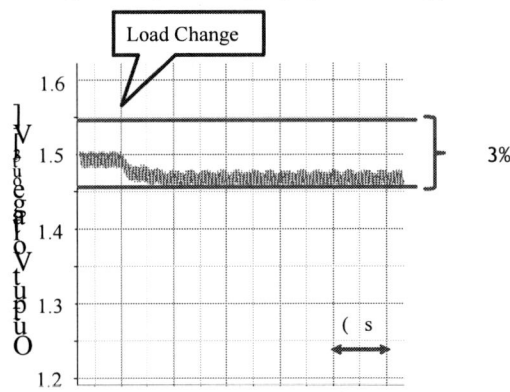

Fig.12 Output Voltage (Light to Heavy)

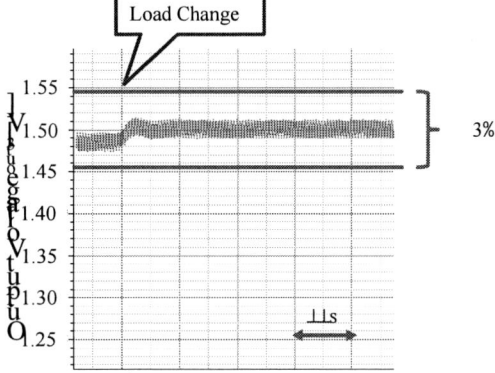

Fig.14 Output Voltage (Heavy to Light)

Fig.15 Output Voltage (Heavy to Light)

V. CONCLUSIONS

In this paper, Matrix-POL architecture for integrated power supply is proposed. From the simulation results, the validity of the Matrix-POL is shown. The results revealed that the fast response to the load current and the voltage change can be done with duty and parallel number control.

REFERENCES

[1] Ministry of Economy, Trade and Industry, Green IT initiative, http://www.meti.go.jp/committee/materials/downloadfiles/g80619b08j.pdf

[2] Mandy Pant and Bill Bowhill "Era of Intelligent Power Delivery," PowerSoC2012, 2012

[3] Ken Shepard , "Scalable integrated voltage regulation technologies, " PowerSoC 2012, 2012

[4] Stephen Kosonocky, "Power Delivery Challenges for Next Generation High Performance SOCs, " PowerSoC 2012, 2012

[5] Gerard Villar Piqué, "Potential Benefits of Integrated Switching Power Converters: Inductive vs. switched-Capacitor , " PowerSoC 2012, 2012

[6] Toke Meyer Andersen, Florian Krismer, Johann W. Kolar, and Thomas Toifl, "High Power Density On- Chip Switched Capacitor Converters for Microprocessor Power Delivery , " PowerSoC 2012, 2012

[7] Wonyoung Kim, David M Brooks, Gu-Yeon Wei, "A Fully-Integrated 3-Level DC/DC Converter for Nanosecond-Scale DVS with Fast Shunt Regulation," ISSCC 2011, 2011.

[8] Tanay Karnik, T6: Power Management Using Integrated Voltage Regulators, IEEE International Solid State Circuit Conference 2012, 2012.2.

On-Chip Buck Converter with Spiral Ferrite Inductor and Reducing IR Drop in 3D Stacked Integration

Hiroshi Fuketa[*], Yasuhiro Shinozuka, Koichi Ishida, Makoto Takamiya, and Takayasu Sakurai

Institute of Industrial Science
University of Tokyo
Tokyo, Japan
[*] fuketa@iis.u-tokyo.ac.jp

Abstract— **In this paper, two topics about an on-chip DC-DC buck converter are described. First, the buck converter with an inductor on interposer is investigated for mobile applications. Simulation results indicate that the efficiency of the buck converter is improved by introducing a ferrite film to the inductor. Next, a circuit technique to reduce IR drop in 3D stacked integration using the buck converter is proposed. In this paper, 3D stacked-die system, which consists of stacked dies and silicon interposer, is fabricated and measurement results show that the proposed technique achieves 78% decrease in IR voltage drop compared with the conventional approach.**

Keywords— *3D stacked integration, Ferrite, Interposer, IR Drop, On-chip DC-DC buck converter*

I. INTRODUCTION

A buck converter is often used as a DC-DC voltage down converter, since the buck converter provides high conversion efficiency at rather high current output among different types of DC-DC voltage down converters.

In this paper, the following two topics about the buck converter for mobile applications and 3D IC's are described. 1) Efficiency improvement of the buck converter by introducing high permeability material to an inductor on interposer is investigated [1]. 2) A circuit technique to reduce IR drop in 3D stacked integration using the buck converter is presented [2].

II. EFFICIENCY INCREASE IN BUCK CONVERTER BY INTRODUCTION OF HIGH PERMEABILITY MATERIAL TO INDUCTOR ON INTERPOSER

A buck converter requires an inductor. Since mobile devices have been currently smaller and thinner, miniaturization of the inductor used for the buck converter is required. Although it is one of promising solutions that inductors are implemented in IC package or interposer, the inductance and Q factor of such inductors are smaller, which results in worsening the efficiency of the buck converter. Therefore, introduction of high permeability material to the inductor is expected to improve the efficiency. In this paper, the efficiency of the buck converter using inductor on interposer with a ferrite film is investigated as shown in Fig. 1.

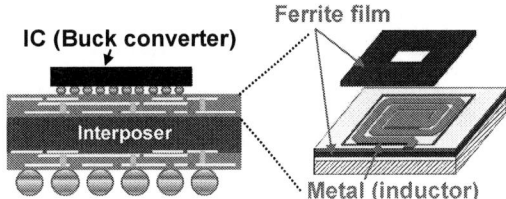

Fig. 1. Inductor on Interposer with ferrite film.

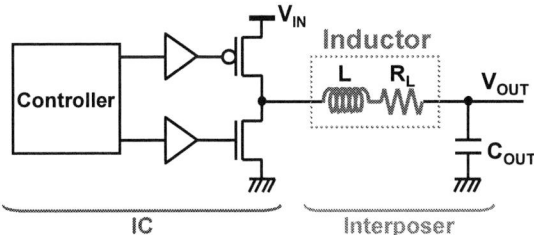

Fig. 2. Circuit schematic of buck converter with inductor on interposer.

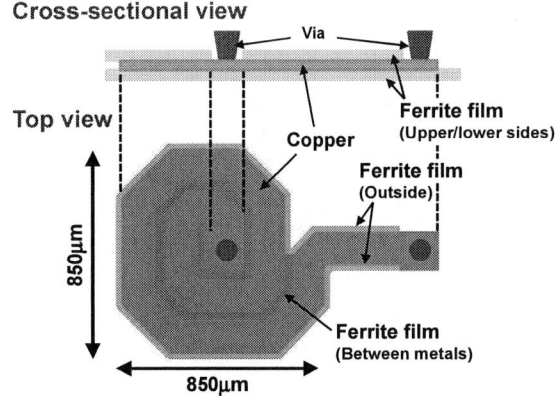

Fig. 3. Layout of spiral ferrite inductor.

Fig. 2 shows a circuit schematic of the buck converter evaluated in this paper. A controller circuit, buffers, and drivers of the buck converter are implemented in the IC, while an inductor is fabricated in the interposer using metal interconnects of the interposer. The layout of the inductor is illustrated in Fig. 3. In this paper, the

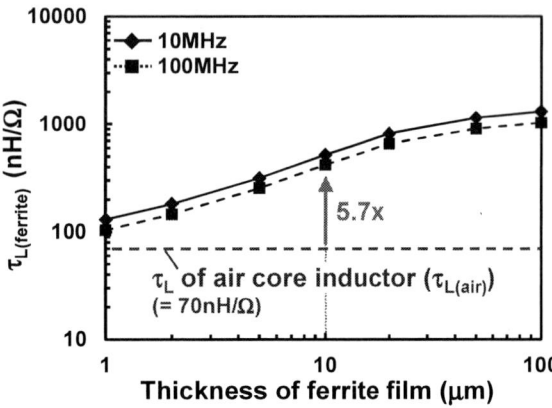

Fig. 4. Dependence of τ_L of ferrite inductor and air core inductor.

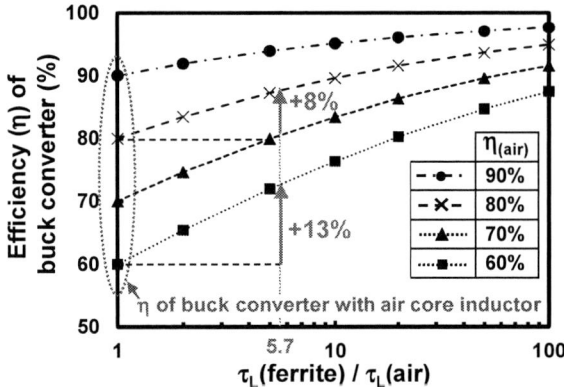

Fig. 5. Efficiency increase of buck converter by using ferrite inductor.

inductance of the inductor is simulated by a 3D electromagnetic simulator (Agilent EMPro).

According to [3], the efficiency of the buck converter depends on τ_L, which is given by

$$\tau_L = L/R_L \,, \qquad (1)$$

where L and R_L are the inductance and the DC resistance of the inductor, respectively. As τ_L increases, the efficiency of the buck converter improves. Therefore, τ_L is evaluated to investigate the efficiency improvement of the buck converter by using the ferrite inductor in this paper.

Fig. 4 shows the dependence of τ_L on the thickness of the ferrite film. $\tau_{L(ferrite)}$ and $\tau_{L(air)}$ are τ_L of the ferrite inductor and the air core inductor, respectively. When the thickness of the ferrite film is 10μm, $\tau_{L(ferrite)}$ is 5.7 times larger than $\tau_{L(air)}$. Fig. 5 shows the efficiency improvement of the buck converter by using the inductor with the ferrite film when the input/output voltage of the buck converter is 1.8/1.0V, the output current is 1.0A, and the switching frequency is 100MHz. By using the inductor with the 10μm-thick ferrite film, the efficiency of buck converter is improved by 13% and 8% when the efficiencies of buck converter with the air core inductor are 60 and 80%, respectively.

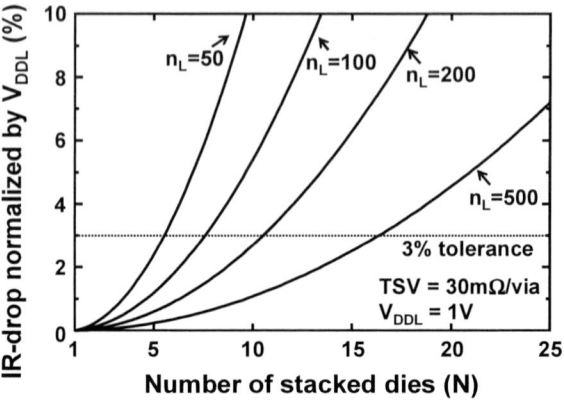

Fig. 6. Calculated IR drop in 3D integration. n_L is the total number of TSV's per die used for power delivery. Each die is assumed to consume 1A.

III. REDUCING IR DROP IN 3D INTEGRATION USING BUCK CONVERTER ON TOP DIE SCHEME

Recently, 3-dimensional integration based on TSV's (through silicon vias) have been explored extensively for achieving low power and high performance to break the limit of single-chip 2D integration. The TSV and associated bump connection structure (TSV link) inherently have parasitic resistance and thus the supply voltage is exposed to IR drop due to the TSV link resistance [4,5]. The increase of IR drop is shown in Fig. 6. The IR drop is calculated as a function of the number of stacked dies (N) and the total number of TSV's per die (n_L) used for power delivery. If allowable IR drop is assumed to be 3% of the nominal power supply voltage (V_{DDL}), it is seen from the figure that hundreds of TSV's are needed to make the IR drop within the tolerance.

In order to reduce the IR drop due to TSV, we proposed a Buck Converter on Top die (BCT) scheme [2]. Figs. 7 and 8 show a typical 3D integration with the conventional power supply scheme and the proposed BCT scheme. In the conventional power supply scheme (Fig. 7), all the current is supplied from a buck converter which is either implemented on the bottom die or is implemented as a separate die residing external to the 3D stacked dies. Since the top die can be far from the power supply circuit, the IR drop gets large as the current must go through many serially-connected TSV's.

In contrast, a buck converter is added on the topmost die (Die N) in the proposed BCT scheme, and the current to an upper half of the 3D integration (Die (N/2)+1 ~ Die N) is delivered from the added BCT (Fig. 8). Each buck converter on the top die and the bottom die stably supply the specified voltage and the current to the furthest die (Die N/2, Die (N/2)+1) from a buck converter through less number of TSV's compared with the conventional case. Thus, the maximum IR drop gets smaller and improved.

In order to evaluate the effectiveness of the proposed BCT scheme, we fabricated 3D stacked-die system, which consists of stacked dies and silicon interposer. Its

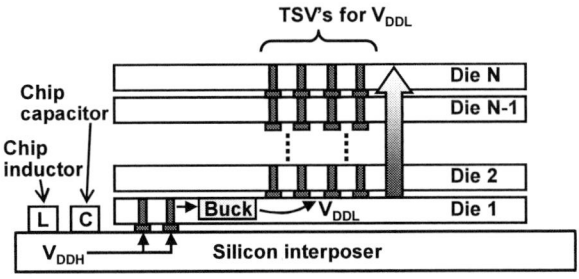

Fig. 7. Conventional power supply scheme.

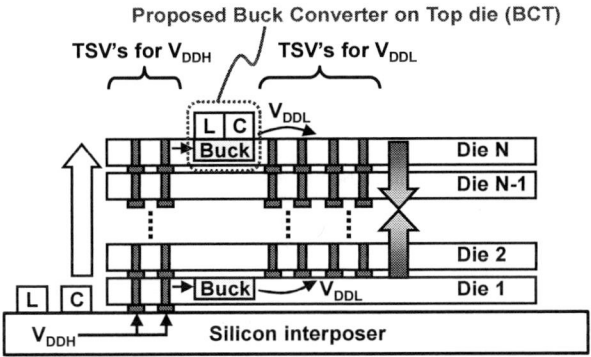

Fig. 8. Proposed Buck Converter on Top die (BCT) scheme.

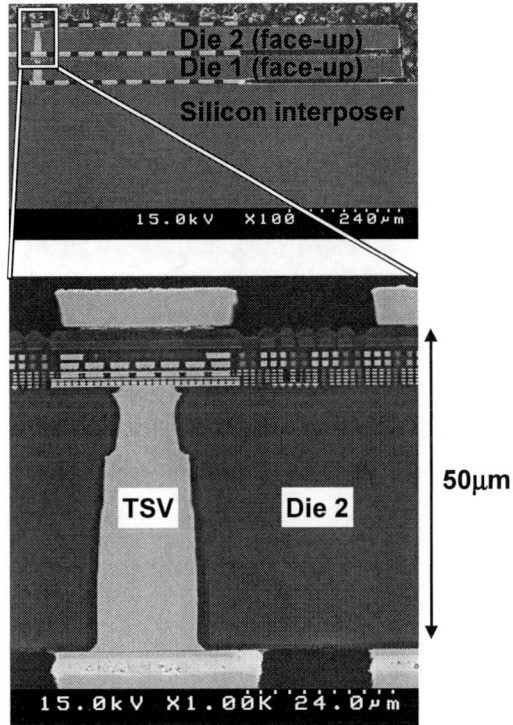

Fig. 9. Cross sectional view of 3D integrated system. The upper photo is a global view and the lower photo is a magnified view around a TSV.

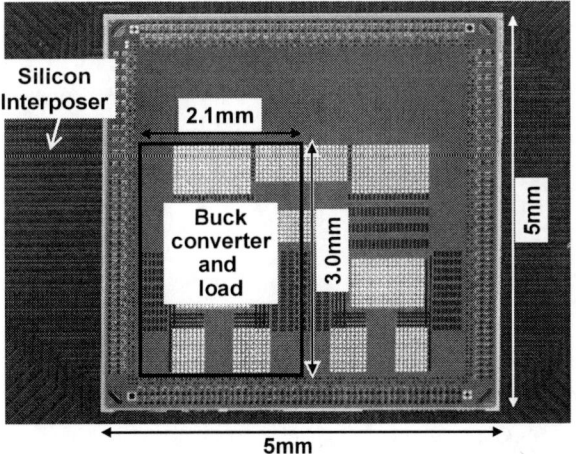

Fig. 10. Two stacked dies using TSV's are mounted on top of a silicon interposer.

cross sectional view is shown in Fig. 9. Two face-up dies are stacked on top of a silicon interposer with a via-last process. The diameter of the TSV is 20μm and the minimum pitch of the TSV's is 50μm. The resistance of a TSV link is 29mΩ (typical). Although only two dies are stacked physically as shown in Fig. 9, 8-die stack case is emulated by the daisy chain of TSV's in two stacked dies in this paper.

The technology used to implement circuits is 8-metal, 90nm CMOS. The typical operation voltage for logic transistors is 1.2V while I/O transistors accept typically 3.3V. Thus, 3.3V is used for V_{DDH} and 1.2V for V_{DDL} in the 3D stacked-die system design.

A microphotograph of the fabricated two stacked dies mounted on top of a silicon interposer is shown in Fig. 10. The silicon interposer is needed to accept 50μm pitch TSV's, which cannot be realized by a PCB board whose design rule is loose. Two identical buck converters are implemented on a die to expect yield improvement in this 3D stacked-die system.

The photo in Fig. 11 shows an assembled 3D stacked-die device. A chip inductor and two chip capacitors are directly mounted on the top die. The blue part in the picture is a silicon interposer. The chip inductor and capacitor for a buck converter is 3.9nH and 1μF, respectively. These values are chosen to optimize the buck converter design following the design theory described in [3].

Fig. 12 shows the measured and simulated IR drop in the fabricated 3D stacked-die system. The simulation results are obtained using SPICE. Although the voltage of only selected locations can be monitored in the manufactured 3D stacked-die system due to TSV constraints, the simulation results can be used to extrapolate the measured points since the simulation results agree very well with the measured results as is seen in Fig. 12. It is shown that the proposed scheme achieves 78% decrease in IR voltage drop compared with the conventional approach. That is, the IR drop is reduced to less than 1/4 by introducing the BCT scheme.

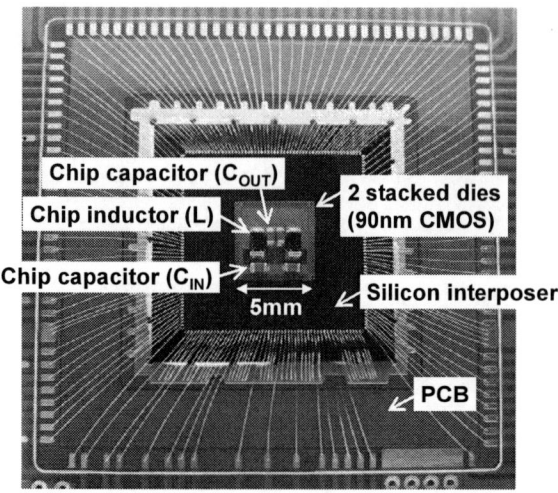

Fig. 11. Photograph of assembled system on a PCB board. Chip inductor and chip capacitor are mounted directly on top of stacked dies at the center of the photo.

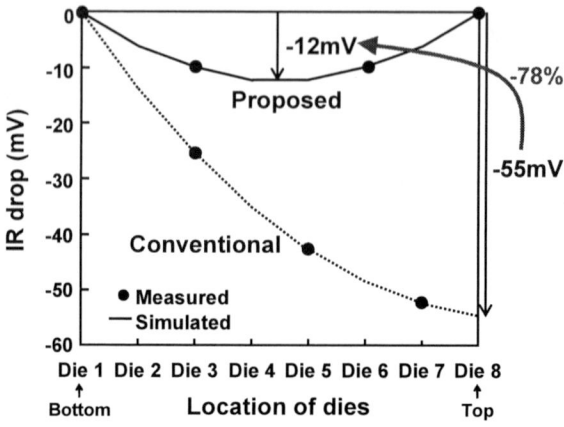

Fig. 12. Measured IR drop and simulated results.

IV. CONCLUSIONS

In this paper, the two topics about the on-chip buck converter were described. First, we investigated the efficiency improvement of the buck converter by introducing high permeability material to an inductor on interposer. The simulation results indicated that the efficiency of the buck converter is improved by 8% by using the inductor with the 10μm-thick ferrite film when the efficiency of the buck converter with the air core inductor is 80%. Next, we proposed the buck converter on top die (BCT) scheme to reduce IR drop in 3D stacked integration. 3D stacked-die system, which consists of stacked dies and silicon interposer, was fabricated and measurement results showed that the proposed BCT scheme achieves 78% decrease in IR voltage drop compared with the conventional approach.

ACKNOWLEDGMENT

The authors would like to thank Prof. Toshiro Sato from Shinshu University, Mr. Tomoharu Fujii, Mr. Hiroshi Shimizu, and Mr. Kazutaka Kobayashi from SHINKO Electric Industries Co., Ltd. for valuable discussions regarding the inductor on interposer with the ferrite film.

The authors also would like to acknowledge Dr. Futoshi Furuta, Dr. Kenichi Osada, and Dr. Kenichi Takeda from Association of Super-Advanced Electronics Technologies (ASET) for fabricating 3D stacked ICs.

REFERENCES

[1] H. Fuketa, Y. Shinozuka, K. Ishida, M. Takamiya, T. Fujii, H. Shimizu, K. Kobayashi, T. Sato, and T. Sakurai, "Efficiency Increase in On-Chip Buck Converter by Introduction of High Permeability Material to Inductor on Interposer," in Proc. *International Conference on Ferrites (ICF)*, p. 75, 2013.

[2] Y. Shinozuka, H. Fuketa, K. Ishida, F. Furuta, K. Osada, K. Takeda, M. Takamiya, and T. Sakurai, "Reducing IR Drop in 3D Integration to Less Than 1/4 Using Buck Converter on Top Die (BCT) Scheme," in Proc. *IEEE International Symposium on Quality Electronic Design (ISQED)*, pp. 210-215, 2013.

[3] G. Schrom, P. Hazucha, F. Paillet, D. S. Gardner, S. T. Moon, and T. Karnik, "Optimal Design of Monolithic Integrated DC-DC Converters," in Proc. *IEEE International Conference on IC Design and Technology (ICICDT)*, pp.1-3, 2006.

[4] P. Singh, R. Sankar, X. Hu, W. Xie, A. Sarkar, and T. Thomas, "Power Delivery Network Design and Optimization for 3D Stacked Die Designs," in Proc. *IEEE International 3D System Integration Conference*, pp. 1-6, 2010.

[5] M. Jung and S. Lim, "A Study of IR-Drop Noise Issues in 3D ICs with Through-Silicon-Vias," in Proc. *IEEE International 3D System Integration Conference*, pp. 1-6, 2010.

The 2014 International Power Electronics Conference

DCM Analysis of a Single SiC Switch Based ZVZCS Tapped Boost Converter

Bo H. Choi[1], Eun S. Lee[1], Ji H. Kim[2], Chun T. Rim[1]

[1]Department of Nuclear and Quantum Engineering   [2]Department of Electrical Engineering

KAIST

Daejeon, Korea

bohwan@kaist.ac.kr

Abstract-The analysis of discontinuous conduction mode (DCM) for a novel single active switch based zero voltage and zero current switching (ZVZCS) tapped boost converter is proposed in this paper. The ZVZCS converter includes a lossless snubber composed of three diodes and two capacitors, and exhibits novel ZVZCS operation for wide operating ranges of duty cycle, load current, and input voltage. The voltage stress of the active switch is always less than load voltage, and soft switching turn-on and off is achieved without cumbersome current or voltage sensing, which could not have been obtained from quasi-resonant converters that also have an active switch. A detailed analysis for the DCM and the design procedures of the proposed converter are presented. Experiments for a 450 W prototype show 99.0% of maximum efficiency with a SiC JFET at the switching frequency of 50 kHz, and the ZVZCS operation is guaranteed even for the DCM under the experiment conditions.

I. INTRODUCTION

'Higher switching frequency' for reducing the size and weight of converters and filters is one of the most important issues in power electronics. In particular, demands for high frequency applications in electric vehicles, renewable energy systems, and light emitting diode (LED) drivers are increasing recently. Another vital issue for reducing the energy loss and generated heat is 'higher power efficiency'. To meet these demands, numerous soft switching techniques achieving zero voltage switching (ZVS) and/or zero current switching (ZCS) of a converter have been presented for last several decades [1]-[3].

Despite of the advantages of soft switching, the ZVS and/or ZCS techniques have demerits such as additional active devices, excessive component stresses, cumbersome voltage and/or current sensing, limited operating range and nonlinearity of a converter, and high frequency ringing after ZCS turn-off. Conventional converters, on the other hand, do not have such problems mentioned above except for the biggest demerit of hard switching characteristic. Therefore, a soft switching converter of an active switch with no larger stresses, having the same operating characteristics as the conventional hard switching converters is strongly required as a countermeasure for the demerits of existing techniques so far.

As a candidate of such soft switching technique, a single switch based zero voltage and/or zero current switching (ZVZCS) for a tapped boost converter example was proposed [4], as shown in Fig. 1. For the ZVZCS operation of the converter, a lossless snubber circuit

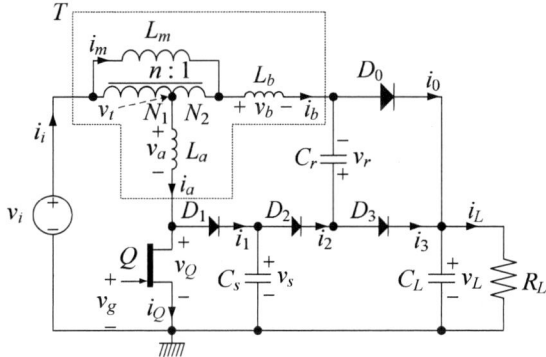

Fig. 1. Single switch based ZVZCS tapped boost converter.

composed of three small size diodes and two capacitors is added to an ordinary tapped boost converter. When the active switch is turned on, it is in ZCS with the help of the leakage inductance of a tapped inductor. When turned off, the energy of the leakage inductance is transferred to a snubber capacitor through a diode, which results in ZVS of the active switch. The energy of the snubber capacitor is then retrieved through a diode and a capacitor connected to the end of the tapped inductor when the active switch is turned on again. The single switch based ZVZCS tapped boost converter has its advantages of a wide ZVZCS range, a low voltage stress, a load independent switching, a low current stress of snubber, a robust DC gain, and no parasitic ringing.

In this paper, the operation principles and design considerations for the discontinuous conduction mode (DCM), which was not covered by the paper [2], is described in detail, and experimentally verified by a SiC JFET based prototype of 50 kHz switching frequency.

II. CIRCUIT ANALYSIS FOR DCM OPERATION

The DCM operating modes of the proposed converter for a switching period T_s can be classified into fourteen, as shown in Fig. 2, where the switching waveforms are shown in Fig. 3. Turning-off transition of the active switch is for *Modes* 1-4, off-mode is *Mode* 5, whereas turning-on transition is for *Modes* 12-13, and on-mode is *Mode* 14, as shown in Fig. 4. Between the off-mode and the turning-on transition, there is DCM transient, which consists of several repetitive LC resonance after the magnetizing current i_m goes to zero, for *Modes* 6-11. The averaged static characteristic of the proposed converter in the DCM is determined simply by an averaged duty cycle D like a conventional tapped-inductor boost converter as

978-1-4799-2706-7/14 $31.00 © 2014 IEEE   2232

The 2014 International Power Electronics Conference

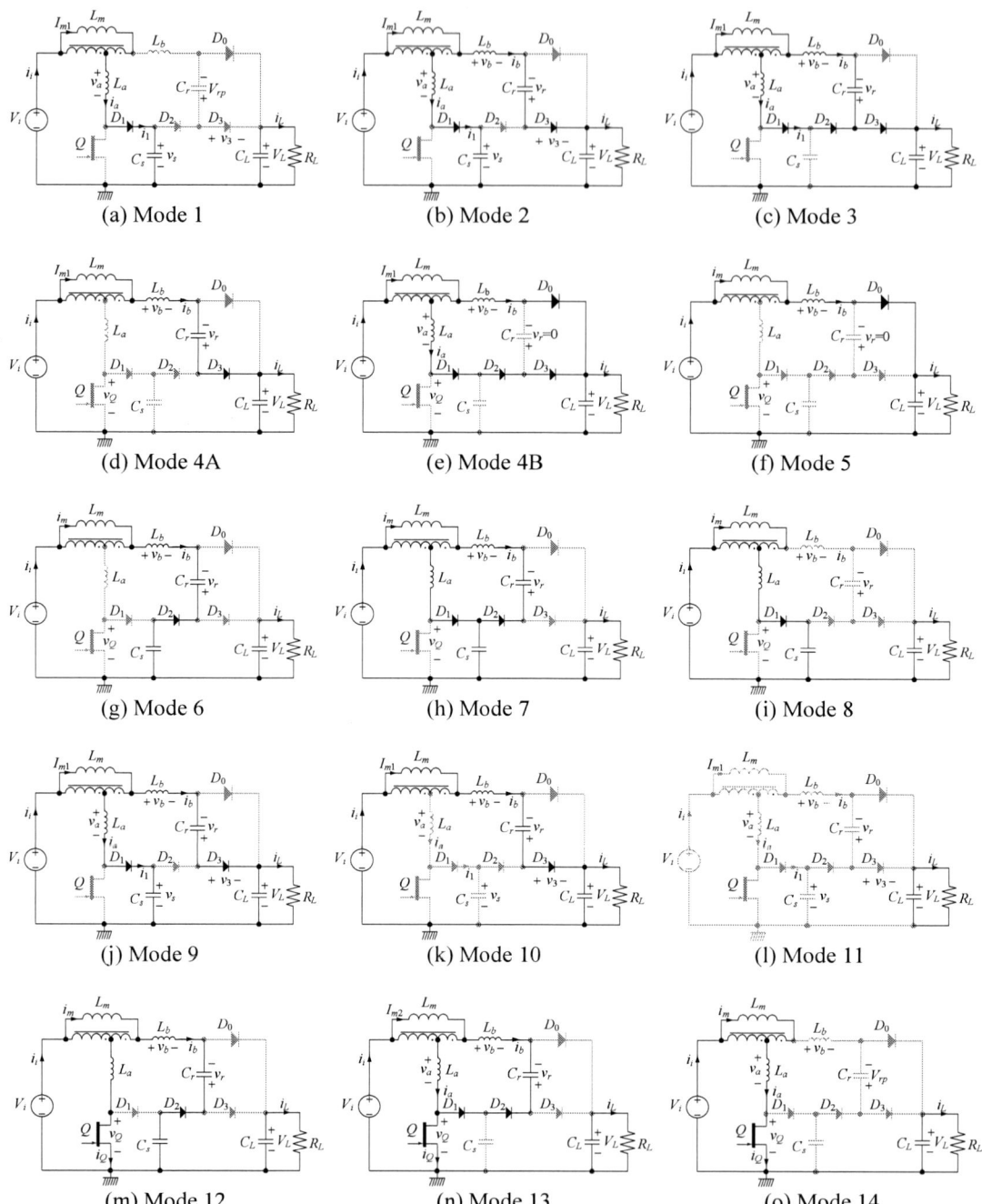

Fig. 2. Operating modes of the proposed converter including DCM.

(1) where D_{off} is defined as (2) which is the time ratio that the output diode D_0 conducts.

$$G_v \equiv \frac{V_L}{V_i} \cong \frac{D(1+1/n)+D_{off}}{D_{off}} = \frac{1+\sqrt{1+4D^2K}}{2} \quad (1)$$

$$D_{off} = \frac{n+1}{n}\frac{D}{G_v-1} \quad \text{where} \quad K = \frac{T_s R_L}{2L_m}\left(\frac{n+1}{n}\right)^2 . \quad (2)$$

Mode 1 $[t_0, t_1]$: At t_0, the switch Q is just turned off, as shown in Fig. 2(a). The leakage current i_a flows to the snubber capacitor C_s and the voltage v_s across C_s increases from zero, which results in ZVS switching of Q.

The voltage v_3 across D_3, when it is turned off, is derived as follows:

$$v_3 = V_i + \frac{n+1}{n}(v_s - V_i) + V_{rp} - V_L \quad (3)$$

where V_{rp} is the recovery capacitor voltage determined from (34), and the magnetizing current i_m is assumed to be constant for a short time period of this mode. The time interval t_{10}, where the voltage slope of v_s is linear as shown in Fig. 3, is determined as follows:

$$t_{10} = \frac{nC_s v_s(t_1)}{(n+1)I_{mp}} = \frac{nC_s}{(n+1)I_{mp}}\left\{\frac{n(V_L-V_{rp})+V_i}{n+1}\right\} \quad (4)$$

978-1-4799-2706-7/14 $31.00 © 2014 IEEE

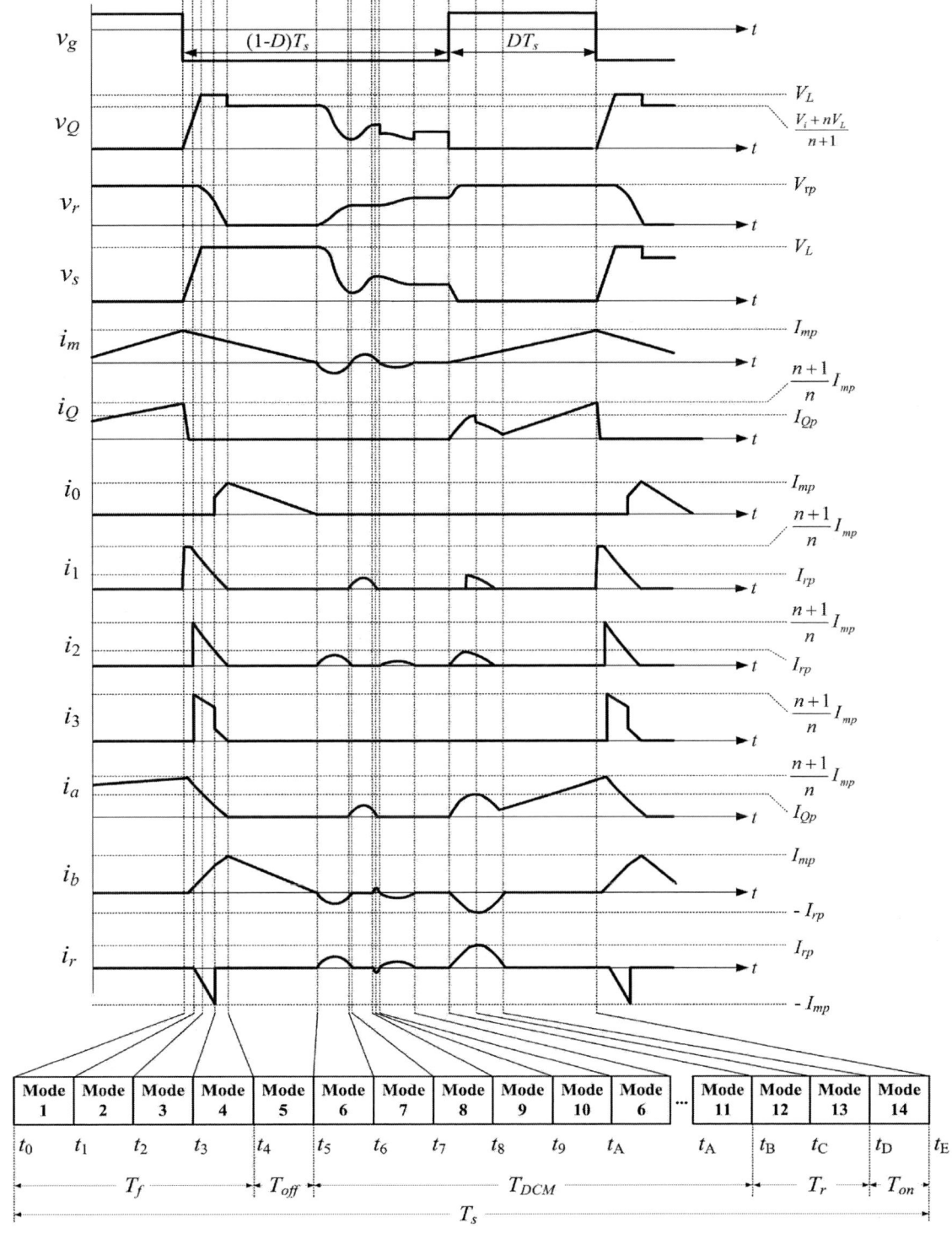

Fig. 3. Waveform diagram of the proposed converter.

where $v_s(t_1)$ is the voltage to turn on the clamp diode D_3 at t_1.

Mode 2 $[t_1, t_2]$: As the clamp diode D_3 is turned on, the recovery capacitor voltage v_r starts to decrease while the snubber capacitor voltage v_s is increasing. Considering the fact that Mode 2 is very short compared to a resonant time period ω_1, which is governed during

this mode as in (5), and that $i_m \cong I_{mp}$, the sinusoidal changes can be neglected and C_s has been charged linearly since Mode 1, as shown in Fig. 3.

$$\omega_1 = 1 \left/ \sqrt{\left\{ \left(\frac{n+1}{n} \right)^2 L_a + L_b \right\} \left\{ \frac{n^2 C_s C_r}{n^2 C_s + (n+1)^2 C_r} \right\}} \right. \quad (5)$$

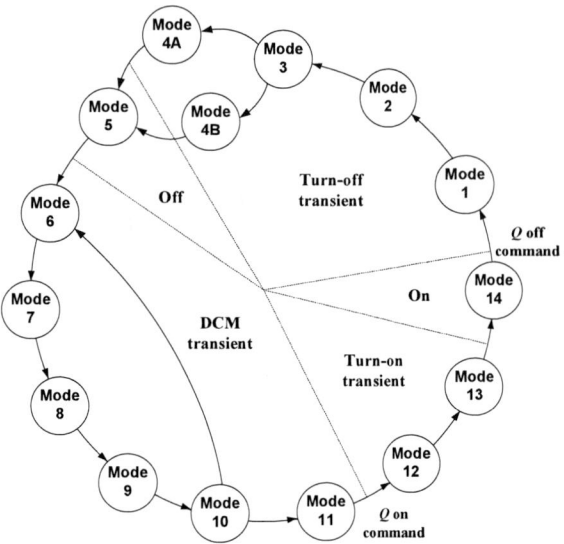

Fig. 4. State diagram of the proposed converter.

The time interval t_{21} to reach $v_s(i_2)=V_L$, therefore, can be straightforwardly determined as follows:

$$t_{21} \cong \frac{nC_s}{(n+1)I_{mp}} \{V_L - v_s(t_1)\}. \qquad (6)$$

Mode 3 [t_2, t_3] : At t_2, the link diode D_2 is turned on as a result of v_s increase up to V_L, as shown in Fig. 2(c). Then, i_b increases, i_a decreases, and v_r decreases. Eventually, either i_a becomes zero, which corresponds to Mode 4A, or v_r becomes zero, which corresponds to Mode 4B, depending on load conditions. Assuming that the magnetizing current is still kept constant, the governing frequency of Mode 3 becomes

$$\omega_2 = 1 \bigg/ \sqrt{\left\{ \left(\frac{n+1}{n}\right)^2 L_a + L_b \right\} C_r} \ . \qquad (7)$$

General solutions of v_r and i_b for Mode 3 can be written, as follows:

$$v_r = \left(V_{rp} + \frac{V_L - V_i}{n}\right) \cos \omega_2 (t - t_2) + \frac{V_i - V_L}{n} \qquad (8a)$$

$$i_b = -C_r \frac{dv_r}{dt} = \omega_2 C_r \left(V_{rp} + \frac{V_L - V_i}{n}\right) \sin \omega_2 (t - t_2). \qquad (8b)$$

This mode ends at t_{3A} when i_b reaches to I_{mp}, at which Mode 4A will start, and the time interval t_{32A} becomes

$$t_{32A} \equiv t_{3A} - t_2 \cong \frac{1}{\omega_2} \sin^{-1}\left\{ \frac{I_{mp}}{\omega_2 C_r V_{rp}\left(1 + \frac{V_L - V_i}{nV_{rp}}\right)} \right\}. \qquad (9)$$

Mode 4A exists only for a light load condition when I_{mp} is small. For a heavier load condition, where v_r becomes zero before i_b reaches to I_{mp}, Mode 3 ends at t_{3B} at which Mode 4B will start, and the time interval t_{32B} becomes

$$t_{32B} \equiv t_{3B} - t_2 \cong \frac{1}{\omega_2} \cos^{-1}\left\{ \frac{1}{1 + \frac{nV_{rp}}{V_L - V_i}} \right\}. \qquad (10)$$

Mode 4A [t_{3A}, t_4] : At t_{3A}, the diodes D_1 and D_2 are just turned off and the magnetizing current i_m, which is still kept constant as I_{mp}, flows to C_r until v_r becomes zero, as shown in Fig. 2(d). Therefore, the time interval t_{43A} can be straightforwardly determined, considering linear charging as follows:

$$
\begin{aligned}
t_{43A} &\equiv t_4 - t_{3A} = C_r \frac{v_r(t_{3A})}{I_{mp}} \\
&= \frac{C_r}{I_{mp}}\left\{ \left(V_{rp} + \frac{V_L - V_i}{n}\right) \cos \omega_2 t_{32A} - \frac{V_L - V_i}{n} \right\}
\end{aligned}, (11)
$$

where t_{32A} is determined from (10).

Mode 4B [t_{3B}, t_4] : At t_{3B}, the recovery capacitor voltage v_r becomes zero and the output diode D_0 is just turned on, as shown in Fig. 2(e). While the magnetizing current i_m is still kept as its maximum value of I_{mp}, i_a and i_b can be solved, as follows:

$$L_l \frac{di_b}{dt} = -L_l \frac{n}{n+1} \frac{di_a}{dt} = \frac{V_L - V_i}{n} \quad \because i_a = \frac{n+1}{n}\left(I_{mp} - i_b\right). \quad (12)$$

The time interval t_{43B}, at which i_a becomes zero, can be determined from (12), using (8b) and (10), as follows:

$$
\begin{aligned}
t_{43B} &\equiv t_4 - t_{3B} = \left\{ \left(\frac{n+1}{n}\right)^2 L_a + L_b \right\} \cdot \frac{n^2}{n+1} \cdot \frac{i_a(t_{3B})}{V_L - V_i} \\
&= n\left\{ \left(\frac{n+1}{n}\right)^2 L_a + L_b \right\} \frac{I_{mp} - \omega_2 C_r \left(V_{rp} + \frac{V_L - V_i}{n}\right) \sin \omega_2 t_{32B}}{V_L - V_i}
\end{aligned}. (13)
$$

Mode 5 [t_4, t_5] : At t_4, all the turn-off transition is finished, and the output diode D_0 is maintained to turn-on, which delivers the magnetizing current i_m to the load side, as shown in Fig. 2(f). This period of Mode 5 is relatively very long, as in (14), compared to previous transition modes; therefore, i_m is no longer constant but linearly decreased to zero like a conventional DCM operated boost converter, which results in soft switching of D_0, as in (15).

$$t_{54} = \frac{I_{mp}L_m}{V_L - V_i} \qquad (14)$$

978-1-4799-2706-7/14 $31.00 © 2014 IEEE

$$i_b = i_m \cong \frac{1}{L_m}\int_{t_4}^{t}(V_i - V_L)dt + I_{mp} \qquad (15)$$

Mode 6 [t_5, t_6] : At t_5, the output diode D_0 is turned off as a result of i_b decrease up to zero, and the link diode D_2 is turned on, which results in a series resonant circuit composed of the magnetizing inductance L_m, leakage inductance L_b and capacitors C_s, C_r, as shown in Fig. 2(g). The governing frequency of Mode 6 becomes

$$\omega_3 = 1 \Big/ \sqrt{\left(L_m + L_b\right)\cdot\left(\frac{C_s C_r}{C_s + C_r}\right)} \;. \qquad (16)$$

The snubber capacitor voltage v_s decreases from V_L and its current i_b increases from zero in sinusoidal waveform, as follows:

$$v_s = \frac{C_r\left(V_L - V_i\right)\cos\omega_3(t-t_5)+C_s V_L + C_r V_i}{C_s + C_r} \qquad (17a)$$

$$i_b = i_m = -C_s\frac{dv_s}{dt} = \frac{\omega_3 C_s C_r\left(V_L - V_i\right)}{C_s + C_r}\sin\omega_3(t-t_5)\cdot \qquad (17b)$$

The recovery capacitor voltage v_r can be determined from (17a) as follows:

$$v_r = \frac{C_s}{C_r}\left(V_L - v_s\right) \quad \because i_r = -i_c \;. \qquad (18)$$

As the voltage v_s increases, the voltage potential of the tap of the tapped inductor v_t is also boosted up to v_s, as in (19), which eventually makes the snubber diode D_1 turned-on at t_6.

$$v_t \equiv \left(v_s - v_r - V_i\right)\frac{L_m}{L_m + L_b}\cdot\frac{n}{n+1}+V_i$$
$$\cong \left(v_s - v_r - V_i\right)\frac{n}{n+1}+V_i \quad \left(\because L_m \gg L_b\right) \qquad (19)$$

The time interval t_{65} is then determined from (17a), using (16) and (19), as follows:

$$t_{65} = \frac{1}{\omega_3}\cos^{-1}\left\{\frac{2}{\alpha(n/\beta-1)}\right\} \qquad (20)$$

where $\alpha \equiv C_r/C_s$ and $\beta \equiv nV_L/V_i$.

Mode 7 [t_6, t_7] : At t_6, the snubber diode D_1 is turned on, while the link diode D_2 is turned on, which results in two series resonant loops are composed in this mode, as shown in Fig. 2(h). Considering the fact that Mode 7 is very short compared to Mode 6 and that the voltage and current changes are insignificant during this mode, the discharge of C_s is assumed to be with the same frequency

of Mode 6. The time interval t_{76}, at which i_b becomes zero, is then determined from (17b) and (20), as follows:

$$t_{76} \cong \frac{1}{\omega_3}\left[\pi - \cos^{-1}\left\{\frac{2}{\alpha(n/\beta-1)}\right\}\right] \to t_{75} \equiv t_{76}+t_{65} \cong \frac{\pi}{\omega_3}.(21)$$

At t_7, Mode 8 starts when $V_i > v_s$, or Mode 11 starts when $V_i < v_s$, depending on load conditions and LC component values.

Mode 8 [t_7, t_8] : At t_7, the link diode D_2 is turned off as a result of i_b decreases to zero, and the snubber diode D_1 is still turned on, which results in a series resonant circuit composed of the magnetizing inductance L_m, leakage inductance L_a and snubber capacitor C_s, as shown in Fig. 2(i). The governing frequency of Mode 6 then becomes

$$\omega_4 = 1 \Big/ \sqrt{\left\{\left(\frac{n}{n+1}\right)^2 L_m + L_a\right\}C_s} \;. \qquad (22)$$

The snubber capacitor voltage v_s increases from $v_s(t_7)$ and its current i_a is assumed to be increased from zero in sinusoidal waveform, as follows:

$$v_s = \frac{1-\alpha}{1+\alpha}\left(V_L - V_i\right)\cos\omega_4(t-t_7)+V_i \qquad (23a)$$

$$i_a = \frac{1-\alpha}{1+\alpha}C_s\omega_4\left(V_L - V_i\right)\sin\omega_4(t-t_7) \qquad (23b)$$

where the particular solution and initial conditions of the circuit are derived from (17) and (21).
The voltage v_3 across D_3, when it is turned off, is derived as follows:

$$v_3 = V_i + \frac{n+1}{n}\left(v_t - V_i\right)+v_r(t_7)-V_L$$
$$\cong V_i + \frac{n+1}{n}\left(v_s - V_i\right)+\frac{2\left(V_L - V_i\right)}{1+\alpha}-V_L \;. \qquad (24)$$
$$\left(\because L_m \gg L_a\right)$$

v_3 becomes zero at $t = t_8$; hence, the time interval t_{87} is then determined from (23a) and (24), as follows:

$$t_{87} = \frac{1}{\omega_4}\cos^{-1}\left\{\frac{n}{n+1}\cdot\frac{\alpha\beta-\beta-n\alpha+n}{(1-\alpha)(\beta-n)}\right\} \;. \qquad (25)$$

Mode 9 [t_8, t_9] : At t_8, the clamp diode D_3 is turned on as a result of v_3 reaches to zero. The recovery capacitor voltage v_r starts to decrease while the snubber capacitor voltage v_s is increasing during Mode 9, as shown in Fig. 2(j). Considering the fact that Mode 9 is very short and insignificant against overall operation compared to Mode 8, the charge of C_s is assumed to be with the same frequency of Mode 8 like the Mode 7 analysis. The time interval t_{98}, at which i_a becomes zero, is then determined from (23b) and (25), as follows:

$$t_{98} \cong \frac{1}{\omega_4}\left[\pi - \cos^{-1}\left\{\frac{n}{n+1}\cdot\frac{\alpha\beta - \beta - n\alpha + n}{(1-\alpha)(\beta - n)}\right\}\right]. \quad (26)$$

$$\rightarrow t_{97} \equiv t_{98} + t_{87} \cong \frac{\pi}{\omega_4}$$

Mode 10 [t_9, t_A] : At t_9, the snubber diode D_1 is turned off as a result of i_a decreases to zero, as shown in Fig. 2(k). The recovery capacitor voltage v_r decreases subtle until the recovery capacitor current $i_r (=i_b)$ becomes zero with a governing frequency of ω_5, as in (27). Due to the fact that the initial conditions of Mode 10 is omitted from the previous analysis, the time interval t_{A9} also can be neglected, considering its unimportance compared to the overall operation.

$$\omega_5 = 1\left/\sqrt{(L_m + L_b)C_r}\right. \quad (27)$$

At t_A, Mode 11 starts when $v_s < v_r + V_i$; otherwise, a cycle from Mode 6 to Mode 10 continues to repeat.

Mode 11 [t_A, t_B] : At t_A, every diode and the switch Q are completely turned off unless there is an external turn-on command for Q, as shown in Fig. 2(l).

Mode 12 [t_B, t_C] : At t_B, the switch Q is turned on by an external turn-on command and the link diode D_2 is turned on, which results in a series resonant circuit composed of the leakage inductances L_a, L_b and capacitors C_s, C_r, as shown in Fig. 2(m). The switch current i_Q, sum of a magnetizing current and a resonant current, increases from zero, which results in ZCS turn-on of Q. The governing frequency of Mode 12 becomes

$$\omega_6 = 1\left/\sqrt{\left\{\left(\frac{n+1}{n}\right)^2 L_a + L_b\right\}\left(\frac{C_s C_r}{C_s + C_r}\right)}\right. \quad (27)$$

Assuming the cycle from Mode 6 to Mode 10 was once occurred before this mode, the snubber capacitor voltage v_s decreases from $v_s(t_B)$ and its current i_b increases from zero in sinusoidal waveform, as follows:

$$v_s = a_1 \cos\omega_6(t - t_B) + a_2 \sin\omega_6(t - t_B) + a_3 \quad (28a)$$

$$i_b = C_s\frac{dv_s}{dt} = \omega_6 C_s\{-a_1 \sin\omega_6(t - t_B) + a_2 \cos\omega_6(t - t_B)\} \quad (28b)$$

where the coefficient $\{a_k\}$ can be determined as follows:

$$a_1 = \frac{\alpha\{4n + 1 + \alpha + \beta(\alpha - 3)\}}{(1+\alpha)^2\beta}V_L \quad (29a)$$

$$a_2 = 0 \quad (29b)$$

$$a_3 = \frac{2n(1-\alpha) - \alpha(1+\alpha) + \beta(3\alpha - 1)}{(1+\alpha)^2\beta}V_L. \quad (29c)$$

This mode ends at t_C when v_s reaches to zero, at which D_1 is turned on and Mode 13 starts. The time interval t_{CB} is then determined as follows:

$$t_{CB} = \frac{1}{\omega_6}\cos^{-1}\left\{-\frac{a_3}{a_1}\right\}. \quad (30)$$

Mode 13 [t_C, t_D] : At t_C, the snubber diode D_1 is turned on and the snubber capacitor C_s is excluded from the resonant loop, as shown in Fig. 2(n). The recovery capacitor voltage v_r increases continuously and its current i_r begins to decrease, and they can be determined using the same governing frequency of (7) as follows:

$$v_r = b_1 \cos\omega_2(t - t_C) + b_2 \sin\omega_2(t - t_C) + b_3 \quad (31a)$$

$$i_r = C_r\frac{dv_r}{dt} = \omega_2 C_r\{-b_1 \sin\omega_2(t - t_C) + b_2 \cos\omega_2(t - t_C)\} \quad (31b)$$

where the coefficient $\{b_k\}$ can be determined as follows:

$$b_1 = \left\{\frac{(3\alpha - 1)\beta + 2n(1-\alpha)}{\alpha\beta(1+\alpha)} - \frac{1}{\beta}\right\}V_L \quad (32a)$$

$$b_2 = \frac{4n + 1 + \alpha + \beta(\alpha - 3)}{(1+\alpha)^{3/2}\beta}V_L \sin\omega_6 t_{CB} \quad (32b)$$

$$b_3 = \frac{V_i}{n}. \quad (32c)$$

This mode ends when the recovery capacitor current i_r becomes zero, and the time interval t_{DC} is then determined as follows:

$$t_{DC} = \frac{1}{\omega_2}\left(\frac{\pi}{2} - \tan^{-1}\frac{b_1}{b_2}\right). \quad (33)$$

The recovery capacitor is fully charged at the time t_D as follows:

$$v_r(t_D) = b_1 \cos\omega_2 t_{DC} + b_2 \sin\omega_2 t_{DC} + b_3 \equiv V_{rp}. \quad (34)$$

$$\rightarrow V_{rp,\max} = \sqrt{b_1^2 + b_2^2} + b_3$$

Mode 14 [t_D, t_E] : At t_D, the resonant current i_r has reached to zero and ZCS has been achieved for D_1 and D_2. Now, Q is solely turned on and all the other diodes are completely turned off for a while, as shown in Fig. 2(o). Considering the fact that $L_a \ll L_m$, the magnetizing current i_m is finally determined as follows:

$$i_Q = i_a = \frac{n+1}{n}i_m$$

$$\cong \frac{n+1}{nL_m}\left(\int_{t_D}^t \frac{n+1}{n}V_i\,dt\right) = \frac{n+1}{nL_m}\left\{\frac{n+1}{n}V_i(t - t_D)\right\}. \quad (35)$$

TABLE I
PEAK COMPONENT RATINGS OF THE PROPOSED CONVERTER

	Peak voltage [V]	Peak current [A]
Switch Q	V_L	$Max\left\{I_{Qp}, I_{mp}\dfrac{n+1}{n}\right\}$
Output diode D_0	$V_L + V_{rp}$	I_{mp}
Snubber diode D_1	V_L	$Max\left\{I_{rp}, I_{mp}\dfrac{n+1}{n}\right\}$
Link diode D_2	$V_{rp} - \dfrac{V_i}{n}$	$Max\left\{I_{rp}, I_{mp}\dfrac{n+1}{n}\right\}$
Clamp diode D_3	V_L	$\dfrac{n+1}{n}I_{mp}$
Snubber capacitor C_s	V_L	$\dfrac{n+1}{n}I_{mp}$
Recovery capacitor C_r	V_{rp}	I_{rp}

Fig. 5. A prototype fabrication of the proposed converter.

TABLE II
SELECTED PARAMETERS OF THE PROPOSED CONVERTER

Parameters	Values
n	2
L_m [μH]	238
L_a [μH]	3.80
L_b [μH]	7.45
C_s [nH]	1
C_r [nH]	4

III. VOLTAGE AND CURRENT STRESS OF EACH COMPONENT

For practical applications of the proposed converter, the voltage and current stress of each component, which is determined based on the circuit analysis results of the previous section, should be within reasonable range. As identified from Figs. 2 and 3, the maximum voltages of the switch Q, diodes D_1, D_2, D_3, and snubber capacitor C_s are always bounded to the load voltage V_L. One exception is the output diode D_0, which suffers from higher voltage stress added by the recovery capacitor voltage V_{rp} during Mode 13 determined by (34), as follows:

$$V_{0,\max} = V_L + V_{rp,\max}. \qquad (36)$$

As identified from Figs. 2 and 3 again, I_{Qp}, which is the peak current of the switch Q, is determined by either I_{Qp} of Mode 12 or I_{mp} contribution of Mode 14. From (28b) and (29a), I_{Qp} is obtained as follows:

$$
\begin{aligned}
I_{Qp} &= \frac{n+1}{n}\left(-I_{b,\min}\right) = \frac{n+1}{n}I_{rp} = \frac{n+1}{n}\omega_6 C_s a_1 \\
&= \omega_6 C_s V_L \frac{n+1}{n}\frac{\alpha\{4n+1+\alpha+\beta(\alpha-3)\}}{(1+\alpha)^2 \beta}
\end{aligned} \qquad (37)
$$

Similarly, the peak current rating of the snubber diode and link diode D_1 and D_2 is determined by either the I_{mp}

contribution of the turn-off transient modes or the I_{rp} of the turn-on transient modes, which is defined in (37).

As identified from Figs. 2(b) and 3, the peak current rating of the clamp diode D_3 is involved in Mode 2; however, the current change during Mode 2 is negligible, so only Mode 3 is examined as follows:

$$
\begin{aligned}
I_{3,\max} &= Max\{i_a + i_b\} = Max\left\{\frac{n+1}{n}(I_{mp} - i_b) + i_b\right\} \\
&= Max\left\{\frac{n+1}{n}I_{mp} - \frac{i_b}{n}\right\} = \frac{n+1}{n}I_{mp} \quad \because i_b > 0
\end{aligned} \qquad (38)
$$

The peak voltage ratings and peak current ratings for all the components of the proposed converter are listed in Table I.

IV. EXPERIMENTAL VERIFICATIONS

The design principle, including the component stress of the previous section, has been applied to a prototype DCM operated ZVZCS tapped boost converter, operating at 50 kHz switching frequency. A SiC JFET (SJEP120R063), having turn-on and turn-off switching times of 20 ns, was selected because it is a common switching device affordable nowadays. The proposed converter was fabricated, as shown in Fig. 5, where circuit parameters are tabulated in Table II.

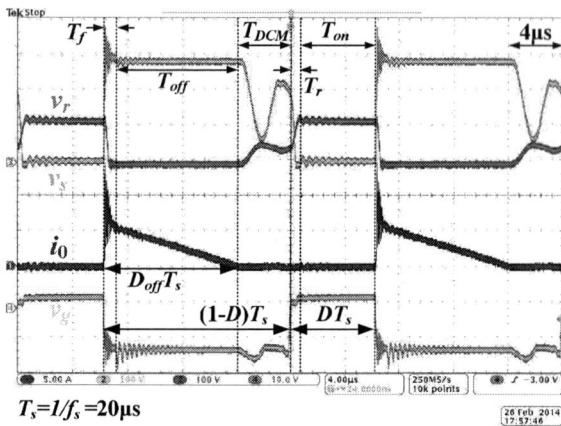

$T_s = 1/f_s = 20\mu s$

Fig. 6. Experimental waveforms for the prototype converter when D=0.35, V_L=300V.

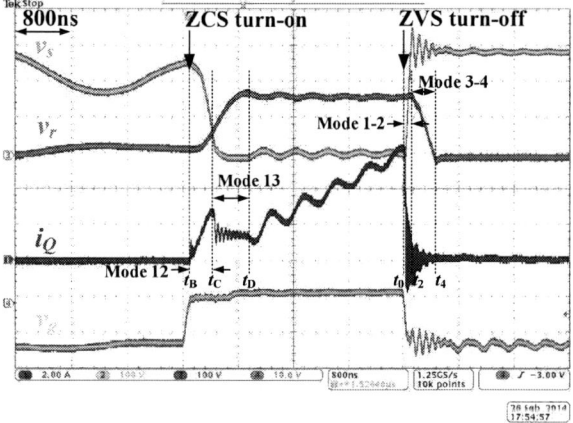

Fig. 7. Experimental waveforms for the ZVZCS operation of the prototype converter when D=0.15, V_L=300V.

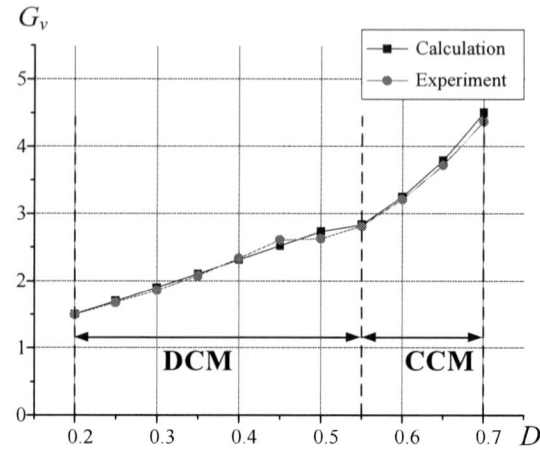

(a) Output voltage gain G_v for the input voltage of 50V

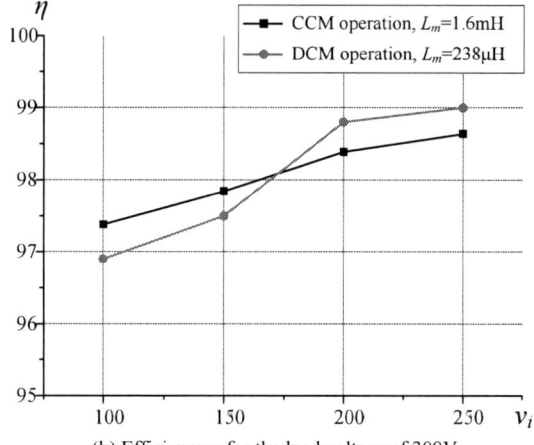

(b) Efficiency η for the load voltage of 300V

Fig. 8. Comparisons of the analysis results with the experiment results.

The experimental waveforms of the prototype are illustrated in Fig 6, where the voltage and current stresses of the components are well mitigated in accordance with the analyses. The ZCS turn-on and ZVS turn-off of the switch Q were also verified for wide operating conditions, for both CCM and DCM operation, as shown in Fig. 7. The input to output voltage gain G_v was measured w.r.t. D and compared with the ideal case of (1), as shown in Fig. 8(a). It is verified that the experimental results coincide with the theory for a wide operating range of D from 0.2 to 0.7. As shown in Fig. 8(b), the maximum efficiency η of the proposed ZVZCS converter for the fixed load voltage of 300 V at the switching frequency of 50 kHz, neglecting the gate driver power consumption, was measured as 99.0 % at DCM operation when L_m=238μH and 98.6 % at CCM operation when L_m=1.6mH, respectively. From temperature measurement, it is found that the efficiency drop is mainly due to the core loss and conduction loss of the inductor. The load power was fixed to 450 W with input voltage range of 100~250 V during the efficiency measurement.

V. CONCLUSION

A novel single SiC switch based ZVZCS tapped boost converter in the DCM has been designed and verified both by theory and experiment. The proposed converter has wide ZVZCS operating range with no severe voltage and current stresses as well as no cumbersome parasitic ringing. Moreover, high efficiencies of 98.6% and 99.0% were achieved by a prototype converter for the CCM and DCM operation, respectively. Due to its wide ZVZCS operating range regardless of DCM or CCM operation with simple circuit structure, the proposed converter can be counted as a suitable candidate for the power factor correction circuits with high efficiency, where the load voltage is fixed and input voltage is rapidly varying in time. A detail design of the tapped inductor of the proposed converter is left for further work.

REFERENCES

[1] H. Do, "A soft switching DC/DC converter with high voltage gain," *IEEE Trans. Power Electronics*, vol. 25, no. 5, pp. 1193-1200, May 2010.

[2] K. Hwu and Y. Yau, "Voltage boosting converter based on charge pump and coupling inductor with passive voltage clamping," *IEEE Trans. Power Electronics*, vol. 57, no. 5, pp. 1719-1727, May 2010.

[3] E. Firmansyah, S. Tomika, S. Abe, M. Shoyama, and T. Ninimiya, "Zero-current-switch quasi-resonant boost converter in power factor correction application," in *2009 APEC conf.*, pp. 1165-1169.

[4] Bo H. Choi and Chun T. Rim, "A novel single SiC switch based ZVZCS tapped boost converter," *IEEE Trans. Power Electronics* (will be published).

Effect of Input and Output Terminal Sources on Dynamic Behavior of Switched-Mode Converters

T. Suntio, J. Viinamäki, J. Jokipii, T. Messo

M. Sitbon, A. Kuperman

Dept. of Electrical Engineering
Tampere University of Technology
Tampere, Finland

Dept. of Electrical Engineering and Electronics
Ariel University
Ariel, Israel

Abstract – **Voltage-type sources such as grid, constant-voltage-regulated power supplies, and storage batteries have dominated as input sources for power electronic converters leading to development of multitude of different power stages dedicated for such input sources. The growing application of renewable energy sources has revealed that there are also current-type input sources such as e.g. photovoltaic generator requiring different kind of power electronic converters for interfacing them to practical usage. There are persistent assumptions that the power stage itself determines the steady-state and dynamic properties of the corresponding converter. This paper introduces the decisive factors determining the properties of specific power stages. The findings are supported with comprehensive experimental measurements.**

Keywords – *Power electronic converter, power stage, dynamic model, stability*

I. INTRODUCTION

Voltage-type input sources such as AC grid, output-voltage-regulated converter, and storage battery have dominated as input sources for power electronic converters leading to the development of multitude of power-stage topologies dedicated for the named applications as introduced e.g. in [1,2]. There are, however, input sources such as photovoltaic (PV) generator [3], superconductive magnetic storage (SMES) devices [4,5], and input-side feedback controlled converters [6], which are basically current sources requiring the use of different kind of converters for utilizing their energy. It is quite common that the input sources are assumed to be voltage sources despite their real nature. The use of the voltage as the base variable of the source is usually justified by means of Norton/Thevenin transformation as e.g. in [7,8]. The transformation does not, however, change the source by no means, and the application of improper converter type would lead easily to severe problems as discussed in [9,10].

It has been observed quite long time ago already that the change of the input and output terminal sources would change profoundly, especially, the dynamic behavior of the converter [11-13] but also its steady-state behavior [9] compared to the conventional application of the corresponding converter: A power stage intended for voltage-fed application requires usually to adding a capacitor at its input terminal for satisfying the terminal constraints stipulated by the current source as discussed in [14]. As a consequence of that the order of the system may increase, and the properties of the converter will resemble the properties of the dual of the original converter (i.e., e.g. buck vs. boost) because of the necessity to interchanging the input and output variables [14]. In addition, the duty ratio has to be decreased for increasing the corresponding output variables [9,14].

This paper presents the results of theoretical and practical investigations of the dynamic properties of a buck-power-stage converter, when the input and output terminal sources are varied but the operating point is maintained exactly the same. The practical evidence provided in this paper explicitly proves the existence of the dynamical changes. The rest of the paper is organized as follows: The general dynamic representations including the non-ideal source and load effects as well as general impedance-based stability assessment method are shortly presented in Section II. The dynamic properties of a buck-power-stage converter in four different terminal-source configurations are introduced in Section III. The experimental evidence is provided in Section IV, and the conclusions are finally drawn in Section V, respectively.

II. GENERAL DYNAMIC REPRESENTATIONS

For facilitating the dynamic analysis, the dynamic representations of the switched-mode converter are developed in general framework including the effect of non-ideal source and load on the internal dynamics of the converter. The set of transfer functions describing the dynamic behavior of the DC-DC converter can be given by

$$\begin{bmatrix} \hat{y}_{in} \\ \hat{y}_{out} \end{bmatrix} = \begin{bmatrix} G_{11-o} & G_{12-o} & G_{13-o} \\ G_{21-o} & -G_{22-o} & G_{23-o} \end{bmatrix} \begin{bmatrix} \hat{u}_{in} \\ \hat{u}_{out} \\ \hat{u}_{c} \end{bmatrix}$$ (1),

where the right-side column vector denotes the input variables and the left-side column vector the output variables. The subscripts 'in' and ''out' denote the corresponding terminal, where the variable physically exists, and the subscript 'c' the control variable, respectively. The negative sign of the element (2,2) in (1) indicates that the direction of the output current is out of

978-1-4799-2706-7/14 $31.00 © 2014 IEEE

the converter in contrast to the current flow in the theoretical two-port network [15].

A. Closed-Loop Dynamics

The feedback variable has to be always one of the output variables (i.e., y_x in (1)). Therefore, there are two different sets of general closed-loop transfer functions as given in (2) (i.e., input-side feedback) and (6) (i.e., output-side feedback), where the input and output-variable vectors correspond to (1). The special parameters denoted by the subscript extension '$-\infty$' describe the low-frequency behavior of the corresponding closed-loop transfer function and can be approximated by

$$G_{11-\infty} \approx -\frac{Y_{in}}{U_{in}} \text{ and } G_{22-\infty} \approx \frac{Y_{out}}{U_{out}}.$$

$$\begin{bmatrix} \dfrac{G_{11-o}}{A} & \dfrac{G_{12-o}}{A} & \dfrac{L_{in}}{G_{se-in}A} \\ \dfrac{G_{21-o}+L_{in}G_{21-\infty}}{A} & -\dfrac{G_{22-o}+L_{in}G_{22-\infty}}{A} & \dfrac{G_{23-0}L_{in}}{G_{se-in}G_{13-o}A} \end{bmatrix} \quad (2)$$

where

$$A = 1 + L_{in} \quad (3)$$

$$G_{21-\infty} = G_{21-o} - \frac{G_{11-o}G_{23-o}}{G_{13-o}} \quad (4)$$

$$G_{22-\infty} = G_{22-o} + \frac{G_{12-o}G_{23-o}}{G_{13-o}} \quad (5)$$

$$\begin{bmatrix} \dfrac{G_{11-o}+L_{out}G_{11-\infty}}{B} & \dfrac{G_{12-o}+L_{out}G_{12-\infty}}{B} & \dfrac{G_{13-o}L_{out}}{G_{se-out}G_{23-o}B} \\ \dfrac{G_{21-o}}{B} & -\dfrac{G_{22-o}}{B} & \dfrac{l_{out}}{G_{se-out}B} \end{bmatrix} \quad (6)$$

where

$$B = 1 + L_{out} \quad (7)$$

$$G_{11-\infty} = G_{11-o} - \frac{G_{21-o}G_{13-o}}{G_{23-o}} \quad (8)$$

$$G_{12-\infty} = G_{12-o} + \frac{G_{21-o}G_{13-o}}{G_{23-o}} \quad (9)$$

B. Source and Load-Affected Dynamics

When assumed that the source (**S**) and load (**L**) subsystems are composed of their ideal source and the corresponding internal impedance or admittance denoted by S_{22} and L_{11}, the source and load-affected sets of converter transfer functions can be given according to (10) and (13), where the input and output-variable vectors correspond to (1).

$$\begin{bmatrix} \dfrac{G_{11-o}}{C} & \dfrac{G_{12-o}}{C} & \dfrac{G_{13-o}}{C} \\ \dfrac{G_{21}}{C} & -\dfrac{1+S_{22}G_{11-xo}}{C}G_{22-o} & \dfrac{1+S_{22}G_{11-\infty}}{C}G_{23-o} \end{bmatrix} \quad (10)$$

where

$$C = 1 + S_{22}G_{11-o} \quad (11)$$

$$G_{11-xo} = G_{11-sco} = G_{11-oco} = G_{11-o} + \frac{G_{12-o}G_{21-o}}{G_{22-o}} \quad (12)$$

$$\begin{bmatrix} \dfrac{1+L_{11}G_{22-xi}}{D}G_{11} & \dfrac{G_{12-o}}{D} & \dfrac{1+L_{11}G_{22-\infty}}{D}G_{13-o} \\ \dfrac{G_{21-o}}{D} & -\dfrac{G_{22-o}}{D} & \dfrac{G_{23-o}}{D} \end{bmatrix} \quad (13)$$

where

$$D = 1 + L_{11}G_{22-o} \quad (14)$$

$$G_{22-xi} = G_{22-sci} = G_{22-oci} = G_{22-o} + \frac{G_{12-o}G_{21-o}}{G_{11-o}} \quad (15)$$

The transfer function G_{11-xo} in (12) denotes the impedance or admittance of the input terminal when the output terminal is either short circuited or open circuit. The transfer function G_{22-xo} in (15) denotes the impedance or admittance of the output terminal when the input terminal is either open circuit or short circuited, respectively.

C. General Stability Assessment

The generalized stability assessment formalism applicable for DC-DC converters are published early in [15]. According to it, the set of transfer functions for the coupled source-load-system can be presented by

$$\begin{bmatrix} \dfrac{S_{21}L_{11}}{1+S_{22}L_{11}} & \dfrac{L_{12}}{1+S_{22}L_{11}} \\ \dfrac{S_{21}}{1+S_{22}L_{11}} & -\dfrac{S_{22}L_{12}}{1+S_{22}L_{11}} \end{bmatrix} \quad (16)$$

and

$$\begin{bmatrix} S_{11}+\dfrac{S_{12}S_{21}L_{11}}{1+S_{22}L_{11}} & \dfrac{S_{12}L_{12}}{1+S_{22}L_{11}} \\ \dfrac{S_{21}L_{21}}{1+S_{22}L_{11}} & -\left(L_{12}+\dfrac{S_{12}S_{22}L_{12}}{1+S_{22}L_{11}}\right) \end{bmatrix} \quad (17)$$

where the matrix elements denoted by S_{ij} correspond to the source-subsystem transfer functions, and L_{ij} to the load-subsystem transfer functions, respectively. If the subsystems are stable as standalone systems, the stability of the coupled system can be analyzed based on the product $S_{22}L_{11}$, which is an impedance ratio known as minor-loop gain [16,17]. The stability exists when the minor-loop gain satisfies *Nyquist* stability criterion. The minor-loop gain has to be determined in such a way that the numerator impedance is the internal impedance of the voltage-type source, and the denominator impedance is the internal impedance of the current-type source, respectively.

III. DYNAMIC MODELS OF BUCK POWER STAGE

Dynamic modeling of the direct-duty-ratio-controlled converter operating in continuous conduction mode (CCM) can be performed by applying e.g. state space averaging (SSA) method introduced in [18,19]. Most essential is to correctly determine the valid input and output variables: It may be obvious that the ideal terminal sources are always the input variables, and the duals of these sources the output variables, respectively. If a resistor is used as a load, then the intended or used feedback variable determines the output variable, and consequently, the corresponding input variable is the dual of the feedback variable.

The power-stage of the buck converter with open input and output-terminal sources is given in Fig. 1. The control scheme of the high and low-side MOSFETs is assumed to be the same as in the conventional application (i.e., the high-side MOSFET conducts during the on time, and the low-side MOSFET during the off time, respectively. The derived small-signal state spaces are given including all the parasitic elements shown in Fig. 1. The corresponding transfer functions can be solved from the given state spaces by applying e.g. Symbolic Toolbox of Matlab™. The explicit transfer functions are given by neglecting the parasitic elements for more convenient comparison, and simplicity.

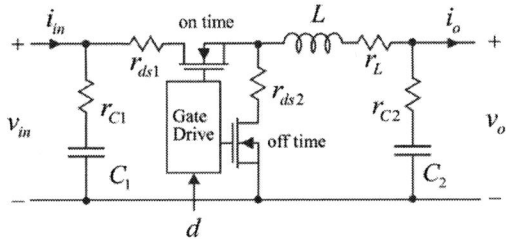

Fig. 1. Buck-power-stage with open input and output terminals

A. Voltage-Fed-Voltage-Output Converter

The small-signal state space of VF/VO converter including the parasitic elements can be given by

$$\frac{d\hat{i}_L}{dt} = -\frac{r_E + r_{C2}}{L}\hat{i}_L - \frac{1}{L}\hat{v}_{C2} + \frac{D}{L}\hat{v}_{in} + \frac{r_{C2}}{L}\hat{i}_o + \frac{V_E}{L}\hat{d}$$

$$\frac{d\hat{v}_{C1}}{dt} = -\frac{1}{r_{C1}C_1}\hat{v}_{C1} + \frac{1}{r_{C1}C_1}\hat{v}_{in}$$

$$\frac{d\hat{v}_{C2}}{dt} = \frac{1}{C_2}\hat{i}_L - \frac{1}{C_2}\hat{i}_o \tag{18}$$

$$\hat{i}_{in} = D\hat{i}_L + I_L\hat{d} - \frac{1}{r_{C1}}\hat{v}_{C1} + \frac{1}{r_{C1}}\hat{v}_{in}$$

$$\hat{v}_o = (1 + r_{C2}C_2\frac{d}{dt})\hat{v}_{C2}$$

where

$$r_E = r_L + Dr_{ds1} + D'r_{ds2} \tag{19}$$
$$V_E = V_{in} + (r_{ds2} - r_{ds1})I_o \tag{20}$$

and the corresponding operating point

$$I_L = I_o \quad I_{in} = DI_o \quad V_o = DV_{in} - r_E I_o \tag{21}$$

as well as the set of transfer functions by

$$\begin{bmatrix} \hat{i}_{in} \\ \hat{v}_{out} \end{bmatrix} = \frac{\begin{bmatrix} C_1 s(s^2 + \frac{C_1 + D^2 C_2}{LC_1 C_2}) & \frac{D}{LC_2} \\ \frac{D}{LC_2} & -\frac{s}{C_2} \end{bmatrix}}{s^2 + \frac{1}{LC_2}} \begin{bmatrix} \hat{v}_{in} \\ \hat{i}_{out} \end{bmatrix} \tag{22}$$

and

$$\begin{bmatrix} \hat{i}_{in} \\ \hat{v}_{out} \end{bmatrix} = \frac{\begin{bmatrix} I_o(s^2 + s\frac{DV_{in}}{I_o L} + \frac{1}{LC_2}) \\ \frac{V_{in}}{LC_2} \end{bmatrix}}{s^2 + \frac{1}{LC_2}} \begin{bmatrix} \hat{d} \\ 0 \end{bmatrix} \tag{23}$$

B. Voltage-Fed-Current-Output Converter

The small-signal state-space of VF/CO converter including the parasitic elements can be given by

$$\frac{d\hat{i}_L}{dt} = -\frac{r_E}{L}\hat{i}_L + \frac{D}{L}\hat{v}_{in} - \frac{1}{L}\hat{v}_o + \frac{V_E}{L}\hat{d}$$

$$\frac{d\hat{v}_{C1}}{dt} = -\frac{1}{r_{C1}C_1}\hat{v}_{C1} + \frac{1}{r_{C1}C_1}\hat{v}_{in}$$

$$\frac{d\hat{v}_{C2}}{dt} = -\frac{1}{r_{C2}C_2}\hat{v}_{C2} + \frac{1}{r_{C2}C_2}\hat{v}_o \tag{24}$$

$$\hat{i}_{in} = D\hat{i}_L + I_L\hat{d} - \frac{1}{r_{C1}}\hat{v}_{C1} + \frac{1}{r_{C1}}\hat{v}_{in}$$

$$\hat{i}_o = \hat{i}_L + \frac{1}{r_{C2}}\hat{v}_{C2} - \frac{1}{r_{C2}}\hat{v}_o$$

and corresponding operating point by

$$I_L = I_o \quad I_{in} = DI_o \quad I_o = \frac{DV_{in} - V_o}{r_E} \tag{25}$$

as well as the set of transfer functions by

$$\begin{bmatrix} \hat{i}_{in} \\ \hat{i}_{out} \end{bmatrix} = \frac{\begin{bmatrix} LC_1(s^2 + \frac{D^2}{LC_1}) & -D \\ D & -LC_2(s^2 + \frac{1}{LC_2}) \end{bmatrix}}{sL} \begin{bmatrix} \hat{v}_{in} \\ \hat{v}_{out} \end{bmatrix} \tag{26}$$

and

$$\begin{bmatrix} \hat{i}_{in} \\ \hat{i}_{out} \end{bmatrix} = \frac{\begin{bmatrix} I_o L(s + \frac{DV_{in}}{LI_o}) \\ \frac{V_{in}}{sL} \end{bmatrix}}{sL} \begin{bmatrix} \hat{d} \\ 0 \end{bmatrix} \tag{27}$$

C. Current-Fed-Current-Output Converter

The small-signal state-space of CF/CO converter including the parasitic elements can be given by

$$\frac{d\hat{i}_L}{dt} = -\frac{r_E + Dr_{C1}}{L}\hat{i}_L + \frac{D}{L}\hat{v}_{C1} + \frac{Dr_{C1}}{L}\hat{i}_{in} - \frac{1}{L}\hat{v}_o$$
$$+ \frac{V_E - D'r_{C1}I_L}{L}\hat{d}$$

$$\frac{d\hat{v}_{C1}}{dt} = -\frac{D}{C_1}\hat{i}_L + \frac{1}{C_1}\hat{i}_{in} - \frac{I_L}{C_1}\hat{d}$$

$$\frac{d\hat{v}_{C2}}{dt} = -\frac{1}{r_{C2}C_2}\hat{v}_{C2} + \frac{1}{r_{C2}C_2}\hat{v}_o \tag{28}$$

$$\hat{v}_{in} = (1 + r_{C1}C_1\frac{d}{dt})\hat{v}_{C1}$$

$$\hat{i}_o = \hat{i}_L + \frac{1}{r_{C2}}\hat{v}_{C2} - \frac{1}{r_{C2}}\hat{v}_o$$

978-1-4799-2706-7/14 $31.00 © 2014 IEEE

and the corresponding operating point by

$$I_L = I_o \quad I_o = \frac{I_{in}}{D} \quad V_{in} = \frac{V_o}{D} + (r_E + DD'r_{C1})I_{in} \qquad (29)$$

as well as the set of transfer functions by

$$\begin{bmatrix} \hat{v}_{in} \\ \hat{i}_{out} \end{bmatrix} = \frac{\begin{bmatrix} \dfrac{s}{C_1} & \dfrac{D}{LC_1} \\[2mm] \dfrac{D}{LC_1} & -C_2 s(s^2 + \dfrac{C_1 + D^2 C_2}{LC_1 C_2}) \end{bmatrix}}{s^2 + \dfrac{D^2}{LC_1}} \begin{bmatrix} \hat{i}_{in} \\ \hat{v}_{out} \end{bmatrix} \qquad (30)$$

and

$$\begin{bmatrix} \hat{v}_{in} \\ \hat{i}_{out} \end{bmatrix} = \frac{\begin{bmatrix} -\dfrac{I_o}{C_1}(s + \dfrac{DV_E}{LI_o}) \\[2mm] \dfrac{V_{in}(s - \dfrac{DI_o}{V_{in}C_1})}{LC_1} \end{bmatrix}}{s^2 + \dfrac{D^2}{LC_1}} \begin{bmatrix} \hat{d} \\ 0 \end{bmatrix} \qquad (31)$$

D. Current-Fed-Voltage Output Converter

The small-signal state-space of CF/VO converter including the parasitic elements can be given by

$$\frac{d\hat{i}_L}{dt} = -\frac{r_E + Dr_{C1} + r_{C2}}{L}\hat{i}_L + \frac{D}{L}\hat{v}_{C1} - \frac{1}{L}\hat{v}_{C2} + \frac{Dr_{C1}}{L}\hat{i}_{in}$$
$$+ \frac{r_{C2}}{L}\hat{i}_o + \frac{V_E - D'r_{C1}I_L}{L}\hat{d}$$
$$\frac{d\hat{v}_{C1}}{dt} = -\frac{D}{C_1}\hat{i}_L + \frac{1}{C_1}\hat{i}_{in} - \frac{I_L}{C_1}\hat{d}$$
$$\frac{d\hat{v}_{C2}}{dt} = \frac{1}{C_2}\hat{i}_L - \frac{1}{C_2}\hat{i}_o \qquad (32)$$
$$\hat{v}_{in} = (1 + r_{C1}C_1\frac{d}{dt})\hat{v}_{C1}$$
$$\hat{v}_o = (1 + r_{C2}C_2\frac{d}{dt})\hat{v}_{C2}$$

and corresponding operating point by

$$I_L = I_o \quad I_o = \frac{I_{in}}{D} \quad V_{in} = \frac{V_o}{D} + (r_E + DD'r_{C1})I_{in} \qquad (33)$$

as well as the set of transfer functions by

$$\begin{bmatrix} \hat{v}_{in} \\ \hat{v}_{out} \end{bmatrix} = \frac{\begin{bmatrix} \dfrac{1}{C_1}(s^2 + \dfrac{1}{LC_2}) & -\dfrac{D}{LC_1 C_2} \\[2mm] \dfrac{D}{LC_1 C_2} & -\dfrac{1}{C_2}(s^2 + \dfrac{D^2}{LC_1}) \end{bmatrix}}{s(s^2 + \dfrac{C_1 + D^2 C_2}{LC_1 C_2})} \begin{bmatrix} \hat{i}_{in} \\ \hat{i}_{out} \end{bmatrix} \qquad (34)$$

and

$$\begin{bmatrix} \hat{v}_{in} \\ \hat{v}_{out} \end{bmatrix} = \frac{\begin{bmatrix} -\dfrac{I_o}{C_1}(s^2 + s\dfrac{DV_{in}}{LI_o} + \dfrac{1}{LC_2}) \\[2mm] \dfrac{V_{in}}{LC_1 C_2}(s - \dfrac{DI_o}{C_1 V_{in}}) \end{bmatrix}}{s(s^2 + \dfrac{C_1 + D^2 C_2}{LC_1 C_2})} \begin{bmatrix} \hat{d} \\ 0 \end{bmatrix} \qquad (35)$$

E. Discussions

The transfer functions given for the buck power stage in different input and output-terminal-source configurations in (22,23), (26,27), (30,31) and (34,35) indicate clearly that there are significant changes taken place in the dynamic representation of the converters such as the change of system order, appearing of right-half (RHP) zero in the control-to-output transfer function, the dependence of the resonant frequency on duty ratio, and change of sign of the control-related transfer functions. All these changes would require careful consideration when designing the control system for ensuring robust stability.

The operating points given in (21), (25), (29), and (33) are basically quite the same if the output and input variables are not taken into account. The operating point in (21) indicates that the high-side MOSFET's conduction time has to be increased for increasing the output voltage and input current. The operating point in (29) indicates that the high-side MOSFET's conduction time has to be decreased for increasing the output current and input voltage, respectively, which is the property of boost converter.

The ideal input ($G_{11-\infty}$) and output ($G_{22-\infty}$) parameters can be computed based on (8) and (5) as well as the stated open-loop transfer functions of each of the converter. This computation yields for all of the converters

$$G_{11-\infty} = Z_{in-\infty} = \frac{1}{C_1(s - \dfrac{DI_o}{C_1 V_{in}})} \qquad (36)$$

$$G_{22-\infty} = Z_{out-\infty} = \frac{s + \dfrac{DV_{in}}{LI_o}}{C_2(s^2 + s\dfrac{DV_{in}}{LI_o} + \dfrac{1}{LC_2})} \qquad (37)$$

which are quite expected results, because these ideal parameters are power-stage dependent, and not affected by any internal or external control method, etc., as discussed in [19] and [20].

The change of input and output terminal sources affects also the impedance-based stability assessment at the input and output terminals of the converters by dictating the numerator and denominator impedances of the minor-loop gain: The numerator impedance has to be always the output/input impedance of the voltage-type source, and the denominator impedance is the output/input impedance of the current-type source, respectively.

The 2014 International Power Electronics Conference

IV. EXPERIMENTAL EVIDENCE

The power stage and its approximated component values and their parasitic elements as well as the used operating point are presented in Fig. 2. The PWM modulator is implemented by using Texas Instruments' DSP TMSF 28335 digital signal processor. Sorenen's DC power supply XHR 100-10 was used as the constant-voltage input source. Agilent technologies' solar array emulator E4360A was used as the constant-current input source. CROMA 63103 electronic load configured either as a constant-current or a constant-voltage sink was used as the load. The frequency responses were measured by using Venable Instruments' frequency-response analyzer Model 3120 with an impedance measurement kit. The predictions given in this section include all the parasitic elements defined in Fig. 2. The duty ratio of the converter equals 0.77. All the given frequency responses indicate that the derived small-signal models are accurate.

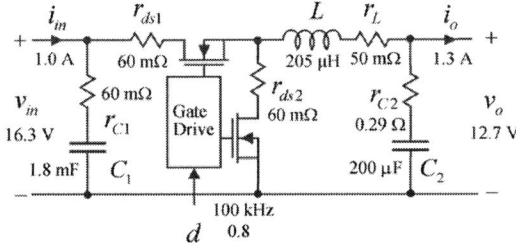

Fig. 2 The power stage with components and operating point

A. Control-to-Output Transfer Functions

The open-loop control-to-output transfer functions denoted in (1) by G_{23-o} for the VF, and CF converters are given in Figs. 3 and 4, respectively, where the output-mode of the converter determines the corresponding output or feedback variable in question.

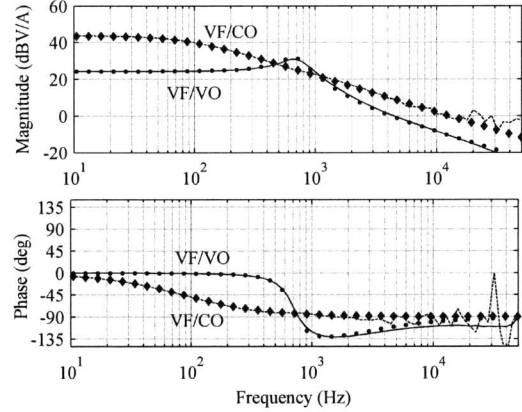

Fig. 3 Measured (solid and dashed lines) and predicted (dots and diamonds) control-to-output transfer functions of VF power stage at voltage (VO) and current (CO) output modes.

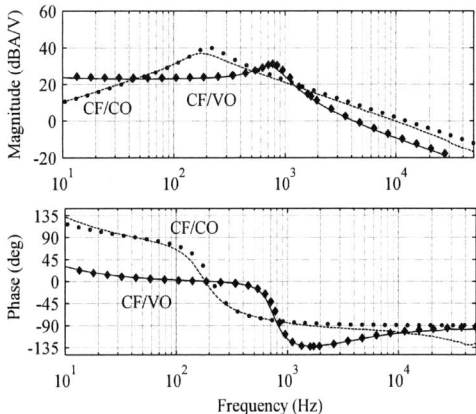

Fig. 4 Measured (solid and dashed lines) and predicted (dots and diamonds) control-to-output transfer functions of CF power stage at current (CO) and voltage (VO) output modes.

Fig. 3 indicates that the change of feedback variable from voltage (i.e., the conventional operation mode) to current (i.e., e.g. storage-battery charging) increases the transfer-function gain significantly, and also reduces the order of the system. The gain increase is earlier shown to lead easily to instability due to the high loop-gain crossover frequency if the models of the conventional buck converter are utilized in the control design [13].

Fig. 4 shows that the change of input source affects especially the phase at low frequencies, which starts at 180 deg. The behavior of the low-frequency phase implies that the sign of the feedback and reference signal has to be interchanged for ensuring stable operation. In addition, the phase behavior implies that a low-frequency RHP zero has appeared which has to be carefully taken into account because of the control bandwidth limitation [19].

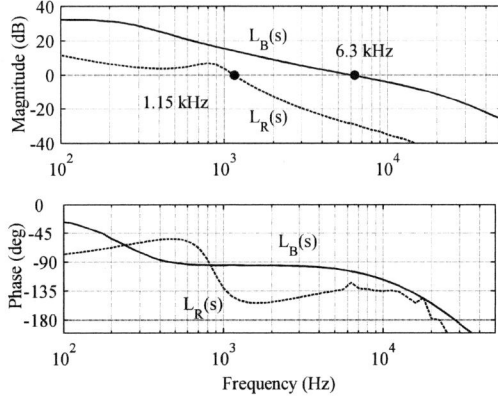

Fig. 5 Measured frequency responses of the output-current loop gains of the VF/CO buck converter with ideal battery load ($L_B(s)$) and resistive load ($L_R(s)$) as reported in [13].

As shown in Fig. 5, the use of resistor as a load, when validating the control design of a current-output

978-1-4799-2706-7/14 $31.00 © 2014 IEEE

converter, is quite problematic, and can lead easily to instability due to the change of crossover frequency by one decade or even more.

B. Audiosusceptibilities

The open-loop audiosusceptibilities or forward transfer functions denoted in (1) by G_{21-o} for VF, and CF converters are given in Figs. 6 and 7 indicating significant changes when the input and output sources vary. As a consequence, the conclusions based on the properties of the conventional buck converter do not hold for the other converters, which is very evident according to Figs. 6 and 7.

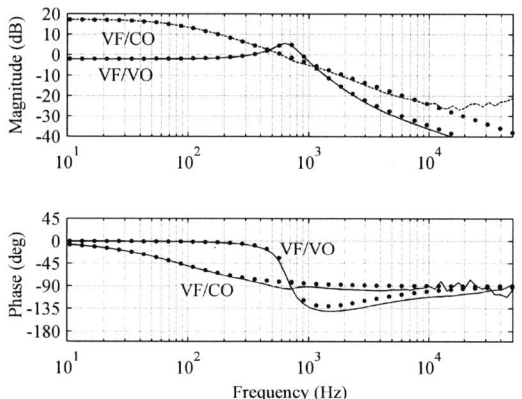

Fig. 6 Measured (solid and dashed lines) and predicted (dots) frequency responses of the audio susceptibilities of the VF power-stage at voltage (VO) and current (CO) output modes.

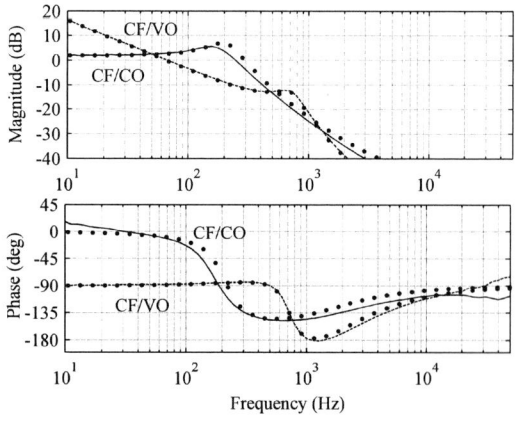

Fig. 7 Measured (solid and dashed lines) and predicted (dots) frequency responses of the audio susceptibilities of the CF power-stage at current (CO) and voltage (VO) output modes.

C. Input and Output Impedances

Figs. 8 and 9 show the measured input and output impedances when the terminal sources are varied. It may be obvious, when the power stage is the same all the time, that the impedances measured through the same terminal are equal if the other terminal source is equal.

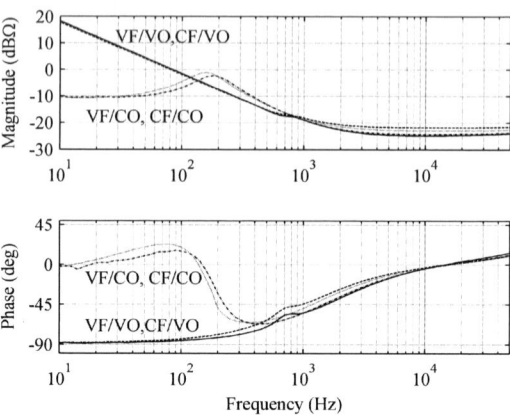

Fig. 8 Measured frequency responses of the buck-power-stage input impedance for all possible source-load combinations.

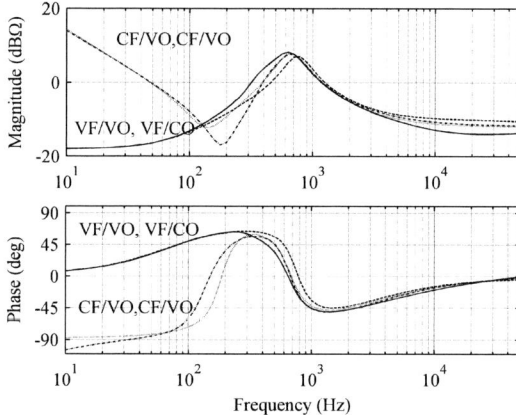

Fig. 9 Measured frequency responses of the buck-power-stage output impedances for all possible source-load combinations.

V. CONCLUSIONS

The main purpose of the paper is to show definitely that the used input and output terminal sources determine the steady-state and dynamic properties of the converter even if the power stage remains the same all the time. The theoretical and experimental investigations reported in this paper prove that definitively. The main reason for this is the fact that the terminal sources actually determine the applicable feedback variables. At open-loop, the power stage is automatically tuned to the behavioral mode, the terminal sources would dictate. If feedback control is used, the used feedback variables would dictate the applicable terminal sources. If resistor is used as the output terminal load either the voltage or current can be used as the feedback variable. If the real internal dynamic behavior is desired to be known then the load effect has to be removed.

When the output terminal is feedback controlled applying the standard negative feedback then the control scheme of the power electronic switches would dictate the usable input source as discussed in detail in [14].

REFERENCES

[1] R. Tymerski, and V. Vorpérian, "Generation, classification, and analysis of switched-mode DC-to-DC converters by the use of converter cells," in *Proc. IEEE INTELEC'86*, 1986, pp. 181-195.

[2] R. Tymerski, and V. Vorpérian, "Generation and classification of PWM DC-to-DC converters," *IEEE Trans. Aerosp. Electron. Syst.*, vol. 24, no. 6, pp. 743-754, 1988.

[3] L. Nousiainen, et al., "Photovoltaic generator as an input source for power electronics converters," *IEEE Trans. Power Electron.*, vol. 28, no. 6, pp. 3028-3038, 2013.

[4] M. H. Ali, B. Wu, and R. A. Dougal, "An overview of SMES applications in power and energy systems," *IEEE Trans. Sustain. Energy*, vol. 1, no. 1, pp. 38-45, 2010.

[5] D. Shmilovitz, and S. Singer, "A switched-mode converter suitable for superconductive magnetic energy storage (SMES) systems," in *Proc. IEEE APEC'02*, 2002, pp. 630-634.

[6] T. Suntio, J. Leppäaho, J. Huusari, and L. Nousiainen, "Issues on solar-generator interfacing with current-fed MPP-tracking converters," *IEEE Trans. Power Electron.*, vol. 25, no. 9, pp. 2409-2419, 2010.

[7] M. G. Villalva, T. G. de Siqueira, and E Ruppert, "Voltage regulation of photovoltaic arrays: Small-signal analysis and control design," *IET Power Electron.*, vol. 3, no. 6, pp. 869-880, 2010.

[8] Y. M. Chen, A. Q. Huang, and X. Yu, "A high step-up three-port DC-DC converter for stand-alone PV/battery power systems," *IEEE Trans. Power Electron.*, vol. 28, no. 11, pp. 5049-5062, 2013.

[9] W. Xiao, W. G. Dunford, P. R. Palmer, and A. Capel, "Regulation of photovoltaic voltage," *IEEE Trans. Ind. Electron.*, vol. 54, no. 3, pp. 1365-1374, 2007.

[10] T. Suntio, J. Huusari, and J. Leppäaho,"Issues on solar-generator interfacing with voltage-fed MPP-tracking converters," *European Power. Electron. and Drives J.*, vol. 20, no. 3, pp. 40-47, 2010.

[11] M. C. Glass, "Advancements in the design of solar array to battery charge current regulators," in *Proc. IEEE PESC'77*, 1977, pp. 346-350.

[12] K. Siri, "Study of system instability in solar-array-based power systems," *IEEE Trans. Aerosp. Electron. Syst.*, vol. 36, no. 3, pp. 957-964, 2000.

[13] M. Hankaniemi, and T. Suntio, "Small-signal models for constant-current regulated converters," in *Proc. IEEE IECON'06*, 2006, pp. 2037-2042.

[14] J. Leppäaho, L. Nousiainen, J. Puukko, and T. Suntio, "Implementing current-fed converters by adding an input capacitor at the input of voltage-fed converter for interfacing solar generator," in *Proc. EPE-PEMC'10*, 2010, pp. 81-88.

[15] C. K. Tse, "*Linear Circuit Analysis*," Harlow, England: Addison Wesley Longham, 1998.

[16] J. Leppäaho, J. Huusari, L. Nousiainen, J. Puukko, and T. Suntio, "Dynamic properties and stability assessment of current-fed converters in photovoltaic applications," *IEEJ Trans. Ind. Appl.*, vol. 131, no. 8, pp. 976-984, 2011.

[17] R. D. Middlebrook, "Input filter considerations in design and applications of switching regulators," in *Proc. IEEE IAS'76*, 1976, pp. 91-107.

[18] R. D. Middlebrook, and S. Ćuk, "A general unified approach to modelling switching-converter power stages," *Int. J. Electron.*, vol. 42, no. 6, pp. 521-550, 1977.

[19] T. Suntio, "*Dynamic Profile of Switched-Mode Converter – Modeling, Analysis, and Control*," Weinheim, Germany: Wiley-VCH, 2009.

[20] S. Vesti, T. Suntio, J. A. Oliver, R. Prieto, and J. A. Cobos, "Effect of control method on impedance-based interactions in a buck converter," *IEEE Trans. Power Electron.*, vol. 28, no. 11, pp. 5311-5322, 2013.

A Fully Soft-Switched Multiphase DC-DC Converter with Reduced Switch Count for High Power Application

Minjae Kim, Daeki Yang, Sewan Choi, *IEEE Senior Member*
Department of Electrical and Information Engineering
Seoul National University of Science and **Tech**nology
Email: schoi@seoultech.ac.kr

Abstract— this paper proposes a fully soft-switched multiphase dc-dc converter for high power application. The proposed multiphase converter exhibits low switch count and simple gate driver circuit. Owing to use of low rated lossless passive clamp and series resonant circuits, ZCS turn-on and ZVS turn-off of switches and ZCS turn-off of diodes are achieved. Further, despite the passive clamp method, operating duty cycle range is not limited, resulting in no additional start-up circuitry required. Experimental results on a 50kHz, 5kW prototype are provided to validate the proposed converter.

Keywords— *Soft switching, high power dc-dc converter, multiphase, lossless snubber*

I. INTRODUCTION

In the application such as fuel cell systems, distributed power generation, hybrid electric vehicles and photovoltaic systems the high power isolated boost dc-dc converter is required. The three-phase dc-dc converter can be used for the high power isolated boost dc-dc converter. They can be classified into voltage-fed and current-fed. The voltage-fed converter [1-2] suffers from high transformer turns ratio, leading to disadvantages associated with large leakage inductance. Compared to the voltage-fed topology, in general, the current-fed topology exhibits lower transformer turns ratio, smaller input current ripple and lower diode voltage rating. The three-phase current-fed topology has two types: passive clamping [3-4] and active clamping [5-8]. The three-phase current-fed converter with passive clamp [3-4] has simple structure and small switch count, but suffers from excessive power losses dissipated in the RCD snubber and associated with hard switching of main switches. The duty cycle of the three-phase current-fed converter with passive clamp is also limited, leading to necessity of the additional start-up circuit and poor dynamic responses to line and load variations. The three-phase current-fed converter with active clamp has three topologies: push-pull[5], full-bridge[6] and half-bridge[7-8]. The three-phase current-fed converter with active clamp is able to achieve ZVS turn on of main switches as well as clamp the voltage surge across the main switch. Also, it does

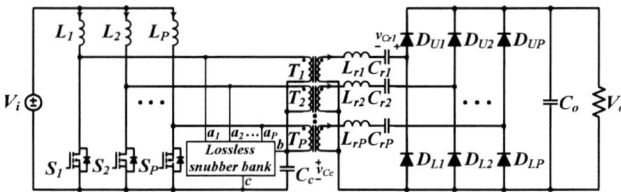

Fig. 1. Proposed multiphase converter

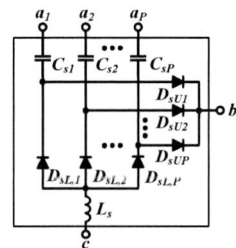

Fig. 2. Lossless snubber bank of the proposed multiphase converter

not have reverse recovery problem due to ZCS turn off of rectifier diodes. Further, unlike the current-fed converter with passive clamp, the whole duty cycle range between 0 and 1 can be used for the switch of the current-fed converter with active clamp. However, a drawback of the current-fed converter with active clamp is to use more switches and gate driver circuits for them.

In this paper, a fully soft switched multiphase converter is proposed for high power application. The proposed multiphase converter has the following features: 1) low switch count and simple gate driver circuit due to the switches with the common ground 2) low rated lossless passive clamp circuit, making it possible to achieve low cost and volume 3) ZCS turn-on and ZVS turn-off of switches regardless of voltage and load variation 4) ZCS turn-off of diodes leading to negligible voltage surge associated with the diode reverse recovery 5) no additional start-up circuitry required since operating duty cycle range is not limited unlike other current-fed converters with passive clamp 6) significantly reduced current rating of the clamp capacitor, compared to other current-fed converters with active clamp. Experimental results on a 50kHz, 5kW prototype are provided to validate the proposed converter.

978-1-4799-2706-7/14 $31.00 © 2014 IEEE

The 2014 International Power Electronics Conference

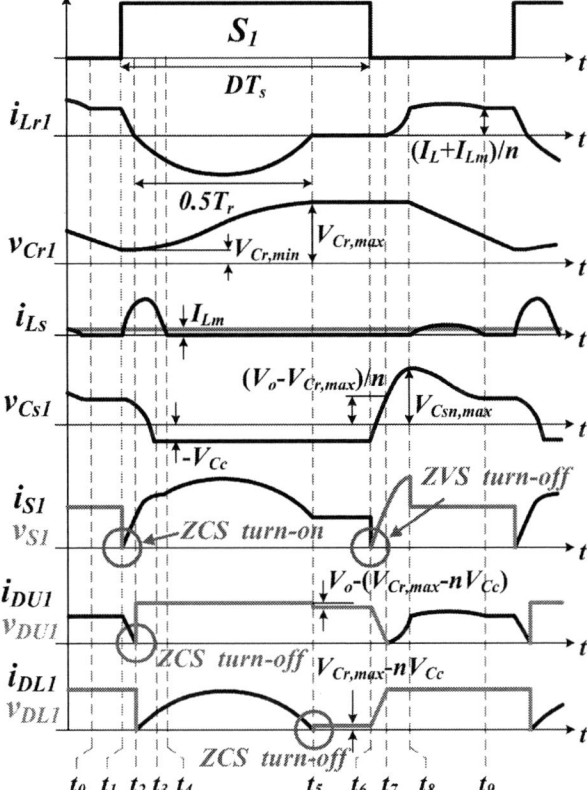

Fig. 3. Key waveform of the proposed multiphase converter (P=1).

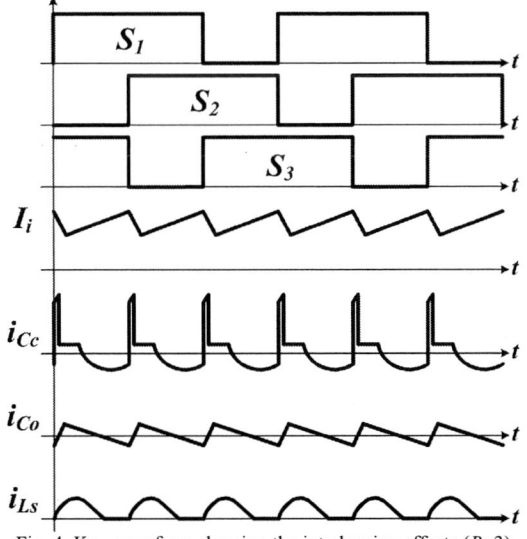

Fig. 4. Key waveform showing the interleaving effects (P=3)

II. PROPOSED CONVERTER

A. Operating principles

Fig. 1 shows the generalized circuit diagram of the proposed multiphase converter with P phases. The proposed P-phase converter consists of P inductor-switch phase legs, a lossless snubber bank and a clamp capacitor at primary side and P L_r-C_r series resonant circuits and P diode phase legs at secondary side. Fig. 2 shows the circuit diagram of the lossless snubber bank of the proposed P-phase converter. Note that the current ratings

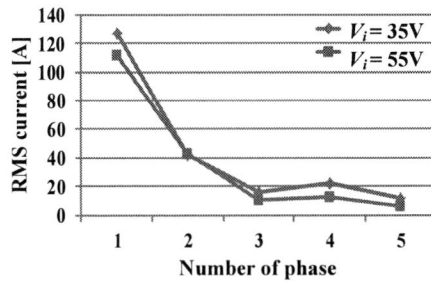

Fig. 5. RMS current of the clamp capacitor as a function of number of phase

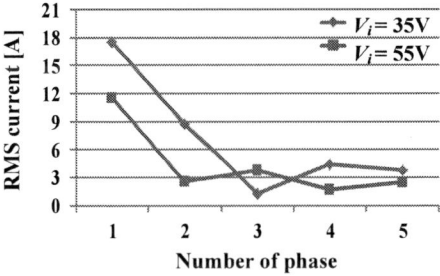

Fig. 6. RMS current of the output capacitor as a function of number of phase

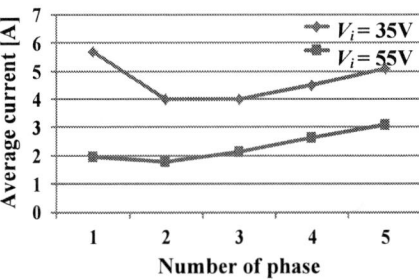

Fig. 7. Average current of the snubber inductor as a function of number of phase

of snubber components are much lower than those of main components. Fig. 3 shows the key waveform of a phase of the proposed multiphase converter. The proposed multiphase converter is able to achieve not only ZCS turn on and ZVS turn off of switches but ZCS turn-off of all diodes owing to the lossless snubber bank and L_r-C_r resonant circuit. It is noted that the turn on loss associated with energy stored in MOSFET's output capacitance is negligible in this low input voltage application [9].

Even though the theoretical range of the duty cycle is $0<D<1$, the actual operating duty cycle of $0.5 \cdot T_r/T_s<D<1$ is used to ensure the aforementioned soft switching switches and diodes over the whole input voltage and load range.

Each leg of the P-phase multi-phase converter is interleaved with a phase difference of $2\pi/P$. Then the effective switching frequencies of the input current, clamp capacitor current and output current become P times the switching frequency, resulting in reduced volume of clamp capacitor C_c and output capacitor C_o. Fig. 4 shows the key waveform showing the interleaving effects of the 3-phase converter.

978-1-4799-2706-7/14 $31.00 © 2014 IEEE

TABLE I.
COMPARISON OF CHARACTERISTICS OF THE PROPOSED AND CONVERTIONAL CONVERTERS (P_o = 5kW, V_i = 35V, V_o = 400V, f_s = 50kHz)

	Clamping method	Duty cycle range	Switching method		Main switch	Reverse recovery problem	Clamp circuit				
			Turn-on	Turn-off			C	L	R	D	Switch
Reference [4]	Passive clamp (RDC snubber)	0.33<D<1	Hard switching	Hard switching	211V_{pk} 63A_{rms} 3EA	Not negligible	211V_{pk} 7.5A_{rms} 3EA	-	50Ω 1.4A_{rms} 1EA	113V_{pk} 1A_{avg} 6EA	-
Reference [7]	Active clamp	0<D<1	ZVS	Hard switching	105V_{pk} 61A_{rms} 3EA	Negligible	105V_{pk} 30A_{rms} 1EA	-	-	-	105V_{pk} 17A_{rms} 3EA
Proposed	Passive clamp (Lossless snubber)	0<D<1	ZCS	ZVS	184V_{pk} 63A_{rms} 3EA	Negligible	135V_{pk} 8.5A_{rms} 3EA	35V_{pk} 6.6A_{rms} 1EA	-	105V_{pk} 1.3A_{avg} 6EA	-

B. Design considerations

The number of phase of the proposed converter can be determined considering filters' volume and components' current ratings. The system specification is as follows:

- V_i=35-55V
- V_o=400V
- P_o=5kW
- f_s=50kHz
- N_p:N_s=1:4
- L_r=5uH
- C_r=1.5uF
- L_s=10uH
- C_s=100nF

Fig. 5 shows the RMS current rating of the clamp capacitor as function of the number of phase. The clamp capacitor current is almost the same as sum of each primary winding current which is referred resonant inductor current consisting of sinusoidal waveform according to L_r-C_r resonance during switch turn on and the sum of input inductor current and snubber inductor current during switch turn off, as shown in Fig. 3. As the number of phase increases, RMS current rating of the clamp capacitor decreases due to the interleaving effect, but the interleaving effect is mitigated for number of phase over 3 in this case. Fig. 6 shows the RMS current of the output capacitor as a function of number of phase. The output capacitor current is sum of upper diode currents. RMS current rating of the output capacitor decreases as the number of phase increases, but the interleaving effect is also mitigated over some number of phase. Fig. 7 shows the average current of the snubber inductor as a function of number of phase. The snubber loss is determined by average current of snubber inductor which is equal to sum of the average current of upper or lower snubber diodes. Considering components' current ratings of Figures 5-7, the number of phase of the proposed multiphase converter is chosen to be 3 in this example.

C. Comparative analysis

In this section, main characteristics and device ratings of the proposed three-phase converter are compared to the conventional three-phase current-fed half-bridge converters with passive clamp [4] and with active clamp [7]. The comparison results are summarized in Table I. In general, the duty cycle less than 0.33 of the conventional three-phase current-fed converter with passive clamp [3-4] is not allowed, requiring an additional start-up circuit. In the meanwhile the proposed converter does not necessitate a start-up circuit since the operating duty cycle range is not limited like the active clamp topology. The switches of the proposed converter are turned ON with ZCS and OFF with ZVS, respectively, while those of the passive clamp topology with RCD snubber [4] are hard switched and those of the active clamp topology [7] are turned ON with ZVS, but tuned OFF with hard switching. Therefore, the switching loss of the proposed converter is lowest.

The rectifier diodes of the passive clamp topology with RCD snubber [4] have switching losses and voltage surge associated with reverse recovery. The proposed converter and the active clamp topology [7] do not have reverse recovery problem due to the ZCS turn OFF of the rectifier diodes.

The losses associated with RCD snubber in the passive clamp topology [4] are significantly large. The clamp circuit of the active clamp topology [7] is more expensive than other two topologies since it has 3 active clamp switches and gate drivers for them and ripple current rating of the clamp capacitor is much higher than those of the passive clamp topology and the proposed converter. Further, the clamp switch of the active clamp topology is turned OFF with hard switching. On the other hand, the clamp circuit of the proposed converter has much lower cost as well as smaller losses than the main circuits since its component ratings are much smaller than those of the main circuit.

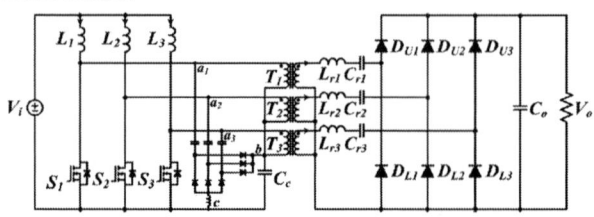

Fig. 8. Circuit diagram of proposed three-phase converter

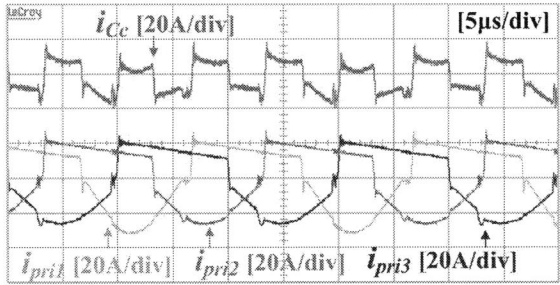

(a) Clamp capacitor current and each primary winding current

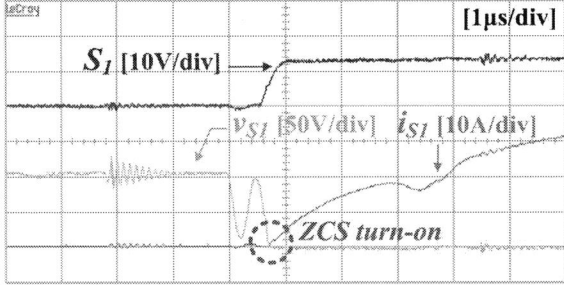

(b) Switch S_1 at turn on

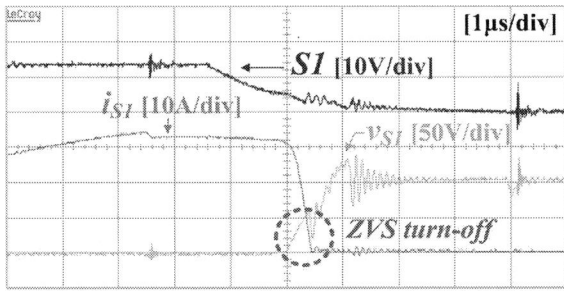

(c) Switch S_1 at turn off

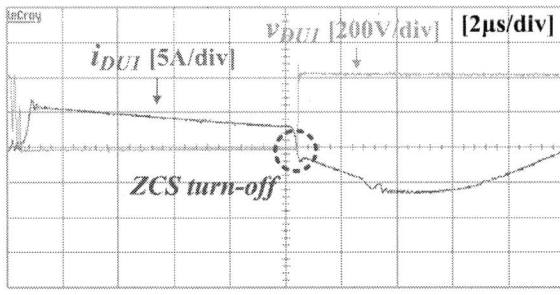

(d) Diode D_{U1} at turn off

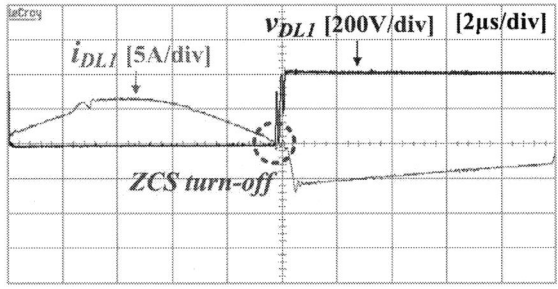

(e) Diode D_{L1} at turn off

Fig. 9. Experimental waveform

Fig. 10. Measured efficiency

Fig. 11. Photograph of the proposed three-phase converter

III. EXPERIMENTAL RESULTS

A 5kW laboratory prototype of the proposed three-phase converter has been built and tested to verify the operating principle. The system specifications used in the experiment are the same as those in Section II-B. In the experimental P=3 is chosen, and is shown in Fig. 8. Fig. 9 shows experimental waveforms at 2.5kW when the input voltage V_i is 55V. Fig. 9(a) shows the clamp capacitor current and three phase primary winding current. We can see that the clamp capacitor current ripple has been greatly reduced due to interleaving effect. Fig. 9(b) and (c) show that switch S_1 is turned ON with ZCS and turned OFF with ZVS, respectively. Fig. 9(d) and (e) show that diode D_{U1} and D_{L1} are turned OFF with ZCS, respectively. The efficiency of the proposed three-phase converter is measured by YOKOGAWA WT3000 and shown in Fig. 10. The maximum efficiency is 96% at 2.5kW. The full load efficiency is 94.5%. Fig. 11 shows the photograph of the proposed three-phase prototype.

IV. CONCLUSIONS

A fully soft-switched multiphase dc-dc converter was proposed in this paper. The proposed converter features low switch count and simple gate driver circuit, fully soft switching of switches and diodes, full operating duty cycle range between 0 and 1. Experimental results on a 50kHz, 5kW prototype are provided to validate the proposed three-phase concept. Enhanced design consideration and loss analysis of proposed converter will be provided in the final paper.

REFERENCES

[1] H. Kim, C. Yoon, and S. Choi, "A three-phase zero-voltage and zero-current switching dc-dc converter for fuel cell applications," *IEEE Trans. on Power Electronics*, vol. 25, no. 2, pp. 391-398, 2010.

[2] C. Lin, A. Johnson, and j. S. Lai, "A novel three-phase high-power soft-switched dc/dc converter for low-voltage fuel cell applications," *IEEE Trans. on Industry Applications*, vol. 41, no. 6, pp. 1691-1697, 2005.

[3] R. L. Andersen, and I. Barbi, "A three-phase current-fed push-pull dc-dc converter," *IEEE Trans. Power Electronics*, vol. 24, no. 2, pp. 358-368, 2009.

[4] S. V. G. Oliveira, and I. Barbi, "a three-phase step-up dc-dc converter with a three-phase high-frequency transformer for dc renewable power source applications," *IEEE Trans. on Industrial Electronics*, vol. 58, no. 8, pp. 3567-3580, 2011.

[5] S. Lee, J. Park, and S. Choi, "A three-phase Current-fed Push-Pull DC-DC Converter with Active Clamp for Feul Cell Applications," *IEEE Trans. on Power Electronics*, vol. 26, no. 8, pp. 2266-2277, 2011

[6] H. Char, J. Choi, and P. N. Enjeti, "A three-phase current-fed dc/dc converter with active clamp for low-dc renewable energy sources," *IEEE Trans. on Power Electronics*, vol. 23, no. 6, pp. 2784-2793, 2008.

[7] J. Choi, H. Char, and B. Han, "A three-phase interleaved dc-dc converter with active clamp for fuel cells," *IEEE Trans. on Power Electronics*, vol. 25, no. 8, pp. 2115-2123, 2010.

[8] C. Yoon, J. Kim, and S. Choi, "Multiphase dc-dc converters using a boost-half-bridge cell for high-voltage and high-power applications," *IEEE Trans. on Power Electronics*, vol. 26, no. 2, pp. 381-388, 2011.

[9] Y. Ren, M. Xu, J. Zhou, and F. C. Lee, "Analytical loss model of power mosfet," *IEEE Trans. on Power Electronics*, vol. 21, no. 2, pp. 310-319, 2006.

A Static Characteristic Analysis of Proposed Bi-Directional Dual Active Bridge DC-DC Converter

Shun Nagata

Nagasaki University
1-14 Bunkyo-machi
Nagasaki-shi, Nagasaki, Japan
bb52113224@cc.nagasaki-u.ac.jp

Mika Takasaki

Nagasaki University
1-14 Bunkyo-machi
Nagasaki-shi, Nagasaki, Japan
bb52112218@cc.nagasaki-u.ac.jp

Yutaka Furukawa

Koga System Works
Saga, Japa

Toshiro Hirose

Nishimu Electronics Industries Co.,Ltd.
700 Tateno, Yoshinogari-cho
Kanzaki-gun, Saga, Japan
hirose@nishimu.co.jp

Yoichi Ishizuka

Nagasaki University
1-14 Bunkyo-machi
Nagasaki-shi, Nagasaki, Japan
sy2@nagasaki-u.ac.jp

Abstract- **Recently, the power supply network with energy storage devices such as battery has been focused. This network topology uses bi-directional isolated DC-DC converter of low or medium capacity is required for the diversification of power supply network. The dual active bridge (DAB) DC-DC converter is one of the effective bi-directional isolated DC-DC converters. However, the circuit has some instinct problems such as degradation of power efficiency and the occurrence of the surge in light-load operation. In this paper, we have been done a static characteristic analysis and highly power-efficient technique for DAB DC-DC Converter at light load. Also the analysis results and the proposed technique are verified with some experimental results.**

I. INTRODUCTION

Recently, the bidirectional dc-dc converter has been focused on because of the huge demand for diversification of power supply network including battery. The DAB dc-dc converter is one of the most popular circuits for bidirectional applications because of its simple structure. Some examples are for UPS [1], for automotive [2]-[4] and for energy storage system [5]. The one of the feature is achieving zero volt switching (ZVS) in natural operation. However, hard switching and/or power efficiency at light load condition is the intrinsic problem [6]. Some research have been done to solve the problem, for instance, use of resonant type converter with snubber circuit [7], silicon carbide (SiC) power device and new magnetic materials [8], and Quasi-ZCS operation with LC filter [9]. Furthermore by applying switching modulation, DAB converter works in wide range of input voltage and load condition [10]-[12]. These objectives of switching modulation controls are to regulate voltage and satisfy load variation [10], to expand soft switching region [11], and minimize the total power losses [12]. However, the problem of switching surges reduction was not addressed. In [13], the novel switching surge reduction technique is proposed and confirmed with some analysis and experimental results. And also, the results of power efficiency improvement of the light load were described.

In this paper, the detailed analysis of the technique is described and confirmed with some experiments.

II. CONVENTIONAL OPERATION OF A DAB DC-DC CONVERTER

Fig. 1 shows the circuit schematic of the basic DAB dc-dc

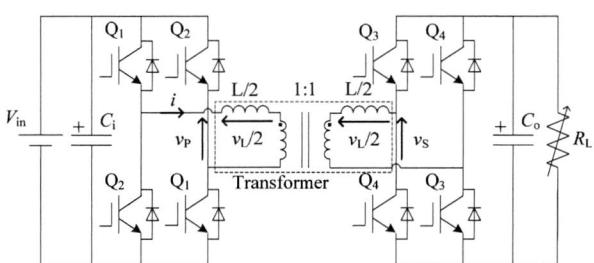

Fig. 1. The circuit schematic of DAB dc-dc converter.

978-1-4799-2706-7/14 $31.00 © 2014 IEEE

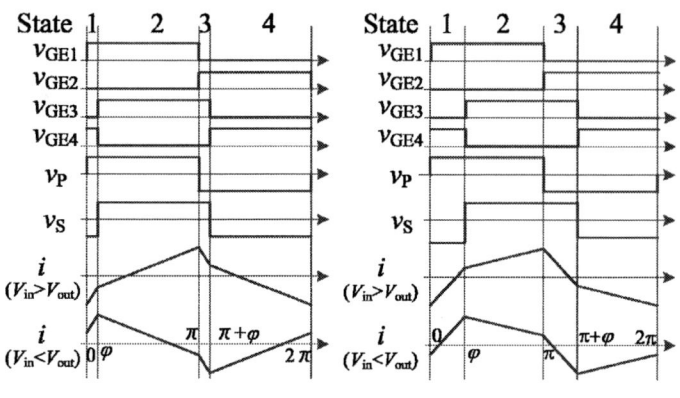

(a) Light load   (b) Heavy load

Fig. 2.   Conventional operating waveform.

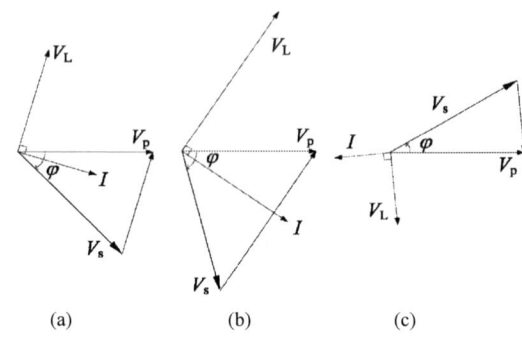

Fig. 3.   Phasor diagram [13]: (a) Forward power flow mode (light load); (b) Forward power flow mode (heavy load); (c) Reverse power flow mode.

converter. Fig. 2 shows the operating waveforms with the conventional operation [14]. In the conventional operation, the output power is operated by the phase-shift shown as φ between the primary voltage v_P and secondary voltage v_S of transformer. Fig. 3 shows the phasor diagram. V_P, V_S, V_L, and I are phasor symbols for v_P, v_S, v_L, i, respectively. When V_S is lagging V_P in forward power flow mode (Fig. 3 (a) and (b)), and when V_S is leading V_P, it is operated in reverse power flow mode (Fig. 3 (c)).

The output power P_o can be obtained as

$$P_o = \frac{V_{in}V_{out}}{\omega L}\varphi\left(1-\frac{\varphi}{\pi}\right). \tag{1}$$

The output power can be controlled with the phase difference φ. The waveform of the current i is changed by the load condition. In this paper, current i crossed the zero line in the state 2 is defined as a light load, and current i crossed the zero line in the state 1 is defined as a heavy load as shown in Fig. 2.

III. INTRINSIC SURGES PROBLEM OF A DAB DC-DC CONVERTER

As mentioned in above, well known problem of a DAB DC-DC converter is hard switching in the light condition. However, previous researches haven't been addressed about the switching surges problem. It is caused by the reverse recovery effect of the diode. Fig. 4 shows φ - P_o. The switching surges occur at light load range of this figure.

Fig. 5 shows the generation mechanism of switching surges when $V_{in} > V_{out}$. The surges voltage occurs in the transition from State 1 (3) to State 2 (4), repeatedly. C_d is the parasitic capacitance of diode which is connected in parallel with the ideal diode, and L_{wire} is parasitic reactance. At the light load condition, the diodes D_4 is conducting in state 1. Then the switches Q_3 is turned on when state changes from State 1 to State 2. At this instantaneous moment, the diode D_4 is switched from a forward bias condition to a reverse bias condition, immediately. And the switching surges are

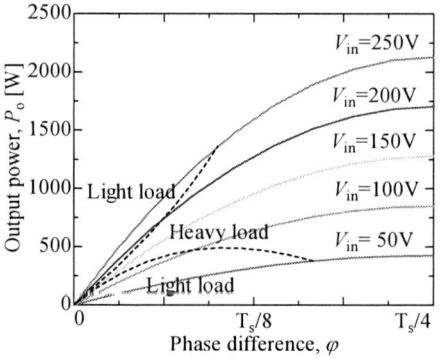

Fig. 4. φ - P_o (V_{in}=150V) [13].

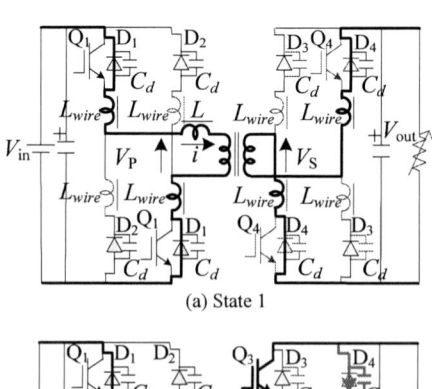

(a) State 1

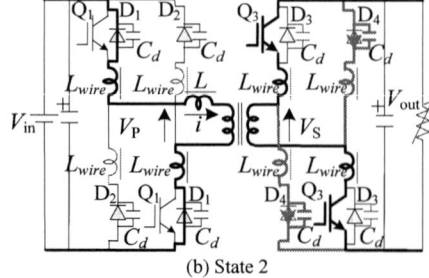

(b) State 2

Fig. 5.   The generation mechanism of switching surge ($V_{in} > V_{out}$).

978-1-4799-2706-7/14 $31.00 © 2014 IEEE   2253

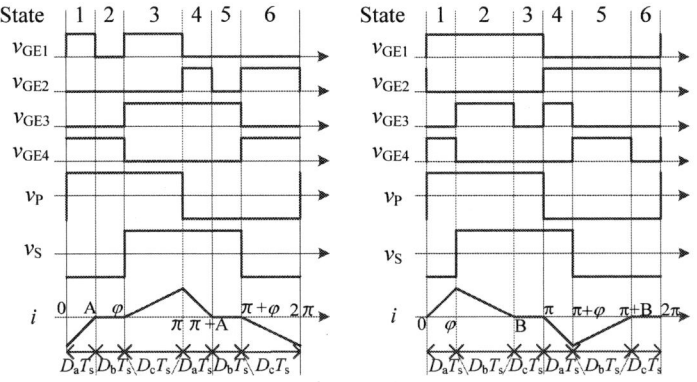

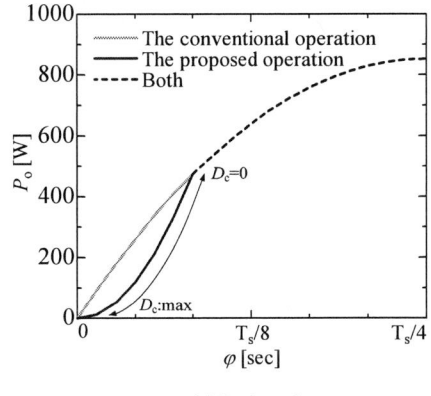

(a) Buck mode   (b) Boost mode

Fig. 6. Idealized waveforms for proposed operating at the light load.

(a) Buck mode

occurred with the resonance of C_d and L_{wire} due to reverse recovery phenomenon. With the same reason, when $V_{in} < V_{out}$, the surges occurs in the transition from State 2 (4) to State 3 (1) on the primary side.

Commonly, to protect the switches from the switching surges, snubber circuit are applied. However, the power loss at the snubber circuit can't be ignored at the light load condition. The other way, the resonant converter type is also popular, but the additional components are needed [9].

IV.˙ PROPOSED OPERATION METHOD

We have proposed the software-based compensation method for basic DAB dc-dc converter topology. It can be reduce the switching surges at the light load, without any of additional circuits such as the snubber circuits or resonant circuits [13]. Fig. 6 shows idealized waveform of the proposed operating method. When $V_{in} < V_{out}$, as it can be seen from the waveforms, the direction of primary side current of transformer i during each on-time of Q_1 and Q_2 is restricted to avoid the crossing the zero line. Due to the restriction, the zero-current-switching can be realized for Q_1 and Q_2, respectively. The ideal static analysis has been done as follows. This converter has six operational states in one switching period for each of the buck and boost mode operation, respectively. The each element is treated as ideal in equivalent circuit.

The detailed description of the ideal circuit is revealed in a previous paper [13]. Therefore, only the results are shown in this paper.

A. Buck Mode Operation in Light Load

In buck mode, the primary side switches Q_1 and Q_2 are turned-on twice in the period. Firstly, Q_1 and Q_2 are turn-on at $t = 0$ and $T_s/2$. Secondly, they are turn-off at $t = A$ and $T_s/2 + A$. Thirdly, they are turn-on at $t = \varphi$ and $T_s/2 + \varphi$. Fourthly, they are turn-off at $t = T_s/2$ and T_s.

A is calculated as

$$A = \frac{V_{in} - V_{out}}{V_{in} + V_{out}} \left(\frac{1}{2} T_s - \varphi \right). \tag{2}$$

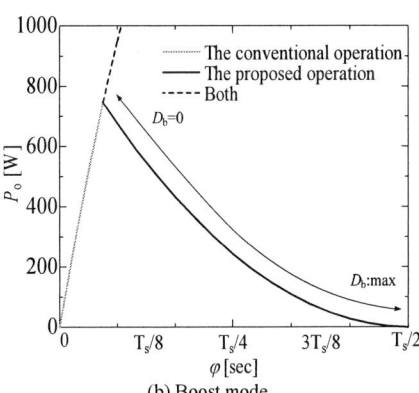

(b) Boost mode

Fig. 7. $\varphi - P_o$ (conventional operation and proposed operation).

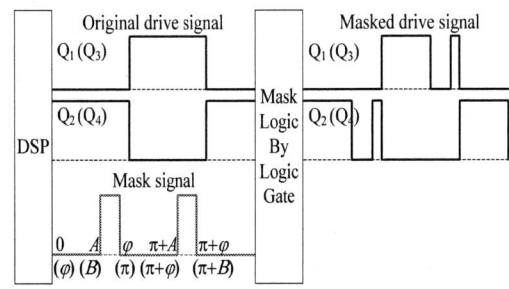

Fig.8.   Masked drive signal generating mechanism.

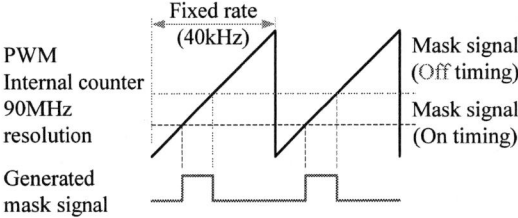

Fig. 9.   Mask signal generating mechanism of PWM in DSP.

B. Boost Mode Operation in Light Load

In boost mode, the secondary side switches Q_3 and Q_4 are turned-on twice in the period. Firstly, Q_3 and Q_4 are turn-on at $t = \varphi$ and $T_s/2 + \varphi$, respectively. Secondly, they are turn-off at $t = B$ and $T_s/2 + B$. Thirdly, they are turn-on at $t = T_s/2$ and 0. Fourthly, they are turn-off at $t = T_s/2 + \varphi$ and φ.

B is calculated as

$$B = \frac{2V_{out}}{V_{out} - V_{in}} \varphi. \qquad (3)$$

C. Output Power Control in Light Load

The ideal analysis for both of buck and boost mode operation can be done for power. For the ideal analysis result, the output power P_o can be obtained as

$$P_o = \frac{2X^2}{T_s L} \cdot \left| \frac{V_{in} + V_{out}}{V_{in} - V_{out}} \right| \cdot V_{in} V_{out}. \qquad (4)$$

In buck mode, $X = A$, and in boost mode, $X = \varphi$.

D. Output Power Control in Heavy Load

In the light load, with the output power increasing, the periods of which all switches turned OFF ($A \sim \varphi$, $\pi + A \sim \pi + \varphi$, $B \sim \pi$, $\pi + B \sim 2\pi$) becomes shorter. A equal to φ or B equal to π is the boundary between light load and heavy load. Therefore, in the heavy load condition, the only conventional phase-shift operation is active. From the results, it can be seen that it is possible to control the output power seamlessly despite of the load condition. Relationship φ and P_o of conventional and proposed operation is shown in Fig. 7.

E. Pulse Generating Method

Fig. 8 and Fig. 9 show the generating mechanism of proposed driving signal. As mentioned above, the gate signal is the combination of the phase shift signal and the masked signal. The mask width is calculated and controlled by (2) and (3), respectively.

V. LOSS INCLUDED ANALYSIS OF CONVENTIONAL OPERATION

Equation of the output power (1) was equation for the ideal state without consideration of the conduction loss of the body diode and switch and the parasitic resistance of the transformer. This chapter will be described analysis of static characteristics in consideration of these losses. In order to analyze and make some definitions, in the operation waveform of Figure 2, both of light load and heavy load can be divided into four states. Since the basic operation of two half cycles are symmetric, only the first half cycle is explained. To analyze the characteristics of the circuit, Extended State-Space Averaging Method [16] is applied.

In order to simplify the loss analysis, loss is defined as r_{loss}. Equivalent circuits corresponding to each state in buck mode operation are shown in Fig.9, where $\hat{v}_o$ is the low-frequency component of V_o.

For analysis, solving for i_L and i_c,

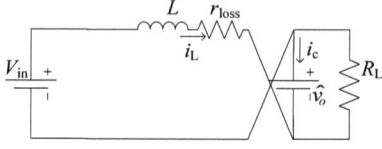

(a) state1.

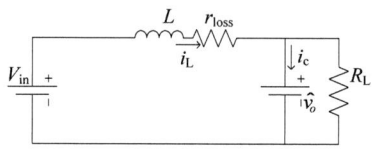

(b) state2.

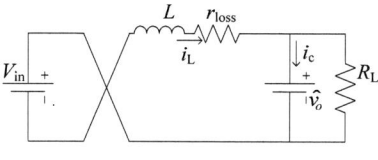

(c) state3.

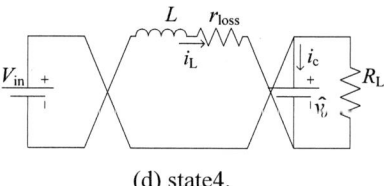

(d) state4.

Figure 9. Equivalent circuit :
(a) state 1; (b) state 2; (c) state 3; (d) state 4;

for $0 \le t \le DT_s$ (state 1)

In Fig. 9(a), Voltage law of the circuit is

$$L \frac{di_{L1}(t)}{dt} = V_{in} + \hat{v}_{out} - i_{L1}(t) r_{loss}. \qquad (5)$$

Integration of eq.(5) is

$$L \cdot i_{L1}(t) = \int (V_i + v_o) dt - r_{loss} \int i_{L1}(t) dt + C. \qquad (6)$$

Linear approximation of eq.(6) is

$$\int i_{L1}(t) dt = \frac{1}{2} \{ i_{L1}(0) + i_{L1}(t) \} t. \qquad (7)$$

Using eq.(5) and (6),

$$i_{L1}(t) \approx \frac{2(V_i + v_o)t}{2L + r_{loss}t} + \frac{2L - r_{loss}t}{2L + r_{loss}t} \cdot i_{L1}(0). \qquad (8)$$

In addition, the current law is expressed as

$$i_{c1}(t) = -i_{L1}(t) - \frac{\hat{v}_{out}}{R_L}. \qquad (9)$$

for $DT_s \le t \le \pi$ (state 2)

Solving for current law and voltage law of circuit in the same way

$$i_{L2}(t) \approx \frac{2(V_i - v_o)(t - DT_s)}{2L + r_{loss}(t - DT_s)} \cdot$$

$$+ \frac{2L - r_{loss}(t - DT_s)}{2L + r_{loss}(t - DT_s)} \cdot \left\{ \frac{2(V_i + v_o)DT_s}{2L + r_{loss}DT_s} + \frac{2L - r_{loss}DT_s}{2L + r_{loss}DT_s} i_{L1}(0) \right\} \quad (10)$$

$$i_{c2}(t) = i_{L2}(t) - \frac{\hat{v}_{out}}{R_L} \cdot \quad (11)$$

In one cycle in the steady state, the current flowing through the leakage inductance is positive and negative symmetry operation. The operation of the state3 and State4 is equivalent to positive and negative symmetry to the operation of the state1 and state2 Therefore, the analysis was performed only for half cycle.

Since the conventional operation of two half cycles are symmetric,

$$i_{L1}(0) = -i_{L2}(\tfrac{1}{2}T_s) \cdot \quad (12)$$

It is possible to determine the initial value of the circuit using the eq. (12).

$$i_{L1}(0) \approx -\frac{2(V_i - v_o) \cdot (2L + r_{loss} \cdot DT_s)(1 - 2D)T_s + 2(V_i + v_o)\{4L - r_{loss} \cdot (1 - 2D)T_s\} \cdot DT_s}{\{4L + r_{loss} \cdot (1 - 2D)T_s\}(2L + r_{loss} \cdot DT_s) + \{4L - r_{loss} \cdot (1 - 2D)T_s\}(2L - r_{loss} \cdot DT_s)} \quad (13)$$

Next, deriving for the average current in the output capacitor of the state1 and state2.The average value of v in each state is calculated with

$$i_{c1\_ave} \approx -\frac{(V_i + v_o) \cdot D^2 T_s^2}{2L + r_{loss} \cdot DT_s} + \left(1 + \frac{2L - r_{loss} \cdot DT_s}{2L + r_{loss} \cdot DT_s}\right) \cdot$$

$$\frac{(V_i - v_o) \cdot (2L + r_{loss} \cdot DT_s)(1 - 2D)DT_s^2 + (V_i + v_o)\{4L - r_{loss} \cdot (1 - 2D)T_s\} \cdot D^2 T_s^2}{\{4L + r_{loss} \cdot (1 - 2D)T_s\}(2L + r_{loss} \cdot DT_s) + \{4L - r_{loss} \cdot (1 - 2D)T_s\}(2L - r_{loss} \cdot DT_s)}$$

$$-\frac{\hat{v}_o}{R_L} \cdot DT_s \quad (14)$$

$$i_{c2\_ave} \approx \frac{1}{2} \frac{(V_i - v_o)(1 - 2D)^2 T_s^2}{4L + r_{loss}(1 - 2D)T_s} + \frac{1}{2} \frac{(V_i + v_o) \cdot (1 - 2D)DT_s^2}{2L + r_{loss} \cdot DT_s}\left[1 + \frac{4L - r_{loss}(1 - 2D)T_s}{4L + r_{loss}(1 - 2D)T_s}\right]$$

$$-\frac{1}{2}\left[1 + \frac{4L - r_{loss}(1 - 2D)T_s}{4L + r_{loss}(1 - 2D)T_s}\right]\frac{2L - r_{loss} \cdot DT_s}{2L + r_{loss} \cdot DT_s} \cdot$$

$$\frac{(V_i - v_o) \cdot (2L + r_{loss} \cdot DT_s)(1 - 2D)^2 T_s^2 + (V_i + v_o)\{4L - r_{loss} \cdot (1 - 2D)T_s\} \cdot (1 - 2D)DT_s^2}{\{4L + r_{loss} \cdot (1 - 2D)T_s\}(2L + r_{loss} \cdot DT_s) + \{4L - r_{loss} \cdot (1 - 2D)T_s\}(2L - r_{loss} \cdot DT_s)} - \frac{\hat{v}_o}{R_L} \cdot \frac{1 - 2D}{2}T_s$$

$$\cdot \quad (15)$$

while $\alpha = 2\omega L + r_{loss}\varphi$, $\beta = 2\omega L - r_{loss}\varphi$, $\gamma = 4\omega L + 2r_{loss}(\pi - \varphi)$ and $\lambda = 4\omega L - 2r_{loss}(\pi - \varphi) \cdot$

The results of static characteristics are obtained by letting $d\hat{v}_o / dt = 0$ (therefore $\overline{i_{c_o}} = C\,d\hat{v}_o/dt = 0$).

The output power Po is

$$P_o \approx VV_o\varphi\left(1 - \frac{\varphi}{\pi}\right)\left[\begin{array}{l}\frac{1}{\pi - \varphi}\left(1 + \frac{V_o}{V_i}\right)\left[\frac{\varphi\lambda}{\alpha\gamma + \beta\lambda} + \left(1 - \frac{\beta\lambda}{\alpha\gamma + \beta\lambda}\right)\left\{\frac{\lambda}{\alpha\gamma}(\pi - \varphi) + \frac{1}{\alpha}(\pi - 2\varphi)\right\}\right] \\ + \left(1 - \frac{V_o}{V_i}\right)\left[\frac{8\omega L}{\alpha\gamma + \beta\lambda} + \frac{2\pi}{\varphi}\left\{\frac{1}{\gamma}\left(1 - \frac{\beta\lambda}{\alpha\gamma + \beta\lambda}\right) - \frac{\beta}{\alpha\gamma + \beta\lambda}\right\}\left(1 - \frac{\varphi}{\pi}\right)\right]\end{array}\right] \quad (16)$$

When the $r_{loss} = 0$ in equation (16), is found to be obtained the same equation as the equation (1). Fig.10 shows the characteristics of changing the value of r_{loss}. We can see the effect of r_{loss} on the output power Po by Fig.10.

VI. LOSS INCLUDED ANALYSIS OF PROPOSED OPERATION

To analyze the characteristics of the circuit, Extended

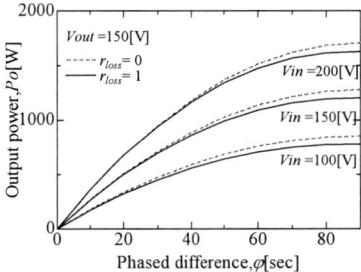

Figure 10. φ - Po
(conventional analysis and proposed analysis).

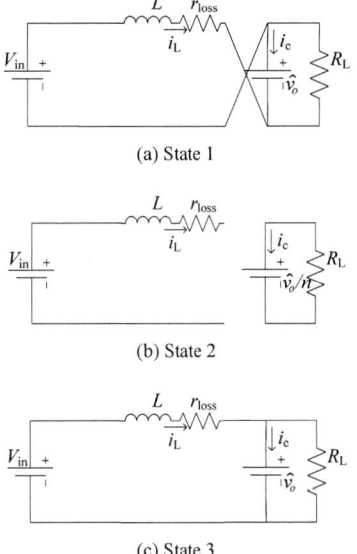

(a) State 1

(b) State 2

(c) State 3

Fig. 11.   Equivalent circuit of buck mode operation.

State-Space Averaging Method [16] is applied again.

The analysis has been done for each of buck mode and boost mode operation, respectively. In order to simplify the loss analysis, loss is defined as r_{loss}.

A.   Buck Mode Operation

Equivalent circuits corresponding to each state in buck mode operation are shown in Fig. 11, where $\hat{v}_o$ is the low-frequency component of V_o. $D_a = (A - 0)/T_s$, $D_b = (\varphi - A)/T_s$, $D_c = (\pi - \varphi)/T_s$ in Fig. 6 (a). For ease of analysis, the calculation has been performed in a half of the switching period because of the symmetric behavior of the circuit topology.

For analysis, solving for i_L and i_c,
for $0 \leq t \leq A$ (state 1)

$$i_L \approx \frac{2(V_{in} + \hat{v}_o)t}{2L + r_{loss}t} - \frac{2L - r_{loss}t}{2L + r_{loss}t}i_L(0) \quad (17)$$

$$i_{C_o} = -i_L - \frac{\hat{v}_o}{R_L} \tag{18}$$

for $A \le t \le \varphi$ (state 2)

$$i_L = 0 \tag{19}$$

$$i_{C_o} = -i_L - \frac{\hat{v}_o}{R_L} \tag{20}$$

for $\varphi \le t \le \pi$ (state 3)

$$i_L \approx \frac{2(V_{in} - \hat{v}_o)\{t - (D_a + D_b)T_s\}}{2L + r_{loss}\{t - (D_a + D_b)T_s\}} \tag{21}$$

$$i_{C_o} = i_L - \frac{\hat{v}_o}{R_L}. \tag{22}$$

From Fig. 6, it is clear that $D_a + D_b + D_c = 1/2$, $i_L(0) = -i_L(\pi)$ and $i_L(A) = 0$. Using the preceding relationships,

$$i_L(0) = -i_L(\pi) = -\frac{2(V_{in} + \hat{v}_o)D_a T_s}{2L - r_{loss}D_a T_s} \tag{23}$$

and

$$D_c = \frac{V_{in} + \hat{v}_o}{V_{in} - \hat{v}_o - 2D_a T_s V_{in} r_{loss} / L} D_a. \tag{24}$$

The average value of i in each state is calculated with

$$i_{c_o\_ave1} = -\frac{1}{2}\frac{2(V_{in} + \hat{v}_o)D_a T_s}{2L - r_{loss}D_a T_s} - \frac{\hat{v}_o}{R_L} \tag{25}$$

$$i_{c_o\_ave2} = -\frac{\hat{v}_o}{R_L} \tag{26}$$

$$i_{c_o\_ave3} = -\frac{1}{2}\frac{2(V_{in} + \hat{v}_o)D_a T_s}{2L - r_{loss}D_a T_s} - \frac{\hat{v}_o}{R_L}. \tag{27}$$

Hence,

$$\overline{i_{c_o}} = 2(i_{c_o\_ave1} \times D_a + i_{c_o\_ave2} \times D_b + i_{c_o\_ave3} \times D_c)$$
$$= \frac{2D_a^2 T_s}{L} \cdot \frac{V_{in}\hat{v}_o(V_{in} + \hat{v}_o)}{V_{in} - \hat{v}_o - V_{in}D_a T_s r_{loss} / L} - \frac{\hat{v}_o}{R_L}. \tag{28}$$

The results of static characteristics are obtained by letting $d\overline{v}_o / dt = 0$. And using $D_a T_s = A$,

$$P_o = \frac{2A^2}{T_s L} \cdot \frac{V_{in}V_{out}(V_{in} + V_{out})}{V_{in} - V_{out} - V_{in}D_a T_s r_{loss} / L} \tag{31}$$

where

$$A = \frac{(V_{in} - V_{out})(T_s / 2 - \varphi)}{V_{in} + V_{out} + V_{in}(T_s / 2 - \varphi)r_{loss} / L}. \tag{32}$$

B. Boost Mode Operation

Equivalent circuits corresponding to each state in boost mode operation are shown in Fig. 12. For analysis, equation is formularized for each state. $D_a = (\varphi - 0)/T_s$, $D_b = (B - \varphi)/T_s$ and $D_c = (\varphi - B)/T_s$ in Fig. 6 (b).

For $0 \le t \le \varphi$ (state 1)

$$i_L \approx \frac{2(V_{in} + \hat{v}_0)t}{2L + r_{loss}t} \tag{33}$$

$$i_{C_o} = -i_L - \frac{\hat{v}_o}{R_L} \tag{34}$$

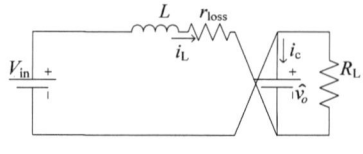

(a) State 1

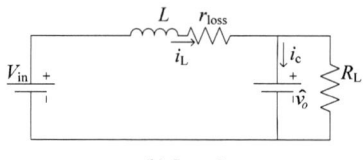

(b) State 2

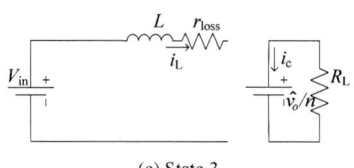

(c) State 3

Fig. 12. Equivalent circuit of boost mode operation.

for $\varphi \le t \le B$ (state 2)

$$i_L \approx \frac{2(V_{in} - \hat{v}_0)(t - D_a T_s)}{L + r_{loss}(t - D_a T_s)} + \frac{2L - r_{loss}(t - D_a T_s)}{2L + r_{loss}(t - D_a T_s)}i(D_a T_s) \tag{35}$$

$$i_{C_o} = i_L - \frac{\hat{v}_o}{R_L} \tag{36}$$

for $B \le t \le \pi$ (state 3)

$$i_L = 0 \tag{37}$$

$$i_{C_o} = -\frac{\hat{v}_o}{R_L}. \tag{38}$$

From Fig. 6, it is clear that $D_a + D_b + D_c = 1/2$, and $i_L(B) = 0$. Using the preceding relationships

$$D_b = -\frac{(V_{in} + \hat{v}_o)}{(V_{in} - \hat{v}_o) - \hat{v}_o D_a T_s r_{loss} / L} D_a. \tag{39}$$

Hence,

$$\overline{i_{c_o}} = -\frac{2D_a^2 T_s}{L}\frac{(V_{out} + V_{in})(2V_{in}V_{out} - V_{out}^2 D_a T_s r_{loss} / 2L)}{(V_{out} - V_{in} + V_{out}r_{loss}D_a T / L)(2 + D_a T_s r_{loss} / L)} - \frac{\hat{v}_o}{R_L}. \tag{40}$$

The results of static characteristics are obtained by letting $d\overline{v}_o / dt = 0$, therefore

$$P_o = \frac{2(D_a T_s)^2}{T_s L}\frac{(V_{out} + V_{in})(2V_{in}V_{out} - V_{out}^2 D_a T_s r_{loss} / 2L)}{(V_{out} - V_{in} + V_{out}r_{loss}D_a T / L)(2 + D_a T_s r_{loss} / L)}. \tag{41}$$

Using $D_a T_s = \varphi$

$$P_o = \frac{2\varphi^2}{T_s L}\frac{(V_{out} + V_{in})}{(V_{out} - V_{in} + V_{out}r_{loss}\varphi / L)}\frac{(2V_{in}V_{out} - V_{out}^2 \varphi r_{loss} / 2L)}{(2 + \varphi_s r_{loss} / L)}. \tag{42}$$

B is calculated as

$$B = (D_a + D_b)T_s = \frac{v_o(2 + r_{loss}D_a T / L)}{(v_o - v_i) + v_o r_{loss}D_a T / L}\varphi \tag{43}$$

978-1-4799-2706-7/14 $31.00 © 2014 IEEE

VII. Experimental Results

In order to select the value of r_{loss}, we perform some experiments with the prototype circuit. The main circuit is DAB dc-dc converter without additional circuits like snubber circuit. We had closed-loop-operation experiments with DSP TI TMS320F28335. And, also the value of A and B are manually supplied in this experiment. Experimental parameters are shown in Table I. Dead time of each switch is set as 1μs.

A. Power Efficiency

Fig. 13 shows the power efficiency results for the both of the conventional and the proposed operation. It can be seen that the power efficiency of buck mode can be apparently improved by up to 37% using the proposed operation at 100W as shown in Fig. 13 (a). It can be seen that the power efficiency of boost mode can be apparently improved by up to 30% at 100W as shown in Fig. 13 (b).

B. Estimating the Value of Loss

Fig. 14 shows φ - P_{o} of analysis and experimental results. The value of r_{loss} for the conventional operation is for the analysis is set to 2.0Ω. The value is the measurement result of series resistance r_{s} of the transformer as shown in Table I measured with LCR meter Agilent 4263B.

The value of r_{loss} for the proposed operation is calculated with averaged equivalent resistance with the averaged power calculation describe below.

$$\overline{P_{loss\_conv}} \approx r_s \frac{\int_0^T \left(i_{Lm}(t)\right)^2 dt}{T} = r_s \frac{V_{in}^2 T^2}{3L^2} \tag{44}$$

$$\overline{P_{loss\_proposed}} \approx \frac{\int_0^{DT} r_s \left(i_{Lm}(t)\right)^2 dt}{T} = \frac{r_s V_{in}^2}{T L^2} \int_0^{DT} t^2 dt$$

$$= \frac{r_s V_{in}^2 D^3 T^2}{3L^2} = r_s D^3 \frac{V_{in}^2 T^2}{3L^2} \tag{45}$$

where D is the conduction time ratio of switching term in no load condition and T is the half of switching term.

From the calculation results, r_{loss} of the proposed operation is calculated as the averaged equivalent resistance as

$$r_{loss} = D^3 r_s . \tag{46}$$

From the result of our optional experiment, D is obtained as 0.5 ohm. Therefore, the r_{loss} for the proposed operation is set as 0.25 ohm.

Comparing loss including analysis and experimental results, the root mean square was in 4% both of boost mode and buck mode.

TABLE I
SPECIFICATION OF DAB DC-DC CONVERTER

Item	Symbol	Specification
Transformer		
1) Turns ratio	A	1:1
2) Leakage inductance (primary-referred)	L	110μH
3) Series resistance (primary-referred)	r_s	2Ω
Converter		
1) Rated output power	P_{o}	1kW
2) Rated input direct voltage	V_{in}	150V
3) Rated output direct voltage	V_{out}	150V
4) Switching frequency	f_{s}	20kHz
5) Absolute maximum ratings of IGBT collector-emitter	v_{CE}	600V
6) On resistance of IGBT	r_{t}	50mΩ
7) Absolute maximum ratings of diode	i_{F}	30A
8) Forward voltage of diode	v_{F}	0.8V
9) Recovery time of diode	t_{rr}	0.1μs

VIII. Conclusion

By the analysis of the circuit operation and the some experiments, the validation of the proposed operation for DAB dc-to-dc converter is revealed. Form the analysis, P_{o} can be calculated with the loss included analysis for both of the conventional and the proposed technique. The analysis results are well matched with the experimental results. Applying the two modes which are proposed operation in light load and conventional operation in heavy load, the circuit can be operated in the full load range. 37% maximum power efficiency improvement can be confirmed at light load.

References

[1] R. Morrison, M. G. Egan, "A new power-factor-corrected single-transformer UPS design," *IEEE Trans. Ind. Appl.*, vol 36, no.1, pp.171-179, Jan./Feb. 2000.

[2] Florian Krismer, and Johann W. Kolar, "Accurate power loss model derivation of a high-current dual active bridge converter for Automotive Application," *IEEE Trans. Ind. Electron.*, vol 57, no.3, pp.881-891, Mar. 2010.

[3] Huang-Jen Chiu and Li-Wei Lin, "A bidirectional dc-dc converter for fuel cell electric vehicle driving system," *IEEE Trans. Power Electron.*, vol. 21, no. 4, pp. 950-958, Jul. 2007.

[4] F. Krismer, J. W. Kolar, "Accurate small-signal model for the digital control of an automotive bidirectional dual active bridge," *IEEE Trans. Power Electron.*, vol. 24, no. 12, pp. 2756-2768, Dec. 2009.

[5] S. Inoue and H. Akagi, "A bidirectional dc-dc converter for an energy storage system with galvanic isolation," *IEEE Trans. Power Electron.*, vol. 22, no. 6, pp. 2299-2306, Nov. 2007.

[6] Toshiro Hirose, Keisuke Nishimura, Takayuki Kimura, and Hirofumi Matsuo, "An AC-link Bidirectional DC-DC Converter with Synchronous Rectifier," *in Proc. IECON*, Nov. 2010.

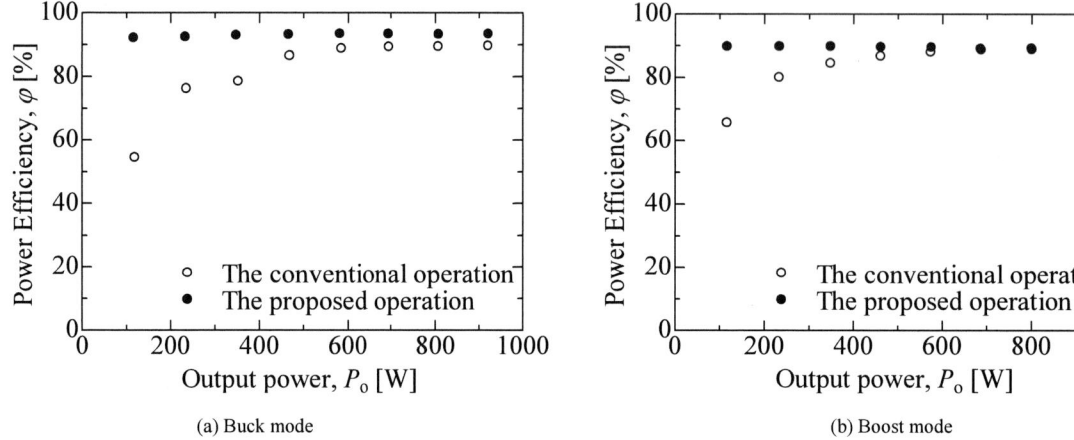

(a) Buck mode (b) Boost mode

Fig. 13. Power efficiency: (a) Buck mode (V_{in}=200V, V_{out}=150V); (b) Boost mode (V_{in}=100V, V_{out}=150V).

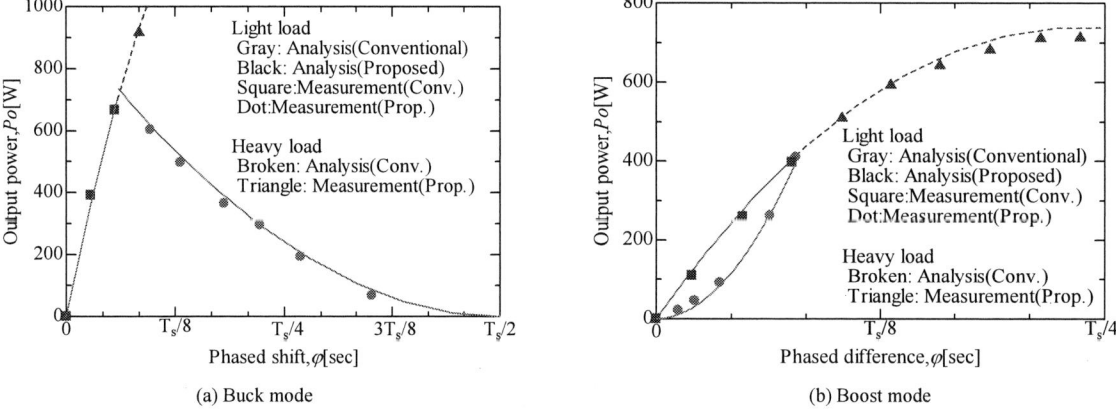

(a) Buck mode (b) Boost mode

Fig. 14. φ - P_o (analysis and experimental results).

[7] Mustansir H. Kheraluwala, Randal W. Gascoigne, Deepakraj M. Divan, and Eric D. Baumann, "Performance characterization of a high-power dual active bridge dc-to-dc converter," *IEEE Trans. Industry Applications*, vol.28, NO.6, pp. 1294-1301, Nov. / Dec. 1992.

[8] Shigenori Inoue and Hirofumi Akagi, "A bidirectional isolated dc-dc converter as a core circuit of the next-generation medium-voltage power conversion system, " *IEEE Trans power Electron.*, vol.22, no.2, pp. 535-542, Mar. 2007.

[9] M. Pavlovsky, S. W. H. de Hann, and J. A. Ferreira, "Concept of 50kW DC/DC converter based on ZVS, quasi-ZCS topology and integrated thermal and electronic design," 2005 European Conference on Power Electronics and Applications.

[10] Haihua Zhou, A. M. Khambadkone, "Hybrid modulation for dual-active-bridge bidirectional converter with extended power range for ultracapacitor application," *IEEE Trans. Ind. Appl.*, vol 45, no.4, pp.1434-1442, July/Aug. 2009.

[11] Yujin Song, P. N. Enjeti, "A new soft switching technique for bidirectional power flow, full-bridge DC-DC converter," in *Proc. 37th IAS Annual Meeting*, pp.2314-2319, Oct. 2002.

[12] German G. Oggier, Guillermo O. Garcia, and Alejandro R. Oliva, "Switching control strategy to mimimize dual active bridge converter losses," *IEEE Trans. Power Electron.*, vol 24, no.7, pp.1826-1838, July. 2009.

[13] Mika Takasaki, Yoichi Ishizuka, Tamotsu Ninomiya, Yutaka Furukawa, and Toshiro Hirose, "A Power Efficiency Improvement Technique for A Bi-Directional Dual Active Bridge DC-DC Converter at light load," EPE-ECCE 2013, pp.1-10, Sept. 2013.

[14] Rik W. A. A. De Doncker, Deepkraj M. Divan, and Mustansir H. Kheraluwala, "A Three-Phase Soft-Switched High-Power-Density dc/dc Converter for High-Power Applications," *IEEE Trans. Industry Applications*, vol.27, NO.1, pp.63-73, Jan. / Feb. 1991.

[15] Shun Nagata, Mika Takasaki, Kazuhide Domoto, Toshiro Hirose , and Yoichi Ishizuka, "A static characteristic analysis and high efficient technique of the light load condition of Bi-Directional Dual Active Bridge DC-DC Converter," IEICE, , vol. 113, no. 155, EE2013-14, pp. 67-72, July 2013.

[16] Tamotsu Ninomiya, Masatoshi Nakahara, Toru Higashi, and Koosuke Harada, "A Unified Analysis of resonant Converters," *IEEE Trans. Industry Applications*, vol.27, NO.1, pp.63-73, Jan. / Feb. 1991.

The 2014 International Power Electronics Conference

Hybrid Battery Charging System Combining OBC with LDC for Electric Vehicles

Seonghye Kim
Dept. of Control & Instrumentation Engineering
Hanbat National University
125 Dongseodero, Yuseong, Dajeon 305-719, Korea
seonghye52@gmail.com

Feel-soon Kang
Dept. of Electronics and Control Engineering
Hanbat National University
125 Dongseodero, Yuseong, Dajeon 305-719, Korea
feelsoon@ieee.org

Abstract— **It presents a hybrid battery charging system unifying an on-board charger (OBC) with a low voltage dc-to-dc converter (LDC) for small and light design of EV charger. The proposed hybrid charger consists of three H-bridge converters including a selective switch, a high frequency transformer, and inductors. Front-end converter connected to the grid by a selective switch has two roles. One is to obtain unity power factor by ac-to-dc converting. It belongs to one of the behavior of OBC. The other is to charge an auxiliary battery by working a step-down converter. Back-end converter employs a transformer to ensure galvanic isolation between the grid and EV. It charges propulsion battery by phase-shift control, and as the occasion demands, energy stored in propulsion battery discharges to dc-link capacitor to supply power to the grid or auxiliary battery. To verify the validity of the proposed approach, we carry out computer-aided simulations and experiments.**

Keywords— Battery chagering system, dc-to-dc converter, inverter, Low Voltage dc-to-dc Converter (LDC), On-board Charger (OBC).

I. INTRODUCTION

Plug-in hybrid electric vehicles (PHEV) or electric vehicles (EV) charge battery from the grid via a battery charging system [1]-[3]. Usually, propulsion battery is charged by on-board charger (OBC), and auxiliary battery used to supply power to electric field of EV is charged by low voltage dc-to-dc converter (LDC) using the propulsion battery. Many researches related to OBC and LDC are carrying out [4]-[6]. However, researches unifying OBC and LDC are not introduced until now. Usually, they are studied separately. In the viewpoint of

compact design of battery charging system, LDC can be merged into OBC.

The motivation of the paper is to reduce the size and weight of on-board battery charging system by means of combining OBC and LDC. To satisfy each indigenous function, the proposed charger is designed to be operated in a bidirectional way. Front-end ac-to-dc converter uses a bridgeless boost PFC converter, and back-end dc-to-dc full-bridge converter employs a circuit configuration of bidirectional phase shifted ZVS converter. This is a basic circuit for OBC. LDC is realized in the same circuit with proper control method sharing the same switching devices. To verify the validity of the proposed approach, we carry out computer aided simulation and experiments.

II. PROPOSED HYBRID BATTERY CHARGING SYSTEM

Fig. 1 shows a circuit configuration of the proposed hybrid battery charger combining OBC and LDC. It consists of three H-bridge converters (Q_1-Q_4, Q_5-Q_8, and Q_9-Q_{12}), a high frequency transformer (Tr), and two inductors (L_1, L_2). The H-bridge converter (Q_1-Q_4) plays important roles to charge an auxiliary battery and grid connection with PFC. By controlling a selective switch (M), it determines whether it works for LDC or OBC. Because EV can be connected to grid during stoppage time, the selective switch can be acceptable.

A. Operational modes for OBC and LDC

1) Grid-connection for charging of propulsion battery
One of the important roles of OBC is to charge a

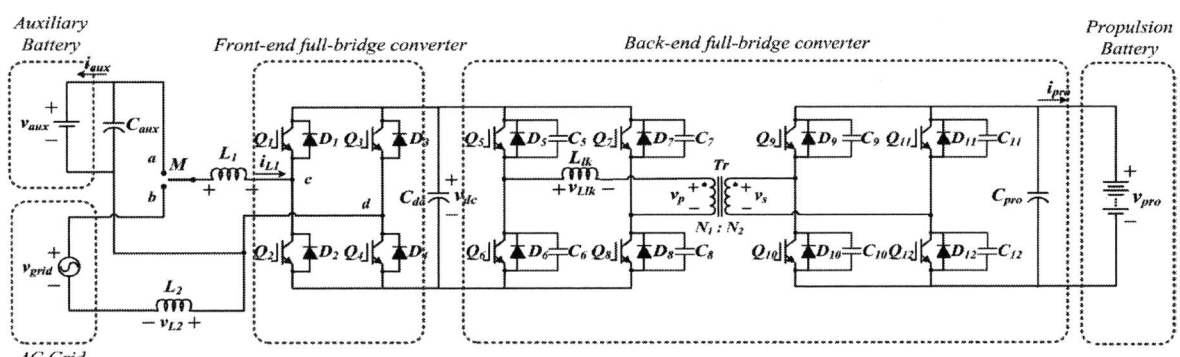

Fig. 1. Circuit configuration of proposed hybrid battery charging system combining OBC with LDC.

978-1-4799-2706-7/14 $31.00 © 2014 IEEE   2260

propulsion battery by connecting selective switch M to terminal b. Here, the proposed converter is connected to the grid. In this time, front-end full-bridge converter does PFC by compensating reactive power resulted in the increase of power density. Rectified and power corrected current is stored to dc-link capacitor (C_{dc}) to be supplied to the propulsion battery. Back-end full-bridge converter steps dc-link voltage up to charge the propulsion battery. To increase energy efficiency, the back-end full-bridge converter uses phase-shift control to supply a continuous current to the propulsion battery.

2) Grid-connection for discharging of propulsion battery

In V2G technology, discharging of the propulsion battery to the grid-connection is also one of the important roles of OBC. Here, the selective switch (M) connects to terminal b. Energy flow is from the propulsion battery to the grid. In this time, back-end full-bridge converter first synthesizes dc-link voltage by phase-shift. And then, front-end full-bridge converter links dc-link to the grid by PWM inverting operation.

3) Charging of auxiliary battery by discharging of propulsion battery

It is a mode of auxiliary battery charging. It belongs to LDC operation receiving energy from the propulsion battery. To do so, the selective switch (M) connects to terminal a. Energy flow is from the propulsion battery to the auxiliary battery. DC-link voltage is generated by back-end phase-shifted converter using a propulsion battery, and it steps down by buck operation of front end converter to charge auxiliary battery. We know that front-end converter works for PFC, inverting, and step-down operations. Back-end phase-shifted full-bridge converter works step up and down regardless of the operation of front-end converter.

B. Operational modes of front-end converter

The operation of front-end converter is different due to charging or discharging of battery. It means that each component plays different roles according to operational mode of front-end converter. In case of L_1 and L_2, they work for current buffer to restrict current in PFC mode. They work as an output filter in inverting mode. And they also work as an output inductor of step down operation.

1) PFC mode

Fig. 2 shows Q_2 operation when ac input voltage is positive. When Q_2 turns on, input current flows through M, L_1, Q_2, D_4, and L_2. Here, L_1 and L_2 store energy from the ac-grid. When Q_2 turns off, both inductor currents start to decrease via M, L_1, D_1, C_{dc}, D_4, and L_2. When ac input voltage is negative, Q_4 works to get high power factor. By iterating of on/off switching, it obtains high power factor. At the same time, it establishes dc-link voltage suitable for charging of propulsion battery.

2) Inverting mode

Fig. 3 shows an inverting mode of front-end full-bridge converter when it supplies a positive voltage to the

grid. In this time, dc-link capacitor discharges energy via Q_1, L_1, M, v_{grid}, L_2, and Q_4 as given in Fig. 3(a). When it supplies a negative voltage to the grid, the current flows through Q_3, L_2, v_{grid}, M, L_1, and Q_2 as shown in Fig. 3(b). Here, L_1 and L_2 work as output filter to obtain high qualified output voltage wave. By this inverting mode, the proposed converter realizes V2G (Vehicle to Grid) operation.

3) Step down (Buck) operation

For a step-down operation of front-end converter, Q_4 maintains on-state, and Q_1 works in PWM. Q_2 and Q_3 are also turned off during this period. Body diode D_2 works for freewheeling like that of conventional Buck converter.

Fig. 4 shows front-end converter in a step-down mode. When Q_1 turns on, dc-link capacitor start to discharge via Q_1, L_1, M, v_{aux}, and Q_4 as shown in Fig. 4(a). During this period, L_1 stores energy, and discharges the energy through M, v_{aux}, Q_4, and D_2 when Q_1 turns off as given in Fig. 4(b). By designing the inductor worked in CCM (Continuous current Conduction Mode), the auxiliary battery can be charged in stable with a low current ripple.

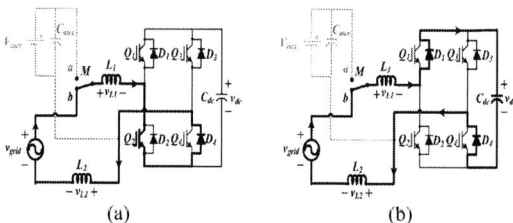

Fig. 2. PFC mode of front-end full-bridge converter when ac input voltage is positive, (a) Q_2=on, (b) Q_2=off.

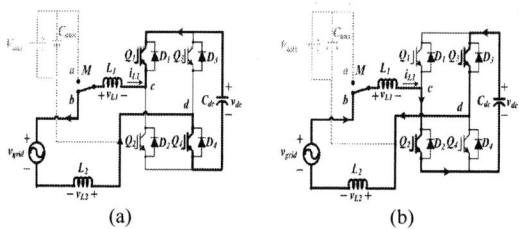

Fig. 3. Inverting mode of front-end full-bridge converter, (a) when supplying a positive voltage to the grid (Q_1 and Q_4=on), (b) when supplying a negative voltage to the grid (Q_2 and Q_3=on)

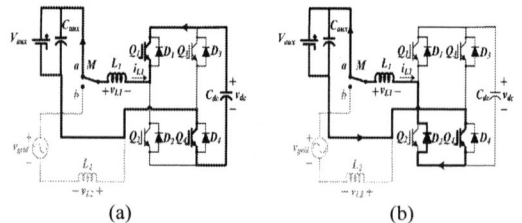

Fig. 4. Step down mode of front-end full-bridge converter, (a) Q_1=on, (b) Q_1=off.

C. Operational mode for back-end full-bridge converter

From Fig. 1, we find a circuit configuration of back-end converter, which consists of a transformer between a full-bridge converter (Q_5-Q_8) and a full-bridge converter (Q_9-Q_{12}). By phase-shift control, it realizes ZVS (zero voltage switching). Fig. 5 shows key waveform of phase-shift. Here, V_{GE5}-V_{GE8} are gate signals for Q_5-Q_8. V_{GE5} controls Q_9 and Q_{12}, and V_{GE6} controls Q_{10} and Q_{11}, respectively. These control signals apply at the same time. We explain a phase-shift operation when charging propulsion battery for a half cycle. Discharging sequence follows the same to the charging.

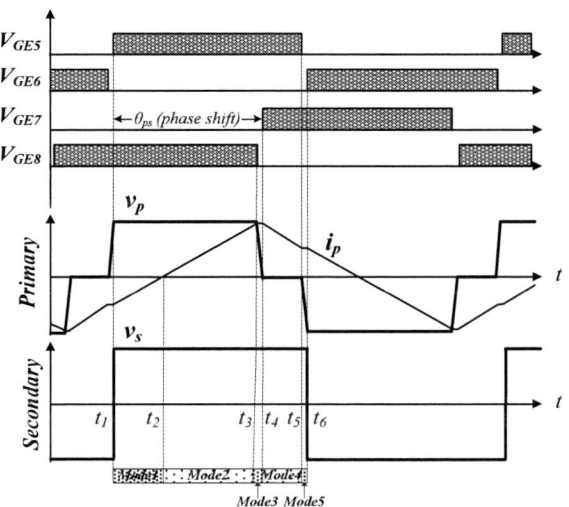

Fig. 5. Phase-shift control of back-end converter for ZVS.

Mode 1 ($t_1 < t \leq t_2$): By control signals of V_{GE5} and VGE8, Q_5 and Q_8 turn on. Voltage across the primary of the transformer applies dc-link voltage saving energy to L_{lk}. During this period, Q_9 and Q_{12} turn on, so voltage across the propulsion battery becomes v_{pro} by induced voltage from the primary. Voltage across the leakage inductance (L_{lk}) is given by (1)

$$v_{Llk} = L_{lk}\frac{d}{dt}i_{Llk} = v_{Llk} - \frac{N_1}{N_2}v_{pro} \qquad (1)$$

Mode2 ($t_2 < t \leq t_3$): At t_2, current of L_{lk} becomes zero. Hence, dc-link capacitor (V_{dc}) supplies energy to the propulsion battery.

Mode3 ($t_3 < t \leq t_4$): Q_5 maintains on-state, and Q_8 turns off. Then the stored energy to L_{lk} charges C_8, and C_7 starts to discharge. When voltage across C_7 becomes zero, D_7 turns on. For realizing ZVS of Q_7, it is essential to ensure proper dead-time.

Mode4 ($t_4 < t \leq t_5$): When Q_7 turns on, the energy stored in L_{lk} discharges. The propulsion battery continuously receives energy from dc-link capacitor. By applying KVL to the close loop of L_{lk}, V_p, Q_7, and Q_5, we obtain (2).

$$v_{Llk} = L_{lk}\frac{d}{dt}i_{Llk} = \frac{N_1}{N_2}v_{pro} \qquad (2)$$

Mode5 ($t_5 < t \leq t_6$): At t_5, Q_7 maintains on-state, and Q_5 turns off. Here, C_5 starts to charge, and C_6 discharges. When voltage across C_6 becomes zero, D_6 turns on for ZVS of Q_6. It also needs a proper dead time to ensure ZVS of Q_6.

III. DESIGN CONSIDERATION OF CONTROLLER

Fig. 6 shows a control block diagram for PFC of front-end converter. It needs feedback for $v_{dc,f}$, $i_{L1,f}$, and $v_{grid,f}$. Because α feeds to Q_2 and Q_4 according to the polarity of input voltage, it needs a polarity detection circuit of $v_{grid,f}$. Fig. 7 shows a control block diagram for phase-shift of back-end converter. It feeds back $v_{pro,f}$, and generates 4 control signals. Here, v_{ratio} means the ratio of dead-time (δ) to a cycle. Fig. 8 shows a control block diagram for step-down operation of front-end converter. It feeds back $v_{aux,f}$, $i_{L1,f}$, and generates PWM signal for Q_1. This algorithm conducts a constant current and constant voltage (CC-CV) control. When it begins to charge an auxiliary battery, it starts on constant current mode (CC), and turns in constant voltage mode (CV) when charging voltage is higher than an arbitrary preset voltage.

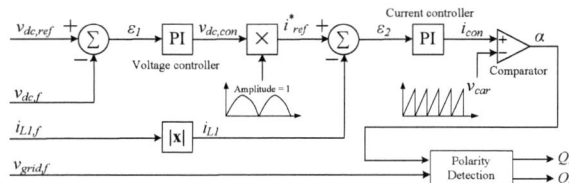

Fig. 6. Control block diagram for PFC of front-end converter.

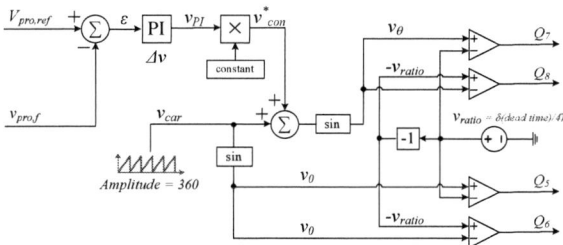

Fig 7 Phase-shift controller for back-end converter.

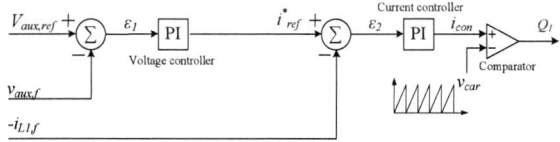

Fig. 8. Control block diagram for step-down operation of front-end converter.

IV. SIMULATION RESULTS

To verify the validity of the proposed approach, we carry out computer-aided simulation based on PSIM. Table I lists up specification of the simulation.

Fig. 9(a) shows input voltage and current waveform, which has a unity power factor. Fig. 9(b) shows the primary voltage and current (v_P and i_P) and the secondary voltage and current (v_S and i_S). When V_{GS5} and V_{GS8} turn on, voltage across the primary of the transformer becomes positive with the increase of i_P. On the other hand, when V_{GS6} and V_{GS7} turn on, the primary voltage becomes negative decreasing i_P. When V_{GS5} and V_{GS6} (or V_{GS7} and V_{GS8}) turn on and during dead-time, the primary current (i_P) increases or decreases. It means that the stored energy of L_{lk} discharges.

Fig. 10 is an inverting mode of the proposed converter. During this mode, propulsion battery discharges to the grid realizing V2G. Fig. 10(a) shows the primary voltage (v_S), the primary current (i_S), voltage across the secondary of the transformer (v_P), and the secondary current (i_P). Fig. 10(b) shows a filtered output voltage (v_{grid}) by inverting operation of front-end converter, pulse width modulated voltage (v_{cd}), and current waveform (i_{grid}).

Fig. 11 shows step-down operation of front-end converter. Fig. 11(a) is phase-shift control of back-end converter to ensure dc-link voltage to charge an auxiliary battery. It shows voltage across the primary of the transformer (v_S), the primary current (i_S), the secondary voltage of the transformer (v_P), and the secondary current (i_P). Fig. 11(b) shows step down operation of front-end converter to charge the auxiliary battery. When Q_1 turns on, the inductor L_1 starts to save energy and the stored energy is discharged to the auxiliary battery when Q_1 turns off.

TABLE I
SPECIFICATION FOR SIMULATION

Specifications	
AC grid	110Vrms / 60Hz
Maximum AC current	14.5A
DC-link voltage	155V
Switching frequency	f = 10kHz @ mode 1
	f = 14kHz @ mode 2
	f = 14kHz @ mode 3
Sampling time	0.5μs
Propulsion battery voltage @Normal	200V
Auxiliary battery voltage @Normal	24V
$L_1 = L_2 = L_m$	2mH
L_{lk}	
$N_1 : N_2$	25 : 30
C_{dc}	4000 μF

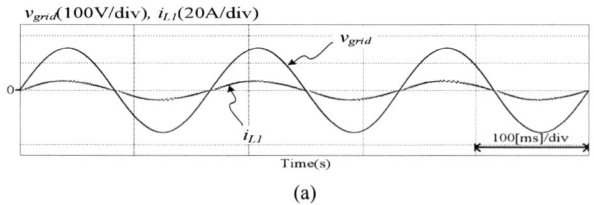

(a)

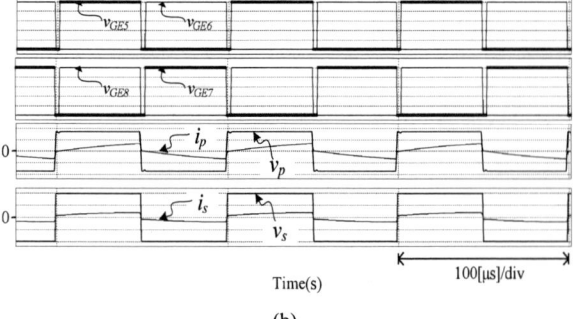

(b)

Fig. 9. Simulation results at ac-to-dc PFC mode, (a) input voltage and current, (b) phase-shift control of back-end converter.

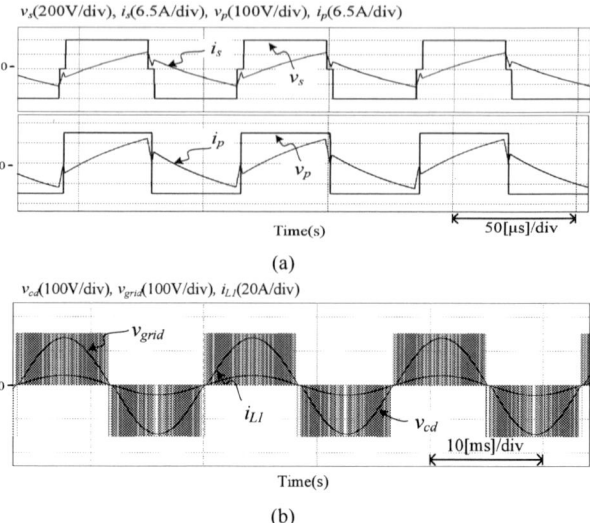

(a)

(b)

Fig. 10. Simulation results at inverting mode, (a) phase-shift control of back-end converter, (b) output voltage and current at inverting operation.

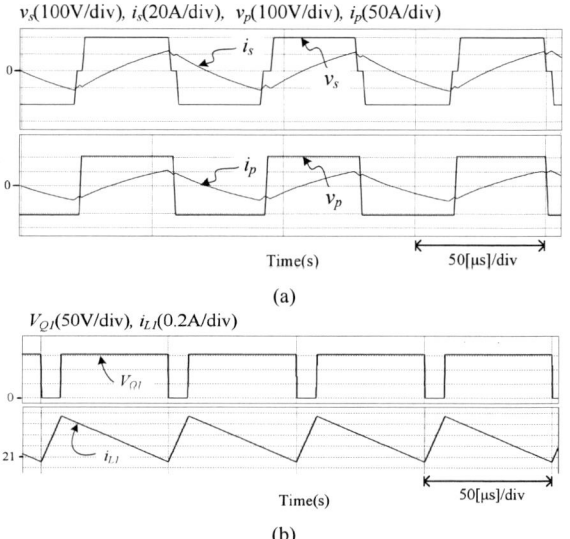

(a)

(b)

Fig. 11. Simulation results at step-down mode, (a) phase-shift control of back-end converter, (b) output voltage and current at step-down operation.

V. EXPERIMENT RESULTS

To verify the validity of the proposed approach, we implement the experiment based on equal specification of the simulation. The experiment targets a 500[W] system, which has the grid voltage of 110[V_{ac}] and 60[Hz]. The propulsion battery is set to 72[V] connected batteries in series, and the auxiliary battery is 24[V].

Fig. 12 shows in the ac-to-dc PFC mode. Fig. 12(a) and (b) are on the front-end converter rolling the power factor correction. Fig. 12(a) shows input voltage (v_{grid}) and current (i_{grid}) in phase condition, which is over 0.95 of PF. Fig. 12(b) is changing the duty ratio along with PWM operation of Q_2. Fig. 12(c) and (d) are on the back-end converter which is for step-up operation. Fig. 12(c) shows primary voltage and current (v_P and i_P), and Fig.

12(d) is about secondary voltage and current (v_S and i_S). It is same with simulation results. Also, for an efficiency increase, switching method of the secondary of the transformer employs the synchronizing rectification.

Fig. 13(a) and (b) are both sides of the transformer voltage and current on the inverting mode. It shows the some phase contrast by phase-shift controller. Also, synchronizing rectification is applied on the secondary of the transformer. Fig. 13(c) shows sinusoidal output voltage (v_{grid}) and current (i_{grid}) from the dc-link to the grid. In this time, THD of the output voltage is about 2.3[%].

Fig. 14 shows each part of the waveform in the step-down mode. Fig. 14(a) is primary voltage (v_S) and current (i_S), and Fig. 14(b) is secondary voltage (v_P) and current (i_P). Fig. 14(c) shows step-down operation in the front-end converter. In this time, Q_1 works into PWM and Q_2 roles wheeling-diode of the conventional buck converter.

Fig. 15(a) shows output voltage (v_{pro}) and current (i_{pro}) when the propulsion battery is charged in the CV mode. Fig. 15(b) shows output voltage (v_{aux}) and current (i_{aux}) when the auxiliary battery is charged in the CV mode.

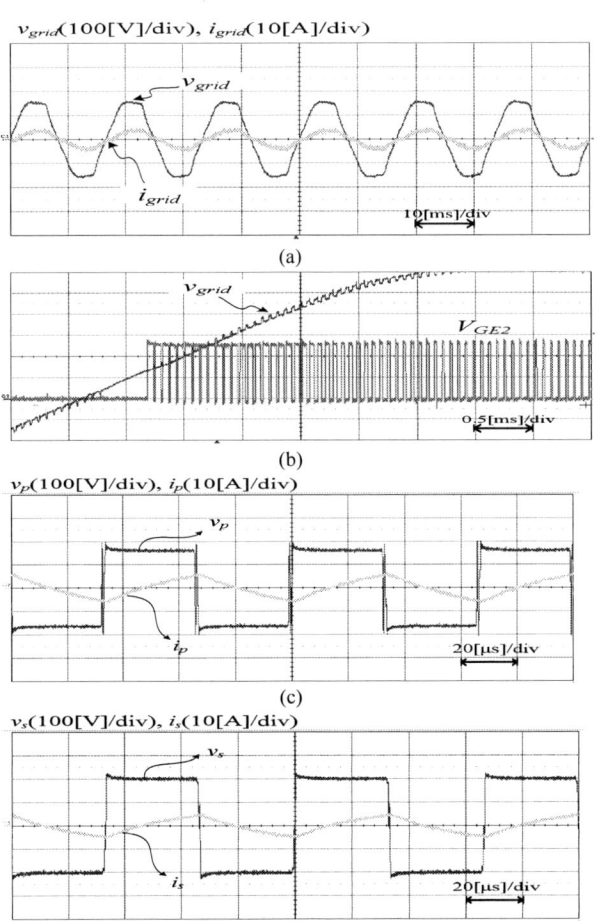

Fig. 12. Experiment results at ac-to-dc PFC mode, (a) input voltage (v_{grid}) and current (i_{grid}), (b) input voltage and PWM of Q_2, (c) primary of the transformer voltage (v_P) and current (i_P), (d) secondary of the transformer voltage (v_S) and current (i_S).

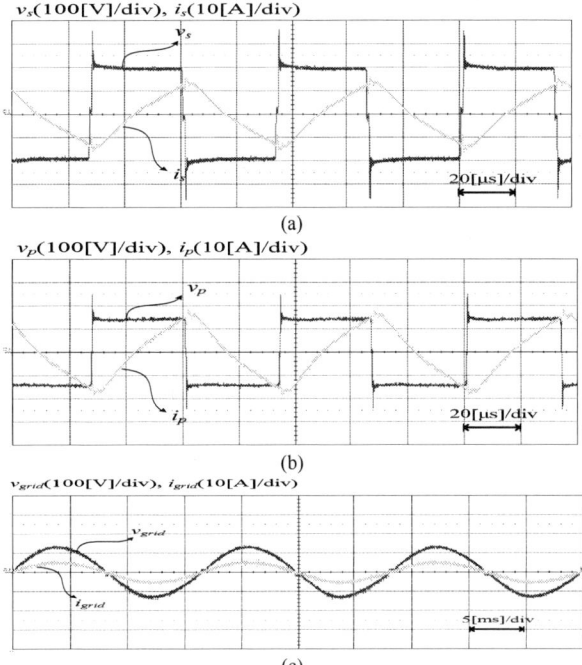

Fig. 13. Experiment results at inverting mode, (a) primary of the transformer voltage (v_S) and current (i_S), (b) secondary of the transformer voltage (v_P) and current (i_P), (c) output voltage and current at inverting operation..

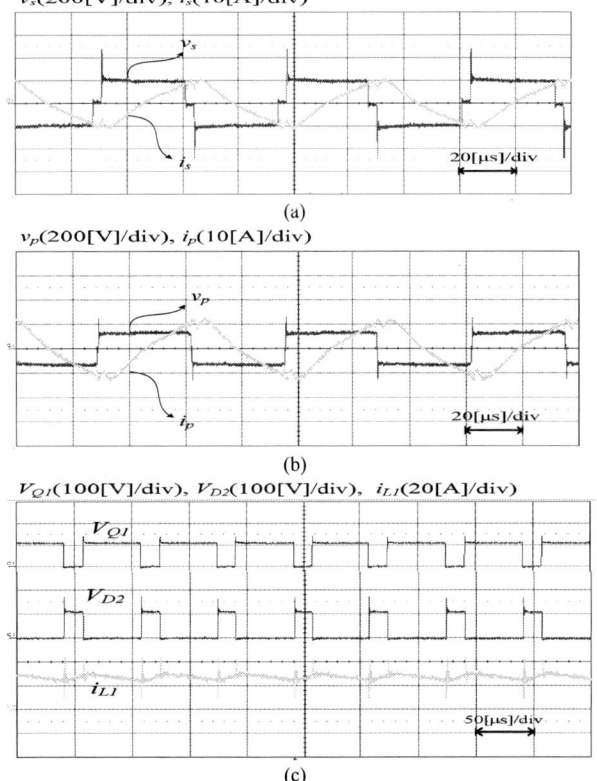

Fig. 14. Experiment results at step-down mode, (a) primary of the transformer voltage (v_S) and current (i_S), (b) secondary of the transformer voltage (v_P) and current (i_P), (c) voltage and current of the front-end converter.

978-1-4799-2706-7/14 $31.00 © 2014 IEEE

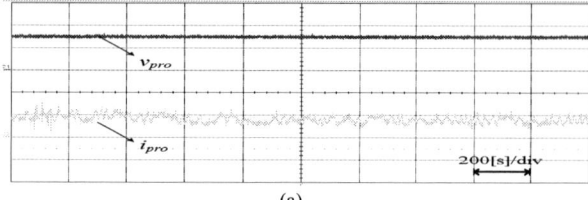

$v_{pro}(50[\text{V}]/\text{div})$, $i_{pro}(1[\text{A}]/\text{div})$

(a)

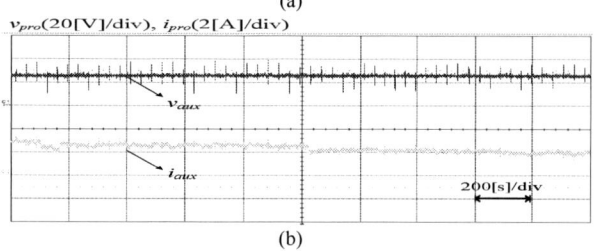

$v_{pro}(20[\text{V}]/\text{div})$, $i_{pro}(2[\text{A}]/\text{div})$

(b)

Fig. 15. Output voltage and current when battery is charged, (a) output voltage (v_{pro}) and current (i_{pro}) by charging the propulsion battery, (b) output voltage (v_{aux}) and current (i_{aux}) by charging the auxiliary battery.

VI. EXPERIMENT RESULTS

We proposed a hybrid battery charging system combining OBC with LDC. It can reduce the system size and weight by sharing a circuit topology. It consists of a selective switch, a high frequency transformer, inductors, and three H-bridge modules. By using this converter, we can implement all behaviors required to OBC and LDC at the same time. To verify the validity of the proposed approach, we carry out computer-aided simulations and experiments. From the results, we would know that the proposed hybrid charger can be a good candidate for battery charging system for EV.

ACKNOWLEDGMENT

This research was supported by Basic Science Research Program through the National Research Foundation of Korea (NRF) funded by the Ministry of Education, Science and Technology (No.2012-006120).

REFERENCES

[1] A. Emadi, Y. J. Lee, and K. Rajashekara, "Power Electronics and motor drives in electric, hybrid electric, and plug-in hybrid electric vehicles," *IEEE Trans. Ind. Electron.*, vol. 55, no. 6, pp. 2237-2245, 2008.

[2] W. D. Jones, "Take this car and plug it [plug-in hybrid vehicles]," *IEEE Spectr.*, vol. 42, no. 7, pp. 10-13, 2005.

[3] K. Morrow, D. Karner, and J. Francfort, "Plug-in hybrid electric vehicle charging infrastructure review," U.S. Dept. Energy-Vehicles Technologies Program, Washington, DC, INL-EXT-08-15058, 2008.

[4] G. Zorpette, " The smart hybrid," *IEEE Spectr.*, vol. 41, no. 1, pp. 44-47, 2004.

[5] A. Y. Saber and G. K. Venayagamoorthy, "Plug-in vehicles and renew-able energy sources for cost and emission reductions," *IEEE Trans. Ind. Electron.*, vol. 58, no. 4, pp. 1229-1238, 2011.

[6] A. Hajimiragha, C. A. cañizares, M. W. Fowler, and A. Elkamel, "Optimal transition to plug-in hybrid electric vehicles in Ontario, Canada, considering the electricity-grid limitations," *IEEE Trans. Ind. Electron.*, vol. 57, no. 2, pp. 690-701, 2011.

Transient Behavior of the Dual Active Bridge Converter in High Efficient Energy Conversion System

Kohei Aoyama
Yokohama National University
Email: aoyama-kohei-bh @ynu.jp

Naoki Motoi
Kobe University
Email: motoi@maritime. kobe-u.ac.jp

Yukinori Tsuruta
Yokohama National University
Email: tsuruta@ynu.ac.jp

Atsuo Kawamura
Yokohama National University
Email: kawamura@ynu.ac.jp

Abstract—This paper shows the transient behavior of a system which boosts a part of the battery voltage for a electric vehicle. This system consists of the powertrain of electric vehicles and a bi-directional isolated DC-DC converter called dual active bridge (DAB). This system can compensate the battery voltage drops with high efficiency compared with the conventional series chopper system. In many studies, the DAB converter operates in the steady state. On the other hand, this paper analyzes the operation of the DAB converter in the transient state. Especially, the operation at the start of the DAB converter is focused on.

I. INTRODUCTION

In recent years, many environmental issues such as depletion of fossil fuels, global warming, and air pollution are treated as great concern. Especially, the burning of oil causes various environmental problems. Hence, various efforts to reduce oil consumption has been conducted. Above all, It is desirable to reduce the dependence on oil in the transportation sector which is primary user of oil. From this view point, in the automobile industry, many researches on hybrid electric vehicles (HEVs) and electric vehicles (EVs) have been reported in recent years. Especially EVs, which run only on electricity, do not emit any pollutants during driving. In addition, Well-to-Wheel efficiency of EVs is more than twice that of the internal combustion engine vehicles. Therefore, total CO_2 emission values of EVs are less than about 70% compared with that of the internal combustion engine vehicles [1].

The powertrain of EV is usually composed of a battery, an inverter, and a motor. In this system, the input voltage of the inverter depends on the output voltage of the battery. However, the output voltage of the battery drops when the discharge rate is increased or when its remaining charge is decreased. In addition, the internal resistance of the battery is increased at low state-of-charge or high output power. Therefore, in this system, the desired output may not be obtained at low state-of-charge or when high output power is required. Besides, at low voltage, the required current to obtain the same amount of power increases compared with normal state. Therefore, the inverter requires large current capacity. Furthermore, in this system, the flux-weakening control is applied at high-speed range [2][3]. Under the flux-weakening control, the total

efficiency may decrease because the extra current flows.

In order to solve these problems, a DC-DC converter is often inserted serially between the battery and the inverter in order to change the input voltage of the inverter [4][5]. Owing to this, the input voltage of the inverter is compensated even if the battery voltage drops. However, in this solution, the total efficiency may be worse because conduction losses in the DC-DC converter always occur. Even if the total efficiency is better, the large-capacity DC-DC converter is heavy. As a result, the cruising range may be shorter. In general, it is difficult to achieve both high efficiency and high power density [6].

In order to solve the above problems, a voltage boosting system called high efficient energy conversion system (HEECS) was reported in [7]. In this system, the Dual Active Bridge DC-DC converter (DAB converter)[8] is used as a buck converter. This system has two operation modes; non-boosting mode, and boosting mode. When the boost is unneeded, the non-boosting mode is selected. In the non-boosting mode, the conduction loss can be minimized. In the boosting mode, only a part of the total power is input to the DAB converter. Thus the loss of the DAB converter is reduced.

Usually, the DAB converter is used to transfer energy between different voltage levels. In such an application, the input/output voltage of the DAB converter is fixed to the voltage of the battery, the fuel cell, and so on. Therefore, the DAB converter performs in the steady state in many cases. However, in this system, the DAB converter is used in the transient state. This is because the switching between powering and regeneration and between on and off state is performed frequently. In addition, the loads fluctuate frequently because the loads are the inverter and the motor. In this paper, among these transient states, the operation at the start/stop (the switching between on and off state) of the DAB converter is focused on.

The rest of this paper is organized as follows. Section II describes HEECS and its validity. Section III introduces the operation of the DAB converter in the transient state and the control method of the DAB converter. Section IV and V perform simulation and experiments. Section VI concludes this paper.

978-1-4799-2706-7/14 $31.00 © 2014 IEEE

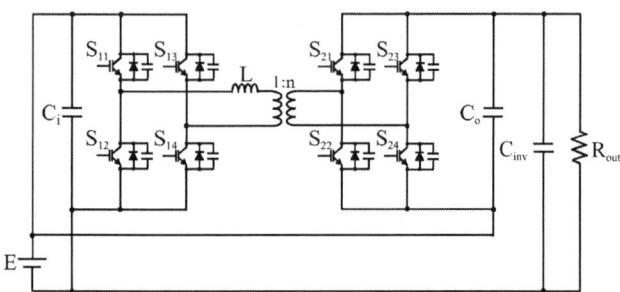

Fig. 1: High efficient energy conversion system (HEECS)

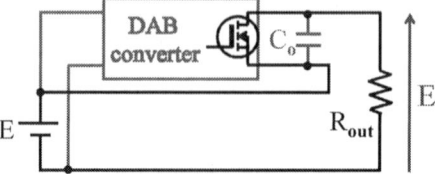

Fig. 2: Non-boosting mode

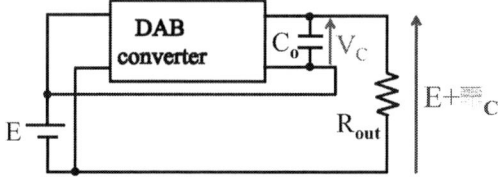

Fig. 3: Boosting mode

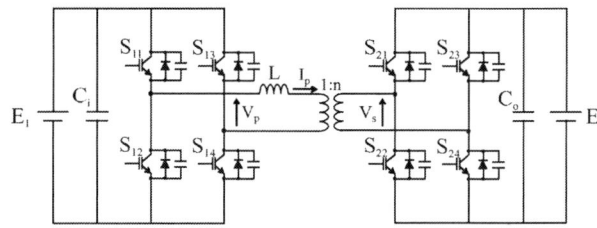

Fig. 4: The DAB converter

II. HIGH EFFICIENT ENERGY CONVERSION SYSTEM

A. Basic circuit configuration

Fig. 1 shows the basic circuit configuration of HEECS. In this circuit, the DAB converter is inserted between a battery and an inverter of an existing powertrain. By inserting the DAB converter in this way, it is possible to boost a part of the battery voltage. The DAB converter has many advantages such as zero voltage switching (ZVS)[9], bidirectionality of power transfer, and low number of components. In HEECS, these advantages are utilized greatly. Specifically, HEECS has the following advantages.
– High efficiency because the current is not passing through the DAB converter during non-boosting.
– High efficiency because only a part of the total power goes through the DAB converter during boosting.
– Relatively small capacity DAB converter can be used because only a part of total power is input to the DAB converter.

B. Operation modes

The operation modes are shown in Fig. 2 and Fig. 3. This system has two operation modes; non-boosting mode(Fig. 2), and boosting mode(Fig. 3).

In the non-boosting mode, the DAB converter is turned off. This mode is selected when the output power is low or when the remaining charge of the battery is enough. In this mode, the input voltage of inverter is approximately equal to the output voltage of the battery. The power is not input to the DAB converter during non-boosting. Thus only conduction losses of the secondary switches occur. When the DAB converter is off, the secondary switches of the DAB converter are on. In this system, the secondary side voltage of the DAB converter is low. Therefore, MOSFETs can be used as the secondary switches. By using MOSFETs which has low on-resistance as the secondary switches, the efficiency almost does not decrease.

In the boosting mode, the DAB converter is turned on. This mode is selected when the output power is high or when the remaining charge of the battery is decreased. In this mode, the input voltage of inverter is the sum of the output voltage of battery E and the output voltage of the DAB converter V_c. Only a part of the total power is input to the DAB converter during boosting. Thus the loss of the DAB converter is reduced

and compact, small capacity DAB converter can be used as well. In [7], the experiments of the non-boosting mode and the boosting mode are performed in the steady state. In these experiments, the total efficiency of 99.86 % with the non-boosting mode and approximately 99 % with the boosting mode were measured.

III. THE OPERATION OF THE DAB CONVERTER

The schematic of the DAB converter is shown in Fig. 4. The DAB converter behaves differently depending on the amount and form of the AC voltage V_p, V_s. Recently, several modulation methods have been proposed to extend ZVS range [10][11]. However, at the moment, the DAB converter in this system may be controlled with square wave AC voltage simply. The slope of the AC current I_p is calculated by equation (1) using parameters shown in Fig. 4.

$$\frac{dI_p}{dt} = \frac{V_p - V_s/n}{L} \qquad (1)$$

The theoretical waveforms are shown in Fig. 5. When V_p is equal to V_s/n, the slope of I_p becomes 0 from equation (1). As a result, the waveforms as shown in Fig. 5(a) are obtained. On the other hand, when V_s is very small or 0, the slope of I_p depends only on V_p. As a result, the waveforms as shown in Fig. 5(b) are obtained. In HEECS, the initial output voltage of the DAB converter is 0 because HEECS operates with non-boosting mode until the DAB converter starts its operation.

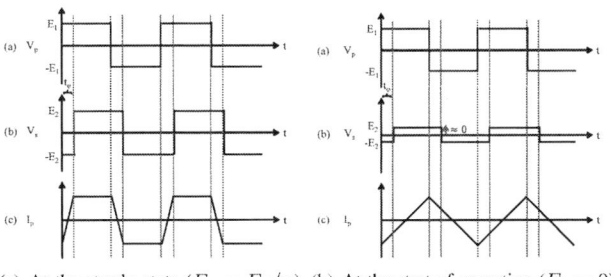

(a) At the steady state ($E_1 = E_2/n$) (b) At the start of operation ($E_2 = 0$)

Fig. 5: Theoretical waveforms

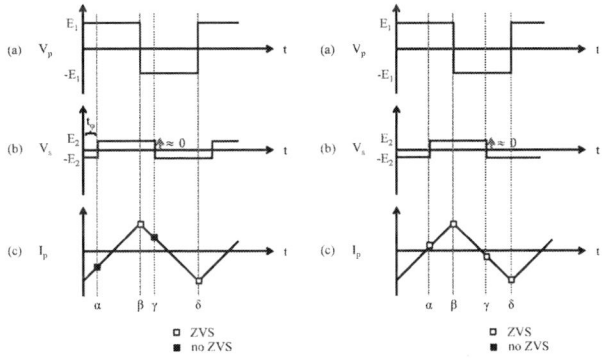

(a) The waveforms with improper t_ϕ (b) The waveforms with proper t_ϕ

Fig. 6: Theoretical waveforms immediately after the start of operation

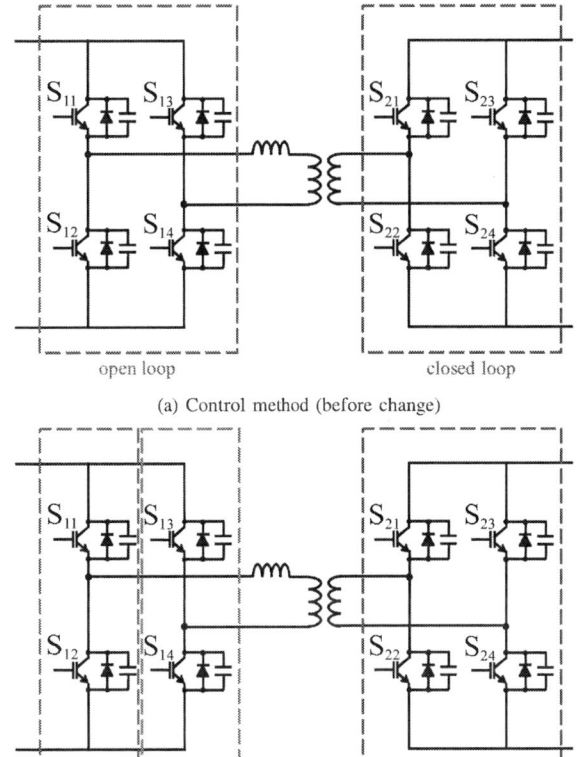

(a) Control method (before change)

(b) Control method (after change)

Fig. 7: Change of the control method

Thus the waveforms like Fig. 5(b) appear immediately after the start of the DAB converter. After the output voltage rises, the waveforms become like Fig. 5(a).

A. Control method of the DAB converter

In [7], a new control method of the DAB converter called hysteresis control was proposed in order to improve the response to the transient state and its effectiveness has been confirmed. Its effectiveness can be confirmed also at the start of the DAB converter in HEECS. The theoretical waveforms are shown in Fig. 6. The following constraints have to be satisfied in order to achieve ZVS. (2) is the constraints of the primary switches and (3) is the constraints of the secondary switches. $I_p\alpha$ means the current I_p at the point α in Fig. 6.

$$I_p(\beta) > 0, I_p(\delta) < 0 \qquad (2)$$
$$I_p(\alpha) > 0, I_p(\gamma) < 0 \qquad (3)$$

At the start of the DAB converter, ZVS cannot be achieved in the secondary switches if the phase-shift t_ϕ is not set properly as shown in Fig. 6(a). By contrast, in hysteresis control, the constraints in (3) are satisfied automatically and the constraints in (2) are satisfied with $E_1 > E_2/n$. Thus at the start, all constraints are satisfied as shown in Fig. 6(b). If the peak

current in Fig. 6 ($I_p(\beta)$, $I_p(\delta)$) is too large for the rated current of switches, there is no choice but to increase the switching frequency as far as the square wave control is used. However, switching losses and the rise time of the output voltage of the DAB converter increase when the switching frequency increases. Then, changing the control method as shown in Fig. 7 in order to suppress the peak current at the start. At first, allowable current of I_s (I_{s_max}) is set. Two of the four switches on the primary side (S_{13}, S_{14}), which was controlled in open loop (Fig. 7(a)), are controlled in closed loop only when I_s exceeds I_{s_max} (Fig. 7(b)). As a result, the peak current is limited as shown in Fig. 8.

IV. SIMULATION

The simulation circuit and the simulation parameters are shown in Fig. 9 and TABLE I. In this simulation, the DAB converter starts its operation with initial input voltage $E_1 = 200[V]$, initial output voltage of the DAB converter $V_{Co} = 0[V]$, and initial output current I_o.

First, change in I_s and the rise time of V_{Co} due to the difference of the control method are confirmed. The simulation results are shown in Fig. 10 - Fig. 12. Compared with Fig. 10

The 2014 International Power Electronics Conference

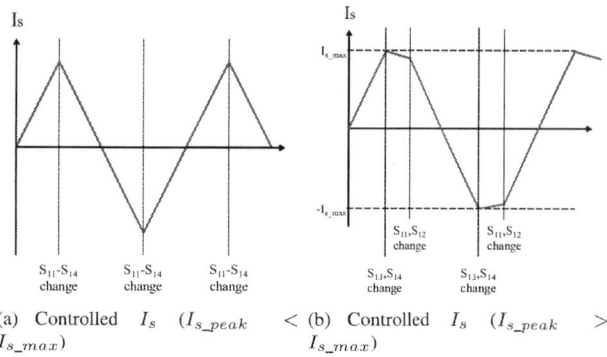

(a) Controlled I_s (I_{s_peak} < I_{s_max}) (b) Controlled I_s (I_{s_peak} > I_{s_max})

Fig. 8: Controlled I_s

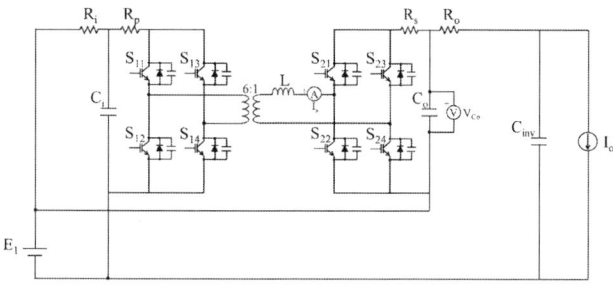

Fig. 9: Simulation circuit

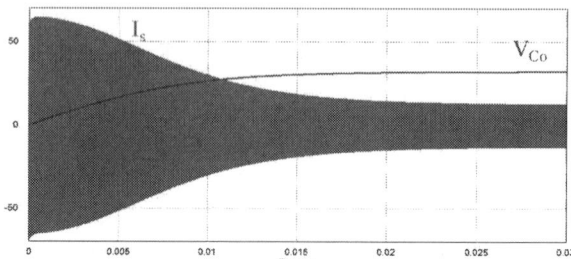

(a) I_s and V_{Co} before the change of control

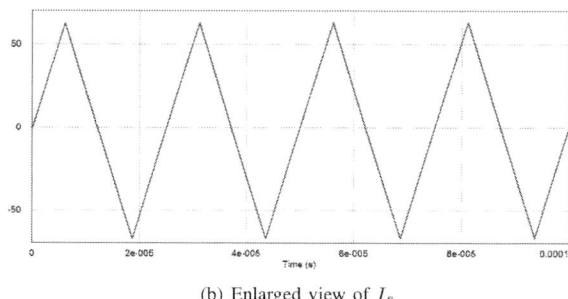

(b) Enlarged view of I_s

Fig. 10: Operation waveforms before the change of control

and Fig. 11, though current is limited in Fig. 11, there is almost no difference in the rise time. However, if the peak current is limited by increasing the switching frequency as shown in Fig. 12, the rise time becomes longer.

Next, change in the rise time of V_{Co} due to the difference of I_o is confirmed. The simulation results are shown in Fig. 13. From Fig. 13, the rise time of the output voltage depends on I_o. The rise time becomes longer under the larger output current. The output voltage no longer rises if the output current is too large. In order to raise the output voltage, the parameters (ex. L, f_{sw}) have to be set properly or the DAB converter has to starts its operation with the low output current.

TABLE I: Simulation parameters

Input voltage	E_1[V]	200
Allowable current of I_s	I_{s_max}[A]	50
AC-link inductor	$L[\mu H]$	3.2
Input capacitor	$C_i[\mu F]$	1500
Output capacitor	$C_o[\mu F]$	4700
Inverter input capacitor	$C_{inv}[\mu F]$	1500
Snubber capacitor (primary)	C_{s1}[nF]	1
Snubber capacitor (secondary)	C_{s2}[nF]	68
Wire resistance	R_i, R_o, R_p, R_s[mΩ]	1
Switching frequency	f_{sw}[kHz]	40
Transformer turn ratio	$n : 1$	6:1

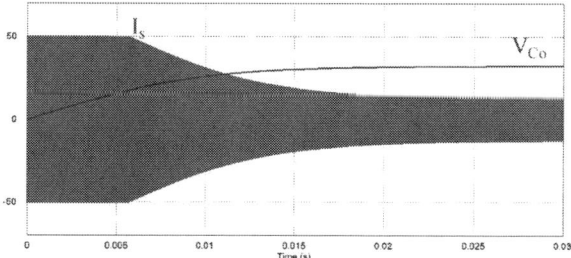

(a) I_s and V_{Co} after the change of control

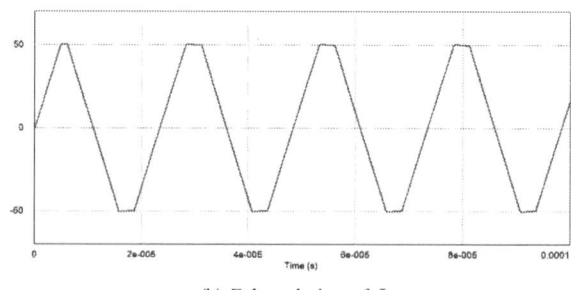

(b) Enlarged view of I_s

Fig. 11: Operation waveforms after the change of control

978-1-4799-2706-7/14 $31.00 © 2014 IEEE   2269

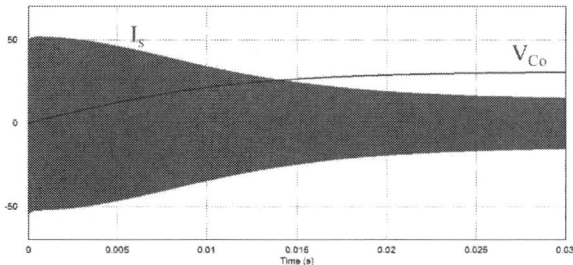

Fig. 12: Suppression of I_{s_peak} by the increase of the switching frequency (50kHz)

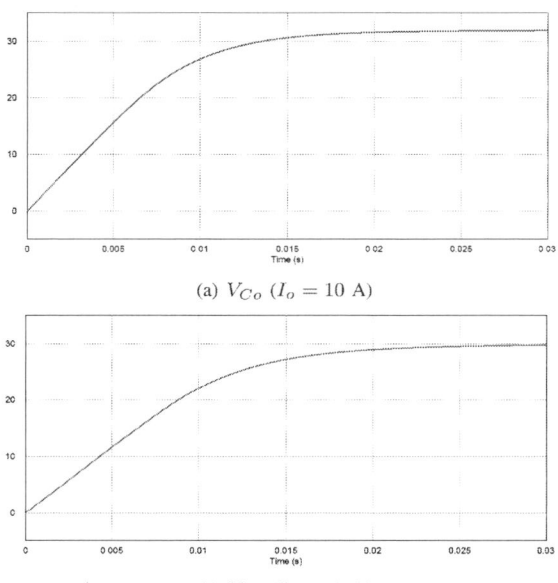

(a) V_{Co} ($I_o = 10$ A)

(b) V_{Co} ($I_o = 15$ A)

Fig. 13: The rise time of V_{Co}

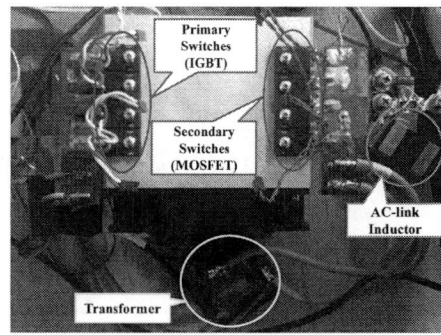

Fig. 14: The DAB converter

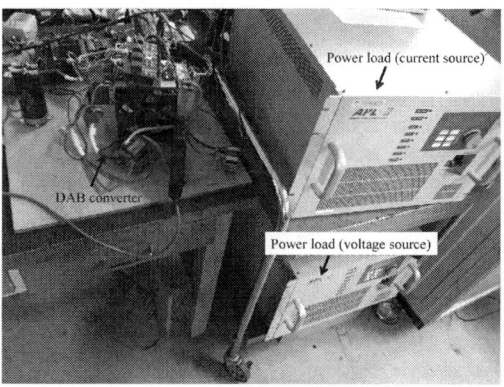

Fig. 15: Whole setup

V. EXPERIMENT

A. Experimental system

The DAB converter used in the experiment is shown in Fig. 14. Each parameter is the same as that used in the simulation. In this experiment, the DAB converter starts its operation under the same conditions in the simulation. Fig. 15 shows the whole setup and Fig. 16 shows the conceptual diagram of the whole setup. Two power loads are used as a voltage source (primary side) and a current source (secondary side). The controller is mainly composed of a DSP board, a FPGA board, and some A/D converters. DSP performs PI control using the measured voltages and currents. FPGA performs hysteresis control using the current command value received from DSP.

B. Experimental results

The output voltage of DAB converter V_{Co} is shown in Fig. 17. Compared with the simulation result under the same

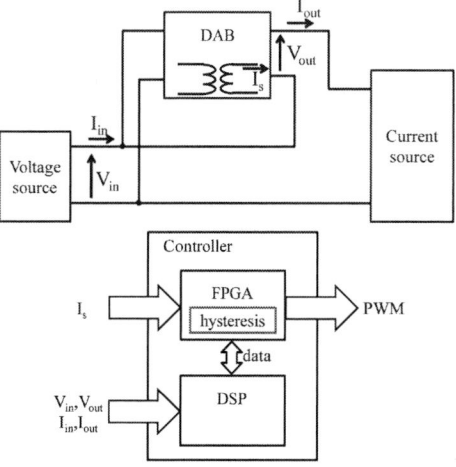

Fig. 16: Conceptual diagram of the whole setup

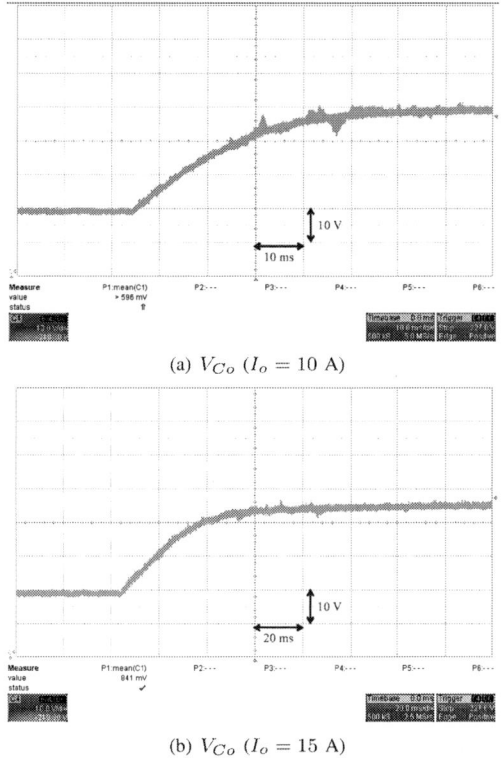

(a) V_{Co} ($I_o = 10$ A)

(b) V_{Co} ($I_o = 15$ A)

Fig. 17: Experimental result

condition (Fig. 13), their ways of rising of the output voltage are similar.

However, the rise time of the output voltage is longer than that of simulation. This may be due to differences in the controller and difference between the real components (switches, transformer, inductor, etc.) and the ideal components.

VI. CONCLUSION

In this paper, the transient behavior of the DAB converter in HEECS was considered. In the simulation and experiment, the DAB converter started its operation under the actual conditions. As a result, the DAB converter showed similar behaviors in the simulation and experiment.

This paper shows the dynamic behavior of the DAB converter only at the start of the operation. In the future, the experiments in the other transient states will be performed. In addition, improvement of the control method may be performed in that process.

REFERENCES

[1] T. Hosokawa, K. Tanihata, H. Miyamoto, "Development of i-MiEV Next-Generation Electric Vehicle(Second Report),h *Mitsubishi Motors Technical Review*, No. 20, pp. 53-60, 2008.

[2] T. M. Jahns, "Flux-Weakening Regime Operation of an Interior Permanent-Magnet Synchronous Motor Drive,h *IEEE Transactions on Industry Applications*, Vol. IA-23, pp. 681-689, 1987.

[3] S. Morimoto, M. Sanada, Y. Takeda, "Effects and Compensation of Magnetic Saturation in Flux-Weakening Controlled Permanent Magnet Synchronous Motor Drives,h *IEEE Transactions on Industry Appllications*, Vol. 30, No. 6, pp. 1632 -1637, 1994.

[4] Jih-sheng(Jason) Lai, D. J. Nelson, "Energy Management Power Converters in Hybrid Electric and Fuel Cell Vehicles,h *Proceedings of the IEEE, Special issue on Electric, Hybrid and Fuel Cell Vehicles*, Vol. 95, No. 4, 2011.

[5] J. O. Estima, A. J. M. Cardoso, "Efficiency Analysis of Drive Train Topologies Applied to Electric/Hybrid Vehicles,h *IEEE Transactions on Vehicular Technology*, Vol. 61, No. 3, pp. 1021-1031, 2012.

[6] J. Biela, U. Badstuebner, J. W. Kolar, "Impact of power density maximization on efficiency of dc-dc converter systems,h *IEEE Transactions on Power Electronics*, Vol. 24, No. 1, pp. 288-300, 2009.

[7] K. Aoyama, N. Motoi, G. Guidi, Y. Tsuruta, A. Kawamura, "Ultra High Efficient Battery Voltage Compensation against Decrease in the Terminal Voltage of Electric Vehicles,h *Proceedings of IECON 2013 - 39th Annual Conference of the IEEE Industrial Electronics Society*, 2013.

[8] R. W. De. Doncker, D. M. Divan and M. H. Kheraluwala, "A Three-Phase Soft-Switched High-Power-Density dc/dc Converter for High-Power Applications,h *IEEE Transactions on Industry Applocations*, Vol. 27, No. 1, pp. 63-73, 1991.

[9] Z. Shen, R. Burgos, D. Broyevich, F. Wang, "Soft-Switching Capability Analysis of a Dual Active Bridge Dc-Dc Converter,h *Proceedings of IEEE-ESTS*, pp. 334-339, 2009.

[10] G. G. Oggier, G. O. Garcia, A. R. Oliva, "Modulation Strategy to Operate the Dual Active Bridge DC-DC Converter Under Soft Switching in the Whole Operating Range,h *IEEE Transactions on Power Electronics*, pp. 1228-1236, 2010.

[11] Y. Wang, S. W. H. de Haan, J. A. Ferreira, "Optimal Operating Ranges of Three Modulation Methods in Dual Active Bridge Converters,h *Proceedings of Power Electronics and Motion Control Conference*, pp. 1397-1401, 2009.

State-of-Charge Estimation for Lithium-ion Battery Pack Using Reconstructed Open-Circuit-Voltage Curve

Chang Yoon Chun, Gab-Su Seo, Sung Hyun Yoon, and Bo-Hyung Cho
Department of Electrical and Computer Engineering
Seoul National University
Seoul, Republic of Korea
wobniw77@snu.ac.kr, bhcho@snu.ac.kr

Abstract— This paper proposes a state-of-charge (SOC) estimation algorithm for a series-connected Li-ion battery pack using a reconstructed open-circuit-voltage (OCV) curve. This method redefines the OCV-SOC relationship based on a cell which has the lowest available capacity and estimates the battery pack SOC and capacity information regardless of the type and existence of balancing circuit. To validate the performance of the proposed estimation method, a constant current profile is used for seven 18650 series-connected Li-ion batteries (7S1P). The experimental results verify the performance of the proposed battery pack SOC estimation algorithm.

Keywords—battery pack; pack capacity; reconstructed open-circuit-voltage (OCV); state-of-charge (SOC).

I. INTRODUCTION

In estimating state-of-charge (SOC) of a Li-ion battery pack with series-connected cells, conventional pack model-based methods experience difficulties providing accurate battery status since they assume all batteries of a pack have the identical characteristics. Even though a screening process may improve the uniformity issue [1]-[3], the cell voltage variations in a pack are still inevitable even in the screened case. Due to unavoidable mismatches resulting from product tolerance and different capacity degradation of individual cells in use [4], it is challenging to estimate accurate battery status.

To overcome the inherent limitations of the pack model-based methods, cell model-based methods have been presented. They use the lowest voltage cell model [5] or averaged cell model [6]-[8] to achieve high estimation accuracy. By utilizing an additional SOC calibration process, the alternative methods obtain better performance. However, although they correct the estimation results considering the cell voltage variances, to achieve high estimation accuracy is still challenging since the battery pack SOC is not determined by a cell with the minimum voltage but by that with the minimum available energy.

The proposed SOC estimation algorithm locates the dominant cell which determines the entire battery pack capacity and SOC. Considering the dominant cell

characteristic, the relation of open-circuit-voltage (OCV) and SOC is reconstructed. As a result, this method is able to accomplish the SOC estimation of a battery pack regardless of existence and type of balancing circuit with simple calculation. This paper is organized as follows. Section II presents discussions on battery pack capacity and pack SOC and the key idea of the SOC estimation algorithm. In Section III, the developed battery pack SOC estimation algorithm is validated using a constant current profile with seven 18650 series-connected Li-ion batteries (7S1P).

II. BATTERY PACK SOC ESTIMATION ALGORITHM

A. Capacity and SOC of a Packed Battery

The definition of the battery pack capacity was introduced in [6]-[7]. The battery pack capacity C_{pack} is expressed as

$$C_{pack} = \begin{cases} \min(SOC_i \cdot C_i) + \min((1-SOC_j) \cdot C_j) & \text{w/o balancing} \\ \min(C_i) & \text{Passive balancing} \\ \text{mean}(C_i) & \text{Active balancing} \end{cases} \quad (1)$$

where SOC_i and C_i are the SOC and capacity of i-th cell in the series-connected battery pack. As (1) describes, the existence and type of the balancing circuit determine the battery pack capacity.

In the balancing circuit-less case, the remaining capacity of a pack is clearly related to that of the lowest remaining capacity cell, $\min(SOC_i \cdot C_i)$, while the chargeable capacity is to one with the smallest chargeable room, $\min((1-SOC_j)C_j)$. If the voltage of the lowest remaining capacity cell (C1) at the moment when the lowest chargeable capacity cell (C2) reaches its upper limit is set as the new upper limit, the newly constructed OCV-SOC curve of C1 would represent the pack SOC.

Every ideally balanced cell in the pack has same chargeable room when the battery pack is charging. This means that C1 and C2 refer to the same cell, and this cell has the lowest cell capacity in the pack. For this reason, the battery pack capacity equals to the minimum cell capacity at passive balancing, and the averaged capacity

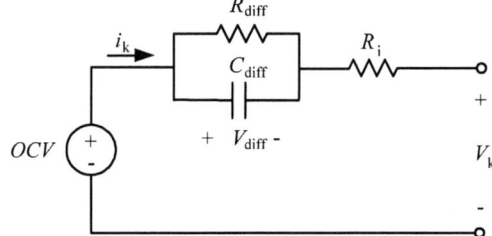

Fig. 1. Battery Thevenin equivalent circuit model.

TABLE I
BATTERY CHARACTERISTICS

Symbol	Meaning	Value
R_i	Series resistance	0.1695 Ω
R_{diff}	Diffusion resistance	0.0249 Ω
C_{diff}	Diffusion capacitance	7000 F

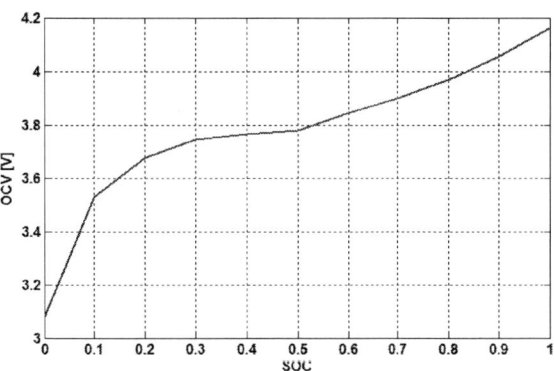

Fig. 2. OCV-SOC curve of a typical 18650 Li-ion battery.

of the entire pack at active balancing because all the cells are adjusted to have same SOC state. According to the relations, it is noted that the status of C1 represents the pack capacity regardless of the type and existence of balancing. As a result, battery pack SOC is derived as

$$SOC_{pack} = \frac{\min(SOC_i \cdot C_i)}{C_{pack}} = \frac{SOC_{C1} \cdot C_{C1}}{\alpha \cdot C_{C1}} \quad (2)$$

where α is the calibration factor which is the capacity ratio coefficient of the newly constructed OCV-SOC curve.

B. Battery Cell Model and OCV Curve Reconstruction

Fig. 1 shows a Thevenin equivalent circuit model for a single battery cell, and Table I presents the characteristic values of a 18650 Li-ion battery including a series resistor (R_i), a R-C branch (R_{diff} and C_{diff}). Fig. 2 illustrates SOC and OCV relationship. As aging proceeds in a battery cell, the battery impedance tends to increase, which leads to the limited available OCV range due to the increased internal energy loss. As another result of aging, the battery capacity also decreases. Since the high internal impedance and the low capacity cause higher and lower cell voltage in charging and discharging, respectively, it limits operating range of the cell and battery pack.

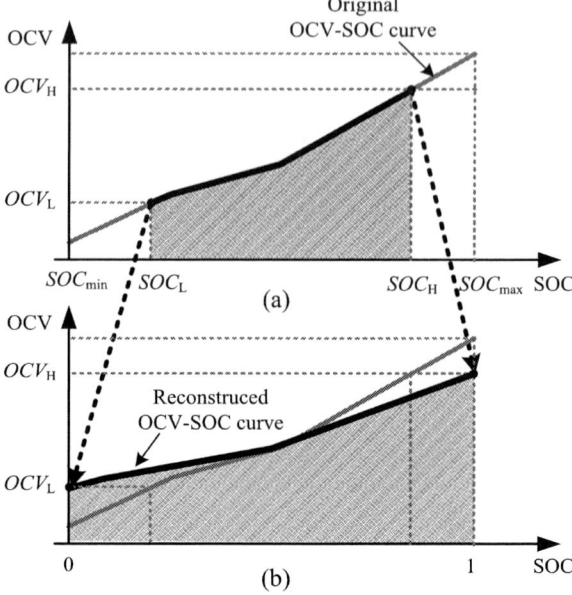

Fig. 3. OCV-SOC curve reconstruction: (a) original OCV-SOC curve with upper and lower boundary identified, (b) reconstructed battery OCV-SOC curve.

This paper uses OCV information for SOC estimation. As demonstrated in Fig. 3, the lowest OCV, OCV_L, is obtained through calculation or measurement with sufficient rest time when the voltage of the lowest remaining capacity cell (C1) meets lower cutoff voltage. The highest OCV value of the lowest remaining capacity cell (C1), OCV_H, is obtained when any cell voltage, which would be the voltage of C1 or not, meets its upper limit. From the OCV-SOC relationship, SOC_H and SOC_L is obtained, and calibration factor α is simply derived from

$$\alpha = \frac{(SOC_H - SOC_L)}{(SOC_{max} - SOC_{min})}. \quad (3)$$

The OCV operating range is reconstructed like the dotted line in Fig. 3(a), and (b).

The relationship between the lowest remaining capacity cell SOC, SOC_k, and the battery pack SOC, SOC_{pack}, can be defined as

$$SOC_{pack} = \begin{cases} 0, & SOC_{pack} \leq 0 \\ \frac{1}{\alpha}(SOC_k - SOC_L) = \frac{(SOC_k - SOC_L)}{(SOC_H - SOC_L)}. \\ 1, & SOC_{pack} \geq 1 \end{cases} \quad (4)$$

In (4), the battery pack SOC is trimmed not to exceed its upper (100%) and lower (0%) boundary.

C. Proposed Battery Pack SOC Estimation

As illustrated in Fig. 4, when the BMS starts its operation, the lowest remaining capacity cell (C1), which has the dominant characteristic, is not identified yet. In the initial operation, the system selects a cell which shows the highest voltage in charging or the lowest in

The 2014 International Power Electronics Conference

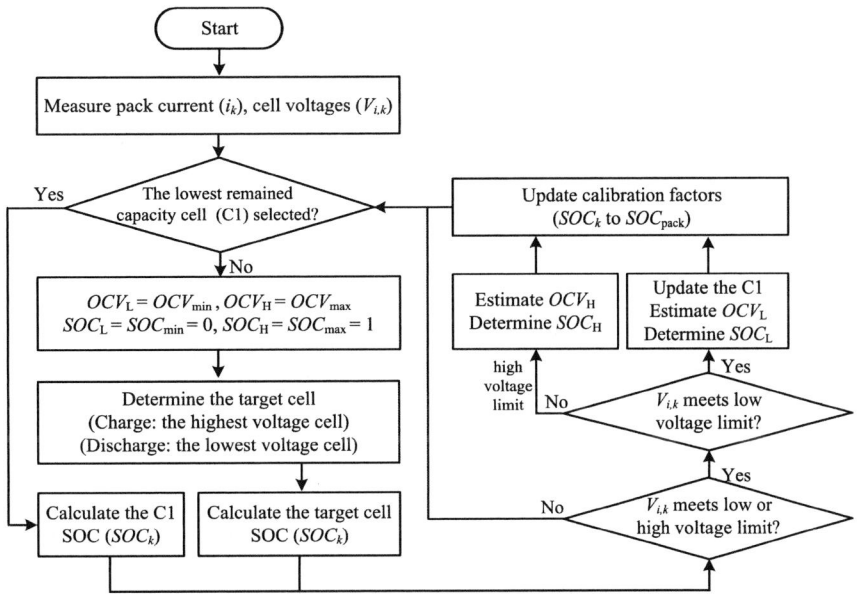

Fig. 4. Flowchart for the calibration factors derivation.

discharging, and lets the cell SOC to represent the pack SOC until C1 is located. OCV_H, OCV_L, SOC_H, and SOC_L are set to the initial values. Because no correction is applied in this state, SOC estimation error would occur.

If a battery voltage of a cell reaches its lowest limit during operation, it is selected as the dominant cell C1 and the BMS sets the OCV of C1 as OCV_L. The SOC estimation algorithm estimates the pack SOC based on the dominant cell characteristic and operating condition. On the other hand, if a battery voltage of a cell reaches its highest limit, it sets the OCV of the cell as OCV_H while the dominant cell is not updated. The management system continuously updates C1 as the $\min(SOC_i \cdot C_i)$, and proceeds the same SOC estimation and updating processes, as shown in Fig. 4.

D. Extended Kalman Filter Algorithm

The extended Kalman filter (EKF) is used to estimate the SOC in many literatures [9]-[11]. The process model and the measurement model, expressed as

$$x_k = \begin{bmatrix} SOC_{k+1} \\ V_{\text{diff},k+1} \end{bmatrix} = \begin{bmatrix} 1 & 0 \\ 0 & 1 - \dfrac{\Delta t}{R_{\text{diff}} C_{\text{diff}}} \end{bmatrix} \begin{bmatrix} SOC_k \\ V_{\text{diff},k} \end{bmatrix} + \begin{bmatrix} -\dfrac{\Delta t}{C_{\text{C1}}} \\ \dfrac{\Delta t}{C_{\text{diff}}} \end{bmatrix} i_k \quad (5)$$

$$V_k = h_k(OCV, V_{\text{diff}}) - R_i i_k = OCV(SOC_k) - V_{\text{diff}} - R_i i_k, \quad (6)$$

defines the SOC and diffusion voltage between over the R-C branch and the estimated terminal cell voltage based on the battery model in Fig. 1. EKF ensures the SOC estimation is within a specified range even though initial error and measurement error occur. The specified range is determined by battery model, especially the OCV. Thus, if the OCV in battery model is more accurate, EKF algorithm guarantees reliable results. Otherwise, SOC information fluctuates in a certain bound.

In this paper, the proposed algorithm uses a reduced order EKF with a noise model and data rejection algorithm from [10].

III. VERIFICATION

In this section, the SOC estimation results for battery pack and the lowest remaining capacity cell are shown in detail. In order to verify the proposed method, a single pack of seven series-connected batteries (7S1P), which are Samsung SDI 2.6Ah 18650 Li-ion batteries, is used. Before the experiment, most of the cell voltages are balanced, and the initial cell voltages in the battery pack stay constant. The battery pack is charged under the constant current-constant voltage (CC-CV) protocol with 0.5 C rating and discharged under the same constant current conditions, and rested for 1.5 h, as shown in Fig. 5(a). For a single cell case, the voltage is held at 4.2 V till the current drops to 130 mA (0.05 C) at CC-CV and the low cutoff voltage is set to be 2.75V at discharging. However, for this battery pack case, 29.05-V/21-V (4.15 V, 3 V each cell) are assigned as the cutoff voltages for charging/discharging case in order to place the margin. In other words, even if the battery balancing circuit exists but the balancing speed is not sufficient, imbalance may occur between cells. For protecting battery cells in the pack, it is necessary to limit the operating range of the battery cells unlike the cell parameters extraction process.

Fig. 5(a) shows seven cell voltages in the battery pack when the battery pack is charging/discharging or rest, and Figs. 5(b), 5(c), 5(d) are the expanded waveforms of Fig. 5(a) at the notable mode changes for further discussions. In the beginning, the lowest remaining capacity cell (C1) is not found yet, the highest voltage cell is a target cell and the SOC of the target cell is estimated. While the charging progresses in Fig. 5(b), the cell #6 has the largest voltage variation. As, the voltage of cell #6 reaches the upper limit, the cell #6 is identified as the

978-1-4799-2706-7/14 $31.00 © 2014 IEEE

2274

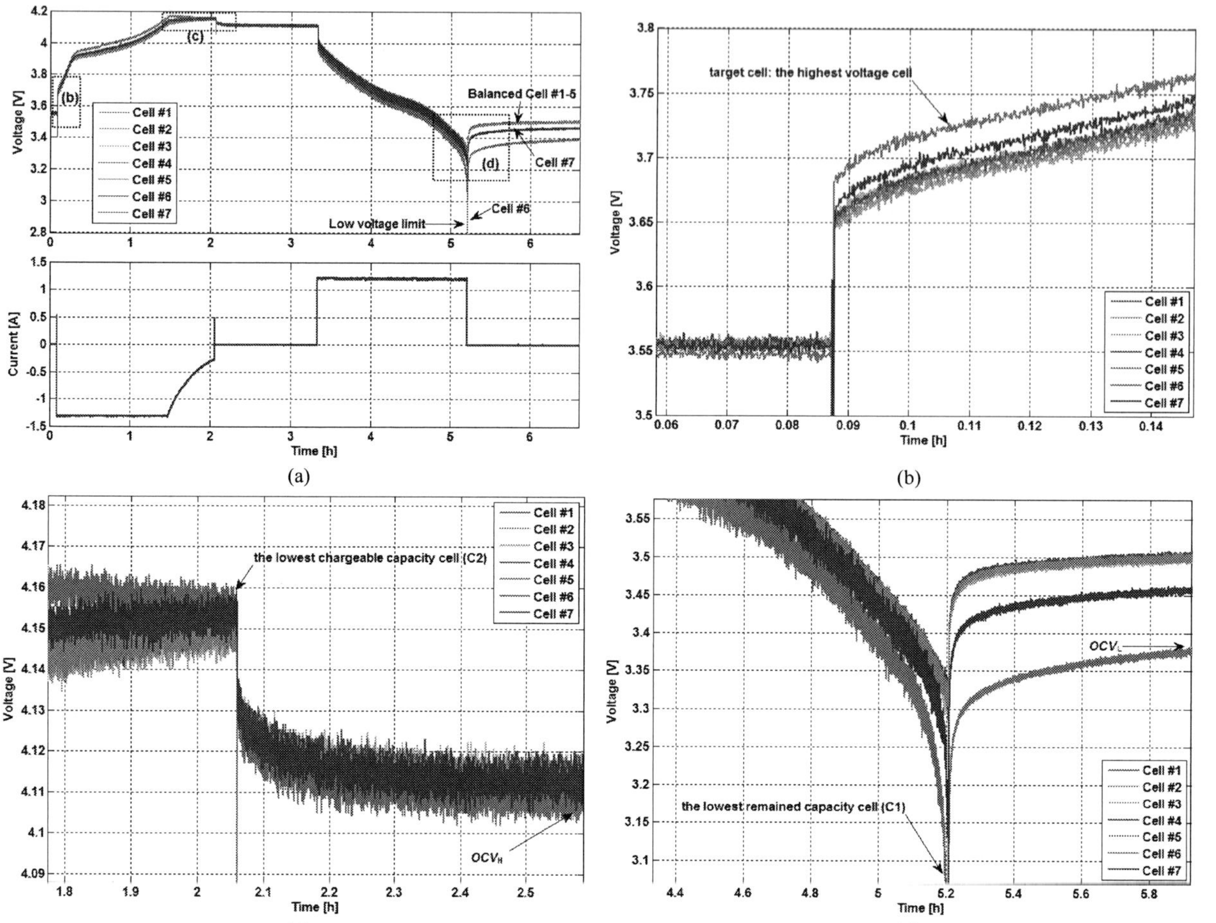

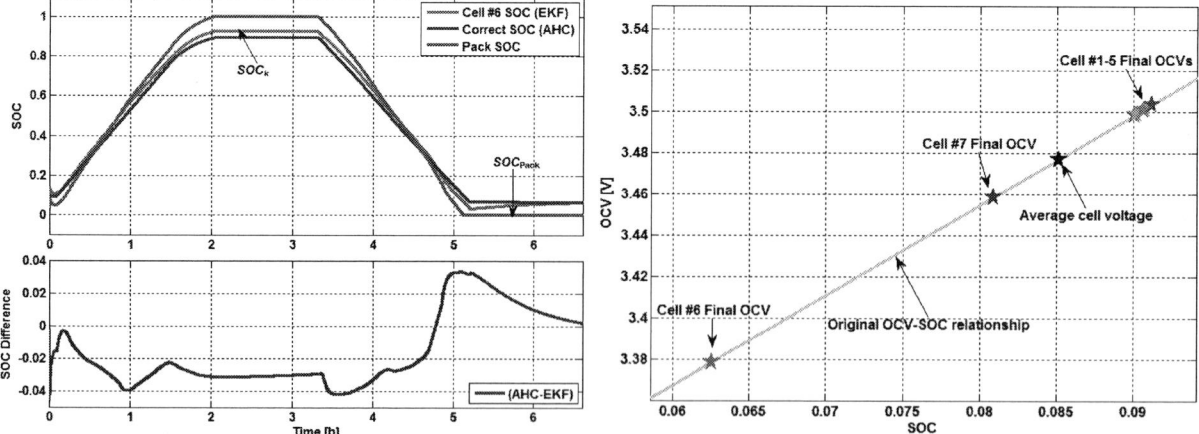

Fig. 5. Experimental result: (a) cell voltages and pack current, (b) expanded waveform at determining the target cell, (c) expanded waveform at the end of charge condition, (d) expanded waveform at the end of discharge condition.

Fig. 6. Estimated SOC (cell # 6), battery pack SOC and SOC difference

Fig. 7. Estimated OCV and SOC values using each cell model and average cell voltage model at $SOC_{pack} = 0$.

lowest chargeable capacity cell (C2) as observed in Fig. 5(c) and the battery pack SOC reaches SOC_{max}. At last, discharging of the pack is halted as shown in Fig. 5(d) as the voltage of cell #6 reaches the lower limit. At the moment, cell #6 is the lowest remaining capacity cell (C1), and the battery pack SOC reaches SOC_{min}. The

SOCs of the C1 and the pack are calculated using the original OCV curve, as shown in Fig. 2, and the updating calibration factors from (3). As a result, the pack capacity comes from the lowest remaining capacity cell (C1) and calibration factor, and the final battery pack SOC is obtained as shown in Fig. 6.

The Fig. 7 is the enlarged version of Fig. 2 in which OCV distribution at the end of discharging is projected on the low SOC range curve. At the beginning of the experiment, all initial OCVs are placed around 9% SOC in Fig. 5. However, the final SOC values are placed on 6.2% to 9% depending on the voltage of the each cell distribution. This slight SOC difference comes from its OCV difference and these results indicate that if the battery pack SOC estimation is based on other cell model or averaged cell model, the SOC estimation error should be larger than the proposed method due to their model error.

IV. CONCLUSIONS

In this paper, a battery pack SOC estimation algorithm is proposed which uses the dominant cell characteristic. The proposed method reconstructs the OCV-SOC relationship based on the lowest available capacity cell since it determines the battery pack SOC regardless of the type and existence of balancing. With the updated calibration factors, the newly defined OCV-SOC relationship allows the battery management system (BMS) to improve its SOC estimate performance.

ACKNOWLEDGMENT

This work was partly supported by the Technological Innovation R&D program (S2058100) funded by the Small and Medium Business Administration (SMBA, Korea); the Human Resources Development program (No. 20124030200030) of the Korea Institute of Energy Technology Evaluation and Planning (KETEP) grant funded by the Korea government Ministry of Trade, Industry and Energy; and the Seoul National University Research Grant.

REFERENCES

[1] J. Kim et al., "High accuracy state-of-charge estimation of Li-ion battery pack based on screening process," in *Proc. IEEE Appl. Power Electron. Conf. Expo. (APEC)*, 2011, pp. 1984-1991.

[2] J. Kim et al., "Stable configuration of a Li-ion series battery pack based on a screening process for improved voltage/SOC balancing," *IEEE Trans. Power Electron.*, vol. 27, no.1, pp. 411-424, 2012.

[3] R. Xiong, "Adaptive state of charge estimator for lithium-ion cells series battery pack in electric vehicles," *J. Power Sources*, vol. 242, pp. 699-713, 2013.

[4] M. Broussely et al., "Main aging mechanisms in Li ion batteries," *J. Power Sources*, vol. 146, pp. 90-96, 2005.

[5] X. Liu et al., "State-of-charge estimation for power Li-ion batteries pack using Vmin-EKF," in *Proc. IEEE Int. Conf. Software Eng. and Data Mining (SEDM)*, 2010, pp. 27-31.

[6] L. Zhong et al., "A method for the estimation of the battery pack state of charge based on in-pack cells uniformity analysis," *Appl. Energy*, vol. 113, pp. 558-564, 2014.

[7] Y. Zheng et al., "LiFePO$_4$ battery pack capacity estimation for electric vehicles based on charging cell voltage curve transformation," *J. Power Sources*, vol. 226, pp. 33-41, 2013.

[8] H. Dai et al., "Online cell SOC estimation of Li-ion battery packs using a dual time-scale Kalman filtering for EV applications," *Appl. Energy*, vol. 95, pp. 227-237, 2012.

[9] G. L. Plett, "Extended Kalman filtering for battery management systems of LiPB-based HEV battery packs: Part 1-3," *J. Power Sources*, vol. 134, pp. 252-292, 2004.

[10] O. Nam et al., "Li-ion battery SOC estimation method based on the reduced order extended Kalman filtering," in *Proc. 4th Int. Energy Convers. Eng. Conf. and Exhibit*, 2006, pp. 1-9.

[11] M. Charkhgard and M. Farrokhi, "State-of-charge estimation for lithium-ion batteries using neural networks and EKF," *IEEE Trans. Ind. Electron.*, vol. 57, no. 12, pp. 4178-4187, 2010.

System Design of Electric Assisted Bicycle using EDLCs and Wireless Charger

Jun-ichi Itoh, Kenji Noguchi and Koji Orikawa
Department of Electrical, Electronics and Information Engineering
Nagaoka University of Technology
Nagaoka, Niigata, Japan
itoh@vos.nagaokaut.ac.jp, k_noguchi@stn.nagaokaut.ac.jp, orikawa@vos.nagaokaut.ac.jp

Abstract— **This paper discusses an electric assisted bicycle which uses electric double layer capacitors (EDLCs) as a power source. EDLCs are charged through a rapid charger by using wireless power transmission. In this paper, first, the energy capacity of EDLCs is designed. Next, the antenna for the wireless power transmission and the charger are investigated. Third, this paper compares the volume and the power loss of the three kinds of DC-DC converters which are step-down type, boost-type and buck-boost type for the charging and discharging of EDLCs. As a result, boost-type is the most compact in the power capacity of the electric assisted bicycle. Finally, the proposed system is experimentally verified as a prototype. As a result, the proposed system can shorten to 1/4 the charging time of the conventional system.**

Keywords— *Electric assisted bicycle, Electric double layer capacitors, Wireless power transfer, System design*

I. INTRODUCTION

Recently, electric assisted bicycles have been attracted because the safety and easy driving on bicycle. The electric assisted bicycles use a lithium-ion battery as energy source. The lithium-ion battery is suitable for a long assist time owing to high energy density. However the lithium-ion battery is the short lifetime and the necessary to the long charging time.

On the other hands, the electric vehicle using electric double layer capacitors (EDLCs) of long cycle life has been developed as city commuters [1].This system uses EDLCs instead of the lithium-ion battery. However, the large vehicle such as a car is afraid of the empty of the energy on the load.

The electric bike with EDLCs has been also proposed [2]. The EDLCs are suitable for the electric bike owing to the small vehicle. However, the electric bike is also afraid of the empty of the energy.

This paper proposes an electric assisted bicycle with EDLCs [3]. The electric assisted bicycle is not afraid of the empty of the energy because it can be driven by pedals on the bicycle even if no electric energy. The concept of this system has two concepts as follows; First, it is assumed that the EDLCs are used as only the assist of starting acceleration and slopes for the electric assisted bicycle in order to suppress the total energy of the EDLCs; Second, the wireless power transfer system is applied to rapid charger for the EDLCs because of small capacity and the short charging time.

In order to maximize the advantages of the proposed system, it is necessary to optimize the design from the perspective of the overall system in consideration of trade-off problem of the time required to fully charge the EDLCs and the total volume of the power conversion system.

Therefore, as the purpose of optimum design of the proposed system, this paper evaluates three items as follows;

1) The energy capacity of EDLCs which is required for assist, based on the running test,

2) The antenna using a print circuit board (PCB) and charger design for the wireless power transmission,

3) The circuit topology of the interface DC-DC converter for EDLCs in the proposed system configuration in terms of the volume and efficiency.

Finally, the proposed system is experimentally verified as the prototype.

II. PROPOSED SYSTEM CONFIGURATION

Fig. 1 shows the configuration of the proposed system. The proposed system uses a Radio Frequency (RF) power supply at the input stage of the transmitting antenna for wireless power transmission [4-8]. Also, the proposed system comprises a rapid rechargeable AC-DC converter and EDLCs in the latter part of the receiving antenna. In addition, the interface DC-DC converter for the assist

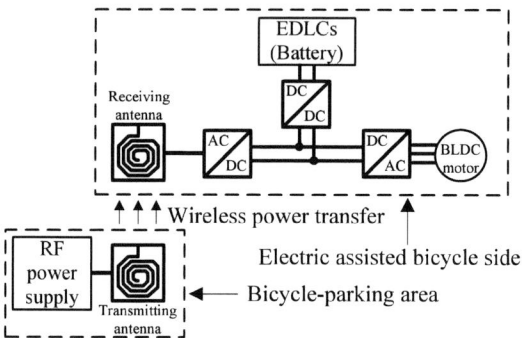

Fig. 1. Configuration of the proposed system. The proposed system can be charged by simply park the electric assisted bicycle to bicycle parking area.

performs discharging control of the EDLCs by assisting the drivers when a brushless DC (BLDC) motor is driven. It is noted that the BLDC motor and DC-AC converter are commercial products [9].

III. DESIGN OF ENERGY CAPACITY OF EDLCs

A. Conditions of Assist

Fig. 2 shows the assist pattern of a driving test. The pattern A is a driving pattern for the evaluation of the assist in the accelerating mode. The pattern B is a driving pattern for demonstration of the assist in the climbing slopes with a gradient (average gradient 6%).

B. Measurement of Cumulative Energy

Fig. 3 shows the method of measuring the cumulative energy of each assist pattern. The energy consumption by pattern A or pattern B is measured using a power meter. The cumulative energy is measured in the conventional system.

Fig. 4 shows the time characteristics of the cumulative energy of each assist pattern. The energy for the assist is 716 J (the energy is 436 J during acceleration time) in the pattern A, and that of the pattern B is 10.6 kJ. From the results in the pattern B, when the distance climbing the distance of the hill was set to 1 km, the necessary energy becomes 53.0 kJ. This paper designs the capacitance of EDLCs that can only assist this energy. Further, the volume of the EDLCs with lithium-ion battery is compared. In addition, the total volume and power loss for three types DC-DC converters using EDLCs as a power source are compared. In addition, the energy of EDLCs is designed in 12.1kJ as a prototype of the system in Section 6.

C. Comparison of Energy

Fig. 5 shows the relationship between the energy and volume of the EDLCs (Nippon Chemi-Con Co, DDLE2R5LGN232KCH2S) and lithium-ion battery in the marketplace [9-10].

The enegy E of EDLCs is calculated by (1) where C_{total} is the total capacitance of EDLCs, V_{max} is the maximum voltage of EDLCs (The rated voltage of EDLC per unit is 2.5 V. In this case, the maximum voltage is designed to 22.5 V), V_{min} is the minimum voltage of EDLCs that is designed by the designer (In this case, the minimum voltage is designed to 8 V). The number and the output voltage range of the EDLCs are designed when the charging and discharging energy of EDLCs meet more than 53.0 kJ.

$$E = \frac{1}{2} C_{total} \left(V_{max}^2 - V_{min}^2 \right) \qquad (1)$$

As shown in Fig. 5, the energy E, which is supplied from the EDLCs is designed to 56.5 kJ. Furthermore, the energy density of the EDLCs in the marketplace is 5.16 Wh/dm^3. When the energy density becomes more than 1.5 times of the present EDLCs, the energy source size of the proposed system can become smaller than that of the conventional systems. In other words, if the distance of 2/3 is considered in order to suppress the total energy

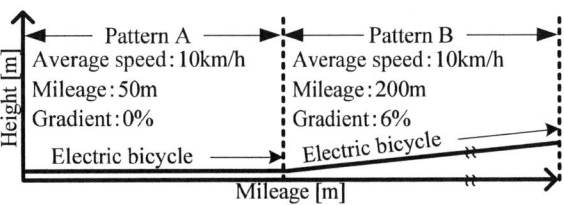

Fig. 2. Assist pattern of driving test.
The assist assume only two patterns.

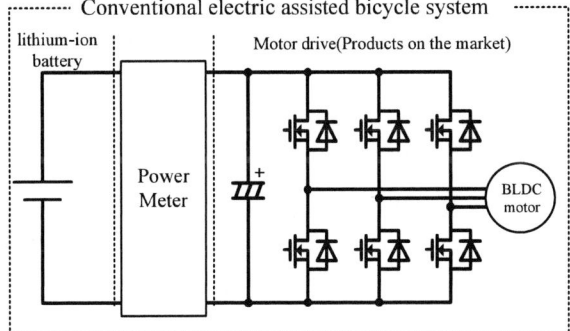

Fig. 3. Method of measuring the cumulative energy of each assist pattern.

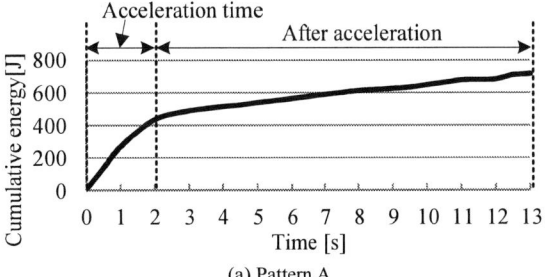

(a) Pattern A.

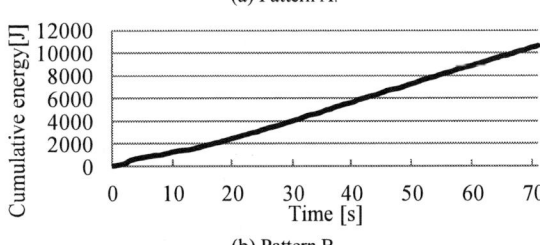

(b) Pattern B.
Fig. 4. Time characteristics of the cumulative energy of each assist pattern. The cumulative energy is measured based on the conditions of Figure 2.

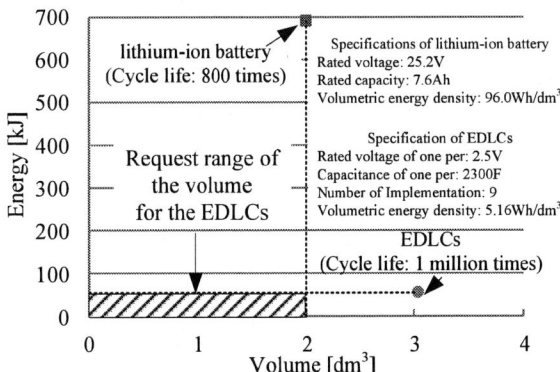

Fig. 5. Relationship between energy and volume of EDLCs and lithium-ion battery.
The request range of volume for EDLCs is shown.

capacity, the proposed system becomes useful. Therefore, operation verification of the system is discussed as a 200m distance in Chapter 6.

IV. DESIGN OF ANTENNA FOR WIRELESS POWER TRANSMISSION AND CHARGER

A small receiving antenna is preferred in order mounting to the electric bicycle. Therefore, the authors have been proposed a formula spiral antenna made of a printed circuit board, which is flat, compact and easy design.

A. Design of Wireless Charger

Fig. 6 shows the structure and equivalent circuits of the transmitting antenna for wireless power transmission. The connection point of the short type and open type is the output of a RF power supply. For the receiving antenna, the equivalent circuit is the same. The antenna has a two-layer structured wiring in order to increase the inductance value.

Fig. 7 shows the design flow chart of the charger and the antenna for the wireless power transmission. From Fig. 7, the charging power is determined by the charging time and the charging energy. The antenna is designed from the specifications of the antenna size and the charging power. The charger considers the short charging time and the volume of power conversion system for the charging power. The target of charging time T is 60 sec, and the charging energy E is supplied to 56.5 kJ. In addition, the average output power of the charger E/T is 1 kW. However, the size of the circuit becomes large when the charging circuit combines with the discharging circuit. Therefore, the charging control should be placed in RF power supply of the power transmission side of the wireless power transfer system is realistic.

Also, the antenna size depends on the mounting position of the power receiving side. The antenna for wireless power transmission is experimentally compared the short type and open type with the same size.

B. Comparison of Transmission Efficiency

Fig. 8 shows the system configuration of the experiments. Arbitrary frequency is output using a function generator. The output power cannot output 1 kW(The average output power of the charger) because of capacity limit of RF power supply. Therefore, the output of the RF power supply is scaled to 100W (10% of 1kW). Reflected power P_R is measured using a power meter in the front stage of the power transmission antenna. Also, the antenna of short type is connected in series with the capacitor of capacitance 500 pF.

Fig. 9 shows the experimental results of the frequency characteristics of the transmission efficiency. The power is amplified by the amplifier. The antenna is designed according to the flowchart of Fig. 7. From Fig. 9, it is confirmed that the short type can realize the low resonance frequency compared to the open type at the same transmission efficiency and antenna size. In general,

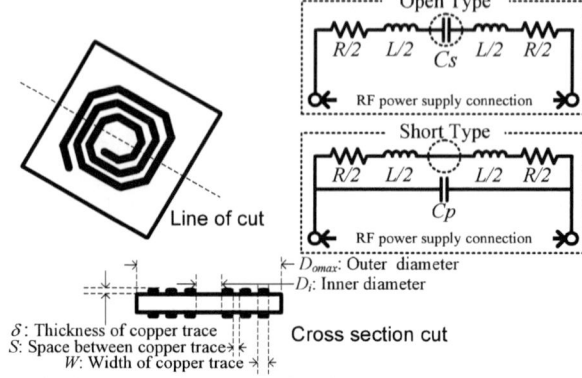

Fig. 6. Structure of the antenna for wireless power transmission. The antenna is a printed circuit board of the two layers.

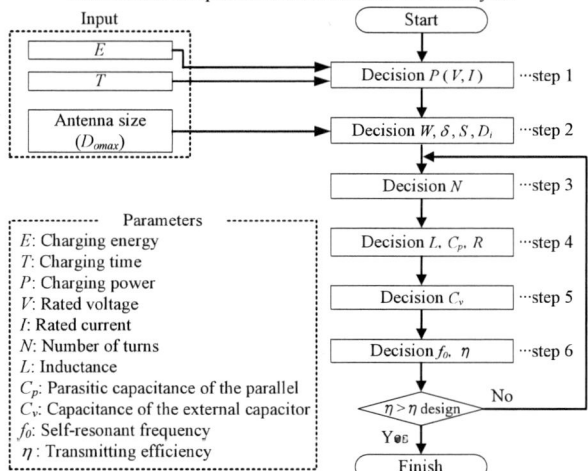

Parameters
E: Charging energy
T: Charging time
P: Charging power
V: Rated voltage
I: Rated current
N: Number of turns
L: Inductance
C_p: Parasitic capacitance of the parallel
C_v: Capacitance of the external capacitor
f_0: Self-resonant frequency
η: Transmitting efficiency

Fig. 7. Design flow chart of charger and antenna for wireless power transmission. Specifications of the antenna is determined by the antenna size and the charging power.

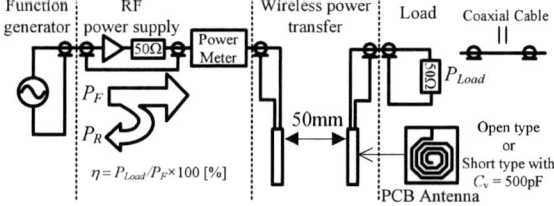

Fig. 8. System configuration of the experiments. The system of the experiment circuit is matched at 50Ω.

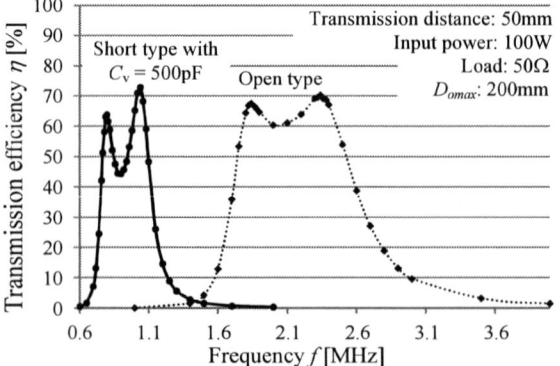

Fig. 9. Experimental results of the frequency characteristics of the transmission efficiency.
It is the result of the experiment under the same conditions.

The 2014 International Power Electronics Conference

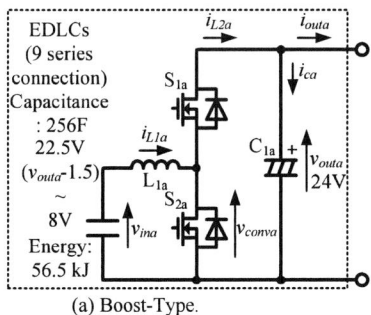 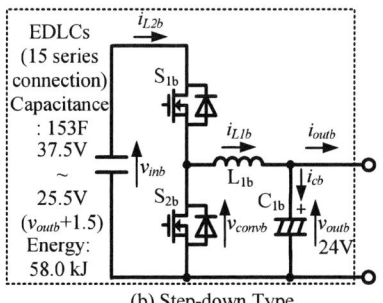 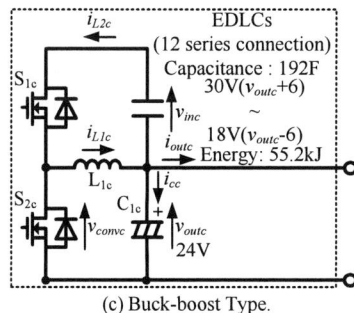

(a) Boost-Type.   (b) Step-down Type.   (c) Buck-boost Type.

Fig. 10. Investigated circuit configuration.
The EDLCs of a circuit of three types are designed with the same energy.

If resonant frequency (Frequency of RF power supply) is the low frequency, the efficiency of RF power supply is high. Therefore, the short type can be high efficiency of the system than the open type. So, the proposed system is adopted the short type.

V. DESIGN OF THE INTERFACE CONVERTER FOR EDLCs

A. Circuit configuration to be compared

Fig. 10 shows the circuit configuration of three kinds of DC-DC converters. This paper investigates the step-down type, boost-type and buck-boost type as the charging and discharging DC-DC converter by using EDLCs. The number and the output voltage range of the EDLCs are designed when the charging and discharging energy of EDLCs meet more than 53.0 kJ. Thus, the DC-DC converters are designed that the charging and discharging energy of EDLCs is as follows: the energy of the boost-type, step-down type and buck-boost type are 56.5 kJ, 58.0 kJ and 55.2 kJ, respectively. In addition, the specification of the input voltage of the DC-AC converter is 24 V (V_{outa}, V_{outb}, V_{outc}), and the output voltage of the DC-DC converter is controlled to 24 V at constant. Further, the reactors of all types are designed to the current ripple value (1.5 A) as same as the step-down type. Therefore, the reactors of all types are designed that the current ripple of the reactor meets less than 1.5A. On the other hand, the smoothing capacitor is designed from the allowable ripple current.

In this system, by the force step on the pedal of a bicycle, the energy, which is used for an assist, and the load of BLDC motor are changed. It is necessary to increase the capacitance of the smoothing capacitor in order to reduce the variation in the output voltage of the DC-DC converter owing to load variations. However, the volume of the smoothing capacitor should be reduced in terms of volume implementation. Therefore, when the load fluctuations occur, it is important to reduce the volume of the smoothing capacitor by implementing the fast control response of the output voltage [11].

B. Comparison of Power Loss

Fig. 11 shows the power loss analysis of three kinds of DC-DC converters when the voltage of the EDLCs V_{in} is changed. The power loss is analyzed under the conditions that the output power of the DC-DC converter is 384 W.

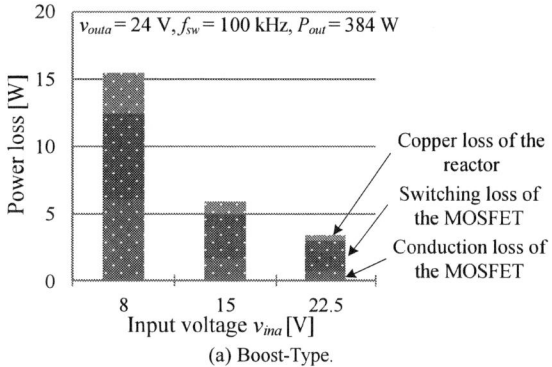

(a) Boost-Type.

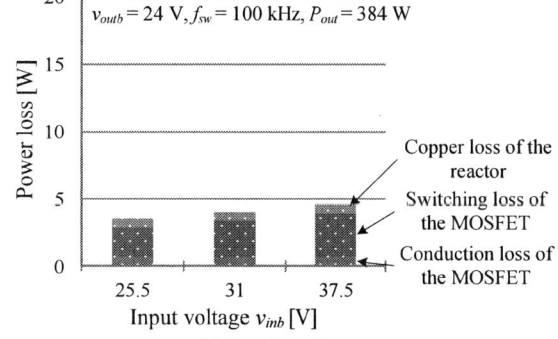

(b) Step-down Type.

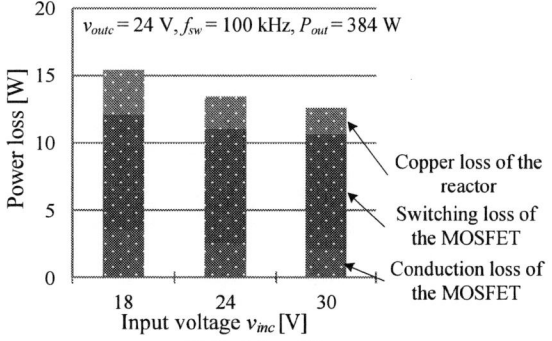

(c) Buck-boost Type.

Fig. 11. Results of power loss analysis of the three kinds of DC-DC converters.
The power loss is neglected iron loss of the reactor,
the conduction loss of the diode, recovery loss.

It is noted that dead time is neglected. As shown in Fig. 11, it is confirmed that the switching loss and the conduction loss of the MOSFET dominates the total loss

978-1-4799-2706-7/14 $31.00 © 2014 IEEE   2280

in the boost-type. Also, it is confirmed that the switching loss of the MOSFET is dominates the total losses in the step-down type and buck-boost type. This is because the input current of the boost-type is increased owing to the input voltage, which is lower than the step-down type and buck-boost type.

C. Comparison of Total Volume

Fig. 12 shows the relationship between the total volume and the output power for the three kinds of DC-DC converters. It is noted that the power loss is maximum for the voltage of EDLCs in Fig. 11. In addition, the volume of the Fig. 12 is the sum of volume of EDLCs and DC-DC converter (Reactor, Heat sink, Electrolytic capacitor).

In this paper, CSPI (Cooling System Performance Index), which is a reciprocal of the product of the volume and the thermal resistance, is introduced to estimate the volume of cooling system. The CSPI indicates the cooling performance per unit volume of the cooling system. It means that a high performance cooling system shows high CSPI. Therefore, the cooling system is miniaturized when CSPI become higher. The volume of the cooling system $vol_{cooling}$ is given by (2) from the relationship between the power loss and the rise in temperature [12].

$$vol_{cooling} = \frac{1}{R_{th} \times CSPI} = \frac{P_{loss}}{(T_j - T_a) \times CSPI} \quad (2)$$

where R_{th} is the thermal resistance of the cooling system, T_j is the junction temperature of the switching device, T_a is the ambient temperature, P_{loss} is the power loss of the switching device.

In this paper, the reactor is designed by the Area Product concept [13] using a window area and a cross sectional area. Then volume of the reactor vol_L is given by (3).

$$vol_L = K_v \left(\frac{2W}{K_u B_m J} \right)^{\frac{3}{4}} \quad (3)$$

where K_v is the constant value depending on the shape of cores, W is the maximum energy of the reactor, K_u is the occupancy of the window, B_m is the maximum flux density of the core, and J is the current density of the wire.

The capacitor volume is calculated based on commercially available electrolytic capacitors [14].The volume of the electrolytic capacitor is proportional to the rms value of the ripple current of the electrolytic capacitor. The volume vol_{CE} of the electrolytic capacitor is given by (4).

$$vol_{CE} = \gamma^{-1}_{VCE} I_{CRMS} \quad (4)$$

where γ^{-1}_{VCE} is the proportionality factor between the rms value of the ripple current and the volume, and I_{CRMS} is the rms value of the ripple current of the electrolytic capacitor.

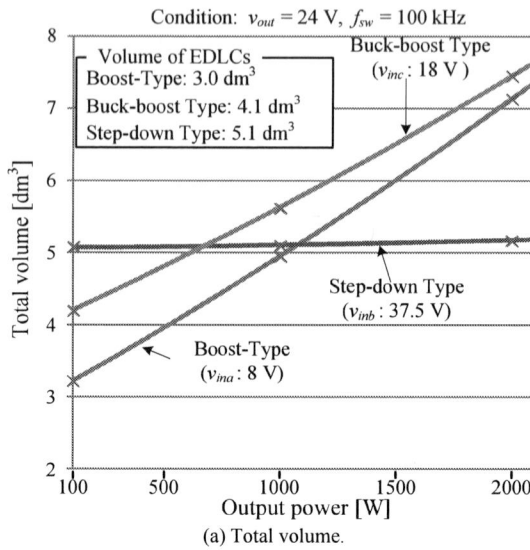

(a) Total volume.

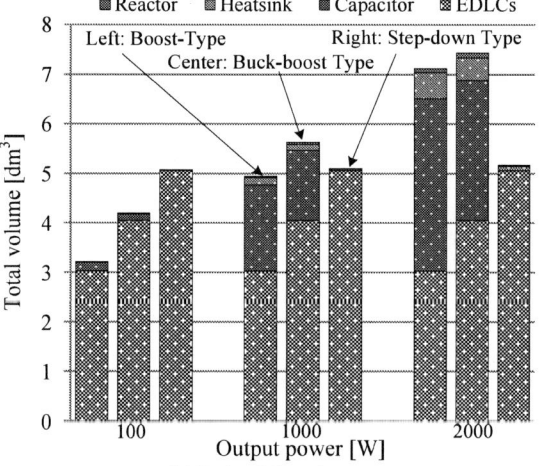

(b) Ratio of the volume.

Fig. 12. Relationship between the total volume and the output power for the three kinds of DC-DC converters.

From Fig. 12 (b), around the output power of 100 W, the volume of the EDLCs is dominant in the converter of all types. The number of EDLCs for the boost-type is the fewest in all of other DC-DC converters. Therefore, the volume of the boost-type is the smallest. However, if the output power is 1.1 kW or more than 675 W, the volume of the step-down type is smaller than that of the boost-type and the buck-boost type. For the boost-type and the buck-boost type, the ripple current of the smoothing capacitor is increased when the output power is increased. Therefore, the ratio of the volume of the smoothing capacitor is increased. On the other hand, for the step-down type, the ripple current of the smoothing capacitor does not depend on the output power. Therefore, when the output power is increased, therefore, when the output power is increased, the volume of heat sink and reactor which is very small compared to the volume of the EDLCs is increased. However, the volume of the converter does not increase too much. Furthermore, if the output voltage of the DC-DC converter is set to over 24V, the cross over point of the volume of three kinds of DC-

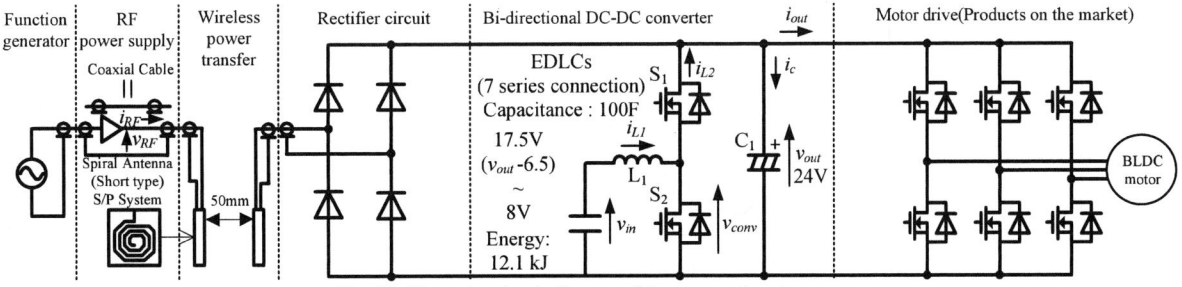

Fig. 13. The entire circuit diagram of the proposed system.
The proposed system is designed based on the system design to chapter 5.

DC converter is shifted toward high output power. The reason is that the volume of the capacitor is decreased depends on its ripple current according to the high output voltage of the DC-DC converter. Also, boost-type is the most compact in the power capacity (384W) of the electric assisted bicycle. So, the proposed system is adopted the boost-type.

VI. CHARGING VERIFICATION OF SYSTEM

A. Design of the whole system

Fig. 13 shows the circuit diagram of the entire system based on the system design to chapter 5. By the system design, the total volume of boost type is most minimum in three kinds of DC-DC converters at the maximum power (384W) of electric assisted bicycle. Therefore, boost type is adopted for the bi-directional DC-DC converter. In the antenna for wireless power transmission, the short type can be high efficiency of the RF power supply than the open type. Therefore, the antenna of short type is adopted. The running distance of a slope is more than 200 m as prototype specifications in this paper. Thus, the energy of the EDLCs (Nippon Chemi-Con Co, DDLE2R5LGN701KAA5S) is designed in 12.1 kJ [10]. In addition, the specification of the charging time is 1 minute. The wireless power transfer system is adopted the S/P system [15].

Fig. 14 shows the equivalent circuit diagram of the S/P system. The transmitting antenna is connected to the RF power supply. The receiving antenna is connected to the rectifier.

Table I shows the specification of antenna for wireless power transmission. The designed coupling factor and the self-inductance are measured by a LCR meter. The capacity C_2 of the parallel capacitor is given by (5). The capacity C_1 of the series capacitor is given by (6).

$$C_2 = \frac{1}{L_2 \omega_0^2} \tag{5}$$

$$C_1 = \frac{1}{(1 - k_0^2) L_1 \omega_0^2} \tag{6}$$

where L_2 is the self-inductance of receiving antenna, ω_0 is the resonance angular frequency, k_0 is the designed coupling factor, L_1 is the self-inductance of transmitting antenna.

Fig. 15 shows the control block diagram of wireless charging. In the S/P system, If the voltage of transmitting

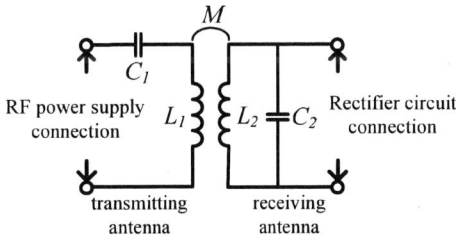

Fig. 14. Equivalent circuit diagram of the S/P system.
The capacitor is connected to an external antenna.

Table I. Specification of antenna.

Items	Values		Remarks
Outline (Length)	250	mm	
Outline (Side)	200	mm	
Number of turns (surface)	10	turn	Short type
Number of turns (reverse face)	10	turn	Short type
Space between copper trace	5	mm	
Width of copper trace	3	mm	
Thickness of PCB	2	mm	FR-4
Thickness of copper trace	70	μm	
Self-inductance L_1	58.27	μH	
Self-inductance L_2	58.70	μH	
Mutual inductance M	21.95	μH	
Designed coupling factor k_0	0.375		
Primary capacitance C_1	586	pF	
Secondary capacitance C_2	500	pF	

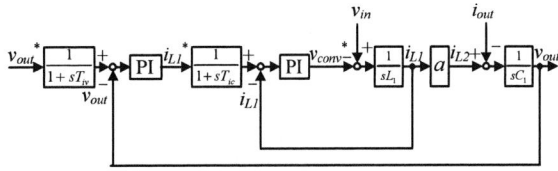

Fig. 15. Control block diagram.
The output voltage of bi-directional DC-DC converter is controlled to a constant 24V.

Table II. Conditions of the experiment.

Items	Values		Remarks
Frequency of the FG	929	kHz	
Resonant frequency	929	kHz	
Voltage of the FG	450	mVrms	
Gain of the RF power source	100		
Transmission distance	50	mm	

978-1-4799-2706-7/14 $31.00 © 2014 IEEE

antenna is not controlled, the voltage of the power receiving antenna is varied by variation of the coupling coefficient and load [16]. In this problem, the breakdown voltage of the switching elements must be designed by taking into account the variation of the coupling coefficient and the load. Therefore, charging control by the combination of two points as below is proposed.

1) The output voltage of the bi-directional DC-DC converter is controlled to be constant.

2) The output voltage of the RF power supply is set higher than the output voltage of the bi-directional DC-DC converter.

Thus, the design of the breakdown voltage of the switching device can be simplified. The charging experiment is based on the experimental conditions shown in Table II.

B. Experimental results of charging

Fig. 16 shows the experimental results of the wireless charging to the EDLCs.

Fig. 16(a) shows the voltage and current waveform of the input and output for the bi-directional DC-DC converter. From Fig. 16 (a), the output voltage of the bi-directional DC-DC converter is controlled to be constant, and the EDLCs are charged.

Fig. 16(b) shows the energy storage capacity of EDLCs from the start of charging to the end of charging. From Fig. 16 (b), the energy capacity of the EDLCs are charged from 0% to 100% in about 6 minutes is confirmed. The reason that charging time is not within 1 minute is for the capacity limit of the RF power supply used in experiments of this paper. Therefore, the charging can be achieved a full charging in 1 minute using the high-capacity RF power supply.

Fig. 16(c) shows the output voltage and current waveforms of the RF power source, input voltage waveform of bi-directional DC-DC converter, voltage waveform of EDLC (The voltage of EDLC is at the 8V). From Fig. 16 (c), the input power factor of the high frequency power source is approximately 1 is confirmed.

Fig. 17 shows the comparison of the charging time for charging the 12.1kJ of the proposed system and conventional system. The charging energy of the conventional system is calculated from specification of the product. From Fig. 17, the proposed system can shorten to 1/4 the charging time of the conventional system.

C. Wireless charging method

Fig. 18 shows the wireless charging method to the electric assisted bicycle. The receiving antenna is connected to the sides of the front basket of the electric assisted bicycle. The proposed system is introduced to a charging side. There are the following advantages by this method.

1) The foreign body (Dust, etc) is less likely to adhere to the power transmitting and receiving antenna, as compared to a system for power transmission from the ground (Therefore, it does not affect the transmission

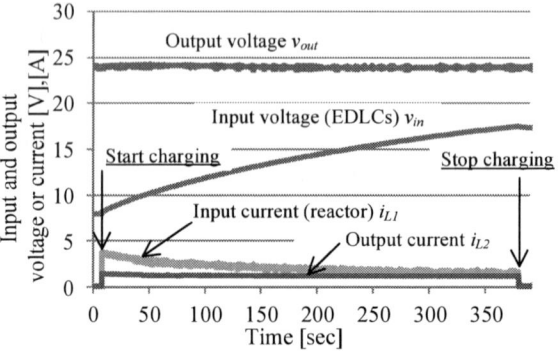

(a) voltage and current waveform of the input and output.

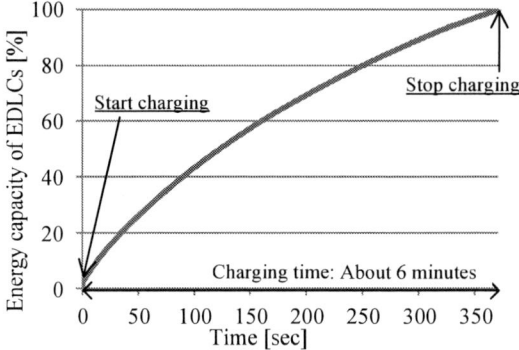

(b) Energy capacity of EDLCs.

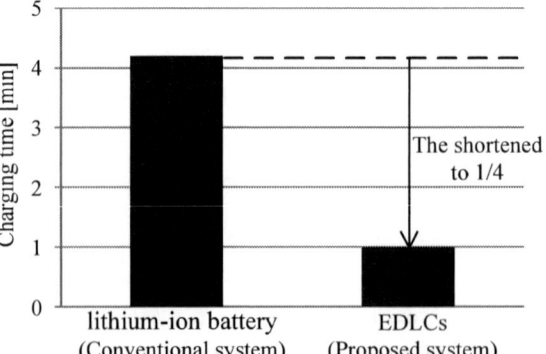

(c) Waveform(Voltage of EDLCs is at the 8V).
Fig. 16. Experimental results of the wireless charging to the EDLCs The energy capacity of EDLCs is the wireless charging in about 6 minutes.

Fig. 17. Comparison of the charging time for charging the 12.1kJ of the proposed system and conventional system. The proposed system can be charged rapidly than conventional system.

efficiency)

2) The variation of transmission distance and the positional deviation of between the transmitting and receiving antenna can be reduced by the front wheel is fixed using bicycle parking stand.

3) It is low cost because the antenna and wiring are not placed into the ground.

VII. CONCLUSION

This paper evaluated the energy capacity of EDLCs, the antenna for wireless power transmission and the charger, the interface converter for EDLCs in proposed system configuration. The proposed system was experimentally verified as prototype.

The results showed that the short type as the antenna for wireless power transmission can be to system of higher efficiency than the open type with the same size. Also, the results showed that the total volume of the boost-type DC-DC converter is the smallest when the output power is less than 1.1 kW. Therefore, in terms of the volume of the system at the minimum point, the boost-type DC-DC converter is suitable for the electric assist bicycle. Also, the results showed that the proposed system can shorten to 1/4 the charging time of the conventional system.

In the future work, the proposed system will be experimentally verified running test in an actual road.

REFERENCES

[1] Y. Hori: "Future vehicle society based on electric motor, capacitor and wireless power supply", *IEEE IPEC2010*, pp.2930 – 2934 (2010)

[2] A. Govindaraj, M. King, S. M. Lukic: "Performance characterization and optimization of various circuit topologies to combine batteries and ultra-capacitors", *IEEE IECON2010*, pp 1850-1857 (2010)

[3] K. Noguchi, K.Orikawa, J. Itoh: "System Design of Electric Assisted Bicycle using the EDLCs as Power Source", *2013 Japan-Korea Joint Technical Workshop on Semiconductor Power Conversion*, No. IEEJ-SPC-P2-21 (2013) (in Japanese).

[4] A. Kurs, A. Karalis, R. Moffatt, J. D. Joannopoulos, P. Fisher, M. Soljacic: "Wireless Power Transfer via Strongly Coupled Magnetic Resonances", *Science*, Vol. 317, pp. 83-86 (2007)

[5] S. Lee, R. D. Lorenz: "Development and Validation of Model for 95%-Efficiency 200-W Wireless Power Transfer Over a 30-cm Air-gap", *IEEE Trans. On Industry Applications*, Vol. 47, No. 6, pp. 2495-2504 (2011)

[6] C.S. Wang, O.H. Stielau, and G.A. Covic: "Design Considerations for a Contactless Electric Vehicle Battery Charger", *IEEE Trans. On Industrial Electronics*, Vol. 52, No. 5, pp. 1308-1314 (2005)

[7] T. Imura, Y. Hori: "Maximizing Air Gap and Efficiency of Magnetic Resonant Coupling for Wireless Power Transfer Using Equivalent Circuit and Neumann Formula", *IEEE Trans. On Industrial Electronics*, Vol. 58, No. 10, pp. 4746-4752 (2011)

[8] A. P. Sample, D. A. Meyer, J. R. Smith: "Analysis, Experimental results, and Range Adaptation of Magnetically Coupled Resonators for Wireless Power Transfer", *IEEE Trans. On Industrial Electronics*, Vol. 58, No. 2, pp. 544-554 (2011)

[9] Yamaha Motor Co., Ltd. PAS wagon. (http://www.yamaha-motor.jp/pas/lineup/wagon/)

[10] Nippon Chemi-Con Co. (http://www.chemi-con.co.jp/catalog/dl.html)

[11] T. Shibuya, J. Itoh: "An Evaluation of optimal Design of Capacitance by High Speed of a Control Response", *SPC Osaka*, No. SPC-12-026 (2012) (in Japanese).

[12] U. DROFENIK, G. LAIMER, and J. W. KOLAR: "Theoretical Converter Power Density Limits for Forced Convection Cooling" *International PCIM Europe Conference*, pp.608-619 (2005)

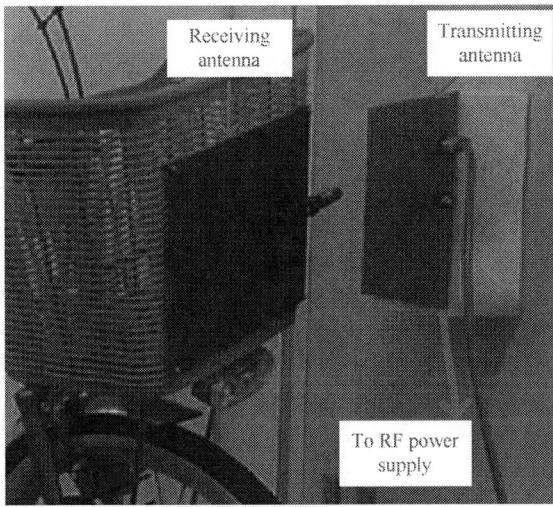

(a) Attached photo of the antenna.

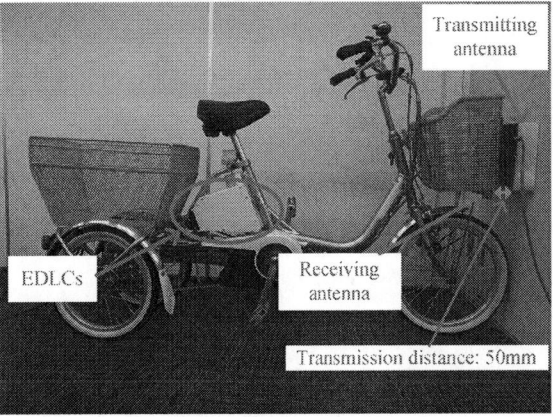

(b) The entire photo.

Fig. 18. Wireless charging method to the electric assisted bicycle.

[13] Wm. T. Mclyman: "Transformer and inductor design handbook", Marcel Dekker Inc. (2004)

[14] Y. Kashihara, J. Itoh: "Parformance Evaluation among Four types of Five-level Topologies using Pareto Front Curves", *IEEE ECCE 2013*, pp.1296-1303 (2013)

[15] S. Lee, R. D. Lorenz: "A Design Methodology for Multi-kW, Large Airgap, MHz Frequency, Wireless Power Transfer Systems", *IEEE ECCE 2011*, pp. 3503-3510 (2011)

[16] K. Takuzaki, N. Hoshi: "Consideration of Operating Condition of Secondary-side Converter of Inductive Power Transfer System for Obtaining High Resonant Circuit Efficiency", *IEEJ Trans. IA*, Vol. 132, No. 10, pp. 966-975 (2012) (in Japanese).

Study on Low-loss Gate Drive Circuit for High Efficiency Server Power Supply using Normally-off SiC-JFET

Kaoru Katoh
Department of Power Electronics Systems Research
Hitachi Research Laboratory, Hitachi, Ltd.
Hitachi, Japan
kaoru.katoh.ry@hitachi.com

Katsumi Ishikawa
Process Designing Dept.
Mito Rail Systems Product Div., Rail Systems Company,
Hitachi, Ltd.
Hitachinaka, Japan
katsumi.ishikawa.zw@hitachi.com

Ayumu Hatanaka
Department of Power Electronics Systems Research
Hitachi Research Laboratory, Hitachi, Ltd.
Hitachi, Japan
ayumu.hatanaka.sv@hitachi.com

Kazutoshi Ogawa
Department of Power Electronics Systems Research
Hitachi Research Laboratory, Hitachi, Ltd.
Hitachi, Japan
kazutoshi.ogawa.ev@hitachi.com

Satoru Akiyama
Department of Energy Electronics Research
Central Research Laboratory, Hitachi, Ltd.
Kokubunji, Japan
satoru.akiyama.ma@hitachi.com

Takashi Ogawa
Department of Communication Electronics Research
Central Research Laboratory, Hitachi, Ltd.
Kokubunji, Japan
takashi.ogawa.xk@hitachi.com

Natsuki Yokoyama
Department of Energy Electronics Research
Central Research Laboratory, Hitachi, Ltd.
Kokubunji, Japan
natsuki.yokoyama.cz@hitachi.com

Naoki Maru
Hardware Technology Development Dept.
IT Platform R & D Management Division, Hitachi, Ltd.
Hadano, Japan
naoki.maru.hr@hitachi.com

Osamu Takahashi
Hardware Technology Development Dept.
IT Platform R & D Management Division, Hitachi, Ltd.
Hadano, Japan
osamu.takahashi.mh@hitachi.com

Koji Nishisu
Hardware Technology Development Dept.
IT Platform R & D Management Division, Hitachi, Ltd.
Hadano, Japan
koji.nishisu.qm@hitachi.com

Abstract— **We investigated how to reduce the energy loss of server power supplies equipped with vertical-trench normally-off Silicon Carbide junction-gate field-effect transistors (SiC-JFETs). High-speed driving circuits consisting of a speed-up capacitor with separated source terminal and timing adjust circuits to ensure a dead time margin are proposed. Applying the developed normally-off SiC-JFETs and the proposed gate driver to PFC circuits and DC/AC circuits resulted in, an increase of server power supply efficiency to 95.10%.**

Keywords— *Silicon Carbide (SiC), junction-gate field-effect transistors (JFETs), normally-off, high-speed driving circuits, server power supply .*

I. Introduction

The amount of electricity use in U.S. data centers increases every year (Figure 1[1][2][3]). Among the power consumption in these data centers, power supply loss typically accounts for 20%. Applying Silicon Carbide (SiC) as a next-generation power semiconductor is expected to reduce this loss. Thus, we have promoted the development of server power supplies equipped with normally-off SiC junction-gate field-effect transistors (JFETs).

SiC-JFETs are promising candidates for low-resistance high-voltage switching devices. JFETs have high electron mobility and no critical problems related to the SiC/oxide interface, which is in contrast to SiC metal-oxide-semiconductor field-effect transistors (SiC MOSFETs), which have inherent problems concerning device reliability [4]. Given this advantageous feature of JFETs, we previously developed 600-V normally-off JFETs [5][6] to enhance the current density and blocking voltage by localized current-path doping and to reduce their inherent gate leakage current by surface oxynitridation.

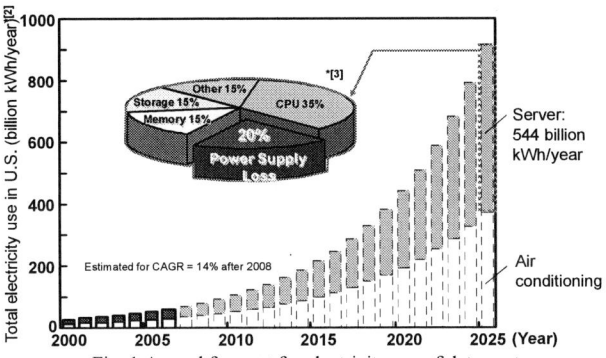

Fig. 1 Annual forecast for electricity use of data center

Further, from the viewpoint of the starting time of the power supply of an inverter and the fail-safe function of a control source, a normally-off type switching device is in strong demand from the system side. However, normally-off SiC-JFETs have several problems that make it difficult to reduce the power loss of the server power supply.

In this work, we carefully designed the optimum gate-drive voltage for reducing the gate driving loss, taking the on-state resistance characteristics and gate leakage current into account. We also developed high-speed driving circuits consisting of a speed-up capacitor and timing adjust circuits for ensuring dead time margins to prevent arm shorts. Applying the normally-off SiC-JFETs with separate source terminals [7][8] and the proposed gate driver for Power Factor Correction (PFC) circuits and DC/AC inverter circuits enabled us to achieve a high-efficiency server power supply.

II. Server Power Supply

A. Overview

Figure 2 shows an overview of a 2-kW server power supply applied with normally-off SiC-JFETs. This power supply consists of filters, AC/DC rectifiers, a PFC, and DC/DC rectifiers. Figure 3 shows a circuit diagram of the PFC and the

DC/AC inverter assuming that the JFETs are used to reduce their conduction loss in the power supply. In the case of the PFC and the DC/AC inverter of the server power supply, zero voltage switching (ZVS) is applied in order to attain high efficiency. Therefore, it is important to operate ZVS normally when the JFETs are used in the PFC and the inverter.

It is also possible to realize high efficiency by higher frequency which is an advantage of SiC. However, server power supplies are significantly affected by the increase of the high frequency core loss of the transformer. Therefore, in this work, the switching frequencies of the PFC and DC/AC inverter are used with their present conditions.

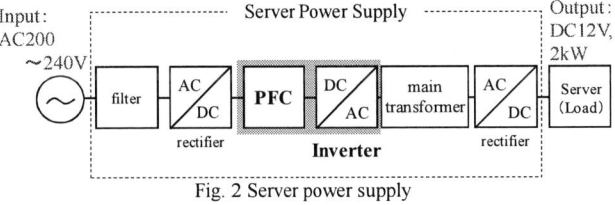

Fig. 2 Server power supply

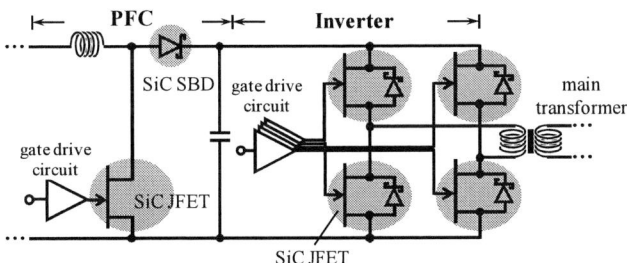

Fig. 3 Block diagram of PFC and DC/AC inverter

B. 600-V normally-off junction-gate field-effect transistor

A schematic cross-sectional view of the developed JFET structure [6] is shown in Figure 4. The substrate is an n-type 4H-SiC wafer (n+ sub) including an 8-µm-thick epilayer with a doping level of $2.0 \times 10^{16} \mathrm{cm}^{-3}$. A gate trench structure was fabricated by dry etching, followed by ion implantation for the p+ gate region. Localized current-path doping (LCD) by nitrogen ion implantation at a tilt was performed to achieve both a high-current density and high blocking voltage at a trench depth D_{CH} of 1.4 µm and source width W_{CH} of 0.9 µm. After dopant activation annealing, surface oxynitridation (SON) was performed to reduce the gate leakage current between the p+ gate and n+ source. Figure 5 shows the dependence of the measured drain current I_D for the developed JFET on drain voltage V_D. Owing to the LCD and SON technique, the fabricated $3.9 \times 3.9 \mathrm{mm}^2$ JFET chip depicted in Fig. 5 has a low on-state resistance of 41 mΩ even at 100°C.

Figure 6 (a) shows the developed JFETs and our 600-V 40-A developed SiC Schottky barrier diode (SBD) packaged in a transfer-mold package with separate source terminals [7][8]. Using the separate source terminals, the influence on the gate

978-1-4799-2706-7/14 $31.00 © 2014 IEEE

drive voltage is small even though the drain current of the SiC-JFET makes fast transitions since the source inductance is divided. The measured electrical characteristics of the JFET and SBD are shown in Fig. 6 (b).

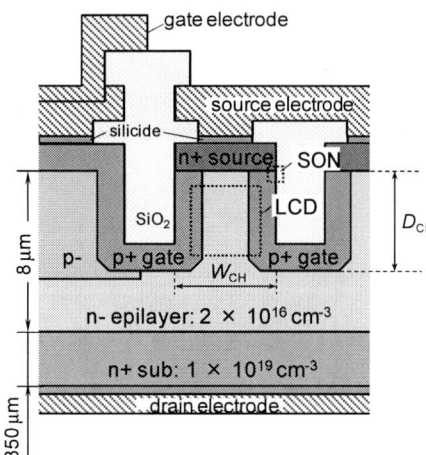

Fig. 4   Schematic cross-sectional view of developed SiC JFET [5, 6]

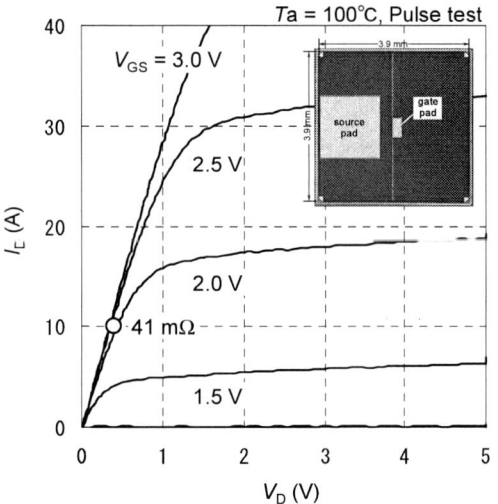

Fig. 5 Dependence of drain current I_D on drain voltage V_D

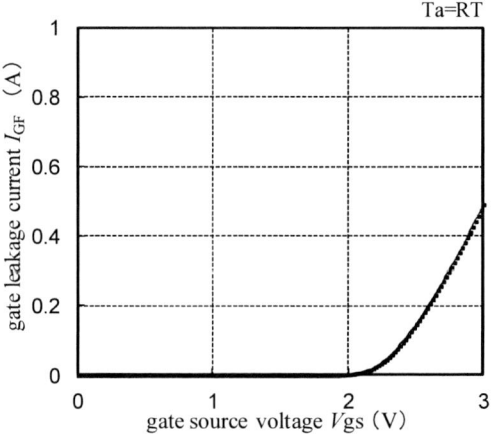

	Item		Unit	Measured data
VJFET	#1	$R_{DS(ON)}$	mΩ	41
	#2	$V_{(BR)DSS}$	V	>700
	#3	I_{GF}	mA	1.3
SBD	#4	V_F	V	1.5
	#5	V_R	V	950

Conditions: #1: I_D = 10 A, V_{GS} = 2.1 V, Ta = 100 °C
#2, #4: I_D = 125 μA, #3: V_G = 2.1 V, Ta = 100 °C
#4, #5: I_F = 40 A

(a)   (b)

Fig. 6 (a) Developed JFET and SBD packaged in transfer-mold package and (b) measured electrical characteristics

C.   Gate drive voltage

The gate-drive voltage must be carefully set to reduce power loss when the JFETs are incorporated into the server power supply because the on-state resistance $R_{DS(ON)}$ of the JFETs depends on the gate-drive voltage and the JFETs have a parasitic pn diode between the p+ gate and n+ source. Figure 7 shows the dependence of the forwarding gate leakage current I_{GF} on the gate voltage V_{gs}. As shown in Figure 7, in addition to the conduction loss, the gate leakage loss (∝ I_{GF}) is induced by a forwarding gate leakage current I_{GF} flowing from the gate of the JFETs to the source when the JFETs are turned on. The gate-drive voltage is thus set by considering the gate leakage characteristics of the JFETs. We estimated the power loss of SiC-JFETs by using the characteristics of $R_{DS(ON)}$ and I_{GF} when the server drives at a 50% load. Figure 8   shows the estimated result of the dependence of the loss on the gate drive voltage V_{DR}. In this case, the switching loss is excluded because the ZVS system reduces switching loss in the PFC and AC/DC inverter. As shown in the graph, the optimum gate-drive voltage was estimated to be 2.2 ± 0.2 V.

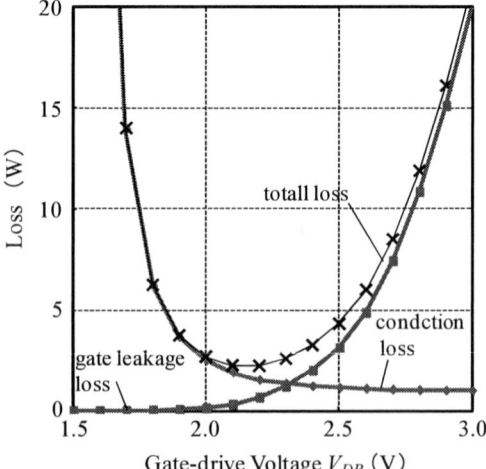

Fig. 7 Dependence of gate leakage current I_{GF} on gate voltage V_{gs}

Fig. 8 Dependence of estimated power loss on gate-drive voltage

978-1-4799-2706-7/14 $31.00 © 2014 IEEE

III. Gate Drive Circuit

When SiC-JFETs with a low threshold voltage (1.0 V) and large input capacitance (Ciss) are applied to a DC/AC inverter, it is difficult to drive fast without arm shorts. To solve this problem, we developed a driving method for the inverter circuit, as shown in Figure 9. The gate drive circuit consists of the timing adjust circuit and speed up circuit. Figure 10 shows the inverter circuit, and a schematic view of a Vgs operation waveform in the inverter circuit is shown in Figure 11. In (a) the Si-MOSFET, the drive voltage VG is generated by using only a positive voltage source, and the pulse voltage that repeats the positive voltage +VG, 0 V, and the negative voltage −VG through a pulse transformer is impressed by a gate. In this case, the inverter can keep a dead time to drive normally without arm shorts because the Si-MOSFET has a much higher threshold voltage (5 V) and much lower gate leak current thus the gate drive voltage can be set to a higher voltage. If (b) the SiC-JFET is applied to such an inverter, its low threshold voltage causes no dead time in the inverter, which results in arm shorts. Therefore, by using the drive circuit shown in Figure 9, the dead time can be kept by shortening the 0-V period of the drive voltage and adjusting the turn-on timing. Figure 11(c) shows a Vgs waveform applied to the gate drive circuit shown in Figure 9. When the SiC-JFET is turned off, by shortening the 0-V period of the drive voltage, Vgs can be changed to a negative voltage quickly. When the SiC-JFET is turned on, the rise timing of Vgs is delayed by the nMOS, C1, and R1 in the timing adjust circuit. At this time, switching loss does not occur since Vds has already fallen to 0 V in order to perform ZVS operation. Moreover, high-speed drive is performed by using a speed up capacitor Csp and a diode D1 with the timing adjust circuit in order to operate ZVS normally.

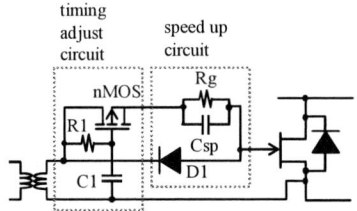

Fig. 9 Gate drive circuit for inverter

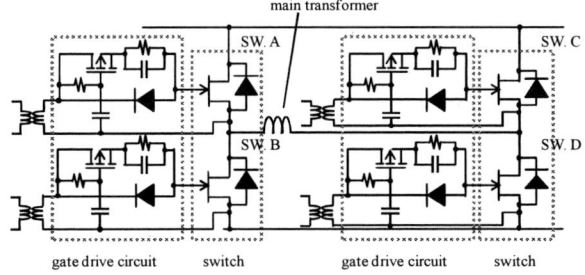

Fig. 10 Inverter circuit

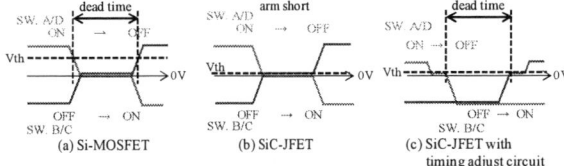

Fig. 11 Vgs waveforms for inverter circuit operation

IV. Results

Table 1 shows the turn-on and turn-off times of PFC with the applied normally-off SiC-JFETs with the proposed and conventional gate drive circuits. For the PFC circuit, the proposed gate drive circuit has a speed-up capacitor Csp and a gate resistance Rg with the separate source terminal while the conventional gate drive circuit has only a Csp and a Rg. The proposed gate-drive circuit reduces turn-on time by 40%.

Table 1 SiC-JFET switching time of PFC circuit

	conventional	**proposed**
ton (ns)	73.4	48.4
toff (ns)	39.5	35.3

Figure 12 shows the measured Vgs waveform for the inverter circuit. In the case of (a) the Si-MOSFET, the inverter could keep the dead time to drive normally. However, in the case of (b) the SiC-JFET using the conventional gate drive circuit, the inverter could not keep the dead time, and thus, arm shorts occurred. In comparison, in the case of (c) the SiC-JFET using the proposed gate drive circuit, including the timing adjust and speed-up circuit, even if Vgs was about 2.2 V during on mode, it was proven to keep dead time the same as the case with the Si-MOSFET.

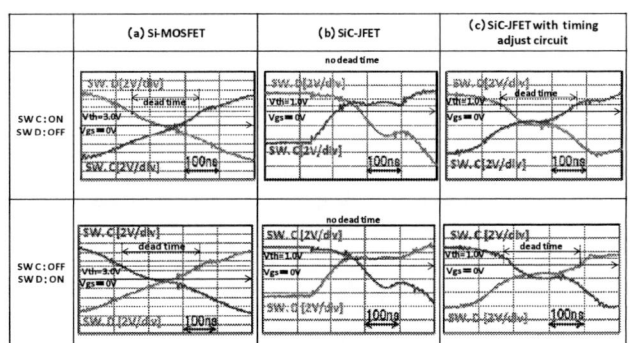

Fig. 12 Measured Vgs waveform for inverter circuit

We used the results above to measure the efficiency of the prototype server power supply by applying the normally-off SiC-JFETs and the proposed gate driver for PFC circuits and DC/AC circuits. Figure 13 shows the measurement results.

The efficiency of the prototype server power supply reached 95.10% at 50% load.

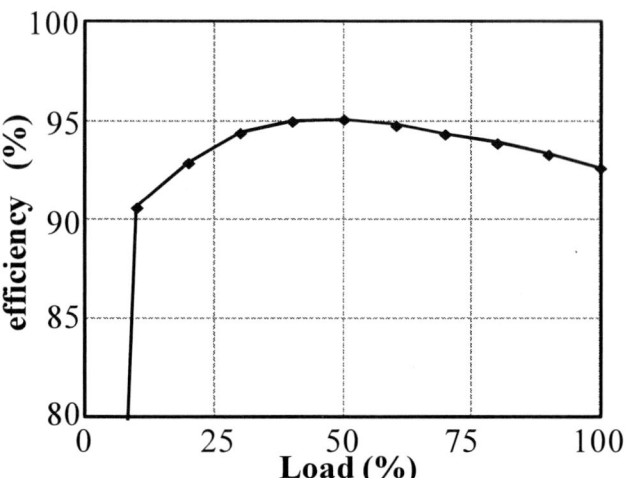

Fig. 13 Efficiency of prototype server power supply

v. Conclusion

We designed a gate-drive voltage for a vertical-trench normally-off SiC-JFET with the aim of fabricating a server power supply with 95.10% efficiency. The conduction loss and gate leakage loss was calculated using the on-state resistance and gate leakage current characteristics of SiC-JFETs. In addition, high-speed driving circuits by a speed-up capacitor and timing adjust circuits to ensure a dead time margin were proposed. A prototype server power supply with high efficiency was achieved by applying the normally-off SiC-JFETs with the proposed gate driver for PFC circuits and DC/AC circuits. These results indicate that precise gate-drive voltage design and high speed gate drive circuits for SiC-JFETs are essential for producing high-efficiency server power supplies.

Acknowledgments

Part of this work was conducted under a joint research contract with the New Energy and Industrial Technology Development Organization (NEDO) and R&D Partnership for Future Power Electronics Technology (FUPET).

References

[1] N. Yokoyama and K. Ishikawa, "R&D of power supply in data center using SiC power devices", *International Symposium on SiC Power Electronics 2012*, 2012.

[2] U.S. Environmental Protection Agency ENERGY STAR Program, Report to Congress on Server and Data Center Energy Efficiency Public Law 109-431.

[3] Denali Memcon '08 San Jose, Micron, "The Quest for Green: Don't Forget the Memory".

[4] M. Gurfinkel, Hao D. Xiong, Kin P. Cheung, John S. Suehle, Joseph B. Bernstein, Yoram Shapira, Aivars J. Lelis, Daniel Habersat, and Neil Goldsman, "Characterization of Transient Gate Oxide Trapping in SiC MOSFETs Using Fast I − V Techniques" , IEEE Trans. on El. Dev., vol. 55, no. 8, pp. 2004−2012, August 2008.

[5] H. Shimizu, Y. Onose, T. Someya, H. Onose, and N. Yokoyama, "Normally-Off 4H-SiC Vertical JFET with Large Current Density" , Material Science Forum, Vol. 600−603, p. 1059, 2008.

[6] H. Shimizu *et al.*, "600 V/27 mΩ Normally-off SiC JFET for High Efficiency Power Supply", *Extended Abstracts of the 2012 International Conference on Solid State Devices and Materials, Kyoto*, pp. 893−894, 2012.

[7] K Ishikawa *et al.*, "High-Speed Drive Circuit with Separate Source Terminal for 600 V/40 A Normally-off SiC-JFET", *Materials Science Forum*, vols. 740−742, pp. 1060−1064, 2013.

[8] S. Akiyama, K. Katoh, H. Shimizu, A. Hatanaka, T. Ogawa, N. Yokoyama, and K. Ishikawa, "Gate-drive Voltage Design for 600-V Vertical-trench Normally-off SiC JFETs toward 94% efficiency Server Power Supply", *The International Conference on Silicon Carbide and Related Materials 2013*, 2013.

A Short Circuit Protection Method Based on a Gate Charge Characteristic

Takeshi Horiguchi, Shin-ichi Kinouchi,
Yasushi Nakayama, and Takeshi Oi

Advanced Technology R&D Center
Mitsubishi Electric Corporation
Amagasaki, Hyogo, Japan
horiguchi.takeshi@aj.mitsubishielectric.co.jp

Hiroaki Urushibata

Department of Electrical and
Electronic Engineering
Kanazawa Institute of Technology
Nonoichi, Ishikawa, Japan

Shoji Okamoto, Shinji Tominaga,
and Hirofumi Akagi

Department of Electrical and
Electronic Engineering
Tokyo Institute of Technology
Meguro, Tokyo, Japan

Abstract—This paper describes a high-speed circuit to protect IGBTs against short-circuit faults. The reverse transfer capacitance depends on a collector–emitter voltage and it produces a significant effect on a switching behavior under short-circuit fault conditions as well as under normal conditions. A gate charge characteristic under short-circuit fault conditions differs from that under normal turn-on conditions. Hence, hard-switching fault (HSF) can be detected by monitoring both a gate–emitter voltage and an amount of gate charge. IGBTs can be rapidly protected from destruction because the protection circuit based on a gate charge characteristic does not require any blanking time. Fault under load can be also detected by almost the same circuit configuration. Simulation and experiment verify the validity of the novel protection circuit based on a gate charge characteristic.

Keywords—Fault under load, Hard-switching fault, Insulated gate bipolar transistor, Protection.

I. INTRODUCTION

Power semiconductor devices such as IGBTs have been used in various industrial applications. The protection of IGBTs against short-circuit failures is of great concern. Especially, a demand for a high-speed protection circuit for IGBTs subjected to a hard-switching fault (HSF) gets higher as a power density of power semiconductor devices is higher.

Many kinds of protection circuits have been reported in [1]–[8]. The collector–emitter-voltage monitoring method (V_{ce}-monitoring method) and the collector-current monitoring method (I_c-monitoring method) are well-known conventional ones [1]. The V_{ce}-monitoring method requires a long blanking time to detect the fault because v_{ce} softly drops toward the steady-state value $V_{ce(sat)}$ under normal conditions. The I_c-monitoring method is not affordable because it uses a current sensor such as a current transformer (CT).

The gate–emitter-voltage monitoring method (V_{ge}-monitoring method) has an advantage in terms of requiring neither high-voltage diode for sensing nor current sensor [2]–[4]. However, it requires a blanking time to distinguish HSF conditions from normal turn-on conditions. A combined of the V_{ge}-monitoring method with the V_{ce}-monitoring method can detect the fault during a turn-on transient period without setting a detection period, although it requires a high-voltage diode for sensing [5], [6].

An amount of gate charge under HSF conditions is smaller than that under normal turn-on conditions. The gate charge monitoring method (Q_g-monitoring method) has been reported [7]–[8]. The Q_g-monitoring method can detect the fault during turn-on transient and it is a cost-effective method. However, the threshold value of the gate charge for detecting HSF is predefined as a function of the gate–emitter voltage.

Since an amount of gate charge under HSF conditions is smaller than that under normal turn-on conditions, a gate charge characteristic under HSF conditions differs from that under normal turn-on conditions. The protection circuit based on a gate charge characteristic can detect HSF.

As for a fault under load (FUL) conditions, a gate–emitter voltage exceeds the gate control voltage, and an amount of gate charge is smaller than that under normal turn-on conditions due to a negative gate current. Hence, FUL can be also detected by the protection circuit based on a gate charge characteristic.

A physics-based IGBT model has high accuracy in a wide range of variations in parameters such as temperature, voltage and current [9]–[11]. Therefore, it is more useful than a conventional behavioral IGBT model in designing a protection circuit for IGBTs subjected to short-circuit faults.

This paper devises a novel high-speed protection circuit based on a gate charge characteristic, and verify the effectiveness of the sophisticated protection circuit by the simulation using the physics-based IGBT model and experiments.

II. NON-PUNCH-THROUGH IGBT MODEL

The IGBT is one of the conductivity-modulated devices and its behavior depends heavily on an excess carrier distribution in the drift region. Fig. 1 shows non-punch-through IGBT (NPT-IGBT) structure and interterminal capacitances [9]. The IGBT consists of pin diode, MOSFET and interterminal capacitances as shown in Fig. 1.

The one-dimensional ambipolar diffusion equation de-

The 2014 International Power Electronics Conference

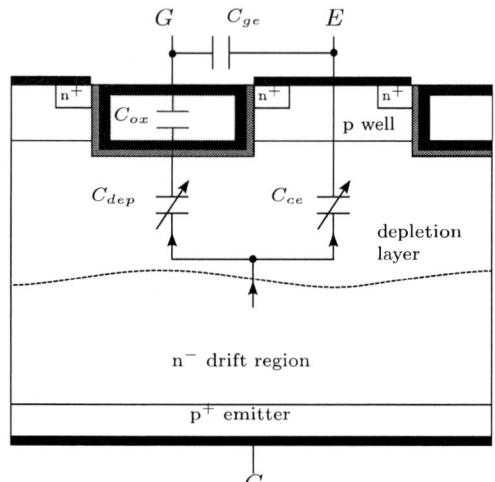

Fig. 1. IGBT structure and interterminal capacitances [9].

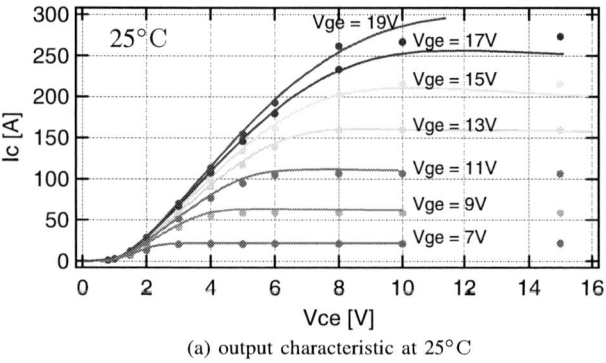

(a) output characteristic at 25°C

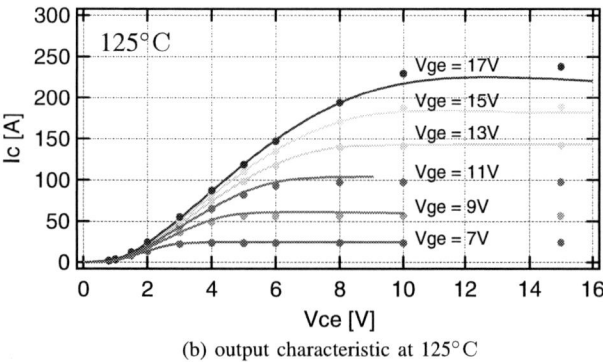

(b) output characteristic at 125°C

Fig. 2. Comparison in output characteristics between simulated results and experimental ones. (line: experiment, symbol: simulation).

scribes the charge dynamics within the drift region [10].

$$D\frac{\partial^2 p(x,t)}{\partial x^2} = \frac{p(x,t)}{\tau_{HL}} + \frac{\partial p(x,t)}{\partial t} \quad (1)$$

where D is the ambipolar diffusion coefficient, τ_{HL} is the high-level carrier lifetime, and $p(x,t)$ is the excess carrier concentration.

The boundary conditions are described by using the carrier

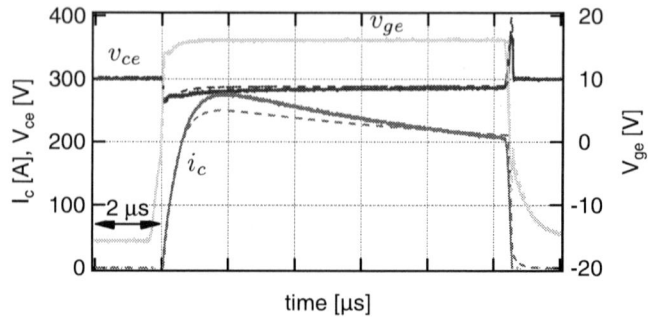

Fig. 3. Simulated results under hard-switching fault conditions compared with experiment ones (broken: simulation, solid: experiment).

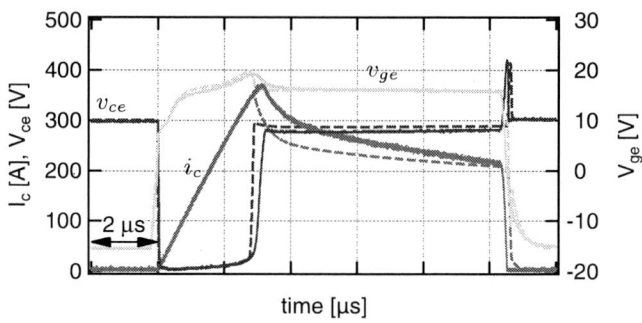

Fig. 4. Simulated results under fault under load conditions compared with experimental ones (broken:simulation, solid:experiment)

concentrations $p(x_1,t)$ at the collector end of the drift region x_1, and $p(x_2,t)$ at the MOS end of the drift region x_2.

$$\left.\frac{\partial p(x,t)}{\partial x}\right|_{x_1} = \frac{1}{2qA}\left(\frac{I_{n1}}{D_n} - \frac{I_{p1}}{D_p}\right)$$
$$\left.\frac{\partial p(x,t)}{\partial x}\right|_{x_2} = \frac{1}{2qA}\left(\frac{I_{n2}}{D_n} - \frac{I_{p2}}{D_p}\right) \quad (2)$$

where q is the electron charge, A is the effective cross-sectional area, D_n and D_p are the electron and the hole diffusion constants, I_{n1} and I_{p1} are the electron and hole currents at x_1, and I_{n2} and I_{p2} are the electron and hole currents at x_2.

For the collector end of the drift region, the electron current I_{n1} is described by

$$I_{n1} = qAh_p p(x_1,t)^2 \quad (3)$$

where h_p is the recombination parameter at the p$^+$ emitter.

For the MOS end of the drift region, the electron current I_{n2} is the MOS channel current. It is given by MOSFET equations in the saturation region

$$I_{n2} = \frac{K_{psat}}{2}(V_{gs} - V_{th})^2 \frac{1 + \lambda V_{mos}}{1 + \theta(V_{gs} - V_{th})}, \quad (4)$$

978-1-4799-2706-7/14 $31.00 © 2014 IEEE

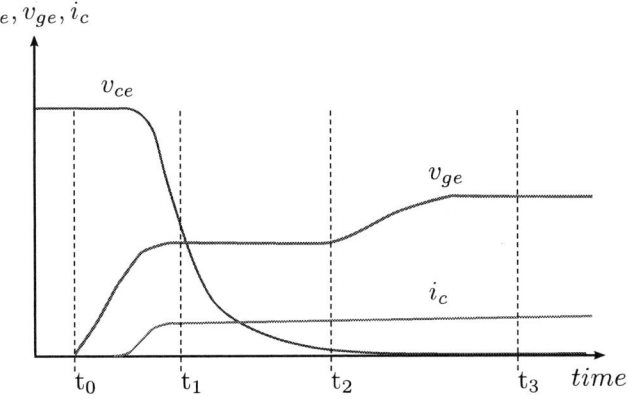

Fig. 5. Switching waveforms under normal conditions.

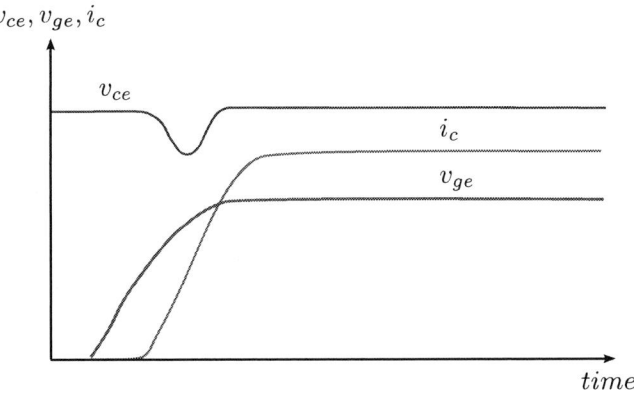

Fig. 6. Switching waveforms under hard-switching fault conditions.

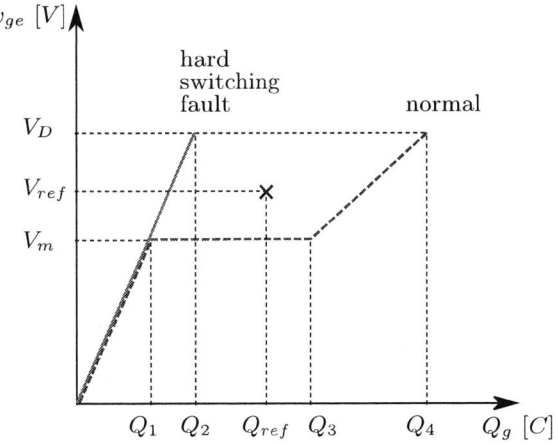

Fig. 7. Gate charge characteristics under normal conditions and hard-switching fault conditions.

while it is given in the linear region

$$I_{n2} = K_{plin}\left\{V_{mos}(V_{gs}-V_{th}) - \frac{K_{plin}}{2K_{psat}}V_{mos}^{2}\right\}$$
$$\times \frac{1+\lambda V_{mos}}{1+\theta(V_{gs}-V_{th})}, \tag{5}$$

where K_{psat} is the transconductance in the saturation region, K_{plin} is the transconductance in the linear region, V_{th} is the threshold voltage, λ is the short-channel parameter, and θ is the transverse field factor [11]. The parameters characterizing a switching property are extracted by using the method described in [9], [12].

A 600-V, 30-A NPT-IGBT is chosen as a test device. Fig. 2 shows comparison in output characteristics between simulated results and experimental ones. The simulated results are in good agreement with the experimental ones.

Figs. 3 and 4 show switching waveforms under HSF and FUL conditions, respectively. The simulated results by using the physics-based IGBT model are in very good agreement with the experimental ones within an error of 9% in terms of collector peak current.

III. HARD-SWITCHING FAULT PROTECTION

A. Protection method

Fig. 5 illustrates the waveforms of the collector–emitter voltage v_{ce}, the collector current i_c and the gate–emitter voltage v_{ge} under normal turn-on conditions.

Phase 1 (t_0 to t_1) : The reverse transfer capacitance C_{cg} is expressed as follows:

$$C_{cg} = \frac{C_{ox}\cdot C_{dep}}{C_{ox}+C_{dep}} = \frac{C_{dep}}{1+C_{dep}/C_{ox}}. \tag{6}$$

The depletion capacitance C_{dep} is much smaller than the MOS oxide capacitance C_{ox} when the collector–emitter voltage is high. Hence, the reverse transfer capacitance C_{cg} corresponds to C_{dep}. Besides, C_{cg} is much smaller than the gate–emitter capacitance C_{ge}. As a consequence, the gate current i_g is charging only C_{ge}.

Phase 2 (t_1 to t_2) : The period of a constant gate–emitter voltage (Miller plateau) is found in v_{ge} under normal conditions. The collector–emitter voltage v_{ce} is dropping toward the steady-state value $V_{ce(sat)}$. The gate current i_g is flowing into only C_{cg}, and i_g is charging only C_{cg}. Hence, the gate–emitter voltage v_{ge} remains at the value of Miller plateau.

Phase 3 (t_2 to t_3) : The IGBT is in the saturation region and the collector–emitter voltage v_{ce} has dropped to the steady-state value $V_{ce(sat)}$. As a consequence, the gate–emitter voltage v_{ge} is increasing to sustain the load current i_c and the gate current i_g is flowing into both C_{ge} and C_{cg}.

Fig. 6 illustrates the waveforms under HSF conditions. Since the collector–emitter voltage v_{ce} remains high under HSF conditions, i_g is charging only C_{ge}. Hence, no Miller plateau appears in the gate–emitter voltage under HSF conditions.

As mentioned above, quite a significant difference exists in gate–emitter voltage waveform between normal conditions and HSF conditions. The difference causes another difference in a gate charge characteristic.

Fig. 7 depicts gate charge characteristics under normal

978-1-4799-2706-7/14 $31.00 © 2014 IEEE

The 2014 International Power Electronics Conference

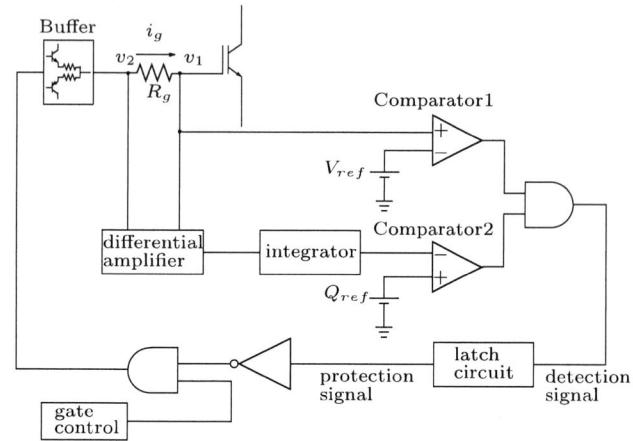

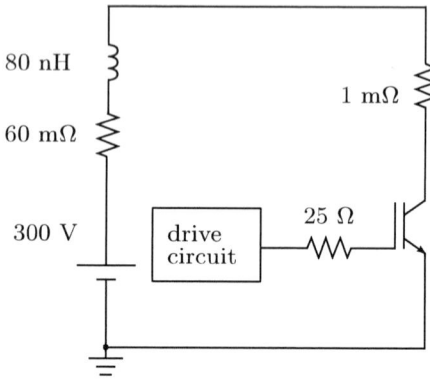

Fig. 9. Test circuit under hard-switching fault conditions.

Fig. 8. Outline of the protection circuit against hard-switching fault.

TABLE I.   LOGIC SIGNALS UNDER NORMAL TURN-ON

v_{ge}	$v_{ge} < V_{ref}$	$V_{ref} \leq v_{ge}$
comparator1 out	0	1
comparator2 out	1	0
detection signal	0	0
protection signal	0	0
gate control	1	1
gate signal	1	1

TABLE II.   LOGIC SIGNALS UNDER HARD-SWITCHING FAULT

v_{ge}	before detection	after detection	
	$v_{ge} < V_{ref}$	$V_{ref} \leq v_{ge}$	$v_{ge} < V_{ref}$
comparator1 out	0	1	0
comparator2 out	1	1	1
detection signal	0	1	0
protection signal	0	1	1
gate control	1	1	1
gate signal	1	0	0

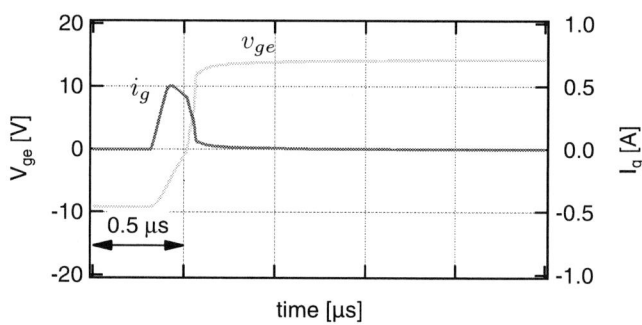

Fig. 10. Gate–emitter voltage and gate current waveforms by simulation.

conditions and HSF conditions. V_D and V_m represent the gate control voltage and the Miller plateau voltage, respectively. The gate charge increases as the gate–emitter voltage increases, so that an amount of gate charge under normal turn-on conditions is larger than that under HSF conditions when v_{ge} is higher than V_m. Hence, HSF is detected by monitoring both a gate–emitter voltage and an amount of gate charge.

The reference charge Q_{ref} and the reference voltage V_{ref} should be set to the value within the trapezoidal area surrounded by points (Q_1, V_m), (Q_3, V_m), (Q_4, V_D), and (Q_2, V_D), as shown in Fig. 7.

Fig. 8 shows an outline of the protection circuit against HSF conditions. It monitors the gate–emitter voltage v_{ge} and the voltage across the on-gate resistor R_g. The detected v_{ge} is compared with V_{ref}. The detected voltage across R_g is transferred to a differential amplifier, and the gate charge Q_g is calculated with an integrator. Q_g is compared with Q_{ref}.

Under normal conditions, Q_g becomes larger than Q_{ref} as v_{ge} becomes higher than V_m, where V_m corresponds to

the voltage of Miller plateau. Under HSF conditions, however, Q_g is always smaller than Q_{ref} because no Miller plateau appears in the waveforms of v_{ge}. Hence, the novel protection circuit can detect the fault as soon as v_{ge} exceeds V_{ref}. After detecting the fault, v_{ge} starts decreasing to protect the IGBT from destruction, so that the detection signal has to be held by using a latch-circuit such as an SRFF (Set-Reset Flip-Flop).

Tables I and II summarize the logic signals in the protection circuit under normal turn-on and HSF conditions, respectively.

The protection circuit does not require any blanking time because it detects the fault based on a gate charge characteristic.

B. Results

Fig. 9 illustrates the test circuit under HSF conditions. A resistor of 1 mΩ is used as a resistive load representing hard-switching fault.

Fig. 10 shows the gate–emitter voltage and the gate current waveforms under hard-switching fault conditions, which are obtained by the simulation using the physics-based IGBT model. No Miller plateau appears and the gate–emitter voltage reaches rapidly the gate control voltage.

978-1-4799-2706-7/14 $31.00 © 2014 IEEE

The 2014 International Power Electronics Conference

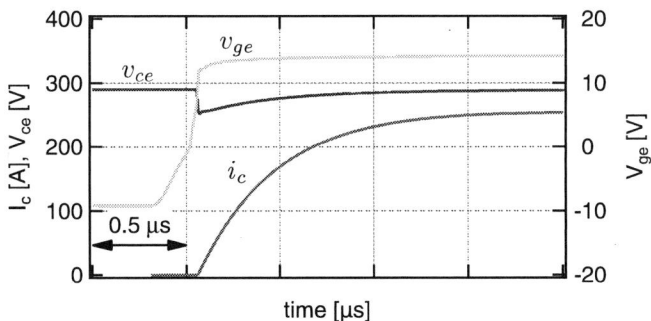

Fig. 11. Simulated waveforms without protection circuit under hard-switching fault conditions.

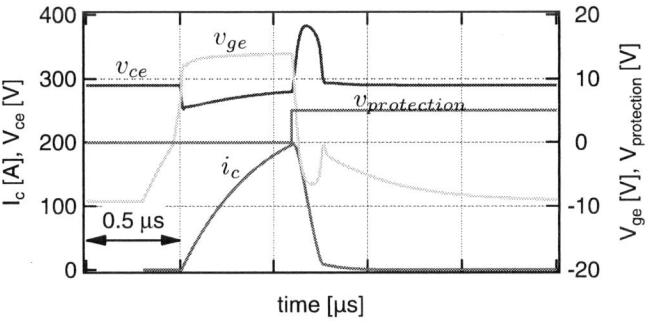

Fig. 12. Simulated waveforms with the protection circuit under hard-switching fault conditions.

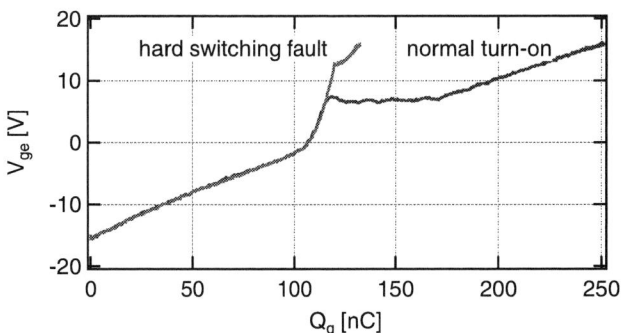

Fig. 13. Gate charge characteristics under normal conditions and under hard-switching fault conditions by experiment.

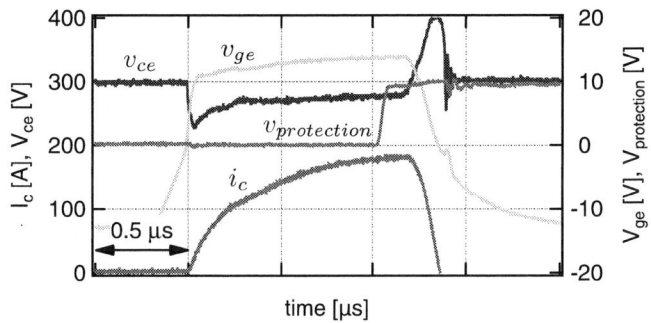

Fig. 14. Experimental waveforms with the protection circuit under hard-switching fault conditions.

Fig. 11 shows simulated waveforms under HSF conditions without protection circuit. No Miller plateau appears and the gate–emitter voltage reaches rapidly the gate control voltage. The collector–emitter voltage remains high and the collector current is rapidly increasing to about 250 A that is about eight times as large as its rated current.

Fig. 12 shows simulated waveforms under HSF conditions with the protection circuit. The fault is detected within about 600 ns later after the collector current starts flowing through the IGBT. As a consequence, the collector peak current is no more than 200 A which is about six times as large as its rated current.

Fig. 13 shows a difference in gate charge characteristics between under normal conditions and under HSF conditions, which are obtained by experiment. An amount of gate charge under HSF conditions is much smaller than that under normal conditions when the gate–emitter voltage exceeds the value of Miller plateau.

Fig. 14 shows experimental results under HSF conditions with the protection circuit. The fault was detected within about 1.0 μs later after the collector current started flowing through the IGBT, and the IGBT was turned-off about 200 ns later after the fault was detected. As a consequence, the collector peak current was about 180 A which is about six times as large as its rated current.

IV. FAULT UNDER LOAD PROTECTION

A. Protection method

Fig. 15 illustrates the waveforms of voltage and current under FUL conditions.

Phase 1 (t_0 to t_1) : The gate–emitter voltage v_{ge} is increasing to the value of Miller plateau, and the gate current i_g is charging only C_{ge} as well as normal turn-on conditions.

Phase 2 (t_1 to t_2) : Miller plateau is found in v_{ge} under FUL conditions. The collector–emitter voltage v_{ce} is dropping toward the steady-state value $V_{ce(sat)}$. The gate current i_g is flowing into only C_{cg}, and the gate–emitter voltage v_{ge} remains at the value of Miller plateau.

Phase 3 (t_2 to t_3) : The IGBT is in the saturation region and the collector–emitter voltage v_{ce} has dropped to the steady-state value $V_{ce(sat)}$. Under FUL conditions, the IGBT is changing from the saturation region to the desaturation region again, so that the collector current i_c is increasing.

Phase 4 (t_3 to t_4) : The collector–emitter voltage v_{ce} rises from the on-state voltage to the dc-link voltage, and the IGBT is in the desaturation region. As v_{ce} is increasing, the reverse transfer capacitance C_{cg} is being small. The collector–emitter voltage v_{ce} becomes higher than the gate–emitter

978-1-4799-2706-7/14 $31.00 © 2014 IEEE

The 2014 International Power Electronics Conference

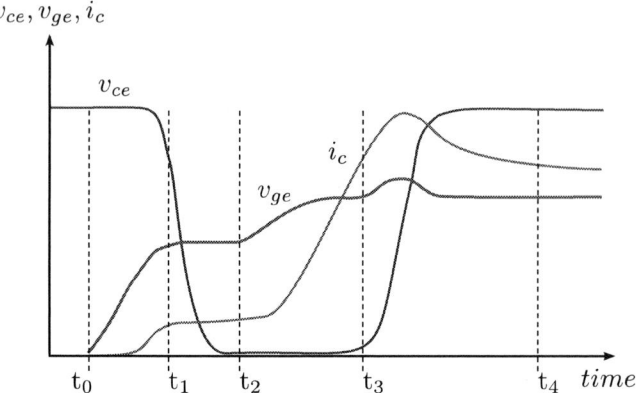

Fig. 15. Switching waveforms under fault under load conditions.

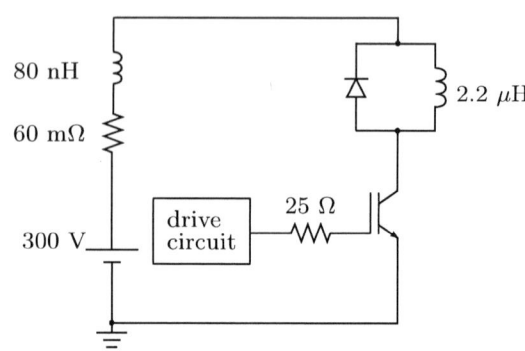

Fig. 17. Test circuit under fault under load conditions.

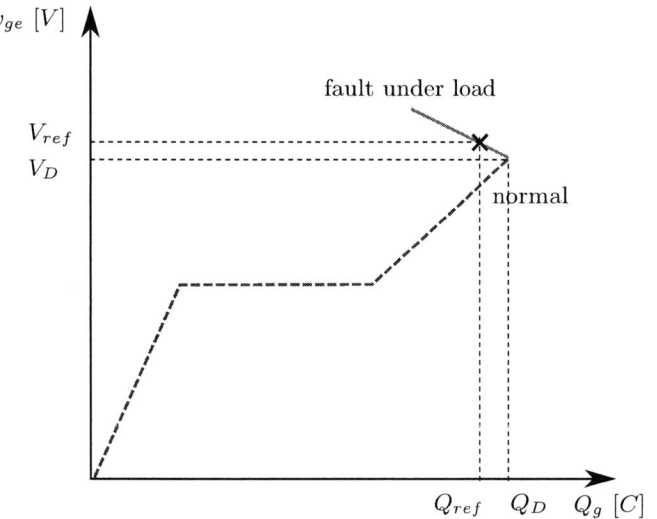

Fig. 16. Gate charge characteristics under normal conditions and fault under load conditions.

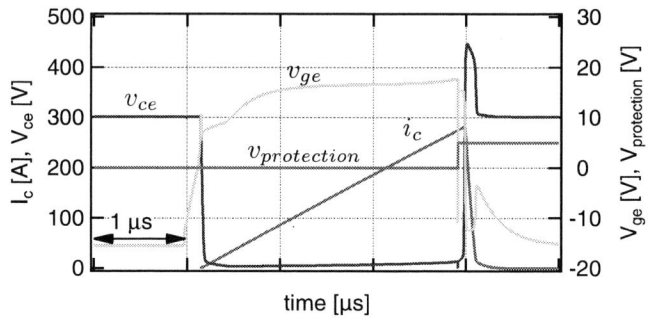

Fig. 18. Simulated waveforms with the protection circuit under fault under load conditions.

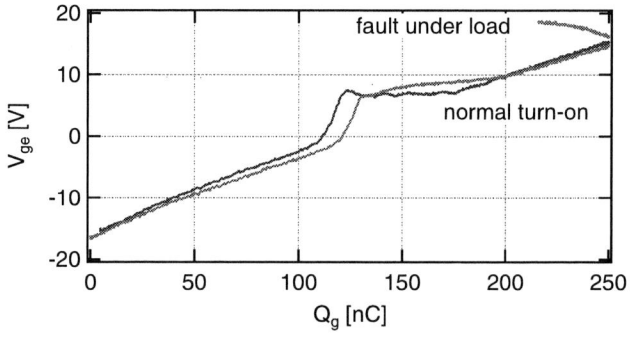

Fig. 19. Gate charge characteristics under normal conditions and fault under load conditions by experiment.

voltage v_{ge}. The gate current (the negative gate current) flows from the collector terminal to the gate drive circuit via the reverse transfer capacitance C_{cg}. The gate–emitter voltage v_{ge} becomes higher than the gate control voltage V_D due to the negative gate current. As a consequence, the gate charge Q_g decreases.

Quite a significant difference appears in gate charge characteristic between normal conditions and FUL conditions. Fig. 16 depicts gate charge characteristics under normal conditions and FUL conditions. Under FUL conditions, v_{ge} becomes higher than V_D, and Q_g becomes smaller than Q_D, where Q_D corresponds to the gate charge in V_D. Hence, FUL can be detected by monitoring the gate–emitter voltage and an amount of gate charge as well as HSF.

The reference charge Q_{ref} should be set to the value lower than Q_D, and the reference voltage V_{ref} should be set to the value higher than V_D. FUL is detected when v_{ge} is higher than V_{ref} and Q_g is lower than Q_{ref}.

B. Results

Fig. 17 illustrates the test circuit under FUL conditions. An inductance of 2.2 μH is used as an inductive load representing fault under load. The protection circuit configuration is the same as that for HSF except for the reference voltage V_{ref} and the reference charge Q_{ref}.

Fig. 18 shows simulated waveforms under FUL conditions. The gate–emitter voltage v_{ge} becomes higher than the gate control voltage V_D due to the negative gate current. The fault is detected within about 3.0 μs after the collector current starts

978-1-4799-2706-7/14 $31.00 © 2014 IEEE   2295

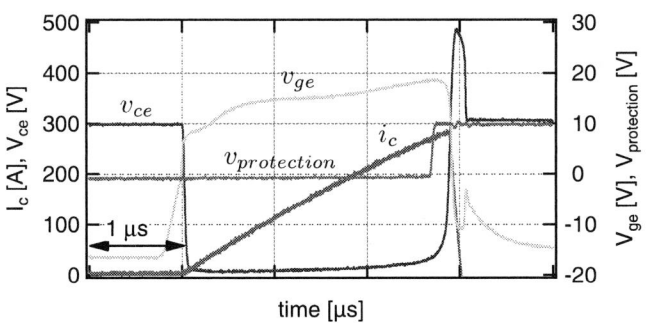

Fig. 20. Experimental waveforms with the protection circuit under fault under load conditions.

flowing through the IGBT. After detecting the fault, v_{ge} starts falling to protect the IGBT, so that the detection signal is held by a latch-circuit.

Fig. 19 shows a difference in gate charge characteristics between under normal conditions and under FUL conditions, which are obtained by experiment. The gate–emitter voltage v_{ge} becomes higher than the gate control voltage V_D and an amount of gate charge decreases under FUL conditions.

Fig. 20 shows experimental waveforms. The fault was detected within about 3.0 μs later after the collector current started flowing through the IGBT. The experimental waveforms are in excellent agreement with the simulated ones.

V. CONCLUSION

This paper has presented a high-speed circuit to protect IGBTs against short-circuit failures such as hard-switching fault and fault under load.

The sophisticated protection circuit is based on a gate charge characteristic, and it can detect both hard-switching fault and fault under load by monitoring a gate–emitter voltage and an amount of gate charge during turn-on transient.

Both simulation and experiment have confirmed the effectiveness of the protection circuit for the IGBTs subjected to hard-switching fault and fault under load.

REFERENCES

[1] R. S. Chokhawala, J. Catt, and L. Kiraly: "A Discussion on IGBT Short-Circuit Behavior and Fault Protection Schemes", *IEEE Trans. Industry Appl.*, Vol. 31, No. 2, pp. 256–263, 1995.

[2] Y. Nakayama, and T. Ohi: "Novel Over Current Protection Methods for IGBT Gate Drivers Using Gate Voltage Monitoring", 10th European Conference on Power Electronics and Applications, Vol. 1, 2003-9.

[3] M. A. Rodríguez-Blanco, A. Claudio-Sánchez, D. Theilliol, L. G. Vela-Valdés, P. Sibaja-Terán, L. Hernández-González, and J. Aguayo-Alquicira: "A Failure-Detection Strategy for IGBT Based on Gate-Voltage Behavior Applied to a Motor Drive System", *IEEE Trans. Industry Electron.*, Vol. 58, No. 5, pp. 1625–1633, 2011.

[4] B. G. Park, J. B. Lee, and D. S Hyun: "A Novel Short-Circuit Detecting Scheme Using Turn-on Switching Characteristics of IGBT", IEEE IAS Annual Meeting, 2008–10.

[5] S. Musumeci, R. Pagano, A. Raciti, G. Belverde, and M. Melito: "A New Gate Circuit Performing Fault Protections of IGBTs During Short Circuit Transients", IEEE IAS Annual Meeting, Vol. 3, pp. 2614–2621, 2002–10.

[6] K. Ishikawa, K. Suda, M. Sasaki, and H. Miyazaki: "A 600 V Driver IC with New Short Circuit Protection in Hybrid Electric Vehicle IGBT Inverter System", Proceedings of the 17th International Symposium on Power Semiconductor Devices and ICs, pp. 59–62, 2005.

[7] K. Yuasa, S. Nakamichi, and I. Omura: "Ultra High Speed Short Circuit Protection for IGBT With Gate Charge Sensing", Proceedings of the 22th International Symposium on Power Semiconductor Devices and ICs, pp. 37–40, 2010.

[8] T. Tanimura, K. Yuasa, and I. Omura: "Full Digital Short Circuit Protection for Advanced IGBTs", Proceedings of the 23th International Symposium on Power Semiconductor Devices and ICs, pp. 60–63, 2011.

[9] A. T. Bryant, L. Lu, E. Santi, P. R. Palmer, and J. L. Hudgins: "Two Step Parameter Extraction Procedure With Formal Optimization for Physics-Based Circuit Simulator IGBT and p-i-n Diode Models", *IEEE Trans. Power Electron.*, Vol. 21, No. 6, pp. 295–309, 2006–3.

[10] A. T. Bryant, L. Lu, E. Santi, J. L. Hudgins, and P. R. Palmer: "Modeling of IGBT Resistive and Inductive Turn-On Behavior", *IEEE Trans. Industry Appl.*, Vol. 44, No. 3, pp. 904–914, 2008.

[11] P. M. Igic, P. A. Mawby, M. S. Towers, W. Jamal, and S. Batcup: "Investigation of the Power Dissipation During IGBT Turn-Off Using a new Physics-Based IGBT Compact Model", *Microelectronics Reliability*, Vol. 42, pp. 1045–1052, 2002.

[12] X. Kang, A. Caiafa, E. Santi, J. L. Hudgins, and P. R. Palmer: "Parameter Extraction for a Physics-Based Circuit Simulator IGBT Model", IEEE Applied Power Electronics Conference and Exposition, Vol. 2, pp.946–952, 2003–2.

Highly reliable 1200-V p-type MOSFET for level-shift circuit used in driver IC

Naoki Sakurai

Hitachi Research Laboratory, Hitachi, Ltd.
Department of Power Electronics Research
7-1-1 Omika-cho, Hitachi-shi, Ibaraki-ken, 319-1292 Japan

Takuma Hakutou

Hitachi Automotive Systems, Ltd.
520 Oaza-takaba, Hitachinaka-shi, Ibaraki-ken,
312-8503 Japan

Masashi Yura

Hitachi Automotive Systems, Ltd.
520 Oaza-takaba, Hitachinaka-shi, Ibaraki-ken,
312-8503 Japan

Abstract— **We developed a highly reliable 1200-V p-type MOSFET for level-shift circuits. A new phenomenon, in which the leakage current increases and decreases irregularly in AC half-wave bias near avalanche voltage and the blocking voltage decreases with the long-term durability, was observed from test results of developed prototype driver ICs at an AC half-wave voltage of 1380 V. We analyzed this phenomenon through device simulation. The reason of this phenomenon was that the avalanche point of a prototype 1200-V p-type MOSFET shifted and the leakage current changed in one cycle of the AC half-wave. The device structure and process were improved on the basis of this analysis, resulting in a highly reliable p-type MOSFET.**

Keywords— *High voltage p-type MOSFET, Driver IC*

Level-shift circuit, Reliability

I. Introduction

The ground voltages of logic circuits in inverter systems that provide control signals are often different from the emitters of Insulated Gate Bipolar Transistors (IGBTs) that are switched with control signals. The upper arm voltages of the inverter systems especially are floating, and the voltages change from a power supply voltage to an on-state voltage of the IGBTs with every switching. Thus, a circuit that communicates between different ground voltages with a high degree of reliability is needed. The inverter systems that use IGBTs of 600 or 1200 V are used with high voltage driver ICs in which level-shift circuits are used for transmission of drive signals from the lower side to the floating high side [1–4]

We developed a new driver IC that consists of up- and down-arm ICs fabricated by fine pattern technology, 1200-V discrete power MOSFETs communicating between the up- and-down arm ICs with a level-shift circuit, and 30-V discrete power MOSFETs for the output stage [5].

In this paper, we report on the level-shift circuit by which a signals transmitted from a upper side to lower side, and the result of the developed 1200-V p-type MOSFET for the level-shift circuit, especially a new phenomenon in which the leakage current increases and decreases in applied AC half-wave bias nearly avalanche voltage and protective measures for this phenomenon.

II. Level-shift circuit

A. Circuit system

Figure 1 shows a block diagram of a motor control system using 2in1 driver ICs. It is a 2-in-1-type, which drives the IGBT with the up-and-down arm of one IC.

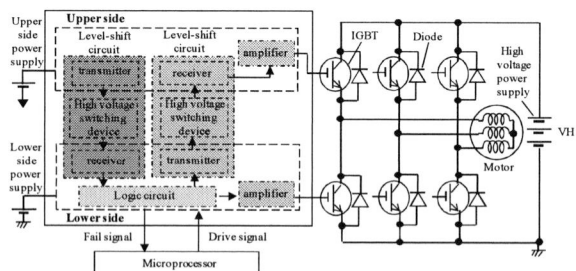

Fig.1 Block diagram of motor control system using 2-in-type driver ICs

The ground voltages of the driver circuits consisting of a driver IC and microprocessor in the inverter systems are often different from the emitters of IGBTs. The upper-arm voltages of the inverter systems, especially, are floating, and the voltages change from a power supply voltage to an on-state voltage of the IGBTs with every IGBT switching. Therefore, a 2-in-1-type driver IC requires a level-shift circuit, which communicates between the lower and upper arms, with floating potential. The high-voltage switching device of the level-shift circuit by which the upper drive signal transmitted to the upper arm from the lower side logic circuit is widely used a high-voltage n-type MOSFET.

If the protection circuit operated, the fail signal can be transmitted from the upper-side to lower-side logic circuit by the level-shift circuit. We considered two types of level-shift circuits by which the signal is transmitted from the upper side to the lower side. In Figure 2, a high-voltage n-type MOSFET is used for the high-voltage switching device. Figure 2(a) shows the details of the level-shift circuit. Figure 2(b) shows the timing charts of fail signal input between the voltage of the upper-side ground voltage (VgndU) and lower-side ground voltage (VgndL) and the fail signal output.

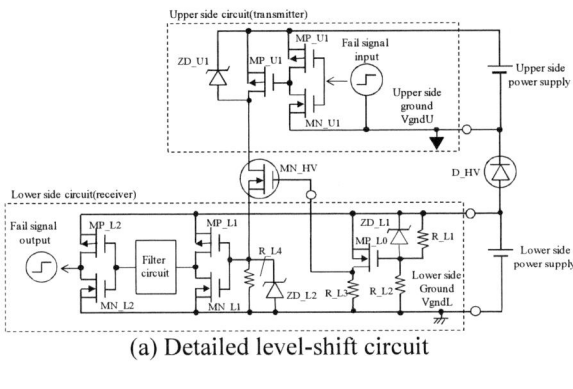

(a) Detailed level-shift circuit

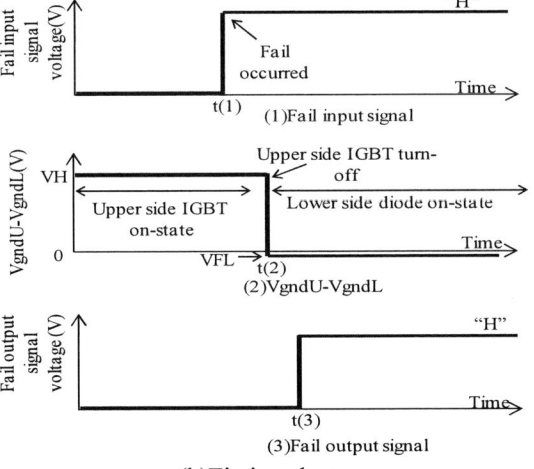

(b)Timing charts

Fig. 2 Level-shift circuit used in high-voltage n-type MOSFETs

This level-shift circuit functions as follows. If the protection circuit of the upper side is operational, the upper-side IGBT is turned off and the fail signal input is "H". By this fail signal, MN_U1 and MP_U1 in Figure 2(a) turn on. The reflex current of the lower-side diode (VFL) flows in the upper-side IGBT turn-off period, and the VgndU is lowered to the forward voltage of the VFL. If the VgndU is lowered to the VFL, the current flows through R_L2, R_L1, and high-voltage diode (D_HV) to the upper-side ground. By this current, MP_L0 turns on, the gate of the high-voltage n-type MOSFET (MN_HV) is "H", and MN_HV turns on. By the current through the

MP_U1, MN_HV and R_L4 to the lower-side ground, the gate of MN_L1 is "H" and MN_L1 turns on. Through a filter circuit, Mp_L2 turns on and the fail signal output is "H".

There are two defects with this circuit. One is that the fail signal is not transmitted until the VFL turns on. The other defect is that if the upper-side IGBT was destroyed, the reflex current of the VFL would not flow and the fail signal would not be transmitted from the upper to lower side.

In Figure 3, a high-voltage p-type MOSFET is used for the high-voltage switching device. Figure 3(a) shows the details of the level-shift circuit, and Figure 3(b) shows the timing charts of fail signal input, VgndU-VgndL, and the fail signal output.

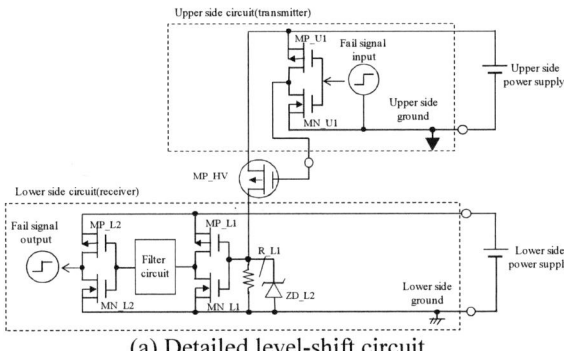

(a) Detailed level-shift circuit

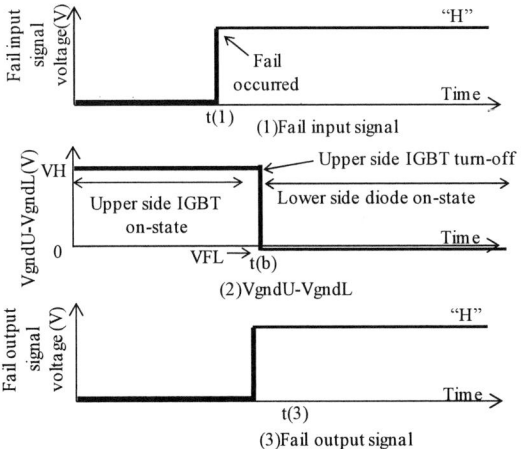

(b) Timing charts

Fig. 3 Level-shift circuit used in high-voltage p-type MOSFET

This level-shift circuit functions as follows. If the protection circuit of the upper side can be operated, the fail signal input is "H", which turns on MN_U1 and the high-voltage p-type MOSFET (MP_HV). By the current through MP_HV, R_L1, and D_HV to the lower-side ground, the gate of MN_L1 is "H" and MN_L1 turns on. Through the filter circuit, Mp_L2 turns on and the fail signal output is "H".

This level-shift circuit can transmit the fail signal from the upper side to the lower side independently of the upper-side IGBT condition. The D_HV is not needed for the level-shift circuit. Therefore, we adopted a high-voltage p-type MOSFET for the level-shift circuit that transmits the fail signal from the upper side to the lower side.

B. Blocking characteristics

Figure 4 shows the blocking characteristics of a prototype IC with applied DC voltage. The avalanche voltage was 1480 V. In a reliability DC blocking test, the avalanche voltage and leakage current did not change for over 1000 hr.

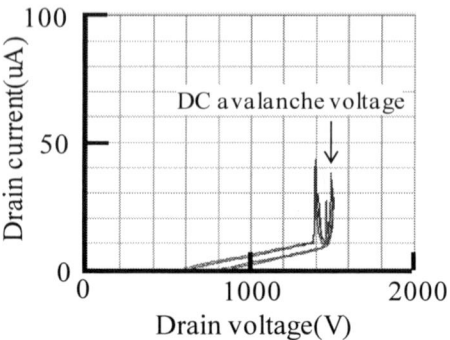

Fig. 5 Blocking characteristics of prototype IC

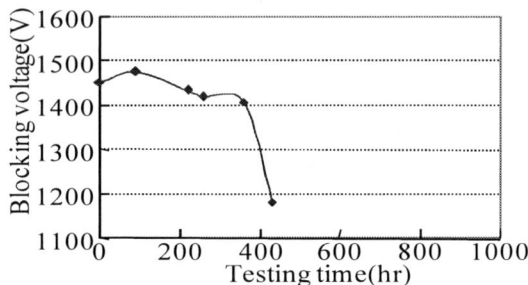

Fig. 6 Long-term durability test results of AC half-wave bias

There were two n-type MOSFETs and one p-type MOSFET as devices to which the high voltage was applied in this driver IC. The device in which the blocking voltage decreased was the p-type MOSFET with specified wire cutting. Therefore, we investigated our p-type high voltage MOSFET

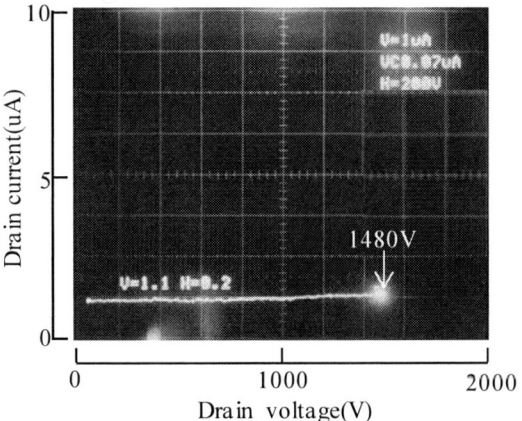

Fig. 4 DC blocking characteristics of IC

Figure 5 shows the blocking characteristics of the prototype IC with applied AC half-wave voltage. The reason there was a linear line in the low-voltage area was that the testing equipment (curve tracer) had parallel resistance for protection. The avalanche voltage was 1480 V, which was the same as that for the DC measurement. The phenomenon in which the leakage current increases and decreases irregularly was observed near the avalanche voltage.

Figure 6 shows long-term durability test results at an AC half-wave voltage of 1380 V in which the leakage current increased and decreased irregularly. The blocking voltage decreased after more than 400 hr and became lower than the standard voltage of 1200 V.

III. High-voltage p-type MOFET

A. Structre

The 1200-V blocking voltage of the p-type discrete MOSFET is the highest in the world. Therefore, we designed it in reference to the structure of the 1200-V n-type IGBT. Figure 7 shows a cross-section of the developed p-type MOSFET. We adopted a termination structure similar to that of the 1200-V IGBT.

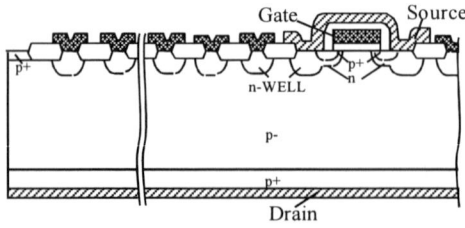

Fig.7 Cross-section of developed p-type MOSFET

Figure 8(a) shows a photograph of the developed p-type MOSFET chip, and Figure 8(b) is a schematic of the mask layout for the active source area. This device had to flow a current with an applied high voltage during level-shift operation. Because the current should be suppressed to reduce loss, the gate width was made to be very narrow (under 10 μm). A planar structure was adopted for the gate structure to suppress the saturation current during the short circuit state with a pinch-off effect.

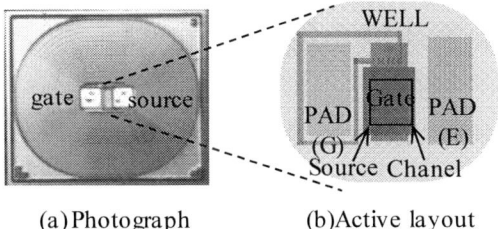

(a) Photograph   (b) Active layout

Fig. 8 Photograph and active layout of the developed p-type MOSFET chip

B. Device simulation

We analyzed the phenomenon in which the leakage current increases and decreases irregularly near the avalanche voltage by device simulation TCAD.

Figure 9 shows the simulation model. The termination area and the wire that connected the source electrode were modeled.

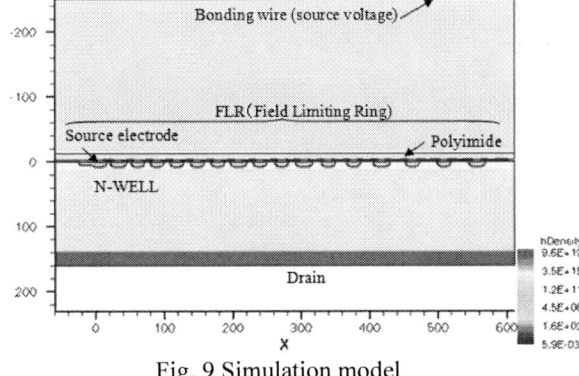

Fig. 9 Simulation model

Figure 10 shows the simulated results of blocking characteristics. The drain currents decreased after having one peak at 1460 V, and the current suddenly increased from 1470 V.

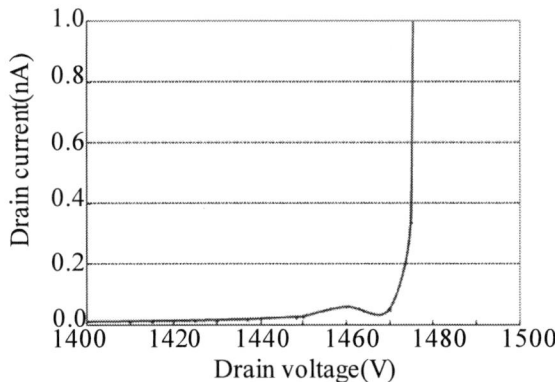

Fig. 10 Blocking characteristics of simulation

Figure 11 shows the avalanche factor for each voltage (1440, 1450, 1460, 1470 V) that was near the avalanche voltage.

Figure 12 shows the hole current density. The avalanche factor was the highest in the 1st N-WELL layer (the source layer) at 1440, 1450, and 1460 V, but the 2nd N-WELL became highest at 1470 V. Like the avalanche factor, the hole currents mainly flowed into the 1st N-WELL layer at 1460 V, but the current mainly flowed into the 2nd N-WELL layer at 1470 V.

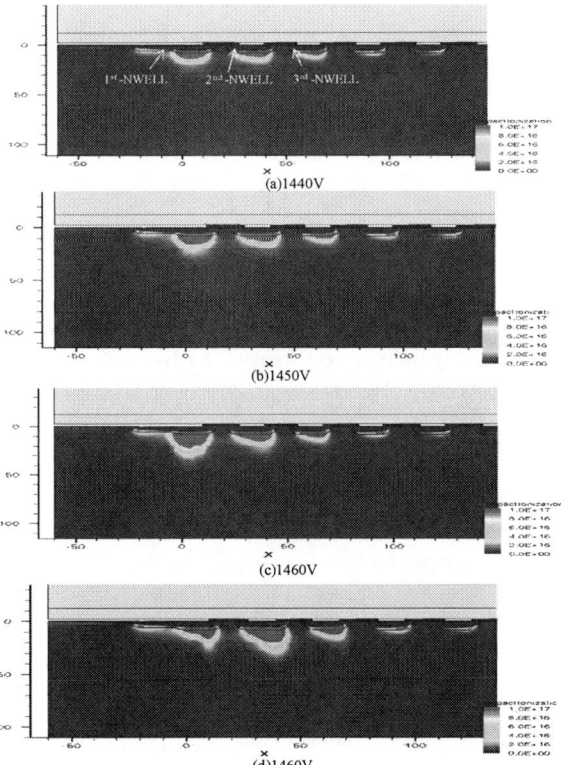

Fig. 11 Simulated avalanche factor of each voltage

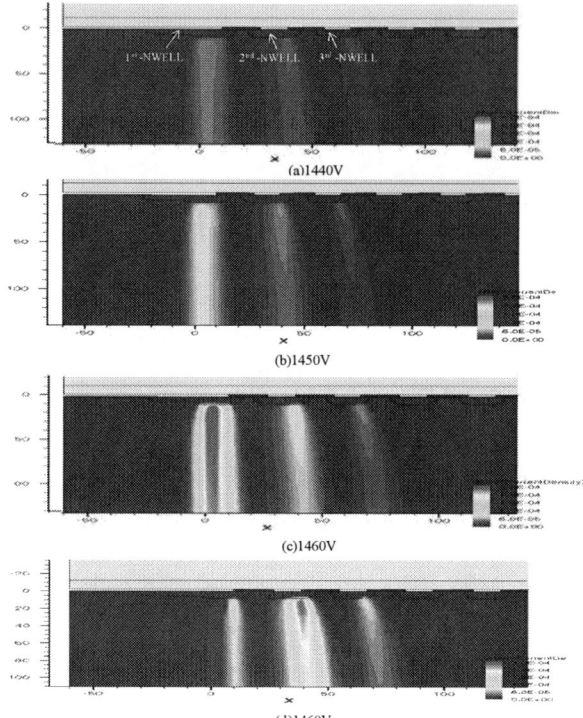

(a)1440V

(b)1450V

(c)1460V

(d)1460V

Fig. 12 Simulated hole current density of each voltage

The simulation showed that the phenomenon, in which the leakage current increases once and decreases again before the avalanche voltage, occurred due to a shift in the avalanche point. This shift may have been caused by changing the distribution of the voltage and electric field strength in the device by generating electrons and holes with avalanche.

Thus, the reason that the measured current increased and decreased irregularly in the AC half-wave voltage added by the curve tracer is that the avalanche point of the p-type MOSFET shifted and the leakage current changed in one cycle of the AC half-wave.

C. Results of improved IC

The simulation also showed that the electric field of the surface neighborhood was strong and that the avalanche occurred there. Therefore, we reviewed the p-layer density and process condition in order to lower the electric field of the surface electric field neighborhood.

Figure 13 shows the blocking characteristics of the 2nd IC used the improved p-type 1200-V MOSFET for level-shift circuit with applied to AC half-wave voltage. The phenomenon in which the leakage current increases and decreases irregularly was not observed.

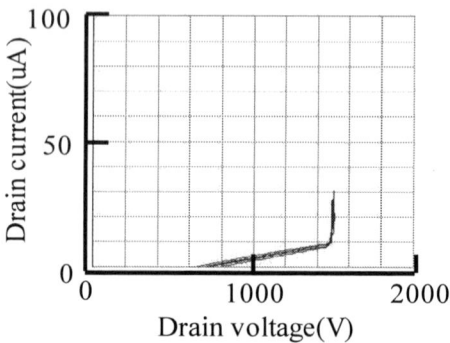

Fig.13 Blocking characteristics of 2nd IC

Figure 14 shows the results of a long-term durability test of the AC half-wave bias of the improved IC. The blocking voltage was constant over 1000 hr.

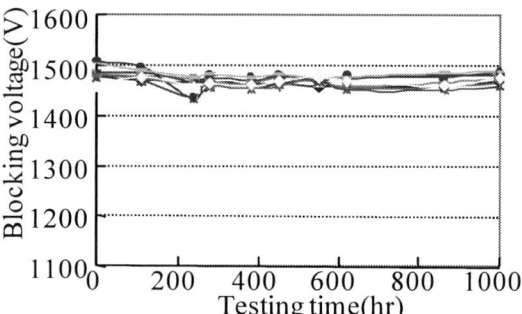

Fig. 14 Long-term durability test of
AC half-wave bias at 2nd IC

IV. Conclusions

We analyzed a new phenomenon in which the leakage current increases and decreases in applied AC half-wave bias near avalanche voltage. The reason of this phenomenon was that the avalanche point of the developed p-type MOSFET shifted and the leakage current changed in one cycle of the AC half-wave. This phenomenon was improved by reducing the electric field of the surface neighborhood and reviewing the p-layer density and process condition.

References

[1] H. Akiyama et al., "A High Breakdown Voltage IC with Lateral Power Device based on SODI Structure," Proceedings, ISPSD 2005, pp. 375–378.

[2] K. Shimizu, S. Rittaku, and J. Moritani, "A 600V HVIC Process with a Built-in EPROM which Enables New Concept Gate Driving," Proceedings, ISPSD 2005, pp. 23–29.

[3] K. Ishikawa et al., "A 600V Driver IC with New Short Protection in Hybrid Electric Vehicle IGBT Inverter System," Proceedings, ISPSD 2005, pp. 59–62.

[4] M. Yoshino, K. Shimizu, T. Terashima, "A new 1200-V HVIC with a novel high voltage Pch-MOS" ,ISPSD 2010, pp. 93–96.

[5] N. Sakurai et al., "A 1200-V High-power Driver IC with Multi-chip Mounting for Strong HEVs," Proceedings of 2nd International Conference on Automotive Power Electronics, 2007.

The 2014 International Power Electronics Conference

A New Level Up Shifter for HVICs with High Noise Tolerance

Masashi Akahane, Akihiro Jonishi, Masaharu Yamaji, Hiroshi Kanno, Takahide Tanaka,
Haruhiko Nishio, and Hitoshi Sumida

Fuji Electric Co.,Ltd.
4-18-1, Tsukama, Matsumoto-city, Nagano, 390-0821, Japan
E-mail: akahane-masashi@fujielectric.co.jp

Abstract — **In this paper, a new level up shifter with high dV/dt and negative transient voltage noise immunities is presented. The proposed level up shifter achieves high noise immunity without an increase in the delay time of high-voltage ICs (HVICs). A fabricated 1200V-class HVIC adopting the proposed level up shifter on a p-type substrate indicates a stable operation under the conditions of dV/dt noise over 50kV/µs and negative transient voltage noise of -150V. And we have confirmed the good performances of a 3-phase inverter driven by the fabricated HVIC.**

Keywords— gate driver, level shift, dV/dt noise, negative transient voltage noise.

I. INTRODUCTION

High-voltage IC (HVICs) [1] is widely used for IGBT gate driver in many applications, such as motor drives, LED lighting and so on because the HVIC provides the cost effective solution for power control system. However, the current rating of the main power devices driven by HVIC is limited to less than one hundred amperes. This is because the HVIC would have unstable operation due to various switching noises which become more severe in higher current applications. Therefore, the HVIC with high tolerance to dV/dt and negative transient voltage noises has been strongly requested.

There are mainly two approaches to improve noise tolerance of the HVICs: one is process approach and the other is circuit approach [2-5]. The process approach includes isolation techniques such as junction isolation using an epitaxial wafer with partial buried layers and dielectric isolation using SOI (Silicon on Insulator). These process techniques lead to an increase in the process cost of the HVIC fabrication. The modified high-side driver circuits including the level shifter have been reported as the circuit approach. Because the circuit approach accompanies additional circuit components, the delay time of the HVIC is an important issue.

Two factors which increase the delay time of the level shifter in the circuit approach are shown below.
 a. An increase in the propagation delay time by the additional logic and passive filters to improve the noise immunity.

 b. Dead time of the latch circuit in the level shifter for preventing the malfunction which is caused by the rapid voltage change at VS node.

However, there are few reports regarding the circuit approach to achieve the higher noise tolerance without an increase in the delay time of the HVICs.

In this paper, we have proposed a new level up shifter with high dV/dt and negative transient voltage noise immunities for the HVICs. The proposed level up shifter can suppress an increase in the delay time caused by the aforementioned two factors. As a result, the high noise immunity without an increase in the delay time of the HVICs can be achieved. And we have fabricated a 1200V-class HVIC, adopting the proposed level up shifter on a p-type substrate, which integrates three low- and high-side drivers to drive a 3-phase inverter. Furthermore, we demonstrated the driving of the 3-phase inverter with the fabricated HVIC. This paper shows the concept of the new level up shifter, the performances of the fabricated 1200V-class HVIC and the demonstration results of the 3-phase inverter driven by the fabricated HVIC.

II. NOISE AFFECTED IN HVIC DUE TO IGBT SWITCHING

Figure 1 shows a half bridge switching system using the HVIC. The block diagram of the HVIC is also indicated in the Fig. 1.

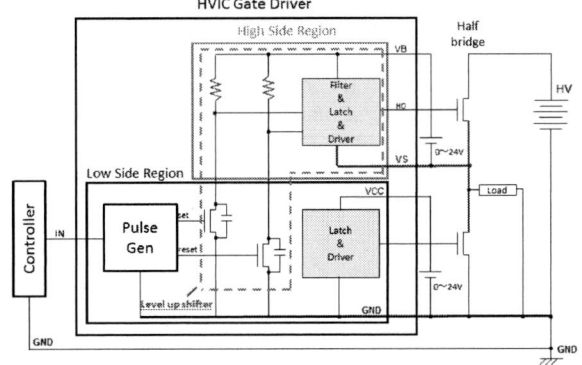

Fig. 1. Half bridge switching system using a HVIC.

978-1-4799-2706-7/14 $31.00 © 2014 IEEE   2302

The HVIC has the low- and high-side gate drivers to drive the IGBTs in the half bride circuit. The level up shifter is to transmit the digital signal from ground to the floating source of the high-side IGBT, which is indicated as VS in Fig. 1. The high-side gate driver operates with the potential of the VS node as a reference, which is changed between GND level and HV level (600V or 1200V) in the high-side IGBT switching.

Figure 2 shows a switching diagram of the HVIC. When the high-side IGBT turns on, the voltage of VS node rises up near HV level. At this time the dV/dt noise is generated by the voltage change and applies to the high-side driver region in the HVIC through the VS line. Meanwhile, when the high-side IGBT turns off, the inductive load causes the current from the ground to VS node to keep on flowing. This current induces –dV/dt noise at the VS terminal, which is applied to the high-side driver region in the HVIC as the negative transient voltage noise.

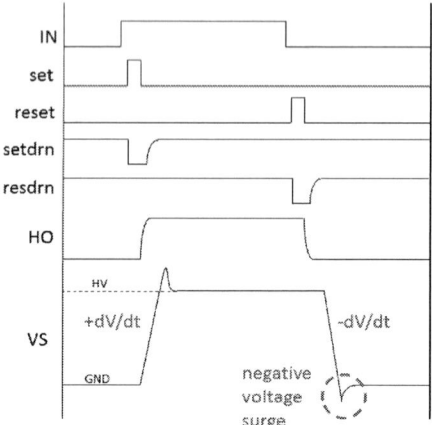

Fig. 2. Switching diagram of a HVIC.

III. DESCRIPTIONS OF THE NEW LEVEL UP SHIFTER

In this section the concept of the new level up shifter is described in detail using the measurement results obtained from the testing of the 1200V-class HVIC.

A. Concept

A-1 Conventional level up shifter features

Figure 3 shows the circuit diagram of the conventional level up shifter. As shown in Fig. 3, the level up shifter consists of two high-voltage MOSFETs, HVN1 and HVN2 for set and reset pulses, respectively. These pulses are made from rising and falling edges of an input signal (IN). The driver output (HO) is controlled by these two short pulses through the RS latch circuit in the high-side driver region.

Table 1 is a truth table of a latch circuit block in the level up shifter. The output, HO, is set to "H" or "L" only when both inputs, setdrn and resdrn, are exclusive. And when the both inputs are the common-mode signal, the output is held because the filter inserted before the latch circuit rejects the common-mode signal. As a result, the common-mode noise, which is generated by the parasitic capacitances of HVN1 and HVN2 when the dV/dt is applied to VS terminal and superimposed to the both inputs, can be rejected and malfunction which is caused by the common-mode noise can be prevented. However, this function causes the delay time of the HVICs.

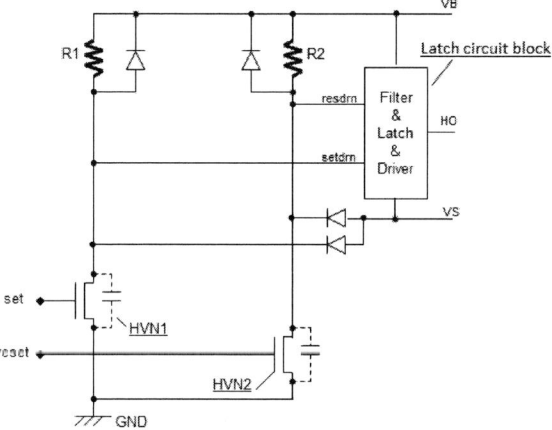

Fig. 3. Circuit diagram of the conventional level up shifter.

Table 1 Truth table of the latch circuit block in Fig. 3.

Input		Output	Behavior
setdrn	resdrn	HO	
H	H	HOLD	Standby
L	H	H	Set
H	L	L	Reset
L	L	HOLD	Common Mode Noise Rejection

Figure 4 shows simulation results of the output pulse. During recovering of setdrn and resdrn to "H", slight difference between setdrn and resdrn waveforms is observed due to the variation of the parasitic capacitances between HVN1 and HVN2. This difference causes malfunction of the latch circuit, which turns the HO to "H" or "L" wrongly. In order to prevent this malfunction, the logic filter and passive low pass filter are added to the level up shifter. However, it increases the delay time of the level up shifter.

A-2. Proposed level up shifter to prevent malfunction

Figure 5 shows the circuit diagram of the proposed level up shifter. The following components exist in the level up shifter.

a. Two feed-back loops with two different logic signals from the latch circuit to the level up shifter.
b. Two p-channel MOSFETs (MVP1 and MVP2) in parallel with level shift resistances (R1 and R2).
c. Serial-connected two resistances (R3-R3' and R4-R4') in each feed-back loop. The drain terminals of HVN1 and HVN2 and the gate terminals of MVP1 and MVP2 connect to the serial-connected two resistances.
d. The resistance values are determined according to following two conditions. Vth is the threshold voltage of the MVP1(or MVP2) based on VS level.

$$R1 < R3' < R3 \tag{1}$$

$$\frac{R3}{R1 + R3' + R3} \times (VB - VS) > Vth \tag{2}$$

Figure 6 shows simulation results of setdrn and resdrn waveforms when the +dV/dt noise is applied to VS terminal. The initial condition of the HO signal was set to a logically high level. Two feedback signals from the latch circuit have different voltage, that is, one feedback signal has VB potential and the other has VS potential. This gives different gate voltages to MVP1 and MVP2. As a result, a larger voltage difference between setdrn and resdrn waveforms can be obtained in comparison with that in Fig. 4. It is found that the malfunction due to the dV/dt noise can be prevented.

In order to estimate an influence of variation in the parasitic capacitance of the proposed level up shifter, extra capacitance is added between source(=GND) and drain terminals of the reset side HVN2.

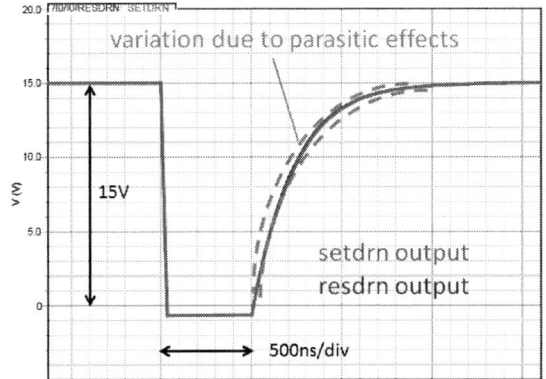

Fig. 4. Outputs of the conventional level up shifter when the +dV/dt noise occurs.

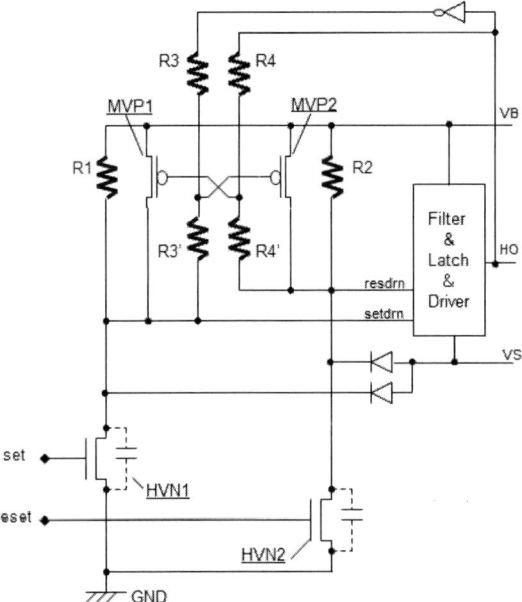

Fig. 5. Circuit diagram of a proposed level up shifter.

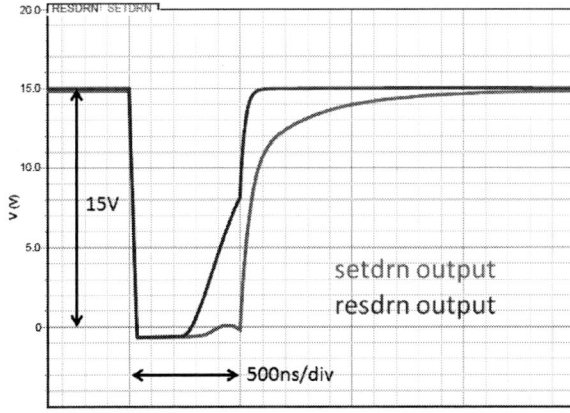

Fig. 6. Outputs of the proposed level up shifter when the +dV/dt noise occurs under the high level HO signal.

978-1-4799-2706-7/14 $31.00 © 2014 IEEE

Figure 7 shows simulation results with additional 1.0pF capacitance between source and drain terminals of the reset side HVN2. The malfunction due to the dV/dt noise occurs in the conventional level up shifter, but does not occur in the proposed level up shifter.

Table 2 shows same simulation results as those in Fig. 7 when the capacitor value is changed between the source and the drain of the HVN2 from 0 to 2.0pF. As a result, an additional 1.0pF capacitance leads to malfunction in the conventional level up shifter. On the other hand, in the proposed level up shifter (M=1), the additional 1.0pF capacitance does not lead to malfunction. Furthermore, in the proposed level up shifter (M=2, which means double channel width of MVP with M=1), no malfunction is observed under an additional 2.0pF capacitance. We have found that the proposed level up shifter improves the dV/dt noise tolerance than the conventional level up shifter. Moreover, the proposed level up shifter improves the dV/dt noise tolerance by increasing the channel width of added p-channel MOSFET (MVP).

Table 2 Comparisons of simulation results of circuit performance between Fig. 3 and Fig. 5.

		added HVN2(reset side) drain-gnd capacitance value				
		none	0.5pF	1.0pF	1.5pF	2.0pF
Conventional		O	O	×	×	×
New	M=1	O	O	O	×	×
	M=2	O	O	O	O	O

M: Multiplier number of p-channel MOSFET MVP1 and MVP2
O : good
× : no good (HO output signal malfunction)

A-3. Proposed level up shifter to improve delay time

The recovery time of the setdrn and resdrn signals disturbed by the dV/dt noise depends on the dV/dt value. The recovery time becomes longer under the smaller dV/dt value. Figure 8 shows simulation results of the delay time due to large and small dV/dt. When the recovery time is longer, the propagation delay time from the input signal (IN shown in Fig. 2) to HO signal becames longer. This is because the level up shifter does not operate due to the lower voltage level of both setdrn and resdrn signals than the logic threshold value.

In order to overcome above delay time issue, we have proposed a circuit component, dV/dt response changing circuit(RCC), and added it to the level up shift circuit as shown in Fig. 9. The dV/dt RCC detects the voltage change of setdrn and resdrn signals due to the dV/dt noise and reduces the impedance of the level up shifter by turning additional p-channel MOSFETs, MVP3 and MVP4, on. By adding the dV/dt RCC, an additional control loop is formed, resulting that the setdrn and resdrn signals indicate a voltage change as shown in Fig. 10 when dV/dt noise occurs.

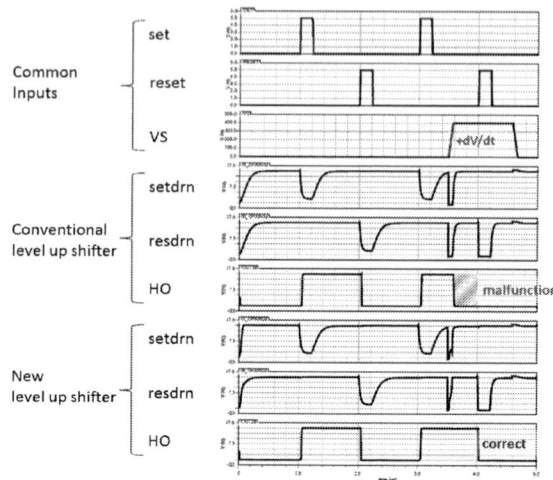

Fig. 7. Simulation results when a 1.0pF capacitance between source and drain terminals of reset side HVN2 is added.

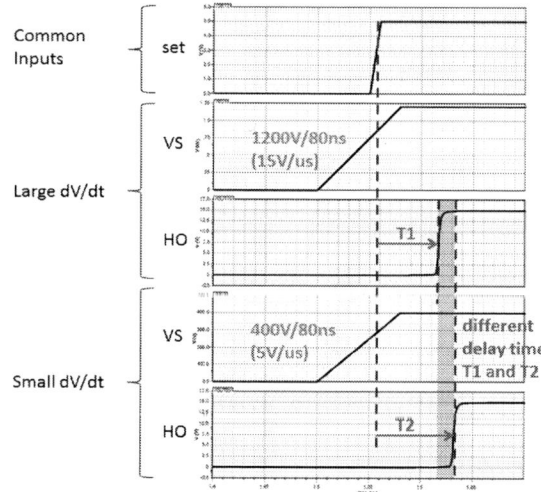

Fig. 8. Delay time due to large and small dV/dt.

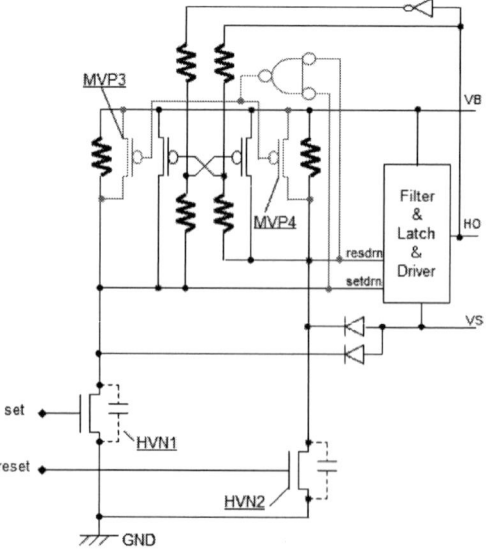

Fig. 9. Proposed level up shifter with the dV/dt RCC.

The signal invalid period (ta) and the recovery time (tb) of the voltage change shown in Fig. 10 can be controlled by adjusting the dV/dt RCC parameters. As a result, following effects are obtained.

a. Faster recovery response of setdrn and resdrn signals can be realized.

b. The smaller capacitance value in the capacitor filter at the internal stage of the level up shifter can be used.

The delay time can be suppressed to be short by above effects.

Figure 11 shows simulation results of the conventional level up shifter and proposed level up shifter with the dV/dt RCC when the rising edge of set terminal appears simultaneously at the dV/dt on VS terminal. Longer propagation delay is observed at the HO output with conventional level up shifter. On the other hand, the delay time is very short with the new level up shifter at the same condition.

A-4. Simulation results of branch and peak currents

Table 3 shows simulation results of branch currents with the conventional and new level up shifters. The standby current of the VB terminal in the new level up shifter is 30μA, it increases by 30μA in comparison with the conventional level up shifter. This current flows through the newly added resistances. This current can be reduced by increasing resistance value. Two HVN-drain currents and VS-diode currents are almost same in the conventional and proposed level up shifters.

Table 4 shows peak currents of several nodes obtained from simulations when the dV/dt noise of ±5kV/μs is applied. The new level up shifter consumes mA order current, which is almost the same level as the conventional level up shifter. When the +dV/dt (rising VS terminal) is applied, the current is flowing to HVN parasitic capacitance through a diode between VS and HVN drain terminals. When the −dV/dt (falling VS terminal) is applied, the current is flowing to HVN parasitic capacitance through VB (power source) terminal.

As a result of the aforementioned reasons, we had conviction that the new level up shifter is definitely useful for the HVIC.

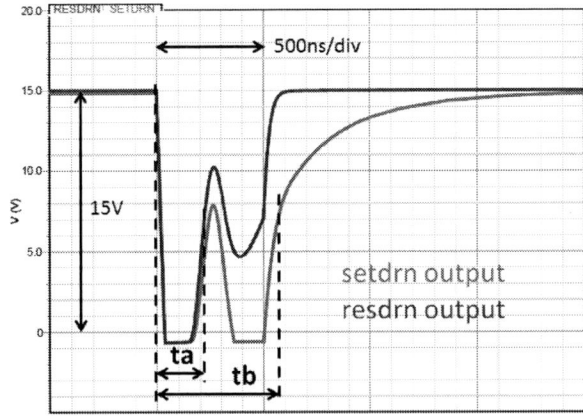

Fig. 10. Waveforms of setdrn and resdrn signals when the HO signal is the logical high level.

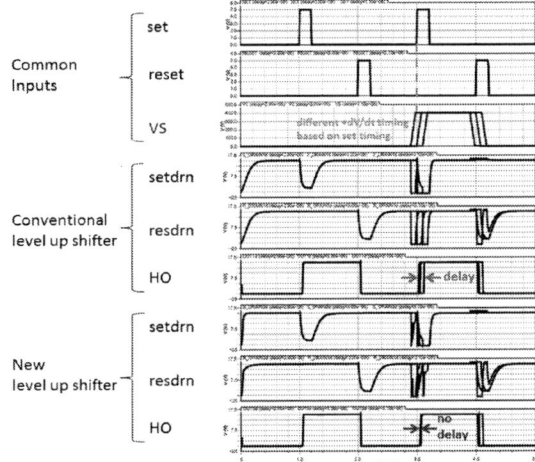

Fig. 11. Simulation results of conventional and proposed level up shifters.

Table 3 Branch currents of the conventional and new level up shifters.

Current Node	Blanch current [μA]			
	Convention	New	New with RCC	condition
VB(power input)	< 0.1	30	30	standby
HVN1-drain	< 0.1	< 0.1	< 0.1	"
HVN2-drain	< 0.1	< 0.1	< 0.1	"
VS-setdrn Diode	< 0.1	< 0.1	< 0.1	"
VS-resdrn Diode	< 0.1	< 0.1	< 0.1	"

Table 4 Peak currents of several nodes when the dV/dt noise of ±5kV/μs is applied.

Current Node	Peak current [mA]			
	Convention	New	New with RCC	condition
VB(power input)	115	115	115	−dV/dt
HVN1-drain	55	55	55	±dV/dt
HVN2-drain	55	55	55	±dV/dt
VS-setdrn Diode	55	50	52	+dV/dt
VS-resdrn Diode	55	50	52	+dV/dt

B. Measurement results

We have fabricated the 1200V-class HVIC adopting the new level up shifter on a p-type substrate. The block diagram is shown in Fig. 12. The HVIC has three low- and high-side drivers to drive a 3-phase inverter on one chip.

Figure 13 shows the waveform of HO under the dV/dt of 72kV/μs. It is confirmed that the proposed level up shifter operates normally under the dV/dt of 50kV/μs in a range of –40 to 150°C.

Measurement results on negative transient voltage tolerance (Vsurge) are shown in Fig. 14 and Table 5. The negative transient voltage of –150V does not affect the HO output.

Temperature dependences of the rising and falling delay time are shown in Figs. 15 and 16, respectively. Both of the rising and falling delay time are below 200ns even under the worst condition with 150°C and VBS (a voltage between VB and VS terminals.) of 12V. It is confirmed that the fabricated HVIC has same switching characteristics as the HVIC adopted the conventional level up shifter.

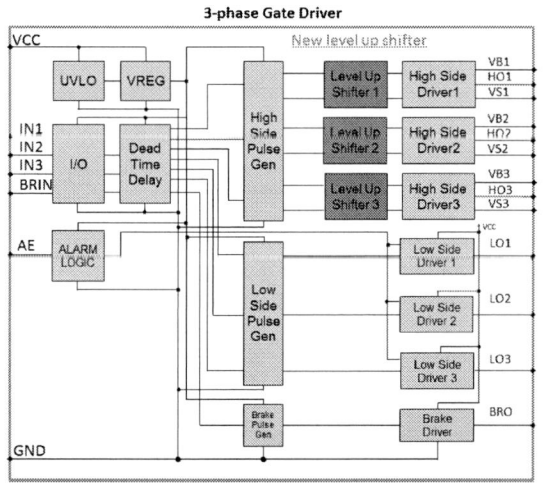

Fig. 12. Block diagram of the fabricated HVIC.

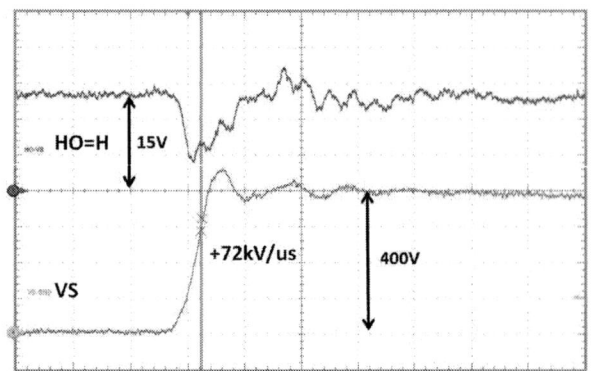

Fig. 13. Waveform of HO under dV/dt of 72kV/μs.

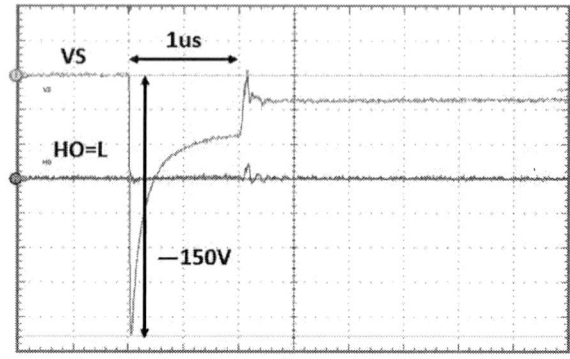

Fig. 14. Waveform of HO under VS of –150V.

Table 5 Measurement results on negative transient.

	Pulse width [μs]			
	0.25	0.5	1	2
Vsurge [V]	< – 150 measurement system limitation			

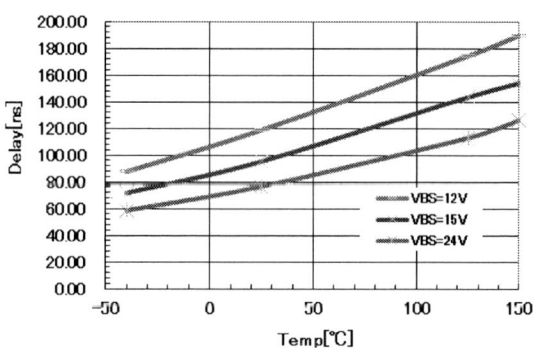

Fig. 15. Temperature dependences of the rising delay time.

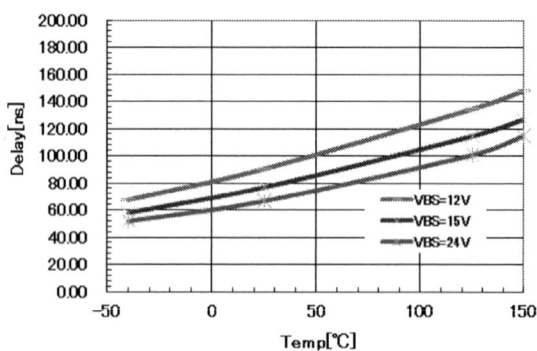

Fig. 16. Temperature dependences of the falling delay time.

C. Comparisons with previous works

Table 6 shows comparisons between our level up shifters and previous works. The maximum dV/dt noise immunity from previous works is 65kV/μs [5]. On the other hand, the malfunction of our proposed level up shifter does not occur until 72kV/μs. As for negative transient voltage surge immunity, our proposed level up shifter has better performance than others, that is −150V. There is no difference of the delay time between the conventional and proposed level up shifters.

IV. APPLICATION TO A 3–PHASE INVERTER

In order to demonstrate the driving operation of the HVIC with the new level up shifter under the actual operating condition, the fabricated HVIC is adopted to the 3-phase inverter. The inverter is Fuji electric FRENIC-mini AC400V/3.7kW. Figure 17 shows the block diagram of inverter system. The input voltage is AC400V, and 3.7kW AC induction motor is connected to the output terminals.

Table 7 shows measurement results on the driving test of the 3-phase inverter. And Fig. 18 shows output current waveforms. We have confirmed that the inverter system operates properly in basic operation sequences such as start-up, driving and shutdown, and also confirmed that the system drives the motor properly in overload conditions up to 200% of the rated output current. These results show that the fabricated HVIC is operating normally under the actual operating condition.

V. Conclusions

In this paper, a new level up shifter with high dV/dt and negative transient voltage noise immunities has been reported. The proposed level up shifter achieves high noise immunity without an increase in the delay time of the HVICs. And the performances of the fabricated 1200V-class HVIC and the demonstration results of the 3-phase inverter driven by the fabricated HVIC have been also shown. The stable operation under the conditions of dV/dt noise over 50kV/μs and negative transient voltage noise of −150V have been observed from the fabricated HVIC. And we have confirmed the good performances of a 3-phase inverter driven by our fabricated HVIC. These results indicate that the HVIC can be applied to higher current applications.

Table 6 Comparisons with previous works.

	Fuji Conventional	Reference [2]	Reference [5]	Fuji New type
dV/dt noise immunity [kV/μs]	50	50*	65	72
Negative surge noise immunity [V]	-20	-75	-13	-150
Level shifter Path Delay [ns]	95	Unknown	Unknown	95

*: dV/dt absolute maximum rating value in M81738FP datasheet.

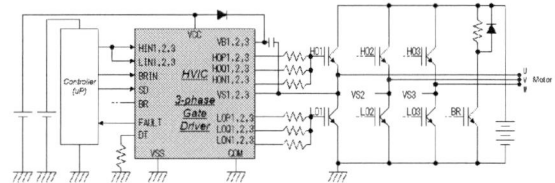

Fig. 17. Block diagram of 3-phase inverter system.

Table 7 Measurement results on driving test.

Test case		judgement result
Basic sequence	etc.	Passed
Load Test	9A(100%)	Passed
	13.5A(150%)	Passed
	18A(200%)	Passed

Passed : Normal functions with no malfunction and destruction.

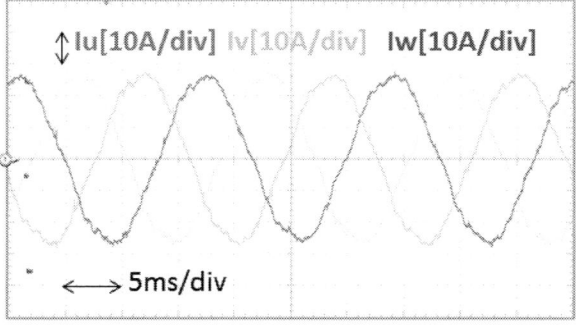

Fig. 18. Output current waveforms of 3-phase inverter. (AC=400V, Fc=15kHz, Fo=60Hz, load=18A [200%])

REFERENCES

[1] T. Fujihira, et al., "Proposal of New Interconnection Technique for Very High-Voltage IC's," Jpn. J. Appl. Phys. , vol. 35, no. 11 , pp. 5635-5663, 1996.

[2] M. Yamamoto, et al., "High reliability 1200V High Voltage Integrated Circuit (1200V HVIC) for half bridge applications," PCIM Europe 2012, 8-10, May 2012, pp. 466-472.

[3] M. Yamaji, et al., "A Novel 600V-LDMOS with HV-Interconnection for HVIC on Thick SOI," Proc. of the 22nd ISPSD, June 2010, pp. 101-104.

[4] J. J. Kim, et al., "The new high voltage level up shifter for HVIC," PESC, 2002, pp. 626-630.

[5] Z. Yunwu, et al., "A noise immunity improved level shift structure for a 600V HVIC," *Journal of Semiconductors,* vol. 34, no. 6, 065008-1, 2013.

The 2014 International Power Electronics Conference

Output Ripple Minimization of Single-Stage Power Factor Corrected Bi-directional Buck AC/DC Converter

Balaji Veerasamy*, Wataru Kitagawa* and Takaharu Takeshita*
*Nagoya Institute of Technology, Gokisocho, Showa-ku, Nagoya, Aichi 466-8555, Japan

Abstract—**This paper presents a PWM control method for a variable-voltage buck single-stage bi-directional AC/DC converter for suppressing the output voltage ripple with high power factor. The proposed method maintains minimum commutations during the control period for realizing three voltage levels close to the output reference voltage on the output of the converter which in turn reduces the voltage ripple. The effectiveness of the proposed PWM strategy is verified by experimental results.**

I. INTRODUCTION

Switching power electronic converters employs Pulse Width Modulation (PWM) method for converting the AC voltage to a desired DC voltage. Efficient single-stage AC/DC converters are replacing the conventional two-stage converters for their compact size and simple control methods for renewable and energy saving technologies. Several single-stage isolated AC/DC converters are developed in [3]-[7] which increases the overall size. Ripple minimization is an important feature of single-stage non-isolated PWM controlled converters [8] with a limited input phase control range is proposed with reduced number of commutations during the control period for reduced DC ripple. As power factor regulations are required to meet the standards additional switching pattens are included in [10] to improve the input source power factor. However the input power factor range is improved by phase shifting, the method lacks to keep the output DC ripple minimum. The main drawback of this control method is the high variation of output voltage, producing an increased voltage stress on the output capacitor leading to a high ripple on the DC voltage.

This paper is organized as follows: Sections II presents the basic control system and analytical model of the single-stage AC/DC buck converter, Section III describes the proposed PWM switching method which is compared with the conventional single-stage PWM switching method. Sections IV discusses the output voltage ripple analysis and finally the experimental results are discussed with comparing the proposed and conventional methods.

II. PROPOSED CIRCUIT AND ITS OPERATING PRINCIPLE

A. Converter Configuration and Analytical Model

Fig.1 shows the buck single-stage bi-directional AC/DC converter system. The three phase input AC voltages e_u, e_v and e_w are converted to a constant desired DC voltage through the six bi-directional IGBT switches $S_{up} - S_{wn}$. The bi-directional operation is achieved by reverse blocking IBGT's

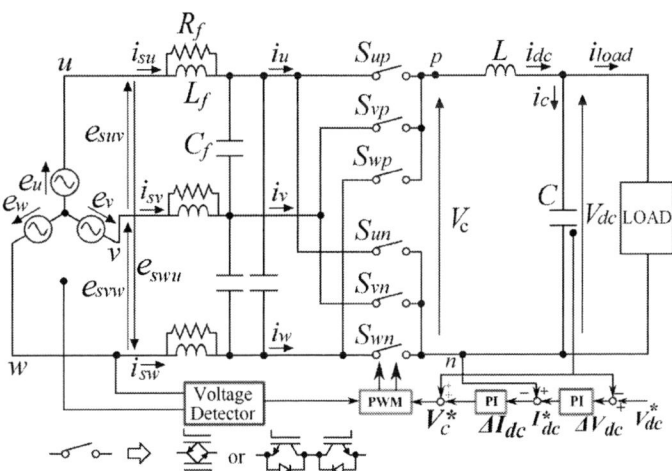

Fig. 1.   Single Stage Bi-directional AC/DC Converter

or connecting two IGBT's in series reverse. The circuit has two LC filters on both the input side and the output side. The input filter reactor L_f and capacitor C_f is connected to suppress the outflow of the harmonic current to the power supply and the output side is connected to the LC filter of the smoothing reactor L and capacitor C for suppressing the output voltage ripple caused by PWM switching. The input current references i_u^*, i_v^* and i_w^*(* denotes reference) of the converter are selected so as to realize unity power factor between the source currents i_{su}, i_{sv} and i_{sw} and corresponding source voltages e_u, e_v and e_w. The switching patterns of switches $S_{up} - S_{wn}$ are controlled so as to obtain a desired output voltage with realizing an unity input source power factor through two PI controllers and the input phase detector as shown in Fig.1.

The input and output instantaneous powers are given in terms of their reference voltages and currents in (1) and (2).

$$p_{in} = e_u i_u^* + e_v i_v^* + e_w i_w^* \qquad (1)$$

$$p_{out} = V_c^* I_{dc} \qquad (2)$$

A symmetrical three-phase source with an effective line voltage E is expressed in (3). The supply voltage is approximately equal to the input phase voltage by ignoring the voltage

978-1-4799-2706-7/14 $31.00 © 2014 IEEE   2310

drop across inductor L.

$$\begin{bmatrix} e_{su} \\ e_{sv} \\ e_{sw} \end{bmatrix} \simeq \begin{bmatrix} e_u \\ e_v \\ e_w \end{bmatrix} = \sqrt{\frac{2}{3}}E \begin{bmatrix} \cos\theta \\ \cos(\theta - 2\pi/3) \\ \cos(\theta + 2\pi/3) \end{bmatrix} \quad (3)$$

The input current references are expressed in terms of its effective value I^* and power factor angle φ^* in (4).

$$\begin{bmatrix} i_u^* \\ i_v^* \\ i_w^* \end{bmatrix} = \sqrt{2}I^* \begin{bmatrix} \cos(\theta + \varphi^*) \\ \cos(\theta + \varphi^* - 2\pi/3) \\ \cos(\theta + \varphi^* + 2\pi/3) \end{bmatrix} \quad (4)$$

The input effective current reference I^* is obtained from (1) - (4).

$$I^* = \frac{V_c^* I_{dc}}{\sqrt{2}\{e_u\cos(\theta+\varphi^*) + e_v\cos(\theta+\varphi^* - \frac{2\pi}{3}) + e_u\cos(\theta+\varphi^* + \frac{2\pi}{3})\}} \quad (5)$$

III. ANALYSIS OF PWM METHODS

A. Basic principle of duty cycles

The input phase pattern changes for every $\pi/3$ of the input phase voltages e_u, e_v and e_w, and therefore six patterns (I-VI) occur for a single cycle of three phase source. The duty cycles of six switches are $d_{\alpha p} - d_{\gamma n}$, respectively. The current references i_α^*, i_β^* and i_γ^* are highest, intermediate and lowest of input current references i_u^*, i_v^* and i_w^*, respectively and they are expressed in terms of duty cycles and output current I_{dc} in (6) - (8).

$$i_\alpha^* = (d_{\alpha p} - d_{\alpha n})I_{dc} \quad (6)$$
$$i_\beta^* = (d_{\beta p} - d_{\beta n})I_{dc} \quad (7)$$
$$i_\gamma^* = (d_{\gamma p} - d_{\gamma n})I_{dc} \quad (8)$$

The time T_s is the control period of the circuit and in order to ensure continuous output current and preventing short circuit of input line voltages only one among the three switches on each output phase should be conducting during this control period. Therefore the following relationship of duty cycles are obtained.

$$d_{\alpha p} + d_{\beta p} + d_{\gamma p} = 1 \quad (9)$$
$$d_{\alpha n} + d_{\beta n} + d_{\gamma n} = 1 \quad (10)$$

Fig.2 shows the principle PWM control of all the patterns for the high and low output voltage reference V_c^*. During the control period T_s only four switches are conducting and the remaining two switches are OFF to reduce the number of commutations. Therefore duty cycles of the switches during each pattern are obtained from (6) - (10). Table.I summarizes the duty cycles for the six patterns. The input current references are in phase with the corresponding phase voltages and the duty cycles are realized through the input current references. The output phase p is connected between the maximum input current phase i_α^* and intermediate phase i_β^* and output phase n is connected between intermediate current phase i_β^* and minimum phase i_γ^*. Thus output phase p is commutated between the input maximum voltage phase and input intermediate phase, and the output phase n is commutated input intermediate phase and minimum phase and

therefore only two commutations appear during the control period which makes three voltage levels to appear on the output side of the converter.

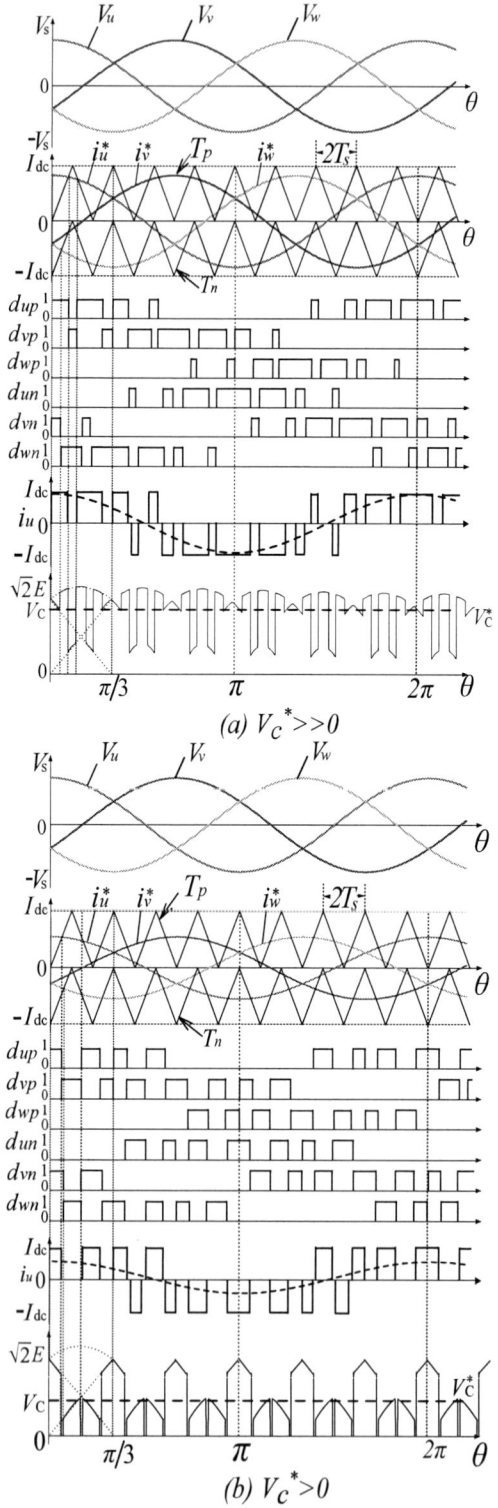

Fig. 2. Principle PWM control

The output voltage reference is realized by the average of the three output voltages of the converter during the control

TABLE I
DUTY CYCLES OF SWITCHES FOR DIFFERENT PATTERNS

Pattern	Voltage Levels			Duty cycles					
	High	Medium	Low	d_{up}	d_{vp}	d_{wp}	d_{un}	d_{vn}	d_{wn}
I	e_u	e_v	e_w	$\dfrac{i_u^*}{I_{dc}}$	$1 - \dfrac{i_u^*}{I_{dc}}$	0	0	$1 + \dfrac{i_w^*}{I_{dc}}$	$-\dfrac{i_w^*}{I_{dc}}$
II	e_v	e_u	e_w	$1 - \dfrac{i_v^*}{I_{dc}}$	$\dfrac{i_v^*}{I_{dc}}$	0	$1 + \dfrac{i_w^*}{I_{dc}}$	0	$-\dfrac{i_w^*}{I_{dc}}$
III	e_v	e_w	e_u	0	$\dfrac{i_v^*}{I_{dc}}$	$1 - \dfrac{i_v^*}{I_{dc}}$	$-\dfrac{i_u^*}{I_{dc}}$	0	$1 + \dfrac{i_u^*}{I_{dc}}$
IV	e_w	e_v	e_u	0	$1 - \dfrac{i_w^*}{I_{dc}}$	$\dfrac{i_w^*}{I_{dc}}$	$-\dfrac{i_u^*}{I_{dc}}$	$1 + \dfrac{i_u^*}{I_{dc}}$	0
V	e_w	e_u	e_v	$1 - \dfrac{i_w^*}{I_{dc}}$	0	$\dfrac{i_w^*}{I_{dc}}$	$1 + \dfrac{i_v^*}{I_{dc}}$	$-\dfrac{i_v^*}{I_{dc}}$	0
VI	e_u	e_w	e_v	$\dfrac{i_u^*}{I_{dc}}$	0	$1 - \dfrac{i_u^*}{I_{dc}}$	0	$-\dfrac{i_v^*}{I_{dc}}$	$1 + \dfrac{i_v^*}{I_{dc}}$

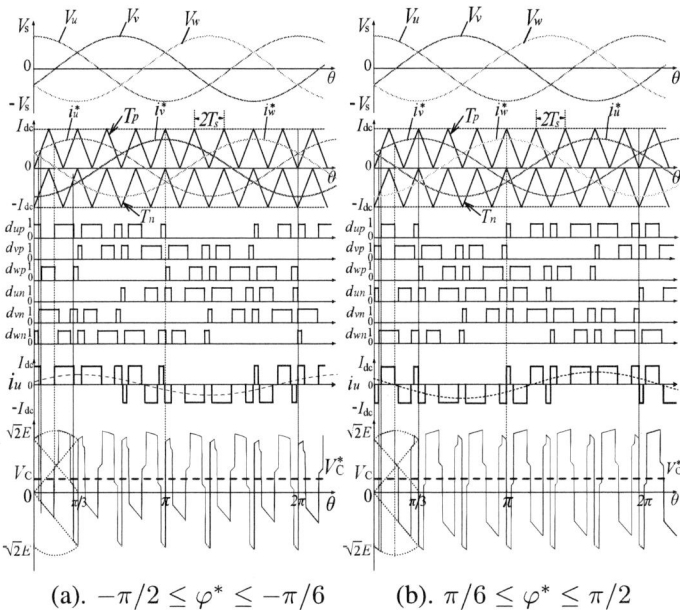

(a). $-\pi/2 \leq \varphi^* \leq -\pi/6$ (b). $\pi/6 \leq \varphi^* \leq \pi/2$

Fig. 3. PWM control for power factor control

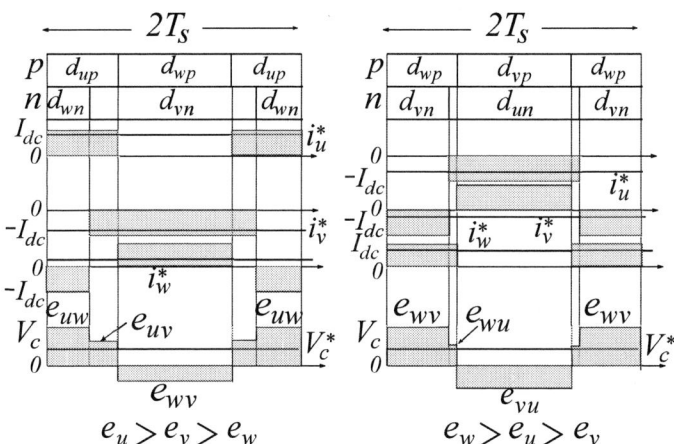

Fig. 4. Conventional switching waveforms

period. From Fig.2(a) for high voltage reference V_c^* the output voltage levels are close to the reference level. From Fig.2(b) for low voltage reference V_c^* the output voltage consists of two low levels and a zero voltage. This PWM control method has restrictions towards the input phase angle φ^* with limits of the control range of $-\pi/6 \leq \varphi^* \leq \pi/6$ and $5\pi/6 \leq \varphi^* \leq 7\pi/6$ as verified in [8].

B. Conventional PWM Strategy

Fig.3 shows the additional two switching strategies for improving the control ranges of the input phase reference φ^* therefore increasing the control range of the input power factor angle from 0 to 2π for low power ranges[10]. Fig.3(a) shows the switching pattern method for a lagging power factor control ranging from $-\pi/2 \leq \varphi^* \leq -\pi/6$ and $\pi/2 \leq \varphi^* \leq 5\pi/6$. Fig.3(a) shows the switching pattern method for a leading power factor control ranging from $\pi/6 \leq \varphi^* \leq \pi/2$ and $7\pi/6 < \varphi^* < 3\pi/2$. Because of this switching methods

the minimum voltage phase is connected to output phase p and maximum voltage phase connected to the output phase n. Therefore during the control period two positive and a negative voltage appears on the output side of the converter to realize a low output reference voltage. The negative voltage gradually increases and reaches maximum when one of the line voltage is zero.

Fig.4 shows the partial waveforms of input reference currents and output voltage in the conventional PWM strategy with the low output voltage reference of V_c^* for a input phase range of $-\pi/2 \leq \varphi^* \leq -\pi/6$ shown in Fig.3(a). During the control period T_s for the condition $e_u > e_v > e_w$ the output phase p is commutated from maximum input phase u and minimum input phase w and output phase n is commutated from minimum input phase w and intermediate input phase v. The duty cycles $(d_{up} - d_{wn})$ of the switches $S_{up} - S_{wn}$ during $0 < \theta < \pi/3$ are given in (11).

$$\left. \begin{array}{lll} d_{up} = \dfrac{i_u^*}{I_{dc}}, & d_{vp} = 0, & d_{wp} = 1 - \dfrac{i_u^*}{I_{dc}} \\[3mm] d_{un} = 0, & d_{vn} = -\dfrac{i_v^*}{I_{dc}}, & d_{wn} = 1 + \dfrac{i_v^*}{I_{dc}} \end{array} \right\} \quad (11)$$

Similarly during $e_w > e_u > e_v$ the output phase p is commutated from maximum input phase w and minimum input phase v and output phase n is commutated from minimum input phase v and intermediate input phase u. In both the conditions, output waveforms have two positive voltages and a negative voltage in both the cases. A large ripple is formed as the negative voltage is approaching maximum when a line voltage is zero.

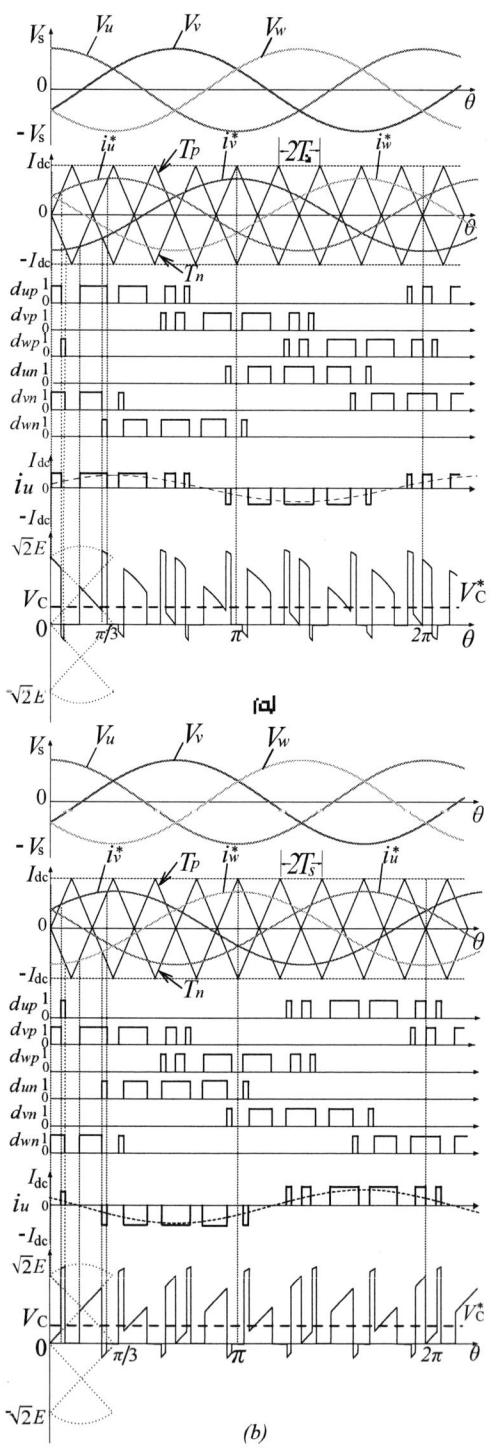

Fig. 5. Proposed PWM control

(a). $-\pi/2 \leq \varphi^* \leq -\pi/6$ (b). $\pi/6 \leq \varphi^* \leq \pi/2$

C. Proposed PWM Strategy

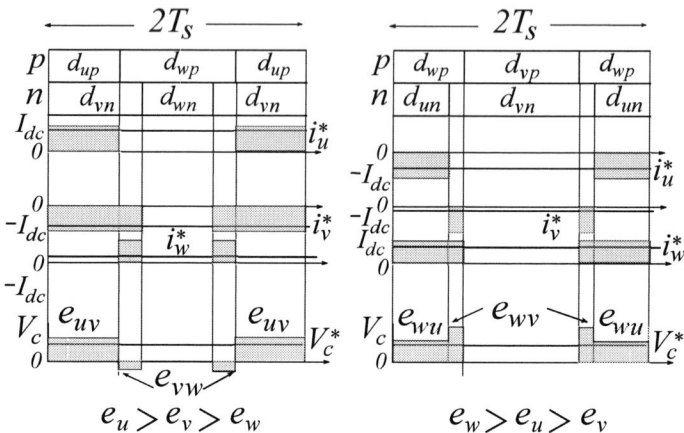

Fig. 6. Proposed switching waveforms

Fig.5 shows the overall duty cycles for all the six patterns by the proposed PWM control method. The input current waveforms consists of current pulses with the same sign as the input current reference which can reduce the harmonics of the switching frequency. In the proposed PWM method the output phase p is commutated from maximum input phase and minimum input phase and output phase n is commutated from intermediate input phase and minimum input phase in order to improve the output waveform of the converter. This is achieved the changing the phase of the carrier wave from which the position of the commutation is decided. In conventional method the two carrier waves T_p and T_n are in phase with each other whereas in the proposed method the carrier waves T_p and T_n are out of phase with each other.

Fig.6 shows the partial waveforms of input references currents and output voltage in the proposed control method for low output voltage reference V_c^* for a input phase range of $-\pi/2 \leq \varphi^* \leq -\pi/6$ shown in Fig.5(a). The input voltage conditions ($e_u > e_v > e_w$ and $e_w > e_u > e_v$) discussed in previous sub-section is considered for comparison. The duty cycles ($d_{up} - d_{wn}$) of the switches $S_{up} - S_{wn}$ during $0 < \theta < \pi/3$ are given in (12).

$$\left. \begin{array}{lll} d_{up} = \dfrac{i_u^*}{I_{dc}}, & d_{vp} = 0, & d_{wp} = 1 - \dfrac{i_u^*}{I_{dc}} \\[2mm] d_{un} = 0, & d_{vn} = -\dfrac{i_v^*}{I_{dc}}, & d_{wn} = 1 + \dfrac{i_v^*}{I_{dc}} \end{array} \right\} \quad (12)$$

The output waveforms have a positive and a negative voltage and a zero voltage in the first case and two positive voltage levels and a zero voltage during the second case. Therefore, the output never reaches a negative voltage when a line voltage is zero and finally the output ripple is reduced. The output voltage of the converter V_c is restricted to reach a maximum negative voltage such that ripple on the DC voltage is minimized when compared to the conventional switching pattern in Fig.4.

The 2014 International Power Electronics Conference

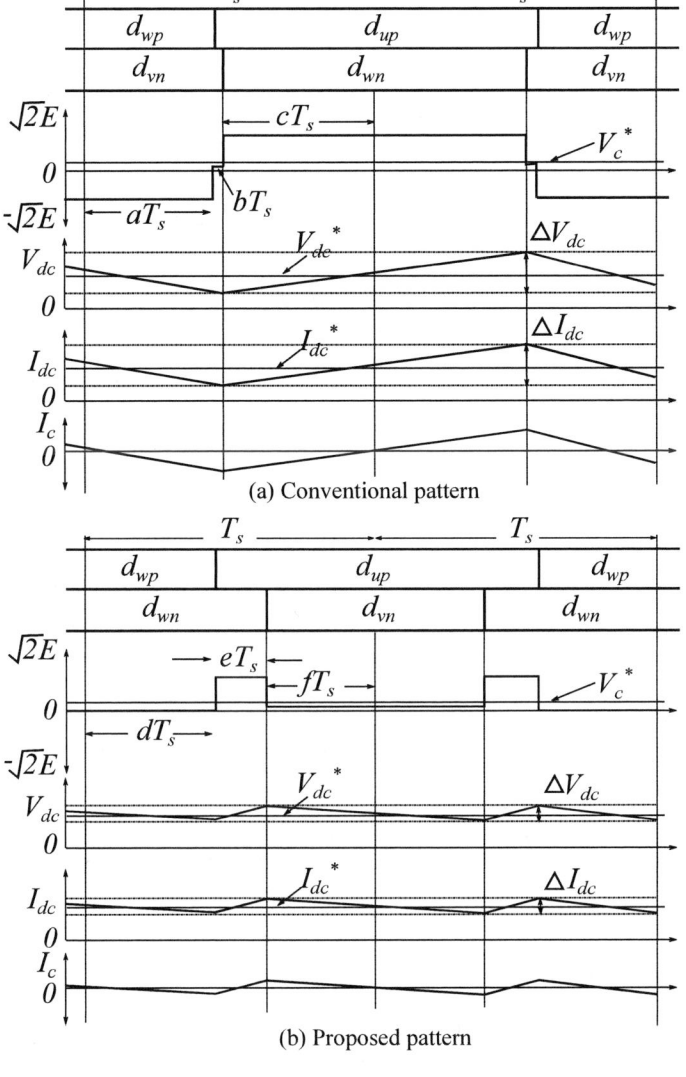

(a) Conventional pattern

(b) Proposed pattern

Fig. 7. Ripple analysis

IV. ANALYSIS OF OUTPUT VOLTAGE RIPPLE

The proposed PWM strategy reduces the output ripple by using voltage levels close to the output voltage reference. When one of the line voltage is zero the absolute value of the other two line voltages is $\sqrt{6}/2E$. In the conventional pattern when one of the line voltage is zero, the small output voltage reference is realized by $\sqrt{6}/2E$, a small positive voltage and $\sqrt{6}/2E$ as shown in Fig.7(a). But in the case of proposed pattern, as shown in Fig.7(b), when one of the line voltage is zero the output voltage reference is realized by $\sqrt{6}/2E$, a small positive voltage and zero voltage. The ripple of the DC voltage and the output current are maximum when ripple on the output voltage is maximum. The DC voltage is controlled at rated voltage V_{dc}^*, and the output side voltage equation is given as:

$$V_c - V_{dc}^* = L\frac{dI_{dc}}{dt} \qquad (13)$$

During the control period three voltage levels appear due to two commutations and the sum of rates of the three voltage levels is equal to the control period T_s. The rates of three voltages for conventional and proposed patterns shown in Fig.7 are given in (14).

$$a + b + c = T_s \quad , \quad d + e + f = T_s \qquad (14)$$

Each rate of the control period is given as per their respective line voltage and is expressed in (15).

$$a = \frac{V_c^*}{E_{la}} \qquad (15)$$

The maximum output ripple ΔI_{dc} of the conventional pattern is given by:

$$\Delta I_{dc} = (a + b + c)T_s \left|\frac{dI_{dc}}{dt}\right| \qquad (16)$$

From (13),(15) and (16) the maximum output ripple ΔI_{dc} is given in (17), where V_{ca}, V_{cb} and V_{cc} are the output voltages at the rates a, b and c respectively.

$$\Delta I_{dc} = \frac{V_c^*}{L}\left[\frac{|V_{ca} - V_{dc}^*|}{E_{la}} + \frac{|V_{cb} - V_{dc}^*|}{E_{lb}} + \frac{|V_{cc} - V_{dc}^*|}{E_{lc}}\right]T_s \qquad (17)$$

The ripple component of the DC voltage is caused by the output current ripple and is maximum when the output current has it s maximum ripple. From Fig.7(a) the DC voltage ripple ΔV_{dc} of the conventional pattern is as follows:

$$\Delta V_{dc} = \frac{1}{C}\left|\int_0^{aT_s} i_c dt + \int_{aT_s}^{bT_s} i_c dt + \int_{bT_s}^{cT_s} i_c dt\right| \qquad (18)$$

Similarly from Fig.7(b) the current ripple of the proposed patten is given in (19) and (20).

$$\Delta I_{dc} = (d + e + f)T_s \left|\frac{dI_{dc}}{dt}\right| \qquad (19)$$

$$\Delta I_{dc} = \frac{V_c^*}{L}\left[\frac{|V_{ce} - V_{dc}^*|}{E_{le}} + \frac{|V_{cf} - V_{dc}^*|}{E_{lf}}\right]T_s \qquad (20)$$

The DC voltage ripple caused by proposed switching patten is stated as:

$$\Delta V_{dc} = \frac{1}{C}\left|\int_0^{dT_s} i_c dt + \int_{dT_s}^{eT_s} i_c dt + \int_{eT_s}^{fT_s} i_c dt\right| \qquad (21)$$

In (17) three ripple components are present causing the ripple to increase but in the case of (20) only two ripple components occur as one of the output voltage is zero so finally reducing the ripple.

The 2014 International Power Electronics Conference

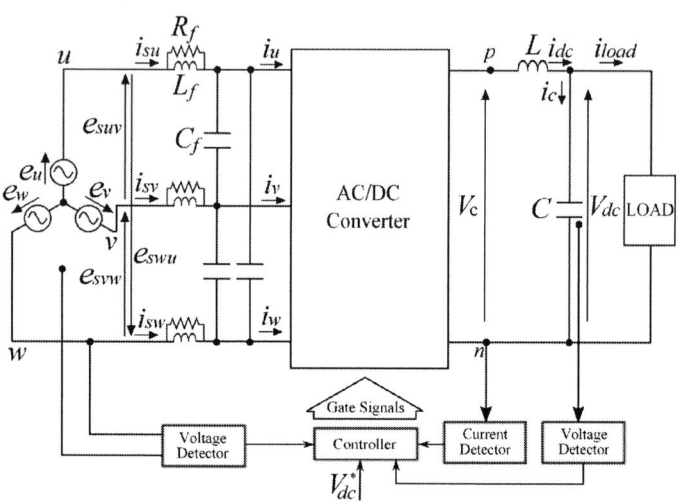

Fig. 8. Experimental configuration

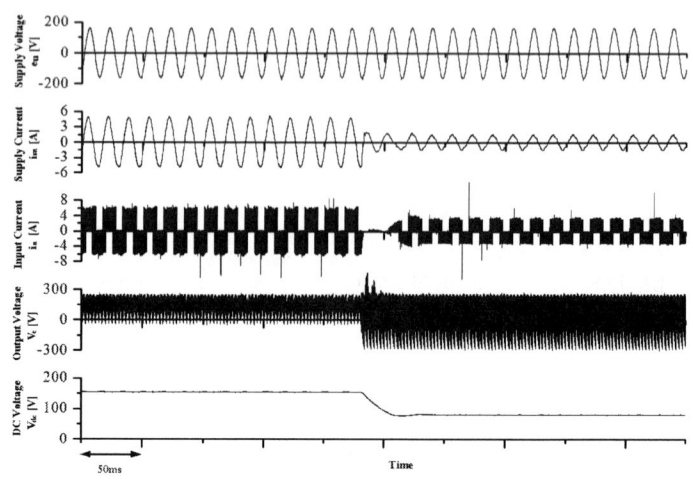

Fig. 9. Overall waveforms by conventional PWM strategy

TABLE II
EXPERIMENTAL CONDITION

Source voltage E, ω	200 V, $2\pi \times 60$ rad/s
DC voltage reference V_{dc}^*	160 V, 80 V
Power factor	−0.43,
angle reference φ^*	−1.11
Output Power P_{out}	1 KW, 240 W
Input filter L_f, C_f, R_f	1.0 mH, 10.47 μF, 47 Ω
Load R	26 Ω
Inductance L	20 mH
Capacitor C	1500 μF
Carrier frequency f_s	10 kHz

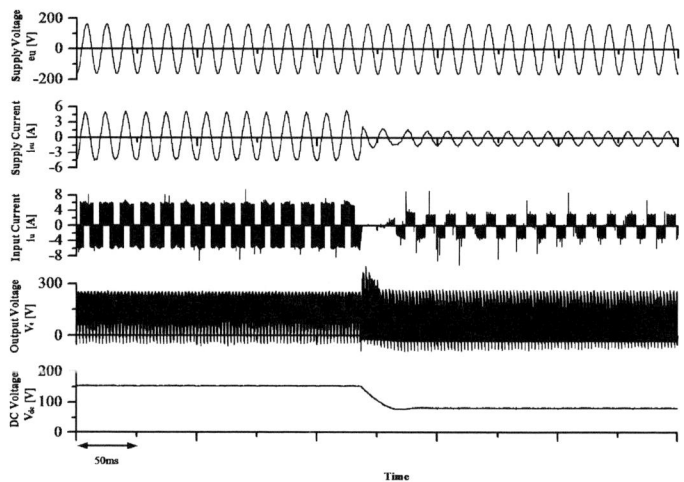

Fig. 10. Overall waveforms by proposed PWM strategy

V. RESULTS AND DISCUSSIONS

A. Experimental Conditions

Fig.8 shows the experimental configuration and Table.II lists the experimental specifications with the supply phase voltage of 200 V and 60 Hz. Fig.9 and Fig.10 shows the overall waveforms of the conventional and the proposed PWM strategies, respectively when the DC voltage reference V_{dc}^* is changed from 160 V to 80 V with a supply phase voltage of 200 V and 60 Hz. In the conventional method when the voltage reference is 80 V the input phase reference shifted by an angle of 64° for maintaining unity power factor so in this condition the output voltage V_c varies between maximum and minimum line voltage. But in the proposed PWM method the output voltage V_c varies between positive line voltage and a small negative line voltage. Fig.11 and Fig.12 shows the partial waveforms of Fig.9 and Fig.10, respectively when the output voltage reference is 80 V. In both the conditions the input supply current is in phase with the corresponding supply voltage with a phase reference φ^* equal to 64°. As derived in Fig.11 the conventional method generates high ripple when the output voltage waveform reaches maximum negative value when one of the line voltage is zero. But in case of the

proposed method the output voltage waveform never reaches maximum and because of this condition the ripple on the DC voltage is low. And also proposed method can reduce the harmonics of the switching frequency component of the input current as the input current waveforms consist of current pulses with the same sign as the input current reference.

Fig.13 shows the partial waveforms of the output voltage V_c and output DC voltage for a output voltage reference of $V_{dc}^* = 80$ V. The proposed PWM strategy reduces the average of the DC voltage by three voltages levels composed of a zero voltage so reducing the ripple to 0.1 V, whereas in the conventional PWM strategy the a negative voltage appearing during every control period and its level increases when one of line voltage is approaching zero, so the ripple on the DC voltage is 0.45 V which is four times the ripple caused by the proposed method. Fig.14 (a) and (b) shows the harmonic spectral analysis of the output voltages of conventional and proposed control methods respectively. When comparing both the frequency analysis the ripple caused by the conventional method is very high at the carrier frequency than the proposed control method. The

978-1-4799-2706-7/14 $31.00 © 2014 IEEE

The 2014 International Power Electronics Conference

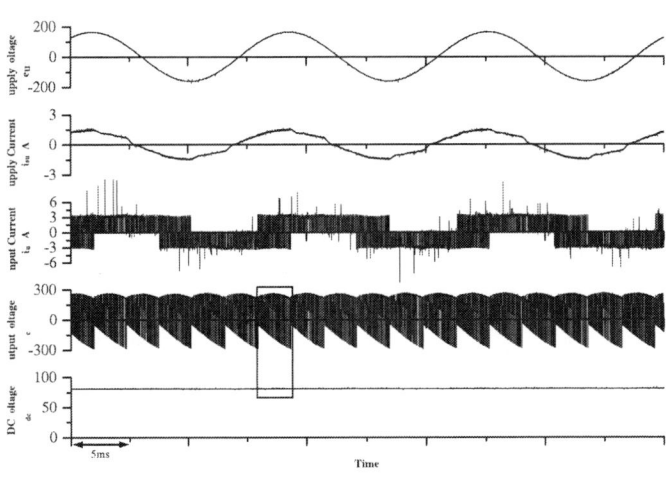

Fig. 11.   Partial waveforms of conventional method

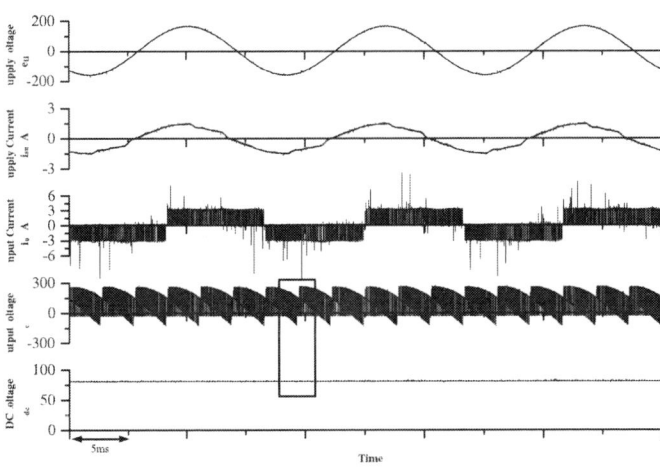

Fig. 12.   Partial waveforms of proposed method

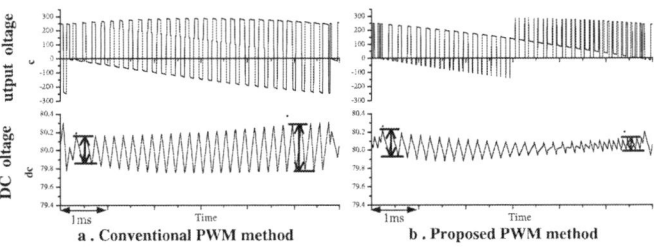

a . Conventional PWM method   b . Proposed PWM method

Fig. 13.   Output waveforms comparison

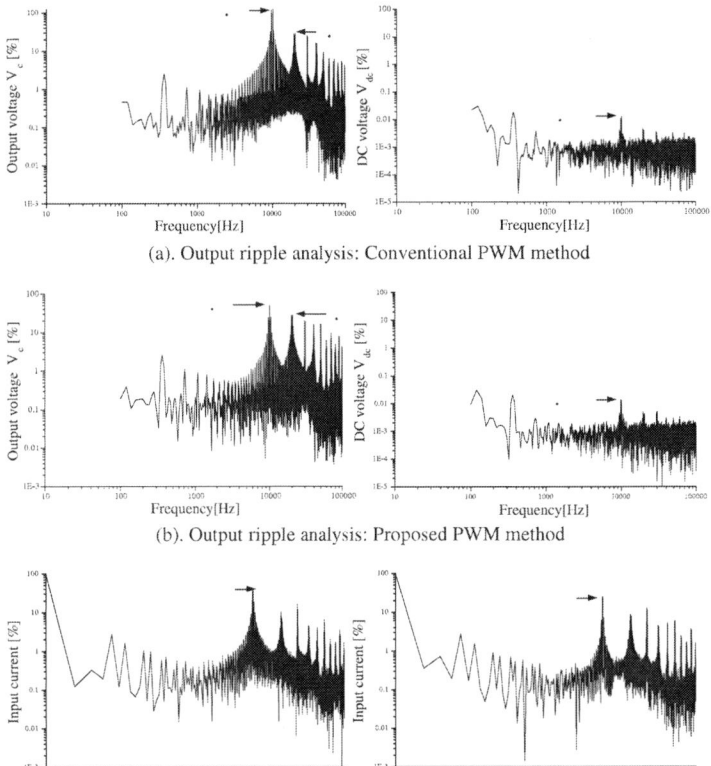

(a). Output ripple analysis: Conventional PWM method

(b). Output ripple analysis: Proposed PWM method

(c). Input current harmonic analysis: (d). Input current harmonic analysis:
Conventional PWM method   Proposed PWM method

Fig. 14.   Ripple analysis of output voltage and input current of converter

output voltage ripple and DC voltage ripple have be been reduced by 58.18% and 21.42%, respectively by proposed PWM strategy by preventing high negative line voltage to appear on the output voltage of the converter. Fig.14 (c) and (d) shows the analysis of the harmonics of the input current and the percentage of each harmonic component when the fundamental wave is considered to be 100%. For the conventional pattern the percentage of input current harmonics at the carrier wave frequency $\omega_c = 2\pi$ x 10 krad/s is 45%. But for the proposed patten the input current harmonics is reduced to 23% as the input current waveforms contains current pulses of same sign as the input current reference.

VI. CONCLUSION

This paper presents a PWM method for a single-stage AC/DC converter for reducing the output voltage ripple by controlling the negative voltage level on the output side of the converter. The converter also maintains the high input power factor range and reduces the input current harmonics. The effectiveness is verified using the experimental results.

ACKNOWLEDGEMENT

This study is supported in a part by Power Academy Japan Grant-in-Aid for Exploratory Research (doctoral student frame) 2013-14, "Development of single stage bi-directional AC/DC converter for battery charging system", and we express our gratitude to all the persons concerned.

978-1-4799-2706-7/14 $31.00 © 2014 IEEE   2316

REFERENCES

[1] M. Daniele, P. K. Jain and G. Joos, "A Single- Stage Power-Factor-Corrected AC/DC Converter", IEEE Trans. Power Electron., Vol. 14, No. 6, pp. 1046-1055, November 1999.

[2] Q.Zhao, F.C.Lee and F.S.Tsai, "Voltage and Current Stress Reduction in Single-Stage Power Factor Correction AC/DC Converters With Bulk Capacitor Voltage Feedback", IEEE Trans. Power Electron., Vol. 17, No. 4, pp. 477-484, July 2002.

[3] W.Y.Choi and J.S. Yoo,"A Bridgeless Single-Stage Half-Bridge AC/DC Converter", IEEE Trans. Power Electron., Vol. 26, No. 12, pp. 3884-95, December 2011.

[4] M. S. Agamy and P. K. Jain,"An Adaptive Energy Storage Technique for Efficiency Improvement of Single-Stage Three-Level Resonant AC/DC Converters", IEEE Trans. Ind. Electron., Vol. 47, No. 1, pp. 176-184, Jan/Feb 2011.

[5] S.K. Ki and D. Lu, "A High Step-Down Transformerless Single-Stage Single-Switch AC/DC Converter", IEEE Trans. On Power Electron., Vol. 28, pp. 36-45, January 2013.

[6] D. S. Wijeratne and G. Moschopoulos, "A Comparative study of Two Buck-Type Three-Phase Single-Stage AC-DC Full-Bridge Converters", IEEE Trans. Power Electron., Vol. 29, No. 4, pp. 1632-1645, Apr. 2014.

[7] Y.Hu, L. Huber and M.M. Jovanovic: "Single-Stage, Universal-Input AC/DC LED Driver with Current-Controlled Variable PFC Boost Inductor", IEEE. Trans. On Power Electronics, Vol. 27, pp.1579-98.

[8] Y.Onoe, W. Kitagawa and T. Takeshita: "PWM Strategy of Bi-directional Buck AC/DC Converter for suppressing Output Voltage Ripple", 2011 Annual Conference of IEEJ, Industry Applications Society (2011 - 9).

[9] S.Ishikawa and T. Takeshita: "Input Power Factor Control of Three-Phase to Three-Phase Matrix Converters", IEEJ Trans IA, Vol. 178, No. 3, pp. 42-52, 2012.

[10] B.Veerasamy, W. Kitagawa and T. Takeshita: "Input Power Factor Control of Bi-directional AC/DC Converter", 2013 IEEE International conference on Power Electronics and Drives(PEDS), pp. 1103 - 1108.

Three-Phase Isolated Full-Bridge Boost PFC with Flyback Passive Auxiliary Converter

Tao Meng, Shuai Yu, Hongqi Ben, Guo Wei, Shaohua Sun
School of Electrical Engineering and Automation
Harbin Institute of Technology
Harbin, China

Abstract—A three-phase isolated full-bridge boost power factor correction (PFC) converter with a flyback passive auxiliary converter is presented and investigated. The PFC converter operates in discontinuous conduction mode (DCM), it can achieve zero voltage turning off for the upper switches and zero voltage turning on for the lower switches. With the adoption of the passive auxiliary converter, the voltage spike of the PFC converter is suppressed efficiently, and during one charging period, the absorbed energy can be transferred to the load by itself, from which the stresses of main circuit can be reduced. The operational process of the PFC converter is analyzed in details, and design considerations of both the PFC converter and the auxiliary converter are given. Finally, the feasibility of the presented method and the validity of the theoretical analysis are verified by the experimental results.

Keywords—single-stage power factor correction (PFC), three-phase, passive auxiliary converter, flyback

I. INTRODUCTION

Power factor correction (PFC) is an important researching orientation in power electronics field. Compared with the two-stage PFC, single-stage PFC integrates the function of PFC and DC/DC conversion, and it has the advantages such as high efficiency, simplicity and low cost [1]. Presently, many low power single-stage PFC converters have been investigated, however, fewer high power schemes, especially very few the high power three-phase schemes [2].

In medium or high power field, the isolated full-bridge topology is attractive in single-stage PFC. However, the topology itself has a serious problem: due to the existence of the transformer leakage inductance, there is a large voltage spike across the bridge leg, which will increase the voltage stress of switches and decrease the reliability [3, 4]. To suppress the voltage spike, a number of techniques have been proposed. For example, the active clamping methods have been introduced in [5-7], A two-switch clamping circuit is presented in [8], and some active auxiliary circuits with single-switch are adopted in [6, 9]. The voltage spike is efficiently suppressed after adoption of each of the active methods above. However, for these methods, one (or two) additional switch is introduced, which increases the complexity of control circuit. Besides the active methods, some passive methods have also been proposed. For example, a *LC* resonance scheme is studied in [10, 11], which can also achieve soft switching of the switches, however, its resonance energy can not be transferred to the load but add to the conduction losses. A *RCD* snubber is used in [12], but its energy is released by the resister. A passive clamping technique is proposed in [13], however, the problem of magnetic bias of the transformer appears after adoption of this method. A passive snubber is investigated in [14], however, after adoption of this snubber, a diode is connected in series with the bridge leg switches, which will increase the losses. In [15], an improved passive snubber is investigated, which can overcome the disadvantage of the snubber in [14]. The two passive snubbers in [14, 15] have a common drawback: the absorbed energy is transferred to the load through the main circuit, which increases the losses of the main circuit.

In this paper, a novel flyback passive auxiliary converter is proposed and investigated in a three-phase single-stage isolated full-bridge PFC converter. Compared with the snubbers investigated in [14, 15], the absorbed energy of this auxiliary converter is transferred to the load by itself, which decreases the stresses of the main circuit.

II. OPERATIONAL PRINCIPLE

The three-phase isolated full-bridge boost PFC converter with flyback passive auxiliary converter is shown in Fig.1 (a), where L_a, L_b, L_c ($L_a=L_b=L_c=L$) are the boost inductors, $D_1\sim D_4$ and $C_1\sim C_4$ ($C_1=C_2=C_3=C_4$) are the parasitic components of $S_1\sim S_4$, L_{lk} and n are the equivalent leakage inductance and turns ratio of transformer T. The auxiliary converter is composed of C_{C1}, C_{C2} ($C_{C1}=C_{C2}$), D_{L1}, D_{L2}, D_C, D_f and the flyback transformer T_f. Magnetic circuit structure of T_f is shown in Fig.1 (b), where two primary windings and one secondary windings are made on a common magnetic core, L_1 and L_2 ($L_1=L_2$) are equivalent inductance of the primary side, L_f is inductance of the secondary side, and n_f is the turns ratio.

The following analysis is in the phase of $0\leq\omega t\leq\pi/6$, where $u_{an}=U\sin\omega t$, $u_{bn}=U\sin(\omega t-2\pi/3)$, $u_{cn}=U\sin(\omega t+2\pi/3)$, and $u_{bn}\leq0\leq u_{an}\leq u_{cn}$. To simplify the analysis, we assumed that: 1) all devices are ideal, 2) the capacitor C is large enough, so the output voltage U_o can be considered as a constant value, 3) during one charging period of the boost inductors (T), the change of u_{an}, u_{bn} and u_{cn} are negligible

978-1-4799-2706-7/14 $31.00 © 2014 IEEE

because T is much shorter than the line period.

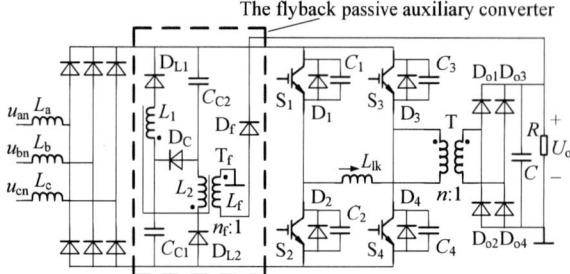

(a) The PFC converter with flyback passive auxiliary converter

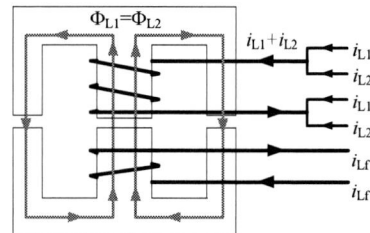

(b) Magnetic circuit structure of the T_f

Fig.1 The proposed configuration

The PFC converter operates in DCM. When the bridge leg are shorted (S_1, S_2 or S_3, S_4 turns on), L_a, L_b, L_c are charged by u_{an}, u_{bn}, u_{cn}, and the input current increases from zero almost linearly. When the bridge diagonal-leg switches turn on (S_2, S_3 or S_1, S_4 turn on), the output current is provided by u_{an}, u_{bn}, u_{cn} and L_a, L_b, L_c, and the input current decreases. The process above is repeated periodically, the discontinuous input current follows the envelopes which are proportional to the input voltage, and both PFC and AC/DC conversion can be achieved.

The following is the operational process analysis of the PFC converter during one charging period. The equivalent circuit of each stage and the theoretical waveforms are shown in Fig.2 and Fig.3 respectively.

Stage 1 (before t_0): S_2, S_3 are turning on and S_1, S_4 are turning off. The PFC converter operates in DCM, i_{La}, i_{Lb}, i_{Lc} has been reduced to zero before t_0, and then the current in both primary and secondary sides of T is zero. The voltage across the primary side of T: $U_{C1}=U_{C4}=nU_o$, $U_{C2}=U_{C3}=0$ and $U_{Cc1}=U_{Cc2}=U_{Cc}$ (It is defined that $2U_{Cc}=anU_o$, $a>1$). In this stage, the output current is only provided by capacitor C.

Stage 2 (t_0-t_1): At t_0, S_1 turns on, and S_3 turns off with zero voltage. i_{La}, i_{Lb}, i_{Lc} increases linearly with the charging of u_{an}, u_{bn}, u_{cn}. In the auxiliary converter, L_1, L_2 are charged by C_{C1}, C_{C2}. The capacitors C_{C1} and C_{C2} are large enough, so the decreasing of their voltage can be ignored in this stage. The current of L_1, L_2 is:

$$i_{L1/L2}(t) = anU_ot/2L_1 \qquad (1)$$

At t_1, i_{L1}, i_{L2} reaches the maximum value in the whole charging period. During this stage, $U_{C1}=U_{C2}=U_{C3}=U_{C4}=0$, $U_{Cc1}=U_{Cc2}=anU_o/2$, and the output current is only provided by capacitor C.

Stage 3 (t_1-t_2): At t_1, S_2 turns off, and S_4 turns on with zero voltage. L_1 is connected in series with L_2, and C_2, C_3 are charged by L_a, L_b, L_c and L_1, L_2. At t_2, $U_{C1}=U_{C4}=0$ and

$U_{C2}=U_{C3}=anU_o$. The value of C_2, C_3 is very small, so the current decreasing in L_a, L_b, L_c, L_1, L_2 and the duration of this stage can be ignored. In this stage, the output current is only provided by capacitor C.

Stage 4 (t_2-t_3): In this stage, the energy in L_1 and L_2 is transferred to L_f entirely, and the current in L_f is decreasing. The expression of i_{Lf} can be obtained:

$$i_{Lf}(t) = anU_oDTn_f/L_1 - U_ot/L_f \qquad (2)$$

Where $D=(t_1-t_0)/T$ is the duty cycle of the PFC converter.

After t_2, C_{C1}, C_{C2} and C_2, C_3 are charged by L_a, L_b and L_c (the capacitance of C_{C1}, C_{C2} is very large, so the increasing of their voltage can be ignored), i_{La}, i_{Lb}, i_{Lc} decreases, i_{Llk} increases from zero, D_{o2}, D_{o3} are turning on, and the input energy is transferred to the load through T. In this stage, the expression of i_{Llk} is:

$$i_{Llk}(t) = (a-1)nU_ot/L_{lk} \qquad (3)$$

At t_3, $i_{Llk}(t_3)=-i_{Lb}$. The value of L_{lk} is very small, so the varying of i_{La}, i_{Lb}, i_{Lc} can be ignored during the this stage. The duration of this stage is:

$$t_{23} = -i_{Lb}L_{lk}/(a-1)nU_o \qquad (4)$$

Stage 5 (t_3-t_4): At t_3, $U_{C2}=U_{C3}=anU_o$, $i_{Llk}(t_3)=-i_{Lb}$, and after t_3, U_{C2}, U_{C3} begins to decrease, U_{Cc1}, U_{Cc2} can not decrease because of the existing of D_C, and i_{Llk} still increases. So the following relationships can be obtained:

$$i_{Llk}(t) + i_{C2}(t) + i_{C3}(t) = -i_{Lb} \qquad (5)$$

$$i_{C2}(t) + i_{C3}(t) = 2C_2d\Delta u_{C2}(t)/dt \qquad (6)$$

$$\Delta u_{C2}(t) + (a-1)nU_o = L_{lk}di_{Llk}(t)/dt \qquad (7)$$

Where $i_{C2}(t)$, $i_{C3}(t)$ is the charging current of C_2, C_3, $\Delta u_{C2}(t)$ is the increasing of the voltage of C_2, C_3 after t_3.

From (5), (6) and (7), it can be obtained that:

$$\Delta u_{C2}(t) + 2L_{lk}C_2d^2\Delta u_{C2}(t)/dt^2 = -L_{lk}i_{Lb}' - (a-1)nU_o \qquad (8)$$

In this stage, i_{La}, i_{Lb}, i_{Lc}, i_{Lf} decreases almost linearly, so the differential of i_{Lb} in (8) can be considered as a constant value. Because $\Delta u_{C2}(t_3)=0$, $i_{Llk}(t_3)=-i_{Lb}$, $i_{C2}(t_3)+i_{C3}(t_3)=0$. Therefore, the solution of (8) is:

$$i_{Llk}(t) = -i_{Lb} + [L_{lk}i_{Lb}' + (a-1)nU_o]\sqrt{2C_2/L_{lk}}\sin\frac{t}{\sqrt{2L_{lk}C_2}}) \qquad (9)$$

It can be seen that after t_3, there is a resonance within C_2, C_3 and L_{lk}, and the input energy is transferred to the output side at the same time. The value of C_2, C_3 is very small, so the resonant amplitude is much lower than the average value, and the resonant frequency is much higher than the charging frequency. It is believed that when i_{La}, i_{Lb}, i_{Lc} reduces to zero, the resonance can be ignored, and at that moment, $U_{C2}=U_{C3}=nU_o$, $i_{Llk}=0$. At t_4, $i_{La}=0$.

Stage 6 (t_4-t_5): In this stage, i_{Lb}, i_{Lc}, i_{Lf} still decreases. At t_5, i_{Lf} reduce to zero.

Stage 7 (t_5-t_6): In this stage, i_{Lb}, i_{Lc} still decreases. At t_6, i_{Lb}, i_{Lc} reduce to zero.

Stage 8 (t_6-t_7): The equivalent circuit of this stage is the same as that of stage 1. The current on the primary side of T is zero and the output current is only provided by capacitor C.

After t_7, the PFC converter operates in the next charging period, and the switching state between S_1 and S_3, S_2 and S_4 are exchanged.

978-1-4799-2706-7/14 $31.00 © 2014 IEEE

The 2014 International Power Electronics Conference

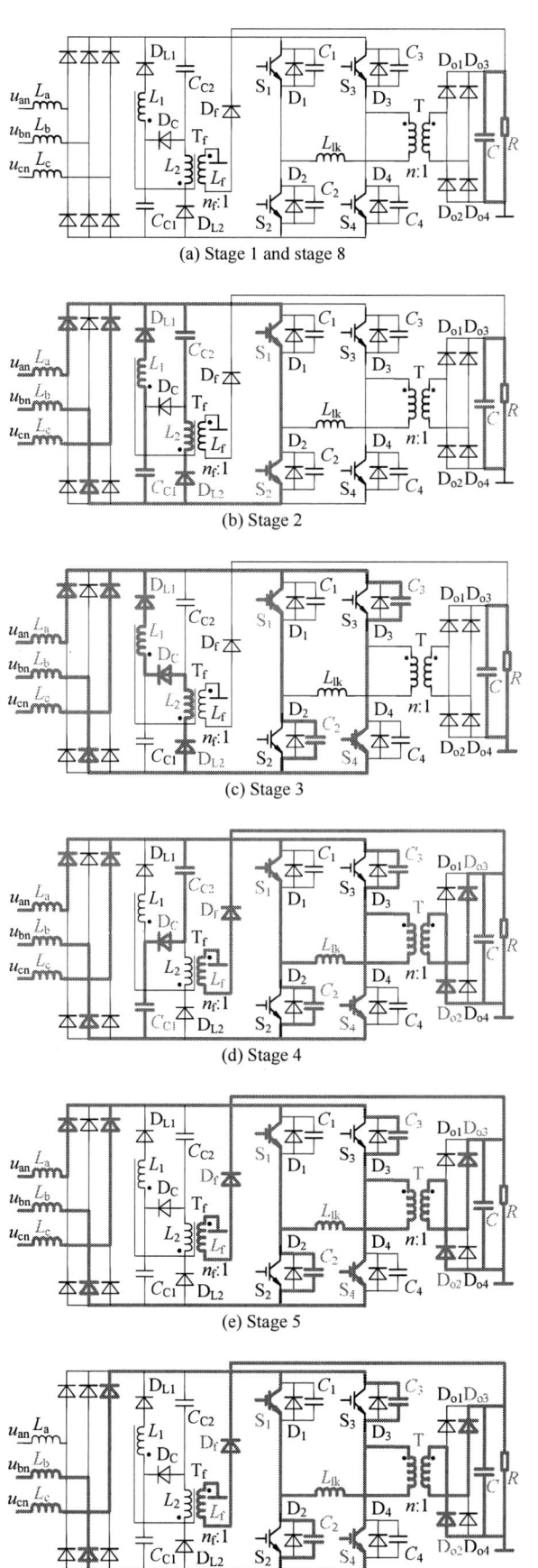

(a) Stage 1 and stage 8

(b) Stage 2

(c) Stage 3

(d) Stage 4

(e) Stage 5

(f) Stage 6

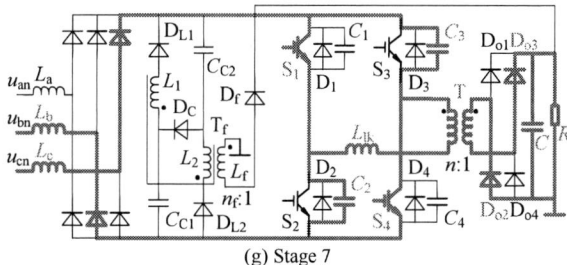

(g) Stage 7

Fig.2 Equivalent circuits

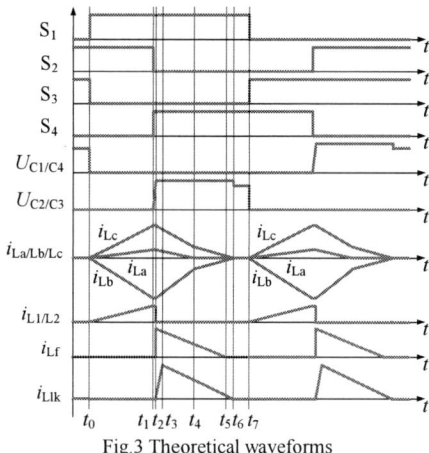

Fig.3 Theoretical waveforms

III. ANSLYSIS AND DESIGN

A. Design Considerations of T_f

T_f can be equivalent to a coupled-inductor, and the basic mathematical model of the coupled-inductor is:

$$\begin{bmatrix} u_{L1} \\ u_{L2} \end{bmatrix} = \frac{d}{dt} \begin{bmatrix} L_{11} & M_{12} \\ M_{12} & L_{22} \end{bmatrix} \begin{bmatrix} i_{L1} \\ i_{L2} \end{bmatrix} \quad (10)$$

Where $L_{11}=L_{22}$ is the self inductance and M_{12} is the mutual inductance.

The two equivalent inductors have a common magnetic circuit, so it can be obtained that: $M_{12}=L_{11}=L_{22}$.

In ideal conditions, it can be obtained: $u_{L1}=u_{L2}$, $i_{L1}=i_{L2}$. Therefore, it can be obtained that:

$$\begin{cases} u_{L1} = L_1 di_{L1}/dt = 2L_{11} di_{L1}/dt \\ u_{L2} = L_2 di_{L2}/dt = 2L_{22} di_{L2}/dt \end{cases} \quad (11)$$

From (11), it can be obtained: $L_1=2L_{11}$, therefore:

$$u_{L1} = L_{11} d(i_{L1} + i_{L2})/dt \quad (12)$$

It can be seen that the coupled-inductor can be equivalent to a single-inductor, of which the inductance is L_{11} and the current is $i_{L1}+i_{L2}$. Furthermore, T_f can be equivalent to a basic flyback transformer with one primary windings (L_{11} is the inductance) and one secondary windings (L_f is the inductance).

B. The Turn Ratio n_f and the Duty Cycle D

In stage 4, the energy of L_1 and L_2 is transferred to L_f entirely, so it can be obtained that:

$$n_f U_o \le U_{Cc} \Rightarrow 2n_f \le an \quad (13)$$

T_f operates in DCM, so it can be obtained:

978-1-4799-2706-7/14 $31.00 © 2014 IEEE

$$anU_oDTn_f/L_1 \le U_o(1-D)T/L_f \qquad (14)$$

From the above analysis, it can be obtained: $n_f^2 = L_1/L_f = 2L_{11}/L_f$. So we can get:

$$2n_f \ge anD/(1-D) \qquad (15)$$

From (13) and (15), the relationship can be obtained:

$$aD/(1-D) \le 2n_f/n \le a \qquad (16)$$

From (16), it can be obtained the related parameters limitation, furthermore, it can be also seen that: $D_{max} \le 0.5$.

C. The Inductance L_1 and the Value of a

C_{C1}, C_{C2} are discharging in stage 2 and charging in stage 4. During one charging period, the average discharging and charging power of C_{C1} can be expressed as followed:

$$P_O = \frac{U_{Cc}}{T}\int_{t_0}^{t_1} i_{L1}(t-t_0)dt = (anU_o)^2 D^2 T/8L_1 \qquad (17)$$

$$P_I = \frac{U_{Cc}}{T}\int_{t_2}^{t_3}[|i_{Lb}| - i_{Llk}]dt = \frac{anU_o}{2T}[|i_{Lb}|t_{23} - \frac{(a-1)nU_o}{2L_{lk}}t_{23}^2] \qquad (18)$$

The charging period is much shorter than the line period, so the average discharging and charging power of C_{C1} in the charging period can be considered as the instantaneous power in the whole line period. Therefore, in the whole line period, the following relationship can be obtained:

$$\int_0^{\pi/6} P_O d\omega t = \int_0^{\pi/6} P_I d\omega t \qquad (19)$$

From (1), (3), (4) and (19), it can be calculated that:

$$L_1 = 1.65a(a-1)M^2 L^2 / L_{lk} \qquad (20)$$

$$M = nU_o / \sqrt{3}U \qquad (21)$$

Where M is the voltage ratio of PFC converter.

From (20), the relationship between L_1, a and M is obtained in Fig.4. It can be seen that L_1 will increase as the value of a or M increases.

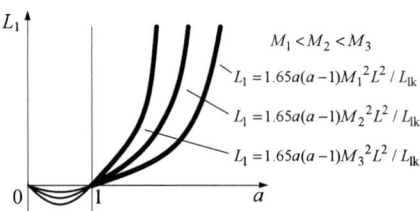

Fig. 4 The relationship between L_1, a and M

According to the analysis in section II, the voltage and current stresses of $S_1 \sim S_4$ in the PFC converter can be expressed:

$$\begin{cases} U_S = 2U_{Cc} = anU_o \\ I_S = -I_{Lbmax} + 2I_{L1max} = UDT/L + anU_oDT/L_1 \end{cases} \qquad (22)$$

It can be seen that U_S will increase as a increases, and if L_1, a increase according to (20), I_S will decrease. So, the parameters of L_1 and a will be designed according to both the voltage and current stresses of the switches selected.

IV. EXPERIMENTAL RESULTS

To verify the proposed method and theoretical analysis, a laboratory-made prototype of PFC converter was built. The basic circuit parameters and the main utilized components' type are: the input voltage: 110Vrms±10%, U_o=200Vdc, $S_1 \sim S_4$: BSM75GB120DN2, the switching frequency is 20kHz, $L_a = L_b = L_c = 76\mu H$, $L_{lk} = 6\mu H$, $n=2$, $C=1000\mu F$. $L_1 = L_2 = 960\mu H$, $C_{C1} = C_{C2} = 10\mu F$.

Fig.5 shows the input waveforms of phase A, it can be seen that the PFC converter operates in DCM, and the discontinuous input current follows the envelopes which are proportional to the input voltage.

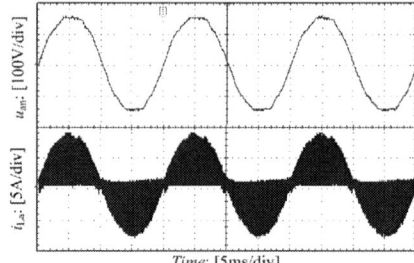

Fig.5 Input voltage and current of phase A

Fig.6 shows the voltage waveform across primary side of T, and Fig.7 shows the driving and voltage waveforms of S_1, S_2. From the voltage waveforms, it can be seen that the voltage spike has been suppressed efficiently. From Fig.7, it can be seen that S_1 turns off with zero voltage, and S_2 turns on with zero voltage. The PFC converter operates in DCM, so the zero current switching can also be achieved in S_1, The principle of which is very easy, so the experimental result isn't presented. The switching state of S_3 and S_4 are the same as that of S_1 and S_2, so the related experimental results are not presented.

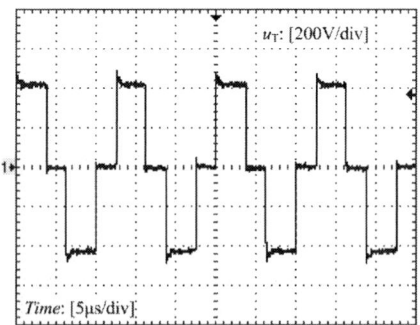

Fig.6 Voltage across primary side of T

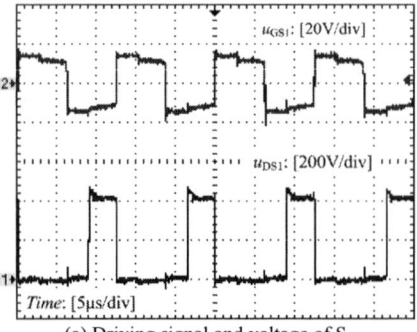

(a) Driving signal and voltage of S_1

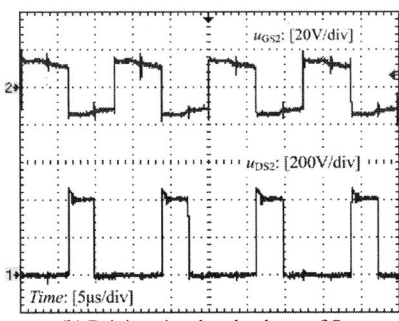

(b) Driving signal and voltage of S_2

Fig.7 Soft switching waveforms

V. CONCLUSIONS

In this paper, a three-phase isolated full-bridge boost PFC converter with flyback passive auxiliary converter is presented and investigated. The theoretical analysis and experimental results show that: 1) It can achieve soft-switching for the switches of the PFC converter, 2) with the adoption of the passive auxiliary converter, the voltage spike of the PFC converter is suppressed efficiently, and 3) the absorbed energy of the passive auxiliary converter can be transferred to the load by itself during one charging period.

ACKNOWLEDGMENT

This work was supported by the National Natural Science Foundation of China (51107017, 51377036), the China Postdoctoral Science Foundation Funded Project (2012M510954) and the Fundamental Research Funds for the Central Universities (HIT.NSRIF.2014014).

REFERENCES

[1] A. A. Badin and I. Barbi, "Unity power factor isolated three-phase rectifier with split dc-bus based on the scott transformer," *IEEE Trans. Power Electron.*, vol. 23, no. 3, pp. 1278-1287, 2008.

[2] D. Wijeratne and G. Moschopoulos, "A three-phase single-stage AC-DC full bridge converter with high power factor and phase-shift," in *Proc. IEEE APEC*. 2009, pp: 977-983.

[3] L. Z. Zhu, K. R. Wang, F. C. Lee and J. S. Lai, "New start-up schemes for isolated full-bridge boost converters," *IEEE Trans. Power Electron.*, vol. 18, no. 4, pp. 946-951, 2003.

[4] L. J. Hang, W. X. Yao, Z. Y. Lu, Z. M. Qian and J. M. Guerrero, "Analysis of flux density bias and digital suppression strategy for single-stage power factor corrector converter" *IEEE Trans. Ind Electron.*, vol. 55, no. 8, pp. 3077-3087, 2008.

[5] E. S. Park, S. J. Choi, J. M. Lee and B. H. Cho, "A soft-switching active-clamp scheme for isolated full-bridge boost converter," in *Proc. IEEE APEC*. 2004, pp: 1067-1070.

[6] A Mousavi, P. Das and G. Moschopoulos, "A comparative study of anew ZCS DC-DCfull-bridge boost converter with a ZVS active-clamp converter," *IEEE Trans. Power Electron.*, vol. 27, no. 3, pp. 1347-1358, 2012.

[7] M. Baei and G. Moschopoulos, "A ZVS-PWM full-bridge boost converter for applications needing high step-up voltage ratio," in *Proc. IEEE APEC*. 2012, pp: 2213-2217.

[8] B. Su and Z. Y. Lu, "An improved single-stage power factor correction converter based on current-fed full-bridge topology," in *Proc. IEEE PESC*. 2008, pp: 472-475.

[9] G. Moschopoulos and P. Jain, "Single-stage ZVS PWM full-bridge converter," *IEEE Trans. Aerosp. Electron. Syst.*, vol. 39, no. 4, pp. 1122-1133, 2003.

[10] J. F. Chen, R. Y. Chen and J. J. Liang, "Study and implementation of a single-stage current-fed boost PFC converter with ZCS for high voltage applications," *IEEE Trans. Power Electron.*, vol. 23, no. 1, pp. 379-386, 2008.

[11] S. Jalbrzykowski and T. Citko, "Current-fed resonant full-bridge boost DC/AC/DC converter," *IEEE Trans. Ind Electron.*, vol. 55, no. 3, pp. 1198-1205, 2008.

[12] L. Z. Zhu, "A novel soft-commutating isolated boost full-bridge ZVS-PWM DC-DC converter for bidirectional high power applications," *IEEE Trans. Power Electron.*, vol. 21, no. 2, pp. 422-429, 2006.

[13] D. Q. Wang, H. Q. Ben, T. Meng and Z. B. Lu, "Single-stage full-bridge PFC technique based on clamp circuit," *Electr. Power Autom. Equip.*, vol. 30, no. 5, pp. 53-56, 2010.

[14] T. Meng, H. Q. Ben, D. Q. Wang and J. M. Zhang, "Research on a novel three-phase single-stage boost DCM PFC topology and the dead zone of its input current," in *Proc. IEEE APEC*. 2009, pp: 1862-1866.

[15] T. Meng, H. Q. Ben and D. Q. Wang, "Novel passive snubber suitable for three-phase single-stage PFC based on an isolated full-bridge boost topology," *J. Power Electron.*, Vol. 11, No. 3, pp. 264-270, 2011.

Control and Experiment of a Modular Push-Pull PWM Converter for a Battery Energy Storage System

Makoto Hagiwara* and Hirofumi Akagi*
*Department of Electrical and Electronic Engineering
Tokyo Institute of Technology, Tokyo, Japan 152–8552
Email: akagi@ee.titech.ac.jp

Abstract—This paper presents a modular push-pull PWM converter (MPC) for a battery energy storage system, which is intended for grid connections to medium- or high-voltage power systems. The converter per phase consists of a center-tapped transformer and two arms based on a cascade connection of multiple bidirectional PWM chopper cells with floating dc capacitors. This paper proposes a new control method for a three-phase MPC, which makes it possible to regulate six average voltages of the floating dc capacitors in each arm independently without any interference. This attractive feature eliminates a need of arm-balancing control, leading to a simpler and more reliable system. The validity of the operating performance and control method developed in this paper is confirmed by experiment using a three-phase, 200-V, 5-kW downscaled system.

I. INTRODUCTION

Massive installation of renewable energy sources such as wind turbine generators and photovoltaics in the recent years has spurred an interest in battery energy storage systems intended for active-power leveling [1]–[8]. A grid-connected converter is typically used for converting a dc-source voltage representing a sodium-sulfur (NaS) battery, a lithium-ion (Li-ion) battery, and so on, into a three-phase ac voltage that is synchronized to the grid. Multilevel converters are promising candidates as a grid-connected converter due to many advantages such as low harmonic voltage/current, low EMI emissions, suitability for high-voltage high-power conversion, and so on.

A modular push-pull PWM converter (MPC) [9]–[11] is one of multilevel converters that is suitable for a battery energy storage system. The MPC can be considered as a push-pull converter equipped with modular arms, the structure of which is identical to that of a modular multilevel cascade converter based on double-star chopper cells (MMCC-DSCC)[1] [12][13]. The authors of this paper has presented a battery energy storage system using the MPC, along with a comparison with the DSCC [11]. It is reported that the dc-source voltage required for the MPC to achieve grid connections is half that required for the DSCC. This means that the MPC is more suitable as a power converter for battery energy storage systems requiring low-voltage large-current power conversion at the dc side.

The same voltage and current equations come into existence in the MPC and the DSCC when an ideal transformer is considered [11]. This means that the same control method can

[1]It is also referred to simply as a modular multilevel converter (MMC) [12].

be applied to both converters for achieving voltage balancing of all the floating capacitors. Reference [11] has carried out experimental verification of a three-phase MPC applying a control method proposed in [14] and [15], which is characterized by a hierarchical structure consisting of three sub-control parts: averaging control, arm-balancing control, and individual-balancing control. The averaging control regulates the average dc-capacitor voltage in each leg, while the arm-balancing control mitigates a difference in average voltages between two arms forming one leg. The individual balancing control makes each dc-capacitor voltage in one arm balance by forming an active power flowing into or out of each chopper cell. However, the use of three controls makes the control system complex and less reliable.

This paper proposes a new control method for an MPC, which can decrease the number of sub-control parts from three to two. The control method is characterized by utilizing six degrees of freedom in the circulating current for a three-phase MPC, whereas the conventional method utilizes three degrees of freedom. As a result, the average voltages of two arms can be regulated independently without any interference. This results in eliminating a need of the arm-balancing control that is indispensable for the DSCC, leading to a simpler and more reliable system. The validity of operating performance and control method proposed in this paper is confirmed by experiment using a three-phase, 200-V, 5-kW downscaled system. Experimental results show that the MPC exhibits high performance not only under steady-state conditions but also under transient-state conditions including a start-up procedure. Moreover, experimental results show that no dc magnetic saturation occurs in the transformer even when a voltage or power command of the MPC is changed under a ramp or step change.

II. CIRCUIT CONFIGURATION FOR THE MPC

A. Circuit Configuration for the Single-Phase MPC

Fig. 1(a) shows the circuit configuration for the single-phase MPC. It consists of a cascade connection of multiple chopper cells, as shown in Fig. 1(b), in which the two arms have the same cell count. The chopper cell consists of a dc capacitor and two IGBTs that form the so-called "bidirectional chopper." Here, v_{CPj} and v_{CNj} are the dc-capacitor voltages and v_{Pj} and v_{Nj} are the low-voltage side voltages of each chopper cell. The two arms are connected via a coupled or center-tapped inductor with three terminals, forming a leg of

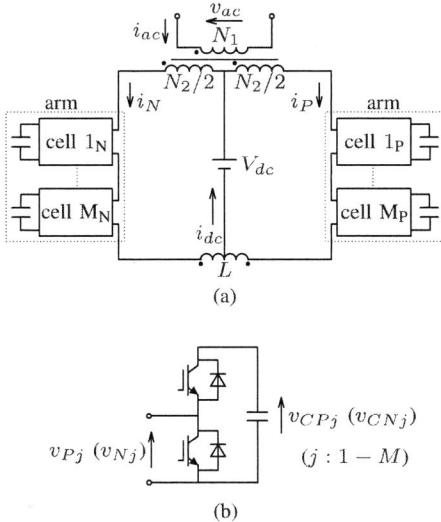

Fig. 1. Circuit configuration for the single-phase MPC. (a) Power circuit. (b) chopper cell.

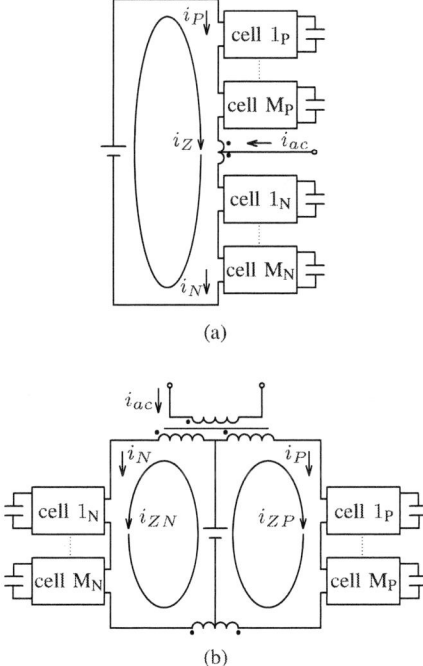

Fig. 2. DC circulating current(s) flowing in the converter. (a) single-phase MMCC-DSCC. (b) single-phase MPC.

the converter. The center terminal of the inductor is connected to the negative side of V_{dc}.

In Fig. 1(a), v_{ac} is the ac voltage (i.e., the primary voltage of the transformer), i_{ac} is the ac current (i.e., the primary current of the transformer), i_P and i_N are the arm currents, and i_{dc} is the dc-source current.

B. DC Magnetic Satuation in the Transformer

A concern of push-pull inverters including the MPC is that dc magnetic saturation may occur in the transformer [16]. The saturation occurs when the dc currents included in the right- and left-arm currents are different in amplitude, resulting in causing a dc flux in the transformer. This means that the difference of the dc currents, i_{ZPN}, should be regulated to zero, in which it is expressed by using i_{ZP} and i_{ZN} as follows:

$$i_{ZPN} = i_{ZP} - i_{ZN}. \tag{1}$$

The MPC can regulate i_{ZPN} to zero by using a circulating-current control method not only under steady-state conditions but also under transient-state conditions as will be shown in experimental waveforms.

III. DC Circulating Current(s) Flowing in the Converter

A. When an MMCC-DSCC is Used

Fig. 2(a) shows a dc circulating current flowing in a single-phase MMCC-DSCC. It has three branch currents: the positive and negative arm currents i_P and i_N, and the ac current i_{ac}. Here, i_P and i_N can be expressed as

$$i_P = -\frac{i_{ac}}{2} + i_Z \tag{2}$$

$$i_N = \frac{i_{ac}}{2} + i_Z, \tag{3}$$

where i_Z is defined in [14] as

$$i_Z = \frac{1}{2}(i_P + i_N). \tag{4}$$

The circulating current i_Z is a current flowing in a dc loop shown in Fig. 2(a). Although i_Z contains a 100-Hz second-order harmonic component and switching-ripple components, they are small enough to be negligible compared to the most dominant dc component. The adjustment of the dc component enables one to exchange an active power flowing into or out of the chopper cells [14], leading to the regulation of an average dc-capacitor voltage in a leg. Fig. 2(a) implies that a three-phase DSCC has three degrees of freedom in the circulating current [13].

On the other hand, an amount of active power is exchanged between positive and negative arms under specific operating conditions, resulting in causing a voltage imbalance between average voltages in both arms [15]. The DSCC cannot regulate voltages of the two arm independently, because they have the common circulating current, as shown in Fig. 2(a). Hence, an additional control that is called "arm-balancing control" [15] is required. However, the use of the arm-balancing control makes the system more complex and less reliable.

B. When an MPC is Used

Fig. 2(b) shows dc circulating currents flowing in a single-phase MPC. It is obvious that the MPC has two dc loops as shown in Fig. 2(b), resulting in yielding six degrees of freedom in the circulating current when a three-phase MPC is used. Let the circulating currents flowing in the two dc loops be i_{ZP} and i_{ZN}, respectively. This paper defines i_{ZP} and i_{ZN} by using

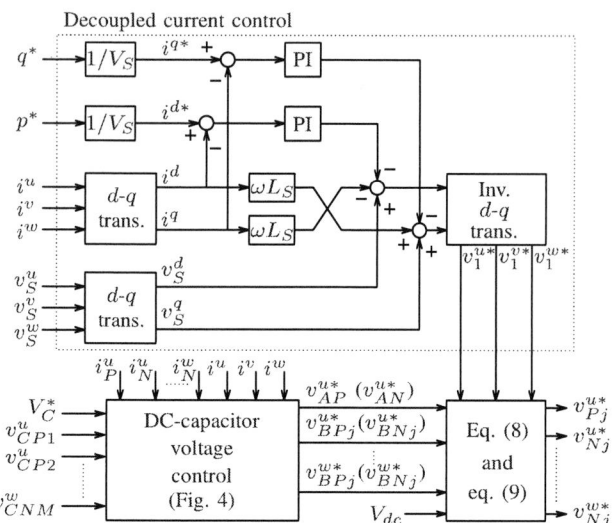

Fig. 3. Block diagrams for the three-phase MPC.

i_P, i_N, and i_{ac} as follows:

$$i_{ZP} = i_P + \frac{N_1}{N_2} i_{ac} \qquad (5)$$

$$i_{ZN} = i_N - \frac{N_1}{N_2} i_{ac}. \qquad (6)$$

Here, the second terms on the right-hand sides in (5) and (6) correspond to the 50-Hz ac components included in the arm currents. This implies that i_{ZP} and i_{ZN} are dc currents without any ac components, ideally. However, they contain 50-Hz fundamental-frequency components and switching-ripple components due to the effects of magnetizing current of the transformer and PWM operation of the converter. These two currents are small enough to be negligible compared to the dc component as will be shown in experimental waveforms.

The MPC can regulate the average voltages of right and left arms independently because they have different circulating currents. This attractive feature eliminates a need of the arm-balancing control from the MPC, giving the following advantages:

1) Communication of voltage signals between arms is unnecessary. It makes a significant contribution to the reduction in a computational burden of the controller, leading to a simpler and more reliable system.

2) The MPC has independent arms in terms of circuit and control, whereas the DSCC has arms with mutual interference between controls. This advantage makes the assembly of the MPC simpler and easier than that of the DSCC.

IV. CONTROL METHOD FOR THE MPC

Fig. 3 shows the control block diagrams for a three-phase MPC. It is applicable to an MPC with any count of cells per arm. Here, p^* and q^* represent the power commands for the instantaneous active and reactive powers [17], [18] at the ac mains, i.e., p and q, as will be shown in Fig. 5,

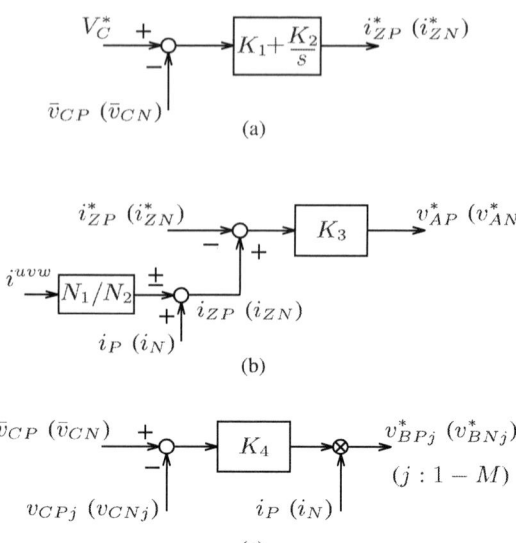

Fig. 4. Block diagrams for dc-capacitor voltage control. (a) Averaging control. (b) Circulating-current control. (c) Individual-balancing control.

respectively. The line-to-neutral voltage commands for the primary windings of the transformer v_1^{u*}, v_1^{v*}, and v_1^{w*} are determined by the decoupled current control of the ac-supply currents i^u, i^v, and i^w.

Fig. 4 shows the block diagrams for the dc-capacitor voltage control of each arm. It should be noted that the block diagrams for the dc-capacitor voltage control of each arm are identical. This control method is characterized by directly controlling the circulating current of each arm, leading to good current regulation of the dc capacitor. Voltage control of the floating dc capacitors shown in Fig. 1(a) can be divided into

1) averaging control,
2) circulating-current control, and
3) individual-balancing control.

A. Averaging Control

Fig. 4(a) shows the block diagram for averaging control. The voltage major loop forces the average voltage of each arm $\bar{v}_{CP}$ or $\bar{v}_{CN}$ to follow the command V_C^*, where they are given by

$$\bar{v}_{CP} = \frac{1}{M} \sum_{j=1}^{M} v_{CPj}$$

$$\bar{v}_{CN} = \frac{1}{M} \sum_{j=1}^{M} v_{CNj}. \qquad (7)$$

In Fig. 4(a), i_{ZP}^* and i_{ZN}^* are the commands for I_{ZP} and I_{ZN}, respectively, which are the dc components in i_{ZP} and i_{ZN}.

B. Circulating-Current Control

Fig. 4(b) shows the block diagram for circulating-current control. The current minor loop forces i_{ZP} and i_{ZN} to follow the commands i_{ZP}^* and i_{ZN}^*, producing the voltage commands v_{AP}^* and v_{AN}^*.

978-1-4799-2706-7/14 $31.00 © 2014 IEEE

The 2014 International Power Electronics Conference

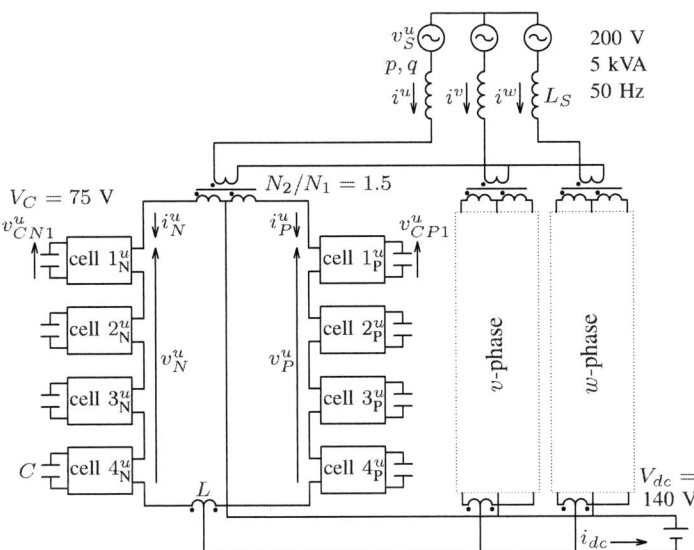

Fig. 5. Circuit configuration for the three-phase MPC used for experiment.

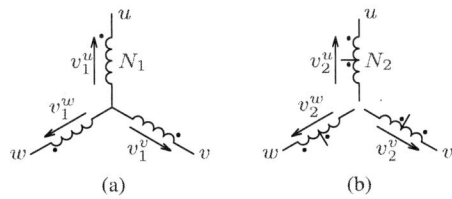

Fig. 6. Circuit configuration for the three-phase converter transformer used in Fig. 5. (a) Primary side. (b) Secondary side.

C. Individual-Balancing Control

Fig. 4(c) shows the block diagram for individual-balancing control. It produces an active power via the voltage at the low-voltage side of each chopper cell (i.e., v_{Pj} or v_{Nj}) and the corresponding arm current, producing the voltage commands v_{BPj}^* and v_{BNj}^*.

Finally, the voltage command for the right-arm chopper cells, v_{Pj}^* for $j = 1 - M$, is given by

$$v_{Pj}^* = v_{AP}^* + v_{BPj}^* - \frac{N_2}{2N_1}\frac{v_{ac}^*}{M} + \frac{V_{dc}}{M}, \qquad (8)$$

and that for the left-arm chopper cells, v_{Nj}^* for $j = 1 - M$, is given by

$$v_{Nj}^* = v_{AN}^* + v_{BNj}^* + \frac{N_2}{2N_1}\frac{v_{ac}^*}{M} + \frac{V_{dc}}{M}, \qquad (9)$$

where v_{ac}^* is the command for v_{ac}. The last term on the right-hand sides in (8) and (9), V_{dc}/M, corresponds to feedforward control. The voltage commands v_{Pj}^* and v_{Nj}^* are normalized by the corresponding dc-capacitor voltages v_{CPj} and v_{CNj}, respectively. They are then compared with a triangular carrier waveform with a maximum value of unity and a minimum value of zero, with a carrier frequency f_C.

TABLE I. CIRCUIT PARAMETERS USED FOR EXPERIMENT.

Capacity		5 kVA
Line-to-line rms voltage	V_S	200 V
Rated rms current	I	15 A
Frequency	f	50 Hz
chopper cell count per arm	M	4
Voltage ratio	N_2/N_1	1.5
DC-link voltage	V_{dc}	140 V
Coupled inductor	L	4 mH (7%*)
DC-voltage command	V_C^*	75 V
DC capacitance	C	3.3 mF
Unit capacitance constant	H	45 ms
Carrier frequency	f_C	2 kHz
AC-link inductor	L_S	2.75 mH (11%**)

*on a three-phase, 5-kVA, 300-V, and 50-Hz base.
**on a three-phase, 5-kVA, 200-V, and 50-Hz base.

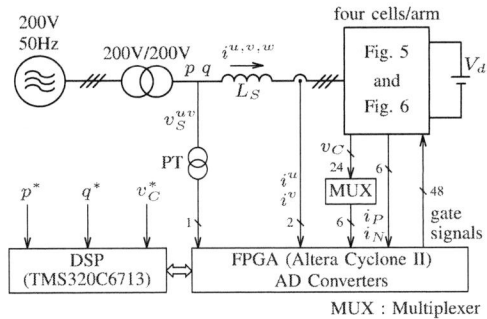

Fig. 7. System configuration of the MPC used for experiment.

V. EXPERIMENT OF THE THREE-PHASE MPC

A. Circuit Configuration for the Three-Phase MPC

Fig. 5 shows the circuit configuration for the three-phase MPC used for experiment, the circuit parameters of which are summarized in Table I. The count of chopper cells per arm was set to $M = 4$, which is the same as that of the experimental system shown in Fig. 7. This results in a total of 24 chopper cells. The carrier frequency is set as $f_C = 2$ kHz. The ac sides of the MPC are connected to a three-phase ac mains of 200 V via ac-link inductors, while the dc sides are connected to a dc power supply (Kikusui PAT160-50T or Myway pCUBE) producing a dc voltage of $V_{dc} = 140$ V.

Fig. 6 shows the circuit configuration of a three-phase converter transformer with a Y-connected primary and an open-Y connection at the secondary. The primary windings are connected to the ac-link inductors, while the secondary windings with center-taps are connected to each converter. The turns ratio (i.e., voltage ratio) of the transformer is set as $N_2/N_1 = 1.5$. As a result, the line-to-line rms voltage of the secondary windings is 300 V ($= 200 \times 1.5$). $v_1^{u,v,w}$ are the line-to-neutral voltages of the primary windings and $v_2^{u,v,w}$ are those of the secondary windings. Note that the center-taps for each phase can be connected to a common terminal as shown in Fig. 5. This means that the number of terminals required for the secondary windings is seven.

978-1-4799-2706-7/14 $31.00 © 2014 IEEE

The 2014 International Power Electronics Conference

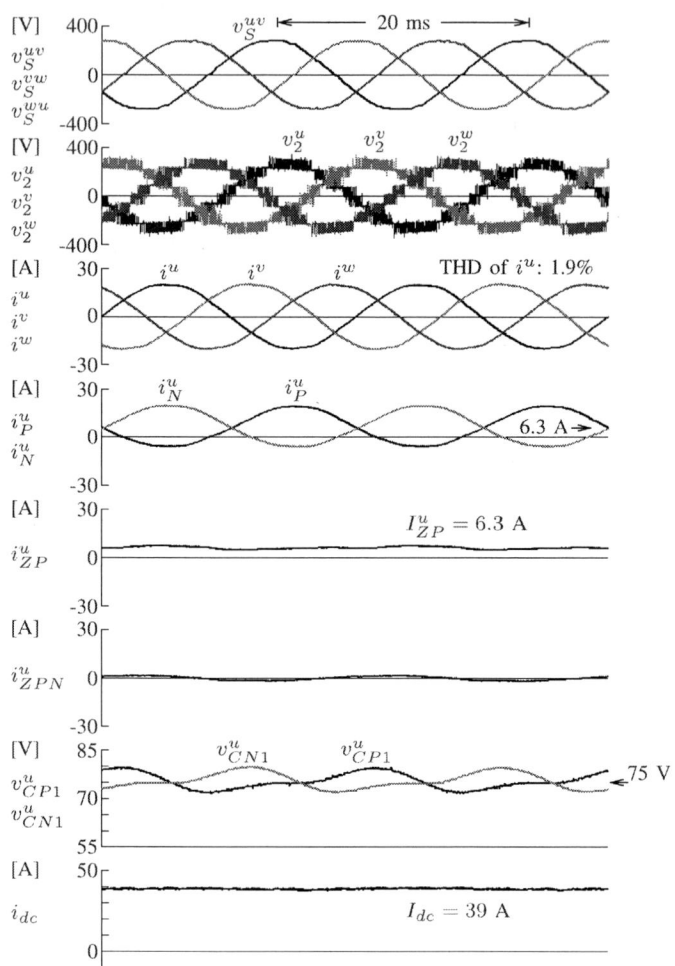

Fig. 8. Experimental steady-state waveforms for the MPC in Fig. 7 during inversion of $p^* = -5$ kW and $q^* = 0$ kVA.

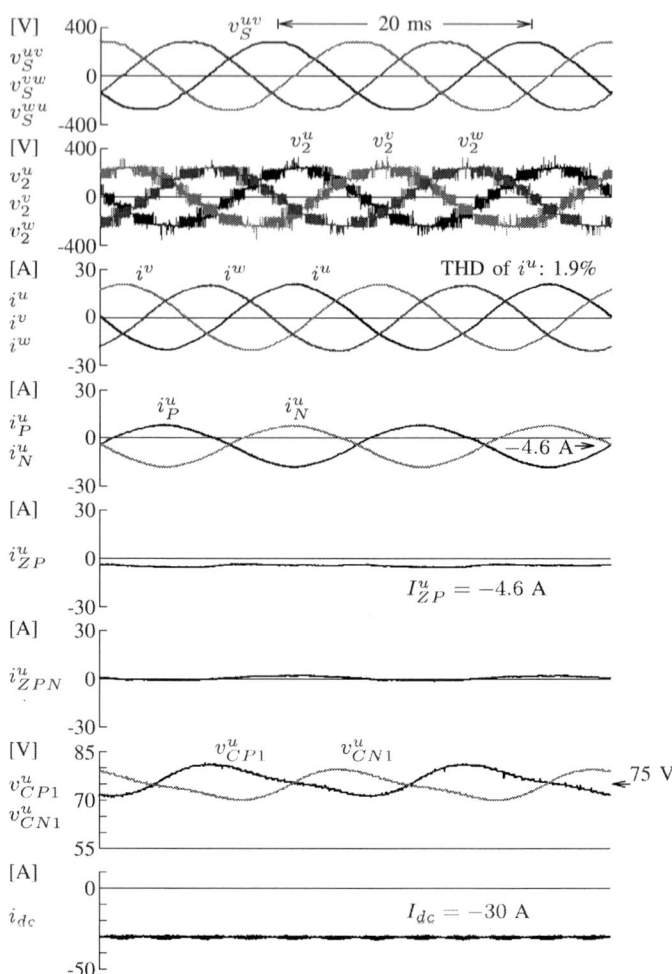

Fig. 9. Experimental steady-state waveforms for the MPC in Fig. 7 during rectification of $p^* = 5$ kW and $q^* = 0$ kVA.

B. Experimental System Configuration

Fig. 7 shows the system configuration of the three-phase, 200-V, 5-kW downscaled MPC system. The control system detects each dc-capacitor voltage v_C, both right-side and left-side arm currents i_P and i_N, the ac supply currents i^u and i^v, and the ac-supply voltage v_S^{uv}. These signals are sent to A/D converters. Here, the multiplexer (MUX) unit is used to reduce the number of A/D converters. A digital signal processor (DSP) unit using a Texas Instruments TMS320C6713 takes in the digital signals from the A/D converters and produces the voltage commands for each chopper cell.

The experimental waveforms were taken by using the Hioki Memory Hicorder 8861-50. The sampling frequency in Fig. 10 is 20 kHz, the sampling frequency in Figs. 8, 9, and 11 is 100 kHz.

C. Experimental Waveforms Under Steady-state Conditions

Fig. 8 shows the experimental waveforms when the MPC acts as an inverter (i.e., $p^* = -5$ kW and $q^* = 0$ kVA). The line-to-neutral voltages of the secondary windings of the transformer, v_2^u, v_2^v, and v_2^w are nine-level PWM waveforms. The THD value of i^u is reduced to 1.9%. The arm currents i_P^u and i_N^u contain both dc and 50-Hz components, in which the dc component is 6.3 A. On the other hand, the amplitudes of 50-Hz components are two thirds ($= N_1/N_2$) of those of the supply currents. The circulating current i_{ZP}^u contains the 50-Hz fundamental-frequency component in addition to the dc component $I_{ZP}^u = 6.3$ A, due to the effect of magnetizing current of the transformer. However, the 50-Hz component is negligible compared to the dc component. i_{ZPN}^u defined by (1) has no dc component from a practical point of view. The dc-capacitor voltages v_{CP1}^u and v_{CN1}^u contain both dc and ac components, as shown in Fig. 8, where the voltage control regulates the dc component at 75 V. The dc component of the dc-source current is $I_{dc} = 39$ A, which is six times as high as that of I_Z^u.

Fig. 9 shows the experimental waveforms when the MPC acts as a rectifier (i.e., $p^* = 5$ kW and $q^* = 0$ kVA). A comparison of Figs. 8 and 9 reveals that the amplitudes of v_2^u, v_2^v, and v_2^w in Fig. 9 are smaller than those in Fig. 8. This is caused by the existence of resistance in the power circuit,

The 2014 International Power Electronics Conference

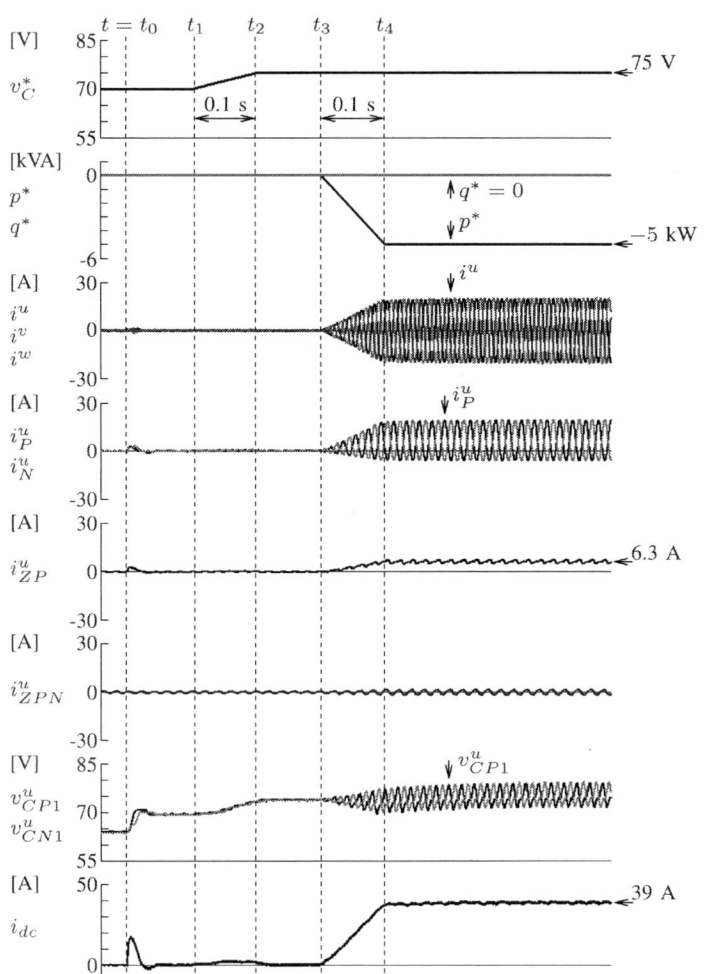

Fig. 10. Experimental start-up performance for the MPC in Fig. 7 when $p^* = -5$ kW and $q^* = 0$ kVA.

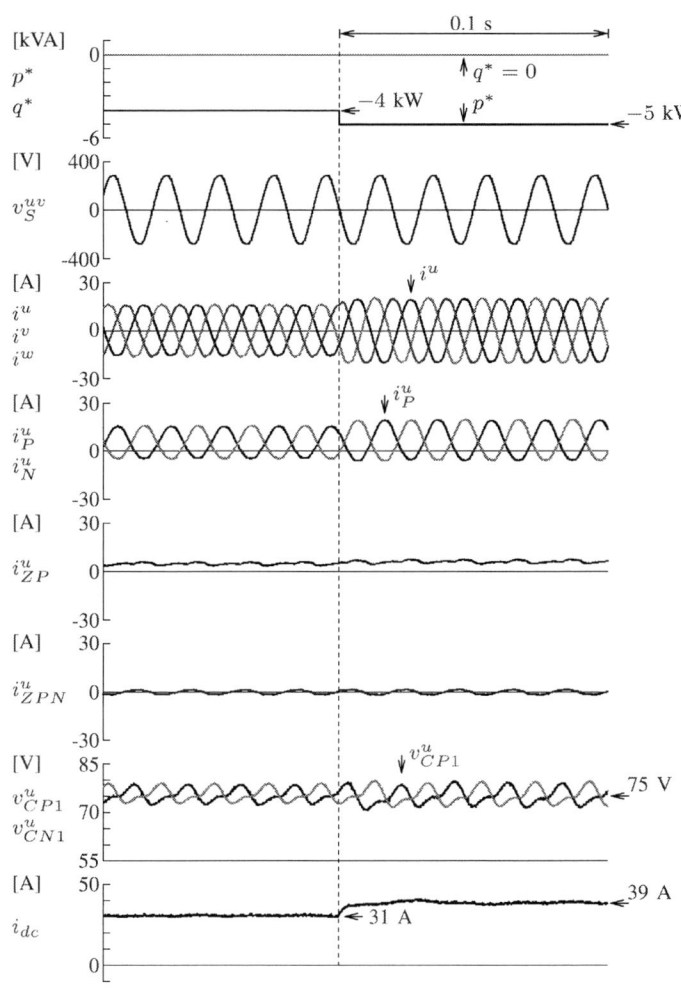

Fig. 11. Experimental transient performance for the MPC in Fig. 7 when p^* was changed from -4 to -5 kW under a step change where $q^* - 0$ kVA.

which decreases their amplitudes during rectification and increases them during inversion vice versa. The polarities of i^u_{ZP} and i_{dc} change to negative due to operation of rectification. The amplitudes of i^u_{ZP} and i_{dc} in Fig. 9 are smaller than those in Fig. 8 due to the converter power loss. The other waveforms are similar to those shown in Fig. 8.

D. Experimental Waveforms Under Transient-state Conditions

Fig. 10 shows the experimental waveforms during a start-up procedure when $V_{dc} = 140$ V, $p^* = -5$ kW, and $q^* = 0$ kVA. From t_1 to t_2, v^*_C was increasing from 70 V to 75 V with a ramp change rate of 5 V/0.1 s. From t_3 to t_4, p^* was decreasing from 0 kW to -5 kW with a ramp change rate of -5 kW/0.1 s while $q^* = 0$ kVA. The amplitudes of the supply currents and the arm currents were increasing without any overcurrent. Moreover, the dc mean voltages of v^u_{CP1} and v^u_{CN1} are regulated to the command value of 75 V without any steady-state error. A dc component included in i^u_{ZPN} was suppressed to zero even during transient-state conditions.

Fig. 11 shows the experimental waveforms when p^* was changed from -4 to -5 kW under a step change. The

experimental waveforms show that the dc-link current i_{dc} exhibits a first-order response against change in p^*, in which time constant estimated from the experimental waveform is as short as 1.5 ms. No dc current occurs in i^u_{ZPN} even when p^* is changed under a step change.

VI. CONCLUSION

This paper has described a modular push-pull PWM converter (MPC) intended for its application to a battery energy storage system. This paper has proposed a new control method for the MPC, which is characterized by utilizing six degrees of freedom in the circulating current for a three-phase MPC. As a result, the average voltages of positive and negative arms can be regulated independently without any interference, leading to a simpler and more reliable system. Experimental results obtained from a three-phase 200-V, 5-kW downscaled system have verified the validity and effectiveness of the converter.

REFERENCES

[1] L. H. Walker, "10-MW GTO converter for battery peaking service," *IEEE Trans. Ind. Appl.*, vol. 26, no. 1, pp. 63–72, 1990.

[2] S. R. Bull, "Renewable energy today and tomorrow," *Proc. IEEE*, vol. 89, no. 8, pp. 1216–1226, 2001.

[3] P. F. Ribeiro, B. K. Johnson, M. L. Crow, A. Arsoy, and Y. Liu, "Energy storage systems for advanced power applications," *Proc. IEEE*, vol. 89, no. 12, pp. 1744–1756, Dec. 2001.

[4] Z. Yang, C. Shen, L. Zhang, M. L. Crow, and S. Atcitty, "Integration of a STATCOM and battery energy storage," *IEEE Trans. Power Sys.*, vol. 16, no. 2, pp. 254–260, 2001.

[5] K. C. Divya and J. Ostergaard, "Battery energy storage technology for power systems - An overview," *Electric Power Systems Research*, vol. 79, no. 4, pp. 511–520, 2009.

[6] L. Maharjan, S. Inoue, and H. Akagi, "A transformerless energy storage system based on a cascade multilevel PWM converter with star configuration," *IEEE Trans. Ind. Appl.*, vol. 44, no. 5, pp. 1621–1630, Sep./Oct. 2008.

[7] L. Maharjan, S. Inoue, and H. Akagi, "SOC (state-of-charge)-balancing control of a battery energy storage system based on a cascade PWM converter," *IEEE Trans. Power Electron.*, vol. 24, no. 6, pp. 1628–1636, June 2009.

[8] N. Wade, P. Taylor, P. Lang, J. Svensson, "Energy storage for power flow management and voltage control on an 11 kV UK distribution network," in *Conf. Rec. CIRED'09*, June 2009.

[9] C. Oates, "A methodology for developing chainlink converters," in *Conf. Rec. EPE* 2009, CD-ROM.

[10] J. A. Ferreira, "The Multilevel Modular DC Converter," *IEEE Trans. Power Electron.*, vol. 28, no. 10, pp. 4460–4465, Oct. 2013.

[11] M. Hagiwara and H. Akagi, "Experiment and simulation of a modular push-pull PWM converter for a battery energy storage system," to be published in *IEEE Trans. Ind. Appl.*

[12] R. Marquardt and A. Lesnicar, "An innovative modular multilevel converter topology suitable for a wide power range," in *Conf. Rec. IEEE Bologna PowerTech* 2003, CD-ROM.

[13] H. Akagi, "Classification, terminology, and application of the modular multilevel cascade converter (MMCC)," *IEEE Trans. Power Electron.*, vol. 26, no. 11, pp. 3119–3130, Nov. 2011.

[14] M. Hagiwara and H. Akagi, "Control and experiment of pulsewidth-modulated modular multilevel converters," *IEEE Trans. Power Electron.*, vol. 24, no. 7, pp. 1737–1746, July 2009.

[15] M. Hagiwara, R. Maeda, and H. Akagi, "Control and analysis of the modular multilevel cascade converter based on double-star chopper-cells (MMCC-DSCC)," *IEEE Trans. Power Electron.*, vol. 26, no. 6, pp. 1649–1658, Jun. 2011.

[16] N. Mohan, T. M. Undeland, and W. P. Robbins, *Power Electronics*: John Willy & Sons, Inc., 2003.

[17] H. Akagi, Y. Kanazawa, and A. Nabae, "Instantaneous reactive power compensators comprising switching devices without energy storage components," *IEEE Trans. Ind. Appl.*, vol. IA-20, no. 3, pp. 625–630, May/Jun 1984.

[18] H. Akagi, E. H. Watanabe, and M Aredes, *Instantaneous Power Theory and Applications to Power Conditioning*: IEEE Press, 2007.

Active front-end topology for 5 level medium voltage drive system with isolated DC bus

Toshiaki Oka, Hironobu Kusunoki,
Masahiko Tsukakoshi
Toshiba Mitsubishi-Electric Industrial Systems Corporation
1 Toshiba-cho, Fuchu-shi, Tokyo, 183-8511, JAPAN

John Kleinecke, Mike Daskalos
TOSHIBA International Corporation
13131 West Little York Road Houston,
Texas, 77041, USA

Abstract— **An active front-end converter that can provide multi-phase isolated DC buses is proposed in this paper. It supplies DC buses for 5 level medium voltage inverters that typically have an isolated transformer and diode rectifiers and it has no regeneration capability. The proposed DC bus control scheme can be used in all four quadrants and has less current harmonics back to the grid. The results of a 746kW (1000HP) load test are also described.**

Keywords— Active front-end, 5 level medium voltage drive, Isolated DC bus

I. INTRODUCTION

The high power motor market is growing worldwide. When higher motor power is required, higher motor voltage has been selected to reduce current and cable size. To get higher output voltage, a 5 level medium voltage drive system that has single bridge NPC (Neutral Point Clamp) leg as a power module, and connects three power modules in star connection as shown in Fig. 1 is often used [1-3]. It has less motor surge and harmonics than a three phase NPC 3 level drive system because it has a lower DC bus. It has isolated DC buses in general, and those voltages are provided through isolated transformers and independent diode rectifier circuits. Using a diode rectifier is very simple and easy, it needs no additional control circuit, but it has no regeneration capability. This is typically not a problem for general Fan and Pump applications. However, some medium voltage motor applications need regeneration capability, such as down-hill conveyers and wind/water turbine generators. Large GD2 fans systems need regeneration if it is required to stop the fan quickly. Sometimes Induced Draft fans (ID fans) can be turning backward because of another fan's draft air, and regeneration power is needed to return to forward rotation.

A three level NPC drive system has a common DC bus. Regeneration systems using identical converter and inverter sections are well known as shown in Fig. 2. These systems have only one DC bus and its control is well known and easily applied to get regeneration capability [4-5].

For the isolated DC bus multilevel drive, it is often suggested that three phase full bridge converters be used for individual power modules as shown in Fig. 3. It has

also individual DC bus controls, that might be PWM control or 120deg fixed gate on the each three phase devices [6]. Since it has multiple DC bus circuits and controls, it is very complicated.

This paper presents development and implementation of an active front-end 5 level medium voltage drive system which has isolated DC buses, as well as test results with a 4.16kV-746kW (1000HP) load.

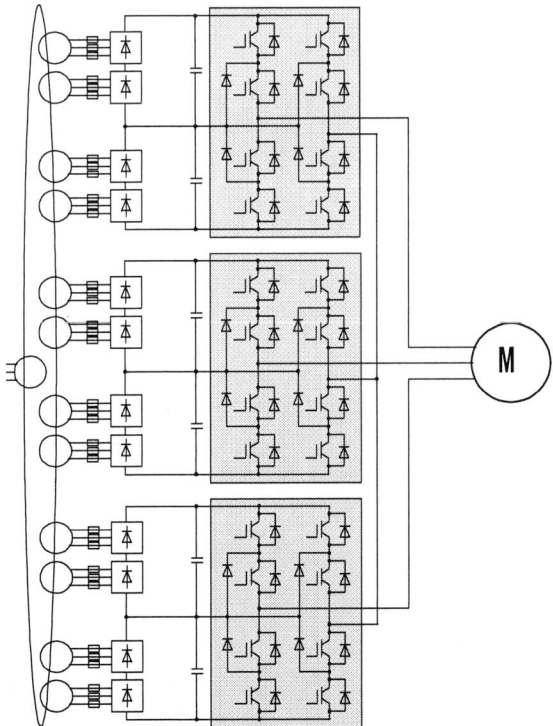

Fig. 1. 5 level medium voltage drive system with isolated transformer and individual diode rectifiers

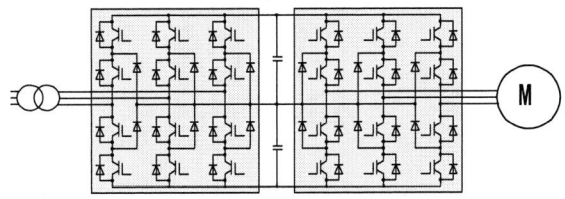

Fig. 2. 3 level NPC drive system with an active front-end converter.

The 2014 International Power Electronics Conference

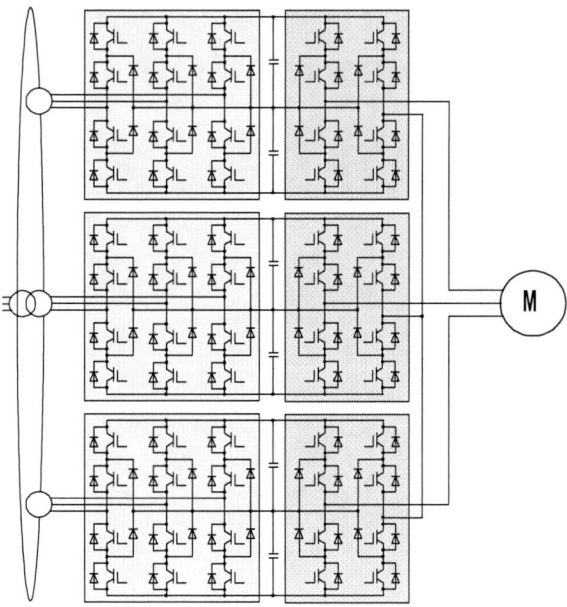

Fig. 3. Conventional active front-end converter for 5 level inverter with isolated DC buses.

II. CIRCUIT TOPOLOGY AND DRIVE SYSTEM OVERVIEW

The proposed active front-end topology is shown in Fig. 4, 4.16V-4474kW (6000HP) regenerative drive system overview is shown in Fig. 5, and its rating is in Table I.

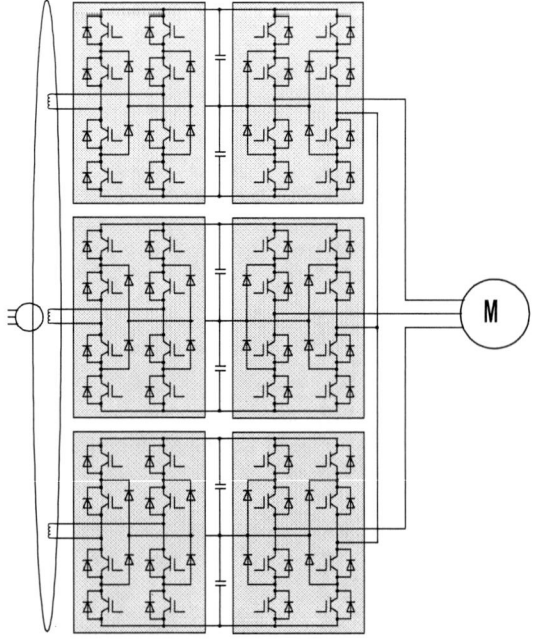

Fig. 4. Proposed active front-end converter for 5 level inverter with isolated DC buses.

The regenerative drive system consists of an input section, an isolation transformer section, a converter/inverter section, and an output section as shown from left to right in Fig. 5.

The input section provides maintenance isolation, main and precharge control, and short-circuit protection for the drive.

The isolation transformer section includes the transformer and associated fan cooling system. Each single phase output from the converter is connected to an open winding secondary of the isolation transformer. The secondary windings are shifted 120deg from each other. The transformer is provided with a delta primary winding to allow the drive to be connected to supply voltages ranging from 2400V to 13,800V.

The converter and inverter power modules are compactly located in a single section, and air cooled by a common fan system. The converter and inverter power modules are each identical single bridge NPC inverters.

The output section contains the drive controls, including the DC bus controls.

Fig. 5. 12.7kVin-4.16kVout 4474kW (6000HP) overview

TABLE I
RATING OF 4474kW (6000HP) REGENERATIVE DRIVE

Input Voltage	12.7kV-60Hz
Drive rating	4.16kV-744A
Device	3.3kV-1200A
DC bus Voltage	1650V
Carrier	2048Hz

The comparison table between Fig. 3 and Fig. 4 is shown in Table II. The proposed circuit is simpler than the conventional active front-end circuit, since there are less devices and transformer secondary windings. The conventional circuit has its own DC bus controllers in the each 3 phase converters, that each control is simple and well-known. On the other hand, the proposed topology has only a single winding for each of the phases. Typically individual single phase control is difficult since its reference is changing in 50Hz or 60Hz utility frequency, and control response needs to be faster.

TABLE II
COMPARISON WITH CONVENTIONAL DRIVE

Item	Conventional Topology	Proposed Topology
Transformer winding	3 phase x 3 phase	Single phase x 3 phase
Switching device	Converter side:12 x 3 phase Inverter side:8 x 3 phase	Converter side:8 x 3 phase Inverter side:8 x 3 phase
DC bus control	Individual DC bus control	Averaged DC bus control and balancing compensation

III. DC BUS CONTROL SCHEME

A. DC bus control for conventional topology

A control scheme of the conventional three phase active front-end shown in Fig. 2 has dq axis separation and individual current regulators as shown in Fig. 6. By controlling q-axis current based upon a DC voltage regulator, it controls active power both motoring and regenerating. Also by controlling d-axis current, it can control reactive power (Static Var control). It is very simple since it has only one DC bus.

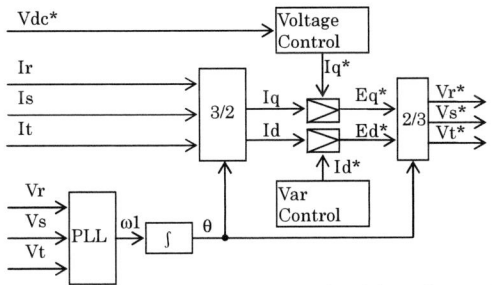

Fig. 6. DC bus control scheme for conventional three phase active front-end topology

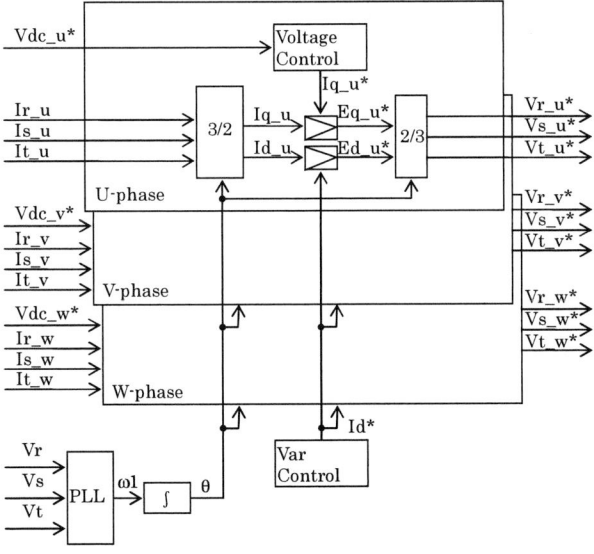

Fig. 7. DC bus control scheme for conventional 5 level active front-end topology

A conventional 5 level medium voltage with active front-end shown in Fig. 3 has individual DC bus controls as shown in Fig. 7. Each control is same as three phase active front-end control, but there are three sets of DC bus controls, and its circuit is much bigger and complex. Descriptions of the variables in Fig. 7 are shown in Table III.

B. DC bus control for proposed topology

The proposed active front-end circuit has isolated DC buses. To use conventional dq axis vector control, the DC bus average is calculated and used for q-axis current control. If inverter section modules output exactly the same power, there would be no voltage unbalance. However, since DC buses are not connected to each other, and there may be some power unbalance, there may be DC voltage unbalance between the phases. To control this unbalance, individual voltage compensation based on the difference from the average is used. For example, when motoring and the DC bus is higher than the average, the converter voltage will be reduced to reduce input motoring power. Control system overview is shown in Fig. 8. Descriptions of the variables in Fig. 8 are shown in Table IV. A detail of the DC bus balancing control is shown in Fig. 9.

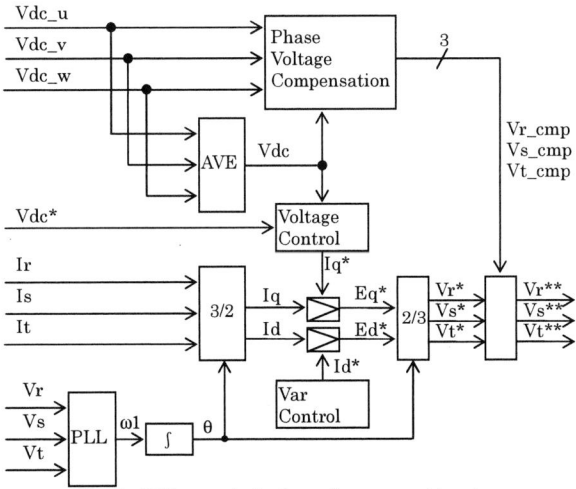

Fig. 8. DC bus control scheme for proposed topology

TABLE III
VARIABLE DESCRIPTIONS FOR CONVENTIONAL TOPOLOGY

Symbol	Description
Vdc_x (x; u, v, w)	Each phase DC bus feedback
Iy_x (x; u, v, w) (y; r, s, t)	Each phase converter current feedback
Iq_x, Iq_x* (x; u, v, w)	Active power current feedback & reference
Id, Id*	Reactive power current feedback & reference
Eq_x*, Ed_x* (x; u, v, w)	dq axis voltage reference
Vy_x* (x; u, v, w) (y; r s, t)	Converter voltage reference
Vy (y; r, s, t)	Utility voltage feedback
ω1	Utility frequency
θ	Utility phase angle

TABLE IV
VARIABLE DESCRIPTIONS FOR PROPOSED TOPOLOGY

Symbol	Description
Vdc_x (x; u, v, w)	Each phase DC bus feedback
Vdc	DC bus voltage average
Vdc*	DC bus voltage reference
Iy (y; r, s, t)	Each phase converter current feedback
Iq, Iq*	Active power current feedback & reference
Id, Id*	Reactive power current feedback & reference
Eq*, Ed*	dq axis voltage reference
Vy_cmp (y; r s, t)	Voltage reference compensation
Vy* (y; r s, t)	Converter voltage reference w/o compensation
Vy** (y; r s, t)	Converter voltage reference with compensation
Vy (y; r, s, t)	Utility voltage feedback
ω1	Utility frequency
θ	Utility phase angle

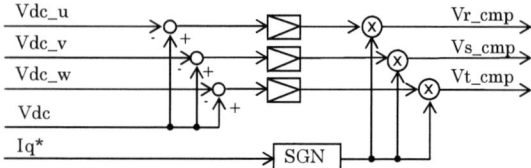

Fig. 9. DC bus balancing control

IV. TEST RESULTS

A. Variable speed test

The variable speed four quadrant test circuit is shown in Fig. 10. A 746kW (1000HP) Thyristor Leonard DC drive is controlling torque on the DC motor, and an active front-end medium voltage drive is controlling 1864kW (2500HP) motor speed. Those motors are coupled together. Other test conditions are shown in Table V.

There are four sample waveforms shown in Fig. 11, (a) Lagging Var, (b) Motoring, (c) Leading Var, and (d) Regenerating. Because of the thyristor commutation, there is a voltage notch every 60 deg in (b) and (d), while the DC drive is on. But there is no voltage notch in (a) and (c) since the DC drive is off.

A load step test is shown for both motoring and regenerating conditions in Fig. 12. The DC bus voltage error is less than 10%.

Four quadrant motor acceleration/deceleration test waveforms are shown in Fig. 13.

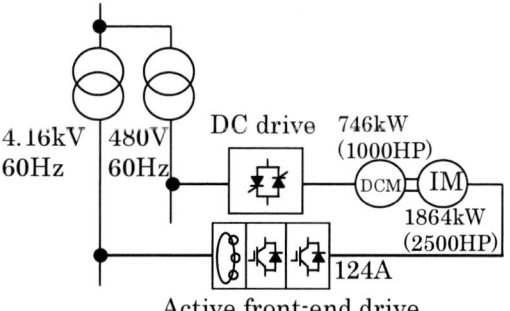

Fig. 10. Test circuit for variable speed four quadrant operation

TABLE V
TEST CONDITION

Utility Voltage	4160V-60Hz
Drive Current	124A
Device	3.3kV-400A
DC bus Voltage	1650V
Carrier	2048Hz

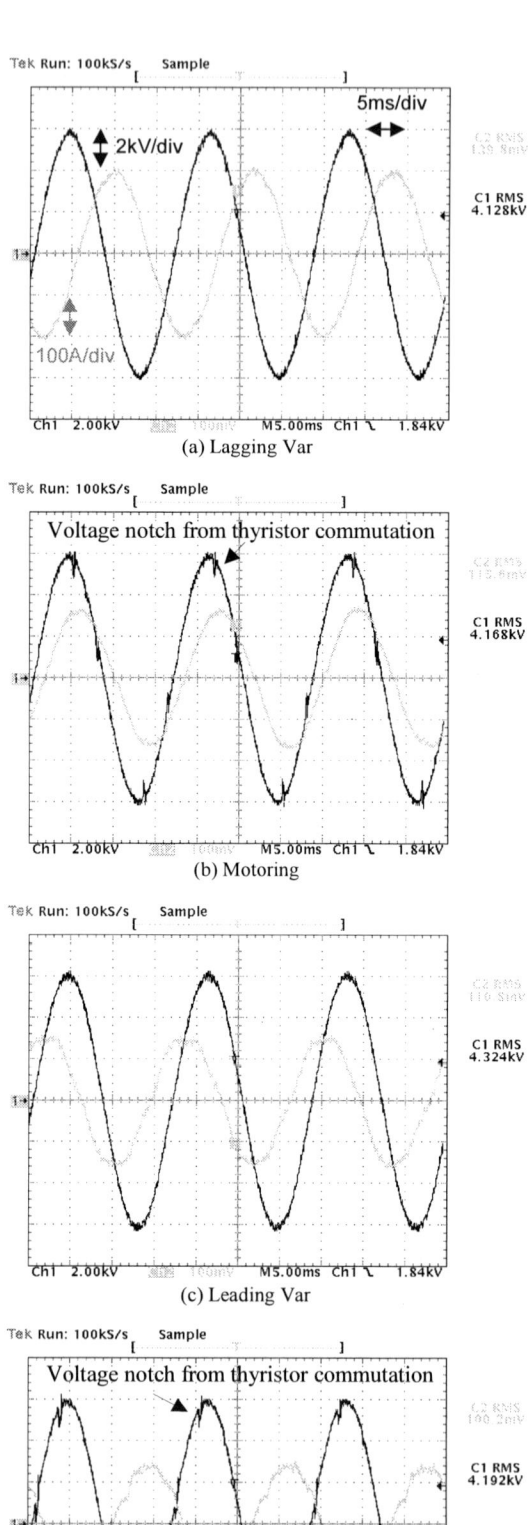

(a) Lagging Var

(b) Motoring

(c) Leading Var

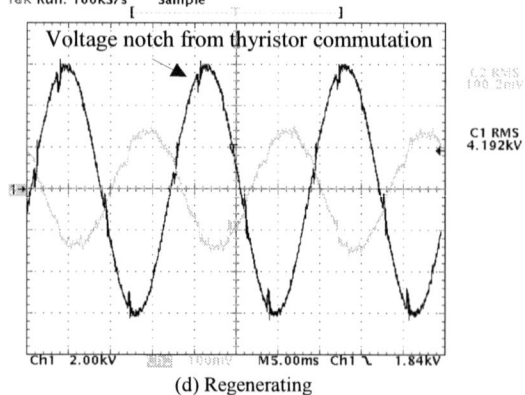

(d) Regenerating

Fig. 11. Voltage and current waveforms

978-1-4799-2706-7/14 $31.00 © 2014 IEEE

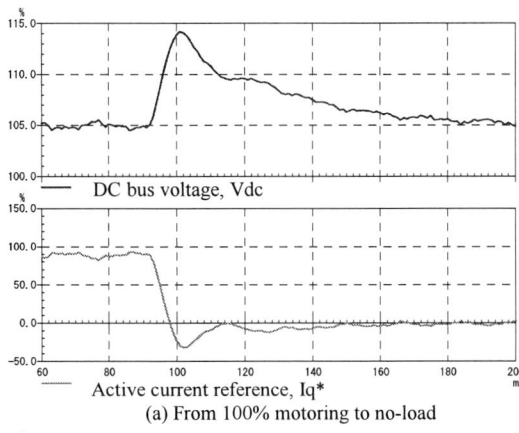

DC bus voltage, Vdc

Active current reference, Iq*

(a) From 100% motoring to no-load

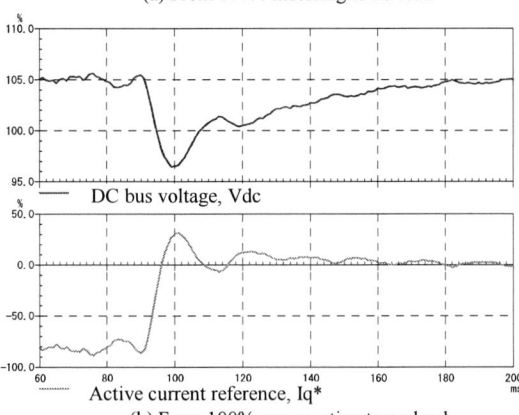

DC bus voltage, Vdc

Active current reference, Iq*

(b) From 100% regenerating to no-load

Fig. 12. Load impact step

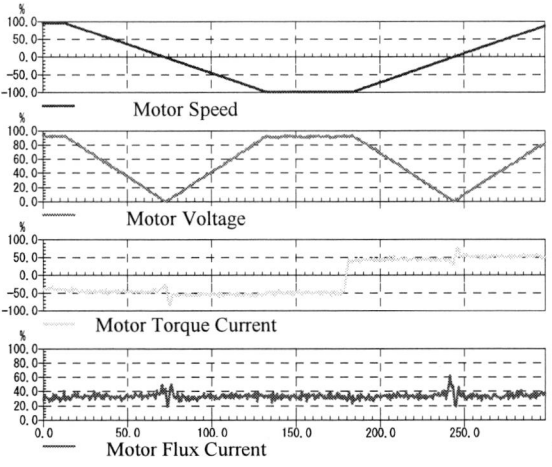

Motor Speed

Motor Voltage

Motor Torque Current

Motor Flux Current

Fig. 13. Four quadrant operation

B. Fixed speed test

There is big voltage distortion from DC drive, and it is difficult to validate the harmonics capability. To remove the utility harmonics from the DC drive, a fixed speed load test was performed as shown in Fig. 14. Two 1864kW (2500HP) motors were coupled together and one was connected to a 4.16kV utility source, and active front-end medium voltage drive is controlling 1864kW (2500HP) motor current, and tested at 746kW (1000HP).

Utility current harmonics results are shown in Fig. 15 and Fig. 16. In the actual test condition, the utility maximum short-circuit (I_{SC}) is as below:

$$I_{sc} = \frac{3750 kVA}{4.16 kV \times \sqrt{3} \times 6.1\%} = 8.5 kA \qquad (1)$$

The fundamental current (I_L) of the drive is determined by 746kW (1000HP) as below:

$$I_l = \frac{746 kW}{4.16 kV \times \sqrt{3}} = 104 A \qquad (2)$$

Then, the I_{SC}/I_L ratio is as below:

$$I_{sc}\Big/I_l = 8.5 kA\Big/104 A = 82 \qquad (3)$$

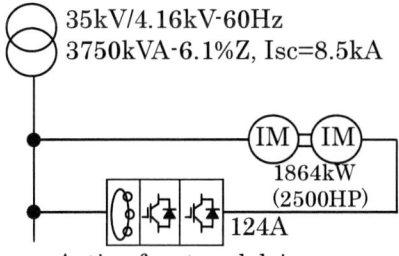

35kV/4.16kV-60Hz
3750kVA-6.1%Z, Isc=8.5kA

IM IM
1864kW
(2500HP)
124A

Active front-end drive

Fig. 14. Test circuit for fixed speed operation

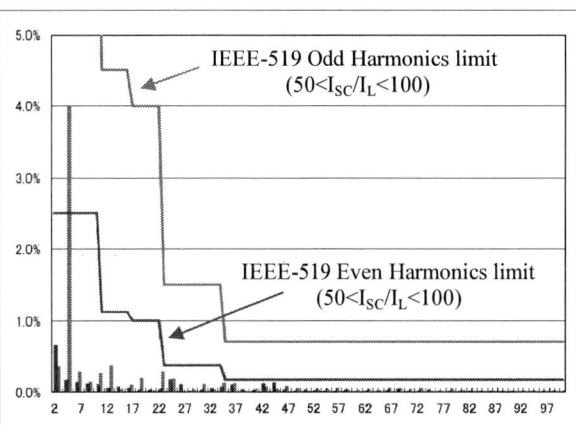

IEEE-519 Odd Harmonics limit
(50<I_{SC}/I_L<100)

IEEE-519 Even Harmonics limit
(50<I_{SC}/I_L<100)

(a) Motoring case with IEEE-519 limits for actual test condition

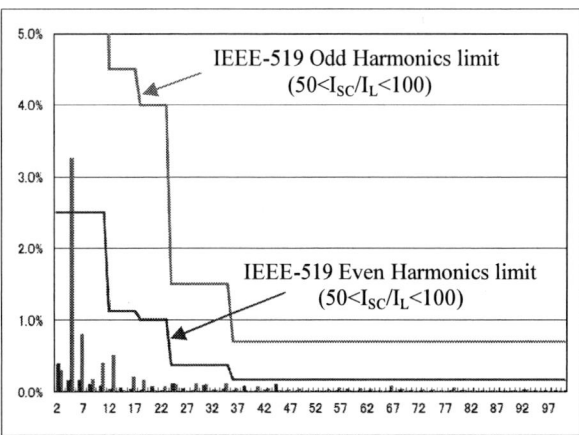

IEEE-519 Odd Harmonics limit
(50<I_{SC}/I_L<100)

IEEE-519 Even Harmonics limit
(50<I_{SC}/I_L<100)

(b) Regenerating case with IEEE-519 limits for actual test condition

Fig. 15. Current harmonics and IEEE-519 limit (50<I_{SC}/I_L<100)

The current harmonics are shown by bars (Red: Odd Harmonics, Blue: Even Harmonics) in (a) and (b) for each motoring and regenerating, and IEEE-519 limits of $50 < I_{SC}/I_L < 100$ are shown, also. All harmonics are lower than IEEE-519 limits.

Most strict harmonics current limit in IEEE-519, when I_{SC}/I_L is less than 20, is shown in Fig. 16. Some of harmonics are bigger than this limit, but almost all harmonics are lower than this limit.

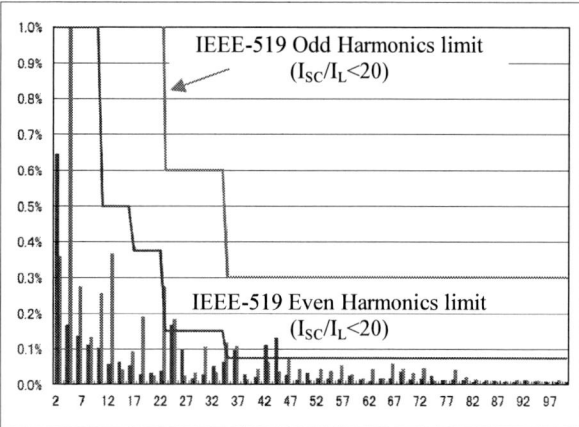

(a) Motoring case with IEEE-519 limits for most strict case

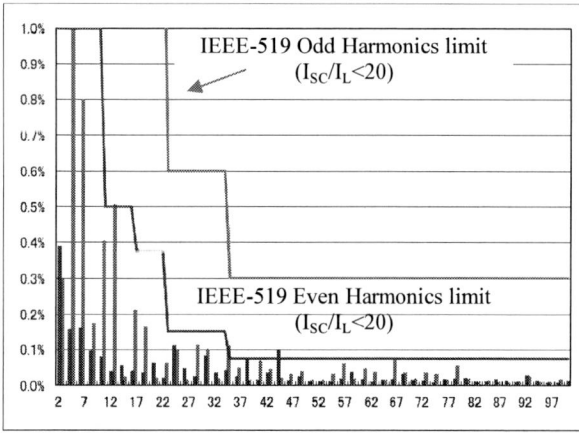

(b) Regenerating case with IEEE-519 limits for most strict case
Fig. 16. Current harmonics and IEEE-519 limit ($I_{SC}/I_L < 20$)

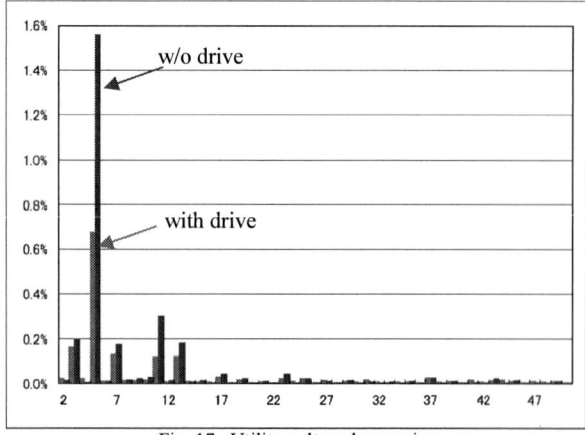

Fig. 17. Utility voltage harmonics

To validate 5th and other low order current harmonics, voltage harmonics is compared in case with the drive and without the drive as shown in Fig. 17. The test utility voltage harmonics is already high even without the drive.

C. Flyback test

Based on the proposed topology, 12.7kVin-4.16kVout 4474kW (6000HP) regenerative drive test is performed in the flyback configuration as shown in Fig. 18. Even this case that the input utility transformer size is smaller than the drive size, the drive could manage its rating current and could have a stable DC bus control.

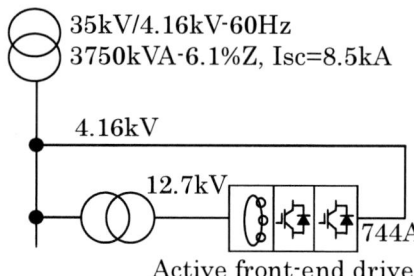

Fig. 18. Test circuit for flyback operation

V. CONCLUSION

A new active front-end converter for a 5 level medium voltage inverter topology is proposed. Compared to a 3 level NPC active front-end drive, there are smaller DC bus steps and less harmonics. Also, voltage surge on the motor is reduced.

Compared to a conventional 5 level active front-end drive, which has a full bridge NPC active front-end converter in each of the DC buses, the proposed drive are has less devices and a simpler DC bus control scheme.

Test results of four quadrant operation and DC bus control capability are shown.

REFERENCES

[1] M. Tsukakoshi, "Performance Evaluation of a Large Capacity VSD System for Oil and Gas Industry", IEEE ECCE, pp.3485-3492, 2009

[2] M. Tsukakoshi, "Novel Torque Ripple Minimization Control for 25MW Variable Speed System Fed by Multilevel Voltage Source Inverter", Proc. Thirty-ninth Turbomachinery Symposium, pp.193-200, 2010.

[3] M.A. Mamun, "Performance Evaluation of a Large Capacity 3-Level IEGT Inverter", IEEE ECCE Asia, pp.201-207, 2013.

[4] A. Nabae, "A New Neutral-Point-Clamped PWM Inverter" IEEE Trans. on Industry Applications, vol. IA-17, no. 5, pp. 518-523, 1981.

[5] F. Hernandez, "A Generalized Control Scheme for Active Front-end Multilevel Converters," Conference Recode in IECON '04, IEEE Industrial Electronics Society, vol.2, pp1446-1451, 2004.

[6] T. Arai, "The power regeneration technology for medium Voltage Inverter" (In Japanese) Conference Recode in the Technical meeting of metal applications IEEJ, MID-07-2, 2007.

A Dual Active Bridge DC-DC Converter with Optimal DC-Link Voltage Scaling and Flyback Mode for Enhanced Low-Power Operation in Hybrid PV/Storage Systems

Shahab Poshtkouhi, and Olivier Trescases

University of Toronto, 10 King's College Road, Toronto, ON, M5S 3G4, Canada

E-mail: shahab.poshtkouhi@utoronto.ca

Abstract—Today's PV micro-inverters (MIVs) provide a modular solution for generation, however the energy storage architectures remain centralized, requiring an additional bi-directional ac-dc converter, with complex cell balancing circuits. Distributing storage capacity within the smart PV panels allows power fluctuations to be locally buffered, while minimizing the need for additional power electronics and balancing circuits. The dual-active-bridge (DAB) topology, which is adopted in this paper, provides bi-directional power flow; however it generally suffers from poor efficiency at low power. It is shown that with a minor modification, the DAB can be operated as a two-transistor flyback converter for improved efficiency. In addition, the dc-link voltage can be dynamically adjusted for the best performance in DAB mode. The proposed control scheme is demonstrated on a 100 W prototype, with up to 8% increase in efficiency at low power.

Index Terms—Photovoltaic (PV) micro-inverters, efficiency, isolated dc-dc stage.

I. INTRODUCTION

The continuous decline of photovoltaic (PV) module prices, compounded with attractive feed-in tariffs in a variety of jurisdictions, is leading to the rapid deployment of PV installations throughout the world [1]. The intermittent nature of PV and other renewable energy resources, and thus the need for energy storage and/or load shedding, is a major challenge in small-scale PV based grids. This is despite the fact that power-quality and other grid strict requirements such as frequency variations are reduced compared to conventional grids. Low-power dc-dc micro-converters (MIC) [2], [3] and ac-dc micro-inverters (MIV) [4], [5] provide high-granularity Maximum Power Point Tracking (MPPT) [6], [7] at the module or sub-string level. This leads to increased robustness to clouds, dirt, and aging effects, as well as irradiance and temperature gradients [7]. A conventional MIV based ac power system is shown in Fig. 1. The Energy Storage System (ESS), which is definitely required for islanded operation on the scale of one or more houses for example, is usually based on a high-power centralized bi-directional ac-dc converter, which is interfaced to a battery bank or a flywheel [8], [9]. Existing MIV architectures satisfy the need for low capital-cost and expandable ac generation, while there is compelling argument to extend this technology to include small-scale distributed storage. A novel topology with distributed storage is proposed

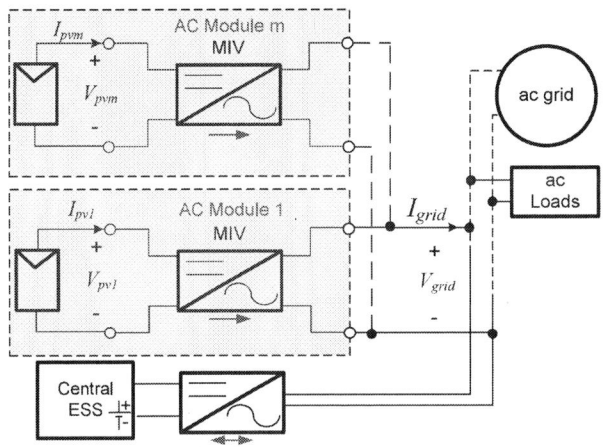

Fig. 1. Conventional micro-inverter based PV system with central ESS.

in [10] for grid stabilization, while saving fuel, and improving the generator lifetime. MIV Integrated storage helps to buffer the frequent irradiance fluctuations, while also providing local back-up power and reactive power support [11], [12]. A low-power single-stage multi-port converter for PV and battery is proposed in [13], while a 3-kW interconnection of a battery pack and a PV module through an isolated dc-dc converter is discussed in [14]. The general architecture of a two-stage MIV with an integrated ESS is shown in Fig. 2. While two-stage MIVs have a slightly lower efficiency than single-stage MIVs, the high-voltage dc link capacitance, C_{bus}, can be used for ac power decoupling in single-phase systems [15], [16].

Interfacing the low-voltage dc storage, either batteries or ultra-capacitors, directly to the PV bus is preferable for high efficiency [14]. Lithium-ion ultra-capacitors [17], which offer $2\text{-}4\times$ higher specific energy than conventional Electric Double Layer Capacitors and can withstand more than 200000 charge/discharge cycles, are an attractive future candidate for short-term MIV integrated storage. The focus of this work is on the front-end dc-dc stage.

The objective of this paper is to 1) discuss a bi-directional isolated dc-dc stage for the PV-bus connection, which is al-

978-1-4799-2706-7/14 $31.00 © 2014 IEEE

ready presented in [10] and 2) demonstrate a novel low-power operating mode and dynamic dc-link optimization scheme to maintain high efficiency over a broad power range. This is crucial in any commercial MIV architecture; for example the European Efficiency index dedicates 32% of the total evaluation weight to operation below 30% of the rated power [18].

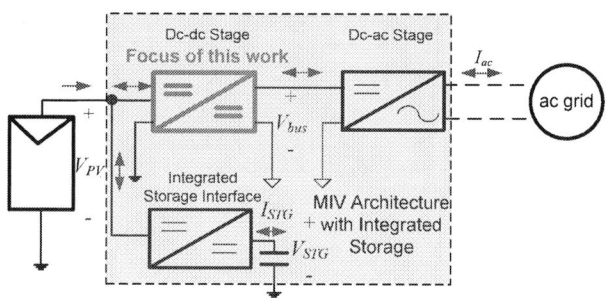

Fig. 2. Two-stage MIV architecture with integrated storage [10].

II. PROPOSED DAB ARCHITECTURE AND PRINCIPLE OF OPERATION

The proposed dc-dc architecture is shown in Fig. 3(a). This converter is a modified Dual-Active-Bridge (DAB) that interfaces V_{pv} with the dc link, V_{bus}.

A. DAB Mode

The DAB topology was selected based on (1) galvanic isolation, (2) soft-switching operation and (3) simple phase-shift power control [19], [20]. In addition, the DAB topology is bi-directional, therefore the storage can be used to transfer energy to/from other elements in the grid. The average power from V_{pv} to V_{bus}, P, is

$$P = \frac{V_{PV} V_{bus}}{n \omega_s L_{DAB}} \phi \left(1 - \frac{\phi}{\pi}\right), \tag{1}$$

where n is the transformer's turns ratio, L_{DAB} is the DAB inductance, which is the sum of transformer's leakage inductance, L_{leak}, and an optional external inductance, L_{ext}. ϕ is the phase-shift between the two bridges, and $\omega_s = 2\pi f_s$, where f_s is the switching frequency.

The switching waveforms of the DAB converter are shown in Fig. 4(a). The slopes of the DAB inductance current, I_{LDAB}, in switching states I and II are respectively calculated as

$$s_1 = \frac{V_{PV} + \frac{V_{bus}}{n}}{L_{DAB}}, \tag{2}$$

$$s_2 = \frac{-V_{PV} + \frac{V_{bus}}{n}}{L_{DAB}}. \tag{3}$$

In two-stage MIV architectures, V_{bus} is generally regulated to a fixed voltage by the inverter stage. The reference voltage, V_{bus}^*, is usually chosen to optimize efficiency at the nominal operating point [7]. It can be shown that the DAB converter achieves turn-on Zero-Voltage-Switching (ZVS) and maximum

efficiency when $V_{bus} = n V_{PV}$, as the reactive circulating current is minimized [19].

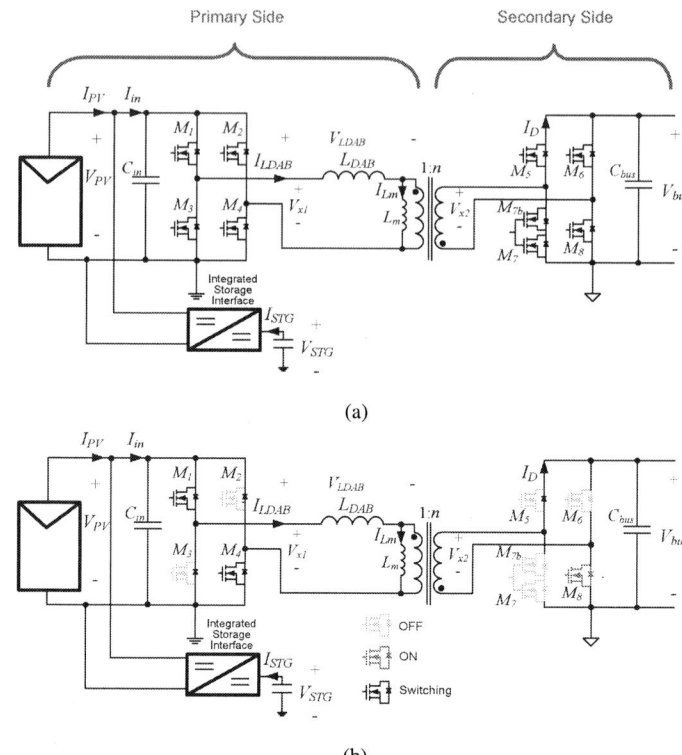

(a)

(b)

Fig. 3. a) Proposed modified DAB dc-dc architecture for improved low power efficiency. b) Switch configuration in Flyback mode.

In order to minimize the losses in the DAB, the reference DC link voltage, V_{bus}^*, is dynamically adjusted in the inverter stage such that $V_{bus}^* = n V_{MPP}$, where V_{MPP} is the PV MPP voltage. It is well-known that V_{MPP} undergoes a relatively low fluctuation of about 30% during the course of a typical day [21]. This is in contrast to the PV current at MPP, I_{MPP}, which is proportional to irradiance and thus has large-scale fluctuations, especially on cloudy days.

B. Flyback Mode

A typical PV generation system spends more than two-thirds of the time operating below 50% of its rated power [7]. The conventional DAB converter suffers from relatively poor efficiency at low power due to high switching and drive losses [19], hence the need for a dedicated low-power mode. By driving M_1 and M_4 simultaneously on the primary side, the converter can be operated as a two-transistor flyback converter (2T-flyback) [22]. The switch configuration in Flyback mode is shown in Fig. 3(b). The switches M_1 and M_4 remain active, M_8 is kept *on* and all other switches are *off*. The secondary-side bridge is slightly modified with the addition of one switch, M_{7b}, to achieve bi-directional blocking capability in Flyback mode. The charging and discharging slopes of the magnetizing

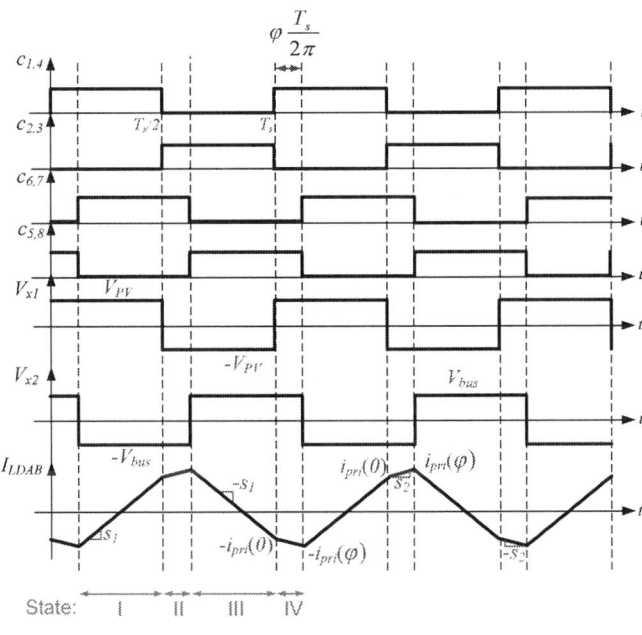

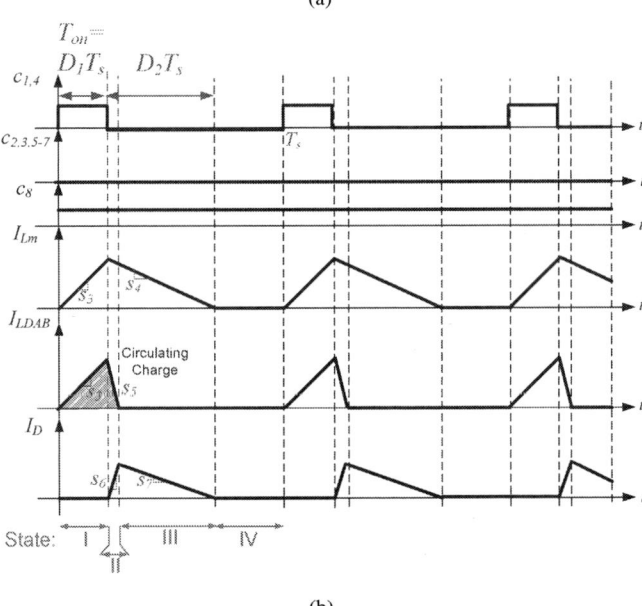

(a)

(b)

Fig. 4. Switching waveforms in (a) DAB mode, and (b) Flyback mode.

inductance current, I_{L_m}, are given by

$$s_3 = \frac{V_{PV}}{L_{DAB} + L_m}, \tag{4}$$

$$s_4 = \frac{-V_{bus}}{nL_m}. \tag{5}$$

The DAB inductance circulates charge in every switching period in this mode. The charging slope of I_{LDAB} is the same as s_3 and the discharging slope is

$$s_5 = \frac{-V_{PV} - \frac{V_{bus}}{n}}{r}. \tag{6}$$

Finally, the output diode's current, I_D, delivers charge to the bus with the following slopes in switching states II and III

$$s_6 = \frac{s_4 + s_5}{n}, \tag{7}$$

$$s_7 = \frac{s_4}{n}. \tag{8}$$

The 2T-flyback topology exhibits several advantages over DAB mode for low power conditions, including less switching and gate-drive losses (two switching devices versus nine in the DAB mode). Unlike the more conventional single transistor flyback topology, the body diodes of M_2 and M_3 clamp the drain voltage on the switching devices M_1 and M_4, which reduces Electromagnetic Interference (EMI) and limits the blocking voltage rating on the primary switches to V_{PV}. The Flyback mode is operated with fixed on-time, T_{on}, in Pulse Frequency Modulation (PFM) mode [23], where T_{on} is given by

$$T_{on} = D_1 T_s, \tag{9}$$

where D_1 is the duty cycle in Flyback mode, and T_s is the switching period.

The corresponding waveforms of the converter are shown in Fig. 4(b). There are two inherent limitations to the 2T-flyback topology: 1) D_1 must be less than 50% in order to avoid transformer saturation, and 2) V_{bus} must be less than nV_{pv}, to ensure that the body diode of M_5 transfers power to V_{bus} when the primary-side switches are *off*. As a result, V_{bus} needs to be reduced in Flyback mode. The presence of L_{DAB} results in additional losses, since it circulates current in a switching period. The energy captured in L_{DAB} is transferred back to the input capacitance, C_{in}, in the 2T-flyback topology, as opposed to a conventional flyback scheme, which docs not provide a return path for the charge in the leakage inductance. In addition, L_{DAB} results in the soft turn-on of the output diode.

The Flyback mode exhibits uni-directional power transfer. The converter can operate with reverse power flow by adding another switch on the primary side. This additional switch is not included in the experimental prototype, as the efficiency in DAB mode is sensitive to conduction losses at the low-voltage, high-current primary-side. While possible, reverse power capability is not strictly needed in low-power Flyback mode; the DAB can be prevented from operating in this condition by adopting burst-mode control instead, albeit at slightly lower efficiency than Flyback mode.

C. Dual Mode Control

The conceptual control diagram of the converter is shown in Fig. 5. c_{1-8} denote the gating voltages for switches M_{1-8}. The DAB mode is adopted if P is higher than a threshold value, P_{thresh}, or if P is negative, in which case the storage is charged directly from the bus. In DAB mode, ϕ is controlled to regulate the power flow to/from the dc-ac stage, while the storage element's State-of-Charge (SOC) and MPPT operation can be controlled by the dedicated interface converter.

In Flyback mode, T_s is adjusted by the controller, $G_{c2}(s)$, in order to regulate P to P^*. Assuming that the magnetizing inductance of the transformer, L_m, is much larger than L_{DAB}, the power flow is given by

$$P = \frac{(V_{PV}D_1)^2 T_s}{2L_m}. \tag{10}$$

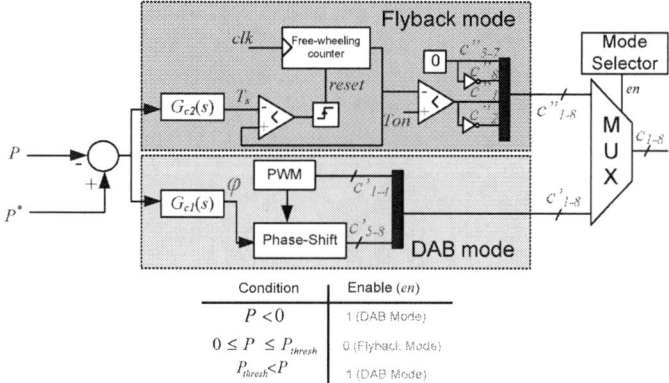

Fig. 5. Simplified conceptual control diagram.

III. EFFICIENCY ANALYSIS

This section discusses the dominant power losses in the DAB and Flyback modes.

A. Conduction Losses

As shown in Fig. 4(a), the current into the leakage inductor and primary winding of the transformer approach a perfect trapezoid when $V_{bus} = nV_{pv}$. The following equation for the RMS current at the primary side of transformer, I_{pri}, can be obtained [24]

$$I_{pri} = \frac{1}{3\pi}(i_{pri}(0)^2\gamma + i_{pri}(\phi)^2(\phi - \gamma)$$
$$+(\pi - \phi)(i_{pri}(0)^2 + i_{pri}(\phi)^2 - i_{pri}(0)i_{pri}(\phi)), \tag{11}$$

where

$$\gamma = \frac{i_{pri}(0)}{i_{pri}(0) - i_{pri}(\phi)}, \tag{12}$$

and $i_{pri}(0)$ and $i_{pri}(\phi)$ are the instantaneous currents of the transformer at the primary side at times $t = 0$ and $t = \frac{\phi}{\omega_s}$, respectively. These can be easily calculated considering the symmetry of the transformer current [24]

$$i_{pri}(0) = \frac{1}{2L_{DAB}\omega_s}(\pi V_{PV} - (\pi - 2\phi)V_{bus}), \tag{13}$$

$$i_{pri}(\phi) = \frac{1}{2L_{DAB}\omega_s}((\pi - 2\phi)V_{PV} - \pi V_{bus}). \tag{14}$$

Similar calculations can be done for the transformer's secondary side RMS current, I_{sec}. Two primary and two secondary switches are conducting at each instance in DAB mode. Thus, the conduction losses in this mode are approximated by

$$P_{DAB,cond} = (2R_{on,pri} + R_{L_{DAB}})I_{pri}^2 + \tag{15}$$
$$2.5R_{on,sec}I_{sec}^2.$$

where $R_{on,pri}$ and $R_{on,sec}$ are the primary and secondary side switches' on-resistances respectively, and $R_{L_{DAB}}$ is the lumped winding resistance of the transformer and inductor. The factor of $2.5\times$ on the secondary side comes from the fact that, there are two back-to-back switches, M_7, and M_{7b} on one leg in the secondary side to support the flyback operation.

Neglecting the clamping diodes' conduction interval, the RMS current for the two switches, I_{M_1,M_4}, and in the output diode, D, in the 2T-flyback converter can be obtained as [25]

$$I_{M_1,M_4} = \frac{nP\sqrt{D_1}}{V_{bus}(1 - D_1)}, \tag{16}$$

$$I_D = \frac{D_2 T_s V_{bus}}{n^2 L_m}\sqrt{\frac{D_2}{3}}, \tag{17}$$

where D_2 is approximated by the following

$$D_2 = \frac{nV_{PV}D_1}{V_{bus}}. \tag{18}$$

The total conduction loss in Flyback mode is

$$P_{Flbk,cond} = (2R_{on,pri} + R_{L_{DAB}})I_{M1}^2 + \tag{19}$$
$$V_F I_D + R_{on,sec}I_D^2,$$

where V_F is the output diode's forward voltage.

B. Switching Losses

The switches in DAB mode can be turned *on* realizing Zero Voltage Switching (ZVS) [24]. However the turn-*off* losses are not eliminated. The total switching losses in this mode can be approximated by

$$P_{DAB,sw} = \frac{1}{2}f_s t_{off}(V_{pv}(i_{pri}(0) + i_{pri}(\phi) + \tag{20}$$
$$V_{bus}(i_{sec}(0) + i_{sec}(\phi)),$$

where t_{off} is the turn-*off* time of the MOSFETs.

Switches M_1 and M_2 exhibit hard switching at turn-*off* in Flyback mode. Thus, the corresponding switching loss in Flyback mode can be approximated as

$$P_{Flbk,sw} = \frac{V_{PV}^2 D_1 t_{off}}{2(L_{DAB} + L_m)}. \tag{21}$$

The switch drive losses are not considered here; However, there are nine switches actively driven in DAB mode compared to only two switches actively switching in Flyback mode. Furthermore, f_s is much lower in Flyback mode then in DAB mode, which further reduces the switch drive losses.

C. Core Losses

Core losses are present in the high-frequency transformer in both DAB and Flyback modes and can be approximated using the Steinmetz equation [26]

$$P_{core} = kf_s^\alpha B_{peak}^\beta, \tag{22}$$

where B_{peak} is the peak flux density, and k, α, and β are the Steinmetz parameters, which depend on the core, and are usually found by curve fitting. Core losses constitute a low percentage of the total losses in DAB mode [24] due to high-frequency ac-ac operation. However, core losses are dominant in Flyback mode. This is due to high peaks in core voltage, and lower frequency operation, which increases B_{peak} in the transformer.

This analysis neglects the skin effect in all conductors, which can be significant, especially in DAB mode due to high frequency operation.

D. Loss Comparison in Two Modes

The calculated loss breakdown for $P = 10$ W and $P = 40$ W is shown in Fig. 6. The conduction losses in all active and passive elements are lumped together. The switching losses also include the drive losses. In Flyback mode, the switching losses are reduced by at least $10\times$, mostly by eliminating the turn-*off* losses on the high voltage side, at a cost of marginally increase in conduction losses. The transformer and inductor core loss is slightly higher in Flyback mode, due to higher B_{peak}. The core losses in Flyback mode increase rapidly with the power due to higher B_{peak} and f_s. This is not the case for the DAB converter, in which the core losses remain almost constant over the power range.

IV. EXPERIMENTAL RESULTS

A prototype of the system shown in Fig. 3(a) was fabricated on a custom Printed Circuit Board, with DAB power rating of 100 W. The specifications of the prototype are listed in Table I. The converters are digitally controlled using an on-

TABLE I
MIV PROTOTYPE SPECIFICATIONS

Parameter	Value	Units
Rated Power, P_{nom}	100	W
Dc-dc Stage Switching Frequency, f_s		
DAB Mode	195	kHz
Flyback Mode	20-50	kHz
Fixed On-Time, T_{on}	8	μs
Input Capacitance, C_{in}	300	μF
Bus Capacitance, C_{bus}	100	μF
DAB Inductance, L_{DAB}	4.2	μH
Magnetizing Inductance, L_m	32	μH
Bus Voltage Range, V_{bus}		
DAB mode	200-270	V
Flyback mode	170	V
Transformer Turns Ratio, n	9	

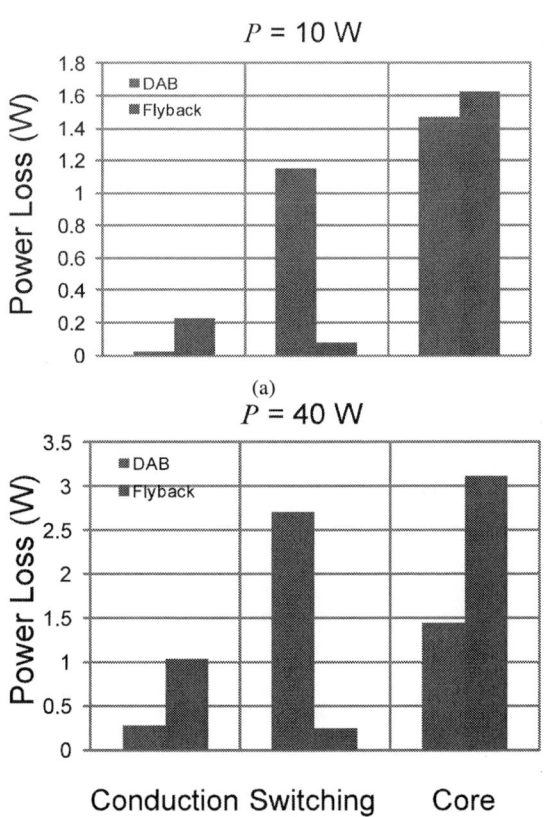

Fig. 6. Simulated power losses for (a) $P = 10$ W, and (b) $P = 40$ W.

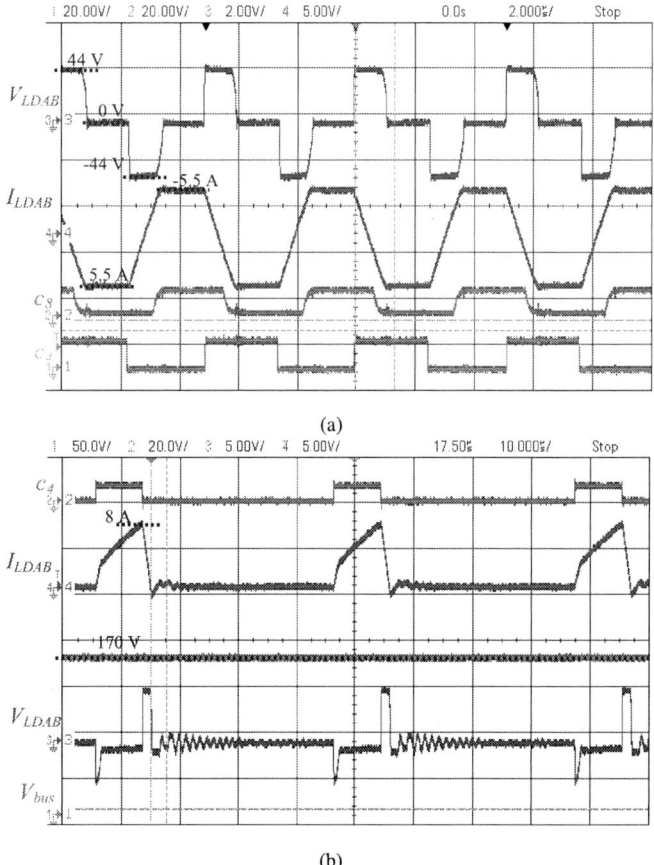

Fig. 7. Steady-state waveforms of the converter in (a) DAB mode at $V_{PV} = 22$ V (I_{LDAB}:5 A/div), and (b) Flyback mode at $V_{PV} = 25$ V (I_{LDAB}:5 A/div).

board FPGA. A custom planar transformer was designed to reduce the weight and profile of the prototype. The steady-state waveforms in DAB and Flyback modes at $P = 70$ W, and $P = 15$ W, are shown in Fig. 7(a) and (b), respectively. V_{bus} is adjusted to nV_{PV} in DAB mode to achieve optimal efficiency, as it is illustrated by the flat portions in I_{LDAB}. The closed-loop dynamic response of Flyback mode for a step

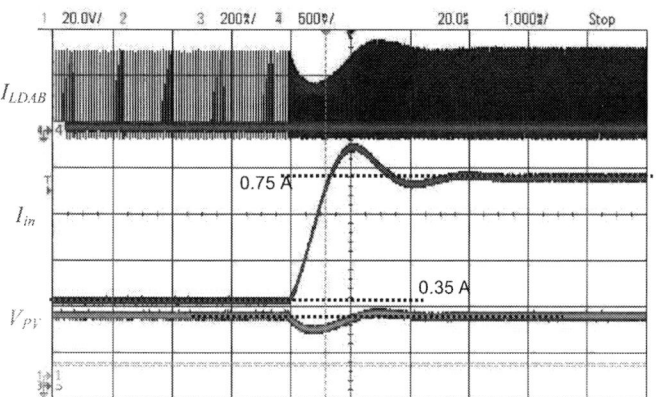

Fig. 8. Step response of Flyback mode with the integrated storage interface off: P: 9.1 W $\rightarrow$ 19.5 W (I_{in}:0.2 A/div, I_{LDAB}: 10 A/div).

change in P^*, while the dedicated integrated storage converter is off, is shown in Fig. 8. f_s is increased in Flyback mode by the controller to accommodate the higher input power.

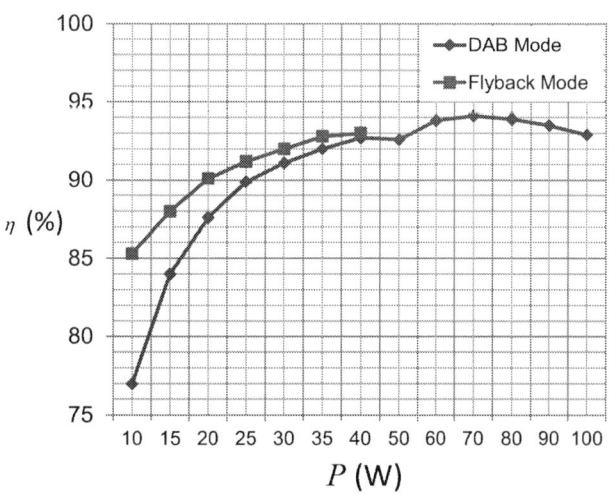

Fig. 9. Measured efficiency, η, of the converter.

The measured efficiency of the converter, η, in both modes is shown in Fig. 9. A peak efficiency of 94% is achieved in DAB mode, while Flyback mode has a superior efficiency up to $P = 40$ W. The power is limited in Flyback mode due to the maximum duty ratio of 50%. However, the design is such that the two efficiency curves intercept at a point close to the maximum transferrable power in Flyback mode. The converter operates at the edge of Discontinuous Conduction Mode (DCM) at the intercept point, such that the operation is switched to DAB mode at higher power.

V. CONCLUSIONS

A novel DAB switching scheme was introduced for the dc-dc stage of module integrated power converters for PV applications. The modified flyback switching scheme exhibits 8% higher efficiency than DAB mode at 10 W, which comes at the cost of an additional switch. While Flyback mode exhibits more core losses and slightly more conduction losses compared to DAB mode, the switching losses are significantly reduced by eliminating most of the switching actions, and reducing frequency. The Flyback mode is operated in Fixed On-Time PFM; further power savings could be achieved in both modes by using Burst-Mode Control.

The bus voltage is adjusted in DAB mode based on MPP voltage in order to achieve maximum efficiency. This voltage is adaptively reduced in Flyback mode by the following dc-ac stage, in order to operate properly.

ACKNOWLEDGEMENT

This work was supported by Solantro Semiconductor, the Ontario Centres of Excellence, the Natural Sciences and Engineering Research Council of Canada, the Canadian Foundation for Innovation and the Ontario Research Fund. The authors also thank Ray Orr, Ben Bacque, Mihai Varlan, and Chris Gerolami for discussions related to nanogrids and micro-inverters.

REFERENCES

[1] P. Denholm, R. Margolis, T. Mai, G. Brinkman, E. Drury, M. Hand, and M. Mowers, "Bright future: Solar power as a major contributor to the u.s. grid," *IEEE Power and Energy Magazine*, vol. 11, no. 2, pp. 22–32, 2013.

[2] R. K. Hester, C. Thornton, S. Dhople, Z. Zhao, N. Sridhar, and D. Freeman, "High efficiency wide load range buck/boost/bridge photovoltaic microconverter," in *IEEE Applied Power Electronics Conference and Exposition*, 2011, pp. 309 –313.

[3] B. York, W. Yu, and J.-S. Lai, "An integrated boost resonant converter for photovoltaic applications," *IEEE Transactions on Power Electronics*, vol. 28, no. 3, pp. 1199–1207, 2013.

[4] R. Erickson and A. Rogers, "A microinverter for building-integrated photovoltaics," in *IEEE Applied Power Electronics Conference and Exposition*, 2009, pp. 911 –917.

[5] "Enphase m190 microninveter," Enphase Datasheet, 2009, available http://enphaseenergy.com.

[6] N. Femia, G. Lisi, G. Petrone, G. Spagnuolo, and M. Vitelli, "Distributed maximum power point tracking of photovoltaic arrays: Novel approach and system analysis," *IEEE Transactions on Industrial Electronics*, vol. 55, no. 7, pp. 2610 –2621, July 2008.

[7] S. Poshtkouhi, V. Palaniappan, M. Fard, and O. Trescases, "A general approach for quantifying the benefit of distributed power electronics for fine grained mppt in photovoltaic applications using 3-d modeling," *IEEE Transactions on Power Electronics*, vol. 27, no. 11, pp. 4656–4666, 2012.

[8] L. Xu and D. Chen, "Control and operation of a dc microgrid with variable generation and energy storage," *IEEE Transactions on Power Delivery*, vol. 26, no. 4, pp. 2513–2522, 2011.

[9] G. Suvire, M. Molina, and P. Mercado, "Improving the integration of wind power generation into ac microgrids using flywheel energy storage," *IEEE Transactions on Smart Grid*, vol. 3, no. 4, pp. 1945–1954, 2012.

978-1-4799-2706-7/14 $31.00 © 2014 IEEE

[10] S. Poshtkouhi, M. Fard, H. Hussein, L. Dos Santos, O. Trescases, M. Varlan, and T. Lipan, "A Dual-Active-Bridge based Bi-Directional Micro-Inverter with Integrated Short-Term Li-Ion Ultra-Capacitor Storage and Active Power Smoothing for Modular PV Systems," in press.

[11] M. Alam, K. Muttaqi, and D. Sutanto, "Mitigation of rooftop solar pv impacts and evening peak support by managing available capacity of distributed energy storage systems," *IEEE Transactions on Power Systems*, vol. PP, no. 99, pp. 1–11, 2013.

[12] L. Liu, H. Li, Z. Wu, and Y. Zhou, "A cascaded photovoltaic system integrating segmented energy storages with self-regulating power allocation control and wide range reactive power compensation," *IEEE Transactions on Power Electronics*, vol. 26, no. 12, pp. 3545–3559, 2011.

[13] Y.-M. Chen, A. Huang, and X. Yu, "A high step-up three-port dc-dc converter for stand-alone pv/battery power systems," *IEEE Transactions on Power Electronics*, vol. 28, no. 11, pp. 5049–5062, 2013.

[14] Z. Wang and H. Li, "An integrated three-port bidirectional dc-dc converter for pv application on a dc distribution system," *IEEE Transactions on Power Electronics*, vol. 28, no. 10, pp. 4612–4624, 2013.

[15] H. Hu, S. Harb, N. Kutkut, Z. Shen, and I. Batarseh, "A single-stage microinverter without using eletrolytic capacitors," *IEEE Transactions on Power Electronics*, vol. 28, no. 6, pp. 2677–2687, June 2013.

[16] S. Kjaer, J. Pedersen, and F. Blaabjerg, "A review of single-phase grid-connected inverters for photovoltaic modules," *IEEE Transactions on Industry Applications*, vol. 41, no. 5, pp. 1292–1306, Sept 2005.

[17] "Lithium-Ion Capacitor," JSR Micro, 2012, available at http://www.jsrmicro.com/index.php/EnergyAndEnvironment/.

[18] "European or CEC Efficiency," available at

http://files.pvsyst.com/help/index.html.

[19] F. Krismer and J. Kolar, "Efficiency-optimized high-current dual active bridge converter for automotive applications," *IEEE Transactions on Industrial Electronics*, vol. 59, no. 7, pp. 2745–2760, 2012.

[20] H. Qin and J. Kimball, "Generalized average modeling of dual active bridge dc-dc converter," *IEEE Transactions on Power Electronics*, vol. 27, no. 4, pp. 2078–2084, 2012.

[21] M. Park and I.-K. Yu, "A study on the optimal voltage for mppt obtained by surface temperature of solar cell," in *30th Annual Conference of IEEE Industrial Electronics Society, 2004*, vol. 3, 2004, pp. 2040–2045 Vol. 3.

[22] D. D. C. Lu, H.-C. Iu, and V. Pjevalica, "A single-stage ac/dc converter with high power factor, regulated bus voltage, and output voltage," *IEEE Transactions on Power Electronics*, vol. 23, no. 1, pp. 218–228, 2008.

[23] R. Erickson and D. Maksimović, *Fundamentals of Power Electronics, Second Ed.* Springer, 2001.

[24] H. Qin and J. Kimball, "Generalized average modeling of dual active bridge dc-dc converter," *IEEE Transactions on Power Electronics*, vol. 27, no. 4, pp. 2078–2084, April 2012.

[25] D. Murthy-Bellur and M. Kazimierczuk, "Two-switch flyback-forward pwm dc-dc converter with reduced switch voltage stress," in *Proceedings of 2010 IEEE International Symposium on Circuits and Systems (ISCAS)*, May 2010, pp. 3705–3708.

[26] J. Reinert, A. Brockmeyer, and R. De Doncker, "Calculation of losses in ferro- and ferrimagnetic materials based on the modified steinmetz equation," *IEEE Transactions on Industry Applications*, vol. 37, no. 4, pp. 1055–1061, Jul 2001.

Novel Modular Multiple-Input Bidirectional DC-DC Power Converter (MIPC)

Andrew Hintz, Udupi. R. Prasanna, *Member IEEE* and Kaushik Rajashekara, *Fellow IEEE*
Electrical Engineering,
University of Texas at Dallas
Richardson, Texas 75080, USA.

Abstract- **This paper proposes a novel multiple-input bidirectional DC-DC power converter to interface more than two dc sources of different voltage levels. This finds applications in distributed energy resources (DER), micro grid, and hybrid electric/fuel cell vehicles, where different dc sources of unequal voltage levels need to be connected with bidirectional power flow capability. Converter can be used to operate in both the buck and boost modes with bidirectional power control. It is also possible to independently control power flow when more than two sources are actively transferring power in either direction. This paper presents a power converter topology based on three switching legs of a standard 3-phase inverter module. The operation, analysis and design of the converter are presented with different modes of power transfer. Proposed converter is demonstrated for fuel cell vehicle application using real-time hardware-in-the-loop system. Results for a 5kW system are presented validating the theoretical analysis.**

Keywords — **Distributed Energy Resources (DER), Micro grid, Hybrid Electric/Fuel Cell Vehicles, Multiple Input Converter, DC-DC converter, Battery, Ultracapacitor.**

I.   INTRODUCTION

In a system like micro grid, different sources like diesel generator, fuel cell, solar photovoltaic (PV), and energy storage systems like battery or ultracapacitor need to be interfaced to three phase grid or to the load. Power conditioning circuitry is necessary to match the differences in voltages from these sources and storage systems. This also serves the purpose of control of power flow to achieve maximum power point tracking and strategies for energy storage system. A block diagram of such a system is shown in Fig. 1. Energy source could be PV, wind, fuel cell, or diesel generator supplying power to the dc link of the inverter through a power converter. Bidirectional power converter is necessary to assimilate energy storage like battery and ultracapacitors with the rest of the system to interface with the grid.

A number of topologies are proposed in literature [1-9] to transfer the power from one source to another in HEV and microgrid applications. In [1], multiple sources are interfaced at a common high-frequency transformer where each source is connected through full-bridge cells utilizing twelve switches for three sources. Both phase shift and duty ratio modulations are suggested. Similar operation is proposed with a half bridge circuit at each source in [2] with half the number of switches and supplementary capacitors. In order to reduce the ripple

current in the battery, current-fed half-bridge topology has been proposed in [3] with the phase shift modulation. Similarly multiple input isolated buck-boost and forward converters along with the stability analysis have been presented in [4]. All these bidirectional converters have galvanic isolation and are coupled together magnetically. Energy sharing between various sources is difficult to control in these types of topologies, although isolation gives better safety and more flexibility in selecting voltage levels. In non-isolated topology presented in [5], the battery and the ultracapacitor are cascaded using a bidirectional converter, which is connected in parallel to the fuel cell supplying a load. This does not provide the flexibility to vary the voltage across battery and ultracapacitor as they are connected directly across the fuel cell. Multiple bidirectional boost converters with various voltage sources are connected in parallel with output connected to common dc link to supply the inverter [6-8]. Voltages from several sources are boosted to dc link voltage with independent control of individual current delivered to the load. Similarly, regenerative energy is shared between the storages during braking. Voltage levels of input auxiliary sources are limited to below dc link voltage because it can only boost the input voltages and number of inductors required is equal to number of sources. Instead of additional dc-dc converter, Z-source (LC) network is used to interface fuel cell and battery in Z-source inverter topology [9]. Although this topology is suitable with optimal devices and components, number of voltage sources is limited to two and unable to extend this topology for multiple sources.

Disadvantage of most of these topologies is the difficulty in independent transfer of power between any two sources with wide variation in voltage levels. Few of the topologies have problems in scaling it to multiple sources. Hence, this paper proposes a multi-input power converter with the following characteristics:

(a) It is possible to interface more than two dc sources of different voltage levels and can be extended to any number of sources,

(b) Ability to transfer power in both directions i.e. bidirectional power flow capability,

(c) It is possible to control power flow between any two of the sources independently, and

(d) It is simpler to design, implement and control.

The topology shown in Fig. 2 can be built by connecting a switching leg with the sources through an inductor and a switching leg for each source. In order to interface a larger number of sources, connecting an

U.S. Government work not protected by U.S. copyright

additional inductor and switching leg for each additional source can extend the topology. Although a similar topology is presented in [10], this paper presents detailed analysis and operation during all possible conditions. In addition, this topology has been extended to operate the converter over wide voltage levels. In [10], power flow between only two of the sources is demonstrated operating in boost mode from the battery/ultracapacitor to the dc link and buck mode in opposite direction.

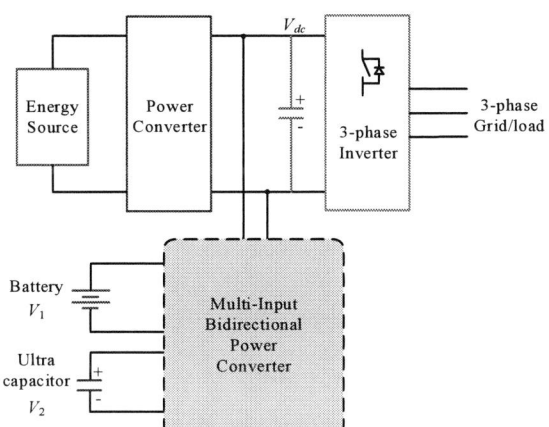

Fig. 1: Block diagram showing application of multiple-input bidirectional converter.

The objective of this paper is to present the operation and steady state analysis of the proposed converter topology. The mode-by-mode operation of the converter is explained. The design procedure of the converter is presented. Analysis and design details for various modes of operation have been verified for a 5kW system using a real-time emulator based on Typhoon HIL-400.

II.   OPERATION AND ANALYSIS OF THE CONVERTER

The proposed topology consists of a standard three-phase inverter modified by adding two high frequency inductors as shown in Fig. 2.

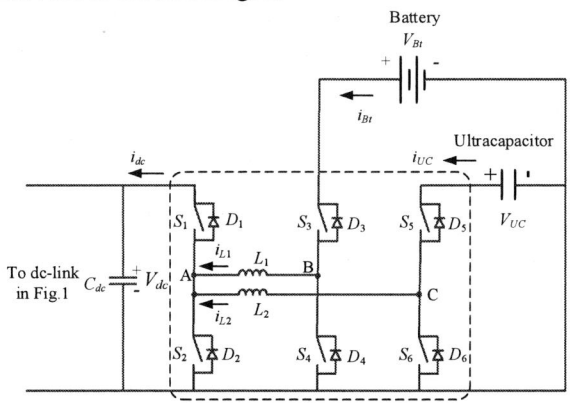

Fig. 2. Schematic of the proposed multiple-input bidirectional converter topology.

In the block diagram, sources V_{Bt} and V_{UC} represents energy storage system like battery and ultracapacitor, which are interfaced with the dc link of inverter V_{dc}. Two

legs of switch modules are connected to the DC voltage sources instead of dc bus. Another leg of the converter is connected to the dc link of the inverter, which is also fed by energy source through a dc-dc converter. If necessary more power/energy sources can be added by adding additional legs to the inverter. Operation of the converter under these modes of operation is discussed in the following sections.

A.   Battery and Dc Link

In this mode of operation, energy stored in battery is transferred to dc link to supply load/grid. Switching sequence in this mode of operation is given in Table I. Inductor L_1, connecting battery and dc link is storing energy from battery at first time interval T_1 by turning-on switches S_2 and S_3. Battery voltage V_{Bt} appears across L_1 resulting in increase of current with a slope of V_{Bt}/L_1. During time interval T_2 both the switches are turned-off. Inductor current i_{L1}, flows to dc link capacitor by forward biasing diodes D_1 and D_4. Energy stored in L_1 during time T_1 is being discharged to C_{dc} in this interval with a negative slope of V_{dc}/L_1. The flow of current through these diodes results in a voltage drop of around 1V-1.2V each. In the next time interval T_3, switches S_1 and S_4 are turned-on. Current continues to flow from L_1 to the dc link through the switches as opposed to the diodes in T_2. Switches S_1 and S_4 are made to operate as synchronous rectifier resulting in a reduced voltage drop to a level of around 0.2V, thereby improving system efficiency.

Under steady state operation, relation between dc link voltage and battery voltage is given by,

$$V_{dc} = \frac{T_1}{T_2 + T_3} \cdot V_{Bt} = \frac{D}{1-D} \cdot V_{Bt} \tag{1}$$

Where, D is the duty ratio defined by T_1/T_S. T_S is the time period of switching cycle, which is total of all three time intervals. If battery voltage V_{Bt} is less than the dc link voltage V_{dc}, the converter boosts V_{Bt} to V_{dc} by operating at $D>0.5$. If battery voltage V_{Bt} is greater than dc link voltage V_{dc}, the converter charges the battery from dc link by operating at $D<0.5$. The switching sequence for this operation is given in mode A(ii) of Table I. Principle of energy transfer remains the same; energy is stored in L_1 in interval T_1 and used to charge the battery in intervals T_2 and T_3. Considering duty ratio D to be T_1/T_S, the relation between two voltages is given in (2).

$$V_{Bt} = \frac{T_1}{T_2 + T_3} \cdot V_{dc} = \frac{D}{1-D} \cdot V_{dc} \tag{2}$$

B.   Ultracapacitor and Dc Link

Energy transfer from ultracapacitor to the dc link is explained in this subsection where voltage level of ultracapacitor needs to be boosted to the higher dc link voltage. Switching sequence for three distinct time intervals of operation is given in mode B(i) in Table I. For the time duration T_1, inductor L_2 is charged by the ultracapacitor by triggering the switches S_2 and S_5 while the remaining switches are turned off. Current i_{L2}, increases linearly with a slope of V_{UC}/L_2. Once inductor L_2 is charged for a predetermined time, both the switches S_2 and S_5 are turned off resulting in conduction of diodes

D_1 and D_6. Energy stored in the inductor L_2 by the ultracapacitor is being utilized to charge the dc link capacitance and hence to supply the load. Since, voltage drop across these diodes are higher, corresponding MOSFETs S_1 and S_6 are turned on in the time interval T_3. This effectively makes it a synchronous rectifier with reduced voltage drop subsequently reducing the losses in the converter. Voltages V_{dc} and V_{UC} are related by duration of these three time intervals and it can be calculated using (3).

$$V_{dc} = \frac{T_1}{T_2 + T_3} \cdot V_{UC} = \frac{D}{1-D} \cdot V_{UC} \qquad (3)$$

In a hybrid vehicle system, ultracapacitor voltage V_{UC} is generally kept smaller than V_{dc}. However, this converter enables to interface two sources without changing the topology even though one voltage is either higher or lower than the other just by varying duty ratio D. Whenever, ultracapacitor needs to be charged from dc link, similar devices are switched as given in mode B(ii) in Table I. Direction of current flow in inductor L_2 is opposite to that of mode B(i), as the direction of power flow is now reversed. In this mode, it is important to observe the changes in switching devices in three different time intervals. Considering duty ratio D to be T_1/T_S, the relation between two voltages is given in (4).

$$V_{UC} = \frac{T_1}{T_2 + T_3} \cdot V_{dc} = \frac{D}{1-D} \cdot V_{dc} \qquad (4)$$

In order to charge the ultracapacitor from dc link that is at higher voltage, converter is operated at $D<0.5$.

C. Battery and Ultracapacitor

Whenever, energy stored in the battery needs to be transferred over to the ultracapacitor or vice versa, switching sequence given in Table I is followed. There are various permutations possible by switching four switches S_3 to S_6, depending on the voltage levels of the two energy sources. While charging the ultracapacitor from the battery in mode C(i), converter can be operated in boost mode, buck mode, or buck-boost mode. Boost mode of operation is given where ripple in the battery current is smaller as compared to other two modes. This is chosen to improve the life of battery by reducing the peak value of charging or discharging current. If the ultracapacitor voltage is lower than the battery voltage, then buck mode can be adopted. During the process of charging of ultracapacitor, if its voltage increases from less than V_{Bt} to above V_{Bt}, then the controller needs to be seamlessly maneuvered from buck mode to boost mode of operation. On the other hand, buck-boost mode can be implemented simplifying the control by changing the value of duty ratio. Switching states given in Table I, the battery voltage is being boosted to charge the ultracapacitor. In this mode of operation, both the inductors L_1 and L_2 are storing energy during interval T_1 from the battery through switches S_3 and S_6. At the end of this interval, switch S_6 is turned-off forcing the inductor current to flow through diode D_5 to charge the ultracapacitor. To decrease the voltage drop across D_5, switch S_5 is gated on in the subsequent time interval T_3

operating as synchronous rectifier. Relation between the two voltages are given as,

$$V_{UC} = \frac{T_S}{T_2 + T_3} \cdot V_{Bt} = \frac{1}{1-D} \cdot V_{Bt} \qquad (5)$$

Where, D is the duty ratio defined by T_1/T_S. In a similar operation, energy can be transferred from the ultracapacitor to the battery by switching same set of active switches S_3, S_5 and S_6 but with different on-time durations resulting in opposite direction of current through the inductors as compared to mode C(i). Energy transferred from the ultracapacitor to the battery depends on the time duration T_1 for which switch S_5 is retained ON given by,

$$V_{Bt} = \frac{T_1}{T_S} \cdot V_{UC} = D \cdot V_{UC} \qquad (6)$$

It is clear from the above equation that the energy can be transferred in this mode only if V_{Bt} is smaller than V_{UC}. If this condition is not met, then different mode of operation like boost or buck-boost can be used to control the switches without changing the circuit topology.

TABLE I CONDUCTION OF DEVICES IN DIFFERENT TIME INTERVALS OF THE MODES

	T_1	T_2	T_3
Mode A(i)	S_2, S_3	D_4, D_1	S_1, S_4
Mode A(ii)	S_1, S_4	D_2, D_3	S_2, S_3
Mode B(i)	S_2, S_5	D_6, D_1	S_1, S_6
Mode B(ii)	S_1, S_6	D_2, D_5	S_2, S_5
Mode C(i)	S_3, S_6	D_5, S_3	S_5, S_3
Mode C(ii)	S_5, S_3	D_6, S_3	S_3, S_6

D. Battery and Ultracapacitor to Dc Link

During peak power demand from the load/grid, battery unit and ultracapacitor provide the peak power demand due to its faster dynamic response as compared with the energy source like fuel cell system. In fuel cell vehicles, the auxiliary sources have to deliver rated power during the process of cold startup of the fuel cell system.

TABLE II. CONDUCTION OF DEVICES IN DIFFERENT TIME INTERVALS WHEN ALL THE THREE SOURCES/LOADS ARE ACTIVE

	T_1	T_2	T_3	T_4	T_5
Mode D	$S_2, S_3,$ S_5	$S_2, D_4,$ S_5	$S_2, S_4,$ S_5	$D_1, S_4,$ D_6	$S_1, S_4,$ S_6
Mode E	$S_1, S_3,$ S_5	$D_2, S_3,$ S_5	$S_2, S_3,$ S_5	$S_2, S_4,$ S_5	$S_1, D_3,$ S_5

Switching states during this mode of operation is given in Table II dividing the switching cycle T_S into five time intervals. Identical gate signals are given to switches S_1 and S_6 with a duty ratio of d_2 and complementary of this signal is being provided to switches S_2 and S_5 as demonstrated in Fig. 3. Gate signal for the switch S_3 is synchronized with that of switch S_1 having duty ratio of

d_1. S_4 is switched complementary to switch S_3 with a dead time.

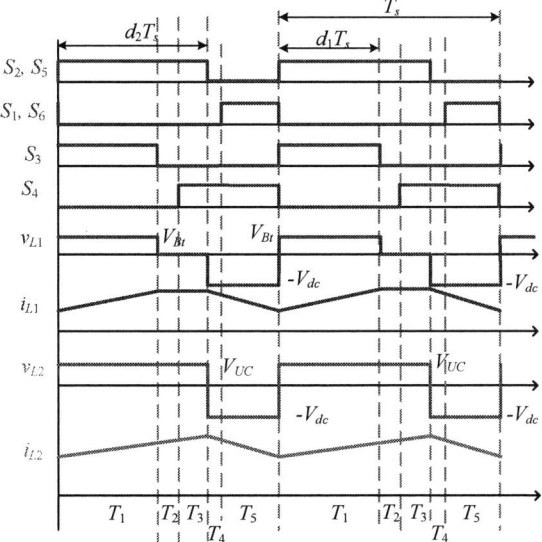

Fig. 3. Steady-state waveforms for mode D.

During interval T_1, switches S_2, S_3 and S_5 are gated-on charging both the inductors L_1 and L_2 by corresponding sources V_{Bt} and V_{UC}. Current increases with a slope of V_{Bt}/L_1 and V_{UC}/L_2 respectively as demonstrated in Fig. 4. At the end of this interval, switch S_3 is turned off providing freewheeling path for i_{L1} through D_4. In interval T_3, switch S_4 is switched-on in order to avoid voltage drop across the diode D_4 while inductor L_2 continues to charge from the ultracapacitor. Switches S_2 and S_5 are turned off at the end of interval T_3, forcing current to flow through anti-parallel diodes of their corresponding complementary switches. Energy stored in the inductors L_1 and L_2 are transferred to the dc link through diode D_1 resulting in a voltage drop of 1V-1.2V. This voltage drop is reduced to the level of around 0.2V by gating on switch S_1 to act as synchronous rectifier. At the end of this interval, gating signals for each switching leg are complemented to start charging the inductors L_1 and L_2 going back to interval T_1.

Current through and voltage across the inductors under steady state operation is given in Fig. 3. Applying volt-sec balance for both the inductors, relation between the three voltages are given by (7) and (8).

$$V_{Bt} = \frac{T_4 + T_5}{T_1} \cdot V_{dc} = \frac{d_2 T_S}{d_1 T_S} \cdot V_{dc} = \frac{d_2}{d_1} \cdot V_{dc} \quad (7)$$

$$V_{UC} = \frac{T_4 + T_5}{T_1 + T_2 + T_3} \cdot V_{dc} = \frac{d_2 T_S}{T_S - d_2 T_S} \cdot V_{dc} = \frac{d_2}{1 - d_2} \cdot V_{dc} \quad (8)$$

Where, d_1 is the ratio of on-time of switch S_3 to total switching period T_S and similarly d_2 corresponds to switch S_2. From (8), it can be observed that the d_2 can be calculated for a required boost ratio of V_{UC} to V_{dc}. Since the ultracapacitor voltage is assumed to be smaller than V_{dc}, the duty ratio d_2 is being functioned always below 0.5 with the voltage gain defined as M_2. Variation in M_2 for change in d_2 between 0 and 0.5 is as shown in Fig. 6. On the other hand, the switch S_3 can be controlled with a duty ratio between 0 to $(1-d_2)$ in order to boost the battery

voltage to dc link. The gain M_1 connecting V_{Bt} and V_{dc} is given in (9).

$$M_1 = \frac{V_{Bt}}{V_{dc}} = \frac{d_2}{d_1}; \qquad M_2 = \frac{V_{UC}}{V_{dc}} = \frac{d_2}{1 - d_2}; \quad (9)$$

For a given value of d_2, the minimum voltage gain M_1 can be obtained considering maximum value of duty ratio d_1. It is clear from Fig. 4 that the maximum value of the boost ratio for the battery to dc link cannot exceed the boost ratio of the ultracapacitor to the dc link. Hence, this switching sequence can be employed only if V_{Bt} is higher than V_{UC}. However, switching sequence corresponds to the switching legs of battery and ultracapacitor needs to be interchanged if V_{UC} is higher than V_{Bt}. With the proposed converter it is possible to transfer energy from two sources of unequal voltage by independently controlling the share from each.

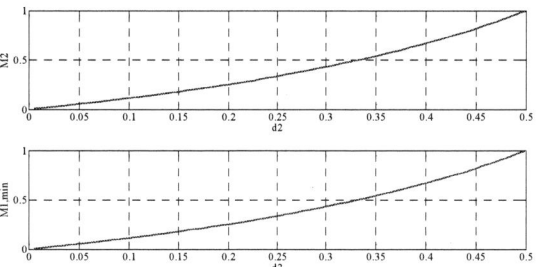

Fig. 4. Relationship between voltage gains between the three sources for mode D.

The proposed converter can be either voltage or current controlled, depending on the role of source in the overall system and their constraints. Total power required from the auxiliary sources can be shared between the battery system and the ultracapacitor bank based on factors like, charging current limitations of battery, state of charge (SOC), dynamics of the converter, etc.

E. Dc Link to Both Battery and Ultracapacitor

Kinetic energy stored in the traction drive is fed back to the source during regenerative braking operation. Since, a fuel cell stack lacks bidirectional power handling capability; battery bank plays an important role. During sudden braking condition, regenerative power can be much higher than the battery can absorb. Additional energy which battery cannot capture is utilized in charging ultracapacitor, which has higher power density as compared to battery. Switching sequence during this mode of operation is given in mode E of Table 2. In this mode of operation, L_2 is always connected to the ultracapacitor by keeping S_5 and S_6 in ON and OFF state respectively throughout the cycle. The switch pairs S_1-S_2 and S_3-S_4 are operated in complementary fashion with duty ratios of d_1 and d_2 as shown in Fig. 8 where d_2 is greater than d_1.

Switches S_1, S_3 and S_5 are turned-on in the time interval T_1 transferring energy from the dc link to charge both the auxiliary sources. Inductor currents i_{L1} and i_{L2} increase as V_{dc} appears at point A. At the end of T_1, switch S_1 is turned-off providing free wheeling path for the inductor currents through diode D_2. In order to reduce the voltage drop across the diode, switch S_2 is gated ON to effectively function as a synchronous rectifier in T_3. In order to control the charging of two sources independently, switch S_3 is turned-off and S_4 is turned-on maintaining the current through inductor L_1 as shown in Fig. 5. During this time interval, the inductor current i_{L2} continues to charge the. Switches S_2 and S_4 are turned-off at the end of the interval T_4. Current in inductors flow from dc link to battery and ultracapacitor through switches S_1, S_5 and diode D_3. One cycle of operation completes when the switch S_3 is turned on providing path for i_{L1} by reducing the voltage drop across it.

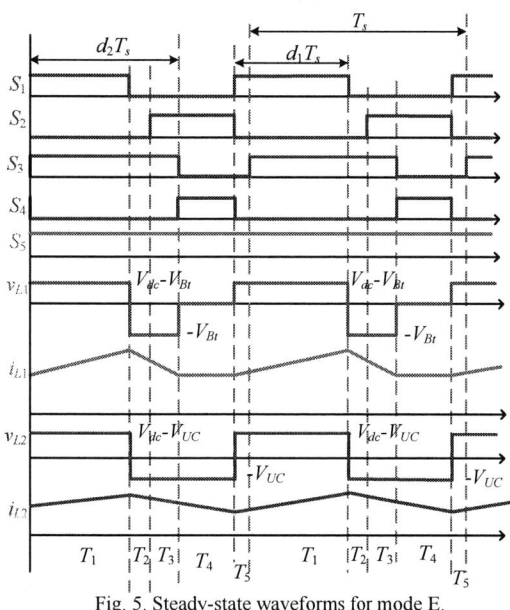

Fig. 5. Steady-state waveforms for mode E.

Relationship between the three voltage sources can be derived by applying volt-sec balance across the inductors, which has been demonstrated in Fig. 5.

$$V_{Bt} = \frac{T_1+T_5}{T_1+T_2+T_3+T_5} \cdot V_{dc} = \frac{T_1+T_5}{T_S-T_4} \cdot V_{dc} = \frac{d_1}{d_2} \cdot V_{dc} \quad (10)$$

$$V_{UC} = \frac{T_1+T_5}{T_S} \cdot V_{dc} = d_1 \cdot V_{dc} \quad (11)$$

The ultracapacitor voltage V_{UC} is controlled by varying d_1, which is related by (11). Energy transferred to the battery is regulated by changing the variable d_2 for a given value of d_1. Voltage gain functions for both the voltage sources are defined and are obtained as,

$$M_1 = \frac{V_{UC}}{V_{dc}} = d_1; \qquad M_2 = \frac{V_{Bt}}{V_{dc}} = \frac{d_1}{d_2}; \quad (12)$$

Fig. 6 shows the variation of voltage gain M_1 with respect to duty ratio d_1, which is a linear function. However, the voltage gain for battery i.e. M_2 is a function of both d_1 and d_2 that can be changed by varying value of d_2 from 0 to $(1-d_1)$. For a given value of d_1, minimum gain is obtained at a duty ratio of $(1-d_1)$ which is shown in Fig. 6. When both the auxiliary source voltages are less than dc link voltage, it is clear that the duty ratio d_1 has to be less than 0.5. Moreover, V_{Bt} needs to be higher than V_{UC} for this specific switching sequence. Otherwise,

S_3 has to be kept on while switching S_5-S_6 at a duty ratio of d_2.

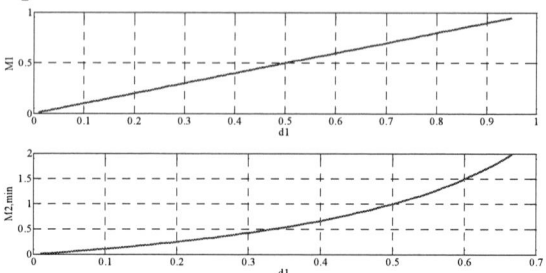

Fig. 6. Relationship between voltage gains between the three sources for mode E.

III. DESIGN OF THE CONVERTER

For grid tied applications, it is assumed that a battery bank is required to support full power until ancillary sources can reach full power and/or peak efficiency. Similarly, during times when the auxiliary source needs to be reduced, excess energy can be stored into the battery bank. For a 5kW power converter, a battery bank of 144V, 17Ah will provide a maximum continuous current of at least 51A, providing a runtime of 30 minutes. Coupling this system with an ultracapacitor bank rated at 125V, 15F provides an additional 20 seconds of high power runtime for quick power bursts where the battery may not need to be engaged. The results of the operation of this grid-tied converter are presented in the following section.

In most practical systems, a traction motor, or grid-tied connection, is driven by a voltage source inverter (VSI) with a constant dc link voltage. A fixed voltage of 300V is chosen. The proposed converter can be built from a standard three-phase inverter by connecting three sources namely V_{dc}, V_{Bt} and V_{UC} to collector terminal of the top switches. Two inductors are inserted between midpoints of the three switching legs. Hence, it is very easy to construct the proposed converter without modifying driver part of the inverter. Values of inductors are designed to limit ripple current to a specific amount for a given switching frequency. Similarly capacitance at the dc link is also designed based on the allowed value of ripple in V_{dc} and switching currents.

IV. RESULTS

The proposed multiple-input dc-dc converter has been tested in various modes of operation discussed in previous section. For validation purpose, the following specifications are taken. V_{dc} = 300 V, output power P_o = 5kW, switching frequency of 20 kHz. The details are as follows:

1. Battery bank: 144V, 17Ah Lithium-ion battery bank, voltage varies from 120V to 140V, series resistance of 30mΩ.

2. Ultracapacitor tank: 125V, 15F, parasitic inductance of 10μH, series resistance of 18mΩ.

3. Dc link capacitor: 3mF

4. Inductors: 0.75mH each.

5. The switches are controlled from DSP320F2808.

The above specified system has been implemented in a 'Typhoon HIL-400' Hardware-In-Loop (HIL) system to validate the analysis and performance of the system. The HIL real-time emulator comprises of an application-specific processing architecture based on a FPGA with fast analog/digital input and output interfaces with a supporting front-end software tool-chain performing the function of power electronics and energy storage systems like battery and ultracapacitors. With the minimum step time of 0.5μs and ultra-low latency of 1μs, most of the experiments of power electronics and drives having switching frequency up to 50 kHz can be emulated. Accuracy and latency of HIL has been proved for various complex systems like grid-connected inverter [11]. Control of the semiconductor devices is being executed externally at a constant switching frequency of 20kHz using DSP.

In order to verify the analysis, the proposed converter is tested for different modes of operation explained in Section II and results are presented in Fig. 7 to Fig. 14. In mode A(i), energy is transferred from the battery to dc link shown in Fig. 7. V_{Bt} of 130V is boosted to V_{dc} of 300V to supply the grid load by transferring energy through inductor L_1. Positive current of i_{dc} indicates the current flowing into dc link capacitor from the MIPC and similarly positive value of i_{Bt} shows discharge current of the battery. A smooth current of around 70A with small ripple is flowing through L_1 maintaining zero current from the ultracapacitor, which can be seen in i_{L2}. For power flow from dc link to the battery, experimental results are shown in Fig. 8. A negative current of approximately 20A through L_1 shows reversal of direction of flow with zero current flowing through L_2 and hence through ultracapacitor. Voltages across the switches S_2 and S_4 show dc link voltage and battery voltage respectively along with their states. Whenever, S_1 is ON, voltage V_{dc} appears across the switch S_2 which can be observed in waveform of V_{S2}. Similarly V_{dc} appears across S_4 when S_3 is gated on for the duration of $(1-D)T_S$. Peak to peak ripple current of 5A in L_1 validates the design selection of the inductors.

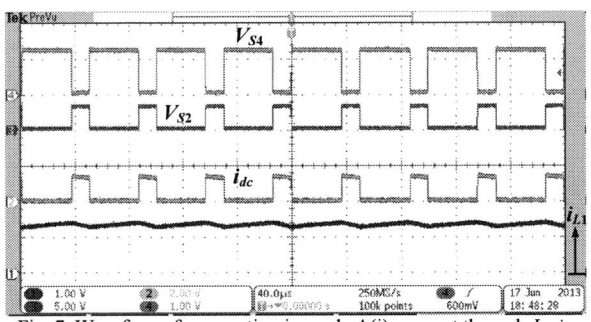

Fig. 7. Waveforms for operation in mode A(i): current through L_1, i_{L1} (scale: 50A/div), current supplied to dc link, i_{dc} (scale: 100A/div), voltage across switch S_2, V_{S2} (scale: 500V/div), and voltage across switch S_4, V_{S4} (scale: 100V/div) (X-axis: 40μs/div).

Likewise, waveforms for transfer of energy from the ultracapacitor to the dc link are presented in Fig. 9. It is clear from the figures that voltage V_{UC} is being boosted to

V_{dc} by transferring current through L_2 maintaining zero current from the battery. Power flow control of each sources are possible without affecting the remaining sources. Since voltage across the ultracapacitor varies during discharge, current drawn from the ultracapacitor needs to be regulated by inclusion of closed loop control to vary the duty ratio based on the feedback information of current. Charging of the ultracapacitor from the dc link is performed by using switching sequence given in mode B(ii) and corresponding results are presented in Fig. 10. Current is fed from the dc link, hence negative waveform is seen in Fig. 10 along with the negative current i_{L2} charging the ultracapacitor tank.

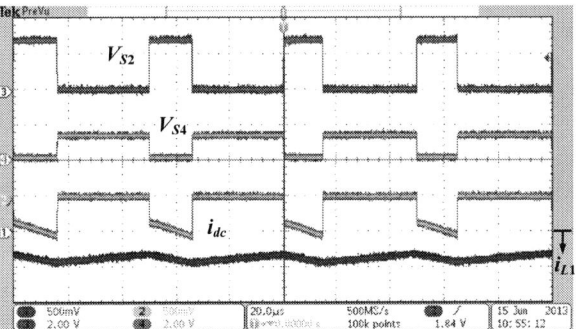

Fig. 8. Waveforms for operation in mode A(ii): (a) current through L_1, i_{L1} (scale: 25A/div), current supplied to dc link, i_{dc} (scale: 25A/div), voltage across switch S_2, V_{S2} (scale: 200V/div), and voltage across switch S_4, V_{S4} (scale: 200V/div) (X-axis: 20μs/div).

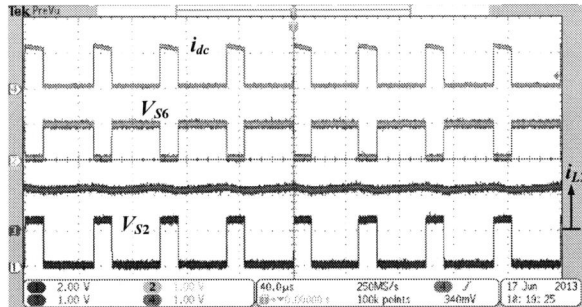

Fig. 9. Waveforms for operation in mode B(i): (a) voltage across switch S_2, V_{S2} (scale: 200V/div), voltage across switch S_6, V_{S6} (scale: 100V/div), current through L_2, i_{L2} (scale: 50A/div), and current flowing into dc link, i_{dc} (scale: 50A/div) (X-axis: 40μs/div).

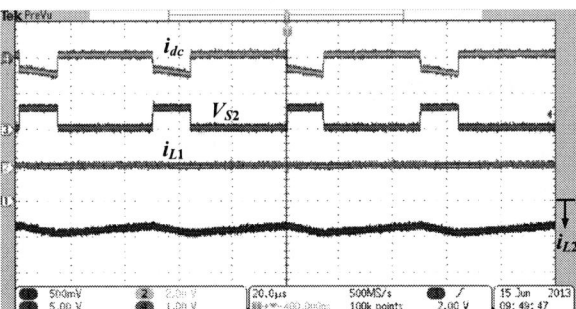

Fig. 10. Waveforms for operation in mode B(ii): (a) current through L_2, i_{L2} (scale: 25A/div), current through L_1, i_{L1} (scale: 100A/div), voltage across switch S_2, V_{S2} (scale: 500V/div), and current flowing from dc link, i_{dc} (scale: 50A/div) (X-axis: 20μs/div).

Fig. 11 shows results for transfer of energy from the battery bank to the ultracapacitor. Waveform of V_{S4}

demonstrates the voltage across the switch S_4. Top and bottom switches in each leg are operated complementary to each other and top switch of one leg is switched with bottom switch of another leg. Whenever switch S_3 and S_6 are ON, battery voltage V_{Bt} appears across switch S_4 and inductor current is getting charged from the battery. Ripple in i_{L1} or i_{L2} have reduced by half since the two inductors are in series in this mode of operation. Dynamic behavior of the converter in mode C(i) is presented in Fig. 11. Operating converter in a fixed duty ratio inductor current increases initially to charge the ultracapacitor and reduces the charging rate once it has reached the specific voltage. Negative current in L_2 implies charging of the ultracapacitor. Pulsating current flowing from the battery is also given which has envelope similar to waveform of i_{L2} but with opposite polarity.

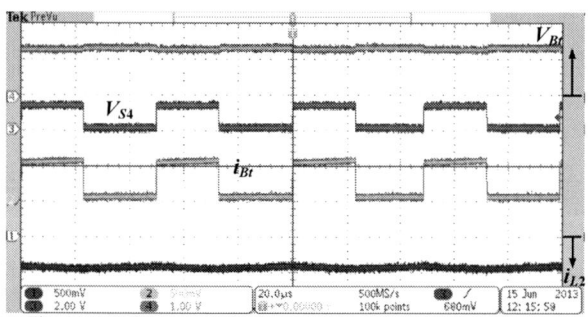

Fig. 11. Steady state results for operation in mode C(i): (a) current flowing through L_2, i_{L2} (scale: 25A/div), current flowing from the battery, i_{Bt} (scale: 25A/div), voltage across switch S_4, V_{S4} (scale: 200V/div), and battery voltage V_{Bt} (scale: 100V/div) (X-axis: 20µs/div).

Functioning of the converter in mode C(ii) is same as that of mode C(i) except variation in duty ratio according to the voltages V_{Bt} and V_{UC} to reverse the direction of current flow from the ultracapacitor to the battery as shown in Fig. 12.

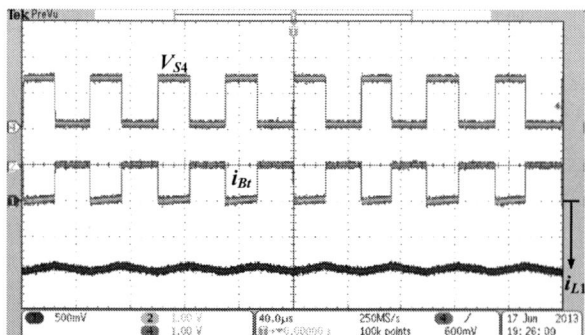

Fig. 12. Steady state results for operation in mode C(ii): current flowing through L_1, i_{L1} (scale: 25A/div), current transferred to the battery, i_{Bt} (scale: 50A/div), and voltage across switch S_4, V_{S4} (scale: 200V/div) (X-axis: 40µs/div).

In mode D, both the auxiliary sources are transferring energy to the dc link to drive the traction motor. Two sources with different voltage levels are boosting voltage to V_{dc} by switching all the devices as explained in Section II. Switch pair S_2-S_5 is gated with the same signal of duty ratio d_2 with complementary signal given to switches S_1-S_6. When S_3 and S_5 are turned on, inductors L_1 and L_2 are

storing energy supplied from the battery and the ultracapacitor respectively as shown in Fig. 13(a). Switch S_3 is turned-off maintaining current i_{L1} for a short time before turning S_1 and S_6 to ON state. Since V_{dc} and V_{UC} are of same order, difference between the duty ratios d_1 and d_2 is very small. Energy is transferred to the dc link whenever switch S_1 is turned-on. The current flowing into the dc link capacitor, i_{dc} can be observed when V_{dc} appears across switch S_2. Fig. 13(b) shows peak value of i_{dc} is sum of i_{L1} and i_{L2} indicating that the energy transferred to dc link is resultant of both the auxiliary sources.

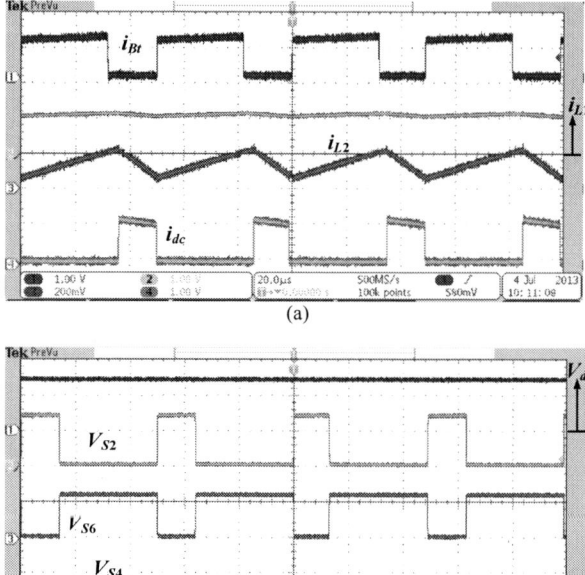

(a)

(b)

Fig. 13. Steady state results for operation in mode D: (a) current flowing from the battery, i_{Bt} (scale: 50A/div), current through L_1, i_{L1} (scale: 50A/div), current through L_2, i_{L2} (scale: 10A/div), and current flowing into dc link, i_{dc} (scale: 50A/div) (X-axis: 20µs/div), (b) dc link voltage, V_{DC} (scale: 200V/div), voltage across switch S_2, V_{S2} (scale: 200V/div), voltage across switch S_6, V_{S6} (scale: 100V/div), and voltage across switch S_4, V_{S4} (scale: 200V/div) (X-axis: 20µs/div).

Energy is being restored back to the battery and the ultracapacitor from dc link during regenerative braking. Since the battery voltage V_{Bt} is higher than the ultracapacitor voltage V_{UC} in this example, switch S_5 is maintained in ON state throughout this mode of operation, while switches S_1 and S_3 are being switched with duty ratio of d_1 and d_2 respectively. Voltage across S_2 and S_4 shown in Fig. 14(a) demonstrates unequal duty ratio in order to pump in regenerative energy to auxiliary sources of uneven voltages. Current through L_2 is also shown in the figure, which is same as current flowing into the ultracapacitor. Fig. 14(b) shows waveform of current supplied from dc link i_{dc} and charging current of battery i_{Bt} along with the voltage across ultracapacitor V_{UC}.

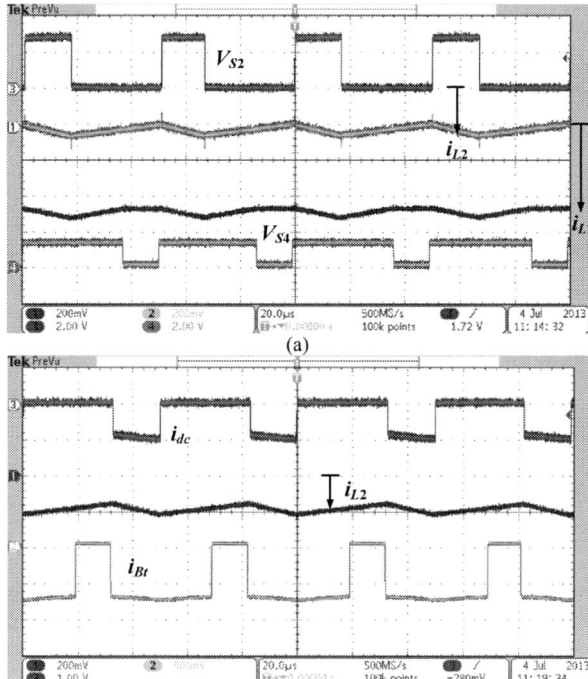

Fig. 14. Steady state results for operation in mode E: (a) current through L_1, i_{L1} (scale: 10A/div), current through L_2, i_{L2} (scale: 10A/div), voltage across switch S_2, V_{S2} (scale: 200V/div), and voltage across switch S_4, V_{S4} (scale: 200V/div) (X-axis: 20µs/div), (b) current flowing through L_2, i_{L2} (scale: 10A/div), charging current flowing into the battery, i_{Bt} (scale: 25A/div), and current transferred from dc link, i_{dc} (scale: 50A/div) (X-axis: 20µs/div).

Emulation results shown here validate the proposed multiple-input DC-DC bidirectional converter topology. It can be observed that the experimental waveforms coincide well with the analytically predicted steady-state operating waveforms. Though the results have been emulated using hardware-in-the-loop real-time system for verification of the proposed converter, this can be applied to practical system finding applications in electric vehicles, smart-grid, and distributed energy resources. In this paper, an application for fuel cell hybrid electric vehicle has been demonstrated. However, the proposed topology can be applied to different applications where more than two dc sources of unequal voltage levels need to transfer energy between themselves. This can also be used in charge balancing of multiple battery banks.

V. SUMMARY AND CONCLUSION

This paper proposes a multiple-input bidirectional dc-dc converter to interface more than two sources of different voltage levels. The converter can be operated either in buck mode or boost mode in either direction of power flow. It is possible to control power flow between each pair of sources independently when more than two sources are active. The major advantage of the proposed topology is that it can be built from a standard three-phase inverter with few changes in connections.

This paper gives detailed analysis, modulation and operation of the converter for various modes. In each mode, relationship between the sources are derived which

assists in implementation of the controller. Design of the converter for a typical hybrid fuel cell vehicle has been explained where dc-link of the inverter, lithium-ion battery bank and ultracapacitor tank are being interfaced together. Typhoon based hardware-in-the-loop (HIL) real-time system has been used to emulate the designed converter to validate the performance. Results obtained from this system have been presented and match very well with the analytically expected waveforms. This converter not only finds application in fuel cell vehicles, this can be utilized in DER, smart and micro grid, battery management systems etc. where more than two dc sources need to be interfaced with bidirectional power flow capability.

REFERENCES

[1] C. Zhao, S. D. Round, and J. W. Kolar, "An isolated three-port bidirectional dc-dc converter with decoupled power flow management", *IEEE Trans. on Power Electronics*, vol. 23, no. 5, pp. 2443-2453, Sept. 2008.

[2] H. Tao, J. L. Duarte, M. A. M. Hendrix, "Three-port triple-half-bridge bidirectional converter with zero-voltage switching", *IEEE Trans. on Power Electronics*, vol. 23, no. 2, pp. 782-792, Mar. 2008.

[3] S. Liu, X. Zhang, H. Guo, and J. Xie, "Multiple input bidirectional dc/dc converter for energy supervision in fuel cell electric vehicles", in *Proc. IEEE International conference on Electrical and Control Engineering*, 2010, pp. 3890-3893.

[4] H. Matsuo, W. Lin, F. Kurokawa, T. Shigemizu, and N. Watanabe, "Characteristics of the multiple-input dc-dc converter", *IEEE Trans. on Industrial Electronics*, vol. 51, no. 3, pp. 625-631, Jun. 2004.

[5] W-S. Liu, J-F. Chen, T-J. Liang, R-L. Lin, and C-H. Liu, "Analysis, design, and control of bidirectional cascoded configuration for a fuel cell hybrid power system", *IEEE Trans. on Power Electronics*, vol. 25, no. 6, pp. 1565-1575, Jun. 2010.

[6] W. Jiang, and B. Fahimi, "Multiport power electronic interface- concept, modeling, and design", *IEEE Trans. on Power Electronics*, vol. 26, no. 7, pp. 1890-1900, Jul. 2011.

[7] L. Solero, A. Lidozzi, and J. A. Pomilio, "Design of multiple-input power converter for hybrid vehicles", *IEEE Trans. on Power Electronics*, vol. 20, no. 5, pp. 1007-1006, Sept. 2005.

[8] L. Solero, A. Lidozzi, and J. A. Pomilio, "Design of multiple-input power converter for hybrid vehicles", in *Proc. IEEE APEC*, 2004, vol. 2, pp. 1145-1151.

[9] F. Z. Peng, M. Shen, and K. Holland, "Application of Z-source inverter for traction drive of fuel cell- battery hybrid electric vehicles", *IEEE Trans. on Power Electronics*, vol. 22, no. 3, pp. 1054-1061, May 2007.

[10] K. Rajashekara, and J. G. Noetzel, "Power system to transfer power between a plurality of power sources", US Patent 7084525 B2, August 1, 2006.

[11] Z. R. Ivanovic, E. M. Adzic, M. S. Vekic, S. U. Grabic, N. L. Celanovic, and V. A. Katic, " HIL evaluation of power flow control strategies for energy storage connected to smart grid under unbalanced condition", *IEEE Trans. on Power Electronics*, vol. 27, no. 11, pp. 4699-4710, Nov. 2012.

Single-Switch PWM Converter Integrating Voltage Equalizer for Photovoltaic Modules Under Partial Shading

Masatoshi Uno and Akio Kukita
Aerospace Research and Development Directorate, Japan Aerospace Exploration Agency
2-1-1 Sengen, Tsukuba, Ibaraki, Japan
uno.masatoshi@jaxa.jp

Abstract - **A photovoltaic (PV) string comprising multiple modules/sub-modules connected in series is well known to suffer from partial shading issues, such as decreased power generation and the occurrence of multiple power-point maxima. Although various kinds of differential power processing converters and voltage equalizers that can preclude partial-shading issues have been proposed, increases in circuit and system complexities are very likely due to the increased switch count and added converters. In this paper, a single-switch PWM converter integrating a voltage equalizer for PV modules under partial shading is proposed. In the proposed integrated converter, since the converter and voltage equalizer are integrated into a single unit without increasing the switch count, both the system level and circuit level simplifications are feasible. Experimental equalization tests emulating partially shaded conditions were performed using a prototype built for the four PV modules connected in series. With the proposed integrated converter, the occurrence of multiple power-point maxima was successfully precluded and the maximum extractable power was significantly augmented, demonstrating the effectiveness and performance of the proposed integrated converter.**

I. INTRODUCTION

Partial shading on a photovoltaic (PV) string comprising multiple PV modules/sub-modules (hereinafter 'modules') connected in series causes a significant mismatch in individual module characteristics; shaded modules are less capable of generating current due to decreased irradiance. In a partially-shaded PV string, a string current circumvents the shaded modules and traverses bypass diodes instead that are connected to the shaded modules in parallel. Accordingly, unshaded

modules generate power, whereas shaded modules are bypassed and operated at virtually zero voltage, producing no power.

This characteristic mismatch is well known to trigger serious issues, such as a significant reduction in power generation and the occurrence of multiple power point maxima that hinder conventional maximum power point tracking (MPPT) algorithms to track a real or global MPP. Although advanced MPPT techniques that sweep string voltage to find a global MPP have been proposed to mitigate the partial shading issues, significant power reduction is inevitable.

In the decentralized power management systems shown in Fig. 1(a), the use of module-level converters, such as micro-inverters/converters that individually control each module is the most typical solution to partial shading issues. However, increased system complexity and cost are very likely because the number of converters required is proportional to the number of PV modules. In addition, efficiency penalty due to the double power conversion of the module-level and central converters is also considered a major drawback [1], [14].

Differential power processing (DPP) converters and voltage equalizers are a very powerful solution [1]−[15]. A fraction of the generated power of unshaded modules is transferred to shaded module(s) via these converters so that all modules in a string operate at virtually the same voltage or even at each MPP. Most topologies of conventional DPP converters and voltage equalizers are

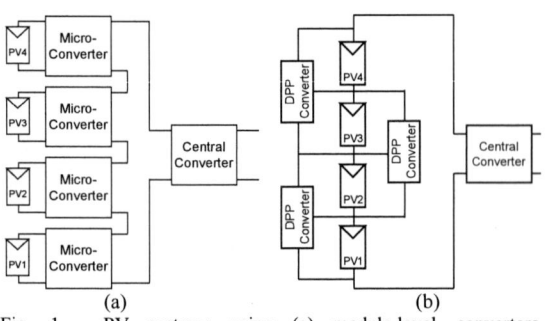

Fig. 1. PV systems using (a) module-level converters, (b) DPP converters or voltage equalizers.

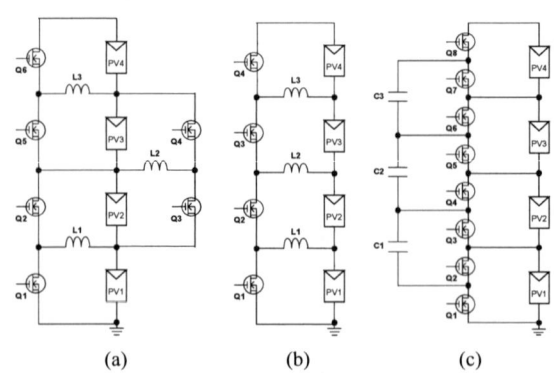

Fig. 2. Differential power processing converters and voltage equalizers based on (a) buck-boost converter, (b) multi-stage buck-boost converter, and (c) switched capacitor converter.

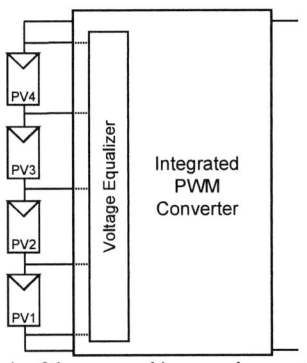

Fig. 3. Schematic of the proposed integrated converter.

based on bidirectional dc-dc converters, such as buck-boost converters [1]–[8] (and some extended topologies [9]–[11]), multi-stage buck-boost converters [12], [13], and switched capacitor converters [14], [15], as listed in Fig. 2. These topologies are basically identical to voltage equalizers that are commonly used to equalize cell voltages of series-connected energy storage devices, such as lithium-ion batteries and supercapacitors [16]–[26], to preclude voltage imbalance and ensure years of safe operation.

With DPP converters or voltage equalizers, as shown in Fig. 1(b), an efficient central converter can be used without suffering from the efficiency penalty of double power conversion. However, most conventional topologies require multiple switches proportional to the number of modules, increasing circuit complexity, as shown in Fig. 2. In addition, there are still more than two converters (i.e. the central converter and DPP converters or the voltage equalizer) in the system. In other words, if the number of converters in the system were to be reduced to one by integrating these converters into a single unit, the PV system would be further simplified.

In this paper, a single-switch PWM converter integrating a voltage equalizer for series-connected PV modules under partial shading is proposed. By integrating a PWM converter and voltage equalizer into a single unit without increasing the switch count, both the system level and circuit level simplifications are feasible. The derivation procedure is initially explained, whereupon fundamental operation analysis for the voltage equalizer is performed in this paper. An experimental test using a prototype of the proposed integrated converter was performed emulating a partially-shaded condition for the four PV modules connected in series.

II. SINGLE-SWITCH INTEGRATED CONVERTER

A. Key Elements for the Single-Switch Integrated Converter

A schematic of the proposed integrated converter, shown in Fig. 3, demonstrates the potential for system-level simplification by integrating a converter and voltage equalizer into a single unit. However, these must be integrated without increasing the circuit complexity, since simplifying the system level at the cost of increased circuit complexity is undesirable.

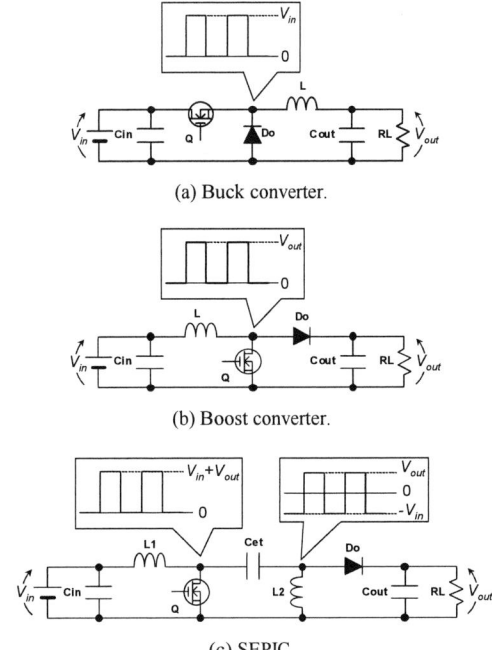

(a) Buck converter.

(b) Boost converter.

(c) SEPIC.

Fig. 4. Representative nonisolated PWM converters; (a) Buck converter, (b) boost converter, and (c) SEPIC.

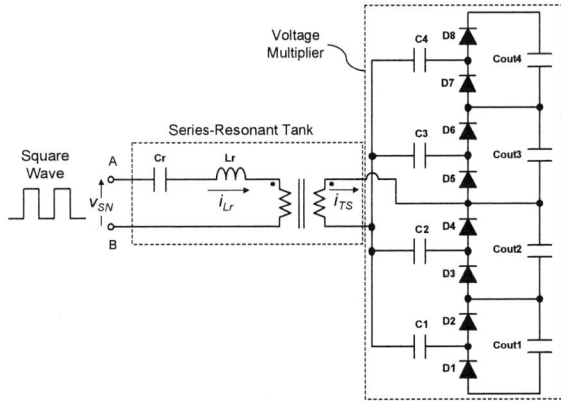

Fig. 5. Series-resonant voltage multiplier.

The proposed integrated converter can be derived based on a combination of a PWM converter and series-resonant voltage multiplier (SRVM). Representative non-isolated PWM converters and SRVM are shown in Figs. 4 and 5, respectively, while the SRVM shown in Fig. 5 can be further divided into a series-resonant tank and voltage multiplier comprising diodes of D_1–D_8, energy transfer capacitors of C_1–C_4, and smoothing capacitors of C_{out1}–C_{out4}. In the series-resonant tank, the leakage inductance of the transformer, L_{kg}, is used as a resonant inductor.

B. Derivation for Single-Switch Integrated Converter

In general, an asymmetric square wave voltage is generated at the switching node of PWM converters, as shown in the insets of Fig. 4. In the proposed integrated converter, this asymmetric square wave voltage is exploited to drive the SRVM. Although the PWM buck converter shown in Fig. 4(a) is taken as an example in the

The 2014 International Power Electronics Conference

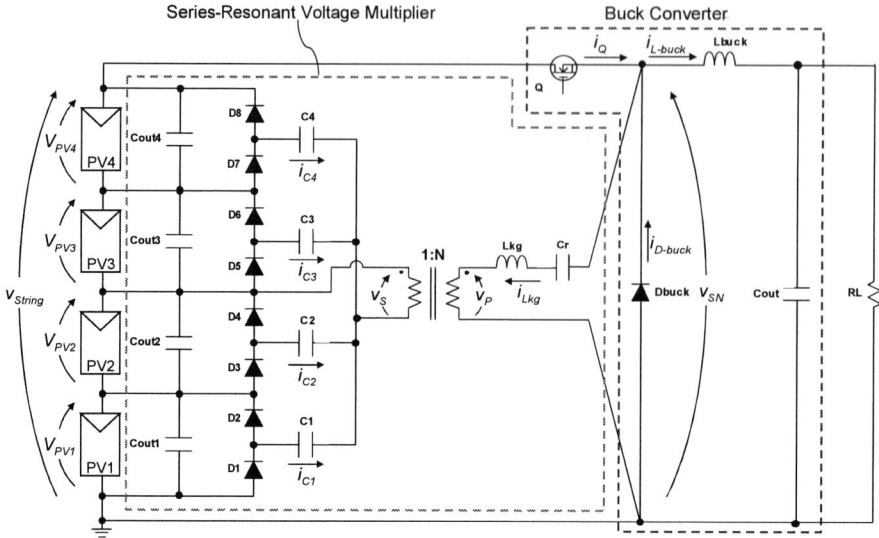

Fig. 6. Single-switch integrated converter based on the combination of a PWM buck converter and series-resonant voltage multiplier for the four PV modules connected in series.

rest of this paper, any converters, including isolated and resonant types, can be used provided there is an asymmetric square wave voltage in the converter and the operational criteria, which will be discussed in Section IV, are ensured.

By combining the PWM buck converter and SRVM shown in Figs. 4(a) and 5, respectively, the proposed single-switch integrated converter, as shown in Fig. 6, can be derived as a representative topology for the four PV modules connected in series. The input of the SRVM is tied to the switching node of the buck converter, while each smoothing capacitor in the SRVM is connected in parallel with each PV module.

The buck converter in the proposed integrated converter operates identically to a conventional one; the input and output voltage conversion ratio is simply determined by a duty cycle D, and any conventional MPPT techniques can be applied. Meanwhile, the SRVM is driven by the asymmetric square wave voltage

generated at the switching node of the buck converter. The SRVM operates so that all the voltages of the series-connected PV modules become virtually uniform. In general, the voltage of a shaded module in a string tends to be lower than that of nonshaded modules, as mentioned in Section I. Accordingly, the SRVM supplies current to the shaded module(s) so that the voltages of all the series-connected modules become virtually uniform. A detailed operation analysis and explanation of the voltage equalization mechanism will be performed in Section III-B.

With the proposed integrated converter, the system-level simplification is feasible by integrating a converter and voltage equalizer into a single unit. In addition, the proposed integrated converter is rather simpler at a circuit level than conventional DPP converters and the voltage equalizers shown in Fig. 2 due to the single-switch circuitry.

III. OPERATIONAL ANALYSIS

A. Waveforms and Current Flow

The key operational waveforms and current flow directions in the case that PV_1 is partially-shaded are shown in Figs. 7 and 8, respectively. The current paths in Fig. 8 are illustrated assuming that the currents in the voltage multiplier are buffered by smoothing capacitors of C_{out1}–C_{out4}.

In Mode 1, the switch Q is turned on, and the voltage at the switching node, v_{SN}, is equal to the input voltage or the string voltage of V_{String}. The inductor current in the buck converter, $i_{L\text{-}buck}$, linearly increases. In the SRVM, as the voltage of V_{String} is applied, the current of L_{kg}, i_{Lkg}, sinusoidally changes. In the voltage multiplier, the diode of D_1, a lower diode connected to a shaded module, starts conducting to charge C_1.

As i_{Lkg} drops to zero, the operation shifts to Mode 2, in which no current flows in the SRVM. This period is

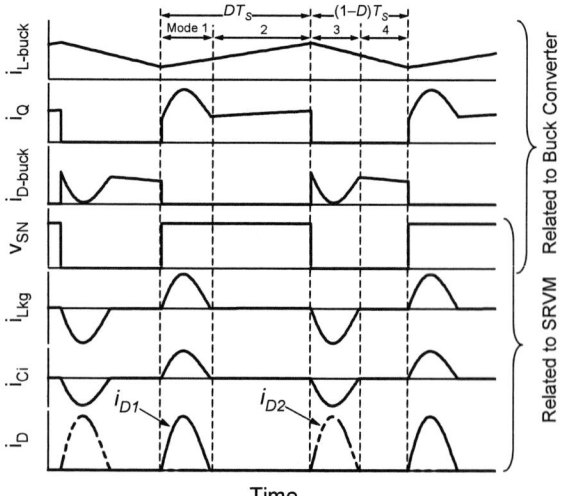

Fig. 7. Operation waveforms when PV_1 is partially-shaded.

978-1-4799-2706-7/14 $31.00 © 2014 IEEE   2353

The 2014 International Power Electronics Conference

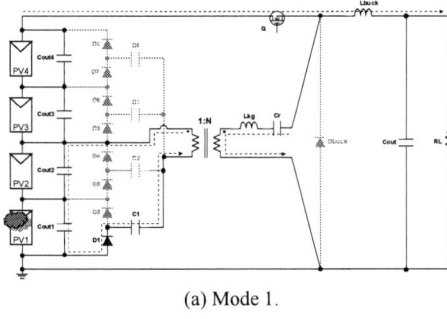

(a) Mode 1.

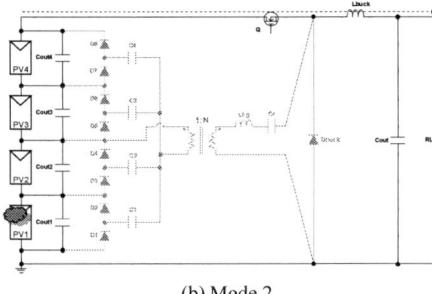

(b) Mode 2.

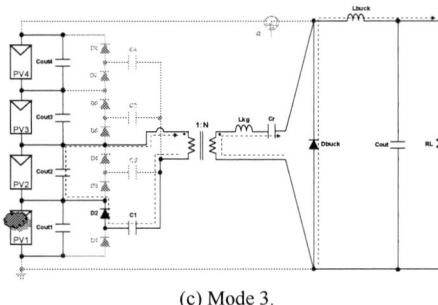

(c) Mode 3.

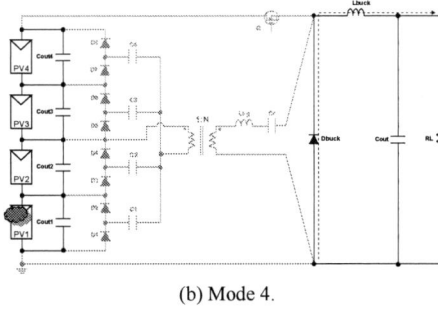

(b) Mode 4.

Fig. 8. Current flow directions when PV₁ is partially shaded: (a) Mode 1, (b) Mode 2, (c) Mode 3, and (d) Mode 4.

exactly identical to an on-period of a conventional buck converter.

Mode 3 begins as Q is turned off, and v_{SN} drops to a low level of nearly 0 V. The diode, D_{buck}, starts conducting and $i_{L\text{-}buck}$ linearly falls. A sinusoidal current of i_{Lkg} flows back to the switching node of the buck converter, while the diode of D_2, an upper diode connected to the shaded module, starts conducting to discharge C_1.

As i_{Lkg} rises and reaches zero, the integrated converter starts operating in Mode 4, in which no current flows in

the SRVM. This mode is equivalent to an off-period in a conventional buck converter.

In the SRVM, odd- and even-numbered diodes that are connected to a shaded module conduct alternately, while other diodes connected to unshaded modules do not. Based on Kirchhoff's Current Law, the average current supplied to a module PV$_i$ (i = 1...4) is equal to that of $D_{(2i-1)}$ or $D_{(2i)}$. Therefore, the SRVM supplies current for the shaded module of PV$_1$ only because only D_1 and D_2 conduct where PV$_1$ is partially-shaded.

B. Analysis for the Series-Resonant Voltage Multiplier

For the sake of clarity, an SRVM for two PV modules, PV$_m$ and PV$_n$, as shown in Fig. 9, is considered for the analysis in this subsection. As shown in Figs. 7 and 8, the SRVM is inactive and no current flows during Modes 2 and 4. By neglecting these inactive operation modes, voltage waveforms in the SRVM can be approximated as symmetrical square waves, and subsequently transformed into sinusoidal waves based on the Fourier Series. Since odd- and even-numbered diodes in the SRVM alternately conduct, voltages at diode pair junctions can also be transformed into sinusoidal waves, as shown in insets of Fig. 9. v_m and v_n are the voltages of $D_{(2m-1)}$ and $D_{(2n-1)}$ relative to ground level. The voltages v_{SN}, v_{TS}, v_m, and v_n, as well as those amplitudes $V_{m\text{-}SN}$, $V_{m\text{-}TS}$, $V_{m\text{-}m}$, and $V_{m\text{-}n}$, are expressed as

$$
\begin{cases}
v_{SN} = V_{m\text{-}SN} \sin \omega_r t = \dfrac{2}{\pi} V_{String} \sin \omega_r t \\[2mm]
v_{TS} = V_{m\text{-}TS} \sin \omega_r t = \dfrac{2}{\pi} V_{TS} \sin \omega_r t \\[2mm]
v_m = V_{m\text{-}m} \sin \omega_r t = \dfrac{2}{\pi} \left(V_{PV\text{-}m} + 2V_D \right) \sin \omega_r t \\[2mm]
v_n = V_{m\text{-}n} \sin \omega_r t = \dfrac{2}{\pi} \left(V_{PV\text{-}n} + 2V_D \right) \sin \omega_r t
\end{cases}
\quad (1)
$$

where ω_r is the resonant angular frequency, $V_{PV\text{-}m}$ and $V_{PV\text{-}n}$ are the voltages of PV$_m$ and PV$_n$, respectively, and V_D is the forward voltage drop of the diodes. The current amplitudes of i_{Lr} and i_{Ci} (i = m or n) can be yielded as

$$
\begin{cases}
I_{m\text{-}Lr} = \dfrac{V_{m\text{-}SN} - N V_{m\text{-}TS}}{|Z_r|} = \dfrac{2}{\pi} \dfrac{V_{String} - N V_{TS}}{R_r} \\[2mm]
I_{m\text{-}TS} = N I_{m\text{-}Lr} = \sum_i I_{m\text{-}Ci} \\[2mm]
I_{m\text{-}Ci} = \dfrac{V_{m\text{-}TS} - V_{m\text{-}i}}{\sqrt{r_i^2 + (1/\omega_r C_i)^2}} = \dfrac{2}{\pi} \dfrac{V_{TS} - (V_{PVi} + 2V_D)}{\sqrt{r_i^2 + (1/\omega_r C_i)^2}}
\end{cases}
\quad (2)
$$

where Z_r and R_r are the impedance and resistance of the series-resonant tank and r_i and C_i are the ESR and capacitance of C_i, respectively.

The average current supplied to the SRVM from the PV string, $I_{SRVM\text{-}ave}$, is expressed as

$$
I_{SRVM\text{-}ave} = \frac{1}{T_S} \int_0^{0.5 T_r} I_{m\text{-}Lr} \sin \omega_r t \, dt = \frac{\omega}{\omega_r} \frac{I_{m\text{-}Lr}}{\pi}, \quad (3)
$$

where T_S is the switching period, and ω is the angular switching frequency. Similarly, the average current flowing over C_i, I_{Ci}, is

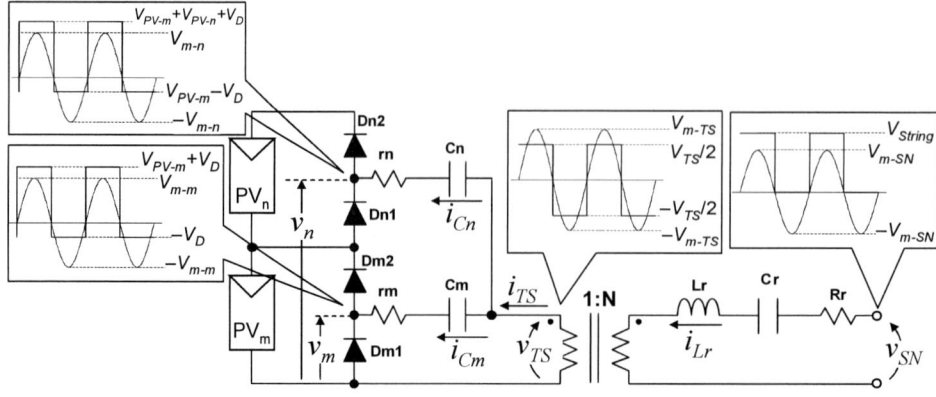

Fig. 9. Square voltage waves and approximated sinusoidal waves in the series-resonant voltage multiplier.

$$I_{Ci} = \frac{1}{T_S} \int_0^{0.5T_r} I_{m-Ci} \sin \omega_r t \, dt = \frac{\omega}{\omega_r} \frac{I_{m-Ci}}{\pi}. \tag{4}$$

Substituting (3) into (4) produces

$$\begin{cases} V_{String} - N V_{TS} = I_{SRVM-ave} R_{in} \\ V_{TS} - (V_{PVi} + 2V_D) = I_{Ci} R_{eq-i} \end{cases}, \tag{5}$$

where

$$\begin{cases} R_{in} = \dfrac{\omega_r}{\omega} \dfrac{\pi^2}{2} R_r \\ R_{eq-i} = \dfrac{\omega_r}{\omega} \dfrac{\pi^2}{2} \sqrt{r_i^2 + \left(\dfrac{1}{\omega_r C_i}\right)^2} \end{cases}. \tag{6}$$

From (5), a dc equivalent circuit of the SRVM can be derived as shown in Fig. 10. Both modules are connected to the secondary winding of the ideal transformer through the respective two diodes and one equivalent resistor R_{eq-i}. This derived dc equivalent circuit reveals an intuitive understanding of how the shaded modules are supported by the SRVM. As mentioned in Section I, the voltage of shaded modules tends to be lower than that of unshaded modules. Since both modules are connected to the common terminal of the transformer secondary winding, current from the SRVM preferentially flows toward the module with the lowest voltage, provided the voltage drop across R_{eq-i} is sufficiently smaller than the voltage difference between V_{PV-m} and V_{PV-n}. Accordingly, the SRVM supplies current for a shaded module so that all the module voltages become virtually uniform.

IV. OPERATIONAL CRITERIA

The SRVM is driven by an asymmetric square wave voltage generated at the switching node of the PWM

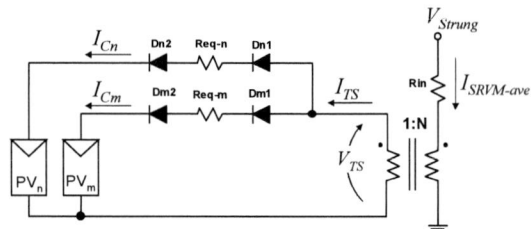

Fig. 10. Derived dc equivalent circuit for the series-resonant voltage multiplier.

buck converter. The shape of the asymmetric square wave varies with the duty cycle of the switch Q. Therefore, the SRVM is better designed so that its characteristic is independent of any duty cycle variation.

The SRVM is powered only during Modes 1 and 3, with no power supplied during Modes 2 and 4. In other words, provided Modes 2 and 4 exist, the power supplied to the SRVM is independent of the duty cycle variation of the PWM buck converter. Therefore, in order for the SRVM to operate independently from duty cycle variation, DT_S and $(1 - D)T_S$ must exceed half the resonant period (where T_S is the switching period). Therefore, the SRVM should be designed to satisfy

$$1 - \frac{f_S}{2f_r} > D > \frac{f_S}{2f_r}, \tag{7}$$

where f_S is the switching frequency, f_r is the resonant frequency given by

$$f_r = \frac{1}{2\pi\sqrt{L_{kg}C_r}}. \tag{8}$$

V. EXPERIMENTAL RESULTS

A. Prototype and Experimental Setup

A prototype of the proposed single-switch integrated converter was built for the four PV modules connected in series, while the buck converter and SRVM were separately built, as shown in Figs. 11(a) and (b), to measure the individual SRVM characteristic. The component values are listed in Table I.

The experimental setup used to measure the characteristics of the SRVM is shown in Fig. 12. A symmetric square wave voltage was generated by Q_a and Q_b at f_S = 50 kHz. PV modules were removed, while a

Table I. Component values

		Component	Value
Synchronous Bidirectional Converter		L	220 µH
		Q	N-Ch MOSFET, FDS86240, R_{on} = 35.3 mΩ
		D	Schottky Diode, D3FJ10, V_D = 0.74 V
		C_{in}, C_{out}	Aluminum Electrolytic Capacitor, 660 µF
Series-Resonant Voltage Multiplier		D_1–D_8	Schottky Diode, CLS01, V_D = 0.47 V, R_D = 20 mΩ
		C_1–C_4	Ceramic Capacitor, 94 µF, 5 mΩ
		C_{out1}–C_{out4}	Ceramic Capacitor, 188 µF
		C_r	Film Capacitor, 660 nF
		Transformer	N_1:N_2 = 24:6, L_{kg} = 2.8 µH, L_{mg} = 1.8 mH

(a) PWM buck converter.

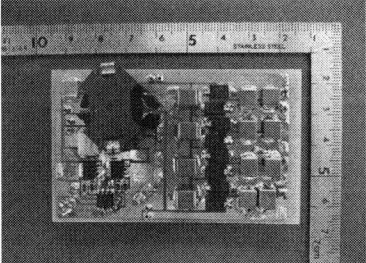

(b) Series-resonant voltage multiplier.

Fig. 11. Photographs of prototype: (a) PWM buck converter, (b) series-resonant voltage multiplier for the four PV modules connected in series.

variable resistor R_{var} with the selectable tap was used to emulate current paths under partially-shaded conditions. The one-module-shaded (PV_1-shaded) condition can be emulated by selecting the tap X, while the tap Y emulates the two-module-shaded condition where both PV_1 and PV_2 are equally shaded.

B. Performance of the Series-Resonant Voltage Multiplier

The efficiencies and output current characteristics measured at various input voltages are shown in Fig. 13. Losses of MOSFET's output capacitance and gate driving are excluded. Measured efficiencies under two-module-shaded conditions were lower than those under one-module-shaded conditions due to the increased losses in the series-resonant circuit, including the transformer; since the resonant circuit is common to all modules, the larger the processed power in the SRVM, the greater the loss in the series-resonant circuit will be.

As the output current of I_{Rvar} increased, the voltage of R_{var} linearly decreased, as implied by the derived dc-

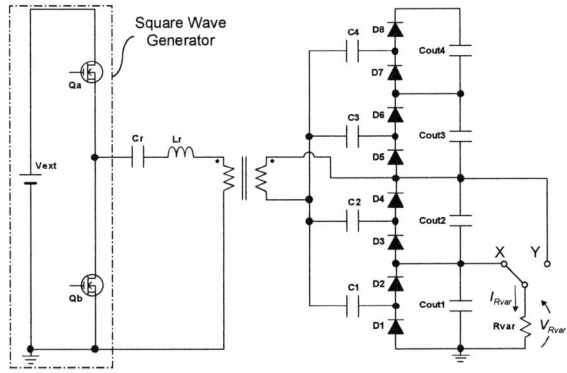

Fig. 12. Experimental setup for efficiency and characteristic measurement for the series-resonant voltage multiplier.

equivalent circuit shown in Fig. 10. By fitting the measured characteristics, the equivalent output resistance R_{out} under one- and two-module-shaded conditions were determined to be approximately 0.94 and 1.62 Ω, respectively. The larger value of R_{out} under the two-module-shaded conditions can be better explained with the derived dc equivalent circuit shown in Fig. 10; since R_{in} is common to all the modules, the larger current traverses R_{in} under the two-module-shaded conditions, increasing the value of R_{out}. These values correlated well with the theoretical ones calculated according to (6).

C. Equalization Emulating Partially-Shaded Conditions

The buck converter was operated at a switching frequency of 50 kHz and an experimental verification test was performed emulating the one-module-shaded (PV_1) condition. The individual PV characteristics used for the experiment are shown in Fig. 14(a). For a constant-voltage load of 24 V, the duty cycle of the integrated converter was varied to sweep the characteristic of the PV string, while the measured string characteristics with and without the integrated converter are compared in Fig. 14(b). Without the integrated converter (i.e. without equalization), two power point maxima, including a local and global MPP, were observed, and the extractable maximum power was as high as 55 W at V_{String} = 30 V. Conversely, with the integrated converter, the local MPP

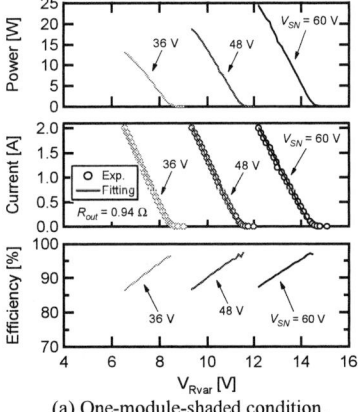

(a) One-module-shaded condition.

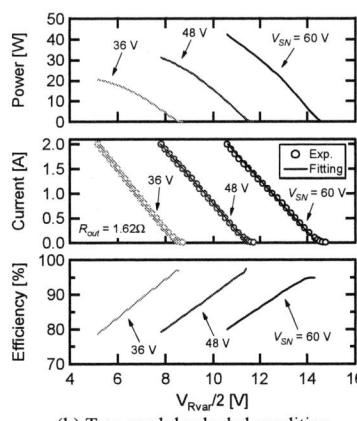

(b) Two-module-shaded condition.

Fig. 13. Measured power conversion efficiencies and output characteristics of the series-resonant voltage multiplier.

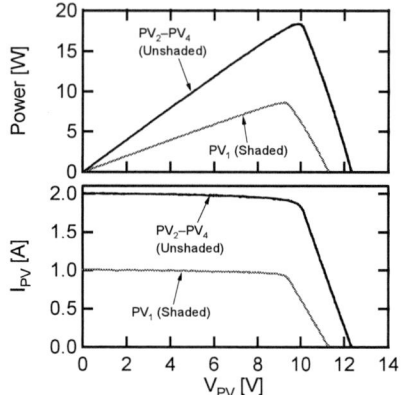

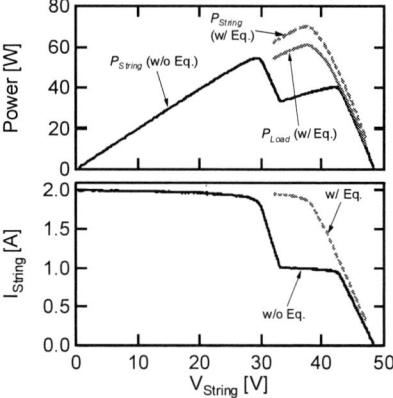

(a) Individual module characteristics used for experiment.

(b) String characteristics with/without equalization.

Fig. 14. Experimental results under one-module-shaded condition: (a) Individual module characteristics used for the experiment, (b) measured string characteristics with/without equalization.

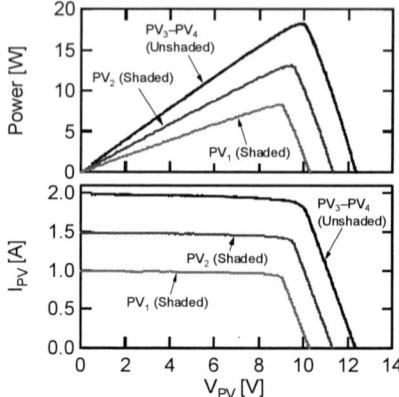

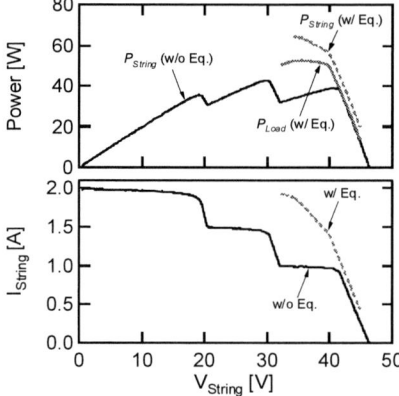

(a) Individual module characteristics used for the experiment.

(b) String characteristics with/without equalization.

Fig. 16. Experimental results under two-module-shaded condition: (a) Individual module characteristics used for the experiment, (b) measured string characteristics with/without equalization.

disappeared and the extractable maximum power soared to approximately 72 W at V_{String} = 36 V.

The measured key operational waveforms at V_{String} = 36 V are shown in Fig. 15. $i_{L\text{-}buck}$ was a triangular wave, while the sinusoidal current of i_{Lkg} flowed as v_{SN} swung. The results correlated well with the theoretical waveforms shown in Fig. 5, verifying the operational

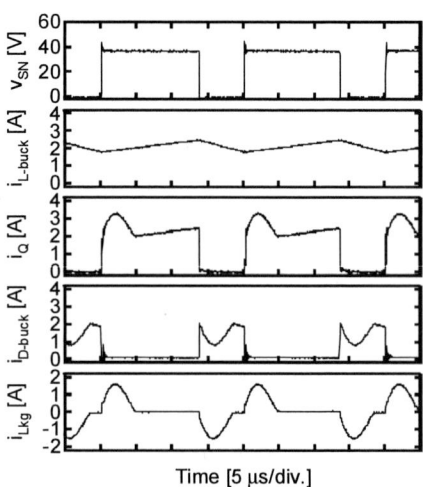

Fig. 15. Measured key waveforms at V_{String} = 36 V under one-module-shaded condition.

principle of the proposed integrated converter.

Another experimental equalization test was performed emulating the two-module-shaded (PV_1 and PV_2) condition. The individual PV characteristics used for the experiment and string characteristics measured with/without equalization are shown in Fig. 16. Similar to the results shown in Fig. 14, local MPPs successfully disappeared with equalization, while the extractable maximum power increased from 40 to 65 W, demonstrating how the proposed integrated converter can effectively prevent the negative impacts of partial shading; even when multiple modules are shaded.

VI. CONCLUSIONS

A single-switch PWM converter integrating a voltage equalizer has been proposed for series-connected PV modules under partial shading. In addition to system-level simplification by integrating a converter and voltage equalizer into a single unit, circuit level simplification is also feasible due to the single-switch circuitry. In the proposed integrated converter that comprises a PWM buck converter and SRVM, the asymmetric square wave voltage generated at the switching node of the buck converter is exploited to drive the SRVM.

Experimental verification tests were performed for the four PV modules connected in series, emulating partially-shaded conditions. With the proposed integrated converter, multiple power point maxima successfully disappeared and the extractable maximum powers soared, demonstrating the effectiveness and performance of the proposed single-switch integrated converter.

ACKNOWLEDGEMENTS

This work was supported in part by the Ministry of Education, Culture, Sports, Science, and Technology through Grant-in-Aid for Young Scientists (B) 25820118.

REFERENCES

[1] P. S. Shenoy, K. A. Kim, B. B. Johnson, and P. T. Krein, "Differential power processing for increased energy production and reliability of photovoltaic systems," *IEEE Trans. Ind. Power Electron.*, vol. 28, no. 6, pp. 2968–2979, Jun. 2013.

[2] H. J. Bergveld, D. Büthker, C. Castello, T. Doorn, A. D. Jong, R. V. Otten, and K. D. Waal, "Module-level dc/dc conversion for photovoltaic systems: the delta-conversion concept," *IEEE Trans. Power Electron.*, vol. 28, no. 4, pp. 2005–2013, Apr. 2013.

[3] S. Qin and R. C. N. P. Podgurski, "Sub-module differential power processing for photovoltaic applications," *IEEE Applied Power Electron. Conf. Expo.*, pp. 101–108, 2013.

[4] S. Qin, S. T. Cady, A. D. D. Garcia, and R. C. N. P. Podgurski, "A distributed approach to MPPT for PV sub-module differential power processing," *IEEE Energy Conversion Conf. Expo.*, pp. 2778–2785, 2013.

[5] R. Kadri, J. P. Gaubert, and G. Champenois, "New converter topology to improve performance of photovoltaic power generation system under shading conditions," *Int. Conf. Power Eng. Energy Electrocal Drives*, pp. 1–7, 2011.

[6] R. Kadri, J. P. Gaubert, and G. Champenois, "Centralized MPPT with string current diverter for solving the series connection problem in photovoltaic power generation system," *Int. Conf. Power Eng. Energy Electrocal Drives*, pp. 116–123, 2011.

[7] R. Giral, C. A. R. Paja, D. Gonzalez, J. Calvente, À. C. Pastpr, and L. M. Salamero, "Minimizing the effects of shadowing in a PV module by means of active voltage sharing," *IEEE Int. Conf. Ind. Technol*, pp. 943–948, 2010.

[8] R. Giral, C. E. Carrejo, M. Vermeersh, A. J. Saavedra-Montes, and C. A. Ramos-Paja, "PV field distributed maximum power-point tracking by means of an active bypass converter," *Int. Conf. Clean Electrical Power*, pp. 94–98, 2011.

[9] L. F. L. Villa, T. P. Ho, J. C. Crebier, and B. Raison, "A power electronics equalizer application for partially shaded photovoltaic modules," *IEEE Trans. Ind. Electron.*, vol. 60, no. 3, pp. 1179–1190, Mar. 2013.

[10] L. F. L. Villa, X. Pichon, F. S. Ardelibi, B. Raison, J. C. Crebier, and A. Labonne, "Toward the design of control algorithms for a photovoltaic equalizer: choosing the optimal switching strategy and the duty cycle," *IEEE Power Electron.*, vol. 29, no. 3, pp. 1447–1460, Mar. 2014.

[11] Z. Salam and M. Z. Ramli, "A simple circuit to improve the power yield of PV array during partial shading," *IEEE Energy Conversion Cong. Expo.*, pp. 1622–1626, 2012.

[12] T. Shimizu, O, Hashimoto, and G. Kimura, "A novel high-performance utility-interactive photovoltaic inverter system," *IEEE Trans. Power Electron.*, vol. 18, no. 2, pp. 704–711, Mar. 2003.

[13] T. Shimizu, M. Hirakata, T. Kamezawa, and H. Watanabe, "Generation control circuit for photovoltaic modules," *IEEE Trans. Power Electron.*, vol. 16, no. 3, pp. 293–300, May 2001.

[14] J. T. Stauth, M. D. Seeman, and K. Kesarwani, "Resonant switched-capacitor converters for sub-module distributed photovoltaic power management," *IEEE Trans. Power Electron.*, vol. 28, no. 3, pp. 1189–1198, Mar. 2013.

[15] S. B. Yaakov, A. Blumenfeld, A. Cervera, and M. Evzelman, "Design and evaluation of a modular resonant switched capacitor equalizer for PV panels," *IEEE Energy Conversion Cong. Expo.*, pp. 4129–4136, 2012.

[16] K. Nishijima, H. Sakamoto, and K. Harada, "A PWM controlled simple and high performance battery balancing system," in Proc. *IEEE Power Electron. Spec. Conf.*, Jun. 2000, pp. 517–520.

[17] P. A. Cassani and S. S. Williamson, "Feasibility analysis of a novel cell equalizer topology for plug-in hybrid electric vehicle energy-storage systems," *IEEE Trans. Veh. Technol.*, vol. 58, no. 8, Oct. 2009, pp. 3938–3946.

[18] P. A. Cassani and S. S. Williamson, "Design, testing, and validation of a simplified control scheme for a novel plug-ion hybrid electric vehicle battery cell equalizer," *IEEE Trans. Ind. Electron.*, vol. 57, no. 12, Dec. 2010, pp. 3956–3962.

[19] J. W. Kimball, B. T. Kuhn, and P. T. Krein, "Increased performance of battery packs by active equalization," in Proc. *IEEE Veh. Power Propulsion Conf.*, Sep. 2007, pp. 323–327.

[20] A. Baughman and M. Ferdowsi, "Double-tiered switched-capacitor battery charge equalization technique," *IEEE Trans. Ind. Appl.*, vol. 55, no. 6, Jun. 2008, pp. 2277–2285.

[21] M. Uno and K. Tanaka, "Influence of high-frequency charge-discharge cycling induced by cell voltage equalizers on the life performance of lithium-ion cells," *IEEE Trans. Veh. Technol.*, vol. 60, no. 4, May 2011, pp. 1505–1515.

[22] Y. Yuanmao, K. W. E. Cheng, and Y. P. B. Yeung, "Zero-current switching switched-capacitor zero-voltage-gap automatic equalization system for series battery string," *IEEE Trans. Power Electron.*, vol. 27, no. 7, Jul. 2012, pp. 3234–3242.

[23] H. S. Park, C. E. Kim, C. H. Kim, G. W. Moon, and J. H. Lee, "A modularized charge equalizer for an HEV lithium-ion battery string," *IEEE Trans. Ind. Electron.*, vol. 56, no. 5, May 2009, pp. 1464–1476.

[24] C. H. Kim, H. S. Park, C. E. Kim, G. W. Moon, and J. H. Lee, "Individual charge equalization converter with parallel primary winding of transformer for series connected lithium-ion battery strings in an HEV," *J. Power Electron.*, vol. 9, no. 3, May 2009, pp. 472–480.

[25] J. Cao, N. Schofield, and A. Emadi, "Battery balancing methods: a comprehensive review," in Proc. *IEEE Veh. Power Propulsion Conf.*, Sep. 2008, pp. 1–6.

[26] K. Z. Guo, Z. C. Bo, L. R. Gui, and C. S. Kang, "Comparison and evaluation of charge equalization technique for series connected batteries," in Proc. *IEEE Power Electron. Spec. Conf.*, Jun. 2006, pp. 1–6.

New DC Rail Side Soft-Switching PWM DC-DC Converter with Voltage Doubler Rectifier for PV Generation Interface

Khairy Sayed[1], Soon-Kurl Kwon[2], Katsumi Nishida[3], Mutsuo Nakaoka[4]

[1]Sohag University, Sohag, Egypt
[2]Kyungnam University, Masn, South Korea
[3]Ube National College of Technology, Ube, Japan
[4]Kyungnam University, South Korea/Yamaguchi University, Yamaguchi, Japan

Abstract: This paper presents a new circuit topology of active edge resonant snubbers assisted half-bridge soft switching PWM inverter type high power dc-dc converter for photovoltaic applications. The proposed step-up DC/DC converters with transformer isolation can obtain high voltage step-up trough turns ratio. The proposed dc-dc power converter is composed of typical voltage source-fed half-bridge high frequency PWM inverter with a high frequency transformer link in addition to input DC busline side power semiconductor switching devices for PWM control scheme and parallel capacitive lossless snubbers. The operating principle of the new DC-DC converter treated here is described by using switching mode equivalent circuits, together with its unique features. All the active power switches in the half-bridge arms and input DC buslines can achieve ZCS turn-on and ZVS turn-off commutation transitions. The total turn-off switching losses of the power switches can be significantly reduced. As a result, a high switching frequency IGBTs can be actually selected in the frequency range of 60 kHz under principle of soft switching.

Keywords- active edge resonant snubbers; DC-DC power converter; dc rail; solar PV; high frequency transformer link; soft switching PWM.

I. INTRODUCTION

Recently, transformer secondary side saturable inductor switch assisted ZVS-PWM dc-dc power converter was proposed [1]-[5]. The total losses associated with such operations may impact the efficiency, size and total cost of the overall design. The stresses and losses upon the electronic devices can be minimized by controlling the switching times that occur at the instants when the current through and/or voltage across the converter switches become zero. This technology is known as soft-switching. In addition to the zero voltage/current switching conditions (ZVS/ZCS), this approach has other advantages, such as: elimination of snubber losses, improvement of device reliability, less dv/dt stress on magnetic device insulation, reduced EMI problems.

These DC-DC power converter circuit topologies are suitable for handling high output power more than about several kW, especially for high or medium voltage and low current dc applications as new energy related power supplies. However, secondary magnetic switches or secondary side semiconductor switching devices as active power rectifier in these converter circuit topologies may cause large conduction loss. Soft-switching permits

higher switching frequency operation, reducing the size, weight and cost of the magnetic components. Therefore, for the low input voltage and large input current applications power supplies, a soft switching DC-DC power converter with active switches in the primary side of high frequency transformer is considered to be more suitable and acceptable. A variety of topologies have been considered for high frequency isolated DC-DC converter, for example Phase shift full bridge. Phase-shifted PWM converter has found many applications in design of high power density DC-Dc converters due its zero voltage switching ZVS [6]. However, the control strategy is more complicated and long freewheeling current affects the circuit efficiency. A new family of ZVT PWM converters was presented in [7]. These converters are derived by adding by adding an LC resonant network to the conventional PWM converters. The proposed ZVT PWM converter has all advantages of the converter presented in [7], and in addition, ZVS is achieved more efficiently, with less additional components. The IGBT's are used to increase power density by increasing switch capacitance with an external capacitor. Compared to the converter presented in [7], the proposed converter has advantage that more output power regulation can be achieved, since the duty cycle can be controlled through switches S1 and S3. The new converter has another advantage: much less circulating energy since no resonant inductor is used in the main power path, soft-switching operation maintained for the entire line range and load range. These features make the new topology attractive for high power applications. Due to increasing demands for high-efficiency high-power-density topologies for future applications, the soft switching is challenging. Voltage doubler rectifier (VDR) is the most promising topology for high step up low voltage input applications [8-11].

This paper presents a novel circuit topology of voltage source soft switching PWM DC-DC power converter. Secondary side of the step-up DC/DC converter is voltage doubler rectifier. Under the newly-proposed power converter circuit; all the active switches in the half-bridge arm and DC bus lines can actively achieve ZCS turn-on and ZVS turn-off commutation operation. According to the electric grid standards which are in place in some countries, some grid-connected PV systems have the transformer for the galvanic isolation between the conversion stage and the grid. Without the isolation transformer, the circulation of ground current through the parasitic capacitance between the PV array and the ground may occur [12]. Appropriate control algorithm including Maximum Power Point Tracking (MPPT) should be implemented in the converter control circuit in order to satisfy high PV energy productivity and better conversion efficiency. The practical effectiveness of the proposed power converter is actually proved on the basis of experimental data.

II. NEW SOFT SWITCHING DC-DC CONVERTER

Fig. 1 shows the proposed voltage source soft switching PWM DC-DC converter circuit using a novel topology of half-bridge soft switching PWM inverter with high frequency transformer link. This DC-DC converter is composed of typical divided voltage source modified type half-bridge inverter with a active PWM switch $Q_3(S_3/D_3)$ in positive dc busline and a active PWM switch $Q_4(S_4/D_4)$ in negative dc bus line and two lossless snubbing capacitors C_1, C_2 and two diodes D_5, D_6. Two centre points of two capacitors C_1, C_2 in parallel with two diodes D_5, D_6 are connected to a mid point of two voltage sources E_1, E_2 and one of terminals of primary winding of high frequency transformer. The voltages represented by E_1, E_2 and capacitance of C_1, C_2 are designed so as to be equal ($E_1=E_2=E$, $C_1 =C_2 =C$). The main active switches Q_1, Q_2 can be periodically turned on and turned off in the same pulse pattern with a certain dead-time as conventional half-bridge type hard switching PWM inverter.

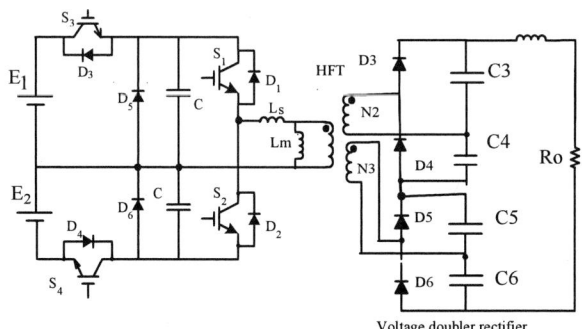

Fig. 1 DC rail side soft switching PWM DC-DC converter with voltage doubler rectifier

Under the proposed DC-DC converter, the switches Q_1, Q_2 in the half-bridge type inverter arms can perform ZVS turn-off transition due to the presence of the active PWM switches Q_3 or Q_4 which are turned off and the snubbing capacitors C_1 or C_2 are completely discharged before the active switches Q_1 or Q_2 in half-bridge type inverter arms are turned off. In addition, all the inverter switches can also perform ZCS at a turn on mode transition with the aid of inductance L_S, as a primary side lumped parasitic leakage inductance of high frequency transformer HF-T.
As for the active PWM switches Q_3 - Q_4 in series with the DC busline side, the PWM controlled switches Q_3, Q_4 can achieve ZVS at a turn-off mode transition due to the lossless snubbing capacitors C_1 or C_2. These active PWM

switches Q_3, Q_4 can also achieve ZVS/ZCS at a turn-on mode transition due to the lossless snubbing capacitors C_1, C_2, which have been charged up to the same voltage as the half voltage E1or E2 of dc power busline voltage source by the energy storage in the leakage inductance L_S after the half-bridge type inverter switches Q_1, Q_2 are turned off completely. Although the conduction power loss of the additional switches Q_3 - Q_4 may increase the total power loss a little, the total turn-off switching loss of half-bridge type PWM high frequency inverter can be significantly decreased with the optimum aid of DC bus line series switches Q3 - Q4 and the lossless snubbing capacitors C1 - C2. Figure 1(b) depicts the modified circuit topology of the proposed PWM DC-DC power converter with high frequency transformer link. The Duty cycle of this converter may theoretically increase to 100%. In practice this is not possible because the series connected transistors, Q_1 and Q_2, have to be switched with a time difference to avoid a short circuit of the input supply. The duty cycle definition can be found from

$$D = \frac{T_{on4}}{T_h} \qquad (1)$$

where T_{on4} is the turning on time of switch Q_4, T_h is the half of switching period T_s.

III. PRINCIPLE OF OPERATION

A. Gate voltage pulse timing sequences

Figure 2 shows the specified timing pattern sequences of switching gate driving pulses. The gate voltage pulse signals for the inverter switches Q_1, Q_2 in voltage source half-bridge inverter arms are the same as PWM signal sequences of conventional half-bridge inverter. Regarding the turn-on gate voltage pulse signals to the DC busline side series switches Q_3, Q_4; the signals are applied to Q_3, Q_4 at the same timing as the turn-on signals to Q_1, Q_2, respectively. As for the turn-off signals to Q_3, Q_4, the signals are delivered to Q_3, Q_4 before the predetermined length t_d of time on the basis of the time when the turn-off signals are applied to Q_1 or Q_2. In other words, the turn-off pulse signals are applied to Q_1, Q_2 after the turn-off gate signals applied to Q_3, Q_4 by time t_d.

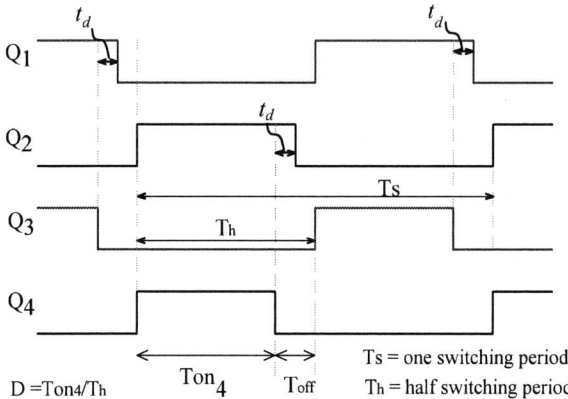

Fig. 2 Timing Pattern sequences of switching gate driving pulses.

B. Switching operation modes and equivalent circuits

Figure 3 illustrates the relevant operating waveforms during a complete switching period for the pattern of gate drive pulse timing sequences shown in Fig. 2. The switching operation modes of this DC-Dc power converter are divided into twelve operation modes from mode 1 to mode 12 in accordance with operating timing sequences, from t_0 to t_{11}. The operation principle is described in the following. The equivalent circuits corresponding to respective operation mode are shown in Fig. 4.

1) Mode 1($t_0 \sim t_1$): Q1, Q3 are turned on at causing the leakage inductance current to start charging from zero at the following slope:

$$\frac{di_{LS}}{dt} = \frac{0.5V_{in}}{L_S} = \frac{E}{L_S} \qquad (2)$$

where L_S is the leakage inductance, E is DC busline voltage, the voltage difference between points a, b will equal; V_{ab}=E (see Fig. 3).

The smaller the leakage inductance is, the faster its current charge-up will be, until it becomes equal to the reflected inductor current, causing the diode D8 to be blocked and the converter to start transferring energy from the primary-side to the secondary-side. This current will continue to charge at a slope of

$$\frac{di_{LS}}{dt} = \frac{E - \dfrac{n_1}{n_2}V_o}{L_S + \left(\dfrac{n_1}{n_2}\right)^2 L_O} \qquad (3)$$

$$i_{LS}(t) = \frac{E - \dfrac{n_1}{n_2}V_o}{L_S + \left(\dfrac{n_1}{n_2}\right)^2 L_O}(t - t_o) + i(t_0) \qquad (4)$$

where $i(t_o) = I_{SW1} = \dfrac{n_2}{n_1}I_O$ (5)

where Io is the load current, L_o is the secondary inductor, n_1 and n_2 are the transformer primary and secondary number of turns, respectively, and Vo is the output voltage. After time t_0, the switches Q1 and Q3 are turned on. At this time, i_{t1} flows through the primary winding of high frequency transformer HF-T. i_{s1} flows through Q1 and i_{s3} flows through Q3. In this period, all currents i_{t1}, i_{s1} and i_{s3} are equal and the voltage v_{C1} across C_1 is the same as the DC busline voltage E_1

2) Mode 2($t_1 \sim t_2$): At time t_1, the turn-off pulse signal is applied to Q3. At this time, the high side series switch Q3 in DC busline can be turned off with ZVS because the current i_{s3} through Q3 is immediately cut off due to the lossless snubbing capacitor C_1. After time t_1, the voltage v_{C1} across the capacitor C_1 discharges constantly toward zero voltage from $E1=E$. At this time, the voltage V_{C1} across the lossless snubber capacitor C_1 is estimated by,

$$v_{C1} = \frac{1}{C_1}\int i_{C1}dt \qquad (6)$$

In this mode the capacitor current is constant

$$v_{C1} = E - \frac{i_{t1}}{C_1}t \qquad (7)$$

$$C_1 = \frac{t_x i_{t1}}{E}, \; t_x = t_2 - t_1 \qquad (8)$$

where, i_{t1} is a primary current of high frequency transformer. From eq. (1), until this voltage v_{C1} becomes zero, the discharging time t_x of the capacitor C_1 is given by;

$$t_x = \frac{C_1 E}{i_{t1}} \qquad (9)$$

From eq. (9), the more the current i_{t1} though the primary winding of high frequency transformer HF-T is large, the more the discharging time for capacitor C_1 is short. On the other hand, the more the current i_{t1} is small, the more the discharging time t_x is long. Under this newly-developed DC-DC power converter circuit (see fig. 1(a)), the delay time t_d indicated in Fig.2 is designed so as to be longer than the time calculated from eq. (9), under the condition of the maximum i_{t1} or the maximum output current. In this case, the switches Q_1 or Q_2 can achieve ZVS turn-off transition completely. To enlarge the complete ZVS operation range at the turn-off commutation for the switches Q_1 or Q_2, the delay time t_d should be varied according to the value of current i_{t1}.

3) Mode 3($t_2 \sim t_3$): At time t_2, the voltage v_{C1} becomes zero. In the interval from t_1 to t_2, the diodes D_5 is turned on and the current i_{t1} through high frequency transformer primary winding flows through the circulation loop; $L_S \rightarrow D_5 \rightarrow S_1 \rightarrow L_S$.

4) Mode 4($t_3 \sim t_4$): At time t_3, the turn-off gate pulse signal (see Fig.2) is applied to Q_1. At this time, the switch Q_1 can be turned off with ZVS because the voltage v_{C2} across C2 was already zero during next half operation cycle and the diodes D_2 of Q_2 are immediately turned on. After that, the capacitor C_2 is charged up to the same voltage as dc busline voltage E_2. During this mode, the capacitor is charged linearly by the reflected output current. Assuming that the output current is constant during this mode, the reflected current through the capacitor is given by

$$I_{C2} = \frac{n_2}{2n_1}I_o \qquad (10)$$

$$v_{C2} = \frac{1}{C_2}\int i_{C2}dt \qquad (11)$$

Then $v_{C2} = \dfrac{i_{C2}}{C_2}(t - t_3)$,   (12)

$$i_{C2} = I_{C2} - C_2 E /(t - t_3) \qquad (13)$$

At t=t_4, i_{c2}=0 $I_{C2} = C_2 E /(t_4 - t_3)$ (14)

Then the capacitor C2 value can be calculated from:

$$C_2 = I_{C2}(t_4 - t_3)/E \qquad (15)$$

At this mode, In order to achieve ZVS, the energy stored in the leakage inductance has to be larger enough to charge and discharge divided capacitance, neglecting the transformer winding capacitance, the energy stored in the leakage inductance has to satisfy the condition that the capacitor C_2 is just charged up to the same voltage as DC bus line voltage E_2, which can be estimated by eq. (15).

$$\frac{1}{2}CE^2 = \frac{1}{2}L_S i_{t1}^2 \qquad (16)$$

where $C1=C2=C$

However, as described after in mode 6, circuit parameters should be designed to meet the condition of;

$$\frac{1}{2}CE^2 \leq \frac{1}{2}L_S i_{t1}^2 \qquad (17)$$

in order to achieve ZVS a turn-on transition of Q_4.

The converter dc voltage gain is assumed to be the same as it is for the conventional half-bridge dc–dc converter, which is given by

$$V_{out} = \frac{n_2 D}{2n_1}V_{in} \qquad (0 < D < 0.5) \qquad (18)$$

5) Mode 5(t_4~t_5): Under a condition of $(1/2)CE^2 < (1/2)L_S(i_{t1})^2$, after the voltage v_{C2} reaches the DC busline voltage E_2, the voltage v_{C2} across the snubber capacitor C_2 is clamped to dc bus line voltage E_2 because the diode D_4 of Q_4 is turned on and the energy stored into the inductor with leakage inductance L_S is back to DC busline voltage source E_2.

6) Mode 6(t_5~t_6):In this mode, all operations are stopped in the primary circuit of high frequency transformer.

7) Mode 7(t_6~t_7):At time t_6, the turn-on gate pulse signals are applied to the switches Q_2 and Q_4. At this time, the switch Q_2 can be turned on with ZCS due to the primary-side lumped parasitic inductance L_S of high frequency transformer. And more the series switch Q_4 in the bus line achieves a complete soft-switching ZVS/ZCS at turn-on transition because the voltage v_{C2} is the same voltage as DC power busline voltage E_2. The voltage Vab=-E (see Fig. 3). This topology generates symmetrical ac waveforms at the primary side of the transformer, in which the core flux is excited bidirectionally. This topology has an advantage of better utilization of the core.

8) Mode 8(t_7~t_8): The turn-off signal is applied to Q_4. At this time, the series switch Q_4 in dc the busline can be turned off with ZVS because the current i_{s4} through Q_4 is immediately cut off due to the lossless snubbing capacitor C_2. The voltage v_{C2} across the capacitor C_2 discharges constantly toward zero.

9) Mode 9(t_8~t_9): The voltage v_{C2} becomes zero, the diodes D_6 are turned on and the current i_{t1} through transformer primary winding flows through the loop; $L_S \rightarrow S_2 \rightarrow D_6 \rightarrow L_S$.

10) Mode 10(t_9~t_{10}): The turn-off gate pulse signal is applied to Q_2. At this time, the switch Q_2 can be turned off with ZVS because the voltage v_{C2} was already zero during the last half operation cycle and the diodes D_1 of

Q_1 are immediately turned on. After that, the capacitor C_1 is charged up to the same voltage as dc busline voltage E_1.

11) Mode 11(t_{10}~t_{11}): After the voltage v_{C1} across the snubber capacitor C_1 reaches the dc busline voltage E_1, it is clamped to dc busline voltage E_1 because the diode D_3 of Q_3 is turned on and the energy stored into leakage inductance L_S is back to dc busline voltage source E_1.

12) Mode 12(t_{11}~t_{12}): At this mode, all operations are stopped in the primary circuit. Thereafter, the operation processes for Q_1, Q_3 and Q_2, Q_4 will be repeated in sequence continuously from mode 1 to 12 as in Fig. 4.

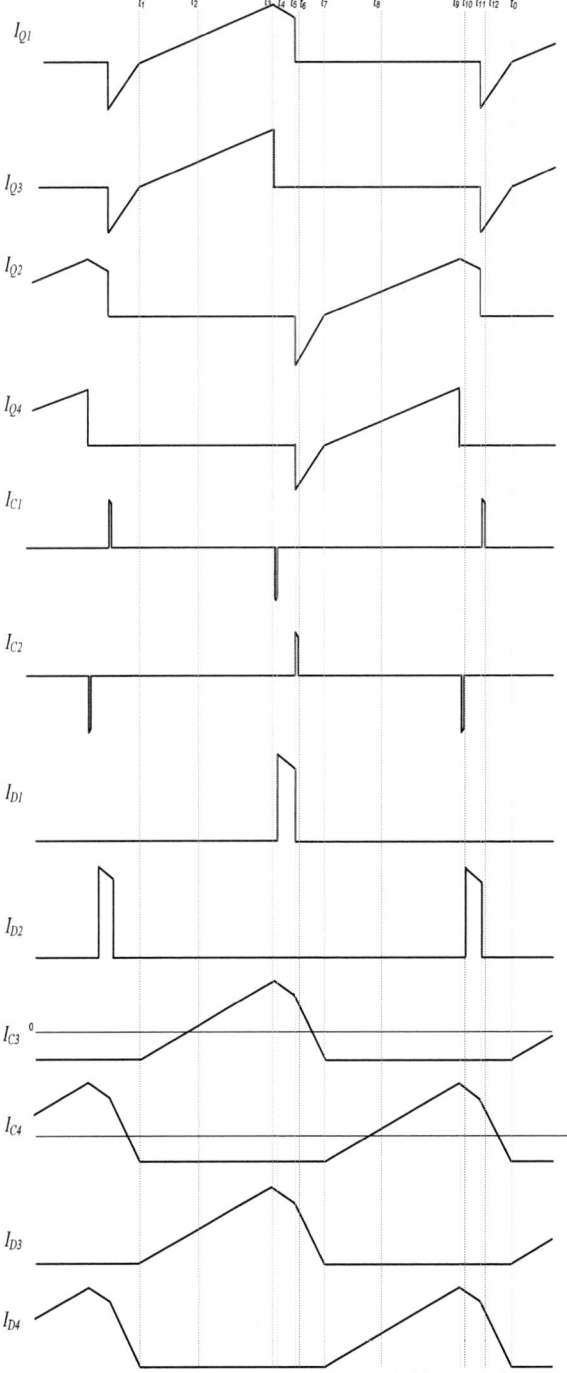

Fig. 3 Operating waveforms during one switching period.

The 2014 International Power Electronics Conference

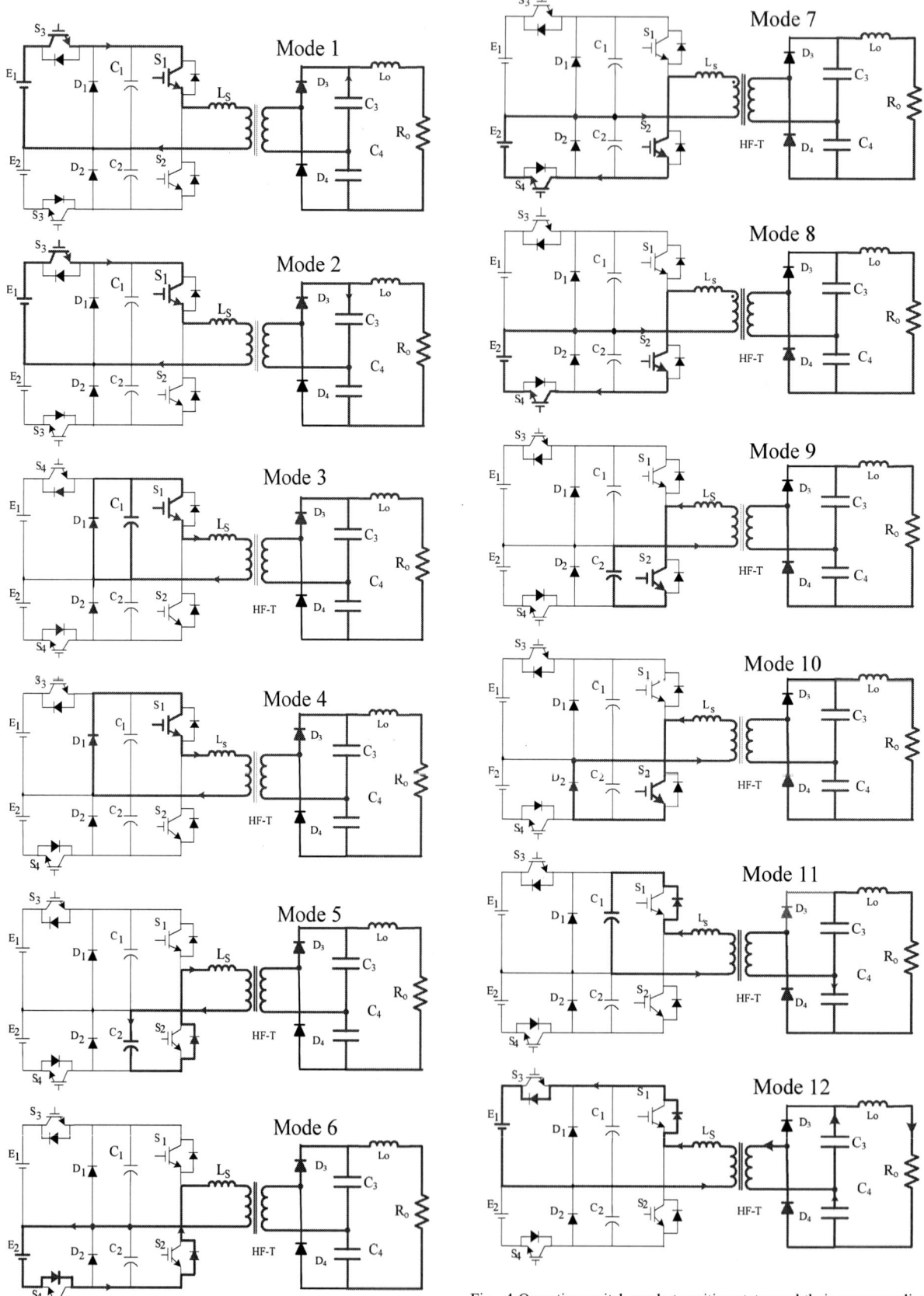

Fig. 4 Operating switch-mode transition states and their corresponding equivalent circuits.

C. Secondary circuit (voltage doubler rectifier)

The voltage doubler rectifier circuit is very suitable for low input voltage applications. Like standard voltage doubler topologies, it reduces the need of rectifier diodes to half but requires output capacitors twice the full bridge rectifier topologies. It has the major advantage of reduction in transformer turns ratio (secondary to primary) to half. Therefore, for the present specifications, voltage doubler topology reduces HF transformer size.

IV. EXPERIMENTAL RESULTS AND DISCUSSION

A. Total system implementations

In the DC-DC converter setup implementation, the maximum output voltage and current are designed for 310V, 6A, respectively. The design specifications and circuit parameters are described in Table. 1. The proposed converter was constructed using IGBTs. These were chosen due to their low on-state losses and their availability in a module, the CM200DU-12NF. This module contains two series connected 600V, 200A IGBTs and two freewheeling diodes in parallel with each IGBT. Its construction reduces the inductance between the two power switches to a minimum and controls the current ringing and voltage overshoot during switching. The resonant components in the circuit were polypropylene capacitors, inductors wound on ferrite cores with air gaps to prevent saturation and the leakage inductance of the isolation transformer. Polypropylene capacitors were used due to their relatively constant value of capacitance over a wide frequency range and low value of inductance. The secondary coil of the isolation transformer was connected to a speed power diode bridge rectifier. The maximum DC output power of this experimental setup is designed for 14.4 kW. The IGBT power modules are mounted on the heat sink and connected by the printed circuit board in which the capacitors C_1 and C_2 are mounted on PCB and the capacitors C_3 and C_4 are directly connected to the output of three phase rectifier. Connecting IGBTs and capacitor C_1, C_2, C_3 and C_4 by the printed circuit board enables to minimize the stray line inductance with optimum connections among IGBTs, C_1, C_2, C_3 and C_4. Actually, the minimum leakage inductance of the high frequency transformer is particularly important on this new soft-switching PWM DC-DC converter, because spike voltage across collector and emitter of IGBTs easily appears at a turn off transition if there is wiring stray inductance between snubbing capacitors and the IGBT switches. Under the experimental setup of power equipment using the proposed power converter, the dynamic performance and power density can be improved by fast control responses in accordance with the high switching frequency.

B. Control circuit design and implementation

An appropriate DC-Dc converter controller by using MPPT with P&O function is used. The MPPT is implemented using P&O algorithm. The controller calculates the power point value from A/D channels for the PV output voltage and PV output current. Usually, the MPPT continuously regulates the duty cycle. The isolated boost DC/DC stage is used to boost the output voltage of the PV module up to about 400V DC and to process the Maximum Power Point Tracking (MPPT). For it, the common and easy Perturb & Observe (P&O) method was used in which the DC-DC converter duty cycle is incremented or decremented by the microcontroller, according to both PV panel power and voltage change (Fig. 5 shows this method).

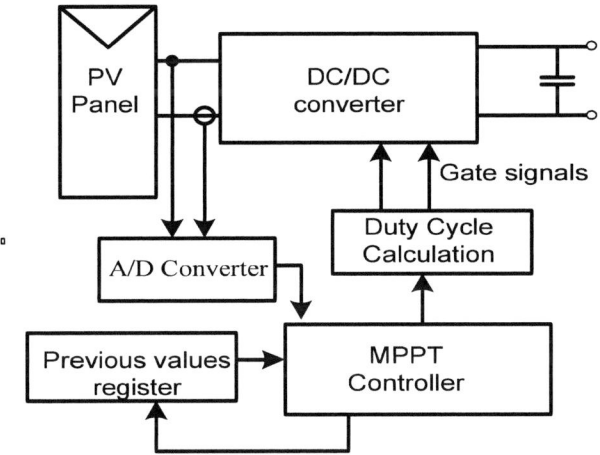

Fig. 5 Block diagram of control system

The galvanic isolation of the DC-DC converter is guaranteed by a high frequency transformer that is connected between the input stage and its output rectifier stage. The controller generates the PWM switching signal of the DC-DC power converter treated here, in which the frequency of gate pulse remain constant, while the pulse width only is varied.

TABLE 1
DESIGN SPECIFICATION AND CIRCUIT PARAMETERS.

Item	Symbol	Value
DC input voltage		40 [V]
Inverter switching frequency	fs	40 [kHz]
Switching period	Ts	25 [μs]
Primary side lumped leakage inductance of HF transformer	Ls	2 [μH]
Quasi resonant capacitors	C_1, C_2	0.2[μF]
DC smoothing filter capacitor	C_3, C_4	20[μF]
Inductance of dc reactor in load side	Lo	60[μH]
Equivalent load resistance	R	50 [Ω]
Maximum load current	Io	5.6 [A]
Turns ratio of hft	N_1:N_2	1:5

C. Measured voltage and current switching waveforms

The switching operating waveforms for voltage and current when the switch Q_1 is turned on and turned off are shown in Fig. 6 (a) and (b). Observing these operating waveforms, the switch Q_1 is turned on with ZCS and also turned off with ZVS. The switching waveforms for voltage and current when the switch Q_3 is turned on and turned off are shown respectively in Fig 6. (a) and (c). Observing the operating waveforms, the switch Q_3 is turned on with ZCS and is turned off with ZVS. However, at the turn-off transition for Q_1 and Q_3, only a few switching power losses still exists due to the inherent tail current characteristic of IGBTs.

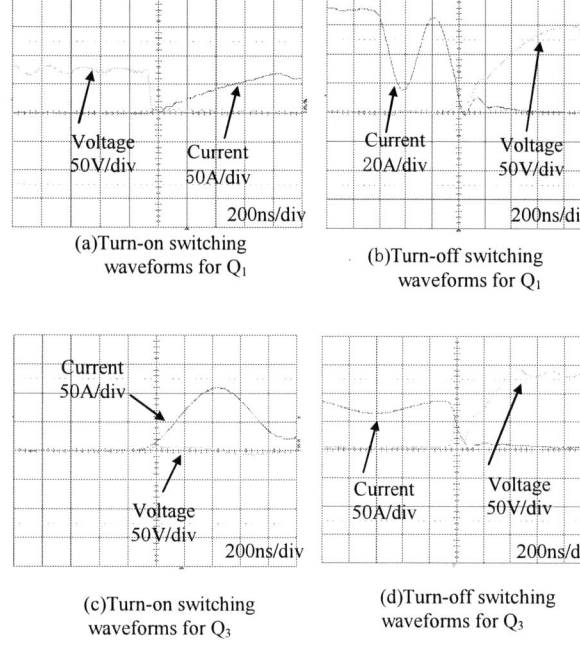

(a)Turn-on switching waveforms for Q_1

(b)Turn-off switching waveforms for Q_1

(c)Turn-on switching waveforms for Q_3

(d)Turn-off switching waveforms for Q_3

Fig. 6 Measured switching voltage and current waveforms for active power switches Q1 and Q3

V. CONCLUSIONS

A novel circuit topology of voltage source modified half-bridge soft switching PWM DC-DC power converter was presented. The operating principle, switching pattern and control strategy of the proposed converter were illustrated and discussed. The power loss analysis of the proposed converter was discussed and evaluated as compared with hard switching PWM DC-DC power converter. Under the simple circuit which has two additional power semiconductor switching devices and two passive circuit components to the typical half-bridge inverter circuit, all the active switching devices incorporated into half bridge inverter arms and DC busline input achieve ZVS turn-off and ZCS turn-on commutation. The more the switching frequency of inverter increases, the more this proposed converter has remarkable advantage as for the power conversion efficiency and power density as compared with the conventional hard switching inverter type DC-DC power converter. Lower conduction loss results in smaller heat sink size. In brief, based on higher efficiency, smaller size (low volume and weight), reduced cost, higher frequency ripples; the proposed converter is more suitable for PV system for high power applications.

REFERENCES

[1] S. Hamada, M. Nakaoka, "Saturable Inductor-Assisted ZVS-PWM Full-Bridge High-Frequency Link DC-DC Power Converter Operating and Conduction Losses", Proceedings of IEE-UK Int. Conf. on Power Electronics and Variable-Speed Drives, October 1994, pp.483-488.

[2] O. D .Patterson and D. M. Divan, "Pseudo-Resonant Full Bridge DC/DC Converter", Records of IEEE-PESC, pp.424-430, June, 1987.

[3] Robert L. Steigerwald, "A Comparison of Half-Bridge Resonant Converter Topologies", IEEE Tran. on Power Electronics, Vol. 3, No. 2, April 1988.

[4] Rajapandian Ayyanar and Ned Mohan, "A Novel Full-Bridge DC–DC Converter for Battery Charging Using Secondary-Side Control Combines Soft Switching Over the Full Load Range and Low Magnetics Requirement", IEEE Transactions on Industry Applications, Vol. 37, No. 2, 2001.

[5] M. Michihira, M. Nakaoka, " A Novel Quasi-Resonant DC-DC converter Using Phase –Shift Modulation in Secondary Side of High Frequency Transformer", Records of IEEE- PESC, Vol. 1, pp 100-105, June 1996.

[6] Kim, Y.J. Maruyama, Y. Nakaoka, M. " Practical evaluations of resonant pole-assisted ZVS-PWM DC/DC converter with series capacitor-connected transformer parasitic parallel resonant tank", IEEE PESC conf. Rec., pp. 795 – 802, 1993

[7] G. Hua, C. S. Leu, and F. C. Lee, "Novel zero-voltagetransition PWM converters," IEEE PESC Rec. 1993, pp. 55-61.

[8] Yuri Panov and Milan M. Jovanovic, "Design and Performance Evaluation of Low-Voltage/High-Current DC/DC On-Board Modules", IEEE Transactions on Power Electronics, Vol. 16, No. 1, pp:26-33, 2001.

[9] W. Chen, G. Hua, D. Sable, and F. C. Lee, "Design of high-efficiency,low-profile, low-voltage converter with integrated magnetics," in Proc. IEEE Appl. Power Electron. Conf., pp. 911–917, 1997.

[10] Hong Mao; Songquan Deng; Abu-Qahouq, J.; Batarseh, I.; "Active-clamp snubbers for isolated half-bridge DC-DC converters ", IEEE Transactions on Power Electronics, Vol. 20, No. 6, pp:1294 – 1302, 2005.

[11] Akshay K. Rathore, "Interleaved soft-switched active-clamped L–L type current-fed half-bridge DC–DC converter for fuel cell applications", international journal of hydrogen energy, 3 4, pp. 9802 – 9815, 2009.

[12] Q. Li and P. Wolfs, "A review of the single phase photovoltaic module integrated converter topologies with three different dc link configurations," IEEE Trans. Power Electron. May (2008), vol. 23, no. 3,1320–1333.

Modeling Method of Stray Magnetic Couplings in an EMC Filter for a SiC Solar Inverter

Takashi Masuzawa
DENSO CORPORATION
1-1, Showa-cho, Kariya-shi, Aichi, 448-8661, Japan
takashi_masuzawa@denso.co.jp

Eckart Hoene, Stefan Hoffmann, Klaus-Dieter Lang
Fraunhofer IZM
Gustav-Meyer-Allee 25, 13355, Berlin, Germany

Abstract— **This paper proposes a remarkably efficient modeling method of stray magnetic couplings in an Electromagnetic Compatibility (EMC) filter focusing on a dominant magnetic field in a power electronic device. The proposed modeling method is applied to a filter performance simulation of an EMC filter for a Silicon Carbide (SiC) solar inverter and its effectiveness is verified by comparing a measurement result and a simulation result. By using the proposed modeling method, the influence of the stray magnetic couplings on the filter performance can be well predicted and matched with the measurement result and the simulation result using the conventional modeling method. Accordingly, the number of the stray magnetic couplings required for the accurate prediction can be dramatically reduced from 325 to just one.**

Keywords—Electromagnetic compatibility, EMC filter, stray magnetic coupling, electromagnetic modeling.

I. INTRODUCTION

An EMC filter plays a key role to comply with the EMC standards. On the other hand, apart from an effect of EMC suppression, an EMC filter can lead to an additional space and cost. Thus, it has to be optimally designed. To realize an optimal filter design, stray magnetic couplings between components should be properly considered in addition to the self stray impedance of the components [1-3]. However, it is not practical to consider all stray magnetic couplings existing in a power electronic device. Because a product designer can not determine which part of an EMC filter has to be modified for better performance by using simulation with consideration of a huge number of stray magnetic couplings. Furthermore, at an early stage of product design process, a product designer has no detailed 3D geometry of a product. Therefore a reduction of the complexity of the modeling, namely an extraction of major couplings, is needed especially in a product design process. With respect to an extraction of major couplings, the idea to identify major couplings has been suggested in [4], but however, since the idea is empirically developed using a specific prototype, it is considerably ambiguous whether the idea is applicable to general EMC filter designs.

In this paper, a highly efficient modeling method of the stray magnetic couplings based on the simplification method [5] is proposed and applied to the filter performance simulation of the EMC filter for the SiC solar inverter [6]. A comparison between the measurement and the simulation is carried out to prove that the proposed modeling method can significantly reduce an effort to accurately predict influence of stray magnetic couplings.

II. SIMPLIFICATION METHOD OF MODELING OF STRAY MAGNETIC COUPLING

A. Complexity of considering stray magnetic coupling

Fig. 1 shows an EMC filter consisting of one coil, two capacitors and stray inductances. Even in this simple filter, there are 10 relevant stray inductances resulting in 45 stray magnetic couplings to be considered. The stray inductances originate from the PCB tracks, the connecting cables, the leads of components, the leakage magnetic flux from the coil, and so on. For example, the three stray inductances in the branch including the coil correspond to the PCB tracks connected to the both sides of the coil and the leakage inductance of the coil respectively. The stray inductance in the branch including the capacitor is the combined stray inductance of the capacitor and the PCB tracks connected to the capacitor.

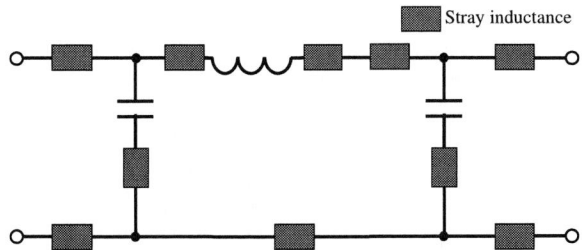

Fig. 1. EMC filter considering stray inductance.

B. Basic idea of simplification method

The EMC filter in Fig. 1 is assumed to be a part of the half-bridge circuit as shown in Fig. 2. Either of the capacitors in the EMC filter is the DC link capacitor, which is a part of the commutation cell. The noise current, the alternating current with high amplitude flowing in the commutation cell, produces a voltage drop along the DC

978-1-4799-2706-7/14 $31.00 © 2014 IEEE

link capacitor. And consequently, this voltage drop mainly produces a conducted noise which spreads to the power supply lines through the EMC filter.

Although most of the conducted noise directly propagates to the power supply, a considerable part of the conducted noise can also indirectly propagate via magnetic couplings. The influence of the indirect propagation will significantly increase, particularly when there is a large difference in amplitude between the conducted noise at the input and the output of the EMC filter. As a result, the filter performance of the EMC filter is severely deteriorated by the indirect propagation.

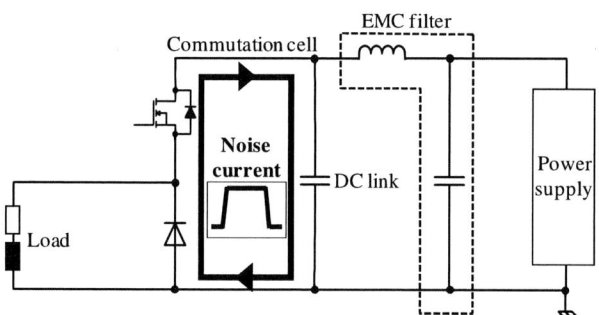

Fig. 2. Noise current in half-bridge circuit.

The simplification method we propose is based on identifying current loops with high amplitude and ones with low amplitude in the EMC filter. Current loops with high amplitude radiate strong magnetic flux which is picked up by current loops with low amplitude.

Fig. 3 illustrates the basic idea of the simplification method. The equivalent circuit of the half-bridge circuit is divided into three parts: the input loop, the high impedance area, and the output loop. The input loop can be, for example, the commutation cell in the half-bridge circuit composed of the current source I_{in}, the impedance of the current source Z_{in}, the impedance of the input capacitor (the DC link capacitor) Z_{Cin}, and the relevant stray inductances [7]. The current I_{in} generates the large magnetic flux Φ_{in}. And the output loop can be, for example, the current loop composed of the impedance of the output capacitor Z_{Cout}, the impedance of the power supply including connectors, cables and Line Impedance Stabilization Network (LISN) Z_{out}, and the relevant stray inductances.

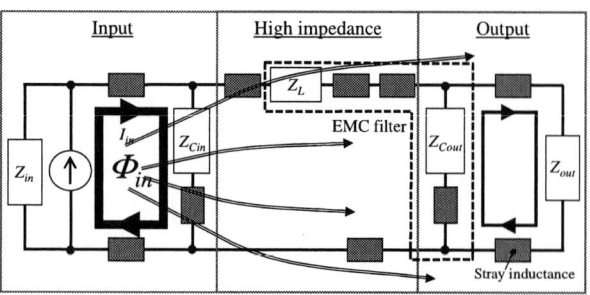

Fig. 3. Basic idea of simplification method.

The magnetic flux Φ_{in} generated from the input loop can cause a significant magnetic coupling between the input loop and the output loop. And also, since the high impedance components Z_L like a filter coil connect the input loop to the output loop, the high impedance area can be sensitive to a magnetic coupling with the input loop as well. An influence of a magnetic coupling can be theoretically defined as an induced voltage. In Fig. 4, e_{high} and e_{out} are the induced voltages caused by the magnetic couplings in the high impedance area and the output loop respectively.

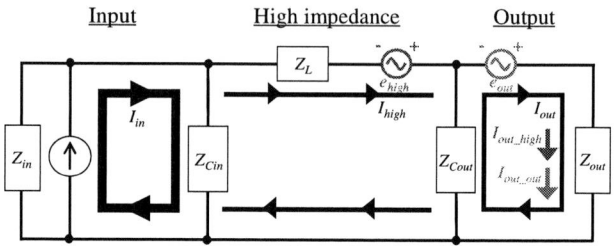

Fig. 4. Influence of magnetic coupling in EMC filter.

These induced voltages e_{high} and e_{out} cause the additional currents in the output loop I_{out_high} and I_{out_out}, which deteriorate the performance of the EMC filter. I_{out_high} and I_{out_out} are given by:

$$
\begin{aligned}
I_{out\_high} &= \frac{Z_{Cout}}{Z_{Cout} + Z_{out}} I_{high} \\[2mm]
&= \frac{Z_{Cout}}{Z_{Cout} + Z_{out}} \frac{e_{high}}{\dfrac{Z_{in} Z_{Cin}}{Z_{in} + Z_{Cin}} + Z_L + \dfrac{Z_{Cout} Z_{out}}{Z_{Cout} + Z_{out}}}
\end{aligned}
\tag{1}
$$

$$
\begin{aligned}
I_{out\_out} &= \frac{e_{out}}{(Z_{in} /\!/ Z_{Cin} + Z_L) /\!/ Z_{Cout} + Z_{out}} \\[2mm]
&= \frac{e_{out}}{\dfrac{\left(\dfrac{Z_{in} Z_{Cin}}{Z_{in} + Z_{Cin}} + Z_L\right) Z_{Cout}}{\dfrac{Z_{in} Z_{Cin}}{Z_{in} + Z_{Cin}} + Z_L + Z_{Cout}} + Z_{out}}
\end{aligned}
\tag{2}
$$

Where I_{out_high} is the current flowing in the output loop caused by e_{high}, I_{out_out} is the current flowing in the output loop caused by e_{out}, and I_{high} is the current flowing in the high impedance area caused by e_{high}.

Since the impedance of capacitors Z_{Cin} and Z_{Cout} and the impedance of the PCB tracks Z_{in} are generally much smaller than the impedance of the high impedance component Z_L, the following approximations can be applied.

$$\frac{Z_{in}Z_{Cin}}{Z_{in}+Z_{Cin}} + Z_L + \frac{Z_{Cout}Z_{out}}{Z_{Cout}+Z_{out}} \approx Z_L \qquad (3)$$

$$\left(\frac{Z_{in}Z_{Cin}}{Z_{in}+Z_{Cin}} + Z_L\right)Z_{Cout} \approx Z_L Z_{Cout} \qquad (4)$$

$$\frac{Z_{in}Z_{Cin}}{Z_{in}+Z_{Cin}} + Z_L + Z_{Cout} \approx Z_L \qquad (5)$$

With the above-mentioned approximations (3)-(5), I_{out_high} and I_{out_out} can be simplified as follows:

$$I_{out\_high} = \frac{Z_{Cout}}{Z_{Cout}+Z_{out}}\frac{e_{high}}{Z_L} \qquad (6)$$

$$I_{out\_out} = \frac{e_{out}}{Z_{Cout}+Z_{out}} \qquad (7)$$

The influence of e_{high} and e_{out} can be compared using a ratio between the additional currents in the output loop I_{out_high} and I_{out_out} expressed by (6) and (7). The ratio r_{Iout} is described as:

$$r_{Iout} = \left|\frac{I_{out\_high}}{I_{out\_out}}\right| = \left|\frac{Z_{Cout}}{Z_L}\right|\frac{e_{high}}{e_{out}} \qquad (8)$$

Unless e_{high} is exceptionally larger than e_{out}, the ratio r_{Iout} is basically much smaller than 1, because Z_L is much larger than Z_{Cout} from the viewpoint of a filter design; otherwise the EMC filter has no effect on noise attenuation. Based on this premise, the stray magnetic couplings in the EMC filter can be simplified as described in Fig. 5, where $M_{in\text{-}out}$ is the major stray magnetic coupling between the input loop and the output loop.

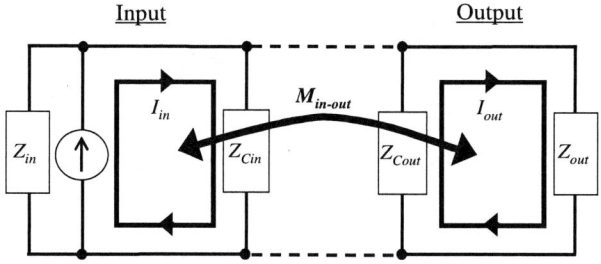

Fig. 5. Major stray magnetic coupling in EMC filter.

III. APPLICATION OF PROPOSED MODELING METHOD

In this chapter, to clarify its applicability and its problems to be solved for application to actual products, we apply the proposed modeling method to an EMC filter for a SiC solar inverter.

A. Tested EMC filter

Fig. 6 depicts the EMC filter for the SiC solar inverter used in the verification.

Fig. 6. Tested EMC filter for SiC solar inverter.

Fig. 7 shows the circuit schematic of the tested EMC filter used for the three-phase grid connection. The filter is composed of the three output inductors L_{out}, the two common mode choke coils L_{CM}, the nine X-capacitors C_X, and the two Y-capacitors C_Y. This EMC filter has 26 stray inductances resulting in 325 stray magnetic couplings.

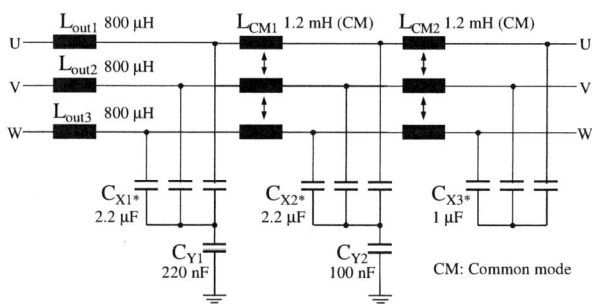

Fig. 7. Circuit schematic of tested EMC filter.

B. Consideration of magnetic material

Since partial impedance is needed for the filter performance simulation of the tested EMC filter, we use the Partial Element Equivalent Circuit (PEEC) method software FastHenry [8].

To estimate an accurate stray magnetic coupling by using simulation, influences of permeability of the magnetic core should be properly considered. However, most of the commercial software including the PEEC software does not consider permeability; simulations considering permeability can lead to a tremendous increase of computational time, even if it is possible with specific software. Thus, in simulation, we have to consider permeability of a core in a simple way only where necessary.

In [9, 10], it is concluded that leakage magnetic flux from a typical common mode choke coil generated by differential mode current is nearly equal to that from a solenoid coil. That is because the magnetic fluxes inside the core, generated by the currents in both of the windings, repel each other and flow outside of the core as shown in Fig. 8. Fig. 9 illustrates the idea of the approximation. The winding of the coil can be described as the solenoid coil with the same effective magnetic length l and the same cross section A, where l is given by:

$$l = 2\pi r \frac{\theta}{360°} \tag{9}$$

Where θ is the winding coverage angle and r is the radius of the coil.

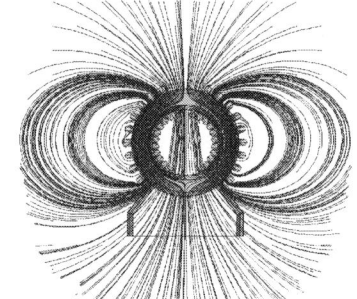

Fig. 8. Magnetic flux line by differential mode current.

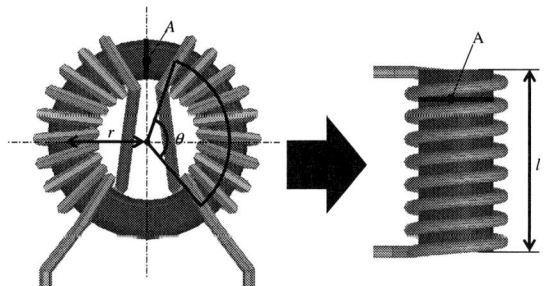

Fig. 9. Idea of approximation of leakage inductance.

Based on the approximation, the effective permeability μ_{eff} related to the leakage magnetic flux from the coil is described as follows:

$$\mu_{eff} = 2.5\left(\sqrt{\frac{\pi}{A}}\frac{l}{2}\right)^{1.45} \tag{10}$$

Whereas it is assumed that the leakage magnetic flux from the coil generated by common mode current is not significantly influenced by the core, because the magnetic fluxes inside the core do not repel each other. Similarly to the common mode choke coil, it is also assumed that the leakage magnetic flux from the output inductor is not significantly influenced by the core for the same reason.

To validate the assumption, a distribution of magnetic flux from the output inductor is compared between ones with and without the core by means of 3D Finite Element Method (FEM) simulation software ANSYS HFSS [11].

The output inductor is built using a Hitachi Metals AMCC-40 core with an air gap of 1.1 mm. The number of turns of the winding is 48. The horizontal and vertical wire sizes of the winding are 6 mm and 2 mm respectively. Fig. 10 shows the simulation model. The input current is 1 A, and the simulating frequency is 1 MHz.

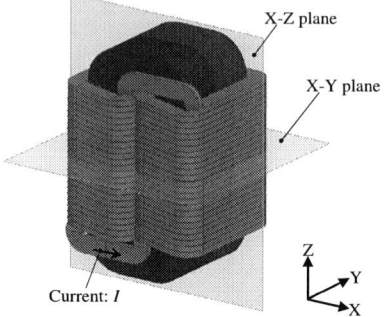

Fig. 10. Model for simulation of magnetic flux distribution.

Fig. 11 presents the comparison of the simulated magnetic flux distribution on X-Y and X-Z planes between the output inductors with and without the core.

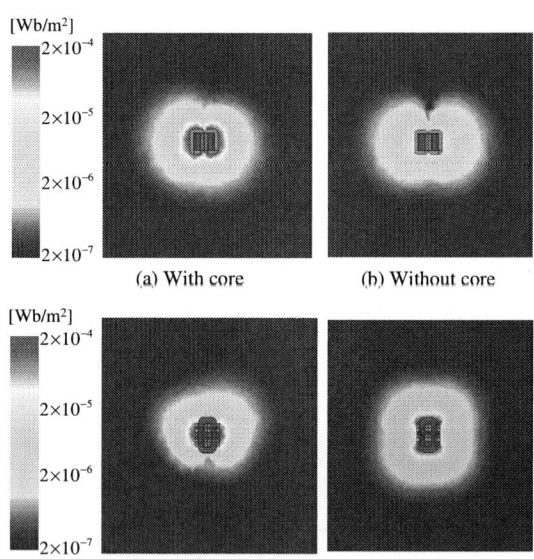

Fig. 11. Distribution of magnetic flux: X-Y plane in (a) and (b); X-Z plane in (c) and (d).

In Fig. 11, there is no significant difference in the distribution of the magnetic flux between the simulation results with the core and without the core. Moreover the common mode choke coil is not the main part of the output loop. Thus we reach a conclusion that it is not necessary to consider μ_{eff} of the output inductor and the common mode choke coils for the filter performance simulation conducted in the next chapter. On the other hand, it is highly likely that μ_{eff} needs to be considered by using the above mentioned approximation for the simulation in differential mode. In future work, the influence of μ_{eff} needs to be investigated in more detail.

978-1-4799-2706-7/14 $31.00 © 2014 IEEE   2369

C. Applicability to the tested EMC filter

As stated in the preceding chapter, a major stray magnetic coupling basically occurs between an input loop and an output loop in an EMC filter. In this section, the applicability of the proposed modeling method to the tested EMC filter is verified.

In Fig. 4, the induced voltage in the output loop e_{out} is generated by the stray magnetic coupling between the input loop and the output loop $M_{in\text{-}out}$. And similarly to e_{out}, the induced voltage in the high impedance area e_{high} is dominantly generated by the stray magnetic coupling between the input loop and the filter coil $M_{in\text{-}L}$. Hence e_{out} and e_{high} can be described as follows:

$$e_{out} = M_{in\text{-}out} \frac{dI_{in}}{dt} \tag{11}$$

$$e_{high} \approx M_{in\text{-}L} \frac{dI_{in}}{dt} \tag{12}$$

By using (11) and (12), the ratio of the output current r_{Iout} is given by:

$$r_{Iout} = \left| \frac{I_{out\_high}}{I_{out\_out}} \right| = \left| \frac{Z_{Cout}}{Z_L} \right| \left| \frac{e_{high}}{e_{out}} \right| = \left| \frac{Z_{Cout}}{Z_L} \right| \left| \frac{M_{in\text{-}L}}{M_{in\text{-}out}} \right| \tag{13}$$

Fig. 12 presents calculated r_{Iout} using (13), where the simulated values of $M_{in\text{-}out}$ and $M_{in\text{-}L}$ are 3.2 nH and 0.25 nH respectively.

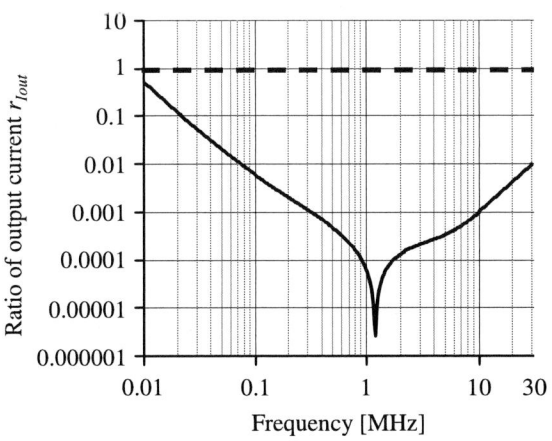

Fig. 12. Calculated ratio of output current r_{Iout}.

It can be seen that r_{Iout} is much smaller than 1 in the frequency range from 0.05 MHz to 30 MHz. Therefore it is concluded that the influence of the induced voltage in the high impedance area e_{high} is negligible in this frequency range. This fact in Fig. 12 corroborates that the major stray magnetic coupling in the tested EMC filter is $M_{in\text{-}out}$.

IV. COMPARISON OF FILTER PERFORMANCE

To verify the effectiveness of the proposed modeling method, the filter performance of the EMC filter is compared between a measurement and a simulation.

Fig. 13 depicts a system configuration for a measurement of a filter performance P_f in common mode. In this measurement, P_f is defined as the ratio of the output voltage A to the reference voltage Ref, measured by means of a Gain-Phase analyzer: Agilent 4395A. The range of measuring frequency is set from 0.01 MHz to 30 MHz. The nanocrystalline cores in Fig. 13 are used to suppress common mode current flowing through the Gain-Phase analyzer.

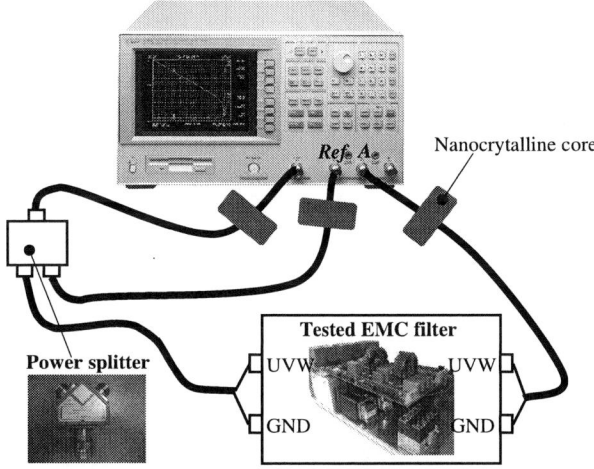

Fig. 13. System configuration for measurement of P_f.

Fig. 14 shows the circuit simulation model with the highlighted input loop, output loop, and major stray magnetic coupling $M_{in\text{-}out}$. The software used for the circuit simulation is Portunus [12].

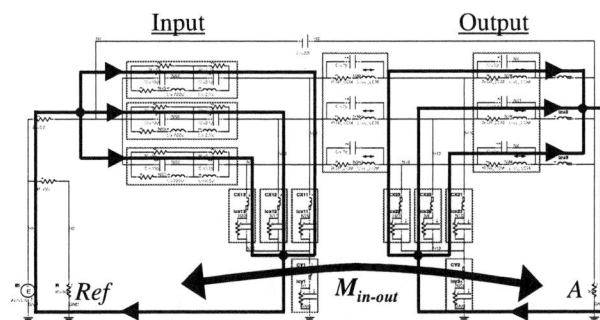

Fig. 14. Circuit simulation model with the highlighted input loop, output loop, and major stray magnetic coupling

The major stray magnetic coupling incorporated into the circuit simulation model described in Fig. 14 is obtained from the 3D geometry by using the PEEC method software. Fig. 15 depicts the 3D simulation model including the output inductor and the common mode choke coil. The inductor and the coil have no core based on the conclusion in the preceding chapter.

The 2014 International Power Electronics Conference

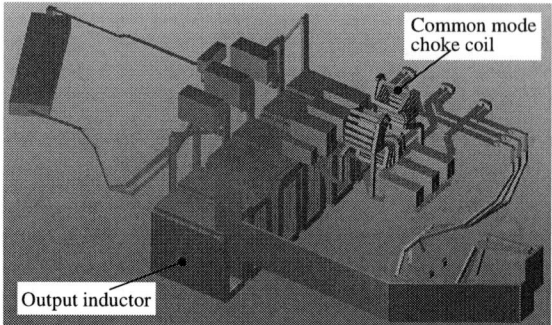

Fig. 15. Simulation model of tested EMC filter

Fig. 16 shows a comparison of the filter performance P_f in common mode between the measurement and the simulation with the following three conditions: considering the classical stray impedances (ESR and ESL of a capacitor and EPR and EPC of a coil), considering the classical stray impedances and all the stray magnetic couplings (325 couplings), considering the classical stray impedances and only the major stray magnetic couplings (1 coupling) derived with the proposed modeling method. Where ESR is an equivalent series resistance, ESL is an equivalent series inductance, EPR is an equivalent parallel resistance, and EPC is an equivalent parallel capacitance.

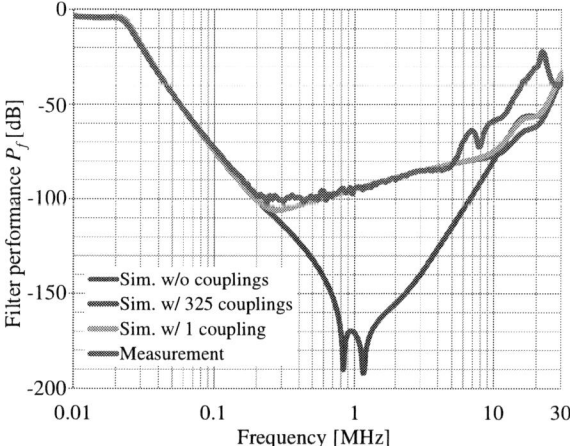

Fig. 16. Comparison of P_f between simulation and measurement.

In Fig. 16, the measurement result and the simulation results considering all the couplings and only the major coupling show a good agreement in the frequency range up to 10 MHz. On the other hand, the simulation without consideration of the stray magnetic coupling inaccurately predicts the filter performance with the difference of more than 100 dB compared to the measurement result. Most importantly, the simulation using the proposed modeling method realizes a comparably accurate prediction in comparison with the simulation using the conventional modeling method, even though the number of the considered stray magnetic couplings is dramatically reduced from 325 to just one. Regarding the difference in the frequency range higher than 10 MHz, it

is assumed that it is caused by the following three elements: 1) the connections between the shielded cables and the tested EMC filter forming a small loop, which can be affected by magnetic flux, 2) the residual common mode current flowing through the measuring instrument, and 3) the changed distribution of magnetic flux of the coils due to the stray capacitances over their self resonance frequencies [13].

V. CONCLUSION

In this paper, the highly efficient modeling method of stray magnetic couplings based on the simplification method was proposed. And its effectiveness was verified by comparing the performance of the EMC filter for the SiC solar inverter, between the measurement and the simulation. Moreover, the simple approximation to consider permeability of a core and its necessity in differential mode and common mode were also outlined. The results of the comparison showed that the proposed modeling method could significantly reduce the number of stray magnetic couplings considered for an accurate simulation.

REFERENCES

[1] E. Hoene, A. Lissner, S. Weber, S. Guttowski, W. John, and H. Reichl, "Simulating Electromagnetic Interactions in High Power Density Converters," in Proc. IEEE 36th Power Electron. Spec. Conf., pp.1665-1670, 2005.

[2] T. De Oliveira, J. Guichon, J. Schanen, and L. Gerbaud, "PEEC-Models for EMC filter layout optimization," in Integrated Power Electronics Systems (CIPS), paper 13.3, 2010.

[3] I. Kovacevic, A. Muesing, T. Friedli, and J. W. Kolar, "Electromagnetic modeling of EMI input filters," in Integrated Power Electronics Systems (CIPS), paper 02.1, 2012.

[4] S. Wang, F.C. Lee, D. Chen, and W. G. Odendaal, "Effects of Parasitic Parameters on EMI Filter Performance," IEEE Trans. Power Electron., vol.19, no.3, pp.869-877, 2004.

[5] T. Masuzawa, and E. Hoene, "A simplification of a modeling of stray magnetic couplings in EMC filters," in PCIM Europe 2013, pp. 779-786, 2013.

[6] S. Hoffmann, E. Hoene, and O. Zeiter, "Electrical, thermal and electromagnetic design of a SiC solar inverter: a case study," in PCIM Europe 2013, pp. 470-477, 2013.

[7] E. Hoene, W. John and H. Reichl, "Simulation of Conducted Electromagnetic Interference of Inverter-Fed Induction Motor," in Proc. EPE'99, 1999.

[8] M. Kamon, M.J. Tsuk, and J.K. White, "FASTHENRY: a multipole-accelerated 3-D inductance extraction program," IEEE Trans. Microw. Theory Tech., vol. 42, no. 9, pp. 1750-1758, 1994.

[9] M. Nave, Power Line Filter Design for Switched-Mode Power Supplies. New York, NY, USA, Springer, 1991.

[10] S. Weber, "Effizienter Entwurf von EMV-Filtern für leistungselektronische Geräte unter Anwendung der Methode der partiellen Elemente," PhD dissertation at Technical University of Berlin, 2007.

[11] ANSYS HFSS, 2014. [Online]. Available: http://www.ansys.com/Products/Simulation+Technology/Electronics/Signal+Integrity/ANSYS+HFSS

[12] Portunus, 2014. [Online]. Available: http://www.adapted-solutions.com/web/AdaptedSolutionsEnglish/ASProductPortunus.html

[13] R. Wang, H. Blanchette, M. Mu, D. Boroyevich, and P. Mattavelli, "Influence of high-frequency near-field coupling between magnetic components on EMI filter design," IEEE Trans. Power Electron., vol.28, no.10, pp.4568-4579, 2013.

978-1-4799-2706-7/14 $31.00 © 2014 IEEE

DC Bus Voltage EMI Mitigation in Three-Phase Active Rectifiers Using a Virtual Neutral Filter

S. G. Parker   D. S. Segaran   D. G. Holmes   B. P. McGrath

School of Electrical and Computer Engineering
RMIT University, Melbourne, Australia

stewart.parker@student.rmit.edu.au
grahame.holmes@rmit.edu.au

dinesh.segaran@ieee.org
brendan.mcgrath@rmit.edu.au

Abstract — **Grid connected PWM inverters/rectifiers inevitably produce a high frequency switched common mode voltage on their DC bus, which can cause substantial EMI problems. The mitigation strategy proposed in this paper is to couple the Y point of the LCL filter capacitors to the DC bus mid-point to connect a virtual neutral to the DC bus, which significantly reduces the high frequency common mode voltages on the bus. However, additional common mode inductances and capacitances are required for this filter, to limit high frequency circulating currents and define the EMI filter breakpoint. Also, the converter current regulator must be split into differential and common mode sections, to eliminate phase coupling between the phase leg active damping current regulators, and to support damping of the common mode EMI filter resonance. The concept is verified by simulations and experimental results.**

Keywords — *EMI, Common Mode, Virtual Neutral, Active Rectifier.*

I.   INTRODUCTION

Three phase active power electronic rectifiers are becoming ever more popular in grid connected applications. However, the PWM switching of these rectifiers produces a high frequency "6 step" common mode voltage on the DC bus with respect to earth, with a rapid dv/dt at each switching transition. This transition causes EMI currents to circulate through parasitic capacitances within the rectifier structure, which can interfere with either external equipment connected to the DC bus or with internal control circuitry within the rectifier. The problem is analogous to common mode voltages that occur with AC motor drive voltage source inverters [1]-[4], and presents a similar level of hazard to reliable converter operation.

The standard approach for EMI mitigation is to include both Differential Mode (DM) and Common Mode (CM) filtering in the conduction path between the switched phase legs of the rectifier, and the external grid connection [5]. The DM filters (i.e. the primary LCL components) reduce the phase leg PWM switching frequency harmonics to acceptable levels, while the CM filters reduce the CM EMI that is injected into the grid. However, neither of these filters have any influence on the CM switched voltage on the DC bus, and thus they do not help to mitigate CM EMI currents that flow from the DC bus components to earth via parasitic capacitances. Hence an additional filter is required to manage this interference.

One approach that has been proposed is to connect a set of star connected capacitors to the rectifier output, and tie their common star point to the DC bus midpoint [5][6]. This clamps a high frequency virtual neutral to the DC bus, which significantly reduces the DC link CM voltage. However, the capacitors must be appropriately sized to limit circulating current, which is challenging if it is desired for convenience to use the capacitors of the primary LCL filter to create the virtual neutral point. In addition, the virtual neutral connection can create an additional LC resonance, which can cause further EMI problems, and can even interfere with the damping operation of the primary rectifier current regulators.

This paper presents an analysis and design methodology to integrate a virtual neutral DC bus clamping filter into a grid connected rectifier. The analysis provides a basis for filter design in terms of harmonic voltages which should be suppressed and identifies the placement of the filter break point to minimise harmonic excitation. The implications of the additional current path on the primary current regulator are then investigated, and a strategy is proposed to eliminate this interference by separating the current regulator into two parts – one to manage the differential LCL filter current, and the second to damp the LC common mode circulating current caused by the virtual neutral filter connection. Simulation and experimental results are presented to validate the theory, filter design and proposed CM active damping current controller.

II.   DC BUS COMMON MODE VOLTAGE IN THREE PHASE GRID CONNECTED CONVERTERS

A.   LCL Filtered Converter Response without CM filter

Fig 1(a) shows the two level Voltage Source Inverter (VSI) topology that is commonly used used for three-phase grid-connected applications. The converter has an LCL differential mode line filter consisting of L_1, C_f and L_2, which provides superior switching frequency harmonic filtering for reduced values of filter components [7]. Note that this topology is also appropriate for an LC filtered converter, if L_2 is understood to represent the external grid impedance. The figure also shows a number of typical parasitic capacitances that couple various sections of the converter physical structure to earth – it is these capacitances that provide a path to earth for EMI currents

978-1-4799-2706-7/14 $31.00 © 2014 IEEE

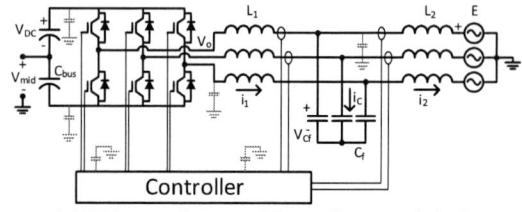

(a) With exemplary parasitic coupling to earth (red).

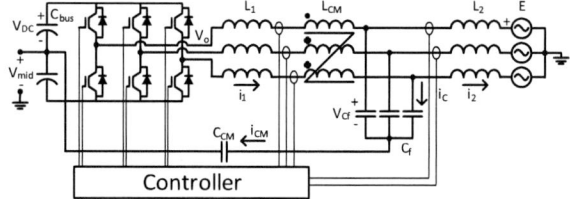

(b) With virtual neutral common mode filter.

Fig. 1: Grid connected converter with LCL filter.

created by CM switching voltages that may develop at various node points within the converter.

Fig. 2(a) shows the ideal CM voltage with respect to neutral/earth that develops at the midpoint of the DC bus. This waveform was developed using simulation without any consideration of parasitic capacitances from the DC bus to earth, and the expected "6-step" waveform with very fast dv/dt transitions can be clearly seen. The peak-to-peak voltage swing of this waveform is $2V_{DC}$, so its potential EMI circulating current influence is quite high.

Fig. 2(c) shows the experimentally measured DC midpoint CM voltage, which is distorted from its ideal counterpoint because of the filtering effect of the parasitic capacitances between various components in the converter and earth.

B. Common Mode DC Bus Harmonic Source Voltage

To develop an effective CM filter it is necessary to identify the CM harmonics that are present on the DC Bus. This can be done by establishing KVL loops (in Laplace form) for each phase of the system without parasitic capacitances, as per Fig. 1(a), to give

$$V_{mid}(s) + V_{DC}S_x(s) = sL_1i_{1x}(s) + sL_2i_{2x}(s) + E_x(s) \quad (1)$$

where $x \in \{a, b, c\}$ represents each phase and the switching function $S_x(s) \in \{1, -1\}$ defines each phase leg switch state. The DC bus capacitance is sufficiently large to be assumed to be a stiff voltage source. For a three-wire three-phase system, the grid phase voltages sum to zero, and the converter, capacitor and grid currents also sum to zero since there is no common mode return path for this circuit. Summing (1) across three phases with these constraints gives

$$V_{mid}(s) = V_{CM}(s) = \frac{-V_{DC}}{3}\{S_a(s) + S_b(s) + S_c(s)\} \quad (2)$$

where $V_{CM}(s)$ must equal the DC bus mid point voltage because there is no path for common mode current to flow from the DC bus to earth unless parasitic capacitances between the two are included into the model.

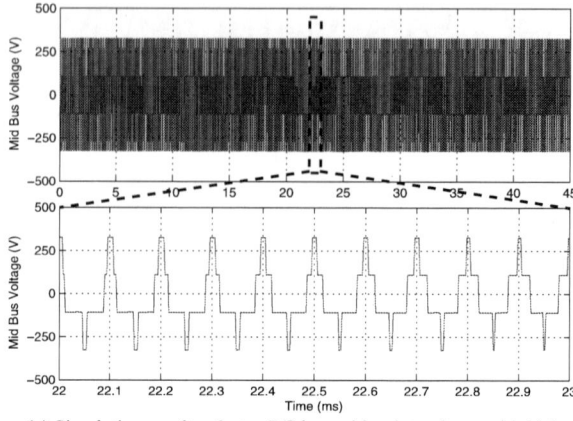

(a) Simulation result – 6-step DC bus mid-point voltage with high dv/dt.

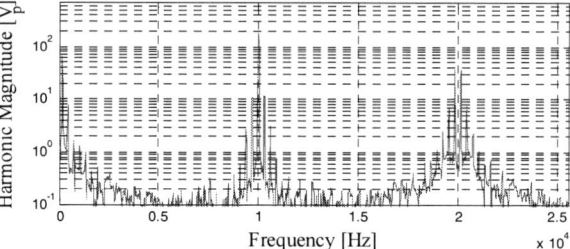

(b) Simulated DC bus mid-point voltage V_{mid} spectrum

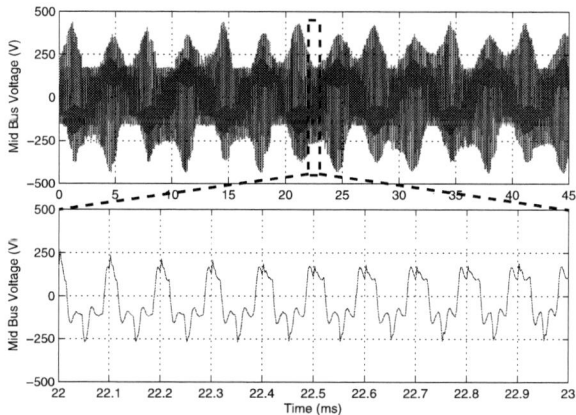

(c) Experimental result – parasitic coupling generates filtering effect on DC bus midpoint voltage.

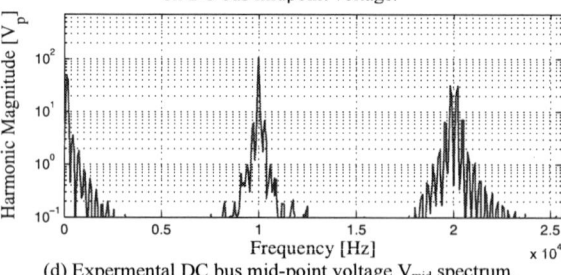

(d) Experimental DC bus mid-point voltage V_{mid} spectrum.

Fig. 2: Converter waveforms without common mode filter showing the high dv/dt voltages present on the DC bus with respect to earth.

The phase leg switching functions can now be replaced with their harmonic summation equivalents, derived using the well known double Fourier series [8]

$$S_x(t) = \frac{A_{00}}{2} + \sum_{n=1}^{\infty}[A_{0n}\cos(n[\omega_o t + \phi_x])]$$

$$\qquad\qquad\qquad(3)$$

$$+ \sum_{m=1}^{\infty}\sum_{n=-\infty}^{\infty}[A_{mn}\cos(m\omega_c t + n[\omega_o t + \phi_x])]$$

where $x \in \{a,b,c\}$, and $\phi_x \in \{0, -2\pi/3, +2\pi/3\}$.

For asymmetrical regular sampled two level modulation of the VSI (with or without third harmonic injection), the non-zero harmonics for each phase leg reduce to:

- Odd baseband harmonics - $m = 0, n \in \{1,3,5\ldots\}$

- Carrier and even sideband harmonics for each odd carrier multiple group - $m = \{1,3,5\ldots\}, n \in \{0, \pm 2, \pm 4\ldots\}$

- odd sideband harmonics for each even carrier multiple group - $m = \{2,4,6\ldots\}, n \in \{\pm 1, \pm 3, \pm 5\ldots\}$

Summing these harmonics across all three phases as per (2) and retaining only those harmonics with any significant magnitude, identifies V_{CM} as the summation of the triplen baseband harmonics of $m = 0, n \in \{3,9\}$ and the triplen carrier and sideband harmonics of $m = \{1,3,5\ldots\}, n \in \{0, \pm 6\}$ and $m = \{2,4,6\ldots\}, n \in \{\pm 3\}$. Fig. 2(b) shows the spectrum of the simulated midpoint voltage of Fig. 2(a), where these significant harmonic components can be readily identified.

In a modelling sense, these common mode harmonics can be represented as a combined harmonic voltage source that connects the DC bus midpoint through an AC side common mode impedance that has a magnitude of one third of the phase leg filter impedances, to the grid supply neutral (earth), as shown in Fig. 3. Note that the LCL filter capacitor neutral point connection is unconnected in this representation, since it is a floating connection between the three phase legs with no path back to earth. From this model, the variation between simulation and experiment for the DC bus midpoint common mode voltage as shown in Fig. 2 is immediately explicable.

For the simulation system, without a parasitic impedance Z_{para} to earth, no common mode current can flow through the AC side CM impedance, and so the AC side of the CM harmonic voltage source stays at earth potential. Consequently the DC bus midpoint V_{CM} waveform w.r.t earth is exactly the sum of the common mode harmonics, i.e. the six-step waveform shown in Fig. 2(a).

For the experimental system, the parasitic impedance Z_{para} creates an impedance divider with the AC side filter impedance. Hence while the DC bus midpoint voltage still has an identifiable six-step response at the switching frequency, as can be seen in Fig 2(c), there are significant harmonic magnitude reductions because of the impedance divider effect. Nevertheless the harmonic spectrum is essentially unchanged, as shown in Fig 2(d).

From this understanding, it becomes clear is that the DC bus common mode harmonic voltages can only be mitigated by providing a low impedance path between the bus midpoint and ground at the common mode harmonic frequencies. However, note that with this DC side connection, the AC side of the harmonic source generator will fluctuate w.r.t earth instead, and this will in turn cause harmonic currents to flow through the AC side filter impedance between the harmonic sources and the grid earth. These harmonic currents have to be managed to achieve an overall satisfactory filtering solution.

I. DYNAMIC MODEL OF VIRTUAL NEUTRAL COMMON MODE FILTER

The approach proposed in this paper derives from [6], and connects an additional capacitor C_{CM} between the LCL filter capacitor star point (a virtual neutral) and the DC bus midpoint, as shown in Fig. 1(b). This creates a harmonic low impedance path from the DC bus midpoint to a virtual neutral point, which limits the high frequency harmonic excusions of the DC bus w.r.t. this node, to low magnitudes. An additional common mode inductor L_{CM} is then added to the converter output inductor path. L_{CM} is sized to limit the circulating current around the virtual neutral circuit, while C_{CM} allows the resonant frequency of the filter circuit to be set to avoid the CM baseband harmonics, while still achieving good harmonic reduction at the PWM switching frequencies. Fig. 4 shows a simplified circuit representation of the differential mode filter circuit for each phase leg of the VSI with this arrangement, and the corresponding common mode equivalent circuit. Note that the common mode inductor does not feature in the differential mode equivalent circuit, since it does not introduce a significant voltage drop into this circuit loop.

The virtual neutral connection can again be analysed with KVL, except that there are now two loops to be considered. <u>Loop #1</u> starts at the DC bus midpoint, follows through the switched phase legs to the converter

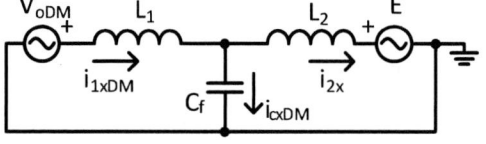

(a) Per phase differential mode circuit.

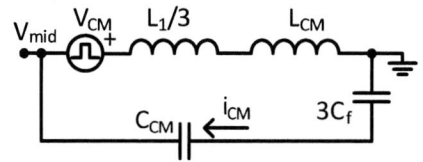

(b) Simplified common mode circuit with mid-point connection.

Fig. 4: Equivalent circuits of grid converter.

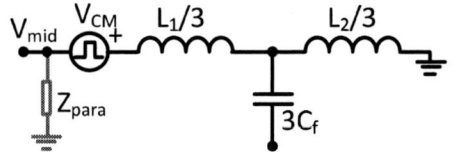

Fig. 3: Simplified CM representation of VSI switching harmonics.

978-1-4799-2706-7/14 $31.00 © 2014 IEEE   2374

side differential and common mode filters, then to the grid side differential mode filter and back through the grid side source voltage. Its KVL loop is given by

$$V_{mid}(s) + V_{DC}S_x(s) = sL_1 i_{1x}(s)$$
$$+ sL_{CM}\{i_{1a}(s) + i_{1b}(s) + i_{1c}(s)\} \quad (4)$$
$$+ sL_2 i_{2x}(s) + E_x(s)$$

Loop #2 flows directly through the switched phase legs to the converter side differential and common mode filter inductors, and immediately returns through the LCL filter capacitor and the common capacitor C_{CM} back to the DC bus midpoint. Its KVL loop is given by

$$V_{DC}S_x(s) = sL_1 i_{1x}(s)$$
$$+ sL_{CM}\{i_{1a}(s) + i_{1b}(s) + i_{1c}(s)\} \quad (5)$$
$$+ \frac{1}{sC_f}i_{cx}(s) + \frac{1}{sC_{CM}}i_{CM}(s)$$

Note that both these loops now have phase current cross-coupling, caused by the common mode inductor.

The CM filtering properties and overall dynamics of these circuits can be analysed by separating (4) and (5) into DM and CM components and considering each mode in turn. The switching functions can be separated, viz:

$$S_x(s) = S_{CM}(s) + S_{xDM}(s) \quad (6)$$

Next, the converter and capacitor currents must be separated, recognizing that each phase carries a third of the total common mode current, viz:

$$i_{1x}(s) = i_{1xDM} + i_{CM}/3 \quad (7)$$
$$i_{cx}(s) = i_{cxDM} + i_{CM}/3 \quad (8)$$

Substituting (6), (7) and (8) into (4) and (5) and separating the DM components gives:

$$V_{oxDM}(s) = sL_1 i_{1xDM}(s) + sL_2 i_{2x}(s) + E_x(s) \quad (9)$$

$$V_{oxDM}(s) = sL_1 i_{1xDM}(s) + \frac{1}{sC_{Cf}}i_{cxDM}(s) \quad (10)$$

Similarly separating the CM components gives:

$$V_{mid}(s) + V_{CM}(s) = s\left(\frac{L_1}{3} + L_{CM}\right)i_{CM}(s) \quad (11)$$

$$V_{CM}(s) = s\left(\frac{L_1}{3} + L_{CM}\right)i_{CM}(s) + \frac{3C_f + C_{CM}}{3sC_f C_{CM}}i_{CM}(s) \quad (12)$$

i_{cxDM} and i_{1xDM} represent the differential mode capacitor and converter currents respectively. Also note that $V_{DC}S_{CM}(s) = V_{CM}(s)$ and $V_{DC}S_{xDM}(s) = V_{oxDM}(s)$, being the CM and DM output voltages respectively.

Upon inspection it is apparent that (9) and (10) describe the dynamics of the standard LCL per-phase DM circuit shown in Fig. 4(a). Equally, (11) and (12) describe the dynamics of the CM circuit shown in Fig. 4(b).

A further result for the CM filtering, relating $V_{mid}(s)$ to $V_{CM}(s)$ can be derived by combining (11) and (12), to give:

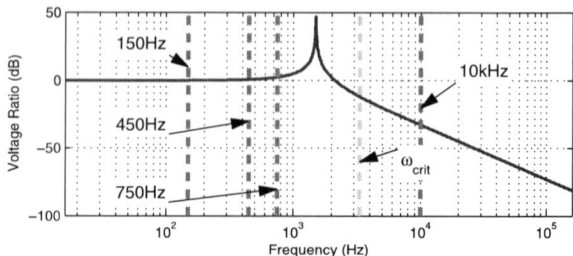

Fig. 5: Frequency response behavior of virtual neutral showing significant common mode voltage harmonics and critical active damping frequency.

$$V_{mid}(s) = \frac{\omega_{resCM}^2}{s^2 + \omega_{resCM}^2}V_{CM}(s) \quad (13)$$

where the CM resonant frequency is

$$\omega_{resCM} = \sqrt{\frac{L_1/3 + L_{CM}}{(L_1/3 + L_{CM})(3C_f C_{CM})/(3C_f + C_{CM})}}.$$

Fig. 5 shows the frequency response relationship between V_{CM} and V_{mid} defined by (13). Note in particular the resonant peak of this response, which is very similar to that of the primary LCL filter. Hence it is likely to similarly resonate unless it is appropriately damped [6]. An active damping approach to resolve this problem would be attractive if it is achievable, as will now be explored.

II. IMPLICATIONS ON CURRENT CONTROLLER DESIGN

A. Differential Mode Controller with Active Damping

Current regulators for an inverter with a LCL filter usually include an active damping loop to control the filter resonance. The feedback variable for this loop is often the filter capacitor current [7], which of course will now contain a CM current component because of the virtual neutral filter connection. This will interact with and degrade the closed loop regulator response for each phase leg unless this CM current is subtracted from the measured capacitor current, i.e. rearranging (8) gives:

$$i_{cxDM}(s) = i_{cx}(s) - i_{CM}(s)/3 \quad (14)$$

Note that in practice the capacitor current is usually indirectly calculated from the converter side and grid side measured currents, but this does not change the need for the CM current subtraction.

With this decoupled capacitor current, each phase leg current can be regulated using the established Proportional plus Resonant (PR) LCL current regulator structure shown in Fig 6(a), with gains calculated according to [7]. The LCL system of Fig. 4(a) can be represented in discrete time using appropriate discrete transformations [7] such as the following equations:

$$G_{icDM}(z) = \frac{\sin(\omega_{resDM}T)}{\omega_{resDM}L_1} \times \frac{z-1}{z^2 - 2z\cos(\omega_{resDM}T) + 1} \quad (15)$$

$$\frac{i_2(z)}{i_{cDM}(z)} = \frac{\gamma_{LCDM}^2 T^2 z}{(z-1)^2} \quad (16)$$

978-1-4799-2706-7/14 $31.00 © 2014 IEEE   2375

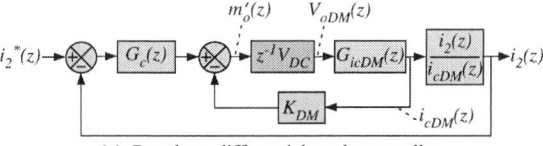

(a) Per phase differential mode controller.

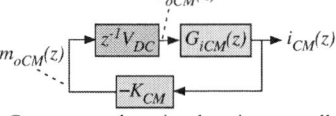

(b) Common mode active damping controller.

Fig. 6: Differential mode and common mode controller structures.

where $\omega_{resDM} = \sqrt{\dfrac{L_1 + L_2}{L_1 L_2 C_f}}$, $\gamma_{LCDM} = \sqrt{\dfrac{1}{L_2 C_f}}$ and $T = 1/2f_{sw}$.

The inverter is represented as a gain and unit delay $z^{-1}V_{DC}$. In order to regulate the fundamental current component a PR controller is incorporated, as [7]:

$$G_c(z) = K_p\left(1 + \frac{1}{T_r}\frac{\sin(\omega_0 T)}{2\omega_0}\frac{z^2 - 1}{(z^2 - 2z\cos(\omega_0 T) + 1)}\right) \quad (17)$$

The PR controller gains are now selected to provide the greatest controller bandwidth, using the equations provided in [7], viz:

$$K_p = \frac{\omega_c(L_1 + L_2)}{V_{DC}} \qquad T_r = 10/\omega_c \quad (18)$$

where $\omega_c = 0.36\omega_{resDM}$. The active damping gain is set by finding the best closed loop pole location, given the system characteristic equation:

$$z + K_{DM}V_{DC}G_{icDM}(z) \\ + K_pV_{DC}G_{icDM}(z)[i_2(z)/i_{cDM}(z)] = 0 \quad (19)$$

as K_{DM} varies. This root locus and the selected gain are shown in Fig. 7(a), for the system parameters given in Table I, with the resultant gains listed in Table II.

B. Common Mode Active Damping Control

A transfer function for the common mode current flowing around the virtual neutral filter circuit shown in Fig. 4(b), is now required to analyse common mode current regulation. This can be derived as:

$$i_{CM}(s) = \frac{1}{L_1/3 + L_{CM}}\frac{s}{s^2 + \omega_{resCM}^2}V_{CM}(s) \quad (20)$$

Since this system is a digital sampled system, the ZOH transformation is used to transform (20) into the discrete time domain, viz:

$$G_{iCM}(z) = \frac{\sin(\omega_{resCM}T)}{\omega_{resCM}(L_1/3 + L_{CM})} \times \\ \frac{z - 1}{z^2 - 2z\cos(\omega_{resCM}T) + 1} \quad (21)$$

Upon inspection one can recognize that this equation has the same dynamics as the capacitor current equation of the LCL filter (15), and thus it must have the same active damping characteristics.

TABLE I: SYSTEM PARAMETERS.

$2V_{DC} = 650V$	$E = 415V_{l-l}$	$f_{sw} = 10kHz$
$L_1 = 3\,mH$	$C_f = 7.5\,\mu F$	$L_2 = 1\,mH$
$L_{CM} = 4.6\,mH$	$C_{CM} = 2.2\,\mu F$	$C_{bus} = 2000\,\mu F$

TABLE II: SYSTEM CONTROLLER GAINS.

$K_p = 0.060A^{-1}$	$T_r = 0.0021s$
$K_{DM} = 0.083A^{-1}$	$K_{CM} = 0.062A^{-1}$

In order to apply feedback active damping the resonant frequency must be well below the critical resonant frequency [7]. The critical frequency represents the resonant frequency where feedback active damping is entirely ineffective, i.e. the closed loop poles are not drawn into stability for any value of gain. The critical frequency is [7][9]:

$$\omega_{crit} = \pi/3T \quad (22)$$

This resonant frequency boundary is also drawn in Fig. 5, and provides the last criteria for the filter design.

A similar control approach to the LCL filter can now be used to actively damp the common mode LC resonance in this circuit. The only difference is that the common mode controller does not need to regulate a fundamental current, but only has to damp any resonance. Hence only a

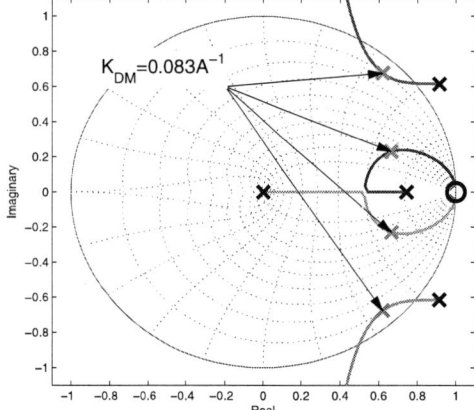

(a) Differential mode controller.

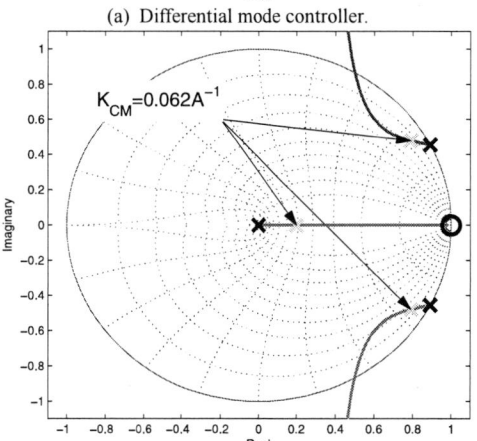

(b) Common mode controller.

Fig. 7: Root locus gain selection and closed loop poles of each controller.

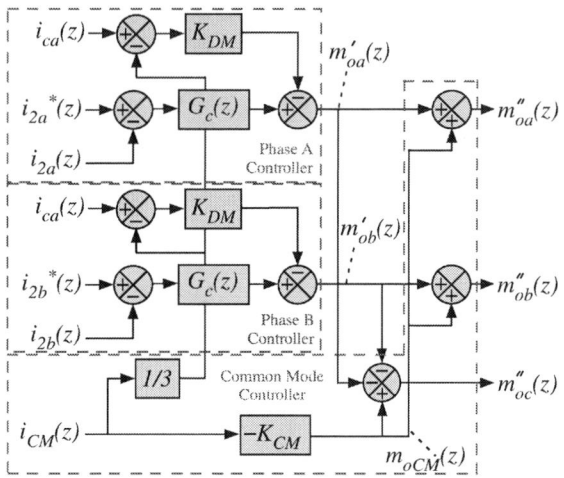

Fig. 8: Combined differential mode and common mode controllers.

proportional feedback controller is required, as shown in Fig. 6(b). The gain of this controller is set using the plant characteristic equation:

$$z + K_{CM} V_{DC} G_{iCM}(z) = 0 \qquad (23)$$

and a similar root locus approach to the LCL current regulator design. The root locus and selected gain are shown in Fig. 7(b). The gain is selected to provide a similar level of damping performance to the differential mode controller, with values listed in Table II.

C. Combining the Controllers

Finally, the differential and common mode controllers are combined to make the complete control system, shown in Fig. 8. Note that as usual for a three phase system, the differential mode PR controller is only used for phases A and B, and the phase C modulation command is the negative sum of phase A and B, viz

$$m'_{oc} = -m'_{oa} - m'_{ob} \qquad (24)$$

The common mode modulation command is then added onto all three differential mode phase leg modulation commands to form the final modulation commands of:

$$m''_{ox} = m'_{ox} + m_{oCM} \qquad (25)$$

III. SIMULATION AND EXPERIMENTAL RESULTS

Simulation and experimental systems have been developed in order to verify the common mode voltage mitigation technique and active damping scheme proposed. The system parameters are those used throughout this paper, as given in Table I, and the gains are those listed in Table II.

The simulation system was developed in PSIM using a full switched PWM model with dead-time. The experimental grid was also characterized and the major harmonics present were included in the simulation. However, no parasitic damping resistances were placed in the circuit to ensure a worst case damping scenario. A step change of current reference from 4A to 8A reactive was used to excite any potential resonances. Three sets of simulation results are presented, demonstrating the system without virtual neutral filter (Fig. 9(a)); with

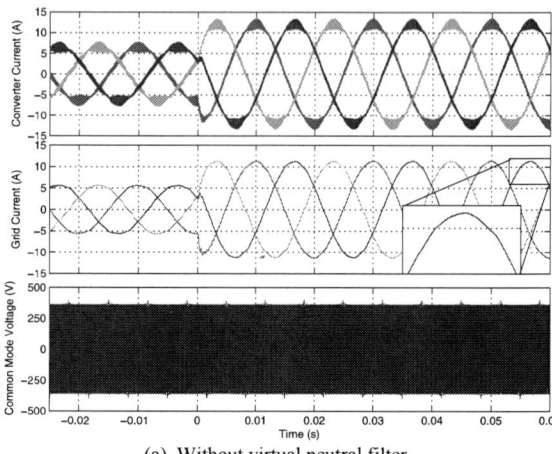

(a) Without virtual neutral filter.

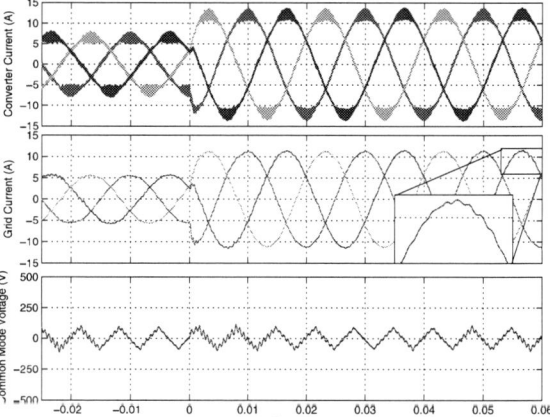

(b) With virtual neutral filter, without common mode active damping.

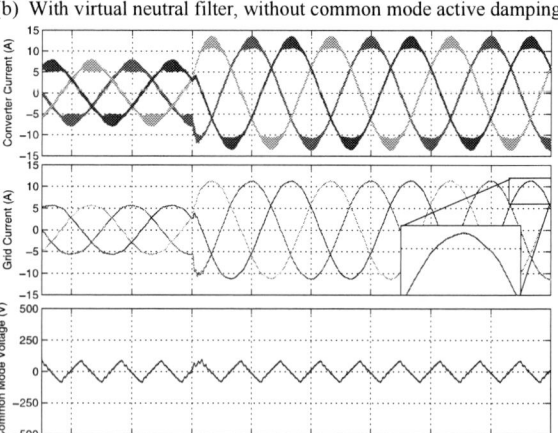

(c) With virtual neutral filter, with common mode active damping.

Fig. 9: Simulation Results.

virtual neutral filter but without common mode damping (Fig.9(b)); and the system with common mode active damping (Fig. 9(c)). NOTE: The system without CM active damping still contains the DM active damping.

Without the virtual neutral common mode filter the DC bus mid-point voltage shows the high frequency voltage ripple predicted by the theory presented in Section II. However, when the virtual neutral filter is incorporated (irrespective of damping) the mid-point ripple is reduced to a triangular waveform caused by the space vector

The 2014 International Power Electronics Conference

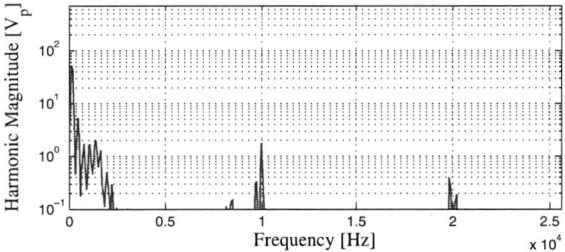

Fig. 11: Experimental DC bus mid-point voltage V_{mid} spectrum, with virtual neutral filter.

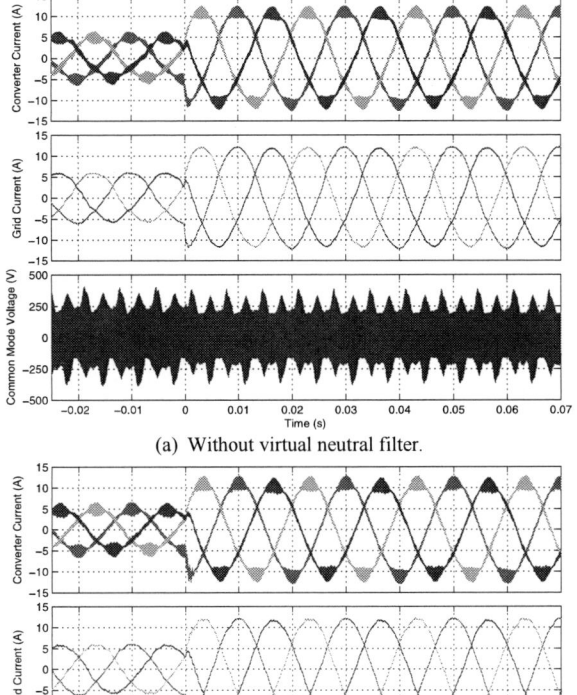

(a) Without virtual neutral filter.

(b) With virtual neutral filter, with common mode active damping.

Fig. 10: Experimental Results.

centering, together with minor levels of switching frequency ripple. This essentially baseband harmonic response is consistent with the low pass frequency characteristic of the virtual neutral network, discussed in Section III.

The system without common mode active damping also shows significant resonant ripple in the DC bus mid-point voltage. This is expected as the closed loop resonant poles are on the stability boundary. This resonant effect also couples into the differential current that enters the grid, demonstrating the need for controller decoupling. This is in contrast to the actively damped system which contains no appreciable resonance except during the transient event, which is quickly damped. Thus the simulation results confirm that active damping of the common mode filter circuit ensures that minimal resonance is present in both the DC bus mid-point voltage and in the target AC output currents.

An experimental system was established to validate these concepts using a grid-connected 5kVA prototype inverter operating as a STATCOM (to avoid the need for a real power source) using a TI TMS320F2810 fixed-point DSP as the controller.

The experimental results (Figs. 10 and 11) show a close match to the simulation results (Fig. 9). The voltage seen at the DC bus mid-point has the same triangular

pattern and the output current quality has not been compromised by the virtual neutral filter. However a small degree of high frequency ripple can be seen in the DC bus measurement. This may be due to two reasons – parasitic coupling in the circuit which has not been modeled and/or non-linear high frequency impedance degradation of the inductors and capacitors.

Note also that a series of harmonic compensators were also added to the differential current control system to overcome baseband harmonic distortion caused by grid voltage harmonics and converter dead-time. These compensators were designed as described in [10], and have no direct contribution to the common mode EMI virtual neutral filter described in this paper. Nevertheless, without them, the grid harmonic voltages from the available grid supply did cause significant distortion of the regulated grid current, which made the experimental results more complex and difficult to interpret.

Fig. 11 shows the experimental spectrum of the DC bus mid-point voltage (with respect to earth) when the virtual neutral filter is in place. When compared to Figs. 2(b) and 2(d), the reduction of switching frequency harmonics is substantial, and clearly validates the filtering concepts developed in this paper. Furthermore, as anticipated, the triplen baseband voltages are still present when the virtual neutral is in place. However, due to the slightly underdamped common mode resonance there is a minor increase in some of these baseband harmonic magnitudes.

IV. CONCLUSION

Common mode voltages on the DC bus midpoint of a grid connected converter can create substantial EMI problems if they are not filtered. This paper has presented a virtual neutral filter strategy that integrates with the main LCL filter components to achieve a very effective filter response. Enhancements are also presented for the converter current regulator, to remove phase cross-coupling caused by the additional filter components, and to provide active damping for the virtual neutral filter circuit. The concepts are validated by simulation and matching experimental results.

REFERENCES

[1] H. Akagi and T. Doumoto, "An approach to eliminating high-frequency shaft voltage and ground leakage current from an inverter-driven motor," IEEE Trans. Ind. Appl., vol. 40, no. 4, pp. 1162-1169, Jul./Aug. 2004.

[2] H. Akagi and T. Doumoto, "A passive EMI filter for preventing high-frequency leakage current from flowing through the

978-1-4799-2706-7/14 $31.00 © 2014 IEEE 2378

grounded inverter heat sink of an adjustable-speed motor drive system," IEEE Trans. Ind. Appl., vol. 41, no. 5, pp. 1215-1223, Sept./Oct. 2005.

[3] H. Akagi and S. Tamura, "A passive EMI filter for eliminating both bearing current and ground leakage current from an inverter-driven motor," IEEE Trans. Power Electron., vol. 21, no. 5, pp. 1459-1469, Sept. 2006.

[4] M. H. Hedayati, A. B. Acharya and V. John, "Common-mode filter design for PWM rectifier-based motor drives," IEEE Trans. Power Electron., vol. 28, no. 11, pp. 5364-5371, Nov. 2013.

[5] Hartmann, H. Ertl and J. W. Kolar, "EMI filter design for a 1MHz, 10kW three-phase/level PWM rectifier," IEEE Trans. Power Electron., vol. 26, no. 4, pp. 1192-1204, Apr. 2011.

[6] J. W. Kolar, U. Drofenik, J. Minibock, et al., "A new concept for minimizing high-frequency common-mode EMI of three-phase PWM rectifier systems keeping high utilization of the output

voltage," in Proc. IEEE Applied Power Electronics Conference (APEC), New Orleans, US, 2000, pp. 519-527.

[7] S. G. Parker, B. P. McGrath and D. G. Holmes, "Regions of active damping control for LCL filters," in Proc. IEEE Energy Conversion Congress and Exposition (ECCE), Raleigh, US, 2012, pp. 53-60.

[8] D.G. Holmes, and T. Lipo, Pulse Width Modulation for Power Converters: Principles and Practice, Wiley, 2003.

[9] J. Yin, S. Duan and B. Liu, "Stability analysis of grid-connected inverter with LCL filter adopting a digital single-loop controller with inherent damping characteristics," IEEE Trans. Ind. Inform., vol. 9, no. 2, pp. 1104-1112, May 2013.

[10] S. G. Parker, B. P. McGrath and D. G. Holmes, "Managing harmonic current distortion for grid connected converters with low per-unit filter impedances," in Proc. IEEE Energy Conversion Congress and Exposition Asia (ECCEAsia), Melbourne, Australia, 2013, pp. 1150-1156.

Effects of Transformer Structures on the Noise Balancing and Cancellation Mechanisms of Switching Power Converters

Hung-I Hsieh
Department of Electrical Engineering
National Chiayi University
Chiayi, Taiwan, R.O.C.
E-Mail: hihsieh@mail.ncyu.edu.tw

Sheng-Fang Shih
Department of Electrical Engineering
National Chiayi University
Chiayi, Taiwan, R.O.C.

Abstract—The focus of this paper is to explore the possibility of minimizing capacitive CdV/dt displacement (common-mode, CM) currents present in the power cord of a switching converter, by use of two types of transformer windings, namely the balancing dV/dt winding (B-Winding) and core cancellation winding (C-Winding). In this paper, techniques for constructing both the B- and C-Windings are first discussed and measurements are given accordingly to verify the study results. Depending on the gradient of the voltage distribution on transformer windings, the noise balancing and/or cancellation conditions (loops) within the transformer, formed by the B- and/or C-Windings, can cause the CM displacement current to be circulated back and forth between the transformer windings and parasitic inter-winding capacitances, so that there should not be any current flowing into the LISN. Therefore, the CM EMI measurements should be zero in theory and extremely small in practice. Design examples are given and experiments are finally carried out to verify the proposed theory.

Keywords—*Capacitive CdV/dt displacement (CM) current, transformer inter-winding capacitance, balancing dV/dt winding (B-Winding), core cancellation winding (C-Winding), transformer structures and winding techniques.*

I. INTRODUCTION

Electromagnetic interference (EMI) noise mitigation is an important issue that should be highlighted when designing switched-mode power supplies (SMPS). In recent years, it has become more and more important because electromagnetic pollution of SMPS systems by switching circuit harmonics is a growing problem due to their fast switching speed at relatively high voltage and current conditions to obtain improved efficiency. As a result, EMI problems of practical power electronic systems are likely to become even more severe in the future. To limit interference with other devices, EMI filters are in general used to meet international regulations limiting conducted emissions [1-4]. Even though EMI filters are the most popular and are intended for power processing industries, large EMI filters normally occupy 15% to 20% of the total size and weight of a power supply system, thus substantial savings in component spaces and costs cannot be achieved. Today, in order to obtain low cost, light weight and high power

density in SMPS applications, a lot of efforts have been made to reduce filter size by reducing the EMI noise emission propagating on switching power circuits [5-7], especially in small (i.e., more components placed in close proximity to each other on devices) and portable electronic apparatus. Hence, transformer construction and winding-techniques are becoming more demanded in recently years for a compact design can be obtained in commercial applications of power electronics circuits.

In this paper, effects of transformer structures on the converter EMI performance are first discussed, and winding techniques are then developed to balance and/or cancel the noise displacement current flowing within the transformer. From the standpoint of EMI noise reduction, incorporating the additional windings into a transformer can effectively reduce the CM noise current driven by the induced dV/dt on the winding of transformer without any conflicts. The theory is developed and refined from existing knowledge. Prototypes are designed to meet EMI standards and noise measured using a spectrum analyzer. Finally, the transformer models are used in practical power supplies for conducted EMI measurements. It can be seen that the proposed techniques give a more physical interpretation in the EMI design of power transformer, and can effectively reduce the CdV/dt displacement currents (CM EMI) by improving the winding techniques and transformer structures.

II. REVIEW OF DV/DT COMMON-MODE COUPLING BETWEEN THE PRIMARY AND SECONDARY WINDINGS OF THE TRANSFORMER

A typical test setup of conducted EMI measurements for an offline power supply based on the transformer-isolated flyback converter is shown in Fig. 1. The ac line voltage is connected through a line impedance stabilization network (LISN) to a diode-bridge rectifier and a power converter. An EMI filter is added to the converter as shown. This setup is just for convenience in representing the noise coupling path caused by the capacitive CdV/dt displacement current (CM EMI) flowing through the two 50-ohm of the LISN resistors in parallel. In offline applications, a switching power converter contains high-frequency transformers; normally, leading to significantly increased switching stress and generated CM EMI. The dominant contributors to the conducted emissions are the parasitic capacitances to ground; hence, the CM EMI noise is driven by the CdV/dt current flowing through these capacitances to power supply's chassis ground. The three dominant contributors to the capacitances are primary side transistor to ground capacitance C_q, transformer core to ground capacitance C_{cg}, and secondary side diode to ground capacitance C_d, as shown in Fig. 1. The other contributors to the parasitic capacitances are the inter-winding capacitance C_w and windings to core capacitance C_{wc}. The transistor and diode, they both perform as the noise sources to cause large dV/dt on the primary and secondary windings of transformer. The magnitudes of the dV/dt may vary over a wide range of values, depending on the converter input and output conditions. The induced dV/dt drives CM noise currents flowing from the transformer inter-winding capacitances C_w, and/or windings to core capacitance C_{wc} into the secondary side ground, and back to the primary side ground as shown in Fig. 1. Although the secondary side of the converter is not grounded, these loops may still exist owing to the parasitic capacitances between the secondary side and ground C_d, and transformer core and ground C_{cg}.

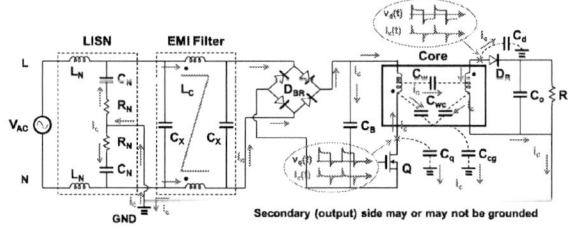

Fig. 1. Test setup of conducted EMI measurements for a two-wire offline flyback converter with an input EMI filter.

III. Effects of Transformer Balancing dV/dt Winding (B-Winding) and Core Cancellation Winding (C-Winding) on EMI Noise Reduction

To meet the requirements on conducted EMI, the conventional approach to accomplish this is to add EMI filters to the converter. In general, however, it is undesirable to allow filter size and weight to become too large, because the space and cost are of practical concern. In addition, the power loss due to filter component nonidealities cannot be neglected when considering the internal power dissipation in EMI filters. In this regard, the effectiveness of incorporating the additional windings (the B- and/or C-Windings) into a transformer is first discussed in this section; the effects of transformer structures on the noise balancing and cancellation mechanisms are investigated to minimize the CM noise of a switching converter.

A. Use of the B-Winding to Balance the Induced dV/dt on the Transformer Windings

Take the flyback converter circuit diagram in Fig. 2 as an example. This figure shows a physical transformer having multiple wound components, including the primary (N_P), secondary (N_S) and balancing (N_B) dV/dt windings. The B-Winding, which is wound on the same core, can be realized using an additional winding with opposite polarity mark of the primary side winding. The copper wire size for the B-Winding can be chosen as the same wire diameter of the primary or secondary windings. The number of turns of this winding can then be chosen to span the entire width of the winding bobbin in the window area of core. It should be pointed out that one side of the coil winding (point D) is not electrically connected within the transformer network, for the purpose of improving the system CM EMI performance without any conflicts.

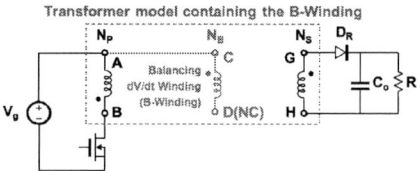

Fig. 2. Transformer design for a flyback converter example, the transformer incorporating an additional winding, namely the balancing dV/dt winding (B-Winding).

As mentioned previously, depending on the polarities of the induced dV/dt on the windings, one of the major contributors to the CM noise are the transformer inter-winding capacitances C_{wp} and C_{ws} shown in Fig. 3. In this case, the switching semiconductor devices act as the noise source generators to cause high dV/dt on the windings of transformer; hence, the dV/dt conducts high-frequency CM noise currents through the inter-winding capacitances of transformer to ground. The CM currents can then be reduced or at least attenuated by providing the noise current balancing conditions (loops) as shown, therefore following Thevenin's theorem, internal to the switching converter for the CdV/dt currents to flow through. This result that the reduced CM currents flow into the LISN can be achieved by adding the B-Winding between the primary and secondary windings inside the transformer. Fig. 3 shows the induced dV/dt of $V_{A/B}$, $V_{D/C}$ and $V_{H/G}$ on the transformer windings. The dV/dt drives the CM displacement currents which follow circular paths around the loops as shown. It can be seen that the B-Winding tend to balance the CM noise currents in the center of the transformer, and minimize the net noise current coupling to ground ($i_{q(CM)} \approx 0$, and $i_{d(CM)} \approx 0$).

978-1-4799-2706-7/14 $31.00 © 2014 IEEE

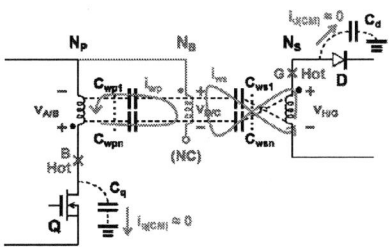

Fig. 3. CM noise-balancing effect of the B-Winding on the converter EMI performance.

Fig. 4 shows an example for this analysis. A flyback transformer contains the primary winding (N_P), B-Winding (N_B) and secondary winding (N_S) as shown. The N_P contains a total of 48 turns of wire arranged in three layers; the N_B and N_S each consist of 16 turns of wire per layer. For convenience in discussion of the geometry of winding structures, only one-half of the windings and core are depicted in this figure. The physical layer to layer capacitances C_{ws} and C_{wp} are represented in the figure. One side of the windings N_P, N_B and N_S can be connected accordingly to the cold points A, C and H of the flyback transformer. The presence of the induced dV/dt, appearing across the windings as shown on the left side of figure, will cause some CdV/dt displacement currents i_{ws} and i_{wp}, flowing from the N_P and N_S into the N_B. Because there is to one turn per layer for each of these windings as shown, the gradient of the voltage distribution on each layer is equal in magnitude but opposite in polarity to each other. It is therefore to be seen as Thevenin-equivalent networks, and the CM noise currents can pass through the parasitic inter-winding capacitances C_{ws} and C_{wp}, without conflicting the basic circuit properties. This will lead to significant reductions in the CM noise emissions of the converter ($i_{cg(CM)} \approx 0$), seen at the LISN side, due to reduced noise coupling via the parasitic capacitance C_{cg} to ground. Hence, the CM noise balancing mechanism is most pronounced in the cases of large dV/dt dominates the overall total noise, particularly in transformer-isolated converters.

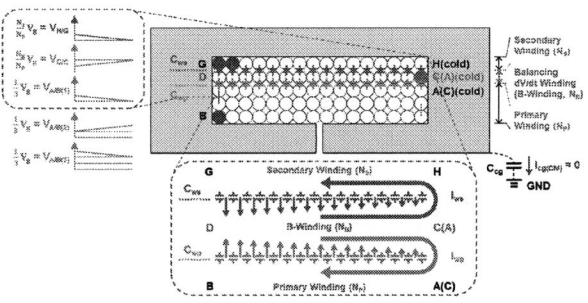

Fig. 4. Transformer winding geometry containing the B-Winding used in noise-balancing applications for CM EMI suppression.

B. Use of the C-Winding to Cancel the CdV/dt Displacement Current through the Transformer Core

Fig. 5 shows a flyback converter that has a different type of transformer structure and winding arrangement.

The transformer core and windings are depicted in this figure. As mentioned previously, the C-Winding can also be realized using an additional winding that is wound on the same core. It should be noted that one side of the C-Winding (point F) is not electrically connected within the transformer. The core cancellation winding (C-Winding), instead of the B-Winding, is used for reducing the CM noise emissions by means of the transformer core.

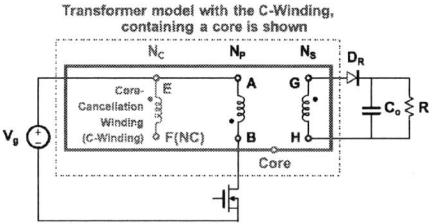

Fig. 5. Transformer design for a flyback converter example, the transformer incorporating an additional winding, namely the core cancellation winding (C-Winding).

Because the circuit designers usually want physically smaller size transformers; the transformer core and windings are placed close together, which maximizes both the parasitic capacitances within the transformer, such as the N_P to N_S capacitance C_{ps}, N_P to N_C capacitance C_{pw}, N_S to core capacitance C_{sc}, and N_C to core capacitance C_{wc}, as shown in Fig. 6. As mentioned previously, the transistor and diode, they both act as the noise voltage generators to produce high-frequency dV/dt on the primary, secondary and core cancellation windings of the transformer. Such induced dV/dt drives CM noise currents through the transformer inter-winding capacitances C_{ps} and C_{pw}, then propagate through the windings to core parasitic capacitances C_{sc} and C_{wc} as shown. The topological structure of the flyback transformer containing the core cancellation winding (C-Winding), which is placed on the bottom adjacent to the core, is illustrated in Fig. 7. Because the voltage gradients along each layer can be seen as identical sketched in Fig. 7, in which the resulting CM noise currents i_{sc} and i_{wc} are shown. It can be seen that the induced CM noise flowing into the transformer core, due to the distribution of voltage gradients on windings of N_S and N_C; hence, the noise currents are then equal in magnitude but opposite in phase. In this way, the CM currents on symmetrical halves of the core can be cancelled each other ($i_{wc} + i_{sc} \approx 0$). For the presence of noise currents ips and ipw within the transformer shown in Figs. 6 and 7, the voltage differences between the N_P and N_C ($V_{A/B(1)}$ and $V_{F/E}$), and between the N_P and N_S ($V_{A/B(3)}$ and $V_{H/G}$), are not obvious; therefore, the effect of these noise currents on the CM EMI performance is insignificant, and can be ignored without any problems. It is demonstrated that using the C-Winding can be effective in canceling the CM noise currents within the transformer, leads to much reduced noise coupling from the core to ground ($i_{cg(CM)} \approx 0$).

978-1-4799-2706-7/14 $31.00 © 2014 IEEE

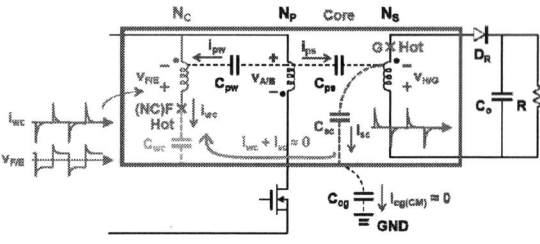

(a)

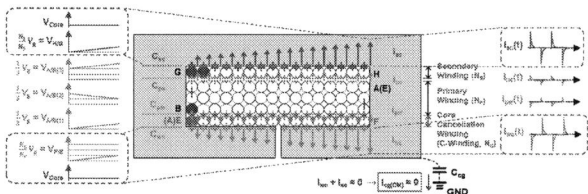

Fig. 6. CM noise-cancellation effect of the C-Winding on the converter EMI performance.

(b)

Fig. 7. Transformer winding geometry containing the C-Winding used in noise-cancellation applications for CM EMI suppression.

IV. EXPERIMENTAL RESULTS

In this paper, the effects of transformer structures on the techniques of noise current balancing and cancellation are first developed. Test setup of the two-wire input offline flyback converter, as shown in Fig. 1, is built to validate the proposed theory. The transformer model is replaced by the B- or C-Windings, which are discussed in Sections III A and III B. For a two-wire (i.e., there is no ground-wire terminal) power supply system, Y capacitors cannot be connected across line (L) and neutral (N) to chassis ground (G) as shown. Therefore, use of the additional windings (the B- and/or C-Windings) into a transformer is an effective way to reduce the CM noise emissions without adding Y capacitors to an EMI filter.

In the test setup of Fig. 1, the converter operates with a switching frequency of 70 kHz. The input ac voltage of the converter is 110 V and the output dc voltage is 5 V. The load power is 10 W. Fig. 8 shows the experimental results for the design of transformer structure containing the B- or C-Windings. It can be seen that, for the measurements of conducted CM EMI emissions of the converter including an input EMI filter shown in Fig. 1, there is up to 8 dB improvement in the frequency range of 150 kHz to 1 MHz, even though an EMI filter is installed. This means that there is still a room for improving the EMI performance by using the B- and/or C-Windings. Fig. 9 shows the CM noise measured results for operation in the frequency range of 1 MHz to 10 MHz. It is also apparent that the CM noise reduced by even more (up to 25 dB) since the B- or C-Windings is used in the transformer-isolated converter.

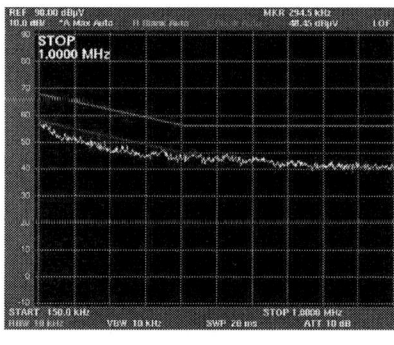

(c)

Fig. 8. Measurements of the CM noise including an EMI filter, in the frequency range of 150 kHz to 1 MHz. (a) Conventional structure of transformer without using an additional winding. (b) Noise measured using the B-Winding. (c) Noise measured using the C-Winding.

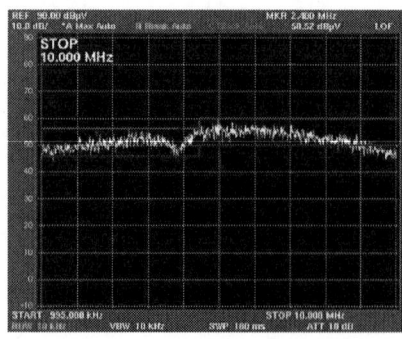

(a)

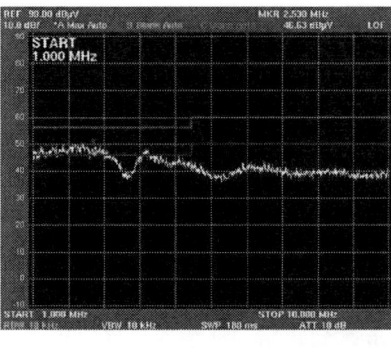

(b)

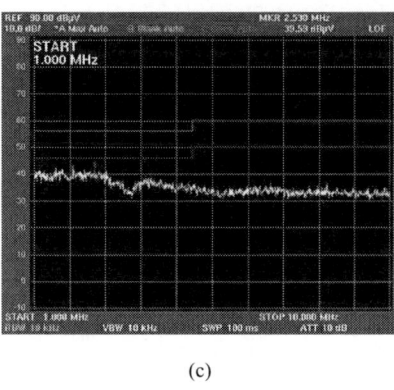

(c)

Fig. 9. Measurements of the CM noise including an EMI filter, in the frequency range of 1 MHz to 10 MHz. (a) Conventional structure of transformer without using an additional winding. (b) Noise measured using the B-Winding. (c) Noise measured using the C-Winding.

V. CONCLUSIONS

In this paper, the effects of transformer structures on the CM noise current balancing and cancellation mechanisms are first discussed. The techniques for constructing both the B- and C-Windings are investigated, and measurements are given to verify the analysis. Depending on the voltage gradients along the transformer windings, the noise current balancing and/or cancellation conditions of the transformer (i.e., the equivalent circuit loops established by the B- and/or C-Windings for CM noise reduction), can provide an effective way to minimize the CM EMI noise measured on the LISN side of the circuit. Based on the analysis, these two winding structures can be integrated simultaneously into one physical transformer to improve EMI performance without any conflicts and problems with test setup.

ACKNOWLEDGMENT

This work was supported primarily by the National Science Council under Award Numbers NSC 100-2221-E-415-001, NSC 101-2221-E-415-010, and NSC 102-2221-E-415-005.

REFERENCES

[1] M. J. Nave, *Power line filter design for switched-mode power supply*. New York: Van Nostrand Reinhild, 1991.

[2] S. Wang, F. Lee, and W. Odendaal, "Characterization and parasitic extraction of EMI filters using scattering parameters," *IEEE Trans. Power Electron.*, vol. 20, no. 2, pp. 502–510, Mar. 2005.

[3] H. Hsieh, J. Li, and D. Chen, "Effects of X capacitors on EMI filter effectiveness," *IEEE Trans. Ind. Electron.*, vol. 55, no. 2, pp. 949–955, Feb. 2008.

[4] H. Hsieh, L. Huwang, T. Lin, and D. Chen, "Use of a Cz common-mode capacitor in two-wire and three-wire offline power supplies," *IEEE Trans. Ind. Electron.*, vol. 55, no. 3, pp. 1435–1443, Mar. 2008.

[5] S. Wang, P. Kong, and F. Lee, "Common mode noise reduction for boost converters using general balance technique," *IEEE Trans. Power Electron.*, vol. 22, no. 4, pp. 1410–1416, Jul. 2007.

[6] S. Wang, F. Lee, and J. van Wyk, "A study of integration of parasitic cancellation techniques for EMI filter design with discrete components," *IEEE Trans. Power Electron.*, vol. 23, no. 6, pp. 3094–3102, Nov. 2008.

[7] J. Biela, A. Wirthmueller, R. Waespe, M Heldwein, K. Raggl, and J. Kolar, "Passive and active hybrid integrated EMI filters," *IEEE Trans. Power Electron.*, vol. 24, no. 5, pp. 1340–1349, May 2009.

A Novel Technique for Reducing Leakage Current by Application of Zero-Sequence Voltage

Hideki Ayano, Kouhei Murakami and Yoshihiro Matsui
Department of Electrical Engineering
Tokyo National College of Technology
1220-2 Kunugida-machi, Hachioji, Tokyo, Japan
e-mail: ayano@tokyo-ct.ac.jp

Abstract—**This paper proposes a technique for reducing the leakage current flowing into a power source as a conductive noise. This leakage current may have an adverse effect on the peripheral equipment, e.g., causing malfunctions. The proposed technique requires us to add a zero-sequence voltage to the output voltage reference of the inverter. The zero-sequence voltage can be controlled without affecting the driving characteristics of the motor. In a motor drive range, a counter EMF increases with an increase in the motor speed. The proposed technique, therefore, controls the zero-sequence voltage as a function of the inverter output frequency; as a result, a stable motor drive and reduction in the common-mode current are realized simultaneously. Simulation and experimental results prove the validity of the proposed technique.**

Keywords—leakage current, zero-sequence voltage, common-mode, EMI

I. INTRODUCTION

PWM inverters and PWM rectifiers contribute greatly to energy saving. However, they produce an electromagnetic interference (EMI) noise[1][2]. The EMI noise includes a conductive noise[1] which flows into a power supply as a leakage current (common-mode current) and a radioactive noise which spreads space as electromagnetic waves. The conductive noise causes malfunctions of peripheral equipment as well as own equipment. As a reduction method of the leakage current, a common-mode choke is generally used. The size of the common-mode choke becomes comparatively large in the converter system because the three coils for each phase are wound around the same choke core. Furthermore, since the common-mode choke requires high-frequency characteristics, it needs a magnetic material with sufficient frequency characteristics such as an amorphous or ferrite. For this reason, the cost of the common-mode choke becomes high.

The way to use a common-mode choke can be classified into two methods. Method (i): a common-mode choke is inserted in the input or output terminals of a converter[2][3], This method can mitigate the unfavorable effect on the peripheral and own equipment because the common-mode choke directly reduce the leakage current. Method (ii): a common-mode choke is used together with grounding capacitors[4][5]. The capacitors provide the leakage current with another route to flow. Since the method (ii) requires less inductance for the choke, it reduces the size of the common-mode choke compared to the method (i).

The three-phase output voltages of an inverter consists of three components: a d-axis voltage that generates a flux current, q-axis voltage that generates a torque current, and zero-sequence voltage. The zero-sequence voltage does not affect the driving characteristics of the motor. The zero-sequence voltage is conventionally used for improving the dc voltage utilization by adding the 3rd harmonic component[6], reducing the number of switching times by employing the two-phase modulation techniques[7], and distributing noises over a wide frequency range[8].

Although the conventional conductive EMI is regulated in the frequency band from 150 kHz to 30 MHz, these days, the reduction in the lower frequency noises has been required. In our previous paper, we proposed a technique for reducing common-mode current by applying an appropriate zero-sequence voltage in an output voltage range nearly zero volts where the EMI noise becomes largest[9][10]. EMI noises are largest when the inverter reference voltage is around zero because the rate of change in the zero-sequence component of the output voltages becomes largest there. This situation occurs in the case of a hill-start of an electric vehicle, a start or stop of an elevator system and a servo lock of servo equipment, where high torque and very low speed are required to the motor.

This paper proposes a leakage current reduction technique not only when the motor is nearly at zero speed but also the motor is driving a load at any speed. In the motor driving range, a counter EMF increases with an increase in the inverter frequency/motor speed. Therefore, this paper proposes a novel technique to control the zero-sequence voltage as a function of the inverter output frequency. With this technique, a stable motor drive and reduction in the leakage current are realized simultaneously. It is noted that the technique is implemented only with software without additional hardware such as noise filters. Simulation and experimental results prove the validity of the proposal technique.

978-1-4799-2706-7/14 $31.00 © 2014 IEEE

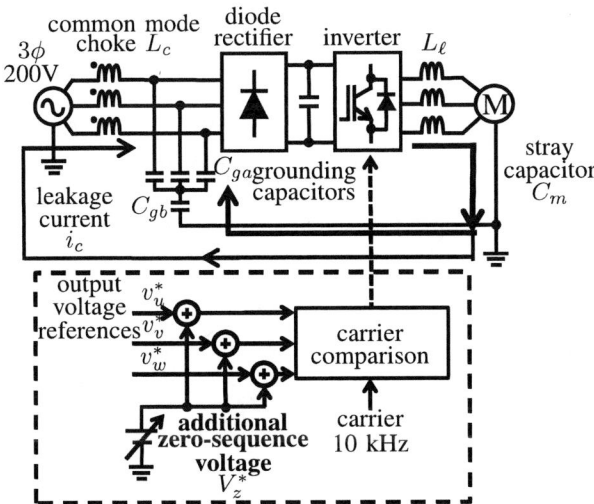

Fig. 1. Configuration of the system.

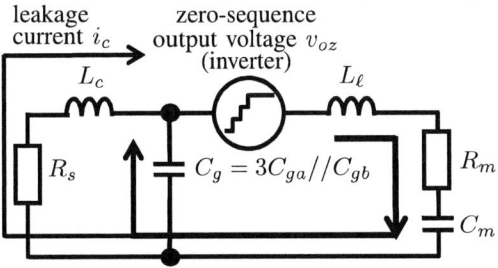

Fig. 2. Zero-sequence equivalent circuit.

TABLE I. PARAMETERS IN EQUIVALENT CIRCUIT.

cable inductance	L_ℓ	8.2 [μH]
load resistance (zero-sequence)	R_m	25 [Ω]
stray capacitor	C_m	2.9 [nF]
common-mode choke inductance	L_c	0.79 [mH]
power source resistance (zero-sequence)	R_s	3.0 [Ω]
grounding capacitor	C_{ga}	0.68 [μF]
	C_{gb}	0.47 [μF]

II. CONFIGURATION OF THE PROPOSED SYSTEM AND PRINCIPLE OF REDUCING THE LEAKAGE CURRENT

A. System configuration and equivalent circuit

Fig. 1 shows the configuration of the inverter system. The common-mode choke L_c is connected between the power supply and the diode rectifier. The grounding capacitors C_{ga} are connected in a Y-connection and the neutral point is connected to the ground through C_{gb}. The inverter drives an induction motor and the line inductance between the inverter and motor is L_ℓ. The switching frequency of the inverter is 10 kHz.

The leakage current, which is caused by the switching of the inverter, flows to the ground through the stray capacitance C_m inside the motor[3][11]. The leakage current has two routes: a part of it flows through the power supply via peripheral equipment and the rest flows through the grounding capacitors. Since the frequency of the leakage current is higher than the inverter switching frequency, the impedance of the latter route is lower than the former one. Thus, the common-mode current mainly flows though the grounding capacitors, and in this way the unfavorable effects on the peripheral equipment can be lowered.

The purpose of this research is to further reduce the leakage current i_c that flows though the power supply. The feature of the proposed technique is to add a dc zero-sequence voltage V_z^* to each of the three-phase output voltage references. Thus, the V_z^* acts as a part of the zero-sequence voltage v_{oz}. The v_{oz}, however, do not affect the motor drive since it is cancelled in the line-to-line voltages.

Fig. 2 shows the equivalent circuit for the leakage current in Fig. 1. R_m denotes the resistance of the load side such as of the motor, and R_s denotes the resistance of the power supply side. The inverter is expressed as a zero-sequence voltage source. The leakage current flows into the ground through L_ℓ, R_m, and C_m. Then, it flows

through two routes: through L_c and R_s connected in the power supply side and through C_g, i.e., the series connection of C_{ga} and C_{gb}. The purpose of this paper is the reduction in the leakage current i_c that flows through the power supply side. Table I shows the specifications of the equivalent circuit. Here, L_c, C_g, L_ℓ, and C_m are measured values in the experimental system, and R_s and R_m are typical values for induction motors of several kWs.

B. Principle of reducing leakage current

Define leakage current gain G_s that is the ratio of i_c to the zero-sequence voltage of the inverter output v_{oz} in Fig. 2 .

$$G_s(s) = \frac{I_c(s)}{V_{oz}(s)} = \frac{C_m \cdot s}{A(s)} \qquad (1)$$

where, $A(s)$ is given by

$$
\begin{aligned}
A(s) = {} & L_\ell \cdot L_c \cdot C_m \cdot C_g \cdot s^4 \\
& + C_m \cdot C_g \cdot (L_\ell \cdot R_s + L_c \cdot R_m) \cdot s^3 \\
& + (L_\ell \cdot C_m + L_c \cdot C_m + L_c \cdot C_g \\
& \quad + C_m \cdot C_g \cdot R_m \cdot R_s) \cdot s^2 \\
& + (R_m \cdot C_m + R_s \cdot C_m + R_s \cdot C_g) \cdot s + 1. (2)
\end{aligned}
$$

Fig. 3 shows the frequency characteristic of $G_s(s)$ and it shows a resonant characteristic. The resonance frequency depends on L_c and C_g, and is 9.2 kHz under the condition of Table I. In the proposed system, L_c and C_g are designed so that the resonance frequency may be lower than the carrier frequency of 10 kHz. Thus, as harmonic order of the carrier frequency becomes higher, gain $|G_s|$ becomes lower. The leakage current i_c is expressed by

$$I_c(s) = G_s(s) \cdot V_{oz}(s). \qquad (3)$$

In order to reduce i_c, it is effective to reduce the carrier frequency component of v_{oz}, namely 10 kHz because it is quite near the resonance frequency of $G_s(s)$ as shown in

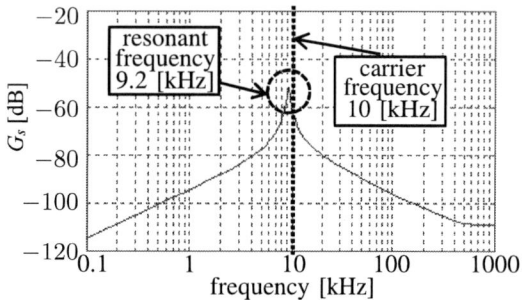

Fig. 3. Admittance gain of the leakage current.

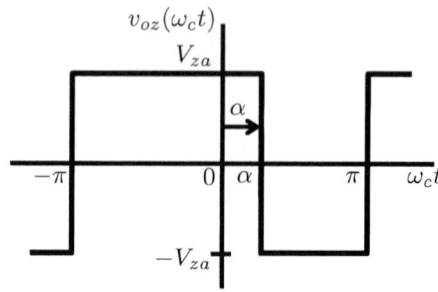

Fig. 4. Inverter output voltage when its reference is nearly zero.

Fig. 3. Let's assume that the inverter reference voltage is nearly zero. Then, the zero-sequence output voltage v_{oz} is rectangular with the carrier frequency of 10 kHz as shown in Fig. 4. This paper proposes to vary the shape of the rectangular inverter output so that its frequency may be raised to a higher harmonic region and the carrier frequency component may be lowered. In order to vary the shape, a dc zero-sequence voltage V_z^* is introduced. If V_z^* is added to the reference, the angle α can be controlled. Thus the form of the the zero-sequence output voltage v_{oz} can be altered. The v_{oz} is expressed by

$$v_{oz}(\omega_c t) = \frac{V_{za}\alpha}{\pi} + \frac{2V_{za}}{\pi}\sum_{k=1}^{\infty}\left(\frac{\sin k\alpha}{k}\cos k\omega_c t \right.$$
$$\left. + \frac{\cos k\pi - \cos k\alpha}{k}\sin k\omega_c t\right) \qquad (4)$$

where, ω_c is the angular frequency of the carrier, and V_{za} is the amplitude of the rectangular zero-sequence voltage v_{oz} and is given by

$$V_{za} = \frac{200\sqrt{2}}{2}. \qquad (5)$$

The duty ratio of v_{oz}, let's call it d hereafter, is 50% when α is 0, and it is 75%, and 95% if α is $\pi/2$ and $9\pi/10$, respectively. From (4), if d is 50%, v_{oz} includes only odd harmonics and has the largest carrier frequency component. With an increase in the duty ratio, the carrier frequency component, which is near the resonance frequency as shown in Fig. 3 , decreases and v_{oz} includes both the odd and even harmonics. If the inverter reference voltage is zero ($v_u^* = v_v^* = v_w^* = 0$ V) with 200 V power supply, α is 0 (the duty ratio is 50%) is equivalent to $V_z^* = 0$ (conventional operation). α is $\pi/2$ ($d = 75\%$) is equivalent to $V_z^* = 50\sqrt{2}$ V, and α is $9\pi/10$ ($d = 95\%$) is equivalent to $V_z^* = 90\sqrt{2}$ V. It is noted that V_z^* increases linearly with α.

C. Reduction in the leakage current at zero speed range of the motor

Consider the leakage current when the motor is at zero speed. Figs. 5 and 6 show the calculation results of the zero-sequence output voltages v_{oz} and their frequency spectra respectively when the dc zero-sequence voltages $V_z^* = 0$ and $V_z^* = 50\sqrt{2}$ V are added. In the case

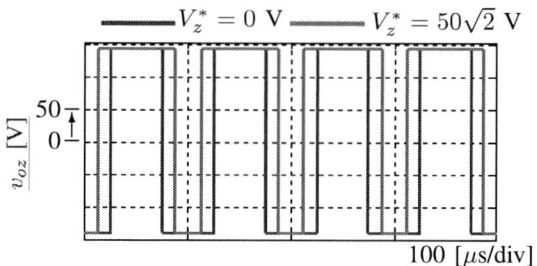

Fig. 5. Calculation results of zero-sequence voltages.

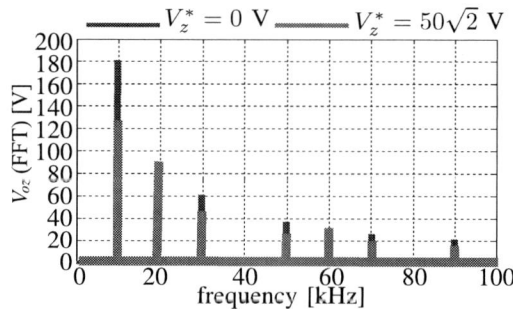

Fig. 6. Calculation results of frequency spectra.

of $V_z^* = 0$, v_{oz} becomes rectangular with a duty ratio of $d = 50\%$. The frequency spectra include only odd harmonics, and the fundamental (carrier frequency, 10 kHz) and 3rd and 5th harmonic components are 180 V, 60 V (1/3 of the fundamental), 35 V (1/5 of the fundamental) respectively. On the other hand, in the case of $V_z^* = 50\sqrt{2}$ V, v_{oz} becomes a rectangular with a duty ratio is 75%. The frequency spectra include both odd and even harmonics, but the fundamental is reduced to 130 V, which results in a 19% reduction in the leakage current i_c.

Fig. 7 shows the peak leakage current vs. duty ratio d characteristics in zero speed condition. The experimental results agree well with the simulation ones. The peak values of i_c decreases with an increase in the duty ratio d. Measured i_c waveforms under $d = 50\%$ (conventional case with $V_z^* = 0$) and $d = 95\%$ are also shown in the figure. The peak values of i_c in conventional case is 150 mA. On the other hand, when the zero-sequence voltage corresponding to $d = 95\%$ is added, it reduces to 40 mA. This is more than a 70% reduction.

978-1-4799-2706-7/14 $31.00 © 2014 IEEE

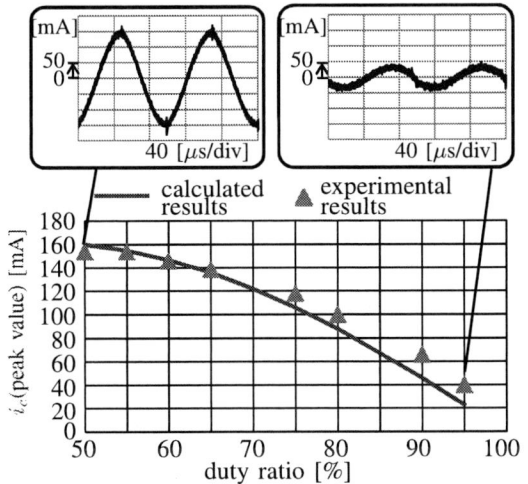

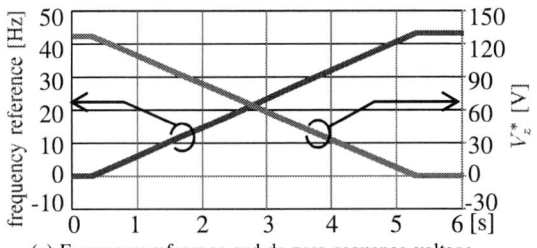

(a) Frequency reference and dc zero-sequence voltage

Fig. 7. Peak values of leakage current when dc zero-sequence voltage added.

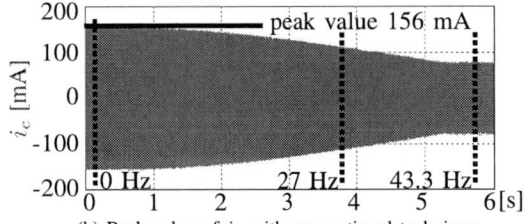

(b) Peak value of i_c with conventional technique

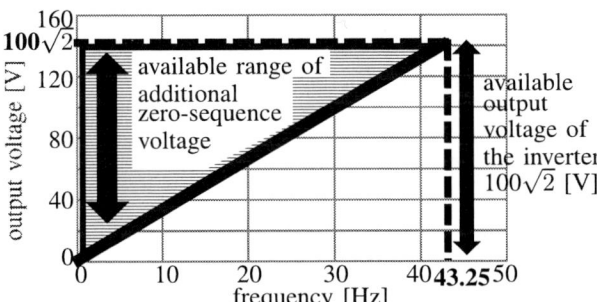

Fig. 8. Available range of dc zero-sequence voltage.

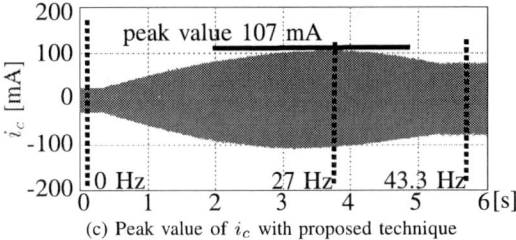

(c) Peak value of i_c with proposed technique

Fig. 9. Simulation results.

III. PROPOSED TECHNIQUE FOR REDUCING LEAKAGE CURRENT IN MOTOR DRIVING RANGE

A. Control of dc zero-sequence voltage

Here, this paper proposes a technique for reducing the leakage current when the motor is driving a load. Conventional constant v/f control is employed for the induction motor drive system shown in Fig. 1. The v/f control fixes a ratio of the output voltage to the output frequency, which realizes a wide speed control range keeping the amount of magnetization constant irrespective of the motor speed.

Fig. 8 shows the relationship between the output voltage of the inverter and available dc zero-sequence voltage V_z^*. Since the rated voltage and frequency of the motor are 200 V and 50 Hz, the ratio of the output phase voltage to the frequency is

$$v/f = v_{max}/f_{max} = 200\sqrt{2/3} \text{ V}/ 50 \text{ Hz}$$
$$= 3.27 \text{ V/Hz.} \tag{6}$$

When a reference of the output phase voltage is sinusoidal, the maximum output voltage that the inverter can output without distortion in the waveform is $100\sqrt{2}$ V. In this case, the maximum output frequency is expressed

$$f = \frac{100\sqrt{2}}{3.27} = 43.3 \text{Hz.} \tag{7}$$

From Fig. 8, it is obvious that the sum of the amplitude of the output voltage reference V_r^* and the added zero-sequence voltage V_z^* must be smaller than the available output voltage of the inverter.

$$V_r^* + V_z^* \leq 100\sqrt{2} \tag{8}$$

This paper proposes to regulate V_z^* linearly with V_r^* as shown in (9).

$$V_z^* = (100\sqrt{2} - V_r^*) \times 0.9 \tag{9}$$

When the motor is at a standstill, V_z^* is 90% of the available inverter output voltage, and V_z^* is 0 when the inverter operates with a maximum frequency (V_r^* is maximum). Substituting (6) to (9), we have another expression of V_z^* as

$$V_z^* = (100\sqrt{2} - 3.27 \times f) \times 0.9. \tag{10}$$

B. Simulation results

Simulation results are shown in Fig. 9. Fig. 9(a) shows how the inverter frequency reference and added dc zero-sequence voltage V_z^* are changed in time. The frequency reference of the motor is given at $t = 0.25$ sec. It is increased linearly with time and reaches the maximum 43.3 Hz at $t = 5.25$ sec.

Fig. 9(b) shows the waveform of i_c with the conventional technique (no dc zero-sequence voltage is applied, $V_z^* = 0$). When the motor is in standstill, i_c takes the maximum, 156 mA. The i_c decreases with an increase

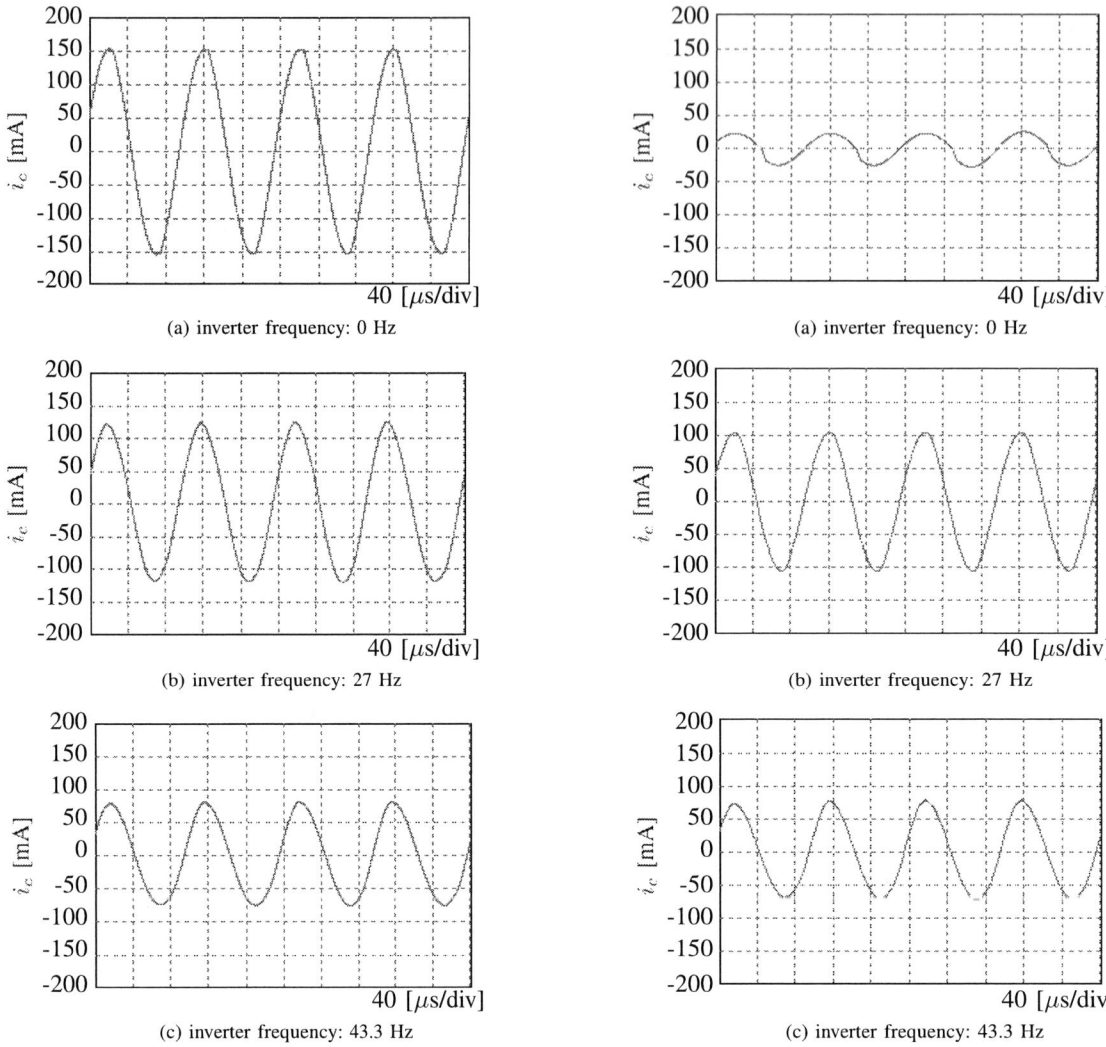

Fig. 10. Enlarged i_c waveforms in Fig. 9(b).

Fig. 11. Enlarged i_c waveforms in Fig. 9(c).

in the inverter output frequency. This is because the carrier frequency component of v_{oz} decreases with an increase in the inverter output frequency since waveform of v_{oz} is getting sinusoidal as the inverter frequency becomes high though it is rectangular when the inverter output frequency is zero. Fig. 10(a)-(c) are the enlarged waveforms of i_c in Fig. 9(b) at the output frequency of 0, 27 and 43.3 Hz, respectively. The fundamental frequency of i_c is 10 kHz, and the amplitudes of it decrease with an increase in the inverter output frequency. Furthermore, the three waveforms are almost sinusoidal.

Fig. 9(c) shows the waveform of i_c when the proposed technique is used. It is noted that i_c is reduced drastically in the low speed range. However, since V_z^* decreases with an increase in the inverter output frequency, i_c increases with the inverter output frequency and takes the maximum at 27 Hz. When the frequency is increased further, the leakage current decreases because the carrier frequency component of v_{oz} decreases with an increase in the inverter output frequency. Fig. 11(a)-(c) shows the enlarged

waveforms of i_c in Fig. 9(c) at the output frequency of 0, 27 and 43.3 Hz, respectively. Fig. 11(a) shows a distorted waveform because the even order harmonics are included in v_{oz} as indicated in (4). Fig. 11(b) shows that i_c takes the maximum of 107 mA. This demonstrates a 31% reduction compared with the maximum of 156 mA in the conventional case as shown in Fig. 10(a). Fig. 11(c) shows the same waveform as that of Fig. 10(c) because the added dc zero-sequence voltage is zero, $V_z^* = 0$.

C. Experimental results

Fig. 12 shows the experimental waveforms in the conventional case with $V_z^* = 0$. When the motor is in standstill, i_c takes the maximum, 150 mA. The i_c decreases with an increase in the output frequency. Each waveform of i_c is almost sinusoidal with the frequency of 10 kHz, and agrees very well with the simulation results in Fig. 10.

Fig. 13 shows the experimental waveform of i_c when the proposed technique is applied. The i_c is reduced

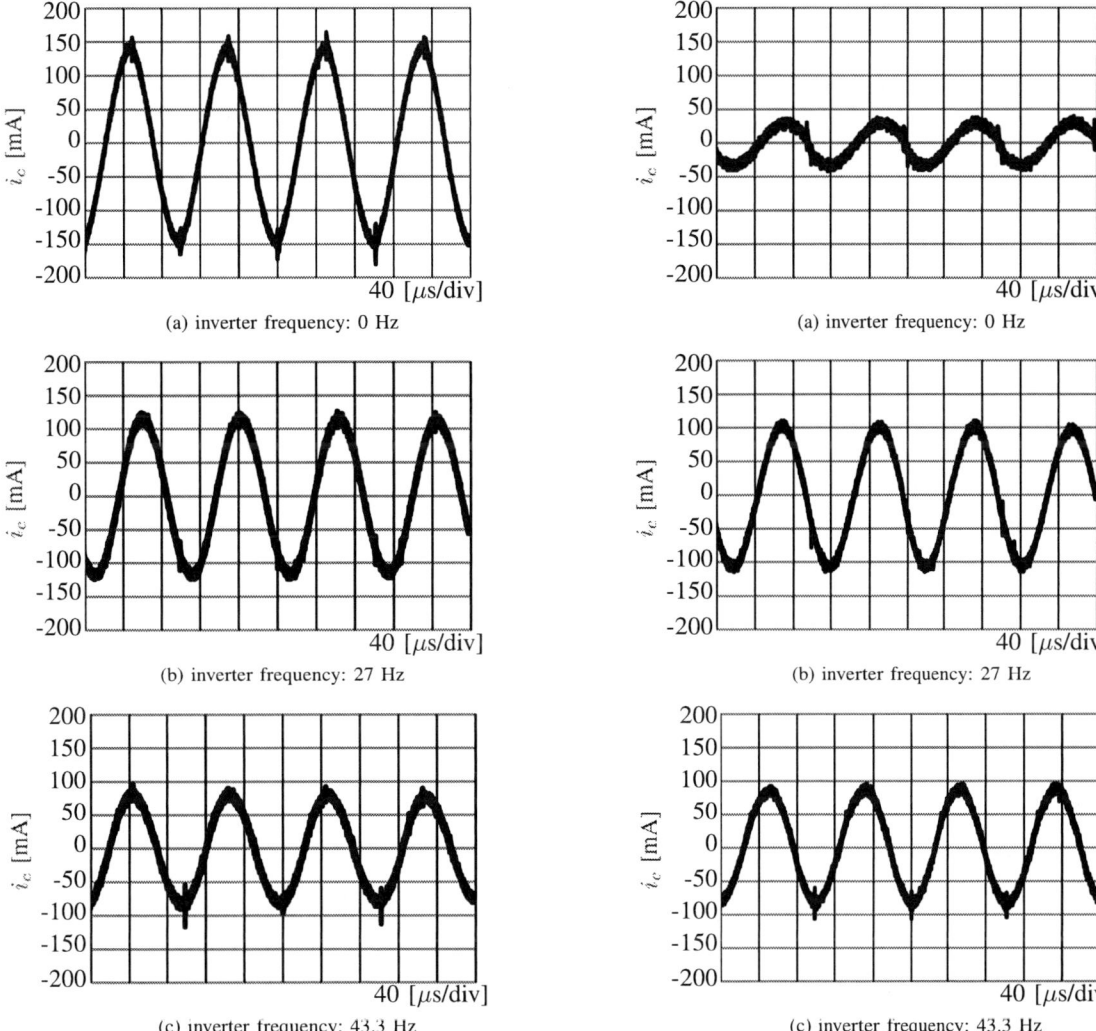

Fig. 12. Experimental results with conventional technique.

Fig. 13. Experimental results with proposed technique.

drastically in the low speed range owing to the additional zero-sequence voltage V_z^*. However, i_c increases with the output frequency since V_z^* decreases with an increase in the output frequency. When the frequency is increased further, i_c decreases because the carrier frequency component of v_{oz} decreases with an increase in the inverter output frequency. Some distortion is observed in i_c in Fig. 13(a) because of the even order harmonics in v_{oz} that is caused by adding large V_z^*. All the waveforms of Fig. 13 agree very well with those in Fig. 11. The i_c in Fig. 13(b) takes the maximum, 110 mA. Comparing with the maximum of 150 mA with the conventional technique in Fig. 12(a), the proposed method enables us to reduce the leakage current i_c by about 30%. This verifies the validity of the simulations and the effectiveness of the proposed method.

IV. CONCLUSIONS

This paper proposed a leakage current reduction technique not only when the motor was nearly at zero speed

but also the motor was driving a load at any speed. The following results were obtained.

1. This paper proposes to add a zero-sequence voltage to an inverter reference which is controlled as a function of the inverter output frequency considering that the counter EMF of the motor is directly proportional to the motor speed.

2. The proposed technique reduced about 30% of the leakage current compared with the conventional technique.

3. With this technique, a stable motor drive and reduction in the leakage current were realized simultaneously. It should be noted that the technique was implemented only with software. No any additional hardware such as noise filters was required.

ACKNOWLEDGEMENT

The authors would like to thank Dr. Shoji Fukuda, retired and was with Hokkaido University, for his suggestions and kind advice on this project. This research is supported by JSPS KAKENHI (25420275).

REFERENCES

[1] P. S. Chen, "Effective EMI Filter Design Method for Three-Phase Inverter Based Upon Software Noise Separation," IEEE Trans. on Power Electronics, vol. 25, no. 11, pp. 2797-2806, 2010.

[2] M. L. Heldwein and J. W. Kolar, "Impact of EMC Filters on the Power Density of Modern Three-Phase PWM Converters," IEEE Trans. on Power Electronics, Vol.24, No.6, p.1577-1588, 2009

[3] S. Ogasawara and H. Akagi, "Modeling and Damping of High-Frequency Leakage Currents in PWM Inverter-fed AC Motor Drive Systems," IEEE Trans. on Industry Applications, Vol.32, No.5, p.1105-1114, 1996

[4] T. Chida, N. Kusuno, A. Mishima, M. Kurita, and S. Ibori, "Filter Design Technique for Inverter Complied with European EMC Standard," Proc. of the 2009 JIASC, Vol.1, No.1-S1-5, pp.163-168 (2009) (in Japanese)

[5] J. Takeda, K. Shimane, T. Tkeda, H. Matsuoka, H. Mochikawa, and J. Tsuda, "The proposal of method reducing the leakage current in an elevator -New method of the passive filter-," Elevator, Escalator and Amusement Rides Conference 2008, pp.23-26 (2009) (in Japanese)

[6] J. W. Kolar, H. Ertl, and F. C. Zach, "Influence of the Modulation method on the Conduction and Switching Losses of a PWM Converter System," IEEE Trans. on Industry Applications, Vol.27, No.6, p.1063-1075, 1991

[7] S. Halasz, B. T. Huu, and A. Zakharov, "Two-phase modulation technique for three-level inverter-fed AC drives," IEEE Trans. on Industrial Electronics, Vol.47, No.6, p.1200-1211, 2000

[8] T. Homma, S. Taniguchi, T. Ogawa, S. Wakao, K. Kondo and T. Yoneyama, "An Experimental Study on Random PWM for Reducing Return Current Harmonics," Annual Conference of IEEJ, No. 5-087, pp.139-140 (2008) (in Japanese)

[9] H. Ayano, Y. Sato and Y. Matsui, "Proposal and Verification of a Technique for Reducing Leakage Current Using Zero-Sequence Voltage," T.IEE Japan, vol.133-D, No.11, pp. 1141-1148 (2012) (in Japanese)

[10] H. Ayano, Y. Sato and Y. Matsui, "Variation and Factor of the Waveform of Leakage Current Depending on the Position of Grounding Capacitors," T.IEE Japan, vol.132-D, No.12, pp. 1048-1056 (2013) (in Japanese)

[11] A. V. Vouanne, P. Enjiti and W. Gray, "Application issues for PWM adjustable speed AC motor drives," IEEE Industry Applications Magazine, Vol.2, No.5, p.10-18 (1996)

AC-Choppers Using Instantaneous Voltage Control Technique to Solve Voltage Sag Problems

Surin Khomfoi

Dept. of Electrical Engineering, Faculty of Engineering
King Mongkut's Institute of Technology Ladkrabang
Bangkok, Thailand
kkhsurin@kmitl.ac.th

Abstract— AC-Choppers using instantaneous voltage control technique to solve voltage sag problems are proposed in this paper. The developed AC-chopper for series voltage sag compensation is applied for three phase four wire system; therefore, three single phase ac chopper circuits and three series transformers are utilized. By using AC-chopper, there are no required for a dc bus; thereupon, the size of proposed voltage sag compensator is smaller. Each AC-chopper cell is independently controlled by using instantaneous voltage control method. With this proposed technique, the AC-chopper can rapidly compensate the related voltage sag and provide constant output voltages. Positive sequence PLL is applied to generate angle velocity (ω) thus the proposed AC-chopper can perform both balance and unbalance voltage sag. MATLAB/Simulink is used to simulate the proposed control algorithm and circuit operation. The 2-kW prototype is also developed for experimental validation. The results illustrate that the developed voltage sag compensator can perform within 3 ms responding time and the total harmonic voltage distortion is less than 2 % with overall system efficiency about 90%.

Keywords— *AC-chopper, series voltage sag compensation, instantaneous value voltage control*

I. INTRODUCTION

Voltage sag is one of the most important issues in power quality. It does cause the problems in the power system and industry. Voltage sag can cause operation downtime in the manufacturing system. This may cause damage in products and components; especially, electronic devices e.g. variable speed drive (VSD), Programmable logic control (PLC). To avoid such problems, the voltage sag compensator should be installed to the system. Several topologies of the voltage sag compensator have been proposed and they can be categorized into two groups: parallel and series compensators. The advantages and disadvantages on both categories have been clearly described in [1-2]. A series compensator can offer a smaller size of VA rating comparing to a parallel compensator. Moreover, a series compensator can provide fault tolerance and fault limiter capabilities to a system. However, a parallel compensator may have less compensator system losses and better dynamic voltage response comparing to a series compensator [1]. Therefore, better system efficiency and fast dynamic voltage response are required in a series compensator.

AC-chopper configuration can have better system efficiency because AC-chopper does not have dc bus. Thereupon, the application of an AC-chopper is widely used in several applications; for instance, a single phase UPS, and a LED driver as shown in [3-4]. A symmetrical PWM AC-chopper voltage controller has been developed in [4]. AC-choppers have been used as AC voltage regulator and voltage sag compensation as proposed in [5-8]. The AC-Chopper can be used to perform like an automatic voltage regulator as presented in [5]. The AC-chopper can offer a minimum energy injection using of micro-SMES has clearly been verified in [6]. The application of an AC-chopper on dynamic voltage restorer has also been reported in [9]. One can see that an AC-chopper can be applied in many applications; however, a few researches are focusing on a series voltage sag compensator.

The PWM line conditioner with fast output voltage control has been developed in [10]. Also, a non-liner feed-forward control technique for single phase UPS has been report in [11]. As can be seen, a developed PWM technique in [10-11] is for fast output voltage control for a conventional converter which can be utilized in AC-chopper with some modification. Hence, fast dynamic voltage response for AC-choppers can be achieved by using instantaneous voltage control technique as presented in [11-12]. The simulation and experimental results have been reported in [11] and have been validated in [12] for a single phase AC-chopper applications. The simulation study of AC-choppers for a series voltage sag application using instantaneous voltage control technique has been reported in [13]. Therefore, the experimental results are required to validate the AC-chopper for a series voltage sag application. A series voltage sag compensator using instantaneous voltage control technique is presented in this paper. AC-chopper can offer low losses and smaller size comparing to a conventional ac-dc-ac configuration.

This paper presents a series voltage sag compensator using ac-chopper as shown in Fig.1. The three single phase ac-choppers are used to compensate voltage sag in three phase four wire system. The advantages of using ac-chopper are without DC-Link, nearly sinusoidal output voltage with small size of input and output low-pass filter, high input power factor. As a result, the size and weight of this compensator is smaller which would lead to cost effectiveness.

978-1-4799-2706-7/14 $31.00 © 2014 IEEE

II. PROPOSED AC-CHOPPER CONFIGURATION AND CONTROL TECHNIQUE

The proposed system consists of three single phase ac-choppers connected in series through the injection transformer between source and load as illustrated in Fig 2. During voltage sag condition, each ac-chopper injects compensated in-phase voltage with the source voltage to maintain the desired load voltage independently. If unbalanced voltage sag occurs (i.e. single phase sag, poly phase sag and unbalanced three phase sag), ac-chopper will generate the compensated voltage to the related phases of the ac-chopper. The other ac-chopper will operate in by-pass mode.

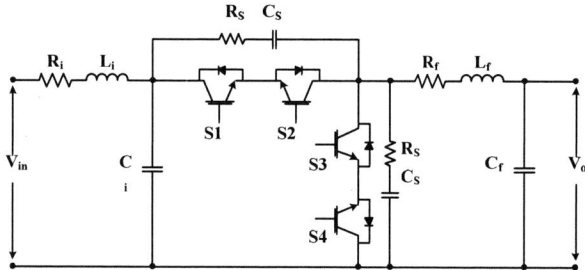

Fig. 1. A single phase ac-chopper.

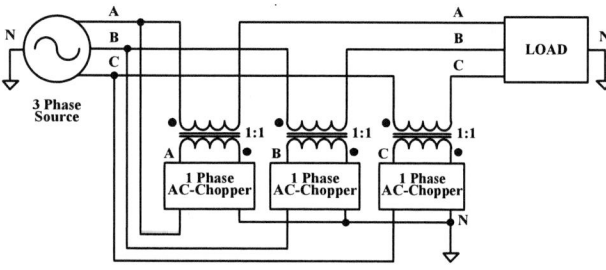

Fig. 2. Proposed system paradigm.

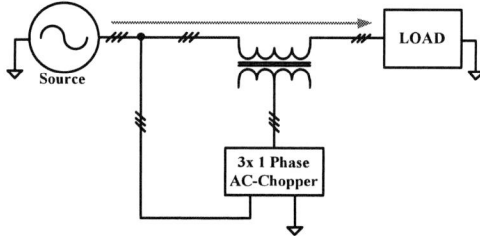

Fig. 3. Power flow diagram during normal condition.

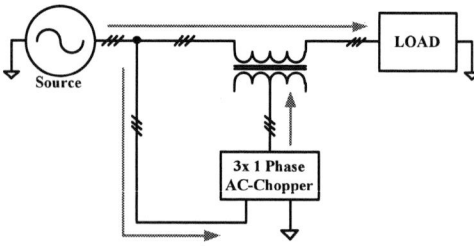

Fig. 4. Power flow diagram during voltage sag condition.

For instance, if voltage sag occurs in phase A and B, both phase A and B ac-choppers are operating but phase C ac-chopper will operate in a by-pass mode. During normal condition, all three single phase ac-chopper will operate in by-pass mode.

In by-pass mode, IGBT switch S1, S2 is turned-off and S3, S4 is turned-on. Power flow diagram from source to load during normal condition is depicted in Fig 3. Fig. 4 shows power flow diagram during voltage sag condition. As can be seen, the ac-chopper will inject power when voltage sag occurs. In this proposed control, the delay ¾ cycles is used after the voltage increases to normal condition. So, the ac-chopper will work for next ¾ cycles to make sure that source voltage is in absolute normal condition. If voltage sag is repeatedly occurred within ¾ cycles, ac-chopper will work until source voltage completely increases to normal condition.

An instantaneous voltage control technique is used to control ac-chopper during voltage sag. This technique provides fast response, robustness against the supply voltage sag, nearly sinusoidal wave form under non-linear load condition [12-13]. The explanation is explained below and the control diagram is illustrated in Fig. 5. Fig. 5 shows the control diagram of ac-chopper. An instantaneous value voltage control is the method that uses sinusoidal control signal comparing with modulated triangle carrier signal. The triangle carrier signal is generated from an integrator operation of source voltage and we will reset the integrator every switching period as shown in Fig.6. As a result, the amplitude of triangle carrier signal will be varied with amplitude of the source voltage. As can be seen in Fig.7, voltage sag occurs at 0.005s: the amplitude of triangle carrier signal will decrease and duty ratio will automatically increase. Therefore, voltage sag of the source is not effect to the load and independent to ac-chopper controller.

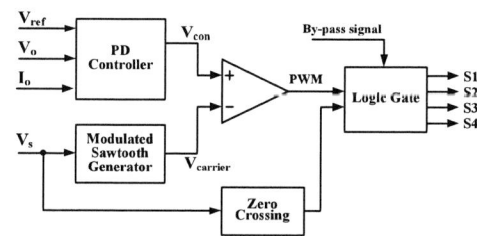

Fig. 5. Control Diagram.

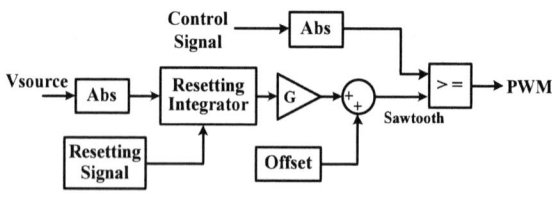

Fig. 6. Simulation model of PWM generator.

$$\xi = \frac{1}{2} \frac{K_D + R_f C_f}{\sqrt{(K_p+1)L_f C_f}} \tag{4}$$

$$s_{1,2} = -\xi\omega_n \pm j\omega_n \sqrt{1 - \xi^2} \tag{5}$$

V_0 = AC-Chopper Output Voltage,

V_{ref} = Reference Input Voltage,

K_p = Proportional Gain,

K_D = Derivative Gain,

L_f = Low-pass filter Inductance,

C_f = Low-pass filter Capacitance,

R_f = Low-pass filter Resistance.

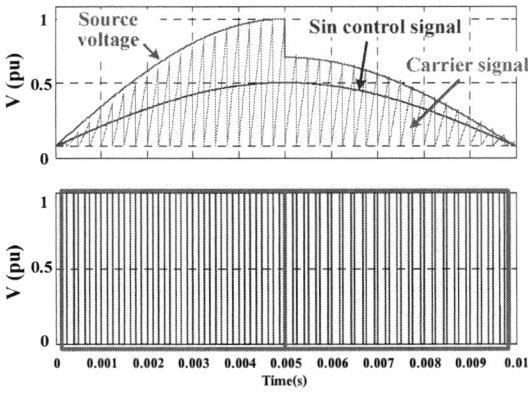

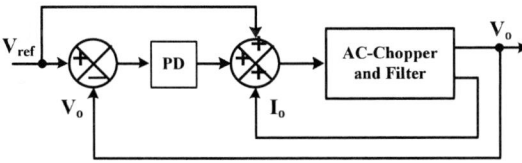

Fig. 7. PWM generation scheme.

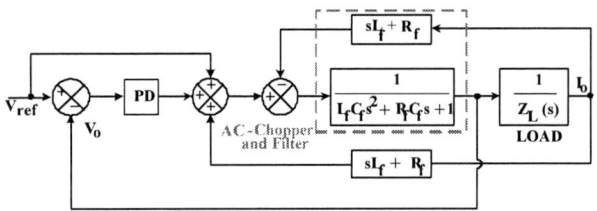

Fig. 8 Proposed PD controller.

As shown in Fig.5, the sinusoidal control signal or V_{con} which is used to compare with triangle carrier signal is obtained from PD controller using output voltage (V_o) feedback and feed-forward control technique as depicted in Fig. 8 [11-12]. The output current of ac-chopper (I_o) is used to compensate current disturbance in which the load is modeled as a current source disturbance. By using the state-variable averaging method, the output voltage of an ac-chopper can be written as (1). Therefore, the control system of the proposed ac chopper can then be represented as Fig. 9.

$$V_o(\text{s}) = \frac{DV_i(\text{s}) - (sL_f + R_f)\,I_o(\text{s})}{L_f C_f \text{s}^2 + R_f C_f \text{s} + 1} \tag{1}$$

D = Duty Ratio

V_0 = AC-Chopper Output Voltage

V_i = AC-Chopper Input Voltage

I_0 = AC-Chopper Output Current

L_f = Low-pass filter Inductance

C_f = Low-pass filter Capacitance

R_f = Low-pass filter Resistance

All control parameters such as a control characteristic equation, natural frequency(ω_n) and damping ratio (ξ) can be calculated as illustrated from (2) to (5).

$$\frac{V_o(\text{s})}{V_{ref}(\text{s})} = \frac{K_D\text{s} + K_p + 1}{L_f C_f \text{s}^2 + (K_D + R_f C_f)\text{s} + K_p + 1} \tag{2}$$

$$\omega_n = \sqrt{\frac{K_p+1}{L_f C_f}} \tag{3}$$

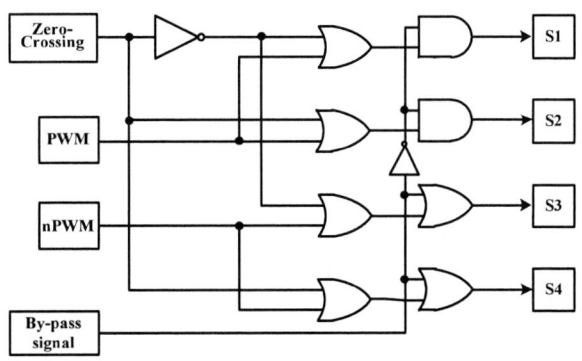

Fig. 9. Proposed control system.

In order to generate all four switching signals, logic gates are used to multiplex among PWM signal, zero-crossing signal and by-pass signal as depicted in Fig. 10. nPWM is the invert of PWM signal with 2µs dead-time. The remaining V_{ref} and by-pass signal will be described in next section.

Fig. 10. Switching signal generation.

III. VOLTAGE SAG DETECTION

The PLL algorithm is shown in Fig. 11 [13]: one can see that source voltage V_a, V_b and V_c are used to calculate V_{qp} (Positive sequence V_q). Then, V_{qp} is applied to estimate the source voltage angle and angular velocity. The estimated angle is used as a commanded signal of voltage transformation to generate the reference voltage $V_{a,ref}$ $V_{b,ref}$

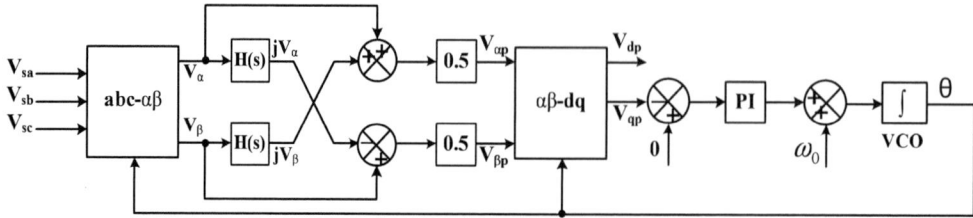

Fig. 11. Three-phase PLL Algorithm.

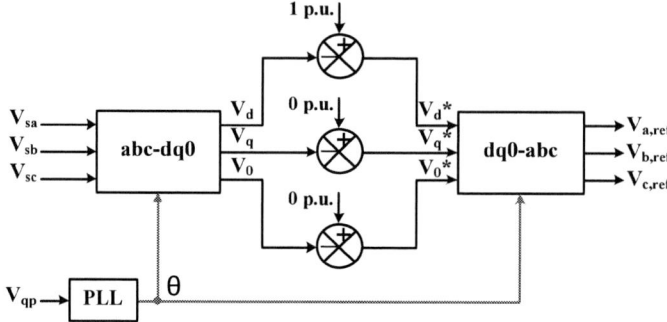

Fig. 12. Voltage sag detection topology.

and $V_{c,ref}$ as shown in Fig.12. H(s) is a phase-shift filter transfer function having $+\pi/2$ output phase-shift at angular frequency which can be expressed as follows:

$$H(s) = -\frac{1 - sT_1}{1 - sT_1}, \qquad (6)$$

where, $T_1 = 1/\omega$ and ω is the angular frequency.

Fig.12 shows the topology to determine the magnitude of voltage sag. 1.0 pu., 0.0 pu., and 0.0 pu. of voltage reference are used to compare with V_d, V_q and V_0 generated from Park's transformation (abc to dq) respectively. Then, calculate V_{ref} from the difference of V_d, V_q, and V_0 by using Inverse Park's transformation (dq to abc) for providing to PD controller as shown in Fig.5. The by-pass signal in each phase is generated by comparing absolute of V_{ref} and 0.05 per unit as shown in Fig. 13. If the absolute of V_{ref} is higher than 0.05, this means that voltage sag occurs, the by-pass signal will deactivate ac-chopper for compensating voltage sag. When the system voltage returns to normal condition, by-pass signal will be delayed for 1 ½ cycles to make sure that the source voltage is completely return to normal condition.

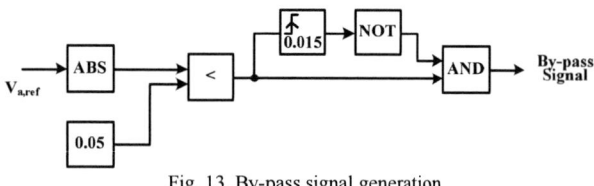

Fig. 13. By-pass signal generation

IV. SIMULATION

MATLAB/Simulink is used to simulate the proposed application. Fig.14 shows the power circuit diagram of the system which has a three single phase ac-chopper connected between the line through the series injection transformer. The source supplies to the 1.5-kW load. The simulation is performed under three different voltage sag conditions: balanced three phase voltage sag, single phase voltage sag, and two phase voltage sag, respectively. Some important parameters used for simulation are shown in Table I.

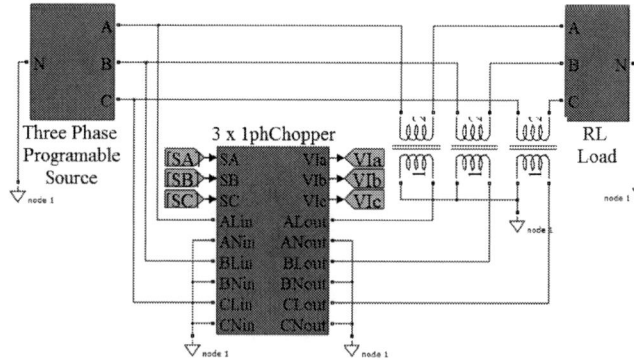

Fig. 14. Power circuit diagram of the system.

The simulation results under source voltage sag to 70% condition are shown in Fig. 15, 16 and 17 which are the source and load voltage waveform under balanced three phase sag, single phase sag and two phase sag at time (t) from t = 0.045s to 0.105s respectively. The results show that an AC-chopper does perform both balanced and unbalanced compensation effectively.

TABLE I.
PARAMETERS FOR SIMULATIONS

AC-chopper	
Supply voltage	220 V
Switching frequency	10 kHz
Low-pass filter	
R_f and R_i	0.5Ω
L_f and L_i	0.5mH
C_f and C_i	25uF
Controller parameters	
K_p	1
K_D	0.0002s
RC Snubber	
R	82Ω
C	22nF

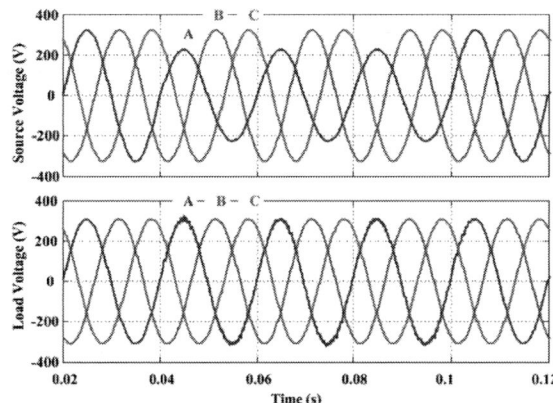

Fig. 17. Two phase sag on phase B and C.

Simulation results on phase B and C voltage sag are illustrated in Fig. 17. Obviously, the voltage sag can be compensated by using the proposed AC-chopper technique. The simulation results indicate that the proposed AC-chopper togather with instantaneous voltage control technique can be applied in a series voltage sag compensator. Fig.18 shows the voltage waveform and the harmonic spectrum of the load of the phase A voltage sag condition. As can be seen, the waveform is nearly sinusoidal which has a total harmonic voltage distortion about 1.9%

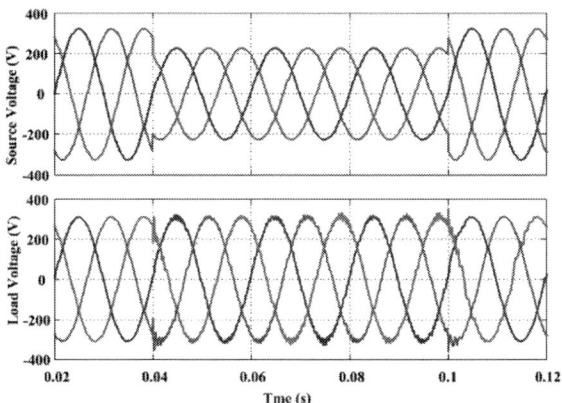

Fig. 15. Balanced three phase sag.

Fig. 15 is a simulation result showing a balance three phase voltage sag to 70% at 0.045 s (top). As can be seen, the proposed AC-chopper series compensator can perform to compensate the voltage sag as the normal voltage at the load (bottom).

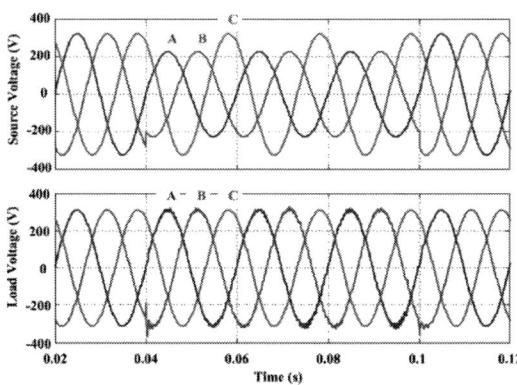

Fig. 16. Single phase sag on phase B.

Fig. 16 shows a simulation result of a phaseB voltage sag to 70% at 0.045 s (top). Clearly, the proposed AC-chopper system can also solve the voltage sag problems.

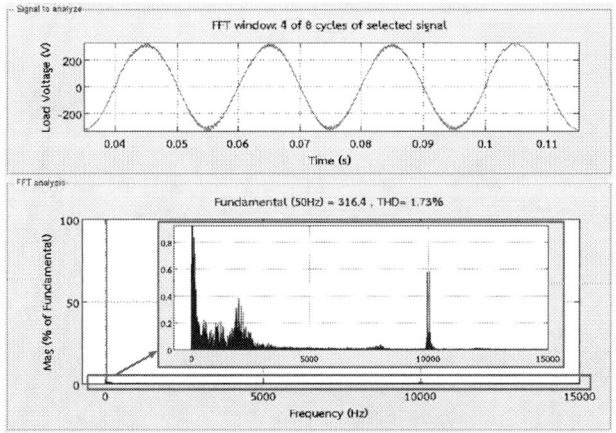

Fig. 18. FFT analysis of the load voltage.

V. EXPERIMENTAL RESULTS

The 2 kW prototype was developed to validate the proposed system as shown in Fig. 19 and Fig. 20. A single chip DSP (TMS320F2808) was used to perform central calculation. During normal operation, the by-passed signal was sent to the AC chopper acting as a standby mode. If the voltage sag occurs at the source voltage, the voltage sag algorithm will detect and then calculate amount of voltage sag. After that, the amount of voltage sag will send to the PWM generator unit via D/A converter to generate desired PWM signals for the AC-choppers as depicted in Fig. 19.

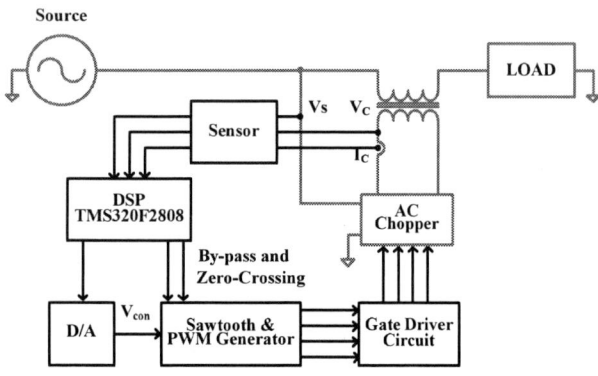

Fig. 19. Prototype diagram of proposed AC-chopper system.

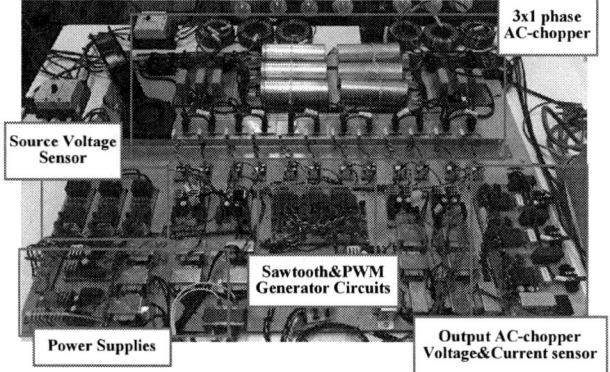

Fig. 20. 2-kW AC-chopper series compensator prototype.

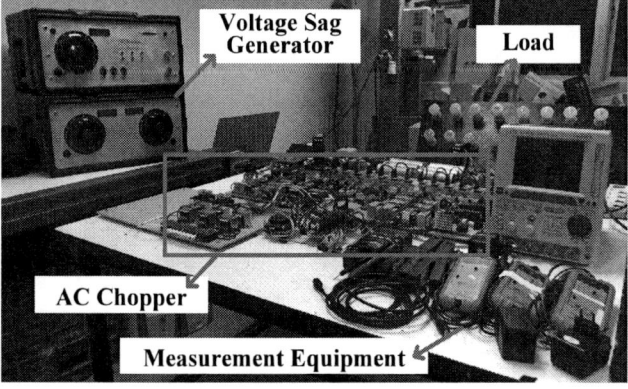

Fig. 21 . Experimental set-up.

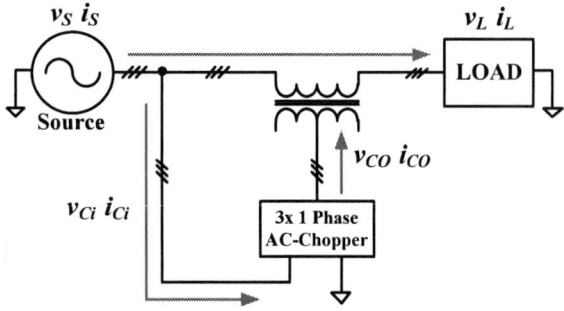

Fig. 22. Measurement points: voltage and current during validation

The experimental set-up is shown in Fig. 21. The voltage sag generator was used to validate the proposed AC- chopper system. The voltage sag generator was used as a voltage source which can perform voltage sag situations. The developed prototype was connected with 1:1 transformer. Normally, AC-chopper can compensate with a fine power quality performance at the source voltage sag to 50 % [13]. Therefore, the source voltage to 90%, 80% and 70% conditions were validated. Also, three phase, two phase and single phase voltage sag categories were examined in this test. The measurement points are illustrated in Fig. 22 to probe the voltage and current data. All parameters of the developed prototype are listed in Table II.

TABLE II.
PARAMETERS OF DEVELOPED PROTOTYPE

AC-chopper	
IGBT-HGTG20N60B3D	600V 40A
Switching frequency	10 kHz
Low-pass filter	
R_f and R_i	0.5Ω
L_f and L_i	0.5mH
C_f and C_i	25uF
Controller parameters	
K_p	0.7
K_D	0.0002s
RC Snubber	
R	82Ω
C	22nF

The experimental results on three phase, two phase and single phase voltage sag conditions are shown in Fig. 23-25, respectively. The (a) is a source voltage from voltage sag generator and the (b) is load voltage. As can be seen, the voltage sag generator was created the voltage sag to 70 % for all categories at 90° magnitude. The proposed AC-chopper series voltage sag compensator can compensate voltage sag for all tested categories. The efficiency evaluation of proposed AC-chopper system is illustrated in Table III

TABLE III.
EFFICIENCY EVALUATION

Magnitude of Voltage Sag to	Chopper Efficiency (%)	System Efficiency (%)
100%	-	96.67
90%	68.12	92.29
80%	78.35	90.32
70%	82.81	90.61

As shown in Table III, the efficiency is about 96.7% at normal condition because of the output filter and transformer losses. Also, the efficiency of AC-chopper will increase if the voltage sag at the source increases. It should be noted that the overall system efficiency is more than 90%; therefore, the proposed AC-chopper series voltage sag compensator is satisfactorily operated under this validation.

978-1-4799-2706-7/14 $31.00 © 2014 IEEE

The 2014 International Power Electronics Conference

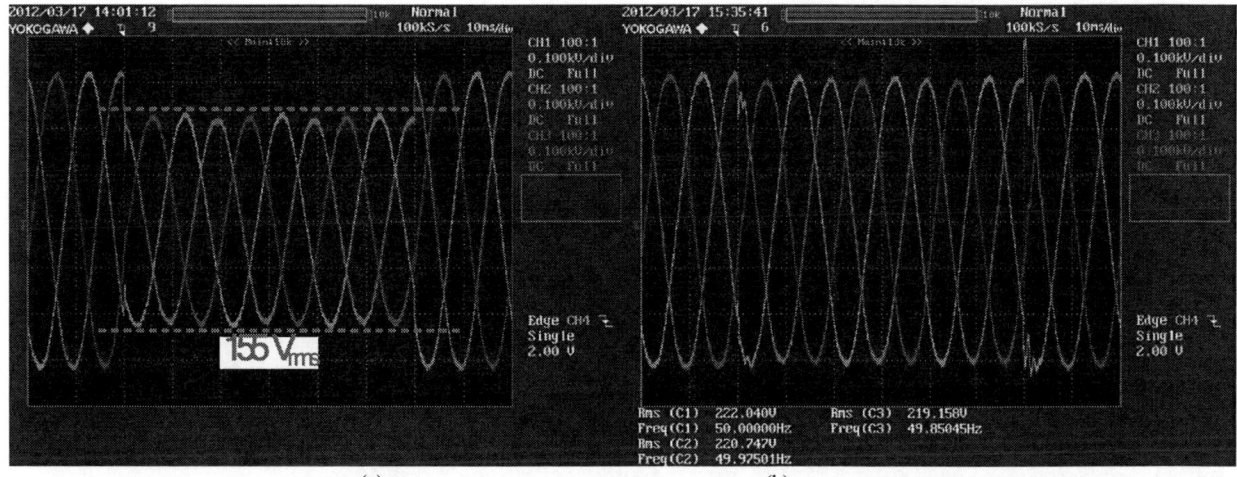

(a)   (b)

Fig. 23. Three phase voltage sag to 70% : (a) showing voltage sag at sources and (b) showing voltage at load terminals

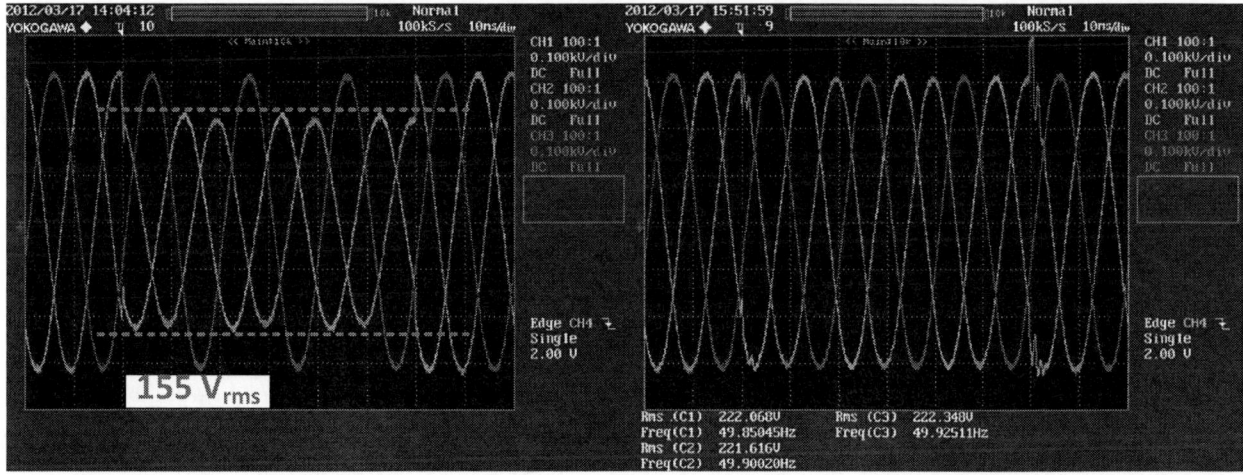

(a)   (b)

Fig. 24. Two phase voltage sag to 70% : (a) showing voltage sag at sources and (b) showing voltage at load terminals

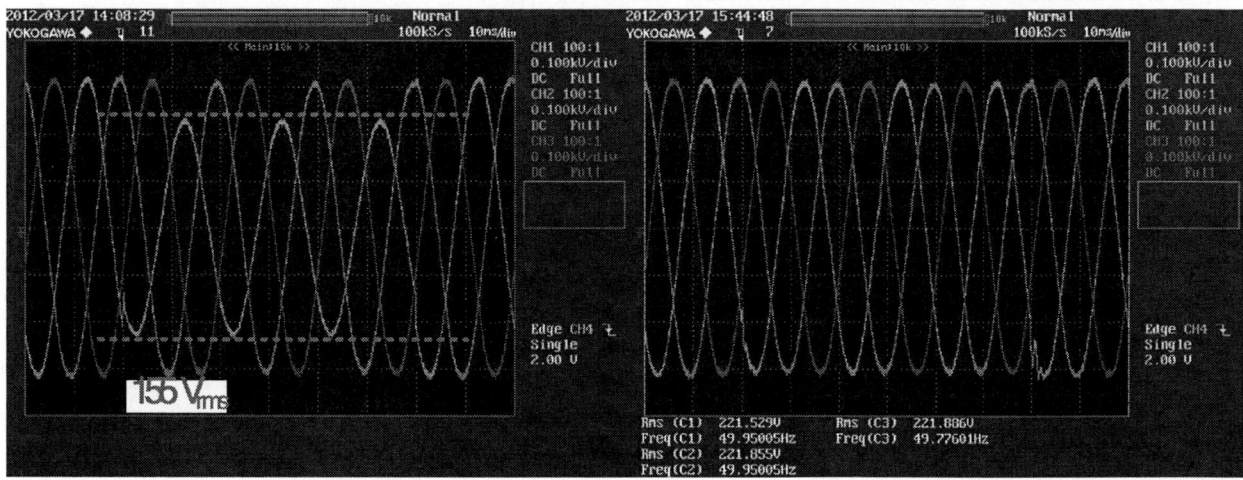

(a)   (b)

Fig.25. Single phase voltage sag to 70% : (a) showing voltage sag at sources and (b) showing voltage at load terminals

978-1-4799-2706-7/14 $31.00 © 2014 IEEE   2398

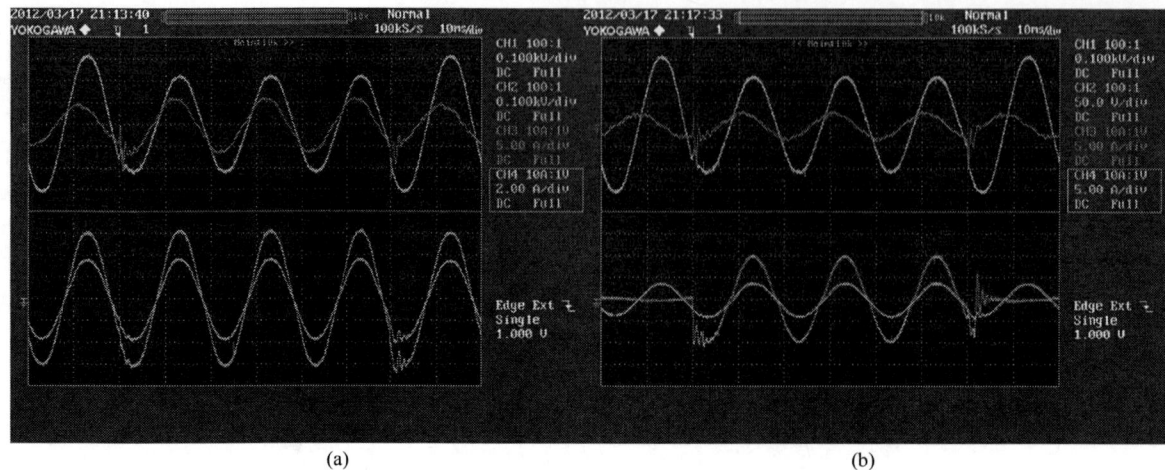

(a)　　　　　　　　　　　　　　　　(b)

Fig. 26. Voltage and current signals of the proposed AC-chopper during voltage sag to 70%:
(a) voltage and current at source (up) and voltage and current at load terminal (bottom),
(b) voltage and current at input ac-chopper (up) and output ac-chopper (bottom).

The voltage and current at the source, load terminal, input AC-chopper and output AC-chopper operating at voltage sag to 70% are illustrated in Fig. 26. Fig. 26 (a) shows the voltage and current at source and load terminals. Fig. 26 (b) illustrates the input/output voltage and current at AC-chopper. The signal both current and voltage at the load terminal is in good power quality. The overshoot current is still under IGBT current limit. The total harmonic voltage distortion at source, load terminal and output AC-chopper are listed in Table IV

TABLE IV.
TOTAL HARMONIC VOLTAGE DISTORTION VALIDATION

Source voltage sag to	THD_{VS} (%) at source	THD_{VL} (%) at load terminal	THD_{VC} (%) at AC-chopper
100%	1.8	1.98	-
90%	1.5	1.68	7.25
80%	1.67	1.90	4.40
70%	1.84	2.02	4.07

VI. CONCLUSION

The AC-chopper to solve voltage sag problem has been proposed. An instantaneous value voltage control technique has been used to control AC-chopper. The simulations for both balanced and unbalanced voltage sag have been performed using MATLAB/Simulink and the 2-kW AC-chopper prototype has been developed and validated with experiments. The system efficiency performance of under test conditions is more than 90% with only 2% total voltage harmonic distortion. The experimental results show that the proposed AC-chopper circuit can be applied in series voltage sag compensation with a fine efficiency and fast dynamic voltage response and total harmonic voltage distortion.

ACKNOWLEDGMENT

The author would prefer to thank Mr. Saradech Varanavin for helping on the experiments, Provincial Electricity Authority (PEA) of Thailand and KMITL for supporting research funds.

REFERENCES

[1] A-A. Edris, "Proposed Terms and Definitions for Flexible ac Transmission System (FACTS)", IEEE Transaction on Power Delivery,Vol 12, No.4, 0ct 1997

[2] Electric Power Research Institute (EPRI), Power quality in commercial buildings, Tech. Rep. BR-105018, 1995

[3] M.J. Ryan; W.E. Brumsickle, and R.D. Lorenz. "Control topology options for single-phase UPS inverters", IEEE Trans. Industry Applications, Vol. 33, No. 2, March-April (1997)

[4] Nabil A. Ahmed, Kenji Amei, and Masaaki Sakui, "A New Configuration of Single-Phase Symmetrical PWM AC Chopper Voltage Controller," IEEE Transactions on Industrial Electronics, Oct 1999, Vol. 46, No.5, pp.942-952.

[5] Steven M. Hietpas and Mark Naden, "Automatic Voltage Regulator Using an AC Voltage–Voltage Converter," IEEE Transactions on Industry Application, Jan/Feb 2000, Vol. 36, No. 1, pp.33-38

[6] S. Polmai, Toshifumi Ise and Sadatoshi Kumagai, "Experiment on Voltage Sag Compensation with Minimum Energy Injection by Use of a Micro-SMES", Proceedings of the Power Conversion Conference-Osaka 2002, Vol. II, pp. 415-420, Osaka, Japan, April 2-5, 2002,

[7] Vazquez, N., Velazquez A., Hernandez C., "AC Voltage Regulator Based on the AC-AC Buck-Boost Converter," IEEE International on Industrial Electronics, ISIE 2007., June 4-7, 2007, pp.533 – 537.

[8] Flores-Arias, J. M., Moreno-Muñoz, A., Domingo-Perez, F., Pallares-Lopez, V. and Gutierrez, D., "Voltage regulator system based on a PWM AC chopper converter," IEEE International Symposium on Industrial Electronics (ISIE), Jun 27-30, 2011, pp.468-473.

[9] Nam-Sup Choi, Byung-Moon Han, Eui-Cheol Nho and Hanju Cha, "Dynamic voltage restorer using PWM ac-ac converter," Proceeding of the Power Electronics Conference (IPEC), June 21-24, 2010, pp. 2690-2695

[10] Josep M. Guerrero, Luis Garcia de Vicuna, Jaume Miret, and Jose Matas. "A nonlinear feed-forward control technique for single-phase UPS inverters", in the Proceeding of the IEEE Conference IECON 02, Vol. 1, pp. 257-261, 2002

[11] S. Polmai, "Instantaneous-value voltage control of a single-phase AC chopper", Proc. International Power Electronics Conference, IPEC-Niigata 2005, vol. 49, No.4, April 4 - 8, 2005

[12] S. Polmai and E. Sugprajun, "Experiment on Instantaneous Value Voltage Control of a Single-Phase AC Chopper", Power Conversion Conference - Nagoya, 2007. PCC '07

[13] Saradech Varanavin and Surin Khomfoi, "An Instantaneous Value Voltage Control AC Chopper for Compensating Voltage Sag in Three Phase Four Wire System," ECTI-CON 2012, Hua-Hin, Thailand, May 16-19, 2012

Voltage Regulation in Distribution System using the Combined DVR

Sota Nakamura, Mutsumi Aoki, Hiroyuki Ukai

Nagoya Institute of Technology
Department of Computer Science and Engineering
Gokiso-cho, Showa-ku, Nagoya 466-8555, Japan
25417584@stn.nitech.ac.jp, aoki.mutsumi@nitech.ac.jp, ukai.hiroyuki@nitech.ac.jp

Abstract- **In recent years, renewable energy system such as Photovoltaic generation system (PV) is widely introduced in distribution system. On the other hand, voltage deviation due to voltage fluctuation and unbalanced voltage caused by PV becomes serious problem. In order to mitigate those problems, the Combined type Dynamic Voltage Regulator (C-DVR) which consists of a step voltage regulator and AC chopper is the effective device. C-DVR can regulate line voltage continuously and quickly. This paper shows the effectiveness of C-DVR by the numerical simulation.**

Keywords— Dynamic voltage regulator, AC chopper, Voltage control, Photovoltaic generation system.

I. Introduction

In recent years, Dispersed Energy Generation systems (DEG) such as Photovoltaic generation system (PVs), Fuel Cell generation (FC) are widely introduced in power systems. Especially, a roof top type PV and FC are attracted and installed in many residential house in Japan, therefore most of DEG are connected via single phase Power Conditioning System (PCS). The much of the unbalanced current flows in a distribution line, therefore the unbalanced voltage may become larger. On the other hand, most of distribution systems are radial system and the voltage of distribution line is controlled by Load Ratio control Transformer (LRT) in a substation (S/S) and Step Voltage Regulator (SVR) in a line and pole transformer to keep within adequate range (Fig.1). Since LRT and SVR are one of a transformer which control the voltage automatically by changing winding ratio, the voltage can be varied discretely and non-phase segregated change. Therefore, the unbalanced voltage in distribution system cannot be regulated by those devices. Furthermore, the output of PV fluctuates frequently and violently depending on weather conditions, amount of unbalanced voltage also may vary frequently and randomly. As a result, single-phase PV system may cause one or two phase voltage deviation from the proper range frequently. However, LRT and SVR has slow speed mechanical operation, and it cannot regulate the voltage effectively. Furthermore, frequently operation of SVR may lead to deterioration of contact for attrition.

In previous research, high speed voltage regulation devices such as Static Var Compensator (SVC), Unified Power Flow Controller (UPFC), are proposed. However, those devices are expensive and it is difficult to install on the distribution line. Thyristor type Step Voltage Regulator (TVR) is also proposed as a faster voltage regulation device, but it cannot regulate the voltage continuously.

From this background, this paper proposes the novel voltage regulator so called the Combined type Dynamic Voltage Regulator (C-DVR). C-DVR consists of a step voltage regulator by thyristor devices (TVR configuration) and AC chopper by IGBT devices and regulates the voltage by injecting the compensating voltage in series to the distribution line. Hence, only the small capacity of AC chopper is combined to TVR configuration, and the cost could be reduced. In this paper, the effectiveness of C-DVR in terms of the suppression of voltage fluctuation due to PV and the reduction of the unbalanced voltage is presented. The configuration of C-DVR is described in detail in Chapter 2. In Chapter 3 the control scheme of the unbalanced voltage compensation is explained. Simulation results are shown in Chapter 4.

II. Configuration of C-DVR

1. Dynamic voltage regulator

C-DVR consists of TVR configuration and AC chopper, regulates the voltage by injecting the compensating voltage in series to the distribution line (Fig. 2). v_{si} (i=u,v,w) is source voltage, R and L are resistance and inductance of the distribution line. v_{fi}(i=u,v,w) is the compensating voltage produced by C-DVR. v_s is the voltage at the source side, and the v_l is the load side voltage.

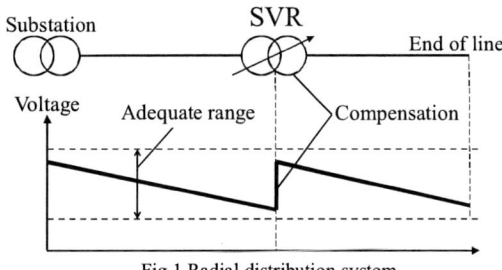

Fig.1 Radial distribution system

978-1-4799-2706-7/14 $31.00 © 2014 IEEE

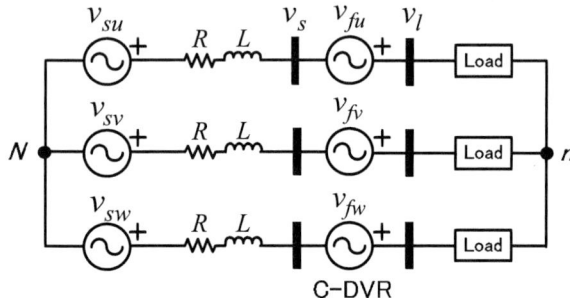

Fig. 2 Schematic diagram of a DVR

2. Combined AC Choppers

C-DVR for three phase line consists of two Combined AC Choppers (CACCs). Fig. 3 shows the configuration of CACC. CACC is a back-boost AC chopper and consists of five bidirectional GTOs (Th0~ Th4) and two bidirectional IGBTs. It has also L-C filters that consist of L_f and C_f in input side and consist of L_o and C_o in output side. GTOs (Th$_1$ ~ Th$_4$) are used for discrete voltage regulation in four levels (V_{in}, $0.5V_{in}$, $-0.5V_{in}$, $-V_{in}$) shown in Fig. 4. When it outputs the zero voltage, Th$_0$ is turned ON. After the discrete voltage regulation by GTOs, AC chopper regulate the output voltage precisely by controlling the duty ratio D in the each range of V$_{RANGE}$ (V_{PH}, V_{PL}, V_{NH} and V_{NL}).

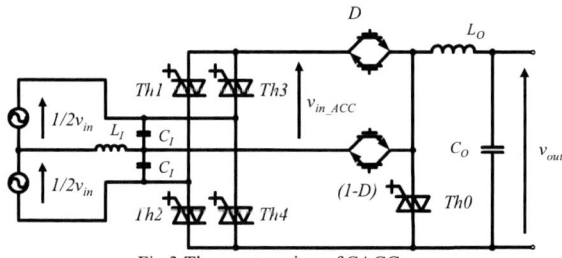

Fig.3 The construction of CACC

Fig. 5 shows current flow diagrams in the each range of V_{RANGE}. v_{out} of the case of V_{PH} (V_{PH} mode) is represented by Eq. (1)

$<V_{PH}$ mode: $0.5V_{in} \leq V_{in\_ACC} \leq V_{in}>$

$$v_{out} = \frac{(1+D)}{2} \cdot V_{in} \sin(\omega t) \qquad (1)$$

In Fig. 5 (a), solid or dotted line shows the change of current flow by switching of the IGBT. Thus HACC can output positive and negative voltage continuously in the range from V_{in}~$0.5V_{in}$.

The control scheme of the other mode is same as V_{PH} mode.

$<V_{PL}$ mode: $0 \leq V_{in\_ACC} \leq 0.5V_{in}>$

$$v_{out} = \frac{(1-D)}{2} \cdot V_{in} \sin(\omega t) \qquad (2)$$

$<V_{NH}$ mode: $-0.5V_{in} \leq V_{in\_ACC} \leq 0>$

$$v_{out} = -\frac{(1-D)}{2} \cdot V_{in} \sin(\omega t) \qquad (3)$$

$<V_{NL}$ mode: $-V_{in} \leq V_{in\_ACC} \leq -0.5V_{in}>$

$$v_{out} = -\frac{(1+D)}{2} \cdot V_{in} \sin(\omega t) \qquad (4)$$

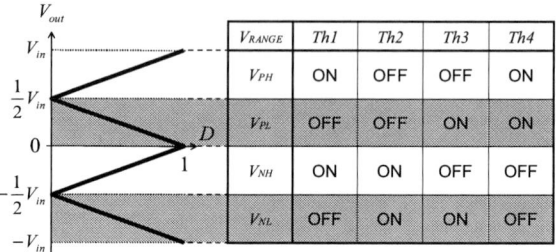

Fig.4 The output of CACC

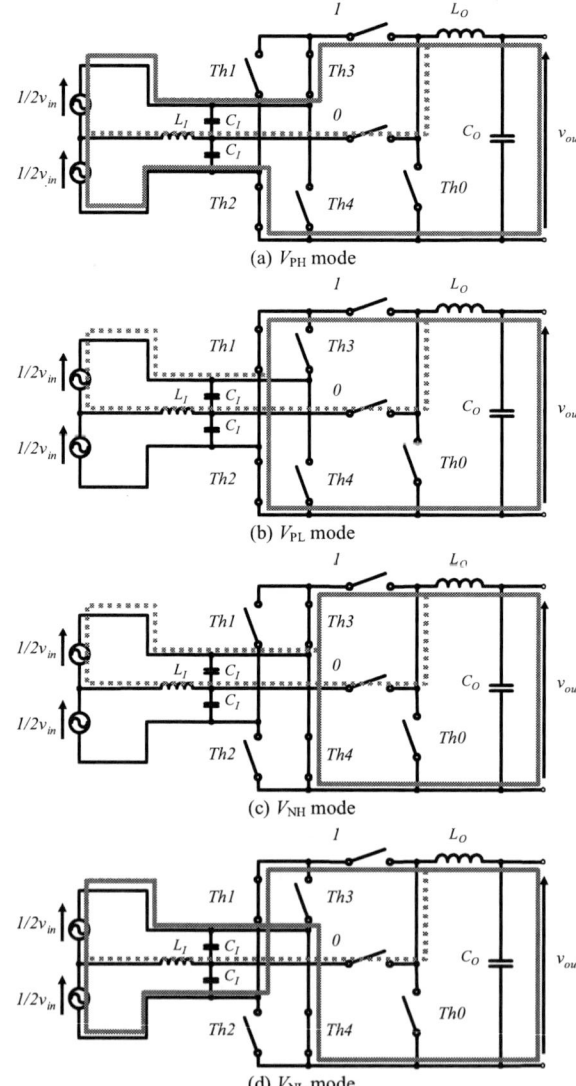

(a) V_{PH} mode

(b) V_{PL} mode

(c) V_{NH} mode

(d) V_{NL} mode

Fig.5 The detail of method to output voltage by CACC

From eq. (1) to eq. (4), V_{in} is determined by Eq. (5) and C-DVR is possible to control the output voltage by determining the coefficient K.

$$v_{out} = K \cdot v_{in} \qquad (-1 \leq K < 1) \qquad (5)$$

Thus, since the voltage applied to the IGBT is half of V_{in}, the capacity of IGBT is small and the cost of CACC may become lower. Furthermore, switching frequency of the small capacity of IGBTs can be high, hence the response of C-DVR can be more quickly.

3. Configuration of C-DVR

The configuration of C-DVR is shown in Fig. 6. V_L stands for the voltage at the load side, and V_S stands for the voltage of the source side. In order to control the three phase voltage, C-DVR has two CACCs. The input voltage of each CACC is provided via the parallel connected transformer from the distribution line and the output voltage of CACC is applied to the distribution line via a series transformer. To reduce the number of switches, two CACCs are connected by V connection.

As described in the last section, CACC can regulate the voltage continuously and quickly, C-DVR can suppress the voltage fluctuation precisely. Therefore the violent voltage fluctuation caused by large amount of PVs can be compensated using C-DVR.

The reference voltage V_{ref} is determined according to the conditions of the distribution line. Then the difference between V_L and V_{ref} is calculated and the compensation voltage which is the output of CACC is determined by conversion from line voltage to phase voltage. Thus, the load voltage can be regulated at constant value.

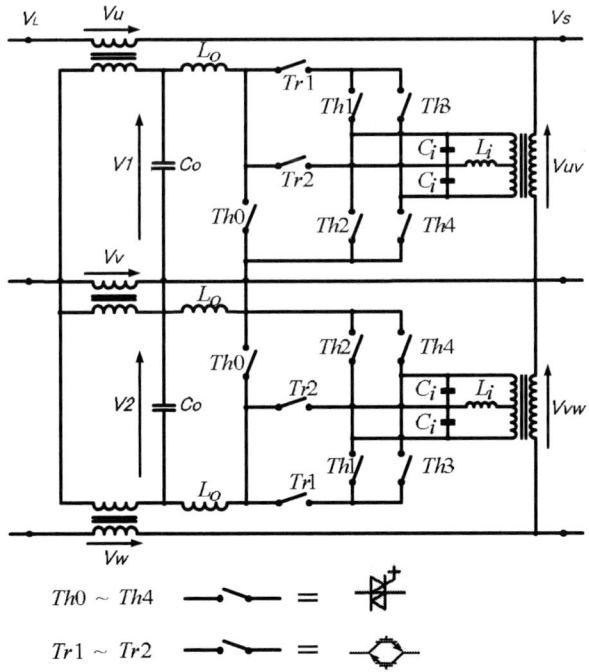

Fig. 6 The configuration of C-DVR

III. Unbalanced voltage compensation by C-DVR

When the large amount of residential PVs is connected to the line in single phase, the problem of unbalanced voltage may become more seriously. The unbalanced voltage causes the problems to an induction motor, such as increasing the loss and torque fluctuation. Furthermore, the control of the voltage in the distribution line by utility company may become more difficult. Therefore, the suppression of unbalanced voltage is very important issue in terms of power quality.

In order to reduce the unbalanced voltage, C-DVR should output the negative phase sequence component of the line voltage. Therefore, the magnitude and phase angle of output voltage should be controlled by the input voltage V_{in} that is the line voltage at the source voltage V_S.

To achieve this aim, the output voltage of CACC is modulated by second harmonic.

The input voltage v_{in} is expressed in Eq. (6).

$$v_{in} = V_{in}\sin(\omega t + \varphi_{in}) \qquad (6)$$

V_{in}: Amplitude of input voltage

φ_{in}: Phase angle

The reference of the output voltage of CACC v_1 is determined by the modulating coefficient K which is given in Eq. (7).

$$v_1 = K \cdot v_{in}$$
$$K = K_0 + K_2\sin(\omega t + 2\varphi_{in}) \qquad (7)$$

K_0: Basic coefficient

K_2: Second harmonic coefficient

Then the output voltage v_1 is expressed in Eq. (8) [3].

$$v_1 = K \cdot v_{in}$$
$$= V_{in}\sqrt{K_0^2 + \left(\frac{K_2}{2}\right)^2}\ \sin(\omega t + \varphi_{in} + \varphi) \qquad (8)$$
$$- \frac{K_2 V_{in}}{2}\cos(3\omega t + 3\varphi_{in})$$

Where, $\varphi = \tan^{-1}\dfrac{K_2}{2K_0}$.

From Eq. (8), the amplitude and phase angle of the output voltage can be controlled by changing K_0 and K_2. Although third harmonic voltage is appeared in the output voltage, it may be small enough for the unbalanced voltage compensation. K_0 and K_2 is determined as follows.

The relation between the output of C-DVR and that of CACCs is represented in Eq. (9).

$$\begin{cases} v_1 = v_u - v_v \\ v_2 = v_v - v_w \end{cases} \qquad (9)$$

v_1: Output voltage of CACC1 (Upper side)

v_2: Output voltage of CACC2 (Lower side)

The reference of the output voltage of C-DVR is defined in Eq. (10) using the reference value of the positive-phase- sequence voltage V_p^* and the negative-phase-sequence voltage V_n^*.

978-1-4799-2706-7/14 $31.00 © 2014 IEEE

$$\begin{cases} v_u{}^* = V_p{}^* \sin(\omega t + \varphi_p) + V_n{}^* \sin(\omega t + \varphi_n) \\ v_v{}^* = V_p{}^* \sin(\omega t + \varphi_p - 2/3\pi) + V_n{}^* \sin(\omega t + \varphi_n - 2/3\pi) \\ v_w{}^* = V_p{}^* \sin(\omega t + \varphi_p + 2/3\pi) + V_n{}^* \sin(\omega t + \varphi_n + 2/3\pi) \end{cases}$$
$$(10)$$

Then the output of CACCs is determined by Eq. (11) from Eq.(9).

$$\begin{cases} v_1 = v_u{}^* - v_v{}^* \\ \quad = \sqrt{(A-C)^2 + (B-D)^2}\, \sin\!\left(\omega t + \tan^{-1}\dfrac{B-D}{A-C}\right) \\ v_2 = 2v_v{}^* + v_w{}^* \\ \quad = \sqrt{(2A+C)^2 + (2B+D)^2}\, \sin\!\left(\omega t + \tan^{-1}\dfrac{2B+D}{2A+C}\right) \\ \quad A = V_p \cos\varphi_p + V_n \cos\varphi_n \\ \quad B = V_p \sin\varphi_p + V_n \sin\varphi_n \\ \quad C = V_p \cos(\varphi_p - 2/3\pi) + V_n \cos(\varphi_n - 2/3\pi) \\ \quad D = V_p \sin(\varphi_p - 2/3\pi) + V_n \sin(\varphi_n - 2/3\pi) \end{cases}$$
$$(11)$$

The reference value of the positive-phase- sequence voltage $V_p{}^*$ is determined according to the conditions of the distribution line and the negative-phase- sequence voltage $V_n{}^*$ is zero in order to suppress the unbalanced voltage. Thus, the coefficient K_0 and K_2 is determined by using Eq. (8) and Eq. (11).

IV. SIMULATION

1. Suppression of the voltage fluctuation

In order to confirm the effectiveness of the C-DVR, the numerical simulation using MATLAB/Simulink is performed. The distribution model is shown in Fig. 7. The large amount of PV (MEGA Solar) is connected at the end of line. The maximum output of PV is 4[MW] and the output of PV is fluctuated as shown in Fig. 8 which is modified form the actual data. C-DVR is installed around the center of the line. The capacity of C-DVR is 180[kVA] that is determined to be able to control the line voltage ± 300 [V] in the steady state. The loads connected in each load is set as R-L load and its capacity is 125[kVA] and power factor is 0.98(lagging). The other simulation conditions are shown in Table 1. The specification of C-DVR is shown in Table 2.

The maximum output voltage V_{out_MAX} at the series transformer of C-DVR is restricted by the line voltage of distribution line. When the effective value of the phase voltage at point A is V_i ($i = u, v, w$), the maximum output voltage of C-DVR is restricted by Eq. (9) using the winding ratio of the shunt transformer and series transformer.

$$V_{out\_MAX} = V_i \times \frac{2000}{6600} \times \frac{100}{500} \tag{9}$$

The output voltage of each CACC is regulated using the modulating coefficient K from Eq. (5). In order to suppress the voltage fluctuation caused by PV, K is determined by Eq. (10).

$$K = \frac{V_i - V_{ref}}{V_i \times 2000/6600 \times 100/500} \tag{10}$$

where, V_i : the phase voltage at the point A

V_{ref} : the reference voltage

The reference voltage V_{ref} is determined in considering of the conditions of a distribution line. In this simulation, the reference voltage V_{ref} is set 6340[V] in considering the voltage drop of the distribution line in case of that the PV is not connected.

The simulation results are shown in Fig. 9. Although the voltage at the point A fluctuates due to the output fluctuation of PV, the voltage at the point B is sufficiently suppressed by using C-DVR.

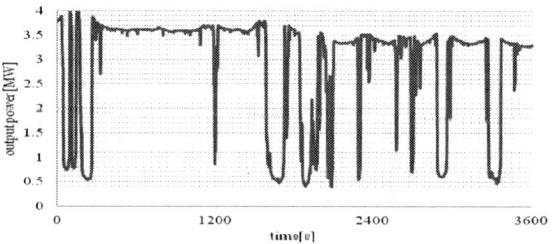

Fig. 8 The output of PV

Table 1
Simulation Conditions

			Value
Sending voltage of S/S			6600 [V]
Line impedance		R	0.139 [Ω]
		L	0.4 [mH]
Load	Capacity	125	[kVA]
	Power factor	0.98	(Lagging)
	impedance	R	118.5 [Ω]
		L	583.7 [H]

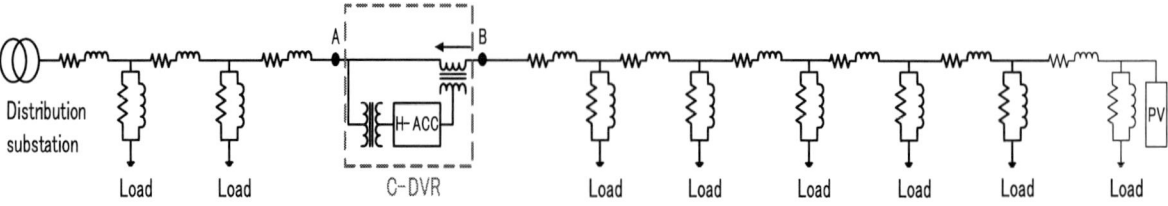

Fig. 7 Simulation model of suppression of voltage fluctuation

Table 2
Specification of C-DVR

	Value
Controllable Voltage Range	$\pm$ 300[V]
Winding ratio of shunt transformer	2000/6600 [V]
Winding ratio of series transformer ratio	100/500 [V]
GTO turn on time	0.05 [ms]
Input filter	L_i 0.5 [mH] C_i 0.05 [mF]
Output filter	L_i 2.0 [mH] C_i 0.01 [mF]
Total AC chopper capacity	90 [kVA]

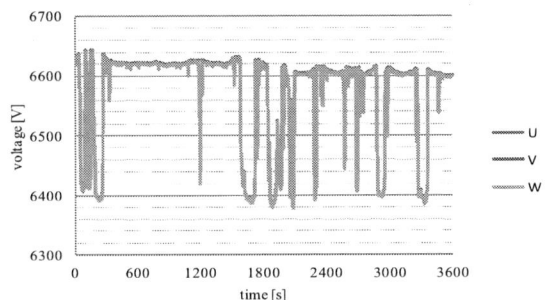

(a) Voltage at point A

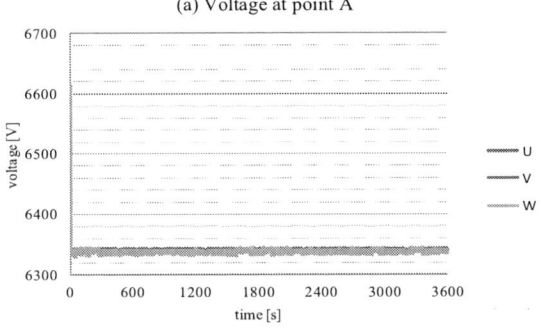

(b) Voltage at point B
Fig. 9 Simulation result of suppression of voltage fluctuation

2. Reduction of the unbalanced voltage

In order to clear the effectiveness of the unbalanced voltage suppression, PV is not considered in this simulation. The simulation model is shown in Fig. 8. In this model, the sending voltage at the distribution substation (S/S) is unbalanced. The amplitude of the positive-phase-sequence component of the sending voltage is set $6600\sqrt{2/3}$ [V] and the negative-phase-sequence component of it is set 100[V] respectively. In this case, the voltage unbalance rate is 1.85%.

The modulating coefficient K_0 and K_2 in Eq.(8) is determined from that the the negative-phase-sequence component of the voltage is zero.

Simulation result of the voltage at the point B is shown in Fig. 9. C-DVR starts the compensation at 0.05[s]. After the suppression of the negative phase sequence component by C-DVR, the unbalanced voltage at the point B is reduced sufficiently and stably. The voltage unbalance rate at the point B becomes 0.08%.

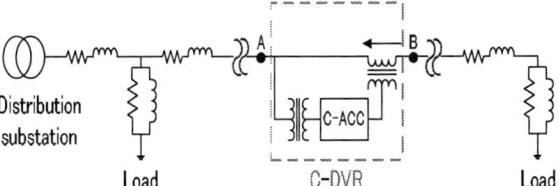

Fig. 8 Simulation model of unbalanced line voltage

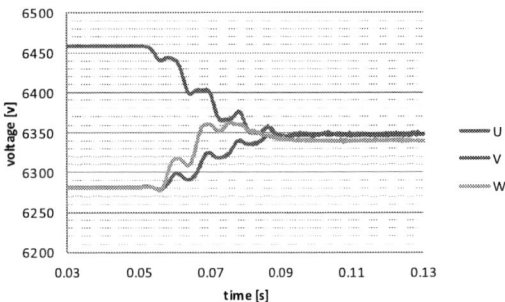

Fig. 9 Simulation result of reduction of unbalanced voltage

V. Conclusion

In this paper, the novel voltage regulator C-DVR is proposed. C-DVR consists of a step voltage regulator by thyristor devices (TVR configuration) and AC chopper by IGBT devices and regulates the voltage by injecting the compensating voltage in series to the distribution line. The feature of C-DVR is fast and continuous regulation of the distribution line and the lower cost as compared with SVC or UPFC.

By the numerical simulation using the distribution model, it is confirmed that C-DVR is able to suppress the voltage fluctuation caused by PV sufficiently.

Furthermore, the control scheme of the unbalanced voltage suppression is also presented, and the unbalanced voltage in the sending voltage at the S/S is reduced at the point of the load side by using C-DVR.

However, the third harmonics is remained in the voltage according to the second harmonic modulation. Therefore, in case of the large negative-phase-component of the voltage compensation, the third harmonic voltage may affects to the distribution line. The reduction method of the third harmonic voltage is the future works. And the experimental verification using a prototype model is also the next challenge.

References

[1] T. Kondo, J. Baba, and A. Yokoyama: "Voltage Control of Distribution Network with a Large Penetration of Photovoltaic Generations using FACTS Devices", IEEJ Trans.PE,Vol.126,No.3,pp347-358,2006

[2] Y. Sasaki: "High Speed TVR for Power Distribution Lines and Test Results", IEEJ Trans.PE,Vol.123,No.9,pp1105-1111,2003

[3] Divan M, J.Sastry: "Voltage Synthesis Using Dual Virtual Quadrature Sources—A New Concept in AC Power Conversion", Power Electronics, IEEE Transactions, Vol.23, pp3004-3013, 2013

[4] Deepak Divan: "Design and Testing of a Medium Voltage Controllable Network Transformer Prototype with an Integrated Hybrid Active Filter", IEEE, pp4035-4042, 2011

[5] Debrup Das, Deepak Divan: "Power Flow Control in Networks Using Controllable Network Transformers", ICEE 2009, pp2224-2231, 2009

[6] Deepak Divan, Jyoti Sastry: "Control of Multilevel Direct AC Converters", ICEE, pp3077-3084, 2009

[7] Hirofumi Akagi, Ryohei Kitada: "Control and Design of a Modular Multilevel Cascade BTB System Using Bidirectional Isolated DC/DC Converters"

[8] Shinya Sekizaki, Mutumi Aoki, Hiroyuki Ukai, Shunsuke Sasaki, Takaya Shigetou: "Control of multistage SVRs using Voltage sensor in Distribution System with Large Amount of Photovoltaic Generations" IEEJ, Vol.133, pp45-55, 2013

Nonlinear Control of Three-Phase Four-Wire Dynamic Voltage Restorers for Distribution System

Seon-Yeong Jeong, Thanh Hai Nguyen, Dong-Choon Lee, and Jang-Mok Kim

Dept. of Electrical Eng., Yeungnam University, 280 Daehak-Ro, Gyeongsan, Gyeongbuk, Korea
E-mail: iamsolinear@gmail.com, nthai@ctu.edu.vn, dclee@yu.ac.kr
Dept. of Electrical Eng., Pusan National University, Busan, Korea
E-mail: jmok@pusan.ac.kr

Abstract-This paper proposes a novel control strategy of the three-phase four-wire dynamic voltage restorers (DVR) under the grid voltage unbalance and nonlinear loads in the distribution system, where a feedback linearization technique is applied. By the input-output feedback linearization for the nonlinear model of DVR with the PWM voltage source inverter (VSI) and the LC filters, the system is linearized and then the controller of the linearized model is designed by a linear control theory. With this control strategy, the control performance of the DVR becomes faster and more stable under the unbalanced source voltages and loads as well as the nonlinear loads. PSIM simulation and experimental results have proved the effectiveness of the proposed control scheme.

Keywords-Dynamic voltage restorers, feedback linearization, nonlinear control, three-phase four-wire VSI.

I. INTRODUCTION

Voltage disturbances are the most common power quality issue in the distribution power system. With the industrial processes based on a large number of electronic devices, the industrial loads are vulnerable to the disturbances of the power supply, which results in the malfunctions, tripping, or even faults of the load system. The voltage disturbances are normally classified into the voltage sags, swells, harmonics, unbalances, and flickers [1], [2]. The main reason for the voltage sag is short-circuit faults such as line-to-ground fault and line-to-line fault, and the startup of high power rating of induction motors. The voltage swell is normally caused by the switching capacitors or the removal of the large load. The extensive use of nonlinear loads increases the current and voltage harmonic issues further [3]-[5].

To improve the power quality in the distribution network, where the sensitive loads require a clean power for their appropriate operation, the DVR is an attractive solution for compensating the voltage quality issues. The basic function of the DVR is to inject a voltage with an appropriate magnitude, phase angle, and frequency in series with the source to maintain the nominal voltage to the load even though the grid voltage is unbalanced or distorted. The DVR is a custom power device consisting of a VSI and output LC filters, which is inserted between the power grid and the loads [6]-[8]. In general, a Δ/Y

transformer is used to connect the industrial loads to the power supply, which prevents the zero-sequence components. However, there are systems using a Y/Y transformer with a grounded neutral [9], in which the zero-sequence components can be propagated to the load during the unbalanced faults. Then, the DVR is required to be able to generate the zero-sequence voltage components. For this requirement, the DVR system based on a four-leg VSI or a three-phase four-wire VSI with split capacitors are preferred [9]-[11].

Control of the DVR is comprised of the derivation of the injected voltage reference and the control of the converter, in which a cascade controller consisting of an outer voltage control loop and d-q inner current control loops has been presented [12]. However, its control dynamic is low due to the limitation of the bandwidth of the voltage control loop [6]. Moreover, under unbalanced sag conditions, the negative- and zero-sequence components may appear in the grid voltage. Then, the *d-q* components of the grid voltage are not DC signals. Under these conditions, the typical PI (proportional integral) controller is not appropriate to regulate the AC signals. Another critical issue considered for controlling the DVR is the nonlinearity of the DVR. To overcome this drawback for controlling the AC signal, a feedback linearization technique has been suggested to control the UPS (uninterruptible power supply) inverter system [13], [14]. Also, for the nonlinear system model, the feedback linearization control gives better transient performance than the PI controller.

In this paper, the feedback linearization control of the three-phase four-wire DVR system is proposed, in which the positive-, negative-, and zero-sequence output voltage components generated from the DVR are injected to the distribution system to keep the load voltage constant. For this control scheme, a nonlinear relationship between the converter currents and the output voltages is derived from a power balance of the converter output and the load sides. Then, the output voltage controller can be designed using the linear control theory. By a pole placement technique, the tracking control law is designed. In addition, for the DVR system based on the three-phase four-wire VSI, the three-dimensional space-vector PWM

978-1-4799-2706-7/14 $31.00 © 2014 IEEE

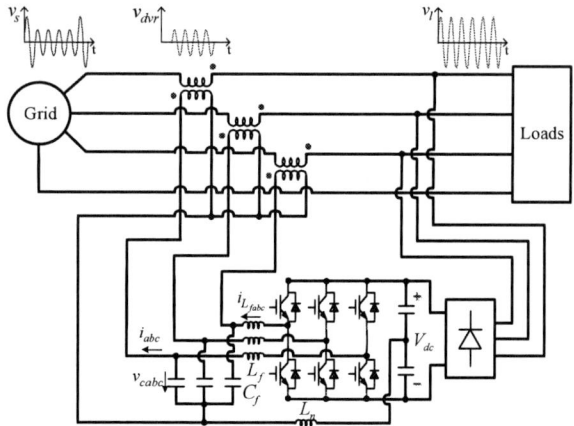

Fig. 1. DVR system based on three-phase four-wire VSI.

is applied. With the nonlinear control scheme, a high dynamic response for the control of the DVR system is obtained under the unbalanced condition of grid voltages and loads as well as the nonlinear loads. Simulation and experimental results for the three-phase four-wire distribution system are shown to verify the validity of the proposed control scheme.

II. MODELING OF THREE-PHASE FOUR-WIRE DVR SYSTEMS

A. System Description

A DVR system is shown in Fig. 1, which consists of a shunt diode rectifier connected at the load side and a series converter connected between the grid and the loads. The series converter uses the three-phase four-wire VSI with the split capacitors, which allows the DVR system to produce the unbalanced voltage with a positive-, negative-, and zero-sequence components for the voltage compensation at the unbalanced sag conditions of the grid. In the configuration of Fig. 1, the DC capacitors of the DVR system do not need a large capacity unlikely in the case of connecting the shunt diode rectifier at the grid side, since the load voltage is compensated to keep at the nominal value even during the grid voltage sag [7], [8].

B. Modeling of DVR Systems

The output voltages and currents of the VSI as shown in Fig. 1 can be expressed in the synchronous d-q-0 reference frame as [14]

$$\dot{i}_{d,L_f} = \frac{1}{L_f}v_d - \frac{1}{L_f}v_{d,c} + \omega i_{q,L_f} \tag{1}$$

$$\dot{i}_{q,L_f} = \frac{1}{L_f}v_q - \frac{1}{L_f}v_{q,c} - \omega i_{d,L_f} \tag{2}$$

$$\dot{i}_{0,L_f} = \frac{1}{\left(L_f + 3L_n\right)}v_0 - \frac{1}{\left(L_f + 3L_n\right)}v_{0,c} \tag{3}$$

$$\dot{v}_{d,c} = \frac{1}{C_f}i_d - \frac{1}{C_f}i_{d,L_f} + \omega v_{q,c} \tag{4}$$

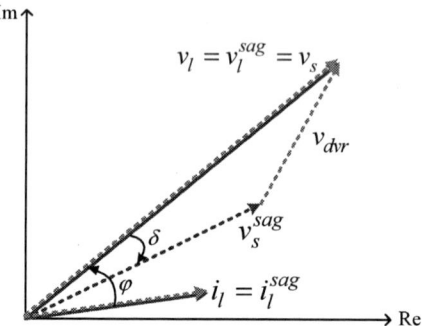

Fig. 2. Phasors of the voltages and currents for pre-sag compensation strategy.

$$\dot{v}_{q,c} = \frac{1}{C_f}i_q - \frac{1}{C_f}i_{q,L_f} - \omega v_{d,c} \tag{5}$$

$$\dot{v}_{0,c} = \frac{1}{C_f}i_{0,L_f} - \frac{1}{C_f}i_0 \tag{6}$$

where i_{dq0,L_f} are the d-q-0 axis inverter output currents, i_{dq0} are the d-q-0 axis output currents of the DVR, v_{dq0} are the d-q-0 axis inverter terminal voltages, $v_{dq0,c}$ are the d-q-0 axis capacitor voltages, ω is the grid angular frequency, L_f, C_f are the filter inductance and capacitance, respectively, and L_n is the neutral-line inductance.

C. Control Requirements for DVR Systems

For the normal operation of the loads, the DVR has to compensate for the grid sags to restore the load voltage at the nominal one. There are three control strategies for the DVR to compensate for the load voltages [8]. The first method is called as a pre-sag compensation, where the load voltage is recovered at the same one as before the sag. The second strategy is the in-phase compensation, in which the load voltage is compensated in-phase with the grid after the sag. This method optimizes the magnitude of the DVR output voltage. However, a phase jump in the load voltage is inevitable. The other strategy is to control the required compensation voltage of the DVR with 90 degrees lagging to the load current, which is able to optimize the compensation energy, called as the energy optimized compensation strategy.

In this work, the pre-sag compensation strategy is adopted, where the amplitude and phase of the load voltage are kept exactly at the same as before the sag. From Fig. 1, the load voltage is expressed as

$$v_{abc,l} = v_{abc,s} - v_{abc,dvr} \tag{7}$$

where $v_{abc,l}$ is the load voltage, $v_{abc,s}$ is the three-phase grid voltage, and $v_{abc,dvr}$ is the voltage injected by the DVR. Fig. 2 shows the voltage and current phasors before and during the sag.

The references of the compensation voltages, $v_{abc,c}^*$, injected by the DVR can be calculated as

$$\begin{bmatrix} v_{a,c}^* = v_{a,l} - v_{a,s} \\ v_{b,c}^* = v_{b,l} - v_{b,s} \\ v_{c,c}^* = v_{c,l} - v_{c,s} \end{bmatrix} \tag{8}$$

Then, the reference voltages expressed in the abc-frame are transformed to the d-q-0 components in the synchronous reference frame used for the DVR voltage controllers.

III. Proposed Feedback Linearization Control of Three-Phase Four-Wire DVR System

A. Feedback Linearization Control Technique

A feedback linearization theory is briefly reviewed for a multi-input multi-output system, which is expressed as [15]

$$\dot{x} = f(x) + g(x)u \tag{9}$$
$$y = h(x) \tag{10}$$

where x is a state vector, u is control inputs, y is outputs, f and g are the smooth vector fields, and h is the smooth scalar function.

To obtain the input-output feedback linearization from the nonlinear model, the outputs in (10) are differentiated until the inputs appear, which are expressed as

$$\begin{bmatrix} y_1^{(r_1)} \\ \cdots \\ \cdots \\ y_m^{(r_m)} \end{bmatrix} = \begin{bmatrix} L_f^{r_1} h_1(x) \\ \cdots \\ \cdots \\ L_f^{r_m} h_m(x) \end{bmatrix} + E(x) \begin{bmatrix} u_1 \\ \cdots \\ \cdots \\ u_m \end{bmatrix} \tag{11}$$

where the $m \times m$ matrix E(x) is defined as

$$E(x) = \begin{bmatrix} L_{g_1} L_f^{r_1-1} h_1 & \cdots & \cdots & L_{g_m} L_f^{r_1-1} h_1 \\ \cdots & \cdots & \cdots & \cdots \\ \cdots & \cdots & \cdots & \cdots \\ L_{g_1} L_f^{r_m-1} h_m & \cdots & \cdots & L_{g_m} L_f^{r_1-1} h_m \end{bmatrix} \tag{12}$$

and $L_g h$ and $L_f h$ are the Lie derivatives of the $h(x)$ with respect to $g(x)$ and $f(x)$, respectively.

From (11), the input transformation can be obtained as

$$u = -E^{-1}(x) \begin{bmatrix} L_f^{r_1} h_1(x) \\ \cdots \\ \cdots \\ L_f^{r_m} h_m(x) \end{bmatrix} + E^{-1}(x) \begin{bmatrix} v_1 \\ \cdots \\ \cdots \\ v_m \end{bmatrix} \tag{13}$$

where $\begin{bmatrix} v_1 & v_2 & \cdots & v_m \end{bmatrix}^T$ is the new inputs, which has a linear differential relation with the output y derived from the (11) and (13) as

$$\begin{bmatrix} v_1 \\ \cdots \\ \cdots \\ v_n \end{bmatrix} = \begin{bmatrix} y_1^{(r_1)} \\ \cdots \\ \cdots \\ y_m^{(r_m)} \end{bmatrix} \tag{14}$$

where $y_1^{(r_1)}, \ldots, y_m^{(r_m)}$ are the outputs differentiated r times, $v_1, \ldots, v_m$ are the new inputs, which have the linear relationship with the control inputs.

B. Proposed Feedback Linearization Control Technique for DVR Systems

Eqs. (1) to (6) can be rewritten so that they are in the form of (9) and (10), which are expressed as

$$f = \begin{bmatrix} \omega i_{q,L_f} - \dfrac{1}{L_f} v_{d,c} \\ -\omega i_{d,L_f} - \dfrac{1}{L_f} v_{q,c} \\ -\dfrac{1}{L_f + 3L_n} v_{n,c} \\ \dfrac{1}{C_f} i_{d,L_f} - \dfrac{1}{C_f} i_{d,c} + \omega v_{q,c} \\ \dfrac{1}{C_f} i_{q,L_f} - \dfrac{1}{C_f} i_q - \omega v_d \\ \dfrac{1}{C_f} i_{0,L_f} - \dfrac{1}{C_f} i_0 \end{bmatrix}, \quad g = \begin{bmatrix} \dfrac{1}{L_f C_f} & 0 & 0 \\ 0 & \dfrac{1}{L_f C_f} & 0 \\ 0 & 0 & \dfrac{1}{(L_f + 3L_n)C_f} \end{bmatrix}$$

$$x = \begin{bmatrix} i_{d,L_f} & i_{q,L_f} & i_{0,L_f} & v_{d,c} & v_{q,c} & v_{0,c} \end{bmatrix}^T$$

$$u = \begin{bmatrix} v_d & v_q & v_0 \end{bmatrix}^T, \quad y = \begin{bmatrix} v_{d,c} & v_{q,c} & v_{0,c} \end{bmatrix}^T.$$

Next, the outputs are differentiated twice to produce the linear relationship with the control inputs, which are expressed as

$$\begin{bmatrix} \ddot{y}_1 \\ \ddot{y}_2 \\ \ddot{y}_3 \end{bmatrix} = A(x) + E(x) \begin{bmatrix} u_1 \\ u_2 \\ u_3 \end{bmatrix} \tag{15}$$

where

$$A(x) = \begin{bmatrix} \dfrac{2}{C_f} \omega i_{q,L_f} - \left(\dfrac{1}{L_f C_f} + \omega^2 \right) v_{d,c} - \dfrac{1}{C_f} \dot{i}_d - \dfrac{1}{C_f} \omega i_q \\ -\dfrac{2}{C_f} \omega i_{d,L_f} - \left(\dfrac{1}{L_f C_f} + \omega^2 \right) v_{q,c} - \dfrac{1}{C_f} \dot{i}_q + \dfrac{1}{C_f} \omega i_d \\ -\dfrac{1}{(L_f + 3L_n)C_f} v_{0,c} - \dfrac{1}{C_f} \dot{i}_0 \end{bmatrix}$$

$$E(x) = \begin{bmatrix} \dfrac{1}{L_f C_f} & 0 & 0 \\ 0 & \dfrac{1}{L_f C_f} & 0 \\ 0 & 0 & \dfrac{1}{(L_f + 3L_n)C_f} \end{bmatrix}.$$

Then, the control law is given as

$$\begin{bmatrix} v_d^* \\ v_q^* \\ v_n^* \end{bmatrix} = \begin{bmatrix} u_1 \\ u_2 \\ u_3 \end{bmatrix} = \left(-A(x) + \begin{bmatrix} \ddot{y}_1 \\ \ddot{y}_2 \\ \ddot{y}_3 \end{bmatrix} \right) E^{-1}(x). \tag{16}$$

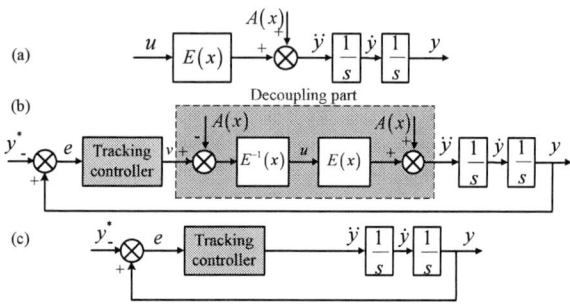

Fig. 3. Process of feedback linearization control. (a) Relation of control input and output. (b) Decoupling process. (c) Simplified control diagram.

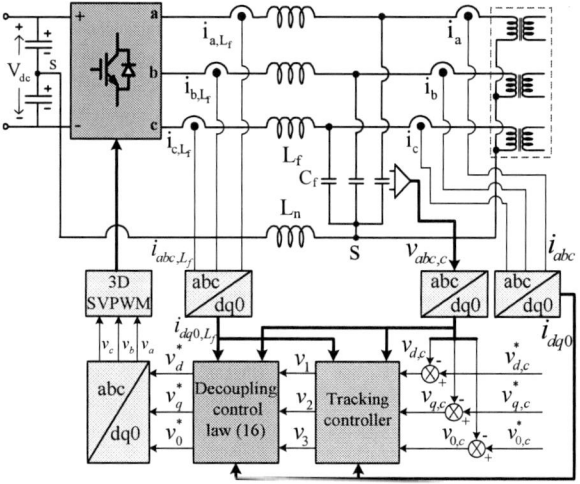

Fig. 4. Control block diagram of DVR.

Fig. 3 shows a process of the feedback linearization control for the DVR, in which Fig. 3(a) shows the input-output relation consisting of a double integrator. A decoupling control law for the DVR system with a tracking controller is shown in Fig. 3(b). Then, the system with decoupling is shown in Fig. 3(c).

The new control inputs for the tracking control are obtained as

$$
\begin{aligned}
v_1 &= \ddot{y}_{1ref} - k_{11}\dot{e}_1 - k_{12}e_1 \\
v_2 &= \ddot{y}_{2ref} - k_{21}\dot{e}_2 - k_{22}e_2 \\
v_3 &= \ddot{y}_{3ref} - k_{31}\dot{e}_3 - k_{32}e_3
\end{aligned}
\tag{17}
$$

where, $e_1 = y_1 - y_{1ref}$, $e_2 = y_2 - y_{2ref}$, $e_3 = y_3 - y_{3ref}$, $k_{ij}(i, j = 1 \sim 3)$ are the tracking controller gains.

For eliminating the error of the tracking controller, the new control inputs including the integral control are given by

$$
\begin{bmatrix} v_1 \\ v_2 \\ v_3 \end{bmatrix} = \begin{bmatrix} \ddot{y}_{1ref} - k_{11}\dot{e}_1 - k_{12}e_1 - k_{13}\int e_1 \\ \ddot{y}_{2ref} - k_{21}\dot{e}_2 - k_{22}e_2 - k_{23}\int e_2 \\ \ddot{y}_{3ref} - k_{31}\dot{e}_3 - k_{32}e_2 - k_{33}\int e_3 \end{bmatrix}.
\tag{18}
$$

Then, the dynamic errors of the tracking controller from (18) are obtained as

TABLE I
PARAMETERS OF GRID, LOAD, AND DVR

	Parameters	Values
Grid	AC grid voltage	220 V / 60 Hz
DVR system	DC input voltage	311 V
	Neutral inductance	0.5 mH
	Filter inductance	3 mH
	Filter capacitance	100 μF
	Transformer ratio	1:1
	Switching frequency	5 kHz
Loads	RL load	50 Ω / 1 mH (3 phases)
	Nonlinear loads	3 single-phase diode rectifiers supply resistive loads: 50 Ω; 50 Ω; 1 kΩ

$$
\begin{aligned}
\dddot{e}_1 + k_{11}\ddot{e}_1 + k_{12}\dot{e}_1 + k_{13}e_1 &= 0 \\
\dddot{e}_2 + k_{21}\ddot{e}_2 + k_{22}\dot{e}_2 + k_{23}e_2 &= 0 . \\
\dddot{e}_3 + k_{31}\ddot{e}_3 + k_{32}\dot{e}_3 + k_{33}e_3 &= 0
\end{aligned}
\tag{19}
$$

From Fig. 3(c) and (19), the closed-loop transfer function for the d-axis DVR voltage can be expressed as

$$
\frac{y}{y^*} = \frac{v_{d,c}}{v_{d,c}^*} = \frac{\left(k_{11}s^2 + k_{12}s + k_{13}\right)}{\left(s^3 + k_{11}s^2 + k_{12}s + k_{13}\right)} .
\tag{20}
$$

which is similar with those of the q-axis and zero-sequence voltages.

The tracking controller gains are chosen by the pole-placement technique, in which the poles of the transfer function, s_1, s_2, s_3, are chosen to be located in the left-half complex plane. Then, the gains of tracking controllers are given as [12]

$$
\begin{aligned}
k_{11} &= k_{21} = k_{31} = -(s_1 + s_2 + s_3) \\
k_{12} &= k_{22} = k_{32} = s_1 \cdot s_2 + s_2 \cdot s_3 + s_3 \cdot s_1 . \\
k_{13} &= k_{23} = k_{33} = k_{21} = k_{31} = -s_1 \cdot s_2 \cdot s_3
\end{aligned}
\tag{21}
$$

Fig. 4 shows the control block diagram of the three-phase four-wire DVR system. The voltage references for the DVR are produced from (8), then transformed to d-q-0 components for the tracking controller. The decoupling control law gives the voltage references for the 3D-SVPWM (three-dimensional SVPWM) [14].

IV. SIMULATION RESULTS

To verify the effectiveness of the feedback linearization control, the PSIM simulation has been carried out for the 3P4W DVR system connected in series with the distribution system as shown in Fig. 1. The parameters of the distribution system and the DVR are listed in Table I.

Fig. 5 shows the performance of the DVR system to restore the load voltages at the rating value of 220 V under the balanced grid sag condition for the balanced and linear loads. Fig. 5(a) shows the grid voltages, where the voltages drop to 50% for 40 ms. In order to avoid the tripping of the loads, the DVR output voltages in Fig. 5(b) are injected to keep the load voltages unchanged as shown in Fig. 5(c). The load voltages after the sag are almost sinusoidal and balanced, which are the same as

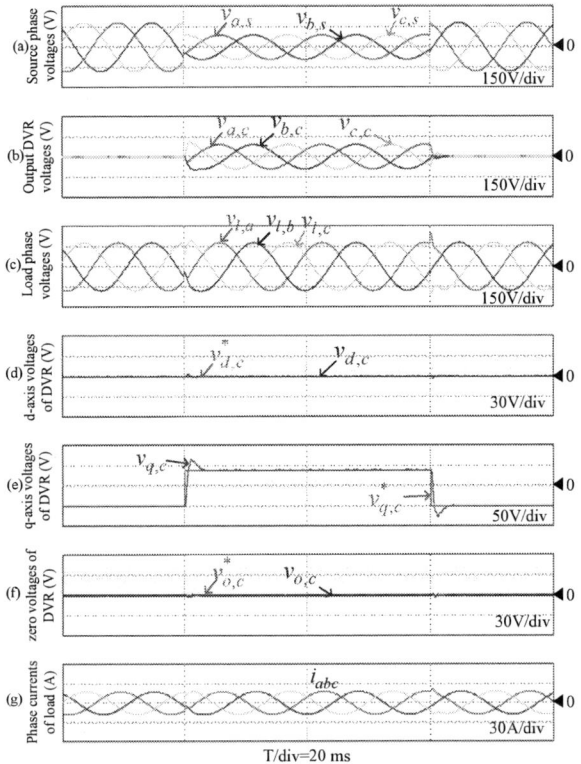

Fig. 5. Performance of DVR by FL control performance under linear load, balanced-voltage sag conditions. (a) Grid voltages (b) DVR output voltages (c) Load voltages (d) d-axis voltages of DVR (e) q-axis voltages of DVR (f) Zero-sequence voltages of DVR (g) Load currents.

TABLE II
CONTROLLER GAINS

Controllers	Gains	
PI method	Voltage controller	$K_P = 0.057$ $K_I = 16$
	Current controller	$K_P = 8.58$ $K_I = 625$
Feedback linearization	$k_{11} = k_{21} = k_{31} = 6.05 \times 10^3$ $k_{12} = k_{22} = k_{32} = 9.3 \times 10^6$ $k_{13} = k_{23} = k_{33} = 4.5 \times 10^9$	

that of the pre-sag. It is seen that the DVR responds very fast less than 3 ms to restore the load voltages. For the DVR control, the d-q-0 components of the DVR voltages are regulated, in which their control performance are satisfactory as shown in Fig. 5(d)-(f), respectively. For the balanced grid sag condition, the d-axis voltage and zero-sequence voltage components are equal to zero as shown in Fig. 5(d) and (f), respectively. The q-axis DVR voltages are shown in Fig. 5(e), which are injected to the system to keep the amplitude of the load voltages. Fig. 5(g) shows the three-phase load currents, which are almost sinusoidal and balanced.

Now, let's investigate the control performances by the PI and feedback linearization methods. With the PI control method, the design method for the controller gains of the outer and inner control loops is described in [17]. Table II lists the controller gains for each control method. At the unbalanced grid sag, the three phase voltages drop to 20%, 50%, and 80%, respectively for 40

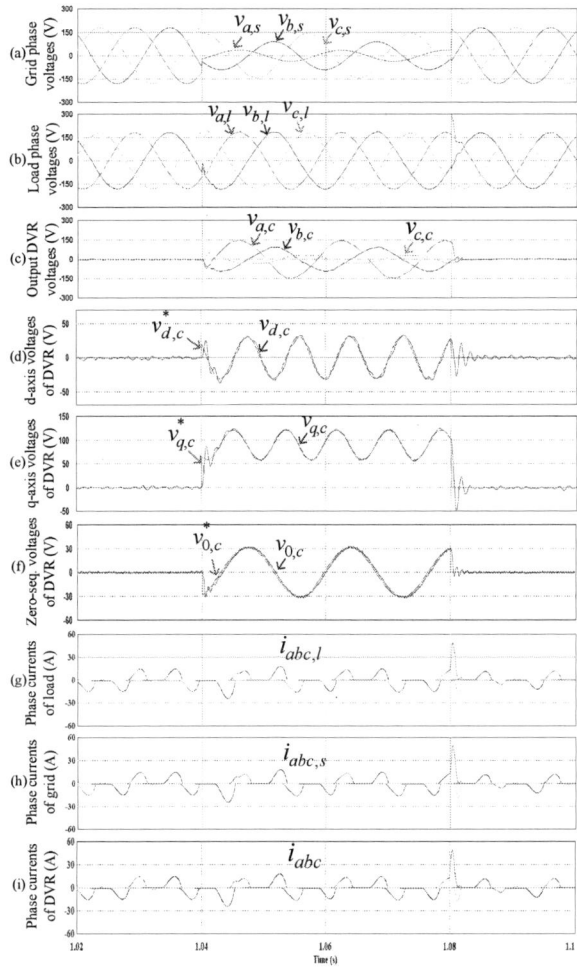

Fig. 6. Performance of DVR using conventional PI control. (a) Grid voltages. (b) Load voltages. (c) DVR voltages. (d) d-axis voltages of DVR. (e) q-axis voltages of DVR. (f) Zero-sequence voltages of DVR. (g) Load currents. (h) Grid currents. (i) DVR currents.

ms. In this condition, the unbalanced and nonlinear loads are connected as shown in Table I.

First, the performance of the DVR with the conventional PI control is shown in Fig. 6, where the unbalanced grid voltage is applied as shown in Fig. 6(a). With compensation, the load voltages still contain the voltage ripple and unbalance, which are shown in Fig. 6(b). Fig. 6(c) shows the DVR voltages. It is seen in Fig. 6(d)-(f) that the control performance of the d-q-0 components of the DVR voltages, respectively, are not satisfied, in which the actual voltages cannot follow their references. Since the references are time-varying, the PI controller cannot give a good performance. Fig. 6(g)-(i) shows the load, grid, and DVR currents, respectively.

Next, with the same simulation condition, the control performance of the DVR with the proposed feedback linearization technique is shown in Fig. 7. Fig. 7(a) shows the AC grid voltages, which are the same as in Fig. 6(a). Fig. 7(b) shows the load voltages, which are kept at the nominal values. It is seen that there are the transients in the load voltages at the beginning and the end of the sag, which can be reduced further by gain tuning of the tracking controller. The output voltages of the DVR to

The 2014 International Power Electronics Conference

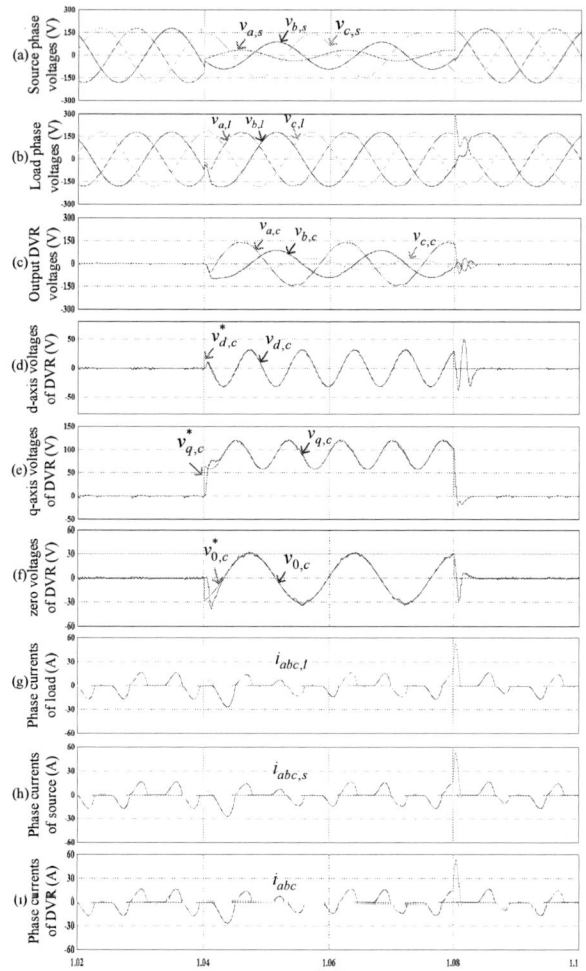

Fig. 7. Performance of DVR using feedback linearization control. (a) Source voltages. (b) Load voltages. (c) DVR voltages. (d) *d*-axis voltages of DVR. (e) *q*-axis voltages of DVR. (f) Zero-sequence voltages of DVR. (g) Load currents. (h) Grid currents. (i) DVR currents.

TABLE III
PARAMETERS OF GRID, LOAD, AND DVR

	Parameters	Values
Source	AC grid voltage	70 V / 60 Hz
DVR system	DC input voltage	100 V
	Neutral inductance	0.5 mH
	Filter inductance	3.5mH
	Filter capacitance	100 µF
	Transformer ratio	1:1
	Switching frequency	5 kHz
Load	Linear load (RL load)	20Ω / 2.5mH (3 phases)

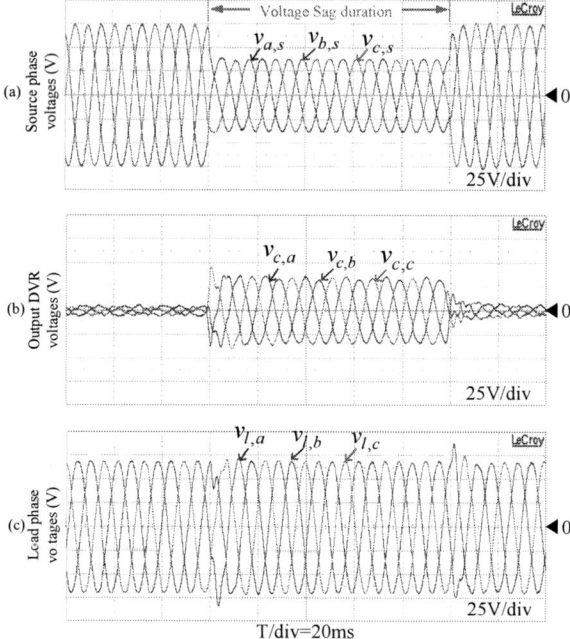

Fig. 8. Performance of the DVR under balanced voltage sag. (a) Grid voltages (b) DVR output voltages (c) Load voltages.

compensate for the voltage sag are shown in Fig. 7(c), which are also unbalanced. Thus, the *d-q* components of the DVR output voltage appear in the AC waveform, which are shown in Fig. 7(d) and (e), respectively. Also, the zero-sequence voltage of the DVR appears as shown in Fig. 7(f). It is obvious in Fig. 7(d)-(f) that the control performances of the d-q-0 component voltages, respectively, are satisfied. Fig. 7(g)-(i) shows the currents of the load, grid, and DVR, respectively, which are unbalanced and nonlinear.

V. EXPERIMENTAL RESULTS

The experimental tests have been performed for a laboratory prototype with the hardware system parameters listed in Table III. The three-phase four-wire split-capacitor VSI is comprised of the intelligent power module (PM75RLA060) and two DC-link capacitors of 1,650 µF. A high-performance DSP chip TMS320C28335 is used as a main digital controller. A 10-kVA grid simulator is used to generate the grid voltage sag.

For investigating the performance of the DVR system, a balanced grid sag is generated by the grid simulator. Fig.

8 shows the three-phase voltages of the grid, load, and the DVR before and after the grid sag. Fig. 8(a) shows the grid voltages, where the three-phase grid voltages simultaneously drop to 50% for 100 ms. In order to keep the load voltages unchanged at the amplitude of 70 V, the DVR output voltages are injected as shown in Fig. 8(b). Then, the load voltages are recovered to their references as shown in Fig. 8(c).

Fig. 9 shows the control performance of the DVR system for the d-q-0 components of the DVR voltages. Under the balanced grid sag condition, the DVR system are also required to generate the balanced voltage to keep the load voltage, in which the d-axis voltage and zero-sequence voltage components of the DVR are equal to zero as shown in Fig. 9(a) and (c), respectively. Fig. 9(b) shows the q-axis voltage of the DVR, which are the amplitude of the DVR output voltage required to compensate for the grid voltage drop. It is seen in Fig. 9 that the control performance is satisfactory, in which the actual voltages are regulated to follow their reference closely.

Fig. 10 shows the load voltages before, during, and after the grid sag. It is seen in Fig. 10(b) that the q-axis

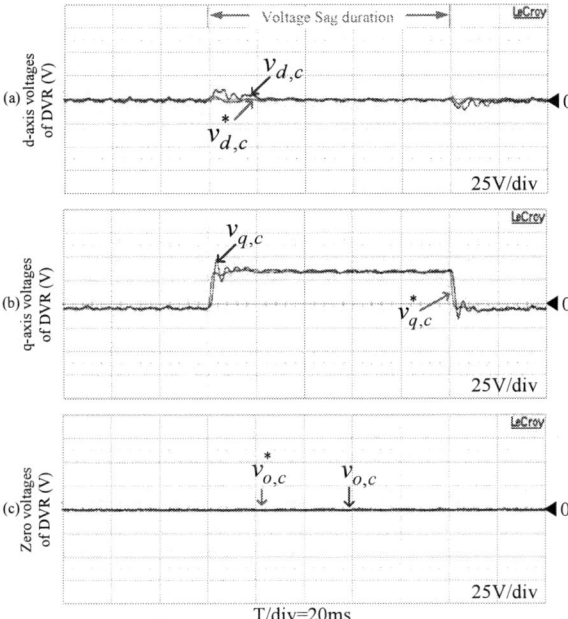

Fig. 9. Control performance of the DVR. (a) d-axis voltages of the DVR (b) q-axis voltages (c) Zero-sequence voltages.

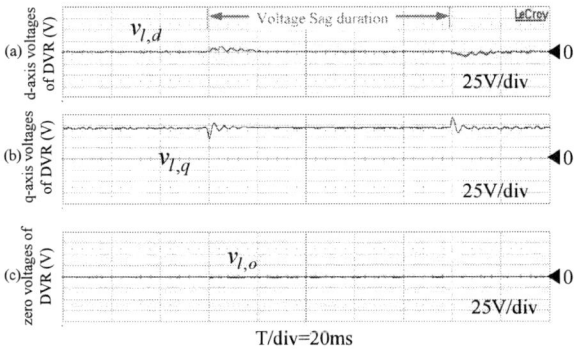

Fig. 10. Load voltages before, during, and after the sag. (a) d-axis voltage (b) q-axis voltage (c) Zero-sequence voltage.

voltage is kept closely to its reference of 70 V even though the grid voltage drops to 50% during the sag. Fig. 10(a) and (c) show the d-axis and zero-sequence voltages of the DVR, which are almost kept at zero. This shows that the load voltages are compensated well, which are balanced.

VI. CONCLUSIONS

In this paper, a novel control scheme for the three-phase four-wire DVR system has been proposed, in which the feedback linearization technique has been applied. The DVR system with the output LC filter is modeled as a nonlinear system. Next, the nonlinearity of the system is eliminated by the feedback linearization. Then, the linear control theory is applied with the pole placement technique for the gain selection. With the proposed scheme, the control performance of the DVR becomes faster and more stable under the unbalanced source voltages and nonlinear loads. The simulation and experimental results have proved the validity of the

proposed control scheme for the three-phase four-wire DVR system.

REFERENCES

[1] E. Babaei, M. F. Kangarlu, and M. Sabahi, "Mitigation of voltage disturbances using dynamic voltage restorer based on direct converters," *IEEE Trans. Power Electron.*, vol. 25, no. 4, pp. 2676-2683, 2010.

[2] Q. N. Trinh, H.-H. Lee, and T. W. Chun, "An enhanced harmonic voltage compensator for general loads in stand-alone distributed generation systems," *Journal of Power Electronics*, vol. 13, no. 6, pp. 1070-1079, Nov. 2013.

[3] V. Khadkikar and A. Chandra, "UPQC-S: a novel concept of simultaneous voltage sag/swell and load reactive power compensations utilizing series inverter of UPQC," *IEEE Trans. Power Electron.*, vol. 26, no. 9, pp. 2414-2425, Sep. 2011.

[4] C. Liu, K. Dai, K. Duan, and Y. Kang, "Application of a C-type filter based LCFC output filter to shunt active power filters," *Journal of Power Electronics*, vol. 13, no. 6, pp. 1058-1069, Nov. 2013.

[5] G. Mahendran, M. Sathikumar, S. Thiruvenkadam, and L. Lakshminarasimman, "Multi-objective unbalanced distribution network reconfiguration through hybrid heuristic algorithm," *Journal of Electric Engineering and Technology*, vol. 8, no. 2, pp. 215-222, Mar. 2013.

[6] H. Kim and S.-K. Sul, "Compensation voltage control in dynamic voltage restorers by use of feed forward and state feedback scheme," *IEEE Trans. Power Electron.*, vol. 20, no. 5, pp. 1169-1177, May 2005.

[7] T. Jimichi, H. Fujita, and H. Akagi, "Design and experimentation of a dynamic voltage restorer capable of significantly reducing an energy-storage element," *IEEE Trans. Ind. Appl.*, vol. 44, no. 3, pp. 817-825, May/Jun. 2008.

[8] C. Meyer, R. W. De Doncker, Y. W. Li, and F. Blaabjerg, "Optimized control strategy for a medium-voltage DVR-theoretical investigations and experimental results," *IEEE Trans. Power Electron.*, vol. 23, no. 6, pp. 2746-2754, Nov. 2008.

[9] S. R. Naidu and D. A. Fernandes, "Dynamic voltage restorer based on a four-leg voltage-source converter," *IET Gener. Transm. Distrib.*, vol. 3, no. 5, pp. 437-447, 2009.

[10] S. B. Karanki, N. Geddada, M. K. Mishra, and B. K. Kumar, "A modified three-phase four-wire UPQC topology with reduced DC-link voltage rating," *IEEE Trans. Ind. Electron.*, vol. 60, no. 9, pp. 3555-3566, Sep. 2013.

[11] V. Khadkikar and A. Chandra, "A novel structure for three-phase four-wire distribution system utilizing unified power quality conditioner (UPQC)," *IEEE Trans. Ind. Appl.*, vol. 45, no. 5, pp. 1897-1902, Sep./Nov. 2009.

[12] S. Lee, Y. Chae, J. Cho, G. Choe, H. Mok, and D. Jang, "A new control strategy for instantaneous voltage compensator using 3-phase PWM inverter," in *Proc. IEEE PESC'88*, 1998, pp. 248-254.

[13] D.-E. Kim and D.-C. Lee, "Feedback linearization control of three-phase UPS inverter system," *IEEE Trans. Ind. Electron.*, vol. 57, no. 3, pp. 963-968, Mar. 2010.

[14] N. Q. T. Vo and D.-C. Lee, "Advanced control of three-phase four-wire inverters using feedback linearization under unbalanced and nonlinear load condition," *Trans. Korean Institute of Power Electronics*, vol. 18, no. 4, pp. 333-341, Aug. 2013.

[15] J. J. E. Slotine and W. Li, *Applied Nonlinear Control*. Englewood Cliffs, NJ:Prentice-Hall, 1991, pp. 207-271.

[16] K.-H. Kim, Y.-C. Jeung, D.-C. Lee, and H.-G. Kim, "LVRT scheme of PMSG wind power systems based on feedback linearization," *IEEE Trans. Power Electron.*, vol. 27, no. 5, pp. 2376-2384, 2012.

[17] J.-I. Jang and D.-C. Lee, "High performance control of three-phase PWM converters under non-ideal source voltage," in Proc. *IEEE ICIT*, India, 2006, pp. 2791-2796.

AUTHOR INDEX

Abe, Kodai 3153
Abe, Seiya 177, 1179, 2216, 2222, 3652
Abe, Shigeru 1109, 1115
Abe, T. 3007
Abe, Takashi...................2183, 2189, 3024
Abe, Tomohiko 1575
Abiko, Hiroshi 634
Achara, Pichetjamroen....................3687
Adachi, Mitsuo 92
Adhikari, Jeevan 1775
Agelidis, Vassilios G................. 1458, 3758, 3764, 3933
Agelidis, Vassilios Georgios.....................640
Aguglia, Davide 3371
Ahmed, Furqan 480, 790
Ahssanuzzaman, S. M. 3582
Aiso, Kohei 1141
Ajima, Toshiyuki..........................383, 682
Akagi, Hirofumi 750, 1586, 1761, 2290, 2323, 3742
Akagi, Masataka 629
Akahane, Masashi.................................2302
Akatsu, Kan....................1128, 1141, 1234, 2673, 3828
Aketa, M. 2074
Akira 3784
Akiyama, Satoru 2285
Alemi, Payam 1201
Alipoor, Jaber.......................................3298
Aljankawey, A. S. 2156
Allen, Scott...3447
Almer, Stefan..3563
Ama, Naji Rajai Nasri 2413, 2988, 3278
Amanci, Adrian Z. 1303
Amano, Yuki 1824
Amma, Ryosuke....................................2027
Anazawa, Yoshihisa 3801
Andersen, Michael A. E. 78, 506, 2842, 3352, 3905
Ando, Itaru 1516
Ando, Masato 1317
Anthon, Alexander 78
An-Yeol Ko 796
Aoki, Mutsumi......................................2400
Aoyagi, Shigehisa 2451
Aoyama, Fumio 2644
Aoyama, Kohei 2266
Aoyama, Masahiro 1405
Aoyama, Tomohiro 3823
Ara, Takahiro..3044
Arai, Haruki 403
Arai, Manabu 3440
Araki, Jun 1728
Araki, Takahiro 1613
Arata, Masanori 1874
Arikawa, S. 3007
Arimatsu, Kenji 415
Arita, Hideaki 2673
Asai, Inami...123

Asakimori, Koki 1567
Asama, Junichi 988
Asano, Katsunori 3440
Asano, Yoshinari 1997
Asano, Yuji 3872
Ashikaga, Tadashi 1886
Athab, Hussain S. 3695
Atsushi, Manabe 2745
Awaji, Sosuke....................................... 3194
Ayano, Hideki 2385
Azuma, M. 1892
Baba, Jumpei 1849
Babasaki, Tadatoshi 1567
Bac Xuan Nguyen 2722
Bafleur, M. 707
Bahman, A. S. 2862
Bahrani, Behrooz 1386
Bak, Claus Leth 3320
Bakran, Mark-M. 2113, 3255
Bang, Deok-Je 2427
Bani Shamseh, Mohammad.................. 2794
Baoquan Liu 1155, 3546
Barater, Davide 433
Barrade, Philippe 1081
Barth, Henry.. 2881
Basari, Amat A. 3194
Basu, Kaushik 3061
Bauer, Florian...................................... 3898
Bauer, Pavol 1193, 3200
Baumgartner, Thomas 1707
Beczckowski, Szymon 2547
Beczkowski, S. M. 2862
Beczkowski, Szymon 2850
Belanger, Jean 2644
Ben Guo... 3129
Ben, Hongqi 2318
Beres, Remus 3320
Berhouet, S. 707
Berkouk, El Madjid.............................. 560
Bessegato, Luca 1087
Besselmann, Thomas............................ 3563
Bhat, Ashoka K. S. 1721
Bhattacharya, Subhashish 651, 656, 758, 1626, 2562, 3225, 3286, 3447, 3726
Bianda, Enea 3432
Biela, J. 868
Biela, Jurgen 1788
Bilal, Akin 230
Bin Wu 3482, 3695
Binbin Li 3680
Bizen, Yosio... 2983
Blaabjerg, F. 548, 1912, 2862
Blaabjerg, Frede..................... 216, 857, 1529, 1634, 1801, 2610, 3320
Blank, Frederic 264

AUTHOR INDEX

Bo Wen...944
Bocker, J...2887
Bocker, Joachim.......................346, 1501, 1508
Boehm, Andreas..283
Boillat, David O...1073
Boitier, V..707
Boroyevich, Dushan.....................944, 2626, 3850
Bortis, D..1291
Bortis, Dominik.........................1309, 2079, 3864
Bosshard, R..2167
Bosshard, Roman...1904
Boyu Wang...2893
Braz Cardoso, F...3225
Burger, Niklaus..1386
Burgos, Rolando.........................944, 2626, 3850
Burkart, Ralph M.....................................891, 3460
Buticchi, Giampaolo..433
Byoungchang Jung..1185
Byung Moon Han...937
Byung-Geuk Cho...2802
Byung-Gyu Yu..3784
Cai, Zheng-Xiu..429
Canales, Francisco..................................1043, 3432
Cao, Guoen...2587
Cao, Wei..567
Cao, Yuan..647
Cardoso, Braz J..3270
Caris, M. L. A...2954
Carvalho, Eden Luiz...1276
Casolari, Ronaldo Pedro....................................1276
Castellazzi, A..2503
Castellazzi, Alberto......................433, 2920, 3718
Ceballos, Salvador...................................3758, 3764
Cha, Honnyong...........................110, 480, 790
Chai Feng..3129
Chang, C.-H..2050
Chang, Chien-Hsuan..................................2523, 3333
Chang, Hsiu-Feng...330
Chang, Kai-Chi...105
Chang, L...2156
Chang, Yuan-Chih.....................................330, 1832
Changsheng Hu...782
Changwoo Kim..1646
Chao Wang..2950
Chao-Fu Wang...2758
Chattopadhyay, Ritwik.......................................3225
Chen, Ching-Guo...1734
Chen, H...1471
Chen, Hsin-Chih...1639
Chen, Hung-Chi...2580
Chen, Jiann-Jong...2910
Chen, Jung-Chieh..677
Chen, Min...485
Chen, Qianhong...1425
Chen, Shen-Li...236

Chen, Wei..66, 72
Chen, Wenjie..2950
Chen, Yaow-Ming..3592
Chen, Ying-Zuo...351
Chen, Zhe..3538
Chen-Feng Chuang...3379
Cheng Deng...782
Cheng, Chun-An.......................................2523, 3333
Cheng, Hung-Liang....................................2523, 3333
Cheng, Po-Tai...1261, 1639
Cheng, Shih-Jen...199, 2593
Cheng, Stone...3425
Cheng-Chieh Yu...2910
Cheng-Wei Chen..3592
Cheol-O Yeon...1738
Cheon, Jun P...3358
Chia-Chi Chu...3379
Chiang, Hsin-Wei...2100
Chiba, Akira...............................982, 988, 3513
Chiba, Yoshinori..634
Chien-Yu Lin...2758
Chih Wei Chen...3938
Ching-Hsiang Yang...1639
Ching-Tasi Pan..3379
Ching-Wei Wang..1639
Chiu, Chian-Song..440
Chiu, Huang-Jen...............172, 199, 2593, 3328
Chiu, Tse-Wei..440
Cho, Bo-Hyung..2272, 2575
Choi, Bo H..2232, 3358
Choi, Byungcho..3638
Choi, Hangseok..2575
Choi, Seong-Chon..409
Choi, Sewan..1394, 2247
Choi, Su Y..1103
Chokchai, Chuenwattanapraniti.........................3789
Chou, Tzu-Han...208, 421
Chow, T. Paul...2208
Chu, Xi...1322
Chun, Chang Yoon...2272
Chung, Tsung-Yuan...2523
Chung-Chuan Hou..2821
Chung-Yi Lin..3185
Chung-Yuen Won..796, 3532
Chunkag, Viboon...694
Chu-Shen Chang..3928
Ciftci, Baris..3734
Coldevin, Grete H..1861
Colmenares, Juan...3712
Concari, Carlo..433
Cortes, Patricio...3864
Cortizio, Porfirio C..3225
Cosovic, Mirsad...1148
Daesu Han..1185
Dahidah, Mohamed S. A.....................................1283

AUTHOR INDEX

Dahono, Pekik Argo 3893
Dai, Wei-Fu 330, 1832
Daikoku, A. 1892
Daikoku, Akihiro 2011, 2673
Dan Chen 3938
Darba, Araz 718
D'Arco, Salvatore 1544
Darus, Rosheila 3758, 3764
Daskalos, Mike 2330
Davoodnezhad, R. 1482
Dawson, Francis P. 1303
De Belie, Frederik 718
De Carvalho, Kelly Caroline
Mingorancia 2413, 2618, 2988, 3278
De Doncker, R. W. 3898
De Doncker, Rik W. 736, 2729, 3145
De Haan, Sjoerd 2787
De Mallac, Louis 3371
De Miranda, Rubens Domingos 1276
De Paula, Helder 3225
De S. Brito, Jose A. 3225
De Vega, Angel Ruiz 2547, 2850
De, Ankan 651, 2562, 3286, 3447
De, D. 2503
De, Dipankar 433
Deguchi, Tadayoshi 3440
Dehong Xu 782
Dekka, Apparao 3468
Demetriades, Georgios D. 1220
Deng, Lirong 465
Dianguo Xu 3174, 3680
Dianguo, Xu 341
Diduch, C. P. 2156
Dilhac, J-M. 707
Diniz, Rogerio Azevedo 3270
Doki, Shinji 907, 2445, 3079, 3823
Domoto, Kazuhide 3652
Dong Le 1837
Dong-Hee Lee 994, 2693
Dong-Jing Lee 1452
Dongkook Son 2914
Dongouk Kim 925
Dongwook Kim 2914
Dou, Qinyun 3604
Dowaki, Kiyoshi 1207
Do-Yun Kim 796
Drofenik, Uwe 1043
Du Yan 2668
Du, Yimian 1721
Duarte, J. L. 2954
Dujic, Drazen 3476
Durand Estebe, P. 707
Dutta, R. 2679
Endo, Takahisa 2541, 2977
Eni, E. 1912

Enomoto, Toshio 2421
Enomoto, Yuji 1997
Erturk, Feyzullah 3734
Eui-Cheol Nho 2763
Fang Zheng 1342
Fang Zhuo 1155, 3546
Fang, Xiaocun 335
Fassler, Lukas 3864
Fei Lin 807
Fei Meng 2815
Fei Zhang 3857
Fernandes, B. G. 2433
Ferrari, Bruno Augusto 3278
Ferreau, Joachim 3563
Ferreira, J. A. 1935
Ferreira, Jan A. 2787
Figueredo, Ricardo Souza 2413, 2618
Fletcher, J. 2679
Fletcher, John 2926, 2932
Foo, Gilbert 2722
Fosso, Olav B. 1861
Foureaux, Nicole C. 3225
Fournier-Bidoz, Sebastien 3496
Franca, Gleisson J. 3270
Franceschini, Giovanni 433
Franke, Toke 78
Fritz, Dominik 3476
Frohleke, N. 2887
Fujii, Junji 1654
Fujii, Kansuke 1748
Fujii, Toshiyuki 2663
Fujimoto, Hiroshi 1671, 2421
Fujimoto, Takafumi 3857
Fujimoto, Yasutaka 1685, 1968
Fujisaki, Keisuke 289, 2856, 2874
Fujisawa, Hiroyuki 3440
Fujita, Hideaki 1006, 1160, 1350, 2027, 2042
Fujitsuna, Masami 3079
Fuketa, Hiroshi 2228
Fukuda, Kenji 3440
Fukuhara, Shuhei 289
Fukumoto, Hisao 724, 730, 3067, 3249
Fukuoka, Hiroki 3341
Fukushima, Kentaro 2189
Fulin Zhou 1050
Funabiki, Shigeyuki 2470
Funaki, Tsuyoshi 1621
Funato, Hirohito 1728, 2517
Furukawa, Kimihisa 383
Furukawa, Tatsuya 724, 730, 3067, 3249
Furukawa, Yutaka 2252
Furuta, R. 2120
Gaing, Zwe-Lee 278
Gao Qiang 3050
Gao, Qiang 614

AUTHOR INDEX

Geng, Hua 543
Gerling, Dieter...774
Ghimire, Pramod 2547, 2850
Giaretta, Antonio Ricardo..........................1276
Goehler, Lutz...2554
Goh Teck Chiang 1028
Gohara, Hiromichi.......................................671
Goto, Akira...130
Goto, Yasuyuki..1490
Goto, Yuichi..1671
Graus, Johannes...270
Grider, Dave..3726
Gruber, Wolfgang 1691, 1701
Gu, Beom W. 1103
Gueldner, Henry 2554
Guidi, Giuseppe...1544
Gunasekaran, Deepak..................................1342
Guo, Wei 160, 475
Gurpinar, Emre 433, 3718
Ha, Jung-Ik...3140
Hafner, Jurgen..3667
Haga, Hitoshi.....................................415, 3153
Haghbin, Saeid..1373
Hagiwara, Makoto 1586, 1761, 2323, 3742
Hahn, Ingo...270, 283
Haining Wang..3702
Haitao Yang..782
Hak-Soo Kim..2763
Hakutou, Takuma..2297
Hama, Ryota...2470
Hamasaki, Shin-Ichi.............................2775, 3674
Hamasaki, Sin Ichi......................................3093
Hamazaki, Yasuhiro.....................................2126
Hanada, T...2074
Hanamoto, Tsuyoshi............................538, 1811
Han-Shin Youn..1743
Hao Huang..2967
Hao Yi 1155, 2960, 3546
Hao, Xiang..2950
Hao-Chien Cheng..3379
Hara, Hidenori 1654, 1898
Harada, Shingo..1671
Harada, Shinsuke..3440
Harakawa, Masaya......................................2638
Hariya, Akinori...3630
Hasegawa, Isamu..1365
Hasegawa, Kohei..3707
Hasegawa, Masaru..........................183, 907, 2445
Hasegawa, Masataka.....................................2212
Hasegawa, Shinya 294, 299, 2972, 3055, 3159, 3162
Hasegawa, Tomonori.....................................2126
Hashimoto, Kento..1974
Hashimoto, S..1471
Hashimoto, Seiji..3194
Hashimoto, Shizuka......................................3018

Hassanpoor, Arman 3667
Hatanaka, Ayumu 2285
Hattori, Fumiya 811
Hatua, Kamalesh 758, 1626
Hau-Chen Yen 3928
Hava, Ahmet M. 498, 2034, 3734
Hayase, Masanori 1207
Hayashi, Makoto 1950
Hayashi, Toshihiko 3440
Hayashi, Yusuke 1560
Hayashiya, Hitoshi 1062
Hazeyama, M. 1892
Hazra, Samir.....................................758, 1626, 3447
He, Guofeng 485
Hee-Jun Lee 3532
Hei, Xinhong 647
Hella, Mona...2208
Hermansson, Willy 1220
Hernandez, Juan C.......................................3352
Heung-Geun Kim 790, 2763
Hibino, Shinya 2638
Hidaka, Akira 2216
Hidayat, Nabil M 573, 2529
Higuchi, Shinichi 1522
Higuchi, T...3007
Higuchi, Tsuyoshi 3024
Hijikata, Hiroki 2673, 3828
Hikita, Masayuki 689
Hinata, Toshifumi 919
Hinkkanen, Marko 2489
Hino, Wataru 3525
Hintz, Andrew 2343
Hira, Yuki 730
Hirahara, Hideaki 3044
Hirai, Junji 1974
Hirakawa, Yuki 3024
Hiraki, Eiji 3292
Hirano, Yosei 1956
Hirao, Kuniaki 191, 1365
Hirase, Yuko 1552
Hirokado, K. 146
Hirose, Toshiro 2252
Hirota, Yukitsugu 1728
Hisada, Yoshihiro 3292
Hisato, Hosoyama 2745
Ho, Kung-Min 3942
Hoene, Eckart 2366
Hoffmann, Stefan 2366
Hofmann, Wilfried 2881
Hojo, Masahide..2152
Hojo, Toshiaki 1276
Hokazono, Hiroaki 2870
Holm, Toni 3432
Holmes, D. G.............................1482, 2019, 2372, 3306
Homma, Hiroshi 1880

AUTHOR INDEX

Hong Li 3314
Hong, Ki-Nam 2598
Hong-Hee Lee 1013
Hongqi Ben 3213
Hori, Yoichi 2421
Horiguchi, Takeshi 2290
Horita, Yasuhisa 1317
Hosaka, Tatsuya 1350
Hoshi, Nobukazu 1242
Hosoyamada, Yu 801
Hou, Chih-Hao 1796
Hou, Jiaxin 526
Hou, Lixiang 577
Hredzak, Branislav 1458
Hsieh, Guan-Cyun 526
Hsieh, Hung-I 526, 2380
Hsieh, Min-Fu 278
Hsieh, Yao-Ching 429, 1796
Hsin-Chih Chen 1261
Hsin-Ping Su 2821
Hu, Jia-Sheng 278
Hu, Shang-Hung 2606
Hu, Sheng 555
Hu, Taiyuan 335
Huang, Hsin-Wei 421
Huang, Jia-Wei 3233
Huang, Lang 2950
Huang, Min 2610
Huang, T. D. 2195
Huang, Wen-Nan 1734
Huang, Zhenhui 647
Huang-Jen Chiu 2758, 2810, 3185, 3913
Huber, Jonas E. 766
Huber, Tobias 1508
Hui Liu 1634
Hui Zhang 1365, 3455
Huisman, H. 2954
Hull, Brett 3447
Huu-Nhan Nguyen 1013
Hwang, Seon-Hwan 2427
Hwang, Yuh-Shyan 2910
Hwu, K. I. 204, 2754, 3190, 3392
Hyoyol Yoo 1646
Ichihara, Junichi 2189
Ichiya, Takahiro 370
Ichiyanagi, Katsuhiro 1490
Ide, Kozo 933
Ieda, Jun 2663
Iga, Yuichi 3341
Igarashi, Kazunori 2983
Igarashi, S. 3702
Iida, Mikiya 2977
Iijima, Ryuji 117
Iijima, Yukihia 2095
Ikawa, O. 2569, 3702

Ikeda, Hidehiro 2476
Ikeda, Masahiro 2183
Ikeda, Tomohiko 1575
Ikeda, Y. 2569
Ikeda, Yoshinari 2870
Il-Kuen Won 796
Ilves, Kalle 1087
Imai, Jun 2470
Imakiire, Akihiro 689
Imamura, Yasutaka 863
Imanishi, Takao 2663
Imaoka, Jun 811, 883, 2497
Inamori, Mamiko 3509
In-Dong Kim 2763
Inomata, Kentaro 1654
Inoue, Kaoru 3872
Inoue, Keita 130
Inoue, M. 1892
Inoue, Tatsuki 363
Inoue, Y. 246, 258, 312, 390
Inoue, Yukinori 324, 356, 363, 370, 2183, 3018, 3519
Irokawa, Shoichi 1357
Ise, Tomofumi 1430
Ise, Toshifumi 1536, 1560, 2632, 3298, 3687
Ishibashi, Makoto 724
Ishida, Koichi 2228
Ishida, M. 146
Ishida, Masaaki 3707
Ishida, Masaki 3162
Ishida, Takahito 634
Ishigami, Takashi 1880
Ishigma, Satoru 403
Ishihara, Chio 1984
Ishihara, Yuji 1135
Ishii, Hirotaka 294
Ishikawa, Hiroki 1135, 2183, 2189
Ishikawa, Katsumi 2140, 2285
Ishikawa, Takeo 252, 1697
Ishimaru, Yusuke 92
Ishimori, Hitoshi 3440
Ishitobi, Manabu 811
Ishizuka, Tomotsugu 2644
Ishizuka, Yoichi 2222, 2252, 2737, 3630, 3652
Isida, Takashi 1950
Itako, Kazutaka 3244
Ito, Yasuhide 3823
Ito, Yoichi 403
Itoh, Hideaki 724, 730, 3067, 3249
Itoh, Jum-Ichi 1943
Itoh, Jun-Chi 1253
Itoh, Junichi 130
Itoh, Jun-Ichi 84, 138, 152, 191, 682, 1021, 1028, 1095, 1613, 2277, 3659, 3815
Itoh, Tomomichi 850
Itoh, Youichi 415

AUTHOR INDEX

Itoh, Yuki 883, 2497
Iwaji, Yoshitaka 2451
Iwakami, Tetsuro 817
Iwasaki, Makoto 1665
Iwasaki, Shinya 2663
Iwata, Tetsuki 403
Iyer, Kartik V 3037, 3061
Iyer, Shivkumar 3482
Izumi, Toru 3440
Jacobson, Bjorn 3667
Jae-Bum Lee 1738
Jaeho Choi 2656
Jae-Hun Jung 2763
Jae-Hyun Kim 1738
Jaesig Kim 2656
Jang, Jinhaeng 3638
Jang, Young-Jin 664
Jang-Hwan Kim 925
Jardini, Jose Antonio 1276
Jauch, Felix 1788
Javed, Riffat 624
Jayoon Kang 1185
Jen-Hao Teng 1452
Jenn-Jong Shieh 3190
Jeon, Jin-Yong 166
Jeong, Seog Y 1103
Jeong, Seon-Yeong 2406
Jcongjoong Kim 1185
Jhen-Yu Jian 3928
Jia Liu 1536
Jia, Y. 1594
Jianfeng Li 3718
Jiang, Dawang 647
Jiang, Maoh-Chin 105
Jiang, W. 1471
Jiang, W. Z. 204, 3190
Jiang, Yongjie 458
Jianhui Su 2668
Jiann-Fuh Chen 2714
Jianwen Zhang 3124
Jih-Hua Yu 2910
Jin Miaoxin 3050
Jin, Miaoxin 614
Jin, Xu 341
Jin, Yasuhiro 3207
Jing Bian 3314
Jing-Hsiao Chen 3233
Jing-Yuan Lin 2758
Jinjun Liu 835
Jinno, Masahito 1781, 3333
Jin-Woo Ahn 994, 2693
Jinyong Zhang 3213
Ji-Shiang Lee 3346
Joebges, Philipp 2729
Jokipii, J. 514, 2240

Jokipii, Juha 1466
Jong Kyou Jeong 937
Jonghyung Park 1185
Jonishi, Akihiro 2302
Jou, Sung-Tak 224
Jung, Hochang 1990
Jung, Jae-Jung 1268
Jung, Sang-Yong 1990
Jung, Yong-Chae 166, 409
Junghum Lee 2656
Junjie Feng 835
Juntao Fei 3168
Juyoung Jang 2656
Kabasawa, Yuichiro 2175
Kabiri, R. 3306
Kadavelugu, Arun 758, 1626, 3726
Kai, Masahiko 1054
Kai-Hui Chen 2750, 3346
Kaipia, T. 587
Kajiwara, Kazuhiro 3950
Kakishima, Takeo 3513
Kalogera, Maria 1193, 3200
Kameshiro, Norifumi 2140
Kamikura, Mamoru 2064
Kamnarn, Uthen 694
Kanagawa, Kinji 2983
Kanai, Yasuyuki 1567
Kanamori, Masaki 2541, 2977
Kaneko, Junji 2745
Kaneko, Yasuyoshi 1109, 1115
Kanematsu, Masato 2421
Kanemoto, Daisuke 2737
Kang, Feel-Soon 2260
Kang, Yong 555
Kanno, Hiroshi 2302
Kano, Yoshiaki 2004, 2457
Kanoda, Akihiko 3920
Kanouda, Akihiko 2058
Kantar, Emre 2034
Kanthaphayao, Yutthana 694
Kari, Mat Nasir 573
Karki, Ujjwal 1342
Karvonen, Andreas 1373
Kasai, Makoto 3194, 3194
Kashihara, Yugo 1943
Kasper, Matthias 2079
Katade, Motohumi 130
Katakami, Shuji 3440
Kataoka, Yasuhiro 3801
Katayama, Noboru 1207, 1227
Kato, Hideaki 2972, 3162
Kato, Koji 403, 415
Kato, Shinji 2175
Kato, Takashi 3828
Kato, Taro 2972

AUTHOR INDEX

Kato, Tomohisa 3440
Kato, Toshiji 2183, 2189, 3872
Kato, Yutaka 2644
Katoh, Kaoru 2285
Katoh, Shuji 850
Katsuki, Akihiko 1575, 3624
Katsura, Seiichiro 1679
Kawachi, Konosuke 863
Kawaguchi, Shinichi 3959
Kawahara, Keiji 1062
Kawakami, Noriko 2095
Kawamura, Atsuo 801, 2266, 2794, 3403
Kawamura, Mitsuhiro 3012
Kawamura, Wataru 3742
Kawano, Daisuke 1671
Kawano, Kenji 883
Kawazoe, Yosuke 2011
Kazuya, Ogura 452
Kempen, S. 2887
Kenji, Matsumoto 3218
Kern, Ansgar 712
Khan, Ashraf Ali 110
Khan, Faisal H. 2161
Khant, Hlaing Kyi Pyar 183
Khomfoi, Surin 2392
Kicin, Slavo 3432
Kihyun Lee 1646
Kikuchi, Takuya 1328
Kim, Bong C. 3358
Kim, Chong-Eun 1738, 1743
Kim, Dong-Hun 790
Kim, Dong-Rak 409
Kim, Hee-Jun 2587
Kim, Heung-Geun 110, 480
Kim, Hyejin 2575
Kim, Jae-Hyun 1743
Kim, Jang-Mok 2406
Kim, Ji H. 2232
Kim, Ji-Won 2427
Kim, Jonghoon 619
Kim, Minjae 2247
Kim, Seonghye 2260
Kim, Su-Han 480
Kim, Sungmin 1268
Kim, Yong-Jae 1990
Kimoto, Tsunenobu 3440
Kimura, Hiroshi 1920
Kimura, Noriyuki...............1299, 1806, 2183, 3341
Kimura, Shota 883, 2497
Kinouchi, Shin-Ichi 750, 2290
Kish, Gregory J. 951
Kitabayashi, Tatsuaki 2517
Kitagawa, Wataru 2310, 3809
Kitajima, Jun 1247
Kitazawa, Satochi..1438

Kiyota, Kyohei 3513
Kleinecke, John 2330
Kluge, Andreas 2554
Knott, Arnold 506
Kobayashi, H. 2569
Kobayashi, Hiroya 2517
Kobayashi, Ryota 1115
Kobayashi, Takenori 1868
Kobayashi, Y. 2569
Kodama, Takashi 1365
Kogi, Ryosuke 2874
Kogoshi, Sumio 1207, 1227
Kohama, Teruhiko 522, 2781
Kohno, Yusuke 2183
Koiwa, Kazuhiro 84, 130, 1028
Kolar, J. W. 1291, 2167
Kolar, Johann W. 766, 821, 891, 899, 975,
        1073, 1309, 1707, 1904, 2079, 2834, 3365,
        3460, 3864
Komada, Satoshi..1974
Komatsu, Wilson 1276, 2413, 2988
Komeda, Shohei 1160
Komiya, Hiroshi 2421
Kon, Saytaro 3263
Kondo, Keiichiro 1438, 2126
Kondo, Seiji 415
Kondo, Takeshi 1365
Kondou, Masahiko 2421
Kono, Y. 2120
Kono, Yasuhiko 2140
Konoto, Masaaki 2189
Konstantinou, Georgios 1458, 3758, 3764
Korner, Olaf. 2113
Kosaka, T. 2438
Kosaka, Takashi 1984, 1997
Koschik, Stefan 3145, 3898
Koseki, Takafumi 1334, 2126
Kotegawa, Ryo 1317
Kotera, Keito 3872
Kouki, Matsuse 3134
Kouno, Yusuke 2189
Kounoto, Masaaki 2175
Koyama, Masato 750
Krafft, Eberhard 2113
Krismer, Florian 2834
Kuan-Hsien Chou 3346
Kubo, H. 395, 1594
Kubo, Hajime..1601
Kubo, Yuji 3134
Kubota, Hisao 919, 1929, 3119, 3134
Kubota, Yutaka 2183
Kudo, Takahiro 1109
Kuga, Shotaro 3955
Kukita, Akio 1444, 2351
Kumagai, Shunji..3194

AUTHOR INDEX

Kumakura, Yoshito.................................1715
Kume, Tsuneo.....................................1898
Kumsuwan, Yuttana 3417
Kun-Hung Chen 3592
Kunomura, Ken 1054
Kuo, Kuan-Yi 278
Kuperman, A. 2240
Kurabayashi, Toshiyuki 1962
Kuribayashi, H. 2569
Kurihara, Takeshi...................................299
Kurihara, Yoshihiro 1874
Kurita, Nobuyuki 252, 1697
Kuroda, Y. 1892
Kurokawa, Fujio 2108, 3611, 3950
Kusaka, Keisuke 191
Kusukawa, Jumpei 2904
Kusunoki, Hironobu 2330
Kutsuki, Tomohiro 2064
Kuwahara, Akinobu 3179
Kuzumaki, Atsuhiko 1929
Kwasinski, Alexis 2649
Kwasinski, Andres 2649
Kwon, Soon-Kurl 2359
Kyungbae Lim 2656
Kyungmin Sung......................................744
Kyungsub Jung 1646
Lai, Yen-Shin 3942
Lamantia, A..2503
Lana, A. 587
Lang, Klaus-Dieter 2366
Larsson, Tomas 1220
Laska, Bernd 2113
Law, Kah Haw 1283
Le Hoai Nam 3659
Lee, Chia-Tse......................................1639
Lee, Dong-Choon 1201, 2406
Lee, Eun S........................1103, 2232, 3358
Lee, Hong-Hee 2826
Lee, Jae-Bum 1743
Lee, June-Hee 493, 532
Lee, June-Seok 493, 532
Lee, Kyo-Beum 224, 493, 532
Lee, Min-Hua 236
Lee, Seong Ryong 3292
Lee, Shiu-Hui 1734
Lee, Sung W. 1103
Lee, Taeck-Kie 595
Lee, Tzung-Lin 2606
Lee, Woo-Cheol 595
Lee, Ya-Ting 440
Lee, Yuang-Shung 208, 421
Lehmann, Oliver 3085
Lehn, Peter W. 951
Lei, Wanjun 160, 475
Leibl, Michael......................................899

Lelie, Markus 2729
Leslie, Scott 3726
Leuenberger, D. 868
Leuer, Michael 346
Li Yan 2899
Li, Ding 341
Li, Haiqing 2095
Li, Hong 2893
Li, Ning 160, 475
Li, Qian 2161
Li, Yanxiang 3002
Lian, K. L. 2195
Liang Hao 3174
Liangyi Tang 3695
Liao, Jhen-Yu 2580
Lie Guo 3489
Lin Cheng 3447
Lin, Chiao-Chien 3072
Lin, Chia-Yu 1832
Lin, Chien-Yu 172
Lin, Chung-Yi 199, 2593
Lin, Fei 335, 1322, 2133
Lin, Jing-Yuan 172
Lin, L.-C. 2050
Lin, Z. Y. 1471
Lindberg-Poulsen, Kristian 2842
Liping Zheng 1837
Liserre, Marco 857, 3320
Liu, Baoquan 577
Liu, Fang 567
Liu, Fangcheng 3604
Liu, Fuxin 458, 2768
Liu, Hanchao 967
Liu, Jianyu 614
Liu, Jilong 66, 72
Liu, Jinjun 624, 2815, 3604
Liu, Kangzhi 3568
Liu, Ning 2156
Liu, Rongqiang 3099
Liu, Tai-Chun 105
Liu, Xiankai 647
Liu, Xiaosheng 3002
Liu, Yi-Hua 3233
Liu, Yu-Chen 199, 2593
Liuchen Chang 1476, 2668, 3842
Lo, Yu-Kang 172, 199, 2593
Lobsiger, Yanick 1309
Loh, Poh Chiang.........216, 1529, 1634, 1801, 2610
Longlong Zhang 782
Lopez-Arevalo, Saul 3718
Lovatt, Howard 2679
Low, K. S. 446
Lu, Dylan D. C. 3553
Lu, Kao-Yi 105
Luo, Guomin 2145

AUTHOR INDEX

Luthardt, Sven 3029
Ma, K. 548, 2862
Ma, Weigang 647
Madawala, Udaya K. 2722
Madhusoodhanan, Sachin 656, 1626
Maekawa, Sari............................919, 1929
Maemura, Akihiko............................1898
Maeyama, Shigetaka............................1575, 3624
Maezono, Paulo Koiti............................1276
Mahdavikhah, Behzad............................3582
Mainali, Krishna 758, 1626
Makaino, Yuki............................914
Makita, Shinji............................3823
Mamun, Mostafa............................97
Manias, Stefanos N............................1606
Mannen, Tomoyuki............................2042
Manolas, Iakovos 1606
Mao, Meiqin............................2156
Maret, C............................3239
Marrero Sosa, Juan Alberto 3476
Martinz, Fernando Ortiz............................2413, 2988, 3278
Maru, Naoki............................2285
Marukawa, Yasuhiro............................1984
Marumori, Hiroki............................3055
Maruta, Hidenori............................3611
Marz, Andreas............................2113
Marzouk, Ahmad Diab............................3496
Masaki, Kenji............................2663
Mashino, Masahiro............................3162
Masic, Semsudin............................1148
Maskell, D. L............................3598
Masuda, Hiroyuki............................92
Masui, Takeshi............................1317
Masutomo, Kazufumi 3624
Masuzawa, Hiroshi............................1054
Masuzawa, Takashi............................2366
Matakas, Lourenco 2413, 2618, 2988, 3278
Matsubara, Masakatsu 1874
Matsuda, Katsuhiro............................415
Matsuhashi, Daiki............................1886
Matsuhashi, Masataka............................1516
Matsui, Hitoshi............................1586
Matsui, Keiju............................183
Matsui, Mikihiko............................3489
Matsui, N. 2438
Matsui, Ryota............................1128
Matsui, Yoshihiro............................2385
Matsui, Yoshinobu............................2745
Matsumoto, Akira............................1560
Matsumoto, Atsushi............................2445
Matsumoto, Kazushi............................3440
Matsumoto, Satoshi............................2216
Matsumoto, Shuhei............................1929
Matsumoto, Yasushi 1920
Matsuo, Hirofumi............................1781

Matsuo, Keisuke 1886
Matsuo, Yusuke 1671
Matsuoka, Kazumasa............................3207
Matsuoka, Yuji............................744
Matsushima, Yoshitarou............................3801
Matsushita, Makoto 3012
Matsuura, Kei 1516
Matsuura, Ken 3630
Matsuzaki, Ryohei 1978
Mattavelli, Paolo............................3850
Mattsson, A. 587
Mauerer, M. 1291
McGrath, B. P............................1482, 2019, 2372, 3306
McLean, Kenneth 3496
Meiqin Mao............................1476, 2668, 3842
Mekhilef, Saad 560, 3574
Melkebeek, Jan 718
Meng, Fei 624
Meng, Tao 2318
Merahi, Farid............................560
Messo, T. 514, 2240
Messo, Tuomas 1466
Mihara, Teruyoshi............................1728
Mii, Kenji 2737
Miiura, Yushi 1430
Mikihiko 3784
Mills, Liam 3718
Ming Yang 3174
Mingfei Wu 3553
Mingyan Wang 3129
Mino, Kazuaki 1920
Minoshima, N............................2438
Minowa, Masanao 3828
Minsoo Jang 3933
Mira, Maria C............................506
Mishima, Tomoakzu 2533
Mishra, Santanu 2707
Mishra, Santanu Kumar 3587
Misu, Daisuke 1874, 3012
Mitterhofer, Hubert 1701
Miura, Yushi 1536, 3298
Miyajima, Hiroki 1054
Miyajima, Takayuki............................2421
Miyakawa, Takayuki............................2421
Miyama, Yoshihiro............................2673
Miyashita, S. 3702
Miyawaki, Satoshi 84
Miyazaki, Hideki 383
Miyazaki, Kensuke 601
Miyazaki, Toshimasa 1956
Miyazaki, Yuji 750
Mizoguchi, Takahiro 1660
Mizukami, Makoto 3440
Mizuki, Tatsuya............................1575, 3624
Mizuma, Takeshi 2126

AUTHOR INDEX

Mizuno, Takayuki 1886
Mizusaki, Hiroshi 3093
Moballegh, Shiva 656
Mochikawa, Hiroshi 1929
Mochizuki, Eiji 671, 2870
Mochizuki, K. 2569
Mohamed, Essam Ebaid 3877
Mohamed, Tarek Hassan 3877
Mohan, Ned 1036, 1412, 3037, 3061, 3750
Mohd Arif, Mohd Johari 573
Molinas, Marta 1861
Momose, Fumihiko 671
Moo, Chin-Sien 1796, 3796, 3928
Moon, Dongok 1394
Moon, Gun-Woo 1738, 1743
Moon, Sang-Ho.......................................224
Moorthy, Radha Sree Krishna 2087, 3616
Moraes, Lenin 3225
Mori, Tomohiro 2983
Morikawa, R. 258
Morimoto, Masayuki 3509
Morimoto, S. 246, 258, 312, 390
Morimoto, Shigeo.................324, 356, 363, 370,
    1997, 3018, 3519
Morimoto, Shinya 1654
Morishita, Shin......................................130
Morita, Hiroshi......................................1490
Morita, Kazunori 191, 582
Morita, Kosuke......................................3624
Morita, M...1892
Morizane, Toshimitsu 1299, 1806
Morizane, Tosimitsu 3341
Moroi, Takayuki 3134
Morozumi, Akira 671
Mory, David...2554
Motizuki, Shun......................................2745
Motoi, Naoki 801, 2266
Motomura, Masashi 3611
Mouri, Masayuki 1728
Mrak, Branimir......................................1701
Mukai, Ryosuke 2775
Mukunoki, Makoto 97, 1950
Munk-Nielsen, Stig........................2547, 2850
Murai, Kensuke 1567
Murai, Toshiaki.....................................1122
Murakami, Daichi 1728
Murakami, Kouhei 2385
Murakami, Toshiyuki 1962
Murata, Koji 2108
Murata, Munehiro 1173
Murata, Yuichiro 2064
Musing, Andreas.....................................821
Mustapa, Rijalul Fahmi.........................2529
Muta, Shoichiro 3067
Nag, Soumya Shubhra 3587

Nagai, Shinichiroh 811
Nagano, Tetsuaki 2638
Nagano, Tsuyoshi 1253
Nagano, Y. 146
Nagashima, Tomohiro 2175
Nagata, Shun 2252
Nagatomo, Yoshinobu 2663
Nagel, Andreas.....................................2113
Nagura, Hirokazu 2451
Nagy, Istvan 2700
Naitoh, Haruo 1135
Nakagawa, Hidehiko 1552
Nakagawa, Yuki 2533
Nakahara, Mizuki 744, 2511
Nakajima, Yoichiro 403
Nakamura, M. 376
Nakamura, Ritaka 92
Nakamura, Sota 2400
Nakamura, T. 2074
Nakamura, Tatsuya 1575
Nakamurame, Fuminori 2632
Nakanishi, Toshiki 1095
Nakano, Y. 2074
Nakao, Hiroshi 2745
Nakao, Noriya 1128, 1141
Nakaoka, Mutsuo 2359, 2533
Nakaoka, Mutuo 3341
Nakashima, Yoshiyasu 2745, 3386
Nakata, Yuki 138
Nakatsu, Kinya 2904
Nakatsugawa, Junnosuke 2451
Nakayama, Koji 3440
Nakayama, Naoyuki 3857
Nakayama, Yasushi 2290
Nakazawa, Yosuke 1357
Nam, Kwang-Hee 664
Narita, Takayoshi.................294, 299, 3055, 3159
Nashida, N. 2569
Nayanasiri, D. R. 3598
Nee, Hans-Peter 3712
Neubert, Markus 3145
Nguyen, D. 2679
Nguyen, Quoc Khanh 318
Nguyen, Thanh Hai 2406
Nha, Quang Trong.................................3913
Nho Van Nguyen 2826
Nian Heng 843
Nicolae, Ileana-Diana 2996
Nicolae, Marian-Stefan 2996
Nicolae, Petre-Marian 2996
Niijima, Koji 1299, 1806
Nilssen, Robert.....................................1412
Nimura, Tomohiro 3079
Ning Liu 2668

AUTHOR INDEX

Ninomiya, Tamotsu 177, 1179, 2216, 2222, 3630, 3652
Nishida, Katsumi 2359
Nishida, Yasuyuki 2189
Nishikata, Shoji 959
Nishimura, T. 3702
Nishimura, Tomohiro 2870
Nishimura, Yoshitaka 671
Nishio, Haruhiko 2302
Nishioka, Tomoya 2152
Nishisu, Koji 2285
Nishiyama, Noriyoshi 2011
Nishizawa, Shinichi 117, 744
Niu, Ruigen 160, 475
Noda, Koji 2541
Noda, Taku 2175
Noguchi, Kenji 2277
Noguchi, S. 2569
Noguchi, Toshihiko 1173, 1405
Noh, Yong-Su 166
Nomura, Naofumi 1522
Nomura, Shinichi 3218
Nonaka, Hirotaka 2737
Norigoe, Isami 117
Noro, Osamu 1552
Norrga, Staffan 1087
Noto, Yasuo 682
Nozaki, Takahiro 1660
Nozawa, Ryosuke 1115
Nussbaumer, Thomas 975, 3365
Nuutinen, P. 587
Oboe, Roberto 1679
O'Byrne, Sean 2926, 2932
Oda, Yoshinori 829
Odawara, Shunya 289, 2856, 2874
Ogasawara, Satoshi 1728, 2977, 3525
Ogashi, Yoshihiro 92
Ogawa, Kazutoshi 2140, 2285
Ogawa, Takashi 2285
Ogura, Kazuya 582, 3455
Ogura, Tsuneo 2068
Oh, Min-Seok 166
Ohara, Shinya 850
Ohashi, Hiromichi 117, 744
Ohashi, Shunsuke 3410
Ohchi, Masashi 724, 730, 3067, 3249
Ohishi, Kiyoshi 1247, 1516, 1956, 3153
Ohnishi, Kouhei 1660, 2483
Ohnuma, Takumi 914
Ohnuma, Yoshiya 84
Ohse, Naoyuki 3440
Ohtake, Asuka 3857
Oi, Kazunobu 452
Oi, Takeshi 2290
Oishi, K. 376

Oishi, Koji 3012
Oiwa, Takaaki 988
Ojika, Satoshi 1430
Oka, T. 376
Oka, Toshiaki 2330
Okamoto, Dai 3440
Okamoto, Masayuki 3292
Okamoto, Shoji 2290
Okamura, Kazuki 3674
Okazaki, Fumihiro 1728
Okazaki, Yuhei 1586
Okitsu, Takashi 1886
Okubo, Toshikazu 2058
Okuma, Jun 1978
Okuma, Yasuhiro 2834
Okumura, Hajime 3440
Okuyama, Yoshihiro 1811
Omata, Shinpei 2944
Omi, Masataro 1317
Omori, Hideki 1299, 1806, 3341
Omote, Kenichiro 1950
Omura, Mototsugu 1685
Ong, Andrew 2722
Onishi, Mitsuru 1054
Ooishi, Eiji 183
Ooshima, Masahide 1715
Orikawa, Koji 191, 1613, 2277, 3659
Ortiz, G. 1291
Ortiz, Gabriel 1309
Oshima, Ryo 1021
Oshinoya, Yasuo 294, 299, 2972, 3055, 3159, 3162
Oso, Hiroshi 629
Ota, Chiharu 3440
Ota, Satoru 2095
Otsuki, Midori 1054
Ouchi, Takayuki 3920
Ouyang, Shaodi 624
Ouyang, Ziwei 2842
Ozaki, Takayuki 2638
Ozkan, Ziya 498
Pala, Vipindas 2208
Palmour, John 3447
Pan, Miao 582
Panda, S K 1775
Panda, Sanjib Kumar 1580
Pansier, F. 1935
Papafotiou, Georgios 1606
Papastergiou, Konstantinos 1220
Park, Gyeong-Jae 1990
Park, Junsung 1394
Park, Yongsoon 2598
Parker, S. G. 2019, 2372
Partanen, J. 587
Patel, Dhaval 758, 1626
Pedersen, Kristian Bonderup 2547

AUTHOR INDEX

Peftitsis, Dimosthenis 3712
Peltoniemi, P. 587
Peng Gao 2926, 2932
Peng Wang 3124
Peng Wen 782
Peng, Han 2208
Peretti, L. 3111
Peters, A. 2887
Peters, Wilhelm 1508
Petersen, Lars P. 3352
Petersen, Lars Press 2842
Petrich, Matthias 318
Pettersson, Sami 3432
Pham Phu Hieu 3913
Pidaparthy, Syam Kumar 3638
Piepenbreier, Bernhard 1816, 3029
Ping-Heng Wu 1261
Pires, Igor A. 3225, 3270
Pittet, Serge 3371
Pittini, Riccardo 3905
Po-Chien Chou 3425
Poh Chiang Loh 857
Po-Jung Tseng 2810, 3328
Popa, Lucian-Dinut 2996
Popova, L. 548
Popovic, J. 1935
Poshtkouhi, Shahab 2336
Pou, Joscp 3758, 3764
Prasanna, I. V. 1580
Prasanna, U. R. 230, 395, 1594
Prasanna, Udupi. R. 2343
Prodic, Aleksandar 3582
Pyrhonen, J. 548
Qi Zhang 3489
Qu, Lizhi 609
Qunzhan Li 1050
Rabkowski, Jacek 3712
Radic, Aleksandar 3582
Radman, Karlo 1691
Rae-Sung Yu 925
Rahman, F. 2679
Rahman, M. A. 982
Rahman, M. F. 2686
Rajashekara, K. 395, 1594
Rajashekara, Kaushik 230, 2343, 3134
Raju, Siddharth 1036
Ramadan, Husam A. 863
Rambetius, Alexander 270, 3029
Rannestad, Bjorn 2547
Rannested, Bjorn 2850
Rathore, Akshay K. 1775
Rathore, Akshay Kumar 2087, 3616
Ray, Olive 2707
Razik, H. 3239
Reiter, Tomas 774

Ren, Kangle 465
Riffat, Javid 2815
Rikitake, Jungo 2216
Rim, Chun T. 1103, 2232, 3358
Rivera, Marco 3574
Robbins, William P 3037, 3061
Rodriguez, Jose 3574
Rongfeng Yang 3680
Rosekeit, Martin 2729
Roth-Stielow, Jorg 264, 318, 3085
Roy, Sudhin 651, 2562, 3286
Ruan, Xinbo 458, 2768
Ruda, Harry E. 1303
Ruderman, Michael 1665
Rufer, Alfred 1081, 1386
Ryo, Mina 3440
Ryu, Sei-Hyung 3726
Saarakkala, Seppo E. 2489
Saga, Yasunao 1748
Sahoo, Ashish Kumar 3750
Saikusa, H. 3007
Saito, Eiichi 1679
Saito, Katsuhiko 2064
Saito, Ryo 3397
Saito, Ryoji 2189
Saito, Takashi 671
Saitoh, Ryoh 914
Sakaba, Kouichi 3159
Sakai, Kazuto 240
Sakai, Tomoyasu 1490
Sakai, Toshifumi 2451
Sakaino, Sho 1978
Sakimoto, Kenichi 1552
Sakurai, Naoki 2297
Sakurai, Takayasu 2228
Sampath, Prasad K. 2722
Sanada, M. 246, 258, 312, 390
Sanada, Masayuki 324, 356, 363, 370, 3018, 3519
Sand, Kjell 1861
Sariyildiz, Emre 2483
Sasaki, Tomotake 2745
Sasongko, Firman 1761
Sato, Daisuke 3815
Sato, Koji 1671
Satria, Andri 3893
Sauer, Dirk Uwe 2729
Sawada, Tadashi 1122
Sayed, Khairy 2359
Sayed, Mahmoud A. 3877
Schob, Reto. T. 1691
Schon, Andre 3255
Schrittwieser, L. 1291
Schupbach, Marcelo 3447
Schuster, Johannes 3085
Segaran, D. S. 2372

AUTHOR INDEX

Segsa, Karl-Heinz....................................2554
Seilmeier, Markus 1816
Sekiba, Yoichi...2175
Sekisue, Takayuki 2175, 2189
Seo, Gab-Su 2272
Seok-Jin Hong 3532
Seunghoo Song 1646
Seung-Ki Sul................................925, 2802
Severson, Eric..1412
Shah, Shahil..................................843, 967
Shao Zhang...1342
Shaodi Ouyang 2815
Shaofeng Xie 1050
Shaohua Sun 3213
Shaohui Zhong 1155
Shen, Na 582
Shen, Zhiyu 3850
Shenghui Cui 1268
Sheng-Kai Kao 2714
Shi, Hongtao...577
Shi, Rongliang..567
Shibahara, Kohei 1575, 3624
Shibahara, Ryota 2222
Shibanuma, Kenichi....................................634
Shibata, Yuichiro......................................3950
Shieh, Hsin-Jang 351
Shieh, Jenn-Jong 204
Shigematsu, Koichi 2183, 2189
Shih, Bing-Jyun 105
Shih, Sheng-Fang 2380
Shih-Jen Cheng 3185
Shimada, Takae...2058
Shimamori, Hiroshl 2745
Shimao, Toshihiro.......................................415
Shimatou, T. 2939
Shimizu, Kyohei 1968
Shimizu, Toshihisa ... 876, 1166, 2944, 2983, 3044, 3771
Shimizu, Toshimasa 1054
Shimode, Daisuke......................................1122
Shimomura, Junichi...................................2183
Shimono, Tomoyuki 1685
Shin Shiung Wang 3938
Shin, Hyunhak..110
Shin, Yesl 493
Shinagawa, Syuhei 252
Shinbo, Mitsuo 634
Shindo, Yuji..1552
Shinnaka, Shinji 1824
Shinohara, Atsushi......................................324
Shinozaki, Ikki 1728
Shinozuka, Yasuhiro..................................2228
Shioda, Masashi 130
Shirakawa, Kazuhiro.......................304, 1379
Shiraki, N..2120
Shirasawa, Koki 3106

Shishida, Yasuhiro 2644
Shiting Weng..1476
Shixi Hou 3168
Shoeiby, B. 1482
Shoji, Hiroyuki 2058
Shoyama, Masahito 863, 3386
Shu-Hung Liao 1452
Shuitao Yang...1342
Shun-Chung Wang 3233, 3778
Shunke Sui 3680
Shuren Wang 3194
Shu-Wei Kuo 3185
Siemaszko, Daniel 3371
Silva, Marcelo 3864
Silva, Sidelmo M. 3225
Silventoinen, P. 587
Sin, Min-Ho..409
Singh, B. N. 3482
Sintamarean, C. 1912
Sitbon, M. 2240
Sivakumar K 1400
Siwakoti, Yam P. 1801
Skuriat, Robert 2920
Smadi, Issam 1968
Smaka, Senad 1148
Smiththisomboon, Somrat 3885
Sogawa, Yuki 522, 2781
Solomon, Adane Kassa 2920
Sone, Kodai 3525
Song Kejian 640
Song, Z. Q. 2686
Sonoda, Hideki...634
Soo-Cheol Shin 3532
Specht, Andreas..1501
Srirattanawichaikul, Watcharin 3417
Steinert, Daniel 975
Steinke, Gina..1081
Stumpf, Peter 2700
Su, Bonan 614
Su, Hong-Wei 1781
Su, Jianhui 2156
Suetsugu, Tadashi.....................................3955
Sugao, Kazumi 403
Sugimoto, Hiroya 982
Sugiura, Makoto 1135
Suh, Yongsug 1185, 1646
Su-Han Kim 790
Sul, Seung-Ki 1268, 2598
Sumida, Hitoshi 2302
Sun, Jian...843, 967, 2202
Sun, Shaohua 2318
Sun, Wei 609
Sunaga, Keita 3162
Sung, Kyungmin 117, 829
Sunsoon Park..1646

AUTHOR INDEX

Suntio, T............514, 2240
Suntio, Teuvo............1466
Suryadevara, Rohit............2433
Suto, Kenji............3194, 3194
Suul, Jon Are............1544
Suwankawin, Surapong............3885
Suzuki, Genri............1697
Suzuki, Hirokazu............3503
Suzuki, Katsumi............959
Suzuki, Michiaki............883
Suzuki, Nobuyuki............919, 2541
Suzuki, Ryosuke............2972
Suzuki, Shun............1166
Suzuki, Takashi............1062
Suzuki, Toshiki............907
Svensson, Jan R............1220
Tabira, K.............2939
Tadano, Yugo............1242
Tadokoro, D.............390
Taeck-Kie Lee............3532
Taekyun Kim............3933
Tae-Won Chun............2763
Taga, Hironori............1328
Tajima, G.............2438
Takada, Hiromu............1697
Takagi, Ryo............1328
Takahashi, Akiko............2470
Takahashi, Hiroki............152, 1021
Takahashi, Hirotaka............1068
Takahashi, Hisashi............3106
Takahashi, K.............2569
Takahashi, Naoya............3920
Takahashi, Nobuhiro............3207
Takahashi, Osamu............2285
Takahashi, Takehiro............3207
Takahashi, Yoshikazu............671, 2870
Takamiya, Makoto............2228
Takao, Kazuto............744, 3440, 3707
Takasaki, Mika............2252
Takashita, Haruomi............3386
Takasu, Shinji............3440
Takatsuka, Yushi............1898
Takayama, Masakazu............3801
Takayanagi, Atsushi............2794
Takeda, Kotaro............1654
Takeda, Masashi............801
Takeda, Takashi............1490
Takei, Manabu............3440
Takemoto, Masatsugu............1000, 3525
Takenaka, Kensuke............3440
Takenami, Fumiaki............2737
Takenoiri, S.............2939
Takeshita, Takaharu............123, 601, 2310, 3809, 3877
Takeuchi, Katsutoku............3012
Takeuchi, Shun............3646

Takezaki, Kenichi............3525
Taki, Hiroshi............876, 1379
Takino, Toshiaki............817
Tam Khanh Tu Nguyen............2826
Tamada, Shunsuke............1357
Tamura, Hiroshi............682
Tan, Nadia M. L.............750
Tanabe, Ryo............1234
Tanai, Masanobu............829
Tanaka, Daiki............3179
Tanaka, Junya............240
Tanaka, Kiminori............2222
Tanaka, Koutaro............1006
Tanaka, Seiyu............982
Tanaka, Takahide............2302
Tanaka, Toshihiko............3292
Tanaka, Yasunori............3440
Tanaka, Yuichiro............1880
Tanifuji, Hikaru............1115
Taniguchi, Shun............2465
Tao Meng............3213
Tatsuta, Fujio............959
Tauchi, Yuki............3119
Teng, Jen-Hao............677
Teodorescu, R.............1912
Tera, Takahiro............304, 876
Terabe, Ryosuke............2638
Terao, Yutaka............2644
Teshima, Masato............1068
Thiringer, Torbjorn............1373
Thogersen, Paul............2547
Thogersen, Paul Bach............2850
Tian, Yanjun............3538
Ting, Pangan............677
Ting, Yeh............2787
Tint Soe Win............3292
Toba, Akio............2011
Toda, Hiroaki............1984
Togashi, Ryo............356
Toi, Takahiro............1109
Tokiwa, Tsuyoshi............2977
Tokuda, Hirokazu............2175
Tokumasu, Akira............1379
Tokuyama, Takeshi............2904
Tominaga, Shinji............2290
Tomioka, Satoshi............3630
Tomita, Mutuwo............907
Tonogi, K.............2438
Toru............3784
Tosaka, Shuhei............1207, 1227
Town, Graham E.............1801
Toyoda, Hajime............1560
Tran, Q. V.............446
Trescases, Olivier............2336, 3496
Trillion Zheng............2893

AUTHOR INDEX

Trintis, Ionut 2547
Tripathi, Awneesh...............................758, 1626
Trompa, Thomas 2554
Tsai, Jiung-Lin 1781
Tsai, Ming-Hsiao 278
Tsan Chen 2810
Tse, Chi K. 1425
Tseng, K. J. 2145
Tsorng-Juu Liang 2750, 3346
Tsubakidani, Takashi 3674
Tsuboi, Yoshiki...1160
Tsuboi, Yuichi 92
Tsuchida, Kazuo...3397
Tsuda, Junichi 1929
Tsuji, Mineo......................2775, 3093, 3674
Tsuji, Satoshi 522, 2781
Tsuji, Toshiaki...1978
Tsukakoshi, Masahiko 92, 2330
Tsuruma, Yoshinori...........................1054, 2644
Tsuruta, Hironori 629
Tsuruta, Ryoji 1350
Tsuruta, Yukinori 2266, 3403
Tsuyoshi, Hanamoto 2476
Tu, Yunwu..465
Tukiman, Rahayu 2529
Turpin, Santiago 1974
Tuysuz, Arda 1904
Tzou, Ying-Yu 3072
Uddin, Muslem 3574
Ueda, K. 312
Ueda, Tetsuya 403
Ueda, Tetsuzo 2075
Ueda, Yoshinobu 1855
Uemura, Hirofumi 891, 2834
Ukai, Hiroyuki...2400
Umeda, Nobuhiro..2183
Umeno, Masayoshi...183
Umesh B S 1400
Umetani, Kazuhiro..304
Undeland, Tore...1412
Uno, Masatoshi 1444, 2351
Urushibata, Hiroaki 2290
Urushibata, Shota 1365, 3455
Ushiro, Nobumasa...3106
Vaisanen, V. 587
Vajk, Istvan 2700
Van Brunt, Edward 3726
Van Wyk, J. D. 1935
Van-Long Tran 1214
Vasiladiotis, Michail 1386
Vasquez-Arnez, Ricardo Leon 1276
Veerasamy, Balaji 2310
Vieto, Ignacio...843
Viinamaki, J. 2240
Vilathgamuwa, D. M...........................2722, 3598

Vogt, T..2887
Wada, Keiji 744, 1379, 2511, 3646
Wahlstroem, Jonas 3476
Wajima, Kiyoshi 1984
Wallscheid, Oliver 1501
Wang Hui 640
Wang, Bin 2133
Wang, Chao-Fu 172
Wang, Fei 470
Wang, Fusheng 465
Wang, H. 1912
Wang, Hengli 72
Wang, Jun 944
Wang, Lingxiang 465
Wang, Lipeng 458
Wang, Xiaojian 2815
Wang, Xinyu 624, 2815
Wang, Xiongfei 216, 1529, 3320, 3538
Wang, Yanbo 3538
Wang, Yong 470
Wang, Yue..160, 475
Wang, Zhao'An 160, 475
Watanabe, Daisuke..988
Watanabe, Hiroki 84
Watanabe, S. 2939
Watanabe, Shoichiro 1334, 2126
Watashima, T..2939
Wei Jiang 3194
Wei Liu 1050
Wei Wang 3680
Wei Yan 3455
Wei, Guo 2318
Wei, Sun 341
Weili Dai 3168
Weirong Chen 3695
Wei-Ting Hsu 2910
Wen, Bo 3850
Wen, Chao-Kai 677
Wen, Huiqing...702
Wen-Chien Hsu 2714
Wenjie Chen 2967
Wen-Tai Li 677
Wheeler, Pat...2920
Won, Chung-Yuen 166, 409
Wong, Siu-Chung 1425
Wonsuk Choi 2914
Woojin Choi 1214
Wu Mingli 640
Wu, Bin 3468
Wu, Chun-Wei 330, 1832
Wu, G. F. 1471
Wu, Gwo-Bin 3796
Wu, T.-F. 2050
Wu, Tsung-Hsi 1796
Wu, Weimin 2610

AUTHOR INDEX

Wu, Weiyang 582
Wu, Wenlong 470
Wu, Wen-Zhe 429
Wunsch, B. 2167
Xia, Huan 2133
Xiangdong Sun 3489
Xiang-Dong Sun 3784
Xiao, D. 2686
Xiao, Fei 66, 72
Xiao, Shuai 543
Xiaojie You 2893
Xiaojie Zhuang 2638
Xiaolong Ma 835
Xiaomei Song 2967
Xie, Ruiliang 2950
Xiong, Li 230
Xiuqin Wei 3955
Xu Cai 1842, 3124
Xu Dianguo 3050
Xu Yang 2967
Xu, David 3099
Xu, Dehong 485
Xu, Dianguo 609, 614, 3002
Xu, Haizhen 567
Xu, Rong 609
Xue, Danhong 3604
Xuling Chen 2768
Yablecki, Jessica 3496
Yachi, Toshiaki 3959
Yakabe, Seichiro 730
Yamada, Hiroaki 538, 1811
Yamada, Kenji 1898
Yamada, Ryuji 1920
Yamada, Takatoshi 2212
Yamada, Tatsuji 3263
Yamagata, Shinichi 829
Yamagishi, Tatsuya 750
Yamaguchi, Shota 3771
Yamaguchi, Takashi 1242
Yamaichi, Katsuya 2517
Yamaji, Masaharu 2302
Yamamoto, Eiji 1654
Yamamoto, Junichi 177, 1179
Yamamoto, Kenji 3106
Yamamoto, Kichiro 689
Yamamoto, Kohei 1438
Yamamoto, Masayoshi 811, 883, 2497
Yamamoto, Shu 3044
Yamamoto, Takashi 3707
Yamamoto, Yasuhiro 1601
Yamamura, N. 146
Yamanaka, Kenji 2152
Yamanaka, Tatsuya 1207, 1227
Yamanoi, Takashi 1062
Yamashita, Nobuyuki 2983

Yamashita, Shigeharu 2745
Yamazaki, Akira 933
Yan Li 3124
Yan Zhang 835
Yanagi, Hiroshige 3630
Yang Chuan 1962
Yang, Cs 199, 2593, 3185
Yang, Daeki 2247
Yang, Geng 543, 582
Yang, Guorun 66
Yang, Hong-Tzer 2100
Yang, Rongfeng 609
Yang, Shih-Sian 2606
Yang, Sihun 863
Yang, Xu 2950
Yang, Zhongping 335, 1322, 2133
Yanhong, Zhang 452
Yano, Yoshihiro 2775
Yanru Zhong 3489
Yaramasu, Venkata 3695
Yashiro, Daisuke 1974
Yashun Li 782
Yasubayashi, Mikio 183
Yasui, Kazuya 2465
Yasumura, Yuji 3568
Yasuno, Takashi 3179
Yau, Y. T. 2754, 3392
Yazdkhasti, Pegah 2156
Yi, Hao 577
Yi-Chun Lin 2714
Yi-Hsun Chiu 3778
Yi-Hua Liu 3778
Yixin Zhu 1155, 3546
Yizhanyi Tang 2977
Yoda, Kazuyuki 1748
Yokoi, Y. 3007
Yokoi, Yuichi 3024
Yokokura, Yuki 1956
Yokoyama, H. 2120
Yokoyama, Natsuki 2285
Yokoyama, Tomoki 3397, 3410
Yonemori, Ryo 689
Yonezawa, Hikaru 3055
Yonezawa, Yoshiyuki 3440
Yonezawa, Yu 2745, 3386
Yong Ding 1476, 3842
Yong, Yu 341
Yong-Cheol Kwon 925
Yongdong Tan 3538
Yongjae Lee 3140
Yoon, Sung Hyun 2272
Yoshida, Morito 3410
Yoshida, S. 2569
Yoshida, T. 2438
Yoshida, Yoshiaki 3503

AUTHOR INDEX

Yoshikawa, Yuichi 2011
Yoshimoto, Kantaro..............................2421
Yoshimura, Eiji 1552
Yoshino, Teruo 2644, 3834
Yoshino, Yukio 2745
Yoshioka, S. 246
Yoshioka, Takashi..............................1956
Yoshizawa, Daisuke 97, 1950
Young-Do Kim 1738, 1743
Youngjoon Choi 1185
Young-Ryul Kim 796
Yu, Changzhou..............................567
Yu, F. Y. 1471
Yu, Ling-Chia 208
Yu, Shuai 2318
Yu, Weikai 647
Yu, Yifan 1458
Yu, Yong 567, 609
Yu-Chen Liu 2810, 3328
Yue Chen 2768
Yue, Xiaolong 2960
Yu-Jen Wang 1420
Yu-Kang Lo 2758, 2810, 3185, 3328, 3913
Yuki, Kazuaki 2465
Yukita, Kazuto 1490
Yukutake, Seigo 2140
Yunchang Kwak 2693
Yun-Chu Chiu 3328
Yung-Ching Huang..............................1452
Yunmei Fang 3168
Yunwei Li 3482
Yura, Masashi 2297
Yu-Shan Cheng 3233, 3778
Yuzurihara, Itsuo 2794
Zaitsu, Toshiyuki 177, 1179
Zanma, Tadanao 3568
Zargari, Navid R. 3468
Zehelein, Matthias..............................3085
Zeliang Shu 1050
Zeljkovic, Sandra 774
Zhang Wei 3050
Zhang Yajing..............................2899
Zhang, Haodong 3604
Zhang, Huiguo 2202
Zhang, Tao..............................485
Zhang, Wei 614, 1425
Zhang, Xing 465, 567
Zhang, Xuning..............................2626
Zhang, Yuzhuo 647
Zhang, Zhe..............................78
Zhao, Wei..............................567
Zhao-Qin Guo 1580
Zhe Wang 3129
Zhe Zhang..............................3905
Zheng Dong 3842

Zheng Li 1842
Zheng, T. Q. 807
Zheng, Trillion Q. 2899, 3314
Zhengzhi Han 3124
Zhenyao Xu 994
Zhongping Yang 807
Zhu, B. 395, 1594
Zhu, Honglin 3002
Zhuo, Fang 577, 2960
Zian Qin 857
Zingerli, Claudius M. 3365
Zitouni, Y. 3239
Zong-Zhen Yang..............................3778
Zou, Xudong 555
Zwyssig, Christof 1707